The Genetic Code (mRNA)

Second position

		U	C	A	G	
U		UUU, UUC \}Phe UUA, UUG \}Leu	UCU, UCC, UCA, UCG \}Ser	UAU, UAC \}Tyr UAA Stop, UAG Stop	UGU, UGC \}Cys UGA Stop UGG Trp	U C A G
C		CUU, CUC, CUA, CUG \}Leu	CCU, CCC, CCA, CCG \}Pro	CAU, CAC \}His CAA, CAG \}Gln	CGU, CGC, CGA, CGG \}Arg	U C A G
A		AUU, AUC \}Ile AUA AUG Met/start	ACU, ACC, ACA, ACG \}Thr	AAU, AAC \}Asn AAA, AAG \}Lys	AGU, AGC \}Ser AGA, AGG \}Arg	U C A G
G		GUU, GUC, GUA, GUG \}Val	GCU, GCC, GCA, GCG \}Ala	GAU, GAC \}Asp GAA, GAG \}Glu	GGU, GGC, GGA, GGG \}Gly	U C A G

First position / Third position

Some Prefixes Used in the International System of Units

10^9	giga	G
10^6	mega	M
10^3	kilo	k
10^{-1}	deci	d
10^{-2}	centi	c
10^{-3}	milli	m
10^{-6}	micro	μ
10^{-9}	nano	n
10^{-12}	pico	p
10^{-15}	femto	f

Conversion Factors

Energy: 1 joule = 10^7 ergs = 0.239 cal
         1 cal = 4.184 joule

Length: 1 nm = 10 Å = 1×10^{-7} cm = 1×10^{-9} m

Mass: 1 kg = 1000 g = 2.2 lb
       1 lb = 453.6 g

Pressure: 1 atm = 760 torr = 1.013 bar = 14.696 psi
           1 bar = 100 kPa = 0.987 atm = 750.1 torr
           1 torr = 1 mm Hg

Temperature: K = °C + 273
              °C = (5/9)(°F − 32)

Volume: 1 L = 1×10^{-3} m^3 = 1000 cm^3

$$A^- + H^+ \rightleftharpoons HA$$

$$HA + OH^- \rightleftharpoons H_2O + A^-$$

Physical Constants

Name	Symbol	SI Units	cgs Units
Avogadro's number	N	6.022137×10^{23}/mol	6.022137×10^{23}/mol
Boltzmann constant	k	1.38066×10^{-23} J/K	1.38066×10^{-16} erg/K
Curie	Ci	3.7×10^{10} d/s	3.7×10^{10} d/s
Electron charge	e	1.602177×10^{-19} coulomb[b]	4.80321×10^{-10} esu
Faraday constant	F	96485 J/V·mol	9.6485×10^{11} erg/V·m
Gas constant[a]	R	8.31451 J/K·mol	8.31451×10^7 erg/K·m
Light speed (vacuum)	c	2.99792×10^8 m/s	2.99792×10^{10} cm/s
Planck's constant	h	6.626075×10^{-34} J·s	6.626075×10^{-27} erg·s

[a]Other values of R: 1.9872 cal/K·mol = 0.082 L·atm/K·mol
[b]1 coulomb = 1 J/V

Biochemistry
CONCEPTS AND CONNECTIONS

SECOND EDITION

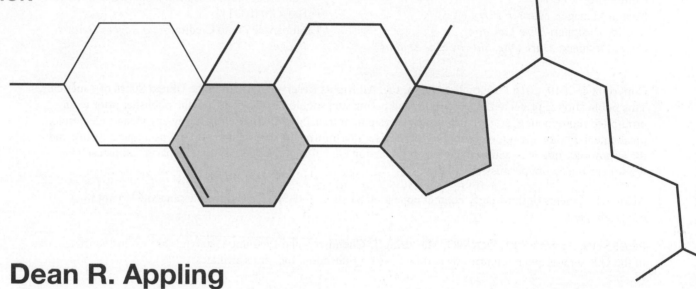

Dean R. Appling
THE UNIVERSITY OF TEXAS AT AUSTIN

Spencer J. Anthony-Cahill
WESTERN WASHINGTON UNIVERSITY

Christopher K. Mathews
OREGON STATE UNIVERSITY

 Pearson

330 Hudson Street, New York, NY 10013

Editor in Chief: *Jeanne Zalesky*
Acquisitions Editor: *Chris Hess*
Director of Development: *Jennifer Hart*
Marketing Manager: *Elizabeth Bell*
Development Editor: *Matt Walker*
Art Development Editor: *Jay McElroy*
Program Manager: *Kristen Flathman*
Content Producer: *Anastasia Slesareva*
Text Permissions Project Manager: *Tim Nicholls*
Text Permissions Specialist: *James Fortney,*
  Lumina Datamatics
Project Management Team Lead: *David Zielonka*
Production Management: *Mary Tindle, Cenveo*
Compositor: *Cenveo*
Design Manager: *Marilyn Perry*
Interior Designer: *Elise Lansdon*
Cover Designer: *Mark Ong, Side by Side Studios*

Illustrators: *ImagineeringArt, Inc.*
Photo Researcher: *Mo Spuhler*
Photo Lead: *Maya Melenchuk / Eric Shrader*
Photo Permissions: *Kathleen Zander / Matt Perry*
Operations Specialist: *Stacey Weinberger*
Cover Background Photo Credit: *The Human*
  Spliceosome
Dr. Berthold Kastner, Max Planck Institute of
  Biophysical Chemisty
Molecular graphics and analyses were performed
  with the UCSF Chimera package. Chimera is
  developed by the Resource for Biocomputing,
  Visualization, and Informatics at the University of
  California, San Francisco (supported by NIGMS
  P41-GM103311)
Chimpanzee Photo Credit: *Fiona Rogers/Getty*
  Images

Acknowledgments of third-party content appear on pages C-1–C-3, which constitutes an extension of this
copyright page.

Library of Congress Cataloging-in-Publication Data
Names: Appling, Dean Ramsay, author. | Anthony-Cahill, Spencer J., author. |
  Mathews, Christopher K., 1937- author.
Title: Biochemistry : concepts and connections / Dean R. Appling, Spencer J.
  Anthony-Cahill, Christopher K. Mathews.
Description: Second edition. | New York : Pearson, [2019] | Includes
  bibliographical references and index.
Identifiers: LCCN 2017047599| ISBN 9780134641621 | ISBN 0134641620
Subjects: | MESH: Biochemical Phenomena
Classification: LCC RB112.5 | NLM QU 34 | DDC 612/.015--dc23
LC record available at https://lccn.loc.gov/2017047599

[Third-Party Trademark] [TM/®] is a [registered] trademark of [Third Party]. Used under license.

1 18

ISBN 10: 0-134-64162-0; ISBN 13: 978-0-134-64162-1
www.pearson.com

Brief Contents

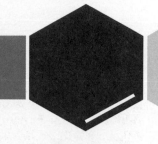

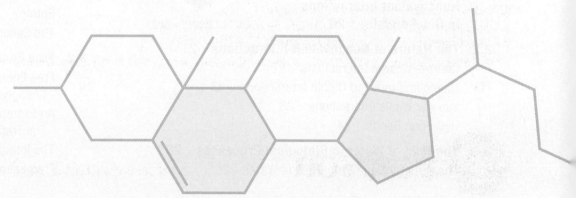

Contents

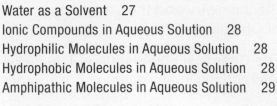

CHAPTER **6**

The Three-Dimensional Structure of Proteins   144

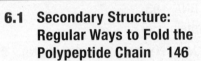
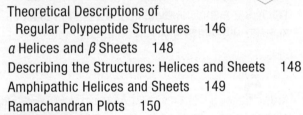

CHAPTER **7**

Protein Function and Evolution   190

CHAPTER **8**

Enzymes: Biological Catalysts 232

HIV

CHAPTER **9**

Carbohydrates: Sugars, Saccharides, Glycans 278

CHAPTER **10**

Lipids, Membranes, and Cellular Transport 304

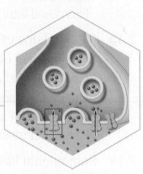

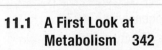

CHAPTER **11**

Chemical Logic of Metabolism   340

CHAPTER **12**

Carbohydrate Metabolism: Glycolysis, Gluconeogenesis, Glycogen Metabolism, and the Pentose Phosphate Pathway   374

CHAPTER **15**

Photosynthesis   486

CHAPTER **16**

Lipid Metabolism   512

CHAPTER **19**

Nucleotide Metabolism   610

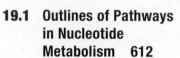

CHAPTER **20**

Mechanisms of Signal Transduction   636

CHAPTER **21**

Genes, Genomes, and Chromosomes   664

CHAPTER **22**

DNA Replication   686

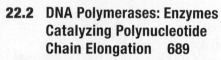

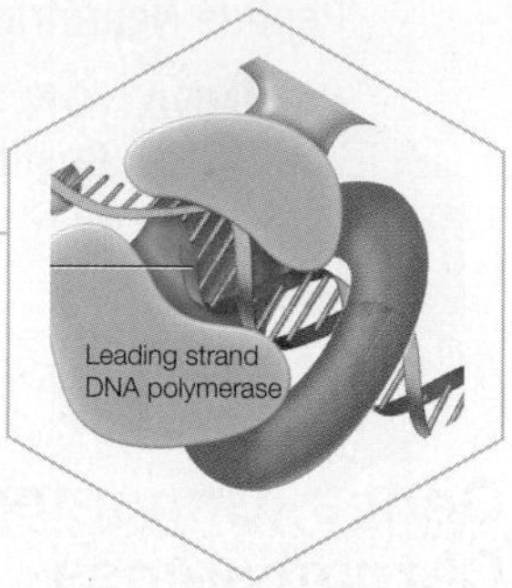

Leading strand DNA polymerase

CHAPTER **23**

DNA Repair, Recombination, and Rearrangement   714

CHAPTER **24**

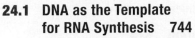

Transcription and Posttranscriptional Processing   742

CHAPTER **25**

Information Decoding: Translation and Posttranslational Protein Processing   766

Preface

Biochemistry: Concepts and Connections

As genomics and informatics revolutionize biomedical science and health care, we must prepare students for the challenges of the twenty-first century and ensure their ability to apply quantitative reasoning skills to the science most fundamental to medicine: biochemistry.

We have written *Biochemistry: Concepts and Connections* to provide students with a clear understanding of the chemical logic underlying the mechanisms, pathways, and processes in living cells. The title reinforces our vision for this book—twin emphases upon fundamental *concepts* at the expense of lengthy descriptive information, and upon *connections,* showing how biochemistry relates to all other life sciences and to practical applications in medicine, agricultural sciences, environmental sciences, and forensics.

Inspired by our experience as authors of the biochemistry majors' text, *Biochemistry, Fourth Edition* and the first edition of this book, and as teachers of biochemistry majors' and mixed-science-majors' courses, we believe there are several requirements that a textbook for the mixed-majors' course must address:

- The need for students to understand the structure and function of biological molecules before moving into metabolism and dynamic aspects of biochemistry.

- The need for students to understand that biochemical concepts derive from experimental evidence, meaning that the principles of biochemical techniques must be presented to the greatest extent possible.

- The need for students to encounter many and diverse real-world applications of biochemical concepts.

- The need for students to understand the quantitative basis for biochemical concepts. The Henderson–Hasselbalch equation, the quantitative expressions of thermodynamic laws, and the Michaelis–Menten equation, for example, are not equations to be memorized and forgotten when the course moves on. The basis for these and other quantitative statements must be understood and constantly repeated as biochemical concepts, such as mechanisms of enzyme action, are developed. They are essential to help students grasp the concepts.

In designing *Biochemistry: Concepts and Connections,* we have stayed with the organization that serves us well in our own classroom experience. The first 10 chapters cover structure and function of biological molecules, the next 10 deal with intermediary metabolism, and the final 6 with genetic biochemistry. Our emphasis on biochemistry as a quantitative science can be seen in Chapters 2 and 3, where we focus on water, the matrix of life, and bioenergetics. Chapter 4 introduces nucleic acid structure, with a brief introduction to nucleic acid and protein synthesis—topics covered in much more detail at the end of the book.

Chapters 11 through 20 deal primarily with intermediary metabolism. We cover the major topics in carbohydrate metabolism, lipid metabolism, and amino acid metabolism in one chapter each (12, 16, and 18, respectively). Our treatment of cell signaling is a bit unconventional, since it appears in Chapter 20, well after we present hormonal control of carbohydrate and lipid metabolism. However, this treatment allows more extended presentation of receptors, G proteins, oncogenes, and neurotransmission. In addition, because cancer often results from aberrant signaling processes, our placement of the signaling chapter leads fairly naturally into genetic biochemistry, which follows, beginning in Chapter 21.

With assistance from talented artists, we have built a compelling visual narrative from the ground up, composed of a wide range of graphic representations, from macromolecules to cellular structures as well as reaction mechanisms and metabolic pathways that highlight and reinforce overarching themes (chemical logic, regulation, interface between chemistry and biology). In addition, we have added two new **Foundation Figures** to the Second Edition, bringing the total number to 10. These novel Foundation Figures integrate core chemical and biological connections visually, providing a way to organize the complex and detailed material intellectually, thus making relationships among key concepts clear and easier to study. The "**CONCEPT**" and "**CONNECTION**" statements within the narrative, which highlight fundamental concepts and real-world applications of biochemistry, have been reviewed and revised for the Second Edition.

In *Biochemistry: Concepts and Connections,* we emphasize our field as an experimental science by including 17 separate sections, called **Tools of Biochemistry,** that highlight the most important research techniques. We also provide students with references (about 12 per chapter), choosing those that would be most appropriate for our target audience, such as links to Nobel Prize lectures.

We consider end-of-chapter problems to be an indispensable learning tool and provide 15 to 25 problems for each chapter. (In the Second Edition we have added 3 to 4 new end-of-chapter problems to each chapter.) About half of the problems have brief answers at the end of the book, with complete answers provided in a separate solutions manual. Additional tutorials in Mastering Chemistry will help students with some of the most basic concepts and operations. See the table of Instructor and Student Resources on the following page.

Producing a book of this magnitude involves the efforts of dedicated editorial and production teams. We have not had the pleasure of meeting all of these talented individuals, but we consider them close colleagues nonetheless. First, of course, is Jeanne Zalesky, our sponsoring editor, now Editor-in-Chief, Physical Sciences, who always found a way to keep us focused on our goal. Susan Malloy, Program Manager, kept us organized and on schedule, juggling disparate elements in this complex project—later replaced by Anastasia Slesareva. Jay McElroy, Art Development Editor, was our intermediary with the talented artists at Imagineering, Inc., and displayed considerable artistic and editorial gifts in his own right. We also worked with an experienced development editor, Matt Walker. His suggested edits, insights, and attention to detail were invaluable. Beth Sweeten, Senior Project Manager, coordinated the production of the main text and preparation of the Solutions Manual for the end-of-chapter problems. Gary Carlton provided great assistance with many of the illustrations. Chris Hess provided the inspiration for our cover illustration, and Mo Spuhler helped us locate much excellent illustrative material. Once the book was in production, Mary Tindle skillfully kept us all on a complex schedule.

Instructor and Student Resources

Resource	Instructor or Student Resource	Description
Solutions Manual ISBN: 0134814800	Instructor	Prepared by Dean Appling, Spencer Anthony-Cahill, and Christopher Mathews, the solutions manual includes worked-out answers and solutions for problems in the text.
Mastering™ Chemistry pearson.com/mastering/chemistry ISBN: 0134787250	Student & Instructor	Mastering™ Chemistry is the leading online homework, tutorial, and assessment platform, designed to improve results by engaging students with powerful content. Instructors ensure students arrive ready to learn by assigning educationally effective content before class, and encourage critical thinking and retention with in-class resources such as Learning Catalytics. Learn more about Mastering Chemistry. **Mastering Chemistry for** *Biochemistry: Concepts and Connections,* **2/e** now has hundreds of more biochemistry-specific assets to help students tackle threshold concepts, connect course materials to real world applications, and build the problem solving skills they need to succeed in future courses and careers.
Pearson eText ISBN: 0134763025	Student	*Biochemistry: Concepts and Connections* 2/e now offers **Pearson eText, optimized for mobile,** which seamlessly integrates videos and other rich media with the text and gives students access to their textbook anytime, anywhere. Pearson eText is available with Mastering Chemistry when packaged with new books, or as an upgrade students can purchase online. The Pearson eText mobile app offers: • Offline access on most iOS and Android phones/tablets. • Accessibility (screen-reader ready) • Configurable reading settings, including resizable type and night reading mode • Instructor and student note-taking, highlighting, bookmarking, and search tools • Embedded videos for a more interactive learning experience
TestGen Test Bank ISBN: 0134814827	Instructor	This resource includes more than 2000 questions in multiple-choice answer format. Test bank problems are linked to textbook-specific learning outcomes as well as MCAT-associated outcomes. Available for download on the Pearson catalog page for *Biochemistry: Concepts and Connections* at www.pearson.com
Instructor Resource Materials ISBN: 0134814843 ISBN: 0134814835	Instructor	Includes all the art, photos, and tables from the book in JPEG format, as well as Lecture Powerpoint slides, for use in classroom projection or when creating study materials and tests. Available for download on the Pearson catalog page for *Biochemistry: Concepts and Connections* at www.pearson.com

The three of us give special thanks to friends and colleagues who provided unpublished material for us to use as illustrations. These contributors include John S. Olson (Rice University), Jack Benner (New England BioLabs), Andrew Karplus (Oregon State University), Scott Delbecq and Rachel Klevit (University of Washington), William Horton (Oregon Health and Science University), Cory Hamada (Western Washington University), Nadrian C. Seaman (New York University), P. Shing Ho (Colorado State University), Catherine Drennan and Edward Brignole (MIT), John G. Tesmer (University of Michigan), Katsuhiko Murakami (Penn State University), Alan Cheung (University College London), Joyce Hamlin (University of Virginia), Stefano Tiziani, Edward Marcotte, David Hoffman, and Robin Gutell (University of Texas at Austin), Dean Sherry and Craig Malloy (University of Texas-Southwestern Medical Center), and Stephen C. Kowalczykowski (University of California, Davis). The cover image, representing in part the structure of the human splicesome, was kindly provided by Karl Bertram (University of Göttingen, Germany).

We are also grateful to the numerous talented biochemists retained by our editors to review our outline, prospectus, chapter drafts, and solutions to our end-of-chapter problems. Their names and affiliations are listed separately.

Our team—authors and editors—put forth great effort to detect and root out errors and ambiguities. We undertook an arduous process of editing and revising several drafts of each chapter in manuscript stage, as well as copyediting, proofreading, and accuracy, reviewing multiple rounds of page proofs in an effort to ensure the highest level of quality control.

Throughout this process, as in our previous writing, we have been most grateful for the patience, good judgment, and emotional support provided by our wives—Maureen Appling, Yvonne Anthony-Cahill, and Kate Mathews. We expect them to be as relieved as we are to see this project draw to a close, and hope that they can share our pleasure at the completed product.

Dean R. Appling
Spencer J. Anthony-Cahill
Christopher K. Mathews

Reviewers

The following reviewers provided valuable feedback on the manuscript at various stages throughout the wiring process:

Paul D. Adams, *University of Arkansas*
Harry Ako, *University of Hawaii–Manoa*
Eric J. Allaine, *Appalachian State University*
Mark Alper, *University of California—Berkeley*
John Amaral, *Vancouver Island University*
Trevor R. Anderson, *Purdue University*
Steve Asmus, *Centre College*
Kenneth Balazovich, *University of Michigan*
Karen Bame, *University of Missouri—Kansas City*
Jim Bann, *Wichita State University*
Daniel Barr, *Utica College*
Moriah Beck, *Wichita State University*
Marilee Benore, *University of Michigan*
Wayne Bensley, *State University of New York—Alfred State College*
Werner Bergen, *Auburn University*

Dean R. Appling is the Lester J. Reed Professor of Biochemistry and the Associate Dean for Research and Facilities for the College of Natural Sciences at the University of Texas at Austin, where he has taught and done research for the past 32 years. Dean earned his B.S. in Biology from Texas A&M University (1977) and his Ph.D. in Biochemistry from Vanderbilt University (1981). The Appling laboratory studies the organization and regulation of metabolic pathways in eukaryotes, focusing on folate-mediated one-carbon metabolism. The lab is particularly interested in understanding how one-carbon metabolism is organized in mitochondria, as these organelles are central players in many human diseases. In addition to coauthoring *Biochemistry, Fourth Edition,* a textbook for majors and graduate students, Dean has published over 65 scientific papers and book chapters.

As much fun as writing a textbook might be, Dean would rather be outdoors. He is an avid fisherman and hiker. Recently, Dean and his wife, Maureen, have become entranced by the birds on the Texas coast. They were introduced to bird-watching by coauthor Chris Mathews and his wife Kate—an unintended consequence of writing textbooks!

Spencer J. Anthony-Cahill is a Professor and chair of the Department of Chemistry at Western Washington University (WWU), Bellingham, WA. Spencer earned his B.A. in chemistry from Whitman College and his Ph.D. in bioorganic chemistry from the University of California, Berkeley. His graduate work, in the laboratory of Peter Schultz, focused on the biosynthetic incorporation of unnatural amino acids into proteins. Spencer was an NIH postdoctoral fellow in the laboratory of Bill DeGrado (then at DuPont Central Research), where he worked on *de novo* peptide design and the prediction of the tertiary structure of the HLH DNA-binding motif. He then worked for five years as a research scientist in the biotechnology industry, developing recombinant hemoglobin as a treatment for acute blood loss. In 1997, Spencer decided to pursue his long-standing interest in teaching and moved to WWU, where he is today.

In 2012, Spencer was recognized by WWU with the Peter J. Elich Award for Excellence in Teaching.

Research in the Anthony-Cahill laboratory is directed at the protein engineering and structural biology of oxygen-binding proteins. The primary focus is on the design of polymeric human hemoglobins with desirable therapeutic properties as a blood replacement.

Outside the classroom and laboratory, Spencer is a great fan of the outdoors—especially the North Cascades and southeastern Utah, where he has often backpacked, camped, climbed, and mountain biked. He also plays electric bass (poorly) in a local blues–rock band and teaches Aikido in Bellingham.

Christopher K. Mathews is Distinguished Professor Emeritus of Biochemistry at Oregon State University. He earned his B.A. in chemistry from Reed College (1958) and his Ph.D. in biochemistry from the University of Washington (1962). He served on the faculties of Yale University and the University of Arizona from 1963 until 1978, when he moved to Oregon State University as Chair of the Department of Biochemistry and Biophysics, a position he held until 2002. His major research interests are the enzymology and regulation of DNA precursor metabolism and the intracellular coordination between deoxyribonucleotide synthesis and DNA replication. From 1984 to 1985, Dr. Mathews was an Eleanor Roosevelt International Cancer Fellow at the Karolinska Institute in Stockholm, and in 1994–1995, he held the Tage Erlander Guest Professorship at Stockholm University. Dr. Mathews has published about 190 research papers, book chapters, and reviews dealing with molecular virology, metabolic regulation, nucleotide enzymology, and biochemical genetics. From 1964 until 2012, he was principal investigator on grants from the National Institutes of Health, the National Science Foundation, and the Army Research Office. He is the author of *Bacteriophage Biochemistry* (1971) and coeditor of *Bacteriophage T4* (1983) and *Structural and Organizational Aspects of Metabolic Regulation* (1990). He was lead author of four editions of *Biochemistry,* a textbook for majors and graduate students. His teaching experience includes undergraduate, graduate, and medical school biochemistry courses.

He has backpacked and floated the mountains and rivers, respectively, of Oregon and the Northwest. As an enthusiastic birder, he is serving as President of the Audubon Society of Corvallis.

Tools of Biochemistry

When an electric field is applied to a solution, solute molecules with a net positive charge migrate toward the cathode, and molecules with a net negative charge move toward the anode. This migration is called **electrophoresis.** Although electrophoresis can be carried out free in solution, it is more convenient to use some kind of *supporting medium* through which the charged molecules move. The supporting medium could be paper or, most typically, a gel composed of the polysaccharide agarose (commonly used to separate nucleic acids; see FIGURE 2A.1) or crosslinked polyacrylamide (commonly used to separate proteins).

The velocity, or **electrophoretic mobility (μ),** of the molecule in the field is defined as the ratio between two opposing factors: the force exerted by the electric field on the charged particle, and the frictional force exerted on the particle by the medium:

$$\mu = \frac{Ze}{f} \qquad (2A.1)^{\ddagger}$$

The numerator equals the product of the negative (or positive) charge (e) times the number of unit charges, Z (a positive or negative integer). The greater the overall charge on the molecule, the greater the force it experiences in the electric field. The denominator f is the **frictional coefficient,** which depends on the size and shape of the molecule. Large or asymmetric molecules encounter more frictional resistance than small or compact ones and consequently have larger frictional coefficients. Equation 2A.1 tells us that the mobility of a molecule depends on its charge and on its molecular dimensions.[‡] Because ions and macroions differ in both respects, electrophoresis provides a powerful way of separating them.

Gel Electrophoresis

In **gel electrophoresis,** a gel containing the appropriate buffer solution is cast in a mold (for agarose gel electrophoresis, shown in Figure 2A.1) or as a thin slab between glass plates (for polyacrylamide gel electrophoresis, shown in FIGURE 2A.2). The gel is placed between electrode compartments, and the samples to be analyzed are carefully pipetted into precast notches in the gel, called wells. Usually, glycerol and a water-soluble anionic "tracking" dye (such as bromophenol blue) are added to the samples. The glycerol makes the sample solution dense, so that it sinks into the well and does not mix into the buffer solution. The dye migrates faster than most macroions, so the experimenter is able to follow the progress of the experiment. The current is turned on until the tracking dye band is near the side of the gel opposite the wells. The gel is then removed from the apparatus and is usually stained with a dye that binds to proteins or nucleic acids. Because the protein

[‡]Equation 2A.1 is an approximation which neglects the effects of the ion atmosphere. See van Holde, Johnson, and Ho in Appendix II for more detail.

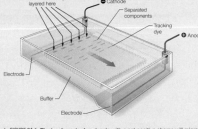

▲ FIGURE 2A.1 **Electrophoresis.** A molecule with a net positive charge will migrate toward the cathode, whereas a molecule with a net negative charge will migrate toward the anode.

or nucleic acid mixture was applied as a narrow band in the well of the gel, components migrating with different electrophoretic mobilities appear as separated bands on the gel. FIGURE 2A.3 shows an example of separation of DNA fragments by this method using an agarose gel. An example of the electrophoretic separation of proteins using a polyacrylamide gel is shown in Chapter 5 (see Figure 5A.9).

▲ FIGURE 2A.2 **Gel electrophoresis.** An apparatus for polyacrylamide gel electrophoresis is shown schematically. The gel is cast between plates. The gel is in contact with buffer in the upper (cathode) and lower (anode) reservoirs. A sample is loaded into one or more wells cast into the top of the gel, and then current is applied to achieve separation of the components in the sample.

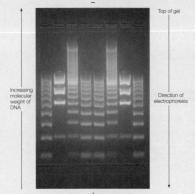

▲ FIGURE 2A.3 **Gel showing separation of DNA fragments.** Following electrophoretic separation of the different-length DNA molecules, the gel is mixed with a fluorescent dye that binds DNA. The unbound dye is then washed off, and the stained DNA molecules are visualized under ultraviolet light.

Polyelectrolytes like DNA or polylysine have one unit charge on each residue, so each molecule has a charge (Ze) proportional to its molecular length. But the frictional coefficient (f) also increases with molecular length, so to a first approximation, a macroion whose charge is proportional to its length has an electrophoretic mobility almost independent of its size. However, gel electrophoresis introduces additional frictional forces that allow the separation of molecules based on size. For linear molecules like the nucleic acid fragments in Figure 2A.3, the relative mobility in an agarose gel is a pproximately a linear function of the logarithm of the molecular weight. Usually, standards of known molecular weight are electrophoresed in one or more lanes on the gel. The molecular weight of the sample can then be estimated by comparing its migration in the gel to those of the standards. For proteins, a similar separation in a polyacrylamide gel is achieved by coating the denatured protein molecule with the anionic detergent sodium dodecylsulfate (SDS) before electrophoresis. This important technique is discussed further in Chapter 5.

Isoelectric Focusing

Proteins are polyamphotytes; thus, a protein will migrate in an electric field like other ions if it has a net positive or negative charge. At its isoelectric point, however, its net charge is zero, and it is attracted to neither the anode nor the cathode. If we use a gel with a stable pH gradient covering a wide pH range, each protein molecule in a complex mixture of proteins migrates to the position of its isoelectric point and accumulates there. This method of separation, called **isoelectric focusing,** produces distinct bands of accumulated proteins and can separate proteins with very small differences in the isoelectric point (FIGURE 2A.4). Since the pH of each portion of the gel is known, isoelectric focusing can also be used to determine experimentally the isoelectric point of a particular protein.

What we have presented here is only a brief overview of a widely applied technique. Additional information on electrophoresis and isoelectric focusing can be found in Appendix II.

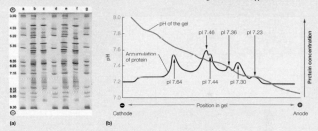

▲ FIGURE 2A.4 **Isoelectric focusing of proteins.** (a) An isoelectric focusing gel with a pH gradient from 3.50 (anode end) to 9.30 (cathode end). (b) A schematic showing where proteins of the indicated pIs would accumulate (peaks shown in red) in a pH gradient gel.

TOOLS OF BIOCHEMISTRY emphasize our field as an experimental science and highlight the most important research techniques relevant to students today.

Foundation Figures

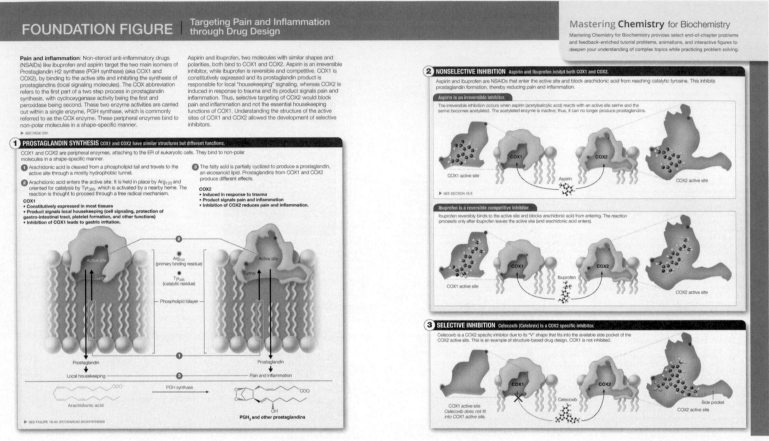

FOUNDATION FIGURES integrate core chemical and biological connections visually and provide a way to organize the complex and detailed material intellectually, thus making relationships among key concepts clear and easier to study.

Enhanced art and media programs
engage students

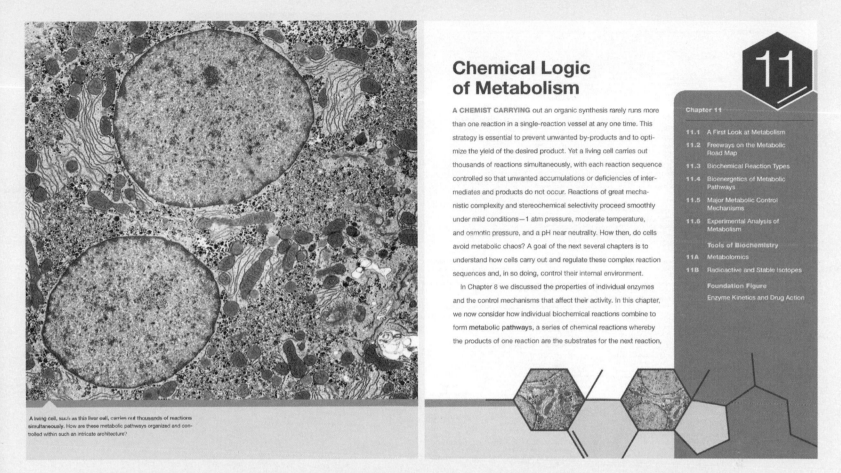

Chemical Logic of Metabolism

A CHEMIST CARRYING out an organic synthesis rarely runs more than one reaction in a single-reaction vessel at any one time. This strategy is essential to prevent unwanted by-products and to optimize the yield of the desired product. Yet a living cell carries out thousands of reactions simultaneously, with each reaction sequence controlled so that unwanted accumulations or deficiencies of intermediates and products do not occur. Reactions of great mechanistic complexity and stereochemical selectivity proceed smoothly under mild conditions—1 atm pressure, moderate temperature, and osmotic pressure, and a pH near neutrality. How then, do cells avoid metabolic chaos? A goal of the next several chapters is to understand how cells carry out and regulate these complex reaction sequences and, in so doing, control their internal environment.

In Chapter 8 we discussed the properties of individual enzymes and the control mechanisms that affect their activity. In this chapter, we now consider how individual biochemical reactions combine to form **metabolic pathways**, a series of chemical reactions whereby the products of one reaction are the substrates for the next reaction,

Chapter 11

11.1 A First Look at Metabolism
11.2 Freeways on the Metabolic Road Map
11.3 Biochemical Reaction Types
11.4 Bioenergetics of Metabolic Pathways
11.5 Major Metabolic Control Mechanisms
11.6 Experimental Analysis of Metabolism

Tools of Biochemistry

11A Metabolomics
11B Radioactive and Stable Isotopes

Foundation Figure

Enzyme Kinetics and Drug Action

A living cell, such as this liver cell, carries out thousands of reactions simultaneously. How are these metabolic pathways organized and controlled within such an intricate architecture?

UPDATED & REVISED! The second edition of Appling, Mathews, & Anthony-Cahill's **Biochemistry: Concepts and Connections** builds student understanding even more with an enhanced art program and a deeper, more robust integration with Mastering Chemistry. This renowned author team's content engages students with visualization, synthesis of complex topics, and connections to the real world resulting in a seamlessly integrated experience.

▲ **FIGURE 12.5 Reaction mechanism for fructose-1,6-bisphosphate aldolase.** The Figure shows the protonated Schiff base intermediate (iminium ion) between the substrate and an active site lysine residue. An aspartate residue facilitates the reaction via general acid–base catalysis.

CONCEPT Aldolase cleaves fructose-1,6-bisphosphate under intracellular conditions, even though the equilibrium lies far toward fructose-1,6-bisphosphate under standard conditions.

Reaction 4 is so strongly endergonic under standard conditions that the formation of fructose-1,6-bisphosphate is highly favored. However, from the actual intracellular concentrations of the reactant and products, ΔG is estimated to be approximately -1.3 kJ/mol, consistent with the observation that the reaction proceeds as written in vivo. Reaction 4 demonstrates the importance of considering the conditions *in the cell* (ΔG) rather than standard state conditions ($\Delta G^{\circ\prime}$) when deciding in which direction a reaction is favored.

Aldolase activates the substrate for cleavage by nucleophilic attack on the keto carbon at position 2 with a lysine ε-amino group in the active site, as shown in **FIGURE 12.5**. This is facilitated by protonation of the carbonyl oxygen by an active site acid (aspartate) ❶. The resulting carbinolamine undergoes dehydration to give an iminium ion, or protonated **Schiff base** ❷. A Schiff base is a nucleophilic addition product between an amino group and a carbonyl group. A retro-aldol reaction then cleaves the protonated Schiff base into an enamine plus GAP ❸. The enamine is protonated to give another iminium ion (protonated Schiff base) ❹, which is then hydrolyzed off the enzyme to give the second product, DHAP ❺.

The Schiff base intermediate is advantageous in this reaction because it can delocalize electrons. The positively charged iminium ion is thus a better electron acceptor than a ketone carbonyl, facilitating retro-aldol reactions like this one and, as we shall see, many other biological conversions. This mechanism also demonstrates why it was important to isomerize G6P to F6P in reaction 2. If glucose had not been isomerized to fructose (moving the carbonyl from C-1 to C-2), then the aldolase reaction would have given two- and four-carbon fragments, instead of the metabolically equivalent three-carbon fragments.

Reaction ❺: Isomerization of Dihydroxyacetone Phosphate
In reaction 5, **triose phosphate isomerase (TIM)** catalyzes the isomerization of dihydroxyacetone phosphate (DHAP) to glyceraldehyde-3-phosphate (GAP) via an enediol intermediate.

Reaction 5: Triose phosphate isomerase (TIM)

$\Delta G^{\circ\prime} = +7.6$ kJ/mol

Like reaction 4, reaction 5 is weakly endergonic under standard conditions, but the intracellular concentration of GAP is low because it is consumed in subsequent reactions. Thus, reaction 5 is drawn toward the right.

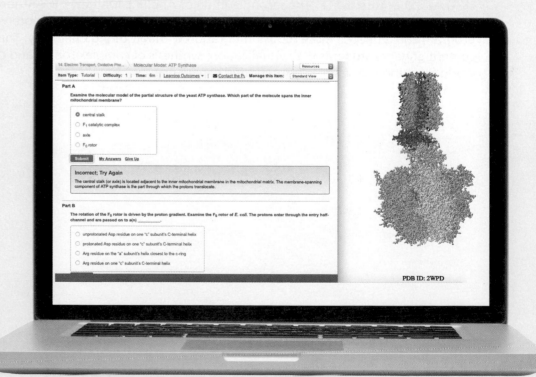

PDB ID: 2WPD

and understand what's happening on the cellular level

Dynamic Study Modules (DSMs) help students study effectively on their own by continuously assessing their activity and performance in real time.

Students complete a set of questions and indicate their level of confidence in their answer. Questions repeat until the student can answer them all correctly and confidently. These are available as graded assignments prior to class and are accessible on smartphones, tablets, and computers.

The DSMs focus on General Chemistry as well as Biochemistry topics.

BioFlix® 3D movie-quality animations help your students visualize complex biology topics and include automatically graded tutorial activities.

Synthesis of information is simplified through features

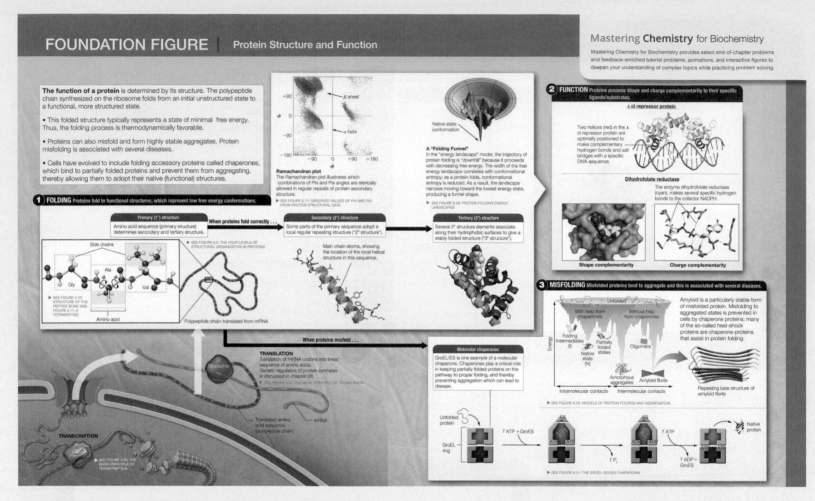

FOUNDATION FIGURE | Protein Structure and Function

Mastering Chemistry for Biochemistry

Mastering Chemistry for Biochemistry provides select end-of-chapter problems and feedback-enriched tutorial problems, animations, and interactive figures to deepen your understanding of complex topics while practicing problem solving.

The function of a protein is determined by its structure. The polypeptide chain synthesized on the ribosome folds from an initial unstructured state to a functional, more structured state.

• This folded structure typically represents a state of minimal free energy. Thus, the folding process is thermodynamically favorable.

• Proteins can also misfold and form highly stable aggregates. Protein misfolding is associated with several diseases.

• Cells have evolved to include folding accessory proteins called chaperones, which bind to partially folded proteins and prevent them from aggregating, thereby allowing them to adopt their native (functional) structures.

UPDATED & REVISED! Foundation Figures integrate core chemical and biological connections visually and provide a way to organize the complex and detailed material intellectually, making relationships among key concepts clear and easier to study. The second edition includes two new foundation figures as well as updated layouts based on learning design principles. Foundation Figures are assignable in Mastering Chemistry and are embedded in the Pearson eText as an interactive part of the narrative.

that help students connect complex concepts

Note that we have expressed concentrations of all solutes in units of molarity, then divided by the proper standard state concentration (also in units of molarity). These steps ensure that the terms in Q are of the proper magnitude and stripped of units:

$$\Delta G = -32.2\,\frac{\text{kJ}}{\text{mol}} + \left(2.478\,\frac{\text{kJ}}{\text{mol}}\right)\ln\left(\frac{(0.0001)(0.035)(0.398)}{(0.005)}\right) \quad (3.31b)$$

or

$$\Delta G = -32.2\,\frac{\text{kJ}}{\text{mol}} + -20.3\,\frac{\text{kJ}}{\text{mol}} = -52.5\,\frac{\text{kJ}}{\text{mol}} \quad (3.31c)$$

Note that the value calculated for ΔG is much more negative (i.e., more favorable) than the standard free energy change $\Delta G^{\circ\prime}$. This last point underscores the fact that it is ΔG and not $\Delta G^{\circ\prime}$ that determines the driving force for a reaction. However, to evaluate ΔG using Equation 3.19, we must be given, or be able to calculate, $\Delta G^{\circ\prime}$ for the reaction of interest. Recall that $\Delta G^{\circ\prime}$ can be calculated from K using Equation 3.22. In the remaining pages of this chapter, we will use examples relevant to biochemistry to illustrate two alternative methods for calculating $\Delta G^{\circ\prime}$.

3.4 Free Energy in Biological Systems

Understanding the central role of free energy changes in determining the favorable directions for chemical reactions is important in the study of biochemistry because every biochemical process (such as protein folding, metabolic reactions, DNA replication, or muscle contraction) must, overall, be a thermodynamically favorable process. Very often, a particular reaction or process that is necessary for life is in itself endergonic. Such intrinsically unfavorable processes can be made thermodynamically favorable by *coupling* them to strongly favorable reactions. Suppose, for example, we have a reaction A → B that is part of an essential pathway but is endergonic under standard conditions:

$$A \rightleftharpoons B \qquad \Delta G^{\circ\prime} = +10 \text{ kJ/mol}$$

C in our previous example) that can undergo reactions with large negative free energy changes. Such substances can be thought of as energy transducers in the cell. Many of these energy-transducing compounds are organic phosphates such as ATP (**FIGURE 3.5**), which can transfer a phosphoryl group ($-PO_3^{2-}$) to an acceptor molecule. You will see many examples of phosphoryl group transfer reactions in this text. As shown in Figure 3.5, we will use a common shorthand notation, Ⓟ, to represent the phosphoryl group when describing these processes.

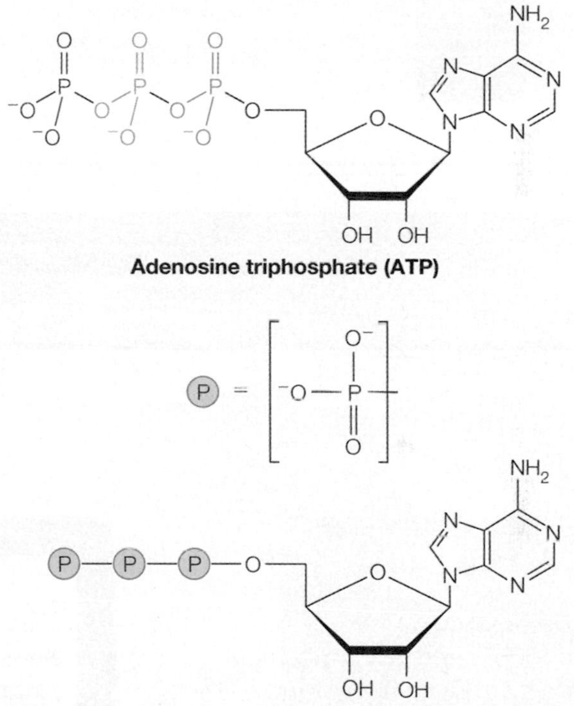
Adenosine triphosphate (ATP)

▲ **FIGURE 3.5 The phosphoryl groups in ATP.** Top: The three phosphoryl groups in ATP are shown in red, blue, and green. Middle: A commonly used shorthand for a phosphoryl group is the symbol Ⓟ. Bottom: The three phosphoryl groups in ATP are represented by this symbol.

NEW! Pearson eText, optimized for mobile, seamlessly integrates videos and other rich media with the text and gives students access to their textbook anytime, anywhere. Pearson eText is available with Mastering Chemistry when packaged with new books, or as an upgrade students can purchase online. The Pearson eText mobile app offers:

- Offline access on most iOS and Android phones/tablets.
- Accessibility (screen-reader ready)
- Configurable reading settings, including resizable type and night reading mode
- Instructor and student note-taking, highlighting, bookmarking, and search tools
- Embedded videos for a more interactive learning experience

Mastering Chemistry now offers double the biochemistry-specific assets

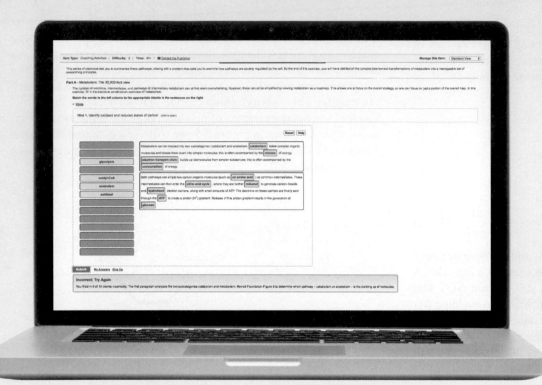

Updated Biochemistry-Specific Tutorial Problems, featuring specific wrong-answer feedback, hints, and a wide variety of educationally effective content guide your students through the most challenging topics.

The hallmark Hints and Feedback offer instruction similar to what students would experience in an office hour, allowing them to learn from their mistakes without being given the answer.

Extended coverage of biochemistry topics and new real-world applications such as non-glucose metabolism have been added to the second edition.

100 NEW! End-of-Chapter Problems from the textbook are assignable within Mastering Chemistry and help students prepare for the types of questions that might appear on an exam. All end-of-chapter problems are automatically graded and include author-written solutions.

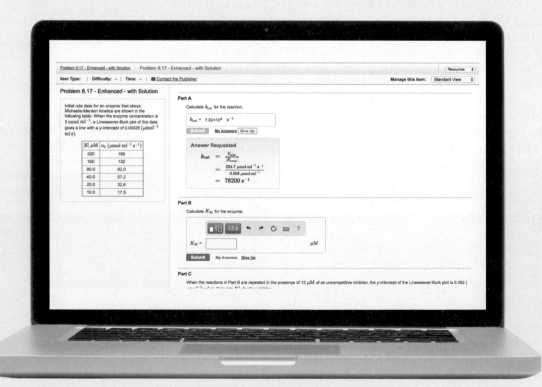

to help students connect course materials to real world applications

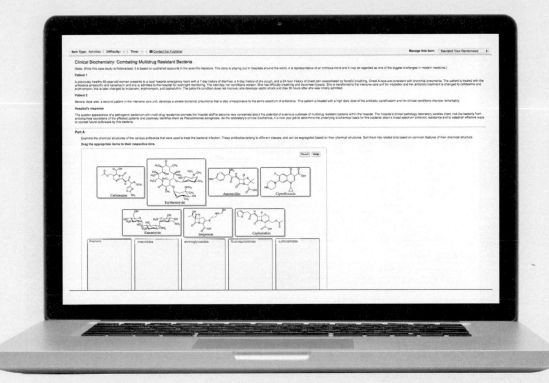

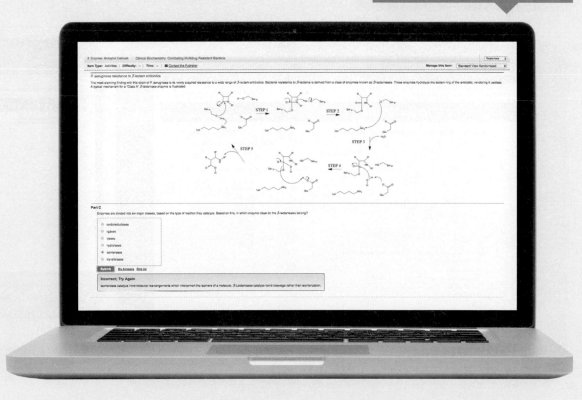

NEW! Interactive Case Studies, assignable in Mastering Chemistry, put students into the role of a biochemist in real-world scenarios, immersing them in topics such as combating multidrug resistant bacteria using Michaelis-Menton enzyme kinetics. Each activity is designed to help students connect the course material to the real world by having them explore actual scientific data from primary literature. Students solve problems that matter to them using a myriad of question types such as multiple choice, drag and drop, and plotting results on graphs.

Students get the threshold concept coverage they need to succeed

NEW! Threshold Concept Tutorials prepare students for success in biochemistry. Much of biochemistry requires foundational knowledge from earlier courses. Unfortunately, many students begin the course either never having truly grasped the important concepts or having forgotten them since they last took the prerequisite.

Based on recent research published in *Biochemistry and Molecular Biology Education*, we have created tutorials and in-class assessment questions for Learning Catalytics to help students assess their own understanding and master these important threshold concepts to prepare them for their biochemistry course.

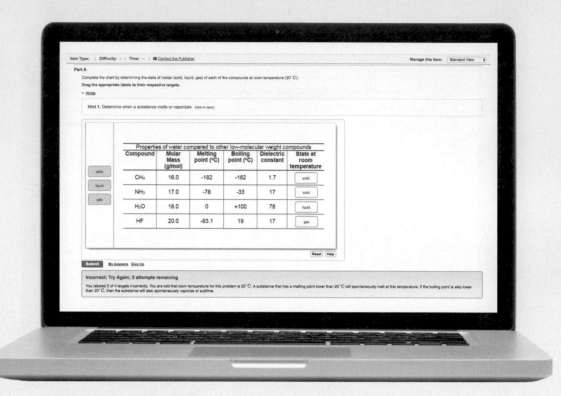

Learning Catalytics™ helps generate class discussion, customize lectures, and promote peer-to-peer learning with real-time analytics.

Learning Catalytics acts as a student response tool that uses students' smartphones, tablets, or laptops to engage them in more interactive tasks and thinking.

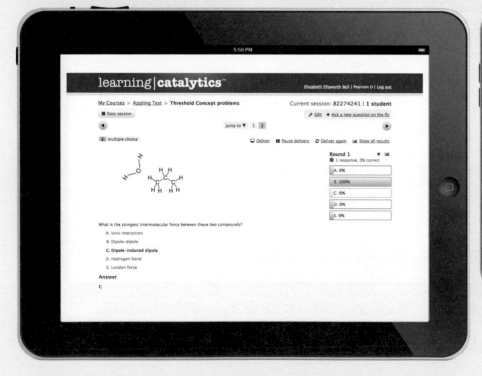

NOW with Biochemistry-specific questions

Biochemistry

CONCEPTS AND CONNECTIONS

SECOND EDITION

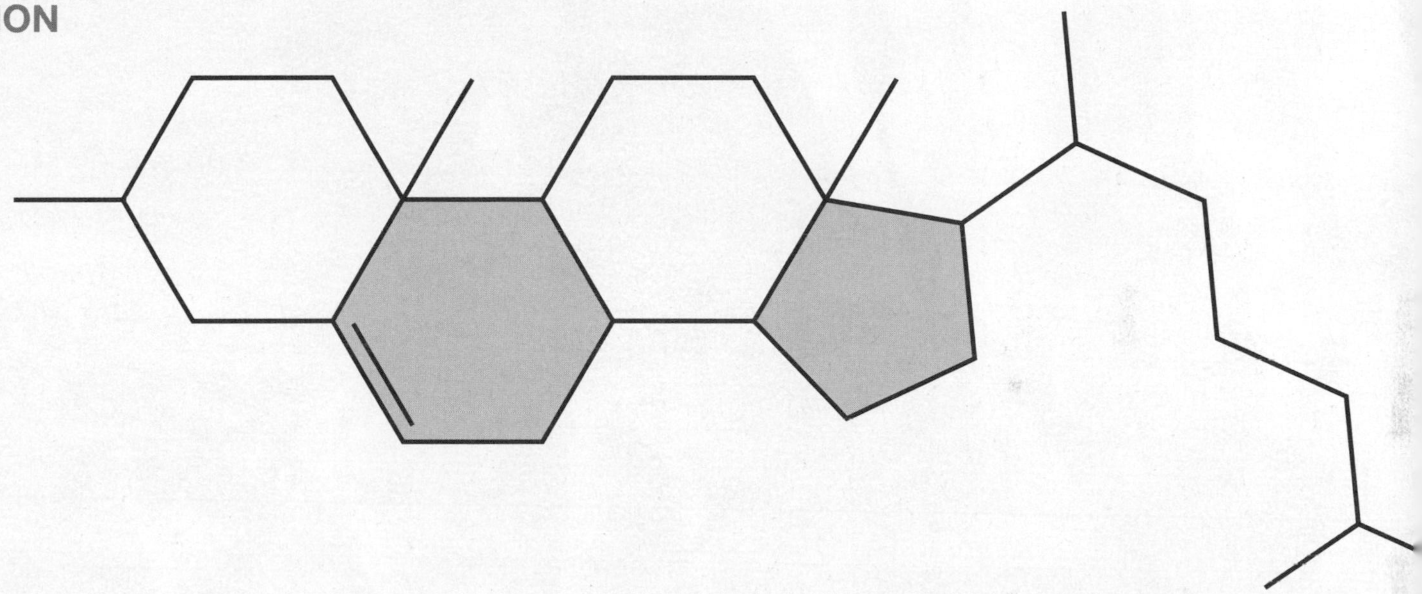

Discovering new medicines requires comprehension of the structure and function of the drug target, whether that be an enzyme, a gene, or a signaling molecule. Success in drug discovery requires deep understanding of biochemistry and its allied disciplines.

Biochemistry and the Language of Chemistry

"**MUCH OF LIFE** can be understood in rational terms if expressed in the language of chemistry. It is an international language, a language for all of time, and a language that explains where we came from, what we are, and where the physical world will allow us to go." These words were written in 1987 by Arthur Kornberg (1918–2007), one of the greatest biochemists of the twentieth century, and they provide a backdrop for our study of biochemistry. Because it seeks to understand the chemical basis for all life processes, biochemistry is at once a biological science and a chemical science. Indeed, all of the traditional disciplines within biology—including physiology, genetics, evolution, and ecology, to name a few—now use the language and techniques of chemistry. Many of you who are using this book are planning careers in life sciences—in teaching, basic research, health sciences, science journalism, drug discovery, environmental science, bioengineering, agriculture, science policy, and more. You will find biochemistry at the heart of all fields within the biological sciences.

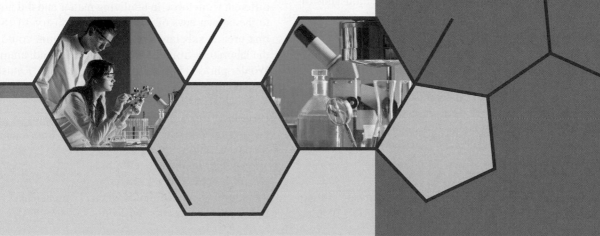

As we proceed through our study of biochemistry, think about "the language of chemistry." To understand a language, we must become familiar with the words and how to incorporate them into sentences. In this text we will be faced with numerous chemical names and structures that must be learned, such as the amino acids in proteins or the sugars in starch or cellulose. These are the *words* in the biochemical language, and learning them will occupy much of the first several chapters of this book. Next, we begin putting these words into *sentences*—chemical reactions—and *paragraphs*—metabolic pathways, which are made up of linked sequences of two or more individual reactions. Reading the sentences and paragraphs will require that we learn about enzymes and catalysis of biochemical reactions. Later we move from paragraphs to pages and chapters, as we explore how metabolic processes in different tissues interrelate to explain, for example, the adaptation of an animal to starvation, or the possible effects of calorie restriction on life-span extension. We will also learn what regulates expression of the biochemical language when we explore chromosomes, genomes, and genes—and how the controlled expression of genes dictates which sentences will be printed and in which cells, and how instructions in the language are transmitted from generation to generation.

As we discuss the biochemical language and its expression, three themes will dominate our discussion—*metabolism, energy,* and *regulation.* What are the chemical reactions? How is metabolic work done? How is expression of the language controlled?

● **CONCEPT** All of the life sciences require an understanding of the language of chemistry.

In order to apply the language of chemistry to learning biochemistry, you will need to recall much of what you learned in organic chemistry—the structures and properties of the principal functional groups, for example. Chapter 2 provides a brief review of the major functional groups, and Chapter 11 describes those reaction mechanisms most directly involved in biochemistry.

Because most of you are learning the biochemical language for the first time, our initial emphasis must be on individual reactions and pathways, operating to some extent in isolation. Be aware, however, that plucking individual reactions out of a cell for investigation is artificial and that a chemical reaction within a cell is but one in a coordinated system of hundreds or thousands of individual reactions, all occurring in the same time and space. In the past two decades, techniques have been developed that allow analysis of a true *systems biology*—chemical reactions as they occur within a complex system rather than in isolation. In time, we will discuss these techniques and what they teach us, but the emphasis in a first course in biochemistry is on elements and expression of the biochemical language.

1.1 The Science of Biochemistry

Humankind has harvested the fruits of biochemistry for thousands of years, perhaps beginning some 8000 years ago with the fermentation of grapes into wine. **FIGURE 1.1** illustrates winemaking as it was carried out in Egypt in about 1500 B.C. However, the science behind winemaking and many other biochemical applications, such as medicinal folk remedies or the tanning of leather, remained obscure until the past three centuries or so, with the birth of biochemistry as a science. With respect to winemaking, see Chapter 12 for a presentation of **glycolysis,** the fundamental process for the breakdown of sugars, which in yeast and other microorganisms converts the sugar to ethanol.

The Origins of Biochemistry

Biochemistry as a science can be said to have originated early in the nineteenth century, with the pioneering work of Friedrich Wöhler (1800–1882) in Germany. Prior to Wöhler's time, it was believed that the substances in living cells and organisms were somehow qualitatively different from those in nonliving matter and did not behave according to the known laws of physics and chemistry. In 1828 Wöhler showed that urea, a substance of biological origin, could be synthesized in the laboratory from the inorganic compound, ammonium cyanate. As Wöhler phrased it in a letter to a colleague, "I must tell you that I can prepare urea without requiring a kidney or an animal, either man or dog." This was a shocking statement in its time, for it breached the presumed barrier between the living and nonliving.

Another landmark in the history of biochemistry occurred in the mid-nineteenth century when the great French chemist Louis

Wöhler's synthesis of urea from ammonium cyanate:

$$\overset{+}{N}H_4NCO^- \longrightarrow H_2N-\overset{\overset{\displaystyle O}{\|}}{C}-NH_2$$

Ammonium cyanate   **Urea**

▲ **FIGURE 1.1 An ancient application of biochemistry.** Manufacture of wine in Egypt, around 1500 B.C.

Pasteur (1822–1895) turned his attention to fermentation in order to help the French wine industry. Pasteur recognized that wine could be spoiled by the accidental introduction of bacteria during the fermentation process and that yeast cells alone possess the ability to convert the sugars in grapes to ethanol in wine. Following this discovery, he devised ways to exclude bacteria from fermentation mixtures.

● **CONCEPT** Early biochemists had to overcome the doctrine of vitalism, which claimed that living matter and nonliving matter were fundamentally different.

Although Pasteur demonstrated that yeast cells in culture could ferment sugar to alcohol, he adhered to the prevailing view known as *vitalism,* which held that biological reactions took place only through the action of a mysterious "life force" rather than physical or chemical processes. In other words, the fermentation of sugar into ethanol could occur only in whole, living cells.

The vitalist dogma was shattered in 1897 when two German brothers, Eduard (1860–1917) and Hans Buchner (1850–1902), found that extracts from broken and thoroughly dead yeast cells could carry out the entire process of fermentation of sugar into ethanol. This discovery opened the door to analysis of biochemical reactions and processes **in vitro** (Latin, "in glass"), meaning in a test tube—or, more generally, outside of a living organism or cell, rather than **in vivo,** in living cells or organisms. In the following decades, other metabolic reactions and reaction pathways were reproduced in vitro, allowing identification of reactants and products and of the biological catalysts, known as **enzymes,** that promoted each biochemical reaction. The name "enzyme," coined in 1878, comes from the Greek *en zyme* (meaning "in yeast"), reflecting the fact that the chemical nature of

these catalysts did not become known until some time later, as described below.

The nature of biological catalysis remained the last refuge of the vitalists, who held that the structures of enzymes were far too complex to be described in chemical terms. But in 1926, James B. Sumner (1887–1955) showed that an enzyme from jack beans, called **urease,** could be crystallized like any organic compound and that it consisted entirely of protein. Although proteins have large and complex structures, they are just organic compounds, and their structures can be determined by the methods of chemistry and physics. This discovery marked the final fall of vitalism.

Although developments in the first half of the twentieth century revealed in broad outline the chemical structures of biological materials, identified the reactions in many metabolic pathways, and localized these reactions within the cell, biochemistry remained an incomplete science. We knew that the uniqueness of an organism is determined by the totality of its chemical reactions. However, we had little understanding of how those reactions are controlled in living tissue or of how the information that regulates those reactions is stored, transmitted when cells divide, and processed when cells differentiate.

What factors determine why yeast cells might ferment sugars to ethanol, while bacteria contaminating a wine culture might convert the sugars to acetic acid and turn the wine culture to vinegar? To answer this question, we must understand expression of **genes,** which control synthesis of the enzymes involved. The idea of the gene, a unit of hereditary information, was first proposed in the mid-nineteenth century by Gregor Mendel (1882–1894), an Austrian monk, from his studies on the genetics of pea plants. By about 1900, cell biologists realized that genes must be found in chromosomes, which are composed of proteins and nucleic acids. Subsequently, the new science of genetics provided increasingly detailed knowledge of patterns of inheritance and development. However, until the mid-twentieth century no one had isolated a gene or determined its chemical composition. Nucleic acids had been recognized as cellular constituents since their discovery in 1869 by Friedrich Miescher (1844–1895). But their chemical structures were poorly understood, and in the early 1900s nucleic acids were thought to be simple substances, fit only for structural roles in the cell. Most biochemists believed that only proteins were sufficiently complex to carry genetic information.

That belief turned out to be incorrect. Experiments in the 1940s and early 1950s proved conclusively that **deoxyribonucleic acid (DNA)** is the primary bearer of genetic information (**ribonucleic acid, RNA,** is also an informational molecule). The year 1953 was a landmark year, when James Watson (1928–) and Francis Crick (1916–2004) described

● **CONCEPT** Biology was transformed in 1953, when Watson and Crick proposed the double-helical model for DNA structure.

the double-helical structure of DNA. This concept immediately suggested ways in which information could be encoded in the structure of molecules and transmitted intact from one generation to the next. The discovery of DNA structure, which we describe more fully in Chapter 4, represents one of the most important scientific developments of the twentieth century (**FIGURE 1.2**).

▲ **FIGURE 1.2** James Watson and Francis Crick with their hand-assembled wire model of the structure of DNA.

Although Watson and Crick made their landmark discovery over six decades ago, the revolution ushered in by that discovery is still underway, as seen by some of the major advances that have occurred since 1953. By the early 1960s, we knew much about the functions of RNA in gene expression, and the genetic code had been deciphered (see Chapters 24 and 25). By the early 1970s, the first recombinant DNA molecules were produced in the laboratory (see Chapter 4), opening the door, as no other discovery had done, to practical applications of biological information in health, agriculture, forensics, and environmental science. By the next decade, scientists had learned how to amplify minute amounts of DNA (see Chapter 21) so that any gene could be isolated by cloning (Chapter 4), allowing any desired change to be made in the structure of a gene. After another decade, by the early 1990s, scientists had learned not only how to introduce new genes into the germ line of plants and animals, but also how to disrupt or delete any gene, allowing analysis of the biochemical function of any gene product (see Chapter 23). A decade later, the nearly complete nucleotide sequence of the human genome was announced—2.9×10^9 base pairs of DNA, representing more than 20,000 different genes. At about the same time came discoveries regarding noncoding properties of RNA, in catalysis and gene regulation (Chapters 7, 25, and 26). The 20-teens saw development of CRISPR (**clustered regularly interspersed short palindromic repeats**) technology, which allowed unprecedented opportunities for editing genes in living organisms (Chapter 23). The wealth of information from genomic sequence analysis and gene regulation by RNA continues to transform the biochemical landscape well into the twenty-first century.

The Tools of Biochemistry

The advances in biochemistry discussed in the previous section and described throughout this book would not have been possible without the development of new technologies for studying biological molecules and processes. Biochemistry is an experimental science—more so, for example, than physics, with its large theoretical component. To understand the key biochemical concepts and processes, we must have some understanding of the experiments that helped us elucidate them. We will describe the experimental basis for much of our understanding of biochemistry in this book. In some cases, the description of experimental techniques will be set apart in end-of-chapter segments called "Tools of Biochemistry."

In the case of DNA structural analysis, the needed technology came from X-ray diffraction. Physicists and chemists had learned that the molecular structures of small crystals could be determined by analyzing patterns showing how X-rays are deflected upon striking atoms in a crystal. Stretched DNA fibers yield comparable data, and these patterns (obtained by Rosalind Franklin, 1920–1958; see Chapter 4), along with the chemical structures of the individual nucleotide units in DNA, led Watson and Crick to their leap of intuition.

FIGURE 1.3 shows a timeline for introduction of methods related to biochemistry beginning at the end of World War II (1945) with the introduction of radioisotopes; these are used to tag biomolecules so that they can be followed through reactions and pathways. Other notable developments include gel electrophoresis (early 1960s), which allows separation and analysis of nucleic acids and proteins. By the early 1970s, restriction enzymes (Chapter 21) had been shown to cut DNA strands at particular sequences in DNA molecules; this finding opened the door to isolating individual genes by recombinant DNA technology. Polymerase chain reaction (Chapter 21) allowed the amplification of selected DNA sequences from minute tissue samples. CRISPR technology (Chapter 23), introduced in 2013, allowed unprecedented opportunities for genome editing in living cells. Throughout this book we will be describing these and other benchmark technologies, and you may wish to refer back to this figure.

● **CONCEPT** Powerful new chemical and physical techniques have accelerated the pace at which biological processes have become understood in molecular terms.

Biochemistry as a Discipline and an Interdisciplinary Science

In trying to define biochemistry, we must consider it both as an interdisciplinary field and as a distinct discipline. Biochemistry shares its major concepts and techniques with many disciplines—with organic chemistry, which describes the properties and reactions of carbon-containing molecules; with physical chemistry, which describes thermodynamics, reaction kinetics, and electrical parameters of oxidation–reduction reactions; with biophysics, which applies the techniques of physics to study the structures of biomolecules; with medical science, which increasingly seeks to understand disease states in molecular terms; with nutrition, which has illuminated metabolism by describing the dietary requirements for maintenance of health; with microbiology, which has shown that single-celled organisms and viruses are ideally suited for the elucidation of many metabolic pathways and regulatory mechanisms; with physiology, which investigates life processes at the tissue and organism levels; with cell biology, which describes the metabolic and mechanical division of labor within a cell; and with genetics, which analyzes mechanisms that give a particular cell or organism its biochemical identity. Biochemistry draws strength from all of these disciplines, and it nourishes them in return; it is truly an interdisciplinary science.

Timeline (left margin)

2015
- Cryo-electron microscopy
- CRISPR-Cas9 technology

2010
- Synthetic biology
- RNA-sequence analysis
- Chromatin immunoprecipitation/ sequencing
- Induced pluripotent cells
- Second generation

2005
- DNA sequence analysis

- Proteomic analysis with mass spectrometry

2000
- Genetic code expansion
- Gene analysis on microchips

- Single-molecule dynamics

1995
- Targeted gene disruption

- In vivo NMR
- Atomic force microscopy

1990
- Scanning tunneling microscopy

1985
- Pulsed field electrophoresis
- Transgenic animals
- Amplification of DNA: polymerase chain reaction
- Automated oligonucleotide synthesis

1980
- Site-directed mutagenesis of cloned genes
- Automated micro-scale protein sequencing
- Rapid DNA sequence determination
- Monoclonal antibodies

1975
- Southern blotting
- Two-dimensional gel electrophoresis
- Gene cloning

1970
- Restriction cleavage mapping of DNA molecules

- Rapid methods for enzyme kinetics

1965

- High-performance liquid chromatography
- Polyacrylamide gel electrophoresis
- Solution hybridization of nucleic acids

1960
- X-ray crystallographic protein structure determination
- Zone sedimentation velocity centrifugation
- Equilibrium gradient centrifugation
- Liquid scintillation counting

1955
- First determination of the amino acid sequence of a protein
- X-ray diffraction of DNA fibers

1950

1945
- Radioisotopic tracers used to elucidate reactions

◀ **FIGURE 1.3 The recent history of biochemistry as shown by the introduction of new research techniques.** The timeline begins with the introduction of radioisotopes as biochemical reagents, immediately following World War II.

You may wonder about the distinction between biochemistry and *molecular biology,* because both fields take as their ultimate aim the complete definition of life in molecular terms. The term *molecular biology* is often used in a narrower sense to denote the study of nucleic acid structure and function and the genetic aspects of biochemistry—an area we might more properly call *molecular genetics* or *genetic biochemistry.*

Regardless of uncertainty in terminology, biochemistry is a distinct discipline, with its own identity. It is distinctive in its emphasis on the structures and reactions of biomolecules, particularly on enzymes and biological catalysis and on the elucidation of metabolic pathways and their control. As you read this book, keep in mind both the uniqueness of biochemistry as a separate discipline and the absolute interdependence of biochemistry and other physical and life sciences.

1.2 The Elements and Molecules of Living Systems

All forms of life, from the smallest bacterial cell to a human being, are constructed from the same chemical elements, which in turn make up the same types of molecules. The chemistry of living systems is similar throughout the biological world; the reactions and pathways that will concern us involve fewer than 200 different molecules. Undoubtedly, this continuity in biochemical processes reflects the common evolutionary ancestry of all cells and organisms. Let us begin to examine the composition of living systems, starting with the chemical elements and then moving to biological molecules.

The Chemical Elements of Cells and Organisms

Life is a phenomenon of the second generation of stars. This rather strange-sounding statement is based on the fact that life, as we conceive it, could come into being only when certain elements—carbon, hydrogen, oxygen, nitrogen, phosphorus, and sulfur (C, H, O, N, P, and S)—were abundant (**FIGURE 1.4**). The primordial universe was made up almost entirely of hydrogen (H) and helium (He), for only these simplest elements were produced in the condensation of matter following the primeval explosion, or "big bang," which we think created the universe. The first generation of stars contained no heavier elements from which to form planets. As these early stars matured over the next seven to eight billion years, they burned their hydrogen and helium in thermonuclear reactions. These reactions produced heavier elements—first carbon, nitrogen, and oxygen, and eventually all the other members of the periodic table. As large stars matured, they became unstable and exploded as novas and supernovas, spreading the heavier elements through the cosmic surroundings. This matter condensed again to form

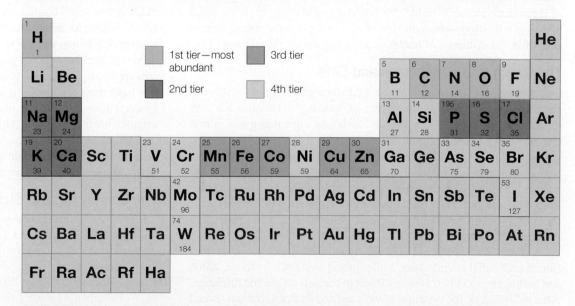

▲ **FIGURE 1.4 Periodic table pertinent to biochemistry.** The four tiers of chemical elements, grouped in order of their abundance in living systems, are highlighted in separate colors.

● **CONCEPT** Life depends primarily on a few elements (C, H, O, N, S, and P), although many others have essential functions as well.

second-generation stars, at least some of which (like our sun) have planetary systems incorporating these heavier elements. Our universe, which is now rich in second-generation stars, has an elemental composition compatible with life as we know it.

Relatively few elements are involved in the creation of living systems. Living creatures on Earth are composed primarily of just four elements—carbon, hydrogen, oxygen, and nitrogen. These are also the most abundant elements in the universe, along with helium and neon. Helium and neon, inert gases, are not equipped for a role in life processes; they do not form stable compounds, and they are readily lost from planetary atmospheres.

The abundance of oxygen and hydrogen in organisms is explained partly by the major role of water in life on Earth. We live in a highly aqueous world, and, as we will see in Chapter 2, the solvent properties of water are indispensable in biochemical processes. The human body, in fact, is about 70% water. The elements C, H, O, and N are important to life because of their strong tendencies to form covalent bonds. In particular, the stability of carbon–carbon bonds and the possibility of forming single, double, or triple bonds give carbon the versatility to be part of an enormous diversity of chemical compounds.

But life is not built on these four elements alone. Many other elements are necessary for organisms on Earth, as you can see in Figure 1.4. A "second tier" of essential elements includes sulfur and phosphorus, which form covalent bonds, and the ions Na^+, K^+, Mg^{2+}, Ca^{2+}, and Cl^-. Sulfur is a constituent of nearly all proteins, and phosphorus plays essential roles in energy metabolism and the structure of nucleic acids. Beyond the first two tiers of elements (which correspond roughly to the most abundant elements in the first two rows of the periodic table), we come to those that play quantitatively minor—but often indispensable—roles. As Figure 1.4 shows, most of these third- and fourth-tier elements are metals, some of which serve as aids to the catalysis of biochemical reactions. In succeeding chapters we shall encounter many examples of the importance of these trace elements to life. Molybdenum, for example, is essential in nitrogen fixation—the reduction of nitrogen gas in the atmosphere to ammonia, for synthesis of nucleic acids and proteins (see Chapter 18).

The Origin of Biomolecules and Cells

Once the chemical elements had formed, during cooling of the second-generation stars, how did the complex molecules that we associate with living systems come into being on Earth? An educated guess is that they arose as part of a "primordial soup" within the oceans. Because the strong oxidant, oxygen, was absent from Earth's atmosphere, scientists hypothesize that a highly reducing environment prevailed within the primordial atmosphere, a condition that tends to promote joining reactions of atoms and molecules. Moreover, high-energy discharges were thought to occur through lightning or volcanic eruptions, providing sufficient energy to drive atoms and small molecules together.

In 1953, Stanley Miller tested this hypothesis by simulating the presumed primordial environment. Miller mixed ammonia, methane, water, and hydrogen in a closed system subject to continuous electric discharge. After several days, the system was analyzed and shown to contain several amino acids, as well as other simple compounds, including carbon monoxide, carbon dioxide, and hydrogen cyanide. Thus, it was established that biological compounds could have been produced abiotically (without living systems). Refinements of the Miller experiment have shown that much more complex organic molecules can also arise under similar conditions.

How we went from the primordial soup, rich in potential biomolecules, to primitive living systems is still a matter of conjecture. Many biochemists believe that the earliest primitive systems, capable of self-replication and some form of metabolism, were based on ribonucleic acid (RNA). RNA is a more versatile molecule than DNA, as we discuss in Chapters 4 and 8, and it is capable of catalyzing chemical reactions as well as storing information. Thus, biochemists speak of an ancient "RNA world," in which simple self-replicating cellular structures, surrounded by crude, lipid-rich membranes, might have existed. Eventually, because DNA is more stable than RNA, this presumed chemical evolution would have led to processes by which RNA or its component nucleotides could give rise to DNA-based life forms.

The earliest living systems would almost certainly have been anaerobic because of the absence of oxygen in the atmosphere. Energy was probably obtained from coupled oxidation–reduction reactions involving inorganic compounds of sulfur and iron. Over time, photosynthetic capability would have arisen, as some organisms evolved the ability to harness light energy from the sun to drive the reduction of inorganic compounds, notably CO_2, to reduced organic compounds. Eventually, organisms would have developed the ability to use water as an electron donor, thereby creating enough oxygen over time to enrich the atmosphere with oxygen. Because much more energy can be derived through complete oxidation of organic compounds than from anaerobic processes (see Chapter 11), aerobic organisms would have had a large evolutionary advantage.

As primitive bacteria underwent the numerous changes leading to characteristic features of eukaryotic cells—condensation of genes into chromosomes, development of intracellular membranous structures—some eukaryotic cells acquired new metabolic capabilities through infection with aerobic bacteria or photosynthetic bacteria. Over time, the intracellular organisms living in this symbiotic relationship underwent their own evolution, eventually becoming what we now recognize as mitochondria and chloroplasts in present-day cells.

How long might this process have taken? Geologists tell us that Earth was formed about 4.6 billion years ago. Rocks containing carbon of likely biological origin have been dated to more than 3.5 billion years ago. Evidence for aerobic bacteria and an oxygen-rich atmosphere dates to about 2.5 billion years ago, with the first eukaryotic microorganisms following about one billion years later. The earliest multicellular eukaryotes are 400 to 500 million years old. Although we understand the forces that have shaped life since it arose—and these will be described as we proceed through our study of biochemistry—our understanding of the origin of life is conjectural. Although the spontaneous generation of self-replicating entities seems highly improbable, the enormous amount of time during which this could have occurred changes the almost impossible to highly likely, and perhaps inevitable.

The Complexity and Size of Biological Molecules

The complexity of life processes requires that many of the molecules governing them be enormous. Consider, for instance, the DNA molecules released from one human chromosome, as shown in **FIGURE 1.5**. The long, looped thread you see corresponds to a small part of a huge molecule, with a molecular mass of about 20 billion daltons. (A dalton, Da, is 1/12 the mass of a carbon-12 atom, 1.66×10^{-24} g.) Even a simple organism such as the single-celled bacterium *Escherichia coli* contains a DNA molecule with a molecular mass of about 2 billion Da—more than one millimeter long. Protein molecules are generally much smaller than DNA molecules,

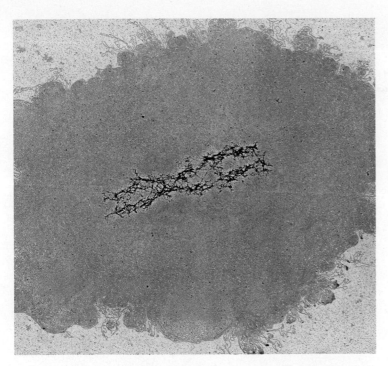

▲ **FIGURE 1.5 Part of the DNA from a single human chromosome.** Most of the chromosomal proteins have been removed in this color-enhanced electron micrograph, leaving only a protein "skeleton" from which enormous loops of DNA emerge.

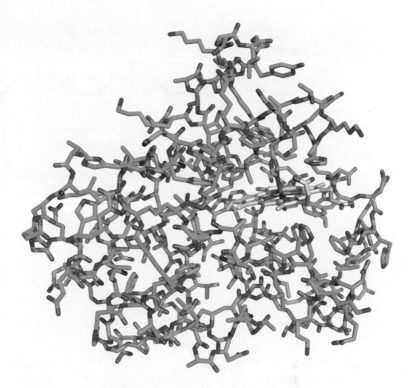

▲ **FIGURE 1.6 The three-dimensional structure of myoglobin.** This computer-generated stick model portrays sperm whale myoglobin, the first protein whose structure was deduced by X-ray diffraction. It depicts, therefore, our first indication of the complexity and specificity of the three-dimensional structure of proteins. PDB ID: 1mbn.

but they are still large, with molecular masses ranging from about 10,000 to one million Da. The complexity of these molecules is seen from the three-dimensional structure of even a fairly small protein. **FIGURE 1.6** illustrates the structure of myoglobin, an oxygen-carrying protein of muscle, which has a molecular mass of about 17,000 Da.

Biological **macromolecules** are giant molecules made up of smaller organic molecule subunits. In living organisms, there are four major classes of macromolecules, all essential to the structure and function of cells: proteins, nucleic acids, carbohydrates, and lipids. As we shall see throughout this text, there are good reasons for some biological materials to be so large. DNA molecules, for example, can be thought of as tapes from which genetic information is read out in a linear fashion. Because the amount of information needed to specify the structure of a multicellular organism is enormous, these tapes must be extremely long. In fact, if the DNA molecules in a single human cell were stretched end to end, they would reach a length of about 2 meters. As revealed in the early twenty-first century through the Human Genome Project, the information encoded in this DNA is sufficient to encode about 100,000 proteins, although the actual number of genes is far smaller.

The Biopolymers: Proteins, Nucleic Acids, and Carbohydrates

The synthesis of such large molecules poses an interesting challenge to the cell. If the cell functioned like an organic chemist carrying out a complex laboratory synthesis bit by bit, millions of different types of reactions would be involved, and thousands of intermediates would accumulate. Instead, cells use a modular approach for constructing large polymeric molecules. These **biopolymers** are made by joining together prefabricated units, or **monomers.** Of the four classes of macromolecules, three of them are biopolymers: proteins, nucleic acids, and carbohydrates. Lipids, the fourth class of macromolecule, are not considered polymers and are discussed in the next section.

The monomers of a given type of macromolecule are of limited diversity and are linked together, or **polymerized,** by identical mechanisms. Each process involves **condensation,** or removal of a molecule of water in the joining reaction. A simple example is the carbohydrate **cellulose** (**FIGURE 1.7(a)**), a major constituent of the cell walls of plants. Cellulose is a polymer made by joining thousands of molecules of glucose, a simple sugar. In this polymer, all of the chemical linkages between the monomers are identical. Covalent links between glucose units are formed by removing a water molecule between two adjoining glucose molecules; the portion of each glucose molecule remaining in the chain is called a glucose **residue.** Because cellulose is a polymer of a simple sugar, or **saccharide,** it is called a **polysaccharide.** This particular polymer is constructed from identical monomeric units, so it is called a **homopolymer.** In contrast, many polysaccharides—and all nucleic acids and proteins—are **heteropolymers,** polymers constructed from a number of different kinds of monomer units.

Nucleic acids (Figure 1.7(b)) are polymers made up of four **nucleotides,** so nucleic acids are also called **polynucleotides.** Similarly, proteins (Figure 1.7(c)) are assembled from combinations of 20 different **amino acids.** Protein chains are called **polypeptides,** a term derived from the **peptide bond** that joins two amino acids together.

● **CONCEPT** Cells use a modular approach for constructing large molecules.

Polymers form much of the structural and functional machinery of the cell. Polysaccharides serve both as structural components, such as cellulose, and as reserves of biological energy, such as **starch,** another type of glucose polymer found in plants. The nucleic acids, DNA and RNA,

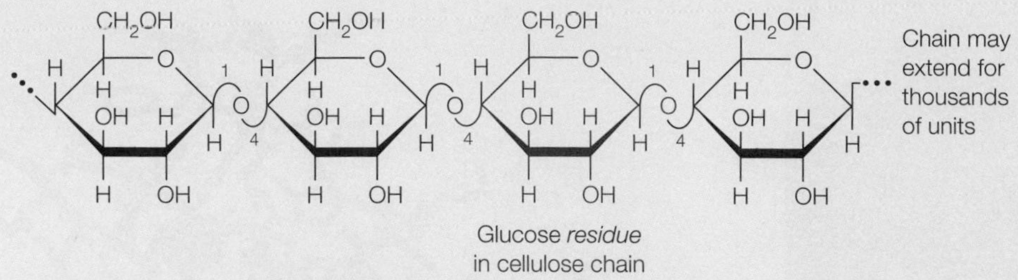

Cellulose, a *polymer* of β-D-glucose

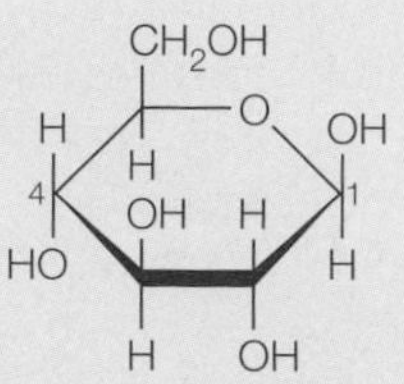

β-D-glucose, the *monomer*

(a) A carbohydrate. The carbohydrate cellulose is a polymer of β-D-glucose monomers, with a molecule of water split out in each joining reaction.

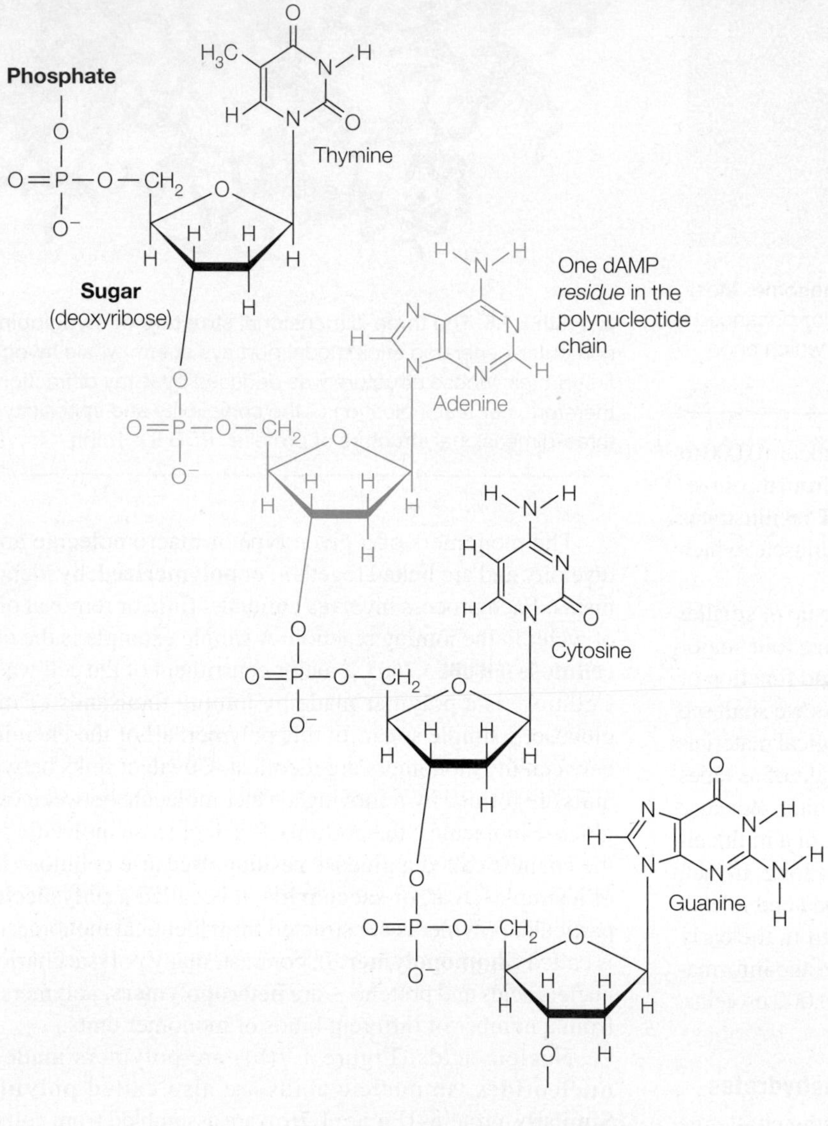

Part of deoxyribonucleic acid (DNA), a polynucleotide

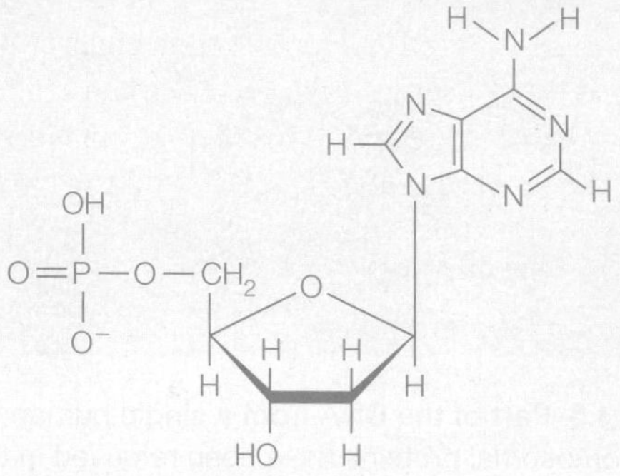

Deoxyadenosine monophosphate (dAMP), one of the four kinds of monomers that make up DNA

(b) A nucleic acid. The nucleic acids, DNA and RNA, are polymers of nucleotides. Part of a DNA molecule is shown, along with one of its monomers, deoxyadenosine monophosphate.

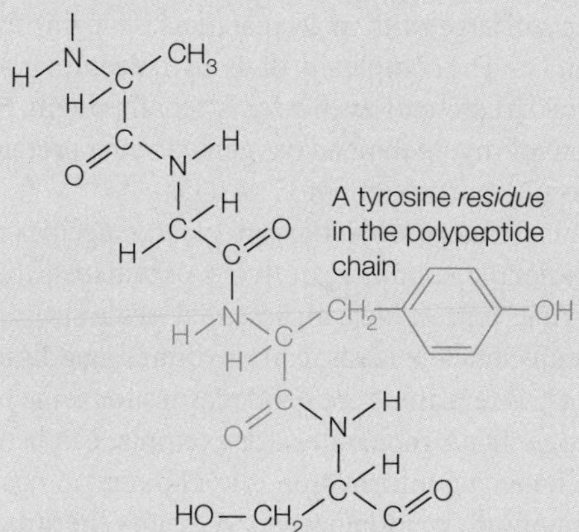

Part of a polypeptide chain in a protein

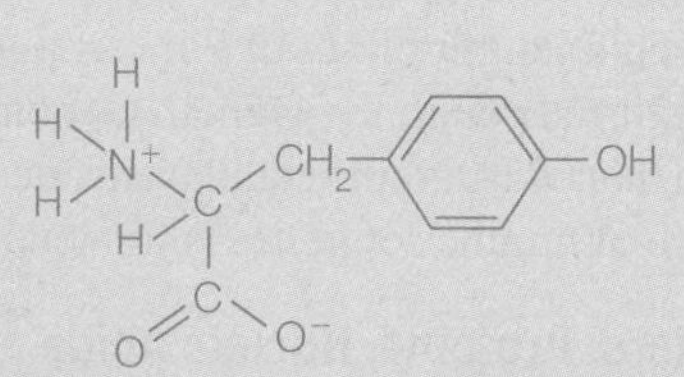

Tyrosine, one of the 20 kinds of monomers that make up polypeptides

(c) A polypeptide. Protein chains, or polypeptides, are polymers assembled from 20 different amino acids. Part of a polypeptide is shown, along with one of its monomers, tyrosine.

▲ FIGURE 1.7 Examples of biological polymers, or biopolymers.

participate in information storage, transmission, and expression. DNA serves principally as a storehouse of genetic information, while the chemically similar RNA is involved in the readout of information stored in DNA.

Proteins, which have far more structural diversity than polysaccharides or nucleic acids, perform a more diverse set of biological functions. Some, such as keratin in hair and skin or collagen in connective tissue, have structural roles in cells. Other proteins act as transport agents, an outstanding example being hemoglobin, the oxygen-carrying protein of blood. Proteins may transmit information between distant parts of an organism, as do protein **hormones** and the cell surface **receptors** that receive the hormone signals, or they may defend an organism against infection, as do the **antibodies.** Most important of all, proteins function as **enzymes,** catalysts for the thousands of chemical reactions that occur within an individual cell. For example, the enzyme **RNA polymerase** catalyzes the joining of nucleotide molecules to synthesize RNA (see Chapter 24). RNA itself is more versatile than suggested by its role as an information carrier; certain RNA molecules, of which the most prominent is the ribosome, act as catalysts, just as do proteins.

Lipids and Membranes

In addition to the biopolymer macromolecules, there is one other major class of cellular constituents—the **lipids;** these exist in large complexes but are not themselves polymeric. Lipids are

● **CONCEPT** Lipids form the major constituents of biological membranes.

complexes but are not themselves polymeric. Lipids are a chemically diverse group of compounds that are classified together because of their hydrocarbon-rich structures, which give them low solubility in the aqueous environment of the cell. This low solubility equips lipids for one of their most important functions—to serve as the major structural element of the **membranes** that surround cells and partition them into various compartments.

FIGURE 1.8 illustrates the chemical diversity of lipid molecules. **Triacylglycerols** (Figure 1.8(a)), formerly called triglycerides, are esters, formed by condensation between the three hydroxyl groups of glycerol and three long-chain organic acids (fatty acids). As discussed in Chapter 16, triacylglycerols store large amounts of energy in their carbon chains and represent the primary energy storehouse, as fat, in most organisms. In many **phospholipids,** one of the fatty acid chains is replaced by a small phosphorus-containing molecule, which has both positive and negative charges. In the example shown in Figure 1.8(b), that molecule is the amino acid **serine** in ester linkage with phosphate. Phospholipids like the one shown are the principal components of membranes.

Shown in Figure 1.8(c) is another type of lipid, called **cholesterol.** Cholesterol is an example of a **sterol,** with a characteristic four-ring structure. Cholesterol is an important constituent of membranes and is also the metabolic precursor in vertebrates to all of the steroid hormones.

Phospholipids, such as the phosphatidylserine shown in Figure 1.8(b), have a distinctive property that enables them to provide the foundation for biological membranes. Such molecules are **amphipathic,** meaning that they have both **hydrophobic** ("water-fearing") and **hydrophilic** ("water-loving") functional groups. The long hydrocarbon chains on the fatty acyl moieties are insoluble in water and tend to cluster together, while the charged portion with the phosphoryl moiety interacts readily with water. When placed in water, phospholipids spontaneously form a **bilayer,** a two-layered structure in which the fatty acid chains associate with one another in the interior, away from the aqueous surroundings, while the charged and hydrophilic phosphate-containing components, also called *polar head groups,* cluster on the outside, in contact with the

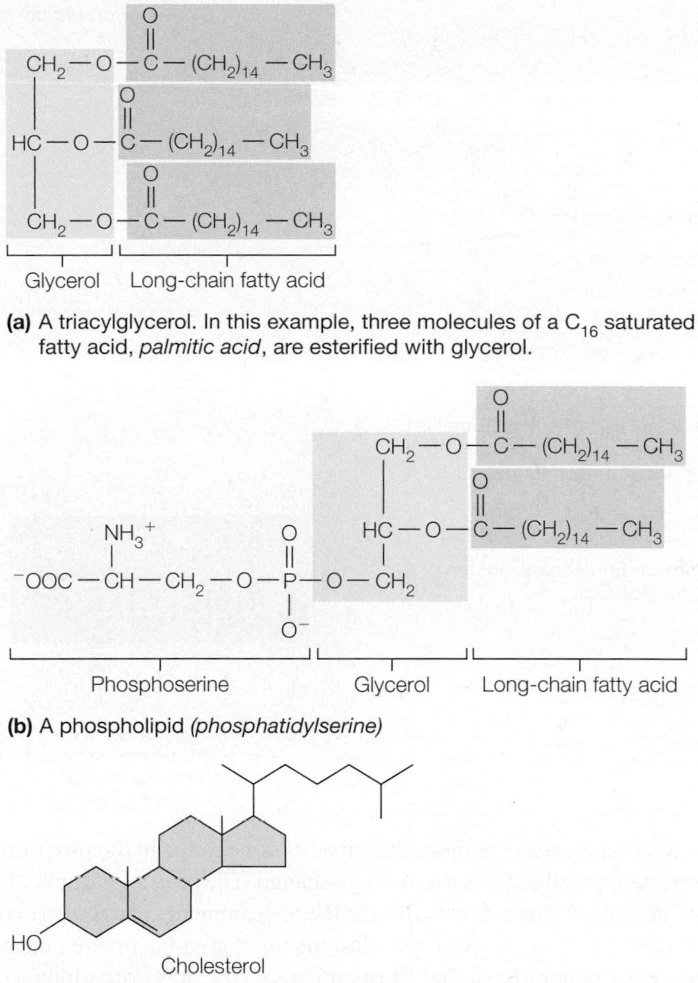

(a) A triacylglycerol. In this example, three molecules of a C_{16} saturated fatty acid, *palmitic acid*, are esterified with glycerol.

(b) A phospholipid *(phosphatidylserine)*

(c) *Cholesterol*, a sterol

▲ **FIGURE 1.8** Structures of typical lipids.

watery milieu. This is the basic structure of cell membranes, which exist largely to separate intracellular from extracellular environments, helping to maintain an internal environment compatible with life. However, the simple lipid bilayer shown in **FIGURE 1.9** is impermeable to nearly all substances, and a critical function of membranes is to control what enters cells and what exits. To this end, phospholipid bilayers in biological membranes are studded with proteins, as discussed in Chapter 10, which serve as conduits controlling the entry and exit of molecules and ions according to the cell's needs.

1.3 Distinguishing Characteristics of Living Systems

We have learned that living cells and organisms are made from the same chemical elements as those found throughout the universe. Yet we know that life and living systems are qualitatively different from nonliving matter. How, then, does one distinguish living from nonliving matter?

Daniel Koshland (1920–2007), another distinguished biochemist, has described seven distinctive "pillars of life"—the essential attributes of all living things. The first pillar is a *program,* or organized plan for constitution and reproduction of an organism. For life on Earth, that program is the information stored in the genome.

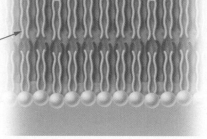

Hydrophobic tail

Polar head group

serine

A phospholipid bilayer

Charged polar head groups lie on the exterior, in contact with the aqueous environment.

Hydrocarbon chains cluster in the interior.

◄ **FIGURE 1.9 Formation of a phospholipid bilayer.** The two hydrocarbon chains on each phospholipid molecule cluster in the interior, while the charged polar head groups lie on the exterior, in contact with the aqueous environment. Although the figure shows a two-dimensional structure, the bilayer is a sheet, which extends in three dimensions. Here, phosphate in the polar head group is linked to the amino acid serine. Substituents other than serine may be attached, as discussed in Chapter 10.

Second is *improvisation,* the capacity for change in the program to promote survival as the surroundings change. The processes of mutation and natural selection ensure that, as the environment changes (aqueous to terrestrial, for example), populations undergo adjustments, usually over many generations, that fit organisms to the new conditions.

The third pillar is *compartmentalization,* the ability of an organism to separate itself from the environment (with membranes, for example), so that the chemistry needed to carry out the program, such as enzyme-catalyzed reactions, can occur under favorable conditions of temperature, pH, and concentrations of reactants and products. The smallest organisms contain just one compartment, the interior of a single cell, while larger organisms contain many cells, specialized to carry out different functions, such as sensory perception or movement. In addition, the cells of multicellular organisms contain subcompartments, the organelles that allow for division of labor within the basic functional unit, the cell itself.

The fourth pillar is *energy.* Thermodynamics tells us that spontaneous processes occur in the direction of simplicity and randomness. Yet living systems must create complexity and order to sustain the program and the other pillars of life. To do this, cells and organisms carry out reactions that yield energy, such as the breakdown of nutrients, and they couple some of that energy to drive energy-requiring reactions that create complexity, such as the synthesis of nucleic acids and proteins or the transmission of nerve impulses. The ultimate source of all energy is the sun, which is used either directly, by photosynthetic plants and bacteria to drive the synthesis of biomolecules, or indirectly, by organisms that consume other organisms and derive their energy from the breakdown of dietary nutrients. An overarching goal of biochemistry, which permeates this book, is to understand the mechanisms by which energy-yielding reactions are coupled in living

● **CONCEPT** Life depends on creating and duplicating order in a chaotic environment. This ordering requires energy.

systems to the energy-consuming reactions needed to maintain the complexity and order required for life to be sustained. We will learn about the energetics of life in Chapter 3, early on in the text because so much of the biochemistry that follows depends on these key concepts.

The fifth pillar is *regeneration,* the ability to compensate for the inevitable wear involved in maintaining a physical state far from equilibrium. For example, all of the proteins in a cell are subjected to continuous degradation, either because they suffer environmental damage or because, like digestive enzymes, they undergo degradation as part of their normal function. The ability to continuously replace damaged molecules of this type is a distinguishing characteristic of life.

The sixth pillar is *adaptability,* the capacity of an organism to respond to environmental changes. For example, when nutrient stores within an animal are depleted, the animal becomes hungry and seeks food. Adaptability, a property of *individual organisms,* is distinguished from improvisation (the second pillar), which is the capacity of *populations* of organisms to respond to environmental change over a time scale of many generations.

The seventh and final pillar is *seclusion,* which means that metabolic processes and pathways must operate in isolation from one another, even though they may take place within the same compartment of a cell. When we digest carbohydrates, a consequence is a rise in intracellular glucose concentration. In liver or muscle, the glucose thus formed can either be consumed to provide energy or be polymerized into glycogen, a glucose polymer, which is stored for later release when there is an energy demand. Glucose polymerization and the initial steps in breakdown occur within the same cell compartment. Yet intracellular control processes and the specificities of the enzymatic catalysts involved ensure that one of the pathways is favored and the other inhibited, in response to the cell's needs.

By these criteria, are viruses living? When isolated in the laboratory, viruses are inert complexes of nucleic acid and protein; some can be crystallized and maintained for years in a dormant state. Viruses can replicate only by invading a living cell and taking over its metabolic machinery. Virus-infected cells display all of the distinctive attributes of living cells and organisms, but virus particles do not.

As we can see, maintaining an orderly intracellular environment in the midst of a chaotic universe is critical to life. The functioning of semipermeable membranes, as discussed in the previous section, plays an essential role. For example, we know that in most cells the concentration of potassium is higher than in the extracellular fluid, while the concentration of sodium is lower. In many cells, the intracellular Na^+ and K^+ concentrations may not change over time. It is important that this **steady state,** or **homeostasis,** not be confused with equilibrium because a continuous input of energy is required to maintain these ion concentrations at different values from the extracellular conditions. If membranes lost their ability to couple energy generation with ion transport, intra- and extracellular concentrations of ions would soon reach equivalence. We will learn about membranes, and the transport of ions and molecules across them, in Chapter 10.

1.4 The Unit of Biological Organization: The Cell

A seminal early discovery in biology was Robert Hooke's (1635–1703) observation in 1665 that plant tissues (in this case, cork) were divided into tiny compartments, which he called *cellulae,* or **cells.** By 1840, improved observations of many tissues led Theodor Schwann (1810–1882) to propose that all organisms exist as either single cells or aggregates of cells. That hypothesis is now firmly established.

Because cells are the universal units of life, let us examine them more closely. Major differences between cell structures define the three great classes of organisms—bacterial, archaeal, and eukaryotic. The **true bacteria** (also called **eubacteria**) and an ancient group of organisms called **archaea** are always unicellular, while **eukaryotes** may be uni- or multicellular.

● **CONCEPT** All living creatures are composed of cells. Most cells are similar in size.

Whereas organisms were originally classified in terms of morphological or structural criteria, current classification is based on biochemical analysis, primarily DNA nucleotide sequence determination. **FIGURE 1.10** shows the three great branches of life as an evolutionary "tree," with relationships based on similarities in ribosomal RNA sequences, which in turn are based on analyses of the DNA sequences that encode these RNAs. The smaller the distance between any two branches, the more closely related are the organisms in those two branches. Note the close relationship between eubacteria and the genomes of mitochondria and chloroplasts. This provides the principal evidence supporting the evolution of these organelles from primitive bacteria, as mentioned on page 8 and discussed in Chapter 14.

A typical prokaryotic organism is shown schematically in **FIGURE 1.11.** Prokaryotic cells are surrounded by a plasma membrane and usually by a rigid outer cell wall as well. Within the membrane is the **cytoplasm,** which contains the **cytosol**—a semiliquid concentrated solution or gel—and the structures suspended within it. In prokaryotes, the cytoplasm is not divided into compartments, and the genetic information is in the form of one or more DNA molecules that exist free in the cytosol. Also suspended in the cytosol are the **ribosomes,** which constitute the molecular machinery for protein synthesis. The surface of a prokaryotic cell may carry **pili,** which aid in attaching the organism to other cells or surfaces, and **flagella,** which enable it to swim.

● **CONCEPT** The two great classes of organisms have different cell types. Prokaryotic cells are uncompartmentalized; eukaryotes have membrane-bound organelles.

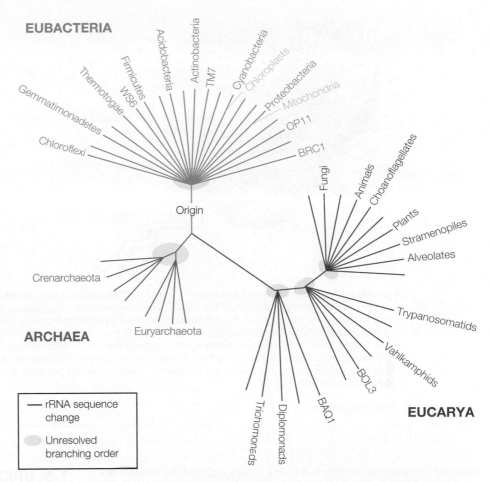

▲ **FIGURE 1.10** Molecular tree of life, based on ribosomal RNA sequence comparisons. Not all lines of descent are shown. The branch points represent common evolutionary origins, and the distance between two branches represents genetic relatedness. Some lineages represent "environmental sequences," that is, sequences detected in the environment from which the relevant organisms have not yet been isolated in culture.

Eukaryotes include the multicellular plants and animals, as well as the unicellular and simple multicellular organisms called protozoans, fungi, and algae. Most eukaryotic cells are larger (by 10- to 20-fold) than bacterial cells, but they compensate for their large size by being **compartmentalized.** Their specialized functions are carried out in **organelles**—membrane-surrounded structures lying within the surrounding cytoplasm.

Schematic views of idealized animal and plant cells are shown in **FIGURE 1.12.** Major organelles common to most eukaryotic cells are the **mitochondria,** which specialize in oxidative metabolism; the **endoplasmic reticulum,** a folded membrane structure rich in ribosomes, where much protein synthesis occurs; the **Golgi complex,** membrane-bound chambers that function in secretion and the transport of newly synthesized proteins to their destinations; and the **nucleus.** The nucleus of a eukaryotic cell contains the cell's genetic information, encoded in DNA that is packaged into **chromosomes.** A portion of this DNA is subpackaged into a dense region within the nucleus called the **nucleolus.** Surrounding the nucleus is a **nuclear envelope,** pierced by pores through which the nucleus and cytoplasm communicate.

There are also organelles specific to plant or animal cells. For example, animal cells contain digestive bodies called **lysosomes,** which are not found in plants. Plant cells have **chloroplasts,** the sites of photosynthesis, and usually a large, water-filled **vacuole,** whose functions may

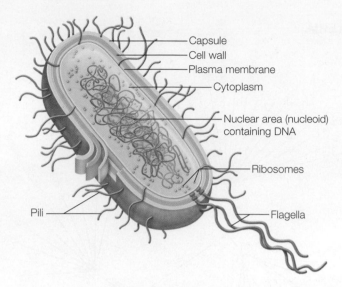

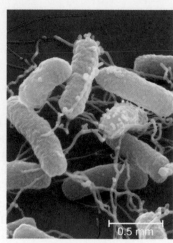

(a) Schematic view of a representative bacterial cell. The DNA molecule that constitutes most of the genetic material is coiled up in a region called the nucleoid, which shares the fluid interior of the cell (the cytoplasm) with ribosomes (which synthesize proteins), other particles, and a large variety of dissolved molecules. The cell is bounded by a plasma membrane, outside of which is usually a fairly rigid cell wall. Many bacteria also have a gelatinous outer capsule. Projecting from the surface may be pili, which attach the cell to other cells or surfaces, and one or more flagellae, which enable the cell to swim through a liquid environment.

(b) Scanning electron micrograph of *Salmonella*: rod-shaped Gram-negative enterobacteria that causes typhoid fever, paratyphoid fever, and food-borne illness.

▲ **FIGURE 1.11** Prokaryotic cells.

include intracellular secretion, excretion, storage, and digestion. Furthermore, whereas most animal cells are surrounded only by a **plasma membrane,** plant cells often have a tough **cell wall** outside the membrane. **Basal bodies** act as anchors for cilia or flagella in animal cells that have those appendages.

● **CONCEPT** The cell can be thought of as a factory, with organelles and compartments specialized to perform different functions.

Having seen the variety of substructures within even a simple bacterial cell, how can we think about how they function in concert to do the work of the cell? It is useful to think of the cell as a factory, an analogy we shall frequently use in later chapters. Membranes enclose the whole structure and separate different organelles, which can be thought of as departments with specialized functions. The nucleus, for example, is the central administration. It contains in its DNA a library of information for cellular structures and processes, and it issues instructions for proper regulation of the business of the cell. The chloroplasts and mitochondria are power generators (the former being solar, the latter fuel-burning). The cytoplasm is the general work area, where protein machinery (enzymes) carries out the formation of new molecules from imported raw materials. Special molecular channels in the membranes between compartments and between the cell and its surroundings monitor the flow of molecules in appropriate directions.

Like factories, cells tend to specialize in function; for example, many of the cells of higher organisms are largely devoted to the production and export of one or a few molecular products. Examples are pancreatic cells, which secrete digestive enzymes, and white blood cells, each of which is specialized to synthesize just one of the several million different antibody molecules that can be produced by humans as part of the immune response.

1.5 Biochemistry and the Information Explosion

● **CONCEPT** Much of today's biochemistry looks at the cell globally, attempting to understand its functions in terms of the expression of all the genes in the genome.

To a great extent, our understanding of biochemistry has been assembled piecemeal, through analysis of single molecules and individual reactions. To understand a metabolic pathway such as the citric acid cycle, the major nutrient oxidation pathway (Chapter 13), we must identify all of the intermediate molecules in the pathway, isolate each of the enzymes that catalyzes a reaction in the pathway (there are eight), and characterize each reaction by identifying substrates, products, stoichiometry, and the means by which it is regulated. Ultimately, we would like to learn how each enzyme works, by determining the atomic structure of the enzyme and learning the molecular mechanism of catalysis.

At the same time, we realize that all of the known metabolic pathways described in this book represent but a small fraction of the total potential of a cell or organism to carry out chemical reactions in support of its existence and function. As mentioned earlier, the amount of DNA in the 23 pairs of human chromosomes was originally estimated to encode 100,000 different proteins, although the number of actual protein-coding genes is about 21,000. The ability of scientists to carry out ultra-large-scale DNA sequence analysis has generated biological data on a scale that far exceeds our capacity to integrate it through analysis of individual reactions and pathways. This has given rise to new fields of science that can be considered areas within biochemistry or molecular biology—*bioinformatics, genomics, proteomics,* and *metabolomics.*

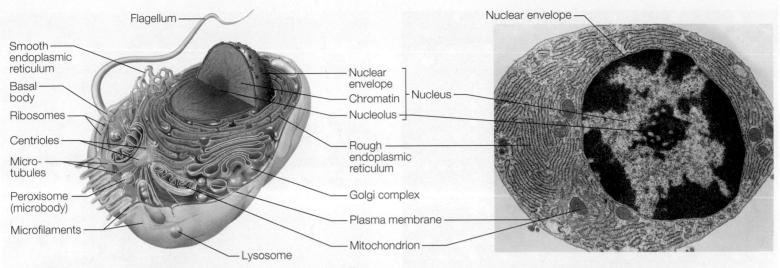

(a) **Typical animal cell.** The accompanying photograph is an electron micrograph of a representative animal cell, a white blood cell.

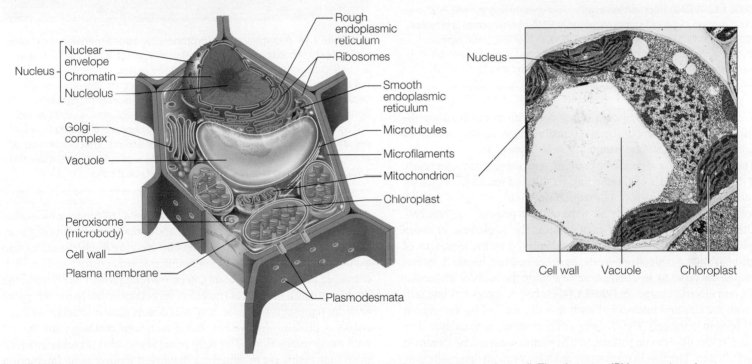

(b) **Typical plant cell.** The accompanying photograph is an electron micrograph of a representative plant cell, Timothy-grass (*Phleum pratense*).

▲ **FIGURE 1.12** Structures of eukaryotic cells.

Bioinformatics can be considered as information science applied to biology. Bioinformatics includes such processes as mathematical analysis of DNA sequence data, computer simulation of metabolic pathways, and analysis of potential drug targets (enzymes or receptors) for structure-based design of new drugs. Each of the "-omics" fields identified below depends heavily on the computational power of bioinformatics.

Whereas genetics concerns itself with the location, expression, and function of individual genes or small groups of genes, **genomics** concerns itself with the entire genome, or the totality of genetic information in an organism. This includes not only determining the nucleotide sequence of the whole genome, but also assessing the expression and function of each gene, as well as evolutionary relationships among genes in the same genome and with genomes of different organisms. The use of "gene chips," or microarrays, allows the collection of enormous amounts of data

concerning genes expressed in a given cell or tissue under a particular set of conditions. For this analysis (see Tools of Biochemistry 24A) minute amounts of thousands of DNA fragments, representing many or most of the genes of an organism, are immobilized in an array on a glass slide. The slide is then incubated with a preparation of total RNA from a cell, under conditions that enable a particular RNA molecule to bind by base-pairing to the corresponding DNA sequences from the gene specifying that RNA. The use of fluorescent reagents to monitor binding allows one to generate a snapshot of the global, or genome-wide, pattern of gene expression within a cell. In the experiment shown in **FIGURE 1.13**, two messenger RNA populations were compared, one from normal cells and one from related tumor cells. One RNA sample was labeled with a red fluorescent dye, the other with green. The color of the fluorescence of each dot represents the relative amount of each individual RNA species binding to the gene

▲ **FIGURE 1.13** A DNA microarray. In this experiment using the MICROMAX detection system, 2 μg of human messenger RNAs from normal and tumor cells were annealed with immobilized cDNAs representing 2400 human genes. Spot intensities and colors indicate the abundance of particular gene-specific mRNAs (for further information, see text and Chapter 24).

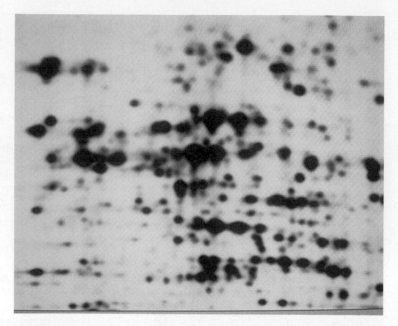

▲ **FIGURE 1.14** A display of the proteome by two-dimensional gel electrophoresis. An extract containing all of the proteins of a cell or tissue is applied to one corner of a slab of gel-like material, and the proteins are resolved along one edge of the gel by *isoelectric focusing,* which separates proteins on the basis of their relative acidity (horizontally in the figure). The gel is rotated by 90 degrees, and resolution is next carried out by denaturing polyacrylamide gel electrophoresis (Chapter 2), which separates proteins on the basis of molecular weight. This is followed by staining with a sensitive silver reagent. In the example shown, more than 1500 proteins were detected in an extract of yeast cells.

represented by the DNA fragment at that dot. With this technology, the experimenter can ask how oncogenic transformation, as shown here, or hormonal stimulation, affects patterns of gene expression.

The products of most genes are proteins, and proteins carry out nearly all of the chemistry of a cell, including catalysis of reactions, signaling between cells, and movement of material both into and out of cells. A goal of **proteomics** is to identify all of the proteins present in a given cell, and the amount and function of each protein. The **proteome,** or entire complement of proteins in a cell, can be displayed by the technique of two-dimensional gel electrophoresis, as described in Chapter 2. In this technique, the proteins in a cell are separated on the basis of molecular weight and electric charge. In **FIGURE 1.14**, each spot represents one protein, with the size and intensity of each spot determined by the amount of that protein in the cell. The challenge of proteomics is to analyze data such as this electrophoretic pattern, so that each protein can be identified and quantitated. Then, as in microarray analysis, one can determine how the proteome changes, for example, in the transformation of a normal cell to a cancer cell, to help scientists understand the chemistry of cancer.

Many of the proteins in a cell are enzymes, and the intracellular rates of enzyme-catalyzed reactions can be estimated by analysis of the intracellular concentrations of the substrates and products of each reaction. A goal of **metabolomics** is to determine the intracellular concentrations of all small molecules, or metabolites, that serve as intermediates in metabolic pathways. Metabolite analysis, of course, is at the heart of much of clinical diagnosis. One example is the measurement of cholesterol in blood to assess risk of cardiovascular disease; another is measurement of glucose levels in the blood to diagnose and monitor diabetes. At present, no single technique would allow complete analysis of the metabolome, or total complement of low-molecular-weight metabolites, but that represents an important goal of much of contemporary biochemistry (see Tools of Biochemistry 11A). Mass spectrometry (Chapter 6) is an approach that has the potential to describe the complete metabolome, but formidable challenges remain.

FIGURE 1.15 illustrates the power of "big data" to generate biological insight. The yeast genome contains about 6000 genes. Investigators examined gene product interactions by creating double mutants and asking whether the gene product interaction was *positive* (cell growth enhanced in the double mutant), *negative* (cell growth inhibited in the double mutant), or *neutral* (no effect on cell growth). In the left panel each dot represents one gene, and lines denote gene interactions. Genes encoding proteins involved in central metabolic pathways are "hubs," with many interactions. The right panel shows that proteins involved in similar pathways or functions displayed similar gene interaction patterns. The data generate a "wiring diagram of cell function," which may prove useful in identifying potential drug targets.

The enzyme and pathway biochemistry that forms the basis of this book has merged with systems biology and bioinformatics to create another new field—**synthetic biology.** The aim of synthetic biologists is to create or modify living systems that can be harnessed to achieve useful goals, such as the production of medicines or biofuels. The potential for this new field was recognized in 2010 when scientists at the J. Craig Venter Institute used chemical synthesis to produce the genome of a known bacterium. Microinjection of this synthetic genome into a bacterial cell from which the native chromosome had been removed led to growth of the recipient cell, under control of the injected genome.

● **CONNECTION** Although metabolomics is a new science, the analysis of individual metabolites in biological material is used routinely to diagnose disease and follow response to treatment.

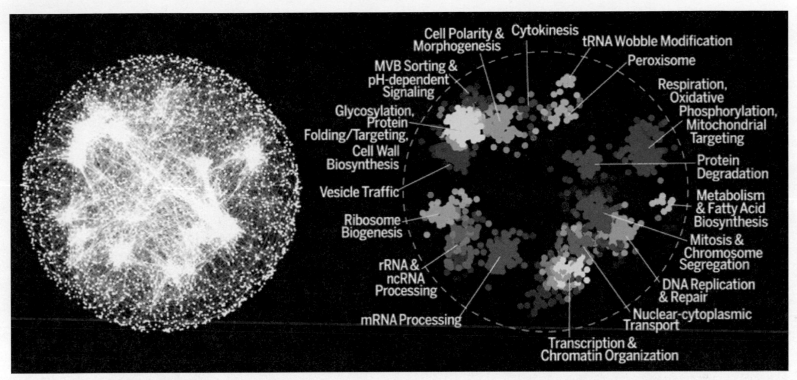

▲ **FIGURE 1.15 The genetic landscape of a yeast cell.** Clustering of genes controlling common functions was established by large-scale analysis of gene interactions. See text for details. From M. Costanza and 53 coauthors (2016) *Science* **353**, 1381.

Bioinformatics and the allied fields have the potential to yield enormous insight into the functions of cells and organisms. However, effective use of that information requires understanding the three themes that we introduced earlier in this chapter—*metabolism,* the individual reactions and pathways in cells, and the functions of those pathways; *information,* the instructions encoded in molecules that assemble and drive metabolic pathways; and *regulation,* the interactions between cells and their environment that modulate flux rates through individual reaction pathways so that processes within a cell are coordinated to fit the functioning of an entire organism.

Summary

- The aim of biochemistry is to explain life in molecular terms because living systems and nonliving matter obey the same laws of physics and chemistry. Modern biochemistry draws on knowledge from chemistry, cell biology, and genetics and uses techniques borrowed from physics. (Section 1.1)

- Although biochemistry deals with organisms, cells, and cellular components, it is fundamentally a chemical science. The chemistry involved is that of carbon, hydrogen, oxygen, nitrogen, phosphorus, and sulfur, but organisms use many other elements in smaller quantities. Many of the important biological substances are giant molecules that are polymers of simpler monomer units. Such **biopolymers** include **polysaccharides, proteins,** and **nucleic acids. Lipids** form the fourth major group of biologically important substances, and they play major roles in energy storage and the formation of membranes. (Section 1.2)

- Distinguishing features of living cells and organisms include the use of energy to create and duplicate orderly structure and information encoded in molecules. (Section 1.3)

- All living organisms are composed of one or more cells, and these cells are similar in size. However, the three great classes of organisms, **eubacteria, archaea,** and **eukaryotes,** have fundamentally different cellular structures: Bacterial and archaeal cells are uncompartmentalized, lacking the membrane-bound **organelles** characteristic of eukaryotic cells. (Section 1.4)

- Biochemistry is an experimental science, and the remarkable recent advances in biochemistry are due in large part to the development of powerful new laboratory techniques. Some of these techniques are generating significant data far more rapidly than they can be integrated and understood without recourse to new approaches in information technology. (Section 1.5)

References

For a list of references related to this chapter, see Appendix II.

As a consequence of noncovalent hydrogen-bonding interactions between water molecules, water possesses remarkable physical and chemical properties. How does hydrogen bonding establish water as the ideal solvent for the biochemistry that is critical to life on Earth?

The Chemical Foundation of Life: Weak Interactions in an Aqueous Environment

2

IN CHAPTER 1, we argued that a deep understanding of biochemistry is based on a strong foundation in chemical principles. In this chapter and the next, we review and expand on a number of key concepts—drawn from several areas of introductory chemistry—that are crucial to the explanation of biochemical phenomena. We focus particularly on bonding interactions and bond polarity, the chemistry of acids, bases and aqueous buffers, and thermodynamics. These topics are of enormous practical relevance to the study of living systems. Throughout this book you'll encounter many examples of biochemical phenomena that illustrate the close relationship between fundamental chemical principles and human physiology. One such example is depicted in **FIGURE 2.1.** The stimulation of muscle tissue growth in response to human growth hormone (hGH) occurs as a result of forming several weak, but highly specific, bonding interactions between the hGH molecule and

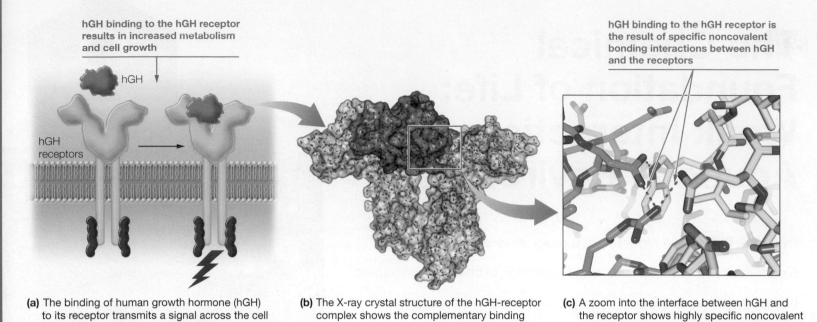

(a) The binding of human growth hormone (hGH) to its receptor transmits a signal across the cell membrane which, in turn, stimulates cell growth (indicated by lightning bolt).

(b) The X-ray crystal structure of the hGH-receptor complex shows the complementary binding surfaces (interfaces) between these proteins.

(c) A zoom into the interface between hGH and the receptor shows highly specific noncovalent bonding interactions (magenta-colored dashed lines). Here, O atoms are colored red, N atoms are blue, and C atoms are either green (on hGH) or yellow (on the receptor).

▲ **FIGURE 2.1** Noncovalent bonding interactions between human growth hormone and its cellular receptor.

the growth hormone receptor protein found on the muscle cell surface.

The macromolecules responsible for cellular structure and function are made up of many thousands of atoms, held together by strong, covalent bonds. Yet covalent bonding alone cannot begin to describe the complexity of molecular structure in biology. Much weaker interactions are responsible for most of the elegant architecture visible in the electron micrographs of bacterial, animal, and plant cells seen in Chapter 1. These intermolecular forces are the noncovalent interactions, also called *noncovalent bonds,* between ions, molecules, and parts of molecules.

We begin this chapter with a review of the major types of noncovalent bonding interactions found in biomolecules such as DNA and proteins. Then we consider the role of water as the solvent in which these interactions occur within cells. Finally, we discuss the need for organisms to tightly control the pH in tissues, the use of buffers to maintain a constant pH, and the relationship between pH and the surface charges of biomolecules.

● **CONCEPT** Noncovalent interactions are critically important determinants of biomolecular structure, stability, and function.

2.1 The Importance of Noncovalent Interactions in Biochemistry

Consider the macromolecules we discussed in Chapter 1. The linear sequence of the nucleotide residues in a strand of DNA is maintained by covalent bonds. But the familiar double helix of DNA is a highly specific three-dimensional structure, which is stabilized by noncovalent interactions between different parts of the molecule. Similarly, every kind of protein is made up of amino acids linked by covalent *peptide bonds;* but it is also folded into a specific molecular conformation that is stabilized by numerous noncovalent interactions (for example, see the structure of sperm whale myoglobin in Figure 1.6). Proteins interact with other protein molecules or with DNA to form still higher levels of organization (for example, the interaction between hGH and its receptor, shown in Figure 2.1), ultimately leading to cells, tissues, and whole organisms. All of this complexity is accounted for by myriad noncovalent interactions within and between macromolecules.

What makes noncovalent interactions so important in biology and biochemistry? A clue is provided by **FIGURE 2.2,** which compares noncovalent and covalent bond energies. The covalent bonds most important in biology (such as C — C and C — H) have bond energies in the range of 150–400 kJ/mol. Biologically important noncovalent bonds are roughly 10 to 100 times weaker. It is their very weakness that makes noncovalent bonds so essential, for it allows them to be

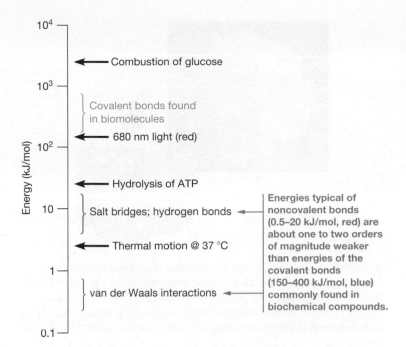

▲ FIGURE 2.2 Covalent and noncovalent bond energies. The energies available from thermal motion, ATP hydrolysis, red light, and aerobic glucose metabolism (all discussed in detail in later chapters) are also shown as reference points. Note that the energy values are plotted on a log scale.

continually broken and re-formed in the dynamic molecular interplay that is life. This interplay depends on rapid exchanges of molecular partners, which could not occur if intermolecular forces were so strong as to lock the molecules in conformation and in place.

The various noncovalent interactions described in the following pages are individually weak, but when many are present within a given macromolecule, or between macromolecules, their energies can sum to a total that is often several hundreds of kilojoules. This amount of energy is sufficient to provide stability to macromolecular structures such as DNA, proteins, and membranes. At the same time, the ease with which *individual* noncovalent bonds can be broken and re-formed gives these structures a dynamic flexibility necessary to their function.

If we are to understand life at the molecular level, we must understand the critical role of noncovalent interactions within and between biomolecules. Furthermore, we must understand how such interactions occur in an aqueous environment, for every cell in every organism on Earth is bathed in and permeated by water. Water is the major constituent of organisms—70% or more of the total weight, in most cases.

2.2 The Nature of Noncovalent Interactions

In this section, we describe the variety of ways in which molecules and ions can interact noncovalently. All of these noncovalent interactions, summarized in **FIGURE 2.3,** are fundamentally *electrostatic;* that is, they all depend on the forces that electrical charges exert on one another.

Type of Interaction	Model	Example	Dependence of Energy on Distance
(a) Charge–charge			$1/r$
(b) Charge–dipole			$1/r^2$
(c) Dipole–dipole			$1/r^3$
(d) Charge–induced dipole			$1/r^4$
(e) Dipole–induced dipole			$1/r^5$
(f) Dispersion (van der Waals)			$1/r^6$
(g) Hydrogen bond	Donor Acceptor		Bond length is fixed

◄ FIGURE 2.3 Types of noncovalent interactions. Each of these interactions is explained in greater detail in the text. The symbols δ^-, δ^+ denote a fraction of an electron or proton charge. The variation in the dependence of bond energy on distance predicts that charge–charge interactions are stronger over much longer distances than are van der Waals interactions.

TABLE 2.1 Energies of some noncovalent interactions in biomolecules

Type of Interaction	Approximate Energy (kJ/mol)
Charge–charge	13 to 17
Hydrogen bond	2 to 21
van der Waals	0.4 to 0.8

Source: Data from S. K. Burley and G. A. Petsko, Weakly polar interactions in proteins, *Advances in Protein Chemistry* (1988) 39:125–189.

TABLE 2.1 lists ranges for the energies of some of the noncovalent interactions prevalent in biomolecules.

Charge–Charge Interactions

The simplest noncovalent interaction is the electrostatic interaction between a pair of ions. Such charge–charge interactions are also referred to as ionic bonds or salt bridges. Many of the molecules in cells, including macromolecules such as DNA and proteins, carry a net electrical charge. In addition to these molecules, the cell contains an abundance of small ions, both cations like Na^+, K^+, and Mg^{2+}, and anions like Cl^- and $HPO_4{}^{2-}$. All of these charged entities exert forces on one another (see Figure 2.3(a)). The predicted force F between a pair of charges q_1 and q_2, separated *in a vacuum* by a distance r, is given by Coulomb's law:

> **CONCEPT** Noncovalent interactions are fundamentally electrostatic and can be described by Coulomb's law.

$$F = k\frac{q_1 q_2}{r^2} \qquad (2.1)$$

where k is a constant whose value depends on the units used.* If q_1 and q_2 have the same sign, F is positive, so a positive value corresponds to repulsion. If one charge is positive and the other is negative, then F is negative, signifying attraction. It is such charge–charge interactions that stabilize a crystal of a salt, like that shown in **FIGURE 2.4**.

The biological environment, of course, is not a vacuum. In a cell, charges are always surrounded, or separated, by water or by other molecules or parts of molecules. The existence of this medium between charges has the effect of screening them from one another, so that the actual force between two charges is always less than that given by Equation 2.1. This screening effect is expressed by inserting a dimensionless number, the *relative permittivity,* or dielectric constant (ε), into Equation 2.1, giving

$$F = k\frac{q_1 q_2}{\varepsilon r^2} \qquad (2.2)$$

Every substance that acts as a dielectric medium has a characteristic value of ε. Equation 2.2 predicts that the higher the value of ε, the weaker the force between the separated charges. The dielectric constant of water is high, approximately 74 at 37 °C, whereas organic liquids usually have much lower values, in the range of 1 to 10. We shall see

*In this book we use the SI, or international, system of units. Here q_1 and q_2 are in coulombs (C), r is in meters (m), and $k = 1/(4\pi\varepsilon_0)$. The quantity ε_0 is the *permittivity of a vacuum* and has the value $8.85 \ 10^{-12} \ J^{-1} C^2 m^{-1}$, where J is the energy unit joules. F is in newtons (N).

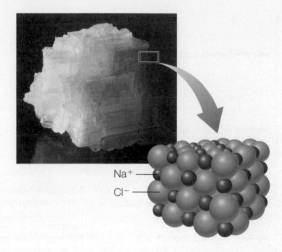

▲ **FIGURE 2.4 Charge–charge interactions in an ionic crystal.** Ionic crystals are held together by charge–charge interactions between positive and negative ions. In a sodium chloride crystal, each positively charged sodium ion is surrounded by six negatively charged chloride ions, and each negatively charged chloride ion is surrounded by six positively charged sodium ions.

presently the reason for the high value of ε in water, but its major consequence is that charged particles interact rather weakly with one another in an aqueous environment unless they are very close together, that is, within 4 to 10 Å (note that 1 Å is 10^{-10} m and that a typical cell has dimensions of $\sim 5 \times 10^6$ m, or 50,000 Å).

Coulomb's law is an expression of *force;* however, every bond formation or cleavage process involves a change in *energy.* Such changes in energy drive all the biochemical processes described in this textbook. Thus, when we consider changes in noncovalent bonding interactions, we are particularly interested in the energy of interaction (E). This is the energy required to separate two charged particles from a distance r to an infinite distance—in other words, to pull them apart by countering the attractive electrostatic force between them. The energy of interaction is given by Equation 2.3, which is similar in form to Equation 2.2:

$$E = k\frac{q_1 q_2}{\varepsilon r} \qquad (2.3)$$

As with force, the energy of an oppositely charged pair q_1 and q_2 is always negative, signifying attraction, but E approaches zero as r (the distance separating the two particles) becomes very large. For charge–charge interactions, the energy of interaction is inversely proportional to the first power of $r;$ so these interactions are relatively strong over greater distances compared to the other noncovalent interactions listed in Figure 2.3. Charge–charge interactions often occur within or between biomolecules—for example, in the attraction between amino and carboxylate groups, as shown in Figure 2.3(a). Charge–charge interactions are also frequently important determinants of binding specificity between proteins and their binding targets, such as between hGH and its receptor (as shown in Figure 2.1) or between an antibody and a bacterial or viral pathogen (discussed in more detail in Chapter 7).

As an illustration of the major concepts summarized in Equation 2.3, let us consider two example problems.

Example Problem 1: The ionic radii of Ca^{2+}, K^+, and Cl^- are, respectively, 1.14 Å, 1.52 Å, and 1.67 Å. Are the ionic bonds stronger in a crystal of KCl or $CaCl_2$?

To solve this problem, we will assume that the ions in each crystal are contacting one another (as suggested in Figure 2.4), and $\varepsilon = 1$. Recall that k is a constant; thus, the terms in Equation 2.3 that we must consider are the ionic charges (q_1 and q_2) and the distance between the ions in the crystal (r). For KCl q_1 = charge on K^+, q_2 = charge on Cl^-, and r = distance separating the K^+ and Cl^-. For $CaCl_2$ q_1 = charge on Ca^{2+}, q_2 = charge on Cl^-, and r = distance separating the Ca^{2+} and Cl^-. The product $q_1 \times q_2$ in the numerator of Equation 2.3 will be a factor of 2 greater for $CaCl_2$ versus KCl; thus, based on differences in ionic charge alone, we would predict that E will be greater for $CaCl_2$. However, we must also consider the effect of charge separation. For KCl r = 1.52 Å + 1.67 Å = 3.19 Å, and for $CaCl_2$ r = 1.14 Å + 1.67 Å, = 2.81 Å. Equation 2.3 predicts E increases as r decreases; thus, E is predicted to be greater for $CaCl_2$ based on charge separation. In this case both factors, r and q, predict a stronger ionic bond in the $CaCl_2$ crystal.

Example Problem 2: Suppose a Ca^{2+} ion and a Cl^- ion are separated in solution by 6 Å. Will the energy of interaction between the ions be greater when the intervening medium is water (with $\varepsilon = 74$), or ethanol (with $\varepsilon = 24$)?

In this case, the only parameter in Equation 2.3 that is changing is ε. Equation 2.3 predicts that as ε increases, the magnitude of E decreases; thus, the energy required to separate the ions will be ~3.1 times greater in ethanol since $1/(24)$ is ~3.1 times greater than $1/(74)$.

The preceding examples have focused on simple cases of ion–ion interactions; however, these same basic concepts apply to the interactions between more complex biomolecules. We will see several applications of Equation 2.3 when we consider protein structures (Chapter 6) and enzyme function (Chapter 8).

Dipole and Induced Dipole Interactions

Molecules that carry no *net* charge may nevertheless have an asymmetric internal distribution of charge. Such a molecule is described as **polar** and is said to have a **dipole moment (μ)**. The dipole moment expresses the magnitude of a molecule's polarity. In a linear molecule with fractional charges δ^+ and δ^-, the dipole moment is a vector directed along the polar bond. We represent the dipole moment as an arrow pointing from δ^+ toward δ^-.[†]

In molecules with a more complex shape, like water, the dipole moment for the entire molecule is a vector sum of the dipole moments along each polar bond (**FIGURE 2.5**). Water has a significant μ because electrons are drawn from the hydrogen atoms toward the oxygen atom, owing to the much greater *electronegativity* of the oxygen atom (electronegativity is the tendency of an atom in a covalent bond to attract electrons to itself). Molecules with large dipole moments are said to be highly polar.

● **CONCEPT** Some molecules interact because they are polar and possess dipole moments.

[†]Chemists use the convention described above. In the Physics convention, the orientation of the arrow is opposite: from δ^- to δ^+.

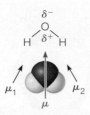

Water: the partial negative charge on O together with the partial positive charge on each H produces two dipole moments, μ_1 and μ_2, directed along the O — H bonds. Their vector sum (μ, shown in blue) represents the net dipole moment of the molecule.

▲ **FIGURE 2.5** The molecular dipole moment of a water molecule.

In the aqueous environment of a cell, a polar molecule can be attracted by a nearby ion (a *charge–dipole interaction*) or by another polar molecule (a *dipole–dipole interaction*). These **dipolar interactions** are depicted in Figure 2.3(b) and 2.3(c). Unlike the simple charge–charge interactions described earlier, the energies of dipolar interactions depend on the relative orientation of the dipoles. Furthermore, they are shorter-range interactions: The energy of a charge–dipole interaction is proportional to $1/r^2$ and that of a dipole–dipole interaction to $1/r^3$. Thus, a pair of polar molecules must be quite close together before the dipolar interaction becomes strong.

Molecules that do not have permanent dipole moments can become polar in the presence of an electric field. This field may be produced by a neighboring charged or polar particle. A molecule in which a dipole can be so induced is said to be **polarizable.** Aromatic rings, for example, are very polarizable, because the electrons can easily be displaced in the plane of the ring, as shown in **FIGURE 2.6(a).** Interactions of polarizable molecules are called **induced dipole interactions.** An anion or a cation may induce a dipole in a polarizable molecule and then be attracted to it (a *charge-induced dipole interaction*, Figure 2.3(d)), or a permanent dipole may do the same (a *dipole-induced dipole interaction*, Figure 2.3(e)). These induced dipole interactions are even shorter in range than permanent dipole interactions, with energies of interaction proportional to $1/r^4$ and $1/r^5$, respectively.

Even two molecules that have neither a net charge nor a permanent dipole moment can attract one another if they are close enough (Figure 2.3(f)). The distribution of electronic charge in a molecule is never static but fluctuates. When two molecules approach very closely, their charge fluctuations tend to localize an area of partial positive charge on one molecule next to an area of partial negative charge on the neighboring molecule, producing a net attractive force. Such intermolecular forces, which can be thought of as mutual dipole induction, are called **van der Waals interactions,** or **dispersion forces.** Their attractive energy varies as $1/r^6$, so van der Waals interactions are significant only at very short range. They can become particularly strong when two planar molecules stack on one another, as shown in Figure 2.6(b) and 2.6(c). We shall encounter many examples of such interactions in the internal packing of molecules such as proteins and nucleic acids. Note that van der Waals interactions are individually weak, yet collectively make significant contributions to the stability of biomolecules, as will be discussed in Chapter 6.

Van der Waals Interactions

When molecules or atoms that do not have covalent bonds between them come so close together that their outer electron orbitals begin to overlap, there is a mutual repulsion. This energy of this repulsion increases very

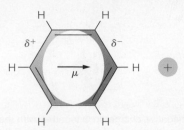

(a) Benzene has neither a net charge nor a permanent dipole moment, but a nearby charge can induce a redistribution of electrons within the benzene ring, producing an induced dipole moment (arrow).

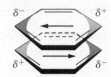

(b) Planar molecules like benzene have a strong tendency to stack because fluctuations in the electron clouds of the stacked rings give rise to mutually attractive induced dipoles (van der Waals interactions).

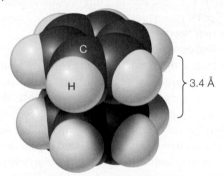

3.4 Å

(c) Although the molecules approach closely, they do not interpenetrate.

▲ **FIGURE 2.6** Induced dipoles and van der Waals interactions.

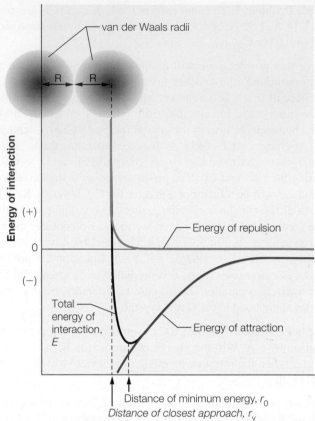

▲ **FIGURE 2.7 Noncovalent interaction energy of two approaching particles.** The interaction energy of two atoms, molecules, or ions is graphed versus the distance between their centers, r. The total interaction energy (E, black curve) at any distance is the sum of the energy of attraction and the energy of repulsion. As the distance between the particles decreases (reading *right to left* along the x-axis), both the attractive energy (<0, red curve) and the repulsive energy (>0, blue curve) increase in magnitude but at different rates. At first the longer-range attraction dominates, but then the repulsive energy increases so rapidly that it acts as a barrier, defining the distance of closest approach (r_v) and the van der Waals radii (R, described by the orange spheres). The position of minimum energy (r_0) is usually very close to (r_v).

rapidly as the distance between their centers (r) decreases; it can be approximated as proportional to r^{-12}. If we combine this repulsive energy with one or more of the kinds of attractive energy described previously, we see that the total energy of noncovalent interaction (E) of a pair of atoms, molecules, or ions will vary with distance of their separation (r) in the manner depicted in **FIGURE 2.7**. Two points on the graph should be noted. First, there is a minimum in the energy curve, at position r_0. This minimum corresponds to the most stable distance between the centers of the two interacting particles. If we allow them to approach each other, this is how close they will come. Second, the repulsive potential rises so steeply at shorter distances that it acts as a "wall," effectively barring an approach closer than the distance r_v. This distance defines the so-called **van der Waals radius, R,** the effective radius for closest molecular packing. For a pair of identical spherical molecules, $r_v = 2R$; for molecules with van der Waals radii R_1 and R_2, $r_v = R_1 + R_2$.

● **CONCEPT** Molecules may attract one another by noncovalent forces but cannot interpenetrate; van der Waals radii determine molecular surfaces.

Real molecules, of course, are not spherical objects like those depicted in Figure 2.7. Because large biological molecules all have complicated shapes, it is useful to extend the concept of the van der Waals radius to atoms or groups of atoms within a molecule. The values for van der Waals radii given in **TABLE 2.2** represent the distances of closest approach for another atom or group. When we depict complex molecules in a so-called space-filling manner, we represent each atom by a sphere with its appropriate van der Waals radius. For example, Figure 2.6(c) shows a space-filling model of two stacked benzene rings. In this case, the van der Waals radii of the carbon atoms (1.7 Å) dictate that the planes of the two stacked benzene rings cannot be closer than 3.4 Å. As we shall see in Chapter 4, this 3.4 Å spacing is also seen between adjacent aromatic bases in double-stranded DNA and RNA.

Hydrogen Bonds

One specific kind of noncovalent interaction, the **hydrogen bond,** is of the greatest importance in biochemistry. The structure and properties of many biological molecules, including water, are

TABLE 2.2 van der Waals radii of some atoms and groups of atoms

	R (Å)
Atoms	
H	1.2
O	1.4
N	1.5
C	1.7
S	1.8
P	1.9
Groups	
—OH	1.4
—NH$_2$	1.5
—CH$_2$—	2.0
—CH$_3$	2.0
Half-thickness of aromatic ring	1.7

determined largely by this type of bond. A hydrogen bond is an interaction between a hydrogen atom covalently bonded to another atom (e.g., —O—H or ≡N—H) and a pair of nonbonded electrons on a separate atom (in biomolecules, this is most often an O or a N atom). Hydrogen bonding is shown in Figure 2.3(g) and **FIGURE 2.8.** The atom to which the hydrogen is covalently bonded is called the *hydrogen-bond donor,* and the atom with the nonbonded electron pair is called the *hydrogen-bond acceptor.* The typical representation of the hydrogen-bonding interaction is a dotted line between the acceptor atom and the shared hydrogen.

The ability of an atom to function as a hydrogen-bond donor depends on its electronegativity. The more electronegative the donor atom, the more electron density it withdraws from the hydrogen to which it is bonded. Thus, the hydrogen gains partial positive charge character and is more strongly attracted to the electron pair of the acceptor. Among the atoms most prevalent in biological compounds, only oxygen and nitrogen are sufficiently electronegative to serve as strong donors. Thus, ≡C—H groups do not form strong hydrogen bonds, but —O—H groups do.

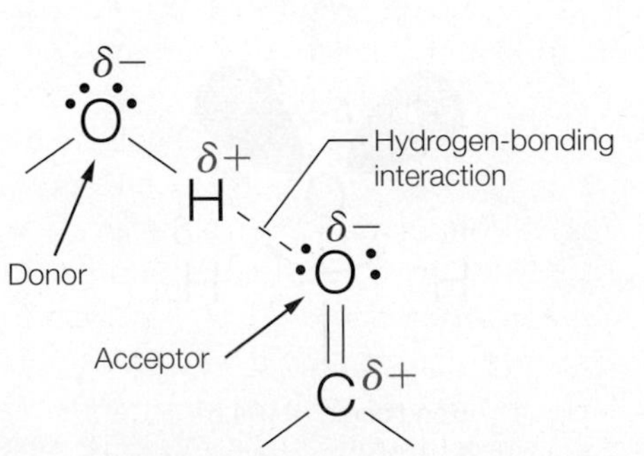

▲ **FIGURE 2.8 The hydrogen bond.** The figure shows an idealized hydrogen bond that might exist, for example, between an alcohol (the donor) and a keto compound (the acceptor). The hydrogen-bonding interaction is indicated by a dotted line between the H and the acceptor atom.

The hydrogen bond has features in common with both covalent and noncovalent interactions. In part, it is like a charge–charge interaction between the partial positive charge on the hydrogen atom and the negative charge of the acceptor electron pair. But it is also true that there is electron sharing (as in a covalent bond) between the hydrogen atom and the acceptor. This dual character is reflected in the bond length of the hydrogen bond. The distance between the hydrogen atom and the acceptor atom in a hydrogen bond is considerably less than would be expected from their van der Waals radii. For example, on the one hand, the distance between H and O in the bond ≡N—H···O=C≡ is only about 1.9 Å, whereas we would predict about 2.6 Å from the sum of the van der Waals radii given in Table 2.2. On the other hand, a *covalent* H—O bond has a length of only 1.0 Å. The distance between the hydrogen-bond donor and acceptor is about 2.9 Å. The donor–acceptor distances of some particularly strong hydrogen bonds are listed in **TABLE 2.3.** Note that these distances are fixed, as they are for covalent (but not for other noncovalent) bonds.

The energy of hydrogen bonds is relatively high compared to that of most other noncovalent bonds, in keeping with their partially covalent character (see Table 2.1). Hydrogen bonds are also like covalent bonds in being highly directional. Computational studies predict that hydrogen bonds are strongest when the angle defined by the donor atom, the shared H atom, and the acceptor atom is 180° (i.e., the three atoms are colinear). The majority of hydrogen-bond angles observed in proteins are within 30° of 180°. The importance of this directionality is seen in the role that hydrogen bonds play in organizing a regular biochemical structure such as the **α helix** in proteins (shown

● **CONCEPT** Hydrogen bonds are among the strongest, most specific noncovalent interactions.

TABLE 2.3 Major types of hydrogen bonds found in biomolecular interactions

Donor . . . Acceptor	Distance between Donor and Acceptor (Å)	Comment
—O—H···O(—H)	2.8 ± 0.1	H bond formed in water
—O—H···O=C	2.8 ± 0.1	Bonding of water to other molecules often involves these
N—H···O(—H)	2.9 ± 0.1	
N—H···O=C	2.9 ± 0.1	Very important in protein and nucleic acid structures
N—H···N	3.1 ± 0.2	
N—H···S	3.7	Relatively rare; weaker than above

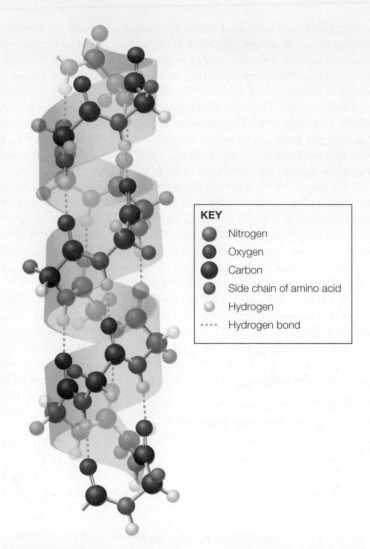

KEY

- Nitrogen
- Oxygen
- Carbon
- Side chain of amino acid
- Hydrogen
- ···· Hydrogen bond

▲ **FIGURE 2.9 Hydrogen bonding in biological molecules.** This example is a portion of a protein in an α-helical conformation. The α helix, a common structural element in proteins, is maintained by $\equiv$N—H $\cdots$ O$=$C$\equiv$ hydrogen bonds between groups in the protein chain. See Chapter 6 for more details on the structure of the α helix.

● **CONCEPT** Hydrogen bonds play a major role in stabilizing the three-dimensional structure of proteins and other important biological macromolecules.

in **FIGURE 2.9** and described in greater detail in Chapter 6). This is but one example of many we shall encounter in which hydrogen bonds stabilize ordered structure in large molecules.

2.3 The Role of Water in Biological Processes

The chemical and physical processes of life require that molecules be able to move about, encounter one another, and change partners. A fluid environment allows molecular mobility, and water, the most abundant fluid on the Earth's surface, is admirably suited to this purpose. To understand why, we must examine the properties of water in some detail.

TABLE 2.4 Properties of water compared to those of some other hydrogen-containing, low-molecular-weight compounds

Compound	Molecular Weight	Melting Point (°C)	Boiling Point (°C)	Heat of Vaporization (kJ/mol)
CH_4	16.04	−182	−164	8.16
NH_3	17.03	−78	−33	23.26
H_2O	18.02	0	+100	40.71
H_2S	34.08	−86	−61	18.66

The Structure and Properties of Water

Water is a curious substance. **TABLE 2.4**, which contrasts H_2O with other hydrogen-rich compounds of comparable molecular weight, reveals a remarkable fact. These other compounds are gases at room temperature and have much lower boiling points than water does. Why is water unique? The answer lies mainly in the strong tendency of water molecules to form hydrogen bonds with other water molecules.

The electron arrangement of a single water molecule is shown in **FIGURE 2.10**. Of the six electrons in the outer orbitals of the oxygen atom, two are involved in covalent bonds to the hydrogen atoms. The other four electrons exist in nonbonded pairs, which are excellent hydrogen-bond acceptors. The —OH groups in water, as we have seen, are strong hydrogen-bond donors. Thus, each water molecule is both a hydrogen-bond donor and a hydrogen-bond acceptor, capable of forming up to *four* hydrogen bonds simultaneously. Ice and liquid water both contain extensive networks of hydrogen-bonded molecules (**FIGURE 2.11**). When water vaporizes, this network of strong hydrogen bonds must be broken as the molecules become completely separated in the gas phase. As a consequence, vaporization of water requires an unusually large amount of energy for a molecule of its size. Both the heat of vaporization and the boiling point of water are therefore remarkably high (see Table 2.4), and water remains in the liquid state at temperatures characteristic of much of the Earth's surface. **Heat capacity** is a measure of the energy that must be added to raise (or removed to lower) the temperature of a substance by 1 °C. Liquid water has a relatively large heat capacity, which accounts for the nearly constant temperatures of large bodies of water. In essence, water is acting as a temperature "buffer" in oceans and large lakes. A physiologically relevant consequence of these physical principles is seen in the effect of perspiration on body temperature.

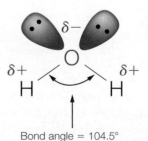

Bond angle = 104.5°

▲ **FIGURE 2.10 Hydrogen-bond donors and acceptors in water.** The two nonbonded electron pairs on oxygen act as hydrogen-bond acceptors, and the two O—H bonds act as hydrogen-bond donors. The angle between the two polar O—H bonds is 104.5°; thus, water has a large net dipole moment, which accounts for its high polarity (see Figure 2.5).

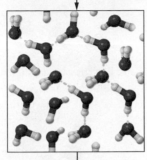

(a) A space-filling model of the structure of ice. Ice is a molecular lattice formed by indefinite repetition of a tetrahedral hydrogen-bonding pattern. Each molecule acts as a hydrogen-bond donor to two others and as an acceptor from two others. Because of the length of the hydrogen bonds, the structure is a relatively open one, which accounts for the low density of ice.

(c) The structure of liquid water. When ice melts, the regular tetrahedral lattice is broken, but substantial portions of it remain, especially at low temperatures. In liquid water, flickering clusters of molecules are held together by hydrogen bonds that continually break and re-form. In this schematic "motion picture," successive frames represent changes occurring in picoseconds (10^{-12} s).

(b) A stick model of the ice lattice. Hydrogen bonds are shown as dashed yellow lines.

▲ **FIGURE 2.11** Water as a molecular lattice.

Property	Water	n-Pentane
Molecular weight (g/mol)	18.02	72.15
Density (g/cm^3)	0.997	0.626
Boiling point (°C)	100	36.1
Dielectric constant	78.3	1.84
Viscosity (g/cm · s)	0.890×10^{-2}	0.228×10^{-2}
Surface tension (dyne/cm)	71.97	17

TABLE 2.5 Important properties of liquid water compared with those of *n*-pentane, a nonpolar, nonhydrogen-bonding liquid[a]

[a]All data are for 25 °C. (Note the higher dielectric constant for water compared to that at 37 °C.)

Perspiration lowers body temperature because heat from the body is transferred to water on the surface of the skin, resulting in evaporation of the water. Thus, thermal energy from the body is transferred from the skin to the environment when the vaporized water molecules leave the skin. The net effect is a cooling of the skin. The cooling effect of vaporization is especially noticeable when a liquid with a lower heat of vaporization, such as ethanol, is applied to the skin.

The hydrogen bonding between water molecules becomes most regular and clearly defined when water freezes to ice, creating a rigid tetrahedral molecular lattice in which each molecule is hydrogen-bonded to four others (Figure 2.11(a, b)). The lattice structure is only partially dismantled when ice melts, and some long-range order persists even at higher temperatures. On average, there are 15% fewer hydrogen bonds in liquid water than in ice (i.e., 3.4 hydrogen bonds per water molecule

in the liquid phase vs. 4 hydrogen bonds per molecule in the solid phase). The structure of liquid water has been described as "flickering clusters" of hydrogen bonds, with remnants of the ice lattice continually breaking and re-forming as the molecules move about (Figure 2.11(c)).

The rather open structure of the ice lattice accounts for another of water's highly unusual properties—the liquid phase of water is more dense than its solid phase because when the lattice breaks down, molecules can move closer together. This seemingly trivial fact is of the utmost importance for life on Earth. If water behaved like most substances and was more dense in the solid phase, the ice formed on lake and ocean surfaces each winter would sink to the bottom. There, insulated by the overlying layers, it would have accumulated over the ages, and most of the water on Earth would by now have become locked up in ice, leaving little liquid water to support life. Another consequence of this remarkable property of water is that the more dense liquid state is favored over the less dense solid state at high pressures. Thus, even at the high pressures in the depths of the ocean, water remains liquid.

Other unusual properties of water, listed in **TABLE 2.5,** are also readily explained by its molecular structure and ability to form strong hydrogen bonds. Relative to most organic liquids, water has a high *viscosity* (resistance to flow)—a consequence of water's interlocked, hydrogen-bonded structure. The cohesive network of hydrogen bonds at the surface of water also accounts for its high *surface tension* (the resistance of a liquid surface to distortion or penetration). The high dielectric constant of water, mentioned in Section 2.2, results from the polar nature of the water molecule. An electric field generated between two dissolved ions causes extensive orientation of intervening water dipoles. These oriented dipoles contribute to a counterfield, reducing the effective electrostatic force between the two ions.

Water as a Solvent

The processes of life require a wide variety of ions and molecules to move about in proximity, that is, to be soluble in a common medium. Water serves as this intracellular and extracellular medium, thanks primarily to the two properties of water we have discussed: its tendency to form hydrogen bonds and its polar character. Substances that can take advantage of these properties so as to readily dissolve in water are described as being **hydrophilic,** or "water loving."

● **CONCEPT** Water is an excellent solvent because of its hydrogen-bonding potential and its polar nature.

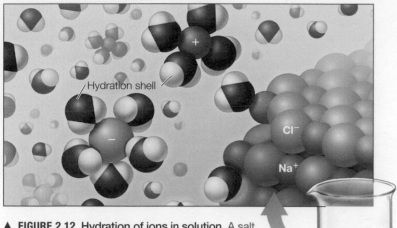

▲ **FIGURE 2.12 Hydration of ions in solution.** A salt crystal is shown dissolving in water. The noncovalent interactions between these ions and the dipolar water molecules produce a sphere of oriented water molecules, or *hydration shell,* around each dissolved ion. To illustrate the formation of hydration shells, only a few water molecules are shown—in fact, there would be very little space between water molecules in the solution.

Ionic Compounds in Aqueous Solution

In contrast to most organic liquids, water is an excellent solvent for ionic compounds. Substances like sodium chloride, which exist in the solid state as stable lattices of ions, dissolve readily in water. The explanation lies in the polar nature of the water molecule. The interactions of the negative ends of the water dipoles with cations (see Figure 2.3(b)) and the positive ends of water dipoles with anions in aqueous solution cause the ions to become **hydrated,** that is, surrounded by shells of oriented water molecules called **hydration shells** (**FIGURE 2.12**). The propensity of many ionic compounds like NaCl to dissolve in water can be accounted for largely by two factors. First, the formation of hydration shells is energetically favorable. The energy released in these favorable hydration interactions compensates for the loss of the charge–charge interactions stabilizing the crystal. Second, the high dielectric constant of water screens and decreases the electrostatic force between oppositely charged ions that would otherwise pull them back together.

Hydrophilic Molecules in Aqueous Solution

Molecules with groups capable of forming hydrogen bonds tend to form them with water. Thus, water tends to dissolve molecules, such as proteins and nucleic acids, which display on their solvent-accessible surfaces groups that can form hydrogen bonds. These include, for example, hydroxyl, carbonyl, and ester groups that are uncharged but polar, as well as charged groups such as protonated amines, carboxylates, and phosphate esters. In addition, when molecules that contain internal hydrogen bonds (such as the α helix shown in Figure 2.9) dissolve in water, some or all of their internal hydrogen bonds may be in dynamic exchange for hydrogen bonds to water (**FIGURE 2.13**).

The polar nature of the water molecule also contributes to water's ability to dissolve such nonionic, but polar, organic molecules as phenols, esters, and amides. These molecules often have large dipole moments, and interaction with the water dipole promotes their solubility in water.

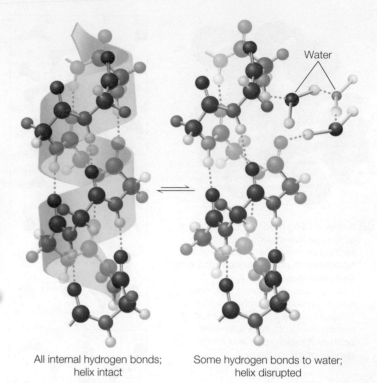

All internal hydrogen bonds;   Some hydrogen bonds to water;
helix intact   helix disrupted

▲ **FIGURE 2.13 Exchange of internal hydrogen bonds for water hydrogen bonds.** A section of a protein molecule like the α helix from Figure 2.9 is shown here, replacing some of its intramolecular hydrogen bonds with hydrogen bonds to solvent water. This dynamic (transient) exchange of hydrogen bonds is observed most often at the ends, rather than in the middle, of the helix.

Hydroxyl group   Carbonyl group   Ester group

Ammonium group   Carboxylate group   Phosphate monoester

Hydrophobic Molecules in Aqueous Solution

The solubility of hydrophilic substances depends on their energetically favorable interaction with water molecules. It is therefore not surprising that substances like hydrocarbons, which are nonpolar, nonionic, and cannot form hydrogen bonds, show only limited solubility in water. Molecules that behave in this way are called **hydrophobic,** or "water fearing." However, the inability to form hydrogen bonds is not the only factor limiting their solubility in water. When hydrophobic molecules do dissolve, they do not form hydration shells as hydrophilic substances do. Instead, the regular water lattice forms ice-like **clathrate** structures, or "cages," about nonpolar molecules (**FIGURE 2.14**). This ordering of water molecules, which extends well beyond the cage, corresponds to a decrease in the **entropy,** or disorder, of the mixture. As is discussed in greater detail in Chapter 3, decreasing entropy is unfavorable

▲ **FIGURE 2.14** One unit of clathrate structure surrounding a hydrophobic molecule (yellow). Oxygen atoms of the water molecules are shown in red. Hydrogens are shown for one pentagon of oxygens (in the rest of the structure the gray bonds represent H–bonds between water molecules). The ordered structure may extend considerably further into the surrounding water.

thermodynamically; thus, the dissolving of a hydrophobic substance in water is entropically unfavorable. This accounts for the well-known tendency of hydrophobic substances to self-associate, rather than dissolve, in water. For example, we have all seen salad oil (a hydrophobic substance) form droplets when we shake it with vinegar (an aqueous substance).

Why do hydrophobic substances self-associate in water? Surrounding two hydrophobic molecules with two *separate* cages requires more ordering of water in clathrates than surrounding both hydrophobes within a *single* cage. Thus, the hydrophobic molecules tend to aggregate because doing so releases some water molecules from the clathrates, thereby increasing the entropy of the solvent. This phenomenon, called the **hydrophobic effect,** plays an important role in the folding of protein molecules (see Chapter 6) as well as in the self-assembly of lipid bilayers, which we discuss briefly in the next section.

Amphipathic Molecules in Aqueous Solution

A most interesting and important class of biomolecules exhibits both hydrophilic and hydrophobic properties simultaneously. Such **amphipathic** substances include fatty acids, detergents, and lipids (**FIGURE 2.15**). This class of amphipathic molecules has a "head" group that is strongly hydrophilic, coupled to a hydrophobic "tail"—usually a hydrocarbon. When we attempt to dissolve them in water, these substances form one or more of the structures shown in **FIGURE 2.16(a)**. For example, they may form a **monolayer** on the water surface, with only the head groups immersed. Alternatively, if the mixture is vigorously stirred, **micelles** (spherical structures formed by a single layer of molecules) or **bilayer vesicles** may form. In such cases,

A simplified representation of an amphipathic lipid molecule

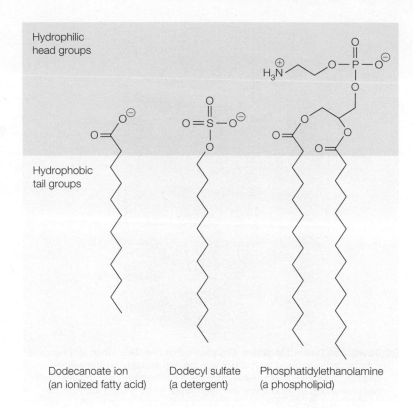

▲ **FIGURE 2.15 Amphipathic molecules.** These three examples illustrate the dual nature of amphipathic molecules, which have a hydrophilic head group attached to a hydrophobic tail.

the hydrocarbon tails of the molecules tend to lie in roughly parallel arrays, which allows them to interact via van der Waals interactions. The polar or ionic head groups are strongly hydrated by the water around them. Most important to biochemistry is the fact that amphipathic molecules are the basis of the biological **membrane bilayers** that surround cells and that form the partitions between cellular compartments (see Figures 1.9, 1.11, and 1.12). These bilayers are made primarily from phospholipids, such as that shown in Figure 2.15. Figure 2.16(b) shows the formation of synthetic membranes from phospholipids. We have much more to say about phospholipids and membranes in Chapter 10.

● **CONCEPT** A molecule is amphipathic if some parts of the molecular surface are significantly hydrophilic and other parts of the surface are significantly hydrophobic.

2.4 Acid–Base Equilibria

Most biochemical reactions occur in an aqueous environment; the exceptions are those that occur within the hydrophobic interiors of membrane bilayers. The many substances dissolved in the aqueous cytosol and extracellular body fluids include free ions like Na^+, K^+, Cl^-, and Mg^{2+}, as well as molecules and macromolecules carrying ionizable groups such as carboxylic acids and amines. The behavior of all these molecules in biochemical processes depends strongly on their state of ionization. Thus, it is important that we review briefly some aspects of ionic equilibrium, particularly acid–base equilibria and the ionization of water.

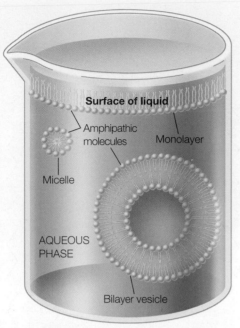

Surface of liquid

Amphipathic molecules

Monolayer

Micelle

AQUEOUS PHASE

Bilayer vesicle

(a) Structures formed in water. Structures that can form when amphipathic substances are mixed with water include a monolayer on the water surface, a micelle, and a bilayer vesicle, a hollow sphere with water both inside and out. In each case, the hydrophilic head groups are in close contact with the aqueous phase, whereas the hydrophobic tails associate with one another.

Vesicle

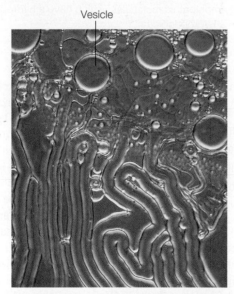

(b) Vesicle formation. When phospholipids are mixed with water, the amphipathic molecules aggregate to form films similar to biological membranes. Agitation causes the film to break up into vesicles.

▲ **FIGURE 2.16** Interactions of amphipathic molecules with water.

Acids and Bases: Proton Donors and Acceptors

A useful definition of acids and bases in aqueous systems is the Brønsted-Lowry definition: that acids are proton (H^+) donors and bases are proton acceptors. A **strong acid** dissociates almost completely into a proton and a weak **conjugate base.** For example, the acid HCl dissociates almost completely in water to yield H^+ and the weak conjugate base Cl^-; thus, the H^+ concentration in the solution is almost exactly equal to the molar concentration of HCl added. Similarly, NaOH is considered a **strong base** because it ionizes entirely, releasing ^-OH ion, which is a powerful proton acceptor.

Most of the acidic and basic substances encountered in biochemistry are **weak acids** or **weak bases,** which dissociate only partially. In an aqueous solution of a weak acid, there is a measurable equilibrium between the acid and its conjugate base. Note that the conjugate base is the substance that can accept a proton to re-form the acid. Examples of weak acids and their conjugate bases are given in **TABLE 2.6.** Note that these bases do not necessarily contain ^-OH groups, but they increase the ^-OH concentration of a solution by extracting a proton from water.

● **CONCEPT** Many biological molecules are weak acids or weak bases.

The weak acids listed in Table 2.6 vary greatly in strength—that is, in their tendency to donate protons. This variation in acid strength is indicated by the range of values of K_a and pK_a, which we will define shortly. The stronger the acid, the weaker its conjugate base. In other words, the greater the tendency of an acid to donate a proton, the less its conjugate base tends to accept a proton and re-form the acid.

Ionization of Water and the Ion Product

Although water is essentially a neutral molecule, it does have a slight tendency to ionize; in fact, it can act as both a very weak acid and a very weak base. The most correct way to understand this ionization reaction is to note that one water molecule can transfer a proton to another to yield a hydronium ion (H_3O^+) and a hydroxide ion (^-OH), so that water is both the proton donor and the proton acceptor:

$$H_2O + H_2O \rightleftharpoons H_3O^+ + {}^-OH \tag{2.4}$$

This is an oversimplification because the H_3O^+ is itself solvated by other water molecules, and protons in aqueous solution are transferred from one water molecule to another with a frequency of about 10^{15} per second.

For the purposes of the following discussion, it suffices to describe the ionization process in an even simpler way,

$$H_2O \rightleftharpoons H^+ + {}^-OH \tag{2.5}$$

as long as we remember that a proton *never* exists in aqueous solution as a free ion—it is always associated with one or more water molecules. Whenever we write a reaction involving aqueous H^+, we are really referring to a *hydrated* proton.

The equilibrium described by Equation 2.5 can be expressed in terms of K_w, called the **ion product** of water, which is 10^{-14} at 25 °C:

$$K_W = \frac{(a_{H^+})(a_{^-OH})}{(a_{H_2O})} = 10^{-14} \tag{2.6}$$

As will be described in greater detail in Chapter 3, the proper terms to use in a mass action expression such as K_w are unitless numbers, called *activities* (designated by a), equal to the effective concentrations of the chemical species. In practice, the distinction between *molar concentration* (i.e., moles of solute per liter of solution, or M) and *activity* is almost always neglected in biochemistry. This approach is appropriate in most biochemical experiments, which are usually conducted in dilute solutions, where activity and molar concentration become nearly equal. The activities of pure liquids and solids have a

TABLE 2.6 Some weak acids and their conjugate bases

Acid (Proton Donor)		Conjugate Base (Proton Acceptor)		pK_a	K_a (M)
HCOOH Formic acid	⇌	HCOO⁻ Formate ion	+H⁺	3.75	1.78×10^{-4}
CH₃COOH Acetic acid	⇌	CH₃COO⁻ Acetate ion	+H⁺	4.76	1.74×10^{-5}
OH | CH₃CH—COOH Lactic acid	⇌	OH | CH₃CH—COO⁻ Lactate ion	+H⁺	3.86	1.38×10^{-4}
H₃PO₄ Phosphoric acid	⇌	H₂PO₄⁻ Dihydrogen phosphate ion	+H⁺	2.14	7.24×10^{-3}
H₂PO₄⁻ Dihydrogen phosphate ion	⇌	HPO₄²⁻ Monohydrogen phosphate ion	+H⁺	6.86	1.38×10^{-7}
HPO₄²⁻ Monohydrogen phosphate ion	⇌	PO₄³⁻ Phosphate ion	+H⁺	12.4	3.98×10^{-13}
H₂CO₃ Carbonic acid	⇌	HCO₃⁻ Bicarbonate ion	+H⁺	6.3*	5.1×10^{-7}*
HCO₃⁻ Bicarbonate ion	⇌	CO₃²⁻ Carbonate ion	+H⁺	10.25	5.62×10^{-11}
C₆H₅OH Phenol	⇌	C₆H₅O⁻ Phenolate ion	+H⁺	9.89	1.29×10^{-10}
NH₄⁺ Ammonium ion	⇌	NH₃ Ammonia	+H⁺	9.25	5.62×10^{-10}

*Apparent pK_a and K_a values (see text for explanation).

value of one, and in biochemical reactions the activity of the solvent water is often assumed to have an activity of one. Thus, an alternative expression for the equilibrium shown in Equation 2.5 is:

$$K_W \cong \frac{[H^+][^-OH]}{1} - 10^{-14} \qquad (2.7)$$

To correctly evaluate Equation 2.7, we must use unitless values for $[H^+]$ and $[^-OH]$ that have the same magnitude as their concentrations expressed in units of *molarity* (moles per liter). Throughout this text we will adopt the convention of indicating molar concentration for a solute with square brackets [ ] around the symbol for that solute.

Because K_w is a constant, $[H^+]$ and $[^-OH]$ cannot vary independently. If we change either $[H^+]$ or $[^-OH]$ by adding acidic or basic substances to water, the other concentration must change accordingly. A solution with a high $[H^+]$ has a low $[^-OH]$, and vice versa. In pure water to which no acidic or basic substances have been added, all the H^+ and ^-OH ions must come from the dissociation of the water itself. Under these circumstances, the concentrations of H^+ and ^-OH must be equal; thus, at 25 °C

$$[H^+] = [^-OH] = 1 \times 10^{-7} M \qquad (2.8)$$

and the solution is said to be *neutral*, that is, neither acidic nor basic. However, the ion product depends on temperature, so a neutral solution does not always have $[H^+]$ and $[^-OH]$ of exactly 10^{-7} M. For example, at human body temperature (37 °C) the concentrations of H^+ and ^-OH ions in a neutral solution are each 1.6×10^{-7} M.

The pH Scale and the Physiological pH Range

To avoid working with negative powers of 10, we almost always express hydrogen ion concentration in terms of pH, defined as

$$pH = -\log(a_{H^+}) \cong -\log[H^+] \qquad (2.9)$$

The higher the $[H^+]$ of a solution, the lower the pH, so a low pH describes an acidic solution. On the other hand, a low $[H^+]$ must be accompanied by a high $[^-OH]$, as indicated by Equation 2.7, so a high pH describes a basic solution.

A diagrammatic scale of pH values is shown in **FIGURE 2.17,** with the values for some well-known solutions indicated. Note that most body fluids have pH values in the range 6.5–8.0, which is often referred to as the **physiological pH range.** Most biochemistry occurs within this region of the scale. Because of the sensitivity of biochemical processes to even small pH changes, controlling and monitoring pH are essential in most biochemical experiments. The control of pH is achieved by using solutions that are *buffered,* that is, solutions containing components to help maintain the solution at a constant pH as biochemical reactions proceed. The composition of buffered solutions is described in detail later in this chapter.

● **CONCEPT** Most biological reactions occur between pH 6.5 and pH 8.0.

As we will see throughout this text, the structures and functions of many biomolecules show a strong dependence on pH. As pH changes, so does the degree of protonation of acidic and basic groups displayed on the surface of a biomolecule. One important consequence

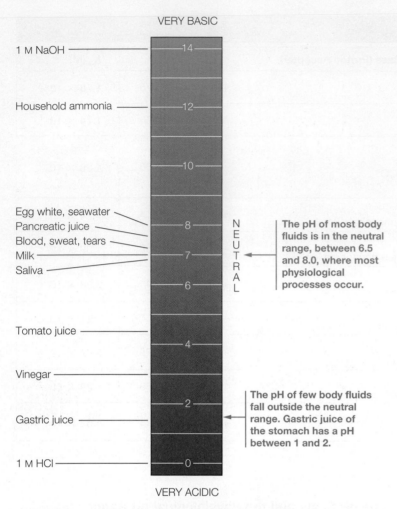

▲ **FIGURE 2.17 The pH scale and the physiological pH range.** The pH values of some common substances and body fluids are listed, with NaOH at the basic end of the range shown here and HCl at the acidic end.

of changing pH is that *the overall charge on the molecule changes.* The overall charge on a protein is determined by the balance of individual positive and negative charges on the surface of the molecule. As **FIGURE 2.18** shows, proteins will carry more positive charge at lower values of pH than at higher values of pH. We will come back to this important point and discuss the basis for this phenomenon later in this chapter and in many subsequent chapters.

Figure 2.18 illustrates another key concept important for understanding specific interactions between biomolecules (such as those shown in Figure 2.1): The charges on the surface of the protein have defined locations. Because surfaces and charge distributions differ between proteins, we can think of the pattern of charges, such as those shown in Figure 2.18, as defining a "fingerprint" for a given protein.

● **CONCEPT** The distribution of charges on the surface of a biomolecule defines its functional characteristics, such as receptor binding or catalysis.

It is this "fingerprint" that allows a receptor, such as the growth hormone receptor shown in Figure 2.1, to identify its specific binding partner from the thousands of other proteins present in cells or in circulation.

In the following sections, we review some principles of acid–base chemistry that allow us to understand and predict the behavior of biomolecules as a function of pH.

Weak Acid and Base Equilibria: K_a and pK_a

Many biologically important compounds contain weak acids and/or weak bases. For example, very large protein molecules carry on their surfaces both acidic (e.g., carboxylic acid) and basic (e.g., amino) groups. As we discussed earlier, the response of such groups to changes in pH is often of considerable importance to the function of a protein. For example, the catalytic efficiency of many enzymes depends critically on the ionization state of certain groups, so these catalysts are effective only in defined pH ranges (enzyme catalysis and the

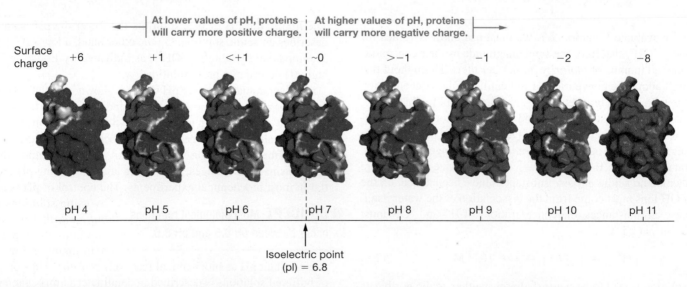

▲ **FIGURE 2.18 The effect of pH on overall surface charge of human ubiquitin.** The charge on the van der Waals surface of human ubiquitin is shown at eight values of pH. The overall charge on ubiquitin at a given pH is listed above each representation of the protein. By convention, negative charge density is shown in red, and positive charge density is shown in blue. This reflects the fact that negative charge is associated with deprotonated carboxylic acid groups and positive charge is associated with protonated amine groups. PDB ID: 1ubq.

importance of pH are discussed in Chapter 8). The dissociation equilibria of weak acids and bases can be used to describe the molecular basis for such effects. We begin this discussion by considering a few examples given in Table 2.6.

Each of the reactions shown in Table 2.6 can be written as the dissociation of an acid. This dissociation may take several forms, depending on the substance involved:

$$HA^+ \rightleftharpoons H^+ + A$$

$$HA \rightleftharpoons H^+ + A^-$$

$$HA^- \rightleftharpoons H^+ + A^{2-}$$

Note that in some cases the conjugate base has a negative charge and in other cases it does not, but in *all* cases it has one proton fewer than the acid. For convenience, we will write such reactions in the generic form: $HA \rightleftharpoons H^+ + A^-$. The equilibrium constant for the dissociation of a weak acid (often called the **acid dissociation constant**) is defined as

$$K_a = \frac{[H^+][A^-]}{[HA]} \qquad (2.10)$$

A larger value of K_a indicates a stronger acid—that is, one with a greater tendency to dissociate. The strength of acids is usually expressed in terms of the pK_a value:

$$pK_a = -\log K_a \qquad (2.11)$$

Because pK_a is the *negative* logarithm of K_a, a numerically *smaller* value of pK_a corresponds to a stronger acid, and a *larger* value corresponds to a weaker acid. Both K_a and pK_a values are given for the acids listed in Table 2.6.

● **CONCEPT** A convenient way to express the strength of an acid is by its pK_a; the lower the value of pK_a, the stronger the acid.

Some acids, such as the phosphoric and carbonic acids in Table 2.6, are capable of losing more than one proton. These acids are called **polyprotic acids.** The successive dissociations involve separate steps, with separate pK_a values; thus, polyprotic acids exist in several different ionization states.

Titration of Weak Acids: The Henderson–Hasselbalch Equation

As shown in Figure 2.18, the overall charge on a protein changes as a function of pH. How does changing the pH of a solution alter the charge on such a molecule? We can derive a relationship that will give us a quantitative answer to this question by taking the negative logarithm of both sides of Equation 2.10 and rearranging:

$$-\log[H^+] = -\log K_a + \log\frac{[A^-]}{[HA]} \qquad (2.12)$$

Substituting pH for $-\log[H^+]$ and pK_a for $-\log K_a$ gives:

$$pH = pK_a + \log\frac{[A^-]}{[HA]} \qquad (2.13)$$

This equation, known as the **Henderson–Hasselbalch equation,** shows the direct relationship between the pH of a solution and the ratio of concentrations of the *deprotonated* form $[A^-]$ to the *protonated*

form $[HA]$ of some ionizable group. In the case of a carboxylic acid, $[A^-]=[RCOO^-]$ and $[HA]=[RCOOH]$. In the case of a primary amine base, $[A^-]=[RNH_2]$ and $[HA]=[RNH_3^+]$. Both of these functional groups are abundant on the surfaces of typical protein molecules; thus, their ionization as a function of pH is largely responsible for the changes in surface charge density illustrated in Figure 2.18.

As an example of the application of the Henderson–Hasselbalch equation to a simple carboxylic acid, let us consider a solution of formic acid, where the R group $=$ H. In this case, Equation 2.13 becomes

$$pH = pK_a + \log\frac{[HCOO^-]}{[HCOOH]} \qquad (2.14)$$

Equation 2.14 allows the pH of the formic acid solution to be calculated for any ratio of the conjugate acid and conjugate base. This ratio—and thus the pH of the solution—changes during a *titration* (i.e., as we add a base or acid to the solution). Suppose we want to titrate a 1 M solution of formic acid with sodium hydroxide. First, we must ask: What is the pH of the solution made by dissolving 1 mol formic acid in sufficient water to make 1 L of solution? Let us calculate the initial concentration of H^+ in the solution, which is determined by how much of the formic acid has dissociated before any titrant is added. This can be calculated from Equation 2.10, if we assume that virtually all of the H^+ in such a solution comes from the formic acid (i.e., the ionization of water is negligible in this case) and that dissociation of one formic acid molecule gives one H^+ and one $HCOO^-$ ion. If we denote their molar concentrations by x, then Equation 2.10 becomes

$$K_a = 1.78 \times 10^{-4} = \frac{[H^+][HCOO^-]}{[HCOOH]} = \frac{x^2}{1-x} \qquad (2.15)$$

We can solve for x using the quadratic equation (see Problem 4 at the end of the chapter). In this example, we find that

$$x = [H^+] = [HCOO^-] = 1.33 \times 10^{-2}\,M.$$

Thus, in a 1 M solution of formic acid, only about 1% of the acid has dissociated.

The pH of the solution can then be calculated using Equation 2.9. In this case, $pH = -\log[H^+] - -\log(1.33 \times 10^{-2}) = 1.88$. Thus, the pH of 1 M formic acid is about 1.9. Now, what happens as a solution of NaOH is added to the formic acid solution? As NaOH is added, it dissociates completely into Na^+ and ^-OH because it is a strong base. However, the hydroxide ions are in equilibrium with protons according to the relation $K_W = [H^+][^-OH]$, so addition of ^-OH reduces the concentration of protons in the solution. In accordance with Le Chatelier's Principle, as $[H^+]$ decreases, more formic acid must dissociate to satisfy the equilibrium relationship given by Equation 2.15. This means that the ratio $[HCOO^-]/[HCOOH]$ increases as NaOH is added. Applying the Henderson–Hasselbalch equation (Equation 2.14), we see that the pH must also increase continuously as the titration proceeds. At the midpoint of the titration, half of the original formic acid has been neutralized. That means that half is still present in the acid form and half is present as the conjugate base, so $[A^-]/[HA] = 1$. The Henderson–Hasselbalch equation then becomes

$$pH = pK_a + \log 1 = pK_a \qquad (2.16)$$

● **CONCEPT** The Henderson–Hasselbalch equation describes the change in pH during titration of a weak acid or a weak base.

Thus, the pH of a weak acid at the midpoint of its titration curve has the same value as its pK_a. This can be confirmed experimentally, as shown by the titration curves of two acids, formic acid and ammonium ion, in **FIGURE 2.19**. The titration curves in Figure 2.19 plot the measured pH against *moles of base added per mole of acid originally present.* Note that over much of the titration curve the pH lies within one pH unit below or above the pK_a.

Buffer Solutions

If we look at Figure 2.19 in a different way, another important point emerges. In the pH range near the pK_a, the pH of the solution changes very little with each increment of base or acid added. In fact, the pH is least changed per increment of acid or base just at the pK_a. This is the principle behind **buffering** of solutions by the use of weak acid–base mixtures, a technique used in virtually every biochemical experiment.

Buffered solutions are able to minimize the change in pH following addition of H^+ or ^-OH because the conjugate acid (HA) and conjugate base (A^-) of the buffering compound (commonly called the **buffer** or **buffer salt**) are present in sufficient concentration to combine with the added H^+ or ^-OH and neutralize them:

Neutralization of ^-OH by conjugate acid: $HA + {}^-OH \rightleftharpoons A^- + H_2O$

Neutralization of H^+ by conjugate base: $A^- + H^+ \rightleftharpoons HA$

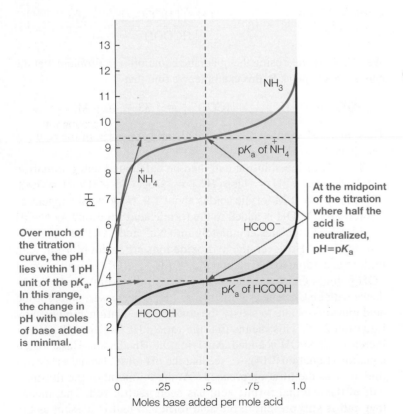

Over much of the titration curve, the pH lies within 1 pH unit of the pK_a. In this range, the change in pH with moles of base added is minimal.

At the midpoint of the titration where half the acid is neutralized, pH=pK_a

▲ **FIGURE 2.19 Titration curves of weak acids.** The curves for titration of 1 M formic acid (HCOOH) and 1 M ammonium ion ($\overset{+}{N}H_4$) show the change in pH as a strong base is added.

Suppose a biochemist wishes to study a reaction at pH 4.00. The reaction may be one that generates or consumes protons. To prevent the pH from drifting too much during the reaction, the experimenter should use a buffer solution consisting of a nearly equal mixture of a weak acid and its conjugate base. Recall that the Henderson–Hasselbalch equation predicts that an equimolar ratio of A^- to HA is achieved when the pH of the solution equals the pK_a of the buffer. In this example, a formic acid–formate ion buffer would be a good choice because the pK_a of formic acid (3.75) is close to the pH value required for the experiment. An acetic acid–acetate mixture would not be as satisfactory because the pK_a of acetic acid (4.76) is nearly 1 pH unit away. The ratio of formate ion to formic acid required to make a buffered solution with a pH of 4.00 can be calculated from the Henderson–Hasselbalch equation:

$$4.00 = 3.75 + \log\frac{[HCOO^-]}{[HCOOH]} \quad (2.17)$$

To calculate the base/acid ratio, we simply subtract 3.75 from 4.00 and take the antilogarithm of both sides of Equation 2.17:

$$\frac{[HCOO^-]}{[HCOOH]} = 10^{(4.00-3.75)} = 10^{0.25} = 1.78 \quad (2.18)$$

This result shows that at pH = 4.00 there will be a ratio of 1.78 mol of $HCOO^-$ in solution to every 1.00 mol of HCOOH in solution. Such a buffer could be made, for example, by mixing equal volumes of 0.1 M formic acid and 0.178 M sodium formate. In practice, it is better to use a calibrated pH meter and prepare the buffer solution by titrating a solution of formic acid to pH 4.00 with sodium hydroxide.

Organisms must maintain the pH inside cells and in most body fluids within the narrow pH range of about 6.5 to 8.0. For example, the normal pH of human blood is 7.4, which is also the pH inside most human cells. When the pH drops below 6.8 or rises above 7.8, animal cells suffer irreparable damage, leading to death. Thus, a drop in the pH of human blood below 7.2 is considered a serious medical emergency. Blood pH may fall, a condition known as **acidemia,** as a result of **ketoacidosis** or **metabolic acidosis.** Ketoacidosis occurs in response to starvation or type I diabetes (described in more detail in Chapters 16 and 17). Metabolic acidosis occurs when the removal of organic acids by the kidneys doesn't keep up with the production of such acids. This condition is associated with several human diseases, as well as binge drinking. Prolonged vigorous exercise can produce a more modest acidemia as a consequence of increased metabolic activity and adenosine triphosphate (ATP) hydrolysis in muscle tissue. Cells have many mechanisms to control pH, the most important of which are biological buffer systems.

● **CONCEPT** Buffer solutions function because the pH of a weak acid–base solution is least sensitive to added acid or base near the pK_a where the conjugate acid and conjugate base of the buffer are both present in nearly equimolar concentrations.

● **CONNECTION** Changes in pH in cells or body fluids can have profound clinical significance, and if untreated can be fatal. Low blood pH (acidemia) is associated with a number of human diseases such as diabetes and kidney failure.

The entries in Table 2.6 include two buffer systems of great importance for biological pH control. The dihydrogen phosphate–hydrogen phosphate system, with a pK_a

of 6.86, plays a major role in controlling intracellular pH because phosphate is abundant in cells. In blood, which contains dissolved CO_2 as a waste product of metabolism, the carbonic acid–bicarbonate system provides buffering capacity. Carbonic acid has a pK_a of 3.8; however, since it readily decomposes into water and dissolved CO_2, the concentration of H_2CO_3 in solution is very low. As a consequence, the *apparent* pK_a (designated pK_a') of carbonic acid is 6.3 (this is the value listed in

● **CONCEPT** Organisms use buffer systems to maintain the pH of cells and body fluids in the appropriate range.

Table 2.6). The relationship between dissolved CO_2 and proton concentration (i.e., pH) can be appreciated by considering the following chemical equations:

$$CO_2 + H_2O \rightleftharpoons H_2CO_3 \quad \text{(reaction to form carbonic acid from } CO_2\text{)}$$

$$H_2CO_3 \rightleftharpoons HCO_3^- + H^+ \quad \text{(first proton dissociation of carbonic acid)}$$

$$CO_2 + H_2O \rightleftharpoons HCO_3^- + H^+ \quad \text{(sum of the two equations)} \quad (2.19)$$

Applied to Equation 2.19, Le Chatelier's Principle predicts that as the concentration of dissolved CO_2 in blood increases, the concentrations of HCO_3^- and H^+ will also increase. Thus, the blood pH will decrease as a result of increasing CO_2 concentration. We will see in Chapter 7 that in actively respiring tissues, which produce dissolved CO_2 as a by-product of metabolism, a drop in pH from 7.4 to 7.2 results in increased oxygen delivery to these tissues by the oxygen-carrying protein hemoglobin (the major protein component inside red blood cells).

From Equation 2.19 we can write an expression that describes the apparent dissociation of carbonic acid to bicarbonate ion and a proton:

$$K_a' = \frac{[H^+][HCO_3^-]}{[CO_2][H_2O]} \quad (2.20)$$

The value of the pK_a' for carbonic acid can be formulated from the combination of the two relevant equilibrium expressions for the chemical equations that were used to derive Equation 2.19. The first of these expressions is for the equilibrium between dissolved CO_2, water, and H_2CO_3:

$$CO_2 + H_2O \rightleftharpoons H_2CO_3$$

$$K_{eq} \cong 3 \times 10^{-3} = \frac{[H_2CO_3]}{[CO_2][H_2O]} \quad (2.21)$$

The second is for the dissociation of H_2CO_3:

$$H_2CO_3 \rightleftharpoons HCO_3^- + H^+$$

$$K_a \cong 1.7 \times 10^{-4} = \frac{[H^+][HCO_3^-]}{[H_2CO_3]} \quad (2.22)$$

When these two equilibrium expressions are multiplied, they yield Equation 2.20:

$$\frac{[H_2CO_3]}{[CO_2][H_2O]} \times \frac{[H^+][HCO_3^-]}{[H_2CO_3]} = \frac{[H^+][HCO_3^-]}{[CO_2][H_2O]} = K_a'$$

$$K_a' = K_{eq} \times K_a = (3 \times 10^{-3}) \times (1.7 \times 10^{-4})$$

$$= 5.1 \times 10^{-7} \text{ and } pK_a' = -\log(5.1 \times 10^{-7}) = 6.3$$

In addition to phosphate and bicarbonate buffers, proteins play a major role in the control of pH in organisms. As mentioned earlier, proteins contain many weakly acidic or basic groups, and some of these have pK_a values near 7.0. Because proteins are abundant both in cells and in body fluids like blood and lymph, pH buffering is very strong.

Molecules with Multiple Ionizing Groups

So far, we have considered molecules containing only one or a few weakly acidic or basic groups. But many molecules contain multiple ionizing groups and display more complex behavior during titration.

A molecule that contains groups with both acidic and basic pK_a values is called an **ampholyte**. Consider, for example, the molecule glycine: H_2N-CH_2-COOH. Glycine is an α-amino acid, one of a group of important amino acids that are the constituents of proteins. The pK_a values of the α-carboxylate and α-amino groups on glycine are 2.3 and 9.6, respectively. If we dissolved glycine in a very acidic solution (e.g., pH 1.0), both the α-amino group and the α-carboxylate group would be protonated, and the molecule would have a net charge of $\sim +1$. If the pH was increased (e.g., by adding NaOH), proton dissociation would occur in the following sequence:

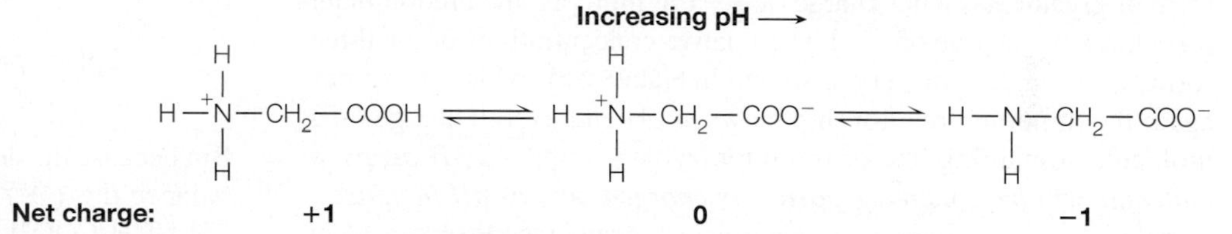

Increasing pH ⟶

H—$\overset{+}{N}$(H)(H)—CH$_2$—COOH ⟷ H—$\overset{+}{N}$(H)(H)—CH$_2$—COO$^-$ ⟷ H—N(H)—CH$_2$—COO$^-$

Net charge: +1   0   −1

Thus, the titration of glycine occurs in two steps as the more acidic α-carboxylate and the less acidic α-amino groups successively lose their protons. Glycine can therefore serve as a good buffer in two quite different pH ranges, as shown in **FIGURE 2.20**. In each range, we can

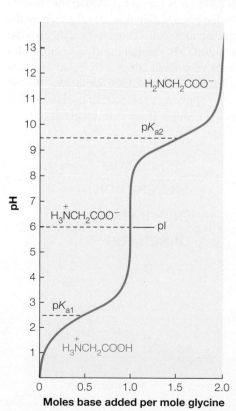

◄ **FIGURE 2.20** Titration of the ampholyte glycine. Since two groups with quite different pK_a values can be titrated, this is a two-step titration curve. The calculated isoelectric point, pI, is shown.

Moles base added per mole glycine

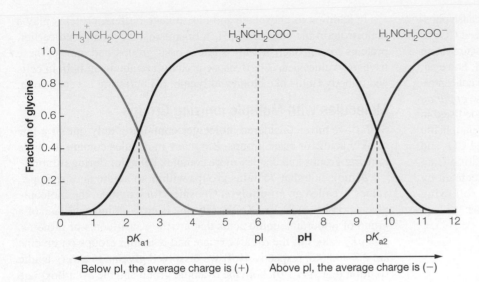

◀ **FIGURE 2.21 The relative concentrations of the three forms of glycine as a function of pH.** The three forms are $H_3\overset{+}{N}CH_2COOH$, shown in blue; $H_3\overset{+}{N}CH_2COO^-$, in black; and $H_2NCH_2COO^-$, in red. The two pK_a values and the isoelectric point (pI) are indicated. As pH increases, the average charge on the molecule becomes more negative, and as pH decreases the average charge on the molecule becomes more positive.

describe the titration curve by applying the Henderson–Hasselbalch equation to the appropriate ionizing group. At low pH, the predominant form of glycine has a net charge of ~ +1 at high pH, the predominant form has a net charge of ~ −1 The relative concentrations of the three forms as a function of pH are shown in **FIGURE 2.21,** which illustrates again the important relationship between pH and overall charge on a molecule containing one or more ionizable groups: *As pH drops, a molecule will become more positively charged, and as pH increases, a molecule will become more negatively charged* (see Problem 14 at the end of the chapter).

The situation near neutral pH is an interesting one. In this region, most glycine molecules are in the form $H_3\overset{+}{N}—CH_2—COO^-$, which has a net charge of zero. An ampholyte in this state, carrying both positive and negative charges, is called a *zwitterion* (German for "double ion"). However, there is only one point within this pH range where the *average* charge on all the glycine molecules sums to exactly zero. At this pH, called the **isoelectric point (pI),** most of the glycine molecules are in the zwitterion form, with very small but exactly equal amounts of $H_3\overset{+}{N}—CH_2—COOH$ and $H_2N—CH_2—COO^-$ molecules. We can calculate the isoelectric point by applying the Henderson–Hasselbalch equation to both of the ionizing groups. If we call the pH at the isoelectric point pI, we have

$$pI = pK_{COOH} + \log\frac{[H_3\overset{+}{N}CH_2COO^-]}{[H_3\overset{+}{N}CH_2COOH]} \qquad (2.23)$$

$$pI = pK^+_{NH_3} + \log\frac{[H_2NCH_2COO^-]}{[H_3\overset{+}{N}CH_2COO^-]} \qquad (2.24)$$

Adding these equations (and remembering that the sum of the logarithms of two quantities is the logarithm of their product) gives

$$2pI = pK_{COOH} + pK^+_{NH_3} + \log\frac{[H_2NCH_2COO^-]}{[H_3\overset{+}{N}CH_2COOH]} \qquad (2.25)$$

But because the definition of pI requires that the $(-)$ charges exactly balance the $(+)$ charges, or in this case that $[H_2NCH_2COO^-] = [H_3\overset{+}{N}CH_2COOH]$, the ratio term to the far right is equal to log 1, or zero, so

$$pI = \frac{pK_{COOH} + pK^+_{NH_3}}{2} \qquad (2.26)$$

The result is simple in this case: For a molecule with only two ionizable groups, the pI is simply the average of the two pK_as. If we insert the pK_a values given previously, we obtain pI = 5.95 for glycine. Actually, as Figure 2.21 shows, glycine will be predominantly in the zwitterion form from about pH 4 to about pH 8. Thus, the average charge on glycine is very close to zero throughout this pH range; but it is exactly zero only when pH = pI = 5.95.

For small molecules that have three ionizable groups, the pI can be calculated by averaging the two pK_as that describe the ionization of the "isoelectric" species. For example, as shown below the amino acid aspartic acid has three ionizable groups with pK_as of 2.1 (α-COOH), 3.9 (β-COOH), and 9.8 (α-NH$_3^+$). The titration of fully protonated aspartic acid with NaOH occurs in three steps (**FIGURE 2.22**), which we can write in the order of *increasing* pK_a values for the ionizable groups.

Increasing pH ⟶

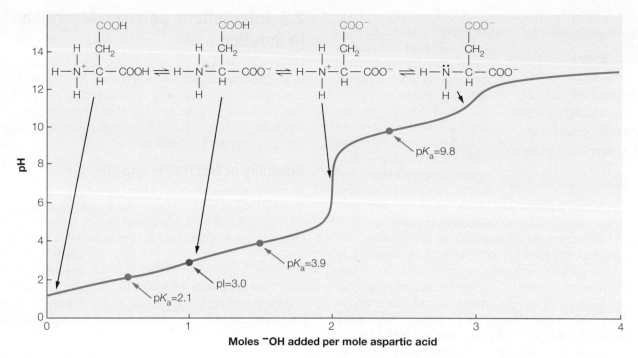

▲ **FIGURE 2.22 Titration curve of aspartic acid.** Since three groups with different pK_a values can be titrated, this is a three-step titration curve. As the titration progresses from lower to higher values of pH, the aspartic acid becomes increasingly deprotonated. Arrows indicate which ionization state is predominant for a particular portion of the titration.

If we sum the charges on each of the four possible ionization states for aspartic acid, we see that one of these, the so-called isoelectric species, carries no net charge. As stated above, the two pK_as that describe the ionization of the isoelectric species, $pK_{\alpha\text{-COOH}}$ and $pK_{\beta\text{-COOH}}$, have values of 2.1 and 3.9, respectively. Thus, the pI for aspartic acid is

$$pI = \frac{pK_{\alpha COOH} + pK_{\beta COOH}}{2} = \frac{(2.1 + 3.9)}{2} = 3.0 \quad (2.27)$$

Careful analysis of the ionization of each functional group at pH = 3.0 would show that the α-amino group is essentially fully protonated and therefore carries a full $(+1)$ charge, whereas the two carboxylic acids are only *partially* deprotonated; thus, each carries a partial $(-)$ charge. At pH = 3.0 the α-COOH is 88.8% deprotonated and therefore carries a net charge of –0.888. The β-COOH is 11.2% deprotonated and therefore carries a net charge of –0.112. The sum of these two partial $(-)$ charges = -1, which balances the +1 charge on the α–amino group; thus, the net charge on the molecule at pH = 3 is zero. For practice performing such an analysis, see Problem 15 at the end of this chapter. It must be noted that the approach described above only works for very simple molecules with three to five ionizable groups. In cases involving many titratable groups, the calculation of pI becomes much more complicated.

Large molecules such as proteins can have *many* acidic and basic groups. Such molecules, called **polyampholytes,** may contain tens to hundreds of individual charged groups. However, as long as the molecule has both positively and negatively charged groups, there is an isoelectric pH, at which the average charge on the molecule is zero. For example, human hemoglobin has 148 ionizable groups on its surface, and its pI is 6.85. Although the theory described above can be used to predict a

pI value from the amino acid composition of a protein, there is some uncertainty associated with such predictions. In practice, pI is determined experimentally using the methods of *electrophoresis* and *isoelectric focusing*. In the *Tools of Biochemistry 2A* at the end of this chapter, we present an overview of these widely used laboratory techniques.

The value of pI for a protein will vary depending on the types of functional groups displayed on its surface. If acidic groups (e.g., $-COOH/-COO^-$) predominate on the protein surface, the pI will be low; if basic groups (e.g., $-NH_3^+/-NH_2$) predominate, the pI will be high. In Chapter 5 we will find that this is an important consideration in working with solutions of proteins.

Three important conclusions can be drawn from this discussion of pI. First, when the pH of a protein solution equals the pI for the protein, there is no net charge on the protein because the sum of all the negative charges, $\Sigma(-)$, is exactly offset by the sum of all the positive charges, $\Sigma(+)$, or,

$$\text{at pH} = pI \quad |\Sigma(-)| = |\Sigma(+)| \quad (2.28)$$

Thus, the net charge of the molecule = 0. Second, when the pH is greater than the pI, the molecule carries a net negative charge because the sum of the negative charges is greater than the sum of the positive charges:

$$\text{at pH} > pI \quad |\Sigma(-)| > |\Sigma(+)| \quad (2.29)$$

Thus, the net charge of the molecule is $(-)$. Third, when the pH is below the pI, the molecule carries a net positive charge because the sum of the negative charges is less than the sum of the positive charges:

$$\text{at pH} < pI \quad |\Sigma(-)| < |\Sigma(+)| \quad (2.30)$$

Thus, the net charge of the molecule is $(+)$.

The behavior described by Equations 2.28–2.30 and illustrated in Figure 2.18 can be understood by considering the acid–base chemistry involved. Ionization by H^+ transfer is a *dynamic* process in which ionizable groups are constantly gaining or losing a proton. Using the Henderson–Hasselbalch equation, we can predict, on average, what fraction of these ionizable groups is protonated.

● **CONCEPT** An ampholyte's isoelectric point, pI, is the pH at which average charge, for all forms of the molecule, is zero. At pH < pI, an ampholyte molecule will be positively charged, and at pH > pI, the molecule will be negatively charged.

As the pH changes, this fraction will also change. For example, protein molecules in aqueous solution become increasingly protonated as the pH decreases. As a consequence, proteins become more positively charged because carboxylic acids, the groups largely responsible for the negative charge density on the surfaces of proteins, become *less negatively charged* as they become more protonated, whereas nitrogenous bases (e.g., amines), which are largely responsible for the positive charge density on the surfaces of proteins, become *more positively charged* as they become more protonated. A similar logic explains the observation that proteins become more negatively charged as pH increases. Under conditions of increasing pH, the ionizable groups become more deprotonated. Thus, acidic groups become more negatively charged, while the basic groups become less positively charged.

Some macromolecules, called **polyelectrolytes,** carry multiples of only positive or only negative charge. Strong polyelectrolytes, like the negatively charged nucleic acids (see Chapter 4), are ionized over a wide pH range. In addition, there are weak polyelectrolytes, like polylysine, a polymer of the amino acid lysine:

Polylysine

When a number of weakly ionizing groups are carried on the same molecule, the pK_a of each group is influenced by the state of ionization of the others. In a molecule like polylysine, the first protons are more easily removed than the last because the strong positive charge on the fully protonated molecule makes deprotonation, and the associated reduction in repulsive charge–charge interactions, more favorable. Conversely, a molecule that develops a strong negative charge as protons are removed gives up the last ones less readily, as reflected in the increasing values of pK_a for successive deprotonations (see the entries for H_3PO_4 and H_2CO_3 in Table 2.6).

Significant fluctuations in the charges on proteins will result in loss of function; thus, the concepts summarized in Figure 2.18 and Equations 2.28–2.30 explain not only the basis for molecular recognition via charge complementarity but also the critical need to maintain intracellular pH within a narrow range such that charges on molecules remain relatively constant.

2.5 Interactions Between Macroions in Solution

Large polyelectrolytes such as nucleic acids and polyampholytes such as proteins are classified as **macroions.** As described in the previous section, such ions may carry a substantial net charge in aqueous solution, depending on the pH. The electrostatic forces of attraction or repulsion between such charged particles play a major role in determining their behavior in solution.

Solubility of Macroions and pH

Because macroions of like net charge repel one another, nucleic acid molecules tend to remain separated in solution (**FIGURE 2.23(a)**). This accounts for the high solubility of DNA in solution. For the same reason, proteins tend to be more soluble when they have a net charge—that is, at pH values above or below their isoelectric points. In contrast, if positively and negatively charged macromolecules are

● **CONCEPT** Interactions between macroions are greatly influenced by pH and the small ions in the solution.

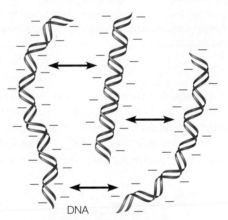

(a) **Repulsion.** DNA molecules, with many negative charges, strongly repel one another in solution.

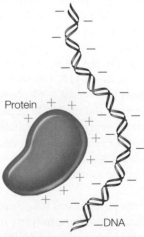

(b) **Attraction.** If DNA is mixed with a positively charged protein, these molecules have a strong tendency to associate.

▲ **FIGURE 2.23** Electrostatic interactions between macroions.

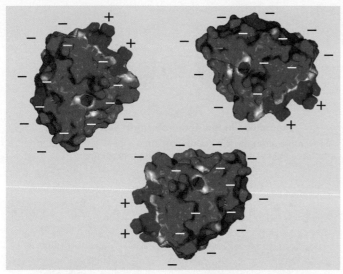

(a) High pH: protein soluble (deprotonated). Most proteins are very soluble at high pH, where the molecules are negatively charged and repel one another.

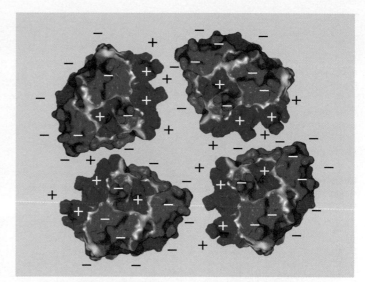

(b) Isoelectric point: protein aggregates. At the isoelectric point, where a protein has no *net* charge, the molecules retain regions of positive and negative charge on their surfaces, resulting in an increased tendency to aggregate and precipitate.

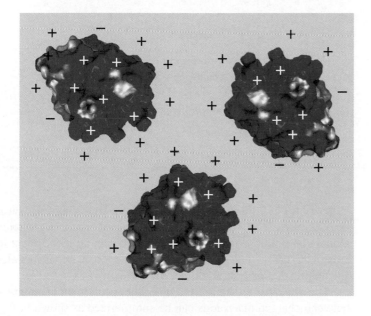

(c) Low pH: protein soluble (protonated). At low pH the proteins are soluble because of repulsions due to their positive charge.

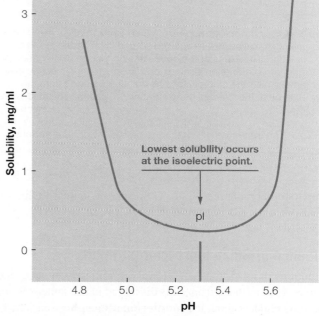

(d) Solubility of β-lactoglobulin with varying pH.

▲ **FIGURE 2.24 Dependence of protein solubility on pH.** Surface charge for β-lactoglobulin was calculated using coordinates from PDB ID: 3blg.

mixed, electrostatic attraction makes them tend to associate with one another (Figure 2.23(b)). Many proteins interact strongly with DNA; most of these proteins turn out to be positively charged. A striking example is found in the chromosomes of higher organisms in which the negatively charged DNA is strongly associated with positively charged proteins called *histones* to form the complex called chromatin (discussed in Chapter 21).

A more subtle type of electrostatic interaction may cause the molecules of a particular protein to self-associate at the isoelectric pH (**FIGURE 2.24**). For example, the common milk protein β-lactoglobulin is a polyampholyte with an isoelectric point of about 5.3. Above or below this pH, the molecules have either negative or positive charges and repel one another. This protein is therefore very soluble at either acidic or basic pH. At the isoelectric point the net charge is zero, but each molecule still carries surface patches of both positive and negative charge (Figure 2.24(b)). The charge–charge interactions, together with other kinds of intermolecular interactions such as van der Waals forces, increase the likelihood that the molecules

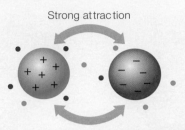

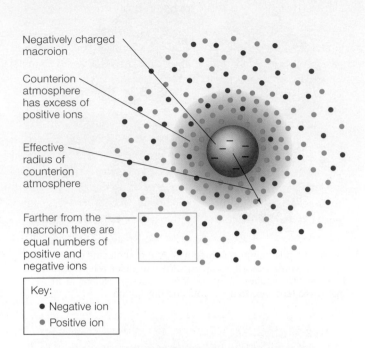

Negatively charged macroion

Counterion atmosphere has excess of positive ions

Effective radius of counterion atmosphere

Farther from the macroion there are equal numbers of positive and negative ions

Key:
● Negative ion
● Positive ion

(a) The counterion atmosphere. When a macroion (in this example, negatively charged) is placed in an aqueous salt solution, small ions of the opposite sign tend to cluster about it, forming a counterion atmosphere. There are more cations than anions near the macroanion shown here; far away from the macroion, the average concentrations of cations and anions are equal.

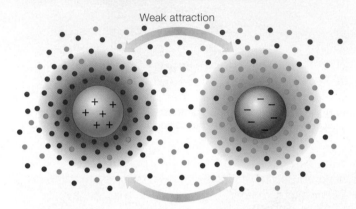

Strong attraction

Macroions in low-ionic-strength salt solution
At low ionic strength, the counterion atmosphere is diffuse and interferes little with the interactions of the macroions.

Weak attraction

Macroions in high-ionic-strength salt solution
At high ionic strength, the counterion atmosphere is concentrated about the macroions and greatly reduces their interactions.

(b) The influence of ionic strength.

▲ **FIGURE 2.25** The influence of small ions on macroion interactions.

will clump together and precipitate. Therefore, β-lactoglobulin, like many other proteins, has minimum solubility at its isoelectric point (Figure 2.24(d)).

The Influence of Small Ions: Ionic Strength

The interactions of macroions are strongly modified by the presence of small ions, such as those from salts dissolved in the same solution. Each macroion collects about it a **counterion atmosphere** enriched in oppositely charged small ions, and this cloud of ions tends to screen the molecules from one another (**FIGURE 2.25(a)**). The larger the concentration of small ions present, the more effective this electrostatic screening will be. However, the precise relationship of screening to concentration is rather complex and is a function of the ion concentration called the **ionic strength (I)**. The ionic strength is defined as

$$I = \tfrac{1}{2}\sum_i M_i Z_i^2 \qquad (2.31)$$

where the sum is taken over all small ions in the solution. For each ion type, M_i is its molarity and Z_i is its charge. For a 1:1 electrolyte like NaCl, we have $Z_{Na^+} = +1$ and $Z_{Cl^-} = +1$, and since $M_{Na^+} = M_{Cl^-} = M_{NaCl}$, we find that $I_{NaCl} = M_{NaCl}$. Thus, ionic strength equals salt

molarity for 1:1 electrolytes, but this is not true if multivalent ions (e.g., Mg^{2+} or SO_4^{2-}) are involved. Multivalent ions make greater individual contributions to the ion atmosphere than do monovalent ions, as reflected in the fact that the *square* of the ion charge is included in calculating the ionic strength. For these electrolytes, $I > M$.

The effects of the ionic strength of the medium on the interaction between charged macroions can be summarized as shown in Figure 2.25(b). At very low ionic strength, the counterion atmosphere is highly expanded and diffuse, and screening is ineffective. In such a solution, macroions attract or repel one another strongly. If the ionic strength is increased, the counterion atmosphere shrinks and becomes concentrated about the macroion, and the attractive interactions between positive and negative groups are effectively screened.

The effects of ionic interactions on the behavior of biological macromolecules mean that the biochemist must pay close attention to both ionic strength and pH. Experimenters usually use a neutral salt (like NaCl or KCl) to control the ionic strength of a solution, as well as a buffer to control the pH. In determining the amount of salt to add, they often try to mimic the ionic strengths of cell and body fluids. Although ionic strengths vary from one cell type or fluid to another, a value of 0.1 to 0.2 M is often appropriate in biochemical experiments.

Summary

- The major concepts discussed in this chapter are summarized in two equations:

  Coulomb's law:

  $$F = k\frac{q_1 q_2}{r^2} \qquad \text{(Eq 2.1; Section 2.1)}$$

  Henderson–Hasselbalch equation:

  $$pH = pK_a + \log\frac{[A^-]}{[HA]} \qquad \text{(Eq 2.13; Section 2.4)}$$

- **Coulomb's law** provides the theoretical foundation for describing the variety of weak, noncovalent interactions that occur between ions, molecules, and parts of molecules in the cell. These interactions, which are 10 to 100 times weaker than most covalent interactions, include charge–charge interactions and the interactions of permanent and induced dipoles. The **hydrogen bond** is a strong noncovalent interaction, yet it shares some features (directionality, specificity) with covalent bonds. Because noncovalent interactions are relatively weak, they can be broken and re-formed in a dynamic fashion during the transient interactions that occur between molecules in living cells. The energies of a large number of such weak interactions can sum to a total that is sufficient to provide stability to macromolecular structures such as DNA, proteins, and membranes. (Section 2.1)

- Water is the essential matrix of life. Most of the unique properties of water as a substance are accounted for by its polarity and hydrogen-bonding properties that also make it an excellent solvent. Polar, hydrogen-bonding, and ionic substances dissolve easily in water and are called **hydrophilic,** whereas other compounds dissolve in water to only a limited extent and are called **hydrophobic. Amphipathic molecules,** which have both polar and nonpolar parts, form distinctive structures such as monolayers, vesicles, and micelles when in contact with water. Such molecules form the membrane bilayers that surround cells and cellular compartments. (Sections 2.2–2.3)

- The ionization of weak acids and bases is of major importance in biochemistry because it establishes the charges on biomolecules. Most processes in living cells occur in the pH range between 6.5 and 8.0, called the *physiological pH range,* which is maintained by buffer compounds present in cells. The behavior of **weak acids** and their **conjugate bases** is described by the **Henderson–Hasselbalch equation,** which relates the conjugate base/undissociated acid ratio to pH and pK_a. Titration curves show that the pH change with added acid or base is most gradual in the range near the pK_a of the acid; this is the basis for preparation of **buffer solutions.** (Section 2.4)

- The behavior of macroions (polyampholytes and polyelectrolytes) in solution depends on pH and on the presence of small ions, which screen the macroions from each other's charges. An **ampholyte** has both acidic and basic ionizing groups; the molecules can have a net positive, zero, or net negative charge, depending on solution pH. A **polyampholyte** has many acidic and basic groups. The **isoelectric point** of an ampholyte or a polyampholyte is the pH at which the average net charge of the molecules is zero. **Polyelectrolytes** have multiple ionizing groups with a single kind of charge. The magnitude of screening due to the presence of small ions depends on the ionic strength of the solution. (Section 2.5)

Problems

Enhanced by Mastering Chemistry for Biochemistry

Mastering Chemistry for Biochemistry provides select end-of-chapter problems and feedback-enriched tutorial problems, animations, and interactive figures to deepen your understanding of complex topics while practicing problem solving.

Answers to red problems are available in the Answer Appendix.

1. Suppose a chloride ion and a sodium ion are separated by a center–center distance of 5 Å. Is the interaction energy (the energy required to pull them infinitely far apart) predicted to be larger if the medium between them is water, or if it is *n*-pentane? (See Table 2.5.)

   If Ca^{2+}, Na^+, and F^- each have ionic radii ~1.16. Which ionic bond is stronger: Ca-F or Na-F?

   If Ca^{2+} is often bound on the surface of a protein by carboxylic acid functional groups. If the pK_a of a particular —COOH group is 4.2, would you predict Ca^{2+} to be most tightly bound at pH 8, pH 4.2, or pH 3? Explain your answer.

2. Draw two different possible hydrogen-bonding interactions between two molecules of formamide ($HCONH_2$). Clearly label the hydrogen-bond donor and acceptor atoms. Which of these two possible hydrogen-bonding interactions is more likely to occur? (*Hint:* Consider resonance structures for formamide.)

3. The accompanying graph depicts the interaction energy between two water molecules situated so that their dipole moments are parallel and pointing in the same direction. Sketch an approximate curve for the interaction between two water molecules oriented with *antiparallel* dipole moments.

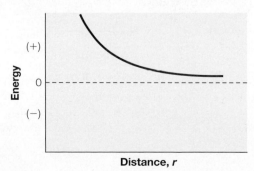

Distance, *r*

4. What is the pH of each of the following solutions? (Note that it may be necessary to use the quadratic formula to solve one or more of these problems.)
   (a) 0.35 M hydrochloric acid
   (b) 0.35 M acetic acid
   (c) 0.035 M acetic acid
   (d) Explain the differences in the pH values between the solutions in parts a–c.

5. (a) Calculate the pH of a 1 M NH_4Cl solution (see Figure 2.19).
   (b) Calculate the pH of the solution that results following addition of 10 mL of 1 M NaOH to 40 mL of 1 M NH_4Cl.
   (c) Calculate the pH of the solution that results following addition of 30 mL of 1 M NaOH to 40 mL of 1 M NH_4Cl.

6. The weak acid HA is 2% ionized (dissociated) in a 0.20 M solution.
   (a) What is K_a for this acid?
   (b) What is the pH of this solution?

7. Calculate the pH values and draw the titration curve for the titration of 500 mL of 0.010 M acetic acid (pK_a 4.76) with 0.010 M KOH.

8. What is the pH of the following buffer mixtures?
   (a) 100 mL 1 M acetic acid plus 100 mL 0.5 M sodium acetate
   (b) 250 mL 0.3 M phosphoric acid plus 250 mL 0.8 M KH_2PO_4

9. (a) Suppose you wanted to make a buffer of exactly pH 7.00 using KH_2PO_4 and Na_2HPO_4. If the final solution was 0.1 M in KH_2PO_4, what concentration of Na_2HPO_4 would you need?
   (b) Now assume you wish to make a buffer at the same pH, using the same substances, but want the total phosphate molarity ($[HPO_4^{2-}] + [H_2PO_4^{-}]$) to equal 0.3. What concentrations of the KH_2PO_4 and Na_2HPO_4 would be required?

10. A 500 mL sample of a 0.100 M formate buffer, pH 3.75, is treated with 5 mL of 1.00 M KOH. What is the pH following this addition?

11. You need to make a buffer whose pH is 7.0, and you can choose from the weak acids shown in Table 2.6. Briefly explain your choice.

12. Describe the preparation of 2.00 L of 0.100 M glycine buffer, pH 9.0, from glycine and 1.00 M NaOH. What mass of glycine is required, and what volume of 1.00 NaOH is required? The appropriate pK_a of glycine is 9.6.

13. Carbon dioxide is dissolved in blood (pH 7.4) to form a mixture of carbonic acid and bicarbonate. Neglecting free CO_2, what fraction will be present as carbonic acid? Would you expect a significant amount of carbonate (CO_3^{2-})?

14. What is the molecular basis for the observation that the overall charge on a protein becomes increasingly positive as pH drops and more negative as pH increases?

15. The amino acid *arginine* ionizes according to the following scheme:

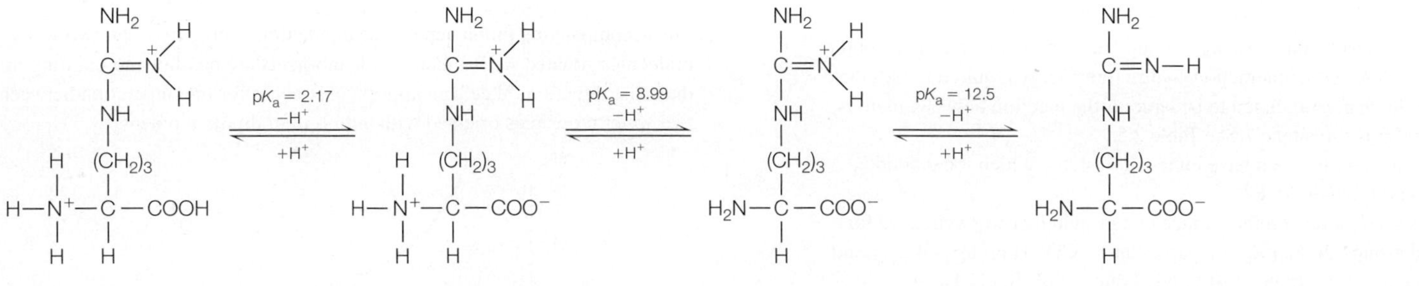

(a) Calculate the isoelectric point of arginine. You can neglect contributions from form I. Why?
(b) Calculate the average charge on arginine when pH = 9.20. (*Hint:* Find the average charge for each ionizable group and sum these together.)
(c) Is the value of average charge you calculated in part b reasonable, given the pI you calculated in part a? Explain your answer.

16. It is possible to make a buffer that functions well near pH 7 using citric acid, which contains only carboxylate groups. Explain.

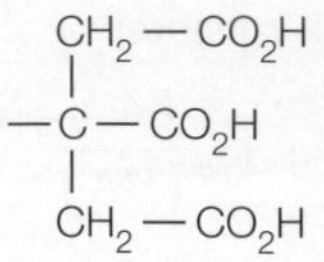

$$CH_2—CO_2H$$
$$|$$
$$HO—C—CO_2H$$
$$|$$
$$CH_2—CO_2H$$

Citric acid

17. A student is carrying out a biological preparation that requires 1 M NaCl to maintain an ionic strength of 1.0. The student chooses to use 1.0 M ammonium sulfate instead. Why is this a serious error?

18. Histidine is an amino acid with three titratable groups: an $-NH_3^+$ group ($pK_a = 9.2$), a $-COOH$ group ($pK_a = 1.8$), and an imidazole (amine-like) group ($pK_a = 6.0$). The titration curve for histidine is shown below with four points highlighted.
   (a) Identify which point on the titration curve corresponds to the pK_a for each of the titratable groups, and which point corresponds to the pI. Explain your choices.
   (b) Calculate the value of pI for histidine.

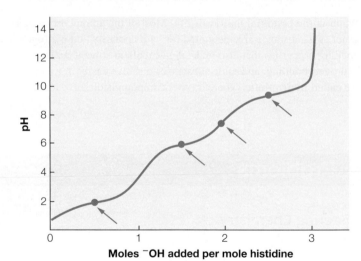

Moles $^-$OH added per mole histidine

19. The activity of an enzyme requires a glutamic acid to display its —COOH functional group in the *protonated* state. Suppose the pK_a of the —COOH group is 4.07.

(a) Will the enzyme be more active at pH 3.5 or 4.5? Explain.

(b) What fraction of the enzymes will be active at pH = 4.07? Explain.

(c) At what pH will the enzyme show 78% of maximal activity?

20. A biochemical reaction takes place in a 1.00 ml solution of 0.0250 M phosphate buffer initially at pH = 7.20 (see Table 2.6 for pK_as of phosphate species).

(a) Are the concentrations of any of the four possible phosphate species negligible? If so, identify them and explain your answer.

(b) During the reaction, 3.80 μmol of HCl are produced. Calculate the final pH of the reaction solution. Assume that the HCl is completely neutralized by the buffer.

21. Is RNA-binding enzyme RNase A more likely to have a pI of 9.2 or 5.0? Briefly explain your reasoning.

22. Consider a protein in which a negatively charged glutamic acid side chain (pK_a = 4.2) makes a salt bridge (ion–ion interaction) with a positively charged histidine side chain (pK_a = 6.5).

(a) Do you predict that this salt bridge will become stronger, become weaker, or be unaffected as pH increases from pH = 7.0 to pH = 7.5?

(b) Justify your answer with calculations of partial charges on these amino acid side chains. (*Hint*: Consider lessons from Coulomb's law, and the Henderson-Hasselbalch equation.)

Problems 23 and 24 refer to material presented in Tools of Biochemistry 2A.

23. What is the optimum pH to separate a mixture of lysine, arginine, and cysteine using electrophoresis? Draw the structures of the three amino acids *in the protonation state that would predominate at the pH you have chosen*. (See Figure 5.3 and Table 5.1.) For each amino acid, indicate the net charge at the chosen pH as well as the direction of migration and relative mobility in the electric field.

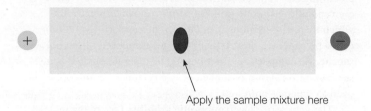

Apply the sample mixture here

24. Suppose you have two genetic variants of a large protein that differ only in that one contains a histidine (side chain pK_a = 6.0) when the other has a valine (uncharged side chain).

(a) Which would be better for separation: gel electrophoresis or isoelectric focusing? Why?

(b) What pH would you choose for the separation?

References

For a list of references related to this chapter, see Appendix II.

When an electric field is applied to a solution, solute molecules with a net positive charge migrate toward the cathode, and molecules with a net negative charge move toward the anode. This migration is called **electrophoresis.** Although electrophoresis can be carried out free in solution, it is more convenient to use some kind of *supporting medium* through which the charged molecules move. The supporting medium could be paper or, most typically, a gel composed of the polysaccharide agarose (commonly used to separate nucleic acids; see **FIGURE 2A.1**) or crosslinked polyacrylamide (commonly used to separate proteins).

The velocity, or **electrophoretic mobility (μ),** of the molecule in the field is defined as the ratio between two opposing factors: the force exerted by the electric field on the charged particle, and the frictional force exerted on the particle by the medium:

$$\mu = \frac{Ze}{f} \qquad (2A.1)^3$$

The numerator equals the product of the negative (or positive) charge (e) times the number of unit charges, Z (a positive or negative integer). The greater the overall charge on the molecule, the greater the force it experiences in the electric field. The denominator f is the **frictional coefficient,** which depends on the size and shape of the molecule. Large or asymmetric molecules encounter more frictional resistance than small or compact ones and consequently have larger frictional coefficients. Equation 2A.1 tells us that the mobility of a molecule depends on its charge and on its molecular dimensions.[‡] Because ions and macroions differ in both respects, electrophoresis provides a powerful way of separating them.

Gel Electrophoresis

In **gel electrophoresis,** a gel containing the appropriate buffer solution is cast in a mold (for agarose gel electrophoresis, shown in Figure 2A.1) or as a thin slab between glass plates (for polyacrylamide gel electrophoresis, shown in **FIGURE 2A.2**). The gel is placed between electrode compartments, and the samples to be analyzed are carefully pipetted into precast notches in the gel, called wells. Usually, glycerol and a water-soluble anionic "tracking" dye (such as bromophenol blue) are added to the samples. The glycerol makes the sample solution dense, so that it sinks into the well and does not mix into the buffer solution. The dye migrates faster than most macroions, so the experimenter is able to follow the progress of the experiment. The current is turned on until the tracking dye band is near the side of the gel opposite the wells. The gel is then removed from the apparatus and is usually stained with a dye that binds to proteins or nucleic acids. Because the protein

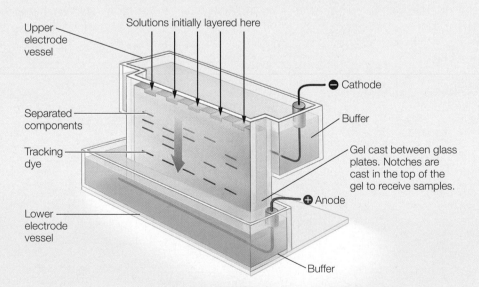

▲ **FIGURE 2A.1 Electrophoresis.** A molecule with a net positive charge will migrate toward the cathode, whereas a molecule with a net negative charge will migrate toward the anode.

or nucleic acid mixture was applied as a narrow band in the well of the gel, components migrating with different electrophoretic mobilities appear as separated bands on the gel. **FIGURE 2A.3** shows an example of separation of DNA fragments by this method using an agarose gel. An example of the electrophoretic separation of proteins using a polyacrylamide gel is shown in Chapter 5 (see Figure 5A.9).

▲ **FIGURE 2A.2 Gel electrophoresis.** An apparatus for polyacrylamide gel electrophoresis is shown schematically. The gel is cast between plates. The gel is in contact with buffer in the upper (cathode) and lower (anode) reservoirs. A sample is loaded into one or more wells cast into the top of the gel, and then current is applied to achieve separation of the components in the sample.

[‡]Equation 2A.1 is an approximation which neglects the effects of the ion atmosphere. See van Holde, Johnson, and Ho in Appendix II for more detail.

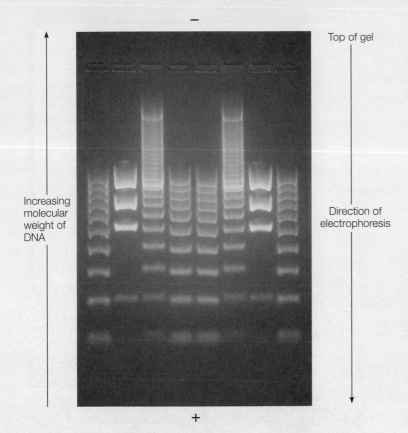

Increasing molecular weight of DNA

− Top of gel

Direction of electrophoresis

+

▲ FIGURE 2A.3 **Gel showing separation of DNA fragments.** Following electrophoretic separation of the different-length DNA molecules, the gel is mixed with a fluorescent dye that binds DNA. The unbound dye is then washed off, and the stained DNA molecules are visualized under ultraviolet light.

Polyelectrolytes like DNA or polylysine have one unit charge on each residue, so each molecule has a charge (Ze) proportional to its molecular length. But the frictional coefficient (f) also increases with molecular length, so to a first approximation, a macroion whose charge is proportional to its length has an electrophoretic mobility almost independent of its size. However, gel electrophoresis introduces additional frictional forces that allow the separation of molecules based on size. For linear molecules like the nucleic acid fragments in Figure 2A.3, the relative mobility in an agarose gel is a pproximately a linear function of the logarithm of the molecular weight. Usually, standards of known molecular weight are electrophoresed in one or more lanes on the gel. The molecular weight of the sample can then be estimated by comparing its migration in the gel to those of the standards. For proteins, a similar separation in a polyacrylamide gel is achieved by coating the denatured protein molecule with the anionic detergent sodium dodecylsulfate (SDS) before electrophoresis. This important technique is discussed further in Chapter 5.

Isoelectric Focusing

Proteins are polyampholytes; thus, a protein will migrate in an electric field like other ions if it has a net positive or negative charge. At its isoelectric point, however, its net charge is zero, and it is attracted to neither the anode nor the cathode. If we use a gel with a stable pH gradient covering a wide pH range, each protein molecule in a complex mixture of proteins migrates to the position of its isoelectric point and accumulates there. This method of separation, called **isoelectric focusing,** produces distinct bands of accumulated proteins and can separate proteins with very small differences in the isoelectric point (**FIGURE 2A.4**). Since the pH of each portion of the gel is known, isoelectric focusing can also be used to determine experimentally the isoelectric point of a particular protein.

What we have presented here is only a brief overview of a widely applied technique. Additional information on electrophoresis and isoelectric focusing can be found in Appendix II.

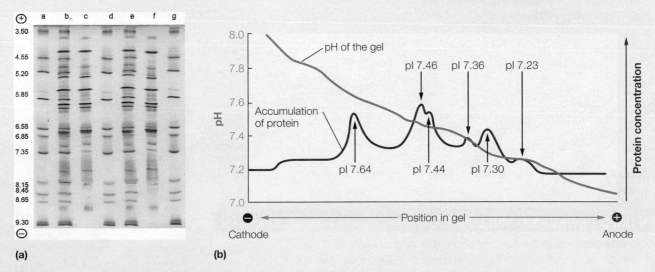

(a)   (b)

▲ FIGURE 2A.4 **Isoelectric focusing of proteins.** **(a)** An isoelectric focusing gel with a pH gradient from 3.50 (anode end) to 9.30 (cathode end). **(b)** A schematic showing where proteins of the indicated pIs would accumulate (peaks shown in red) in a pH gradient gel.

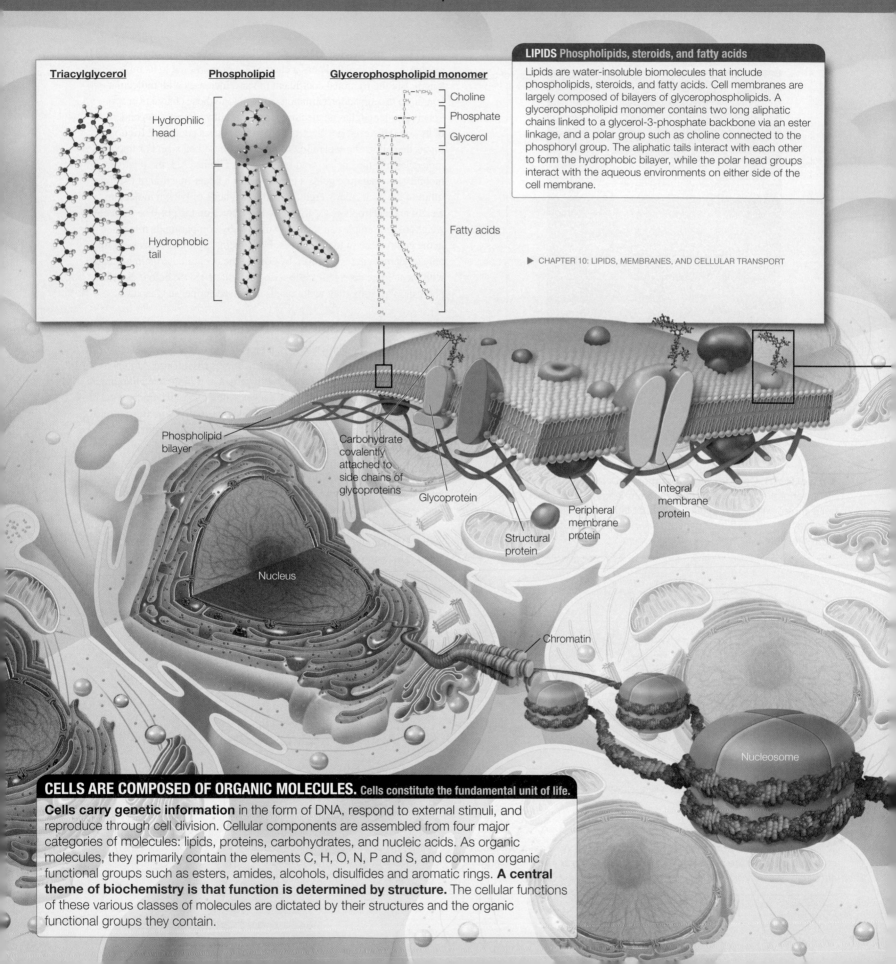

Triacylglycerol   **Phospholipid**   **Glycerophospholipid monomer**

Hydrophilic head

Hydrophobic tail

Choline

Phosphate

Glycerol

Fatty acids

LIPIDS Phospholipids, steroids, and fatty acids

Lipids are water-insoluble biomolecules that include phospholipids, steroids, and fatty acids. Cell membranes are largely composed of bilayers of glycerophospholipids. A glycerophospholipid monomer contains two long aliphatic chains linked to a glycerol-3-phosphate backbone via an ester linkage, and a polar group such as choline connected to the phosphoryl group. The aliphatic tails interact with each other to form the hydrophobic bilayer, while the polar head groups interact with the aqueous environments on either side of the cell membrane.

▶ CHAPTER 10: LIPIDS, MEMBRANES, AND CELLULAR TRANSPORT

Phospholipid bilayer

Carbohydrate covalently attached to side chains of glycoproteins

Glycoprotein

Structural protein

Peripheral membrane protein

Integral membrane protein

Nucleus

Chromatin

Nucleosome

CELLS ARE COMPOSED OF ORGANIC MOLECULES. Cells constitute the fundamental unit of life.

Cells carry genetic information in the form of DNA, respond to external stimuli, and reproduce through cell division. Cellular components are assembled from four major categories of molecules: lipids, proteins, carbohydrates, and nucleic acids. As organic molecules, they primarily contain the elements C, H, O, N, P and S, and common organic functional groups such as esters, amides, alcohols, disulfides and aromatic rings. **A central theme of biochemistry is that function is determined by structure.** The cellular functions of these various classes of molecules are dictated by their structures and the organic functional groups they contain.

Mastering **Chemistry** for Biochemistry

Mastering Chemistry for Biochemistry provides select end-of-chapter problems and feedback-enriched tutorial problems, animations, and interactive figures to deepen your understanding of complex topics while practicing problem solving.

PROTEINS Macromolecules containing one or more polypeptide chains

Proteins perform a wide range of cellular functions. These polypeptides are polymers of α-amino acids linked together via amide, or peptide, bonds. Three of the 20 common α-amino acids are shown at right; each amino acid contains an amine group and a carboxylic acid functional group. The order of amino acid residues in a polypeptide dictates its folding into a specific and highly complex structure, such as the protein ubiquitin shown here.

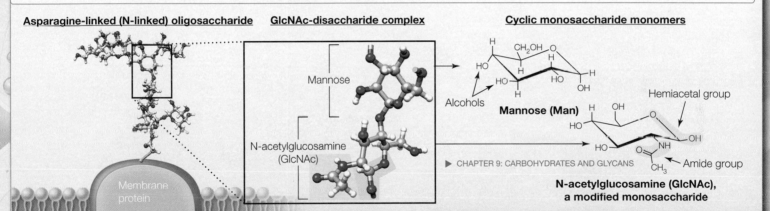

Protein (ubiquitin)

Polypeptide chain

Gly · Ala · Val

Amino acid monomers

Valine (Val)

Carboxylic acid group (–)

Amine group (+)

Glycine (Gly)

Alanine (Ala)

▶ CHAPTERS 5 AND 6: INTRODUCTION TO PROTEINS

CARBOHYDRATES Sugars, saccharides, and glycans

Also called saccharides, they have the simple empirical formula $(CH_2O)_n$. Carbohydrates are used diversely in cells, from energy storage to structure to cell recognition. The ring forms of monosaccharides are polyalcohols with hemiacetal or hemiketal functional groups. Cyclic monosaccharides can be modified in a variety of ways and covalently linked via glycosidic bonds to form highly complex structures. Here an N-linked polysaccharide is shown attached to an integral membrane protein.

Asparagine-linked (N-linked) oligosaccharide

GlcNAc-disaccharide complex

Cyclic monosaccharide monomers

Membrane protein

Mannose

N-acetylglucosamine (GlcNAc)

Alcohols

Mannose (Man)

Hemiacetal group

Amide group

▶ CHAPTER 9: CARBOHYDRATES AND GLYCANS

N-acetylglucosamine (GlcNAc), a modified monosaccharide

NUCLEIC ACIDS Polymers of nucleotides linked via phosphodiester bonds

Nucleotides are made up of a ribose or deoxyribose sugar with an aromatic nitrogenous base attached. These bases display hydrogen bond donors and acceptors which interact in a complementary fashion between the two strands of the double helix. This base-pairing complementarity establishes the mechanism to maintain the fidelity of genetic information. Individual nucleotides, such as ATP, can also function as energy transfer molecules.

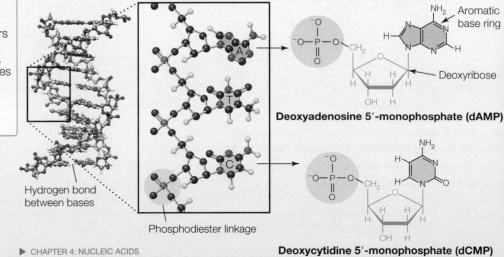

DNA

DNA double helix

DNA single strand

Nucleotide monomers

Aromatic base ring

Deoxyribose

Deoxyadenosine 5′-monophosphate (dAMP)

Hydrogen bond between bases

Phosphodiester linkage

Deoxycytidine 5′-monophosphate (dCMP)

▶ CHAPTER 4: NUCLEIC ACIDS

Luciferin + ATP + $O_2 \rightarrow$ oxyluciferin + AMP + PP_i + light. Fireflies expend significant metabolic energy to attract mates via biolumines-cence. How do organisms extract energy from their environment and use it to carry out critical cellular functions?

The Energetics of Life

A LIVING CELL is a dynamic structure. It grows, it moves, it synthesizes complex macromolecules, and it selectively shuttles substances in and out and between membrane-bound compartments. Because all of this activity requires energy, every cell and every organism must obtain energy from its surroundings and

● **CONCEPT** Bioenergetics describes how organisms capture, transform, store, and utilize energy.

expend it as efficiently as possible. Plants gather radiant energy from sunlight, whereas animals use the chemical energy stored in plants or other animals that they consume. Much of the elegant molecular machinery that exists in every cell is dedicated to the production and utilization of the energy necessary for an organism to maintain the living state.

How much food must an animal eat every day to maintain its health? Why does the brain consume energy even when "resting"? Why is it important to maintain the proper balance of ions (electrolytes) in cells? These are just a few of the questions that can be answered with a basic understanding of bioenergetics—the quantitative analysis of the capture, transformation, storage, and utilization of energy in organisms.

Bioenergetics may be regarded as a special part of the general science of energy transformations, which is called **thermodynamics.** In this chapter, we review a few fundamental concepts in that field—such as enthalpy, entropy, and free energy—that provide the biochemist or biologist with the conceptual framework to provide quantitative answers to questions such as those listed on the previous page.

3.1 Free Energy

The central purpose of this chapter is to illustrate how the **free energy change (ΔG)** of a process tells us whether that process will require energy or release energy—and if the latter is the case, how much energy it will make available to do useful work. In a cell, a favorable free energy change drives many critical processes such as protein synthesis, the folding of proteins into their functional conformations, the transport of ions across membranes, and the extraction of metabolic energy from nutrients to make **adenosine triphosphate (ATP),** which plays a central role in cellular bioenergetics.

Various thermodynamic equations are used to calculate the changes in free energy for biochemical processes. In this chapter, we introduce several of these equations and the associated fundamental concepts in cellular bioenergetics. These topics will then be developed further in subsequent chapters where the molecular details of biochemical processes are described.

Thermodynamic Systems

In any discussion involving thermodynamics, it is important to distinguish clearly between the *system* and the *surroundings.* In this context, a **system** is any part of the universe that we choose for study. It can be a single bacterial cell, a Petri dish containing nutrients and millions of cells, the whole laboratory in which this dish rests, or the entire Earth. A system must have defined boundaries, but otherwise there are few restrictions. Anything not defined as part of the system is considered to be the **surroundings.** The system may be *isolated* and thus unable to exchange energy and matter with its surroundings; it may be *closed,* able to exchange energy but not matter; or it may be *open,* so that both energy and matter can pass between the system and surroundings. For example, our planet displays the essential features of a closed system: Earth can exchange energy (e.g., in the form of electromagnetic radiation) with its surroundings, but except for a few human artifacts (spacecraft and satellites) and some astronomical debris (such as meteorites), material is not exchanged between the planet and its surroundings. In contrast, organisms are open systems, as they can exchange both energy (e.g., heat) and material (e.g., nutrients and excreted wastes) with their environments.

The First Law of Thermodynamics and Enthalpy

The **first law of thermodynamics** states that the total energy of an isolated system is constant. When extended to closed or open systems, the first law requires that energy be *conserved.* In other words, although energy can be transferred between the system and the surroundings in different ways (i.e., in the form of *heat* or *work*), in the chemical processes we will consider here, energy can be neither created nor destroyed. For the total energy of a closed system to remain constant as it loses energy by doing some work on its surroundings (e.g., as happens during muscle contraction), an equal amount of energy must be absorbed by the system. The potential to transfer matter in open systems complicates the accurate accounting of energy transfers between the system and surroundings. To a first approximation, we will assume that the principle of conservation of energy can be used to evaluate energy transfers in open systems. For example, during muscle contraction, the energy to do mechanical work comes from the energy released from the hydrolysis of ATP, which in turn, is produced by metabolic reactions in cells that extract energy from nutrients such as carbohydrates, fats, or proteins.

● **CONCEPT** According to the first law of thermodynamics, energy can be converted from one form to another; but it is conserved in a closed system.

To appreciate how much metabolic energy can be extracted from nutrients, let us consider a specific chemical reaction—the complete oxidation of 1 mole of a fatty acid, palmitic acid:

$$CH_3(CH_2)_{14}COOH \text{ (solid)} + 23\,O_2\text{(gas)} \longrightarrow$$
$$16\,CO_2\text{ (gas)} + 16\,H_2O \text{ (liquid)}$$

This is a combustion reaction that results in the complete oxidation of palmitic acid to CO_2 and H_2O. This process also releases a tremendous amount of energy in the form of heat. We can measure the heat released by igniting the mixture of O_2 and palmitic acid in a sealed vessel (a "bomb" calorimeter) immersed in a water bath (**FIGURE 3.1**). If these measurements are made under conditions of constant pressure, the heat released in the reaction is equal to the change in **enthalpy** (ΔH). For the oxidation of palmitic acid $\Delta H = -9977.6$ kJ/mol. The negative sign indicates that the reaction *releases* energy. The energy within the system *decreased* as this energy was transferred as heat to the surroundings. To put this energy change in perspective, the average human requires the expenditure of about 6000 kJ per day (roughly 1500 kcal or 1500 of the "calories" used in dietetics) just to sustain basal metabolic function. With moderate exercise, this need for metabolic energy may easily double.

● **CONCEPT** The heat evolved in a reaction at constant pressure is equal to the change in enthalpy, ΔH.

The value of any **state function** depends only on the initial and final states of the system it describes; it is therefore independent of the pathway taken to get from the initial to the final state. Examples of state functions include temperature, pressure, free energy, enthalpy, and *entropy* (defined later in this chapter). The change in enthalpy for

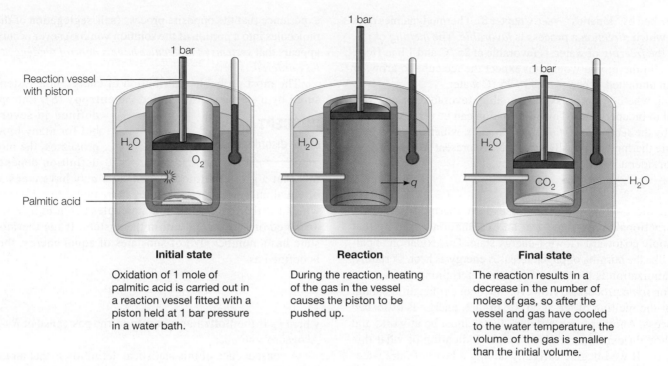

▲ FIGURE 3.1 Measuring the heat of reaction. The total amount of heat ("q" in the middle panel) delivered to the bath at constant pressure is equal to the change in enthalpy (ΔH). By using this apparatus, the calorie content of foodstuffs can be determined.

any system is defined as the difference in the enthalpy between the final and initial states of the system:

$$\Delta H = H_{final} - H_{initial} \qquad (3.1)$$

The pathway independence of state functions is critically important for the analysis of bioenergetics. For example, the complete oxidation of palmitic acid is an important biochemical reaction that takes place in our bodies in a much more indirect way than that described in Figure 3.1. Even so, measuring changes in ΔH in a calorimeter is of practical use to biochemists and dieticians because the values of ΔH for the oxidation of palmitic acid are *exactly* the same in both pathways. Because ΔH depends only on the final and initial states, the calorimeter provides an exact measurement of the energy available to a human from each mole of palmitic acid oxidized completely to CO_2 and H_2O in a mitochondrion.

● **CONNECTION** The calorie content of food can be determined by measuring ΔH of combustion in a calorimeter.

The Driving Force for a Process

However useful the first law may be for keeping track of energy changes in processes, it cannot give us one very important piece of information: What is the thermodynamically favored direction for a process? The first law cannot answer questions like the following:

- We place an ice cube in a glass of water at room temperature. It melts. Why doesn't the rest of the water freeze instead?

- We place an ice cube in a jar of supercooled water. All the water freezes. Why?

- We touch a lit match to a piece of paper. The paper burns to carbon dioxide and water. Why can't we mix carbon dioxide and water to form paper?

One characteristic of such processes is their *irreversibility* under the conditions described above. An ice cube in a glass of room-temperature water at 1 bar will continue to melt—there is no way to turn that process around without making major changes in the conditions. But there is a *reversible* way to melt ice—to have it in contact with water at 0 °C and 1 bar. Under these conditions, adding a bit of heat to the glass will result in a small amount of ice melting, whereas removing a little heat will cause a small amount of the water to freeze. A **reversible** process like melting ice at 0 °C is always near a state of **equilibrium.** Two defining features of the equilibrium state for a system undergoing a reaction or process are that (1) it is the lowest energy state for the system, and (2) the forward and reverse rates for the process are equal. As discussed below, lower-energy states are favored over those of higher energy; thus, systems tend to adopt states of lower energy. **Irreversible** processes (such as burning paper) happen when systems are set up far from an equilibrium state. They then drive *toward* a state of equilibrium.

● **CONCEPT** Reversible processes always occur near a state of equilibrium; irreversible processes drive toward equilibrium.

In the jargon of thermodynamics, an irreversible process is often called a "spontaneous" process. However, we prefer the word *favorable* instead because *spontaneous* implies that the process is rapid, yet thermodynamics has nothing to say about how fast a process will be (this is described by "kinetics"—see Chapter 8). Thermodynamics does indicate in which *direction* a process is *favorable*. The *melting* of ice, rather than the *freezing* of water, is favorable at 25 °C and 1 bar. Here, the result is intuitive; you would not expect the ice cube to grow, or even remain unmelted, when placed in 25 °C water.

Knowing whether a process is reversible, favorable, or unfavorable is vital to bioenergetics. This information can be expressed most succinctly by the *second law of thermodynamics,* which tells us which processes are thermodynamically favorable. To present the second law, we must consider a new concept—entropy.

Entropy

Why do chemical and physical processes have thermodynamically favored directions? A first guess at an explanation might be that systems simply go toward a lowest-energy state. The oxidation of palmitic acid, like the burning of paper, releases energy as heat. Certainly, energy minimization is the major factor in determining the favored direction for *some* processes. However, such an explanation cannot account for the melting of ice at 25 °C. In fact, energy is *absorbed* in that process. Another, very different factor must be at work, and a simple thought-experiment gives a clear indication of what this factor may be. If we imagine carefully adding a layer of pure water on top of a sucrose solution, we would observe as time passes that the solution becomes more and more uniform (**FIGURE 3.2**). Eventually, the sucrose molecules will be evenly distributed throughout the solution. Although there is practically no energy change, in terms of heat and work, the process is clearly a favorable one. We know from experience that the opposite process (self-segregation of the sucrose molecules into a portion of the solution volume) never occurs. Thus, it appears that *systems of molecules have a natural tendency to become less ordered.*

The most probable distribution of energy in a system is measured by a state function called the **entropy** (S). Entropy can be

● **CONCEPT** Entropy is a measure of the disorder in a system.

defined in several ways, but for many biochemical processes, the most useful definition depends on the fact that a given thermodynamic state may have many *substates* of equal energy. Those substates correspond to different ways in which the thermal energy and molecules (as in Figure 3.2) can be arranged or distributed within the system. If the thermodynamic state has a number (W) of substates of equal energy, the entropy is defined as

$$S = k_B \ln W \tag{3.2}$$

where k_B is the **Boltzmann constant,** the gas constant R divided by Avogadro's number.

A consequence of this statistical definition is that an increase in entropy is associated with an increase in disorder for a system. There will always be many more ways of putting a large number of molecules into a disorderly arrangement than into an orderly one; therefore, the entropy of an ordered state is lower than that of a disordered state of the same system. In fact, the minimal value of entropy (zero) is

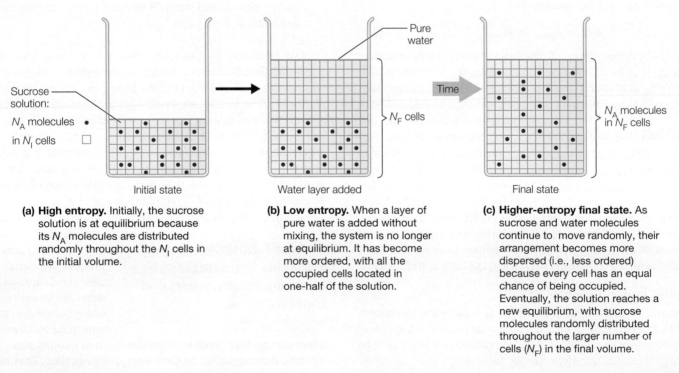

(a) **High entropy.** Initially, the sucrose solution is at equilibrium because its N_A molecules are distributed randomly throughout the N_I cells in the initial volume.

(b) **Low entropy.** When a layer of pure water is added without mixing, the system is no longer at equilibrium. It has become more ordered, with all the occupied cells located in one-half of the solution.

(c) **Higher-entropy final state.** As sucrose and water molecules continue to move randomly, their arrangement becomes more dispersed (i.e., less ordered) because every cell has an equal chance of being occupied. Eventually, the solution reaches a new equilibrium, with sucrose molecules randomly distributed throughout the larger number of cells (N_F) in the final volume.

▲ **FIGURE 3.2 Diffusion as an entropy-driven process.** The gradual mixing of a dilute sucrose solution and pure water is the result of random movement of their molecules. We can visualize the increase in entropy if we imagine the volume of the two liquids to be made up of cells, each big enough to hold one sucrose molecule. The drive toward equilibrium is a consequence of the tendency for entropy to increase. The likelihood that this system would go from state **(c)** to state **(b)** is improbably low. In other words, going from state **(c)** to state **(b)** is extremely unfavorable.

TABLE 3.1 Examples of lower-entropy and higher-entropy states

Lower Entropy	Higher Entropy
Ice, at 0 °C	Water, at 0 °C
Water, at 10 °C	Water vapor, at 10 °C (e.g., fog)
An unblended mixture of yogurt, whole bananas, honey, and whole strawberries	A fruit smoothie (i.e., the same ingredients *after* blending)

predicted *only* for a perfect crystal at the absolute zero of temperature (0 K or −273.15 °C). In Figure 3.2, the process of diffusion disperses the sucrose molecules in the solution simply because there are more ways to distribute the molecules over a larger volume than over a smaller one. In Figure 3.2, the state shown in panel (c) has more substates of equal energy (i.e., different random arrangements of the sucrose molecules in the entire volume of the solution) than does the state shown in panel (b), where sucrose molecules would be restricted to a much smaller volume. Thus, the value of *W* is greater for state (c), and this accounts for the greater entropy of state (c) compared to state (b).

In Chapter 6 we will apply the definition of entropy given in Equation 3.2 to the analysis of protein structure and stability. Meanwhile, to make the concept of entropy a bit more familiar, consider the examples given in **TABLE 3.1**.

The Second Law of Thermodynamics

It can be shown that the driving force toward equilibrium for the sucrose solution in Figure 3.2 is determined by the increase in entropy (see Appendix II). This observation can be generalized as the **second law of thermodynamics.** *The entropy of an isolated system will tend to increase to a maximum value.* The entropy of such a system will not decrease—sucrose will never "de-diffuse" into a corner of the solution. This simply reflects our commonsense understanding that things, if left alone, will not become more ordered.

● **CONCEPT** The fact that the entropy of an isolated system will tend to increase to a maximum value explains the thermodynamic driving force for a favorable process.

3.2 Free Energy: The Second Law in Open Systems

The form of the second law as stated in the previous section is not very useful to biologists or biochemists because we never deal with isolated systems. Every biological system (cell, organism, or population, for example) can exchange energy and matter with its environment. Thus, to predict whether or not a process is likely to be favorable in an open system, we must consider how entropy changes in both the system and the surroundings. This overall entropy change is commonly referred to as *the entropy of the universe* (recall that the "universe" is, by definition, an isolated system that comprises the system under study and its surroundings):

$$\Delta S_{universe} = \Delta S_{system} + \Delta S_{surroundings} \qquad (3.3)$$

Equation 3.3 allows a favorable process in an open system to be defined as one for which $\Delta S_{universe} > 0$. In other words, the entropy of the universe must increase as a consequence of favorable processes occurring in open systems such as living cells.

● **CONCEPT** For a favorable process in an open system, such as a living cell, $\Delta S_{universe}$ must increase.

Free Energy Defined in Terms of Enthalpy and Entropy Changes in the System

It is more practical to measure parameters for the system than for the surroundings. Thus, it would be more useful to define $\Delta S_{universe}$ solely in terms of the system. That can be done using the classical thermodynamic definition of entropy change at constant pressure:

$$\Delta S_{surroundings} = -\frac{\Delta H_{system}}{T} \qquad (3.4)$$

In Equation 3.4, ΔH_{system} represents the heat transferred from the system to the surroundings at constant pressure and some temperature *T*. The negative sign is included to account for the fact that any *loss* of energy by the system represents a *gain* of energy by the surroundings (and vice versa). With $\Delta S_{surroundings}$ defined as in Equation 3.4, we can reformulate Equation 3.3 solely in terms of thermodynamic parameters of the system:

$$\Delta S_{universe} = \Delta S_{system} - \frac{\Delta H_{system}}{T} \qquad (3.5)$$

Multiplying Equation 3.5 by −*T* gives:

$$-T\Delta S_{universe} = -T\Delta S_{system} + \Delta H_{system} = \Delta H_{system} - T\Delta S_{system} \qquad (3.6)$$

J. W. Gibbs recognized that at constant pressure and temperature $-TS_{universe}$ is a state function, which is defined as the **free energy** and represented by the symbol *G* in honor of Gibbs. If we define the free energy change for a process as

$$\Delta G = -T\Delta S_{universe} \qquad (3.7)$$

we can derive an important equation for predicting the favorability of a process:

$$\Delta G = \Delta H - T\Delta S \qquad (3.8)$$

● **CONCEPT** The free energy change for a process at constant temperature and pressure is $\Delta G = \Delta H - T\Delta S$.

where *T* is the absolute temperature measured in kelvin. The Gibbs free energy is relevant to bioenergetics because most cells and organisms live under conditions of constant pressure and temperature.

We can gain insight into the meaning of free energy by considering the factors that make a process favorable. A decrease in energy (i.e., $\Delta H < 0$) and/or an increase in entropy ($\Delta S > 0$) are typical of favorable processes. As shown by Equation 3.8, either of these conditions will tend to make ΔG negative. In fact, another way to state the second law of thermodynamics is this: *The criterion for a favorable process in*

TABLE 3.2 Free energy rules

If ΔG is . . .	Free energy is . . .	The process is . . .
Negative	Available to do work	Thermodynamically favorable (and the reverse process is unfavorable)
Zero	Zero	Reversible; the system is at equilibrium
Positive	Required to do work	Thermodynamically unfavorable (and the reverse process is favorable)

● **CONCEPT** A thermodynamically favored process tends in the direction that minimizes free energy (results in a negative ΔG). This is one way of stating the second law of thermodynamics.

a nonisolated system, at constant temperature and pressure, is that ΔG be negative. Conversely, a positive ΔG means that a process is not favorable; rather, the reverse of that process is favorable.

The importance of quantifying free energy lies in this *predictive* power of ΔG. Given a set of conditions for a process (e.g., temperature, concentrations of reactants and products, pH, etc.), we can calculate the value of ΔG and thereby determine whether or not the process is favorable. Processes accompanied by negative free energy changes are said to be **exergonic;** those for which ΔG is positive are **endergonic.**

Suppose that the ΔH and $T\Delta S$ terms in Equation 3.8 are equal. In this case $\Delta G = 0$, and the process is not favored to go either forward or backward. In fact, the system is at equilibrium. Under these conditions, the process is reversible; that is, it can be displaced in either direction by an infinitesimal push one way or the other. These simple but important rules about free energy changes are summarized in **TABLE 3.2.**

An Example of the Interplay of Enthalpy and Entropy: The Transition Between Liquid Water and Ice

To illustrate these concepts, let us consider in detail a process we mentioned before, the transition between liquid water and ice. This familiar example demonstrates the interplay of enthalpy and entropy in determining the state of a system. In an ice crystal, there is a maximum number of hydrogen bonds between the water molecules (see Figure 2.11(b)). When ice melts, some of these bonds must be broken. The enthalpy difference between ice and water corresponds almost entirely to the energy that must be put into the system to break these hydrogen bonds. Because energy is added to break these hydrogen bonds, the enthalpy change for the transition ice $\rightarrow$ water is positive ($\Delta H > 0$).

The entropy change in melting arises primarily because liquid water is a more disordered structure than ice. On one hand, in an ice crystal, each water molecule has a fixed place in the lattice and binds to its neighbor in the same way as every other water molecule. On the other hand, molecules in liquid water are continually moving, exchanging hydrogen-bond partners as they go (compare Figures 2.11(b) and 2.11(c)). For this reason, the entropy change for melting ice to liquid water is a positive quantity ($\Delta S > 0$). If we use Equation 3.8 to calculate

the free energy change for the ice $\rightarrow$ water transition, we find the following: At low temperatures, ΔH dominates and ΔG is positive. For example, at 263 K ($-10\,°C$), where $\Delta H = +5630$ J/mol and $\Delta S = +20.6$ J/K · mol, we find:

$$\Delta G = \Delta H - T\Delta S = 5630\,\frac{J}{mol} - (263\ K)\left(20.6\,\frac{J}{K \cdot mol}\right)$$
$$= +212\,\frac{J}{mol} \tag{3.9}$$

Because $\Delta G > 0$, the transition ice $\rightarrow$ water is *not* favorable under these conditions. In fact, the opposite transition (water $\rightarrow$ ice) is favorable (and irreversible) at this temperature.

At a temperature above the melting point of ice, say 283 K ($+10\,°C$), an ice cube will irreversibly melt. Again, we would predict this outcome from a calculation of ΔG at 283 K, where $\Delta H = +6770$ J/mol and $\Delta S = +24.7$ J/K · mol:

$$\Delta G = \Delta H - T\Delta S = 6770\,\frac{J}{mol} - (283\ K)\left(24.7\,\frac{J}{K \cdot mol}\right)$$
$$= -220\,\frac{J}{mol} \tag{3.10}$$

The sign of ΔG is now negative because the $T\Delta S$ term dominates when T becomes large enough.

At the melting temperature, 273 K (0 °C), the ΔH and $T\Delta S$ terms are both equal to 6010 J/mol; thus, $\Delta G = 0$, and we know that ice and liquid water are in equilibrium at 273 K. The change is now reversible; when ice and liquid water are together at 273 K, we can melt a bit more ice by adding an infinitesimal amount of heat. Alternatively, we can take a minute amount of heat away from the system and freeze a bit more water. At this temperature, the enthalpically favored process of freezing is in balance with the entropically favored process of melting. The melting point of any substance is simply the temperature at which the values for ΔH and $T\Delta S$ are equal. Neither ΔH nor ΔS alone can tell us what will happen, but their combination, expressed as $\Delta H - T\Delta S$, defines exactly which form of water is stable at any temperature.

● **CONCEPT** At the melting point of any substance, enthalpy and entropy contributions to ΔG balance, and $\Delta G = 0$.

The Interplay of Enthalpy and Entropy: A Summary

For all chemical and physical processes, it is the relationship between enthalpy and entropy terms that determines the favorable direction. As **FIGURE 3.3** shows, in some processes the enthalpy change drives the overall process, whereas in others the entropy change is the dominant factor. Furthermore, because ΔS is multiplied by T in Equation 3.8, the favorable direction may change as a function of temperature, depending on the signs of ΔH and ΔS. **TABLE 3.3** lists the possibilities. Note that when ΔH is negative and ΔS is positive, ΔG must always be negative, so the reaction is favorable at all temperatures. The reverse is true when ΔH is positive and ΔS is negative; ΔG is always positive, and the reaction is not favorable at any temperature.

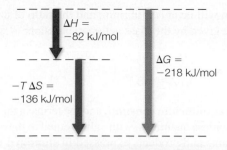

(a) Fermentation of glucose to ethanol

$C_6H_{12}O_6(s) \longrightarrow 2C_2H_5OH(l) + 2CO_2(g)$

ΔH negative, $-T\Delta S$ negative (because ΔS is positive). Thus, both the enthalpy and the entropy changes favor this reaction.

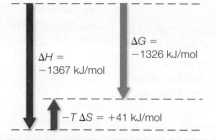

(b) Combustion of ethanol

$C_2H_5OH(l) + 3O_2(g) \longrightarrow 2CO_2(g) + 3H_2O(l)$

ΔH negative, $-T\Delta S$ positive (because ΔS is negative). The entropy decrease arises because there are fewer moles of gases in the products compared to the reactants. The negative ΔH favors this reaction, but the negative ΔS opposes it. If water vapor were the product, the associated entropy increase would favor the reaction as well.

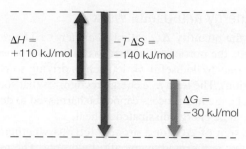

(c) Decomposition of nitrogen pentoxide

$N_2O_5(s) \longrightarrow 2NO_2(g) + \frac{1}{2}O_2(g)$

ΔH and ΔS both positive. The positive ΔH opposes this reaction, but the positive ΔS favors it. The large entropy increase results from the formation of gaseous products.

▲ **FIGURE 3.3 Contributions of enthalpy and entropy to favorable processes.** Each of the processes shown in panels **(a)–(c)** has a negative free energy change, but the change is produced in different ways. (Note that the arrows in the diagrams are not to scale.)

TABLE 3.3 The effect of temperature on ΔG for a process depends on the signs of ΔH and ΔS

ΔH	ΔS	Low T	High T
+	+	ΔG positive; not favored	ΔG negative; favored
+	−	ΔG positive; not favored	ΔG positive; not favored
−	+	ΔG negative; favored	ΔG negative; favored
−	−	ΔG negative; favored	ΔG positive; not favored

Two matters that frequently cause confusion should be cleared up at this point. First, we must emphasize something we have mentioned already: The thermodynamic favorability of a process does not determine its *rate*. A reaction may have a large negative free energy change but still proceed at a slow rate (for reasons discussed in Chapter 8). A surprising example of this situation is the simple reaction C (diamond) → C (graphite). The free energy change for this transformation, at room temperature, is −2880 J/mol. Thus, relative to graphite, diamond is unstable. If you have a diamond ring, it should be turning into graphite—and in fact it is doing so as you read this sentence. Yet the reaction is imperceptibly slow because it is very difficult for the rigid crystal lattice to change its form. A **catalyst** may increase the rate for some reactions, but the favored direction is always dictated by ΔG and is independent of whether or not the reaction is catalyzed. We will see in Chapter 8 that the protein catalysts called **enzymes** selectively increase the rates for specific, *thermodynamically favorable* reactions in cells.

● **CONCEPT** Favorable processes are not necessarily rapid.

Second, the entropy of an open system can *decrease*. We have just seen that this happens whenever water freezes. More important to biochemists is that decreases in entropy happen all the time in living organisms. An organism takes in nutrients, often in the form of high-entropy small molecules, and from them it builds enormous, complex, highly ordered macromolecules like proteins and nucleic acids. From these macromolecules, it constructs elegantly structured cells, tissues, and organs. All of this activity involves a tremendous entropy decrease. The implication of Equation 3.8 is that entropy can decrease in a favored process, but only if this change is accompanied by a large enthalpy decrease. *Energy must be expended to pay the price of organization.* This exchange is what life is all about. Living organisms spend energy to overcome entropy. For these life processes to occur, the *overall* free energy changes in the organism must be negative; thus, life is an irreversible process. An organism that comes to equilibrium with its surroundings is dead.

● **CONCEPT** Life involves a temporary decrease in entropy, paid for by the expenditure of energy.

Bioenergetics has an even deeper philosophical implication. The universe as a whole is an isolated system. Thus, *the entropy of the whole universe must be increasing.* It follows that each of us, as a living organism that locally and temporarily decreases entropy, must produce an increase in entropy somewhere in the world around us. As we metabolize food, for example, we give off heat and increase random molecular motion around us. Thus, life is sustained in exchange for increasing the entropy of the universe.

Free Energy and Useful Work

Why is the quantity ΔG called *free* energy? The reason is that ΔG represents the portion of an energy change in a process that is *available,* or *free,* to do useful work (such as driving some process toward completion). The term *free* energy recognizes that some portion of the energy change in a process cannot be harnessed to do work; typically, this would be energy dissipated as heat.

The sign of ΔG for a process tells us whether that process, or its reverse, is thermodynamically favorable. The magnitude of ΔG is an indication of how far the process is from equilibrium and how much useful work may be obtained from it. Thus, ΔG is a quantity of fundamental importance in determining which processes will or will not occur in a cell and for what purposes they may be used. Knowing that ΔG measures the maximum amount of useful work that can be obtained from a chemical process is of great importance to biochemistry because quantifying free energy changes allows us to answer questions such as: Which chemical reactions are capable of achieving muscle contraction or cell motility? How much energy is required to transport ions across membranes in neurons to support the proper function of nervous tissues?

To determine how much work a typical biological system can do, we need to answer this question: How does the free energy of a system depend on the concentrations of various components in the system? In the next two sections of this chapter, we derive an alternative expression for ΔG that allows us to answer that question.

● **CONCEPT** The free energy change, ΔG, is a measure of the maximum useful work obtainable from any reaction.

3.3 The Relationships Between Free Energy, the Equilibrium State, and Nonequilibrium Concentrations of Reactants and Products

An important distinction between reactions carried out in a laboratory and those carried out in the cells of a living organism is that the reactions in the laboratory come to equilibrium, whereas those in a living organism do not—until the organism dies. In the following pages, we will see that to properly evaluate ΔG for a reaction in a cell we must know how far the reaction conditions are from the equilibrium state. For many biochemical reactions, this is determined primarily by the relative concentrations of reactants and products.

Equilibrium, Le Chatelier's Principle, and the Standard State

To begin, let us consider the general reaction scheme

$$aA + bB \rightleftharpoons cC + dD \tag{3.11}$$

Even though the reaction can proceed in either direction, we have written A and B on the left as *reactants* and C and D on the right as *products.* Here the symbols *a, b, c,* and *d* represent the coefficients that are needed to balance the chemical equation. The **Law of Mass Action** states that when the rates of the forward and reverse reactions

are equal, the system will be at equilibrium, and the ratio of products and reactants will be given by the mass action expression:

$$K = \left(\frac{[C]^c [D]^d}{[A]^a [B]^b} \right)_{eq} \tag{3.12}$$

Here K represents an **equilibrium constant,** and the terms in the square brackets are dimensionless quantities, with values equal to the concentrations of the reactants and products expressed in **standard units.*** In solutions, the standard concentration unit for each solute is **mol per liter** (M). Thus, if A, B, C, and D are solutes, we could calculate K using Equation 3.12 and their respective *molar* concentrations.

Le Chatelier's Principle states that a system at equilibrium will respond to any perturbation to the equilibrium by moving to reestablish the equilibrium state as defined by Equation 3.12. What would happen if the reaction described by Equation 3.11 were at equilibrium, and we perturbed that equilibrium by increasing the concentration of solute A? Le Chatelier predicts that the system would act to restore the equilibrium ratio of reactants and products by increasing the concentrations of solutes C and D, and decreasing the concentrations of solutes B and A.

● **CONCEPT** Le Chatelier's Principle states that for any system not at equilibrium, there is a thermodynamic driving force that favors reestablishing the equilibrium state.

Another way to state Le Chatelier's Principle is that *for any system not at equilibrium, there is a driving force toward the equilibrium state.* The important concept for this discussion is that the magnitude of this driving force is described by ΔG, which is determined, in turn, by the relative concentrations of reactants and products. How can we use this concept to evaluate ΔG for a given process? First we must define a point of reference, called the **standard free energy,** or $\Delta G°$, which is related to the *chemical standard state.*

The **chemical standard state** is defined by the following conditions: pressure $=1$ bar, $T = 25\,°C$ (273.15 K), and solutes at concentrations of 1 M each. For example, let us assume that for the reaction above (Equation 3.11), $K \neq 1$. If solutes A, B, C, and D were mixed together at 25 °C such that they were each at their standard state concentration of 1 M, then the system would not be at equilibrium. This is evident by comparing the value of the mass action expression for these conditions to K. We use the symbol Q to distinguish the value of the mass action expression for a system that is not at equilibrium from K (which is a *constant* that only describes mass action for a system at equilibrium). Thus, under standard state conditions:

$$Q = \frac{[C]^c [D]^d}{[A]^a [B]^b} = \frac{(1)^c (1)^d}{(1)^a (1)^b} = 1 \tag{3.13}$$

Recall that we assumed $K \neq 1$ for this reaction; thus, in this case the standard state $\neq$ the equilibrium state. Because this system is not at equilibrium, Le Chatelier's Principle predicts a driving force that

---

*As mentioned in Chapter 2, the proper terms to use in a mass action expression are *activities* rather than *concentrations.* However, in most biochemical reactions the differences between the values of activities and *molar* concentration are not significant. Thus, we can frequently substitute molar concentration values for activities without introducing significant error in the evaluation of the relevant mass action expression.

would move the system from the standard state (i.e., the "reference state") toward the equilibrium state. The magnitude of that driving force is defined to be $\Delta G°$. Before we can use this definition of $\Delta G°$ to evaluate the overall driving force, ΔG, for a reaction that is not at equilibrium, we must know how changes in concentration alter the value of free energy.

Changes in Concentration and ΔG

Classical thermodynamics shows that the free energy of any component of the system increases from its standard value with a logarithmic dependence on its *activity*.[†] For example, the free energy, G_A, of 1 mol of solute A can be described by Equation 3.14

$$G_A = G_A° + RT \ln [A]$$ (3.14)

where $G_A°$ is the standard free energy for solute A, R is the gas constant, and T is temperature in units of kelvin. We can now use Equation 3.14 and the fact that free energy is a state function to relate the concentrations of reactants and products to ΔG for the reaction. For any chemical reaction, we can equate the reactant state with the "initial state" and the product state with the "final state." Therefore, the free energy change for any reaction will equal the free energy of the products minus that of the reactants:

$$\Delta G = G_{final} - G_{initial} = G_{(products)} - G_{(reactants)}$$ (3.15)

From the example given in Equation 3.11, the products are c mol of C and d mol of D, and the reactants are a mol of A and b mol of B. The driving force for the reaction, ΔG, is the total free energy of the products minus that of the reactants, which is derived by combining Equations 3.11 and 3.15, and multiplying the free energy term for each component by the proper coefficient given in Equation 3.11:

$$\Delta G = cG_C + dG_D - aG_A - bG_B$$ (3.16)

Substitution of the appropriate expressions for G_A, G_B, and so forth from Equation 3.14 into Equation 3.16 yields

$$\Delta G = cG°_C + cRT \ln [C] + dG°_D + dRT \ln [D]$$
$$- aG°_A - aRT \ln [A] - bG°_B - bRT \ln [B]$$ (3.17a)

or

$$\Delta G = (cG°_C + dG°_D - aG°_A - bG°_B) + (RT \ln [C]^c$$
$$+ RT \ln [D]^d - RT \ln [A]^a - RT \ln [B]^b)$$ (3.17b)

or

$$\Delta G = \Delta G° + RT \ln \left(\frac{[C]^c [D]^d}{[A]^a [B]^b} \right)$$ (3.18)

In going from Equation 3.17a to 3.18, we have done two things: grouped the $G°$ terms for each solute into $\Delta G°$ and made use of rearrangements like: $aRT \ln [A] = RT \ln [A]^a$. $\Delta G°$ represents the **standard state free energy** for the reaction. As described in the preceding section, it represents the driving force from a reference state (i.e., the standard state—where all solute concentrations equal 1 M) toward the equilibrium state.

You will recognize that the ratio of concentrations given inside the large parentheses in Equation 3.18 is a mass action expression, which can be represented by the symbol Q (as was done in Equation 3.13). Making this substitution gives the general form of Equation 3.18:

$$\Delta G = \Delta G° + RT \ln Q$$ (3.19)

Equation 3.19 shows that ΔG depends both on the standard free energy change ($\Delta G°$), which is a reference value, and a term ($RT \ln Q$) that accounts for the *actual* concentrations of the reactants and products relative to that reference value. Thus, we can use Equation 3.19 to calculate ΔG for a reaction at *any* concentration of reactants and products that may occur in a cell.

● **CONCEPT** The free energy change in a chemical reaction depends on the standard state free energy change ($\Delta G°$) and on the concentrations of reactants and products (described by $RT \ln Q$).

ΔG versus $\Delta G°$, Q versus K, and Homeostasis *versus* Equilibrium

If a reaction comes to equilibrium, two things must be true. First, the concentrations in the mass action expression must be equilibrium concentrations. Thus, at equilibrium, the factor Q in Equation 3.19 has a value that is identical to the *equilibrium constant K* for the reaction (see Equation 3.12). Second, if the system is at equilibrium, there is no driving force in either direction, so ΔG must equal zero. In this case, Equations 3.18 and 3.19 reduce to

$$0 = \Delta G° + RT \ln \left(\frac{[C]^c [D]^d}{[A]^a [B]^b} \right)_{eq} = \Delta G° + RT \ln K$$ (3.20)

or

$$\Delta G° = -RT \ln K$$ (3.21)

This can be rearranged as

$$K = e^{-\Delta G°/RT}$$ (3.22)

Equations 3.21 and 3.22 express an important relationship between the standard free energy change, $\Delta G°$, and the equilibrium constant, K, that we shall use many times throughout this textbook. These equations make it possible, for example, to use data from tables of standard state free energy changes to calculate the value of the equilibrium constant for a reaction.

● **CONCEPT** The equilibrium constant K can be calculated from the standard state free energy change ($\Delta G°$) and vice versa.

To illustrate an application of these thermodynamic concepts, let us consider a simple but important biochemical reaction—the isomerization of glucose-6-phosphate (G6P) to fructose-6-phosphate (F6P) shown in **FIGURE 3.4**:

Glucose-6-phosphate $\rightleftharpoons$ Fructose-6-phosphate
$$\Delta G° = +1.7 \text{ kJ/mol}$$

which may be written more compactly as

$$\text{G6P} \rightleftharpoons \text{F6P} \qquad \Delta G° = +1.7 \text{ kJ/mol}$$

[†]Again, we will approximate the value of the activity by the molar concentration of the solute A.

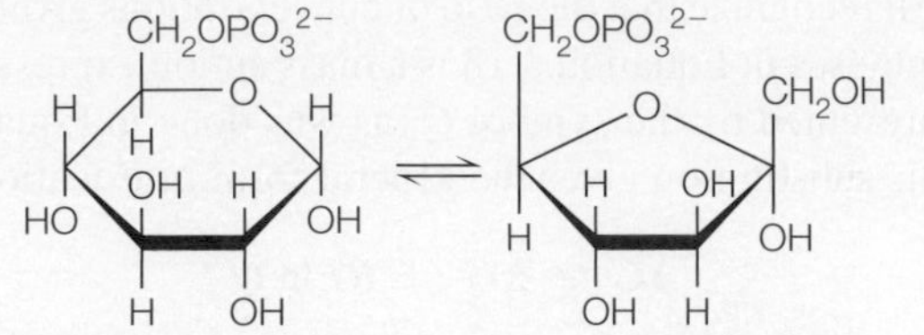

▲ **FIGURE 3.4** Isomerization of glucose-6-phosphate (G6P) to fructose-6-phosphate (F6P).

This is the second step in the *glycolytic pathway*, which is discussed in Chapter 12. The reaction is clearly endergonic under standard conditions. In other words, the system is not at equilibrium when G6P and F6P are both at 1 M because $\Delta G°$ is positive (+1.7 kJ/mol), and the reverse reaction is favored under standard conditions. Therefore, the equilibrium must lie to the left, with a higher concentration of G6P than F6P. We can express this quantitatively by calculating the equilibrium constant using Equation 3.22. Using the given value for $\Delta G°$ and the standard state temperature of 25 °C (298 K), we obtain:

$$K = e^{\left(\frac{-\Delta G°}{RT}\right)} = e^{\left(\frac{-\left(1700 \frac{J}{mol}\right)}{\left(8.314 \frac{J}{mol \cdot K}\right)(298 \text{ K})}\right)} = 0.504 = \left(\frac{[\text{F6P}]}{[\text{G6P}]}\right)_{eq} \quad (3.23)$$

where $([\text{F6P}]/[\text{G6P}])_{eq}$ is the equilibrium ratio of the concentrations of fructose-6-phosphate and glucose-6-phosphate. The fact that $K < 1$ shows that the equilibrium lies to the left or favors the reactants. In this case, G6P will have a concentration about twice that of F6P at equilibrium.

However, the concentrations of metabolites for any process in *living* cells are not typically the equilibrium concentrations defined by $\Delta G°$ for that process. The state of chemical equilibrium is characteristic of dead cells, not living ones. As stated at the beginning of this chapter, characteristics of the living state include the capture, transformation, storage, and utilization of energy. Such processes can occur only when each has a favorable thermodynamic driving force, that is, when $\Delta G < 0$.

Life occurs within relatively narrow ranges of temperature, pH, and concentrations for ions and metabolites. This set of conditions is referred to as **homeostasis** or the **homeostatic condition.** Because the concentrations of certain solutes inside cells remain relatively constant, homeostasis is frequently confused with true thermodynamic equilibrium; however, *homeostasis must not be confused with equilibrium!* The critical distinction between the two is that, under homeostatic conditions, ΔG for numerous vital processes is < 0, whereas, at equilibrium, ΔG for any process $= 0$. In addition, *energy is required to maintain homeostasis*—hence, the need to capture, transform, and store energy.

● **CONCEPT** The homeostatic condition, which is far from equilibrium and a characteristic of living cells, must not be confused with true thermodynamic equilibrium.

We know from Le Chatelier's Principle that any system that is not at equilibrium has a driving force ($\Delta G < 0$) to move in the direction that reestablishes the equilibrium state. We can understand the magnitude of these driving forces by comparing equilibrium and nonequilibrium systems. As stated earlier, for the reaction shown in Equation 3.11

$$a\text{A} + b\text{B} \rightleftharpoons c\text{C} + d\text{D}$$

at equilibrium, Le Chatelier's Principle predicts that a perturbation of the equilibrium by increasing the concentration of *either* reactant, [A] or [B], would result in an increase in the concentrations of *both* products, [C] and [D], as the system moves back to the equilibrium state. Similarly, if the equilibrium were perturbed by an increase in either [C] or [D], the system would respond by increasing [A] and [B]. We can now express Le Chatelier's Principle in terms of the thermodynamic arguments developed above. Let us reconsider Equation 3.19

$$\Delta G = \Delta G° + RT \ln Q$$

and substitute $-RT \ln K$ for $\Delta G°$ (see Equation 3.21) to give:

$$\Delta G = -RT \ln K + RT \ln Q \quad (3.24)$$

For a system at equilibrium $Q = K$; thus

$$\Delta G = -RT \ln K + RT \ln K = 0 \quad (3.25)$$

Because $\Delta G = 0$, there is no driving force in either direction. On the other hand, for a system that is not at equilibrium $Q \neq K$; thus

$$\Delta G = -RT \ln K + RT \ln Q \neq 0 \quad (3.26)$$

In this case, $\Delta G \neq 0$, so there is a driving force for the reaction, which Le Chatelier predicts will favor the direction that reestablishes equilibrium. We can predict which direction, toward reactants or toward products, will be favored by comparing the relative values of K and Q. This is most easily seen if we rearrange Equation 3.24:

$$\Delta G = RT(\ln Q - \ln K) = RT \ln\left(\frac{Q}{K}\right) \quad (3.27)$$

Equation 3.27 predicts that a reaction will proceed as written whenever the ratio $Q/K < 1$ (see **TABLE 3.4**). We can imagine two ways that a living cell could maintain $Q/K < 1$ for a particular reaction. The first is by consuming products as fast as they are formed, such that the homeostatic concentrations of the products are always relatively low compared to the reactant concentrations. This is a common strategy employed in multistep metabolic pathways. For example, for the hypothetical pathway

$$\text{A} \longrightarrow \text{B} \longrightarrow \text{C} \longrightarrow \text{D} \longrightarrow \text{E}$$

B is the "product" of the first step; but it is also the "reactant" for the second step. The concentration of B in the cell will be low if B is rapidly converted to C. By Le Chatelier's Principle, the reaction $\text{A} \rightarrow \text{B}$ will be driven to the right as a result of removing B from the system. The second way to maintain $Q/K < 1$ is by keeping the concentrations of the "reactant" species relatively high. Again, Le Chatelier's Principle

TABLE 3.4 Relationships between K, Q, and ΔG for a reaction

Value of Q	Value of ΔG	Favored Direction
$<K$	<0	Forward reaction (formation of products)
$=K$	$=0$	Neither (system at equilibrium)
$>K$	>0	Reverse reaction (formation of reactants)

predicts that as [A] increases, there is a greater tendency to drive the reaction A → B to the right. The magnitude and sign of that driving force are succinctly expressed as ΔG and can be readily calculated using Equation 3.19.

Equation 3.19 allows us to calculate ΔG under the homeostatic conditions (i.e., nonequilibrium conditions) that are relevant to the myriad biochemical reactions that support the living state. Tabulations of $\Delta G°$ values for different reactions are common, but in applying such data to biochemical problems, we must always keep in mind that it is ΔG as determined by the actual concentrations of reactants and products found in the cell, rather than $\Delta G°$, that determines whether or not a reaction is favorable under the homeostatic conditions found in vivo.

● **CONCEPT** It is ΔG as determined by the actual concentrations of reactants and products in the cell, rather than $\Delta G°$, that determines whether or not a reaction is favorable in vivo.

In summary, whenever a system is displaced from equilibrium, it will proceed in the direction that moves toward the equilibrium state because the direction leading toward equilibrium will have a $\Delta G < 0$. Because living cells are not at equilibrium, most cellular reactions are driven to proceed in either the forward or reverse direction.

We are now prepared to answer a problem of fundamental importance in bioenergetics: "How are unfavorable reactions driven forward in living cells?" Thermodynamically unfavorable reactions can be driven by either, or both, of the following strategies:

1. Maintaining $Q < K$
2. Coupling an unfavorable reaction to a highly favorable reaction

The theoretical basis for strategy number 1 has been described above. In the following sections, we will describe the theoretical basis for strategy number 2. In later chapters, we will illustrate the implementation of these strategies in numerous biochemical reactions where highly favorable processes, such as ion transport across membranes and/or hydrolysis of ATP, are commonly used to drive unfavorable reactions. The processes of ATP consumption and ATP production feature prominently in biochemistry; thus, we introduce here the features of ATP (and similar compounds) that make it suitable as the central energy-transduction molecule in cells. However, we must first recognize that because standard conditions for reactions in cells are not completely compatible with the *chemical standard state* described earlier in this chapter, it is appropriate at this point to define a *biochemical standard state* that is used as the reference state in bioenergetics.

● **CONCEPT** Thermodynamically unfavorable reactions become favorable when $Q < K$ and/or when coupled to strongly favorable (i.e., highly exergonic) reactions.

Water, H⁺ in Buffered Solutions, and the "Biochemical Standard State"

In many reactions in cells, H^+ and H_2O appear as reactants or products. For example, consider the hydrolysis of adenosine triphosphate (ATP) to adenosine diphosphate (ADP), phosphate ion (HPO_4^{2-}), and a proton:

$$ATP + H_2O \rightleftharpoons ADP + HPO_4^{2-} + H^+ \qquad (3.28)$$

Because H^+ appears in the chemical equation, the hydrogen ion concentration (i.e., the pH) will affect ΔG for the reaction, as shown in Equation 3.29.

$$\Delta G = \Delta G° + RT \ln \left(\frac{[ADP][HPO_4^{2-}][H^+]}{[ATP][H_2O]} \right) \qquad (3.29)$$

Because biochemical reactions typically occur in a relatively dilute aqueous solution, buffered near pH 7, it is appropriate to treat the mass action terms for H_2O and H^+ differently than is done for thermodynamic calculations based on the standard state conditions described previously. Thus, we define them relative to a somewhat different set of conditions known as the **biochemical standard state:**

- In a dilute solution, the concentration of water is ~55 M. If we assume that the concentration of water inside cells remains constant and high, the activity of water is not changed significantly by reactions that consume or produce H_2O; thus, *for the biochemical standard state, the activity of water is defined as unity.*

- For chemical reactions, the standard state for solutes is defined as 1 M; however, in living cells the concentration of H^+ is roughly 10^{-7} M, much lower than the standard value of 1 M. It is therefore appropriate to define the reference concentration of H^+ in biochemical reactions relative to the H^+ concentration found in the living state (i.e., 10^{-7} M), rather than the value 1 M defined by the chemical standard state. Recall that when a solute in a dilute solution has a concentration of 1 M, the activity of that solute is unity. *For the biochemical standard state, we define the activity of H^+ to be unity when $[H^+] = 10^{-7}$ M.*

● **CONCEPT** Standard free energy changes for biochemical reactions are specified by $\Delta G°'$ where the water concentration is assumed to be constant (thus, a value of "1" assigned to the activity of water is defined as unity), and the activity of H^+ is defined as unity at pH 7.0.

We distinguish values of $\Delta G°$ that are referenced to the chemical standard state from those referenced to the biochemical standard state by a *superscript prime: $\Delta G°'$.* If we assume the activity of $H_2O = 1$ and we calculate the activities of the other solutes[‡] relative to their biochemical standard state concentrations, Equation 3.29 can be written as follows:

$$\Delta G = \Delta G°' + RT \ln \left(\frac{\dfrac{[ADP]}{(1\,M)} \dfrac{[HPO_4^{2-}]}{(1\,M)} \dfrac{[H^+]}{(10^{-7}\,M)}}{\dfrac{[ATP]}{(1\,M)} (1)} \right) \qquad (3.30)$$

Equation 3.30 illustrates two key points regarding the calculation of ΔG for the reactions you will encounter throughout this text:

[‡]For many reactions described in this text, the assumption that activity ≅ molar concentration will hold; but students should be aware that a rigorous calculation of chemical potential requires the use of more sophisticated definitions of activity and the biochemical standard state. Both are described in the references cited in Appendix II.

1. $\Delta G^{\circ\prime}$ is used, signifying the biochemical standard state.

2. The mass action expression Q is unitless. We strip the units from each concentration term in Q by dividing each by its proper standard concentration (e.g., 1 M for all solutes except H^+; 10^{-7} M for H^+; 1 bar for gases, etc.).

The significance of these two points is illustrated in the following example. Let us calculate ΔG for the hydrolysis of ATP at pH 7.4, 25 °C, where the concentrations of ATP, ADP, and HPO_4^{2-} are, respectively, 5 mM, 0.1 mM, and 35 mM. As we will see in the next section of this chapter, $\Delta G^{\circ\prime} = -32.2$ kJ/mol for ATP hydrolysis. Under these conditions, Equation 3.30 can be written as

$$\Delta G = -32.2\,\frac{kJ}{mol} + \left(0.008314\,\frac{kJ}{mol\cdot K}\right)(298\text{ K})$$

$$\ln\left(\frac{\dfrac{(0.0001\text{ M})}{(1\text{ M})}\dfrac{(0.035\text{ M})}{(1\text{ M})}\dfrac{(10^{-7.4}\text{ M})}{(10^{-7}\text{ M})}}{\dfrac{(0.005\text{ M})}{(1\text{ M})}(1)}\right) \tag{3.31a}$$

Note that we have expressed concentrations of all solutes in units of molarity, then divided by the proper standard state concentration (also in units of molarity). These steps ensure that the terms in Q are of the proper magnitude and stripped of units:

$$\Delta G = -32.2\,\frac{kJ}{mol} + \left(2.478\,\frac{kJ}{mol}\right)$$

$$\ln\left(\frac{(0.0001)(0.035)(0.398)}{(0.005)}\right) \tag{3.31b}$$

or

$$\Delta G = -32.2\,\frac{kJ}{mol} + -20.3\,\frac{kJ}{mol} = -52.5\,\frac{kJ}{mol} \tag{3.31c}$$

Note that the value calculated for ΔG is much more negative (i.e., more favorable) than the standard free energy change $\Delta G^{\circ\prime}$. This last point underscores the fact that it is ΔG and not $\Delta G^{\circ\prime}$ that determines the driving force for a reaction. However, to evaluate ΔG using Equation 3.19, we must be given, or be able to calculate, $\Delta G^{\circ\prime}$ for the reaction of interest. Recall that $\Delta G^{\circ\prime}$ can be calculated from K using Equation 3.22. In the remaining pages of this chapter, we will use examples relevant to biochemistry to illustrate two alternative methods for calculating $\Delta G^{\circ\prime}$.

3.4 Free Energy in Biological Systems

Understanding the central role of free energy changes in determining the favorable directions for chemical reactions is important in the study of biochemistry because every biochemical process (such as protein folding, metabolic reactions, DNA replication, or muscle contraction) must, overall, be a thermodynamically favorable process. Very often, a particular reaction or process that is necessary for life is in itself endergonic. Such intrinsically unfavorable processes can be made thermodynamically favorable by *coupling* them to strongly favorable reactions. Suppose, for example, we have a reaction $A \rightarrow B$ that is part of an essential pathway but is endergonic under standard conditions:

$$A \rightleftharpoons B \qquad \Delta G^{\circ\prime} = +10\text{ kJ/mol}$$

At the same time, suppose another process is highly exergonic:

$$C \rightleftharpoons D \qquad \Delta G^{\circ\prime} = -30\text{ kJ/mol}$$

If the cell can manage to *couple* these two reactions, the $\Delta G^{\circ\prime}$ for the overall process will be the algebraic sum of the values of $\Delta G^{\circ\prime}$ for the individual reactions:

$$
\begin{array}{ll}
A \rightleftharpoons B & \Delta G^{\circ\prime} = +10\text{ kJ/mol} \\
C \rightleftharpoons D & \Delta G^{\circ\prime} = -30\text{ kJ/mol} \\
\hline
\text{Overall: } A + C \rightleftharpoons B + D & \Delta G^{\circ\prime} = -20\text{ kJ/mol}
\end{array}
$$

Equilibrium for the overall process now lies far to the right, with the consequence that, in the coupled process, B is more favorably produced from A. Many critical reactions in cells are driven forward by coupling an unfavorable reaction to a highly favorable one.

Organic Phosphate Compounds as Energy Transducers

In cells, driving an unfavorable process by coupling it to a favorable one requires the availability of compounds (like the hypothetical C in our previous example) that can undergo reactions with large negative free energy changes. Such substances can be thought of as energy transducers in the cell. Many of these energy-transducing compounds are organic phosphates such as ATP (**FIGURE 3.5**), which can transfer a phosphoryl group ($-PO_3^{2-}$) to an acceptor molecule. You will see many examples of phosphoryl group transfer reactions in this text. As shown in Figure 3.5, we will use a common shorthand notation, Ⓟ, to represent the phosphoryl group when describing these processes.

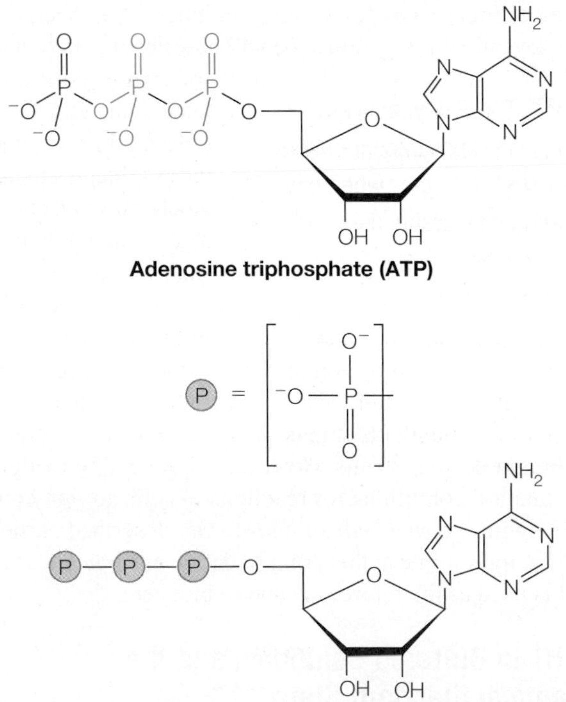

Adenosine triphosphate (ATP)

▲ **FIGURE 3.5 The phosphoryl groups in ATP.** Top: The three phosphoryl groups in ATP are shown in red, blue, and green. Middle: A commonly used shorthand for a phosphoryl group is the symbol Ⓟ. Bottom: The three phosphoryl groups in ATP are represented by this symbol.

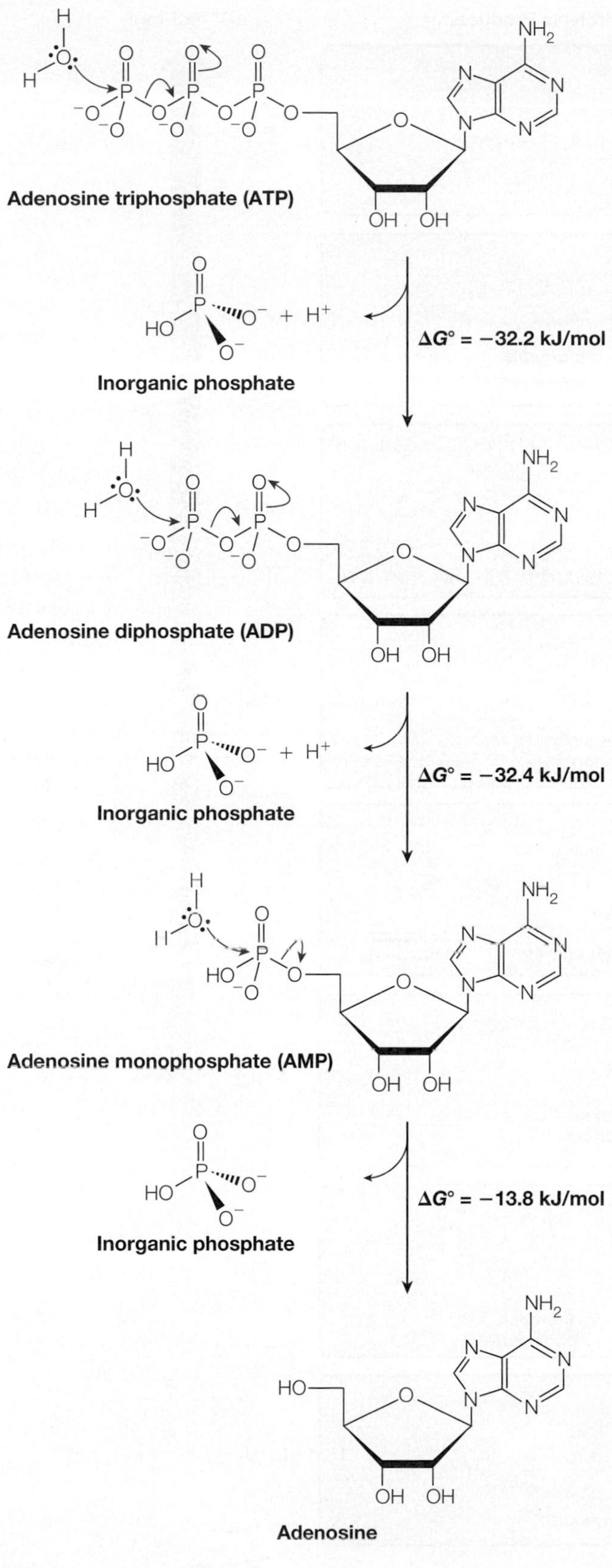

Adenosine triphosphate (ATP)

Inorganic phosphate

$\Delta G^\circ = -32.2$ kJ/mol

Adenosine diphosphate (ADP)

Inorganic phosphate

$\Delta G^\circ = -32.4$ kJ/mol

Adenosine monophosphate (AMP)

Inorganic phosphate

$\Delta G^\circ = -13.8$ kJ/mol

Adenosine

▲ **FIGURE 3.6 Hydrolysis of ATP, ADP, and AMP.** Hydrolysis of ATP or ADP cleaves a phosphoanhydride bond, whereas hydrolysis of AMP cleaves a phosphate ester bond. Inorganic phosphate is produced in each of these hydrolysis reactions.

In many cases, the phosphoryl group is transferred to a water molecule via hydrolysis. Three such hydrolysis reactions that would convert ATP to adenosine are illustrated in **FIGURE 3.6**, along with the corresponding values of $\Delta G^{\circ\prime}$ for each reaction. A phosphate ion (HPO_4^{2-}) is produced when water is the phosphoryl group acceptor. HPO_4^{2-} is also called *inorganic phosphate* and is commonly represented by the shorthand notation P_i. Note that the two reactions involving cleavage of a phosphoanhydride bond are significantly more exergonic than the reaction that cleaves the phosphate ester bond.

The hydrolysis reactions for several important phosphate compounds are shown in **FIGURE 3.7**. You will encounter all of these substances in later chapters on metabolism. Some of these substances, like ATP, **phosphoenolpyruvate (PEP), creatine phosphate (CP),** and **1,3-bisphosphoglycerate (1,3-BPG),** have large negative standard free energies of hydrolysis (Figure 3.7 and **TABLE 3.5**). For example, hydrolysis of ATP to ADP is highly exergonic, with a $\Delta G^{\circ\prime}$ of -32.2 kJ/mol. This value corresponds to an equilibrium constant greater than 10^5. This equilibrium lies so far to the right that ATP hydrolysis can be considered essentially irreversible.

Figure 3.7 shows a wide range of $\Delta G^{\circ\prime}$ values for the hydrolysis reactions listed. Some are highly exergonic processes; others are not. Hydrolysis of phosphoanhydride (as in ATP, ADP, and pyrophosphate) and mixed anhydride (1,3-BPG) bonds is much more exergonic than is hydrolysis of phosphate esters (AMP, glycerol-1-phosphate). These differences in reactivity are expected based on the observed reactivities of analogous carboxylic acid anhydrides and esters. However, $\Delta G^{\circ\prime}$ for the hydrolysis of PEP, -61.9 kJ/mol, is significantly more favorable than one would predict for the hydrolysis of a simple phosphate ester (e.g., compare to $\Delta G^{\circ\prime} = -13.8$ kJ/mol for AMP hydrolysis). The explanation for this unexpected high reactivity of PEP lies in the structural isomerization, in this case called *tautomerization,* of the pyruvate product that occurs following the release of P_i from PEP. The direct product of phosphate hydrolysis from PEP is the *enol* form of pyruvate, which rapidly tautomerizes

TABLE 3.5 $\Delta G^{\circ\prime}$ for hydrolysis of some phosphate compounds	
Hydrolysis Reaction	**$\Delta G^{\circ\prime}$ (kJ/mol)**
Phosphoenolpyruvate + $H_2O \longrightarrow$ pyruvate + P_i	-61.9
1, 3-Bisphosphoglycerate + $H_2O \longrightarrow$ 3-phosphoglycerate + P_i + H^+	-49.4
ATP + $H_2O \longrightarrow$ AMP + PP_i + H^+	-45.6
Acetyl phosphate + $H_2O \longrightarrow$ acetate + P_i + H^+	-43.1
Creatine phosphate + $H_2O \longrightarrow$ creatine + P_i	-43.1
ADP + $H_2O \longrightarrow$ AMP + P_i + H^+	-32.4
ATP + $H_2O \longrightarrow$ ADP + P_i + H^+	-32.2
PP_i + $H_2O \longrightarrow 2P_i$	-19.2
Glucose-1-phosphate + $H_2O \longrightarrow$ glucose + P_i	-20.9
Glucose-6-phosphate + $H_2O \longrightarrow$ glucose + P_i	-13.8

Based on R. A. Alberty (1994) Recommendations for nomenclature and tables in biochemical thermodynamics. *Pure & Appl. Chem.* 66:1641–1666; P. Frey and A. Arabshahi (1995) Standard free energy change for the hydrolysis of the α, β-bridge in ATP. *Biochemistry* 34:11307–11310.

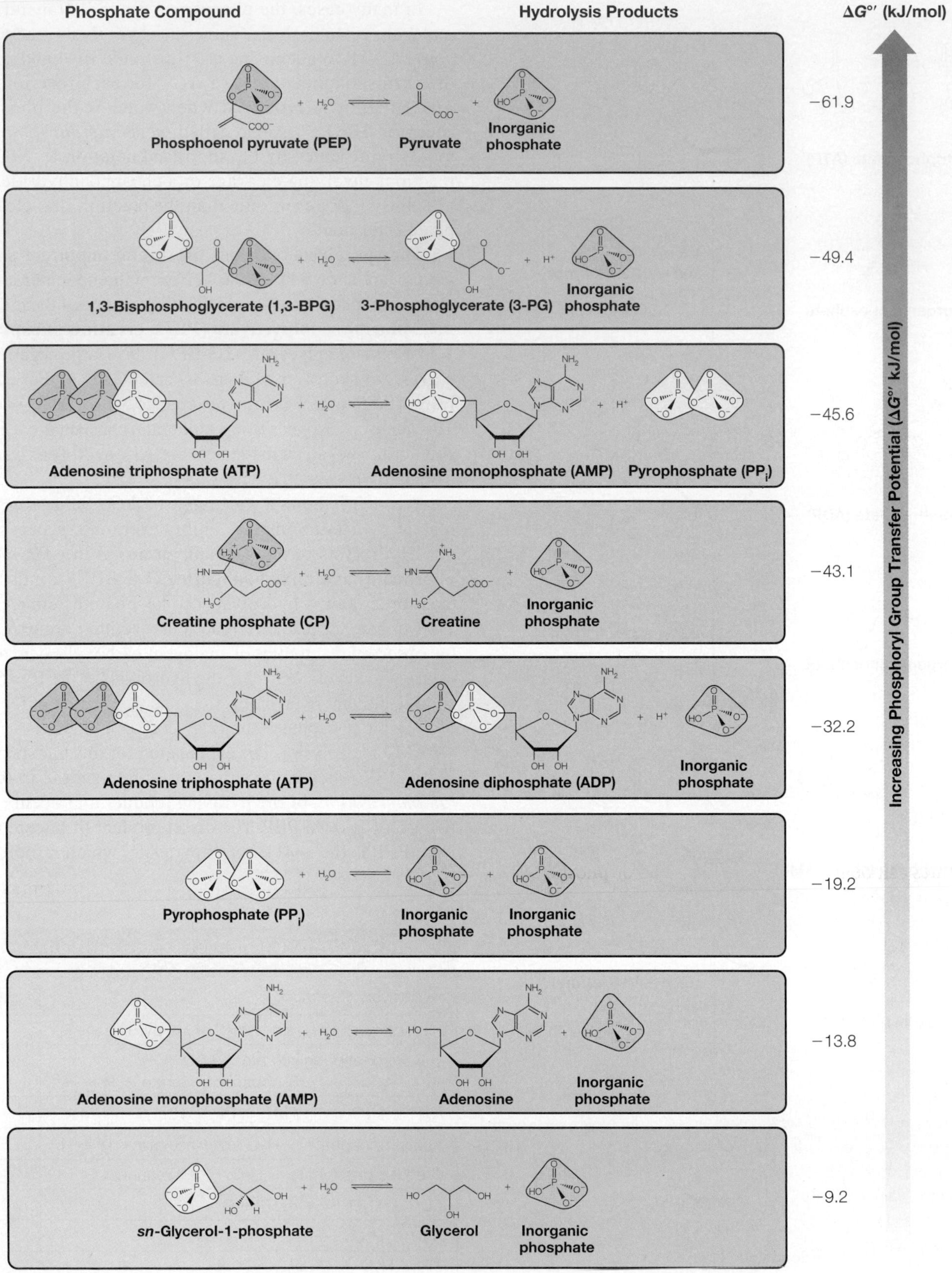

▲ **FIGURE 3.7 Hydrolysis reactions for some biochemically important phosphate compounds.** The reactive phosphoryl groups in each compound are highlighted (with darker color corresponding to groups with a greater phosphoryl group transfer potential). The stable inorganic phosphate ion, HPO_4^{2-}, is shown in gray.

(1) Hydrolysis of phosphoenolpyruvate	$PEP + H_2O \rightleftharpoons pyruvate + P_i$	$\Delta G^{\circ\prime} = -61.9$ kJ/mol
(2) phosphorylation of adenosine diphosphate	$ADP + P_i + H^+ \rightleftharpoons ATP + H_2O$	$\Delta G^{\circ\prime} = +32.2$ kJ/mol
(1) + (2): Coupled phosphorylation of ADP by PEP	$PEP + ADP + H^+ \rightleftharpoons ATP + pyruvate$	$\Delta G^{\circ\prime} = -29.7$ kJ/mol

to the thermodynamically favored *keto* form. The overall $\Delta G^{\circ\prime}$ for PEP hydrolysis is the result of a moderately favorable ester hydrolysis (with $\Delta G^{\circ\prime} \approx -16$ kJ/mol) coupled to a highly favorable tautomerization (with $\Delta G^{\circ\prime} \approx -46$ kJ/mol):

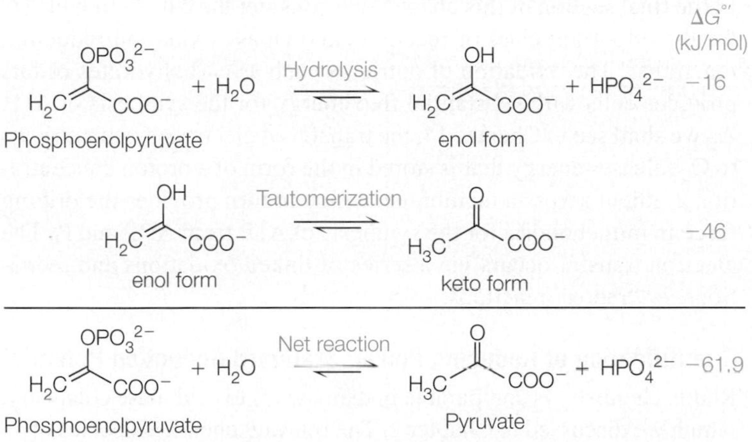

At 37 °C, the equilibrium for this tautomerization lies far toward the keto form ($K_{eq} \approx 6 \times 10^7$). Thus, it is this essentially irreversible isomerization reaction that provides most of the driving force for the overall hydrolysis of PEP to pyruvate and P_i.

Phosphoryl Group Transfer Potential

There is another useful way in which we can think about the $\Delta G^{\circ\prime}$ values for hydrolysis of organic phosphate compounds. As Figure 3.7 shows, these values can be arranged to form a scale of **phosphoryl group transfer potentials.** The phosphoryl group transfer potential is correlated with $-\Delta G^{\circ\prime}$ of hydrolysis; thus, of the compounds listed in Figure 3.7, PEP has the highest phosphoryl group transfer potential and glycerol-1-phosphate has the lowest. A given compound in Figure 3.7 is capable of driving the phosphorylation of compounds lower on the scale, provided that a suitable coupling mechanism is available. Consider, for example, the reactions shown above.

Note that reaction (2) above is the *reverse* of the ATP hydrolysis reaction listed in Table 3.5; thus, the value of $\Delta G^{\circ\prime}$ has the opposite sign to reflect the fact that the free energy change for any reaction in the forward direction is equal in magnitude but opposite in sign when compared to the reaction in the reverse direction.

Thus, PEP, which has a greater phosphoryl group transfer potential than ATP, is capable of adding a phosphoryl group to ADP in a thermodynamically favored process. ATP, in turn, can phosphorylate

● **CONCEPT** The phosphoryl group transfer potential shows which compounds can phosphorylate others under standard conditions.

glucose because the phosphoryl group transfer potential of glucose-6-phosphate (G-6-P) lies still farther down the scale as shown below.

These examples emphasize how ATP can act as a versatile phosphoryl group transfer agent through coupled reactions. In each case, the coupling is accomplished by having the reactions take place on the surface of a large protein molecule (i.e., an *enzyme*). For example, the phosphorylation of glucose by ATP is catalyzed by the enzyme *hexokinase*. We shall discuss in Chapter 8 how enzymes both facilitate such coupling *and* accelerate reactions.

These examples also show how $\Delta G^{\circ\prime}$ can be calculated for a reaction of interest by summing the $\Delta G^{\circ\prime}$ values for two (or more) reactions that, when added together, yield an overall chemical equation that is *identical* to the reaction of interest. Again, when the chemical equation for a reaction is reversed, the sign of $\Delta G^{\circ\prime}$ must also be reversed. (For more practice with the calculation of $\Delta G^{\circ\prime}$ by this method, see Problems 17–19 at the end of this chapter.)

Free Energy and Concentration Gradients: A Close Look at Diffusion Through a Membrane

The reactions of the compounds listed in Figure 3.7 are not the only sources of free energy in cells. We will consider two more in this chapter: concentration gradients across biological membranes and electron transfer reactions. In this section we examine the concentration gradients.

We know from experience that if a substance can diffuse across a permeable membrane, it will do so in such a direction as to make the concentrations on the two sides equal. Now we will see how the thermodynamic arguments presented earlier explain this behavior.

Suppose there are two solutions of substance A separated by a membrane through which A can pass freely (**FIGURE 3.8**). Assume that in region 1 the concentration is initially $[A]_1$ and in region 2 the concentration is $[A]_2$. We must define the direction of transfer for the process to evaluate ΔG. Let us consider transferring some quantity of A from region 1 (the "initial" state of A) to region 2 (the "final" state of A). By the definition of ΔG given in Equation 3.15, the overall free energy change is determined by the difference in free energies of A in regions 1 and 2, or

$$\Delta G = G_{A_2} - G_{A_1} \qquad (3.32)$$

We expect these free energies to be different because G for a solute depends on the concentration of that solute (see Equation 3.14). The free energy change for moving A from region 1 to region 2 can be

(1) Hydrolysis of ATP	$ATP + H_2O \rightleftharpoons ADP + P_i + H^+$	$\Delta G^{\circ\prime} = -32.2$ kJ/mol
(2) phosphorylation of glucose	$glucose + P_i \rightleftharpoons G\text{-}6\text{-}P + H_2O$	$\Delta G^{\circ\prime} = +13.8$ kJ/mol
(1) + (2): Coupled phosphorylation of glucose by ATP	$glucose + ATP \rightleftharpoons ADP + G\text{-}6\text{-}P + H^+$	$\Delta G^{\circ\prime} = -18.4$ kJ/mol

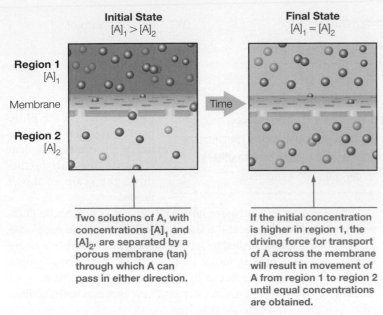

Initial State
$[A]_1 > [A]_2$

Final State
$[A]_1 = [A]_2$

Region 1
$[A]_1$

Membrane

Time

Region 2
$[A]_2$

Two solutions of A, with concentrations $[A]_1$ and $[A]_2$, are separated by a porous membrane (tan) through which A can pass in either direction.

If the initial concentration is higher in region 1, the driving force for transport of A across the membrane will result in movement of A from region 1 to region 2 until equal concentrations are obtained.

▲ **FIGURE 3.8 Equilibration across a membrane.**

calculated by substituting expressions for G_{A_1} and G_{A_2} from Equations 3.14 into Equation 3.32:

$$\Delta G = G_{A_2} - G_{A_1}$$
$$G_{A_1} = G_A^\circ + RT \ln[A]_1 \qquad \text{Equation 3.14}$$
$$G_{A_2} = G_A^\circ + RT \ln[A]_2 \qquad \text{Equation 3.14}$$
$$\Delta G = (G_A^\circ + RT \ln[A]_2) - (G_A^\circ + RT \ln[A]_1) \quad (3.33)$$

After canceling G_A° terms, Equation 3.33 simplifies to

$$\Delta G = RT \ln \frac{[A]_2}{[A]_1} \qquad (3.34)$$

Equation 3.34 predicts the following:

1. If $[A]_2 < [A]_1$, ΔG is negative; thus, transfer from region 1 to region 2 is favorable (this is the situation described in the "initial state" of Figure 3.8).

2. Conversely, for a system where $[A]_2 > [A]_1$, ΔG is positive; thus, transfer from region 1 to region 2 would not be favorable (but transfer in the opposite direction would be favored).

3. If $[A]_2 = [A]_1$, ΔG is zero; thus, there is no net driving force in either direction for the transfer of A. The system is at equilibrium (this is the situation described in the "final state" of Figure 3.8).

The difference in concentrations across the membrane establishes a **concentration gradient.** From this analysis, we can conclude that if a substance can pass through a membrane, the direction of favorable transfer will always be from the region of higher concentration to the region of lower concentration. In the example of Figure 3.8, the concentration gradient provides the *driving force* (with $\Delta G < 0$) to move molecules of A from region 1 to region 2. This process is of fundamental importance in the transmission of neural signals, which requires rapid movements of Na^+ and K^+ ions across the membranes surrounding nerve cells (discussed further in Chapter 10).

There are cases in which substances are transported from regions of lower concentration to regions of higher concentration; but in such circumstances the necessary free energy price is paid by coupling the unfavorable transport process to one or more thermodynamically favorable chemical reactions (for example, ATP hydrolysis). The favorable ΔG for transport of some substance along its concentration gradient can also be coupled to some other unfavorable process to drive the coupled process forward. Thus, concentration gradients across membranes represent important free energy stores in cells.

ΔG and Oxidation/Reduction Reactions in Cells

In the final section of this chapter, we consider the calculation of ΔG for the important class of reactions known as "oxidation/reduction reactions." The oxidation of nutrients such as carbohydrates or fats provides cells with substantial free energy for the synthesis of ATP. As we shall see in Chapter 14, the transfer of electrons from nutrients to O_2 releases energy that is stored in the form of a proton concentration gradient across a membrane, which in turn provides the driving force in mitochondria for the synthesis of ATP from ADP and P_i. The electron transfer occurs via a series of linked oxidations and reductions, or "redox" reactions.

Quantification of Reducing Power: Standard Reduction Potential

Redox chemistry is comparable in many ways to acid–base chemistry, which we discussed in Chapter 2. The relevant chemical species in an acid–base equilibrium are an acid (HA) and its conjugate base (A⁻), which represent a proton donor and a proton acceptor, respectively:

$$HA \rightleftharpoons H^+ + A^-$$

For example, $\quad CH_3COOH_{(aq)} \rightleftharpoons H^+_{(aq)} + CH_3COO^-_{(aq)}$

Similarly, in a redox reaction a donor and an acceptor of *electrons* are paired:

reduced compound (e^- donor) $\rightleftharpoons$
$\qquad\qquad$ oxidized compound (e^- acceptor) + e^-

For example, $\qquad Fe^{2+} \rightleftharpoons Fe^{3+} + e^-$

Free protons and free electrons exist at negligible concentrations in aqueous media, so these equilibrium expressions are merely half-reactions in an overall acid–base or redox reaction scheme. A complete redox reaction must show one reactant as an electron acceptor, which becomes *reduced* by gaining electrons, and another reactant as an electron donor, which becomes *oxidized* by losing electrons. One well-known mnemonic for the definitions of **reduction** and **oxidation** is "OILRIG": <u>O</u>xidation <u>I</u>s <u>L</u>oss (of electrons); <u>R</u>eduction <u>I</u>s <u>G</u>ain (of electrons). Of the two reactants in a redox reaction, the electron donor is the **reductant** (or **reducing agent**), which becomes oxidized while transferring electrons to the other reactant, the **oxidant** (or **oxidizing agent**), which becomes reduced. The general form of a redox reaction is then:

reductant + oxidant $\rightleftharpoons$ oxidized reductant + reduced oxidant

or: $\qquad A_{(red)} + B_{(ox)} \rightleftharpoons A_{(ox)} + B_{(red)}$

for example, $\quad Cu^{1+} + Fe^{3+} \rightleftharpoons Cu^{2+} + Fe^{2+}$

Cu^{1+} is the reductant in this reaction because it is the electron donor, and Fe^{3+} is the oxidant because it accepts an electron from Cu^{1+}.

Critical to our understanding of acid–base chemistry is the concept of pK_a, which represents a quantitative measure of the tendency of an acid to lose a proton. In the same sense, our understanding of biological oxidations demands a comparable measure of the tendency of a reductant to lose electrons (or of an oxidant to gain electrons). Such an index is provided by the **standard reduction potential,** or $E°$. In acid–base equilibria, water, with a pK_a of 7.0, is arbitrarily defined as neutral. Redox chemistry also employs a reference standard: the *standard hydrogen electrode* in an electrochemical cell.

● **CONCEPT** $E°$ is the tendency of a reductant to lose an electron, in the same sense that pK_a is the tendency of an acid to lose a proton.

An electrochemical cell consists of two **half-cells,** each containing an electron donor and its conjugate acceptor. In **FIGURE 3.9**, the right-hand beaker constitutes the reference half-cell, a standard hydrogen electrode, with $[H^+]$ at 1 M and H_2 at 1 bar (the standard unit of pressure, equal to 100 kPa or ~750 torr). The left-hand beaker is the test half-cell, with the solution containing the test electron donor and its conjugate acceptor, each at 1 M concentration. In this example, the solution contains Fe^{2+} and Fe^{3+} each at 1 M. The half-cells are connected with an agar salt bridge, which maintains charge neutrality by allowing ions to flow between the cells as electrons move through the completed circuit. A voltmeter placed between the two half-cells measures the **electromotive force,** or **emf,** in volts. Electromotive force is a measure of the potential, or "pressure," for electrons to flow from one half-cell to the other. The electrons may flow either toward or away from the reference half-cell, depending on whether H_2 or the

test electron donor has the greater tendency to lose electrons. Because H_2 loses electrons more readily than Fe^{2+}, the electrons in our example will flow from the reference half-cell to the test half-cell, oxidizing H_2, reducing Fe^{3+}, and causing the voltmeter to record a positive emf. If the test electron donor loses electrons more readily than H_2, then electrons flow in the reverse direction, reducing $2H^+$ to H_2 in the reference half-cell and causing a negative emf to be recorded. The stronger oxidant, whether H^+ or the test electron acceptor, will draw electrons away from the other half-cell and become reduced.

By convention, $E°$ for the standard hydrogen electrode is set at 0.00 volts. Any redox couple that tends to donate electrons to the standard hydrogen electrode has a negative value of $E°$. A positive $E°$ means that electrons from H_2 are flowing toward the test cell and reducing the electron acceptor, or that the test cell acceptor is oxidizing H_2. *The higher the value of $E°$ for a redox couple, the stronger an oxidant is the electron acceptor of that couple.*

● **CONCEPT** The greater the standard reduction potential, the greater the tendency of the oxidized form of a redox couple to attract electrons.

As discussed earlier, standard conditions for biochemists include a pH value of 7.0, a condition far from that seen in the standard hydrogen electrode, which contains H^+ at 1.0 M. Therefore, biochemists use a modified term, $E°'$, which is the standard reduction potential for a half-reaction measured at 10^{-7} M H^+. These are the values used in this book and most other biochemical references. $E°'$ values for a few biochemically important redox pairs are recorded in **TABLE 3.6** (a more complete table of $E°'$ values is given in Chapter 14). This table

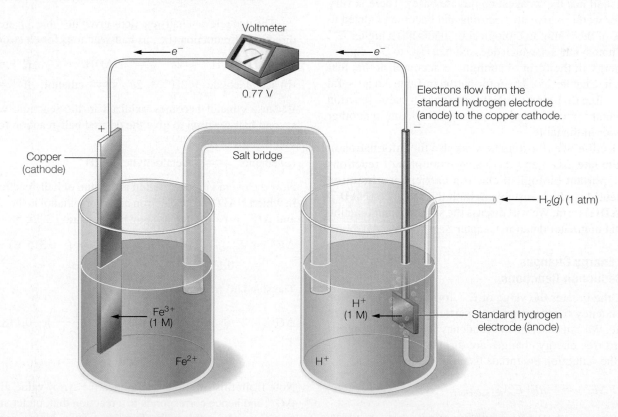

▲ **FIGURE 3.9 A galvanic cell to measure $E°$.** The standard hydrogen reference cell is shown on the right, and a test cell containing Fe^{2+}/Fe^{3+} ions is shown on the left.

TABLE 3.6 A few standard reduction potentials ($E^{\circ\prime}$) of interest in biochemistry

Oxidant (e^- acceptor)		Reductant (e^- donor)	n	$E^{\circ\prime}$ (V)
$H^+ + e^-$	$\rightleftharpoons$	$\frac{1}{2}H_2$	1	-0.421
$NAD^+ + H^+ + 2e^-$	$\rightleftharpoons$	NADH	2	-0.315
1,3-Bisphosphoglycerate $+ 2H^+ + 2e^-$	$\rightleftharpoons$	Glyceraldehyde-3-phosphate $+ P_i$	2	-0.290
$FAD + 2H^+ + 2e^-$	$\rightleftharpoons$	$FADH_2$	2	-0.219
Acetaldehyde $+ 2H^+ + 2e^-$	$\rightleftharpoons$	Ethanol	2	-0.197
Pyruvate $+ 2H^+ + 2e^-$	$\rightleftharpoons$	Lactate	2	-0.185
$Fe^{3+} + e^-$	$\rightleftharpoons$	Fe^{2+}	1	$+0.769$
$\frac{1}{2}O_2 + 2H^+ + 2e^-$	$\rightleftharpoons$	H_2O	2	$+0.815$

Note: $E^{\circ\prime}$ is the standard reduction potential at pH 7 and 25 °C, n is the number of electrons transferred, and each potential is for the partial reaction written as follows:

$$\text{Oxidant} + ne^- \rightarrow \text{reductant}$$

The entry for the H^+/H_2 couple $E^{\circ\prime} = -0.421$ V is not zero because it is measured with $[H^+] = 1$ M in the reference cell (i.e., the standard hydrogen electrode) and $[H^+] = 10^{-7}$ M in the test cell.

is organized with the strongest oxidizing agents listed at the bottom of the "Oxidant" column and the strongest reducing agents listed at the top of the "Reductant" column. Thus, O_2 in the O_2/H_2O couple is the strongest oxidant, and H_2 in the H^+/H_2 couple is the strongest reducing agent. Another way to express this relationship is to say that H_2O in the O_2/H_2O couple is the weakest reducing agent (just as the strongest acid has the weakest conjugate base). There is very little tendency for water to give up electrons and become oxidized to O_2 because none of the common biological oxidants has a higher $E^{\circ\prime}$ than O_2/H_2O. Photosynthesis, which does oxidize H_2O to O_2, requires considerable energy in the form of sunlight to accomplish this feat (more about this in Chapter 15). The information in Table 3.6 is useful because the favorable direction of electron flow in a redox reaction is from the reductant in one redox couple to the oxidant in another couple listed lower in the table.

In the next section, we illustrate how we use the information in Table 3.6 to calculate $\Delta G^{\circ\prime}$ and ΔG using examples of reactions involving the important biological electron carrier **nicotinamide adenine dinucleotide,** which is stable in both the oxidized (NAD^+) and reduced (**NADH**) form. We will discuss the structure and function of NAD^+/NADH in greater detail in Chapter 8.

Standard Free Energy Changes in Oxidation–Reduction Reactions

To recapitulate, the greater the value of $E^{\circ\prime}$ for a redox couple, the greater is the tendency for that couple to participate in oxidation of another substrate. We can describe this tendency in quantitative terms because standard free energy changes are directly related to differences between the reduction potentials listed in Table 3.6:

$$\Delta G^{\circ\prime} = -nF\Delta E^{\circ\prime} = -nF\left[E^{\circ\prime}_{(e^- \text{ acceptor})} - E^{\circ\prime}_{(e^- \text{ donor})}\right] \quad (3.35)$$

where n is the number of electrons transferred in the balanced half-reactions, F is the Faraday constant (96.5 kJ mol^{-1}V^{-1}), and $\Delta E^{\circ\prime}$ is the difference in standard reduction potentials between the two redox

couples. Note that $\Delta E^{\circ\prime}$ is to ΔE° as $\Delta G^{\circ\prime}$ is to ΔG°. The "prime" has the same significance in both cases: Each reactant and product (except H^+) is at 1 M, and the reaction is carried out at a pH of 7.0.

For example, consider the oxidation of ethanol by NAD^+. This reaction is catalyzed by the enzyme alcohol dehydrogenase:

$$\text{ethanol} + NAD^+ \rightleftharpoons \text{acetaldehyde} + \text{NADH} + H^+$$

The two relevant half-reactions given in Table 3.6 are written in the direction of reduction (i.e., as half-reactions for electron acceptors):

(a) $NAD^+ + H^+ + 2e^- \rightleftharpoons$ NADH $\qquad E^{\circ\prime} = -0.315$ V

(b) acetaldehyde $+ 2H^+ + 2e^- \rightleftharpoons$ ethanol $\quad E^{\circ\prime} = -0.197$ V

Because ethanol becomes oxidized in the reaction, we reverse the second half-reaction to give the desired half-reaction for the electron donor:

(c) ethanol $\rightleftharpoons$ acetaldehyde $+ 2H^+ + 2e^-$

Now the overall redox reaction is the sum of half-reactions (a) and (c), in which NAD^+ is the electron acceptor, ethanol is the electron donor, and $\Delta E^{\circ\prime}$ is calculated as shown in Equation 3.35:

$$\Delta E^{\circ\prime} = E^{\circ\prime}_{(e^- \text{ acceptor})} - E^{\circ\prime}_{(e^- \text{ donor})} = (-0.315 \text{ V}) - (-0.197 \text{ V})$$
$$= -0.118 \text{ V} \quad (3.36)$$

The standard free energy change is thus:

$$\Delta G^{\circ\prime} = -nF\Delta E^{\circ\prime} = -(2)\left(96485 \frac{\text{J}}{\text{mol} \cdot \text{V}}\right)(-0.118 \text{ V})$$
$$= +22.8 \frac{\text{kJ}}{\text{mol}} \quad (3.37)$$

Note from this example that a negative $\Delta E^{\circ\prime}$ value gives a positive $\Delta G^{\circ\prime}$ and hence corresponds to a reaction that, under standard conditions, is *not* favorable in the direction written. Note also that if we were to calculate ΔG° for the reverse reaction (reduction of acetaldehyde by NADH), $\Delta E^{\circ\prime}$ would be $+0.118$ V, and $\Delta G^{\circ\prime}$ would be -22.8 kJ/mol.

Calculating Free Energy Changes for Biological Oxidations under Nonequilibrium Conditions

The values given in Table 3.6 (and also in Table 14.1) allow calculation of the biochemical standard state free energy changes. To calculate ΔG for a redox reaction under nonstandard conditions, we must use Equation 3.19. Let us consider the free energy change associated with the transfer of electrons from NADH to O_2 which occurs in mitochondria. The overall electron transport process, which actually occurs in several steps, is given by the following equation:

$$O_2 + 2NADH + 2H^+ \rightleftharpoons 2H_2O + 2NAD^+$$

According to Equation 3.19, ΔG for the reaction is given by

$$\Delta G = \Delta G^{\circ\prime} + RT \ln\left(\frac{[H_2O]^2[NAD^+]^2}{[O_2][NADH]^2[H^+]^2}\right) \quad (3.38)$$

Let us assume that inside the mitochondrion the temperature is 37 °C, pH = 8.1 (the mitochondrial matrix is a relatively alkaline environment compared to the cellular cytosol), the partial pressure of oxygen is 2 torr, and the concentrations of NAD^+ and NADH are 10 mM and 100 μM, respectively. The value of the $RT\ln Q$ term can be calculated as described previously; however, in this example, we will enter the concentration term for oxygen relative to the standard concentration for a gas, which is 1 bar (or 750 torr), rather than as a molar concentration:

$$\Delta G = \Delta G^{\circ\prime} + \left(8.314 \frac{J}{mol \cdot K}\right)(310\,K)$$

$$\ln\left(\frac{(1)^2\left(\dfrac{1 \times 10^{-2}\,M}{1\,M}\right)^2}{\left(\dfrac{2\,torr}{750\,torr}\right)\left(\dfrac{1 \times 10^{-4}\,M}{1\,M}\right)^2\left(\dfrac{10^{-8.1}\,M}{10^{-7.0}\,M}\right)^2}\right) \quad (3.39a)$$

or

$$\Delta G = \Delta G^{\circ\prime} + \left(2.58 \frac{kJ}{mol}\right)$$

$$\ln\left(\frac{(1)(1 \times 10^{-4})}{(2.66 \times 10^{-3})(1 \times 10^{-8})(6.31 \times 10^{-3})}\right) \quad (3.39b)$$

$$\Delta G = \Delta G^{\circ\prime} + \left(2.58 \frac{kJ}{mol}\right)\ln(5.96 \times 10^9)$$

$$\Delta G = \Delta G^{\circ\prime} + \left(52.1 \frac{kJ}{mol}\right) \quad (3.39c)$$

Now we turn our attention to the calculation of $\Delta G^{\circ\prime}$. Because this is a biochemical redox reaction, we can use the information in Table 3.6 and Equation 3.35. Which values of $E^{\circ\prime}$ should be used to calculate $\Delta E^{\circ\prime}$ (and thereby $\Delta G^{\circ\prime}$)? To answer that question, the relevant "half-reactions" must be identified. In this case, it is reasonable to assume that the electron acceptor is O_2 (since O_2 is among the strongest oxidizing agents). The reduction of oxygen in the presence of protons yields water

$$\frac{1}{2}O_2 + 2H^+ + 2e^- \rightleftharpoons H_2O \qquad E^{\circ\prime} = +0.815\,V$$

Protons can't act as electron donors; thus, the electron donor in the overall reaction must be NADH. Oxidation of NADH yields NAD^+, and these two species are related by the following half-reaction:

$$NAD^+ + H^+ + 2e^- \rightleftharpoons NADH \qquad E^{\circ\prime} = -0.315\,V$$

which, when reversed, shows NADH giving up two electrons:

$$NADH \rightleftharpoons NAD^+ + H^+ + 2e^-$$

The half-reactions above are adjusted for stoichiometry, then added together to re-create the overall reaction of interest:

$$O_2 + 4H^+ + 4e^- \rightleftharpoons 2H_2O$$

$$\underline{2NADH \rightleftharpoons 2NAD^+ + 2H^+ + 4e^-}$$

$$\text{net: } O_2 + 2H^+ + 2NADH \rightleftharpoons 2NAD^+ + 2H_2O$$

Note that when we adjust stoichiometry (i.e., by multiplying each half-reaction by a factor of 2), we do not make any adjustments to the value of $E^{\circ\prime}$ because the adjustment for changes in stoichiometry is accounted for in the value of n used in Equation 3.35. When we multiply each half-reaction by 2, we also double the number of electrons transferred; thus, the value of n is also doubled (in this case, from 2 to 4), and

$$\Delta G^{\circ\prime} = -nF\Delta E^{\circ\prime}$$

$$= -(4)\left(96485 \frac{J}{mol \cdot V}\right)((+0.815\,V) - (-0.315\,V))$$

$$= -436.2 \frac{kJ}{mol} \quad (3.40)$$

We can now use this value of $\Delta G^{\circ\prime}$ in Equation 3.39c to calculate ΔG for the reaction

$$\Delta G = \left(-436.2 \frac{kJ}{mol}\right) + \left(+52.1 \frac{kJ}{mol}\right) = -384.1 \frac{kJ}{mol} \quad (3.41)$$

This calculation shows that under the specified conditions, the oxidation of NADH by O_2 is highly favorable. As will be described in greater detail in Chapter 14, a significant portion of this energy is stored in the form of a proton concentration gradient across the inner mitochondrial membrane. The free energy available in the proton gradient (see Equation 3.34) is sufficient to drive the synthesis of ATP from ADP and P_i in mitochondria.

A Brief Overview of Free Energy Changes in Cells

Some of the processes discussed earlier in this chapter are used in **FIGURE 3.10** to illustrate major themes in bioenergetics. Conversion of ATP to ADP (yellow arrows) provides energy that is used to drive many processes forward. The regeneration of ATP from ADP (green arrow) requires an input of free energy, which is supplied by the proton concentration gradient across the inner mitochondrial membrane (magenta arrow). This proton gradient is formed during the energetically favorable redox reactions that ultimately transfer electrons from NADH to O_2. NADH, in turn, is a product of the multistep oxidation of glucose to CO_2. These reactions are all carried out by enzymes that fold into their active forms via energetically favorable processes. The various equations used to determine changes in free energy for these processes are shown in boxes in Figure 3.10. In the preceding sections of this chapter, we have presented the basic thermodynamics that describes free energy changes for these intracellular processes. These concepts will appear again in subsequent chapters which describe the details of these and many more diverse, yet interconnected, processes devoted to the capture and utilization of energy in cells.

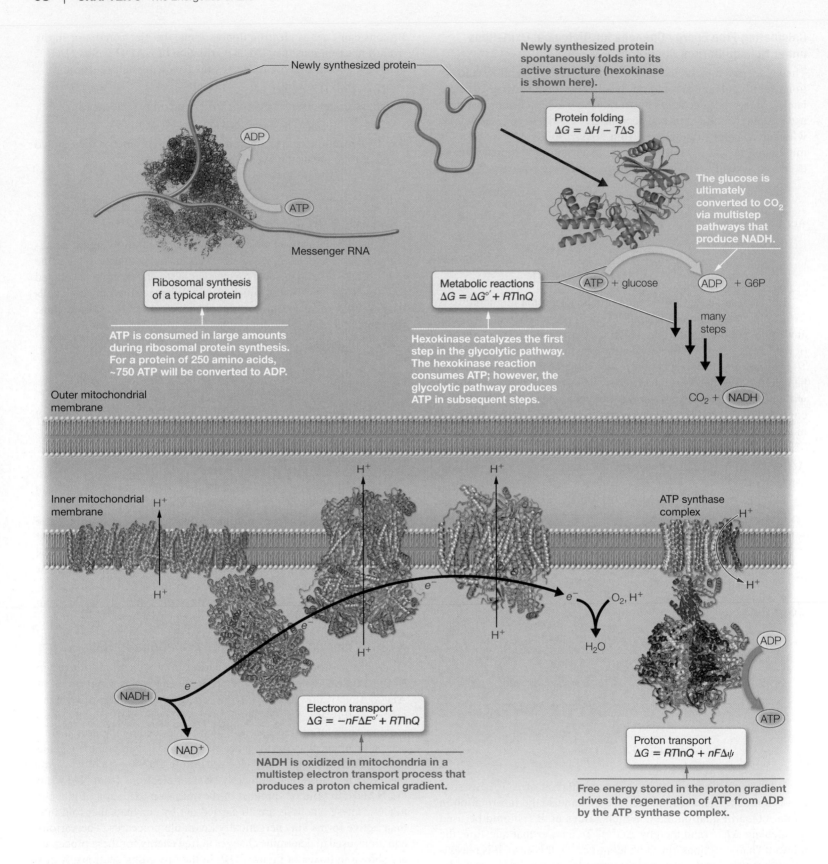

▲ **FIGURE 3.10 Examples of bioenergetic calculations applied to cellular processes.** The processes shown here are all accompanied by a favorable change in free energy (ΔG) in vivo. See text for brief descriptions of these processes. The equation for proton transport includes a term ($nF\Delta\Psi$), which is introduced in Chapter 10, where transport across membranes is described in greater detail. Protein DataBank (PDB) IDs for structure coordinates used in this figure: the ribosome 3o30, 3o5h; (see PDB ID: 4v7r) hexokinase 2yhx; electron transport proteins 3m9s, 1ppj, 2eij; ATP synthase 1c17, 1e79.

Summary

Key equations:

$$\Delta G = \Delta H - T\Delta S \qquad \text{(Equation 3.8; Section 3.2)}$$

$$\Delta G = \Delta G^{\circ\prime} + RT \ln Q \qquad \text{(Equation 3.19; Section 3.3)}$$

$$\Delta G^{\circ\prime} = -RT \ln K \qquad \text{(Equation 3.21; Section 3.3)}$$

$$\Delta G^{\circ\prime} = -nF\Delta E^{\circ\prime} \qquad \text{(Equation 3.35; Section 3.4)}$$

- Thermodynamic principles explain how organisms extract energy from their environment and use it to drive critical processes in cells. (Sections 3.1–3.4)

- The thermodynamically favored direction of a reaction (the direction that leads toward equilibrium) is determined by changes in both the **enthalpy** (H) and the **entropy** (S). The **free energy**, $G = H - TS$, takes both into account. (Section 3.2)

- The criterion for a favorable process is that the free energy change, $\Delta G = \Delta H - T\Delta S$, be negative ("**exergonic**"), rather than positive ("**endergonic**"); this is one statement of the second law of thermodynamics. The ice-to-water transition demonstrates the importance of temperature (T) in determining reaction direction. At the melting point, solid and liquid are in equilibrium ($\Delta G = 0$). The entropy of an open system can decrease, as in freezing of water, but only if the enthalpy decreases. Thus, organisms must constantly expend energy to maintain organization. (Section 3.2)

- Because living cells operate under **homeostatic conditions** that are far from equilibrium, we must consider the intracellular concentrations of each substance in a system to calculate the total free energy of the system. Intracellular conditions of temperature, pH, and concentrations of metabolites are maintained within narrow ranges that are far from equilibrium values. This state of homeostasis is distinct from true chemical equilibrium in that $\Delta G \neq 0$ for many processes. The driving force for a process under homeostatic conditions can be quantified using Equation 3.19: $\Delta G = \Delta G^{\circ\prime} + RT \ln Q$. To evaluate the mass action term Q, biochemists commonly use molar concentrations for dilute aqueous solutes, which approximate the *activities* of the solutes in the reaction and are expressed relative to the "biochemical" standard concentrations listed in the text. (Section 3.3)

- $\Delta G^{\circ\prime}$ represents the free energy change for a process under standard biochemical conditions. There are three common methods for the evaluation of $\Delta G^{\circ\prime}$:

  1. $\Delta G^{\circ\prime}$ can be calculated from the **equilibrium constant, K,** using $\Delta G^{\circ\prime} = -RT \ln K$.

  2. $\Delta G^{\circ\prime}$ can be calculated from tables of standard reduction potentials, ($E^{\circ\prime}$) using $\Delta G^{\circ\prime} = -nF\Delta E^{\circ\prime}$.

  3. $\Delta G^{\circ\prime}$ for a reaction of interest can be calculated from the values of $\Delta G^{\circ\prime}$ for two or more reactions that sum to give the chemical equation of interest. (Sections 3.3 and 3.4)

- Reactions that are not thermodynamically favored may nevertheless be driven forward if coupled to reactions or processes that have large negative ΔG values. In living systems, the hydrolysis of certain organic phosphate compounds is frequently used for this purpose. The **phosphoryl group transfer potential** ranks these compounds according to their ability to phosphorylate other compounds under standard conditions. **ATP,** the most important of these compounds, is generated in the energy-producing metabolic pathways and is used to drive many reactions. **Concentration gradients** across membranes also represent important free energy stores in cells. The favorable direction of movement of some substance through its concentration gradient is from an area of higher concentration to one of lower concentration. (Section 3.4)

Problems

Enhanced by
Mastering Chemistry
for Biochemistry

Mastering Chemistry for Biochemistry provides select end-of-chapter problems and feedback-enriched tutorial problems, animations, and interactive figures to deepen your understanding of complex topics while practicing problem solving.

Answers to red problems are available in the Answer Appendix.

1. The process of a protein folding from an inactive unfolded structure to the active folded structure can be represented by the following equation:

$$\textit{unfolded protein} \rightleftharpoons \textit{folded protein}$$

The values of ΔH° and ΔS° for the folding of the protein lysozyme are:

$$\Delta H^{\circ} = -280 \text{ kJ/mol}$$

$$\Delta S^{\circ} = -790 \text{ J/mol} \cdot \text{K}$$

(a) Calculate the value of ΔG° for the folding of lysozyme at 25 °C.

(b) At what temperature would you expect the unfolding of lysozyme to become favorable?

(c) At what temperature would the ratio of unfolded protein to folded protein be 1:5?

2. Given the following reactions and their enthalpies:

	ΔH (kJ/mol)
$H_2(g) \longrightarrow 2H(g)$	+436
$O_2(g) \longrightarrow 2O(g)$	+495
$H_2(g) + \frac{1}{2}O_2(g) \longrightarrow H_2O(g)$	−242

(a) Devise a way to calculate ΔH for the reaction

$$H_2O(g) \longrightarrow 2H(g) + O(g)$$

(b) From this, estimate the H—O bond energy.

3. The decomposition of crystalline N_2O_5

$$N_2O_5(s) \longrightarrow 2NO_2(g) + \frac{1}{2}O_2(g)$$

is an example of a reaction that is thermodynamically favored, even though it absorbs heat. At 25 °C we have the following values for the standard state enthalpy and free energy changes of the reaction:

$$\Delta H° = +109.6 \text{ kJ/mol}$$

$$\Delta G° = -30.5 \text{ kJ/mol}$$

(a) Calculate $\Delta S°$ at 25 °C.

(b) Why is the entropy change so favorable for this reaction?

4. The oxidation of glucose to CO_2 and water is a major source of energy in aerobic organisms. It is a reaction favored mainly by a large negative enthalpy change.

$$C_6H_{12}O_6(s) + 6O_2(g) \longrightarrow 6CO_2(g) + 6H_2O(l)$$

$$\Delta H° = -2816 \text{ kJ/mol} \qquad \Delta S° = +181 \text{ J/mol·K}$$

(a) At 37 °C, what is the value for $\Delta G°$?

(b) In the overall reaction of aerobic metabolism of glucose, 32 moles of ATP are produced from ADP for every mole of glucose oxidized. Calculate the standard state free energy change for the *overall* reaction when glucose oxidation is coupled to the formation of ATP at 37 °C.

(c) What is the *efficiency* of the process in terms of the percentage of the available free energy change captured in ATP?

5. The first reaction in glycolysis is the phosphorylation of glucose:

$$P_i + \text{glucose} \longrightarrow \text{glucose-6-phosphate} + H_2O$$

This is a thermodynamically unfavorable process, with $\Delta G°' = +13.8 \text{ kJ/mol}$.

(a) In a liver cell at 37 °C the concentrations of both phosphate and glucose are normally maintained at about 5 mM each. What would be the *equilibrium* concentration of glucose-6-phosphate, according to the above data?

(b) This very low concentration of the desired product would be unfavorable for glycolysis. In fact, the reaction is coupled to ATP hydrolysis to give the overall reaction

$$\text{ATP} + \text{glucose} \longrightarrow \text{glucose-6-phosphate} + \text{ADP} + H^+$$

What is $\Delta G°'$ for the coupled reaction?

(c) If, in addition to the constraints on glucose concentration listed previously, we have in the liver cell ATP concentration = 3 mM and ADP concentration = 1 mM, what is the theoretical concentration of glucose-6-phosphate at equilibrium at pH = 7.4 and 37 °C? The answer you will obtain is an absurdly high value for the cell and is never approached in reality. Explain why.

6. In another key reaction in glycolysis, dihydroxyacetone phosphate (DHAP) is isomerized into glyceraldehyde-3-phosphate (GAP):

DHAP   **GAP** $\Delta G°' = +7.5 \text{ kJ/mol}$

Because $\Delta G°'$ is positive, the equilibrium lies to the left.

(a) Calculate the equilibrium constant and the equilibrium fraction of GAP from the above, at 37 °C.

(b) In the cell, depletion of GAP makes the reaction proceed. What will ΔG be if the concentration of GAP is always kept at 1/100 of the concentration of DHAP?

7. Assume that some protein molecule, in its folded native state, has *one* favored conformation. But when it is denatured, it becomes a "random coil," with many possible conformations.

(a) If we only consider the change in entropy for the protein, what must be the sign of ΔS for the change: native → denatured? (*Note*: As suggested in the next problem, this does not include solvent effects, which also make significant contributions to ΔS.)

(b) How will the contribution of ΔS for native → denatured affect the favorability of the process? What *apparent* requirement does this impose on ΔH if proteins are to be stable structures?

8. When a hydrophobic substance like a hydrocarbon is dissolved in water, a clathrate cage of ordered water molecules is formed about it (see Figure 2.14). If we consider only the effects on water, what do you expect the sign of ΔS to be for this process? Explain your answer.

9. It is observed that as temperature is increased, most protein molecules go from their defined, folded state into a random-coil, denatured state that exposes more hydrophobic surface area than is exposed in the folded state.

(a) Given what you have learned so far about ΔH and ΔS, explain why this is reasonable. (*Hint*: Consider Problem 7.)

(b) Sometimes, however, proteins denature as temperature is *decreased*. How might this be explained? (*Hint*: Consider Problem 8.)

10. Suppose a reaction has $\Delta H°$ and $\Delta S°$ values independent of temperature (i.e., assume these values are *constant* over some range of temperature). Show from this, and the equations given in this chapter, that

$$\ln K = \frac{-\Delta H°}{RT} + \frac{\Delta S°}{R}$$

where K is the equilibrium constant. How could you use values of K determined at different temperatures to determine $\Delta H°$ for the reaction? (*Hint*: Recall that you are treating $\Delta H°$ and $\Delta S°$ as constants.)

11. The following data give the ion product, K_w (see Equation 2.7, for water at various temperatures):

$T(°C)$	K_w
0	1.14×10^{-15}
25	1.00×10^{-14}
30	1.47×10^{-14}
37	2.56×10^{-14}

(a) Using the results from Problem 10, calculate $\Delta H°$ for the ionization of water.

(b) Use these data, and the ion product at 25 °C, to calculate $\Delta S°$ for water ionization.

12. The phosphoryl group transfer potentials for glucose-1-phosphate and glucose-6-phosphate are 20.9 kJ/mol and 13.8 kJ/mol, respectively.

(a) What is the equilibrium constant for the reaction shown below at 25 °C?

Glucose-1-phosphate   **Glucose-6-phosphate**

(b) If a mixture was prepared containing 1 M glucose-6-phosphate and 1×10^{-3} M glucose-1-phosphate, what would be the thermodynamically favored direction for the reaction?

13. Wheeler and Mathews (*J. Biol. Chem.* 287:31218–31222 (2012)) reported the NAD^+ and NADH concentrations in yeast mitochondria as 20 mM and 0.3 mM, respectively. Consider the malate dehydrogenase reaction, which is part of the citric acid cycle (Chapter 13):

$$malate + NAD^+ \longrightarrow oxaloacetate + NADH + H^+$$
$$\Delta G^{\circ\prime} = +29.7\,kJ/mol$$

If malate concentration in yeast mitochondria is 0.4 mM, what is the maximum concentration of oxaloacetate needed to make the reaction exergonic at pH 7.0 and 37 °C?

14. Undergoing moderate activity, an average person will generate about 350 kJ of heat per hour. Using the heat of combustion of palmitic acid ($\Delta H = -9977.6\,kJ/mol$) as an approximate value for fatty substances, estimate how many grams of fat would be required per day to sustain this level, if all were burned for heat.

15. The major difference between a protein molecule in its native state and in its denatured state lies in the number of conformations available. To a first approximation, the native, folded state can be thought to have one conformation. The unfolded state can be estimated to have three possible orientations about each bond between residues.
    (a) For a protein of 100 residues, estimate the entropy change per mole upon denaturation.
    (b) What must be the enthalpy change accompanying denaturation to allow the protein to be half-denatured at 50 °C?
    (c) Will the fraction denatured increase or decrease with increasing temperature?

16. Suppose the concentration of glucose inside a cell is 0.1 mM and the cell is suspended in a glucose solution of 0.01 mM.
    (a) What would be the free energy change, in kJ/mol, for the transport of glucose from the medium into the cell? Assume $T = 37$ °C.
    (b) What would be the free energy change, in kJ/mol, for the transport of glucose from the medium into the cell if the intracellular and extracellular concentrations were 1 mM and 10 mM, respectively?
    (c) If the processes described in parts (a) and (b) were coupled to ATP hydrolysis, how many moles of ATP would have to be hydrolysed, per mole of glucose transported, in order to make *each* process favorable? (Use the standard free energy change for ATP hydrolysis.)

17. For part (b) of this problem, use the following standard reduction potentials, free energies, and nonequilibrium concentrations of reactants and products:

    ATP = 3.10 mM P_i = 5.90 mM   ADP = 220 μM
    glucose = 5.10 mM   pyruvate = 62.0 μM
    NAD^+ = 350 μM   NADH = 15.0 μM CO_2 = 15.0 torr

half reaction	$E^{\circ\prime}$ (V)
$NAD^+ + H^+ + 2e^- \longrightarrow NADH$	-0.315
$2\text{Pyruvate} + 6H^+ + 4e^- \longrightarrow \text{glucose}$	-0.590

$$pyruvate + NADH + 2H^+ \longrightarrow ethanol + NAD^+ + CO_2$$
$$\Delta G^{\circ\prime} = -64.4\,kJ/mol$$
$$ATP + H_2O \longrightarrow ADP + P_i + H^+ \qquad \Delta G^{\circ\prime} = -32.2\,kJ/mol$$

Consider the last two steps in the alcoholic fermentation of glucose by brewer's yeast:

$$pyruvate + NADH + 2H^+ \rightarrow ethanol + NAD^+ + CO_2$$

(a) Do you predict that ΔS° for this reaction is > 0 or < 0?
(b) Calculate the nonequilibrium concentration of ethanol in yeast cells, if $\Delta G = -38.3\,kJ/mol$ for this reaction at pH = 7.4 and 37 °C when the reactants and products are at the concentrations given above.
(c) How would a drop in pH affect ΔG for the reaction described in part (b)?
(d) How would an increase in intracellular CO_2 levels affect ΔG for the reaction in part (b)?
(e) How would an increase in intracellular CO_2 levels affect $\Delta G^{\circ\prime}$ for the reaction in part (b)?

18. Consider the degradation of glucose to pyruvate by the glycolytic pathway:

$$glucose + 2ADP + 2P_i + 2NAD^+ \longrightarrow 2\,pyruvate + 2ATP$$
$$+ 2H_2O + 2NADH + 2H^+$$

Calculate ΔG for this reaction at pH = 7.4 and 37 °C when the reactants and products are at the concentrations given in Problem 17.

19. (a) Consider the malate dehydrogenase reaction, which is part of the citric acid cycle:

$$malate + NAD^+ \rightarrow oxaloacetate + NADH + H^+$$

In yeast mitochondria, where the pH = 8.1, this reaction is exergonic only at low oxaloacetate concentrations. Assuming a pH = 8.1, a temperature of 37 °C, and the steady-state concentrations given below, calculate the maximum concentration of oxaloacetate at which the reaction will still be exergonic.

    malate = 410 μM   pyruvate = 3.22 mM   lactate = 1.10 mM
    NAD^+ = 20.0 mM   NADH = 290 μM CO_2 = 15.5 torr

half reaction	$E^{\circ\prime}$ (V)
$\text{Pyruvate} + CO_2 + H^+ + 2e^- \longrightarrow \text{malate}$	-0.330
$\text{Pyruvate} + 2H^+ + 2e \longrightarrow \text{lactate}$	-0.190

$$oxaloacetate + H^+ \longrightarrow pyruvate + CO_2 \qquad \Delta G^{\circ\prime} = -30.9\,kJ/mol$$
$$lactate + NAD^+ \longrightarrow pyruvate + NADH + H^+ \qquad \Delta G^{\circ\prime} = +25.1\,kJ/mol$$

(b) How would a drop in pH affect ΔG for the reaction described in part (a)?
(c) How would an increase in intracellular malate levels affect ΔG for the reaction in part (a)?
(d) How would an increase in intracellular pyruvate levels affect ΔG for the reaction in part (a)?

20. Bovine ribonuclease folds with $\Delta H^\circ = -280\,kJ\,mol^{-1}$ and $\Delta S^\circ = -0.79\,kJ\,mol^{-1}\,K^{-1}$. Assume ΔH° and ΔS° are independent of temperature. What fraction of bovine ribonuclease is unfolded at 42 °C?

References

For a list of references related to this chapter, see Appendix II.

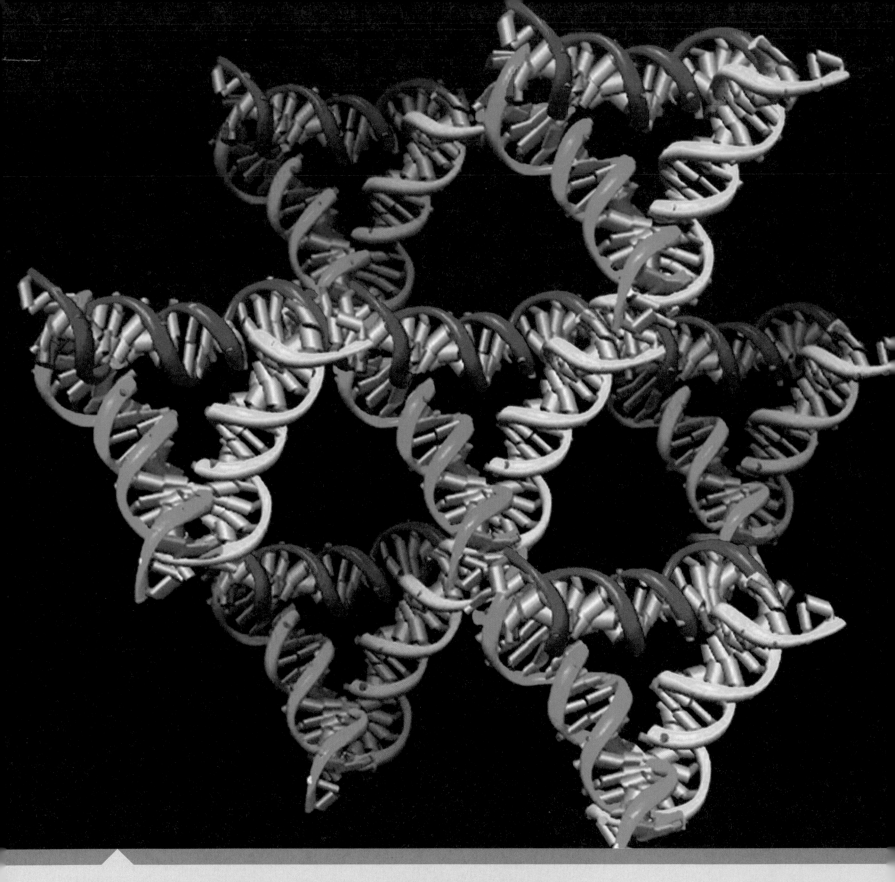

The base-pairing rules for DNA, together with the ability to chemically synthesize short DNA molecules of a known base sequence, have made it possible to design DNA molecules that fold into predetermined three-dimensional shapes—an exercise called structural DNA nanotechnology. Designer DNA structures have applications in the design of minute motors and crystalline arrays and other useful constructs. (Image courtesy of Nadrian C. Seeman, New York University.)

Nucleic Acids

WE BEGIN OUR treatment of biomolecular architecture with nucleic acids because, with their roles in the storage and transmission of biological information, these can be considered the ultimate determinants of form and function in all organisms. It seems probable that life began its evolution with nucleic acids, for only they, of all biological substances, carry the potential for self-duplication and hence, for passing information from one generation to the next. The blueprint for an organism is encoded in its nucleic acid, in gigantic DNA molecules such as that shown in **FIGURE 4.1**. Much of an organism's physical development throughout life is programmed in these molecules.

In this chapter, we first describe the nucleic acids structurally and then provide a brief introduction to the ways in which they preserve and transmit genetic information. We briefly review events in DNA replication, transcription, and translation, processes that are covered in more detail later in this book. Because of the tremendous impact of recombinant DNA techniques on biomolecular science and technology, some of the most fundamental and informative techniques are presented in Tools of Biochemistry 4A.

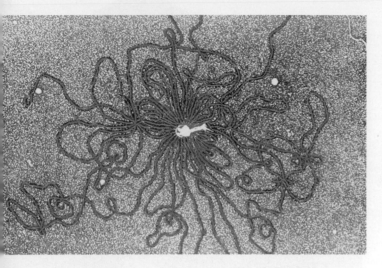

▲ **FIGURE 4.1** A nucleic acid molecule visualized by electron microscopy. The DNA of the bacteriophage T2, a virus that infects *Escherichia coli*. The virus particle (center) was subjected to osmotic shock, which ruptured the head, allowing release of a linear DNA molecule, about 0.05 mm in length.

4.1 Nucleic Acids—Informational Macromolecules

The Two Types of Nucleic Acid: DNA and RNA

DNA was discovered in 1869 by Friedrich Miescher, a military surgeon during the Franco-Prussian War. Miescher found quantities of an acidic substance in the pus from discarded surgical dressings. Because this material was predominantly in the nuclei of the white blood cells that constituted much of this material, he named it nucleic acid. Chemically, nucleic acid was found to consist of organic nitrogenous bases, a pentose sugar, and phosphate. Later, it was recognized that there are two chemical species of nucleic acid, differing in the nature of the sugar component. Miescher had discovered **DNA, deoxyribonucleic acid,** while the other major form, discovered later, is **RNA, or ribonucleic acid.** As **FIGURE 4.2** shows, each is a polymeric chain in which the monomer units are connected by covalent bonds. The monomer units of RNA and DNA have the following structures:

Repeating unit of ribonucleic acid (RNA)

Repeating unit of deoxyribonucleic acid (DNA)

RNA

DNA

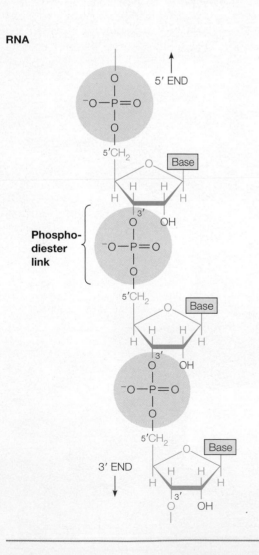

◄ **FIGURE 4.2** Chemical structures of ribonucleic acid (RNA) and deoxyribonucleic acid (DNA). The ribose–phosphate or deoxyribose–phosphate backbone of each chain is shown. The bases shown schematically here are detailed in Figure 4.3.

In each case, the monomer unit contains a five-carbon sugar, **ribose** in RNA and **2-deoxyribose** in DNA (shown in blue in the structures). The carbon atoms of the sugars are designated by primes (1′, 2′, etc.) to distinguish them from atoms of the bases. The difference between the two sugars lies solely in the 2′ hydroxyl group on ribose in RNA, which is replaced by hydrogen in DNA. The connection between successive monomer units in nucleic

● **CONCEPT** Both DNA and RNA are polynucleotides. RNA has the sugar ribose; DNA has deoxyribose.

acids is through a phosphate group attached to the hydroxyl on carbon 5′ of one unit and the hydroxyl on carbon 3′ of the next one. This forms a **phosphodiester link** between two sugar residues (Figure 4.2). The name refers to the fact that hydrolysis of this phosphate diester link yields one acid (phosphoric acid) and two alcoholic sugar hydroxyls. Through these links, long nucleic acid chains, containing up to hundreds of millions of units, are built up. The phosphate group is a strong acid, with a pK_a of about 1; this is why DNA and RNA are acidic. Every **residue,** or polymerized monomer unit, in a DNA or RNA molecule carries a negative charge at physiological pH.

The phosphodiester-linked sugar residues form the backbone of the nucleic acid molecule. By itself, the backbone is a repetitious structure, incapable of encoding information. The importance of the nucleic acids in information storage and transmission derives from their being **heteropolymers.** Each monomer in the chain carries a heterocyclic base, or **nucleobase,** which is always linked to the 1′ carbon of the sugar (see Figure 4.2). The structures of the

Purine

Pyrimidine

major nucleobases found in nucleic acids are shown in **FIGURE 4.3**. There are two types of heterocyclic bases, derivatives of **purine** and **pyrimidine.**

DNA has two purine bases, **adenine (A)** and **guanine (G),** and two pyrimidines, **cytosine (C)** and **thymine (T).** RNA has the same bases except that **uracil (U)** replaces thymine. RNA, particularly the species called **transfer RNA** (page 89), also contains several chemically modified bases, which serve in part to stabilize the molecule, as discussed in Chapter 25. Most eukaryotic DNAs contain a small proportion of cytosines methylated at carbon 5. The biological significance of DNA cytosine methylation is discussed in Chapter 26.

DNA and RNA can each be regarded as a polymer made from four kinds of monomers. The monomers are phosphorylated ribose or deoxyribose molecules, with

● **CONCEPT** The nucleic acid bases are of two kinds: the purines, adenine and guanine, and the pyrimidines, cytosine, thymine, and uracil. RNA and DNA use the same bases, except that RNA contains uracil, whereas DNA contains thymine.

purine or pyrimidine bases attached to their 1′ carbons. In purines, the attachment is at nitrogen 9; in pyrimidines, the attachment is at nitrogen 1. The bond between carbon 1′ of the sugar and the base nitrogen is referred to as a **glycosidic bond.** These

monomers are called **nucleotides.** Each nucleotide can be considered the 5′-monophosphorylated derivative of a **nucleoside (FIGURE 4.4),** which comprises just a base and a sugar. Thus, these nucleotides are also called *nucleoside 5′-monophosphates.* You have already encountered one of these molecules in Chapter 3: adenosine 5′-monophosphate, or AMP.

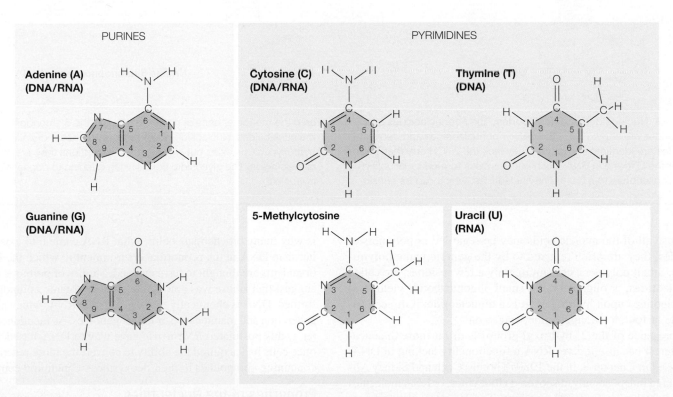

▲ **FIGURE 4.3 Purine and pyrimidine bases found in DNA and RNA.** DNA always contains the bases A, G, C, and T, whereas RNA always contains A, G, C, and U. Thymine is simply 5-methyluracil. Many DNAs contain small amounts of 5-methylcytosine.

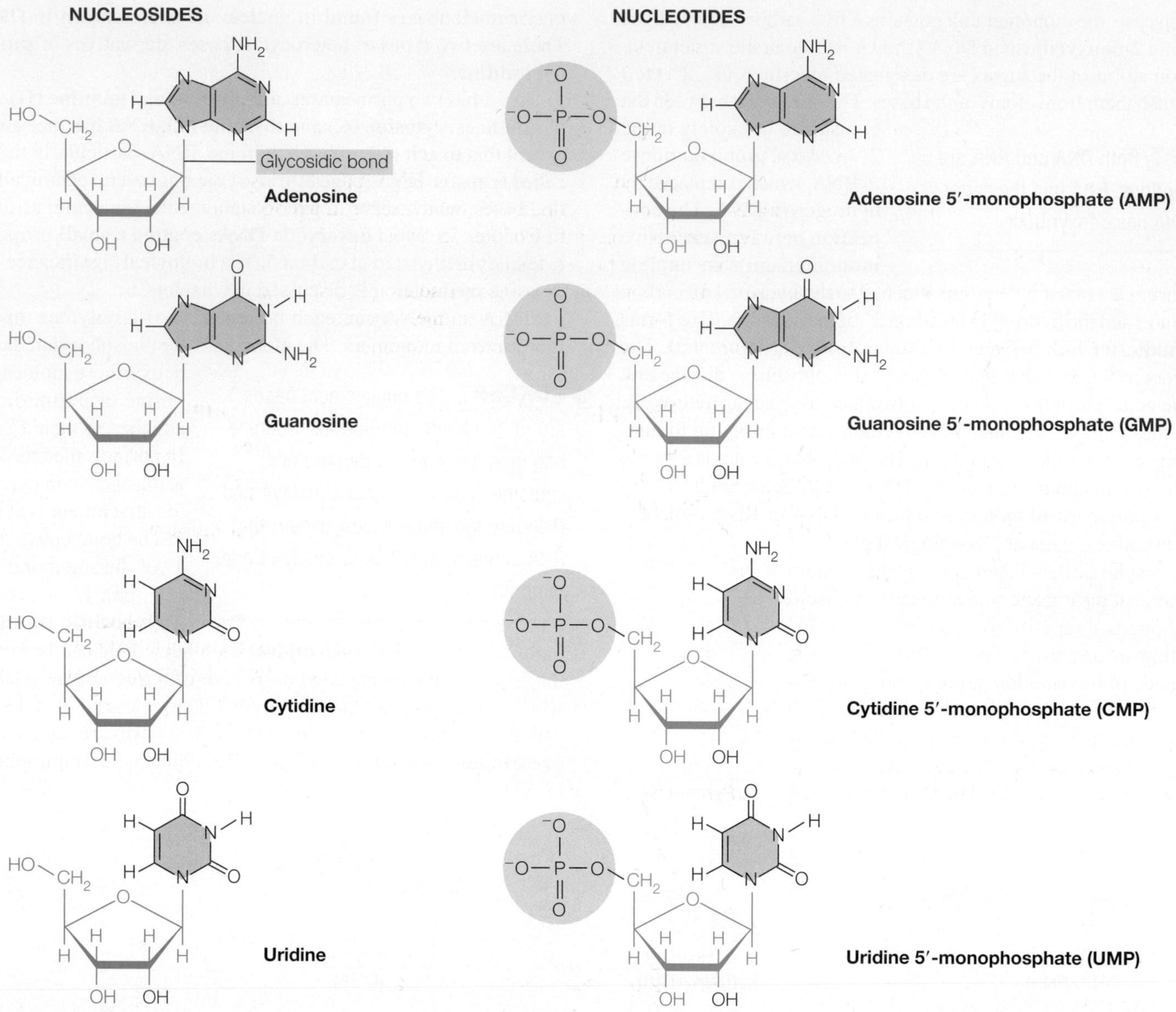

NUCLEOSIDES

NUCLEOTIDES

Adenosine

Glycosidic bond

Adenosine 5′-monophosphate (AMP)

Guanosine

Guanosine 5′-monophosphate (GMP)

Cytidine

Cytidine 5′-monophosphate (CMP)

Uridine

Uridine 5′-monophosphate (UMP)

▲ **FIGURE 4.4 Nucleosides and nucleotides.** The ribonucleosides and ribonucleotides are shown here; the deoxyribonucleosides and deoxyribonucleotides are identical except that they lack the 2′ OH and that T substitutes for the U found in RNA. Each nucleoside is formed by coupling ribose or deoxyribose to a base. The nucleotides, which can be considered the monomer units of nucleic acids, are the 5′-monophosphates of the nucleosides. Nucleoside phosphates with phosphorylation on other hydroxyl groups exist, but with very few exceptions they are not found in nucleic acids. The glycosidic bond linking adenine to ribose in adenosine is labeled.

Because all of the nucleic acids may be regarded as polymers of nucleotides, they are often referred to by the generic name **polynucleotides.** Small polymers, containing only a few residues, are called **oligonucleotides,** or oligomers. If a small oligonucleotide yields two mononucleotides upon hydrolysis, it is a dinucleotide; if three, a trinucleotide; if four, a tetranucleotide; and so on.

The presence of the 2′ hydroxyl groups is of far more than academic interest because it gives RNA a functionality lacking in DNA. As discussed in Chapter 8, in the 1980s Thomas Cech and Sidney Altman, working independently, discovered **ribozymes,** or RNA molecules capable of catalyzing chemical reactions. The 2′ hydroxyl groups are critically involved in the catalytic mechanisms. Hence, RNA molecules have the capacity both for information storage and for catalysis. This

is why many biochemists believe that RNA came into existence earlier than DNA in the primordial environment in which the first living organisms are thought to have evolved. Studies of prebiotic chemistry suggest that ribose was among the earliest organic compounds to be formed. DNA is chemically more stable than RNA, which permits the generation and maintenance of longer genomes. As mentioned in Chapter 1, this postulated *RNA world* gave way to DNA-based organisms once cells had acquired the ability to convert the more reactive ribose-containing compounds to their deoxyribose-containing counterparts.

Properties of the Nucleotides

As noted earlier, nucleotides are strong acids; the primary ionization of the phosphate occurs with a pK_a of approximately 1.0. Both secondary ionization of the phosphate and protonation or deprotonation of the amino groups on the bases within the nucleotides can be observed at pH values closer to neutrality (**TABLE 4.1**). The bases are also capable of conversion between **tautomeric** forms. **Tautomers** are structural isomers differing only in the location of their hydrogen atoms and double bonds. The major forms are those shown in Figure 4.2, but G, T, and U can partially isomerize to enol forms, and A and C to imino forms, as shown in **FIGURE 4.5**.

As a consequence of the conjugated double-bond systems in the purine and pyrimidine rings, the bases and all of their derivatives (nucleosides, nucleotides, and nucleic acids) absorb light strongly in the ultraviolet region of the spectrum. This absorption depends somewhat on pH because of the ionization reactions in the bases; representative spectra at neutral pH for ribonucleotides are depicted in **FIGURE 4.6**. This strong absorbance is often used for quantitative determination of nucleic acids because it allows measurement of nucleic acid concentrations by measurement of light absorption at 260 nm in a spectrophotometer.

Stability and Formation of the Phosphodiester Linkage

If we compare the structures of the nucleotides shown in Figure 4.4 with the polynucleotide chains depicted in Figure 4.2, we see that, in principle, a polynucleotide could be generated from its nucleotide monomers by eliminating a water molecule between each pair of monomers. However, the free energy change in this hypothetical reaction is quite positive, about +25 kJ/mol; therefore,

▲ FIGURE 4.5 Tautomerization of the bases. The most stable (and therefore common) forms are shown at the left. The less common imino and enol forms, shown on the right, are found in some special base interactions.

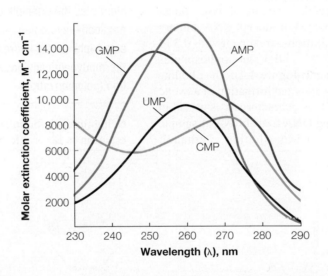

◄ FIGURE 4.6 Ultraviolet absorption spectra of ribonucleotides. Absorbance is a measure of the relative amount of light absorbed at a given wavelength.

TABLE 4.1 Ionization constants of ribonucleotides expressed as pK_a values				
Phosphate			Base	
Primary Ionization		Secondary Ionization		
	pK_{a1}	pK_{a2}	pK_a	Reaction (as Loss of Proton from)
5′ AMP	0.9	6.1	3.8	N-1
5′ GMP	0.7	6.1	2.4	N-7
			9.4	N-1
5′ UMP	1.0	6.4	9.5	N-3
5′ CMP	0.8	6.3	4.5	N-3

equilibrium lies far to the side of hydrolysis of the phosphodiester bond in the aqueous environment of the cell. Hydrolysis of polynucleotides to nucleotides is the thermodynamically favored process.

We encounter here the first of many examples of the **metastability** of biologically important polymers. Metastable compounds are thermodynamically favored to break down, but they do so only slowly unless the reaction is catalyzed. On the basis of the free energy change involved, polynucleotides should undergo hydrolysis under conditions existing in living cells, but this hydrolysis is exceedingly slow in the absence of a catalyst. This characteristic is of great importance, for it ensures that the DNA in cells is sufficiently stable to serve as a repository of genetic information through successive generations. In dehydrated conditions, DNA is so stable that fragments of DNA molecules can be recovered from bones as old as 430,000 years. Indeed, the complete genome sequence of a Neanderthal, an archaic human, was reported in 2010, and "ancient DNA" has become an important adjunct to paleontology.

When catalysts *are* present, however, hydrolysis can be exceedingly rapid in aqueous solution. Acid treatment catalyzes hydrolysis of the phosphodiester bonds in nucleic acids, yielding a mixture of nucleotides. In both RNA and DNA, the glycosidic bond between the base and the sugar is also hydrolyzed; a mixture of bases, phosphoric acid, and ribose (or deoxyribose) is produced. RNA, but not DNA, is also labile in mild alkaline solution—treatment with 0.3 M alkali yields a mixture of nucleoside 2′- and 3′-monophosphates (unlike the 5′-monophosphates shown in Figure 4.4). The alkaline hydrolysis of RNA involves the base-catalyzed formation of a cyclic intermediate, as shown in **FIGURE 4.7**. This characteristic is useful in the laboratory, for example, in preparing DNA free of RNA. A sample can be treated with mild alkali followed by addition of ethanol;

DNA precipitates, while the nucleotides derived from RNA remain in solution.

Both DNA and RNA can be broken down by enzymes called **nucleases,** which catalyze the hydrolysis of phosphodiester bonds in both RNA and DNA. Your body can break down and utilize polynucleotides in food you eat because your digestive system contains nucleases. These enzymes are described in Chapter 19.

The unfavorable thermodynamics of the hypothetical formation of phosphodiester bonds via dehydration leads us to ask: If polynucleotides cannot be synthesized in vivo by the direct elimination of water, how are they actually made? The answer is that their synthesis involves the energy-rich ribonucleoside or deoxyribonucleoside *triphosphates*. The nucleoside monophosphate being added to a growing chain is presented as a *nucleoside triphosphate,* like ATP or deoxy ATP (dATP), and pyrophosphate is released in the reaction (**FIGURE 4.8**).

Recall from Chapter 3 our discussion of the energetics of the phosphate bonds in adenine nucleotides. The innermost, or α, phosphate is linked to ribose by a low-energy ester bond. The central and outermost phosphates (β and γ, respectively) are connected to each other and to the α phosphate by energy-rich anhydride bonds. Nucleic acid biosynthesis is driven, as shown in Figure 4.8, by the cleavage of the energy-rich anhydride bond linking the β and γ phosphates. The coupled reaction is favorable because the *net* $\Delta G°'$ is negative. The reaction is further favored because the hydrolysis of the pyrophosphate product (PP_i) to orthophosphate, or inorganic phosphate (P_i) has a $\Delta G°' = -19$ kJ/mol. Thus, the pyrophosphate is readily removed, driving the synthesis reaction even further to the right and yielding an overall $\Delta G°'$ of -25 kJ/mol. Polynucleotide synthesis is an example of a principle we emphasized in Chapter 3—the use of favorable reactions to drive thermodynamically unfavorable ones.

● **CONCEPT** Polynucleotide synthesis illustrates the principle of a thermodynamically favorable reaction (nucleoside triphosphate hydrolysis) driving a thermodynamically unfavorable reaction (polymerization).

It is important to appreciate how the energetics of such processes fit into the overall scheme of life. An organism obtains energy from the metabolism of nutrients and stores part of this energy by generating ATP, GTP, dATP, and similar energy-rich compounds. It uses these compounds in turn as energy sources to drive the synthesis of macromolecules such as DNA, RNA, and proteins. All four common ribonucleoside triphosphates, but particularly ATP, are used as energy donors in metabolic reactions. Because ATP is the most abundant and most frequently used energy donor, it is often referred to as the energy currency of the cell. You encountered this theme in Chapter 3 and will see it repeated throughout your study of biochemistry.

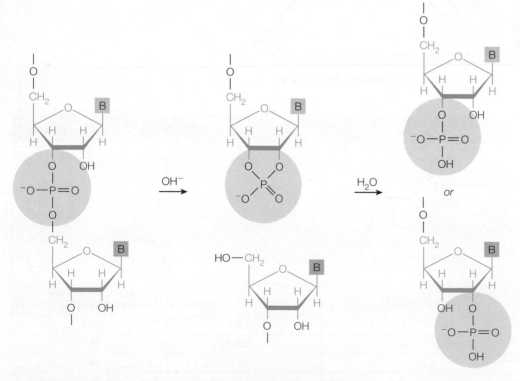

▲ **FIGURE 4.7** Alkaline hydrolysis of RNA via a cyclic intermediate.

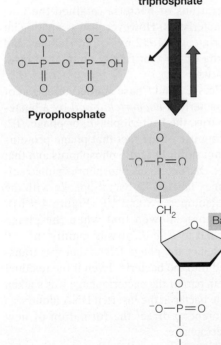

4.2 Primary Structure of Nucleic Acids

Our treatment of the structure of nucleic acids begins with the linear order of nucleotides in polymeric nucleic acids—what we call the **primary structure**—and continues with consideration of the three-dimensional architecture of these molecules—the **secondary** and **tertiary** structures.

The Nature and Significance of Primary Structure

A closer examination of Figure 4.2 reveals two important features of all polynucleotides:

1. A polynucleotide chain has a *sense* or *directionality*. The phosphodiester linkage between monomer units is between the 3′ carbon of one monomer and the 5′ carbon of the next. Thus, the two ends of a linear polynucleotide chain are distinguishable. One end normally carries an unreacted 5′ phosphate, and the other end an unreacted 3′ hydroxyl group.

2. A polynucleotide chain has *individuality*, determined by the sequence of its bases—that is, the *nucleotide sequence*. This sequence is called the **primary structure** of that particular nucleic acid.

● **CONCEPT** Every naturally occurring polynucleotide has a defined sequence, its primary structure.

If we want to describe a particular polynucleotide sequence (either DNA or RNA), it is awkward and unnecessary to draw the molecule in its entirety as in Figure 4.2. Accordingly, some compact conventions have been devised. If we state that we are describing a DNA molecule or an RNA molecule, then most of the structure is understood. We can then abbreviate a small DNA molecule as follows, where the vertical lines represent sugars and the diagonals are phosphodiester bonds:

This notation shows (1) the sequence of nucleotides, by their letter abbreviations (A, C, G, T); (2) that all phosphodiester links are between 3′ hydroxyls and 5′ phosphates; and (3) that this particular molecule has a phosphate group at its 5′ end and an unreacted hydroxyl at its 3′ end. It also tells us it is a DNA sequence, not RNA, because it has T, not U.

If all of the phosphodiester bonds can be assumed to link a 3′ hydroxyl to a 5′ phosphate (as is usually the case), a more compact notation is possible for the same molecule:

$$pApCpGpTpT$$

The 3′ –OH group is understood to be present and unreacted. Were there a phosphate on the 3′ end and an unreacted hydroxyl on the 5′ end, we would write

$$ApCpGpTpTp$$

Finally, if we are concerned *only* with the sequence of bases in the molecule, as will often be the case, we can write it still more compactly as

$$ACGTT$$

Note that the sequence of a polynucleotide chain is usually written, by convention, with the 5′ end to the left and the 3′ end to the right.

The main importance of primary structure, or sequence, is that *genetic information is stored in the primary structure of DNA. A gene is nothing more than a*

▲ **FIGURE 4.8 Use of a metabolically activated nucleotide in nucleic acid biosynthesis.** Cleavage of the anhydride bond in the triphosphate provides the energy to drive incorporation of the nucleotide into a growing polynucleotide chain.

● **CONCEPT** The primary structure of DNA encodes genetic information.

___

of the bases. Later in this chapter we describe how the primary structure—the nucleotide sequence of DNA—is actually determined.

particular DNA sequence, encoding information in a four-letter language in which each "letter" is one

DNA as the Genetic Substance: Early Evidence

The search for the chemical nature of genes started in the late 1800s, shortly after Friedrich Miescher had first isolated DNA. Some scientists suspected that DNA might be the genetic material. However, the fact that DNA contained only four kinds of monomers seemed to argue against such a complicated role. Early researchers thought it more likely that genes were made of proteins, for proteins were beginning to be recognized as extremely complex molecules. For most of the first half of the twentieth century, nucleic acids were considered to be merely some kind of structural material in the cell nucleus.

Between 1944 and 1952, two crucial experiments identified DNA as the genetic material. In 1944, Oswald Avery, Colin MacLeod, and Maclyn McCarty found that DNA from pathogenic strains of the bacterium *Pneumococcus* could be transferred into nonpathogenic strains, making them pathogenic (**FIGURE 4.9(a)**). This **transformation** was genetically stable; succeeding generations of bacteria retained the new characteristics. However, it was an elegant experiment in 1952 by Alfred Hershey and Martha Chase that finally convinced most scientists.

Hershey and Chase studied infection of the bacterium *Escherichia coli* by a bacterial virus, the bacteriophage, or phage, T2. Making use of the fact that phage proteins contain sulfur but little phosphorus and that phage DNA contains phosphorus but no sulfur, they labeled T2 bacteriophage with the radioisotopes ^{35}S and ^{32}P (Figure 4.9(b)). They then showed that when the phage attached to *E. coli*, it was mainly the ^{32}P (and hence the phage DNA) that was transferred into the bacteria. Even if the residual protein part of the bacteriophage was shaken off the bacteria, the inserted DNA alone was sufficient to direct the formation of new bacteriophage.

Through these and similar experiments, it was generally recognized by 1952 that DNA must be the genetic substance. But how could it carry the enormous amount of information that a cell needed, how could it transmit that information to the cell, and, above all, how could it be accurately replicated in cell division? The answers to these questions came only after 1953, when James Watson and Francis Crick proposed a structure for DNA that opened a whole new world of molecular biology.

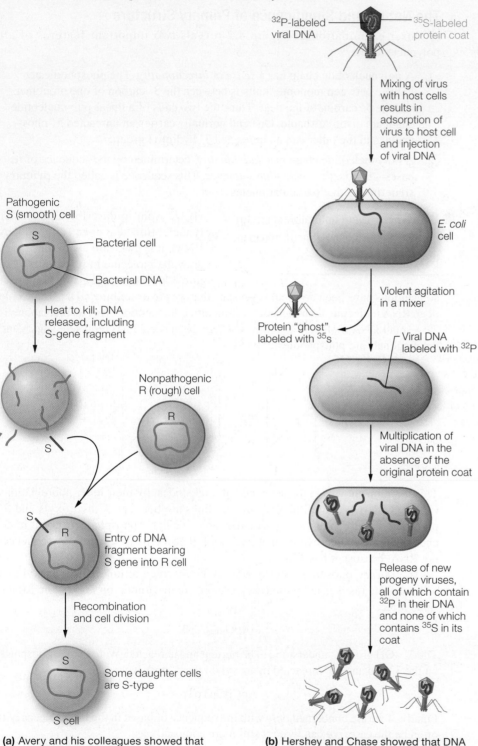

(a) Avery and his colleagues showed that nonpathogenic pneumococci could be made pathogenic by transfer of DNA from a pathogenic strain.

(b) Hershey and Chase showed that DNA transferred from a virus to a bacterium contains sufficient information to direct the synthesis of new viruses.

▲ **FIGURE 4.9 Experiments that showed DNA to be the genetic substance.**

4.3 Secondary and Tertiary Structures of Nucleic Acids

The DNA Double Helix

We mentioned earlier that the primary structure of nucleic acids refers to the sequence of nucleotides in the linear molecules. The **secondary structure** refers to the three-dimensional arrangements of the nucleotide residues with respect to one another—in the case of DNA, to the double helix of native DNA molecules and related structures, to be discussed later in this chapter. The **tertiary structure** refers to longer-range interactions in three dimensions, also discussed in this chapter, such as supercoiling. Note that these terms have comparable significance with respect to protein structure, as we will see in Chapter 6.

Data Leading Toward the Watson–Crick Double-Helix Model

In Chapter 1, we mentioned the leap of intuition that led Watson and Crick to propose the double-helical model for DNA structure, a leap that represents one of the great scientific achievements of the twentieth century. We also pointed out that biochemistry is an experimental science, with concepts based on experimental data. Watson and Crick relied on data gathered by others to make their proposal—data from chemical analysis and from X-ray diffraction.

The chemical data came from Erwin Chargaff, who in the late 1940s subjected DNAs from various organisms to acid hydrolysis, conditions that released the purine and pyrimidine bases. Analysis of the DNA base composition from numerous sources showed a striking regularity. Within experimental error, the mole percent adenine was always the same as the mole percent thymine, while the mole percent guanine was always the same as that of cytosine. DNAs from different organisms varied considerably in their abundance of (A + T) relative to (G + C). Human DNA, for example, contains about 30% each of A and T, and 20% each of G and C, making it relatively AT-rich. By contrast, the common bread mold *Neurospora crassa* is GC-rich, with about 23% each of A and T, and 27% each of G and C. Surveying his data, Chargaff wrote, "There exist a number of regularities. Whether these are merely accidental cannot yet be decided."

X-Ray Analysis of DNA Fibers

The other key data came from X-ray diffraction analysis of stretched DNA fibers, data obtained by Rosalind Franklin and Maurice Wilkins. We discuss the principle of this technique in Tools of Biochemistry 4B. In a crystal, even one of a macromolecule such as a protein, the position of every atom is fixed. Therefore, the structure of the molecule can be determined from X-ray analysis at atomic resolution, provided that one can collect sufficient data and apply sufficient computing power. However, because of its length and flexibility, DNA cannot be crystallized. Instead, Franklin and Wilkins carried out their analyses on DNA fibers, which, when stretched, produce approximate alignment of the long helical molecules with the fiber axis.

FIGURE 4.10 shows a diffraction pattern of wet DNA fibers. Analysis of patterns like this, using diffraction theory, indicated that the DNA molecules within the fibers were helical in structure, with the following characteristics:

1. A *repeat* distance of 10 nucleotide residues, meaning that the structure repeats itself at intervals of 10 nucleotides.

2. A *pitch* of 3.4 nm, meaning that the distance parallel to the helix axis at which the helix makes one turn is 3.4 nm.

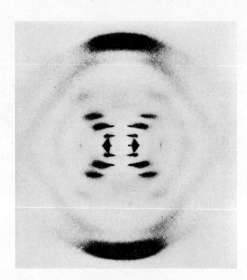

▲ **FIGURE 4.10 Evidence for the structure of DNA.** This photograph, taken by Rosalind Franklin, shows the X-ray diffraction pattern produced by wet DNA fibers. The cross pattern indicates a helical structure, and the strong spots at the top and bottom correspond to a helical rise of 0.34 nm. The pattern suggested further that there are 10 bases per repeat.

3. A *rise* of 0.34 nm, meaning that the distance parallel to the helix axis between two successive nucleotide residues is 0.34 nm.

If we think of a spiral staircase as an example of a helix, the rise is the height of each step and the pitch is the distance from where one is standing to the corresponding spot directly overhead.

Watson was a biologist and Crick was a physicist. To formulate their model, they needed help from a chemist, Jerry Donohue, who informed them that for those bases that underwent keto-enol tautomerization (see Figure 4.5), the keto form was likely to predominate under intracellular conditions. This information was essential to the Watson–Crick proposal that the two strands in the double helix were stabilized by hydrogen bonding between A and T and between G and C, as shown in **FIGURE 4.11(a)**. With this pairing, distances between the 1′ carbons of the deoxyribose moieties of A–T and of G–C are the same—1.08 nm in each case. This pairing arrangement meant that the double helix could be uniform in diameter, an impossibility if purines were paired with purines, or pyrimidines with pyrimidines.

In the Watson–Crick model, the hydrophilic phosphate–deoxyribose backbones of the helix were on the outside, in contact with the aqueous environment, and the base pairs were stacked on one another with their planes perpendicular to the helix axis. Two views of such a structure are shown in Figure 4.11(b) and (c). (The figure shows a recent, refined model, based on better data than Watson and Crick had available: The bases are not exactly perpendicular to the helix axis, and the sugar conformation is slightly different from that in the original model.) Stacking of the bases, as shown in Figure 4.11(b), allows strong interactions between them, probably of van der Waals type (see Chapter 2); this is usually referred to as a "stacking interaction." Each base pair is rotated by 36°, that is, 1/10 of a 360° rotation with respect to the next, to accommodate 10 base pairs in each turn of the helix (though later structural studies have shown the number to be closer to 10.5). The diffraction pattern showed the repeat distance to be about 3.4 nm, so the helix *rise*—that is, the distance between base pairs—had to be about 0.34 nm (Figure 4.11(c)). This distance is just twice the van der Waals thickness of a planar ring

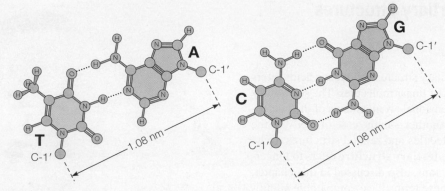

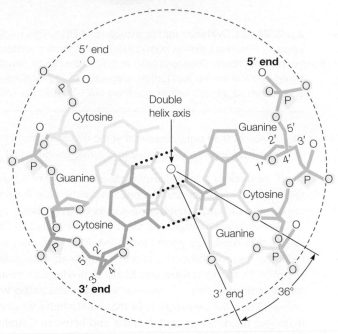

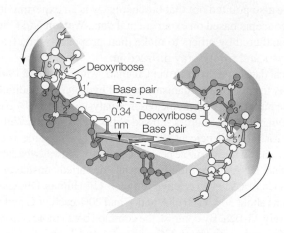

(a) Base pairing. A-T and G-C are the base pairs in the Watson–Crick model of DNA. This pairing allows the C-1′ carbons on the two strands to be exactly the same distance apart for both pairs.

(b) Stacking of the base pairs. This view down the helix axis shows how the base pairs stack on one another, with each pair rotated 36° with respect to the next.

(c) Distance between the base pairs. A side view of the base pairs shows the 0.34-nm distance between them. This distance is the rise of the helix.

▲ **FIGURE 4.11** Fundamental elements of structure in the DNA double helix.

(see Table 2.2), so the bases are closely packed within the helix, as shown in a space-filling model (**FIGURE 4.12**).

The model also shows that, although the bases are inside, they can be approached through two deep spiral grooves called the *major* and *minor* grooves. The major groove gives more direct access to the bases; the minor groove faces the sugar backbone. Building molecular models of two-stranded DNA structures soon convinced Watson and Crick that the DNA strands must run in opposite directions. This arrangement can be seen in Figure 4.11(c). The model Watson and Crick presented was for a right-hand helix, although at that time evidence for the sense (direction of the turn) of the helix was weak. Their guess proved to be correct.

● **CONCEPT** The Watson–Crick model for DNA is a two-stranded, antiparallel double helix with 10 base pairs per turn. Pairing is A-T and G-C.

The structural relationships between A and T, and between G and C, explained the regularity in DNA base compositions observed by Chargaff that we mentioned previously. What was not explained was one instance of a DNA that did not obey Chargaff's rule. That DNA, from bacteriophage ΦX174, contained 24.0% A, 23.3% G, 21.5% C, and 31.2% T. The mystery was solved when it was learned that this particular DNA is single-stranded. When it replicates in infected bacteria, that single strand directs synthesis of a complementary strand, and this double-stranded **replicative form** DNA does follow Chargaff's rule (Chapter 22).

The Watson–Crick model not only explained the structure of DNA and Chargaff's rule, but also carried implications that went to the heart of biology. Since A always pairs with T, and G always pairs with C, the

● **CONCEPT** The complementary, two-stranded structure of DNA explains how the genetic material can be replicated.

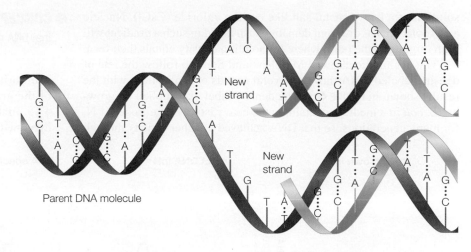

▲ **FIGURE 4.13 A model for DNA replication.** Each strand acts as a template for a new, complementary strand. When copying is complete, there will be two double-stranded daughter DNA molecules, each identical in sequence to the parent molecule. The point at which the parental DNA strands unwind is called the *replication fork,* as discussed later in this chapter.

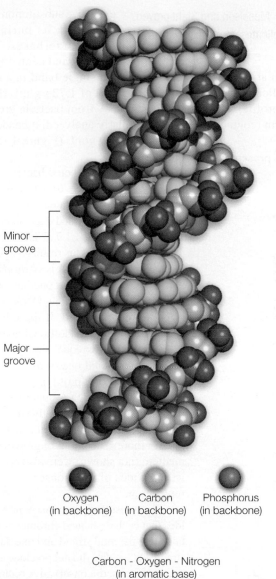

▲ **FIGURE 4.12 A space-filling model of DNA.** The DNA molecule as modeled by Watson and Crick is shown here, with each atom given its van der Waals radius. This model shows more clearly than Figure 4.11 how closely the bases are packed within the helix. The major and minor grooves are indicated.

two strands are *complementary.* If the strands could be separated and new DNA synthesized along each, following the same base-pairing rule, two double-stranded DNA molecules would be obtained, each an *exact* copy of the original (**FIGURE 4.13**). This **self-replication** is precisely the property that the genetic material must have: When a cell divides, two complete copies of the genetic information carried in the original cell must be produced. In their 1953 paper announcing the model (see Appendix II), Watson and Crick expressed this idea in a masterpiece of understatement: "It has not escaped our notice that the specific pairing we have postulated immediately suggests a possible copying mechanism for the genetic material."

Semiconservative Nature of DNA Replication

The DNA copying mechanism we have mentioned involves unwinding the two strands of a parental DNA duplex, with each strand serving

as template for the synthesis of a new strand, complementary to and wound about that parental strand. Complete replication of a DNA molecule would yield two "daughter" duplexes, each consisting of one-half parental DNA (one strand of the original duplex) and one-half new material. This mode of replication is called **semiconservative** because half of the original material is conserved in each of the two copies (**FIGURE 4.14**). It is distinguished from two other possible modes: **conservative,** in which one of the two daughter duplexes is the conserved parental duplex while the other is synthesized de novo, and **dispersive,** in which parental material is scattered through the structures of the daughter duplexes.

The first experimental test of this model came in 1958, when Matthew Meselson and Franklin Stahl realized that molecules differing in density by very small amounts could be separated from each other by centrifugation in density gradients. In this method, a density gradient is created by centrifuging to equilibrium a concentrated

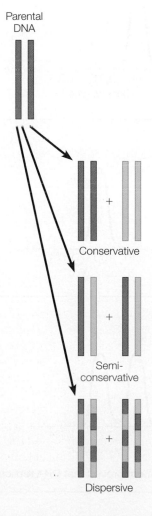

◀ **FIGURE 4.14 Three models of DNA replication.** Experimental evidence supports the semiconservative model. Purple, parental DNA, blue, new DNA.

solution of a heavy metal salt like cesium chloride (CsCl). Nucleic acid molecules of different densities suspended in such a gradient will each migrate to the point where the solution density equals their own.

This technique allowed Meselson and Stahl to follow the fate of density-labeled DNA through several rounds of replication, with the results shown in **FIGURE 4.15**. The density label was applied by growing *E. coli* in a medium containing the heavy isotope of nitrogen, ^{15}N, for many generations, so that DNA achieved a higher density through

● **CONCEPT** Meselson and Stahl proved that DNA replicates semiconservatively.

extensive substitution of ^{15}N for ^{14}N in its purine and pyrimidine bases. When isolated and centrifuged to equilibrium at pH 7.0, this DNA formed a single band in a region of the gradient corresponding to a density of 1.724 g/mL (Figure 4.15, first profile). By contrast, when DNA from bacteria grown in light medium (containing ^{14}N) was similarly analyzed, it banded at a density of 1.710 g/mL (Figure 4.15, second profile).

When density-labeled bacteria grown in heavy medium were transferred to light medium, the DNA isolated after one generation of growth banded exclusively at an intermediate density, 1.717 g/mL (Figure 4.15, third profile). This result is expected if the newly replicated DNA is a *hybrid* molecular species, consisting of one-half parental material and one-half new DNA (synthesized in light medium). If these bacteria were cultured for an additional generation in light medium, two equal-sized bands were seen, one light and one of hybrid density (Figure 4.15, fourth profile), as expected if the hybrid-density DNA underwent a second round of semiconservative replication. Analysis of DNA molecules isolated after three or more generations of growth in light medium (not shown) were also consistent with the idea of semiconservative DNA replication.

These results were consistent with the idea that each replicated chromosome contains one parental strand and one daughter strand, but the data did not preclude alternative forms of semiconservative replication, involving the breaking of DNA strands. These models were ruled out by centrifugal analysis of the density-labeled DNAs at pH 12, where DNA strands separate. The inescapable conclusion was that the replicative hybrid contains one complete strand of parental DNA and one complete strand of newly synthesized DNA.

Alternative Nucleic Acid Structures: B and A Helices

At the time Watson and Crick proposed their model, two quite different X-ray diffraction patterns had already been obtained for DNA, indicating that the molecule can exist in more than one form. The **B form,** which is seen in DNA fibers prepared under conditions of high humidity, is shown in Figures 4.11 and 4.12. Watson and Crick chose to study the B form because they correctly expected

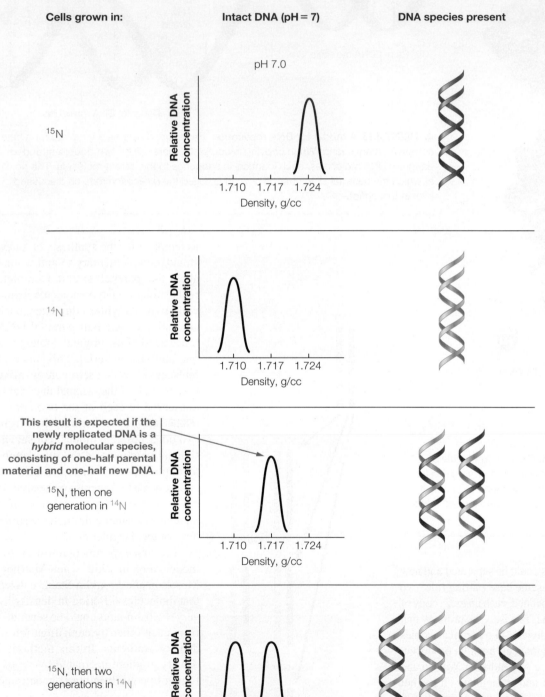

▲ **FIGURE 4.15** The Meselson–Stahl experiment proved that DNA replicates semiconservatively.

it to be the predominant form in the aqueous milieu of the cell. DNA fibers prepared under conditions of low humidity have a different structure, the so-called **A form** (**FIGURE 4.16(c)** and (d)). Although a B helix is indeed the form of DNA found in cells, the A helix is also important biologically. Double-stranded RNA molecules always form the A structure, and so do **DNA–RNA hybrid molecules,** which are formed by the pairing of one DNA strand with an RNA strand of complementary sequence. Thus, two major kinds of secondary structures exist in polynucleotides. As we shall see later in this chapter, other kinds of secondary structures are possible under special circumstances.

As Figure 4.16 shows, the A and B forms are quite different, although both are right-hand helices (turning upward and to the right). In the B helix, the bases lie close to the helix axis, which passes between the hydrogen bonds (note the end-on views of the helices in Figure 4.16(a) and (c)). In the A helix, the bases lie farther to the outside and

● **CONCEPT** The two major forms of polynucleotide secondary structure helices are called A and B. Most DNA adopts the B form; double-stranded RNA and DNA-RNA hybrids adopt the A form.

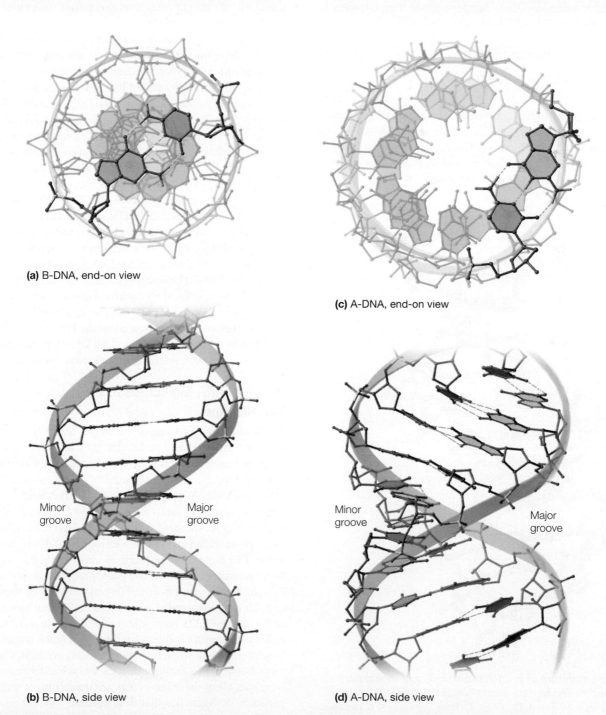

(a) B-DNA, end-on view

(c) A-DNA, end-on view

Minor groove   Major groove

Minor groove   Major groove

(b) B-DNA, side view

(d) A-DNA, side view

▲ **FIGURE 4.16 Comparison of the two major forms of DNA.** The structures of B-DNA and A-DNA, as deduced from fiber diffraction studies, are shown here in both end-on and side views.

are strongly tilted with respect to the helix axis. The surfaces of the helices are also different. In the B helix the major and minor grooves are quite distinguishable, whereas in the A helix the two grooves are more nearly equal in width. The B helix has 10.5 residues per turn and a rise of 0.34 nm, as noted earlier, while the A helix has 11 residues per turn and a rise of 0.255 nm.

As mentioned earlier, a crystal gives far more information from X-ray diffraction than a fibrous molecule, because every atom within a crystal is in a fixed position. Therefore, a major advance took place in 1981, when Richard Dickerson and his colleagues succeeded in *crystallizing* a small double-stranded DNA fragment, made from chemically synthesized oligonucleotides, which had the sequence

$$5'CGCGAATTCGCG 3'$$
$$3'GCGCTTAAGCGC 5'$$

Molecular crystallography of this fragment and other small DNA fragments has given us much more detailed information concerning the secondary structure of polynucleotides. The results of such a study of B-DNA are shown in **FIGURE 4.17**. Here we can show DNA molecules, with the position of every atom clearly specified.

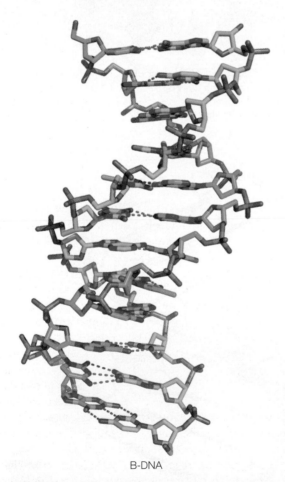

B-DNA

▲ **FIGURE 4.17 The structure of B-DNA from studies of molecular crystals.** The structure shows local distortions of the idealized structure presented in Figures 4.11 and 4.16. PDB: ID 1bna. Data from R. E. Dickerson, *Sci. Am.* December 1983, pp. 100–104.

A major point emerging from the molecular crystal studies is that the models drawn from the fiber patterns represent oversimplifications of the structures. The real structure of B-DNA involves local variations in the angle of rotation between base pairs, the sugar conformation, the tilt of the bases, and even the rise distance. If you examine Figure 4.17 carefully, you can see many distortions from the idealized structures. Nucleic acid secondary structure is not homogeneous. It varies in response to the local sequence, and it can be changed by interaction with other molecules. The parameters given for various forms of DNA–10.5 base pairs per turn for B-DNA, for example—should therefore be thought of as *average* values, from which considerable local deviation is possible.

If one closely examines structures like that shown in Figure 4.17, another departure from the original Watson–Crick model becomes apparent. Many DNA molecules are slightly *bent;* that is, the helix axis does not follow a straight line. The degree and directions of bending depend in a complicated way on DNA sequence. They can also be strongly influenced by the interaction of DNA with various protein molecules; we will see examples of this in later chapters.

The molecular crystallographic studies also provide a possible explanation for why B-DNA is favored in an aqueous environment. The B form of DNA, but not A-DNA, can accommodate a spine of water molecules lying in the minor groove. The hydrogen bonding between these water molecules and the DNA may confer stability on the B form. According to this hypothesis, when this water is removed (as in fibers at low humidity), the B form becomes less stable than the A form.

Why, then, do double-stranded RNA and DNA–RNA hybrids always adopt the A form? The answer probably lies in the extra hydroxyl group on the ribose in RNA. This hydroxyl interferes sterically in the B form by lying too close to the phosphate and carbon 8 on an adjacent purine base. Therefore, RNA *cannot* adopt the B form, even under conditions in which hydration might favor it. In DNA the hydroxyl is replaced by hydrogen, and such steric hindrance does not occur.

DNA and RNA Molecules in Vivo
DNA Molecules

We have described some of the major features of nucleic acids. But in what forms do these molecules exist in the living cell? Most of the DNA in most organisms is double-stranded, with the two strands being complementary, although some DNA viruses carry single-stranded DNA molecules. We have mentioned the bacteriophage ΦX174, whose genome consists of one single-stranded DNA molecule 5386 nucleotides in length. Another "ssDNA phage," M13, has 6407 nucleotides in its single-stranded genome. Bacteriophage T4, which is closely related to the T2 phage used by Hershey and Chase, has a double-stranded linear genome of 168,899 base pairs, giving it a length of about 57.4 μm. *E. coli,* the bacterial host for infection by all of these phages, has a circular double-stranded genome some 20-fold larger than that of T2 or T4, at 4,639,221 base pairs and a length of 1.57 mm (**FIGURE 4.18**). Eukaryotic chromosomes, also linear, are much larger—typically, centimeters in length when fully extended. By contrast, the DNA in human mitochondria is a circular duplex, only 16,569 base pairs in length. Note that all of these molecules are much longer than the cells, virus particles, or organelles that they inhabit.

The proportions of B and A forms of polynucleotides in vivo are as you might expect from the conditions under which these conformations

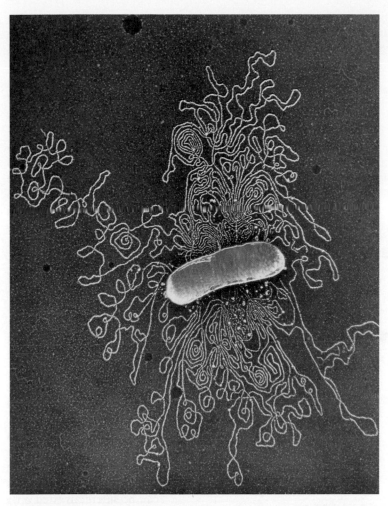

▲ **FIGURE 4.18** A DNA molecule as seen by the electron microscope. The large circular double-stranded DNA of *E. coli* exists as a number of supercoiled loops bound to a protein matrix.

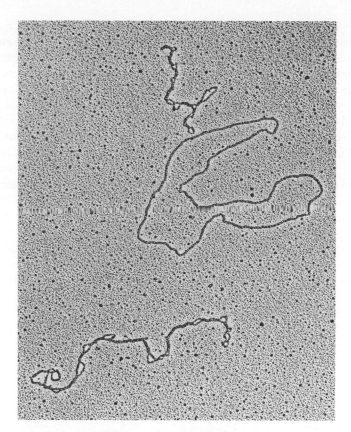

▲ **FIGURE 4.19** Relaxed and supercoiled DNA molecules. Electron micrograph showing three human mitochondrial DNA molecules. All three are identical in sequence and contain 16,569 bp each. However, the molecule in the center is relaxed, whereas the molecules at top and bottom are tightly supercoiled.

are stable. Since cells contain much water, we expect most of the double-stranded DNA to be in the B form or something very like it. There is evidence that B-DNA dissolved in solution is only a little different in conformation from the B form seen in fiber preparations, having about 10.5 base pairs per turn instead of the expected 10.0. Double-stranded RNA, as mentioned earlier, always exists in the A form.

Circular DNA and Supercoiling

Another feature of naturally occurring DNA molecules is illustrated in **FIGURE 4.19**: Many of these molecules are *circular;* that is, they do not have free 5′ or 3′ ends. The circles may be small, as in bacteriophage ΦX174 DNA, or immense, as in *E. coli* DNA (Figure 4.18), and they may involve either a single strand or two strands intertwined in a B-form double helix. Not all DNA molecules are circular, however. We have mentioned

● **CONNECTION** Some viruses that cause disease in humans contain linear double-stranded DNA genomes.

● **CONCEPT** Most DNA molecules in vivo are double-stranded; many are closed circles. Most double-stranded circular DNA molecules are supercoiled.

bacteriophages T2 and T4. Adenoviruses, which cause respiratory infections in animals and humans, contain linear DNA, as do poxviruses, including the virus that causes smallpox. Eukaryotic chromosomes also contain giant linear DNA molecules.

Circular DNA molecules have another important property, which can be seen in Figure 4.19. This electron micrograph shows three human mitochondrial DNA molecules, one of which has an open circular structure and the other two that appear to have a strained conformation. All three molecules are circular, and the two that appear strained are said to be **supercoiled.** Three-dimensional structure, such as supercoiling, that involves a higher-order folding of elements of regular secondary structure, is referred to as the **tertiary structure** of a polymer.

● **CONCEPT** The higher-order folding of a biopolymer's secondary structure is called its tertiary structure.

Why is supercoiling important? As you can see from Figure 4.19, the supercoiled DNA molecules are more compact than the relaxed circular molecule. In general, DNA molecules are far longer than the cells that they inhabit. As we discuss further in Chapter 21, supercoiling is an essential feature of the compaction of DNA that allows them to fit within cells or virus particles.

To understand supercoiling, consider a double-stranded DNA molecule 105 base pairs in length, as shown in **FIGURE 4.20(a)**, with the ends brought together to form a circle. Because B-form DNA has 10.5 base pairs per turn, the molecule has 10 complete helical turns. Now suppose that the ends are sealed covalently together as shown in Figure 4.20(b); enzymes called **DNA ligases** carry out this reaction in cells. The molecule still has 10 complete helical turns; we say that it has a **Twist** (T) of 10. We can also note that each strand crosses the other in this circle exactly 10 times; we say that the molecule has a **Linking Number** (L) of 10.

Now suppose that before joining the ends we grasp the two ends and rotate one end counterclockwise by one turn (360°). This creates a strained structure upon closure because the most stable form of the DNA B helix has 10.5 base pairs per turn, and this structure has 11.67 base pairs per turn. If we force the molecule to lie in a plane, it assumes the structure shown in Figure 4.20(c), with a twist of 9 and a linking number of 9. Now if the molecule is allowed to reach its most stable conformation in solution, it **writhes,** re-creating a helix with 10.5 base pairs per

turn. We say that this structure is **underwound** by one turn, or that it has a writhe (W) of −1. We can also say that this molecule is **negatively supercoiled.** If we were to rotate the two ends by two complete turns before rejoining, our molecule would have a writhe, W, of −2.

The same considerations apply to *overwound* DNA. Suppose that instead of unwinding by one turn in Figure 4.20, we rotated the end clockwise by two turns. The resultant DNA would be overwound, or positively supercoiled. Most naturally occurring circular DNAs, however, are underwound, or negatively supercoiled. The degree of supercoiling can be defined in terms of the **superhelix density, σ**.

$$\sigma = \Delta L / L_0$$

where L_0 is the linking number for the relaxed DNA and ΔL is the change in linking number caused by supercoiling. For the structure in Figure 4.21(d) ΔL is −1 and L is 9, so σ is −0.11. Most naturally occurring superhelical DNA molecules are underwound, with superhelix densities of about −0.06. Note that if one DNA strand is broken at one site, the entire structure will relax to the most stable conformation, and supercoiling will no longer

● **CONCEPT** Most circular DNA molecules found in vivo are underwound supercoils.

─────────────────────

● **CONNECTION** Topoisomerases, which interconvert circular DNA topoisomers, are targets for several antibacterial and anticancer drugs.

─────────────────────

● **CONCEPT** The superhelix density, σ, is a quantitative measure of the degree of supercoiling.

─────────────────────

be possible. **FIGURE 4.21** shows schematically the difference between overwinding and underwinding.

The DNA molecules shown in Figures 4.20 and 4.21 differ only in their topology; therefore, they are called **topoisomers.** Topoisomers can be interconverted *only* by cutting and resealing the DNA. Cells have enzymes capable of doing this. These enzymes are called **topoisomerases,** and they regulate the superhelicity of natural DNA molecules. Topoisomerases play essential roles in DNA replication and gene expression, and many antibiotics and anticancer drugs act as topoisomerase inhibitors. We discuss topoisomerases in more detail in Chapter 22.

Differences in supercoiling can be detected by gel electrophoresis. As described in Chapter 2, the rate at which a molecule migrates through a gel matrix in an electric field depends on its dimensions; hence, the more compact superhelical forms will move faster than the relaxed form. **FIGURE 4.22** shows electrophoretic patterns for supercoiled DNA molecules that are being progressively relaxed by the action of a topoisomerase enzyme. Thus, gel electrophoresis allows us to separate topoisomers of a given DNA.

Single-Stranded Polynucleotides

Most DNAs found within cells are double-stranded, although the two DNA strands in a duplex molecule can be separated. Most naturally occurring RNA molecules are single-stranded (with the exception of some viruses

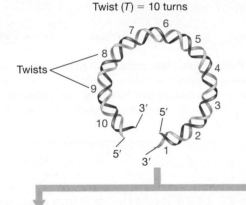

(a) Double-stranded, linear DNA of 105 bp, and 10.5 bp/turn (as for DNA in solution) bent in a circle Twist (T) = 10 turns

Twists

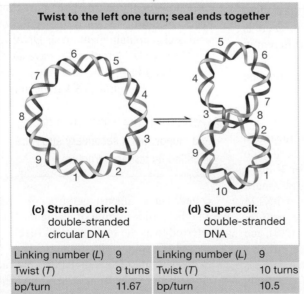

Seal ends together		Twist to the left one turn; seal ends together			
(b) Unstrained circle: double-stranded circular DNA		**(c) Strained circle:** double-stranded circular DNA		**(d) Supercoil:** double-stranded DNA	
Linking number (L)	10	Linking number (L)	9	Linking number (L)	9
Twist (T)	10 turns	Twist (T)	9 turns	Twist (T)	10 turns
bp/turn	10.5	bp/turn	11.67	bp/turn	10.5
Writhe (W)	0	Writhe (W)	0	Writhe (W)	−1

▲ **FIGURE 4.20 Creating supercoiled circular DNA.** For details, see the text.

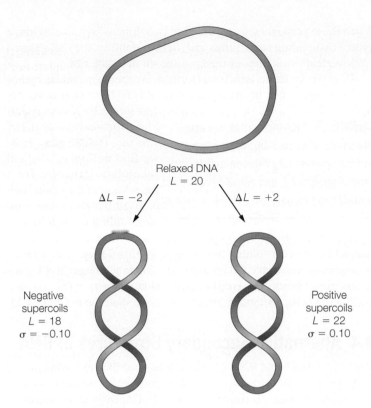

Relaxed DNA
$L = 20$

$\Delta L = -2$ $\Delta L = +2$

Negative supercoils
$L = 18$
$\sigma = -0.10$

Positive supercoils
$L = 22$
$\sigma = 0.10$

▲ **FIGURE 4.21 Negative and positive supercoils.** Note that in underwinding, the DNA axis turns to the left, and in overwinding, it turns to the right.

Relaxation by topoisomerase

1   2   3   4   5   6

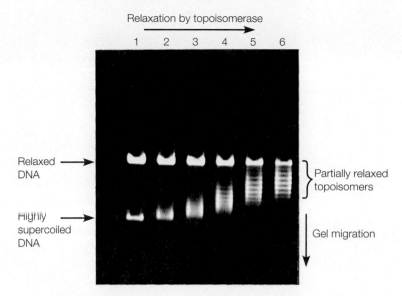

Relaxed DNA

Highly supercoiled DNA

Partially relaxed topoisomers

Gel migration

▲ **FIGURE 4.22 Gel electrophoresis demonstrating DNA supercoiling.** Lane 1: A mixture of relaxed and highly supercoiled DNA. Lanes 2 to 6: The progress of relaxation catalyzed by the enzyme topoisomerase. Samples have been taken at successive times after adding the enzyme. Individual topoisomers are resolved as individual bands on the gel. The highly supercoiled material, which forms a densely packed series of overlapping bands at the bottom, gradually disappears. The DNA species are resolved by electrophoresis in an agarose gel and then visualized by adding the dye ethidium bromide to the gel, a treatment that makes each DNA species fluorescent.

that have double-stranded RNA genomes). Single-stranded molecules, either RNA or DNA, can exist in solution in a random coil form, as depicted in **FIGURE 4.23(a)**. Such a structure is characterized by flexibility and freedom of rotation about backbone bonds, which leads to a floppy, constantly changing form. However, under conditions closer to those found in vivo, the stacking interactions will tend to form regions of single-chain, stacked-base helix (Figure 4.23(b)). Furthermore, most naturally occurring nucleic acid sequences contain regions of self-complementarity between which base pairing is possible. Here the

molecule can loop back upon itself to form a double-stranded structure, as diagrammed in Figure 4.23(c).

Many single-stranded RNA molecules have extensive regions of internal self-complementarity, so that defined regions of structure (in this case, tertiary structure) are formed by *intramolecular* base pairing. The **transfer RNA (tRNA)** molecules involved in protein synthesis are among the smaller RNA molecules and were the first to have their nucleotide sequences determined. All of the first molecules sequenced had four short regions of self-complementarity.

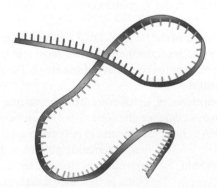

Self-complementary stem–loop

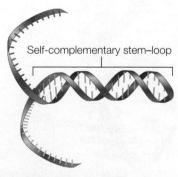

(a) The random coil structure of denatured single strands. There is flexibility of rotation of residues and no specific structure.

(b) Stacked-base structure adapted by non–self-complementary single strands under "native" conditions. Bases stack to pull the chain into a helix, but there is no H-bonding.

(c) "Hairpin" structures formed by self-complementary sequences (green and orange regions of the single strand); the chain folds back on itself to make a stem–loop structure.

▲ **FIGURE 4.23 Conformations of single-stranded nucleic acids.**

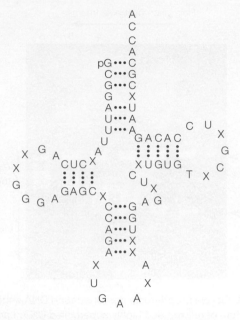

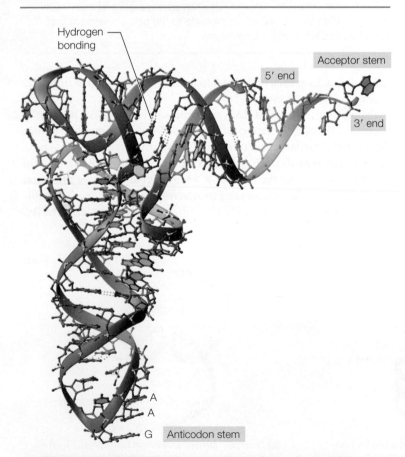

▲ **FIGURE 4.24 How self-complementarity dictates the tertiary structure of a tRNA molecule.** The nucleotide sequence of a yeast transfer RNA used in incorporation of the amino acid phenylalanine into protein. The molecule is folded upon itself to maximize the stabilization provided by intramolecular base pairing. Xs represent unusual or modified bases.

▲ **FIGURE 4.25 The tertiary structure of a transfer RNA as determined by X-ray diffraction.** This molecule is the same phenylalanine tRNA whose nucleotide sequence was shown in Figure 4.24. The path of the phosphate-ribose backbone is traced by the purple ribbon. Note that there are some regions of triple-base bonding.

When these sequences were folded in two dimensions, to maximize hydrogen-bonding capabilities and, hence, stability, they all showed a "cloverleaf" structure, of the kind shown in **FIGURE 4.24**.

Does this structure, which was formed on paper to maximize hydrogen bonding, in fact have biological validity? The answer is yes, as was shown by X-ray crystallographic analysis of tRNA structure (**FIGURE 4.25**). Here we find not only A-helical secondary structure from the folding of the chain back on itself but also more complex folding of such helices together. Thus, the tRNA molecule possesses a defined *tertiary* structure, a higher-order folding that gives it a defined shape and internal arrangement necessary for its function. The much larger RNA molecules from ribosomes have similar secondary structure features, but much more complex tertiary structures, as we shall see in Chapter 25.

● **CONCEPT** RNA molecules are usually single-stranded, but most have self-complementary regions that form hairpin structures, and some have well-defined tertiary structures.

4.4 Alternative Secondary Structures of DNA

Most of the DNA and RNA in cells can be described as having one of the three secondary structures—random coil (which is really a lack of secondary structure), B form, or A form. But these three structures do not nearly exhaust the conformational possibilities of nucleic acids. Here we consider some of the alternative structures that have been recognized in recent years.

Left-Handed DNA (Z-DNA)

Since both the A and B forms of polynucleotide helices are right-handed, the discovery of a left-handed form in 1979 caused considerable surprise. Alexander Rich and his colleagues carried out X-ray diffraction studies of crystals of the small oligodeoxynucleotide

$$5'\text{CGCGCG}\,3'$$
$$3'\text{GCGCGC}\,5'$$

and determined that it is a double-stranded helix with G-C base pairing, as they had expected. However, the data were consistent only with a peculiar *left-handed* structure they called **Z-DNA** (Z for zig-zag). A model for a long DNA molecule in the Z conformation is shown in **FIGURE 4.26**.

In addition to the reverse sense of the helix, Z-DNA exhibits other structural peculiarities. In polynucleotides there are two most stable orientations of the bases with respect to their deoxyribose rings. They are called *syn* and *anti;* see **FIGURE 4.27**.

In both A- and B-form polynucleotides, all bases are in the *anti* orientation. In Z-DNA, however, the *pyrimidines* are always *anti,* but the *purines* are always *syn.* Because Z-DNA is most often found in polynucleotides with alternating purines and pyrimidines in each strand (such as that shown earlier), the base orientations will alternate. Parameters for Z-DNA reflect this characteristic in that the repeating unit is not one base pair but two base pairs.

● **CONCEPT** Z-DNA is a left-handed helix with purine/pyrimidine bases in alternate *syn/anti* conformation.

There is now abundant evidence that Z-DNA exists in living cells. However, the exact role played by Z-DNA in vivo is still an open

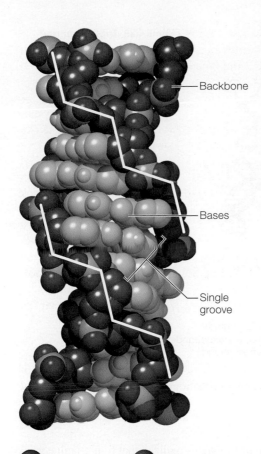

◀ **FIGURE 4.26 Z-DNA.** The structure of Z-DNA as determined by single-crystal X-ray diffraction studies. Compare this left-handed form of DNA with the similar, space-filling model of B-DNA in Figure 4.12. The single groove of Z-DNA is shown in green. The white line follows the zig-zag phosphate backbone.

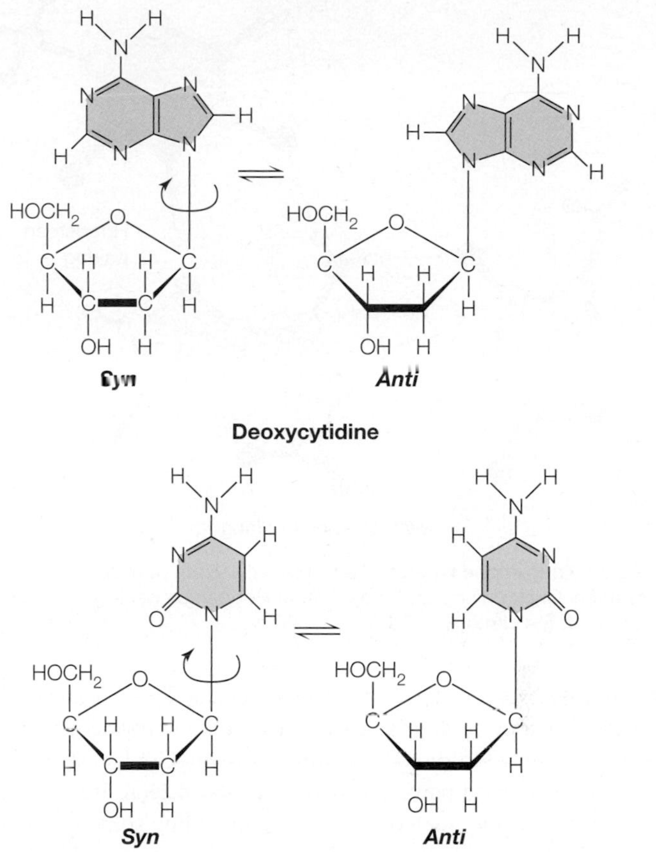

- Backbone
- Bases
- Single groove

● Oxygen ● Carbon
◐ Phosphorus ○ Aromatic base

Deoxyadenosine

Syn   *Anti*

Deoxycytidine

Syn   *Anti*

▲ **FIGURE 4.27** *Syn* and *anti* conformations for two nucleosides.

question. It is perhaps significant that methylation of cytosines on carbon 5, a fairly common modification in vivo, favors Z-DNA formation.

Hairpins and Cruciforms

We have already encountered examples of "hairpin" structures—notably in the transfer RNA shown in Figure 4.25. In these single-stranded molecules, self-complementarity in base sequence allows the chain to fold back on itself and form a base-paired, antiparallel helix. The schematic structure of a tRNA shown in Figure 4.24 depicts how a particular base sequence accomplishes this folding.

Double hairpins, often called **cruciform** (cross-like) structures, can be formed in some DNA sequences (see **FIGURE 4.28**). To create this structure, the sequence must be **palindromic.** As mentioned elsewhere, the word *palindrome* is of literary origin and refers to a statement that reads the same backward and forward, such as "Marge lets Norah see Sharon's telegram." As used in descriptions of DNA, the word refers to segments

of complementary strands that are the exact (or almost exact) reverse of one another. Such sequences are called **inverted repeats,** as shown in Figure 4.28. In most instances, formation of the cruciform structure leaves a few bases unpaired at the ends of the hairpins. Thus, under normal circumstances, the cruciform will be less stable than the extended structure. One effect of superhelical strain is to stabilize cruciforms.

Triple Helices

It has long been recognized that certain homopolymers can form *triple* helices. The first such structure to be discovered was the synthetic RNA structure

$$poly(U) \cdot poly(A) \cdot poly(U)$$

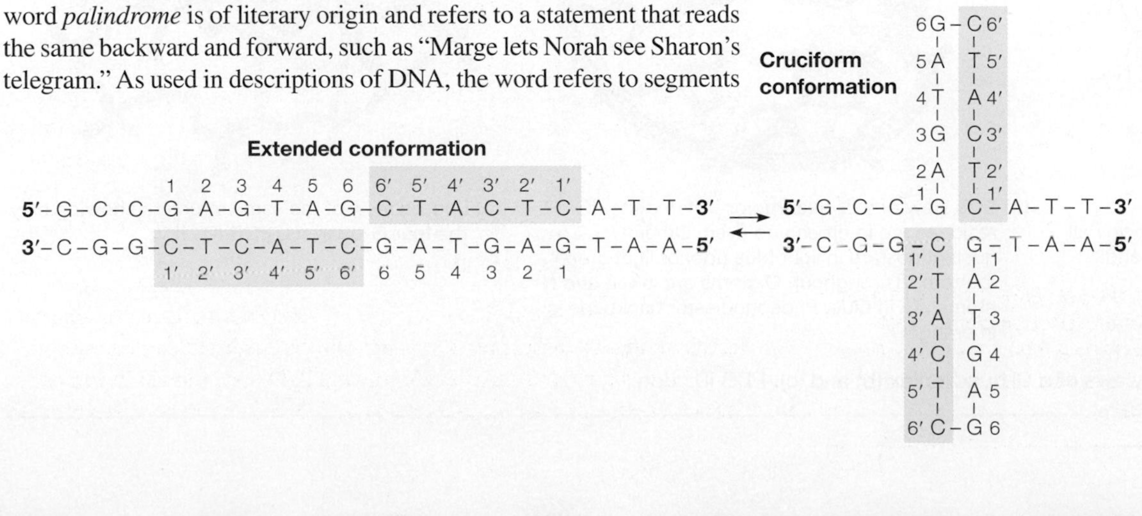

Extended conformation

```
        1   2   3   4   5   6   6'  5'  4'  3'  2'  1'
5'– G – C – C – G – A – G – T – A – G – C – T – A – C – T – C – A – T – T –3'
3'– C – G – G – C – T – C – A – T – C – G – A – T – G – A – G – T – A – A –5'
        1'  2'  3'  4'  5'  6'  6   5   4   3   2   1
```

Cruciform conformation

```
          6 G – C 6'
          5 A   T 5'
          4 T   A 4'
          3 G   C 3'
          2 A   T 2'
          1 G   C 1'
5'– G – C – C – G       C – A – T – T –3'
3'– C – G – G – C       G – T – A – A –5'
          1' C   G 1
          2' T   A 2
          3' A   T 3
          4' C   G 4
          5' T   A 5
          6' C – G 6
```

◀ **FIGURE 4.28 A palindromic DNA sequence.** A palindrome is symmetrical with respect to nucleotide sequence about a center of symmetry. Note that both segments in blue read the same 5′→3′, as do both segments in tan. The sequence is shown in both extended and cruciform conformations, along with numbers that identify base-pairing partners in both conformations.

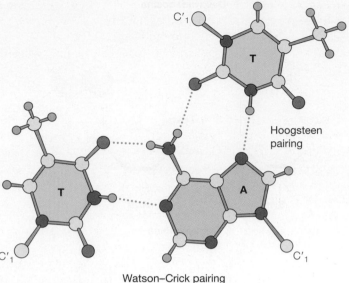

Watson–Crick pairing

(a) Base pairing in one type of DNA triple helix. Both normal Watson–Crick pairing and the unusual Hoogsteen pairing occur on the same A residue.

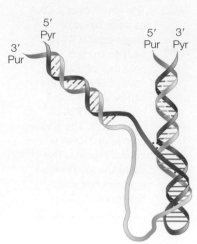

(b) Self-complementarity within a single DNA molecule, showing creation of a triple-stranded structure and an unpaired single strand.

Later, it was observed that deoxy triplets such as $T \cdot A \cdot T$ and $C^+ \cdot G \cdot C$ (where C^+ is a protonated cytosine) can also be formed. Such structures involve, in addition to the normal Watson–Crick base pairing, the *Hoogsteen-type* base pairing shown in **FIGURE 4.29(a)**. It is now recognized that many polynucleotide strands, including some of nonrepeating sequence, can enter into such triple helices, or triplex structures, with conversion of part of the DNA to single-stranded conformation, as shown in Figure 4.29(b). Normally, the loss of base pairing makes this structure unstable, except in situations where the strain in the DNA molecule favors it.

G-Quadruplexes

As long ago as 1962, David R. Davies realized that four guanine molecules could fit together in a planar hydrogen-bonded structure; he proposed that such structures might form naturally in guanine-rich sections of DNA. **FIGURE 4.30(a)** shows such a **G-quartet.** The structure

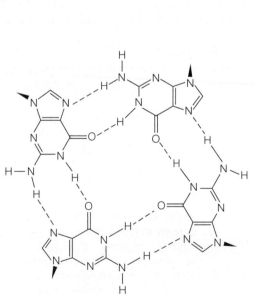

(a) Arrangement of bases in a G-quartet surrounding a central metal ion (not shown); all of the G bases are in the same DNA strand.

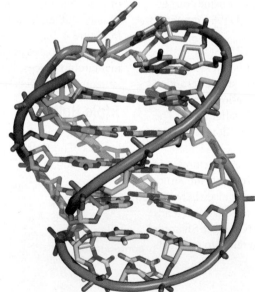

(b) Side view of a G-quadruplex. Alternating dGMP residues are in green and cyan. Other nucleotides are in light blue (top) or light green (bottom). Throughout, O atoms are in red and N atoms are in blue. Phosphodiester backbone is in orange.

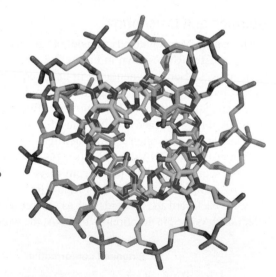

(c) End-on view of the same structure showing only the four G-quartets. PDB ID: 1jpq.

▲ **FIGURE 4.30** A G-quartet (a) and two views of a G-quadruplex (b) and (c). PDB ID: 1jpq.

containing several G-quartets, as shown in Figure 4.30(b) and (c), is called a **G-quadruplex.** Such structures actually form in living cells and probably exist in **telomeres**—special sequences at the ends of linear eukaryotic chromosomes (see Chapter 21). More recent studies show that G-quadruplexes exist in the transcriptional control sites, or **promoters,** of several biologically important genes, including the **oncogene** cMYC. Current efforts are aimed at targeting such structures with anticancer drugs.

● **CONNECTION** G-Quadruplex structures may be target sites for anticancer drugs.

4.5 The Helix-to-Random Coil Transition: Nucleic Acid Denaturation

The major polynucleotide secondary structures (the A and B forms) are relatively stable for RNA and DNA, respectively, under physiological conditions. Yet, they must not be *too* stable because DNA replication and transcription require that the double-helix structure be opened up.

When it extends over large regions, this loss of secondary structure is called **denaturation** (**FIGURE 4.31**). Competing factors create a balance between structured and unstructured forms of nucleic acids.

DNA strand separation, like any process, obeys thermodynamic laws (see Chapter 3).

$$\Delta G = \Delta H - T\Delta S \qquad (\text{helix} \leftrightarrows \text{random coil}) \qquad (4.1)$$

At low temperature, ΔG is positive. The DNA double-stranded structure is stabilized both by base pairing and by base stacking interactions. This contribution is largely enthalpic, meaning that it appears as a positive ΔH in Equation 4.1. The entropy of the double-stranded structure is low because it is much more highly organized than the two separate DNA strands, which can adopt a limitless number of conformations. Thus, ΔS for the transition from double-strand to random coil is also positive. At low temperature, the term $T\Delta S$ is smaller than ΔH; therefore, $\Delta G > 0$, and the helix is stable. But as the temperature is increased, $T\Delta S$ becomes greater than ΔH, and ΔG becomes negative. Hence, at higher temperatures the double-stranded structure becomes unstable and falls apart (Figure 4.31(a)).

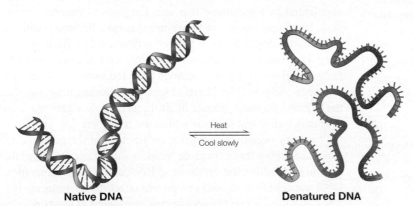

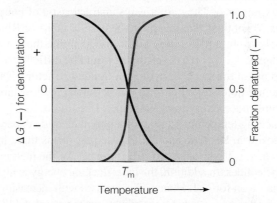

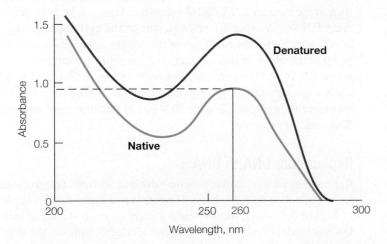

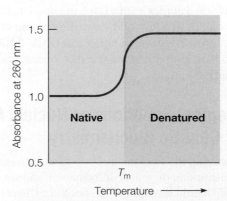

(a) When native (double-stranded) DNA is heated above its "melting" temperature, it becomes denatured (separating into single strands). The two random-coil strands have a higher energy than the double helix.

(b) As T increases, $-T\Delta S$ overcomes a positive ΔH, making ΔG negative and denaturation favorable. The midpoint of the curve marks the "melting" temperature, T_m, of DNA.

(c) Absorption spectra of native and denatured DNA show that native DNA absorbs less light than denatured DNA.

(d) This change in absorbance can be used to follow the denaturation of DNA as temperature increases. An abrupt increase in absorbance, corresponding to the sudden "melting" of DNA, is seen at T_m.

▲ **FIGURE 4.31** Denaturation of DNA.

One can follow denaturation by observing the absorption of ultraviolet light with a wavelength of about 260 nm in a DNA solution as the solution is slowly heated (Figure 4.31(d)). When the nucleotides are polymerized into a polynucleotide and the bases are packed into a helical structure, the absorption of light is reduced (Figure 4.31(c)). This phenomenon, called *hypochromism,* results from close interaction of the light-absorbing purine and pyrimidine rings. If secondary structure is lost, the absorbance increases and becomes closer to that of a mixture of the corresponding free nucleotides. Therefore, raising the temperature of a DNA solution, with accompanying breakdown of the secondary structure, will result in an absorbance change of about 50%, as shown in Figure 4.31(d).

The striking feature of this helix-to-random-coil transition is that it is so sharp. It occurs over a small temperature range, almost like the melting of ice into water, as described in Chapter 3. Therefore, nucleic acid denaturation is sometimes referred to as a *melting* of the polynucleotide double helix, even though the term is not technically correct.

● **CONCEPT** AT-rich regions melt more easily than GC-rich regions.

We shall encounter similar abrupt changes in the configuration of proteins in Chapter 6. They are always characteristic of **cooperative transitions.** What this term means in the case of DNA or RNA is that a double helix cannot melt bit by bit. If you examine the double helix shown in Figure 4.12, you can see that it would be difficult for a single base to pop out of the stacked, hydrogen-bonded structure. Rather, the whole structure holds together until it is at the verge of instability and then denatures over a very narrow temperature range.

The "melting temperature" (T_m) of a polynucleotide depends on its base composition. Because each G-C base pair forms three hydrogen bonds and each A-T pair only two, ΔH is greater for the melting of GC-rich polynucleotides. In addition, the base-stacking energy is higher for a G-C base pair than for A-T. Hence, T_m increases with increasing G + C content. However, base stacking contributes more toward stabilization of the double helix than does H-bonding.

DNA denaturation is reversible. For example, when heat-denatured DNA is cooled, DNA duplexes can re-form. The rate of cooling must be slow, allowing time for complementary strands to find one another and pair up, or **renature** (a process also called **annealing**). Similarly, an RNA molecule can form a duplex with a DNA of complementary base sequence, creating a DNA–RNA hybrid, consisting of one strand each of DNA and RNA. DNA–DNA renaturation and DNA–RNA hybridization are at the root of several important research techniques, as we shall see later in this book.

4.6 The Biological Functions of Nucleic Acids: A Preview of Genetic Biochemistry

We have emphasized that the fundamental role of the nucleic acids is the storage and transmission of genetic information, and we will continue to develop this theme throughout the book. In Chapters 21–26 we describe in detail how nucleic acids are passed from parent cell to daughter cell (or from an organism to its descendants) and how they direct biochemical processes—specifically, the synthesis of proteins. In this chapter and the next, we present a preliminary overview of these nucleic acid functions. Even though this may not be your first exposure to these topics, you should gain some appreciation of the relationships between nucleic acid and protein structures, of evolution at the molecular level, and of our ability to modify microbes, plants, and animals through genetic engineering.

Genetic Information Storage: The Genome

Every organism carries in each of its cells at least one copy of the total genetic information possessed by that organism. This is referred to as the *genome.* Usually, the genomic information is coded in the nucleotide sequence of double-stranded DNA, but some viruses use single-stranded DNA or even RNA. Genomes vary enormously in size; the smallest viruses need only a few thousand bases (b) or base pairs (bp), whereas the human genome consists of about 3×10^9 bp of DNA, distributed in 23 chromosomes.

Recent years have seen remarkable advances in our ability to determine DNA or RNA sequences. In 1976, Allan Maxam and Walter Gilbert devised a chemical sequencing method involving selective cleavage at A, T, G, or C residues, followed by separation of the fragments on the basis of fragment length, using gel electrophoresis. This ingenious technique allowed researchers to begin the exploration of genomic information. The Maxam–Gilbert method has been largely supplanted by a technique that uses enzymes to generate oligonucleotide fragments started and terminated at specific base positions. This method, developed by Fred Sanger, is described in Tools of Biochemistry 4A. Automation of this technique and development of much faster "high-throughput" methods have led to the complete genome sequence analysis of hundreds of species, including near completion of the human genome sequence in 2001 and complete genome sequences for many individual humans within the following few years.

In every organism, a significant fraction of the genomic DNA is capable of being transcribed, or "read," to allow expression of its information in directing the synthesis of RNA and protein molecules. Each DNA segment that encodes one protein or one RNA molecule is a gene. The DNA in each cell of every organism contains at least one copy (and sometimes several) of the gene carrying the information to make each protein that the organism requires. In addition, there are genes (often reiterated manyfold) for the several specific functional RNA molecules, such as the transfer RNA (tRNA) shown in Figure 4.24. Like proteins, these RNAs play specific roles as part of the cell's machinery. We briefly summarize the functions of messenger RNA, ribosomal RNA, and transfer RNA in this chapter, and the more recently discovered micro RNAs and their role in *RNA interference* in Chapter 26. RNA plays a greater number of different functions in gene regulation than was once thought; in Chapter 26 we also mention *long noncoding RNAs* and *riboswitches*.

Replication: DNA to DNA

Replication passes on the genetic information from cell to cell and from generation to generation. The essence of the process is depicted in Figure 4.13—a complementary copy of each strand of duplex DNA is made, usually resulting in two identical copies of the original duplex. The process is highly accurate—making less than 1 error in 10^8 bases—but occasionally mistakes are made. These contribute to the mutations that have allowed the evolution of life to ever more complex forms.

The replication of DNA is accomplished by a complex of enzymes, acting in concert like a finely tuned machine. These enzymes are

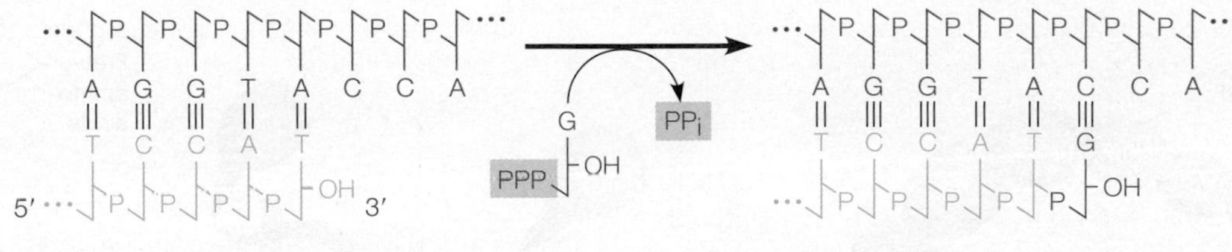

▲ **FIGURE 4.32 The DNA polymerase reaction.** DNA polymerase fits a deoxyribonucleoside triphosphate molecule to its complementary nucleotide in the template strand (purple) and catalyzes formation of a phosphodiester link between the incoming nucleotide and the 3′-hydroxyl of the 3′-terminal nucleotide in the growing daughter strand (blue), with pyrophosphate being split out.

described in detail in Chapter 22. Each enzyme complex, or **replisome,** centered on the enzyme *DNA polymerase,* has multiple functions. As parental DNA strands unwind, forming a *replication fork* (see Figure 4.13), DNA polymerase guides the pairing of incoming deoxyribonucleoside triphosphates, each with its complementary partner on the strand being copied. It then catalyzes the formation of the phosphodiester bond to link this residue to the new growing chain. Thus, each of the parental DNA strands serves as a **template,** specifying the sequence of a daughter strand. DNA polymerase adds nucleotides, one at a time, to the lengthening daughter strand as it grows from its 5′ end toward its 3′ end (**FIGURE 4.32**). In most cases, the enzyme complex also checks or "proofreads" the addition before proceeding to add the next residue, which contributes to the high overall accuracy of replication. Because the two DNA strands run in opposite directions, one daughter strand is elongated in the direction that the replication fork is moving, while the other is formed in the opposite direction. This and other aspects of DNA replication are discussed in more detail in Chapter 22.

● **CONCEPT** Replication is the copying of both strands of a duplex DNA to produce two identical DNA duplexes.

Transcription: DNA to RNA

Expression of genetic information always involves as a first step the **transcription** of genes into RNA molecules of nucleotide sequence complementary to that of the gene being expressed. This production of specific RNA molecules is easy to visualize. Just as a DNA strand can direct replication, it can equally well direct transcription, the formation of a complementary RNA strand (**FIGURE 4.33**). Of course, the monomers required in transcription are different from those required in replication. Instead of the deoxyribonucleoside triphosphates (dNTPs), the ribonucleoside triphosphates, ATP, GTP, CTP, and UTP, are needed to make RNA. (Note that U in the new RNA pairs with A in the DNA template.) DNA transcription, like DNA replication, requires the enzyme catalysts known as *RNA polymerases.* Another distinction from DNA replication is that only one of the two DNA strands, the **template strand,** is copied. Transcription is presented in detail in Chapter 24.

● **CONCEPT** Transcription is the copying of a DNA strand into a complementary RNA molecule.

Translation: RNA to Protein

Transcription is the central process in producing the many functional RNA molecules of the cell: messenger RNAs, transfer RNAs, ribosomal RNAs, and the more recently discovered microRNAs and noncoding RNAs. The synthesis of specific proteins, under the direction of specific genes, is a more complex matter. The problem, as we will see in Chapter 5, is that proteins are polymers made from 20 different kinds of amino acid monomers. Because there are only four different nucleotide monomer types in DNA, there cannot be a one-to-one relationship between the sequence of nucleotides in a DNA molecule and the sequence of amino acids in a protein. Rather, the linear sequence of bases that constitutes the protein-coding information is "read" by the cell in blocks of three nucleotides, or **codons,** each of which specifies a different amino acid. The set of rules that specifies which nucleic acid codons correspond to which amino acids

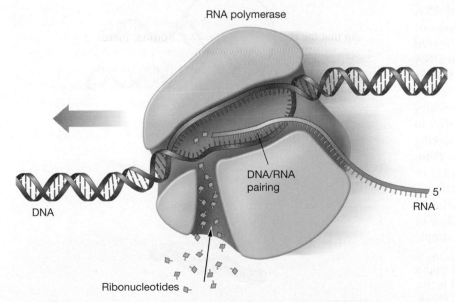

▲ **FIGURE 4.33 The basic principle of transcription.** An enzyme (RNA polymerase) travels along a DNA molecule (right to left as shown), opening the double strand and making an RNA transcript by adding one ribonucleotide at a time. It copies the oligonucleotide sequence from only one of the two DNA strands. After the enzyme passes, the DNA rewinds.

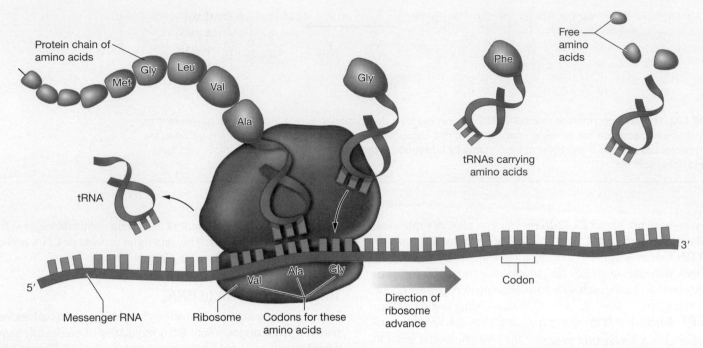

▲ **FIGURE 4.34 The basic principle of translation.** A messenger RNA molecule is bound to a ribosome, and transfer RNA molecules bring amino acids to the ribosome one at a time. Each tRNA identifies the appropriate codon on the mRNA and adds this amino acid to the growing protein chain. The ribosome travels along the mRNA, so that the genetic message can be read and translated into a protein.

is the **genetic code.** We discuss this code in Chapter 5 after we have described the amino acids and the structure of proteins.

Although the information for all protein sequences is coded in DNA, the production of proteins does not proceed directly from DNA. For the information to be converted from the DNA sequences of the genes into amino acid sequences of proteins, special RNA molecules are needed as intermediates. Complementary copies of the genes to be expressed are transcribed from DNA in the form of **messenger RNA (mRNA)** molecules, so called because they carry information from DNA to the protein-synthesizing machinery of the cell. The protein-making machinery includes tRNA molecules, special enzymes, and **ribosomes,** which are RNA–protein complexes where the assembly of new proteins takes place. This **translation** of RNA information is outlined in **FIGURE 4.34**. We present the main features of translation in Chapter 5 and describe it in detail in Chapter 25. The flow of genetic information in the cell can be summarized by the simple schematic diagram shown in **FIGURE 4.35**.

As we show in the following chapters, proteins are the major structural and functional molecules in most cells. What a cell is like and what it can do depend largely on the proteins it contains. These, in turn, are dictated by the information stored in the cell's DNA, transcribed and processed into mRNA, and expressed by the protein-synthesizing machinery. Overlaid on these processes are recently

● **CONCEPT** In translation, an RNA nucleotide sequence dictates a protein amino acid sequence.

described small RNA molecules, barely 20 nucleotides long, which help regulate the process. These *microRNAs* are discussed further in Chapter 26.

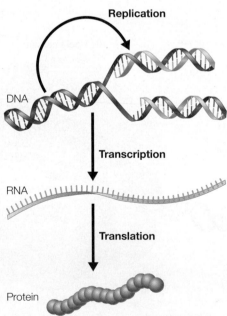

▲ **FIGURE 4.35 The flow of genetic information in a typical cell.** DNA can both replicate and be transcribed into RNA. Messenger RNAs are translated into protein amino acid sequences.

Summary

- There are two kinds of nucleic acids, **DNA** and **RNA.** Each is a polynucleotide, a polymer of four kinds of nucleoside 5′-phosphates, connected by links between 3′ hydroxyls and 5′ phosphates. RNA has the sugar **ribose;** DNA has **deoxyribose.** The **phosphodiester linkage** is inherently unstable, but it undergoes hydrolysis only slowly in the absence of catalysts. Each naturally occurring nucleic acid has a defined sequence, or **primary structure.** Early evidence indicated that DNA is the genetic material, but it was not until Watson and Crick elucidated its two-strand secondary structure in 1953 that it became obvious how DNA might direct its own replication. The structure they proposed involved specific pairing between **A** and **T** and between **G** and **C.** The helix is right-handed, with about 10.5 base pairs (bp) per turn. Such a structure can replicate in a **semiconservative** manner, as demonstrated by Meselson and Stahl in 1958. Other forms of polynucleotide structures exist, of which the most important is the **A form,** found in RNA–RNA and DNA–RNA double helices. In vivo, most DNA is double-stranded; many molecules are circular. Most of the circular DNA molecules found in nature are **supercoiled.** Topoisomerases are enzymes that can either supercoil DNA or relax supercoils by breaking and re-forming phosphodiester bonds. Most RNA is single-stranded, but it may fold back to form hairpins and other well-defined **tertiary structures.** (Sections 4.1–4.3)

- Polynucleotides can form a number of unconventional structures. These structures include left-handed DNA (**Z-DNA**), **cruciforms,** in some cases triple helices, and **G-quadruplexes.** The secondary structures of polynucleotides can be changed in various ways. The helix can "melt," which involves strand separation. This change is easiest for regions rich in A-T pairs. Energy stored in superhelical DNA may promote local DNA melting or changes to a variety of alternative structures, such as Z-DNA, cruciforms, or a particular triple-helical structure. (Sections 4.4, 4.5)

- The biological functions of nucleic acids may be summarized as follows: DNA contains stored genetic information, which is transcribed into RNA. Some of these RNA molecules act as messengers to direct protein synthesis. The **messenger RNA** is translated on a ribosome, using the genetic code, to produce proteins. (Section 4.6)

- Modern molecular biological techniques allow us to manipulate DNA for increased protein expression and directed changes in protein structure, through gene cloning, rapid DNA sequence analysis, polymerase chain reaction, **site-directed mutagenesis,** and other technologies. (Tools of Biochemistry 4.A, 4.B)

Problems

Enhanced by
Mastering Chemistry
for Biochemistry

Mastering Chemistry for Biochemistry provides select end-of-chapter problems and feedback-enriched tutorial problems, animations, and interactive figures to deepen your understanding of complex topics while practicing problem solving.

Answers to red problems are available in the Answer Appendix.

1. Fill in the blanks to identify each substance as
   (a) A nucleobase
   (b) A ribonucleoside
   (c) A ribonucleotide
   (d) A deoxyribonucleoside
   (e) A deoxyribonucleotide
   (f) A dinucleotide

   _____Adenosine
   _____Cytidine 5′-monophosphate
   _____Guanine
   _____Thymine
   _____Deoxyguanosine
   _____Uridine
   _____Deoxycytidine 5′-triphosphate
   _____Deoxyuridine
   _____Guanosine 5′-diphosphate
   _____Thymidine triphosphate

2. What is the difference between a nucleoside triphosphate and a trinucleotide?

3. pppApCpCpUpApGpApU-OH
   (a) Using the straight-chain sugar convention shown on page 79, write the structure of the DNA strand that encoded this short stretch of RNA.
   (b) Using the simplest convention for representing the DNA base sequence (page 79), write the structure of the nontemplate DNA strand.

4. Shown is a representation of a DNA molecule being transcribed.
   (a) Identify every 3′ end and every 5′ end in the picture.
   (b) Identify the template strand.

(c) The nontemplate strand is also called the "sense strand." Explain.

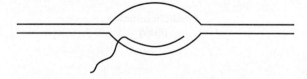

5. Base analysis of DNA from maize (corn) shows it to have 23 mole percent cytosine (moles per 100 moles total nucleotide). What are the percentages of the other three bases?

6. Using the pK_a data in Table 4.1 and the Henderson-Hasselbalch equation, calculate the approximate net charge on each of the four common ribonucleoside 5′-monophosphates (rNMPs) at pH 3.8. If a mixture of these rNMPs was placed in an electrophoresis apparatus, as shown, draw four bands to predict the direction and relative migration rate of each.

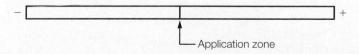

Application zone

7. For some DNAs, it is possible to separate the two strands, after denaturation, in a CsCl gradient.
  (a) What property of any DNA determines where it will band in a CsCl gradient?
  (b) What kind of DNA might have two strands that differ sufficiently in this property that they could be separated after denaturation?

8. Refer to Figure 4.15, which presents the Meselson–Stahl experiment. DNA molecules can be denatured by high pH, as well as by heat. Suppose that the CsCl gradient centrifugations were run at pH 12, conditions under which DNA strands separate. Sketch the gradient profiles expected for each of the four samples depicted in the figure.

9. Suppose that you centrifuged a transfer RNA molecule to equilibrium in an alkaline CsCl gradient, as described in Problem 7. What result would you expect?

10. Predict the structure of a cruciform that could be formed from this oligonucleotide.

5′ GCAATCGTACGATTAGGGC

3′ CGTTAGCATGCTAATCCCG

11. DNA from a newly discovered virus was purified, and UV light absorption was followed as the molecule was slowly heated. The absorbance increase at the melting temperature was only 10%. What does this result tell you about the structure of the viral DNA?

12. Would you expect *Neurospora crassa* DNA to have a higher or lower T_m than human DNA (see page 81)? Explain.

13. A circular double-stranded DNA molecule contains 4200 base pairs. In solution, the molecule is in a B-form helix, with about 10.5 base pairs per turn. The DNA circle has 12 superhelical turns. What is its superhelix density σ?

14. The gel electrophoresis pattern in Figure 4.23 was determined by soaking the gel in a solution of ethidium bromide (EtBr). This is a fluorescent molecule with a planar structure:

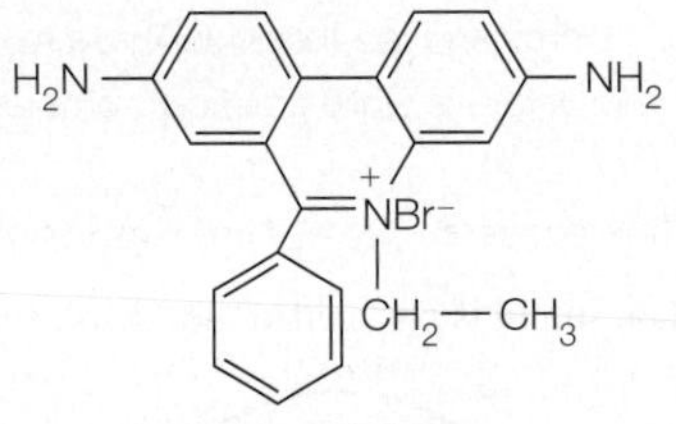

Ethidium bromide
(EtBr)

The flat molecule *intercalates,* or fits directly between two adjacent base pairs in a double helix. In doing so, it unwinds the double helix by 26° for each ethidium molecule bound.
  (a) If EtBr was added to relaxed, closed circular DNA, would you expect positive or negative supercoiling to occur? Explain.
  (b) If the circular DNA were nicked (had a single-stranded break) on one strand, what would be the effect on supercoiling?
  (c) If negatively supercoiled DNA is titrated with EtBr, the electrophoretic mobility decreases at first but then increases at higher EtBr concentrations. Explain.

15. DNA polymerase requires both a template, to be copied, and a primer, which provides a 3′ hydroxyl from which polymerase can extend. Yet, this molecule supports DNA polymerase activity. Explain.

pTGACACAGGTTTAGCCCATCGATGGG−OH

16. Draw the gel electrophoretic pattern that would be seen in dideoxy sequence analysis of the DNA molecule in Problem 10.

17. Early gene-cloning experiments involved insertion at one restriction site in the vector; for example, the insert would have an EcoRI site at each end, and the vector would be opened at an EcoRI site prior to ligation. Under what circumstances would asymmetric cloning be desirable, with the insert having a different restriction site at each end (see Tools of Biochemistry 4A)?

18. (a) What two enthalpic factors stabilize DNA in double-helical form at low temperature?
  (b) What entropic factor destabilizes helical DNA at high temperature?
  (c) Why is the double-helical structure of DNA stabilized at moderate to high ionic strength?

19. (a) The plasmid pBR322 (4362 base pairs) was isolated and the circular DNA was found to be underwound, with a superhelix density (σ) of −0.05. How many superhelical turns does this molecule have?
  (b) Suppose that the molecule was overwound, also with a superhelix density of 0.05. How many superhelical turns would this molecule have?

20. In studying the mechanism of a particular enzyme, for which the cloned gene is available, you wish to change a putative active site histidine residue to a proline by site-directed mutagenesis. The His codon to be changed is 5′-CAC. You wish to change it to 5′-CCC, one of four Pro codons. The nucleotide sequence surrounding the His codon is 5′-CTGGAATCT-**CAC**TTTATCTGG-3′. Write the nucleotide sequence of an oligonucleotide (a 21-mer) that could force the conversion of the CAC codon to CCC in a site-directed mutagenesis operation.

21. What DNA sequence feature is required for a G-quadruplex to form?

References

For a list of references related to this chapter, see Appendix II.

The biological sciences have been revolutionized by laboratory techniques that allow the researcher to isolate any desired gene, amplify the gene for determination of its nucleotide sequence, express the gene at a high level for production and analysis of its protein product, and introduce any desired mutation into the gene, either for analysis of structure–function relationships within the protein or to create some desirable property, such as increased stability. The key discoveries, made in the 1970s and early 1980s, spawned a large biotechnology industry, which has created hundreds of new drugs and genetically modified organisms, such as drought-resistant crop plants.

Because these methods have taught us so much about the structure and function of the proteins and enzymes we will be discussing in the next several chapters, we introduce some of the benchmark techniques here, even though much of the genetic biochemistry upon which they are based is presented later in this book. Within this family of **recombinant DNA** techniques, we present here outlines of (1) gene cloning, (2) chemical synthesis of oligonucleotides of defined sequence, (3) DNA sequence analysis, and (4) site-directed mutagenesis. A fifth technique, *polymerase chain reaction* (PCR), which allows amplification of any DNA sequence, starting from minute amounts of material, is presented in Chapter 21, and CRISPR technology, which revolutioned gene modification *in vivo*, is presented in Chapter 23.

Gene Cloning

In 1973, Stanley Cohen and Herbert Boyer realized that two recent developments had set the stage for them to **clone,** or isolate, a single gene. In classical biology, a clone is a population of organisms that are genetically homogeneous because they were derived from a single ancestor. For example, all of the bacterial cells in a colony represent a clone because they were derived from a single cell that was deposited at that location on a Petri plate.

The first development leading to cloning of single genes was the characterization of **plasmids** as small circular DNA molecules capable of independent replication within bacterial cells. Clinical resistance to antibiotics is often caused by mutations in genes carried on plasmid DNA molecules. For example, resistance to penicillin results from a plasmid-encoded enzyme called β-lactamase, which cleaves penicillin to produce an inactive derivative.

The second development was the discovery of a class of bacterial enzymes called **restriction endonucleases,** many of which catalyze cleavage of DNA at specific sites (Chapter 21). For example, the enzyme *Eco*RI cleaves DNA whenever it recognizes the sequence

$$5'\ldots\text{GAATTC}\ldots3'$$
$$3'\ldots\text{CTTAAG}\ldots5'$$

Notice that the complement of the base sequence on one strand is exactly the same sequence running in the opposite direction. This section of DNA is therefore a palindrome (Section 4)—it reads the same in both directions. The enzyme cuts between G and A on both strands. Thus, each of the cleavage products has a short (four-base) single-stranded end, $3' \ldots \text{TTAA}5'$. What this means is that *any* two DNA molecules that have been cut by *Eco*RI can rejoin end to end when

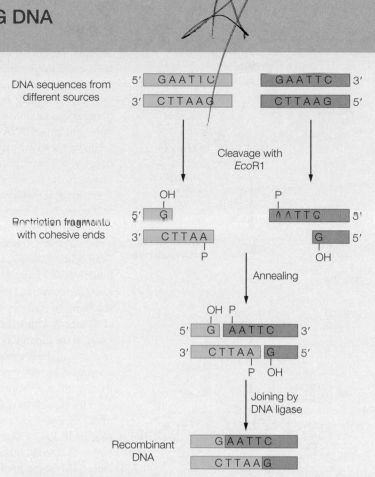

▲ **FIGURE 4A.1** Creation of a recombinant DNA molecule in vitro.

annealed, or subjected to DNA renaturation conditions, by base pairing between the two 5'-terminated AATT sequences (also called *cohesive ends* or *sticky ends.*)

Once this pairing has occurred, an enzyme called **DNA ligase** (**FIGURE 4A.1**) catalyzes formation of a covalent bond in the openings between 3'-terminal G and 5'-terminal A. If the two DNA sequences being rejoined come from different sources, the product is called a recombinant DNA molecule, after the fact that genetic recombination in vivo involves cutting and rejoining DNA from different chromosomes (Chapter 23).

Cohen and Boyer also devised a technique for high-frequency *transformation,* so that recombinant DNA molecules created could be introduced into living bacterial cells. This technique originally involved treating bacteria with calcium chloride followed by heat shock, but it has now been supplanted by more efficient methods, including application of a high-voltage electric current. The **vector** is a DNA molecule into which the gene to be cloned is inserted. As schematized in **FIGURE 4A.2**, the vector is a plasmid containing a gene that specifies resistance to a particular antibiotic, such as ampicillin. The plasmid contains just one site for cleavage by the restriction enzyme to be used. Both the plasmid and a DNA molecule containing the gene of interest are cleaved with the same restriction enzyme, such as *Eco*RI. Annealing, or renaturation, followed by enzymatic ligation, yields recombinant DNA molecules in which the gene of interest has been

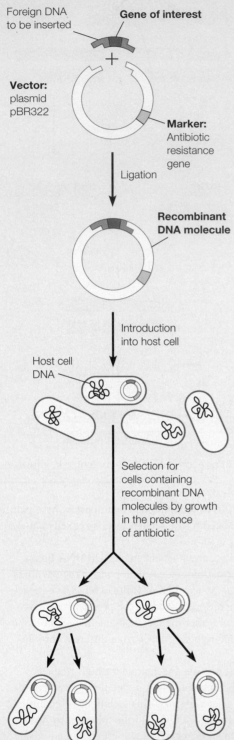

Foreign DNA to be inserted

Gene of interest

Vector: plasmid pBR322

Marker: Antibiotic resistance gene

Ligation

Recombinant DNA molecule

Introduction into host cell

Host cell DNA

Selection for cells containing recombinant DNA molecules by growth in the presence of antibiotic

▲ FIGURE 4A.2 Cloning a fragment of DNA into a plasmid vector and introducing the recombinant molecule into bacteria.

spliced into the vector. Note that the vector need not be a plasmid. Any DNA capable of independent replication within a cell, such as a virus genome, can serve as a vector.

After annealing and ligation, the DNA, containing a mixture of recombinant and nonrecombinant DNAs, is introduced into recipient bacteria by transformation. Any transformed bacteria can be identified by plating in the presence of ampicillin. Only those bacteria that have been transformed will grow. The investigator must then carry out additional experiments to establish that a particular transformed cell contains the gene being cloned. A variety of methods are available, including additional antibiotic-resistance screens, DNA–DNA renaturation reactions, or analysis of expression of the cloned gene with an antibody or activity assay. If the cloning is designed for protein expression, the presence of the protein can be confirmed by mass spectrometry or by reactivity with an appropriate antibody.

FIGURE 4A.3 shows the structure of pBR322, a plasmid engineered for gene cloning in 1977 and still in occasional use today. This small recombinant DNA molecule, 4361 base pairs in length, contains two antibiotic resistance genes, one for ampicillin resistance (amp^R) and one for tetracycline resistance (tet^R). Note that each of these genes contains restriction cleavage sites

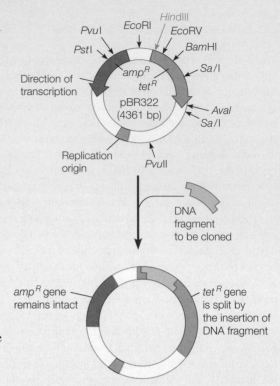

▲ FIGURE 4A.3 pBR322, one of the earliest cloning vectors. Some of the restriction sites are shown, as well as the direction of transcription of the ampicillin and tetracycline resistance genes. The bottom diagram shows the effect of cloning a novel sequence into the HindIII site.

within its sequence. A site for the restriction enzyme HindIII lies within the tet^R sequence. Hence, if this site is used for cloning, the tet^R gene is split and, therefore, inactive. Because the amp^R gene is intact, all bacteria that have acquired a plasmid, whether recombinant or not, can be selected on the basis of their resistance to ampicillin. But now, recombinant plasmids can be identified because these bacteria are sensitive to tetracycline and are also ampicillin-resistant, whereas bacteria that acquired the original plasmid, without an insert, are resistant to both drugs.

Many variations on this original approach have been devised. Blunt-ended DNA molecules, containing no sticky ends, can now be cloned. Many cloning vectors have been created, usually as recombinant DNA molecules themselves. Particularly widely used are **expression vectors,** in which signals to regulate and activate high-level expression of the cloned gene are built into the vector. Other modifications include sequences that aid in purification of the recombinant protein (see Tools of Biochemistry 5A). In the biotechnology industry, techniques such as these led to the introduction as early as 1982 of human insulin as a recombinant gene product, used for diabetes treatment. Other recombinant proteins later approved for clinical use include blood-clotting factors, clot-dissolving enzymes used to treat heart attack victims, pituitary growth hormone, and interferons.

● **CONNECTION** Recombinant DNA technology has allowed the generation of many new drugs and organisms modified in useful ways.

Automated Oligonucleotide Synthesis

In 1976, the British biochemist Fred Sanger introduced **dideoxy DNA sequencing,** the rapid DNA sequence analysis that led to completion of the 3 billion-base-pair human genome sequence in 2001. In 1978, the Canadian biochemist Michael Smith devised a method for site-directed

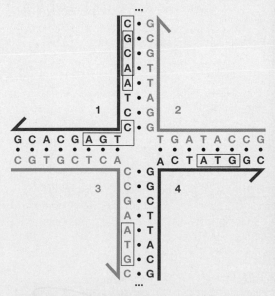

▲ FIGURE 4A.4 The coupling and oxidation steps in the solid-phase synthesis of oligonucleotides by the phosphoramidite method.

mutagenesis, which allows an investigator to introduce any desired mutation into a cloned gene. Both researchers received the Nobel Prize for their contributions.

Both dideoxy DNA sequencing and site-directed mutagenesis depend on the availability of oligodeoxyribonucleotide molecules of precisely known sequence to serve as *primers* in DNA polymerase-catalyzed reactions. The oligonucleotide synthesis process most widely used at present is the **phosphoramidite** method. This method is popular because it can be carried out in automated fashion, with the oligonucleotide linked to a solid support as nucleotides are added in stepwise fashion, one at a time.

The key reaction in the first step is shown in **FIGURE 4A.4**. The first nucleotide is attached via its 3′-hydroxyl group to a solid support. The second nucleotide is introduced with its 5′-hydroxyl blocked with a dimethyltrityl (DMTr-) group and its 3′-hydroxyl linked to a phosphoramidate group, with the phosphorus in the reactive trivalent form. After coupling (the reaction shown), the phosphorus is oxidized to the

pentavalent form. The cycle is completed with demethylation of the phosphate and removal of the dimethyltrityl group, creating a free 5′-hydroxyl for the second round of nucleotide addition.

This process is repeated in stepwise fashion until up to 150 nucleotides have been added, in precisely determined sequence. Finally, the blocking groups are removed, and the finished chain is removed from the solid support. Each step proceeds with about 98% efficiency, meaning that a 20-mer can be produced at about 80% final yield. Note that this method proceeds from the 3′ end of the polymer toward the 5′ end, whereas the enzymatic synthesis of DNA by DNA polymerase proceeds in the opposite direction.

Because of the ease with which oligonucleotides of defined sequence can be synthesized, and because of the regularity of DNA secondary structure, scientists have devised numerous methods for creating defined DNA-based nanostructures. **FIGURE 4A.5** illustrates examples of this "DNA nanotechnology." This approach may generate materials useful in biomedicine or electronics.

◀ FIGURE 4A.5 Design and synthesis of three-dimensional nanostructures. Source: Nadrian C. Seeman, Nanomaterials Based on DNA, *Annu. Rev. Biochem.* 2010. 79:65–87.

(a) A four-arm junction made from design, synthesis, and annealing of four 16-mers with the sequences shown.

(b) A DNA-truncated octahedron, with each edge consisting of two turns of helical DNA.

Dideoxynucleotide Sequence Analysis

We have mentioned the Maxam–Gilbert method for DNA sequence analysis, which involves treating DNA with reagents that cleave at specific nucleotides. The product is a population of molecules that can be resolved on the basis of molecular weight by gel electrophoresis. The enzymatic method introduced by Sanger similarly involves electrophoretic analysis of DNA fragments terminated at specific nucleotides. However, it allows analysis of longer stretches of DNA, and it lends itself more readily to automation. The method begins with an oligonucleotide *primer* that is complementary in sequence to the 3′ end of one strand of the DNA being analyzed (**FIGURE 4A.6**). Extension of this primer by DNA polymerase will copy the insert. The polymerase reactions are run in the presence of deoxyribonucleoside triphosphate analogs, the **2′, 3′-dideoxyribonucleoside triphosphates** (ddNTPs) which serve as terminators of chain extension because they lack 3′ hydroxyl termini. The dideoxy analog (ddATP) of deoxyadenosine triphosphate is shown here.

2′,3′-Dideoxyadenosine triphosphate

To generate a series of A-terminated fragments, one runs the DNA polymerase reactions in the presence of equal concentrations of dATP, dCTP, dGTP, and dTTP, plus one-tenth that concentration of ddATP. When T is in the template strand, DNA polymerase occasionally inserts ddAMP instead of dAMP. When that happens, DNA replication stops and the fragment is released from the enzyme. Thus, a series of fragments of varying lengths accumulates, with a common 5′ end (the primer) and variable 3′ ends, and with each 3′ end identifying a T residue in the insert sequence that is being analyzed. Similarly, one identifies sites terminated by C, G, and T simply by running polymerase reactions with the other three dideoxy analogs, one at a time. Inclusion of a radioactive nucleotide in the polymerization mixture and gel electrophoresis followed by radioautography (exposure of the gel to X-ray film) yields four "sequencing ladders," as shown in Figure 4A.6. Each band in the radioautographic image of the electrophoretic gel identifies one of the four bases at that site.

Sanger sequencing is now done automatically. Each ddNTP is modified with a fluorescent dye, each of a different color. Thus, each fragment has a distinct color based on the identity of the ddNTP that terminated the sequencing reaction. This allows all four reaction mixtures to be resolved in one lane of a sequencing gel, permitting analysis of far more DNA in one sequencing operation. The gel is scanned for its fluorescence, and a computer reads the DNA sequence directly from the resulting pattern of differently colored peaks (**FIGURE 4A.7**).

Further refinements of this method have greatly increased its "throughput," or the amount of sequencing information derived from one operation. These methods yielded complete genomic sequences for several bacteria in the mid-1990s and the near completion of the human genome sequence in 2001. Since then, additional modifications have yielded several approaches that greatly expand the speed and accuracy of DNA sequencing operations. Several of the most prominent of these "second-generation" sequencing technologies are described in a review cited in Appendix II. Remarkably, the most recent approaches allow DNA sequence determination at the single-molecule level. As detailed later in this book, enormous amounts of information about health and disease, biological individuality, and evolutionary relationships have come from these developments.

Site-Directed Mutagenesis

Analysis of the function of a protein involves altering the protein's structure and then determining how its biological function has been affected. Until the 1980s, investigators treated proteins with chemical reagents that would react with particular amino acids. Such studies were hampered by lack of specificity and incomplete yields, among other problems.

Once it became possible to clone the gene encoding a protein of interest, one could systematically alter the gene at specific sites to

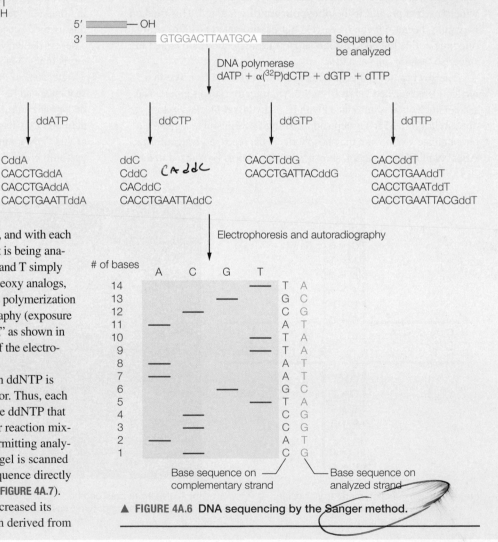

▲ **FIGURE 4A.6** DNA sequencing by the Sanger method.

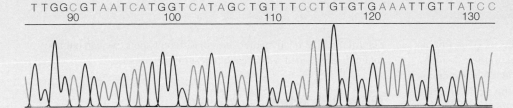

T T G G C G T A A T C A T G G T C A T A G C T G T T T C C T G T G T G A A A T T G T T A T C C
90   100   110   120   130

◄ **FIGURE 4A.7** Data from a DNA sequencing gel. Data from Dr. Robert H. Lyons, The University of Michigan DNA Sequencing Core.

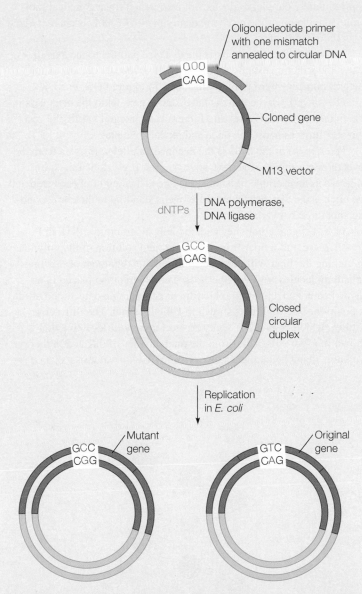

Oligonucleotide primer with one mismatch annealed to circular DNA

GGG
CAG

Cloned gene

M13 vector

dNTPs

DNA polymerase, DNA ligase

GCC
CAG

Closed circular duplex

Replication in *E. coli*

GCC
CGG

Mutant gene

GTC
CAG

Original gene

▲ **FIGURE 4A.8** Use of a mismatched synthetic oligonucleotide primer to introduce mutations in a gene cloned into a single-stranded vector.

generate virtually any desired mutation, a technique known as **site-directed mutagenesis.** Introduction of the cloned mutant gene into a host cell, followed by its expression, could then yield the mutant protein for study of its altered function.

The original method for site-directed mutagenesis, conceived by Michael Smith, allows the introduction of practically any mutation at any site, including single-base substitutions, short deletions, or insertions. The approach, illustrated in **FIGURE 4A.8**, requires that the gene first be cloned into a single-stranded vector, such as phage M13. This is a single-stranded DNA phage, like ΦX174, that replicates via a double-stranded intermediate.

The next step is to synthesize an oligodeoxynucleotide, about 20 nucleotides long, that is complementary in sequence to the cloned gene at the site of the desired mutation, *except in the center of the sequence.* Here the oligonucleotide sequence contains one or two deliberate mistakes—either single nucleotides that do not pair with the template or insertions or gaps of a few nucleotides. Upon annealing to the cloned gene, these alterations create either non-Watson–Crick base pairs or bases that have no partners and therefore form a "looping out." The correctly matched bases on both sides of the mismatch cause it to remain annealed, despite the mismatch.

DNA polymerase is then used to synthesize around the circular vector from this primer. The enzyme adds nucleotides stepwise from the 3′ end of the oligonucleotide, copying the vector around to the 5′ end of the primer. This is followed by enzymatic ligation to create a closed circular duplex. After this circular molecule is introduced into bacteria by transformation, both strands replicate and yield phage. In principle, 50% of the phage should contain the desired mutation within the inserted sequence. In practice, that percentage is considerably less, but contemporary approaches, using polymerase chain reaction (PCR; see Chapter 21), eliminate this difficulty. The resulting mutant gene, whether prepared using phage M13 or PCR, is then recloned into an expression vector for large-scale preparation and subsequent isolation of the mutant protein.

References

For a list of references related to this chapter, see Appendix II.

As discussed in this chapter, X-ray diffraction analysis of DNA fibers provided the crucial insight leading Watson and Crick to their double-helix model. Similar analysis of protein crystals has allowed structural determination of thousands of individual proteins at the atomic level, and new structures, which provide unparalleled insight into protein function, are being revealed on a weekly basis. We provide a brief introduction to the technique here.

When radiation of any kind passes through a regular, repeating structure, *diffraction* is observed. This means that radiation scattered by the repeating elements in the structure shows reinforcement of the scattered waves in certain specific directions and weakening of the waves in other directions. A simple example is given in **FIGURE 4B.1**, which shows radiation being scattered from a row of equally spaced atoms. Only in certain directions will the scattered waves be in phase and therefore constructively interfere with (reinforce) one another. In all other directions, they will be out of phase and destructively interfere with one another. Thus, a **diffraction pattern** is generated. For that pattern to be sharp, it is essential that the wavelength of the radiation used be somewhat shorter than the regular spacing between the elements of the structure. This is why X-rays are used in studying molecules, for X-rays typically have a wavelength of only a few Å. If the regular spacing in the object being studied is large (as in a window screen), we can observe exactly the same phenomenon with visible light, which has a wavelength many times longer than X-rays. We will find that a point source, seen through a window screen, gives a rectangular diffraction pattern of spots.

The rule relating the periodic spacings in object and diffraction pattern is simple: Short spacings in the periodic structure correspond to large spacings in the diffraction pattern, and vice versa. In addition, by determining the relative intensities of different spots, we can tell how matter is distributed within each repeat of the structure.

Fiber Diffraction

We consider first the diffraction from helical molecules, aligned approximately parallel to the axis of a stretched fiber. A helical molecule, like the one shown schematically in **FIGURE 4B.2**, is characterized by certain parameters:

The *repeat* (c) of the helix is the distance parallel to the axis in which the structure exactly repeats itself. The repeat contains some integral number (m) of polymer residues. In Figure 4B.2, $m = 4$.

The *pitch* (p) of the helix is the distance parallel to the helix axis in which the helix makes one turn. If there is an integral number of residues per turn (as here), the pitch and repeat are equal.

The *rise* (h) of the helix is the distance parallel to the axis from the level of one residue to the next, so $h = c/m$. If we think of a spiral staircase as an example of a helix, the rise is the height of each step and the pitch is the distance from where one is standing to the corresponding spot directly overhead.

To investigate a polymer with the helical structure shown in Figure 4B.2, we pull a fiber from a concentrated solution of the polymer. Stretching the fiber will produce approximate alignment of the long helical molecules with the fiber axis. The fiber is then placed in an X-ray beam, and a photographic film or comparable detection medium is positioned behind it, as shown in **FIGURE 4B.3(a)**. The diffraction pattern, which consists of spots or short arcs, will look like that shown in Figure 4B.3(b). From the mathematics of diffraction theory, we know that a helix always gives rise to this kind of cross-shaped

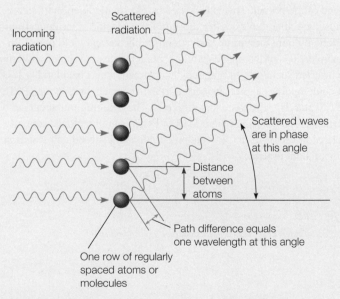

▲ **FIGURE 4B.1** Diffraction from a very simple structure—a row of atoms or molecules.

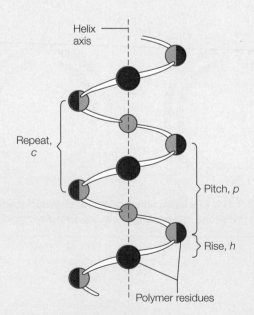

▲ **FIGURE 4B.2** A simple helical molecule.

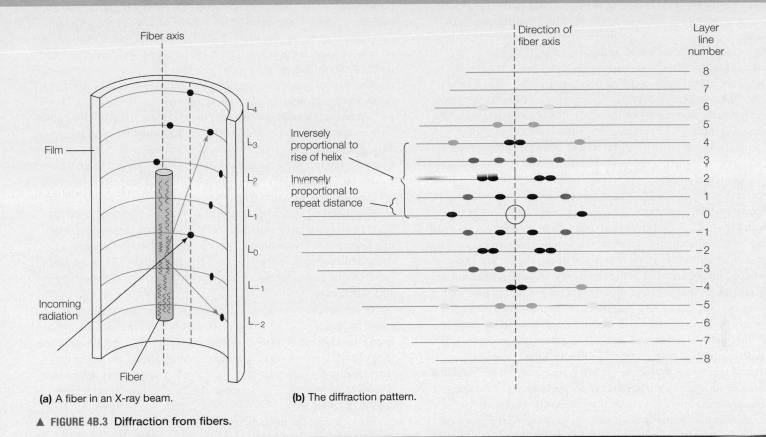

Fiber axis

Film

Incoming radiation

Fiber

L_4

L_3

L_2

L_1

L_0

L_{-1}

L_{-2}

Direction of fiber axis

Layer line number

8
7
6
5
4
3
2
1
0
−1
−2
−3
−4
−5
−6
−7
−8

Inversely proportional to rise of helix

Inversely proportional to repeat distance

(a) A fiber in an X-ray beam.

(b) The diffraction pattern.

▲ **FIGURE 4B.3** Diffraction from fibers.

pattern. Therefore, we know we are dealing with a helical structure. The spots all lie on lines perpendicular to the fiber axis; these are called **layer lines.** The spacing between these lines is inversely proportional to the repeat of the helix, c, which in this case equals the pitch. Note that the cross pattern repeats itself on every fourth layer line. This repetition pattern tells us that there are exactly 4.0 residues per turn in the helix. Thus, the rise in the helix is $c/4$. This is the kind of evidence telling Watson and Crick that B-DNA was a helix with 10 residues per turn.

The preceding information is given directly by the pattern. To find out exactly how all of the atoms in each residue are arranged in each repeat, a more detailed analysis is necessary. Usually, a model is made using the indicated repeat, pitch, and rise. Model making is simplified because we know approximate bond lengths and the angles between many chemical bonds. The model must also be inspected to see that no two atoms approach closer than their van der Waals radii. From such a model, the intensities of the various spots can be predicted. These predictions are compared with the observed intensities, and the model is readjusted until a best fit is obtained. The initial determination of the structure of DNA was done in just this way. As you can see from Figure 4.10, real fiber diffraction patterns are not as neat as the idealized example, mainly because of incomplete alignment of the molecules.

Crystal Diffraction

To study molecular crystals such as those formed by small oligonucleotides, molecules like tRNA, and globular proteins, the experimenter

faces a different problem from the study of helical fibers and so proceeds in a quite different way. A schematic drawing of such a crystal is shown in **FIGURE 4B.4.**

The repeating unit is now the **unit cell,** which may contain one, two, or more molecules. The unit cell may be thought of as the basic building block of the crystal. Repetition of the unit cell in three

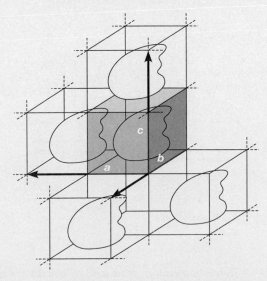

▲ **FIGURE 4B.4** Schematic drawing of a molecular crystal.

dimensions (marked by arrows on the figure) creates the whole crystal. A simple two-dimensional analog of the crystal unit cell is the repeating pattern in wallpaper. No matter how random a wallpaper pattern may seem, if you stare at it long enough, you can always find a unit that, by repetition, fills the entire wall.

Just as in fiber diffraction, passing an X-ray beam through a molecular crystal produces a diffraction pattern. The pattern shown in **FIGURE 4B.5** was obtained from a crystal of a small synthetic DNA molecule. Again, the spacing of the spots allows us to determine the repeating distances in the periodic structure—in this case, the *x, y,* and *z* dimensions of the unit cell labeled *a, b,* and *c* in Figure 4B.4. But the important information in crystal diffraction studies is just how the atoms are arranged *within* each unit cell, for that arrangement describes the molecule. Again, this information is contained in the relative intensities of the diffraction spots in a pattern like that shown in Figure 4B.5. But in crystal diffraction, more exact information can be extracted than from a fiber diffraction pattern because the corresponding molecules in each unit cell of the crystal are of the same shape and are oriented in the same way. In fiber diffraction, the helical molecules may all have their long axes pointed in the same direction, but they are rotated randomly about these axes. This difference in exactness of arrangement can be appreciated by comparing the sharpness of the crystal diffraction pattern shown in Figure 4B.5 with the fiber pattern depicted in Figure 4.10.

After obtaining the diffraction pattern from a molecular crystal, the experimenter measures the intensities of a large number of the spots. If the molecule being studied is small, one can proceed in much the same manner as with fiber patterns. A structure is guessed at, and expected intensities are calculated and compared with the observed intensities. The structure is refined until the relative intensities of all spots are correctly predicted. However, such a procedure won't work with a molecule as complex as the tRNA shown in Figure 4.25—there is simply no way to guess such a structure.

Why not proceed directly from spot intensities to the structure? The difficulty is that some of the information contained in the spot intensities is hidden. To greatly simplify a complex problem, we may say that it is as if the quantities that the experimenter needed in order to deduce the structure (which are called **structure factors**) were the square roots of the intensities. If the intensity has a value of, say, 25, the investigator knows that the number needed is +5 or −5. But which? This is the essence of the *phase problem,* which prevented progress in large-molecule crystallography for many years.

One way of solving the problem was discovered in the early 1950s. Suppose a heavy metal atom, such as mercury, can be introduced into some point in the molecule such that the molecule and crystal are otherwise unchanged. This process is called an **isomorphous replacement.** Now suppose the heavy metal contributes a value of +2 to the structure factor for the spot we were just discussing. If the original value was +5, its new value is +7 and its square is 49. If the original value was −5 the value now becomes 3 and its square is 9. The investigator takes a diffraction photograph of the crystal with the heavy metal inserted. If the new crystal has an intensity of 9 for this spot, the original structure factor must have been −5, not +5. Although this example is an oversimplification, it gives the essence of the method. Usually, multiple isomorphous replacements are necessary to determine the phases of the structure factors.

Given structure factors for all of the spots, the investigator can calculate the positions of all atoms in the unit cell. What is actually calculated is an **electron density** distribution (**FIGURE 4B.6**), but this amounts to the same thing, for regions of high electron density are where the atoms are. In the particular view shown in Figure 4B.6, we are looking at a two-dimensional "slice" through the three-dimensional electron density distribution.

To determine the three-dimensional structure of a macromolecule from crystal diffraction studies, one must first purify the material to homogeneity (see Tools of Biochemistry 5A) and then obtain crystals of good quality, at least a few tenths of a millimeter in the minimum dimension. A good deal of trial and error is needed to obtain satisfactory crystals. One then obtains a diffraction pattern and measures the intensities of many of the spots. Next, and again by trial and error, the investigator finds a way to make isomorphous replacements in the molecule, usually two or more. Each isomorphous replacement must

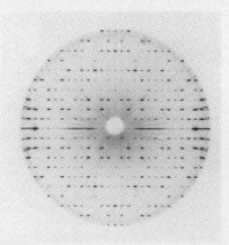

▲ **FIGURE 4B.5 Diffraction pattern produced by a crystal of a small DNA molecule.** Courtesy of P. Shing Ho, Colorado State University.

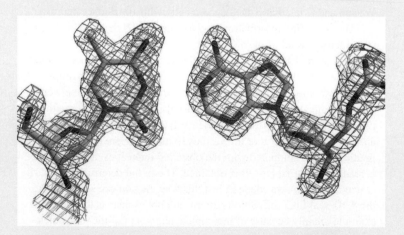

▲ **FIGURE 4B.6 Part of an electron density map derived from the crystal diffraction pattern shown in Figure 4B.5.** Courtesy of P. Shing Ho, Colorado State University.

be crystallized and its diffraction pattern analyzed, as are done for the native macromolecule. From these data the structure factors and then the electron density distribution can be calculated. These calculations are done on a large computer.

Usually, the investigator will first carry out this analysis with a small number of spots. This procedure will give a *low-resolution* structure. If all is going well, more spots will be measured and the calculations refined to give higher resolution (**FIGURE 4B.7**). With the best crystals, one can now obtain resolutions of below 1 Å. This resolution is sufficient to identify individual groups and even some atoms and to show how they interact.

Most of the detailed three-dimensional structures of biological macromolecules shown in this book have been determined by X-ray diffraction studies of crystals. Tens of thousands of such structures are known. The results allow us to understand macromolecular function at a level that would have been unbelievable only a few decades ago.

Within the past few years a new technique, called cryo-electron microscopy or cryo-EM, has emerged and has yielded spectacular images of complexes too large to be crystallized, such as mitochondria, and improved images of complexes that have been crystalllized, such as ribosomes. In this technique, the sample is frozen to liquid nitrogen temperature, and a beam from an electron microscope is passed through it, to yield the image. The technique has also provided images of crystallized proteins superior to those previously obtained from X-ray diffraction.

References

For a list of references related to this chapter, see Appendix II.

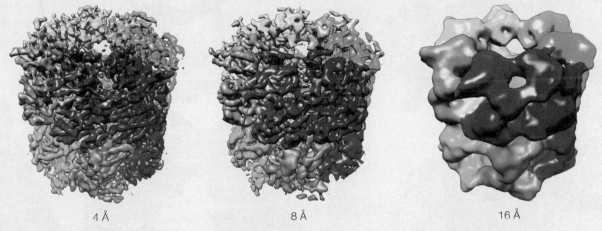

4 Å   8 Å   16 Å

▲ **FIGURE 4B.7** Models of the structure of GroEL (a chaperone protein; see Chapter 6) at resolutions (left to right) of 4, 8, and 16 Å.

Polymerized sickle hemoglobin

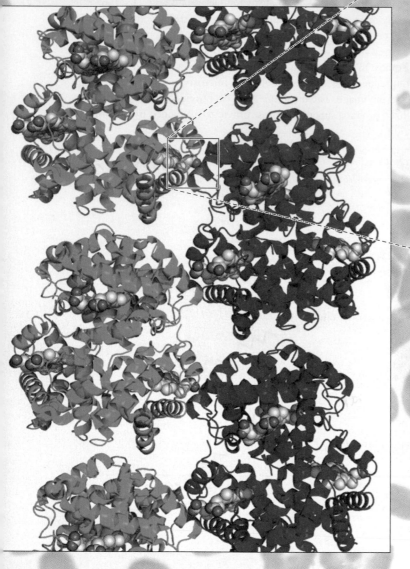

Zoom of contact surface

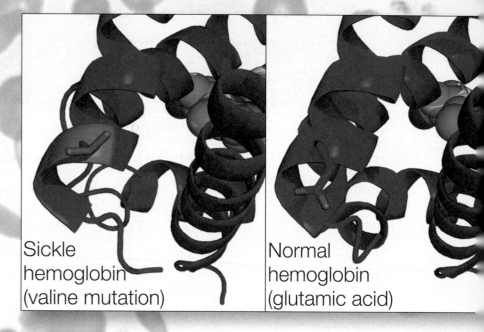

Sickle hemoglobin (valine mutation)

Normal hemoglobin (glutamic acid)

The structures of biomolecules define their functions. The elaborate structures of proteins are determined by the structures and chemical properties of their constituent amino acids. The relationship between structure and function in biochemistry is dramatically illustrated by diseases such as cystic fibrosis and sickle-cell anemia, which arise as a result of a single amino acid mutation that leads to loss of normal protein function. A change in amino acid structure at a specific location in the oxygen-carrying protein hemoglobin results in the undesired polymerization of hemoglobin in sickle-cell patients. Mutation of the normal, highly polar, glutamic acid in this position to the mutant, nonpolar, valine creates a hydrophobic surface that leads to association of the mutant hemoglobins. What are the features of amino acids that determine the structures of proteins? How do mutations affect protein structure and function?

Introduction to Proteins: The Primary Level of Protein Structure

5

ONE CLASS OF biopolymers, the nucleic acids, stores and transmits the genetic information of the cell. Much of that information is expressed in another class of biopolymers, the proteins. Proteins play an enormous variety of roles: Some transport small molecules or ions across membranes, whereas others make up much of the structural framework of cells and tissues. Proteins also play central roles in processes such as muscle contraction, the immune response, and blood clotting. Many critically important proteins are enzymes—catalysts that promote the tremendous variety of reactions that are required to support the living state. Each type of cell in every organism has several thousand kinds of proteins to serve these many diverse functions.

A fundamental concept in biochemistry is that the *structure* of a given biomolecule is directly related to its *function.* This is a concept we develop and illustrate throughout this textbook. The structures of most protein molecules are highly complex. For example, **FIGURE 5.1** shows the molecular structure of myoglobin, a relatively

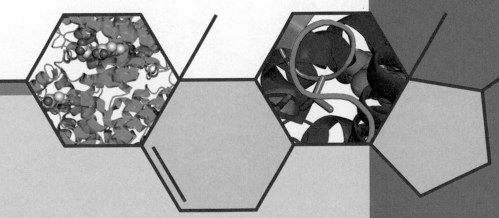

▶ **FIGURE 5.1** **The three-dimensional structure of the globular protein myoglobin.** Most proteins in cells are linear polymers of amino acids that are synthesized on the ribosome by the translation of mRNA. The resulting amino acid polymer then folds upon itself and adopts a generally well-defined, stable, three-dimensional structure (note that this process is much more complex in cells and involves many more cellular components than are shown in this highly simplified scheme). Shown in the lower left are all the atoms (except hydrogens) in the protein myoglobin from sperm whale. This molecular model was generated from the X-ray crystal structure determined by H. C. Watson and J. C. Kendrew (PDB ID: 1mbn). The amino acid atoms are shown as sticks, where carbon atoms are gray, oxygen atoms are red, nitrogen atoms are blue, and sulfur atoms are yellow.

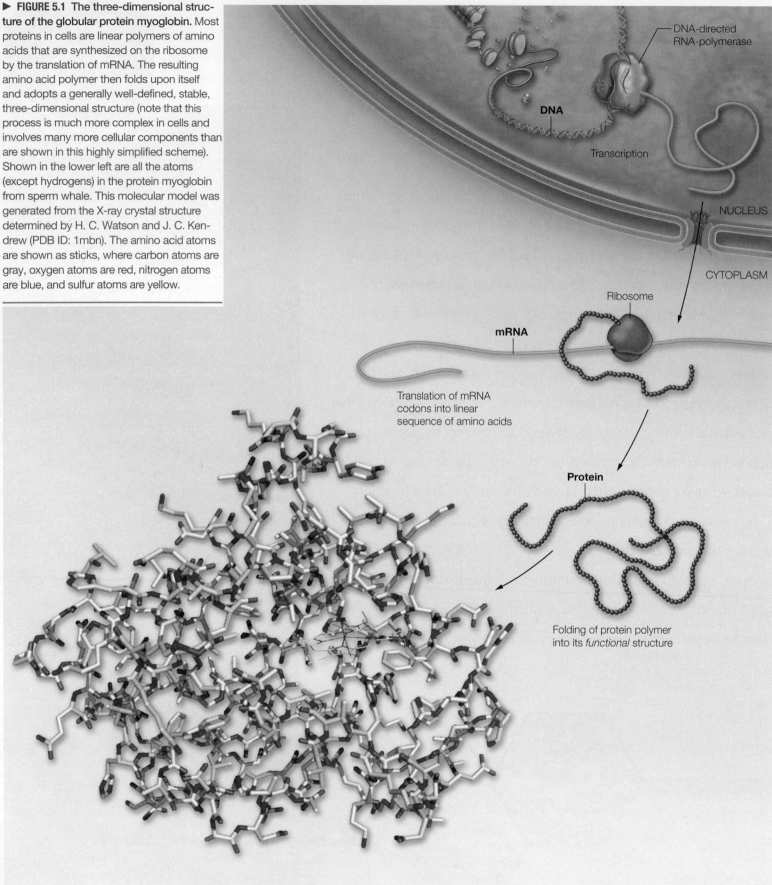

DNA-directed RNA-polymerase

DNA

Transcription

NUCLEUS

CYTOPLASM

Ribosome

mRNA

Translation of mRNA codons into linear sequence of amino acids

Protein

Folding of protein polymer into its *functional* structure

small protein that functions primarily in oxygen binding and storage in animal tissues.

In Chapters 6–8, we analyze in detail the structures and functions of a handful of proteins, including myoglobin. Although most proteins share some common structural features, we will see that each protein has a distinct structure that is optimally suited to its function. Protein structures may appear at first glance to be hopelessly complex, but there is an elegant and readily comprehensible chemical logic to protein structure, which we describe here and in Chapter 6. The exploration of protein structure begins with a description of the structures and chemical properties of amino

● **CONCEPT** Protein function is determined by protein structure, which is determined, in turn, by the structures and properties of the various amino acids that make up the protein.

acids, the simple "building blocks" that are found in all proteins.

One consequence of the correlation between protein structure and protein function is that even modest changes in protein structure can lead to unstable or nonfunctional proteins. Many human diseases are the result of a single amino acid substitution, or *mutation,* in a protein molecule. Among these diseases are cystic fibrosis, sickle-cell anemia, and familial amyotrophic lateral sclerosis (ALS). Amino acid mutations in certain tumor suppressor proteins are also correlated with increased risk for some cancers. Thus, an understanding of the relationship between amino acid structure and protein function is critical to our understanding of the molecular basis for many diseases.

● **CONNECTION** Many diseases are a consequence of amino acid mutations that render a protein unstable or non-functional.

5.1 Amino Acids

Structure of the α-Amino Acids

All proteins are polymers, and the monomers that combine to make them, are **α-amino acids,** so-named because the amino group is attached to the α-carbon (the carbon next to the carboxylic acid group). A general representation of an α-amino acid is shown in **FIGURE 5.2(a)**. Notice that a *hydrogen atom* and a *side chain* (R group) are attached to the α-carbon of every amino acid, too. Different α-amino acids are distinguished by their different side chains. Although the representation shown in Figure 5.2(a) is chemically correct, it does not convey the conditions in vivo. As pointed out in Chapter 2, most biochemistry occurs in the physiological pH range near 7 (neutrality). The pK_as of the carboxylic acid and amino groups of the α-amino acids are about 2 and 10, respectively. Near neutral pH, then, the carboxylic acid group will be deprotonated ($-COO^-$), and the amino group will be protonated ($-NH_3^+$), to yield the **zwitterion** form shown in Figure 5.2(b). This is the form in which we will customarily write amino acid structures.

Twenty different kinds of amino acids are commonly incorporated into proteins during the process of translation (see Figure 4.34). The complete structures of these amino acids are shown in **FIGURE 5.3**, and other important data are given in **TABLE 5.1**. At least two additional amino acids, selenocysteine and pyrrolysine, are encoded genetically and incorporated into proteins, but they are found in only a relatively small number of proteins. For the purposes of this introductory discussion, we will focus our attention on the 20 common amino acids shown in Figure 5.3.

● **CONCEPT** Proteins are polymers of α-amino acids. There are 20 common α-amino acids that are the major building blocks of proteins.

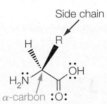

Side chain

(a) A general representation of a nonionized amino acid showing the carboxylic acid group, the α-amino group, the hydrogen bonded to the α-carbon, and the side chain (R group) that gives the amino acid its unique properties. The stereochemistry shown in this figure is that for the α-amino acids found in biosynthetic proteins, with lines representing bonds in the plane of the page, a solid wedge projecting out of the plane toward the viewer, and the dashed wedge projecting behind the plane of the page.

zwitterion

(b) An amino acid shown as a zwitterion at neutral pH, where the α-carboxylic acid group is deprotonated and the α-amino group is protonated. The negative charge on the α-carboxylate is delocalized over the two oxygen atoms.

▲ **FIGURE 5.2** The chemical structure of an α-amino acid.

Stereochemistry of the α-Amino Acids

The asymmetry of biomolecules plays a critical role in determining their structures and functions. Thus, we need to understand the basic stereochemistry of amino acids to understand the biochemistry of proteins.

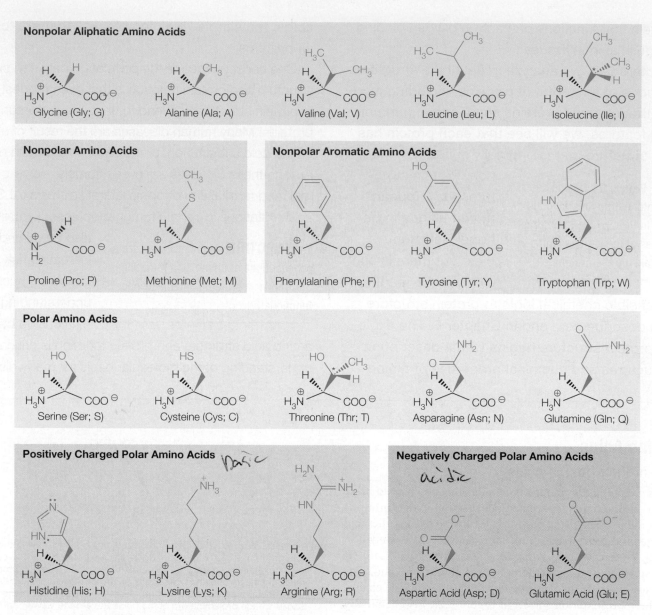

▲ **FIGURE 5.3 The 20 common amino acids found in proteins.** The chemical structures of the 20 common α-amino acids that are incorporated into proteins are shown and arranged here in the order in which they are discussed in the text. The "side chain" or "R group" of each amino acid is highlighted in green. Below each amino acid are its name, its three-letter abbreviation, and its one-letter abbreviation.

The central α-carbon in all 20 common amino acids is sp^3 hybridized, so the four groups bonded to it in Figure 5.2(a) are arranged in a tetrahedral geometry. In Figures 5.2 and 5.3, the projection of these groups around the α-carbon ($C_α$) is drawn such that lines represent bonds in the plane of the page, solid wedges represent bonds projecting forward from the page, and dashed wedges represent bonds projecting behind the page. When a carbon atom has four different substituents attached to it, it is said to be **chiral,** or a **stereocenter,** or (preferably) an asymmetric carbon. The stereochemistry of the amino acids can also be represented by a convention known as the **Fischer projection** (**FIGURE 5.4(b)**). In a Fischer projection, the bonds are all represented as solid lines, where the horizontal bonds are understood to project forward from the page and the vertical bonds project behind the page.

The spatial orientation of the four groups bound to the $C_α$ is the same in Figures 5.2, 5.3, and 5.4.

If a molecule contains one asymmetric carbon, then two distinguishable **stereoisomers** exist; these are nonsuperimposable mirror images of one another, or **enantiomers.** The stereoisomers of **alanine** shown in **FIGURE 5.5**, called the L and D enantiomers,* can be distinguished from one another experimentally because their solutions rotate plane-polarized light in opposite directions. All

---

*Those who are familiar with organic chemistry will know that there are two commonly used systems for distinguishing stereoisomers—the older D–L system and the newer, more comprehensive R–S system (the Cahn–Ingold–Prelog system). Both are discussed in more detail in Chapter 9.

TABLE 5.1 Properties of the common amino acids found in proteins

Name	Abbreviations: 1- and 3-letter codes	pK_a of α-COOH Group[a]	pK_a of α-NH_3^+ Group[a]	pK_a of Ionizing Side Chain[a]	Residue[b] Mass (daltons)	Occurrence[c] in Proteins (mol %)
Alanine	A, Ala	2.3	9.7	—	71.08	8.7
Arginine	R, Arg	2.2	9.0	12.5	156.20	5.0
Asparagine	N, Asn	2.0	8.8	—	114.11	4.2
Aspartic acid	D, Asp	2.1	9.8	3.9	115.09	5.9
Cysteine	C, Cys	1.8	10.8	8.3	103.14	1.3
Glutamine	Q, Gln	2.2	9.1	—	128.14	3.7
Glutamic acid	E, Glu	2.2	9.7	4.2	129.12	6.6
Glycine	G, Gly	2.3	9.6	—	57.06	7.9
Histidine	H, His	1.8	9.2	6.0	137.15	2.4
Isoleucine	I, Ile	2.4	9.7	—	113.17	5.5
Leucine	L, Leu	2.4	9.6	—	113.17	8.9
Lysine	K, Lys	2.2	9.0	10.0	128.18	5.5
Methionine	M, Met	2.3	9.2	—	131.21	2.0
Phenylalanine	F, Phe	1.8	9.1	—	147.18	4.0
Proline	P, Pro	2.0	10.6	—	97.12	4.7
Serine	S, Ser	2.2	9.2	—	87.08	5.8
Threonine	T, Thr	2.6	10.4	—	101.11	5.6
Tryptophan	W, Trp	2.4	9.4	—	186.21	1.5
Tyrosine	Y, Tyr	2.2	9.1	10.1	163.18	3.5
Valine	V, Val	2.3	9.6	—	99.14	7.2

[a]Approximate values found for side chains on the *free* amino acids. W. P. Jencks and J. Regenstein (1976) Ionization constants of acids and bases in *Handbook of Biochemistry and Molecular Biology,* 3rd ed., G. Fasman (ed.), CRC Press, Boca Raton, FL.

[b]To obtain the mass of the amino acid itself, add the mass of a molecule of water, 18.02 daltons. The values given are for neutral side chains; slightly different masses are observed at pH values where protons have been gained or lost from the side chains.

[c]Average for a large number of proteins. Individual proteins can show large deviations from these values. Data from J. M. Otaki, M. Tsutsumi, T. Gotoh, and H. Yamamoto, Secondary structure characterization based on amino acid composition and availability in proteins (2010) *Journal of Chemical Information and Modeling* 50:690–700 © 2010 American Chemical Society.

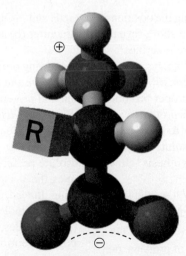

(a) This ball-and-stick model shows the three-dimensional arrangement of the atoms, with C atoms in black, H atoms in gray, the N atom in blue, and O atoms in red. The α-carbon is asymmetric, with tetrahedral bonding geometry. The variable R group is shown as an orange cube.

(b) In a Fischer projection (left), the horizontal bonds project toward the viewer, and the vertical bonds project away from the viewer. This orientation of bonds in the Fischer projection is represented on the right by solid and dashed wedges, respectively.

▲ **FIGURE 5.4** Three-dimensional representations of α-amino acids.

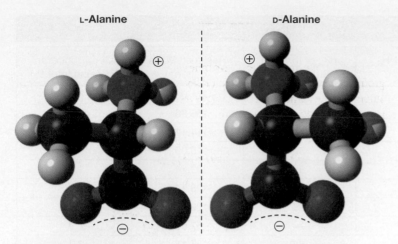

L-Alanine   **D-Alanine**

(a) L-Alanine and its enantiomer D-alanine are shown as ball-and-stick models. The alanine side chain is –CH$_3$. The two models are nonsuperimposable mirror images. The mirror plane is represented by the vertical dashed line.

(b) The same two enantiomers in Fischer projections.

▲ **FIGURE 5.5** Stereoisomers of α-amino acids.

amino acids except glycine can exist in D and L forms because in each case the α-carbon is asymmetric. Glycine is the sole exception because two of its four groups bonded to the α-carbon are the same (they are H atoms), thus eliminating the asymmetry. Two amino acids, isoleucine and threonine, have an additional stereocenter at the β-carbon in their respective side chains (indicated by red asterisks in Figure 5.3).

As a self-test of your skill in properly visualizing three-dimensional stereochemistry from two-dimensional representations, consider the following problem: Which of the following Fischer projections (a)–(d) represents L-serine?

First, recall that in a Fischer projection the vertical bonds project behind the plane of the page, and the horizontal bonds project forward:

Second, a frame of reference is needed. For example, the image of L-serine in Figure 5.3 could serve as the reference to determine which of the four structures represents the L enantiomer of serine. Thus, each structure above will be rotated such that the —NH$_3$^+ and —COO^- groups are drawn in the plane of the page, with the —NH$_3$^+ on the left (as it is in Figure 5.3):

Comparing the positions of the α-H and —CH$_2$OH groups in Figure 5.3 with those above reveals that structures (b) and (c) are L-serine and (a) and (d) are D-serine.

Chemical analysis of naturally occurring proteins shows that *nearly all of their constituent amino acids have the L form.* Cell viability depends on proper protein function, and protein function depends on the ability of a protein to adopt a well-defined *active* structure. Thus, the need for cells to produce many structurally identical copies of a given protein is absolute. For this reason, nature uses only L-amino acids in protein biosynthesis. How the absolute preference for the L-isomer over the D-isomer evolved is puzzling. Indeed, each of the three major classes of biological macromolecules (polypeptides, polysaccharides, and polynucleotides) exhibits a strong preference for one stereoisomer class or the other. Most naturally occurring polysaccharides prefer D-sugars, as do DNA and RNA. It may be that productive interaction between these substances was established early in the evolution of life, but why was a particular set of enantiomers chosen at all? It is hard to see how L-amino acids have any inherent selective advantage over the D-isomers

● **CONCEPT** All α-amino acids except glycine contain an asymmetric α-carbon. In the vast majority of proteins, only the L-enantiomers are found.

for biological function. Indeed, D-amino acids exist in nature, and some play important biochemical roles, but they are only rarely found in large protein molecules.

The preference for L-amino acids in natural proteins has two important consequences, which we discuss further in subsequent chapters:

● **CONCEPT** The stereochemistry of the amino acids is a major determinant of the structure of proteins.

● **CONNECTION** The asymmetry of protein binding sites determines which small molecules they will bind. Thus, the efforts of medicinal chemists to design new drugs include matching the shape of the drug to that of the binding site in the target protein.

1. The surface of any given protein, which is where the interesting biochemistry occurs, is asymmetric. This asymmetry is the basis for the highly specific molecular recognition of binding targets by proteins, including enzyme–substrate binding, neurotransmitter binding to a specific receptor, and a drug binding to an enzyme or receptor.

2. The stereochemistry of the amino acids promotes the formation of so-called secondary structure in proteins (i.e., α helices and β strands), which, in turn, is a major determinant of the overall structure of proteins.

Properties of Amino Acid Side Chains: Classes of α-Amino Acids

The 20 common amino acids contain, in their 20 different side chains, a variety of chemical groups. This diversity allows proteins to exhibit a great variety of

● **CONCEPT** The variety of side chains on amino acids allows proteins enormous versatility in structure and function.

structures and properties. Figure 5.3 organizes the 20 amino acids into five different classes based on the dominant chemical features of their side chains. These features include aliphatic or aromatic character, polarity of the side chain, and presence or absence of ionizable groups.

Amino Acids with Nonpolar Aliphatic Side Chains

Glycine, alanine, valine, leucine, and **isoleucine** have aliphatic side chains. As we progress from left to right along the top row of Figure 5.3, the R group becomes more extended and more hydrophobic. Isoleucine [R = —CH(CH₃)CH₂CH₃], for example, has a much greater tendency to transfer from water to a hydrocarbon solvent than alanine (R = —CH₃) does. As we illustrate in Chapter 6, the more hydrophobic amino acids such as isoleucine are usually found *within the core* of a folded protein molecule, where they are shielded from water.

Glycine (R = —H) is difficult to fit into any category. Although its side chain has aliphatic character, it is frequently found on the *surfaces* of proteins due to its ability to form tight **turns** in protein structures. Such turns are sites in the protein structure where the protein polymer folds back on itself to maintain a compact structure (see Figure 5.1).

Proline, which has a secondary α-amino group, is also difficult to classify. It is the only amino acid in this group in which the side chain forms a covalent bond with the α-amino group. The proline side chain has a primarily aliphatic character, but, like glycine, it is frequently found on the surfaces of proteins because the rigid ring of proline is well suited to turns.

Finally, we include **methionine** (R = —CH₂CH₂SCH₃) in this group, even though it is not technically an aliphatic amino acid. Because sulfur has the same electronegativity as carbon, the thioether side chain is quite hydrophobic; thus, methionine displays many of the same properties as the other nonpolar amino acids.

Amino Acids with Nonpolar Aromatic Side Chains

Three amino acids—**phenylalanine, tyrosine,** and **tryptophan**—have nonpolar aromatic side chains. Phenylalanine ((R = —CH₂C₆H₅), together with the aliphatic amino acids valine, leucine, and isoleucine, is one of the most hydrophobic amino acids. Tyrosine and tryptophan have hydrophobic character as well, but it is tempered by the polar groups in their side chains. In addition, tyrosine can ionize at high pH:

The aromatic amino acids, like most highly conjugated compounds, absorb light in the near-ultraviolet region of the electromagnetic spectrum (**FIGURE 5.6**). This characteristic is frequently used for the detection and/or quantification of proteins, by measuring their absorption at 280 nm.

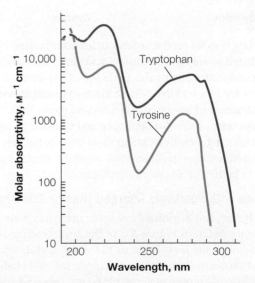

▲ **FIGURE 5.6 Absorption spectra of two aromatic amino acids in the near-ultraviolet region.** Tryptophan (red; λ_{max} = 278 nm) and tyrosine (blue; λ_{max} = 274 nm) account for most of the UV absorbance by proteins in the region around 280 nm. Phenylalanine does not absorb at 280 nm, and it absorbs only weakly at 258 nm. Note that the absorptivity scale is logarithmic. Compared with nucleic acids, amino acids absorb only weakly in the UV (see Figure 4.6 for comparison).

Amino Acids with Polar Side Chains

Serine, threonine, asparagine, and **glutamine** have polar side chains, and they can form multiple H bonds with water molecules and/or other good H bond donors and acceptors. As described in the following paragraph, cysteine also has polar character (though it is a poor H bond donor or acceptor). As a result, these five amino acids are most often found on the surfaces of proteins, where they can contact the aqueous environment in cells or in circulation. As we will see in Chapter 8, the —OH group of serine and the —SH group of cysteine are good nucleophiles and often play key roles in enzyme activity.

Cysteine is noteworthy in two respects. First, the side chain can ionize at moderately high pH:

$$\text{SH-CH}_2\text{-C(H)(}^+\text{NH}_3\text{)-COO}^- \underset{pK_a = 8.3}{\rightleftharpoons} \text{S}^-\text{-CH}_2\text{-C(H)(}^+\text{NH}_3\text{)-COO}^- + \text{H}^+$$

Second, the oxidation of two cysteine side chains yields a **disulfide bond:**

$$\text{Cysteine} \underset{\text{reduction}}{\overset{\text{oxidation}}{\rightleftharpoons}} \text{Cystine} + 2\text{H}^+ + 2e^-$$

Cysteine   **Cystine**

The covalently bonded product of this oxidation is called **cystine.** Cystine is not listed among the 20 amino acids in Figure 5.3 because it is formed posttranslationally by the oxidation of two cysteine side chains and it is not coded for by DNA. These kinds of disulfide bonds stabilize the active structure of a protein that contains them.

Although we classified methionine and tyrosine as hydrophobic, because we do not generally consider them to be polar, they can display more hydrophilic character than their aliphatic analogs due to the S and O atoms in their respective side chains.

Amino Acids with Positively Charged (Basic) Side Chains

Histidine, lysine, and **arginine** have basic groups in their side chains. They are represented in Figure 5.3 in the form that predominates at pH 7. Histidine is the least basic of the three, and its titration curve in **FIGURE 5.7** shows that the imidazole ring in the side chain of the free amino acid loses its proton at about pH 6 (see Table 5.1 for pK_a values of the side chains of *free* amino acids). When histidine is incorporated into proteins, the pK_a typically ranges from 6.5 to 7.4 (**TABLE 5.2**). As we describe later in this chapter, the pK_a value for an ionizable side chain depends on its electrostatic environment. Thus, when other charged groups are in proximity, the pK_as of ionizable side chains can be altered significantly from the values given in Table 5.1. For example, in the folded structures of proteins, the local electrostatic

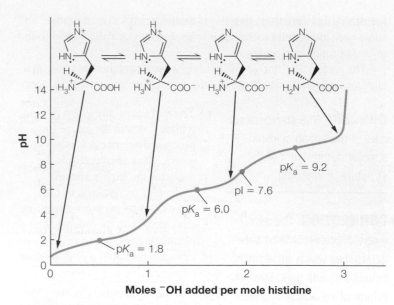

▲ **FIGURE 5.7 Titration curve of histidine.** The dots correspond to pK_a values (orange) or the pI (green). The ionic forms of histidine predominating at different pH values are shown above the curve. Titratable hydrogens are shown in red. The starting solution was adjusted to pH < 2 by the addition of strong acid to the dissolved amino acid. See also Figure 2.22 for the titration curve of aspartic acid.

environment can perturb the pK_a of an ionizable side chain by ± 3 pH units. Because the histidine side chain has a pK_a near physiological pH, it is often involved in proton transfers during enzymatic catalysis. Lysine and arginine are more basic than histidine, and their pK_a values (Tables 5.1 and 5.2) indicate that their side chains are almost always positively charged under physiological conditions. The guanidinyl group of arginine is a particularly strong base due to the resonance stabilization of the protonated side chain.

The basic amino acids are strongly polar, so they are usually found on the exterior surfaces of proteins, where they can be hydrated by the surrounding aqueous environment, or in substrate binding clefts of enzymes, where they can interact with polar groups on the substrate that binds to the enzyme.

TABLE 5.2 Typical ranges observed for pK_a values of ionizable groups in proteins

Group Type	Typical pK_a Range[a]
α-Carboxyl	3.5–4.0
Side-chain carboxyl (aspartic and glutamic acids)	4.0–4.8
Imidazole (histidine)	6.5–7.4
Cysteine (—SH)	8.5–9.0
Phenolic (tyrosine)	9.5–10.5
α-Amino	8.0–9.0
Side-chain amino (lysine)	9.8–10.4
Guanidinyl (arginine)	~12

[a]Values outside these ranges are observed. For example, side-chain carboxyls have been reported with pK_a values as high as 7.3.

Amino Acids with Negatively Charged (Acidic) Side Chains

Aspartic acid and **glutamic acid** typically carry negative charges at pH 7 (see their anionic forms in Figure 5.3). The titration curve of aspartic acid is shown in Figure 2.22. The side chain pK_a values of these two acidic amino acids are so low (see Table 5.2) that the negatively charged form of the side chain typically predominates under physiological conditions, even when they are incorporated into proteins. Hence, aspartic acid and glutamic acid are often referred to as **aspartate** and **glutamate,** respectively (i.e., as the conjugate bases rather than as the acids). Like the basic and polar amino acids, Asp and Glu are hydrophilic and tend to be on the surface of a protein molecule, in contact with the surrounding water or a bound substrate.

Aspartic acid and glutamic acid share steric similarities with asparagine and glutamine, their respective amide analogs. In contrast to their acidic analogs, however, asparagine and glutamine have nonionizable, *uncharged* polar side chains.

● **CONCEPT** Nonpolar amino acids are typically found in the interiors of soluble proteins, whereas polar and charged amino acids are typically found on the surfaces of proteins.

Rare Genetically Encoded Amino Acids

Besides the 20 common amino acids that are coded for in DNA and are incorporated directly into proteins by ribosomal synthesis, there are two other amino acids encoded in gene sequences —namely, selenocysteine, which is widely distributed but found in few proteins, and pyrrolysine, which is restricted to a few archaea and eubacteria (**FIGURE 5.8**). Selenocysteine ("Sec") and pyrrolysine ("Pyl") are sometimes referred to as the 21st and 22nd amino acids. Selenocysteine is a structural analog of cysteine in which the sulfur atom is replaced by a selenium atom. Pyrrolysine is a derivative of lysine in which a 4-methyl-pyrroline-5-carboxylic acid forms an amide bond with the ε-amino group of the lysine side chain.

Modified Amino Acids

The variety of side-chain groups in proteins can be expanded beyond the 20 canonical structures in Figure 5.3 by chemical modification after the amino acids have been assembled into proteins. The structures of four such *posttranslationally modified amino acids* are as follows, where the modifying group is shown in red:

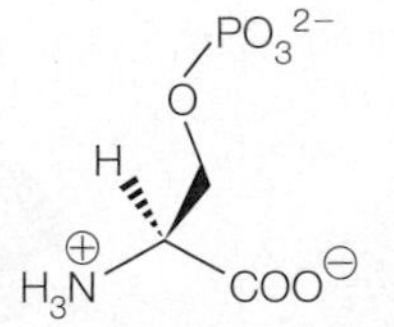

Phosphoserine 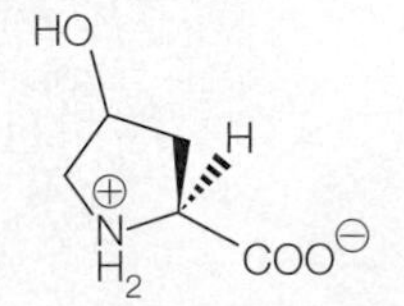   **4-Hydroxyproline**

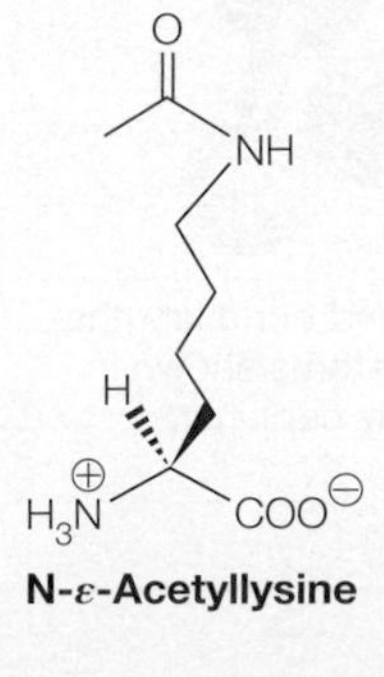

N-ε-Acetyllysine   **γ-Carboxyglutamate**

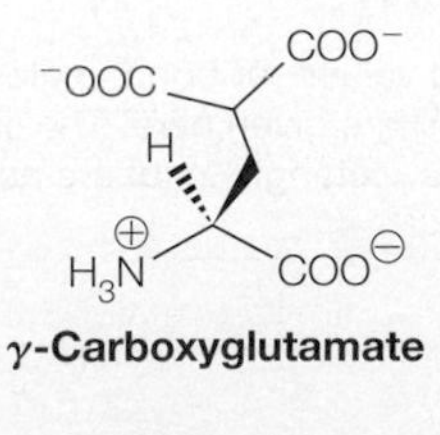

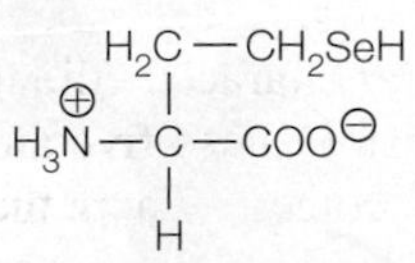

▲ **FIGURE 5.8 Structures of selenocysteine and pyrrolysine.** The selenium atom in selenocysteine and the 4-methyl-pyrroline-5-carboxylic acid in pyrrolysine are highlighted in red.

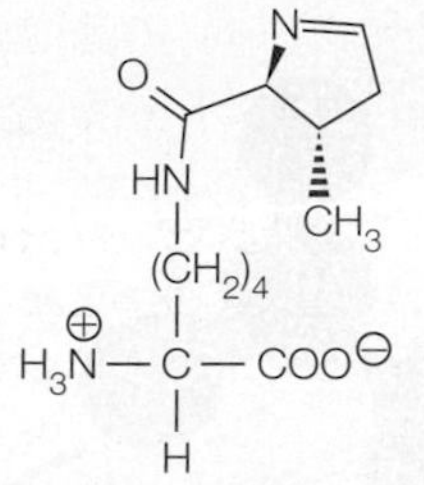

Posttranslational modification of amino acids serves many functions. The hydroxylation of proline provides an additional hydrogen bonding site to stabilize collagen fibers. The carboxylation of glutamate introduces Ca^{2+} binding sites into blood-clotting proteins. Calcium ion binding to these proteins is required for clot formation. As described in later chapters, protein phosphorylation is critical in many cell signaling pathways, as well as in the "fight-or-flight" response. Acetylation of lysine plays a role in gene expression or suppression.

● **CONCEPT** Sometimes amino acid side chains are modified after being incorporated into a protein—a process called posttranslational modification.

The amino acids found in biosynthetic proteins are by no means the only ones that occur in living organisms. Many other amino acids play important roles in metabolism.

5.2 Peptides and the Peptide Bond

Amino acids can be covalently linked together by the formation of an **amide bond** between the α-carboxylic acid group on one amino acid and the α-amino group on another. This bond is often referred to as a **peptide bond,** and the products formed by such a linkage are called **peptides.** The formation of a peptide bond between glycine and alanine is shown in **FIGURE 5.9**. The product in this case is called a **dipeptide** because two amino acids have been combined by the formal elimination of a water molecule between the carboxylic acid of one amino acid and the amino group of the other. During ribosomal protein synthesis, amino acids are added to only one end of the growing peptide (the so-called C-terminal end, which is defined later in this chapter). As each amino acid is added to the chain, another molecule of water is formally eliminated to create the new peptide bond. The portion of each amino acid remaining in the chain is called an **amino acid residue.** When specifying an amino acid residue in a peptide, the suffix –**yl** may be used to replace –ine or –ate in the name of the amino acid (e.g., glycyl for glycine and aspartyl for aspartate; tryptophanyl and cysteinyl are exceptions to this general rule). Thus, an *alanyl residue* in a protein sequence would have this structure:

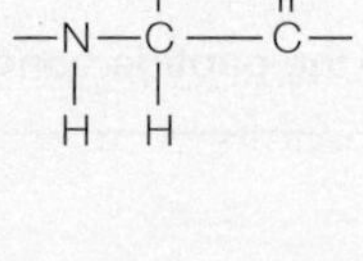

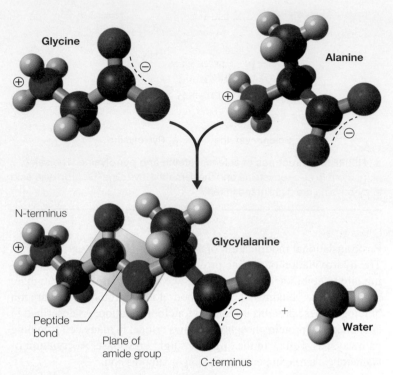

▲ **FIGURE 5.9 A peptide bond between amino acids.** The dipeptide glycylalanine (Gly–Ala) is produced when a peptide bond forms between the —COO⁻ group of glycine and the —NH₃⁺ group of alanine (formally eliminating a water molecule). The condensation of free amino acids to form a peptide bond is not thermodynamically favorable under physiological conditions. In fact, the reverse reaction-peptide bond *hydrolysis*-is favored. Peptide bond *formation* during translation is coupled to ATP hydrolysis (see Figure 3.10).

The Structure of the Peptide Bond

In the Gly–Ala dipeptide shown in Figure 5.9, the shaded blue rectangle highlights the peptide bond. The features of the peptide bond are shown in greater detail in **FIGURE 5.10**. This substituted amide bond, which is found between every pair of residues in a protein, has properties that are important for defining the structures of proteins. For example, almost invariably the amide carbonyl (C=O) and amide N—H bonds are nearly parallel; in fact, the six atoms shown lying in the blue shaded rectangle in Figure 5.10 are usually coplanar. There is little twisting possible around the peptide bond because the C—N bond has a substantial fraction of double-bond character. The peptide bond can be considered a resonance hybrid of two forms:

A schematic depiction of the electron density about the peptide bond is shown in Figure 5.10(a), and bond lengths and angles are given in Figure 5.10(b).

X-ray crystallography data of proteins show that the spatial arrangement of the two α-carbon atoms on either side of a peptide bond is described by two possible configurations, trans or cis:

trans   **cis**

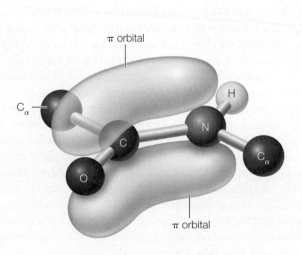

(a) Delocalization of the π-electron orbitals over the three atoms O—C—N accounts for the partial double-bond character of the C—N bond.

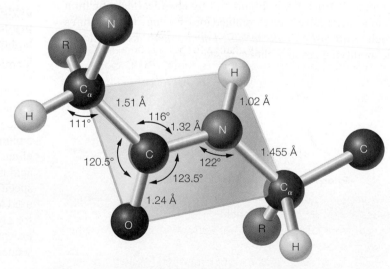

(b) The presently accepted values for bond angles and bond lengths (in Angstrom, Å or m⁻¹⁰) are given here. The six atoms shown in (a) and those in the blue rectangle in (b) are nearly coplanar.

▲ **FIGURE 5.10** Structure of the peptide bond.

TABLE 5.3 The sequence specificities of some proteolytic enzymes and cyanogen bromide

$$\text{N-terminal} \cdots -\overset{\displaystyle |}{\underset{\displaystyle |}{N}}-\overset{\displaystyle R_1}{\underset{\displaystyle |}{C}}-\overset{\displaystyle O}{\underset{}{\overset{\|}{C}}}-\overset{\displaystyle |}{\underset{\displaystyle |}{N}}-\overset{\displaystyle R_2}{\underset{\displaystyle |}{C}}-\overset{\displaystyle O}{\overset{\|}{C}}-\cdots \text{C-terminal}$$

Enzyme	Preferred Site[a]	Source
Trypsin	R_1 = Lys, Arg	From digestive systems of animals, many other sources
Chymotrypsin	R_1 = Tyr, Trp, Phe, Leu	Same as trypsin
Thrombin	R_1 = Arg	From blood; involved in coagulation
V-8 protease	R_1 = Asp, Glu	From *Staphylococcus aureus*
Prolyl endopeptidase	R_1 = Pro	Lamb kidney, other tissues
Subtilisin	Very little specificity	From various bacilli
Carboxypeptidase A	R_2 = C-terminal amino acid	From digestive systems of animals
Thermolysin	R_2 = Leu, Val, Ile, Met	From *Bacillus thermoproteolyticus*
Cyanogen bromide	R_1 = Met	

[a]The residues indicated are those next to which cleavage is most likely. In some cases, preference is determined by the residue on the N-terminal side of the cleaved bond (R_1) and sometimes by the residue to the C-terminal side (R_2). Generally, proteases do not cleave where proline is on the other side of the bond. Even prolyl endopeptidase will not cleave if R_2 = Pro.

● **CONCEPT** The peptide bond is nearly planar, and the trans form is favored.

The trans form is highly favored because the R groups on adjacent α-carbons can sterically interfere in the cis configuration. The major exception is the bond in the sequence X—Pro, where X is any other amino acid. In this bond the cis configuration is sometimes allowed, although the trans configuration is still favored by about 4:1.

Stability and Formation of the Peptide Bond

Figure 5.9 implies that a peptide bond could be formed by the elimination of a water molecule between two amino acids. In fact, this process is thermodynamically disfavored in an aqueous environment like the one found in cells. The free energy change for peptide bond formation at room temperature in aqueous solution is about +10 kJ/mol, so the favored reaction under these conditions is the *hydrolysis* of the peptide bond, with equilibrium lying well to the right:

$$\text{(structure with } +H_2O / -H_2O \text{)}$$

The uncatalyzed reaction, however, is exceedingly slow at physiological pH and temperature. Like polynucleotides, polypeptides are metastable, hydrolyzing rapidly only under extreme conditions or when catalysts are present.

Peptide bonds can be hydrolyzed in several ways. A general method,

● **CONCEPT** The peptide bond is metastable. Peptide bonds hydrolyze in aqueous solution when a catalyst is present.

which cleaves all peptide bonds, is boiling in strong mineral acid (e.g., 6 M HCl), a harsh treatment that also destroys side chains of Asn,

Gln, and Trp (and to a lesser extent Ser, Thr, and Tyr). More specific cleavage under milder conditions is provided by **proteolytic enzymes** or **proteases.** Many of these enzymes cleave only specific peptide bonds (see **TABLE 5.3**). Some of these enzymes are secreted into the digestive tracts of animals, where they catalyze the initial breakdown of dietary proteins to smaller peptides, which are then further digested to amino acids by other enzymes. Others, such as papain, are found in certain plant tissues. Biochemists use proteolytic enzymes with different sequence specificities to cleave polypeptides in well-defined ways. In Tools of Biochemistry 5B, we describe how these specific cleavage reactions can aid in determining the sequence of residues in a protein. Specific cleavage of peptide bonds can also be achieved using certain chemical reagents. The most useful of these is **cyanogen bromide** (Br-C≡N), which cleaves amide bonds on the C-terminal side of methionine residues.

If polypeptides are thermodynamically unstable (like polynucleotides), then how can they be synthesized in the aqueous medium of the cell? The unfavorable synthetic reaction is coupled to the favorable hydrolysis of certain organophosphate compounds. In fact, every amino acid must be activated by an ATP-driven reaction before it can be incorporated into proteins. The activation of amino acids for protein biosynthesis is discussed in detail in Chapter 25. General features of translation are described in Figure 4.34 and later in this chapter.

● **CONCEPT** Peptides can be cleaved at specific sites by proteolytic enzymes or by treatment with chemical agents such as cyanogen bromide.

Peptides

Chains containing only a few amino acid residues, like the tetrapeptide shown in **FIGURE 5.11**, are collectively referred to as **oligopeptides.** If the chain is longer (i.e., >~15–20 residues), then it is called a **polypeptide.** Polypeptides greater than ~50 residues are generally referred

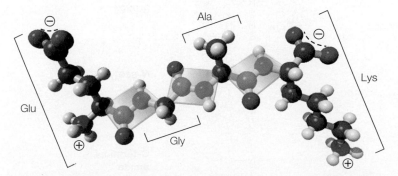

▲ **FIGURE 5.11 A tetrapeptide.** The tetrapeptide Glu–Gly–Ala–Lys (EGAK in one-letter code) is shown in ball-and-stick rendering. The planar peptide bonds that make up the main chain, or "backbone," are highlighted within shaded blue rectangles.

to as **proteins** (most globular proteins contain 250–600 amino acid residues). Figure 5.11 shows that most oligopeptides and polypeptides retain an unreacted amino group at one end (called the **amino terminus** or **N-terminus**) and an unreacted carboxylic acid group at the other end (the **carboxyl terminus** or **C-terminus**). Exceptions are certain small cyclic oligopeptides in which the N- and C-termini have been linked covalently. In addition, many proteins have N-termini blocked by *N*-formyl or *N*-acetyl groups, and a few have C-terminal carboxylates that have been modified to amides (**FIGURE 5.12**).

● **CONCEPT** Oligopeptides and polypeptides are formed by the polymerization of amino acids. The amino acids in oligopeptides and polypeptides are held together by peptide bonds. All proteins are polypeptides.

In writing the sequence of an oligopeptide or polypeptide, it becomes awkward to spell out all of the amino acid residue names. Therefore, biochemists usually write these sequences using either the three-letter abbreviations or the one-letter abbreviations given in Figure 5.3. For example, the oligopeptide shown in Figure 5.11 could be written as either

Glu-Gly-Ala-Lys

or

EGAK

The convention is to *always* write the N-terminal residue to the left and the C-terminal residue to the right. The need to follow this convention when reporting protein sequence arises because amino acids are asymmetric and there is only one free N-terminal amino group (and one free C-terminal carboxylate). Thus, the sequences represented as EGAK and KAGE are not equivalent structurally and could be expected to show very different functional properties in cells.

In the structure of a peptide, we distinguish between the side chains (i.e., the R groups in Figure 5.3) and the **main chain** (or peptide backbone), which is composed of the atoms that make up the peptide bonds—namely, the α-NH, the C_α, and the α-C=O groups of each amino acid residue in the peptide. The N-terminal amino group and C-terminal carboxylate are also part of the main chain.

As a self-test of your skill in recognizing side chains and main-chain atoms in a peptide, consider the following problem: What is the sequence of the following pentapeptide in one-letter code? Here, N atoms are shown in blue, O atoms are red, C atoms are green, and H atoms are white:

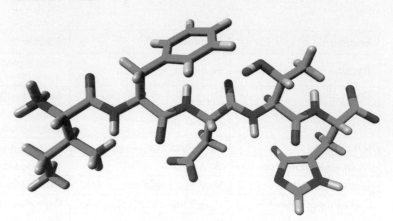

First, it is important to identify the N- and C-termini. Since they are part of the peptide main chain (or backbone), we will distinguish the main chain from the side chains by coloring main-chain C atoms in yellow:

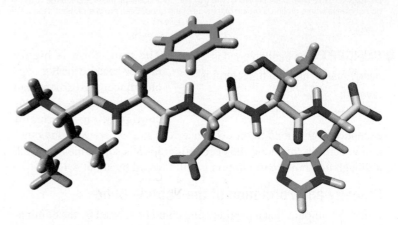

This makes the repeating pattern of $(\alpha$-NH)-$(\alpha$-C)-$(\alpha$-CO) more evident. Following the main chain to the left, we see it terminates in —NH$_3^+$; thus, the N-terminus is on the left. Likewise, following the main chain to the right, we see it terminates in —COO$^-$, thus, the C-terminus is to the right. Since the proper sequence is written starting with the N-terminal residue, we need to compare the side chains to those in Table 5.3 starting, in this case, with the leftmost residue. The N-terminal residue is isoleucine (I), and working to the right we can identify the peptide sequence as:

IFDTH

Polypeptides as Polyampholytes

In addition to the free amino group at the N-terminus and the free carboxylate group at the C-terminus, polypeptides usually contain some amino acids that have ionizable groups on their side chains. These various groups have a wide range of pK_a values, as shown in Table 5.2, but all are weakly acidic or basic groups. As a result, polypeptides are excellent examples of the polyampholytes described in Chapter 2. Recall that amino acid side chains display a range of pK_a values in different proteins due to differences in the local electrostatic

▲ **FIGURE 5.12 Groups that may block N- or C-termini in proteins.** Blocking of the N-terminus with a formyl or acetyl group is more common than modifying the C-terminus to an amide.

environment. For example, compared to free glutamic acid, a glutamic acid side chain near another negatively charged group (i.e., Glu or Asp) is *less* likely to lose a proton because the resulting unfavorable charge–charge repulsion disfavors proton dissociation (**FIGURE 5.13(b)**). Thus, the apparent pK_a of the glutamic acid *increases* (it becomes less acidic). If the glutamic acid is near a positively charged group (i.e., Lys, His, or Arg), then it is *more* likely to lose a proton because the resulting favorable charge–charge interaction favors proton dissociation (Figure 5.13(c)). Thus, the apparent pK_a of the glutamic acid *decreases* (it becomes more acidic).

The interactions between charged side chains in proteins form the basis for considerable interesting physiology. For example, the **Bohr effect** in hemoglobin results in the release of additional oxygen to tissues that have increased need for O_2 to support ATP production (e.g., for muscle contraction during exercise; described in Chapters 7 and 14). Thus, we can gain greater understanding of physiology by understanding the relationships between intracellular pH, side chain pK_a, and the overall charge on proteins.

> ● **CONNECTION** Favorable charge–charge interactions between amino acid side chains in the protein hemoglobin result in increased oxygen delivery to respiring tissues such as active muscles.

To illustrate how the ionization of side chains affects the molecular charge of a protein, consider the titration of Glu-Gly-Ala-Lys, the tetrapeptide first shown in Figure 5.11. We begin with the tetrapeptide in a very acidic solution with pH = 0. At this pH, which is below the pK_a of any of the groups present, all of the ionizable residues will be in their protonated forms:

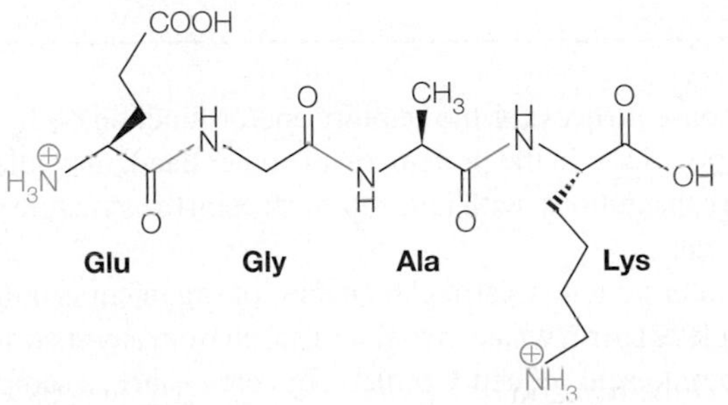

| Glu | Gly | Ala | Lys |

That is, both amino groups will be positively charged, and both carboxyl groups will have zero charge. As a result, the tetrapeptide has an overall charge of +2 at pH = 0. If we raise the pH of the solution (e.g., by titrating with NaOH), then the various ionizable groups will lose protons at pH values in the vicinity of their pK_a values. The progress of this titration is shown in **FIGURE 5.14(a)**. As the pH of the solution increases, more groups become deprotonated; thus, the net positive charge decreases and passes through zero at the isoelectric point (Figure 5.14(b)). As more base is added, the tetrapeptide becomes increasingly negatively charged, ultimately reaching a net charge of −2 at very high pH, because both amino groups are neutral and both carboxyl groups are negatively charged.

The acid–base properties of side chains are of considerable importance in biochemistry. For example, even a small shift in pH can alter the constellation of charges displayed on the surface, or in the active site, of a protein and will thereby significantly affect its stability and/or functional properties (see Figure 2.18). Also, the fact that

> ● **CONCEPT** Amino acids, peptides, and proteins are ampholytes; each has an isoelectric point.

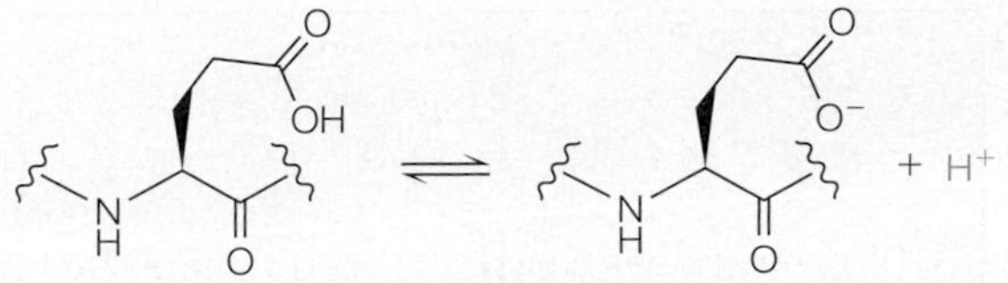

(a) The carboxylic acid proton on the side chain of free glutamic acid, or glutamic acid in an unperturbed electrostatic environment, will dissociate with a pK_a near 4.2.

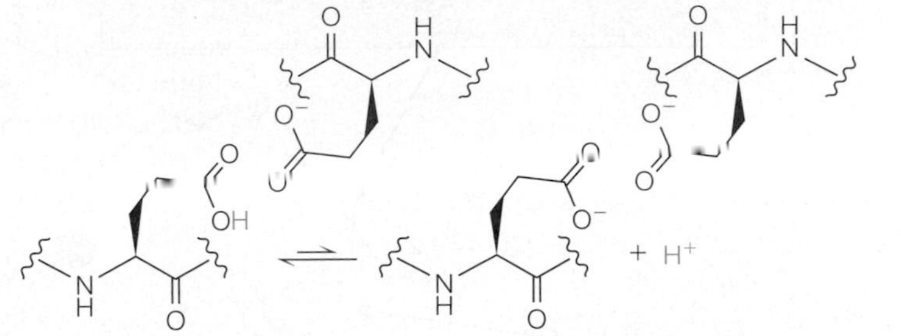

(b) When the side-chain environment is perturbed by the proximity of a (−) charged group, then proton dissociation is disfavored because deprotonation of the glutamic acid side chain results in unfavorable electrostatic interactions with the (−) charged group. The pK_a of this glutamic acid is expected to be > 4.2, since proton dissociation is less likely under these conditions.

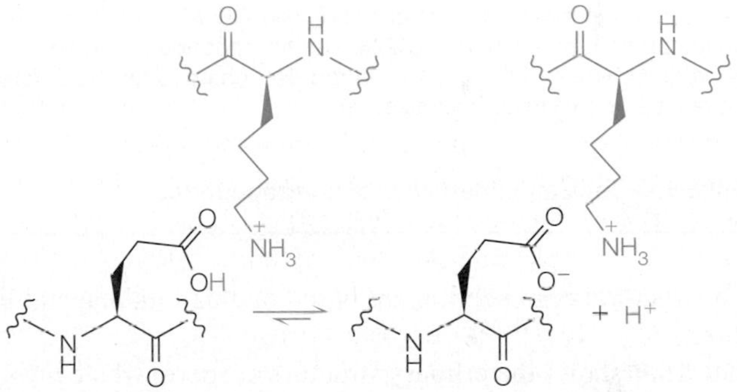

(c) When the side-chain environment is perturbed by the proximity of a (+) charged group, then proton dissociation is favored because deprotonation of the glutamic acid side chain results in favorable electrostatic interactions with the (+) charged group. The pK_a of this glutamic acid is expected to be < 4.2, since proton dissociation is more likely under these conditions.

▲ **FIGURE 5.13** The effect of the local electrostatic environment on side chain pK_a.

---

different proteins and oligopeptides have different net charges at a given pH (Figures 2.18 and 5.14(b)) can be used to separate them, either by electrophoresis (see Tools of Biochemistry 2A) or by ion exchange chromatography (see Tools of Biochemistry 5A).

5.3 Proteins: Polypeptides of Defined Sequence

Proteins are polypeptides of defined sequence. Every protein has a defined number and order of amino acid residues. As with the nucleic acids, this sequence is referred to as the *primary structure* of the protein. In later chapters we shall see that all higher levels of

> ● **CONCEPT** Every protein has a unique, defined amino acid sequence—its primary structure.

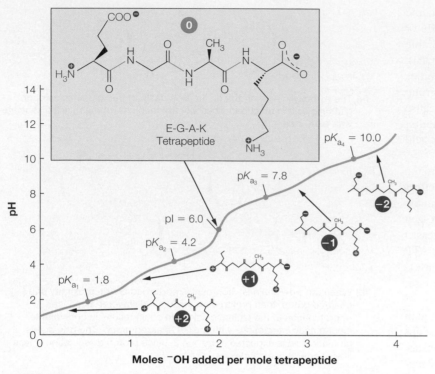

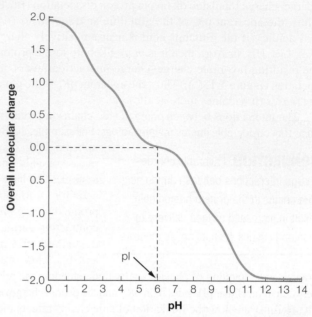

(a) This titration curve for the tetrapeptide Glu-Gly-Ala-Lys shows the major ionization states present as a function of pH. The tetrapeptide is shown schematically with (+) charges in blue and (−) charges in red. Net charges for the different ionization states are shown in solid circles.

(b) The net charge on Glu-Gly-Ala-Lys as a function of pH is not a step function. Rather, between pH 1 and 11, it changes gradually and smoothly from +2 to −2. At values of pH > pI, the overall charge on the tetrapeptide is negative. At values of pH < pI, the net charge is positive.

▲ **FIGURE 5.14** Ionization behavior of a tetrapeptide.

protein structural organization are based on this fundamental level of structure.

FIGURE 5.15 shows the primary structure of sperm whale *myoglobin,* the protein whose three-dimensional structure is shown in Figure 5.1. The primary structure of human myoglobin, the protein that serves the same O_2 binding function in humans, is also provided. The two amino acid sequences are aligned to show that they share a great deal of similarity. Comparing these two amino acid sequences highlights two important points about protein structures. First, proteins are *long* polypeptides. Sperm whale and human myoglobin both contain 153 amino acids, but these are among the smaller proteins found in nature. Some proteins have sequences extending for many hundreds or even thousands of amino acid residues. Second, although the two myoglobin sequences are similar, they are not identical. Their similarity is sufficient for each to serve the same biochemical function (which is why we call each a *myoglobin*). They are not quite the same, though, because many millions of years have passed since sperm whales and humans had a common evolutionary ancestor.

Over extended periods of time, populations evolve, and such evolution is reflected in changes in the amino acid sequences of proteins. Some of these are called **conservative** changes, because they conserve the chemical properties and/or size of the side chain (e.g., Asp for Glu). Other, **nonconservative** changes (e.g., Asp for Ala) may have more serious consequences. Structural context is of great importance in determining the effects of changing the identity of an amino acid at a given position in the sequence of the protein. If, on one hand, the amino acid side chain is in the interior of the protein, even a "conservative" substitution could have

dramatic consequences for the stability and/or function of the protein. A solvent exposed site in the protein, on the other hand, may tolerate many side-chain substitutions, with little or no effect on the structure or function of the protein.

A given protein in a particular species of organism is unique. Thus, every sample of sperm whale myoglobin, taken from any sperm whale, has the same amino acid sequence (unless, by rare chance, a sample is taken from a whale that carries a mutated myoglobin gene).

Biochemists have come to understand the complex structure of the myoglobin molecule by analyzing successively higher levels of complexity in the protein structure. To begin this kind of study of a protein, the protein must be prepared in a pure form, free of contamination by other proteins or other cellular substances. Methods for protein purification are described in Tools of Biochemistry 5A. Historically, the next step after purifying a protein would be to determine its amino acid composition—that is, the relative amounts of the different amino acids in the protein—and then the sequence of the amino acids in the polypeptide chain. These steps have been superseded by the routine mass determination of purified proteins by mass spectrometry. Modern mass spectrometers can also determine the amino acid sequence of a protein. These procedures are described in Tools of Biochemistry 5B.

● **CONCEPT** Some proteins contain two or more polypeptide chains held together by noncovalent interactions or by covalent disulfide bonds.

A complication arises in the sequence determination if the protein being studied contains more than one polypeptide chain. These chains may be held together by noncovalent

| | | Identical amino acids | | Conservative substitutions | | Nonconservative substitutions |

Number	1	2	3	4	5	6	7	8	9	10	11	12	13	14	15
Human	G	L	S	D	G	E	W	Q	L	V	L	N	V	W	G
Whale	V	L	S	E	G	E	W	Q	L	V	L	H	V	W	A

Number	16	17	18	19	20	21	22	23	24	25	26	27	28	29	30
Human	K	V	E	A	D	I	P	G	H	G	Q	E	V	L	I
Whale	K	V	E	A	D	V	A	G	H	G	Q	D	I	L	I

Number	31	32	33	34	35	36	37	38	39	40	41	42	43	44	45
Human	R	L	F	K	G	H	P	E	T	L	E	K	F	D	K
Whale	R	L	F	K	S	H	P	E	T	L	E	K	F	D	R

Number	46	47	48	49	50	51	52	53	54	55	56	57	58	59	60
Human	F	K	H	L	K	S	E	D	E	M	K	A	S	E	D
Whale	F	K	H	L	K	T	E	A	E	M	K	A	S	E	D

Number	61	62	63	64	65	66	67	68	69	70	71	72	73	74	75
Human	L	K	K	H	G	A	T	V	L	T	A	L	G	G	I
Whale	L	K	K	H	G	V	T	V	L	T	A	L	G	A	I

Number	76	77	78	79	80	81	82	83	84	85	86	87	88	89	90
Human	L	K	K	K	G	H	H	E	A	E	I	K	P	L	A
Whale	L	K	K	K	G	H	H	E	A	E	L	K	P	L	A

Number	91	92	93	94	95	96	97	98	99	100	101	102	103	104	105
Human	Q	S	H	A	T	K	H	K	I	P	V	K	Y	L	E
Whale	Q	S	H	A	T	K	H	K	I	P	I	K	Y	L	E

Number	106	107	108	109	110	111	112	113	114	115	116	117	118	119	120
Human	F	I	S	E	C	I	I	Q	V	L	Q	S	K	H	P
Whale	F	I	S	E	A	I	I	H	V	L	H	S	R	H	P

Number	121	122	123	124	125	126	127	128	129	130	131	132	133	134	135
Human	G	D	F	G	A	D	A	Q	G	A	M	N	K	A	L
Whale	G	N	F	G	A	D	A	Q	G	A	M	N	K	A	I

Number	136	137	138	139	140	141	142	143	144	145	146	147	148	149	150	151	152	153
Human	E	L	F	R	K	D	M	A	S	N	Y	K	E	L	G	F	Q	G
Whale	E	L	F	R	K	D	I	A	A	K	Y	K	E	L	G	Y	Q	G

▲ **FIGURE 5.15 The amino acid sequences of sperm whale myoglobin and human myoglobin.** Single-letter abbreviations are used here for the amino acids, and they are numbered beginning at the N-terminus. Of the 153 amino acid residues, 128 (84%) are identical in humans and whales. If we include the 13 conservative substitutions (e.g., isoleucine for leucine), then the two proteins are 92% similar.

interactions, as in the protein *hemoglobin,* which is made up of four myoglobin-like chains (see Chapter 7). Alternatively, covalent bonds such as disulfide bonds may link the chains in a protein that contains multiple polypeptides. An example is the hormone **insulin** (**FIGURE 5.16**). Tools of Biochemistry 6B describes methods for detecting such multi-chain proteins and how to separate them.

Although direct sequencing of purified proteins provided much of the early protein sequence information, biochemists are turning increasingly to gene sequencing for such information. As described in Chapter 4, the primary structure of every protein is dictated by a particular gene. Because we now know the code that relates DNA sequence to protein sequence, determination of the nucleotide sequence of a gene (or more often, the sequence of the mRNA transcribed from that gene) allows us to translate the nucleic acid sequence into the corresponding protein sequence. Keep in mind that gene sequencing tells us only the amino acid sequence of the protein *as synthesized by ribosomal translation.* There are often posttranslational modifications of the polypeptide chain that are not revealed in this way and can be found only by analyzing the protein purified from its natural source.

The techniques for identifying protein-encoding sequences in the genome and for retrieving, cloning, and sequencing the genes are discussed in Chapters 21–24 of this text (see also Tools of Biochemistry 4A).

5.4 From Gene to Protein

The Genetic Code

Recall from Chapter 4 that the DNA sequences of genes are transcribed into mRNA molecules (Figure 4.33), which are translated, in turn, into proteins (Figure 4.34). But there are only four kinds of nucleotides in DNA, each of which is transcribed to a particular nucleotide in mRNA, and there are 20 kinds of amino acids. A 1:1 correspondence between nucleotide and amino acid, then, is impossible. In fact, triplets of nucleotides (**codons**) are used to code for each amino acid, allowing 4^3, or 64, different combinations. This number is more than enough to code for 20 amino acids, so most amino acids have multiple codons. We can consider the information shown in **FIGURE 5.17** to be the "standard" genetic code because nearly all organisms use the same codons to translate their genomes into proteins. The few exceptions are scattered throughout the biological kingdoms. We discuss these and other details of the genetic code in Chapter 25.

● **CONCEPT** The genetic code specifies RNA nucleotide triplets that correspond to each amino acid residue.

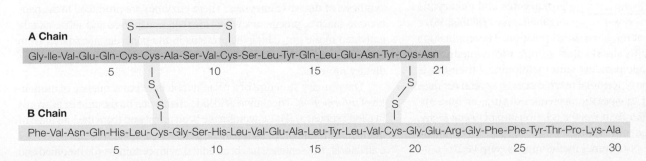

A Chain
Gly-Ile-Val-Glu-Gln-Cys-Cys-Ala-Ser-Val-Cys-Ser-Leu-Tyr-Gln-Leu-Glu-Asn-Tyr-Cys-Asn
5   10   15   21

B Chain
Phe-Val-Asn-Gln-His-Leu-Cys-Gly-Ser-His-Leu-Val-Glu-Ala-Leu-Tyr-Leu-Val-Cys-Gly-Glu-Arg-Gly-Phe-Phe-Tyr-Thr-Pro-Lys-Ala
5   10   15   20   25   30

◀ **FIGURE 5.16 The primary structure of bovine insulin.** This protein is composed of two polypeptide chains (A and B) joined by disulfide bonds. The A chain also contains an internal disulfide bond.

Base in second position

Base in first position (5′ end)	U	C	A	G	Base in third position (3′ end)
U	UUU Phe	UCU Ser	UAU Tyr	UGU Cys	U
U	UUC Phe	UCC Ser	UAC Tyr	UGC Cys	C
U	UUA Leu	UCA Ser	UAA Stop	UGA Stop	A
U	UUG Leu	UCG Ser	UAG Stop	UGG Trp	G
C	CUU Leu	CCU Pro	CAU His	CGU Arg	U
C	CUC Leu	CCC Pro	CAC His	CGC Arg	C
C	CUA Leu	CCA Pro	CAA Gln	CGA Arg	A
C	CUG Leu	CCG Pro	CAG Gln	CGG Arg	G
A	AUU Ile	ACU Thr	AAU Asn	AGU Ser	U
A	AUC Ile	ACC Thr	AAC Asn	AGC Ser	C
A	AUA Ile	ACA Thr	AAA Lys	AGA Arg	A
A	AUG Met/start	ACG Thr	AAG Lys	AGG Arg	G
G	GUU Val	GCU Ala	GAU Asp	GGU Gly	U
G	GUC Val	GCC Ala	GAC Asp	GGC Gly	C
G	GUA Val	GCA Ala	GAA Glu	GGA Gly	A
G	GUG Val	GCG Ala	GAG Glu	GGG Gly	G

▲ **FIGURE 5.17 The genetic code.** The table is arranged so that users can quickly find any amino acid from the three bases of the mRNA codon (written in the 5′ → 3′ direction). For example, to find the amino acid corresponding to the codon 5′ AUC 3′, look first in the A row, then in the U column, and finally in the C space, which indicates that the amino acid is Ile.

Figure 5.17 depicts the genetic code in terms of the mRNA triplets that correspond to the different amino acid residues. Three triplets—UAA, UAG, and UGA—typically do not code for any amino acids but serve as "stop" signals to end translation at the C-terminus of the chain. In rare cases, the codon UGA codes for selenocysteine and UAG codes for pyrrolysine. The codon AUG, which normally codes for methionine, also serves as a translation "start" signal that directs the placement of *N*-formylmethionine (in prokaryotes and eukaryotic organelles; see **FIGURE 5.18**) or methionine (in eukaryotes) at the N-terminal position. The implication is that all prokaryotic proteins should start with *N*-formylmethionine ("fMet"), and most eukaryotic proteins with methionine. Often this is true, though in many cases the N-terminal residue or even several residues will be removed in the cell by specific proteases during, or immediately after, translation. **FIGURE 5.19** shows the relationship between DNA, mRNA, and polypeptide sequences for the N-terminal portion of human myoglobin. In this case, the N-terminal methionine is removed.

● **CONCEPT** Translation of mRNA into a protein sequence begins with an AUG codon, which encodes Met in eukaryotes and fMet in prokaryotes and eukaryotic organelles.

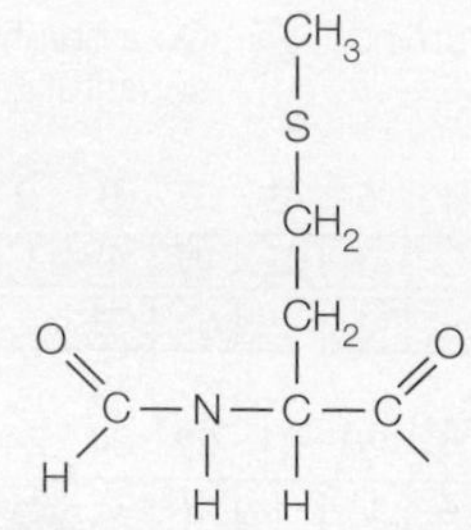

▲ **FIGURE 5.18** ***N*-Formylmethionine.** This amino acid residue initiates translation in prokaryotes and in eukaryotic organelles. It is coded by AUG when that triplet appears at the start of a protein coding region in a gene.

Posttranslational Processing of Proteins

When a polypeptide chain is released from a ribosome following translation, its synthesis is not necessarily finished. It must fold into its correct three-dimensional structure, and in some cases disulfide bonds must form. Certain amino acid residues may be acted upon by enzymes in the cell to produce, for example, the kinds of posttranslational modifications shown earlier in this chapter (page 117).

Many proteins are further modified by specific proteolytic cleavage, which shortens the chain length. A remarkable example is found in the synthesis of insulin (**FIGURE 5.20**). Insulin is a two-chain protein held together by disulfide bonds (Figure 5.16), but it is actually synthesized as a single, much longer polypeptide chain, called **preproinsulin,** step ❶. The residues at the N-terminus of the molecule (the exact number varies with the species) serve as a "signal peptide" (also called a **leader sequence**), which targets the preproinsulin molecule to cellular machinery that will transport it across the hydrophobic cell membranes. This transport is essential because insulin is a protein hormone that functions outside the cells in which it is synthesized. The leader sequence is then removed by a specific protease, leaving **proinsulin,** step ❷. Proinsulin folds into a specific three-dimensional structure, which helps it form the correct disulfide bonds, step ❸. The connecting sequence between the A chain and the B chain is then removed by further protease action, yielding the mature, active form, of the insulin molecule, step ❹. This process provides an important physiological advantage. Because proinsulin is not an active hormone, it can be produced and stored in the pancreas at high concentrations, whereas similar high levels of active insulin would be toxic. The proinsulin can then be converted to insulin by proteolysis to rapidly secrete active insulin when needed by the body. As described in Chapter 8, a similar strategy is used for the synthesis of digestive enzymes. These enzymes are produced in the pancreas as inactive precursors, which are then transported and subsequently activated in the gut. Thus, the proteins in the pancreas are spared from destruction by the digestive enzymes it produces to aid in the uptake of dietary proteins.

● **CONCEPT** After being translated from mRNA, a protein may be modified in many ways—including cleavage of specific peptide bonds.

The primary structure of a protein molecule is a sequence of atomic-level *information.* The amino acid side chains can be thought of as words in a long sentence. These words have been translated from the language of nucleic acid sequences stored in the genes and copied into mRNA. After translation, the sentence has been edited, with certain words modified and

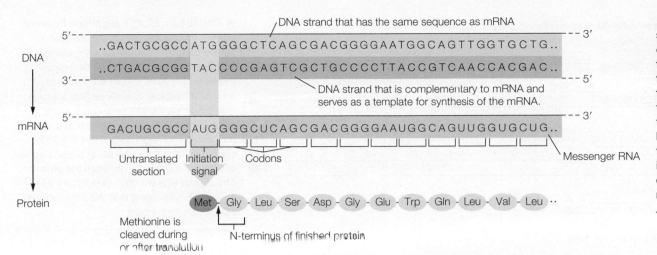

DNA strand that has the same sequence as mRNA

DNA

5′── ..GACTGCGCC ATG GGG CTC AGC GAC GGG GAA TGG CAG TTG GTG CTG.. ──3′

3′── ..CTGACGCGG TAC CCC GAG TCG CTG CCC CTT ACC GTC AAC CAC GAC.. ──5′

DNA strand that is complementary to mRNA and
serves as a template for synthesis of the mRNA.

mRNA

5′── GACUGCGCC AUG GGG CUC AGC GAC GGG GAA UGG CAG UUG GUG CUG.. ──3′

Untranslated   Initiation   Codons
section   signal

Messenger RNA

Protein

Met ─ Gly ─ Leu ─ Ser ─ Asp ─ Gly ─ Glu ─ Trp ─ Gln ─ Leu ─ Val ─ Leu ··

Methionine is
cleaved during
or after translation

N-terminus of finished protein

◀ **FIGURE 5.19 Relationships between DNA encoding a gene, the transcribed mRNA, and the translated polypeptide chain.** These relationships are shown for the first 12 residues of human myoglobin. Note that the DNA strand that is transcribed is the strand complementary to the final mRNA message.

others deleted in the posttranslational processing. In Chapter 6 we shall see that the information contained in the "sentence" of a protein sequence dictates how that protein folds in three dimensions. This folding, in turn, determines the function of the protein—how it interacts with small molecules and ions, with other proteins, and with substances like nucleic acids, carbohydrates, and lipids. The information expressed in protein sequences plays an essential role in determining how cells and organisms function.

5.5 From Gene Sequence to Protein Function

The development of **systems biology** has created a revolution in the molecular biosciences. The impact of systems biology on biochemistry is described in much greater detail in later chapters of this textbook; however, we briefly introduce the topic here because vast amounts of protein primary structure information have been generated as a result of the genomics projects that have spurred the rapid growth of systems biology.

Historically, biochemistry has been a discipline in which the functions and/or structures of highly purified macromolecules are studied in isolation. Systems biology is directed at understanding the functions of biomolecules in the context of intact cells where, for example, several thousand different proteins, nucleic acids, and smaller molecules can potentially interact and thereby affect cellular processes. Systems biology is a relatively new discipline that has grown out of the development of recent high-throughput gene sequencing and protein analysis technologies. Mapping cellular interaction networks is a complex task that relies on large-scale computation resources to manage information, as well as on laboratory instruments and techniques for generating large amounts of data.

To map possible interactions between biomolecules in a cell, it is necessary to know what molecules are present in the cell. For this reason, one of the earliest goals of systems biology was the sequencing of an entire genome for an organism. In 1995, a team led by J. Craig Venter at The Institute for Genome Research (TIGR) sequenced the first genome of a free-living organism, *Haemophilus influenzae* (1.8 million base pairs), and in 2001 two large teams, one led by Venter (then at the privately held company Celera) and the other by Francis Collins (at the National Institutes of Health), jointly announced the draft sequence of the human genome (3 billion base pairs). The sequencing of the human genome is one of the most significant achievements in science. Since then, an exponential growth in genome sequencing has occurred. As of November 2016, the Kyoto Encyclopedia of Genes and Genomes (KEGG) database had reported complete sequences for ~4200 microbial, ~320 viral, and ~350 eukaryotic genomes.

❶ Preproinsulin is synthesized as a random coil on membrane-associated ribosomes

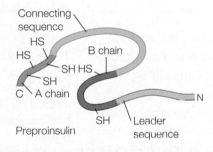

Connecting
sequence

HS
HS
B chain
SH HS
SH
C A chain
N

SH   Leader
      sequence

Preproinsulin

❷ After membrane transport, the leader sequence is cleaved and the resulting proinsulin folds into a stable conformation

❸ Disulfide bonds form

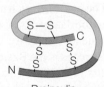

S—S
C
S   S
S   S
N

Proinsulin

❹ The connecting sequence is cleaved to form the mature insulin molecule

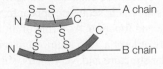

S—S   A chain
N   C
S   S
S   S   C
N   B chain

Insulin

▲ **FIGURE 5.20 Structure of preproinsulin and its conversion to insulin.**

Score = 30.8 bits (68), **Expect = 6e-06**, Method: Compositional matrix adjust.
Identities = 32/133 (25%), Positives = 48/133 (37%), Gaps = 40/133 (30%)

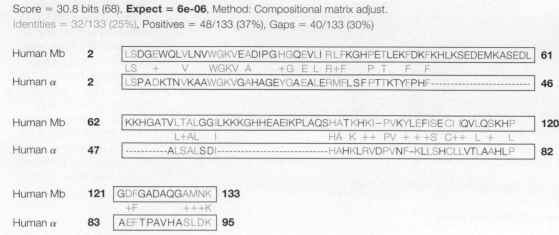

Human Mb 2 LSDGEWQLVLNVWGKVEADIPGHGQEVLI RLFKGHPETLEKFDKFKHLKSEDEMKASEDL 61
             LS +   V WGKV A   +G E LR+F   PT F   F
Human α   2 LSPADKTNVKAAWGKVGAHAGEYGAEALERMFLSFPTTKTYFPHF------------------------ 46

Human Mb 62 KKHGATVLTALGGILKKKGHHEAEIKPLAQSHATKHKI−PVKYLEFISECI IQVLQSKHP 120
              L+AL   I   HA K ++ PV + ++S C++ L +   L
Human α   47 -----------ALSALSDI--------------------HAHKLRVDPVNF−KLLSHCLLVTLAAHLP 82

Human Mb 121 GDFGADAQGAMNK 133
               +F   +++K
Human α   83 AEFTPAVHASLDK 95

◀ **FIGURE 5.21** BLAST alignment between human myoglobin and human α globin sequences. The sequence of myoglobin (Mb) appears above that for the α globin. Between them are shown the identical amino acids (blue) and those that are considered similar (green "+"). Gaps in the alignment are shown by red text and red dashes. In this alignment, there is a 25% sequence identity, suggesting a high degree of structural similarity between the proteins. The expect score in this case is very low (6×10^{-6}), indicating that the alignment is statistically significant.

Nearly all DNA sequences are deposited in public databases such as GenBank or KEGG, and researchers can use several web-based tools to search these databases (some of these are cited in Appendix II). In addition, the Protein Data Bank (PDB) is a repository of protein and nucleic acid structures.[†] Throughout this text you will see representations of biomolecular structures that have been generated from the data in the PDB. The exponential growth of gene sequence information has had a profound effect on many areas of biochemistry. One such area is the prediction of protein function from translated gene sequences.

Because the technology to determine DNA sequence has developed to the point where entire genomes can be determined in a relatively short period of time, there is tremendous interest in identifying how much of a given genome encodes functional properties. For example, how much of the human genome encodes metabolic function, or tissue differentiation and growth, or cell signaling?

To attempt to answer these questions, researchers can translate putative gene sequences into protein sequences, and then search protein sequence databases to find similar sequences. In most cases, proteins with a similar amino acid sequence possess similar structural and functional properties (e.g., the human and sperm whale myoglobins in Figure 5.15). Thus, it is reasonable to use sequence similarity to propose functional properties for an uncharacterized protein.

What does it mean for two proteins to have "similar" amino acid sequences? We must first distinguish **sequence identity** from **sequence similarity.** In this context, "identity" refers to those parts of the amino acid sequence that are an exact match (e.g., the amino acids highlighted in blue in Figure 5.15). The definition of "similarity" is less clear-cut, but, as discussed earlier in this chapter, it is based on the classification of the chemical properties of the side chains, such as hydrophobicity, polarity, and charge. The classification of some amino acids is ambiguous because they possess more than one of these properties. The side chain of lysine, for example, has a charged amino group at the end of four hydrophobic methylene groups. In one structural context in a particular protein, lysine might be best characterized as "hydrophobic," whereas in a different structural context in a different protein it would be considered "charged."

Once a definition of "similarity" has been adopted, how is sequence similarity assessed? Sequence similarity is based on finding the best **alignment** of the sequences that are to be compared. For example, imagine a process in which two (or more) sequences are aligned and an amino acid similarity score for the alignment is calculated based on some scoring rubric (e.g., a "high" score for a perfect amino acid match, a "medium" score for a chemically similar amino acid, and a "low" score for a chemically dissimilar amino acid). The registry of the alignment is then changed and the similarity score is calculated again. This iterative process continues until the alignment is found that gives the highest similarity score. The alignment of the two myoglobin sequences in Figure 5.15 gives the maximum value for sequence similarity. This example is rather straightforward given the high degree of identity between these two sequences. In many cases, it is necessary to allow gaps in the alignment to get the best overall alignment score.

● **CONCEPT** The degree of similarity between protein sequences is determined from a procedure called alignment.

● **CONCEPT** Primary sequence analysis can be used to predict protein function because similarities in aligned sequences are correlated with similarities in protein structure and function.

The alignment optimization process occurs many times during a protein database search. The output of such a search typically lists several hundred proteins from the database that are best matches based on a statistical analysis of sequence similarity. One widely used protocol for conducting database searches is the Basic Local Alignment Search Tool (BLAST). The results of a BLAST search can be used to infer potential function for an uncharacterized protein sequence, although any proposed function must be validated experimentally.

To summarize, given a newly discovered gene sequence, it is possible to make a best guess as to the function of the gene product based on a search of protein sequence databases using BLAST. Statistically significant similarities in protein sequence appear to be correlated with similarities in structure and function. As a rule of thumb, if two aligned protein sequences share at least a 25% amino acid sequence identity, then they will have similar structure and, very likely, similar function. An example of a BLAST alignment between human myoglobin and human α globin is shown in **FIGURE 5.21.** The statistical significance of the sequence alignment is indicated by the "expect" score (top line in Figure 5.21), which predicts

---

[†] As of November 2016, there were more than 39,000 distinct protein structures deposited in the PDB and over 220 billion nucleotide bases from more than 190 million individual sequences (obtained from more than 300,000 different organisms) deposited in GenBank (about 11 million of the sequences in GenBank are of human origin).

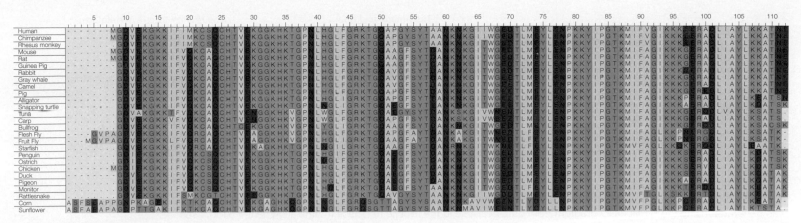

(a) Alignment of cytochrome *c* sequences from 27 organisms, where hydrophobic amino acids are highlighted in gray, basic amino acids in blue, acidic amino acids in red, and polar uncharged amino acids in green (except for Asn and Gln, which are magenta).

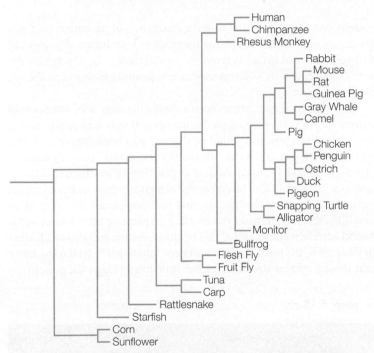

(b) A phylogenetic tree for the sequences shown in (a). Branches indicate points of evolutionary divergence based on differences in the amino acid sequences of the aligned proteins.

▲ **FIGURE 5.22 Sequence alignment and a phylogenetic tree for cytochromes *c* from different organisms.** Both the alignment and the phylogenetic tree were created using CLUSTALW2 and CINEMA (both available via the ExPASY website; see Appendix II).

---

the basis for so-called personalized medicine. Some controversy exists around whether or not it is desirable to have access to such detailed genetic information. Whereas it could be invaluable in the treatment of individual patients, it could also lead to discriminatory practices and stigmatization.

5.6 Protein Sequence Homology

Protein sequence similarity is also used to map evolutionary relationships between organisms. Two organisms that have a common evolutionary ancestry are likely to have gene sequences, and therefore protein sequences, that are related. Protein sequences are classified as homologous in cases where any sequence similarity is thought to be the result of a common evolutionary ancestry. Sequence similarity can also arise without common ancestry via convergent evolution (e.g., two proteins may have evolved independently to bind to the same peptide sequence found on a particular transcription factor). Two such protein sequences would be *similar* but not *homologous*. Hence, "sequence similarity," which is based on sequence alignment, is distinct from "sequence homology," which is used only to indicate an evolutionary relationship between two sequences.

For organisms that have a common ancestor, a greater degree of protein sequence similarity indicates a closer evolutionary relationship, whereas a lesser degree of similarity indicates greater divergence. **FIGURE 5.22(a)** shows an alignment of lesser degree of similarity sequences for cytochromes *c* from 27 different organisms. In this figure, the hydrophobic (gray), positively charged (blue), negatively charged (red), and polar uncharged (green and magenta) amino acids are highlighted to illustrate how these similar amino acid properties cluster in this alignment. Figure 5.22(b) shows a phylogenetic tree for the sequences, where branches show points

● **CONCEPT** Protein sequence analysis can be used to propose evolutionary relationships between organisms.

---

the likelihood that the alignment is due to chance. An expect score close to one indicates that the match is possible due to random chance, whereas a score closer to zero indicates the match is less likely to be random.

Given recent advancements in genome sequencing technology, determining genome sequences for individuals may become routine. Such information would make it possible to identify gene sequences associated with specific diseases or resistance to certain treatments. For example, some cancer treatments target a specific receptor that is expressed on the surface of the cancer cells. If a patient lacks the gene for that receptor, the treatment will be ineffective. This is

● **CONNECTION** Gene sequence analysis can be used to identify protein variants associated with human disease or resistance to certain treatments.

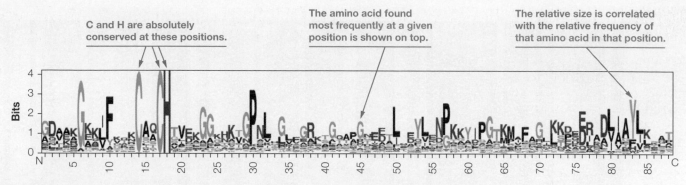

C and H are absolutely conserved at these positions.

The amino acid found most frequently at a given position is shown on top.

The relative size is correlated with the relative frequency of that amino acid in that position.

▲ **FIGURE 5.23** **Sequence logo from the alignment of 412 sequences from the cytochrome c family.** The sequence logo was generated using the WEBLOGO program from data available in the Prosite database (matrix alignment # 51007; see the ExPASY website).

of proposed evolutionary divergence based on differences in the amino acid sequences of the aligned proteins. The construction of phylogenetic trees involves complex analyses of gene mutation rates and genetic polymorphism within populations—topics beyond the scope of this introduction to protein sequence analysis. The important point for this discussion is that protein sequence analysis can be used to propose evolutionary relationships between organisms, such as those shown in Figure 5.22(b).

Finally, when several homologous protein sequences are aligned, then the so-called **consensus sequence** can be determined. A common representation of the consensus sequence lists the amino acid that is most frequently found at a given position in the sequence, but this representation can be an oversimplification. If a certain amino acid is found at a specific position in 80% of the aligned sequences, that means that 20%, a significant percentage, of the sequences do not have that amino acid at that position. A more information-rich representation is the **sequence logo** in which the type size used to represent the one-letter

code is correlated with the relative frequency of an amino acid at a given position. **FIGURE 5.23** shows a sequence logo for an alignment of 412 sequences of proteins from the cytochrome c family, where the amino acid found most frequently at any position is shown at the top of each column.

The sequence logo more clearly shows the degree of **amino acid conservation** within the aligned sequences. If only one amino acid is found at a given position within an alignment of homologous proteins, that amino acid is said to be "absolutely conserved" (e.g., cysteine at positions 14 and 17, and histidine at position 18 in Figure 5.23). An amino acid type can also be conserved at a given position. For example, only leucine, valine, isoleucine, and methionine are found at position 80 in the 412 aligned cytochrome c sequences; thus, hydrophobic amino acids are conserved at this position. As we will discuss further in Chapter 6, the most highly conserved amino acids tend to be those that serve a critical structural and/or functional role in the protein.

Summary

- Proteins are polymers of L-α-amino acids. Oligopeptides and polypeptides are produced by polymerization of amino acids via peptide bond formation. The peptide bond is nearly planar, and the trans form is favored. This bond is metastable and can be readily hydrolyzed in the presence of catalysts. Twenty common amino acids and two rare amino acids are coded for in genes and incorporated into proteins by the process of translation. Other, less common, amino acid structures are found in nature (e.g., in bacterial cell walls, antibiotics, and venoms). The variety of side chains—hydrophilic, hydrophobic, acidic, basic, or neutral—allows much functional complexity in proteins. Additional variation is made possible by posttranslational modification of some amino acids. The presence of both positive and negative charges on side chains makes proteins polyampholytes. (Sections 5.1–5.3)

- The unique, defined sequence of amino acids in each protein constitutes its primary structure and is dictated by its gene. Some proteins contain more than one polypeptide chain, held together by either covalent (disulfide) bonds or noncovalent interactions. Proteins are synthesized in the cell by an ATP-dependent process called translation. The genetic code is made up of a standard set of codons of three nucleotides, each of which specifies a particular amino acid. Specific "start" and "stop" triplets in the genetic code specify the chain length for a particular protein. (Sections 5.4–5.5)

- Genome sequencing has generated vast amounts of gene sequence information. The functions of most of these genes are unknown, but sequence similarity analysis can be used to predict the function, and in some cases the structure, of a gene product. (Section 5.6)

Problems

Enhanced by
Mastering Chemistry
for Biochemistry

Mastering Chemistry for Biochemistry provides select end-of-chapter problems and feedback-enriched tutorial problems, animations, and interactive figures to deepen your understanding of complex topics while practicing problem solving.

Answers to red problems are available in the Answer Appendix.

Note that some of these problems refer to information presented in the Tools of Biochemistry 5A and 5B.

1. Using the data in Table 5.1, calculate the *average* amino acid residue weight in a protein of typical composition. This is a useful number to know for approximate calculations.

2. Draw the structure of the peptide DTLH, showing the backbone and side-chain atoms in the ionization states favored at pH = 7.0.
   (a) Draw a water molecule making a hydrogen bond to a side-chain H-bond *donor*.
   (b) Draw a water molecule making a hydrogen bond to a main-chain H-bond *acceptor*.
   (c) Using the values of pK_as given in Table 5.1, calculate the pI for DTLH.

3. Identify each amino shown below in Fischer projection. Indicate whether the D- or L-enantiomer is shown.

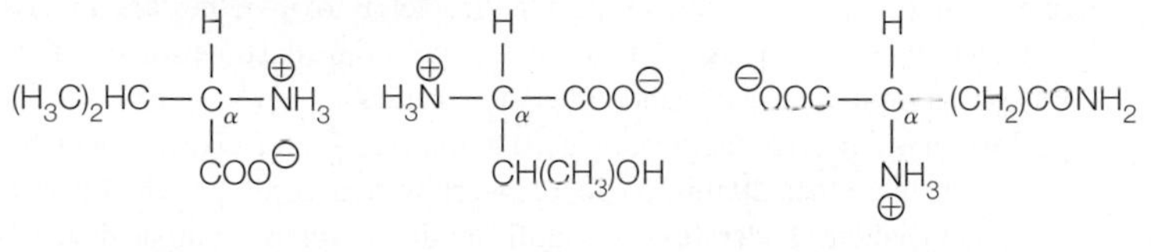

$(H_3C)_2HC-\overset{\overset{\displaystyle H}{|}}{\underset{\underset{\displaystyle COO^\ominus}{|}}{C_\alpha}}-\overset{\oplus}{N}H_3$ $H_3\overset{\oplus}{N}-\overset{\overset{\displaystyle H}{|}}{\underset{\underset{\displaystyle CH(CH_3)OH}{|}}{C_\alpha}}-COO^\ominus$ $^\ominus OOC-\overset{\overset{\displaystyle H}{|}}{\underset{\underset{\displaystyle \overset{\oplus}{N}H_3}{|}}{C_\alpha}}-(CH_2)CONH_2$

   (a)   **(b)**   **(c)**

4. The melanocyte-stimulating peptide hormone α-*melanotropin* has the following sequence:

   Ser–Tyr–Ser–Met–Glu–His–Phe–Arg–Trp–Gly–Lys–Pro–Val

   (a) Write the sequence using the one-letter abbreviations.
   (b) Calculate the molecular weight of α-melanotropin, using data in Table 5.1.

5. (a) Sketch the titration curve you would expect for α-melanotropin (Problem 4). Assume the pK_as of the N- and C-termini are 7.9 and 3.8, respectively. For side chains, assume the pK_a values given in Table 5.1.
   (b) Calculate to three decimal places the charge on α-melanotropin at pH values of 11, 5, and 1.
   (c) Calculate the pI (isoelectric point) of α-melanotropin. For parts (b) and (c) of this problem, refer to the logic used to solve Problem 2.15. You may find a spreadsheet useful to solve part (c).

6. What peptides are expected to be produced when α-melanotropin (Problem 4) is cleaved by (a) trypsin, (b) cyanogen bromide, or (c) thermolysin? (Refer to Table 5.3.)

7. There is another melanocyte-stimulating hormone called β-*melanotropin*. Cleavage of β-melanotropin with trypsin produces the following peptides plus free aspartic acid.

   WGSPPK   DSGPYK   MEHFR

   If you assume maximum sequence similarity between α-melanotropin and β-melanotropin, then what must the sequence of the latter be?

8. Given the following peptide

   SEPIMAPVEYPK

   (a) Estimate the net charge at pH 7 and at pH 12. Assume the pK_a values given in Table 5.1.

   (b) How many peptides would result if this peptide were treated with (1) cyanogen bromide, (2) trypsin, or (3) chymotrypsin?
   (c) Suggest a method for separating the peptides produced by chymotrypsin treatment.

9. A mutant form of polypeptide hormone angiotensin II has the amino acid *composition*

   (Asp, Arg, Ile, Met, Phe, Pro, Tyr, Val)

   The following observations are made:
   • Trypsin yields a dipeptide containing Asp and Arg and a hexapeptide with all the rest.
   • Cyanogen bromide cleavage yields a dipeptide containing Phe and Pro and a hexapeptide containing all the others.
   • Chymotrypsin cleaves the hormone into two tetrapeptides, of composition

   (Asp, Arg, Tyr, Val) and (Ile, Met, Phe, Pro)

   • The dipeptide of composition (Pro, Phe) cannot be cleaved by either chymotrypsin or carboxypeptidase.

   What is the sequence of angiotensin II?

10. A protein has been sequenced after cleavage of disulfide bonds. The protein is known to contain 3 Cys residues, located as shown here. Only one of the Cys has a free —SH group, and the other two are involved in an —S—S— bond.

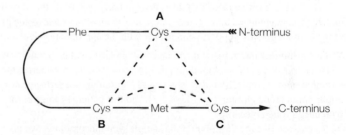

   The only methionine and the only aromatic amino acid (Phe) in this protein are in the positions indicated. Cleavage of the *intact* protein (i.e., with disulfide bonds intact) by either cyanogen bromide or chymotrypsin does *not* break the protein into two peptides. Where is the —S—S— bond (i.e., AB, BC, or AC)?

11. *Apamine* is a small protein toxin present in the venom of the honeybee. It has the sequence

   CNCKAPETALCARRCQQH

   (a) If apamine does not react with iodoacetate (see Tools of Biochemistry 5B), then how many disulfide bonds are present?
   (b) If trypsin cleavage gave two peptides, then where is(are) the S—S bond(s)?

12. (a) Write a possible sequence for an mRNA segment coding for apamine.
   (b) Do you think apamine is synthesized in the form shown in Problem 11, or is it more likely a product of proteolytic cleavage of a larger peptide? Explain.

13. Assume the following portion of an mRNA. Find a start signal, and write the amino acid sequence that is coded for.

   5′...GCCAUGUUUCCGAGUUAUCCCAAAGAUAAAAAAGAG...3′

14. Suppose you had separated the A and B chains of insulin by disulfide reduction. What chromatographic method should make it possible to isolate pure A and B chains? Explain your choice of separation method. You will find some useful information in Tools of Biochemistry 5A.

15. Acetylating agents such as acetic anhydride react preferentially with primary amines, iodoacetate reacts preferentially with sulfhydryl groups (see Tools of Biochemistry 5B), and ATP-dependent kinases preferentially add a phosphoryl group to side-chain hydroxyl or phenolic —OH groups. Which amino acid side chains, or main-chain groups, in a polypeptide are most likely to be modified by treatment with:
    (a) acetic anhydride
    (b) iodoacetate
    (c) a kinase + ATP

16. Consider the chemical modifications proposed in Problem 15. For each modification, consider:
    (a) Would the chemical modification change the overall charge on the peptide at pH = 7?
    (b) If so, is the pI of the modified peptide higher or lower than that of the untreated peptide?

17. Sickle-cell disease is caused by a so-called point mutation in the human β-globin gene. A point mutation is the result of a single base substitution in the DNA encoding a gene. The sickle-cell mutation results in the substitution of Val for Glu at position 6 in the β-globin protein.
    (a) Using the information in Figure 5.17, explain how a point mutation could change a codon for Glu to a codon for Val.
    (b) Do you expect the pI for the sickle-cell β-globin to be higher or lower than the pI for wild-type β-globin? Explain.

18. You have discovered a novel protein that has a pI = 5.5. To study the functional properties of this new protein, your research group has made a mutant that contains two amino acid changes—namely, a surface Phe residue in the normal protein has been replaced by His (side chain pK_a = 6.1) and a surface Gln has been replaced by Glu (side chain pK_a = 6.0). Is the pI of the mutant protein predicted to *be greater than, less than,* or *the same as* the pI of the normal protein? Support your answer with the appropriate calculation.

19. You are a summer intern in a clinical hematology lab. The lab director gives you a sample of a patient's blood proteins and asks you to characterize the thrombin in the sample. She also tells you that thrombin is a serine protease important in blood clotting (see Table 5.3), and this patient is a newborn with uncontrolled bleeding.
    (a) To characterize the thrombin in the sample, you must remove two proteins that interfere with the thrombin activity assay: cytochrome *c* and lactoglobin. You find some CM-cellulose (see Figure 5A.5) and a phosphate buffer (pH = 6.4) on the shelf in your lab. You decide to load the protein sample onto a column of CM-cellulose equilibrated in the pH = 6.4 buffer. Predict the order of elution for the three proteins shown in the accompanying table. At pH = 6.4, which protein(s) do you predict will remain bound to the column?

Protein	pl
Cytochrome *c*	10.6
Lactoglobin	5.2
Thrombin (wild type)	7.1

    (b) List two different ways you could change the buffer to elute the bound protein(s) and achieve proper separation of the proteins.
    (c) You are surprised to observe that the patient's thrombin flows through the CM-cellulose column at pH = 6.4 and does not bind. Confident in your technique, you suspect the patient's thrombin is different from wild-type thrombin. Using a different buffer system, you manage to purify some of the patient's thrombin, and you submit the purified sample for amino acid sequencing. The sequence analysis shows that the patient's thrombin contains a mutation in the enzyme active site. A lysine residue in the wild type has been mutated to an asparagine in the patient's thrombin. Does this mutation explain the anomalous CM-cellulose binding behavior you observed?
    (d) How many nucleotide changes are required at the level of the gene for a lysine to asparagine mutation? Do you think Lys-to-Asn (or Asn-to-Lys) is likely to be a frequently observed mutation in proteins?
    (e) Based on their side-chain structures, compare and contrast the potential of Lys and Asn to form noncovalent interactions. In other words, can each form H bonds and/or salt bridges and/or van der Waals contacts?

20. Suppose you performed two BLAST searches, one on the peptide sequence FIDPWE and another on the sequence KRTIAAVNSPLLEVATY. Which search do you predict will give you alignments with "expect scores" closer to one? Explain your reasoning.

21. Despite the fact that many peptides have critical physiological functions (e.g., as hormones, neurotransmitters, antibacterial toxins), they are not considered to be ideal as drugs. The following questions illustrate some of the issues that limit the use of peptides as therapeutics.
    (a) Insulin is a peptide therapeutic used to manage Type 1 diabetes, which affects more than 20 million people worldwide according to the International Diabetes Federation. A significant limitation to the broad distribution and use of insulin to treat Type 1 diabetes is the fact that it must be administered by injection rather than orally. Why is insulin administered by injection and not orally?
    (b) Many of the D-amino acids found in nature have been discovered in bacterially produced peptides that have antibiotic properties. Bacteria secrete these peptides into their environments to kill competitor bacteria and thereby gain a selective advantage. Given your answer to part (a) of this question, what potential advantages might D-amino acids confer to a secreted peptide toxin?
    (c) As a class of biomolecule, peptides have low membrane permeability (i.e., few peptides readily cross the membrane bilayer). This limits most peptide-based drugs to targets that are on the surfaces of cells (rather than in the cytoplasm or the nucleus). Review the information in Figures 2.15 and 2.16 and propose a reason that most peptides are not likely to cross the membrane bilayer.
    (d) Does Figure 5.12 suggest a strategy for increasing the membrane permeability of peptides?

22. Based on the information in Figure 5.17, which single-nucleotide mutation event is more likely: Arg-to-His, or Arg-to-Ser? Explain.

23. If you want to purify a DNA-binding protein from a crude mixture of proteins at pH 7, should you use a DEAE-cellulose or a CM-cellulose column (see Figure 5.A5)? Briefly explain your reasoning.

References

For a list of references related to this chapter, see Appendix II.

Much of the information presented in this text was gained through the study of highly purified protein molecules. To determine the structure and/or functional properties of a specific protein, it is necessary to separate that protein from the other biomolecules in the cell, which include lipids, nucleic acids, saccharides, and all other proteins. Typically, the protein of interest is a minor component of this mixture, so the isolation of that protein from several hundred other proteins presents a challenge. Modern methods of gene expression and protein purification have simplified the problem by increasing the concentration of the desired protein within the cell and exploiting specific interactions between the protein and materials used in the purification process.

Recombinant Protein Expression

To begin, let's consider the problem of protein abundance. For example, a typical enzyme may represent only 0.01% of the soluble protein in a cell. Thus, a 10,000-fold enrichment is necessary to purify that protein to homogeneity. If recombinant DNA technology (see Tools of Biochemistry 4A) can increase the intracellular abundance of that protein to 1%, then only a 100-fold enrichment is required; if the increase is to 10%, then a 10-fold enrichment will suffice. Historically, proteins were purified from natural sources, such as animal or plant tissues. The first proteins to be studied in detail were those with high abundance in particular tissues (e.g., hemoglobin in red blood cells). For proteins that are present in low concentration in their natural tissues, it is necessary to harvest large amounts of the tissue to isolate a useful amount of the desired protein. This unfortunate scenario was a fact of life for biochemists for many years until the late 1970s and early 1980s when the tools of recombinant protein expression were developed and widely adopted. Recombinant protein expression allows researchers to produce proteins of interest at relatively high concentrations within cells and also enables the production of so-called **site-directed mutants,** which are protein variants with designed amino acid sequence alterations (see Tools of Biochemistry 4A). Frequently, the mutant proteins exhibit changes in structure, function, and/or stability relative to the "wild-type" (i.e., naturally occurring) protein; thus, the mutants provide great insight into the relationships between protein structure and function.

Another important feature of recombinant technology is the ability to express a wide variety of foreign proteins in host cells. For example, many proteins of animal or plant origin have been successfully expressed in *Escherichia coli* cells. Because *E. coli* can be easily programmed to produce foreign proteins and the recombinant bacteria grow quickly compared to most plants and animals, *E. coli* cells can be viewed as convenient "factories" for protein production; however, *E. coli* cells are limited in the types of posttranslational modifications they carry out. It is usually necessary to express proteins in eukaryotic systems if some posttranslational modification (e.g., glycosylation) is required for activity.

Recombinant protein expression technology is based on the observation that the amino acid sequence for a protein is determined by the sequence of the DNA in the gene that encodes that protein, as described

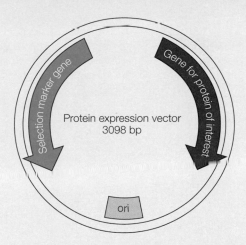

▲ **FIGURE 5A.1 Schematic representation of a generic protein expression vector.** The circle represents the double-stranded DNA sequence for the entire vector. The box marked "ori" is the "origin of replication," which determines how many copies of the vector will be made in the cell. The arrows represent the locations of protein genes in the vector DNA sequence. This vector will express two proteins: the recombinant protein of interest (red arrow) and a so-called selection marker (green arrow), which allows the bacterium to survive in a growth medium that contains antibiotic.

in Chapters 4 and 5 (see Figures 4.34 and Figure 5.17) . In theory, any protein sequence can be expressed in a cell that contains a copy of the gene encoding that protein. *E. coli* can be made to take up small circular DNA molecules, called expression vectors, that are on the order of 2–10 kilobase pairs in length. An expression vector is a modified form of a natural extrachromosomal DNA, such as a plasmid, which is capable of autonomous replication in a bacterial cell (similar technology exists to express proteins in yeast and other eukaryotic cells). Recombinant DNA technology allows a researcher to cleave that plasmid at a desired site and splice in a gene encoding the protein of interest. As shown in **FIGURE 5A.1**, the gene encoding the wild-type or mutant protein to be expressed is within each vector, along with a gene encoding a so-called **selection marker.** The selection marker is usually a protein that confers resistance to a toxic antibiotic that is included in the cell growth medium; thus, only those cells that have taken up the vector, and are thereby capable of expressing the desired protein, will be resistant and survive in the growth medium. With many tens to hundreds of copies of the vector in each cell, production of the desired protein is maximized. Using this approach, we find that even those proteins that occur in low intracellular concentrations in nature can be produced in sufficient yield in *E. coli* to allow biochemical characterization and/or commercial production.

The Purification Process

Although recombinant technology can increase the concentration of a specific protein inside the cell, the problem of separating the desired protein from all the other cellular components remains. The sequence

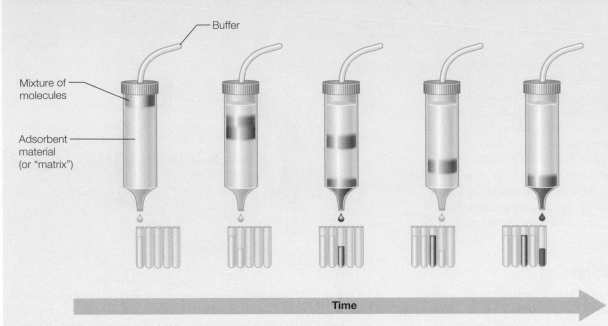

Buffer

Mixture of molecules

Adsorbent material (or "matrix")

Time

◄ **FIGURE 5A.2** The principle of column chromatography. A mixture of proteins is separated as a result of differential interactions with the column matrix. The more a protein interacts with the matrix, the later it will elute from the column.

of steps taken to purify a given protein will be unique to that protein because proteins vary in chemical properties; however, many features of the purification process are common. For example, most purification steps begin with lysis, or rupture, of the cells. Cell lysis can be achieved by sonic disruption ("sonication"), by mechanical rupture using a homogenizer, or by enzymatic digestion of the cell wall. This is followed by centrifugation to remove unbroken cells and insoluble cell parts (e.g., membranes) to yield an extract, called the cell lysate, which contains the soluble proteins and other biomolecules in a cell. The desired protein is then purified from the other proteins in the lysate by one or more of the following commonly used steps: (1) affinity chromatography, (2) ion exchange chromatography, or (3) size exclusion chromatography. As illustrated in **FIGURE 5A.2**, purification of the desired protein by chromatography is the result of differential interactions between the various proteins in the mixture loaded onto the chromatography column and the adsorbent material, or "matrix," within the column. In general, the more strongly some protein interacts with the matrix, the later it will elute from the column. Proteins are generally detected by UV absorbance at 280 nm (λ_{max} for Trp and Tyr), or 220 nm (amide bond absorbance), as they elute from the column.

Affinity Chromatography

Affinity chromatography relies on selective adsorption of a protein to a natural or synthetic ligand, typically a substrate or inhibitor, which is immobilized by covalent attachment to an inert solid support. The support that displays the bound ligand is called the affinity matrix. The interactions between the desired protein and the affinity matrix are expected to be highly specific; thus, most of the contaminants in the mixture will not interact with the affinity matrix, whereas the desired protein is expected to bind tightly. As shown in **FIGURE 5A.3**, when a complex mixture flows through the affinity matrix, the desired protein binds tightly and remains bound until most contaminants are washed

through the column. The bound protein can then be eluted using a variety of methods that preserve the structure and activity of the protein (for example, addition of the free ligand to the elution buffer). In some exceptional cases, the elution methods require extreme conditions to disrupt the bonds between the protein and the matrix, including denaturation of the protein.

Specific, complementary, noncovalent interactions are the basis for much interesting biological chemistry, including antibody binding, enzyme–substrate recognition, enzymatic catalysis, gene regulation, cell signaling, and muscle contraction, just to name a few examples. Early in the development of affinity purification methods, many investigators immobilized the natural binding targets ("ligands") for a protein on a solid chromatography support to take advantage of such specific binding interactions. A related technique, called immunoaffinity chromatography, takes advantage of the high binding specificity of antibodies for their ligands. Antibodies raised against a given protein can be covalently attached to a solid support to make an affinity matrix that binds selectively and reversibly to that protein. Although immunoaffinity columns are efficient, the costs and time required to produce the antibodies are significant; thus, this method has largely given way to faster and less expensive techniques.

One of the most prevalent affinity methods, called immobilized metal affinity chromatography (IMAC), takes advantage of the strong interactions between a Ni^{2+}, Zn^{2+}, or Co^{2+} ion and a string of six sequential histidine residues. The amino acid sequence $(His)_6$ is called a hexahistidine-tag (or "His-tag"), and it is quite rare in nature; thus, few naturally occurring proteins will bind to the IMAC matrix. Using recombinant DNA technology, one can append the His-tag to the gene encoding the desired protein, as shown in **FIGURE 5A.4**. When this protein is expressed, it will include the His-tag sequence, so it will bind tightly to the IMAC matrix. The bound protein can be eluted from the IMAC column with a buffer containing imidazole (an analog of the His side chain), or a low pH buffer (which protonates the His side chains

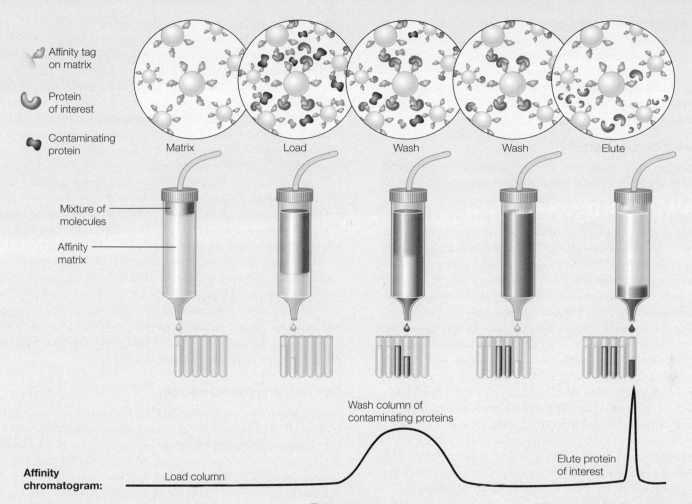

| Matrix | Load | Wash | Wash | Elute |

Affinity tag on matrix

Protein of interest

Contaminating protein

Mixture of molecules

Affinity matrix

Affinity chromatogram:

Load column

Wash column of contaminating proteins

Elute protein of interest

The proteins that don't bind the affinity matrix are washed off and elute early, giving rise to a large protein absorbance (broad peak).

After the contaminants are washed off, the desired protein is eluted, giving a smaller protein absorbance (narrow peak).

▲ **FIGURE 5A.3 An overview of affinity chromatography.** Top: Specific binding of the desired protein (blue shapes) to the affinity matrix is shown. The contaminating proteins (red) wash through the column without binding, resulting in a significant purification of the desired protein. The bound protein can then be eluted by any one of several methods discussed in the text. Bottom: A schematic representation of an affinity chromatogram, where each peak represents protein absorbance at 280 nm.

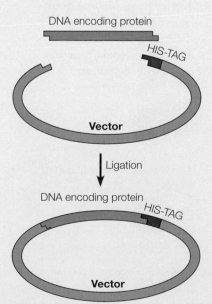

DNA encoding protein

HIS-TAG

Vector

Ligation

DNA encoding protein

HIS-TAG

Vector

◄ **FIGURE 5A.4** Insertion of a protein gene into one of several commercially available expression vectors results in addition of the affinity His-tag (i.e., a sequence of six His residues) to the protein sequence.

and reduces metal ion binding), or with a buffer containing the metal chelator EDTA. EDTA effectively removes the metal ion from the column and thereby disrupts the bonding interactions between the protein and the affinity matrix.

The His-tag affinity method just described is used extensively to produce purified proteins for biochemical characterization. This example illustrates the power of recombinant technology not only for increasing protein concentration in cells, but also for improving the protein-purification process. One possible limitation of this method is that the resulting protein sequence carries some modification compared to the wild type (in this case, an extra six histidine amino acids), but certain expression vectors are designed to allow removal of the His-tag by treatment with specific proteases (e.g., thrombin; see Table 5.3).

Affinity chromatography is so efficient that in many cases no further purification steps are required; however, further steps may be required to achieve separation of the desired protein from those contaminants that are closely related (e.g., proteins that differ by posttranslational modifications).

Ion Exchange Chromatography

Ion exchange chromatography (IEC) is used to separate molecules on the basis of their surface charge. The strength of interaction between a protein molecule and an ion exchange matrix depends on (1) the charge density on the protein and (2) the ionic strength of the mobile phase, which is always a buffered solution. The charge density on the protein is modulated by altering the pH of the solution. Recall that the overall charge on a protein is zero when the pH of a protein solution is equal to the isoelectric point (pI) for the protein and that the charge on the protein molecules becomes increasingly negative as the pH of the protein solution increases (see Figures 2.18 and 5.14(b)). Conversely, the charge on the protein molecules becomes increasingly positive as the pH of the solution decreases. This behavior is a consequence of the fact that the ionizable groups on the surface of a protein are either carboxylic acids or amines (see Problem 14 in Chapter 2).

The two main types of ion exchange matrices, shown in **FIGURE 5A.5**, are anion exchangers and cation exchangers. *Anion* exchangers, such as diethylaminoethyl (DEAE) cellulose and quaternary ammonium ("Q") resins, carry a *positive* charge, so negatively charged proteins bind to these resins. *Cation* exchangers, such as carboxymethyl (CM) cellulose and sulfonic acid ("S") resins, carry a *negative* charge, so positively charged proteins will bind to them. DEAE and CM exchangers are considered "weak" ion exchangers because they carry functional groups that can lose their charges at pH $> \sim 10$ (DEAE) or pH $< \sim 4$ (CM). The "Q" and "S" resins are effectively always charged in aqueous buffers, so they are considered to be "strong" ion exchangers. It is critical to match the IEC resin and buffer to the pI of the protein of interest. For example, a protein will bind to a column of DEAE-cellulose when the pH of the mobile phase is above the pI for that protein, but it will not bind the column if buffer pH $<$ pI.

The essential features of IEC are shown in **FIGURE 5A.6**, using a strong anion exchange resin as an example. If a protein mixture is loaded onto the column using a mobile phase that is buffered at a relatively high pH (e.g., pH $= 8.5$), then those proteins with pIs above 8.5 will carry a $(+)$ charge and will not bind to the anion exchange matrix. Thus, under these conditions, proteins with high pIs will wash off the column, while those with lower pIs will carry a $(-)$ charge and will bind to the matrix. Decreasing the pH of the mobile phase reduces the $(-)$ surface charge on the proteins, thereby weakening the attractive interactions between the proteins and the matrix. The proteins with the lowest pI values will retain a $(-)$ surface charge over a wider range of pH, so they are retained the longest and elute latest.

Elution of a protein bound to an IEC matrix can also be achieved by increasing the ionic strength of the mobile phase (e.g., by the addition of some salt to the elution buffer). Soluble ions compete for binding to the charged functional groups on the matrix. As the concentration of soluble ions increases, the ions will outcompete, and thereby displace, the protein from the matrix.

In summary, IEC allows for the separation of proteins based on differences in charge density. The theoretical basis for this separation technique is presented in Chapter 2 (Section 2.4) and includes the following concepts: (1) relative strengths of electrostatic interactions, as described by Coulomb's law, (2) the isoelectric point or pI for a given protein, and (3) modulation of charge density on a protein as a function of pH, as described by the Henderson–Hasselbalch equation.

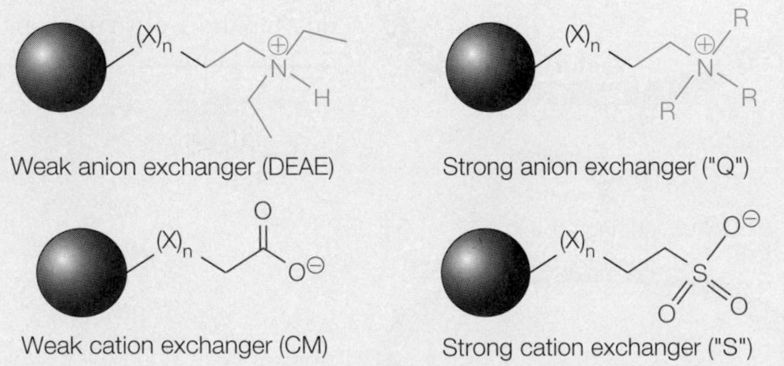

Size Exclusion Chromatography

Size exclusion chromatography (SEC), which is also known as **gel filtration chromatography,** differs from affinity chromatography and ion exchange chromatography in that noncovalent interactions between the protein and support are negligible. As shown in **FIGURE 5A.7**, SEC separates proteins on the basis of apparent size, or hydrodynamic radius. The apparent size of a protein molecule is approximately correlated with the length of the protein amino acid sequence. This rule of thumb assumes, to a first approximation, that soluble folded proteins behave as spheres. The distinguishing feature of the SEC stationary phase is the porous structure of the matrix. These matrices are generally spherical beads with pores in the surface of the bead that are on a size scale close to that of protein molecules. Different SEC matrices have larger- or smaller-sized pores in the beads.

The principles at work in SEC are diffusion by Brownian motion and excluded volume. The total volume in an SEC column includes the volume of the porous beads and the volume of the mobile phase, which is in the space between the beads and also inside the beads. The volume of buffer solution required to elute a protein from the column depends on what fraction of the column volume that protein can occupy: The more volume the protein can occupy, the more buffer will be needed to elute the protein, and the later it will elute from the column. The size of the pores in the beads determines which proteins can occupy volume inside the bead and which proteins cannot. Because smaller proteins are more likely to diffuse into the interior of a bead, they will occupy more of the total column volume than a large protein, which can only occupy the volume between—but not inside—the beads. In other words, larger proteins are "excluded" from the volume found inside the beads, so larger proteins elute from an SEC column earlier than smaller proteins. Note that this order of elution by size is the reverse of the order of migration in a typical electrophoresis experiment (see Tools of Biochemistry 2A and 6B).

▲ **FIGURE 5A.5** The chemical structures of commonly used ion exchange media.

Weak anion exchanger (DEAE)

Strong anion exchanger ("Q")

Weak cation exchanger (CM)

Strong cation exchanger ("S")

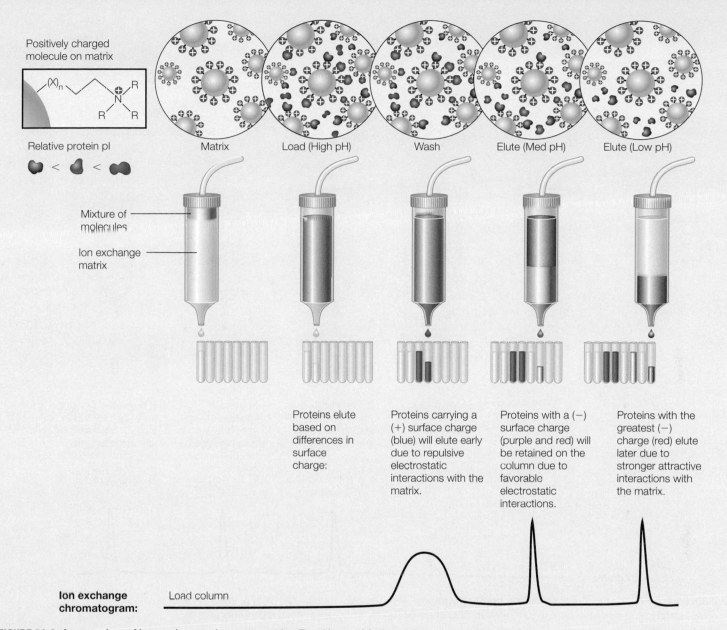

Positively charged
molecule on matrix

Relative protein pI

Matrix   Load (High pH)   Wash   Elute (Med pH)   Elute (Low pH)

Mixture of
molecules

Ion exchange
matrix

Proteins elute
based on
differences in
surface
charge:

Proteins carrying a
(+) surface charge
(blue) will elute early
due to repulsive
electrostatic
interactions with the
matrix.

Proteins with a (−)
surface charge
(purple and red) will
be retained on the
column due to
favorable
electrostatic
interactions.

Proteins with the
greatest (−)
charge (red) elute
later due to
stronger attractive
interactions with
the matrix.

**Ion exchange
chromatogram:**   Load column

▲ **FIGURE 5A.6 An overview of ion exchange chromatography.** Top: Here, a (+)-charged matrix is shown. As the pH
of the mobile phase decreases, or as the ionic strength increases, interactions between the proteins and the matrix
are weakened, and the proteins elute. Bottom: A schematic representation of an ion exchange chromatogram, where
each peak represents protein absorbance at 280 nm. In this example, proteins elute in the order of decreasing pI.

SEC works best as a purification technique when the differences
in size between the desired protein and the contaminants are a factor
of two or greater; thus, for SEC to be effective, the complexity of the
protein mixture must be relatively low. For this reason, SEC is often
the last step of a purification process. In addition, SEC is a convenient
way to "desalt" or change the buffer composition of a protein solution because the salts that make up a buffer are low molecular weight
and elute well after the desired protein. For example, it is convenient
to run affinity columns and IEC columns in nonvolatile buffers such
as phosphate-buffered saline (PBS); however, nonvolatile buffers are
not compatible with lyophilization (freeze-drying) and many mass

spectrometry techniques. Mass spectrometry is a powerful tool for
protein analysis and is often used to confirm the identity of a purified
protein (see Tools of Biochemistry 5B). To prepare a protein for lyophilization or mass spectrometry, an SEC column is equilibrated with
a mobile phase that uses a volatile buffer, such as ammonium acetate.
A protein that is loaded onto the SEC column in a small volume of
PBS becomes separated from the phosphate buffer salts as it moves
through a column equilibrated with ammonium acetate. In this way,
a "buffer exchange" from PBS to ammonium acetate occurs, and the
protein is now ready for lyophilization and/or mass spectrometric
analysis.

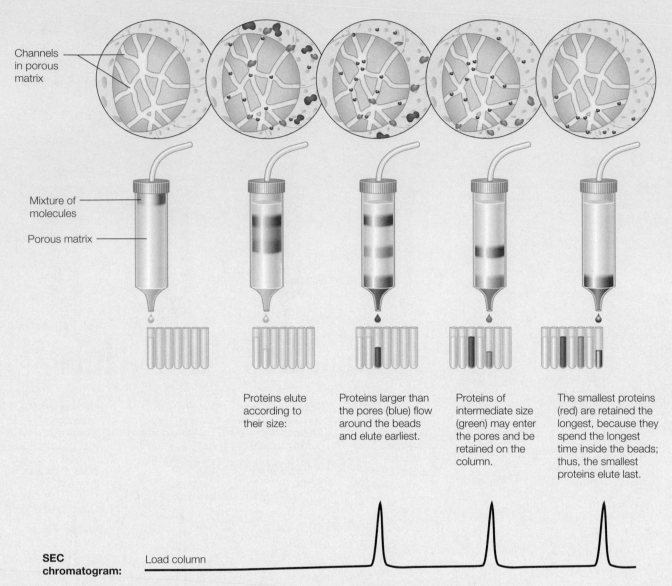

Channels in porous matrix

Mixture of molecules

Porous matrix

Proteins elute according to their size:

Proteins larger than the pores (blue) flow around the beads and elute earliest.

Proteins of intermediate size (green) may enter the pores and be retained on the column.

The smallest proteins (red) are retained the longest, because they spend the longest time inside the beads; thus, the smallest proteins elute last.

SEC chromatogram:

Load column

▲ FIGURE 5A.7 An overview of size exclusion chromatography. Top: The SEC matrix (gray with white channels) is porous. Bottom: A schematic representation of an SEC chromatogram, where each peak represents protein absorbance at 280 nm. In this example, proteins elute in the order of decreasing molecular weight.

Recombinant Hemoglobin Purification Scheme

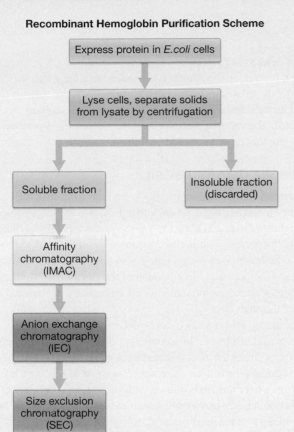

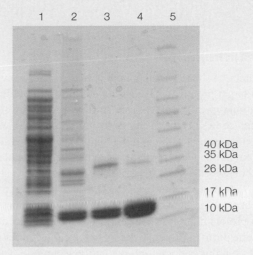

▲ **FIGURE 5A.9** **Gel electrophoresis of samples taken at different stages of protein purification.** The hemoglobin mutant is purified to greater than 95% homogeneity by a three-column procedure (see Figure 5A.8). A Coomassie-stained SDS-PAGE gel is shown. Lane 1: *E. coli* lysate (after centrifugation). Lane 2: IMAC-purified proteins; the prominent bands are the α–globin and β–globin chains which make up the mutant hemoglobin. Lane 3: IEC-purified protein. Lane 4: Overloaded sample after SEC. Lane 5: Protein molecular weight markers. Courtesy of Cory Hamada, Western Washington University.

▲ **FIGURE 5A.8** Flowchart for the purification of recombinant hemoglobin.

Example: Purification of a Recombinant Hemoglobin Mutant

FIGURE 5A.8 shows a scheme for the purification of a mutant form of human hemoglobin expressed in *E. coli* bacteria. Hemoglobin contains two different types of polypeptide chains: α-globin and β–globin. This particular mutant includes an eight-amino acid insertion into the β-globin sequence. Following the period of protein production in bacterial cells, the cells are lysed so that the contents of the cytoplasm are released into a buffered solution. The soluble material can then be separated from the insoluble material (e.g., membranes and precipitated protein aggregates) by centrifugation. The resulting supernatant

is a complex mixture of nucleic acids and proteins, as shown in Lane 1 of **FIGURE 5A.9.** An efficient purification scheme will significantly reduce the complexity of the mixture in the early steps of the process, so affinity chromatography is a good choice for the first chromatography step. In this case, an IMAC purification step significantly purifies the mutant hemoglobin, as shown in Lane 2 of Figure 5A.9. After IMAC, several contaminants are removed by anion exchange on Q resin (Lane 3). Finally, SEC is used to separate the hemoglobin from the contaminant at 35 kDa (Lane 4). Lane 4 is overloaded in order to detect the presence of impurities in the final preparation. In this case, the hemoglobin mutant is greater than 95% of the total protein in the purified sample.

References

For a list of references related to this chapter, see Appendix II.

Mass Determination

Once a protein has been purified, how is a researcher convinced that the correct target protein has been obtained in a purified form? The first indication typically comes from an SDS-PAGE gel (see Figure 5A.9), which shows (1) the purity of the protein and (2) an approximate molecular weight estimated by comparing the migration of the target protein in the gel to protein molecular weight standards. For example, the molecular weights of the α- and β–globins shown in Figure 5A.9 are predicted to be ~16 kDa each, based on the amino acid sequences of the translated genes. Figure 5A.9 shows that the mutant protein migrates between 10 and 17 kDa as would be expected. The purified protein appears to have the correct mass by SDS-PAGE, but this is not a high-resolution technique. The actual mass could differ from the expected mass by several hundred daltons and still appear by SDS-PAGE to be reasonably close to the expected mass. Mass spectrometry (MS) provides the most accurate mass measurements of large biomolecules. For this reason, it is desirable to obtain high-resolution mass data via MS to confirm that the protein has no unexpected posttranslational modifications (e.g., proteolytic cleavage and/or covalent modifications).

FIGURE 5B.1 shows a simplified diagram of a mass spectrometer that contains a single mass analyzer, sufficient for the routine determination of accurate protein masses using **electrospray ionization (ESI)** or **matrix-assisted laser desorption/ionization** (MALDI) techniques. In ESI, a fine mist of protein solution is accelerated toward a mass analyzer. By the time the mist reaches the analyzer, most of the solvent has evaporated, leaving protein molecules with a varying number of charges to be separated in the mass analyzer. The detector records the ratio of mass to charge (m/z, where m = mass and z = charge). In the MALDI technique, the protein is embedded in a large excess (~10,000-fold) of some matrix that absorbs UV light. When a laser pulse hits the matrix, it absorbs the energy of the laser light and is vaporized. The vaporized matrix carries intact, charged protein molecules into the gas phase and toward the mass analyzer.

Sequence Determination

An accurate protein mass is usually sufficient to confirm the identity of a known protein, but if an unknown protein is the target of some purification scheme, then the mass alone is not typically sufficient to identify the protein. In this case, sequence information is also desirable. If the function of the protein is also unknown, then the sequence will allow the potential identification of the function by similarity searching. Posttranslational modifications can also be detected at this stage of the analysis.

The amino acid sequence can be determined in several ways. As mentioned in Chapter 5 (Section 5-4), determination of the gene sequence is one of the easiest methods. Because the entire genomes of many organisms have been determined, amino acid sequence information is known for hundreds of thousands of proteins. For many of them, the function remains unknown. Protein sequences translated from cloned genes do not provide information concerning the modification of amino acids or the existence of intramolecular cross-links such as disulfide bonds. To find these, the protein itself must be sequenced. Two methods can be used to obtain peptide sequences—namely, tandem MS, which was developed in the mid-1980s and is now the method of choice for most labs, and Edman sequencing, which was developed 20 years earlier

Electrospray Ionization (ESI)

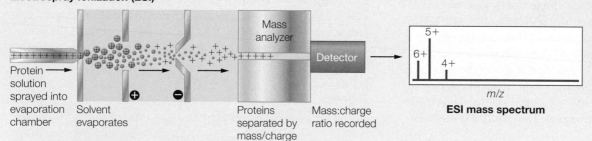

Protein solution sprayed into evaporation chamber | Solvent evaporates | Proteins separated by mass/charge | Mass:charge ratio recorded

ESI mass spectrum

◀ **FIGURE 5B.1** Electrospray ionization (ESI) and matrix-assisted laser desorption/ionization (MALDI) mass spectrometry techniques.

Matrix-Assisted Laser Desorption/Ionization (MALDI)

Laser pulse (orange) vaporizes matrix containing embedded protein

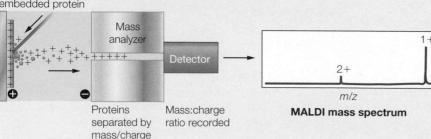

Proteins separated by mass/charge | Mass:charge ratio recorded

MALDI mass spectrum

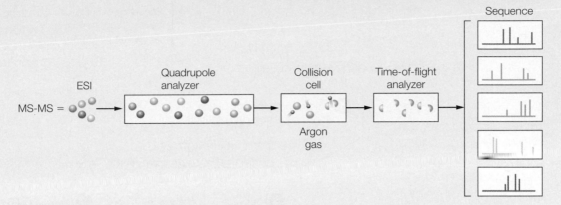

Sequence

ESI   Quadrupole analyzer   Collision cell   Time-of-flight analyzer

MS-MS =

Argon gas

▲ **FIGURE 5B.2 Peptide sequencing by MS-MS.** Each peptide fragment that enters the quadrupole analyzer is directed into the collision cell one at a time. The fragments generated in the collision cell give rise to mass spectra (shown on the right). Thus, each fragment that enters the quadrupole analyzer will yield a distinct mass spectrum in the time-of-flight analyzer.

by Pehr Edman and is still in use. We present the procedure for the tandem MS method since it is now the dominant technology.

To obtain peptide-sequence information, the mass spectrometer must have a collision cell and two mass analyzers rather than the single mass analyzer shown in Figure 5B.1. **FIGURE 5B.2** shows a schematic diagram of a tandem mass spectrometer (MS-MS) capable of peptide sequencing, where the two mass analyzers are labeled quadrupole analyzer and time-of-flight analyzer.

MS-MS sequencing works best when the protein to be sequenced is first fragmented into smaller peptides. The peptide fragments can be generated using proteases or in the mass spectrometer itself. Consider the case in which the fragments are generated using a protease and then introduced into the MS-MS instrument using electrospray. As depicted in Figure 5B.2, the mixture of peptide fragments is introduced by electrospray into the first mass analyzer (the quadrupole analyzer). The quadrupole analyzer can be tuned to select a specific peptide (i.e., a specific m/z range) for introduction into the collision cell. In the course of the complete sequencing analysis, each peptide fragment of the full-length protein will be directed from the quadrupole analyzer into the collision cell one at a time.

In the collision cell, the selected peptide is fragmented further by collisions with argon atoms. To a large extent, the fragmentations in the collision cell result in cleavage of the peptide backbone as shown in **FIGURE 5B.3**, where cleavage of the first peptide bond gives two subfragments: the N-terminal subfragment that includes the residue R_1, and the C-terminal subfragment that includes residues R_2–R_4. By convention, the N-terminal subfragments are called *b* ions and the C-terminal subfragments are called *y* ions.

In the collision cell, two series of subfragments are generated simultaneously—a series of *b* ions and the corresponding set of *y* ions. Within each series, the masses of the ions differ from one another by the mass of a single amino acid residue (i.e., in Figure 5B.3, the masses of b_2 and b_3 differ by the mass of residue R_3). The m/z ratio for each

ion is determined in the time-of-flight analyzer and recorded to generate a complex spectrum with peaks for each ion. Because the masses for amino acid residues are known (see Table 5.1), the amino acids present in each fragment can be reliably identified. Modern MS-MS instruments include software that can rapidly identify fragmentation patterns consistent with a specific amino acid sequence. This process is repeated until every fragment that enters the quadrupole analyzer is sequenced (a matter of a few minutes), thereby generating a set of peptide sequences for the protein. To find the order of these peptides fragments in the sequence of the entire protein, a second MS-MS analysis is performed on a series of different fragments generated by a different protease (e.g., see **FIGURE 5B.4**).

We will use the sequencing of bovine insulin to illustrate this process. This choice is appropriate because it was the first protein ever sequenced, by Frederick Sanger and his coworkers in the early 1950s (work for which Sanger won his *first* Nobel Prize). The example is also more complicated than most because we must deal with two covalently connected chains and locate the disulfide bonds. The steps of the procedure are outlined in Figure 5B.4.

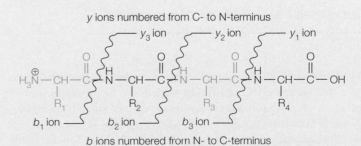

y ions numbered from C- to N-terminus

▲ **FIGURE 5B.3 Principal ions generated by low-energy collision-induced fragmentation.** The wavy red lines indicate sites of peptide bond cleavage in the collision cell (see Figure 5B.2).

139

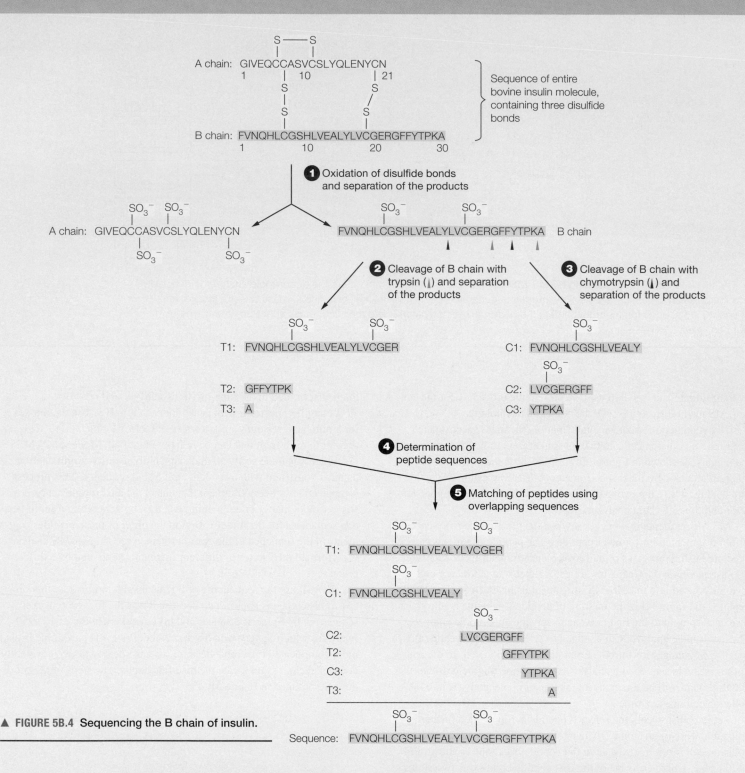

▲ FIGURE 5B.4 Sequencing the B chain of insulin.

The researcher intending to sequence a protein must first make sure that the material is pure. The protein can be separated from other proteins by some combination of the methods described in Tools of Biochemistry 5A and checked for purity by means of electrophoresis and/or isoelectric focusing (see Tools of Biochemistry 2A). Next, it must be determined whether the material contains more than one polypeptide chain because in some cases, disulfide bridges covalently bond chains together. SDS-PAGE in the presence and absence of reducing agents can answer this question (see Tools

of Biochemistry 6B). In the insulin example, there are two chains, A and B, as shown in Figure 5B.4. The MS-MS technique does not require that chains be separated, but sequence determination is simplified if the chains are separated and analyzed individually. To break disulfide bonds and thus separate the chains, several reactions are available.

Performic acid oxidation is the technique used in Figure 5B.4, step 1. The strong oxidizing agent performic acid will *irreversibly* react with cystine to yield cysteic acid residues:

Reduction with β-mercaptoethanol is a milder, and *reversible,* technique.

Performic acid

β-Mercaptoethanol

Reduction leaves free sulfhydryl groups, often positioned so that reoxidation to reform the disulfide bond is likely. Therefore, the sulfhydryls are usually blocked to prevent this. A common blocking reagent is iodoacetate:

Cysteine residue   Iodoacetate   Carboxymethyl-cysteine residue

If either of these methods is carried out with insulin, then the intact protein is cleaved into A and B chains. These chains can then be separated by chromatographic methods.

In bovine insulin, the A and B chains are so short that modern instrumentation could sequence either without prior fragmentation; however, to demonstrate the methods needed for larger proteins, we assume that the investigator must cleave the insulin chains into shorter polypeptides (this was indeed the case in Sanger's pioneering sequencing studies on insulin). Suppose the insulin B chain is to be sequenced. A first step would be to cleave separate aliquots of the chain with two or more of the specific cleavage reagents described in Table 5.3. Trypsin and chymotrypsin, for instance, would yield the sets of peptides shown in steps 2 and 3, respectively, of Figure 5B.4. The mixture of peptide fragments generated by one protease would be analyzed by MS-MS; then the fragments from the second protease digestion would be analyzed (Figure 5B.4, step 4).

Although the tryptic peptides alone cover the whole sequence, they are insufficient to allow us to write down the sequence of the insulin B chain because we do not know the order in which they appear in the intact chain. To overcome this problem, we also have the chymotryptic peptides, which *overlap* the tryptic peptides; therefore, all ambiguity is removed. Only one arrangement of the whole chain is consistent with the sequences of these two sets of peptides, as can be seen by matching overlapping sequences (Figure 5B.4, step 5).

Finally, a complete characterization of the covalent structure of a protein requires that the positions of any disulfide bonds be determined. In preparation for sequencing, these bonds would have been destroyed, but the positions of all cysteines, some of which *might* have been involved in bonding, would have been determined. How can we determine which cysteines are linked via disulfide bonds in the native protein?

To determine the arrangement of disulfide bonds, the experimenter again starts with the native protein—insulin in the example shown in **FIGURE 5B.5**. Reaction with a sulfhydryl-specific reagent (such as

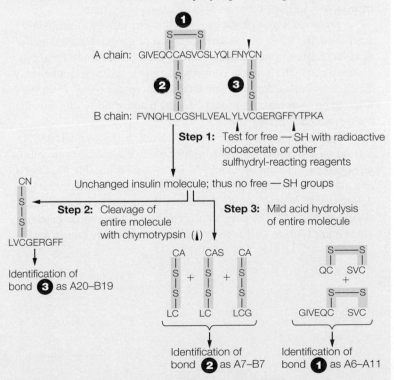

▲ FIGURE 5B.5 Locating the disulfide bonds in insulin.

141

radioactively labeled iodoacetate) marks any free cysteine residues, and fragmentation of the protein into the same peptides used in sequencing allows the positions of these nonbonded cysteines to be identified (step 1). Then samples of the intact protein are cut with various cleavage reagents, but now without first cleaving the disulfide bonds (steps 2 and 3). Some peptides, which are connected by these bonds, are attached to one another. These can then be isolated and their disulfide bonds cleaved to map the location of each disulfide-bonded cysteine in the protein.

We have described how the entire amino acid sequence, or primary structure, of a protein can be determined. Such analyses have been carried out on several thousand different proteins in the years since Sanger first determined the sequence of insulin. Today it is rare that an entire protein sequence would be determined using these methods because the sequencing of genes is much more rapid (and frequently precedes the isolation of the protein of interest); however, MS-MS is often used to determine a portion of the sequence of an unknown protein. With a sequence of only 6–10 amino acids, one can often identify a protein by searching databases of protein sequences. This use of protein sequence information is the basis for the field of **proteomics,** which we introduce briefly here.

The complement of proteins present in a given cell makes up the so-called **proteome** of that cell. As mentioned in Chapter 1, proteomics is the field of study that attempts to understand the complex relationships between proteins and cell function through global analysis of the entire proteome rather than investigating the properties of individual proteins in isolation. Proteomics includes, among other things, efforts to understand how protein expression and/or posttranslational modification levels change in cells and the consequences of such changes. For example, how does the proteome of a normal pancreatic cell differ from the proteome of a cancerous pancreatic cell? How might a researcher begin to address this question since every cell has the potential to express thousands of different proteins at any given time?

A proteomics experiment can, for example, rapidly identify differences in the levels of cellular expression or posttranslational modification (e.g., phosphorylation) between the proteomes of normal and diseased cells. The key is to correctly identify the affected proteins. The best technique for achieving this is mass spectrometry. Once the affected proteins have been identified, the putative role of the protein in the disease can be researched.

A typical proteomics experiment includes the following steps: (1) separation and isolation of proteins, or protein fragments, from cells or an organism; (2) identification by MS-MS sequencing (Figures 5B.2 and 5B.3) of a particular protein within the complex mixture; and (3) database searching to identify the target protein and its putative function. In practice, 2-D electrophoresis can be used to separate peptides (see Figure 1.14) as well as to identify potential protein targets for proteomic analysis. However, the extraction of peptides from 2-D gels is laborious, and direct analysis of complex peptide mixtures using tandem mass spectrometry is preferred for large-scale proteomic analyses.

A basic proteomics experiment is illustrated in **FIGURE 5B.6**. A complex mixture of proteins is digested—in this case with the protease trypsin. This mixture gives rise to a complex mixture of peptides that can be separated by chromatography. Here, the chromatography effluent is injected directly into a mass spectrometer. The complexity of this peptide mixture is represented by the many peaks in the total ion current (TIC) chromatogram [panel (a)]. If a very complex sample (e.g., a cellular extract) is the starting material for this experiment, then it is likely that each of the peaks in the TIC chromatogram will contain several peptides; however, these peptides can be separated within the mass spectrometer based on differences in the mass-to-charge ratio between the peptides. These separated peptides appear as "parent ions" in a much less complex mass spectrum [panel (b)]. Each parent ion can be analyzed by MS-MS to yield the amino acid sequence (see Figures 5B.2 and 5B.3). Panel (c) of Figure 5B.6 shows the MS-MS spectrum of one of the parent ions from the mass spectrum in panel (b). The experimentally determined amino acid sequence is then used as input to search a protein sequence database. In the example here, the sequence of the peptide is a fragment of bovine serum albumin [panel (d)].

MS-MS is compatible with the analysis of complex peptide mixtures because only one peptide fragment from the mixture is selected for sequencing by the mass analyzer. Thus, it is still possible to make a positive identification of a protein from complex mixtures of many proteins. In theory, all the proteins in the mixture can be identified, assuming the sequences are listed in some database.

Mass spectrometry is particularly suited to the detection of posttranslational modifications. A common modification is protein phosphorylation, which confers changes in both mass and charge on the phosphorylated protein. The presence of some enzymatic activity can also be detected by mass analysis by adding a chemical labeling reagent that is either covalently attached to some substrate by the target enzyme or is converted to a lower molecular weight product. These types of proteomics experiments have been used to detect metabolic disorders in newborns (see Appendix II).

There are many challenges to proteomic analysis. For example, the proteins present at low levels in a cell lysate can be difficult to detect. In some eukaryotic cells, the concentration differences between the most and least abundant proteins can be 10^6-fold, and many proteins that are interesting targets (e.g., for drug development) are low-abundance proteins. For these reasons a purification step, such as affinity chromatography, prior to mass analysis may be included to either remove highly expressed proteins and/or increase the concentrations of low-level proteins.

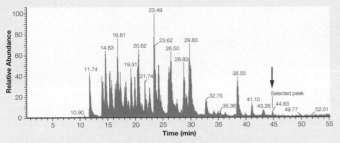

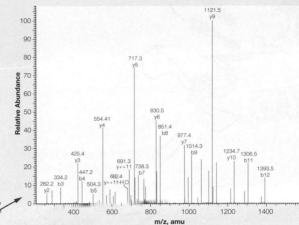

(a) A total ion current (TIC) chromatogram for a complex mixture of peptides separated by HPLC. The peptide ions from any portion of the TIC chromatogram can be selected within the mass spectrometer and subjected to further analysis by tandem mass spectroscopy (e.g., the peak under the magenta arrow).

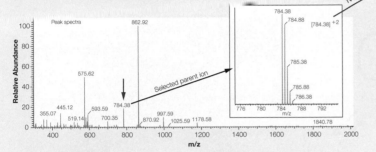

(c) An MS-MS spectrum for one of the parent ions in panel (b) [magenta arrow in panel (b)]. The 13 amino acid sequence obtained from MS-MS analysis is used in a database search.

Sequence identification

DAFLGSFLYEYSR

Database search

P02769 Serum albumin precursor (Allergen Bos d 6) (BSA)
MKWVTFISLLLLFSSAYSRGVFRRDTHKSEIAHRFKDLGEEHFKGL
VLIAFSQYLQQCPFDEHVKLVNELTEFAKTCVADESHAGCEKSLHT
LFGDELCKVASLRETYGDMADCCEKQEPERNECFLSHKDDSPDL
PKLKPDPNTLCDEFKADEKKFWGKYLYEIARRHPYFYAPELLYYAN
KYNGVFQECCQAEDKGACLLPKIETMREKVLASSARQRLRCASIQ
KFGERALKAWSVARLSQKFPKAEFVEVTKLVTDLTKVHKECCHGD
LLECADDRADLAKYICDNQDTISSKLKECCDKPLLEKSHCIAEVEK
DAIPENLPPLTADFAEDKDVCKNYQEAK**DAFLGSFLYEYSR**RHPEY
AVSVLLRLAKEYEATLEECCAKDDPHACYSTVFDKLKHLVDEPQNL
IKQNCDQFEKLGEYGFQNALIVRYTRKVPQVSTPTLVEVSRSLGKV
GTRCCTKPESERMPCTEDYLSLILNRLCVLHEKTPVSEKVTKCCTE
SLVNRRPCFSALTPDETYVPKAFDEKLFTFHADICTLPDTEKQIKKQ
TALVELLKHKPKATEEQLKTVMENFVAFVDKCCAADDKEACFAVEG
PKLVVSTQTALA

(d) A database search using the sequence DAFLGSFLYEYSR (magenta) shows that the peptide is part of the amino acid sequence of bovine serum albumin precursor protein.

(b) Fragmentation of the ions selected in panel (a) gives a set of ions that can be further separated by mass-to-charge ratio in the mass spectrometer.

▲ **FIGURE 5B.6 Identification of a protein of interest using proteomics methods.** The 13 amino acid sequence obtained from MS-MS analysis is used in a database search.

References

For a list of references related to this chapter, see Appendix II.

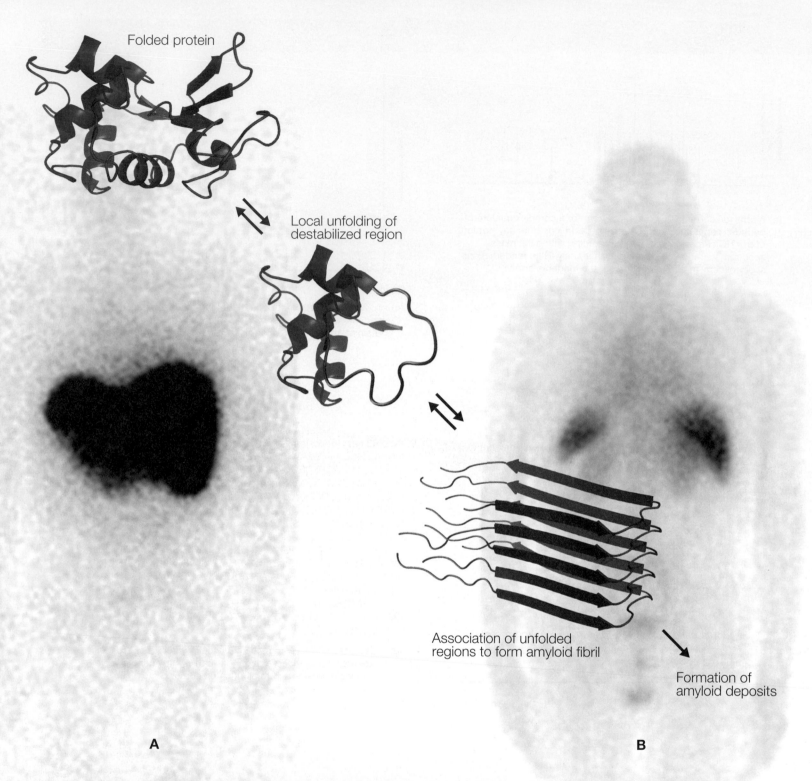

Folded protein

Local unfolding of destabilized region

Association of unfolded regions to form amyloid fibril

Formation of amyloid deposits

A

B

Whole-body scan of a patient with amyloidosis (dark areas) at diagnosis (A), after treatment (B).

Amyloidosis is caused by an abnormal deposition and accumulation of protein aggregates in tissues. How do proteins adopt and maintain a stable folded structure? What features of the protein amino acid sequence determine the stability of the folded structure? How does disruption of that structure lead to protein deposition diseases such as amyloidosis, Alzheimer's disease, and Parkinson's disease?

The Three-Dimensional Structure of Proteins

IN CHAPTER 5 we introduced the concept of protein primary structure. We emphasized that this first level of organization, the amino acid sequence, is dictated by the DNA sequence of the gene for the particular protein. However, nearly all proteins exhibit higher levels of structural organization as well. It is the three-dimensional structure of each protein that specifies its function in a particular biological process. Thus, an understanding of protein structure is the key to understanding how proper protein function promotes normal cellular functions, as well as how perturbations to protein structure are linked to pathology in diseased cells.

Figure 5.1 includes a representation of sperm whale myoglobin, showing a well-defined spatial location for every atom (except hydrogen atoms) in the protein. FIGURE 6.1 is another representation of the three-dimensional conformation of the myoglobin molecule, which illustrates that there exist two distinguishable levels of structural organization in the folded polypeptide chain. First, the chain appears to be locally coiled into regions of helical structure. Such local *regular* folding is called the secondary structure of the

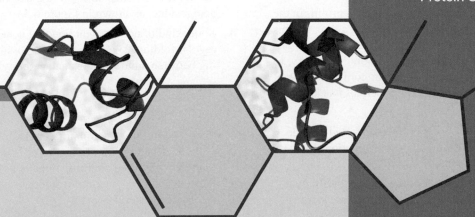

molecule. The helically coiled regions themselves are, in turn, folded into a specific compact structure for the entire polypeptide chain. We call this further level of three-dimensional organization the **tertiary structure** of the molecule. Later in this chapter, we shall find that some proteins consist of several folded polypeptide chains, called subunits, arranged in a well-defined manner. The arrangement of subunits defines the **quaternary structure** of a multisubunit protein.

● **CONCEPT** Protein molecules have four levels of structural organization: primary (sequence), secondary (local folding), tertiary (overall folding), and quaternary (subunit association).

FIGURE 6.2 shows the relationships between these four levels of protein structure and features a common representation of protein structure—the "cartoon" rendering. A cartoon rendering shows the location of the protein main-chain atoms as a wide ribbon and does not show side chains. The advantage of a cartoon rendering over "stick" or all-atom renderings is that it allows the viewer to more easily identify the essential features of protein secondary and tertiary structure.

In this chapter, we examine the structural features of amino acids that promote the formation of secondary and tertiary structures in stably-folded proteins. We

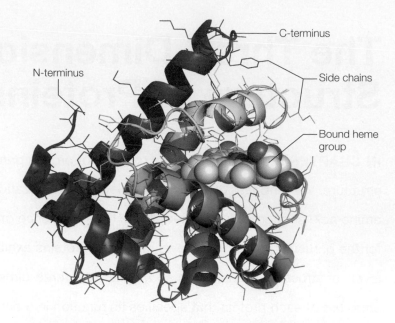

▲ **FIGURE 6.1 Three-dimensional folding of the protein myoglobin.** This rendering generated from the X-ray crystal structure, shows the polypeptide main chain as helices (flat ribbons) connected by thick lines. Amino acid side chains are shown as sticks. Individual helical regions are color-coded, with the peptide N-terminus shown in blue and the C-terminus shown in red. Myoglobin binds a heme group (shown in space-filling display). The orientation of the protein in this figure is the same as that shown in Figure 5.1. PDB ID: 1mbn.

will then consider some factors that impart stability to folded proteins, and the mechanisms by which protein folding is thought to occur.

6.1 Secondary Structure: Regular Ways to Fold the Polypeptide Chain

Theoretical Descriptions of Regular Polypeptide Structures

Our understanding of protein secondary structure has its origins in the remarkable work of Linus Pauling. In the early 1950s, Pauling and his collaborators used X-ray diffraction data of amino acids and peptides, together with a deep understanding of chemical bond geometry, to begin a systematic analysis of the possible regular conformations of the polypeptide chain. They postulated several principles that any such structure must obey:

1. The bond lengths and bond angles should be distorted as little as possible from those found through X-ray diffraction studies of amino acids and peptides, as shown in Figure 5.10(b).

2. No two atoms should approach one another more closely than is allowed by their van der Waals radii; thus, there are steric restrictions to rotations around the bonds that make up the peptide backbone (see **FIGURE 6.3**).

3. The six atoms in the peptide amide group must remain coplanar with the associated α-carbons in the *trans* configuration, as shown in Figure 5.10(b). Consequently, rotation is possible only about the two bonds adjacent to the α-carbon in each amino acid residue, as shown in Figure 6.3.

4. Some kind of noncovalent bonding is necessary to stabilize a regular folding. The most obvious possibility is hydrogen bonding between amide protons and carbonyl oxygens:

$$\diagdown N - H \cdots O = C \diagdown$$

Such a concept was familiar to Pauling, who had much to do with the development of the idea of hydrogen bonds. In summary, the preferred conformations of polypeptides must be those that allow a maximum amount of hydrogen bonding, yet also satisfy criteria 1–3.

...KEFTPPVQAAYQKVVAGVANALAHKYH...

(a) Primary structure (amino acid sequence):
A portion of the amino acid sequence of human beta globin is shown. The sequence highlighted in cyan adopts a helical conformation, and is shown in the same orientation in parts (b–d).

Some parts of the primary sequence adopt a local regular repeating structure ("2° structure")

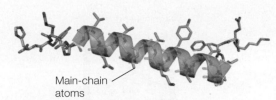

Main-chain atoms

(b) Secondary structure:
A stick representation of the amino acid sequence from part (a) is shown. Superimposed on the stick structure is a cartoon rendering of the main-chain atoms, showing the location of the local helical structure in this sequence.

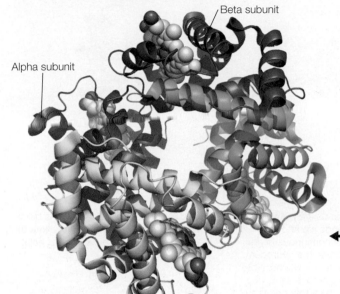

Beta subunit

Alpha subunit

Several 2° structure elements associate along their hydrophobic surfaces to give a stably folded structure ("3° structure")

Quaternary structure ("4° structure") arises when two or more proteins folded into tertiary structures interact to form well-defined multisubunit complexes.

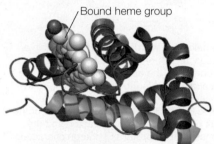

Bound heme group

(d) Quaternary structure:
Four separate protein subunits, two alpha subunits (magenta and green) and two beta subunits (yellow and blue/cyan) associate to form the fully assembled hemoglobin protein. The four subunits are shown in cartoon rendering with hemes in space-filling display (PDB ID: 2hhb).

(c) Tertiary structure:
The entire beta globin chain is shown in its well-defined folded structure. As in myoglobin (Figure 6.1), the helical regions interact to define the folded structure, which binds a heme (shown in space-filling display).

▲ **FIGURE 6.2** The four levels of structural organization in proteins.

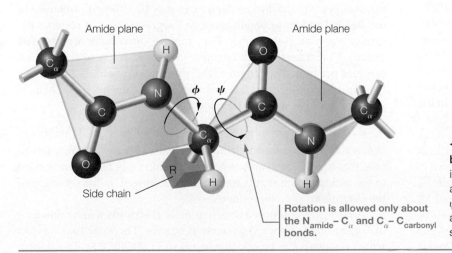

Amide plane

Amide plane

ϕ ψ

Side chain

Rotation is allowed only about the $N_{amide}-C_\alpha$ and $C_\alpha-C_{carbonyl}$ bonds.

◄ **FIGURE 6.3** Rotation around the bonds in a polypeptide backbone. Two adjacent amide planes are shown in light blue. Rotation is allowed only about the $N_{amide}-C_\alpha$ and $C_\alpha-C_{carbonyl}$ bonds. The angles of rotation about these bonds are defined as ϕ (phi) and ψ (psi), respectively, with directions defined as positive rotation as shown by the arrows. The extended conformation of the chain shown here corresponds to $\phi = +180°$, $\psi = +180°$.

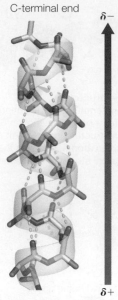

C-terminal end

$\delta-$

N-terminal end

$\delta+$

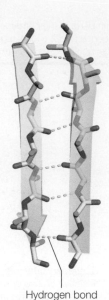

Hydrogen bond

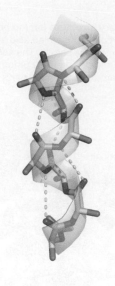

(a) In the α helix, the hydrogen bonds are within a contiguous stretch of amino acids and are almost parallel to the helix axis. This orientation of the amide bonds in the helix gives rise to a helical macrodipole moment shown by the arrow (see Figure 2.5). The N-terminal end of the helix has partial (+) charge character, and the C-terminal end has partial (−) charge character.

(b) In the β sheet, the hydrogen bonds are between adjacent strands (only two strands are shown here), which are not necessarily contiguous in the primary sequence. In this structure, the hydrogen bonds are nearly perpendicular to the chains. Note that in the cartoon rendering a strand is shown by a flat arrow, where the head of the arrow points to the C-terminus of the strand.

(c) The 3_{10} helix is found in proteins but is less common than the α helix. Note that, compared to the α helix, the 3_{10} helix forms a tighter spiral.

▲ **FIGURE 6.4 The right-handed α helix, β sheet, and 3_{10} helix.** The right-handed α helix and β sheet are the two most frequently observed regular secondary structures of polypeptides. Backbone atoms are shown in stick rendering (N atoms blue, O atoms red, and C atoms gray) superimposed on a transparent cartoon rendering. For clarity, the side chains have been deleted at the beta carbons (green). Hydrogen bonds are shown as yellow dashed lines. **(a)** The α helix. **(b)** A β sheet containing two strands. **(c)** The 3_{10} helix. Panels (a) and (b) from PDB ID: 2kl8, and panel (c) from PDB ID: 1lb0.

α Helices and β Sheets

Working mainly with molecular models, Pauling and his associates were able to arrive at a small number of regular conformations that satisfied all of these criteria. Some were helical structures formed by a single-polypeptide chain, and some were sheet-like structures formed by adjacent chains. The two structures they proposed as most likely—the right-handed **α helix,** and the **β sheet**—are shown in **FIGURE 6.4(a)** and (b). These two structures are, in fact, the most commonly observed secondary structures in proteins. Figure 6.4(c) shows the so-called **3_{10} helix,** which is observed in some proteins but is not as common as the α helix. All of the protein secondary structures shown in Figure 6.4 satisfy the criteria listed earlier: In each structure the amide group is planar, and all amide protons and carbonyl oxygens (except a few near the ends of helices, or edges of sheets) are involved in hydrogen bonding. The arrangement of the main-chain hydrogen bonds in the α helix orients the amide N—H and C=O groups such that the dipole moments (see Figure 2.5) for each of these polar bonds align and give rise to a **helical dipole moment** (also called a "macrodipole"). In effect, the N-terminus of the helix has partial (+) charge character, and the C-terminus has partial (−) charge character, as shown by the arrow in Figure 6.4(a).

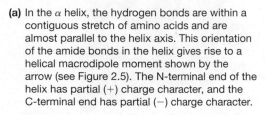

● **CONCEPT** Of the several possible secondary structures for polypeptides, the most frequently observed are the α helix and the β sheet.

Describing the Structures: Helices and Sheets

In Chapter 4, we discussed some terms that define features of the helical structure of DNA: the pitch (p), the rise (h), and the number of residues per turn (n). Molecular helices may be either right-handed or left-handed and may contain either an integral number of residues per turn or a nonintegral number. For example, the α helix repeats after exactly 18 residues, which amounts to 5 turns. It has, therefore, 3.6 residues per turn. The length of the 18 residue α helix is 27 Å; thus, the rise of the α helix is 1.5 Å/residue. Because the pitch of a helix is given by $p = nh$, we have for the α helix:

$$p = (3.6 \text{ residues/turn}) \times (1.5 \text{ Å/residue}) = 5.4 \text{ Å/turn}$$

When you examine the model for the α helix (Figure 6.4(a)), you will note that a given carbonyl oxygen, on residue i, is hydrogen-bonded to the amide proton that is four residues removed, in the direction of the C-terminus (i.e., on residue $i + 4$).

A β sheet is composed of two or more **β strands** with main-chain hydrogen bonds between adjacent β strands. There are two ways in which β strands can be oriented in a β sheet (**FIGURE 6.5**). The β sheet shown in Figure 6.4(b) shows two β strands arranged such that the N-terminus to C-terminus orientations of the two strands are in opposite directions. Such an arrangement of strands is called **antiparallel,** whereas the arrangement with both strands oriented in the same direction is called **parallel** (see Figure 6.5(b)).

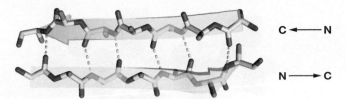

(a) An antiparallel arrangement of β strands.

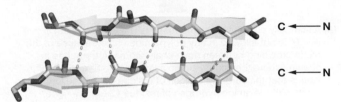

(b) A parallel arrangement of β strands.

▲ **FIGURE 6.5 β Sheets.** Backbone atoms are shown in stick rendering (N atoms blue, O atoms red, and C atoms gray) superimposed on a transparent cartoon rendering. For clarity, the side chains have been deleted at the beta carbons (green). Hydrogen bonds are shown as yellow dashed lines. Images are generated from PDB files 2kl8 (panel a) and 2lvb (panel b).

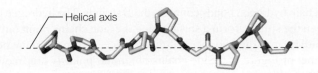

▲ **FIGURE 6.6 The polypeptide II helix.** A PPGPPGPPG sequence is shown. The peptide backbone is extended in a left-handed helical twist. This image rendered from PDB ID: 1k6f.

designation "polyproline II helix." Glycine is also often found in this helix, as are, albeit to a much lesser extent, several other amino acids. Because this conformation is not restricted to proline residues we will refer to this secondary structure motif by the more general term **polypeptide II helix** (**FIGURE 6.6**).

Parameters for the secondary structures described above are listed in **TABLE 6.1**.

Amphipathic Helices and Sheets

In an α helix, the side chain of each amino acid residue points outward, away from the center of the helix (**FIGURE 6.7**). In a β sheet, a network of

In addition to the helices and sheets described above, there is one more regularly repeating conformation that is commonly observed in protein structures—the so-called **polyproline II helix.** This particular conformation was not predicted on theoretical grounds because it does not satisfy Pauling's requirement for hydrogen bonding. Nevertheless, it is a common motif in protein structures. Unlike the α and 3_{10} helices, this structure does not have stabilizing hydrogen bonds between main-chain groups, and it is left-handed. Roughly a third of the amino acid residues found in this conformation are prolines, leading to the

TABLE 6.1 Parameters of some polypeptide secondary structures

Structure Type	Residues per Turn	Rise (h) per Residue	Pitch (p)
β Strand (antiparallel)	2.0	3.4 Å	6.8 Å
β Strand (parallel)	2.0	3.2 Å	6.4 Å
α helix	3.6	1.5 Å	5.4 Å
3_{10} helix	3.0	2.0 Å	6.0 Å
Polypeptide II helix	3.0	3.1 Å	9.3 Å

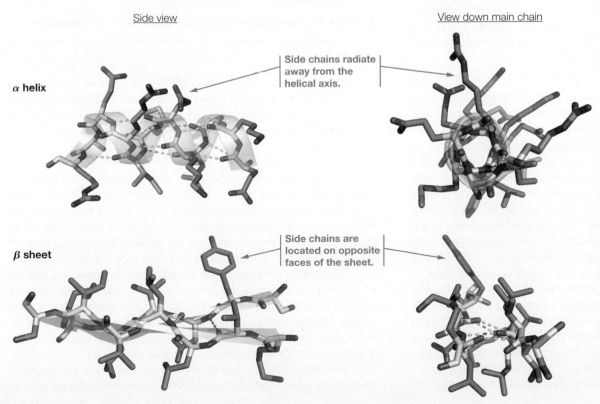

Side view

View down main chain

α helix

β sheet

Side chains radiate away from the helical axis.

Side chains are located on opposite faces of the sheet.

◄ **FIGURE 6.7 The positions of side chains in the α helix and β sheet.** A right-handed α helix and a two-stranded β sheet are shown in stick rendering superimposed on a backbone cartoon (gray ribbons). The O atoms are shown in red, N atoms in blue, S atoms in yellow, with C atoms in the backbone in gray and C atoms in the side chains in green. Hydrogen bonds are shown as yellow dashes. In the α helix, the side chains radiate away from the center of the helix (upper right panel, looking along the helical axis). In the β sheet, the side chains are located on opposite faces of the sheet defined by the H-bonded main-chain amides. This is shown in the lower left panel, looking at the main-chain strands from the side (one strand is behind the other), and the lower right panel, looking down the main chains of both strands. Images are generated from PDB files 2kl8 (helix) and 2lvb (sheet).

main-chain hydrogen bonds connects the β strands. The hydrogen bonds between the strands form a surface, and the side chains point outward, perpendicular to the two faces of this surface, as shown in the bottom left panel of Figure 6.7. Side chains of similar polarity are frequently found together on one side of a helix or a sheet, forming extended hydrophilic or hydrophobic surfaces, or "faces." Secondary structures that display a predominantly hydrophobic face opposite a predominantly hydrophilic face are said to be **amphipathic** (or **amphiphilic**). Many α helices and β sheets have this characteristic because it allows two or more secondary structures to associate via contacts between the

● **CONCEPT** An amphiphilic α helix will have side chains of similar polarity every 3–4 residues, whereas a β strand in an amphiphilic β sheet will have alternating polar and nonpolar side chains.

hydrophobic surfaces (see Figure 6.2), while projecting the hydrophilic surfaces toward the aqueous solvent. Based on the number of residues per turn listed in Table 6.1, an amphiphilic α helix will have side chains of similar polarity every 3–4 residues, whereas a β strand in an amphiphilic β sheet will have alternating polar and nonpolar side chains. These distinct patterns of side-chain polarity are the basis for many secondary structure prediction programs, discussed later in the chapter.

The parameters listed in Table 6.1 distinguish the common secondary structures shown in Figures 6.4, 6.5, and 6.6. Another way to describe the regular repeat for any of these secondary structural motifs is to specify the values of the angles around the N_{amide} — C_α and C_α — $C_{carbonyl}$ bonds. These angles are designated, respectively, ϕ and ψ (see Figure 6.3). As illustrated in **FIGURE 6.8(a)**, some combinations of ϕ and ψ are not allowed due to steric crowding. These steric constraints on peptide conformation can be appreciated by considering space-filling models for helices and sheets. For example, as shown in Figure 6.8(b), the main-chain atoms in the α helix are closely packed, with R groups projecting away from the helical axis. The combinations of ϕ and ψ angles that are most favorable (or "allowed")—because they relieve steric crowding—are shown in a systematic description of polypeptide backbone conformation called a **Ramachandran plot.**

Ramachandran Plots

As shown in Figure 6.3, each residue in a polypeptide chain has two backbone bonds about which rotation is permitted. The angles of rotation about these bonds, defined as ϕ (phi) and ψ (psi), describe the backbone conformation of any particular residue in any protein. To make the definition meaningful, we must specify what we mean by a positive direction of rotation and the zero-angle conformation of each. The conventions chosen for directions of positive rotation about ϕ and ψ are given by the arrows in Figure 6.3—that is, clockwise when looking in either direction from the α-carbon. The conformation shown in that figure corresponds to $\phi = +180°$ and $\psi = +180°$, the fully extended form of the polypeptide chain. Compare that to the (nonallowed) zero-angle conformation, $\phi = 0°$ and $\psi = 0°$, shown in Figure 6.8(a).

With these conventions, the backbone conformation of any particular residue in a protein can be described by a point on a map (**FIGURE 6.9**) with coordinates ϕ and ψ. Such maps are called Ramachandran plots, after the biochemist G. N. Ramachandran, who first made extensive use of them. For any regularly repeating secondary structure (e.g., α helix, β sheet), all the residues that are part of the structure are in nearly equivalent conformations and therefore have nearly equivalent ϕ and

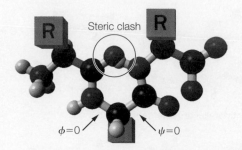

(a) A sterically nonallowed conformation. The conformation $\phi = 0°$, $\psi = 0°$ is not allowed in any polypeptide chain because of the steric crowding between main-chain atoms. A tripeptide is shown, where the central amino acid has $\phi = 0°$ and $\psi = 0°$. Notice that the carbonyl oxygen of residue #1 (on the left) would clash with the amide hydrogen of residue #3 (on the right).

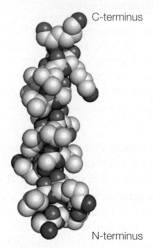

(b) The atoms in a helix are closely packed but do not clash sterically. Here, a segment of an α helix in sperm whale myoglobin is shown as a space-filling model (this is the longer green helix in Figure 6.1; PDB ID: 1mbn).

▲ **FIGURE 6.8 Steric interactions determine peptide conformation.**

ψ angles; thus, the points on a Ramachandran plot that correspond to those residues will cluster within a narrow range of sterically allowed ϕ and ψ angles for a given secondary structure. **TABLE 6.2** lists the ranges of ϕ and ψ angles that correspond to the various helices and β sheets described above.

One of the most useful features of Ramachandran plots is that they allow us to describe very simply which structures are sterically possible and which are not. For many of the conceivable combinations of ϕ and ψ values, some atoms in the chain would approach closer than their van der Waals radii would allow (an example is shown in Figure 6.8(a)). Such conformations are unfavorable because they are sterically crowded. The theoretically allowed ϕ, ψ combinations for poly-L-alanine correspond to the white regions in Figure 6.9, and the dots correspond to ϕ, ψ angles observed for ~10,000 individual alanine residues in high-resolution crystal structures of proteins. Clearly, only a small fraction of the possible conformations is actually favorable. All of the regular secondary structures we have discussed fall within or very close to these regions of the Ramachandran plot.

The theoretically allowed ϕ, ψ combinations shown in Figure 6.9 were drawn with the assumption that all residues are L-alanine (that is,

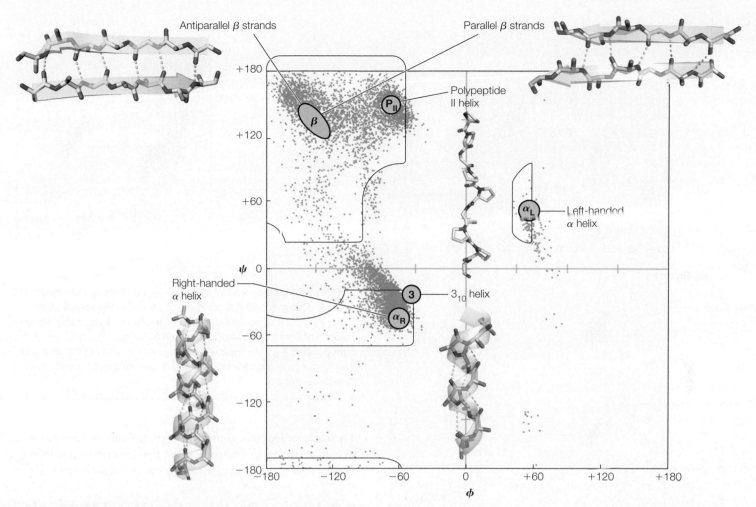

▲ **FIGURE 6.9 A Ramachandran plot for poly-L-alanine.** A map of this type can be used to describe the backbone conformation of any polypeptide residue as well as the secondary structures of proteins. The coordinates are the values of the bond angles ϕ and ψ, defined as in Figure 6.3. The white areas correspond to sterically allowed conformations for poly-L-alanine (i.e., a peptide consisting of only L-Ala residues) based on theoretical predictions. The gray dots in the background show the observed ϕ and ψ angles for over 10,000 alanine residues found in protein X-ray crystal structures. Although a left-handed α helix is theoretically allowed, it is very rare; thus, we do not show a model for the structure. The circles with the following symbols correspond to the secondary structures discussed in the text: α_R, right-handed α helix; 3, 3_{10} helix; β, β strand; P_{II}, polypeptide II helix; α_L, left-handed α helix. Data plot, courtesy of P. A. Karplus (Oregon State University).

● **CONCEPT** The Ramachandran plot illustrates which combinations of ϕ and ψ angles are sterically allowed in regular repeats of secondary structure.

all have —CH_3 side chains). If bulkier side chains were considered, the "allowed" regions would shrink due to increasing steric clash between side chains and main-chain atoms. Conversely, glycine, with its —H side chain, allows more conformations, as shown in **FIGURE 6.10(a)**. Proline has fewer combinations of allowed ϕ and ψ angles due to restricted rotation around ϕ (Figure 6.10(b)).

The foregoing analysis of protein structure is based on a relatively simple consideration of probable steric interactions. How does it compare with observations of actual protein structures? In this case, the correspondence between theory and observation is remarkably good, as shown by comparing the white regions of Figure 6.9 (values of ϕ and ψ allowed based on prediction of steric interactions) to the distribution of gray dots (actual values of ϕ and ψ for alanine residues observed in protein crystal structures).

The predictions of Ramachandran apply equally well to other amino acid residues in proteins. **FIGURE 6.11** shows a plot of ϕ, ψ pairs for 30,692 residues found in 209 different polypeptide chains for which high-resolution X-ray crystallography data exist. As is evident in this figure, the majority of the observed ϕ, ψ pairs (gray dots in Figure 6.11) cluster in the regions of the Ramachandran plot that are predicted

TABLE 6.2 Ranges of allowed ϕ and ψ angles for some polypeptide secondary structures

Structure Type	ϕ	ψ
β strand	−150° to −100°	+120° to +160°
α helix	−70° to −60°	−50° to −40°
3_{10} helix	−70° to −60°	−30° to −10°
Polypeptide II helix	−80° to −60°	+130° to +160°

Data from *Protein Science* 18:1321–1325 (2009), S. A. Hollingsworth, D. S. Berkholz, and P. A. Karplus, On the occurrence of linear groups in proteins.

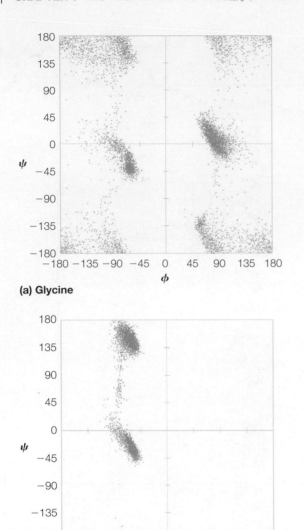

(a) Glycine

(b) Proline

▲ **FIGURE 6.10** Ramachandran plots for glycine and proline. The data shown are for glycine and proline residues found in high-resolution X-ray crystal structures of proteins. The favorable combinations of the bond angles ϕ and ψ are shown by the clustering of several points in the darker regions. Glycine has the greatest number of allowed ϕ, ψ angle combinations, whereas proline has the fewest. Courtesy of S. A. Hollingsworth and P. A. Karplus (Oregon State University).

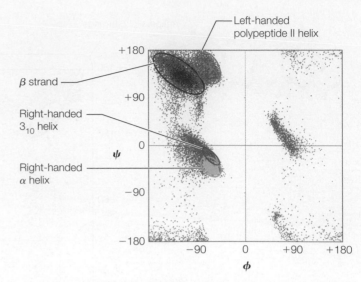

▲ **FIGURE 6.11** Observed values of ϕ and ψ from protein structural data. ϕ, ψ pairs for 30,692 residues observed in high-resolution ($\leq$1.2 Å) crystal structures of proteins are shown (gray dots). Different colors highlight those residues defined to be right-handed α helix (cyan), right-handed 3_{10} helix (purple), β strand (blue), and left-handed polypeptide II helix (orange). Courtesy of S. A. Hollingsworth and P. A. Karplus (Oregon State University).

to give the most favorable combinations of ϕ and ψ (white spaces in Figure 6.9). Those residues classified by H-bonding patterns and geometry to be helix or strand are identified by color in Figure 6.11 and illustrate the range of ϕ and ψ values that correspond to each of the secondary structure types shown in Table 6.2.

Although most of the points fall in "allowed" regions, a few lie in "nonallowed" regions. These are mainly glycines, for which a much wider range of ϕ and ψ angles is allowed because the side chain is so small (see Figure 6.10(a)).

Our discussion so far provides a background for understanding the observation of stable and widespread local protein secondary structures. We now examine the structures of some well-characterized proteins. We begin with the observation that two major classes of proteins exist.

These are called *fibrous* and *globular* proteins and are distinguished by major structural differences. We first describe a few examples of fibrous proteins, before considering the globular proteins.

6.2 Fibrous Proteins: Structural Materials of Cells and Tissues

Fibrous proteins are distinguished from globular proteins by their filamentous, or elongated, form. Most of them play structural roles in animal cells and tissues—they hold things together. Fibrous proteins include the major proteins of skin and connective tissue, and of animal fibers like hair and silk. The amino acid sequence of each of these proteins favors a particular kind of secondary structure, which confers on each protein a particular set of appropriate mechanical properties. **TABLE 6.3** lists the amino acid composition of three examples of fibrous proteins: α-keratin, fibroin, and collagen. Compared to the typical distributions of the 20 amino acids in globular proteins (see the "All Proteins" column in Table 6.3), each of these fibrous proteins is significantly enriched in 3–4 particular amino acids, which stabilize the extended secondary structures typical of the fibrous proteins.

● **CONCEPT** Fibrous proteins are elongated molecules with well-defined secondary structures. They usually play structural roles in the cell.

The Keratins

Two important classes of fibrous proteins that have similar amino acid sequences and biological function are called α- and β-keratins. The **α-keratins** are the major proteins of hair and fingernails, and comprise a major fraction of animal skin. The α-keratins are members of a broad group of **intermediate filament proteins,** which play important

TABLE 6.3 Amino acid compositions of some fibrous proteins

Amino Acid	α-Keratin (wool)	Fibroin (silk)	Collagen (bovine tendon)	All Proteins[c]
Gly	8.1	44.6	32.7	7.9
Ala	5.0	29.4	12.0	8.7
Ser	10.2	12.2	3.4	5.8
Glu + Gln	12.1	1.0	7.7	6.6 (3.7)
Cys	11.2	0	0	1.3
Pro	7.5	0.3	22.1[a]	4.7
Arg	7.2	0.5	5.0	5.0
Leu	6.9	0.5	?.1	8.9
Thr	6.5	0.9	1.6	5.6
Asp + Asn	6.0	1.3	4.5	5.9 (4.2)
Val	5.1	2.2	1.8	7.2
Tyr	4.2	5.2	0.4	3.5
Ile	2.8	0.7	0.9	5.5
Phe	2.5	0.5	1.2	4.0
Lys	2.3	0.3	3.7[b]	5.5
Trp	1.2	0.2	0	1.5
His	0.7	0.2	0.3	2.4
Met	0.5	0	0.7	2.0

Note: The three most abundant amino acids in each protein are indicated in magenta. Values are given in mole percent.

[a]About 39% of this is hydroxyproline.

[b]About 14% of this is hydroxylysine.

[c]Data from *Journal of Chemical Information and Modeling* (2010) 50:690–700, J. M. Otaki, M. Tsutsumi, T. Gotoh, and H. Yamamoto, Secondary structure characterization based on amino acid composition and availability in proteins.

(a) Side view of two monomers interacting via a parallel coiled-coil. N- and C-termini are indicated.

(b) View looking down the axis of the coiled-coil from the C-terminal end of the two monomers (side chains removed for clarity).

▲ **FIGURE 6.12** The coiled-coil structure of α-keratin intermediate filaments. PDB ID: 3ntu.

structural roles in the nuclei, cytoplasm, and surfaces of many cell types. All of the intermediate filament proteins are predominantly α-helical in structure; in fact, it was the characteristic X-ray diffraction pattern of α-keratins that Pauling and his colleagues sought to explain by their α helix model.

Individual α-keratin molecules contain long sequences—over 300 residues—that are wholly α-helical. Pairs of these right-handed helices twist about one another in the left-handed **coiled-coil** structure shown in **FIGURE 6.12**. This pairing of α helices occurs because of the amino acid sequence of α-keratin. Every third or fourth amino acid has a nonpolar, hydrophobic side chain. Because the α helix has 3.6 residues/turn, this means there is a strip of contiguous hydrophobic surface area along one face of each helical chain. This strip of hydrophobic surface area makes a shallow spiral around the helix, rather than a straight strip down one side of the helix, because the nonpolar residues occurring every third or fourth residue are slightly offset from the 3.6 residue repeat of the helix. As we noted in Chapter 2, hydrophobic surfaces tend to associate in aqueous medium. Thus, to maximize contact between the hydrophobic surfaces of two α-keratin helices, they entwine with a left-handed helical twist (Figure 6.12).

The α-keratin molecule includes a small globular region that is covalently attached to the coiled-coil sequence. In keratin intermediate filaments, pairs of coiled coils tend to associate into stretchy and flexible fibrils. In some tissues α-keratin is hardened, to differing degrees, by the introduction of disulfide cross-links within the several levels of fiber structure (note that α-keratin has an unusually high content of cysteine—see Table 6.3). The α-keratin in fingernails has many cross-links, whereas that in hair has relatively few. The process of introducing a "permanent wave" into human hair involves reduction of these disulfide bonds, rearrangement of the fibers, and reoxidation to "set" the tight curls thus introduced.

● **CONCEPT** α-Keratin is built on a coiled-coil α-helical structure.

The **β-keratins**, as their name implies, contain much more β-sheet structure. Indeed, they represented the second major structural class described by Pauling and his coworkers. The β-keratins are found mostly in birds and reptiles in structures like feathers and scales.

Fibroin

The β-sheet structure is most elegantly utilized in the fibers spun by silkworms and spiders. Silkworm fibroin (**FIGURE 6.13**) contains long regions of stacked antiparallel β sheets, with the polypeptide chains

Fibers are very flexible because bonding between the sheets involves only the weak van der Waals interactions between the side chains.

Fiber axis

Fiber axis

Fiber axis

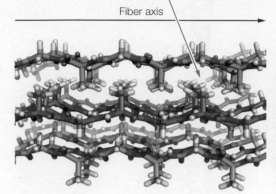

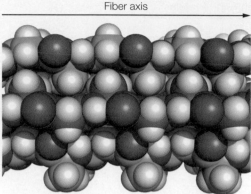

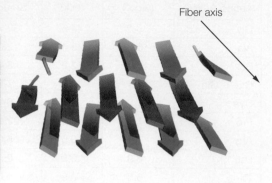

(a) A three-dimensional model of the stacked β sheets of fibroin showing alternating layers of interdigitating alanine or glycine side chains. Three sheets are shown with H atoms in gray, N atoms in blue, O atoms in red, and C atoms in either green or magenta.

(b) Space-filling rendering of the image in panel (a), showing the close contact between sheets.

(c) Cartoon model of the image in panel (a). This image is rotated 90° with respect to panel (a); thus, the fiber axis would be coming out of the page toward the viewer.

▲ **FIGURE 6.13 Theoretical model for the structure of silk fibroin.** These images were generated from PDB ID: 2slk.

running parallel to the fiber axis. The stacked sheets are held together by noncovalent interactions between the interdigitated side chains. The β-sheet regions comprise, almost exclusively, multiple repetitions of the sequence:

[Gly-Ala-Gly-Ala-Gly-Ser-Gly-Ala-Ala-Gly-(Ser-Gly-Ala-Gly-Ala-Gly)$_8$]

In silkworm fibroin, almost every other residue is glycine, which is usually followed by alanine or serine residues. In other species, the residues following the glycine residues are different, leading to differences in physical properties. This alternating pattern of residues allows the sheets to fit together and pack on top of one another in the manner shown in Figure 6.13. This arrangement of sheets results in a fiber that is strong and relatively inextensible (incapable of being stretched) because the covalently bonded chains are already stretched to nearly their maximum possible length. Yet the fibers are very flexible because bonding between the sheets involves only the weak van der Waals interactions between the side chains, which provide little resistance to bending.

● **CONCEPT** Fibroin is a β-sheet protein. Almost half of its residues are glycine.

With the fibroin protein, there are also compact folded regions that periodically interrupt the β segments, and they probably account for the limited amount of stretchiness that silk fibers do have. In fact, different species of silkworms produce fibroins with different extents of such non–β-sheet structure and corresponding differences in elasticity. The overall fibroin structure is a beautiful example of a protein molecule that has evolved to perform a particular function—to provide a tough, yet flexible, fiber for the silkworm's cocoon or the spider's web.

Collagen

Because it performs such a wide variety of functions, **collagen** is the most abundant single protein in most vertebrates. In large animals, it may make up a third of the total protein mass. Collagen fibers form the matrix material in bone, on which the mineral constituents precipitate; collagen fibers constitute the major portion of tendons; and a network of collagen fibers is an important constituent of skin. In effect, collagen holds most animals together.

The basic unit of the collagen fiber is the **tropocollagen** molecule, a *triple helix* of three polypeptide chains, each about 1000 residues in length. This triple helical struc-

● **CONCEPT** Collagen fibers are built from triple helices of polypeptides rich in glycine and proline.

ture, shown in **FIGURE 6.14(a)** and (b), is unique to collagen. The individual chains are left-handed helices, with ~3.3 residues/turn. Three of these chains wrap around one another in a right-handed sense, with hydrogen bonds extending between the chains. Examination of the model reveals that every third residue, which lies near the center of the triple helix, *must* be glycine (see Figure 6.14(a) and Table 6.3). Any side chain other than —H would be too bulky to fit within the tropocollagen triple helix. The conformation of the left-handed collagen helix is favored by the presence of proline or hydroxyproline in the tropocollagen molecule. A repetitive motif in the sequence is Gly–X–Y, where X is often proline and Y is proline or hydroxyproline (see Table 6.3); however, other residues are often tolerated in these positions. Due to its limited conformational flexibility, proline favors an extended structure, which fits well within the collagen triple helix. Like silk fibroin, collagen is a good example of how a particular kind of repetitive sequence dictates a particular structure. To properly serve

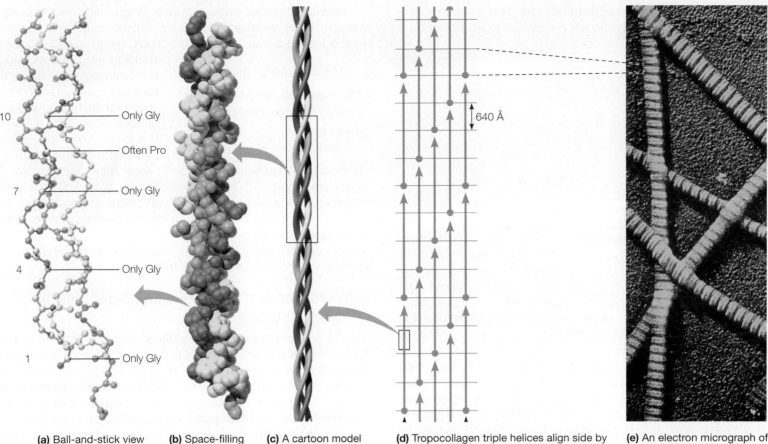

(a) Ball-and-stick view

(b) Space-filling view

(c) A cartoon model emphasizes the interwoven triple-helical secondary structure.

(d) Tropocollagen triple helices align side by side in a staggered fashion to form the collagen fiber. This regular arrangement leads to a periodic pattern of bands separated by 640 Å (blue lines).

(e) An electron micrograph of collagen shows the crisscrossing of fibers, with the 640 Å periodic pattern clearly visible in each.

▲ **FIGURE 6.14 The structure of collagen fibers.** The protein collagen is made up of tropocollagen molecules packed together to form fibers. The tropocollagen molecule is a triple helix. In panel (a), strands are color coded: green, yellow, multicolor. The multicolor strand shows N atoms in blue, O atoms in red, and C atoms in white. This strand is shown in red in panels (b) and (c).

---

the multiple functions it does, collagen exists in a large number of genetic variants in higher organisms.

(2S,4R)-4-hydroxyproline

(2S,5R)-5-hydroxylysine

Collagen is also unusual in its widespread modification of proline to hydroxyproline. Most of the hydrogen bonds between chains in the triple helix are from amide protons to carbonyl oxygens, but the —OH groups of hydroxyproline also seem to participate in stabilizing the fiber. Hydroxylation of lysine residues in collagen also occurs but is much less frequent. It plays a different role, serving to form attachment sites for polysaccharides.

The enzymes that catalyze the hydroxylations of proline and lysine residues in collagen require **vitamin C,** L-ascorbic acid. A symptom of extreme vitamin C deficiency, called **scurvy,**

● **CONNECTION** Scurvy is caused by vitamin C deficiency, which leads to failure to hydroxylate prolines and lysines in collagen.

---

is the weakening of collagen fibers caused by the failure to hydroxylate these side chains, which results in reduced hydrogen bonding between chains in the collagen fiber. Consequences are as might be expected: Lesions develop in skin and gums, and blood vessels weaken. The condition quickly improves with administration of vitamin C.

The individual tropocollagen molecules pack together in a collagen fiber in a specific way (Figure 6.14(d)). Each molecule is about 300 nm long and overlaps its neighbor by about 640 Å, producing the characteristic banded appearance of the fibers shown in Figure 6.14(e). This structure contributes remarkable strength: Collagen fibers in tendons have a tensile strength comparable to that of hard-drawn copper wire.

Part of the toughness of collagen is due to the cross-linking of tropocollagen molecules to one another via a reaction involving lysine side chains. Some of the lysine side chains are oxidized to aldehyde derivatives (allysine), which can then react with either a lysine residue, or with one another. Two allysine side chains react to produce a cross-link with this structure:

Cross-linking makes collagen less elastic and more brittle

Allysine residues

● **CONNECTION** Collagen becomes more crosslinked, and therefore more brittle, with age.

This process continues through life, and the accumulating cross-links make the collagen steadily less elastic and more brittle. As a result, bones and tendons in older individuals are more easily snapped, and the skin loses much of its elasticity. Many of the signs we associate with aging are consequences of this simple cross-linking process.

This brief overview of a few of the fibrous proteins brings out several points. First, proteins have evolved to serve a diversity of functions. Second, the fibrous proteins do this by taking advantage of the propensities of particular repetitive sequences of amino acid residues to favor one kind of secondary structure or another. Finally, posttranslational modification of proteins, including cross-linking, is an important adjunct in tailoring a protein to its function. We say more about the cellular sites for such modification in Chapter 25, when we consider the whole process of protein synthesis in detail.

These few examples do not exhaust the list of structural proteins. There are other important ones, such as actin and myosin of muscle and tubulin in microtubules. These proteins are constructed in a quite different way. Actin and myosin are discussed in Chapter 7.

6.3 Globular Proteins: Tertiary Structure and Functional Diversity

Different Folding for Different Functions

Abundant and essential as the fibrous structural proteins may be in any organism, they constitute only a small fraction of the *kinds* of proteins an organism possesses. Most of the chemical work of the cell—the synthesizing, transporting, and metabolizing—is mediated by the enormously varied **globular proteins.** These proteins are so named because their polypeptide chains are folded into compact structures very unlike the extended, filamentous forms of the fibrous proteins. Myoglobin (see Figure 6.1) is a typical globular protein. A glance at its compact three-dimensional structure, when compared with the extended structure of collagen, illustrates this qualitative difference.

Within the protein molecule, the polypeptide chain is often locally folded into one or another of the kinds of secondary structure (α helix, β sheet, and so forth) that we have already discussed. But to make the structure globular and compact, these regions of secondary structure must themselves be folded on one another. This packing together of the protein chain defines the tertiary structure of the protein (see Figure 6.2).

In nature, structure defines function. Of particular interest to biochemists is the observation that the *tertiary* structure of a globular protein is the major determinant of its functional properties. We have

● **CONCEPT** Globular proteins not only possess secondary structures but also are folded into compact tertiary structures.

learned a great deal about the biochemistry that occurs in cells as a result of the detailed structural analysis of protein molecules, largely through the use of X-ray diffraction methods (see Tools of Biochemistry 4B) and **nuclear magnetic resonance spectroscopy** or **NMR** (see Tools of Biochemistry 6A). Both of these methods provide structural information with atomic-level details. In this chapter, we identify important general features of globular protein structure. In the remaining chapters of this book, we will describe the critical functions of many different globular proteins and highlight the structural features that impart specific functional properties.

Different Modes of Display Aid Our Understanding of Protein Structure

To set the stage for a discussion of the structural features of globular proteins, let us consider the common, yet varied, ways that protein structural information is illustrated. We begin with one of the smallest globular proteins—human ubiquitin, which contains only 76 amino acid residues. This protein, synthesized in nearly all eukaryotic cells, plays a critical role in targeting other proteins for degradation. It serves here as an example of a globular protein with a relatively simple tertiary structure.

In spite of being a very small protein, ubiquitin has a complex molecular structure due to the 1231 atoms distributed among its 76 amino acids. Although all of the atoms in a protein molecule are important in defining its folded structure, showing all of them can obscure important structural features of the protein. Thus, protein scientists use various simplified representations of proteins that allow the viewer to focus on particular structural features.

FIGURE 6.15(a) shows a cartoon model of ubiquitin. Cartoon models emphasize the position of the main chain and the associated secondary structure elements. Here, helices are shown as cyan ribbons, and β strands are shown as red arrows (with the arrowhead pointing to the C-terminus of the strand). A cartoon rendering allows easy identification of the helices and sheets, and how they are packed together; thus, these models are useful for classification of proteins based on similarities in the main-chain folding (discussed in the next section). Figure 6.15(b) shows a stick model of ubiquitin. In this stick model, all the atoms (except H atoms) are shown. Even with just 76 amino acids in the ubiquitin protein, it is difficult to distinguish the main chain from the side chains in this model. Such models are useful when we zoom in to examine specific interactions, such as the hydrogen bond shown in the inset. Thus, a stick model is informative when it is important to identify a set of specific noncovalent interactions within the protein that serve to stabilize the folded structure (as in Figure 6.15(b)), or those between a protein and its binding target (as in Figure 2.1(c)).

In most cases, it is the outer surface of the protein that is of greatest importance in defining its function. We can think of this surface as the protein's "skin"—it is the part of the molecule that is in direct contact with its environment. Figure 6.15(c) shows three representations of the surface of ubiquitin. This surface is defined by rolling a water molecule over the van der Waals radii of the atoms on the surface of the protein.

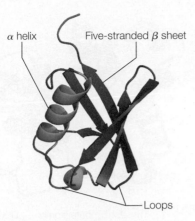

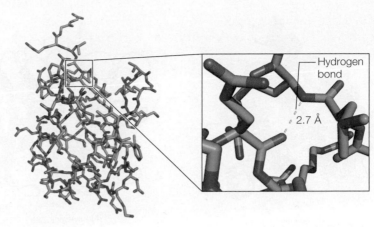

(a) A cartoon model of the protein backbone. An α helix (cyan) is packed against a five-stranded β sheet (red) composed of parallel and antiparallel strands. Loops are shown in magenta.

(b) A stick model showing the locations of all atoms (excluding H atoms). C atoms are green, N atoms are blue, and O atoms are red. The inset shows a hydrogen bond of 2.7 Å between main-chain atoms.

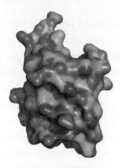

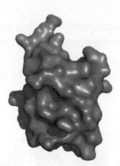

Atom coloring is the same as in panel (b).

Monochrome to emphasize the irregularities of the protein surface.

Distribution of positive (blue) and negative (red) charge density on the protein surface at pH = 7.

(c) Three surface models, showing the solvent-accessible surface of the molecule.

▲ **FIGURE 6.15** **The structure of human ubiquitin.** All images were generated from PDB ID: 1ubq and show the ubiquitin molecule in the same orientation.

In essence, Figure 6.15(c) shows all the places a water molecule could contact the outermost surface of ubiquitin, which is why this surface is also called the "solvent-accessible surface."

The leftmost image in Figure 6.15(c) was made by superimposing a surface on the ubiquitin model in panel (b). This shows the locations of oxygen (red), nitrogen (blue), and carbon (green) atoms on the surface of the molecule. Areas of red and blue indicate the presence of polar and/or charged groups, whereas larger patches of green would indicate a nonpolar surface. Ubiquitin does not show a hydrophobic patch on its surface; however, such nonpolar patches are typical of protein–protein interfaces in multiprotein complexes, or in the portions of proteins that span a membrane bilayer (discussed in Chapter 10).

The middle image in Figure 6.15(c) is monochrome, which more clearly highlights the irregularities of the surface compared to the leftmost image. Deeper depressions and clefts are frequently sites of ligand binding, or "active sites" where an enzyme promotes a specific chemical transformation on its binding target.

Finally, the rightmost image in Figure 6.15(c) shows the distribution of (+) and (−) charges on the surface of ubiquitin at pH = 7.0 (see also Figure 2.18). By convention, (+) charge is shown in blue and (−) charge in red. In Chapter 7 we will discuss the binding of antibodies to

their targets, and we will see that the binding surface of the antibody and its target are complementary with respect to both shape and charge. Thus, a representation of the charge distribution on the surface of a protein allows identification of the portion of the protein that is likely to bind to an oppositely charged molecule (e.g., proteins that bind DNA are predominantly (+) charged at pH = 7.0).

The surface renderings in Figure 6.15(c) also illustrate the important point that globular proteins are densely packed structures rather than the open structures suggested by cartoon or stick renderings. In short, there is no open space in the interior of a typical globular protein.

Varieties of Globular Protein Structure: Patterns of Main-Chain Folding

As you study more biochemistry and molecular biology, you will appreciate the enormous variety of folded structures found in proteins. At first glance, it might seem that there would be an almost infinite number of ways in which the helices and strands in globular proteins could associate to yield a stably folded structure. Yet when a large number of protein structures is examined, certain common motifs and principles emerge. For example, the secondary structures of some proteins are composed almost solely of either α helix or β sheet, whereas others

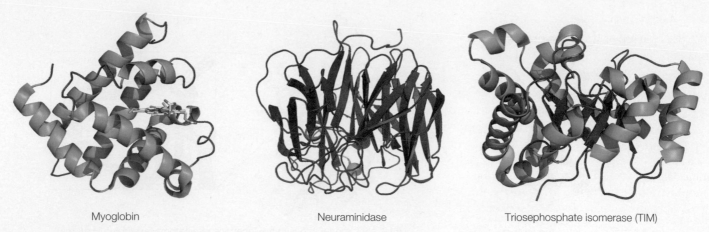

Myoglobin   Neuraminidase   Triosephosphate isomerase (TIM)

▲ **FIGURE 6.16** **Examples of diversity in protein tertiary structures.** Cartoon images were generated from the following PDB IDs: 1mbn (myoglobin), 1nn2 (neuraminidase), and 1n55 (triosephosphate isomerase). The α helices are shown in cyan, β sheets are red, and loops are magenta.

---

have both in roughly equal measure. **FIGURE 6.16** shows a few examples: Myoglobin is mostly α helix, neuraminidase (an enzyme important for viral pathogenesis) is mostly β sheet, and triosephosphate isomerase (TIM, an enzyme important for the utilization of glucose as a nutrient) has both helix and sheet structures. Thus, the first common principle is

that we can classify a protein based on the dominant secondary structural motifs. We develop this idea further in the next section.

The second common principle is that many proteins are made up of more than one **domain.** A domain is a compact, locally folded region of tertiary structure of roughly 150–250 amino acids. Domains are interconnected by the polypeptide strand that runs through the whole molecule. Multiple domains are especially common in the larger globular proteins, whereas relatively small proteins like ubiquitin, and those shown in Figure 6.16, tend to be single folded domains. Different domains often perform differing functions, and a given domain type may be found in several different proteins. For example, an ATP-binding domain may appear in many different proteins, each of which will utilize the bound ATP to perform a function specific to that protein.

A third common principle is that domains may themselves be composed of repeating secondary structure motifs, sometimes called supersecondary structures. For example, a β-strand sequence linked to an α-helix sequence is referred to as a "β/α motif," and a β strand linked to two α-helical sequences is known as a $\beta/\alpha/\alpha$ motif. The enzyme TIM is a good example of a single-domain protein composed of repeating structural motifs. The TIM main chain is composed of four β/α motifs and four $\beta/\alpha/\alpha$ motifs (**FIGURE 6.17(a)**) linked together by loops of indeterminate structure (Figure 6.17(b)). The eight β strands in these eight structural motifs associate in the center of the molecule to form a so-called β-barrel (or TIM-barrel). The β-barrel is one of the most common domain folds found in proteins. Figure 6.17(c) shows

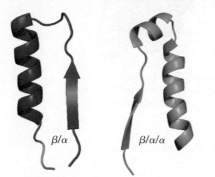

β/α $\beta/\alpha/\alpha$

(a) The N-terminal β/α supersecondary structure motif is shown in blue, and one of four $\beta/\alpha/\alpha$ supersecondary structure motifs is shown in green.

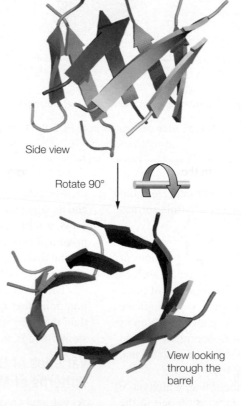

Side view

Rotate 90°

N-terminus

View looking through the barrel

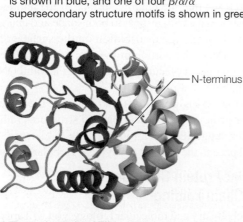

(b) The main chain of TIM, which is an example of a "β-barrel" domain. The cartoon is colored in a "rainbow" fashion where the N-terminus is blue, and the C-terminus is red. The color scheme matches those in panels (a) and (c).

(c) The association of eight β strands to form the β-barrel. Here, the associated helices have been stripped away for clarity. The upper image shows the β-barrel from the side, and the lower image is a view looking through the barrel.

▲ **FIGURE 6.17** **Two examples of repeated supersecondary structures within a protein domain.** Features of the TIM backbone are shown in cartoon renderings (PDB ID: 1n55).

● **CONCEPT** A protein domain is a compact, locally folded region of tertiary structure. Smaller proteins typically contain a single domain. Larger proteins may contain several domains.

Over 1400 distinct protein domain structures (or "folds") have been classified and organized into searchable online databases (see Appendix II). These databases allow identification of potential functional and evolutionary relationships among proteins that share similar structural domains. To illustrate the immense structural variation observed in globular proteins, let us consider how one of these databases—CATH (**C**lass, **A**rchitecture, **T**opology, and "**H**omologous superfamily")——sorts the ~170,000 protein domain structures found in the PDB (Protein Data Bank) into one of these 1400 categories.

FIGURE 6.18 shows seven distinct domain folds out of the ~1400 cataloged in CATH. However, as suggested by the first common principle discussed earlier in this section, this immense variation can be distilled down to four basic folding patterns in globular proteins: (1) those that are built about a packing of α helices, (2) those that are constructed on a framework of β sheets, (3) those that include both helices and sheets, and (4) those that contain little helix or sheet structure. In the CATH classification system, "class" refers to one of these four categories based on the prevalent secondary structure in the domain fold (top row, Figure 6.18). If we look more closely at the "$\alpha + \beta$" class, we find 14 different general shapes, or "architectures," for domain folds that include α-helix and β-sheet structure. Two of these architectures, "3-layer $\alpha/\beta/\alpha$ sandwich" and "α/β barrel," are shown in the second row of Figure 6.18. The next level of classification is called topology. In the context of protein structure, **topology** refers to the order in which the secondary structural features are connected in the protein main chain. For example, in different proteins, two α helices and one β strand might be linked "helix-helix-strand" or "helix-strand-helix" or "strand-helix-helix." Each of these represents a different topology for these three secondary structural elements. Thus, the β/α and $\beta/\alpha/\alpha$ motifs shown in Figure 6.17(a) would be examples of different main-chain topologies. Row three of Figure 6.18 shows two examples of the 126 different topologies for the "3-layer $\alpha/\beta/\alpha$ sandwich" architecture: "aminopeptidase" and the "Rossman fold."

● **CONCEPT** The majority of globular protein structures can be broadly classified as "mainly α," "mainly β," and "$\alpha + \beta$." A small number of globular proteins has little α or β secondary structure.

the barrel-like structure formed by the eight β strands (the associated helices have been stripped away for clarity). In the case of TIM-barrels, a repeat of a relatively simple motif (β/α or $\beta/\alpha/\alpha$) led to the evolution of a very stable domain structure.

The **Rossman fold** is a common domain structure for an important class of enzymes that bind a nicotinamide adenine dinucleotide (NAD) cofactor and carry out

reactions critical to the production of ATP in metabolic processes (see Figure 3.10). There are 130 distinct "homologous superfamilies" of proteins that include a Rossman fold topology. The proteins within each superfamily appear to be homologous (i.e., they have a common evolutionary ancestor). There are 8249 different domain entries within

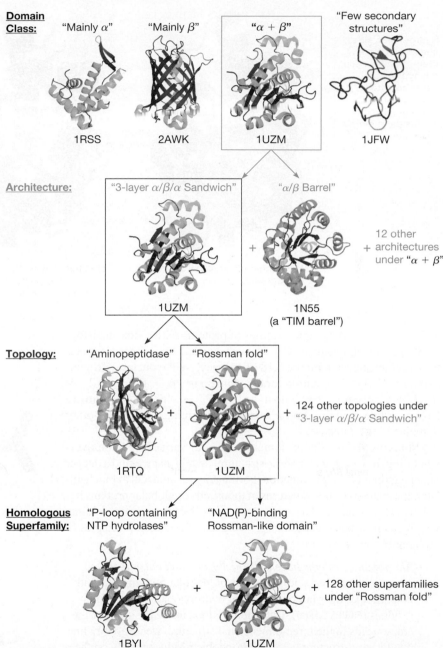

▲ **FIGURE 6.18 Structural diversity in globular protein domains.** Top row: Representatives of the four main classes of domain structure are shown in cartoon rendering with α helices in cyan, β strands in red, and connecting loops in magenta. Second row: two representatives of the 14 architectures within the "$\alpha + \beta$" class. Third row: two representatives of the 126 topologies found in the "3-layer $\alpha/\beta/\alpha$ sandwich" architecture. Fourth row: two representatives of the 130 homologous superfamilies found in the "Rossman fold" topology. PDB IDs: Ribosomal protein S7 (1rss); green fluorescent protein (2awk); β-ketoacyl ACP reductase (1uzm); HIV1 transactivator protein (1jfw); triosephosphate isomerase (1n55); leucyl aminopeptidase (1rtq); dethiobiotin synthase (1byi).

VLSEGEWQLV LHVWAKVEAD VAGHGQDILI RLFKSHPETL EKFDRFKHLK
TEAEMKASED LKKHGVTVLT ALGAILKKKG HHEAELKPLA QSHATKHKIP
IKYLEFISEA IIHVLHSRHP GDFGADAQGA MNKALELFRK DIAAKYKELG
YQG

(a) The amino acid sequence of sperm whale myoglobin. Hydrophobic (green), hydrophilic (magenta), and ambivalent (black) residues appear to be scattered throughout the sequence.

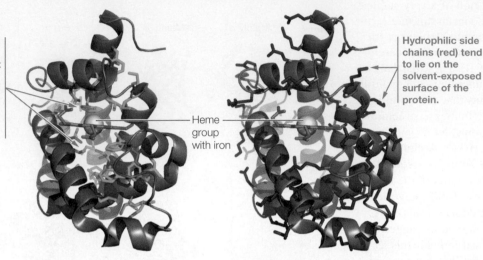

Hydrophobic side chains (shown in green) cluster about the hydrophobic heme cofactor (orange with iron ion in gray) and on the inside of the molecule.

Hydrophilic side chains (red) tend to lie on the solvent-exposed surface of the protein.

Heme group with iron

(b) The three-dimensional structure of the same protein.

(c) In this view of the myoglobin structure, the hydrophilic side chains are shown in red.

◀ **FIGURE 6.19** The distribution of hydrophilic and hydrophobic residues in globular proteins. PDB ID: 1mbn.

the "NAD(P)-binding superfamilies of proteins with a Rossman-fold (~3% of total entries in the CATH database), and one of these includes the structure shown in Figure 6.18 (β-ketoacyl ACP reductase). In summary, the CATH designation for this domain is: "$\alpha + \beta$" → "3-layer $\alpha/\beta/\alpha$ sandwich" → "Rossman fold" → "NAD(P)-binding Rossman-like domain," and it is structurally related to ~8200 other domains in the database.

Studying the details of domain structural variation can be overwhelming; however, it is in these structural details that we find a deeper understanding of protein function. Fortunately for students of biochemistry, the analysis of the structures of thousands of globular proteins has led to the formulation of a few general rules governing tertiary folding, which can now be added to our growing list of common principles of globular protein structure:

- *All globular proteins have a defined inside and outside.* If we examine the amino acid sequences of globular proteins, we find no particular distribution pattern of hydrophobic or hydrophilic residues (**FIGURE 6.19(a)**). But when we look at the positions of the amino acids in the three-dimensional structure, we invariably find that tertiary structure places hydrophobic residues mostly on the inside (Figure 6.19(b)), whereas the hydrophilic residues are on the surface, in contact with water (Figure 6.19(c)).

- *β sheets are usually twisted, or wrapped into barrel structures.* Examples can be seen in Figures 6.16–6.18. It is probable that the structure of silk fibroin is not exactly planar, as depicted in the model in Figure 6.13, but slightly twisted.

- *The polypeptide chain can turn corners in a number of ways,* to go from one β segment or α helix to the next. One kind of compact turn is called a β turn (**FIGURE 6.20**). There are several

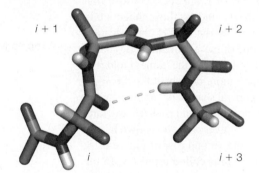

Type I

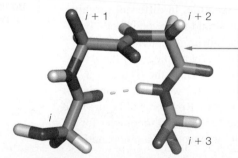

In the type II turn, residue $i+2$ is usually glycine, presumably because a bulky R group would clash with the carbonyl oxygen of residue $i+1$.

Type II

▲ **FIGURE 6.20 Examples of β turns.** Each of these turn types allows an abrupt change in polypeptide chain direction. Here H atoms are gray, N atoms blue, O atoms red, main-chain C atoms green, and side-chain C_β atoms orange. In a β turn, the carbonyl O of residue i is hydrogen-bonded (yellow dashed line) to the amide H of residue $i + 3$. The type I turn here is from residues 67–70 of myohemerythrin (PDB ID: 2mhr), and the type II turn is from residues 114–117 of the same protein.

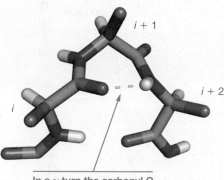

In a γ turn the carbonyl O of residue *i* is hydrogen-bonded (yellow dashed line) to the amide H of residue *i*+2.

▲ **FIGURE 6.21** A γ turn. Here H atoms are gray, N atoms blue, O atoms red, main-chain C atoms green, and side-chain C$_\beta$ atoms orange. Proline is often found at residue *i* + 1 in γ turns. The turn here is from residues 25–27 of thermolysin (PDB ID: 4d9w).

varieties of β turn, each able to accomplish a complete reversal of the polypeptide chain direction in only four residues; in each case the carbonyl of residue *i* hydrogen-bonds to the amide hydrogen of residue *i* + 3. In the even tighter γ turn, bonding is to residue *i* + 2 (**FIGURE 6.21**). Proline promotes bending of the main chain; thus, it often is found in turns, and also as a "breaker" of α helices, because this proline amide is not an H bond donor (see Figure 6.4(a)). Bends and turns most often occur at the surface of proteins.

- *Not all parts of globular proteins can be conveniently classified as helix, β sheet, or turns.* Examination of Figures 6.16 and 6.18 reveals many strangely contorted loops and folds in the chains (the regions shown in magenta). These have sometimes been referred to as "random coil" regions, but this term is a misnomer because such sections of the chain are not flexible in the same way that a true random coil is. Rather, X-ray diffraction and NMR data indicate that such regions do, in fact, possess a well-defined fold. We might call these *irregularly structured regions.* Several proteins also have intrinsically unstructured regions, a feature that is particularly important in a class of signaling proteins that bind to several different targets.

6.4 Factors Determining Secondary and Tertiary Structure

The Information for Protein Folding

What ultimately determines the complex mixture of secondary and tertiary folding that characterizes each globular protein? Much evidence indicates that *most of the information for determining the three-dimensional structure of a protein is carried in the amino acid sequence of that protein.* This can be demonstrated by experiments in which the *native,* or folded, three-dimensional structure is perturbed by changing the temperature or pH of a protein solution. If we raise the temperature sufficiently, or make the pH extremely acidic or alkaline, or add to the solvent certain kinds of organic molecules like

alcohols or urea, the protein structure will unfold (**FIGURE 6.22**). As with nucleic acids, this process is called **denaturation** because the natural structure of the protein has been lost, along with many of its specific functional properties.

In diagrams such as Figure 6.22(a), the unfolded chain is often drawn as a random coil, with freedom of rotation about bonds in both the polypeptide backbone and the side chains. Much recent evidence suggests that this is a gross oversimplification. In many cases the unfolded state for a protein is a dynamic ensemble of largely unstructured, extended conformations; however, even in the "unfolded" state, regions of structure may persist.

● **CONCEPT** Amino acid sequence (primary structure) determines secondary and tertiary structure.

In the classic experiment by Chris Anfinsen depicted in Figure 6.22(a), and recognized with the Nobel Prize in 1972, the enzyme ribonuclease A (RNase A) was denatured by addition of urea, and the four native disulfide bonds were reduced by addition of β-mercaptoethanol (BME). RNase A catalyzes the hydrolysis of ribonucleic acids. When RNase A is denatured, its tertiary and secondary structures are lost, and it can no longer catalyze RNA cleavage. The denaturation process can be tracked by various physical measurements, as shown in Figure 6.22(b). Remarkably, the total disruption of RNase A structure was shown by Anfinsen to be reversible. If the urea is removed by dialysis, the reduced RNase A will spontaneously refold into its native structure. Oxidation of the refolded protein in air restores the native disulfide bonds and full enzymatic activity. However, if the unfolded RNase A is first oxidized, followed by removal of the urea, a mixture of protein molecules with randomly formed disulfide bonds results. This mixture has ~1% of the original enzymatic activity of native RNase A, suggesting that molecules with native disulfide bonds account for ~1% of the mixture.

Random formation of four disulfides from eight cysteines predicts 105 possible different combinations. The first —SH group to pick a partner will have seven choices for random disulfide bond formation. That leaves the second —SH group with five possible bonding partners, the third —SH with three possible bonding partners, and the last —SH with only one, so there are $7 \times 5 \times 3 \times 1$, or 105, equally probable combinations for the formation of four random disulfide bonds. One of these combinations is the set of four native disulfide bonds, which accounts for the observed 1% activity when oxidation of disulfides precedes refolding ($1/105 \approx 1\%$). The observation that refolding, followed by disulfide bond formation, restores *full* enzymatic activity indicates that the folding of RNase A to its thermodynamically favored native conformation positions the —SH groups for correct disulfide bond formation.

Anfinsen's work showed that a protein can self-assemble into its functional conformation, and it needs no information to guide it other than that contained in its amino acid sequence. The same phenomenon has been observed experimentally for many other proteins; thus, we might expect a newly synthesized polypeptide in a cell to spontaneously fold into the proper active conformation. As we shall see later in this chapter, the actual process in vivo is more complicated. To avoid misfolding or aggregation, some proteins interact with cellular machinery that guides the folding process. Nonetheless, the basic principle of self-assembly of secondary and tertiary structure seems to be the general rule.

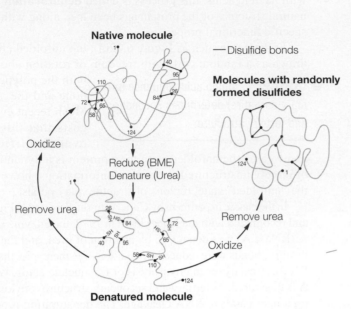

(a) In the classic RNase A refolding experiment of Anfinsen, renaturation before disulfide bond formation yields the active, native conformation, but disulfide bond formation before renaturation yields multiple conformations with little recovery of enzymatic activity.

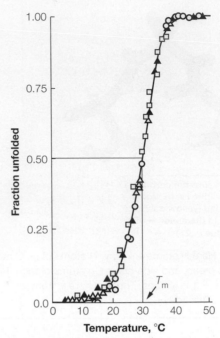

(b) Thermal denaturation of RNase A monitored by various physical methods. Differences between the native and denatured conformations can be detected by several of the spectroscopic methods discussed in Tools of Biochemistry 6A. All three techniques indicate the same fraction unfolding as a function of increasing temperature. Measurements of a second denaturation after cooling (▲) produce the same curve, showing that the process is reversible with a melting temperature (T_m) of 30.0 °C.

▲ **FIGURE 6.22 The denaturation and refolding of ribonuclease A.**
(a) This schematic drawing depicts the classic RNase A refolding experiment of Anfinsen. **(b)** Thermal denaturation of RNase A monitored by various physical methods. This graph shows the fraction of protein that is denatured, as measured by the increase in solution viscosity (□), change in optical rotation at 365 nm (o), or change in UV absorbance at 287 nm (Δ). The experiments were conducted at pH 2.1, ionic strength 0.019 M. Under physiological conditions, RNase A is much more stable, not denaturing until about 70–80 °C.

The Thermodynamics of Folding

Since proteins fold spontaneously in buffer solutions that mimic the intracellular environment (pH ~ 7, 150 mM NaCl), the folding of a globular protein is clearly a thermodynamically favorable process under physiological conditions. In other words, the overall free energy change for folding must be negative. But this negative free energy change is achieved by a balance of several thermodynamic factors.

Conformational Entropy

The folding process, which involves going from a multitude of "random-coil" conformations to a *single* folded structure,* involves a

*For the sake of simplicity, we will present an idealized case for which there is a single well-defined native state conformation. Because most proteins require some conformational flexibility to perform their function, it is more proper to acknowledge that the native state comprises a limited number of well-defined conformations with similar values of free energy. In any event, the entropy of the native state ensemble of conformations is much less than that of the denatured state ensemble.

decrease in randomness and thus a decrease in entropy ($\Delta S < 0$). This change is termed the **conformational entropy** of folding:

random coil (higher entropy) $\longrightarrow$ folded protein (lower entropy)

As we learned in Chapter 3, the free energy equation, $\Delta G = \Delta H - T\Delta S$, shows that a negative ΔS makes a *positive* contribution to ΔG. In other words, the conformational entropy change works *against* folding. To seek the explanation for an overall negative ΔG, we must seek features of protein folding that yield either a large negative ΔH or some other *increase* in entropy upon folding. Both can be found.

● **CONCEPT** The decrease in conformational entropy when a protein folds disfavors folding. This is compensated in part by energy stabilization through internal noncovalent bonding.

The major source for a negative ΔH is energetically favorable interactions between groups within the folded molecule. These include many of the noncovalent interactions described in Chapter 2.

Charge–Charge Interactions

Charge–charge interactions can occur between positively and negatively charged side-chain groups. For example, a lysine side-chain ε-amino group may be close to the γ-carboxyl group of some glutamic acid residue in the folded protein. At neutral pH, one group will be charged positively and the other negatively, so an electrostatic attractive force exists between them. Such interactions are also called **salt bridges.** These ionic bonds are broken if the protein is taken to pH values high enough or low enough that either side chain loses its charge. This loss of salt bridges is a partial explanation for acid or base denaturation of proteins. The mutual repulsion between pairs of the numerous similarly charged groups that are present in proteins in very acidic or basic solutions contributes further to the instability of the folded structure under these conditions (see Figure 2.18).

Internal Hydrogen Bonds

Many of the amino acid side chains carry groups that are either good hydrogen bond donors or good acceptors. Examples are the hydroxyl groups of serine or threonine, the amino groups and carbonyl oxygens of asparagine or glutamine, and the ring nitrogens in histidine. Furthermore, if amide protons or carbonyls in the polypeptide backbone are not involved in secondary structure formation, they are potential candidates for interaction with side-chain groups. A network of several types of internal hydrogen bonds is seen in the portion of the enzyme lysozyme shown in **FIGURE 6.23**. As we have seen before, hydrogen bonds are relatively weak in aqueous solution, but their large number can add considerably to stability.

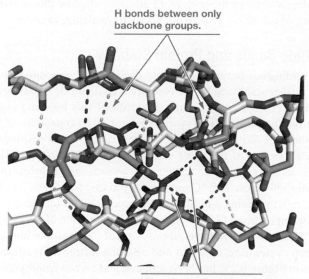

H bonds between only backbone groups.

H bonds that include a side-chain group to either a backbone group or another side-chain group.

▲ **FIGURE 6.23 Details of hydrogen bonding in a typical protein.** A network of hydrogen bonds within the enzyme lysozyme is illustrated. O atoms are red, N atoms blue, main-chain C atoms gray, and side-chain C atoms green. Two kinds of hydrogen bonds are shown: H bonds between only backbone groups are shown as yellow dashes, and H bonds that include a side-chain group to either a backbone group or another side-chain group are shown as magenta dashes. Image generated from PDB ID: 2lyz. The region of the protein spanning residues 40–72 is shown.

Van der Waals Interactions

As discussed in Chapter 2, the weak induced dipole–induced dipole interactions between nonpolar groups can also make significant contributions to protein stability because in the folded protein, nonpolar groups are densely packed and thus make a large number of van der Waals contacts.

The change in enthalpy for folding, $\Delta H_{U \rightarrow F}$, is dominated by the differences in noncovalent bonding interactions between the unfolded and folded states:

$$\Delta H_{U \rightarrow F} = H_{\text{folded}} - H_{\text{unfolded}} \tag{6.1}$$

where the unfolded state is characterized by noncovalent interactions between the extended protein chain and solvent water molecules, and the folded state includes many fewer interactions with solvent and many more intramolecular interactions instead. Typically, the only new *covalent* bonds that form upon folding are disulfide bonds (note, however, that most proteins do not contain disulfide bonds).

Each individual noncovalent interaction can contribute only a small amount (at most only a few kilojoules per mole) to the overall negative enthalpy of interaction. But the sum of the contributions of many interactions can yield significant stabilization to the folded structure. Examples of the total enthalpy changes for folding are given for some representative proteins in **TABLE 6.4**. In many cases, a favorable energy contribution from the sum of intramolecular interactions more than compensates for the unfavorable entropy of folding.

The Hydrophobic Effect

Yet another factor makes a major contribution to the thermodynamic stability of many globular proteins. Recall from Chapter 2 that hydrophobic substances in contact with solvent water cause the water molecules to form clathrate, or cagelike, structures around them (see Figure 2.14). This ordering corresponds to a loss of randomness in the solvent; thus, the entropy of the solvent is decreased. Suppose a protein contains, in its amino acid sequence, a substantial number of residues with hydrophobic side chains (for example, leucine, isoleucine, and phenylalanine). When the polypeptide chain is in an unfolded form, these residues are in contact with water and cause ordering of the surrounding water structure into clathrates. When the chain folds into a globular structure, the hydrophobic residues become buried within the molecule (see Figure 6.19(b)), and the water molecules that were ordered around the solvent-exposed hydrophobic surfaces in the denatured protein are now released from the clathrates, thereby gaining

TABLE 6.4 Thermodynamic parameters for folding of some globular proteins at 25 °C in aqueous solution

Protein	ΔG (kJ/mol)	ΔH (kJ/mol)	ΔS (kJ/K·mol)
Ribonuclease A	−46	−280	−0.79
Chymotrypsin	−55	−270	−0.72
Lysozyme	−62	−220	−0.53
Cytochrome *c*	−44	−52	−0.027
Myoglobin	−50	0	+0.17

Note: Data adapted from *Journal of Molecular Biology* (1974) 86:665–684, P. L. Privalov and N. N. Khechinashvili, A thermodynamic approach to the problem of stabilization of globular protein structure: A calorimetric study. Each dataset has been taken at the pH value where the protein is maximally stable; all are near physiological pH. Data are for the folding reaction: Denatured (unfolded) ⇌ native (folded).

freedom of motion. Thus, the randomness of the solvent is *increased* by internalizing hydrophobic groups via folding.

For the process of protein folding, the *overall* change in entropy is the sum of the change in conformational entropy for the polypeptide chain *and* the change in entropy of the solvent water molecules:

$$\Delta S_{U \to F} = \Delta S_{\text{protein}} + \Delta S_{\text{solvent}} \qquad (6.2)$$

In Equation 6.2, $\Delta S_{\text{protein}}$ is the conformational entropy change for folding, which is *negative;* however, this unfavorable entropy change is counteracted by the favorable entropy increase for the solvent ($\Delta S_{\text{solvent}}$). In summary, the burial of hydrophobic surface area in the solvent-inaccessible core of the protein acts to stabilize the folded state by making the value of $\Delta S_{U \to F}$ more positive. This source of protein stabilization is referred to as the **hydrophobic effect.** Examples of the importance of the hydrophobic effect can be seen in Table 6.4. The very small negative ΔS for cytochrome *c* and the positive value for myoglobin are a consequence of the hydrophobic effect in these proteins. Indeed, the stability of myoglobin comes mainly from the hydrophobic effect.

● **CONCEPT** The burying of hydrophobic groups within a folded protein molecule produces a stabilizing entropy increase known as the hydrophobic effect.

In summary, the stability of the folded structure of a globular protein depends on the interplay of three factors:

1. The conformational entropy change, $\Delta S_{\text{protein}}$, which favors the unfolded state

2. The enthalpy change, ΔH (associated with changes in noncovalent bonding interactions), which generally favors the folded state

3. The favorable entropy change of the solvent due to the release of water from clathrates, $\Delta S_{\text{solvent}}$, which occurs when solvent-exposed hydrophobic groups become buried within the molecule.

Thus, factor 1 works against folding, whereas factors 2 and 3 favor folding. A picture of the way these components might contribute to the free energy of folding for some protein is shown in **FIGURE 6.24**.

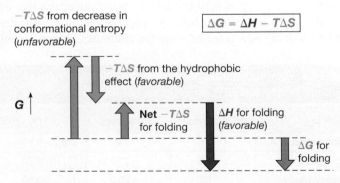

▲ **FIGURE 6.24 Contributions to the free energy of folding of globular proteins.** The conformational entropy change works against folding, but the enthalpy of internal interactions and the entropy change from the hydrophobic effect favor folding. Summing these three quantities makes the total free energy of folding negative (favorable); thus, the folded structure is stable.

The relative contributions to the stability of the folded protein from enthalpic interactions and the hydrophobic effect differ between different proteins (see Table 6.4), but the overall consequence is the same: Some particular folded structure corresponds to a free energy minimum for the polypeptide under physiological conditions. This is why the protein folds spontaneously.

Examination of the data in Table 6.4 reveals another important aspect of protein stability. On the whole, proteins are relatively unstable molecules. The free energy difference between the native and denatured states is modest—typically, 20–60 kJ/mol. If we compare that value to the strength of a hydrogen bond (5–15 kJ/mol; see Figure 2.2), we see that the free energy stabilizing a typical folded protein is equivalent to the energy change associated with formation, or disruption, of a few noncovalent bonds. Thus, the native, *functional*, conformations of many proteins can be easily perturbed by changes in intracellular conditions, such as increased temperature (~5 °C) or fluctuations in pH (typically in the range of +/−2 pH units). This underscores the need for tight control of pH and temperature within cells. This relatively low thermodynamic stability of proteins plays a role in the regulation of intracellular processes. For many proteins there is a well-established relationship between the activity of the protein and its native (folded) conformation (e.g., as in the case of RNase A in Anfinsen's denaturation studies). Thus, the activity of certain proteins can be regulated by disrupting the native conformation. In later chapters, the various mechanisms for achieving such disruption of the active conformation will be described in detail. Some of these are reversible—allowing switching between active and inactive conformations in response to cellular conditions—and others are irreversible—resulting in the complete degradation of the protein. In all cases, the low thermodynamic stability of the native protein allows such regulation to occur.

Disulfide Bonds and Protein Stability

Once folding has occurred, the three-dimensional structure is in some cases further stabilized by the formation of disulfide bonds between cysteine residues. An extreme example of this bonding is found in the bovine pancreatic trypsin inhibitor (BPTI) protein, depicted in **FIGURE 6.25**. With three —S—S— bridges in 58 residues, this molecule is one of the most stable proteins known. The BPTI is quite inert to unfolding reagents, such as urea, and exhibits thermal denaturation below 100 °C only in very acidic solutions. At pH 2.1, where most proteins would be denatured at room temperature, the melting temperature (T_m) for reversible denaturation of BPTI is about 80 °C (**FIGURE 6.26**). However, if only *one* of the disulfide bonds (that between cysteine residues 14 and 38—the leftmost disulfide bond shown in Figure 6.25) has been reduced and carboxymethylated, the T_m is decreased to 60 °C. When all the disulfide bonds in BPTI are reduced, the protein is unfolded at room temperature; yet, upon reoxidation of the sulfhydryls, native protein with the three correct disulfide pairings is efficiently formed. Many studies of this and other proteins containing disulfide bonds (e.g., the RNAse A described in the previous section) indicate that correct pairing is regained in almost 100% of the molecules if sufficient time is allowed. As discussed for RNase A refolding, this finding must mean that it is the preferred folding of the

● **CONCEPT** Some folded proteins are stabilized by internal disulfide bonds, in addition to noncovalent forces.

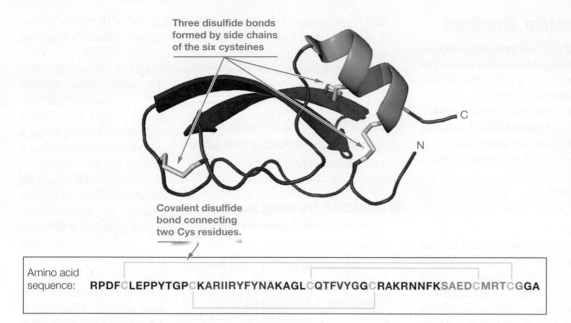

Three disulfide bonds formed by side chains of the six cysteines

Covalent disulfide bond connecting two Cys residues.

Amino acid sequence: **RPDF**C**LEPPYTGP**C**KARIIRYFYNAKAGL**C**QTFVYGG**C**RAKRNNFK**S**AED**C**MRTC**GG**A**

▲ **FIGURE 6.25 Disulfide bonds in bovine pancreatic trypsin inhibitor (BPTI).** The main chain of BPTI is shown in cartoon rendering, and the side chains of the six cysteines that form the three disulfide bonds are shown as yellow sticks. The amino acid sequence of BPTI is shown with the strands colored red, the helix cyan, and the cysteines yellow. The covalent disulfide bonds are shown schematically as yellow lines connecting two cysteine residues.

protein that places the —SH groups in position for correct pairing. The corollary of this statement is that the disulfide bonds are not themselves essential for correct refolding. They do, however, contribute to the stability of the structure once it is folded.

The apparent advantage of disulfide-bond formation raises a question: Why don't *most* proteins have disulfide bonds? In fact, such bonds are relatively rare and are found primarily in proteins that are exported from cells, such as ribonuclease, BPTI, and insulin. One explanation is that the environment inside most cells is reducing and tends to keep sulfhydryl groups in the reduced state. External environments, for the most part, are oxidizing and stabilize —S—S— bridges.

Prosthetic Groups, Ion-Binding, and Protein Stability

The folded conformation of a protein may also be stabilized by the binding of a metal ion, as is the case for a class of DNA-binding proteins called **zinc finger proteins** (**FIGURE 6.27(a)**), or by binding to a **cofactor** or a **prosthetic group** (i.e., a small molecule that is required for the protein to be active). When the ion or cofactor is absent, the "stripped" protein is called an **apoprotein,** and when the ion or cofactor is bound, the resulting complex is called a **holoprotein** (Figure 6.27(b)). Formation of the holoprotein stabilizes the active conformation because favorable noncovalent interactions are formed between the apoprotein and the bound ion or prosthetic group.

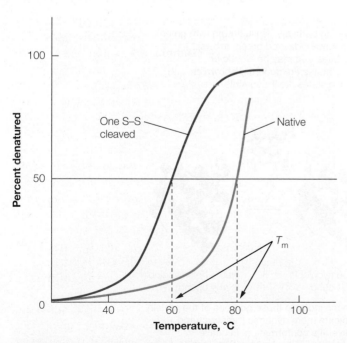

▲ **FIGURE 6.26 Thermal denaturation of BPTI.** The percent denaturation as a function of temperature at pH 2.1 is indicated for the native protein and for the protein in which the Cys 14–Cys 38 disulfide bond has been reduced and carboxymethylated.

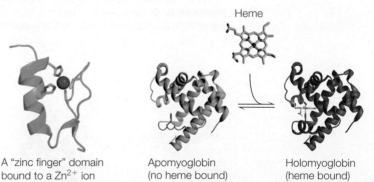

A "zinc finger" domain bound to a Zn²⁺ ion

Apomyoglobin (no heme bound)

Holomyoglobin (heme bound)

Heme

(a) The side chains from two histidines and two cysteines bind specifically to a Zn^{2+} ion (red sphere) in a zinc finger domain. The Zn finger domain is a common structure among certain DNA-binding proteins (PDB ID: 1tf6).

(b) Heme binding to apomyoglobin stabilizes the folded structure of myoglobin and gives the holoprotein its red color. Holomyoglobin includes both the myoglobin protein and the heme prosthetic group.

▲ **FIGURE 6.27 Ion or prosthetic group binding increases protein stability.** Images in panel (b) courtesy of J. S. Olson (Rice University)

6.5 Dynamics of Globular Protein Structure

The structural and thermodynamic descriptions we have developed so far tend to give too static a picture of globular proteins. The models of folded proteins produced from structural studies often give the impression of a rigid structure. Likewise, the thermodynamic analysis of folding concentrates on the initial (unfolded) and final (folded) states. However, it is now recognized that globular proteins fold via complex *kinetic* pathways and that even the folded structure, once attained, is a dynamic structure. The following section explores some of these dynamic aspects of globular proteins.

Kinetics of Protein Folding

The folding of globular proteins from their denatured conformations is a remarkably rapid process, often complete in less than a second under physiological conditions. This observation has been of profound interest to biochemists because at first glance, proper folding to reach the well-defined structure typical of globular proteins would seem to be very difficult. This point of view was dramatically expressed in "Levinthal's paradox," first enunciated by Cyrus Levinthal in 1968: Assuming just two possible values for each angle ϕ and ψ in a polypeptide chain such as RNase A (124 residues), there are $\sim 10^{77}$ different conformations possible for the protein. Even if the protein could adopt a new conformation every 10^{-13} seconds (roughly the time needed to rotate a covalent bond to generate a "new" conformation), it would still take $\sim 10^{56}$ years to sample a significant fraction of them! Yet RNase A is observed to fold, in vitro, in about 1 minute. Clearly, something is wrong with such a calculation.

Levinthal's paradox assumes that folding is a random process; thus, a protein must sample a vast number of *possible* conformations to find the desired native conformation. Years of experimental and theoretical studies have shown that protein folding is not a completely random search through the vast number of all possible conformations. Rapid kinetic studies, using a variety of physical techniques to monitor different aspects of protein structure, show that folding takes place through a series of partially structured intermediate states. One particularly well-studied intermediate is the so-called **molten globule.** The molten globule is a compact, partially folded intermediate state that has native-like secondary structure and backbone folding topology but lacks the defined tertiary structure interactions of the native state. These observations led initially to the classical "pathway" model of folding, depicted in **FIGURE 6.28**.

● **CONCEPT** Protein folding can be rapid but seems to involve well-defined intermediate states.

● **CONCEPT** The molten globule is a compact, partially folded intermediate state that has native-like secondary structure and backbone folding topology but lacks the defined tertiary structure interactions of the native state.

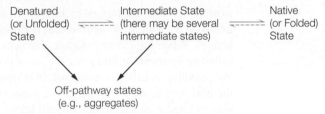

(a) A simple schematic for protein folding is shown. Here, the "folding pathway" is illustrated with green arrows; the "unfolding pathway" is the reverse of the folding pathway (dashed green arrows). Off-pathway states include aggregates and other non-native states that may be kinetic or thermodynamic "dead-ends" (red text); thus, the paths to these states are generally shown as irreversible (red arrows). In fact, not all pathways leading to such states are irreversible.

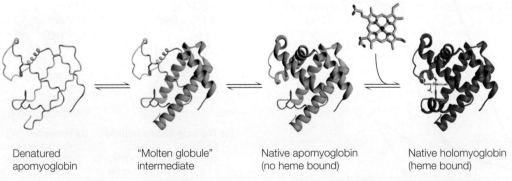

(b) A pathway model for the folding of myoglobin. The wider ribbons indicate those portions of the protein that adopt native-like conformation, while the thin ribbons indicate more dynamic ("unstructured") regions. In the molten globule state, three helices are in a native-like conformation. In native apomyoglobin, most of the protein is folded; however, heme binding is required to stabilize the fully folded conformation of holomyoglobin.

▲ **FIGURE 6.28** A simplified representation of the folding pathway for a protein.
Images courtesy of J. S. Olson (Rice University).

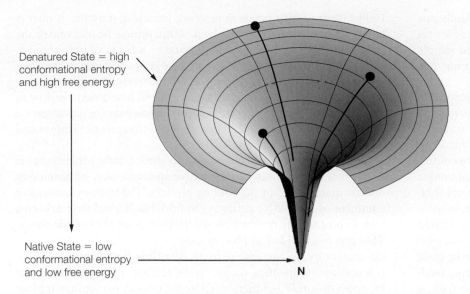

Denatured State = high conformational entropy and high free energy

Native State = low conformational entropy and low free energy

N

(a) Highly idealized energy landscape for which all trajectories lead to productive folding of the native state.

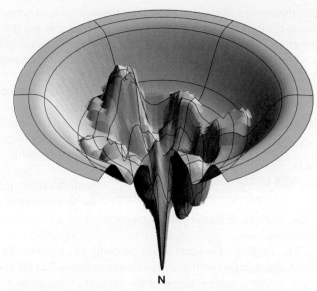

N

(b) A more "rugged" energy landscape. Here, many different paths are possible, some of which lead "downhill" with no local energy minima and give rapid folding. Others may lead to conformations corresponding to local energy minima (i.e., stable intermediate states), which may slow folding (see Figure 6.30).

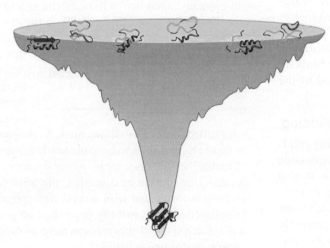

(c) A relatively smooth energy landscape with no significant energy barriers between the denatured and native states. This landscape describes so-called *two-state folding* because the only significantly populated states are the denatured and native states.

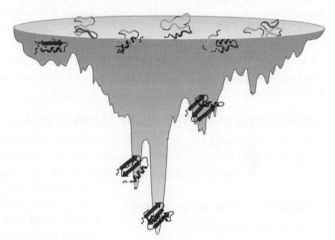

(d) A more rugged energy landscape showing local minima for metastable intermediate states. This landscape describes a multistate folding process.

▲ **FIGURE 6.29 Protein folding energy landscapes.** For all folding funnels shown, the widest part of the funnel corresponds to the unfolded denatured state ensemble of conformations, and the deepest and narrowest part of the funnel corresponds to the folded native state ensemble (many fewer conformations).

The "Energy Landscape" Model of Protein Folding

Several proteins are known to fold via well-characterized intermediate states, as shown in Figure 6.28; yet this simple model of protein folding is not sufficient to resolve Levinthal's paradox. That can only be achieved if the number of conformational states that the protein samples during folding is significantly restricted. The so-called **energy landscape,** or **folding funnel** model, described in the mid-1990s by José Onuchic and

Peter Wolynes, explains how conformational restriction can be achieved during folding. Imagine that the energetics of folding is described by a funnel-shaped landscape (**FIGURE 6.29**) for which the depth of the funnel corresponds to free energy, and the width of the funnel corresponds to the number of conformational states at a given value of free energy. The funnel becomes narrower, and the number of conformations accessible to the protein decreases as a protein molecule follows a "downhill" (i.e., thermodynamically favorable) trajectory on the energy landscape toward the folded conformation (which is presumably the state of lowest free energy). According to this model, any one molecule need sample only an infinitesimal fraction of the total conformations possible, and thereby, Levinthal's paradox is averted.

The classical pathway model and the folding funnel are compatible if we consider the denatured and intermediate states in Figure 6.28 to be made up of vast ensembles of conformations, rather than unique conformations. Recent studies have shown that, in some cases, the native state is also an ensemble of a few closely related conformations.

> ● **CONCEPT** In the "energy landscape" model, the trajectory of protein folding is "downhill"—it proceeds with a decrease in free energy.

The folding of several small proteins has been studied in great detail, and much experimental evidence suggests that the energy landscape model is robust. Some of the earliest events in protein folding are "nucleation" of secondary structure (typically α helix; see Problem 6.18 at the end of the chapter) and "hydrophobic collapse." Nucleation events are critical because it is much more difficult to initiate an α helix than to extend it (note that at least four residues must fold properly to make the first stabilizing H bond). Hydrophobic collapse is driven by the hydrophobic effect and typically involves several highly hydrophobic side chains rapidly associating to create a desolvated hydrophobic core. Whether secondary structure nucleation or hydrophobic collapse (or both) is the first event in protein folding, the formation of *any* partially folded structure imposes significant constraints on the conformational entropy of the peptide chain, as indicated by the landscape model.

Intermediate and Off-Pathway States in Protein Folding

On any folding pathway leading to the "global" free energy minimum, there may be "local" energy minima corresponding to metastable intermediates, just as in the classical pathway model. There is also evidence for "off-pathway" states—those in which some key element is incorrectly folded. Such states also correspond to local free energy minima in the funnel and may temporarily, or permanently, trap the protein. As described below, cells contain specialized proteins and protein complexes to assist incorrectly folded proteins in finding their proper conformations.

> ● **CONCEPT** Folding can be delayed by the trapping of molecules in "off-pathway" states.

One of the most common folding errors results from the incorrect *cis–trans* isomerization of the amide bond adjacent to a proline residue:

trans   *cis*

Unlike other peptide bonds in proteins, for which the *trans* isomer is highly favored (by a factor of about 1000), proline residues favor the *trans* form in the preceding peptide bond by a factor of only about 4. Thus, there is a significant chance that the conformationally incorrect *cis* isomer will form first. Conversion to the *trans* configuration may involve a significant chain rearrangement and hence may be slow in vitro. Cells have an enzyme, called **prolyl isomerase** (or peptideprolyl isomerase, or PPIase), to catalyze this *cis–trans* isomerization and thereby accelerate folding in vivo.

Similarly, for proteins containing disulfide bonds, some disulfide bonds that are not found in the native structure may be formed in intermediate stages of the folding process. The protein can utilize a number of alternative pathways to fold, but it ultimately achieves both its proper tertiary structure and the correct set of disulfide bonds. This process is aided in vivo by catalysis of —S—S— bond rearrangement by the enzyme **protein disulfide isomerase** (or PDI). If a non-native disulfide forms during folding, PDI will reduce the incorrect disulfide and thereby allow the protein to continue folding toward its native structure.

Chaperones Facilitate Protein Folding in Vivo

The fact that a protein can, by itself, find its proper folded state in vitro does not necessarily mean that the same events occur in vivo. We have already noted two of the catalytic aids to folding that are present in cells: the enzymes that accelerate the *cis–trans* isomerization at proline residues and those that catalyze disulfide bond rearrangement. However, to achieve proper folding, some proteins require the action of specialized proteins called **molecular chaperones.** As the name implies, the function of these chaperones is to keep the newly formed protein out of trouble. "Trouble," in this case, means either improper folding or aggregation. As described in the next section, the misfolding or aggregation of proteins is associated with several widespread diseases. Normal cellular functions are critically dependent on proper protein folding, so it is not surprising that organisms have evolved systems to ensure that protein misfolding is avoided.

> ● **CONCEPT** Protein folding and assembly in vivo are sometimes aided by chaperone proteins.

Improper folding may correspond to being trapped in a deep local minimum on the energy landscape (**FIGURE 6.30**). Aggregation is often a danger because the protein, released from the ribosome in an unfolded state, will have hydrophobic groups exposed. These can form *intermolecular* hydrophobic contacts with other polypeptide strands and thereby lead to aggregation.

A molecular chaperone can be defined as an accessory protein that binds to and stabilizes a non-native protein against aggregation and/or helps it achieve folding to its native structure, but is not part of the final functional structure of the correctly folded protein. Several chaperone systems have been discovered. As an example, consider the best studied of all chaperones: the GroEL-GroES complex from *E. coli* (**FIGURE 6.31**). GroEL is made of two rings, each consisting of seven protein molecules, or *subunits*. The center of each ring is an open cavity, accessible to the solvent at the ends. Either cavity can be "capped" with GroES, a seven-membered ring of smaller protein subunits. Such double-ringed chaperone complexes are also known as **chaperonins.**

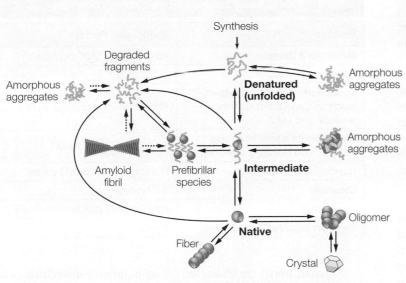

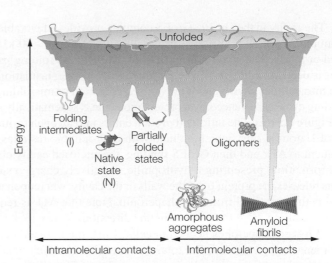

(a) A pathway model showing various protein conformations and their interconversions.

(b) The same information shown in an energy landscape model. The purple portion of the folding funnel shows trajectories leading to the native conformation. The orange portion of the funnel shows trajectories leading to amorphous and ordered aggregates.

▲ **FIGURE 6.30** Models of protein folding and aggregation.

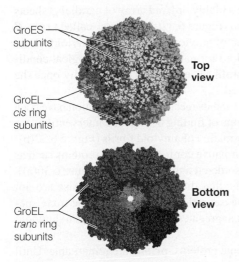

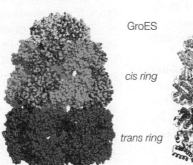

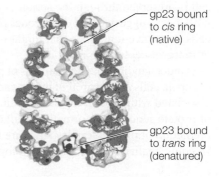

(a) Space-filling top and bottom views of the X-ray diffraction structure of the GroEL-GroES(ADP)₇ complex (PDB ID: 1aon). GroES subunits are highlighted in gold or orange, the subunits in the *trans* ring of GroEL (i.e., the ring opposite the GroES cap) are highlighted in magenta and red, and those in the *cis* GroEL ring (bound by GroES) are highlighted in green.

(b) Space-filling side view of the chaperonin, colored as in (a).

(c) Cartoon view showing the enclosed cavity formed by the *cis* ring of GroEL and GroES (top), as well as the compaction of the *trans* ring compared to the *cis* ring (bottom).

(d) Electron density map obtained from cryoelectron microscopy of a chaperonin complexed with bacteriophage T4 coat protein gp23. The gp23 bound to the *trans* ring of GroEL is shown in red and appears to be denatured. The gp23 bound to the *cis* ring is shown in green and appears to have a native-like conformation.

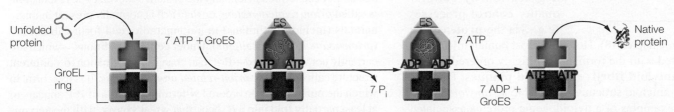

(e) A schematic of chaperonin function. Coloring matches that in panel (c).

▲ **FIGURE 6.31** The GroEL-GroES chaperonin.

The cavity in the GroEL-GroES complex provides a favorable environment in which non-native protein chains of up to ~60 kDa can fold properly. The chaperonin does not stipulate the folding pattern; that is determined by the protein sequence. However, insulation from the intracellular environment prevents aggregation or misfolding. The folding cycle experienced by a protein molecule is schematically shown in Figure 6.31(e). The unfolded protein enters into an open form of the GroEL double ring, which is lined with hydrophobic residues. Subsequent to ATP and then GroES binding, the enclosed cavity changes conformation, presenting a hydrophilic, negatively charged surface. This releases the protein from the walls of the cavity, whereupon it folds and is then released from the chaperonin. Note that ATP is required, presumably to drive the process in one direction.

Why are chaperones needed in cells? First, the intracellular environment is very crowded. The total concentration of proteins and other macromolecules inside cells is on the order of 300–400 grams per liter, or roughly 1000 times more concentrated than is typical for in vitro studies of folding. Intermolecular interactions are more likely to occur in such a concentrated intracellular "stew"; thus, aggregation of misfolded or incompletely folded proteins is problematic in vivo. Second, chaperones play a critical role in protecting cellular proteins during times of stress. In fact, many chaperones were discovered during studies of the heat shock response. When the temperature inside cells is raised by a few degrees Celsius, so-called **heat-shock proteins** (Hsp proteins) are produced. For example, GroEL is also known as Hsp60. We now know that the Hsp proteins are preventing irreversible denaturation of cellular proteins. Once the temperature is restored to normal, temperature-sensitive proteins can be refolded; thus, the viability of the cell is preserved through the action of molecular chaperones. Finally, it has been suggested that chaperones provide a critical mechanism for cells to survive mutations that might render a protein less stable and/or more susceptible to misfolding.

Understanding the mechanisms of protein folding and how folding occurs in cells is of profound importance given the number of diseases associated with protein misfolding. In addition, with the vast amount of protein sequence information available from genomics studies, the prediction of protein tertiary structure from sequence remains an outstanding challenge in molecular biology and biophysics.

Protein Misfolding and Disease

A broad range of diseases is associated with protein misfolding. A few misfolding diseases, such as cystic fibrosis, are the consequence of mutations that reduce the stability of a critical protein, thereby leading to its misfolding and subsequent degradation by cellular quality control processes (e.g., via the **proteasome,** which is discussed in Chapter 18). However, most human misfolding diseases are associated with the formation of highly ordered protein aggregates called **amyloid fibrils** or **amyloid plaques.** Proteins that misfold to form amyloid structures are said to be *amyloidogenic.* **TABLE 6.5** lists a few examples of amyloidogenic proteins associated with human disease states. In many cases, amyloid disease is correlated with inherited genetic mutations that yield a destabilized variant of the protein in question.

● **CONNECTION** Protein misfolding is the basis for several diseases, including Alzheimer's disease and Parkinson's disease.

TABLE 6.5 Examples of amyloid-related human diseases

Disease	Associated Protein
Alzheimer's disease	Amyloid β or "Aβ" peptide
Parkinson's disease	α-Synuclein
Spongiform encephalopathies (such as Creutzfeldt-Jakob disease and kuru)	Prion protein
Amyotrophic lateral sclerosis (ALS, or Lou Gehrig's disease)	Superoxide dismutase I
Huntington's disease	Huntingtin with polyQ tracts
Cataracts	γ-Crystallin
Type II diabetes	Islet amyloid polypeptide (IAPP)
Injection-localized amyloidosis	Insulin

Amyloid fibrils are characterized by highly organized arrays of β sheet structure. **FIGURE 6.32(a)** shows the globular structure of native human insulin—note that it is predominantly α-helical, with no β-sheet secondary structure. Figure 6.32(b) shows two electron microscopy images of insulin amyloid fibrils. These fibrils are formed from a left-handed helix of four "protofibrils" (Figure 6.32(c)). Each protofibril, in turn, is formed from a highly ordered array of parallel β sheets separated by roughly 1.0 nm (Figure 6.32(d)). This so-called **cross-β structure** is a characteristic of amyloids. Studies on amyloidogenic peptides indicate that amyloid formation under physiological conditions is slow; however, protofibrils and fibrils form rapidly once the cross-β structure is nucleated.

The structure of amyloid and its mechanism of formation have been studied extensively in the hope of finding therapeutic interventions for misfolding diseases. The formation of amyloid fibrils (Figure 6.32(b)) is thought to occur from non-native conformations of proteins such as folding intermediates or disordered aggregate states (Figure 6.30(a)). Recent experiments suggest that the amyloid fibrils per se are not responsible for causing disease; rather, it appears that the toxic species may be the amorphous aggregates that are precursors to amyloid formation.

Among the amyloidogenic proteins, **prions** are remarkable. Until recently, virtually all researchers believed that the only ways in which diseases could be transmitted from one organism to another were via viruses or microorganisms. However, there is now evidence that a class of diseases is transmitted by a protein and nothing more. The most notorious of these diseases is *bovine spongiform encephalopathy,* or "mad cow disease," but they also include *scrapie* in sheep and certain neuropathologies in humans (see entries in Table 6.5). The infectious agent has been termed *prion,* and the protein believed to be responsible is called *prion-related protein,* or *PrP.* PrP is normally present in many animals (including humans) in a nonpathological form called PrP^c (*prion-related protein cellular*). Under certain conditions—which are currently not well defined—PrP^c can change conformation to a different structure called PrP^{sc}, or *prion-related protein scrapie.* It is this form, in which the intrinsically disordered N-terminal portion of PrP appears to at least partially fold into a β sheet, that wreaks havoc with the nervous system. Even more remarkable is the fact that ingesting PrP^{sc} can induce conversion of PrP^c in the recipient to PrP^{sc}; thus, the condition is transmitted. How this conversion is catalyzed is unknown, but it strongly

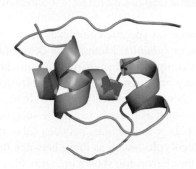

(a) Human insulin in its native conformation.

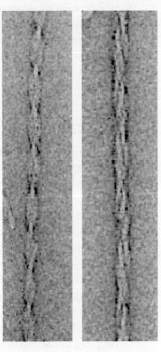

(b) Electron micrographs of amyloid fibrils of insulin.

(c) A model of insulin amyloid fitted to electron density data from cryo-electron microscopy. Four protofibrils wind together to form a left-handed helix.

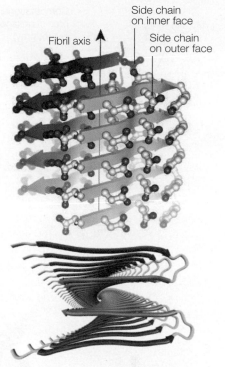

Side chain on inner face

Fibril axis

Side chain on outer face

(d) The cross-β structure of the protofibril. The top figure shows the X-ray crystal structure of the protofibril formed from the GNNQQY peptide from the yeast prion protein Sup35. The bottom figure shows a model of the protofibril formed from residues 1–40 of the Aβ peptide. Solid-state NMR data were used to derive this model.

▲ **FIGURE 6.32** Amyloid fibril structure.

suggests that PrP^{sc} represents an especially stable off-pathway conformation of the type hypothesized in the preceding section. All known prion diseases are untreatable and fatal. In 1997, Stanley Prusiner was awarded the Nobel Prize in Physiology or Medicine for his work in defining the relationship of PrP to these diseases.

6.6 Prediction of Protein Secondary and Tertiary Structure

Can protein structure be predicted? In one sense, the answer to this question must surely be yes because the molecular information necessary to determine the secondary and tertiary structures is carried in the amino acid sequence itself; thus, the gene "predicts" the structure. The implication of this fact is that prediction of the entire three-dimensional conformation of any protein, starting with nothing more than the knowledge of its sequence, is a stringent test of our current understanding of the thermodynamics and kinetics of protein folding. This kind of prediction cannot yet be done completely. Secondary structure can be predicted

● **CONCEPT** Protein secondary structure can now be predicted with good accuracy. The *de novo* prediction of more complex tertiary structure is not yet as accurate.

with some reliability, but prediction of tertiary folding, while improving, is not as robust.

Prediction of Secondary Structure

Although several approaches have been applied to the problem of predicting secondary structure, the most successful method is entirely empirical. From analysis of the *known* structures of a number of proteins, data have been compiled to show the relative frequency with which a particular kind of amino acid residue lies in α helices, β sheets, or turns. For example, Ala, Leu, Met, and Glu are all strong helix formers; Gly and Pro do not favor helices. Similarly, Ile, Val, and Phe are strong β-sheet formers, whereas Pro does not fit well into β sheets. Gly is frequently found in β turns, whereas Val is not. We have already mentioned that Pro tends to lie in turns.

In addition to such statistical data, patterns of side-chain polarity are used to predict secondary structure. A stretch of alternating polar and nonpolar residues is diagnostic for a β strand that is part of an amphiphilic β sheet. Similarly, a pattern of side chains with similar polarity every 3–4 residues indicates an amphiphilic α helix.

As the amount of structural data stored in public databases grows, so too does the sophistication of the computational methods used to make structural predictions from sequence data. Many different algorithms exist online to make rapid predictions of secondary structure

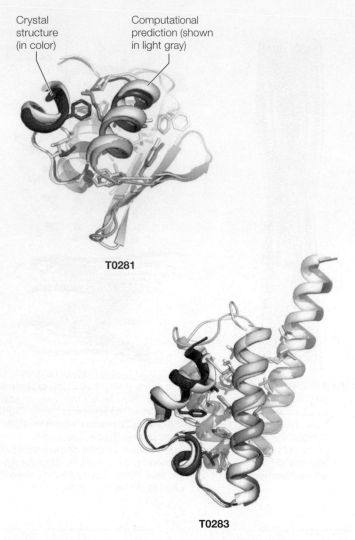

Crystal structure (in color)

Computational prediction (shown in light gray)

T0281

T0283

▲ **FIGURE 6.33 Comparison of de novo predictions to X-ray crystal structures.** Top panel: Computational prediction (shown in light gray) for the Critical Assessment of Techniques for Protein Structure Prediction (CASP) 6 target, T0281, superimposed on the *subsequently* released crystal structure (PDB ID: 1whz; shown with rainbow coloring from the N-terminus [blue] to the C-terminus [red]). The agreement between the backbones of the two structures is within 1.6 Å over 70 residues. Bottom panel: The prediction for the CASP7 target, T0283, (light gray) agrees with the subsequently released crystal structure (PDB ID: 2hh6; shown in rainbow coloring) with a backbone accuracy of 1.6 Å over 90 residues.

from amino acid or gene sequence information (see Appendix II). The accuracy of the best prediction methods is ~80%.

Tertiary Structure Prediction: Computer Simulation of Folding

The prediction of tertiary structure has proved much more difficult, probably because the higher-order folding depends so critically on specific noncovalent interactions, often between residues far removed from one another in the sequence. In spite of the difficulty of the problem, recent efforts in de novo structure prediction (i.e., prediction from the amino acid sequence alone) have met with spectacular success. **FIGURE 6.33** shows the results of two recent de novo predictions that

were achieved at high resolution *before* the X-ray crystal structures for the protein sequences were released publicly. The close agreement between the predicted structures and the crystal structures is remarkable and bodes well for the future of de novo structure prediction. There is tremendous need to develop reliable methods of de novo structure prediction, given the fact that structures have been determined for only 1% of all known protein sequences.

6.7 Quaternary Structure of Proteins

In Chapter 5 and in this chapter, we have explored increasingly complex levels of protein structure, from primary to secondary to tertiary. Functional protein organization can reach at least one more level—**quaternary structure** (see Figure 6.2). Many proteins exist in the cell (and in solution, under physiological conditions) as specific complexes of two or more folded polypeptide chains, or *subunits*. Methods for determining whether or not a protein is composed of multiple subunits are described in Tools of Biochemistry 6B. Such quaternary organization can be of two kinds—association between identical or nearly identical polypeptide chains (**homotypic**—such as those between the subunits in hemoglobin shown in Figure 6.2) or interactions between subunits of very different structures (**heterotypic**—such as those between the GroES and GroEL subunints in the chaperonin shown in Figure 6.31). In either case, *multisubunit proteins* are formed.

The interactions between the folded polypeptide chains in multisubunit proteins are of the same kinds that stabilize tertiary structure—salt

● **CONCEPT** Association of polypeptide chains to form specific multisubunit structures is the quaternary level of protein organization.

bridges, hydrogen bonding, van der Waals forces, and sometimes disulfide bonding. Such interactions, along with the hydrophobic effect, stabilize the multisubunit structure.

Symmetry in Multisubunit Proteins: Homotypic Protein–Protein Interactions

Each polypeptide chain in a multisubunit complex is an asymmetric unit, but the overall quaternary structure may exhibit a wide variety of symmetries, depending on the geometry of the interactions. For purposes of illustration, we shall use an asymmetric object familiar to everyone—a right shoe. Think of this shoe as a polypeptide chain folded into a compact three-dimensional form. We can stick shoes together in many ways. If the interacting surfaces (A and B) were at toe and heel, a linear complex could form:

Interactions between subunits that can give rise to indefinite growth of a polymeric complex by addition of more subunits to either end of the polymer are known as **heterologous** interactions. Heterologous interactions must be specially oriented to give a truly linear complex. More often, the interaction is such that each unit is twisted through some angle with respect to the preceding one. This twisting gives rise to a helical structure. **FIGURE 6.34(a)** shows an arrangement of shoes that forms a right-handed helix with *n* units per turn. The top of the toe of each shoe is attached to the sole of the toe of the next, with a rotation of 360/*n* degrees. Two biological examples, both

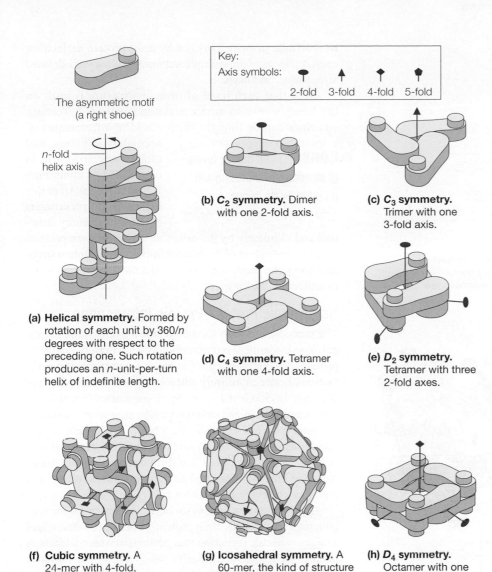

Key:
Axis symbols:

2-fold   3-fold   4-fold   5-fold

The asymmetric motif
(a right shoe)

n-fold helix axis

(a) Helical symmetry. Formed by rotation of each unit by $360/n$ degrees with respect to the preceding one. Such rotation produces an n-unit-per-turn helix of indefinite length.

(b) C_2 symmetry. Dimer with one 2-fold axis.

(c) C_3 symmetry. Trimer with one 3-fold axis.

(d) C_4 symmetry. Tetramer with one 4-fold axis.

(e) D_2 symmetry. Tetramer with three 2-fold axes.

(f) Cubic symmetry. A 24-mer with 4-fold, 3-fold, and 2-fold axes.

(g) Icosahedral symmetry. A 60-mer, the kind of structure found in the protein coat of a number of viruses. This structure has 5-fold, 3-fold, and 2-fold axes.

(h) D_4 symmetry. Octamer with one 4-fold axis and two 2-fold axes.

▲ **FIGURE 6.34 Symmetries of protein quaternary structures.** Although composed of asymmetric polypeptides, proteins adopt many symmetrical patterns in forming quaternary structures. In this figure, a right shoe represents the asymmetric structural unit.

A special situation of great importance arises when two subunits are related to one another by a 2-fold axis (also called a **dyad** axis) to give C_2 symmetry. That is, each subunit is rotated by 180° about this axis with respect to the other (this arrangement is also shown in Figures 6.34b and 6.36a):

Dyad axis perpendicular to paper

Imagine that there are interacting groups at A and B. For example, A could be a hydrogen-bond donor, B an acceptor. Note that in this case the 2-fold symmetry means that *two* identical interactions occur, symmetrically placed about the dyad axis. Such a symmetric interaction is called **isologous.** Dimers are the most common of all quaternary structures, and they almost always display C_2 symmetry. An example is shown in **FIGURE 6.36(a)**. Further isologous interactions can easily give rise to more complex quaternary structures of higher symmetry. An example is **dihedral symmetry** (D_2, Figures 6.34(c), and 6.36(b)), the most common structure for tetrameric proteins. It has three mutually perpendicular 2-fold axes and, therefore, involves three pairs of isologous interactions.

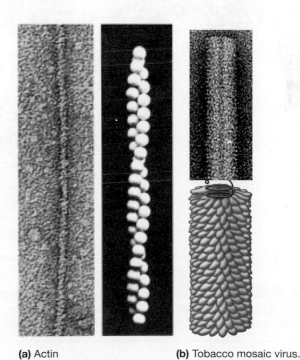

(a) Actin

(b) Tobacco mosaic virus. In the virus, protein subunits form a helical array about a helically coiled RNA (red).

▲ **FIGURE 6.35 Two helical proteins.** Beside each electron micrograph is a diagrammatic representation of the helical aggregate structure.

of **helical symmetry,** are shown in **FIGURE 6.35**: the helix of the muscle protein actin and the helical protein coat of tobacco mosaic virus. Note that both linear and helical arrays are potentially capable of indefinite growth as described above. For example, actin filaments can be thousands of units in length.

Most assemblies of protein subunits are based not on helical symmetry but on one of the classes of **point-group symmetry** (Figure 6.34(b–h)). The classes of point-group symmetry involve a defined number of subunits arranged about one or more **axes of symmetry.** An n-fold axis of symmetry corresponds to rotation of each subunit by $360/n$ degrees with respect to its neighbor; thus, a 2-fold axis corresponds to 180° rotation. The simplest kinds of point-group symmetry are the **cyclic symmetries** C_n, shown in Figure 6.34(b–d). These rings of subunits involve heterologous interactions where $n = 3$ or greater.

● **CONCEPT** Two general classes of symmetry—helical and point-group—characterize most quaternary structures.

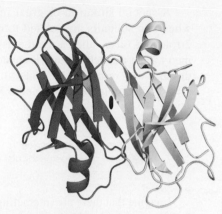

(a) In the transthyretin dimer, the two monomers combine to form a complete β sandwich, or flattened β barrel. The dimer has 2-fold symmetry about the C_2 axis perpendicular to the paper (black oval). The isologous interactions are mostly hydrogen bonds between specific β sheet strands.

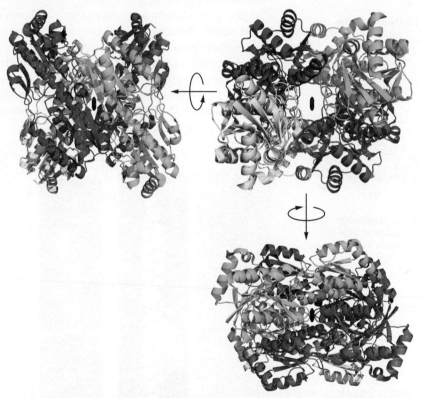

(b) Three views of the tetrameric enzyme phosphofructokinase. Each view is down one of the three mutually perpendicular C_2 axes.

▲ **FIGURE 6.36 Examples of common point-group symmetries in multisubunit proteins.** (a) The transthyretin dimer displays C_2 symmetry. PDB ID: 1gko. (b) The phosphofructokinase tetramer displays D_2 symmetry. PDB ID: 4pfk.

Other, more complex point-group symmetries exist in some protein quaternary structures; examples are shown in Figure 6.34(f–h). Note that the more complex structures involve both 2-fold axes and axes with $n > 2$. Most important is that molecules exhibiting any of the point-group symmetries are always constrained to a definite number of subunits. Most multisubunit proteins exhibit this kind of association geometry, rather than the linear or helical aggregation that can lead

to indefinite growth. This is why most protein molecules, even if they contain multiple subunits, have a well-defined size, shape, and mass.

Note that each level of protein structure is built on the lower levels, as summarized in Figure 6.2. Tertiary structure can be thought of as a folding of elements of secondary structure, and quaternary structure is established by combining folded subunits. *All* of this higher-level structuring is dictated by primary structure and ultimately by the gene that codes for each protein. This understanding of the relationship between protein structure and gene sequence is one of the most important ideas in molecular biology.

● **CONCEPT** All higher levels of structure of a protein are dictated by its genes.

Heterotypic Protein–Protein Interactions

The preceding section focused on the association of identical or near-identical protein subunits. However, the range of protein–protein interactions is much greater; specific associations between entirely different protein molecules are common (as in GroEL-GroES). Sometimes these associations lead to organized structures containing several different subunit types. For example, the RNA polymerase complex (described in Chapter 24) includes five different types of subunits. The interactions that form these assemblies are of the same kind we described earlier—noncovalent forces at complementary protein surfaces.

Complementary protein surfaces also determine the specific interactions between a protein, or protein complex, and a target molecule (often another protein) to which it binds. A simple example involves bovine pancreatic trypsin inhibitor (BPTI), which is so named because it forms a tight, specific complex with the enzyme trypsin, thereby inhibiting trypsin proteolytic activity in the pancreas (**FIGURE 6.37**).

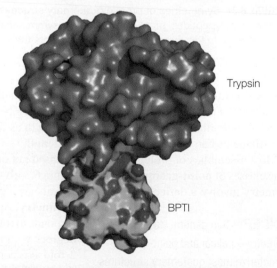

Trypsin

BPTI

▲ **FIGURE 6.37 Interaction of BPTI with trypsin.** The BPTI molecule fits snugly onto the surface of the trypsin molecule, blocking the active site of trypsin. PDB ID: 2ra3.

Summary

- Protein molecules typically have several levels of organization. The first, or primary, level is the amino acid sequence, dictated by the gene. In turn, primary sequence dictates local folding (secondary structure), global folding (tertiary structure), and organization into multisubunit structures (quaternary structure). Ultimately, all of the levels of protein structure are determined by the gene sequence. (Section 6.1)

- Of many conceivable secondary structures, only a limited number are sterically allowed and can be stabilized by hydrogen bonds. These include the **α helix,** the **β sheet,** and the **3_{10} helix.** There are also specific structures that allow a polypeptide chain to make sharp turns. **Ramachandran plots** provide a way to visualize the possibilities and describe various secondary structures in terms of allowed ϕ and ψ angles. (Section 6.1)

- Proteins can be grouped into two broad categories—fibrous and globular. **Fibrous proteins** are elongated, usually of regular secondary structure, and perform structural roles in the cell and organism. Important examples include the **keratins** (α helix), the **fibroins** (β sheet), and **collagen** (triple helix). **Globular proteins** have more complex tertiary structures and fold into compact shapes that often contain defined **domains.** Globular proteins exhibit a wide variety of structures and associated functions. Several classes of folding motifs have been recognized, such as "$\alpha/\beta/\alpha$ sandwiches" and "β barrels." (Sections 6.2 and 6.3)

- A number of factors determine globular protein stability. The conformational entropy of the polypeptide disfavors the folded state. The enthalpy from internal noncovalent bonding as well as the **hydrophobic effect,** which is a solvent entropy effect, favor the folded state. For some proteins, disulfide bonds stabilize the folded state. (Section 6.4)

- The folding of many globular proteins occurs spontaneously and rapidly under "native" conditions. The folding of proteins is a thermodynamically favorable process. The folding of proteins may include marginally stable intermediate states (e.g., the "**molten globule**"). In the cell, proteins called **chaperones** help prevent formation of incorrectly folded structures or undesired intermolecular interactions leading to aggregation. Even when folded, globular proteins are dynamic structures undergoing several kinds of internal motions. Protein secondary structure can be predicted with good accuracy, but prediction of tertiary structure, though improving, remains challenging. (Sections 6.5 and 6.6)

- Many (perhaps most) globular proteins exist and function as multisubunit assemblies forming a quaternary level of structure. A few of these proteins are elongated structures with **helical symmetry.** Most have a small number of subunits (often 2, 4, or 6) and exhibit **point-group symmetry.** (Section 6.7).

Problems

Enhanced by
Mastering Chemistry
for Biochemistry

Mastering Chemistry for Biochemistry provides select end-of-chapter problems and feedback-enriched tutorial problems, animations, and interactive figures to deepen your understanding of complex topics while practicing problem solving.

Answers to red problems are available in the Answer Appendix

1. Polyglycine, a simple polypeptide, can form a helix with $\phi = -80°$, $\psi = +150°$. From the Ramachandran plot (see Figure 6.9), describe this helix with respect to handedness.

2. Bovine pancreatic trypsin inhibitor (BPTI; Figure 6.25) contains six cysteine residues that form three disulfide bonds in the native structure of BPTI. Suppose BPTI is reduced and unfolded in urea (as illustrated for RNase A in Figure 6.22). If the reduced unfolded protein were oxidized *prior* to the removal of the urea, what fraction of the resulting mixture would you expect to possess native disulfide bonds?

3. A schematic structure of the subunit of hemerythrin (an oxygen-binding protein from invertebrate animals) is shown to the right.
   (a) It has been found that in some of the α-helical regions of hemerythrin, about every third or fourth amino acid residue is a hydrophobic one. Suggest a structural reason for this finding.
   (b) What would be the effect of a mutation that placed a proline residue at point A in the structure?

4. In the protein *adenylate kinase*, the C-terminal region has the sequence

   Val-Asp-Asp-**Val**-**Phe**-Ser-Gln-**Val**-Cys-Thr-His-
   Leu-Asp-Thr-**Leu**-**Lys**-

   The hydrophobic residues in this sequence are presented in boldface type. Suggest a possible reason for the periodicity in their spacing.

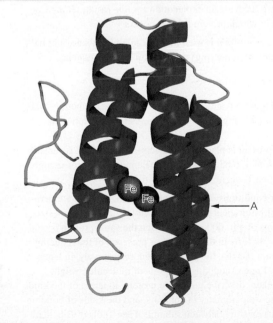

5. Give two reasons to explain why a proline residue in the middle of an α helix is predicted to be destabilizing to the helical structure.

6. Consider a small protein containing 101 amino acid residues. The protein backbone will have 200 bonds about which rotation can occur. Assume that three orientations are possible about each of these bonds.
   (a) Based on these assumptions, about how many *random-coil* conformations will be possible for this protein?
   (b) The estimate obtained in (a) is surely too large. Give one reason why.

7. (a) Based on a more conservative answer to Problem 6 (2.7×10^{92} conformations), estimate the conformational entropy change on folding a mole of this protein into a native structure with only one conformation. (*Hint*: Consider Equation 3.2.)
   (b) If the protein folds *entirely* into α helix with H bonds as the only source of enthalpy of stabilization, and each mole of H bonds contributes -5 kJ/mol to the enthalpy, estimate $\Delta H_{folding}$. Note that the ends of helices contain fewer hydrogen bonds per residue than in the middle (see Figure 6.4).
   (c) From your answers to (a) and (b), estimate $\Delta G_{folding}$ for this protein at 25 °C. Is the folded form of the protein stable at 25 °C?

8. The following sequence is part of a globular protein. Predict the secondary structure in this region.

   $$\ldots RRPVVLMAACLRPVVFITYGDGGTYYHWYH \ldots$$

9. (a) A protein is found to be a tetramer of identical subunits. Name two symmetries possible for such a molecule. What kinds of interactions (isologous or heterologous) would stabilize each?
   (b) Suppose a tetramer, like hemoglobin, consists of two each of two types of subunits, α and β. What is the highest symmetry now possible?

10. Under physiological conditions, the protein hemerythrin exists as an octamer of eight chains of the kind shown in Problem 3.
    (a) Name two symmetries possible for this molecule.
    (b) Which do you think is more likely? Explain.
    (c) For the more likely symmetry, what kinds of interactions (isologous, heterologous, or both) would you expect? Why?

11. Theoretical and experimental measurements show that in many cases, the contributions of ionic and hydrogen-bonding interactions to ΔH for protein folding are close to zero. Provide an explanation for this result. (*Hint:* Consider the environment in which protein folding occurs.)

12. The peptide hormone *vasopressin* is used in the regulation of saltwater balance in many vertebrates. Porcine (pig) vasopressin has the sequence

    Asp-Tyr-Phe-Glu-Asn-Cys-Pro-Lys-Gly

    (a) Using the data in Figure 5.6 and Table 5.1, estimate the extinction coefficient ε (in units of cm^2/mg) for vasopressin, using radiation with $\lambda = 280$ nm.
    (b) A solution of vasopressin is placed in a 0.5-cm-thick cuvette. Its absorbance at 280 nm is found to be 1.3. What is the concentration of vasopressin, in mg/cm^3? (See Tools of Biochemistry 6A.)

13. A protein gives, under conditions of buffer composition, pH, and temperature that are close to physiological conditions, a molecular weight by size exclusion measurements of 140,000 g/mol. When the same protein is studied by SDS gel electrophoresis in the absence or presence of the reducing agent β-mercaptoethanol (BME), the patterns seen, respectively, in lanes A and B are observed. Lane C contains standards of molecular weight indicated. From these data, describe the native protein, in terms of the kinds of subunits present, the stoichiometry of subunits, and the kinds of bonding (covalent, noncovalent) existing between subunits. (See Tools of Biochemistry 5A, 6B.)

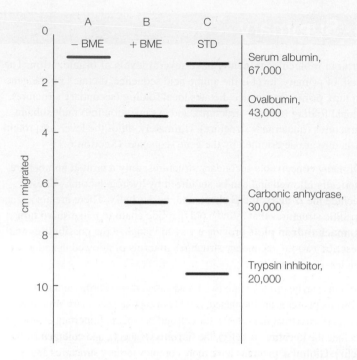

14. A protein gives a single band on SDS gel electrophoresis, as shown in lanes 1 and 2 below. There is little, if any, effect from adding β-mercaptoethanol (BME) to the sample; if anything, the protein runs a little bit slower. When treated with the proteolytic enzyme thrombin (see Chapter 5) and electrophoresis in the absence of BME, the protein migrates a bit more rapidly (lane 3). But if BME is present, two much more rapidly migrating bands are found (lane 4). Explain these results in terms of a model for the protein.

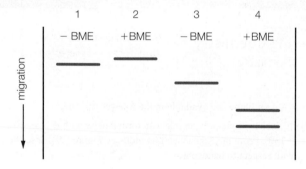

15. It has been postulated that the normal (noninfectious) form of prion differs from the infectious form only in secondary/tertiary structure.
    (a) How might you show that changes in secondary structure occur?
    (b) If this model is correct, what are the implications for structural prediction schemes?

16. Below are shown two views of the backbone representation of the Myc-Max complex binding to DNA (PDB ID: 1npk).

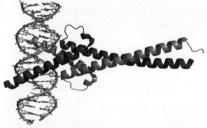

Side view of Myc-Max bound to DNA

Top view of Myc-Max bound to DNA (looking down the DNA helical axis)

Myc and Max are members of the basic helix-loop-helix (bHLH) class of DNA-binding proteins. Myc (red) and Max (blue) associate via a coiled-coil interaction and bind DNA as a dimer.

(a) Are the helices bound to the DNA likely to be amphiphilic? Explain.

(b) Where do you predict the N- and C-termini are located for Max? Explain your rationale.

17. Do you expect a Pro → Gly mutation in a surface-loop region of a globular protein to be stabilizing or destabilizing? Assume the mutant folds to a native-like conformation. Explain your answer in terms of the predicted *enthalpic* and *entropic* effects of the mutation on the ΔG for protein folding compared to ΔG of folding for the wild-type protein.

18. Rank the following in terms of predicted rates: the nucleation of an α helix; the nucleation of a *parallel* β sheet; the nucleation of an *antiparallel* β sheet. Justify your predictions.

19. Shown below are two cartoon views of the small globular protein StrepG in which an α helix is packed against a four-strand β sheet. The sheet is made up of two "β-hairpins" (a β-hairpin is a "β-turn-β" structure). Refer to the images and answer the questions that follow:

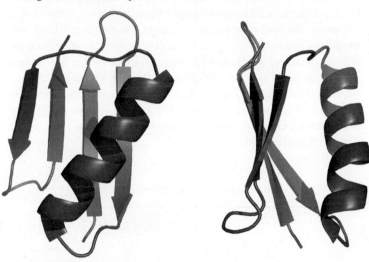

(a) Identify the locations of the N- and C-termini of StrepG.

(b) Indicate the orientation of the helical macrodipole, showing the ($\delta+$) and ($\delta-$) ends of the macrodipole.

(c) How many residues are in the helix?

(d) Do you predict that the α helix and β sheet are amphiphilic or not? Briefly explain.

(e) The following two peptides are part of the primary sequence of StrepG. Based on your answer to part (d), which one is more likely to correspond to the α helix? Which is most likely to be part of a β-hairpin? Explain your choice.

Peptide #1: **DAATAEKVFKQYAND** or Peptide #2: **VDGEWTYDDATKTFTV**

20. Why does it make biochemical sense that chaperones recognize hydrophobic surface area? What catastrophic event are chaperones meant to prevent in cells?

21. In most cases, mutations in the core of a protein that replace a smaller nonpolar side chain in the wild-type (e.g., Ala, Val) with a larger nonpolar side chain (e.g., Leu, Ile, Phe, Trp) in the mutant, result in significant destabilization and misfolding of the mutant. What feature of the protein core explains this observation? Why would such a mutation prevent a protein from folding properly?

22. A Leu → Ala mutation at a site buried in the core of the enzyme lysozyme is found to be destabilizing. Explain the observed effect of this mutation on lysozyme stability by predicting how enthalpy ($\Delta H°$), conformational entropy ($\Delta S°_{peptide}$), and the hydrophobic effect ($\Delta S°_{solvent}$) are expected to change for the mutant compared to wild-type lysozyme. Explain how $\Delta G°$ for unfolding is affected by your predicted changes in enthalpy or entropy.

23. Disulfide bonds have been shown to stabilize proteins (i.e., make them less likely to unfold). Consider the cases shown schematically below for two variants of the same protein. In case #1 the disulfide forms between Cys residues that have been introduced near the protein N- and C-termini, and in case #2 the disulfide forms between Cys residues that have been introduced in the middle of the protein sequence. Which protein is likely to be more stable? (*Note*: Assume the disulfide bond is intact in both the unfolded and folded states). Explain your reasoning.

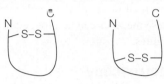

Case #1:
Disulfide forms
near termini

Case #1:
Disulfide forms in
middle of protein

24. Cartoon renderings of the proteins Top7 and adaH2 are shown below. Both are soluble, densely packed proteins of roughly 96 residues, and each has a topology of 2 α helices packed onto a 5-stranded β sheet.

adaH2

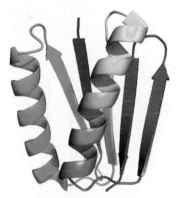

Top7

The accompanying table lists some information about the amino acid composition and values of $\Delta S°$ for the folding of these proteins (i.e., for: Unfolded → Folded) at 25 °C. Based on the information in the table, which protein do you predict buries the greater hydrophobic surface area upon folding? Assume 2-state folding (i.e., no intermediates), and that the Unfolded state for both proteins is 100% solvent-exposed. Explain your answer in terms of expected contributions from $\Delta S°_{peptide}$ and $\Delta S°_{solvent}$.

Protein	Number of Gly in Primary sequence	Number of Pro in Primary sequence	$\Delta S°(J/(mol\ K)$
adaH2	10	0	−53
Top7	5	0	−58

 References

For a list of references related to this chapter, see Appendix II.

X-ray diffraction (see Tools of Biochemistry 4B) is a very powerful method for determining the details of the three-dimensional structure of globular proteins and other biopolymers; however, this technique has the fundamental limitation that it can be employed only when the molecules are crystallized—and crystallization can be challenging. For example, proteins containing significant regions of sequence that are intrinsically unstructured are notoriously difficult to crystallize. Furthermore, X-ray diffraction cannot easily be used to study conformational changes in response to changes in the molecules' environment. Other methods, however, allow us to study molecules in the solution state. A number of these methods can be grouped in the category of **spectroscopic techniques.**

Absorption Spectroscopy

Proteins, carbohydrates, and nucleic acids are complex molecules and can absorb electromagnetic radiation over a wide spectral range. The basic principles of such absorption phenomena, however, can be explained in terms of the simplest kind of molecule, a diatomic molecule.

When two atoms interact to form a molecule, the potential energy curve for the lowest-energy electronic state (the **ground state**) will look like the lower curve in **FIGURE 6A.1(a)**. **Excited electronic states** will have similar curves for energy versus interatomic distance, but at higher energies. For each electronic state of the molecule, there will

be a series of allowed **vibrational states,** with energy levels indicated by horizontal lines in the figure. The basics of molecular absorption spectroscopy can be understood by two simple rules: (1) Transitions are possible only between allowed energy states of the molecule (that is, energy levels are **quantized**), and (2) the energy (ΔE) that is absorbed (or emitted) in any transition between allowed states determines the wavelength (λ) of the radiation that is absorbed (or released) to accomplish that transition. The energy in a **quantum** (or **photon**) of radiation is inversely proportional to λ

$$E_{\text{final state}} - E_{\text{initial state}} = \Delta E = \frac{hc}{\lambda} \qquad (6A.1)$$

Here h is Planck's constant (6.626×10^{-34} J s), and c is the velocity of light in a vacuum (2.998×10^8 m s^{-1}). According to Equation 6A.1, transitions with smaller energy differences between states correspond to absorption (or release) of longer wavelength radiation, and transitions with larger energy differences correspond to absorption (or release) of shorter wavelength radiation. This relationship is in accord with Figure 6A.1(b), which indicates that the high-energy transitions between electronic states of a molecule lead to absorption in the visible or ultraviolet region of the spectrum, whereas the low-energy transitions between different vibrational energy levels correspond to absorption of infrared energy.

▼ **FIGURE 6A.1** The principles of absorption spectroscopy.

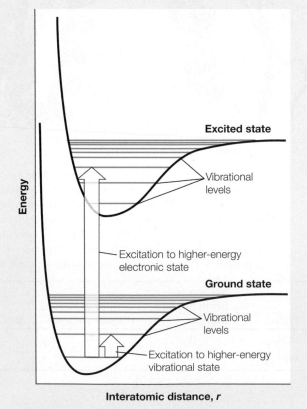

(a) Electronic and vibrational transitions between allowed states in a diatomic molecule.

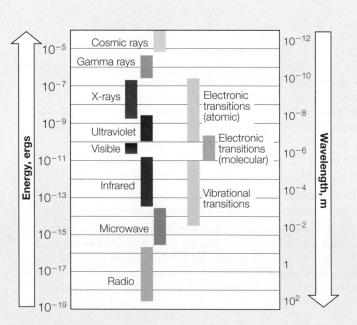

(b) The electromagnetic spectrum.

Complex biopolymers like proteins and nucleic acids can undergo many kinds of molecular vibrations and oscillations. **Infrared spectroscopy** can therefore provide direct information concerning macromolecular structure. For example, the exact positions of infrared absorbance bands corresponding to vibrations in the polypeptide backbone are sensitive to the conformational state of the protein backbone (α helix, β sheet, and so forth). Thus, studies in this region of the spectrum can be used to investigate the secondary structural features in protein molecules.

Most biopolymers do not absorb visible light to a significant extent. Some proteins are colored, but they invariably contain prosthetic groups (such as the heme in myoglobin) or metal ions (such as copper) that confer the visible absorption. Blood and red meat owe their color to the heme groups carried by hemoglobin, myoglobin, and other heme proteins. Such absorption can often be exploited to investigate changes in the molecular environment of the prosthetic group. An example is the use of absorption spectroscopy in the visible spectrum to follow the oxygenation of myoglobin or hemoglobin (see Figure 7.17).

The most common uses of spectroscopic techniques in biochemistry involve **ultraviolet spectroscopy.** In the ultraviolet region of the electromagnetic spectrum, both proteins and nucleic acids absorb strongly (**FIGURE 6A.2**). The strongest protein absorbances are found in two wavelength ranges within the ultraviolet (UV) region, at approximately 280 and 220 nm. In the near UV range 270–290 nm, we see absorbance by the aromatic side chains of tyrosine and tryptophan. Because this region of the spectrum is easy to study, absorbance at 280 nm is used routinely to measure protein concentrations. The second region of strong absorbance in the protein spectrum lies in the far UV range (180–220 nm). Absorbance at such wavelengths arises from electronic transitions in the polypeptide backbone itself. Thus, spectroscopy at these wavelengths is used to monitor changes in the secondary structure of proteins.

Spectroscopic measurements of protein concentration use a **spectrophotometer,** in which a cuvette with a light path of length l containing a solution of the protein is placed in a beam of monochromatic radiation of intensity I_0 (**FIGURE 6A.3**). The intensity of the emerging beam will be decreased to a value I because the solution absorbs some of the radiation. The **absorbance** at wavelength λ is defined as

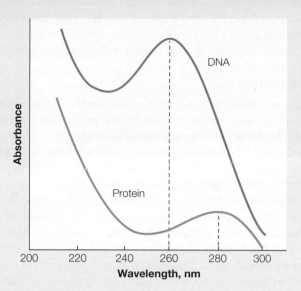

▲ **FIGURE 6A.2 The near-ultraviolet absorbance spectra of a typical protein and of DNA.** Absorbance at 280 nm is commonly used to measure protein concentrations, whereas absorbance at 260 nm is more sensitive for nucleic acids.

$A_\lambda = \log(I_0/I)$ and is related to l and the concentration c by the **Beer-Lambert law:**

$$A = \varepsilon_\lambda l c \qquad (6A.2)$$

Here ε_λ is the **extinction coefficient** (or **molar absorptivity**) at wavelength λ for the particular substance being studied. The dimensions of ε_λ depend on the concentration units employed. When protein concentration is measured in molarity (M) and l in cm, then ε_λ must have the dimensions $M^{-1}cm^{-1}$, because A is a dimensionless quantity. Note that the molar absorptivities of the aromatic amino acids differ in the order: Trp > Tyr $\gg$ Phe (see Figure 5.6. Note that Cys also absorbs at 280 nm, albeit 10-fold less than Tyr).

Once the molar absorptivity for a particular protein has been determined (for example, by measuring the absorbance of a solution containing

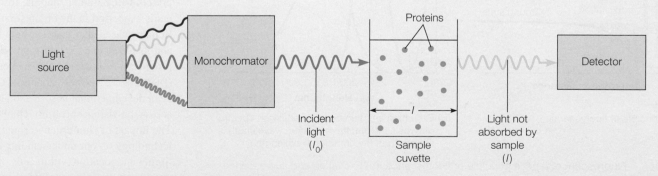

▲ FIGURE 6A.3 Measurement of light absorption with a spectrophotometer.

a known mass of the protein), the concentration of any other solution of that protein can be calculated from a simple absorbance measurement, using Equation 6A.2. The same method is routinely used with nucleic acids, but in that case a wavelength of 260 nm is usually employed because nucleic acids absorb most strongly in that spectral region.

Fluorescence

In most cases, molecules raised to an excited electronic state by absorption of radiant energy return, or "relax," to the ground state by **radiationless transfer** of the excitation energy to surrounding molecules. In short, the energy of relaxation most often is manifested as heat rather than an emitted photon. As shown in **FIGURE 6A.4(a)**, an excited state molecule can lose a small part of its energy of excitation by radiationless transfer (yellow arrow) and will lose the larger part in the form of an emitted photon (red arrow). This gives rise to the phenomenon called **fluorescence.** Because, as Figure 6A.4(a) shows, the quantum of energy reemitted as fluorescence is always of lower energy than the quantum that was initially absorbed (blue arrow), the wavelength of the emitted light will be longer than the wavelength of the light used for excitation. The **fluorescence emission spectrum** of tyrosine is contrasted with its absorbance (or **excitation**) spectrum in Figure 6A.4(b). In proteins, tyrosine and tryptophan are the major fluorescent groups. The local environment of these residues can greatly modify the intensity and wavelength of maximum fluorescence (the so-called λ_{max}). For example, the fluorescence λ_{max} shifts to shorter wavelength (a "blue-shift"), and the intensity of the fluorescence signal increases as the polarity of the solvent surrounding a tryptophan residue decreases. Tryptophan residues buried in the hydrophobic cores of proteins can have values for λ_{max} that are blue-shifted by 10 to 20 nm compared to tryptophans in solvent-accessible locations; thus, fluorescence spectroscopy of tryptophan can be used to monitor changes in protein conformation, such as the transition from a folded to a denatured state (see Figure 6.22).

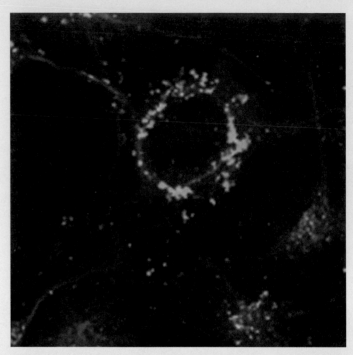

▲ **FIGURE 6A.5 Localization of a GFP-fusion protein.** Fibroblast growth factor receptor-3 fused to GFP (yellow-green) localizes to the periphery of endosomes. The cytoskeletal protein vimentin is stained red.

Because fluorescence spectroscopy is a sensitive technique that can detect a small number of fluorescent molecules, it has become widely used as a tool for precisely locating proteins in cells or subcellular organelles. Confocal microscopy allows such location, if the protein can be specifically labeled with a fluorescent tag. Sometimes this can be done by covalently attaching fluorescent dyes, but this process is challenging in vivo. A powerful modern technique uses a highly fluorescent protein found in some jellyfish, called **green fluorescent protein (GFP).** The intense fluorescence is due to an unusual chromophore, generated by oxidation of the amino acid sequence Ser-Tyr-Gly. GFP is used most effectively as a **fusion protein;** the gene for GFP is fused to the gene for the protein being studied, and the fused product is expressed in the organism of interest. In many cases, the fusion protein functions and localizes like the native protein, and provides a brilliant marker in microscopy (**FIGURE 6A.5**). Several variants of GFP have been developed that absorb and fluoresce across the visible spectrum (**FIGURE 6A.6**). The impact of fluorescent–protein fusion technology on our understanding of the timing and location within cells of gene expression, protein movement, and changes in pH and/or Ca^{2+} levels was recognized

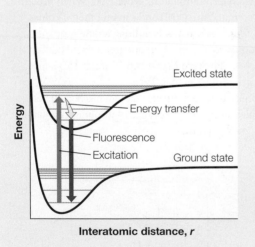

(a) The principle of fluorescence.

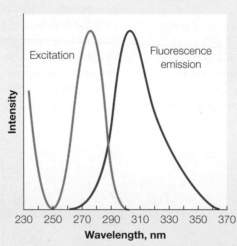

(b) Excitation and fluorescence emission spectra of tyrosine (*Note*: the excitation wavelength is shorter than the emission wavelength.)

▲ **FIGURE 6A.4 Fluorescence. (a)** The principle of fluorescence. **(b)** Excitation and fluorescence emission spectra of tyrosine.

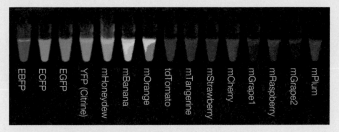

(a) Fifteen different fluorescent proteins are shown.

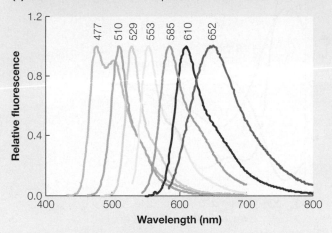

(b) Emission spectra for some of the proteins in (a) are shown. The emission λ_{max} (nm) is listed above each peak. The spectra were generated from data available at: http://www.tsienlab.ucsd. edu/Documents/REF-FluorophoreSpectra.xls.

▲ **FIGURE 6A.6 Fluorescence emission of several fluorescent proteins. (a)** Examples of several fluorescent variants of GFP. (Roger Tsien Lab/Composite by Paul Steinbach). **(b)** Emission spectra for some of the proteins in (a) are shown.

by the 2008 Nobel Prize in Chemistry, shared by Osamu Shimomura, Martin Chalfie, and Roger Tsien.

Circular Dichroism

Although fluorescence spectroscopy, and in some cases visible absorption spectroscopy, are useful for monitoring significant changes in protein *tertiary* conformation due to local or global unfolding, such measurements are difficult to interpret directly in terms of changes in *secondary* structure. For this purpose, infrared spectroscopy (see above) and techniques involving polarized light are more informative.

Light can be polarized in various ways. Most familiar is **plane polarization** (**FIGURE 6A.7(a)**, top), in which the oscillating electric field of the radiation has a fixed orientation in a single plane. In contrast, *unpolarized* light consists of waves vibrating in *all* planes perpendicular to the direction of travel. Less familiar, but equally important, is **circular polarization,** in which the direction of polarization *rotates* with the frequency of the radiation (Figure 6A.7(a), bottom). If you observe a circularly polarized beam coming toward you, the electric field can be rotating in either a clockwise or counterclockwise direction; thus, as the light moves toward you, the oscillation of the electric field describes a

right-handed or a left-handed helix. The former is called *right circularly polarized light,* the latter, *left circularly polarized light.*

Most of the molecules studied by biochemists are asymmetric—for example, L- and D-amino acids, right- and left-handed protein helices, and right- and left-handed nucleic acid helices. Such molecules exhibit a preference for the absorption of either left or right circularly polarized light. For example, a right circularly polarized beam interacts differently with a right-handed α helix than does a left circularly polarized beam. This difference in absorption, called **circular dichroism,** is defined as

$$\Delta A = \frac{A_L - A_R}{A} \qquad (6A.3)$$

where A_L is the absorbance for left circularly polarized light, A_R is the corresponding quantity for right circular polarized light, and A is the absorbance for unpolarized light. Because ΔA can be either positive or negative, a **circular dichroism spectrum** (or **CD spectrum**) is unlike a normal absorption spectrum in that both (+) and (−) values are allowed.

Figure 6A.7(b) shows CD spectra for polypeptides in the α helix, β sheet, and random-coil conformations. The three spectra in the figure are very different, so circular dichroism is a sensitive technique for following conformational changes of proteins in solution. For example, if a protein is denatured so that its native structure, containing α-helix and β-sheet regions, is transformed into an unfolded, random-coil structure, this transformation will be reflected in a dramatic change in its CD spectrum. An example is shown in **FIGURE 6A.8**, which shows the denaturation of myoglobin as a function of increasing urea concentration. Myoglobin possesses significant helical structure in the absence of denaturant (black data points in Figure 6A.8(a)); however, in the presence of 8 M urea (red data points) the helical structure is disrupted. The transition from the folded helical structure to the unfolded ("random coil") structure can be easily monitored by recording the CD signal at 222 nm (vertical blue arrow in Figure 6A.8). Changes in the CD signal at 222 nm as a function of increasing urea concentration are plotted in Figure 6A.8(b) (compare this unfolding curve to that in Figure 6.22).

Although circular dichroism is an extremely useful technique for monitoring global changes in protein or nucleic acid structures, it is not a high-resolution one. That is, it cannot provide insight into biomolecular structure at the level of atomic detail.

Two methods that can provide the details of structure at the atom level are X-ray crystallography and **nuclear magnetic resonance spectroscopy** (NMR). The influence of these two structure-determination methods on the field of biochemistry is profound. Our detailed understanding of the functions of biomolecules is enhanced by, and in some cases based on, the knowledge of the high-resolution structures obtained by these methods. The first report of an X-ray crystal structure of a protein (myoglobin), published in 1958 by Nobel laureate John Kendrew and coworkers, established crystallography as the standard technique for the determination of high-resolution structures of biomolecules. As of 2011, more than 12 X-ray crystallographers had received Nobel prizes for their significant achievements in the determination of protein and/or nucleic acid structures and the insights these structures provided to our understanding of molecular

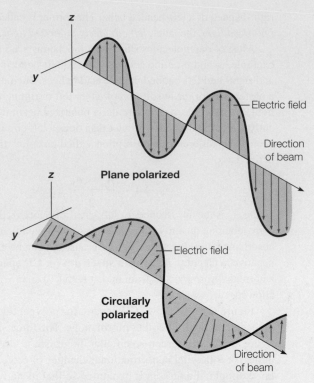

Plane polarized

Circularly polarized

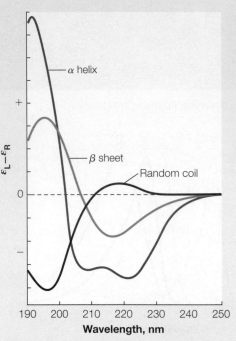

(a) Polarization of light. Above: Plane polarized light, in which the amplitude of the electric field oscillates in a single plane. Below: In circularly polarized light, the oscillation of the electric field follows a helical path around the axis describing the direction of the beam.

(b) Circular dichroism spectra for polypeptides in various conformations. Here the y-axis records differences in molar absorptivity (ε) between left and right circularly polarized light.

▲ **FIGURE 6A.7** Circular dichroism.

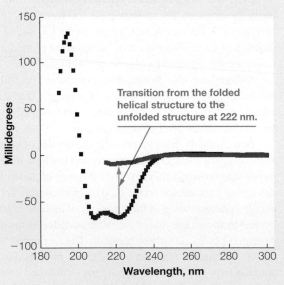

Transition from the folded helical structure to the unfolded structure at 222 nm.

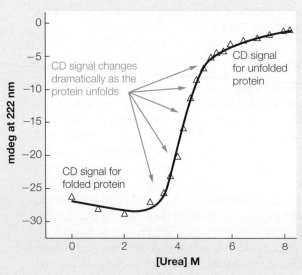

CD signal changes dramatically as the protein unfolds

CD signal for unfolded protein

CD signal for folded protein

(a) Denaturation of myoglobin as a function of increasing urea concentration. The black spectrum was obtained under native conditions. The red spectrum was obtained in the presence of 8 M urea.

(b) Changes in the CD signal at 222 nm as a function of increasing urea concentration.

▲ **FIGURE 6A.8** Monitoring protein unfolding (or refolding) by circular dichroism.

TABLE 6A.1 Nuclei most often used in biochemical NMR experiments

Isotope	Spin	Natural Abundance[a] (%)	Relative Sensitivity[b]	Applications
^{1}H	$\frac{1}{2}$	99.98	(1.000)	Almost every kind of biochemical study
^{2}H	1	0.02	0.0096	Studies of selectively deuterated compounds; structure determination of proteins > 20 kDa
^{13}C	$\frac{1}{2}$	1.11	0.0159	Multidimensional NMR; residue assignment
^{15}N	$\frac{1}{2}$	0.37	0.0104	Multidimensional NMR; residue assignment; protein backbone dynamics
^{19}F	$\frac{1}{2}$	100.00	0.834	Substituted for H (e.g., ^{19}F-Tyr) to probe local structure
^{31}P	$\frac{1}{2}$	100.00	0.0664	Studies of nucleic acids and phosphorylated compounds

[a]The number represents the percentage of this isotope in the naturally occurring mix of isotopes of each element. Isotopes with figures that are close to 100% can be studied directly in the naturally occurring biopolymers. Rare isotopes, such as ^{2}H (deuterium), ^{13}C, and ^{15}N, usually must be artificially enriched in the biomolecules to be studied. This is achieved by including one or more of these isotopes in the medium used to grow the organism containing the molecule of interest (usually a recombinant protein).

[b]Indicates the sensitivity (relative to ^{1}H) of conventional NMR instruments to each isotope, when that isotope has been enriched to 100%. Low values mean that the experiment will be more difficult or time consuming.

biology. The first structure of a protein determined by NMR methods (proteinase inhibitor IIa) was published in 1985 by Kurt Wüthrich and coworkers. It took some time for NMR to become widely accepted as a reliable method for high-resolution structure determination. However, that is no longer in doubt, and in 2002 Wüthrich was awarded the Nobel Prize for his contributions to the development of NMR methods for the determination of protein structures in solution. The recent development of powerful methods for the analysis of protein structure, dynamics, and function by NMR justifies a brief introduction to some of the basic experiments in this important field.

Nuclear Magnetic Resonance Spectroscopy (NMR)

The nuclei of certain elemental isotopes have a property referred to as **spin,** which makes these nuclei behave like microscopic magnets. A limited number of isotopes have this property; some that are particularly useful

to biochemists are listed in **TABLE 6A.1**. The most useful nuclei for NMR have spin states of $-\frac{1}{2}$ or $+\frac{1}{2}$. If an external magnetic field is applied to a sample containing such nuclei, the different nuclear spin states will align with or against the magnetic field and therefore have different energies. As **FIGURE 6A.9(a)** shows, the difference in energy (ΔE) between these two spin states increases as the strength of the external magnetic field increases. For most NMR spectrometers, a pulse of radio frequency (RF) radiation can change the orientation of the nuclear spin (or "magnetization") in the external field, a phenomenon called nuclear magnetic resonance. Such reorientation of the nuclear magnetization is analogous to the "excited" electronic states described above. Whereas the techniques of absorption spectroscopy described earlier record the wavelengths of light absorbed during "allowed" *electronic* transitions, NMR spectroscopy records the frequencies of RF radiation that are absorbed when the spin states of different nuclei in a molecule become reoriented in the external field.

The energy of a nuclear spin in a magnetic field is very sensitive to the chemical environment surrounding the atom in question.

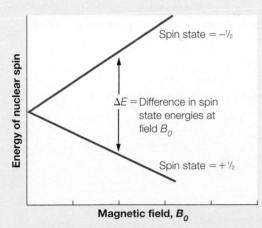

(a) The effect of magnetic field strength on the energies of nuclear spin states (e.g., ^{1}H, ^{13}C).

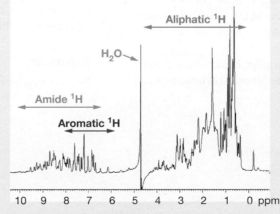

(b) A 500 MHz ^{1}H NMR spectrum of human ubiquitin (1 mM ubiquitin in 25 mM sodium phosphate, 150 mM NaCl, pH 7.0). This protein has 76 residues, which give rise to ~600 peaks in the ^{1}H NMR spectrum. The *x*-axis is the chemical shift, δ in parts per million (ppm).

◄ **FIGURE 6A.9** Nuclear magnetic resonance spectroscopy.
(b) courtesy of S. Delbecq and R. Klevit (University of Washington).

The chemical environment of a nucleus is defined by factors such as the polarity, hydrophobicity, and charge state of the surroundings. For example, due to differences in chemical environments, different hydrogen nuclei in a compound will reach resonance at different field strengths. These differences are recorded in the **NMR spectrum** and are expressed in terms of **chemical shifts (δ)** defined with respect to a reference material:

$$\delta = \frac{B_{ref} - B}{B_{ref}} \qquad (6A.4)$$

Here B is the field strength at which the nucleus in question reaches resonance, and B_{ref} is that for a reference nucleus. The differences between B_{ref} and B are quite small, and this fact is reflected in the units for chemical shift, which are *parts per million* or ppm. A common reference compound for protein NMR is 4,4-dimethyl-4-silapentane-1-sulfonic acid (DSS), which has a strong ^{1}H resonance for its nine methyl protons. This resonance is assigned the value of 0 ppm. The electronic absorption spectra described earlier in this section are obtained by plotting signal intensity versus wavelength. Similarly, an NMR spectrum is obtained by plotting signal intensity (for each resonance) versus chemical shift. For example, a ^{1}H NMR spectrum will record a peak (or "line") for every ^{1}H nucleus in the molecule.

As illustrated in Figure 6A.9(b), the ^{1}H NMR spectrum of a protein is extremely complex, containing many peaks because even relatively small proteins have several hundred protons with many overlapping resonances (e.g., human ubiquitin, with only 76 amino acid residues, contains > 600 protons). Most of the ^{1}H resonances for aliphatic side chains occur between 0.5 and 5 ppm, whereas the backbone amide proton resonances generally occur between 6.5 and 10 ppm, and aromatic side-chain resonances occur between 6 and 8 ppm. In cases where specific ^{1}H resonances in a protein can be identified, it becomes possible to monitor changes in the chemical environment of a specific amino acid residue. An example of such an experiment is given in **FIGURE 6A.10**, which uses NMR to generate titration curves of individual histidine residues in the protein ribonuclease A. The figure also graphically illustrates a principle described in Chapter 5: Individual side chains of a given amino acid type can show quite different pK_a values because of their different chemical environments within the protein molecule.

Changes in chemical shift are also correlated with conformational changes in protein structure; thus, NMR can be used to monitor dynamic motions of proteins, such as those that occur upon local or global unfolding of the polypeptide.

With modern NMR instruments, it is possible to use advanced techniques to resolve resonances for many of the ^{1}H nuclei in proteins as large

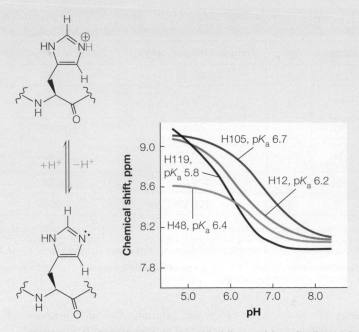

▲ **FIGURE 6A.10 Titration of the four histidine residues in ribonuclease A by NMR.** The *y*-axis is the ^{1}H chemical shift, and each curve follows the titration of an individual histidine group, as detected by the NMR chemical shift of a ^{1}H bound to either one of the two C atoms in the imidazole ring (red H atoms on ring). The labels such as H12 and H48 indicate the positions of the histidines in the primary sequence. The two histidines with lowest pK_a values (H12 and H119) are involved directly in the catalytic process.

as 30 kDa. These data are then used to generate a 3-D model of the protein. Structure determination by solution phase NMR for proteins larger than 50 kDa is challenging. This is the major limitation on structure determination by NMR compared to X-ray crystallography; however, NMR is the more powerful technique for studying dynamic processes in solution.

This introduction to NMR has only scratched the surface of describing applications of this versatile tool for the study of protein structure, dynamics, and ligand binding. More detail is given in Appendix II. A brief description of multidimensional NMR for the determination of protein structures in solution is given in the online supplements to this text.

References

For a list of references related to this chapter, see Appendix II.

When a new protein has been identified and purified, three questions immediately arise:

1. Does the protein exist under physiological conditions as a single-polypeptide chain, or is it made up of multiple subunits?

2. If the functional protein has more than one subunit, are the subunits identical, or are there several kinds?

3. If the functional protein has more than one subunit, are the subunits covalently linked by disulfide bonds or not?

The answers to these questions can usually be obtained by first determining the **molecular mass** (sometimes called the molecular weight, M_W) of the protein under native conditions (i.e., in a buffer that approximates the relevant physiological pH and ionic strength) and then subjecting it to conditions under which dissociation into subunits should occur. If subunits are held together by noncovalent interactions, changing the solvent environment will often promote dissociation. For example, the pH might be raised or lowered well outside the physiological range. Alternatively, non-native (i.e., denaturing) solvents like concentrated solutions of urea or guanidine hydrochloride (GnHCl) might be used. These compounds, which are excellent hydrogen-bond formers, disrupt the regular water structure (see Figure 2.11). For this reason they are also called **chaotropic** ("chaos-forming") agents. Disruption of the regular water structure decreases the hydrophobic effect and thereby promotes the unfolding and dissociation of protein molecules. Detergents like sodium dodecyl sulfate (SDS), which form micelle-like structures about individual polypeptide chains, are even more effective at unfolding proteins. By determining the molecular masses of the dissociated subunits under non-native solvent conditions and comparing them with the native M_W, we can tell how many subunits are present in the native protein.

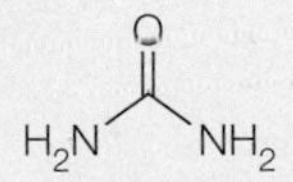

Urea

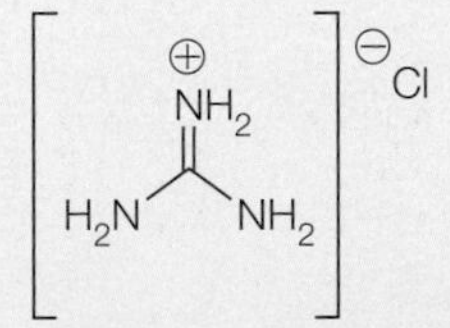

Guanidinium chloride
(guanidine hydrochloride)

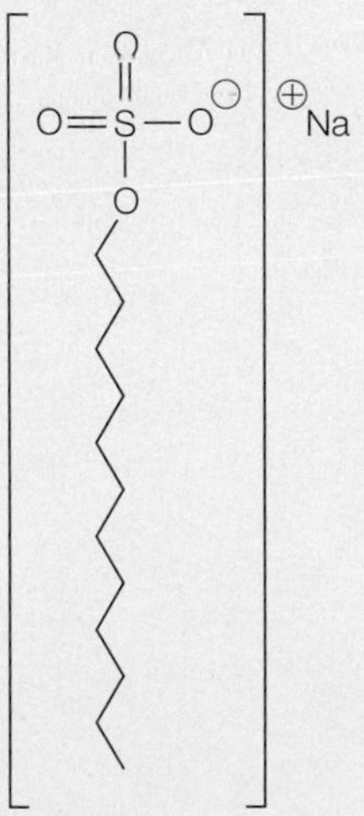

Sodium dodecyl sulfate
(SDS)

Determining the Molecular Mass of the Native Structure

To determine the molecular masses of proteins in their physiological states, several techniques are available. Recall from Tools of Biochemistry 5A that size exclusion chromatography (SEC) separates proteins in a mixture based on differences in hydrodynamic radius. The hydrodynamic radius is related to the M_W; thus, SEC can be used to estimate the native M_W of a protein. The shape of the protein (i.e., rod-like vs. spherical) has a significant impact on hydrodynamic radius, which can skew SEC data. Nevertheless, SEC provides a simple means to estimate, with reasonable accuracy, the M_W of a protein under native conditions. This is done by making a calibration curve of log M_W versus retention time for a series of protein standards of known M_W.

A more exact, but less commonly available, way to measure M_W uses the technique of **sedimentation equilibrium.** If a protein solution is sedimented for many hours at low rotor speed in an analytical ultracentrifuge, an equilibrium will be established between the tendency of the molecules to sediment and their tendency to diffuse back into the solution. Details of sedimentation equilibrium and other physical techniques used to determine the molecular masses of native proteins are given in van Holde et al. (see Appendix II).

Mass spectrometry (see Tools of Biochemistry 5B) is routinely employed to determine masses of protein subunits with great accuracy. In most cases, it is possible to obtain masses for each subunit of a multisubunit complex following the isolation of the native complex by SEC.

Determining the Number and Approximate Masses of Subunits: SDS Gel Electrophoresis

Once the native M_W has been determined, the easiest way to estimate the molecular masses of the subunits is to use gel electrophoresis in the presence of SDS. Under these conditions, quaternary, tertiary, and secondary structures of proteins are all disrupted. The chain is unfolded and coated by SDS molecules. The numerous negative charges carried by the many SDS molecules bound noncovalently to the protein mask any charges associated with the side chains or protein termini. The folded polypeptide chain is therefore transformed into an unstructured elongated object, the length and charge of which are each proportional to the number of amino acid residues in the chain (and hence the polypeptide M_W). As pointed out in Tools of Biochemistry 2A and 5A, such particles will migrate during gel electrophoresis, with relative mobilities depending only on their lengths. This phenomenon is demonstrated by the graph shown in **FIGURE 6B.1**. If electrophoresis of an unknown protein chain is carried out on the same gel as a set of standards, the M_W of the unknown can be measured by interpolation using a graph like that shown in Figure 6B.1. In practice, the stained SDS-PAGE gel is scanned using image analysis software (e.g., ImageJ) that automatically creates the calibration curve and calculates an apparent M_W for the unknown protein.

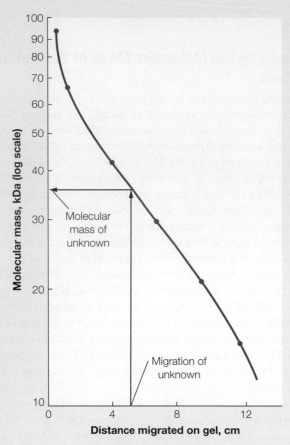

▲ **FIGURE 6B.1** Estimation of protein molecular weight by SDS gel electrophoresis. The graph plots log M_W versus relative electrophoretic mobility for a series of proteins dissolved in a solution containing the detergent SDS. The resulting curve is used to estimate the apparent M_W for an unknown protein.

In investigating the subunits of a protein by this technique, it is advisable to do two experiments: one in the presence of a disulfide-bond–reducing agent like β-mercaptoethanol (HSCH$_2$CH$_2$OH, or BME) and one in its absence. This will distinguish between subunits that are bonded by —S—S— bridges and those that are held together only by noncovalent forces. If a single band is found on each of these SDS gels, corresponding to the M_W of the native protein, we may conclude that the protein exists under physiological conditions as a single-polypeptide chain. If the band or bands observed are of much lower molecular mass, a multisubunit structure is indicated. Even though the M_W values obtained from gel electrophoresis may be approximate, they should be sufficiently accurate for a good guess as to the number of subunits (see Problem 13). For example, **FIGURE 6B.2** shows the pattern of bands expected on an SDS gel when a sample of an antibody is analyzed in the absence or presence of a disulfide reducing agent like BME. When the disulfides are intact ("–BME"), the antibody runs as a single band on the gel, with an apparent M_W of ∼150 kDa. Treatment with BME ("+BME") cleaves the disulfide bonds and yields a mixture of higher (∼53 kDa) and lower (∼23 kDa) M_W chains, which appear as two distinct bands on the gel.

Assuming that multiple subunits are indicated, is there only one kind, or are there several? More than one band on the SDS gel is a clear indication of multiple types of subunits. But finding only one band does not prove that subunits are identical. There may, in fact, be several kinds of subunits with distinct amino acid sequences but nearly identical molecular masses; these different subunits usually cannot be resolved on SDS gels. To be satisfied that only one type of chain is present, the researcher must turn to other methods. Again, mass spectrometry is the method of choice for such determinations.

References

For a list of references related to this chapter, see Appendix II.

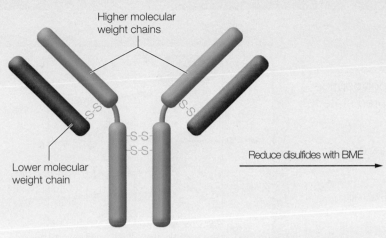

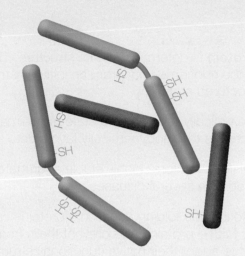

Reduce disulfides with BME

(a) The intact antibody has a *Mw* of ~150 kDa and is composed of two higher *Mw* heavy chains (~53 kDa, blue) and two lower *Mw* light chains (~23 kDa, red). When BME is added to the sample, the disulfide bonds between the chains are broken.

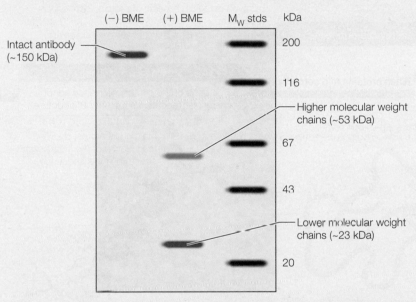

(b) A schematic representation of the stained SDS gel. With no added BME, a single band of ~150 kDa is observed (leftmost lane of the gel). When BME is added to the sample, the two chains are separated by electrophoresis (middle lane). The apparent *Mw*s of the bands can be estimated by comparing them to a set of protein *Mw* standards (rightmost lane of the gel). The colors of the bands here correspond to the colors of the heavy and light chains shown in panel (a). In practice, the protein bands in the gel are colorless until stained with some dye that binds to proteins.

◄ **FIGURE 6B.2** SDS-PAGE analysis of an antibody.

The function of a protein is determined by its structure. The polypeptide chain synthesized on the ribosome folds from an initial unstructured state to a functional, more structured state.

• This folded structure typically represents a state of minimal free energy. Thus, the folding process is thermodynamically favorable.

• Proteins can also misfold and form highly stable aggregates. Protein misfolding is associated with several diseases.

• Cells have evolved to include folding accessory proteins called chaperones, which bind to partially folded proteins and prevent them from aggregating, thereby allowing them to adopt their native (functional) structures.

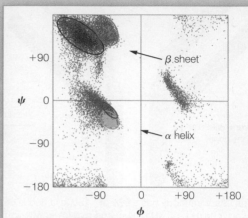

Ramachandran plot
The Ramachandran plot illustrates which combinations of Phi and Psi angles are sterically allowed in regular repeats of protein secondary structure.

▶ SEE FIGURE 6.11: OBSERVED VALUES OF PHI AND PSI FROM PROTEIN STRUCTURAL DATA

1 FOLDING Proteins fold to functional structures, which represent low free energy conformations.

Primary (1°) structure
Amino acid sequence (primary structure) determines secondary and tertiary structure.

When proteins fold correctly . . .

Secondary (2°) structure
Some parts of the primary sequence adopt a local regular repeating structure ("2° structure").

Side chains

Ala

Gly ϕ ψ Val

▶ SEE FIGURE 5.10: STRUCTURE OF THE PEPTIDE BOND AND FIGURE 5.11: A TETRAPEPTIDE

Amino acid

▶ SEE FIGURE 6.2: THE FOUR LEVELS OF STRUCTURAL ORGANIZATION IN PROTEINS

Polypeptide chain translated from mRNA

Main chain atoms, showing the location of the local helical structure in this sequence.

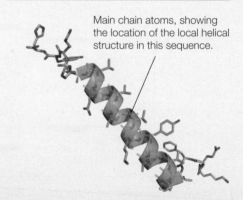

When proteins misfold . . .

TRANSLATION
Translation of mRNA codons into linear sequence of amino acids.
Genetic regulation of protein synthesis is discussed in chapter 26.

▶ SEE FIGURE 4.34: THE BASIC PRINCIPLE OF TRANSLATION

Ribosome

Translated amino acid sequence (polypeptide chain)

mRNA

TRANSCRIPTION

▶ SEE FIGURE 4.33: THE BASIC PRINCIPLE OF TRANSCRIPTION

Mastering **Chemistry** for Biochemistry

Mastering Chemistry for Biochemistry provides select end-of-chapter problems and feedback-enriched tutorial problems, animations, and interactive figures to deepen your understanding of complex topics while practicing problem solving.

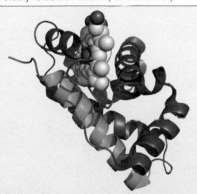

Native state conformation

A "Folding Funnel"

In the "energy landscape" model, the trajectory of protein folding is "downhill" because it proceeds with decreasing free energy. The width of the free energy landscape correlates with conformational entropy; as a protein folds, conformational entropy is reduced. As a result, the landscape narrows moving toward the lowest energy state, producing a funnel shape.

▶ SEE FIGURE 6.29: PROTEIN FOLDING ENERGY LANDSCAPES

Tertiary (3°) structure

Several 2° structure elements associate along their hydrophobic surfaces to give a stably folded structure ("3° structure").

② FUNCTION Proteins possess shape and charge complementarity to their specific ligands/substrates.

λ cl repressor protein

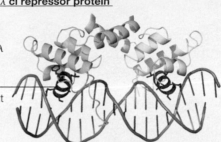

Two helices (red) in the λ cl repressor protein are optimally positioned to make complementary hydrogen bonds and salt bridges with a specific DNA sequence.

Dihydrofolate reductase

The enzyme dihydrofolate reductase (cyan), makes several specific hydrogen bonds to the cofactor NADPH.

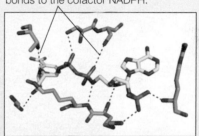

Shape complementarity　　　　**Charge complementarity**

③ MISFOLDING Misfolded proteins tend to aggregate and this is associated with several diseases.

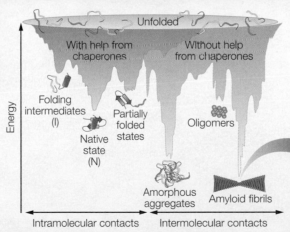

Unfolded

With help from chaperones　　Without help from chaperones

Energy

Folding intermediates (I)

Partially folded states

Oligomers

Native state (N)

Amorphous aggregates　　Amyloid fibrils

Intramolecular contacts　　Intermolecular contacts

▶ SEE FIGURE 6.30: MODELS OF PROTEIN FOLDING AND AGGREGATION

Amyloid is a particularly stable form of misfolded protein. Misfolding to aggregated states is prevented in cells by chaperone proteins; many of the so-called heat-shock proteins are chaperone proteins that assist in protein folding.

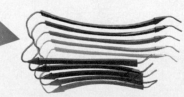

Repeating beta structure of amyloid fibrils

Molecular chaperones

GroEL/ES is one example of a molecular chaperone. Chaperones play a critical role in keeping partially folded proteins on the pathway to proper folding, and thereby preventing aggregation which can lead to disease.

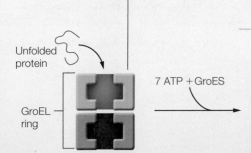
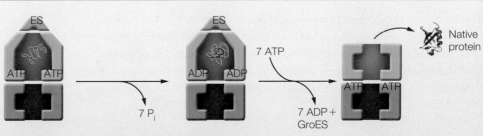

Unfolded protein

GroEL ring

ES
ATP ATP

7 ATP + GroES

7 P_i

ES
ADP ADP

7 ATP

7 ADP + GroES

ATP ATP

Native protein

▶ SEE FIGURE 6.31: THE GROEL-GROES CHAPERONIN

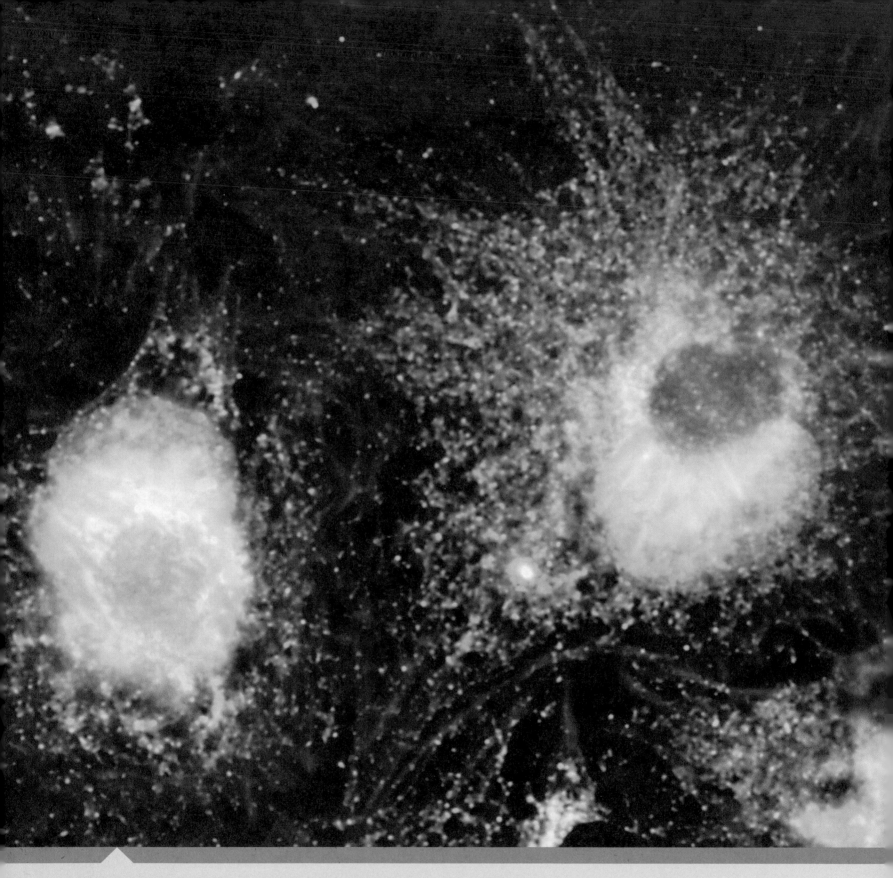

Fluorescence microscopy image of a mouse immortalized cell line. The cells have been modified, and different human genes have been transfected into the cells so that they can be used as a tool for study of the human immune system. The cells have been stained with fluorescent antibodies for different cell structures. The structures in red are the cell cytoskeleton, those in green are cell organelles, while those in blue are nuclei.

Protein Function and Evolution

7

WITH AN UNDERSTANDING of the complex, folded structures of globular proteins presented in Chapter 6, we can now look more closely at how such structures are related to the proteins' functions and how they may have evolved to fulfill those functions. In this chapter, we examine three groups of proteins: (1) antibodies of the immune system, (2) the oxygen-binding proteins myoglobin (found primarily in muscle) and hemoglobin (the major constituent of red blood cells), and (3) the motility proteins actin and myosin. These proteins illustrate several features of protein function common to many of the other proteins discussed throughout this text. The structure–function relationships for another diverse and important group of proteins, the enzymes, are discussed in Chapter 8.

We start with a description of the structural features of antibodies that account for their highly selective binding to specific targets. We will see that binding specificity is due to interactions between *complementary binding surfaces* on the antibody and its target ligand. Such complementary interactions play critical roles in many biochemical processes, including enzyme catalysis, recognition of

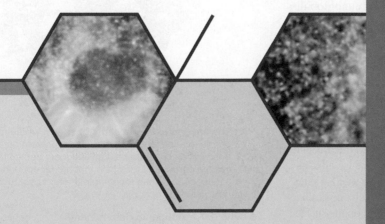

specific nucleic acid sequences by transcription factors, and hormone-receptor binding.

We then move from simple ligand binding by antibodies to a ligand-binding event in hemoglobin that results in conformational change in that protein. This change in protein conformation alters the oxygen-binding properties of hemoglobin and leads into a discussion of **allostery,** which is modulation of protein activity via conformational change. Allostery is an important mechanism for the regulation of key enzymes in the metabolic pathways discussed in later chapters.

We will also use the globin family of proteins to illustrate many basic principles of protein evolution and the relationship between mutation of a protein and disease.

Finally, we will investigate the molecular basis of muscle contraction, which requires a multistep cycle driven by ATP hydrolysis that is coupled to several protein conformation changes. We will see more examples of protein conformation change linked to ATP hydrolysis when we discuss membrane transport in Chapter 10.

7.1 Binding a Specific Target: Antibody Structure and Function

Antibodies, also called **immunoglobulins,** are large proteins produced in a multitude of amino acid sequence variations—each displayed on a common structural framework. Each variant binds with exquisite specificity and, essentially nonreversibly, to a unique target. Our primary defenses against infectious disease depend on the ability of immunoglobulins to recognize and bind to "foreign" (or "non-self") molecules, such as bacterial or viral pathogens. This section focuses on the structural features of antibodies that allow them to recognize non-self molecules and target them for destruction.

7.2 The Adaptive Immune Response

When a foreign substance—a virus, a bacterium, or even a foreign molecule—invades the tissues of a higher vertebrate, the organism defends itself by the **immune response.** We start with a brief description of the **adaptive immune response,** which includes humoral and cellular components. In the **humoral immune response,** lymphatic cells called **B lymphocytes** synthesize specific immunoglobulin molecules that are secreted and bind to the invading substance. This binding causes aggregation of the foreign substance and marks it for destruction by cells called **macrophages.** In the **cellular immune response,** lymphatic cells called **T lymphocytes** recognize and destroy foreign cells.

● **CONCEPT** The immune response involves the defense of the body against foreign substances or pathogens and operates via many different cellular mechanisms.

● **CONCEPT** In the humoral immune response, B lymphocytes secrete antibodies (immunoglobulins) that bind tightly to specific antigens.

The substance that elicits an immune response is called the **antigen,** and a specific immunoglobulin that binds to this substance is called the **antibody.** If the invading particle is sufficiently large—like a cell, a virus, or a protein—many different antibodies may be elicited, each one binding specifically to a given **antigenic determinant** (or **epitope**) on the surface of the particle (**FIGURE 7.1(a)**). Such antigenic determinants may be, for

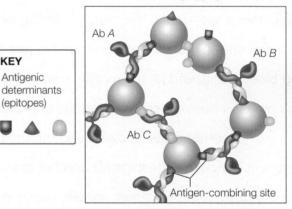

(a) **Precipitation.** A foreign object, or antigen (such as a virus, a bacterial cell, or a foreign protein molecule), may elicit the production of antibodies (shown in blue and light blue) to several different antigenic determinants on its surface. When the antigen is mixed with this collection of antibodies, precipitation occurs because each antibody molecule has two binding sites for its antigenic determinant. Thus, a crosslinked network is formed.

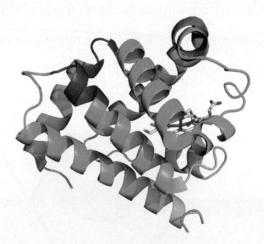

(b) **The antigenic determinants of sperm whale myoglobin.** The purple, orange, and cyan portions represent different segments of the polypeptide chain that act as antigens. Some antigenic determinants involve portions of the chain that are far apart in the primary sequence but close together in the tertiary structure—a so-called discontinuous epitope.

▲ **FIGURE 7.1** Antigenic determinants.

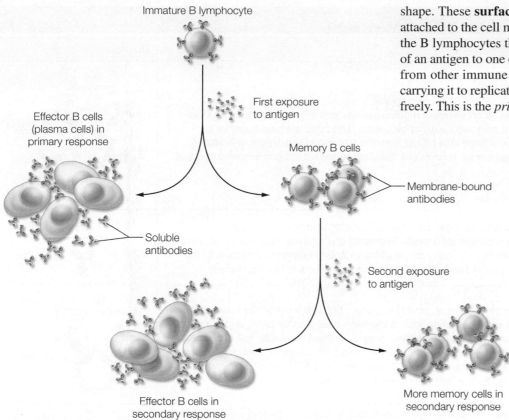

Immature B lymphocyte

First exposure
to antigen

Effector B cells
(plasma cells) in
primary response

Memory B cells

Membrane-bound
antibodies

Soluble
antibodies

Second exposure
to antigen

Effector B cells in
secondary response

More memory cells in
secondary response

▲ **FIGURE 7.2 Two developmental paths for stimulated B lymphocytes.** Exposure to antigen causes two kinds of cells to develop from immature B lymphocytes. Effector B cells, or plasma cells, synthesize soluble antibody (see Figure 7.3). Memory cells carry membrane-bound antibody to allow a rapid and enhanced response to a second exposure of the same antigen. Note that these processes require the involvement of several immune system components (e.g., T helper cells and interleukins), which we have omitted in this highly simplified diagram.

shape. These **surface immunoglobulins,** or **surface antibodies,** are attached to the cell membrane and are exposed on the outer surfaces of the B lymphocytes that circulate in blood plasma (**FIGURE 7.2**). Binding of an antigen to one of these surface antibodies, in the presence of help from other immune system cells (T helper cells), stimulates the cell carrying it to replicate and produce *soluble* antibodies that can circulate freely. This is the *primary immune response.*

As shown in Figure 7.2, two classes of B cells are produced in the primary response. **Effector B cells,** or **plasma cells,** produce soluble antibodies. These antibodies have the same antigen-binding sites as the surface antibodies of the B lymphocyte from which the effector cells arose, but they lack a hydrophobic tail that binds the surface antibodies to the lymphocyte membrane. The other class of cells—**memory cells**—will persist for some time, even after antigen is no longer present. This persistence constitutes the immune memory: It allows a rapid *secondary response* to a second stimulation by the same antigen, as shown in Figure 7.2.

At this point in our discussion of the adaptive immune response, a critical question may have occurred to you: Why do we not find B cells producing antibodies against *our own* proteins and tissues? Immature B lymphocytes in bone marrow will encounter self antigens. Those immature B lymphocytes that bind self antigens are *not* stimulated to replicate. Rather, these B lymphocytes are destroyed before they can mature. Thus, B lymphocytes producing antibodies against all of the potential "self" antigens are eliminated before release to circulation.

Normally, the only B lymphocytes that mature are those that produce antibodies against "non-self" substances. Occasionally, the immune system goes awry and produces antibodies against the normal tissues of an adult. The causes for such **autoimmunity** are not wholly understood, but the resulting diseases can be devastating. For example, in *lupus erythematosus* the individual's own nucleic acids become the object of attack. Other autoimmune diseases include rheumatoid arthritis, multiple sclerosis, type 1 diabetes mellitus, and psoriasis.

● **CONNECTION** In autoimmune diseases, such as rheumatoid arthritis, multiple sclerosis, and type 1 diabetes mellitus, the immune system attacks normal tissues.

instance, groups of sugar residues in a carbohydrate or groups of amino acids on a protein surface; an example is shown in Figure 7.1(b).

The adaptive immune response has some remarkable features. First, it is incredibly versatile, being able to respond to an enormous number of different foreign substances. These foreign substances range from cells of another individual of the same species (e.g., rejection of tissue grafts or transplanted organs) to synthetic molecules that could never have been encountered in nature. Second, the adaptive immune response has a so-called *memory:* After an initial exposure to a given antigen, a second exposure at a later date will result in rapid and more massive production of the antigen-specific antibodies. Vaccines are developed using modified pathogens that have been treated to reduce their ability to cause disease, but that still contain antigens that elicit an immune response against the virulent form of the pathogen. Thus, the immune system of a vaccinated individual should prevent the onset of disease when exposed to the virulent pathogen, owing to the rapid production of neutralizing antibodies from memory cells.

The immune system of mammals has an inherent ability to produce an immense diversity of antibodies with different amino acid sequences that are able to bind an enormous range of antigens. The generation of antibodies begins in the bone marrow, where immature **B lymphocytes** are produced. Every B lymphocyte produces a single type of immunoglobulin molecule, with a binding site that recognizes a specific molecular

7.3 The Structure of Antibodies

To see how the adaptive immune response works at the molecular level, we must explore the structure of the immunoglobulin molecules that constitute the antibody arsenal. There are five classes of immunoglobulin molecules which carry out various functions in the immune system (**TABLE 7.1**). However, all are built from the same basic immunoglobulin structure, which is shown schematically in **FIGURE 7.3**. Different kinds of antibodies may contain from one to five immunoglobulin molecules;

TABLE 7.1 The five classes of immunoglobulins

Class	Definition	Diagram
IgM	IgM is produced during the early response to an invading microorganism. It is the largest immunoglobulin, containing five Y-shaped units of two light and two heavy chains each. The units are held together by a component called a J chain. The relatively large size of IgM restricts it to the bloodstream. It is also effective in triggering an important mechanism for foreign cell destruction, called the complement system.	IgM (pentamer) — J chain
IgG	IgG molecules, also known as *γ-globulin*, are the most abundant circulating antibodies. A variant is attached to B-cell surfaces. IgG molecules consist of a single Y-shaped unit and can traverse blood vessel walls rather readily; they also cross the placenta to carry some of the mother's immune protection to the developing fetus. Specific receptors allow such passage. IgG also triggers the complement system.	IgG (monomer)
IgA	IgA is found in body secretions, including saliva, sweat, and tears, and along the walls of the intestines. It is the major antibody of colostrum, the initial secretion from a mother's breasts after birth, and of milk. IgA occurs as a monomer or as double-unit aggregates of the Y-shaped protein molecule. IgA molecules tend to be arranged along the surface of body cells and to combine there with antigens, such as those on a bacterium, thus preventing the foreign substance from directly attaching to the body cell. The invading substance can then be swept out of the body together with the IgA molecule.	IgA (monomer or dimer) — J chain
IgE/ IgD	Less is known about the IgD and IgE immunoglobulins. IgD molecules are found on the surface of B cells. IgE is associated with some of the body's allergic responses, and its levels are elevated in individuals who have allergies. The constant regions of IgE molecules can bind tightly to mast cells, a type of epithelial and connective tissue cell that releases histamines as part of the allergic response. Both IgD and IgE consist of single Y-shaped units.	IgD (monomer) IgE (monomer)

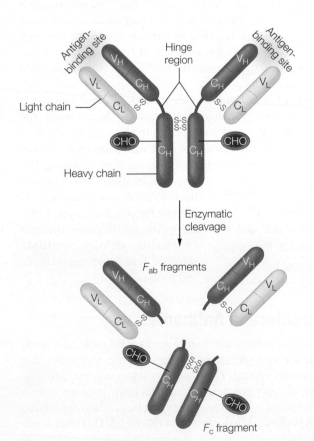

when more than one is present, the monomers are linked by an additional polypeptide, called a J chain (see Table 7.1).

Each immunoglobulin monomer consists of four polypeptide chains, two identical **heavy chains** ($M_W = 53{,}000$ Da each) and two identical **light chains** ($M_W = 23{,}000$ Da each) held together by disulfide bonds. Each chain contains **constant domains** (identical in all antibodies of a given class) and a **variable domain.** It is variation in the amino acid sequence (and therefore the tertiary structure) of the variable domains of the light and heavy chains that establishes the binding specificity of a given antibody to its unique antigen. Note that the four variable domains are carried at the ends of the Y-like fork of the molecule, where they form two identical binding sites for antigens.

A large protein, a virus, or a bacterial cell has many different potential antigenic determinants on its surface. Antibodies may be generated against several of these determinants, binding many antigen molecules together and thereby aggregating the antigen (see Figure 7.1(a)). If the antigen is so small that it has only one determinant, binding will occur but aggregation will not. Antibody-mediated aggregation, also called **immunoprecipitation,**

◀ **FIGURE 7.3 Schematic models of an IgG antibody molecule and an F_{ab} fragment.** The IgG is made from two identical heavy chains (dark blue) and two identical light chains (light blue), all held together by disulfide bonds. Each chain contains both *constant* domains (C) and *variable* domains (V). Constant domains are the same in all antibody molecules of a given class (see Table 7.1), whereas variable domains confer specificity to a given antigenic determinant. Cleavage by certain proteolytic enzymes such as papain at the *hinge regions* allows production of two identical monovalent F_{ab} fragments and one F_c fragment (see Figure 7.4). The carbohydrate (CHO; red) attached to the heavy chains helps determine the destinations of antibodies in the tissues and in stimulating secondary responses such as phagocytosis. The crystal structure of an immunoglobulin molecule is shown in Figure 7.4.

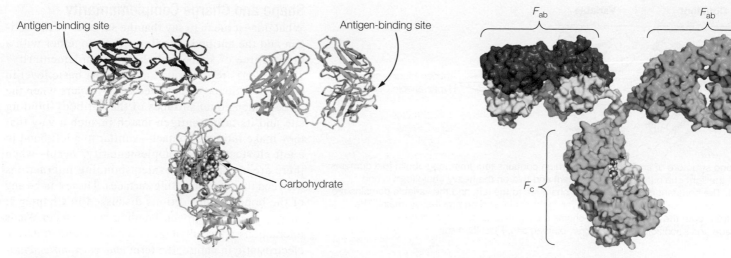

▲ **FIGURE 7.4   The crystal structure of an IgG molecule from mouse.** The identical heavy chains are colored yellow and cyan; the identical light chains are magenta and green. A cartoon model, illustrating the high degree of β secondary structure, is shown on the left. On the right is a surface rendering showing the intimate contact between the chains. The carbohydrate attached to the heavy chain is indicated in the leftmost figure. Note that the two F_{ab} fragments have identical structures; one is rotated relative to the other in this image. PDB ID: 1igt.

● **CONCEPT**   Immunoglobulin molecules contain both constant and variable regions. The variable regions form the antigen-binding sites.

requires the antibody to be *bivalent* (to have two binding sites). In the laboratory, it is possible to cleave antibodies at the *hinge* regions (see Figure 7.3) with specific protein-cleaving enzymes, to produce a single F_c **fragment,** which has no antigen-binding site, and two F_{ab} **fragments,** which have only one binding site each. The F_{ab} fragments will bind, but not precipitate, antigen.

The constant domains of the heavy chains in the base of the Y-shaped molecule serve to hold the chains together. More important, these regions also function as effectors that signal other cells of the immune response, such as T cells or macrophages, to attack particles or cells that have been marked for destruction by antibody binding. Macrophages are large, white blood cells that are specially adapted to engulf and digest foreign particles, a process called phagocytosis. In addition, differences in heavy chains identify immunoglobulin types for delivery to different tissues or for secretion (see Table 7.1).

7.4 Antibody:Antigen Interactions

The antigen-binding sites lie at the extreme ends of the variable domains (Figure 7.3 and **FIGURE 7.4**) and involve amino acid residues from the variable regions of both heavy and light chains. Different sequences in these variable regions give rise to different local secondary and tertiary structure; they can thereby define binding sites to fit different antigens with exquisite specificity.

● **CONCEPT**   The diversity as well as the exquisite specificity of antigen-binding sites is determined by the hypervariable complementarity determining regions from both the light and the heavy chains.

The domains in immunoglobulins are built on a common motif—the *immunoglobulin domain,* or *Ig domain*—in which two antiparallel β sheets lie face to face (**FIGURE 7.5**). The Ig domain probably represents the primitive structural element

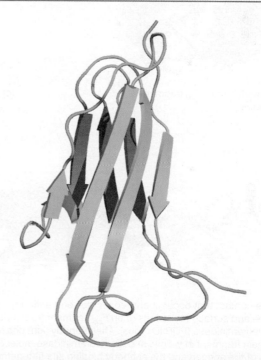

▲ **FIGURE 7.5 The immunoglobulin domain.** The immunoglobulin domain is a common structural motif among proteins in the immunoglobulin superfamily (see text). Two antiparallel β sheets (cyan and orange) are stacked face to face and covalently bonded by a disulfide bond (not shown). This folding motif is found 12 times in the IgG molecule (see the leftmost panel of Figure 7.4) and 4 times within an F_{ab}. PDB ID: 1igt.

in the evolution of the adaptive immune response. Indeed, the Ig domain is also found in a number of other proteins that are involved in cell recognition. The various proteins that contain Ig domains are classified as members of the **immunoglobulin superfamily.** We show some examples later in this chapter.

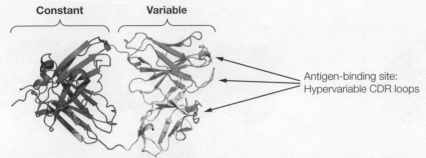

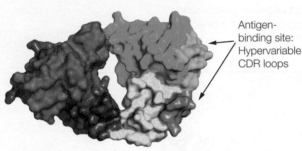

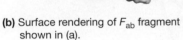

(a) Backbone structure of an F_{ab} fragment, which contains four immunoglobulin fold domains: two on the light chain (magenta and green) and two on the heavy chain (brown and yellow). The constant domains on each chain are to the left, and the variable domains are on the right. The CDRs are shown in cyan (light chain) and orange (heavy chain). The CDRs from both the heavy and light chains are hypervariable in sequence and determine the shape and specificity of the antigen-binding site. PDB ID: 1aqk.

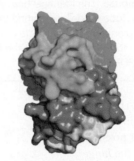

(b) Surface rendering of F_{ab} fragment shown in (a).

(c) Same rendering as in (b), but rotated 90 degrees to view the surface of the antigen-binding site formed by the CDR loops.

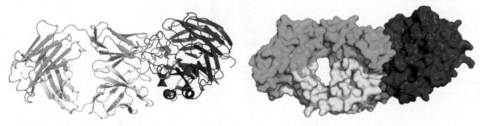

(d) The close contact that occurs between antigen and antibody surfaces is shown in the backbone and surface renderings of an F_{ab} fragment from mouse bound to the viral protein neuraminidase (PDB ID: 1nca). The antibody light chain is shown in green, the heavy chain fragment in yellow, and the neuraminidase molecule in purple. The surfaces of the antigen and the antibody binding site fit together with a high degree of shape and charge complementarity.

▲ **FIGURE 7.6** Antigen binding by an F_{ab} fragment.

The Ig domain is a stable scaffold on which to display the hypervariable loops that determine the *shape* and *charge* complementarity of the antigen-binding site (**FIGURE 7.6**). These hypervariable loops are known as **complementarity determining regions** or **CDRs.** Figure 7.6 shows the results of X-ray diffraction studies of the interaction of an F_{ab} fragment with a viral protein antigen, neuraminidase, an enzyme that facilitates viral infection (discussed in Chapter 9). Note that the antigen and antibody surfaces fit together in a highly complementary fashion.

Shape and Charge Complementarity

What does it mean to say that the surfaces of the antigen and the antibody binding site fit together with a high degree of shape and charge complementarity? These ideas are illustrated at the most basic level in **FIGURE 7.7**. "Shape complementarity" occurs when the three-dimensional surfaces of the antibody-binding site and its target antigen match in such a way that they make intimate contact—similar to a left hand in a left glove. "Charge complementarity" occurs when there are specific noncovalent-bonding interactions between the two contacting surfaces. These can be any of the bonding interactions discussed in Chapter 2: charge–charge, hydrogen-bonding, and van der Waals interactions. Since all of these bonding interactions are electrostatic in nature, the term *charge complementarity* applies, even though it includes interactions that may lack formal charge.

Shape complementarity and charge complementarity are important concepts to understand because they provide a fundamental explanation for many biochemical phenomena. As Figure 7.7 indicates, in addition to explaining the specificity of antibody–antigen binding, shape and charge complementarity explain receptor–ligand binding, enzyme–substrate binding, and some features of enzyme catalysis.

● **CONCEPT** Most of the biochemical processes in cells involve complementary binding interactions between biomolecules and their ligands.

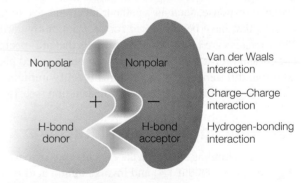

Antibody–Antigen binding site ⟶ Target 5 antigen
Receptor–Ligand binding site ⟶ Target 5 ligand
Enzyme–Substrate binding site ⟶ Target 5 substrate

▲ **FIGURE 7.7** A simplified illustration of shape and charge complementarity. Shape complementarity is achieved when two surfaces have significant close contact. Charge complementarity occurs when favorable noncovalent-binding interactions occur in multiple locations between the two surfaces. These concepts apply to a wide variety of binding interactions between biomolecular receptors and their target ligands.

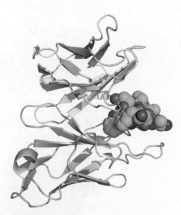

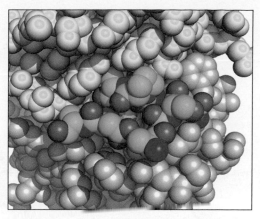

(a) Cartoon rendering of the F_v fragment and space-filling rendering of the viral peptide. The CDR loops of the F_v make several contacts with the peptide.

(b) Surface rendering of the F_v fragment shown in (a) illustrates the intimate contact between the CDRs and the peptide.

(c) Same rendering as in panel (b), but rotated 90 degrees and zoomed in on the antigen-binding site.

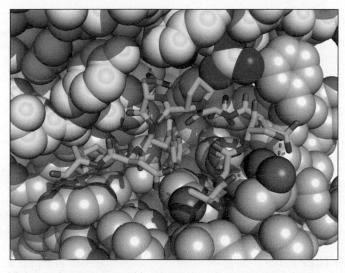

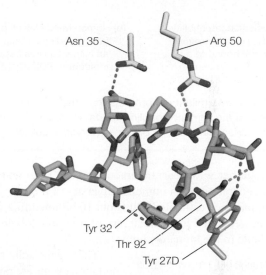

(d) Same view as in panel (c) with the peptide rendered in sticks to show that the shape of the antigen-binding site matches the shape of the peptide antigen.

(e) Same view as in panel (d) showing selected side chains from the CDRs in stick rendering. A few hydrogen bonds between the F_v and the antigen are shown as dotted lines (magenta lines show side chain–side chain H bonds; green lines show side chain–main chain H bonds).

▲ **FIGURE 7.8 Shape and charge complementarity between an antibody and its binding target.** This figure shows the molecular details of binding between a peptide antigen derived from the small envelope protein of human hepatitis B virus (cyan) and the variable domains of a neutralizing antibody (yellow and gray). This variable portion of an IgG is also called an F_v fragment. In all panels, the F_v light chain is yellow, the heavy chain is gray, and the C atoms of the viral peptide are cyan. Selected N atoms are blue, and O atoms are red. PDB ID: 2eh8.

When an antibody not only binds to the antigen, but also prevents the antigen from exerting its physiological effects (i.e., infection), it is said to have "neutralized" the antigen. **FIGURE 7.8** illustrates shape and charge complementarity between a neutralizing antibody and a peptide antigen from the small envelope protein found on the surface of the human hepatitis B virus. Panels (a)–(d) show the intimate contact between the antigen and the CDR loops (i.e., "shape" complementarity), and panel (e) shows "charge" complementarity

● **CONNECTION** Detailed knowledge of receptor–ligand binding interactions can be useful in the design of new therapeutics and vaccines.

in the form of several specific hydrogen-bonding interactions between the peptide and the CDR loops. Understanding the molecular details of the binding between this antigen and the antibody can help efforts to develop an effective vaccine against hepatitis B, which affects ~ 400 million people worldwide.

Generation of Antibody Diversity

In antibody-producing B cells, the heavy and light chain gene sequences are rearranged and spliced to create many different sequence combinations. The details and mechanism of this process are described in Chapter 23. In addition, within B cells, the gene sequences corresponding to the IgG CDR loops mutate at an unusually high rate.

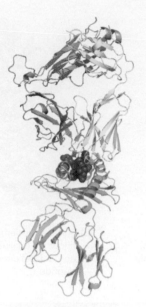

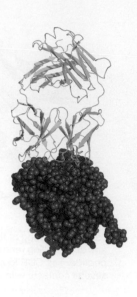

(a) Human major histocompatibility complex (MHC) class I protein (cyan and green) bound to a fragment of an HIV protein (magenta). PDB ID: 1a1m.

(b) Human MHC class II protein (green and cyan) bound to an influenza virus peptide (magenta). PDB ID: 1dlh.

(c) Human T-cell receptor (rose and yellow) binding to an MHC class I molecule (cyan and green) displaying a viral peptide (magenta). PDB ID: 1bd2.

(d) Murine F_{ab} fragment (cyan and green) of IgG bound to neuraminidase (magenta; see Figure 7.6, PDB ID: 1nca).

▲ **FIGURE 7.9 Structural similarity of proteins from the immunoglobulin superfamily.** This figure shows crystal structures for a few members of the immunoglobulin superfamily of proteins, which includes not only the immunoglobulin family but many related cell-surface and soluble proteins involved in cell recognition and binding.

This process, together with rearrangement of gene fragments, can account for the generation of an immense diversity of immunoglobulin molecules. It has been estimated that about 10 billion different IgG sequences can be generated from the library of immunoglobulin gene fragments available in the human genome.

● **CONCEPT** Through gene recombination and rapid mutation, a human can generate over 10 billion different antibodies.

These gene splicing and mutation events are random; thus, antibody-binding specificity is not preprogrammed. Accordingly, it is possible to elicit an immune response to synthetic substances as long as there is a B lymphocyte that displays a complementary antibody.

7.5 The Immunoglobulin Superfamily

Whereas the humoral immune response of the adaptive immune system is based on antibody-produced aggregation, usually followed by digestion by macrophages, the cellular immune response involves a quite different mechanism for killing foreign cells. The cellular response plays a major role in tissue rejection and in destruction of virus-infected cells. It can also destroy potential cancer cells before they have a chance to propagate. Although the mechanisms of the humoral and cellular processes are quite different, both require specific recognition between a receptor and its ligand. This *function* (i.e., target recognition) is carried out in both pathways by protein molecules from the immunoglobin superfamily (**FIGURE 7.9**), which have similar *structures*. Notice the occurrence of several immunoglobulin domains in all these proteins.

The major effectors in the cellular immune response are **cytotoxic T cells,** also referred to as **killer T cells.** These cells carry on their surfaces receptor molecules that are structurally similar to the F_{ab} fragments of antibody molecules (Figure 7.9(c)). Like antibodies, these fragments have a wide range of binding specificities, mostly directed toward short oligopeptide sequences. The T-cell receptor recognizes foreign oligopeptides that are presented on the surface of the infected cell by another immunoglobulin-like molecule known as **major histocompatibility complex,** or **MHC** (see Figure 7.9(a–c)). When a killer T cell identifies (via its receptor) a foreign antigen carried on the surface of another cell by an MHC protein, it releases a protein called **perforin.** Perforin forms pores in the plasma membrane of the cell being attacked, allowing critical ions to diffuse out and thereby killing the cell.

● **CONCEPT** The cellular immune response uses killer T cells to destroy foreign or infected cells.

7.6 The Challenge of Developing an AIDS Vaccine

AIDS (acquired immune deficiency syndrome) is a disease of the immune system. It is caused by the **human immunodeficiency virus,** or **HIV** (**FIGURE 7.10**), which attacks a number of kinds of cells but is particularly virulent toward a class of T cells that activate B cells to produce neutralizing antibodies. HIV infects rapidly replicating T cells, and eventually the rate of T-cell destruction exceeds the rate of replication. The consequence is a deterioration of the whole immune response. Most AIDS patients succumb either to diseases they could have easily resisted before contracting AIDS or to certain kinds of cancer. AIDS is so deadly because it attacks our most fundamental defenses against all disease.

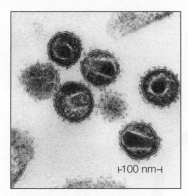

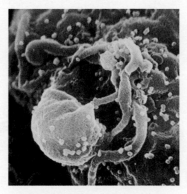

(a) Electron micrograph of the human immunodeficiency virus (HIV) that is responsible for AIDS.

(b) False-color scanning electron micrograph of budding HIV-1 virus particles (green spheres) on the surface of a human lymphocyte (red).

▲ **FIGURE 7.10** The human immunodeficiency virus, or HIV.

● **CONCEPT** In AIDS, the causal virus attacks T cells, destroying the body's immunological defense system.

Since 1983, more than 60 million people, predominantly in the developing world, have been infected with HIV, and nearly half of those have died. Because AIDS poses such a grave threat to world health, efforts to develop a vaccine are being pursued intensely. Such research entails unusual problems because HIV has an unparalleled capacity to mutate and thus develop strains resistant to any vaccine. Mutations occur in the HIV genome at a rate many times higher than in the human genome. This results in a high level of amino acid mutation in the viral coat proteins that would be the typical immunological targets in a vaccine development strategy. The magnitude of the problem can be grasped by considering our experience with the influenza virus. We have never been able to produce a lifelong "flu" vaccine because of the great variability of the influenza virus. HIV mutates about 60 times faster than the influenza virus.

Unfortunately, no effective vaccine has yet been developed for HIV. Antiviral drugs have been developed that target virus-specific enzymes involved in replication of the viral genome or processing of the gene products (see Chapter 8 opener). These therapies can slow the progression of AIDS, but they do not constitute a cure for the disease. One may hope that improved therapeutics, and an eventual vaccine, will halt the AIDS pandemic.

7.7 Antibodies and Immunoconjugates as Potential Cancer Treatments

Current cancer treatments frequently include radiation and/or chemotherapy with highly toxic drugs. Chemotherapy is associated with many undesirable side effects, owing in large part to the broad cytotoxicity of the drugs. The efficacy of chemotherapeutics would be increased if they could be delivered specifically to cancer cells rather than systemically. To that end, hybrid drugs, called **immunoconjugates,** have been designed which link a cytotoxic agent to an antibody that has been raised against a tumor-specific antigen (**FIGURE 7.11**). In principle, the antibody specifically delivers the drug to the target tumor cell where it is taken up only by that cell (and not a healthy cell). Once inside the cancer cell, the covalent link between the antibody and drug is cleaved,

Stable upon storage
Stable in circulation in vivo
Cleaved inside target cell

| Antibody | Linker | Drug |

High tumor specificity
Tight binding to target antigen
Multiple sites for attachment of drug

High potency
Linkable
Water soluble

(a) The desirable features for each component of the immunoconjugate: targeting antibody, linker, and cytotoxic drug.

(b) Common sites of attachment of drugs to the antibody constant regions are shown as orange spheres. The tumor-specific antigen binding sites are shown in green and yellow.

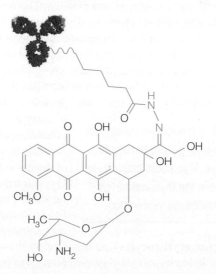

mAb-doxorubicin (hydrazone)

(c) A schematic (not to scale) of the immunoconjugate. An acid-labile hydrazone linker is highlighted in green. The linker is stable in circulation in blood but is cleaved in the acidic environment of the endosome following endocytosis into the tumor cell.

▲ **FIGURE 7.11** Immunoconjugate drugs for targeted chemotherapy.

releasing the drug to exert its cell-killing effect. In addition to cytotoxic drugs, such as taxol, radioactive isotopes have also been conjugated to antibodies. A few such immunoconjugates have been approved for clinical use, with more in clinical trials.

Because antibody binding is the first step in the recruitment of a cytotoxic response, antibodies that specifically recognize tumor antigens are under development as anticancer drugs that can selectively target tumors for destruction. At the beginning of 2017, roughly 35 antibodies or immunoglobulin derivatives had been approved for use as human therapeutics, primarily as anticancer and anti-inflammatory agents.

● **CONNECTION** Several antibody-based therapeutics have been approved for use in targeted cancer treatment.

7.8 Oxygen Transport from Lungs to Tissues: Protein Conformational Change Enhances Function

We now turn our attention to two proteins that play key roles in one of the most important aspects of animal metabolism—the acquisition and utilization of molecular oxygen (O_2). These proteins are **myoglobin (Mb)** and its molecular relative **hemoglobin (Hb),** members of a family of proteins collectively termed **globins.**

The most efficient energy-generating mechanisms in animal cells require O_2 for the oxidation of nutrients; therefore, proteins that transport O_2 to respiring cells are essential for any higher organism (this is described in detail in Chapter 14). Myoglobin is an O_2-binding protein found primarily within the muscle tissue of animal species; hemoglobin circulates in the blood and is used for oxygen transport in all vertebrates and some invertebrates. Hemoglobin also plays a role in removing CO_2 from tissues. CO_2 is a major product of metabolite oxidation and must be continually removed and exhaled. The role of hemoglobin in O_2 and CO_2 transport is schematically illustrated in **FIGURE 7.12**. Later in this chapter we consider important insights into the evolution of protein function, which emerged from analysis of the close structural relationship between hemoglobin and myoglobin.

The evolution of all higher organisms has been accompanied by the development of **oxygen transport proteins,** which allow the blood to carry a 100-fold higher concentration of O_2 than would be possible based on the solubility of the gas in blood plasma alone. Oxygen transport proteins may be either dissolved in the blood (as in some invertebrates) or concentrated in specialized cells, like the human **erythrocytes** (red blood cells) shown in **FIGURE 7.13**.

● **CONCEPT** Hemoglobin transports O_2 from lungs or gills to tissues fed by capillaries. Myoglobin binds O_2 in tissues, where the O_2 is ultimately consumed in cellular respiration.

In deep-diving mammals, the concentration of myoglobin in skeletal muscle is 10–30 times higher than that of terrestrial mammals. Thus, myoglobin also acts as an O_2 storage molecule, to provide a substantial reserve of O_2 to support the demand for ATP production while the animal is submerged. In larger whales, this can be 30 minutes or longer!

Myoglobin and hemoglobin are perhaps the most completely studied proteins on Earth and were important in defining the early history of biochemistry and molecular biology. Indeed, the properties of hemoglobin have been investigated extensively since the protein was successfully crystallized in the first half of the nineteenth century. Myoglobin holds the distinction of being the first protein for which

◄ **FIGURE 7.12 Role of the globins in oxygen transport and storage.** Vertebrate animals use hemoglobin and myoglobin to provide their tissues with a continuous O_2 supply.

LUNGS or GILLS

CO_2 O_2

HCO_3^-

VEINS

$NHCOO^-$

Fe Fe
Fe Fe

$NHCOO^-$

Deoxyhemoglobin

CO_2 produced by oxidative processes in the tissues is carried back to the lungs or gills by hemoglobin, or in the plasma as HCO_3^-, and released.

ARTERIES

O_2 O_2

Fe Fe
Fe Fe

O_2 O_2

Oxyhemoglobin

Hemoglobin transports O_2 from the lungs or gills to the respiring tissues, where it is used for aerobic metabolism in the mitochondria.

HCO_3^-

CO_2 O_2

Fe
Myoglobin
Fe

O_2

O_2

TISSUES

Inside cells, dissolved O_2 diffuses freely or is bound to myoglobin, which aids transport of O_2 to the mitochondria. Myoglobin can also store O_2 for later use (as in deep-diving mammals).

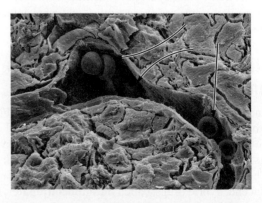

▲ **FIGURE 7.13 Human erythrocytes.** Arrows point to red blood cells, or erythrocytes, shown moving in a capillary. Each erythrocyte contains about 300 million hemoglobin molecules.

the structure was determined at the atomic level using X-ray crystallography (work for which John Kendrew received the Nobel Prize in Chemistry in 1962). Thus, both of these proteins have served for many years as primary models of protein structure and function.

The binding of small molecules to hemoglobin can alter its O_2-binding affinity. This is an example of an **allosteric effect,** whereby ligand binding results in a structural change in the protein and thereby alters its functional properties. This is an important feature in the regulation of many enzyme reactions, and also accounts for the mechanism of action of many therapeutic drugs.

7.9 The Oxygen-Binding Sites in Myoglobin and Hemoglobin

Myoglobin and hemoglobin are built on a common structural motif, as shown in **FIGURE 7.14**. In myoglobin, a single polypeptide chain is folded about a prosthetic group, the **heme** (**FIGURE 7.15**), which contains the O_2-binding site. Hemoglobin is a tetrameric protein, made up of four polypeptide chains, each of which binds a heme group, and resembles myoglobin in structure. We begin the discussion of how these structures enable O_2 transport from lungs to mitochondria with a description of O_2 binding in myoglobin.

An oxygen storage or transport molecule must be able to bind O_2 reversibly and protect it from reaction with any other substance before it is used for ATP production in mitochondria. How do the globins achieve this? To answer this question, we must consider how the peptide and the prosthetic group interact. The peptide portion of any protein without its prosthetic group bound is called the **apoprotein,** whereas with its prosthetic group bound, it is called a **holoprotein.** The apoglobins are unable to bind O_2 themselves; however, certain transition metals in their lower oxidation states—particularly Fe(II) and Cu(I)—have a strong tendency to bind O_2. The globin proteins have evolved such that Fe(II) is bound to the proteins to produce a site at which O_2 binds reversibly.

Various iron-containing proteins can hold Fe(II) in a number of possible ways. Throughout the myoglobin–hemoglobin family, the

● **CONNECTION** The iron–porphyrin in globin proteins is responsible for the red color of blood and meat, while the magnesium–porphyrin in chlorophyll is responsible for the green color of plants.

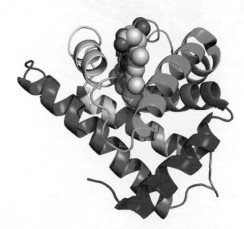

(a) Myoglobin shown in cartoon rendering with the heme in space-filling rendering (PDB ID: 1mbn). The protein backbone is shown in rainbow coloring, with the N-terminus in blue and the C-terminus in orange.

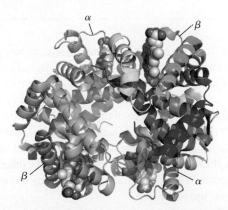

(b) Each of the four chains in hemoglobin (PDB ID: 2dn2) has a folded structure similar to that of myoglobin, and each carries a heme (shown in space-filling rendering). Hemoglobin contains two identical α chains (light pink) and two identical β chains (light blue and rainbow coloring). The β chain shown in rainbow coloring is oriented to show the high degree of structural similarity to the myoglobin in panel (a).

▲ **FIGURE 7.14** Comparison of myoglobin and hemoglobin.

(a) The structure of protoporphyrin IX.

—Fe²⁺

→ 2 H⁺

(b) Heme, which is protoporphyrin IX complexed with Fe(II), is the prosthetic group of hemoglobin and myoglobin. There is resonance delocalization of the electrons in the tetrapyrrole ring.

▲ **FIGURE 7.15** The structures of protoporphyrin IX and heme.

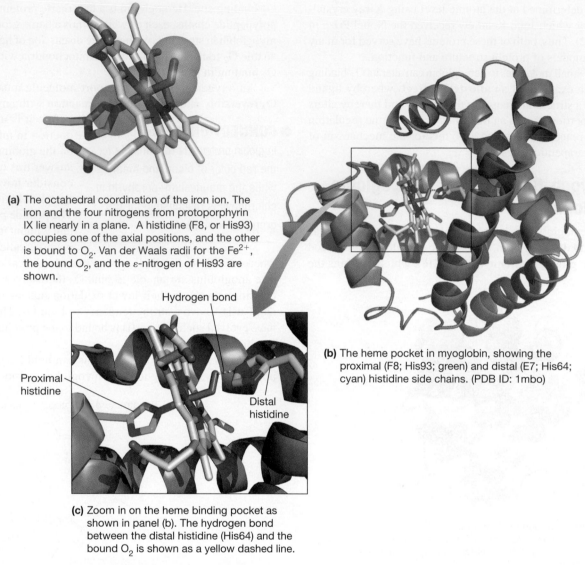

(a) The octahedral coordination of the iron ion. The iron and the four nitrogens from protoporphyrin IX lie nearly in a plane. A histidine (F8, or His93) occupies one of the axial positions, and the other is bound to O_2. Van der Waals radii for the Fe^{2+}, the bound O_2, and the ε-nitrogen of His93 are shown.

Hydrogen bond

Proximal histidine

Distal histidine

(b) The heme pocket in myoglobin, showing the proximal (F8; His93; green) and distal (E7; His64; cyan) histidine side chains. (PDB ID: 1mbo)

(c) Zoom in on the heme binding pocket as shown in panel (b). The hydrogen bond between the distal histidine (His64) and the bound O_2 is shown as a yellow dashed line.

▲ **FIGURE 7.16** The geometry of iron coordination in oxymyoglobin.

iron is chelated by a tetrapyrrole ring system called **protoporphyrin IX** (Figure 7.15(a)), one of a large class of **porphyrin** compounds. We will encounter other porphyrins in chlorophyll (Chapter 15), the cytochrome proteins (Chapter 14), and some natural pigments. Like most compounds with large conjugated ring systems, the porphyrins are strongly colored.

The complex of protoporphyrin IX with Fe^{2+} is called heme (Figure 7.15(b)). This prosthetic group is bound in a hydrophobic crevice in the myoglobin or hemoglobin molecule (see Figure 7.14). The binding of oxygen to heme is illustrated in **FIGURE 7.16**, which shows the oxygenated form of myoglobin. Normally, ferrous iron (Fe^{2+}) is octahedrally coordinated, which means it should have six **ligands,** or binding groups, attached to it. As shown in Figure 7.16(a), the nitrogen

atoms of the porphyrin ring account for only four of these ligands. Two remaining coordination sites are available, and they lie along an axis perpendicular to the plane of the heme. In both the deoxygenated and the oxygenated forms of

The ε-tautomer of histidine

myoglobin, one of these remaining coordination sites is occupied by the ε-nitrogen of histidine residue number 93. The eight helical segments in the globins are designated A through H, and residue 93 is located in the F helix. Using a nomenclature that allows for meaningful comparisons between the homologous sequences of different globins, this residue is called histidine F8 (it is the eighth residue in the F helix). Because it is in direct contact with the Fe^{2+}, it is also called the **proximal histidine.** In **deoxymyoglobin,** the remaining coordination site, on the other side of the iron ion, is unoccupied. When oxygen is bound, making **oxymyoglobin,** the O_2 molecule occupies this site.

The Fe^{2+} — O_2 complex is stabilized by a hydrogen bond between the bound O_2 and another important His residue located in the O_2-binding pocket—the so-called **distal histidine** (His 64, or E7; see Figure 7.16(c)). The hydrogen bond between His E7 and O_2 selectively increases the affinity of myoglobin for O_2 versus CO, which doesn't make a similar

● **CONCEPT** Coordination of Fe(II) in a porphyrin (heme) within a hydrophobic globin pocket allows reversible O_2 binding without iron oxidation.

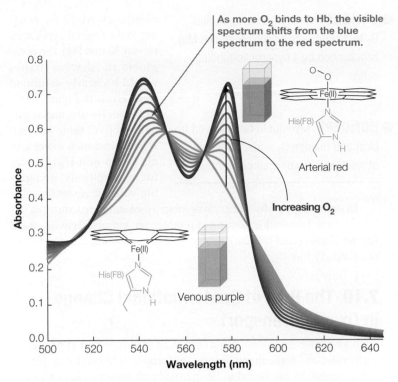

As more O_2 binds to Hb, the visible spectrum shifts from the blue spectrum to the red spectrum.

▲ FIGURE 7.17 Changes in the visible spectrum of hemoglobin. Spectra for hemoglobin in the deoxygenated state (blue trace) and the O_2-bound state (red trace) are shown. Hemoglobin in the deoxygenated state is a venous purple, whereas completely oxy-Hb is bright red. As more O_2 binds to Hb, the visible spectrum shifts from the blue to the red trace (several spectra for partially bound Hb are shown). Thus, the ligand-binding behavior of the globins is easily monitored by visible spectroscopy (see Tools of Biochemistry 6A) due to the distinctive spectral differences between the various forms of the globins. Courtesy of John Olson (Rice University).

hydrogen bond to His E7. Even so, CO binds ~200 times more tightly to myoglobin than does O_2. However, without the E7 hydrogen bond, that ratio would be ~6000:1 in favor of CO. Thus, the distal histidine plays a critical role in promoting O_2 binding over other ligands.

A similar mode of oxygen binding, with histidines at the homologous F8 and E7 positions, is found in each subunit of hemoglobin.

Although myoglobin and hemoglobin are ideally adapted to reversibly bind an O_2 molecule, they also bind other diatomic gases such as carbon monoxide (CO) and nitric oxide (NO). The toxicity of CO is due to its ability to block respiration by binding tightly to the Fe^{2+}-hemes in globins as well as those in other critical respiratory proteins called cytochromes. Nitric oxide also inhibits respiratory proteins (primarily cytochrome-c oxidase; discussed in Chapter 14) and is released by macrophages to destroy invading organisms as part of the immune response. At low levels, NO is also a cell-signaling molecule (see Chapter 20).

Analysis of Oxygen Binding by Myoglobin

The binding of O_2 by myoglobin must meet certain physiological requirements. As Figure 7.12 shows, myoglobin located inside muscle cells binds O_2, which diffuses into cells from hemoglobin circulating in capillaries. The myoglobin then delivers the O_2 to the mitochondria. To understand these functions on a quantitative basis, we must examine how the binding of a ligand like O_2 depends on its concentration in the surroundings.

First, a way to quantify the concentration of dissolved O_2 is needed. Because the concentration of any gas dissolved in a fluid is proportional to the *partial pressure* of that gas above the fluid, we can conveniently express O_2 concentration as this partial pressure: P_{O_2}.

To study ligand binding, we must have a way of measuring the fraction of myoglobin molecules carrying O_2. When myoglobin (or hemoglobin) is oxygenated, it changes color (the absorption spectrum changes due to alteration of the electronic structure of the heme iron—see Tools of Biochemistry 6A). This allows a spectrophotometric determination of the fraction of binding sites that are oxygenated (**FIGURE 7.17**). The results of such analysis, using myoglobin in solution at neutral pH, are shown in **FIGURE 7.18**. Such a graph is called a *binding curve* because it describes how the fraction of the myoglobin sites that have O_2 bound to them (Y_{O_2}) depends on the concentration of free O_2 (P_{O_2}).

We can describe the binding of a ligand (in this case: O_2) to myoglobin (Mb) by the following reaction

$$\text{Mb} + \text{O}_2 \xrightarrow{k_{\text{on}}} \text{MbO}_2 \tag{7.1}$$

and ligand dissociation by

$$\text{MbO}_2 \xrightarrow{k_{\text{off}}} \text{Mb} + \text{O}_2 \tag{7.2}$$

where k_{on} and k_{off} are the rate constants for binding and dissociation, respectively. Thus, reversible ligand binding is described by the following equilibrium

$$\text{MbO}_2 \underset{k_{\text{on}}}{\overset{k_{\text{off}}}{\rightleftharpoons}} \text{Mb} + \text{O}_2$$

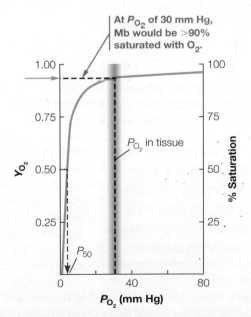

At P_{O_2} of 30 mm Hg, Mb would be >90% saturated with O_2.

▲ FIGURE 7.18 Oxygen-binding curve for myoglobin. The free oxygen concentration is expressed as P_{O_2}, the partial pressure of oxygen. The proportion of myoglobin O_2-binding sites that are occupied is expressed as a fraction (Y_{O_2}, on the left) or as percent saturation (on the right). As P_{O_2} becomes large, 100% saturation is approached asymptotically, as described by Equation 7.6. The value of P_{50}, the partial pressure of oxygen at 50% saturation, is indicated on the graph (magenta arrow).

and

$$K_d = \frac{[Mb][O_2]}{[MbO_2]} \qquad (7.3)$$

where the equilibrium constant K_d is called a **dissociation constant.** The fraction of myoglobin sites occupied is defined as follows:

$$Y_{O_2} = \frac{\text{sites occupied}}{\text{total sites available}}$$

Each myoglobin molecule has only one site, so the total number of potentially available sites is proportional to the total concentration of myoglobin species $= [MbO_2] + [Mb]$. Therefore,

$$Y_{O_2} = \frac{[MbO_2]}{[Mb] + [MbO_2]} = \frac{\frac{[Mb][O_2]}{K_d}}{[Mb] + \frac{[Mb][O_2]}{K_d}} \qquad (7.4)$$

where we have used $[MbO_2] = [Mb][O_2]/K_d$ from Equation 7.3 to obtain the expression on the right. The concentration of deoxymyoglobin, $[Mb]$, can be factored out of the numerator and denominator to give

$$Y_{O_2} = \frac{\frac{[O_2]}{K_d}}{1 + \frac{[O_2]}{K_d}} = \frac{\frac{[O_2]}{K_d}}{\frac{K_d}{K_d} + \frac{[O_2]}{K_d}} = \frac{[O_2]}{K_d + [O_2]} \qquad (7.5)$$

Equation 7.5 shows that K_d describes the oxygen concentration at which half the Mb molecules have O_2 bound. You can check this relationship by setting $Y_{O_2} = 1/2$ in Equation 7.5. Because oxygen concentration is proportional to oxygen partial pressure, Equation 7.5 can also be written as

$$Y_{O_2} = \frac{P_{O_2}}{P_{50} + P_{O_2}} \qquad (7.6)$$

where P_{50} is the oxygen partial pressure required for 50% O_2 saturation. The value of P_{50} is an indicator of the relative binding affinity for a ligand. In the case of myoglobin, when $P_{O_2} = P_{50}$, $[Mb] = [MbO_2]$.

● **CONCEPT** P_{50} is an indicator of the relative binding affinity of a globin for a ligand: For a globin with higher O_2-binding affinity, the value of P_{50} is lower. For a globin with lower O_2-binding affinity, the value of P_{50} is higher.

For a globin protein with high O_2-binding affinity, half-saturation occurs at *low* P_{O_2}; thus, the value of P_{50} is *low*. For a globin that has a low O_2-binding affinity, half-saturation would occur at *high* P_{O_2}; thus, the value of P_{50} would be *high*.

Equation 7.6 allows us to predict how many ligand-binding sites (in this case: heme Fe^{2+} sites) will be bound to the ligand as the concentration of ligand varies. We will see a similar expression used to describe the effect of reactant concentration on the rate of enzyme-catalyzed reactions in Chapter 8.

We see that Equation 7.6 describes the *hyperbolic*-binding curve shown in Figure 7.18; Y_{O_2} starts at zero at $P_{O_2} = 0$ and approaches 1 as P_{O_2} increases. The P_{50} for myoglobin is very low (3–4 mm Hg), signifying that myoglobin has a high affinity for O_2. This characteristic is appropriate for a protein that must extract O_2 from the blood in

● **CONCEPT** Binding of a ligand like O_2 to a single site on a protein (like Mb) is described by a hyperbolic-binding curve.

● **CONCEPT** Myoglobin has evolved to bind and release O_2 under conditions of relatively low oxygen concentration.

capillaries. At the P_{O_2} existing in the arterial capillaries (about 30 mm Hg), the myoglobin in adjacent tissues would be nearly saturated (blue arrow in Figure 7.18). When cells are metabolically active, their internal P_{O_2} falls to much lower levels (3–18 mm Hg). Under these conditions, myoglobin will release its O_2.

In summary, we observe in myoglobin a molecular structure that has been selected through evolution to produce an optimized environment for the binding *and* release of oxygen under physiological conditions of relatively low P_{O_2}.

7.10 The Role of Conformational Change in Oxygen Transport

All higher animals contain some kind of an oxygen transport protein. In vertebrates and some invertebrates, this protein is hemoglobin. Nearly all hemoglobins are found to be multisubunit proteins, in contrast to the single-subunit myoglobins. Why should this be so? Investigating this question reveals new aspects of globin protein function.

Cooperative Binding and Allostery

FIGURE 7.19 illustrates the physiological demands placed on an O_2 transport protein. It must bind O_2 efficiently at the partial pressure found

● **CONCEPT** Higher animals use O_2-binding proteins to transport oxygen from lungs or gills to respiring tissues where it is needed to support metabolism.

in lungs or gills (approximately 100 mm Hg; red lines in Figure 7.19) and then release a significant fraction of the O_2 to tissues. At rest, the P_{O_2} in capillaries is about 30 mm Hg (light blue lines in Figure 7.19). The amount of O_2 released in capillaries is calculated as the difference in fractional saturation between the lungs and capillaries: $\Delta Y_{O_2} = Y_{O_2}(\text{lungs}) - Y_{O_2}(\text{capillaries})$. In this example, $\Delta Y_{O_2} = Y_{O_2}(@100\text{ mmHg}) - Y_{O_2}(@30\text{ mmHg})$. Panels (a–c) in Figure 7.19 show how different O_2-binding characteristics for a transport protein would affect ΔY_{O_2}.

Consider a transport protein that had a hyperbolic-binding curve like that of myoglobin. If the transport protein had a high O_2-binding affinity, it would achieve $\sim 100\%$ O_2 binding in the lungs but would release very little O_2 in the capillaries (panel (a)). Alternatively, if the transport protein had a low O_2-binding affinity, it would release O_2 in capillaries but not achieve saturation in the lungs (panel (b)). Neither of these situations results in efficient delivery of O_2 between lungs and capillaries (i.e., ΔY_{O_2} is small in each case).

To achieve optimum O_2 delivery to tissues, an ideal oxygen transport protein would be nearly saturated at 100 mm Hg, deliver sufficient O_2 to tissues at rest to support basal metabolism, yet maintain a significant O_2 reserve for periods of high demand such as chasing prey or flight from predators. Panel (c) shows that the O_2-binding curve for hemoglobin meets the criteria for an efficient O_2 transporter. In lungs, where $P_{O_2} \sim 100$ mm Hg, hemoglobin is nearly saturated with oxygen, and in capillaries of

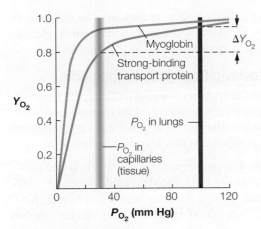

(a) Transport protein efficient in binding but inefficient in unloading (hyperbolic-binding curve).

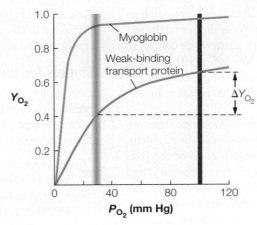

(b) Transport protein efficient in unloading but inefficient in binding (hyperbolic-binding curve).

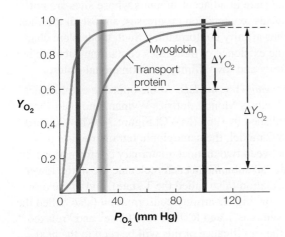

(c) Transport protein efficient in both binding and unloading because it can switch between higher- and lower-affinity states (sigmoidal-binding curve).

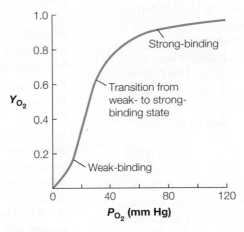

(d) Switch from lower- to higher-affinity states yields the sigmoidal curve.

◄ **FIGURE 7.19 Cooperative versus non-cooperative O_2-binding curves.** These graphs show why an O_2 transport protein such as hemoglobin is more efficient if it switches cooperatively between lower- and higher-affinity states. The vertical blue and red bars represent P_{O_2} in capillaries and lungs, respectively. The difference in O_2 saturation of Hb in lungs versus capillaries (ΔY_{O_2}) is represented by the gap between the values of Y_{O_2} in lungs and capillaries (horizontal dashed lines). **(a)** If the transport protein were noncooperative and had a high O_2 affinity, ensuring saturation in the lungs, transfer of O_2 to the tissues would be inefficient. In this case ΔY_{O_2} is very small. **(b)** If a noncooperative transport protein had a lower O_2 affinity, it would efficiently transfer O_2 to tissues; but it would not be saturated in the lungs. In this case, ΔY_{O_2} is somewhat greater than it is for the high-affinity noncooperative Hb. **(c)** A transport protein that binds efficiently in the lungs and unloads efficiently in the tissues requires a sigmoidal-binding curve and delivers the largest fraction of its bound O_2 (i.e., it has the largest value of ΔY_{O_2}). The curve shown here is that for hemoglobin. The dark blue line represents intracellular P_{O_2} under conditions of high metabolic demand (e.g., extreme muscular exertion). **(d)** The sigmoidal-binding curve illustrates hemoglobin's switch from a lower-affinity state at low oxygen pressures (in tissues) to a higher-affinity state at high oxygen pressures (lungs or gills).

resting muscles (where $P_{O_2} \sim 30$ mm Hg), hemoglobin is $\sim 60\%$ saturated. In other words, hemoglobin in capillaries will release $\sim 40\%$ of the O_2 it carried from the lungs to support basal metabolism. Under conditions of extreme metabolic demand for O_2, P_{O_2} in tissues can drop to 10 mm Hg (dark blue line in panel (c)), and ΔY_{O_2} increases from 0.4 to roughly 0.85.

Optimized delivery of O_2 to tissues has been achieved through the evolution of O_2 transport proteins, such as hemoglobin, that have the *sigmoidal*-binding curve shown in Figure 7.19(c). This sigmoidal curve is observed because hemoglobin switches between states of higher and lower affinity for O_2 binding (panel (d)). This binding behavior is efficient because it allows nearly full saturation of the protein in the lungs or gills, as well as optimal release in the capillaries. Examination of the O_2-binding curve for hemoglobin shows that at low oxygen pressures hemoglobin binds O_2 with low affinity (i.e., it looks similar to the curve in panel (b) at low P_{O_2}); but, as more oxygen is bound, the affinity for oxygen becomes greater (i.e., the binding curve is similar to the curve in panel (a) at high P_{O_2}).

● **CONCEPT** Efficiency in O_2 transport is achieved by cooperative binding in multisite proteins, described by a sigmoidal-binding curve.

● **CONCEPT** Hemoglobin switches between conformational states with lower and higher O_2-binding affinities. In the O_2-rich environment of the lungs or gills, the higher-affinity state is favored, and oxygen binds to hemoglobin. In the O_2-poor environment of respiring tissues, the lower-affinity state is favored, and oxygen is released from hemoglobin.

● **CONCEPT** Cooperativity in binding requires communication between binding sites.

Such behavior means that a *cooperative interaction* must exist among the O_2-binding sites in hemoglobin. Ligand binding to the first empty site somehow increases the O_2-binding affinity of the remaining sites, thus promoting complete saturation of the protein with O_2. We can also express this idea the other way around, by saying that dissociation of one O_2 from the protein makes it easier for the remaining O_2 to dissociate, thus promoting the fully deoxygenated state.

Such behavior requires some kind of intramolecular communication between the binding sites. A single-subunit protein, such as myoglobin, cannot achieve this sort of modulation of ligand-binding affinity; however, such communication *is* possible between the subunits of a multisubunit protein, such as hemoglobin. Thus, the answer to the question "Why are hemoglobins multisubunit proteins?" lies in the evolutionary advantages conferred by cooperative switching between high- and low-affinity states. The advantages of enhanced aerobic capacity translate to greater survival for both predators and prey and provide the selective pressure for the evolution of optimized O_2 transporters.

Vertebrate hemoglobin has evolved from the *monomeric* structure of myoglobin into the *tetrameric* structure shown in Figure 7.14(b). Hemoglobin can bind four O_2 molecules—one in each of the four subunits. Although each of the subunits has primary, secondary, and tertiary structures similar to those of myoglobin, the amino acid side chains in hemoglobin also provide other necessary interactions—salt bridges, hydrogen bonds, and hydrophobic interactions—to stabilize a particular quaternary structure.

The functional difference between hemoglobin and myoglobin lies in the cooperativity exhibited by the ligand-binding sites in hemoglobin. This cooperativity is possible because the oxygenation state (filled or empty) of one site can be communicated to another.

As described later in this chapter, there is a structural basis for the cooperative ligand binding in hemoglobin: The lower-affinity state has a protein conformation that is distinct from that of the higher-affinity state. We can replicate the shape of the binding curve in Figure 7.19 using an equation derived by Archibald Hill, who in 1910 proposed that the binding of O_2 by hemoglobin could be empirically described by the function

$$Y_{O_2} = \frac{P_{O_2}^h}{P_{50}^h + P_{O_2}^h} \tag{7.7}$$

The parameter **h,** which is called the **Hill coefficient,** is related to the number of ligand-binding sites (n) and the energy of their interaction, such that the value of h approaches the value n as the interaction between sites increases. Note that in the case of myoglobin, $h = 1$, and Equation 7.7 is then identical to Equation 7.6. Fitting Equation 7.7 to an oxygen-binding curve yields values for P_{50} and h that allow for useful comparison of the ligand-binding behaviors of different hemoglobins (e.g., between normal and mutant hemoglobins). Once values for P_{50} and h are known, Equation 7.7 can be used to calculate ΔY_{O_2} for a mutant to determine whether it would deliver more or less O_2 to tissues than normal hemoglobin.

The value of h indicates whether or not ligand binding is cooperative. We will consider two cases for a molecule with n ligand-binding sites:

> Case 1. $h = 1$: There is no interaction between the sites; thus, the molecule binds ligands noncooperatively (e.g., as for myoglobin in Figure 7.19(c)). This situation may also be observed for a multisite protein if the binding sites do not interact with one another.

> Case 2. $1 < h < n$: There is interaction between the sites. This situation is the usual one for a protein that binds ligands with so-called positive cooperativity, as depicted for hemoglobin in Figure 7.19(c). The Hill coefficient measured under physiological conditions for hemoglobin ($n = 4$) is ~3.

The cooperative binding of oxygen by hemoglobin is one example of what is referred to as an **allosteric** effect. In allosteric binding, the uptake of one ligand by a protein influences the affinities of remaining unfilled binding sites. The ligands may be of the same kind, as in the case of O_2 binding to hemoglobin, or they may be different. As discussed in Chapter 8, allostery is also an important mechanism for regulating the activity of enzymes.

Models for the Allosteric Change in Hemoglobin

How do allosteric transitions from lower-affinity binding states to higher-affinity binding states actually occur? A number of theories have been developed to describe the allosteric transitions in hemoglobin. These theories can be grouped into four classes.

1. *Sequential models:* The prototype for such models is one proposed by Gilbert Adair in 1925, then further developed in 1966 by Daniel Koshland, George Némethy, and David Filmer (KNF; **FIGURE 7.20(a)**). The Adair/KNF model assumes that the subunits can change their tertiary conformation one at a time in response to the binding of oxygen. Positive cooperativity arises because the binding of O_2 in one subunit favors the higher-affinity conformational state in adjacent subunits whose sites are not yet filled. Thus, as oxygenation progresses, almost all the sites adopt the higher-affinity conformation. Such models are characterized by the existence of molecules containing subunits in both high-affinity and low-affinity conformational states.

2. *Concerted, or symmetry, models:* At the opposite extreme lies the theory of Jacques Monod, Jeffries Wyman, and Jean-Pierre Changeux published in 1965 (MWC; Figure 7.20(b)). According to the MWC model, the hemoglobin tetramer exists in an equilibrium between two distinct quaternary conformations. In the deoxy state, all subunits in each molecule are in the lower-affinity conformation (also called the T state), and in the oxy state, all are in the higher-affinity conformation (also called the R state). The symbols T and R stand for "tense" and "relaxed," respectively; the significance of this will be seen in the next section. An equilibrium between these states is presumed to exist, and ligand binding shifts that equilibrium toward the R state. The shift is a *concerted* one, so that molecules with some subunits in the T state and some in the R state are specifically excluded. Historically, this has been the most widely accepted model for allostery in hemoglobin.

3. *Multistate models:* It has become clear since the early 1990s that neither the KNF nor the MWC model completely explains the allosteric behavior of proteins, including hemoglobin. Consequently, more complex models have been devised, though most retain some elements of the KNF and/or MWC models.

4. *Dynamics models:* A completely different proposal for allostery attributes changes in functional properties to changes in the dynamic behavior of proteins rather than to conformational changes per se.

We will focus on the symmetry (MWC) model for the remaining discussion of hemoglobin allostery.

Changes in Hemoglobin Structure Accompanying Oxygen Binding

To understand allosteric behavior in hemoglobin, it is necessary to examine the protein structure in more detail. The hemoglobins of higher vertebrates are made up of two types of subunits, α and β. The primary structures of human hemoglobins are compared with that of

(a) KNF (sequential) Model

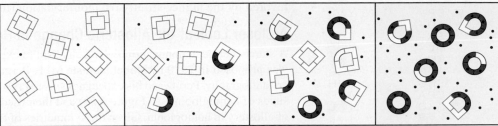

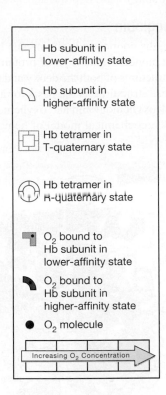

Hb subunit in lower-affinity state

Hb subunit in higher-affinity state

Hb tetramer in T-quaternary state

Hb tetramer in R-quaternary state

O_2 bound to Hb subunit in lower-affinity state

O_2 bound to Hb subunit in higher-affinity state

O_2 molecule

Increasing O_2 Concentration

No O_2 bound. Almost all subunits are in the lower-affinity state. Few are in the higher-affinity state.

At low O_2 concentrations, most subunits are in the lower-affinity state. O_2 binding at one subunit favors the transition to the higher-affinity state in an adjacent subunit.

As O_2 concentration increases, more of the higher-affinity state subunits bind ligand and promote the transition to the higher-affinity state in adjacent subunits.

At high O_2 concentrations, the Hb approaches saturation. Almost all ligand-binding sites are filled and in the higher-affinity conformational state.

(b) MWC (symmetry) Model

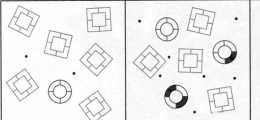
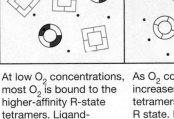

No O_2 bound. Almost all Hb tetramers are in the T state. Only a few are in the R state.

At low O_2 concentrations, most O_2 is bound to the higher-affinity R-state tetramers. Ligand-binding shifts the T $\rightleftharpoons$ R equilibrium toward the R state.

As O_2 concentration increases, more of the tetramers are in the R state. Note that the T state tetramers can also bind O_2, but do so with lower affinity.

At high O_2 concentrations, the Hb approaches saturation. Almost all Hb tetramers are in the higher-affinity R state.

▲ **FIGURE 7.20 Two classical models for the cooperative ligand binding by hemoglobin.** (a) The Adair/Koshland, Némethy, and Filmer (KNF) model. As each subunit binds a ligand, it promotes a change in an adjacent subunit to the higher-affinity conformation. (b) The Monod, Wyman, and Changeux (MWC) model. The entire molecule has two different quaternary states—tense (T) and relaxed (R)—which are in equilibrium. Binding of ligands shifts the equilibrium toward the higher-affinity (R) state.

sperm whale myoglobin in **FIGURE 7.21**. The human α and β sequences have 44% amino acid identity to one another and 18% identity to the sequence of whale myoglobin. Essential residues, like the proximal and distal histidines (F8 and E7, respectively), are strictly conserved, and those that stabilize the tertiary structure appear to be conserved as well, as the hemoglobin subunits and myoglobin all have very similar tertiary structures (see Figure 7.14). As a rule of thumb, sequences with greater than 30% amino acid identity are predicted to have the same tertiary structure. The hemoglobin molecule contains two copies of each type of subunit, so the whole molecule can be described as an ($\alpha_2\beta_2$) tetramer. The subunits are placed in a roughly tetrahedral arrangement, as shown in Figure 7.14. When hemoglobin is dissolved in mildly denaturing solutions, it dissociates into $\alpha\beta$ dimers; this suggests that the strongest subunit contacts are between α and β chains, rather than between two α subunits or between two β subunits. In other words, the molecule can be thought of as a dimer of $\alpha\beta$ dimers. Figure 7.14 also shows that the hemes, with their O_2-binding sites, are all close to the surface but *not* close to one another. Therefore,

● **CONCEPT** Vertebrate hemoglobins are tetramers ($\alpha_2\beta_2$) made up of two kinds of myoglobin-like chains.

we cannot seek the source of cooperative binding in anything as simple as direct heme–heme interaction.

General features of the protein structural changes that occur upon ligand binding/release can be seen in **FIGURE 7.22**, which illustrates two views of the differences between the crystal structures of the T state and the R state conformations of hemoglobin. Early X-ray diffraction studies suggested that these differences arise primarily as a result of changes in the quaternary structure, accompanied by much smaller tertiary structure changes. One $\alpha\beta$ dimer rotates 15 degrees and slides with respect to the other, as seen in Figure 7.22(a). Upon O_2 binding, this movement brings the β subunits closer together and narrows a central cavity in the molecule, as can be seen in Figure 7.22(b). To a first approximation, then, we can regard the hemoglobin molecule as having two quaternary

● **CONCEPT** Oxygenation causes hemoglobin quaternary structure to change: One $\alpha\beta$ dimer rotates and slides with respect to the other.

● **CONCEPT** R-state hemoglobin has a higher O_2-binding affinity (lower P_{50}). T-state hemoglobin has a lower O_2-binding affinity (higher P_{50}).

	Mb Hbβ Hbα			Mb Hbβ Hbα			Mb Hbβ Hbα	
1	VVV		52	EP–		103	YNN	
	–H–			AD–			LFF	
	LLL			EA–			ERK	
	STS			MV–			FLL	
	EPP			KM–			ILL	
	GEA			AGG			SGS	
	EED			SNS			ENH	
	WKK			EPA			AVC	
	QST			DKQ			ILL	
	LAN			LVV			IVL	
	VVV			KKK			HCV	
	LTK			KAG			VVT	
	HAA			HHH	E7		LLL	
	VLA			GGG			HAA	
	WWW			VKK			SHA	
	AGG			TKK			RHH	
	KKK			VVV			HFL	
	VVV			LLA		120	PGP	
	E–G			TGD			GKA	
	A–A			AAA			DEE	
	DNH			LFL			FFF	
	VVA			GST			GTT	
	ADG			ADN			APP	
	GEE			IGA			DPA	
	HVY			LLV			AVV	
	GGG			KAA			QQH	
	QGA			KHH			GAA	
	DEE			KLV			AAS	
	IAA		80	GDD			MYL	
	LLL			HND			NQD	
	IGG			HLM			KKK	
	RRR			EKP			AVF	
	LLM			AGN			LVL	
	FLF			ETA			EAA	
	KVL			LFL			LGS	
	SVS			KAS			FVV	
	HYF			PTA			RAS	
	PPP			LLL			KNT	
	EWT			ASS			DAV	
	TTT			QED			ILL	
40	LQK			SLL			AAT	
	ERT			HHH	F8		AHS	
	KFY			ACA			KKK	
	FFF		FG corner	TDH			YYY	
	DEP		FG corner	KKK			KHR	
	RSH		FG corner	HLL			E	
	FFF		FG corner	KHR			L	
	KG–		FG corner	IVV	FG5		G	
	HDD			PDD			Y	
	LLL			IPP			Q	
	KSS			KEV		153	G	
	TTH							

▲ **FIGURE 7.21 Comparison of sequences of myoglobin and the α and β subunits of hemoglobin.** The aligned sequences are those of whale myoglobin and the two human hemoglobin subunits. Gaps (indicated by dashes) have been inserted where necessary to provide maximum alignment of the sequences; the residue numbers to the left of the chains are for the myoglobin sequence. A residue critical to the functioning of these proteins is indicated to the right of the sequences; F8 and E7 are the proximal and distal histidines, respectively (see Figure 7.16). Yellow indicates the residues that are identical in all three sequences, and purple indicates the residues identical in both hemoglobin sequences. "FG corner" refers to residues in the loop between the F and G helices.

states, one characteristic of the lower-affinity deoxy conformation (the T state) and the other favored by the higher-affinity oxy conformation (the R state).

A Closer Look at the Allosteric Change in Hemoglobin

A stereochemical, nonquantitative mechanism to explain the cooperativity in oxygen binding was proposed in 1970 by Nobel laureate M. F. Perutz, a pioneer in the field of protein X-ray crystallography. Perutz and his colleagues solved crystal structures of both the deoxy and the oxy states of hemoglobin (see Figure 7.22) and then used these structures as the basis for a model of allostery in hemoglobin. Given the symmetries of the deoxyhemoglobin and oxyhemoglobin structures, it was reasonable for Perutz to describe his stereochemical model of hemoglobin

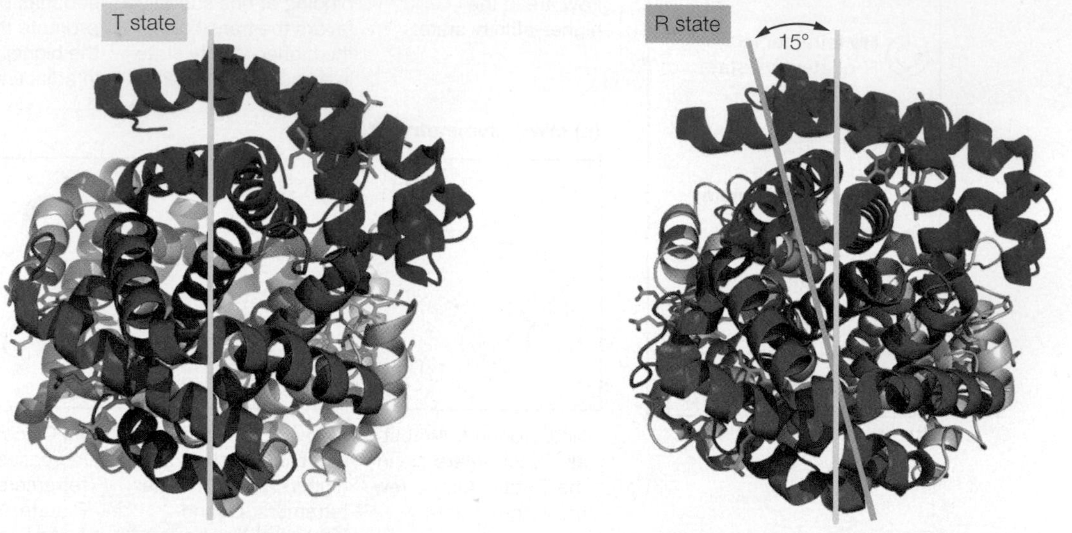

(a) The transition viewed with the $\alpha_1\beta_1$ dimer (darker blue and red cartoons) in front of the $\alpha_2\beta_2$ dimer (light blue and pink cartoons). Hemes are shown in stick rendering (green). T-state hemoglobin is shown on the left, and R-state hemoglobin on the right. Note the 15° rotation of $\alpha_1\beta_1$ with respect to $\alpha_2\beta_2$. The rotation of roughly 15° is accompanied by sliding because the center of rotation is not centrally located (note intersection of yellow and orange lines).

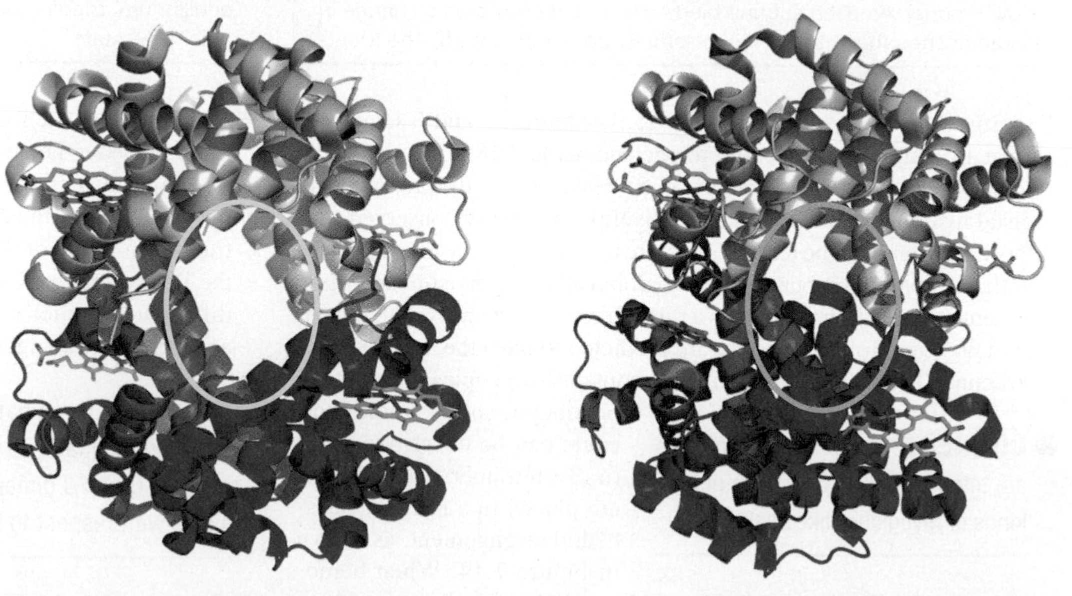

(b) Top views of hemoglobin, achieved by rotating the tops of molecules in panel (a) toward the viewer by 90°. The two β subunits are in the foreground; α subunits are in the background. Note that the central cavity in T-state hemoglobin is broad (yellow oval). The shift from the T to the R state is evident by the shrinkage of the central cavity (compare the open space within the orange and yellow ovals).

▲ **FIGURE 7.22 The change in hemoglobin quaternary structure during oxygenation.** PDB IDs: T state, 1hga; R state, 1bbb.

allostery in terms of the symmetric MWC model. Before we present the Perutz model of allostery, let us consider some important differences between the T and R crystal structures.

The transition from the T to the R conformation involves significant disruption of several noncovalent interactions. As shown in **FIGURE 7.23**, in the T state each of the β-globin C-termini makes two salt bridges: an intersubunit interaction between the His 146 C-terminal — COO^- and the — NH_3^+ side chain of Lys 40 on a nearby α-globin, and an intrasubunit

interaction between the *protonated* His 146 side chain and the — COO^- of Asp 94. The latter interaction plays a major role in stabilizing the T state under conditions of low pH, where protonation of the His 146 side chain is favored (see the section "Response to pH Changes: The Bohr Effect"). In the T state each α-globin C-terminus is involved in four critical intersubunit interactions. The C-terminal — COO^- of Arg 141 of one α-globin interacts with the side chain — NH_3^+ of Lys 127 on the other α-globin chain. The guanidinium side chain of Arg 141 forms a

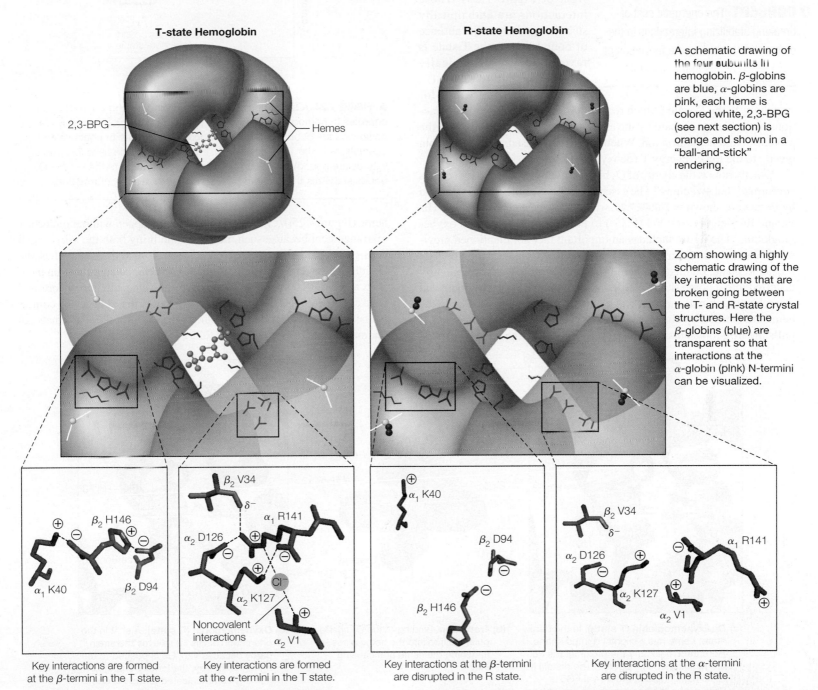

T-state Hemoglobin

R-state Hemoglobin

A schematic drawing of the four subunits in hemoglobin. β-globins are blue, α-globins are pink, each heme is colored white, 2,3-BPG (see next section) is orange and shown in a "ball-and-stick" rendering.

2,3-BPG — Hemes

Zoom showing a highly schematic drawing of the key interactions that are broken going between the T- and R-state crystal structures. Here the β-globins (blue) are transparent so that interactions at the α-globin (pink) N-termini can be visualized.

Key interactions are formed at the β-termini in the T state.

Key interactions are formed at the α-termini in the T state.

Key interactions at the β-termini are disrupted in the R state.

Key interactions at the α-termini are disrupted in the R state.

▲ **FIGURE 7.23 Key noncovalent interactions disrupted during the switch between T- and R-state hemoglobin quaternary structures.** Zoom showing crystal structure data of the key interactions at the β-termini and α-termini. See the text for a detailed description of the residues involved. Noncovalent interactions in the T state are indicated by black dashed lines. All of these interactions are disrupted during the

transition to the R state. Side chains of β-globins are highlighted in blue, and those of α-globins are highlighted in magenta. The chloride ion that bridges the C-terminus of one α-globin (α_1 R141) to the N-terminus of the other α-globin (α_2 V1) is shown in orange. Note that identical sets of symmetry-related interactions occur at the other α-globin and β-globin termini. PDB IDs: T state, 1hga; R state, 1bbb.

salt bridge with the side chain—COO^- of Asp126 and a hydrogen bond with the amide carbonyl group of Val34 on a nearby β-globin. Finally, the Arg 141 side chain makes a bridging intersubunit interaction with a chloride ion and the N-terminal —NH_3^+ of Val1. As described in the next section, increasing $[Cl^-]$ also stabilizes the T state by promoting this bridging interaction.

All of the interactions shown in the lower left of Figure 7.23 are disrupted when hemoglobin switches from the T to the R state (see lower

● **CONCEPT** The energetic cost of breaking stabilizing interactions in the deoxy state is paid by the formation of Fe^{2+}—O_2 bonds in the oxy state.

right of Figure 7.23). These interactions are enthalpically stabilizing; thus, in the absence of bound ligand, the T state is favored thermodynamically over the R state (**FIGURE 7.24**). The thermodynamic price for

switching to the R state (which requires breaking the bonding interactions in the T state) is paid by the energy provided by binding O_2 to the protein-bound heme iron ion. When the O_2 is released, the hemoglobin reverts to the lower-energy T (deoxy) conformation.

Exactly how is the energy of O_2 binding communicated to induce this conformational switching? The essential features of the model proposed by Perutz are shown in **FIGURE 7.25**, which shows the relationship of His F8 and the neighboring Val (FG5*) to the heme in deoxy- and oxyhemoglobins. The figure includes an important fact not mentioned previously: Not only is the iron ion in the deoxy conformation a bit outside the heme plane, but also the heme itself is not quite flat—it is slightly distorted into a dome shape. Furthermore, in both deoxymyoglobin and deoxyhemoglobin, the axis of His F8 is not exactly perpendicular to the heme but is tilted by about 8°. When oxygen binds to the other side, it pulls the iron ion a short distance down into the heme and flattens the

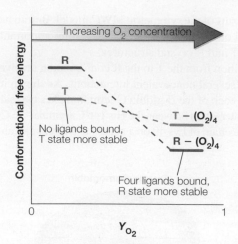

▲ **FIGURE 7.24 A simplified view of ligand binding and conformation energies in hemoglobin.** The deoxy (T) conformation is favored when no ligands are bound, owing to the increased number of noncovalent interactions in the T state (see Figure 7.23). As Y_{O_2} increases (i.e., more ligands are bound), the energy provided by formation of the Fe^{2+}—O_2 bonds stabilizes the R conformation relative to the T conformation.

heme (Figure 7.25(b,c)). This change cannot happen without molecular rearrangement because such motion would bring both the ε-hydrogen of His F8 and the side chain of Val FG5 too close to the heme. Thus, the histidine shifts its orientation toward the perpendicular, pulling on the F helix and the FG corner as it does so. This movement in turn distorts and weakens the complex of H-bonds and salt bridges that connect FG corners of one subunit with C helices of another. Consequently, the conformational rearrangements shown in Figures 7.22 and 7.23 occur.

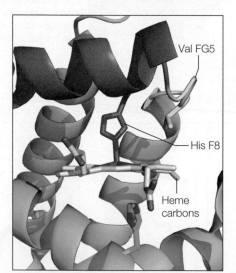 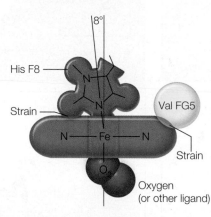 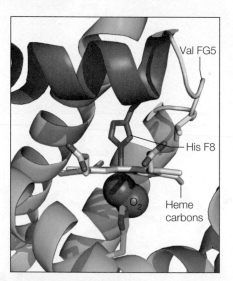

(a) Deoxyhemoglobin (T state). In the deoxy state, heme has a slightly domed shape. Here the F helix and the proximal His F8 are highlighted in magenta. The distal His E7 is highlighted in green, Val FG5 is yellow, and the heme carbons are white.

(b) Transition. Binding of the O_2 ligand pulls the iron into the heme plane, flattening the heme and causing strain.

(c) Oxyhemoglobin (R state). A shift in the orientation of His F8 relieves the strain, partly because Val FG5 is pushed to the right. In this way, the tertiary change in heme is communicated to the FG corner.

▲ **FIGURE 7.25 The essential features of the "Perutz mechanism" for the T → R transition in hemoglobin.** The binding of oxygen to deoxyhemoglobin causes conformational changes in the heme. PDB IDs: T state, 2dn2; R state, 2dn1.

*Note: "FG5" refers to the fifth residue in the loop connecting the F and G helices.

● **CONCEPT** In the Perutz mechanism, a small movement of the heme iron upon O_2 binding is translated into a larger movement of the F helix by the covalent connection between the F helix and the proximal histidine.

In the simplest terms, the binding of O_2 pulls the Fe^{2+} 0.2 Å into the heme, which levers the F helix and thereby produces a much larger shift in the surrounding protein structure, particularly at the critical $\alpha-\beta$ interfaces (which are 17–23 Å distant from the heme). This remodeling of the $\alpha-\beta$ interfaces provides the physical link between the ligand-binding sites, which explains the observed cooperativity.

This mechanism to explain the cooperativity in oxygen binding was proposed by Perutz based on the insights he gained from identifying differences between the deoxy- and oxyhemoglobin crystal structures. As such, it represents a brilliant example of the application of structural biology to explain the physiologically relevant behavior of a protein. But does it correspond to reality? Site-directed mutagenesis studies (see Tools of Biochemistry 4A) and rapid ligand-binding studies (see Tools of Biochemistry 8A) have allowed many features of the Perutz mechanism to be tested. Although it is not entirely correct, much experimental data support the general features of the model, and it remains widely accepted. For example, site-directed mutagenesis was used to replace the proximal histidine residues with glycines in α and β chains. The mutant protein was then studied in the presence of 10 mM imidazole; the imidazole molecule can substitute for the proximal histidine side chain and bind the heme iron *but is not covalently linked to the F helix* (see **FIGURE 7.26**). As a consequence, although oxygen binding can still

flatten the heme, it does not move the F helix. Cooperativity in oxygen binding is largely lost in the glycine mutant, and although the subunits appear to remain in the T state, the ligand-binding affinity of the mutant is *increased* compared to wild-type hemoglobin. These findings support the main features of the Perutz model and demonstrate that the proximal histidine plays a significant role in hemoglobin cooperativity. The observation that cooperativity is not completely abolished in these hemoglobin mutants suggests that other features of hemoglobin structure, perhaps residues in the distal heme pocket, also contribute to cooperative ligand binding.

The mechanism of hemoglobin allostery remains a matter of vigorous debate. However, to discuss the effects of specific allosteric effectors on hemoglobin function, we will continue to use the central ideas of the Perutz and MWC models we presented earlier because the general features of these models explain much of the allosteric behavior in hemoglobin.

7.11 Allosteric Effectors of Hemoglobin Promote Efficient Oxygen Delivery to Tissues

As we shall see throughout this text, allostery is a general mechanism for the regulation of many important proteins. Here we give some general definitions of terms related to allostery. A given protein may bind specifically to several different molecules in more than one location on that protein's surface. The **active site** of a protein is where the protein must bind one or more **substrate** molecules to carry out its primary function. Since the primary function of hemoglobin is oxygen transport, we can view the heme pocket as the "active site" of hemoglobin because that is where the O_2 ligand is bound. In addition to the active site, there may be other sites, called **regulatory sites,** which bind specifically to molecules that regulate the function of the protein. The molecules that regulate protein function in this way are called **effectors,** and they typically exert their effects through an allosteric mechanism.

Allosteric effectors can be characterized in many ways. Those that increase protein activity are called **positive effectors,** and those that decrease activity are **negative effectors.** Effectors can also be differentiated by the site on the protein to which they bind. We have seen that O_2 binding *at the heme* has a positive cooperative effect on subsequent O_2-binding events. Because O_2 affects its own binding, by binding to the protein active site, it is called a **homotropic** effector. Effectors that bind at regulatory sites (which are typically distant from the active site) are called **heterotropic** effectors.

We now discuss the effects of four heterotropic negative effectors of hemoglobin: H^+, CO_2, Cl^-, and 2,3-bisphosphoglycerate (2,3-BPG).

Response to pH Changes: The Bohr Effect

As oxygen is consumed for respiration in aerobic tissues, carbon dioxide is produced. As described in Chapter 2, accumulation of CO_2 also lowers the pH in erythrocytes through the *bicarbonate reaction,*

$$CO_2 + H_2O \rightleftharpoons HCO_3^- + H^+ \tag{7.8}$$

This reaction in erythrocytes is catalyzed by the enzyme *carbonic anhydrase.* At the same time, the high demand for oxygen, especially in muscle involved in vigorous activity, can result in oxygen deficit, or **hypoxia.** As we shall see in Chapter 12, a consequence of this deficit is

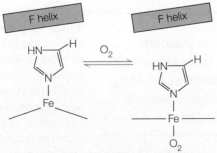

(a) The effect of O_2 binding according to the Perutz model: The F helix is drawn toward the heme.

(b) Now lacking a covalent connection to the heme, the F helix is not disturbed by O_2 binding, and there is significantly reduced cooperativity.

▲ **FIGURE 7.26 The effect of replacing the proximal histidine in hemoglobin with a glycine residue and adding a noncovalently bonded imidazole.**

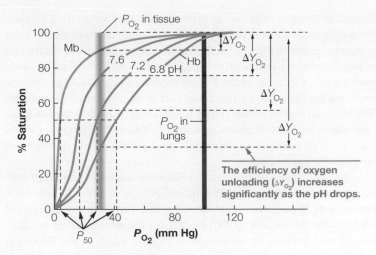

▲ **FIGURE 7.27 The Bohr effect in hemoglobin.** Oxygen-binding curves for hemoglobin (green) are shown for pH 7.6, 7.2, and 6.8. As the hemoglobin circulates from lungs to tissues, the lower pH favors the lower-affinity conformation (this is also reflected in the increase in P_{50} values as pH drops). Myoglobin displays little Bohr effect, so its oxygen-binding curve (orange) is approximately the same at all three pH values.

the production of lactic acid, which also lowers the pH. The falling pH in tissue and venous blood signals a demand for more oxygen delivery. Hemoglobin functions efficiently to meet the need for increased O_2 delivery through its allosteric transition between high-affinity oxy (R) and low-affinity deoxy (T) states.

Blood plasma has a pH that is normally 7.4. As shown in **FIGURE 7.27**, a pH drop initially has the effect of raising the P_{50} of hemoglobin (i.e., lowering O_2-binding affinity), thereby facilitating greater release of O_2. This response of hemoglobin to pH change is called the **Bohr effect** after Christian Bohr (father of physicist Niels Bohr), who reported it in 1904. The overall reaction may be written as

$$Hb \cdot 4O_2 + nH^+ \rightleftharpoons Hb \cdot nH^+ + 4O_2$$

where n has a value somewhat greater than 2. Physiologically, this reaction has two consequences. First, in the capillaries, H^+ ions promote the release of O_2 by driving the reaction to the right. Then, when the venous blood recirculates to the lungs or gills, the oxygenation has the effect of releasing the H^+ by shifting the equilibrium to the left. This, in turn, releases CO_2 from bicarbonate dissolved in the blood plasma by the reversal of the bicarbonate reaction (Equation 7.8). The free CO_2 can then be exhaled.

A stereochemical mechanism to explain the Bohr effect was first proposed by Perutz and coworkers in 1970. Perutz argued that certain proton-binding sites in hemoglobin are of higher affinity in the deoxy form than in the oxy form, and he predicted that a major contribution comes from histidine residue 146 at the C-terminus of each β chain. A change in proton affinity for some ionizable group is manifested as a change in pK_a. How can the pK_a of an amino acid side chain be altered? This can be achieved by altering the chemical environment of the ionizable side chain (see Figure 5.13). Histidine $\beta146$ in the

● **CONCEPT** A decrease in blood pH results in stabilization of the deoxy state and thereby favors greater O_2 released from hemoglobin.

R-state tetramer has a pK_a of roughly 6.4 and is therefore predominantly deprotonated at the normal pH of blood, 7.4. As shown in Figure 7.23, when hemoglobin is in the T state, the side chain of β Asp 94 moves close enough to β His 146 to make a salt bridge *if the histidine is protonated*. Because this salt bridge stabilizes the proton against dissociation, the pK_a of β His 146 is increased to 7.9 in the T state. Thus, as the proton concentration increases, protonation of β His 146 is favored, which in turn favors the *deoxy* conformation and thereby promotes the release of oxygen.

The overall effect of lowering pH on the O_2-binding affinity of hemoglobin is illustrated in Figure 7.27. Note that a decrease in pH of only 0.8 unit shifts the P_{50} from less than 20 mm Hg to over 40 mm Hg, greatly increasing the amount of O_2 released to respiring tissues (compare ΔY_{O_2} for all three values of pH).

Carbon Dioxide Transport

Release of carbon dioxide from respiring tissues lowers the oxygen-binding affinity of hemoglobin in two ways. First, as mentioned above, most of the CO_2 is rapidly converted to bicarbonate in erythrocytes, releasing protons that contribute to the Bohr effect. Most of this bicarbonate is transported out of the erythrocytes and is carried dissolved in the blood plasma. A small portion of the CO_2 (estimated to be 5–13%) reacts directly with hemoglobin, binding to the N-terminal amino groups of the chains to form **carbamates:**

$$-\overset{+}{N}H_3 + HCO_3^- \rightleftharpoons -\overset{\overset{\displaystyle H}{|}}{N}-COO^- + H^+ + H_2O$$

This *carbamation reaction* allows hemoglobin to aid in the transport of CO_2 from tissues to lungs or gills, and the protons released by carbamate formation contribute to the Bohr effect.

We may summarize the effects of H^+ and CO_2 in terms of the respiratory cycle shown in Figure 7.12: In the lungs or gills of an animal, O_2 is abundant. Oxygenation favors the oxy conformation of hemoglobin, which stimulates the release of CO_2. As the blood then travels via arteries into the tissue capillaries, the lower pH and high CO_2 content favor the deoxy form, promoting O_2 release and binding of CO_2. Carbon dioxide, both in forming bicarbonate and in reacting with hemoglobin to form carbamate, contributes to decreasing the pH, further stimulating O_2 release.

The role of increasing CO_2 in stimulation of O_2 release is seen in hyperventilation. If a person breathes too rapidly, plasma CO_2 concentration is significantly reduced, and consequently, release of oxygen into the tissues is impaired. This condition leads to dizziness and, in extreme cases, unconsciousness. Hyperventilation can be easily corrected by breathing into a paper bag—this brings exhaled CO_2 back into the blood, thereby increasing its concentration.

● **CONCEPT** Hemoglobin also transports CO_2 (in the form of carbamates) from tissues to gills or lungs. CO_2 acts as a negative allosteric effector of O_2 binding.

Response to Chloride Ion at the α-Globin N-Terminus

The bicarbonate ions formed inside erythrocytes are transported across the erythrocyte membrane into the surrounding plasma. To maintain charge neutrality within the red cell, Cl^- ions are exchanged for HCO_3^- ions (ion transport is discussed in detail in Chapter 10).

Chloride ion binds to deoxyhemoglobin, between the terminal residues in each α-globin, forming a bridge between the positively charged amino-terminus of Val1 and the side chain of Arg 141. This is possible due to the proximity of these groups in the deoxy conformation (see Figure 7.23). Chloride ion binding favors protonation of the N-terminal amino group of Val1, thereby increasing its pK_a. Val1 and Arg 141 don't interact in the oxy conformation; thus, both the bound Cl^- and the H^+ are released. In this way, chloride ion binding augments the Bohr effect.

2,3-Bisphosphoglycerate

H^+ and CO_2 are the effectors that function rapidly to facilitate the exchange of O_2 and CO_2 in the respiratory cycle. One other major effector operates over longer periods to permit organisms like humans to adapt to gradual changes in oxygen availability. It is a common observation that people who move to high altitudes at first experience some distress but gradually acclimate to the lower oxygen pressure. In the short term (1–2 days), this acclimation results from increased concentration of red cells in plasma due to a reduction in the volume of the blood plasma; but a more significant short-term adaptive effect is due to changes in the concentration in red cells of the allosteric effector **2,3-bisphosphoglycerate** (or **2,3-BPG; FIGURE 7.28(a)**). Within 2 days of moving to higher altitude, the concentration of 2,3-BPG in red cells nearly doubles (from 4.5 mM to ~7.6 mM), resulting in increased binding of this effector to Hb. Longer-term adaptation to higher altitude, which requires 2–3 months, is the result of increased red cell production.

● **CONCEPT** 2,3-Bisphosphoglycerate (2,3-BPG) is found inside red blood cells and is a potent allosteric effector that lowers the O_2 affinity of hemoglobin.

● **CONNECTION** Short-term adaptation to higher altitude is achieved by increasing the concentration of 2,3-BPG inside red blood cells.

Like the effects of H^+ and CO_2, the binding of 2,3-BPG acts to lower the oxygen affinity of hemoglobin. At first glance, this may seem a strange way to adapt to lower O_2 pressure, but in fact, the more efficient unloading of oxygen in the tissues compensates for the slight decrease in loading efficiency in the lungs (see end-of-chapter Problem 11). The action of 2,3-BPG

is illustrated in **FIGURE 7.29**. 2,3-BPG binds via ionic interactions to positively charged groups lining the cavity between the β chains in the deoxy state. Comparison of the two hemoglobin conformations shown in Figure 7.22(b) shows that this cavity is much narrower in oxyhemoglobin than in deoxyhemoglobin. In fact, 2,3-BPG cannot be bound in the oxy form. The higher the 2,3-BPG content in red blood cells, the more the deoxy structure is favored. Once again, a decrease in O_2 affinity is explained by preferential stabilization of the

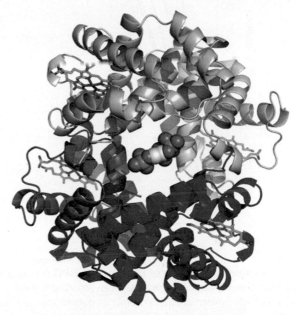

(a) The 2,3-BPG binding site is located in the central cavity of the adult hemoglobin (HbA) tetramer between the two β-globin chains (see also Figure 7.22).

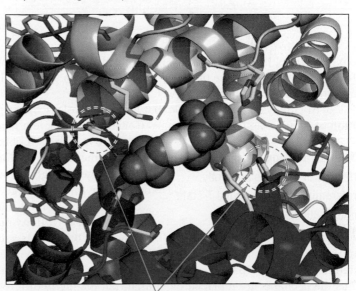

His 143 → Ser 143 in fetal Hb

(b) Zoom in on the central cavity, which is lined with eight positively charged groups (highlighted in yellow) that favorably bind the negatively charged 2,3-BPG molecule. Note the His residues (β143) that are replaced by Ser in fetal hemoglobin (cyan arrows and dashed ovals).

▲ **FIGURE 7.29** Binding of 2,3-bisphosphoglycerate to deoxyhemoglobin. Human hemoglobin colored as in Figure 7.22. PDB ID: 1b86.

◀ **FIGURE 7.28** Two anionic compounds that bind to deoxyhemoglobin.

$$
\begin{array}{c}
COO^{\ominus} \\
| \\
H-C-O-PO_3{}^{2-} \\
| \\
H-C-O-PO_3{}^{2-} \\
| \\
H
\end{array}
$$

(a) 2,3-Bisphosphoglycerate (2,3-BPG), found in mammals.

(b) myo-Inositol-1,3,4,5,6-pentaphosphate (IPP), found in birds.

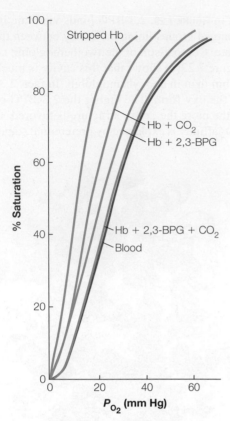

▲ **FIGURE 7.30 Combined effects of CO_2 and 2,3-BPG on oxygen binding by hemoglobin.** Hemoglobin that has been stripped of both CO_2 and 2,3-BPG has a high oxygen affinity. When both substances are added to hemoglobin at the levels found in blood emerging from the capillaries, the hemoglobin displays almost exactly the same binding curve as that observed for whole blood.

deoxy structure. Increased 2,3-BPG levels are also found in the blood of smokers, who—because of the carbon monoxide in smoke—also suffer from limitation in O_2 transport.

2,3-BPG plays one other subtle, but important, role in the respiration of humans and other mammals. Consider the problem faced

● **CONCEPT** Negative allosteric effectors of hemoglobin act by preferentially stabilizing the T-state conformation.

by a fetus, which must obtain O_2 from the mother's blood by exchange through the placenta. For this exchange to work well, fetal blood must have a higher O_2 affinity than the mother's blood. In fact, the human fetus has a hemoglobin different from the adult form. Whereas adult hemoglobin (HbA) has two α and two β chains ($\alpha_2\beta_2$), in the fetus the β chains are replaced by similar, but distinctly different, polypeptides. These are called γ chains; thus, fetal hemoglobin (HbF) has an $\alpha_2\gamma_2$ structure. The intrinsic oxygen affinity of HbF is very similar to that of HbA, but HbF has a much lower affinity for 2,3-BPG than does HbA. This difference is largely due to the replacement of His 143 in the adult β chain by a Ser in the fetal γ chain (see Figure 7.29(b)). Loss of the positively charged His 143 in HbF reduces the binding affinity for 2,3-BPG. The concentration of 2,3-BPG is about the same in the circulatory systems of mother and fetus. Under these

conditions, HbF will have less 2,3-BPG bound than will HbA and therefore tends to favor the R state; thus, HbF will have a higher oxygen affinity and efficiently extracts O_2 from the lower-affinity maternal HbA.

The use of effectors that facilitate oxygen release is not restricted to mammals. The blood of birds contains **inositol pentaphosphate** (IPP; see Figure 7.28(b)), and fish use ATP for a similar purpose. All of these molecules have a high negative charge and bind in the central cleft of deoxyhemoglobin. All of these allosteric effectors, including H^+, CO_2, Cl^-, and 2,3-BPG, act in the same general manner—by biasing the conformational equilibrium in hemoglobin toward the deoxy form. However, they interact at distinctly different sites, and therefore their effects can be additive, as illustrated for CO_2 and 2,3-BPG in **FIGURE 7.30**.

7.12 Myoglobin and Hemoglobin as Examples of the Evolution of Protein Function

Each protein sequence produced by an organism is encoded by a gene. The nucleotide sequence in that gene dictates the amino acid sequence of the protein, which in turn defines the protein's secondary, tertiary, and quaternary structures. Evolution of proteins occurs through accumulated changes in the nucleotide sequences of genes. To explore this process, we will use the evolutionary development of the myoglobin–hemoglobin family of proteins as an example. First, however, we must examine in a bit more detail the structure of eukaryotic genes and the mechanisms by which mutation can occur.

The Structure of Eukaryotic Genes: Exons and Introns

In previous chapters, we stated that a direct correspondence exists between the nucleotide sequence in a gene and (via mRNA) the amino

● **CONCEPT** Eukaryotic genes are discontinuous, containing regulatory and protein-encoding sequences (exons) and intervening sequences (introns).

acid sequence of the polypeptide chain it encodes. For most genes in prokaryotic organisms, this concept is true. But investigation of the genomes of higher organisms has produced a surprising result: Within most eukaryotic genes are DNA sequences that are never expressed in the polypeptide chain. These noncoding regions, called **introns,** alternate with regions called **exons** that include the DNA encoding the translated polypeptide sequence and untranslated regions at the 5′ and 3′ ends that are important in regulating transcription and translation. **FIGURE 7.31** shows how the exon–intron structure of the β-globin gene is related to the structure of β-globin protein.

Clearly, this remarkable situation means that mRNA production in eukaryotes must be a more complex process than the process in prokaryotes. As step ❶ in Figure 7.31 shows, transcription first produces a primary transcript, or **pre-mRNA,** corresponding to the whole gene—exons, introns, and flanking regions. The pre-mRNA, while still in the cell nucleus, is cut and spliced in step ❷ to remove the intron regions, thereby producing an mRNA that codes correctly for expression of the polypeptide chain in step ❸. We describe the details of this process in Chapter 24. For now, keep in mind that most eukaryotic genes are "patchwork" structures containing extensive regions that do not correspond to any part of the protein sequence.

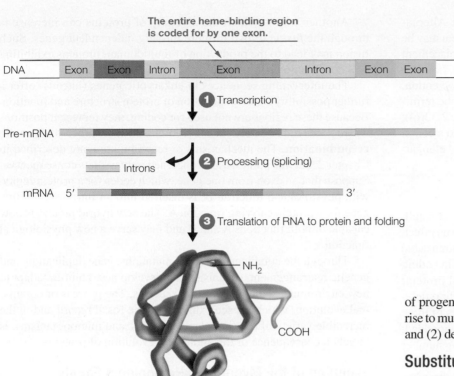

The entire heme-binding region is coded for by one exon.

◀ **FIGURE 7.31 Coding and noncoding regions of the β-globin gene.** The gene for the human β-globin chain has regulatory (blue boxes) and coding (purple, orange, and green boxes) regions, or exons, alternating with noncoding regions, or introns (gray boxes). This figure follows the transcription and translation of the gene to yield the final β-globin chain. ❶ **Transcription:** A primary transcript (pre-mRNA) containing complementary copies of the exons and introns is produced from the gene. ❷ **Splicing:** The intron sequences are removed and the exons spliced together to yield the final mRNA.) ❸ **Translation:** The coding regions of the spliced mRNA produce a β-globin chain, which adopts its favored three-dimensional structure and incorporates a heme group. Note that the entire heme-binding region (orange) is coded for by one exon.

of progeny. Two basic kinds of changes in the DNA sequence may give rise to mutations in proteins: (1) substitution of DNA nucleotides by others and (2) deletion or insertion of nucleotides in the gene sequence.

Substitution of DNA Nucleotides

Substitution of one nucleotide for another can have several possible consequences. First, the change may not affect the protein sequence at all. The substitution may occur in an intron, for example. But even if the nucleotide substitution occurs in a protein-coding exon, it will not change the amino acid sequence if the new codon codes for the same amino acid as the original codon. This kind of mutation is called a **silent,** or **synonymous mutation** (**FIGURE 7.32(a)**). The redundancy of the genetic code (see Figure 5.17) is such that

7.13 Mechanisms of Protein Mutation

When organisms reproduce, they copy their DNA, and occasionally mistakes are made. These mistakes may be random errors that occur during copying, or they may be results of damage the DNA has sustained from radiation or chemical **mutagens,** substances that produce mutations (discussed in greater detail in Chapter 23). In any event, these alterations will appear as **mutations** in the DNA of the next and subsequent generations

● **CONCEPT** Mutations result from changes in the DNA sequence of genes, including base substitutions, deletions, or additions.

Residue number	1	2	3	4	5	6	7	8	9	10
Normal β gene ...A T G	G T G	C A **C**	C T G	A C **T**	C C T	G **A** G	G A G	**A** A G	T C T	G C C ...
	Val	His	Leu	Thr	Pro	Glu	Glu	Lys	Ser	Ala
(a) Silent, or synonymous mutation	G T G	C A T	C T G	A C T	C C T	G A G	G A G	A A G	T C T	G C C ...
	Val	His	Leu	Thr	Pro	Glu	Glu	Lys	Ser	Ala
(b) Missense, or nonsynonymous mutation	G T G	C A C	C T G	A C T	C C T	G T G	G A G	A A G	T C T	G C C ...
	Val	His	Leu	Thr	Pro	Val	Glu	Lys	Ser	Ala
(c) Nonsense mutation	G T G	C A C	C T G	A C T	C C T	G A G	G A G	T A G	T C T	G C C ...
	Val	His	Leu	Thr	Pro	Glu	Glu	Stop		
(d) Frameshift mutation by deletion	G T G	C A C	C T G	A C □	C C T	G A G	G A G	A A G	T C T	G C C ...
	Val	His	Leu	Thr	Leu	Arg	Arg	Ser	Leu	

▲ **FIGURE 7.32 Mutation types.** Some of the ways in which mutations can occur in the β-globin chain are shown here. The first 10 residues of the normal human β chain, together with their DNA codons, are shown at top. **(a)** A silent, or synonymous, mutation has occurred in the codon for residue 2 (CAC to CAT). **(b)** A missense, or nonsynonymous, mutation has occurred in the codon for residue 6 (GAG to GTG). This is the sickle-cell mutation. **(c)** A nonsense mutation has introduced a stop signal after the codon for residue 7 (AAG to TAG), terminating the chain prematurely. **(d)** A frameshift mutation has occurred by deletion of a single T residue. The rest of the chain, with a completely altered sequence, will continue to be produced until a stop signal is encountered in the new frame.

frequently a base change does not alter the protein product. Alternatively, the codon for an amino acid residue in the original protein may be changed to a codon for a different amino acid; this type of replacement is called a **missense,** or **nonsynonymous mutation** (Figure 7.32(b)). Occasionally, a codon for an amino acid is changed to a *stop* codon. We call this a **nonsense mutation** because the protein will be terminated prematurely and will usually be nonfunctional (Figure 7.32(c)). Sometimes the opposite happens—a stop codon mutates into a codon for an amino acid residue. In this case translation continues, elongating the peptide chain.

Nucleotide Deletions or Insertions

Nucleotide deletions or insertions in the gene may be large or small. Such mutations outside the coding regions will generally have no effect, unless they modify sites of transcriptional control (e.g., the untranslated regions of the spliced mRNA). Large insertions or deletions in coding regions almost invariably prevent the production of useful protein; sometimes even whole genes may be deleted. The effect of short deletions or insertions depends on whether they involve multiples of three nucleotides. If one, two, or more *whole codons* are removed or added, the consequence is the deletion or addition of a corresponding number of amino acid residues. However, a deletion or insertion in a coding region of any number of nucleotides *other* than a multiple of three has a much more profound effect: It causes a shift in the reading frame during translation. Such **frameshift mutations** result in a complete change in the amino acid sequence in the C-terminal direction from the point of mutation (Figure 7.32(d)).

The effects of these kinds of mutations on the functionality of the protein product, and therefore on the organism itself, can be quite varied. Base substitutions may, in some cases, be neutral in effect, either not changing the amino acid coded for or changing it to another that functions equally well at that position in the protein. Frequently, however, the result is deleterious. Occasionally, such mutations confer some functional advantage to the protein, and the mutated organisms may be selected for in future generations. Nonsense mutations and frameshift mutations, by contrast, almost always result in loss of protein function. If the protein function is critical to the life of the organism, such mutations are strongly selected against in the course of evolution—those who inherit them do not survive to reproduce.

Gene Duplications and Rearrangements

By accumulating many small mutational changes over millions of years, proteins gradually evolve. The diversity of functions that they can collectively perform is increased by two other phenomena: **gene duplication** and **exon recombination.**

Occasionally, replication of the genome occurs in such a way that some DNA sequence, containing a particular gene, is copied twice. Initially, the only result of such duplication is that the descendants of the organism have two copies of the same gene. This mutation may be advantageous if the protein is needed in large amounts because the capacity for its production will be increased. In such cases, there will be selective pressure to maintain two or even more copies of the same gene. Alternatively, the two copies may evolve independently. One copy may continue to express the protein fulfilling the original function, but the other may evolve through mutations into an entirely different protein with a new function. Recall from Chapter 5 that proteins related by a common evolutionary origin are called **homologs.**

Another way in which the diversity of proteins can increase is through the *fusion* of two or more initially independent genes. Such fusion may lead to the production of multidomain proteins exhibiting new combinations of functions.

The intervening sequences in eukaryotic genes (introns) offer a further possibility for diversification of protein structure and function. Because these regions are not used for coding, they represent positions where genes can be safely cut and recombined in the process of **genetic recombination.** The mechanisms of recombination are described in Chapter 23; at this point we are concerned only with its consequences. Suppose that an exon from one gene, which codes for a protein region with physiological function B, is inserted into an intron region in a gene for a protein carrying function A. The new hybrid protein is now capable of both functions A and B and may serve a new physiological function.

Through the combined effects of mutations, gene duplication, and genetic rearrangement, organisms can develop new abilities, adapt to new environments, and become new species. The process of organismal evolution, which we see exhibited in the fossil record and in the incredible variety of existing plants, animals, and microorganisms, is largely a consequence of this molecular evolution of proteins.

Evolution of the Myoglobin–Hemoglobin Family of Proteins

Myoglobin and hemoglobin represent sophisticated molecular machines, each finely tuned to carry out its functions. After learning about the various mechanisms of protein mutation, we are ready to explore how these structures might have evolved. We will focus on the O_2 transport functions of vertebrate myoglobin and hemoglobin; however, it should be noted that globins are ancient proteins that are found in greater structural and functional variety in bacteria and archaea. In these lower organisms, globin genes seem to be involved primarily in interactions with the gas nitric oxide (NO). Thus, O_2 transport appears to be a relatively recent functional adaptation to the ancestral globin fold, which has been present on Earth since atmospheric levels of O_2 were much lower than they are at present.

We have already seen an example of the process of protein evolution. If we compare the sequences of sperm whale and human myoglobin (see Figure 5.15), we find 25 amino acid changes. Because fossil evidence indicates that the evolutionary lines that led to sperm whales and humans diverged from a common mammalian ancestor about 100 million years ago, we can gain an idea of the rate of this process. If the rate was uniform, there has been an average of one amino acid replacement every 8 million years.

If we compare human myoglobin with that of the shark, we find about 88 differences. Because these evolutionary lines diverged about 400 million years ago, the accumulated differences are about what we would expect from the preceding example. In other words, the number of amino acid substitutions in two related proteins is roughly proportional to the evolutionary time that has elapsed since the proteins (and the species) had a common ancestor. Using this principle, we can compare the sequences of both hemoglobins and myoglobins and attempt to construct a "family tree" of globin proteins. The tree is complicated by the fact that higher eukaryotes, including humans, carry genes for both myoglobin and several *different* hemoglobin chains. These different genes are expressed at different times in human development (**FIGURE 7.33**). The α and β chains, as mentioned earlier, are normally present in adults. But in the early embryo, the

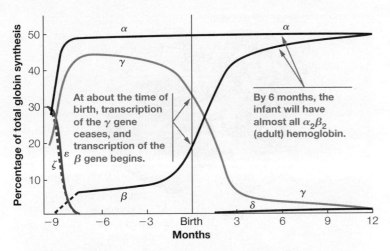

▲ FIGURE 7.33 Expression of human globin genes at different stages in development. The human ζ and ε genes make the $\zeta_2\varepsilon_2$ hemoglobin found in the very early embryo. This is soon supplanted by the $\alpha_2\gamma_2$ hemoglobin of the fetus. The δ gene is never transcribed at high rates. There are two copies of the α gene: α_1 and α_2. Both contribute to the production of α chains.

hemoglobin genes expressed are those for the embryonic chains, ζ (zeta) and ε. As the fetus develops, these chains are replaced by α and γ chains, ensuring efficient oxygen transfer from mother to fetus. Finally, at about the time of birth, the γ chains are replaced by β chains. In addition, after birth a small amount of a δ chain is produced. These developmental types of hemoglobin chain are slightly different, and each is coded for by a separate gene in the human genome.

Comparison of the sequences of many globins from many different species yields the evolutionary tree shown in **FIGURE 7.34**. According

● CONCEPT Myoglobin and hemo-globin evolved from an ancestral myoglobin-like protein.

to these results, very primitive animals had only a myoglobin-like, single-chain ancestral globin for oxygen storage. Most of these animals, like protozoans and flatworms, were so small that they did not require a transport protein. Roughly 500 million years ago, an important event occurred: The ancestral myoglobin gene was duplicated. One of the copies became the ancestor of the myoglobin genes of all higher organisms; the other evolved into the gene for an oxygen transport protein and gave rise to the hemoglobins.

Along the evolutionary line leading to vertebrates and mammals, the most primitive animals to possess hemoglobin are the lampreys. Lamprey hemoglobin can form dimers but not tetramers and is only weakly cooperative; it represents a first step toward allosteric binding. But subsequently a *second* gene duplication occurred, giving rise to the ancestors of the present-day α- and β-hemoglobin chain families. Reconstruction, from sequence comparison, indicates that this must have happened about 400 million years ago, at about the time of divergence of the sharks and bony fish. The evolutionary line of the bony fish led to the reptiles, and eventually to the mammals, all carrying genes for both α- and β-globins and capable of forming tetrameric $\alpha_2\beta_2$ hemoglobins. Further gene duplications have occurred in the hemoglobin line, leading to the embryonic forms ζ and ε and the fetal γ. As Figure 7.34 shows, the duplications that led to a distinction between adult and embryonic subtypes coincide fairly well with the development of placental mammals, about 200 million years ago. This concurrence is functionally appropriate because in these mammals the later stages of embryo development occur within the mother, and a special hemoglobin, adapted to promote oxygen transfer through the placenta from mother to fetus, is essential (see Figure 7.29).

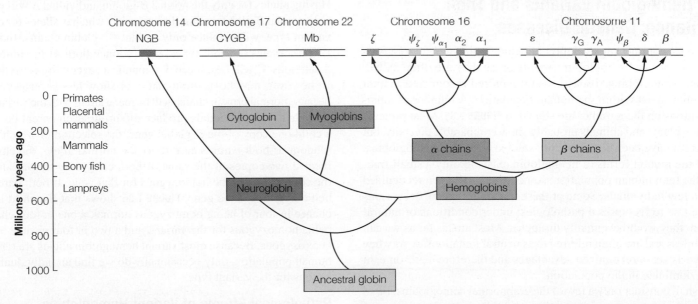

▲ FIGURE 7.34 Evolution of the globin genes. The arrangement of human globin genes is shown at the top. Note that they are found in five different chromosomes. Functional genes are shown in color; pseudogenes, which are nontranscribed variants of a gene, are in gray. The diagram underneath shows the probable evolution of the globin gene family, based on sequence differences among the various globin genes in humans and other animals. The times at which gene duplications occurred are inferred from a combination of sequence and fossil evidence and are only approximate. The two α genes (and the two γ genes) are too similar in sequence to allow us to judge the time of their divergence. We know only that it must have happened relatively recently.

During the long evolution of the myoglobin–hemoglobin family of proteins, only a few amino acid residues have remained invariant. These *conserved residues* may mark the truly essential structural features of the molecule. As Figure 7.21 shows, they include the histidines proximal and distal to the heme iron (F8 and E7; see Figure 7.16(b)). Interestingly, Val FG5, which has been implicated in the hemoglobin deoxy–oxy conformation change described earlier, is invariant in hemoglobins, replacing the isoleucine found at this position in most myoglobins. Other regions highly conserved in hemoglobins are those near the $\alpha_1 - \beta_2$ and $\alpha_2 - \beta_1$ contacts. These contacts are most directly involved in the allosteric conformational change.

Despite the major changes that have occurred in the primary structure of the myoglobin–hemoglobin family over hundreds of millions of years, the secondary and tertiary structures of these proteins have remained surprisingly unchanged, particularly in the region that binds the heme. At first glance, this similarity seems inconsistent with our earlier statements that primary structure determines secondary and tertiary structure. However, careful examination of many sequences shows that many of the replacements have been *conservative*—that is, an amino acid has been replaced by another of the same general class (e.g., polar replaces polar, or nonpolar replaces nonpolar). Obviously, evolution of these proteins has proceeded not at random but under the constraint of maintaining a physiologically functional structure. Survival of mutant proteins in the globin family has been restricted to those that maintain the basic "globin fold" and function.

● **CONCEPT** Evolution of globins has retained the common "globin fold" that holds the heme. Evidence for continuing evolution is found in the many variant proteins in existing species.

7.14 Hemoglobin Variants and Their Inheritance: Genetic Diseases

Evidence for the ongoing evolution of hemoglobin genes can be seen in the existence of hemoglobin variants or, as they are often called, abnormal hemoglobins. Today, several hundred recognized mutant hemoglobins exist within the human population. A number of mutation positions on the tetramer are shown in **FIGURE 7.35**. Most proteins in existing plants and animals probably show comparable diversity, but few of them have been as thoroughly studied as human hemoglobins. Each of the mutant forms of hemoglobin exists in only a small fraction of the total human population; some forms have been recognized in only a few individuals. Some of these mutant forms are deleterious and give rise to recognized pathologies; under conditions of natural selection, they would eventually disappear. Most are, as far as we can tell, harmless and are often referred to as neutral mutations. A very few may have as yet unrecognized advantages and therefore may come, in time, to dominate in the population.

We shall consider only a few of these abnormal hemoglobins. First, it is necessary to review a bit of genetics. All human cells, except for the germ-line cells (sperm and ova), are **diploid;** that is, they carry two copies of each chromosome. Therefore, they carry two copies of each gene, one on each of the paired chromosomes. A gene may exist in the genome in variant forms, which we refer to as **alleles.** Suppose we consider a gene such as the adult β-globin gene, which can exist in two allelic forms—the "normal" type, β, and a variant (mutant) type,

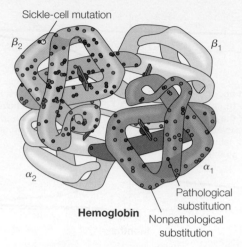

Sickle-cell mutation

β_2 β_1

α_2 α_1

Hemoglobin   Pathological substitution / Nonpathological substitution

▲ **FIGURE 7.35 Distribution of mutations in human hemoglobins.** The blue and red dots represent all positions at which amino acid substitutions have been found in the α and β chains (only one pair is illustrated, for clarity). Those substitutions that have known pathological effects are shown in red. At many of these positions, more than one substitution has been observed. Position 6 in the β chain, at which the sickle-cell mutation occurs, is shown in yellow.

β^*. Individuals can have three possible combinations of these alleles in their paired chromosomes:

A. $\beta + \beta$: **homozygous** (identical alleles) in the normal type

B. $\beta + \beta^*$: **heterozygous** (different alleles), one copy of each gene type

C. $\beta^* + \beta^*$: homozygous in the variant type

Having alleles for only the normal β-globin, individual A will produce only normal β-globin chains. Individual C, who has alleles for only the variant type, will produce only variant β^*-globin chains. Individual B, with alleles for both types, will produce both. If the mutation is deleterious, C will be expected to manifest severe disease symptoms. B, however, may be asymptomatic, or show less severe symptoms because normal protein chains will be made along with the variant ones.

When two individuals produce offspring, each parent donates to a child one copy of the β-globin gene, the selection of which will be random. If both parents carry only the normal allele, the child will receive two copies of the same. If both carry only the variant gene, the child must also be homozygous for that gene. If both parents are heterozygous for the gene, **FIGURE 7.36** shows that the child has one chance in four of being homozygous normal, a one in four chance of being homozygous for the variant, and a two in four chance of being heterozygous. Because most variant hemoglobin alleles are rare in the human population, only occasionally do we find an individual homozygous for the variant type.

Pathological Effects of Variant Hemoglobins

Of the large number of hemoglobin mutations, a significant fraction have deleterious effects. As Figure 7.35 shows, the known deleterious mutations are mostly clustered about the heme pockets and in the vicinity of the $\alpha - \beta$ contact region that is so important in the allosteric transition. A few of the well-studied pathological missense mutations are listed in **TABLE 7.2**.

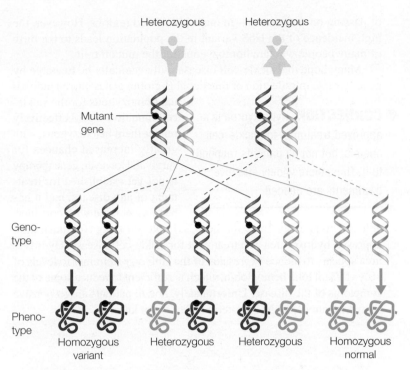

Heterozygous   Heterozygous

Mutant gene

Geno-type

Pheno-type

Homozygous variant   Heterozygous   Heterozygous   Homozygous normal

▲ **FIGURE 7.36 Inheritance of normal and variant proteins in a heterozygous cross.** Diploid organisms can exist as one of three types with respect to any gene: homozygous normal, homozygous variant, or heterozygous. The offspring of a heterozygous pair may be any of the three types, as shown here: homozygous normal, heterozygous, or homozygous variant, with a probability ratio of 1:2:1. We leave it as an exercise for you to work out other possibilities—for instance, the offspring of one homozygous normal parent and one homozygous variant parent.

(a) Typical sickled cells, together with some normal, rounded red blood cells.

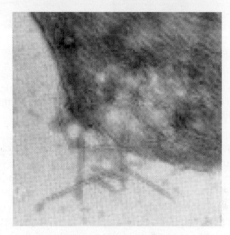

(b) Scanning electron micrograph of a sickled cell that has ruptured, with hemoglobin fibers spilling out.

▲ **FIGURE 7.37 Erythrocytes in sickle-cell disease.**

The most infamous of all variant hemoglobins, *sickle-cell hemoglobin* (HbS), is a source of misery and early death to many humans. The variant has gained its name because it causes red blood cells to adopt an elongated, sickle shape at low oxygen concentrations (**FIGURE 7.37(a)**). This "sickling" is a consequence of the tendency of the mutant hemoglobin, in its deoxygenated state, to aggregate into long, rodlike structures (**FIGURE 7.38**; see also the opening to Chapter 5). The elongated cells tend to block capillaries, causing inflammation and considerable pain. Even more serious is that the sickled cells are fragile (Figure 7.37(b)); their breakdown leads to an anemia that leaves the victim susceptible to infections and diseases. Individuals who are homozygous for the sickle-cell mutation often do not survive into adulthood, and those who do are seriously debilitated. Heterozygous individuals, who can still produce some normal hemoglobin, usually suffer distress only under conditions of prolonged oxygen deprivation. For example, flying may be dangerous for HbS heterozygotes because of the lower oxygen level in the cabin of the aircraft.

TABLE 7.2 Selected list of missense mutations in human hemoglobins

Effect	Residue Changed	Change	Name	Consequences of Mutation	Explanation
Change in erythrocyte shape	$\beta 6$ (A3)	Glu ⟶ Val	S	Sickling	Val fits into EF pocket in chain of another hemoglobin molecule.
Change in O_2 affinity	$\alpha 87$ (F8)	His ⟶ Tyr	M Iwate	Decreases O_2 affinity due to oxidation of heme to Fe^{3+} state	The His normally ligated to Fe^{2+} has been replaced by Tyr.
	$\alpha 141$ (HC3)	Arg ⟶ His	Suresnes	Increases O_2 affinity by favoring R state	Replacement eliminates bond between Arg 141 and Asn 126 in deoxy state.
	$\beta 146$ (HC3)	His ⟶ Asp	Hiroshima	Increases O_2 affinity, reduced Bohr effect	Disrupts salt bridge in deoxy state and removes a His that binds a Bohr effect proton.

  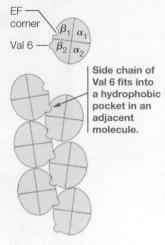

(a) An electron micrograph of one sickle-cell fiber.

(b) A computer-graphic depiction of one fiber.

(c) A schematic model of fiber formation.

EF corner

Val 6

Side chain of Val 6 fits into a hydrophobic pocket in an adjacent molecule.

▲ **FIGURE 7.38 Sickle-cell hemoglobin.** Molecules of sickle-cell hemoglobin tend to aggregate, forming long fibers. Deoxyhemoglobin S molecules lock together to form a two-stranded cluster because the side chain of Val 6 in the β chain of one hemoglobin molecule fits into a hydrophobic pocket in an adjacent molecule. Interaction of these two-stranded structures with one another produces the multistrand fibers shown in (a) and (b).

Linus Pauling first suggested in 1949 that sickle-cell disease was a "molecular disease" resulting from a mutation in the hemoglobin molecule. Remarkably, sickling stems from what we might expect to be an innocuous mutation. The glutamic acid residue normally found at position 6 in β chains is replaced by a valine (see Figure 7.32(b)). This hydrophobic valine can fit into a pocket at the EF corner of a β chain in another hemoglobin molecule, and thus, as shown in Figure 7.38(c), adjacent hemoglobin molecules can fit together into a long, rodlike helical fiber. Why sickling occurs with deoxyhemoglobin, but not with the oxygenated form, is simply explained: In the oxy form, the rearrangement of subunits makes the EF pocket inaccessible to Val 6.

Sickle-cell disease is confined largely to populations originating in tropical areas of the world. At first glance, this distribution seems unexpected. Why should a *genetic* disease be climate-related? The answer tells us something about the persistence of what seem to be unfavorable traits. A high incidence of sickle-cell disease in a population generally coincides with a high incidence of malaria, a parasitic disease carried by a tropical mosquito. Individuals *heterozygous* for sickle-cell hemoglobin have a higher resistance to malaria than those who do not carry the sickle-cell mutation. The malarial parasite spends a portion of its life cycle in human red cells, and the increased fragility of the sickled cells, even in heterozygous individuals, tends to interrupt this cycle. In addition, the distortion of the cell membrane of intact sickled cells leads to a loss of potassium ions from these cells, providing a less favorable environment for the parasite. Heterozygous individuals have a higher survival rate—and therefore a better chance

● **CONCEPT** Sickle-cell disease results from a single-base substitution in the β chain.

of passing on their genes—in malaria-infested regions. However, the high incidence of the HbS variant in the population leads to the birth of many people who are homozygous for the mutant trait.

Many hope that sickle-cell disease will eventually be treatable by gene therapy. Introduction of functional β-globin genes into an individual homozygous for the sickle-cell mutation would effectively render them heterozygous, with greatly increased chances for survival. However, gene therapy is not yet established for treatment of any disease, and it has been associated with considerable risk to patients. In 1998, the U.S. Food and Drug Administration approved hydroxyurea as a treatment for sickle-cell disease. Hydroxyurea appears to induce expression of the HbF $\alpha_2\gamma_2$ tetramer to levels of 10%–15% of total hemoglobin, which is sufficient to reduce some of the symptoms of the disease. Unfortunately, not all patients are responsive to hydroxyurea, and its long-term safety is not known.

● **CONNECTION** Hydroxyurea is an approved treatment for sickle-cell anemia, but not all patients respond to it. Thus, more widely effective treatments are needed.

$$H_2N \overset{O}{\underset{H}{\overset{\|}{C}}} N \overset{}{\underset{H}{}} OH$$

Hydroxyurea

7.15 Protein Function Requiring Large Conformational Changes: Muscle Contraction

We have seen how proteins, either as monomers or in defined multi-subunit structures, carry out a variety of functions. We now turn to an example in which proteins organize into larger, more complex structures involving many kinds of polypeptide chains. Such supramolecular structures perform many cellular functions; the one we will consider here is the mechanical work of motion carried out by **motor proteins.** This motion may involve the whole organism or parts thereof, individual cells, or subcellular constituents. We will see that protein conformational change mediated by the binding, hydrolysis, and release of ATP is a key feature of motor protein function.

Of the many kinds of motion exhibited by living systems, the one we are most familiar with is the muscle contraction required for bodily movement. However, muscle contraction accomplishes a remarkable variety of other things as well. Even the emission of sound is a muscular action, as is the injection of venom by an insect or a snake. Equally important muscular motions maintain the animal's internal world, including the beating of its heart, the breathing motions of its lungs or gills, and the peristaltic motions of its digestive system. Each of these kinds of movements is produced by a specific muscular tissue.

All muscles, as well as some other contractile systems, are based on the interaction of two major proteins, **actin** and **myosin.** However, certain kinds of directed motions exist—motions of individual cells and parts of cells—that do not depend on the actin–myosin system at all but use other protein mechanisms. For example, the beating of cilia and flagella and the movement of chromosomes and organelles within cells are accomplished by different classes of motor proteins.

● **CONCEPT** Certain proteins act as energy transducers, using free energy from ATP hydrolysis or free energy stored in ion gradients to do the mechanical work of motion.

The biological systems that produce movement share one common feature—they hydrolyze ATP. The energy released by the hydrolysis of ATP is converted into work by producing motions in parts of protein molecules. Thus, proteins can act as **energy transducers.** That is, some proteins can convert the free energy of ATP hydrolysis into mechanical work. When the motions of proteins are properly coordinated, directed macroscopic motion occurs.

7.16 Actin and Myosin

The major proteins in muscle tissue are actin and myosin; however, these proteins are also found in many other types of cells and are involved in several kinds of cellular and intracellular motions. To understand how muscles and other actin–myosin systems work, we must consider the properties of these two proteins.

Actin

Under physiological conditions, actin exists as a long, helical polymer (filamentous actin, or **F-actin**) of a globular protein monomer (**G-actin**). The G-actin monomer, shown in **FIGURE 7.39**, is a four-domain molecule with a molecular weight (M_W) of 42 kDa. The binding of ATP by a G-actin monomer leads to its polymerization; the ATP is subsequently hydrolyzed, but the ADP is held in the actin filament. In F-actin filaments, the G-actin monomers are arranged in a two-strand helix (**FIGURE 7.40**). Because of the asymmetry of the subunits, the F-actin filament has a defined directionality, and the two ends have been called the **"barbed"** or **"plus" end,** and the **"pointed"** or **"minus" end.** The polymerization reaction exhibits a preferred direction, and the plus end is defined as that end that grows more rapidly under physiological conditions. The sites on the F-actin filament that bind to myosin are located on domain 1 of each actin subunit.

Myosin

Six of the 20 or so forms of myosin found in cells, and their various functions, are listed in **TABLE 7.3**. The most studied myosin molecule

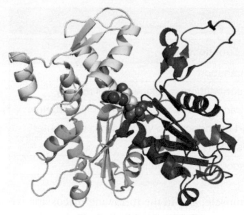

G-actin with ATP bound
(PDB ID: 1atn)

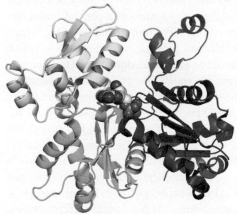

G-actin with ADP bound
(PDB ID: 1j6z)

◄ **FIGURE 7.39 G-actin.** Cartoon representations of the X-ray crystal structures of the G-actin monomer are shown. Domain 1 is red, domain 2 is blue, domain 3 is green, and domain 4 is yellow. Nucleotides are shown as spheres. The "plus" end is down in each panel.

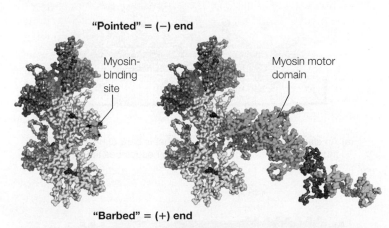

"Pointed" = (−) end

Myosin-binding site

Myosin motor domain

"Barbed" = (+) end

The α-carbon backbones of five G-actin monomers are shown. Individual monomers are distinguished by colors, with bound ATP shown in red. The green residues on domain 1 of the gray actin monomer show the myosin-binding site.

A model of the myosin S1 fragment (see text) bound to F-actin. The myosin motor domain is shown in green, and the essential and regulatory light chains are, respectively, magenta and cyan (PDB ID: 1alm).

▲ **FIGURE 7.40 A model for F-actin filaments.**

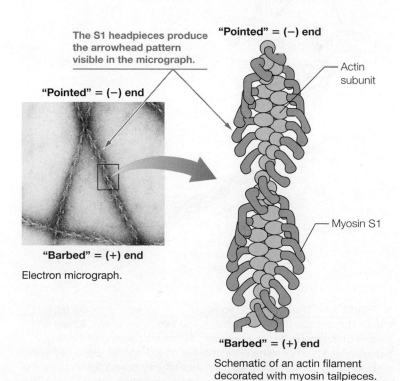

The S1 headpieces produce the arrowhead pattern visible in the micrograph.

"Pointed" = (−) end

"Pointed" = (−) end

Actin subunit

Myosin S1

"Barbed" = (+) end

Electron micrograph.

"Barbed" = (+) end

Schematic of an actin filament decorated with myosin tailpieces.

TABLE 7.3 Myosin types and their functions

Myosin Type	Primary Functions
I	Vesicle transport
II	Muscle contraction; cytokinesis; cell motility
V	Vesicle transport and localization at cell periphery
VI	Transport of endocytotic vesicles into cell center
VIII	Cell division in plants
XV	Part of acoustic sensor in inner ear

is myosin II from striated muscle, and in the following discussion we will refer to myosin II simply as "myosin." The functional myosin molecule (**FIGURE 7.41**) is composed of six polypeptide chains: two identical heavy chains ($M_W = 230$ kDa) and two each of two kinds of light chains ($M_W = 20$ kDa). Together, they form a complex of molecular weight 540 kDa. The heavy chains have long α-helical tails, which form a two-stranded coiled coil, and globular head domains to which the light chains are bound. Between each head domain and tail domain, the heavy chain acts as a flexible stalk. The coiled-coil structure of the tails is reminiscent of the structure of α-keratin (see Figure 6.12).

Just as IgGs can be cleaved specifically by proteases to yield F_{ab} and F_c fragments, proteolytic cleavage of myosin yields a coiled-coil tail and two **S1 fragments,** each consisting of a head domain carrying the light chains (see **FIGURE 7.42**). The ability to cleave the myosin molecule in this way has helped researchers understand the functions of its several parts. Myosin exhibits aspects of both fibrous and globular proteins, and its functional domains play quite different roles.

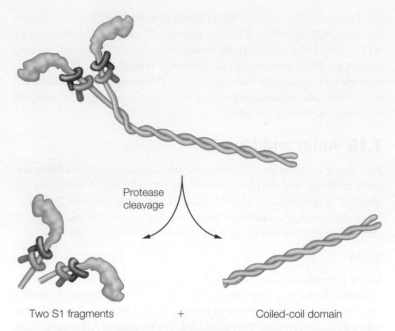

Two S1 fragments   +   Coiled-coil domain

▲ **FIGURE 7.42 Dissection of myosin II by proteases.** Cleavage by proteases cuts the myosin tail to yield two S1 fragments and a coiled-coil domain.

The tail domains have a pronounced tendency to aggregate, causing myosin molecules to form the kind of thick bipolar filaments shown in **FIGURE 7.43**. Each S1 fragment, or **headpiece,** includes the globular **motor domain,** which binds ATP and actin and two myosin light chains—the **essential** and the **regulatory** light chains. The crystal structure of an S1 fragment is shown in **FIGURE 7.44**.

● **CONCEPT** The major muscle systems of animals are based on the proteins actin and myosin.

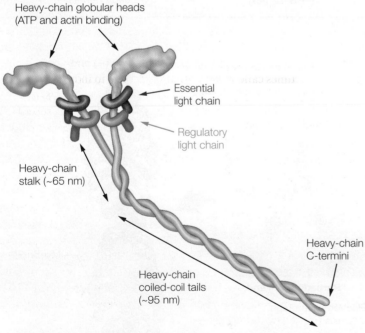

Heavy-chain globular heads
(ATP and actin binding)

Essential light chain

Regulatory light chain

Heavy-chain stalk (~65 nm)

Heavy-chain C-termini

Heavy-chain coiled-coil tails (~95 nm)

▲ **FIGURE 7.41 The myosin II molecule.** This schematic model depicts the six polypeptide chains of myosin. The two heavy chains (green and yellow) of the molecule are connected by the intertwining of the two α helices of the heavy chains in the rod-like coiled-coil tail. Each of the two globular head domains carries two noncovalently bound light chains: the essential light chain (red) and the regulatory light chain (blue).

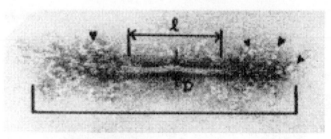

(a) An electron micrograph. The zone that is bare of headpieces is indicated by the ℓ, and some myosin headpieces are indicated by arrowheads.

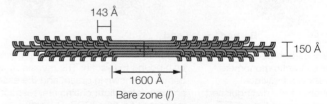

143 Å

150 Å

1600 Å

Bare zone (l)

(b) A schematic drawing of the filament structure, showing dimensions in Ångstrom. The projections are the pairs of headpieces on each myosin molecule.

▲ **FIGURE 7.43 A thick filament of myosin II molecules.**

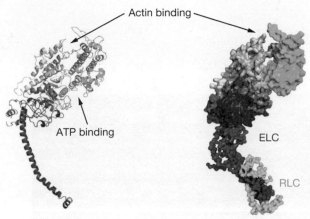

A cartoon rendering of the heavy chain in rainbow coloring. The "neck" of the S1 fragment is shown by the extended red helix.

The heavy chain in a surface representation with the α-carbon backbone of the essential (ELC, magenta) and regulatory (RLC, cyan) light chains. The position of ATP binding is shown, as well as the point of contact with actin.

▲ **FIGURE 7.44** The X-ray crystal structure of an S1 fragment of myosin II. PDB ID: 2mys.

● **CONCEPT** Myosins are motor proteins that move toward the (+) end of the actin filament.

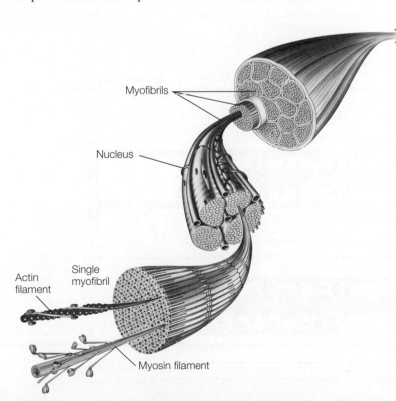

▲ **FIGURE 7.45** Levels of organization in striated muscle.

7.17 The Structure of Muscle

In muscle tissue, the actin and myosin filaments interact to produce the contractile structure. Vertebrates have three morphologically distinct kinds of muscle. *Striated muscle* is the kind most often associated with the term *muscle,* because it is the striated muscles in arms, legs, eyelids, and so forth that make voluntary motions possible. *Smooth muscle* surrounds internal organs such as the blood vessels, intestines, and gallbladder, which are capable of slow, sustained contractions that are not under voluntary control. *Cardiac muscle* can be considered a specialized form of striated muscle, adapted for the repetitive, involuntary beating of the heart. The following discussion considers only the structure of striated muscle.

FIGURE 7.45 shows successive levels of organization of a typical vertebrate striated muscle. The individual muscle fibers, or **myofibers,** are very long (1–40 mm) multinucleate cells, formed by the fusion of muscle precursor cells. Each myofiber contains a bundle of protein structures called **myofibrils.** A myofibril exhibits a periodic structure when examined under the light microscope. Dark **A bands** alternate with lighter **I bands.** The I bands are divided by thin lines called **Z disks** (or sometimes **Z lines**). At the center of the A band is found a lighter region called the **H zone.** The repeating unit of muscle structure can be taken as extending from one Z disk to the next. It is called the **sarcomere** and is about 2.3 μm long in relaxed muscle.

The molecular basis for this periodic structure of the myofibril can be seen by electron microscopic studies of thin sections of muscle, as shown in **FIGURE 7.46**. **Thin filaments** of actin extend in both directions from the Z disks, interdigitating with myosin **thick filaments.** The regions in which the thick and thin filaments overlap form the dark areas of the A band. The I bands contain only thin filaments, which extend to the edges of the H zones. Within the H zones, only thick filaments are found. The dark line at the center of the H zone (sometimes called the M band) is believed to indicate positions where thick filaments associate with one another.

● **CONCEPT** The sarcomere is the basic repeating unit of a muscle myofibril.

If we look closely at electron micrographs of myofibrils (see Figure 7.46(d)), we can see small projections extending from the thick (myosin) filaments, often contacting the thin (actin) filaments. The projections correspond to the headpieces of the myosin molecules. These *cross-bridges* between myosin and actin filaments are the key to muscle contraction.

● **CONCEPT** Thin filaments are mainly actin; thick filaments are mainly myosin. They are connected by noncovalent cross-bridges.

The organization of actin, myosin, and other muscle proteins into this elaborate yet specific structure found in the sarcomere is a remarkable example of how several kinds of protein can combine in a specific way to form a functional structure. We will now examine how this structure works.

7.18 The Mechanism of Contraction

Our understanding of the mechanism of muscle contraction has come from observation of both the fine details of muscle structure and changes in the sarcomere-banding pattern during contraction. The muscle sections shown in Figure 7.46 and at the top of **FIGURE 7.47**

Most myosin types are motor proteins that move toward the (+) end of the actin filament. As we shall see, myosin's **ATPase activity** (i.e., ATP binding and hydrolysis), as well as its ability to bind and release F-actin, are essential parts of the multistep mechanism of muscle contraction.

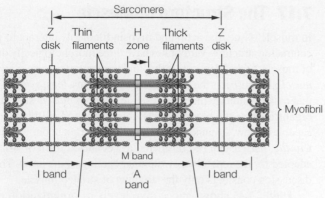

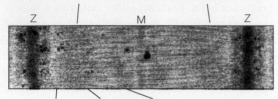

(a) A model of the sarcomere, the repeating unit in striated muscle. The I bands, A bands, and Z disks shown in panel (b) of this figure are identified, and structural elements of the sarcomere are indicated.

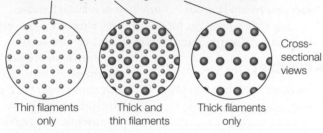

(b) An electron micrograph showing the same features.

Cross-sectional views

Thin filaments only

Thick and thin filaments

Thick filaments only

(c) A schematic drawing of cross sections of a sarcomere in the various regions shown in (a) and (b). Thick filaments are indicated by heavy red dots, thin filaments by small purple dots.

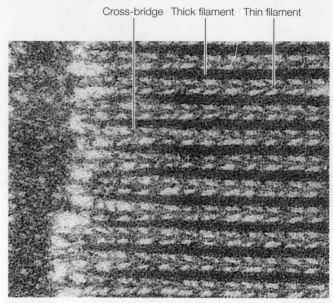

(d) A higher magnification within an A band showing cross-bridges between actin and myosin filaments.

▲ **FIGURE 7.46 Muscle structure seen at the electron microscopy level.**

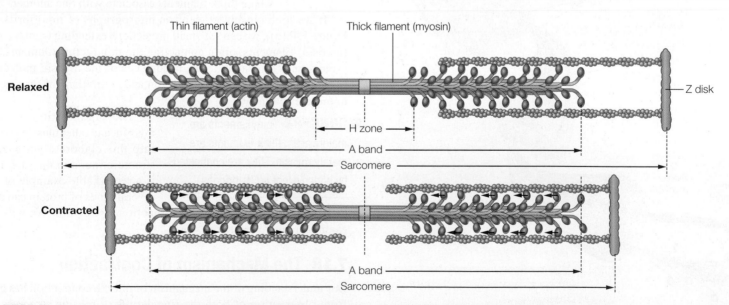

▲ **FIGURE 7.47 The sliding filament model of muscle contraction.** Contraction of striated muscle occurs when the myosin headpieces pull the actin filaments toward the center of the sarcomere.

are in the relaxed, or extended, state. In a fully contracted muscle, each sarcomere shortens from a length of about $2.3 \, \mu m$ to $2.0 \, \mu m$. During this process, the I bands and H zones disappear, and the Z disks move right up against the A bands (bottom of Figure 7.47). Such observations led two independent (and unrelated) investigators, Hugh Huxley and Andrew Huxley, to propose in the 1950s the **sliding filament model** of muscle contraction depicted in Figure 7.47. According to this model, the myosin headpieces are presumed to "walk" along the interdigitated actin filaments, pulling them past, and thereby shortening the sarcomere.

To produce such directed motion against an opposing force on the muscle, energy must be expended. You might expect energy to be derived in some way from ATP hydrolysis, and our previous mention of the ATPase activity of the actin–myosin complex alludes to how this energy could be gained: According to a refinement of the sliding filament model called the **swinging cross-bridge model,** each myosin headpiece takes part in a repetitive cycle of making and breaking cross-bridges to an adjacent actin thin filament. We imagine the cycle starting with the myosin attached to F-actin, as shown at the top of **FIGURE 7.48**. Binding of ATP leads to release of the myosin cross-bridge in step ❶. Hydrolysis of ATP then causes a conformational change, "cocking" the headpiece in step ❷. Myosin, because it has now cocked, binds to the thin filament at a site that is closer to the Z disk—where the (+) end of the actin filament is anchored (step ❸). This initial binding is weak. Phosphate release in step ❹ results in strong binding to the thin filament prior to the power stroke in step ❺, which pulls the thin filament toward the center of the sarcomere. Release of ADP in step ❻ and binding a new ATP will then restart the cycle.

It is the "neck region" of the S1 fragment (Figure 7.44), where the light chains are bound, that undergoes the greatest conformational change during the power stroke. Based on comparisons of X-ray crystal structures of myosin in the ADP-bound and nucleotide-free states, it is thought that the C-terminal part of the myosin neck moves about 10 nm during the power stroke in step ❺.

● **CONCEPT** In the swinging cross-bridge model, the periodic attachment and release of cross-bridges, with a cross-bridge conformational change, slide the thin and thick filaments past one another.

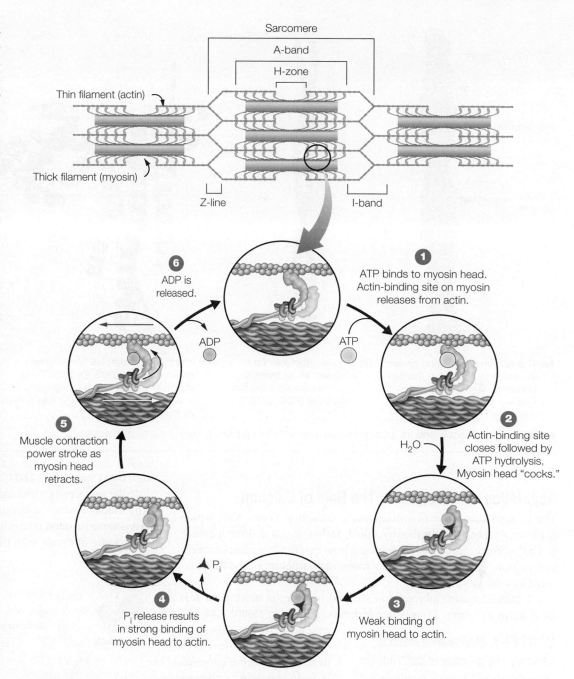

▲ **FIGURE 7.48 A model of the ATP cycle in muscle contraction.** For clarity, only one of the two myosin headpieces is shown going through a binding, power stroke, and release cycle. The second myosin headpiece is shown in light green. The individual steps of the cycle are described in the text.

At the end of each cycle, the actin filament has been moved with respect to the myosin so that each headpiece makes successive steps along the thin filament. The "walking" is rather like that of a millipede—there is always contact between some of the thick filaments' "legs" (i.e., the many S1 headpieces) and the thin filament. Thus, the thick filament doesn't slip back during muscle contraction. A thick filament displays several hundred S1 fragments, and each contacts the thin filament 5 times per second during muscle contraction. The distance the actin filament is moved per power stroke is about 100–200 Å at high resistance. When working against weaker resistance, the multiple power strokes can propel the actin for as much as 1000 Å per ATP hydrolyzed.

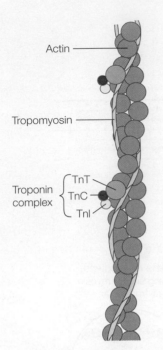

Actin

Tropomyosin

Troponin
complex {
  TnT
  TnC
  TnI
}

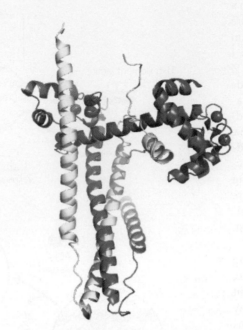

(a) This schematic drawing shows the proteins present in the thin filaments of striated muscle: F-actin, tropomyosin, and troponins (Tn) I, C, and T.

(b) Crystal structure of a fragment of rat skeletal tropomyosin showing the coiled coil (PDB ID: 2b9c).

(c) The troponin complex from chicken skeletal muscle (PDB ID: 1ytz). TnI is shown in yellow, TnT in orange, and TnC in red, with four bound Ca^{2+} ions shown as blue spheres.

▲ **FIGURE 7.49** F-actin and its associated proteins. Panels (b) and (c) are not drawn to scale.

Regulation of Contraction: The Role of Calcium

The critical substance in stimulating contraction is not ATP, which is generally available in the myofibril; rather, it is a sudden increase in Ca^{2+} concentration. To understand how calcium regulates muscle contraction, we must examine the molecular structure of the thin filament in a bit more detail.

A thin filament, as found in striated muscle, is more than just an F-actin polymer. Four other proteins, shown in **FIGURE 7.49**, are essential to the thin filament's contractile function. One of these proteins is **tropomyosin,** a fibrous coiled-coil protein that overlaps the myosin-binding site in the F-actin helix. Bound to each tropomyosin molecule are three small proteins called **troponins I, C, and T.** Troponin C (TnC) is homologous to the Ca^{2+}-binding protein **calmodulin,** and like calmodulin, TnC undergoes a conformational change upon binding Ca^{2+}. The structure and function of calmodulin are presented in greater detail in Chapter 12 (see Figure 12.28). The presence of tropomyosin and the troponins inhibits the binding of myosin heads to actin *unless calcium is present* at a concentration of about 10^{-5} M. In resting muscle, the Ca^{2+} concentration is approximately 10^{-7} M, so new cross-bridges cannot be formed. An influx of Ca^{2+} stimulates contraction because the ion is bound by troponin C, causing a rearrangement of the troponin–tropomyosin complex, which moves the tropomyosin ~10 Å closer to the central

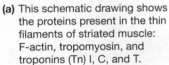

● **CONCEPT** Muscle contraction is stimulated by an influx of Ca^{2+} into the sarcomere. Ca^{2+} binding by troponin C causes a rearrangement of the troponin–tropomyosin–actin complex, allowing actin–myosin cross-bridges to form.

groove of the F-actin helix. This shift reveals the sites on actin to which the myosin headpieces bind. The postulated mechanism, shown in **FIGURE 7.50**, permits step ❸ and the subsequent steps in the cycle of Figure 7.48 to take place.

We have now traced the activation of muscle contraction to the influx of calcium into the myofibrils. But why does this influx occur? In particular, how can it be brought about by the nerve impulses that excite muscles to contract? The answer can be found from a closer examination of the myofiber, or muscle cell (**FIGURE 7.51**). Within the cell, each myofibril is surrounded by a structure called the **sarcoplasmic reticulum,** formed of membranous tubules. In resting muscles, the level of Ca^{2+} in the myofibrils is maintained at about 10^{-7} M, whereas the Ca^{2+} level within the lumen of the sarcoplasmic reticulum may be 10,000-fold higher. Impulses from motor nerves stimulate the opening of Ca^{2+} channels (see Chapter 10), and Ca^{2+} pours out of the sarcoplasmic reticulum into the myofibrils, stimulating muscle contraction. The signal is rapidly transmitted to the entire sarcoplasmic reticulum of a myofiber via **transverse tubules,** invaginations of the plasma membrane that connect at periodic intervals with the reticulum. Following the contraction, a

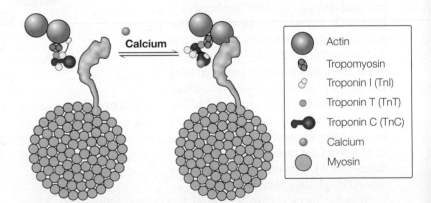

Calcium

Actin
Tropomyosin
Troponin I (TnI)
Troponin T (TnT)
Troponin C (TnC)
Calcium
Myosin

Relaxed muscle. At low Ca^{2+} levels, the configuration of actin, tropomyosin, and the troponin complex in the thin filament blocks most myosin headpieces from contracting the thin filament.

The binding of Ca^{2+} to TnC opens a binding site on TnI for the inhibitory domain of TnI, which also binds actin (see left panel). The release of TnI from actin allows the tropomyosin–troponin complex to slide closer to the central groove in F-actin, thereby exposing the myosin-binding site. Cross-bridge formation (step 3 of Figure 7.48) can then occur, and the muscle contracts.

▲ **FIGURE 7.50** The regulation of muscle contraction by calcium.

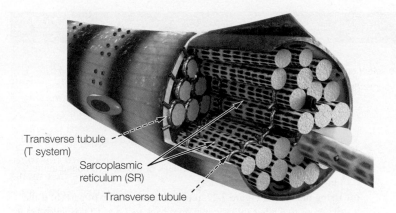

Transverse tubule
(T system)

Sarcoplasmic
reticulum (SR)

Transverse tubule

◀ **FIGURE 7.51 Structure of a myofiber (muscle cell).** The sarcoplasmic reticulum (SR, white; see solid arrows) is a network of specialized endoplasmic reticulum tubules that surrounds the myofibrils within the myofiber. In resting muscle, the SR accumulates Ca^{2+}, which it discharges into the myofibrils when a neural signal reaches the plasma membrane. The transverse tubules (T system; see dashed arrows) are invaginations of the plasma membrane that make contact with the SR at many points, ensuring uniform response to the signal.

Ca^{2+}-specific transport protein pumps Ca^{2+} ions out of the sarcomere to restore the resting $[Ca^{2+}]$ to 10^{-7} M. The activities of such ion transporters are described in Chapter 10.

The functions of actin and myosin are not limited to muscle contraction. Indeed, members of the actin and myosin families are found in most eukaryotic cells, even those that are in no way involved in muscular tissues (see Table 7.3). Actin and myosin play significant roles in cell motility, changes of cell shape, and directed movements of materials in cells. We will not present the details of these other actin–myosin systems here; however, they function via mechanisms that are closely related to the swinging bridge mechanism shown in Figure 7.48.

Summary

- The major theme of this chapter is that protein function is directly related to protein structure. The function of many proteins can be modulated by altering the structure of the protein (e.g., hemoglobin; Section 7.9). Modulation of function can be achieved by the binding of an allosteric effector to a regulatory site on the protein (Section 7.10). The functions of some proteins require large conformational changes, which may be driven by ATP hydrolysis (e.g., myosin; Section 7.18).

- The innate and adaptive immune responses are among the body's main defenses against infection. In the **humoral response** of the adaptive immune system, antibodies (specific **immunoglobulin** molecules) that will bind with specific **antigens** are generated and secreted from B cells (Section 7.2). Antibodies are composed of two identical **light chains** and two identical **heavy chains** (Section 7.3). Each **antibody** displays two identical antigen-binding sites (Sections 7.3 and 7.4). The antigen-binding sites are made up of loops, known as the **complementarity determining regions** (CDRs), displayed by the variable regions of the light and heavy chains (Section 7.4). Immense antibody binding diversity is achieved through amino acid variations in the CDRs (Section 7.4). The exquisite binding specificity of antibodies has been exploited to develop targeted anticancer drugs (Section 7.7).

- Most organisms need oxygen to support cellular respiration. Vertebrates use **hemoglobin** for oxygen transport between lungs/gills and respiring tissues (Section 7.8). In the heme globins, O_2 is bound at an Fe^{2+}-porphyrin (heme); the heme is carried in a hydrophobic pocket, inhibiting oxidation of the iron (Section 7.9). **Myoglobin** carries a single oxygen-binding site and consequently exhibits a noncooperative, hyperbolic-binding curve, whereas hemoglobin binds O_2 cooperatively, with a sigmoidal-binding curve (Sections 7.9 and 7.10). Binding O_2 to hemoglobin sites causes tertiary structure changes. When strain from these changes accumulates, a quaternary $(T \rightarrow R)$ transition occurs, shifting the molecule from the lower-affinity T state to the higher-affinity R state (Section 7.10). The allosteric effectors H^+, CO_2, Cl^-, and **2,3-BPG** stabilize the T state and result in increased oxygen release in respiring tissues (Section 7.11).

- Myoglobin and hemoglobin, like other proteins, are evolving via mutations, duplications, and recombinations in their genes. Both types of globin evolved from a myoglobin-like ancestral protein, with the development of a true hemoglobin coinciding approximately with the emergence of vertebrates (Sections 7.12 and 7.13). Evolution of these proteins continues, as evidenced by the existence of a multitude of variant hemoglobins in the human population. Most nucleotide substitution (missense) mutations are neutral, but some, like the sickle-cell hemoglobin mutation, are deleterious (Section 7.14).

- A number of macromolecular protein systems exist in cells to convert the free energy of ATP hydrolysis into mechanical work. A major example is the actin–myosin contractile system of muscle (Section 7.16). In muscle, interdigitating filaments of actin and myosin are driven past one another by attachment, motion, and detachment of myosin cross-bridges (Sections 7.17 and 7.18). Muscle contraction is stimulated by the influx of calcium ions, which causes rearrangement of actin-associated proteins. The direct source of contractile energy is ATP hydrolysis (Section 7.18).

Problems

Enhanced by
Mastering Chemistry
for Biochemistry

Mastering Chemistry for Biochemistry provides select end-of-chapter problems and feedback-enriched tutorial problems, animations, and interactive figures to deepen your understanding of complex topics while practicing problem solving.

Answers to red problems are available in the Answer Appendix.

1. What physiological effect would you predict from a mutation that replaced with serine the cysteine in the constant part of the immunoglobulin light chain that is involved in disulfide-bond formation with the heavy chain? (See Figure 7.3.)

2. Certain antibodies have been shown to bind only to the folded structure of their target protein ligand; yet others will bind to both the folded and denatured states of the same protein ligand. Explain this observation.

3. Antibodies raised against a macromolecular antigen (e.g., a protein) generally form an antigen-Ab precipitate when mixed with the antigen at roughly equimolar concentrations. However, little to no precipitate forms if the antibody is added in great excess (e.g., 20–fold molar excess) compared to the target antigen. Explain this observation.

4. The following data describe the binding of oxygen to human myoglobin at 37 °C.

P_{O_2} (mm Hg)	Y_{O_2}	P_{O_2} (mm Hg)	Y_{O_2}
0.5	0.161	6	0.697
1	0.277	8	0.754
2	0.434	12	0.821
3	0.535	20	0.885
4	0.605		

From these data, estimate (a) P_{50} and (b) the fraction saturation of myoglobin at 30 mm Hg, the partial pressure of O_2 in venous blood.

5. What qualitative effect would you expect each of the following to have on the P_{50} of hemoglobin?
(a) Increase in pH from 7.2 to 7.4
(b) Increase in P_{CO_2} from 20 to 40 mm Hg
(c) Dissociation into monomer polypeptide chains
(d) Decrease in 2,3-BPG concentration from 7 mM to 5 mM in red cells.

6. Measurements of oxygen binding by whole human blood, at 37 °C, at pH 7.4, and in the presence of 40 mm Hg of CO_2 and normal physiological levels of 2,3-BPG (5 mmol/L of cells), give the following:

P_{O_2} (mm Hg)	% Saturation ($=100 \times Y_{O_2}$)
10.6	10
19.5	30
27.4	50
37.5	70
50.4	85
77.3	96
92.3	98

(a) From these data, construct a binding curve, and estimate the percent oxygen saturation of blood at (1) 100 mm Hg, the approximate partial pressure of O_2 in the lungs, and (2) 30 mm Hg, the approximate partial pressure of O_2 in venous blood.
(b) Under these conditions, what percentage of the oxygen bound in the lungs is delivered to the tissues?
(c) Using the data in Figure 7.27, repeat the calculation of part (b) if the pH drops to 6.8 in capillaries but goes back to 7.4 as CO_2 is unloaded in the lungs.

7. Crocodile hemoglobin does not bind 2,3-BPG. Instead, it binds bicarbonate ion, which is a strong negative allosteric effector. Why might crocodiles have a hemoglobin that is responsive to HCO_3^- instead of 2,3-BPG? Recall that crocodiles hold their prey underwater to kill them.

8. Suggest probable consequences of the following real or possible hemoglobin mutations. [*Note*: Consult Figures 7.23, 7.25.]
(a) At β146 (HC3) His $\rightarrow$ Asp
(b) At β92 (F8) His $\rightarrow$ Leu
In each case, indicate whether a single-nucleotide change is sufficient for the mutation.

9. Suppose each of the mutants listed in Problem 7 was electrophoresed in comparison with native hemoglobin (pI = 7.0) at pH 8.0. Which would move faster toward the anode than native protein, and which would move more slowly?

10. In the experiments of Barrick et al. (see Figure 7.26), it was observed that replacement of histidine by a noncovalently bonded imidazole not only reduced cooperativity but also increased the oxygen affinity of the hemoglobin. Suggest an explanation.

11. Suppose you visit the Dalai Lama in Dharamsala, India (elevation 1460 m), and you begin to ponder the "big questions," such as "What is the fractional saturation of the Dalai Lama's hemoglobin?"
(a) Assuming the Dalai Lama's hemoglobin has a Hill coefficient = 3.2, and a P_{50} = 31 mm Hg, calculate the change in fractional O_2 saturation of his hemoglobin going from his lungs (where P_{O_2} = 85 mm Hg) to his capillaries (where P_{O_2} = 25 mm Hg).
(b) Why do you suppose the Dalai Lama's hemoglobin has a P_{50} higher than normal (where "normal" = 27 mm Hg)?

12. Assume that a new oxygen transport protein has been discovered in certain invertebrate animals. X-ray diffraction of the deoxy protein reveals that it has the dimeric structure shown in panel (a) of the accompanying figure, with a salt bridge between residues histidine 13 and aspartic acid 85. The two monomers interact by salt bridges between the C- and N-termini. The O_2-binding site lies *between* the two iron atoms shown, which are rigidly linked to helices A and C (see panel (b)). In the deoxy form, the space between the iron atoms is too small to hold O_2, and so the Fe atoms must be forced apart when O_2 is bound.

Answer the following questions, explaining your answer in each case in terms of the structure shown below.
(a) Is this molecule likely to show cooperative oxygen binding?
(b) Is this molecule likely to exhibit a Bohr effect?
(c) Predict the likely effect of a mutation that replaced aspartic acid 85 by a lysine residue.

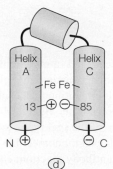

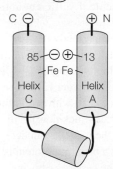

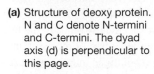

(a) Structure of deoxy protein. N and C denote N-termini and C-termini. The dyad axis (d) is perpendicular to this page.

(b) Detail of the oxygen-binding site in the oxy form.

Oxygen molecule

13. The mutation in hemoglobin at β82 Lys → Asp results in lowered O_2-binding affinity compared to normal hemoglobin. β82 is one of the residues that lines the 2,3-BPG binding site (see Figure 7.29; β82 is adjacent to His 143). Based on the location of this residue and the differences between Lys and Asp, suggest a rationale for the observed reduction in O_2-binding affinity.

14. Suppose your biking partner claims that hyperventilating at the bottom of steep hill climbs is a good idea because it will "increase the O_2 saturation of the blood" and thereby provide "more O_2 for leg muscles during the climb." Hyperventilation reduces the CO_2 content of blood. Given this fact, is hyperventilation likely to increase O_2 delivery to muscles (as your biking partner claims) or not?

15. Hemoglobin *Rainier* (HbR) is a mutant in which Tyr 145 of the β-globin is replaced by Cys. This Cys forms an intramolecular disulfide with β Cys93 in the *oxy* state but cannot form the intramolecular disulfide in the *deoxy* state.
(a) How do you expect O_2-binding affinity of disulfide-bonded HbR to compare to that of normal adult Hb (HbA)?
(b) Do you predict that the Bohr effect for disulfide-bonded HbR is greater than, less than, or the same as the Bohr effect in HbA? In other words, will changes in pH likely have a greater, lesser, or the same effect on O_2 transport by these two Hbs?
(c) Is a person with disulfide-bonded HbR likely to adapt to higher elevation more slowly or more quickly compared to a person with HbA? (i.e., would HbR be an asset or a liability for someone trying to climb tall mountains?)
(d) HbR is reported to have $P_{50} = 14$ torr and a Hill coefficient = 1.2. Calculate ΔY_{O_2} for a climber with HbR assuming that, at 14,000 ft ($\sim$ 4300 m), $P_{O_2} = 48$ mm Hg in lungs and $P_{O_2} = 15$ mm Hg in muscle capillaries. Compare the calculated value of ΔY_{O_2} for a climber with HbR to ΔY_{O_2} for

a climber with HbA under the same conditions of lung and capillary P_{O_2}. (For HbA: $P_{50} = 28$ mm Hg and a Hill coefficient = 3.2.)

16. An antibody has been isolated that binds to F-actin but not to G-actin. What structural feature(s) of F-actin do you suppose the antibody binds (i.e., how is the antibody able to distinguish between these two forms of actin)?

17. A typical relaxed sarcomere is about 2.3 μm in length and contracts to about 2 μm in length. Within the sarcomere, the thin filaments are about 1 μm long and the thick filaments are about 1.5 μm long.
(a) Describe the overlap of thick and thin filaments in the relaxed and contracted sarcomere.
(b) An individual "step" by a myosin head in one cycle pulls the thin filament about 15 nm. How many steps must each actin fiber make in one contraction?

18. Each gram of mammalian skeletal muscle consumes ATP at a rate of about 1×10^{-3} mol/min during contraction. To bridge the short interval between the moderate demand for ATP met by aerobic metabolism and the high demand met by anaerobic ATP production, muscles carry a small reserve of the compound *creatine phosphate* which, due to its high phosphoryl group transfer potential (see Figure 3.7), is capable of phosphorylating ADP very efficiently. The reaction is catalyzed by the enzyme *creatine kinase*:

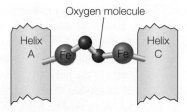

$$^-OOC-CH_2-N-\overset{+}{C}=NH_2 + ADP + H^+ \underset{\text{kinase}}{\overset{\text{Creatine}}{\rightleftharpoons}}$$

with CH_3 on the N and $H-N-PO_3^{2-}$

Creatine phosphate

$$^-OOC-CH_2-N-\overset{+}{C}=NH_2 + ATP \qquad \Delta G^{\circ\prime} = -12.6 \text{ kJ/mol}$$

with CH_3 on N and NH_2 below C

Creatine

Because the equilibrium lies well to the right, virtually all of the muscle ADP or AMP is converted to ATP as long as creatine phosphate is available. Concentrations of ATP and creatine phosphate in muscle are about 4 mM and 25 mM, respectively, and the density of muscle tissue can be taken to be about 1.2 g/cm^3.
(a) How long could contraction continue using ATP alone?
(b) If all creatine phosphate were converted into ATP and utilized as well, how long could contraction continue?
(c) What do these answers tell you about the role of ATP in providing energy to cells?

19. A few hours after the death of an animal, the corpse will stiffen as a result of continued contraction of muscle tissue (this state is called *rigor mortis*). This phenomenon is the result of the loss of ATP production in muscle tissue.
(a) Consult Figure 7.48 and describe, in terms of the six-step model of muscle contraction, how a lack of ATP in sarcomeres would result in *rigor mortis*.
(b) The Ca^{2+} transporter in sarcomeres that keeps the $[Ca^{2+}] \sim 10^{-7}$ M requires ATP to drive transport of Ca^{2+} ions across the membrane of the sarcoplasmic reticulum. How would a loss of this Ca^{2+} transport function result in the initiation of *rigor mortis*?
(c) *Rigor mortis* is maximal at $\sim$12 hrs after death and by 72 hrs is no longer observed. Propose an explanation for the disappearance of *rigor mortis* after 12 hrs.

 References

For a list of references related to this chapter, see Appendix II.

Because of the ease with which antibodies against biological materials can be prepared in the laboratory and because of their great specificity, antibodies form the core of many important analytical and preparative biochemical procedures.

Experiments in which an antigen is injected into an animal show that one antigen can elicit formation of several different antibodies. Each of these antibodies recognizes one particular portion of the antigen molecule, called an *antigenic determinant*, or *epitope*. Figure 7.1(b) shows the epitopes that have been identified in sperm whale myoglobin. Each epitope is a region encompassing

five or six residues in the myoglobin sequence. Thus, a myoglobin **antiserum** (i.e., serum from an animal immunized against myoglobin) carries at least three different antimyoglobin antibodies, each directed against one of the three epitopes shown in the figure. Most antibodies that are useful in biochemistry are of the IgG type (see Table 7.1 and Figure 7.4). Each of these Y-shaped monomers has two antigen-combining sites. In an antigen–antibody reaction, each site usually binds to a different antigen molecule if sufficient antigen is present.

A technique widely used to quantify antigen–antibody reactions is the **enzyme-linked immunosorbent assay (ELISA).** The ELISA can be carried out in 96-well plates for high-throughput applications. We illustrate the widely used **indirect ELISA** in **FIGURE 7A.1**. In ❶, a 96-well plate is coated with a solution containing the sample mixture. Proteins stick nonspecifically to the polystyrene surfaces of the well via van der Waals forces. Next, as shown in ❷, a blocking protein is added to bind to any bare polystyrene surfaces (this prevents nonspecific binding of the antibody to the well). In ❸ a **primary antibody** that specifically binds to the target antigen is added. This is the "immunosorbent" part of the assay in which the detection of a specific antibody–antigen interaction occurs. The detection of this antigen–antibody interaction is achieved using a **secondary antibody** that recognizes the F_c region of the primary antibody (❹ ❺ ❻). If, for example, a murine antibody was used as the primary antibody, an antibody from goats or rabbits raised against murine antibodies could be used as the secondary antibody. The secondary antibody is covalently crosslinked to an enzyme (typically, horseradish peroxidase) whose activity can be easily assayed spectrophotometrically using a substrate that changes color when it is converted to the product of the enzyme-catalyzed reaction—this is the "enzyme-linked" part of the assay. As shown in ❻, the development of color in the wells of the plate indicates the presence of the target antigen. Because the readout is based on the activity of the enzyme, it is *indirect* detection; however, it is also *general* because the secondary antibody will react with *any* murine primary antibody (so the same equipment can be used to run assays for several different target antigens by altering the primary antibody in ❸. Although this method has many variations, the principle is to assay for bound antibody by analyzing for the activity of the conjugated enzyme. This technique forms the basis of many clinical diagnostic tests, such as the most widely

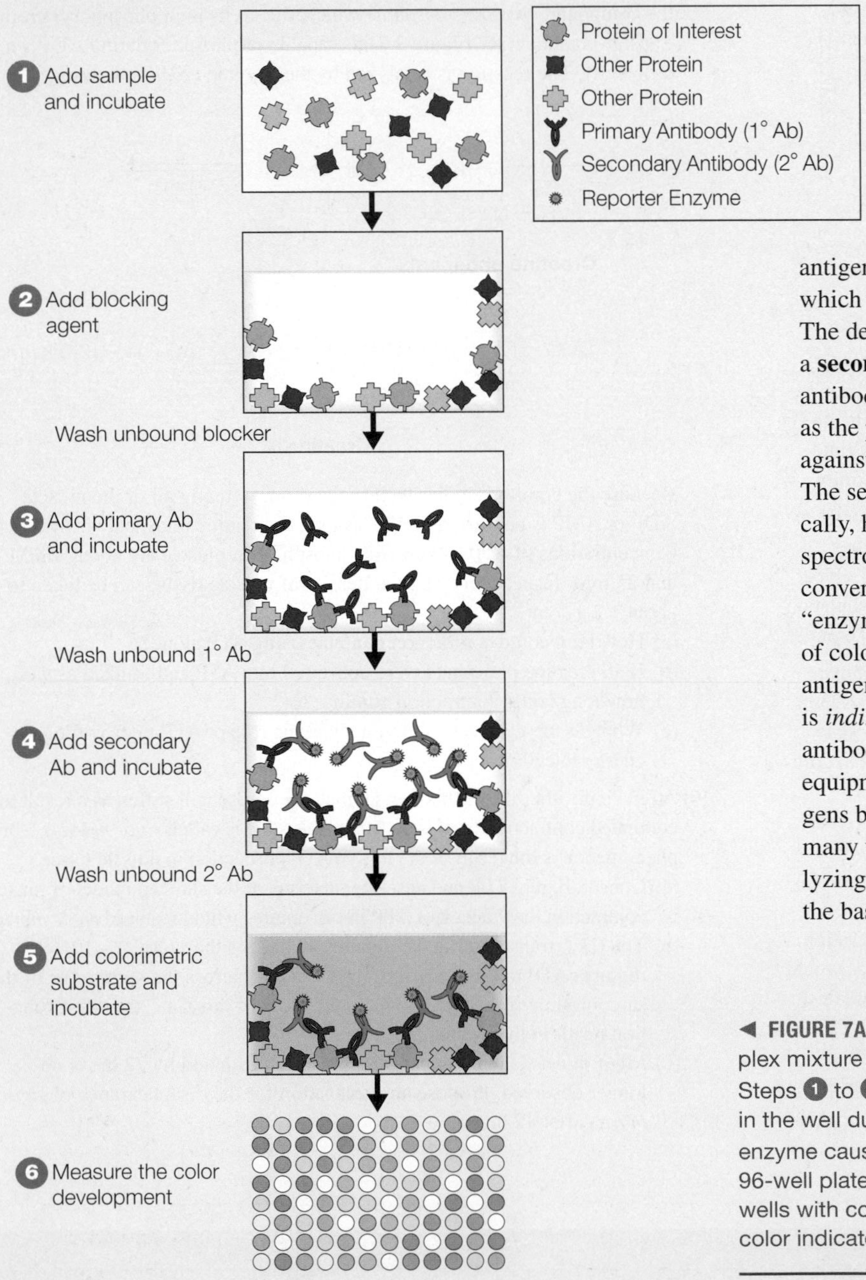

❶ Add sample and incubate

Legend:
- Protein of Interest
- Other Protein
- Other Protein
- Primary Antibody (1° Ab)
- Secondary Antibody (2° Ab)
- Reporter Enzyme

❷ Add blocking agent

Wash unbound blocker

❸ Add primary Ab and incubate

Wash unbound 1° Ab

❹ Add secondary Ab and incubate

Wash unbound 2° Ab

❺ Add colorimetric substrate and incubate

❻ Measure the color development

◄ **FIGURE 7A.1 The indirect ELISA assay.** Detection of one protein in a complex mixture is shown. The assay is typically carried out in a 96-well plate. Steps ❶ to ❺ show a close-up view of a single well. In ❺, color develops in the well due to the action of the enzyme linked to the 2° antibody. The enzyme causes a chromogenic substrate to change color. In ❻, the entire 96-well plate is analyzed (each circle represents one of the 96 wells). Those wells with color are presumed to contain the target antigen, where a darker color indicates a higher concentration of the antigen in the sample well.

used current test for HIV infection. An enzyme-linked antibody to one of the surface proteins of the human immunodeficiency virus is used to detect the presence of the viral antigen in human blood samples. The presence of the enzyme activity on antigen–antibody complexes can be detected with great sensitivity.

Another analytical technique useful to characterize proteins is **western blotting** (also called **immunoblotting**), which is used to detect, in a complex mixture of proteins, those that react with a specific antibody. In this technique, the antibody-reactive proteins in a mixture are analyzed by first resolving the proteins in that mixture by denaturing gel electrophoresis. Often, 2-D gels are used. After electrophoresis, the gel is placed in contact with a sheet of nitrocellulose, and the proteins are transferred (or "blotted") to the nitrocellulose by an electric current. The proteins are bound irreversibly to the nitrocellulose sheet, so the antigen–antibody reactions can be visualized after treatment of the sheet with primary and secondary antibodies, as we have described for indirect ELISA. Alternatively, the target can be detected by autoradiography using a primary (or secondary) antibody labeled with radioactive ^{125}I. An example is shown in **FIGURE 7A.2**.

Biochemists and cell biologists must be concerned with intracellular organization and with the location of enzymes that catalyze reactions of interest. An array of techniques with the generic term **immunocytochemistry** uses antibodies to help localize particular antigens in cytological preparations. In the simplest form, an antibody is conjugated with a fluorescent dye such as fluorescein. A thin section of cell or tissue is then immersed in a solution of the fluorescent antibody. After the excess is washed off, the bound antibody can be visualized by fluorescence microscopy. It is possible to visualize different antigens within a cell simultaneously using multiple dye-linked antibodies. **FIGURE 7A.3** shows a cell stained with three different antibodies—each linked to a different fluorescent dye. Each fluorescent antibody binds to a different macromolecular complex—in this case chromatin, actin filaments, and microtubules. Alternatively, the antibody can be linked to the iron-binding protein ferritin, and the bound iron can be visualized from its high electron density in the electron microscope.

Because of their high specificity in protein binding, antibodies can also be used to purify proteins. In this technique, called immunoaffinity chromatography (see Tools of Biochemistry 5A), the antibody is coupled to a chromatographic support, and a column of this material is used to adsorb selectively the protein being purified. The protein is then desorbed, usually by a pH adjustment in the eluting solution (typically as low as pH 2.5 or as high as pH 11.5) and often in a state close to homogeneity.

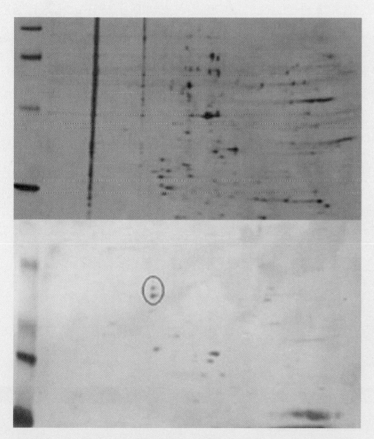

▲ **FIGURE 7A.2 Western blotting.** On the top is a 2-D gel of total protein from tobacco leaf. On the bottom is the same gel, blotted with an antibody against proteins containing phosphothreonine residues.

▲ **FIGURE 7A.3 Immunofluorescent light micrograph of a rat kangaroo kidney epithelial cell during mitotic cell division.** The chromosomes (blue, center) are condensed, after replication. The actin microfilaments (red) and tubulin microtubules (green) of the cytoskeleton maintain the structure of the cell. Antibodies have been used to attach different fluorescent dyes to the chromatin, actin, and tubulin.

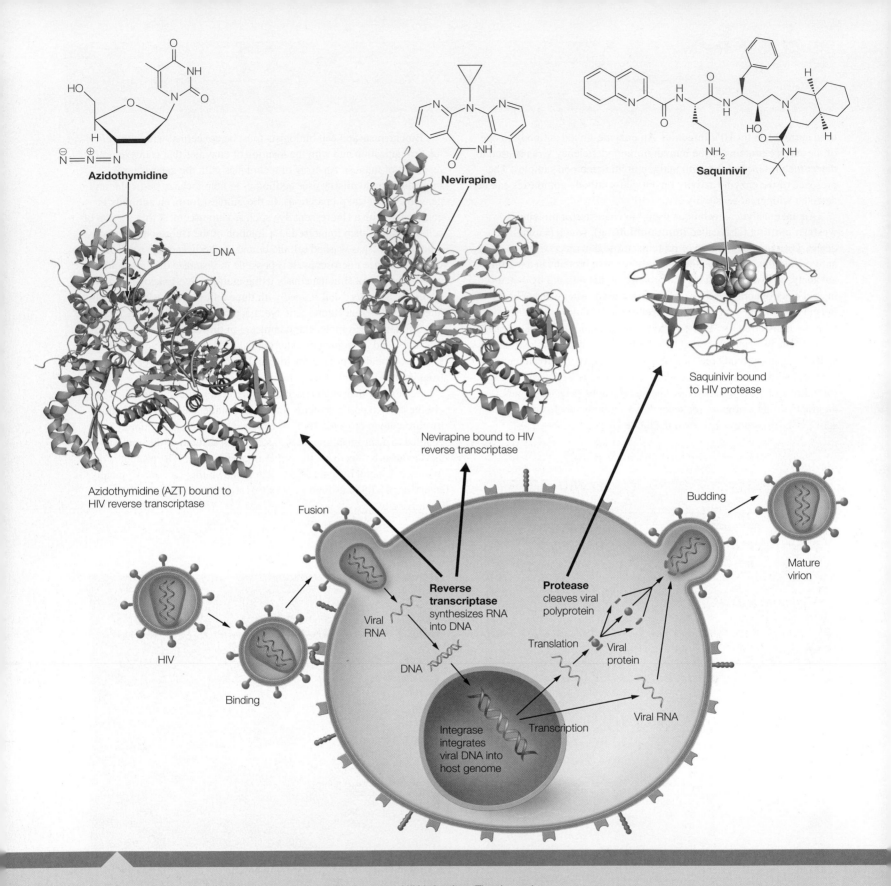

Azidothymidine

Nevirapine

Saquinivir

DNA

Azidothymidine (AZT) bound to
HIV reverse transcriptase

Nevirapine bound to HIV
reverse transcriptase

Saquinivir bound
to HIV protease

Budding

Mature
virion

Fusion

**Reverse
transcriptase**
synthesizes RNA
into DNA

Protease
cleaves viral
polyprotein

Viral
RNA

Translation

Viral
protein

HIV

DNA

Viral RNA

Binding

Transcription

Integrase
integrates
viral DNA into
host genome

Many different therapeutic drugs have been developed to treat HIV infection. The three shown here—azidothymidine (AZT), nevirapine, and saquinivir—all have different modes of action against two enzymes critical to the HIV life cycle. AZT is a substrate analog that causes premature DNA chain termination. Nevirapine distorts the enzyme active site and thereby inhibits reverse transcription. Saquinivir blocks the active site of the HIV protease. How does knowledge of enzyme structure and mechanism help researchers to develop new drugs?

Enzymes: Biological Catalysts

EARLIER WE ALLUDED to the importance of specific catalysts called enzymes in regulating the chemistry of cells and organisms. Catalysis is essential to making most critical biochemical reactions proceed at useful rates under physiological conditions. A reaction that takes many hours to approach completion cannot be metabolically useful to a bacterium that must reproduce in 20 minutes or to a human nerve cell that must respond instantly to a stimulus.

In the complex milieu of the cell, countless thermodynamically favorable reactions are possible. The cell employs catalysts to direct reactive substances into useful pathways rather than into unproductive side reactions. Each reaction we will encounter in subsequent chapters is catalyzed by a specific enzyme that has been optimized by evolution to perform its required task.

Enzymes are large molecules. As such, their reactivity is based on the physical-chemical principles presented in the first few sections of this chapter. We focus next on the biochemical features of enzymes that make them such efficient catalysts. The most important feature is the ability to lower the free energy of the transition

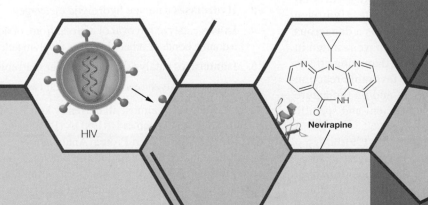

HIV

Nevirapine

state for the reaction being catalyzed. Close inspection of enzyme structures has revealed some common strategies: covalent catalysis, stabilization of the transition state by shape and charge complementarity, and cofactor catalysis.

In normal cells, the activities of enzymes are regulated to modulate the production of different substances in response to cellular and organismal needs. Many human diseases are caused by the loss of enzyme activity or by aberrant regulation of that activity. To appreciate how enzyme activity is related to cellular function, we will introduce analysis of enzyme rates and interpretation of kinetic data. The development of many therapeutic drugs relies on our ability to measure enzyme kinetics and compare how a potential drug affects the activity of the target enzyme. This leads into the last topic of the chapter, which is a more detailed discussion of reversible and irreversible modes of enzyme regulation.

8.1 Enzymes As Biological Catalysts

In general terms, a **catalyst** is a substance that increases the rate, or velocity, of a chemical reaction without itself being changed in the overall process. Most biological catalysts are proteins. We have already encountered a few enzymes in earlier chapters. For example, trypsin catalyzes hydrolysis of peptide bonds in proteins and polypeptides. The substance that is acted on by an enzyme is called the **substrate** of that enzyme. Thus, polypeptides are the natural substrates for trypsin.

We can see the power of enzyme catalysis in a familiar example—the decomposition of hydrogen peroxide (H_2O_2) into water and oxygen:

$$2H_2O_2 \longrightarrow 2H_2O + O_2$$

This reaction, though strongly favored thermodynamically, is very slow unless catalyzed. A bottle of H_2O_2 solution can be kept for many months before it breaks down. If, however, a bit of Fe^{2+} ion (as $FeSO_4$ for example) is added, the reaction rate increases about 1000-fold. The iron-containing protein hemoglobin is even better at increasing the rate of this reaction. When hydrogen peroxide solution is applied to a cut finger, bubbles from released O_2 appear immediately—the reaction is now proceeding about 1 million times faster than the uncatalyzed process. However, even higher rates can be achieved with an enzyme specific for this reaction. Catalase, an enzyme present in many cells, increases the rate of H_2O_2 decomposition about 1 billion-fold over the uncatalyzed rate. Hydrogen peroxide is produced in some cellular reactions and is a dangerous oxidant (see Chapter 14); thus, selective pressures have resulted in the evolution of catalase to defend cells against the damaging effects of H_2O_2. This example shows that the rate of a favorable reaction depends greatly on whether a catalyst is present and on the nature of the catalyst. Enzymes are among the most efficient and specific catalysts known.

Two features of catalysts deserve emphasis. First, although a true catalyst participates in the reaction process, it is unchanged by the process. For example, after catalyzing the decomposition of an H_2O_2 molecule, catalase is found again in exactly the same state as before, ready for another round of reaction. In contrast, although hemoglobin accelerates the rate of H_2O_2 decomposition, it is oxidized in the process, from the active Fe^{2+} to the inactive Fe^{3+} form; thus, hemoglobin is not a true catalyst for this reaction. Second, catalysts change *rates* of processes but do not affect the thermodynamic favorability of a reaction. A thermodynamically favorable process is not made more favorable, nor is an unfavorable process made favorable, by the presence of a catalyst. The equilibrium state is just approached more quickly in the presence of a catalyst.

● **CONCEPT** Catalysts increase the velocity of chemical reactions. Enzymes are biological catalysts.

● **CONCEPT** Catalysts accelerate the approach to equilibrium for a given reaction. Catalysts do not change the thermodynamic favorability of a reaction.

8.2 The Diversity of Enzyme Function

An enormous number of different proteins act as enzymes. Many of these enzymes were given common names, especially during the earlier years of enzymology. Some enzyme names, like *triose phosphate isomerase*, are descriptive of the enzyme's substrate and its function; others, like *trypsin*, are not. To reduce confusion, a rational naming and numbering system has been devised. Enzymes are divided into six major classes, with subgroups and sub-subgroups to define their functions more precisely. The major classes are as follows:

1. **Oxidoreductases** catalyze oxidation–reduction reactions.
2. **Transferases** catalyze transfer of functional groups from one molecule to another.
3. **Hydrolases** catalyze hydrolytic cleavage.
4. **Lyases** catalyze removal of a group from, or addition of a group to, a double bond, or other cleavages involving electron rearrangement.
5. **Isomerases** catalyze intramolecular rearrangement.
6. **Ligases** catalyze reactions in which two molecules are joined.

Information for almost all currently known enzymes can be found in online databases such as BRENDA (BRaunschweig ENzyme DAtabase) or ExPASy (Expert Protein Analysis System). The ~6500 entries in these databases do not include all enzymes; more are being discovered all the time. Indeed, it has been estimated that the typical cell contains many thousands of different kinds of enzymes. **TABLE 8.1** lists one example enzyme and reaction from each of the major classes. We will discuss each of these reactions later in this book.

TABLE 8.1 Examples of each of the major classes of enzymes

Class	Example (reaction type)	Reaction Catalyzed
1. Oxidoreductases	Alcohol dehydrogenase (oxidation with NAD$^+$)	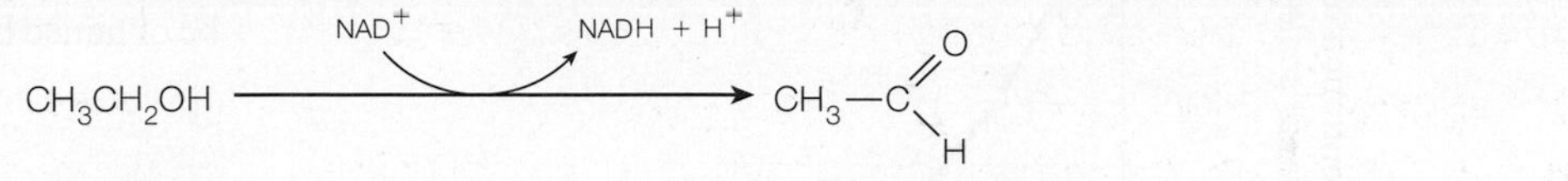
2. Transferases	Hexokinase (phosphorylation)	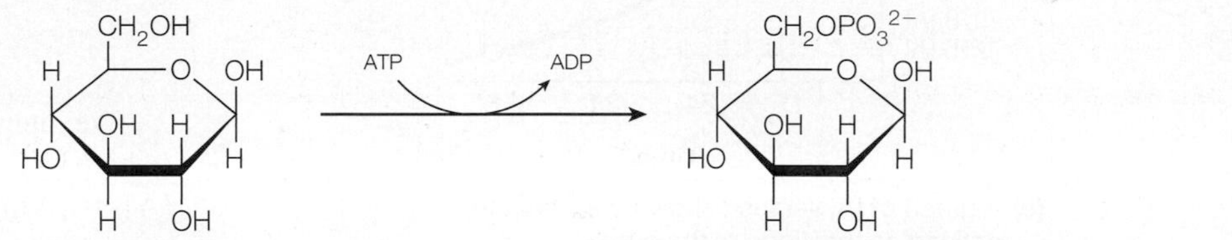
3. Hydrolases	Carboxypeptidase A (peptide bond cleavage)	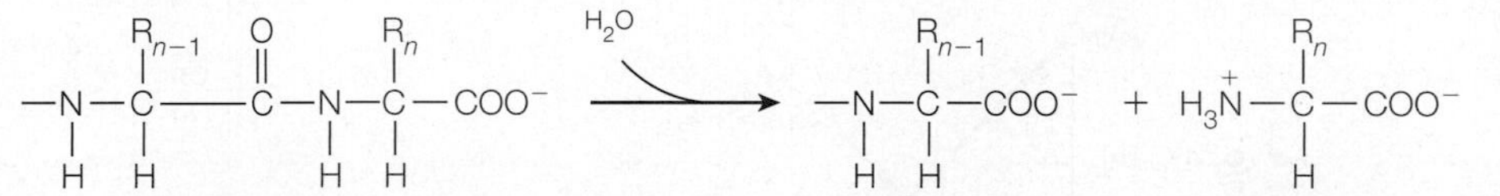
4. Lyases	Pyruvate decarboxylase (decarboxylation)	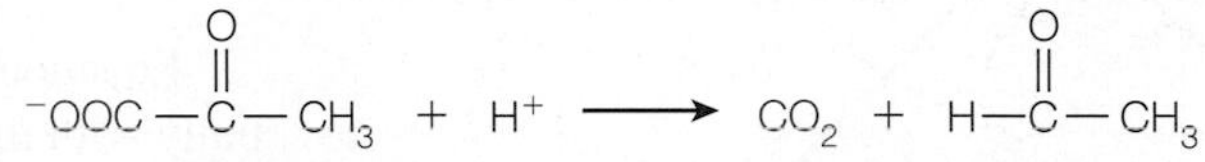
5. Isomerases	Maleate isomerase (*cis–trans* isomerization)	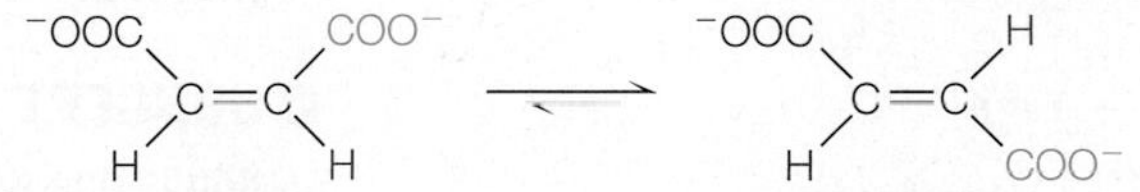
6. Ligases	Pyruvate carboxylase (carboxylation)	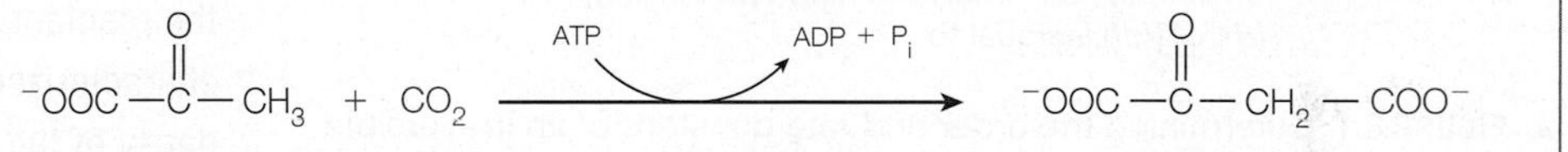

8.3 Chemical Reaction Rates and the Effects of Catalysts

To understand how enzymes achieve their remarkable rate accelerations, it is necessary to understand a few basic principles of catalysis of chemical reactions.

Reaction Rates, Rate Constants, and Reaction Order
First-Order Reactions

To understand what is meant by a reaction rate and how it might be measured, let us first consider the simplest possible reaction, the *irreversible* conversion of substance A to substance B:

$$A \longrightarrow B$$

The single arrow here means that the reverse reaction ($B \longrightarrow A$) is negligible; that is, the equilibrium state lies far to the right.

We can define the **reaction rate,** or **velocity (v),** at any instant as the rate of formation of the product. In this case, the rate is the *increase* in concentration of B with time:

$$v = \frac{d[B]}{dt} \tag{8.1}$$

The units of v are *concentration per unit time* (e.g., molar per second: $M \cdot s^{-1}$, where [B] symbolizes molar concentration of B). If we note that, for every B molecule formed, an A molecule must disappear, it is clear that v can be written equally well as

$$v = -\frac{d[A]}{dt} \tag{8.2}$$

where the negative sign indicates [A] is decreasing with time. Over time, as molecules of A are consumed, the number of molecules left to change is diminished, and the rate decreases as the reaction proceeds

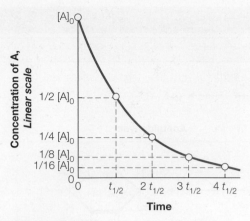

(a) A graph of [A] versus *t* shows that the rate, defined as the slope of the curve, decreases as the reaction continues.

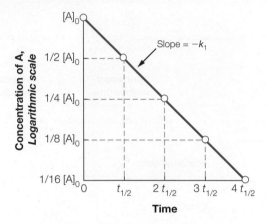

(b) A graph of ln [A] versus *t*, when linear, indicates that the reaction follows Equation 8.7 and is first-order. The slope of this line (*d* ln[A]/*dt*) is equal to −k_1.

▲ **FIGURE 8.1 Determining the order and rate constant of an irreversible first-order reaction.** Graphs **(a)** and **(b)** analyze the rate of a single reaction, with time expressed as multiples of the half-life ($t_{1/2}$) of the reactant. Note that for each interval of $t_{1/2}$ the reactant concentration is halved.

(**FIGURE 8.1(a)**). Mathematically, we state this by saying that the rate is proportional to [A]:

$$v = \frac{d[B]}{dt} = -\frac{d[A]}{dt} = k_1[A]^n \tag{8.3}$$

The factor *n* in Equation 8.3 describes the dependence of the observed rate on the concentration of A. In the case where $n = 1$, the reaction rate depends on the first power of the reactant concentration, and the reaction is called a **first-order reaction.** The constant k_1 is called the **rate constant** and for a first-order reaction has units of 1/(time), typically, s^{-1} or min^{-1}. The rate constant provides a direct measure of how fast this reaction is. The larger the value of k_1, the more rapid the rate and vice versa. The most common example of a first-order reaction is the decay of radioactive elements (see Tools of Biochemistry 11B).

Initially, the value of *n* is not known. It may be 1 (first order), or 2 (second order), or 3 (third order), and so on. The order of a reaction is determined experimentally by comparing the kinetic data (i.e., a plot of [A]

vs. time) to mathematical models, which describe the predicted change in [A] for each type of reaction. We illustrate this idea for a first-order reaction.

Let us begin with an equation that describes how the concentration of A changes with time during a first-order reaction. This description can be obtained by rearranging, then integrating, Equation 8.3, where $n = 1$:

$$\frac{d[A]}{[A]} = -k_1 dt \tag{8.4}$$

$$\int_{[A]_0}^{[A]_t} \frac{d[A]}{[A]} = -k_1 \int_0^t dt \tag{8.5}$$

We integrate both sides of Equation 8.5 over limits from time $= 0$ (where [A] equals the initial concentration "$[A]_0$") to time $= t$ (where $[A] = [A]_t$), yielding:

$$\ln\frac{[A]_t}{[A]_0} = -k_1 t \tag{8.6}$$

or

$$\ln[A]_t = \ln[A]_0 - k_1 t \tag{8.7}$$

Or raising each side of the equation to be a power of *e*

$$[A]_t = [A]_0 e^{-k_1 t} \tag{8.8}$$

Equation 8.8 tells us that the concentration of A decreases exponentially with time, as shown in Figure 8.1(a). A characteristic of such an exponential decay is the **half-life ($t_{1/2}$),** which is the time needed for [A] to decrease by one-half. For a first-order reaction, the half-life is inversely proportional to k_1 (see Problem 1 at the end of this chapter). To test whether a reaction is first-order, we need only make a graph of ln [A] versus *t,* as shown in Figure 8.1(b). A straight line with a slope of −k_1 and an intercept of $[A]_0$, as predicted by Equation 8.7, is consistent with a first-order reaction.

● **CONCEPT** A first-order reaction is one whose rate is directly proportional to the first power of the reactant concentration and is characterized by single exponential decay of the reactant.

Most biochemical processes cannot be described over their full course by equations as simple as Equation 8.6. One reason is that many of the reactions and processes we encounter are *reversible,* and as product accumulates, the reverse reaction becomes important. For example, we may have a reaction like the following:

$$A \underset{k_{-1}}{\overset{k_1}{\rightleftharpoons}} B$$

Because A is being consumed in the reaction to the right and formed by the reaction to the left, the corresponding rate equation is

$$v = -\frac{d[A]}{dt} = k_1[A]^n - k_{-1}[B]^m = \frac{d[B]}{dt} \tag{8.9}$$

In the case where *n* and *m* both equal 1, k_1 and k_{-1} are, respectively, the rate constants for the first-order forward and reverse reactions. Such a reaction approaches a state of equilibrium, at which point the rates of the forward and reverse reactions become equal, and so the observed rate becomes zero (i.e., there is no apparent change in [A] or [B] over time). Thus, at equilibrium

$$k_1[A] = k_{-1}[B] \tag{8.10}$$

Note that the equilibrium constant, *K,* for a reversible reaction can be written as the ratio of the forward and reverse rate constants, which is evident when we rearrange Equation 8.10:

$$K = \frac{[B]}{[A]} = \frac{k_1}{k_{-1}} \qquad (8.11)$$

Second-Order Reactions

The reactions we have described so far are first-order—they involve changes happening in individual molecules; but many biochemical reactions are more complex, involving encounters between molecules. A **second-order reaction** occurs typically when two molecules must come together to form products:

$$A + B \longrightarrow C$$

This is illustrated by the binding of oxygen to myoglobin:

$$Mb + O_2 \xrightarrow{k_1} MbO_2$$

The rate for this process is given by

$$v = -\frac{d[Mb]}{dt} = -\frac{d[O_2]}{dt} = k_1[Mb]^n[O_2]^m \qquad (8.12)$$

where *n* and *m* both equal 1 and k_1 is the **second-order rate constant,** with dimensions of $M^{-1}s^{-1}$. Equation 8.12 predicts that the rate of formation of the product, MbO_2, is dependent on both the concentration of free Mb *and* the concentration of free O_2. In this case, we say the reaction is first-order in [Mb] ($n = 1$) and first-order in [O_2] ($m = 1$) and second-order overall ($n + m = 2$). The binding of substrate (S) to an enzyme to form an **enzyme-substrate complex** is formally a second-order process:

$$Enz + S \longrightarrow [Enz \cdot S] \longrightarrow Enz + Product$$

Later in this chapter, we will explore this process further when we consider simple kinetic models for enzyme-catalyzed reactions.

Enzyme-catalyzed reactions, when analyzed in detail, are generally more complicated than those described earlier. Many include complex, multistep processes. Often, however, the analysis of complex multistep reaction schemes can be simplified by the recognition of a **rate-limiting step.** The rate-limiting step is the slowest step in a multistep process. As such, it determines the experimentally observed rate for the entire process.

● **CONCEPT** A first-order rate constant has units of $(time)^{-1}$, whereas a second-order rate constant has units of $(concentration)^{-1}(time)^{-1}$.

Transition States and Reaction Rates

The observed rate of a chemical reaction depends on one or more of the following: (1) the order of the reaction, (2) the concentrations of the reactants and products, (3) temperature, and (4) the value of the rate constant for the reaction. We next consider the factors that affect the value of the rate constant. The thermodynamic concepts we presented in Chapter 3 allow us to determine whether or not a reaction is favorable. However, such information, on its own, does not explain reaction rates. A free energy diagram for a favorable reaction, drawn on the basis of thermodynamics alone, will look like **FIGURE 8.2(a).** It shows the free energy of the system versus the **reaction coordinate,** a generalized measure of the progress of the reaction through intermediate states. For a favorable reaction, the free energy of the products is lower than that

of the reactants. However, what may be most important in determining the reaction rate is what happens in the *transition* from reactants to product. Equilibrium measurements, which pertain to final and initial states, do not reveal any information about the transition between these states or the energetic barrier(s) between them.

A molecule in a first-order reaction must only *occasionally* reach an energy state in which the process can occur; otherwise, all molecules

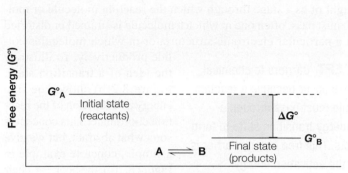

(a) Information provided by thermodynamic studies of the equilibrium: Only the free energy difference between the initial state and the final state is revealed. $G°_A$ and $G°_B$ represent the standard free energies per mole, respectively, of A and B molecules. $\Delta G°$ is the standard state free energy change for the reaction.

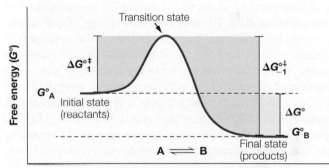

(b) Free energy diagram filled in to include the transition state through which the molecule must pass to go from A to B or vice versa. $\Delta G°^{\ddagger}_1$ is the energy of activation for the A → B transition, and $\Delta G°^{\ddagger}_{-1}$ is for the B → A transition.

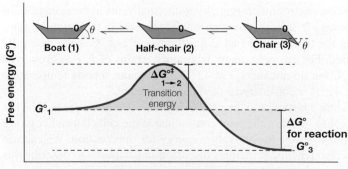

(c) A reasonable path for the transition of a pyranose (such as glucose) from boat (1) to chair (3) conformation. The highest energy state—the transition state—will look something like (2).

▲ **FIGURE 8.2** Free energy diagrams for the simple reaction A → B.

would already have reacted. This observation suggests that only a fraction of the molecules—those that are sufficiently energetic—can undergo reaction. Similarly, in a second-order reaction, not all encounters between reactants can be productive because some collisions may not be sufficiently energetic, or the colliding molecules may not be properly oriented with respect to each other. Such considerations have given rise to the idea of a **free energy barrier** to reaction and the concept of a **transition state** (symbolized by $\ddagger$). The transition state is thought of as a stage through which the reacting molecule or molecules must pass, often one in which a molecule is strained or distorted or has a particular electronic structure, or in which molecules collide productively. To illustrate

● **CONCEPT** Barriers to chemical reactions occur because a reactant molecule must pass through a high-energy transition state to form products. This free energy barrier is called the activation energy.

the idea of a transition state, Figure 8.2(b) shows the free energy as a function of the reaction coordinate. This concept is somewhat abstract, but we give a simple, concrete example in Figure 8.2(c), the boat → chair conversion of a pyranose ring, such as glucose. Both the initial state (boat) and the final state (chair) are of lower energy than the most strained, flattened state (the half-chair state). To make the conversion, the ring must go through the half-chair state, which represents the high-energy transition state in this example.

We now consider two strategies for increasing reaction rates: (1) raising the temperature of the reaction, and (2) lowering the free energy of the transition state. The **standard free energy of activation, $\Delta G^{\circ\ddagger}$,** represents the additional free energy (above the average free energy of reactant molecules) that molecules must have to attain the transition state. If the activation barrier to the reaction is high, only a small fraction of the molecules will have enough energy to surmount it, or only a small fraction of collisions will be energetic enough for the reaction to occur. We know that in any sample or solution, not all molecules have the same energy at any instant. As shown in **FIGURE 8.3(a)**, at a given temperature some molecules will have lower kinetic energies and others will have much higher kinetic energies. As temperature increases, the average kinetic energy of the molecules in a sample increases. Thus, as the temperature of a sample increases, we expect more molecules to have sufficient energy to attain the transition state (compare the shaded portions of Figure 8.3(a)), and the rate of the reaction should increase. Except for the organisms that have adapted to extreme environments (e.g., hydrothermal vents in the ocean), most organisms are sensitive to even small increases in temperature; thus, raising the temperature inside a cell beyond 2–3 °C is not generally tolerated. For example, many laboratory strains of *E. coli* grow well at 37 °C but do not survive at 42 °C. In humans, a body temperature above 41 °C is considered a medical emergency.

Nature's response to the need to increase reaction rates in cells has been not to increase the temperature inside the cells but rather to use enzymes to lower the activation energy for the reaction. We can use the thermodynamic principles discussed in Chapter 3 to describe this effect in qualitative terms.

If we consider the simple reaction A ⇌ B, and assume that molecules of the reactant A are in equilibrium between the initial state and the activated state, the concentration of activated molecules at any instant will be given by Equation 3.22 ($K = e^{-\Delta G^{\circ}/RT}$). It is important to recognize that a true thermal equilibrium is not possible between the reactant state and something as fleeting and dynamic as the transition

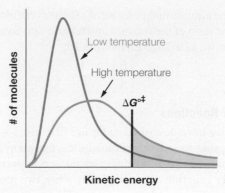

(a) At higher temperature, more molecules have this energy (compare the shaded area under the light blue curve versus the shaded area under the orange curve).

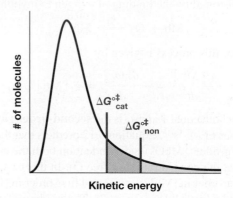

(b) Lowering the value of $\Delta G^{\circ\ddagger}$ also increases the number of molecules with sufficient energy to attain the transition state (compare shaded areas under the curve for each value of $\Delta G^{\circ\ddagger}$). The subscripts "cat" and "non" indicate, respectively, the catalyzed and noncatalyzed processes.

▲ **FIGURE 8.3 Effect of increasing temperature or lowering $\Delta G^{\circ\ddagger}$ on the rates of reactions. The curves show Maxwell–Boltzmann distributions of kinetic energies for systems of molecules at lower and higher temperatures.** For a given system, the rates of reactions are proportional to the number of molecules possessing sufficient energy to overcome the activation barrier $\Delta G^{\circ\ddagger}$.

state. Instead, we will assume that $A^{\ddagger}$ represents an average structure that is a reasonable approximation of the transition state. With that caveat in mind, if we let $[A^{\ddagger}]$ represent the fraction of molecules with sufficient energy to attain the transition state and $[A]$ the fraction of A that remains in the reactant state, we can use Equation 3.22 to describe the ratio $[A^{\ddagger}]/[A]$ as a function of the activation energy, $\Delta G^{\circ\ddagger}$:

$$\frac{[A^{\ddagger}]}{[A]} = e^{\left(\frac{\Delta G^{\circ\ddagger}}{RT}\right)} \tag{8.13}$$

where T is temperature and R is the gas constant. Equation 8.13 and Figure 8.3(b) provide us with a framework to understand the most important feature of enzyme catalysis: Enzymes achieve faster rates by lowering the activation energy for a reaction. As $\Delta G^{\circ\ddagger}$ decreases, a larger fraction of molecules will possess sufficient energy to attain the transition state, and the rate of reaction will increase (compare shaded portions of Figure 8.3(b)).

In 1921, Michael Polanyi proposed that a reaction catalyst preferentially binds the transition state structure and thereby stabilizes it, relative to the ground state, leading to a reduction in activation energy. Twenty-five years later, Linus Pauling extended this idea to biological catalysts,

● **CONCEPT** Catalysts increase reaction rates by lowering the activation energy.

suggesting that specific, complementary binding interactions between the transition-state structure and the enzyme active site would account for the extraordinary rate enhancements achieved by enzymes. Indeed, **transition state theory** has proved to be widely applicable to the study of enzyme catalysts, both in the laboratory and in computer simulations.

Transition State Theory Applied to Enzymatic Catalysis

In its simplest formulation, transition state theory assumes that a reactant molecule that attains the transition state rapidly decomposes to a lower-energy state, such as the product state, or to an intermediate state (see below). Because bonds are in the process of breaking and/or forming in the transition state, the lifetime of the transition state is similar to the vibrational frequencies of covalent bonds—on the order of a picosecond (10^{-12} s). These concepts are summarized in the following general expression for the rate constant k:

$$k = Ae^{\left(\frac{\Delta G^{\circ\ddagger}}{RT}\right)} \tag{8.14}$$

where T is temperature, R is the gas constant, and A is a factor that accounts for the vibrational frequency of bonds in the transition state. At 310 K (37 °C), A has a value of $\sim 6.4 \times 10^{12}$ s^{-1}.

The **rate enhancement** is the ratio of the rate constants for the catalyzed (k_{cat}) and the noncatalyzed (k_{non}) reactions for a given set of conditions (e.g., temperature, pH). The rate enhancement indicates how much faster the reaction occurs in the presence of the enzyme. For example, we can compare the rate of amide bond hydrolysis by the enzyme *carboxypeptidase A* versus the noncatalyzed rate at pH = 8 and 23 °C:

$$\text{rate enhancement} = \frac{k_{cat}}{k_{non}} = \frac{238 \text{ s}^{-1}}{1.8 \times 10^{-11} \text{s}^{-1}} = 1.3 \times 10^{13} \tag{8.15}$$

How significant is this rate enhancement? The noncatalyzed peptide-bond hydrolysis has a half-life of ~ 2500 years, whereas the enzyme-catalyzed reaction has a half-life of ~ 0.005 second! Without the catalyst, this reaction would not occur on a physiologically useful time scale. **FIGURE 8.4** illustrates the enormous rate enhancements that are characteristic of enzyme-catalyzed reactions.

● **CONCEPT** The rate enhancement for an enzyme-catalyzed reaction is the ratio of the rate constants for the catalyzed (k_{cat}) and the noncatalyzed (k_{non}) reactions. The rate enhancement indicates how much faster the reaction occurs in the presence of the enzyme.

We can combine Equations 8.14 and 8.15 to evaluate by how much an enzyme must stabilize the transition state to achieve these observed rate enhancements:

$$\text{rate enhancement} = \frac{k_{cat}}{k_{non}} = \frac{A_{cat}e^{\left(\frac{-\Delta G^{\circ\ddagger}_{cat}}{RT}\right)}}{A_{non}e^{\left(\frac{-\Delta G^{\circ\ddagger}_{non}}{RT}\right)}}$$

$$= \left(\frac{A_{cat}}{A_{non}}\right)e^{\left(\frac{\Delta\Delta G^{\circ\ddagger}}{RT}\right)} \tag{8.16}$$

where $\Delta\Delta G^{\circ\ddagger} = (\Delta G^{\circ\ddagger}_{non} - \Delta G^{\circ\ddagger}_{cat})$, or the difference in activation energies between the noncatalyzed and catalyzed reactions (**FIGURE 8.5**). $\Delta\Delta G^{\circ\ddagger}$ indicates by how many kJ/mol the transition state

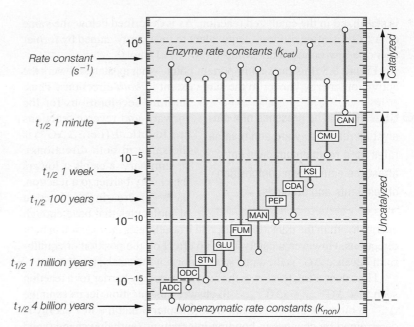

▲ **FIGURE 8.4 Enzymatic rate enhancements.** Logarithmic scale of k_{cat} and k_{non} values (white circles) for some representative enzyme-catalyzed reactions at 25 °C. The length of each vertical bar represents the rate enhancement achieved by the enzyme. ADC = arginine decarboxylase; ODC = orotidine 5′-phosphate decarboxylase; STN = staphylococcal nuclease; GLU = sweet potato α-amylase; FUM = fumarase; MAN = mandelate racemase; PEP = carboxypeptidase B; CDA = *E. coli* cytidine deaminase; KSI = ketosteroid isomerase; CMU = chorismate mutase; CAN = carbonic anhydrase. Data from Accounts of Chemical Research 34:938–945, R. Wolfenden and M. J. Snyder, The depth of chemical time and the power of enzymes as catalysts. © 2001 American Chemical Society.

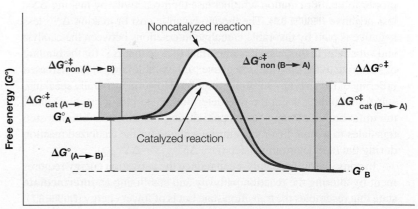

▲ **FIGURE 8.5 Effect of a catalyst on activation energy.** A free energy diagram (blue curve) similar to that in Figure 8.2(b) is shown, along with an alternative catalyzed path (red curve) for the reaction. The catalyst lowers the standard free energy of activation, $\Delta G^{\circ\ddagger}$, and thereby accelerates the rate because more of the reactant molecules have the energy needed to reach this lowered transition state. The rate enhancement is related to $\Delta\Delta G^{\circ\ddagger}$. Note that the values of $\Delta G^{\circ}_{A \to B}$ for both the catalyzed and noncatalyzed reactions are the same; thus, the reaction equilibrium is not perturbed by the presence of the catalyst.

is stabilized in the catalyzed reaction. As is described below, the value of $\Delta\Delta G^{\circ\ddagger}$ is often modest—equivalent to the energy gained by formation of a few noncovalent bonds ($\sim$ 30–90 kJ/mol).

Figure 8.5 illustrates an important point—that a catalyst lowers the activation energy barrier to the same extent in *both* directions. Thus,

● **CONCEPT** The presence of a catalyst increases forward and reverse rates for a reaction but does not affect the equilibrium composition of reactants and products.

the rate accelerations for the forward and reverse reactions are identical (i.e., $\Delta\Delta G^{\circ\ddagger}$ is the same in both directions). In summary, a catalyst lowers the energy barrier to a reaction, thereby increasing the fraction of molecules that have enough

energy to attain the transition state, and it accelerates the reaction in both directions. However, a catalyst has no effect on the position of equilibrium because ΔG° is the same whether or not the catalyst is present.

How does a catalyst lower the activation energy barrier for a reaction such that $\Delta G^{\circ\ddagger}_{cat} < \Delta G^{\circ\ddagger}_{non}$? To answer this question, let us examine $\Delta G^{\circ\ddagger}$ a bit more closely. Recall from Chapter 3 that the free energy is determined by changes in bonding interactions (enthalpy change) and changes in the ordering of particles, or the energetic microstates within the system (entropy change). We can describe the activation energy in similar terms: $\Delta G^{\circ\ddagger} = \Delta H^{\circ\ddagger} - T\Delta S^{\circ\ddagger}$. Thus, we can imagine two possibilities for making $\Delta G^{\circ\ddagger}_{cat} < \Delta G^{\circ\ddagger}_{non}$: either make $\Delta H^{\circ\ddagger}$ more negative (such that $\Delta H^{\circ\ddagger}_{cat} < \Delta H^{\circ\ddagger}_{non}$) or increase $\Delta S^{\circ\ddagger}$ (such that $\Delta S^{\circ\ddagger}_{cat} > \Delta S^{\circ\ddagger}_{non}$). $\Delta H^{\circ\ddagger}$ can be made more negative (i.e., favorable) by increasing the number of bonding interactions between the catalyst and the transition state. As described below, this appears to be the dominant effect in most enzyme-catalyzed reactions. The $\Delta S^{\circ\ddagger}$ term reflects the fact that a particular *orientation* between reactants or parts of a molecule may be necessary to achieve the transition state. For example, when collisions between molecules occur (as in a second-order reaction), most encounters are unproductive just because the molecules happen to be pointed the wrong way when they hit. A catalyst that can bind two reacting molecules in proper mutual orientation will increase their reactivity by making $\Delta S^{\circ\ddagger}$ less negative (**FIGURE 8.6**). The thermodynamic cost of making $\Delta S^{\circ\ddagger}$ less negative is paid by favorable enthalpic interactions between the catalyst and substrate. In this case, it is important to note that ΔS° for the binding of substrate to the enzyme is negative (unfavorable) due to the increased ordering of the enzyme–substrate complex. However, generally speaking, substrate binding is not the rate-determining step for a reaction. Thus, the unfavorable entropy "cost" paid during the substrate binding step translates to a more favorable entropy change for the catalyzed reaction during the rate-determining step (i.e., $\Delta S^{\circ\ddagger}_{cat} > \Delta S^{\circ\ddagger}_{non}$).

In some cases, the catalyst can reduce the activation energy requirement by altering the reaction pathway and stabilizing an **intermediate** state that resembles the transition state but is of lower energy (**FIGURE 8.7**). The result is that two lower-activation energy barriers replace the single higher barrier. We distinguish such an intermediate state from a transition state by the fact that the former corresponds to a local free energy minimum and the latter to a free energy maximum.

8.4 How Enzymes Act as Catalysts: Principles and Examples

We have seen that the role of a catalyst is to decrease $\Delta G^{\circ\ddagger}$ by facilitating the formation of the transition state. In this section, we describe some common strategies used by enzymes to achieve a reduction in

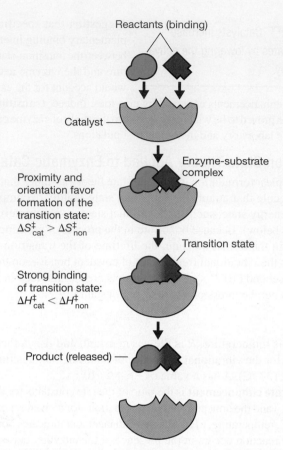

Reactants (binding)

Catalyst

Enzyme-substrate complex

Proximity and orientation favor formation of the transition state: $\Delta S^{\ddagger}_{cat} > \Delta S^{\ddagger}_{non}$

Transition state

Strong binding of transition state: $\Delta H^{\ddagger}_{cat} < \Delta H^{\ddagger}_{non}$

Product (released)

▲ **FIGURE 8.6 Entropic and enthalpic factors in catalysis.** In this example, two reactants are bound to sites on the catalyst, which ensures their correct mutual orientation and proximity, and binds them most strongly when they are in the transition state conformation. Note that the entropy change to form the enzyme–substrate complex, ΔS°, is unfavorable; but the entropy change for the rate-determining formation of the transition state, $\Delta S^{\circ\ddagger}$, is more favorable for a catalyzed process compared to the noncatalyzed process.

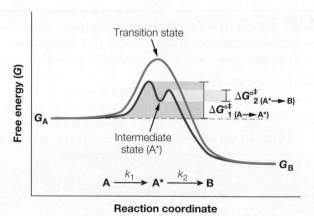

▲ **FIGURE 8.7 Importance of intermediate states.** An enzyme may alter the reaction pathway to one that includes one or more intermediate states that resemble the transition state but have a lower free energy (red curve). In the case of a single intermediate, the activation energies for formation of the intermediate state and for conversion of the intermediate to product ($\Delta G^{\circ\ddagger}_{1}$ and $\Delta G^{\circ\ddagger}_{2}$, respectively) are lower than the activation energy for the uncatalyzed reaction (blue curve). In this figure, only activation energies for the enzyme-catalyzed forward reaction (A → B) are shown.

$\Delta G^{\circ\ddagger}$, and then illustrate those strategies with two examples: lysozyme and chymotrypsin.

Models for Substrate Binding and Catalysis

An enzyme binds a molecule of substrate (or in many cases, multiple substrates) into a region of the enzyme called the **active site,** as shown schematically in **FIGURE 8.8**. The active site is often a pocket or cleft surrounded by amino acid residues that help bind the substrate, and other residues that play a role in catalysis. The extraordinary specificity of enzyme catalysis is due in part to the complex tertiary structure of an enzyme, which enables the active site to recognize the substrate through *complementary* binding interactions (like those we described for the binding between an antibody and its cognate antigen in Chapter 7). This possibility was realized as early as 1894 by Emil Fischer, who proposed a *lock-and-key hypothesis* for enzyme action. According to this model, the enzyme accommodates the specific substrate as a lock fits its specific key (Figure 8.8(a)).

Although the lock-and-key model explained enzyme specificity, it did not increase our understanding of the catalysis itself because a lock does nothing to its key. This understanding came from an elaboration of Fischer's idea: What fits the enzyme active site best is a substrate molecule induced to take up a configuration approximating the transition state. In other words, the enzyme does not simply accept the substrate—the enzyme also demands that the substrate be distorted into something close to the transition state. This **induced fit hypothesis,** proposed by Daniel Koshland in 1958, is still an important model for enzymatic catalysis; however, it has been extended to include the idea that the substrate can also induce conformational changes in the enzyme that lead to stabilization of the transition state.

The structural changes induced upon substrate binding may be local distortions or may involve major changes in enzyme conformation (Figure 8.8(b)). A conformational change of this kind can be seen when substrate binds the enzyme *hexokinase,* which catalyzes the phosphorylation of glucose to glucose-6-phosphate in the first step in the metabolic pathway called glycolysis (see Chapter 12). The structure of this enzyme has been determined by X-ray diffraction in both the presence and absence of bound glucose. As **FIGURE 8.9** shows, the binding of glucose causes two domains of the enzyme to fold toward each other, closing the binding-site cleft about the substrate.

● **CONCEPT** An enzyme active site is complementary in shape, charge, and polarity to the transition state for the reaction and, to a lesser extent, its substrate. Such complementarity is the basis for the specificity of enzyme-catalyzed reactions.

Mechanisms for Achieving Rate Acceleration

Enzymes do more than simply distort or position their substrates. Often, we find specific amino acid side chains poised in exactly the right places to aid in the catalytic process itself. In many cases, these side chains are acidic or basic groups that can promote the addition or removal of protons. In other instances, the enzyme holds a metal ion in exactly the right position to participate in catalysis. Thus, an enzyme (1) binds the substrate or substrates, (2) lowers the energy of the transition state, and (3) directly promotes the catalytic event. When the catalytic process has been completed, the enzyme must be able to release the product or products and return to its

● **CONCEPT** An enzyme binds substrate(s), preferentially stabilizes the transition state, and releases product(s)

(a) Lock-and-key model
In this early model, the active site of the enzyme fits the substrate as a lock does a key.

(b) Induced fit model
In this elaboration of the lock-and-key model, both enzyme and substrate are distorted on binding. The substrate is forced into a conformation approximating the transition state; the enzyme keeps the substrate under strain.

Distortion to transition state conformation

▲ **FIGURE 8.8 Two models for enzyme–substrate interaction.** In this example, the enzyme catalyzes a cleavage reaction.

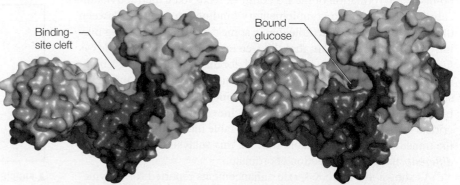

(a) Before glucose binding

(b) After glucose binding

Binding-site cleft

Bound glucose

▲ **FIGURE 8.9 The induced conformational change in hexokinase.** The binding of glucose to hexokinase induces a significant conformational change in the enzyme. The enzyme is a single polypeptide chain, with two major domains. A surface representation is shown with rainbow coloring (N-terminus is blue; C-terminus is red). Notice how the obvious cleft between the domains (panel a) closes around the glucose molecule (magenta spheres in panel b). PDB IDs: panel **(a)** 2yhx; panel **(b)** 3b8a.

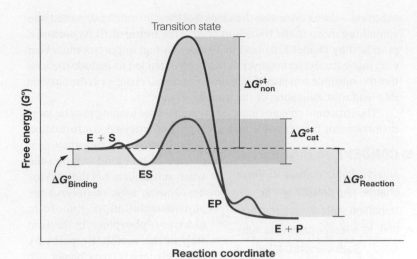

▲ **FIGURE 8.10 Reaction coordinate diagram for a simple enzyme-catalyzed reaction.** The blue curve shows the free energy profile for the noncatalyzed reaction; the red curve shows one of several possible free energy profiles for a simple enzyme-catalyzed reaction that follows the scheme shown in Equation 8.17 (i.e., the free energy of the ES complex could lie above that of free enzyme and substrate).

---

original state, ready for another round of catalysis. For an enzyme (E) that catalyzes the conversion of a single substrate (S) into a single product (P), the expression for the reaction includes three steps:

$$E + S \underset{k_{-1}}{\overset{k_1}{\rightleftharpoons}} ES \underset{k_{-2}}{\overset{k_2}{\rightleftharpoons}} EP \underset{k_{-3}}{\overset{k_3}{\rightleftharpoons}} E + P \qquad (8.17)$$

Here ES represents the **enzyme–substrate complex,** and EP represents the enzyme bound to the product. For many enzyme-catalyzed reactions the first step, binding of substrate, is reversible (i.e., $k_{-1} \gg k_2$); the second step, conversion of ES to EP, lies far to the right (i.e., $k_2 \gg k_{-2}$); and the third step, release of product, is rapid compared to the catalytic step (i.e., $k_3 \gg k_2$).

A reaction coordinate diagram based on Equation 8.17 is shown by the red curve in **FIGURE 8.10**. This figure illustrates several important points. First, formation of the ES complex tends to be thermodynamically favorable, due to complementary binding interactions between the enzyme and substrate (note that in some cases, E + S binding may introduce sufficient strain that the free energy of the ES complex is *higher* than that of free E + S). Second, for maximum efficiency, binding to product should be less favorable than its release from the active site. Third, the enzyme active site must bind the transition state more favorably than it binds the substrate to achieve $\Delta G^{\circ\ddagger}_{cat} < \Delta G^{\circ\ddagger}_{non}$. It is critical to distinguish the ES complex (a stable intermediate state) from the transition state (an unstable state) in this analysis—they represent *different* states on the reaction coordinate.

As shown in Figure 8.4, rate enhancements reported for enzyme-catalyzed reactions range from 10^7 to 10^{19}. If we consider these observations in terms of transition state theory, assuming 10^3 as an upper limit for the value of (A_{cat}/A_{non}), we expect $e^{(\Delta\Delta G^{\circ\ddagger}/RT)}$ to contribute at least 10^4 to 10^{16} to the rate enhancement (see Equation 8.16). This means the enzyme must provide a reduction in the activation energy (i.e., $\Delta\Delta G^{\circ\ddagger}$) on the order of 24–95 kJ/mol at 37 °C. This is equivalent to the energy of a few noncovalent bonds. Thus, it is reasonable to propose that specific noncovalent interactions between the enzyme

active site and the transition state can account for the stabilization that gives rise to the observed rate enhancements.

We have discussed several means by which an enzyme might achieve such significant rate enhancements:

1. Preferential binding to the transition state through complementary noncovalent bonding interactions (H-bonds, charge–charge interactions, etc.). Recall from Chapter 2 that noncovalent bonds are electrostatic in nature; thus, this is referred to as **electrostatic catalysis.**

2. Distortion of the substrate and/or active site, which promotes reduction of the activation energy (induced fit).

3. Binding of substrates to optimize proximity and orientation (making $\Delta S^{\circ\ddagger}$ more favorable).

4. Altering the reaction pathway to include intermediate states (see Figure 8.7). This is typical of **covalent catalysis.**

In addition to these four, we will add two other widely observed mechanisms for achieving rate enhancements in enzymes: **general acid/base catalysis (GABC)** and **metal ion catalysis.**

General acid/base catalysis (GABC) is important in reactions involving proton transfer. An active-site amino acid residue is classified as a **general acid** if it donates H^+ to an atom that develops negative charge in the transition state (**FIGURE 8.11**). A **general base** removes H^+

Enzyme-catalyzed ester cleavage

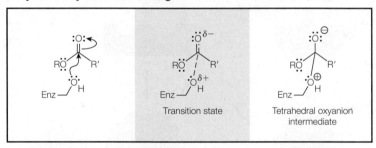

▲ **FIGURE 8.11 Enthalpic stabilization of the transition state in an enzyme-catalyzed reaction.** The upper panel shows the transition state and tetrahedral intermediate for an enzyme-catalyzed ester cleavage. This transition state might be stabilized by electrostatic interactions with active site amino acids and/or metal ions (lower left panel), or it might be stabilized by general acids or bases (lower right panel). The direction of proton transfer from the proton donor to the proton acceptor is indicated by the placement of the wide end of the dashed bond near the proton acceptor. The GAC is a proton donor and the GBC is a proton acceptor.

from an atom that develops positive charge in the transition state. Thus, GABC may be viewed as a specialized case of electrostatic catalysis involving the transfer of a positive charge (H^+).

In general, the catalytically important residues in the enzyme active site are polar. Histidine is a very common GABC catalyst in active sites, due to its ability to accept or donate protons at physiological pH. Residues like Glu, Asp, Lys, and Arg are also common participants in proton transfers, as well as frequently serving to make electrostatic bonds with substrate molecules. A number of other residues, such as Ser, Tyr, and Cys, are also found to play important roles in the active sites of enzymes, either as hydrogen bond donors/acceptors or as nucleophiles.

Over one-third of enzymes characterized to date contain metal ions in their active sites; thus, **metalloenzymes** are an important class of enzymes, and much current research is devoted to understanding the roles of the metal ions in catalysis. If the metal ion behaves as a Lewis acid, by accepting electron density from an electron-rich atom (e.g., an atom that develops negative charge in the transition state), it is acting as an electrostatic catalyst. Metal ions can also promote the formation

of hydroxide ion (^-OH) in the enzyme active site. This species is an important nucleophile in many hydrolytic reactions, such as the cleavage of peptide bonds in proteins or phosphodiester bonds in DNA and RNA.

To illustrate these general principles in a more concrete way, we consider the mechanisms of two specific enzyme-catalyzed reactions for which the details of the catalytic cycle are well understood.

Case Study #1: Lysozyme

Lysozyme is an enzyme that cleaves the peptidoglycan layer of the cell walls of bacteria (see the opening figure of Chapter 9 and Figure 9.23), resulting in lysis, and subsequent death, of the bacterial cell. As such, it defends against bacterial infection, and it is found in the secretions of tissues that contact the external environment, such as tears, saliva, and mucus. The first X-ray crystal structure of an enzyme was that of lysozyme from hen egg white, reported in 1965 by the laboratory of David Phillips at the Royal Institution in London.

The active site of lysozyme is located in a deep cleft (**FIGURE 8.12**) where six glycosyl residues bind in subsites labeled A–F. The

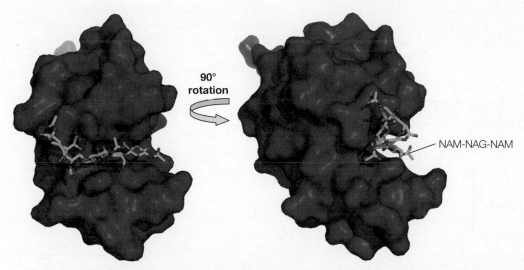

The solvent-accessible surface of hen lysozyme (PDB ID: 2war) is shown in blue. The trisaccharide NAM-NAG-NAM is shown in stick representation bound to the active site.

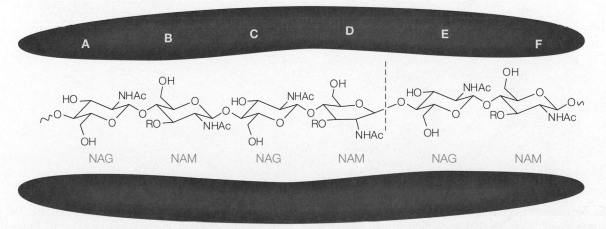

A schematic drawing of (NAG-NAM)₃ bound to the A–F subsites in lysozyme. The site of glycosidic bond cleavage, between the D and E subsites, is shown by the red dashed line. NAM = N-acetylmuramic acid; NAG = N-acetylglucosamine.

▲ **FIGURE 8.12** The active site-cleft of lysozyme.

glycosidic bond cleavage occurs between residues bound in the D and E subsites. Based on the crystal structure of the enzyme, Phillips proposed a stereochemical mechanism in which glutamic acid 35 (E35) acts as a general acid and aspartate 52 (D52) acts as an electrostatic catalyst during the generation of an oxocarbenium ion in the transition state (**FIGURE 8.13**). Phillips also proposed that the residue in the D subsite is distorted into a half-chair conformation upon binding in the active site (see Figure 8.2(c)). This conformation approximates that of the oxocarbenium ion and is an example of the substrate distortion ("strain") discussed earlier.

Additional experimental evidence for the proposed mechanism includes the observation of *kinetic isotope effects* (discussed in Tools of Biochemistry 8A) and the results of mutagenesis studies in which E35 and D52 were changed to nonionizing residues such as glutamine (Q), asparagine (N), and alanine (A). Amino acid mutations are often described using the single-letter code and the following convention: [wild-type amino acid][wild-type residue number] [mutant amino acid]. For example, a mutant in which glutamine replaces

glutamic acid at residue 35 would be "E35Q." In kinetics experiments using defined substrates, the E35Q and D52N mutants each showed <0.1% of wild-type activity, suggesting that both residues play critical roles in catalysis. As shown in **FIGURE 8.14**, lysozyme has optimum activity at a pH of ~5 and is dependent on a base with a pK_a of ~4 and an acid with a pK_a of ~6. The pK_as of all the Glu and Asp residues in lysozyme were determined by NMR titration (see Tools of Biochemistry 6A), and the pK_as of E35 and D52 were found to be, respectively, 6.2 and 3.7. Thus, at values of pH between 3.7 and 6.2, E35 would be predominantly protonated and D52 would be predominantly deprotonated, as required by the Phillips mechanism.

The Phillips mechanism proposes that D52 stabilizes the oxocarbenium ion via electrostatic stabilization. This aspect of the mechanism was challenged in 2001 when the Withers laboratory reported an X-ray crystal structure (PDB ID: 1h6m) of an enzyme-glycosyl intermediate that showed the D52 side chain covalently bound to C1 (see right side of Figure 8.13). The covalent mechanism, initially proposed by Daniel Koshland in 1953, is now favored.

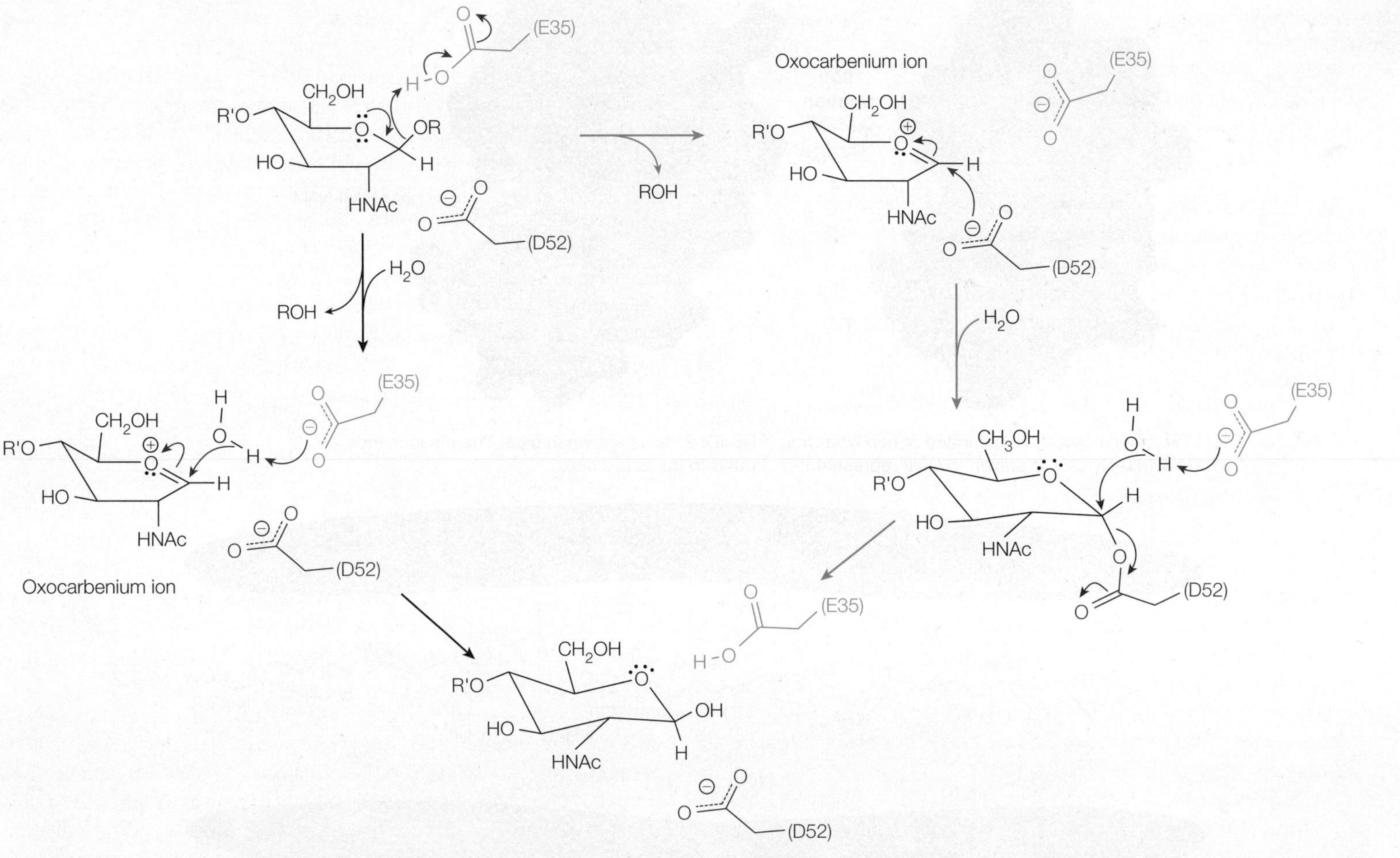

▲ **FIGURE 8.13 Two proposed mechanisms of action for lysozyme.** The Phillips mechanism is illustrated by the black reaction arrows along the left side of the diagram. In the first step, E35 acts as a general acid to promote cleavage of the glycosidic bond and concomitant formation of the oxocarbenium ion (which is stabilized electrostatically by D52). In the second step, E35 acts as a general base, deprotonating a water molecule, which then attacks C1 of the substrate. The pathway that includes the covalent intermediate reported by Steve Withers follows the green reaction arrows along the right side of the diagram. In this case, the second step involves covalent bond formation between C1 of the substrate and D52. Attack of the water displaces D52 in the subsequent step.

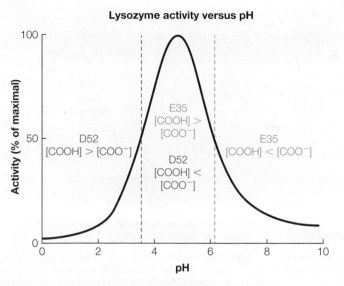

Lysozyme activity versus pH

▲ **FIGURE 8.14 The effect of pH on the activity of lysozyme.** E35 must be protonated to act as a general acid catalyst in the first step of the mechanism; thus, at pH values *below* 6.2 (blue line) the ratio of [COOH]/[COO⁻] is greatest, favoring catalysis. D52 must be deprotonated to interact with the oxocarbenium ion; thus, at pH values *above* 3.7 (red line) the ratio of [COO⁻]/[COOH] is greatest, favoring catalysis. These two boundary requirements give rise to the observed pH optimum (~5) where both protonated E35 and deprotonated D52 are abundant.

For the reaction to be catalytic, the enzyme active site must be restored to its initial state; in this case, the side chain of E35 must be protonated, and that of D52 must be free and deprotonated. Both of these requirements are achieved in the second half of the mechanism shown in Figure 8.13, in which a water molecule attacks the intermediate. Note that in this step E35 acts as a general base, removing a proton from the attacking water molecule.

In summary, lysozyme employs GABC, substrate distortion, and covalent catalysis to achieve its rate enhancement.

Case Study #2: Chymotrypsin, a Serine Protease

As a second example, let us consider the catalysis of peptide-bond hydrolysis by one of the **serine proteases.** This important class of enzymes includes trypsin and chymotrypsin, which we first encountered in Chapter 5. These enzymes are called proteases because they catalyze the hydrolysis of peptide bonds in polypeptides and proteins. Many kinds of proteases exist, exhibiting a wide range of substrate specificities and utilizing a variety of catalytic mechanisms. The *serine* proteases are distinct because they all have a critical serine nucleophile in the active site.

Catalysis of peptide-bond hydrolysis by a serine protease (chymotrypsin, in this example) proceeds as shown in **FIGURE 8.15**. First (as shown in step ➊), the polypeptide chain to be cleaved is bound to the enzyme surface. Most of the polypeptide binds nonspecifically, but the side chain of the residue to the N-terminal side of the peptide bond to be cleaved must fit in the specificity pocket within the enzyme active site. This pocket defines not only the position of the bond cleavage but also the *specificity* of serine proteases. Each of the serine proteases preferentially cleaves the amide bond immediately C-terminal to a

specific kind of amino acid side chain. Examples of sites of preferential cleavage are shown in Table 5.3. For example, trypsin cleaves preferentially to the carboxylate side of basic amino acid residues like lysine or arginine, whereas chymotrypsin prefers a large hydrophobic residue like phenylalanine, tyrosine, tryptophan, and leucine in this position. On the one hand, chymotrypsin has a narrow specificity pocket lined with small glycine residues, which can accommodate bulky nonpolar side chains. Trypsin, on the other hand, has an aspartate side chain in the bottom of the specificity pocket. This negatively charged residue provides complementary binding interactions to the positive charge on an arginine or lysine side chain. This very specific binding of a particular type of amino acid also serves to place the active site serine very close to the carbonyl group of the bond to be cleaved—also called the **scissile bond.**

A common feature of serine proteases is the so-called **catalytic triad** of a *nucleophile,* a *general base,* and an *acid.* In many of the serine proteases that have been studied in detail, this catalytic triad is composed of serine, histidine, and aspartic acid residues presented in a similar 3-D orientation in the active site. As shown in **FIGURE 8.16**, in chymotrypsin the catalytic triad residues are Ser 195 (S195), His 57 (H57), and Asp 102 (D102).

Because of the high pK_a of alcohol groups, serine side chains are usually in the protonated (i.e., —OH) form and are therefore not very reactive nucleophiles. However, S195 is in an environment that optimizes its reactivity. The S195 proton is transferred to the imidazole ring of H57, leaving a negative charge on the serine. Normally, this transfer would be unlikely (due to the lower pK_a of a His side chain vs. a typical Ser side chain), but it appears to be facilitated by D102, which, by its negative charge, stabilizes the protonation of the adjacent H57 side chain (see Figure 8.16). These interactions make S195 an unusually reactive nucleophile that attacks the amide carbonyl of the scissile bond.

Attack by the serine nucleophile on the amide carbonyl results in the formation of a **tetrahedral oxyanion,** which then collapses to an **acyl–enzyme intermediate** (Figure 8.15 steps ➊ and ➋). The formation of the oxyanion intermediate requires that the planar sp^2 hybridized amide carbon adopt a tetrahedral sp^3 configuration. Modeling of these species in the X-ray crystal structure of a serine protease active site led Jon Robertus and Joseph Kraut to propose the existence of an "**oxyanion hole**" that stabilizes the tetrahedral intermediate through specific H-bonding interactions to the negatively charged oxygen atom in the oxyanion (**FIGURE 8.17**). In chymotrypsin, the H-bond donors are two backbone amide protons from residues S195 and G193. These H-bonds are stronger when the configuration of the amide carbon atom is tetrahedral rather than planar. Indeed, one of the H-bonds (from G193) appears to form only *after* the oxyanion is generated by attack of S195. Presumably, the enzyme also stabilizes the transition state between the ES complex and this oxyanion intermediate because both states have significant negative charge on the oxygen atom and sp^3-like geometry. The oxyanion hole is a clear example of an enthalpic interaction that preferentially stabilizes high-energy states along the reaction coordinate and thereby lowers the overall energies of those states.

The protonated H57 acts as a general acid in the collapse of the oxyanion that yields the acyl–enzyme intermediate (Figure 8.15, steps ➋ and ➌). This leaves the N-terminal part of the polypeptide substrate covalently bound to the enzyme and allows the C-terminal portion to diffuse from the active site. A water molecule can then enter the active site

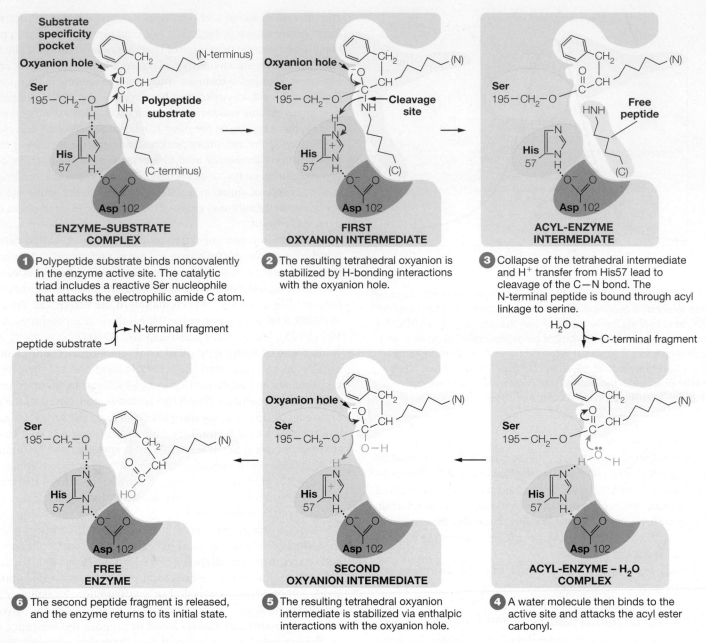

1 Polypeptide substrate binds noncovalently in the enzyme active site. The catalytic triad includes a reactive Ser nucleophile that attacks the electrophilic amide C atom.

2 The resulting tetrahedral oxyanion is stabilized by H-bonding interactions with the oxyanion hole.

3 Collapse of the tetrahedral intermediate and H⁺ transfer from His57 lead to cleavage of the C—N bond. The N-terminal peptide is bound through acyl linkage to serine.

6 The second peptide fragment is released, and the enzyme returns to its initial state.

5 The resulting tetrahedral oxyanion intermediate is stabilized via enthalpic interactions with the oxyanion hole.

4 A water molecule then binds to the active site and attacks the acyl ester carbonyl.

▲ **FIGURE 8.15 Catalysis of peptide bond hydrolysis by chymotrypsin.** The figure shows the steps in the cleavage of a polypeptide chain as catalyzed by chymotrypsin. The figure is highly schematic and does not represent the actual spatial arrangement of atoms (see Figure 8.16).

to cleave the acyl–enzyme intermediate (Figure 8.15, steps **3** and **4**). H57 acts as a general base, deprotonating the H_2O to make it a more potent nucleophile, which attacks the acyl-enzyme ester to form a second tetrahedral oxyanion intermediate (Figure 8.15, steps **4** and **5**). Finally, H57, acting as a general acid, facilitates the collapse of the oxyanion to regenerate S195 and release the remainder of the cleaved polypeptide chain (Figure 8.15, steps **5** and **6**). The enzyme is back in its original state, ready to catalyze the hydrolysis of another amide bond.

Mutagenesis studies have been used to confirm the critical contributions of each residue in the catalytic triad toward the $\sim 10^{10}$ rate

● **CONCEPT** The catalysis of peptide-bond cleavage by serine proteases involves stabilization of transition states and tetrahedral intermediate states.

enhancement observed for serine proteases. S195A or H57A mutations reduced the rate enhancement by $\sim 10^6$ compared to the wild-type enzyme, and the D102A mutant showed a reduction of $\sim 10^4$. These results suggest that S195 and H57 contribute 100-fold more to the rate enhancement than does D102. The triple mutant, S195A:H57A:D102A, also shows a $\sim 10^6$ reduction in rate enhancement, which suggests that other features of the enzyme active site, such as the oxyanion hole, contribute $\sim 10^3$–10^4

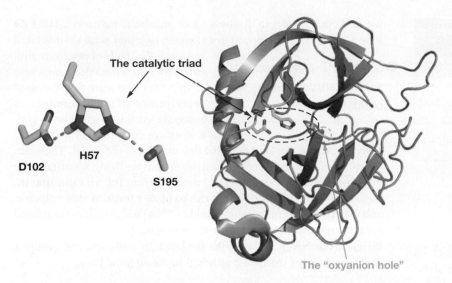

▲ FIGURE 8.16 The structure of chymotrypsin and the serine protease catalytic triad. The backbone of bovine chymotrypsin determined by X-ray crystallography (PDB ID: 4cha) is shown in cartoon representation (blue) with the side chains of the catalytic triad, S195, H57, and D102, shown in stick form within the dotted red oval. The location of the oxyanion hole (see text) is indicated by the dotted orange circle. To the left is shown a high-resolution neutron diffraction structure of the catalytic triad from porcine elastase (PDB ID: 3hgn). The neutron diffraction data show the positions of H atoms with greater resolution than do the X-ray data. Here, H57 is protonated, S195 and D102 are deprotonated, and H-bonds are shown by orange dashes.

to the overall rate enhancement (see Problem 12 at the end of this chapter).

Additional support for the mechanism shown in Figure 8.15 comes from a large number of crystal structures of serine proteases bound to reaction intermediates as well as to **transition-state analogs.** Transition-state analogs are compounds designed to mimic the transition state for a reaction. As such, they are meant to possess greater shape and electrostatic complementarity to the enzyme active site than do the natural substrates. In fact, transition-state analogs typically bind to enzymes with affinities that are at least 10^2–10^4 greater than those of the natural substrates. In some cases the tight-binding analog is formed in the enzyme active site. An example is shown in Figure 8.17, where the tetrapeptide Ala–Ala–Pro–Ala bearing a C-terminal boronic acid is converted to a mixed borate ester following attack by S195 on the boron atom. This converts the planar boronic acid to the tetrahedral borate ester, which can then interact with the amides in the oxyanion hole.

In summary, serine proteases employ covalent catalysis and electrostatic stabilization of the transition state to achieve rate enhancement (**TABLE 8.2**). It is important to keep in mind that many enzyme-catalyzed reactions are thought to work in ways not listed in Table 8.2. Two such alternate catalytic strategies are discussed in the next section.

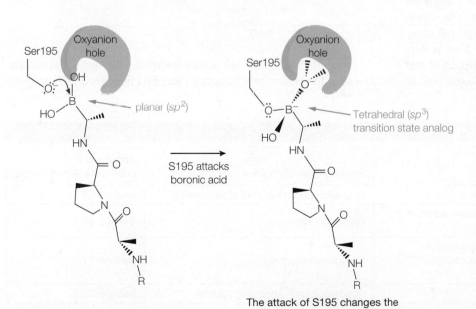

The attack of S195 changes the geometry of the boron atom. This places an O atom in the oxyanion hole, where H-bonds to the main chain form.

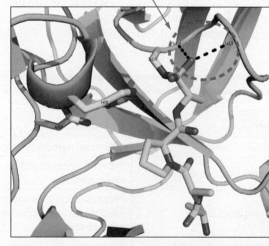

The active site of α-lytic protease from the bacterium *Lysobacter enzymogenes* is shown (PDB ID: 1gbb). The backbone is shown in cartoon form (green) with the catalytic triad in stick form (C atoms yellow, N atoms blue, O atoms red) and the peptide boronic ester in stick form (C atoms light blue, B atom pink). H-bonds from the main chain amides of S195 and G193 in the oxyanion hole are shown by black dashed lines.

▲ FIGURE 8.17 Binding in the oxyanion hole of a serine protease.

TABLE 8.2 The strategies used by lysozyme and serine proteases to lower $\Delta G^{\circ\ddagger}$

Catalytic Strategy	Lysozyme	Ser Proteases
GABC	E35	H57 (catalytic triad)
Covalent	D52	S195 (catalytic triad)
Electrostatic		oxyanion hole
Other	strain of D ring	

8.5 Coenzymes, Vitamins, and Essential Metals

The complexity of globular protein structure and the variety of side-chain structures in a protein allow the formation of many kinds of catalytic sites. This variability allows enzymes to act as efficient catalysts for many reactions. However, for some kinds of biological processes, the chemical potential of amino acid side chains alone is not sufficient. A protein may require the help of some other small molecule or ion to carry out the reaction. The ions or molecules that are bound to enzymes for this purpose are called **cofactors** or **coenzymes.** Like enzymes, coenzymes are not irreversibly changed during catalysis; they are either unmodified or regenerated, as described below.

● **CONCEPT** Many enzymes use ions or small bound molecules called cofactors or coenzymes to aid in catalysis.

Coenzyme Function in Catalysis

Coenzymes often have complex organic structures that cannot be synthesized by some organisms—mammals in particular. The water-soluble vitamins, those usually referred to as the vitamin-B complex, are metabolic precursors of several coenzymes. This is why such vitamins are so important in metabolism. Our approach in this book is to introduce the detailed biochemistry of each coenzyme as we first encounter it in discussions of metabolic pathways. **TABLE 8.3** lists a number of important coenzymes, together with their related vitamins, the kinds of reactions with which they are associated, and where in this text you will find detailed descriptions of their function in catalysis. At this point, just to give a more concrete idea of how coenzymes function, we will describe one class in some detail. These are the nicotinamide nucleotides, a major example being **nicotinamide adenine dinucleotide (NAD$^+$)** derived from the vitamin **niacin.** Note that "NAD$^+$" represents the oxidized form of this cofactor, and the reduced form is represented as "NADH." The closely related cofactor **nicotinamide adenine dinucleotide phosphate (NADP$^+$)** differs from NAD$^+$ only by the presence of a phosphoryl group at the 2' position of the ribose attached to the adenine base.

● **CONCEPT** Many essential vitamins are constituents of coenzymes.

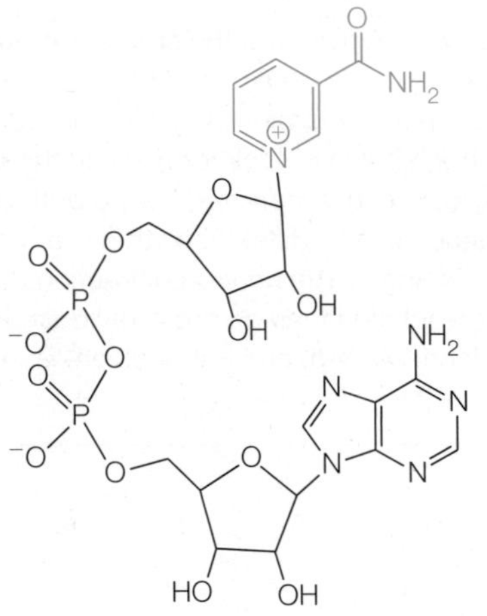

NAD$^+$

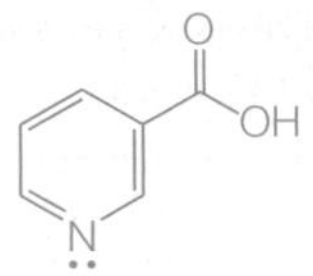

Niacin

Vitamin deficiencies are often associated with nutritional disease. In fact, the concept of a "vitamin" was first described following the

TABLE 8.3 Some important coenzymes and related vitamins

Vitamin	Coenzyme	Reactions involving the coenzyme	Section in which coenzyme is introduced
Thiamine (vitamin B$_1$)	Thiamine pyrophosphate	Activation and transfer of aldehydes	Section 12.3
Riboflavin (vitamin B$_2$)	Flavin mononucleotide; flavin adenine dinucleotide	Oxidation-reduction	Section 13.2
Niacin (vitamin B$_3$)	Nicotinamide adenine dinucleotide; nicotinamide adenine dinucleotide phosphate	Oxidation-reduction	Section 11.3
Pantothenic acid (vitamin B$_5$)	Coenzyme A	Acyl group activation and transfer	Section 13.2
Pyridoxine (vitamin B$_6$)	Pyridoxal phosphate	Various reactions involving amino acid activation	Section 18.4
Biotin (vitamin B$_7$)	Biotin	CO_2 activation and transfer	Section 12.5
Lipoic acid	Lipoamide	Acyl group activation; oxidation-reduction	Section 13.2
Folic acid (vitamin B$_9$)	Tetrahydrofolate	Activation and transfer of single-carbon functional groups	Section 18.4
Vitamin B$_{12}$	Adenosyl cobalamin; methyl cobalamin	Isomerizations and methyl group transfers	Section 18.4

● **CONNECTION** Vitamin deficiency is associated with a variety of nutritional diseases.

observation that beriberi could be treated by thiamine, which is present in unpolished but not refined rice. Similarly, deficiency in niacin gives rise to pellagra. Although pellagra is less widespread than it was 100 years ago, it is still common in populations where malnutrition or chronic alcoholism is prevalent.

The nicotinamide portion is the part of NAD^+ responsible for its metabolic functions because it can be reduced and thus can serve as an *oxidizing* agent to which two electrons and a proton are added to the nicotinamide ring: $NAD^+ + 2e^- + H^+ \longrightarrow NADH$. The reaction is reversible (i.e., NADH acts as a *reducing* agent in several reactions) and is formally a **hydride ion** transfer: $NAD^+ + H^- \rightleftharpoons NADH$.

Hydride donor: $H^\ominus$ Hydride acceptor

NAD^+ NADH

Here, R stands for the remainder of the molecule.

A typical reaction in which NAD^+ acts as an oxidizing agent is the conversion of alcohols to aldehydes or ketones (for example, the conversion of ethanol to acetaldehyde by the *alcohol dehydrogenase* of liver):

H_3C—OH $\rightleftharpoons$ H_3C—CHO $+ H^- + H^+$

$NAD^+ + H^- \rightleftharpoons NADH$

$CH_3CH_2OH + NAD^+ \rightleftharpoons CH_3CHO + H^+ + NADH$
Ethanol   Acetaldehyde

It is the C-linked H, not the O-linked H, that is transferred to NAD^+, as can be demonstrated by studies using deuterated compounds. Furthermore, these reactions are stereospecific. Even when the hydroxyl carbon has *two* hydrogens attached (as with ethanol), a particular one of the hydrogens is transferred to NAD^+. This specificity may seem surprising because the hydroxyl carbon of ethanol is not a chiral center. How can a particular hydrogen be favored when the substrate molecule has a plane of symmetry? The answer lies in the asymmetric nature of the enzyme surface, to which both NAD^+ and the alcohol are bound. If a symmetrical molecule like ethanol is bound by at least *three* points to an asymmetric object, the two H atoms are no longer equivalent; they are said to be *prochiral* (**FIGURE 8.18**). Furthermore, although the nicotinamide ring is planar, the transfer of hydrogen in a particular reaction is always to a specific face of the ring, as the two faces are *not* equivalent in the asymmetric active site of an enzyme. Such considerations lie behind the high stereospecificity of many enzyme-catalyzed reactions, in contrast to nonenzymatic catalysis.

Sometimes it is difficult to make a clear distinction between a true coenzyme and a second substrate in a reaction. The reaction we have just discussed is a good example of this problem. The dehydrogenase enzymes, such as alcohol dehydrogenase, each have a strong binding

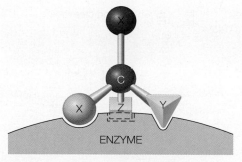

▲ **FIGURE 8.18 Stereospecificity conferred by an enzyme.** This figure shows how the asymmetric surface of an enzyme can confer stereospecificity in the reaction of a symmetric substrate. If the substrate molecule X_2CYZ makes at least three contacts with unique complementary groups on the enzyme, its two X atoms are no longer equivalent. Only a specific one of the two X atoms can contact the surface properly. In many cases, a minimum of *four* contacts is required to distinguish prochiral functional groups. This is discussed in greater detail in Chapter 13.

site for the oxidized form of the cofactor, NAD^+. After oxidation of the substrate, the reduced form, NADH, leaves the enzyme and is reoxidized by other electron-acceptor systems in the cells. The NAD^+ so formed can now bind to another enzyme molecule and repeat the cycle. In such cases, NAD^+ is acting more like a second substrate than a true cofactor. Yet NAD^+ and NADH differ from most substrates in that they are continually recycled in the cell and are used over and over again by the hundreds of different enzymes that require this cofactor. Because of this behavior, we consider NAD^+ and NADH to be coenzymes.

Metal Ions in Enzymes

Many enzymes contain one or more metal ions, usually coordinated by certain amino acid side chains but sometimes bound in a prosthetic group like heme. Such enzymes are called **metalloenzymes.** The bound ion acts in much the same way as a coenzyme, conferring on the metalloenzyme a property it would not possess in its absence. As **TABLE 8.4** shows, these ions play diverse roles. For example, the zinc ion in

TABLE 8.4 Metals and trace elements important as enzymatic cofactors

Metal	Example of Enzyme	Role of Metal
Fe	Cytochrome oxidase	Oxidation–reduction
Cu	Ascorbic acid oxidase	Oxidation–reduction
Zn	Alcohol dehydrogenase	Helps bind NAD^+
Mn	Histidine ammonia lyase	Aids in catalysis by electron withdrawal
Co	Glutamate mutase	Co is part of cobalamin coenzyme
Ni	Urease	Catalytic site
Mo	Xanthine oxidase	Oxidation–reduction
V	Nitrate reductase	Oxidation–reduction
Se	Glutathione peroxidase	Replaces S in one cysteine in active site
Mg	Many kinases	Helps bind ATP

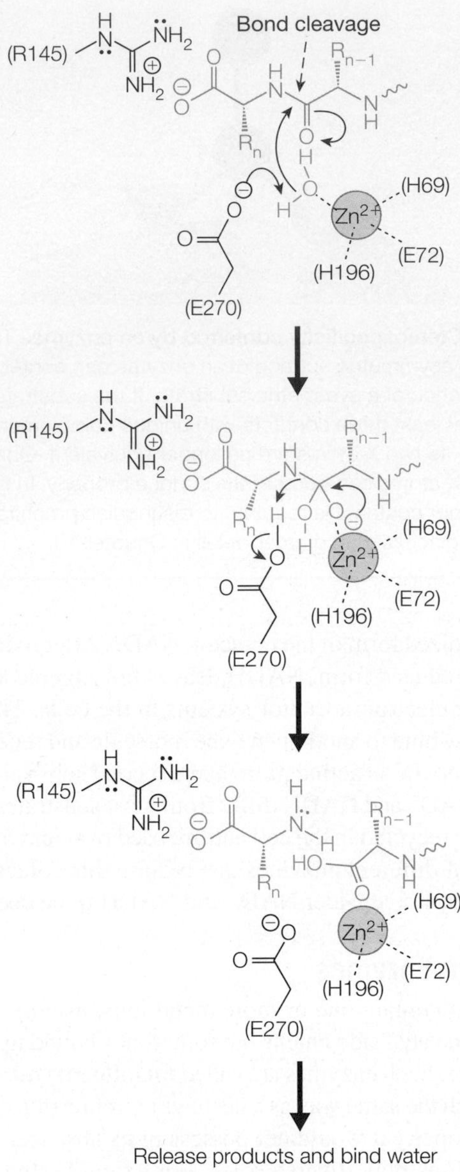

▲ **FIGURE 8.19 The mechanism of the protease carboxypeptidase A.** The zinc ion (orange circle) binds a water molecule (blue) and serves as an electrostatic catalyst to promote hydrolysis of the C-terminal amino acid from a peptide substrate (green). It does so by stabilizing the negative charge on the oxygen in the tetrahedral transition state. Enzyme active site residues are indicated by black coloring. The bond cleaved is indicated by the dashed red arrow.

● **CONCEPT** Some enzymes require metal ions for their catalytic function.

carboxypeptidase A binds the water molecule that attacks the carbonyl of the scissile bond and also acts as an electrostatic catalyst (**FIGURE 8.19**). The zinc ion stabilizes the tetrahedral oxyanion in the transition state and the intermediate state, in much the same way as the oxyanion hole serves this function in chymotrypsin (see Figure 8.17 for comparison).

In other cases, the metal in a metalloenzyme serves as a redox reagent. We have mentioned the example of the heme-iron–containing enzyme *catalase,* which catalyzes the breakdown of hydrogen peroxide,

a potentially destructive agent in cells. Because the reaction involves both reduction and oxidation of H_2O_2, the Fe^{2+} is reversibly oxidized and reduced, acting as an electron exchanger. Such redox activity requires metals like Fe or Cu with multiple stable oxidation states.

In many other enzymatic reactions, certain ions are necessary for catalytic efficiency, even though they may not remain permanently attached to the protein nor play a direct role in the catalytic process. For example, a number of enzymes that couple ATP hydrolysis to other processes require Mg^{2+} for efficient function. In most cases, Mg^{2+} is necessary because the Mg–ATP complex is a better substrate than ATP itself.

8.6 The Kinetics of Enzymatic Catalysis

Earlier, we described details of the mechanisms for two well-studied enzyme-catalyzed reactions. Hundreds more are known, and such knowledge is essential for understanding and treating many human diseases. For example, as we will see in Chapters 18 and 19, compounds that inhibit the activity of the enzyme dihydrofolate reductase have found wide use in treating cancer and infectious diseases. Where does this mechanistic knowledge come from? X-ray crystallography and solution NMR studies have provided key structural insights into enzyme activity; however, much of our understanding of enzyme mechanisms comes from the careful mathematical analysis of enzyme kinetics. In 1913, long before enzymes had been purified—or even classified as proteins—pioneers in the analysis of enzyme kinetics, Leonor Michaelis and Maude Menten, developed expressions that explained enzyme–substrate affinity and some modes of enzyme inhibition. It is to their seminal analysis that we now turn.

Reaction Rate for a Simple Enzyme-Catalyzed Reaction: Michaelis–Menten Kinetics

Earlier in this chapter, we introduced Equation 8.17 as an expression for a simple reaction involving a single substrate and product:

$$E + S \underset{k_{-1}}{\overset{k_1}{\rightleftharpoons}} ES \underset{k_{-2}}{\overset{k_2}{\rightleftharpoons}} EP \underset{k_{-3}}{\overset{k_3}{\rightleftharpoons}} E + P \qquad (8.17)$$

If we analyze the **initial rate** of an enzyme-catalyzed reaction (i.e., before a significant concentration of P appears) and we assume that the chemical transformation of ES to EP is rate-limiting (i.e., k_1, k_{-1}, and $k_3 \gg k_2$), Equation 8.17 simplifies to the following:

$$E + S \underset{k_{-1}}{\overset{k_1}{\rightleftharpoons}} ES \overset{k_{cat}}{\longrightarrow} E + P \qquad (8.18)$$

where k_{cat} is the apparent rate constant for the rate-determining conversion of substrate to product.*

We have assumed that initial reaction conditions are such that the reverse reaction between E and P is negligible. The catalytic formation of the product, with enzyme regeneration, will then be a simple first-order reaction, and its rate will be determined solely by the concentration of ES and the value of k_{cat}. Therefore, the reaction rate, or velocity, defined as the observed rate of formation of products, can be expressed as

$$v = k_{cat}[ES] \qquad (8.19)$$

---

*k_{cat} is an aggregate rate constant $= k_2 k_3 / (k_2 + k_3)$. In the limiting case where $k_3 \gg k_2$, $k_{cat} \approx k_2$.

If v and [ES] can be measured for a specific enzyme and substrate, the rate constant, k_{cat} for that particular reaction can be derived. In practice, [ES] is difficult to measure in kinetics experiments. The easily measured parameters are: (1) the observed reaction rate, (2) the initial substrate concentration, and (3) the *total* concentration of enzyme added to the reaction, which must be the sum of free enzyme and enzyme bound to substrate:

[E]$_t$	=	[E]	+	[ES]	
Total enzyme		Free enzyme		Enzyme in ES complex	(8.20)

Thus, it is desirable to express the rate, v, in terms of the substrate concentration [S] and the total enzyme concentration [E]$_t$.

The way we have written Equation 8.18 suggests that E and S should be in equilibrium with ES, with an equilibrium dissociation constant K_S:

$$K_S = \frac{k_{-1}}{k_1} = \frac{[E][S]}{[ES]} \qquad (8.21)$$

This is usually an incorrect assumption, but under certain circumstances (i.e., $k_{cat} \ll k_{-1}$), this approximation is valid. This assumption was used in early attempts to solve the problem of expressing the reaction rate. It doesn't apply in general because E, S, and ES are not truly in equilibrium; some ES is continually being converted to P, albeit slowly. An analysis that avoids the assumption of equilibrium was presented by G. E. Briggs and J. B. S. Haldane in 1925. The Briggs–Haldane model is based on the following argument: The more ES that is present, the faster ES will dissociate either to products (k_{cat}) or back to reactants (k_{-1}). Therefore, when the reaction is started by mixing enzymes and substrates, the ES concentration builds up at first but quickly reaches a **steady state,** in which it remains almost constant. This steady state will persist until almost all of the substrate has been consumed (**FIGURE 8.20**). Because the steady state accounts for nearly all the reaction time, we can calculate the reaction velocity by assuming steady-state conditions. Normally, we measure rates only after the steady state has been established and before [ES] has changed much. We can then express the velocity as follows.

In the steady state, the rates of formation and breakdown of ES are equal. Therefore,

k_1[E][S]	=	k_{-1}[ES]	+	k_{cat}[ES]	
Formation of ES complex		Dissociation of ES complex		Breakdown to E + P	(8.22)

which can be rearranged to give

$$[ES] = \left(\frac{k_1}{k_{-1} + k_{cat}}\right)[E][S] \qquad (8.23)$$

Combining the ratio of rate constants in Equation 8.23 gives a single constant, K_M:

$$K_M = \frac{k_{-1} + k_{cat}}{k_1} \qquad (8.24)$$

> **CONCEPT** The steady-state assumption proposes that the concentration of enzyme–substrate complex remains nearly constant through much of the reaction.

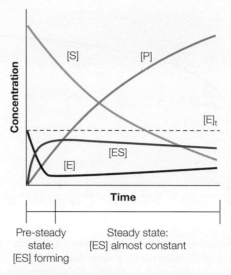

▲ **FIGURE 8.20 The steady state in enzyme kinetics.** The figure shows how the concentrations of substrate [S], free enzyme [E], enzyme–substrate complex [ES], and product [P] vary with time for a simple enzyme-catalyzed reaction described by E + S $\rightleftharpoons$ ES $\longrightarrow$ E + P. After a very brief initial period, [ES] reaches a steady state in which ES is consumed approximately as rapidly as it is formed, so $d[ES]/dt \approx 0$. The concentrations of E and ES are greatly exaggerated for clarity. Note that $[E] + [ES] = [E]_t$, or total enzyme concentration, and that [ES] actually falls very slowly as substrate is consumed, while [E] accordingly rises.

Equation 8.23 can now be rewritten as

$$K_M[ES] = [E][S] \qquad (8.25)$$

At this point, [ES] is expressed in terms of [E] and [S]. To get [E]$_t$ into the equation, rather than [E], recall from Equation 8.20 that [E] = [E]$_t$ − [ES]. Putting this into Equation 8.25 yields

$$K_M[ES] = [E]_t[S] - [ES][S] \qquad (8.26)$$

This rearranges to

$$[ES] = \frac{[E]_t[S]}{K_M + [S]} \qquad (8.27)$$

Finally, inserting this result into Equation 8.19 gives an expression for v in terms of [E]$_t$ and [S]:

$$v = \frac{k_{cat}[E]_t[S]}{K_M + [S]} \qquad (8.28)$$

Equation 8.28 is called the **Michaelis–Menten equation,** and K_M the **Michaelis constant.** We will discuss the meaning of K_M shortly; in the meantime, there are two important points to keep in mind. First, because K_M is a ratio of the rate constants for a specific reaction (see Equation 8.24), it is a characteristic of that reaction. Thus, a given enzyme acting upon a given substrate has a defined K_M. Second, you can see from Equations 8.24 and 8.25 that K_M has units of concentration.

Now consider the graph of v versus [S] for the Michaelis–Menten equation shown in **FIGURE 8.21**. At high substrate concentrations, where [S] is much greater than K_M, the reaction approaches a **maximum velocity, V$_{max}$,** because the enzyme molecules are *saturated;* every enzyme molecule is bound by substrate. Thus, [ES] = [E]$_t$, and

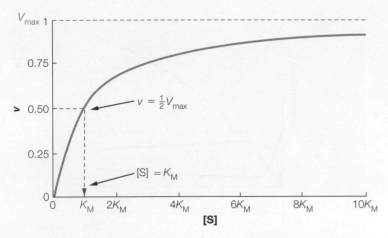

▲ **FIGURE 8.21 Reaction velocity as a function of substrate concentration.** This graph, a plot of Equation 8.30, shows the variation of reaction velocity with substrate concentration according to the Michaelis–Menten model of enzyme kinetics. Here, [S] is given in terms of K_M. At the point where [S] = K_M, the reaction has exactly half its maximum velocity. The values of v plotted here are determined from the *initial rates* of the reaction (see Figure 8.22) and are expressed as a fraction of V_{max}. Note that V_{max} is approached asymptotically.

Equation 8.19 will reach its maximum value. When [S] ≫ K_M, $K_M + [S] \approx [S]$, and Equation 8.28 simplifies to the expression for V_{max}:

$$v = \frac{k_{cat}[E]_t[S]}{[S]} = k_{cat}[E]_t = V_{max} \qquad (8.29)$$

Thus, $k_{cat}[E]_t$ in Equation 8.28 is equivalent to V_{max}, and the Michaelis–Menten rate equation is

$$v = \frac{V_{max}[S]}{K_M + [S]} \qquad (8.30)$$

This is the most familiar form of the Michaelis–Menten equation.

Interpreting K_M, k_{cat}, and k_{cat}/K_M

The two quantities that characterize an enzyme obeying Michaelis–Menten kinetics are K_M and k_{cat}. What do they signify? The Michaelis constant, K_M, is often associated with the affinity of enzyme for substrate. However, this relationship is true only in the limiting case: a two-step reaction in which $k_{cat} \ll k_{-1}$ where Equation 8.24 then yields $K_M \approx k_{-1}/k_1 = K_S$, the equilibrium constant defined in Equation 8.21. For more complex kinetic schemes, K_M is a ratio of several rate constants. For any reaction that follows the Michaelis–Menten equation, K_M is numerically equal to the substrate concentration at which the reaction velocity has attained *half* of its maximum value (see Figure 8.21). Thus, K_M is a measure of the substrate concentration required for effective catalysis to occur. **TABLE 8.5** lists K_M values for a number of important enzymes.

● **CONCEPT** The Michaelis constant, K_M, indicates the substrate concentration at which the reaction rate is $\frac{1}{2}V_{max}$ Enzyme activity will be high when substrate concentration is above K_M.

The second constant, k_{cat}, gives a direct measure of the rate of product formation under optimum conditions (saturated enzyme). The units of k_{cat} are usually given as s^{-1} or min^{-1}, so the reciprocal of k_{cat} can be thought of as a time—the time required by an enzyme molecule to "turn over" one substrate molecule. Alternatively, k_{cat} measures the number of substrate molecules turned over per enzyme molecule per unit time. Thus, k_{cat} is sometimes called the **turnover number.** Some typical values of k_{cat} are listed in Table 8.5.

● **CONCEPT** The turnover number, k_{cat} measures the rate of the catalytic process.

The enzymes listed in Table 8.5 are arranged in order of increasing value of the ratio k_{cat}/K_M. This ratio is often thought of as a measure of enzyme efficiency. Note that either a large value of k_{cat} (i.e., rapid turnover) or a small value of K_M (i.e., $\frac{1}{2} V_{max}$ occurs at relatively low [S]) will make k_{cat}/K_M large. We can gain another insight into the meaning of k_{cat}/K_M by considering the situation at very low substrate

TABLE 8.5 Michaelis–Menten parameters for selected enzymes, arranged in order of increasing efficiency as measured by k_{cat}/K_M

Enzyme	Reaction Catalyzed	K_M(mol/L)	$k_{cat}(s^{-1})$	$k_{cat}/K_M[(mol/L)^{-1}s^{-1}]$
Chymotrypsin	Ac–Phe–Ala $\xrightarrow{H_2O}$ Ac–Phe + Ala	1.5×10^{-2}	0.14	9.3
Pepsin	Phe–Gly $\xrightarrow{H_2O}$ Phe + Gly	3×10^{-4}	0.5	1.7×10^3
Tyrosyl-tRNA synthetase	Tyrosine + tRNA ⟶ tyrosyl-tRNA	9×10^{-4}	7.6	8.4×10^3
Ribonuclease	Cytidine 2′, 3′ cyclic phosphate $\xrightarrow{H_2O}$ cytidine 3′-phosphate	7.9×10^{-3}	7.9×10^2	1.0×10^5
Carbonic anhydrase	$HCO_3^- + H^+ \longrightarrow H_2O + CO_2$	2.6×10^{-2}	4×10^5	1.5×10^7
Fumarase	Fumarate $\xrightarrow{H_2O}$ malate	5×10^{-6}	8×10^2	1.6×10^8

concentrations. In this case, $[S] \ll K_M$, and most of the enzyme is free, so $[E]_t \approx [E]$. Then Equation 8.30 becomes

$$v = \frac{k_{cat}}{K_M}[E][S] \qquad (8.31)$$

Therefore, under these circumstances, the ratio k_{cat}/K_M behaves as a second-order rate constant for the reaction between substrate and free enzyme, and it provides a direct measure of enzyme efficiency and specificity. It allows direct comparison of the effectiveness of an enzyme toward different substrates. Suppose an enzyme has a choice of two substrates, A or B, present at equal concentrations. Then, under conditions in which both substrates are dilute and are competing for the enzyme, we find

$$\frac{v_A}{v_B} = \frac{\left(\dfrac{k_{cat}}{K_M}\right)_A [E][A]}{\left(\dfrac{k_{cat}}{K_M}\right)_B [E][B]} = \frac{\left(\dfrac{k_{cat}}{K_M}\right)_A}{\left(\dfrac{k_{cat}}{K_M}\right)_B} \qquad (8.32)$$

TABLE 8.6 lists values of k_{cat}/K_M for cleavage of various amino acid esters by chymotrypsin. Within the group shown, k_{cat}/K_M varies 1 million-fold, showing the range of preference the enzyme has for different peptide substrates. These data clearly show the preference to cleave next to the most hydrophobic residues.

N-acetyl amino acid methyl ester
(see Table 8.6)

We have just seen that the ratio k_{cat}/K_M corresponds to the second-order rate constant for the enzyme–substrate combination under circumstances of low-substrate concentration. Such a rate constant has a maximum possible value, which is determined by the frequency with which enzyme and substrate molecules can collide in solution. A reaction that attains such a velocity is said to be "diffusion-limited"; every encounter leads to reaction, so nothing but the rate of molecular encounters limits the velocity. If *every* collision results in formation of an enzyme–substrate complex, diffusion theory predicts that k_{cat}/K_M will attain a maximum value of about 10^8 to 10^9 $(mol/L)^{-1}s^{-1}$. Thus, an enzyme that approaches maximum possible efficiency will demonstrate this by having a value of k_{cat}/K_M in this range. As Table 8.5 shows, enzymes such as carbonic anhydrase and fumarase actually approach this limit.

● **CONCEPT** The ratio k_{cat}/K_M is a convenient measure of enzyme efficiency.

TABLE 8.6 Preferences of chymotrypsin in the hydrolysis of several *N*-acetyl amino acid methyl esters, as measured by k_{cat}/K_M

Amino Acid in Ester	Amino Acid Side Chain	$k_{cat}/K_M[(mol/L)^{-1}s^{-1}]$
Glycine	—H	1.3×10^{-1}
Norvaline	—$CH_2CH_2CH_3$	3.6×10^2
Norleucine	—$CH_2CH_2CH_2CH_3$	3.0×10^3
Phenylalanine	—CH_2—⟨⟩	1.0×10^5

Enzyme Mutants May Affect k_{cat} and K_M Differently

As discussed earlier for lysozyme and chymotrypsin, site-directed mutagenesis has been used extensively to test mechanistic models for most enzymes. A particular mutation may change the apparent value of k_{cat} or K_M, or both. The simplest interpretation of such data is that a mutation that affects only k_{cat} changes an amino acid side chain involved solely in catalysis (e.g., the active site nucleophile or a GABC) but not involved in binding to the substrate prior to the rate-determining catalytic step. Conversely, a mutation that affects only K_M alters a side chain that binds to the substrate but is not involved in stabilizing the transition state. Frequently, both k_{cat} and K_M are affected by mutations to active site residues. This would occur for a residue that binds substrate and then makes stronger interactions with the transition state as a result of an altered geometry in the transition state that optimizes the interaction.

● **CONCEPT** The observed effects of an amino acid mutation in an enzyme active site on K_M and k_{cat} can be used to identify the role of the amino acid in substrate binding (K_M effects) and transition-state stabilization (k_{cat} effects).

Analysis of Kinetic Data: Testing the Michaelis–Menten Model

The measurement of reaction velocity as a function of substrate concentration is used to determine whether an enzyme-catalyzed reaction follows the Michaelis–Menten model (Equation 8.30) and, if so, to determine the constants K_M and k_{cat}.

A few analytical methods for the measurement of rates are described in Tools of Biochemistry 8A. One general point should be noted: In principle, one could simply mix enzyme and substrate and follow the change in substrate concentration with time, as shown in Figure 8.1(a). As substrate is consumed, the reaction velocity decreases, until equilibrium is eventually reached. But measuring the instantaneous velocity at specific times during the reaction is difficult and usually inaccurate. It is usually easier to set up a series of experiments, all at the same enzyme concentration but at different substrate concentrations, and measure the *initial* rates (**FIGURE 8.22**). Because we know the initial $[S]$ precisely, and the change in $[S]$ versus t is almost linear in the initial stages, accurate data for v as a function of $[S]$ can be obtained. An enzyme that obeys the Michaelis–Menten kinetic model will yield a plot of initial velocity versus substrate concentration that is hyperbolic—compare Figures 8.21 and 8.22(b).

Given such data for concentrations and initial rates, how are K_M and k_{cat} calculated? In practice, modern nonlinear curve-fitting software can fit the data plotted in Figure 8.22(b) directly to provide these parameters. Before such data analysis became widely available, a different method was (and still is) used. Equation 8.30 can be rearranged to give the expression for a linear graph. Several kinds of graphs are possible, but it is most common to use a **double reciprocal plot**, also called a **Lineweaver–Burk plot** (**FIGURE 8.23**). A Lineweaver–Burk plot provides a quick test for adherence to Michaelis–Menten kinetics and allows easy evaluation of the critical constants. It also allows discrimination between different kinds of enzyme inhibition and regulation. Taking the reciprocal of both sides of Equation 8.30, we find

$$\frac{1}{v} = \frac{K_M + [S]}{V_{max}[S]} = \frac{K_M}{V_{max}[S]} + \frac{[S]}{V_{max}[S]} \qquad (8.33)$$

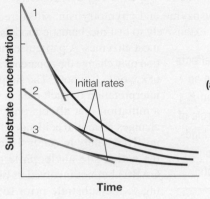

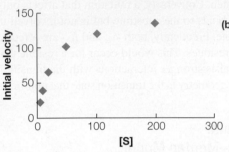

(a) Several reactions are performed with varying concentrations of substrate, and the values of the initial rates are determined from the slopes of the curves in the early phase for each reaction.

(b) Initial rate data, determined as described in (a), are plotted as a function of the substrate concentration. The enzyme appears to obey the Michaelis–Menten kinetic model (compare this plot to Figure 8.21).

▲ **FIGURE 8.22** Analysis of initial rates.

or

$$\frac{1}{v} = \left(\frac{K_M}{V_{max}}\right)\frac{1}{[S]} + \frac{1}{V_{max}} \qquad (8.34)$$

Thus, plotting $1/v$ versus $1/[S]$ should yield a straight line. At $1/[S] = 0$, [S] is infinitely large and the reaction velocity is at its maximum. Therefore, the y-intercept at $1/[S] = 0$ is equal to $1/V_{max}$. Given V_{max} and $[E]_t$ (from the initial conditions of the kinetics experiment), k_{cat} can be calculated from $V_{max} = k_{cat}[E]_t$. In similar fashion, K_M is

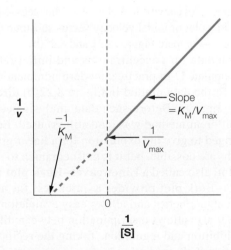

▲ **FIGURE 8.23** A Lineweaver–Burk plot. In this double reciprocal plot, $1/v$ is graphed versus $1/[S]$, according to Equation 8.34. Note that a linear extrapolation of the data gives both V_{max} and K_M.

● **CONCEPT** Lineweaver–Burk plots provide convenient ways to determine K_M and k_{cat} from initial-rate data.

calculated using the slope of the plot, which gives K_M/V_{max}, and the value of V_{max} obtained from the y-intercept. A disadvantage of a Lineweaver–Burk plot is that a long extrapolation is often required to determine K_M, which introduces corresponding uncertainty in the result. Performing nonlinear curve fitting to the raw data with readily available software is now preferred; however, as shown in the next section, the double-reciprocal plot is useful in distinguishing different modes of enzyme inhibition.

8.7 Enzyme Inhibition

Studies of enzyme inhibition are important because they provide critical insights into the mechanisms of enzyme catalysis, and the therapeutic action of many drugs depends on their acting as enzyme inhibitors. An understanding of various mechanisms of inhibition will provide insights into the mode of action of such drugs. Many different kinds of molecules inhibit enzymes, and they act in a variety of ways. We will present specific, detailed examples of such therapeutics in subsequent chapters.

A major distinction must be made between **reversible inhibitors** and **irreversible inhibitors.** Reversible inhibition involves *noncovalent* binding of the inhibitor and can always be reversed, at least in principle, by removal of the inhibitor. In some cases, noncovalent binding may be so strong as to appear irreversible under physiological conditions. Trypsin inhibitor binding to trypsin (see Figure 6.37) is one such example. In contrast, irreversible inhibition results when a molecule is *covalently* bound to the enzyme and inactivates it. Irreversible inhibition is frequently encountered in the action of specific toxins and poisons, many of which kill by incapacitating key enzymes.

● **CONCEPT** Inhibition of enzymes can be either reversible or irreversible.

Reversible Inhibition

The various modes of reversible inhibition all involve the noncovalent binding of an inhibitor to the enzyme, but they differ in the mechanisms by which they decrease the enzyme's activity and in how they affect the kinetics of the reaction.

Competitive Inhibition

Consider a molecule that can bind to an enzyme active site because it is structurally similar to the true substrate for the enzyme. If this molecule can also be processed by the enzyme, it is merely a competing alternative substrate. However, if the molecule binds to the active site but *cannot* undergo the chemical conversion step, it effectively reduces the enzyme's availability to carry out chemistry on true substrates. Such a molecule is called a **competitive inhibitor** because it competes with the substrate for binding to the same site on the enzyme (**FIGURE 8.24**).

For whatever fraction of time a competitive inhibitor molecule is occupying the active site, the enzyme is unavailable for catalysis. The overall effect is as if the enzyme cannot bind substrate as well when the inhibitor is present. Thus, the enzyme is predicted to act as if its

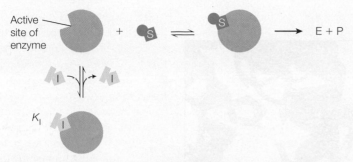

▲ **FIGURE 8.24 Competitive inhibition.** Both substrate and inhibitor can fit the active site. Substrate can be processed by the enzyme, but the inhibitor cannot.

K_M were increased by the presence of the inhibitor. These ideas can be expressed by writing the reaction scheme as

$$E + S \underset{k_{-1}}{\overset{k_1}{\rightleftharpoons}} ES \xrightarrow{k_{cat}} E + P$$
$$+$$
$$I$$
$$K_I \updownarrow$$
$$EI$$

Here I stands for the inhibitory substance, and K_I is a dissociation constant for inhibitor binding, defined as $K_I = [E][I]/[EI]$ where $[I]$ = concentration of free inhibitor. The rate equations can be solved as shown above (see Equations 8.20–8.28), but note that now

$[E]_t$	=	$[E]$	+	$[ES]$	+	$[EI]$	
Total enzyme		Free enzyme		Enzyme bound to substrate		Enzyme bound to inhibitor	(8.35)

Upon analysis of this case, the expression for v is found to be

$$v = \frac{k_{cat}[E]_t[S]}{K_M\left(1 + \dfrac{[I]}{K_I}\right) + [S]} \tag{8.36}$$

or

$$v = \frac{V_{max}[S]}{\alpha K_M + [S]} = \frac{V_{max}[S]}{K_M^{app} + [S]} \tag{8.37}$$

where $\alpha = (1 + [I]/K_I)$ and $K_M^{app} = \alpha K_M$. Thus, increasing [I] causes an apparent increase in the value observed for K_M because the factor α is always >1. Because the formation of EI depends on [I] just as the formation of ES depends on [S],

● **CONCEPT** A competitive inhibitor competes with substrate for the enzyme active site. It increases the apparent K_M, but does not change the observed V_{max}.

the rate of a competitively inhibited reaction is strictly dependent on the relative concentrations of I and S. Note that K_M for the substrate is not changing per se; rather, the presence of a competitive inhibitor increases the value of [S] required to reach ½ V_{max} (**FIGURE 8.25**). We use the symbol K_M^{app} to distinguish this *apparent* K_M, measured in the presence of the inhibitor, from the true K_M (measured without the inhibitor present).

For competitive inhibition, the value of V_{max} is unchanged because as [S] becomes very large, v approaches V_{max} (just as in the absence of inhibition), and $V_{max} = k_{cat}[E]_t$. Physically, this simply means that when [S] is very large at a given [I], the more numerous substrate molecules will outcompete the inhibitor for binding to the enzyme active site. The effect of competitive inhibition on a graph of v versus [S] is shown in Figure 8.25(a). Because the system, at a given [I], still obeys an equation of the Michaelis–Menten form,

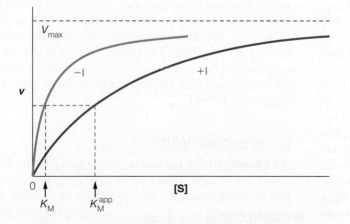

(a) The effect of a competitive inhibitor (I) on reaction velocity at different substrate concentrations. Two sets of substrate–velocity experiments were carried out, one with (red line) and one without (blue line) inhibitor present. Addition of the inhibitor decreases the velocity but not the V_{max}. The apparent K_M is higher in the presence of inhibitor.

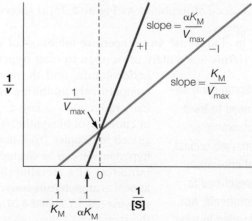

(b) Lineweaver–Burk plots of the reactions shown in (a). The lines cross the $1/v$ axis at the same V_{max}, showing that I is a competitive inhibitor.

▲ **FIGURE 8.25 Effects of competitive inhibition on enzyme kinetics.**

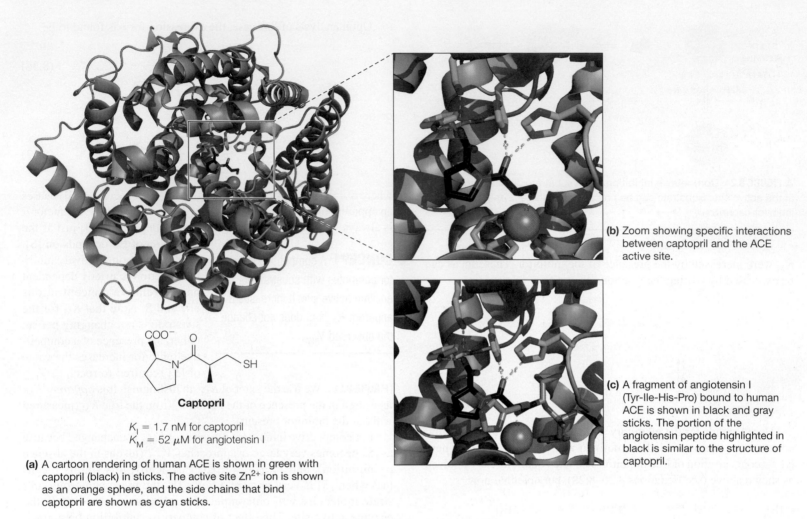

Captopril

K_I = 1.7 nM for captopril
K_M = 52 μM for angiotensin I

(a) A cartoon rendering of human ACE is shown in green with captopril (black) in sticks. The active site Zn^{2+} ion is shown as an orange sphere, and the side chains that bind captopril are shown as cyan sticks.

(b) Zoom showing specific interactions between captopril and the ACE active site.

(c) A fragment of angiotensin I (Tyr-Ile-His-Pro) bound to human ACE is shown in black and gray sticks. The portion of the angiotensin peptide highlighted in black is similar to the structure of captopril.

▲ **FIGURE 8.26** Captopril is a competitive inhibitor of angiotensin-converting enzyme.

the Lineweaver–Burk plots should be linear, with K_M (but not V_{max}) changed by the presence of inhibitor. As Figure 8.25(b) shows, this is exactly what happens.

Several drugs in clinical use are competitive inhibitors of a particular enzyme. Statins are widely prescribed to treat hypercholesterolemia, and they act by competitively inhibiting HMG-CoA reductase—a key enzyme in cholesterol biosynthesis (discussed in Chapter 16). Likewise, hypertension can be treated with inhibitors of a metalloprotease known as **angiotensin-converting enzyme** (ACE). **FIGURE 8.26** shows the structure of the competitive inhibitor captopril bound in the active site of human ACE. Captopril is a structural analog of the normal ACE substrate **angiotensin I,** which is a short peptide: Asp-Arg-Val-Tyr-Ile-His-Pro-Phe—His-Leu. Cleavage of the Phe-His peptide bond, highlighted in red, releases **angiotensin II,** which

● **CONNECTION** Some competitive inhibitors are used to treat human disease. For example, *statins* inhibit an enzyme critical for the synthesis of cholesterol; thus, statins are prescribed to treat hypercholesterolemia. ACE inhibitors are prescribed to treat hypertension.

causes elevated blood pressure. As shown in Figure 8.26, captopril binds tightly to several side chains in the ACE active site, as well as the Zn^{2+} ion that is required for activity, and thereby blocks the binding of angiotensin I to the enzyme active site. In general, for an enzyme inhibitor to be an effective therapeutic, it must bind its target specifically and with a K_I that is nanomolar or lower (note that K_I for captopril is 1.7 nM).

Uncompetitive Inhibition

An **uncompetitive inhibitor** binds tightly to the ES complex but shows low or zero affinity for the free enzyme. A simple uncompetitive inhibitor binds at a *second* site on an enzyme surface (not the active site) and prevents conversion of the bound substrate to product (**FIGURE 8.27(a)**). Alternatively, in the case of bisubtrate reactions (described later in this chapter), an inhibitor that binds at the active site *after* one of the substrates has bound will show

● **CONNECTION** The drugs methotrexate (an anticancer agent) and mycophenolic acid (an antirejection, or immunosuppressive, agent) are uncompetitive inhibitors of their target enzymes.

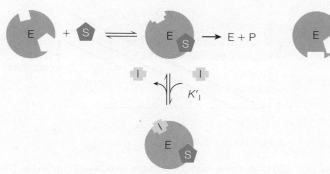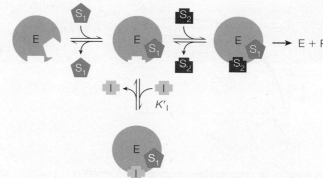

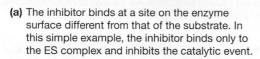

(a) The inhibitor binds at a site on the enzyme surface different from that of the substrate. In this simple example, the inhibitor binds only to the ES complex and inhibits the catalytic event.

(b) In a bisubstrate reaction with ordered binding shown, the inhibitor is uncompetitive with respect to substrate S1 and competitive with respect to substrate S2. The anticancer drug methotrexate shows uncompetitive inhibition of dihydrofolate reductase with respect to NADPH, and competitive inhibition with respect to dihydrofolate.

▲ **FIGURE 8.27 Modes of action for uncompetitive inhibition.** The ordered binding mechanism shown in part (b) is an example of a multisubstrate reaction, which is described later in this chapter.

uncompetitive kinetic behavior (Figure 8.27(b)). This mode of inhibition of dihydrofolate reductase is shown by the anticancer drug methotrexate (**FIGURE 8.28**).

An uncompetitive inhibitor typically affects both k_{cat} and K_M. The simplest case to consider is one in which the inhibitor molecule binds only to the ES complex and completely prevents the catalytic step (Figure 8.27(a)). Here we distinguish the equilibrium constant for inhibitor binding to free enzyme K_I from K'_I, the equilibrium constant for inhibitor binding to the ES complex.

$$E + S \underset{k_{-1}}{\overset{k_1}{\rightleftharpoons}} ES \overset{k_{cat}}{\longrightarrow} E + P$$

$$+$$
$$I$$
$$\updownarrow K'_I$$
$$EIS \xrightarrow{\times\!\!\!\!\times} \text{No reaction}$$

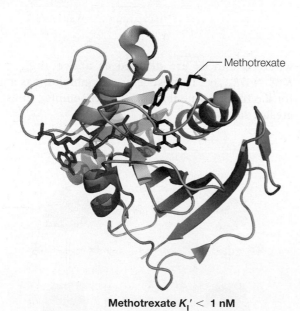

Methotrexate $K_I' < 1$ nM

(a) Methotrexate (black sticks) binds tightly to the complex of dihydrofolate reductase (green cartoon) and NADPH (magenta sticks). NADPH is a phosphorylated derivative of NADH. PDB ID: 1u72.

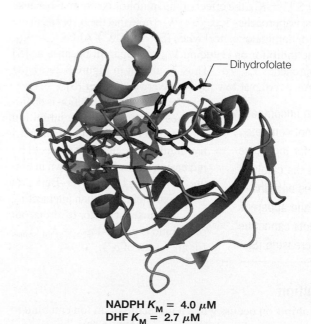

NADPH $K_M = 4.0\ \mu$M
DHF $K_M = 2.7\ \mu$M

(b) Dihydrofolate reductase bound to both its normal substrates NADPH (magenta sticks) and dihydrofolate (black sticks). PDB ID: 2w3m.

▲ **FIGURE 8.28 Methotrexate inhibition of dihydrofolate reductase.**

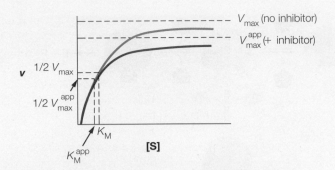

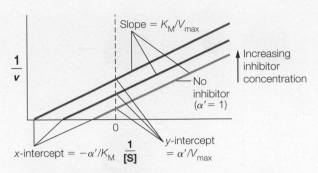

(a) The effect of an uncompetitive inhibitor (**I**) on reaction velocity at different substrate concentrations. In this simple example, both K_M and V_{max} are decreased by a factor of $1/\alpha'$.

(b) Lineweaver–Burk plots of the reactions shown in (a). The lines are parallel and cross the $1/v$ axis at different points, clearly distinguishing this situation from competitive inhibition (see Figure 8.25(b)).

▲ **FIGURE 8.29** Effects of uncompetitive inhibition on enzyme kinetics.

It can be shown that the Michaelis–Menten equation for the uncompetitive inhibition scheme is

$$v = \frac{\dfrac{V_{max}}{\alpha'}[S]}{\dfrac{K_M}{\alpha'} + [S]} = \frac{V_{max}^{app}[S]}{K_M^{app} + [S]} \tag{8.38}$$

where $\alpha' = (1 + [I]/K'_I), V_{max}^{app} = V_{max}/\alpha'$, and $K_M^{app} = K_M/\alpha'$. In the presence of an uncompetitive inhibitor, both V_{max} and K_M appear to be *reduced* by a factor of $1/\alpha'$ (**FIGURE 8.29(a)**). By binding only to the ES complex, an uncompetitive inhibitor increases the effective S binding, which reduces the apparent K_M. How can this observation be explained? At $[S] < K_M$, the effect of the inhibitor is minimal because as [S] decreases, v approaches $V_{max}[S]/K_M$ (note that the α' terms in the numerator and denominator cancel when $[S] \ll K_M^{app}$). At $[S] > K_M$, the effect of the inhibitor on reducing V_{max} is apparent because as [S] increases, v approaches V_{max}/α'. Thus, as shown in Figure 8.29(a), the v versus [S] plots overlap at low [S] but diverge at higher [S].

● **CONCEPT** An uncompetitive inhibitor does not compete with substrate for the active site but affects the catalytic event. It reduces both the apparent V_{max} and apparent K_M. These effects cannot be reversed by increasing [S].

Uncompetitive inhibition is distinguished from competitive inhibition by two observations: (1) uncompetitive inhibition cannot be reversed by increasing [S] and (2) as shown in Figure 8.29(b), the Lineweaver–Burk plot yields parallel rather than intersecting lines (because the factor α' drops out of the ratio to give a slope $= K_M/V_{max}$ for all values of α').

Mixed Inhibition

This form of inhibition occurs when a molecule or an ion can bind to both the free enzyme and the ES complex (**FIGURE 8.30**):

$$
\begin{array}{c}
\mathrm{E} + \mathrm{S} \underset{k_{-1}}{\overset{k_1}{\rightleftharpoons}} \mathrm{ES} \xrightarrow{k_{cat}} \mathrm{E} + \mathrm{P} \\
+ \qquad\qquad + \\
K_I \updownarrow \qquad\qquad \updownarrow K'_I \\
\mathrm{EI} + \mathrm{S} \dashleftarrow\dashrightarrow \mathrm{EIS} \xrightarrow{\;\times\;} \text{No reaction}
\end{array}
$$

Again, we distinguish the equilibrium constants K_I and K'_I, respectively, for inhibitor binding to E and ES. In the case where $K_I = K'_I$, the mixed mode of inhibition is also called **noncompetitive inhibition.** Substrate binding to the EI complex (green arrows in the scheme above) is typically significantly reduced compared to that for binding to free enzyme. Thus, the process $\mathrm{EI} + \mathrm{S} \rightleftharpoons \mathrm{EIS}$ will not be considered here, although it is part of a complete thermodynamic analysis; in doing so, the derivation of the Michaelis–Menten equation for mixed inhibition is greatly simplified without significantly altering the conclusions of the analysis. For this simplified case:

$$v = \frac{\dfrac{V_{max}}{\alpha'}[S]}{\dfrac{\alpha K_M}{\alpha'} + [S]} = \frac{V_{max}^{app}[S]}{K_M^{app} + [S]} \tag{8.39}$$

where α and α' are defined as above, $V_{max}^{app} = V_{max}/\alpha'$, and $K_M^{app} = \alpha K_M/\alpha'$. In most cases, the inhibitor has a greater affinity for the free enzyme than for the ES complex; thus, α is typically greater than α'.

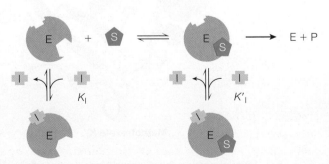

▲ **FIGURE 8.30 A model for mixed inhibition.** The inhibitor binds at a site on the enzyme surface different from that of the substrate. In this simplified example, the inhibitor binds to both free enzyme and the ES complex. EI has reduced substrate binding affinity compared to free enzyme. The EIS complex cannot carry out the catalytic event. As noted in the text, the process $\mathrm{EI} + \mathrm{S} \rightleftharpoons \mathrm{EIS}$ is not considered here.

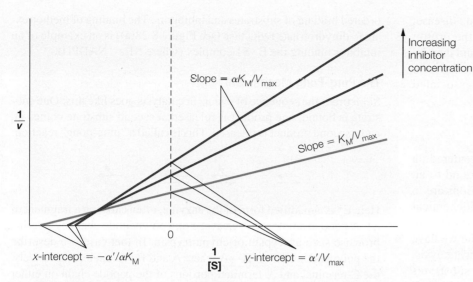

▲ **FIGURE 8.31 Lineweaver–Burk plot for mixed inhibition kinetics.** V_{max} is decreased by a factor of $1/\alpha'$, and K_M is increased by a factor of α/α'. Compare this plot with those in Figures 8.25(a) and 8.29 (b) to find the features that distinguish competitive, uncompetitive, and mixed modes of inhibition.

This mode of inhibition is "mixed" because the denominator of Equation 8.39 contains terms found in the equations for competitive inhibition (αK_M) and uncompetitive inhibition (V_{max}/α' and K_M/α'). As in competitive inhibition, K_M appears to be *increased* (by a factor of α/α'), and as in uncompetitive inhibition, V_{max} appears to be *reduced* (by a factor of $1/\alpha'$). **FIGURE 8.31** shows that the Lineweaver–Burk plot for mixed inhibition reflects this decreased V_{max} and increased K_M, and is distinct from similar plots for competitive and uncompetitive modes of inhibition.

● **CONCEPT** A mixed inhibitor does not compete with substrate for the active site but affects the catalytic event. It reduces the apparent V_{max} at all [S] and increases the apparent K_M.

Mixed inhibitors effectively reduce v at low and high values of [S]. At $[S] \ll K_M$, v approaches $V_{max}[S]/\alpha K_M$, and at $[S] \gg K_M$, v approaches V_{max}/α'. Thus, V_{max}^{app} will be less than V_{max} for all values of [S].

Examples of mixed inhibitors include the approved AIDS treatment nevirapine (shown on the opening page of this chapter) as well as several protein kinase inhibitors, which are currently in clinical trials for cancer treatment. The target of nevirapine, HIV reverse transcriptase, is discussed in greater detail in Chapter 22.

Irreversible Inhibition

Some substances combine *covalently* with enzymes to inactivate them irreversibly. Almost all **irreversible enzyme inhibitors** are toxic substances, either natural or synthetic. In most cases, such substances react with some functional group in the active site to leave it catalytically inactive or to block substrate binding.

● **CONCEPT** Many irreversible inhibitors bind covalently to the active sites of enzymes.

A typical example of an irreversible competitive inhibitor is found in *diisopropyl fluorophosphate (DFP)*. This compound reacts rapidly and irreversibly with serine hydroxyl groups to form a covalent *adduct,* as shown in **FIGURE 8.32**. Therefore, DFP acts as an irreversible inhibitor of enzymes that contain an essential serine in their active site. These enzymes include, among others, the serine proteases and the enzyme *acetylcholinesterase.* It is the inhibition of acetylcholinesterase that makes DFP such an exceedingly toxic substance to animals. Acetylcholinesterase is essential for nerve conduction (see Chapter 20), and its inhibition causes rapid paralysis of vital functions. Many insecticides and nerve gases resemble DFP and are potent acetylcholinesterase inhibitors.

For such irreversible inhibitors to react selectively with a critical residue, they must bind strongly to the active site. Many do so because they are transition-state analogs. Examples include DFP and the nerve gas *sarin,* which have a tetrahedral structure surrounding the phosphorus atom that is similar to the tetrahedral oxyanion transition states for the substrates in many hydrolytic enzymes.

The *penicillin* antibiotics also act as irreversible inhibitors of serine-containing enzymes used in bacterial cell wall synthesis (see Chapter 9). Not all irreversible inhibitors are toxins; some are therapeutic drugs. For example, omeprazole, which is widely prescribed to treat

● **CONNECTION** Several pesticides and some nerve gases irreversibly inactivate acetyl-cholinesterase, leading to inhibition of nerve conduction and the associated paralysis.

▲ **FIGURE 8.32 Irreversible inhibition by adduct formation.** Diisopropyl fluorophosphate (DFP) reacts with a serine group on a protein to form a covalent adduct. The covalent bond renders the catalytically important serine ineffective in catalysis. The adduct also may block substrate binding to the active site.

● **CONNECTION** Omeprazole is an irreversible inhibitor of the proton pump in the stomach and is widely prescribed to treat gastroesophageal reflux disease.

gastroesophageal reflux disease, irreversibly inhibits the proton pump in the stomach and thereby reduces the acid concentration in the stomach and the associated "heartburn."

Multisubstrate Reactions

Our discussions of enzyme kinetics have, to this point, centered on simple reactions in which one substrate molecule is bound to an enzyme and undergoes reaction there. In fact, such reactions are in the minority. Most biochemical reactions involve two or more substrates. An example we have already discussed is proteolysis, which involves two substrates (the polypeptide and water) and two products (the two fragments of the cleaved polypeptide chain). Phosphorylation of glucose, as catalyzed by hexokinase, is another such case: The two substrates are glucose and ATP, and the products are glucose-6-phosphate and ADP.

● **CONCEPT** Multisubstrate reactions fall into several classes, depending on the order of substrate binding: random, ordered, or ping-pong.

When an enzyme binds two or more substrates and releases multiple products, the order of the steps becomes an important feature of the enzyme mechanism. Knowledge of the order of substrate binding is used in the design of inhibitors that are used as therapeutic drugs. Several major classes of mechanisms for multisubstrate reactions are recognized. We shall briefly illustrate them using two substrates, S1 and S2, and two products, P1 and P2. Many of the enzymes that follow one of these schemes obey Michaelis–Menten kinetics, although the definitions of k_{cat} and K_M are more complex in these cases.

Random Substrate Binding

In random substrate binding, either substrate can be bound first, although in many cases one substrate will be favored for initial binding, and its binding may promote the binding of the other. The general pathway is

$$\begin{array}{c} \text{either} \quad S1 \quad \rightleftharpoons \quad E\cdot S1 \quad \searrow^{S2} \\ E \qquad\qquad\qquad\qquad\qquad E\cdot S1\cdot S2 \longrightarrow E + P1 + P2 \\ \text{or} \quad S2 \quad \rightleftharpoons \quad E\cdot S2 \quad \nearrow_{S1} \end{array}$$

The phosphorylation of glucose by ATP, with hexokinase as the enzyme, appears to follow such a mechanism, although there is some tendency for glucose to bind first.

Ordered Substrate Binding

In some cases, one substrate *must* bind before a second substrate can bind significantly:

$$E \underset{\longleftarrow}{\overset{S1}{\rightleftharpoons}} E\cdot S1 \underset{\longleftarrow}{\overset{S2}{\rightleftharpoons}} E\cdot S1\cdot S2 \longrightarrow E + P1 + P2$$

This mechanism is often observed in oxidations of substrates by the cofactor $NAD^+/NADH$, discussed earlier as a coenzyme for alcohol dehydrogenase. The actions of therapeutic drugs can also depend on

ordered binding of substrates and inhibitors. The binding of methotrexate to dihydrofolate reductase (see Figure 8.28(a)) is an example of an inhibitor binding the $E\cdot S1$ complex (where $S1 = NADPH$).

The Ping-Pong Mechanism

Sometimes the sequence of events in catalysis goes like this: One substrate is bound, one product is released, a second substrate comes in, and a second product is released. This is called a "ping-pong" reaction:

$$E \underset{\longleftarrow}{\overset{S1}{\rightleftharpoons}} E\cdot S1 \overset{P1}{\underset{\longleftarrow}{\rightleftharpoons}} E^* \underset{\longleftarrow}{\overset{S2}{\rightleftharpoons}} E^*\cdot S2 \overset{P2}{\longrightarrow} E$$

Here E* is a modified form of the enzyme, often carrying a fragment of S1. A good example is the cleavage of a polypeptide chain by a serine protease such as trypsin or chymotrypsin. In that case, we describe the polypeptide as S = B—A where A and B designate, respectively, the C-terminal and N-terminal portions of the peptide chain on either side of the scissile bond:

$$E \underset{\longleftarrow}{\overset{S}{\rightleftharpoons}} E\cdot S \overset{A}{\underset{\longleftarrow}{\rightleftharpoons}} E^*\cdot B \underset{\longleftarrow}{\overset{H_2O}{\rightleftharpoons}} E^*\cdot B\cdot H_2O \overset{B}{\longrightarrow} E$$

Here $E^*\cdot B$ and $E^*\cdot B\cdot H_2O$ indicate the covalent intermediates described earlier (see Figure 8.15).

Qualitative Interpretation of K_M and V_{max}: Application to Multisubstrate Reaction Mechanisms

As described in the previous section, the parameters V_{max} and $\frac{V_{max}}{K_M}$ are directly obtained from a Lineweaver–Burk plot by taking the reciprocals of the y-intercept and the slope, respectively. These parameters can provide a qualitative prediction of the order of substrate (or inhibitor) binding for multisubstrate reactions. This qualitative assessment is based on the observation that at saturating concentrations of substrate (i.e., $[S] \gg K_M$), the Michaelis–Menten equation simplifies to $v = V_{max}$. Thus, changes in V_{max} can be interpreted to provide information on the effects of an added second substrate (or inhibitor) *when substrate is bound to the enzyme* (i.e., when $[ES] \gg [E]$). Likewise, at very low concentrations of substrate (i.e., $[S] \ll K_M$), the Michaelis–Menten equation simplifies to $v = \frac{V_{max}}{K_M}[S]$. Thus, changes in $\frac{V_{max}}{K_M}$ can be interpreted to provide information on the effects of an added second substrate (or inhibitor) *when substrate is not bound to the enzyme* (i.e., when $[ES] \ll [E]$). Qualitatively, changes in only V_{max} reflect competitive binding modes for the second substrate (or inhibitor), changes in only $\frac{V_{max}}{K_M}$ reflect uncompetitive modes of binding, and changes in both parameters reflect mixed modes of binding. These principles are illustrated in **FIGURE 8.33**, which shows initial velocity experiments for the interaction of NADPH and DHF with **dihydrofolate reductase** (DHFR). Both Lineweaver–Burk plots show changes in V_{max} and $\frac{V_{max}}{K_M}$. This is consistent with a random substrate binding model as either substrate appears to bind the free enzyme or the ES complex.

Panel (a) of **FIGURE 8.34** shows the Lineweaver–Burk analysis for inhibition studies of DHFR. In this case, the experiments shown in Figure 8.33 are repeated except that the concentration of one of the substrates is held constant for a series of experiments where the concentration of inhibitor is varied. When tested against dihydrofolate (Figure 8.34(a)), the addition of increasing amounts of the inhibitor

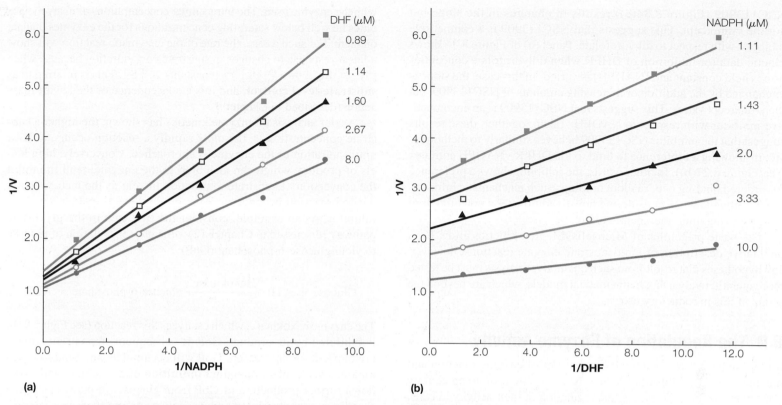

(a)

(b)

▲ **FIGURE 8.33** **Predicting order of binding of substrates to *E. coli* dihydrofolate reductase (DHFR).** Initial velocities at various concentrations of NADPH were recorded as a function of increasing dihydrofolate (panel (a)). In Panel (b), the concentration of NADPH is varied, while [dihydrofolate] is held constant. In both panels the reciprocal of concentration ($M^{-1} \times 10^{-5}$) is plotted.

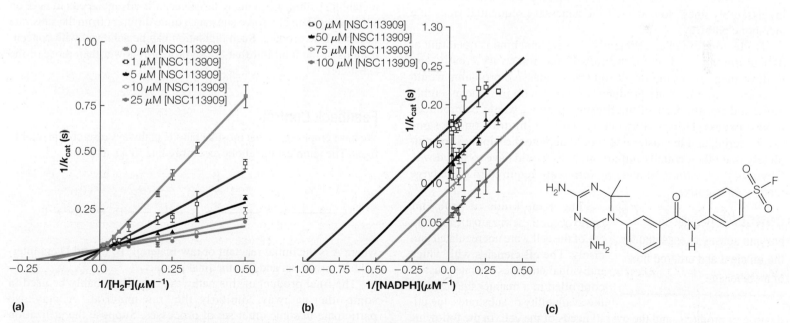

(a)

(b)

(c)

▲ **FIGURE 8.34** **Predicting mechanism of binding of an inhibitor to *E. coli* dihydrofolate reductase (DHFR).** Lineweaver–Burk plots of kinetic data from inhibition experiments show a change in slope when DHFR is treated with increasing concentrations of inhibitor NSC113909, while the concentration of dihydrofolate is varied and the co-substrate NADPH is held constant (panel (a)). Only the y-intercept changes when increasing concentrations of the inhibitor are added to reactions containing various [NADPH] with [dihydrofolate] held constant (panel (b)). The structure of the inhibitor is shown in panel (c).

NSC113909 (Figure 8.34(c)) results in changes in the slope but not the *y*-intercept. This suggests that NSC113909 is a competitive inhibitor with respect to dihydrofolate. Panel (b) of Figure 8.34 shows kinetic data for inhibition of DHFR when dihydrofolate concentration is held constant and [NADPH] is varied. In this case, the slope is unchanged by the addition of increasing amounts of NSC113909, but the *y*-intercept varies. This suggests that NSC113909 is an uncompetitive inhibitor with respect to NADPH. Taken together, these results suggest that the inhibitor NSC133909 behaves similarly to methotrexate, in showing a preference to bind to the DHFR · NADPH complex (see Figure 8.27(b)). In other words, the inhibitor shows a preference for ordered binding over random binding. Such mechanistic information is valuable when considering how to administer a potential drug to achieve maximal therapeutic effect.

The basic principles of Michaelis–Menten kinetics allow for a qualitative description of multisubstrate enzyme reactions; however, any hypotheses that result from such qualitative analysis must be tested more quantitatively using mathematical models, which are beyond the scope of this introductory text.

8.8 The Regulation of Enzyme Activity

So far we have described the basic features of enzyme function and mechanisms of inhibition. We now turn our attention to an equally important feature of enzymes: the regulation of their activity in cells. In Chapter 1 we made an analogy between a living cell and a factory. That analogy is especially appropriate when we consider the roles that enzymes play in living cells. We note that a cell has certain raw materials available to it and must produce specific products from them. Enzymes make up the majority of the machines that facilitate these transformations in the cell. Often, as we shall see, enzymes are arranged in "assembly lines" to carry out the necessary sequential steps in a metabolic pathway.

No factory operates efficiently if every machine is operating at its maximum rate. The capabilities of machines vary greatly, and if all of them were running at top speed, massive problems would soon arise. Intermediate products would pile up in some assembly lines, and certain parts of the finished product would be produced in vast excess. Different assembly lines might draw on the same raw material, and the faster ones could deplete the supplies so completely that other, equally important, lines would have to shut down. Obviously, *coordination* and *regulation* are required to run a large factory efficiently.

● **CONCEPT** Regulation of enzyme activity is essential for the efficient and ordered flow of metabolism.

The same kinds of problems could occur if the enzymatic machinery of the cell were not regulated precisely. The efficiencies with which individual enzymes operate must be controlled in a manner that reflects the availability of substrates, the utilization of products, and the overall needs of the cell. In the following chapters we will see many examples of such regulation.

Substrate-Level Control

Some enzyme regulation occurs in a simple way, through direct interaction of the substrates and products of each enzyme-catalyzed reaction with the enzyme itself. The intracellular concentrations of many metabolites lie well below saturating concentrations for the enzymes that act on them. For such a case, the rate of the enzymatic reaction will show a linear response to changes in substrate concentration because when $[S] \ll K_M, v = \frac{V_{max}}{K_M}[S] = (constant) \times [S]$. This is referred to as **substrate-level control,** and it is a consequence of the law of mass action (described in Chapter 3).

As our analysis of enzyme kinetics has shown, the higher a substrate concentration is, the more rapidly a reaction occurs, at least until saturation of the enzyme is approached. Conversely, high levels of product, which can also bind to the enzyme, tend to inhibit the conversion of substrate to product. Insofar as the metabolically desired reaction is concerned, the product can therefore act as an inhibitor. As an example, consider the first step in the glycolytic pathway (discussed in Chapter 12)—the phosphorylation of glucose to yield glucose-6-phosphate (G6P):

$$Glucose + ATP \xrightarrow{\text{Hexokinase}} glucose\text{-}6\text{-}phosphate + ADP$$

The enzyme hexokinase, which catalyzes this reaction (see Figure 8.9), is inhibited by its product, G6P. If subsequent steps in glycolysis are blocked for any reason, G6P will accumulate and bind to hexokinase. This results in **product inhibition** of hexokinase and slows down further production of G6P from glucose. In many cases, the reaction product binds the enzyme active site and therefore acts as a competitive inhibitor. Hexokinase is an interesting example because its product, G6P, can act both as a competitive inhibitor (by binding to the active site) and as an uncompetitive inhibitor (by binding at another site).

Substrate-level control is not sufficient for the regulation of many metabolic pathways. In many instances, it is advantageous to have an enzyme regulated by some substance quite different from the substrate or immediate product. Such regulation can be achieved with concentrations of the inhibitor that are significantly lower than those of the substrates.

Feedback Control

We have emphasized that most metabolic pathways resemble assembly lines. The simplest metabolic assembly line looks like this:

A →(Enzyme 1)→ B →(Enzyme 2)→ C →(Enzyme 3)→ D →(Enzyme 4)→ E

where A is the initial reactant or raw material; B, C, and D are intermediate products; and E is the final product.

The final product of this pathway, E, will probably be used in some other pathway. Similarly, the "raw material," A, may also participate in some other set of processes. Suppose the utilization of E suddenly slows down. If everything kept going as before, E would accumulate, and consumption of A would continue. But this process is inefficient. A more efficient process would solve this problem by closely monitoring the concentration of E and, as E accumulated, sending a signal back to inhibit its production. The cell can control generation of the final product through activation (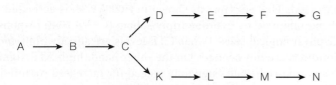) or inhibition (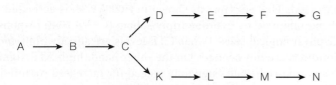) of a key step in the pathway. It would be most efficient to slow the *first step*—the conversion of A to B. So the A ⟶ B "machine" should be regulated by the concentration of E.

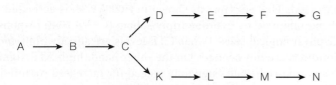

This type of **feedback control** is called **feedback inhibition** because an *increase* in the concentration of E leads to a *decrease* in its rate of production. Note that by inhibiting the first step, we prevent both unwanted utilization of A and accumulation of E. Furthermore, because most biochemical processes are reversible to some extent, generation of a large quantity of E will tend to build up the concentration of intermediate products. The feedback control mechanism visualized above prevents accumulation of any intermediates, which might have undesired effects on metabolism.

● **CONCEPT** Feedback control is important in the efficient regulation of complex metabolic pathways.

Other metabolic situations require more complicated patterns, in which **activation** as well as inhibition may be useful. For example, consider a slightly more complex case, in which A is fed into two pathways, which lead to two products needed in roughly equivalent amounts. Then a scheme like the following emerges:

To control the pathways so that G and N remain at their proper homeostatic concentrations, higher concentrations of G might *inhibit* the C → D enzyme and/or *activate* the C → K enzyme. Conversely, increasing the concentration of N might inhibit the C → K enzyme and/ or activate the C → D enzyme. Finally, it might be useful to have G and N act *together* to inhibit the A → B enzyme, to provide overall regulation. An example of this kind of control is found in the synthesis of the purine and pyrimidine monomers that go into making DNA because approximately equal quantities of all four deoxyribonucleotides are required for DNA replication.

Both inhibition and activation of enzymes are essential to regulate metabolism. Furthermore, control of pathways by their end products means that the necessary inhibitions and activations *must* be produced by molecules that come from far down the assembly line and therefore bear little or no structural resemblance to either the substrates or the direct products of the enzymes to be regulated. Activation or inhibition of enzymes that catalyze reactions at key control points in a pathway typically occurs through an allosteric mechanism. We have already studied an example of allosteric

control of protein function. Hemoglobin (see Chapter 7) is a four-subunit protein that has four binding sites for its "substrate," oxygen. The binding of oxygen is cooperative and is influenced by other molecules and ions. The basic ideas presented in Chapter 7 for the analysis of hemoglobin function apply equally well to allosteric enzymes.

Allosteric Enzymes

The term *allostery* has Greek roots that translate to "other solid" or "other structure." An allosteric effect occurs when the binding of a ligand results in a conformational change ("other structure") in the enzyme. As described in greater detail in the next section for aspartate carbamoyltransferase, such conformational change is associated with increased, or decreased, enzyme activity. Allosteric enzymes are frequently multisubunit proteins, with multiple active sites. They exhibit cooperativity in substrate binding (**homoallostery**) and regulation of their activity by other, effector molecules (**heteroallostery**).

Homoallostery

Let us first consider the homoallosteric effects (cooperative substrate binding). In Chapter 7, we contrasted O_2 binding by the single-subunit protein myoglobin with binding by the multisubunit hemoglobin. Myoglobin gives a hyperbolic binding curve (Figure 7.18); hemoglobin, with its cooperative binding, gives a sigmoidal curve (Figure 7.19(d)). We find *exactly the same contrast* when we compare the v versus [S] curve of a single-site enzyme obeying Michaelis–Menten kinetics with that of a multisite enzyme showing cooperative binding (**FIGURE 8.35**). The same kind of reasoning applies: An enzyme that binds substrate cooperatively will behave, at low-substrate concentration, as if it were poor at substrate binding (that is, as if it had a large K_M). But as the substrate levels are

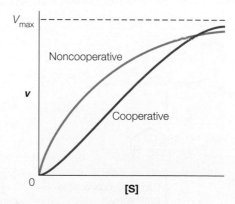

▲ **FIGURE 8.35 Effect of cooperative substrate binding on enzyme kinetics.** Comparison of v versus [S] curves for a noncooperative enzyme and an allosteric enzyme with cooperative binding. The two enzymes are assumed to have the same V_{max}. Compare this plot with the myoglobin and hemoglobin oxygen binding curves shown in Figure 7.19.

increased and more substrate is bound, the enzyme becomes more and more effective because it binds substrate more avidly in the last sites to be filled. We imagine this happening, as with hemoglobin, because as more substrate is bound, the enzyme undergoes a transition from a lower affinity state (T state) to a higher affinity state (R state). The kinds of models that have been used to describe O_2 binding by hemoglobin (see Figure 7.20) can account equally well for the kinetics exhibited by enzymes that show cooperative substrate binding.

Heteroallostery

The major advantage of allosteric control is found in the role of **heteroallosteric effectors,** which may be either inhibitors or activators. These effectors are the analogs, in enzyme kinetics, of the CO_2, BPG, and H^+ that so elegantly regulate O_2 binding by hemoglobin.

● **CONCEPT** Allosteric enzymes show cooperative substrate binding and can respond to a variety of inhibitors and activators.

The activation and inhibition of enzymes by allosteric effectors are the keys to the kind of complex feedback control described above. If an enzyme molecule can exist in two conformational states (T and R) that differ dramatically in the strength with which substrate is bound or in the catalytic rate, then its kinetics can be controlled by *any* other substance that, in binding to the protein, alters the T $\rightleftharpoons$ R equilibrium. Allosteric *inhibitors* shift the equilibrium toward T, and *activators* shift it toward R (**FIGURE 8.36**). Some enzymes are regulated by multiple inhibitors and activators, allowing extremely subtle and complex patterns of metabolic control.

Aspartate Carbamoyltransferase: An Example of an Allosteric Enzyme

An excellent example of allosteric regulation is provided by the enzyme **aspartate carbamoyltransferase** (also known as aspartate transcarbamoylase, or ATCase), a key enzyme in pyrimidine synthesis (Chapter 19). As can be seen from **FIGURE 8.37**, ATCase stands at

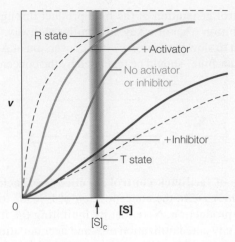

▲ **FIGURE 8.36 Heteroallosteric control of an enzyme.** In the absence of activation or inhibitors, the *v* versus [S] curve is sigmoidal. Activators shift the system toward the R state; inhibitors stabilize the T state. $[S]_c$ represents the homeostatic concentration range for S. Note that effectors significantly alter the activity of the enzyme over this range of [S].

a crossroads in biosynthetic pathways. Glutamine, glutamate, and aspartate are also used in protein synthesis; but once aspartate has been carbamoylated to form **N-carbamoyl-L-aspartate** (CAA), the molecule is committed to pyrimidine biosynthesis. Thus, the enzyme that controls this step must be sensitive to pyrimidine need. In bacteria like *E. coli*, the activity of ATCase is regulated to respond to this need. This enzyme, as shown in **FIGURE 8.38**, is activated by ATP and inhibited by cytidine triphosphate (CTP). Both responses make physiological sense; when CTP levels are already high, more pyrimidines are not needed. On the other hand, high ATP signals both a purine-rich state (signaling a need for increased pyrimidine synthesis) and an energy-rich cell condition under which DNA and RNA synthesis will be active. Recent analysis of ATCase activity suggests that in vivo, a complex of CTP, uridine triphosphate (UTP,

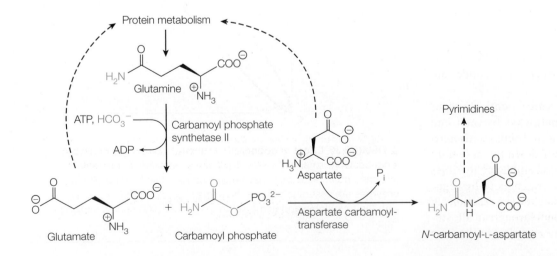

Glutamate

Carbamoyl phosphate

N-carbamoyl-L-aspartate

◄ **FIGURE 8.37 Control points in pyrimidine synthesis.** This figure shows the formation of *N*-carbamoyl-L-aspartate from carbamoyl phosphate and aspartate. This reaction is the first step in a series of reactions committed to synthesis of pyrimidine nucleotides, so control at or near this point is essential. In prokaryotes, the aspartate carbamoyltransferase is regulated; in most eukaryotes, regulation is on the preceding step, catalyzed by carbamoyl phosphate synthetase II.

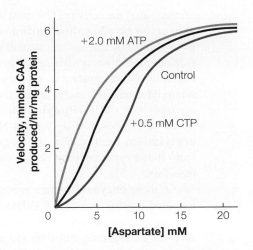

▲ **FIGURE 8.38 Regulation of aspartate carbamoyltransferase by ATP and CTP.** ATP is an activator of aspartate carbamoyltransferase, and CTP is an inhibitor (when no divalent cations are present). The curve marked as "control" shows the behavior of the enzyme in the absence of both regulators. *N*-Carbamoyl-L-aspartate (CAA) is the product of the reaction.

another product of the pyrimidine biosynthesis pathway) and Mg^{2+} is required to inhibit ATCase.

Like most allosteric enzymes, ATCase is a multisubunit protein. Its quaternary structure has been examined in some detail and is depicted schematically in **FIGURE 8.39**. There are six *catalytic*

subunits, in two tiers of three, held together by six *regulatory* subunits. Pairs of regulatory subunits appear to connect catalytic subunits in the two tiers. The three-dimensional structure of ATCase has been solved to high resolution, and a detailed representation of two catalytic subunits linked to two regulatory subunits is shown in **FIGURE 8.40**. The catalytic subunit comprises two domains, one binding aspartate and the other carbamoyl phosphate, and the active site lies between them. The regulatory subunit likewise has two parts; the so-called zinc domain and the allosteric domain. The zinc domain binds a structurally necessary zinc ion; the allosteric domain contains the ATP/CTP-UTP binding site. ATP and the two pyrimidines thus compete for the same site, so that the activity of ATCase is regulated by the *ratio* of ATP to pyrimidines in the cell.

As in the case of hemoglobin, the allosteric regulation of ATCase involves changes in the quaternary structure of the molecule. Conformations of the R and T states have been determined by X-ray diffraction. As Figure 8.40 shows, a rearrangement of the relative subunit positions occurs in the T ⇌ R transition.

Virtually every metabolic pathway we shall encounter in the following chapters is subject to complex feedback control, and in almost all cases multisubunit, allosteric enzymes are employed. The pattern of control, even in a given pathway, is not the same in every organism. For example, whereas ATCase is the major control point in the pyrimidine biosynthesis pathway in bacteria, eukaryotes regulate at the preceding step—the synthesis of carbamoyl phosphate (see Figure 8.37).

It should be clear at this point that organisms can regulate metabolism in complex and subtle ways through allosteric enzymes; however, this kind of regulation is not sufficient for all needs. We turn now to covalent modification, an entirely different kind of regulatory mechanism.

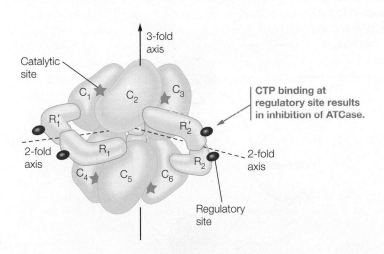

(a) Quaternary structure of ATCase in the T state. This schematic view of the enzyme shows the six catalytic subunits (C) and six regulatory subunits (R). Six catalytic sites (blue stars) lie in or near the grooves between the catalytic subunits. Regulatory sites (red ellipses) lie on the outer surfaces of the regulatory subunits. The molecule has 1 three-fold axis (solid arrow) and 3 two-fold axes (dashed lines; D_3 symmetry). This is a side view of the molecule with the three-fold axis in the plane of the paper.

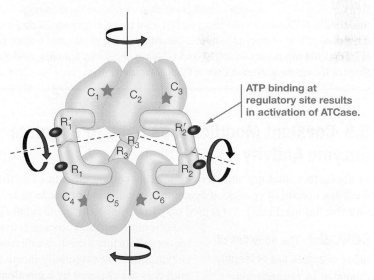

(b) Transition of ATCase to the R state. The transition involves a rotation of the regulatory subunits, which pushes the two tiers of catalytic subunits apart and rotates them slightly about the three-fold axis.

▲ **FIGURE 8.39 Quaternary structure of aspartate carbamoyltransferase (ATCase).**

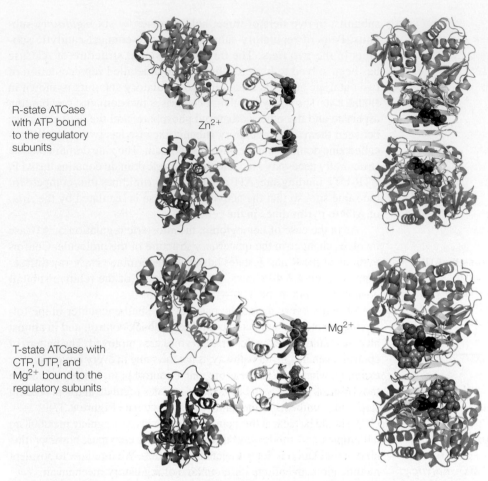

▲ FIGURE 8.40 The structure of the catalytic and regulatory subunits of ATCase in R and T conformations. Only two catalytic (green and dark green) and two regulatory domains, (yellow and light yellow) are shown. The three-fold axis (see Figure 8.39) is in the plane of the page. The location of the Zn^{2+} in each regulatory domain is shown by an orange sphere. The top row shows the R-state ATCase with ATP bound. The right panel is rotated 90 degrees with respect to the left panel. ATP is shown as spheres. The bottom row shows the T-state with CTP, UTP and Mg^{2+} bound to the regulatory site. CTP and UTP are shown as spheres, and the Mg^{2+} that bridges them is shown as a cyan sphere. PDB ID: 4kgv and 4fyy.

R-state ATCase with ATP bound to the regulatory subunits

Zn^{2+}

T-state ATCase with CTP, UTP, and Mg^{2+} bound to the regulatory subunits

Mg^{2+}

8.9 Covalent Modifications Used to Regulate Enzyme Activity

In the factory analogy, allosteric regulation can be thought of as the feedback control of continuously running machines. But any large factory also has machinery that is used only from time to time and is left on

● **CONCEPT** The activities of many enzymes are reversibly regulated by phosphorylation (and dephosphorylation). ATP-dependent *kinases* add a phosphoryl group to the target enzyme, whereas *phosphatases* remove the added phosphoryl group via hydrolysis.

standby until needed. The same is true for the cell. In this section, we discuss enzymes that are essentially inactive until they are changed by a **covalent modification** and then begin to function. In some cases, such modification acts in the opposite direction, to inactivate otherwise active enzymes. Some such modifications can be reversed; others cannot.

A number of kinds of covalent modification are commonly used to

regulate enzyme activity. The most widespread appears to be **phosphorylation** or **dephosphorylation** of various amino acid side chains (serine, threonine, tyrosine, and histidine, for example). Other covalent modifications include **adenylylation,** the transfer of an adenylate moiety from ATP; **ADP-ribosylation,** the transfer of an ADP-ribosyl moiety from NAD^+; and **acetylation,** the transfer of an acetyl group from the enzyme cofactor acetyl-coenzyme A (see Table 8.3).

Of those enzymes that are subject to covalent modification, most are regulated by reversible phosphorylation. **Protein kinases** are ATP-dependent enzymes that add a phosphoryl group to the —OH group of a Tyr, Ser, or Thr on some target protein (**FIGURE 8.41**). This process is made reversible by a second class of enzymes, called **phosphatases,** which hydrolyze the resulting side-chain phosphate esters, releasing P_i. Much research activity has been devoted to understanding the roles of various kinases and phosphatases in cell signaling and regulation of metabolism. Protein phosphorylation can affect enzyme activity in several ways. The addition of a highly charged phosphoryl group to an uncharged hydroxyl can result in repulsive (or attractive) electrostatic interactions that lead to enzyme conformational change. Binding affinity for a substrate or co-regulatory protein can also be dramatically altered by phosphorylation or dephosphorylation.

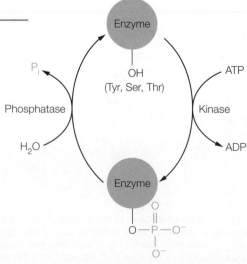

▲ FIGURE 8.41 Reversible covalent modification by kinases/phosphatases. The target residues for ATP-dependent phosphorylation by kinases are serine, threonine, or tyrosine. The phosphoprotein is dephosphorylated via a hydrolysis reaction catalyzed by a phosphatase.

Such covalent modifications are part of complex regulatory pathways, frequently under hormonal control. Several of these pathways are discussed in great detail in subsequent chapters, where the significance of regulation—and the consequences of aberrant regulation—will be presented. At this point, we shall concentrate instead on a third kind of covalent modification—the irreversible activation of certain enzymes by proteolytic cleavage of precursor forms.

Pancreatic Proteases: Activation by Irreversible Protein Backbone Cleavage

An important example of covalent enzyme activation, **proteolytic cleavage,** is found in the maturation of **pancreatic proteases.** These include a number of enzymes—for example, trypsin, chymotrypsin, elastase, and carboxypeptidase—some of which we have already discussed. All are synthesized in the pancreas. They are secreted through the pancreatic duct into the duodenum of the small intestine in response to a hormone signal generated when food passes from the stomach. They are not, however, synthesized in their final, active form because a battery of potent proteases free in the pancreas would digest the pancreatic tissue. Rather, they are made as slightly longer, catalytically inactive enzyme precursor molecules, called **zymogens.** The names given to the zymogens of these enzymes are *trypsinogen, chymotrypsinogen, proelastase,* and *procarboxypeptidase,* respectively. The zymogens must be cleaved proteolytically in the intestine to yield the active enzymes. The cleavage of zymogens to active enzymes is diagrammed in **FIGURE 8.42**.

● **CONCEPT** Some enzymes, such as pancreatic proteases, are synthesized in an inactive precursor form, called a *zymogen,* to mask an activity that would otherwise be toxic to the parent cell.

The first step in the activation of these four enzymes is the activation of trypsin in the duodenum. A hexapeptide is removed from the N-terminal end of trypsinogen by *enteropeptidase,* a protease secreted by duodenal cells. This action yields the active trypsin, which then activates the other zymogens by specific proteolytic cleavages. In fact, once some active trypsin is present, it will activate other trypsinogen molecules to make more trypsin; thus, its activation is *autocatalytic.* This is an example of the kind of **cascade** process frequently observed when enzymes are activated by covalent modification. The production of just a few trypsin molecules leads quickly to many more, as each enzyme molecule, when activated, can process many more every minute.

● **CONCEPT** Pancreatic zymogens are converted to fully active enzymes by irreversible proteolytic cleavage.

The activation of chymotrypsinogen to chymotrypsin is one of the most complex and best-studied examples of proteolytic activation of an enzyme; it is illustrated in **FIGURE 8.43**. In the first step, trypsin cleaves the bond between arginine 15 and isoleucine 16. The N-terminal peptide remains attached to the rest of the molecule because of the disulfide bond between residues 1 and 122. The product, called *π-chymotrypsin,* is an active enzyme.

Just how the cleavage of one peptide bond transforms an essentially inactive protein into an active one can now be understood as a result of detailed X-ray diffraction studies of the zymogen and the active enzyme. Cleavage of the peptide bond between residues 15 and 16 creates a new, positively charged N-terminal residue at Ile 16. This residue shifts its position and forms a salt bridge with Asp 194, the neighbor of the active site Ser 195 (see Figure 8.16). This change in turn triggers further conformational rearrangements in the active site. These changes result in the formation of a catalytically competent active site pocket, including the movement of main-chain

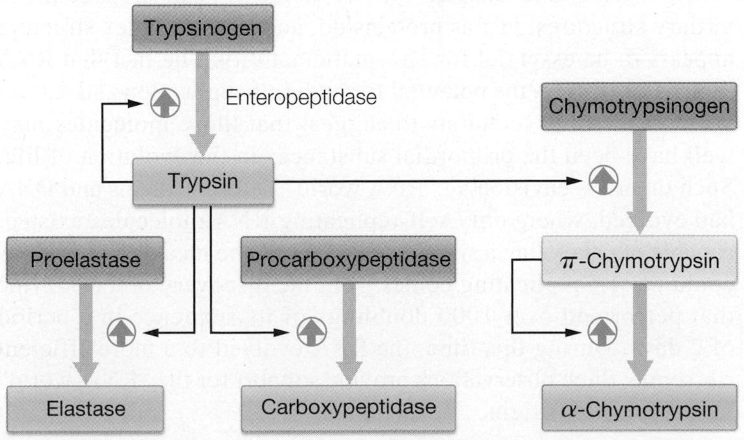

▲ **FIGURE 8.42 Zymogen activation by proteolytic cleavage.** This schematic view shows the activation of pancreatic zymogens, molecules that become catalytically active when cleaved. Zymogens are shown in orange and active proteases in yellow or green. The difference between π-chymotrypsin and α-chymotrypsin is shown in Figure 8.43.

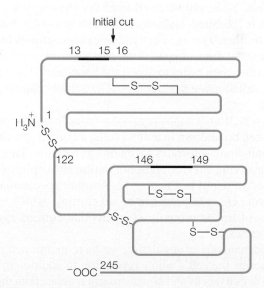

▲ **FIGURE 8.43 Activation of chymotrypsinogen.** The figure is a schematic rendition of the primary sequence of chymotrypsinogen. A series of peptide bond cleavages produces the enzyme chymotrypsin, with the disulfide bonds continuing to hold the structure together. The initial cleavage between residues 15 and 16 (arrow) results in the formation of π-chymotrypsin. Subsequent proteolytic removal of the segments shown in black yields α-chymotrypsin.

amino groups of residues 193 and 195 to form the oxyanion hole. Thus, both the substrate-binding pocket and the catalytic site are formed correctly only after the peptide bond between Arg 15 and Ile 16 has been cleaved.

π-Chymotrypsin is not the most active form of chymotrypsin. More autocatalytic cleavages remove residues 14–15 and 147–148 from the molecule, to produce the final α-*chymotrypsin,* which is the principal and fully active form found in the digestive tract.

This battery of enzymes, trypsin, chymotrypsin, elastase, and carboxypeptidase, together with the *pepsin* of the stomach and other proteases secreted by the intestinal wall cells, is capable of ultimately digesting most ingested proteins into free amino acids, which can be absorbed by the intestinal epithelium. The enzymes themselves are continually subjected to mutual digestion and auto-digestion so that high levels of these enzymes never accumulate in the intestine.

Even inactive zymogens are a potential source of danger to the pancreas. Because trypsin activation can be autocatalytic, the presence of even a single active trypsin molecule could set the activation cascade in motion prematurely. Therefore, the pancreas protects itself further by synthesizing a protein called the *secretory pancreatic trypsin inhibitor* (to be distinguished from the pancreatic trypsin inhibitor shown in Figure 6.37, which is an intracellular protein found only in ruminants). This competitive inhibitor binds so tightly to the active site of trypsin that it effectively inactivates it even at very low concentration. The bonding between trypsin and its inhibitor is among the strongest noncovalent associations known in biochemistry. Only a tiny amount of trypsin inhibitor is present—far less than needed to inhibit all of the potential trypsin in the pancreas. Thus, only a fraction of the trypsin generated in the duodenum is inhibited, and the rest can be activated. Because protection is limited, zymogen activation can sometimes be triggered in the pancreas—for example, if the pancreatic duct is blocked. The active enzymes then begin to digest the pancreatic tissue itself. This condition, called *acute pancreatitis,* is extremely painful and sometimes fatal.

● **CONNECTION** Premature activation of zymogens in the pancreas can result in acute pancreatitis.

The first well-understood regulatory cascade was the one controlling glycogen breakdown in animal cells, a critical process that provides carbohydrate substrates for energy generation. This regulatory cascade, involving enzyme phosphorylation and dephosphorylation, is described in detail in Chapter 12. Another spectacular example of an enzyme cascade occurs in blood clotting, which is mediated by a series of linked proteolysis events that convert zymogens to active proteases.

The mechanisms of regulation we have mentioned here by no means describe the cell's whole repertoire. In addition to regulation of enzyme activity, cells and organisms can regulate both the synthesis and degradation of enzymes, as well as the compartmentalization of enzymes within specific organelles or multienzyme complexes. However, description of these processes is more appropriate in the broader framework of Chapter 11.

8.10 Nonprotein Biocatalysts: Catalytic Nucleic Acids

Throughout this chapter, we have described how the proteins called enzymes function as biocatalysts. Indeed, for many years, it was assumed that *all* biochemical catalysis was carried out by proteins. But biochemistry is full of surprises, and research performed in the 1980s revealed something wholly unexpected: Some RNA molecules, called **ribozymes,** can act as enzymes.

The first hint that RNA might have catalytic activity came from studies of **ribonuclease P,** an enzyme that cleaves the precursors of tRNAs to yield the functional tRNAs (see Chapter 24). It had been known for some time that active ribonuclease P contained both a protein portion and an RNA "cofactor," but it was widely assumed that the active site resided on the protein portion. However, careful studies of the isolated components revealed an astonishing fact: Whereas the protein component alone was wholly inactive, the RNA by itself, if provided with either a sufficiently high concentration of magnesium ion or a small amount of magnesium ion plus the small basic molecule spermine, was capable of catalyzing the specific cleavage of pre-tRNAs. Furthermore, the RNA acted like a true enzyme, being unchanged in the process and obeying Michaelis–Menten kinetics. Addition of the protein portion of ribonuclease P does enhance the activity (k_{cat} is markedly increased) but is in no way essential for either substrate binding or cleavage. At high salt concentrations, the RNA itself becomes a very efficient catalyst; K_M becomes very low, and k_{cat}/K_M approaches $10^7 \text{ M}^{-1}\text{s}^{-1}$. This example is by no means unique; we discuss other RNA-catalyzed reactions in Chapters 24 and 25.

● **CONCEPT** Ribozymes, a class of ribonucleic acids, function as biological catalysts.

As discussed in Chapter 4, RNA molecules can adopt complex tertiary structures, just as proteins do, and such complex structure appears to be essential for enzymatic activity. The fact that RNA molecules possess the potential for both self-replication and catalysis has led some scientists to suggest that these molecules may well have been the primordial substances in the evolution of life. Such theorists envision an "RNA world" before proteins and DNA had evolved, where only self-replicating RNA molecules existed, capable of catalyzing a simple metabolism. The idea that ribozymes could be self-replicating comes from the discovery of a ribozyme that performed over 1000 doublings of its sequence in a period of 2 days. During this time, the RNA evolved to a more efficient ribozyme. Such observations provide support for the "RNA-world" model for life's origin.

Summary

- Enzymes are biological **catalysts** that increase the rates of biochemical processes. Like all catalysts, enzymes are regenerated by the reactions they catalyze. Most (but not all) enzymes are proteins. Some nucleic acid molecules function as enzymes. (Sections 8.1, 8.2, and 8.10).

- The rate of a chemical reaction is determined by reactant concentrations and by the rate constant. The **rate constant** in turn depends on the activation energy needed to reach the transition state. All catalysts function by lowering the activation energy for a reaction. In doing so, they increase the reaction rate but do not affect the chemical equilibrium. (Section 8.3)

- In enzyme catalysis, one or more substrates are bound at the **active site** of an enzyme, to form the **enzyme–substrate complex;** products are then formed and released. (Section 8.4)

- The **induced fit hypothesis** states that enzymes induce bound substrates to adopt conformations close to the transition state, though binding may also bring about conformational change in the enzyme. (Section 8.4)

- Many enzymes utilize **coenzymes** in their function; others require specific metal ions. A number of coenzymes are closely related to vitamins required in the human diet. (Section 8.5)

- Most simple enzymatic reactions can be described by the **Michaelis–Menten equation,** with two parameters, the **Michaelis constant,** K_M, and the **turnover number,** k_{cat}. (Section 8.6)

- Enzymes can be inhibited reversibly or irreversibly. **Reversible inhibition** can be competitive, uncompetitive, or mixed. Competitive inhibition increases the apparent K_M; uncompetitive inhibition reduces the apparent V_{max} and apparent K_M; and mixed inhibition reduces the apparent V_{max} and increases the apparent K_M. **Irreversible inhibition** usually involves covalent binding to the active site. (Section 8.7)

- Regulation of enzymatic activity takes many forms. **Substrate-level regulation** simply depends on ambient concentrations of reactants and products. Allosteric regulation provides sensitive **feedback control** for complex metabolic pathways. For more drastic changes in activity, some enzymes are switched on or off (or both) by covalent modification. (Sections 8.8 and 8.9)

Problems

Enhanced by
Mastering Chemistry
for Biochemistry

Mastering Chemistry for Biochemistry provides select end-of-chapter problems and feedback-enriched tutorial problems, animations, and interactive figures to deepen your understanding of complex topics while practicing problem solving.

Answers to red problems are available in the Answer Appendix.

1. Show that the half-life for a first-order reaction is inversely proportional to the rate constant, and determine the constant of proportionality.

2. The enzyme urease catalyzes the hydrolysis of urea to ammonia plus carbon dioxide. At 21 °C the uncatalyzed reaction has an activation energy of about 125 kJ/mol, whereas in the presence of urease the activation energy is lowered to about 46 kJ/mol. By what factor does urease increase the velocity of the reaction?

3. An enzyme contains an active site aspartic acid with a pK_a = 5.0, which acts as a *general acid catalyst*. On the accompanying template, draw the curve of enzyme activity (reaction rate) versus pH for the enzyme (assume that the protein is stably folded between pH 2–12 and that the active site Asp is the only ionizable residue involved in catalysis). Briefly explain the shape of your curve.

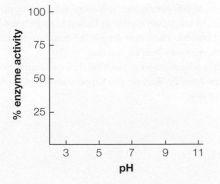

4. The folding and unfolding rate constants for a myoglobin mutant have been determined. The unfolding rate constant $k_{F \to U} = 3.62 \times 10^{-5} \text{s}^{-1}$ and the folding rate constant $k_{U \to F} - 255 \text{ s}^{-1}$, where F is the folded protein and U is the unfolded (denatured) protein. For wild-type myoglobin, $\Delta G^{\circ\prime}_{F \to U} = +37.4$ kJ/mol. Which myoglobin is more thermodynamically stable, the mutant or the wild-type?

5. In some reactions, in which a protein molecule is binding to a specific site on DNA, a rate *greater* than that predicted by the diffusion limit is observed. Suggest an explanation. [*Hint:* The protein molecule can also bind weakly and nonspecifically to *any* DNA site.]

6. Would you expect an "enzyme" designed to bind to its target substrate as tightly as it binds the reaction transition state to show a rate enhancement over the uncatalyzed reaction? In other words, would such a protein actually be a catalyst? Explain why or why not.

7. The initial rate for an enzyme-catalyzed reaction has been determined at a number of substrate concentrations. Data are as follows:

[S](μmol/L)	v[(μmol/L)min^{-1}]
5	22
10	39
20	65
50	102
100	120
200	135

(a) Estimate V_{max} and K_M from a direct graph of v versus [S]. Do you find difficulties in getting clear answers?

(b) Now use a Lineweaver–Burk plot to analyze the same data. Does this work better?

8. (a) If the total enzyme concentration in Problem 7 was 1 nmol/L, how many molecules of substrate can a molecule of enzyme process in each minute?

(b) Calculate k_{cat}/K_M for the enzyme reaction in Problem 7. Is this a fairly efficient enzyme? (See Table 8.5.)

9. Figure 8.19 shows a proposed mechanism for carboxypeptidase A.

(a) What is the role of Glu 270 in catalysis?

(b) What is the role of Arg 145 in catalysis?

10. The catalytic efficiency of many enzymes depends on pH. Chymotrypsin shows a maximum value of k_{cat}/K_M at pH 8. Detailed analysis shows that k_{cat} increases rapidly between pH 6 and 7 and remains constant at higher pH. K_M also increases rapidly between pH 8 and 10. Suggest explanations for these observations.

11. The following data describe the catalysis of cleavage of peptide bonds in small peptides by the enzyme elastase.

Substrate	K_M(mM)	k_{cat}(s^{-1})
PAPA↓G	4.0	26
PAPA↓A	1.5	37
PAPA↓F	0.64	18

The arrow indicates the peptide bond cleaved in each case.

(a) If a mixture of these three substrates was presented to elastase with the concentration of each peptide equal to 0.5 mM, which would be digested most rapidly? Which most slowly? (Assume enzyme is present in excess.)

(b) On the basis of these data, suggest what features of amino acid sequence dictate the specificity of proteolytic cleavage by elastase.

(c) Elastase is closely related to chymotrypsin. Suggest two kinds of amino acid residues you might expect to find in or near the active site.

12. At 37 °C, the serine protease subtilisin has $k_{cat} = 50$ s^{-1} and $K_M = 1.4 \times 10^{-4}$ M. It is proposed that the N155 side chain contributes a hydrogen bond to the oxyanion hole of subtilisin. J. A. Wells and colleagues reported (1986, *Phil. Trans. R. Soc. Lond. A* 317:415–423) the following kinetic parameters for the N155T mutant of subtilisin: $k_{cat} = 0.02$ s^{-1} and $K_M = 2 \times 10^{-4}$ M.

(a) Subtilisin is used in some laundry detergents to help remove protein-type stains. What unusual kind of stability does this suggest for subtilisin?

(b) Subtilisin does have a problem in that it becomes inactivated by oxidation of a methionine close to the active site. Suggest a way to make a better subtilisin.

(c) Is the effect of the N155T mutation what you would expect for a residue that makes up part of the oxyanion hole? How do the reported values of k_{cat} and K_M support your answer?

(d) Assuming that the T155 side chain cannot H-bond to the oxyanion intermediate, by how much (in kJ/mol) does N155 appear to stabilize the transition state at 37 °C?

(e) The value you calculated in part (d) represents the strength of the H-bond between N155 and the oxyanion in the transition state. This value is higher than typical H-bonds in water. How might this observation be rationalized? *Hint:* Consider Equation 2.2 (Coulomb's Law).

13. The accompanying figure shows three Lineweaver–Burk plots for enzyme reactions that have been carried out in the presence, or absence, of an inhibitor. Indicate what type of inhibition is predicted based on each Lineweaver–Burk plot. For each plot indicate which line corresponds to the reaction without inhibitor and which line corresponds to the reaction with inhibitor present.

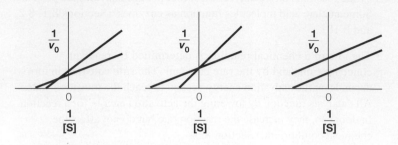

14. The steady-state kinetics of an enzyme are studied in the absence and presence of an inhibitor (inhibitor A). The initial rate is given as a function of substrate concentration in the following table:

	v[(mmol/L)min^{-1}]	
[S] (mmol/L)	No inhibitor	Inhibitor A
1.25	1.72	0.98
1.67	2.04	1.17
2.50	2.63	1.47
5.00	3.33	1.96
10.00	4.17	2.38

(a) What kind of inhibition (competitive, uncompetitive, or mixed) is involved?

(b) Determine V_{max} and K_M in the absence and presence of inhibitor.

15. The same enzyme as in Problem 14 is studied in the presence of a different inhibitor (inhibitor B). In this case, two different concentrations of inhibitor are used. Data are as follows:

	v[(mmol/L)min^{-1}]		
[S] (mmol/L)	No inhibitor	3 mM inhibitor B	5 mM inhibitor B
1.25	1.72	1.25	1.01
1.67	2.04	1.54	1.26
2.50	2.63	2.00	1.72
5.00	3.33	2.86	2.56
10.00	4.17	3.70	3.49

(a) What kind of inhibitor is inhibitor B?

(b) Determine the apparent V_{max} and K_M at each inhibitor concentration.

(c) Estimate K_I from these data.

16. Enalapril is an anti-hypertension "pro-drug" (i.e., a drug precursor) that is inactive until the ethyl ester (arrow in figure) is hydrolyzed by esterases present in blood plasma. The active drug is the dicarboxylic acid ("enalaprilat") that results from this hydrolysis reaction.

(a) Enalapril is administered in pill form, but enalaprilat must be administered intravenously. Why do you suppose enalapril works as a pill but enalaprilat does not?

Enalapril

(b) Enalaprilat is a *competitive* inhibitor of the angiotensin-converting enzyme (ACE), which cleaves the blood-pressure regulating peptide angiotensin I. ACE has a $K_M = 52 \mu M$ for angiotensin I, which is present in plasma at a concentration of 75 μM. When enalaprilat is present at 2.4 nM, the activity of ACE in plasma is 10% of its uninhibited activity. What is the value of K_I for enalaprilat?

17. Initial rate data for an enzyme that obeys Michaelis–Menten kinetics are shown in the following table. When the enzyme concentration is 3 nmol ml^{-1}, a Lineweaver–Burk plot of this data gives a line with a y-intercept of 0.00426 (μmol^{-1} ml s).

$[S] \mu M$	$v_0 (\mu mol\ ml^{-1} s^{-1})$
320	169
160	132
80.0	92.0
40.0	57.2
20.0	32.6
10.0	17.5

(a) Calculate k_{cat} for the reaction.
(b) Calculate K_M for the enzyme.
(c) When the reactions in part (b) are repeated in the presence of 12 μM of an uncompetitive inhibitor, the y-intercept of the Lineweaver–Burk plot is 0.352 (μmol^{-1} ml s). Calculate K'_I for this inhibitor.

18. TPCK and TLCK are irreversible inhibitors of serine proteases. One of these inhibits trypsin and the other chymotrypsin. Which is which? Explain your reasoning.

TPCK

TLCK

19. Suggest the effects of each of the following mutations on the physiological role of chymotrypsinogen:
(a) R15S
(b) C1S
(c) T147S

20. The inhibitory effect of an uncompetitive inhibitor is greater at high [S] than at low [S]. Explain this observation.

21. The allosterically regulated enzyme ATCase binds aspartic acid as a substrate and acylates the α–amino group. Succinate acts as a competitive inhibitor of ATCase because it binds the active site but can't be acylated. The dependence of v_0 on [aspartic acid] for ATCase is shown in panel (a) of the accompanying figure. Panel (b) shows the effect of increasing [succinate] on v_0 when [Asp] is held at a low concentration (see thick vertical arrow in panel (a)). Note that in panel (b), v_0 is not zero when [succinate] = 0 (see thin horizontal arrow). Explain the shape of the curve in panel (b). Why does v_0 increase initially, before decreasing at higher [succinate]?

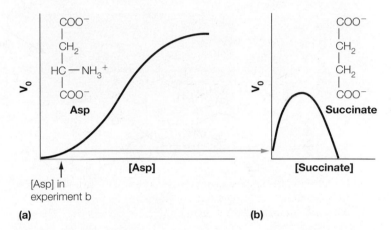

(a)
(b)

22. ATP is a (+) allosteric effector, and CTP is a (−) allosteric effector of the enzyme ATCase. Both of these heterotropic effectors bind to the regulatory subunits on ATCase. The substrates of ATCase, aspartate and carbamoyl phosphate, bind the enzyme active site with positive cooperativity (i.e., they exert a "+" homotropic effect on activity). As the concentrations of the substrates change from values where $[S] \ll K_M$ to values where [S] is saturating ($[S] \gg K_M$), how will the binding constants for each of the two allosteric effectors change? In other words, does ATP bind ATCase with higher affinity when [S] is low or high? Does CTP bind ATCase with higher affinity when [S] is low or high?

23. Shown below is a proposed mechanism for the cleavage of sialic acid by the viral enzyme neuraminidase. The k_{cat} for the wild-type enzyme at pH = 6.15, 37 °C is 26.8 s^{-1}.
(a) Describe the roles of the following amino acids in the catalytic mechanism: Glu117, Tyr409, and Asp149. List all of the following that apply: general acid/base catalysis (GABC), covalent catalysis, electrostatic stabilization of transition state.
(b) Based on the information shown in the scheme, would you expect mutation of Glu 117 to Ala to have a greater effect on K_M or k_{cat}?
(c) For the R374N mutant at pH = 6.15, 37 °C, k_{cat} is 0.020 s^{-1}, and K_M is relatively unaffected. Based on this result, it seems that R374 is more critical for catalysis than for substrate binding. Explain how R374 stabilizes the reaction transition state more than the substrate (i.e., what feature of this reaction would explain tighter binding to the transition state *vs.* substrate?).

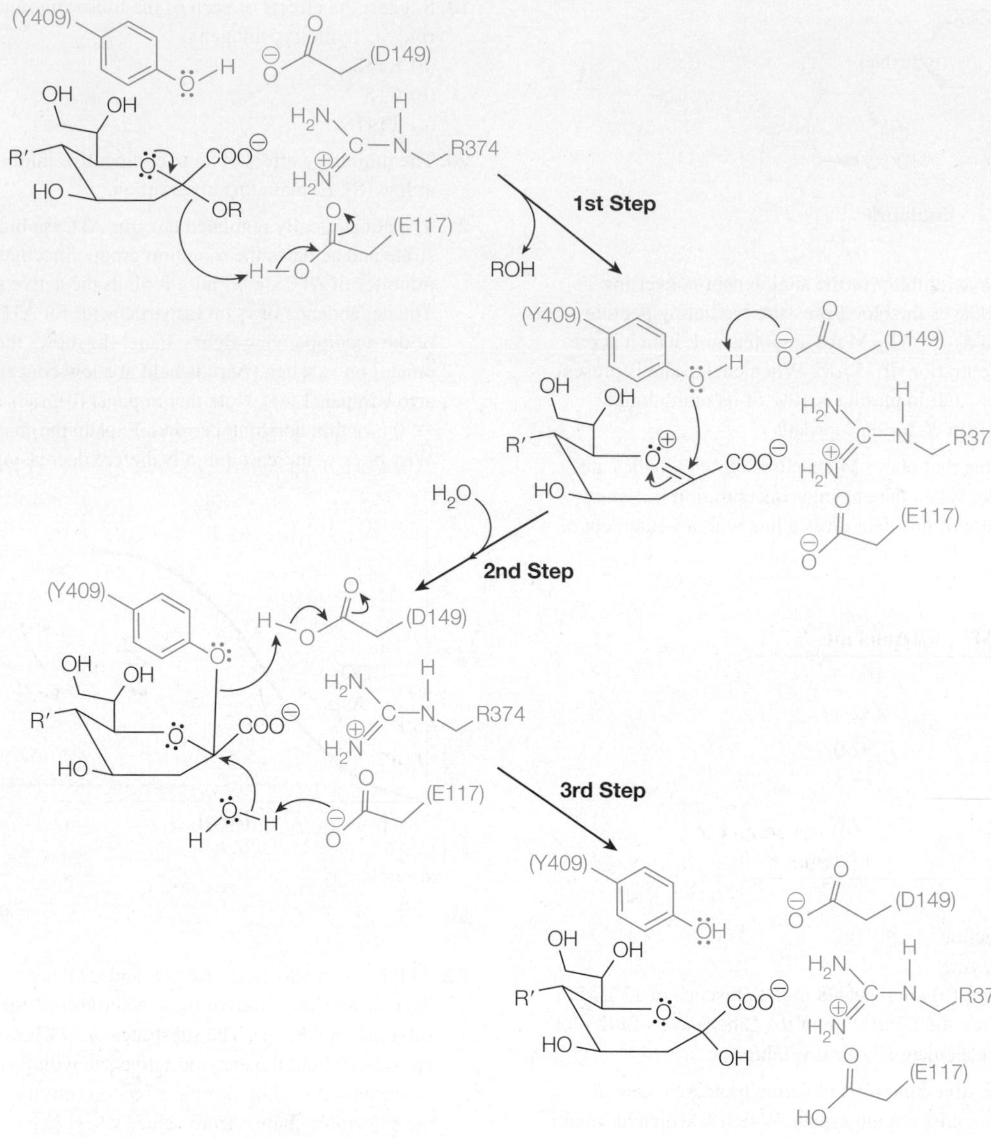
(Y409)
(D149)
R374
(E117)
1st Step
ROH
H2O
2nd Step
3rd Step

24. In kinetics experiments, the hydrolysis of the substrate sialic acid by neuraminidase appears to obey Michaelis–Menten kinetics. Neuraminidase activity is critical for viral infectivity; thus, this enzyme is the target of much work by pharmaceutical companies to develop a drug to treat influenza virus infection. The drug "Tamiflu" is a competitive inhibitor of neuraminidase. Initial rate data collected at pH $= 6.15$, 37 °C with 0.021 μM neuraminidase and 25.0 μM sialic acid gives a Lineweaver–Burk plot with a slope of 51.2 s.

(a) Recall from Problem 23 that the k_{cat} for neuraminidase at pH $= 6.15$, 37 °C is 26.8 s^{-1}. Calculate K_M for the hydrolysis of sialic acid.

(b) When the reactions in part (a) are repeated in the presence of 0.040 μM of Tamiflu, the slope of the Lineweaver–Burk plot is 198.8 s. Calculate the value of K_I for Tamiflu.

References

For a list of references related to this chapter, see Appendix II.

There are essentially two approaches to enzyme kinetic analysis. The first and simplest is to make measurements of rates under conditions in which the steady-state approximation holds (see Figure 8.20). Under these conditions, the Michaelis–Menten equation is often applicable, and determination of the reaction velocity as a function of substrate and enzyme concentrations will yield K_M and k_{cat}. Almost all enzymatic studies at least start in this way. But if the experimenter wishes to learn more of the details of the mechanism, it is often important to carry out studies before the steady state has been attained. Such *pre-steady-state* experiments require the use of special rapid techniques. Here we describe some of the experimental techniques that can be employed.

Analysis in the Steady State

The steady state in most enzymatic reactions is established within seconds or a few minutes and persists for many minutes or even hours thereafter. Therefore, extreme rapidity of measurement is not important, and many techniques are available to the experimenter wishing to follow the reaction. Descriptions of the most commonly used techniques follow.

Spectrophotometry Spectrophotometric methods are simple and accurate (see Tools of Biochemistry 6A). However, an obvious requirement is that either a substrate or a product of the reaction must absorb light in a spectral region where other substrates or products do not. Classic examples are reactions that generate or consume NADH. NADH absorbs quite strongly at 340 nm, but NAD^+ does not absorb in this region. Thus, we could, for example, follow the oxidation of ethanol to acetaldehyde, as catalyzed by alcohol dehydrogenase, by measuring the formation of NADH spectrophotometrically. Even if the reaction being studied does not involve a light-absorbing substance, it may be possible to couple this reaction to another, very rapid reaction that does.

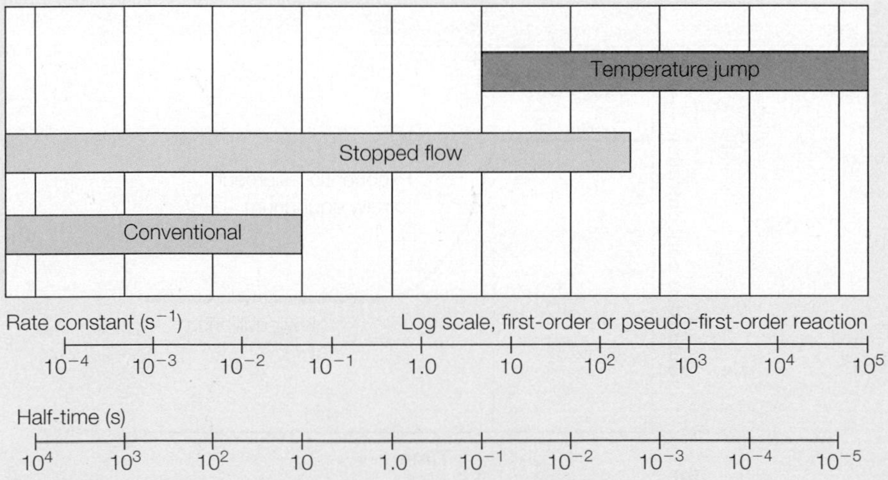

Fluorescence The applications of fluorescence are similar to those of spectrophotometry, and the problems are also similar: A substrate or a product must have a distinctive fluorescence emission spectrum (see Tools of Biochemistry 6A). However, fluorescence often has the advantage of high sensitivity, so extremely dilute solutions may be employed, enabling an experimenter to greatly extend the concentration range (i.e., [S]) over which studies are practicable.

Radioactivity Assays If a substrate is labeled with a radioactive isotope that will be lost or transferred during the reaction to be studied, measurement of changes in radioactivity can be an extremely sensitive kinetic method. This procedure requires that the labeled compound can be separated quickly at different, precisely defined times during the reaction. An example is a method often used with radioactive ATP. The ATP can be adsorbed on charcoal-impregnated filter disks by very fast filtration of aliquots from the reaction mixture. The radioactivity can then be measured in a scintillation counter (see Tools of Biochemistry 11B). Another example of the use of radioisotopes comes from measuring the rates of peptide-bond cleavage (by a protease) or protein biosynthesis (e.g., ribosomal protein synthesis). Peptides are most commonly labeled with radioactive amino acids that contain ^3H, ^{14}C, or ^{35}S. The rate of a peptide cleavage or synthesis reaction can be monitored by rapidly precipitating the peptide (or peptide fragments) from the reaction solution using cold trichloroacetic acid and collecting the precipitate on filter paper. The radioactivity present on the filter paper can be quantified using a scintillation counter.

Analysis of Very Fast Reactions

Reactions that are extremely rapid require special techniques to investigate the pre-steady-state processes. Two major methods are currently employed to cover the rapid time scales shown in **FIGURE 8A.1**.

Stopped Flow. **FIGURE 8A.2** shows a **stopped-flow apparatus,** first described by Quentin Gibson in the 1950s. Enzyme and substrate are initially in separate syringes. The syringes are driven, within a few milliseconds, to deliver their contents through a mixing chamber and into a third, "stopping" syringe. This step triggers a detector to begin observing (for example, by light absorption or fluorometry) the solution in the tube connecting the mixer to the stopping syringe. Flow rates can easily be made as high as 1000 cm/s. If the mixture was moving at this rate when the flow was stopped, and if the observation point is 1 cm from the mixer, the detection system first sees a mixture that is 1 ms "old." The reaction can then be followed for as long as desired—often for a period of only a few seconds. The limitations of the method are imposed only by the initial "dead time" (i.e., the time it takes the mixed solutions to arrive at the detector—in the example above, 1 ms) and the rapidity of the detection system.

▲ FIGURE 8A.1 Time scales for kinetic techniques described here.

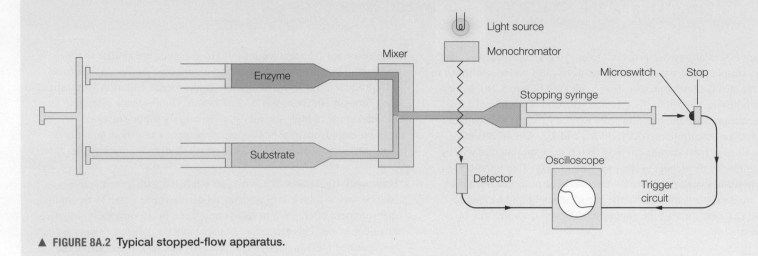

▲ FIGURE 8A.2 Typical stopped-flow apparatus.

Stopped-flow is used to measure rates of rapid enzymatic reactions as well as ligand binding events, such as O_2 binding to, or release from, hemoglobin (see Chapter 7).

Temperature Jump Some processes are so fast that they are essentially completed in the dead time of a stopped-flow apparatus. The experimenter may then turn to temperature jump (T-jump) methods. The basic apparatus and principle of the method are shown in FIGURE 8A.3 (a and b), respectively. A reaction mixture that is at equilibrium at a temperature T_1 is suddenly jumped to a temperature T_2. Because chemical equilibria are typically temperature-dependent, the position of equilibrium will shift, and the system must now react to attain this new equilibrium. A rapid jump in temperature (5–10 °C in 1 μs) can be obtained by passing a large burst of electrical current between electrodes immersed in the reaction mixture. Even more rapid jumps (10–100 ns) can be obtained if a pulsed infrared laser is used to heat the mixture. The relaxation (approach) to a new equilibrium, monitored by absorption or fluorescence measurements, is an exponential process related to the rate constants for the reaction.

Although a number of other techniques are employed for even faster reactions, including some newly developed NMR methods and pulsed laser techniques, the methods described here are widely used. If we consider the variety of techniques available to the experimenter, we can see that they cover a wide time range. Altogether, time scales from nanoseconds to hours can be studied.

Relating Kinetics to a Mechanism: Kinetic Isotope Effects

Kinetic data are the basis for proposing a detailed chemical mechanism such as those illustrated in Figures 8.13, 8.15, and 8.19. The determinations of rates for the turnover of isotopically labeled substrates are among the more useful data for distinguishing one possible mechanism from another (e.g., see the discussion of the lysozyme mechanism).

The rates of bond formation/cleavage depend on the masses of the atoms involved because the vibrational frequencies of bonds are sensitive to the masses of the bonded atoms. Bond cleavage/formation reactions involving heavier isotopes proceed with slower

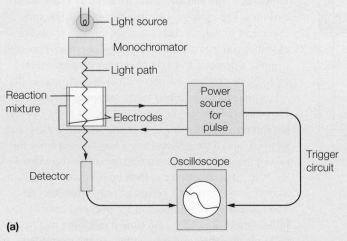

(a)

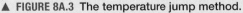

▲ FIGURE 8A.3 The temperature jump method.

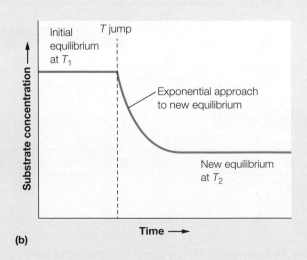

(b)

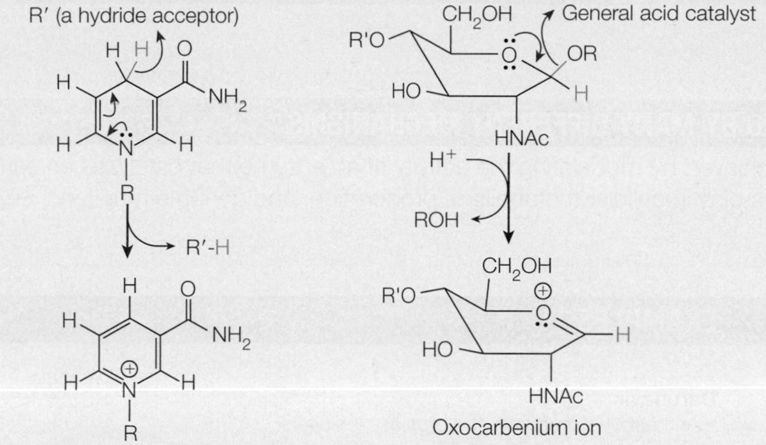

Primary Kinetic Isotope Effect:
- the bond *to* the isotope is broken/formed
- $k_H/k_D = 3-5$ for hydride transfer

Secondary Kinetic Isotope Effect:
- the bond *adjacent* to the isotope is broken/formed
- $k_H/k_D = 1.05-1.12$

▲ **FIGURE 8A.4** Examples of primary and secondary kinetic isotope effects.

---

rates, and this effect is known as a **kinetic isotope effect** (or KIE). Chemists have developed methods to synthesize substrates for enzyme-catalyzed reactions with atom-specific isotopic substitutions. In other words, at some specific location in the substrate molecule, a deuterium (^{2}H or D) or a tritium (^{3}H or T) might be substituted for the common isotope of hydrogen, ^{1}H (sometimes referred to as "protium"). The KIE is largest for hydrogen isotopes because the change in mass is more significant for ^{1}H versus ^{2}H (or ^{3}H) than it is for heavier elements (e.g., ^{12}C vs. ^{13}C, or ^{16}O vs. ^{18}O).

A so-called **primary KIE** is observed when the bond including the atom in question is broken/formed in the rate-determining step, whereas a **secondary KIE** is observed when the bond adjacent to the atom in question is broken/formed (**FIGURE 8A.4**). The KIE is recorded as the ratio of rates for the reactions of a substrate labeled with two different isotopes, for example, k_H/k_D. For primary KIE, the range of observed values for hydrogen transfer reactions is 2–15. For secondary KIE, the range is closer to 1 (1.05–1.12). These differences in magnitude allow primary and secondary KIEs to be distinguished.

If researchers suspect that a particular bond is broken during the slow step of a reaction, they can synthesize a labeled substrate and compare the kinetics of the reaction with unlabeled and labeled substrates. If the expected KIE is observed, the proposed mechanism may be correct. If the expected KIE is not observed, the proposed mechanism is likely incorrect, or the bond cleavage/formation is more rapid than some other rate-determining step in the mechanism. In this way, KIEs are useful tools for the elucidation of mechanistic detail in enzyme-catalyzed reactions.

References

For a list of references related to this chapter, see Appendix II.

ENZYME ACTIVITIES ARE REGULATED BY MANY DIFFERENT MECHANISMS

Regulation of metabolism is frequently achieved by modulating the activity of an enzyme that catalyzes an early step in a given metabolic pathway. Given the central roles enzymes play in cellular metabolism, proliferation, and genome maintenance, they are the targets of several therapeutic drugs.

MECHANISM OF REGULATION

Reversible inhibition

• Inhibitor binds noncovalently and inhibition can be reversed, at least in principle, by removal of the inhibitor.

• The various modes of reversible inhibition involve the noncovalent binding of an inhibitor to the enzyme, but they differ in the mechanisms by which they decrease the enzyme's activity.

▶ FOUNDATION FIGURE 4: ENZYME KINETICS AND DRUG ACTION

Irreversible inhibition

• Irreversible inhibitor binds covalently to an enzyme and irreversibly inactivates it.

• Almost all irreversible enzyme inhibitors are toxic substances, either natural or synthetic. In most cases, such substances react with a functional group in the active site to leave it catalytically inactive or to block substrate binding.

Substrate level control: Product inhibition

• High levels of product bind to enzyme and inhibit conversion of substrate to product.

EXAMPLES

Darunavir
• Antiretroviral drug from the protease inhibitor class used to treat HIV infection and AIDS.

• Inhibits the HIV protease, and thereby prevents maturation of new viruses. This results in a reduced viral load and reduced symptoms in AIDS patients.

Captopril is another example of reversible inhibition.

▶ FIGURE 8.26: CAPTOPRIL IS A COMPETITIVE INHIBITOR OF ANGIOTENSIN-CONVERTING ENZYME.

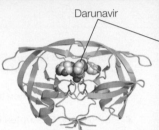
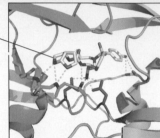

HIV protease with darunavir bound

Zoom showing extensive interactions between inhibitor and active site.

Omeprazole
• Widely prescribed to treat gastroesophageal reflux disease (GERD)

• Irreversibly inhibits the proton pump in the stomach.

• Acts by covalently modifying a key cysteine (Cys 813) in the H^+/K^+ ATPase (or, more commonly, the gastric proton pump) of the gastric parietal cells. The proton pump is structurally related to the sodium/potassium pump discussed in detail in Chapter 10. The covalent modification can be reversed by a reducing agent, but this process is quite slow in the stomach, making it effectively irreversible.

▶ FIGURE 10.32: THE STRUCTURE OF THE Na^+–K^+ ATPase. FIGURE 10.33: A SCHEMATIC DIAGRAM OF THE FUNCTIONAL CYCLE OF THE Na^+–K^+ PUMP.

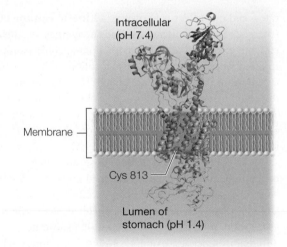

Intracellular (pH 7.4)

Membrane

Cys 813

Lumen of stomach (pH 1.4)

The first step in the glycolytic pathway—the phosphorylation of glucose—yields glucose-6-phosphate (G6P):

$$\text{Glucose} + \text{ATP} \xrightarrow{\text{Hexokinase}} \text{Glucose-6-phosphate} + \text{ADP}$$

• The enzyme hexokinase I is inhibited by its product, G6P. If subsequent steps in glycolysis are blocked for any reason, G6P will accumulate and bind to hexokinase. This results in product inhibition of hexokinase and slows down further production of G6P from glucose.

• G6P can act both as a competitive inhibitor (by binding to the active site), as well as an uncompetitive inhibitor (by binding at another site).

▶ FIGURE 12.14: MAJOR CONTROL MECHANISMS AFFECTING GLYCOLYSIS AND GLUCONEOGENESIS

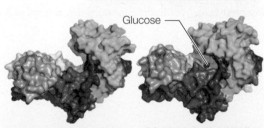

Glucose

(a) Before glucose binding **(b)** After glucose binding

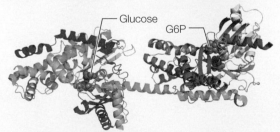

Glucose

G6P

Human hexokinase bound to two molecules each of G6P (gray) and glucose (magenta)

MECHANISM OF REGULATION

Feedback Control

- In an enzymatic pathway, a cell can control generation of the final product through activation or inhibition of a key step in the pathway—often the first step.

- Control of pathways by their end products means that the necessary inhibition or activation must be produced by molecules that come from far down the assembly line and therefore bear little or no structural resemblance to either the substrates or the direct products of the enzymes to be regulated. Activation or inhibition of enzymes at key control points in a pathway typically occurs through an allosteric mechanism.

Covalent Modifications

- Enzymes are activated or inactivated by covalent changes. Often these modifications lead to an enzyme conformational change, affecting binding sites for substrates or cofactors.

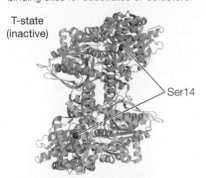

T-state
(inactive)

Ser14

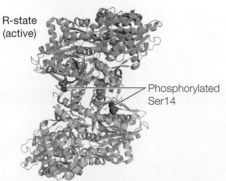

R-state
(active)

Phosphorylated
Ser14

X-ray crystal structures of rabbit muscle glycogen phosphorylase, showing the effect of Ser14 phosphorylation on enzyme conformation (and thereby on activity).

EXAMPLES

Replication of DNA requires that the purine and pyrimidine deoxyribonucleotides are available in approximately equal concentrations. This is achieved by feedback control.

▶ SECTION 19.5

Allostery—An allosteric effect occurs when the binding of a ligand results in a conformational change in the enzyme which results in increased, or decreased, enzyme activity. Allosteric enzymes are frequently multisubunit proteins, with multiple active sites. They exhibit cooperativity in substrate binding (homoallostery) and regulation of their activity by other, effector molecules (heteroallostery).

The activity of *E. coli* aspartate carbamoyltransferase (also known as aspartate transcarbamoylase, or ATCase), a key enzyme in pyrimidine synthesis (Chapter 19), is controlled by allosteric regulation. See section 8.8 for details on ATCase structure and function.

▶ SECTION 8.8: FIGURES 8.35–8.38

Phosphorylation—Many enzymes, and their associated metabolic and signaling pathways, are regulated by reversible phosphorylation.

- Glycogen phosphorylase catalyzes the rate-limiting step in glycogenolysis in animals. It is regulated by both allosteric control and by phosphorylation.

- The regulation of phosphorylase activity is itself an enzyme-regulated process, mediated by specific kinases (which add a phosphoryl group to Ser-14), and phosphatases (which remove phosphoryl groups).

▶ SECTION 12.9: COORDINATED REGULATION OF GLYCOGEN METABOLISM

Proteolytic Cleavage—Certain enzymes become irreversibly activated by proteolytic cleavage of a precursor form.

- Pancreatic proteases (trypsin, chymotrypsin, elastase, and carboxypeptidase)

▶ SECTION 8.9: COVALENT MODIFICATIONS USED TO REGULATE ENZYME ACTIVITY

- The blood coagulation pathway is another example of a regulatory cascade involving a series of linked proteolysis events.

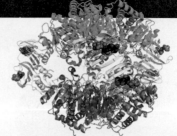

R-state ATCase. Six regulatory domains (yellow) are shown with bound ATP.

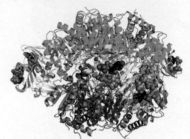

T-state ATCase. CTP, UTP, and Mg^{2+} are bound to the regulatory domains.

Blood coagulation pathway

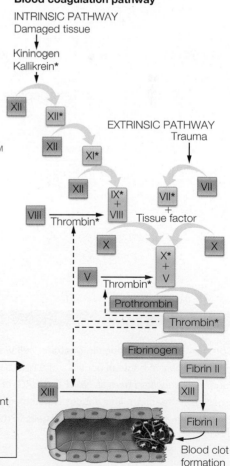

Factor key:

V	Proaccelerin	X*	Stuart factor
VII*	Proconvectin	XI*	Thromboplastin antecedent
VIII	Antihemophilic factor	XII*	Hageman factor
IX*	Christmas factor	XIII	Fibrin stabilizing factor

Asterisk (*) denotes serine proteases

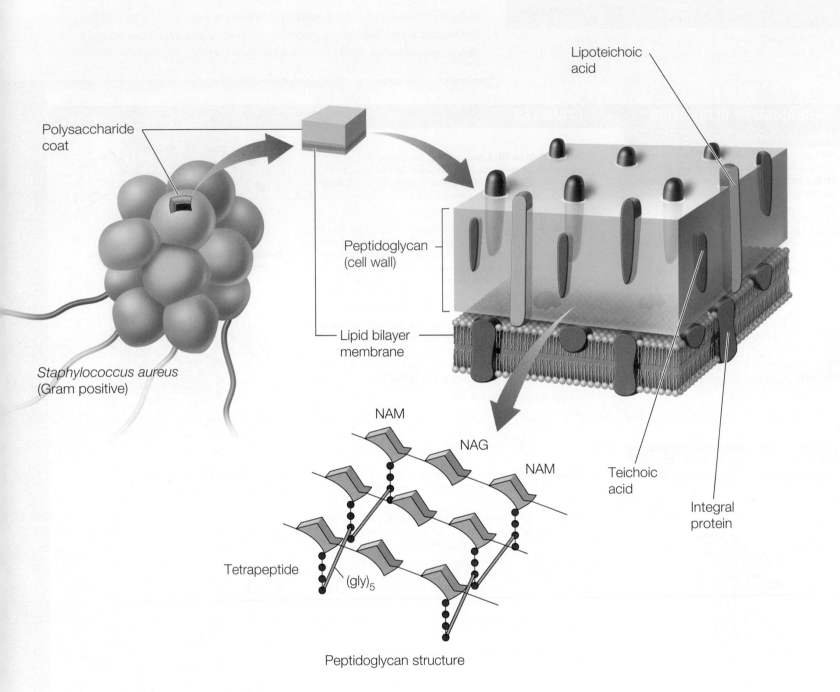

Polysaccharide coat

Staphylococcus aureus (Gram positive)

Lipoteichoic acid

Peptidoglycan (cell wall)

Lipid bilayer membrane

Teichoic acid

Integral protein

NAM

NAG

NAM

Tetrapeptide

(gly)$_5$

Peptidoglycan structure

Complex polysaccharides in bacterial cell walls, the site of action of antibiotics. The cell surface of Gram-positive bacteria consists of a lipid bilayer membrane surrounded by complex polymers of sugars (abbreviated here as *NAM* and *NAG*) crosslinked by short peptide chains. Penicillin inhibits the crosslinking reaction. This causes weakening of the cell wall, which, coupled with continued growth of the bacteria, makes it impossible for cells to resist internal pressure, and eventually they rupture.

Carbohydrates: Sugars, Saccharides, Glycans

9

WE TURN NOW to the third great class of biological molecules, the carbohydrates, or saccharides. Like the nucleic acids and proteins, carbohydrates play metabolic roles both as their constituent monomeric units, such as glucose or ribose, and as polymers, such as starch or glycogen. Unlike proteins and nucleic acids, which are

● **CONCEPT** Carbohydrate formation in photosynthesis and its oxidation in metabolism together constitute the major energy cycle of life.

strictly linear polymers, macromolecular polysaccharides also exist as branched polymers. For much of this book we concern ourselves with the roles of carbohydrates in generating and storing biological energy.

FIGURE 9.1 illustrates the indispensable functions of carbohydrates in the major energy cycle of the biosphere—the light-driven synthesis of sugars and oxygen from carbon dioxide in photosynthetic organisms and the metabolism of those sugars by most organisms to generate the energy needed to sustain life.

However, carbohydrates play far more functions than just in energy metabolism—notably, functions as diverse as molecular recognition (as in the immune system), cellular protection (as in bacterial and plant cell walls), cell signaling, cell adhesion, biological

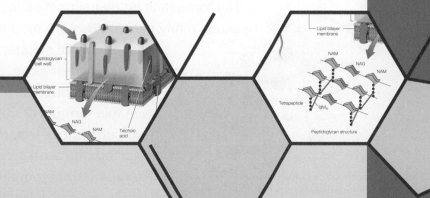

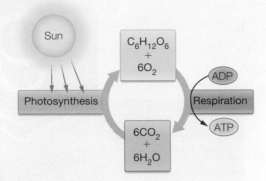

◀ **FIGURE 9.1 The major energy cycle of life.** In photosynthesis, plants and photosynthetic microorganisms use the energy of sunlight to combine carbon dioxide and water into carbohydrates, releasing oxygen in the process. In respiration, all aerobic organisms oxidize the carbohydrates made by plants, releasing energy and reforming CO_2 and H_2O.

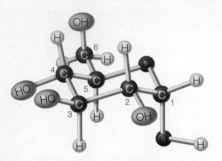

(a) Glucose, a monosaccharide.

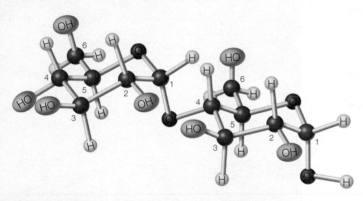

(b) Maltose, a disaccharide containing two glucose units.

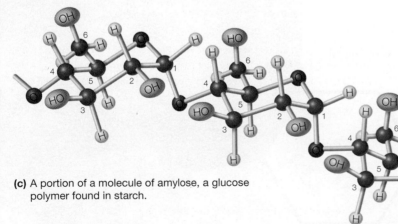

(c) A portion of a molecule of amylose, a glucose polymer found in starch.

▲ **FIGURE 9.2 Representative carbohydrates.** The three compounds shown here are composed entirely of C, H, and O, with glucose (a) forming the monomer for the oligomer (b) and the polymer (c).

lubrication, control of protein trafficking, and maintenance of biological structure.

Many carbohydrates are already familiar to you. The simplest carbohydrates are small, monomeric molecules—the monosaccharides, typically containing from three to nine carbon atoms, which include simple sugars such as *glucose* (**FIGURE 9.2(a)**). Other important carbohydrates are formed by linking such monosaccharides together. If only a few monomer units are involved, we call the molecule an oligosaccharide. An example is *maltose* (Figure 9.2(b)), a disaccharide made by linking two glucose molecules together. Long polymers of the monosaccharides, like the starch *amylose* (Figure 9.2(c)), are called polysaccharides. Many kinds of polysaccharides exist, some of which are complex polymers made from many types of sugar monomers. Oligosaccharides and polysaccharides are also referred to as glycans.

Saccharides are often referred to by the more familiar name *carbohydrates* because many of them can be represented by the simple stoichiometric formula $(CH_2O)_n$. The name was first given when chemists knew only the stoichiometry of saccharides and thought of them as "hydrated carbon."

● **CONCEPT** Carbohydrates are compounds with the empirical formula $(CH_2O)_n$, while saccharides include carbohydrates and all of their derivatives.

The formula is an oversimplification, however, because many saccharides are modified, and some contain amino, sulfate, or phosphate groups. Nevertheless, all of the compounds described in this chapter either have this formula or can be derived from substances that do. Strictly speaking, the term *carbohydrate* is reserved for compounds with the $(CH_2O)_n$ empirical formula, while the term **saccharide** covers both these compounds and all derivatives of carbohydrates. We will occasionally stray from strict usage and use the terms *carbohydrate* and *saccharide* interchangeably. The term *sugar* generally refers to underivatized monosaccharides and small oligosaccharides, such as *sucrose,* a disaccharide containing glucose and fructose. As noted earlier, an oligosaccharide or polysaccharide is also called a glycan.

Also as noted earlier, carbohydrates play many roles in addition to their functions in energy storage and generation, particularly as structural elements (e.g., cell walls) and in molecular recognition. Examples of the latter role include highly specific processes such as the binding of viruses or antibodies on particular cells. Thus, like proteins, carbohydrates are extremely versatile molecules, essential to all organisms.

9.1 Monosaccharides

We begin our discussion of carbohydrates with the simple, monomeric sugars—the monosaccharides. The simplest compound with the empirical formula of the class $(CH_2O)_n$ is found when $n = 1$. However, *formaldehyde,* $H_2C{=}O$, has little in common with our usual concept of sugars; indeed, it is a noxious, poisonous gas. The smallest molecules usually regarded as monosaccharides are the **trioses,** with $n = 3$. (The suffix *ose* is commonly used to designate compounds as saccharides.) Monosaccharides are generally characterized by the presence of one carbonyl group (aldehyde or ketone) and one or more hydroxyl groups.

Aldoses and Ketoses

There are two trioses: *glyceraldehyde* and *dihydroxyacetone* (**FIGURE 9.3**). These molecules, as simple as they are, exhibit certain features that we shall encounter again and again in discussing sugars. In fact, they represent the two major classes of monosaccharides. Glyceraldehyde is an aldehyde, one of a class of monosaccharides called **aldoses.** Dihydroxyacetone is a ketone; such monosaccharides are called **ketoses.** Note that glyceraldehyde and dihydroxyacetone each has one carbonyl carbon and that both have the same atomic composition. They are *tautomers* (see Chapter 4) and can undergo interconversion. Such tautomeric interconversions occur to a certain extent

● **CONCEPT** The two major classes of monosaccharides are aldoses and ketoses.

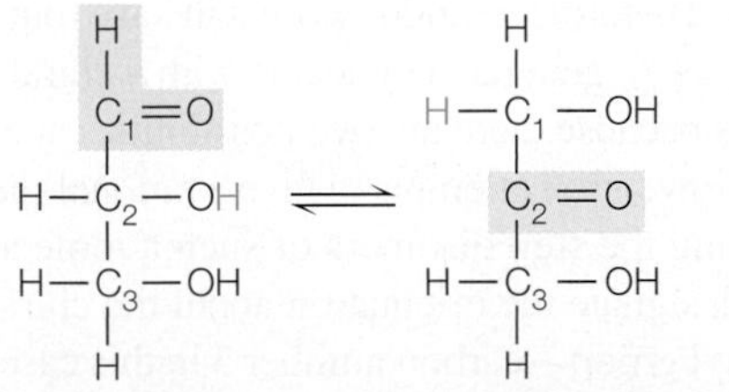

between all such pairs of aldose and ketose monosaccharides, but the reactions are usually very slow unless catalyzed. Thus, glyceraldehyde and dihydroxyacetone can each exist as a stable compound.

Enantiomers

An essential feature of monosaccharide structure can be seen by examining the formula for glyceraldehyde. The second carbon atom carries four different substituents, so it is *chiral,* like the α-carbon in most α-amino acids. Therefore, glyceraldehyde has two stereoisomers of the type called **enantiomers,** which are nonsuperimposable mirror images. Three-dimensional drawings of the two forms, designated as D- and L-glyceraldehyde, are shown in the accompanying figure, using the same bond convention that we used for amino acids in Chapter 5.

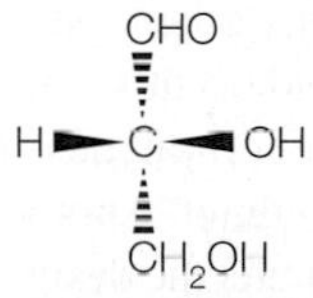

D-Glyceraldehyde 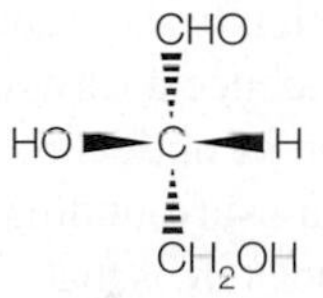   **L-Glyceraldehyde**

● **CONCEPT** D and L forms of a monosaccharide are nonsuperimposable mirror images and are called enantiomers.

Note that we do not need to draw the spatial orientation of atoms about carbons 1 or 3 because these carbons are not chiral centers.

The most compact way to represent enantiomers is to use a **Fischer projection.** Again, as described in Chapter 5, the bonds that are drawn horizontally are imagined as coming toward you; those drawn vertically are receding. Thus, for D-glyceraldehyde and L-glyceraldehyde, we have

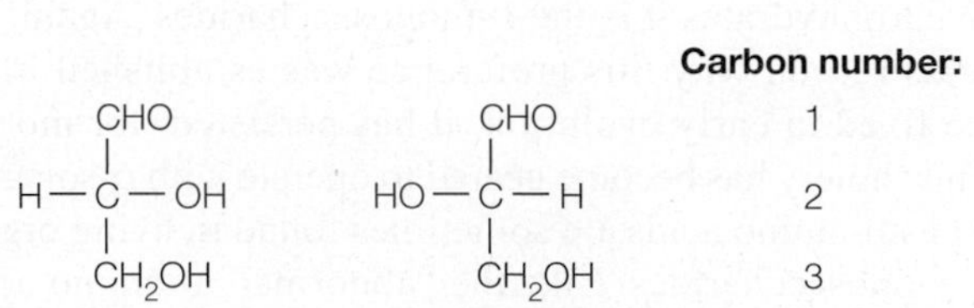

D-Glyceraldehyde   **L-Glyceraldehyde**

Alternative Designations for Enantiomers: D–L and *R–S*

Originally, the terms *D* and *L* were meant to indicate the direction of rotation of the plane of polarization of polarized light: D for right (dextro), L for left (levo). It is true that a solution of D-glyceraldehyde does rotate the

▲ **FIGURE 9.3 Trioses, the simplest monosaccharides.** The two triose tautomers illustrate the difference between aldose and ketose monosaccharides, also called more descriptively, aldotriose and ketotriose, respectively. Carbon numbering begins in all aldoses with the aldehyde carbon and in ketoses with the end carbon closest to the ketone group. (Because dihydroxyacetone has only three carbons, the two end carbons are equivalent and either of them could be designated number one.)

Rotate molecule so group of lowest priority
(H) faces away:

If priority of remaining groups decreases in *clockwise* direction, configuration is *R*

D-Glyceraldehyde
= *R*-Glyceraldehyde

If priority decreases in *counterclockwise* direction, configuration is *S*

L-Glyceraldehyde
= *S*-Glyceraldehyde

▲ **FIGURE 9.4** *R–S* nomenclature. The *R–S* system describes absolute stereochemical configuration, as shown in this example. Each type of group attached to a chiral carbon (gray) is given a priority, according to a set of defined rules. Priorities for groups common in carbohydrate chemistry are $SH > OR > OH > NH_2 > CO_2H > CHO > CH_2OH > CH_3 > H$. We view the molecule with the group of lowest priority away from us (H in our example). If the priority of the remaining three groups *decreases clockwise*, the absolute configuration is called *R* (from Latin *rectus*, meaning "right"). If priority *decreases counterclockwise*, the configuration is *S* (from Latin *sinister*, meaning "left"). In this notation, D-glyceraldehyde is *R*-glyceraldehyde, and L-glyceraldehyde is *S*-glyceraldehyde.

plane of polarization to the right, as do many other D-monosaccharides, but this correspondence does not always hold because the magnitude and even the direction of optical rotation are a complicated function of the electronic structure surrounding the chiral center. Another disadvantage of the D–L nomenclature is that it is not absolute; the designation is always with respect to some reference compound. Accordingly, an *absolute* convention has been developed that allows us to assign a stereochemical designation to any compound from examination of its three-dimensional structure. This *R–S* convention, shown in **FIGURE 9.4**, is also called the Cahn–Ingold–Prelog convention, after its inventors. Although the *R–S* convention is more general and gives the absolute configuration about a chiral center, the D–L convention is still in common use by biochemists and will be used in this chapter.

Monosaccharide Enantiomers in Nature

Just as in the case of amino acids, one enantiomeric form of monosaccharides dominates in living organisms. In proteins it is the L-amino acids; in carbohydrates it is the D-monosaccharides. Again, there is no obvious reason why this preference was established in nature. But once fixed in early evolution, it has persisted, for most of the cellular machinery has become geared to operate with D-sugars. However, just as D-amino acids are sometimes found in living organisms, so are L-monosaccharides. Like the "abnormal" D-amino acids, the L-monosaccharides play rather specialized roles.

Diastereomers

When we consider monosaccharides with more than three carbons, a further structural issue appears. Such a monosaccharide may have more than one chiral carbon, which results in its having two types of stereoisomers. These types are *enantiomers* (mirror-image isomers),

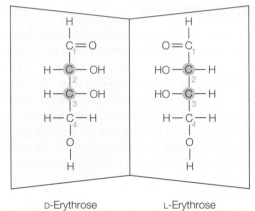

D-Threose   L-Threose

D-Erythrose   L-Erythrose

▲ **FIGURE 9.5 Stereochemistry of aldotetroses.** These molecules have two chiral carbons (2 and 3) and thus have two diastereomeric forms, threose and erythrose, each with a pair of enantiomers. Note that the threose enantiomers have the *opposite* configuration about carbons 2 and 3, whereas erythrose enantiomers have the *same* configuration about these two carbons.

which we have already discussed, and *diastereomers*, which we first encounter in the tetrose monosaccharides.

Tetrose Diastereomers

Tetroses, with the empirical formula $(CH_2O)_4$, have two chiral carbons in the aldose forms. Therefore, an aldotetrose will have four stereoisomers, as shown in **FIGURE 9.5**. In general, a molecule with *n* chiral centers will have 2^n stereoisomers because there are two possibilities at each chiral center. The following convention attempts to give a rational method for naming and distinguishing the stereoisomers of such a molecule: The prefix D or L is used to designate the orientation about the chiral carbon *farthest* from the carbonyl group—carbon number 3 in this case. Molecules with different orientations about the carbons preceding this reference carbon are given separate names. Thus, *threose* and *erythrose* are two aldotetroses with opposite orientations about carbon 2. Stereoisomers of this kind, which are *not* mirror images, are called **diastereomers.** Threose and erythrose are diastereomers, and each has two enantiomers (D and L) that are nonsuperimposable mirror images. Unfortunately, there is no general logical rule for forming the specific names (such as threose and erythrose); they must simply be learned, like the names of the amino acids.

The four-carbon ketose, which is called *erythrulose*, has only one pair of enantiomers because this monosaccharide has only one chiral

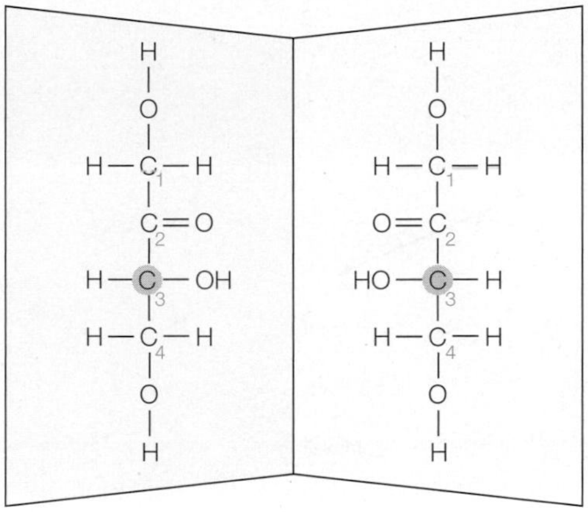

▲ **FIGURE 9.6** The two enantiomers of erythrulose. Unlike the four-carbon aldoses (see Figure 9.5), the four-carbon ketose has only one chiral carbon (C3) and only one pair of enantiomers.

● **CONCEPT** When monosaccharides contain more than one chiral carbon, the prefix D or L designates the configuration about the carbon farthest from the carbonyl group. Isomers differing in orientation about other carbons are called diastereomers and are given different names.

carbon (**FIGURE 9.6**). Another naming convention appears at this point: Usually, the ketose name is derived from the corresponding aldose name by insertion of the letters *ul*. Thus, *erythrose* becomes *erythrulose*. As with glyceraldehyde (and other monosaccharides), the ketose and aldose forms are interconvertible via tautomerization in dilute alkali. The aldose–ketose conversion also provides a route for interconversion of aldose diastereomers, using the ketose as an intermediate.

Pentose Diastereomers

Adding one more carbon, we obtain the **pentoses**. The **aldopentoses** have three chiral centers; therefore, we expect 2^3, or eight, stereoisomers—in four pairs of enantiomers. The D forms of the pentoses are shown in **FIGURE 9.7(a)**, which provides a summary of the aldoses containing three to six carbons. Note that each of the aldopentoses shown has the D orientation about carbon 4 and that all possible combinations of orientations about carbons 2 and 3 are included. (From here on, in our illustration of carbohydrate structure we will show only the D forms; you can easily draw the L forms from the rules given above.) **Ketopentoses,** as shown in Figure 9.7(b), have two chiral carbons, so four isomers (two pairs of enantiomers) must exist. The D diastereomers are called D-*ribulose* and D-*xylulose*.

Hexose Diastereomers

Monosaccharides containing six carbon atoms are called **hexoses**. As you might imagine, there is a large number of possible hexoses. To keep their structures in mind, it is useful to relate them to the simpler pentoses, tetroses, and trioses. Figure 9.7 summarizes these relationships. The hexoses we will most frequently encounter are *glucose* and *fructose*. However, *mannose* and *galactose* are also widespread in nature. In fact, almost all of the hexoses play some significant biological role.

Aldose Ring Structures

With the pentoses and hexoses, another feature of monosaccharide chemistry assumes critical importance. Having five or six carbons in the chain gives these compounds the potential to form stable ring structures via internal *hemiacetal* formation. A hemiacetal results from reaction of an aldehyde with an alcohol.

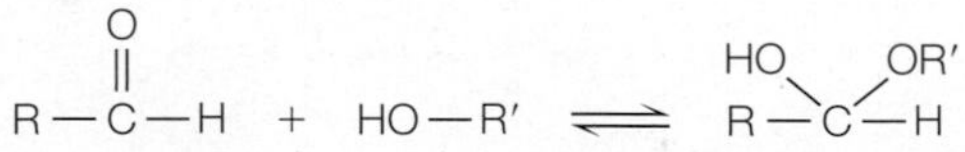

The bond angles characteristic of carbon and oxygen bonding are such that rings containing fewer than five atoms are strained to a considerable extent, whereas five- or six-membered rings are easily formed. In principle, aldotetroses can also form five-membered ring structures, but they rarely do.

Pentose Rings

Consider this hemiacetal ring formation in an aldopentose, such as D-ribose (see Figure 9.7(a)). Two modes of ring closure are possible, as shown in **FIGURE 9.8**. Reaction of the C-1 of D-ribose with the C-4 hydroxyl produces a five-membered ring structure called a **furanose;** the name reflects its structural similarity to the heterocyclic compound furan. Alternatively, a six-membered ring is obtained if the reaction occurs with the C-5 hydroxyl. Such a six-membered ring is called a **pyranose,** to indicate its relation to the heterocyclic compound pyran.

Both of the reactions shown in Figure 9.8 have equilibria that highly favor the cyclic structures for pentoses or larger sugars. Under

● **CONCEPT** Monosaccharides with five or more carbons exist preferentially in five- or six-membered ring structures, resulting from internal hemiacetal formation.

physiological conditions in solution, monosaccharides with five or more carbons typically exist more than 99% of the time in the ring forms. The distribution between pyranose and furanose forms depends on the particular sugar structure, the pH, the solvent composition, and the temperature. When the monomers are incorporated into polysaccharides, the structure of the polymer may also influence the ring form chosen. For example, D-ribose exists in solution as a mixture of the two ring forms. But in biological compounds, specific forms are stabilized. RNA, for example, contains ribose exclusively as ribofuranose, whereas some plant cell wall polysaccharides have pentoses entirely in the pyranose form.

Let us look more closely at the ring structures shown in Figure 9.8. Cyclization has created a new asymmetric center at carbon 1. That is why we have drawn two stereoisomers of D-ribofuranose, referred to as α-D-ribofuranose and β-D-ribofuranose, as well as a corresponding pair of ribopyranoses. Like other kinds of stereoisomers, these α and β forms rotate the plane of polarized light differently and can be distinguished in that way. Such isomers, differing in configuration only at the carbonyl carbon, are called **anomers,** and carbon 1 is often referred to as the *anomeric carbon atom*. The monosaccharides can undergo interconversion between the α and β forms, using the open-chain structure as an intermediate. This process is referred to as **mutarotation.** A purified anomer, dissolved in aqueous solution, will approach the equilibrium mixture, with an accompanying change in the optical rotation of the

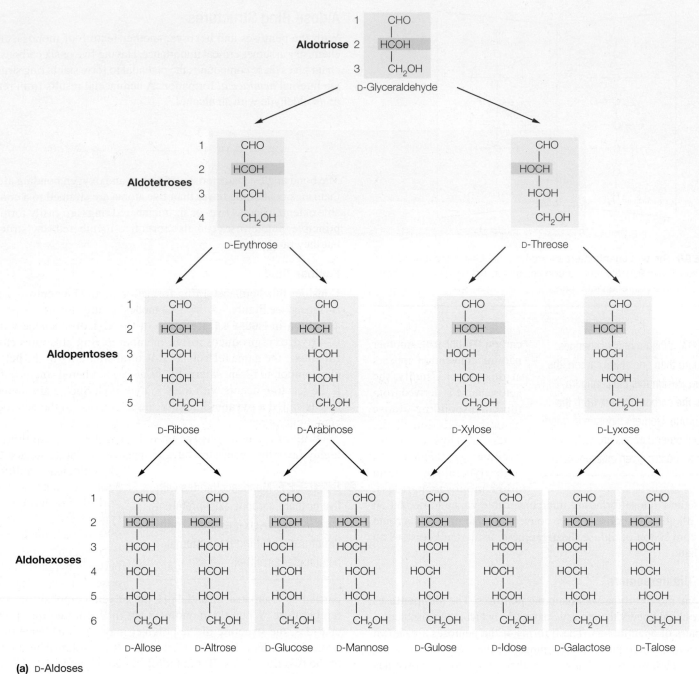

(a) D-Aldoses

▲ FIGURE 9.7 Stereochemical relationships of the D-aldoses and D-ketoses. This figure shows the relationships between pairs of diastereomers in the D-aldose series (a) and the D-ketose series (b). Each series is generated by successive additions of one CHOH group (shaded) just below the carbonyl carbon. In each case, the two possible orientations of the added group generate a pair of diastereomers. The L forms are not shown; they are just the mirror images of the D forms.

solution. Enzymes called **mutarotases** catalyze this process in vivo. Anomers exist only for aldoses, not ketoses. Can you tell why?

The representation of a cyclic sugar structure we have used in Figure 9.8 is called a **Haworth projection.** Imagine that you are seeing the ring in perspective, and the groups attached to the ring carbons (H, OH, CH₂OH) are pictured as being above or below the ring. In all D-monosaccharides, the —CH₂OH is above the ring. The relationship between hydroxyl orientations in a Fischer and a Haworth projection is straightforward. Those represented to the right of the chain in a Fischer projection are shown below the ring in a Haworth. For example, Fischer projections of α-D-ribofuranose and β-D-ribofuranose would look like this:

α-D-Ribofuranose   β-D-Ribofuranose

Even Haworth projections do not accurately depict the three-dimensional structure of molecules like ribofuranose or ribopyranose.

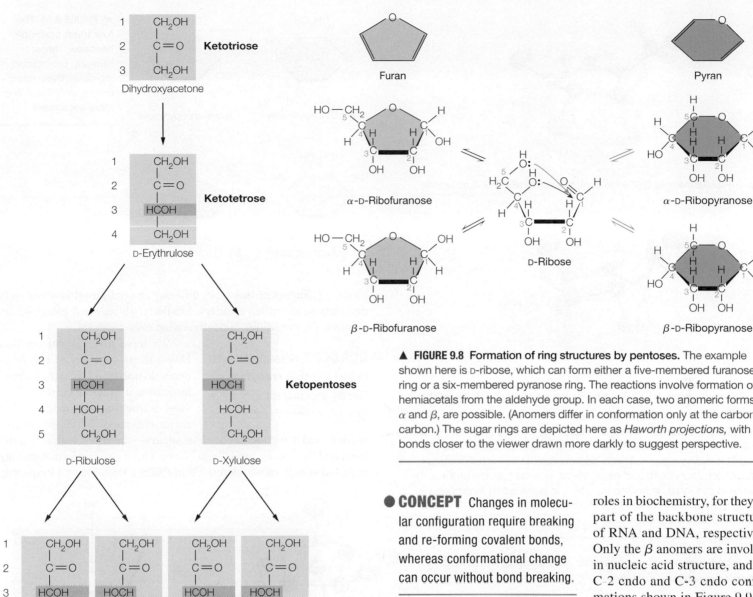

(b) D-Ketoses

▲ **FIGURE 9.7** *(Continued)*

▲ **FIGURE 9.8 Formation of ring structures by pentoses.** The example shown here is D-ribose, which can form either a five-membered furanose ring or a six-membered pyranose ring. The reactions involve formation of hemiacetals from the aldehyde group. In each case, two anomeric forms, α and β, are possible. (Anomers differ in conformation only at the carbonyl carbon.) The sugar rings are depicted here as *Haworth projections,* with bonds closer to the viewer drawn more darkly to suggest perspective.

● **CONCEPT** Changes in molecular configuration require breaking and re-forming covalent bonds, whereas conformational change can occur without bond breaking.

roles in biochemistry, for they are part of the backbone structures of RNA and DNA, respectively. Only the β anomers are involved in nucleic acid structure, and the C-2 endo and C-3 endo conformations shown in Figure 9.9 are favored. However, there is some variation in ring conformation, even locally, along DNA and RNA chains, with resulting changes in secondary structure. This flexibility points up a fundamental difference between *conformation* and *configuration.* Conformational isomers can interchange by a simple deformation of the molecule. But *configurational* isomers, such as the various kinds of stereoisomers described earlier, can interconvert only through the breaking and re-formation of covalent bonds.

Like the aldopentoses, the ketopentoses exist almost entirely in ring form under physiological conditions. However, only the furanose form is possible for ketopentoses. An example is *α-D-ribulose,* which is an intermediate in the carbon fixation processes in photosynthesis.

α-D-Ribulose

Hexose Rings

The hexoses also exist primarily in ring forms under physiological conditions. As with the aldopentoses, two kinds of rings are found: five-membered furanoses and six-membered pyranoses. In each

Saturated five- and six-membered rings cannot be planar because the C—C—C bond angles are about 109° and the C—O—C angle is about 118°. Furthermore, the ring can pucker out of plane in many different ways. The different ring conformations produced by slightly different bond angles are called **conformational isomers.** Ball-and-stick models of two of the several possible conformational isomers of β-D-ribofuranose are shown in **FIGURE 9.9**.

We have already encountered β-D-ribofuranose (and its close relative β-D-2-deoxyribofuranose) in Chapter 4. These sugars play critical

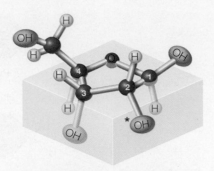

(a) β–D–Ribofuranose in C–2 endo conformation where C–2 is above the plane.

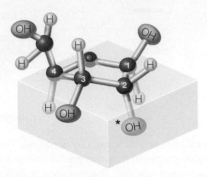

(b) β–D–Ribofuranose in C–3 endo conformation where C–3 is above the plane.

▲ **FIGURE 9.9 Conformational isomers.** These models show two of the possible ring conformations for β-D-ribofuranose. In both of them, C-1, O, and C-4 define a plane. These isomers are the two most common conformations for ribose and deoxyribose in nucleic acids. (In DNA the hydroxyl at carbon 2, indicated here by *, is replaced by hydrogen.) A C-3 exo conformation would look like the figure in (b), but C-3 would be flipped below the plane.

case, α and β anomers are possible. An example, illustrated by Haworth projections, follows:

α-D-Glucopyranose   **β-D-Glucopyranose**

FIGURE 9.10 shows Haworth projections of the structures of the four most common hexoses in their usual configurations. Forms that are favored depend greatly on the structure of the particular sugar and its environment, although in general hexoses prefer the pyranose ring structure when in aqueous solution. This preference is also true of fructose, but we have depicted D-fructose in Figure 9.10 in its furanose configuration because that is how it is found in its most common biological source, the disaccharide sucrose. Elucidation of the distribution of anomeric and tautomeric forms of the sugars existing in solutions has been greatly facilitated by nuclear magnetic resonance spectroscopy (see Tools of Biochemistry 6A), which allows determination of molecular conformation in solution.

There is something else to note from Figure 9.10. Glucose and mannose differ from each other only in the configuration

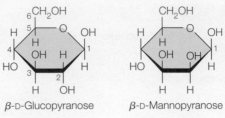

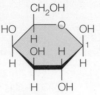

β-D-Glucopyranose   β-D-Mannopyranose

β-D-Galactopyranose   β-D-Fructofuranose

◀ **FIGURE 9.10 The four most common hexoses.** These Haworth projections represent the D enantiomers; only the β anomers are shown.

about C2. Sugars of this type, differing in configuration about only one carbon, are called **epimers.** Similarly, glucose and galactose are epimers, for they differ in configuration only about C4.

● **CONCEPT** Hexoses can exist in boat and chair conformations. Usually, the chair is more stable.

We have already shown that Haworth projections of the furanoses do not depict the actual three-dimensional structure correctly. The same is true for the pyranoses. Two major classes of pyranose conformations exist for the six-carbon sugars—the more stable "chair" form and the less favored "boat" form. These two conformations are depicted as ball-and-stick models in **FIGURE 9.11(a)**. We will frequently

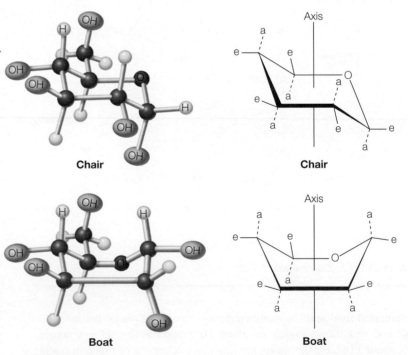

Chair   **Chair**

Boat   **Boat**

(a) Ball-and-stick models.   **(b)** Skeletal diagrams of the bonding. Axial bonds (a) and equatorial bonds (e) are indicated.

▲ **FIGURE 9.11 The pyranose ring in chair and boat conformations.** Three-dimensional representations of α-D-glucopyranose in the chair form (left) and the boat form (right).

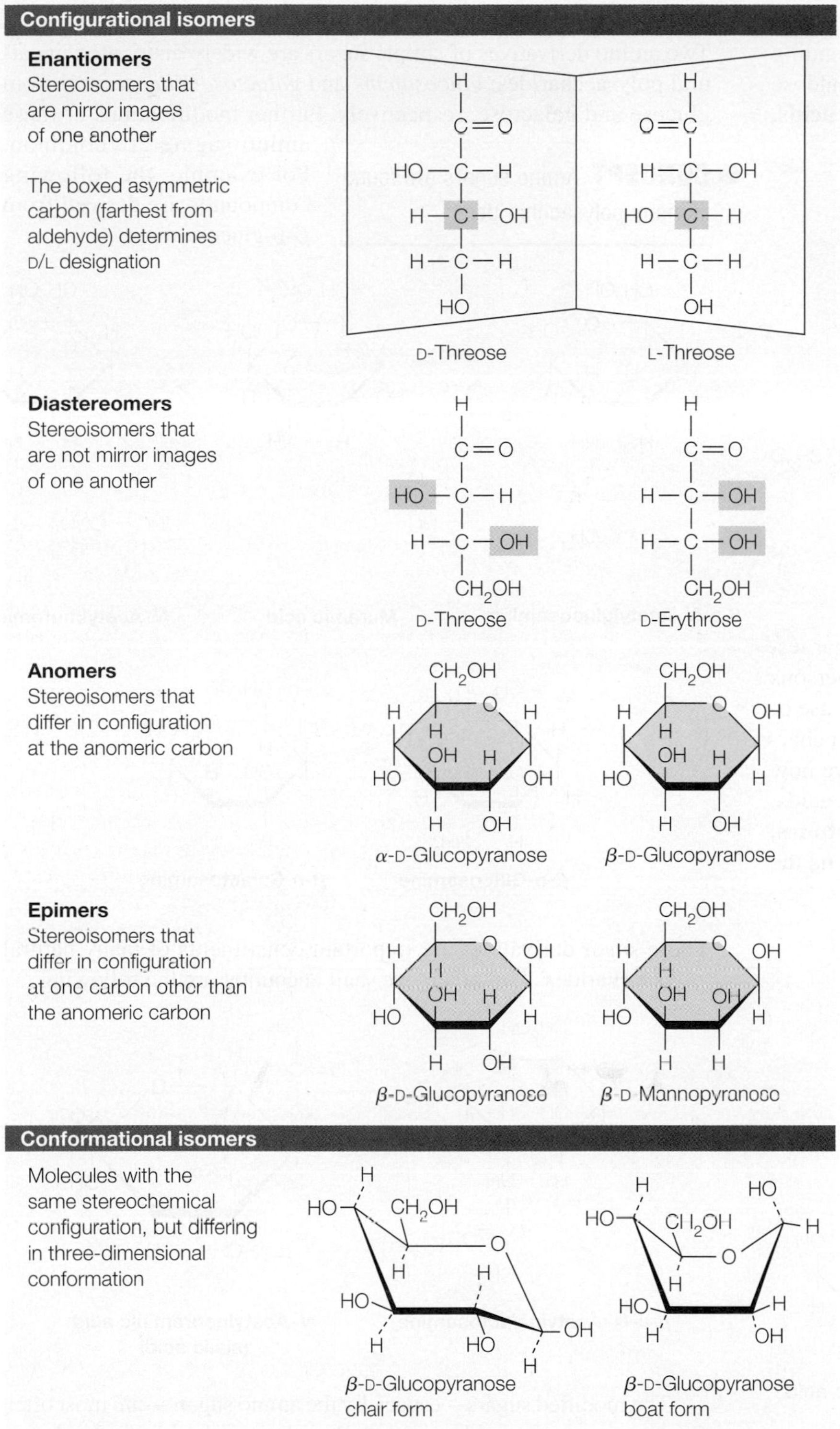

Configurational isomers

Enantiomers
Stereoisomers that are mirror images of one another

The boxed asymmetric carbon (farthest from aldehyde) determines D/L designation

D-Threose   L-Threose

Diastereomers
Stereoisomers that are not mirror images of one another

D-Threose   D-Erythrose

Anomers
Stereoisomers that differ in configuration at the anomeric carbon

α-D-Glucopyranose   β-D-Glucopyranose

Epimers
Stereoisomers that differ in configuration at one carbon other than the anomeric carbon

β-D-Glucopyranose   β-D-Mannopyranose

Conformational isomers

Molecules with the same stereochemical configuration, but differing in three-dimensional conformation

β-D-Glucopyranose chair form   β-D-Glucopyranose boat form

▲ **FIGURE 9.12 Terminology for carbohydrate stereochemistry.** Conformational isomers are distinguished from configurational isomers in that the former can interconvert without breaking and re-forming bonds.

more stable because substituents on axial bonds tend to be more crowded in the boat form.

Sugars with More Than Six Carbons

Monosaccharides with seven or even more carbons exist in nature, but most are of minor importance. However, one heptose, called *sedoheptulose,* plays a major role in the fixation of CO_2 in photosynthesis (see Chapter 15) as well as in the pentose phosphate pathway (see Chapter 12).

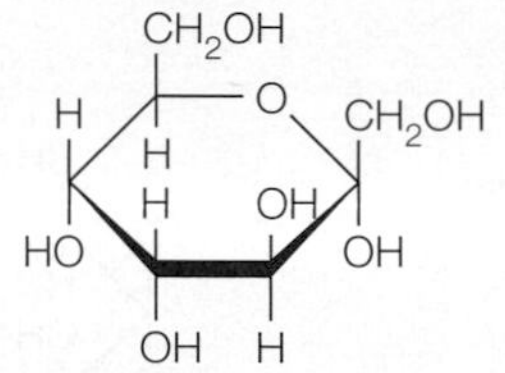

α-D-**Sedoheptulopyranose**

At this point, we have introduced several terms used to describe the structures of sugar molecules—enantiomers, diastereomers, anomers, epimers, and ring conformations. For review, this terminology is summarized in **FIGURE 9.12**.

9.2 Derivatives of the Monosaccharides

Each monosaccharide carries a number of hydroxyl groups to which substituents might be attached or which could be replaced by other functional groups. An enormous number of sugars are in fact modified in this way. In this section, we describe only a few of them—primarily those that play biologically important roles.

Phosphate Esters

We have already encountered sugar phosphorylation in compounds like AMP, ATP, and the nucleic acids. As we see in later chapters, the phosphate esters of the monosaccharides themselves are major participants in metabolic pathways. One such compound, β-D-**glucose-1-phosphate,** is shown here.

● **CONCEPT** Sugar phosphates are important intermediates in metabolism, functioning as activated compounds in syntheses.

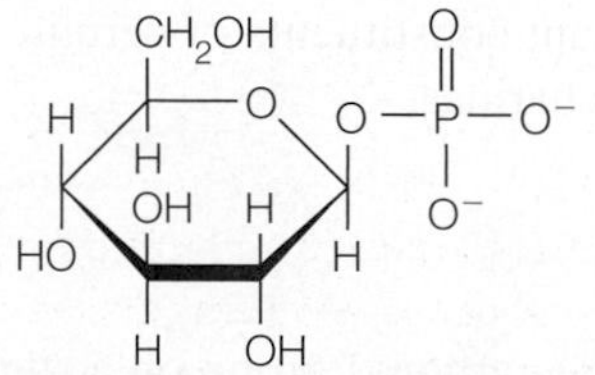

β-D-**Glucose-1-phosphate**

depict them in the ways shown in Figure 9.11(b). For both the boat and chair forms of pyranose rings, a molecular axis can be defined perpendicular to the central plane of the molecule. Bonds to substituents on ring carbons can then be classified as *axial (a)* or *equatorial (e),* depending on whether they are approximately parallel or perpendicular to the axis (Figure 9.11(b)). For most sugars, the chair form is

Sugar phosphate esters are quite acidic, with pK_a values for the two stages of phosphate ionization of about 1–2 and 6–7, respectively. Consequently, these compounds exist under physiological conditions as a mixture of the monoanions and dianions.

Lactones and Acids

Oxidation of monosaccharides can proceed in several ways, depending on the oxidizing agent used. For example, mild oxidation of an aldose with alkaline Cu(II) (Fehling's solution) produces the **aldonic acids,** as shown in the following example:

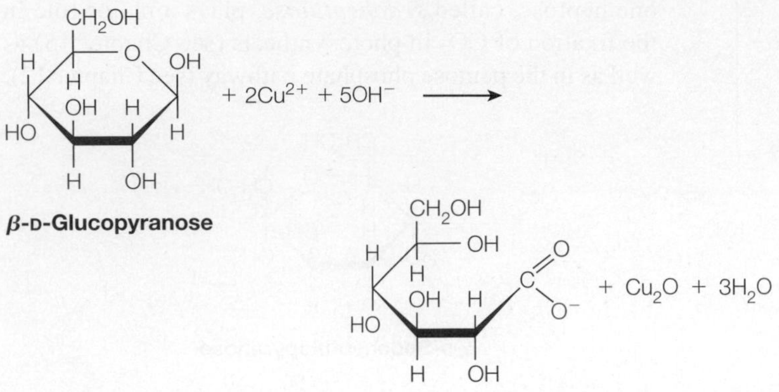

β-D-**Glucopyranose**

$+ 2Cu^{2+} + 5OH^- \longrightarrow$

$+ Cu_2O + 3H_2O$

D-**Gluconic acid**

The production of a red precipitate of Cu_2O is a classic sugar test and was used formerly to test for excess sugar in the urine of persons thought to have diabetes. Another, similar reaction involves the use of Ag^+ ion as an oxidant; its reduction to metallic silver leaves a characteristic "mirror" on the glassware. These older methods have now been replaced by more specific enzyme assays. Free aldonic acids, such as gluconic acid, are in equilibrium in solution with **lactones,** which are cyclic esters, in this case involving the C1 carboxyl and the C5 hydroxyl.

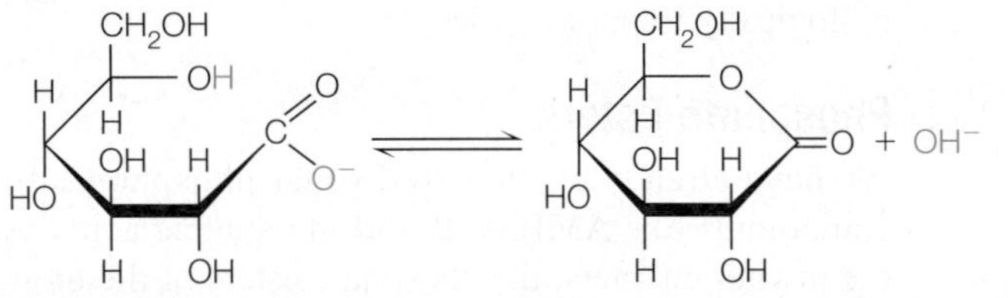

D-**Gluconic acid** ⇌ D-δ-**Gluconolactone** $+ OH^-$

Enzyme-catalyzed oxidation of monosaccharides gives other products, including **uronic acids** such as **glucuronic acid,** in which oxidation has occurred at carbon 6. Uronic acids are, as we see later in this chapter, important constituents of certain natural polysaccharides.

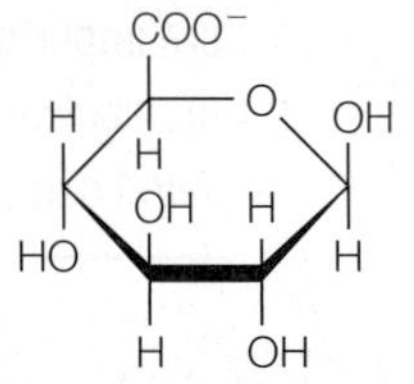

β-D-**Glucuronic acid**

Alditols

Reduction of the carbonyl group on a sugar gives rise to the class of polyhydroxy compounds called **alditols.** The reduced form of glucose, D-glucitol, is also called *sorbitol.*

Each is named from the corresponding monosaccharide. When sorbitol accumulates in the lens of a diabetic's eye, it can lead to the formation of cataracts.

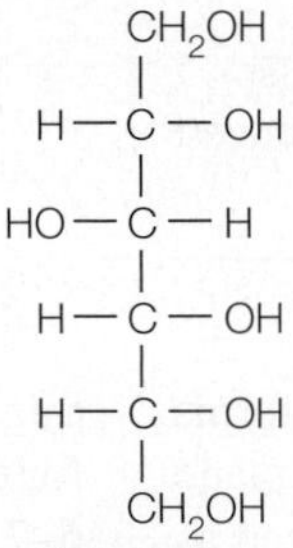

D-**Glucitol (sorbitol)**

Amino Sugars

Two amino derivatives of simple sugars are widely distributed in natural polysaccharides: *glucosamine* and *galactosamine,* derived from glucose and galactose, respectively. Further modifications of these amino sugars are common. For example, the following compounds are derived from β-D-glucosamine:

● **CONCEPT** Amino sugars are found in many polysaccharides.

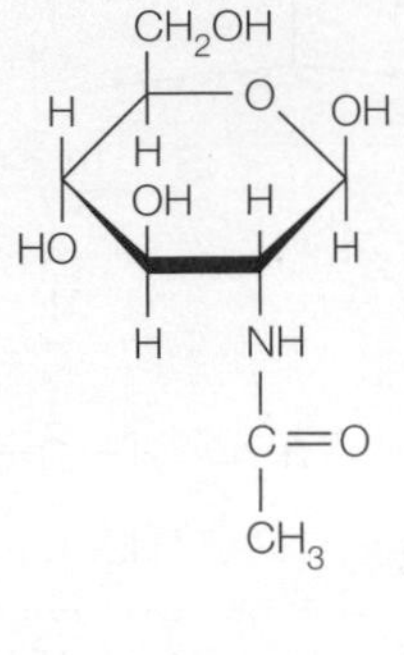

β-D-*N*-**Acetylglucosamine** 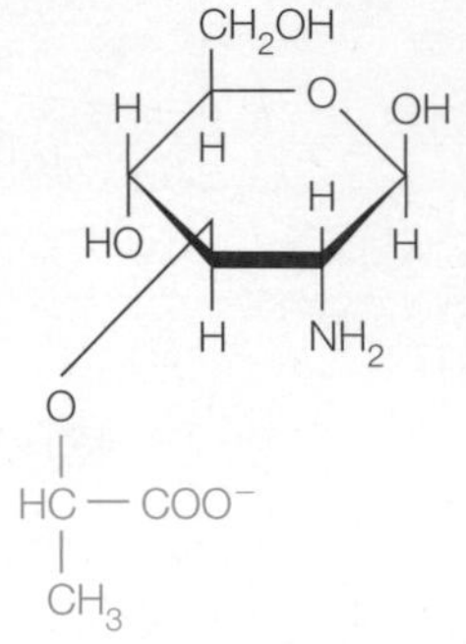   **Muramic acid** 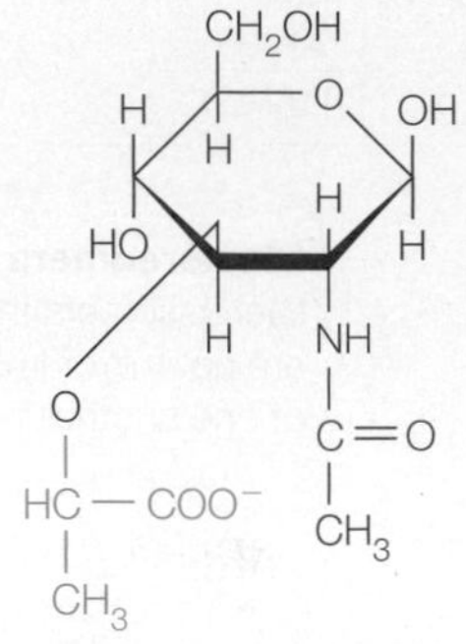   *N*-**Acetylmuramic acid**

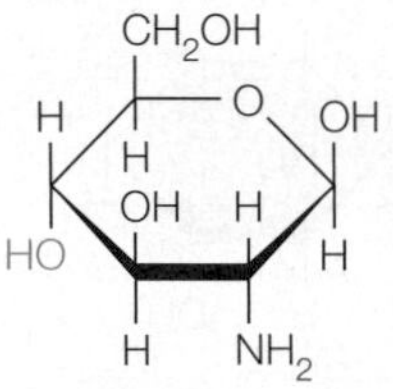

β-D-**Glucosamine** 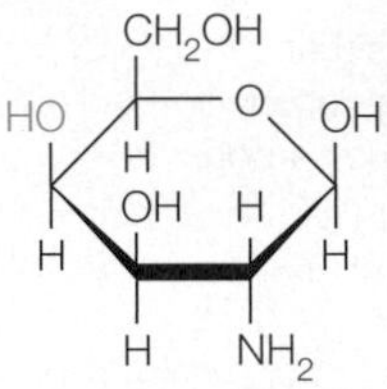   β-D-**Galactosamine**

These sugar derivatives are important constituents of many natural polysaccharides. Two others we shall encounter are the following:

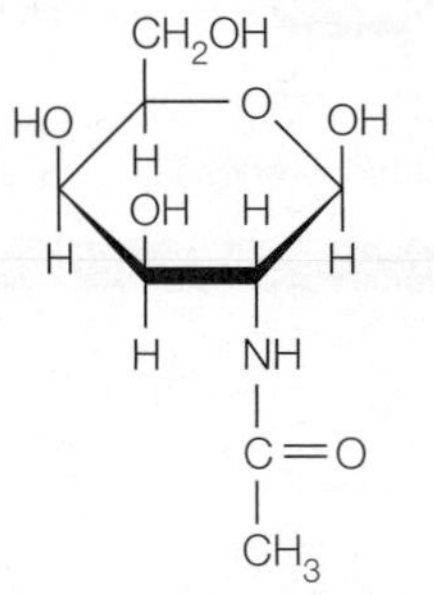

β-D-*N*-**Acetylgalactosamine** 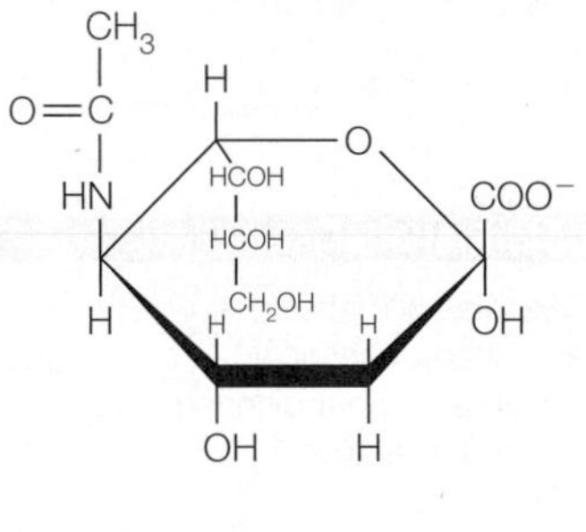   *N*-**Acetylneuraminic acid (sialic acid)**

The modified sugars—especially the amino sugars—are most often found as monomer residues in complex oligosaccharides and polysaccharides. To aid in writing the structures of such molecules, it is useful to have a shorthand notation, as is used in describing nucleic acid and protein structure. Therefore, a set of abbreviations has been defined for the simple sugars and their derivatives. A number of the most important ones are listed in **TABLE 9.1**.

Glycosides

Elimination of water between the anomeric hydroxyl of a cyclic monosaccharide and the hydroxyl group of another compound yields an **O-glycoside** (the O signifying attachment at a hydroxyl). The ether-like

TABLE 9.1 Abbreviations for some common monosaccharide residues

Monosaccharides	Abbreviation
Arabinose	Ara
Fructose	Fru
Fucose	Fuc
Galactose	Gal
Glucose	Glc
Lyxose	Lyx
Mannose	Man
Ribose	Rib
Xylose	Xyl
Monosaccharide derivatives	**Abbreviation**
Gluconic acid	GlcA
Glucuronic acid	GlcUA
Galactosamine	GalN
Glucosamine	GlcN
N-Acetylgalactosamine	GalNAc
N-Acetylglucosamine (or NAG)	GlcNAc
Muramic acid	Mur
N-Acetylmuramic acid (or NAM)	MurNAc
N-Acetylneuraminic acid (or sialic acid)	NeuNAc (or Sia)

Ouabain

Amygdalin

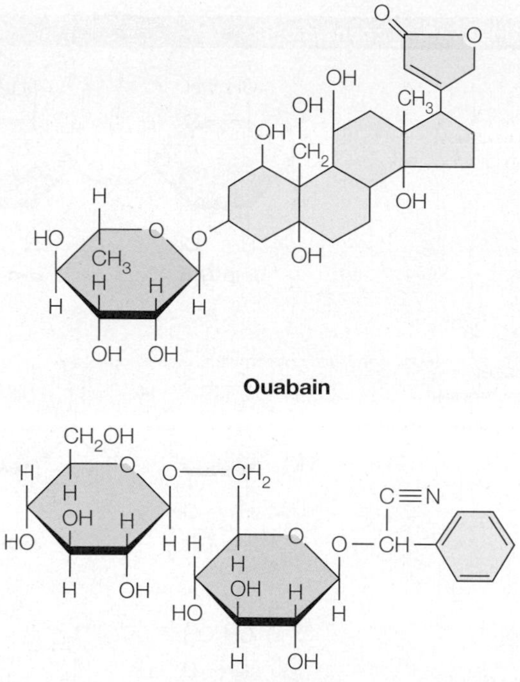

▲ **FIGURE 9.13 Two naturally occurring glycosides.** Ouabain and amygdalin are highly toxic glycosides produced by plants.

bond formed is referred to as a **glycosidic bond.** A simple example is the formation of methyl-α-D-glucopyranoside:

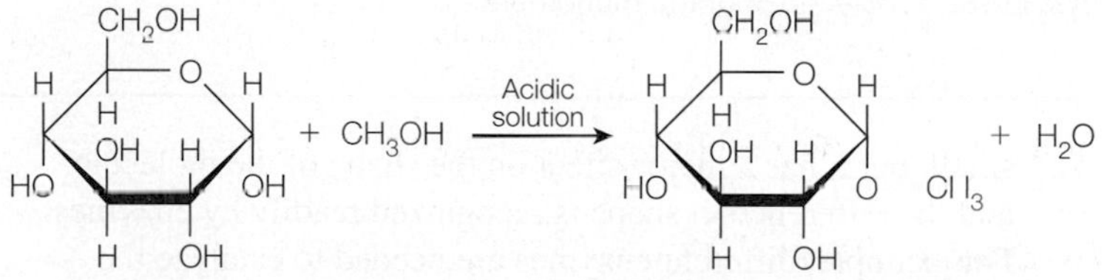

α-**D-Glucopyranose**　　　　　　　　**Methyl-α-D-glucopyranoside**

Unlike the anomers of the sugars themselves, the anomeric glycosides (e.g., methyl-α-D-glucopyranoside in the example shown, and methyl-β-D-glucopyranoside) do not interconvert by mutarotation in the absence of an acid catalyst. This property makes them useful in determining sugar configurations.

Many glycosides are found in plant and animal tissues. Some are toxic substances, in most cases because they act as inhibitors of enzymes involved in ATP utilization. Two toxic glycosides, *ouabain* and *amygdalin,* are shown in **FIGURE 9.13**. Ouabain inhibits the action of the enzymes that pump Na$^+$ and K$^+$ ions across cell membranes to maintain necessary electrolyte balance. It comes from an African shrub and was discovered when it was observed that Somali hunters dipped arrowheads in an extract

● **CONCEPT** O-glycosides are formed by elimination of a water molecule between a hydroxyl group on a saccharide and a hydroxyl on another compound.

from the plant. Ouabain now finds use in treatment of some cardiac conditions. Amygdalin is toxic for a quite different reason. Found in the seeds of bitter almonds, this glycoside yields hydrogen cyanide (HCN) upon hydrolysis. It is for this reason that HCN gas is said to have the odor of bitter almonds.

9.3 Oligosaccharides

Just as monosaccharides can form glycosidic bonds with other kinds of hydroxyl-containing compounds, they can do so with one another. Such bonding gives rise to glycans—the oligosaccharides and polysaccharides.

Oligosaccharide Structures

The simplest and biologically most important oligosaccharides are the *disaccharides,* made up of two residues. The disaccharides play many biological roles. Some, like *sucrose, lactose,* and *trehalose,* are soluble energy stores in plants and animals. Others, like *maltose* and *cellobiose,* can be regarded primarily as intermediate products in the degradation of much longer polysaccharides. Still others, like *gentiobiose,* are found principally as constituents of more complex, naturally occurring substances. The structures of these disaccharides are depicted in **FIGURE 9.14**.

Distinguishing Features of Different Disaccharides

Four major features distinguish disaccharides from one another:

1. *The two specific sugar monomers involved and their stereoconfigurations.* The monomers may be of the same kind, as the two D-glucopyranose residues in maltose, or they may be different, as the D-glucopyranose and D-fructofuranose residues in sucrose.

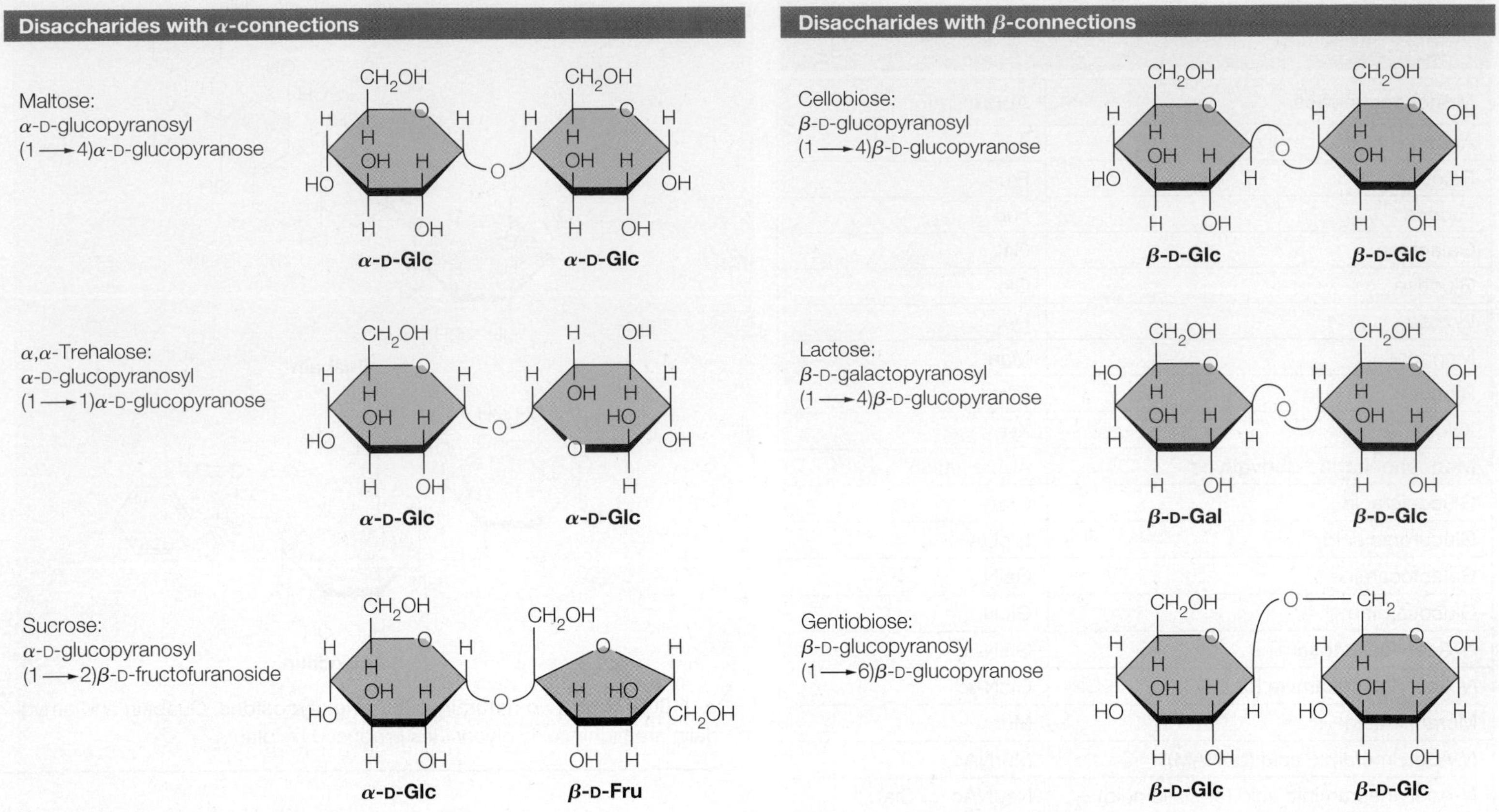

Disaccharides with α-connections

Maltose:
α-D-glucopyranosyl
(1 ⟶ 4)α-D-glucopyranose

α-D-Glc   α-D-Glc

α,α-Trehalose:
α-D-glucopyranosyl
(1 ⟶ 1)α-D-glucopyranose

α-D-Glc   α-D-Glc

Sucrose:
α-D-glucopyranosyl
(1 ⟶ 2)β-D-fructofuranoside

α-D-Glc   β-D-Fru

(a) Disaccharides linked through the C-1 of the α anomer: maltose, trehalose, and sucrose.

Disaccharides with β-connections

Cellobiose:
β-D-glucopyranosyl
(1 ⟶ 4)β-D-glucopyranose

β-D-Glc   β-D-Glc

Lactose:
β-D-galactopyranosyl
(1 ⟶ 4)β-D-glucopyranose

β-D-Gal   β-D-Glc

Gentiobiose:
β-D-glucopyranosyl
(1 ⟶ 6)β-D-glucopyranose

β-D-Glc   β-D-Glc

(b) Disaccharides with β linkage: cellobiose, lactose, and gentiobiose.

▲ **FIGURE 9.14 Structures of some important disaccharides.** Shown are Haworth projections of the same molecules, with color-coded monomers: blue = glucose, pink = fructose, teal = galactose. Note the convention used to draw glycosidic bonds between monomers in disaccharides. The "curved bonds" allow the Haworth projections of the monomers to be drawn on the same line.

2. *The carbons involved in the linkage.* Although many possibilities exist, the most common linkages are 1 → 1 (as in trehalose), 1 → 2 (as in sucrose), 1 → 4 (as in lactose, maltose, and cellobiose), and 1 → 6 (as in gentiobiose). Note that all of these disaccharides involve the anomeric hydroxyl of at least one sugar as a participant in the bond.

3. *The order of the two monomer units, if they are different kinds.* The glycosidic linkage involves the anomeric carbon on one sugar, but in most cases the other is free. Thus, the two ends of the molecule can be distinguished by their chemical reactivity. For example, the glucose residue in lactose, having a free anomeric carbon and thus a potential free aldehyde group, could be oxidized by Fehling's solution; the galactose residue could not be. Lactose is therefore a reducing sugar, and the glucose residue is at its *reducing end*. The other end is called the *nonreducing end*. In sucrose, neither residue has a potential free aldehyde group; both anomeric carbons are involved in the glycosidic bond. Therefore sucrose is a nonreducing sugar.

4. *The configuration of the anomeric hydroxyl group of each residue.* This feature is especially important for the anomeric carbon(s) involved in the glycosidic bond. The configuration may be either α (as in the disaccharides shown in Figure 9.14(a) or β (as in those in Figure 9.14(b)). This difference may seem

small, but it has a major effect on the shape of the molecule, and the difference in shape is recognized readily by enzymes. For example, different enzymes are needed to catalyze the hydrolysis of maltose and cellobiose, even though both are dimers of D-glucopyranose. Furthermore, we shall see that in polysaccharides the anomeric orientation plays a critical role in determining the secondary structures adopted by these polymers.

Writing the Structure of Disaccharides

A convenient way to describe the structures of these and more complex oligosaccharides has been devised. The rules are as follows:

1. The sequence is written starting with the nonreducing end at the left, using the abbreviations defined in Table 9.1.

2. Anomeric and enantiomeric forms are designated by prefixes (e.g., α-, D-).

3. The ring configuration is indicated by a suffix (*p* for pyranose, *f* for furanose).

4. The atoms between which glycosidic bonds are formed are indicated by numbers in parentheses between residue designations (e.g., (1 → 4) means a bond from carbon 1 of the residue on the left to carbon 4 of the residue on the right).

TABLE 9.2 Occurrence and biochemical roles of some representative disaccharides

Disaccharide	Structure	Natural Occurrence	Physiological Role
Sucrose	Glcα(1 ⟶ 2)Fruβ	Many fruits, seeds, roots, honey	A final product of photosynthesis, used as a primary energy source in many organisms
Lactose	Galβ(1 ⟶ 4)Glc	Milk, some plant sources	A major animal energy source
α,α-Trehalose	Glcα(1 ⟶ 1)Glcα	Yeast, other fungi, insect blood	A major circulatory sugar in insects; used for energy
Maltose	Glcα(1 ⟶ 4)Glc	Plants (starch) and animals (glycogen)	The dimer derived from the starch and glycogen polymers
Cellobiose	Glcβ(1 ⟶ 4)Glc	Plants (cellulose)	The dimer of the cellulose polymer
Gentiobiose	Glcβ(1 ⟶ 6)Glc	Some plants (e.g., gentians)	Constituent of plant glycosides and some polysaccharides

As an example, we can write the structure of sucrose as

$$\alpha\text{-D-Glc}p(1 \longrightarrow 2)\text{-}\beta\text{-D-Fru}f$$

In many cases, the nomenclature is further shortened by omitting the D and L designations (except in the unusual cases in which L enantiomers are encountered) and by omitting the p and f suffixes when the monomers have their usual ring forms. Thus, we would more likely write sucrose as Glcα(1 → 2)Fruβ. The system can be applied to oligosaccharides of any length and can include branched structures, as we will see in the discussion of starch later in this chapter. If only one carbon involved in the linkage between two residues is anomeric, the representation can be even more condensed because the anomeric configuration at the reducing end will equilibrate in solution. For example, maltose can be represented as Glcα(1 → 4)Glc. **TABLE 9.2** shows the abbreviated names for the most common disaccharides.

The list of biologically important oligosaccharides is by no means restricted to dimeric structures. Many trimers, tetramers, and even larger, yet specifically constructed, molecules are known. Examples of these compounds will be encountered later in this chapter.

Stability and Formation of the Glycosidic Bond

Formation of the glycosidic bond between two monomers in an oligosaccharide is a condensation involving the elimination of a molecule of water. Thus, we might expect the synthesis of lactose to proceed as follows:

β-D-Galactose   β-D-Glucose

Lactose

This reaction is analogous to the elimination of water between amino acids in the formation of polypeptides or between nucleotides in the formation of nucleic acids. As in those cases, the reaction as written is thermodynamically unfavored. Instead, the hydrolysis of

oligosaccharides and polysaccharides is favored under physiological conditions by a standard free energy change of about 15 kJ/mol, corresponding to an equilibrium constant of about 800 in favor of the hydrolysis products. Nevertheless, like peptides and oligonucleotides, saccharide polymers are sufficiently metastable to persist for long periods unless their hydrolysis is catalyzed by enzymes or acid. So the situation is the same as the ones we have encountered with the other important biopolymers.

The breakdown of oligosaccharides and polysaccharides in vivo is controlled by the presence of specific enzymes. Furthermore, *synthesis* of these sugar polymers never proceeds in living cells by reactions like the one we have just shown. As in protein or nucleic acid synthesis, activated monomers are required. For glycan biosynthesis, those activated monomers are usually nucleotide-linked sugars. The activated sugar molecule in lactose biosynthesis is **uridine diphosphate galactose** (UDP-galactose or UDP-Gal), a nucleotide-linked sugar formed by reaction of uridine triphosphate with galactose-1-phosphate.

β-D-Galactose-1-phosphate   **Uridine triphosphate**

Uridine diphosphate galactose   **Pyrophosphate**

The formation of UDP-galactose illustrates a common mechanism that activates the anomeric carbon for transfer to a carbohydrate acceptor. The details of this mechanism are shown in Chapter 12 for the biosynthesis of glycogen. In the reaction catalyzed by the enzyme **lactose synthase,** that acceptor is glucose, as shown in the final reaction in **FIGURE 9.15**, and the product is lactose (Glc β(1 → 4)Glc). Note the specificity of the enzyme. Although glucose has five different hydroxyl groups to which the galactosyl moiety could be transferred, only the hydroxyl at carbon 4 serves as an acceptor for this enzyme. Also note that the synthesis of UDP-Gal from Gal-1-P and UTP has a standard free energy change close to zero (there is no net loss or gain of phosphoanhydride bonds in that reaction); thus, it is PP$_i$ hydrolysis that ultimately drives the reaction forward.

● **CONCEPT** Like the phosphodiester bond in nucleic acid and amide bond in proteins, the glycosidic bond is metastable. Enzymes control its hydrolysis.

Because different disaccharides (and oligosaccharides and polysaccharides) are distinguished both by the kinds of monomers involved and by the precise glycosidic linkages between them, the enzymes needed for their breakdown must also be specific. For example, hydrolysis of the common nutritional disaccharides maltose, lactose, and sucrose, which takes place in cells lining the wall of the small intestine, requires three different and specific enzymes. None will substitute for another.

Lactose synthase is an example of a **glycosyltransferase.** These reactions all involve transfer of an activated glycosyl moiety to an acceptor. The first known glycosyltransferase was the enzyme responsible for synthesis of **glycogen,** a carbohydrate storage polymer in animals (glycogen metabolism is discussed in Chapter 12). This enzyme uses **UDP-glucose** (UDPG or UDP-Glc) as the activated glycosyl donor. Although both glycosyltransferases mentioned so far use uridine nucleotides for activation, there are exceptions. For example, the biosynthesis of starch in plants uses **adenosine diphosphate glucose** (ADP-glucose) as the activated nucleotide.

Another glycosyltransferase is responsible for the synthesis of sucrose in plants.

$$\text{UDP-glucose} + \text{fructose-6-phosphate} \longrightarrow \text{sucrose-6-phosphate} + \text{UDP}$$

The product, sucrose-6-phosphate, is subsequently hydrolyzed to sucrose plus phosphate. One reason to think about sucrose metabolism is sucrose's involvement in a glycan biosynthetic reaction that does not involve nucleotide-linked sugars. Some bacteria carry out the synthesis of **dextran,** an $\alpha(1 \rightarrow 6)$-linked polymer of glucose with $\alpha(1 \rightarrow 2)$, $\alpha(1 \rightarrow 3)$, or $\alpha(1 \rightarrow 4)$ branch points. The polymerization, catalyzed by **dextran sucrase,** uses sucrose itself as the substrate.

$$n \text{ sucrose} \longrightarrow \text{glucose}_n(\text{dextran}) + n \text{ fructose}$$

● **CONNECTION** Bacterial synthesis of polymeric dextran from dietary sucrose is an important factor in the generation of dental plaque.

Several bacteria growing in the human oral cavity synthesize large quantities of dextran, which contributes to formation of dental plaque—hence, one concern nutritionists have, in addition to obesity, about excessive sucrose consumption.

● **CONCEPT** Glycan biosynthesis is carried out by glycosyltransferases, enzymes that transfer an activated glycosyl moiety, such as UDP-glucose, to a specific position on a carbohydrate acceptor.

As we see later in this chapter, the larger oligosaccharides and the polysaccharides can exhibit complex structures. There is one important way in which oligosaccharide and polysaccharide synthesis differs from synthesis of nucleic acids and proteins. These sugar polymers are never copied from template molecules. Instead, in the formation of glycans, a different enzyme is employed to catalyze the addition of each kind of monomer unit. Clearly, a vast array of plant and animal enzymes must be devoted to the synthesis and degradation of saccharide polymers.

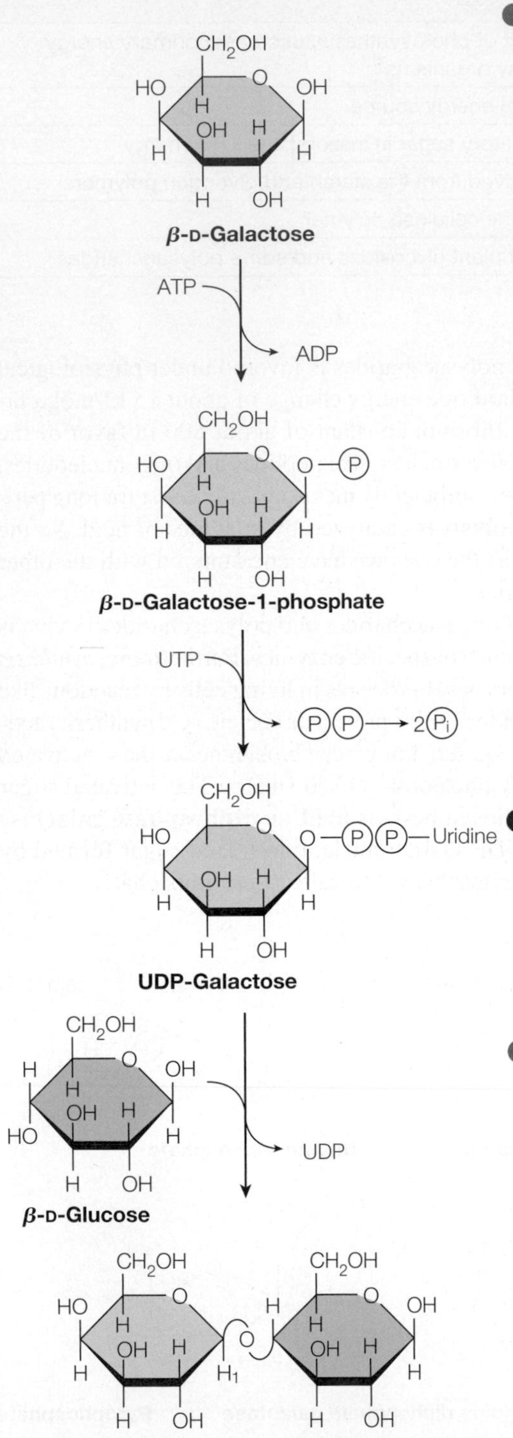

9.4 Polysaccharides

Polysaccharides fulfill numerous biological functions. Some, like starch and glycogen, serve mainly to store sugars for energy in plants and animals. Others, like **cellulose, chitin,** and the polysaccharides of bacterial cell walls, are structural materials analogous to the fibrous proteins described in Chapter 6. It is simplest to consider these molecules in terms of their functional categories.

As with polypeptides and polynucleotides, the sequence of monomer residues in a polysaccharide defines its primary structure. Whereas proteins usually have complex sequences, polysaccharides often have rather simple primary structures. In some cases (e.g., cellulose), the polymer is made from only one kind of monomer residue (β-D-glucose for cellulose); these kinds of polymers are referred to as **homopolysaccharides.** If two or more different monomers are involved, the polymer is called a **heteropolysaccharide.** Even those storage and structural polysaccharides that are heteropolymers are rarely complex; usually, no more than two kinds

β-D-Galactose

β-D-Galactose-1-phosphate

UDP-Galactose

β-D-Glucose

Lactose

▲ **FIGURE 9.15 Enzymatic formation of lactose.** The reaction shown occurs in the formation of milk in mammary tissue. Galactose is phosphorylated by ATP, then transferred to uridine diphosphate (UDP). UDP-galactose transfers galactose to glucose, with the accompanying cleavage of a phosphate ester bond. The reaction is catalyzed by the enzyme *lactose synthase.*

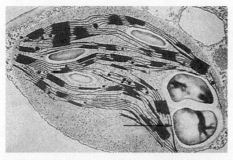
(a) Starch granules in a plant leaf chloroplast.

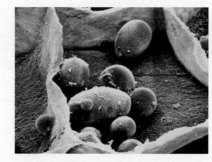
(b) Starch granules in potato tuber cells.

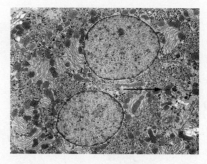
(c) Glycogen granules in liver.

▲ **FIGURE 9.16 Storage of starch and glycogen in granules.** In each case, a representative granule is indicated by an arrow.

of residues are involved. In further contrast to protein and nucleic acid molecules, which are almost always of defined length, polysaccharide chains grow to random lengths. And, as mentioned earlier, glycans are distinctive in that they can form branched chains. The ability to branch, as well as the number of functional groups on each monomer that can participate in bond formation, gives polysaccharides amazing structural diversity, and this diversity undoubtedly contributes to the large number of roles played by polysaccharides.

The functional reasons for the distinctions from proteins and nucleic acids are not hard to find. A storage material, such as starch, needs neither to convey information nor to adopt a complicated three-dimensional form. It is simply a bin in which to put away glucose molecules for future use. Many structural polysaccharides (like fibrous proteins) form extended, regular secondary structures, well suited to the formation of fibers or sheets. Often a regular repetition of some simple monosaccharide or disaccharide motif will serve this function. (Recall, for comparison, the simple and repetitive amino acid sequences of collagen and silk fibroin described in Chapter 6.) The only glycan polymers in which well-defined and complex sequences are found are some of the oligosaccharides attached to cell surfaces or those attached to specific glycoproteins. Because these oligomers serve to identify cells or molecules, they must convey information. This function requires precisely defined "words" in the polysaccharide language, just as nucleic acid sequences spell out information in their own language.

Storage Polysaccharides

The principal storage polysaccharides are **amylose** and **amylopectin,** which together constitute starch in plants, and **glycogen,** which is stored in animal and microbial cells. Both starch and glycogen are stored in granules within cells (**FIGURE 9.16**). Starch is found in almost every kind of plant cell, but grain seeds, tubers, and unripe fruits are especially rich in this material. Glycogen is deposited in the liver, which acts as a central energy storage organ in many animals. Glycogen is also abundant in muscle tissue, where it is more immediately available for energy release.

Amylose, amylopectin, and glycogen are all polymers of α-D-glucopyranose. They are homopolysaccharides of the class called **glucans,** the polymers of glucose. The three polymers differ only in the kinds of linkages between glucose residues. Amylose is a linear polymer, involving exclusively $\alpha(1 \rightarrow 4)$ links between adjacent glucose residues. Amylopectin (**FIGURE 9.17**) and glycogen are both branched polymers because they contain, in addition to the $\alpha(1 \rightarrow 4)$ links, some $\alpha(1 \rightarrow 6)$ links as well. The branches in glycogen are somewhat

Amylopectin

(a) The primary structure of amylopectin. Nonreducing ends (N) and reducing ends (R) are indicated.

▲ **FIGURE 9.17 Amylopectin, a branched glucan.** To simplify the figure, some ring hydroxyls are not shown. $\alpha(1 \rightarrow 6)$ branch points are shown in red.

(b) Detailed structure of a branch point.

▲ **FIGURE 9.18 The secondary structure of amylose.** The orientation of successive glucose residues favors helix generation. Note the large interior core. Hydrogen bonds (not shown) stabilize the helix.

more frequent and shorter than those in amylopectin, and glycogen is usually of higher molecular weight, but in most respects the structures of these two polysaccharides are very similar.

The regular and simple primary structure of amylose allows a regular secondary structure for this molecule. As with polynucleotides and polypeptides, the details of this structure initially came from X-ray diffraction studies. In fact, amylose was the first biopolymer whose structure was elucidated by this method. Because of the $\alpha(1 \rightarrow 4)$ link, each residue is angled with respect to the preceding residue, favoring a regular helical conformation (**FIGURE 9.18**). The branched nature of amylopectin and glycogen inhibits the formation of helices because the helix requires 6 residues for each turn; there is a branch point about every 10–20 residues in amylopectin and about every 8 in glycogen.

● **CONCEPT** The starches—amylose, amylopectin, and glycogen—are storage polysaccharides. Amylose is linear; amylopectin and glycogen are branched.

The storage polysaccharides are admirably designed to serve their function. Glucose and even maltose are small, rapidly diffusing molecules, which are difficult to store. Were such small molecules present in large quantities in a cell, they would give rise to a high osmotic pressure, which would be deleterious. Therefore, most cells build the glucose into long polymers, so that large quantities can be stored in a manner that prevents its diffusion and loss. Whenever glucose is needed, it can be obtained by selective degradation of the polymers by specific enzymes. These processes are discussed in detail in Chapter 12, but one aspect should be mentioned now. Most of the enzymes employed attack the chains at their nonreducing ends, releasing one glucose residue at a time. Such "end-nibbling" (as opposed to internal cutting) prevents the continual breakup of the long polymers, which would lead to their complete solubilization. The branched structure of both amylopectin and glycogen is such that each molecule has *many* nonreducing ends that can be attacked simultaneously (see Figure 9.17), allowing rapid mobilization of glucose when it is needed. However, the linear chain of amylose with its single nonreducing end is used mainly for long-term storage of glucose.

Structural Polysaccharides

Plants do not seem to synthesize or use fibrous structural *proteins* (like keratin and collagen) but instead rely entirely on special polysaccharides. Animals use both kinds of materials. Because each structural use requires different properties, a great variety of structural polysaccharides exists. We begin by considering those from plants.

Cellulose

The major polysaccharide in woody and fibrous plants (like trees and grasses), cellulose is the most abundant single polymer in the biosphere. Like amylose, cellulose is a linear polymer of D-glucose (and hence is also a glucan), but in cellulose the sugar residues are connected by $\beta(1 \rightarrow 4)$ linkages, not α (**FIGURE 9.19**). This seemingly small difference from starch (i.e., amylose) has remarkable structural consequences. Cellulose can exist as fully extended chains, with each glucose residue flipped by 180° with respect to its neighbor in the chain. In this extended form, the chains can form ribbons that pack side by side with a network of hydrogen bonds within and between them. This arrangement is reminiscent of the β-sheet structure in silk fibroin, and as in fibroin, the fibrils of cellulose have great mechanical strength but limited extensibility.

The same small difference between cellulose and starch has another important consequence: Animal enzymes that are able to catalyze cleavage of the $\alpha(1 \rightarrow 4)$ link in starch cannot cleave cellulose. For this reason, humans, even if they are starving, are unable to utilize the enormous quantities of glucose all around them in the form of cellulose. Ruminants such as cows can digest cellulose only because their digestive tracts contain symbiotic bacteria that produce the necessary **cellulases.** Termites manage to eat woody substances in a somewhat more complicated fashion—their guts harbor protozoans capable of cellulose digestion, but their salivary glands also produce a cellulase. Many fungi also produce such enzymes, which is why some mushrooms can live on wood as a carbon source.

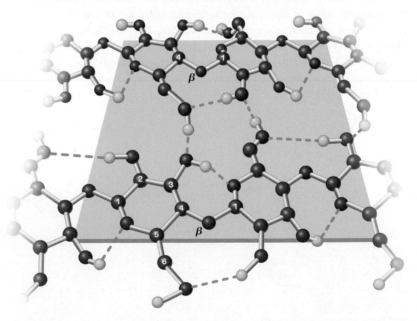

▲ **FIGURE 9.19 Cellulose structure.** The β $(1 \rightarrow 4)$ linkages of cellulose generate a planar structure. The parallel cellulose chains are linked together by a network of hydrogen bonds. Hydrogens involved in such bonds are shown in blue. For clarity, all of the carbons are numbered in only one glucose residue. Not all hydrogen atoms are shown.

Although humans don't digest cellulose, high-fiber foods containing cellulose are important nutritionally. The bulk in fiber produces a feeling of satiety, or fullness, signaling when we have probably had enough to eat. Insoluble fiber increases the rate of transport of digestion products through the alimentary tract, by increasing its bulk, and is thought to reduce exposure to potential toxins or carcinogens in the diet.

● **CONNECTION** Cellulose is indigestible in humans, but it provides an important source of dietary fiber, essential for gastrointestinal health.

A major goal of the biofuels industry is to develop efficient means for industrial-scale conversion of cellulose in plant waste materials to glucose or other substrates that can be fermented to ethanol or other potential fuels. A particular challenge is the close association of cellulose in plant tissues with *lignin,* a complex polymer derived from phenylalanine, primarily in woody tissue, that resists breakdown much more strongly than cellulose. So cellulose must be separated from lignin, or else means must be found to degrade lignin to potential fuel molecules.

Fibrous parts of plants are not made exclusively from cellulose. A variety of other polysaccharides are present in plant cell walls. These include the **xylans,** which are polymers with $\beta(1 \rightarrow 4)$-linked D-xylopyranose, often with substituent groups attached; the **glucomannans;** and many other polymers. Often these polysaccharides are grouped together under the term **hemicellulose.**

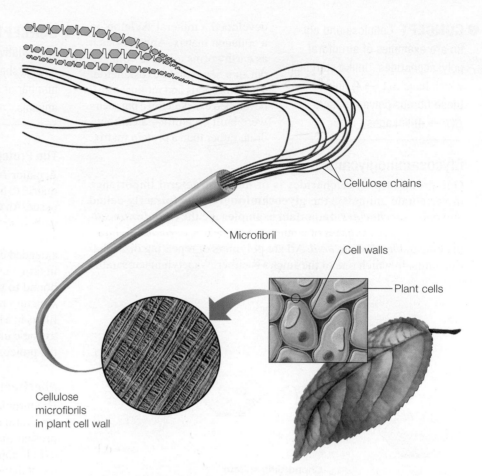

▲ **FIGURE 9.20 Organization of plant cell walls.** Microfibrils of cellulose are embedded in a matrix of hemicellulose. Note that the fibers are laid down in a crosshatched pattern to give strength in all directions.

$$\cdots \beta\text{-D-Xyl}p(1 \longrightarrow 4)[\beta\text{-D-Xyl}p(1 \longrightarrow 4)]\text{-}\beta\text{-D-Xyl}p(1 \longrightarrow 4)\text{-}\beta\text{-D-Xyl}p(1 \longrightarrow 4) \cdots$$

Acetyl at C-2 or C-3   4-O-Me-α-D-Glc$p$(1 $\longrightarrow$ 2)

A typical xylan structure

$$\cdots \beta\text{-D-Glc}p(1 \longrightarrow 4)\text{-}\beta\text{-D-Man}p(1 \longrightarrow 4)\text{-}\beta\text{-D-Man}p(1 \longrightarrow 4)\text{-}\beta\text{-D-Man}p(1 \longrightarrow 4) \cdots$$

β-D-Gal$p$(1 $\longrightarrow$ 6)   Acetyl at C-2 or C-3

A typical glucomannan structure

The cell wall of a plant is a complex structure made up of several layers. Microfibrils of cellulose are laid down in a crosshatched pattern (**FIGURE 9.20**) and impregnated with a matrix of the other polysaccharides and some proteins. The same principle is used when glass fibers are embedded in a tough resin to produce strong, durable sheets of fiberglass.

Cellulose is not confined exclusively to the plant kingdom. The marine invertebrates called *tunicates,* such as the sea squirt, contain considerable quantities of cellulose in their hard outer mantle. There are even reports of small amounts of cellulose in human connective tissue. However, as a structural material, cellulose seems to have been largely passed over in animal evolution.

Chitin

A homopolymer of *N*-acetyl-β-D-glucosamine, chitin has a structure similar to that of cellulose, except that the hydroxyl on carbon 2 of each residue is replaced by an acetylated amino group.

Chitin

Chitin is widely distributed among all kingdoms. It is a minor constituent in most fungi and some algae, where it often substitutes for cellulose or other glucans. In dividing yeast cells, chitin is found in the septum that forms between the separating cells. The best known role of chitin, however, is in invertebrate animals; it constitutes a major structural material in the exoskeletons of many arthropods and mollusks. In many of these exoskeletons, chitin forms a matrix on which mineralization takes place, much as collagen acts as a matrix for mineral deposition in vertebrate bones. The evolutionary implications are interesting. As animals evolved to the size that made rigid body parts essential, quite different paths were taken. The ancestors of the vertebrates

● **CONCEPT** Cellulose and chitin are examples of structural polysaccharides. Unlike starches, which have $\alpha(1 \rightarrow 4)$ links, these fibrous polymers have $\beta(1 \rightarrow 4)$ linkages.

developed a mineral skeleton on a collagen matrix. Annelids such as earthworms also use collagen but in a segmented exoskeleton. The arthropods and mollusks also developed exoskeletons, but theirs were built on chitin—a carbohydrate rather than a protein matrix.

Glycosaminoglycans

One group of polysaccharides is of major structural importance in vertebrate animals—the **glycosaminoglycans,** formerly called *mucopolysaccharides*. Important examples are the *chondroitin sulfates* and *keratan sulfates* of connective tissue, the *dermatan sulfates* of skin, and *hyaluronic acid*. All are polymers of repeating disaccharide units, in which one of the sugars is either *N*-acetylgalactosamine

● **CONCEPT** Glycosaminoglycans are negatively charged heteropolysaccharides that serve a number of structural functions in animals.

or *N*-acetylglucosamine or one of their derivatives. All are acidic, through the presence of either sulfate or carboxylate groups. Representative structures of glycosaminoglycans are shown in **FIGURE 9.21**.

The Proteoglycan Complex

A major function of the glycosaminoglycans is the formation of a matrix to hold together the protein components of skin and connective tissue. An example is given in **FIGURE 9.22**, which illustrates the protein–carbohydrate, or **proteoglycan,** complex in cartilage. The filamentous structure is built on a single long hyaluronic acid molecule, to which extended core proteins are attached noncovalently. The core proteins, in turn, have chondroitin sulfate and keratan sulfate chains covalently bound to them through serine side chains. In cartilage, this kind of structure binds collagen (see Chapter 6) and helps hold the collagen fibers in a tight, strong network. The binding apparently involves electrostatic interactions between the sulfate and/or carboxylate groups of the proteoglycan complex and the basic side chains in collagen.

Nonstructural Roles of Glycosaminoglycans

Hyaluronic acid has other functions in the body besides its use as a structural component. The polymer is highly soluble in water and is present in synovial fluid of joints and in the vitreous humor of the eye. It appears to act as a viscosity-increasing agent or lubricating agent in these fluids, possibly related to electrostatic repulsion among the many carboxylate groups in the polymer.

Chondroitin sulfate

Keratan sulfate

Hyaluronic acid

▲ **FIGURE 9.21 Repeating structures of some glycosaminoglycans.** In each case, the repeating unit is a disaccharide, of which two are shown for each structure. Abbreviations of residues (6s means sulfonated at carbon 6) are presented in Table 9.1. To simplify the figure, hydrogens and nonreacted hydroxyls are not shown.

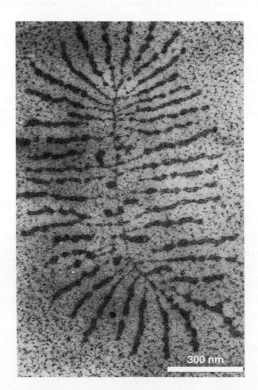

▲ **FIGURE 9.22** An electron micrograph of a proteoglycan aggregate in bovine cartilage.

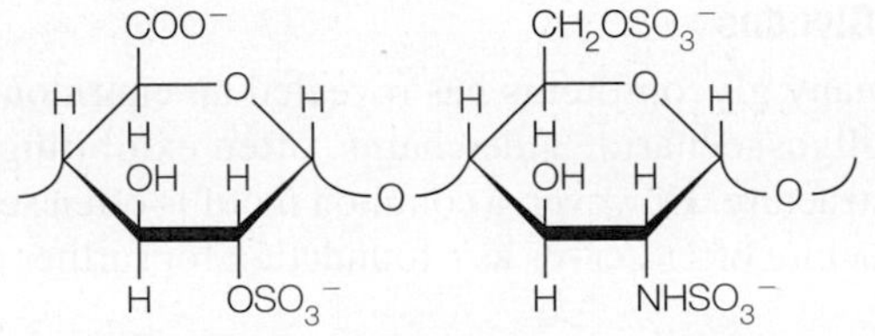

Heparin

Another highly sulfated glycosaminoglycan is **heparin.** One repeat unit of its complex chain is shown above. Heparin appears to be a natural anticoagulant and is found in many body tissues. It binds strongly to a blood protein, antiprothrombin III, and the complex inhibits enzymes of the blood-clotting process. Therefore, heparin is used medicinally to inhibit clotting in blood vessels.

● **CONNECTION** Heparin finds medical use as an anticoagulant.

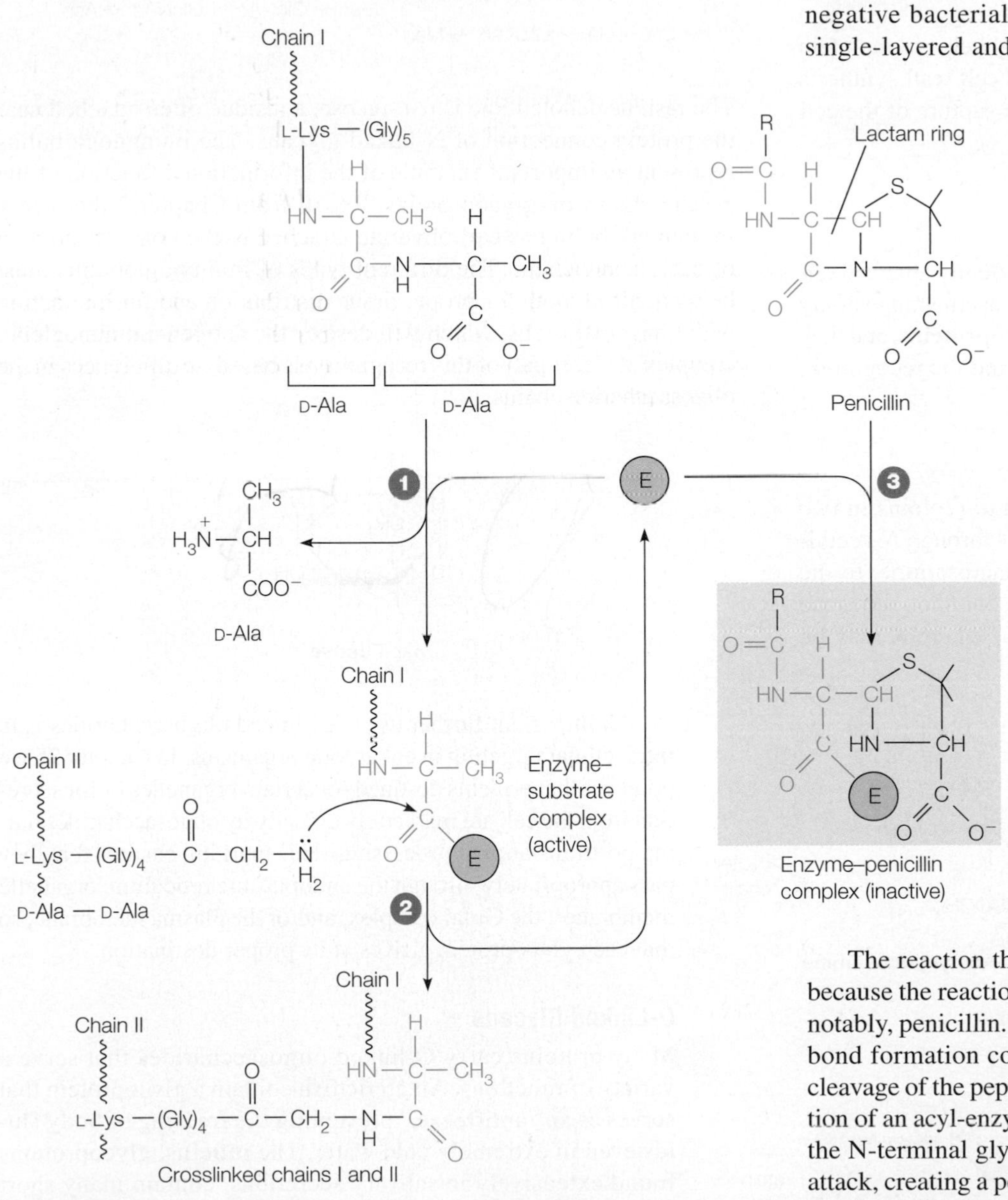

The glycosaminoglycans are interesting examples of how sugar residues can be modified to provide polymers with a wide variety of properties and functions.

Bacterial Cell Wall Polysaccharides; Peptidoglycan

In Chapter 1 we noted that plants, bacteria, and most other unicel-lular organisms possess a *cell wall*. In microorganisms, the nature of

● **CONCEPT** The cell walls of many bacteria are constructed of pepti-doglycans, composite polymers of polysaccharides and oligopeptides.

this cell wall is the basis for cat-egorizing bacteria into two major classes: those that retain the Gram stain (a dye–iodine com-plex), which are called *Gram-positive* bacteria, and those that do not, known as *Gram-negative*

bacteria. Gram-positive bacteria have a cell wall (page 278) with a crosslinked, multilayered polysaccharide–peptide complex called **peptidoglycan** at the surface, outside the lipid cell membrane. Gram-negative bacterial cell walls also contain peptidoglycan, but it is single-layered and covered by an outer lipid membrane layer. This difference allows the Gram stain to be washed from Gram-negative bacteria.

The chemical structure of the peptidoglycan of a Gram-positive bacterium was schematized on page 278. Long polysaccharide chains, which are strictly alternating copolymers of *N*-acetyl-glucosamine (NAG) and *N*-acetylmuramic acid (NAM), are crosslinked through short pep-tides. These peptides have unusual structures. Attached to a NAM residue is a tetrapeptide with the sequence

$$(\text{L-Ala})–(\text{D-Glu})–(\text{L-Lys})–(\text{D-Ala})$$

This peptide is unusual in two respects: It con-tains some D-amino acids, and the glutamate residue is linked into the chain through its γ-carboxyl instead of the usual α-carboxyl link-age. To the ε-amino group of each lysine residue is attached a glycine pentapeptide, which is bonded at its other end to the terminal D-Ala residue of an adjacent chain. The result is the formation of a covalently crosslinked structure that envelops the bacterial cell. The entire cell wall can be regarded as a single enormous molecule made up of mul-tiple layers of crosslinked peptidoglycan strands. The cell wall protects bacteria from lysis when they are in the blood of host animals.

The reaction that generates the cross-link is of particular interest because the reaction is the target site for several important antibiotics, notably, penicillin. As shown in **FIGURE 9.23**, reaction ❶, the energy for bond formation comes from a **transpeptidation** reaction, in which cleavage of the peptide bond between two D-Ala residues drives forma-tion of an acyl-enzyme intermediate. In reaction ❷, the amino group of the N-terminal glycine in a Gly pentapeptide initiates a nucleophilic attack, creating a peptide bond between D-Ala and the N-terminal Gly. As shown in reaction ❸, penicillin resembles the two C-terminal D-Ala residues linked to Chain I. The penicillin molecule contains a ther-modynamically unstable four-membered **lactam** ring. Spontaneous

▲ **FIGURE 9.23** The crosslinking reaction in peptidoglycan synthesis and inhibition of the transpeptidase reaction by penicillin.

● **CONNECTION** Biosynthesis of the peptidoglycan cell wall presents target sites for several important antibiotics, including penicillin, which attacks the transpeptidase activity involved in the crosslinking reaction.

rupture of the ring leads to formation of a covalently bonded ternary complex linking the enzyme and Chain 1 to the penicillin molecule. Because inhibition of peptidoglycan crosslinking weakens the cell wall, the bacterial cell eventually ruptures because of internal turgor pressure. This action of penicillin illustrates an important principle in using biochemistry to treat disease; the most effective drugs are those that exploit a biochemical feature specific to the disease process. In this case, that feature is the presence of a peptidoglycan cell wall in the pathogen that is not present in the treated human or animal patient.

Peptidoglycan synthesis is also the target for natural antibiotics, notably the enzyme **lysozyme,** which cleaves the glycosidic links between GlcNAc and MurNAc residues in the polysaccharide chain (see Chapter 8, section 8.4). Lysozyme is abundant in egg white and in tears in the eye. In both cases, this enzyme activity helps to maintain a sterile environment. Although lysozyme attacks cell wall synthesis at a different site, the result is the same—osmotic rupture of the cell caused by weakening of the cell wall.

9.5 Glycoproteins

More than half of all eukaryotic proteins carry covalently attached oligosaccharide or polysaccharide chains. There is an astonishing variety of these modified proteins, which are known as **glycoproteins,** and they serve multiple functions, including cell adhesion and the recognition of eggs by sperm cells.

N-Linked and O-Linked Glycoproteins

The saccharide chains, or glycans, can be linked to proteins in two major ways. *N-linked glycans* are attached, usually through *N*-acetylglucosamine, or sometimes through *N*-acetylgalactosamine, to the side-chain amide group in an asparagine residue. A common sequence surrounding the asparagine is –Asn–X–Ser/Thr–, where X may be any amino acid residue. *O-linked glycans* are usually attached by an O-glycosidic bond between *N*-acetylgalactosamine and the hydroxyl group of a threonine or serine residue, although in a few cases—collagen, for example (Chapter 6)—hydroxylysine or hydroxyproline is employed.

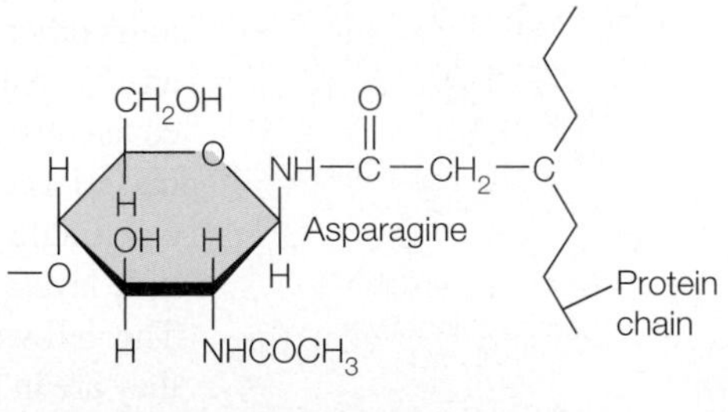

N-Acetylglucosamine

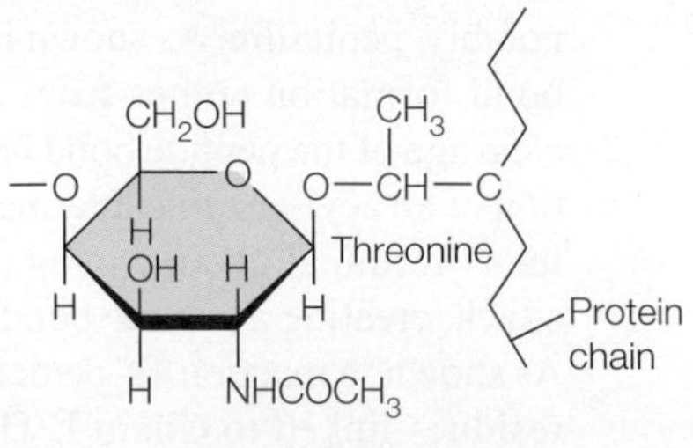

● **CONCEPT** Oligosaccharides and proteins can be linked to form glycoproteins in two ways: O-linked glycans are attached via threonine or serine hydroxyls, and N-linked glycans via asparagine amide groups.

N-Acetylgalactosamine

N-Linked Glycans

Study of many glycoproteins has revealed an enormous variety of N-linked oligosaccharide side chains, often exhibiting a complex branched structure. However, a common motif is often seen. The following structure often serves as a foundation for further elaboration:

This motif can be seen, for example, in the glycan moieties of ovalbumin in egg whites and in the immunoglobulins. The structure of the oligosaccharide found attached to a human immunoglobulin G (IgG) is shown in simplified form below.

The residue denoted Fuc is α-L-fucose, a residue often attached near the protein connection of N-linked glycans. The immunoglobulins represent an important example of the informational function of the glycan chains on glycoproteins. Recall from Chapter 7 that every immunoglobulin has carbohydrate attached to the constant domain of each heavy chain. The different types of immunoglobulins must be recognized, both for proper tissue distribution and for interaction with phagocytic cells, which will destroy the antigen–immunoglobin complex. At least part of this recognition is based on differences in the oligosaccharide chains.

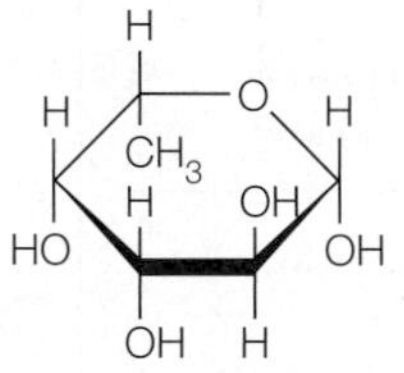

α-L-**Fucose**

An important further use of N-linked oligosaccharides is in intracellular targeting in eukaryotic organisms. In Chapter 25 we point out that proteins destined for certain organelles or for secretion from the cell are marked specifically by oligosaccharides during posttranslational processing. This marking ensures that they pass appropriately through the endoplasmic reticulum, organelle membranes, the Golgi complex, and/or the plasma membrane, so that each glycoprotein arrives at its proper destination.

O-Linked Glycans

Many proteins carry O-linked oligosaccharides that serve a variety of functions. Antarctic fish contain a glycoprotein that serves as an "antifreeze," preventing the freezing of body fluids even in extremely cold water. The **mucins,** glycoproteins found extensively in salivary secretions, contain many short O-linked glycans. The highly extended and highly hydrated mucins increase the viscosity of the fluids in which they are dissolved. Some O-linked glycans also function in intracellular targeting and molecular and cellular identification. An example significant to human biology is found in the *blood group antigens*.

Blood Group Antigens

On some cells, the blood group antigens are attached as O-linked glycans to membrane proteins. Alternatively, the oligosaccharide may be linked to a lipid molecule to

● **CONCEPT** The blood group substances are a set of antigenic oligosaccharides attached to the surfaces of red cells.

form a **glycolipid** (see Chapter 10). The lipid portion of the molecule helps anchor the antigen to the outside surface of erythrocyte membranes. It is these oligosaccha-

rides that determine the blood group types in humans. Their presence in a blood sample is detected by blood typing—determining whether antibodies to a particular antigen cause the red cells of that blood sample to clump, or agglutinate. Although the system consisting of blood types A, B, AB, and O is probably most familiar to you, it is just one of 14 genetically characterized blood group systems, with more than 100 different blood group antigens. These substances are also present in many cells and tissues other than blood, but we often focus on blood because of the widespread use of typing in establishing familial relationships and selecting blood for transfusion.

For simplicity, we will take the ABO system as an example. **FIGURE 9.24** depicts cell-surface oligosaccharides corresponding to each of these blood types. Almost all humans can produce the type O trisaccharide, but addition of either galactose (to make type B) or *N*-acetylgalactosamine (to make type A) requires an additional enzyme. Type B individuals carry a glycosyltransferase that uses UDP-galactose to transfer Gal to the O substance, yielding the B antigen. Type A individuals carry a similar enzyme, which uses UDP-*N*-acetylglucosamine to transfer GalNAc to the O substance, yielding the A antigen. Type AB individuals possess both glycosyltransferases and carry both antigens on their cell surfaces; type O individuals possess neither.

Humans can produce antibodies against the A and B oligosaccharides, but the O type is nonantigenic. Normally, a person does not produce antibodies against his or her own antigen but does produce them against the other antigen type. Thus, an individual with type A blood carries antibodies directed against the B polysaccharide. If he or she accepts blood from a type B donor, these antibodies will cause clumping and precipitation of the donated blood cells. Nor can a type

● **CONNECTION** The presence or absence of two specific glycosyltransferase enzymes is the basis for typing blood as A, B, AB, or O, and hence, of governing the ability of blood to be transfused into certain individuals.

B individual safely accept type A blood. People with type O blood normally have antibodies against both A and B and thus can receive from neither. Those with AB type, since they carry both A and B antigens themselves, have antibodies against neither.

In donating blood, an inverse relationship holds. Those with type O blood, which carries no antigenic

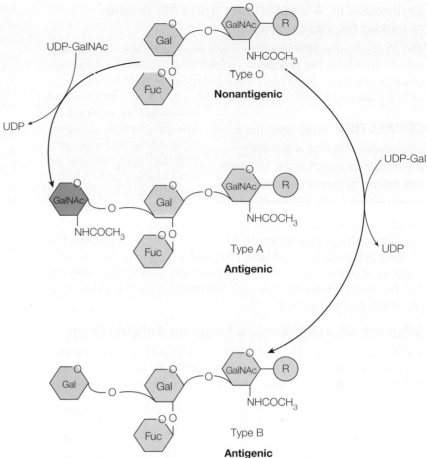

▲ **FIGURE 9.24 The ABO blood group antigens.** The O oligosaccharide (top) does not elicit antibodies in most humans. The A and B antigens are formed by addition of GalNAc or Gal, respectively, to the O oligosaccharide. Each of the A and B antigens can elicit a specific antibody. In this figure, R can represent either a protein molecule or a lipid molecule to which the respective oligosaccharide is added.

determinants, can safely donate to any other person—they are the "universal donors." Type AB individuals can donate *only* to other ABs; a person of any other type will carry antibodies to A or B, or both. These relationships are summarized in **TABLE 9.3**.

Person Has Blood Type:	Makes Antibodies Against:	Can Safely Receive Blood from:	Can Safely Donate Blood to:
O	A, B	O	O, A, B, AB
A	B	O, A	A, AB
B	A	O, B	B, AB
AB	None	O, A, B, AB[a]	AB

TABLE 9.3 Transfusion relationships among ABO blood types

[a]In principle, this relationship is true. However, ABs are not given donations from other types because the donor's antibodies could react with the recipient's antigens, unless the red blood cells are first separated from the plasma.

Erythropoetin: A Glycoprotein with Both O- and N-Linked Oligosaccharides

The glycoprotein **erythropoetin** is a hormone synthesized in the kidney, which stimulates the production of red blood cells. The hormone, often called EPO, is a 165-residue polypeptide with N-linked oligosaccharides (13-mers) at Asn 24, 38, and 83, and an O-linked trisaccharide at Ser-126. The carbohydrate stabilizes EPO within the blood, preventing rapid removal by the kidney. Anemia—low red blood cell count—is a common side effect of cancer chemotherapy, and EPO is often administered along with anticancer drugs to counteract this effect. EPO is also subject to misuse, particularly by athletes who wish to improve their aerobic capacity. Recombinant EPO, which is usually used for this purpose, can be identified by drug testing laboratories on the basis of its abnormal glycosylation pattern.

● **CONNECTION** Erythropoetin is a glycoprotein that acts to stimulate erythrocyte formation and is an important adjunct to cancer chemotherapy.

Influenza Neuraminidase, a Target for Antiviral Drugs

Influenza virus, an RNA virus, carries on its surface a virus-coded enzyme, **neuraminidase.** As shown in **FIGURE 9.25**, the spherical virus particle has two kinds of spikes on its exterior surface. One such spike, made up of a protein called *hemagglutinin,* binds to *N*-acetylneuraminic acid (also called sialic acid; see page 288). The virus attaches to host cells through binding of the hemagglutinin to sialic acid residues in cell-surface glycoproteins or glycolipids. At the conclusion of the virus infection cycle, release of virus from infected cells requires cleavage of the sialic acid from the rest of each oligosaccharide chain, and this action is carried out by the viral neuraminidase.

The crystal structure of the neuraminidase complex with sialic acid was solved in the 1980s, leading to synthesis of sialic acid analogs that might inhibit the enzyme and, hence, block release of virus particles from infected cells. One such analog, *oseltamivir,* is shown here. Oseltamivir, marketed as *Tamiflu,* assumed importance during the H1N1 influenza epidemic of 2009, before an effective vaccine had become available. Tamiflu acts, as predicted, by blocking release of newly formed virus particles from infected cells, but to be effective it must be administered very soon after the onset of flu symptoms.

Sialic acid **Oseltamivir**

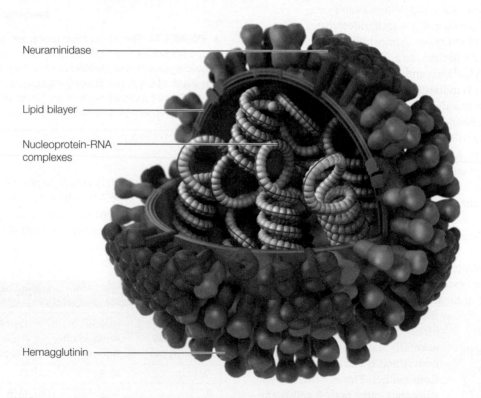

▲ **FIGURE 9.25 The structure of influenza virus.** The 13,600-nucleotide RNA genome is packaged within the sphere, about 120 nm in diameter. The spikes on the virion exterior include the hemagglutinin molecule and a spike that terminates in four neuraminidase molecules. http://www.cdc.gov/h1n1flu/images.htm.

Neuraminidase

Lipid bilayer

Nucleoprotein-RNA complexes

Hemagglutinin

Summary

- **Carbohydrates** (or **saccharides**) are compounds with the stoichiometric formula $(CH_2O)_n$ or derivatives of such compounds. They are the primary product of photosynthesis, and their oxidation provides a major energy source for both plants and animals. Because of their multiple chiral centers, these saccharides exist as enantiomeric pairs (D and L mirror images) of multiple **diastereomers.** **Monosaccharides** may be either **aldoses** or **ketoses.** Those containing five or more carbons exist mainly in the form of rings of five (**furanose**) or six (**pyranose**) atoms, resulting from internal hemiacetal formation. Such rings exist as α or β anomers and exhibit multiple conformations (e g , boat and chair) as well. (Section 9.1)

- Derivatives of monosaccharides include phosphate esters, acids and **lactones, alditols,** amino sugars, and glycosides. Phosphate esters are important as metabolic intermediates; glycosides represent a large class of compounds formed by elimination of water between a sugar and another hydroxy compound. (Section 9.2)

- **Oligosaccharides** and **polysaccharides** are formed by making glycosidic links between monosaccharides. The glycosidic linkage is metastable, so enzymes control its hydrolysis in vivo. Polysaccharides serve multiple functions—energy storage (starch and glycogen), structural roles (**cellulose, xylans, chitin, glycosaminoglycans,** cell wall polysaccharides), and identification tags. The blood group antigens are examples of the identification function. (Sections 9.3, 9.4)

- Complex glycan chains are assembled by stepwise transfer of monosaccharide units from nucleotide linked sugars, through the action of glycosyltransferases. In Gram-positive bacteria, the crosslinking of peptidoglycan chains is the site of action of penicillin and related antibiotics. The complexity of many important glycoconjugates is being revealed through selective degradation techniques coupled with nuclear magnetic resonance spectroscopy (NMR). (Section 9.5, Tools of Biochemistry 9A)

Problems

Answers to red problems are available in the Answer Appendix.

1. Draw Haworth projections for the following:

   (a)

   ```
            CHO
             |
      H — C — OH
             |
     HO — C — H
             |
      H — C — OH
             |
           CH₂OH
   ```

   in α-furanose form. Name the sugar.
   (b) The L isomer of (a)
   (c) α-D-GlcNAc
   (d) α-D-Fructofuranose

2. α-D-Galactopyranose rotates the plane of polarized light, but the product of its reduction with sodium borohydride (galactitol) does not. Explain the difference.

3. Provide an explanation for the fact that α-D-mannose is more stable than β-D-mannose, whereas the opposite is true for glucose.

4. Why is a type O individual considered a universal blood donor? Why is a type AB individual considered a universal acceptor?

5. The disaccharide α,β-trehalose differs from the α, α structure in Figure 9.14(a) by having an (α1 → β1) linkage. Draw its structure as a Haworth projection.

6. A *reducing* sugar will undergo the Fehling reaction, which requires a (potential) free aldehyde group. Which of the disaccharides shown in Figure 9.14 are reducing and which are nonreducing?

7. *Dextrans* are polysaccharides produced by certain species of bacteria. They are glucans, with primarily α(1 → 6) linkages and with frequent α(1 → 3) branching. Draw a Haworth projection of a portion of a dextran, including one (1 → 3) branch point.

8. What is the natural polysaccharide whose repeating structure can be symbolized by GlcUAβ(1 → 3)GlcNAc, with these units connected by β(1 → 4) links?

9. Indicate whether the structures shown are R or S in the absolute system.

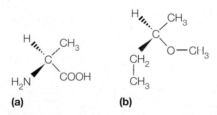

(a)   (b)

10. The reagent periodate (IO_4^-) oxidatively cleaves the carbon–carbon bonds between two adjacent carbons carrying hydroxyl groups. Explain how periodate oxidation might be used to distinguish between methyl glycosides of glucose in the pyranose and furanose forms.

11. Draw (using Haworth projections) the fragments of xylan and glucomannan structures shown on page 295.

12. One or more of the compounds shown below will satisfy each of the following statements. Not all compounds may be used; some may be used twice. Put the number(s) in the blank.
    (1) Found in chitin. _____
    (2) An L-saccharide. _____

(3) The first residue attached to asparagine in N-linked glycans. _____

(4) A uronic acid. _____

(5) A ketose. _____

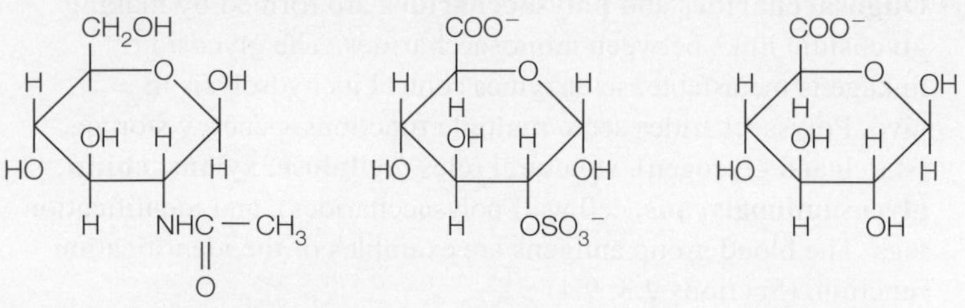

(a)

(b)

(c)

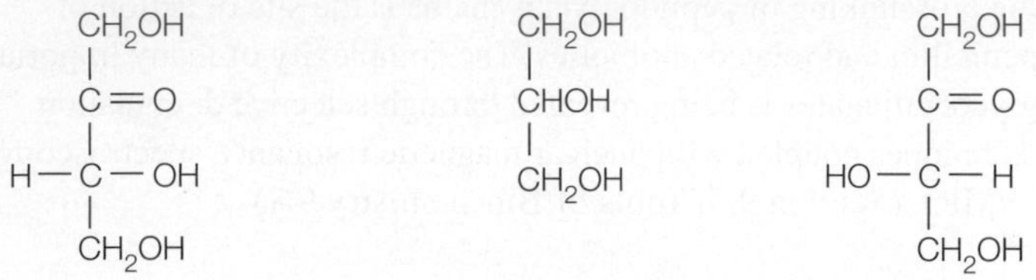

(d)

(e)

(f)

13. Why do you suppose that the influenza virus protein that binds the virus to an infected cell is called hemagglutinin? Hemagglutination is the clumping together of red blood cells.

14. The diversity of functional groups on sugars that can form glycosidic bonds greatly increases the information content of glycans relative to oligopeptides. Consider three amino acids, A, B, and C. How many tripeptides can be formed from one molecule of each amino acid? Now consider three sugars—glucose, glucuronic acid, and N-acetylglucosamine. Use shorthand (e.g., Glcα(1 → 4)GlcUAβ(1 → 4)GlcNAc to represent 10 trisaccharides with the sequence Glc-GlcUA-GluNAc. Is your list exhaustive?

15. Are mannose and galactose epimers? Allose and altrose? Gulose and talose? Ribose and arabinose? Consider only D-sugars. Explain your answers.

16. Which of the pairs in Problem 15 are diastereoisomers?

17. Explain in about one sentence why it is important to animals for the major carbohydrate storage polymer, glycogen, to be branched rather than unbranched.

18. Write the structure of UDP-N-acetylglucosamine and UDP-galactosamine.

19. Consider the dextran sucrase reaction. Why do you suppose there is not an ATP requirement to energetically drive the creation of glycosidic bonds in the dextran product?

20. Indicate whether each of the following disaccharides is a reducing (R) or nonreducing (NR) sugar by the criterion of reaction with Fehling's solution.
    (a) Glcα(1 → 2)Frucβ
    (b) Galβ(1 → 4)Glc
    (c) Glcα(1 → 1)Glcα
    (d) Glcα(1 → 4)Glc
    (e) Glcβ(1 → 6)Glc

21. Briefly describe the function of uridine triphosphate (UTP) in carbohydrate metabolism.

22. Explain how oseltamivir (Tamiflu) interferes with influenza virus replication.

23. Why is a person with type AB blood able to receive a blood transfusion from a donor with any of the major blood types (A, B, AB, and O) but is able to donate blood only to another type AB individual?

References

For a list of references related to this chapter, see Appendix II.

Biologists realize that such molecules as the blood group antigens represent only a special case of a much more general phenomenon—cell marking by glycans. In a multicellular organism, it is essential that different kinds of cells be marked on their surfaces so that they can interact properly with other cells and molecules and so that an organism can recognize its own cells as immunologically distinct from foreign cells. In accord with this view is the growing appreciation that the surfaces of many cells are nearly covered with polysaccharides, which are attached to either proteins or lipids in the cell membrane. The nearly limitless variety of possible glycan structures, which can mark specific cell types, has made carbohydrates major players in cell recognition.

Why do oligosaccharides so often play the role of cellular markers? Certain possibilities suggest themselves. First, oligosaccharides can present a seemingly limitless variety of structures in relatively short chains. The multiple choices of monomers (including modified sugars), linkages, and branching patterns allow a vast but specific vocabulary. Second, oligosaccharides are especially potent antigens, which means that specific antibodies can be elicited swiftly against them, and these can be used as analytical tools. Whether this interaction is the result of some intrinsic property of sugar molecules or of the antibody molecules is unclear. It is possible that antibodies evolved as a defense against bacteria, which have polysaccharide-rich walls, and thus have always favored glycans as targets.

At least partly because of the great variety of monomeric units and branching possibilities, the structural analysis of polysaccharides has lagged behind that of nucleic acids or proteins, which are almost exclusively linear polymers. However, two developments have brought complex polysaccharides within reach and created an emerging field of **glycomics** (by analogy with genomics, proteomics, and so forth). First has been the characterization of enzymes that digest specific links in polysaccharides. Hence, complex polysaccharides can be selectively degraded, allowing analysis of smaller degradation products. An example is shown in **FIGURE 9A.1**. Second is the development of improved analytical techniques, notably, nuclear magnetic resonance, which permits structural determination of minute samples. Third is the fact that oligosaccharides tend to be highly antigenic, and this property allows the development of specific antibodies that can be used as analytical reagents.

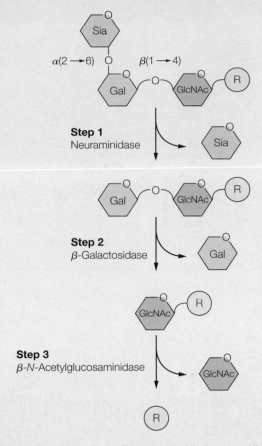

▲ **FIGURE 9A.1 Cleavage of an oligosaccharide by specific glycosidases.** R represents the glycoprotein to which the oligosaccharide is attached.

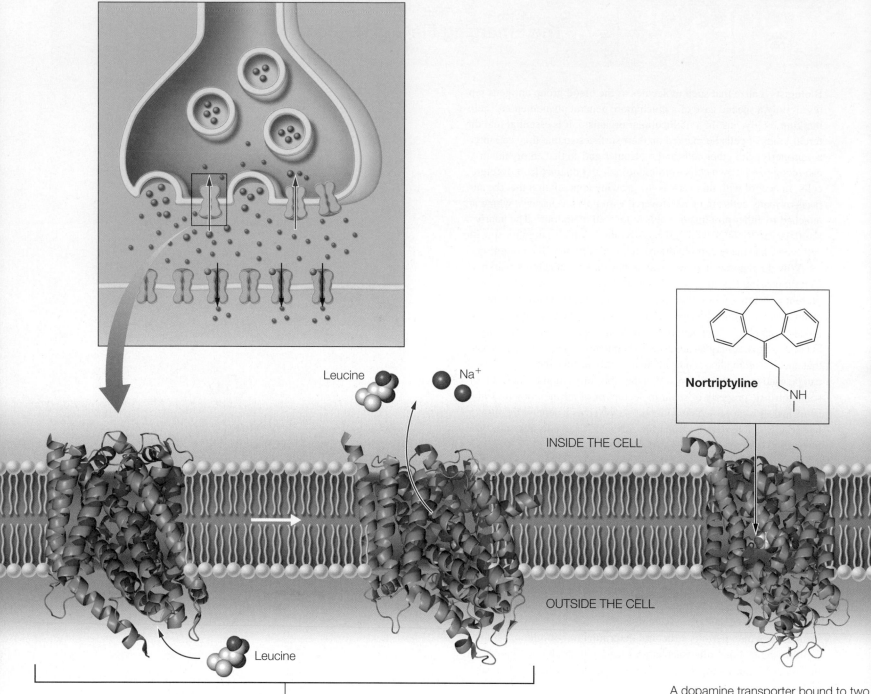

Leucine

Na$^+$

INSIDE THE CELL

OUTSIDE THE CELL

Nortriptyline

NH

Leucine

A bacterial Leucine/Na$^+$ transporter
(model for dopamine transport across membranes)

A dopamine transporter bound to two
Na$^+$ ions and the neurotransmitter
reuptake inhibitor Nortriptyline.

The depletion of neurotransmitters at neuron synapses is associated with some types of depression. Several anti-depressant drugs act by blocking the reuptake of neurotransmitters into presynaptic neurons (see red arrows in figure above). An example is shown by the membrane-bound dopamine transporter (cyan cartoon) bound to two Na$^+$ ions (purple spheres) and the tricyclic antidepressant nortriptyline (gray spheres). The reuptake of dopamine requires the cotransport of Na$^+$ ions via a mechanism that is similar to that used to transport amino acids across bacterial membranes (as shown by the bacterial LeuT transporter in green cartoon). How is transport of material across membranes regulated? What are the driving forces for transport? How does knowledge of the structures and mechanisms of transporters aid in the development of selective inhibitors as useful drugs?

Lipids, Membranes, and Cellular Transport

10

THE LIPIDS ARE a structurally and functionally diverse group of molecules. They are generally hydrophobic, although many also possess polar or charged groups in addition to a hydrocarbon core. Given their structural diversity, lipids carry out multiple functions. Primary among these functions are energy storage, signaling, and formation of membrane structures. In this chapter we focus on the structures of lipid molecules, the general features of membranes formed from lipids, and selective transport across membranes. The roles of lipids as energy stores and signaling molecules are described in detail in Chapters 16 and 20. In those chapters, we also discuss lipid biosynthesis as well as lipid transport and the lipid–protein complexes, or *lipoproteins*, involved in that process.

The largest fraction of the lipids in most cells is used to form membranes, the partitions that divide cellular compartments from one another and separate the cell from its surroundings. Cellular membranes are not just passive barriers. Rather, they contain highly selective gates that control the passage of materials in specific directions. It is this property of *selective membrane permeability*

that allows each part of the cell to carry out its specific operations. It is difficult to imagine the organization and evolution of complex multicellular organisms without membranes. Membranes establish order within a cell (by compartmentation), and they also allow free energy to be stored in the form of concentration gradients (e.g., proton gradients in mitochondria or ion gradients in nerve cells). Such concentration gradients were first described in Chapter 3 (Section 3.4). Thus, membranes have a vital role in maintaining the living state, which is essentially an ongoing campaign against entropy and the tendency to move toward chemical equilibrium.

10.1 The Molecular Structure and Behavior of Lipids

In contrast to the other classes of biomolecules we have considered in earlier chapters, lipids are not generally water-soluble, owing to the large portion of their structure that is hydrocarbon. This means that lipids are rarely found free in solution. Rather, lipids are either in complex with soluble protein transporters or part of higher-order assemblies that sequester the hydrophobic surface area from the surrounding aqueous environment.

Unlike the amino acids in proteins, the nucleotides in nucleic acids, and the monosaccharides in complex carbohydrates, lipids do not form large covalent polymers. Instead, they tend to associate with each other through noncovalent interactions. For example, the lipids that make up membranes are usually characterized by the kind of structure shown in the accompanying figure: a polar, hydrophilic "head" connected to a larger nonpolar, hydrophobic hydrocarbon "tail."

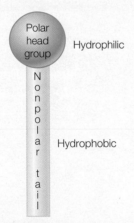

**A simplified representation
of an amphipathic lipid molecule**

Such lipids in an aqueous environment tend to associate for two fundamental reasons. Just as nonpolar groups in proteins associate via an entropy-driven hydrophobic effect, so do the nonpolar tails of lipids. A second stabilizing force comes from the van der Waals interactions between the hydrocarbon regions of the molecules.

The polar, hydrophilic head groups of membrane lipids tend to associate with water. Such lipids are prime examples of the kinds of amphipathic substances described in Chapter 2. The amphipathic nature of membrane lipids has a number of consequences, including the formation of surface monolayers, bilayers, micelles, and vesicles by lipids in contact with water (see Figure 2.16). From a biological point of view, the most important of these consequences is the tendency of lipids to

● **CONCEPT** Lipid molecules tend to be insoluble in water, but they can associate to form water-soluble structures, such as micelles, vesicles, and bilayers.

form micelles and membrane bilayers. Exactly what kind of structure is formed when a lipid is in contact with water depends on the specific molecular structure of the hydrophilic and hydrophobic parts of that lipid molecule. Thus, it is appropriate that we now examine the structure of some of the major types of lipids.

Fatty Acids

The simplest lipids are the **fatty acids,** which are also constituents of many more complex lipids. Their basic structure exemplifies the amphipathic lipid model described earlier: A hydrophilic carboxylate group is attached to one end of the hydrocarbon chain, which contains typically 12 to 24 carbons. An example is *stearic acid,* which is widely distributed in organisms. We show it in **FIGURE 10.1** as the ionized form, the *stearate* ion. Stearic acid is an example of a **saturated** fatty acid, one in which the carbons of the tail are all saturated with hydrogen atoms (i.e., no $C{=}C$ double bonds). Several biologically important saturated fatty acids are listed in **TABLE 10.1**. Note that each has a common name (such as stearic acid) and a systematic name (in this case, octadecanoic acid).

Many important naturally occurring fatty acids are **unsaturated**—that is, they contain one or more double bonds (see Table 10.1). One such example is *oleic acid* (or oleate), which is found in many animal fats (Figure 10.1). In most of the naturally occurring unsaturated fatty acids, the orientation about double bonds is *cis* rather than *trans*. This orientation has an important effect on molecular structure because each *cis* double bond inserts a bend into the hydrocarbon chain. Keep in mind, however, that although Figure 10.1 depicts the molecules as extended structures, there is freedom of rotation about each single bond in the hydrocarbon chain. Thus, many conformations are possible.

● **CONCEPT** Most naturally occurring fatty acids contain an even number of carbon atoms. If double bonds are present (unsaturation), they are usually *cis*.

Most of the naturally occurring fatty acids have an even number of carbon atoms because they are synthesized by sequential additions of a two-carbon precursor (see Chapter 16). Although the hydrocarbon chains are linear in most fatty acids, some fatty acids (found primarily in bacteria) contain branches or even cyclic structures.

Table 10.1 includes a set of abbreviations developed to provide a systematic description of fatty acid structure. By convention, the number before the colon gives the total number of carbons, and the number after the colon gives the count of double bonds. The configurations and

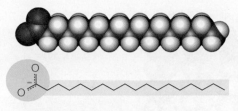

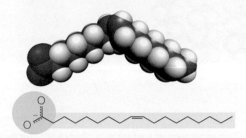

(a) Stearate ion. Stearate (the anionic, deprotonated form of stearic acid) is a saturated fatty acid.

(b) Oleate ion. Oleate is an unsaturated fatty acid with one *cis* double bond.

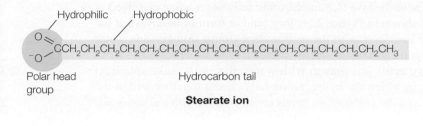

Hydrophilic   Hydrophobic

O
$\overset{\ominus}{\underset{O}{\parallel}}CCH_2CH_2CH_2CH_2CH_2CH_2CH_2CH_2CH_2CH_2CH_2CH_2CH_2CH_2CH_2CH_2CH_3$

Polar head group   Hydrocarbon tail

Stearate ion

O
$\overset{\ominus}{\underset{O}{\parallel}}CCH_2CH_2CH_2CH_2CH_2CH_2CH_2\overset{H}{\underset{}{C}}=\overset{H}{\underset{}{C}}CH_2CH_2CH_2CH_2CH_2CH_2CH_2CH_3$

Oleate ion

(c) Formulas. Alternate representations for (a) and (b).

▲ **FIGURE 10.1 Structures of the ionized forms of two representative fatty acids.** Hydrophilic portions (head groups) of the molecules are indicated by a pale blue background in the models, hydrophobic portions (tails) by a yellow background.

positions of double bonds are indicated by c (*cis*) or t (*trans*) followed by Δ and one or more numbers. These numbers denote the carbon atom (with the carboxylic carbon designated as "1") at which each double bond starts. Thus, oleic acid is designated by 18:1cΔ9 and linolenic acid by 18:3cΔ9,12,15. Because it contains multiple double bonds, linolenic acid is an example of a **polyunsaturated fatty acid** or **PUFA.**

Animals generally do not produce via biosynthesis all the PUFAs they need for proper cellular function; thus, these fatty acids must be part of their diet. The nutritional value and consequences of ingesting various fats, including the so-called trans fats and omega-3 fats, are discussed in Chapter 16 in the context of fuel metabolism.

TABLE 10.1 Some biologically important fatty acids

Common Name	Systematic Name	Abbreviation	Structure	Melting Point (°C)
Saturated Fatty Acids				
Capric acid	Decanoic acid	10:0	$CH_3(CH_2)_8COOH$	31.6
Lauric acid	Dodecanoic acid	12:0	$CH_3(CH_2)_{10}COOH$	44.2
Myristic acid	Tetradecanoic acid	14:0	$CH_3(CH_2)_{12}COOH$	53.9
Palmitic acid	Hexadecanoic acid	16:0	$CH_3(CH_2)_{14}COOH$	63.1
Stearic acid	Octadecanoic acid	18:0	$CH_3(CH_2)_{16}COOH$	69.6
Arachidic acid	Eicosanoic acid	20:0	$CH_3(CH_2)_{18}COOH$	76.5
Behenic acid	Docosanoic acid	22:0	$CH_3(CH_2)_{20}COOH$	81.5
Lignoceric acid	Tetracosanoic acid	24:0	$CH_3(CH_2)_{22}COOH$	86.0
Cerotic acid	Hexacosanoic acid	26:0	$CH_3(CH_2)_{24}COOH$	88.5
Unsaturated Fatty Acids				
Palmitoleic acid	*cis*-9-Hexadecenoic acid	16:1cΔ9	$CH_3(CH_2)_5CH=CH(CH_2)_7COOH$	0
Oleic acid	*cis*-9-Octadecenoic acid	18:1cΔ9	$CH_3(CH_2)_7CH=CH(CH_2)_7COOH$	16
Linoleic acid	*cis,cis*-9,12-Octadecenoic acid	18:2cΔ9,12	$CH_3(CH_2)_4CH=$ $CHCH_2CH=CH(CH_2)_7COOH$	5
Linolenic acid	all-*cis*-9,12,15-Octadecenoic acid	18:3cΔ9,12,15	$CH_3CH_2CH=CHCH_2CH=$ $CHCH_2CH=CH(CH_2)_7COOH$	−11
Arachidonic acid	all-*cis*-5,8,11,14-Eicosatetraenoic acid	20:4cΔ5,8,11,14	$CH_3(CH_2)_4CH=$ $CHCH_2CH=CHCH_2CH=$ $CHCH_2CH=CH(CH_2)_3COOH$	−50

The fatty acids are weak acids, with pK_a values averaging about 4.5:

$$RCOOH \xrightleftharpoons{pK_a \cong 4.5} RCOO^- + H^+$$

Thus, these acids exist in the anionic form ($RCOO^-$) at physiological pH. Because of this hydrophilic charge and the long hydrophobic tail, fatty acids behave like amphipathic substances when dissolved in water. As shown in Figure 2.16, they tend to form **monolayers** at the air–water interface, with the carboxylate groups immersed in water and the hydrocarbon tails out of water.

If fatty acids are shaken with water, they will make spherical **micelles,** in which the hydrocarbon tails cluster together within the structure and the carboxylate heads are in contact with the surrounding water.

Although the fatty acids play important roles in metabolism, large quantities of the free acids or their anions are not found in living cells. Instead, these compounds almost always occur as constituents of more complex lipids. We now turn to consideration of some of these classes of biologically important lipid molecules.

Triacylglycerols: Fats

The long hydrocarbon chains of fatty acids are extraordinarily efficient for energy storage because they contain carbon in a reduced form and will therefore yield a large amount of energy on oxidation. For these reasons, lipids are used by many organisms, including humans, for storage of metabolic energy. A typical human stores sufficient calories in fats to survive for several weeks (assuming access to potable water).

● **CONCEPT** Fats, or triacylglycerols, are triesters of fatty acids and glycerol. They are the major long-term energy storage molecules in many organisms.

Storage of fatty acids in organisms is largely in the form of **triacylglycerols,** or **triglycerides,** or simply, **fats.** These substances are *triesters* of fatty acids and **glycerol;** the general formula is shown here.

Triacylglycerol

Here R_1, R_2, and R_3 correspond to the hydrocarbon tails of various fatty acids. We have depicted the structure with the hydrophobic chains to the right, according to a commonly used convention. This convention does not indicate stereochemical configuration (the correct stereochemical configuration around each C atom in the glycerol moiety is depicted later in this chapter in Figure 10.6(a)).

As a particular example of a triglyceride, if $R_1 = R_2 = R_3 = (CH_2)_{16}CH_3$, the hydrocarbon tail of stearic acid, the molecule is *tristearin* (**FIGURE 10.2**). Triacylglycerols with the same fatty acid esterified at each position are called "simple fats." Most triacylglycerols, however, are "mixed fats" that contain a mixture of different fatty acids. **TABLE 10.2** lists the fatty acid composition of some naturally occurring fats.

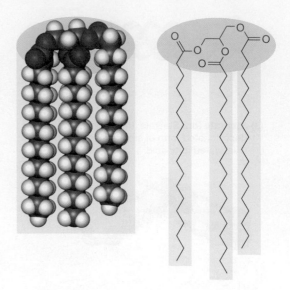

▲ **FIGURE 10.2 The structure of tristearin, a simple fat.** Tristearin is a triacylglycerol (fat) composed of glycerol and three stearate molecules.

An interesting correlation exists between the fatty acid composition of fats and their physical state as a function of temperature. Fats rich in unsaturated fatty acids (like olive oil) are liquid at room temperature, whereas those with a higher content of saturated fatty acids (like butter) are more solid. Indeed, a wholly saturated fat yields a firm solid, especially if the hydrocarbon chains are long. This is shown by the melting point data in Table 10.1. This observation can be explained by considering the strength of noncovalent interactions between fat molecules. Long saturated chains can pack closely together, thereby increasing the number of van der Waals contacts to form regular, semicrystalline structures. In contrast, the kind of bend imposed by one or more *cis* double bonds (see Figure 10.1(b)) makes molecular packing less regular and therefore more dynamic. Indeed, partial **hydrogenation** of unsaturated fat oils (like corn oil) is used commercially to convert them to firmer fats, which can be used as butter substitutes such as margarine or to stabilize them against spoilage.

Esterification with glycerol diminishes the hydrophilic character of the head groups of the fatty acids. As a consequence, triacylglycerols

TABLE 10.2 Composition of some natural fats in percent of total fatty acids

Number of C Atoms in Chain	Percent Present in:		
	Olive Oil	Butter[a]	Beef Fat
Saturated			
4–12	2	11	2
14	2	10	2
16	13	26	29
18	3	11	21
Unsaturated			
16–18	80	40	46

[a]Numbers do not total 100% because the substance contains small amounts of other fatty acids.

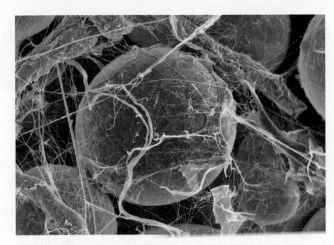

▲ **FIGURE 10.3 Adipocytes.** Adipocytes, or animal fat storage cells, make up a large part of adipose tissue.

are water-insoluble. Fats accumulated in plant and animal cells therefore form as oily droplets in the cytoplasm. In **adipocytes,** animal cells specialized for fat storage, almost the entire volume of each cell is filled by a fat droplet (**FIGURE 10.3**). Such cells make up most of the adipose (fatty) tissue of animals.

Fat storage in animals serves three distinct functions:

1. *Energy production.* As described in Chapter 16, triacylglycerols in most animals are oxidized for the generation of ATP, to drive metabolic processes.

2. *Heat production.* Some specialized cells (in "brown fat" of warm-blooded animals, for example) oxidize triacylglycerols for heat production rather than to make ATP.

3. *Insulation.* In animals that live in a cold environment, layers of fat cells under the skin serve as thermal insulation. The blubber of whales is one obvious example.

Soaps and Detergents

If fats are hydrolyzed with strong bases such as NaOH or KOH (in earlier times, wood ashes were used), a *soap* is produced. This process is called **saponification.** The fatty acids are released as either sodium or potassium salts, which are fully ionized. If fatty acid salts are mixed with water *and* an oily or greasy substance (for example, a hydrocarbon), micelles will form around the oil droplets, emulsifying them. In this way, soaps and synthetic detergents solubilize grease. However, as cleansers, soaps have the disadvantage that the fatty acids are precipitated by the calcium or magnesium ions present in "hard" water, forming a scum and destroying the emulsifying action. Synthetic detergents have been devised that do not have this defect. One class is exemplified by *sodium dodecyl sulfate* (*SDS*):

$$Na^+ \ ^-O_3SO(CH_2)_{11}CH_3$$

The salts of dodecyl sulfate with divalent cations (i.e., Ca^{2+} and Mg^{2+}) are more soluble. Recall that SDS is widely used in forming micelles about proteins for gel electrophoresis (see Tools of Biochemistry 6B).

Waxes

In the natural **waxes,** a long-chain fatty acid is esterified to a long-chain alcohol (**FIGURE 10.4**). This yields a head group that is only weakly

◀ **FIGURE 10.4 Structure of a typical wax.** Waxes are formed by esterification of fatty acids and long-chain alcohols. The small head group can contribute little hydrophilicity, in contrast to the significant hydrophobic contribution of the two long tails.

hydrophilic, attached to two long hydrocarbon chains. As a consequence, the waxes are completely water-insoluble. In fact, they are so hydrophobic that they often serve as water repellents, as in the feathers of some birds and the leaves of some plants. In some marine microorganisms, waxes are used instead of other lipids for energy storage. In beeswax, they serve a structural function. As with the triacylglycerols, the firmness of waxes increases with chain length and degree of hydrocarbon saturation.

10.2 The Lipid Constituents of Biological Membranes

Lipids are major constituents of all biological membranes. The lipid molecules that play the dominant roles in membrane formation all have highly polar head groups and, in most cases, two hydrocarbon tails. This composition promotes formation of **bilayer** membranes: If a large head group is attached to a *single* hydrocarbon chain, the molecule is wedge-shaped and will tend to form spherical micelles (**FIGURE 10.5(a)**). A tail of two fatty acids yields a roughly cylindrical molecule (Figure 10.5(b)); such cylindrical molecules pack in a parallel array to form extended sheets of bilayer membranes, with the hydrophilic head groups facing outward into the aqueous regions on either side (Figure 10.5(c)).

● **CONCEPT** Membrane lipids are amphipathic. They tend to form surface monolayers, bilayers, or vesicles when in contact with water.

(a) Fatty acids are wedge-shaped and tend to form spherical micelles.

(b) Phospholipids are more cylindrical and pack together to form a bilayer structure.

(c) A computer simulation of a phospholipid bilayer showing approximate boundaries for the hydrophobic core, the interface regions, and the bulk water (above and below the interfacial regions). Water is shown in white and red. The hydrocarbon portions of the bilayer lipids are shown in green and cyan.

▲ **FIGURE 10.5 Phospholipids and membrane structure.** Panel (c) generated from data published in Heller, H., Schaefer, M., and K. Schulten (1993) "Molecular dynamics simulation of a bilayer of 200 lipids in the gel and the liquid-crystal phases." *J. Phys. Chem.* 97:8343–8360.

The membrane bilayer is roughly 60 Å thick, with ~15 Å of interface on either side of the ~30 Å hydrophobic core (Figure 10.5(c)). The interface region is composed of the lipid head groups and associated water molecules. Beyond the interface is the bulk water.

The major classes of membrane-forming lipids include the glycerophospholipids, sphingolipids, glycosphingolipids, and glycoglycerolipids. As shown in the next section, they differ principally in the nature of the head group.

● **CONCEPT** The major lipid components of biological membranes are glycerophospholipids, sphingolipids, glycosphingolipids, and glycoglycerolipids.

Glycerophospholipids

Glycerophospholipids (also called *phosphoglycerides*) are the major class of naturally occurring **phospholipids,** lipids with phosphate-containing head groups. These compounds make up a significant fraction of the membrane lipids throughout the bacterial, plant, and animal kingdoms. Like the other biomolecular building blocks we have considered, lipids possess a specific stereochemistry. Glycerol does not possess any stereocenters; however, derivatization of one or the other of the equivalent —CH_2OH groups bonded to the central carbon atom will generate an asymmetric center with a defined stereochemistry at C2. Thus, glycerol is an example of a **prochiral** molecule—it has no stereocenter until it is derivatized. The general structure and

stereochemistry for a glycerophospholipid is shown in **FIGURE 10.6(a)**. Because the groups at C1 and C2 are generally hydrophobic and make up the interior of the bilayer, and the group at C3 is polar and on the outside face of the bilayer, we will draw these groups in the manner shown in Figure 10.6(b), with the hydrophobic tails drawn to the right and the hydrophilic head group to the left.

Typically, the groups R_1 and R_2 are acyl side chains derived from the fatty acids; often R_1 is saturated, and R_2 is unsaturated. The hydrophilic R_3 head group varies greatly and confers the greatest variation in properties among the glycerophospholipids. A gallery of the most common glycerophospholipid head groups is shown in **TABLE 10.3**, and their relative abundances in some membranes are given in **TABLE 10.4**. The simplest member of the group, **phosphatidic acid,** is only a minor membrane constituent; its principal role is as an intermediate in the synthesis of other glycerophospholipids or triglycerides (described in Chapter 16). The names of glycerophospholipids are derived from phosphatidic acid: *phosphatidylcholine, phosphatidylethanolamine,* and so on. As Table 10.3 shows, the glycerophospholipids have very polar head groups. The net charge on a phospholipid is a function of the charge (if any) on the R_3 group in combination with the negatively charged phosphodiester in the head group. Because the hydrocarbon tails are derived from the naturally occurring fatty acids in various combinations, an enormous variety of

(a) Stereochemical view of a generalized glycerophospholipid.

(b) The same structure represented in the convention used in this text, with hydrophobic groups to the right, hydrophilic to the left. R_1 and R_2 are generally long-chain nonpolar groups, and R_3 is a hydrophilic group (see Table 10.3).

▲ **FIGURE 10.6 Glycerophospholipid structure.**

TABLE 10.3 The hydrophilic groups[a] that distinguish common glycerophospholipids

Name of Glycerophospholipid	R₃ (in Figure 10.6)
Phosphatidic acid	H— (ionized at neutral pH)
Phosphatidylethanolamine (PE)	$H_3\overset{+}{N}$—CH_2—CH_2—
Phosphatidylcholine (PC)	$(CH_3)_3\overset{+}{N}$—CH_2—CH_2—
Phosphatidylserine (PS)	
Phosphatidyl inositol (PI)	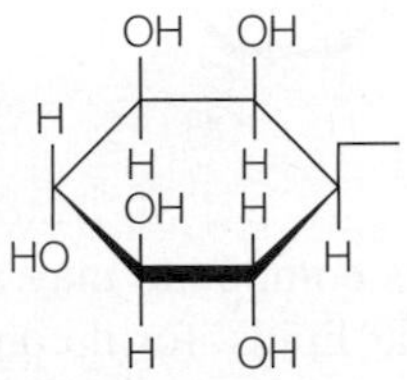

[a] These are the R₃ groups in Figure 10.6. In addition to this variation, there is also a great deal of variation in the hydrocarbon tails (R₁ and R₂ groups).

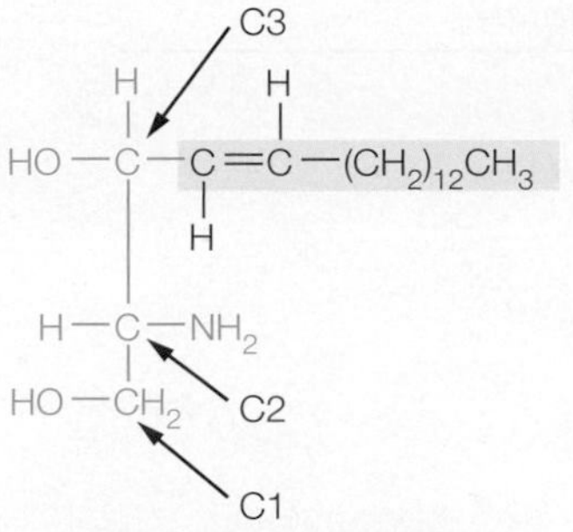

Sphingosine = (2S,3R)-2-aminooctadec-4-ene-1,3-diol

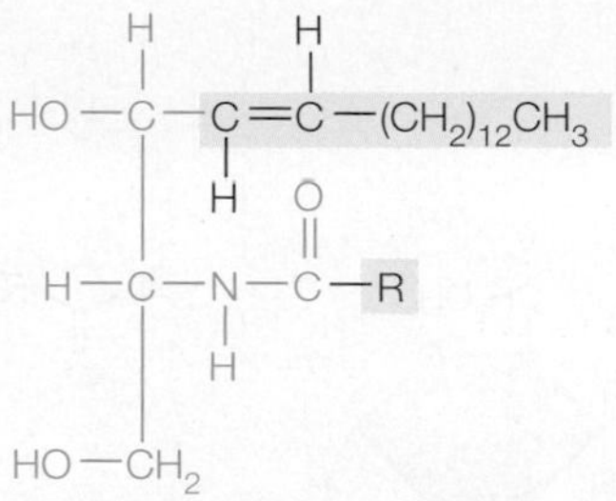

General structure of a ceramide (R = hydrocarbon)

Ceramides consist of sphingosine and a fatty acid. Further modification, by addition of groups to the C-3 hydroxyl of sphingosine, leads to a variety of other sphingolipids. An especially important example is *sphingomyelin,* in which a *phosphocholine* group is attached to the C-3 hydroxyl.

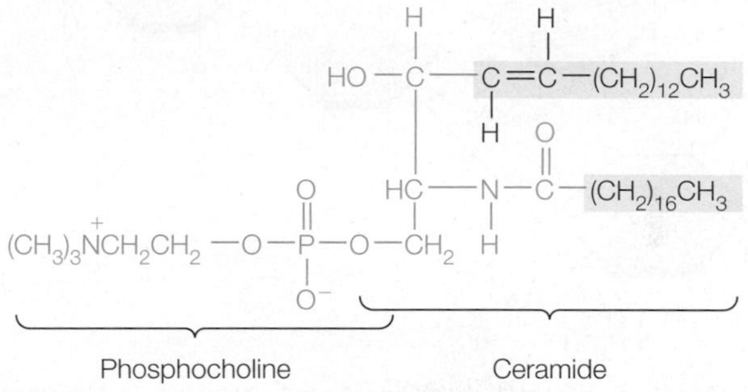

Phosphocholine · Ceramide

Sphingomyelin

glycerophospholipids exists. For example, the erythrocyte membrane contains molecules with hydrocarbon chains of 16 to 24 carbons, with 0 to 6 double bonds. Such variation in membrane composition allows "fine-tuning" of membrane properties for the diverse functions that different membranes must perform.

Sphingolipids and Glycosphingolipids

A second major class of membrane constituents is built on the amino alcohol **sphingosine** rather than on glycerol. The structure of sphingosine includes a long-chain hydrophobic tail, so it requires the addition of only one fatty acid to make it suitable as a membrane lipid. If a fatty acid is linked via an amide bond to the —NH₂ group, the class of **sphingolipids** referred to as **ceramides** is obtained:

● **CONCEPT** A wide variety of sphingolipids is built upon a sphingosine core. These include the ABO blood group antigens.

In some of the membrane lipids built on sphingosine, the head group contains saccharides. Lipids containing saccharide groups go under the general name of **glycolipids.** The **glycosphingolipids** constitute the third major class of membrane

TABLE 10.4 Lipid composition of some biological membranes

Lipid	Percentage of Total Composition in			
	Human Erythrocyte Plasma Membrane	Human Myelin	Bovine Heart Mitochondria	*E. coli* Cell Membrane
Phosphatidic acid	1.5	0.5	0	0
Phosphatidylcholine	19	10	39	0
Phosphatidylethanolamine	18	20	27	65
Phosphatidylglycerol	0	0	0	18
Phosphatidylinositol	1	1	7	0
Phosphatidylserine	8.0	8.0	0.5	0
Sphingomyelin	17.5	8.5	0	0
Glycolipids	10	26	0	0
Cholesterol	25	26	3	0
Others	0	0	23.5	17

Data from C. Tanford (1973) *The Hydrophobic Effect.* Wiley, New York.

Sugar   Ceramide

(a) Galactosylceramide

(b) GalNAcβ(1 → 4)Galβ(1 → 4)Glcβ(1 → 1)ceramide or, Ganglioside GM2

$$\begin{pmatrix} 3 \\ \uparrow \\ \alpha 2 \end{pmatrix}$$
Sia

▲ **FIGURE 10.7 Examples of glycosphingolipids. (a)** Galactosylceramide, a cerebroside, is an important constituent of brain cell membranes. **(b)** An example of a ganglioside. This particular ganglioside, called GM$_2$ or the Tay–Sachs ganglioside, accumulates in neural tissue of infants with Tay–Sachs disease. The defect responsible for this inherited condition is the lack of an enzyme that normally cleaves the terminal GalNAc (see Chapter 16).

● **CONNECTION** Several human diseases—including Tay-Sachs, Gaucher's, and Fabry's diseases—result from incomplete enzymatic processing of the glycan portion of glycosphingolipids.

lipids. In addition to being constituents of the ABO blood group antigens (described in Chapter 9), they include such molecules as the **cerebrosides** (monoglycosyl ceramides) and **gangliosides,** anionic glycosphingolipids containing one or more sialic acid residues. Examples of these are shown in **FIGURE 10.7**. As the names of these compounds suggest, they are especially common in the membranes of brain and nerve cells.

Glycoglycerolipids

Another class of lipids, less common in animal membranes but widespread in plant and bacterial membranes, are the **glycoglycerolipids,** exemplified by **monogalactosyl diglyceride:**

This compound may actually be the most abundant of all polar lipids, for it constitutes about half the lipid in chloroplast membranes (see Chapter 15). Such lipids are also abundant in archaea, where they are the major membrane components.

Cholesterol

One important lipid constituent of many membranes bears little superficial resemblance to the compounds we have studied so far. This substance is **cholesterol,** the structure of which is shown in **FIGURE 10.8**. Cholesterol is the biosynthetic precursor for a large group of substances called **steroids,** which include a number of important hormones, among them the sex hormones of higher animals. Its role in these syntheses is discussed in Chapter 16, along with a detailed description of other steroids and their functions in cell signaling. Another important and diverse class of signaling molecules derived from lipids is the **eicosanoids,** which are derived from arachidonic acid (Table 10.1). These are potent activators of a wide range of physiological functions, including inflammation, blood clotting, blood pressure regulation, and reproduction (see Chapter 16).

(a) Structural formula.

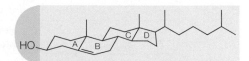

(b) Conformational model.

(c) Space-filling model.

▲ **FIGURE 10.8 Cholesterol.**

● **CONCEPT** Cholesterol, a component of many animal membranes, influences membrane fluidity by its bulky, rigid structure.

Cholesterol is a weakly amphipathic substance because of the hydroxyl group at one end of the molecule. As the conformational structure in Figure 10.8(b) shows, the fused cyclohexane rings in cholesterol are all in the chair conformation. This makes cholesterol a bulky, rigid, structure as compared with other hydrophobic membrane components such as the fatty acid tails; thus, the cholesterol molecule tends to disrupt regular packing of fatty acid tails in membrane structure. This property can have a major effect on membrane fluidity because cholesterol constitutes 25% or more of the lipid content in some membranes (see Table 10.4). Such changes in membrane structure can also have profound effects on such properties as membrane stiffness and permeability.

The molecules that have been described in this section constitute the major portion of membrane lipids in most organisms. However, one of the "three kingdoms" of organisms—the archaea—are unique in having glycoglycerolipids as their major membrane lipids.

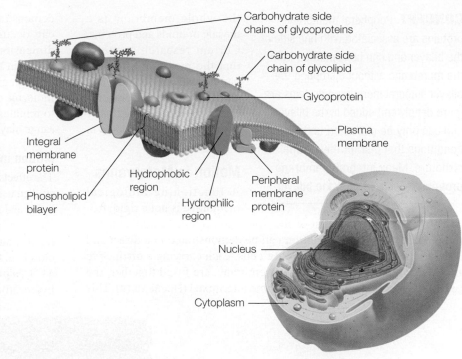

▲ **FIGURE 10.9 Structure of a typical cell membrane.** In this schematic view, a strip of the plasma membrane of a eukaryotic cell has been peeled off. Proteins are embedded in and on the phospholipid bilayer; some of them are glycoproteins, carrying oligosaccharide chains. The membrane is about 60 Å thick. Most membranes are more densely packed with proteins than is shown here.

10.3 The Structure and Properties of Membranes and Membrane Proteins

The membranes of living cells are remarkable bits of molecular architecture, with many functions. To say that a membrane is essentially a phospholipid bilayer is a gross oversimplification. To be sure, the phospholipid bilayer, as depicted in Figure 10.5(b) and (c), forms the basic structure, but there is much more to the membranes found in living cells. Some of the complex features of a typical eukaryotic cell membrane are shown schematically in **FIGURE 10.9**. An important feature of cellular membranes is the wide variety of specific proteins contained within the lipid bilayer or bound to its surface. Many of these proteins carry oligosaccharide groups that project into the surrounding aqueous medium. Other oligosaccharides are carried by glycolipids, with the lipid portions inserted in the membrane. The two sides of the bilayer are usually different, both in lipid composition and in the placement and orientation of proteins and oligosaccharides.

The protein content varies greatly among different kinds of membranes (see **TABLE 10.5**) and appears to be directly related to the functions a particular membrane must carry out. Mitochondrial inner membranes and bacterial cell wall membranes, which carry out many metabolic and transport functions, are about 75% protein. The myelin of nerve fibers, which acts primarily as an electrical insulator, has a much lower protein content (~20%). As a rule of thumb, a typical membrane is roughly 60% protein and 40% lipid by mass.

● **CONCEPT** According to the fluid mosaic model, a membrane is a fluid mixture of lipids and proteins.

Much of our current understanding concerning biological membranes is based on the **fluid mosaic model** proposed by S. J. Singer and G. L. Nicolson in 1972. This is the model depicted in Figure 10.9. The asymmetric lipid bilayer is fluid and carries within it many proteins. Some of them, called **peripheral membrane proteins,** are exposed at only one membrane face or the other. They are

held to the membrane by noncovalent interaction with lipid heads or integral membrane proteins. The **integral membrane proteins** are largely buried within the membrane but are usually exposed on both faces. Integral proteins are frequently involved in transporting specific substances, or transducing chemical signals, through the membrane. Thus,

TABLE 10.5 Protein, lipid, and carbohydrate content of some membranes

Membrane	Percent by Weight		
	Protein	Lipid	Carbohydrate
Myelin	18	79	3
Human erythrocyte (plasma membrane)	49	43	8
Mitochondria (outer membrane)	52	48	0
Sarcoplasmic reticulum (muscle cells)	67	33	0
Chloroplast lamellae	70	30	0
Gram-positive bacteria	75	25	0
Mitochondria (inner membrane)	76	24	0

Adapted from *Annual Review of Biochemistry* 41:731, G. Guidotti, Membrane proteins. © 1972 Annual Reviews.

● **CONCEPT** Peripheral membrane proteins are associated with one side of the bilayer and can be separated from the membrane without disrupting the bilayer. Integral membrane proteins are more deeply embedded in the bilayer and can only be extracted under conditions that disrupt membrane structure. Many integral membrane proteins extend through the bilayer.

zen structure. In fact, most of the lipid and protein components are in constant motion. This motion can be demonstrated in a direct and dramatic way. If human and mouse cells, each carrying a distinctive fluorescent marker in its plasma membrane, are fused together, the two kinds of markers gradually become intermixed (**FIGURE 10.10**). This

the whole membrane is a mosaic of lipids and proteins. Current research suggests that the membrane surface is even more crowded, and the distribution of proteins in membranes is more highly organized, than is depicted in Figure 10.9.

Motion in Membranes

A functioning biological membrane is not a rigid, frozen structure.

demonstrates that *lateral diffusion* (parallel to the membrane surface) can occur in the membrane. The rapidity with which such two-dimensional diffusion can occur depends on the membrane fluidity, which in turn depends on temperature and lipid composition. Under physiological conditions, the average time required for a phospholipid molecule to wander completely around a cell is on the order of seconds to minutes; membrane proteins also move, but more slowly, and their range may be constrained by other structural features of the membrane.

Motion in Synthetic Membranes

The effects of temperature and composition on fluidity can be most simply studied using artificial membranes containing only one or a few kinds of lipids and no proteins. **FIGURE 10.11(a)** depicts the behavior of a membrane made entirely from phosphatidylcholine carrying two 16-carbon saturated chains. At low temperatures, the hydrocarbon tails pack together closely to form a nearly solid *gel* state. If the temperature is raised above 41 °C, a **phase change** occurs in which this regular order is lost, and the hydrocarbon tails become free to move about. The membrane "melts" to

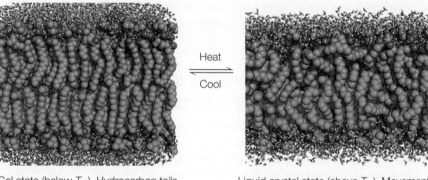

Gel state (below T_m). Hydrocarbon tails are packed together in a highly ordered gel state.

Liquid crystal state (above T_m). Movement of the chains becomes more dynamic, and the interior of the membrane resembles a liquid hydrocarbon.

(a) Transition from gel to liquid crystal states. A computational model of the change in bilayer structure at the transition temperature (T_m). Lipids are shown as green or cyan spheres, and water molecules (above and below the lipid bilayer) are shown as sticks.

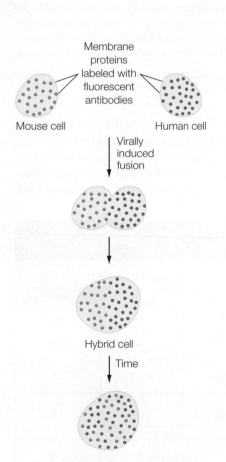

Membrane proteins labeled with fluorescent antibodies

Mouse cell — Human cell

Virally induced fusion

Hybrid cell

Time

▲ **FIGURE 10.10 Experimental demonstration of membrane fluidity.** When cells with surface membrane protein marked by fluorescent tags are induced to fuse, the proteins gradually mix over the fused surface.

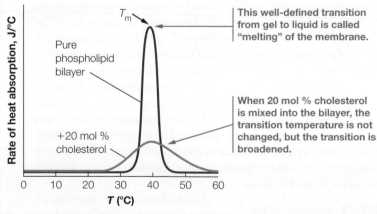

Pure phospholipid bilayer

T_m

This well-defined transition from gel to liquid is called "melting" of the membrane.

+20 mol % cholesterol

When 20 mol % cholesterol is mixed into the bilayer, the transition temperature is not changed, but the transition is broadened.

Rate of heat absorption, J/°C

T (°C)

0 10 20 30 40 50 60

(b) Transition with and without cholesterol. Measurement of the heat absorbed by a membrane as the temperature is raised each degree shows a sharp spike at the T_m for a pure dipalmitoylphosphatidylcholine bilayer.

▲ **FIGURE 10.11 The gel–liquid crystalline phase transition in a synthetic lipid bilayer.** (a) generated from computational modeling data published in Heller, H., Schaefer, M., and K. Schulten (1993) "Molecular dynamics simulation of a bilayer of 200 lipids in the gel and the liquid-crystal phases." *J. Phys. Chem.* 97:8343–8360.

adopt a semifluid *liquid crystalline* state. The temperature at which this happens is called the **transition temperature** or **melting temperature** (T_m). Figure 10.11(b) shows that the transition temperature becomes broadened when cholesterol is added to the pure phospholipid bilayer.

The T_m is very sensitive to the nature of the hydrocarbon tails. Compared to the bilayer described in the previous paragraph, if a phosphatidylcholine with 14-carbon saturated tails is used to make the bilayer, the T_m drops to 23 °C. If a single *cis* double bond is incorporated into each 16-carbon tail, melting occurs at 36 °C. As explained earlier, *cis* double bonds put bends in the chains, disrupting their close packing; thus, such chains must be cooled to a lower temperature to produce the gel phase. Changing the head group can also make a big difference: If phosphatidylethanolamine (Table 10.3) is substituted for phosphatidylcholine,

● **CONCEPT** The transition temperature for a membrane depends on its lipid composition. Lipids with longer, saturated tails tend to increase the transition temperature, whereas those with more *cis* double bonds and/or shorter tails will reduce the transition temperature.

the thermal transition shown in Figure 10.11(b) is raised to 63 °C. The sensitivity of the transition to lipid composition is shown dramatically by the fact that combining several of the changes described above can change the membrane transition temperature over a range of 100 °C.

Motion in Biological Membranes

Biological membranes, which contain complex mixtures of lipid components plus protein, exhibit much broader and more complex phase transitions than those observed for synthetic bilayers of the kind described above. Indeed, there is now evidence for the existence of quite stable "domains" of different composition in different parts of a cell membrane. Biological membranes must be fluid (not gel-like) to allow their associated proteins to move within the bilayer, to interact with binding partners or substrates, and to change conformation. Because

● **CONCEPT** Under physiological conditions, biological membranes exist in a semifluid liquid crystalline state.

it is essential that the membranes in living cells be fluid, the membrane composition is regulated so as to keep the transition temperature below the body temperature of the organism. One example is found in bacteria, which will alter the saturated/unsaturated fatty acid ratio in their membranes in response to a change in the temperature at which they are grown. A remarkable case in the animal kingdom is that of the reindeer's leg. Its cell membranes show an increase in relative amount of unsaturated fatty acids near the hoof, which is usually cooler than the rest of the body.

Cholesterol has a specific and complex effect on membrane fluidity. As Figure 10.11(b) shows for a synthetic membrane, cholesterol does not influence the transition temperature markedly, but it does broaden the transition. It has been hypothesized that this broadening occurs because cholesterol can both stiffen the membrane above the transition temperature and inhibit regularity in structure formation below the transition temperature. Thus, it blurs the distinction between the gel and the fluid state. There is evidence that variations in cholesterol content are used to regulate membrane behavior in some organisms. As we discuss later in this chapter, cholesterol also appears to play a role in organizing smaller regions of the bilayer into functional units referred to as "lipid rafts."

The Asymmetry of Membranes

Every biological membrane has two distinct faces, each encountering a different environment. The plasma membrane of a cell faces the external environment on the outside and the cytoplasm on the inside, whereas the membrane around a chloroplast faces the photosynthetic apparatus on the inside and the cytoplasm on the outside. Because the two faces of a membrane must deal with different surroundings, the faces are usually quite different in composition and structure. This

● **CONCEPT** The two leaflets of a membrane usually differ in lipid composition.

difference extends even to the level of phospholipid composition. Recall that all phospholipid membranes are bilayers; the two individual layers are called *leaflets*. The compositions of the two leaflets in the plasma membranes of several kinds of cells are shown in **FIGURE 10.12**. Not only are the individual lipids distributed very asymmetrically, but the distribution also varies considerably among cell types.

Membranes are highly specialized structures that serve diverse functions in different cells/tissues. The lipid and protein content of a given membrane is tailored to the specific function of that membrane. For example, differences in charged groups between bilayer leaflets lead to differences in the electrical potentials across various membranes (discussed later in this chapter). Glycoproteins and glycolipids carried in the outer leaflet of a plasma membrane contribute to identification of cells via their oligosaccharide chains (e.g., the ABO blood group antigens discussed in Chapter 9).

In contrast to the ease of lateral movement, the "flip-flop" of lipid molecules *across* synthetic lipid bilayers, from one leaflet to the other, is much slower. The reason is not hard to see: When a phospholipid molecule turns from one face to the other, it must pass its very hydrophilic head through the hydrophobic medium of the hydrocarbon tails. Such an event has a significant activation energy; thus, the process is slow. There exist enzymes (*translocases* and *flippases*) that catalyze membrane lipid flipping.

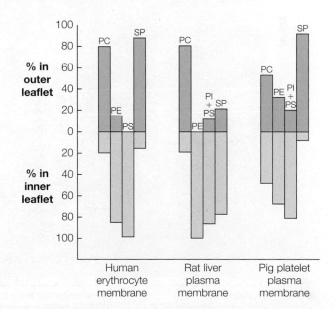

▲ **FIGURE 10.12 Phospholipid asymmetry in plasma membranes.** Lipid composition in the outer leaflet (green) and inner leaflet (gold) of the plasma membrane is graphed for three cell types. PC = phosphatidylcholine; PE = phosphatidylethanolamine; PS = phosphatidylserine; PI = phosphatidylinositol; SP = sphingomyelin.

Much of our knowledge of membrane asymmetry comes from studies of **vesicles,** fragments of membrane that have resealed to form hollow shells, with an inside and an outside. Reagents can be either captured inside the vesicle or added only to the surrounding solution so that they can react specifically with either outward-facing or inward-facing proteins or lipids. A membrane protein in a vesicle may be reacted covalently with a radioactively labeled reagent, isolated, and cleaved into peptides by proteases. Identification of which peptides are labeled by "inside" or "outside" reactants can reveal which portions of the protein were on the inner face and which were on the outer face. In a similar way, lipids can be tested, using enzymes or other reagents that cleave off or otherwise modify the head groups. Experiments of this kind, performed inside or outside vesicles, have provided much of the information shown in Figure 10.12.

Characteristics of Membrane Proteins

Membrane proteins possess special characteristics that distinguish them from other globular proteins. They often contain a high proportion of hydrophobic amino acids in the parts of the protein molecules that are embedded in the membrane (see **FIGURE 10.13**). The segments of proteins that span membranes are often α-helical, although β-barrels

Examples of β-barrel transmembrane structures.

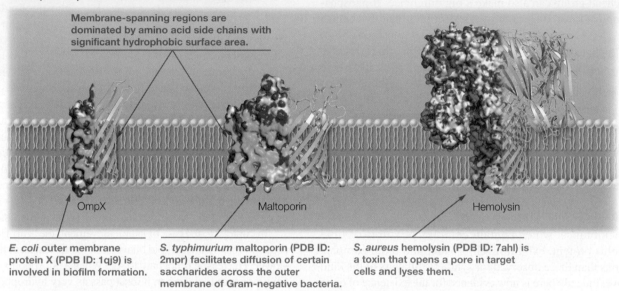

Membrane-spanning regions are dominated by amino acid side chains with significant hydrophobic surface area.

OmpX

Maltoporin

Hemolysin

E. coli outer membrane protein X (PDB ID: 1qj9) is involved in biofilm formation.

S. typhimurium maltoporin (PDB ID: 2mpr) facilitates diffusion of certain saccharides across the outer membrane of Gram-negative bacteria.

S. aureus hemolysin (PDB ID: 7ahl) is a toxin that opens a pore in target cells and lyses them.

Examples of α-helical transmembrane proteins.

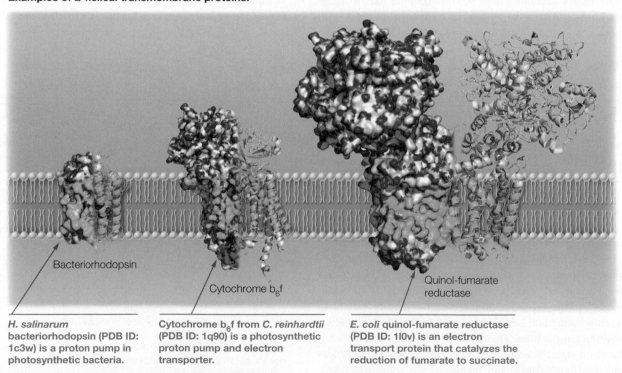

Bacteriorhodopsin

Cytochrome b₆f

Quinol-fumarate reductase

H. salinarum bacteriorhodopsin (PDB ID: 1c3w) is a proton pump in photosynthetic bacteria.

Cytochrome b₆f from *C. reinhardtii* (PDB ID: 1q90) is a photosynthetic proton pump and electron transporter.

E. coli quinol-fumarate reductase (PDB ID: 1l0v) is an electron transport protein that catalyzes the reduction of fumarate to succinate.

▲ **FIGURE 10.13 Examples of structures for several integral membrane proteins.** Each protein is rendered half in a space-filling model (left half of each structure) and half in a cartoon (right half). The cartoon rendering shows the differences in membrane-spanning structure, whereas the space-filling models show similar distributions of hydrophilic (red and blue) and hydrophobic (green and gray) surface areas for both classes of proteins.

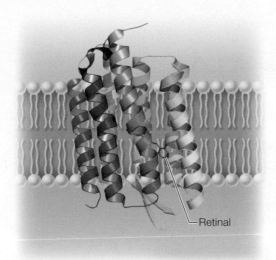

▲ **FIGURE 10.14** Bacteriorhodopsin—an integral membrane protein. Bacteriorhodopsin (PDB ID: 1c3w) functions as a light-driven proton pump in certain bacteria. Seven helices span the membrane and hold a molecule of the light-absorbing pigment retinal (magenta).

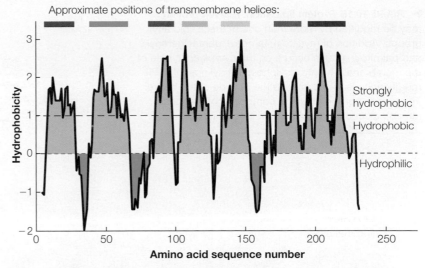

▲ **FIGURE 10.15** Hydrophobicity plot for the bacteriorhodopsin molecule. The hydrophobicity index has been calculated at each residue by the method of Kyte and Doolittle using the ProtScale Web tool (http://web .expasy.org/protscale/). Values greater than zero are considered hydrophobic, and those below zero are hydrophilic. The dashed line at 1 is provided as a reference to highlight the sequences of strong hydrophobic character. The colored horizontal bars above the plot show the approximate positions of the transmembrane helices shown in Figure 10.14.

are also common membrane-spanning motifs. **FIGURE 10.14** depicts bacteriorhodopsin, an integral membrane protein whose structure has been solved to high resolution. Like many such proteins, it contains a bundle of seven α-helical segments that pass back and forth through the membrane. The presence of such transmembrane segments can sometimes be inferred from the kind of **hydrophobicity plot** shown in **FIGURE 10.15**. This plot has been calculated according to the relative hydrophobicities of the 20 amino acid side chains. It reveals maxima in regions of the sequence corresponding to the transmembrane helices. Transmembrane helices typically contain 20 to 25 residues, which are predominantly hydrophobic.

Another class of membrane-associated proteins is covalently modified with lipids as shown in **FIGURE 10.16**. Many proteins involved in signaling are modified in this way, via reactions that are formally acyl group transfers, or **acylations.** Specific sequences at the N-terminus or C-terminus of the protein are required for the addition of **geranylgeranyl, farnesyl,** and **myristoyl** groups. For example, myristoylation occurs at N-terminal Gly residues following cleavage of the N-terminal Met, and farnesylation requires a C-terminal "CaaX" motif (where C = Cys, a = an aliphatic amino acid, and X = any amino acid except Pro). These processes are presented in greater detail in Chapters 16 and 20. Another common protein modification is the addition of one or more palmitoyl groups to a Cys side chain (or to a lesser extent, the N-terminal amino group).

As discussed below, the attached lipid may serve primarily to target the modified protein to a particular subcellular organelle or to a specific location in the plasma membrane. For example, proteins linked to **glycosylphosphatidylinositol** (GPI) are frequently associated with regions of the outer leaflet rich in cholesterol and sphingolipids. Although the lipid moiety may insert directly into the bilayer, several membrane proteins with lipid-binding sites are known. Thus, the association of a lipid-linked protein with the membrane may be mediated by protein-lipid binding.

Insertion of Proteins into Membranes

Membrane proteins must be properly oriented in the bilayer to ensure they can carry out their function. Thus, the necessary asymmetry of the orientation of integral proteins must be properly established. For example, a hormone receptor must display its ligand-binding site on the *extracellular* side of the membrane if it is to detect the presence of the hormone in circulation. Likewise, many transport proteins must move their substrates *in only one direction* across the bilayer to create electrochemical gradients that are critical for several cell functions. Thus, inserting transmembrane peptide sequences into the membrane with the correct topology is a matter of survival. There appear to be special components of the cellular protein synthesis machinery that direct the placement of proteins in membranes and ensure their asymmetric orientation (see Chapter 25).

Greater than 30% of proteins must cross, or integrate into, a cellular membrane. How do hydrophilic, globular proteins make it across a hydrophobic membrane bilayer during, or following, their ribosomal synthesis? The ribosomes are either free in the cytosol or bound to the rough endoplasmic reticulum (see Chapter 25 for a discussion of protein **secretion**). Given the cytosolic location of protein synthesis and the asymmetric orientation of integral membrane proteins, two fundamental questions arise: (1) How do integral membrane proteins get inserted into the membrane? and (2) how are they inserted into the membrane in the correct orientation?

One solution to the problem of protein insertion into a membrane is illustrated in **FIGURE 10.17**. The membrane-spanning regions of several integral membrane proteins are inserted into the bilayer cotranslationally (i.e., *during* ribosomal protein synthesis), where they then fold. This process is facilitated by a protein-conducting channel called a **translocon.** These protein channels are multisubunit complexes called

▶ **FIGURE 10.16 Protein lipidation.** A given protein may be modified by more than one of these lipid acyl groups. Addition of myristoyl, farnesyl, geranylgeranyl, and palmitoyl groups occurs on the cytoplasmic side of the membrane. The glycosylphosphatidylinositol (GPI) modification is found only on the outside of the bilayer. The glycan portion of GPI is composed of mannose (green), glucosamine (blue), and inositol (red).

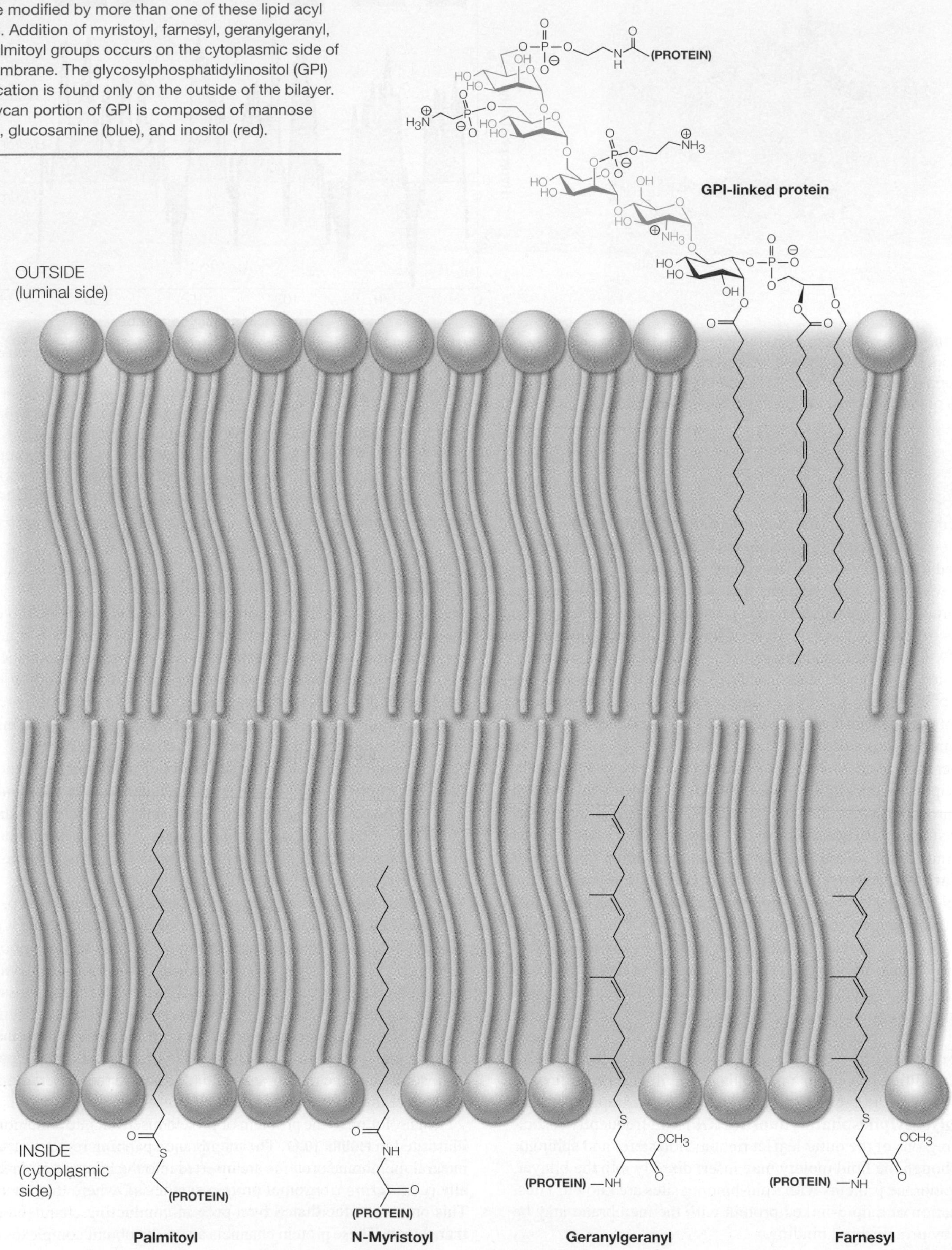

GPI-linked protein

OUTSIDE
(luminal side)

INSIDE
(cytoplasmic side)

Palmitoyl   N-Myristoyl   Geranylgeranyl   Farnesyl

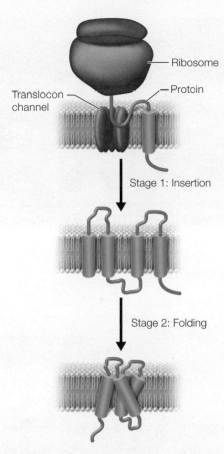

▲ **FIGURE 10.17 Cotranslational insertion and folding of transmembrane helices in an integral membrane protein.** The bilayer lipids are indicated in tan, the ribosome in brown, the translocon channel in purple, and the peptide in green. Some parts of the protein (in this case, the loops) are conducted across the bilayer through the translocon; transmembrane portions exit the translocon and remain embedded in the bilayer. The transmembrane helices fold after they are inserted.

SecY in prokaryotes and **Sec61** in eukaryotes, where **Sec** refers to proteins involved in secretion.

How are the hydrophobic transmembrane segments of the inserted peptide recognized, and how do they get out of the translocon channel and into the bilayer? One plausible explanation, supported by the crystal structure of SecY from *P. furiosus,* is that the translocon opens along its length to allow the peptide to slide out laterally. Hydrophobic peptides will be more likely to partition into the bilayer core, whereas polar sequences will remain in the channel and partition into the polar interface region of the membrane. In essence, the translocon is acting like a molecular separatory funnel, exposing the peptide sequence to polar and nonpolar phases. The peptide will move to the phase that best matches its polarity.

Given this model for insertion of transmembrane helices, how are they inserted in the correct orientation? Several factors are likely

● **CONCEPT** A protein channel called the "translocon" facilitates the insertion of integral membrane proteins into the membrane bilayer.

at work to assure the proper topology for the inserted protein. Work by Gunnar von Heijne and colleagues has supported an "inside positive" rule that explains the enrichment in Lys and Arg residues in the cytoplasmic portions of proteins with multiple membrane-spanning regions. Due to the electrical potential across the membrane, the cytoplasmic side of the bilayer is more negatively charged compared to the outside; thus, a cluster of positively charged side chains would tend to migrate to the cytoplasmic side of the membrane. This "inside positive" orientation of transmembrane proteins is also aided by the preponderance of acidic (i.e., negatively charged) phospholipids on the inside of membranes.

Evolution of the Fluid Mosaic Model of Membrane Structure

Many features of the Singer–Nicolson fluid mosaic model have been confirmed since it was proposed in 1972; however, with improvements in imaging technology, more details have emerged and prompted reinterpretation of the model. Specifically, most membranes appear to be very crowded with proteins and show significant variation in both thickness of the bilayer and distribution of proteins and lipids within a leaflet.

We mentioned previously that GPI-modified proteins are frequently localized in regions of the membrane that are enriched in cholesterol and sphingolipids. This observation led to the proposal that these three membrane components can coalesce to form separate membrane domains called **lipid rafts** (or **membrane rafts**). These membrane rafts are small (~10 nm), short-lived dynamic structures that, in response to certain stimuli, can transiently associate with each other to form larger "raft platforms" (**FIGURE 10.18**). Actin fibers may act to stabilize and/or initiate formation of the rafts. These raft platforms are thought to play significant roles in cell signaling and the sorting of proteins into specific organelles within a cell. GPI-anchored proteins are frequently involved in cell signaling, and the clustering of GPI proteins in raft platforms may accelerate signal transduction across the membrane—particularly in cases where dimerization of a signal receptor is required. More recently, rafts have been implicated in facilitating bacterial entry into host cells.

● **CONCEPT** Membrane rafts are rich in cholesterol, sphingolipids, and GPI-linked proteins. The bilayer is thicker in the raft domains than in the surrounding membrane.

From studies in model membranes it is known that the bilayer thickness is a function of the lipid and protein composition in membrane domains (**FIGURE 10.19**). This may be due to the effect of cholesterol on the phase behavior of lipids. At certain concentration ratios of sterols to lipids, membrane lipids form more ordered and elongated structures. For example, the membrane rafts described in Figure 10.18 are thicker than the surrounding nonraft membrane. Membrane thickness is also influenced by the proteins in the bilayer. This raises the interesting question illustrated in **FIGURE 10.20**: Do proteins change conformation to optimize interactions with the bilayer as membrane thickness changes, or do lipids change their structure to accommodate proteins? Current evidence suggests that it is more common for lipids to adjust to proteins than vice versa; however, both adaptations are observed in cell membranes.

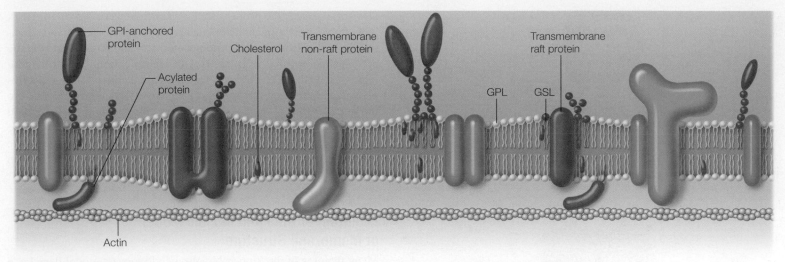

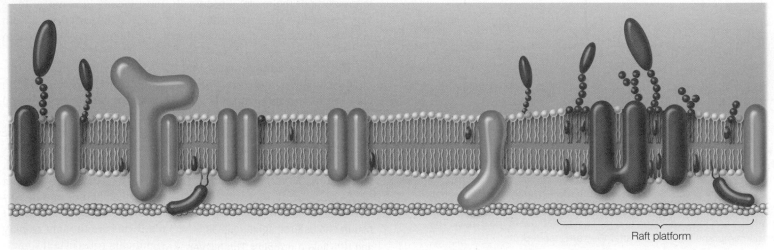

(a) Cholesterol, sphingolipids, and GPI-anchored proteins coalesce and form nanometer-sized dynamic raft domains, which may be stabilized by interactions with actin fibers.

(b) Rafts can associate to form larger structures ("platforms"). Certain proteins interact preferentially with rafts (pink shading), while others do not (green shading) or are excluded.

▲ **FIGURE 10.18 Membrane rafts.** Abbreviations: GPL, glycerophospholipid; GSL, glycosyl sphingolipid; GPI, glycosyl phosphatydylinositol.

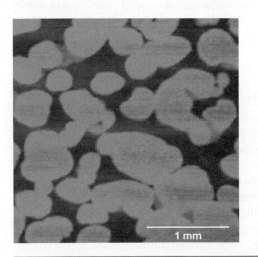

◀ **FIGURE 10.19 Atomic force microscopy image of membrane domains in a model membrane.** Top view of a bilayer composed of a mixture of 1,2-dilauroylphosphatidylcholine (DLPC) and 1,2-distearoylphosphatidylcholine (DSPC) separates into two phases as the temperature is lowered from 70 °C to 25 °C. The DSPC molecules coalesce and form thicker gel-like "islands" or domains (lighter gray spots) in a "sea" of fluid DLPC (darker gray background). Such simple systems have been used to model the more complex lipid rafts described in Figure 10.18. Courtesy of Marjorie Longo (University of California, Davis).

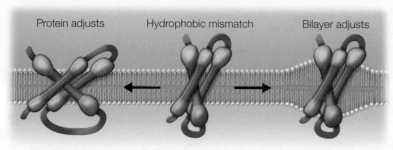

▲ FIGURE 10.20 Adaptation to hydrophobic mismatch in a membrane. If the thickness of the bilayer core and the hydrophobic surface area of an embedded protein do not match (middle image), either the protein will undergo conformational change (left) or the bilayer will change composition (right) until the dimensions of these hydrophobic regions match.

10.4 Transport Across Membranes

A cell or an organelle can be neither wholly open nor wholly closed to its surroundings. Its interior must be protected from certain toxic compounds, and metabolites must be taken in and waste products removed. A few specific examples are shown in **FIGURE 10.21**. Because the cell must contend with the proper transport of thousands of substances, it is not surprising that much of the complex structure of membranes is devoted to the regulation of transport.

▲ FIGURE 10.21 Specific transport processes. This composite plant–animal cell illustrates some of the most important specific transport processes. All of the substances shown here, and many more, are transported in specific directions across cellular membranes. Magenta dots signify known transport proteins.

In this section, we describe various ways in which molecules are transported across membranes. These include the actions of small molecules that act as ion carriers, larger proteins that are highly specific transporters, and proteins that promote the formation of membrane vesicles. Before considering specific mechanisms of transport, we review the underlying thermodynamics of transport processes.

The Thermodynamics of Transport

In Chapter 3 (Section 3.4; Figure 3.8) we discussed the general thermodynamic principles governing the diffusion of substances across membranes. It was shown that the free energy change, ΔG, for transporting one mole of a substance from a location in which its concentration is C_1 to a different location where its concentration is C_2 is given by

$$\Delta G = \Delta G^{\circ\prime} + RT \ln Q = -RT \ln K_{eq} + RT \ln \frac{[C_2]}{[C_1]} \quad (10.1)$$

For a process that only involves transport of some substance across a membrane, the term $\Delta G^{\circ\prime}$ equals zero. Why is this so? Recall that $\Delta G^{\circ\prime}$ describes the *equilibrium state* for a process, and for the case of transport across a membrane, the equilibrium state is reached when the concentrations of the substance are the same on both sides of the membrane (i.e., at equilibrium $[C_1] = [C_2]$). In this case $K_{eq} = [C_2]/[C_1] = 1$. Since $\Delta G^{\circ\prime} = -RT \ln K_{eq}$, if $K_{eq} = 1$, $\Delta G^{\circ\prime} = 0$ for the process. Note that, when $[C_2] = [C_1]$, the term $Q = 1$; thus, $\Delta G = 0$, as expected for a system at equilibrium.

For a transport process that is not at equilibrium $[C_2] \neq [C_1]$; thus, $RT \ln Q \neq 0$, and ΔG for the process is given by

$$\Delta G = \Delta G^{\circ\prime} + RT \ln Q = 0 + RT \ln Q$$
$$= RT \ln \frac{[C_2]}{[C_1]} \quad (10.2)$$

A more detailed derivation of this equation is given in Chapter 3 (see Equation 3.34). According to this equation, if $[C_2] < [C_1]$, ΔG is negative, and the process is thermodynamically favorable. As more and more substance is transferred between the two locations, $[C_1]$ decreases and $[C_2]$ increases, until $[C_2] = [C_1]$. At this point $\Delta G = 0$, and the system is at equilibrium. *Unless other factors are involved,* this equilibrium is the ultimate state approached by transport across any membrane. In short, a substance that can traverse the membrane will eventually reach the same concentration on both sides. We can describe the same process in kinetic terms. If the molecules are colliding with the membrane at random, the number entering from any side will be proportional to the concentration on that side. When the concentrations become equal, the rates of transport in the two directions will be the same, and no net transport will occur.

● CONCEPT For a substance that can pass through a membrane, the normal state of equilibrium is achieved when the concentrations of the substance are equal on both sides of the membrane.

There are three circumstances under which this equilibrium state can be circumvented, and each is important in the behavior of real membranes:

1. A substance may be preferentially bound by macromolecules confined to one side of the membrane, or it may be chemically modified once it crosses. We may find that compound A is more concentrated inside a cell (in terms of total moles of A per unit volume) than outside. But much of A may be bound to some cellular macromolecules or may have been modified; that portion is not accounted for in Equation 10.2, which simply states that the concentrations of *free* A on the two sides must be equal at equilibrium. An appropriate example is oxygen in erythrocytes. If we were to measure the *total* oxygen concentration in an erythrocyte, we would find it higher than the concentration of O_2 in the surrounding blood plasma. But the total concentration inside the cell includes oxygen bound to hemoglobin. The *free* oxygen concentration in the fluids inside and outside an erythrocyte is the same at equilibrium.

2. A **membrane electrical potential** may be maintained across a membrane that influences the distribution of ions. This tendency can be expressed quantitatively in the following way. For an ion of charge Z, the free energy change for transport across a cell or organelle membrane now involves two contributions: the normal concentration term, as given in Equation 10.2, plus a second term describing the energy change (or work involved) in moving a mole of ions across the membrane electrical potential. We consider a process in which one mole of ions is transported from *outside* to inside.

$$\Delta G = RT \ln \frac{[C_{in}]}{[C_{out}]} + ZF\Delta\psi \qquad (10.3)$$

Here F is the Faraday constant (96.5 kJ mol^{-1}V^{-1}), and $\Delta\psi$ is the membrane potential in volts. We define $\Delta\psi$ in terms of the initial and final locations of the transported ion ($\Delta\psi = \psi_{final} - \psi_{initial}$: in this case, $\Delta\psi = \psi_{in} - \psi_{out}$). In this example, $\Delta\psi$ will be negative if the inside of the membrane is negatively charged compared to the outside. Under these conditions, if Z is positive, the $ZF\Delta\psi$ term in Equation 10.3 is negative and makes ΔG more exergonic (favorable). That is, the transport of cations *into* this hypothetical cell (and toward increasing negative charge) is favored. For anions, of course, the opposite is true; they will be driven out. In the presence of a nonzero membrane potential, the equilibrium state ($\Delta G = 0$) will *not* correspond to equal concentrations of ions on the two sides of the membrane. However, energy must be expended continually to keep up the potential difference; otherwise migration of ions would neutralize it. Conversely, Equation 10.3 may be interpreted to mean that if a difference in ionic concentration is maintained, an electrical potential will be produced across the membranes (see Problems 9 and 10 at the end of this chapter).

3. If some thermodynamically favored process is *coupled* to the transport, then the $\Delta G^{\circ\prime}$ and $RT \ln Q$ for this favorable process must be included in the free energy equation. This is the general case of *active transport,* for which we can write

$$\Delta G = \Delta G^{\circ\prime} + RT \ln \frac{Q[C_{in}]}{[C_{out}]} \qquad (10.4)$$

where the term Q includes activities for the species in the favorable reaction that is coupled to the transport of C. If the substance C being transported is an ion, we must also include the $ZF\Delta\psi$ term

$$\Delta G = \Delta G^{\circ\prime} + RT \ln \frac{Q[C_{in}]}{[C_{out}]} + ZF\Delta\psi \qquad (10.5)$$

In this case, the quantities $\Delta G^{\circ\prime}$ and Q correspond to a thermodynamically favored reaction (e.g., hydrolysis of one mol ATP) that is coupled to the process of transport. This equation is a generalization of Equation 10.3, now allowing a variety of processes—not just those that maintain an electrical potential difference—to participate in the transport. In the case where hydrolysis of one mol of ATP results in transport of n mol of an ion C, we must modify Equation 10.5 to include the proper stoichiometry:

$$\Delta G = \Delta G^{\circ\prime} + RT \ln \frac{Q[C_{in}]^n}{[C_{out}]^n} + nZF\Delta\psi \qquad (10.6)$$

where $n = $ mol of C transported per mol of ATP. We will show examples of such calculations in the discussion that follows.

● **CONCEPT** Equalization of the concentrations of some substance across a membrane can be circumvented (1) by the binding of the substance to macromolecules, (2) by maintaining a membrane potential (if the substance is ionic), or (3) by coupling transport to an exergonic process.

With this background, we turn now to the mechanisms whereby substances are passed through membranes. We introduce the problem by asking two questions: (1) Does the process approach a state in which there are equal concentrations of the free substance on both sides, or is it maintained far from equilibrium? (2) How fast does the transport occur? Some molecules that are not actively transported against a concentration gradient can still traverse some membranes very rapidly, whereas others are transported so slowly as to be effectively excluded.

Nonmediated Transport: Diffusion

Nonmediated diffusion across membranes is accomplished by the random wandering of molecules through membranes. The process is the same as the Brownian motion of molecules in any fluid, which is termed **molecular diffusion.** Nonmediated transport ultimately results in the concentration of the diffusing substance being the same on both sides of the membrane.

The rate of nonmediated transport, J, can be described as follows:

$$J = -P(C_2 - C_1) \qquad (10.7)$$

where C_1 and C_2 are the concentrations of the substance on either side of the membrane (thus, $C_2 - C_1 = $ the concentration difference across the membrane) and P is the **permeability coefficient,** which has

● **CONCEPT** The rate of nonmediated transport across lipid membranes is faster for nonpolar substances.

units of (distance the substance migrates)/(time). The value of P increases with increasing solubility of the substance in the lipid phase. Equation 10.7 predicts that the rate of nonmediated

TABLE 10.6 Permeability coefficients (cm/s) for some ions and molecules through membranes

Ion/molecule	Synthetic Membrane (Phosphatidylserine)	Biological Membrane (Human Erythrocyte)
K^+	$< 9 \times 10^{-13}$	2.4×10^{-10}
Na^+	$< 1.6 \times 10^{-13}$	10^{-10}
Cl^-	1.5×10^{-11}	$1.4 \times 10^{-4*}$
Glucose	4×10^{-10}	$2 \times 10^{-5*}$
Water	5×10^{-3}	5×10^{-3}

Data from M. K. Jain and R. C. Wagner (1980) *Introduction to Biological Membranes.* Wiley, New York.

*Facilitated transport. Note that whenever facilitated transport is encountered, the permeability coefficient rises dramatically.

transport is dependent on the concentration difference across the membrane and the polarity of the material. Nonmediated diffusion tends to increase as the hydrophobic character of the substance increases. For ions and other hydrophilic substances, nonmediated diffusion through lipid membranes is extremely slow.

TABLE 10.6 lists permeability coefficients for a number of small molecules and ions in membranes. The low *P* values of the ions are as expected because ions have low solubility in the nonpolar lipid layer. However, the relatively large permeability value for water is surprising. Despite their hydrophobicity, biological membranes are not, in fact, very good barriers against water. Although the reasons for this are not entirely clear, it is probably fortunate for life, for it allows cells to exchange water, albeit slowly, with their surroundings. When water loss is to be strenuously avoided, as in the leaves of desert plants, waxy substances, with their much more hydrophobic structures, provide a nearly impermeable barrier. In some cells, very rapid transport of water is required. As described in the next section, such rapid transport of water is achieved by specific membrane-spanning channels called **aquaporins.**

Facilitated Transport: Accelerated Diffusion

For many substances, the slow transport provided by nonmediated diffusion is insufficient for the functional and metabolic needs of cells, and the means must be found to increase transport rates. For example, exchange of Cl^- and HCO_3^- is essential to erythrocyte function. If we examine the permeability of erythrocyte membranes to chloride or bicarbonate ions, we find permeability coefficients of about 10^{-4} cm/s.

● **CONCEPT** Facilitated transport, via pores, permeases, or carriers, can increase the rate of diffusion across a membrane by many orders of magnitude.

This value is about 10 million times greater than the permeability coefficient for ions in pure lipid bilayers like the artificial phosphatidylserine membrane listed in Table 10.6. Clearly, some special mechanism is required to account for this difference. Three general types of **facilitated transport,** or **facilitated diffusion,** are known to occur: transport through pores or channels formed by transmembrane proteins (**FIGURE 10.22(a)**); transport by carrier molecules (Figure 10.22(b)); and transport by **permeases** (Figure 10.22(c)).

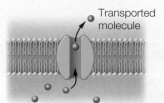

(a) Protein pores

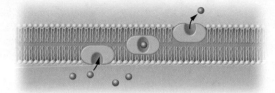

(b) Carrier molecules

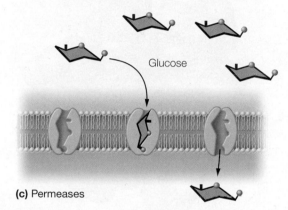

Glucose

(c) Permeases

▲ **FIGURE 10.22** The three major mediators of facilitated transport.

Carriers

Ionophores increase the permeability of a cell membrane to ions. Thus, many bacteria secrete ionophores that act as chemical warfare agents, or *antibiotics*, to kill other bacteria with which they compete for nutrients. The ionophores kill the neighboring bacteria because unregulated ion transport destroys the electrochemical gradients that store free energy needed to drive vital processes in living cells. Some ionophores create ion-conducting pores in the membrane, whereas others carry ions from one side of the bilayer to the other (see Figure 10.22(b)). For example, *valinomycin*, produced by a *Streptomyces*, is an ion carrier and has the structure shown in **FIGURE 10.23**. When complexed with K^+, it is a cyclic polypeptide-like molecule, involving three repeats of the sequence (D-valine)–(L-lactate)–(L-valine)–(D-hydroxyisovalerate). Its folded conformation presents an outside surface rich in hydrophobic $—CH_3$ groups and an interior cluster of polar nitrogen and oxygen atoms that is well suited to chelating cations. The dimensions of the interior cavity nicely accommodate a K^+ ion but do not match other cations as well. This structure is exactly what is needed for a cation carrier: The outer surface is hydrophobic, making the molecule soluble in the lipid bilayer, whereas the inside mimics in some ways the hydration shell that the cation would have in aqueous solution. A number of other ion-carrier antibiotics have the same kind of structure. These molecules are either cyclic or linear chains that can fold into cage-like

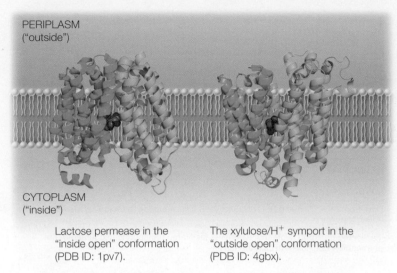

Lactose permease in the "inside open" conformation (PDB ID: 1pv7).

The xylulose/H$^+$ symport in the "outside open" conformation (PDB ID: 4gbx).

▲ **FIGURE 10.24 A model for permease function.** Cartoon renderings of two transporters from *E. coli* with homology to the GLUT family of glucose transporters are shown. The switch from one conformation to the other involves the movements of two bundles of helices. One helical bundle is shown in cyan, with the other in green. The bound substrate is shown in red spheres.

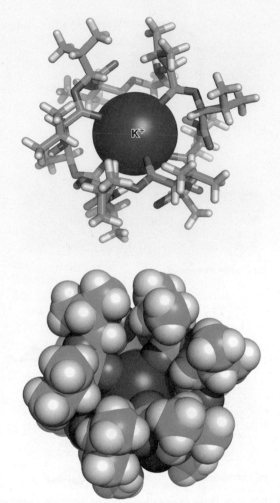

▲ **FIGURE 10.23 Valinomycin, an antibiotic that acts as an ion carrier.** Top: a sticks rendering of valinomycin with carbon atoms colored green, H atoms white, N atoms blue, O atoms red, and the K$^+$ ion shown as a purple sphere. Bottom: a space-filling model of valinomycin. The outside of this roughly spherical cyclic polypeptide is hydrophobic. The central cavity surrounded by oxygens complexes a K$^+$ ion.

structures. Their relative affinities for different ions vary greatly. For example, valinomycin has nearly a 20,000-fold preference for K$^+$ over Na$^+$, whereas the antibiotic *monensin* prefers Na$^+$ by only 10-fold.

● **CONCEPT** Facilitated transport can be passive or active. Passive transport can only achieve net transport of a substrate along its concentration gradient. Active transport can move a substrate against its concentration gradient, but this requires an input of free energy.

A molecule like valinomycin can diffuse to one surface of a membrane, pick up an ion, and then diffuse to the other surface and release it. There is no *directed* flow, but the carrier in effect increases the solubility of the ion in the membrane. We could say that it increases the value of the factor P in Equation 10.7. For such ion carriers, the net transport of ions will be in the direction that equalizes concentration of the ion on both sides of the membrane. Such facilitated transport is also called **passive transport** to distinguish it from **active transport**—a strictly directional process, which requires an input of free energy.

Permeases

Membrane-spanning proteins that recognize specific molecules for transport are called permeases or **transporters.** Like the carriers described earlier, some permeases act in a passive fashion—transporting their substrates in both directions, with a net flow toward the side of the membrane with the lower substrate concentration. The glucose transporter in erythrocytes (**GLUT1**) is thought to operate in this way. The small energy demands of an erythrocyte are met by glucose, which is readily available in the surrounding blood plasma. However, as Table 10.6 shows, the nonmediated transport of glucose through artificial phospholipid membranes is exceedingly slow: $P = 4 \times 10^{-10}$ cm/s. GLUT1, a 492-residue protein with 12 membrane-spanning helices, increases the glucose diffusion rate 50,000-fold. GLUT1 is quite discriminating; for example, D-glucose is transported orders of magnitude more rapidly than L-glucose.

The key feature of permease function is shown schematically in Figure 10.22(c) and **FIGURE 10.24**, namely, the permease shifts between two conformations: one open only to the "outside," and the other open only to the "inside." Thus, the permease never forms a pore that allows unrestricted flow of the transported substrate. Transport requires both binding of the substrate and conformational change by the permease.

Some permeases couple the transport of more than one substrate or ion. When transport of the two molecules, or ions, is in the same direction, the transporter is referred to as a **symport;** when the substrates move in opposite directions, the transporter is called an **antiport** (**FIGURE 10.25**). This cotransport strategy allows the thermodynamically unfavorable transport of some substrate *against* its concentration gradient, when coupled to the favorable transport of the cosubstrate. We will return to this topic later, after we present the general features of active transport.

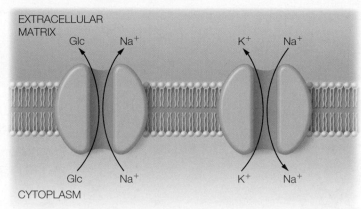

A model glucose/Na$^+$ symport   A model Na$^+$/K$^+$ antiport

▲ **FIGURE 10.25 Models for symports and antiports.** A symport transports two substances across the bilayer in the same direction. An antiport transports two substances in opposing directions.

Pore-Facilitated Transport

Many pathogenic bacteria synthesize and secrete protein toxins that act as ionophores by creating pores in the plasma membranes of cells of the receiving organism. An example, shown in **FIGURE 10.26** (and Figure 10.13), is α-hemolysin from *Staphylococcus aureus*. This protein is made up of seven subunits, which associate to produce a membrane-spanning ion channel. Likewise, the toxin *gramicidin A* produced by the bacterium

Bacillus brevis acts as a cation-specific ion pore, allowing a breakdown in the unequal ratio of [K$^+$] and [Na$^+$] normally maintained between the inside and outside of living cells. Gramicidin A is a 15-residue polypeptide, containing alternating L- and D-amino acids (**FIGURE 10.27**). Gramicidin adopts an open helical conformation when dissolved in the membrane, but one molecule of the antibiotic is long enough to traverse only half the thickness of the membrane. An open pore forms only when two gramicidin molecules line up to form a head-to-head dimer (Figure 10.27). Potassium ions (and, to a lesser extent, sodium ions) can then pass through the channel.[†]

In addition to channels that do damage to cells, many channels facilitate transport processes that are critical to cell survival. Many types of eukaryotic cells must move large amounts of water rapidly across their membranes as part of their physiological function. These include erythrocytes (which experience a wide range of solution osmolarity as they transit through lungs, capillaries, and kidneys), secretory cells in salivary glands, and epithelial cells in the kidney. Although water can cross membranes, it does so relatively slowly; thus, the inherent permeability of membranes toward water (see Table 10.6) is not sufficient to support the rapid transport observed in many cell types.

Such rapid transport is achieved by water-specific channels called **aquaporins.** The aquaporins function as tetramers of identical monomers. Each monomer contains six membrane-spanning helices and two shorter helices that contain a conserved N-terminal Asn-Pro-Ala (NPA) motif.

[†]The helix formed by gramicidin is very different from the α-helix. Recall from Figure 6.8b that there is no pore in an α-helix- that space is filled by backbone atoms.

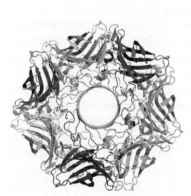

(a) Looking down the sevenfold axis.

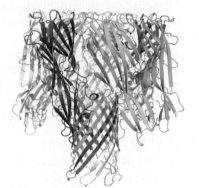

(b) Perpendicular to the sevenfold axis.

(c) One protomer extracted from the heptamer structure. The heptamer is 100 Å in diameter and 100 Å in length, as measured along the sevenfold axis. The β-barrel stem, which penetrates the membrane, is about 60 Å long.

▲ **FIGURE 10.26 The channel-forming hemolysin from *Staphylococcus aureus*.** Ribbon drawings of the α-hemolysin heptamer. PDB ID: 7ahl.

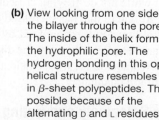

(a) Side view, from within the plane of the bilayer, of two molecules of gramicidin A (one shown with carbon atoms in cyan, the other in green) forming a pore through the membrane with their hydrophobic side chains in contact with the lipid. The N-termini are inside and the C-termini are outside the bilayer core.

(b) View looking from one side of the bilayer through the pore. The inside of the helix forms the hydrophilic pore. The hydrogen bonding in this open helical structure resembles that in β-sheet polypeptides. This is possible because of the alternating D and L residues in the gramicidin sequence.

▲ **FIGURE 10.27 Gramicidin A, an antibiotic that acts as an ion pore.**

EXTRACELLULAR MATRIX

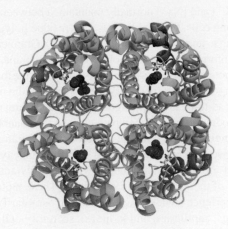

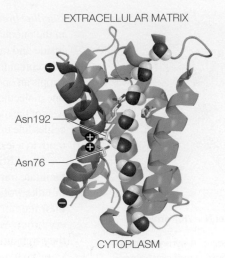

CYTOPLASM

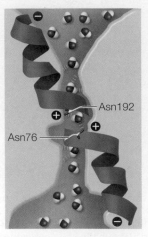

(a) Cartoon rendering of human aquaporin-5 tetramer (PDB ID: 3d9s), looking along the four water channels. Water molecules are shown as red spheres. The two short helices containing the NPA sequence are shown in blue. Side chains for Asn76, Asn192, His180, and Arg195 are highlighted in yellow.

(b) A cutaway view of the water channel in one of the monomers. The narrowest part of the channel is where the two short helices meet. Note the location of Asn76 and Asn192 at this restriction. The two helical macrodipoles and Arg195 provide an electrostatic barrier to the H_3O^+ passage.

(c) Schematic view of the aquaporin channel, showing the electrostatic repulsion of H_3O^+ and the reorientation of the water molecules as they pass through the central restriction.

▲ **FIGURE 10.28** **The aquaporin water channel.** (c) Based on Federation of the European Biochemical Societies, FEBS Letters 555:72–78, P. Agre and D. Kozono, Water channels: Molecular mechanisms for human disease.

Crystal structures of the aquaporins reveal that selectivity for water is achieved by three means (**FIGURE 10.28**). First, the channel is quite narrow (~2.8 Å) and excludes anything larger than a water molecule (including hydrated ions). Second, H_3O^+ is excluded by electrostatic repulsion. A conserved Arg places a (+) charge at this constriction, effectively repelling any cations. In addition, the two shorter helices are oriented with their N-termini pointing into the narrowest part of the channel; thus, the positive ends of the helical macrodipoles provide additional repulsion to H_3O^+ Third, water molecules can pass through the channel only in single file. As they do so, main-chain carbonyl groups as well as the conserved Asn72 and Asn192 side chains in the conserved NPA motifs form H-bonds with the individual water molecules, thereby reorienting the water molecules and disrupting H-bonding between the water molecules in the channel. This is a critical feature of the transport mechanism because it prevents protons from traversing the membrane via an H-bonding network of water molecules. As will become clear later in this chapter, and in subsequent chapters, many membranes must maintain ion gradients to carry out critical processes (e.g., bacterial flagellar motion, ATP synthesis, firing of neurons, etc.). The aquaporins elegantly solve the problem of maintaining the osmotic balance in a cell while not destroying critical ion gradients.

Research that has increased our understanding of transport across membranes was recognized by the 2003 Nobel Prize in Chemistry, awarded to Peter Agre for the discovery of aquaporins and to Roderick MacKinnon for his work elucidating the structure and function of potassium ion channels.

Ion Selectivity and Gating

Two outstanding features of ion channels that we explore here are their selectivity for a particular ion and the control of ion transport through the channel. Many protein ion channels and transporters display high selectivity for the biologically relevant ions K^+, Na^+, Ca^{2+}, or Cl^-. These include the bacterial K^+ channel, KcsA (shown in **FIGURE 10.29(a–c)**), and the Na^+/Leu transporter, LeuT. It is instructive to compare and contrast the binding of K^+ and Na^+ by these two proteins to see how nature has achieved such exquisite selectivity. As shown in **FIGURE 10.30**, the Na^+ and K^+ bound in the **selectivity filter** are completely desolvated and chelated by multiple oxygen atoms. In LeuT, the two Na^+ binding sites provide either five or six chelating oxygen atoms (from main-chain carbonyl or side-chain carboxylate, hydroxyl, or amide groups) with a mean Na^+ — O distance of 2.3 Å. In KcsA, each K^+ is bound by eight oxygen atoms (from main-chain carbonyl or side-chain —OH groups) at a mean K^+ — O distance of 2.8 Å. These chelating groups replace solvating interactions the ions would make with water molecules; thus, there is no enthalpic penalty for desolvating the ions as they transit the channel. Based on the structures shown in Figure 10.30, the major determinant of discrimination between Na^+ and K^+ appears to be the geometry of the chelating groups in the selectivity filter.

● **CONCEPT** Ion selectivity is achieved by optimal geometry of chelating groups in ion channels.

It is critical that the activity of any ion channel is regulated to maintain proper cell function; thus, there should be "open" (i.e., ion-conducting) and "closed" conformations for the channel. The switching between conductive and nonconductive conformations is called **gating.** Figure 10.29 compares open and closed conformations for the K^+ channel. Note that the K^+ channel pore structure is highly conserved and is defined by two membrane-spanning helices and the shorter selectivity filter described above. In the closed conformation, one transmembrane helix from each subunit extends into the channel cavity on the cytoplasmic side of the channel. The convergence of these four helices occludes the channel and prevents K^+ transport

Closed conformation

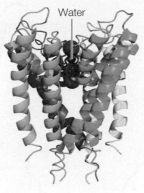

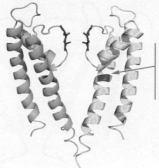

Bending of the helices at the hinge Gly residues opens the channel gate.

Selectivity filter

Water

(a) A view looking along the pore axis from the extracellular face. The four identical transmembrane subunits are shown in different colors. The K$^+$ ions (purple spheres) are bound in the "selectivity filter" (highlighted in red) by amide carbonyl groups from each protein subunit.

(b) A view from within the plane of the bilayer. Here, the periplasmic side is on top and the cytoplasmic side is on the bottom. Three K$^+$ ions (purple spheres) and one water molecule (smaller red sphere) are shown in the selectivity filter (see text and Figure 10.30(b)).

(c) Here, two of the four subunits have been removed to show better the convergence of the pore helices that close the channel. The selectivity filter and the "hinge Gly" are highlighted in red.

Open conformation

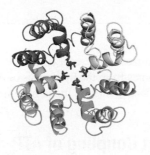

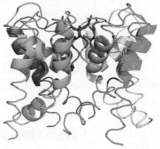

(d)

(e)

(f)

▲ **FIGURE 10.29 The structure of the potassium channel pore.** The transmembrane pore region of the potassium channel KcsA (PDB ID: 1bl8) from the bacterium *Streptomyces lividans* is shown in a "closed" conformation in panels (a–c). The pore region of the potassium channel MthK

(PDB ID: 1lnq) from the bacterium *M. thermautotrophicus* is shown in an "open" conformation in panels (d–f). The views in panels (d–f) are the same as those in (a–c), except that K$^+$ ions were not crystallized in the selectivity filter.

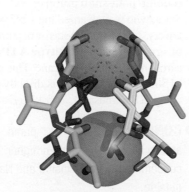

(a) Sodium. Two Na$^+$ binding sites make up the selectivity filter in the transmembrane region of LeuT (PDB ID: 2a65).

(b) Potassium. Two K$^+$ are bound in the filter of the KcsA K$^+$ channel (PDB ID: 1k4c).

▲ **FIGURE 10.30 Selective binding of Na$^+$ and K$^+$ in ion channels.**

(Figure 10.29(b,c)). In response to some gating stimulus (e.g., a change in pH, or membrane potential, or binding of some ligand to the extracellular portion of the channel protein, etc.), the conformations of these helices change. Near the middle of each helix is a so-called hinge Gly (highlighted in red in Figure 10.29(c,f)). Bending of the transmembrane helices around this Gly moves the C-terminal ends of the helices apart, thereby opening the channel (Figure 10.29d-f).

To get a better understanding of the mechanism of channel gating, let us consider the current model for the opening and closing of voltage-gated K$^+$ channels. The voltage-gated channels include six helical transmembrane segments labeled S1–S6 (**FIGURE 10.31**), where the K$^+$ pore is formed by the S5 and S6 helices along with the intervening selectivity filter sequence. This pore is structurally similar to the channels shown in Figure 10.29. Sequences S1–S4 make up the voltage-sensing domain, where S4 contains a sequence in which every third residue is lysine or arginine, separated by two hydrophobic residues. The position of the S4 helix in the channel is thought to change as a function of the membrane potential ($\Delta\psi$). The cytoplasmic side of the membrane is more negatively charged; thus, in the resting state

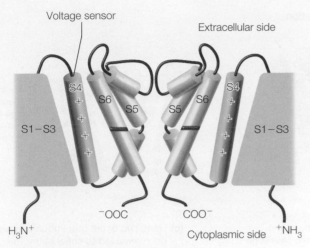

Voltage sensor

Extracellular side

(a) Closed conformation. The helices are closer to the cytosolic side of the bilayer, and this position seals the channel.

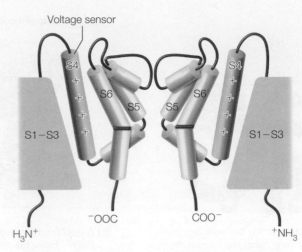

Voltage sensor

(b) Open conformation. The channel opens when the charged helices move toward the extracellular side of the bilayer.

▲ **FIGURE 10.31 A model for voltage-gating in the K$^+$ channel.** The channel portion of the voltage-gated K$^+$ channel is structurally homologous to the KcsA channel (compare to Figure 10.29). The Arg- and Lys-rich S4 helices are highlighted in blue. The depth of these helices in the bilayer changes as a function of the membrane potential. The red stripe in helix S6 indicates the location of the "hinge Gly" residue.

of the membrane, with the channel closed, the S4 helix is positioned closer to this side of the membrane. When the potential across the membrane changes, as in nerve signal conduction, the cytosolic side of the membrane becomes less negatively charged and the S4 helix moves toward the other side of the membrane. As it does so, it pulls S5 along, which in turn allows S6 to bend and open the channel. In this model, S6 corresponds to the helix with the hinge Gly (Figure 10.29). The channel returns to the closed conformation upon restoration of the resting-state membrane potential by active ion transporters.

In conclusion, we must emphasize that even though facilitated transport is sometimes very fast and very selective, it is still only a special form of diffusion. Transporters effectively increase the solubility of the substance in the membrane. The equilibrium state for a system exhibiting facilitated transport is the same as that for nonmediated transport—the substance will be transported down its concentration gradient until the concentrations on both sides of the membrane are equal.

Active Transport: Transport Against a Concentration Gradient

Many cells or cellular compartments need to transport substances *against* concentration gradients, even very unfavorable ones. To take an extreme example, under some circumstances a [Ca$^+$] ratio of 30,000:1 must be established across membranes of the sarcoplasmic reticulum in muscle fibers (see Chapter 7). According to Equation 10.1, ΔG for Ca^{2+} transport against this concentration gradient is unfavorable by $^+$26.6 kJ/mol—a formidable barrier. Nevertheless, this ratio of Ca^{2+} ions is built up and maintained in living cells. Such transport against a concentration gradient is called **active transport.** Clearly, to

● **CONCEPT** In active transport, substances are moved across a membrane against a concentration gradient. Direct or indirect coupling of transport to ATP hydrolysis provides the required free energy.

pump ions against a gradient requires a free energy source of some kind. In most cases, this energy comes from the hydrolysis of ATP. It is estimated that most cells spend 20–40% of total metabolic energy just on active transport.

10.5 Ion Pumps: Direct Coupling of ATP Hydrolysis to Transport

The best known physiological example of active transport is the maintenance of sodium and potassium gradients across the plasma membranes of cells. The fluid surrounding cells in most animals is about 145 mM in Na$^+$ and 4 mM in K$^+$ Yet, animal cells maintain a Na$^+$ concentration of about 12 mM and a K$^+$ concentration of about 155 mM in their cytosol.

These concentration differences are maintained by the action of the **sodium–potassium pump** or **Na$^+$ — K$^+$ ATPase,** first described by Jens Skou, who was awarded the 1997 Nobel Prize in Chemistry for this discovery. The sodium–potassium pump is only one member of a large class of structurally related **P-type ATPases** that function in active transport across the plasma membrane. This molecular machine consists of a large 113-kDa α subunit that is directly involved in ATP hydrolysis and the coupled ion transport, a 55-kDa β subunit that acts as a chaperone and is

● **CONCEPT** The Na$^+$ — K$^+$ pump acts in all cells to maintain higher concentrations of K$^+$ inside and Na$^+$ outside.

required to target the α subunit to the plasma membrane, and a much smaller regulatory subunit, γ. The α subunit traverses the membrane 10 times, forming a multihelix channel, and has three cytoplasmic domains (**FIGURE 10.32**). These include an ATP binding domain, a phosphorylation domain, and an actuator domain. During the ion pumping cycle, the γ-phosphoryl group of ATP is transferred to the side chain of aspartic acid 376 in the phosphorylation domain (**FIGURE 10.33**).

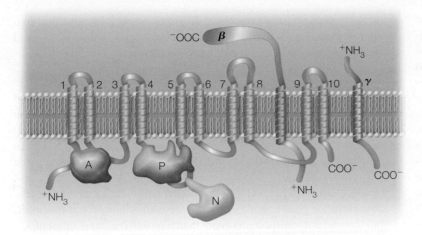

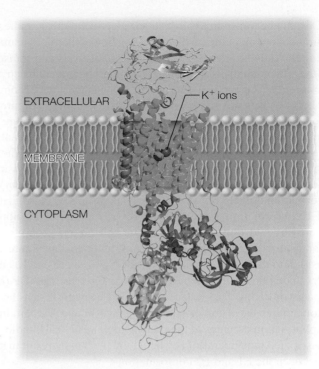

(a) Schematic representation of the domain and subunit structure of the Na^+—K^+ ATPase. The α subunit is shown with the ten transmembrane helices in green, the ATP-binding domain (N) in cyan, the phosphorylation domain (P) in orange, and the actuator domain (A) in purple. The β subunit is beige and the small regulatory protein (γ or FXYD) is shown in gray.

(b) Crystal structure of the Na^+—K^+ ATPase with K^+ bound. The domains and subunits are colored as in (a) with K^+ ions shown as purple spheres. The actuator domain translates conformational changes in the cytoplasmic domains to the transmembrane ion channel (see scheme in Figure 10.33). The two K^+ are bound in the transmembrane helical bundle and will be transported across the membrane. A phosphate ion analog, MgF_4, is shown in red spheres at the site where reversible phosphorylation of the ATPase occurs (Asp 376, which is shown in blue spheres). PDB ID: 2zxe.

▲ **FIGURE 10.32** The structure of the Na^+—K^+ ATPase.

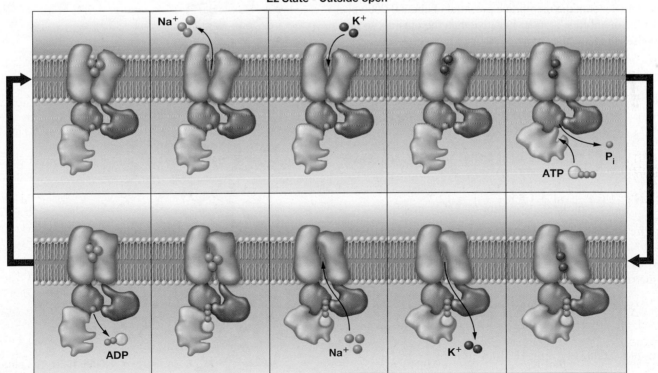

▲ **FIGURE 10.33** A schematic diagram of the functional cycle of the Na^+—K^+ pump. The α subunit is believed to have two major conformational states, one open only to the outside ("E2"), the other open only to the inside ("E1"). See text for a description of this cycle.

Despite the transport against strong electrochemical gradients, the sodium–potassium pump involves no violation of thermodynamic principles. The only requirement is that ATP hydrolysis and transport be *coupled*. This coupling is apparently accomplished in a multistep process reminiscent of the cycle of conformational changes that accompanies ATP hydrolysis in the myosin headpiece (see Chapter 7; Figure 7.48). A model for the entire process is shown schematically in Figure 10.33. It is proposed that the pump can adopt two major conformational states, one open only to cytosol ("E1"), the other open only to the cell's surroundings ("E2"). Several minor conformation changes are postulated within each major conformational state. Completion of one conformational cycle requires hydrolysis of one ATP. Transition to the E1 conformation, which allows K^+ release and Na^+ uptake, is triggered by hydrolytic cleavage of the phosphoryl group from Asp 376 and subsequent binding of ATP (top right panel of Figure 10.33). Transition to the E2 state, which permits Na^+ release and K^+ uptake, occurs upon phosphorylation of Asp 376 and release of ADP (bottom left panel of Figure 10.33). Current evidence suggests that the actuator domain translates conformational changes in the ATP-binding and phosphorylation domains to the transmembrane helical bundle, where the associated movements of key helices in the bundle result in switching between the E1 and E2 conformations.

Two K^+ ions are pumped into the cell, and three Na^+ ions are pumped out for every ATP hydrolyzed. Is this estimate reasonable from a thermodynamic point of view? To answer this question we calculate the free energy required to transport 3 moles of Na^+ outward from 12 mM to 145 mM, and 2 moles of K^+ inward from 4 mM to 155 mM at 37 °C. First, let us use Equation 10.3 to calculate the free energy required to transport 3 moles of Na^+ from within the cell to outside. We must take into account the membrane potential of about 0.060 volt. The inside of the membrane is more negative than the outside, so this potential opposes the flow. Per mole of Na^+ we have

$$\Delta G = RT \ln \frac{[C_{Na^+}]_{out}}{[C_{Na^+}]_{in}} + Z_{Na^+}F\Delta\psi_{in\to out}$$

$$\Delta G = \left(0.008314 \frac{kJ}{mol\ K}\right)(310\ K)\left(\ln \frac{(0.145)}{(0.012)}\right)$$

$$+ (+1)\left(96.48 \frac{kJ}{mol\ V}\right)(+0.060\ V)$$

$$\Delta G = \left(6.4 \frac{kJ}{mol}\right) + \left(5.8 \frac{kJ}{mol}\right) = 12.2 \frac{kJ}{mol}$$

Thus, for the transport of 3 moles of Na^+, we calculate $\Delta G = 3\ mol\ Na^+ \times 12.2\ kJ/(mol\ Na^+) = +36.6\ kJ$.

When K^+ is transported inward, the membrane potential is working in favor of the flow. Per mole of K^+ we have

$$\Delta G = RT \ln \frac{[C_{K^+}]_{in}}{[C_{K^+}]_{out}} + Z_{K^+}F\Delta\psi_{out\to in}$$

$$\Delta G = \left(0.008314 \frac{kJ}{mol\ K}\right)(310\ K)\left(\ln \frac{(0.155)}{(0.004)}\right)$$

$$+ (+1)\left(96.48 \frac{kJ}{mol\ V}\right)(-0.060\ V)$$

$$\Delta G = \left(9.4 \frac{kJ}{mol}\right) + \left(-5.8 \frac{kJ}{mol}\right) = 3.6 \frac{kJ}{mol}$$

or, for the transport of 2 moles of K^+, $\Delta G = +7.2$ kJ. The total free energy requirement for the outward transport of 3 moles of Na^+ and the inward transport of 2 moles of K^+ is then

$$\Delta G_{total} = 36.6\ kJ + 7.2\ kJ = +43.8\ kJ$$

At first glance, it would appear that the hydrolysis of 1 mole of ATP would not provide the necessary free energy, because $\Delta G^{\circ\prime}$, the standard-state free energy change for ATP hydrolysis under physiological conditions, is about -32 kJ/mol. In most cells, however, ATP is in much higher concentration than ADP, so the actual free energy change per mole is typically -45 to -50 kJ/mol (see Problem 5 at the end of this chapter). Thus, ATP hydrolysis is sufficient to maintain these concentration gradients under the observed stoichiometry of transport, but it could not transport any more than 3 mol of Na^+ and 2 mol of K^+ per mol ATP hydrolyzed.

10.6 Ion Transporters and Disease

The E2 state has an especially high affinity for cardiotonic steroids, like **digitoxin** (digitalis) and **ouabain**. These agents inhibit the Na^+–K^+ pump by locking it in the E2 conformation (**FIGURE 10.34**). Such inhibition has major effects on muscles, especially in the heart. The resulting buildup of Na^+ in cells leads to measures to reduce its concentration, including

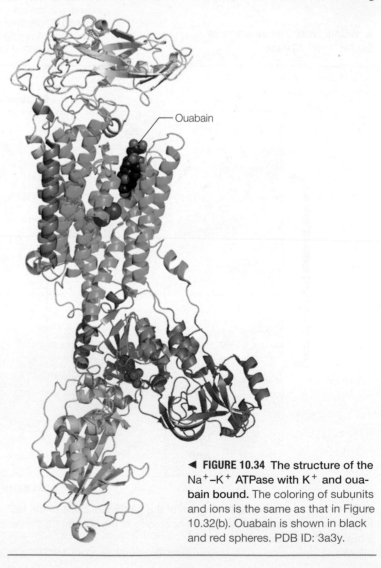
Ouabain

◀ **FIGURE 10.34 The structure of the Na^+–K^+ ATPase with K^+ and ouabain bound.** The coloring of subunits and ions is the same as that in Figure 10.32(b). Ouabain is shown in black and red spheres. PDB ID: 3a3y.

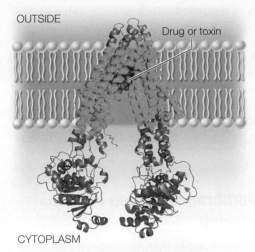

OUTSIDE

Drug or toxin

ATP

ADP + P$_i$

Bound ATP

CYTOPLASM

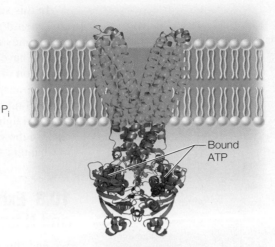

P-glycoprotein from mouse is shown in the conformation open to the cytoplasm (PDB ID: 4m2t). The transmembrane region is colored green. This is the conformation which binds drugs (magenta and gold spheres) or toxins present in the cytoplasm. ATP binding to the cytoplasmic domains results in conformational change that expels the drug/toxin.

The conformation open to the outside is illustrated by the homologus ABC transporter, MsbA from *S. typhimurium*, bound to an ATP analog, which is shown as red spheres (PDB ID: 3b60). Hydrolysis of the bound ATP and release of ADP and P$_i$ results in the conformational switch to the "inside open" state.

◀ **FIGURE 10.35** Proposed mechanism of action for ABC transporters.

● **CONNECTION** Cardiotonic steroids inhibit the Na$^+$–K$^+$ pump, resulting in increased Ca^{2+} ion concentration in heart muscle, which, in turn, leads to stronger contractions of the heart muscle. In small doses, digitoxin improves cardiac function.

As described later in this chapter, the Na$^+$–K$^+$ ATPase plays a critical role in maintaining the resting state of neurons. Thus, mutations that impair proper function of the Na$^+$–K$^+$ ATPase are associated with neurological disorders such as early-onset Parkinson's disease.

● **CONNECTION** Mutations that impair ion transport by the Na$^+$—K$^+$ ATPase are associated with early-onset Parkinson's disease.

● **CONNECTION** Proton pump inhibitors reduce the acidity of the stomach and are widely prescribed for treatment of gastroesophageal reflux disease (GERD) and peptic ulcer disease.

A large class of active transporters is the **ATP-binding cassette transporters,** or **ABC transporters.** These include transporters that are responsible for multiple drug resistance in bacteria (and the human homolog **P-glycoprotein**) and the **cystic fibrosis transmembrane regulator (CFTR),** which is a chloride ion channel. Many mutations

a Ca^{2+}–Na$^+$ exchange process catalyzed by another ion pump. The resulting increase in Ca^{2+} in the sarcoplasmic reticulum of heart muscle cells leads to much stronger contractions (see Chapter 7). This is why substances like digitoxin and ouabain, in low doses, act as heart stimulants.

In some cases, it is desirable to interrupt the activity of ion transporters through the action of therapeutic drugs. For example, so-called **proton pump inhibitors** or **PPIs,** such as omeprazole, inhibit the gastric H$^+$–K$^+$ ATPase and thereby reduce the transport of H$^+$ into the stomach. PPIs are widely prescribed to treat peptic ulcer disease and gastroesophageal reflux disease (GERD).

in CFTR which disrupt Cl$^-$ transport (primarily in the lungs) are associated with cystic fibrosis; however, the deletion of Phe 508 accounts for ~70% of cases worldwide. P-glycoprotein binds to small lipophilic molecules and transports them out of cells (**FIGURE 10.35**). This is one mechanism whereby toxins are removed from cells; thus, the function of P-glycoprotein is beneficial. However, P-glycoprotein is known to bind to small molecule drugs—most of which are lipophilic, to facilitate their transport across membranes into cells—and remove them from their intended sites of action. This activity of P-glycoprotein reduces the effectiveness of many anticancer drugs (e.g., taxol) because it removes these drugs from malignant cells, particularly those types of cancer cells that express high levels of P-glycoprotein.

● **CONNECTION** Multiple-drug resistance is linked to the activity of certain ABC transporters (e.g., P-glycoprotein). Mutation of the ABC transporter CFTR is associated with cystic fibrosis.

10.7 Cotransport Systems

There are other kinds of active transport that do not depend directly on ATP as an energy source but employ ATP hydrolysis in an indirect way. The ATP-driven ion pumps described earlier can generate large ion concentration gradients across membranes. These ion gradients are far from equilibrium and therefore represent a potential source of free energy (as described by Equation 10.3). The **sodium–glucose cotransport system** of the small intestine (**FIGURE 10.36**) is an example of how an ion gradient is used in driving transport of glucose from a region of low [glucose] (the intestinal lumen) to one where [glucose] is higher (inside epithelial cells of the intestine wall). The *unfavorable* transport of each glucose molecule from within the intestinal lumen into the epithelial cells is accompanied by the simultaneous *favorable* transport of one Na$^+$ ion in the same direction.

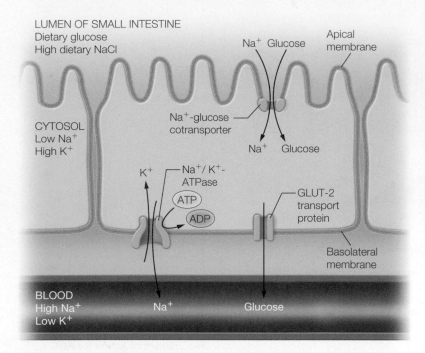

LUMEN OF SMALL INTESTINE
Dietary glucose
High dietary NaCl

Apical membrane

Na⁺ Glucose

Na⁺-glucose cotransporter

CYTOSOL
Low Na⁺
High K⁺

Na⁺ Glucose

K⁺ — Na⁺/K⁺-ATPase

ATP

ADP

GLUT-2 transport protein

Basolateral membrane

BLOOD
High Na⁺
Low K⁺

Na⁺ Glucose

▲ **FIGURE 10.36 A schematic model for the sodium–glucose cotransport (symport) system.** As in the case of the sodium–potassium pump, the sodium–glucose cotransport channel is presumed to have two possible states—one open only to the outside, the other open only to the inside of the cell. See Figure 10.24 for a model of the glucose transporter. The sodium gradient from inside to outside provides the driving force for the unfavorable transport of glucose. That gradient must be maintained by the sodium–potassium pump.

● **CONCEPT** In cotransport, the unfavorable movement of a substance through the membrane is coupled to the favorable transport of another substance.

Because the Na⁺ gradient is maintained by the ATP-driven Na⁺−K⁺ pump of these cells, glucose can be continuously transported against the unfavorable gradient in glucose concentration.

A large number of such cotransport systems are known, and many of them are utilized to move nutrients into cells. A few examples are listed in **TABLE 10.7**. Many use the Na⁺ gradient as a driving force, but some, like the *lactose permease system* in *E. coli*, depend on an H⁺ gradient. As we shall see in later chapters, generation of H⁺ gradients is a central step in energy production by most cells.

TABLE 10.7 Some cotransport systems		
Molecule Transported	Ion Gradient Used	Organism or Tissue
Glucose	Na⁺	Intestine, kidney of many animals
Amino acids	Na⁺	Mouse tumor cells
Glycine	Na⁺	Pigeon erythrocytes
Alanine	Na⁺	Mouse intestine
Lactose	H⁺	*E. coli*

In this section, we have described only a few examples of specific membrane transport. This phenomenon is encountered many more times in later chapters on metabolism. The modes of transport we have described for small molecules or ions typically involve the action of a single protein or carrier. Larger volumes of solution can be transported across various membranes in bulk as a result of remodeling membranes. This is achieved by the actions of proteins that bind tightly to the bilayer and deform it, forming pits or sacs that enclose the volume to be transported. These processes are discussed in Chapters 16 and 25.

10.8 Excitable Membranes, Action Potentials, and Neurotransmission

We end this chapter with an example that demonstrates the enormous variety of properties that membranes can exhibit, through their ability to regulate ion transport. The conduction of neural impulses in animals is a remarkable process, but it depends on very simple physical principles.

Neurons, the cells responsible for conduction of electrical impulses and thus nervous system communication, have specialized thin cell projections called dendrites and axons that act as the "wires" of the nervous system (**FIGURE 10.37**). Neurons are truly remarkable cells that must meet unusual requirements. They must be able to conduct impulses over relatively long distances without significant signal loss (e.g., from the spinal cord to the tip of the toe), and they must conduct impulses on millisecond time scales to control rapid and coordinated thought and behavior. Nerve conduction is accomplished not by electron flow, as in wires, but by waves in the membrane electrical potential on the surface of the membrane.

● **CONCEPT** Neurons conduct electrical impulses by membrane potential changes in regions of the plasma membrane cell.

The Resting Potential

To understand changes in membrane potentials, we must first understand the source and nature of the cell's resting membrane potential. We begin with a simplified model, which draws on our earlier discussion of the electrochemical potential difference across a semipermeable membrane. Suppose we have an ion (M^Z) of charge Z that is present outside the membrane at concentration $[M^Z]_{out}$ and inside at concentration $[M^Z]_{in}$.

If the system is at equilibrium, ΔG for transport will be zero. Then, from rearranging Equation 10.3 when $\Delta G = 0$, we find

$$\frac{RT}{ZF} \ln \frac{[M^Z]_{out}}{[M^Z]_{in}} = \Delta\psi \qquad (10.8)$$

where $\Delta\psi$ is defined as $\psi_{out} - \psi_{in}$. Equation 10.8 is one form of the **Nernst equation.** For monovalent ions ($Z = \pm 1$) at 20 °C, this form of the Nernst equation reduces to

$$\Delta\psi = \pm 59 \log_{10} \frac{[M]_{out}}{[M]_{in}} \qquad (10.9)$$

when $\Delta\psi$ is expressed in millivolts (mV).

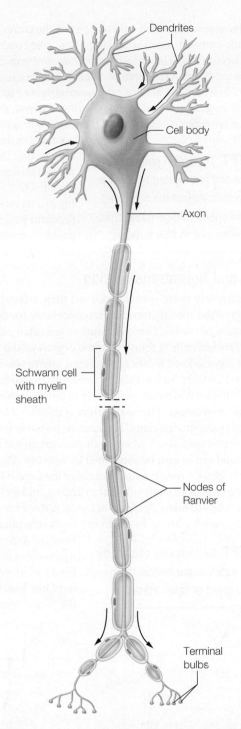

▲ **FIGURE 10.37 Structure of a typical mammalian motor neuron.** A motor neuron transmits nerve impulses to muscles. The cell body contains the nucleus and most of the cellular machinery. The dendrites receive signals from the axons of other neurons; the axon transmits signals via the synaptic termini, which communicate to the dendrites of other neurons or to muscle cells. Along the axon are Schwann cells, which envelop the axon in layers of an insulating myelin membrane. The Schwann cells are separated by nonmyelinated regions called the nodes of Ranvier.

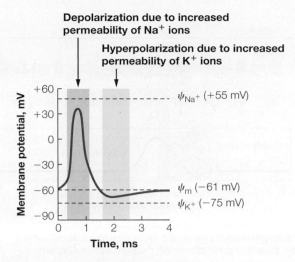

▲ **FIGURE 10.38 The action potential.** Changes in membrane potential accompanying the permeability changes for Na^+ and K^+ ions. As Na^+ rushes into the neuronal axon, the potential increases and becomes more positive ("depolarization"). As K^+ efflux increases, the potential decreases, undershooting the resting potential ("hyperpolarization"), before returning to the resting potential $\Delta\psi_m$.

$\Delta\psi = -59$ mV. Alternatively, if Cl^- ($Z = -1$) is unevenly distributed in this way, the potential will be $+59$ mV.

The major mechanisms creating ionic imbalance across cellular membranes are the specific ion pumps (e.g., the Na^+-K^+ ATPase) that continually act to concentrate certain ions on one side or the other. This imbalance gives rise to the resting potential across the membrane of a nerve axon. The value of $\Delta\psi$ in neurons is dependent on concentration gradients of several ions, each of which can pass through the membrane to some degree. If we used the Nernst equation to calculate the potential from the distribution of K^+ alone (375 mM inside and 20 mM outside the neuron), we would predict a value of $\Delta\psi_{K^+} = -75$ mV. On the other hand, using the Nernst equation with the Na^+ concentrations inside (50 mM) and outside (430 mM), we would find $\Delta\psi_{Na^+} = +55$ mV. When we measure the potential across a resting axonal membrane, we find a value of about -61 mV.

● **CONCEPT** The resting potential of a nerve fiber is determined primarily by the permeabilities of the membrane to K^+, which has a high permeability due to K^+ leak channels.

Because -61 mV lies much closer to -75 mV than to $+55$ mV, it means that the potential existing across the axon membrane is determined primarily by K^+, which has a 25-fold higher permeability than Na^+ due to the presence of K^+ leak channels. If the membrane were to become fully permeable to Na^+ ions, the resulting influx of Na^+ ions would shift the membrane potential to more positive values (approaching $\Delta\psi_{Na^+}$) That is just what happens when an action potential is transmitted along an axon (**FIGURE 10.38**).

The Action Potential

The action potential is a controlled and rapidly propagated change in membrane potential that is transmitted down the length of an axon. This process involves a unique set of voltage-gated potassium and voltage-gated sodium channels (Figure 10.31). Because of the biochemical

According to Equation 10.9, if we maintain an ion concentration difference across a membrane, an electrical potential, $\Delta\psi$, will be produced. For example, if an ion such as K^+ ($Z = +1$) is kept 10 times as concentrated inside as outside, the membrane will be polarized, with

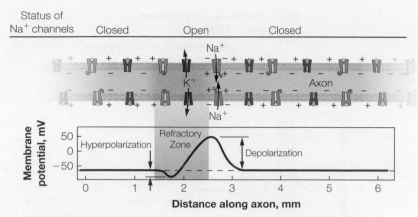

Status of Na$^+$ channels   Closed   Open   Closed

(a) At time = 0, action potential is occurring at the 2.5-mm position. The depolarization spreads down the axon, triggering development of the action potential downstream (to the right in this diagram).

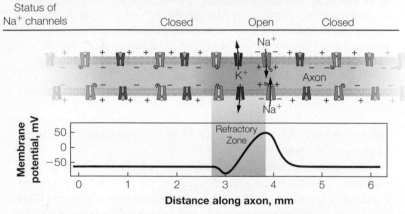

Status of Na$^+$ channels   Closed   Open   Closed

(b) At time = 1 ms, the action potential peak has moved to the 3.8-mm position. The potential can move in only one direction because after it has passed, the region behind the potential becomes refractory for a few milliseconds due to the rapid efflux of K$^+$.

▲ **FIGURE 10.39 Transmission of the action potential.** Shown are two "snapshots," taken 1 ms apart, of potential along the axon. Arrows show Na$^+$ influx and K$^+$ efflux.

properties of the ion channels, the action potential is self-propagating and travels in a single direction down an axon.

As described previously, voltage-gated channels are open or closed, depending on the membrane potential. For example, they are closed in the resting state. In response to some excitatory stimulus at the dendrites, ion channels will open up where the axon and cell body meet (Figure 10.37). This initiates a wave of depolarization/hyperpolarization that travels along the axon toward the terminal bulbs, which form **synapses,** or junctions, with other neuronal cells. The steep changes in membrane potential shown in Figure 10.38 would be a localized effect were it not for the sequential and highly regulated opening and closing of Na$^+$ and K$^+$ channels as shown in **FIGURE 10.39**. As Na$^+$ channels open and Na$^+$ ions rush in, they diffuse away from the region of stimulus and trigger the same round of depolarization in a "downstream" section of the axon. At the same time, "upstream" K$^+$ channels open and counteract the depolarization—this ensures that the wave of depolarization only travels in one direction: *down* the axon toward the terminal bulb (and the associated synapses). The action potential moves rapidly—typical values for propagation along an axon range from 1 to 100 meters per second.

We have described here only one part of the complex phenomenon of the transmission of neural impulses—the conduction along a single nerve fiber. The equally important problem of how these impulses are transmitted from one cell to another is discussed in Chapter 20, where the actions of *neurotransmitter* substances are described in greater detail. Briefly, at the synapse, the arrival of the action potential in the presynaptic neuron stimulates an action potential in the postsynaptic neuron. Thus, the signal that initiated the action potential (e.g., ligand binding to a membrane surface receptor) can be propagated over many neurons and long distances.

● **CONCEPT** The action potential is generated and propagated because a small depolarization of the nerve cell membrane opens voltage-gated channels, allowing ions to flow through.

Toxins and Neurotransmission

Many extremely toxic substances exert their effect by blocking the action of the specific ion channels necessary for development of the action potential. These substances are often called **neurotoxins.** *Tetrodotoxin* is found in some organs of the puffer fish. This fish is considered a delicacy in Japan, where special chefs are trained and certified for their ability to remove the toxin-containing organs. Tetrodotoxin binds specifically to the Na$^+$ channel, blocking all ion movement. The same effect is produced by *saxitoxin,* contained in the marine dinoflagellates responsible for "red tide." These microscopic algae, along with their toxin, are ingested by shellfish and can in turn be consumed by humans. These two toxins, which attack a fundamental process of the nervous system, are among the most poisonous substances known, and their accidental ingestion leads to many deaths every year. A third very poisonous substance, *veratridine,* is found in the seeds of a plant of the lily family, *Schoenocaulon officinale.* This toxin also binds to the Na$^+$ channels but blocks them in the "open" configuration.

● **CONCEPT** Neurotoxins can act by blocking gates in the axonal membrane in closed or open states.

Tetrodotoxin

Saxitoxin

Veratridine

Summary

- Many of the important properties of lipids stem from the fact that these substances are largely hydrophobic. Some are amphipathic, containing both hydrophobic and hydrophilic regions. (Section 10.1)

- Most naturally occurring fatty acids contain an even number of carbon atoms. When they are unsaturated, the double bonds are usually *cis*. Fatty acids are present in fats (triacylglycerols), where they serve for energy storage and insulation, and in membranes, where they are constituents of phospholipids, sphingolipids, glycosphingolipids, and glycoglycerolipids. (Sections 10.1 and 10.2)

- Membranes are bilayer structures containing proteins and lipids in a fluid mosaic. The two leaflets differ in protein and lipid composition. The fluidity of the membrane is increased as the number of *cis*-double bonds in the constituent fatty acids increases and/or the chain length of the fatty acid decreases. Peripheral proteins are confined to one face or the other, whereas integral proteins extend through the membrane, with hydrophobic α helices common in the transmembrane region. (Section 10.3)

- Transport through membranes may be achieved by nonmediated diffusion, or facilitated by pores, permeases, or carriers, or actively

driven by exergonic reactions. Only in the last case can transport against a concentration gradient occur. An example is the Na^+-K^+ ATPase, which maintains the ionic imbalance and membrane potential found between cells and their surroundings. Active transport may be indirect, as in cotransport. (Sections 10.4 and 10.5)

- Disruption of ion transport is associated with disease and/or disease treatment. (Section 10.6)

- Several cotransport systems exist. These depend on the action of the Na^+/K^+ ATPase. (Section 10.7)

- Neural conduction of impulses depends on a moving wave of depolarization (an action potential) in the membrane potential of a neural cell. This depolarization is produced by the flow of ions through voltage-gated channels in the membrane. (Section 10.8)

*Note: An Appendix titled *Guidelines for Evaluating the Thermodynamics of Ion Transport* can be downloaded by your professor from the Pearson Instructor Resource Center.

Problems

Enhanced by
Mastering Chemistry
for Biochemistry

Mastering Chemistry for Biochemistry provides select end-of-chapter problems and feedback-enriched tutorial problems, animations, and interactive figures to deepen your understanding of complex topics while practicing problem solving.

Answers to red problems are available in the Answer Appendix.

1. Give structures for the following, based on the data in Table 10.1.
   (a) *cis*-9-Dodecenoic acid
   (b) 18:1cΔ11
   (c) A saturated fatty acid that should melt below 30 °C

2. Given these molecular components—glycerol, fatty acid, phosphate, long-chain alcohol, and carbohydrate—answer the following:
   (a) Which two are present in both waxes and sphingomyelin?
   (b) Which two are present in both fats and phosphatidylcholine?
   (c) Which are present in a ganglioside but not in a fat?

3. The classic demonstration that cell plasma membranes are composed of bilayers depends on the following kinds of data:
   - The membrane lipids from 4.74×10^9 erythrocytes will form a monolayer of area 0.89 m^2 when spread on a water surface.
   - The surface of one erythrocyte is approximately 100 μm^2 in area.

   Show that these data can be accounted for only if the erythrocyte membrane is a bilayer.

4. The lipid portion of a typical bilayer is about 30 Å thick.
   (a) Calculate the minimum number of residues in an α-helix required to span this distance.
   (b) Calculate the minimum number of residues in a β-strand required to span this distance.

(c) Explain why α helices are most commonly observed in transmembrane protein sequences when the distance from one side of a membrane to the other can be spanned by significantly fewer amino acids in a β-strand conformation.

(d) The epidermal growth factor receptor has a single transmembrane helix. Find it in this partial sequence:

   . . . RGPKIPSIATGMVGALLLLVVALGIGILFMRRRH . . .

5. In the following situations, what is the free energy change if 1 mole of Na^+ is transported across a membrane from a region where the concentration is 1 μM to a region where it is 100 mM? (Assume $T = 37$ °C.)
   (a) In the absence of a membrane potential.
   (b) When the transport is opposed by a membrane potential of 70 mV.
   (c) In each case, will hydrolysis of 1 mole of ATP suffice to drive the transport of 1 mole of ion, assuming pH 7.4 and the following cytoplasmic concentrations: ATP = 4.60 mM, P$_i$ = 5.10 mM, ADP = 310 μM?

6. Propose an experiment that would distinguish pore-mediated diffusion (e.g., by gramicidin) from carrier-mediated diffusion (e.g., by valinomycin).

7. In contrast to phospholipids, the transport of fatty acids across membranes is much more rapid (less than a second). Propose an explanation for this observation.

8. Peptide hormones (such as insulin) must bind to receptors on the outside surfaces of their target cells before their signal is transmitted to the inside of the cell. In contrast, the receptors for steroid hormones (such as estradiol, shown in the figure) are found inside cells. What features of these two different hormones explain the locations of their receptors?

Estradiol

9. Suppose calcium ion is maintained within an organelle at a concentration 1000 times greater than outside the organelle ($T = 37\,°C$). Assuming the membrane is permeable to Ca^{2+}, what is the contribution of Ca^{2+} to the membrane potential? Which side of the organelle membrane is positive, and which is negative?

10. Calculate the equilibrium membrane potentials to be expected across a membrane at $37\,°C$, with a NaCl concentration of 0.10 M on the "right side" and 0.01 M on the "left side", given the following conditions. In each case, state which side is $(+)$ and which is $(-)$.
(a) Membrane permeable only to Na^+
(b) Membrane permeable only to Cl^-
(c) Membrane equally permeable to both ions.

11. In each of a, b, and c of Problem 10, will any appreciable transport of material take place in establishing the membrane potential? Briefly explain each answer.

12. List two differences you would expect to see in the composition of lipids in the *E. coli* membrane when the cells are incubated at $25\,°C$ compared to incubation at $37\,°C$.

13. Many transmembrane proteins are oligomeric, with several identical subunits. The oligomers are usually found to have some form of C_n symmetry, rather than D_n or any higher order. Suggest a reason for this observation.

14. The average human generates approximately his or her weight in ATP every day. A resting person uses about 25% of this in ion transport—mostly via the Na^+–K^+ ATPase. About how many grams of Na^+ and K^+ will a sedentary 70-kg person pump across membranes in a day?

15. The concentration of glucose in your circulatory system is maintained near 5.0 mM by the actions of the pancreatic hormones glucagon and insulin. Glucose is imported into cells by protein transporters that are highly specific for binding glucose. Inside the liver cells the imported glucose is rapidly phosphorylated to give glucose-6-phosphate (G-6-P). This is an ATP-dependent process that consumes 1 mol ATP per mol of glucose.
(a) The process of phosphorylating the glucose after it has been transported into the cell is considered a form of active transport—called "transport by modification"—even though ATP is not bound by the transporter protein, nor is ATP hydrolysis directly involved in the movement of glucose across the membrane. Explain the thermodynamic basis for this form of active transport. (*Hint:* Consider Le Chatelier's principle.)
(b) Given ATP = 4.7 mM; ADP = 0.15 mM; P_i = 6.1 mM, calculate the theoretical maximum concentration of G-6-P inside a liver cell at $37\,°C$, pH = 7.2 when the glucose concentration outside the cell (i.e., $[\,glucose\,]_{outside}$) is 5.0 mM:

$$ATP + glucose_{inside} \rightarrow ADP + glucose\text{-}6\text{-}phosphate + H^+$$

For $ATP + H_2O \rightarrow ADP + P_i + H^+$ $\Delta G^{\circ\prime} = -32.2\ kJ/mol$ and for

$G\text{-}6\text{-}P + H_2O \rightarrow Glucose + P_i$ $\Delta G^{\circ\prime} = -13.8\ kJ/mol$

16. ATP is synthesized from ADP, P_i, and a proton on the *matrix side* of the inner mitochondrial membrane. We will refer to the matrix side as the "inside" of the inner mitochondrial membrane (IMM).
(a) H^+ transport from the outside of the IMM into the matrix drives this process. The pH inside the matrix is 8.2, and the outside is more acidic by 0.8 pH units. Assuming the IMM membrane potential is 168 mV (inside negative), calculate ΔG for the transport of 1 mol of H^+ across the IMM into the matrix at $37\,°C$: $H^+_{(outside)} \rightarrow H^+_{(inside)}$.
(b) Assume three mol H^+ must be translocated to synthesize one mol ATP by coupling of the following reactions:

$$ADP + P_i + H^+_{(inside)} \rightarrow ATP + H_2O \quad \text{(ATP synthesis)}$$
$$3H^+_{(outside)} \rightarrow 3H^+_{(inside)} \quad \text{(proton transport)}$$

Write the overall reaction for ATP synthesis coupled to H^+ transport [and use this equation for part (c)]:
(c) Assume three mol H^+ must be translocated to synthesize one mol ATP as described in part (b) above. Given the following steady-state concentrations: ATP = 2.70 mM and P_i = 5.20 mM, the membrane potential $\Delta\psi$ = 168 mV (inside negative), and the pH values in part (a), calculate the steady-state concentration of ADP at $37\,°C$ when for the coupled process (ATP synthesis + H^+ transport), $\Delta G = -11.7\ kJ/mol$.

17. The Na^+/glucose symport transports glucose from the lumen of the small intestine into cells lining the lumen. Transport of 1 glucose molecule is directly coupled to the transport of 1 Na^+ ion into the cell.

$$1\ Na^+_{out} + 1\ glucose_{out} \rightarrow 1\ Na^+_{in} + 1\ glucose_{in}$$

Assume the following conditions at $37\,°C$: $[Na^+]_{in}$ = 12 mM, $[Na^+]_{out}$ = 145 mM, $[glucose]_{out}$ = 28 μM, and $\Delta\psi$ = -72 mV (inside negative).
(a) What is ΔG for transport of Na^+ from outside to inside under these conditions?
(b) What is the upper limit for $[glucose]_{in}$ under these conditions?
(c) Which of the two hypothetical symports shown below (**A** or **B**) would achieve the highest concentration of $[glucose]_{in}$ under the conditions described above? Briefly explain your choice.

A: $1\ Na^+_{out} + 2\ glucose_{out} \rightarrow 1\ Na^+_{in} + 2\ glucose_{in}$
B: $2\ Na^+_{out} + 1\ glucose_{out} \rightarrow 2\ Na^+_{in} + 1\ glucose_{in}$

18. Shown below is a schematic diagram of the *E. coli* leader peptidase (Lep), which has several basic amino acids in a cytoplasmic loop. Propose a mutant of Lep that would be a test of the "inside positive" rule for the orientation of proteins in membranes.

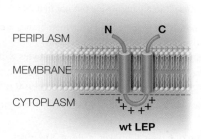

wt LEP

19. One of the curves in this graph describes nonmediated diffusion, and the other describes facilitated transport. Which is which? Explain your choices.

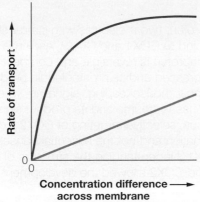

20. The transport of aspirin ($pK_a = 3.5$, structure shown here) from the digestive tract to the circulation occurs by nonmediated absorption into cells lining the stomach (where pH = 0.8) and the small intestine (where pH = 6.0). Do you expect absorption to be faster in the stomach or in the small intestine?

21. The sarcoplasmic reticulum Ca^{2+}-ATPase, pumps 2 mol Ca^{2+} *out* of sarcomeres per mol ATP hydrolyzed.

(a) Given the following steady-state concentrations and a membrane potential of 65 mV (inside negative), calculate ΔG for the following active transport process at 37 °C and pH = 7.4:

$$2Ca^{2+}_{(in)} + ATP + H_2O \rightarrow 2Ca^{2+}_{(out)} + ADP + P_i + H^+$$

ATP = 2.6 mM, ADP = 210 μM, P_i = 5.1 mM, $Ca^{2+}_{(in)}$ = 32 μM, $Ca^{2+}_{(out)}$ = 2.2 mM

(b) In active muscle the pH can drop below 7.4. Is the reaction above *more* or *less* favorable under these conditions?

(c) The activity of the Ca^{2+}-ATPase is regulated reversibly under normal conditions to maintain homeostatic concentrations of Ca^{2+} inside the sarcomere. However, in a rare genetic disorder, irreversible activation of the Ca^{2+}-ATPase can occur. Assuming 37 °C, pH = 7.4, and the steady-state concentrations for ATP, ADP P_i, and $Ca^{2+}_{(out)}$ given in part (a), calculate the minimum $[Ca^{2+}]$ *inside* a sarcomere that has irreversibly activated Ca^{2+}-ATPase (i.e., the Ca^{2+}-ATPase activity is always "on").

References

For a list of references related to this chapter, see Appendix II.

Pain and inflammation: Non-steroid anti-inflammatory drugs (NSAIDs) like ibuprofen and aspirin target the two main isomers of Prostaglandin H2 synthase (PGH synthase) (aka COX1 and COX2), by binding to the active site and inhibiting the synthesis of prostaglandins (local signaling molecules). The COX abbreviation refers to the first part of a two step process in prostaglandin synthesis, with cyclooxygenase activity being the first and peroxidase being second. These two enzyme activities are carried out within a single enzyme, PGH synthase, which is commonly referred to as the COX enzyme. These peripheral enzymes bind to non-polar molecules in a shape-specific manner.

Aspirin and ibuprofen, two molecules with similar shapes and polarities, both bind to COX1 and COX2. Aspirin is an irreversible inhibitor, while ibuprofen is reversible and competitive. COX1 is constitutively expressed and its prostaglandin product is responsible for local "housekeeping" signaling, whereas COX2 is induced in response to trauma and its product signals pain and inflammation. Thus, selective targeting of COX2 would block pain and inflammation and not the essential housekeeping functions of COX1. Understanding the structure of the active sites of COX1 and COX2 allowed the development of selective inhibitors.

▶ SEE PAGE 550

1 **PROSTAGLANDIN SYNTHESIS** COX1 and COX2 have similar structures but different functions.

COX1 and COX2 are peripheral enzymes, attaching to the ER of eukaryotic cells. They bind to non-polar molecules in a shape-specific manner.

1 Arachidonic acid is cleaved from a phospholipid tail and travels to the active site through a mostly hydrophobic tunnel.

2 Arachidonic acid enters the active site. It is held in place by Arg_{120} and oriented for catalysis by Tyr_{385}, which is activated by a nearby heme. The reaction is thought to proceed through a free radical mechanism.

COX1
• Constitutively expressed in most tissues
• Product signals local housekeeping (cell signaling, protection of gastro-intestinal tract, platelet formation, and other functions)
• Inhibition of COX1 leads to gastric irritation.

3 The fatty acid is partially cyclized to produce a prostaglandin, an eicosanoid lipid. Prostaglandins from COX1 and COX2 produce different effects.

COX2
• Induced in response to trauma
• Product signals pain and inflammation
• Inhibition of COX2 reduces pain and inflammation.

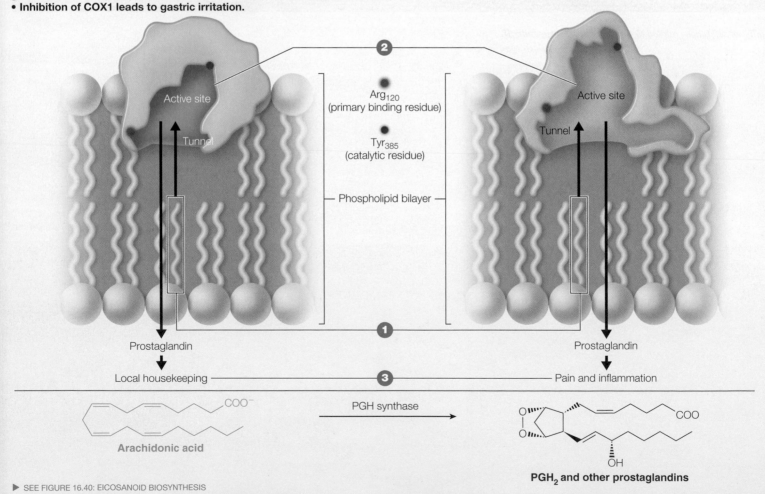

Arg$_{120}$
(primary binding residue)

Tyr$_{385}$
(catalytic residue)

Active site

Tunnel

Active site

Tunnel

— Phospholipid bilayer —

Prostaglandin

Prostaglandin

Local housekeeping ——— **3** ——— Pain and inflammation

PGH synthase

Arachidonic acid

PGH$_2$ and other prostaglandins

▶ SEE FIGURE 16.40: EICOSANOID BIOSYNTHESIS

2 | NONSELECTIVE INHIBITION Aspirin and ibuprofen inhibit both COX1 and COX2.

Aspirin and ibuprofen are NSAIDs that enter the active site and block arachidonic acid from reaching catalytic tyrosine. This inhibits prostaglandin formation, thereby reducing pain and inflammation.

Aspirin is an irreversible inhibitor.

The irreversible inhibition occurs when aspirin (acetylsalicylic acid) reacts with an active site serine and the serine becomes acetylated. The acetylated enzyme is inactive; thus, it can no longer produce prostaglandins.

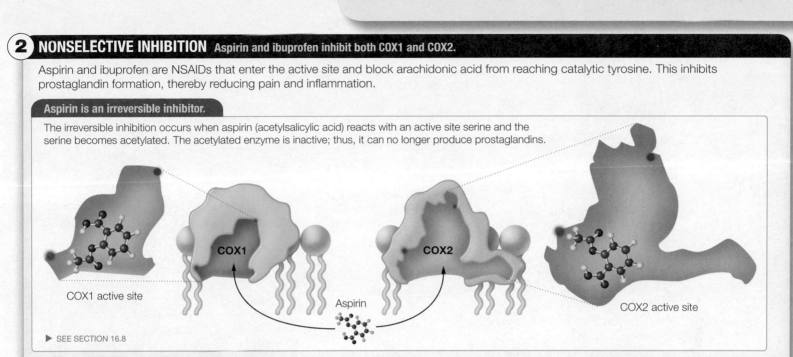

COX1 active site

COX1 COX2 Aspirin

COX2 active site

▶ SEE SECTION 16.8

Ibuprofen is a reversible competitive inhibitor.

Ibuprofen reversibly binds to the active site and blocks arachidonic acid from entering. The reaction proceeds only after ibuprofen leaves the active site (and arachidonic acid enters).

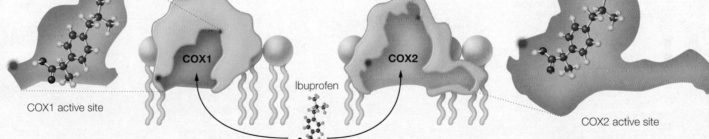

COX1 active site

COX1 COX2 Ibuprofen

COX2 active site

3 | SELECTIVE INHIBITION Celecoxib (Celebrex) is a COX2 specific inhibitor.

Celecoxib is a COX2 specific inhibitor due to its "V" shape that fits into the available side pocket of the COX2 active site. This is an example of structure-based drug design. COX1 is not inhibited.

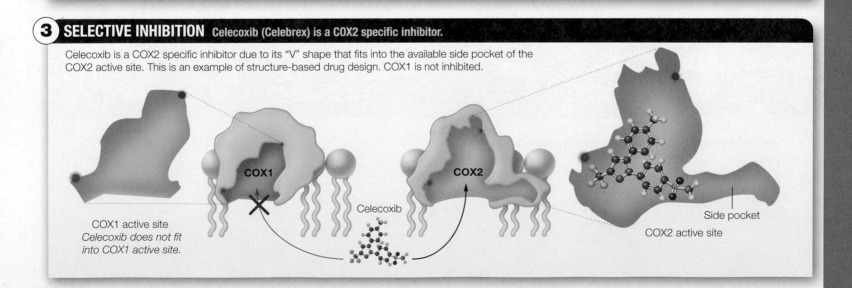

COX1 active site
Celecoxib does not fit into COX1 active site.

COX1 COX2 Celecoxib

Side pocket

COX2 active site

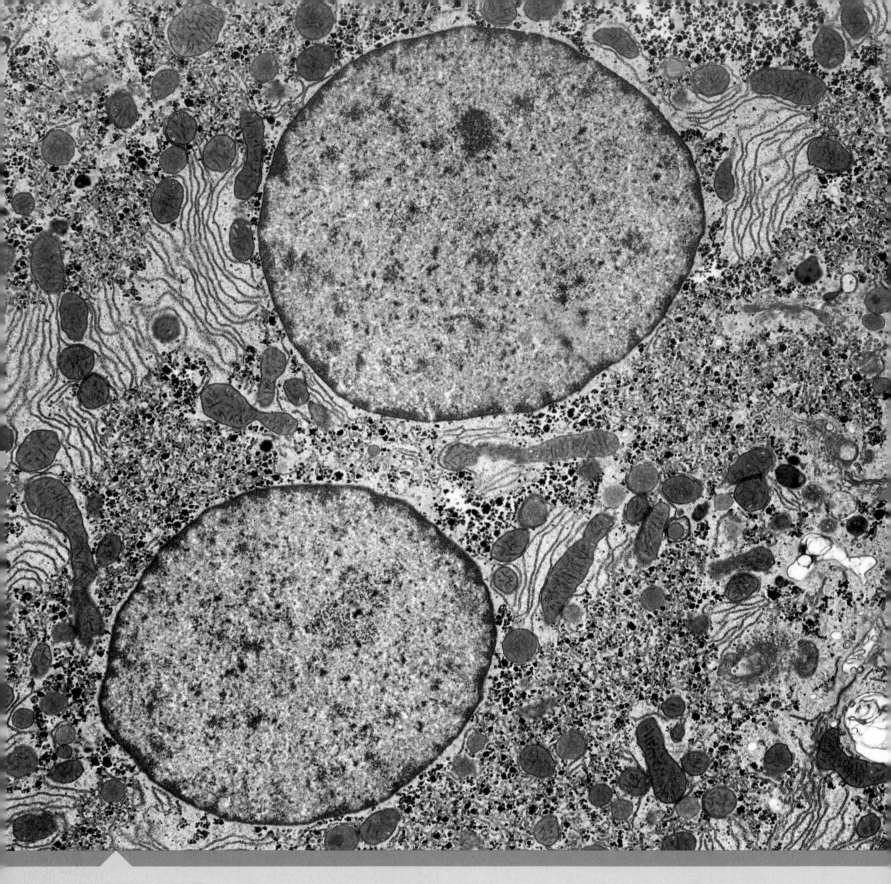

A living cell, such as this liver cell, carries out thousands of reactions simultaneously. How are these metabolic pathways organized and controlled within such an intricate architecture?

Chemical Logic of Metabolism

11

A CHEMIST CARRYING out an organic synthesis rarely runs more than one reaction in a single-reaction vessel at any one time. This strategy is essential to prevent unwanted by-products and to optimize the yield of the desired product. Yet a living cell carries out thousands of reactions simultaneously, with each reaction sequence controlled so that unwanted accumulations or deficiencies of intermediates and products do not occur. Reactions of great mechanistic complexity and stereochemical selectivity proceed smoothly under mild conditions—1 atm pressure, moderate temperature, and osmotic pressure, and a pH near neutrality. How then, do cells avoid metabolic chaos? A goal of the next several chapters is to understand how cells carry out and regulate these complex reaction sequences and, in so doing, control their internal environment.

In Chapter 8 we discussed the properties of individual enzymes and the control mechanisms that affect their activity. In this chapter, we now consider how individual biochemical reactions combine to form **metabolic pathways,** a series of chemical reactions whereby the products of one reaction are the substrates for the next reaction,

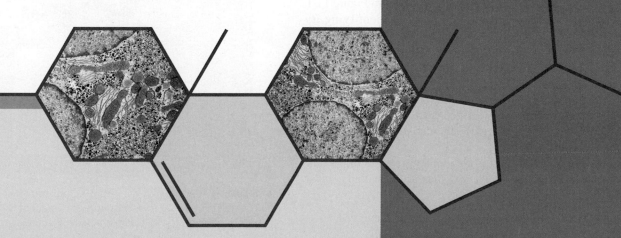

and so on until the end product is generated. We briefly review the organic chemistry mechanisms of the most common chemical transformations found in cells. We also discuss **compartmentation,** the spatial relationship between metabolic pathways and cellular architecture. We conclude the chapter by considering the control mechanisms that regulate **flux,** or intracellular reaction rate, and by reviewing some of the main experimental methods used to investigate metabolism.

11.1 A First Look at Metabolism

FIGURE 11.1 is a simplified overview of the metabolic processes we consider. This figure illustrates two important principles:

- First, metabolism can be thought of as having two major divisions—**catabolism,** those processes in which complex substances are degraded to simpler molecules, and **anabolism,** those processes concerned primarily with the synthesis of complex organic molecules. Catabolism is generally accompanied by the net release of chemical energy, whereas anabolism requires a net input of chemical energy. As Figure 11.1(a) shows, these two sets of reactions are coupled together by adenosine triphosphate (ATP).

- Second, both catabolic and anabolic pathways can be subdivided into three stages, based on the complexity of the **metabolites** (the intermediates and products of metabolism) involved (Figure 11.1(b)):
  - Stage 1: the interconversion of polymers and complex lipids with monomeric intermediates
  - Stage 2: the interconversion of monomeric sugars, amino acids, and lipids with still simpler organic compounds
  - Stage 3: the ultimate degradation to, or synthesis from, inorganic compounds, including CO_2, H_2O, and NH_3

As we proceed through this chapter, we will add detail to Figure 11.1 by introducing each major metabolic process and identifying its functions.

During our discussion, we will see that energy-yielding (catabolic) pathways also generate intermediates used in biosynthetic (anabolic) processes. Thus, although we will focus first on the degradation of organic compounds to provide energy, you should keep in mind that metabolism is really an integrated system, with many of the same metabolites playing roles in both degradative and biosynthetic processes.

In these chapters, we will refer frequently to three important concepts: *intermediary metabolism, energy metabolism,* and *central pathways.*

- **Intermediary metabolism** includes all reactions involved in generating and storing metabolic energy (catabolic reactions), and with using that energy in the biosynthesis of

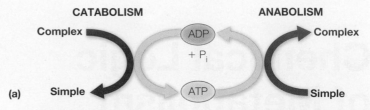

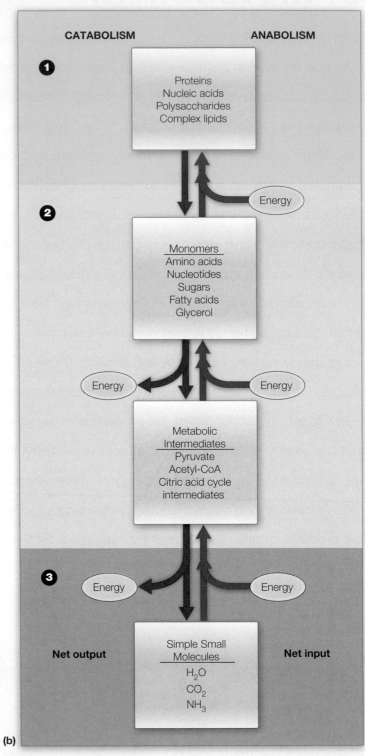

▲ **FIGURE 11.1 A brief overview of metabolism.** The three stages are numbered and colored according to a scheme that will be used in many figures throughout the following chapters: Catabolic pathways are in red, and anabolic pathways are in blue.

low-molecular-weight compounds, or intermediates, as well as energy-storage compounds (anabolic reactions). Chapters 12–19 are devoted to intermediary metabolism. Not included in intermediary metabolism is the biosynthesis of nucleic acids and proteins from monomeric precursors, which are discussed in Chapters 22–25.

- **Energy metabolism** comprises those pathways of intermediary metabolism that generate or store metabolic energy. These pathways are a major focus of Chapters 12–17.

- The **central pathways** of metabolism are those that carry the heaviest traffic, accounting for most of the mass transfer and energy generation within a cell. These pathways, which are substantially the same in many different organisms, include those of energy metabolism, such as glycolysis (Chapter 12) and the citric acid cycle (Chapter 13).

Most organisms derive both the raw materials and the energy for biosynthesis from organic fuel molecules such as glucose. The central pathways involve the oxidation of fuel molecules and the synthesis of small biomolecules from the resulting fragments. But while these pathways are found in most organisms, we can draw a fundamental distinction between two different ways organisms can make a living based on the source of their fuel molecules. **Autotrophs** (from the Greek word for "self-feeding") synthesize glucose and all of their other organic compounds from inorganic carbon, supplied as carbon dioxide. By contrast, **heterotrophs** ("feeding on others") can synthesize their organic metabolites only from other organic compounds, which they must therefore consume. A primary difference between plants and animals is that plants are autotrophs and animals are heterotrophs. With the exception of rare insect-eating plants, such as the Venus flytrap, green plants obtain all of their organic carbon through photosynthetic fixation of CO_2. Animals feed on plants, or on other animals that have fed on plants, and synthesize their metabolites by transforming the organic molecules they consume.

Microorganisms are either autotrophic or heterotrophic, exhibiting a wide range of biosynthetic capabilities and sources of metabolic energy. Microorganisms also show adaptability with respect to their ability to survive in the absence of oxygen. Virtually all multicellular organisms and many bacteria are strictly **aerobic** organisms—that is, they depend absolutely upon **respiration,** the coupling of energy generation to the oxidation of nutrients by oxygen. By contrast, some microorganisms either can or must grow in **anaerobic** environments, deriving their metabolic energy from processes that do not involve oxygen.

Nature has evolved a tremendous variety of metabolic strategies in response to unique or specialized ecological niches. But these seemingly disparate strategies are generally just variations on a few common themes. We introduce those themes in the next section.

● **CONCEPT** Intermediary metabolism refers primarily to those reactions involved in generating and storing metabolic energy, and with using that energy in the biosynthesis of low-molecular-weight compounds (intermediates) and energy-storage compounds.

11.2 Freeways on the Metabolic Road Map

You have probably seen metabolic charts—those wall hangings like giant road maps that adorn biochemistry laboratories and offices. Figure 11.1 is a highly simplified example of such a chart, but a detailed metabolic chart could include hundreds or even thousands of reactions. Faced with the bewildering variety of individual reactions that constitute metabolism, how do we approach this vast topic?

FIGURE 11.2, which presents metabolism in slightly greater detail than Figure 11.1, is the basic road map for this section of the book. This figure will reappear in subsequent chapters, with the specific pathways presented in those chapters highlighted. Our first concerns here are with central pathways and with energy metabolism. Therefore, in the next several chapters we consider the processes that are most important in energy generation—the catabolism of carbohydrates and lipids. We shall also consider how these substances are biosynthesized. These reactions are typically located in the middle of metabolic charts and are illustrated with the biggest arrows—freeways, so to speak, on the metabolic road map.

The road map analogy is also useful when we consider directional flow in metabolism. Just as most traffic flows from the suburbs to downtown in the morning and from downtown back to the suburbs in the evening, so also will we see that some conditions favor biosynthesis, whereas others favor catabolism, and that parts of the same highways are used in both processes.

Central Pathways of Energy Metabolism

The first pathway that we present in detail (in Chapter 12) is **glycolysis,** a stage 2 pathway for the degradation of carbohydrates in both aerobic and anaerobic cells. As schematized in **FIGURE 11.3**, the major input to glycolysis is glucose, which is usually derived from either energy-storage polysaccharides or dietary carbohydrates. Glycolysis leads to the three-carbon intermediate pyruvate, coupled to the production of a small amount of ATP. In the absence of oxygen, pyruvate is reduced to a variety of products, such as lactate or ethanol plus carbon dioxide. These anaerobic processes, called *fermentations*, are described later in this chapter.

● **CONCEPT** In aerobic organisms, all catabolic pathways converge at the citric acid cycle.

When oxygen is present, pyruvate undergoes **oxidative metabolism,** or respiration. Diagrammed in **FIGURE 11.4**, pyruvate is oxidized to **acetyl-coenzyme A** (acetyl-CoA). The two carbons in the acetyl group then undergo oxidation in the **citric acid cycle.** This series of reactions, discussed in Chapter 13, is the principal stage 3 pathway for aerobic organisms. The citric acid cycle accepts acetyl-CoA derived not only from carbohydrates but also from lipids or proteins, and completes the oxidation of these fuels to CO_2. Using the freeway analogy again (Figure 11.2), we see that numerous on-ramps from the highways and byways of stage 1 and stage 2 metabolism lead to the citric acid cycle. In fact, all catabolic pathways converge at this point.

Oxidative reactions of the citric acid cycle generate reduced electron carriers that are then reoxidized to drive the synthesis of ATP. This synthesis occurs primarily through **electron transport** and **oxidative phosphorylation,** also shown in Figure 11.4. Because this process oxidizes the organic fuels completely to CO_2 and H_2O

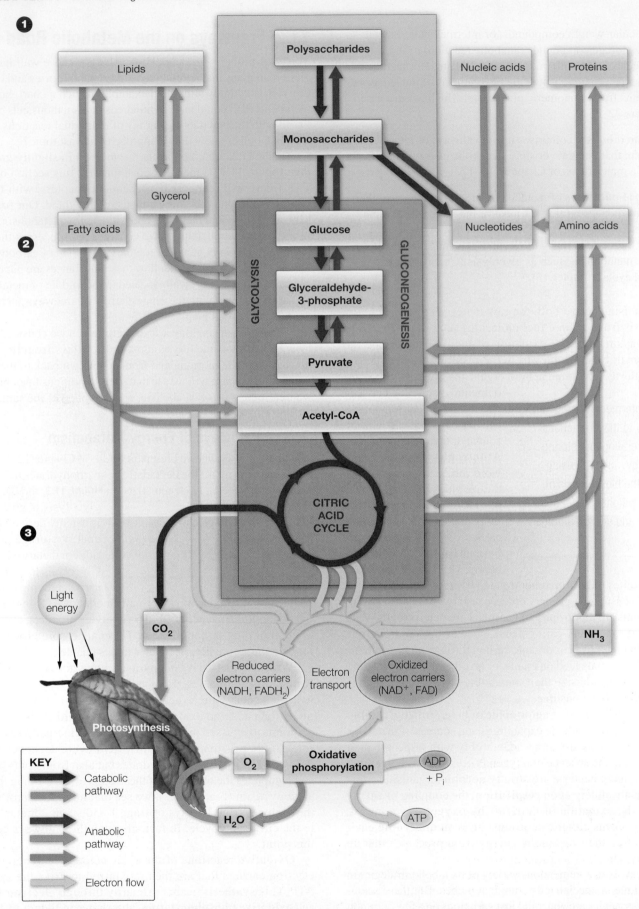

▲ **FIGURE 11.2 Overview of metabolism.** Shown here are the central metabolic pathways and some key intermediates. Catabolic pathways (red) proceed downward, and anabolic pathways (blue and green) proceed upward. The three stages of metabolism are identified as in Figure 11.1.

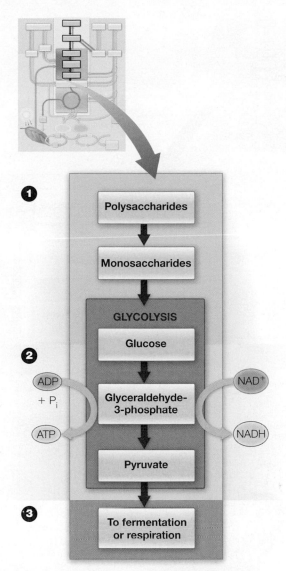

▲ **FIGURE 11.3** The initial phase of carbohydrate catabolism: glycolysis. Pyruvate either undergoes reduction in fermentation reactions or enters oxidative metabolism (respiration) via conversion to acetyl-CoA, as shown in Figure 11.4.

by oxygen, far more energy is generated than from the fermentations of pyruvate. The mechanism by which aerobic cells use oxidative electron transport to drive the synthesis of ATP is explored in Chapter 14.

As Figure 11.2 shows, other stage 2 pathways in addition to glycolysis deliver fuel to the citric acid cycle. Acetyl-CoA comes not only from pyruvate oxidation but also from the oxidation of fatty acids (Chapter 16) and from some amino acid oxidation pathways (Chapter 19). If the two carbons of acetyl-CoA are not oxidized in the citric acid cycle, they can be used in an anabolic pathway, providing substrates for synthesis of fatty acids and steroids (Chapter 16).

Carbohydrates are central to several important biosynthetic processes presented in this book (**FIGURE 11.5**). Chapter 12 discusses **gluconeogenesis**—the synthesis of glucose from noncarbohydrate precursors—and polysaccharide biosynthesis, particularly the biosynthesis of glycogen in animal cells. In Chapter 15 we present **photosynthesis**

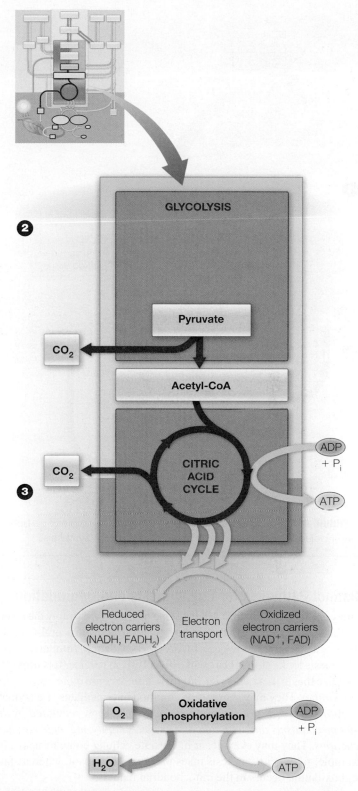

▲ **FIGURE 11.4 Oxidative metabolism.** Oxidative metabolism includes pyruvate oxidation, the citric acid cycle, electron transport, and oxidative phosphorylation. Pyruvate oxidation supplies acetyl-CoA to the citric acid cycle.

(**FIGURE 11.6**), the process by which green plants capture light energy to drive the generation of energy (ATP) and reducing power (NADPH), both of which are used for carbohydrate synthesis.

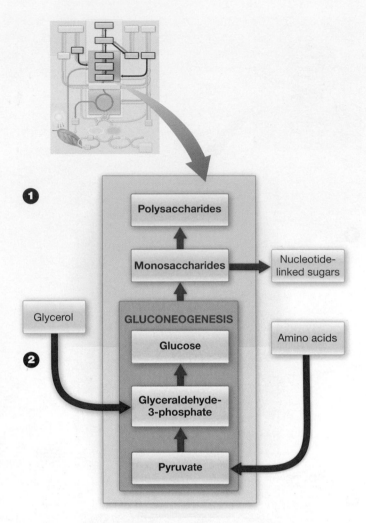

▲ **FIGURE 11.5 Carbohydrate anabolism.** The biosynthesis of carbohydrates includes gluconeogenesis and polysaccharide synthesis.

Distinct Pathways for Biosynthesis and Degradation

It may appear from Figure 11.2 that some pathways are simply the reverse of others. For example, fatty acids are synthesized from acetyl-CoA, but they are also converted to acetyl-CoA by β-oxidation. Similarly, gluconeogenesis looks at first glance like glycolysis in reverse. It is important to realize that in these cases *the opposed pathways are quite distinct from one another.* They may share some common intermediates or enzymatic reactions, but *they are separate reaction sequences, regulated by distinct mechanisms and with different enzymes catalyzing their regulated reactions.* They may even occur in separate cellular compartments. For example, fatty acid synthesis takes place in the cytosol, whereas fatty acid oxidation occurs in the mitochondrial matrix.

Biosynthetic and degradative pathways, even though they often begin and end with the same metabolites, require separate unidirectional pathways for two reasons. First, a pathway cannot proceed in a particular direction unless it is exergonic in that direction—that is, $\Delta G_{pathway} < 0$ (Section 3.2). If a pathway is strongly exergonic, then the reverse of that pathway is just as strongly endergonic (i.e., $\Delta G_{pathway} > 0$), and therefore impossible, under the same conditions. Opposed biosynthetic and degradative pathways must *both* be exergonic, and thus unidirectional, in their respective directions.

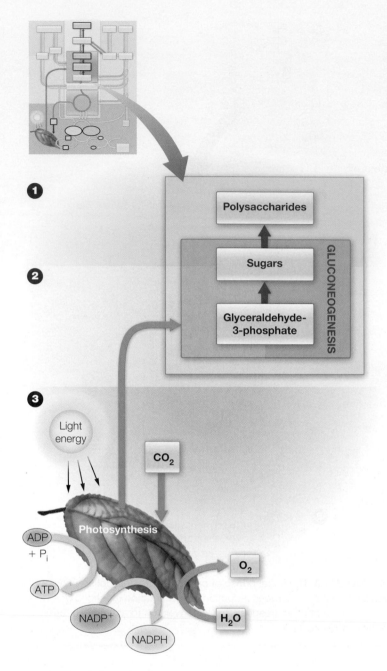

▲ **FIGURE 11.6 Photosynthesis.**

Second, and equally important, is the need to control the flow of metabolites, so that the cell makes only what it needs in the amounts it needs. When ATP levels are high, for example, there is less need for carbon to be oxidized in the citric acid cycle. The cell can thus store carbon as fats and carbohydrates, so fatty acid synthesis, gluconeogenesis, and related pathways come into play. When ATP levels are low, the cell must mobilize stored carbon to generate substrates for the citric acid cycle, so carbohydrate and fat breakdown must occur. Using separate pathways for the biosynthetic

● **CONCEPT** Degradative and biosynthetic pathways are distinct for two reasons: A pathway can be exergonic in only one direction, and pathways must be separately regulated to avoid futile cycles.

and degradative processes is crucial for control; therefore, conditions that activate one pathway tend to inhibit the opposed pathway, and vice versa.

Consider what would happen, for example, if fatty acid synthesis and oxidation (blue and red arrows in Figure 11.2) took place in the same cell compartment in an uncontrolled fashion. Acetyl-CoA released by oxidation would be immediately used for the resynthesis of fatty acids, a process called a **futile cycle.** No useful work would be done, and the net result would be the consumption of more ATP in the endergonic reactions of fatty acid synthesis than was produced in the oxidation reactions.

Futile cycle

A similar futile cycle would result from the interconversion of fructose-6-phosphate and fructose-1,6-bisphosphate in carbohydrate metabolism.

Fructose-6-phosphate $+$ ATP $\rightarrow$ fructose-1,6-bisphosphate $+$ ADP
Fructose-1,6-bisphosphate $+$ H_2O $\rightarrow$ fructose-6-phosphate $+$ P_i

Net: $ATP + H_2O \rightarrow ADP + P_i$

The first reaction is part of glycolysis (Figure 11.3), whereas the second occurs in gluconeogenesis (Figure 11.5); both reactions are energetically favorable. The net effect of carrying out these two reactions simultaneously would be the wasteful hydrolysis of ATP to ADP and P_i. This does not happen in living cells: although both of these processes occur in the cytosol, the enzymes catalyzing the two reactions respond differently to allosteric effectors (Section 8.8), such that one enzyme is inhibited by conditions that activate the other. This reciprocal control prevents the futile cycle from occurring, even though the two enzymes occupy the same cell compartment. Such an arrangement—two seemingly opposed cellular reactions that are independently controlled—is called a **substrate cycle.** Studies of metabolic control suggest that a substrate cycle represents an efficient regulatory mechanism because a small change in the activity of either or both enzymes can have a much larger effect on the flux of metabolites in one direction or the other. You can test this idea in Problem 7 at the end of the chapter.

● **CONCEPT** Compartmentation and reciprocal control of anabolic and catabolic processes prevent futile cycles, which simply waste energy.

11.3 Biochemical Reaction Types

There is nothing mysterious in biochemistry—the chemistry of living systems follows the same chemical and physical laws as the rest of nature. The complexity of these biochemical pathways may seem overwhelming at first, but only five general types of chemical transformations are commonly found in cells: nucleophilic substitutions; nucleophilic additions; carbonyl condensations; eliminations,

and oxidations/reductions. Although all of these transformations are catalyzed by cellular enzymes, the reactions proceed by straightforward organic chemistry mechanisms. We consider these reaction types briefly here, but you should consult your organic chemistry textbook if you need a more detailed review.

Nucleophilic Substitutions

Much of the chemistry of biological molecules is the chemistry of the carbonyl group $(C{=}O)$ because the vast majority of biological molecules contain one or more carbonyl groups. The chemistry of carbonyl groups typically involves **nucleophiles** (represented by "**Nu**") and **electrophiles.** A nucleophile (a "nucleus-loving" substance) is a negatively polarized, electron-rich atom that can form a bond by donating a pair of electrons to an electron-poor atom. An electrophile (an "electron-loving" substance) is a positively polarized, electron-poor atom that can form a bond by accepting a pair of electrons from an electron-rich atom.

Carbonyl groups are polar, with the electron-poor C atom bearing a partial positive charge $(\delta+)$ and the electron-rich O atom bearing a partial negative charge $(\delta-)$.

Aldehyde   **Ketone**   **Carboxylic acid**   **Ester**

Carbonyl carbons are thus very common electrophiles in biochemical reactions. Other common electrophiles are protonated imines, phosphate groups, and protons.

Carbonyl   Protonated imine   Phosphate   Proton

Electrophiles

Common nucleophiles in biochemical reactions include oxyanions (e.g., hydroxide ion, alkoxides, or ionized carboxylates), thiolates (deprotonated sulfhydryls), carbanions, deprotonated amines, and the imidazole side chain of histidine.

Alkoxide   Hydroxide ion   Carbanion

Carboxylate   Thiolate   Amine   Imidazole

Nucleophiles

In a *nucleophilic substitution reaction,* one nucleophile replaces a second nucleophile (the leaving group) on an sp^3-hybridized carbon atom of an alkyl group. The leaving group develops a partial negative charge in the transition state, and the best leaving groups are those that are stable as anions. Thus, halides (e.g., Cl^-, Br^-, and I^-) and the conjugate bases of strong acids (e.g., PO_4^{3-}) are good leaving groups.

Nucleophilic alkyl substitutions proceed by either S_N1 or S_N2 mechanisms. In the S_N1 (*Substitution, Nucleophilic, Unimolecular*) mechanism, the leaving group (Y: $^-$) departs with the bonding electrons, generating a carbocation intermediate, *before* the attacking nucleophile (X: $^-$) arrives. The carbocation intermediate is planar, so X: $^-$ can attack from either side of the plane; in this case, half the product retains the original configuration at the reacting center, and half has an inverted configuration.

S_N1 reaction

Carbocation
intermediate

In the S_N2 (*Substitution, Nucleophilic, Bimolecular*) mechanism, the attacking nucleophile approaches one side of the electrophilic center, while the leaving group remains partially bonded to the other side, resulting in a transient pentavalent intermediate. Departure of the leaving group from the opposite side results in a substituted product with inverted configuration.

S_N2 reaction

Pentavalent
intermediate

A very important class of substitution reaction in biochemistry is the **nucleophilic *acyl* substitution** of carboxylic acid derivatives. Acyl substitutions occur most readily when the carbonyl carbon is bonded to an electronegative atom (such as O or N) or a highly polarizable atom (such as S) that can stabilize a negative charge and thereby act as a good leaving group. Thus, carboxylic acids and their derivatives (e.g., esters, amides, thioesters, and acyl phosphates) are common substrates in acyl substitution reactions.

Common substrates in acyl substitutions

In contrast to the S_N1 and S_N2 mechanisms of *alkyl* substitutions, *acyl* substitutions involve a tetrahedral oxyanion reaction intermediate. The planar carbonyl group is converted to a tetrahedral geometry as the carbonyl carbon rehybridizes from sp^2 to sp^3.

Nucleophilic acyl substitution

Oxyanion intermediate

As the electron pair moves from the oxyanion back toward the central carbon, the leaving group is expelled and the C=O bond is regenerated. Many enzymes, such as carboxypeptidase A, catalyze nucleophilic acyl substitutions in which an activated water molecule serves as the attacking nucleophile (see Figure 8.19). This mechanism is also the basis of a variety of **group transfer reactions,** in which an acyl, glycosyl, or phosphoryl group is transferred from one nucleophile to another.

Nucleophilic Additions

Unlike the corresponding atom in carboxylic acids and their derivatives, the carbonyl carbon in aldehydes and ketones is bonded to atoms (C and H) that cannot stabilize a negative charge; thus, these are not good leaving groups. These carbonyl groups typically undergo **nucleophilic addition reactions** instead of substitution reactions. Like the nucleophilic acyl substitution mechanism, addition of a nucleophile leads to a tetrahedral oxyanion intermediate, as the electron pair from the C=O bond moves onto the oxygen. (Recall from Chapter 8, for example, that an oxyanion intermediate is formed in the initial steps of peptide bond hydrolysis by the serine proteases.)

The oxyanion intermediate then has several fates, depending on the nucleophile (**FIGURE 11.7**). When the attacking nucleophile is a hydride ion (H: $^-$), the oxyanion intermediate undergoes protonation to form an alcohol. Alcohols also result when the attacking nucleophile is a carbanion (R_3C^-), and this is one of the mechanisms that yield new C—C bonds. When an oxygen nucleophile adds, such as an alcohol (ROH), the oxyanion intermediate undergoes proton transfer to yield a hemiacetal (or hemiketal). This reaction is the basis of ring formation in monosaccharides (Chapter 9). Reaction with a second alcohol gives an acetal (or ketal). Glycosidic bonds (Chapter 9) are acetals. When the attacking nucleophile is a primary amine ($R'NH_2$), the oxyanion intermediate picks up a proton from the amino group, giving a carbinolamine, which loses water to form an **imine** (R_2C=NR'). Imines (called **Schiff bases**) are common reaction intermediates in many enzyme-catalyzed reactions because of their ability to spread electrons over several adjacent atoms. This **delocalization** stabilizes the reaction intermediates, facilitating the reactions.

Carbonyl Condensations

Formation of new C—C bonds is a critical element of metabolism, and we've already seen one strategy involving nucleophilic addition by a carbanion. The condensation of two carbonyl compounds is another common strategy used in many biosynthetic pathways. A carbonyl

► **FIGURE 11.7** Nucleophilic addition reactions of aldehydes (at least one R=H) and ketones (both R=C).

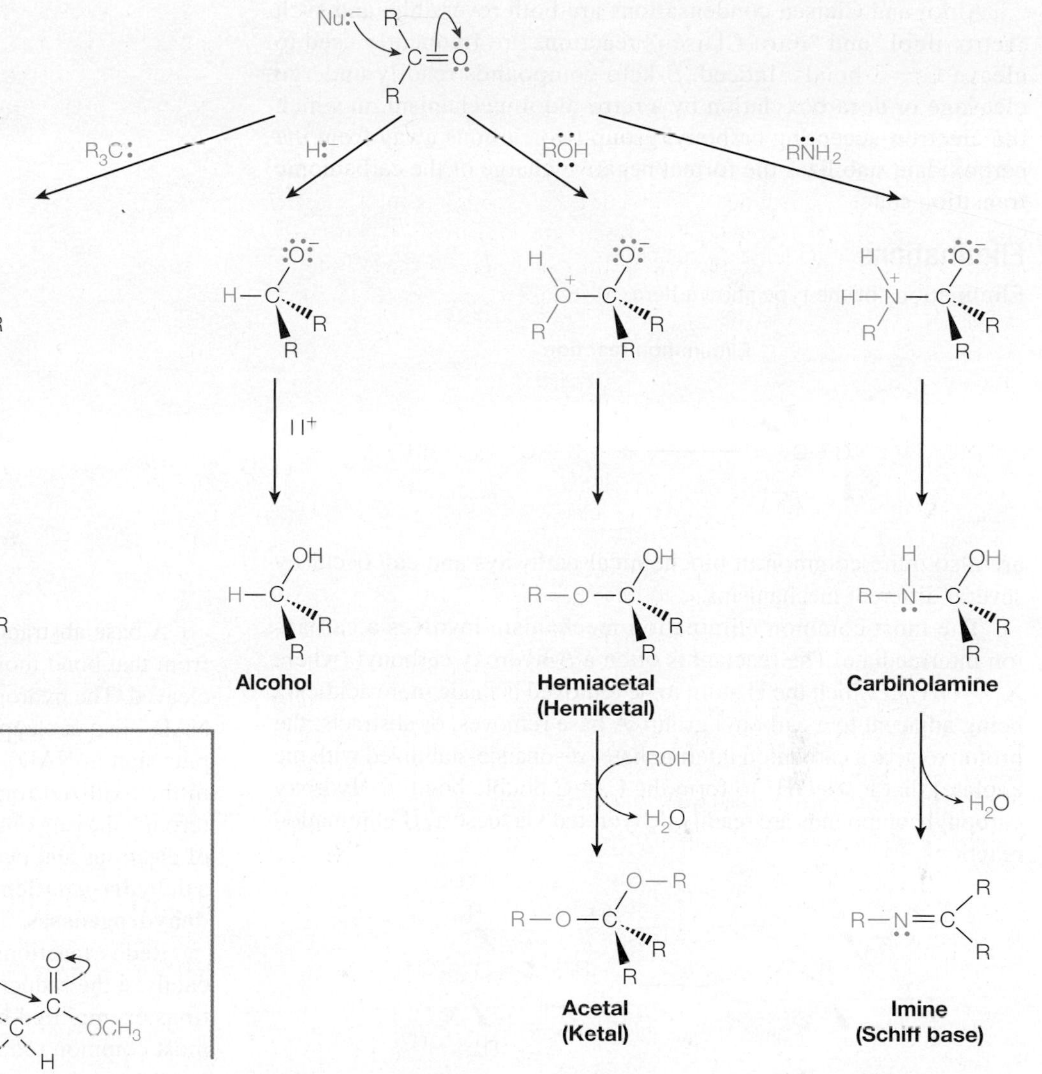

Alcohol   **Alcohol**   **Hemiacetal (Hemiketal)**   **Carbinolamine**

Acetal (Ketal)   **Imine (Schiff base)**

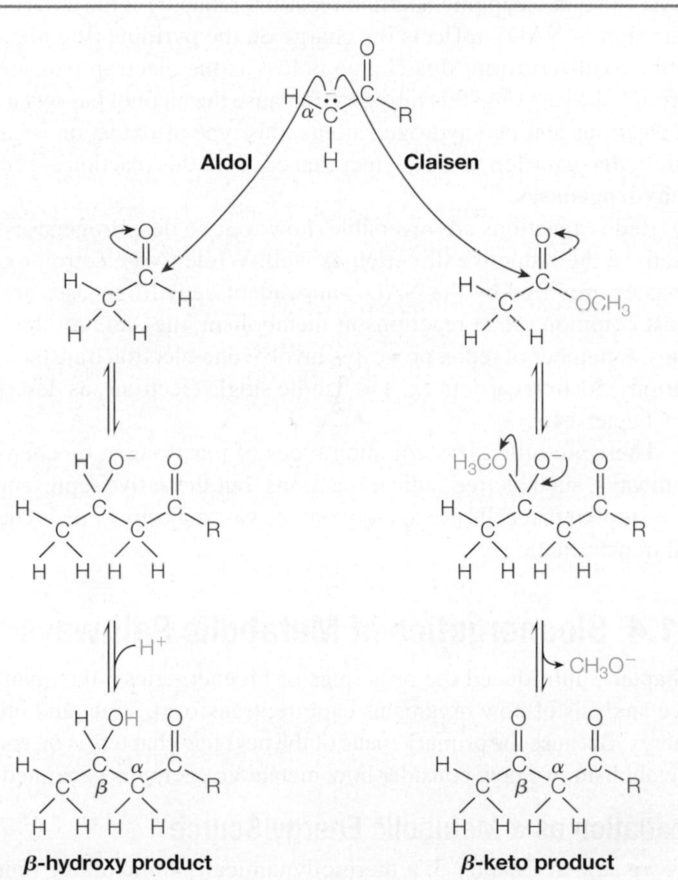

Aldol   **Claisen**

β-hydroxy product   **β-keto product**

▲ **FIGURE 11.8** Carbonyl condensation reactions. These reactions are initiated by deprotonation of the weakly acidic α hydrogen to give a resonance-stabilized enolate ion (top). In an aldol condensation (left side), the enolate adds to an aldehyde or ketone, yielding a β-hydroxy carbonyl product. In a Claisen condensation (right side), the enolate adds to an ester, yielding a β-keto product.

condensation reaction relies on the weak acidity of the carbonyl α hydrogen, producing a carbanion, which is in resonance with a nucleophilic enolate ion.

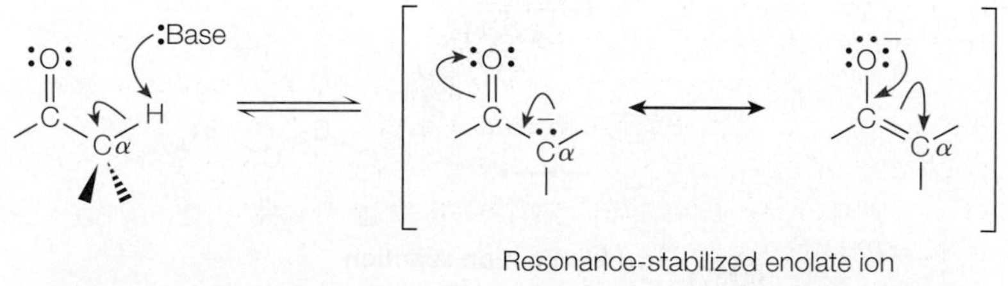

Resonance-stabilized enolate ion

The enolate ion, stabilized by resonance, nucleophilically adds to the electrophilic carbon of a second carbonyl, forming a new C—C bond (**FIGURE 11.8**). If the second carbonyl is an aldehyde or ketone (an **aldol condensation**), this nucleophilic addition produces an oxyanion intermediate, which is protonated to give a β-hydroxy carbonyl product. If the second carbonyl is an ester (a **Claisen condensation**), the intermediate oxyanion expels the ester alkoxide (RO—) as the leaving group, giving a β-keto product. Carbonyl condensations thus result in a new bond formed between the carbonyl carbon of one reactant with the α carbon of the other.

Aldol and Claisen condensations are both reversible, and such "retro-aldol" and "retro-Claisen" reactions are frequently used to cleave C—C bonds. Indeed, β-keto compounds readily undergo cleavage or decarboxylation by a retro-aldol mechanism, in which the electron-accepting carbonyl group two carbons away from the carboxylate stabilizes the formal negative charge of the carbanionic transition state.

Eliminations

Eliminations of the type shown here

Elimination reaction

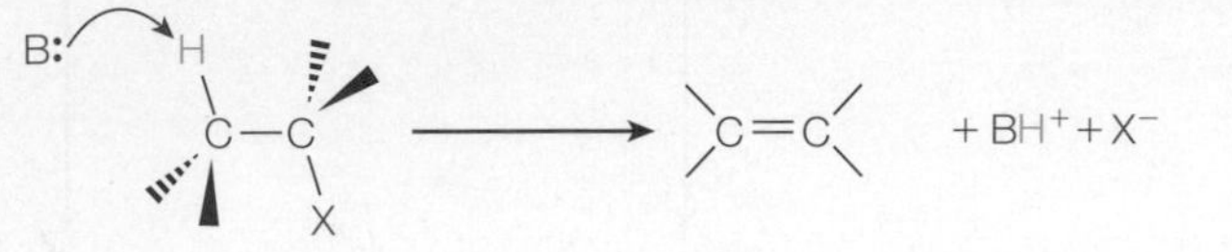

are also quite common in biochemical pathways and can occur by several different mechanisms.

The most common elimination mechanism involves a carbanion intermediate. The reactant is often a β-hydroxy carbonyl (where X = OH) in which the H atom to be removed is made more acidic by being adjacent to a carbonyl group. A base removes, or abstracts, the proton to give a carbanion intermediate (resonance-stabilized with the enolate) that loses OH⁻ to form the C=C double bond. β-Hydroxy carbonyl compounds are readily dehydrated via these α, β-elimination reactions.

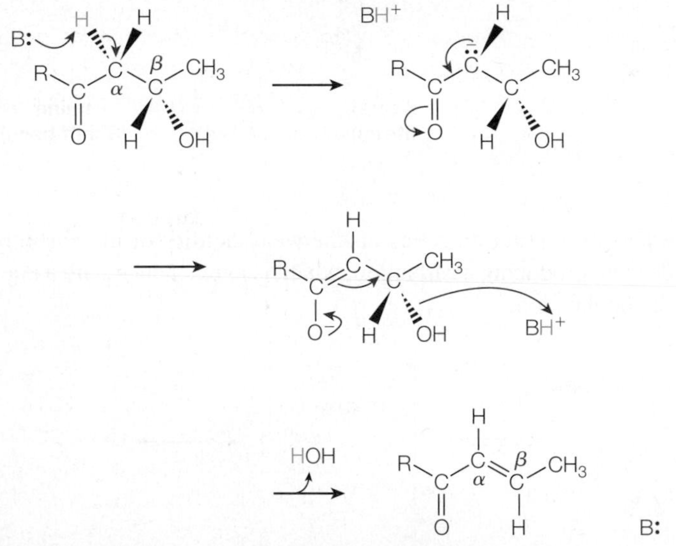

α,β-elimination reaction

Oxidations and Reductions

Energy production in most cells involves the oxidation of fuel molecules such as glucose. Oxidation–reduction, or **redox,** chemistry thus lies at the core of metabolism. Redox reactions involve reversible electron transfer from a donor (the **reductant**) to an acceptor (the **oxidant**). Cells have evolved a number of electron carriers, such as the NAD⁺ (nicotinamide adenine dinucleotide) coenzyme introduced in Section 8.5. Oxidations involving coenzymes such as NAD⁺ occur by a reversible hydride (H⁻) transfer mechanism, illustrated by the oxidation of an alcohol to a carbonyl compound.

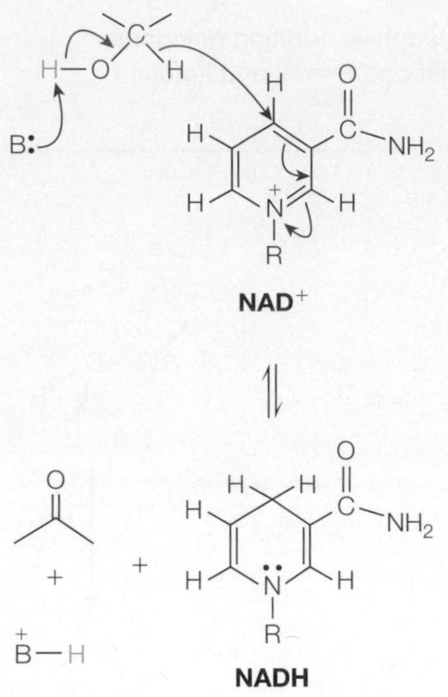

NAD⁺

NADH

A base abstracts the weakly acidic O—H proton, the electrons from that bond move to form a C=O bond, and the C—H bond is cleaved. The hydrogen *and the electron pair* (i.e., hydride, H⁻) add to NAD⁺ in a nucleophilic addition reaction, reducing it to NADH. The plus sign in NAD⁺ reflects the charge on the pyridine ring nitrogen in the oxidized form; this charge is lost as the electron pair moves through the ring onto this nitrogen. Because the alcohol has lost a pair of electrons and two hydrogen atoms, this type of oxidation is called a **dehydrogenation,** and enzymes that catalyze this reaction are called **dehydrogenases.**

Redox reactions are reversible, however, so dehydrogenases can catalyze the reductive direction as well. While two-electron oxidations exemplified by the NAD⁺-dependent dehydrogenases are the most common redox reactions in metabolism, they are not the only ones. A number of redox processes involve one-electron transfers, and various electron carriers exist to handle single electrons, as described in Chapter 14.

There are other, less common types of reactions in biochemical pathways, such as free radical reactions, but these five represent the basic toolkit that cells use to carry out the vast majority of their chemical transformations.

11.4 Bioenergetics of Metabolic Pathways

Chapter 3 introduced the principles of bioenergetics—the quantitative analysis of how organisms capture, transform, store, and utilize energy. Because the primary focus of the next few chapters is on energy metabolism, we now consider how metabolic energy is generated.

Oxidation as a Metabolic Energy Source

As we saw in Chapter 3, a thermodynamically unfavorable (endergonic) reaction will proceed readily in the unfavored direction if it can be coupled to a thermodynamically favorable (exergonic) reaction. In principle, any exergonic reaction can serve this purpose, provided that it releases more free energy than the unfavorable reaction consumes; that is, the ΔG of the combined reactions is negative. In living systems, most of the energy needed to drive biosynthetic reactions is derived from the *oxidation* of organic substrates. Oxygen, the ultimate electron acceptor for aerobic organisms, is a strong oxidant; it has a marked tendency to attract electrons, becoming reduced in the process. Given this tendency and the abundance of oxygen in our atmosphere, it is not surprising that living systems have gained the ability to derive energy from the oxidation of organic substrates.

Biological Oxidations: Energy Release in Small Increments

In a thermodynamic sense, the biological oxidation of organic substrates like glucose is comparable to nonbiological oxidations such as the burning of wood. The free energy release is the same whether we are talking about oxidation of the glucose polymer cellulose in a wood fire, combustion of glucose in a calorimeter, or the metabolic oxidation of glucose:

$$C_6H_{12}O_6 + 6O_2 \longrightarrow 6CO_2 + 6H_2O \quad \Delta G^{\circ\prime} = -2870 \text{ kJ/mol} \tag{11.1}$$

Equation 11.1 reveals the **reaction stoichiometry** of glucose combustion. Biological oxidations, however, are far more complex processes than combustion. When wood is burned, all of the energy is released as heat; useful work cannot be performed except through the action of a device such as a steam engine. In biological systems, by contrast, oxidation reactions occur without a large increase in temperature and in ways that capture some of the free energy as chemical energy, largely through the synthesis of ATP. Recall from Chapter 3 that the hydrolysis of ATP can be coupled to many processes to provide energy for biological work. In the catabolism of glucose, about 40% of the released energy is used to drive the synthesis of ATP from ADP and P_i.

Unlike the oxidation of glucose by oxygen shown in Equation 11.1, most biological oxidations do not involve direct transfer of electrons from a reduced substrate to oxygen (Figure 11.4). Instead, a series of coupled oxidation–reduction reactions occurs, with the electrons passed to intermediate electron carriers such as NAD^+ and FAD (flavin adenine dinucleotide, discussed in Chapter 13), before they are finally transferred to oxygen. Thus, the biological oxidation of glucose might be more accurately represented by the following coupled reactions:

$$C_6H_{12}O_6 + 10NAD^+ + 2FAD + 6H_2O \longrightarrow$$
$$6CO_2 + 10NADH + 10H^+ + 2FADH_2 \tag{11.2}$$
$$10NADH + 10H^+ + 2FADH_2 + 6O_2 \longrightarrow$$
$$10NAD^+ + 2FAD + 12H_2O \tag{11.3}$$
Net: $$C_6H_{12}O_6 + 6O_2 \longrightarrow 6CO_2 + 6H_2O \tag{11.4}$$

The net reaction of the biological oxidation process (Equation 11.4) is identical to that of direct combustion (Equation 11.1). Equations 11.2 and 11.3 are examples of **obligate-coupling-stoichiometry**—stoichiometric relationships fixed by the chemical nature of the process. Thus, the complete oxidation of glucose requires the transfer of 12 pairs of electrons from glucose to molecular oxygen, whether as a direct process, as in combustion, or via intermediate electron carriers, as in the biological process. In the biological process, 12 moles of electron carriers (NAD^+ and FAD) are obligately coupled to the oxidation of 1 mole of glucose to 6 moles of CO_2.

● **CONCEPT** Most biological energy derives from the oxidation of reduced metabolites in a series of reactions, with oxygen as the final electron acceptor.

The transfer of electrons from these intermediate electron carriers to oxygen is catalyzed by the **electron transport chain,** or **respiratory chain** (Figures 11.2 and 11.4), and oxygen is called the

terminal electron acceptor. Because the potential energy stored in the organic substrate is released in small increments, it is easier to control oxidation and capture some of the energy as it is released—small energy transfers waste less energy than a single large transfer does.

Not all metabolic energy comes from oxidation by oxygen. Substances other than oxygen can serve as terminal electron acceptors. Many microorganisms either can or must live anaerobically (in the absence of oxygen). For example, *Desulfovibrio* carry out anaerobic respiration using sulfate as the terminal electron acceptor:

$$SO_4^{2-} + 8e^- + 8H^+ \longrightarrow S^{2-} + 4H_2O$$

Most anaerobic organisms, however, derive their energy from **fermentations,** which are energy-yielding catabolic pathways that proceed with no net change in the oxidation state of the products as compared with that of the substrates. A good example is the production of ethanol and CO_2 from glucose (discussed in Chapter 12). Other anaerobic energy-yielding pathways are seen in some deep-sea hydrothermal vent bacteria, which reduce sulfur to sulfide as the terminal electron transfer reaction, and in other bacteria that reduce nitrite to ammonia. These organisms oxidize the substrates that sustain them, but they use terminal electron acceptors other than oxygen.

Energy Yields, Respiratory Quotients, and Reducing Equivalents

If metabolic energy comes primarily from oxidative reactions, then the more highly reduced a substrate, the greater its potential for generating biological energy. We can use a calorimeter to measure the heat output (enthalpy) from the oxidation of fat, carbohydrate, or protein. The combustion of fat provides more heat energy than the combustion of an equivalent mass of carbohydrate because they differ in the relative proportions of carbon, hydrogen, and oxygen they contain. In other words, fat has a higher **energy density** than carbohydrate. Compare, for example, the oxidation of glucose with the oxidation of a typical saturated fatty acid, palmitic acid:

$$C_6H_{12}O_6 + 6O_2 \longrightarrow 6CO_2 + 6H_2O$$
$$\Delta G^{\circ\prime} = -3.81 \text{ kcal/g} = -15.94 \text{ kJ/g}$$
$$C_{16}H_{32}O_2 + 23O_2 \longrightarrow 16CO_2 + 16H_2O$$
$$\Delta G^{\circ\prime} = -9.30 \text{ kcal/g} = -38.90 \text{ kJ/g}$$

● **CONNECTION** 1 kJ is equal to 0.239 kcal; 1 kcal (10^3 cal) is equal to 1 large calorie, or Cal, the unit favored by nutritionists and found on our food labels.

The caloric content of food is not identical to the Gibbs free energy of combustion because of corrections for efficiency of digestion and absorption. Thus, fat has an energy density of about 9 Cal/g compared to ~4 Cal/g for protein and carbohydrate.

When these substrates are oxidized, electrons are released in a particular obligate-coupling stoichiometry. As we will see in Chapter 14, it is the passage of these electrons down the respiratory chain that generates the electrochemical potential energy used to drive ATP synthesis. Because biological redox reactions rarely involve free electrons, we refer instead to **reducing equivalents,** defined as an amount of a reducing compound that donates the equivalent of 1 mole of electrons. A hydrogen atom (one proton and one electron)

CH$_2$OH

Glucose

$CH_3(CH_2)_{14}COOH$

Palmitic acid

has one reducing equivalent. The complete oxidation of glucose (Equations 11.2 and 11.3) produces 24 reducing equivalents (12 pairs of electrons). By comparison, the complete oxidation of palmitic acid produces 92 reducing equivalents (46 pairs of electrons). The carbons in palmitic acid are more highly reduced (92 reducing equivalents/16C = 5.75) than those in carbohydrate (24 reducing equivalents/6C = 4.0).

Another measure of the energy content of a substrate is the number of moles of CO_2 produced per mole of O_2 consumed during oxidation, a ratio called the **respiratory quotient (RQ).** In general, because oxygen is the final electron acceptor, the more highly reduced a substrate is, the more oxygen will be consumed during its oxidation. Let's compare the RQ for glucose to that for palmitic acid, using the preceding reaction stoichiometries. Oxidation of 1 mole of glucose requires 6 moles of O_2 and produces 6 moles of CO_2, giving an RQ of 1.0 ($6CO_2/6O_2$). Oxidation of 1 mole of palmitic acid requires 23 moles of O_2 and produces 16 moles of CO_2, giving an RQ of 0.70 ($16CO_2/23O_2$). Thus, the lower the RQ for a substrate, the greater the potential per mole of substrate for generating ATP.

Just as the breakdown of complex organic compounds yields both energy and reducing equivalents, the biosynthesis of such compounds utilizes both. For example, we know that both carbons of acetate are used for fatty acid biosynthesis:

$$8CH_3COO^- \longrightarrow \longrightarrow \longrightarrow CH_3(CH_2)_{14}COO^-$$
Acetate   **Palmitate**

Fifteen of the 16 carbon atoms of palmitate are highly reduced—14 at the methylene level ($-CH_2-$) and one at the methyl level ($-CH_3$). Therefore, many reducing equivalents are required to complete this biosynthesis.

The major donor of electrons for reductive biosynthesis is **NADPH—nicotinamide adenine dinucleotide phosphate (reduced).** $NADP^+$ and NADPH are identical to NAD^+ and NADH, respectively, except that $NADP^+$ and NADPH have an additional phosphate esterified at C2′ on the adenylate moiety. NAD^+ and $NADP^+$ are equivalent in their thermodynamic tendency to accept electrons; they have equal standard reduction potentials (Chapter 3). However, nicotinamide nucleotide-linked enzymes that act primarily in a catabolic direction usually use NAD^+/NADH, whereas those acting primarily in anabolic pathways use $NADP^+$/NADPH. In other words, as shown in **FIGURE 11.9**, nicotinamide nucleotide-linked enzymes that oxidize substrates (dehydrogenases) usually use NAD^+, and those enzymes that reduce substrates, called **reductases,** usually use NADPH.

Both dehydrogenases and reductases catalyze reversible reactions—the directions of the reactions are determined by the **redox state,** or the ratio of the oxidized and reduced forms of each cofactor pair that prevail in the cell. Thus, in a healthy cell, NAD^+ dominates the NAD^+/NADH ratio, whereas NADPH dominates the $NADP^+$/NADPH ratio. These ratios generally drive NAD^+-linked reactions in the oxidative direction and $NADP^+$-linked reactions in the reductive direction, for example,

● **CONCEPT** NAD^+ is the cofactor for most dehydrogenases that oxidize metabolites. NADPH is the cofactor for most reductases.

L-Malate + NAD^+ $\longrightarrow$ Oxaloacetate + NADH + H^+

Pyruvate + HCO_3^- + NADPH + H^+ $\longrightarrow$
L-Malate + $NADP^+$ + H_2O

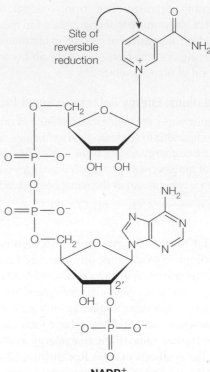

NADP⁺
Nicotinamide adenine
dinucleotide phosphate
(oxidized)

ATP as a Free Energy Currency

Why is ATP commonly referred to as a "free energy currency."? Currency is a medium of exchange. For example, a $20 bill has a generally recognized value that can be readily exchanged for various goods or services, such as lunch at a moderately priced restaurant or about 15 minutes of labor by a skilled auto mechanic. In the same sense, cells exchange the energy released from the breakdown of ATP for biologically useful work. Typically, they do this by converting the chemical energy released in ATP hydrolysis to other forms of energy involved in the performance of vital functions, such as mechanical energy in muscle contraction (Chapter 7), electrical energy in conducting nerve impulses (Chapter 10), or osmotic energy in transporting substances across membranes against a concentration gradient (Chapter 10).

Thus ATP serves as an immediate donor of free energy, continuously being formed and consumed. It is estimated that a resting human turns over as much as 65 kg of ATP every 24 hours, about equal to the entire body weight! (The Na^+/K^+–ATPase you learned about in Chapter 10 accounts for approximately 25% of this resting ATP consumption.) During strenuous exercise, ATP turnover can be as high as 0.5 kg/min.

Figures 11.3 and 11.4 show that the regeneration of all this ATP is coupled to fermentative or oxidative processes. In fact, as we will see in Chapter 14, the oxidation of 1 mole of glucose during respiration is coupled to the phosphorylation of ~ 32 moles of ADP:

$$C_6H_{12}O_6 + 6O_2 + 32ADP + 32P_i \longrightarrow$$
$$6CO_2 + 38H_2O + 32ATP \quad (11.5)$$

The stoichiometries for ADP, P_i, and ATP in Equation 11.5 are fundamentally different from the simple reaction and obligate-coupling

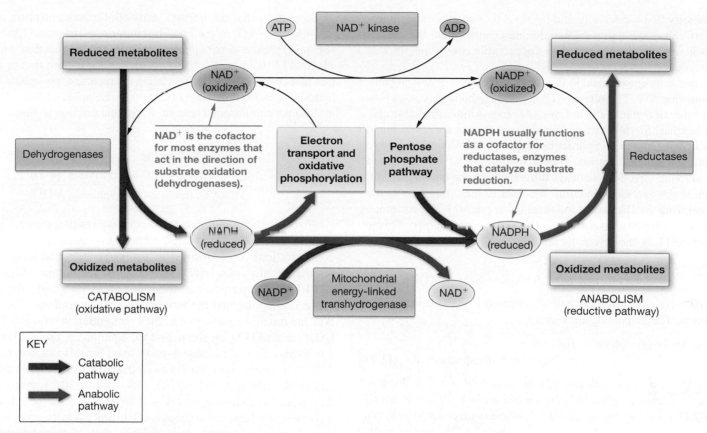

▲ **FIGURE 11.9 Nicotinamide nucleotides in catabolism and biosynthesis.** NAD^+ is the cofactor for most enzymes that act in the direction of substrate oxidation (dehydrogenases), whereas NADPH usually functions as a cofactor for enzymes that catalyze substrate reduction (reductases). NADPH is regenerated either from $NADP^+$ in the pentose phosphate pathway (Chapter 12) or from NADH through the action of mitochondrial energy-linked transhydrogenase (Chapter 14). $NADP^+$ is synthesized from NAD^+ by an ATP-dependent kinase reaction.

stoichiometries we have encountered thus far (see Equations 11.2–11.4). There is no chemical necessity for the production of 32 moles of ATP; in fact, there is not even a direct chemical connection between glucose oxidation and ATP synthesis. We could not predict a stoichiometry of 32 ATP solely from chemical considerations. The energy released by this oxidation could in theory be used to produce a different number of ATP molecules—or even to drive some other altogether different reaction. This is instead an **evolved-coupling stoichiometry.** Evolved-coupling stoichiometries are biological adaptations, phenotypic traits that are acquired during evolution—they are the result of compromise.

To understand the nature of that compromise, we can ask, why did evolution settle on a stoichiometry of ~32 ATP? Recall from Chapter 3 that the free energy changes of coupled reactions are additive. Thus, Equation 11.5 can be broken into its free energy yielding and free energy requiring processes:

$$\Delta G\,(\text{kJ/mol glucose})$$

$$C_6H_{12}O_6 + 6O_2 \longrightarrow 6CO_2 + 6H_2O \qquad -2900$$

$$32ADP + 32P_i \longrightarrow 32H_2O + 32ATP \quad \Delta G = +50\ \text{kJ/mol}$$

$$\times 32\ \text{mol ATP} = +1600$$

Sum: $C_6H_{12}O_6 + 6O_2 + 32ADP + 32P_i \longrightarrow$

$$6CO_2 + 38H_2O + 32ATP \qquad -1300$$

Using estimates of the Gibbs free energy changes under physiological conditions (ΔG) of -2900 kJ/mol for glucose oxidation and $+50$ kJ/mol for the synthesis of each mole of ATP (Equation 3.31), we can see that the process, assuming an evolved-coupling stoichiometry of 32 ATP, occurs with a net ΔG of -1300 kJ/mol glucose. A ΔG of this magnitude represents such an immense driving force that respiration is emphatically favorable under virtually any physiological condition and thus goes to completion.

● **CONCEPT** The stoichiometry for ATP production in respiration is an evolved-coupling stoichiometry, a phenotypic trait acquired during evolution.

Imagine, instead, an ancient cell that mutated in such a way that it acquired an ATP-coupling stoichiometry for respiration of 58 instead of 32. Repeating the calculations above, the energy-requiring process would have a ΔG of $+2900$ kJ/mol glucose (58 ATP X $+50$ kJ/mol), giving an overall ΔG of 0 for respiration. Thus, there would be no net driving force for the reaction. Although this hypothetical cell might have had a higher yield of ATP per glucose oxidized, it would come to equilibrium before much glucose was metabolized. This cell would probably have been a poor competitor, especially if glucose concentrations were limiting, and the mutation(s) that led to this stoichiometry would likely have been selected against. The ATP-coupling stoichiometry that we actually find (~32) is thus an evolved compromise between maximizing ATP yield and ensuring that the overall process is unidirectional under any conceivable conditions the cell might encounter.

Of course, as emphasized in Figure 11.1, many biochemical pathways consume ATP. The role of ATP in these anabolic processes is to convert a thermodynamically unfavorable process into a favorable one. Remember that biosynthetic and degradative pathways are never simple reversals of one another. In particular, opposed pathways always have different ATP stoichiometries. That is, the number of ATP produced in the catabolic direction is always different from the number of ATP required in the anabolic direction. We refer to these numbers as the **ATP-coupling coefficient** of the reaction or pathway. For example, in the fructose-6-phosphate/fructose-1,6-bisphosphate substrate cycle (see Section 11.2), the glycolytic reaction,

$$\text{Fructose-6-phosphate} + \text{ATP} \longrightarrow$$
$$\text{fructose-1,6-bisphosphate} + \text{ADP} \quad (11.6)$$

has an ATP-coupling coefficient of –1, because 1 ATP is consumed in the process. The gluconeogenic reaction,

$$\text{Fructose-1,6-bisphosphate} + \text{H}_2\text{O} \longrightarrow$$
$$\text{fructose-6-phosphate} + \text{P}_i \quad (11.7)$$

has an ATP-coupling coefficient of 0 because ATP is not involved in the process. Similarly, glycolysis (as we will see in Chapter 12) has an overall ATP-coupling coefficient of +2, whereas gluconeogenesis has an overall ATP-coupling coefficient of −6. Such differences in ATP-coupling coefficients between opposed reactions or pathways ensure that each pathway is thermodynamically favorable (that is, has a negative ΔG) in its respective direction.

> **● CONCEPT** The fundamental biological role of ATP as an energy-coupling compound is to convert thermodynamically unfavorable processes into favorable ones.

What do we really mean when we say that ATP converts a thermodynamically unfavorable process into a favorable one? Coupling ATP hydrolysis to a pathway provides a new chemical route, using different reactions with different stoichiometries, resulting in a different overall equilibrium constant for the process. In fact, coupling ATP hydrolysis to a process changes the equilibrium ratio of certain reactant to product concentrations by a factor of 10^8! For example, consider how the equilibrium ratio of [fructose-1,6-bisphosphate] to [fructose-6-phosphate] changes when we compare a direct addition of P_i to fructose-6-phosphate to the reaction pathway that is coupled to ATP hydrolysis. The direct addition of P_i (Equation 11.8) is simply the reverse of Equation 11.7:

$$\text{Fructose-6-phosphate} + \text{P}_i \rightleftharpoons$$
$$\text{fructose-1,6-bisphosphate} + \text{H}_2\text{O} \quad (11.8)$$

In the forward direction, the standard free energy change, $\Delta G°'$, for Equation 11.8 is +16.3 kJ/mol. Using Equation 3.22, we can calculate the equilibrium constant for this reaction:

$$K = e^{\left(\frac{-\Delta G°'}{RT}\right)} = e^{\left(\frac{-16300 \text{ J/mol}}{(8.315 \text{ J/mol·K})(298 \text{ K})}\right)} = 0.0014$$
$$= \frac{[\text{fructose-1,6-bisphosphate}]}{[\text{fructose-6-phosphate}][\text{P}_i]} \quad (11.9)$$

In many cells, the normal intracellular concentration of P_i is ~1 mM (10^{-3} M), so the equilibrium ratio of [fructose-1,6-bisphosphate]/[fructose-6-phosphate] achieved by this reaction would be $(0.0014)(10^{-3}) = 1.4 \times 10^{-6}$. Now let's compare that to a reaction that couples the phosphorylation of fructose-6-phosphate to ATP hydrolysis, as in Equation 11.6. $\Delta G°'$ for Equation 11.6 is −14.2 kJ/mol, so the equilibrium constant, K, for this reaction is 308:

$$K = e^{\left(\frac{14200 \text{ J/mol}}{(8315 \text{ J/mol · K})(298 \text{ K})}\right)} = 308$$
$$= \frac{[\text{fructose-1,6-bisphosphate}][\text{ADP}]}{[\text{fructose-6-phosphate}][\text{ATP}]} \quad (11.10)$$

To compare the resulting equilibrium ratio of [fructose-1,6-bisphosphate]/[fructose-6-phosphate], we need to estimate the intracellular concentrations of adenosine diphosphate (ADP) and ATP. In most normal, healthy cells, [ATP] is several-fold (3–10 times) higher than [ADP]. For the purposes of this calculation, let's make the simplifying assumption that the intracellular concentrations of ADP and ATP are roughly equal—that is, their concentration ratio is ~1. Thus, [ADP] and [ATP] cancel out, and the equilibrium ratio of [fructose-1,6-bisphosphate]/[fructose-6-phosphate] for this reaction becomes 308. This equilibrium ratio is more than 10^8 times higher than that achieved without ATP coupling: $308/(1.4 \times 10^{-6}) = 2.2 \times 10^8$. This effect is completely general—it depends only on the existence of a chemical mechanism that couples ATP hydrolysis to the process at hand (e.g., an enzyme like phosphofructokinase that catalyzes Equation 11.6 in glycolysis).

Note that ATP is not directly hydrolyzed in this reaction. Instead, it is converted to ADP as its phosphate is transferred to fructose-6-phosphate.

> **● CONCEPT** Any process that couples the conversion of ATP to ADP gains the thermodynamic equivalent of the free energy of ATP hydrolysis.

However, *if a process is coupled to the conversion of ATP to ADP, it gains the thermodynamic equivalent of the free energy of hydrolysis of ATP.*

Metabolite Concentrations and Solvent Capacity

The use of ATP as the fundamental coupling agent in biochemical systems has had far-reaching implications for the evolutionary design of cell metabolism. Evolution has produced a complex cell metabolism comprising thousands of enzymes and metabolic intermediates, each at very low individual concentrations. A cell must maintain its components at very low concentrations for at least two reasons. First, even a simple bacterial cell contains several thousand different metabolites dissolved in the aqueous compartment. This aqueous compartment has a finite capacity for the total amount of dissolved substances (metabolites and macromolecules); this is its **solvent capacity.** Thus, individual metabolites must exist at low concentrations (10^{-3}–10^{-6} M, or even lower) to avoid exceeding the solvent capacity of the cell.

Second, low metabolite concentrations minimize unwanted side reactions. For example, imagine two metabolites, A and B, that can react nonenzymatically to form C. Assume that this unwanted side reaction is first-order (see Section 8.3) with respect to each metabolite, so that the reaction rate is directly proportional to the product of [A] and [B] ($\nu = k[A][B]$). We can compare the amount of C that would be produced under two different metabolite concentrations: 1 M each versus 10^{-5} M each. Whatever the rate constant k is, the velocity of the unwanted side reaction at the two different concentrations will differ by a factor of 10^{10} (i.e., $10^{-5} \times 10^{-5}$). Thus, the same amount of C that would be produced in 1 second if each metabolite were at 1 M would take ~317 years (10^{10} seconds) to produce if each metabolite were at 10^{-5} M instead.

How does ATP coupling help avoid high-metabolite concentrations? It does so by activating metabolic intermediates. Let's return to our previous example, comparing the phosphorylation of fructose-6-phosphate using inorganic phosphate to phosphorylation using ATP. Assume that the cell needs to maintain a [fructose-1,6-bisphosphate]/[fructose-6-phosphate] ratio of 10 to thrive. What concentration of P_i is required to ensure this nonequilibrium ratio using Equation 11.8? We can solve for the required $[P_i]$ by recognizing that $\Delta G < 0$ for this reaction and by rearranging the relevant form of Equation 3.18:

$$\Delta G = \Delta G^{\circ\prime} + RT \ln\left(\frac{[\text{F-1,6-BP}][\text{H}_2\text{O}]}{[\text{F-6-P}][\text{P}_i]}\right) < 0$$

Thus,

$$\ln\left(\frac{[\text{F-1,6-BP}][\text{H}_2\text{O}]}{[\text{F-6-P}][\text{P}_i]}\right) < \frac{-\Delta G^{\circ\prime}}{RT} \qquad (11.11)$$

$$\ln\left(\frac{[10][1]}{[1][\text{P}_i]}\right) < \frac{\dfrac{-16.3 \text{ kJ}}{\text{mol}}}{\left(\dfrac{0.008314 \text{ kJ}}{\text{mol K}}\right)(298 \text{ K})} \qquad (11.12)$$

Solving for $[P_i]$ gives

$$[\text{P}_i] > \frac{10}{0.0014} = 7143 \text{ M}$$

There is no way to fit 7000+ moles of phosphate into a liter of aqueous solution. And even if the cell could survive with a [fructose-1,6-bisphosphate]/[fructose-6-phosphate] ratio of 1, it would still need $[P_i] > 714$ M to make the reaction thermodynamically favorable (i.e., to make $\Delta G < 0$)—still an impossibly high concentration.

Now consider the process coupled to ATP hydrolysis, via the phosphofructokinase reaction (Equation 11.6). Here, the relevant calculation is

$$\frac{[\text{F-1,6-BP}][\text{ADP}]}{[\text{F-6-P}][\text{ATP}]} < e^{\left(\frac{-\Delta G^{\circ\prime}}{RT}\right)} \qquad (11.13)$$

where $\Delta G^{\circ\prime} = -14.2$ kJ/mol. If we assume an [ADP]/[ATP] ratio of ~1 under physiological conditions, then the maximum ratio of [fructose-1,6-bisphosphate]/[fructose-6-phosphate] that still makes the reaction favorable is

$$\frac{[\text{fructose-1,6-bisphosphate}]}{[\text{fructose-6-phosphate}]} < 308$$

Thus, when the reaction is coupled to ATP hydrolysis, it becomes thermodynamically favorable as long as the ratio of [F-1,6-BP] to [F-6-P] remains under 308—far above the desired ratio of 10. This scenario completely avoids the involvement of impossible concentrations of inorganic phosphate. In this example, ATP can be thought of as an activated form of phosphate. Cells have evolved a number of activated intermediates besides ATP, including acetyl-CoA,

● **CONCEPT** Activated intermediates, such as ATP, allow reactions to occur under physiologically relevant concentrations of metabolic intermediates.

acyl-lipoate, and nucleoside diphosphate sugars (NDP-sugars). We discuss each of these intermediates in later chapters, but suffice it to say here that they all function by allowing reactions to occur under physiologically relevant concentrations of intermediates.

Thermodynamic Properties of ATP

What factors equip ATP for its special role as energy currency? First, there is nothing unique about the chemistry of ATP; the phosphoanhydride bonds whose breakdown can be coupled to drive endergonic reactions are shared by all other nucleoside di- and triphosphates and by several other metabolites as well. When we call ATP a "high-energy compound," as we did in Chapter 3, we use that term within a defined context: a high-energy compound is one containing at least one bond with a sufficiently favorable ΔG° of hydrolysis. ATP has two phosphoanhydride bonds. Cleavage of one yields ADP and inorganic phosphate (P_i), whereas cleavage of the other gives adenosine monophosphate (AMP) and pyrophosphate (PP_i). Either reaction proceeds with a large negative $\Delta G^{\circ\prime}$ of hydrolysis (Figure 3.7).

Calling a substance a high-energy compound does *not* mean that it is chemically unstable or unusually reactive. In fact, ATP is *kinetically* stable—its spontaneous hydrolysis is slow at physiological pH and temperature. When hydrolysis does occur, however, whether spontaneously or enzyme-catalyzed, substantial free energy is released. Keep in mind, though, that utilizing that energy to drive endergonic reactions usually does *not* involve hydrolysis. Rather, ATP breakdown (as we saw previously) is usually coupled with a thermodynamically unfavorable reaction, such as the synthesis of glucose-6-phosphate from glucose. In this case, the phosphate released from ATP does not become P_i. Instead, it is transferred directly to glucose, forming the esterified phosphate of glucose-6-phosphate. Thus, it is more accurate to say that ATP has a "high phosphoryl group transfer potential" than to call it a high-energy compound. As long as you understand the context in which the term is used, though, the concept of a high-energy compound is quite useful.

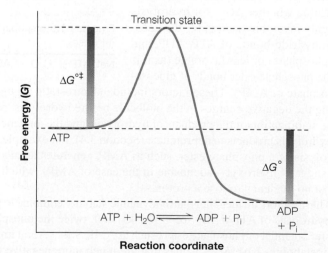

Phosphoanhydride bonds are *thermodynamically* unstable, but *kinetically* stable—large free energies of activation require enzymes to lower the activation barrier.

The product of ATP hydrolysis (top) has a much greater tendency to lose a proton than the product of AMP hydrolysis (bottom).

intermediate on the scale of "energy wealth," which means that the breakdown of a compound such as phosphoenolpyruvate can be coupled to drive the synthesis of ATP from ADP and P_i. In fact, these kinds of coupled reactions, called **substrate-level phosphorylation** reactions, are how ATP is synthesized in glycolysis, as we will see in Chapter 12.

The Important Differences Between ΔG and ΔG°′

You might be wondering where the energy needed for the synthesis of compounds with a much *higher* phosphate transfer potential than that of ATP itself comes from. Much of the answer lies in the fact that ΔG values for reactions under intracellular conditions are quite different from their standard ($\Delta G°′$) values. This is mainly because intracellular concentrations are far different from the 1 M concentrations used to compute standard free energies. If you work Problem 6 at the end of this chapter, you will see that ATP hydrolysis has a considerably more negative ΔG value at intracellular concentrations of ATP, ADP, AMP, and P_i than it has under standard conditions. In fact, the physiological ratio of [ATP]/[ADP] is ~10^8 times higher than the equilibrium ratio.

This value is not coincidental—if the physiological [ATP]/[ADP] ratio is 10^8 times larger than its equilibrium ratio, then the equilibrium ratio of any other reaction that is coupled to ATP hydrolysis will be altered by this same order of magnitude, as we calculated earlier. Maintaining the physiological ratio of [ATP]/[ADP] so far away from equilibrium is accomplished by *kinetic* control—that is, by the regulation of enzymes. The key point is that maintaining the physiological ratio so far from equilibrium provides the thermodynamic driving force for nearly every biochemical event in the cell.

Kinetic Control of Substrate Cycles

Substrate cycles beautifully illustrate the advantages of ATP as a coupling agent and the role of kinetic control over the direction, as well as the rate, of these opposing pathways. Let's return once again to the fructose-6-phosphate/fructose-1,6-bisphosphate substrate cycle:

	$\Delta G°′$	K	ATP coupling coefficient
Fructose-6-phosphate + ATP ⟶ fructose-1,6-bisphosphate + ADP (PFK)	−14.2	308	−1
Fructose-1,6-bisphosphate + H_2O ⟶ fructose-6-phosphate + P_i (FBPase)	−16.3	719	0

Net: ATP + H_2O → ADP + P_i

Recall from Chapter 3 that several factors contribute to the thermodynamic stability of a hydrolyzable bond and determine whether $\Delta G°′$ for hydrolysis is highly favorable (as it is for the phosphoanhydride bonds of ATP, ADP, and pyrophosphate) or less favorable (as it is for the phosphate ester bonds of glucose-6-phosphate or AMP). These factors include electrostatic repulsion among the negative charges in the molecule before hydrolysis, resonance stabilization of the products of hydrolysis, and the tendency of the hydrolysis products to deprotonate (Section 3.4). For example, the hydrolysis of a phosphate ester, such as AMP, generates an alcohol (the sugar 5′ hydroxyl—adenosine in the case of AMP), which has almost no tendency to lose a proton.

These factors influencing thermodynamic stability combine to give the hydrolysis of ATP a $\Delta G°′$ of −32.2 kJ/mol, twice the phosphate transfer potential of phosphate esters such as AMP. Still, several important metabolites have $\Delta G°′$ values that are much more negative than that of ATP. Examples shown in Figure 3.7 include phosphoenolpyruvate (−61.9 kJ/mol), 1,3-bisphosphoglycerate (−49.4 kJ/mol), and creatine phosphate (−43.1 kJ/mol). ATP, therefore, is actually

For the phosphofructokinase (PFK) reaction under physiological conditions, where [ATP] ≈ [ADP], we calculated a value of 308 for the equilibrium ratio of [fructose-1,6-bisphosphate]/[fructose-6-phosphate] (Equation 11.10). Thus, this reaction is favorable in the direction of fructose-1,6-bisphosphate until its concentration approaches a level 308 times that of fructose-6-phosphate. Likewise, we can calculate the equilibrium ratio of [fructose-6-phosphate]/[fructose-1,6-bisphosphate] for the fructose-1,6-bisphosphatase (FBPase) reaction using the physiological concentration of P_i (~10^{-3} M):

$$K = 719 = \frac{[\text{fructose-6-phosphate}]\,[P_i]}{[\text{fructose-1,6-bisphosphate}]}$$

$$\frac{719}{0.001} = \frac{[\text{fructose-6-phosphate}]}{[\text{fructose-1,6-bisphosphate}]} = 719{,}000$$

Thus, under any likely cellular condition, the FBPase reaction is thermodynamically favorable in the direction of fructose-6-phosphate.

It follows, then, that *both* the PFK and FBPase reactions are thermodynamically favorable in their respective directions as long as the [fructose-1,6-bisphosphate]/[fructose-6-phosphate] ratio is between 0.00000139 (1/719,000) and 308. In fact, the ratio will *always* be in this wide range in a healthy cell, so both reactions are *always* thermodynamically favorable. What, then, prevents these two reactions from occurring simultaneously, accomplishing nothing more than the net hydrolysis of ATP (i.e., a futile cycle)? The answer is the independent, but coordinated, kinetic control of the two opposed enzymes. As we will see in the next chapter, enzymes in substrate cycles are kinetically regulated by the levels of allosteric effectors.

Thus, substrate cycles illustrate the important metabolic design feature mentioned previously: ATP-coupling coefficients of opposing reactions or pathways always differ. This difference allows both sequences to be thermodynamically favorable at all times. The choice of which pathway operates, however, is determined entirely by the metabolic needs of the cell (via allosteric effectors), not by thermodynamics (since both pathways are favorable). Why is it so important for both pathways to be thermodynamically favorable at all times? *Because regulation can be imposed only on reactions that are displaced far from equilibrium.* Consider the analogy illustrated in the accompanying figure, in which a dam separates two bodies of water. If the water level is the same on both sides of the dam, the system is at equilibrium. What happens if we now open the floodgate? Water may move back and forth through the floodgate, but there will be no net movement of water or change in water levels. Thus, regulation (opening the floodgate) has no impact on a system at equilibrium. Now imagine a system in which the water level on the left side of the dam is much higher than the level on the right side. This system is far from equilibrium. If we now open the floodgate, water will rush from the left side to the right side and will continue flowing until it reaches equilibrium, or until we impose regulation (closing the floodgate).

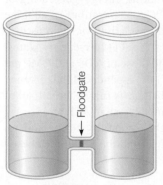

Equilibrium

Far from equilibrium

The fact that each pathway of an opposing pair is thermodynamically favorable (made possible by the different ATP-coupling coefficients for each) results in each pathway being unidirectional (because the reverse direction is highly unfavorable thermodynamically). These opposing pathways (e.g., glycolysis vs. gluconeogenesis) are like dams holding back high waters, each poised to flow downhill (thermodynamics), just

● **CONCEPT** Regulation can be imposed only on reactions that are displaced far from equilibrium.

waiting for the signal to open the floodgates (kinetics). The signal (e.g., allosteric effectors) will be different for each floodgate (regulatory enzyme), so that both pathways never flow at the same time (though both could if allowed).

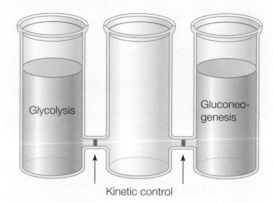

Kinetic control

The theoretical basis of this concept was developed in Section 3.3, where we introduced the relationship of Q (the mass action ratio) to K (the equilibrium constant). Recall that the free energy is at a minimum when the Q/K ratio $= 1$ (i.e., when the actual concentrations of reactants and products equal their equilibrium concentrations). In other words, when the system is at equilibrium, $Q = K$ and $\Delta G = 0$. The magnitude of ΔG is greatest when the mass action ratio is far from equilibrium concentrations, and the further the reaction is from equilibrium, the greater the driving force:

- When the concentration of reactant is high and the concentration of product is low ($Q/K < 1$), ΔG will be large and negative, and the reaction will proceed toward product due to the large driving force.

- When the concentration of product is high and the concentration of reactant is low ($Q/K > 1$), ΔG will be large and positive, and the reaction will proceed toward reactant in response to an equally large driving force in the opposite direction.

Earlier we estimated that ΔG for the oxidation of one mole of glucose during respiration is approximately -1300 kJ/mol. We can calculate the corresponding Q/K ratio at 298 K using Equation 3.27:

$$\Delta G = RT \ln\left(\frac{Q}{K}\right)$$

$$\frac{Q}{K} = e^{\left(\frac{\Delta G}{RT}\right)} = e^{\left(\frac{-1300 \times 10^3 \text{ J/mol}}{(8315 \text{ J/mol}) \cdot (298 \text{ K})}\right)} \approx 10^{-228}$$

This Q/K ratio, 10^{-228}, is based on approximations of ΔG, so it might be off by a few orders of magnitude. Whatever the actual value, however, it is extremely far from equilibrium and thus represents an immense driving force.

Other High-Energy Phosphate Compounds

This principle, that the mass action ratio can make thermodynamically unfavorable reactions favorable, allows ATP to drive the synthesis of compounds of even higher phosphate transfer potential, such as creatine phosphate (CrP). CrP shuttles phosphate bond energy from ATP in mitochondria to myofibrils, where that bond energy is transduced to the mechanical energy of muscle contraction. CrP is produced from creatine by the enzyme **creatine kinase:**

Creatine

creatine kinase

Creatine phosphate

Based on the respective $\Delta G°'$ values for CrP and ATP (see Figure 3.7), this reaction is endergonic under standard conditions ($\Delta G°' = +10.9 \text{ kJ/mol}$). However, because ATP levels are high within mitochondria and CrP levels are relatively low ($Q/K \ll 1$), the reaction is exergonic as written and proceeds toward creatine phosphate in mitochondria.

● **CONCEPT** ATP can drive the synthesis of higher-energy compounds, if nonequilibrium intracellular concentrations make the reactions exergonic.

Other High-Energy Nucleotides

As discussed earlier, there is nothing unique about the properties endowing ATP with its special role as energy currency. All other nucleoside triphosphates, as well as more complex nucleotides, such as NAD^+, have $\Delta G°'$ values close to −31 kJ/mol and could have been selected instead of ATP to be the energy currency in cells. However, evolution has created an array of enzymes that preferentially bind ATP and use its free energy of hydrolysis to drive endergonic reactions. There are exceptions, such as the use of guanosine triphosphate (GTP) as the primary energy-providing nucleotide in protein synthesis, but phosphate–bond energy is created almost exclusively as ATP, through oxidative phosphorylation in aerobic cells, photosynthesis in plants, and substrate-level phosphorylation during glycolysis in virtually all organisms. As a result, ATP is usually the most abundant nucleotide.

In most cells, ATP levels, at 2–8 mM, are severalfold higher than those of the other nucleoside triphosphates and also severalfold higher than the levels of ADP or AMP. These factors give ATP a strong tendency to distribute its γ (outermost) phosphate in the synthesis of other nucleoside triphosphates. This is accomplished through the action of **nucleoside diphosphate kinase,** which synthesizes cytidine triphosphate (CTP) from cytidine diphosphate (CDP) in the following example.

$$\text{ATP} + \text{CDP} \rightleftharpoons \text{ADP} + \text{CTP}$$

Nucleoside diphosphate kinase is active with a wide variety of phosphate donors and acceptors. Because its equilibrium constant is close to unity and because ATP is the most abundant nucleotide within cells, the enzyme normally uses ATP to drive the synthesis of the other common ribo- and deoxyribonucleoside triphosphates from their respective diphosphates.

Some metabolic reactions, such as the activation of amino acids for protein synthesis, cleave ATP, not to ADP and P_i, but to AMP and PP_i. Not only does this hydrolysis yield a bit more free energy ($\Delta G°' = -45.6 \text{ kJ/mol}$; Figure 3.7), but the subsequent hydrolysis of PP_i to $2P_i$ provides an additional driving force ($\Delta G°' = -19.2 \text{ kJ/mol}$). The conversion of AMP to ATP, allowing reuse of the nucleotide, involves another enzyme, **adenylate kinase** (also called myokinase because of its abundance in muscle).

$$\text{AMP} + \text{ATP} \rightleftharpoons 2\text{ADP}$$

ADP is reconverted to ATP by substrate-level phosphorylation, oxidative phosphorylation, or photosynthetic energy (in plants). Because the reaction is readily reversible, it can also be used for the resynthesis of ATP when ADP levels rise, as after a burst of energy consumption. This function is particularly important in muscle metabolism.

Adenylate Energy Charge

Many enzymes that participate in regulating energy-generating or storage pathways are acutely sensitive to concentrations of adenine nucleotides. In general, energy-generating pathways, such as glycolysis and the citric acid cycle, are activated at low-energy states, when levels of ATP are relatively low and those of ADP and AMP are relatively high. Daniel Atkinson (1977) in Appendix II, has likened the cell to a battery. When the cellular battery is fully charged, all of the adenine ribonucleotides are present in the form of ATP, and we say that the **adenylate energy charge** is high. When fully discharged, all of the ATP has been broken down to AMP, and the adenylate energy charge is low.

11.5 Major Metabolic Control Mechanisms

The living cell uses a remarkable array of regulatory devices to control its functions. These include:

- Control of enzyme *concentration* through the regulation of enzyme synthesis (Chapters 24–26) and degradation (Chapter 18).

- Mechanisms that act primarily to control enzyme *activity*, such as substrate concentration, allosteric control, and covalent modification of enzymes (Chapter 8).

- *Compartmentation* (in eukaryotic cells), which makes it possible to regulate the concentration of metabolites by controlling their flow through membranes (Chapter 10).

- Overlying all of these mechanisms are the actions of *hormones*, chemical messengers that act at all levels of regulation (Chapter 17).

Control of Enzyme Levels

If you were to prepare a cell-free extract of a particular tissue and determine intracellular concentrations of several different enzymes, you would find tremendous variations. Enzymes of the central energy-generating pathways are present at many thousands of molecules per cell, whereas enzymes that have limited or specialized functions might

● **CONCEPT** Enzyme levels in a cell may change in response to changes in metabolic needs.

be present at fewer than a dozen molecules per cell. Two-dimensional gel electrophoresis of a cell extract (see Figure 1.14) gives an impression, from the varying spot intensities, of the wide variations in amounts of individual proteins in a particular cell.

The level of a particular enzyme can also vary widely from tissue to tissue and under different environmental conditions. For example, when a usable substrate is added to a bacterial culture, the abundance of the enzymes needed to process the substrate may increase, through synthesis of new enzymes, from less than one molecule per cell to many thousands of molecules per cell. This phenomenon is called enzyme **induction.** Similarly, the presence of the end product of a pathway may turn off the synthesis of enzymes needed to generate that end product, a process called **repression.**

For some time it was thought that controlling the intracellular level of a protein was primarily a matter of controlling the *synthesis* of that protein—in other words, through genetic regulation. We now know that intracellular protein *degradation* is also important in determining enzyme levels, as we will see in Chapter 18.

Control of Enzyme Activity

The *catalytic activity* of an enzyme molecule can be controlled in two ways: by reversible interaction with ligands (such as substrates, products, and allosteric modifiers) or by covalent modification of the protein molecule.

● **CONCEPT** Enzyme activity is regulated by interaction with substrates, products, and allosteric effectors or by covalent modification of the enzyme.

Enzyme activity is most commonly controlled by low-molecular-weight ligands, principally substrates and allosteric effectors. Substrates are usually present within cells at concentrations lower than the K_M values for the enzymes that act on them, but generally within an order of magnitude of these values. In other words, substrate concentrations usually lie within the first-order (linear) ranges of substrate concentration–velocity curves for the enzymes that act on them (recall Figure 8.21). Therefore, reaction velocities respond to small changes in substrate concentration. Ligands that control enzyme activity can also be polymers. For example, protein–protein interactions can affect enzyme activity, and several enzymes of nucleic acid metabolism are activated by binding to DNA.

We saw in Chapter 8 that allosteric activation or inhibition by effectors usually acts on committed steps of a metabolic pathway, often initial reactions. The effectors function by binding at specific regulatory sites, thereby affecting subunit–subunit interactions in the enzyme. This effect, in turn, either facilitates or hinders the binding of substrates. Such a mechanism controls product formation if a pathway is unidirectional and unbranched. Some substrates, however, are involved in numerous pathways, so many branch points exist. Therefore, some of the allosteric enzymes that we will describe display somewhat more complicated regulation than the examples presented in Chapter 8.

Covalent modification of enzyme structure represents another efficient way to control enzyme activity. Chapter 8 introduced several types of covalent modification that are used to regulate enzyme activity, including *phosphorylation*, *acetylation*, and *adenylylation*. Many other less common covalent modifications are now known. Phosphorylation, though, is by far the most widespread.

Control through covalent modification is often associated with regulatory cascades. Modification activates an enzyme, which in turn acts on a second enzyme, which may activate yet a third enzyme, which finally acts on the substrate. Because enzymes act catalytically, this cascading provides an efficient way to *amplify* the original biological signal. Suppose that the original signal modifying enzyme A activates it 10-fold, that modified enzyme A then activates enzyme B by 1000-fold, and that B activates enzyme C by 100-fold. Thus, with the involvement of relatively few molecules of enzyme, a pathway can be activated by a million-fold ($10^1 \times 10^3 \times 10^2$).

The first well-understood regulatory cascade was the one controlling glycogen breakdown in animal cells, a critical process that provides carbohydrate substrates for energy generation. This regulatory cascade, involving enzyme phosphorylation and dephosphorylation, is described in detail in Chapter 12. Blood clotting, described in Chapter 8, is another well-understood regulatory cascade.

Compartmentation

We have already described the physical division of labor that exists in a eukaryotic cell, in the sense that enzymes participating in the same process are localized to a particular *compartment* within the cell. For example, RNA polymerases are found in the nucleus and nucleolus, where DNA transcription occurs, and the enzymes of the citric acid cycle are all found in mitochondria. **FIGURE 11.10** presents the locations of a number of metabolic pathways within eukaryotic cells.

Compartmentation increases the efficiency of cell function. The creatine–creatine phosphate shuttle discussed in the previous section is a good example of this efficiency. In addition, compartmentation has an important regulatory function. This function derives largely from the selective permeability of membranes to different metabolites, which allows the passage of intermediates from one compartment into another. Typically, intermediates of a pathway remain trapped within an organelle, while specific carriers allow substrates to enter and products to exit. Therefore, the flux through a pathway can be regulated by controlling the rate at which a substrate enters the compartment. One way in which the hormone insulin stimulates carbohydrate utilization, for example, is by moving glucose transporters into the plasma membrane, so that glucose is more readily taken into cells for catabolism or for the synthesis of glycogen.

Compartmentation is more than a matter of sequestering enzymes in specific organelles. If enzymes that catalyze sequential reactions are in close physical proximity to one another, the local concentrations of their substrates will tend to be high even if they are not isolated within membrane-bound organelles. Intermediates are less likely to diffuse away because the product of one reaction is released close to the active site of the enzyme catalyzing the next reaction. The enzymes in a pathway may be bound to one another in a membrane, as are the enzymes of mitochondrial electron transport. Alternatively, they may be part of a highly organized multiprotein complex, such as the pyruvate dehydrogenase complex discussed in Chapter 13, a major entry point to the citric acid cycle.

Compartmentation can also result from weak interactions among enzymes that do not remain complexed when they are isolated. For example, conversion of glucose to pyruvate by glycolysis is catalyzed by enzymes that interact quite weakly in solution. However, there is evidence that these enzymes interact more strongly within the cytosol, forming a supramolecular structure that facilitates the multistep

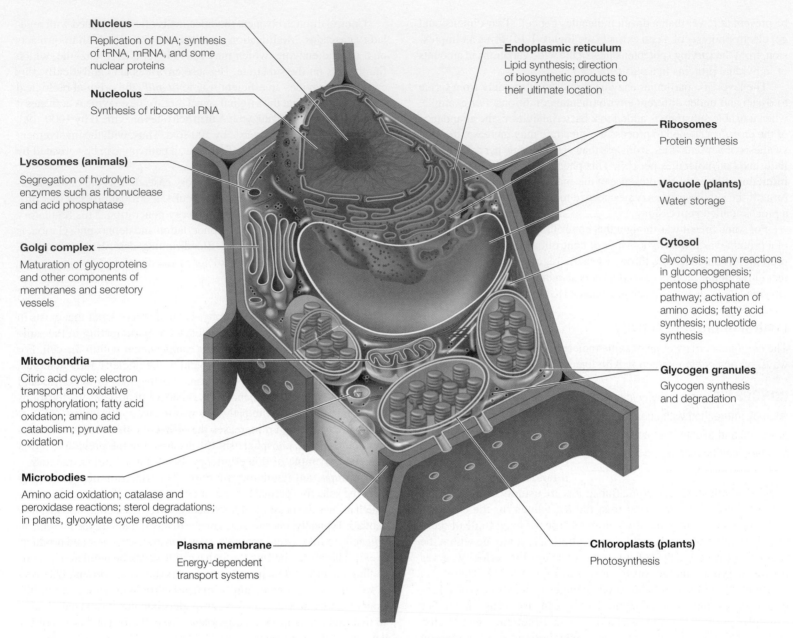

Nucleus
Replication of DNA; synthesis of tRNA, mRNA, and some nuclear proteins

Nucleolus
Synthesis of ribosomal RNA

Lysosomes (animals)
Segregation of hydrolytic enzymes such as ribonuclease and acid phosphatase

Golgi complex
Maturation of glycoproteins and other components of membranes and secretory vessels

Mitochondria
Citric acid cycle; electron transport and oxidative phosphorylation; fatty acid oxidation; amino acid catabolism; pyruvate oxidation

Microbodies
Amino acid oxidation; catalase and peroxidase reactions; sterol degradations; in plants, glyoxylate cycle reactions

Plasma membrane
Energy-dependent transport systems

Endoplasmic reticulum
Lipid synthesis; direction of biosynthetic products to their ultimate location

Ribosomes
Protein synthesis

Vacuole (plants)
Water storage

Cytosol
Glycolysis; many reactions in gluconeogenesis; pentose phosphate pathway; activation of amino acids; fatty acid synthesis; nucleotide synthesis

Glycogen granules
Glycogen synthesis and degradation

Chloroplasts (plants)
Photosynthesis

▲ **FIGURE 11.10** **Locations of major metabolic pathways within a eukaryotic cell.** This hypothetical cell combines features of a plant cell and an animal cell.

glycolytic pathway. The concept of intracellular interactions among readily solubilized enzymes developed as scientists began to realize that the cytosol is much more highly structured than was formerly thought. High-resolution electron micrographs of mammalian cytosol reveal the outlines of an organized structure that has been called the **cytomatrix** (**FIGURE 11.11**). It is likely that these structures form as a result of the extremely high concentrations of proteins inside cells, which decrease the concentration of water and drive weakly interacting proteins to associate. Furthermore, soluble enzymes may be bound within the cell to the structural elements of the cytomatrix.

● **CONCEPT** Enzymes catalyzing sequential reactions are often associated, even in the cytosol, where organized structures are difficult to visualize.

Whether highly structured or loosely associated, multienzyme complexes can efficiently control reaction pathways. Enzyme complexes restrict the diffusion of intermediates, thereby keeping the average concentrations of intermediates low but their local concentrations at enzyme catalytic sites high. Thus, the flux through a pathway can change more quickly in response to a change in the concentration of the first substrate for that pathway, without having to depend on changes in the substrate concentration in the surrounding bulk solution.

Hormonal Regulation

On top of, but integrated with, the regulatory mechanisms operating within a eukaryotic cell are messages dispatched from other tissues and organs. The process of transmitting and receiving such

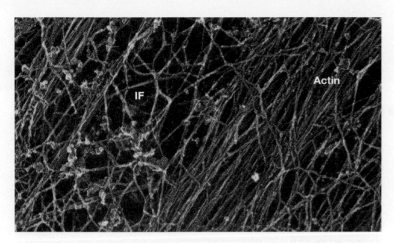

▲ **FIGURE 11.11 Structural organization of the cytomatrix.** This electron micrograph of the cytoplasmic matrix of a cultured mammalian fibroblast reveals networks of filaments anchored to the plasma membrane. MT = microtubules; IF = intermediate filaments. Approximate magnification, 150,000.

messages and then responding with metabolic changes is called **signal transduction.** The extracellular messengers include hormones, growth factors, neurotransmitters, and pheromones, which interact with specific receptors, resulting in specific metabolic changes in the target cell.

Metabolic responses to hormones can involve changes in gene expression, leading to changes in enzyme levels. This type of response typically operates on a time scale of hours to days, resulting in a reprogramming of the metabolic capability of the cell. On a shorter time scale (seconds to hours), some hormones stimulate the synthesis of intracellular **second messengers** that control metabolic reactions. One of the most important second messengers is **adenosine 3′,5′-cyclic monophosphate** commonly known as **cyclic AMP** or **cAMP.** The hormone (the first messenger) binds to the extracellular portion of the receptor protein embedded in the plasma membrane. In response to extracellular binding of the first messenger, the receptor acts as a conduit that transmits the signal inside the cell, where the intracellular portion of the receptor stimulates the formation of a second messenger. These signal transduction systems can effect precise control over metabolic pathways, often through reversible covalent modification of critical enzymes in the pathway.

Adenosine-3′,5′-cyclic monophosphate
(cyclic AMP)

Signal transduction systems are modular in structure, allowing for a diversity of metabolic responses based on a common set of operating principles. Thus, secretion of one hormone can have quite diverse

● **CONCEPT** Second messengers transmit information from hormones bound at the cell surface, thereby controlling intracellular metabolic processes.

effects in different tissues, depending on the nature of the receptors, second messengers, and other signaling components in different target cells. Moreover, a single second messenger may have diverse effects within a single cell. cAMP, for example, which activates glycogen degradation, also activates a cascade that inhibits the synthesis of glycogen. This dual effect is an example of a coordinated metabolic response, where glycogen synthesis is *inhibited* under the same physiological conditions that *promote* glycogen breakdown. These signal transduction systems are described in Chapters 12, 17, and 20.

Distributive Control of Metabolism

With the discovery in the 1950s and 1960s of allosterically regulated enzymes, the metabolic pathway flux was thought to be regulated primarily through control of the intracellular activity of one or a few key enzymes in that pathway. When it was realized that allosteric enzymes often catalyze committed reactions—that is, the first reaction in a pathway that leads to an intermediate with no other known function—these enzymes were assumed to catalyze the "rate-limiting reactions" in metabolic pathways.

We now realize, however, that metabolic regulation is more complex and that all of the enzymes in a pathway contribute toward control of pathway flux. An approach called **metabolic control analysis** assigns to each enzyme in a pathway a **flux-control coefficient,** a value that can vary between zero and one. For a given enzyme, the flux-control coefficient is the relative increase in flux divided by the relative increase in enzyme activity that brought about that flux increase. For a true rate-limiting enzyme, the flux-control coefficient is 1; a 20% increase in the activity of that enzyme would increase that flux rate by 20%. But metabolic control theory predicts that all enzymes in a pathway contribute to regulation, meaning that all enzymes have flux-control coefficients greater than zero, but none has a value as high as 1 (which would be the case if flux were truly controlled solely by one rate-limiting enzyme).

The predictions of metabolic control theory can be tested, for example, by using mutations that affect the activity of a specific enzyme in vivo and then measuring the change in flux rate of a pathway in which that enzyme is involved. Such analyses confirm that enzymes catalyzing committed reactions do play large roles in regulation; that is, they have high flux-control coefficients. More important, however, these analyses confirm that *every* enzyme in a pathway—not just one—plays a part in the control of that pathway. Thus, regulation of a pathway is *distributed* among all of the enzymes involved in the pathway, giving rise to the concept of **distributive control of metabolism.** In retrospect, this concept should have been predicted simply from the complexity of metabolism. Many intermediates participate in more than one pathway, making different pathways interdependent and interlocking. Regulatory schemes that depend on control of just one or two enzymes in each pathway lack the flexibility and subtlety to account for the ability of cells to maintain homeostasis under widely varying nutritional and energetic conditions. Nevertheless, for each pathway or process, one or a few regulatory enzymes of primary importance have been identified, and these will be pointed out as we present the individual pathways involved.

11.6 Experimental Analysis of Metabolism

Given that metabolism consists of all the chemical reactions in living matter, how does a biochemist approach metabolism in the laboratory? The particular metabolic process must first be subdivided into experimentally attainable goals and then investigated at multiple levels of organization and with a wide array of biochemical methods.

Goals of the Study of Metabolism

To fully understand a metabolic process, the biochemist seeks (1) to identify reactants, products, and cofactors, as well as the stoichiometry, for each reaction involved; (2) to understand how the rate of each reaction is controlled in the tissue of origin; and (3) to identify the physiological function of each reaction and control mechanism. These three goals necessitate isolating and characterizing the enzyme catalyzing each reaction in a pathway.

This third task—extrapolating from test tube biochemistry to the intact cell—is especially challenging. For instance, given that most enzymes catalyze reactions that can proceed in either direction, what is the direction of an enzymatic reaction in vivo? Many reactions originally found to proceed one way in vitro have been shown to proceed in the opposite direction in vivo. The mitochondrial enzyme that synthesizes ATP from ADP in cells, for example, was originally characterized as an ATPase—an enzyme that hydrolyzes ATP to ADP and P_i. Therefore, it is not sufficient to isolate an enzyme and to demonstrate that it catalyzes a particular reaction in the test tube. We must also show that the same enzyme catalyzes the same reaction in intact tissue—usually a more difficult task.

To achieve the stated goals, a biochemist must perform analyses at several levels of biological organization, from living organisms and intact cells to broken-cell preparations and ultimately to purified components. Cell-free (in vitro) preparations can be manipulated in ways that intact cells cannot—for example, by adding substrates and cofactors that will not pass through cell membranes. The researcher attempts to duplicate in vitro the process that is known to occur in vivo.

Levels of Organization at Which Metabolism Is Studied

We have seen why it is necessary to study metabolism at various levels of biological organization, from intact organism to purified chemical components. Here we discuss what can be learned at each level.

Whole Organisms

Biochemists must investigate metabolism in whole organisms because their ultimate aim is to understand chemical processes in intact living systems. Radioisotopic tracers have long been used to characterize metabolic pathways in whole organisms (see Tools of Biochemistry 11B). A classic example, described in Chapter 16, is the elucidation of cholesterol synthesis in the 1940s. Konrad Bloch injected ^{14}C-labeled acetate into rats and followed the flow of label into intermediates by sacrificing rats at intervals and analyzing the radioactive compounds in their livers. The use of radioisotopes and other **metabolic probes** to study metabolism is described in more detail shortly.

● **CONNECTION** The glucose tolerance test involves giving a large oral dose of glucose and monitoring the glucose level in the blood over the next several hours. This test is used to diagnose diabetes and other disorders of carbohydrate metabolism.

Many diagnostic tests in clinical medicine are in vivo metabolic experiments. Instead of using radioisotopes, we sample tissue at intervals and carry out biochemical assays. In the **glucose tolerance test,** for example, used to diagnose diabetes and other disorders of carbohydrate metabolism, a subject consumes a large oral dose of glucose, and the glucose level in the blood is then determined at intervals over several hours.

Isolated or Perfused Organs

Some of the difficulties in transporting a precursor or an inhibitor to the desired organ when studying whole organisms can be circumvented by using an isolated organ. A researcher usually **perfuses** the isolated organ during the experimental manipulations. Perfusion involves pumping a buffered isotonic solution containing nutrients, drugs, or hormones through the organ. The solution partly takes the place of the normal circulation, delivering nutrients and removing waste products. This technique is widely used with rodent hearts to study the biochemistry of **myocardial ischemia,** the decrease in blood flow and oxygen delivery to the heart muscle caused by a partial or complete blockage of the coronary arteries. The researcher can also perfuse an organ within a living animal, following appropriate surgical procedures. Perfusion is much less efficient than circulation, however, so experiments at this level must be of limited duration.

Whole Cells

Any plant or animal organ contains a complex mixture of different cell types. Several means of fractionating the cells after disaggregation of an organ are used to obtain preparations enriched in one cell type. The most common method is centrifugation, which separates cells on the basis of size or density. **Fluorescence-activated cell sorting** is also widely used. In a typical application, a cell suspension is treated with a fluorescent-tagged antibody that recognizes a cell-surface antigen, which is present in varying amounts among different cell types. Cells pass in single file through a laser beam and are physically separated according to the amount of fluorescence recorded from each cell. These machines, which can sort several thousand cells per second, result in fractionation based on the abundance of the selected surface antigen.

Uniformity in a cell population can also be achieved by growing cells in **tissue culture.** Disaggregated cells of an organ or a tissue can, with special care, be induced to grow in a medium containing cell nutrients and protein growth factors. The cells grow and divide independently of one another, much like the cells in a bacterial culture. Although animal cells usually cease to grow after a certain number of divisions in culture, variant lines arise that are capable of indefinite

● **CONNECTION** Transgenic animal models provide powerful experimental systems to study metabolic diseases. A gene encoding a specific metabolic enzyme can be knocked out in mice to create a whole animal model of the corresponding human metabolic disorder.

growth, as long as they are adequately nourished. In such cultures, **clonal** cell lines can be generated in which all of the cells in a line are derived from a single cell, so that they are genetically and metabolically uniform. This uniformity is a boon for many biochemical investigations. For example, much of our understanding of virus replication depends on the ability to infect a large number of identical cells in culture simultaneously and then follow the metabolic changes by sampling the cell culture at various times after infection.

One problem with tissue culture is that cells adapted to long-term growth in culture take on characteristics different from those of their parent cells, which were originally embedded in plant or animal tissue. Maintaining specialized cell characteristics in culture always presents a challenge.

Cell-Free Systems

Problems of transport through membranes are avoided by working with broken-cell preparations. Animal cells are easily lysed by mild shear forces, suspension in hypotonic medium, or freezing and thawing. Bacterial cells have a rigid cell wall that requires more vigorous treatment, such as sonication. Enzymatic digestion with lysozyme is often used to rupture bacterial cells under relatively mild conditions. Breaking open the especially tough cell walls of yeast and plant cells usually requires combinations of enzymatic and mechanical treatments.

Initial metabolic experiments are usually carried out in unfractionated cell-free homogenates. However, localizing a metabolic pathway within a particular cell compartment requires fractionating the homogenate to separate the organelles. Using **differential centrifugation,** morphologically intact organelles can be partially separated from each other. Nuclei, mitochondria, chloroplasts, lysosomes, and **microsomes** (artifactual membrane vesicles formed from disrupted endoplasmic reticulum) sediment at the bottom of the centrifuge tube at different rates due to their slightly different densities. The contents of the cytosol remain in the soluble **supernatant fraction** after the final centrifugation step. Much of our understanding of DNA replication and transcription in eukaryotic cells comes from investigations with isolated nuclei, whereas purified mitochondria have yielded much of what we know about respiratory electron transport and oxidative phosphorylation.

The advantages of cell-free systems must be balanced against the fact that the preparation of cell-free components usually destroys biological organization. Components that were in separate compartments in the intact cell become mixed, making it possible to misinterpret data obtained from in vitro systems.

Purified Components

To understand a biological process at the molecular level, an investigator must first purify to homogeneity all of the factors thought to be involved, and then determine their intrinsic properties and their interactions. Purification techniques were introduced in Tools of Biochemistry 5A. Often, as with the citric acid cycle, this process is simply a matter of purifying the individual enzymes involved, determining the substrate and cofactor requirements of each, recombining the purified enzymes, and showing that the entire process can be catalyzed by purified components. This process is called **reconstitution.** Some pathways require cell constituents other than enzymes, such as the ribosomes and tRNAs needed for protein synthesis.

In purifying individual components, biochemists continually risk losing factors that are essential for normal control or for some other aspect of the process under study. Avoiding such pitfalls requires painstaking experiments in which criteria for biological activity are defined and fractions are continually examined to ensure that each activity is retained through each fractionation step. A good example of this approach is presented in Chapter 22, where we discuss the enzymes and proteins that must function at a DNA replication fork.

Systems Level

Systems biology approaches are now being applied to the study of metabolism. Systems biologists try to catalogue the full set of components in a system, and then they study the interactions between these components and how these interactions give rise to the specific function and behavior of that system. Metabolomics is an example of this kind of approach (see Tools of Biochemistry 11A).

Metabolic Probes

Metabolic probes are agents that allow a researcher to follow a metabolic process or to interfere specifically with one or a small number of reactions in a pathway. We have already discussed one type, radioisotopes, which can be injected into an organism and traced throughout the body as metabolism proceeds. Two other kinds of probes that are also widely used are *metabolic inhibitors* and *mutations*.

● **CONCEPT** By inactivating individual enzymes, mutations and enzyme inhibitors help identify the metabolic roles of enzymes.

By blocking a specific reaction in vivo and determining the results of the blockade, inhibitors and mutations help identify the metabolic role of a reaction. For example, respiratory poisons such as carbon monoxide and cyanide block specific steps in respiration; in addition, the use of metabolic inhibitors helped researchers to identify the order of electron carriers in the respiratory electron transport chain (see Chapter 14). However, inhibitors can present difficulties for the researcher—they may be difficult to transport into cells, or they may have multiple sites of action in addition to the process under study. It is often easier to interfere with a pathway by selecting mutant strains deficient in the enzyme of interest.

In the 1940s, George Beadle and Edward Tatum were the first to use mutations as biochemical probes in their work with the bread mold *Neurospora crassa*. Beadle and Tatum isolated a number of X-ray-induced mutants that required arginine for growth (something that is normally synthesized by *Neurospora* itself), in addition to the constituents of minimal medium. Furthermore, Beadle and Tatum discovered that different mutations affected different enzymes in the arginine biosynthetic pathway. In each case, the intermediate that was the substrate for the deficient enzyme would accumulate in the culture medium. This observation allowed Beadle and Tatum to order the enzymes according to the reactions they controlled, by the rationale illustrated in **FIGURE 11.12**. If a culture filtrate (containing an accumulated intermediate) from one mutant allowed a second mutant to grow without arginine, the researchers concluded that the first mutation blocked an enzymatic step *later* in the pathway than the step blocked by the second mutation. Ultimately, the accumulating intermediates were identified, and the pathway was elucidated.

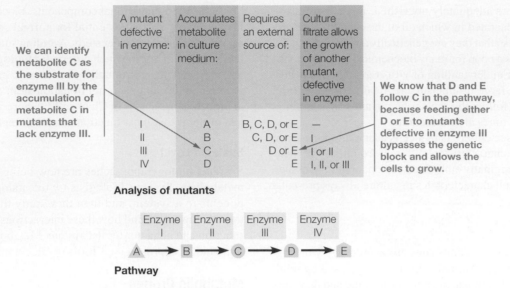

We can identify metabolite C as the substrate for enzyme III by the accumulation of metabolite C in mutants that lack enzyme III.

A mutant defective in enzyme:	Accumulates metabolite in culture medium:	Requires an external source of:	Culture filtrate allows the growth of another mutant, defective in enzyme:
I	A	B, C, D, or E	—
II	B	C, D, or E	I
III	C	D or E	I or II
IV	D	E	I, II, or III

We know that D and E follow C in the pathway, because feeding either D or E to mutants defective in enzyme III bypasses the genetic block and allows the cells to grow.

Analysis of mutants

Enzyme I Enzyme II Enzyme III Enzyme IV

A → B → C → D → E

Pathway

▲ **FIGURE 11.12 Using mutations as biochemical probes.** The steps of a hypothetical metabolic pathway are identified by analyzing mutants defective in individual steps of the pathway.

The impact of this "biochemical genetics" approach was much wider than simply working out the details of a metabolic pathway. Beadle and Tatum recognized that there was a one-to-one correspondence between a genetic mutation and the loss of a specific enzyme, leading them to propose the **one gene–one enzyme hypothesis,** years before the chemical nature of the gene was known.

In addition to identifying enzymatic pathways, mutants have been used to elucidate regulatory mechanisms. The earliest successes came from studies in the 1960s by François Jacob and Jacques Monod, who isolated dozens of *Escherichia coli* mutants with defects in the regulation of lactose catabolism or with abnormalities in virus–host relationships. These data led ultimately to the discovery of mRNA and the repressor–operator mechanism of genetic regulation (see Chapter 26). Recombinant DNA technologies (Tools of Biochemistry 4A) have allowed even more sophisticated perturbations of metabolic regulatory systems. For example, a researcher might use site-directed mutagenesis to disable the allosteric regulatory sites on an enzyme in a particular substrate cycle. The mutated enzyme is then introduced back into the cell, replacing the wild-type version, and the effects of loss of regulation can be examined in vivo.

A single investigation can use both metabolic inhibitors and mutations. This combined approach helped to elucidate the function of DNA gyrase, one of the DNA topoisomerases mentioned in Chapter 4. This enzyme is inhibited by nalidixic acid. When nalidixic acid was administered to bacteria, DNA replication was inhibited, suggesting that DNA gyrase plays an essential role in DNA replication. However, this observation alone was not enough to cement DNA gyrase's role in replication—it is possible that nalidixic acid was inhibiting DNA replication by also blocking some other enzyme, in addition to DNA gyrase. Thus, stronger evidence was needed. This evidence was obtained when mutants resistant to nalidixic acid were found to contain an altered form of DNA gyrase that was resistant to nalidixic acid. Thus, a single mutation abolished nalidixic acid sensitivity for both the DNA gyrase enzyme and the ability of the cells to replicate their DNA, strongly supporting an essential role for DNA gyrase in DNA replication.

In this chapter, we have described the general strategy of metabolism, identified the major pathways, outlined how pathways are regulated, and identified experimental approaches to understanding metabolism. We are now prepared for detailed descriptions of metabolic pathways, which we begin in Chapter 12 with carbohydrates.

 # Summary

• Metabolism is the totality of chemical reactions occurring within a cell. **Catabolic pathways** break down substrates to provide energy, largely through oxidative reactions, whereas **anabolic pathways** synthesize complex biomolecules from small molecules. Catabolic and anabolic pathways with the same end points are actually different pathways, not simple reversals of each other, so both pathways can be thermodynamically favorable. Regulation can be imposed only on reactions that are displaced far from equilibrium. Most metabolic energy comes from the oxidation of substrates, with energy release coming in a series of small steps as the electrons released are transferred from carrier to carrier, and ultimately to oxygen. The more highly reduced a substrate, the more energy is released when it is catabolized. (Sections 11.1–11.3)

• **Flux** through **metabolic pathways** is controlled by an array of regulatory processes. These processes include regulating enzyme concentration (through the control of enzyme synthesis and degradation), enzyme activity (through concentrations of substrates, products, and allosteric effectors, and covalent modification of enzyme proteins), compartmentation, and hormonal control. Hormonal regulation may involve the control of enzyme synthesis at the genetic level or the regulation of enzyme activity. In the latter case, intracellular second messengers are formed in response to hormonal signals. (Section 11.5)

• To understand metabolic processes, each reaction in the pathway must be identified and the reaction's function and control must be established. This understanding requires experimentation at all levels of biological organization, from the living organism to the purified enzyme. The ability to block specific enzymes, either with inhibitors or mutations, can be used to identify the functions of those enzymes. (Section 11.6)

Problems

Enhanced by

Mastering Chemistry for Biochemistry

Mastering Chemistry for Biochemistry provides select end-of-chapter problems and feedback-enriched tutorial problems, animations, and interactive figures to deepen your understanding of complex topics while practicing problem solving.

Answers to red problems are available in the Answer Appendix.

1. Write a balanced equation for the complete oxidation of each of the following, and calculate the respiratory quotient for each substance.
   (a) Ethanol
   (b) Acetic acid
   (c) Stearic acid
   (d) Oleic acid
   (e) Linoleic acid

2. Given what you know about the involvement of nicotinamide nucleotides in oxidative and reductive metabolic reactions, predict whether the following intracellular concentration ratios should be 1, > 1, or < 1. Explain your answers.
   (a) $[NAD^+]/[NADH]$
   (b) $[NADP^+]/[NADPH]$
   (c) Since NAD^+ and $NADP^+$ are essentially equivalent in their tendency to attract electrons, discuss how the two concentration ratios might be maintained inside cells at greatly differing values.

3. (a) NAD^+ kinase catalyzes the ATP-dependent conversion of NAD^+ to $NADP^+$. How many reducing equivalents are involved in this reaction?
   (b) How many reducing equivalents are involved in the conversion of ferric ion to ferrous ion?
   (c) How many reducing equivalents are involved in reducing one molecule of oxygen gas to water?

4. On page 351 we showed that the oxidation of glucose and palmitic acid yields 15.94 kJ/g and 38.90 kJ/g, respectively. Calculate these values in terms of kJ/mol and kJ per carbon atom oxidized for both glucose and palmitic acid.

5. The Nutrition Facts label on a full-fat yogurt lists the following amounts per 200 g serving: 10 g fat; 8 g sugars; 18 g protein. Calculate the calories (Cal) from each nutrient, and the total calories in the 200 g serving. Compare these values to a low-fat yogurt (0 g fat; 8.2 g sugars; 21 g protein per 200 g serving).

6. Free energy changes under intracellular conditions differ markedly from those determined under standard conditions. $\Delta G°' = -32.2$ kJ/mol for ATP hydrolysis to ADP and P_i. Calculate ΔG for ATP hydrolysis in a cell at 37 °C that contains $[ATP] = 3$ mM, $[ADP] = 1$ mM, and $[P_i] = 1$ mM.

7. (a) Consider the following hypothetical metabolic pathway:

$$A \rightleftharpoons B \xrightarrow{\quad X \quad} C \rightleftharpoons D$$

Under intracellular conditions, the activity of enzyme X is 100 pmol/10^6 cells/s. Calculate the effect on the metabolic flux rate (pmol/10^6 cells/s) of B $\rightarrow$ C of the following treatments. Calculate as % change (increase or decrease).

Treatment	% Change in flux
Inhibitor that reduces activity of X by 10%	
Activator that increases activity of X by 10%	

(b) Now consider a substrate cycle operating with enzymes X and Y in the same hypothetical metabolic pathway:

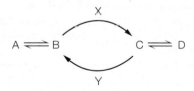

$$A \rightleftharpoons B \qquad C \rightleftharpoons D$$

Under intracellular conditions, the activity of enzyme X is 100 pmol/10^6 cells/s, and that of enzyme Y is 80 pmol/10^6 cells/s. What are the direction and rate (pmol/10^6 cells/s) of metabolic flux between B and C?

(c) Calculate the effect on the direction and metabolic flux rate of the following treatments. Calculate as % change (increase or decrease).

Treatment	Direction of flux (B $\longrightarrow$ C or C $\longrightarrow$ B)	% Change in flux
Inhibitor that reduces activity of X by 10%		
Activator that increases activity of X by 10%		
Doubling the activity of enzyme Y		

(d) Briefly summarize the regulatory advantage(s) of a substrate cycle in a pathway.

8. Consider the simple bimolecular reaction $A + B \underset{k_{-1}}{\overset{k_{+1}}{\rightleftharpoons}} C + D$, where k_{+1} and k_{-1} are the rate constants for the forward and reverse reactions. ΔG is related to the concentration of the reactants and products as well as the equilibrium constant, K, as described by Equation 3.27:

$$\Delta G = RT \ln\left(\frac{Q}{K}\right)$$

Since the law of mass action governs reaction kinetics, the forward and reverse fluxes, J, can be defined as:

$$J^+ = k_{+1}[A][B] \,(\text{forward flux}); J^- = k_{-1}[C][D] \,(\text{reverse flux})$$

ΔG can be defined in terms of the ratio of forward and reverse fluxes:

$$\Delta G = -RT \ln\left(\frac{J^+}{J^-}\right)$$

Derive this equation.

9. Assume that you have a solution of 0.1 M glucose 6-phosphate. To this solution you add the enzyme phosphoglucomutase, which catalyzes the reaction:

glucose-6-P $\rightleftharpoons$ glucose-1-P $\Delta G°' = +1.7$ kJ/mol

(a) Does this reaction proceed at all as written at 25 °C, and if so, what are the final concentrations of glucose 6-P and glucose 1-P?
(b) What effect would omitting the enzyme have on the reaction. Be specific.
(c) Under what cellular conditions, if any, would this reaction continuously produce glucose 1-P at a high rate?

10. The glucose/glucose-6-phosphate substrate cycle involves distinct reactions of glycolysis and gluconeogenesis that interconvert these two metabolites. Assume that under physiological conditions, [ATP] = [ADP] and [Pi] = 1 mM.

Consider the following glycolytic reaction catalyzed by hexokinase:

ATP + glucose $\rightleftharpoons$ ADP + glucose-6-phosphate $\Delta G°' = -16.7$ kJ/mol

(a) Calculate the equilibrium constant (K) for this reaction at 298 K, and from that, calculate the maximum [glucose-6-phosphate]/[glucose] ratio that would exist under conditions where the reaction is still thermodynamically favorable.
(b) The reverse of this interconversion in gluconeogenesis is catalyzed by glucose-6-phosphatase:

glucose-6-phosphate + H$_2$O $\rightleftharpoons$ glucose + P$_i$ $\Delta G°' = -13.8$ kJ/mol

$K = 262$ for this reaction. Calculate the maximum ratio of [glucose]/[glucose-6-phosphate] that would exist under conditions where the reaction is still thermodynamically favorable.
(c) Under what cellular conditions would both directions in the substrate cycle be strongly favored?
(d) What ultimately controls the direction of net conversion of a substrate cycle such as this in the cell?

11. Wheeler and Mathews (*J. Biol. Chem.* 286:16992–16996, 2011) reported the concentrations of adenine nucleotides in rat liver mitochondria as follows: ATP, 5.5 mM; ADP, 5.1 mM; AMP, 1.8 mM.
(a) Calculate the adenylate energy charge within the mitochondrion

$$\text{(adenylate energy charge} = \frac{[\text{ATP}] + 0.5[\text{ADP}]}{[\text{ATP}] + [\text{ADP}] + [\text{AMP}]})$$

(b) Most measurements of adenylate energy charge in whole cells or cytosol give values close to 0.9. Speculate on reasons why it might be advantageous for mitochondria to have an ADP concentration almost as high as that of ATP.
(c) succinyl-CoA + ADP + P$_i$ $\longrightarrow$ succinate + ATP + CoA-SH $\Delta G°'$ = −2.9 kJ/mol
If [P$_i$] within the mitochondrion is 0.05 M and succinate and succinyl-CoA are present at equimolar concentrations, what is the maximum mitochondrial concentration of CoA-SH at which the reaction can be exergonic?

12. (a) A mitochondrion can be modeled as a cylinder 1.5 μm (microns) in length and 0.6 μm in diameter. Calculate the number of citrate synthase enzyme molecules contained in a single mitochondrion, assuming the mitochondrial citrate synthase concentration is 1 μM.
(b) Is this a reasonable answer? Assume that a single citrate synthase molecule is roughly spherical with a diameter of 100 Å (10 nm). Show your reasoning.

13. Multiprotein complexes are formed by weak noncovalent interactions between the proteins. Suppose proteins A and B form a heterodimer with a $K_d = 10^{-6}$ M:

$$AB \rightleftharpoons A + B \qquad K_d = \frac{[A][B]}{[AB]} = 10^{-6}M$$

(a) Calculate the fraction of monomers that exist in the heterodimer complex if the total concentration of A + B inside the cell is 5μM. Assume [A]$_{\text{total}}$ = [B]$_{\text{total}}$.
(b) If the cell is now disrupted in buffer, the cell contents are diluted 5-fold. What fraction of monomers would exist in the complex under these conditions?
(c) What does this calculation tell you about detecting multiprotein complexes in cell-free homogenates? Should they be readily detectable? If not, what methods might you use to aid in their detection?

Read Tools of Biochemistry 11B before attempting Problems 14–17.

14. Two-dimensional gel electrophoresis of proteins in a cell extract provides a qualitative way to compare proteins with respect to intracellular abundance. Describe a quantitative approach to determine the number of molecules of an enzyme per cell.

15. Mammalian cells growing in culture were labeled with [^{3}H]-thymidine to estimate the rate of DNA synthesis. The thymidine administered had a specific activity of 3000 cpm/pmol. At intervals, samples of culture were taken and acidified to precipitate nucleic acids. The rate of incorporation of isotope into DNA was 1500 cpm/10^6 cells/min. A portion of culture was taken to determine the specific activity of the intracellular dTTP pool, which was found to be 600 cpm/pmol.
(a) What fraction of the intracellular dTTP is synthesized from the exogenous precursor?
(b) What is the rate of DNA synthesis, in molecules per minute per cell of thymine nucleotides incorporated into DNA?
(c) How could you determine the specific activity of the dTTP pool?

16. Suppliers of radioisotopically labeled compounds usually provide each product as a mixture of labeled and unlabeled material. Unlabeled material is added deliberately as a *carrier,* partly because the specific activity of the carrier-free product is too high to be useful and partly because the product is more stable at lower specific activities. Using the radioactive decay law, calculate the following.
(a) The specific activity of carrier-free [^{32}P]-orthophosphate, in mCi/mmol.
(b) The fraction of H atoms that are radioactive in a preparation of uniform-label [^{3}H]-leucine, provided at 10 mCi/mmol.

17. Predict the product(s) of the following reactions:

Substrate	Nucleophile	Reaction Type	Product(s)
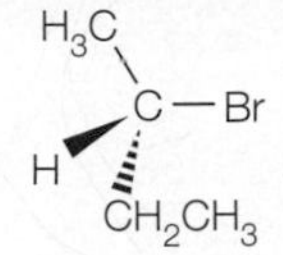	OH⁻	S$_N$1	
	OH⁻	S$_N$2	
	OH⁻	Elimination	
	R—NH$_2$	Addition	

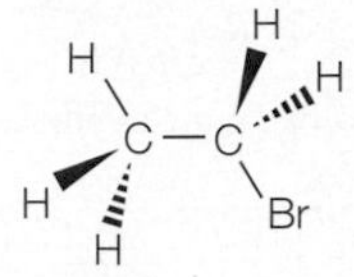

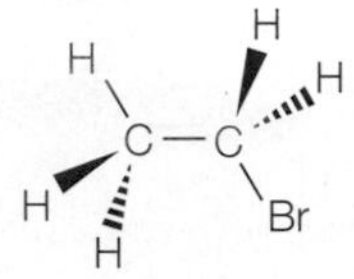

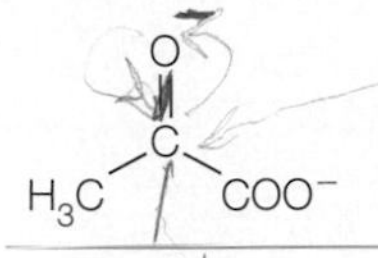

References

For a list of references related to this chapter, see Appendix II.

TOOLS OF BIOCHEMISTRY | 11A Metabolomics

Virtually all the facts in this textbook came from experiments in which a biochemist measured something in a cell extract: the level of a particular mRNA or protein, the activity of a particular enzyme, or the concentration of a particular metabolite. Indeed, a large part of experimental biochemistry involves the development of specific and sensitive *assays* for a particular cellular component. For example, Hans Krebs's elucidation of the citric acid cycle (see Chapter 13) depended on his ability to accurately measure the concentrations of potential substrates such as pyruvate, citrate, succinate, or oxaloacetate in muscle or liver slices. A specific assay had to be developed to measure each metabolite. To fully analyze a pathway such as the citric acid cycle or glycolysis might require 10 or more separate metabolite assays on each sample. Furthermore, these assays could measure only metabolites that were already known—they could not be used to discover new or missing metabolic intermediates.

In recent years, however, new technologies have made it possible to move beyond the measurement of single mRNAs, single proteins, or single metabolites. The advent of these new technologies has driven the "-omics" revolution—the development of the new fields of *genomics, transcriptomics, proteomics,* and **metabolomics**—in which hundreds or even thousands of specific components are measured simultaneously in a biological sample. Thus, it is now feasible to measure the full set of RNA transcripts (transcriptome), proteins (proteome), or metabolites (metabolome) in a particular cell or tissue. The **metabolome** represents the ultimate molecular phenotype of a cell under a given set of conditions, since all the changes in gene expression and enzyme activity eventually lead to changes in cellular metabolite levels (the metabolic state or profile). The metabolic state of a cell or a whole organism is a sensitive indicator of the physiological status of the organism. Changes in metabolic state can be used to understand and diagnose disease, study the effects of drugs, and even predict the effectiveness of a drug in a particular patient.

There are many analytical approaches to determining the metabolic state of a cell, a process called metabolic profiling, but they all follow the same basic steps (**FIGURE 11A.1**): sample extraction; metabolite identification and quantitation; and data analysis (informatics).

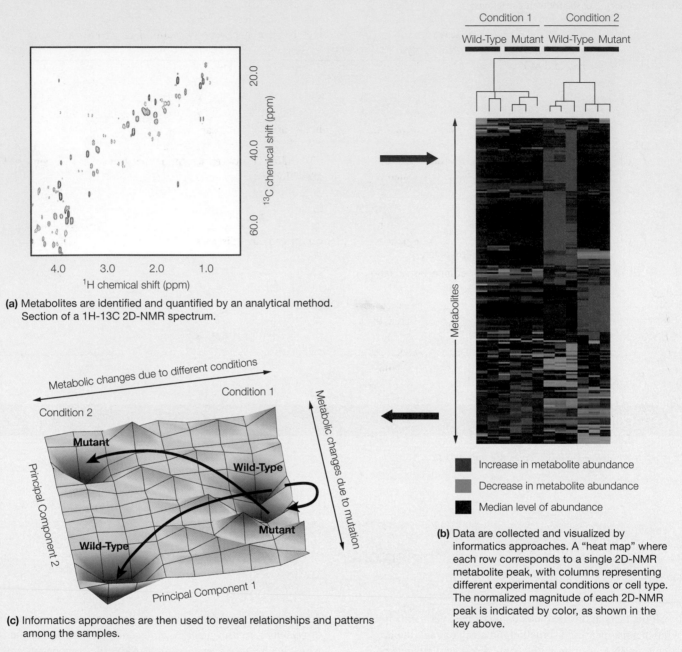

(a) Metabolites are identified and quantified by an analytical method. Section of a 1H-13C 2D-NMR spectrum.

(b) Data are collected and visualized by informatics approaches. A "heat map" where each row corresponds to a single 2D-NMR metabolite peak, with columns representing different experimental conditions or cell type. The normalized magnitude of each 2D-NMR peak is indicated by color, as shown in the key above.

Increase in metabolite abundance

Decrease in metabolite abundance

Median level of abundance

(c) Informatics approaches are then used to reveal relationships and patterns among the samples.

▲ FIGURE 11A.1 Basic process of metabolic profiling.

Metabolic Profiling

In the first step of metabolic profiling, samples of interest are collected—for example, drug-treated versus untreated cells, serum from a patient, or a tissue biopsy from a tumor. The small molecules are extracted from the samples by methods that are matched to the analytical technique to be used. No single method will extract all metabolites from a sample. Different extraction procedures can be used to select specific subsets of the metabolome (e.g., lipophilic metabolites, amino acids, or carbohydrates).

While the number of distinct metabolites in a cell is not known, it is certainly on the order of a few thousand. Therefore, the second critical step, metabolite identification, requires an analytical method that has both a powerful separation component and a sensitive detection component. Mass spectrometry (MS) and nuclear magnetic resonance (NMR) are the most widely used detection methods. The MS-based techniques (see Tools of Biochemistry 5B) use liquid chromatography (LC), gas chromatography (GC), or capillary electrophoresis (CE) to separate the metabolites based on some chemical or physical property (size, charge, hydrophobicity, etc.). Effluent from the LC, GC, or CE is introduced into the mass spectrometer, where the metabolites are detected and quantified. NMR (see Tools of Biochemistry 6A) can be applied directly to samples without a separation step by obtaining spectra in two dimensions (e.g., ^{1}H vs. ^{13}C), a technique known as 2D-NMR (Figure 11A.1(a)).

Both MS and NMR methods are capable of detecting and quantifying hundreds of distinct metabolites in a single run, and both have advantages and disadvantages. NMR can detect many different classes of metabolites and offers great power in metabolite identification and quantitation. MS-based methods are much more sensitive than NMR, but quantitation is not as straightforward. Note, however, that both methods can detect and quantify unidentified metabolites, which can facilitate the discovery of new intermediates or even new pathways.

The third critical step in metabolic profiling is data analysis. Because the datasets are generally quite large, and sophisticated statistical methods are required to visualize and compare metabolic profiles, data analysis uses modern informatics tools. These informatics algorithms attempt to identify patterns in the data that are reproducibly characteristic of a particular metabolic state. These patterns, or fingerprints, can then be compared between samples. The most common informatics methods for analyzing and comparing metabolome data are hierarchical clustering (the partitioning of a dataset into subsets or clusters) (Figure 11A.1(b)) and principal component analysis (PCA) (Figure 11A.1(c)). PCA is used to reduce multidimensional datasets to lower dimensions (principal components) and can often reveal the internal structure of the data in a way that best explains the variance in the data. In the example shown, PCA indicates that the most important variables in this experiment were growth media (Principal Component 1) and the presence or absence of the specific mutation (Principal Component 2).

Applications

Metabolomics has many applications, such as the identification of new biomarkers to predict or diagnose disease. A recent example involves acetaminophen (e.g., Tylenol), a widely used over-the-counter analgesic. Excess acetaminophen is toxic to the liver because it causes oxidative stress, but individuals vary in their susceptibility to liver damage. To identify potential biomarkers for the oxidative stress caused by acetaminophen, CE-MS was used to study changes in liver metabolites in mice treated with excess acetaminophen. Metabolic profiling identified 132 compounds, including a new

metabolite whose levels rise in response to the drug. This new metabolite may serve as a useful biomarker for acetaminophen-induced liver toxicity in humans.

Metabolomics is also proving useful in the diagnosis of disease at earlier stages. For example, survival from oral cancer is better than 90% at 5 years, whereas late-stage disease survival is only 30%. To identify a metabolic signature for oral cancer that might be detected early in the disease, blood samples of oral cancer patients were analyzed using NMR spectroscopy. As illustrated in **FIGURE 11A.2**, PCA readily discriminated between serum samples from cancer patients and from a healthy control group. This method could also discriminate between different stages of disease, and could detect early-stage disease and identify patients with relatively small tumors. Clearly, metabolic profiling has the potential to revolutionize the diagnosis and treatment of cancer and other diseases.

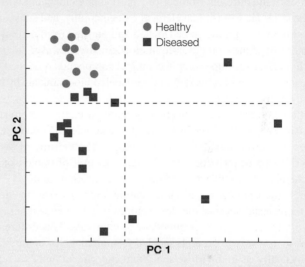

▲ **FIGURE 11A.2 Principal component analysis of 1H NMR spectra of human blood sera from oral cancer patients.** Scatter plot of first principal component (PC1) versus second principal component (PC2). Red squares are oral cancer patients. Blue triangles are healthy subjects. Courtesy of Stefano Tiziani, University of Texas at Austin.

Radioisotopes revolutionized biochemistry when they became available to investigators shortly after World War II. Radioisotopes extend—by orders of magnitude—the sensitivity with which chemical species can be detected. Traditional chemical analysis can detect and quantify molecules in the micromole (10^{-6} mole) or nanomole (10^{-9} mole) range. A compound that is "labeled," containing one or more atoms of a radioisotope, can be detected in picomole (10^{-12} mole) or even femtomole (10^{-15} mole) amounts. Radiolabeled compounds are called **tracers** because they allow an investigator to follow specific chemical or biochemical transformations in the presence of a huge excess of nonradioactive material.

Isotopes are different forms of the same element, so they have different atomic weights but the same atomic number. Thus, the chemical properties of the different isotopes of a particular element are virtually identical. Isotopic forms of an element exist naturally, and substances enriched in rare isotopes can be isolated and purified from natural sources. Most of the isotopes used in biochemistry, however, are produced in nuclear reactors. Simple chemical compounds produced in such reactors are then converted to radiolabeled biochemicals by chemical and enzymatic synthesis.

Although radioisotopes are still commonly used in biochemistry, *stable isotopes* are also used as tracers. For example, the two rare isotopes of hydrogen include a stable isotope (**deuterium,** ^{2}H) and a radioactive isotope (**tritium,** ^{3}H). Of the many uses of stable isotopes in biochemical research, we mention three applications here.

- First, incorporation of a stable isotope often increases the density of a material because the rare isotopes usually have higher atomic weights than their more abundant counterparts. This difference presents a way to separate labeled from nonlabeled compounds physically, as in the Meselson–Stahl experiment on DNA replication (see Chapter 4).

- Second, compounds labeled with stable isotopes, particularly ^{13}C, are widely used in nuclear magnetic resonance studies of molecular structure and dynamics (see Tools of Biochemistry 6A).

- Third, stable isotopes are used to study reaction mechanisms. The "isotope rate effect" refers to the effect on the reaction rate of replacing an atom by a heavy isotope. As discussed in Chapter 8, this effect helps to identify rate-limiting steps in enzyme-catalyzed reactions. **TABLE 11B.1** lists information about the isotopes, both stable and radioactive, that have found the greatest use in biochemistry.

The Nature of Radioactive Decay

The atomic nucleus of an unstable element can decay, giving rise to one or more of the three types of ionizing radiation: α-, β-, and γ-rays. Only β- and γ-emitting radioisotopes are used in biochemical research; the most useful are listed in Table 11B.1. A β-ray is an emitted electron, and a γ-ray is a high-energy photon. Most biochemical uses of radioisotopes involve β emitters.

Radioactive decay is a first-order kinetic process. The probability that a given atomic nucleus will decay is affected neither by the number of preceding decay events that have occurred nor by interaction with other radioactive nuclei. Rather, it is an intrinsic property of that nucleus. Thus, the number of decay events occurring in a given time interval is related only to the number of radioactive atoms present. This phenomenon gives rise to the **Law of Radioactive Decay:**

$$N = N_0 e^{-\lambda t}$$

where N_0 is the number of radioactive atoms at time zero, N is the number remaining at time t, and λ is a radioactive decay constant for a particular isotope, related to the intrinsic instability of that isotope.

TABLE 11B.1 Some Useful Isotopes in Biochemistry

Isotope	Stable or Radioactive	Emission	Half-Life	Maximum Energy (MeV*)
^{2}H	Stable			
^{3}H	Radioactive	β	12.3 years	0.018
^{13}C	Stable			
^{14}C	Radioactive	β	5730 years	0.155
^{15}N	Stable			
^{18}O	Stable			
^{24}Na	Radioactive	β (and γ)	15 hours	1.39
^{31}P	Stable			
^{32}P	Radioactive	β	14.3 days	1.71
^{35}S	Radioactive	β	87 days	0.167
^{45}Ca	Radioactive	β	163 days	0.254
^{59}Fe	Radioactive	β (and γ)	45 days	0.46, 0.27
^{131}I	Radioactive	β (and γ)	8 days	0.335, 0.608

*MeV = million electron volts

According to this equation, the *fraction* of nuclei in a population that decays within a given time interval is constant. For this reason, a more convenient parameter than the decay constant λ is the **half-life,** $t_{1/2}$, the time required for half of the nuclei in a sample to decay. The half-life is equal to $-\ln 0.5/\lambda$ or $+0.693/\lambda$. The half-life, like λ, is an intrinsic property of a given radioisotope (see Table 11B.1).

The basic unit of radioactive decay is the **curie** (Ci). This unit is defined as an amount of radioactivity equivalent to that in 1 g of radium—specifically, 2.22×10^{12} disintegrations per minute (dpm). The most widely used method for measuring β-emissions is **liquid scintillation counting.** The sample is dissolved or suspended in an organic solvent containing one or two fluorescent organic compounds, or *fluors*. A β-particle emitted from the sample has a high probability of hitting a molecule of the solvent. This contact excites the solvent molecule, boosting an electron to a higher-energy level. When that electron returns to the ground state, a photon of light is emitted. The photon is absorbed by a molecule of the fluor, which in turn becomes excited. A photomultiplier detects the fluorescence and for each disintegration converts it to an electrical signal, which is recorded and counted.

Nuclear Magnetic Resonance

In recent years, *nuclear magnetic resonance* (NMR) spectroscopy has become widely available for noninvasive monitoring of intact cells and organs. As explained in Tools of Biochemistry 6A, compounds containing certain atomic nuclei can be identified from an NMR spectrum, which measures shifts in the frequency of absorbed electromagnetic radiation. A researcher can determine an NMR spectrum of whole cells, or of organs or tissues in an intact plant or animal. NMR has even become a powerful noninvasive diagnostic tool, referred to as magnetic resonance imaging (MRI) in the medical arena.

For the most part, macromolecular components do not contribute to the spectrum, nor do compounds that are present at less than about 0.5 mM. The nuclei most commonly used in this in vivo technique are ^{1}H, ^{31}P, and ^{13}C (Table 11B.1). **FIGURE 11B.1** shows ^{31}P NMR spectra that represent components in the human forearm muscle. The five major peaks correspond to the phosphorous nuclei in orthophosphate (P_i), creatine phosphate, and the three phosphates of ATP. Because peak area is proportional to concentration, the energy status of intact cells can be determined. For example, an energy-rich muscle has lots of creatine phosphate, whereas a fatigued muscle uses up most of its creatine phosphate in order to maintain ATP levels (note also the accumulation of AMP—peak 6—in the third scan). NMR is finding wide applicability in monitoring recovery from heart attacks, in which cellular ischemia (insufficient oxygenation) damages cells by reducing ATP content. NMR can also be used to study metabolite compartmentation, flux rates through major metabolic pathways, and intracellular pH.

▶ **FIGURE 11B.1 The effect of anaerobic exercise on ^{31}P NMR spectra of human forearm muscle.** Peak areas are proportional to intracellular concentrations. See Tools of Biochemistry 6A for interpretation of NMR spectra. Courtesy of Dean Sherry, Craig Malloy and Jimin Ren of University of Texas-Southwestern Medical Center.

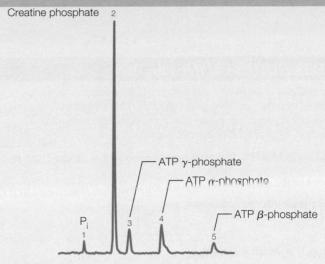

(a) Before exercise.

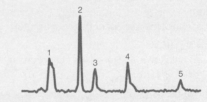

(b) 30 seconds into a 2-minute exercise period.

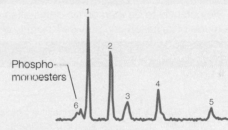

(c) At the end of exercise.

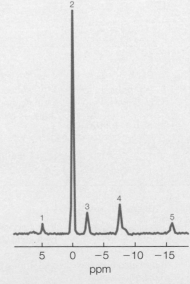

(d) Ten minutes after exercise.

FOUNDATION FIGURE | Enzyme Kinetics and Drug Action

MICHAELIS-MENTEN ENZYME KINETICS

• Kinetic analysis of enzyme reactions provides essential clues to mechanisms of catalysis. Reaction velocity (v) is defined in terms of experimentally measurable parameters—specifically, substrate concentration [S] and total enzyme concentration.

• Michaelis and Menten derived a rate equation for a simple enzyme reaction with a single substrate:

Michaelis-Menten equation: $v = \dfrac{V_{max}[S]}{K_M + [S]}$

K_M: [S] at which $v = 1/2\ V_{max}$
K_M is a measure of the substrate concentration required for effective catalysis to occur.

$k_{cat} = V_{max}/[\text{enzyme}]$
Reactions catalyzed per unit time (per second, or per minute)

Enzyme efficiency $= k_{cat}/K_M$
An efficient enzyme can catalyze many reactions per second at a low substrate concentration.

• Applying modified forms of this equation to more complex systems provides great insight into enzyme function and regulation, and the effectiveness of drugs that function as enzyme inhibitors.

◀ SECTION 8.6

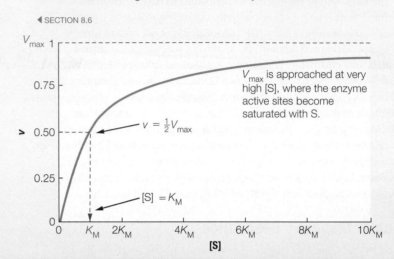

V_{max} is approached at very high [S], where the enzyme active sites become saturated with S.

$v = \frac{1}{2}V_{max}$

$[S] = K_M$

REVERSIBLE ENZYME INHIBITORS

Small molecules that have the potential to be therapeutic drugs often reversibly inhibit enzyme activity in one of three ways. Analysis of the mode of inhibition can provide insight into catalytic mechanisms and guide drug design.   ▶ SECTION 8.7

Mode of inhibition	Apparent K_M	Apparent V_{max}	Lineweaver-Burk plot
Competitive • Inhibitor competes with substrate for binding to the active site of the free enzyme. • Substrate can be processed by the enzyme, but inhibitor cannot.	αK_M (K_M appears to increase) $\alpha = 1 + [I]/K_i$	V_{max} In competitive inhibition, V_{max} is unchanged.	Slope = apparent K_M'/V_{max} $\dfrac{1}{V_{max}}$, +inhibitor, −inhibitor Slope = K_M/V_{max} $-\dfrac{1}{K_M}$, $-\dfrac{1}{K_M^{app}}$, $\dfrac{1}{[S]}$
Uncompetitive • Molecule or an ion can bind to a second site on the enzyme-substrate complex to form an ESI complex which is catalytically inactive.	K_M/α' (K_M appears to decrease) $\alpha' = 1 + [I]/K_i'$	V_{max}/α' V_{max} appears to decrease. Both K_M and V_{max} are reduced by a factor of $1/\alpha'$. $\alpha' = 1 + [I]/K_i'$	Increasing inhibitor concentration Slope = apparent K_M/apparent V_{max} No inhibitor Slope = K_M/V_{max} ($\alpha' = 1$) x-intercept $= -\alpha'/K_M$, y-intercept $= \alpha'/V_{max}$
Mixed • Molecule or an ion can bind to both the free enzyme or the ES complex.	$\alpha K_M/\alpha'$ (typically, K_M appears to increase, because α is usually greater than α') $\alpha = 1 + [I]/K_i$	V_{max}/α' V_{max} appears to decrease by a factor of $1/\alpha'$. $\alpha' = 1 + [I]/K_i'$	Increasing inhibitor concentration Slope = apparent K_M/apparent V_{max} Slope = K_M/V_{max} x-intercept $= -\alpha'/\alpha K_M$, y-intercept $= \alpha'/V_{max}$

Mastering Chemistry for Biochemistry

Mastering Chemistry for Biochemistry provides select end-of-chapter problems and feedback-enriched tutorial problems, animations, and interactive figures to deepen your understanding of complex topics while practicing problem solving.

ENZYME REGULATION AND METABOLIC CONTROL

Organisms control the utilization and production of metabolites through manipulation of enzyme activity. This can be done through:

• Changing enzyme concentration in a cell by altering levels of gene expresson.

• Changing the catalytic activity of an enzyme.

The two most common methods for modulating enzyme activity are:

• Allosteric regulation (mediated by the binding of effector molecules)

• Reversible covalent modification.   ▶ SECTIONS 8.7–8.9

Competitive Inhibition by a Drug

Atorvastatin

• Competitive inhibitor of HMG-CoA reductase, one of the main control points in the cholesterol biosynthesis pathway.

• Pathway inhibition of HMG-CoA reductase leads to a decrease in production of cholesterol and has been used to treat patients with atherosclerosis and high serum cholesterol.

• HMG-CoA reductase catalyzes the committed step in cholesterol biosynthesis (conversion of HMG-CoA to mevalonate):

HMG-CoA reductase (no inhibitor)

HMG-CoA ➡ Mevalonate ➡ Cholesterol

HMG-CoA reductase with Atorvastatin

HMG-CoA → Mevalonate --→ Cholesterol

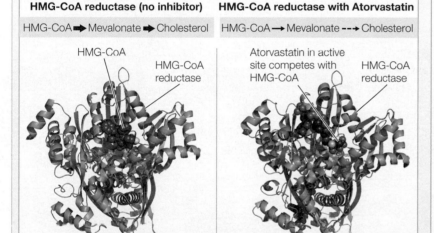

Reversible Covalent Modification: Phosphorylation

• Phosphorylation/dephosphorylation is the most common mechanism of reversible covalent modification.

• For glycogen synthase, phosphorylation stabilizes the T state conformation and thereby decreases enzyme activity. Dephosphorylation favors the active R state conformation.

• The activities of many key regulatory enzymes are modulated by both allosteric regulation and reversible covalent modification. The inactivation of glycogen synthase by covalent modification can be overcome by the binding of glucose-6-phosphate (G6P) which allosterically activates the enzyme.

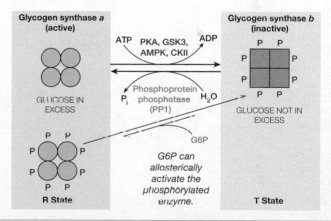

RECIPROCAL REGULATION OF OPPOSING PATHWAYS

• Cells possess pathways both for energy generation and energy utilization. In liver, the catabolic glycolysis pathway converts glucose to pyruvate, producing ATP, whereas the anabolic gluconeogenesis pathway converts pyruvate to glucose and consumes ATP. If both pathways operated simultaneously, the result would be the net consumption of ATP. This futile cycle is prevented by reciprocal regulation of key enzymes. In this case AMP, a sensor of low energy status, activates PFK and inhibits FBPase.

• Reciprocal regulation of key enzymes prevents catastrophic ATP depletion. Generally these enzymes are regulated by the same compound (or the same reversible covalent modification) but with opposite results. For example, both PFK and FBPase are regulated by changes in the AMP concentration. As [AMP] rises—a signal that [ATP] is low—PFK is activated and FBPase is inhibited.

▶ SECTION 12.6

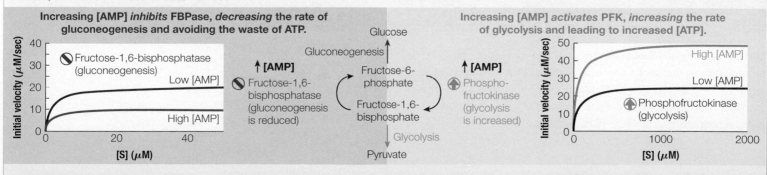

The bubbles and alcohol in this glass of beer are produced by fermentation, a metabolic process in which yeast cells convert the sugar in the grains to CO_2 and ethanol. Fermentation is a way of life for anaerobic microorganisms, but it is also central to the metabolism of air-breathing animals, in the form of glycolysis.

Carbohydrate Metabolism: Glycolysis, Gluconeogenesis, Glycogen Metabolism, and the Pentose Phosphate Pathway

IN CHAPTER 11, we introduced the general principles of metabolic pathways and their regulation. Here, we begin the detailed study of metabolism in earnest with the first phases of carbohydrate metabolism (**FIGURE 12.1**). Much of this chapter is devoted to **glycolysis,** the initial pathway in the catabolism of carbohydrates. *Glycolysis* is derived from the Greek words meaning "sweet" and "splitting." These terms are literally correct because glycolysis is the pathway by which six-carbon sugars are split, yielding a three-carbon compound, pyruvate. During glycolysis, some of the potential energy stored in the hexose structure is used to drive the synthesis of ATP from ADP and inorganic phosphate (P_i). Glycolysis can proceed under *aerobic* or *anaerobic* conditions. *Anaerobes,* microorganisms that live in oxygen-free environments, can derive all of their metabolic energy

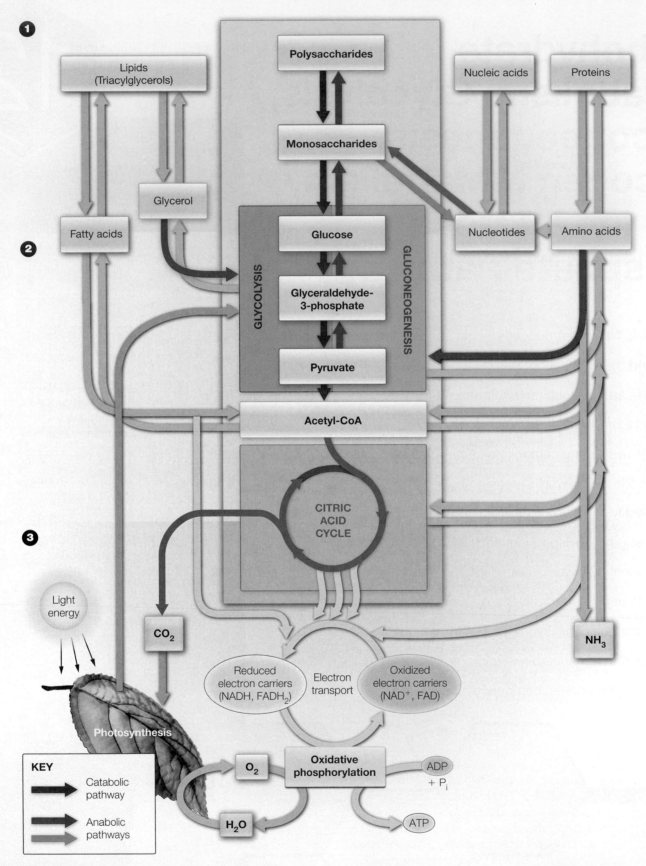

▲ **FIGURE 12.1 Catabolic and anabolic processes in anaerobic carbohydrate metabolism.** The red arrows show the glycolytic pathway and the breakdown of polysaccharides that supply this pathway. Glycolysis generates ATP anaerobically and provides fuel for the aerobic energy-generating pathways. The blue arrows show the gluconeogenesis pathway, and the synthesis of polysaccharides such as glycogen. The green arrow shows the pentose phosphate pathway, an alternative carbohydrate oxidation pathway needed for nucleotide synthesis. The numbers 1, 2, and 3 identify the three stages of metabolism (see Section 11.1).

from glycolysis. Indeed, carbohydrate is the only fuel whose catabolism can produce ATP in the absence of oxygen.

This chapter also introduces **gluconeogenesis,** the synthesis of glucose from noncarbohydrate precursors, as well as the synthesis of storage polysaccharides (which, in animals, is primarily glycogen). We discuss these two energy-requiring biosynthetic processes here because their regulation is so intimately coordinated with that of glycolysis. We end the chapter with a discussion of the multipurpose pentose phosphate pathway, an alternative way to catabolize glucose.

In contrast to most other biochemistry textbooks, we consider these four carbohydrate-metabolizing pathways (glycolysis, gluconeogenesis, glycogen metabolism, and the pentose phosphate pathway) in a single chapter because they are so intricately linked in animals. Discussing these pathways together in a single, albeit long, chapter allows us to present them as they occur in cells—an integrated metabolic network. These pathways beautifully illustrate the fundamental metabolic concepts introduced in Chapter 11, including the elegant coordinate regulation of these pathways, and how that regulation depends on thermodynamic principles (bioenergetics).

12.1 An Overview of Glycolysis

Glycolysis is an appropriate place to begin a detailed study of metabolism for several reasons: (1) It is nearly universal in living cells, (2) the regulation of glycolysis is particularly well understood, and (3) glycolysis plays the central metabolic role in generating both energy and metabolic intermediates for other pathways. It is one of the busiest freeways on the metabolic road map, but it is also connected to many less traveled roads.

Glucose is the major carbohydrate fuel for many cells. Indeed, some animal tissues, such as the brain, normally use glucose as the only energy source. Most cells, however, can utilize other sugars, and later in this chapter we will explore how those sugars are converted to intermediates in glycolysis. In addition, we will consider processes by which stored carbohydrate in the form of polysaccharides is made available for use in glycolysis.

Relation of Glycolysis to Other Pathways

● **CONCEPT** The 10 reactions of glycolysis occur in two phases: energy investment (first five reactions) and energy generation (last five reactions).

Glycolysis is a 10-step pathway that converts one molecule of glucose to two molecules of pyruvate, generating two molecules of ATP in the process. We will focus initially on the pathway as it begins with glucose, and we will discuss the routes for the entry of other carbohydrates in Section 12.7.

The 10 reactions that lead from glucose to pyruvate can be divided into two distinct phases, shown in **FIGURE 12.2.** The first five reactions constitute the *energy investment phase,* in which sugar phosphates are synthesized at the expense of two equivalents of ATP (converted to ADP), and the six-carbon substrate is split into 2 three-carbon sugar phosphates (triose phosphates). The last five reactions make up the *energy generation phase,* during which the two triose phosphates are converted to energy-rich compounds. Two moles of these compounds transfer 4 moles of phosphate to ADP, producing 4 moles of ATP. The net yield, per mole of glucose metabolized, is thus 2 moles of ATP and 2 moles of pyruvate. Two reducing equivalents (see Section 11.4) are generated as well, in the form of NADH.

In aerobic organisms, glycolysis is just the first process in the complete oxidation of glucose to CO_2 and water; the second process is the oxidation of pyruvate to acetyl-CoA; and the final process is the oxidation of the acetyl group carbons in the citric acid cycle (see Figure 12.1). These last two processes are discussed in detail in Chapter 13. In eukaryotic cells, glycolysis occurs in the cytosol, and the further oxidation of pyruvate occurs in mitochondria. Glycolysis also provides biosynthetic intermediates for amino acids and other important metabolites. Thus, glycolysis is both an anabolic and a catabolic pathway, with an importance that extends beyond the synthesis of ATP and substrates for the citric acid cycle.

Anaerobic and Aerobic Glycolysis

Glycolysis is an ancient metabolic pathway that probably evolved before the earliest known photosynthetic organisms began contributing O_2 to the Earth's atmosphere. Thus, initially glycolysis had to function under anaerobic conditions—with no net change in the oxidation state of the substrates as they are converted to products. Recall from Chapter 11

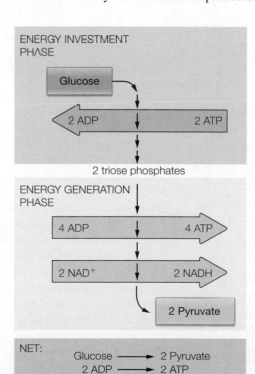

◀ **FIGURE 12.2** The two phases of glycolysis and the products of glycolysis.

that oxidation involves the loss of electrons from a substrate, and the electrons are transferred to an *electron acceptor,* which thereby becomes reduced. For example, as shown in Figure 12.2, the conversion of glucose to pyruvate oxidizes the carbons of glucose, but it also involves the reduction of two equivalents of NAD^+ to NADH. For this pathway to operate anaerobically, NADH must be reoxidized to NAD^+ by transferring its electrons to an electron acceptor so that *a steady-state concentration of NAD^+ is maintained.*

For example, lactic acid bacteria reoxidize NADH to NAD^+ by transferring the electrons to pyruvate, reducing it to lactate via the enzyme **lactate dehydrogenase (LDH).**

Lactate dehydrogenase reaction

$$
\begin{array}{ccc}
COO^- & & COO^- \\
| & & | \\
C{=}O + NADH + H^+ \rightleftharpoons & HO{-}C{-}H + NAD^+ \\
| & & | \\
CH_3 & & CH_3 \\
\textbf{Pyruvate} & & \textbf{L-Lactate}
\end{array}
$$

$$\Delta G^{\circ\prime} = -25.1 \text{ kJ/mol}$$

Lactate formation is strongly favored under standard conditions, as indicated by the large negative standard free energy change.

● **CONCEPT** A fermentation is an energy-yielding metabolic pathway with no net change in the oxidation state of the products compared to that of the substrates.

Glycolysis is therefore part of a *fermentation,* an energy-yielding metabolic pathway that involves no net change in oxidation state of the carbon substrate as it is converted to product (Section 11.4). *Homolactic fermentation,* in which all six carbons of glucose are converted to lactate,* is an important step in the manufacture of cheese. In *alcoholic fermentation,* pyruvate is cleaved into acetaldehyde and CO_2, and the acetaldehyde is then reduced to ethanol by **alcohol dehydrogenase (ADH):**

Alcohol dehydrogenase reaction

$$
\begin{array}{ccc}
H \quad O & & OH \\
\diagdown // & & | \\
C & + NADH + H^+ \rightleftharpoons & CH_2 + NAD^+ \\
| & & | \\
CH_3 & & CH_3 \\
\textbf{Acetaldehyde} & & \textbf{Ethanol}
\end{array}
$$

$$\Delta G^{\circ\prime} = -23.7 \text{ kJ/mol}$$

● **CONNECTION** Dozens of different fermentation processes have been discovered in different organisms, and many of them play major roles in the food and chemical industries. Alcoholic fermentation by yeasts is an important industrial process in brewing and baking. It generates the ethanol and bubbles of CO_2 in alcoholic beverages. In baking, the CO_2 causes bread to rise, while the ethanol merely evaporates. Other useful fermentations lead to acetic acid (in the manufacture of vinegar) and propionic acid (in the manufacture of Swiss cheese).

All fermentations are simply variations on the same theme: The NADH produced earlier in the pathway must be reoxidized to NAD^+ by transferring its electrons to some abundant electron acceptor. This ensures that there is enough NAD^+ available for the cell to continue to produce ATP via glycolysis under anaerobic conditions. **FIGURE 12.3** illustrates the common strategy used by virtually all fermentations.

Animal cells also carry out homolactic fermentation whenever pyruvate is produced faster than it can be oxidized through the citric acid cycle. During strenuous exertion, skeletal muscle cells derive most of their energy from this fermentation process, called **anaerobic glycolysis**—glycolysis occurring under anaerobic conditions.

By contrast, consider a cell undergoing active *respiration,* the oxidative breakdown and release of energy from nutrient molecules by reaction with oxygen.

● **CONCEPT** Anaerobic glycolysis (like aerobic glycolysis) leads to pyruvate, but the pyruvate is then reduced, so no net oxidation of glucose occurs.

In these cells, pyruvate is oxidized to acetyl-CoA, which enters the citric acid cycle (see Figure 12.1). The NADH produced during glycolysis is reoxidized through the mitochondrial electron transport chain for additional energy production (discussed in Chapter 14), with the electrons transferred ultimately to O_2, the terminal electron acceptor. The conversion of glucose to pyruvate in a respiring cell is called **aerobic glycolysis.**

(a) Homolactic fermentation

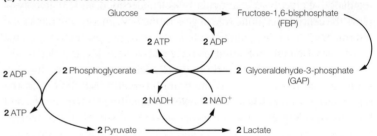

(b) Alcoholic fermentation

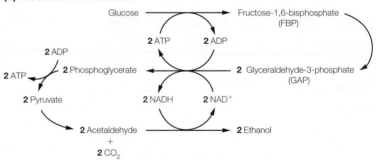

(c) Butanediol fermentation

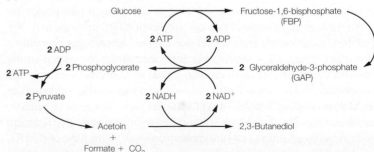

▲ **FIGURE 12.3** Fermentations use a common strategy to regenerate oxidized NAD^+: (a) Homolactic, (b) Alcoholic, and (c) Butanediol.

---

*Some organisms carry out a *heterolactic fermentation* in which only one lactate is produced, with the other three carbons converted to one ethanol plus one CO_2.

Chemical Strategy of Glycolysis

Glycolysis is such an important pathway that we shall examine each of its 10 reactions in some detail. Before doing so, it is helpful to have an overview of how glycolysis uses the potential energy in glucose to drive the synthesis of ATP. Three basic processes are involved:

Process 1. (Priming) Adding phosphoryl groups to glucose, yielding compounds with low phosphoryl group transfer potential.

Process 2. Chemically converting these low phosphoryl group transfer potential intermediates into compounds with high phosphoryl group transfer potential.

Process 3. (Substrate-level phosphorylation) Chemically coupling the energy-yielding hydrolysis of these high phosphoryl group transfer potential compounds to the synthesis of ATP by direct transfer of the phosphoryl group to ADP.

FIGURE 12.4 presents an abbreviated look at the conversion of glucose to pyruvate. In the energy investment phase (reactions ①–⑤), the sugar is metabolically activated by phosphorylation (Process 1). This priming process yields a six-carbon doubly phosphorylated sugar, **fructose-1,6-bisphosphate,** which undergoes cleavage to yield two equivalents of triose phosphate: one **glyceraldehyde-3-phosphate** and one **dihydroxyacetone phosphate.** Both of these compounds have a phosphoryl group transfer potential *lower* than that of ATP.

In the energy generation phase (reactions ⑥–⑩), the triose phosphates undergo further activation to yield two compounds containing high phosphoryl group transfer potential—first **1,3-bisphosphoglycerate** and then **phosphoenolpyruvate** (Process 2). Recall from Figure 3.7 that each of these compounds has a *higher* $\Delta G°'$ of hydrolysis than ATP.

The high phosphoryl group transfer potential of 1,3-bisphosphoglycerate and phosphoenolpyruvate then allows them to phosphorylate ADP, yielding ATP (Process 3). This third process is called **substrate-level phosphorylation**—the direct transfer of a phosphoryl group from a donor compound to ADP, yielding ATP.

12.2 Reactions of Glycolysis

Now let us consider in sequence the 10 reactions leading from glucose to pyruvate, numbering each reaction as indicated in Figure 12.4. The complete names of substrates and products are given when each reaction is presented, but in the text these names are shortened for simplicity. Thus, glucose-6-phosphate (G6P) is the same as α-D-glucose-6-phosphate.

Reactions 1–5: The Energy Investment Phase

The first five reactions, which constitute the energy investment phase, are summarized in the top half of Figure 12.4.

Reaction ①: The First ATP Investment

The first reaction is the ATP-dependent phosphorylation of glucose to form **glucose-6-phosphate (G6P),** catalyzed by **hexokinase.** This reaction involves nucleophilic attack of the C6—OH of glucose on the electrophilic terminal (γ) phosphate of ATP.

Reaction 1: Hexokinase

α-D-**Glucose** + **Mg^{2+}•ATP** →

α-D-**Glucose-6-phosphate** + ADP + H$^+$ $\Delta G°' = -18.4$ kJ/mol

Magnesium ion is required because the reactive form of ATP is its chelated complex with Mg^{2+}. This is true for virtually all ATP-requiring enzymes. Mg^{2+} partially neutralizes the negative charges on the oxygen atoms, making the γ-phosphorous atom more accessible for nucleophilic attack, and thus a better electrophile.

Hexokinase exists in various forms in different organisms, but is generally characterized by broad specificity for hexose sugars and low K_M for the sugar substrate (0.01 to 0.1 mM). The broad specificity allows various hexose sugars to be phosphorylated, including fructose and mannose, making it possible to use them in glycolysis. Besides serving the priming function, phosphorylating glucose helps retain it in the cell because phosphorylated compounds cross the plasma membrane very poorly.

Mammals possess several molecular forms of hexokinase. Different molecular forms of an enzyme catalyzing the same reaction are called **isoenzymes, isozymes,** or **isoforms.** Most tissues express hexokinase I, II, or III, all of which are low-K_M enzymes. As the accompanying graph of enzyme activity versus substrate concentration shows, hexokinase I is active, and thus glycolysis can proceed, even at very low blood glucose levels. Because intracellular glucose levels (2–15 mM) are usually far higher than the K_M value for hexokinase, the enzyme often functions in vivo at saturating substrate concentrations. Vertebrate liver expresses a distinctive isozyme, hexokinase IV, characterized by a much lower affinity for glucose and an insensitivity to inhibition by physiological concentrations of glucose-6-phosphate. More importantly, hexokinase IV exhibits a sigmoidal concentration dependence on glucose (see the graph below), requiring 5–10 mM glucose

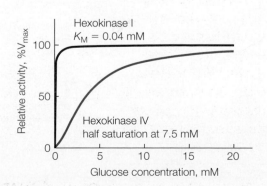

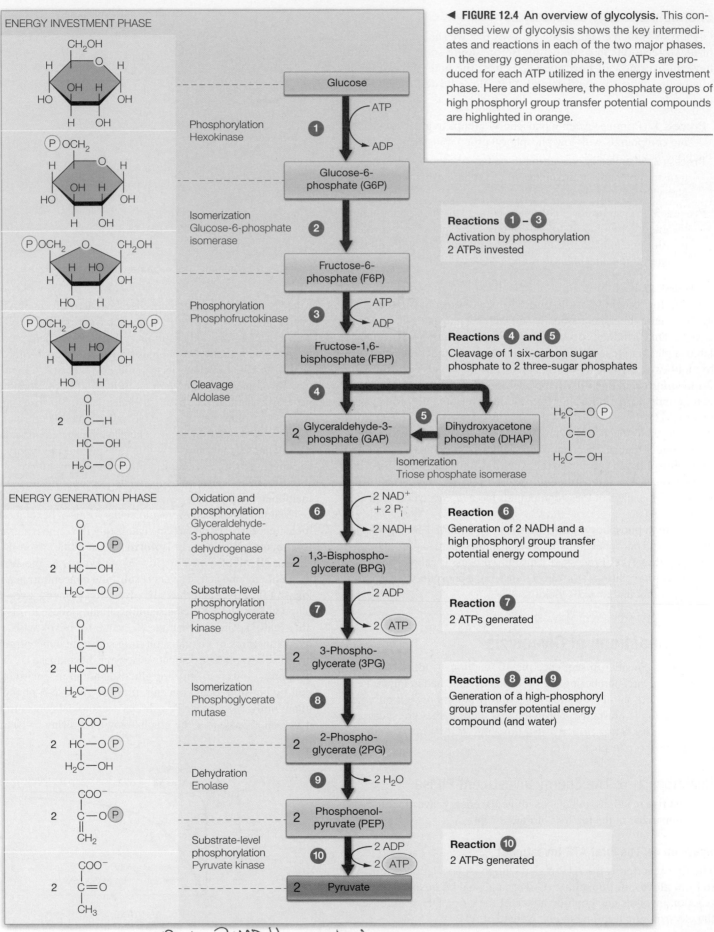

◀ **FIGURE 12.4 An overview of glycolysis.** This condensed view of glycolysis shows the key intermediates and reactions in each of the two major phases. In the energy generation phase, two ATPs are produced for each ATP utilized in the energy investment phase. Here and elsewhere, the phosphate groups of high phosphoryl group transfer potential compounds are highlighted in orange.

Net 2ATP + 2NADH incytosol

● **CONCEPT** A low-affinity isozyme of hexokinase in liver has a sigmoidal dependence on glucose concentration; thus, the liver is able to adjust its glucose use depending on the supply of glucose available in the blood.

for half-saturation. This special hexokinase isozyme allows the liver to adjust its rate of glucose utilization in response to variations in blood glucose levels. In fact, a major role of the liver is to regulate blood glucose levels (as we will see in Chapter 17), and hexokinase IV represents one of the principal mechanisms by which it does so.

Reaction ❷: Isomerization of Glucose-6-Phosphate

The second reaction, catalyzed by **glucose-6-phosphate isomerase** (also called G6P isomerase or phosphoglucoisomerase), is the readily reversible isomerization of glucose-6-phosphate (G6P), an aldose, to **fructose-6-phosphate (F6P),** a ketose.

Reaction 2: Glucose-6-phosphate isomerase

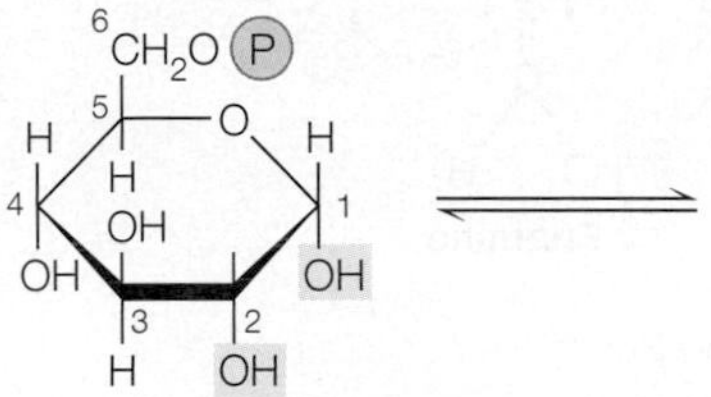

α-D-**Glucose-6-phosphate (G6P)**

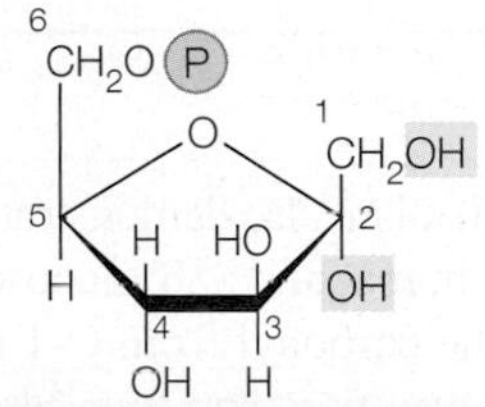

$\Delta G^{\circ\prime} = +1.7$ kJ/mol

D-**Fructose-6-phosphate (F6P)**

Transferring the carbonyl oxygen from carbon 1 (aldose) to carbon 2 (ketose) generates a hydroxyl group at carbon 1 that can be readily phosphorylated in reaction 3. It also sets up the sugar for a symmetric aldol cleavage later in reaction 4.

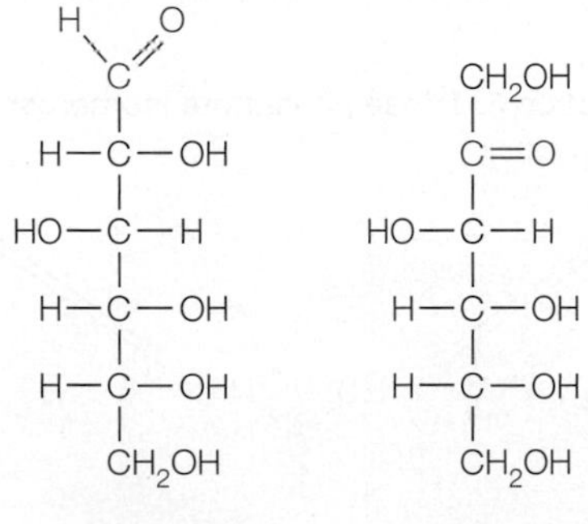

D-**Glucose**
(aldose)

D-**Fructose**
(ketose)

Reaction ❸: The Second Investment of ATP

In reaction 3, **phosphofructokinase (PFK)** catalyzes a second ATP-dependent phosphorylation, converting fructose-6-phosphate (F6P) to **fructose-1,6-bisphosphate (FBP),** a hexose derivative phosphorylated at both carbons 1 and 6.

Reaction 3: Phosphofructokinase

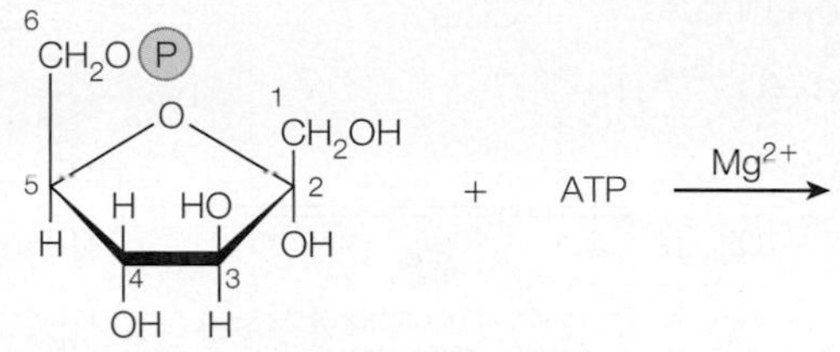

D-**Fructose-6-phosphate (F6P)**

$+$ ADP + H$^+$

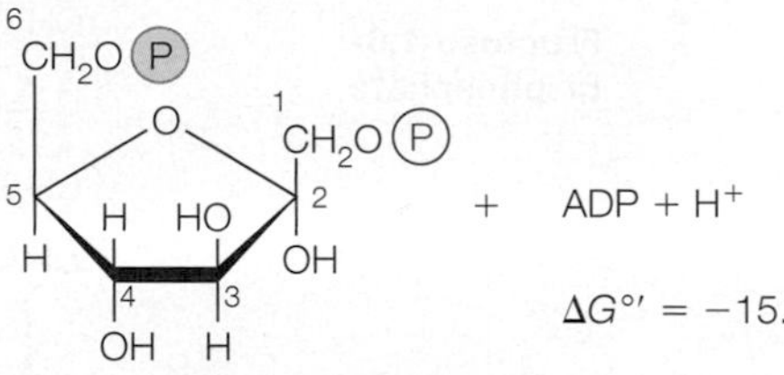

$\Delta G^{\circ\prime} = -15.9$ kJ/mol

D-**Fructose-1,6-bisphosphate (FBP)**

The reaction involves the same nucleophilic substitution chemistry we saw in reaction 1. Here in reaction 3, however, the C1—OH of F6P is the nucleophile that attacks the electrophilic γ-phosphate of ATP. Like phosphorylation at the 6 position, this reaction is sufficiently exergonic

● **CONCEPT** The phosphofructokinase reaction is the primary step at which glycolysis is regulated.

to be essentially irreversible in vivo. Irreversibility is important because PFK is the primary site for regulation of the flow of carbon through glycolysis. PFK is an allosteric enzyme whose activity is acutely sensitive to the energy status of the cell. Interactions with allosteric effectors, discussed later in Section 12.6, activate or inhibit PFK. This allosteric regulation increases carbon flux through glycolysis when there is a need to generate more ATP and inhibits it when the cell contains ample stores of ATP or oxidizable substrates.

Reaction ❹: Cleavage to Two Triose Phosphates

In reaction 4, fructose-1,6-bisphosphate (FBP) is cleaved to give **glyceraldehyde-3-phosphate (GAP)** and **dihydroxyacetone phosphate (DHAP),** each of which has three carbons. The enzyme is **fructose-1, 6-bisphosphate aldolase,** usually called **aldolase,** because reaction 4 is similar to the reverse of an aldol condensation.

Reaction 4: Aldolase

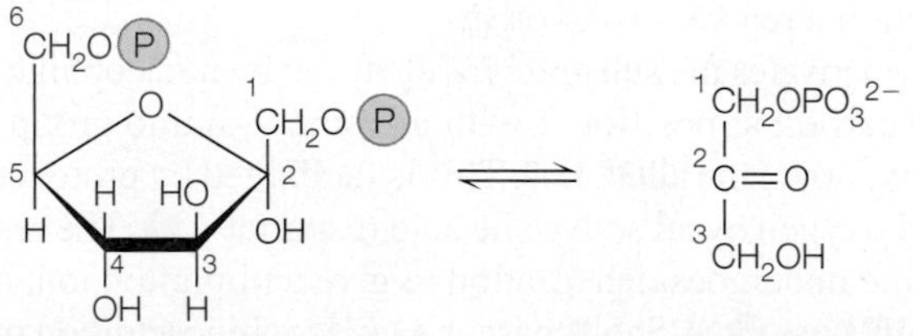

D-**Fructose-1,6-bisphosphate**
(FBP)

Dihydroxyacetone phosphate
(DHAP)

$+$

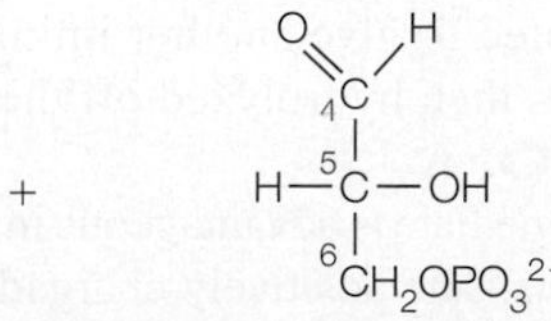

$\Delta G^{\circ\prime} = +23.9$ kJ/mol

D-**Glyceraldehyde-3-phosphate**
(GAP)

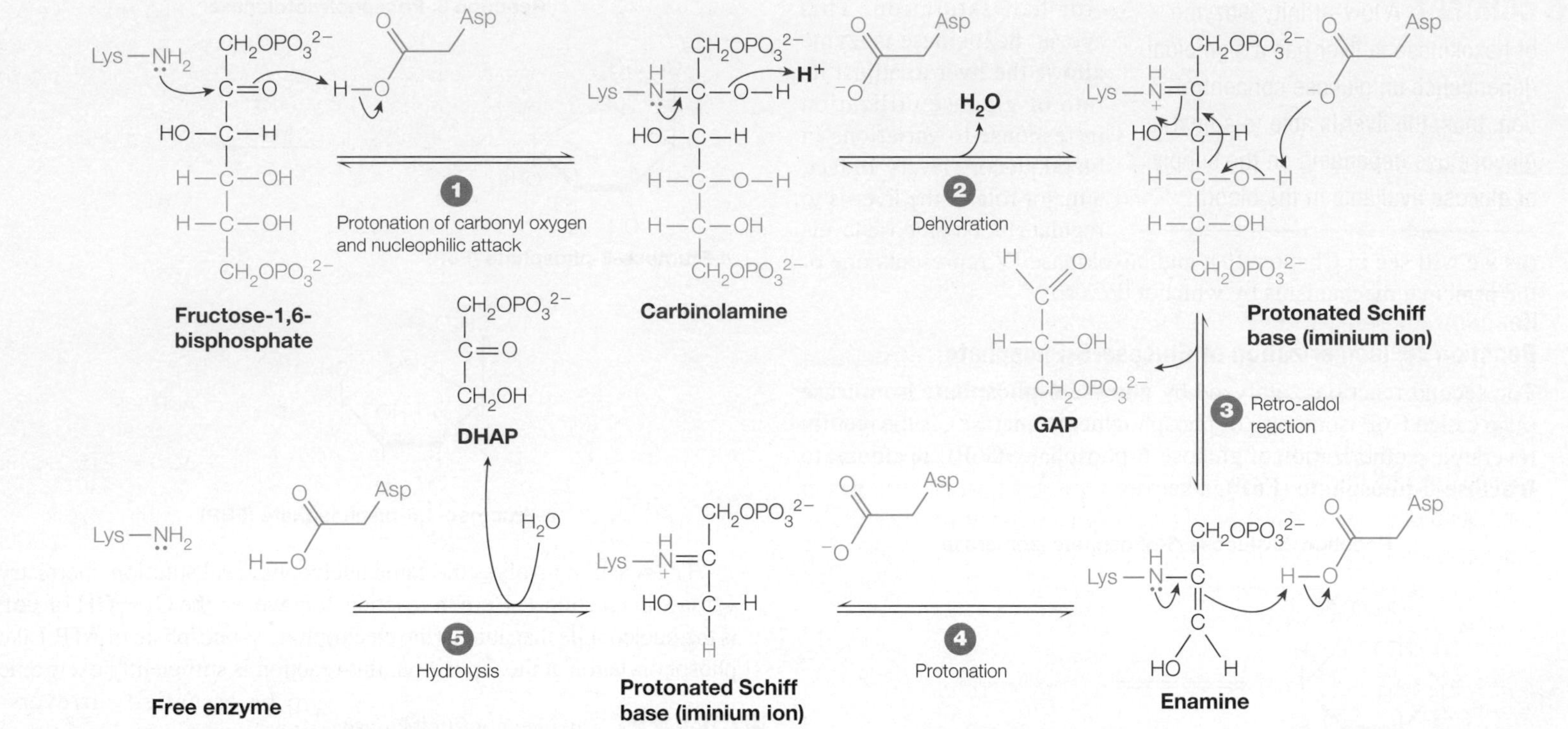

Fructose-1,6-bisphosphate

① Protonation of carbonyl oxygen and nucleophilic attack

DHAP

Carbinolamine

② Dehydration

GAP

Protonated Schiff base (iminium ion)

③ Retro-aldol reaction

Free enzyme

⑤ Hydrolysis

Protonated Schiff base (iminium ion)

④ Protonation

Enamine

▲ **FIGURE 12.5 Reaction mechanism for fructose-1,6-bisphosphate aldolase.** The figure shows the protonated Schiff base intermediate (iminium ion) between the substrate and an active site lysine residue. An aspartate residue facilitates the reaction via general acid–base catalysis.

● **CONCEPT** Aldolase cleaves fructose-1,6-bisphosphate under intracellular conditions, even though the equilibrium lies far toward fructose-1,6-bisphosphate under standard conditions.

Reaction 4 is so strongly endergonic under standard conditions that the formation of fructose-1,6-bisphosphate is highly favored. However, from the actual intracellular concentrations of the reactant and products, ΔG is estimated to be approximately -1.3 kJ/mol, consistent with the observation that the reaction proceeds as written in vivo. Reaction 4 demonstrates the importance of considering the conditions *in the cell* (ΔG) rather than standard state conditions ($\Delta G^{\circ\prime}$) when deciding in which direction a reaction is favored.

Aldolase activates the substrate for cleavage by nucleophilic attack on the keto carbon at position 2 with a lysine ε-amino group in the active site, as shown in **FIGURE 12.5**. This is facilitated by protonation of the carbonyl oxygen by an active site acid (aspartate) ①. The resulting carbinolamine undergoes dehydration to give an iminium ion, or protonated **Schiff base** ②. A Schiff base is a nucleophilic addition product between an amino group and a carbonyl group. A retro-aldol reaction then cleaves the protonated Schiff base into an enamine plus GAP ③. The enamine is protonated to give another iminium ion (protonated Schiff base) ④, which is then hydrolyzed off the enzyme to give the second product, DHAP ⑤.

The Schiff base intermediate is advantageous in this reaction because it can delocalize electrons. The positively charged iminium ion is thus a better electron acceptor than a ketone carbonyl, facilitating retro-aldol reactions like this one and, as we shall see, many other biological

conversions. This mechanism also demonstrates why it was important to isomerize G6P to F6P in reaction 2. If glucose had not been isomerized to fructose (moving the carbonyl from C-1 to C-2), then the aldolase reaction would have given two- and four-carbon fragments, instead of the metabolically equivalent three-carbon fragments.

Reaction ⑤: Isomerization of Dihydroxyacetone Phosphate

In reaction 5, **triose phosphate isomerase (TIM)** catalyzes the isomerization of dihydroxyacetone phosphate (DHAP) to glyceraldehyde-3-phosphate (GAP) via an enediol intermediate.

Reaction 5: Triose phosphate isomerase (TIM)

Dihydroxyacetone phosphate (DHAP)

Enediol intermediate

D-Glyceraldehyde-3-phosphate (GAP)

$\Delta G^{\circ\prime} = +7.6$ kJ/mol

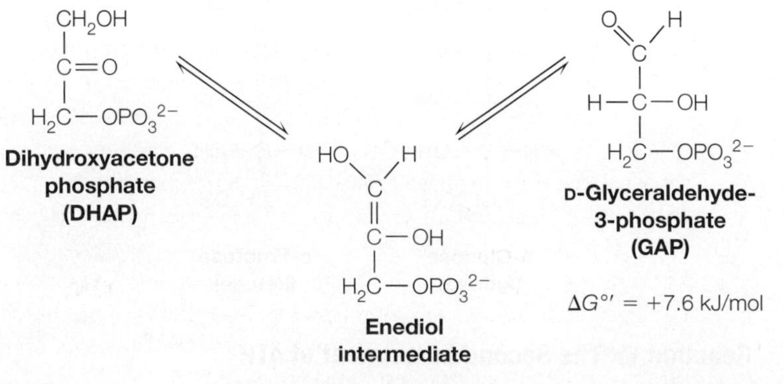

Like reaction 4, reaction 5 is weakly endergonic under standard conditions, but the intracellular concentration of GAP is low because it is consumed in subsequent reactions. Thus, reaction 5 is drawn toward the right.

Taking stock, the first five reactions of glycolysis expend two ATP molecules and convert one molecule of glucose to two molecules of GAP. And because GAP is the substrate for reaction 6, the isomerization of DHAP to GAP in reaction 5 means that all six carbon atoms of glucose are utilized. Each of these GAP molecules will next be metabolized to give compounds with high phosphoryl group transfer potential that can drive the synthesis of ATP. The energy investment phase of the cycle is complete, and the energy generation phase is about to begin.

Reactions 6–10: The Energy Generation Phase

Reaction ❻: Generation of the First Energy-Rich Compound

Reaction 6, catalyzed by **glyceraldehyde-3-phosphate dehydrogenase (GAPDH),** is particularly important because it generates the first intermediate with high phosphoryl group transfer potential and because it generates a pair of reducing equivalents. The overall reaction is as follows:

● **CONCEPT** Glyceraldehyde-3-phosphate dehydrogenase creates a compound with high phosphoryl group transfer potential (BPG) and generates a pair of reducing equivalents (NADH).

Reaction 6: Glyceraldehyde-3-phosphate dehydrogenase (GAPDH)

D-**Glyceraldehyde-3-phosphate (GAP)**

$\Delta G^{\circ\prime} = +6.3$ kJ/mol

1,3-Bisphosphoglycerate (BPG)

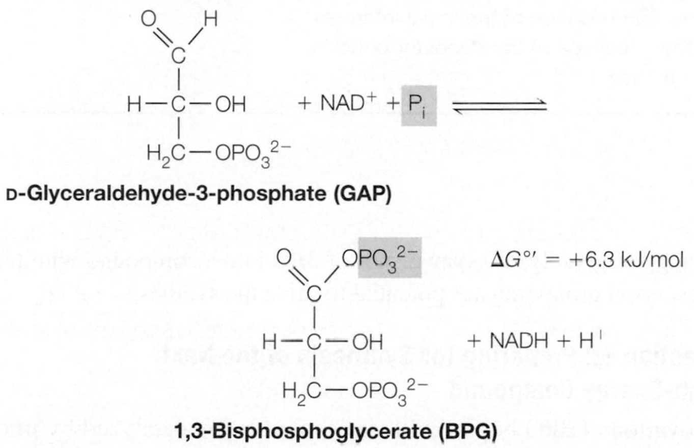

Reaction 6 involves a two-electron oxidation of the carbonyl carbon of GAP to the carboxyl level, a reaction that is normally quite exergonic. However, the overall reaction is slightly *endergonic* under *standard* conditions because the enzyme utilizes most of the energy released by the oxidation to drive the synthesis of **1,3-bisphosphoglycerate (BPG),** a compound with high phosphoryl group transfer potential. BPG contains a carboxylic acid–phosphoric acid mixed anhydride, or an **acyl-phosphate group,** at position 1, a functional group with a very high standard free energy of hydrolysis (-49.4 kJ/mol). This enzyme also requires a coenzyme, NAD^+, to accept electrons from the substrate being oxidized.

Because the acyl-phosphate group has a greater phosphoryl group transfer potential than ATP, BPG can drive the synthesis of ATP from ADP. In fact, it does so in reaction 7, the first of two substrate-level phosphorylations in glycolysis. Because it is important to understand how ATP is synthesized, much attention has been focused on understanding how the high phosphoryl group transfer potential compounds in substrate-level phosphorylation are synthesized.

For glyceraldehyde-3-phosphate dehydrogenase, the reaction proceeds as outlined in **FIGURE 12.6**, starting in ❶ with formation of a **thiohemiacetal** group involving the substrate carbonyl group and a cysteine

thiol group on the enzyme. The thiohemiacetal is next oxidized by NAD^+ in ❷ to give an acyl-enzyme intermediate, or thioester. Thioesters are energy-rich compounds; cleavage of this thioester by P_i preserves much of the energy as the acyl phosphate, which is the product.

This mechanism conserves the energy of oxidation by coupling an exergonic reaction to an endergonic reaction:

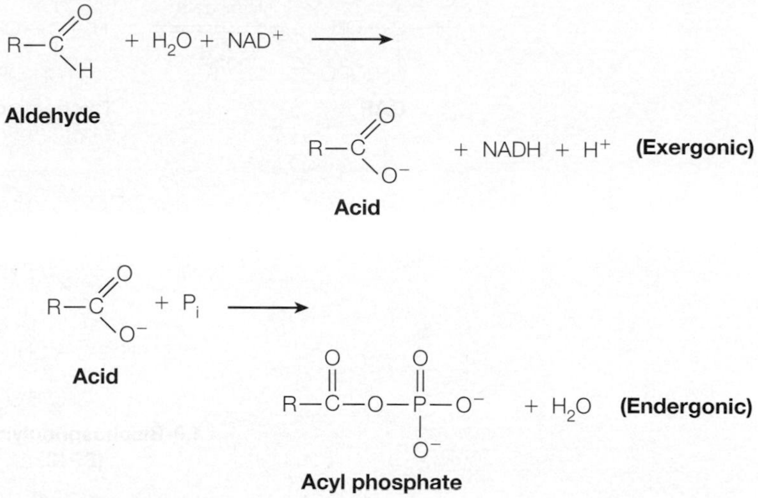

Aldehyde

$+ NADH + H^+$ **(Exergonic)**

Acid

Acid

Acyl phosphate

$+ H_2O$ **(Endergonic)**

In step ❷ of Figure 12.6, an aldehyde (the thiohemiacetal) is oxidized to the level of an acid (the thioester), a two-electron process. But rather than releasing the acid, the enzyme incorporates P_i using the oxidation energy to create a "high-energy" acyl-phosphate compound (BPG), ❸, ❹, possessing higher phosphoryl group transfer potential than ATP. This reaction thus accomplishes the second of the three processes in the chemical strategy of glycolysis outlined previously.

The overall stoichiometry of reaction 6 involves the reduction of 1 mole of NAD^+ to $NADH + H^+$ per mole of glyceraldehyde-3-phosphate. This reaction is the source of the NADH formed in glycolysis, which was first identified in Figures 12.2 and 12.3.

Reaction ❼: The First Substrate-Level Phosphorylation

Reaction 7 accomplishes the third process in the chemical strategy of glycolysis, the synthesis of ATP by transfer of the phosphoryl group of 1,3-bisphosphoglycerate to ADP. After phosphoryl group transfer gives ATP, the remaining product is **3-phosphoglycerate (3PG).** This substrate-level phosphorylation reaction is catalyzed by **phosphoglycerate kinase** as follows:

Reaction 7: Phosphoglycerate kinase

$+ ADP$

Mg^{2+}

1,3-Bisphosphoglycerate (BPG)

$\Delta G^{\circ\prime} = -17.2$ kJ/mol

$+ ATP$

3-Phosphoglycerate (3PG)

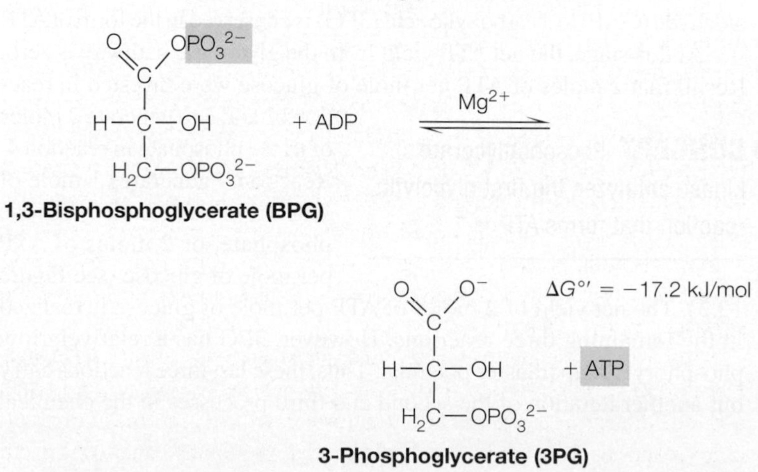

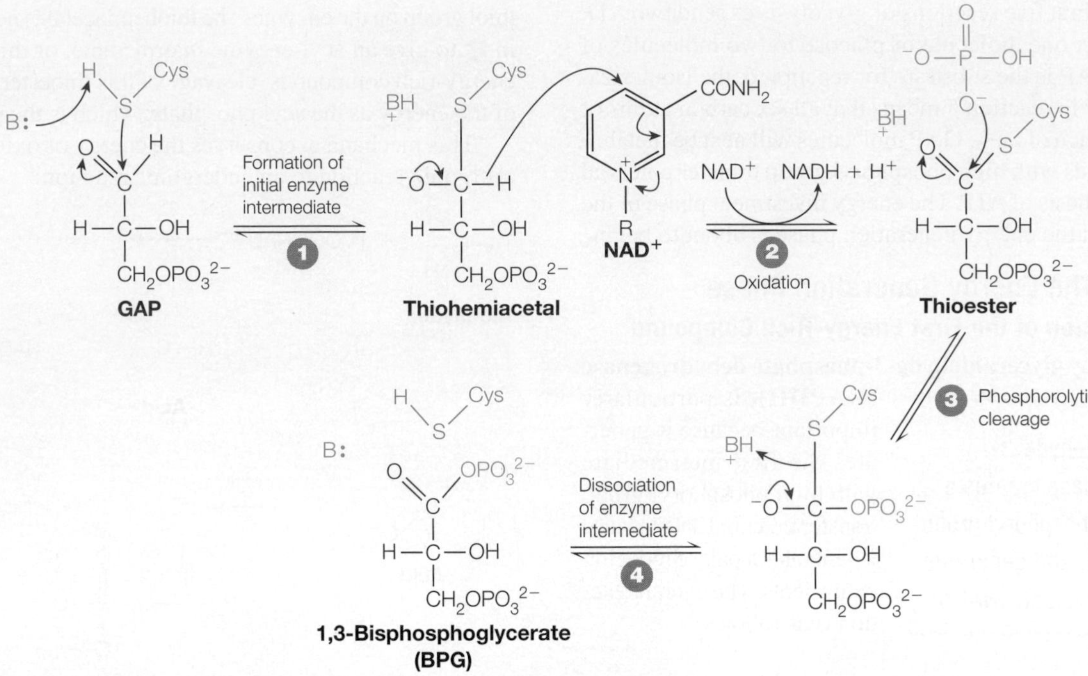

▲ **FIGURE 12.6 Reaction mechanism for glyceraldehyde-3-phosphate dehydrogenase.** ❶ Formation of the initial thiohemiacetal intermediate between glyceraldehyde-3-phosphate (GAP) and the enzyme. ❷ Oxidation of the initial intermediate by NAD^+ to give an acyl-thioester enzyme intermediate. ❸ and ❹ Phosphorolytic cleavage of the thioester bond in the acyl-enzyme intermediate. B: represents an active site histidine acting as a general base.

Glyceraldehyde-3-phosphate dehydrogenase (reaction 6) and phosphoglycerate kinase (reaction 7) are thermodynamically coupled:

$$\Delta G^{o'} \ (kJ/mol)$$

glyceraldehyde-3-P + P_i + $NAD^+ \rightarrow$
1,3-bisphosphoglycerate + NADH + H^+ +6.3

1,3-bisphosphoglycerate + ADP $\rightarrow$
3-phosphoglycerate + ATP   −17.2

glyceraldehyde-3-P + P_i + ADP + $NAD^+ \rightarrow$
3-phosphoglycerate + ATP + NADH + H^+ $\Delta G^{o'}_{Sum} = -10.9$

Thus, through two consecutive reactions, the energy of oxidation of an aldehyde (GAP) to a carboxylic acid (3PG) is conserved in the form of ATP.

At this stage, the net ATP yield from the glycolytic pathway is zero. Recall that 2 moles of ATP per mole of glucose were invested in reactions 1 and 3 to generate 2 moles of triose phosphate in reaction 4. Reaction 7 generates 1 mole of ATP from each mole of triose phosphate, or 2 moles of ATP per mole of glucose (see Figure 12.3). The net yield of 2 moles of ATP per mole of glucose is realized in the remaining three reactions. However, 3PG has a relatively low phosphoryl group transfer potential. Thus, these last three reactions carry out another iteration of the second and third processes in the chemical

● **CONCEPT** Phosphoglycerate kinase catalyzes the first glycolytic reaction that forms ATP.

strategy of glycolysis—conversion of 3PG into a compound with high phosphoryl group transfer potential to drive the synthesis of ATP.

Reaction ❽: Preparing for Synthesis of the Next High-Energy Compound

Activation of 3PG begins with an isomerization catalyzed by **phosphoglycerate mutase.** The enzyme transfers the phosphoryl group from position 3 to position 2 of the substrate to yield **2-phosphoglycerate (2PG).** Mg^{2+} is required.

Reaction 8: Phosphoglycerate mutase

3-Phosphoglycerate (3PG)

$\Delta G^{o'} = +4.4$ kJ/mol

2-Phosphoglycerate (2PG)

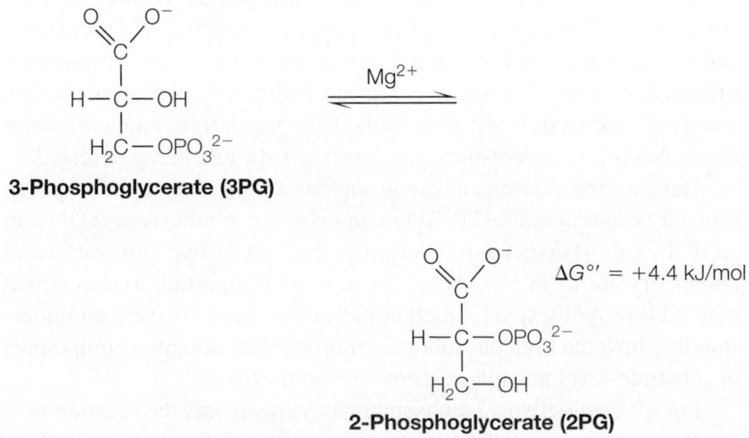

The reaction is slightly endergonic under standard conditions, but the intracellular level of 3PG is high relative to that of 2PG, so reaction 8 proceeds to the right in vivo. The enzyme contains a phosphohistidine residue in the active site.

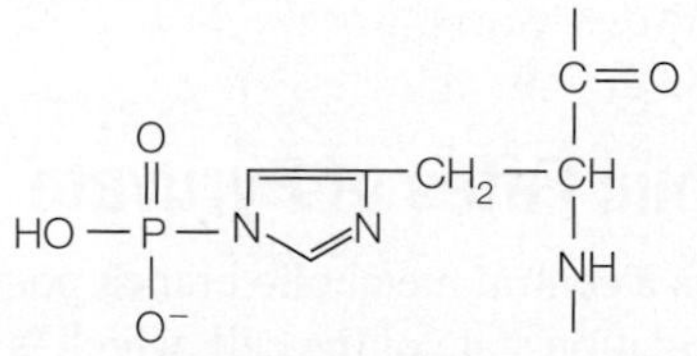

N-Phosphohistidine residue

In the first step of the reaction, this phosphoryl group is transferred from the enzyme to the substrate to give 2,3-bisphosphoglycerate. Transfer of the phosphoryl group from C-3 to the active site of the enzyme regenerates the phosphorylated enzyme and forms the product, which is released.

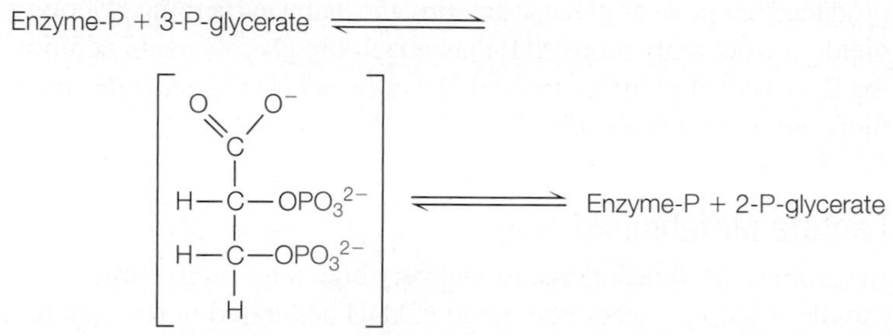

Enzyme-P + 3-P-glycerate ⇌

Enzyme-bound 2,3-bis-P-glycerate ⇌ Enzyme-P + 2-P-glycerate

Enzyme-bound 2,3-bis-P-glycerate

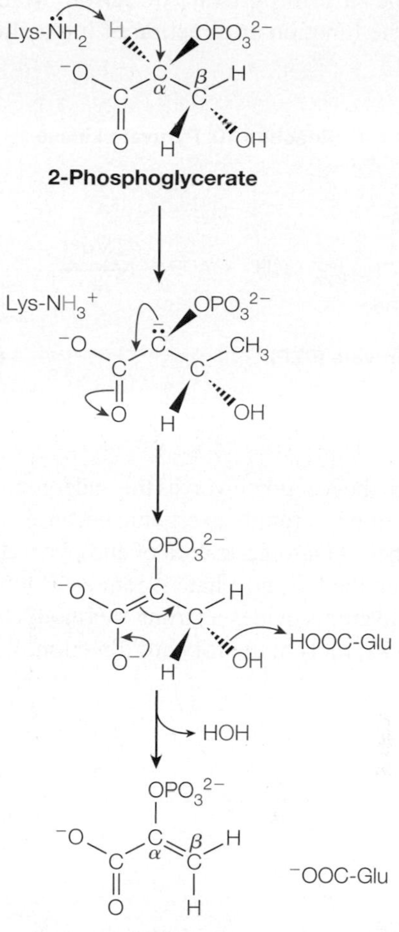

Reaction ⑨: Synthesis of the Second High-Energy Compound

Reaction 9, catalyzed by **enolase,** converts 2PG, which has a low phosphoryl group transfer potential, to **phosphoenolpyruvate (PEP),** which has a very high phosphoryl group transfer potential. This allows PEP to participate in the second substrate-level phosphorylation of glycolysis in reaction 10.

Reaction 9: Enolase

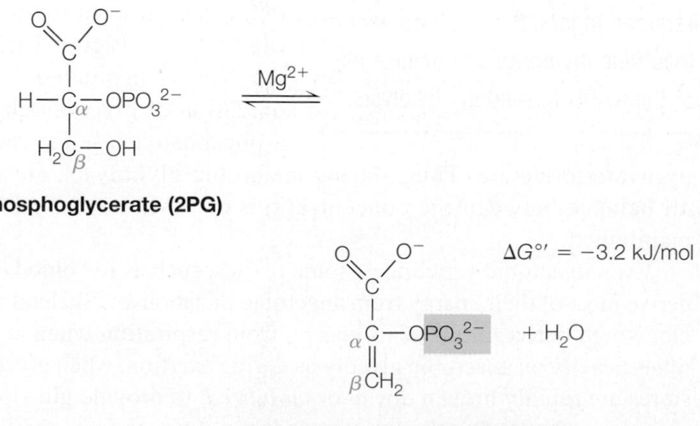

$\Delta G^{\circ\prime} = -3.2 \text{ kJ/mol}$

2-Phosphoglycerate (2PG)

Phosphoenolpyruvate (PEP)

2-Phosphoglycerate

PEP

Like most β-hydroxy carbonyl compounds, 2PG is readily dehydrated via an $\alpha\beta$-elimination reaction (see Section 11.3). An active site Lys, acting as a general base, abstracts a proton from C-2 (α carbon) of 2PG, generating a carbanionic intermediate (see above). An active site Glu facilitates the elimination, protonating the leaving group (OH–) by general acid catalysis. Despite its small overall free energy change, this seemingly simple transformation increases enormously the standard free energy of hydrolysis of the phosphate ester bond, from -15.6 kJ/mol for 2PG to -61.9 kJ/mol for PEP. Phosphorylation of the C2—OH prevents the otherwise highly favorable tautomerization to the keto form. As discussed in Section 3.4, the great thermodynamic instability of enolpyruvate is chiefly responsible for the large negative free energy of hydrolysis of phosphoenolpyruvate.

Reaction ⑩: The Second Substrate-Level Phosphorylation

● **CONCEPT** Pyruvate kinase catalyzes the second ATP-forming reaction in the glycolytic pathway.

In the last reaction, catalyzed by **pyruvate kinase (PK),** PEP transfers its phosphoryl group to ADP in another substrate-level phosphorylation, forming **pyruvate** and ATP. (Note that pyruvate kinase is named as if it were acting in the pyruvate to PEP direction, even though it is strongly exergonic in the PEP to pyruvate direction. Many enzymes were named before the function or direction of intracellular catalysis had been identified.)

Reaction 10: Pyruvate kinase

Phosphoenolpyruvate (PEP)

Pyruvate

$$\Delta G°' = -29.7 \text{ kJ/mol}$$

Even though the reaction involves the endergonic synthesis of ATP, the overall reaction is strongly exergonic because, as noted in Section 3.4, the spontaneous tautomerization of enolpyruvate (the initial product formed when the C2 phosphate is removed) to the highly favored keto form (pyruvate) provides a strong thermodynamic driving force ($\Delta G°' = -46 \text{ kJ/mol}$) in the forward direction.

PEP

Enolpyruvate

Pyruvate

Reaction 10 is another site for metabolic regulation. As we will discuss in more detail in Section 12.6, pyruvate kinase (PK) is allosterically activated by fructose-1,6-bisphosphate, the product of reaction 3 of glycolysis, and is inhibited by high ATP levels. These allosteric mechanisms allow inhibition of glycolysis when ample energy is already available and ensure that flux through the early part of the pathway is coordinated with that at the end of the pathway.

The pyruvate kinase reaction assures a net synthesis of ATP over the 10 reactions of glycolysis. Two ATP per hexose are generated here in reaction 10, to go with the two generated by phosphoglycerate kinase in reaction 7 (see Figure 12.3). Subtracting the two ATPs invested at the hexokinase and phosphofructokinase steps (reactions 1 and 3, respectively) gives a net yield of 2 moles of ATP per mole of glucose—not a high yield, to be sure, but glycolysis is fast and can meet the metabolic energy requirements of many anaerobes. Moreover, subsequent metabolism of pyruvate through aerobic pathways generates much additional ATP.

TABLE 12.1 summarizes the reactions of glycolysis, showing free energy changes and ATP yields at each step. Note the difference between standard ($\Delta G°'$) and estimated (ΔG) Gibbs free energy changes, and the three values in red. We will discuss the importance of these shortly.

12.3 Metabolic Fates of Pyruvate

Pyruvate represents a central metabolic branch point. Its fate depends crucially on the oxidation state of the cell, which is related to reaction 6 of glycolysis. Recall that in this reaction glyceraldehyde-3-phosphate dehydrogenase converts 1 mole of NAD^+ per mole of triose phosphate to NADH. The cytoplasm has a finite supply of NAD^+, so *this NADH must be reoxidized to NAD^+* for glycolysis to continue. During aerobic glycolysis, the NADH is oxidized by the mitochondrial electron transport chain, with the electrons transferred ultimately to molecular oxygen. This oxidation of NADH, which we consider in detail in Chapter 14, yields about 2.5 moles of ATP (synthesized from ADP) per mole of NADH oxidized. Because 2 moles of NADH are produced per mole of glucose entering the pathway, aerobic glycolysis yields considerably more ATP than anaerobic glycolysis. In addition, the oxidation of pyruvate through the citric acid cycle generates much more energy, via respiration.

Lactate Metabolism

In aerobic cells that are undergoing very high rates of glycolysis (e.g., rapidly dividing cancer cells), the NADH generated in reaction 6 of glycolysis cannot all be reoxidized to NAD^+ at comparable rates in the mitochondrion. In such cases, or in anaerobes (which lack mitochondria), NADH must be used to drive the reduction of an organic substrate in order to maintain sufficient NAD^+ levels for continued glycolysis. As noted earlier, that substrate is pyruvate itself, both in eukaryotic cells and in lactic acid bacteria, and the product is lactate. The enzyme catalyzing this reaction is lactate dehydrogenase (LDH) (see p. 378).

$$\text{Pyruvate} + \text{NADH} + \text{H}^+ \rightleftharpoons \text{L-Lactate} + \text{NAD}^+$$
$$\Delta G°' = -25.1 \text{ kJ/mol}$$

● **CONCEPT** Pyruvate must be reduced to lactate when tissues are insufficiently aerobic to oxidize all of the NADH formed in glycolysis.

The equilibrium for this reaction lies far toward lactate. As illustrated in Figure 12.3(a), the NADH produced in the oxidation of glyceraldehyde-3-phosphate is used to reduce pyruvate to lactate. Thus, during anaerobic glycolysis, an overall balance between the concentrations of NAD^+ and NADH is maintained.

Even in aerobic vertebrates, some tissues, such as red blood cells, derive most of their energy from anaerobic metabolism. Skeletal muscle, which derives most of its energy from respiration when at rest, relies heavily on anaerobic glycolysis during exertion, when glycogen stores are rapidly broken down, or *mobilized,* to provide glucose for glycolysis. Normally, the lactate produced diffuses from the tissue and is transported through the bloodstream to highly aerobic tissues, such as the heart and liver. The aerobic tissue can catabolize lactate

TABLE 12.1 Summary of Glycolysis

Reaction	Enzyme	ATP yield	$\Delta G^{\circ\prime}$	ΔG
Glucose (G) → [1] ATP/ADP → Glucose-6-phosphate (G6P)	Hexokinase (HK)	−1	−18.4	−33.5
Glucose-6-phosphate (G6P) → [2] → Fructose-6-phosphate (F6P)	Glucose-6-phosphate isomerase		+1.7	−2.5
Fructose-6-phosphate (F6P) → [3] ATP/ADP → Fructose-1,6-bisphosphate (FBP)	Phosphofructokinase (PFK)	−1	−15.9	−22.2
Fructose-1,6-bisphosphate (FBP) → [4] → Glyceraldehyde-3-phosphate (GAP) + dihydroxyacetone phosphate (DHAP)	Aldolase		+23.9	−1.3
→ [5] → Two glyceraldehyde-3-phosphate (GAP)	Triose phosphate isomerase (TIM)		+7.6	~0
Two glyceraldehyde-3-phosphate (GAP) → [6] $NAD^+ + P_i$ / $NADH + H^+$ → 1,3-Bisphosphoglycerate (BPG)	Glyceraldehyde-3-phosphate dehydrogenase (GAPDH)		+6.3 (+12.6)	−1.7 (−3.4)
1,3-Bisphosphoglycerate (BPG) → [7] ADP/ATP → 3-Phosphoglycerate (3PG)	Phosphoglycerate kinase (PGK)	+1(+2)	−17.2 (−34.4)	~0
3-Phosphoglycerate (3PG) → [8] → 2-Phosphoglycerate (2PG)	Phosphoglycerate mutase		+4.4 (+8.8)	~0
2-Phosphoglycerate (2PG) → [9] H_2O → Phosphoenolpyruvate (PEP)	Enolase		−3.2 (−6.4)	−3.3 (−6.6)
Phosphoenolpyruvate (PEP) → [10] ADP/ATP → Pyruvate (Pyr)	Pyruvate kinase (PK)	+1(+2)	−29.7 (−59.4)	−16.7 (−33.4)
Net: Glucose + 2ADP + 2P$_i$ + 2NAD$^+$ → 2 pyruvate + 2ATP + 2NADH + 2H$^+$ + 2H$_2$O		+2	−79.9	−102.9

Note: $\Delta G^{\circ\prime}$ and ΔG values in kJ/mol. The values in parentheses are based on doubling all the reactions past reaction 5, since the energy generation phase involves 2 three-carbon substrates per glucose molecule. ΔG values are estimated from the approximate intracellular concentrations of glycolytic intermediates in rabbit skeletal muscle.

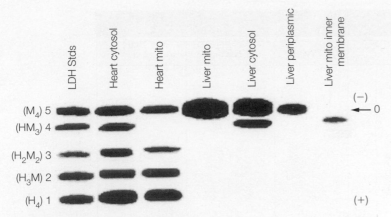

(M_4) 5
(HM_3) 4
(H_2M_2) 3
(H_3M) 2
(H_4) 1

▲ FIGURE 12.7 Structural basis for the existence of isozymes of lactate dehydrogenase (LDH). Proteins were subjected to electrophoresis in a nondenaturing starch gel, which was then treated to reveal bands containing enzymatically active protein. H4 is a tetramer containing only the H subunit, whereas M4 contains only M subunits. Heart tissue contains all five LDH isozymes, whereas liver contains primarily the M4 isozyme.

reaction, the NADH-dependent reduction of acetaldehyde to ethanol, catalyzed by alcohol dehydrogenase (ADH).

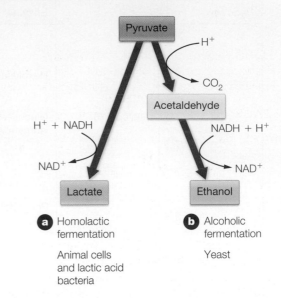

a Homolactic fermentation

Animal cells and lactic acid bacteria

b Alcoholic fermentation

Yeast

further, through respiration, or can convert it back to glucose, through gluconeogenesis.

Isozymes of Lactate Dehydrogenase

Lactate dehydrogenase (LDH) was the first enzyme that established the structural basis for the existence of isozymes. Most tissues contain five isozymes of LDH. They can be resolved electrophoretically, as shown in **FIGURE 12.7**.

LDH is a tetrameric protein consisting of two types of subunits, called M and H, which have small differences in amino acid sequence. M subunits predominate in skeletal muscle and liver, whereas H subunits predominate in heart. M and H subunits combine randomly with each other, so that the five major isozymes have the compositions M_4, M_3H, M_2H_2, MH_3, and H_4. Because of this random subunit reassortment, the isozymic composition of a tissue is determined primarily by the expression levels of the genes specifying the two subunits. The physiological need for the existence of different forms of this enzyme is not yet clear.

● **CONNECTION** The tissue specificity of isozyme patterns is useful in clinical medicine. Such pathological conditions as myocardial infarction (heart attack), infectious hepatitis, and muscle diseases all involve the death of cells in the affected tissue and the release of their contents to the blood. The pattern of LDH isozymes in the blood serum is representative of the tissue that released the isozymes. This information can be used to diagnose such conditions and to monitor the progress of treatment.

Ethanol Metabolism

Pyruvate has numerous alternative fates in anaerobic microorganisms. As we have seen, lactic acid bacteria reduce pyruvate to lactate in a single step (see next column). By contrast, yeasts convert pyruvate to ethanol in a two-step pathway. This alcoholic fermentation starts with the nonoxidative decarboxylation of pyruvate to acetaldehyde, catalyzed by **pyruvate decarboxylase.** NAD^+ is regenerated in the next

The first reaction requires **thiamine pyrophosphate (TPP)** as a coenzyme. This coenzyme is derived from thiamine, the first of the B vitamins to be identified, and so it is also called vitamin B_1. TPP is the coenzyme for all decarboxylations of α-keto acids, and the mechanism of these reactions is described in Chapter 13.

Animal tissues also contain alcohol dehydrogenase, even though ethanol is not a major metabolic product in animal cells. Some of the major metabolic consequences of ethanol intoxication result from ethanol oxidation by this enzyme in the liver. First, there is massive reduction of NAD^+ to NADH, which greatly lowers NAD^+ levels. This depletion decreases flux through glyceraldehyde-3-phosphate dehydrogenase (reaction 6), thereby reducing the amount of energy generated via glycolysis in reactions 7–10. Second, acetaldehyde is quite toxic, and many of the unpleasant effects

● **CONNECTION** Acetaldehyde is quite toxic, and many of the unpleasant effects of hangovers, such as headache and nausea, result from an accumulation of acetaldehyde. Antabuse (disulfiram) is a drug used to treat alcoholism. It is an inhibitor of acetaldehyde dehydrogenase and causes acetaldehyde to accumulate even faster, resulting in the unpleasant side effects after just one drink.

of hangovers, such as headache and nausea, result from an accumulation of acetaldehyde and its metabolites.

12.4 Energy and Electron Balance Sheets

By writing a balanced chemical equation for glycolysis, we can compute the ATP yield accompanying conversion of 1 mole of glucose. For homolactic fermentation, we can write the following balanced equation:

$$\text{Glucose} + 2\text{ADP} + 2\text{P}_i + 2\text{H}^+ \longrightarrow 2 \text{ lactate} + 2\text{ATP} + 2\text{H}_2\text{O}$$

Similarly, we can write a balanced equation for alcoholic fermentation:

$$\text{Glucose} + 2\text{ADP} + 2\text{P}_i + 4\text{H}^+ \longrightarrow$$
$$2 \text{ ethanol} + 2\text{CO}_2 + 2\text{ATP} + 2\text{H}_2\text{O}$$

Note first that both processes involve no net change in oxidation state. NAD^+ and NADH, both of which participate in the reaction pathways, do not appear in the overall reactions. This is an example of the obligate-coupling stoichiometry that was first introduced in Chapter 11. The

● **CONCEPT** Glycolysis, which yields 2 ATP per glucose, is fast but releases only a small fraction of the energy available from glucose.

metabolism of glucose to lactate represents a nonoxidative process because there is no change in the overall oxidation state of the carbons in the empirical formulas for glucose $(\text{C}_6\text{H}_{12}\text{O}_6)$ and lactate $(\text{C}_3\text{H}_6\text{O}_3)$—the numbers of hydrogens and oxygens bound per carbon atom are identical for both compounds. The same is true for ethanol plus CO_2, when the atoms in both are counted, so alcoholic fermentation is likewise nonoxidative. While some individual carbon atoms of lactate (and of ethanol plus CO_2) undergo oxidation, others become reduced, so there is no net oxidation in the process as a whole. By contrast, pyruvate $(\text{C}_3\text{H}_4\text{O}_3)$ is more highly oxidized than glucose.

Next, note that both of these balanced equations represent an exergonic process coupled to an endergonic process. In alcoholic fermentation, for example:

	$\Delta G^{\circ\prime}$
Exergonic:	
Glucose → 2 ethanol + 2CO$_2$	−228 kJ/mol glucose
Endergonic:	
2ADP + 2P$_i$ → 2ATP + 2H$_2$O	$2(32.2) = -64.4$ kJ/mol glucose
	$\Delta G^{\circ\prime}{}_{\text{Sum}} = -163.6$ kJ/mol glucose

Thus, the efficiency of the process—the amount of available free energy actually captured in ATP—is $64.4/228 = 28.2\%$ under standard conditions. The rest of the free energy released ensures that the process goes to completion. The equilibrium constant K for this process can be calculated from Equation 3.22:

$$K = e^{\left(\frac{-\Delta G^{\circ\prime}}{RT}\right)} = 4.7 \times 10^{28}$$

Is glycolysis favorable under physiological conditions? Table 12.1 lists ΔG values for the individual steps, estimated from the approximate intracellular concentrations of glycolytic intermediates in rabbit

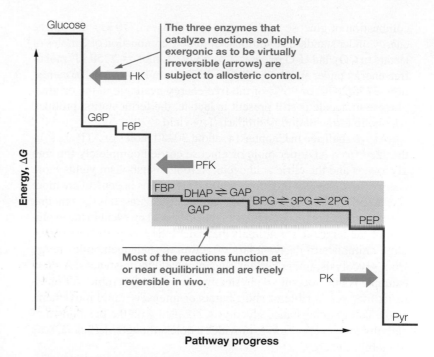

▲ **FIGURE 12.8 Energy profile of anaerobic glycolysis.** The graph shows the change in actual free energy for each reaction in the pathway, based on estimated ΔG values calculated in Table 12.1. Metabolite and enzyme abbreviations are as defined in Table 12.1.

skeletal muscle. These estimated ΔG values are plotted in **FIGURE 12.8** to illustrate the actual free energy changes in the pathway. All but three of the reactions function at or near equilibrium and are freely reversible in vivo. The three exceptions are the reactions catalyzed by hexokinase (HK, reaction 1), phosphofructokinase (PFK, reaction 3), and pyruvate kinase (PK, reaction 10), which take place with large decreases in free energy. These three nonequilibrium reactions are irreversible in vivo, and they make the entire glycolytic pathway unidirectional. As we will see in the next section, these are the steps that must be bypassed in gluconeogenesis. Not coincidentally, these three nonequilibrium reactions are also the sites of regulation of glycolysis because, as we discussed in Section 11.4, regulation can be imposed *only* on reactions displaced far from equilibrium.

This 10-step pathway is fast—anaerobic glycolysis can produce ATP at rates 100 times higher than that of aerobic oxidative phosphorylation. When the oxygen supply of muscle cells cannot keep up with their demand for ATP during strenuous exercise, anaerobic glycolysis satisfies their energy needs. Cancer cells also take advantage of the high rate of ATP production afforded by glycolysis to support their abnormally fast proliferation. Most rapidly dividing cancer cells metabolize glucose by glycolysis, producing lactate even though oxygen is abundant. This phenomenon, described in greater detail in Chapter 17, was first noted by Otto Warburg in 1925 and is known as the Warburg effect. The high rate of ATP production via glycolysis means that glucose must be utilized at a high rate, too, since only two ATPs are produced per glucose.

This brings us to a final point about the bioenergetics of glycolysis. Glycolysis releases but a small fraction of the potential energy stored in the glucose molecule. As noted earlier in Chapters 3 and 11, the complete combustion of glucose to CO_2 and H_2O releases 2870 kJ/mol of free energy under standard conditions. Complete combustion of 2 moles of lactate to CO_2 and H_2O releases 2×1379 kJ/mol $= 2758$ kJ/mol of free energy under standard conditions. Thus, in homolactic fermentation, 2758/2870, or 96%, of the free energy available in the original glucose molecule is still present in lactate, the fermentation product. Alcoholic fermentation is similarly low yield.

As we shall see in Chapter 14, about 30–32 moles of ATP are synthesized from ADP per mole of glucose carried completely through glycolysis and the citric acid cycle. Aerobic metabolism yields more energy from glucose; therefore, aerobic organisms in general are more successful and widespread than anaerobic organisms. It is thought that the early invention of aerobic metabolism was a key event in the evolution of the large, active animals that exist today. Nevertheless, many large animals still derive a large fraction of their metabolic energy from glycolysis, under certain physiological circumstances. A good example is the crocodile—physically inactive (and aerobic) for much of its life, yet capable of short bursts of intensely rapid movement. In the latter circumstance glycolysis, coupled with the breakdown of carbohydrate energy stores, provides a quick, though inefficient, way to mobilize energy.

12.5 Gluconeogenesis

Gluconeogenesis—literally, the production of new glucose—is the synthesis of glucose from noncarbohydrate precursors. Here we encounter the first instance of a principle presented in Chapter 11—that *biosynthetic processes are never simply the reversal of the corresponding catabolic pathways.* Superficially, gluconeogenesis looks very much like glycolysis in reverse, but different enzymatic reactions are used for crucial steps. These steps are strongly exergonic reactions that are controlled largely in reciprocal fashion, so that physiological conditions that activate glycolysis inhibit gluconeogenesis and vice versa. Much the same picture will emerge later in this chapter from our discussion of glycogen synthesis as compared with glycogen mobilization.

Physiological Need for Glucose Synthesis in Animals

Most animal organs can metabolize a variety of carbon sources—lipids, various sugars, pyruvate, and amino acids—to generate energy. The brain and central nervous system, however, require glucose as the sole or primary carbon source. The same is true for some other tissues, such as muscle, kidney medulla, testes, and erythrocytes (**FIGURE 12.9**). Consequently, animal cells must be able to synthesize glucose from other precursors and also to maintain blood glucose levels within narrow limits—both for proper functioning of the brain and central nervous system and for providing precursors for glycogen storage in other tissues. The glucose requirements of the human brain are relatively enormous—120 grams per day, out of about 160 grams needed by the entire body.

● **CONCEPT** Synthesis of glucose from noncarbohydrate precursors is essential for the maintenance of blood glucose levels within acceptable limits.

The amount of glucose that can be generated from the body's glycogen reserves at any time is about 190 grams, and the total amount of glucose in body fluids is

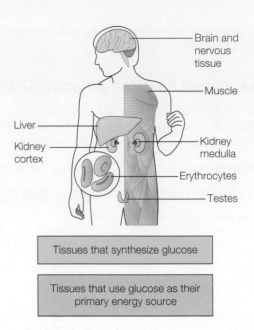

▲ **FIGURE 12.9 Synthesis and use of glucose in the human body.** Liver and kidney cortex are the primary gluconeogenic tissues. Brain, skeletal muscle, kidney medulla, erythrocytes, and testes use glucose as their sole or primary energy source, but they lack the enzymatic machinery to synthesize it.

little more than 20 grams. Thus, the readily available glucose reserves amount to about one day's supply. During periods of fasting for more than one day, glucose must be formed from other precursors. During long intervals between meals (e.g., during overnight sleep), glycogenolysis (the breakdown of glycogen) and gluconeogenesis both contribute to overall glucose production. If you skip breakfast, however, your glycogen stores become depleted, and gluconeogenesis becomes the predominant process for maintaining blood glucose levels. The same thing occurs during intense exertion, such as a marathon run. Initially, glycogenolysis in the liver is the primary source of extra glucose for skeletal muscle, but hepatic gluconeogenesis becomes gradually more important as the glycogen stores are depleted.

The biosynthetic process of gluconeogenesis starts with three-carbon and four-carbon precursors, generally noncarbohydrate in nature. Like glycolysis, gluconeogenesis occurs primarily in the cytosol, although some precursors are generated in mitochondria and must be transported to the cytosol to be utilized. The primary gluconeogenic organ in animals is the liver, with the kidney cortex contributing in a lesser but still significant way (see Figure 12.9). The major fates of glucose formed by gluconeogenesis are catabolism by nervous tissue and utilization by skeletal muscle. In addition, glucose is the primary precursor for all other carbohydrates, including amino sugars, complex polysaccharides, and the carbohydrate components of glycoproteins and glycolipids. The need for glucose as a biosynthetic intermediate means that gluconeogenesis is an important pathway in plants, microorganisms, and animals, and the pathway is essentially identical in all organisms. The wealth of information on the control of gluconeogenesis in animals, however, leads us to concentrate on animal metabolism.

Enzymatic Relationship of Gluconeogenesis to Glycolysis

Gluconeogenesis closely resembles glycolysis in reverse, but there are some important differences that allow the pathway to run in the direction of glucose *synthesis* in the cell.

● **CONCEPT** Gluconeogenesis uses specific enzymes to bypass three irreversible reactions of glycolysis.

Glycolysis proceeds from glucose to pyruvate because it is strongly exergonic in that direction; under typical intracellular conditions, ΔG is about -103 kJ/mol (see Table 12.1). How, then, can the conversion of pyruvate to glucose be made exergonic in gluconeogenesis? Recall that reactions 1, 3, and 10 of the glycolytic pathway (catalyzed by hexokinase, phosphofructokinase, and pyruvate kinase, respectively) are so strongly exergonic as to be essentially irreversible (see Figure 12.8). In gluconeogenesis, different chemistry and enzymes are used at each of these steps. For example, the conversion of fructose-1,6-bisphosphate to fructose-6-phosphate in gluconeogenesis is not simply a reversal of the phosphofructokinase reaction. In essence, the three irreversible reactions of glycolysis are bypassed by enzymes specific to gluconeogenesis, which catalyze quite different reactions that are favorable in the direction of glucose synthesis. This biosynthetic process involves a substantial energy cost, which must be paid if the overall process is to be thermodynamically favored.

The remaining seven reactions of gluconeogenesis are reversible because they are all near equilibrium under cellular conditions (Figure 12.8). These reactions are catalyzed by the same enzymes used in glycolysis—they are driven in either direction by mass action (i.e., the relative concentrations of product and reactant). Another way to relate glycolysis to gluconeogenesis is to say that they differ at only three steps, those controlled by *substrate cycles* (see Section 11.4).

The entire gluconeogenic pathway, from pyruvate to glucose, is summarized in **FIGURE 12.10**. We focus here on the reactions that bypass the three irreversible steps in glycolysis.

Bypass 1: Conversion of Pyruvate to Phosphoenolpyruvate

The bypass of pyruvate kinase (glycolysis reaction 10) begins in the mitochondrion. **Pyruvate carboxylase** catalyzes the ATP- and biotin-dependent carboxylation of pyruvate to oxaloacetate. The enzyme requires acetyl-CoA as an allosteric activator:

Pyruvate carboxylase

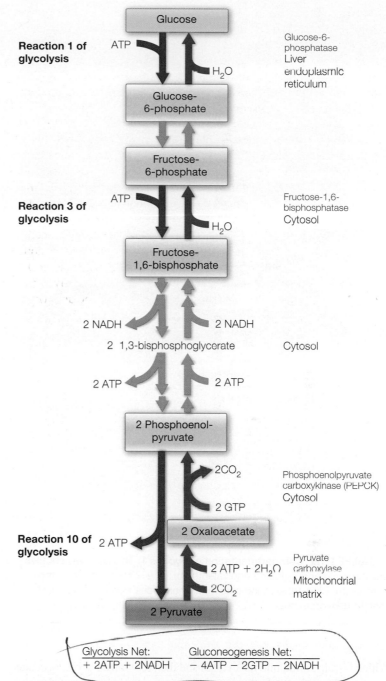

▲ **FIGURE 12.10 Reactions of glycolysis and gluconeogenesis.** Irreversible reactions of glycolysis are shown in dark purple. The opposed reactions in gluconeogenesis, which bypass these steps, are shown in dark blue, with the gluconeogenic enzyme names shown in magenta. Pale arrows identify reversible reactions used in both pathways.

As we will see in Chapter 13, this is one of the *anaplerotic* ("filling up") reactions used to maintain levels of citric acid cycle intermediates. Pyruvate carboxylase generates oxaloacetate in the mitochondrial

matrix, where it can be oxidized in the citric acid cycle (**FIGURE 12.11**). To be used for gluconeogenesis, oxaloacetate must move out of the mitochondrion into the cytosol, where the remainder of the pathway occurs. The mitochondrial membrane does not have an effective transporter for oxaloacetate, however, so oxaloacetate is reduced by mitochondrial malate dehydrogenase (mitochondrial MDH) to malate,

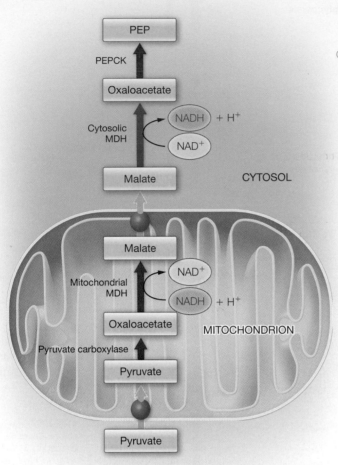

▲ FIGURE 12.11 Compartmentation of bypass 1. This pathway shuttles oxaloacetate and reducing equivalents from the mitochondrion to the cytoplasm. Blue spheres indicate inner membrane transporters. MDH = malate dehydrogenase.

which is transported into the cytosol and then reoxidized to oxaloacetate by cytosolic malate dehydrogenase. This process also accomplishes the transfer of reducing equivalents (as NADH) from mitochondrion to cytoplasm, which will be used later in the glyceraldehyde-3-phosphate dehydrogenase (GAPDH) reaction.

Once in the cytosol, oxaloacetate is acted on by **phosphoenolpyruvate carboxykinase (PEPCK)** to give phosphoenolpyruvate:

$$\text{Oxaloacetate} + \text{GTP} \rightleftharpoons$$

$$\text{phosphoenolpyruvate} + CO_2 + \text{GDP} \quad \Delta G^{\circ\prime} = +1.2 \text{ kJ/mol}$$

The PEPCK reaction requires Mg^{2+} or Mn^{2+} and is readily reversible. Note the use of GTP, rather than ATP, as an energy donor. Note also that the same CO_2 that was fixed by pyruvate carboxylase in the conversion of pyruvate to oxaloacetate is released in this reaction, so that no net fixation of CO_2 occurs. In the reaction, the carboxyl group

formed from the transferred CO_2 provides electrons to facilitate O—P bond formation:

GTP   OAA   GDP   PEP   CO_2

The overall reaction for the bypass of pyruvate kinase is as follows:

$$\text{Pyruvate} + \text{ATP} + \text{GTP} \longrightarrow \text{phosphoenolpyruvate} + \text{ADP}$$

$$+ \text{ GDP} + P_i + H^+ \quad \Delta G^{\circ\prime} = -2.6 \text{ kJ/mol}$$

Although $\Delta G^{\circ\prime}$ for the two reactions combined is only slightly negative, the sequence is strongly exergonic ($\Delta G \approx -25 \text{ kJ/mol}$) under intracellular conditions. As shown in the summary reaction, two high-energy phosphates must be invested for the synthesis of one phosphoenolpyruvate. After this bypass, phosphoenolpyruvate is converted to fructose-1,6-bisphosphate by glycolytic enzymes acting in reverse (glycolysis reactions 9 through 4). The glyceraldehyde-3-phosphate dehydrogenase (GAPDH) reaction requires NADH when operating in the direction of glucose synthesis (Figure 12.10), and those reducing equivalents are provided by the shuttle described in Figure 12.11.

Bypass 2: Conversion of Fructose-1,6-bisphosphate to Fructose-6-phosphate

The phosphofructokinase (PFK) reaction of glycolysis (reaction 3) is essentially irreversible, but only because it is driven by phosphoryl group transfer from ATP. The bypass reaction in gluconeogenesis involves a simple hydrolytic reaction, catalyzed by **fructose-1,6-bisphosphatase.**

$$\text{Fructose-1,6-bisphosphate} + H_2O \xrightarrow{Mg^{2+}} \text{fructose-6-phosphate} + P_i$$

$$\Delta G^{\circ\prime} = -16.3 \text{ kJ/mol}$$

The negative $\Delta G^{\circ\prime}$ favors the reaction in the direction shown. Fructose-6-phosphate formed in this reaction is then isomerized by phosphoglucoisomerase to glucose-6-phosphate (the reverse of reaction 2 in glycolysis).

Bypass 3: Conversion of Glucose-6-phosphate to Glucose

Glucose-6-phosphate cannot be converted to glucose by reverse action of hexokinase because of the high positive $\Delta G^{\circ\prime}$ of that reaction; the phosphoryl group transfer from ATP makes reaction 1 of glycolysis virtually irreversible. Another enzyme specific to gluconeogenesis, **glucose-6-phosphatase,** comes into play instead. This bypass reaction, like the previous one, also involves a simple hydrolysis.

$$\text{Glucose-6-phosphate} + H_2O \xrightarrow{Mg^{2+}} \text{glucose} + P_i$$

$$\Delta G^{\circ\prime} = -13.8 \text{ kJ/mol}$$

TABLE 12.2 Summary of gluconeogenesis, from pyruvate to glucose

Reaction	$\Delta G^{\circ\prime}$(kJ/mol)
Pyruvate + HCO_3^- + ATP $\longrightarrow$ oxaloacetate + ADP + P_i	−3.8 (−7.6)
Oxaloacetate + GTP $\rightleftharpoons$ phosphoenolpyruvate + CO_2 + GDP	+1.2 (+2.4)
Phosphoenolpyruvate + H_2O $\rightleftharpoons$ 2-phosphoglycerate	+6.4(+12.8)
2-Phosphoglycerate $\rightleftharpoons$ 3-phosphoglycerate	−4.4 (−8.8)
3-Phosphoglycerate + ATP $\rightleftharpoons$ 1,3-bisphosphoglycerate + ADP	+17.2 (+34.4)
1,3-Bisphosphoglycerate + NADH + H^+ $\rightleftharpoons$ glyceraldehyde-3-phosphate + NAD^+ + P_i	−6.3 (−12.6)
Glyceraldehyde-3-phosphate $\rightleftharpoons$ dihydroxyacetone phosphate	−7.6
Glyceraldehyde-3-phosphate + dihydroxyacetone phosphate $\rightleftharpoons$ fructose-1,6-bisphosphate	−23.9
Fructose-1,6-bisphosphate + H_2O $\rightleftharpoons$ fructose-6-phosphate + P_i	−16.3
Fructose-6-phosphate $\rightleftharpoons$ glucose-6-phosphate	−1.7
Glucose-6-phosphate + H_2O $\rightleftharpoons$ glucose + P_i	−13.8
Net: 2 Pyruvate + 4ATP + 2GTP + 2NADH + $2H^+$ + $4H_2O$ $\longrightarrow$ glucose + 4ADP + 2GDP + $6P_i$ + $2NAD^+$	−42.7

Note: The reactions in red are those that bypass irreversible glycolytic reactions; the remaining reactions are reversible reactions of glycolysis. The $\Delta G^{\circ\prime}$ values in parentheses are based on doubling the first six reactions because 2 three-carbon precursors are required to make one molecule of glucose. The individual reactions are not necessarily balanced for H^+ and charge.

Glucose-6-phosphatase is expressed predominantly in the liver and kidney. Muscle has little or no glucose-6-phosphatase activity. As a result, the liver is uniquely positioned to synthesize glucose for export to the tissues via the bloodstream.

Stoichiometry and Energy Balance of Gluconeogenesis

Catabolic pathways generate energy, whereas anabolic pathways carry an energy cost. What is the energy cost for gluconeogenesis? $\Delta G^{\circ\prime}$ for the overall conversion of 2 moles of pyruvate to 1 mole of glucose is about −43 kJ/mol (**TABLE 12.2**).

Gluconeogenesis

2 Pyruvate + 4ATP + 2GTP + 2NADH + $2H^+$ + $4H_2O$ $\longrightarrow$

glucose + 4ADP + 2GDP + $6P_i$ + $2NAD^+$

$$\Delta G^{\circ\prime} = -42.7 \text{ kJ/mol}$$

● **CONCEPT** The equivalent of 11 high-energy phosphates are consumed per mole of glucose synthesized by gluconeogenesis.

The synthesis of glucose is energetically expensive in order to make the overall process exergonic. Six high-energy phosphate groups are consumed (four ATPs and two GTPs), as well as 2 moles of NADH, which is the energetic equivalent of five more ATPs (because mitochondrial oxidation of 1 mole of NADH generates ~2.5 moles of ATP).

If glycolysis could operate in reverse, the net equation would show an input of only 2 moles of ATP:

Reversal of Glycolysis

2 Pyruvate + 2ATP + 2NADH + $2H^+$ + $2H_2O$ $\longrightarrow$

glucose + 2ADP + $2P_i$ + $2NAD^+$ $\Delta G^{\circ\prime} = +79.9$ kJ/mol

This process would be highly endergonic, however, with a $\Delta G^{\circ\prime}$ of +79.9 kJ/mol. These are standard free energy changes, but four additional high-energy phosphate bonds must be invested if the net synthesis of glucose is to occur as an irreversible process in vivo.

Substrates for Gluconeogenesis

Besides pyruvate, the principal substrates for gluconeogenesis are lactate and amino acids. Glycerol and propionate, derived from the breakdown of fat, are also important gluconeogenic substrates, but we will save discussion of those until Chapter 16.

Lactate

In quantitative terms, lactate is the most significant gluconeogenic precursor. Recall that skeletal muscle derives much of its energy from glycolysis, particularly during intense exertion, when respiration cannot deliver sufficient oxygen to the tissues for the complete oxidation of glucose. Under these conditions, glycolysis produces pyruvate more rapidly than it can be further metabolized via the citric acid cycle. Lactate dehydrogenase is abundant in muscle, and the equilibrium strongly favors the reduction of pyruvate to lactate. Thus, lactate from working muscle is released to the blood, whence it is readily taken up by the heart and oxidized as fuel.

Some of the lactate produced in muscle enters the liver and is reoxidized to pyruvate by liver LDH. This pyruvate can then undergo gluconeogenesis to give glucose, which is returned to the bloodstream

● **CONNECTION** The acidosis that occurs during prolonged exertion is a significant factor limiting athletic performance. Since 1920, lactic acid production has been taken as the cause of this acidosis; however, more recent research indicates that ATP hydrolysis is the major source of protons acidifying muscle tissue during strenuous exercise.

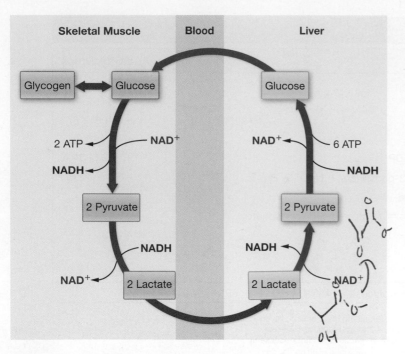

▲ **FIGURE 12.12 The Cori cycle.** Lactate produced in glycolysis during muscle exertion is transported to the liver, for resynthesis of glucose by gluconeogenesis. Transport of glucose back to muscle for synthesis of glycogen, and its reutilization in glycolysis, complete the cycle.

and taken up by muscle to regenerate the glycogen stores. This process, described originally by Carl and Gerti Cori and appropriately called the **Cori cycle,** is schematized in **FIGURE 12.12.** The pathway is particularly active during recovery from intense muscular exercise. During this time the breathing rate is elevated, and the increased oxidative metabolism generates more ATP, much of which is used to rebuild glycogen stores via gluconeogenesis.

In a parallel process called the **glucose–alanine cycle,** pyruvate in peripheral tissues undergoes transamination to alanine, which is returned to the liver and used for gluconeogenesis. This pathway, which is presented in detail in Chapter 18, helps tissues to dispose of toxic ammonia formed during protein degradation.

Amino Acids

Like alanine, many other amino acids can readily be converted to glucose, primarily through degradative pathways that generate citric acid cycle intermediates, which can be converted to oxaloacetate. As we will see in Chapter 18, such amino acids are called **glucogenic** (that is, able to be converted to glucose), although *gluconeogenic* is probably a more accurate term. Among the 20 amino acids found in proteins, only the catabolic pathways for leucine and lysine do not generate gluconeogenic precursors. During fasting, when insufficient carbohydrate is ingested, the catabolism of muscle proteins is the major source of intermediates needed to maintain normal blood glucose concentrations. The same is true in the disease diabetes mellitus, as discussed further in Chapter 17.

Ethanol Consumption and Gluconeogenesis

Although it is possible to visualize pathways by which ethanol could be converted to glucose, ethanol is actually a poor gluconeogenic

precursor. In fact, ethanol strongly inhibits gluconeogenesis and can bring about **hypoglycemia,** a potentially dangerous decrease in blood glucose levels.

Ethanol is metabolized primarily in the liver, by reversal of the alcohol dehydrogenase reaction (Section 12.3):

$$\text{Ethanol} + \text{NAD}^+ \rightleftharpoons \text{acetaldehyde} + \text{NADH} + \text{H}^+$$

This reaction elevates the [NADH]/[NAD$^+$] ratio in liver cytosol, which in turn shifts the equilibrium of the lactate dehydrogenase and glyceraldehyde-3-phosphate dehydrogenase reactions, inhibiting glycolysis. The same mechanism shifts the equilibrium of the cytosolic malate dehydrogenase reaction (Figure 12.11), so that oxaloacetate tends to be reduced to malate and hence becomes unavailable for gluconeogenesis. The resultant hypoglycemia can affect the parts of the brain concerned with temperature regulation. This response, in turn, can lower the body temperature by as much as 2 °C. Therefore, the time-honored practice of feeding brandy or whiskey to those rescued from cold or wet conditions is counterproductive. To be sure, alcohol creates a sense of warming through vasodilation, but this peripheral vasodilation causes further heat loss. Metabolically speaking, glucose would be far more effective in raising body temperature.

12.6 Coordinated Regulation of Glycolysis and Gluconeogenesis

Glycolysis and gluconeogenesis are closely coordinated with other major pathways of energy generation and utilization, notably the synthesis and breakdown of glycogen (or starch), the pentose phosphate pathway (both described later in this chapter), the citric acid cycle (Chapter 13), and fatty acid metabolism (Chapter 16). Metabolic factors that control glycolysis and gluconeogenesis tend to regulate these other processes in a coordinated fashion. Thus, it is difficult to consider regulation in isolation from these other processes. We return to this topic again after we have presented the other major pathways in energy metabolism. However, glycolysis and gluconeogenesis provide a very useful introduction to the principles of coordinated metabolic regulation, and so we will describe here the key enzymes that serve as regulatory targets in these two opposing pathways.

The Pasteur Effect

Long before anything was known about pathways of glucose utilization, much less control mechanisms, Louis Pasteur (1822–1895) observed that when anaerobic yeast cultures metabolizing glucose were exposed to air, the rate of glucose utilization decreased dramatically. This phenomenon, known as the **Pasteur effect,** involves the inhibition of glycolysis by oxygen. This effect makes biological sense because far more energy is derived from the complete oxidation of glucose than from glycolysis alone. What is the mechanism of this effect if oxygen is not an active participant in glycolysis?

The needed insight came much later from analyses of the intracellular contents of glycolytic intermediates in aerobic and anaerobic cells. Experiments revealed that when oxygen is introduced to anaerobic cells, the levels of all the glycolytic intermediates from fructose-1,6-bisphosphate (the product of reaction 3) onward *decrease,* while all of the *earlier* intermediates accumulate at higher levels (see next page). This finding is consistent with the idea that the metabolic flux through phosphofructokinase (the enzyme for reaction 3) is specifically decreased in the presence of O_2, probably because of changes in the concentration of allosteric effectors.

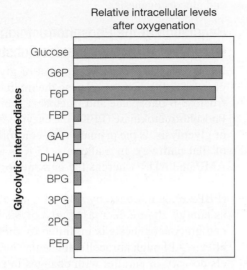

Relative intracellular levels after oxygenation

(Y-axis: Glycolytic intermediates — Glucose, G6P, F6P, FBP, GAP, DHAP, BPG, 3PG, 2PG, PEP)

Other important conclusions emerged from the discovery that when glycolysis is activated, the intracellular levels of ADP, AMP, and NADH are high, whereas the level of ATP is low. Conversely, when the pathway is turned off, ATP concentration is high and ADP, AMP, and NADH concentrations are low. This pattern revealed that the activity of glycolysis depends in some way on the adenylate energy charge (Section 11.4) and suggested that an enzyme regulated by energy charge must be a major control point. Phosphofructokinase is just such an enzyme.

Regulation of glycolysis is crucial because glycolysis not only generates ATP and provides pyruvate for oxidation via the citric acid cycle but it also provides intermediates for other pathways. Intermediates in glycolysis are precursors for a number of compounds, particularly lipids and amino acids. Many pathways lead into glycolysis, and many pathways diverge from it, creating a substantial flux through the pathway (see Figure 12.1).

Regulation of gluconeogenesis is equally crucial for many physiological functions, but particularly so for proper functioning of nervous tissue. Although other organs can use a variety of energy sources, the well-being of the central nervous system requires that blood glucose levels be maintained within narrow limits. Gluconeogenic control is important also as an animal adjusts to muscular exertion or to cycles of feeding and fasting. Flux through the pathway increases and decreases, depending on the availability of lactate produced by the muscles, of glucose from the diet, or of other gluconeogenic precursors.

Gluconeogenesis is controlled in large part by the diet. Animals fed a high-carbohydrate diet show low rates of gluconeogenesis, whereas fasting animals or those fed carbohydrate-poor diets show high flux through this pathway. As we noted in Section 11.5, these responses are mediated by hormones (primarily insulin and glucagon) and involve

● **CONCEPT** Gluconeogenic flux rates are inversely related to the carbohydrate content of the diet. This effect is mediated by hormones.

both control of the synthesis of critical enzymes and regulation through the control of cAMP levels. Our discussion here focuses on these cAMP-mediated effects as well as other mechanisms affecting enzyme activities. We discuss hormonal effects upon enzyme synthesis in Chapter 20, where we present hormone action in detail.

Reciprocal Regulation of Glycolysis and Gluconeogenesis

Gluconeogenesis and glycolysis both proceed largely in the cytosol. Because gluconeogenesis synthesizes glucose and glycolysis catabolizes glucose, *gluconeogenesis and glycolysis must be controlled in reciprocal fashion.* In other words, intracellular conditions that activate one pathway tend to inhibit the other. Without such reciprocal control, glycolysis and gluconeogenesis would operate together as a giant futile cycle. **Reciprocal regulation** is related in large part to the adenylate energy charge. Conditions of low energy charge (i.e., low ATP levels) tend to activate the rate-controlling steps in glycolysis while inhibiting carbon flux through gluconeogenesis. Conversely, gluconeogenesis is stimulated at high energy charge, under conditions where catabolic flux rates are low but are adequate to maintain sufficient ATP levels.

● **CONCEPT** Conditions that promote glycolysis inhibit gluconeogenesis, and vice versa.

The regulatory demands placed on the cell are too complex to meet with a single rate-controlling reaction, and thus both glycolysis and gluconeogenesis are controlled at multiple points. Glycolysis is controlled primarily by regulation of the three strongly exergonic, nonequilibrium reactions of the pathway—those catalyzed by hexokinase, phosphofructokinase, and pyruvate kinase (reactions 1, 3, and 10, respectively; see Figure 12.8). The opposed reactions in gluconeogenesis—those catalyzed by glucose-6-phosphatase, fructose-1,6-bisphosphatase, and the combination of pyruvate carboxylase and phosphoenolpyruvate carboxykinase—are also strongly exergonic and represent the chief targets for control of this pathway. In other words, the three substrate cycles that differentiate glycolysis from gluconeogenesis (**FIGURE 12.13**) represent the primary sites for reciprocal regulation of these pathways. This illustrates the principle (introduced in Section 11.4) that regulation can be imposed *only* on reactions displaced far from equilibrium, such as those that comprise these three substrate cycles.

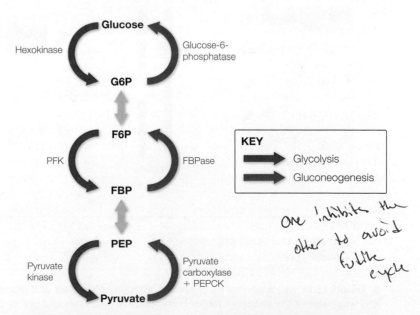

KEY
→ Glycolysis
→ Gluconeogenesis

(Handwritten note: One inhibits the other to avoid futile cycle)

▲ **FIGURE 12.13** Substrate cycles in glycolysis/gluconeogenesis.

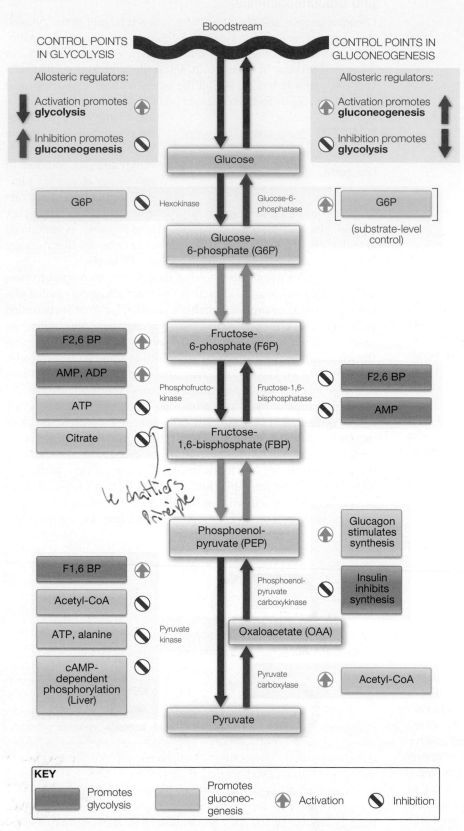

▲ **FIGURE 12.14 Major control mechanisms affecting glycolysis and gluconeogenesis.** The figure shows the strongly exergonic reactions of glycolysis and gluconeogenesis and the major activators and inhibitors of these reactions.

FIGURE 12.14 identifies the major allosteric activators and inhibitors of the key exergonic reactions in glycolysis and gluconeogenesis.

Regulation at the Phosphofructokinase/Fructose-1,6-Bisphosphatase Substrate Cycle

Energy charge affects the control of glycolysis and gluconeogenesis by regulating the interconversion of fructose-6-phosphate and fructose-1,6-bisphosphate. Phosphofructokinase (PFK), the enzyme for reaction 3 of glycolysis, is the primary flux-controlling enzyme of that pathway. In mammals, PFK is activated by AMP and ADP, whereas the enzyme for the reverse step in gluconeogenesis, fructose-1,6-bisphosphatase (FBPase), is inhibited by AMP (Figure 12.14). Thus, as energy charge decreases, glycolysis is activated and gluconeogenesis is inhibited by these opposing effects. Although intracellular adenine nucleotide levels do vary in parallel with changes in flux through glycolysis and gluconeogenesis, as expected if energy charge is a major regulatory factor, the correspondence is not absolute. These observations, suggesting additional control mechanisms, led to the discovery of **fructose-2,6-bisphosphate (F2,6BP),** an even more important physiological regulator.

Fructose-2,6-Bisphosphate and the Control of Glycolysis and Gluconeogenesis

Phosphofructokinase (PFK) is a homotetramer that interconverts between two conformational states, R and T (recall the discussion of R and T states in Section 8.8. The R state is the more active conformation). In addition to the catalytic sites that bind substrates (ATP and fructose-6-phosphate), mammalian PFK has binding sites for several allosteric effectors, including AMP, ADP, ATP, citrate, and fructose-2,6-bisphosphate. (Bacterial PFKs have only a single allosteric site that can bind either an inhibitor, phosphoenolpyruvate, or an activator, ADP.) Fructose-2,6-bisphosphate is considered to be the major regulator controlling carbon flux through glycolysis and gluconeogenesis in mammalian liver—it is active at much lower concentrations than the other physiological regulators we have discussed. As shown in **FIGURE 12.15(a)**, a very low concentration of fructose-2,6-bisphosphate *activates* PFK. AMP and ADP also activate PFK.

The most significant *inhibitors* of mammalian PFK, from a biological standpoint, are ATP (Figure 12.15(b)) and citrate. The effect of ATP may seem anomalous because ATP is a substrate and hence essential for the reaction. As an inhibitor, ATP binds to a site on the enzyme separate from the catalytic site and with lower affinity (Figure 12.15(c)). At low ATP concentrations, the substrate saturation curve for fructose-6-phosphate is nearly hyperbolic because the regulatory site is unoccupied, and the enzyme is almost all in the R state. At high ATP levels, the T state predominates, causing the

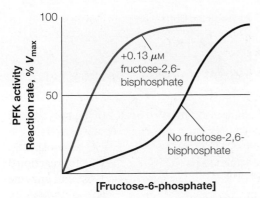

(a) Activation of PFK by fructose-2,6-bisphosphate

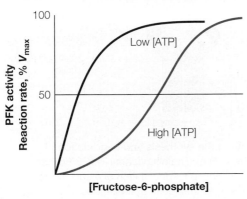

(b) How ATP increases the apparent K_M for substrate fructose-6-phosphate, inhibiting PFK

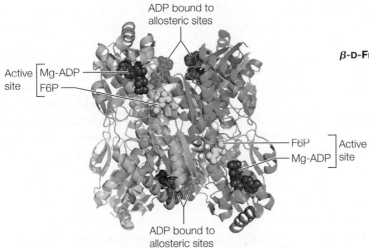

(c) X-ray structure of phosphofructokinase homotetramer from *Bacillus stearothermophilis* (PDB ID: 4pfk). F6P, fructose-6-phosphate

▲ **FIGURE 12.15** Allosteric control of liver phosphofructokinase and structure of a bacterial homolog.

curve to become sigmoidal and shift far to the right (Figure 12.15(b)). Thus, inhibition is achieved because the apparent affinity for fructose-6-phosphate is greatly reduced. Activators such as AMP, ADP, and fructose-2,6-bisphosphate bind to allosteric sites but stabilize the R state, thus increasing the apparent affinity for the substrate fructose-6-phosphate.

The control of PFK by adenine nucleotides represents a way in which energy metabolism responds to the adenylate energy charge. At high energy charge, the relative abundance of ATP signals that the

energy-yielding glycolytic pathway should diminish in activity, so PFK is inhibited. Conversely, a high AMP or ADP level signals that energy charge is low and that flux through glycolysis should increase. Inhibition by citrate represents another energy-level sensor. At high energy charge, flux through the citric acid cycle diminishes, via mechanisms that are discussed in Chapter 13. Under these conditions, citrate accumulates and is transported out of mitochondria. Interaction with PFK in the cytosol can signal that energy generation is adequate, and hence the production of citric acid cycle precursors via glycolysis can be diminished.

Fructose-2,6-bisphosphate also regulates the gluconeogenic side of this substrate cycle. But here it is a potent inhibitor of fructose-1,6-bisphosphatase, at least in vitro. Thus, accumulation of the same regulatory molecule has the effect of simultaneously activating glycolysis and inhibiting gluconeogenesis (Figure 12.14).

Fructose-2,6-bisphosphate is formed from fructose-6-phosphate by **6-phosphofructo-2-kinase.** It is called PFK-2 to distinguish it from the well-known PFK of glycolysis. Another enzymatic activity, **fructose-2,6-bisphosphatase,** cleaves fructose-2,6-bisphosphate back to fructose-6-phosphate. This activity is abbreviated FBPase-2 to distinguish it from the FBPase of gluconeogenesis.

These two activities are localized on separate domains of a 100-kilodalton enzyme, PFK-2/FBPase-2 (**FIGURE 12.16**). This bifunctional enzyme thus catalyzes the opposing reactions of a substrate cycle that determines the level of fructose-2,6-bisphosphate, an important regulatory molecule. The velocities of the two reactions of this substrate cycle are controlled, in turn, by phosphorylation/dephosphorylation of the bifunctional enzyme. Conformational changes caused by phosphorylation increase the activity of one domain while decreasing the activity of the other domain, thereby altering the ratio of kinase to bisphosphatase activity.

Mammals express several different tissue-specific PFK-2/FBPase-2 isozymes, each with different regulatory properties that fit the metabolic needs of various tissues. The activities of the liver isozyme are controlled by the pancreatic hormones insulin and glucagon, and also by glucose. All of these regulatory molecules act through signaling cascades that lead to the reversible phosphorylation of a specific serine residue, which results in conformational changes in the protein. Phosphorylation *decreases* the

● **CONCEPT** Fructose-2,6-bisphosphate, the most important regulator of glycolysis and gluconeogenesis, is synthesized and degraded by different active sites on the bifunctional enzyme PFK-2/FBPase-2.

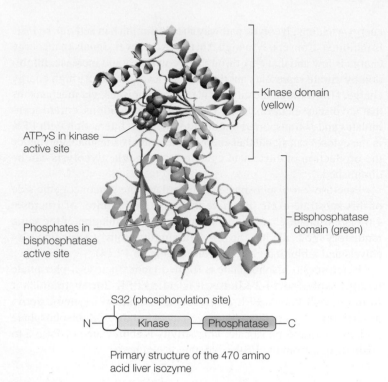

S32 (phosphorylation site)

N—[ ]—[Kinase]—[Phosphatase]—C

Primary structure of the 470 amino
acid liver isozyme

▲ **FIGURE 12.16 Bifunctional PFK-2/FBPase-2.** The crystal structure of one subunit of the homodimeric human liver enzyme (PDB ID: 1k6m), from X-ray analysis. The kinase active site is marked by a bound nonhydrolyzable ATP analog ATPγS. The bisphosphatase active site is marked by bound phosphates. Residues 1–38, including Ser-32, are not visible in the X-ray structure, suggesting that this segment is highly flexible.

ATPγS
(nonhydrolyzable ATP analog)

activity of PFK-2 and *increases* the activity of FBPase-2 (**FIGURE 12.17**). Dephosphorylation reverses this effect.

Phosphorylation of PFK-2/FBPase-2 is catalyzed by cAMP–dependent protein kinase. This important regulatory kinase, also known as protein kinase A (PKA), is discussed in greater detail later in this chapter. As pointed out in Chapter 11, cAMP plays numerous roles in regulating metabolism, both in eukaryotes and in prokaryotes. In eukaryotes, it functions as a second messenger, receiving hormonal messages originating outside the cell and transmitting them within the cell. This transmission involves the activation of some metabolic processes and the inhibition of others. Glucagon, released by the pancreas in response to low blood glucose levels, is the primary hormone whose action raises cAMP levels in liver.

Dephosphorylation of PFK-2/FBPase-2 is catalyzed by one or more specific protein phosphatases that are activated by extracellular

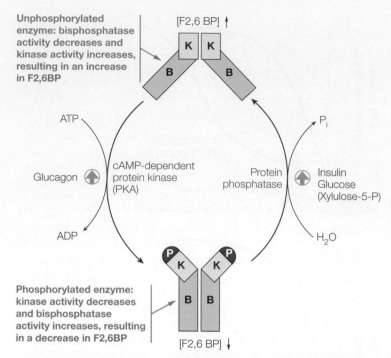

▲ **FIGURE 12.17 Regulation of the synthesis and degradation of fructose-2,6-bisphosphate in liver.** The bifunctional PFK-2/FBPase-2 enzyme is controlled by reversible phosphorylation of a specific serine residue near the N-terminus of each subunit of the homodimeric protein. In the unphosphorylated form, the 6-phosphofructo-2-kinase domain (K) is active, and fructose-2,6-bisphosphate (F2,6BP) is synthesized. In the phosphorylated form, the fructose-2,6-bisphosphatase domain (B) is active, and F2,6BP is degraded.

signals (Figure 12.17). Insulin, released in response to high blood glucose levels, stimulates dephosphorylation and thus activation of the 6-phosphofructo-2-kinase domain. The precise signaling pathway is not known. The glucose signal is mediated by **protein phosphatase 2A (PP2A).** An increase in liver glucose concentration leads to increased synthesis of intermediates in the pentose phosphate pathway (discussed later in this chapter). One of these metabolites, xylulose-5-phosphate, is a specific activator of PP2A.

We discuss the hormonal control of energy metabolism in much greater detail in Chapter 17, but we can begin to grasp some of the mechanistic details here:

- Glucagon, released by the pancreas in response to low blood glucose levels, binds to its plasma membrane receptors on liver cells.

- Binding activates the receptor, which then initiates the cAMP cascade, resulting in an active protein kinase A.

- Protein kinase A catalyzes the phosphorylation of PFK-2/FBPase-2, stimulating its fructose-2,6-bisphosphatase activity.

- The resultant drop in fructose-2,6-bisphosphate levels causes this regulatory molecule to dissociate from PFK, thereby increasing the sensitivity of PFK to the allosteric inhibitors citrate and ATP.

- Inhibition of PFK in turn *reduces* flux through glycolysis and *stimulates* gluconeogenesis by relieving the inhibition of fructose-1,6-bisphosphatase (refer to Figure 12.14). The newly synthesized glucose is then exported from the liver to the bloodstream.

This is one mechanism by which glucagon increases blood glucose concentration. Insulin, released by the pancreas in response to high blood glucose levels, or glucose itself, stimulates dephosphorylation of PFK-2/FBPase-2, activating its 6-phosphofructo-2-kinase activity. The resultant rise in fructose-2,6-bisphosphate levels *inhibits* gluconeogenesis (glucose is in excess, so there is no need to synthesize more) and *stimulates* glycolysis and the storage of glucose in the form of glycogen or fat.

Regulation at the Pyruvate Kinase/Pyruvate Carboxylase + PEPCK Substrate Cycle

Earlier, we identified pyruvate kinase (reaction 10) as a control point for glycolysis. At least four separate mechanisms are involved. Pyruvate kinase (PK), like many enzymes, exists in animal tissues as multiple isozymes. Mammals carry two PK genes, which, through alternative exon splicing, produce four different PK isozymes. As we will see in Chapter 24, alternative splicing is a common mechanism used by eukaryotes to greatly expand the number of different protein products produced from a single gene. The PK-L and PK-R isozymes are expressed specifically in liver and red blood cells, respectively. PK-M1 is expressed in muscle and brain and other terminally differentiated tissues. PK-M2 is expressed during embryonic development, but also in tumor cells.

● **CONNECTION** The pyruvate kinase M2 isozyme (PK-M2) is expressed in essentially all human cancers. Cancer cells are able to allosterically attenuate its activity so that upstream glycolytic intermediates accumulate and become available for biosynthetic pathways necessary to support their rapid proliferation. This metabolic reprogramming is part of the Warburg effect.

The L (liver) and R (erythrocyte) isozymes of pyruvate kinase, like those of PFK, are allosterically inhibited at high ATP concentrations (**FIGURE 12.18**), in a kinetically similar fashion: High ATP levels reduce the apparent affinity of pyruvate kinase for phosphoenolpyruvate (PEP), its other substrate. A second allosteric effect is the **feedforward activation** of pyruvate kinase by fructose-1,6-bisphosphate. In fact, all but the M1 isozyme require allosteric activation by fructose-1,6-bisphosphate

for full activity. This effect, the converse of feedback inhibition, ensures that carbon passing the first regulated step in the pathway (PFK; reaction 3) will be able to complete its passage through glycolysis and that undesirable accumulation of intermediates will not occur.

A third feedback control effect is inhibition of pyruvate kinase by acetyl-CoA, the major product of fatty acid oxidation. This inhibition allows the cell to reduce glycolytic flux when ample substrates for ATP production are available from fat breakdown. Finally, pyruvate kinase is inhibited by some amino acids, particularly alanine, the major gluconeogenic precursor among the amino acids. This relationship makes it possible to inhibit glycolysis, with consequent activation of gluconeogenesis, specifically in gluconeogenic tissues, when ample energy and substrates are available. Control at the pyruvate kinase step allows high phosphate transfer potential to be conserved in the PEP molecule.

The liver pyruvate kinase isozyme is also regulated by reversible phosphorylation/dephosphorylation, with the dephosphorylated form far more active than the phosphorylated form. Phosphorylation is stimulated by glucagon, acting through the same cAMP-dependent protein kinase pathway that phosphorylates PFK-2/FBPase-2. Thus, when blood glucose is low, glucagon secretion inactivates pyruvate kinase in liver and inhibits glycolysis. PEP is diverted instead to gluconeogenesis to support the export of glucose to the blood. Muscle pyruvate kinase, by contrast, is not regulated by covalent modification, and virtually all of the PEP produced in muscle is converted to pyruvate for the production of ATP to support muscle contraction.

Acetyl-CoA can also be seen as a reciprocal regulator of glycolysis and gluconeogenesis, acting on the enzymes that interconvert pyruvate and PEP (Figure 12.14). In addition to its role as an inhibitor of pyruvate kinase (and thus an inhibitor of glycolysis), acetyl-CoA is a required activator of pyruvate carboxylase (and thus an activator of gluconeogenesis) (Figure 12.18). It can therefore signal, when its levels rise, that adequate substrates are available to provide energy through the citric acid cycle and that more carbon can instead be shuttled into gluconeogenesis and ultimately stored as glycogen.

On top of these allosteric mechanisms, the synthesis of the liver pyruvate kinase isozyme (PK-L) is under dietary control. Intracellular activity may increase as much as 10-fold from increased enzyme synthesis, or induction, as a result of high carbohydrate ingestion.

● **CONCEPT** Dietary carbohydrate induces the biosynthesis of pyruvate kinase and increases the ability of the body to obtain energy from glycolysis.

Finally, glucagon controls levels of the key gluconeogenic enzyme phosphoenolpyruvate carboxykinase (PEPCK) by activating transcription of the structural gene for PEPCK (Figure 12.14). Insulin has the opposite effect. By inhibiting PEPCK gene transcription, insulin tends to depress gluconeogenic flux rates. Glucagon has an additional action at the genetic level—it represses synthesis of pyruvate kinase, thereby contributing to increased gluconeogenic flux from pyruvate to PEP.

Regulation at the Hexokinase/Glucose-6-Phosphatase Substrate Cycle

Recall that mammals possess several different isozymes of hexokinase that differ in their kinetic and regulatory properties. The hexokinase isoenzymes expressed in most tissues (HK-I, HK-II, and HK-III) are

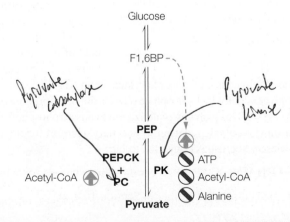

▲ **FIGURE 12.18 Allosteric regulation at the pyruvate kinase/pyruvate carboxylase + PEPCK substrate cycle.** PK, pyruvate kinase; PEPCK, phosphoenolpyruvate carboxykinase; PC, pyruvate carboxylase; F1,6BP, fructose-1,6-bisphosphate; PEP, phosphoenolpyruvate.

inhibited by their product, glucose-6-phosphate (G6P), a mechanism that controls the influx of substrates into the glycolytic pathway. The liver isozyme, hexokinase IV (HK-IV), is not subject to feedback inhibition by G6P. HK-IV exhibits a sigmoidal concentration dependence on glucose, allowing the liver to adjust its rate of glucose utilization in response to variations in blood glucose levels (see p. 379).

Glucose-6-phosphatase is not known to be allosterically controlled, but its K_M for G6P is far higher than intracellular concentrations of this substrate. Thus, intracellular activity is largely controlled in first-order fashion by the concentration of G6P.

In summary, glycolysis and gluconeogenesis are controlled in large part by the energy charge of the cell and by fuel status. Regulation is distributed over multiple steps and is highly coordinated so that the two pathways never operate simultaneously in the same cell. The liver has additional control systems that reflect its special role in maintaining glucose homeostasis for the entire animal. The other major control point of glucose metabolism, at least in animals, is the breakdown and synthesis of glycogen. This extremely important process will be discussed shortly.

12.7 Entry of Other Sugars into the Glycolytic Pathway

Thus far, our discussion of glycolysis has focused on glucose as a source of carbon for this pathway. Many other sources of carbohydrate energy are available, whether through the digestion of foodstuffs or the utilization of endogenous metabolites. This section outlines the utilization of monosaccharides other than glucose, of disaccharides, and of glycerol derived from fat metabolism. These pathways are summarized in **FIGURE 12.19**. We conclude the section with a discussion of the metabolism of polysaccharides, including those from dietary carbohydrates as well as the body's own glycogen reserves.

Monosaccharide Metabolism

As stated earlier, hexokinases I, II, and III have broad substrate specificities. Thus, they can utilize hexoses other than glucose, such as fructose and mannose. A separate enzyme, galactokinase, converts galactose to galactose-1-phosphate.

Galactose Utilization

D-Galactose is derived principally from hydrolysis of the disaccharide lactose $[\mathrm{Gal}\beta(1\rightarrow4)\mathrm{Glc}]$, which is particularly abundant in milk. The main route for galactose utilization is conversion to glucose-6-phosphate via glucose-1-phosphate by a pathway not covered here.

A variety of genetic disorders in humans go by the generic name **galactosemia.** They all involve a failure to metabolize galactose, so that galactose, galactose-1-phosphate, or both, accumulate in the blood and tissues. Clinical consequences include mental retardation, visual cataracts, and enlargement of the liver and other organs. These disorders result from a hereditary deficiency of any one of three enzymes involved in galactose utilization. Because the major dietary source of galactose is lactose in milk, the symptoms usually occur in infants. The condition can be alleviated by eliminating milk and milk products from the diet.

Fructose Utilization

Fructose is present as the free sugar in many fruits, and it is also derived from the hydrolysis of the disaccharide sucrose (see Figure 12.19).

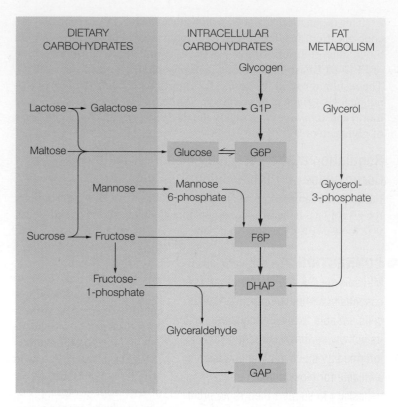

▲ **FIGURE 12.19 Routes for utilizing substrates other than glucose in glycolysis.** In animals, most of the carbohydrate other than glucose and glycogen comes from the diet, and most of the glycerol is derived from lipid catabolism. G1P, glucose-1-phosphate.

Phosphorylation of fructose in most tissues yields fructose-6-phosphate, the product of reaction 2 of glycolysis. A different pathway is involved in vertebrate liver, where the enzyme **fructokinase** phosphorylates fructose to **fructose-1-phosphate (F1P).** F1P is then cleaved by a specific enzyme, **aldolase B.** The cleavage products are dihydroxyacetone phosphate (DHAP), one of the products of reaction 4 of glycolysis, and D-glyceraldehyde.

● **CONNECTION** Fructose is the major component in the "high-fructose corn syrup" used to sweeten soft drinks as well as many other foods, and there is mounting evidence linking consumption of high-fructose corn syrup to an increased risk of obesity and type 2 diabetes.

D-Glyceraldehyde is then phosphorylated in an ATP-dependent reaction to give glyceraldehyde-3-phosphate (GAP), the other product of reaction 4 of glycolysis. This pathway of utilization bypasses phosphofructokinase regulation (glycolysis reaction 3) and may account for the ease with which dietary sucrose is converted to fat:

$$\mathrm{F1P} \rightarrow \mathrm{DHAP} + \mathrm{GAP} \rightarrow \text{glycerol-3-phosphate} \rightarrow \text{triacylglycerols}$$

(This topic is discussed further in Section 16.4.)

Disaccharide Metabolism

The three disaccharides most abundant in foods are maltose, lactose, and sucrose. Maltose is available primarily as an artificial sweetener, derived from starch, while lactose and sucrose are abundant natural products. In animal metabolism, they are hydrolyzed in cells lining the small intestine to give the constituent hexose sugars:

$$\text{Maltose} + H_2O \xrightarrow{\text{Maltase}} 2 \text{ D-glucose}$$

$$\text{Lactose} + H_2O \xrightarrow{\text{Lactase}} \text{D-galactose} + \text{D-glucose}$$

$$\text{Sucrose} + H_2O \xrightarrow{\text{Sucrase}} \text{D-fructose} + \text{D-glucose}$$

The hexose sugars pass via the portal vein to the liver, where they are catabolized, as described in the previous section.

Lactase is secreted in the intestines of infants to digest the lactose in their mothers' milk. Because most mammals do not ingest milk after weaning, lactase secretion decreases in adults as part of a normal developmental program. Humans are unusual in the animal kingdom in that many of us continue to drink milk into adulthood. Lactase deficiency in adult humans ranges from 5 to 20% in whites, to 75% in blacks, to almost 90% in Asians. This causes **lactose intolerance,** a condition in which ingestion of milk or lactose-containing milk products causes intestinal distress because the gut bacteria ferment the lactose that accumulates.

● **CONNECTION** Lactose intolerance, the inability to digest lactose, is caused by a deficiency in intestinal lactase.

Glycerol Metabolism

The digestion of neutral fat (triacylglycerols) and most phospholipids generates glycerol as one product. In animals, glycerol is phosphorylated by the action of **glycerol kinase** in liver:

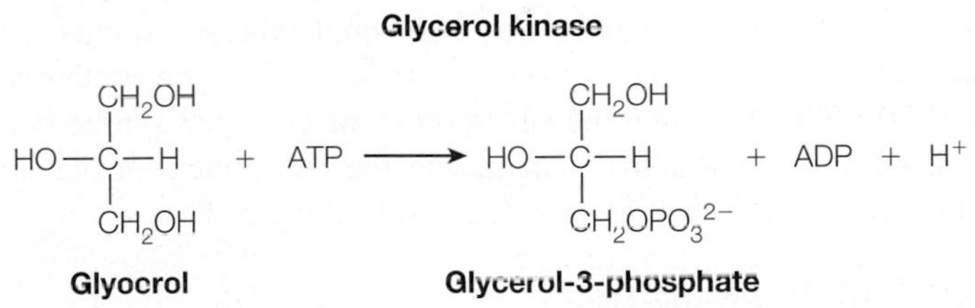

Glycerol kinase

$$\begin{array}{c} CH_2OH \\ | \\ HO-C-H \\ | \\ CH_2OH \end{array} + ATP \longrightarrow \begin{array}{c} CH_2OH \\ | \\ HO-C-H \\ | \\ CH_2OPO_3^{2-} \end{array} + ADP + H^+$$

Glycerol   **Glycerol-3-phosphate**

The product is then oxidized by **glycerol-3-phosphate dehydrogenase** to yield dihydroxyacetone phosphate, which is catabolized by glycolysis (see Figure 12.19).

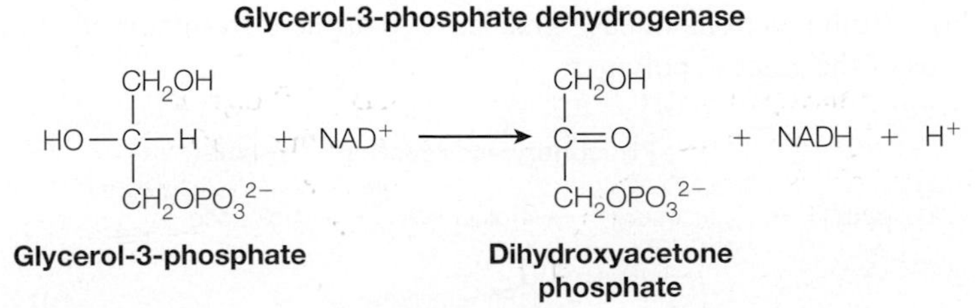

Glycerol-3-phosphate dehydrogenase

$$\begin{array}{c} CH_2OH \\ | \\ HO-C-H \\ | \\ CH_2OPO_3^{2-} \end{array} + NAD^+ \longrightarrow \begin{array}{c} CH_2OH \\ | \\ C=O \\ | \\ CH_2OPO_3^{2-} \end{array} + NADH + H^+$$

Glycerol-3-phosphate   **Dihydroxyacetone phosphate**

Polysaccharide Metabolism

In animal metabolism, glucose is derived from two primary polysaccharide sources: (1) the digestion of dietary polysaccharides, chiefly starch from plant foodstuffs and glycogen from meat; and (2) the mobilization of the animal's own glycogen reserves. Recall from Chapter 9 that starch, the major nutrient polysaccharide of plants, consists of the unbranched

glucose polymer amylose and the branched polymer amylopectin. Glucose residues in both polymers are linked by $\alpha(1 \rightarrow 4)$ glycosidic bonds, but amylopectin also has $\alpha(1 \rightarrow 6)$ linkages, which provide branch points in the otherwise linear polymer. Glycogen is chemically similar to amylopectin, except that it is more highly branched and is of higher molecular weight. Many microorganisms, like animals, store carbohydrate as glycogen.

Hydrolytic and Phosphorolytic Cleavages

Polysaccharide digestion and glycogen mobilization both involve sequential cleavage of monosaccharide units from nonreducing ends of glucose polymers (the anomeric carbon of the terminal residue is involved in the glycosidic bond at nonreducing ends). The first of these processes occurs via *hydrolysis* and the second via *phosphorolysis*. These processes are chemically similar, involving either water or inorganic phosphate as the nucleophile (**FIGURE 12.20**). Hydrolysis is the cleavage of a bond by addition of the elements of water across that bond, and a phosphorolytic cleavage occurs by addition of the elements of phosphoric acid. An enzyme catalyzing a phosphorolysis is often called a **phosphorylase,** to be distinguished from a *phosphatase*, which catalyzes the hydrolytic cleavage of a phosphate ester bond.

Energetically speaking, the advantage of a phosphorolytic mechanism is that mobilization of glycogen yields most of its monosaccharide units in the form of sugar phosphates. These units can be converted directly to intermediates in glycolysis, without the investment of additional ATP. By contrast, starch digestion via hydrolysis yields glucose

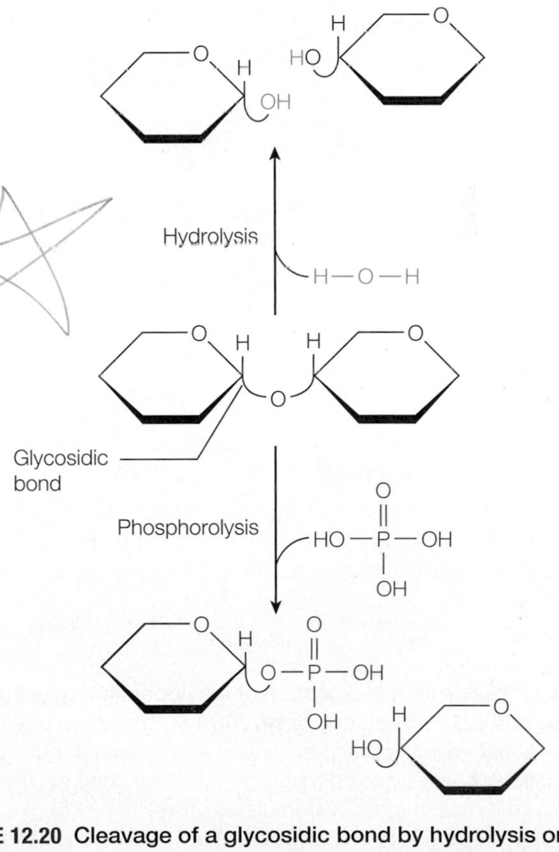

▲ **FIGURE 12.20 Cleavage of a glycosidic bond by hydrolysis or phosphorolysis.** This formal diagram shows how the elements of water or phosphoric acid, respectively, are added across a glycosidic bond.

plus some maltose, so that ATP and the hexokinase reaction are necessary to initiate glycolytic breakdown of these sugars.

● **CONCEPT** Dietary polysaccharides are metabolized by hydrolysis to monosaccharides. Intracellular carbohydrate stores, as glycogen, are mobilized as phosphorylated monosaccharides by phosphorolysis.

● **CONNECTION** In the brewing of beer, the controlled germination of cereal seeds such as barley releases hydrolytic enzymes that break starch down to mono- and disaccharides for later fermentation by yeast. This process is called malting.

The hydrolytic mechanism is useful, however, for the digestion of dietary carbohydrate, which occurs largely in the intestine. Digestion products must be absorbed and transported to the liver, where they are converted into glucose. Because sugar phosphates, like other charged compounds, are inefficiently transported across cell membranes, the hydrolytic digestion of polysaccharides to yield hexose sugars facilitates their uptake by tissues.

internal $\alpha(1 \rightarrow 4)$ linkages of both polymers. In the intestine, digestion continues, aided by α-amylase secreted by the pancreas. α-Amylase degrades amylose to maltose and a little glucose. However, it only partially degrades amylopectin and glycogen, as shown in **FIGURE 12.21**, because it cannot cleave the $\alpha(1 \rightarrow 6)$ linkages found at branch points.

The product of exhaustive digestion of amylopectin or glycogen by α-amylase is called a **limit dextrin;** its continued degradation requires the action of a "debranching enzyme," $\alpha(1 \rightarrow 6)$-*glucosidase* (also called *isomaltase*). This action exposes a new group of $\alpha(1 \rightarrow 4)$-linked branches, which can be attacked by α-amylase until a new set of $\alpha(1 \rightarrow 6)$-linked branches is reached. The end result of the sequential action of these two enzymes is the complete breakdown of starch or glycogen to maltose and some glucose. Maltose is cleaved hydrolytically by **maltase,** yielding two molecules of glucose, which is then absorbed into the bloodstream and transported to various tissues for utilization.

12.8 Glycogen Metabolism in Muscle and Liver

Before describing the enzymology and regulation of animal glycogen metabolism, we should have some idea of the different functions of the glycogen stores in muscle and liver. Glycogen is the major energy source for the contraction of skeletal muscle. Because liver derives most of its own metabolic energy from fatty acid oxidation, however, liver glycogen instead plays a very different role: as a source for blood glucose, to be transported to other tissues for catabolism. Liver serves primarily as a "glucostat," adjusting the synthesis and breakdown of glycogen to maintain appropriate blood glucose levels. As befits this role, the liver contains relatively large glycogen stores, from 2% to 8% of the weight of the organ. In liver the maximal rates of glycogen synthesis and degradation are about equal, whereas in muscle the maximal rate of glycogenolysis exceeds that of glycogen synthesis by about 300-fold. Although the enzymology of glycogen synthesis and breakdown is similar in liver and muscle, the endocrine control in liver is quite different, as we discuss here and in Chapter 17.

Glycogen Breakdown

The principal glycogen stores in vertebrates are in skeletal muscle and liver. Breakdown of these stores to provide usable energy—that is, the **mobilization** of glycogen—involves sequential phosphorolytic cleavages of $\alpha(1 \rightarrow 4)$ bonds, catalyzed by **glycogen phosphorylase.** In plants, starch is similarly mobilized by the action of **starch phosphorylase.** Both reactions release glucose-1-phosphate from nonreducing ends of the glucose polymer:

Starch and Glycogen Digestion

In animals, the digestion of starch and glycogen begins in the mouth, with the action of α-amylase secreted in saliva. α-Amylase cleaves

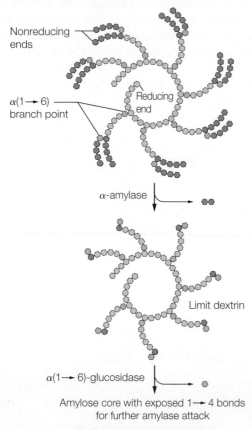

▲ FIGURE 12.21 Sequential digestion of amylopectin or glycogen by α-amylase and $\alpha(1 \rightarrow 6)$-glucosidase. (Top) α-Amylase in saliva cleaves $1 \rightarrow 4$ bonds between the maltose units of amylopectin (or glycogen). It cannot cleave $1 \rightarrow 6$ glycosidic bonds in the branched polymer, however, and a limit dextrin (gray) accumulates unless $\alpha(1 \rightarrow 6)$-glucosidase (debranching enzyme) is present. **(Bottom)** $\alpha(1 \rightarrow 6)$-Glucosidase in the intestine cleaves the branch points, exposing the amylose core to further digestion by amylase.

Phosphorylase reaction

Glucose$\alpha(1 \longrightarrow 4)$glucose$\alpha(1 \longrightarrow 4)$glucose$\alpha(1 \longrightarrow 4)$glucose $\cdots$

P_i ⟩ Phosphorylase

α-D-Glucose-1-(P) + glucose$\alpha(1 \longrightarrow 4)$glucose$\alpha(1 \longrightarrow 4)$glucose $\cdots$

The cleavage reaction is slightly disfavored under standard conditions ($\Delta G^{\circ\prime} = +3.1$ kJ/mol), but the relatively high intracellular levels of inorganic phosphate cause this reaction to operate in vivo almost exclusively in the degradative, rather than the synthetic, direction.

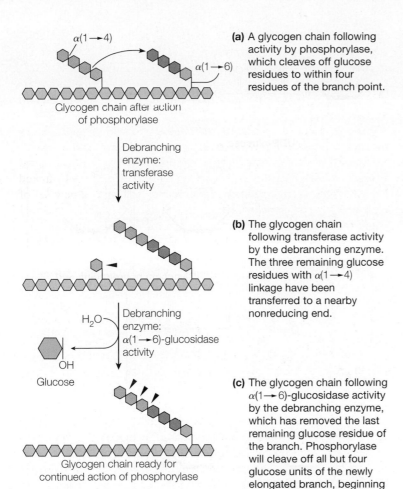

(a) A glycogen chain following activity by phosphorylase, which cleaves off glucose residues to within four residues of the branch point.

Glycogen chain after action of phosphorylase

Debranching enzyme: transferase activity

(b) The glycogen chain following transferase activity by the debranching enzyme. The three remaining glucose residues with $\alpha(1\to4)$ linkage have been transferred to a nearby nonreducing end.

H_2O

Debranching enzyme: $\alpha(1\to6)$-glucosidase activity

Glucose

(c) The glycogen chain following $\alpha(1\to6)$-glucosidase activity by the debranching enzyme, which has removed the last remaining glucose residue of the branch. Phosphorylase will cleave off all but four glucose units of the newly elongated branch, beginning the debranching process again. The new cleavage points are indicated by wedges.

Glycogen chain ready for continued action of phosphorylase

▲ **FIGURE 12.22 The debranching process in glycogen catabolism.**

Note that the reaction proceeds with retention of configuration at carbon 1 (i.e., the phosphate is in α linkage in the glucose-1-phosphate product).

Like the α-amylase used to digest dietary starch and glycogen, phosphorylases cannot cleave past $\alpha(1\to6)$ branch points. In fact, cleavage stops four glucose residues from a branch point. The debranching process involves the action of a second enzyme, as shown in **FIGURE 12.22**. This glycogen "debranching enzyme," $(\alpha1,4\to\alpha1,4)$ **glucantransferase,** catalyzes two reactions. First is the transferase activity, in which the enzyme removes three of the remaining glucose residues and transfers this trisaccharide moiety intact to the end of some other outer branch via a new $\alpha(1\to4)$ linkage. Next, the remaining glucose residue, which is still attached to the chain by an $\alpha(1\to6)$ bond, is cleaved by the $\alpha(1\to6)$-glucosidase activity of the same debranching enzyme. This yields one molecule of free glucose and a branch extended by three $\alpha(1\to4)$-linked glucose residues. This newly exposed branch is now available for further attack by phosphorylase. The end result of the action of these two enzymes is the complete breakdown of glycogen to glucose-1-phosphate ($\sim$90%) and glucose ($\sim$10%).

Why has the glycogen breakdown scheme evolved to include this complex debranching process? The importance of storing carbohydrate energy in the form of a highly branched polymer may well lie in an animal's need to generate energy very quickly following appropriate stimuli. Glycogen phosphorylase cleaves sequentially from nonreducing ends. The more of these ends that exist in a polymer, the faster the polymer can be mobilized.

To be metabolized via glycolysis, the glucose-1-phosphate produced by phosphorylase action must be converted to glucose-6-phosphate. This isomerization is accomplished by *phosphoglucomutase*. The reaction is mechanistically similar to that of phosphoglycerate mutase (Section 12.2), except that in phosphoglucomutase a phosphoserine residue on the enzyme reacts with substrate instead of phosphohistidine:

E-Ser-P + Glucose-1-P $\rightleftharpoons$ E-Ser-P + Glucose-6-P

Enzyme-bound glucose-1,6-bisphosphate

Most of the glycogen in vertebrate animals is stored as granules in cells of liver and skeletal muscle. The liver provides glucose to other tissues for metabolism through glycogen mobilization and gluconeogenesis. Both processes yield phosphorylated forms of glucose, which cannot exit from liver cells. Conversion to free glucose requires the action of glucose-6-phosphatase, the same enzyme used in gluconeogenesis.

Glycogen Biosynthesis

A major fate of glucose in animals is the synthesis of glycogen. The mechanisms that are used to form glycosidic bonds in glycogen are general mechanisms used in the synthesis of all polysaccharides.

Biosynthesis of UDP-Glucose

The immediate substrate for glycogen biosynthesis is **uridine diphosphate glucose (UDP-Glc).**

Uridine diphosphate glucose (UDP-Glc)

UDP-Glc is synthesized from blood glucose, which is transported into cells by a plasma membrane **glucose transporter.** As shown in **FIGURE 12.23**, glucose is then phosphorylated by hexokinase to give glucose-6-phosphate, which is isomerized to glucose-1-phosphate by phosphoglucomutase (the same enzyme used for the reverse reaction in glycogen breakdown). The enzyme **UDP-glucose pyrophosphorylase**

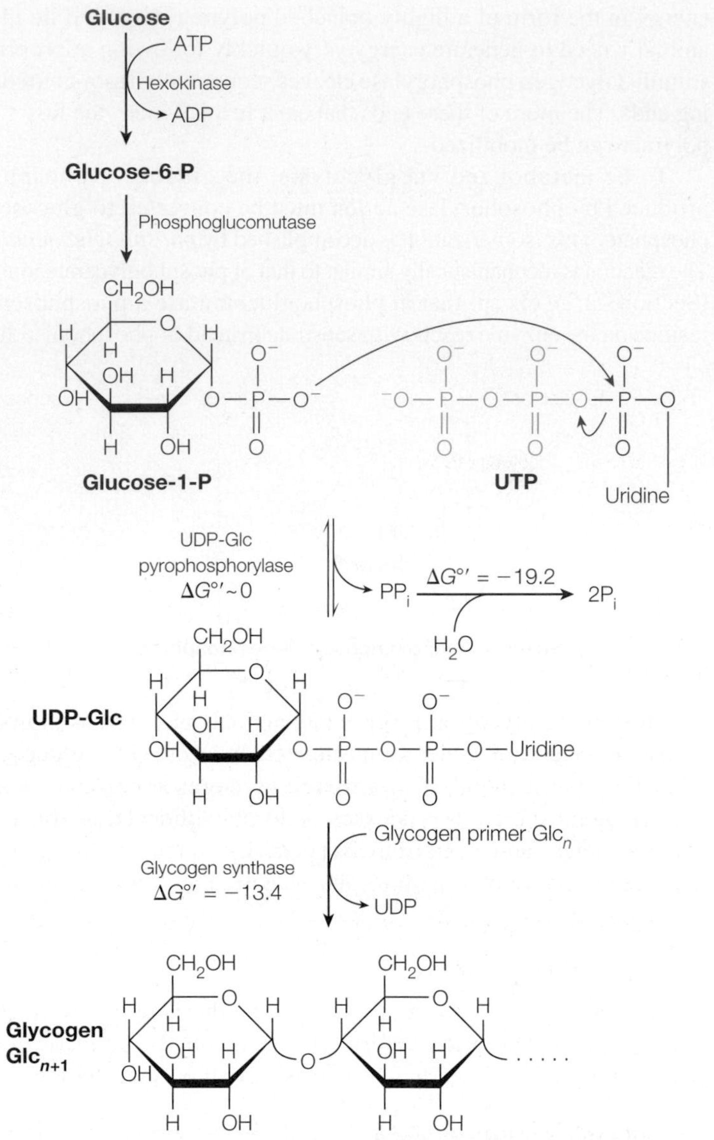

▲ **FIGURE 12.23** Pathway for the conversion of glucose monomers to polymeric glycogen.

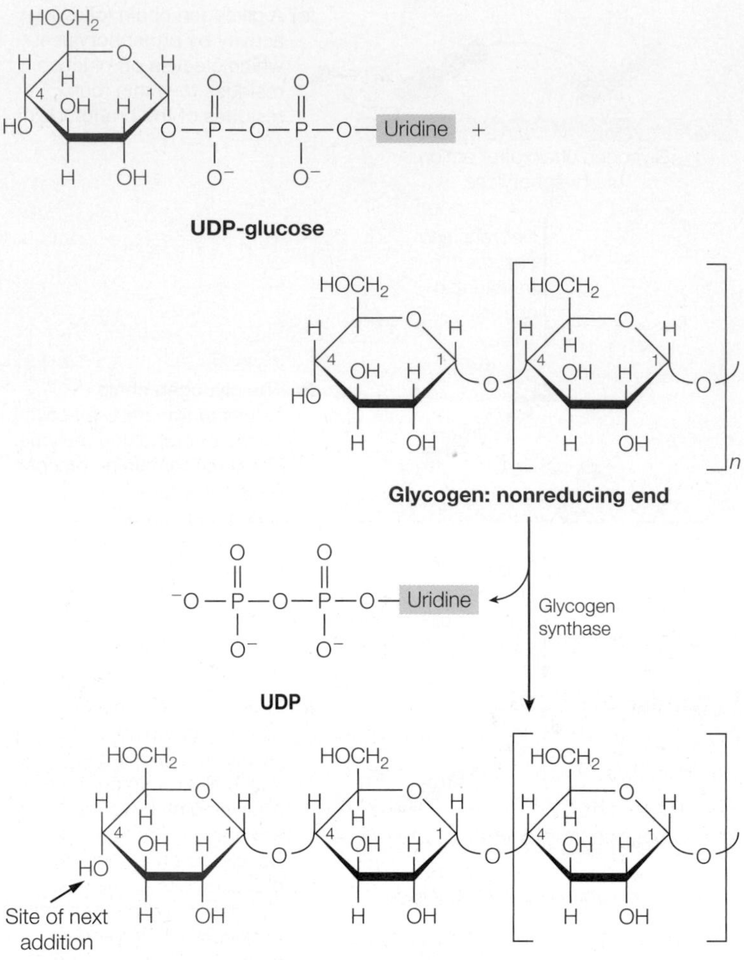

▲ **FIGURE 12.24** The glycogen synthase reaction.

● **CONCEPT** UDP-glucose is the metabolically activated form of glucose for glycogen synthesis.

then catalyzes the synthesis of UDP-glucose. The free energy change of this phosphoanhydride exchange reaction is negligible, but it is drawn forward by rapid enzymatic cleavage of pyrophosphate to orthophosphate, catalyzed by pyrophosphatase. The $\Delta G^{\circ\prime}$ for the hydrolysis of pyrophosphate is $\sim -19\ \text{kJ/mol}$.

The Glycogen Synthase Reaction

The enzyme involved in converting UDP-glucose to glycogen, **glycogen synthase,** is bound tightly to intracellular glycogen granules. Glycogen synthase is a **glycosyltransferase**—an enzyme that transfers an activated sugar unit to a nonreducing sugar hydroxyl group (Section 9.3). Here it catalyzes the immediate donation of a glucosyl residue from UDP-Glc to the nonreducing end of a glycogen branch,

which must be at least four glucose residues in length. The reaction, depicted in **FIGURE 12.24**, generates an $\alpha(1 \rightarrow 4)$ glycosidic linkage between C-1 of the incoming glucosyl moiety and C-4 of the glucose residue at the terminus of the glycogen chain. This transfer involves nucleophilic attack by the 4—OH of the incoming glucosyl residue on C-1 of UDP-glucose. C-1 is rendered electrophilic by elimination of UDP, an excellent leaving group, but the precise mechanism remains unsettled.

The enzyme continues to add glucose residues successively to the 4-hydroxyl groups at the nonreducing ends of the glycogen polymer. Because UDP-Glc is a high-energy compound, the glycogen synthase reaction is exergonic, with a $\Delta G^{\circ\prime} \approx -13.4\ \text{kJ/mol}$. Glycogen synthase catalyzes the rate-limiting step of glycogen biosynthesis and is the site where this anabolic pathway is regulated.

The primer for glycogen synthase is a short chain of glucose residues assembled by a $\sim$37,000 Da protein called **glycogenin,** which transfers glucose from UDP-Glc to a tyrosine residue on the protein itself. Glycogenin then transfers additional glucosyl units from UDP-Glc, to give $\alpha(1 \rightarrow 4)$ linked primers up to eight residues long. These primers are extended by glycogen synthase. Glycogenin thus forms the core of the mature glycogen particle that eventually consists of up to 60,000 glucose residues. These particles are stored in the granules of

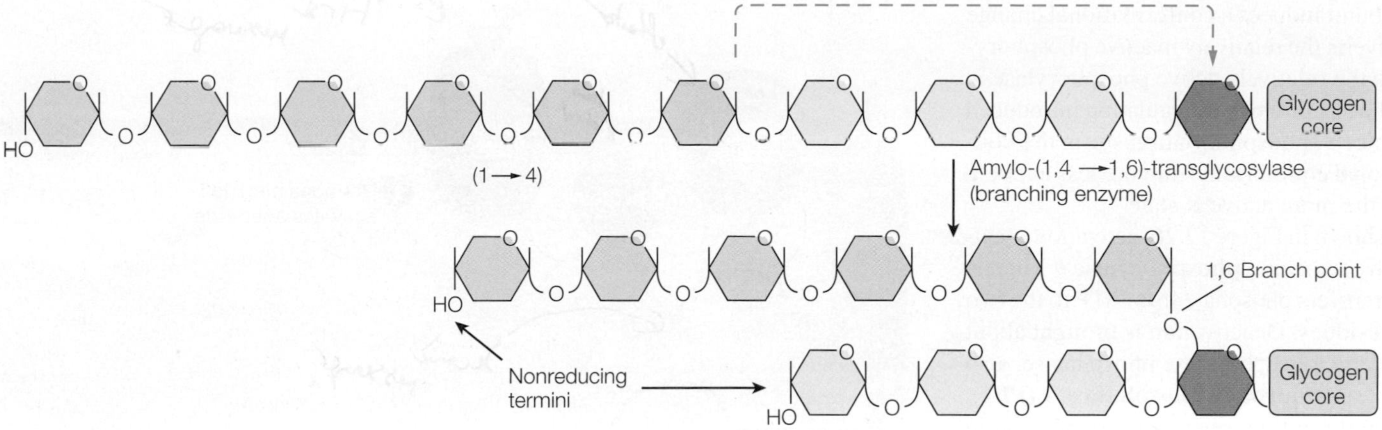

▲ **FIGURE 12.25 The branching process in glycogen synthesis.** Branching is brought about by the action of amylo-(1,4 → 1,6)-transglycosylase.

liver and muscle cells (see Figure 9.16). These granules contain all of the enzymes that metabolize glycogen as well.

Formation of Branches

Glycogen synthesis requires both the polymerization of glucose units and branching from $\alpha(1 \rightarrow 6)$ linkages. These branches are important because they increase the solubility of the polymer and the number of nonreducing ends from which glucose-1-phosphate can be derived during glycogen mobilization. However, these branches cannot be introduced by glycogen synthase. Another enzyme, called **branching enzyme,** but more accurately called **amylo-(1,4 → 1,6)-transglycosylase,** comes into play, as shown in **FIGURE 12.25**. This branching enzyme transfers a terminal fragment, some 6 or 7 residues long, from a branch terminus at least 11 residues long to a hydroxyl group at the 6-position of a glucose residue in the interior of the polymer. The reaction involves nucleophilic attack of the C-6 hydroxyl on C-1 of the oligosaccharide that will form the branch. The reaction thus creates two nonreducing termini for continued action by glycogen synthase, whereas just one existed before. The branching process does not involve a large free energy change because of the chemical similarity of $(1 \rightarrow 4)$ and $(1 \rightarrow 6)$ linkages.

> ● **CONCEPT** Glycogen biosynthesis requires glycogen synthase for polymerization and a transglycosylase to create branches.

12.9 Coordinated Regulation of Glycogen Metabolism

In Chapter 8, we mentioned the control of glycogen breakdown, or glycogenolysis, as a particularly well-understood example of a regulatory cascade, a process in which the intensity of an initial regulatory signal is amplified manyfold through a series of enzyme activations. This amplification is particularly important in the case of glycogenolysis because fright, for example, or the need to catch prey, can trigger an instantaneous need for increased energy generation and utilization. Glycogen represents the most immediately available *large-scale* source of metabolic energy; hence, it is important that animals be able to activate glycogen mobilization rapidly.

The hormonal regulation of glycogen breakdown as it might occur in a muscle cell after stimulation by epinephrine, or in a liver cell after stimulation by glucagon or epinephrine, is summarized in **FIGURE 12.26**:

❶ Binding of a hormone (glucagon or epinephrine) to its plasma membrane receptor on the outside of the cell triggers an interaction between the receptor and a G protein on the inside of the cell, which in turn activates adenylate cyclase (a process described in more detail in Chapter 20).

❷ Adenylate cyclase catalyzes the formation of cAMP, which binds to the R (regulatory) subunits of the protein kinase A (PKA) R_2C_2 tetramer, causing their dissociation from the C (catalytic) subunits.

❸ The active C monomer of PKA catalyzes phosphorylation of specific serine residues on inactive phosphorylase *b* kinase, activating the enzyme.

❹ This active kinase phosphorylates a serine residue on each of the two subunits of the homodimeric glycogen phosphorylase, converting the inactive phosphorylase *b* to the active phosphorylase *a*.

❺ Active phosphorylase *a* then catalyzes glycogen breakdown.

Each reaction in the regulatory cascade amplifies the hormonal signal, so that binding of very few hormone molecules at the cell surface triggers an enormous release of glucose-1-phosphate from intracellular glycogen stores. Inactivation of the pathway involves the action of a phosphatase, which removes the phosphates from phosphorylase *b* kinase and from phosphorylase *a*.

Structure of Glycogen Phosphorylase

To fully comprehend the regulation of glycogen metabolism, we must first understand the structure of the glycogen phosphorylase. In skeletal muscle, glycogen phosphorylase is a dimer containing two identical polypeptide chains, each of 97,400 daltons. The enzyme exists in two interconvertible forms—the relatively *active* phosphorylase *a* and the relatively *inactive* phosphorylase *b*.* Phosphorylation of serine 14 on

---

*For enzymatically interconvertible enzyme systems such as glycogen phosphorylase, *a* and *b* refer to the more active and less active forms, respectively.

each subunit induces a conformational change that converts the relatively inactive phosphorylase *b* to the relatively active phosphorylase *a*. Using the terminology of regulation introduced in Chapter 8, phosphorylation shifts the conformational equilibrium from the less active T state to the more active R state.

As shown in Figure 12.26, activation is catalyzed by a specific **phosphorylase *b* kinase,** which transfers phosphate from ATP to the two serine residues. Deactivation is brought about by a specific phosphorylase phosphatase, also called **phosphoprotein phosphatase 1 (PP1).** PP1 is a ubiquitous serine/threonine protein phosphatase in eukaryotes that regulates many cellular processes through the dephosphorylation of dozens of substrates. PP1 is directed to these various functions through its association with a diverse set of targeting proteins. The activity of PP1 is also subject to hormonal control.

Control of Phosphorylase Activity

Phosphorylase *b* kinase is also converted by phosphorylation from an inactive to an active form (Figure 12.26). This reaction is catalyzed by PKA, the same cAMP–dependent protein kinase that phosphorylates the bifunctional PFK-2/FBPase-2 in glycolysis and gluconeogenesis. In glycogenolysis, cAMP exerts a rapid and efficient activation. At the same time, it inhibits glycogen synthesis through a separate regulatory cascade.

The primary hormone promoting glycogenolysis in muscle is epinephrine (formerly called adrenaline), which is secreted from the adrenal medulla and binds to specific receptors on muscle cell membranes. Mobilization of liver glycogen is stimulated largely by the pancreatic peptide hormone glucagon, although liver can also respond to epinephrine. Figure 12.26 illustrates how the secretion of

● **CONCEPT** Glycogen mobilization is controlled hormonally by a metabolic cascade that is activated by cAMP formation and involves successive phosphorylations of enzyme proteins.

● **CONCEPT** The rapid mobilization of muscle glycogen triggered by epinephrine is one of several components of the "fight-or-flight" response.

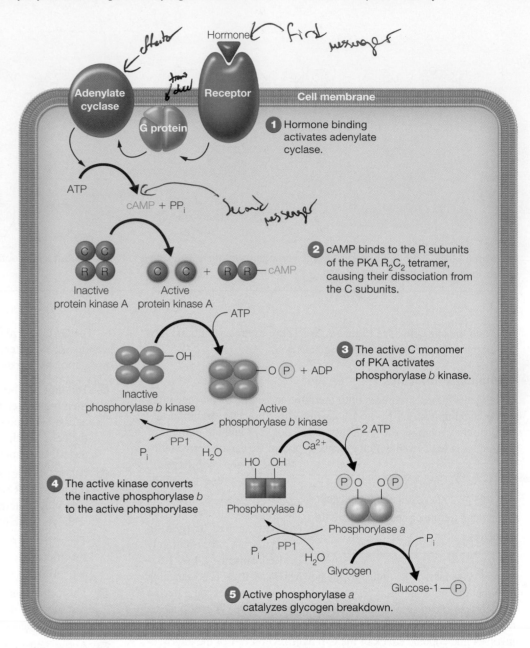

▲ **FIGURE 12.26** The regulatory cascade controlling glycogen breakdown.

relatively few molecules of hormone, such as epinephrine, can, within just a few moments, trigger a massive conversion of glycogen to glucose-1-phosphate.

Epinephrine is the principal hormone governing the "fight-or-flight" response to various stimuli. In addition to stimulating glycogenolysis, epinephrine triggers a variety of physiological events, such as increasing the strength and frequency of heartbeats. These cardiac effects, triggered by increased intracellular

Ca^{2+} concentrations, are also mediated via cAMP, as discussed further in Chapter 20. Cyclic AMP also regulates other metabolic processes, including the stimulation of fat breakdown and the inhibition of glycogen synthesis.

Proteins in the Glycogenolytic Cascade

Our presentation of the glycogenolytic cascade started with the phosphorylase reaction and then worked backward to the initial hormonal signal. Now let us start with the hormone and work forward, with emphasis on the proteins involved (again, refer to Figure 12.26). The hormone binds to a specific receptor located on the outside of the cytoplasmic membrane. This binding activates adenylate cyclase, which is bound to the inside of the membrane, in a process that is mediated by a G protein. (G proteins and adenylate cyclase are discussed in greater detail in Chapter 20.)

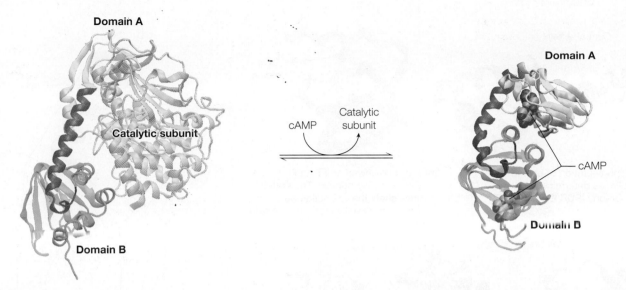

Domain A

Catalytic subunit

Domain B

cAMP

Catalytic
subunit

Domain A

cAMP

Domain B

C subunit bound. When a regulatory (R) subunit is bound to a catalytic (C) subunit, it has an extended dumbbell shape, with the two cAMP-binding domains (A and B) wrapped around the catalytic subunit (blue).

cAMP bound. Upon binding cAMP, Domain B of the R subunit undergoes a conformational change, rotating ~125° toward Domain A to adopt a compact globular structure and releasing the active catalytic subunit.

▲ **FIGURE 12.27** **Activation of cyclic AMP-dependent protein kinase (protein kinase A) by cAMP.** X-ray structures of the cAMP binding domains of the regulatory subunit of protein kinase A bound to the catalytic subunit (left, PDB ID: 2qcs) or to cAMP (right, PDB ID: 1rgs).

Cyclic AMP–Dependent Protein Kinase

Cyclic AMP–dependent protein kinase, also called protein kinase A (PKA), is a tetramer consisting of two *catalytic* subunits, C, and two *regulatory* subunits, R. The tetramer, R_2C_2, is catalytically inactive. Each R subunit possesses two cAMP-binding sites (domains A and B). When bound to the C subunits, the R subunits have an extended dumbbell shape, with the two cAMP-binding domains of each R subunit wrapped around each catalytic subunit (**FIGURE 12.27**). Binding of cAMP to the R subunits causes them to undergo a dramatic conformational change that packs the two cAMP-binding domains of each R subunit together in a compact globular structure. This conformational change causes the tetramer to dissociate, releasing the catalytically active C subunits to catalyze the phosphorylation of target proteins, including phosphorylase *b* kinase.

Phosphorylase *b* Kinase

Phosphorylase *b* kinase is a complex multisubunit protein of ~1.3 MDa, composed of four copies each of α, β, γ, and δ subunits. The γ subunit contains the catalytic site, and the regulatory α and β subunits contain the sites of phosphorylation by PKA. The δ subunit is a protein called **calmodulin,** or calcium-modulating protein. Calcium ion has long been known as an important physiological regulator, particularly of processes related to nerve conduction and muscle contraction. Most of these effects are mediated through the binding of Ca^{2+} to calmodulin, which is sensitive to small changes in intracellular Ca^{2+} concentration.

Calmodulin

Calmodulin is a small protein (~17,000 Da) of highly conserved amino acid sequence. It contains four calcium ion binding sites (**FIGURE 12.28(a)**). Each Ca^{2+}-binding site (Figure 12.28(b)) is composed of a helix–loop–helix motif known as an EF hand (**FIGURE 12.29**). This motif is found in a large number of Ca^{2+}-binding proteins. EF hand domains bind Ca^{2+} with a K_D of about 10^{-6} M, consistent with observations that calcium can effect intracellular metabolic changes in concentrations as low as 1 μM. Binding stimulates a major conformational change in the protein, leading to a more compact and more highly helical structure, which augments the affinity of calmodulin for a number of regulatory target proteins (Figure 12.28(c)).

In the case of phosphorylase *b* kinase, calmodulin plays a special role as an integral subunit of the enzyme. Hence, the glycogenolysis cascade depends on intracellular calcium concentration as well as on cAMP levels. This dependence is particularly important in muscle, where contraction is stimulated by calcium release. Thus, Ca^{2+} plays a dual role, in provision of the energy substrates needed to support muscle contraction and in contraction itself.

Nonhormonal Control of Glycogenolysis

Glycogen breakdown is under nonhormonal as well as hormonal control. Recall that phosphorylase *b* is relatively inactive, existing primarily in its T state. This form of the enzyme is activated allosterically by 5′-AMP (but not by cAMP). This activation does not usually occur in the cell because ATP, which is far more abundant and does not activate phosphorylase *b*, competes with AMP for binding to the enzyme. However, under energy-deprived conditions, AMP may accumulate at the expense of ATP breakdown. AMP binding shifts the conformational equilibrium of phosphorylase *b* to the more active R state (**FIGURE 12.30**). ATP and glucose-6-phosphate, signs of adequate energy status, shift the equilibrium of phosphorylase *b* back to the less active T state.

Once glycogen phosphorylase *b* is phosphorylated to its more active *a* state, it exists primarily in the R form and is unresponsive to most metabolite effectors. However, glucose and glucose-6-phosphate act

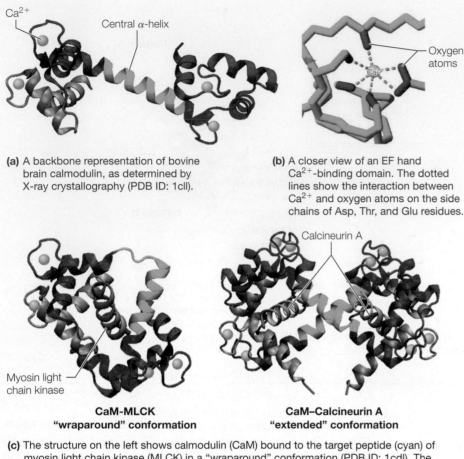

(a) A backbone representation of bovine brain calmodulin, as determined by X-ray crystallography (PDB ID: 1cll).

(b) A closer view of an EF hand Ca^{2+}-binding domain. The dotted lines show the interaction between Ca^{2+} and oxygen atoms on the side chains of Asp, Thr, and Glu residues.

Calcineurin A

Myosin light chain kinase

CaM-MLCK "wraparound" conformation

CaM–Calcineurin A "extended" conformation

(c) The structure on the left shows calmodulin (CaM) bound to the target peptide (cyan) of myosin light chain kinase (MLCK) in a "wraparound" conformation (PDB ID: 1cdl). The structure on the right shows a calmodulin dimer bound to the target peptides (cyan) of a calcineurin A in an "extended" conformation (PDB ID: 2w73).

▲ **FIGURE 12.28 Calmodulin structure.** Calmodulin contains four Ca^{2+} binding domains (colored orange, purple, red, and blue) connected by a long central α-helix (green).

E helix

Ca^{2+}

F helix

EF hand

▲ **FIGURE 12.29 The EF hand Ca^{2+}-binding domain.** This common helix–loop–helix motif is found in many Ca^{2+}-binding proteins.

synergistically on phosphorylase *a,* shifting its equilibrium slightly back toward the T state. In the T state, the phosphoserine side chains are more accessible to phosphoprotein phosphatase 1 (PP1), so that the T state is more readily dephosphorylated than the R state.

Thus, the mobilization of energy reserves from glycogen can be brought about either by hormonal stimulation, reflecting a physiological need for increased ATP production, or by an allosteric mechanism triggered when the energy level is deficient for the maintenance of normal functions. The nonhormonal mechanism, which does not involve a metabolic cascade, stimulates glycogenolysis in response to a low-energy charge, whereas the hormonally induced cascade predominates when the need is to rapidly augment energy generation. In both cases, the phosphorolysis of glycogen to glucose-1-phosphate is enhanced. If, however, the cell has a high-energy charge, signaled by high ATP and/or glucose-6-phosphate levels, then glycogenolysis is turned off.

Control of Glycogen Synthase Activity

Earlier we noted that epinephrine secretion inhibits glycogen synthesis in muscle at the same time that it promotes glycogen mobilization. Glucagon has similar effects in liver. Control of both the synthesis and degradation of glycogen is mediated by distinct regulatory cascades involving cAMP–dependent protein kinase and reversible

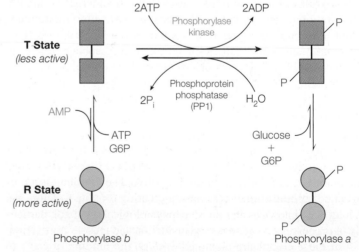

T State *(less active)*

2ATP   2ADP

Phosphorylase kinase

AMP

$2P_i$

Phosphoprotein phosphatase (PP1)

H_2O

ATP
G6P

Glucose + G6P

R State *(more active)*

Phosphorylase *b*

Phosphorylase *a*

▲ **FIGURE 12.30 Control of glycogen phosphorylase activity.** The enzyme exists in an equilibrium between a less active T state and a more active R state. The unphosphorylated form, phosphorylase *b,* exists largely in the T state, and its T ⇌ R equilibrium is controlled by allosteric effectors AMP, ATP, and glucose-6-phosphate (G6P). Glucose and G6P synergistically inhibit phosphorylase *a* by shifting its equilibrium to the T state, which is then rapidly dephosphorylated by phosphoprotein phosphatase 1 (PP1).

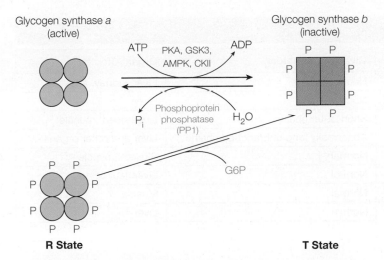

▲ FIGURE 12.31 Control of glycogen synthase activity. The enzyme can be phosphorylated by several different protein kinases, including the catalytic subunit of cAMP-dependent protein kinase (PKA), AMP-activated protein kinase (AMPK), glycogen synthase kinase 3 (GSK3), and casein kinase II (CKII). Dephosphorylation is catalyzed by phosphoprotein phosphatase 1 (PP1). Glucose-6-phosphate (G6P) can allosterically activate the phosphorylated enzyme.

● **CONCEPT** Conditions that activate glycogen breakdown inhibit glycogen synthesis, and vice versa.

protein phosphorylations. However, whereas the cascade controlling glycogenolysis *activates* glycogen phosphorylase (see Figure 12.26), the cascade controlling glycogen synthesis *inhibits* glycogen synthase (**FIGURE 12.31**).

Glycogen synthase from vertebrate tissues is a tetrameric protein consisting of four identical subunits. Its activity is controlled by covalent modification and allosteric activation. Like phosphorylase, glycogen synthase exists in phosphorylated and dephosphorylated states, with up to nine serine residues on each subunit subject to this modification. Several different protein kinases are known to act on glycogen synthase

● **CONCEPT** Glycogen synthase activity is controlled by phosphorylation, through mechanisms comparable to those controlling glycogen breakdown by phosphorylase, but having reciprocal effects on enzyme activity.

(Figure 12.31). Dephosphorylation is catalyzed by PP1, the same phosphatase that acts on glycogen phosphorylase and phosphorylase *b* kinase.

In contrast to glycogen phosphorylase, it is the unphosphorylated enzyme, glycogen synthase *a*, that is the active form. Glycogen synthase *a* is active even in the absence of G6P, whereas the phosphorylated forms (glycogen synthase *b*) depend on allosteric activation by G6P. Binding of this effector shifts the equilibrium back toward the R state, overriding the inhibition caused by phosphorylation (Figure 12.31). In addition, G6P binding induces a conformational change that makes the enzyme a better substrate for dephosphorylation by PP1.

Let's examine the consequences of hormone release upon glycogen synthase (see Figure 12.31). Just as depicted in Figure 12.26, the activation of adenylate cyclase by epinephrine (in muscle) or glucagon (in liver) promotes the dissociation of cAMP-dependent protein kinase

(PKA) to give free catalytic C subunits. These C subunits phosphorylate active glycogen synthase *a* to inactive glycogen synthase *b*. To complicate the picture somewhat, several additional protein kinases, including AMP-activated protein kinase (AMPK), glycogen synthase kinase 3 (GSK3), and casein kinase II, can act on glycogen synthase *a*. The most important of these is GSK3 (see Section 20.3). Each of these kinases phosphorylates different serine residues, but in a hierarchical manner, so that there are several different forms of glycogen synthase *b*, and it is an oversimplification to speak of just two forms.

In general, as more sites are phosphorylated, the activity of the enzyme progressively decreases because of the following changes: (1) decreased affinity for UDP-glucose (the substrate); (2) decreased affinity for glucose-6-phosphate (the allosteric activator); and (3) increased affinity for ATP and P_i, both of which tend to antagonize the activation by glucose-6-phosphate. Thus, there is a graded series of responses to changing metabolic conditions, involving a series of different protein kinases. Whichever kinase is used, the net effect of phosphorylating glycogen synthase is to inhibit the enzyme, with consequent inhibition of glycogen synthesis.

These regulatory mechanisms explain the observation that when plasma glucose levels rise after a meal, the liver gradually takes up glucose and stores it as glycogen. Insulin is released from the pancreas under these conditions, stimulating glycogen synthesis (via inactivation of GSK3; see Section 20.3). Conversely,

● **CONCEPT** The liver regulates blood glucose levels partly by control of its glycogen synthase and phosphorylase.

when blood glucose concentrations fall during fasting or exercise, glucagon secretion inhibits hepatic glycolysis (Figure 12.17) and stimulates glycogen breakdown (Figure 12.26). The liver thus maintains circulating blood glucose levels by mobilizing its glycogen stores.

Skeletal muscle, which does not carry out gluconeogenesis, must rely on blood glucose supplied from the liver and its own glycogen stores for fuel. Because muscle cells lack receptors for glucagon and express an isoform of pyruvate kinase (PK-M1) that is not regulated by covalent modification, muscle glycolysis is not inhibited when blood glucose concentrations are low. Muscle cells respond instead to epinephrine as part of the fight-or-flight response. Epinephrine stimulates glycogen breakdown (Figure 12.26), producing glucose-6-phosphate for the generation of ATP via glycolysis.

Congenital Defects of Glycogen Metabolism in Humans

A number of inherited human diseases involve mutations in genes encoding enzymes of glycogen metabolism. The clinical symptoms of these conditions, called **glycogen storage diseases,** can be quite severe and usually result from the storage of abnormal quantities of glycogen or the storage of glycogen with abnormal properties. Accu-

● **CONNECTION** Human mutations affecting enzymes of glycogen metabolism can have mild or profound clinical consequences.

mulation of abnormal glycogen results from its failure to be broken down. Studies on these conditions have helped identify the roles of the enzymes involved in glycogen metabolism.

Among the earliest glycogen storage diseases to be described was *von Gierke disease,* named for a German physician who studied an 8-year-old girl with a chronically enlarged liver. After her death in 1929 from influenza, her liver was found to contain 40% glycogen. The

TABLE 12.3 Human congenital defects of glycogen metabolism

Type	Common Name	Enzyme Deficiency	Glycogen Structure	Organ Affected
Ia	von Gierke disease	Glucose-6-phosphatase (ER)	Normal	Liver, kidney, intestine
Ib		Glucose-6-phosphate transporter (ER)	Normal	Liver
III	Cori or Forbes disease	Debranching enzyme	Short outer chains	Liver, heart, muscle
IV	Andersen disease	Branching enzyme	Abnormally long unbranched chains	Liver and other organs
V	McArdle disease	Muscle glycogen phosphorylase	Normal	Skeletal muscle
VI	Hers disease	Liver glycogen phosphorylase	Normal	Liver, leukocytes
VII	Tarui disease	Muscle phosphofructokinase	Normal	Muscle
IX		Liver phosphorylase kinase	Normal	Liver
—		Glycogen synthase	Normal	Liver

glycogen appeared normal but could not be degraded by extracts of the girl's liver, only by extracts of other livers. Today we know that these symptoms can result from deficiency of either glucose-6-phosphatase or the debranching enzyme. When the debranching enzyme is deficient (Cori or Forbes disease), phosphorylase can degrade glycogen only until branch points are reached and no farther.

TABLE 12.3 provides information on several of the glycogen storage diseases that have been characterized. Among the most serious clinically is the type I disease, resulting from functional lack of glucose-6-phosphatase. Individuals with this condition can break down glycogen normally, but they are chronically hypoglycemic because they cannot cleave G6P to glucose for release from the liver to the bloodstream. In a less severe form of this disease, blood glucose levels are normal except after stress, when the normal hyperglycemic response is inhibited. One form of this disease (type Ia) results from a deficiency of glucose-6-phosphatase itself. The type Ib disease involves a deficiency of the specific transporter for glucose-6-phosphate into the lumen of the endoplasmic reticulum (ER). This transporter is part of a multiprotein complex, which includes glucose-6-phosphatase itself, located on the lumenal face of the ER.

Other forms of glycogen storage diseases involve abnormalities that can be understood in terms of the known enzymatic defect. In type III individuals, who have a defective debranching enzyme, glycogen with very short outer branches accumulates, leading to enlargement of the liver. By contrast, type IV disease, which is associated with a defective branching enzyme, involves accumulation of glycogen with very long outer branches. Early death from liver failure is often observed in type IV individuals. Type III, V, VI, VII, and IX diseases have less severe symptoms. Individuals with type V disease, for instance, who have a deficiency of muscle glycogen phosphorylase, usually show no symptoms until about age 20. Once symptoms appear, the principal ones are severe muscle cramps upon exercising and the failure of lactate to accumulate in blood after exercise. There are even some rare cases of hepatic glycogen synthase deficiency, in which affected patients have severely decreased liver glycogen stores.

12.10 A Biosynthetic Pathway That Oxidizes Glucose: The Pentose Phosphate Pathway

The predominant pathway for glucose catabolism is glycolysis to yield pyruvate, followed by oxidation to CO_2 in the citric acid cycle (Chapter 13). An alternative process, the **pentose phosphate pathway,** is a remarkable, multipurpose pathway that operates to varying extents in

different cells and tissues. The role of this pathway is primarily anabolic rather than catabolic, but we present it here because it does involve the catabolism of glucose. The pathway, which operates exclusively in the cytosol, is summarized in **FIGURE 12.32**.

The pentose phosphate pathway has two primary functions:

- to provide reducing equivalents (in the form of NADPH) for reductive biosynthesis and for dealing with oxidative stress
- to provide ribose-5-phosphate for nucleotide and nucleic acid biosynthesis.

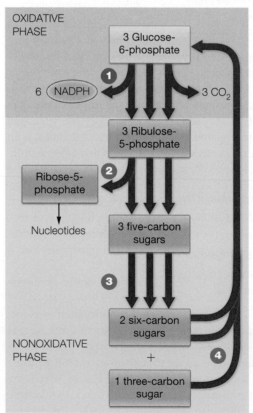

Stage 1: G6P oxidized to Ribulose-5-P + CO_2, with production of NADPH

Stage 2: Ribulose-5-P converted to ribose-5-P and other 5C sugars

Stage 3: 3 5C sugars converted to 2 6C sugars + 1 3C sugar

Stage 4: Production of G6P from 3C and 6C sugars and repeat of cycle

▲ **FIGURE 12.32 Overall strategy of the pentose phosphate pathway.** The oxidative phase, composed of the first three reactions, produces NADPH. The remaining stages constitute the nonoxidative phase of the pathway.

● **CONCEPT** The pentose phosphate pathway converts glucose to various other sugars that can be used for energy, but its most important products are NADPH for reductive biosynthesis and ribose-5-phosphate for nucleotide biosynthesis.

In addition, the pathway metabolizes dietary pentose sugars, derived primarily from the digestion of nucleic acids. In plants, a variant of the pentose phosphate pathway operates in reverse as part of the carbon fixation process of photosynthesis (Chapter 15).

Recall from Section 11.4 that $NADP^+$ is identical to NAD^+ except for the additional 2′ phosphate on one of the ribose moieties of $NADP^+$. Metabolically, the difference between NAD^+ and $NADP^+$ is that nicotinamide nucleotide-linked enzymes whose primary function is to *oxidize* substrates use the NAD^+/NADH pair, whereas enzymes functioning primarily in a *reductive* direction use $NADP^+$/NADPH (see Figure 11.9). Because NADPH is used for fatty acid and steroid biosynthesis, tissues such as adrenal gland, liver, adipose, and mammary gland are rich in enzymes of the pentose phosphate pathway. NADPH is also the ultimate electron source for reducing ribonucleotides to deoxyribonucleotides for DNA synthesis, so rapidly proliferating cells generally have high activity of pentose phosphate pathway enzymes, for the production of both NADPH and ribose-5-phosphate.

The Oxidative Phase: Generating Reducing Power as NADPH

The pentose phosphate pathway operates in two phases—oxidative and nonoxidative. Two of the first three reactions in this pathway are oxidative, each involving reduction of one $NADP^+$ to NADPH. As shown in **FIGURE 12.33**, the first reaction, catalyzed by **glucose-6-phosphate dehydrogenase**, oxidizes glucose-6-phosphate to **6-phosphogluconolactone.** (A **lactone** is an internal ester, in this case linking carbons 1 and 5.) Phosphogluconolactone is hydrolyzed by **6-phosphogluconolactonase** to **6-phosphogluconate,** which undergoes an oxidative decarboxylation to yield CO_2, another NADPH, and **ribulose-5-phosphate** (a pentose phosphate). The net result of the oxidative phase is the generation of 2 molecules of NADPH, the oxidation of one carbon to CO_2, and the synthesis of 1 molecule of pentose phosphate per 1 molecule of glucose-6-phosphate (G6P).

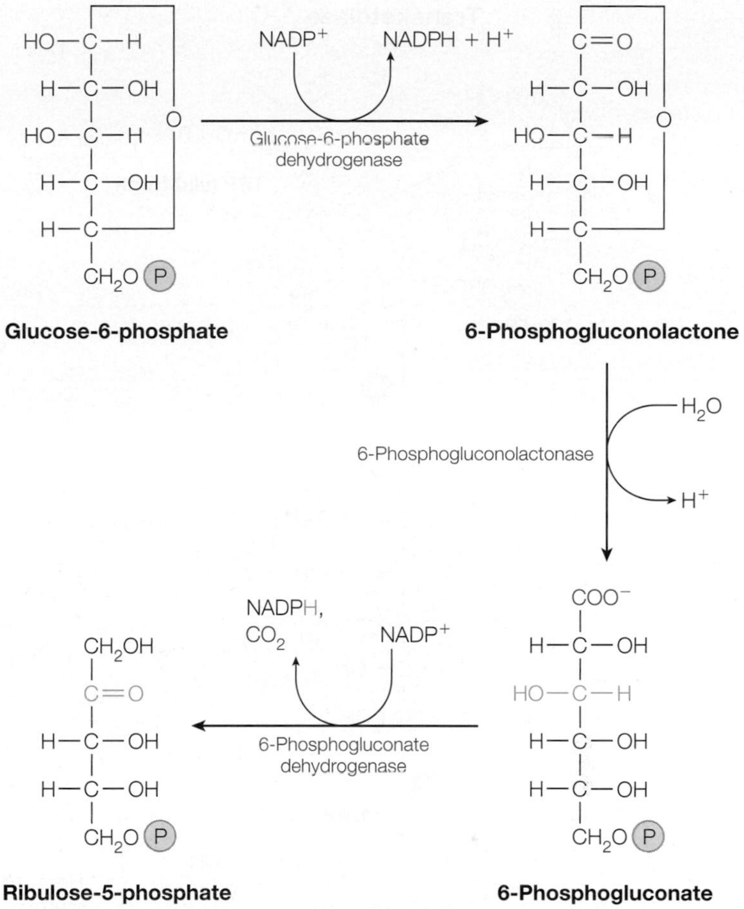

The Nonoxidative Phase: Alternative Fates of Pentose Phosphates

In the nonoxidative phase, some of the ribulose-5-phosphate produced in the oxidative phase is converted to ribose-5-phosphate by **phosphopentose isomerase.**

Phosphopentose isomerase reaction

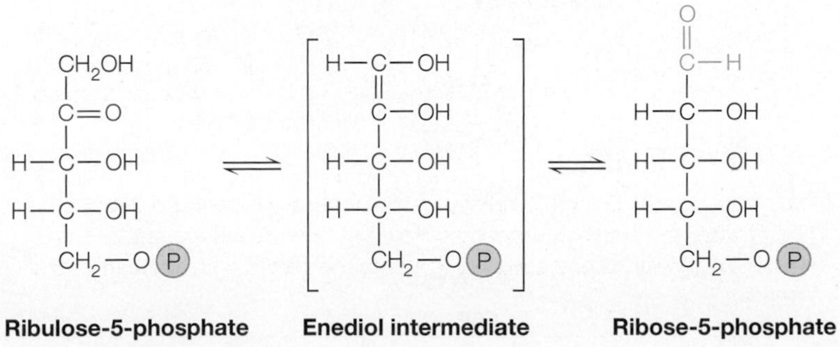

Ribulose-5-phosphate   **Enediol intermediate**   **Ribose-5-phosphate**

Glucose-6-phosphate   **6-Phosphogluconolactone**

Ribulose-5-phosphate   **6-Phosphogluconate**

▲ **FIGURE 12.33 Oxidative phase of the pentose phosphate pathway.** The three reactions of the oxidative phase include two oxidations, each of which produces NADPH.

This reaction, like the isomerization in step 5 of glycolysis, proceeds via an enediol intermediate.

Production of Six-Carbon and Three-Carbon Sugar Phosphates

At this stage, the primary functions of the pentose phosphate pathway—the generation of NADPH and ribose-5-phosphate—have been accomplished. The balanced equation for what has transpired thus far is:

$$\text{Glucose-6-phosphate} + 2NADP^+ \longrightarrow$$
$$\text{ribose-5-phosphate} + CO_2 + 2NADPH + 2H^+$$

Many cells need the NADPH for reductive biosynthesis but do not need the ribose-5-phosphate in such large quantities. How, then, is this ribose-5-phosphate catabolized? The process involves three enzymes in a series of sugar phosphate transformations that may look complicated but that have a simple result. *The reaction sequence converts 3 five-carbon sugar phosphates to 2 six-carbon sugar phosphates and 1 three-carbon sugar phosphate.* The hexose phosphates formed can be catabolized either by recycling through the pentose phosphate pathway or by glycolysis. The triose phosphate is glyceraldehyde-3-phosphate, an intermediate in glycolysis.

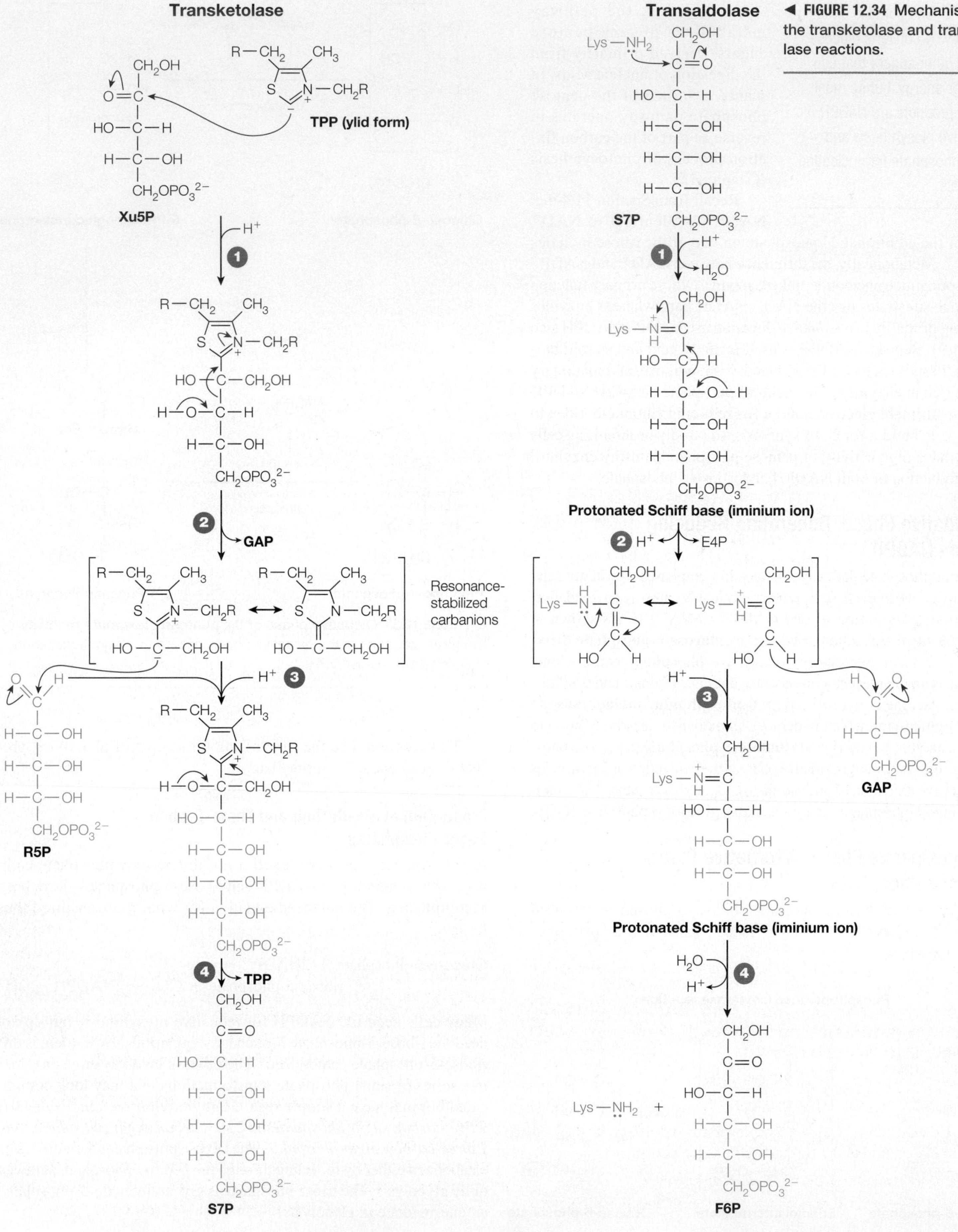

◀ **FIGURE 12.34** Mechanism of the transketolase and transaldolase reactions.

Two of the enzymes that participate in this sequence, *transketolase* and *transaldolase,* catalyze the transfer of two-carbon or three-carbon fragments from one sugar phosphate to another. These two- or three-carbon transfers require cleavage of C—C bonds—a process involving carbanion intermediates that must be stabilized—but transketolase and transaldolase use quite different mechanisms.

Thiamine pyrophosphate (TPP)

Transketolase uses a cofactor, thiamine pyrophosphate (TPP). TPP, which is also used in enzymatic decarboxylations, contains two heterocylic rings: a substituted pyrimidine and a thiazole. The thiazolium (an iminium ion) can be deprotonated at C2 with the aid of a general base in the enzyme active site to form a reactive carbanion at C2, termed an **ylid** (left side of **FIGURE 12.34**). This carbanion can attack the carbonyl carbon of a ketose phosphate, such as xylulose-5-phosphate (Xu5P), giving an addition compound **1**. The addition compound undergoes bond cleavage between C2 and C3, releasing glyceraldehyde-3-phosphate (GAP) **2**, with the thiazolium ring acting as an electron sink because it forms a resonance-stabilized carbanion. This carbanion then attacks the aldehyde carbon of an acceptor aldose phosphate, such as ribose-5-phosphate (R5P), to give a seven-carbon TPP adduct **3**. This intermediate undergoes an elimination reaction to yield sedoheptulose-7-phosphate (S7P) and the TPP carbanion **4**.

The right side of Figure 12.34 illustrates the transaldolase reaction. The enzyme activates the ketose substrate (e.g., S7P) by forming a Schiff base with a lysine residue on the enzyme **1**. Protonation of the Schiff base leads to carbon–carbon bond cleavage, similar to what occurs in reaction 4 of glycolysis (Figure 12.5), with release of the four-carbon erythrose-4-phosphate (E4P) **2**. The dihydroxyacetone unit remains bound as a resonance-stabilized carbanion, which then adds to the carbonyl carbon of glyceraldehyde-3-phosphate in an aldol condensation reaction **3**. Hydrolysis of the protonated Schiff base (an iminium ion) yields the six-carbon product, fructose-6-phosphate (F6P) **4**.

Thus, the transketolase and transaldolase mechanisms represent two different chemical strategies to effect C—C bond cleavage. Both mechanisms use an iminium ion as an electron sink (**FIGURE 12.35**). In addition, both mechanisms involve carbanion intermediates that must be stabilized by resonance. Transketolase uses a cofactor (TPP), while transaldolase uses a protonated Schiff base with an active site lysine to solve the same problem.

In order to write a balanced equation for the pentose phosphate pathway, we must start with three molecules of glucose-6-phosphate passing through the oxidative phase:

$$3 \text{ Glucose-6-phosphate} + 6\text{NADP}^+ + 3\text{H}_2\text{O} \longrightarrow$$
$$3 \text{ pentose-5-phosphate} + 6\text{NADPH} + 6\text{H}^+ + 3\text{CO}_2$$

The nonoxidative phase results in the conversion of three pentose phosphates to 2 six-carbon and 1 three-carbon sugar phosphates:

$$2 \text{ Xylulose-5-phosphate} + \text{ribose-5-phosphate} \longrightarrow$$
$$2 \text{ fructose-6-phosphate} + \text{glyceraldehyde-3-phosphate}$$

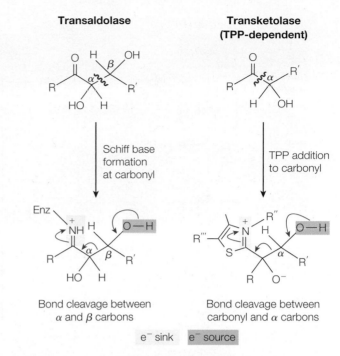

▲ **FIGURE 12.35** C—C bond cleavage in transketolase and transaldolase reactions. Yellow, electron sink; green, electron source.

Thus, the balanced equation for the entire pathway is as follows:

$$3 \text{ Glucose-6-phosphate} + 6\text{NADP}^+ + 3\text{H}_2\text{O} \longrightarrow$$
$$2 \text{ fructose-6-phosphate} + \text{glyceraldehyde-3-phosphate}$$
$$+ 6\text{NADPH} + 6\text{H}^+ + 3\text{CO}_2$$

Tailoring the Pentose Phosphate Pathway to Specific Needs

In the equation for the overall reaction, three hexose phosphates yield two hexose phosphates, one triose phosphate, and three molecules of CO_2. In a formal sense, therefore, the pathway can be seen as a means to oxidize the six carbons of glucose-6-phosphate to CO_2, just as occurs when the products of glycolysis enter the citric acid cycle.

The pentose phosphate pathway, however, is *not* primarily an energy-generating pathway. The actual fate of the sugar phosphates depends on the metabolic needs of the cell in which the pathway is occurring. If the primary need is for nucleotide and nucleic acid synthesis, then the major product is ribose-5-phosphate, and most of the rearrangements of the nonoxidative phase do not take place (**FIGURE 12.36(a)**). If the primary need is for the generation of NADPH (for fatty acid or steroid synthesis), then the nonoxidative phase generates compounds that can easily be reconverted to glucose-6-phosphate, for subsequent passage through the oxidative phase (Figure 12.36(b)). In this mode, repeated turns of the cycle result ultimately in the complete oxidation of glucose-6-phosphate to CO_2 and water, while generating the maximum number of reducing equivalents.

Finally, in a cell with moderate needs for both NADPH and pentose phosphates, the fructose-6-phosphate and glyceraldehyde-3-phosphate produced in the nonoxidative phase can be further catabolized by glycolysis and the citric acid cycle (Figure 12.36(c)). Because of the cell's multiple metabolic needs for biosynthesis, it is unlikely that any one of these three modes operates exclusively in any one cell.

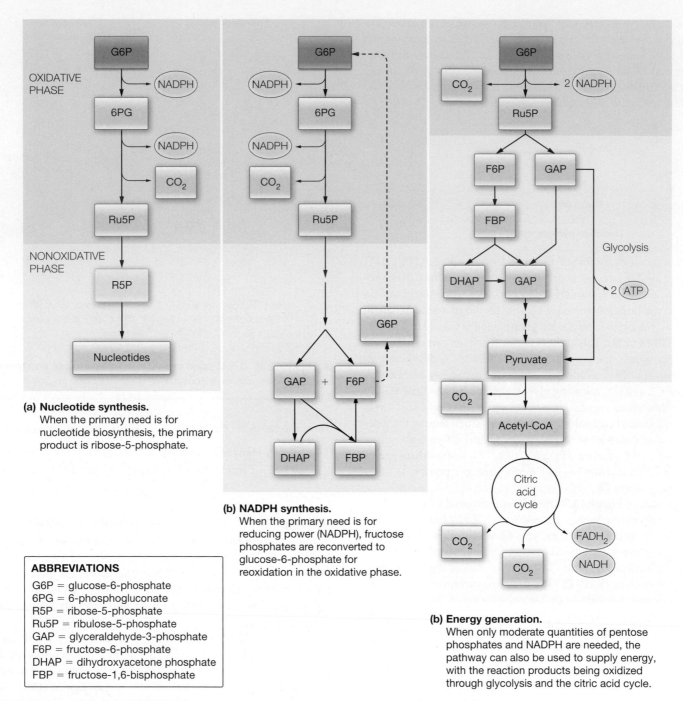

(a) Nucleotide synthesis.
When the primary need is for nucleotide biosynthesis, the primary product is ribose-5-phosphate.

(b) NADPH synthesis.
When the primary need is for reducing power (NADPH), fructose phosphates are reconverted to glucose-6-phosphate for reoxidation in the oxidative phase.

(b) Energy generation.
When only moderate quantities of pentose phosphates and NADPH are needed, the pathway can also be used to supply energy, with the reaction products being oxidized through glycolysis and the citric acid cycle.

ABBREVIATIONS

G6P = glucose-6-phosphate
6PG = 6-phosphogluconate
R5P = ribose-5-phosphate
Ru5P = ribulose-5-phosphate
GAP = glyceraldehyde-3-phosphate
F6P = fructose-6-phosphate
DHAP = dihydroxyacetone phosphate
FBP = fructose-1,6-bisphosphate

▲ **FIGURE 12.36 Alternative pentose phosphate pathway modes.** The pentose phosphate pathway has different modes of operation to meet varying metabolic needs.

Regulation of the Pentose Phosphate Pathway

The pentose phosphate pathway competes with glycolysis for glucose-6-phosphate. Whereas glycolysis is regulated primarily by energy charge and fuel availability, flux through the pentose phosphate pathway is sensitive to the $NADP^+/NADPH$ ratio of the cell. The first enzyme of the pathway, glucose-6-phosphate dehydrogenase, represents the committed step, and its activity controls flux through the entire pentose phosphate pathway. Glucose-6-phosphate dehydrogenase is regulated by the availability of $NADP^+$. If the $NADP^+/NADPH$ ratio is low, indicating that the cell has plenty of reducing power, glucose-6-phosphate dehydrogenase activity

● **CONNECTION** Cancer cells are especially dependent on the pentose phosphate pathway because it generates ribose-5-phosphate, to support their high rate of nucleic acid synthesis, and NADPH, which is required for the synthesis of fatty acids and for cell survival under stress conditions.

will be low and the pathway will not divert glucose-6-phosphate from glycolysis. If the cell needs more reducing equivalents, however, the high $NADP^+/NADPH$ ratio will stimulate flux through glucose-6-phosphate dehydrogenase, regenerating the necessary NADPH.

Human Genetic Disorders Involving Pentose Phosphate Pathway Enzymes

The pentose phosphate pathway is particularly active in the generation of reducing power in the red blood cells of vertebrates. The importance of this activity became apparent through investigation of a fairly widespread human genetic disorder: a deficiency of glucose-6-phosphate dehydrogenase.

During World War II, the antimalarial drug primaquine was prophylactically administered to members of the armed forces. As a result, a significant proportion of servicemen suffered a severe hemolytic anemia (massive destruction of red blood cells). They were also sensitive to a variety of compounds that, like primaquine, generate oxidative stress, as manifested by the appearance of hydrogen peroxide and organic peroxides in red cells. These individuals were later found to be deficient in glucose-6-phosphate dehydrogenase.

● **CONNECTION** Abnormal sensitivity to antimalarial drugs was shown to result from mutations affecting glucose-6-phosphate dehydrogenase.

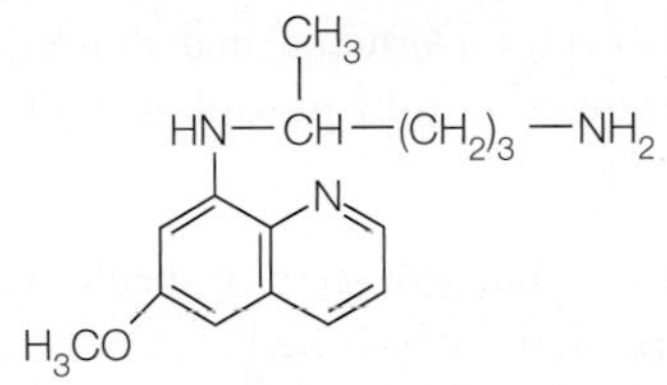

Primaquine

Normally, peroxides are inactivated via reduction by **glutathione,** which is the tripeptide γ-glutamylcysteinylglycine:

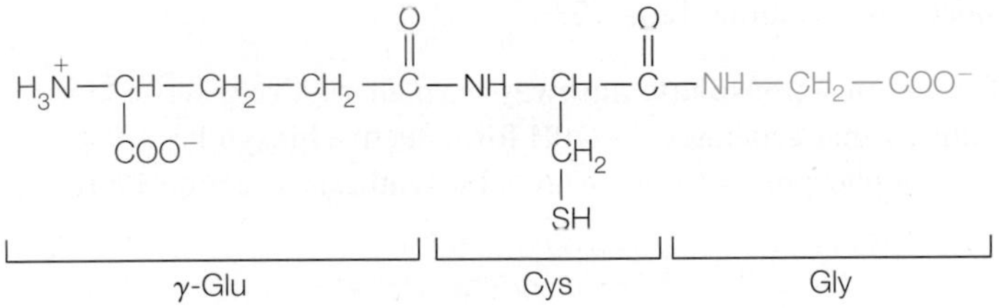

Glutathione

Glutathione (GSH) is abundant in most cells, and because of its free thiol group, it represents a major protective mechanism against oxidative stress. For example, it helps keep cysteine thiol groups in proteins in the reduced state. If two thiol groups become oxidized, they can be reduced nonenzymatically by glutathione.

● **CONCEPT** Glutathione, an abundant thiol-containing tripeptide, is a major intracellular reductant.

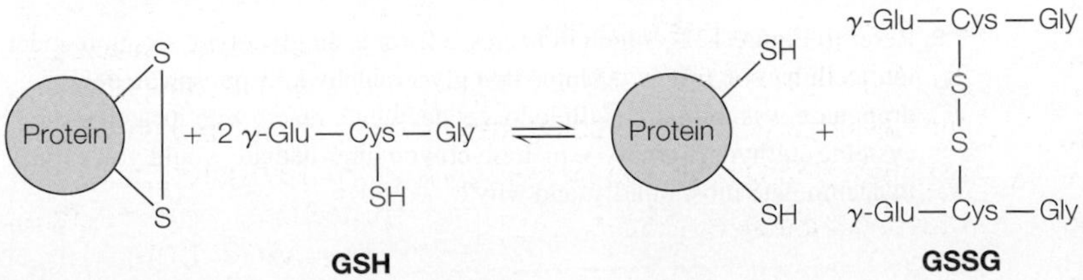

As noted, glutathione also reduces peroxides; this is an enzymatic reaction, catalyzed by **glutathione peroxidase** (Section 14.9).

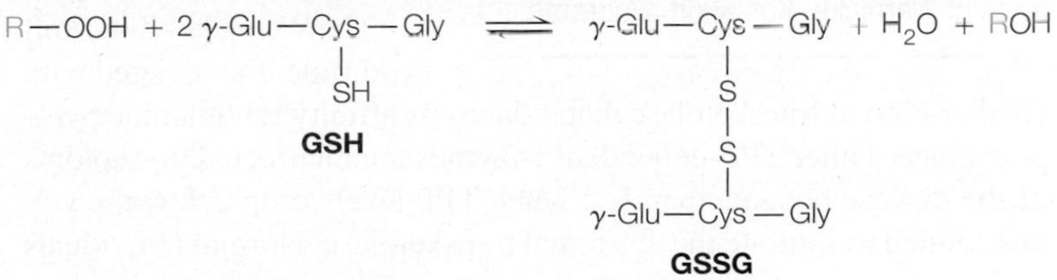

In both of these processes, glutathione becomes oxidized. Oxidized glutathione (GSSG) is reduced by the NADPH-dependent enzyme **glutathione reductase.**

$$\text{GSSG} + \text{NADPH} + \text{H}^+ \longrightarrow 2\text{GSH} + \text{NADP}^+$$

In healthy cells, where the NADPH/NADP$^+$ ratio is high, this reaction is essentially unidirectional, so that the ratio of reduced glutathione (GSH) to oxidized glutathione (GSSG) in most cells is about 500 to 1.

In the erythrocyte, a particularly important role of glutathione is to maintain hemoglobin in the reduced (Fe^{2+}) state because methemoglobin (Fe^{3+}) cannot bind O$_2$ (Table 7.2). Therefore, the erythrocyte is especially sensitive to the depletion of reduced glutathione. And because the pentose phosphate pathway is the major pathway for the generation of NADPH, the erythrocyte is especially vulnerable to conditions that impair flux through this pathway and thereby lower intracellular NADPH levels. Thus, the individuals who were deficient in glucose-6-phosphate dehydrogenase were the ones most sensitive to oxidative stress caused by primaquine.

● **CONNECTION** Glucose-6-phosphate dehydrogenase deficiency, like the sickle-cell trait, confers resistance to malaria caused by *Plasmodium falciparum* (Section 7.14). Thus, the deficiency has a positive survival value in tropical and subtropical regions of the world, where malaria is common. This explains why glucose-6-phosphate dehydrogenase deficiency is seen most frequently among individuals of African or Mediterranean descent.

In most cases of glucose-6-phosphate dehydrogenase deficiency, the enzyme in red cells is not totally inactive but instead is decreased in activity by about 10-fold. Individuals with this deficiency are asymptomatic until stressed. That is, they are asymptomatic until primaquine or a related agent generates enough peroxides that the available GSH becomes depleted. Reduction of the resultant GSSG back to GSH is impaired because NADPH levels are inadequate to allow glutathione reductase to function. This causes methemoglobin (Fe^{3+}) to accumulate at the expense of hemoglobin (Fe^{2+}), which in turn changes the structure of the cell, weakening the membrane and rendering it sensitive to rupture, or hemolysis.

Another disorder related to the pentose phosphate pathway is the **Wernicke–Korsakoff syndrome.** This mental disorder is coupled with loss of memory and partial paralysis and develops when affected individuals suffer a moderate thiamine

● **CONNECTION** A defect in transketolase that increases its apparent K_M for TPP is responsible for the neurological symptoms of Wernicke–Korsakoff syndrome.

an alteration of transketolase that reduces its affinity for thiamine pyrophosphate. Other TPP-dependent enzymes are unaffected. Symptoms of the disease become manifest when TPP levels drop below the values needed to saturate the abnormal transketolase. Normal individuals

deficiency. The symptoms often appear in alcoholics, whose diets are apt to be vitamin deficient.

The Wernicke–Korsakoff syndrome is associated with

contain a transketolase that binds TPP strongly enough that no change in enzyme function occurs as a result of these slight to moderate thiamine deficiencies.

Both glucose-6-phosphate dehydrogenase deficiency and the Wernicke–Korsakoff syndrome, like sickle-cell disease (Chapter 7), illustrate the interdependence of genetic and environmental factors in the onset of clinical disease. Symptoms of the hereditary change express themselves only after some kind of moderate stress that does not affect normal individuals.

Summary

• **Glycolysis** is the central pathway by which energy is extracted from carbohydrates. A two-stage pathway leads from glucose to pyruvate in both aerobic and anaerobic cells. Under anaerobic conditions, the overall pathway proceeds as a **fermentation,** with no net change in oxidation state of the products compared to that of the substrate. Pyruvate is reduced by NADH, so that NAD^+ is regenerated. Under aerobic conditions, NAD^+ can be regenerated by mitochondrial oxidation of NADH, and pyruvate can be further oxidized to yield much more ATP (**respiration**). In the energy investment phase, ATP is used to synthesize a six-carbon sugar phosphate that is split to yield two triose phosphates; in the energy generation phase, the energy of two high-energy compounds is used to drive ATP synthesis from ADP. The reactions catalyzed by **phosphofructokinase, pyruvate kinase,** and **hexokinase** are the major sites for control of the pathway. Control is related to the energy needs of the cell and occurs via allosteric mechanisms. Conditions of low-energy charge stimulate the pathway, and conditions of energy abundance inhibit the pathway. (Sections 12.1–12.4)

• **Gluconeogenesis** is the synthesis of carbohydrate from noncarbohydrate three-carbon and four-carbon compounds.

Although gluconeogenesis uses many of the same enzymes used in glycolysis, it is not a simple reversal of glycolysis. To ensure thermodynamic favorability, specific gluconeogenesis reactions evolved to bypass the three irreversible steps in glycolysis. Coordinate regulation of gluconeogenesis and glycolysis occurs at the sites of these three substrate cycles. Hormonal and allosteric mechanisms are involved, with **fructose-2,6-bisphosphate** being a key regulator. (Sections 12.5–12.6)

• **Glycogen,** the intracellular storage polysaccharide in animals, is mobilized by a hormonally controlled metabolic cascade involving cAMP and protein kinases that set in motion events that activate the breakdown of glycogen, providing glucose-1-phosphate. Glycogen phosphorylase is the rate-limiting step of glycogenolysis. Glycogen synthase is the rate-limiting step for glycogen synthesis. These enzymes are reciprocally regulated by hormonal and nonhormonal processes. (Sections 12.8–12.9)

• The **pentose phosphate pathway** is an alternative glucose oxidative pathway that generates NADPH for reductive biosynthesis and pentose phosphates for nucleotide biosynthesis. (Section 12.10)

Problems

Enhanced by
Mastering Chemistry
for Biochemistry

Mastering Chemistry for Biochemistry provides select end-of-chapter problems and feedback-enriched tutorial problems, animations, and interactive figures to deepen your understanding of complex topics while practicing problem solving.

Answers to red problems are available in the Answer Appendix.

1. Intracellular concentrations in resting muscle are as follows: fructose-6-phosphate, 1.0 mM; fructose-1,6-bisphosphate, 10 mM; AMP, 0.1 mM; ADP, 0.5 mM; ATP, 5 mM; and P_i, 10 mM. Is the phosphofructokinase reaction in muscle *more* or *less* exergonic than under standard conditions? By how much?

2. Methanol is highly toxic, not because of its own biological activity but because it is converted metabolically to formaldehyde, through action

of alcohol dehydrogenase. Part of the medical treatment for methanol poisoning involves administration of large doses of ethanol. Explain why this treatment is effective.

3. Refer to Figure 12.8, which indicates ΔG for each glycolytic reaction under intracellular conditions. Assume that glyceraldehyde-3-phosphate dehydrogenase was inhibited with iodoacetate, which reacts with its active site cysteine sulfhydryl group. Which glycolytic intermediate would you expect to accumulate most rapidly, and why?

4. In different organisms, sucrose can be cleaved either by hydrolysis or by phosphorolysis. Calculate the ATP yield per mole of sucrose metabolized by anaerobic glycolysis starting with (a) hydrolytic cleavage and (b) phosphorolytic cleavage.

5. Suppose it were possible to label glucose with ^{14}C at any position or combination of positions. For yeast fermenting glucose to ethanol, which form or forms of labeled glucose would give the *most* radioactivity in CO_2 and the *least* in ethanol?

6. Write balanced chemical equations for each of the following: (a) anaerobic glycolysis of 1 mole of sucrose, cleaved initially by sucrose phosphorylase; (b) aerobic glycolysis of 1 mole of maltose; (c) fermentation of one glucose residue in starch to ethanol, with the initial cleavage involving α-amylase.

7. Because of the position of arsenic in the periodic table, arsenate (AsO_4^{3-}) is chemically similar to inorganic phosphate and is used by phosphate-requiring enzymes as an alternative substrate. Organic arsenates are quite unstable, however, and spontaneously hydrolyze. Arsenate is known to inhibit ATP production in glycolysis. Identify the target enzyme, and explain the mechanism of inhibition.

8. As early as the 1930s, it was known that frog muscles could still contract when glycolysis was inhibited. Where did the ATP come from to drive these contractions?

9. Suppose that you made some wine whose alcohol content was 10% w/v (i.e., 10 g of ethanol per 100 mL of wine). The initial fermentation mixture would have had to contain what molar concentration of glucose or its equivalent to generate this much ethanol? Is it likely that an initial fermentation mixture would contain that much glucose? In what other forms might the fermentable carbon appear?

10. Briefly discuss why each of the three common forms of galactosemia involves impaired utilization of galactose. Which metabolic process is blocked in each condition?

11. Some anaerobic bacteria use alternative pathways for glucose catabolism that convert glucose to acetate rather than to pyruvate. Shown below is one possible metabolic pathway. The first part of this pathway (glucose to fructose-1,6-bisphosphate) is identical to the glycolytic pathway. In the second part of the alternative pathway, Enzymes 1–6 all have mechanisms/activities analogous to enzymes in glycolysis. Note that there are two C—C bond cleavage reactions in this new pathway: $A \rightarrow B + C$ (Enzyme 1) and $C \rightarrow B + D$. All the steps where ATP is consumed or generated have been shown; however, the addition or loss of NAD^+/NADH, P_i, H_2O, or H^+ has not been shown explicitly. Draw the structures for the intermediates B, F, G, H, and I, and include other reaction participants as needed.

12. Write a pathway leading from glucose to lactose in mammary gland, and write a balanced equation for the overall pathway.

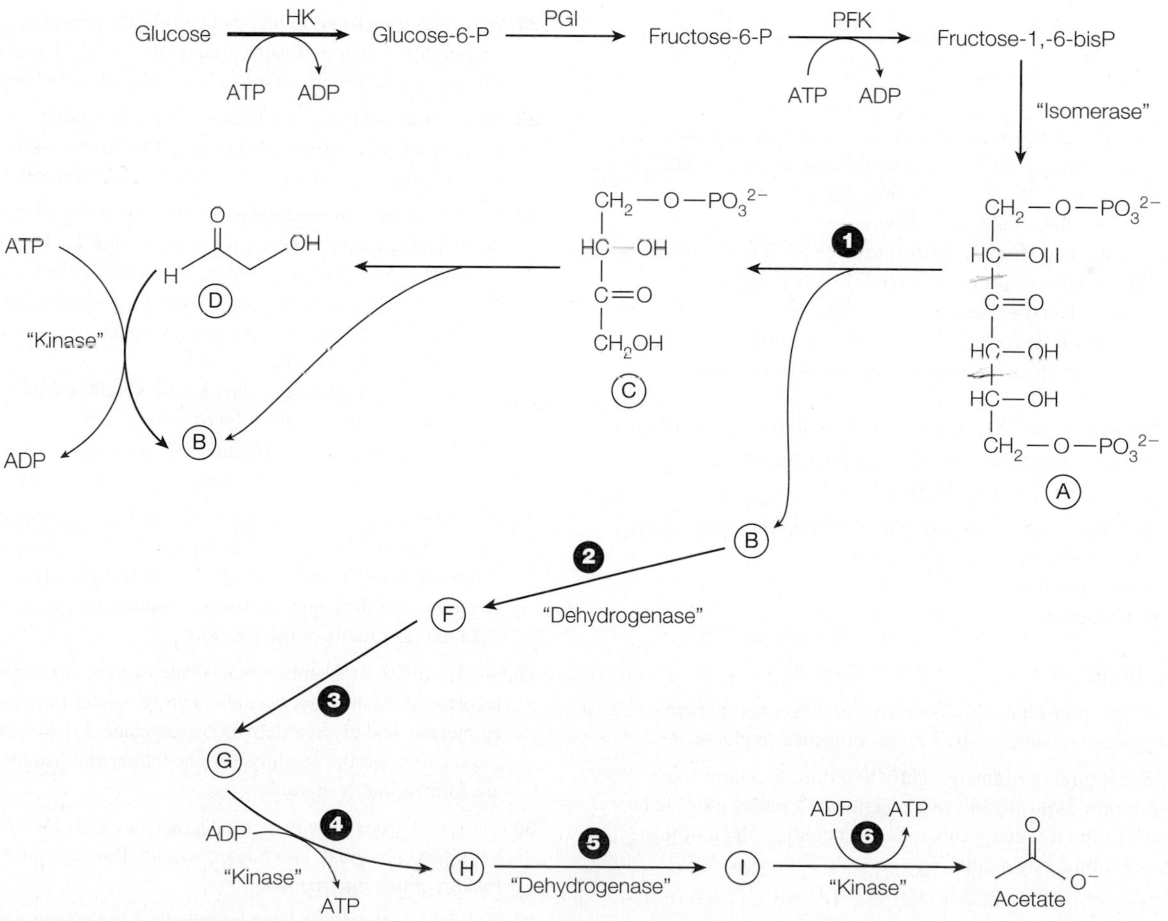

13. Sketch a curve that would describe the expected behavior of phosphofructokinase activity as a function of the adenylate energy charge.

14. Enolase has a strict requirement for two Mg^{2+} ions in its active site. Propose a role for these ions in the catalytic mechanism of the enzyme.

15. The muscle isozyme of lactate dehydrogenase is inhibited by lactate. Steady-state kinetic analysis yielded the following data, with lactate either absent or present at a fixed concentration.

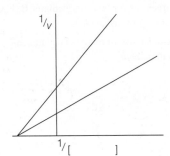 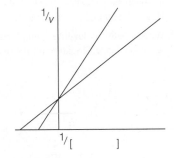

(a) Pyruvate is the substrate whose concentration is varied in one plot, NADH in the other. Identify each. Use an arrow and the appropriate letter (b, c, d, or e) to identify each of the following.
(b) Reciprocal of V_{max} for the uninhibited enzyme.
(c) The line representing data obtained in the presence of lactate acting as a competitive inhibitor with respect to the variable substrate.
(d) The line representing data obtained in the presence of lactate acting as a noncompetitive inhibitor with respect to the variable substrate.
(e) Reciprocal of K_M in the presence of lactate acting as a competitive inhibitor.
(f) If K_M for NADH is 2×10^{-5} M, then which of the following is the most appropriate NADH concentration to use when determining K_M for pyruvate: 10^{-7} M, 10^{-6} M, 10^{-5} M, 10^{-4} M, or 10^{-3} M?

16. How many ATP equivalents are consumed in the conversion of each of the following to a glucosyl residue in glycogen?
(a) Dihydroxyacetone phosphate
(b) Fructose-1,6-bisphosphate
(c) Pyruvate
(d) Glucose-6-phosphate

17. How many high-energy phosphates are generated or consumed in converting (a) 1 mole of glucose to lactate? (b) 2 moles of lactate to glucose?

18. Avidin is a protein that binds extremely tightly to biotin, so avidin is a potent inhibitor of biotin-requiring enzyme reactions. Consider glucose biosynthesis from each of the following substrates and predict which of these pathways would be inhibited by avidin.
(a) Lactate
(b) Oxaloacetate
(c) Malate
(d) Fructose-6-phosphate
(e) Phosphoenolpyruvate

19. $^{14}CO_2$ was bubbled through a suspension of liver cells that was undergoing gluconeogenesis from lactate to glucose. Which carbons in the glucose molecule would become radioactive?

20. Write a balanced equation for each of the following reactions or reaction sequences.
(a) The reaction catalyzed by PFK-2
(b) The conversion of 2 moles of oxaloacetate to glucose
(c) The conversion of glucose to UDP-Glc
(d) The conversion of 2 moles of glycerol to glucose
(e) The conversion of 2 moles of malate to glucose-6-phosphate

21. Sketch curves for reaction velocity versus [fructose-6-phosphate] for the phosphorylated *and* nonphosphorylated forms of PFK-2 in liver.

22. Predict the effect of each of the following mutants on the *rate of glycolysis* in liver cells (increase, decrease, no change):
(a) Loss of the allosteric site for ATP in PFK-1
(b) Loss of the binding site for citrate in PFK-1
(c) Loss of the phosphatase domain of PFK-2/FBPase-2
(d) Loss of the binding site for fructose-1,6-bisphosphate in pyruvate kinase
(e) Loss of acetyl-CoA binding site in pyruvate carboxylase

23. Based on information presented on pages 408–409, sketch curves relating glycogen synthase reaction velocity to [UDP-glucose], for both the *a* and *b* forms of the enzyme, in the presence and absence of glucose-6-phosphate.

24. Glycogen synthesis and breakdown are regulated primarily at the hormonal level. However, important *nonhormonal* mechanisms also control the rates of synthesis and mobilization. Describe these nonhormonal regulatory processes.

25. Why does it make good metabolic sense for phosphoenolpyruvate carboxykinase, rather than pyruvate carboxylase, to be the primary target for regulating gluconeogenesis at the level of control of enzyme synthesis?

26. What is the metabolic significance of the following observations? (a) Only the liver form of pyruvate kinase is inhibited by alanine, and (b) only gluconeogenic tissues contain appreciable levels of glucose-6-phosphatase.

27. Write a one-sentence explanation for each of the following statements.
(a) In liver, glucagon stimulates glycogen breakdown via cAMP. Although you might expect glucagon to stimulate catabolism of the glucose formed as well, glucagon *inhibits* glycolysis and stimulates gluconeogenesis in liver.
(b) An individual with a glucose-6-phosphatase deficiency suffers from chronic hypoglycemia.
(c) The action of phosphorylase kinase simultaneously activates glycogen breakdown and inhibits glycogen synthesis.
(d) The presence in liver of glucose-6-phosphatase is essential to the function of the liver in synthesizing glucose for use by other tissues.

28. Write a balanced chemical equation for the pentose phosphate pathway in the first two modes depicted in Figure 12.36, where (a) ribose-5-phosphate synthesis is maximized and (b) NADPH production is maximized, by conversion of the sugar phosphate products to glucose-6-phosphate for repeated operations of the pathway.

29. [1-^{14}C]Ribose-5-phosphate is incubated with a mixture of purified transketolase, transaldolase, phosphopentose isomerase, phosphopentose epimerase, and glyceraldehyde-3-phosphate. Predict the distribution of radioactivity in the erythrose-4-phosphate and fructose-6-phosphate that are formed in this mixture.

30. Pyruvate carboxylase is thought to activate CO_2 by ATP, through formation of carboxyphosphate as an intermediate. Propose a mechanism for the formation of this intermediate.

31. Xylulose-5-phosphate is an intermediate in the pentose phosphate pathway (see Figure 12.34). As xylulose-5-phosphate levels rise in response to excess glucose shunting through the pentose phosphate pathway, does flux through glycolysis increase or decrease?

32. Although most enzymes are quite specific, they can catalyze side reactions with compounds that are structurally similar to their physiological substrates, but usually at much slower rates. For example, glyceraldehyde-3-phosphate dehydrogenase (GAPDH), which normally catalyzes the oxidative phosphorylation of glyceraldehyde-3-phosphate, can slowly convert erythrose-4-phosphate, an intermediate in the pentose phosphate pathway, to 1,4-bisphosphoerythronate:

Erythrose-4-phosphate   **1,4-Bisphosphoerythronate**

Draw a plausible mechanism for this side reaction of GAPDH.

33. Many of the mechanisms involved in the regulation of glycogen metabolism involve covalent modification of enzymes. In particular, reversible phosphorylation/dephosphorylation plays a central role.
(a) Estimate the $\Delta G^{\circ\prime}$ for the dephosphorylation of phosphorylated Ser-14 of glycogen phosphorylase (catalyzed by phosphoprotein phosphatase 1). Explain your reasoning.
(b) Estimate the $\Delta G^{\circ\prime}$ for the phosphorylation of Ser-14 of glycogen phosphorylase by ATP (catalyzed by phosphorylase kinase). Explain your reasoning.
(c) Would you expect both reactions to be favorable under physiological conditions (i.e., to possess a negative ΔG)?

References

For a list of references related to this chapter, see Appendix II.

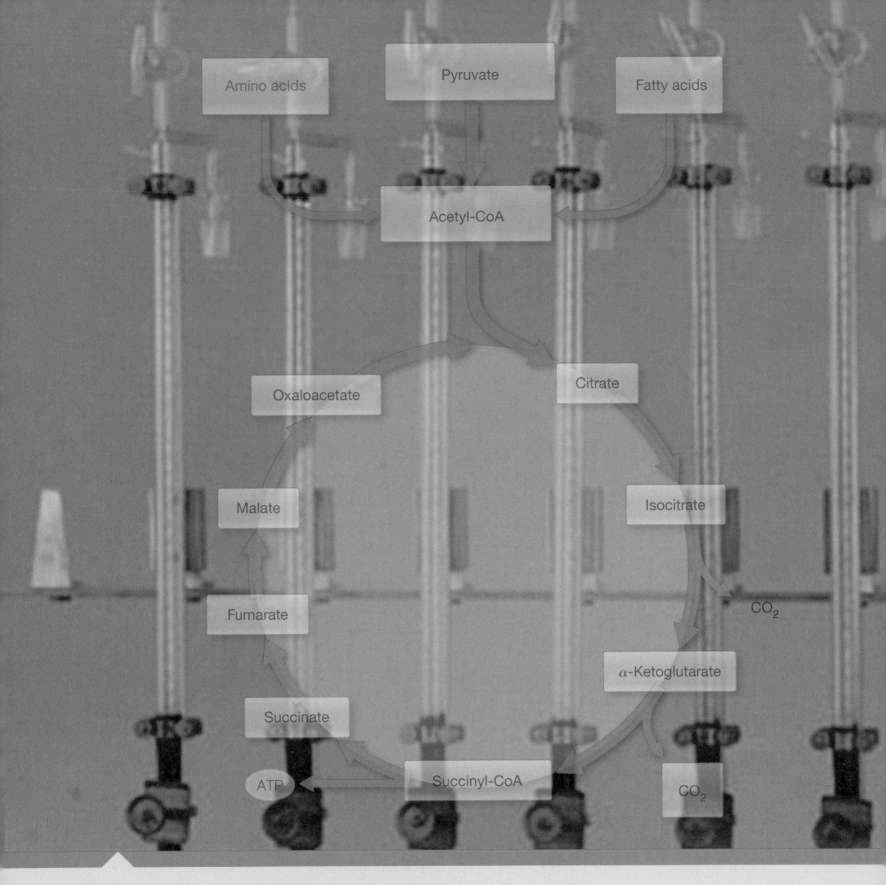

The citric acid cycle is the central oxidative pathway in cellular respiration, in which organic fuels are completely oxidized by molecular oxygen. Much of the early work on this pathway was performed using a manometer such as this to measure the oxygen consumption of tissue homogenates or tissue slices supplied with various organic acids as fuel substrates.

The Citric Acid Cycle

IN CHAPTER 12 we explored the initial, fermentative phase of carbohydrate utilization. The glycolytic pathway leads from glucose to pyruvate. In the absence of oxygen, pyruvate undergoes reductive reactions, and the overall pathway proceeds with no net change in oxidation state—that is, fermentation. Here in Chapter 13, we follow the aerobic reactions by which carbohydrates are ultimately oxidized to carbon dioxide and water via the citric acid cycle. These

● **CONCEPT** The citric acid cycle is a pathway for oxidizing all metabolic fuels.

reactions and the areas of intermediary metabolism that are discussed in this chapter are highlighted in **FIGURE 13.1**.

As we shall see, the citric acid cycle is the central oxidative pathway in respiration, the process by which all metabolic fuels—carbohydrate, lipid, and protein—are catabolized in aerobic organisms and tissues.

We saw in Chapter 12 that a fermentation process such as glycolysis releases only a fraction of the energy available in glucose. This is because ethanol and lactate, like other fermentation products of carbohydrate catabolism, are at an oxidation level similar to that of the starting material, glucose. We can see how inefficient glycolysis is by considering how much energy is actually retained

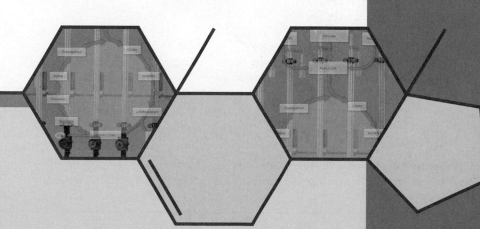

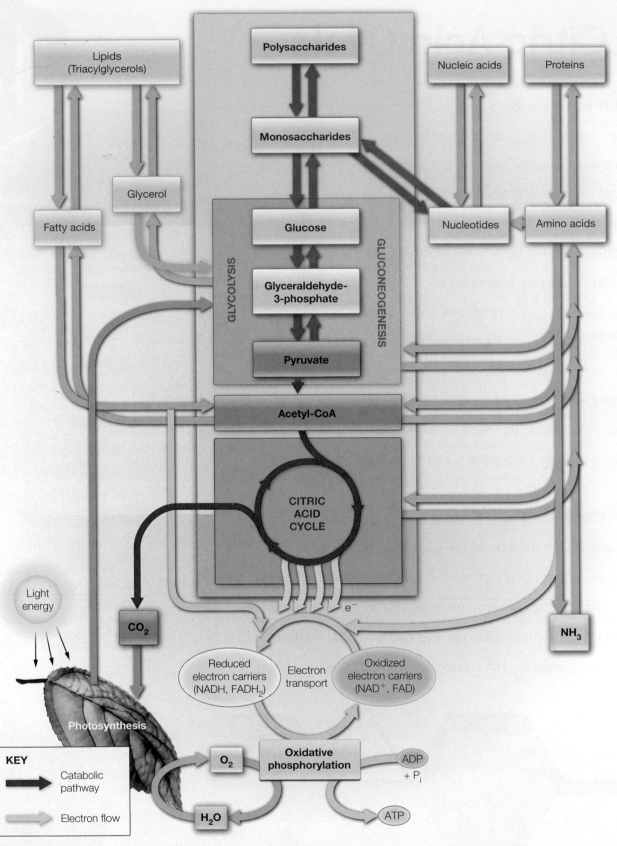

▲ **FIGURE 13.1 Oxidative processes in the generation of metabolic energy.** This overview of intermediary metabolism highlights the citric acid cycle and the entry of fuel from glycolysis. Other pathways that deliver fuel to the citric acid cycle for oxidation (in gray) will be discussed in subsequent chapters.

in the ethanol produced by fermentation of glucose. The complete combustion of ethanol to CO_2 and H_2O releases 1326 kJ/mol of free energy under standard conditions. Because 2 moles of ethanol and 2 moles of CO_2 are produced per mole of glucose, the actual energy yield from the combustion of ethanol generated from one mole of glucose would be $2 \times 1326 = 2652$ kJ/mol. By comparison, the complete combustion of glucose to CO_2 and H_2O releases 2870 kJ/mol of free energy under standard conditions. Thus, 2652/2870, or 92%, of the potential energy stored in the original glucose molecule remains in the fermentation product (ethanol), and only 8% is released. Indeed, all fermentations are characterized by a similarly low-energy yield.

Far more energy is generated if organic fuels like glucose are completely oxidized to CO_2 in the presence of an inorganic electron acceptor such as molecular oxygen. This process is termed **cellular respiration** to distinguish it from respiration in the sense of breathing. Much of the energy released is used to drive dehydrogenation reactions that generate reduced electron carriers, primarily NADH. These carriers are then reoxidized in the mitochondrial respiratory (electron transport) chain, providing the energy for ATP synthesis through oxidative phosphorylation. The electrons released from the carriers are ultimately transferred to oxygen, which becomes reduced to water. We will learn about the chain of electron carriers and the synthesis of ATP in Chapter 14.

This chapter focuses on the fates of the oxidizable substrates. We'll compare the net energy yield of the citric acid cycle to that of glycolysis alone and see how much more efficient oxidative metabolism is compared to fermentative metabolism. We'll discover important control points for regulation of the citric acid cycle. We'll also consider the evolution of the citric acid cycle and its role in human disease, as well as cellular mechanisms for replacing cycle intermediates. Finally, we conclude with a discussion of the glyoxylate cycle, a variation of the citric acid cycle found in plants that allows for the net synthesis of carbohydrates from fat.

● **CONCEPT** Most of the energy yield from oxidation of substrates in the citric acid cycle is used to generate reduced electron carriers such as NADH. The subsequent reoxidation of these carriers provides the energy for the synthesis of ATP.

13.1 Overview of Pyruvate Oxidation and the Citric Acid Cycle

Let's start with a bird's eye view of pyruvate oxidation and the citric acid cycle.

The Three Stages of Respiration

It is convenient to think of the metabolic oxidation of organic substrates as a three-stage process, schematized in **FIGURE 13.2**.

- Stage ❶ is the generation of a chemically activated two-carbon fragment—the acetyl group of acetyl-coenzyme A, or acetyl-CoA—and a pair of electrons.

- Stage ❷ is the oxidation of those two carbon atoms in the citric acid cycle to form two CO_2 molecules and four pairs of electrons.

- Stage ❸ comprises electron transport and oxidative phosphorylation, in which the reduced electron carriers generated in stages 1 and 2 become reoxidized, providing the energy for the synthesis of ATP.

Stage ❶ is a family of pathways that operate separately on carbohydrate, fat, and protein. Carbon from carbohydrate enters the stage as pyruvate, and the oxidation of pyruvate to acetyl-CoA is described in this chapter. Acetyl-CoA produced in other pathways, such as amino acid and fatty acid catabolism, is oxidized by the same stage ❷ reactions (Figure 13.2). We will explore the other ways in which acetyl-CoA is produced in subsequent chapters.

In bacteria, the enzymes for pyruvate oxidation and the citric acid cycle are located in the cytoplasm and on the plasma membrane. In eukaryotic cells, which contain specialized organelles with different metabolic functions, respiration takes place in mitochondria. Reactions of stages ❶ and ❷ occur within the interior, **matrix** compartment of the mitochondrion (**FIGURE 13.3(a)**). Reactions of stage ❸, electron transport and oxidative phosphorylation, are catalyzed by membrane-bound enzymes in the inner mitochondrial membrane. This inner membrane is extensively stacked and folded into projections, called **cristae,** that greatly expand its surface area (Figure 13.3(b)). Most of the enzymes of the citric acid cycle are soluble proteins in the matrix, but one is a membrane protein bound to the matrix side of the inner membrane. These structural and biochemical relationships are discussed further when we explore the third stage of respiration in Chapter 14.

❶ In stage 1, carbon from metabolic fuels is incorporated into acetyl-CoA.

❷ In stage 2, the citric acid cycle oxidizes acetyl-CoA to produce CO_2, reduced electron carriers, and a small amount of ATP.

❸ In stage 3, the reduced electron carriers are reoxidized, providing energy for the synthesis of additional ATP.

Amino acids · Pyruvate · Fatty acids

CO_2

Acetyl-CoA

CO_2 · NH_3

CITRIC ACID CYCLE

Oxaloacetate · Citrate · Isocitrate · α-Ketoglutarate · CO_2 · Succinyl-CoA · CO_2 · Succinate · ATP · Fumarate · Malate

Reduced electron carriers (NADH, $FADH_2$) · Electron transport · Oxidized electron carriers (NAD^+, FAD)

O_2 · **Oxidative phosphorylation** · ADP + P_i · H_2O · ATP

▲ **FIGURE 13.2** The three stages of respiration.

Chemical Strategy of the Citric Acid Cycle

To fully understand the chemical strategy that underlies substrate oxidation in the citric acid cycle, we first briefly review the oxidation and reduction of organic compounds as introduced at the end of Chapter 3 and touched upon in Chapter 11. Quantitative aspects of biological oxidations are presented in Chapter 14.

Oxidation involves the loss of electrons from a substrate; that substrate is the *electron donor,* and the electrons are transferred to an *electron acceptor,* which thereby becomes reduced. Free electrons cannot exist in the cell; electrons released in an enzyme-catalyzed oxidation must be transferred to specialized electron carriers (such as NAD^+ or FAD). The oxidized substrate and the electron acceptor will have different affinities for electrons. This affinity difference drives an exergonic electron flow, releasing free energy that can be captured by the cell—ultimately in the form of ATP.

Carbon atoms become oxidized either through loss of a hydride ion (H^-) or through combination with oxygen. The latter process removes electrons from the shell around a carbon nucleus because the electronegativity of the oxygen draws shared electrons toward its own nucleus. Similarly, when an organic compound loses a hydride ion, it loses both shared electrons associated with that C—H bond. Thus, either process involves a loss of electrons from the carbon atom undergoing oxidation. Formally, the two processes are equivalent.

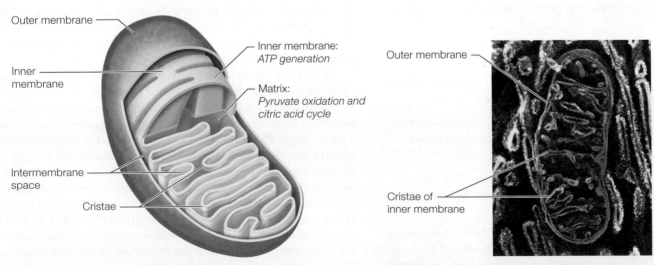

(a) Schematic of a mitochondrion.

Outer membrane · Inner membrane · Intermembrane space · Cristae · Inner membrane: *ATP generation* · Matrix: *Pyruvate oxidation and citric acid cycle*

Outer membrane · Cristae of inner membrane

(b) Colored scanning electron micrograph of a single mitochondrion in the cytoplasm of an intestinal epithelial cell. The stacked and folded cristae are clearly extensions of the inner membrane.

▲ **FIGURE 13.3** Structure of the mitochondrion.

● **CONCEPT** Dehydrogenases catalyze the most common substrate oxidations, which involve the loss of hydrogen.

A point of potential confusion arises in naming enzymes that catalyze oxidation reactions. Because most metabolic oxidations involve loss of hydrogen (typically a hydride ion, H^-, plus a proton, H^+) from the electron donor, we call enzymes that catalyze those reactions **dehydrogenases.**

Referring to **FIGURE 13.4**, let's closely examine the citric acid cycle, focusing on the metabolic fates of the two carbons that enter the cycle (at reaction ❶). These carbons, the acetyl group of acetyl-coenzyme A, are transferred to a four-carbon dicarboxylic acid, **oxaloacetate,** to yield a six-carbon tricarboxylic acid, **citrate** (reaction ❶). Citrate enters into a series of reactions during which two carbons are released as CO_2 (reactions ❸ and ❹) and the remaining four carbons are converted back to oxaloacetate (reactions ❺, ❻, ❼, and ❽), which is ready to begin the process again. Hence, the cyclic nature of the pathway: oxaloacetate

is present at the beginning, to react with an activated two-carbon fragment, and at the end, after two carbons have been oxidized to CO_2. Thus oxaloacetate, and indeed the entire cycle, acts catalytically in the oxidation of acetyl-CoA to CO_2. Note also that of the eight reactions shown in Figure 13.4 , four are oxidations (reactions ❸, ❹, ❻, and ❽), which together generate eight reducing equivalents (all derived from the acetyl-CoA) in the form of three $NADH/H^+$ and one $FADH_2$.

The oxidation of acetyl-CoA to 2 CO_2 would seem to be a relatively simple transformation. Why, then, do cells use such a complicated pathway? The answer lies in the chemistry: oxidation of acetyl-CoA requires C—C bond cleavage, a difficult reaction for the two-carbon acetyl group. As discussed in Chapter 12, C—C bond cleavage is much easier if a carbonyl group is nearby to stabilize the carbanionic transition state, as we saw in the cleavage of fructose-1,6-bisphosphate in glycolysis (see Figure 12.5). The strategy used in the citric acid cycle is to metabolize acetyl-CoA through a series of β-keto acid and α-keto acid intermediates, which are readily

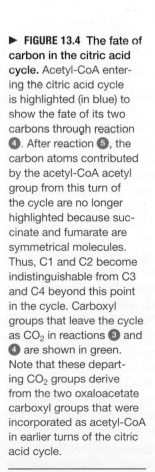

▶ **FIGURE 13.4 The fate of carbon in the citric acid cycle.** Acetyl-CoA entering the citric acid cycle is highlighted (in blue) to show the fate of its two carbons through reaction ❹. After reaction ❺, the carbon atoms contributed by the acetyl-CoA acetyl group from this turn of the cycle are no longer highlighted because succinate and fumarate are symmetrical molecules. Thus, C1 and C2 become indistinguishable from C3 and C4 beyond this point in the cycle. Carboxyl groups that leave the cycle as CO_2 in reactions ❸ and ❹ are shown in green. Note that these departing CO_2 groups derive from the two oxaloacetate carboxyl groups that were incorporated as acetyl-CoA in earlier turns of the citric acid cycle.

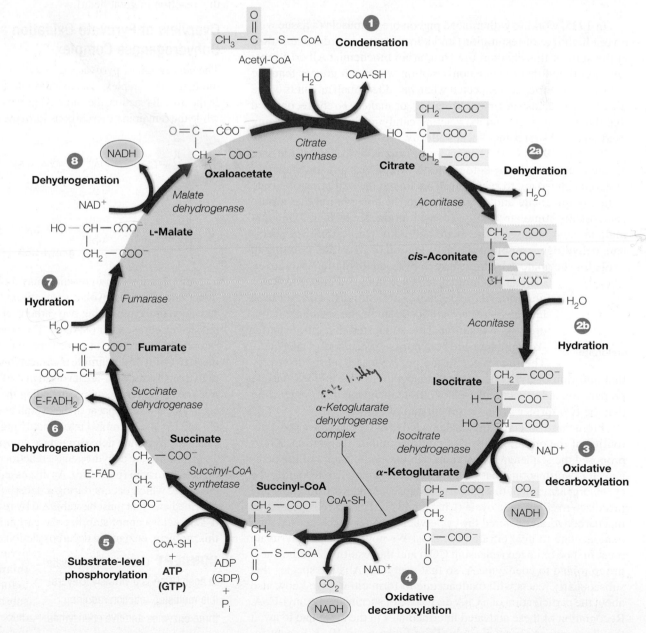

decarboxylated in enzyme-catalyzed reactions, thereby achieving the necessary C—C bond cleavage.

Discovery of the Citric Acid Cycle

The idea that organic fuels are oxidized via a cyclic pathway was proposed in 1937 by Hans Krebs, on the basis of studies for which he later shared a Nobel Prize with Fritz Lipmann (the discoverer of coenzyme A). Beginning in 1932, Krebs tested the ability of various organic acids to be oxidized by following the rate of oxygen consumption in liver and kidney slices. He found that citrate, succinate, fumarate, malate, and acetate were readily oxidized in these tissues.

Citrate **Succinate** **Fumarate** **Malate** **Acetate**

In 1935, working with minced pigeon breast muscle (a tissue with a very high rate of respiration) and a manometer like the one shown at the start of this chapter, the Hungarian biochemist Albert Szent-Györgyi found that oxygen consumption was much greater than the 1:1 stoichiometric ratio expected when he added small amounts of the dicarboxylic acids succinate, fumarate, or malate. Krebs recognized that these acids were somehow acting catalytically rather than being used up as substrates in a linear pathway.

Then in 1937, Carl Martius and Franz Knoop discovered that citrate is converted to α-ketoglutarate, which was already known to undergo conversion to succinate. This linked the oxidation of citrate to that of succinate and helped explain the observation that citrate catalytically stimulated oxygen consumption. Krebs then found that malonate, an analog of succinate and a known inhibitor of succinate dehydrogenase, blocked the oxidation of pyruvate, pointing to a role for succinate dehydrogenase in pyruvate oxidation. Moreover, malonate-inhibited cells accumulated citrate, α-ketoglutarate, and succinate, indicating that these three intermediates are all upstream of the succinate dehydrogenase reaction. This suggested that citrate and α-ketoglutarate are both normal precursors of succinate.

Malonate

The final piece to the puzzle was Krebs's discovery that addition of pyruvate plus oxaloacetate to this minced muscle preparation led to accumulation of citrate in the medium, suggesting that the two former acids are precursors of citrate.

From these observations and knowledge of the structures and reactivities of the organic acids that could stimulate respiration, Krebs proposed the sequence of reactions and the cyclic nature of the pathway. It is worth noting that Krebs was already primed to recognize the cyclic organization of this pathway because he and Kurt Henseleit had just discovered the urea cycle (Chapter 18) in 1932. Krebs postulated that carbohydrates entered the cycle via pyruvate, which reacted with oxaloacetate to give citrate plus CO_2. We now know that pyruvate must first be oxidized (releasing CO_2) and the resulting 2-carbon acyl group joined to coenzyme A, so that acetyl-CoA is the species that subsequently reacts with oxaloacetate to form citrate. We know also about the participation of a CoA derivative of succinate, succinyl-CoA. Recognition of these activated intermediates in the cycle had to await the discovery of coenzyme A by Fritz Lipmann in 1947. Except for

these changes, the pathway as proposed by Krebs was correct. The reactions are outlined in Figure 13.4.

The citric acid cycle is also known by other names: the Krebs cycle, after its discoverer, and the tricarboxylic acid (TCA) cycle because it was apparent from the outset that tricarboxylic acids are involved as intermediates. Only later did it become clear that citrate was one of those intermediates. The following discussion of the citric acid cycle will focus on mammalian biochemistry, but essentially identical versions of this pathway are found in virtually all aerobic organisms, and it is clearly an evolutionarily ancient pathway.

13.2 Pyruvate Oxidation: A Major Entry Route for Carbon into the Citric Acid Cycle

As noted earlier, pyruvate derived from carbohydrate oxidation is one of the major suppliers of acetyl-CoA for oxidation in the citric acid cycle. The chemistry of this transformation and the enzyme that catalyzes it illustrate several important metabolic themes, and we will thus study this reaction in great detail.

Overview of Pyruvate Oxidation and the Pyruvate Dehydrogenase Complex

The conversion of pyruvate to acetyl-CoA, catalyzed by a specialized multienzyme complex, is an oxidative decarboxylation of an α-keto acid. In the overall reaction, the carboxyl group of pyruvate is released as CO_2, while the remaining two carbons form the acetyl moiety of acetyl-CoA.

$$H_3C-\overset{O}{\overset{||}{C}}-COO^- + NAD^+ + CoA-SH \longrightarrow$$

Pyruvate

$$H_3C-\overset{O}{\overset{||}{C}}-S-CoA + NADH + CO_2$$

Acetyl-CoA $\Delta G^{\circ\prime} = -33.5$ kJ/mol

Although the overall reaction may look straightforward, it is in fact rather complicated, involving decarboxylation of pyruvate, metabolic activation of the remaining two carbons of pyruvate, and generation of a reduced electron carrier (NADH). The process is highly exergonic and is essentially irreversible in vivo. Three enzymes—**pyruvate dehydrogenase (E_1)**, **dihydrolipoamide transacetylase (E_2)**, and **dihydrolipoamide dehydrogenase (E_3)**—are involved in the five-step sequence. In addition, five coenzymes are required, including the two coenzymes—NAD^+ and coenzyme A—that appear in the overall reaction. The three enzymes (E_1, E_2, and E_3) are assembled into a highly organized multienzyme complex called the **pyruvate dehydrogenase complex,** or **PDH complex.**

The pyruvate dehydrogenase reaction involves the decarboxylation of an α-keto acid (pyruvate). As discussed in Section 12.3, C—C bond cleavage, which occurs during a decarboxylation, involves carbanion intermediates that must be stabilized by resonance. Unlike β-keto acids, α-keto acids cannot stabilize the carbanion transition state. To solve this problem, enzymatic decarboxylations of α-keto acid substrates like pyruvate all utilize the coenzyme **thiamine pyrophosphate (TPP)** to form a covalent adduct with the substrate and provide the electron delocalization required to stabilize the carbanion intermediate. This

● **CONCEPT** Oxidation of pyruvate to acetyl-CoA is a virtually irreversible multistep reaction requiring three enzymes and five coenzymes.

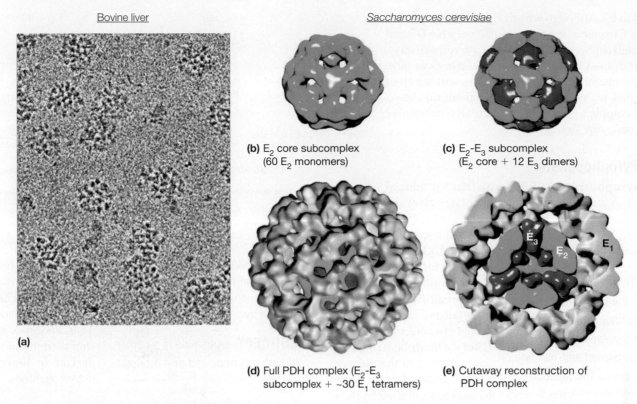

Bovine liver

Saccharomyces cerevisiae

(b) E_2 core subcomplex
(60 E_2 monomers)

(c) E_2-E_3 subcomplex
(E_2 core + 12 E_3 dimers)

(a)

(d) Full PDH complex (E_2-E_3
subcomplex + ~30 E_1 tetramers)

(e) Cutaway reconstruction of
PDH complex

▲ **FIGURE 13.5 Structure of the pyruvate dehydrogenase complex.**
(a) Electron micrograph of the purified pyruvate dehydrogenase complex from bovine liver. **(b–d)** A model for the eukaryotic complex, based on cryoelectron microscopy of the yeast (*Saccharomyces cerevisiae*) PDH complex and its subcomplexes. E_1, E_2, and E_3 are shown in yellow,
green, and red, respectively. Source: (b–e) *The Journal of Biological Chemistry* 276:38329–38336, L. Reed, A trail of research from lipoic acid to α-keto acid dehydrogenase complexes. Reprinted with permission. © 2001. The American Society for Biochemistry and Molecular Biology. All rights reserved.

same strategy is used in the pyruvate decarboxylase reaction of alcohol fermentation (discussed in Section 12.3), and, indeed, the chemistry of these two decarboxylations is nearly identical.

FIGURE 13.5 shows an electron micrograph of the PDH complex as purified from bovine liver and a model of the eukaryotic complex, based on cryoelectron microscopy of the complex in yeast (*Saccharomyces cerevisiae*). The eukaryotic complex (Figure 13.5(b–d)) is composed of a core of 60 E_2 monomers arranged as a pentagonal dodecahedron; it exhibits icosahedral symmetry (shown in Figure 6.34). This core structure also contains 12 E_3 homodimers and is surrounded by approximately 30–45 E_1 $\alpha_2\beta_2$ heterotetramers. In addition, the mammalian PDH complex contains about 12 copies of E_3 binding protein (E_3BP) and small amounts of two regulatory enzymes as well. (We will discuss regulation of the PDH complex later in this chapter.)

Bacteria possess a smaller PDH complex, typically composed of 24 polypeptide chains of E_1, 24 chains of E_2, and 12 chains of E_3. Although only about half the size of the eukaryotic complex, bacterial PDHs are still large protein complexes with a mass of about 4.6 million daltons—larger than a ribosome.

Coenzymes Involved in Pyruvate Oxidation and the Citric Acid Cycle

To understand how the three enzymes (E_1, E_2, and E_3) in the pyruvate dehydrogenase complex interact, we must understand the functions of the five coenzymes that participate in the reaction; the five cofactors are given in **TABLE 13.1**. TPP is tightly bound to E_1, lipoic acid is covalently bound to E_2, and flavin adenine dinucleotide (FAD) is

TABLE 13.1 Coenzymes of the pyruvate dehydrogenase reaction

Cofactor	Location	Function
Thiamine pyrophosphate (TPP)	Tightly bound to E_1	Decarboxylates pyruvate, yielding hydroxyethyl-TPP
Lipoic acid (lipoamide)	Covalently bound to E_2 via lysine ("swinging arm")	Accepts hydroxyethyl carbanion from TPP as acetyl group
Coenzyme A (CoA)	Dissociable substrate for E_2	Accepts acetyl group from lipoamide
Flavin adenine dinucleotide (FAD)	Tightly bound to E_3	Accepts pair of electrons from reduced lipoamide
Nicotinamide adenine dinucleotide (NAD^+)	Dissociable substrate for E_3	Accepts pair of electrons from reduced $FADH_2$

tightly bound to E_3. Although we briefly introduced all five coenzymes in Chapter 8, we have studied only NAD^+ and TPP in any detail (in Sections 12.2 and 12.10, respectively). Here, we will discuss each of the five coenzymes in turn, focusing on the chemistry they facilitate. We will see all of these coenzymes many more times throughout this book, but the PDH complex provides a particularly instructive example of coenzyme function.

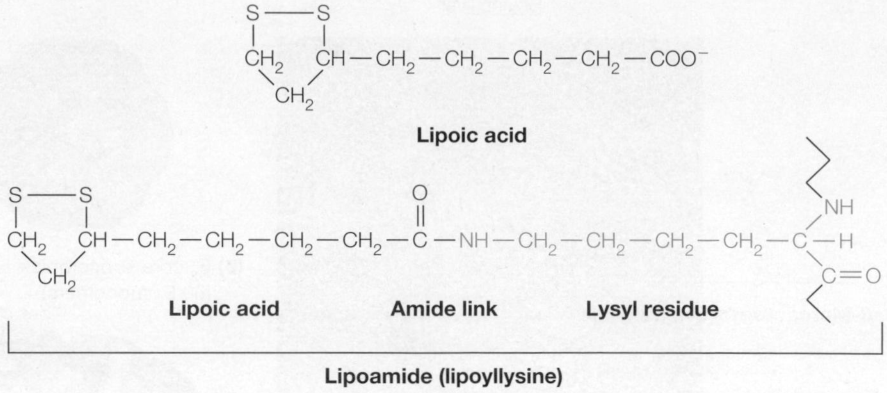

Thiamine Pyrophosphate

Thiamine pyrophosphate (TPP) was first introduced in Chapter 12 as a coenzyme which facilitates cleavage of C—C bonds by stabilizing carbanion intermediates (see Figures 12.34 and 12.35). TPP is also the coenzyme used for all decarboxylations of α-keto acids. Unlike β-keto acids, α-keto acids cannot stabilize the carbanion transition state that develops during decarboxylation, so they require the aid of a cofactor (TPP). The C2 of the thiazolium ring—the carbon between the nitrogen and the sulfur—can be deprotonated to form a nucleophilic carbanion. This carbanion can attack the carbonyl carbon of α-keto acids such as pyruvate. During the subsequent decarboxylation, the thiazolium ring stabilizes the carbanion intermediate through resonance. We will take a look at the details of this mechanism shortly.

● **CONCEPT** Thiamine pyrophosphate (TPP) is the coenzyme used for all decarboxylations of α-keto acids, which require a cofactor to help stabilize the carbanion intermediate that develops during the reaction.

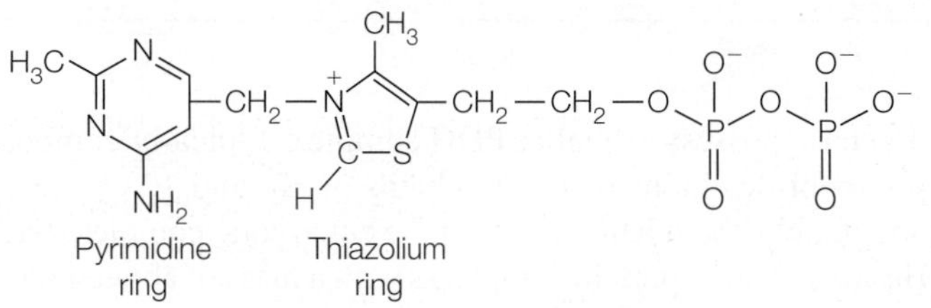

Thiamine pyrophosphate (TPP)

Lipoic Acid (Lipoamide)

This coenzyme participates in the transfer of acyl groups in several α-ketoacid dehydrogenases. Lipoic acid was identified in 1951 by Lester J. Reed and I. C. Gunsalus. Nearly 10 tons of pork and beef liver had to be processed to isolate $\sim$30 mg of crystalline substance! The coenzyme is joined to E_2 via an amide bond linking the carboxyl group of lipoic acid to a lysine ε-amino group. Thus, the reactive species is an amide, called **lipoamide,** or **lipoyllysine.** Each lipoyllysine side chain is $\sim$14 Å long and is located within a flexible lipoyl domain of E_2,[1] allowing it to function as a "swinging arm" that can interact with the active sites of both the E_1 and E_3 components of the PDH complex, as we will see shortly.

In pyruvate oxidation, lipoamide serves as the acceptor of a two-carbon aldehyde fragment carried by TPP. This transfer involves simultaneous oxidation of the aldehyde, coupled to reduction of the disulfide of lipoamide. This generates an acyl group, which, in pyruvate dehydrogenase, is transferred next to coenzyme A. The pair of electrons is transferred to lipoamide to form dihydrolipoamide (**FIGURE 13.6**). Thus, lipoamide is both an electron carrier and an acyl group carrier in these α-ketoacid dehydrogenases.

● **CONCEPT** Lipoamide is a carrier of both electrons and acyl groups.

Coenzyme A: Activation of Acyl Groups

Coenzyme A (A for acyl) participates in activation of acyl groups in general, including the acetyl group derived from pyruvate. The coenzyme is derived metabolically from ATP, the vitamin **pantothenic acid,** and β-mercaptoethylamine.

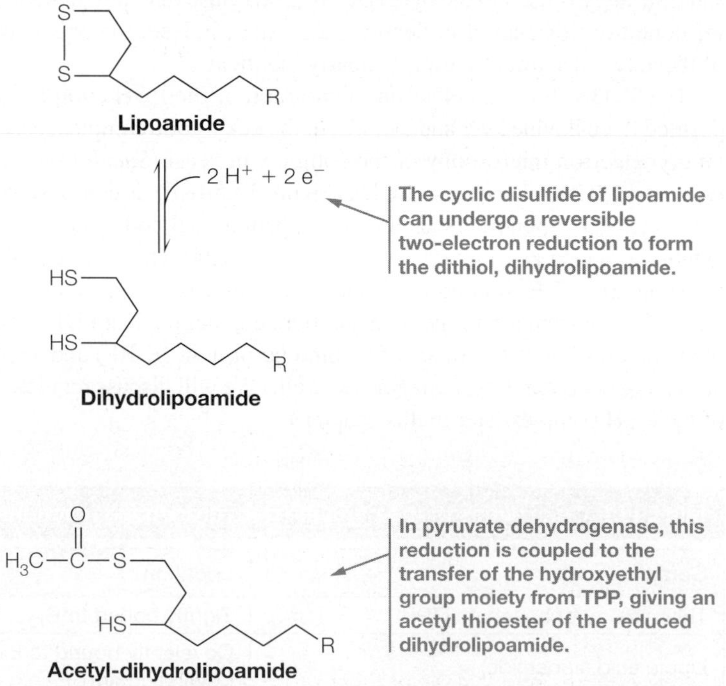

Lipoamide

The cyclic disulfide of lipoamide can undergo a reversible two-electron reduction to form the dithiol, dihydrolipoamide.

Dihydrolipoamide

In pyruvate dehydrogenase, this reduction is coupled to the transfer of the hydroxyethyl group moiety from TPP, giving an acetyl thioester of the reduced dihydrolipoamide.

Acetyl-dihydrolipoamide

▲ **FIGURE 13.6** Oxidized and reduced forms of lipoamide. Lipoamide is both an electron carrier and an acyl group carrier.

Lipoic acid

Lipoamide (lipoyllysine)

---

[1]The number of lipoyl domains in E_2 is species-specific. Mammalian E_2s have two lipoyl domains per subunit; *E. coli* E_2s have three.

β-Mercaptoethylamine   **Pantothenic acid**   **Adenosine 3′-phosphate 5′-diphosphate**

Coenzyme A structure

A free thiol on the β-mercaptoethylamine moiety is the business end of the coenzyme molecule; the rest of the molecule provides enzyme-binding sites. In acylated derivatives, such as acetyl-coenzyme A, the acyl group is linked to the thiol group to form an energy-rich thioester.

$$CoA-SH + HO-\overset{O}{\underset{||}{C}}-CH_3 \underset{H_2O}{\rightleftharpoons} CoA-S-\overset{O}{\underset{||}{C}}-CH_3$$

Coenzyme A   Acetic acid   Acetyl-CoA

The equilibrium of this reaction as written lies far to the left, indicating that acylation of coenzyme A requires an input of energy. The energy required to activate substrates for transfer to CoA can come from ATP hydrolysis or, as in the case of the pyruvate dehydrogenase reaction, by the oxidative decarboxylation of pyruvate. We designate the acylated forms of coenzyme as acyl-SCoA, and the unacylated form as HS-CoA.

The high chemical potential of thioesters, as compared with ordinary esters, is related primarily to resonance stabilization (**FIGURE 13.7**). Most esters have two resonance forms. Stabilization involves π-electron orbital overlap, giving partial double-bond character to the C—OR link.

● **CONCEPT** Thioesters such as acetyl-CoA are energy rich because thioesters are destabilized relative to ordinary oxygen esters.

In thioesters, the larger atomic size of S (relative to O) reduces the π-electron overlap between C and S so that the C—SR bond does not contribute significantly to resonance stabilization. Thus, the thioester is *destabilized*—it has a higher potential energy of acyl group transfer relative to that of an oxygen ester, reflected by a more favorable ΔG of hydrolysis.

The lack of double-bond character in the C—SR bond of acyl-CoAs makes this bond weaker than the corresponding C—OR bond in ordinary esters, and the thioalkoxide ion (R—S⁻) is a good leaving group in nucleophilic displacement reactions. Thus, the acyl group is readily transferred to other metabolites, which is what occurs in the first reaction of the citric acid cycle.

Flavin Adenine Dinucleotide

Flavin adenine dinucleotide, or **FAD,** is one of two coenzymes derived from vitamin B_2, or **riboflavin.** The other is the simpler **flavin mononucleotide (FMN),** or riboflavin phosphate (**FIGURE 13.8**). The functional part of both coenzymes is the **isoalloxazine ring** system, which serves

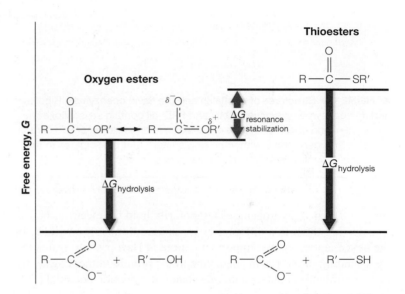

▲ **FIGURE 13.7 Comparison of free energies of hydrolysis of thioesters and oxygen esters.** Lack of resonance stabilization in thioesters is the basis for the more favorable ΔG of hydrolysis of thioesters, relative to that of ordinary oxygen esters. The free energies of the hydrolysis products are similar for the two classes of compounds.

as a two-electron acceptor. Compounds containing such a ring system are called **flavins.** In riboflavin and its derivatives the ring system is attached to **ribitol,** an open-chain version of ribose with the aldehyde carbon reduced to the alcohol level. The five-carbon of ribitol is linked to phosphate in FMN. FAD is a *dinucleotide,* having an adenosine moiety attached to the ribitol via a pyrophosphate linkage. Thus, these compounds are somewhat analogous to nicotinamide mononucleotide and nicotinamide adenine dinucleotide (NAD⁺), respectively.

Enzymes that use a flavin coenzyme are called **flavoproteins,** or **flavin dehydrogenases.** FMN and FAD undergo virtually identical electron transfer reactions. Flavoprotein enzymes preferentially bind either FMN or FAD. In a few cases, that binding is covalent. In most cases, however, the flavin is bound tightly, though noncovalently, so that the coenzyme cannot easily dissociate from the enzyme. Thus, flavins do not transfer electrons by diffusing from one enzyme to another, as the nicotinamide coenzymes do.

● **CONCEPT** Flavin coenzymes participate in two-electron oxidoreduction reactions that can proceed in 2 one-electron steps.

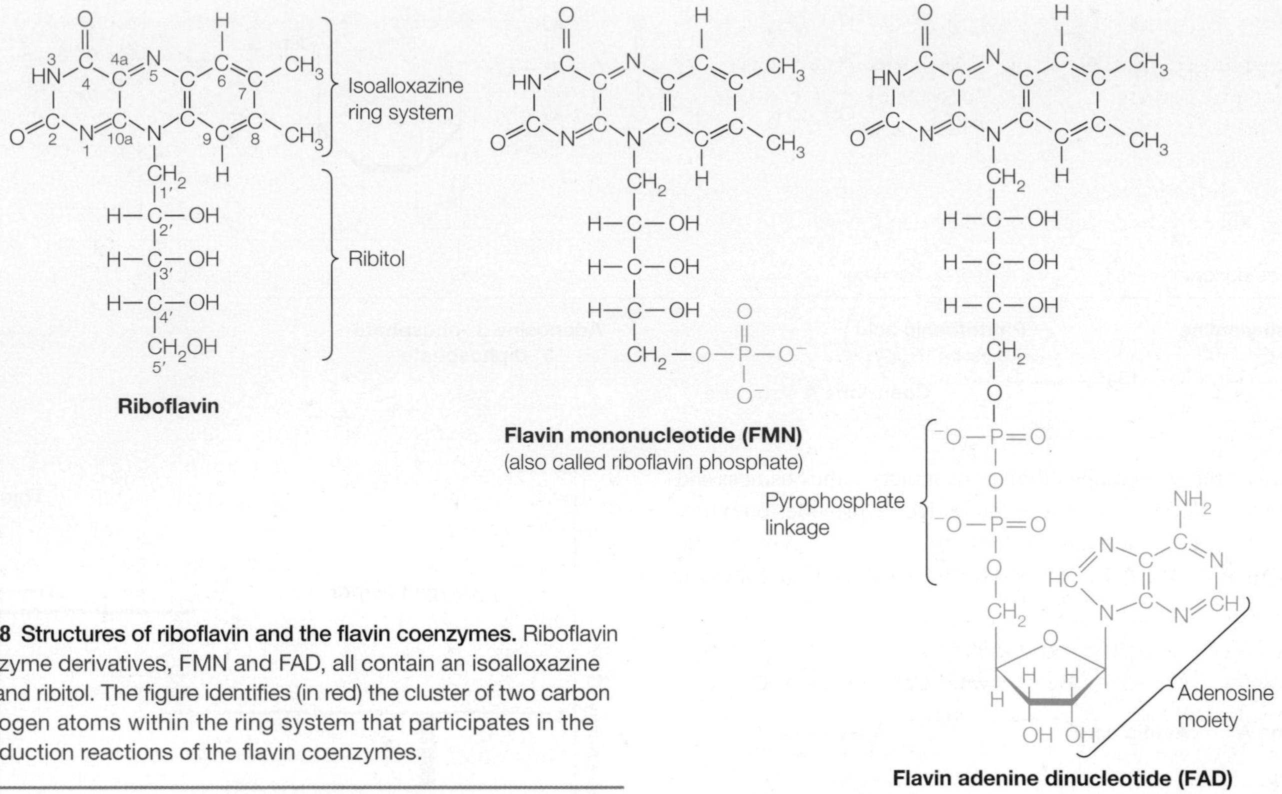

Riboflavin

Flavin mononucleotide (FMN)
(also called riboflavin phosphate)

Flavin adenine dinucleotide (FAD)

▲ **FIGURE 13.8 Structures of riboflavin and the flavin coenzymes.** Riboflavin and its coenzyme derivatives, FMN and FAD, all contain an isoalloxazine ring system and ribitol. The figure identifies (in red) the cluster of two carbon and two nitrogen atoms within the ring system that participates in the oxidation–reduction reactions of the flavin coenzymes.

Instead, flavin dehydrogenases temporarily hold the electrons before transferring them to a different electron acceptor. As we will see in the next chapter, another important feature of flavoproteins is that the tight binding of the flavin coenzyme to the protein (either covalently or noncovalently) confers a unique standard reduction potential ($E^{0'}$) on the flavin ring.

Like the nicotinamide coenzymes (NAD^+, $NADP^+$), the flavins undergo two-electron oxidation and reduction reactions. The flavins, however, are distinctive in having a stable one-electron-reduced species,

a **semiquinone** free radical, as shown in **FIGURE 13.9**. This free radical can be detected spectrophotometrically; whereas oxidized FAD and FMN are bright yellow, and fully reduced flavins are colorless, the semiquinone intermediate is either red or blue, depending on pH. The stability of the semiquinone intermediate gives flavins a catalytic versatility not shared by nicotinamide coenzymes in that flavins can interact with either two-electron or one-electron donor–acceptor pairs. Moreover, some (though not all) flavoproteins can interact directly with oxygen and are thus classified as oxidases.

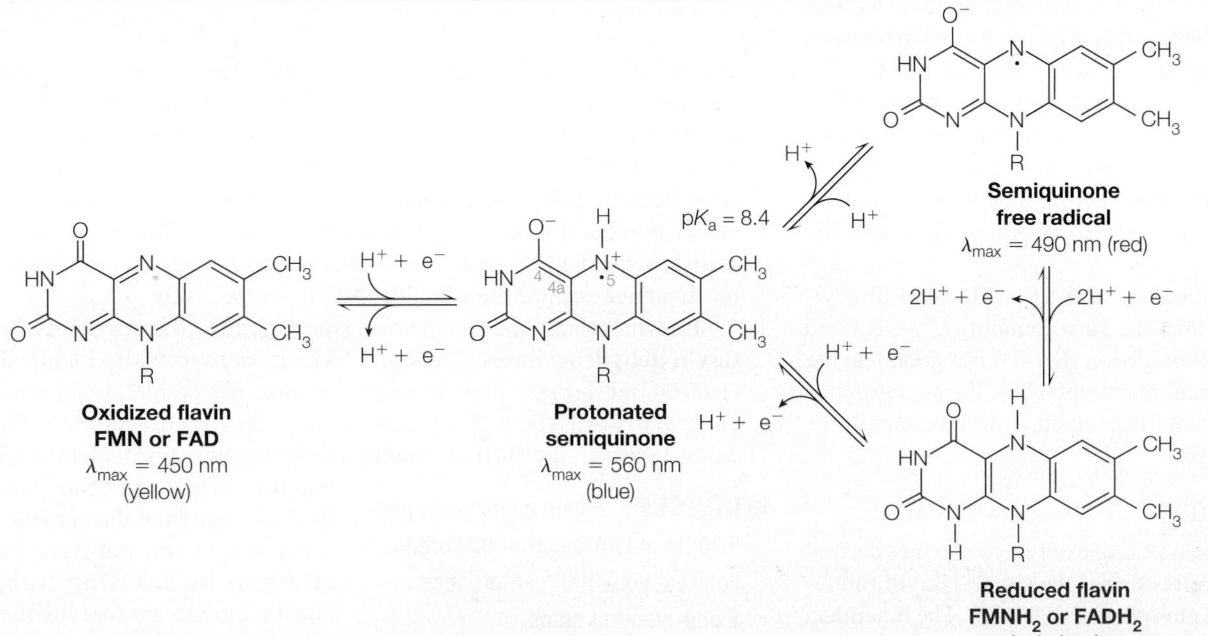

Oxidized flavin
FMN or FAD
λ_{max} = 450 nm
(yellow)

Protonated
semiquinone
λ_{max} = 560 nm
(blue)

pK_a = 8.4

Semiquinone
free radical
λ_{max} = 490 nm (red)

Reduced flavin
FMNH$_2$ or FADH$_2$
(colorless)

◀ **FIGURE 13.9 Oxidation and reduction reactions involving flavin coenzymes.** Flavins participate in two-electron reactions, but the existence of the stable semiquinone free radical intermediate allows these reactions to proceed one electron at a time. Thus, reduced flavins can readily be oxidized by one-electron acceptors. Spectral maxima (λ_{max}) are indicated for the oxidized flavin and the protonated and deprotonated forms of the semiquinone intermediate. In both semiquinone forms, the unpaired electron is delocalized between N-5 and C-4a.

Nicotinamide Adenine Dinucleotide

Nicotinamide adenine dinucleotide (NAD$^+$) was first introduced in Chapter 8 as an electron carrier. NAD$^+$ participates in reversible two-electron redox reactions that occur by a hydride (H$^-$) transfer mechanism—both the hydrogen and the electron pair are transferred. After oxidation of the substrate, the reduced form, NADH, leaves the enzyme and is reoxidized by other electron–acceptor systems in the cell.

Hydride donor: H$^-$ Hydride acceptor

NAD$^+$ **NADH**

Action of the Pyruvate Dehydrogenase Complex

As noted earlier, the oxidation of pyruvate to acetyl-CoA involves the coenzymes TPP, lipoic acid, CoA-SH, FAD, and NAD$^+$, acting in concert with three enzymes in the pyruvate dehydrogenase complex. Now we are ready to see how all of these components function together to effect the conversion of pyruvate to acetyl-CoA. The overall process is summarized in **FIGURE 13.10**.

The entire process, beginning with the decarboxylation of pyruvate, reaction ❶, and ending with the transfer of a pair of electrons to NAD$^+$, reactions ❺ₐ/❺ᵦ, requires five steps. A central feature of this reaction sequence is that in the three middle steps ❷, ❸, and ❹, reaction intermediates are formed and transferred between active sites by covalent attachment to the lipoamide moieties on the flexible lipoyl domains of E$_2$. The reaction sequence begins at the active site of E$_1$ (pyruvate dehydrogenase), which catalyzes a nucleophilic addition of the TPP carbanion to the ketone carbonyl group of pyruvate. The resulting addition product immediately undergoes decarboxylation to give hydroxyethyl-TPP, ❶. The hydroxyethyl group is next transferred from TPP on E$_1$ to a lipoamide moiety on E$_2$, ❷. This reaction, also catalyzed by E$_1$, occurs by an S$_N$2-like attack of the hydroxyethyl carbanion on the lipoamide disulfide (refer to Section 11.3 to review the S$_N$2 mechanism). TPP is eliminated to form an acetyl thioester on dihydrolipoamide and regenerate E$_1$, ❷. Through the combination of steps ❶ and ❷, pyruvate undergoes a two-electron oxidation to an acetyl group, with simultaneous two-electron reduction of the lipoamide disulfide to acetyl-dihydrolipoamide.

The acetyl group is now covalently bound to the active site of E$_2$, via the flexible lipoyllysine "swinging arm." E$_2$ (dihydrolipoamide transacetylase) next catalyzes transfer of the acetyl group to CoA. This nucleophilic acyl substitution reaction simply exchanges one thioester for another, giving acetyl-CoA and dihydrolipoamide, ❸.

The last two steps of the process, ❹ and ❺, are required to reoxidize the dihydrolipoamide of E$_2$ and transfer the pair of electrons to a

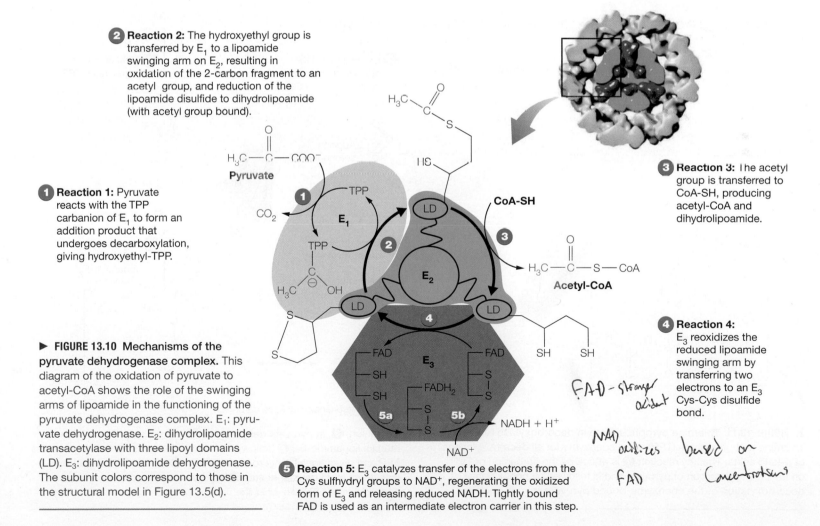

❷ **Reaction 2:** The hydroxyethyl group is transferred by E$_1$ to a lipoamide swinging arm on E$_2$, resulting in oxidation of the 2-carbon fragment to an acetyl group, and reduction of the lipoamide disulfide to dihydrolipoamide (with acetyl group bound).

❶ **Reaction 1:** Pyruvate reacts with the TPP carbanion of E$_1$ to form an addition product that undergoes decarboxylation, giving hydroxyethyl-TPP.

❸ **Reaction 3:** The acetyl group is transferred to CoA-SH, producing acetyl-CoA and dihydrolipoamide.

❹ **Reaction 4:** E$_3$ reoxidizes the reduced lipoamide swinging arm by transferring two electrons to an E$_3$ Cys-Cys disulfide bond.

CoA-SH

Pyruvate

Acetyl-CoA

FAD - stronger oxidant

NAD oxidizes based on concentrations FAD

▶ **FIGURE 13.10 Mechanisms of the pyruvate dehydrogenase complex.** This diagram of the oxidation of pyruvate to acetyl-CoA shows the role of the swinging arms of lipoamide in the functioning of the pyruvate dehydrogenase complex. E$_1$: pyruvate dehydrogenase. E$_2$: dihydrolipoamide transacetylase with three lipoyl domains (LD). E$_3$: dihydrolipoamide dehydrogenase. The subunit colors correspond to those in the structural model in Figure 13.5(d).

❺ **Reaction 5:** E$_3$ catalyzes transfer of the electrons from the Cys sulfhydryl groups to NAD$^+$, regenerating the oxidized form of E$_3$ and releasing reduced NADH. Tightly bound FAD is used as an intermediate electron carrier in this step.

● **CONCEPT** Lipoamide is tethered to one enzyme (E_2) in the pyruvate dehydrogenase complex, but it interacts with all three enzymes via a flexible swinging arm.

active site of E_3 contains a redox active Cys–Cys disulfide bond and a tightly bound FAD. The electrons are first transferred from dihydrolipoamide to the E_3 Cys–Cys disulfide to give the oxidized lipoamide and two sulfhydryl groups on E_3. The sulfhydryl groups are then reoxidized by the E_3-bound FAD, ❺a. The final step, ❺b, is transfer of the electron pair from $FADH_2$ on E_3 to NAD^+, regenerating the oxidized form of E_3 and releasing reduced NADH.

Because TPP is used in so many decarboxylation reactions, it is worth discussing the chemistry of step 1 of the pyruvate dehydrogenase reaction (Figure 13.10) in a bit more detail. The thiazolium moiety of TPP (an iminium ion) can be deprotonated at C2, with the aid of a general base in the enzyme active site, to form a nucleophilic carbanion at C2, termed an **ylid** (❶ in **FIGURE 13.11**). This carbanion can attack the carbonyl carbon of α-keto acids, such as pyruvate, giving an addition compound,

dissociable carrier (NAD^+). In reaction ❹, E_3 (dihydrolipoamide dehydrogenase) catalyzes the transfer of the electron pair from dihydrolipoamide to NAD^+. The

❷. The addition compound undergoes nonoxidative decarboxylation in ❸, with the thiazolium ring acting as an electron sink in forming a resonance-stabilized carbanion. This hydroxyethyl-TPP carbanion then nucleophilically attacks the lipoamide disulfide to achieve transfer of the acetyl moiety from TPP to the lipoyl group on E_2, ❹.

Recall from Chapter 12 that the pyruvate decarboxylase reaction used in ethanol metabolism also uses TPP (Section 12.3). The chemistry of that reaction is identical to the first three steps of Figure 13.11. However, in pyruvate decarboxylase, the resonance-stabilized carbanion undergoes protonation, followed by an elimination reaction to release free acetaldehyde and the TPP carbanion. In pyruvate dehydrogenase, the nucleophilic carbanion attacks lipoamide, ❹, prior to the elimination step that regenerates TPP, ❺. Thus, in general terms, TPP functions in two major types of reaction: (1) C—C bond cleavage (as in transketolase, discussed in Section 12.10) and (2) the transfer of an activated aldehyde species to an acceptor (in this case, lipoamide). We will come across this coenzyme several times in other pathways.

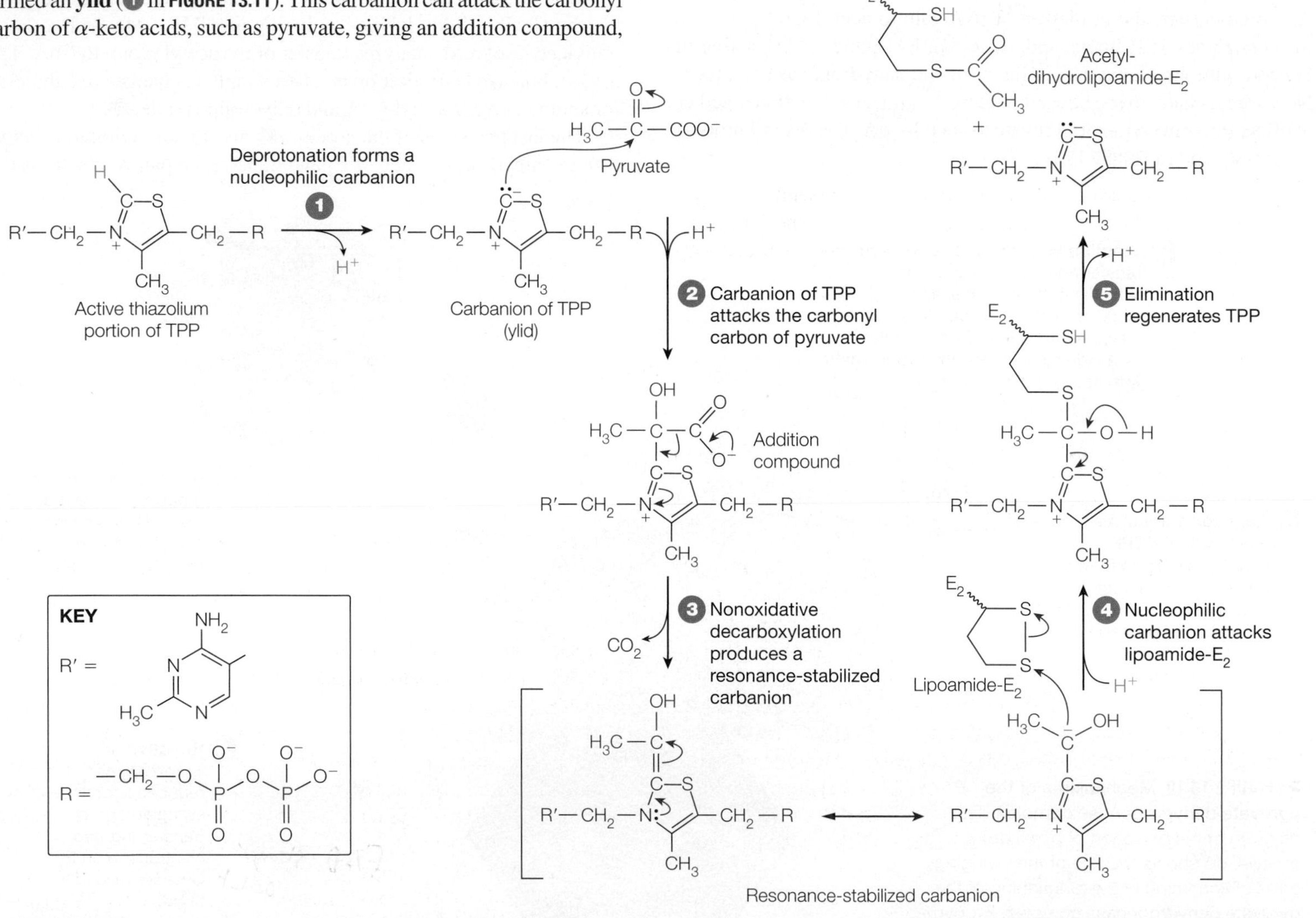

▲ **FIGURE 13.11 Thiamine pyrophosphate in decarboxylase reactions.** Thiamine pyrophosphate (TPP) is the coenzyme for all decarboxylations of α-keto acids. The key reaction ❷ is attack by the carbanion of TPP on the carbonyl carbon of pyruvate and is followed by nonoxidative decarboxylation of the coenzyme-bound pyruvate to give another carbanion, ❸. In pyruvate dehydrogenase, the nucleophilic carbanion attacks lipoamide-E_2, ❹, followed by an elimination step that regenerates TPP and gives acetyl-dihydrolipoamide-E_2, ❺. In pyruvate decarboxylase, there is no transfer at step ❹, and the acetyl group is released as acetaldehyde.

Physical juxtaposition of the enzymes of the complex and covalent attachment of reaction intermediates via the lipoamide swinging arm provide a number of advantages. The five-step reaction sequence in Figure 13.10 nicely illustrates the concept of **substrate channeling:** intermediates of a multistep pathway are "handed off" from one active site to the next without diffusing from the complex. Channeling allows the overall reaction to proceed smoothly, without unwanted side reactions or diffusion of intermediates from catalytic sites. The local concentration of substrates can be very high, allowing greater flux through the pathway, analogous to an assembly line in a factory.

The pyruvate dehydrogenase complex represents one of the best-understood examples of how cells can achieve economy of function by formation of multienzyme complexes catalyzing sequential reactions in a pathway. In fact, this same E_1 E_2 E_3 multienzyme structure is used to oxidize several other α-keto acids. Other examples include the branched-chain α-ketoacid dehydrogenase complex (discussed in Chapter 18) and the α-ketoglutarate dehydrogenase complex involved in the citric acid cycle, which we will discuss shortly. Indeed, the E_3 subunits are identical in all three complexes from a given species. It is clear from Figure 13.10 that each complex must have unique E_1 and E_2 subunits specific to its particular α-ketoacid substrate. But once the acyl group has been transferred to CoA-SH by E_2, the resulting dihydrolipoamide, the substrate of the E_3 subunits, is identical in each complex. Thus, as organisms evolved new α-ketoacid dehydrogenase complexes, the same E_3 subunit could be reused.

Trivalent arsenic-containing compounds such as arsenite (AsO_3^{3-}) and organic arsenicals react readily with thiols. They are especially reactive with dithiols, such as dihydrolipoamide, forming bidentate adducts:

Formation of arsenic bidentate adducts

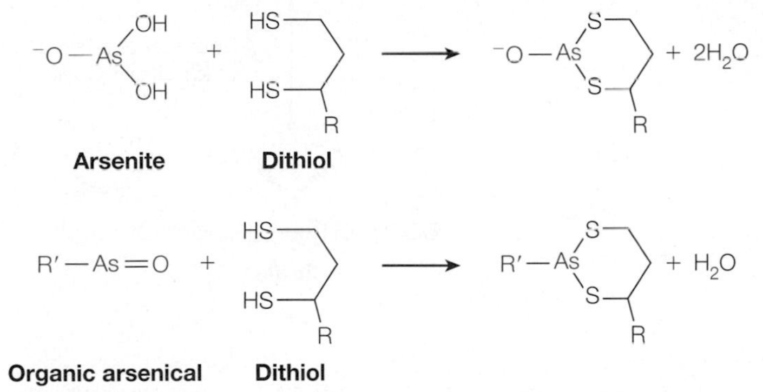

This covalent modification of lipoamide groups inactivates E_1–E_2–E_3 multienzyme complexes, including the pyruvate dehydrogenase and α-ketoglutarate dehydrogenase complexes of the citric acid cycle, thereby inhibiting respiration and accounting for the toxicity of these compounds. In fact, arsenic poisoning, both intentional and unintentional, has had a long history, dating back to at least the eighth century. Arsenic became a favorite poison during the Middle Ages and the Renaissance, often being employed by impatient heirs to "get a jump"

● **CONNECTION** Arsenic compounds inhibit respiration by covalently modifiying lipoamide groups. This toxicity led to their discontinued use as medical treatments once penicillin and other antibiotics were developed.

on their inheritances. Several prominent historical figures, including Francisco de' Medici, King George III, and Napoleon Bonaparte may have been poisoned by arsenic. Arsenic was also an ingredient in many "tonics"

that were popular during the Victorian Era. Organic arsenicals were even used in the early twentieth century to treat syphilis and trypanosomiasis because the lipoamide-containing enzymes of the pathogens are often more sensitive than those of the host.

13.3 The Citric Acid Cycle

We have discussed the chemical logic behind the citric acid cycle, and we have described how pyruvate is converted to acetyl-CoA by the PDH complex. Now we are ready to follow the fate of that acetyl-CoA through the citric acid cycle. The cycle is composed of eight steps, beginning with addition of a two-carbon moiety (acetyl-CoA) to a four-carbon compound (oxaloacetate) to give a six-carbon tricarboxylic acid (citrate), followed by loss of two carbons as CO_2, and finally the regeneration of oxaloacetate. During this process, four pairs of electrons are removed from the substrates and passed to carriers (NAD^+ and FAD) on their way to the respiratory chain. Figure 13.4 presented the entire citric acid cycle, showing the structure of each intermediate. We will now discuss the chemistry and enzymology of each reaction.

Step 1: Introduction of Two Carbon Atoms as Acetyl-CoA

The initial reaction of the citric acid cycle, catalyzed by **citrate synthase,** is akin to an aldol condensation.

Step 1: The citrate synthase reaction

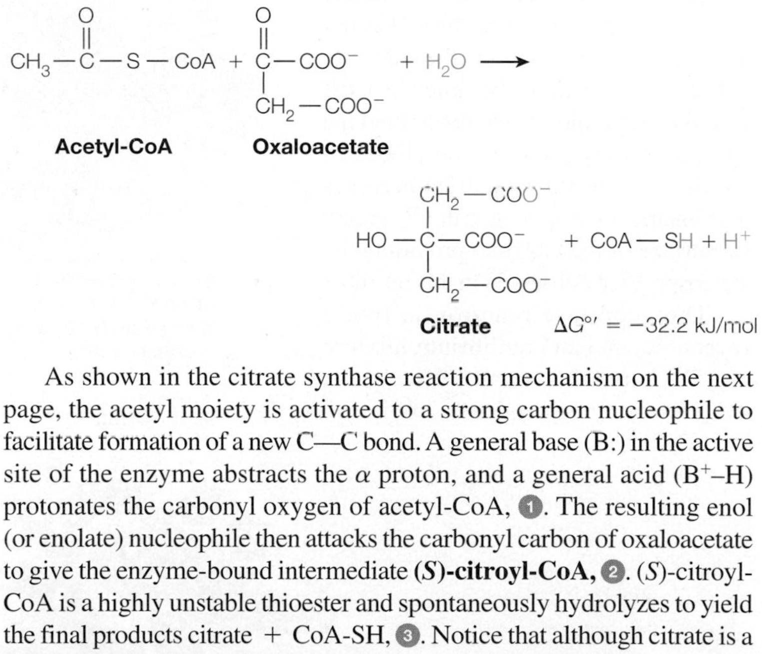

As shown in the citrate synthase reaction mechanism on the next page, the acetyl moiety is activated to a strong carbon nucleophile to facilitate formation of a new C—C bond. A general base (B:) in the active site of the enzyme abstracts the α proton, and a general acid (B^+–H) protonates the carbonyl oxygen of acetyl-CoA, ❶. The resulting enol (or enolate) nucleophile then attacks the carbonyl carbon of oxaloacetate to give the enzyme-bound intermediate **(S)-citroyl-CoA,** ❷. (S)-citroyl-CoA is a highly unstable thioester and spontaneously hydrolyzes to yield the final products citrate + CoA-SH, ❸. Notice that although citrate is a symmetrical molecule, the two carboxymethyl (—CH_2COO^-) groups occupy different positions relative to the OH and COO^- groups on C3. Thus, citrate is **prochiral**—it can become chiral by substitution of one of its carboxymethyl groups (see Section 10.2). The two chemically equivalent groups of a prochiral molecule are designated *pro-R* and *pro-S*. As we shall see shortly, the next enzyme in the pathway, aconitase, catalyzes chemistry specifically on the *pro-R* arm of citrate.

The thioester hydrolysis in ❸ makes the forward reaction highly exergonic—the K_{eq} for the citrate synthase reaction is about 3×10^5, which ensures continued operation of the cycle even when the oxaloacetate concentration is low. Finally, crystallographic analysis of citrate synthase (**FIGURE 13.12**) gives excellent evidence for the induced fit model of enzyme catalysis, as described in Section 8.4.

Step 2: Isomerization of Citrate

Citrate is a tertiary alcohol, meaning that the carbon atom bearing the hydroxyl group is bonded to three other carbon atoms. This presents yet another chemical problem: Tertiary alcohols cannot be oxidized without breaking a carbon–carbon bond. This is because the carbon atom bearing the hydroxyl group, already bonded to three other carbons, cannot form a carbon–oxygen double bond. To set up the next oxidation in the pathway, citrate is first converted to isocitrate, a chiral secondary alcohol, which can be more readily oxidized. This isomerization reaction, catalyzed by **aconitase,** involves successive dehydration and hydration, through *cis*-aconitate as a dehydrated intermediate, which remains enzyme-bound.

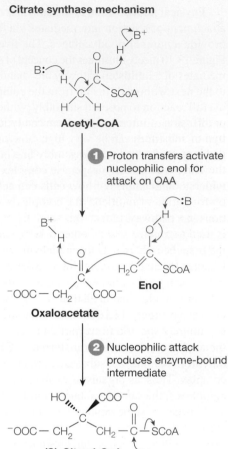

Citrate synthase mechanism

① Proton transfers activate nucleophilic enol for attack on OAA

Acetyl-CoA

Enol

Oxaloacetate

② Nucleophilic attack produces enzyme-bound intermediate

Step 2: The aconitase reaction

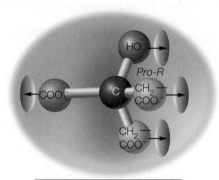

Citrate
(3° alcohol)   *cis*-Aconitate   D-Isocitrate
(2° alcohol)

$\Delta G°' = +6.3$ kJ/mol

This reaction is highly stereospecific: Of the four possible diastereomers of isocitrate, only one, the *2R,3S* diastereomer, is produced. How can citrate, a symmetric molecule, react asymmetrically with aconitase? If the enzyme binds the substrate at four points, then the binding site itself would be asymmetric and could bind the substrate in only one way. Thus, the symmetrical, but prochiral, citrate becomes asymmetric upon binding to the asymmetric surface of the enzyme, providing the stereospecificity observed in this reaction.

The aconitase reaction is freely reversible, and an equilibrium mixture at 25 °C contains about 90% citrate, 4% cis-aconitate, and 6% isocitrate.

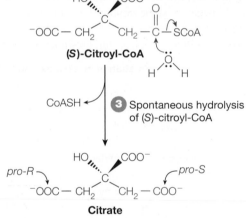

(S)-Citroyl-CoA

③ Spontaneous hydrolysis of (S)-citroyl-CoA

Citrate

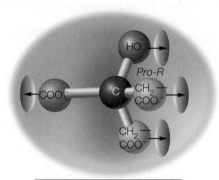

Four-point binding to aconitase enzyme confers asymmetry to the symmetrical (but prochiral) citrate substrate.

► **FIGURE 13.12** Three-dimensional structure of citrate synthase. In cells, citrate synthase functions as a homodimer. Shown here are two conformations of pig heart citrate synthase, as determined by crystal-lographic methods, that support the induced fit model of enzyme catalysis (Chapter 8).

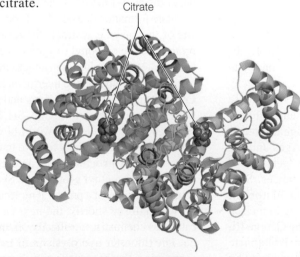

(a) In the absence of substrate, the enzyme crystallizes in an "open" form (PDB ID: 1cts).

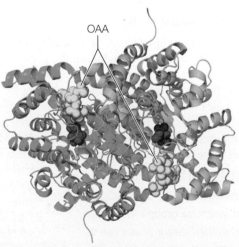

(b) Binding of OAA (yellow) causes the enzyme to adopt a "closed" conformation, which is required for acetyl-CoA binding (PDB ID: 2cts).

● **CONNECTION** The plant product fluoroacetate is an example of a pro-drug that is converted into 2-fluorocitrate, a mechanism-based, or suicide, inhibitor of aconitase.

However, the exergonic nature of the next reaction draws the aconitase reaction to the right as written.

Aconitase is the target site for the toxic action of **fluoroacetate,** a plant product that was originally employed as a rodenticide. Its subsequent use by ranchers in the West to control coyote populations also led to the death of eagles and other endangered animals. Fluoroacetate blocks the citric acid cycle by its metabolic conversion to **2-fluorocitrate,** which is a potent **mechanism-based inhibitor,** or **suicide inhibitor,** of aconitase. As described in Chapter 8, a mechanism-based inhibitor undergoes the first few chemical steps of the enzymatic reaction but is then converted into a compound that binds tightly, often irreversibly, thereby inactivating the enzyme. In other words, the inhibitor requires the normal enzyme mechanism to inactivate the enzyme. At the same time, fluoroacetate can be considered a **pro-drug:** though not toxic to cells by itself, it resembles a normal metabolite closely enough that it undergoes metabolic transformation to a product that does inhibit a crucial enzyme. The cell "commits suicide" by transforming the analog to a toxic product. In this case, fluoroacetate is first converted to fluoroacetyl-CoA by acetate thiokinase (an enzyme discussed later in this chapter), and then to 2-fluorocitrate by citrate synthase. 2-Fluorocitrate then inhibits aconitase, halting the entire citric acid cycle.

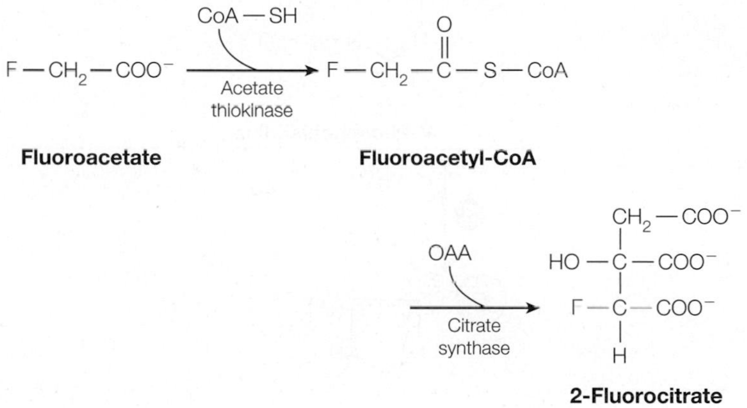

Fluoroacetate   **Fluoroacetyl-CoA**

2-Fluorocitrate

Step 3: Conservation of the Energy Released by an Oxidative Decarboxylation in the Reduced Electron Carrier NADH

The first of two oxidative decarboxylations in the cycle is catalyzed by **isocitrate dehydrogenase.** Isocitrate is oxidized to a ketone, **oxalosuccinate,** an unstable enzyme-bound intermediate that spontaneously decarboxylates to give the product, α-ketoglutarate. The strategy here is to oxidize isocitrate's secondary alcohol to a keto group that is β to the carboxyl group to be removed. The β-keto group acts as an electron sink to stabilize the carbanionic transition state, facilitating decarboxylation, as described in Section 11.3.

This reaction occurs with a $\Delta G^{\circ\prime}$ of -11.6 kJ/mol, and under physiological conditions it is essentially irreversible and sufficiently exergonic to pull the aconitase reaction forward. A mitochondrial form of isocitrate dehydrogenase that is specific for NAD^+ is probably the chief participant in the citric acid cycle in most cells. The NADH produced in this

● **CONNECTION** Mutations in the active site of isocitrate dehydrogenase (IDH) are common features in some brain cancers. These mutations alter the activity of IDH, causing the enzyme to reduce α-ketoglutarate to 2-hydroxyglutarate (2HG). 2HG is an "oncometabolite"—its accumulation causes cancer by inhibiting the demethylation of DNA and histones.

reaction, carrying two reducing equivalents, is the first link between the citric acid cycle and the electron transport process of respiration. As such, this enzyme is an important regulatory site for controlling flux through the cycle. Most cells also contain an $NADP^+$-specific form of isocitrate dehydrogenase, found in both cytosol and mitochondria; its likely role is the generation of NADPH for reductive biosynthetic processes.

Step 3: The isocitrate dehydrogenase reaction

Isocitrate   **Oxalosuccinate**

$\Delta G^{\circ\prime} = -11.6$ kJ/mol

α-**Ketoglutarate**

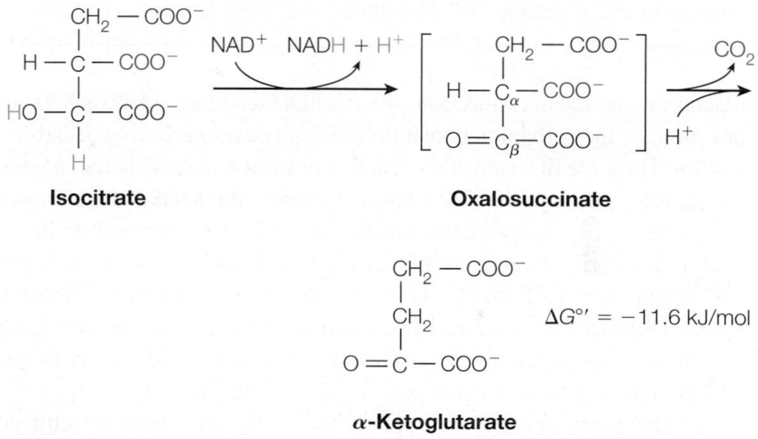

Step 4: Conservation of Energy in NADH by a Second Oxidative Decarboxylation

The fourth reaction of the citric acid cycle is a multistep reaction entirely analogous to the pyruvate dehydrogenase reaction detailed in Section 13.2. An α-keto acid substrate undergoes oxidative decarboxylation, with concomitant formation of an acyl-CoA thioester.

Step 4: The α-ketoglutarate dehydrogenase reaction

α-**Ketoglutarate**

Succinyl-CoA

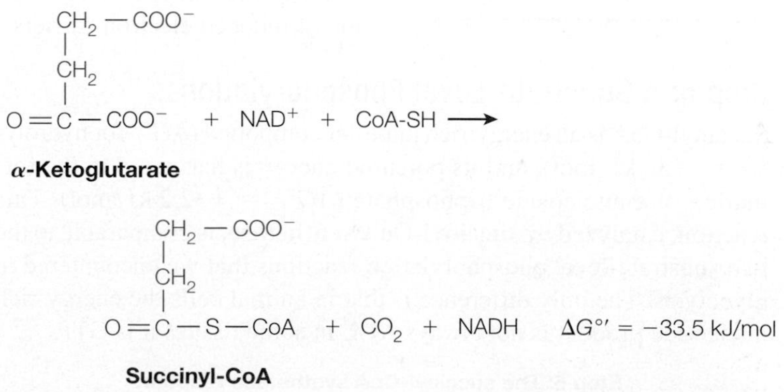

This reaction is catalyzed by the α-**ketoglutarate dehydrogenase complex,** an enzyme cluster similar to the pyruvate dehydrogenase complex, with three analogous enzyme activities and the same five coenzymes—TPP, lipoic acid, CoA-SH, FAD, and NAD^+. Indeed, the E_3 subunits are shared by both complexes. TPP is required for the same

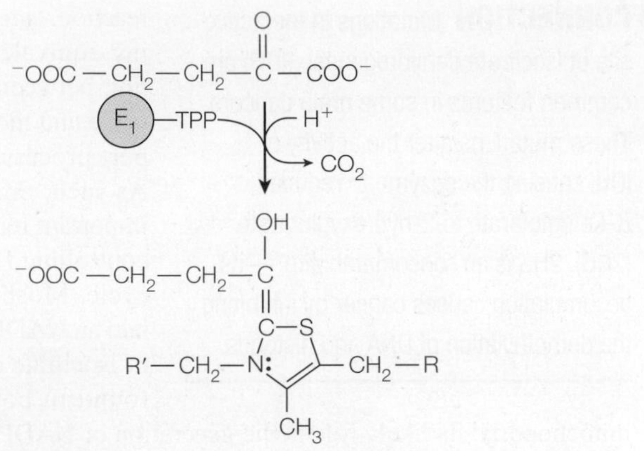

▲ **FIGURE 13.13 Decarboxylation of α-ketoglutarate.** The first step carried out by the α-ketoglutarate dehydrogenase complex is a decarboxylation catalyzed by α-ketoglutarate decarboxylase (E_1 of the complex), producing a four-carbon TPP derivative.

reason we learned in connection with the PDH reaction: α-keto acids cannot stabilize the carbanion transition state that develops during decarboxylation. Thus, the first step of the reaction decarboxylates α-ketoglutarate, producing a four-carbon TPP addition compound (**FIGURE 13.13**; compare to Figure 13.11). Subsequent transfer of the four-carbon unit to lipoic acid, transesterification of the dihydrolipoamide thioester with CoA-SH, and oxidation by FAD and NAD^+ are analogous to the reactions shown in Figure 13.10 for the pyruvate dehydrogenase complex. The succinyl-CoA product is a high potential energy thioester of succinic acid—much of this energy will be extracted in the next reaction of the pathway.

At this point in the cycle, two carbon atoms have been introduced as acetyl-CoA (at the citrate synthase step), and two have been lost as CO_2. Because of the stereochemistry of the aconitase reaction, the two carbon atoms lost are not the same as the two carbons introduced at the beginning of the cycle. In the remaining reactions, the four-carbon oxaloacetate is regenerated from the four-carbon succinyl-CoA, with two of the four steps involving oxidations and conservation of that energy in the generation of reduced electron carriers.

● **CONCEPT** Two carbon atoms enter the citric acid cycle as acetyl-CoA, and two are lost as CO_2 in the oxidative decarboxylations of steps 3 and 4 (Figure 13.4).

Step 5: A Substrate-Level Phosphorylation

Succinyl-CoA is an energy-rich thioester compound ($\Delta G°'$ for hydrolysis ≈ -36 kJ/mol), and its potential energy is harvested in the formation of a nucleoside triphosphate ($\Delta G°' = +32.2$ kJ/mol). This reaction, catalyzed by **succinyl-CoA synthetase,** is comparable to the two substrate-level phosphorylation reactions that we encountered in glycolysis. The only difference is that in animal cells the energy-rich nucleotide product is not always ATP; in some tissues it is GTP.

Step 5: The succinyl-CoA synthetase reaction

Succinyl-CoA + P_i + ADP(GDP) $\rightleftharpoons$
succinate + ATP(GTP) + CoA-SH $\Delta G°' = -3.8$ kJ/mol

Succinyl-CoA synthetase is a heterodimer consisting of an α and a β subunit, with the β subunit determining the substrate specificity (ADP or GDP). In animals, tissues that are dependent on oxidative metabolism, such as brain, heart, and skeletal muscle, contain

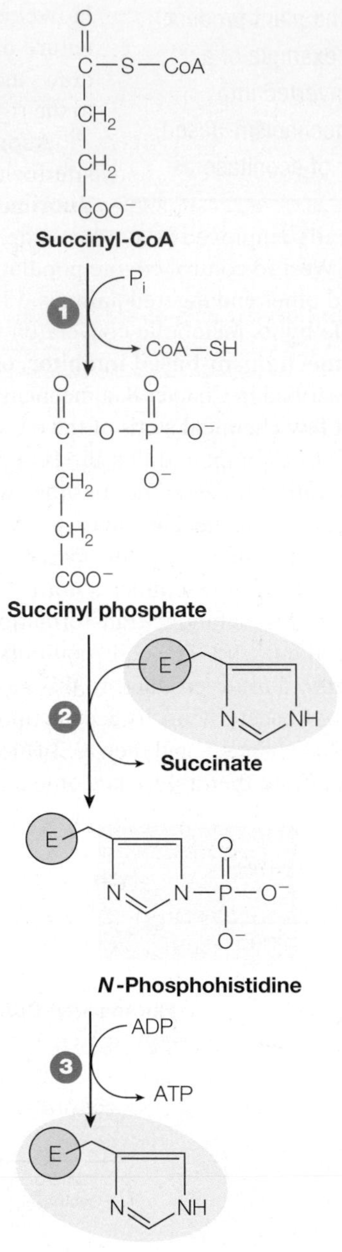

▲ **FIGURE 13.14 Covalent catalysis by the succinyl-CoA synthetase reaction.** Three successive nucleophilic substitution reactions conserve the energy of the thioester of succinyl-CoA in the phosphoanhydride bond of ATP (or GTP). A histidine side chain in the active site of the enzyme is transiently phosphorylated to form *N*-phosphohistidine during the reaction.

the ATP-linked enzyme, while in kidney and liver ("biosynthetic," or anabolic, tissues) the GTP-linked succinyl-CoA synthetase predominates. The two isozymes of succinyl-CoA synthetase presumably serve different metabolic roles in the various tissues.

Whichever nucleotide is used, the reaction occurs via a phosphorylated enzyme intermediate, an example of covalent catalysis. As shown in **FIGURE 13.14**, an initial mixed anhydride between inorganic phosphate and a carboxyl group of succinate is formed in a nucleophilic acyl substitution reaction, displacing CoA-SH, step ❶. In a second nucleophilic substitution, a histidine in the active site of the enzyme attacks the phosphorus atom of succinyl phosphate, displacing succinate, step ❷. The resulting *N*-phosphohistidine residue then transfers its phosphate to the nucleoside diphosphate substrate (ADP or GDP) in a final nucleophilic substitution reaction, step **❸**.

Step 6: A Flavin-Dependent Dehydrogenation

Completion of the cycle involves conversion of the four-carbon succinate to the four-carbon oxaloacetate and requires three more reactions. The first of these reactions, catalyzed by **succinate dehydrogenase,** is the FAD-dependent dehydrogenation of two saturated carbons to a double bond.

Step 6: The succinate dehydrogenase reaction

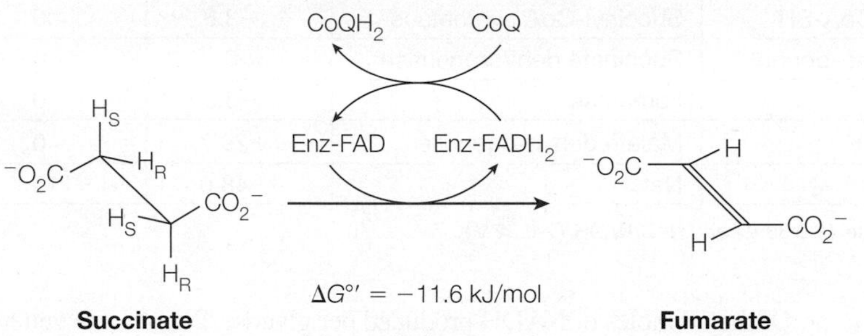

$\Delta G°' = -11.6$ kJ/mol

Succinate　　　　　　　　　　　**Fumarate**

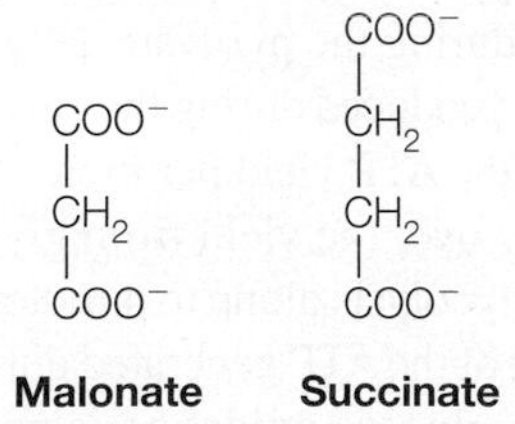

Malonate　　**Succinate**

Succinate dehydrogenase is competitively inhibited by malonate, a structural analog of succinate. Recall that malonate inhibition of pyruvate oxidation was one of the clues that led Krebs to propose the cyclic nature of this pathway. Note that succinate dehydrogenase is stereoselective, removing the *pro-S* hydrogen from one carbon and the *pro-R* hydrogen from the other, producing only the *trans* isomer, fumarate. The *cis* isomer, maleate, is not formed.

A C—C single bond is more difficult to oxidize than a C—O bond. Therefore, the redox coenzyme for succinate dehydrogenase is not NAD^+ but the more powerful oxidant FAD. The flavin is bound covalently to the enzyme protein, designated E, through a specific histidine residue.

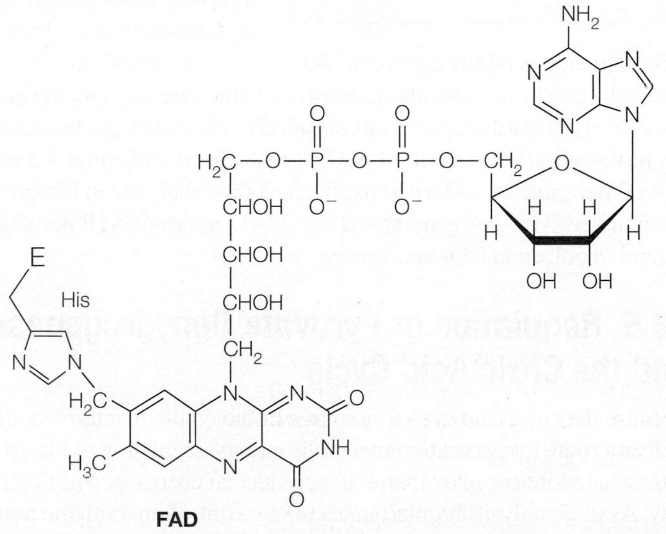

FAD

The importance of this covalent binding is that the reduced flavin must be reoxidized for the enzyme to act again. The two electrons from the reduced flavin are transferred, through three iron–sulfur centers in the enzyme molecule, to Coenzyme Q, another electron carrier in the

mitochondrial electron transport system. Thus, the reaction catalyzed by succinate dehydrogenase can be summarized as:

$$\text{Succinate} + Q \rightleftharpoons \text{fumarate} + QH_2$$

In fact, succinate dehydrogenase is so tightly coupled to the other electron carriers that it is referred to as **Complex II** of the respiratory chain (more about this in Chapter 14). This also explains the fact that, unlike the other enzymes of the citric acid cycle, succinate dehydrogenase is an integral membrane protein found embedded in the mitochondrial inner membrane.

The succinate dehydrogenase reaction is followed by hydration of the fumarate double bond (Figure 13.4, step **❼**) and dehydrogenation of the resulting α-hydroxy acid to give the α-keto acid oxaloacetate (Figure 13.4, step **❽**).

Step 7: Hydration of a Carbon–Carbon Double Bond

The stereospecific *trans* hydration of the carbon–carbon double bond is catalyzed by **fumarate hydratase,** more commonly called **fumarase.**

Step 7: The fumarase reaction

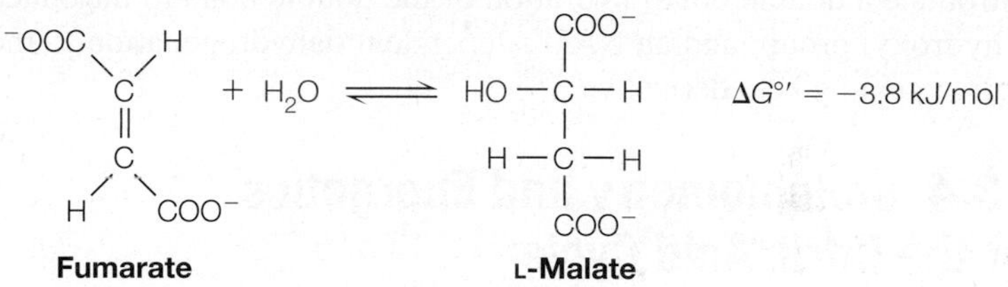

$\Delta G°' = -3.8$ kJ/mol

Fumarate　　　　　　　　　　**L-Malate**

Isomers of fumarate and L-malate

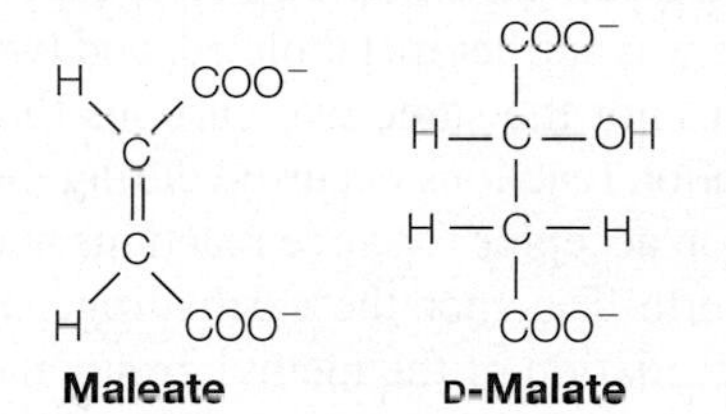

Maleate　　　**D-Malate**

The reaction is mechanistically similar to the addition of water to *cis*-aconitate in the aconitase reaction (Figure 13.4 step **❷**) and produces exclusively the *S* enantiomer (L-malate). The *cis* isomer of fumarate, namely, maleate, is not a substrate for the forward reaction, nor can the enzyme act on D-malate in the reverse direction.

Step 8: An Oxidation that Regenerates Oxaloacetate

Finally, the cycle is completed with the NAD^+-dependent dehydrogenation of malate to oxaloacetate, catalyzed by **malate dehydrogenase.**

Step 8: The malate dehydrogenase reaction

$\Delta G°' = +29.7$ kJ/mol

L-Malate　　　　　　　**Oxaloacetate**

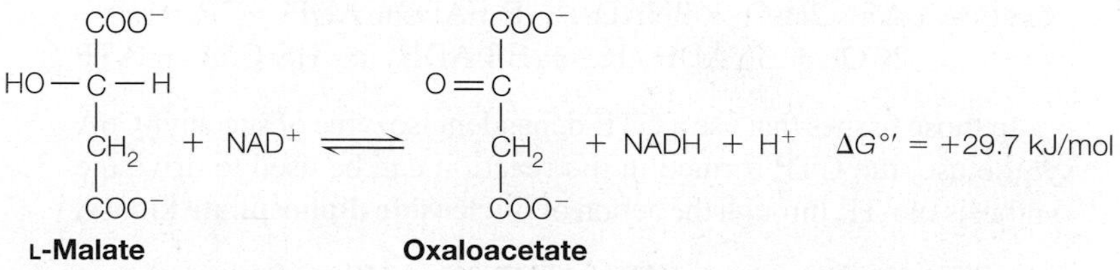

Despite the large positive standard free energy change $(\Delta G°' = +29.7 \text{ kJ/mol})$, this reaction proceeds to the right in mitochondria because the highly exergonic citrate synthase reaction (the next, and first, reaction in the cycle) keeps intramitochondrial oxaloacetate levels exceedingly low (below 10^{-6} M).

These last three steps of the citric acid cycle (Figure 13.4 **❻**, **❼**, and **❽**) use a chemical strategy that we will see several times again in other pathways: an FAD-dependent acyl-CoA dehydrogenation to

TABLE 13.2 Reactions of the citric acid cycle

	Reaction	Enzyme	$\Delta G°'$ (kJ/mol)	ΔG (kJ/mol)
1.	Acetyl-CoA + oxaloacetate + H_2O $\longrightarrow$ citrate + CoA-SH + H^+	Citrate synthase	−32.2	∼−55
2a.	Citrate $\rightleftharpoons$ cis-aconitate + H_2O	Aconitase	+6.3	∼0
2b.	cis-Aconitate + H_2O $\rightleftharpoons$ isocitrate	Aconitase		
3.	Isocitrate + NAD^+ $\rightleftharpoons$ α-ketoglutarate + CO_2 + NADH	Isocitrate dehydrogenase	−11.6	∼−20
4.	α-Ketoglutarate + NAD^+ + CoA-SH $\rightleftharpoons$ succinyl-CoA + CO_2 + NADH	α-Ketoglutarate dehydrogenase complex	−33.5	∼−40
5.	Succinyl-CoA + P_i + ADP (GDP) $\rightleftharpoons$ succinate + ATP (GTP) + CoA-SH	Succinyl-CoA synthetase	−3.8	∼0
6.	Succinate + FAD (enzyme-bound) $\rightleftharpoons$ fumarate + $FADH_2$ (enzyme-bound)	Succinate dehydrogenase	0	∼0
7.	Fumarate + H_2O $\rightleftharpoons$ L-malate	Fumarase	−3.8	∼0
8.	L-Malate + NAD^+ $\rightleftharpoons$ oxaloacetate + NADH + H^+	Malate dehydrogenase	+29.7	∼0
		Net	−48.0	∼−115

Note: $\Delta G°'$ value for reaction 3 was calculated from the $E°'$ values for α-ketoglutarate/isocitrate (−0.38 V) and NAD/NADH (−0.32 V).

introduce a double bond; hydration of the double bond to introduce a hydroxyl group; and an NAD^+-dependent dehydrogenation to the corresponding keto derivative.

13.4 Stoichiometry and Energetics of the Citric Acid Cycle

Now let us review what has been accomplished in one turn of the citric acid cycle, summarized in **TABLE 13.2**. The cycle started when a two-carbon fragment (acetyl-CoA) combined with a four-carbon acceptor (oxaloacetate). The resulting citrate was further metabolized, and two carbons were removed as CO_2 (but not the same two carbons that entered as acetyl-CoA). Four oxidation reactions occurred during the cycle, with NAD^+ serving as electron acceptor for three reactions and FAD as electron acceptor for the fourth. Together, these dehydrogenations accomplish the six-electron oxidation of the methyl group and the two-electron oxidation of the carbonyl carbon of acetyl-CoA (all eight electrons derived from acetyl-CoA). ATP (or GTP) was generated directly in only one reaction (catalyzed by succinyl-CoA synthetase). Finally, oxaloacetate was regenerated; it is now ready to start the cycle again by condensation with another molecule of acetyl-CoA.

We can write a chemical equation representing the sum of the eight reactions involved in one turn of the cycle:

$$\text{Acetyl-CoA} + 2H_2O + 3NAD^+ + \text{E-FAD} + \text{ADP} + P_i \longrightarrow$$
$$2CO_2 + 3NADH/H^+ + \text{E-FADH}_2 + \text{HS-CoA} + \text{ATP}$$

In those tissues that use a GTP-dependent isozyme of succinyl-CoA synthetase, the GTP formed in the reaction can be used to drive the synthesis of ATP, through the action of **nucleoside diphosphate kinase.**

$$\text{GTP} + \text{ADP} \rightleftharpoons \text{ATP} + \text{GDP} \qquad \Delta G°' = 0 \text{ kJ/mol}$$

Now if we take into account glycolysis and the pyruvate dehydrogenase reaction, and if we recall that each molecule of glucose generates two molecules of pyruvate, we can write the following equation for catabolism of glucose through glycolysis and the citric acid cycle.

$$\text{Glucose} + 2H_2O + 10NAD^+ + 2FAD + 4ADP + 4P_i \longrightarrow$$
$$6CO_2 + 10NADH/H^+ + 2FADH_2 + 4ATP$$

Of the 10 moles of NADH produced per glucose, 2 moles are generated in the cytoplasm at the glyceraldehyde-3-phosphate dehydrogenase step of glycolysis, and 2 moles are generated during the pyruvate dehydrogenase reaction. The remaining 6 moles are produced during the citric acid cycle. By the end of the citric acid cycle, the ATP yield per mole of glucose metabolized has not increased greatly over the yield from glycolysis alone: 2 moles of ATP per glucose in glycolysis alone to 4 moles as shown in the chemical equation above. Most of the ATP generated during glucose oxidation is not formed directly from reactions of glycolysis and the citric acid cycle (substrate-level phosphorylations), but rather from the reoxidation of reduced electron carriers in the respiratory chain. These electron carriers, NADH and $FADH_2$, have a high redox potential, in the sense that their oxidation is highly exergonic. As electrons are transferred from the reduced carriers to molecular oxygen, in a stepwise fashion, the energy released is used to drive the synthesis of ATP from ADP, producing about 2.5 moles of ATP per mole of NADH reoxidized and about 1.5 moles of ATP per mole of $FADH_2$ reoxidized. As we shall see in Chapter 14, this coupled synthesis generates about 30–32 moles of ATP per mole of glucose oxidized to CO_2 and water.

● **CONCEPT** One turn of the citric acid cycle generates one high-energy phosphate through substrate-level phosphorylation, plus three NADH and one $FADH_2$. The energy released as these electron carriers are reoxidized in the electron transport chain is used to drive the synthesis of ATP from ADP + P_i.

13.5 Regulation of Pyruvate Dehydrogenase and the Citric Acid Cycle

Because the citric acid cycle is a source of biosynthetic intermediates as well as a route for generating metabolic energy, regulation of the cycle is somewhat more complex than if it were just an energy-generating pathway. As in glycolysis, regulation occurs by controlling both the point of entry of fuel into the cycle (pyruvate dehydrogenase complex and citrate synthase) and key irreversible reactions within the cycle (isocitrate dehydrogenase and α-ketoglutarate dehydrogenase). **FIGURE 13.15** summarizes the major factors involved in regulation at both levels.

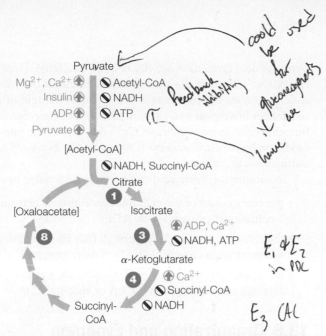

▲ **FIGURE 13.15 Major regulatory factors controlling pyruvate dehydrogenase and the citric acid cycle.** Red brackets indicate concentration dependence. Blue step numbers refer back to Figure 13.4 (page 425). NADH can inhibit through allosteric interactions, but reduced NAD$^+$ availability also slows the citric acid cycle.

Control of Pyruvate Oxidation

As we have seen, fuel enters the cycle primarily as acetyl-CoA, which arises from carbohydrates via pyruvate dehydrogenase and also from the β-oxidation of fatty acids (to be discussed in Chapter 16). The activity of pyruvate dehydrogenase is controlled in two ways: by feedback inhibition and, as noted earlier, by a covalent modification (phosphorylation) that is in turn controlled by the energy state of the cell.

Feedback inhibition operates on two of the three components of the complex. E_2, the dihydrolipoamide transacetylase component (see Figure 13.10), is competitively inhibited by acetyl-CoA. E_3, the dihydrolipoamide dehydrogenase component, is competitively inhibited by NADH. Thus, if the products of the reaction (acetyl-CoA and NADH) are not being continuously removed by subsequent metabolic processes, feedback inhibition by these products will shut down further pyruvate oxidation (**FIGURE 13.16(a)**).

In the mammalian pyruvate dehydrogenase complex, however, the primary mechanism by which enzyme activity is controlled is covalent modification of E_1, the pyruvate dehydrogenase component. As shown schematically in Figure 13.16(b), this involves phosphorylation and dephosphorylation of serine residues in E_1. Mammals possess four different isozymes of pyruvate dehydrogenase kinase that phosphorylate several specific E_1 serine residues, deactivating the complex. Two isozymes of pyruvate dehydrogenase phosphatase that hydrolytically remove the bound phosphate and reactivate the PDH complex are also known. These protein kinase and phosphatase isozymes are expressed in different tissues, and together they mediate tissue-specific regulation of the PDH complex. Recall that these regulatory enzymes are both integrated within the PDH complex, and PDH activity in different tissues is a subtle balance between the relative activities of the kinases and the phosphatases.

PDH kinase is activated by both NADH and acetyl-CoA (products of the PDH reaction) and by ATP. Thus, the kinase turns off PDH activity when products of the reaction accumulate. PDH kinase is inhibited

by ADP and pyruvate. These inhibitors of the kinase, which indicate low cellular energy charge ([ATP]/[ADP] ratio) and available substrate, respectively, result in an increase in the ratio of active to inactive PDH, increasing flux through the complex.

PDH phosphatase is activated by Ca^{2+} and Mg^{2+} and also as a result of insulin secretion. The Ca^{2+} activation of PDH phosphatase mediates stimulation of PDH activity during muscle contraction and in response to epinephrine. Recall from Chapter 7 that Ca^{2+} is a critical signaling molecule for contraction in vertebrate muscle. Using this same signaling molecule to regulate PDH flux provides an elegant mechanism to match ATP demand to its production by the citric acid cycle and subsequent oxidative phosphorylation. The Mg^{2+} activation of PDH phosphatase regulates flux through PDH in response to the energy charge. Because ATP binds Mg^{2+} more tightly than does ADP, the concentration of free Mg^{2+} reflects the [ATP]/[ADP] ratio within the mitochondrion. That is, free Mg^{2+} accumulates at low [ATP]/[ADP] ratios, and it increases PDH activity by stimulating PDH phosphatase activity (and thus dephosphorylation of the complex). When ATP is abundant, however, and further energy production is not needed, pyruvate dehydrogenase is turned off due to activation of PDH kinase (which phosphorylates the complex).

Together, these short-term regulatory mechanisms allow the cell to manage the utilization of fuels for entry into the citric acid cycle. As we will see in Chapter 16, oxidation of fatty acids is another important source of acetyl-CoA and NADH, and both of these activate the

(a) Regulation by feedback inhibition. The products of the pyruvate dehydrogenase reaction, acetyl-CoA and NADH, inhibit pyruvate oxidation if allowed to accumulate.

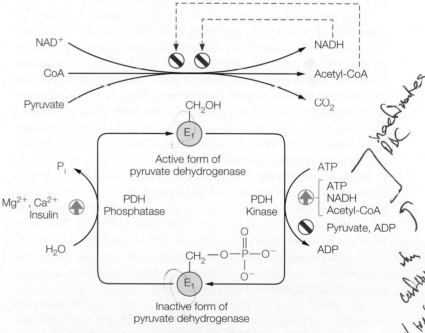

(b) Regulation by covalent modification of E_1. A kinase and a phosphatase inactivate and activate the first component (E_1) of the PDH complex by phosphorylating and dephosphorylating, respectively, three specific serine residues (depicted as —CH$_2$OH).

▲ **FIGURE 13.16 Regulation of the mammalian pyruvate dehydrogenase complex by feedback inhibition and by covalent modification of E_1.**

● **CONCEPT** Activity of the mammalian pyruvate dehydrogenase complex is regulated primarily by reversible phosphorylation of the E_1 subunit.

PDH kinase. Thus, when metabolic conditions favor fatty acid oxidation as the primary fuel source (e.g., during fasting or long-term exercise), carbohydrate reserves are conserved by turning off PDH activity. Once the carbohydrate supply is replenished, PDH can be quickly turned back on.

Control of the Citric Acid Cycle

Flux through the citric acid cycle is controlled by allosteric regulation of key enzymes (Section 8.8), but the concentrations of substrates also play a critical role. Though details of regulation vary among different cells and tissues, the major effects are as summarized in Figure 13.15. As we saw in glycolysis, most of the reactions of the citric acid pathway operate near equilibrium under cellular conditions ($\Delta G \sim 0$) (see Table 13.2). Thus, the key sites for allosteric regulation are those enzymes that catalyze reactions that occur with large free energy decreases: citrate synthase, isocitrate dehydrogenase, and α-ketoglutarate dehydrogenase (Table 13.2).

The most important factor controlling citric acid cycle activity is the ratio of [NAD$^+$] to [NADH] within the mitochondrion. NAD$^+$ is a substrate for three cycle enzymes (Figure 13.15, steps ❸, ❹, and ❽ as well as for pyruvate dehydrogenase. Under conditions

● **CONCEPT** The citric acid cycle is controlled primarily by the relative mitochondrial concentrations of NAD$^+$ and NADH.

that decrease the [NAD$^+$]/[NADH] ratio, such as limitation of the oxygen supply, the low concentration of NAD$^+$ can limit the activities of these dehydrogenases.

Flux through the citrate synthase reaction (Figure 13.15, step ❶) is controlled primarily by substrate availability, since oxaloacetate and acetyl-CoA are generally at concentrations below their K_Ms for the enzyme. Recall from Section 8.6 that at substrate concentrations below the K_M, that is, in the linear region of the substrate–velocity plot, enzyme velocity is quite responsive to changes in substrate concentration. Succinyl-CoA is a competitive inhibitor of citrate synthase, providing a form of feedback inhibition for the cycle. In some mammalian tissues, notably liver, the levels of citrate vary as much as 10-fold, and citrate competes with oxaloacetate binding to citrate synthase (an example of product inhibition). Recall that in some animal tissues citrate is also a prime regulator of flux through glycolysis via allosteric regulation of phosphofructokinase (PFK) (Section 12.6), helping to match the rate of glycolysis to that of the citric acid cycle. This is not true for all tissues. Heart cells, for example, cannot transport citrate out of mitochondria, so citrate interaction with cytosolic PFK probably does not occur to a significant extent. However, citrate levels still control the citric acid cycle in heart muscle.

The other important regulatory sites are the reactions catalyzed by isocitrate dehydrogenase (Figure 13.15, step ❸) and α-ketoglutarate dehydrogenase (Figure 13.15, step ❹). In many cells, isocitrate dehydrogenase is allosterically activated by ADP and allosterically inhibited by NADH and ATP. This control is in addition to the indirect reduction of activity seen at low [NAD$^+$]/[NADH] ratios. The activity of α-ketoglutarate dehydrogenase is inhibited by accumulation of its products succinyl-CoA and NADH. The mechanisms are comparable to the mechanisms by which levels of acetyl-CoA and NADH control

pyruvate dehydrogenase activity (Figure 13.16(a)). Finally, in vertebrates, Ca^{2+} allosterically stimulates both isocitrate dehydrogenase and α-ketoglutarate dehydrogenase. Ca^{2+} can be thought of as a second messenger in a signal transduction pathway, but one that can cross the mitochondrial inner membrane. Ca^{2+} thus allows the rate of substrate oxidation by the citric acid cycle to respond to increased ATP demand during muscle contraction.

To summarize, citric acid cycle flux is responsive to

- the energy state of the cell, through allosteric activation of isocitrate dehydrogenase by ADP;
- the redox state of the cell, through flux rate limitation caused when intramitochondrial [NAD$^+$] decreases; and
- the availability of energy-rich compounds, through inhibition of relevant enzymes by acetyl-CoA or succinyl-CoA.

13.6 Organization and Evolution of the Citric Acid Cycle

The matrix of mitochondria, where the citric acid cycle enzymes are all located, is not the simple aqueous solution we might imagine from the cartoons of mitochondria found in textbooks (including this one—see Figure 13.3). Indeed, it is estimated that the protein concentration of the mitochondrial matrix approaches 500 mg/mL or more, so the matrix is more like a viscous gel. Consistent with this incredibly high protein concentration, there is now considerable evidence that the enzymes of the citric acid cycle are organized in a supramolecular multienzyme complex, or **metabolon,** that is associated with the matrix side of the inner membrane where succinate dehydrogenase is anchored. Physical association of enzymes catalyzing sequential steps in a metabolic pathway could provide significant kinetic advantages through substrate channeling, as was discussed for the pyruvate dehydrogenase complex. It is likely that many, if not most, multistep metabolic pathways are organized into metabolons in intact cells.

As we discussed in Chapter 11, metabolic pathways are the products of evolution, being built from enzymes and pathways that may initially have had other functions. Indeed, as we shall see shortly, even the citric acid cycle as we know it in extant aerobic organisms is used for more than just oxidation of acetyl-CoA. Genomic analyses have revealed the existence of genes for citric acid cycle enzymes in organisms from all three domains of life (Bacteria, Archaea, and Eukarya), including anaerobic chemotrophs. These latter organisms harvest energy by fermenting glucose, using an incomplete citric acid cycle composed of reductive and oxidative branches (see below). The reductive branch reverses the last three enzymes of the traditional cycle (oxaloacetate to succinate) in order to regenerate the oxidized cofactor NAD$^+$ from the NADH produced in the glyceraldehyde-3-phosphate dehydrogenase step of glycolysis. This is the essence of a fermentative pathway (Section 12.1). In the oxidative branch, the first three steps of the citric acid cycle produce α-ketoglutarate, an important biosynthetic precursor. However, these anaerobic chemotrophs lack the enzymes necessary for conversion of α-ketoglutarate to succinate. We can imagine that these reductive and oxidative branches of an incomplete citric acid cycle existed in organisms that evolved before the appearance of atmospheric oxygen some 2.5 billion years ago. Once oxygen levels rose to a level that would support the much more efficient aerobic energy metabolism, it is not difficult to see how a complete citric acid cycle could evolve by recruitment of just a couple of new enzymes (α-ketoglutarate dehydrogenase and succinyl-CoA synthetase).

Branched incomplete citric acid cycle used by anaerobic chemotrophs

Reductive branch:
Oxaloacetate
NADH → NAD$^+$
Malate
Fumarate
NADH → NAD$^+$
Succinate

Oxidative branch:
Acetyl-CoA
Citrate
Isocitrate → NAD$^+$
CO_2 → NADH
α-Ketoglutarate

Reductive branch   **Oxidative branch**

13.7 Citric Acid Cycle Malfunction as a Cause of Human Disease

Given the ubiquitous nature of the citric acid cycle and the critical role it plays in the energy metabolism of a cell, one might expect that a deficiency in any of the enzymes in the cycle would be lethal. However, this is not always the case—we now know that defects in certain citric acid cycle enzymes are indeed linked to a number of rare neurodegenerative diseases and tumors in humans. For example, inherited deficiency of succinate dehydrogenase leads to formation of neuroendocrine tumors. Mutations in the fumarase gene are linked to uterine and/or renal cell cancer. Defects in α-ketoglutarate dehydrogenase, succinate dehydrogenase, and fumarase cause neurodegenerative diseases (Leigh Syndrome or other encephalopathies). Mutations in isocitrate dehydrogenase are found in a majority of several types of malignant gliomas, the most common type of brain tumors in humans.

Although we do not fully understand the mechanisms underlying these processes, a link between organic acid accumulation and abnormal cell proliferation, resulting in tumor formation, is suspected. It is hypothesized that accumulation of certain citric acid cycle metabolites leads to activation of hypoxia-inducible factor 1 (HIF-1), a transcription factor that regulates tumor *angiogenesis* (growth of new blood vessels) and tumor–cell energy metabolism. Despite the wealth of knowledge we

● **CONNECTION** Defects in certain citric acid cycle enzymes are linked to a number of rare neurodegenerative diseases and tumors in humans.

have acquired since Krebs first elucidated the citric acid cycle, we still have a lot to learn about this pathway and its role in cellular physiology.

13.8 Anaplerotic Sequences: The Need to Replace Cycle Intermediates

So far, our discussion of the citric acid cycle has focused on its role in catabolism and energy generation. The cycle also serves as an important source of biosynthetic intermediates and is thus considered an **amphibolic** pathway, or a pathway involving both catabolic and anabolic processes. **FIGURE 13.17** summarizes the most important anabolic roles of citric acid cycle components. These pathways tend to draw carbon from the cycle. Succinyl-CoA, for example, is used in the synthesis of heme and other porphyrins. Oxaloacetate and α-ketoglutarate are the α-keto acid analogs of the amino acids aspartate and glutamate, respectively, and are used in the synthesis of these and other amino acids. In some tissues, citrate is transported from mitochondria to the cytosol, where it is cleaved to provide acetyl-CoA for fatty acid biosynthesis.

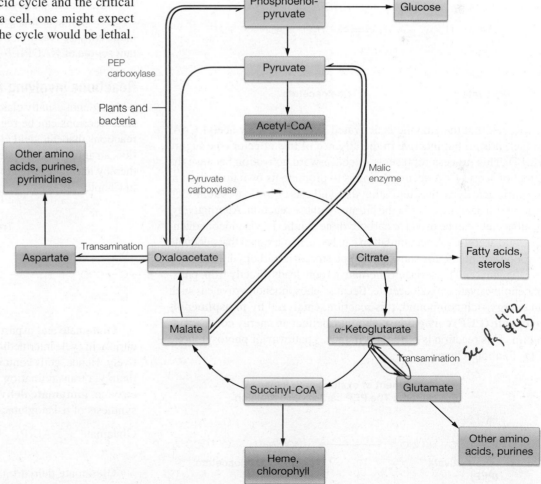

▲ **FIGURE 13.17 Major biosynthetic roles of some citric acid cycle intermediates.** Anaplerotic pathways for replenishment of these intermediates are shown with red arrows.

● **CONCEPT** Citric acid cycle intermediates used in biosynthetic pathways must be replenished to maintain flux through the cycle. Anaplerotic pathways serve this purpose.

Because these and other reactions tend to deplete citric acid cycle intermediates by drawing carbon away, operation of the cycle would be impaired were it not for other processes that replenish the stores of citric acid cycle intermediates. These processes are called anaplerotic pathways, from a Greek word that means "filling up." In most cells, the flow of carbon out of the cycle is balanced by these anaplerotic reactions so that the intramitochondrial concentrations of citric acid cycle intermediates remain constant with time. The anaplerotic processes are summarized by the red arrows in Figure 13.17.

Reactions that Replenish Oxaloacetate

In animals, the most important anaplerotic reaction, particularly in liver and kidney, is the reversible, biotin-dependent carboxylation of pyruvate to give oxaloacetate. This reaction is catalyzed by the biotin-dependent *pyruvate carboxylase*, which we first saw in gluconeogenesis (Chapter 12).

**Replenishment of oxaloacetate in animals:
The pyruvate carboxylase reaction**

Pyruvate　　　　　　**Oxaloacetate**

Recall that the enzyme is activated allosterically by acetyl-CoA; in fact, it is all but inactive in the absence of this effector (see Figure 12.14). This process represents a feedforward activation because the effect of acetyl-CoA accumulation is to promote its own utilization in the citric acid cycle by stimulating the synthesis of oxaloacetate, which reacts with acetyl-CoA, via the citrate synthase reaction. Alternatively, oxaloacetate can be used for carbohydrate synthesis, via gluconeogenesis, and acetyl-CoA accumulation can be seen as a signal that adequate carbon is available for some of it to be stored as carbohydrate.

In plants and bacteria, an alternative route leads directly from phosphoenolpyruvate to oxaloacetate. Because phosphoenolpyruvate is such an energy-rich compound, this reaction, catalyzed by **phosphoenolpyruvate (PEP) carboxylase,** requires neither an energy cofactor nor biotin. This reaction is important in the C_4 pathway of photosynthetic CO_2 fixation (Chapter 15).

**Replenishment of oxaloacetate in plants
and bacteria: The PEP carboxylase reaction**

**Phosphoenolpyruvate
(PEP)**　　　　　　**Oxaloacetate**

A related enzyme, *phosphoenolpyruvate carboxykinase* (PEPCK), interconverts phosphoenolpyruvate and oxaloacetate in animals. The primary role of this enzyme in gluconeogenesis is to produce phosphoenolpyruvate from oxaloacetate, as we discussed in Chapter 12. However, in heart and skeletal muscle, this same enzyme can catalyze the reverse reaction and thereby act as an anaplerotic enzyme, producing oxaloacetate for the citric acid cycle from phosphoenolpyruvate.

The Malic Enzyme

In addition to pyruvate carboxylase and phosphoenolpyruvate carboxylase, a third anaplerotic process is provided by an enzyme commonly known as malic enzyme but more officially as **malate dehydrogenase (decarboxylating:NADP$^+$).** The malic enzyme catalyzes the reductive carboxylation of pyruvate to give L-malate.

Replenishment of malate: The malic enzyme reaction

Pyruvate　　　　　　**L-Malate**

Note that this enzyme uses NADPH, rather than NADH, as the electron donor in this reduction. As we shall see in Chapter 16, the malic enzyme reaction, running in the opposite direction, is an important source of NADPH for fatty acid synthesis.

Reactions Involving Amino Acids

Although not usually classified as anaplerotic pathways, **transamination** reactions can be regarded that way because they are reversible reactions that can yield citric acid cycle intermediates. In transamination, an amino acid transfers its α-amino group to an α-keto acid and is thereby itself converted to an α-keto acid. The mechanism is discussed in Chapter 18.

Transamination reaction

Glutamate and aspartate undergo transamination to generate the citric acid cycle intermediates α-ketoglutarate and oxaloacetate, respectively. Hence, cells containing amino acids in abundance can convert them via transamination to citric acid cycle intermediates. Another enzyme, **glutamate dehydrogenase,** presents an additional route for synthesis of α-ketoglutarate from glutamate.

$$\text{Glutamate} + \text{NAD(P)}^+ + H_2O \rightleftharpoons$$
$$\alpha\text{-ketoglutarate} + \text{NAD(P)H} + NH_4^+$$

Glutamate dehydrogenase, which we discuss in more detail in Chapter 18, uses either NAD$^+$ or NADP$^+$. Being reversible, transaminations and the glutamate dehydrogenase reaction can be used either for amino acid synthesis or for replenishment of citric acid cycle intermediates, depending on the needs of the cell.

Transamination pairs

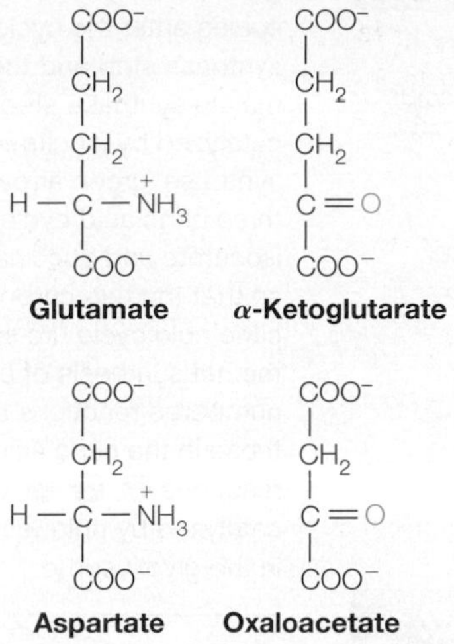

Glutamate   α-Ketoglutarate

Aspartate   Oxaloacetate

Finally, many plants and bacteria can convert two-carbon fragments to four-carbon citric acid cycle intermediates via the glyoxylate cycle, as described in the final section of this chapter.

13.9 The Glyoxylate Cycle: An Anabolic Variant of the Citric Acid Cycle

Metabolically, plant and animal cells differ in many important respects. Of particular concern here is that plant cells, along with some microorganisms, can carry out the net synthesis of carbohydrate from fat. This conversion is crucial to the sprouting of seeds, which generally require a great deal of energy stored in the form of triacylglycerols. (In fact, most vegetable oils available in grocery stores, such as peanut oil, olive oil, or corn oil, are mixtures of triacylglycerols derived from seeds or their accompanying fruit.) When the seeds germinate, triacylglycerols are broken down and converted to sugars, which provide energy and raw material needed for growth of the plant. By contrast, animal cells cannot carry out the net synthesis of carbohydrate from fat.

● **CONCEPT** Unlike plant cells, animal cells cannot carry out the net synthesis of carbohydrate from fat.

● **CONNECTION** Most vegetable oils used in cooking are mixtures of triacylglycerols derived from seeds.

Plants synthesize sugars from fat by using the **glyoxylate cycle,** which can be considered an anabolic variant of the citric acid cycle. To understand the importance of this cycle, consider first the two primary fates of acetyl-CoA in animal metabolism—oxidation through the citric acid cycle, and the synthesis of fatty acids. Because of the virtual irreversibility of the pyruvate dehydrogenase reaction, acetyl-CoA cannot undergo net conversion to pyruvate and hence cannot participate in the net synthesis of carbohydrate. To be sure, the two carbons of acetyl-CoA can be incorporated into oxaloacetate, which is an efficient gluconeogenic precursor. However, because two carbons are lost in this part of the citric acid cycle, there is no *net* accumulation of carbon in carbohydrate. The glyoxylate cycle, however, permits the net synthesis of oxaloacetate by bypassing the reactions in which CO_2 is lost.

The glyoxylate cycle is a cyclic pathway that results in the net conversion of two acetyl units, as acetyl-CoA, to one molecule of succinate. This process occurs in the **glyoxysome,** a specialized organelle that carries out both β-oxidation of fatty acids to acetyl-CoA and utilization of that acetyl-CoA in the glyoxylate cycle. The succinate generated is then transported from the glyoxysome to the mitochondrion, where it is converted, via reactions ❻, ❼, and ❽ of the citric acid cycle (see Figure 13.4), to oxaloacetate. The oxaloacetate is readily utilized for carbohydrate synthesis via gluconeogenesis. Alternatively, some of this oxaloacetate can condense with triacylglycerol-derived acetyl-CoA for oxidation in the mitochondrial citric acid cycle.

● **CONCEPT** The glyoxylate cycle allows plants and bacteria to carry out net conversion of fat to carbohydrate, bypassing CO_2-releasing reactions of the citric acid cycle.

In the glyoxysome, the pathway of the glyoxylate cycle uses some of the same enzymes as the citric acid cycle, but the glyoxysome lacks reactions 3–5 of the citric acid cycle, and thus the glyoxylate cycle bypasses the reactions in which CO_2 is lost during the citric acid cycle. The second mole of acetyl-CoA is brought in during this bypass. Thus, each turn of the cycle involves incorporation of 2 two-carbon fragments (two acetyl groups) and results in the net synthesis of a four-carbon molecule (succinate). The glyoxylate cycle is illustrated in **FIGURE 13.18**.

FIGURE 13.19 illustrates the relationships of the intracellular compartments in a plant cell and the exchange of metabolites between those compartments.

The glyoxylate cycle also allows many microorganisms to metabolize two-carbon substrates, such as acetate. *E. coli*, for example, can grow in a medium that provides acetate as the sole carbon source, as can many fungi, protozoans, and algae. These cells synthesize acetyl-CoA, which is used both for energy production, via the citric acid cycle, and for synthesis of gluconeogenic precursors, via the glyoxylate cycle.

Now let us examine the individual reactions of the glyoxylate cycle in Figure 13.18. As noted, acetyl-CoA is provided from fatty acid oxidation. Alternatively, acetate itself is converted to acetyl-CoA by acetate thiokinase, an enzyme found in nearly all organisms, including those lacking the glyoxylate cycle.

$$\text{Acetate} + \text{CoA-SH} + \text{ATP} \rightleftharpoons \text{acetyl-CoA} + \text{AMP} + \text{PP}_i$$

Next, acetyl-CoA condenses with oxaloacetate to give citrate, ❶, just as in the citric acid cycle, and citrate reacts with aconitase to give isocitrate, ❷. At this point, the glyoxylate cycle diverges from the citric acid cycle. The next reaction, catalyzed by isocitrate lyase, cleaves the 6-carbon isocitrate to glyoxylate and succinate, ❸G.

● **CONNECTION** Isocitrate lyase plays a critical role in the survival of *Mycobacterium tuberculosis* in macrophages during chronic infection; this enzyme is thus a potential drug target for treating tuberculosis.

The isocitrate lyase reaction

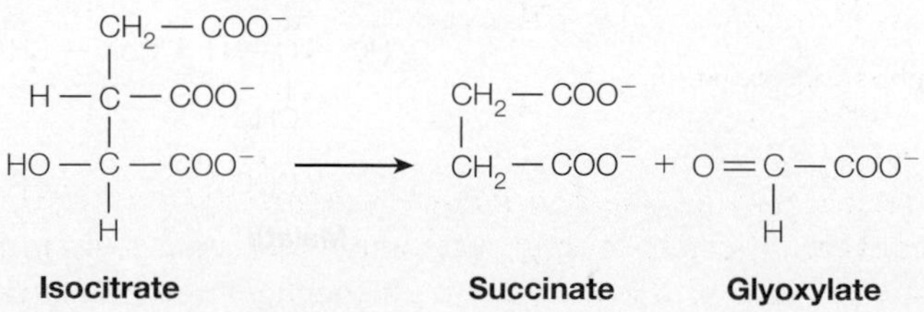

Isocitrate   Succinate   Glyoxylate

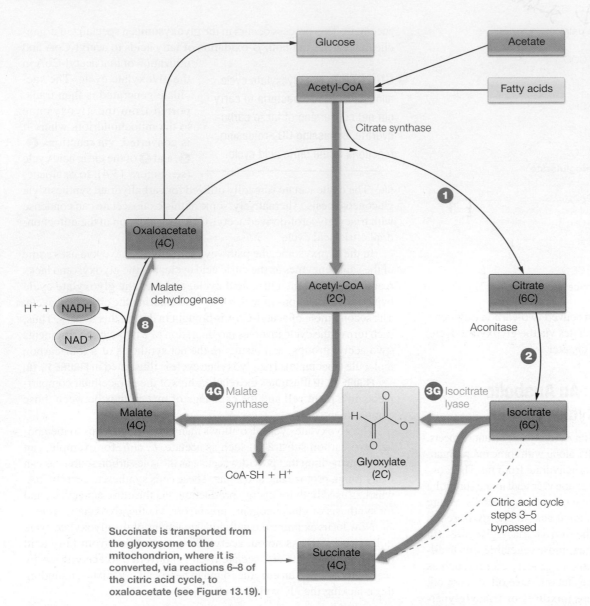

◀ **FIGURE 13.18 Reactions of the glyoxylate cycle.** Two acetyl-CoA molecules enter the cycle, one at the citrate synthase step and the second at the malate synthase step. The reactions catalyzed by isocitrate lyase and malate synthase (green arrows) bypass the three citric acid cycle steps between isocitrate and succinate (blue dashes) so that the two carbons lost in the citric acid cycle are saved, resulting in the net synthesis of oxaloacetate. The numbered reactions are identical to those in the citric acid cycle; however, reactions ❶, ❷, ❸G, ❹G, and ❽ are catalyzed by unique isozymes located in the glyoxysome.

Succinate is transported from the glyoxysome to the mitochondrion, where it is converted, via reactions 6–8 of the citric acid cycle, to oxaloacetate (see Figure 13.19).

Isocitrate lyase uses a reversible aldol cleavage mechanism, similar to that of aldolase in glycolysis. Glyoxylate then accepts two carbons from another acetyl-CoA, in a reaction catalyzed by malate synthase, ❹G.

The malate synthase reaction

$$O = C - COO^- + CH_3 - \overset{\overset{\displaystyle O}{\|}}{C} - S - CoA + H_2O \longrightarrow$$
$$\underset{|}{\overset{}{}}$$
$$H$$

Glyoxylate   **Acetyl-CoA**

$$HO - \underset{\underset{\displaystyle COO^-}{\underset{\displaystyle |}{\underset{\displaystyle CH_2}{|}}}}{\overset{\overset{\displaystyle COO^-}{|}}{C}} - H + CoA - SH + H^+$$

Malate

Mechanistically, this reaction is comparable to that catalyzed by citrate synthase, involving nucleophilic attack of the carbanion form of acetyl-CoA on a carbonyl carbon, in this case the aldehyde carbon of glyoxylate. The malate is then dehydrogenated to regenerate oxaloacetate, ❽, completing the cycle. The enzyme involved here, malate dehydrogenase, is localized in glyoxysomes and is distinct from the mitochondrial form of the enzyme, which is involved in the citric acid cycle. The same is true for citrate synthase and aconitase isozymes used in the glyoxylate cycle (Figure 13.18).

As noted earlier, the glyoxylate cycle results in the net conversion of 2 two-carbon fragments, acetyl-CoA, to a four-carbon compound, succinate, as shown by the following balanced equation.

$$2 \text{ Acetyl-CoA} + NAD^+ + 2H_2O \longrightarrow$$
$$\text{succinate} + NADH + H^+ + 2CoA\text{-}SH$$

The primary fate of succinate is its entry into gluconeogenesis via its conversion to oxaloacetate (Figure 13.19).

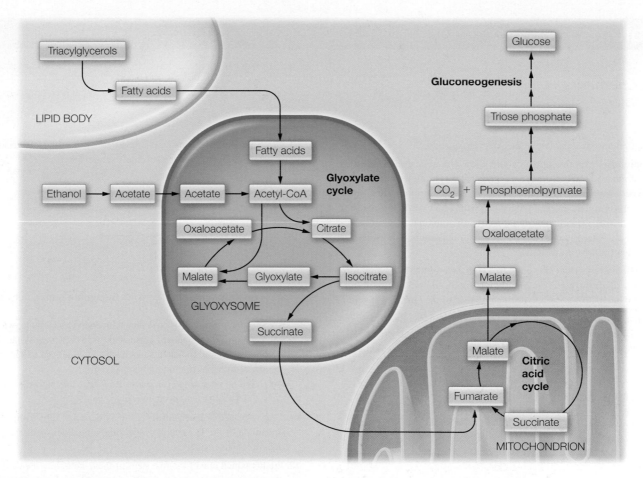

▲ **FIGURE 13.19 Intracellular relationships involving the glyoxylate cycle in plant cells.** Fatty acids released in lipid bodies are oxidized in glyoxysomes to acetyl-CoA, which can also come directly from acetate. Acetyl-CoA is then converted to succinate in the glyoxylate cycle, and the succinate is transported to mitochondria. There it is converted in the citric acid cycle to oxaloacetate, which is readily converted to sugars by gluconeogenesis.

Summary

- **The citric acid cycle** is a central pathway for oxidation of carbohydrates, lipids, and proteins. A principal entrant to this cyclic pathway is pyruvate produced in glycolysis, which undergoes oxidation to acetyl-CoA by the **pyruvate dehydrogenase complex.** Each turn of the citric acid cycle involves entry of two carbons as the acetyl group of acetyl-CoA and loss of two carbons as CO_2. During the cycle, reduced electron carriers, primarily NADH, are generated, and their reoxidation in mitochondria provides the energy for ATP synthesis. (Sections 13.1–13.4)

- Regulation of the citric acid cycle occurs both at the level of entry of fuel into the cycle and at the level of control of key

irreversible reactions within the cycle. In mammals, activity of the pyruvate dehydrogenase complex is regulated by phosphorylation/dephosphorylation, catalyzed by specific protein kinases and phosphatases. (Section 13.5)

- **Anaplerotic reactions** replace citric acid cycle intermediates that are consumed in biosynthetic pathways. (Section 13.8)

- In plants and bacteria the **glyoxylate cycle** bypasses the two decarboxylation reactions of the citric acid cycle, allowing acetyl-CoA to undergo net conversion to carbohydrate. (Section 13.9)

Problems

Enhanced by

Mastering Chemistry
for Biochemistry

Mastering Chemistry for Biochemistry provides select end-of-chapter problems and feedback-enriched tutorial problems, animations, and interactive figures to deepen your understanding of complex topics while practicing problem solving.

Answers to red problems are available in the Answer Appendix.

1. Design a radiotracer experiment that would allow you to determine which proportion of glucose catabolism in a given tissue preparation occurs through the pentose phosphate pathway and which proportion occurs through glycolysis and the citric acid cycle. Assume that you can synthesize glucose labeled with ^{14}C in any desired position or combination of positions. Assume also that you can trap CO_2 after administration of labeled glucose and determine its radioactivity.

2. Consider the fate of pyruvate labeled with ^{14}C in each of the following positions: carbon 1 (carboxyl), carbon 2 (carbonyl), and carbon 3 (methyl). Predict the fate of each labeled carbon during one turn of the citric acid cycle.

3. Suppose that aconitase did not bind its substrate asymmetrically. What fraction of the carbon atoms introduced in one cycle as acetyl-CoA would be released in the first turn of the cycle? What fraction of the carbon atoms that entered in the first cycle would be released in the second turn?

4. [methyl-^{14}C]Pyruvate was administered to isolated liver cells in the presence of sufficient malonate to block succinate dehydrogenase completely. After a time, isocitrate was isolated and found to contain label in both carbon 2 and carbon 5:

$$^{14}CH_2-COO^-$$
$$HC-COO^-$$
$$H^{14}C-COO^-$$
$$OH$$

How do you explain this result?

5. Considering the evidence that led Krebs to propose a cyclic pathway for oxidation of pyruvate, discuss the type of experimental evidence that might have led to realization of the cyclic nature of the glyoxylate pathway.

6. Which carbon or carbons of glucose, if metabolized via glycolysis and the citric acid cycle, would be most rapidly lost as CO_2?

7. Assume you have a solution containing the pyruvate dehydrogenase complex and all the enzymes of the citric acid cycle, but none of the metabolic intermediates. When you supplement this solution with 5 μmoles each of pyruvate, oxaloacetate, coenzyme A, NAD$^+$, FAD, GDP, and P$_i$, you find that 5 μmoles of CO_2 are evolved and then the reaction stops. When you add alcohol dehydrogenase and its substrate acetaldehyde, additional CO_2 is produced.
   (a) How do you explain this result?
   (b) How many μmoles of acetaldehyde are required to allow complete oxidation of the pyruvate to 15 μmoles of CO_2?

8. Outline the mechanism of the conversion of α-ketoglutarate to succinyl-CoA catalyzed by α-ketoglutarate dehydrogenase complex. Include all products, coenzymes, and reactions in your discussion.

9. Briefly describe the biological rationale for each of the following allosteric phenomena: (a) activation of pyruvate carboxylase by acetyl-CoA; (b) activation of pyruvate dehydrogenase kinase by NADH; (c) inhibition of isocitrate dehydrogenase by NADH; (d) activation of isocitrate dehydrogenase by ADP; (e) inhibition of α-ketoglutarate dehydrogenase by succinyl-CoA; (f) activation of pyruvate dehydrogenase phosphatase by Ca^{2+}.

10. Certain microorganisms with a modified citric acid cycle decarboxylate α-ketoglutarate to produce succinate semialdehyde:

$$\begin{array}{cc} COO^- & COO^- \\ | & | \\ CH_2 & CH_2 \\ | & \xrightarrow{CO_2} & | \\ CH_2 & CH_2 \\ | & | \\ C=O & C=O \\ | & | \\ COO^- & H \end{array}$$

α-Ketoglutarate   **Succinate semialdehyde**

   (a) Succinate semialdehyde is then converted to succinate, which is further metabolized by standard citric acid cycle enzymes. What kind of reaction is required to convert succinate semialdehyde to succinate? Show any coenzymes that might be involved.
   (b) Based on your answer in part a, how does this pathway compare to the standard citric acid cycle in energy yield?

11. Draw a plausible mechanism for the oxidation of dihydrolipoamide to lipoamide by the E$_3$ subunit (dihydrolipoamide dehydrogenase) of pyruvate dehydrogenase complex.

12. Predict which one of the five steps of the α-ketoglutarate dehydrogenase complex reaction is metabolically irreversible under physiological conditions and explain why.

13. Draw a plausible mechanism for the oxidative decarboxylation of isocitrate by isocitrate dehydrogenase.

14. Aconitase catalyzes the reaction: citrate ⇌ isocitrate

   The standard free energy change, $\Delta G°'$, for this reaction is +6.3 kJ/mol. However, the observed free energy change (ΔG) for this reaction in mammalian mitochondria at 25°C is ~0 kJ/mol (Table 13.2).
   (a) Calculate the ratio of [isocitrate]/[citrate] in mitochondria.
   (b) Is this reaction likely to be a control point for the citric acid cycle? Why or why not?

15. The same E$_1$–E$_2$–E$_3$ multienzyme structure found in the pyruvate dehydrogenase and the α-ketoglutarate dehydrogenase complexes is also used in the branched-chain α-ketoacid dehydrogenase complex, which participates in the catabolism of branched-chain amino acids. Draw the reaction product when the following substrate is acted on by the branched-chain α-keto acid dehydrogenase complex.

$$\begin{array}{c} H_3C \\ \diagdown \\ \diagup \\ H_3C \end{array} \begin{array}{c} O \\ \| \\ C-COO^- \end{array}$$

16. Given what you know about the function of the glyoxylate cycle and the regulation of the citric acid cycle, propose control mechanisms that might regulate the glyoxylate cycle.

17. Write a balanced equation for the conversion in the glyoxylate cycle of two acetyl units, as acetyl-CoA, to oxaloacetate.

18. Isocitrate lyase cleaves the 6-carbon isocitrate to glyoxylate and succinate. Draw a plausible mechanism for this reaction.

19. Construct a table that lists all of the metabolic reactions from Chapters 12 and 13 that form or break C—C bonds. Arrange them according to whether the C—C bond is α or β to a carbonyl. Which reactions use thiamine pyrophosphate (TPP) as a cofactor? Explain the chemical basis of your observations.

20. Given the roles of NAD^+/NADH in dehydrogenation reactions and NADPH/$NADP^+$ in reductions, as discussed in Chapter 11 (Section 11.4), would you expect the intracellular ratio of NAD^+ to NADH to be high or low? What about the ratio of $NADP^+$ to NADPH? Explain your answers.

21. When the O_2 supply from blood fails to meet the demand of O_2-consuming cells, oxygen deprivation (hypoxia) occurs. This is common, for example, in exercising muscle. It has been recognized for over 100 years that O_2-deprived cells show increased conversion of glucose to lactate, known as the Pasteur effect. Activation of the Pasteur effect during hypoxia is mediated by hypoxia-inducible factor-1 (HIF-1). HIF-1 is a transcription factor that upregulates the expression of several glycolytic enzymes that support the increased glycolytic ATP production as mitochondria become starved for O_2. At the same time glycolysis is increasing, the rate of mitochondrial respiration decreases. New research reveals that in addition to upregulating enzymes in the glycolytic pathway, HIF-1 also induces the expression of cytoplasmic lactate dehydrogenase (LDH) and mitochondrial pyruvate dehydrogenase kinase (PDK).

    (a) Explain why glucose consumption must increase in hypoxic tissues to provide the same amount of ATP that could be produced from glucose in normoxic (normal O_2 levels) tissues.

    (b) How would increasing LDH expression increase the rate of glycolysis?

    (c) How would increasing PDK expression decrease the rate of mitochondrial respiration?

22. Some bacteria use the citric acid cycle intermediate, α-ketoglutarate, plus acetyl-CoA, as the starting point for lysine biosynthesis. The first part of this biosynthetic pathway uses the same chemical strategy found in the citric acid cycle. Propose a four-step pathway for the conversion of α-ketoglutarate to 2-oxoadipate. Draw the three missing intermediates, and indicate the chemistry involved in each reaction. Include any cofactors that you think might be required for specific steps.

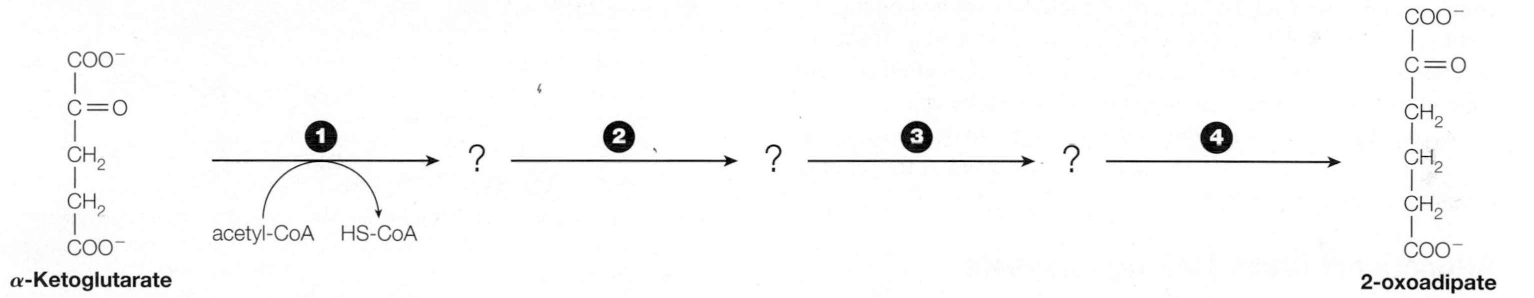

References

For a list of references related to this chapter, see Appendix II.

The enzymes of glycolysis or the citric acid cycle are readily isolated as soluble proteins, but several lines of evidence suggest that they are physically associated within living cells. For years, biochemistry students have been told, "A cell is not a bag of enzymes," implying that enzymes are organized into functional supramolecular units in intact cells. Often these organized units are stabilized by weak, noncovalent forces that are easily disrupted when cells are broken open, as must occur if the enzymes within are to be isolated and characterized. Even when cells are gently lysed, most protein extraction processes dilute intracellular contents by several orders of magnitude, and that alone can disrupt associations that are highly concentration-dependent. Biochemists are trying to define how the organization of functionally related enzymes facilitates the flow of metabolites and the control and coordination of metabolic pathways.

If enzymes functioned as part of a multienzyme complex that could be isolated intact, its properties could be explored by the methods for molecular weight determination, described in Tools of Biochemistry 6B. However, because of the difficulties often encountered in isolating enzyme complexes held together by ephemeral forces, scientists usually use multiple approaches to demonstrate and characterize the protein–protein interactions involved. A few of those techniques are described here.

Bifunctional Cross-Linking Reagents

These are reagents containing two functional groups capable of forming covalent bonds with specific amino acid residues in closely associated proteins. For instance, *dimethylsuberimidate* (DMS) reacts with lysine ε-amino groups and N-terminal amino groups, cross-linking two proteins in a form that can be detected by gel electrophoresis due to the increase in molecular weight.

Some reagents have cleavable cross-links, such as a disulfide bond that can be reductively cleaved, allowing analysis of the separate cross-linked partners. Although the technique can be very informative, experimentation with many reagents is often required to find the right combination of functional groups and distance between the reactive partners for the cross-linking reaction to proceed to a measurable extent. Also, care must be taken not to overinterpret results, for even transient contacts between molecules, which may occur nonspecifically, sometimes lead to cross-linking.

Affinity Chromatography

In this technique, described in Tools of Biochemistry 5A, one protein is immobilized on a chromatographic support, and a mixture of proteins is passed through a column of this material. Proteins that are retained can be identified after elution, by biological activity or by electrophoretic techniques, such as immunoblotting or two-dimensional electrophoretic analysis. The chief limitations of affinity chromatography are the need to have one of the test proteins available in pure form for immobilization and the fact that interactions occur in a rather artificial environment. Again, controls are essential because of nonspecific retention of some proteins on affinity columns.

Immunoprecipitation

Antibody to a purified protein can be added to a protein mixture, often with immunoprecipitation (see Tools of Biochemistry 7A) of both the antigenic protein and any interactive proteins bound to it (co-immunoprecipitation). Although this technique is qualitative, like the approaches described earlier, it is simple to do, and it needs only small amounts of material. Because multiple assays can be run simultaneously, co-immunoprecipitation can be used, for example, to study the effects of the binding of small molecules (substrates or effectors) upon protein associations.

Kinetic Analysis

If enzymes catalyzing sequential reactions interact, the interactions can facilitate the flow of metabolites through multistep pathways (metabolic channeling), and this can be detected in vitro in several ways. Generally, a channeled pathway will display one or more of the following characteristics: (1) reduced *transient time,* the interval after initiating a multistep pathway needed for the formation of final product to reach its maximal rate; (2) steady-state levels of intermediates much lower than expected if they must seek the next enzyme acting on them by diffusion rather than by direct or facilitated transfer to a nearby enzyme molecule; and (3) restricted ability of an exogenous intermediate to equilibrate with the same intermediate in a channeled pathway, usually determined by radio-isotope experiments.

Library-Based Methods

These methods allow the screening of a large number, or library, of cloned genes. They allow tentative identification of interacting partners without first purifying and identifying one of the partners. A popular method called the two-hybrid system uses a transcriptional activation system in yeast that requires two proteins to interact in order to initiate transcription at an appropriate gene site (see Chapters 24 and 26). One of these proteins binds at the DNA site, and the other activates transcription. Two hybrid, or fusion, proteins are generated by recombinant DNA techniques (see Tools of Biochemistry 4A); the gene for one test protein (X) is fused to the DNA-binding protein gene, and the gene for another test protein (Y), or a library of cloned genes, is fused to the gene for the transcriptional activation domain. The recombinant genes are transferred into yeast cells, where the interaction of proteins X and Y can form a fully functional transcriptional activator (assuming that the functional domains of the fusion proteins fold as they do in their native state). Transcription of the target gene is then monitored by assays for the activity of a reporter gene, a gene cloned downstream of the promoter and whose biological activity is easily assayed. Once a specific protein association has been detected, it becomes essential to isolate the interactive partners as full-length proteins and to ascertain that the interactions detected by this somewhat qualitative method are indeed biologically significant.

Biosensor Analysis

In recent years, a new kind of instrumentation has been developed that allows both qualitative and quantitative analysis of protein–protein interactions, using rather small quantities of purified proteins. One such instrument, BIACORE, measures an optical property called *surface plasmon resonance,* which is related to minute changes in the refractive index that occur when a protein in solution interacts with a protein immobilized on a chip. The signal measured is proportional to total protein concentration, over a wide range. Thus, the kinetics

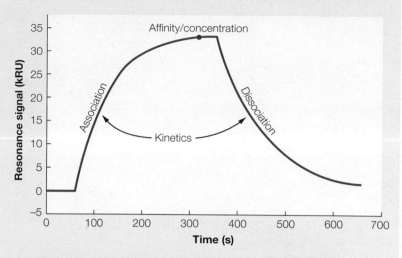

▲ **FIGURE 13A.1** **Biacore analysis of a protein–protein association.** Test protein flows past the immobilized protein in the association phase and is replaced by buffer in the dissociation phase. The height of the plateau response, as compared with standards, is related to the stoichiometry of the association.

of a protein association reaction can be monitored by following the increase in signal as a protein in solution passes over a chip containing immobilized protein. The amount of protein bound at equilibrium gives the affinity constant for the interaction; the kinetics of dissociation can then be followed by passing buffer over the chip and following the decrease in signal, as indicated in **FIGURE 13A.1**. Limitations of this useful technique, which can be controlled for, include the possibility that immobilization alters the protein in a way that affects the interaction and the fact that the two interacting proteins are in different phases (solid, or immobilized, and liquid).

Mitochondria synthesize ATP using energy discharged from an electrochemical gradient across the inner membrane. The electrical charge across this membrane is similar to the voltage of a lightning bolt.

Electron Transport, Oxidative Phosphorylation, and Oxygen Metabolism

14

THE AVERAGE ADULT human synthesizes ATP at a rate of nearly 10^{21} molecules per second, equivalent to producing his or her own weight in ATP *every day.* How is this massive amount of energy extracted from nutrients? As we saw in Chapters 12 and 13, glycolysis and the citric acid cycle by themselves generate relatively little ATP directly. However, under aerobic conditions, six substrate oxidation steps—one in glycolysis, another in the pyruvate dehydrogenase reaction, and four more in the citric acid cycle—collectively reduce 10 moles of NAD^+ to NADH and 2 moles of FAD to $FADH_2$ per mole of glucose. Reoxidation of these reduced electron carriers in cellular respiration generates most of the energy that is then used for ATP synthesis. Respiration constitutes the third stage of the metabolic oxidation of substrates (**FIGURE 14.1**).

In eukaryotic cells, NADH and $FADH_2$ are reoxidized by electron transport proteins bound to the inner mitochondrial membrane.

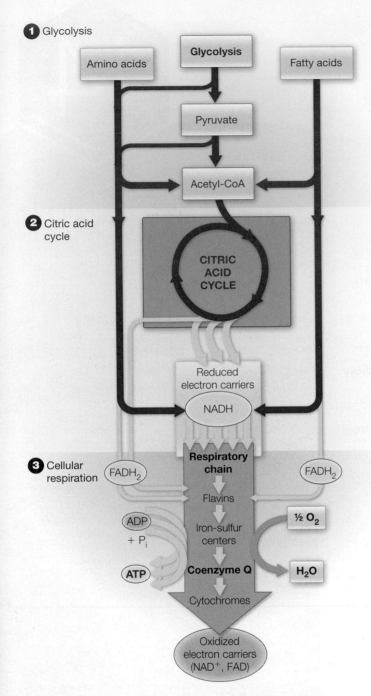

1 Glycolysis

2 Citric acid cycle

3 Cellular respiration

▲ **FIGURE 14.1 Overview of oxidative energy generation.** Stage 1, conversion of carbon from metabolic fuels into an activated two-carbon fragment—acetyl-CoA. Stage 2, oxidation of acetyl-CoA in the citric acid cycle. Stage 3, electron transport and oxidative phosphorylation.

● **CONCEPT** Oxidation of 1 mole of NADH by the respiratory chain provides sufficient energy for synthesis of ~2.5 moles of ATP from ADP.

between 2 and 3 moles of ATP from ADP and P_i by a process called **oxidative phosphorylation.** How is the free energy from the oxidative reactions of the respiratory chain harnessed, or coupled, to drive the synthesis of ATP? The mechanism of this coupling will concern us throughout this chapter. In addition, we will consider a number of other metabolic roles oxygen plays in aerobic cells.

In Chapter 12, we saw that the free energy of glucose breakdown in glycolysis is used to convert low-energy phosphorylated intermediates into high-energy phosphorylated intermediates. These then transfer their phosphates to ADP, forming ATP in a process called *substrate-level phosphorylation.* It was thus assumed, during the early years of research into cellular energetics, that a similar mode of direct chemical coupling must underlie the formation of ATP in respiration. For a long time, therefore, researchers sought a high-energy intermediate that could link electron transport to ATP synthesis.

However, several nagging facts argued against a direct chemical coupling process like the substrate-level phosphorylation reactions in glycolysis. Perhaps most troubling was the realization that the number of ATP molecules produced per glucose during respiration was not integral. That is, the passage of a pair of electrons down the respiratory chain generates approximately 2.5 ATPs, not an integral number, as would be predicted by a direct chemical coupling mechanism. Furthermore, respiration was shown to require an intact inner mitochondrial membrane. If the membrane is disrupted, the passage of electrons down the respiratory chain becomes *uncoupled* from ATP synthesis: Electrons still flow, but ATP production ceases.

As we shall soon see, cells long ago evolved an elegantly simple mechanism that explains all of these observations, a mechanism that turns out to have much broader significance than just the synthesis of ATP. In short, the passage of electrons down the respiratory chain creates a proton gradient across the inner mitochondrial membrane, and the energy in this proton gradient provides the driving force for ATP synthesis.

A series of linked oxidation and reduction reactions occurs, with electrons being passed along a series of electron carriers known as the **electron transport chain,** or **respiratory chain** (Figure 14.1).

The final step in the respiratory chain is reduction of O_2 to water. One pair of reducing equivalents, generated from 1 mole of NADH, suffices to drive the synthesis of

14.1 The Mitochondrion: Scene of the Action

Our comprehension of biological oxidations requires an understanding of both the chemistry of oxidation–reduction reactions and the cell biology of the mitochondrion. Before reviewing the chemistry, let us describe the intracellular sites where these reactions occur. Cellular metabolism generates reduced compounds in all of the major compartments of a eukaryotic cell. As noted earlier, glycolysis takes place in the cytosol of eukaryotic cells, whereas pyruvate oxidation, fatty acid β-oxidation, amino acid oxidation, and the citric acid cycle occur within the mitochondrial matrix. Individual cells vary widely in the abundance and structure of their mitochondria. Most vertebrate cells contain from a few hundred to a few thousand mitochondria, but the number can be as low as 1 or as high as 100,000.

The mitochondrion consists of four distinct subregions, shown in **FIGURE 14.2(a)**—the outer membrane, the intermembrane space, the inner membrane, and the matrix, enclosed by the inner membrane. The inner membrane is highly folded into *cristae* that project into— often very deeply into—the interior of the mitochondrion. Because respiratory proteins are embedded in the inner membrane, the density of cristae is related to the respiratory activity of a cell. For example, heart muscle cells, which have high rates of respiration, contain mitochondria with densely packed cristae. By contrast, liver cells have much lower respiration rates and mitochondria with more sparsely distributed cristae.

Whatever the compartment in which biological oxidations occur, all of these processes generate reduced electron carriers, primarily NADH. For example, the citric acid cycle intermediates isocitrate, α-ketoglutarate, and malate are all NAD^+-linked substrates, whose oxidation generates NADH. Most of this NADH is reoxidized by the enzymes of the respiratory chain, which are firmly embedded in the inner membrane. The inner membrane itself consists of about 70% protein and 30% lipid, making it perhaps the most protein-rich of all biological membranes. About half of the inner membrane proteins in bovine heart mitochondria are directly involved in electron transport and oxidative phosphorylation. Most of the remaining proteins are involved in transport of substances into and out of mitochondria. By contrast, a completely different set of proteins is bound to the outer membrane, including enzymes of amino acid oxidation, fatty acid elongation, membrane phospholipid biosynthesis, and enzymatic hydroxylations.

The inner membrane proteins that constitute the respiratory chain are assembled into five multiprotein enzyme complexes, named I, II, III, IV, and V (Figure 14.2(b)). Complex I and complex II receive electrons from the oxidation of NADH and succinate, respectively, and pass them along to a lipid-soluble electron carrier, coenzyme Q, which moves freely in the membrane. Complex III catalyzes the transfer of electrons from the reduced form of coenzyme Q to cytochrome *c*, a soluble protein electron carrier that is mobile within the intermembrane space. Finally, complex IV catalyzes the oxidation of cytochrome *c*, reducing O_2 to water. The energy released by these exergonic reactions is used to pump protons from the matrix into the intermembrane space, creating a proton gradient across the inner membrane. Protons then reenter the matrix through a specific channel in complex V. The free energy of this exergonic process drives the endergonic production of ATP from ADP and inorganic phosphate (P_i). Throughout this chapter we develop the structural and functional basis for our understanding of these energy-coupling processes.

Critical to comprehension of these processes was the isolation of physiologically intact mitochondria, using differential centrifugation of cell homogenates (see Section 11.6). This feat was accomplished in the late 1940s by Eugene Kennedy and Albert Lehninger, who demonstrated that isolated mitochondria could synthesize ATP from ADP and P_i in vitro, but only if an oxidizable substrate was present as well. Fractionation and analysis of these functional, isolated mitochondria then allowed identification and characterization of the protein complexes described in the previous paragraph.

The situation in bacterial and archaeal cells is comparable, although different electron carriers are involved. However, because these cells lack organelles, all of the electron carriers and enzymes of oxidative phosphorylation are bound to the inner side of the plasma membrane. Therefore, electron transport and oxidative phosphorylation occur at the cell periphery.

● **CONCEPT** Most electron carriers in the respiratory chain are protein complexes embedded in the mitochondrial inner membrane. The exceptions are coenzyme Q, a lipid-soluble electron carrier that moves freely in the membrane, and cytochrome *c*, a soluble protein electron carrier that is mobile within the intermembrane space.

14.2 Free Energy Changes in Biological Oxidations

Biological electron transport consists of a series of linked oxidations and reductions, also called redox reactions or oxidoreduction reactions. To understand the logic behind the sequence of reactions in the respiratory chain, as well as the mechanisms by which metabolic energy is generated from these reactions, recall our discussion of the thermodynamics of redox reactions in Section 3.4. We saw that the higher the value of the *standard reduction potential, $E°'$*, for a redox couple, the greater the tendency for that couple to participate in oxidation of another substrate. We can describe this tendency in quantitative terms because free energy changes are directly related to differences in reduction potential:

$$\Delta G°' = -nF\Delta E°' = -nF(E°'_{acceptor} - E°'_{donor}) \quad (14.1)$$

where n is the number of electrons transferred in the half-reactions, F is Faraday's constant (96485 J $mol^{-1}V^{-1}$), and $\Delta E°'$ is the difference in standard reduction potentials between the two redox couples. $E°'$ values for a number of biochemically important redox pairs are recorded in **TABLE 14.1**.

The values given in Table 14.1 allow calculation of free energy changes only under standard conditions (including, by convention, a pH of 7.0). For nonstandard conditions such as those that might exist in a cell, recall that we learned to use Equation 3.18 to evaluate ΔG for a redox reaction such as

$$A_{ox} + B_{red} \rightarrow A_{red} + B_{ox}$$

$$\Delta G = \Delta G°' + RT \ln\left(\frac{[A_{red}][B_{ox}]}{[A_{ox}][B_{red}]}\right) \quad (14.2)$$

In this example, A is the electron acceptor and B is the donor.

(a) A mitochondrion from a pancreatic cell, shown as a thin section in a color-enhanced transmission electron micrograph. The major mitochondrial compartments are shown, along with principal enzymes and pathways localized to each compartment.

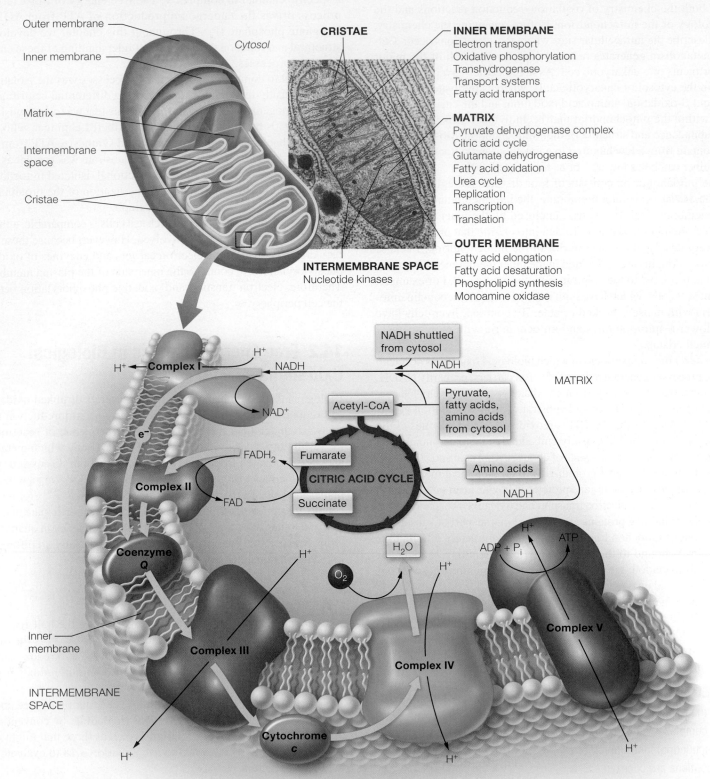

Outer membrane

Cytosol

Inner membrane

Matrix

Intermembrane space

Cristae

CRISTAE

INNER MEMBRANE
Electron transport
Oxidative phosphorylation
Transhydrogenase
Transport systems
Fatty acid transport

MATRIX
Pyruvate dehydrogenase complex
Citric acid cycle
Glutamate dehydrogenase
Fatty acid oxidation
Urea cycle
Replication
Transcription
Translation

INTERMEMBRANE SPACE
Nucleotide kinases

OUTER MEMBRANE
Fatty acid elongation
Fatty acid desaturation
Phospholipid synthesis
Monoamine oxidase

NADH shuttled from cytosol

H^+

Complex I

H^+

NADH

NADH

MATRIX

e^-

NAD^+

Acetyl-CoA

Pyruvate, fatty acids, amino acids from cytosol

$FADH_2$

Fumarate

Complex II

CITRIC ACID CYCLE

Amino acids

FAD

Succinate

NADH

H^+

H_2O

H^+

$ADP + P_i$

ATP

Coenzyme Q

O_2

Complex V

Inner membrane

Complex III

Complex IV

INTERMEMBRANE SPACE

Cytochrome c

H^+

H^+

H^+

H^+

(b) Overview of oxidative phosphorylation. Reduced electron carriers, produced by cytosolic dehydrogenases and mitochondrial oxidative pathways, become reoxidized by enzyme complexes bound in the inner mitochondrial membrane. These complexes actively pump protons outward from the matrix into the intermembrane space, creating an energy gradient whose discharge through complex V drives ATP synthesis.

▲ **FIGURE 14.2** Localization of respiratory processes in the mitochondrion.

TABLE 14.1 Standard reduction potentials of interest in biochemistry

Oxidant		Reductant	n	$E^{\circ\prime}$ (V)
Acetate + CO_2 + $2H^+$ + $2e^-$	$\rightleftharpoons$	Pyruvate + H_2O	2	−0.70
Succinate + CO_2 + $2H^+$ + $2e^-$	$\rightleftharpoons$	α-Ketoglutarate + H_2O	2	−0.67
Acetate + $3H^+$ + $2e^-$	$\rightleftharpoons$	Acetaldehyde + H_2O	2	−0.60
Ferredoxin (oxidized) + e^-	$\rightleftharpoons$	Ferredoxin (reduced)	1	−0.43
$2H^+$ + $2e^-$	$\rightleftharpoons$	H_2	2	−0.42
α-Ketoglutarate + CO_2 + $2H^+$ + $2e^-$	$\rightleftharpoons$	Isocitrate	2	−0.38
Acetoacetate + $2H^+$ + $2e^-$	$\rightleftharpoons$	β-Hydroxybutyrate	2	−0.35
Pyruvate + CO_2 + H^+ + $2e^-$	$\rightleftharpoons$	Malate	2	−0.33
NAD^+ + H^+ + $2e^-$	$\rightleftharpoons$	NADH	2	−0.32
$NADP^+$ + H^+ + $2e^-$	$\rightleftharpoons$	NADPH	2	−0.32
Lipoate (oxidized) + $2H^+$ + $2e^-$	$\rightleftharpoons$	Lipoate (reduced)	2	−0.29
1,3-Bisphosphoglycerate + $2H^+$ + $2e^-$	$\rightleftharpoons$	Glyceraldehyde-3-phosphate + P_i	2	−0.29
Glutathione (oxidized) + $2H^+$ + $2e^-$	$\rightleftharpoons$	2 Glutathione (reduced)	2	−0.23
FAD (free coenzyme) + $2H^+$ + $2e^-$	$\rightleftharpoons$	$FADH_2$	2	−0.22
Acetaldehyde + $2H^+$ + $2e^-$	$\rightleftharpoons$	Ethanol	2	−0.20
Pyruvate + $2H^+$ + $2e^-$	$\rightleftharpoons$	Lactate	2	−0.19
Oxaloacetate + $2H^+$ + $2e^-$	$\rightleftharpoons$	Malate	2	−0.17
O_2 + e^-	$\rightleftharpoons$	O_2^- (superoxide)	1	−0.16
α-Ketoglutarate + NH_4^+ + $2H^+$ + $2e^-$	$\rightleftharpoons$	Glutamate + H_2O	2	−0.14
FAD (enzyme-bound) + $2H^+$ + $2e^-$	$\rightleftharpoons$	$FADH_2$ (enzyme-bound)	2	~0 to −0.30
Methylene blue (oxidized) + $2H^+$ + $2e^-$	$\rightleftharpoons$	Methylene blue (reduced)	2	0.01
Fumarate + $2H^+$ + $2e^-$	$\rightleftharpoons$	Succinate	2	0.03
Q + $2H^+$ + $2e^-$	$\rightleftharpoons$	QH_2	2	0.04
Dehydroascorbate + $2H^+$ + $2e^-$	$\rightleftharpoons$	Ascorbate	2	0.06
Cytochrome b (+3) + e^-	$\rightleftharpoons$	Cytochrome b (+2)	1	0.07
Cytochrome c_1 (+3) + e^-	$\rightleftharpoons$	Cytochrome c_1 (+2)	1	0.23
Cytochrome c (+3) + e^-	$\rightleftharpoons$	Cytochrome c (+2)	1	0.25
Cytochrome a (+3) + e^-	$\rightleftharpoons$	Cytochrome a (+2)	1	0.29
O_2 + $2H^+$ + $2e^-$	$\rightleftharpoons$	H_2O_2	2	0.30
Ferricyanide + $2e^-$	$\rightleftharpoons$	Ferrocyanide	2	0.36
NO_3^- (Nitrate) + $2H^+$ + $2e^-$	$\rightleftharpoons$	NO_2^- (Nitrite) + H_2O	2	0.42
Cytochrome a_3(+3) + e^-	$\rightleftharpoons$	Cytochrome a_3 (+2)	1	0.55
Fe (+3) + e^-	$\rightleftharpoons$	Fe (+2)	1	0.77
$\frac{1}{2}O_2$ + $2H^+$ + $2e^-$	$\rightleftharpoons$	H_2O	2	0.82

Note: $E^{\circ\prime}$ is the standard reduction potential at pH 7 and 25 °C, n is the number of electrons transferred, and each potential is for the partial reaction written as follows: Oxidant + $ne^- \rightleftharpoons$ reductant.

Each of the coupled redox reactions in biological electron transport involves the transfer of electrons from one redox couple to another couple of higher (more positive) reduction potential. Thus, each individual redox reaction in the sequence is exergonic under standard conditions. For electrons entering the respiratory chain as NADH, the overall reaction sequence is given by the following equation:

$$\text{NADH} + \text{H}^+ + \tfrac{1}{2}\text{O}_2 \rightleftharpoons \text{NAD}^+ + \text{H}_2\text{O}$$

This sequence is strongly exergonic under standard conditions:

$$\Delta G^{\circ\prime} = -nF\,\Delta E^{\circ\prime} = -2(96485\ \text{J mol}^{-1}\text{V}^{-1})(0.82\ \text{V} - (-0.32\ \text{V}))$$

$$= -220\ \text{kJ/mol}$$

As discussed later in this chapter, the oxidation of 1 mole of NADH in the respiratory chain can drive the synthesis of about 2.5 moles of ATP from ADP and P_i.

14.3 Electron Transport

The preceding sections have provided a glimpse of the respiratory process and a review of the thermodynamic principles that underlie this process. Now we are ready to take a closer look at the chemistry and structure of the various electron carriers and the respiratory complexes that make up this elegant energy transformation system.

Electron Carriers in the Respiratory Chain

If you use Table 14.1 to look up the standard reduction potentials of the electron carriers in the respiratory chain (shown in sequence in **FIGURE 14.3**), you will find that their $E°'$ values increase in a sequence that corresponds precisely to their position in the chain. This order suggests that each individual oxidoreduction reaction in electron transport is exergonic under standard conditions and that electrons flow in continuous fashion from low-potential to high-potential carriers. This is a very neat arrangement, but is it real? After all, we have seen that glycolysis and the citric acid cycle both proceed smoothly despite the inclusion of some reactions with large positive $\Delta G°'$ values. We shall explore some of the lines of evidence by which the currently accepted pathway of electron transport was determined. First, however, let us become better acquainted with the participants—the electron carriers involved.

● **CONCEPT** The respiratory chain catalyzes the transport of electrons from low-potential carriers to high-potential carriers.

Flavoproteins

Flavoproteins, introduced in Section 13.2, contain tightly bound flavin mononucleotide (FMN) or flavin adenine dinucleotide (FAD) as redox cofactors. Each flavoprotein provides a different microenvironment for the isoalloxazine ring, conferring a unique standard reduction potential on the flavin. Flavin nucleotides can link two-electron and one-electron processes because they are stable in both a one-electron reduced semi-quinone form and a two-electron fully reduced form (see Figure 13.8). Complexes I and II are both flavoproteins.

Iron–Sulfur Proteins

Iron–sulfur proteins consist of nonheme iron complexed with the thiol sulfurs of cysteine residues in the protein. The simplest iron–sulfur arrangement, designated **FeS,** involves one iron atom complexed tetrahedrally with four sulfur atoms from cysteine residues (**FIGURE 14.4**). Another common form (**Fe₄S₄**) contains a cluster of four irons, four sulfides (inorganic sulfur), and four cysteine residues. In all of these centers, the iron can gain or lose an electron to alternate between the Fe^{+2} and Fe^{+3} oxidation states. The standard reduction potential of the iron in these **iron–sulfur clusters** varies dramatically depending on the type of cluster and the microenvironment provided by the protein to which it is attached. Complexes I, II, and III all contain iron–sulfur clusters and can thus be considered iron–sulfur proteins.

Coenzyme Q

The respiratory electron carrier **coenzyme Q** is an extremely lipophilic electron carrier comprising a benzoquinone linked to a number of isoprene units, usually 10 in mammalian cells and 6 in bacteria

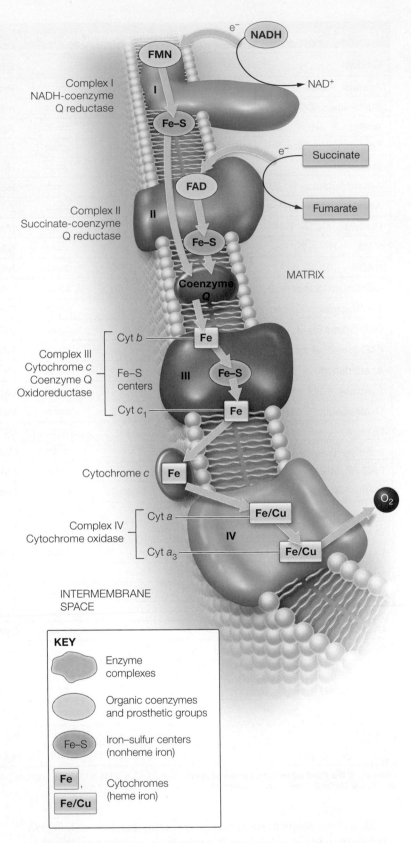

▲ **FIGURE 14.3 Respiratory electron carriers in the mitochondrion.** This figure shows the sequence of electron carriers that oxidize NADH and FAD-linked substrates such as succinate in the inner membrane.

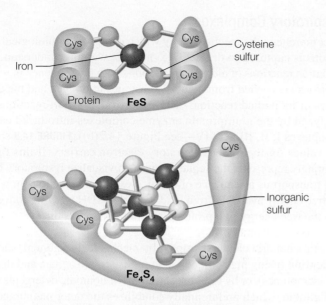

▲ **FIGURE 14.4** Structures of iron–sulfur clusters.

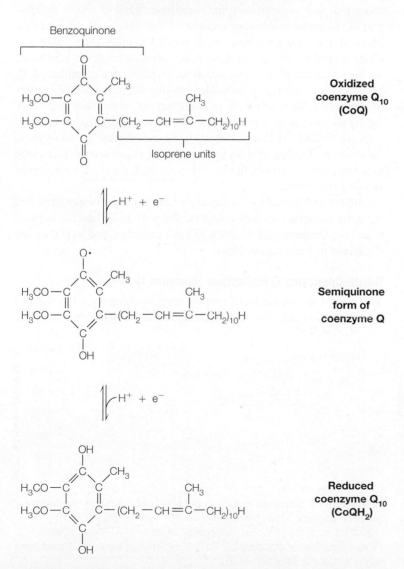

▲ **FIGURE 14.5** Structure and oxidoreduction chemistry of coenzyme Q.

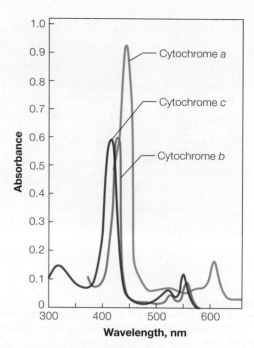

▲ **FIGURE 14.6** Absorption spectra of cytochromes. The plots show the absorption spectra of cytochromes b, c, and a in their reduced states.

(**FIGURE 14.5**). Because the substance is ubiquitous in living cells, it is sometimes called ubiquinone, but we will use coenzyme Q, or simply Q. The long tail of nonpolar isoprene units gives the molecule its hydrophobic character, which allows coenzyme Q to diffuse rapidly through the inner mitochondrial membrane. As with the flavins FMN and FAD, oxidation or reduction of this coenzyme proceeds one electron at a time through a stable semiquinone intermediate, so that coenzyme Q provides another link between two-electron carriers such as NADH and the one-electron carriers such as the cytochromes, which we turn to now.

Cytochromes

Finally, we come to the **cytochromes,** a group of red or brown heme proteins having distinctive visible-light spectra. The major respiratory cytochromes are classified as b, c, or a, depending on the wavelengths of the spectral absorption peaks (**FIGURE 14.6**). The unique spectra of each class result from differences in the groups attached to the heme and in how the heme is ligated to the protein.

Among the respiratory electron carriers are three b-type cytochromes (found in complexes II and III), cytochrome c_1 (found in complex III), cytochromes a and a_3 (found in complex IV), and cytochrome c. Cytochromes b, c, and c_1 all contain the same heme found in hemoglobin and myoglobin—iron complexed with protoporphyrin IX (**FIGURE 14.7(a)**). Cytochromes a and a_3 contain a variant form of heme, called heme A, in which one of the side chains is modified with a hydrophobic tail composed of three isoprene units (Figure 14.7(b)). Although cytochromes a and a_3 represent two identical heme A moieties attached to the same polypeptide chain in complex IV, the hemes exist in different environments in the inner membrane and thus have different reduction potentials. In cytochromes a and a_3, each of the hemes is associated with a copper ion, located close to the heme iron.

(a) General structure of cytochromes *c* and *c*₁.
Covalent bonds join the heme and the protein component in cytochromes *c* and *c*₁. Two vinyl groups on heme are linked to the thiol groups of two cysteine residues (red).

(b) Heme A in cytochromes *a* and *a*₃.
Heme A, the form found in cytochromes *a* and *a*₃, has two modified side chains–a formyl group (red) and an isoprenoid side chain (blue).

▲ **FIGURE 14.7** The hemes found in cytochromes.

A wide range of standard reduction potentials are available to the cytochromes, owing to differences in heme environment. Cytochromes undergo oxidoreduction through the complexed metal, which cycles between $+2$ and $+3$ states of heme iron and $+1$ and $+2$ states for the copper in cytochromes *a* and *a*₃. Thus, like the iron–sulfur clusters, the cytochromes are one-electron carriers.

Cytochrome *c* is a protein of about 100 amino acids, which is associated with the inner membrane but is readily extracted in soluble form. The small size and relative abundance of this molecule have allowed its structure to be studied in detail. Recall from Chapter 5 that the amino acid sequence of cytochrome *c* has been highly conserved in evolution, with nearly 50% identity between residues at corresponding positions of cytochromes *c* in organisms as diverse as yeast and human.

Respiratory Complexes

To comprehend the mechanism by which energy from biological oxidations is captured to drive ATP synthesis, we must understand the oxidation reactions of electron transport—both the sequence in which electrons are carried from reduced substrates to oxygen and the energetics of individual reactions. These electron transport reactions are catalyzed by the multiprotein enzyme complexes introduced earlier (complexes I, II, III, and IV—see Figure 14.2(b)). **FIGURE 14.8** shows $E^{\circ\prime}$ values for the major respiratory electron carriers. If this figure accurately represents the sequence, we can visualize respiratory electron transport as a sequence of coupled exergonic reactions in which the free energy available from oxidation of NADH by O_2 is converted into the free energy of a proton gradient that powers the synthesis of ATP from ADP and P_i.

Mitochondria can be disrupted by mechanical treatment, such as **sonication** (using high-frequency sound energy to agitate and disrupt cell membranes), or by low concentrations of nonionic detergents such as digitonin, which preferentially solubilizes the outer membrane but leaves many protein–protein associations intact. Using combinations of these techniques, one can fractionate the mitochondrial respiratory chain into four separate enzyme complexes, each of which contains part of the entire respiratory sequence (complexes I, II, III, and IV), plus a fifth (complex V), which catalyzes ATP synthesis from ADP. The electron transfer activity of these multisubunit complexes is retained during solubilization and fractionation, revealing that complexes I, II, III, and IV are membrane-embedded enzymes that catalyze the transfer of electrons from a relatively mobile electron carrier (one that is not tightly membrane-bound) to another mobile carrier. These mobile carriers are NADH, succinate, coenzyme Q, cytochrome *c*, and oxygen. Analysis of each complex for the presence of electron carriers, as well as for reactions catalyzed, has helped to establish the currently accepted sequence of carriers.

FIGURE 14.9 provides a summary of the protein composition and catalytic activities of each complex. We will now examine in more detail the structure and function of each complex, and how they are organized in a respiratory chain.

NADH–Coenzyme Q Reductase (Complex I)

The main donor of electrons into the respiratory chain is the reduced nicotinamide nucleotide NADH. Numerous dehydrogenases in the

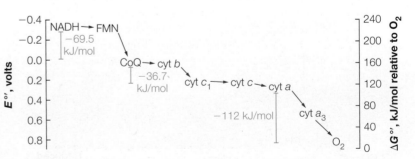

▲ **FIGURE 14.8** Standard reduction potentials of the major respiratory electron carriers. Three reactions have $\Delta G^{\circ\prime}$ values greater than -32.2 kJ/mol, the standard free energy change for ATP hydrolysis. (These data are for standard conditions, which may be very different from those within the mitochondrion.)

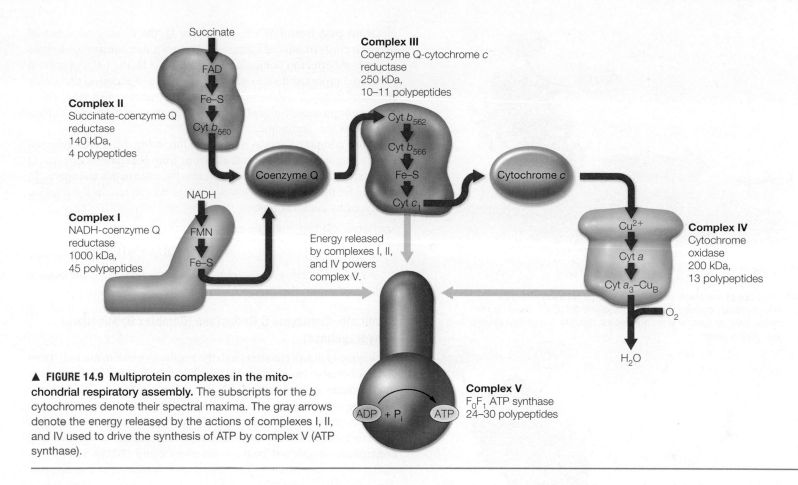

▲ **FIGURE 14.9 Multiprotein complexes in the mitochondrial respiratory assembly.** The subscripts for the *b* cytochromes denote their spectral maxima. The gray arrows denote the energy released by the actions of complexes I, II, and IV used to drive the synthesis of ATP by complex V (ATP synthase).

cell catalyze the oxidation of substrates using NAD^+ as the electron acceptor:

$$\text{reduced substrate} + NAD^+ \rightleftharpoons$$
$$\text{oxidized substrate} + NADH + H^+$$

As described in Section 8.5, these reversible dehydrogenations involve the transfer of two electrons, in the form of a hydride ion (H^-), from the substrate to NAD^+ to give NADH. The nicotinamide nucleotides readily dissociate from their enzymes and thus act as soluble redox cofactors that carry electrons between different enzymes and pathways. In mitochondrial respiration, NADH from many different dehydrogenases becomes oxidized in the first step of electron transport by complex I, which catalyzes the following reaction.

$$NADH + H^+ + CoQ \rightleftharpoons NAD^+ + CoQH_2$$

The mitochondrial enzyme is a large, membrane-embedded multisubunit complex ($\sim$1000 kDa) with about 45 separate polypeptide chains. The bacterial complex is much smaller, composed of just 14 "core" subunits that are conserved from bacteria to humans. The larger mitochondrial enzyme evolved by gradual recruitment of additional subunits to this core. The complex contains FMN as a tightly bound prosthetic group as well as eight iron–sulfur clusters, which transfer electrons from reduced flavin to coenzyme Q.

The overall reaction catalyzed by the NADH dehydrogenase complex is shown below. Because the electrons from NADH are used in the last step to reduce coenzyme Q, the name for complex I is **NADH–coenzyme Q reductase.**

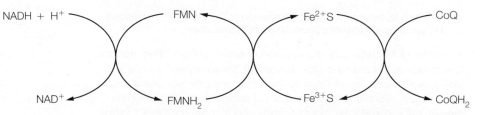

Notice that this process begins with a two-electron donor (NADH) but uses several iron–sulfur clusters, which can carry only single electrons. FMN acts as a link between the two-electron and one-electron carriers.

FIGURE 14.10 illustrates our current understanding of how these various electron carriers function in complex I. Electron microscopy of complex I from several different eukaryotic, archaeal, and bacterial sources reveals an L-shaped structure with two arms: a hydrophobic membrane arm embedded in the mitochondrial inner membrane (or bacterial plasma membrane) and a hydrophilic peripheral arm extending into the mitochondrial matrix (or bacterial cytoplasm). Complex I catalyzes a hydride transfer (two electrons) from NADH to FMN, which

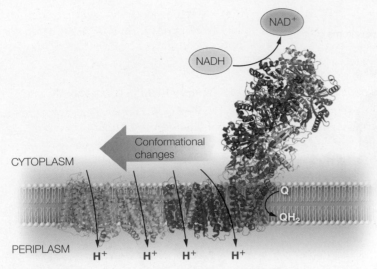

(a) Structure of the entire complex I from the archaea *Thermus thermophilus*, derived from X-ray analysis (PDB ID: 4hea). In the hydrophilic peripheral arm, iron–sulfur clusters are shown in blue, and the FMN is green.

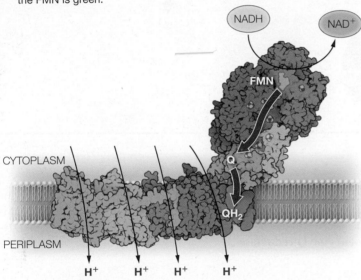

(b) The path of electron transport from NADH to CoQ and the direction of H⁺ pumping are shown schematically in this cartoon.

▲ **FIGURE 14.10 Structure and function of complex I (NADH–coenzyme Q reductase).** Regarding proton movement, the cytoplasm and periplasm in archaea are equivalent to the matrix and intermembrane space, respectively, in mitochondria. The path of electron transport from FMN through the iron–sulfur clusters to coenzyme Q is shown by the red arrows.

then transfers the electrons, one at a time, to the series of iron–sulfur clusters. In the final step of the reaction, coenzyme Q is reduced, one electron at a time, to $CoQH_2$.

Complex I catalyzes a second important process, which is tightly coupled to the electron flow through the complex: the transfer of four protons (H^+) from the matrix side of the inner membrane to the inter-membrane space side (Figure 14.10(b)). The X-ray structure of the *T. thermophilus* enzyme suggests a possible mechanism to explain how electron transfer drives proton pumping. As

● **CONCEPT** Electron transport through complex I is coupled to the pumping of protons from the matrix to the intermembrane space.

electrons pass from FMN to coenzyme Q, the membrane subunits undergo conformational changes. Through a mechanism analogous to the Bohr effect in hemoglobin (Section 7.11), the pK_as of several conserved protonated side chains are altered, and access to those side chains switches from the matrix to the intermembrane space. These changes in proton affinity effectively pump protons through the channels across the membrane.

Thus, complex I acts as a proton pump, using the energy released in the exergonic transfer of two electrons from NADH to coenzyme Q to transport four H^+ from the matrix into the intermembrane space. To account for these four translocated protons, we can rewrite the reaction catalyzed by complex I as:

$$NADH + 5H^+_{matrix} + Q \rightleftharpoons$$
$$NAD^+ + QH_2 + 4H^+_{intermembrane\ space}$$

As we will see shortly, this obligately coupled process is the essence of the mechanism that drives ATP synthesis.

Succinate–Coenzyme Q Reductase (Complex II; Succinate Dehydrogenase)

Coenzyme Q draws electrons into the respiratory chain, not only from NADH but also from succinate (as shown in Figure 14.3) and from intermediates in fatty acid oxidation. In step 6 of the citric acid cycle, succinate dehydrogenase uses an FAD coenzyme to extract electrons from succinate (see Section 13.3). Unlike the other citric acid cycle enzymes, succinate dehydrogenase is an inner membrane protein and constitutes complex II of the respiratory chain (**FIGURE 14.11**). The

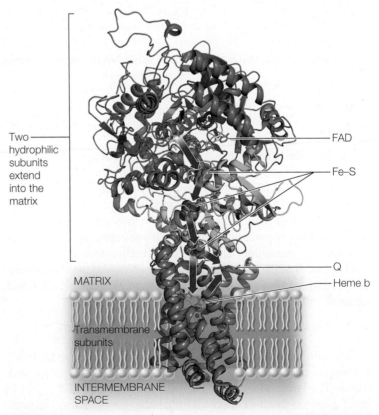

▲ **FIGURE 14.11 Structure of complex II (succinate–coenzyme Q reductase) from pig heart mitochondria (PDB ID: 1zoy).** Succinate binds near FAD in the cyan subunit.

enzyme thus transfers electrons directly from its bound $FADH_2$ to the other membrane-bound respiratory carriers. Like NADH–coenzyme Q reductase (complex I), succinate dehydrogenase transfers electrons via a series of iron–sulfur centers to coenzyme Q and is thus more completely named **succinate–coenzyme Q reductase.** Complex II does not pump protons, so the reaction it catalyzes can be summarized as:

$$\text{succinate} + \text{CoQ} \rightleftharpoons \text{fumarate} + \text{CoQH}_2$$

Thus far, we have seen two ways in which electrons are delivered to coenzyme Q: from NADH via complex I and from succinate via complex II. At least two other flavoprotein dehydrogenases, including the *electron-transferring flavoprotein (ETF):ubiquinone oxidoreductase* and *glycerol-3-phosphate dehydrogenase*, also deliver electrons to coenzyme Q (**FIGURE 14.12**). These enzymes transfer electrons not from citric acid cycle intermediates, but rather from other oxidation pathways, briefly described below.

Like succinate dehydrogenase, **ETF:coenzyme Q oxidoreductase (ETF-QO)** is bound to the matrix side of the inner membrane and uses FAD and an iron–sulfur cluster as electron carriers. **ETF (electron-transferring flavoprotein)** is a small, soluble protein in the mitochondrial matrix that receives electrons from at least 12 different mitochondrial FAD-containing dehydrogenases involved in fatty acid and amino acid oxidations. ETF, with its bound $FADH_2$, is then reoxidized by ETF:ubiquinone oxidoreductase, transferring the electrons to coenzyme Q in the respiratory chain.

Glycerol-3-phosphate dehydrogenase, located on the *intermembrane face* of the inner membrane, catalyzes the oxidation of glycerol-3-phosphate (glycerol-3-P) to dihydroxyacetone phosphate (DHAP), reducing coenzyme Q to $CoQH_2$ (Figure 14.12). We shall see later in this chapter that this reaction plays an important role in shuttling electrons from cytoplasmic NADH into the mitochondrial matrix. Because coenzyme Q is subsequently oxidized by complex III, it can be seen as gathering electrons from several flavoprotein dehydrogenases and delivering them down the respiratory chain, ultimately to O_2 (see Figure 14.9).

Coenzyme Q:Cytochrome *c* Oxidoreductase (Complex III)

The oxidation of reduced coenzyme Q is mediated by another large multisubunit complex embedded in the inner membrane, complex III of the respiratory chain. This enzyme catalyzes the transfer of electrons from $CoQH_2$ to cytochrome *c* and is thus called **coenzyme Q: cytochrome *c* oxidoreductase.** Mammalian complex III functions as a dimer, with each monomer composed of 10 or 11 protein chains ($\sim$250 kDa), including cytochrome *b,* cytochrome c_1, and a protein called the **Rieske iron–sulfur protein (ISP). FIGURE 14.13** shows the X-ray crystal structure of the bovine mitochondrial complex III and the locations of the redox centers.

The path of electrons from $CoQH_2$ to cytochrome *c* through complex III is more complicated than indicated in Figure 14.9 because at

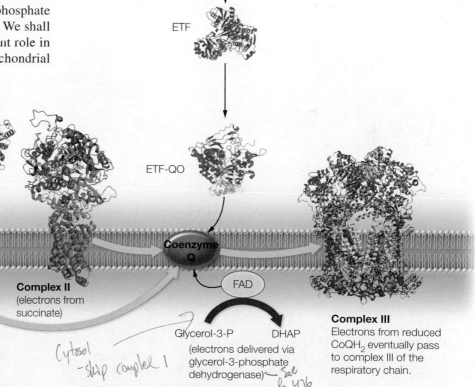

▲ **FIGURE 14.12 Coenzyme Q collects electrons from multiple flavo-proteins.** This lipid-soluble mobile redox cofactor serves as the electron acceptor for at least four mitochondrial flavoprotein dehydrogenases. Electrons from NADH and succinate are delivered via complex I and complex II, respectively. Electron-transferring flavoprotein (ETF):coenzyme Q oxidoreductase (ETF-QO) catalyzes the transfer of electrons from reduced ETF to Q. ETF receives electrons from many mitochondrial oxidations, including fatty acid β-oxidation. Glycerol-3-phosphate dehydrogenase, located on the intermembrane face of the inner membrane, delivers electrons from glycerol-3-phosphate (glycerol-3-P) to coenzyme Q. Electrons from reduced $CoQH_2$ eventually pass to complex III of the respiratory chain.

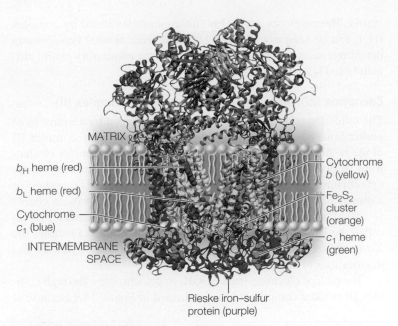

b_H heme (red)

b_L heme (red)

Cytochrome c_1 (blue)

MATRIX

Cytochrome b (yellow)

Fe_2S_2 cluster (orange)

c_1 heme (green)

INTERMEMBRANE SPACE

Rieske iron–sulfur protein (purple)

▲ **FIGURE 14.13 Structure of complex III (coenzyme Q:cytochrome c oxidoreductase).** X-ray structure of the dimeric complex from bovine mitochondria (PDB ID: 1ppj). The core subunits are colored and identified on the figure; the noncore subunits are shown in silver.

this point, a two-electron donor, $CoQH_2$, is transferring electrons to one-electron acceptors, the cytochromes. The so-called **Q cycle** has been proposed to account for this stoichiometry (**FIGURE 14.14**). Complex III has two binding sites for Q: Q_0 and Q_1. QH_2 is oxidized at the Q_0 site, where the two electrons take two different paths. The first electron is passed to the ISP, then on to cytochrome c_1 and then finally cytochrome c. The resultant QH semiquinone then transfers its second electron to the low-potential b_L heme component of cytochrome b, and this electron next passes to the high-potential b_H heme component. The b_H heme is located at the Q_1 site near the matrix side of the membrane, where it reduces a molecule of oxidized Q to a QH semiquinone.

This process is repeated, with a second molecule of QH_2 being oxidized at the Q_0 site. One electron is passed to the ISP and onto cytochrome c, and another is passed from b_L heme to b_H heme as before. However, this time the electron from the b_H heme reduces a QH semiquinone—not a molecule of oxidized Q—to QH_2 at the Q_1 site (Figure 14.14). The result is that two molecules of QH_2 become oxidized and one molecule of Q becomes reduced, for a net transfer of two electrons to reduce two molecules of cytochrome c. Because the proton-consuming reactions occur within the matrix, while proton release takes place in the intermembrane space, the Q cycle contributes to the proton gradient needed to drive ATP synthesis, as we will describe shortly.

● **CONCEPT** Electron transport through complex III is coupled to the pumping of protons from the matrix to the intermembrane space.

Cytochrome c Oxidase (Complex IV)

The final stage of electron transport is carried out by **cytochrome c oxidase** (complex IV). The mitochondrial enzyme exists as a large homodimeric complex in the inner membrane. Each monomer is composed of 13 subunits, with 28 transmembrane helices separating hydrophilic domains facing the matrix and intermembrane space compartments (**FIGURE 14.15**). Bacterial cytochrome c oxidase is much simpler, composed of just four subunits. However, homologs of three of the bacterial subunits form the core of the mitochondrial complex, revealing the evolutionary origin of this enzyme. Indeed, subunits

MATRIX

H^+ Q

$QH·$ e^- b_H

e^-

e^- b_L

QH_2 e^- Q

Fe_2S_2

e^-

c_1

e^-

Stage 1

cyt c $2H^+$

H^+ $QH·$

QH_2 e^- b_H

e^-

e^- b_L

QH_2 e^- Q

Fe_2S_2

e^-

c_1

e^-

Stage 2

INTERMEMBRANE SPACE

cyt c $2H^+$

Stage 1: $QH_2 + Q + H^+_{Mat} + cyt\ c_{ox} \longrightarrow$
$QH· + Q + 2H^+_{IMS} + cyt\ c_{red}$

Stage 2: $QH_2 + QH· + H^+_{Mat} + cyt\ c_{ox} \longrightarrow$
$QH_2 + Q + 2H^+_{IMS} + cyt\ c_{red}$

Net: $QH_2 + 2cyt\ c_{ox} + 2H^+_{Mat} \longrightarrow Q + 2cyt\ c_{red} + 4H^+_{IMS}$

◀ **FIGURE 14.14 The Q cycle.** The spatial arrangement of the redox centers in one monomer of the dimeric bovine mitochondrial complex III (shown in Figure 14.13) is depicted with the protein components in gray (PDB ID: 1ppj). The path of electrons in each stage is indicated by red arrows and the path of protons by blue arrows. The Q_0 site is in the middle of the diagram; the Q_1 site is at the top (the matrix side of the inner membrane). The stoichiometry of each stage of the cycle is shown below, as well as the net stoichiometry of the full cycle. H^+_{Mat}, matrix protons; H^+_{IMS}, intermembrane space protons.

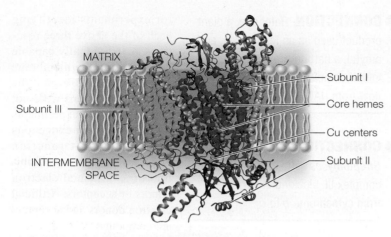

MATRIX

Subunit III

INTERMEMBRANE SPACE

Subunit I

Core hemes

Cu centers

Subunit II

▲ **FIGURE 14.15 Structure of cytochrome *c* oxidase (complex IV).** X-ray structure of one monomer of the bovine mitochondrial enzyme (PDB ID: 2eij). Subunits I, II, and III are colored; the remaining subunits are gray. The two core hemes are orange, and the two Cu centers are green. Cyt *c* binds near the Cu_A center of subunit II.

I, II, and III of cytochrome *c* oxidase are encoded by the mitochondrial genome and synthesized on mitochondrial ribosomes in the matrix. Cytochrome *c* oxidase contains two hemes, *a* and a_3, bound by subunit I in the interior of the membrane, and two copper centers (Cu_A and Cu_B). The Cu_B atom sits within 5 Å of heme a_3 on subunit I. The iron of heme a_3 and Cu_B thus constitute a "binuclear center," functioning as a single unit in electron transfer.

Complex IV catalyzes the transfer of electrons from reduced cytochrome *c* to oxygen. The initial oxidation of cytochrome *c* is carried out by Cu_A, with the electron transferred to heme *a* and then to the binuclear heme a_3–Cu_B site. The binuclear center is the catalytic site where O_2 undergoes its four-electron reduction to water. The redox reaction catalyzed by complex IV is as follows:

$$4 \text{ cyt } c_{red} \text{ (Fe}^{2+}\text{)} + O_2 + 4H^+_{matrix} \rightleftharpoons 4 \text{ cyt } c_{ox} \text{ (Fe}^{3+}\text{)} + 2H_2O$$

The four substrate protons consumed in the oxidation of O_2 to H_2O come from the matrix. In addition, for each electron transferred, complex IV pumps approximately one proton from the matrix side to the intermembrane space side of the membrane. Thus, each turnover of the enzyme transports eight positive charges across the inner membrane. The net reaction catalyzed by complex IV can be summarized as follows:

$$4 \text{ cyt } c_{red} \text{ (Fe}^{2+}\text{)} + O_2 + 8H^+_{matrix} \rightleftharpoons$$
$$4 \text{ cyt } c_{ox}\text{(Fe}^{3+}\text{)} + 2H_2O + 4H^+_{intermembrane space}$$

How are these four protons per molecule of oxygen reduced pumped from the matrix to the intermembrane space? The mechanism probably involves conformational changes in the protein driven by changes in redox state of the binuclear heme a_3–Cu_B center as electrons flow through the complex. These conformational changes in turn alter the pK_as of proton-binding residues (e.g., Glu or Asp), allowing movement of protons across the membrane, as described for complex I.

● **CONCEPT** Electron transport through complex IV is coupled to the pumping of protons from the matrix to the intermembrane space.

We have now followed a pair of electrons through the entire respiratory chain. We've described the structure and chemistry of the electron carriers, how they are arranged in protein complexes in the inner membrane, and how electrons flow from one to the other. We have also seen how, in three of the complexes, protons are pumped from the matrix to the intermembrane space to generate a proton concentration gradient. Now, we are ready to understand how the energy in this proton gradient is harnessed for ATP production through the process of oxidative phosphorylation.

14.4 Oxidative Phosphorylation

We turn now to the question of how the free energy of the proton gradient is used for ATP synthesis—in short, the mechanism of oxidative phosphorylation. Before considering the mechanism, it is important to recall the energetics of the process.

The P/O Ratio: Energetics of Oxidative Phosphorylation

Earlier in this chapter, we calculated that for a pair of electrons entering the respiratory chain as NADH and traversing the entire chain to O_2, the standard free energy change, $\Delta G°'$, is −220 kJ/mol:

$$NADH + H^+ + \tfrac{1}{2}O_2 \rightleftharpoons NAD^+ + H_2O$$
$$\Delta G°' = -nF\,\Delta E°' = -2(96485)(0.82 - (-0.32)) =$$
$$-220 \text{ kJ/mol}$$

How much of this free energy is actually conserved as ATP in oxidative phosphorylation? This question could be answered if we could measure the quantity of ATP synthesized per mole of substrate oxidized in isolated mitochondria. What we usually measure is the **P/O ratio,** which is the number of molecules of ATP synthesized per pair of electrons carried through electron transport. ATP synthesis is quantitated as phosphate incorporation into ATP, and electron pairs are quantitated as oxygen uptake, in μmol of O atoms (not O_2 molecules) reduced to water.

As you might imagine, precise measurements of oxygen consumption and ATP synthesis in a mitochondrial preparation are difficult to obtain, and many experimental pitfalls can lead to inaccurate estimates of the P/O ratio. Early experiments suggested that the mitochondrial oxidation of NADH proceeds with a P/O ratio of 3 and oxidation of succinate proceeds with a P/O ratio of 2. However, as researchers got better at preparing intact mitochondria and measuring oxygen consumption and ATP synthesis, it became clear that the P/O ratios were *not integers*. The general consensus now is that the P/O ratio is ~2.5 for oxidation of NADH and ~1.5 for oxidation of succinate. Indeed, these noninteger P/O values contributed to the realization that phosphorylation and oxidation are not directly coupled, as we mentioned earlier. As we will see next, this mechanism of indirect coupling between oxidation and phosphorylation does not require an integral stoichiometric relationship between reducing equivalents consumed and ATP synthesized.

With these ratios in mind, we can write a balanced equation for the mitochondrial oxidation of NADH coupled to the synthesis of ATP:

$$NADH + H^+ + \tfrac{1}{2}O_2 + 2.5ADP + 2.5P_i \rightleftharpoons$$
$$NAD^+ + H_2O + 2.5ATP$$

ΔG for ATP hydrolysis under intracellular conditions is estimated to be -50 kJ/mol or more, so the synthesis of 2.5 mol ATP requires at least 2.5×50 kJ $= 125$ kJ. As discussed in Section 3.4, ΔG for the oxidation of NADH by O_2 under intracellular conditions is estimated to be -384 kJ/mol O_2, or -192 kJ/mol electron pairs (see Equation 3.41)—clearly enough to drive ATP synthesis. The excess free energy ensures that the process is essentially irreversible under physiological conditions.

Oxidative Reactions That Drive ATP Synthesis

A glance at Figure 14.9 reveals that the transfer of reducing equivalents from NADH to O_2 involves about a dozen consecutive, linked oxidoreduction reactions. Which of these reactions actually drive ATP synthesis? This question was of paramount concern in the early days of bioenergetics research, when it was thought that ATP synthesis was directly coupled to individual exergonic reactions, as it is in substrate-level phosphorylation. The most straightforward interpretation of the P/O ratio of 3 initially estimated for NADH oxidation was that three of the individual reactions of the respiratory chain are sufficiently exergonic to drive the synthesis of one ATP molecule each.

Indeed, three of these reactions do have $\Delta G°'$ values exceeding -32.2 kJ/mol, the minimum barrier that must be overcome (under standard conditions) to make the synthesis of each ATP exergonic (see Figure 14.8). Those three reactions are the oxidation of NADH by coenzyme Q, catalyzed by complex I ($\Delta G°' = -69.5$ kJ/mol); the oxidation of $CoQH_2$ by cytochrome c, catalyzed by complex III ($\Delta G°' = -36.7$ kJ/mol); and the oxidation of reduced cytochrome c by O_2, catalyzed by cytochrome c oxidase ($\Delta G°' = -112$ kJ/mol). Each of these reactions was thus considered to be a **"coupling site"** for ATP synthesis; that is, each was thought to be a reaction in which ATP synthesis was driven directly by the energy released from that reaction.

Because we now know that coupling between oxidation and ATP synthesis is indirect, the concept of coupling sites is an oversimplification. Nevertheless, the idea was useful because it provided a framework

● **CONNECTION** Rotenone, a plant product used as an insecticide, and amytal, a barbiturate drug, both inhibit complex I, blocking electron flow from NADH to coenzyme Q.

● **CONNECTION** Antimycin A, a *Streptomyces* antibiotic, inhibits complex III, blocking electron flow from cytochrome b to c_1.

for experiments identifying each of the above three reactions as individually capable of energizing the membrane for ATP synthesis, even if the other two reactions were not operating. These experiments relied heavily on exogenous compounds that function either as respiratory inhibitors or as artificial electron donors or acceptors. Artificial electron donors and acceptors are compounds that can either feed electrons into or draw electrons away from the respiratory chain in spontaneous nonenzymatic redox reactions. The sites of action of several useful compounds are shown in **FIGURE 14.16**.

Let us now examine some of the evidence for these "coupling sites," which is summarized in Figure 14.16. First, succinate is oxidized with a P/O ratio of ~1.5, not 2.5, suggesting that one of the three coupling sites is not used in electron transport when succinate is the donor. This was confirmed by blocking electron transport past cytochrome b in complex II with **antimycin A,** an antibiotic that blocks electron flow from cytochrome b to c_1. **Ferricyanide** was added as an artificial electron acceptor so that electrons could continue to flow from low potential components of the transport chain. Under these conditions, NAD^+-linked substrates, such as β-hydroxybutyrate, were oxidized with a P/O ratio of ~1, confirming the existence of one coupling site before cytochrome b. We now know this is complex I.

Another approach involved an artificial electron donor, **ascorbate.** In the presence of an intermediate electron acceptor (TMPD), electrons could be supplied to the respiratory chain at cytochrome c. These electron carriers reduced cytochrome c nonenzymatically, and

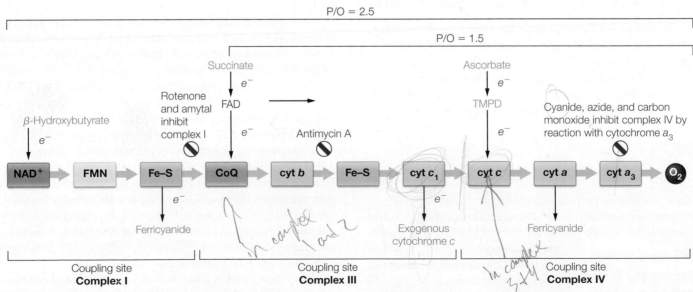

▲ **FIGURE 14.16 Experimental identification of "coupling sites."** Electron transport is restricted to particular parts of the chain by use of selected electron donors (blue), electron acceptors (green), and respiratory inhibitors (red), as indicated. For each segment of the chain that is thus isolated, P/O ratios are determined, allowing identification of coupling sites. TMPD, tetramethyl-*p*-phenylene diamine.

its subsequent oxidation via cytochrome oxidase proceeded with a P/O ratio of ~1, thus localizing one coupling site beyond cytochrome c.

Finally, purified exogenous oxidized cytochrome c, when added to disrupted mitochondria, can act as an electron acceptor, withdrawing electrons from the electron transport chain. The further addition of a cytochrome oxidase inhibitor, such as **cyanide (CN⁻),** forces electrons to exit from the respiratory chain at cytochrome c. Under these conditions, succinate is oxidized with a P/O ratio of ~0.5, which localizes a site between cytochromes b and c. However, if antimycin A is then added, no ATP is synthesized, showing that complex II (succinate dehydrogenase) is not a coupling site.

In sum, these experiments demonstrate that complexes I, III, and IV are each capable of driving ATP synthesis but that complex II is not (Figure 14.10). We now know that these three "coupling sites" indirectly drive ATP synthesis by pumping protons across the membrane as electrons flow through the complexes.

Mechanism of Oxidative Phosphorylation: Chemiosmotic Coupling

What is the actual mechanism by which energy released from electron transport through the respiratory chain is harnessed to drive the synthesis of ATP? For many years, researchers sought a high-energy intermediate that could link electron transport to ATP synthesis. However, such activated intermediates have never been demonstrated in oxidative phosphorylation.

Although other models have been considered, there is now widespread acceptance of a model involving **chemiosmotic coupling,** proposed in 1961 by British biochemist Peter Mitchell. Although this model was resisted at first, overwhelming evidence has now accumulated in its support, and Mitchell's achievements were recognized in 1978 with a Nobel Prize. In its most basic form, this model proposes that *the free energy from electron transport drives an active transport system, which pumps protons out of the mitochondrial matrix into the intermembrane space. This action generates an electrochemical gradient for protons. The protons on the outside have a thermodynamic tendency to flow back in, down their electrochemical gradient, and this flow provides the driving force for ATP synthesis.* Put another way, free energy must be expended to maintain the proton gradient. When protons do flow back into the matrix, that energy is dissipated, and some of it is harnessed to drive the synthesis of ATP.

To understand the chemiosmotic theory in more detail, recall that some (though not all) of the reactions of electron transport transfer hydrogen ions (protons) as well as electrons. These reactions include the dehydrogenations of NADH, FADH₂, FMNH₂, and reduced coenzyme Q. Mitchell proposed that the enzymes catalyzing these dehydrogenations are asymmetrically oriented in the inner membrane so that protons are always taken up from inside the matrix and released in the intermembrane space. **FIGURE 14.17** shows how this process occurs.

This **proton pumping** by respiratory proteins results in conversion of the energy of electron transport to osmotic energy, in the form of an **electrochemical gradient**—or a gradient of chemical concentration that also establishes an electrical potential (as discussed in

● **CONCEPT** The F_0F_1 complex contains a proton channel (F_0) and the enzyme that synthesizes ATP (F_1).

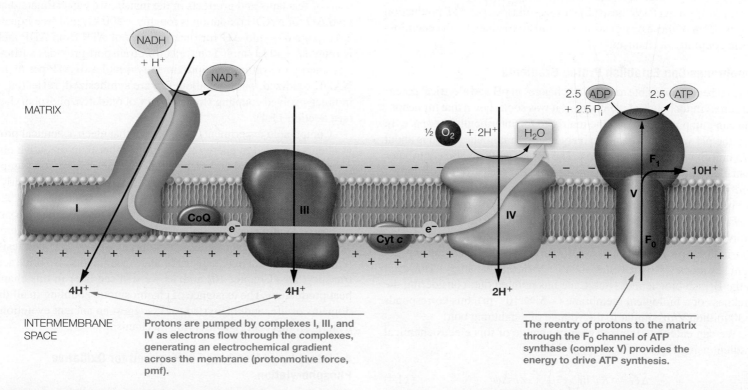

▲ **FIGURE 14.17 Chemiosmotic coupling of electron transport and ATP synthesis.** This depiction of protein complexes in the mitochondrial inner membrane shows the path of electron flow from NADH to O₂ (yellow arrow), and the proton pumping that is driven by the electron transport. Complex II is not included because it does not contribute to the proton gradient.

Labels within figure: NADH + H⁺, NAD⁺, MATRIX, I, CoQ, e⁻, III, Cyt c, e⁻, ½ O₂ + 2H⁺, H₂O, IV, 2.5 ADP + 2.5 Pᵢ, 2.5 ATP, F₁, 10H⁺, V, F₀, 4H⁺, 4H⁺, 2H⁺, INTERMEMBRANE SPACE

Protons are pumped by complexes I, III, and IV as electrons flow through the complexes, generating an electrochemical gradient across the membrane (protonmotive force, pmf).

The reentry of protons to the matrix through the F₀ channel of ATP synthase (complex V) provides the energy to drive ATP synthesis.

detail in Chapter 10). The energy released from discharging this gradient can be coupled with the phosphorylation of ADP to ATP, with no isolatable "high-energy" intermediates being formed. This process involves **ATP synthase** (complex V). ATP synthase is composed of two domains, named F_0 and F_1. The F_0 portion of the complex spans the inner membrane and contains a specific channel for the return of protons to the mitochondrial matrix. The free energy released as protons traverse this channel to return to the matrix is harnessed to drive the synthesis of ATP, catalyzed by the F_1 component of the complex.

A Closer Look at Chemiosmotic Coupling: The Experimental Evidence

Before we describe the structure and function of ATP synthase, let's look at the experimental evidence for the chemiosmotic coupling mechanism in greater detail—partly because of its importance to an understanding of oxidative phosphorylation and partly because

● **CONCEPT** Chemiosmotic coupling refers to the use of a transmembrane proton gradient to drive endergonic processes like ATP synthesis.

it provides insight into other biological processes, including active transport across membranes and photosynthesis. As we have already mentioned, several nagging facts argued against ATP synthesis by a direct chemical coupling process like that in glycolysis. The first was the realization that the number of ATP molecules produced per pair of electrons is approximately 2.5, not an integer as would be predicted by a direct chemical coupling mechanism. Second, oxidative phosphorylation was shown to require an intact membrane. If the membrane is disrupted, the passage of electrons down the respiratory chain becomes *uncoupled* from ATP synthesis: Electrons still flow, but ATP production ceases. Both of these observations are fully explained by the chemiosmotic coupling mechanism.

Membranes Can Establish Proton Gradients

When it became possible to measure changes in pH and electrical potential across mitochondrial membranes, it was soon shown that mitochondria can pump protons from the matrix to the intermembrane space. In fact, the pH value outside an actively respiring mitochondrion is about 0.75 units lower than in the matrix. The proton gradient also contributes to an electrical potential across the membrane because of the net movement of positively charged protons outward across the inner membrane. The pH gradient and the membrane potential both contribute to an *electrochemical H^+ gradient*, or **protonmotive force (pmf)** of 150 to 200 millivolts (mV), although the electrical component is by far the major contributor. An electrical charge of 150 mV across a membrane may not seem like much (1/10 the voltage of a 1.5 V flashlight battery). However, as the British biochemist Nick Lane has pointed out, considering the thickness of a biological membrane ($\sim 5 \times 10^{-9}$ m), this corresponds to 30 million V/m, similar to the voltage of a lightning bolt!

We can calculate the free energy change of this electrochemical gradient using Equation 10.3 (page 322):

$$\Delta G = RT \ln\left(\frac{C_2}{C_1}\right) + ZF\Delta\psi \qquad (14.4)$$

Here C_2 and C_1 are the concentrations of the ion in the two compartments, Z is the charge on the ion ($+1$ for H^+), and $\Delta\psi$ is the membrane potential, in volts. ΔG in this equation is equivalent to $\Delta\mu_H$, the proton electrochemical gradient. For a proton electrochemical gradient, because pH is a logarithmic function of $[H^+]$, the term $\ln\left(\frac{C_2}{C_1}\right)$ can be replaced by 2.3ΔpH. $Z = 1$ for protons, so Equation 14.4 can be simplified to:

$$\Delta\mu_H = 2.3\ RT\ \Delta\text{pH} + F\Delta\psi \qquad (14.5)$$

$\Delta\mu_H$ is also called Δp, the protonmotive force, or pmf. ΔpH has a positive value ($+0.75$) because it is defined as the pH in the matrix minus the pH in the intermembrane space (recall that pH $= -\log[H^+]$). Thus, the contribution of the pH gradient is $2.3RT(+0.75) = +4.5$ kJ/mol at 37 °C (310 K). The membrane potential across the inner membrane of an actively respiring mitochondrion is 0.15 to 0.20 V, so the contribution of the electrical component is $+14.5$ to $+19.3$ kJ/mol. Thus, the total free energy change of transporting a proton from the matrix to the intermembrane space is on the order of $+21$ kJ per mole of protons.

Because the formation of this protonmotive force is an endergonic process (positive ΔG), discharge of the gradient is an exergonic process: It is this free energy that is used to drive the phosphorylation of ADP. As summarized in Figure 14.17, approximately 10 protons are pumped per pair of electrons transferred from NADH to O_2. Thus, the protonmotive force conserves approximately 210 kJ (21 kJ per mole of protons $\times$ 10 moles protons) of free energy for ATP synthesis per pair of electrons. Recall that the free energy change for the oxidation of NADH by O_2 is -220 kJ/mol under standard conditions (Equation 14.3), most of which is conserved in the electrochemical proton gradient. If we make some reasonable assumptions regarding the concentrations of reactants and products in the matrix, we can estimate that in vivo ΔG for NADH oxidation is roughly -200 kJ/mol (see Equation 3.41, page 67), and ΔG for the synthesis of ATP from ADP and P_i is roughly $+50$ kJ/mol. Thus, electron transport provides sufficient free energy to synthesize approximately 4 moles of ATP per mole of NADH oxidized, but only about 2.5 are synthesized, reflecting the indirect, evolved coupling stoichiometry of oxidative phosphorylation (see Section 11.4).

Comparable experiments have shown that electrochemical proton gradients are used in energy transactions other than oxidative phosphorylation. Bacterial and archaeal membranes use proton pumping to transduce energy both for oxidative phosphorylation and for driving flagellar motors that allow movement of the cell. Bacteria and archaea also have many membrane solute transporters that use the protonmotive force to pump solutes into or out of their cells. Proton pumping across the chloroplast thylakoid membrane drives ATP synthesis in photophosphorylation (Chapter 15). Proton gradients also drive active transport (see Chapter 10) and, as we will see later in this chapter, heat production. The existence of chemiosmotic coupling in all three domains of life, and its myriad uses, suggest an ancient evolutionary history for this energy-conserving mechanism.

An Intact Inner Membrane Is Required for Oxidative Phosphorylation

When the physical continuity of the membrane is interrupted—for example, by mechanical disruption—the resultant particles can carry out electron transport but not ATP synthesis. The necessity of a structurally intact membrane for maintenance of a membrane potential is consistent with the idea that a proton gradient is essential for oxidative phosphorylation.

Key Electron Transport Proteins Span the Inner Membrane

If the respiratory proteins are to serve as proton pumps, then the electron carriers that pump protons should be in contact with both the inner and the outer sides of the membrane. Moreover, these carriers should be asymmetrically oriented in the membrane to account for proton transport in only one direction—outward, from the matrix to the intermembrane space.

Asymmetric orientation has been demonstrated by the use of agents that react with respiratory proteins but cannot themselves traverse the membrane, such as antibodies, proteolytic enzymes, or labeling reagents. Treatment of intact mitochondria with such reagents allows detection of proteins located at the outer surface of the inner membrane, whereas reaction with membrane vesicles allows access to the inner, or matrix, side. Such approaches have shown, for example, that the cytochrome oxidase complex (complex IV) binds to cytochrome *c* only on the intermembrane space side (see Figure 14.15). Moreover, 9 of the 13 subunits of the complex can be labeled from only one side or the other, indicating asymmetric placement of the complex in the membrane. Similar findings have now been made for the subunits of complexes I, II, and III.

Uncouplers Act by Dissipating the Proton Gradient

A class of compounds, exemplified by the lipophilic weak acid **2,4-dinitrophenol (DNP),** are called uncoupling agents, or **uncouplers.** Uncoupling agents, when added to mitochondria, permit electron transport along the respiratory chain to O_2 to occur without ATP synthesis. That is, they *uncouple* the process of electron transport from the process of ATP synthesis.

The pK_a of the phenolic hydroxyl group in DNP is such that it is normally dissociated at intracellular pH. However, a DNP molecule that approaches the inner membrane from the outside becomes protonated because of the lower pH value in this vicinity. The protonated DNP diffuses into and through the inner mitochondrial membrane. Once inside the matrix, the higher pH causes the phenolic hydroxyl to deprotonate. The deprotonated dinitrophenolate ion is still lipophilic enough to pass back across the membrane, where it can bind another proton and repeat the cycle. Thus, the uncoupler has the effect of transporting H^+ into the matrix, bypassing the F_0 proton channel and thereby preventing ATP synthesis.

Generation of a Proton Gradient Permits ATP Synthesis Without Electron Transport

Andre Jagendorf (1926–2017), while studying photosynthetic ATP production, provided important evidence for chemiosmotic coupling in the chloroplast. As we will see in Chapter 15, the chloroplast couples light energy to ATP synthesis. Jagendorf showed that ATP synthesis can proceed in the chloroplast in the absence of electron transport, as long as a proton gradient is present. Chloroplasts were incubated at pH 4 for several hours and then quickly transferred to a buffer at pH 8. Thus, like the situation in intact cells, the inside of the organelle was at a lower pH than the outside (chloroplast membranes pump protons *inward*, not outward). Addition of ADP and P_i to these chloroplasts

generated a burst of ATP synthesis, simultaneous with dissipation of the pH gradient. Similar results have now been observed with mitochondria. These experiments show that the establishment of a proton gradient, even without a corresponding energy input, suffices to drive the synthesis of ATP.

Complex V: The Enzyme System for ATP Synthesis

That complexes I, III, and IV could individually drive ATP synthesis was confirmed in elegant reconstitution studies involving complex V, which catalyzes the actual synthesis of ATP from ADP. Let us review the discovery and nature of complex V and then discuss the reconstitution experiments.

Discovery and Reconstitution of ATP Synthase

Electron microscopy of mitochondria reveals that the cristae are covered with knoblike projections on the matrix side, each attached to the inner membrane by a short stalk (**FIGURE 14.18(a)**). The knobs are known as **F_1 spheres.** Disruption of mitochondria by sonication generates fragments of inner membrane, which reseal in the form of closed vesicles, such that the matrix face of the membrane is on the outside of the vesicles. Thus, the knoblike F_1 spheres are likewise found on the outside (Figure 14.18(b)). These **submitochondrial particles** respire and synthesize ATP, just as intact mitochondria do.

Efraim Racker (1913–1991) and his colleagues showed in the 1960s that treatment of these vesicles with trypsin or urea caused the knobs to dissociate from the vesicles. After centrifugation to separate the "stripped" vesicles from the knobs, the vesicles could still oxidize substrates and reduce oxygen, but no ATP was synthesized. When knobs were added back to the vesicles, there was substantial reconstitution of particles that could then catalyze ATP synthesis as a result of the oxidation of exogenous substrates. Thus, readdition of the knobs recoupled ATP synthesis to electron transport.

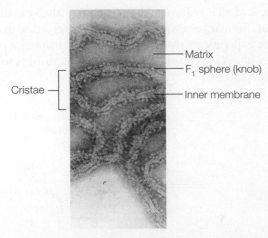

(a) A negatively stained portion of bovine heart mitochondrial inner membrane, showing knoblike projections along the matrix side of the membrane. The knob is attached by a short stalk to a base, which is embedded in the inner membrane of intact mitochondria.

(b) Surface rendering of tubular vesicles from rat liver mitochondria imaged by a cryoelectron microscopy technique. The length of the tube is 280 nm. The F_1 knobs (yellow) appear to exist in double rows, suggesting that ATP synthase is organized as dimers.

▲ **FIGURE 14.18 Fine structure of mitochondrial cristae.** Preparations of the inner membrane show the F_1 spheres as "knobs" projecting from the cristae.

Purification of the F_1 knobs attached to the underlying stalks revealed a large multiprotein aggregate consisting of more than a dozen polypeptide chains, as schematized in **FIGURE 14.19**. The entire structure of ATP synthase, called the **F_0F_1 complex,** consists of the knob, the central stalk to which it is attached, and a complex that is embedded in the inner membrane. The stalked knob is called F_1, and the base is called F_0. The F_1 complex consists of five proteins, designated α, β, γ, δ, and ε, with a subunit stoichiometry of $\alpha_3\beta_3\gamma\delta\varepsilon$. The F_0 complex consists of an oligomer of c subunits (10 copies in yeast mitochondrial ATP synthase; 8 copies in higher eukaryotes) plus one subunit a, and single copies of subunits b, d, F_6, and OSCP (oligomycin sensitivity conferral protein) that form a peripheral stalk.

The entire F_0F_1 complex has an ATP hydrolysis activity in vitro, as does factor F_1 alone; this ATPase activity was assumed to represent the reverse of the true physiological reaction, namely, ATP synthesis. The OSCP subunit in the F_0 complex is the binding site of the ATP synthesis inhibitor oligomycin (discussed further shortly). The ATPase activity, sensitivity to oligomycin, and results of the reconstitution experiments confirm that the role of the F_0F_1 complex (also called complex V) is to synthesize ATP. The action of this amazing structure, a molecular "rotary engine," will be discussed after we describe its structure in more detail.

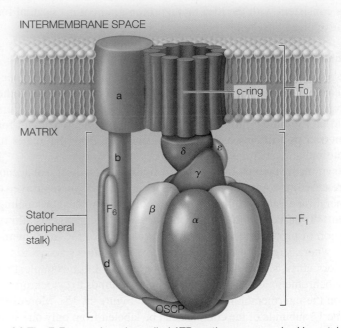

(a) The F_0F_1 complex, also called ATP synthase or complex V, contains an F_1 knob that projects into the mitochondrial matrix and is connected by a central stalk to the F_0 base. The globular F_1 knob contains three $\alpha\beta$ dimers, arranged about the central stalk composed of the γ, δ, and ε subunits, also part of the F_1 complex. The central stalk and the c-ring of the F_0 complex compose the "rotor" of ATP synthase. The remainder of the F_0 subunits (a, b, d, F_6 and OSCP) make up the "stator," a structure that prevents the rotation of the three $\alpha\beta$ dimers of F_1. This model is based on the X-ray crystal structures of the yeast and bovine mitochondrial F_0F_1 complex.

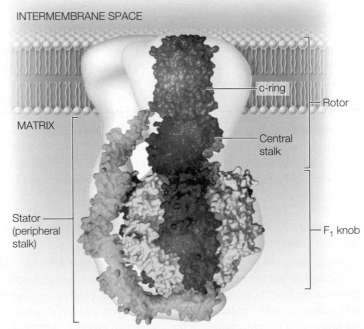

(b) The yeast mitochondrial F_0F_1 X-ray structure has been superimposed on cyroelectron microscopy reconstructions of the bovine complex. The subunits are colored as in panel (a).

▲ **FIGURE 14.19 Structure of the F_0F_1 complex.**

Continuing with his reconstitution experiments, Efraim Racker found that isolated submitochondrial respiratory complexes (I, II, III, or IV) could be reconstituted into artificial membranes (liposomes) containing purified phospholipids by sonication. When F_0F_1 complex was included in the sonication mixture, it was also incorporated into the vesicles. In this case, each preparation had the electron transport properties of one original complex, plus a phosphorylation activity. For reconstituted complexes I, III, or IV, the P/O ratio was ~1 in each case. For example, reconstituted complex III could transfer electrons from coenzyme Q to cytochrome c, with concomitant synthesis of ~1 mole of ATP per pair of reducing equivalents. This type of evidence showed that respiratory complexes I, III, and IV each contain one pumping site. By contrast, complex II (succinate dehydrogenase) showed no ATP synthesis, confirming the absence of protein pumping in this complex.

Structure of the Mitochondrial F_1 ATP Synthase Complex

In 1994, the publication by John Walker's group of the crystal structure of the 371-kDa F_1 component of complex V helped to clarify the mechanism by which the passage of protons through the F_0F_1 complex drives the synthesis of ATP. As was suggested in Figure 14.19, the knob of the F_1 complex contains three identical $\alpha\beta$ dimers, forming a flattened sphere (**FIGURE 14.20(a)**). Each monomer can bind an adenine nucleotide at the interfaces between the α and β subunits, but the catalytic sites are located in the β subunits. The α and β subunits have nearly identical tertiary structures, composed of an N-terminal β barrel, a central domain containing the nucleotide-binding site, and a C-terminal helical domain (Figure 14.20(b)). An α-helical coiled-coil domain of the γ subunit of the central stalk penetrates the interior of the three $\alpha\beta$ dimers, where it interacts with the central and C-terminal domains of the α and β subunits.

Careful analysis of the crystal structure showed important structural differences among the three $\alpha\beta$ dimers. Most significant, the nucleotide-binding site of one contained ADP, while another contained AMP-PNP (a nonhydrolyzable ATP analog added during crystallization) and the third was empty (Figure 14.20(c)). Thus, each $\alpha\beta$ dimer can exist in three alternating conformations. What would cause otherwise identical $\alpha\beta$ dimers to adopt different conformations? The key is the interaction of the γ subunit in the interior of the three $\alpha\beta$ dimers. Because the γ subunit is itself asymmetric, it makes unique contacts with each $\alpha\beta$ dimer. In response, each $\alpha\beta$ dimer assumes a different conformation.

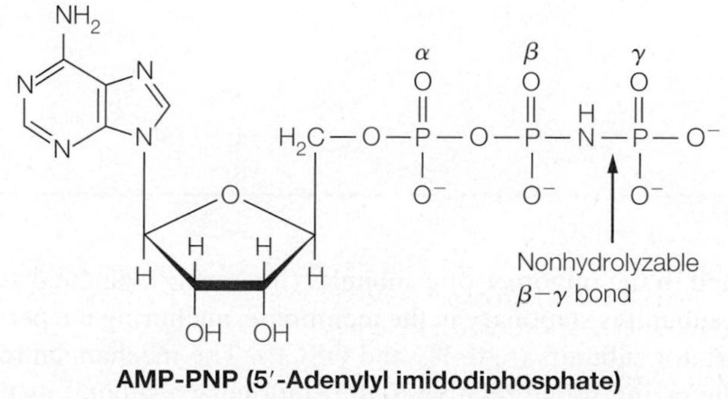

AMP-PNP (5′-Adenylyl imidodiphosphate)

Mechanism of ATP Synthesis

The structure of the ATP synthase complex supported a mechanism proposed by Paul Boyer in the 1970s, called the **binding-change model.** The essential idea of the model is that *rotation of the γ subunit, driven by the passage of protons through channels in F_0, causes sequential conformational changes in the three $\alpha\beta$ dimer assemblies that alter their substrate-binding abilities.*

In this model, the ATP synthase complex functions essentially as a three-cylinder engine. As shown in **FIGURE 14.21**, the γ subunit rotates in 120° steps, while the $\alpha\beta$ dimer assemblies remain stationary (held in place by the F_0 subunits a, b, d, F_6, and OSCP, acting as a stator). Thus, the dimers interact sequentially with different parts of the asymmetric γ subunit as it rotates, causing their nucleotide-binding sites, in turn, to adopt three different conformations, termed loose (L), tight (T), and open (O). The L conformation has loosely bound ADP and P_i; the T conformation has tightly bound ATP; and the O conformation has an empty nucleotide-binding site. In step ❶, passage of protons through channels in F_0 causes γ to rotate 120° counterclockwise. This leads to a conformational change in all three $\alpha\beta$ dimers simultaneously:

- The conformation of the site at lower left changes from T to O, causing it to release its bound ATP.

- The conformation of the site at upper left changes from O to L, allowing it to loosely bind ADP and P_i.

- The site at right changes from an L conformation to a T conformation, in which the substrates are bound much more tightly.

This last change initiates step ❷, in which the substrates react in the T site to form ATP.

Steps ❸ and ❹, and ❺ and ❻, are just repeats of steps ❶ and ❷, forming and releasing two additional ATP molecules, except that the three conformations (and thus the sites of ATP synthesis) migrate around the three $\alpha\beta$ dimers as the γ subunit rotates in 120 steps.

There is now good evidence for additional conformational intermediates. For example, P_i binds to the O site before ADP. This prevents the unwanted binding of ATP instead of P_i + ADP and explains how the enzyme is able to make ATP under cellular conditions where the ATP concentration is 10 to 50 times that of ADP.

● **CONCEPT** ATP synthase functions as a three-cylinder rotary engine driven by the passage of protons through channels in F_0.

Walker and Boyer's research on the structure determination of the F_0F_1 complex and the mechanism of ATP synthesis suggested by that structure earned them the Nobel Prize in Chemistry in 1997.

Does the γ subunit really rotate within the $\alpha\beta$ hexameric ring, as implied by the model of Figure 14.21? Several experimental approaches indicate that it does. The most graphic evidence comes from experiments in which the β subunits were immobilized to a glass coverslip and a fluorescent actin filament was attached to the γ subunit (see **FIGURE 14.22**). α subunits were added, and F_1 complexes were allowed to assemble on the glass slide. When this structure was examined by fluorescence microscopy, the addition of ATP could be seen to stimulate the rotation of the fluorescent probe. Analysis showed that the rotation occurred in discrete steps of 120°, just as predicted from the binding-change model. Rotation in a fully coupled F_0F_1 ATP synthase has now been observed in both

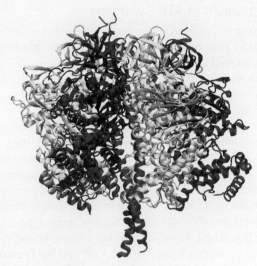

(a) X-ray structure of the bovine mitochondrial F₁ complex (PDB ID: 1bmf). α subunits (red) and β subunits (yellow) are arranged in a hexameric ring around the central γ subunit (blue). The membrane and F₀ complex would be at the bottom of the structure in this view.

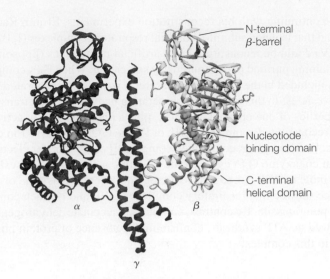

(b) One α subunit and one β subunit, from opposite sides of the hexameric ring, and the γ subunit are shown.

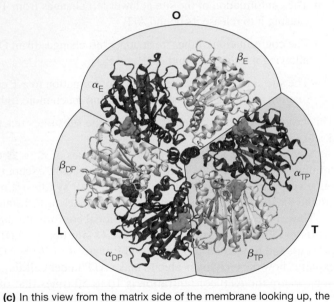

(c) In this view from the matrix side of the membrane looking up, the arrangement of the α and β subunits as three αβ dimers surrounding the central γ subunit is emphasized. Only the central nucleotide binding domain of each subunit is shown. The γ subunit is asymmetric, and makes unique contacts with each αβ dimer. Although not obvious from this figure, each αβ dimer has a slightly different conformation, especially in the central nucleotide-binding domain of each β subunit. The ATP analog AMP-PNP (green) can be seen bound in the catalytic site of the "T" β subunit. ADP (purple) can be seen bound in the catalytic site of the "L" β subunit, and the catalytic site of the "O" β subunit is empty. AMP-PNP molecules can also be seen at the interfaces of α and β subunits, but these are not catalytic sites.

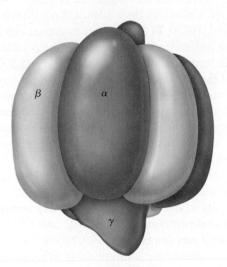

(d) Schematic of the F₁ complex.

▲ **FIGURE 14.20 Structure of the mitochondrial F₁ ATP synthase complex.**

synthesis and hydrolysis modes. The rotor turns counterclockwise in the ATP hydrolysis direction and clockwise in the ATP synthesis direction, when viewed from the membrane, and performs up to 700 revolutions per second (~100 revolutions per second in vivo).

How do protons flowing through the F₀ membrane component drive rotation of the γ subunit? As illustrated in Figure 14.19, the γ subunit is attached to the oligomer of c subunits (the c-ring) via the δ subunit. The a subunit is stationary in the membrane, anchoring the peripheral stalk stator subunits (b, d, F₆, and OSCP). The mechanism requires that the c-ring rotate relative to the stationary a subunit as protons flow from the intermembrane space through the complex into the matrix. Biochemical and crystallographic studies have identified

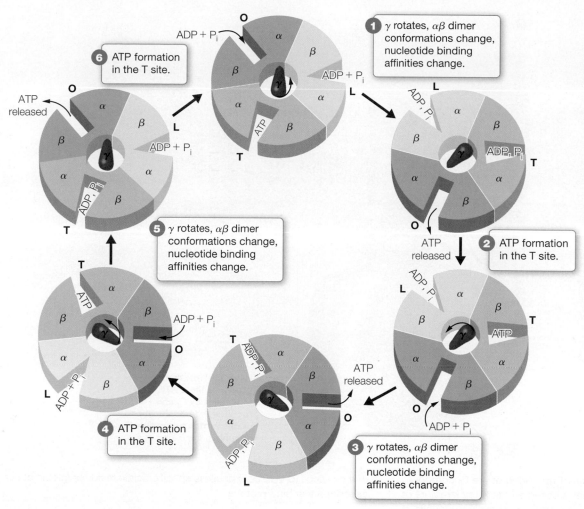

6 ATP formation in the T site.

1 γ rotates, αβ dimer conformations change, nucleotide binding affinities change.

5 γ rotates, αβ dimer conformations change, nucleotide binding affinities change.

2 ATP formation in the T site.

4 ATP formation in the T site.

3 γ rotates, αβ dimer conformations change, nucleotide binding affinities change.

◄ **FIGURE 14.21 Binding-change model for ATP synthase.** The nucleotide binding site (catalytic site) of the three αβ dimers exists in three different conformations, termed loose (L), tight (T), and open (O). In this scheme, the γ subunit rotates counterclockwise, driven by the passage of protons through channels in F₀, while the αβ dimer assemblies are held stationary by a stator. Each αβ dimer is distinguished by a different color, with the catalytic sites shown at the interface of α and β subunits. Step **1** depicts a 120° rotation of the γ subunit, causing a conformational change in all three αβ dimers simultaneously (see text). The result, in step **2**, is a conformation of the catalytic site that favors ATP formation from tightly bound ADP and Pᵢ. Steps **3** and **4**, and steps **5** and **6**, are just repeats of steps **1** and **2** but with the three conformations migrating around the three αβ dimers.

access channels for proton movement through the a and c subunits (**FIGURE 14.23**). The a subunit possesses two half-channels for protons—one leading to the matrix and the other to the intermembrane space. The high proton concentration in the intermembrane space drives protons into the half-channel of the a subunit facing the intermembrane space. The proton binds to a specific acidic residue (glutamate or aspartate) on the c subunit in contact with the a subunit. Upon protonation, the c subunit undergoes a conformational change that causes rotation of the c-ring. As the c subunit rotates all the way around to contact the a subunit again, the protonated carboxyl is exposed to a more hydrophobic environment, lowering its pK_a. The proton is released into the half-channel of the a subunit facing the matrix, favored by the low proton concentration in the matrix. Thus, protonation/deprotonation reactions cause rotation of the c-ring versus a. And since the c-ring is attached to the γ subunit, net proton-driven rotation of the γ subunit results. Each 360° rotation produces three ATP molecules by F₁ and requires the translocation of one proton per each c subunit in the ring of F₀.

▶ **FIGURE 14.22 An experimental system for observing rotation of the F₁ component of F₀F₁ ATP synthase.**

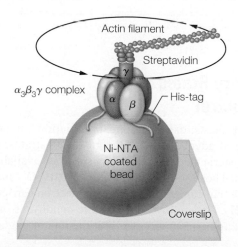

(a) The cloned gene encoding the F₁ β subunit was modified by adding a sequence coding for a polyhistidine tag (His-tag) that allows binding to a nickel-coated bead (Ni-NTA coated bead). After in vitro assembly with α and γ subunits, the F₁ complex was immobilized on the bead and attached to a glass coverslip. Streptavidin is a protein used to couple fluorescent-tagged actin to the γ subunit.

Labeled actin filament rotates counterclockwise

(b) Fluorescence microscopic examination showed that, following addition of ATP and its hydrolysis by the αβ catalytic subunits, the actin molecule rotated, which proved that the γ subunit itself was rotating.

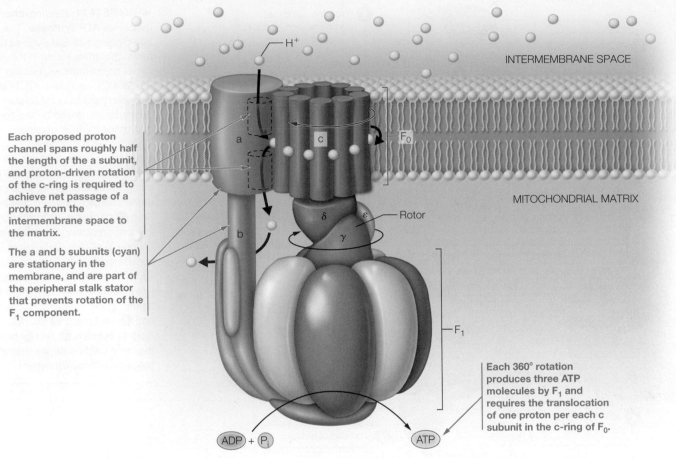

Each proposed proton channel spans roughly half the length of the a subunit, and proton-driven rotation of the c-ring is required to achieve net passage of a proton from the intermembrane space to the matrix.

The a and b subunits (cyan) are stationary in the membrane, and are part of the peripheral stalk stator that prevents rotation of the F_1 component.

INTERMEMBRANE SPACE

MITOCHONDRIAL MATRIX

Each 360° rotation produces three ATP molecules by F_1 and requires the translocation of one proton per each c subunit in the c-ring of F_0.

▲ **FIGURE 14.23 Proton-driven rotation of the c-ring of the F_0 component of F_0F_1 ATP synthase.** This model is based on the X-ray structure of the *E. coli* ATP synthase. The membrane-spanning α-helical c subunits of the c-ring (10 in *E. coli*) possess specific acidic residues (glutamate or aspartate) that bind protons.

14.5 Respiratory States and Respiratory Control

Like any metabolic process, oxidative phosphorylation can occur only in the presence of adequate quantities of its substrates. It is controlled not by allosteric mechanisms but by substrate availability and thermodynamics. Those substrates include ADP, P_i, O_2, and an oxidizable metabolite that can generate reduced electron carriers—NADH and/or $FADH_2$. Depending on metabolic conditions in the cell, any one of these four substrates can limit the rate of oxidative phosphorylation.

The dependence of oxidative phosphorylation on ADP reveals an important general feature of this process: *Respiration is tightly coupled to the synthesis of ATP.* Not only is ATP synthesis absolutely dependent on continued electron flow from substrates to oxygen, but the reverse is true as well—electron flow in normal mitochondria occurs only when ATP is being synthesized. This regulatory phenomenon, called **respiratory control,** makes biological sense because it ensures that substrates will not be oxidized wastefully. Their utilization is controlled by the physiological need for ATP.

In most aerobic cells, the concentration of ATP is 4 to 10 times greater than that of ADP. Thus, it is convenient to think of respiratory control as a dependence of respiration on ADP as a substrate for phosphorylation. If the energy demands on a cell cause ATP to be consumed at high rates, the resultant accumulation of ADP will stimulate respiration and the resynthesis of ATP. Conversely, in a relaxed and well-nourished cell, ATP accumulates at the expense of ADP, and the depletion of ADP limits the rate of both electron transport and its own phosphorylation to ATP. Thus, the energy-generating capacity of the cell is closely attuned to its energy demands.

Experimentally, respiratory control is demonstrated by following oxygen utilization in isolated mitochondria (**FIGURE 14.24**). In the absence of added substrate or ADP, very little oxygen is used because there is very little substrate available for oxidation. Addition of an oxidizable substrate, such as glutamate or malate, by itself has little effect on oxygen uptake. However, if ADP is then added, oxygen uptake proceeds at an enhanced rate until all of the added ADP has been converted to ATP, at which point oxygen uptake returns to the basal rate. This stimulation of respiration is stoichiometric; that is, addition of twice as much ADP causes twice the amount of oxygen

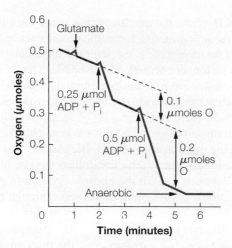

▲ **FIGURE 14.24 Experimental demonstration of respiratory control.** Oxygen uptake is monitored in isolated mitochondria. At the start of the experiment, the reaction chamber contains 0.5 μmol oxygen. As the mitochondria take up oxygen during respiration, the oxygen content of the chamber decreases, until all of the oxygen is used up, and the chamber becomes anaerobic. The slow oxygen uptake at the beginning reflects oxidation of endogenous substrates in the mitochondria. The addition of an exogenous oxidizable substrate (glutamate) stimulates respiration only slightly unless ADP + P_i is added as well. Both ADP additions represent limiting amounts; the second addition is twice the amount of the first, to show that the magnitude of oxygen uptake is stoichiometric. ADP stimulates respiration only until all of the ADP + P_i has been converted to ATP. (Oxygen uptake is recorded in μmoles O because one pair of electrons reduces one atom of O, not one molecule of O_2.)

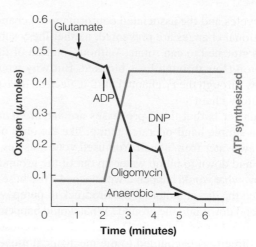

▲ **FIGURE 14.25 Effects of an inhibitor and an uncoupler on oxygen uptake and ATP synthesis.** The plot shows the results of an experiment in which an inhibitor (oligomycin) and an uncoupler (dinitrophenol, DNP) of oxidative phosphorylation were added to a mixture of isolated mitochondria, an oxidizable substrate (glutamate), and excess ADP + P_i. The red trace shows oxygen uptake; the blue trace represents ATP synthesis. The addition of oligomycin inhibits phosphorylation and consequently slows respiration. DNP uncouples respiration from phosphorylation so that O_2 uptake is stimulated even in the presence of oligomycin, but ATP synthesis remains blocked.

Oligomycin

uptake at the enhanced rate. If excess ADP is present instead of oxidizable substrate, the addition of substrate in limiting amounts will stimulate oxygen uptake until the substrate is exhausted. Thus, electron flow and oxygen utilization are tightly coupled to ATP synthesis in these mitochondria.

Maintenance of respiratory control depends on the structural integrity of the inner mitochondrial membrane. Disruption of the organelle causes electron transport to become uncoupled from ATP synthesis. Under these conditions, oxygen uptake proceeds at high rates even in the absence of added ADP. ATP synthesis is inhibited, even though electrons are being passed along the respiratory chain and used to reduce O_2 to water. Thus, electron flow and oxygen utilization are *uncoupled* from ATP synthesis. Uncoupling of respiration from phosphorylation can also be achieved with chemicals such as 2,4-dinitrophenol (**FIGURE 14.25**) or carbonyl cyanide 4-(trifluoromethoxy)phenylhydrazone (FCCP), which act by dissipating the proton gradient (see Section 14.4).

Another group of compounds, exemplified by the antibiotic oligomycin, act as inhibitors of oxidative phosphorylation. Addition of oligomycin to actively respiring, well-coupled mitochondria inhibits both oxygen uptake and ATP synthesis, as shown in Figure 14.25. However, no direct inhibition of electron transport occurs, as shown by the fact that subsequent addition of an uncoupler such as DNP greatly stimulates oxygen uptake.

Figures 14.24 and 14.25 show that blocking ATP synthesis either by depletion of substrate (ADP) or inhibition by oligomycin also blocks electron transport in coupled mitochondria. It is easy to visualize how inhibition of electron transport hampers ATP synthesis because no proton gradient is being formed. But how does inhibition of ATP synthesis block electron transport? Oligomycin inhibits ADP phosphorylation by binding to specific sites in the F_0 complex, blocking the flow of protons through the F_0 proton channel and thus blocking rotation of the c-ring. Likewise, depletion of ADP inhibits ATP synthesis by preventing the $\alpha\beta$ dimers from going through their catalytic cycles and the associated conformational changes. If these conformational changes are prevented, neither the γ subunit nor the c-ring it is attached to can rotate. Without rotation of these components, proton flow through F_0 is blocked. But why would blocking proton flow through the F_0 channel inhibit electron transport through the respiratory chain?

The answer is that these processes are *mechanically* coupled. Consider a simple hand-operated pump, like the ones our ancestors used to draw water from a well. You used your muscles to move the handle up and down to pump water up out of the ground and out the faucet. Now what would happen if we attached a fire hose to the faucet and we used the diesel motor on the fire truck to pump water through the pump and down into the ground? What would happen to the pump handle? It would move (rapidly!) up and down because water flow (in either direction) is coupled to the mechanical movement of the pump handle. The pump, in other words, is reversible. What if we now held onto the pump handle, preventing it from moving? What would happen to the flow of water from the fire truck through the pump? It would stop, unless its pressure was greater than the force holding the handle (or the hose bursts!).

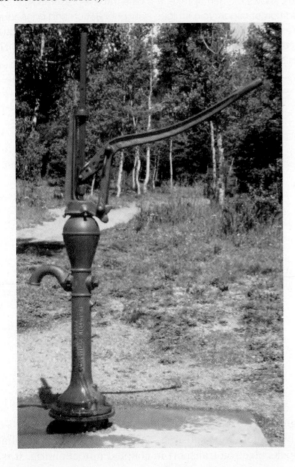

Like the pump we have been describing, ATP synthase is reversible. In fact, as mentioned earlier, ATP synthase was originally discovered as an ATPase. Using an inhibitor of ATP synthase (such as oligomycin) is like holding the pump handle. If we block the ATP synthase from going through its catalytic cycle, there can be no rotation of the c subunits in the F_0 complex in the membrane, and thus no flow of protons through the F_0 due to the mechanical coupling of these two processes. If there

is no flow of H^+ from the intermembrane space back into the matrix to dissipate the protonmotive force, the proton pressure builds up on the outside of the membrane to the point where pumping protons out of the matrix is no longer energetically favorable. That is, the thermodynamically favorable electron transport through these complexes cannot provide sufficient free energy to overcome the high opposing protonmotive force. Under these conditions, $\Delta G_{e^-\text{transport}} + \Delta G_{H^+\text{pumping}} = 0$. So electron transport stops because it too is *mechanically* coupled to proton pumping through complexes I, III, and IV.

Thus, the fundamental factor controlling the rate of respiration is the balance between ΔG for the phosphorylation of ADP (the ATP synthase reaction), ΔG for electron transport through the respiratory complexes, and ΔG for H^+ pumping (defined by the protonmotive force). If the free energy required to synthesize ATP is exactly balanced by the free energy available from the proton gradient, there will be no proton flow and no ATP synthesis (i.e., $\Delta G_{\text{ATP synthesis}} + \Delta G_{H^+\text{ reentry to matrix}} = 0$). Now let's consider what happens when the cell begins to consume ATP to drive biosynthetic processes. As ATP is consumed, levels of ATP in the matrix will fall, making ΔG for phosphorylation of ADP more favorable (i.e., $\Delta G_{\text{ATP synthesis}} + \Delta G_{H^+\text{ reentry to matrix}} < 0$). ATP synthesis will then proceed, as will the coupled flow of protons through F_0. This increased proton reentry will reduce the thermodynamic barrier to electron transport (i.e., $\Delta G_{e^-\text{ transport}} + \Delta G_{H^+\text{ pumping}} < 0$). This in turn leads to increased electron transport and, hence, increased respiration rate.

> ● **CONCEPT** The rate of electron transport is limited by the availability of ADP for conversion to ATP.

Under some natural conditions, the ability to uncouple respiration from phosphorylation is highly desirable. Many mammals, particularly those that are born hairless (including human infants), those that hibernate, and those that are cold-adapted, have special needs for maintenance of core body temperature. Such animals have a special tissue, called brown adipose tissue (BAT), in the neck and upper back. Mitochondria in this tissue are especially rich in respiratory electron carriers, particularly cytochromes, which give BAT its brown color. These mitochondria are specialized to generate heat from the oxidation of acetyl-CoA derived from the breakdown of fatty acids. The inner membranes of BAT mitochondria are rich in **uncoupling protein 1 (UCP1),** a 33-kDa channel that allows protons to return to the matrix in a process that bypasses ATP synthase, thereby uncoupling electron transport from ATP synthesis. Thus, the energy that is derived from acetyl-CoA oxidation in the citric acid cycle is dissipated as heat.

A comparable phenomenon is seen in the plant world among species that emerge in early spring, often when the ground is still covered with snow. The skunk cabbage is a particularly dramatic example: It uses the heat generated by uncoupling oxidation from phosphorylation to melt its way through the frozen snow. The floral spike of this plant can maintain a temperature some 10° to 25° above ambient temperature.

> ● **CONNECTION** Mitochondrial uncoupling proteins play several important roles besides heat generation. In mammals, uncoupling protein 2 (UCP2) limits the production of reactive oxygen species (ROS) that would otherwise contribute to endothelial cell dysfunction in cardiovascular diseases.

14.6 Mitochondrial Transport Systems

Whereas the mitochondrial outer membrane is freely permeable to molecules up to ~5000 Da, the permeability of the inner membrane is severely limited. The importance of this selective permeability can be seen from our discussions of electrochemical gradients and the shuttle systems used to transport reducing equivalents into the mitochondrion. We must also consider substrate transport, including the inward transport of intermediates for oxidation in the citric acid cycle, the export of intermediates used for biosynthesis in other cell compartments, and the exit of newly synthesized ATP. Properties of the principal mitochondrial transport systems are outlined in **FIGURE 14.26**. Because the matrix is negatively charged relative to the cytoplasm, it is energetically unfavorable for a negatively charged solute to enter the matrix. Consequently, mitochondrial carriers use cotransport with protons, or exchange with OH^-, to import these metabolites. Thus, some of the free energy of the proton gradient is used for transport.

Transport of Substrates and Products into and out of Mitochondria

First, let us consider ATP, ADP, and P_i, the participants in oxidative phosphorylation. Two systems are involved, an **adenine nucleotide translocase (ADP/ATP carrier,** or **ANT)** and a **phosphate translocase.** The ADP/ATP carrier spans the inner membrane, and it binds ADP at a specific site on the outer surface of the inner membrane. The protein couples the efflux of free ATP from the matrix to the influx of an equivalent amount of free ADP from the intermembrane space. Because this antiporter exchanges ATP, with a charge of -4, for ADP, with a charge of -3, its action is driven by the membrane potential (outside positive). It is generally true of the mitochondrial transport systems that at least one of the participants is moving down a concentration gradient or is coupled to the proton gradient so that no further energy source is required.

● **CONNECTION** As might be expected from the critical role the ADP/ATP and phosphate carriers play in providing substrates for the ATP synthase, their deficiency in humans is usually characterized by exercise intolerance, muscular hypotonia (decreased muscle tone), hypertrophic cardiomyopathy (thickened heart muscle), and elevated plasma lactate levels with decreased pH (lactic acidosis).

The phosphate translocase acts in either the *antiport* or the *symport* mode (Section 10.4) to transport inorganic phosphate, depending on the ionization state. As an antiporter, it transports dihydrogen phosphate ion ($H_2PO_4^-$) into the matrix, coupled with the efflux of a hydroxide ion. In the alternative symport mode, it transports monohydrogen phosphate ion (HPO_4^{2-}) into the matrix along with two protons. Both modes of transport maintain electrical neutrality and are driven by the proton concentration difference (ΔpH) component of the protonmotive force generated by electron transport. The net effect of the adenine nucleotide and phosphate transport systems is to couple the inward transport of the substrates of oxidative phosphorylation, ADP and P_i, to the efflux of the product, ATP.

Next, let us consider the substrates for oxidation. The major substrate from carbohydrate catabolism is pyruvate, which, like phosphate, is exchanged for OH^-, using the pyruvate transport system. Dicarboxylic acid substrates—succinate, fumarate, and

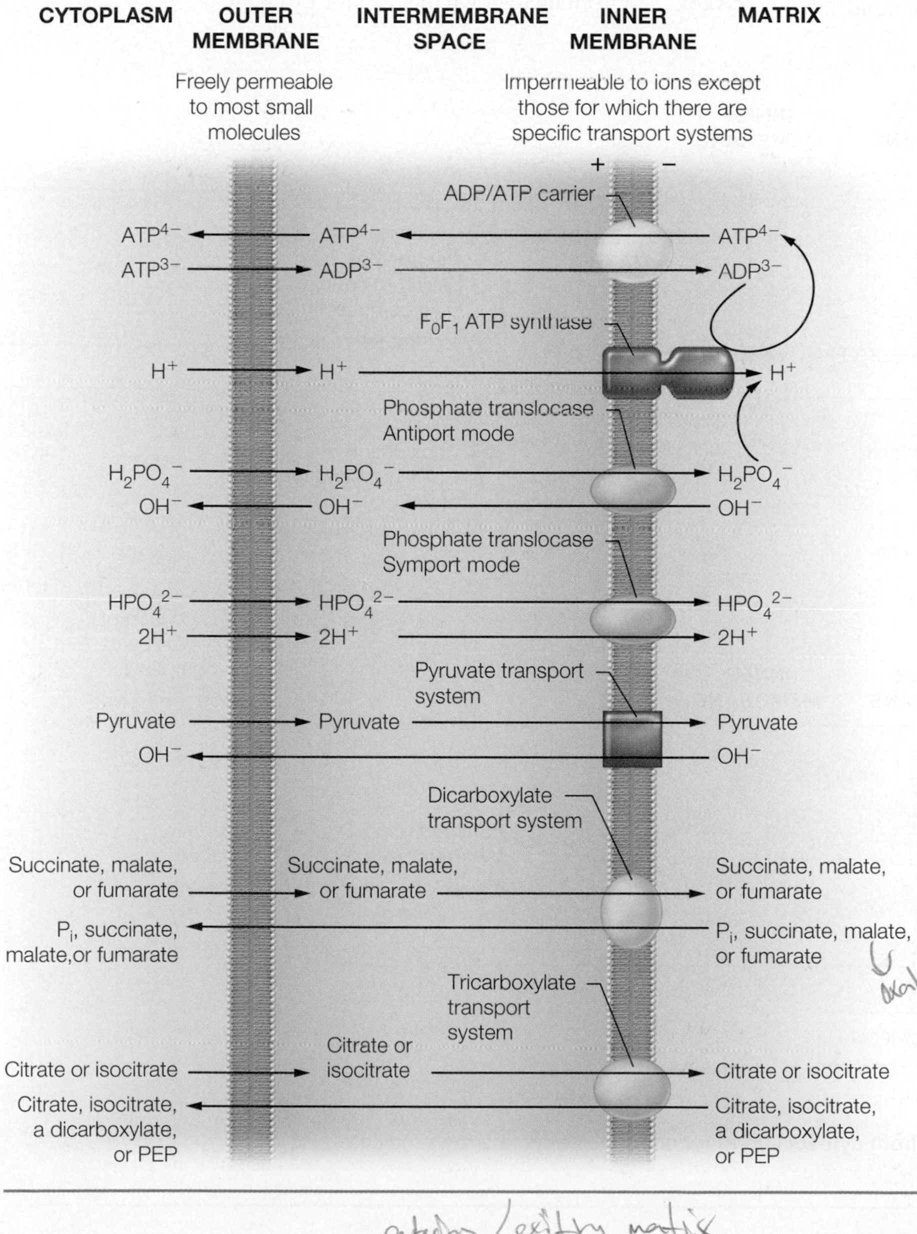

FIGURE 14.26 Major inner membrane transport systems for respiratory substrates and products.

malate—can be exchanged for each other or for orthophosphate in the dicarboxylate transport system. Similarly, the tricarboxylate transport system carries either citrate or isocitrate, coupled each with the other or with a dicarboxylic acid or phosphoenolpyruvate.

Shuttling Cytoplasmic Reducing Equivalents into Mitochondria

Another important role of mitochondrial carriers is to shuttle cytoplasmic reducing equivalents into the matrix for subsequent reoxidation by the respiratory chain. Recall that during aerobic glycolysis the NADH produced at the glyceraldehyde-3-phosphate dehydrogenase step is not reoxidized because pyruvate can be further oxidized in the citric acid cycle (Section 12.1). To extract the energy from this NADH and to regenerate oxidized NAD^+ for continued glycolysis, the reducing equivalents must be transferred into the mitochondrion. However, the NADH generated by a cytosolic NAD-linked dehydrogenase cannot itself traverse the mitochondrial

● **CONCEPT** Electrons are transported into mitochondria by metabolic shuttles.

membrane to be oxidized by the respiratory chain. Therefore, the reducing equivalents must be shuttled to respiratory assemblies in the inner mitochondrial membrane, without physical movement of NADH. This process involves the reduction of a substrate by NADH in the cytoplasm, passage of the reduced substrate into the mitochondrial matrix via a specific transport system, reoxidation of that compound inside the matrix, and passage of the oxidized substrate back to the cytoplasm, where it can undergo the same cycle again.

The dihydroxyacetone phosphate/glycerol-3-phosphate shuttle is particularly active in brain and skeletal muscle. As shown in **FIGURE 14.27(a)**, dihydroxyacetone phosphate (DHAP) is first reduced by NADH in the cytosol (step ❶). The resulting glycerol-3-phosphate is then reoxidized (step ❷) by a flavin-dependent glycerol-3-phosphate dehydrogenase, bound at the outer face of the inner mitochondrial membrane. This process involves reduction of FAD, followed by transfer of an electron pair from $FADH_2$ to coenzyme Q (step ❸), just as intramitochondrial NADH transfers electrons to coenzyme Q (see Figure 14.12). Once dihydroxyacetone phosphate has returned to the cytosol (step ❹), the net effect has been to transfer two reducing equivalents from cytosolic NADH to mitochondrial $FADH_2$ and from there down the respiratory chain.

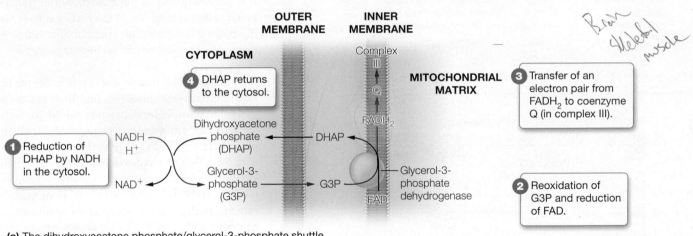

(a) The dihydroxyacetone phosphate/glycerol-3-phosphate shuttle.

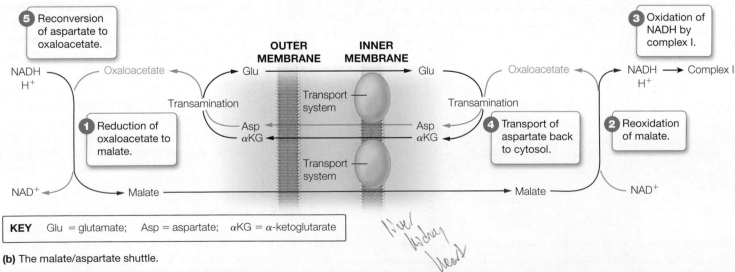

(b) The malate/aspartate shuttle.

▲ **FIGURE 14.27** Shuttles for transfer of reducing equivalents from cytosol into mitochondria. Magenta arrows indicate flow of reducing equivalents.

A different shuttle system, particularly active in liver, kidney, and heart, is the malate/aspartate shuttle, shown in Figure 14.27(b). Here, a cytosolic isozyme of malate dehydrogenase, together with NADH, reduces oxaloacetate to malate (step ❶), which passes into the matrix via a specific α-ketoglutarate/malate exchanger in the inner mitochondrial membrane. The malate is reoxidized by the malate dehydrogenase of the citric acid cycle, which also uses NAD^+ (step ❷). The resulting matrix NADH is then oxidized by complex I (step ❸). Because oxaloacetate cannot cross the inner membrane, it is transaminated to aspartate (step ❹), which is then transported out via a specific aspartate/glutamate exchanger. Once in the cytoplasm, aspartate is reconverted to oxaloacetate by transamination (step ❺), to begin the cycle anew. Because of the transaminations involved, this process requires that α-ketoglutarate be continuously transported out of mitochondria and that glutamate be continuously transported in. This balance is ensured by the substrate specificity of the two exchangers.

In tissues that use the malate/aspartate shuttle, approximately 2.5 moles of ATP are generated per mole of cytoplasmic NADH. In tissues that use the dihydroxyacetone phosphate/glycerol-3-phosphate shuttle, only about 1.5 moles of ATP are generated per mole of NADH because those cytoplasmic reducing equivalents enter the respiratory chain at complex III (via $CoQH_2$) rather than at complex I.

14.7 Energy Yields from Oxidative Metabolism

Much of the past three chapters has been devoted to the pathways by which carbohydrates are oxidized to CO_2 and water. Finally, we are in a position to calculate the total energy yield and metabolic efficiency of these combined pathways. Let us review how much energy is recovered in the form of ATP from the entire oxidative catabolism of glucose. First, we present a balanced equation for each of the three pathways involved, and then we estimate the amount of ATP that can be derived through oxidative phosphorylation from the reduced electron carriers.

Glycolysis:

Glucose $+$ 2ADP $+$ $2P_i$ $+$ $2NAD^+$ $\longrightarrow$

$\qquad$ 2 pyruvate $+$ 2ATP $+$ 2NADH $+$ $2H_2O$ $+$ $2H^+$

Pyruvate dehydrogenase complex:

2 pyruvate $+$ $2NAD^+$ $+$ 2 CoA-SH $\longrightarrow$

$\qquad$ 2 acetyl-CoA $+$ 2NADH $+$ $2CO_2$

Citric acid cycle (including conversion of GTP to ATP):

2 acetyl-CoA $+$ $4H_2O$ $+$ $6NAD^+$ $+$ 2FAD $+$ 2ADP $+$ $2P_i$ $\longrightarrow$

$\qquad$ $4CO_2$ $+$ 6NADH $+$ $2FADH_2$ $+$ 2 CoA-SH $+$ 2ATP $+$ $4H^+$

Net:

Glucose $+$ $10NAD^+$ $+$ 2FAD $+$ $2H_2O$ $+$ 4ADP $+$ $4P_i$ $\longrightarrow$

$\qquad$ $6CO_2$ $+$ 10NADH $+$ $6H^+$ $+$ $2FADH_2$ $+$ 4ATP

These three processes generate 4 moles of ATP directly, plus 10 moles of NADH and 2 moles of $FADH_2$. Using P/O ratios for oxidation of NADH and $FADH_2$ of 2.5 and 1.5, respectively, we find that the total realizable ATP yield is 4 $+$ (2.5 $\times$ 10) $+$ (1.5 $\times$ 2), or about 32 per mole of glucose oxidized. In prokaryotes and in cells using the malate/aspartate shuttle (Figure 14.27(b)), the reducing equivalents from cytoplasmic glycolysis are carried into the mitochondrion with no energy cost. However, cells using the glycerol phosphate shuttle (Figure 14.27(a)) incur an energy cost because the electrons from cytosolic NADH enter the respiratory chain as $FADH_2$. Therefore, the ATP yield from each of these two NADHs is $\sim$1.5, not 2.5. This decreases the overall ATP yield to 30 per mole of glucose. For the following discussion, we shall use 32 as the best estimate of total moles of ATP derived per mole of glucose oxidized. Recalling that $\Delta G°'$ for glucose oxidation is -2870 kJ/mol (Section 11.4) and that $\Delta G°'$ for ATP hydrolysis is -32.2 kJ/mol, we can calculate the efficiency for operation of this biochemical machine: (32 $\times$ 32.2)/2870, or $\sim$36%, under standard conditions. Because ΔG for ATP hydrolysis under intracellular conditions is worth 50–60 kJ/mol, the actual efficiency in the cell is considerably higher.

● **CONCEPT** The complete oxidation of 1 mole of glucose generates about 30–32 moles of ATP synthesized from ADP.

14.8 The Mitochondrial Genome, Evolution, and Disease

Mitochondria possess a double-stranded circular genome of approximately 16,500 base pairs. In humans, mitochondrial DNA (mtDNA) contains 37 genes that code for 13 proteins, all of which are subunits of the respiratory chain complexes (**FIGURE 14.28(a)**). Complexes I, III, IV, and V all contain some subunits encoded by mtDNA, whereas complex II and cytochrome c are encoded by nuclear DNA. The remaining mtDNA genes encode 22 tRNAs and 2 rRNAs. These tRNAs and rRNAs are part of the mitochondrial protein synthesis machinery, required to translate the 13 proteins encoded in the mtDNA. Of course, most of the proteins found in mitochondria (more than 900) are encoded by the nucleus, translated on cytoplasmic ribosomes, and then transported into the mitochondria.

One of the biggest questions in evolutionary biology concerns the origin of mitochondria. Where did they come from? Although many theories have been proposed, the most widely accepted one is the *endosymbiont hypothesis.* In this scenario, over a billion years ago an archaea, perhaps a methanogen that used H_2, entered into a symbiotic relationship with a facultative anaerobic α-proteobacterium, perhaps one that produced H_2. Eventually, the archaeal host completely engulfed the α-proteobacterium without killing it. In this endosymbiotic relationship, the host supplied the endosymbiont with oxidizable substrates, and the endosymbiont in turn supplied the host with energy (ATP). This relationship was sealed forever as most of the genes of the endosymbiont were transferred to the host genome, giving rise to the first eukaryotic cell, with mitochondria representing the modern descendants of the original α-proteobacterium.

Diseases that affect mitochondrial function (mitochondrial diseases) are probably quite common in humans—perhaps as much as 2% of the population has some mitochondrial defect. However, only a small number of these disorders are caused by mutations in mtDNA ($\sim$1 in 10,000 individuals). mtDNA mutations cause defects in the respiratory chain, either directly through mutations in one of the 13 structural genes (Figure 14.28(a)) or indirectly through mutations in a tRNA or rRNA gene. Many of these diseases involve

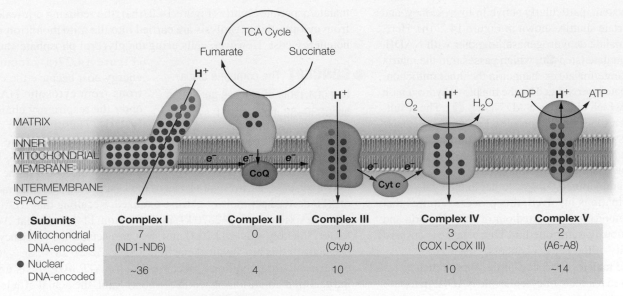

Subunits	Complex I	Complex II	Complex III	Complex IV	Complex V
● Mitochondrial DNA-encoded	7 (ND1-ND6)	0	1 (Ctyb)	3 (COX I-COX III)	2 (A6-A8)
● Nuclear DNA-encoded	~36	4	10	10	~14

(a) The five multisubunit complexes of the respiratory chain are shown with mtDNA-encoded subunits in orange and nuclear-encoded subunits in purple.

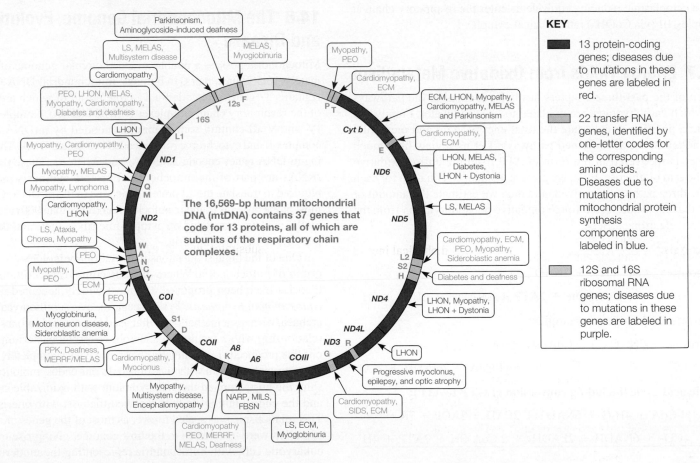

(b) A morbidity map of the human mitochondrial genome.

▲ **FIGURE 14.28 Mitochondrial DNA and mitochondrial diseases.** Data for (b) from *Biochimica et Biophysica Acta* 1658: 80–88, S. DiMauro, Mitochondrial diseases.

brain and skeletal muscle, and are thus known as *mitochondrial encephalomyopathies.*

The "morbidity map" in Figure 14.28(b) shows the locations of mtDNA mutations and the mitochondrial encephalomyopathies they cause. Although the clinical features of these disorders are quite variable, patients with mtDNA diseases often exhibit progressive and disabling neurological problems, muscle weakness, chronic progressive weakness of the eye muscles (ophthalmoplegia), and exercise-induced fatigue, all of which stem from defective respiration and ATP production.

As mentioned above, the majority of inherited mitochondrial diseases are due to mutations in nuclear genes. This is because efficient assembly and functioning of the respiratory chain require a large number of nuclear-encoded proteins. For example, both the mitochondrial DNA polymerase and RNA polymerase are encoded by nuclear genes, as are all of the protein subunits of the mitochondrial ribosomes. Mutations in the gene for the catalytic subunit of mitochondrial DNA polymerase (POLG) are linked with a wide variety of mitochondrial diseases, including Parkinsonism and male infertility. Finally, there is a group of late-onset neurodegenerative diseases that all involve mitochondrial dysfunction, including Huntington's disease, Alzheimer's disease, and amyotrophic lateral sclerosis (ALS, also known as Lou Gehrig's disease). The precise role of mitochondria in the pathogenesis of these hereditary neurodegenerative diseases is not yet understood but is an area of intense investigation.

● **CONNECTION** Huntington's disease, Alzheimer's disease, and amyotrophic lateral sclerosis are all late-onset neurodegenerative diseases that involve mitochondrial dysfunction.

Recall the location of cytochrome c (Figure 14.2). It is on the intermembrane space side of the inner membrane. Cytochrome c plays another important role in the cell besides that of electron carrier in respiration—it is also a central signaling component in the mitochondrial pathway of **apoptosis,** or programmed cell death. This pathway is initiated in response to oxidative damage, DNA damage, and many other chemical and physical insults that damage the cell directly. These diverse stresses all trigger the same response: The mitochondrial outer membrane becomes more permeable, and cytochrome c is released from the intermembrane space into the cytoplasm, where it binds with several other proteins in a complex (the **apoptosome**). Formation of this complex leads to activation of a proteolytic cascade, eventually resulting in cell death. Thus, this integral component of the respiratory chain is also an integral component of a cell death pathway, hinting at the relationship between mitochondria and the evolution of the eukaryotic cell.

● **CONNECTION** In mammalian cells, mitochondria play a central signaling role in apoptosis, or programmed cell death. Multiple stimuli, including oxidative damage, DNA damage, metabolic disruptions, and oncogene activation all trigger the release of cytochrome c into the cytoplasm, initiating this ancient cell death pathway.

14.9 Oxygen as a Substrate for Other Metabolic Reactions

In most cells, at least 90% of the molecular oxygen consumed is utilized for oxidative phosphorylation. The remaining O_2 is used in a wide variety of specialized metabolic reactions. At least 200 known enzymes use O_2 as a substrate. Because O_2 is rather unreactive, virtually all of these 200 enzymes use a metal ion to enhance the reactivity of oxygen, just as cytochrome oxidase does. In this final section of the chapter, we briefly categorize these enzymes, and we consider the metabolism of partially reduced forms of oxygen, which arise continually in all cells and are highly toxic because of their great reactivity.

Oxidases and Oxygenases

As we saw in Chapter 13, the term *oxidase* is applied to enzymes that catalyze the oxidation of a substrate *without* incorporation of oxygen from O_2 into the product. *Oxygenases* are enzymes that incorporate oxygen atoms from O_2 into the oxidized products. There are two classes of oxygenases—monooxygenases and dioxygenases. **Dioxygenases** incorporate both atoms of O_2 into one substrate. **Monooxygenases** incorporate one atom from O_2 into a product and reduce the other oxygen atom to water. Because one substrate usually becomes hydroxylated by this class of oxygenase, the term **hydroxylase** is also used. An example of this type of reaction is the hydroxylation of steroids:

● **CONCEPT** Oxidases are enzymes that catalyze the oxidation of a substrate *without* incorporation of oxygen from O_2 into the product. Oxygenases are enzymes that incorporate oxygen atoms from O_2 into the oxidized products.

$$RH + NADPH + H^+ + O_2 \longrightarrow R\!-\!OH + NADP^+ + H_2O$$

Cytochrome P450 Monooxygenase

The most numerous hydroxylation reactions involve a superfamily of heme proteins with the collective name **cytochrome P450,** found in nearly all organisms, from bacteria to mammals. The human genome contains 57 different structural genes for cytochromes P450, making this a large and diverse protein family. These proteins resemble hemoglobin and mitochondrial cytochrome oxidase in being able to bind both O_2 and carbon monoxide. When complexed with carbon monoxide, cytochromes P450 absorb light strongly at 450 nm.

● **CONCEPT** Cytochromes P450 catalyze hydroxylations of numerous unreactive substrates, making them easier to metabolize.

Cytochromes P450 are involved in hydroxylating a large variety of compounds, including the hydroxylations in steroid hormone biosynthesis mentioned above. In addition, these monooxygenases act upon thousands of **xenobiotics** (foreign compounds), including drugs such as phenobarbital and environmental carcinogens such as benzo[a]pyrene, a constituent of tobacco smoke, and aflatoxin B, a carcinogenic compound produced by a mold and found in peanuts that have not been properly screened. Hydroxylation of foreign substances usually increases their solubility and is an important step in their detoxification, or metabolism and excretion. However, some of these reactions result in activation of potentially carcinogenic substances to more reactive species, as with aflatoxin B.

Cytochrome P450 systems participate in a wide variety of additional reactions, including epoxidation, desulfuration, dealkylation, deamination, and dehalogenation. These reactions are particularly active in liver.

Reactive Oxygen Species, Antioxidant Defenses, and Human Disease

Formation of Reactive Oxygen Species

As we have seen, the terminal step in electron transport is the four-electron reduction of O_2 to water. Cytochrome oxidase, like most oxidases, transfers electrons to oxygen from metal ions that change their valence states by one electron at a time—such as the heme iron and copper in cytochrome oxidase. Because the interactions of one-electron carriers with two-electron carriers are rarely 100% efficient, oxidases often generate incompletely reduced oxygen species—**superoxide** ($O_2^{\cdot -}$), formed from a one-electron reduction of O_2; hydrogen peroxide (H_2O_2), formed from a two-electron reduction; and hydroxyl radical ($OH^{\cdot}$), formed via a three-electron reduction (**FIGURE 14.29**). In addition, some enzymes, such as xanthine oxidase (Section 19.3), generate hydrogen peroxide as their ordinary products. Superoxide, hydrogen peroxide, and hydroxyl radical are all more reactive than oxygen and are referred to collectively as **reactive oxygen species (ROS).**

> ● **CONCEPT** Partially reduced oxygen species—superoxide, hydrogen peroxide, and hydroxyl radical—are extremely toxic. Their toxicity is counteracted by both enzymatic and nonenzymatic mechanisms.

Hydroxyl radical is particularly reactive and is responsible for damage to other biological molecules. Hydroxyl radical alters proteins in various ways and degrades membranes by initiating the oxidation of fatty acids in membrane lipids, a process termed **lipid peroxidation.** Hydroxyl radical also damages nucleic acids, both by causing polynucleotide strand breakage (double-stranded DNA breaks are lethal) and by changing the structure of DNA bases. About 20 different base changes, or DNA lesions, are known to result from reactions of hydroxyl radical with DNA. Some lesions are mutagenic because the altered base created forms non-Watson–Crick base pairs during DNA replication. These lesions, together with their repair processes, will be discussed in Chapter 23.

Superoxide per se is relatively nontoxic. However, because it contains an unpaired electron, it is a free radical, and combines readily with another free radical, nitric oxide ($NO^{\cdot}$) (Figure 14.29), a biological signaling agent that is produced in many animal tissues (Chapters 7, 18, and 20). The product is peroxynitrite ($OONO^-$), also considered a reactive oxygen species. Peroxynitrite causes lipid peroxidation and also causes nitration of tyrosyl hydroxyl groups in proteins, a reaction particularly damaging to membrane proteins.

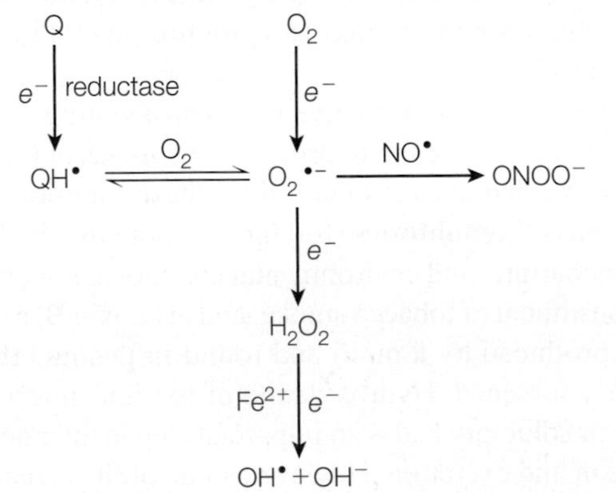

Normal cellular metabolism produces ROS, and mitochondria are an important source of these ROS. It is estimated that no more than 0.02% of all the electrons that start down the respiratory chain leak from complexes I, II, and III, and bring about one-electron reductions of oxygen to superoxide. There is growing evidence that such ROS production constitutes a second messenger system that fine-tunes cellular energy metabolism in response to changes in redox status and mitochondrial function. There are even some cases where overproduction of reactive oxygen species is a normal part of the functioning of a cell. For example, certain white blood cells contribute to defense against infectious agents by **phagocytosis** (from the Greek word meaning "cell eating"). Such cells can engulf a bacterial cell. This event is followed by a **respiratory burst,** a rapid increase in oxygen uptake. In a deliberate and controlled process, much of this oxygen is reduced to superoxide ion and to H_2O_2. The hydrogen peroxide is then converted to more reactive oxidants such as hypochlorous acid (HOCl), which help to kill the engulfed bacterium.

However, uncontrolled overproduction of reactive oxygen species has the potential to inflict considerable damage on the tissues in which they are produced, a situation called **oxidative stress.** Because oxidative stress can damage many biomolecules—lipids, proteins, and nucleic acids—the tissue injury that results can, in principle, lead to a variety of disease states. Oxidative damage has been implicated in many different human disorders, including cardiovascular disease, cancer, stroke, neurodegenerative diseases, chronic inflammatory diseases, and even aging. Not surprisingly, a series of elaborate mechanisms has evolved to minimize its harmful consequences.

Dealing with Oxidative Stress

As discussed in Chapter 15, the Earth had an anaerobic atmosphere for its first billion years, and oxygen was intensely toxic to all life forms existing at that time. With the evolution of oxygen in our atmosphere, life forms developed both enzymatic and nonenzymatic defenses against oxidative stress. The nonenzymatic protection is afforded by **antioxidant** compounds, including glutathione (Section 12.10), vitamins C and E, and uric acid, an end product of purine metabolism (Chapter 19). These compounds can scavenge ROS before they can cause damage, or they can prevent oxidative damage from spreading.

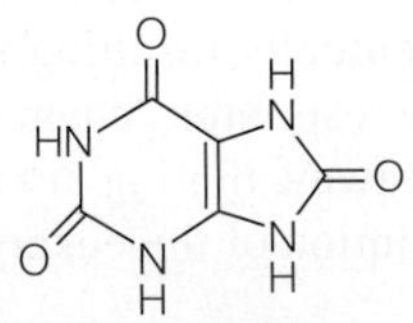

▲ **FIGURE 14.29 Reactive oxygen species (ROS).** The generation and interconversion of the most common reactive oxygen species are shown. $O_2^{\cdot -}$, superoxide; $OH^{\cdot}$, hydroxyl radical; $NO^{\cdot}$, nitric oxide; $OONO^-$, peroxynitrite; Q, oxidized coenzyme Q; $QH^{\cdot}$, semiquinone radical; H_2O_2, hydrogen peroxide.

Uric acid

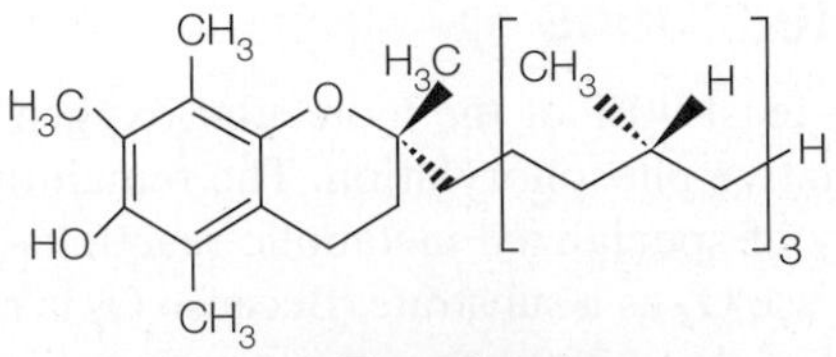

***α*-Tocopherol (vitamin E)**

Among enzymatic mechanisms, the first line of defense is **superoxide dismutase (SOD),** a family of metalloenzymes that catalyze a **dismutation** (a reaction in which two identical substrate molecules have different fates). Here, one molecule of superoxide is oxidized, and one is reduced.

$$O_2^{\cdot-} + O_2^{\cdot-} + 2H^+ \longrightarrow H_2O_2 + O_2$$

Hydrogen peroxide is metabolized either by **peroxiredoxins,** a ubiquitous family of thiol proteins, by **catalase,** another widely distributed enzyme, or by a more limited family of **peroxidases.** Peroxiredoxins, abundant enzymes found in all domains of life, play an antioxidant role through their peroxidase activity:

$$ROOH + 2H^+ + 2e^- \longrightarrow ROH + H_2O$$

These enzymes can reduce and detoxify not only hydrogen peroxide, but also peroxynitrite and a wide range of organic hydroperoxides (ROOH).

Catalase is a heme protein with an extremely high turnover rate (>40,000 molecules per second). It catalyzes the following reaction:

$$2H_2O_2 \longrightarrow 2H_2O + O_2$$

Peroxidases reduce H_2O_2 to water at the expense of oxidation of an organic substrate. An example of a peroxidase is found in erythrocytes, which are especially sensitive to peroxide accumulation. (See Section 12.10 for a discussion of the consequences of peroxide accumulation in glucose-6-phosphate dehydrogenase deficiency.) Erythrocytes are protected from peroxide accumulation by *glutathione peroxidase*, which catalyzes the reduction of peroxide, along with the oxidation of glutathione (GSH):

$$2GSH + H_2O_2 \longrightarrow GSSG + 2H_2O$$

Summary

- Most of the energy captured from oxidative reactions in cells is used for ATP synthesis by means of mitochondrial **oxidative phosphorylation.** Reduced electron carriers, both NADH and $FADH_2$, shuttle reducing equivalents in the mitochondrial matrix. Enzyme complexes bound to the inner mitochondrial membrane pass these electrons through the **respiratory chain,** a series of electron carriers of ever-increasing reduction potential (complexes I–V). Electrons are eventually transferred to O_2, which is reduced to water. (Sections 14.3–14.4)

- The redox reactions of complexes I, III, and IV provide energy to pump protons from the matrix across the inner membrane, generating an electrochemical gradient termed the **protonmotive**

force. Dissipation of the resultant proton gradient, as protons pass back into the matrix through a specific channel on ATP synthase, provides the energy that is necessary to drive ATP synthesis. (Sections 14.4–14.5)

- Although respiration accounts for about 90% of the total oxygen uptake in most cells, dozens of enzymes use O_2 as a substrate— oxygenases, oxidases, and hydroxylases. Some reactions generate partially reduced oxygen species—as hydroxyl radical, **superoxide,** and peroxide—termed reactive oxygen species, which are toxic and mutagenic. Cells possess numerous mechanisms for detoxification of these reactive oxygen species. (Section 14.9)

Problems

Enhanced by
Mastering Chemistry
for Biochemistry

Mastering Chemistry for Biochemistry provides select end-of-chapter problems and feedback-enriched tutorial problems, animations, and interactive figures to deepen your understanding of complex topics while practicing problem solving.

Answers to red problems are available in the Answer Appendix.

1. Referring to Table 14.1 for E'_0 values, calculate $\Delta G°'$ for oxidation of malate by malate dehydrogenase.

2. When pure reduced cytochrome c is added to carefully prepared mitochondria along with ADP, P_i, antimycin A, and oxygen, the cytochrome c becomes oxidized, and ATP is formed, with a P/O ratio approaching 1.0.
 (a) Indicate the probable flow of electrons in this system.
 (b) Why was antimycin A added?
 (c) What does this experiment tell you about the location of coupling sites for oxidative phosphorylation?
 (d) Write a balanced equation for the overall reaction (including cyt c oxidation and ATP synthesis).
 (e) Calculate $\Delta G°'$ for the above reaction, using E'_0 values from Table 14.1 and a $\Delta G°'$ value for ATP hydrolysis of -32.2 kJ/mol.

3. Freshly prepared mitochondria were incubated with β-hydroxybutyrate, oxidized cytochrome c, ADP, P_i, and cyanide. β-hydroxybutyrate is oxidized by an NAD^+-dependent dehydrogenase.

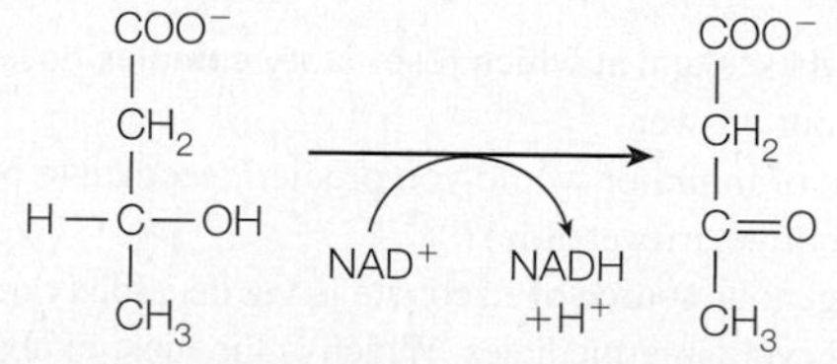

$$
\begin{array}{ccc}
\text{COO}^- & & \text{COO}^- \\
| & & | \\
\text{CH}_2 & & \text{CH}_2 \\
| & \longrightarrow & | \\
\text{H}-\text{C}-\text{OH} & & \text{C}=\text{O} \\
| & \text{NAD}^+ \quad \text{NADH} & | \\
\text{CH}_3 & \qquad\quad +\text{H}^+ & \text{CH}_3 \\
\end{array}
$$

The experimenter measured the rate of oxidation of β-hydroxybutyrate and the rate of formation of ATP.
 (a) Indicate the probable flow of electrons in this system.
 (b) How many moles of ATP would you expect to be formed per mole of β-hydroxybutyrate oxidized in this system?
 (c) Why is β-hydroxybutyrate added rather than NADH?
 (d) What is the function of the cyanide?
 (e) Write a balanced equation for the overall reaction occurring in this system (electron transport and ATP synthesis).

(f) Calculate the net standard free energy change ($\Delta G^{\circ\prime}$) in this system, using E'_0 values from Table 14.1 and a $\Delta G^{\circ\prime}$ value for ATP hydrolysis of -32.2 kJ/mol.

4. If you were to determine the P/O ratio for oxidation of α-ketoglutarate, you would probably include some malonate in your reaction system. Why? Under these conditions, what P/O ratio would you expect to observe?

5. Inflammatory stimuli cause macrophages to undergo dramatic metabolic reprogramming, including a switch from oxidative phosphorylation to aerobic glycolysis. Cordes et al. (*J. Biol. Chem.* 291:14274–14284, 2016) explored the role of itaconate in this metabolic reprogramming. Itaconate is an antimicrobial metabolite that is also produced in mammalian immune cells. Permeabilized RAW264.7 macrophages were treated with various oxidizable substrates in the presence or absence of 10 mM itaconate. Maximal uncoupled oxygen consumption rate (OCR) in response to four different oxidizable substrates (maximal uncoupled OCR) were then measured, and the results are shown below.

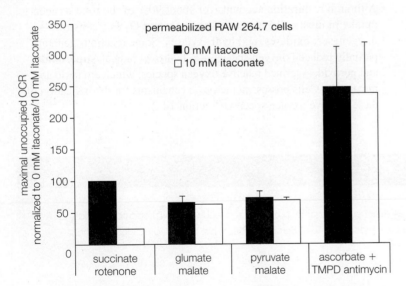

itaconate

(a) Based on these data, at which respiratory complex does itaconate act? Explain your answer.
(b) What kind of inhibitor would you predict itaconate to be (competitive, noncompetitive, irreversible)?
(c) The endogenous source of itaconate is via decarboxylation of one of the citric acid cycle intermediates. Which is the most likely substrate for this decarboxylation reaction?

6. Of the various oxidation reactions in glycolysis and the citric acid cycle, the only one that does not involve NAD^+ is the succinate dehydrogenase reaction. What would $\Delta G^{\circ\prime}$ be for an enzyme that oxidizes succinate with NAD^+ instead of FAD? If the intramitochondrial concentration of succinate was 10-fold higher than that of fumarate, what minimum $[NAD^+]/[NADH]$ ratio in mitochondria would be needed to make this reaction exergonic at 37 °C?

7. Intramitochondrial ATP concentrations are about 5 mM, and phosphate concentration is about 10 mM. If ADP is five times more abundant than AMP,

calculate the molar concentrations of ADP and AMP at an energy charge of 0.85. Calculate ΔG for ATP hydrolysis at 37 °C under these conditions. The energy charge is the concentration of ATP plus half the concentration of ADP divided by the total adenine nucleotide concentration:

$$\frac{[ATP] + 1/2[ADP]}{[ATP] + [ADP] + [AMP]}$$

8. From $E^{\circ\prime}$ values in Table 14.1, calculate the equilibrium constant for the glutathione peroxidase reaction at 37 °C.

9. In the early days of "mitochondriology," P/O ratios were determined from measurements of volume of O_2 taken up by respiring mitochondria and chemical assays for disappearance of inorganic phosphate. Now, however, it is possible to measure P/O ratios simply with a recording oxygen electrode. How might this be done?

10. Years ago there was interest in using uncouplers such as dinitrophenol as weight control agents. Presumably, fat could be oxidized without concomitant ATP synthesis for re-formation of fat or carbohydrate. Why was this a bad (i.e., fatal) idea?

11. Referring to Figure 14.16, predict the P/O ratio for oxidation of ascorbate by isolated mitochondria.

12. As a representation of the respiratory chain, what is wrong with this picture? There are four deliberate errors.

Malate → NAD^+ → $FADH_2$ → $CoQH_2$ → cyt b^{2+}
Oxalo-acetate → NADH → FAD → CoQ → cyt b^{3+}

cyt c^{3+} → cyt c_1^{2+} → cyt a-a_3^{3+} → H_2O_2
cyt c^{2+} → cyt c_1^{3+} → cyt a-a_3^{2+} → $½O_2$

13. GSSG + NADPH + H^+ → 2GSH + $NADP^+$
(a) Calculate $\Delta G^{\circ\prime}$ for the glutathione reductase reaction in the direction shown, using $E^{\circ\prime}$ values from Table 14.1.
(b) Suppose that a cell contained an isoform of glutathione reductase that used NADH instead of NADPH as the reductive coenzyme. Would you expect $\Delta G^{\circ\prime}$ for this enzyme to be higher, lower, or the same as the corresponding value for the real glutathione reductase? Briefly explain your answer.
(c) Given what you know about the metabolic roles and/or intracellular concentration ratios of NAD^+/NADH and $NADP^+$/NADPH, would you expect ΔG (not $\Delta G^{\circ\prime}$) for this enzyme to be higher, lower, or the same as ΔG for the real enzyme under intracellular conditions? Briefly explain your answer.

14. To carefully prepared mitochondria were added succinate, oxidized cytochrome *c*, ADP, orthophosphate, and sodium cyanide. Referring to Figure 14.16, answer the following.
(a) List the sequence of electron carriers in this system.
(b) Write a balanced equation for the overall reaction occurring in this system, showing oxidation of the initial electron donor, reduction of the final acceptor, and synthesis of ATP.
(c) Calculate $\Delta G^{\circ\prime}$ for the overall reaction. $\Delta G^{\circ\prime}$ for ATP hydrolysis is -32.2 kJ/mol.
(d) Why was cyanide added in this experiment?
(e) What would the P/O ratio be if the same experiment were run with addition to the mitochondria of 2,4-dinitrophenol?

15. In order to function as an oxidative phosphorylation uncoupler (page 467), 2,4-dinitrophenol must act catalytically, not stoichiometrically. What does this mean? Identify and discuss an important implication of this conclusion.

16. (a) Calculate the standard free energy change as a pair of electrons is transferred from succinate to molecular oxygen in the mitochondrial respiratory chain.

(b) Based on your answer in part a, calculate the maximum number of protons that could be pumped out of the matrix into the intermembrane space as these electrons are passed to oxygen. Assume 25 °C, $\Delta pH = 1.4$; $\Delta\psi = 0.175$ V (matrix negative).

(c) At which site(s) are these protons pumped?

17. Four electron carriers, a, b, c, and d, whose reduced and oxidized forms can be distinguished spectrophotometrically, are required for respiration in a bacterial electron transport system. In the presence of substrates and oxygen, three different inhibitors block respiration, yielding the patterns of oxidation states shown below. What is the order of the carriers in the chain from substrates to O_2, and where do the three inhibitors act?

	Carriers			
Inhibitor	**a**	**b**	**c**	**d**
1	O	O	R	O
2	R	R	R	O
3	O	R	R	O

O and R indicate fully oxidized and fully reduced, respectively.

18. Biochemists working with isolated mitochondria recognize five energy "states" of mitochondria, depending on the presence or absence of essential substrates for respiration—O_2, ADP, oxidizable substrates, and so forth. The characteristics of each state are:

state 1: mitochondria alone (in buffer containing P_i)

state 2: mitochondria + substrate, but respiration low due to lack of ADP

state 3: mitochondria + substrate + limited amount of ADP, allowing rapid respiration

state 4: mitochondria + substrate, but all ADP converted to ATP, so respiration slows

state 5: mitochondria + substrate + ADP, but all O_2 used up (anoxia), so respiration stops

(a) On the graph, identify the state that might predominate in each stage of the trace indicated with a letter.

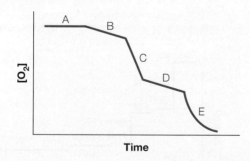

Time

(b) To determine whether isolated mitochondria exhibit respiratory control, one determines the ratio of rates of oxygen uptake in two different states. Which states?

(c) Which state probably predominates in vivo in skeletal muscle fatigued from a long and strenuous workout?

(d) Which state probably predominates in resting skeletal muscle of a well-nourished animal?

(e) Which state probably predominates in heart muscle most of the time?

19. FAD is a stronger oxidant than NAD^+; FAD has a higher standard reduction potential than NAD^+. Yet in the last reaction of the pyruvate dehydrogenase complex (Chapter 13), $FADH_2$ bound to the E_3 subunit is oxidized by NAD^+. Explain this apparent paradox.

20. The antibiotic valinomycin (see Section 10.4) is an ionophore that forms a specific complex with potassium ion. Because the complex is lipophilic and can diffuse into the membrane, valinomycin brings about the transport of K^+ through the inner membrane. Valinomycin acts by decreasing the $\Delta\psi$ (membrane potential) component of the pmf, without a direct effect on the pH gradient. Another antibiotic, nigericin, acts as a K^+/H^+ antiporter; it carries H^+ in one direction, coupled with the reverse transport of K^+. Thus, nigericin dissipates the pH component of the pmf, with little effect on $\Delta\psi$. Which antibiotic, nigericin or valinomycin, do you predict would have the greater effect on oxidative phosphorylation when administered to respiring mitochondria? Assume the antibiotics are added to a suspension of mitochondria in equimolar amounts. Briefly explain your reasoning.

21. As discussed in Chapter 13, calcium is an important regulator of the citric acid cycle. Calcium is transported across the mitochondrial inner membrane by a Ca^{2+} uniporter that is driven by the negative potential inside the matrix.

(a) Assuming a membrane potential across the inner membrane of 180 mV (inside negative), calculate the ratio of the $[Ca^{2+}]$ in the matrix to that in the cytoplasm $\left(\dfrac{[Ca^{2+}]_m}{[Ca^{2+}]_c}\right)$ that would exist at equilibrium (i.e., $\Delta G = 0$).

(b) Cytoplasmic $[Ca^{2+}]$ is on the order of 10^{-7} M in a healthy cell. Based on your answer in (a), calculate the $[Ca^{2+}]$ that would exist in the matrix at equilibrium. Is this a physiologically reasonable answer? If not, provide an explanation.

References

FOUNDATION FIGURE | Intermediary Metabolism

METABOLISM IS COMPOSED OF TWO MAJOR PROCESSES-catabolism and anabolism.

- **Catabolism** (complex substances degraded to simpler molecules) is accompanied by the net release of chemical energy.
- **Anabolism** (biosynthesis of complex organic molecules from small molecules) requires a net input of chemical energy.
- These two sets of reactions are coupled together by ATP.
- Catabolic and anabolic pathways with the same end points are not simple reversals of each other. Each pathway must be thermodynamically favorable in its own direction.

Metabolic pathways are regulated by several mechanisms, including control of enzyme concentration, enzyme activity, and enzyme compartmentation.
- Regulation (kinetic control) can be imposed only on reactions that are displaced far from equilibrium.

Amino acid synthesis   **Pentose phosphate pathway**   **Nucleotide synthesis**

NADPH

PATHWAYS

1a

ENERGY UTILIZATION (REDUCTIVE BIOSYNTHESIS)

Lipid synthesis ← Fatty acids ← **Fatty acid synthesis** ← Met — NADPH — Acetyl-CoA ←
← Glycerol ←

Glycogen synthesis — Glucose-(P) — **Gluconeogenesis** ← – PEP ← Met — NADPH — OAA ←

+/−   BG   Met   Hor   +/−   BG   Met   Hor

CYTOSOL

1b

ENERGY PRODUCTION (OXIDATION)

Glycogenolysis → Glucose-(P) → **Glycolysis** – → PEP → Pyruvate →

ATP   +/−   CYTOSOL

Lipids ● — Glycerol + Fatty acids

Hor

1a ANABOLISM (reduction)

Synthesis of metabolic fuels from simple inorganic compounds– CO_2 and H_2O.

Anabolism generally requires a net *input* of energy.

The Gibbs free energy change (ΔG) of a metabolic pathway can be estimated from the intracellular concentrations of pathway intermediates. For glycolysis, all but three of the reactions function at or near equilibrium, and are freely reversible in vivo. The three exceptions are the reactions catalyzed by hexokinase (HK), phosphofructokinase (PFK), and pyruvate kinase (PK), which take place with large decreases in free energy. These three nonequilibrium reactions are irreversible in vivo, and they make the entire glycolytic pathway unidirectional. These are the steps that must be bypassed in gluconeogenesis. Not coincidentally, these three nonequilibrium reactions are also the sites of regulation of glycolysis, because regulation can be imposed only on reactions displaced far from equilibrium.

▶ SECTION 12.4

1b CATABOLISM (oxidation)

Conversion of carbon from metabolic fuels into an activated two-carbon fragment– acetyl-CoA.

Cytosol

Catabolism is generally accompanied by a net *release* of energy.

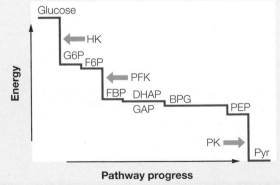

▶ FIGURE 12.8

COMPARTMENTATION

ENERGETICS

REGULATION

 Energy charge (ATP/ADP) "Energy charge" is defined as
$$\frac{[ATP] + 1/2[ADP]}{[ATP] + [ADP] + [AMP]}$$

 Redox state   Blood glucose   Metabolite   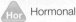 Hormonal

Mastering **Chemistry** for Biochemistry

Mastering Chemistry for Biochemistry provides select end-of-chapter problems and feedback-enriched tutorial problems, animations, and interactive figures to deepen your understanding of complex topics while practicing problem solving.

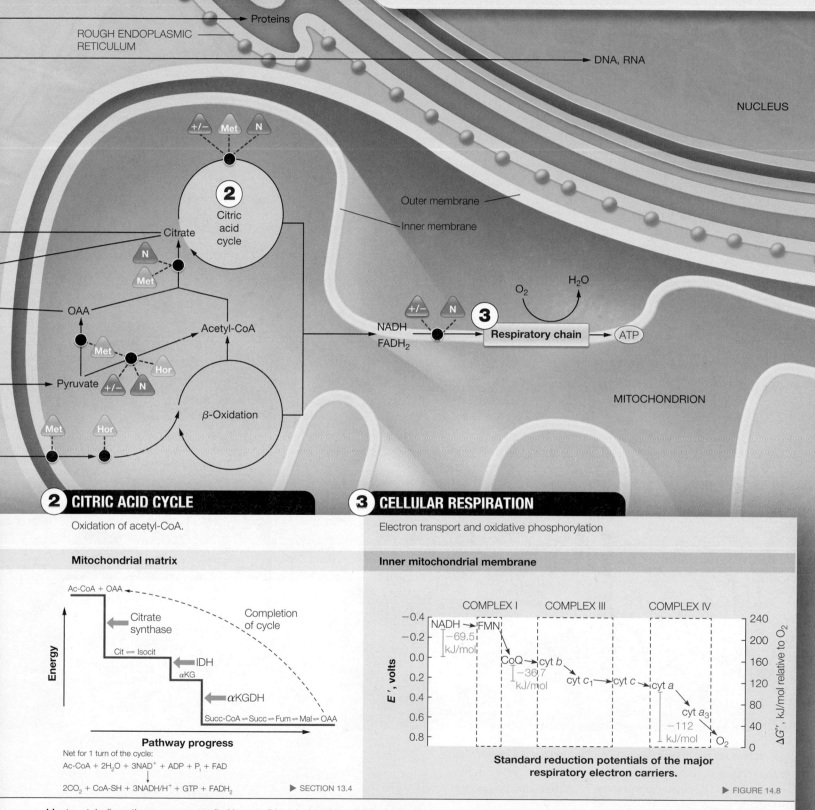

Proteins

ROUGH ENDOPLASMIC RETICULUM

DNA, RNA

NUCLEUS

+/−   Met   N

2

Citric acid cycle

Citrate

N

Met

Outer membrane

Inner membrane

OAA

Acetyl-CoA

Met

Hor

Pyruvate   +/−   N

β-Oxidation

Met   Hor

O₂ H_2O

+/−   N   **3**

NADH
FADH₂

Respiratory chain   ATP

MITOCHONDRION

---

② CITRIC ACID CYCLE

Oxidation of acetyl-CoA.

Mitochondrial matrix

Ac-CoA + OAA

Completion of cycle

Citrate synthase

Cit ⇌ Isocit

IDH

αKG

αKGDH

Succ-CoA ⇌ Succ ⇌ Fum ⇌ Mal ⇌ OAA

Energy

Pathway progress

Net for 1 turn of the cycle:

Ac-CoA + 2H₂O + 3NAD⁺ + ADP + Pᵢ + FAD

2CO₂ + CoA-SH + 3NADH/H⁺ + GTP + FADH₂

▶ SECTION 13.4

③ CELLULAR RESPIRATION

Electron transport and oxidative phosphorylation

Inner mitochondrial membrane

COMPLEX I   COMPLEX III   COMPLEX IV

NADH → FMN

−69.5 kJ/mol

CoQ → cyt b

−36.7 kJ/mol

cyt c_1 → cyt c → cyt a

cyt a_3

−112 kJ/mol

O_2

E', volts

$\Delta G''$, kJ/mol relative to O_2

Standard reduction potentials of the major respiratory electron carriers.

▶ FIGURE 14.8

Most metabolic pathways are controlled by conditions inside the cell (where the pathway operates), and by conditions outside the cell (in the entire organism). External conditions are usually transmitted via a hormone, while internal conditions can be signaled in at least three ways:

1. Energy charge of the cell
2. Redox state within the cell
3. Concentrations of one or more metabolites, either in the same pathway or a related one (This includes feed-forward activators and feed-back inhibitors).

Tropical rainforests are often referred to as the lungs of the planet, breathing in CO_2 and exhaling O_2 through the process of photosynthesis. Deforestation is a major threat to the ability of these rainforests to regulate global CO_2 levels.

Photosynthesis

IN THE PRECEDING chapters, we described the ways in which organisms extract a substantial portion of the energy available from the oxidation of carbohydrates, such as glucose:

$$C_6H_{12}O_6 + 6O_2 \longrightarrow 6CO_2 + 6H_2O \qquad \Delta G^{\circ\prime} = -2870 \text{ kJ/mol.}$$

We noted that as much as 40% of this energy could be recovered for useful biochemical work through the production of ATP.

But life cannot depend on oxidative metabolism as its ultimate source of energy, and it cannot continue indefinitely returning organic carbon to the atmosphere as CO_2. The reaction above is only half of the great energy–carbon cycle of nature (**FIGURE 15.1**). The reverse of the carbohydrate oxidation reaction—the production of hexose carbohydrates and oxygen from carbon dioxide and water— is a *reductive* process, requiring a steady supply of reducing equivalents (electrons). This reverse process is accomplished by plants, algae, and some bacteria, using the energy from sunlight to provide the enormous amount of free energy required.

$$6CO_2 + 6H_2O \xrightarrow{\text{Light energy}} C_6H_{12}O_6 + 6O_2$$
$$\Delta G^{\circ\prime} = +2870 \text{ kJ/mol}$$

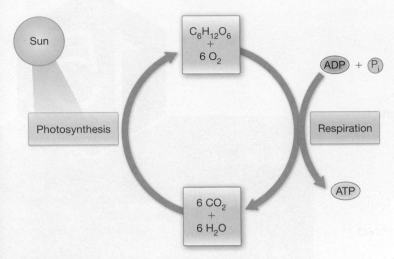

▲ **FIGURE 15.1 The carbon cycle in nature.** Carbon dioxide and water are combined through photosynthesis to form carbohydrates. In both photosynthetic and nonphotosynthetic organisms, these carbohydrates can be reoxidized to regenerate CO_2 and H_2O. Part of the energy obtained from both photosynthesis and fuel oxidation is captured in ATP.

This process is called **photosynthesis.** Not only does it provide carbohydrates for energy production in virtually all organisms, it is also the major path through which carbon reenters the biosphere—that is, the principal means of carbon fixation, the conversion of carbon dioxide into organic carbon molecules. Furthermore, photosynthesis is the major source of oxygen in the Earth's atmosphere.

Prior to the evolution of photosynthetic organisms, the Earth's atmosphere was probably devoid of oxygen (though rich in carbon dioxide). The earliest organisms must have used other hydrogen/electron donors, such as H_2S, NH_3, Fe^{2+}, to drive the conversion of organic acids to the necessary carbohydrates. All of these were in limited supply compared to water. Without the advent of photosynthesis, these energy sources would have eventually been wholly consumed, and life would have perished. Photosynthesis provided life a way to take advantage of the unlimited supply of reducing equivalents (the oceans of water)

needed to convert carbon dioxide into carbohydrates and the other organic molecules necessary for life. The fossil and geochemical records suggest that photosynthetic organisms first appeared approximately 3.4 billion to 2.3 billion years ago. Their gradual conversion of the primitive, nonoxidizing atmosphere of the Earth (O_2 absent) to an oxidizing atmosphere (O_2 present in abundance) paved the way for aerobic metabolism and the evolution of the diversity of life across all domains. Today, photosynthesis represents the ultimate source of energy for almost all life.* Photosynthesis—the fixation of inorganic carbon into organic molecules—is performed by plants, algae, and a wide variety of bacteria, all of which are food sources for other organisms. A comprehensive view of the relationship of photosynthesis to other pathways we have studied is shown in **FIGURE 15.2.**

In the following sections, we will look at the structure of the chloroplast and the light-harvesting systems. We will learn how electrons from water flow through these structures to generate a proton gradient that is used to generate ATP. And finally, we will describe the pathways by which CO_2 is converted into carbohydrate.

● **CONCEPT** Photosynthesis provides carbohydrates for energy production, fixes CO_2, and is the major source of atmospheric O_2.

---

* Recent developments make it necessary to qualify the statement that all life derives its energy directly from photosynthesis. It has been found that some bacteria, such as those associated with submarine "black smoker" hydrothermal vents, use the oxidation of substances like H_2S, formate, or H_2 as an alternate energy source for the synthesis of organic molecules, in the complete absence of light. This energy cycle represents, however, only a small fraction of the energy flow in the biosphere.

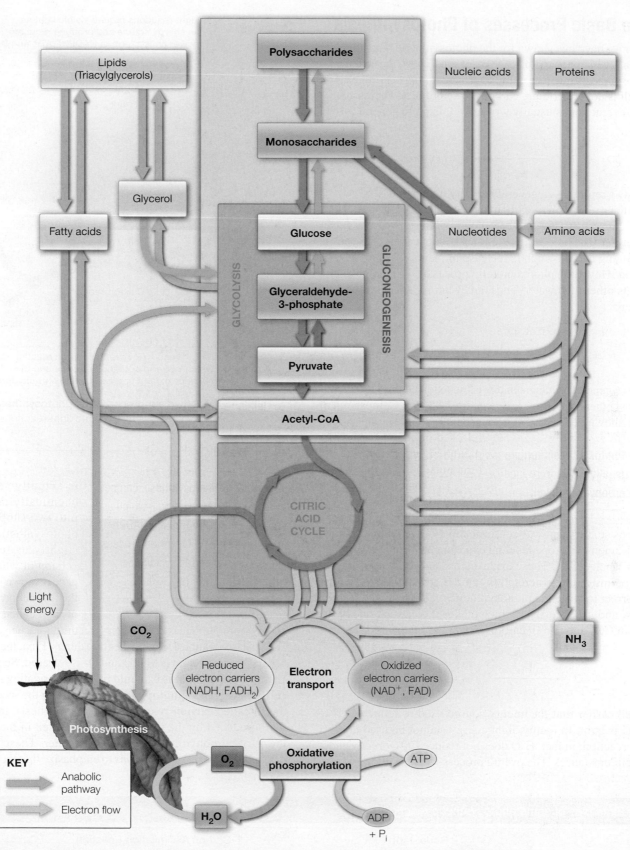

▲ FIGURE 15.2 The role of photosynthesis in metabolism. The major biosynthetic pathways leading from carbon dioxide and water to polysaccharides are highlighted in green. Oxygen derived from the water is released as a by-product of photosynthesis.

15.1 The Basic Processes of Photosynthesis

The equation we have just shown for the photosynthetic reaction is, of course, a great oversimplification. As you might expect, the actual process of photosynthesis involves many intermediate steps. Furthermore, a hexose itself is not the primary carbohydrate product. Therefore, the photosynthetic reaction is usually written in this more general form:

$$CO_2 + H_2O \xrightarrow{\text{Light energy}} [CH_2O] + O_2$$

where $[CH_2O]$ represents a general carbohydrate.

Because the catabolism of carbohydrates to form CO_2 is an *oxidative* process, converting CO_2 into carbohydrate must involve a *reduction* of the carbon. The preceding reaction statement shows H_2O as the ultimate reducing agent, which is the case in plants, most algae, and cyanobacteria. However, photosynthetic processes in many bacteria use reductants other than water. Thus, an even more general reaction can be written:

$$CO_2 + 2H_2A \xrightarrow{\text{Light energy}} [CH_2O] + H_2O + 2A$$

where H_2A is a general reductant and A is the oxidized product. Examples of photosynthetic reactions are given in **TABLE 15.1**. Comparison of the reactions shown in the table suggests that the source of the oxygen released in photosynthesis by plants, algae, and cyanobacteria must be H_2O rather than CO_2. This source was predicted in the early 1930s by C. B. van Niel, one of the pioneers in photosynthesis studies, and was confirmed in 1941 by Samuel Ruben and Martin Kamen in isotope labeling experiments using ^{18}O-labeled water and unlabeled CO_2. These experiments showed that neither of the oxygen atoms in the generated O_2 comes from CO_2. Therefore, it is more correct to write the photosynthetic reactions in the fashion shown below and in Table 15.1, which makes it clear that one of the oxygens from CO_2 ends up in carbohydrate, the other in water:

$$CO_2 + 2H_2O \xrightarrow{\text{Light energy}} [CH_2O] + H_2O + O_2$$

We stated earlier that the energy source used to reduce CO_2 to carbohydrate is light. In reality, light energy cannot be used *directly* to drive this reaction; in fact, H_2O does not reduce CO_2 *directly* under any known circumstances. The overall process we have just described

● **CONCEPT** Photosynthesis requires a reductant, usually H_2O, to reduce CO_2 to the carbohydrate level.

Light reactions require visible light as an energy source and produce reducing power (in the form of NADPH), ATP, and O_2.

NADPH and ATP drive the carbon reactions, which occur in both the presence and the absence of light and fix CO_2 into carbohydrates.

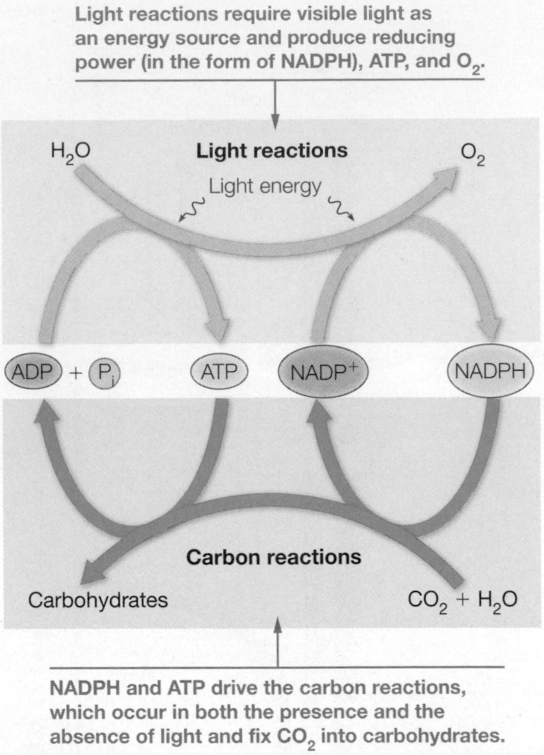

▲ **FIGURE 15.3 The two subprocesses of photosynthesis.** The overall process of photosynthesis is divided into light reactions and carbon reactions.

● **CONCEPT** Photosynthesis can be divided into light reactions, which use sunlight energy to produce NADPH and ATP, releasing O_2 in the process, and carbon reactions, which use NADPH and ATP to fix CO_2.

is actually separated, both chemically and physically, into two subprocesses in all photosynthetic organisms. A slightly more sophisticated version of what actually happens is shown in **FIGURE 15.3**. In the first subprocess, in a series of steps called **light reactions,** energy from sunlight is used to carry out the photochemical oxidation of H_2O. This oxidation accomplishes two things. First, the oxidizing agent $NADP^+$ is reduced to NADPH and O_2 is released. Second, part of the energy from sunlight is captured by phosphorylating ADP to produce ATP. This is called **photophosphorylation.** In the second subprocess, the so-called **carbon reactions** of photosynthesis, the NADPH and ATP produced by the light reactions are used in the reductive synthesis of carbohydrate from CO_2 and water. These reactions were originally termed *dark reactions* to emphasize that they do not require

TABLE 15.1 Examples of some photosynthetic reactions

Organisms	Reductant	Carbon Assimilation Reaction
Plants, algae, cyanobacteria	H_2O	$CO_2 + 2H_2O \longrightarrow [CH_2O] + H_2O + O_2$
Green sulfur bacteria	H_2S	$CO_2 + 2H_2S \longrightarrow [CH_2O] + H_2O + 2S$
Purple bacteria	SO_3^{2-}	$CO_2 + 2SO_3^{2-} + 2H_2O \longrightarrow [CH_2O] + 2SO_4^{2-} + H_2O$
Nonsulfur photosynthetic bacteria	H_2 or many other reductants	$CO_2 + 2H_2 \longrightarrow [CH_2O] + H_2O$

the direct participation of light energy. In fact, the carbon reactions are essentially inactive in the dark. As we shall see, the carbon reactions and the light reactions operate together and are coordinately regulated by light.

15.2 The Chloroplast

In all higher plants and algae, photosynthetic processes are localized in **chloroplasts.** In plants, most of these organelles are found in cells just under the leaf surface (mesophyll cells). Each mesophyll cell may contain 20 to 50 chloroplasts (**FIGURE 15.4**). Eukaryotic algae also have chloroplasts, but often only one very large one is found in each cell.

● **CONCEPT** Photosynthesis in plants and algae occurs in chloroplasts.

Like mitochondria, chloroplasts are semiautonomous, carrying their own DNA to code for some of their proteins, as well as the ribosomes necessary for translation of the appropriate messenger RNAs. The internal structure of a chloroplast, as shown in Figure 15.4(b) and (c), bears some resemblance to that of a mitochondrion (see Figure 14.2(a)), except that chloroplasts have a third set of membranes. There is an outer, freely permeable membrane and an inner membrane that is selectively permeable. The inner membrane encloses a compartment called the **stroma** that is analogous to the mitochondrial matrix. Immersed in the stroma are many flat, saclike membrane structures

called **thylakoids,** which are often stacked like coins to form units called *grana* (see Figure 15.4(c)). Individual grana are irregularly interconnected by thylakoid extensions called *stroma lamellae.* This third membrane, the thylakoid membrane, encloses an interior space, the **lumen** of the thylakoid. In most plants, much of carbohydrate produced by the carbon reactions is stored in the chloroplasts as starch (amylose). These starch grains are located in the stroma.

Photosynthetic bacteria do not contain chloroplasts, but they have membrane structures that play the same roles as chloroplast membranes. We will briefly describe the light reaction machinery in these photosynthetic bacteria, but this chapter will focus primarily on the process as it occurs in higher plants and algae.

The division of labor within a chloroplast is simple. Absorption of light and all of the light reactions occur within or on the thylakoid membranes. The ATP and NADPH produced by these reactions are released into the surrounding stroma where all of the biosynthetic carbon reactions occur. Thus, there are analogies in structure and role between the mitochondrial matrix and chloroplast stroma and between

● **CONCEPT** Absorption of light and the light reactions occur in the chloroplast membranes. The carbon reactions occur in the stroma.

the inner membrane of the mitochondrion and the thylakoid membrane of the chloroplast. Indeed, we shall find that a similar kind of chemiosmotic ATP generation is carried out

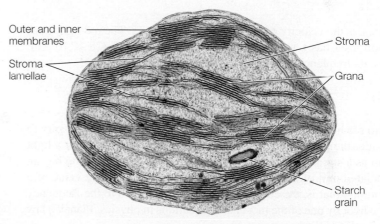

(b) Enlarged view of a single chloroplast from a leaf of Nitella (stonewort).

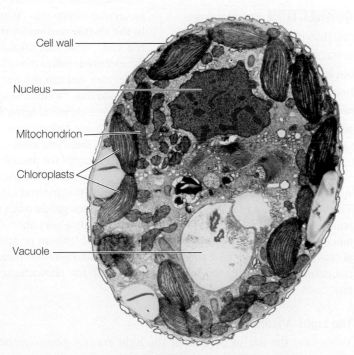

(a) Several chloroplasts and other structures are shown in this color-enhanced transmission electron micrograph (TEM) of a plant cell.

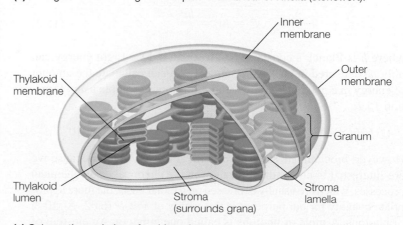

(c) Schematic rendering of a chloroplast.

▲ **FIGURE 15.4** Chloroplasts, the photosynthetic organelles of higher plants and algae.

across these membranes in both mitochondria and chloroplasts. To see how this ATP generation occurs, we must first examine the light reactions in detail, beginning with the process of light absorption.

15.3 The Light Reactions

The light reactions represent the process by which energy from sunlight is used to carry out the photochemical oxidation of H_2O to O_2. The electrons from this oxidation are used to reduce $NADP^+$ to NADPH, and part of the energy from sunlight is captured by phosphorylating ADP to produce ATP.

Absorption of Light: The Light-Harvesting System

The Energy of Light

To understand how energy from sunlight can be captured and utilized, we must first review the nature of electromagnetic radiation. The quantum-mechanical theory of radiation states that light (and all other electromagnetic radiation) has two aspects: wave-like and particle-like. We can characterize a particular kind of radiation by its wavelength (λ) or frequency (ν); these parameters characterize the *wave* aspects of the light. If waves with a length of λ are passing an observer at a velocity c, the number of waves passing per second is the frequency, ν. Thus,

$$\nu = \frac{c}{\lambda} \tag{15.1}$$

where c is the velocity of light, 2.998×10^8 m/s. The red light from a neon laser has a wavelength, $\lambda = 632.8$ nm, or 6.328×10^{-7} m. Thus, its frequency, ν, is

$$\nu = \frac{2.998 \times 10^8 \text{ m/s}}{6.328 \times 10^{-7} \text{ m}} = 4.74 \times 10^{14} \text{ s}^{-1}$$

But to see how *energy* might be obtained from light, it is necessary to consider the particulate aspect of radiation. We must think of a light beam as a stream of light particles, or **photons.** Each photon has an associated unit of energy called a **quantum.** The energy value of a quantum—that is, the energy per photon—is related to the frequency of the light by one of the most basic equations in physics, Planck's law (see Tools of Biochemistry 6A).

$$E = h\nu = \frac{hc}{\lambda} \tag{15.2}$$

where h is Planck's constant, 6.626×10^{-34} J·s. Light energy can only be delivered in packets, or quanta. We can calculate the amount of energy in a photon from the neon laser in our example using Equation 15.2:

$$(6.626 \times 10^{-34} \text{ J·s}) \times (4.74 \times 10^{14} \text{ s}^{-1}) = 3.14 \times 10^{-19} \text{ J}$$

However, biochemists rarely deal with single photons. Because we are interested in how radiation can promote chemical or biochemical processes, which are usually expressed on a molar basis, the more appropriate quantity for our purposes is the energy of a *mole* (6.02×10^{23}) of photons. A mole of photons is called one **einstein.** An einstein of the neon laser light contains 189 kJ:

$$(3.14 \times 10^{-19} \text{ J/photon}) \times (6.02 \times 10^{23} \text{ photons/mol})$$
$$= 189028 \text{ J/mol photons} = 189 \text{ kJ/einstein}$$

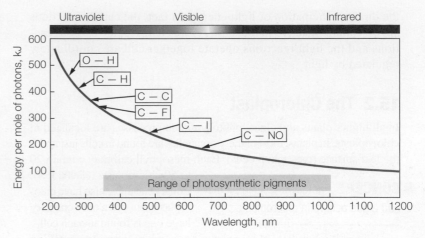

▲ **FIGURE 15.5 The energy of photons.** The graph shows energy per mole of photons as a function of wavelength, compared with energies of several chemical bonds.

FIGURE 15.5 shows a graph of energy per mole of photons as a function of wavelength, through the infrared, visible, and ultraviolet parts of the electromagnetic spectrum. For comparison, the energies associated with various covalent bonds are indicated. When photons of infrared radiation are absorbed by a molecule, they can do little except stimulate molecular vibrations, which we perceive as heat. Visible light can break some weak bonds. Photons of far-ultraviolet radiation, however, have energies capable of breaking covalent bonds.

Photosynthesis depends primarily on light in the visible and near-infrared regions of the spectrum, lying between the extremes of covalent bond-breaking and stimulating molecular vibrations. Photons in the visible and near infra-red can cause transitions in the electronic states of organic molecules that can drive reactions and thus capture the light energy in a chemical form. The ability to use radiation in this range has had clear evolutionary advantages for photosynthetic organisms. Most of the sun's energy that reaches the Earth's surface lies in this spectral range. The small amount of ultraviolet radiation that does get through the atmosphere to the Earth's surface can penetrate only a very short distance into water and thus would have been unavailable to primitive photosynthetic organisms living in the sea. The photons of far-infrared radiation have energies too low to be useful for any photochemical processes.

● **CONNECTION** Far-ultraviolet radiation is chemically destructive to humans and to other organisms, but fortunately most of it is screened from the Earth's surface by the ozone layer. This is one of the reasons depletion of the ozone layer is of such serious concern.

The Light-Absorbing Pigments

To capture the useful portion of the light energy, photosynthetic organisms have evolved a set of pigments that efficiently absorb visible and near-infrared light. Structures of a few of the most important photosynthetic pigments are shown in **FIGURE 15.6**. In **FIGURE 15.7**, the absorption spectra of these photosynthetic pigments are compared with the distribution of solar radiation in the spectrum. Together,

(a) Chlorophylls a and b

CHO in chlorophyll b
CH₃ in chlorophyll a

Phytol side chain

(b) β-Carotene

(c) Lutein

▲ **FIGURE 15.6 Some photosynthetic pigments.** Chlorophylls a and b are the most abundant plant and algal pigments, whereas β-carotene and lutein are examples of accessory pigments. There are also bacteriochlorophylls, which differ slightly in structure. The colored highlights indicate the extended conjugated double-bond systems, which are responsible for light absorption and color reflection.

the chromophores "blanket" the visible spectrum; scarcely a visible photon can reach the Earth's surface that cannot be absorbed by one chromophore or another.

The most abundant pigments in higher plants are chlorophyll a and chlorophyll b. As you can see by comparing Figure 15.6(a) with Figure 7.15(b), these molecules are related to the protoporphyrin IX found in globins and cytochromes. However, the bound metal in the chlorophylls is Mg^{2+} rather than Fe^{2+}. In Figure 15.6(b) and (c), two *accessory pigments* are also shown. All of these pigment molecules absorb light in the visible region of the spectrum because they have large conjugated double-bond systems. Because chlorophylls a and b absorb strongly in both the deep blue and red, the light that is *not* absorbed but *reflected* from them is green, the color we associate with most growing plants. The other observed colors, such as the red, brown, or purple of algae and photosynthetic bacteria, are accounted for by differing amounts and types of accessory pigments. Phycocyanin and phycoerythrin

● **CONNECTION** Loss of chlorophylls in autumn leaves allows the colors of the accessory pigments, as well as nonphotosynthetic pigments, to dominate, resulting in the brilliant yellows, oranges, and reds of fall foliage.

are open-chain tetrapyrroles that are abundant in aquatic photosynthetic organisms. These pigments absorb strongly in the 500–600 nm range, wavelengths that can efficiently pass through water (Figure 15.7). Carotenoids are the most abundant accessory pigments in plants. These include the red-orange β-carotene (Figure 15.6(b)) and xanthophylls (oxygen-containing carotenoids), such as the yellow lutein (Figure 15.6(c)). Some photosynthetic bacteria use pigments that absorb wavelengths up to about 1000 nm, in the near infrared.

The Light-Gathering Structures

Chlorophyll and some of the accessory pigments are contained in the **thylakoid membranes** of the chloroplast. The composition of these membranes is rather unusual. They contain only a small fraction of the common phospholipids but are rich in glycolipids. They also contain much protein, and some of the photosynthetic pigments are attached to certain of these proteins. Other photosynthetic pigments, including chlorophylls a and b, are not covalently bound but interact with both proteins and membrane lipids. These pigments interact with membrane lipids through their hydrophobic phytol tails (see Figure 15.6(a)).

The assemblies of light-harvesting pigments in the thylakoid membrane, together with their associated proteins, are organized into well-defined **photosystems**—structural units dedicated to the task of absorbing visible photons and recovering some of their energy in a chemical form. As we shall see, plants use two distinct photosystems, I and II. Each photosystem is a

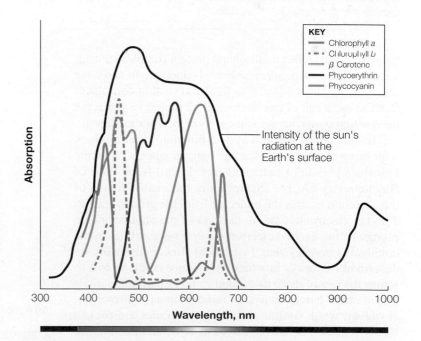

KEY
— Chlorophyll a
- - - Chlorophyll b
— β Carotene
— Phycoerythrin
— Phycocyanin

Intensity of the sun's radiation at the Earth's surface

Absorption

Wavelength, nm

▲ **FIGURE 15.7 Absorption spectra and light energy.** The absorption spectra of various plant pigments are compared with the spectral distribution of the sunlight that reaches the Earth's surface.

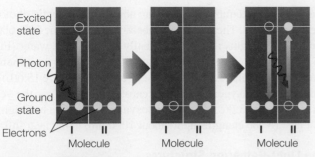

(a) Resonance transfer
Molecule I transfers its excitation energy to an identical molecule II, which rises to its higher-energy state as molecule I falls back to the ground state.

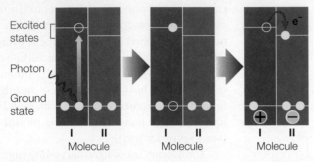

(b) Electron transfer
An excited electron in molecule I is transferred to the slightly lower excited state of molecule II, making molecule I a cation and molecule II an anion.

▲ **FIGURE 15.8 Two modes of energy transfer following photo-excitation.** For each of the two types of energy transfer that occur in a photosystem, the left-hand illustration shows a molecule being excited to a higher-energy state by absorption of a photon of radiation. The right-hand illustration shows how the energy is transferred to an adjacent molecule.

membrane-embedded multisubunit protein complex containing many **antenna** pigment molecules (chlorophylls and some accessory pigments) bound to **light-harvesting complexes (LHCs)** and a pair of special chlorophyll molecules that act as the **reaction center,** trapping the energy of the absorbed photons.

To understand how this system functions, we must look a bit more closely into what can happen when a molecule absorbs a photon of radiant energy. Recall from Tools of Biochemistry 6A that absorption in the visible region of the spectrum excites the molecule from the ground state to a higher electronic state. In the case of the photosynthetic pigments, the excited electron occupies a π orbital in the conjugated bond system. In Tools of Biochemistry 6A we described two ways in which the energy could be lost to return the molecule to its ground state: radiationless dissipation of the energy as heat or reradiation as fluorescence. However, when similar absorbing molecules are packed tightly together, as in a photosystem, two other possibilities arise. First, the excitation energy may be passed from one molecule to an adjacent one—a process called **resonance transfer** or **exciton transfer** (**FIGURE 15.8(a)**). Alternatively,

the excited electron itself may be passed to a nearby molecule with a slightly lower excited state—an **electron transfer** reaction (Figure 15.8(b)). Both of these processes are important in photosynthesis.

The clue that eventually led to the recognition that resonance transfer played a role in photosynthesis came from measurements by Robert Emerson and William Arnold in the 1930s. They showed that even when the photosynthetic system of the alga *Chlorella* was operating at maximum efficiency, only one O_2 molecule was produced for every 2500 chlorophyll molecules. As we now realize, most of the chlorophyll molecules are not directly engaged in the photochemical process itself but act, instead, as antenna molecules of the light-harvesting complexes. The structure of a light-harvesting complex (LHCII) that is part of photosystem II of higher plants is shown in **FIGURE 15.9**. Antenna molecules absorb photons, and the energy is passed

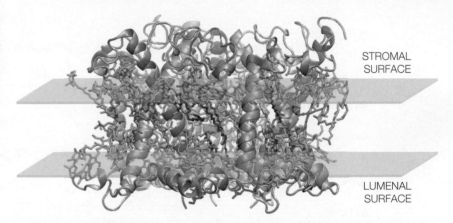

STROMAL SURFACE

LUMENAL SURFACE

(a) A side view of the trimer (approximate locations of the stromal and lumenal faces of the membrane are indicated). Each trimer (grey) contains 24 chlorophyll *a* molecules (green), 18 chlorophyll *b* molecules (cyan), and 12 carotenoids (luteins and xanthophylls; orange), all serving as antenna molecules. Bound lipids are shown in purple.

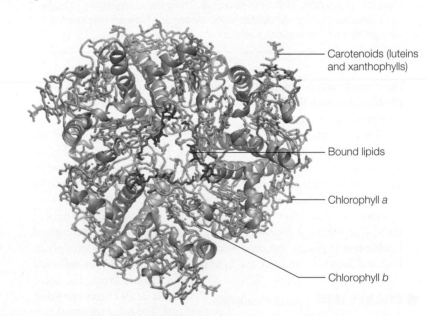

Carotenoids (luteins and xanthophylls)

Bound lipids

Chlorophyll *a*

Chlorophyll *b*

(b) A top view from the lumenal side of the membrane.

▲ **FIGURE 15.9 Three-dimensional structure of the trimeric light-harvesting complex II of plants.** The X-ray structure of the pea LHCII (PDB ID: 2bhw) shows that the protein exists as a homotrimer buried in the thylakoid membrane.

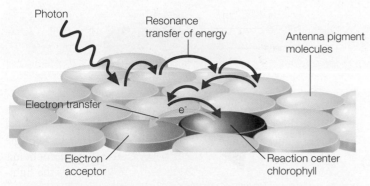

▲ **FIGURE 15.10 Resonance transfer of energy in a light-harvesting complex.** The excitation energy originating in a photon of light wanders from one antenna molecule to another until it reaches a reaction center. There an electron is transferred to a primary electron acceptor molecule, and the energy is trapped.

● **CONCEPT** Most chlorophyll molecules are used as antennae to catch photons and pass their energy on to chlorophyll molecules within reaction centers.

by resonance transfer to specific chlorophyll molecules in a relatively few reaction centers. In other words, the energy of a photon absorbed by any antenna molecule in a photosystem wanders about the system randomly until the energy finds its way to a chlorophyll molecule in the reaction center (**FIGURE 15.10**). Resonance transfer is extraordinarily fast, and the energy is funneled from antenna molecules to reaction centers on a 10–100 picosecond time scale. The reaction center chlorophyll is like the other chlorophylls, except it is in a different protein microenvironment, so that its excited-state energy level is a bit lower. Thus, it acts as a trap for quanta of energy absorbed by any of the other pigment molecules. It is the excitation of this reaction center that begins the actual photochemistry of the light reactions, for it starts a series of electron transfers.

Photochemistry in Plants and Algae: Two Photosystems in Series

Our understanding of the photochemical light reactions has developed from many elegant experiments in many different laboratories. In a pioneering study in 1939, Robert Hill at the University of Cambridge made the seminal observation that isolated chloroplasts could carry out redox chemistry when illuminated in the presence of any of a variety of electron acceptors. For example, when ferricyanide was used as the electron acceptor, the following reaction proceeded efficiently:

$$4\text{Ferricyanide [Fe(CN)}_6^{3-}] \ + \ 2H_2O \ \xrightarrow{\text{Light energy}}$$

$$4\text{Ferrocyanide [Fe(CN)}_6^{4-}] \ + \ 4H^+ \ + \ O_2$$

Thus, the chloroplasts catalyzed the oxidation of water to O_2 and protons, using the energy of absorbed photons. A number of such reactions involving different inorganic oxidants are known and are now referred to collectively as *Hill reactions*. Such reactions, in the absence of light energy, are very unfavorable. Ferricyanide, for example, is a much weaker oxidant than O_2; its $E^{\circ\prime}$ is 0.36 V compared to 0.82 V for O_2 (Table 14.1). $\Delta G^{\circ\prime}$ for the ferricyanide reaction as written can be calculated from Equation 14.1:

$$\Delta G^{\circ\prime} = -4(96485 \text{ J mol}^{-1}\text{V}^{-1})(0.36 \text{ V} - 0.82 \text{ V})$$

$$= +178 \text{ kJ/mol } O_2$$

Clearly, the equilibrium should lie far to the left. Hill's discoveries showed that *chloroplasts irradiated by light are capable of driving thermodynamically unfavorable reactions*. The Hill reactions also demonstrated that the *photosynthetic system can oxidize water to O_2 without any involvement of* CO_2 (see Figure 15.3). This observation was the first clear indication that the light and carbon reactions are separate processes.

Further studies revealed that *two* kinds of photosystems must be involved in photosynthesis in plants and algae. The first hint came from experiments that measured the quantum efficiency of photosynthesis in algae, using light of different wavelengths. The *quantum efficiency* (Q) is the ratio of oxygen molecules released to photons absorbed. As the wavelength of the monochromatic light used was raised above 680 nm (far red), an abrupt drop in Q was noted. This "red drop" was a strange observation, for chlorophylls in plants still show appreciable absorbance even at higher wavelengths. Somehow the energy was not being used efficiently above 680 nm. However, if the chloroplasts were simultaneously illuminated with yellow light (650 nm), there was a marked increase in the quantum efficiency from the light at 700 nm. Even if the yellow light was switched off a few minutes before the measurement, the quantum efficiency remained high. The only reasonable explanation for these results is that two complementary photosystems exist, one absorbing most strongly at wavelengths around 700 nm and the other at shorter wavelengths. The action of *both* must be required for photosynthesis to proceed with maximal efficiency.

The two photosystems predicted by early experimenters have now been identified and characterized. They are both localized in the thylakoid membrane. Each photosystem is a multisubunit, transmembrane protein complex, carrying antenna and reaction center chlorophyll molecules and electron transport agents. The photosystems have been named according to the order in which they were discovered (but not the order in which they pass electrons, as we shall see). The one that absorbs far red light (up to 700 nm) is called **photosystem I (PSI),** and the one that absorbs only to a wavelength of about 680 nm is called **photosystem II (PSII).** In algae, cyanobacteria, and all higher plants, these two photosystems are linked in series to carry out the complete sequence of the light reactions. The basic sequence is illustrated in **FIGURE 15.11(a)**, which depicts the path of electrons through the two systems. Figure 15.11(b) emphasizes the energetics of the electron flow and places the major participants in the light reactions on a scale of reduction potential.

● **CONCEPT** Two photosystems, linked in series, are involved in the photosynthetic light reactions in algae, cyanobacteria, and higher plants.

▼ **FIGURE 15.11 The two-photosystem light reactions.** In the two-photosystem mode of photosynthesis, the light reactions are carried out by two photosystems linked in series.

Key: OEC = oxygen-evolving complex; Y_Z = donor to P680; P680 = photosystem II reaction center chlorophyll; Ph = pheophytin acceptor; Q_A, Q_B = protein-bound plastoquinones; QH_2 = plastoquinol (reduced plastoquinone) in membrane; PC = plastocyanin; P700 = photosystem I reaction center chlorophyll; A_0 = chlorophyll acceptor; A_1 = protein-bound phylloquinone; F_A, F_B, F_X = iron–sulfur clusters; Fd = ferredoxin; FNR = ferredoxin: $NADP^+$ oxidoreductase.

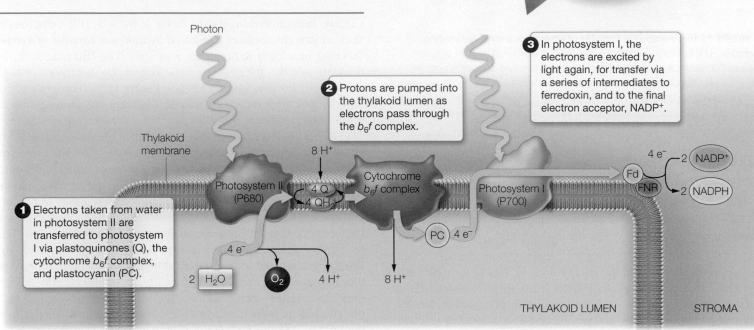

(a) A schematic view of the path of electrons through the two photosystems. The two systems and the cytochrome complex are embedded in the thylakoid membrane.

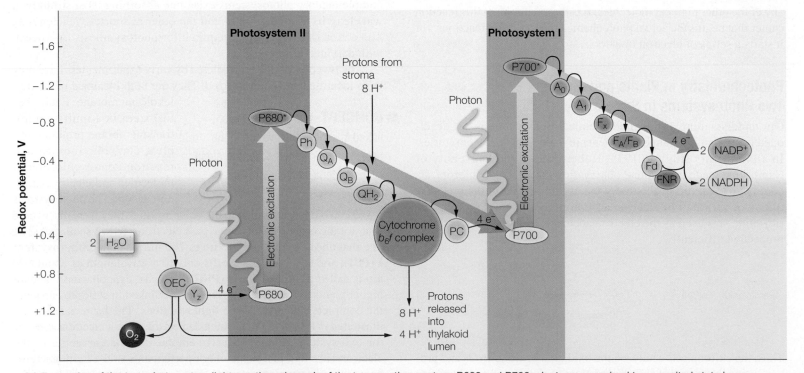

(b) Energetics of the two-photosystem light reactions. In each of the two reaction centers, P680 and P700, electrons are raised to an excited state by absorption of photons and then passed through an electron transport chain. The two-photosystem mode has historically been called the Z-scheme because of the pattern of energy changes shown here, but N-scheme would be more accurate.

● **CONCEPT** Thylakoid membranes capture light energy to drive protons into the thylakoid lumen, establishing a proton gradient for ATP synthesis.

● **CONCEPT** ATP and reducing power in the form of NADPH are the products of the light reactions.

In each of the two photosystems, the primary step is transfer of a light-excited electron from a reaction center (P680 or P700) into an electron transport chain. The ultimate source of the electrons is the water molecules shown at the left in both parts of Figure 15.11. The final destination of the electrons is the molecules of NADP$^+$ at the right, which are thereby reduced to NADPH. At two stages in the electron transport process, protons are released into the thylakoid lumen. Some of the protons come from the H_2O that is broken down, and some come from the stroma. This transfer of protons into the thylakoid lumen produces a pH gradient across the thylakoid membrane. It should not surprise you to find that this proton gradient drives ATP production, just as a proton gradient drives ATP synthesis in mitochondria (see Chapter 14). Thus, ATP and reducing power in the form of NADPH are the products of the light reactions. These compounds are exactly what is needed to drive the carbohydrate syntheses carried out in the carbon reactions. To examine ATP and NADPH generation in detail, we begin with photosystem II, for it is where electrons enter the scheme.

Photosystem II: The Splitting of Water

Each of the photosystems contains an electron transport chain, which extracts energy when an excited electron loses its energy of excitation in a stepwise fashion. The photosystem carries out a series of oxidation–reduction reactions. It is easiest to follow the events in a photosystem by starting with the absorption of a photon picked up by the light-harvesting system of photosystem II. The photon is funneled to a reaction center chlorophyll, designated P680. Excitation of P680 raises the molecule from the ground state to an excited state at –0.8 volts (Figure 15.11(b)). Thus, the excited P680 has become a strong reducing agent, able to quickly transfer an electron from P680 to a lower-energy primary electron acceptor—*pheophytin a (Ph)*, as shown in Figure 15.11(b). The pheophytins are molecules identical to chlorophylls, except that two protons substitute for the centrally bound magnesium ion. We can consider this excited electron as a low-redox-potential electron and thus a strong reducing agent.

The electron is then transferred to a series of *plastoquinone* molecules (Q_A and Q_B) associated with photosystem II proteins.

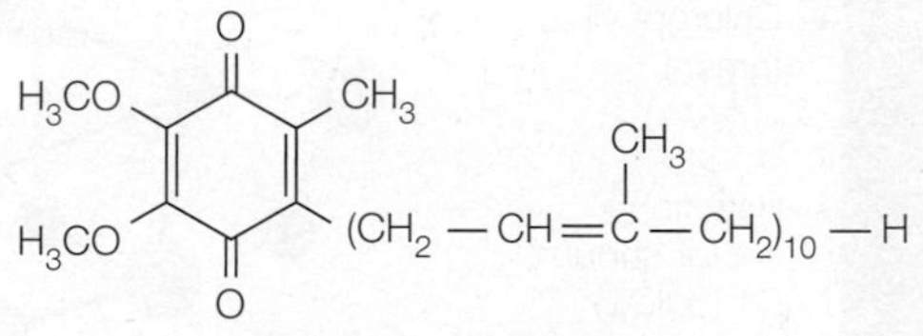

Ubiquinone (Coenzyme Q$_{10}$)

Plastoquinones are structurally and functionally similar to the coenzyme Q (ubiquinone) found in the mitochondrial respiratory chain. Ultimately, two electrons and two protons are picked up by the plastoquinone Q_B; the protons come from the stroma. The reduced plastoquinone, QH_2 (*plastoquinol*), is then released into the lipid portion of the thylakoid membrane. The overall reduction of plastoquinone to plastoquinol can be written as follows:

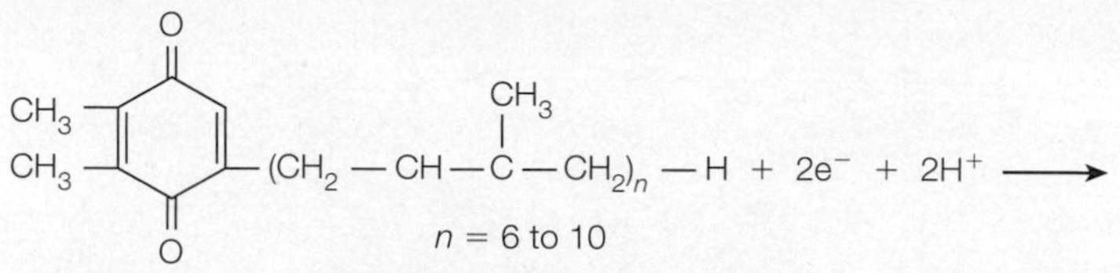

Plastoquinone

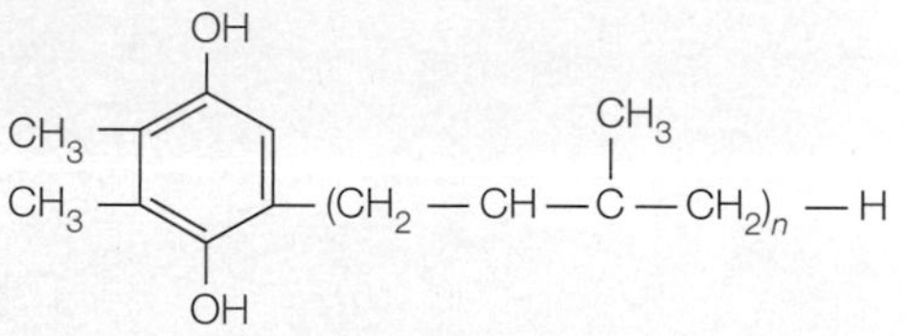

Plastoquinol

● **CONCEPT** Photosystem II extracts electrons from water, passing them to photosystem I and releasing O_2.

Plastoquinol then interacts with a membrane-bound complex of cytochromes and iron–sulfur proteins, the cytochrome $b_6 f$ complex. This complex catalyzes the transfer of the electrons to a copper protein, *plastocyanin* (PC). In doing so, the $b_6 f$ complex serves two purposes. First, it transfers electrons from photosystem II to photosystem I. At the same time, it pumps protons from the stroma into the thylakoid lumen (2 H$^+$ per electron). The major components of this complex are cytochrome f (which contains one c-type heme), cytochrome b_6 (which contains two b-type hemes), and a Rieske iron–sulfur protein (**FIGURE 15.12(a)**). The cytochrome $b_6 f$ complex is thus analogous to complex III of the mitochondrial respiratory chain (Figure 15.12(b)), and catalyzes a similar Q cycle (as shown in Figure 14.14). As plastoquinol is oxidized back to plastoquinone, the two protons it has taken from the stroma are released into the thylakoid lumen. Plastocyanin, a mobile protein in the thylakoid lumen, passes the electrons on to P700 reaction centers. In this process, the copper in plastocyanin is first reduced to Cu (I) and then reoxidized to Cu (II). We shall consider the fate of the electrons passed to P700 shortly, when we discuss photosystem I.

So far, we have described how light energy excites an electron in the P680 reaction center, which then gets passed into an electron transport chain. Note that this process leaves P680 reaction centers deficient in electrons. In other words, each P680 reaction center is oxidized to a strong oxidant, known as P680$^+$. P680$^+$ is probably the most powerful oxidant in nature; it is such a strong oxidant that it can remove electrons from water, generating oxygen in the process. New insight into the mechanism of this water-splitting process has recently been provided by X-ray structures of photosystem II complexes from several cyanobacteria. **FIGURE 15.13(a)** shows the core of photosystem II as a large multisubunit complex embedded in the thylakoid membrane. Each monomer of the homodimer comprises 20 subunits, including antenna proteins with their numerous antenna chlorophylls surrounding two subunits (D1 and D2) that contain the P680 reaction center chlorophylls and the water-splitting catalytic components. The light-harvesting complex II component (Figure 15.9) is not visible in this X-ray structure but would make contact with the core PSII homodimer in the membrane.

As indicated in Figure 15.11(b) (left), the electron acceptor is a subunit of PSII referred to as the oxygen-evolving complex (OEC). The OEC contains a cube-shaped cluster of four oxygen-bridged manganese ions and one calcium ion (Figure 15.13(b)). This metal cluster can

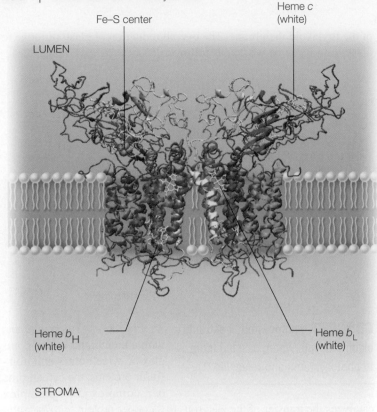

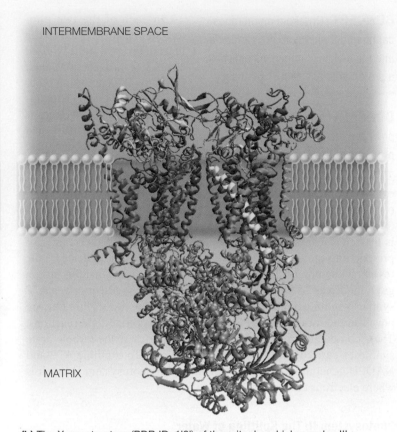

(a) The X-ray structure (PDB ID: 2d2c) of the cytochrome b_6f complex from the thermophilic cyanobacterium, *Mastigocladus laminosus*, shows that the protein exists as a homodimer buried in the thylakoid membrane. This view is in the plane of the membrane, with the thylakoid lumen at the top. Cytochrome *f* subunits are red, with the *c*-type hemes shown in white; the Rieske iron–sulfur protein subunits are yellow, with the Fe_2S_2 iron–sulfur centers shown in orange; cytochrome b_6 subunits are blue, with the b_H and b_L hemes shown in white; and subunits IV are purple. Four additional small subunits per monomer are shown in green.

(b) The X-ray structure (PDB ID: 1l0l) of the mitochondrial complex III illustrates the conservation of the hydrophobic heme-binding transmembrane domain of the cytochrome *b* subunit (blue) between b_6f and bc_1 complexes, revealing the evolutionary relationship of these cytochrome complexes. Homologous or analogous subunits have the same colors, based on the color scheme in panel (a).

▲ **FIGURE 15.12** Structure of the cytochrome b_6f complex.

◀ **FIGURE 15.13** Structure of photosystem II core.

(a) The X-ray structure (PDB ID: 3bz1, 3bz2) of PSII from the thermophilic cyanobacterium, *Thermosynechococcus elongatus*, shows that the protein exists as a homodimer buried in the thylakoid membrane. This view is in the plane of the membrane, with the thylakoid lumen at the top. Each monomer is composed of 20 subunits, including the antenna proteins CP47 and CP43, the reaction center subunits D1 and D2, and two cytochrome b559 subunits. Each monomer contains 35 chlorophyll *a* molecules and 12 carotenoids bound to CP43 and CP47. The Mn_4Ca cluster (shown in blue; structure in panel b) of the oxygen-evolving center is bound by the D1 subunit of each monomer.

(b) The structure of the Mn_4Ca cluster from the PSII of *Thermosynechococcus vulcanus* (PDB ID: 3arc) is shown in greater detail. Calcium atom is green, manganese atoms are purple, and oxygen atoms are red.

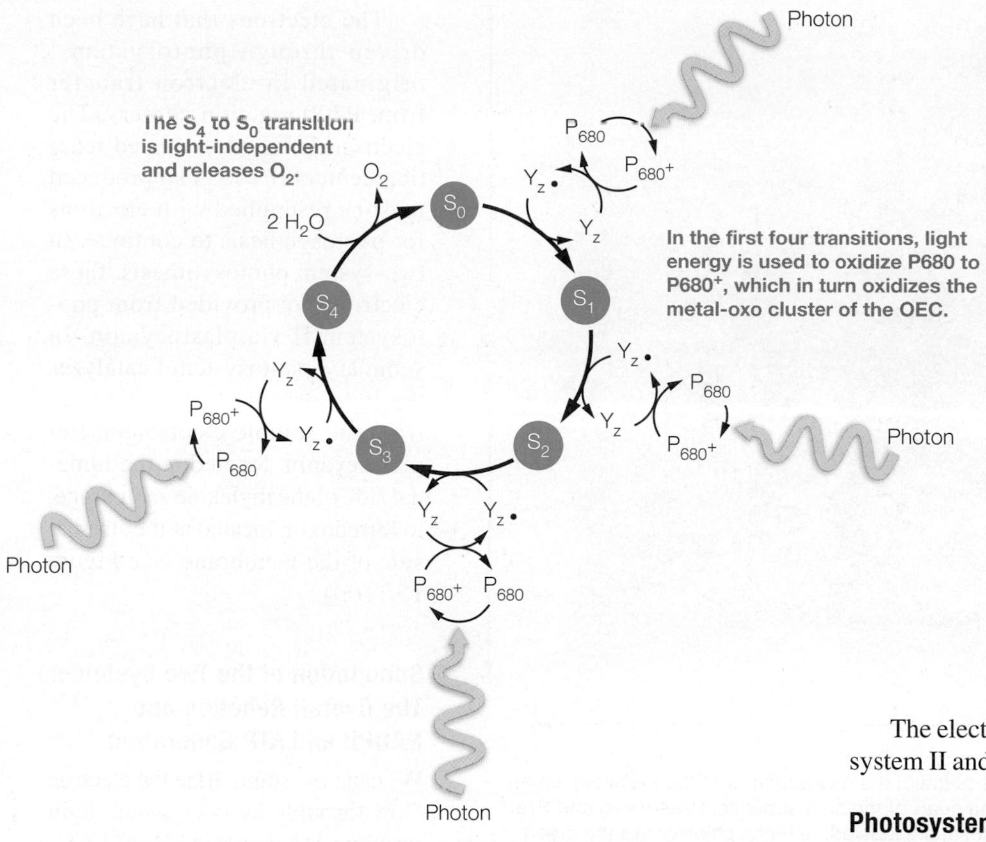

◀ **FIGURE 15.14 A model for the catalytic function of the oxygen-evolving complex (OEC) cluster in PSII.** S_0-S_4 represent the different oxidation states that the ligated metal cluster cycles through as e^- and H^+ are abstracted from the H_2O molecules. Four photons are required to oxidize two H_2O to one O_2. $Y_Z \cdot$ = tyrosine radical.

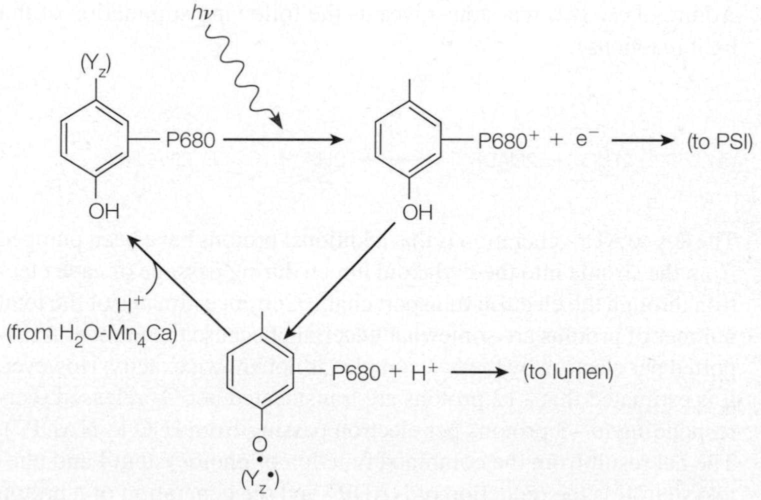

The system has in effect stripped four electrons from the four hydrogen atoms in two water molecules. The oxygen produced diffuses out of the chloroplast. The four protons that are produced from the two water molecules are released into the thylakoid lumen, helping to generate a pH difference between the lumen and stroma. We may summarize the reaction carried out by photosystem II as follows:

$$2H_2O \xrightarrow{4h\nu} 4H^+ + 4e^- + O_2$$

The electrons produced travel through the transport chain of photosystem II and are passed on to photosystem I through the b_6f complex.

Photosystem I: Production of NADPH

We have seen that in plants and algae, which utilize two photosystems, photosystem II accomplishes the splitting of water with evolution of O_2 and helps generate a proton gradient across the thylakoid membrane. However, the electrons from the water molecules have not yet reached their final destination in NADPH. This process is the task of photosystem I, in which electrons are again released from a reaction center by light excitation and passed through a second electron transport chain. These electrons are replaced by those passed on from photosystem II.

Photosystem I (PSI) is another multiprotein complex that spans the thylakoid membrane (**FIGURE 15.15**). Plant PSI consists of two membrane complexes, the core complex and the light-harvesting complex I (LHCI). Like the LHCII described earlier (Figure 15.9), LHCI serves as an additional antenna system that collects photons and transmits the energy to the core complex. The heart of the core complex is formed by the PsaA and PsaB subunits, which bind all of the electron transport chain components, including the P700 reaction center chlorophylls, as well as 80 chlorophylls that function as light-harvesting antennae. The entire plant PSI-LHCI supercomplex is ~600 kDa and contains 45 transmembrane helices and 168 antenna chlorophylls.

As was shown in Figure 15.11(b), excitation by a photon absorbed by antenna chlorophylls raises electrons in P700 from a ground state to an excited state at about −1.3 V—probably the most powerful

● **CONCEPT** Photosystem I receives electrons from photosystem II and transfers them to $NADP^+$ to make NADPH.

reductant in nature. Each excited electron then passes through a transport chain. It is first taken up by a special chlorophyll acceptor (designated A_0), then transferred to a molecule of *phylloquinone* (A_1—also known as vitamin K_1—see p. 551) and finally passed through a series of three iron–sulfur proteins (F_X, F_B, and F_A). These proteins contain Fe_4S_4 clusters of the

exist in a series of oxidation states (S_0-S_4), as indicated in **FIGURE 15.14**; light-driven cycling through these oxidation states allows the cluster to dismantle two water molecules, passing four electrons back to P680 and releasing the four accompanying protons into the thylakoid lumen. Exactly at which points in the cycle individual electrons and protons are released is still a matter of debate, with several models proposed. We know that the electron donor returning electrons to $P680^+$ is a redox-active tyrosine in the D1 subunit of photosystem II. This yields a tyrosine radical ($Y_Z\bullet$) that releases a proton. It is proposed, then, that each of the steps in Figure 15.14 in which a hydrogen atom ($H^+ + e^-$) is extracted involves the following cycle:

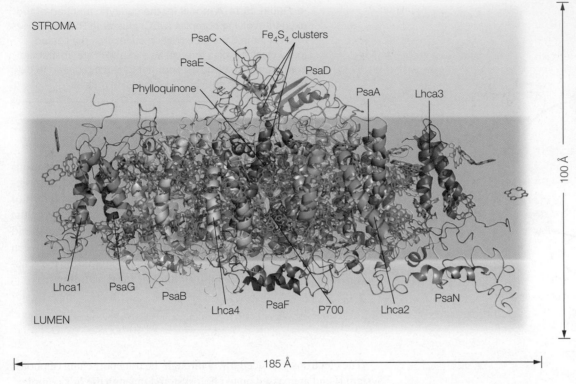

A view of the pea PSI structure (PDB ID: 2o01) as seen from the plane of the membrane, with the thylakoid lumen at the bottom. The four LHCI subunits (Lhca1–4) are in front. The heart of the core subunits, PsaA (pink) and PsaB (silver), are behind the LHCI subunits. The other core subunits (Psa) are labeled. Antenna chlorophylls are green, the phylloquinones are blue, and the Fe₄S₄ clusters are shown as yellow-red spheres. The P700 reaction center chlorophylls are red.

▲ **FIGURE 15.15** The structure of plant photosystem I.

Phylloquinone

kinds depicted in Figure 14.4. Finally, the electron is transferred to another iron–sulfur protein, *soluble ferredoxin* (Fd), which is present in the stroma. The enzyme ferredoxin: NADP⁺ oxidoreductase catalyzes the transfer of electrons to NADP⁺ after ferredoxin has been reduced by photosystem I:

$$2Fd\ (red) + H^+ + NADP^+ \xrightarrow[\text{Fd–NADP}^+\text{ Reductase}]{} 2Fd\ (ox) + NADPH$$

In a sense, ferredoxin, rather than NADP⁺, can be considered the *direct* recipient of electrons from the pathway. Although much of the reduced ferredoxin is used to reduce NADP⁺, some is used for other reductive reactions, which we discuss later in this chapter. In fact, we may consider reduced ferredoxin a source of low-potential electrons for many reductive processes. The NADPH produced by ferredoxin oxidation is released into the stroma, where it will be used in the carbon reactions.

The electrons that have been driven through photosystem I originated in electron transfer from P700 reaction centers. The electron-deficient, oxidized reaction centers (P700⁺) so produced must be resupplied with electrons for photosynthesis to continue. In two-system photosynthesis, these electrons are provided from photosystem II via plastocyanin. In summary, photosystem I catalyzes the light-driven electron transfer from the soluble electron carrier plastocyanin, located at the lumenal side of the thylakoid membrane, to ferredoxin, located at the stromal side of the membrane (see Figure 15.11(a)).

Summation of the Two Systems: The Overall Reaction and NADPH and ATP Generation

We can now summarize the electron flow through the two-system light reactions. As shown in Figure 15.11, electrons are taken from water and end up in NADPH. We wrote for the overall reaction in photosystem II:

$$2H_2O \xrightarrow{4h\nu} 4H^+ + 4e^- + O_2$$

The reactions of photosystem I, if written for four electrons and with all intermediates eliminated, are

$$4e^- + 2H^+ + 2NADP^+ \xrightarrow{4h\nu} 2NADPH$$

Adding these two reactions gives us the following summation of the light reactions:

$$2H_2O + 2NADP^+ \xrightarrow{8h\nu} 2H^+ + O_2 + 2NADPH$$

The key to ATP generation is that additional protons have been pumped from the stroma into the thylakoid lumen during passage of each electron through the electron transport chain. Current estimates of the total number of protons are somewhat uncertain because the number transported per electron by the $b_6 f$ complex is not known exactly. However, it is estimated that ~12 protons are translocated per O₂ released (corresponding to ~3 protons per electron passing from H₂O to NADP⁺). The net result from the combined function of photosystem I and photosystem II is the reduction of NADP⁺ and the generation of a proton gradient across the thylakoid membrane, with the lumen becoming more acidic than the stroma. Recall from Chapter 14 that the free energy available in a proton gradient ($\Delta\mu_H$, the protonmotive force) is composed of both a chemical component (the proton concentration gradient, ΔpH) and an electrical component (the membrane potential, $\Delta\Psi$):

$$\Delta\mu_H = 2.3RT\Delta pH + F\Delta\Psi \qquad (15.3)$$

Unlike the mitochondrial inner membrane, the thylakoid membrane of chloroplasts is permeable to ions such as Mg^{2+} and Cl^-. Movement of these ions across the thylakoid membrane maintains electrical neutrality, dissipating much of the membrane potential. (Light-induced translocation of Mg^{2+} into the stroma also has a regulatory role, as we shall see later.) Thus, in illuminated chloroplasts, the protonmotive force is dominated by the H^+ gradient. The pH difference produced across the thylakoid membrane can become very large—as much as 3.5 pH units in brightly illuminated chloroplasts. This pH gradient across the membrane corresponds to more than a 3000-fold difference in $[H^+]$ and a free energy change of about -20 kJ per mole of protons at 25 °C (see end-of-chapter problem 3). Based on the estimated stoichiometry of 12 mole H^+ per mole O_2 produced, this corresponds to roughly 240 kJ/mol O_2 energy available to drive the synthesis of ATP.

As in the case of ATP generation in mitochondria, these protons can pass back through the thylakoid membrane only through membrane-bound ATP synthase complexes. In chloroplasts these complexes are called CF_0-CF_1 complexes, and they exhibit considerable resemblance to the F_0-F_1 complexes of mitochondria (see Chapter 14). It has been estimated that one ATP is produced for each three protons passing through the CF_0-CF_1 complex. Three moles of protons would supply ~ 60 kJ (240/4) to drive the synthesis of 1 mole of ATP, a thermodynamically reasonable result. Because ~ 12 moles of H^+ are transported per mole of O_2 produced, $\sim 12/3 \approx 4$ moles of ATP are generated for each mole of O_2 evolved.

● **CONCEPT** Both photosystems transport protons from the stroma into the thylakoid lumen. The return of protons, through CF_0-CF_1 complexes, is used to generate ATP.

A summary view of the whole set of light reactions is shown in **FIGURE 15.16**. It should be noted that photosystem I and photosystem II, the cytochrome $b_6 f$ complex, and ATP synthase (CF_0-CF_1) are all individual entities embedded in the thylakoid membrane but are not necessarily contiguous. The components that link the photosystems and the $b_6 f$ complex are mobile—plastoquinone in the lipid phase of the

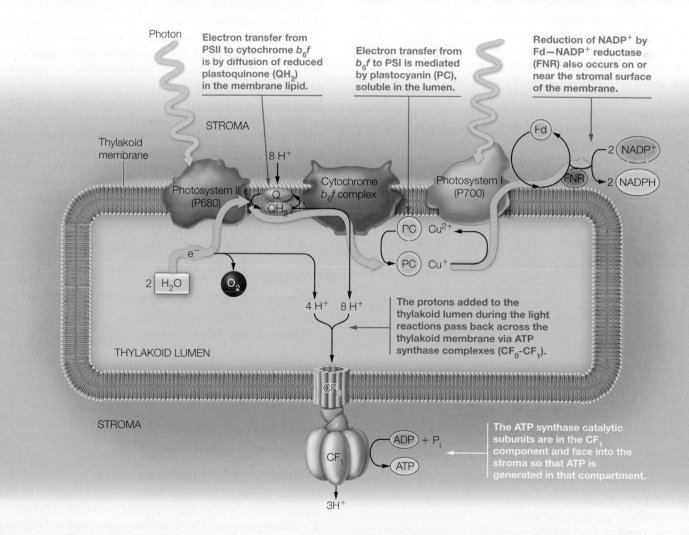

▲ **FIGURE 15.16 Summary view of the light reactions as they occur in the thylakoid.** Photosystems I and II and the cytochrome $b_6 f$ complex are physically separate protein complexes embedded in the thylakoid membrane.

membrane and plastocyanin in the thylakoid lumen. Thus, electrons can be moved over long distances in this system.

Such long-range transport is facilitated by the arrangement of components in the thylakoid membrane. Careful analysis of the composition of grana (the stacked membrane regions) indicates that the interior membrane layers of the grana are rich in photosystem II; by contrast, the stroma lamellae (the unstacked regions) are rich in photosystem I. Cytochrome b_6f complex is found in both regions. This physical segregation of photosystems is possible because of the high mobilities of plastoquinone and plastocyanin. ATP synthase is distributed in both the stroma lamellae and the top and bottom surfaces of the granum, allowing $NADP^+$ reduction and ATP generation to occur at or near these stroma-facing surfaces.

An Alternative Light Reaction Mechanism: Cyclic Electron Flow

In the two-system light reactions just described, the electrons displaced from photosystem I by excitation are replaced by photosystem II, which receives them from water. The entire process is called **noncyclic electron flow,** and generation of ATP by this process is called **noncyclic photophosphorylation.** An alternative pathway for the light reactions, called **cyclic electron flow,** utilizes the components of photosystem I, plus plastocyanin and the cytochrome b_6f complex (**FIGURE 15.17**). Whether or not this pathway is used depends on the levels of $NADP^+$ in the chloroplast stroma. When $NADP^+$ is present in only small amounts, electrons excited in the P700 center are not transferred to $NADP^+$. Instead, they are passed from ferredoxin back to the cytochrome b_6f complex, and from there they are returned via plastocyanin to the P700 ground state. One way to look at this cyclic electron flow is to consider the b_6f complex and $NADP^+$ as competitors for the electrons from Fd. The b_6f complex pumps protons across the thylakoid membrane during this cyclic process, thereby ensuring the generation of ATP. Approximately one ATP is generated for every two electrons that complete the cycle, a process

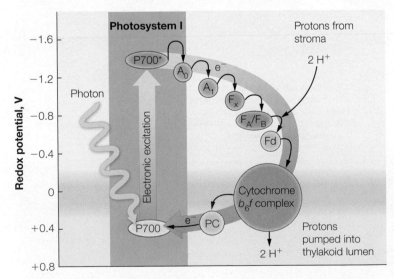

▲ **FIGURE 15.17 Cyclic electron flow.** When levels of $NADP^+$ are low and levels of NADPH are high, electrons from the P700 center are returned to it via the cytochrome b_6f complex. There is no $NADP^+$ reduction, but protons are pumped across the membrane and therefore ATP is generated. Symbols are as in Figure 15.11.

called **cyclic photophosphorylation.** However, in this process, no O_2 is released and no NADPH is produced.

Cyclic electron flow apparently serves to generate ATP in situations when the reductant NADPH is abundant and little $NADP^+$ is

● **CONCEPT** Cyclic electron flow, an alternative to two-system (noncyclic) electron flow, generates extra ATP when NADPH is plentiful.

available as an electron acceptor. It may also play a more fundamental role. As we shall see, the requirements for ATP in the photosynthetic carbon reactions are substantial and may not always be fully met by noncyclic electron flow. Cyclic photophosphorylation, which produces ATP but no NADPH, helps maintain the necessary balance between ATP and NADPH production.

Reaction Center Complexes in Photosynthetic Bacteria

The two-photosystem light reactions just described are those that occur in plants, algae, and cyanobacteria. The other photosynthetic bacteria, however, possess either a photosystem I or a photosystem II, but never both types in the same organism. The purple bacteria (named for the colored bacteriochlorophyll and carotenoid pigments in their membranes) have a reaction center complex that most closely resembles photosystem II in plants. It contains pheophytins (bacteriopheophytin), quinones, and a cytochrome bc_1 complex that is homologous to the cytochrome bc_1 complex in mitochondria. Electrons transferred from the reaction center to the cytochrome bc_1 complex are returned to the reaction center via a mobile c-type cytochrome. Protons are pumped from the cytoplasm into the periplasmic space as electrons flow through the cytochrome bc_1 complex, generating a proton gradient. Return of protons is through ATP synthase complexes spanning the plasma membrane, resulting in generation of ATP (**FIGURE 15.18**). Because the electron flow is cyclic, there is no net oxidation–reduction, and thus no need for an external reductant, such as water. Consequently, this process does not produce O_2 and is referred to as *anoxygenic* photosynthesis. These bacteria produce the NADPH necessary to carry out the carbon reactions of photosynthesis (CO_2 assimilation) by transferring electrons from various substrates such as H_2S, $S_2O_3^{2-}$ (thiosulfate), or H_2 to $NADP^+$.

● **CONNECTION** A number of scientists are trying to design artificial solar cells that can convert sunlight into chemical fuel. "Artificial photosynthesis," patterned on a process invented by nature several billion years ago, could provide a major energy source in the future.

Other anoxygenic photosynthetic bacteria, such as the anaerobic green bacterium *Chlorobium*, use a photosystem that more closely resembles photosystem I in plants. These type-I systems catalyze a linear light-driven electron transport process in which electrons are extracted from H_2S to replenish the oxidized reaction center and S_2 is released.

Evolution of Photosynthesis

How did photosynthesis arise? It is generally accepted that anoxygenic photosynthesis appeared first, probably soon after life began more than 3.8 billion years ago. These earliest photosynthetic bacteria possessed a single photosystem, the predecessor of modern type-I and type-II photosystems, and relied on substances like hydrogen sulfide as an electron source. Oxygenic photosynthesis, such as that seen in modern

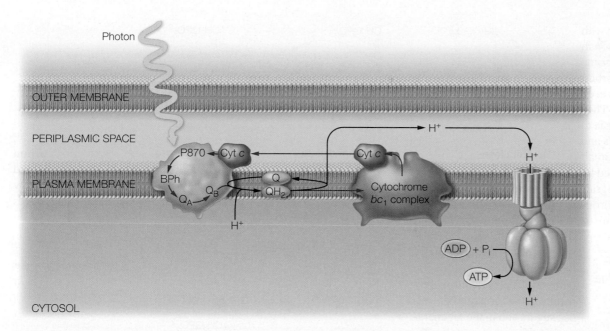

◄ **FIGURE 15.18 Postulated mechanism for purple bacterial photosynthesis.** This process somewhat resembles the photosystem II reactions in the thylakoid (see Figure 15.16), with a reaction center and a membrane-bound cytochrome complex. However, there is only one kind of reaction center, water is not split, O_2 is not produced, and electron flow is cyclic. BPh = bacteriopheophytin; P870 = reaction center; Q_A, Q_B = protein-bound plastoquinones; QH_2 = plastoquinol.

cyanobacteria, requires two photosystems, working in series, and a catalyst that splits water, producing O_2 and electrons—the manganese-dependent oxygen-evolving complex (OEC) of photosystem II (Figure 15.13(b)). Whether a gene duplication event or a lateral gene transfer brought the two photosystems together in one ancestral "protocyano-bacterium" is the subject of much speculation.

Now we come to the question of the invention of oxygenic photosynthesis. Manganese ions are readily photooxidized by ultraviolet light, which would have been abundant before the ozone layer formed in the early atmosphere. Upon oxidation, a Mn ion ejects an electron, an electron that could be used to replenish an oxidized photosystem. The oxidized Mn would then grab an electron from the most abundant source available— water. If our ancestral protocyanobacterium bound a few Mn ions to the surface of its photosystem II, it could extract electrons from H_2O and pass them through the proton-pumping complex, no longer requiring a cyclic pathway. Photosystem I could now be coupled to photosystem II. Over time, the soluble manganese ions were replaced by the more stable Mn_4Ca cluster, and the photosystems became more efficient, eventually leading to the oxygenated, green earth that exists today. There would have been substantial selective pressure for this evolutionary scenario, since oxygenic photosynthesis released our ancestral protocyanobacterium from scarce inorganic or organic electron donors. As biochemist John F. Allen of the University of London has said, "Water is everywhere, so the organisms never ran out of electrons. They were unstoppable."

15.4 The Carbon Reactions: The Calvin Cycle

The carbon reactions occur in the stroma of the chloroplast (or in the cytoplasm of photosynthetic bacteria). Their function is to fix atmospheric carbon dioxide into carbohydrates, utilizing ATP energy and reducing power (NADPH) generated by the light reactions.

Carbon dioxide fixation is accomplished by adding one CO_2 at a time to an acceptor molecule and passing the molecule through a cyclic series of reactions, shown schematically in **FIGURE 15.19**. The whole series is called the **Calvin cycle,** after the biochemist Melvin Calvin, who in 1961 received the Nobel Prize in Chemistry for his

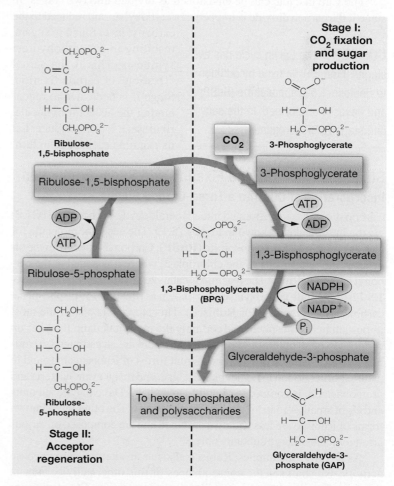

▲ **FIGURE 15.19 Schematic view of the Calvin cycle.** The cycle may be divided into two stages. In stage I, CO_2 is fixed and glyceraldehyde-3-phosphate (GAP) is produced. Part of this GAP is used to make hexose phosphates and eventually polysaccharides. Another fraction of the GAP is used in stage II to regenerate the acceptor molecule, ribulose-1,5-bisphosphate.

work in this field. Calvin, working with James Bassham and Andrew Benson in the 1940s and 1950s, carried out some of the earliest experiments on carbon fixation with the newly discovered radioactive isotope of carbon, ^{14}C. They exposed cultures of unicellular algae to ^{14}C-labeled CO_2 for a few seconds and then rapidly killed the cells to stop all metabolism. Extracts were analyzed by two-dimensional paper chromatography, and radioactive compounds were detected by exposing the chromatogram to X-ray film (autoradiography; see Tools of Biochemistry 4A). These methods allowed the researchers to discover the first product of CO_2 fixation (3-phos-phoglycerate) and eventually to elucidate the entire cyclic pathway. The cycle ultimately results in the formation of hexose sugars and in the regeneration of the acceptor molecule.

● **CONCEPT** The Calvin cycle uses the ATP and NADPH generated in the light reactions to fix atmospheric CO_2 into carbohydrate.

The Calvin cycle can be envisioned as divided into two stages. In stage I, the carbon dioxide is trapped as a carboxylate and reduced to the carbonyl level found in sugars, resulting in net carbohydrate synthesis. This is the actual "fixation" of carbon fixation. Stage II is dedicated to regenerating the acceptor molecule, ribulose-1,5-bisphosphate. Let us examine each stage in turn.

● **CONCEPT** The Calvin cycle has two stages. First, CO_2 is fixed by addition to ribulose-1,5-bisphosphate (RuBP) and hexoses are formed. In the second stage RuBP is regenerated.

Stage I: Carbon Dioxide Fixation and Sugar Production
Incorporation of CO_2 into a Three-Carbon Sugar

Carbon dioxide is incorporated into glyceraldehyde-3-phosphate (GAP) via the intermediates shown in Figure 15.19. The acceptor molecule for CO_2 is **ribulose-1,5-bisphosphate (RuBP).** Carbon dioxide from the air diffuses into the stroma of the chloroplast, where it is added at the carbonyl carbon of RuBP. The reaction is catalyzed by the enzyme **ribulose-1,5-bisphosphate carboxylase,** also known as ribulose-1,5-bisphosphate carboxylase/oxygenase (or **Rubisco**). This enzyme is one of the most important in the biosphere and certainly the most abundant. It makes up about 15% of all chloroplast proteins, and there are an estimated 40 million tons of it in the world. Four different forms of Rubisco are found in nature. Form I, found in higher plants, algae, and many cyanobacteria and proteobacteria, is composed of eight large (~50 kDa) catalytic subunits and eight small (~15 kDa) noncatalytic subunits. The other three known forms of Rubisco are less widely distributed and are composed of various arrangements of large subunits only.

As its full name implies, Rubisco also has an alternative oxygenase activity. We shall see the consequences of this other activity later in this text. For the moment we will concentrate on its CO_2-fixing, or carboxylase, function. The carboxylase reaction is a complex series of steps in which the actual CO_2 acceptor is proposed to be the five-carbon enediolate intermediate:

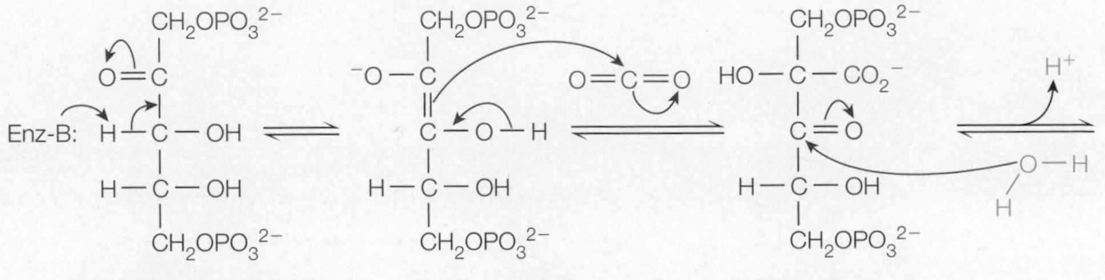

The chemistry of this reaction is the subject of intense research, and several mechanisms have been proposed. The Rubisco enzyme is first activated by *carbamation* of an active site lysine residue. This carbamate, formed by nonenzymatic reaction of a nonsubstrate CO_2 with the

Ribulose-1,5-bisphosphate　　**Enediolate intermediate**　　**Carboxy-β-keto intermediate**

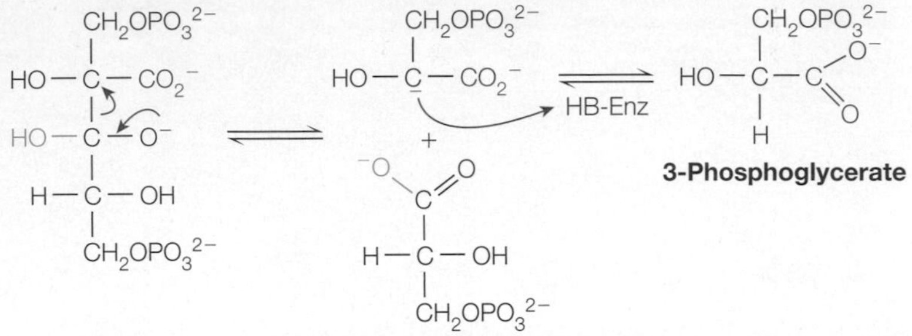

3-Phosphoglycerate

3-Phosphoglycerate

ε-amino group of lysine, is analogous to the carbamates on the N-termini of hemoglobin (Chapter 7). The negatively charged carbamate forms part of the binding site for an essential Mg^{2+} ion, which is involved in binding RuBP and activating the H_2O molecule that hydrates the carboxy-β-keto intermediate. Once the enediolate intermediate is carboxylated, the product is hydrated and then cleaved to yield two molecules of 3-phosphoglycerate (3PG). Cleavage is facilitated by protonation by an active site acid (HB-Enz). The reaction is essentially irreversible, with $\Delta G^{\circ\prime} = -35.1$ kJ/mol. At this point CO_2 has already been fixed into a carbohydrate. The remainder of the Calvin cycle reactions are dedicated to producing hexoses from the triose and regenerating RuBP.

Each molecule of 3PG is phosphorylated by ATP, in a reaction catalyzed by *phosphoglycerate kinase*. The two 1,3-bisphosphoglycerates so produced are then reduced to two glyceraldehyde-3-phosphates (GAP), with accompanying loss of one phosphate from each. The reducing agent is NADPH, which was produced in the light reaction, and the reaction is catalyzed by the enzyme *glyceraldehyde-3-phosphate dehydrogenase:*

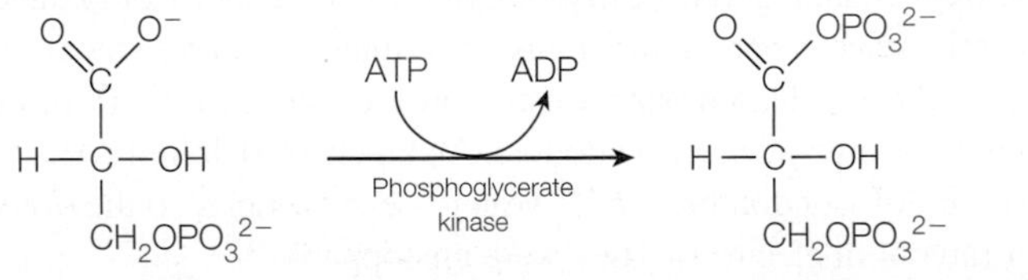

3-Phosphoglycerate　　**1,3-Bisphosphoglycerate**

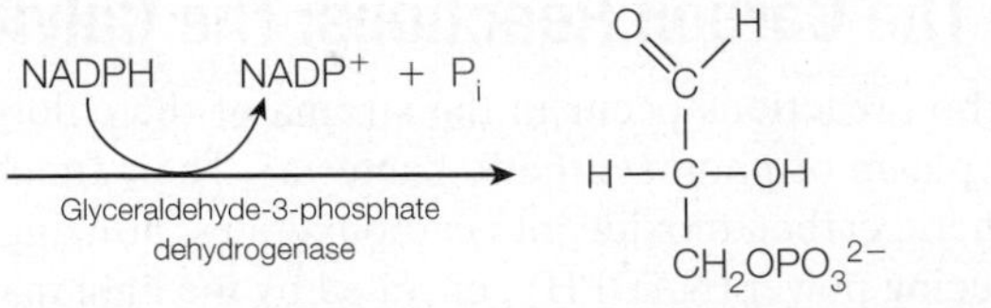

Glyceraldehyde-3-phosphate

We have encountered these enzymes earlier, in connection with their roles in glycolysis (see Chapter 12).

At this stage of the cycle, one molecule of CO_2 has been fixed into a hexose, which is then cleaved into two three-carbon monosaccharides.

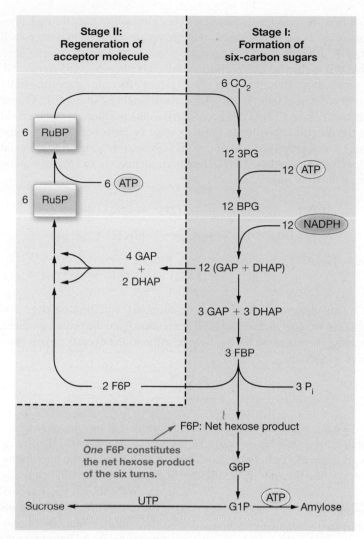

▲ **FIGURE 15.20 Stoichiometry of the Calvin cycle.** In six turns of the Calvin cycle, six CO_2 molecules will have entered and are bound to six molecules of ribulose-1,5-bisphosphate (RuBP) to yield 12 molecules of glyceraldehyde-3-phosphate (GAP). BPG, 1,3-bisphosphoglycerate; DHAP, dihydroxyacetone phosphate; F6P, fructose-6-phosphate; FBP, fructose-1,6-bisphosphate; G6P, glucose-6-phosphate; G1P, glucose-1-phosphate; UTP, uridine triphosphate.

It is informative to note the requirements in ATP and NADPH up to this point. For each CO_2 molecule that has passed through these steps, two molecules of ATP have been hydrolyzed and two molecules of NADPH have been oxidized. However, it is more appropriate to keep accounts on a "per glucose" basis because we want to see what must happen to account for the generation of one hexose molecule from CO_2. **FIGURE 15.20** provides a schematic view of the stoichiometry of the whole Calvin cycle. Six molecules of CO_2 have to enter the cycle to provide the six carbons needed for every new molecule of hexose produced. That requires formation of 12 GAP, and therefore 12 ATP and 12 NADPH are needed.

At this point, the pathway splits so as to satisfy the two essential goals—to make hexoses and to regenerate the acceptor. Of the 12 molecules of GAP that have been produced, 2 will be used to make a molecule of a hexose. The remaining 10 will be utilized to regenerate the 6 molecules of ribulose bisphosphate necessary to maintain the cycle. That is, 10 three-carbon molecules will be converted to 6 five-carbon molecules.

Formation of Hexose Sugars

This is actually familiar ground, for it follows a portion of the gluconeogenic pathway described in Chapter 12. The reactions are shown schematically in Figure 15.20. Recall that glyceraldehyde-3-phosphate can be isomerized to dihydroxyacetone phosphate (DHAP) by triose phosphate isomerase (Section 12.2). Thus, the 12 molecules of GAP produced can be considered to be an interconvertible equilibrium pool of GAP and DHAP. A molecule of GAP and a molecule of DHAP can be combined, via the enzyme *fructose bisphosphate aldolase*, to yield fructose-1,6-bisphosphate (FBP). As Figure 15.20 shows, six of the GAP molecules follow this path, to yield three molecules of FBP. The FBP is dephosphorylated by fructose-1,6-bisphosphatase to yield three molecules of fructose-6-phosphate (F6P). Of these, two will be employed in the regeneration pathway, but one is available as a net product of the Calvin cycle; it is then isomerized to glucose-6-phosphate (G6P) and finally to glucose-1-phosphate (G1P).

Glucose-1-phosphate is, in plants as in animals, the precursor to oligosaccharide and polysaccharide formation. Formation of plant starch (amylose) follows a path similar to that used by animals in glycogen synthesis. However, instead of using uridine triphosphate (UTP) to activate the glucose monomer, as in glycogen formation, ATP is employed in the polymerization of amylose:

$$\text{Glucose-1-phosphate} + \text{ATP} \longrightarrow \text{ADP-glucose} + \text{PP}_i$$
$$\text{ADP-glucose} + (\text{glucose})_n \longrightarrow (\text{glucose})_{n+1} + \text{ADP}$$

Amylose, which is not very soluble, is a storage carbohydrate. However, much of the saccharide synthesized in plant leaves is exported to other parts of the plant, mostly in the form of sucrose. Sucrose is synthesized in the cytosol of plant leaves by the following sequence of reactions:

$$\text{UTP} + \text{glucose-1-P} \longrightarrow \text{UDP-glucose} + \text{PP}_i$$
$$\text{UDP-glucose} + \text{fructose-6-P} \longrightarrow \text{UDP} + \text{sucrose-6-P}$$
$$\text{Sucrose-6-P} + \text{H}_2\text{O} \longrightarrow \text{sucrose} + \text{P}_i$$

The UDP produced is then converted back to UTP by phosphate transfer from ATP.

Stage II: Regeneration of the Acceptor

The reactions we have considered to this point can account for the introduction of one carbon into one molecule of hexose, with subsequent formation of oligosaccharides or polysaccharides. But to complete the Calvin cycle, it is necessary to regenerate enough ribulose-1,5-bisphosphate to keep the cycle going. This means we

● **CONNECTION** For years, tomato breeders have selected for varieties that produce pale green fruit to facilitate harvests of evenly ripened fruit. It turns out that this selection process mutated a gene encoding a transcription factor that increases the fruit's photosynthetic capacity, resulting in higher sugar content. This random selection process inadvertently traded fruit quality for production benefits.

● **CONNECTION** The sequence of the dog genome suggests that domestication of dogs was accompanied by adaptation from the carnivorous diet of their wolf ancestors to a starch-rich diet. Scavenging in waste dumps near human settlements during the dawn of the agricultural revolution may have been a major driving force behind dog domestication.

● **CONCEPT** The six ribulose-1,5-bisphosphate pentoses (30 carbons) are regenerated from two hexoses and six trioses (30 carbons).

● **CONNECTION** The fixation of atmospheric carbon dioxide through photosynthesis represents a potential large-scale source of sustainable energy. Biofuels can be produced from agricultural crops or dedicated cellulosic feedstocks. Cellulosic feedstocks, such as switchgrass, may have greater potential for reducing greenhouse gases and cause less damage to ecosystems compared with agricultural crops, such as corn.

To make five-carbon molecules from six-carbon and three-carbon molecules, several rearrangements are required. These rearrangements are accomplished by *transketolases* and *transaldolases*, whose chemistry was described in our discussion of the pentose phosphate pathway (Section 12.10). What is important here is that two hexoses and six trioses (30 carbons) have been rearranged and recombined to form six pentoses (30 carbons):

2 Fructose-6-P + 2 dihydroxyacetone phosphate
  + 4 glyceraldehyde-3-phosphate ⟶ 6 ribulose-5-phosphate

The final step in the regeneration of ribulose-1,5-bisphosphate is phosphorylation, catalyzed by the enzyme *ribulose-5-phosphate kinase* and utilizing ATP. For six rounds of the cycle, this step will require 6 ATPs in addition to the 12 already accounted for. Therefore, the requirements for synthesizing 1 mole of hexose from CO_2 are 12 moles of NADPH and 18 moles of ATP:

$6CO_2 + 12NADPH + 12H^+ \longrightarrow C_6H_{12}O_6 + 12NADP^+ + 6H_2O$
$18ATP + 18H_2O \longrightarrow 18ADP + 18P_i + 18H^+$

[H is balanced in the second equation when the H in P_i (HPO_4^{2-}) is included.] Summing these two equations, we can write the overall carbon reaction as

$6CO_2 + 18ATP + 12NADPH + 12H_2O \longrightarrow$
$\qquad C_6H_{12}O_6 + 18ADP + 18P_i + 12NADP^+ + 6H^+$

15.5 A Summary of the Light and Carbon Reactions in Two-System Photosynthesis

The Overall Reaction and the Efficiency of Photosynthesis

The ATP and NADPH needed for the carbon reactions are released into the stroma by the light reactions of photosynthesis. If we recall

will need to regenerate 6 moles of RuBP for every 6 moles of CO_2 taken up. This is accomplished by the regenerative phase of the cycle schematized in Figures 15.19 and 15.20. Note that the *input* molecules in this complex reaction pathway are as follows:

1. Two molecules of DHAP and four molecules of GAP, from the six GAP that were diverted to the regeneration pathway in Figure 15.20.

2. Two of the three molecules of fructose-6-phosphate (F6P) that were produced from the remaining three GAP and three DHAP.

that two photons are required for every electron to pass through photosystems I and II, and that two electrons are required to reduce each $NADP^+$, then four photons are necessary for the production of each NADPH molecule. This corresponds to eight photons per O_2 produced, a number in agreement with the quantum efficiency experimentally observed when both photosystems are operating—about 0.12 O_2 per photon. For the 12 NADPH needed in the dark reaction, as summarized in the previous section, 48 photons must be absorbed. If we assume that these photons will also pump enough protons across the thylakoid membrane to yield the 18 ATP required, we may, as an approximation, write the light reactions as

$$12H_2O + 12NADP^+ \longrightarrow 12H^+ + 12NADPH + 6O_2$$
$$18ADP + 18P_i + 18H^+ \longrightarrow 18ATP + 18H_2O$$
$$\text{Sum:} 12NADP^+ + 18ADP + 18P_i + 6H^+ \longrightarrow$$
$$18ATP + 6H_2O + 12NADPH + 6O_2$$

This equation differs from the summary equation on page 500 because we now include the ATP generation from the proton gradient. Adding this equation for the light reaction to the overall carbon reaction, we obtain

$$6H_2O + 6CO_2 \xrightarrow{\text{48 photons}} C_6H_{12}O_6 + 6O_2$$

This estimate of 48 photons assumes that noncyclic photophosphorylation provides enough ATP for the carbon reactions. If, as many workers in the field believe, additional ATP from cyclic photophosphorylation is required, the number of photons needed will be greater.

● **CONCEPT** The overall energy efficiency of photosynthesis can approach 36%.

We can, on the basis of these calculations, estimate the energy efficiency of photosynthesis. As we saw at the start of this chapter, forming a mole of hexose from CO_2 and water requires 2870 kJ. The energy input per photon depends on the wavelength of light used. Assuming that light of 650 nm wavelength is used, 48 einsteins of such light correspond to about 8000 kJ (see p. 492). From this value, we calculate a theoretical efficiency of 2870/8000 ≈ 36%. Direct experimental measurements of the efficiency under optimal conditions give results in the same range or slightly lower.

Regulation of Photosynthesis

It should be evident that the carbon reactions of photosynthesis, which result in the production of sugars, require careful regulation. Because the carbon reactions depend on the reductive power and ATP supplied by the light reactions, it is not surprising that they are stimulated by the light reactions. This stimulation is accomplished in three major ways. First, the central enzyme in the carbon reactions, ribulose-1,5-bisphosphate carboxylase (Rubisco), is stimulated by high pH and by both CO_2 and Mg^{2+} (**FIGURE 15.21**). Recall that the pumping of protons from the stroma into the thylakoid lumen by the light reactions increases the stromal pH; at the same time, Mg^{2+} ions enter the stroma to compensate for the positive charge of the H^+ ions that have been lost.

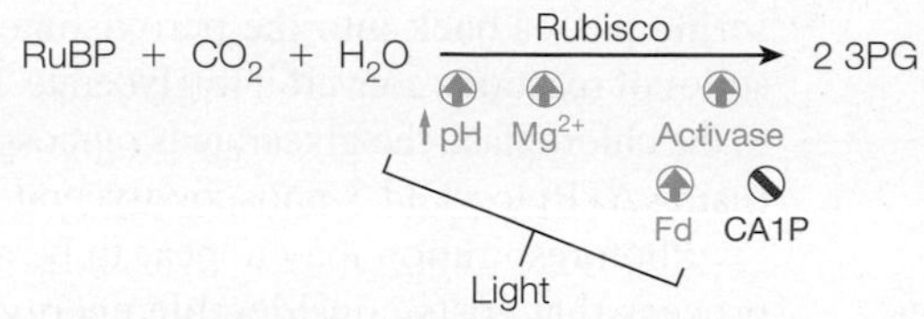

$$RuBP + CO_2 + H_2O \xrightarrow{\text{Rubisco}} 2\ 3PG$$

↑pH   Mg²⁺   Activase

Fd   CA1P

Light

▲ **FIGURE 15.21 Regulation of Rubisco by light.** High pH, Mg^{2+}, and Rubisco activase stimulate Rubisco activity. Reduced ferredoxin (Fd) stimulates activase via the thioredoxin system. High pH, Mg^{2+}, and reduced Fd all increase in response to light. 2-Carboxy-D-arabinitol-1-phosphate (CA1P) is an inhibitor.

Second, Rubisco is sensitive to a number of naturally occurring sugar phosphates that act as tight-binding inhibitors, including xylulose-1,5-bisphosphate, 2-carboxy-D-arabinitol-1-phosphate (CA1P), as well as the ribulose-1,5-bisphosphate substrate itself. These inhibitors resemble transition-state intermediates and cause the active site of Rubisco to adopt a closed conformation, preventing carbamation and/or substrate binding. Genetic experiments with the model flowering plant *Arabidopsis* led to the discovery of a regulatory enzyme, termed **Rubisco activase,** that is required for activation and maintenance of Rubisco activity in vivo. Rubisco activase promotes the ATP-dependent dissociation of the inhibitor, facilitating the carbamation reaction and the binding of Mg^{2+} and substrate.

One of the most potent Rubisco inhibitors, CA1P, is synthesized in chloroplasts in the dark and is responsible for the characteristic decline in Rubisco activity during darkness. When the chloroplasts are again illuminated, Rubisco activase removes CA1P and reactivates Rubisco. Rubisco activase is itself activated in the light by another light-dependent mechanism, which will be described next.

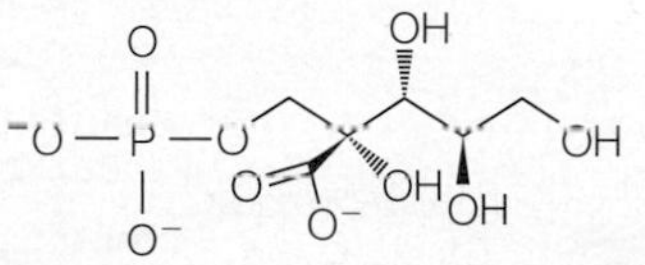

2-Carboxy-D-arabinitol-1-phosphate (CA1P)

The third way by which the carbon reactions are stimulated by the light reactions depends on the redox state of critical disulfide bonds in several enzymes of the Calvin cycle. Rubisco activase, fructose-1,6-bisphosphatase and glyceraldehyde-3-phosphate dehydrogenase (see page 504), and ribulose-5-phosphate kinase (see page 506) are all activated by reduction of disulfides to sulfhydryls in the enzymes. Reduction is promoted by a disulfide exchange reaction with the protein *thioredoxin* (**FIGURE 15.22**). Thioredoxin, a small protein carrying two reversibly oxidizable Cys-SH groups, is used in a wide variety of redox reactions. The reduction of thioredoxin, in turn, is promoted

● **CONCEPT** The photosynthetic carbon reactions are regulated by the amount of light available to the organism through the activation of key enzymes.

by oxidation of the reduced form of ferredoxin via a reaction catalyzed by the enzyme *ferredoxin–thioredoxin reductase*. In strongly irradiated chloroplasts, in which $NADP^+$ stores are depleted, reduced

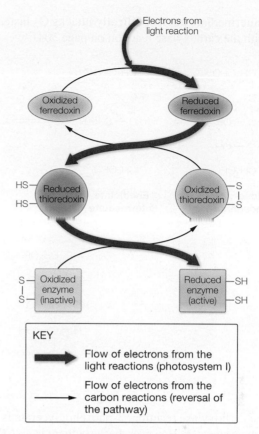

▲ **FIGURE 15.22 Light-dependent activation of carbon-reaction enzymes.** Several enzymes of the Calvin cycle are activated by disulfide reduction, which is mediated by reduced thioredoxin. Thioredoxin is reduced by reduced ferredoxin, which accumulates in irradiated chloroplasts. Additional functions of thioredoxin are presented in Chapter 20.

ferredoxin accumulates. High levels of reduced ferredoxin thereby lead to activation of the Calvin cycle enzymes, stimulating the Calvin cycle reactions when the light reactions are very active. The same compound, reduced thioredoxin, also stimulates the CF_0-CF_1 complexes, ensuring a high rate of ATP generation when illumination is intense.

In the dark, the plant "turns into an animal" in terms of its biochemistry. It begins to draw on its energy reserves, using pathways familiar from our studies of animal catabolism: glycolysis, the citric acid cycle, and the pentose phosphate pathway. In general, these pathways are inhibited in plants in the presence of sunlight and become more active in the dark. The key light-inhibited enzymes are phosphofructokinase (in glycolysis) and glucose-6-phosphate dehydrogenase (in the pentose phosphate pathway). Glucose-6-phosphate dehydrogenase is inhibited by the same reduced form of thioredoxin that *activates* Calvin cycle enzymes.

15.6 Photorespiration and the C₄ Cycle

Ribulose bisphosphate carboxylase is a peculiar enzyme. Under normal environmental conditions, it can behave as an *oxygenase* as well as a carboxylase. In the oxygenase reaction, it is proposed that the enediolate intermediate nucleophilically attacks O_2 instead of CO_2 (compare with the carboxylase reaction on page 504):

Ribulose-1,5-bisphosphate

Enediolate intermediate

2-Phosphoglycolate

3-Phosphoglycerate

As in the carboxylase reaction, 3-phosphoglycerate is a product of the final cleavage step. However, because there is no CO_2 fixation in the oxygenase reaction, the other cleavage product is a 2-carbon compound, **2-phosphoglycolate.** The relative rates of the carboxylase and oxygenase reactions are determined by the concentrations of CO_2 and O_2 at the active site of the enzyme and its K_M values for those two gases. Oxygenation occurs under conditions of high O_2 and low CO_2 concentrations (such as is found in normal air) because the K_M for O_2 is about 10-fold higher than that for CO_2. When the oxygenase reaction becomes significant, it initiates a reaction pathway known as **photorespiration,** with production of 3-phosphoglycerate and phosphoglycolate in the chloroplast. As **FIGURE 15.23** shows, the phosphoglycolate is then dephosphorylated and passed to organelles called **peroxisomes.** Here, it is further oxidized, yielding glyoxylate and hydrogen peroxide. The toxic H_2O_2 is broken down by catalase, and the glyoxylate is transaminated, producing glycine. The glycine enters mitochondria, where *two* molecules are converted into *one* molecule of serine, plus one molecule each of CO_2 and NH_4^+. This process, involving the multiprotein glycine cleavage system and serine hydroxymethyltransferase, is described in detail in Chapter 18. The gases CO_2 and ammonia are released. The

● **CONCEPT** Under conditions of low CO_2 and high O_2, plants exhibit photorespiration, in which O_2 is consumed and CO_2 released.

serine passes back into the peroxisome, where a series of reactions convert it to glycerate. Returning to the chloroplast, the glycerate is rephosphorylated (using ATP) to yield 3-phosphoglycerate.

Photorespiration may appear to be a wasteful process that costs considerable energy and limits the carbon assimilation by the plant. Note the following:

1. Ribulose-1,5-bisphosphate is lost from the Calvin cycle.

2. The fixation of CO_2 is reversed: O_2 is consumed and CO_2 is released.

3. Only a part (75%) of the carbon is returned to the chloroplast.

4. ATP is expended.

So why does photorespiration exist? Rubisco evolved as a carboxylase ~3 billion years ago, when the atmosphere was characterized by high CO_2 and low O_2 levels. Under these conditions, the

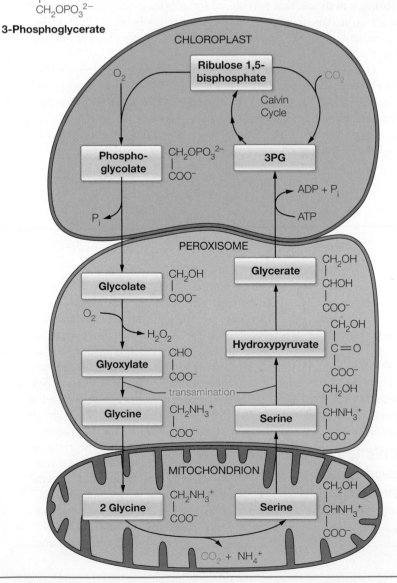

▶ **FIGURE 15.23 Photorespiration.** Ribulose-1,5-bisphosphate can be diverted from the Calvin cycle, especially when the concentration of CO_2 is low. RuBP carboxylase/oxygenase (Rubisco) catalyzes the oxidation of RuBP to form phosphoglycolate. In the reactions that follow, more O_2 is used, CO_2 is generated, and ATP is hydrolyzed, as metabolites pass from the chloroplast to nearby peroxisomes and mitochondria and then back into the chloroplast in the form of 3-phosphoglycerate (3PG) for reentry into the Calvin cycle.

oxygenation reaction was negligible. However, as O_2 levels increased in the atmosphere, its oxygenase activity became significant. Plants then had to evolve strategies to deal with the inefficiency introduced by the oxygenase activity of Rubisco. You might expect they would modify the ribulose bisphosphate carboxylase/oxygenase enzyme to suppress the oxygenase function. Surprisingly, that is not the case. This enzyme, despite (or perhaps because of) its vital importance, has changed little over long ages. It remains a relatively inefficient catalyst (with $k_{cat} \cong 2\ s^{-1}$) and has never lost its oxygenase function. Stuck with this "design flaw" in Rubisco, plants instead evolved a pathway to reclaim the two-carbon fragment (phosphoglycolate) released by the oxygenase activity. This pathway, photorespiration, can thus be thought of as an evolutionary compromise, albeit an inelegant one.

Certain plants, which are called **C₄ plants,** have evolved an additional photosynthetic pathway that helps conserve CO_2 released by photorespiration. This pathway is called the **C₄ cycle** because it involves incorporation of CO_2 into a C_4 intermediate (oxaloacetate). This cycle is distinguished from the Calvin cycle, which utilizes a three-carbon intermediate and is hence sometimes called the C_3 cycle (and plants lacking the C_4 cycle are called C_3 plants). The C_4 cycle is found in several crop species (maize and sugarcane, for example) and is important in tropical plants, which are exposed to intense sunlight and high temperatures. Although photorespiration occurs to some extent at all times in all plants, it is most active under conditions of high illumination, high temperature, and CO_2 depletion.

● **CONCEPT** Some plants (called C₄ plants) minimize the wastefulness of photorespiration by utilizing an alternative CO_2 concentrating mechanism.

C_4 photosynthesis is really just a CO_2 concentrating mechanism, providing a higher CO_2/O_2 ratio at the active site of Rubisco that favors carboxylation. C_4 plants concentrate their Calvin cycle (C_3) photosynthesis in specialized *bundle sheath cells,* which lie below a layer of mesophyll cells (**FIGURE 15.24(a)**). In contrast, the mesophyll cells, which are most directly exposed to external CO_2, contain the enzymes for the C_4 cycle. This pathway, as it operates in most C_4 plants, is shown in Figure 15.24(b). It is essentially a mechanism for trapping CO_2 into a four-carbon compound and passing it on to bundle sheath cells for decarboxylation and use of the resulting CO_2 in their Calvin (C_3) cycle.

The key to the efficiency of C_4 plants is that the CO_2-fixing enzyme used in this pathway, *phosphoenolpyruvate carboxylase,* lacks the oxygenase activity shown by ribulose bisphosphate carboxylase and has a much lower K_M for CO_2. Thus, even under conditions of high O_2 concentration and low CO_2 concentration in the

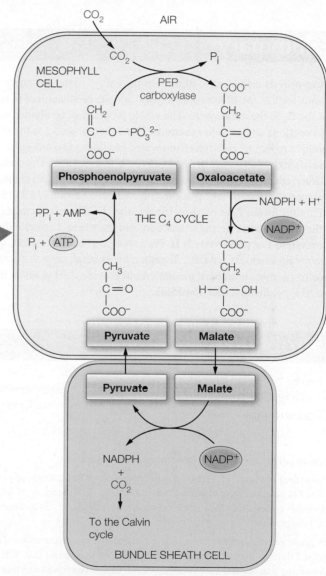

(a) In C₄ plants, the mesophyll cells (light green) trap CO_2 in C_4 intermediates. The C_4 compounds are then delivered to the bundle sheath cells (dark green), where most of the Calvin cycle (C_3) photosynthesis takes place.

(b) CO_2 is transported from mesophyll cells to the bundle sheath cells by coupling it to phosphoenolpyruvate, forming oxaloacetate. Oxaloacetate is then reduced to malate, which is passed to the bundle sheath cells and decarboxylated. The pyruvate product is returned to the mesophyll cells, where it is phosphorylated to regenerate phosphoenolpyruvate.

▲ **FIGURE 15.24** Reactions of the C_4 cycle.

atmosphere, the mesophyll cells continue to pump CO_2 to the photosynthesizing bundle sheath cells, where Rubisco is localized. This process helps maintain high enough CO_2 levels in the bundle sheath cells so that CO_2 fixation, rather than photorespiration, is favored. Furthermore, if photorespiration *does* occur, the CO_2 that is released in that process can be largely salvaged in the surrounding mesophyll cells and returned to the Calvin cycle.

As Figure 15.24 shows, the C_4 cycle costs the plant energy in the form of ATP. In fact, because ATP is hydrolyzed to AMP and inorganic phosphate in regenerating phosphoenolpyruvate, the expense is equivalent to *two* extra ATPs for every C_4 molecule fixed.

The inefficiency of Rubisco as an enzyme and its participation in photorespiration greatly reduce the efficiency of plants as food producers. Not only the very large amounts of Rubisco that must be synthesized but also the energy expended in photorespiration

● **CONNECTION** C_4 crops are more efficient than C_3 crops in hot and arid climates. The threat of climate change has stimulated intense efforts to engineer the C_4 photosynthesis pathway into C_3 crops such as wheat and rice.

place seemingly unnecessary demands on plant metabolism. If a more efficient enzyme could be developed, crop yields could be significantly increased and nitrogen demands reduced. There were intensive attempts to engineer a more efficient Rubisco into crop plants, but these efforts were unsuccessful and have been largely abandoned. This failure is perhaps not surprising, since evolution has had a few billion years to eliminate the oxygenase activity and has not managed to do so. Alternative approaches that are currently being pursued include introducing C_4 photosynthesis into C_3 crop plants and improving the activity of Rubisco activase.

Summary

• **Photosynthesis** is the source of most of the energy in the biosphere and accounts for fixation of atmospheric CO_2 and the production of most or all of the O_2 in the atmosphere. The whole process can be divided into **light reactions** and **carbon reactions.** The light reactions use the energy of sunlight to extract electrons from water, producing O_2, reductive potential, and a proton gradient that drives ATP formation. The carbon reactions reduce CO_2 into carbohydrates. In higher plants and algae, both types of reactions take place in chloroplasts. (Sections 15.1, 15.2)

Photons for the light reaction are absorbed by antenna pigments, and the energy is transferred to reaction centers, where it enters either **photosystem I** or **photosystem II.** Photosystem II oxidizes water, and photosystem I reduces $NADP^+$. Together, the systems drive the transport of protons across chloroplast membranes to provide a pH gradient to drive ATP production. (Section 15.3)

• The carbon reactions are accomplished by the **Calvin cycle**, which may be divided into two stages. In the first stage, CO_2 is added to ribulose-1,5-bisphosphate (RuBP), which is then cleaved and reduced to form trioses that can then be combined to form hexose. The second stage of the cycle uses most of the trioses and hexoses to regenerate RuBP. The carbon reactions are regulated by several mechanisms that all respond to light intensity. (Section 15.4)

• Under conditions of low CO_2 and high O_2, plants undergo an oxidative process called **photorespiration.** The process is essentially inefficient, and some tropical plants compensate for it via the C_4 cycle, which is less sensitive to high O_2 levels. (Section 15.6)

Problems

Enhanced by
Mastering Chemistry
for Biochemistry

Mastering Chemistry for Biochemistry provides select end-of-chapter problems and feedback-enriched tutorial problems, animations, and interactive figures to deepen your understanding of complex topics while practicing problem solving.

Answers to red problems are available in the Answer Appendix.

1. According to Figure 15.11(b), upon excitation, the P700 reaction center is raised in potential from about +0.4 to −1.3 volts. To what value of $\Delta G°'$ does this correspond? How does it compare with the energy in an einstein of 700 nm photons?

2. In cyclic photophosphorylation, it is estimated that two electrons must be passed through the cycle to pump enough protons to generate one ATP. Assuming that the ΔG for hydrolysis of ATP under conditions existing in the chloroplast is about −50 kJ/mol, what is the corresponding percent efficiency of cyclic photophosphorylation, using light of 700 nm?

3. Assume a pH gradient of 4.0 units across a thylakoid membrane, with the lumen more acidic than the stroma.
   (a) What is the standard free energy change per mol of protons associated with this gradient at 25 °C?

   (b) What is the *longest* wavelength of light that could provide enough energy per photon to pump one proton against this gradient, assuming 20% efficiency in photosynthesis and $T = 25°C$?
   (c) What is the standard free energy change per mol O_2 produced? How does this compare to the energy required to drive the synthesis of ATP?
   (d) The intensity of natural sunlight drops off dramatically beyond 650 nm. Calculate the maximum number of moles of protons that could be pumped against the gradient by the energy in a mole of photons of 650 nm wavelength. Assume 100% efficiency and $T = 25$ °C for this calculation.

4. Suppose a brief pulse of $^{14}CO_2$ is taken up by a green plant.
   (a) Trace the ^{14}C label through the steps leading to fructose-1, 6-bisphosphate synthesis, showing which carbon atoms in each compound should carry the label during the first cycle.
   (b) Will all molecules of fructose-1,6-bisphosphate carry two ^{14}C atoms? Explain.

5. The flux of solar energy reaching the Earth's surface is approximately $7\ J/cm^2$. Assume that *all* of this energy is used by a green leaf ($10\ cm^2$ in area), with maximal efficiency of 35%. How many moles of hexose could the leaf theoretically generate in an hour? Use 600 nm for an average wavelength.

6. The substance dichlorophenyldimethylurea (DCMU) is an herbicide that inhibits photosynthesis by blocking electron transfer between plastoquinones in photosystem II.
(a) Would you expect DCMU to interfere with cyclic photophosphorylation?
(b) Normally, DCMU blocks O_2 evolution, but addition of ferricyanide to chloroplasts allows O_2 evolution in the presence of DCMU. Explain.

7. Suppose a researcher is carrying out studies in which she adds a nonphysiological electron donor to a suspension of chloroplasts. Illumination of the chloroplasts yields oxidation of the donor. How could she tell whether photosystem I, II, or both are involved?

8. Suppose ribulose-5-phosphate, labeled with ^{14}C in carbon 1, is used as the substrate in carbon reactions. In which carbon of 3PG will the label appear?

9. The following data, presented by G. Bowes and W. L. Ogre in *J. Biol. Chem.* (1972) 247:2171–2176, describe the relative rates of incorporation of CO_2 by Rubisco under N_2 and under pure O_2. Decide whether O_2 is a competitive or uncompetitive inhibitor.

$[CO_2]$ (mM)	Under N_2	Under O_2
0.20	16.7	10
0.10	12.5	5.6
0.067	8.3	4.2
0.050	7.1	3.2

10. J. C. Servaites, in *Plant Physiol.* (1985) 78:839–843, observed that Rubisco from tobacco leaves collected before dawn had a much lower specific activity than the enzyme collected at noon. This difference persisted despite extensive dialysis, gel filtration, or heat treatment. However, precipitation of the predawn enzyme by 50% $(NH_4)_2SO_4$ restored the specific activity to the level of the noon-collected enzyme. Suggest an explanation.

11. It is believed that the ratio of cyclic photophosphorylation to noncyclic photophosphorylation changes in response to metabolic demands. In each of the following situations, would you expect the ratio to increase, decrease, or remain unchanged?
(a) Chloroplasts carrying out both the Calvin cycle and the reduction of nitrite (NO_2^-) to ammonia (This process does not require ATP.)
(b) Chloroplasts carrying out not only the Calvin cycle but also extensive active transport
(c) Chloroplasts using both the Calvin cycle and the C_4 pathway

12. If a photosynthetic organism is illuminated in a closed, sealed environment, it is observed that the CO_2 and O_2 levels in the surrounding atmosphere reach a constant ratio.
(a) Suggest an explanation.
(b) What factor would you think primarily determines the value of this ratio?

13. If algae are exposed to $^{14}CO_2$ for a brief period while illuminated, the labeled carbon is initially found almost entirely in the carboxyl group of 3-phosphoglycerate. However, if illumination is continued after the label pulse, other carbon atoms become labeled. Explain.

14. Daniel Arnon and coworkers carried out experiments with intact, salt-washed chloroplasts to study photophosphorylation. When the chloroplasts were illuminated in the presence of ADP + P_i, ATP was produced, but oxygen was not produced nor consumed. ATP formation was not accompanied by a measurable electron transport involving any external electron donor or acceptor. The overall reaction for this result is:

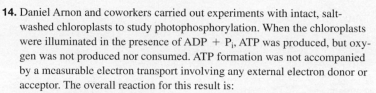

$$ADP + P_i \xrightarrow{h\nu} ATP$$

When $NADP^+$ was included in addition to the ADP + P_i, illumination of the intact chloroplasts again resulted in the photophosphorylation of ADP to ATP. In addition, the $NADP^+$ was reduced to NADPH + H^+, and O_2 was produced. Moreover, the light-induced reduction of $NADP^+$ was greatly decreased if ADP + P_i was omitted. The equation for this reaction is:

$$NADP^+ + H_2O + ADP + P_i \xrightarrow{h\nu} NADPH + H + ATP + \tfrac{1}{2}O_2$$

Briefly describe the mechanism(s) of these two types of photophosphorylation that explain all of these results.

15. Macías-Rubalcava et al., in *J. Photochem. Photobiol.* (2017) *B* 166:35–43, investigated the mechanism of action of several compounds that might serve as new herbicides. One of these new compounds (#7) inhibited photophosphorylation. The data below show the effects of compound 7 on uncoupled electron transport in isolated illuminated chloroplasts. The scheme below is a linear version of the Z-scheme in Figure 15.11(b), showing the sequence of electron transport through PSII and PSI. Arrows show the sites of electron donation and acceptance for various artificial donors/acceptors. Vertical dashed lines show sites of inhibition by Tris and DCMU.

Based on these data, predict the site of action of the new compound. Explain your reasoning.

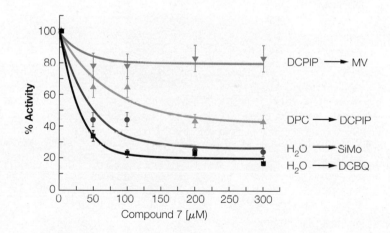

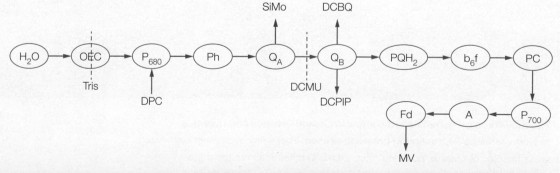

For a list of references related to this chapter, see Appendix II.

The ruby-throated hummingbird winters in Central America and migrates to North America for the spring breeding season. These tiny birds must cross the Gulf of Mexico, covering more than 500 miles in a nonstop marathon. Prior to migration, the hummingbirds fatten up, accumulating about 2 g of fat in a total body weight of 4.5 g. The energy needed for a typical 20-hour crossing is supplied by the oxidation of 1.5 g of this stored fat.

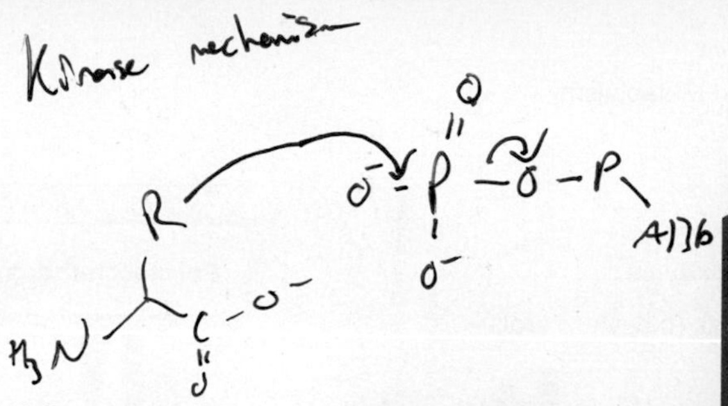

Lipid Metabolism

WE FIRST INTRODUCED the structures and general features of various lipid molecules in Chapter 10. In this chapter, we discuss the metabolism and functions of lipids. Like the carbohydrates we have discussed in previous chapters, lipids play roles in energy metabolism as well as in a variety of other processes. For lipids, those other processes include their roles as components of membranes, hormones, fat-soluble vitamins, thermal insulators, and signaling molecules. Reflecting these distinct roles, this chapter is organized into two parts.

Part I discusses the synthesis and breakdown of energy storage lipids—the triacylglycerols—as well as fatty acid oxidation and biosynthesis (**FIGURE 16.1**), processes that are quite similar among plants, animals, and microorganisms. We also present topics related more directly to animal metabolism—fat digestion, absorption, storage, and mobilization.

Part II focuses on some of the nonbioenergetic roles of lipids. We will outline the metabolism of membrane lipids, sterolds (including cholesterol), and a few of those lipids that function as signaling molecules.

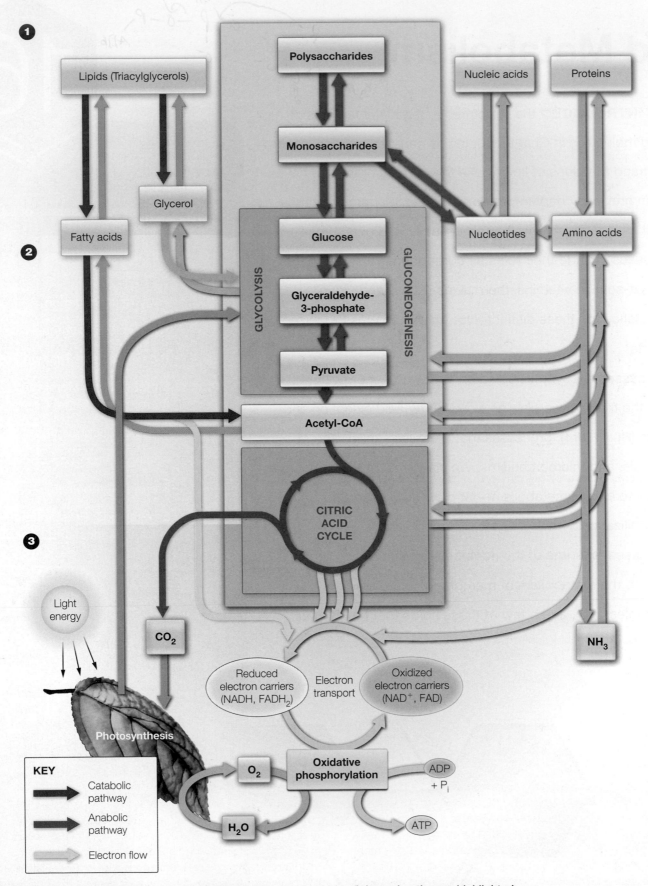

▲ **FIGURE 16.1** Overview of intermediary metabolism with fatty acid and triacylglycerol pathways highlighted.

PART I: BIOENERGETIC ASPECTS OF LIPID METABOLISM

Fat is the major energy storage form in most organisms. In Part I of this chapter, we will discuss how fat is obtained from the diet and distributed to the tissues, how fatty acids are oxidized to yield energy, and how fatty acids are synthesized and incorporated into triacylglycerols for storage. Although cholesterol is not used as a fuel molecule, we include it in this part because its transport is intertwined with that of the other lipids.

16.1 Utilization and Transport of Fat and Cholesterol

As discussed in Chapter 10, the great bulk of the lipid in most organisms is in the form of *triacylglycerols* (formerly called *triglycerides*). The term *fat,* or *neutral fat,* also refers to this most abundant class of lipids.

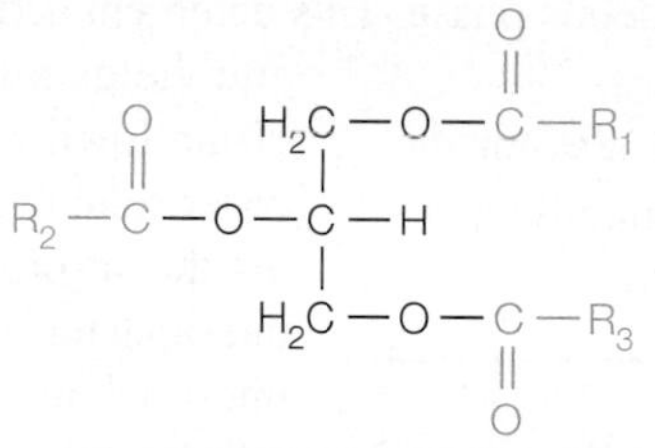

Triacylglycerol (TG)

A mammal contains 5% to 25% or more of its body weight as lipid, with as much as 90% of this lipid in the form of triacylglycerols. Most of this fat is stored in adipose tissue and constitutes the primary energy reserve. Mammals (including humans) are "fat burners"—we eat in pulses (i.e., meals), convert the excess carbohydrate not utilized immediately into fat, and store it. The fat is then burned at a later time as needed. Indeed, certain tissues, such as heart and liver, obtain as much as 80% of their energy needs from fat oxidation. In animal systems, fat is stored in specialized cells called *adipocytes,* where giant fat globules occupy most of the intracellular space (see Figure 10.3). Plant seeds also store great quantities of fat to provide energy to the developing plant embryo. Because plant lipids contain mostly unsaturated fatty acids, the triacylglycerols of seeds are largely in the form of liquid oils.

Triacylglycerols play roles other than that of energy storage. Fat serves to cushion organs against shock, and it provides an efficient thermal insulator, particularly in marine mammals, which must maintain a body temperature far higher than that of the seawater in which they live.

Fats as Energy Reserves

Recall from Chapter 10 that most of the carbon in triacylglycerols is more highly reduced than the carbon in carbohydrates. To be sure, the carboxyl carbons of fatty acids are highly oxidized, but most of the fatty acid carbons are at the highly reduced methyl or methylene level. As we learned in Chapter 11, metabolic oxidation of fat consumes more oxygen, on a weight basis, than oxidation of carbohydrate, with correspondingly larger metabolic energy release. The complete metabolic oxidation of triacylglycerols yields 37 kJ/g or more, whereas that of carbohydrates and proteins yields about 17 kJ/g. Contributing to this difference in energy storage capacity between fat and carbohydrate is

the hydrophilic nature of glucose polymers. Glycogen binds about 2 g of water per gram of carbohydrate. Fat, being extremely nonpolar, is anhydrous. Thus, because 1 g of intracellular glycogen contains only 1/3 g of anhydrous glucose polymer, intracellular fat contains about six times as much potential metabolic energy, on a mass basis, as intracellular glycogen. This is an obvious advantage in many situations, such as in hibernating animals, which must store several months' worth of food, or in the flight muscles of small birds, in which weight is at a premium. Incredibly, some small land birds prepare for migration by increasing their body weight about 15% per day, with all of the weight gain being triacylglycerols. Such obese birds can then fly nonstop for 60 hours or more. In addition, the insolubility of fat allows it to be stored in cells without affecting intracellular osmotic pressure.

● **CONCEPT** Stored fat has six times more caloric content by weight than stored carbohydrate because fat is more highly reduced and is anhydrous.

Little wonder, then, that fat is the major form of energy storage in most organisms. A typical 70 kg human may have fuel reserves of 500,000 kJ or more in total body fat and about 100,000 kJ in total protein (mostly muscle protein). By contrast, the glycogen stores amount to just 6800 kJ of available energy and the total free glucose to about 300 kJ. Fat stores are maintained from the diet; 35%–50% of the caloric value of Western diets comes from fat. Most nutritionists recommend that this value be closer to 25%–35% for cardiovascular health. In addition, carbohydrate ingested in excess of its ability to be catabolized or stored as glycogen is readily converted to fat.

Tissue Fuel Stores for Average 70 kg Human

Fuel	Weight (g)	Energy Content (kJ/g)	Total Energy (kJ)
Triacylglycerols	~15,000	37	555,000
Protein	~6,000	17	100,000
Glycogen	~400	17	6,800
Glucose	~20	17	340
Total fuel stores			662,140

Most of the energy derived from fat breakdown comes from oxidation of the constituent fatty acids. Fatty acid oxidation provides the major energy source for many animal tissues. The brain is distinctive in being unable to use fatty acids as a significant energy supply; it has a highly specific requirement for glucose. However, under conditions of starvation, when blood glucose levels decrease, the brain can adjust to use a class of lipid-related compounds called *ketone bodies,* as we will learn later in this chapter when we turn to the subject of fatty acid oxidation.

Fat Digestion and Absorption

The triacylglycerols that mammals use as fuel are derived from three primary sources: (1) the diet; (2) de novo biosynthesis, particularly in liver; and (3) storage depots in adipocytes. The processes by which these sources are utilized in animals are summarized in **FIGURE 16.2**. The major problem that animals must cope with in the digestion, absorption, and transport of dietary lipids is their insolubility in aqueous media. The action of **bile salts,** detergent substances synthesized in liver and stored in the gallbladder, is essential to the digestion of lipids and their absorption through the intestinal mucosa.

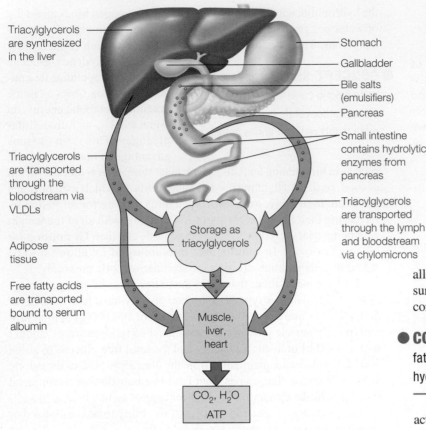

Triacylglycerols are synthesized in the liver

Triacylglycerols are transported through the bloodstream via VLDLs

Adipose tissue

Free fatty acids are transported bound to serum albumin

Stomach

Gallbladder

Bile salts (emulsifiers)

Pancreas

Small intestine contains hydrolytic enzymes from pancreas

Triacylglycerols are transported through the lymph and bloodstream via chylomicrons

Storage as triacylglycerols

Muscle, liver, heart

CO_2, H_2O
ATP

◄ **FIGURE 16.2 Overview of fat digestion, absorption, storage, and mobilization in the human.** Triacylglycerols (fats) are ingested, synthesized in the liver, or mobilized from storage in adipose tissue. Dietary triacylglycerols are enzymatically hydrolyzed in the lumen of the small intestine. Hydrolysis products absorbed by the intestinal mucosa are resynthesized into triacylglycerols, which combine with proteins to form the lipoproteins called chylomicrons. This process solubilizes the lipids and permits their transport out of the intestinal mucosa and through lymph and blood. Triacylglycerols synthesized in liver are combined with carrier proteins to form very low-density lipoproteins (VLDLs) for transport. Lipoproteins transported to peripheral tissues are hydrolyzed at the inner surfaces of capillaries. Hydrolysis products entering cells are either catabolized for energy or recombined into triacylglycerols for storage.

allows bile salts to orient at an oil–water interface, with the hydrophobic surface in contact with the apolar phase and the hydrophilic surface in contact with the aqueous phase. This detergent action emulsifies lipids and yields micelles (see Chapter 10), allowing digestive attack by water-soluble enzymes and facilitating the absorption of lipid through intestinal mucosal cells. Most of the intestinal digestion occurs through action of **pancreatic lipase,** an unusual calcium-requiring enzyme that catalyzes reactions at an oil–water interface. The substrate being cleaved is in an apolar phase, and the other substrate, of course, is water. Pancreatic lipase also functions in a 1:1 complex with **colipase,** a 90-amino acid protein that aids in the binding to the lipid surface.

The products of fat digestion in the intestine are a mixture of glycerol, free fatty acids, monoacylglycerols, and diacylglycerols.

● **CONCEPT** Bile salts emulsify fats, thereby promoting their hydrolysis in digestion.

A bile salt molecule is made up of a **bile acid,** such as cholic acid, and an associated cation. Bile acids are derived from cholesterol, as we will see later in Section 16.7. A bile salt molecule has both hydrophobic and hydrophilic surfaces (**FIGURE 16.3**). This amphipathic character

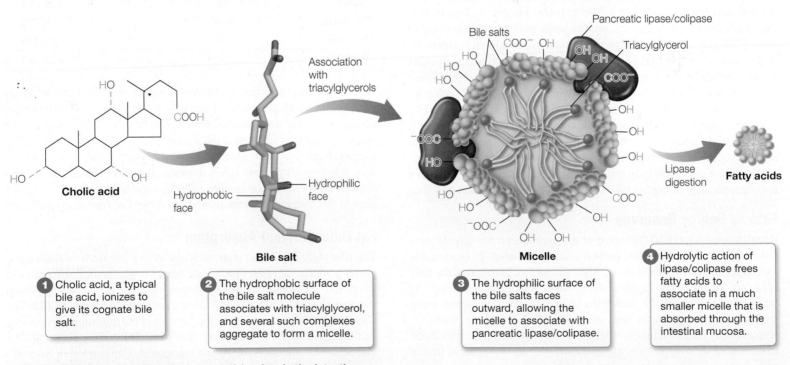

Cholic acid

Bile salt

Hydrophobic face

Hydrophilic face

Association with triacylglycerols

Bile salts

Pancreatic lipase/colipase

Triacylglycerol

Micelle

Lipase digestion

Fatty acids

1. Cholic acid, a typical bile acid, ionizes to give its cognate bile salt.

2. The hydrophobic surface of the bile salt molecule associates with triacylglycerol, and several such complexes aggregate to form a micelle.

3. The hydrophilic surface of the bile salts faces outward, allowing the micelle to associate with pancreatic lipase/colipase.

4. Hydrolytic action of lipase/colipase frees fatty acids to associate in a much smaller micelle that is absorbed through the intestinal mucosa.

▲ **FIGURE 16.3 Action of bile salts in emulsifying fats in the intestine.**

During absorption through intestinal mucosal cells, much resynthesis of triacylglycerols occurs from the hydrolysis products. This resynthesis occurs in the endoplasmic reticulum and Golgi complex of mucosal cells.

Transport of Fat to Tissues: Lipoproteins

The problem of transport of these hydrophobic triacylglycerols through the blood and lymph is dealt with in part by interaction of the lipids

● **CONCEPT** Lipoproteins are noncovalent lipid–protein complexes that allow movement of apolar lipids through aqueous environments like blood and lymph.

with proteins to form soluble aggregates called **lipoproteins.** Triacylglycerols emerge into the lymph system complexed with protein to form the lipoproteins called **chylomicrons.** The chylomicron is essentially an oil droplet coated with more polar lipids and a skin of protein, which help disperse and partially solubilize the fat for transport to tissues. The chylomicron is also a transport vessel for cholesterol obtained through the diet.

Chylomicrons constitute just one class of lipoproteins found in the bloodstream. These noncovalent complexes play essential roles in the transport of lipids to tissues, either for energy storage or for oxidation. Free lipids are all but undetectable in blood. The polypeptide components of lipoproteins are called **apolipoproteins.** These are synthesized mainly in the liver, though about 20% are produced in intestinal mucosal cells.

Classification and Functions of Lipoproteins

Distinct families of lipoproteins have been described, each of which plays defined roles in lipid transport. These families are classified in terms of their density, as determined by centrifugation (**TABLE 16.1**). Lipoproteins in each class contain characteristic apolipoproteins and have distinctive lipid compositions. A total of 10 major apolipoproteins are found in human lipoproteins.

Because lipids are of much lower density than proteins, the lipid content of a lipoprotein class is inversely related to its density: The

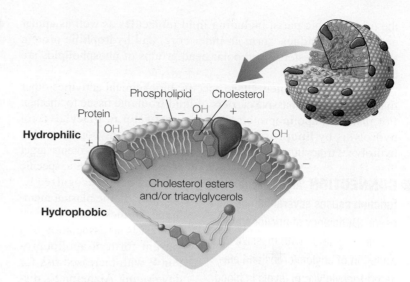

▲ **FIGURE 16.4** Generalized structure of a plasma lipoprotein. The spherical particle, part of which is shown, has a hydrophobic inner core (yellow) composed of cholesterol esters and triacylglycerols surrounded by a hydrophilic surface formed by the polar head groups of phospholipids and free cholesterol. The apolipoproteins orient with their hydrophobic regions in the inner core and their hydrophilic regions at the surface.

higher the lipid abundance, the lower the density. The standard lipoprotein classification includes, in increasing order of density: *chylomicrons*, **very low-density lipoprotein (VLDL), intermediate-density lipoprotein (IDL), low-density lipoprotein (LDL),** and **high-density lipoprotein (HDL).** Some classification schemes recognize two classes of HDL; in addition, there is a quantitatively minor lipoprotein called very high-density lipoprotein (VHDL).

Despite their differences in lipid and protein composition, all lipoproteins share common structural features, notably a spherical shape that can be detected by electron microscopy. As shown in **FIGURE 16.4,**

TABLE 16.1 Properties of major human plasma lipoprotein classes

	Chylomicron	VLDL	IDL	LDL	HDL
Density (g/mL)	< 0.95	0.950–1.006	1.006–1.019	1.019–1.063	1.063–1.210
Diameter (Å)	10^3–10^4	300–800	250–350	180–250	50–120
Components (% dry weight)					
Protein	2	8	15	22	40–55
Triacylglycerol	86	55	31	6	4
Free cholesterol	2	7	7	8	4
Cholesterol esters	3	12	23	42	12–20
Phospholipids	7	18	22	22	25–30
Apolipoprotein composition	A-I, A-II,	B-100,	B-100,	B-100, E	A-I, A-II,
	A-IV, B-48,	C-I, C-II,	C-I, C-II,		C-I,
	C-I, C-II,	C-III, E	C-III, E		C-II,
	C-III, E				C-III,
					D, E

Data from A. Jonas (2002) Lipoprotein structure. In *Biochemistry of lipids, lipoproteins and membranes,* 4th ed., D. E. Vance and J. E. Vance, eds., Ch. 18, pp. 483–504, Elsevier, Amsterdam; and R. J. Havel and J. P. Kane (2001) Introduction: Structure and metabolism of plasma lipoproteins. In *The Metabolic and Molecular Bases of Inherited Disease,* C. R. Scriver, A. L. Beaudet, W. S. Sly, D. Valle, B. Childs, K. W. Kinzler, and B. Vogelstein, eds., Vol. II, Ch. 114, pp. 2705–2716, McGraw-Hill, New York.

the hydrophobic parts, including lipid molecules as well as apolar amino acid residues, form an inner core, and hydrophilic protein structures, cholesterol, and polar head groups of phospholipids are on the outside.

Some apolipoproteins have specific biochemical activities other than their roles as passive carriers of lipid from one tissue to another. For instance, apolipoprotein C-II is an activator of triacylglycerol hydrolysis by **lipoprotein lipase,** a cell-surface glycoprotein that hydrolyzes triacylglycerols in lipoproteins. Other apolipoproteins target specific lipoproteins to specific cells by being recognized by receptors in the plasma membranes of these cells. Of great interest is an association of a variant form of apolipoprotein E with increased risk for developing Alzheimer's disease. The mechanism underlying this association is not yet understood, but there is a solid epidemiological link between high serum cholesterol at midlife and Alzheimer's disease in later life, and apolipoprotein E is the most abundant cholesterol transport protein in the central nervous system. There are three common allelic forms of apolipoprotein E (E2, E3, and E4), and possessing at least one E4 allele is the major known genetic risk factor for Alzheimer's disease.

● **CONNECTION** Altered lipoprotein function causes several human diseases. Deficiency of apolipoprotein C-II is associated with massive accumulation of chylomicrons and elevated triacylglycerol levels in blood. A variant form of apolipoprotein E is associated with increased risk for developing Alzheimer's disease.

Following digestion and absorption of a meal, the lipoproteins help maintain in emulsified form some 500 mg of total lipid per 100 mL of human blood. Of this 500 mg, typically about 120 mg is triacylglycerol, 220 mg is cholesterol (one-third free and two-thirds esterified with fatty acids—which renders them even more hydrophobic), and 160 mg is phospholipids, principally phosphatidylcholine and phosphatidylethanolamine. Indeed, following a high-fat meal, chylomicrons are so abundant in blood that they give the plasma a milky appearance.

Transport and Utilization of Lipoproteins

As noted previously, chylomicrons represent the form in which dietary fat is transported from the intestine to peripheral tissues, notably heart, muscle, and adipose tissue (see Figure 16.2). VLDL plays a comparable role for triacylglycerols synthesized in liver. The triacylglycerols in both of these lipoproteins are hydrolyzed to glycerol and fatty acids at the inner surfaces of capillaries in the peripheral tissues. This hydrolysis involves activation of the extracellular enzyme lipoprotein lipase by apolipoprotein C-II, a component of both chylomicron and VLDL (**FIGURE 16.5**). Lipoprotein lipase is a member of the serine esterase family, which includes pancreatic lipase and *hormone sensitive lipase* or *HSL* (an enzyme involved in the regulated mobilization of stored fat from adipose tissue—discussed shortly). This family is characterized by use of a catalytic triad of serine, histidine, and aspartate residues as well as an acyl-enzyme intermediate, similar to the serine proteases described in Chapter 8.

● **CONNECTION** A major consequence of liver dysfunction is an inability to synthesize apolipoproteins and, hence, to transport fat out of the liver.

into the cell, the fatty acids derived from lipoprotein lipase action can be either catabolized to generate energy or, in adipose cells, used to resynthesize triacylglycerols. **FIGURE 16.6** summarizes the overall aspects of lipoprotein transport and metabolism.

As a consequence of triacylglycerol hydrolysis in the capillaries, both chylomicrons and VLDL are degraded to protein- and cholesterol-rich remnants. The IDL class of lipoprotein is derived from VLDL, and chylomicrons are degraded to what are simply called chylomicron remnants. Both classes of remnants are taken up by the liver through interaction with specific receptors and further degraded in liver lysosomes. Apolipoprotein B-100 is reused for synthesis of LDL (via IDL). As described in the next section, LDL is the principal form in which cholesterol is transported to tissues, and HDL plays the primary role in returning excess cholesterol from tissues to the liver for metabolism or excretion. The importance of lipoproteins as transport vehicles is evident from the fact that a major consequence of chronic liver cirrhosis is fatty liver degeneration, where the liver becomes engorged with fat.

Some of the released fatty acids are absorbed by nearby cells, while others, still rather insoluble, become complexed with serum albumin for transport through the blood to more distant cells. After absorption

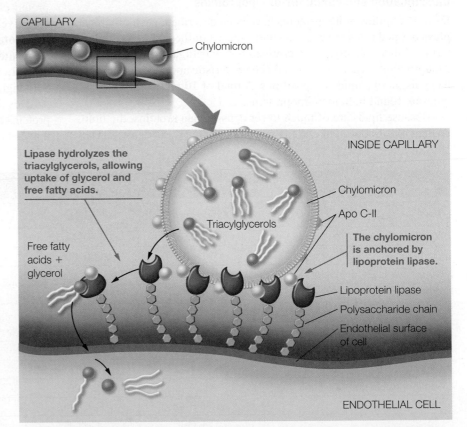

▲ **FIGURE 16.5 Binding of a chylomicron to lipoprotein lipase on the inner surface of a capillary.** The chylomicron is anchored by lipoprotein lipase, which is linked by a polysaccharide chain to the lumenal surface of the endothelial cell. When activated by apolipoprotein C-II (Apo C-II), the lipase hydrolyzes the triacylglycerols in the chylomicron, allowing uptake into the cell of the glycerol and the free fatty acids.

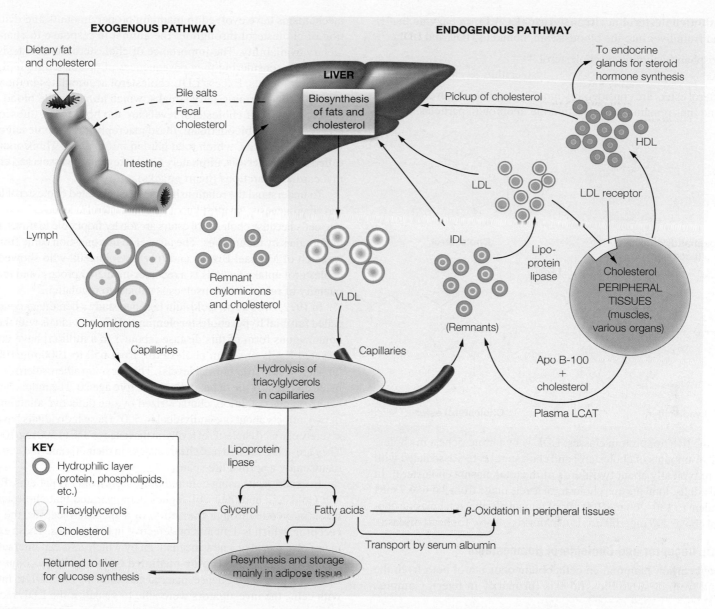

▲ **FIGURE 16.6 Liver orchestrates the lipoprotein transport pathways that redistribute fat and cholesterol throughout the body.** The liver processes and remodels the lipoproteins as they transport lipids obtained from a meal or biosynthesized to peripheral tissues for energy or storage.

Because the liver is the major site of apolipoprotein synthesis, damage to this organ causes endogenously synthesized fat to accumulate there because it cannot be transported to peripheral tissues.

Cholesterol Transport and Utilization in Animals

Cholesterol plays a number of essential roles in cells. As discussed in Section 10.3, cholesterol is critical for maintaining proper membrane fluidity. It is also the precursor to bile acids and steroid hormones. At the same time, an abnormally elevated level of cholesterol in the blood is a primary risk factor for heart disease. Prolonged cholesterol accumulation in the blood contributes to the development of

● **CONNECTION** Cholesterol accumulation in the blood is correlated with development of atherosclerotic plaque and heart disease.

atherosclerotic plaques, fatty deposits that line the inner surfaces of coronary arteries.

As shown in Figure 16.6, cholesterol is obtained both from diet and from de novo biosynthesis, and lipoproteins are involved in the transport of cholesterol from both sources. Dietary cholesterol arrives at the liver via the remnant chylomicrons after they have had most of their triacylglycerols removed in the capillaries (Figure 16.6). The liver then repackages this dietary cholesterol, along with de novo synthesized cholesterol, into VLDL. The VLDL particles are then converted to IDL, and eventually to LDL particles during their transit through the bloodstream.

Recall from Table 16.1 that cholesterol in plasma lipoproteins exists both as the free sterol and as cholesterol esters. Esterification occurs at the cholesterol hydroxyl position with a long-chain fatty acid. Cholesterol esters are synthesized from cholesterol and an acyl chain on phosphatidylcholine (lecithin). This reaction is catalyzed by **lecithin:cholesterol acyltransferase (LCAT),** an enzyme that is secreted from liver into the bloodstream, bound to HDL and LDL:

Phosphatidylcholine + cholesterol ⇌

lysolecithin + cholesterol ester

Cholesterol esters are considerably more hydrophobic than cholesterol itself and thus remain trapped within the lipoprotein particles.

Phosphatidylcholine (Lecithin)

Cholesterol

LCAT

Lysolecithin

Cholesterol ester

Of the five lipoprotein classes, LDL is by far the richest in cholesterol. The amounts of cholesterol and cholesterol esters associated with LDL are typically about two-thirds of the total plasma cholesterol. In normal adults, total plasma cholesterol levels range from 3.5 to 6.7 mM (equivalent to 130–260 mg/100 mL of human plasma; total plasma cholesterol above 200 mg/100 mL is a major risk factor for heart disease).

The LDL Receptor and Cholesterol Homeostasis

As noted earlier, mammalian cells obtain cholesterol both from the diet and from de novo biosynthesis (primarily in liver). Complex mechanisms have evolved to maintain proper amounts and distribution of cholesterol throughout the animal in response to changing dietary availability. The importance of cholesterol homeostasis can be seen by considering the consequences of prolonged high plasma cholesterol levels. Excess LDL cholesterol accumulates in the inner arterial walls, forming fatty streaks, which attract white blood cells (macrophages). If cholesterol levels are too high for its subsequent removal into the bloodstream, these macrophages become engorged with fatty deposits, which then harden into plaque. This condition, called **atherosclerosis,** ultimately blocks key blood vessels and causes myocardial infarctions (heart attacks).

To understand the relationship between elevated cholesterol levels and atherogenesis, we must know how cholesterol is taken up from LDL into cells because cholesterol esters are too hydrophobic to traverse cell membranes by themselves. The answer to this question came from the research of Michael Brown and Joseph Goldstein, who showed that cholesterol uptake by cells is a receptor-mediated process and that the quantity of receptors themselves is subject to regulation.

In 1972, Brown and Goldstein began to study a hereditary condition called **familial hypercholesterolemia,** or **FH.** Individuals with the rare homozygous form of this disease (about 1 in a million) have grossly elevated levels of serum cholesterol, from 650 to 1000 mg/100 mL (about five-fold over normal levels). They develop atherosclerosis early in life and usually die of heart disease before age 20. The more common heterozygous condition, characterized by one defective allele instead of two, affects about one individual in 500. These individuals have less severely elevated cholesterol levels, in the range of 350 to 500 mg/100 mL. They are at high risk to have heart attacks in their 30s and 40s, although many enjoy a normal life span.

In experiments using cultured fibroblasts from FH patients, Brown and Goldstein and their colleagues demonstrated that cholesterol is taken into cells through the action of a specific receptor, the **LDL receptor,** which is deficient or defective in FH patients. These experiments also revealed a new mechanism by which cells can interact with their environment—**receptor-mediated endocytosis.** By conjugating LDL with an electron-dense material and allowing this LDL to interact with cells, the investigators were able to visualize the LDL receptor on cell surfaces (**FIGURE 16.7**). These experiments showed that the receptors are clustered in a structure called a **coated pit,** an invagination whose most abundant intracellular protein is **clathrin,** a self-interacting protein capable of forming a cage-like structure (**FIGURE 16.8**).

Endocytosis is a process by which cells take up large molecules from the extracellular environment. Although LDL uptake involves a cell-surface receptor, the interaction of LDL with its receptor is unlike the interaction of hormones such as epinephrine with their receptors. As discussed in Chapters 11 and 12, the binding of epinephrine at its receptor in the plasma membrane triggers intracellular metabolic changes, but the hormone itself does not enter the cell. By contrast, when LDL binds to its receptor, the entire LDL–receptor complex is engulfed and taken into the cell (endocytosed), as schematized in **FIGURE 16.9**. Binding to the LDL receptor occurs through recognition of the B-100 apolipoprotein, the primary protein

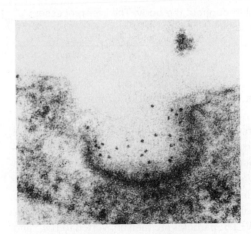

(a) The LDL–ferritin (dark dots) binds to a coated pit on the surface of a cultured human fibroblast (a type of connective tissue cell).

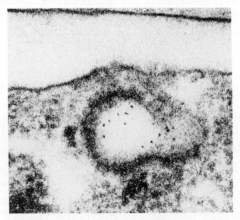

(b) The plasma membrane closes over the coated pit, forming an endocytic vesicle.

▲ **FIGURE 16.7 Receptor-mediated endocytosis of LDL.** Low-density lipoprotein (LDL) was conjugated with ferritin to permit electron microscopic visualization.

(a) Clathrin, the major protein in coated pits, forms triskelions (named after the symbol of three legs radiating from the center), which assemble into polyhedral lattices composed of hexagons and pentagons, such as the barrel shown in the next panel.

(b) Image reconstruction from electron cryomicroscopy of a clathrin barrel formed from 36 triskelions. A single clathrin triskelion is highlighted in yellow.

(c) A coated pit on the inner surface of the plasma membrane of a cultured mammalian cell is visualized by freeze-fracture electron microscopy. The cagelike structure of the pit is due to the clathrin lattice.

100 nm

▲ **FIGURE 16.8 Structure of a clathrin-coated pit.**

[handwritten annotations:] omega-3 lowers cholestrol serum — cholestrol goes into cell   Saturated fats ↑ serum cholestrol

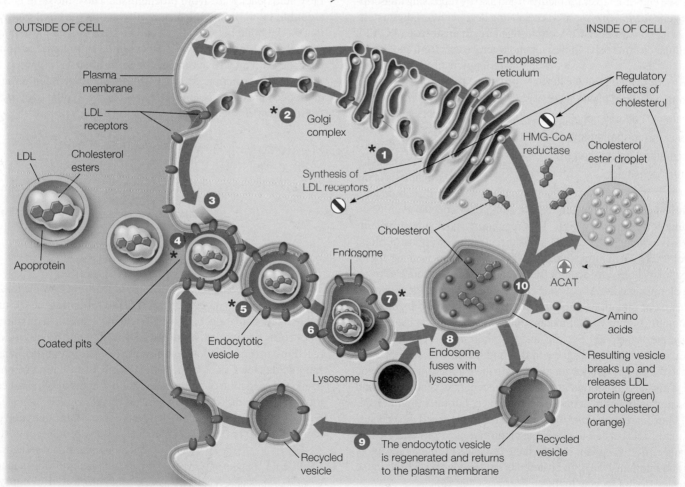

OUTSIDE OF CELL

INSIDE OF CELL

Plasma membrane

LDL receptors

LDL

Cholesterol esters

Apoprotein

Coated pits

Endocytotic vesicle

Lysosome

Recycled vesicle

* ❷ Golgi complex

* ❶

Synthesis of LDL receptors

❸

❹ *

* ❺

❻

❼ *

Endosome

❽ Endosome fuses with lysosome

❾ The endocytotic vesicle is regenerated and returns to the plasma membrane

Endoplasmic reticulum

HMG-CoA reductase

Cholesterol

❿

ACAT

Regulatory effects of cholesterol

Cholesterol ester droplet

Amino acids

Resulting vesicle breaks up and releases LDL protein (green) and cholesterol (orange)

Recycled vesicle

▲ **FIGURE 16.9 Involvement of LDL receptors in cholesterol uptake and metabolism.** LDL receptors are synthesized in the endoplasmic reticulum ❶ and mature in the Golgi complex ❷. They then migrate to the cell surface, where they cluster in clathrin-coated pits ❸. LDL, made up of cholesterol esters and apolipoprotein, binds to the LDL receptors ❹ and is internalized in endocytotic vesicles ❺. Several such vesicles fuse to form an organelle called an endosome ❻. Proton pumping in the endosome membrane causes the pH to drop, which in turn causes LDL to dissociate from the receptors ❼. The endosome fuses with a lysosome, ❽ and the receptor-bearing clathrin coat dissociates and returns to the membrane ❾. The LDL particle is degraded in the lysosomes, ❿ and cholesterol has various fates. Regulatory targets of cholesterol are shown in red. ACAT, acyl-CoA:cholesterol acyltransferase. Red asterisks indicate steps affected by mutations in the LDL receptor gene.

● **CONCEPT** Uptake of cholesterol from the blood occurs at the LDL receptor via receptor-mediated endocytosis.

component of the LDL particle. The plasma membrane fuses in the vicinity of the LDL–receptor complex, and the coated pit becomes an endocytic vesicle. Several of these clathrin-lined vesicles fuse to form an **endosome.** The endosome then fuses with a lysosome, putting the LDL–receptor complex in contact with the hydrolytic enzymes of the lysosome. The LDL apolipoprotein is hydrolyzed to amino acids, and the cholesterol esters are hydrolyzed to give free cholesterol. The receptor itself is recycled, moving back to the plasma membrane to pick up more LDL. About 10 minutes is required for each round trip. The discovery of the LDL receptor and receptor-mediated endocytosis earned Brown and Goldstein the Nobel Prize for Physiology or Medicine in 1985.

Much of the cholesterol released in the lysosome moves to the endoplasmic reticulum, where it is used for membrane synthesis. The internalized cholesterol exerts three regulatory effects on the cell (Figure 16.9). (1) It reduces endogenous cholesterol synthesis, by inhibiting HMG-CoA reductase (see end-of-chapter Problem 30) and also by repressing transcription of the gene for this enzyme and accelerating degradation of the enzyme protein. (2) It activates **acyl-CoA:cholesterol acyltransferase (ACAT),** an intracellular enzyme that synthesizes cholesterol esters from cholesterol

● **CONCEPT** Intracellular cholesterol regulates its own level inside the cell by controlling (1) de novo cholesterol biosynthesis, (2) formation and storage of cholesterol esters, and (3) LDL receptor density.

and a long-chain acyl-CoA. This promotes the storage of excess cholesterol in the form of droplets of cholesterol esters. (3) It regulates the synthesis of the LDL receptor itself by decreasing transcription of the receptor gene. Decreased synthesis of the receptor ensures that cholesterol will not be taken into the cell in excess of the cell's needs, even when extracellular levels are very high. This regulatory mechanism explains why excessive dietary cholesterol leads directly to elevations of blood cholesterol levels. With intracellular cholesterol levels so well regulated, the extracellular cholesterol accumulates because it has nowhere else to go.

Gene cloning and DNA sequence analysis have allowed identification of five classes of mutations affecting the LDL receptor and its metabolism in humans (Figure 16.9). First are mutations that lead to insufficient receptor synthesis (step ❶). Second, and most common, are mutations in which the receptor is synthesized but fails to migrate from the endoplasmic reticulum to the Golgi complex for transport to the cytoplasmic membrane (step ❷). Third are mutations in which the receptor is synthesized and processed normally and reaches the cell surface, but fails to bind LDL (step ❹). Fourth are mutations in which the receptors reach the cell surface and bind LDL but fail to cluster in clathrin-coated pits and thus do not internalize LDL (step ❺). Finally, there is a class of mutant receptors that bind and internalize LDL in coated pits but fail to release LDL in the endosome and do not recycle to the cell surface (step ❼).

Receptor-mediated endocytosis is now known to be a widely used pathway for internalization of extracellular substances, including other lipoproteins, cell growth factors, the iron-binding protein transferrin, some vitamins, and even viruses.

Cholesterol, LDL, and Atherosclerosis

Thanks principally to the work of Brown and Goldstein, we now know a great deal about the genetic and biochemical factors that control

serum cholesterol levels, and overwhelming epidemiological evidence now links prolonged hypercholesterolemia to the development of atherosclerotic plaque. Much, however, remains to be learned. We don't know, for example, why diets rich in saturated fatty acids tend to elevate serum cholesterol levels. Nor do we know why a particular class of naturally occurring polyunsaturated fatty acids (PUFAs) called ω-3 (**omega-3**) **fatty acids** tends to depress levels of both serum cholesterol and triacylglycerols and slow the buildup of atherosclerotic

● **CONNECTION** ω-3 polyunsaturated fat (PUFA) ingestion is correlated with low plasma cholesterol levels. The mechanisms involved are not completely understood.

plaques. But nutritionists have found that adding to a Western diet fish or fish oils, which are abundant in ω-3 fatty acids, does indeed have this effect. That is why it has been recommended that fish be consumed in place of red meat, which tends to be rich in both saturated fatty acids and cholesterol. The most prominent ω-3 fatty acid is linolenic acid, which is an 18:3cΔ9,12,15 fatty acid (see Chapter 10; as discussed there, nutritionists number fatty acids backward from biochemists. Thus, the term ω-3 refers to a double bond on the third carbon from the terminal methyl group, that is, the bond between C-15 and C-16 in this 18-carbon molecule; see below). Linolenic acid is an essential fatty acid for humans; we lack the enzymes to synthesize it, so it must be obtained from the diet. Two other important ω-3 PUFAs—eicosapentaenoic acid (EPA) and docosahexaenoic acid (DHA)—can be synthesized by humans from dietary linolenic acid.

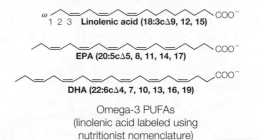

Omega-3 PUFAs
(linolenic acid labeled using nutritionist nomenclature)

Progress is being made, however, in learning how elevated cholesterol levels lead to atherogenesis (plaque formation). LDL undergoes rather ready oxidation, both in cells and in plasma, to a mixture of molecules collectively called **oxidized LDL.** Although the specific oxidation reactions are not well defined, they include peroxidation of unsaturated fatty acids (see Chapter 14, page 480), hydroxylation of cholesterol itself, and oxidation of amino acid residues in the apolipoprotein. Accumulation of oxidized LDL on the vessel wall triggers an inflammation response in which the endothelial cells express adhesion molecules that recruit monocytes and T lymphocytes. Some of these cells differenti-

● **CONCEPT** Uptake of oxidized LDL by scavenger receptors on macrophages is a key event in atherogenesis.

ate into macrophages that take up the lipids that accumulate at sites of arterial injury. Uptake occurs through one of a family of **scavenger receptors;** these receptors take up many substances in addition to oxidized LDL. Unlike the LDL receptor, the scavenger receptor is not downregulated by cholesterol, so cholesterol uptake into these cells is virtually unlimited, which converts these cells to a cholesterol-engorged species called a **foam cell.** These events have a chemotactic effect, causing more white cells to migrate to the site and leading them to accumulate more cholesterol, which ultimately becomes one of the chief chemical constituents of the plaque that forms at such a site.

TV advertisements talk about "bad cholesterol" and "good cholesterol." These are actually inappropriate terms because cholesterol itself is a natural metabolite, an essential component of all membranes, and the precursor to all steroid hormones and bile acids. However, cholesterol present in LDL is considered "bad" because prolonged elevation of LDL levels is what leads to atherosclerosis. By contrast, cholesterol in HDL is called "good" because high levels of HDL counteract atherogenesis. Cholesterol cannot be metabolically degraded, and excess cholesterol is returned from peripheral cells to the liver, for passage through the bile to the intestine, for ultimate excretion (Figure 16.6). As the agent for this transport back to the liver, HDL plays a role in lowering total serum cholesterol levels, which is "good."

Mobilization of Stored Fat for Energy Generation

In general, the capacity of animal storage depots to store fat is virtually unlimited. Whatever appears in the body from the diet is absorbed, and most of it is transported to adipose tissue for storage. The lack of control of this storage process is sadly evident from the prevalence of obesity among humans. By contrast, the release of fat from storage depots in adipose tissue is controlled hormonally to meet the needs of the organism for energy generation.

The catabolism of fat (**lipolysis**) begins with the hydrolysis of triacylglycerol to yield glycerol plus free fatty acid (often abbreviated FFA). About 95% of the energy derived from subsequent oxidation of the fat comes from the fatty acids, with only 5% coming from glycerol. All of the carbons from fatty acids are catabolized to two-carbon fragments, as acetyl-coenzyme A, except for the small proportion of fatty acids that contain odd-numbered chains.

The release of metabolic energy stored in triacylglycerols is comparable to the mobilization of carbohydrate energy stored in animal glycogen, in that the first step of fat breakdown—its hydrolysis to glycerol and fatty acids—is hormonally regulated. Three lipolytic enzymes are now known to participate in this process: **triacylglycerol lipase,** also called **hormone-sensitive lipase (HSL), adipose triglyceride lipase (ATGL),** and **monoacylglycerol lipase (MGL).** All three of these enzymes are serine esterases, catalyzing the hydrolysis of the ester linkage between the glycerol backbone and a fatty acid using a serine nucleophile to form the acyl-enzyme intermediate. HSL and MGL each possess the classic catalytic triad of serine, aspartate, and histidine, whereas ATGL has a catalytic dyad of serine and aspartate. Although these enzymes catalyze the same chemical reaction, their preferences for hydrolyzing triacylglycerol (TG), diacylglycerol (DG), and monoacylglycerol (MG) differ. ATGL catalyzes the first step in TG mobilization, generating DG and FFA. HSL catalyzes hydrolysis of DG, generating MG and FFA, and MGL releases the third FFA from the glycerol backbone (**FIGURE 16.10**).

How is lipolysis regulated? HSL and ATGL activities are controlled by hormones that bind to plasma membrane receptors, leading to activation of adenylate cyclase, as described in Chapter 12. This cyclic AMP/protein kinase A-dependent signal transduction cascade results in the phosphorylation of several target proteins in the adipose cell, causing

● **CONCEPT** Fat mobilization in adipose cells is hormonally controlled, via the cyclic AMP-dependent phosphorylation of lipolytic enzymes and lipid droplet-associated proteins.

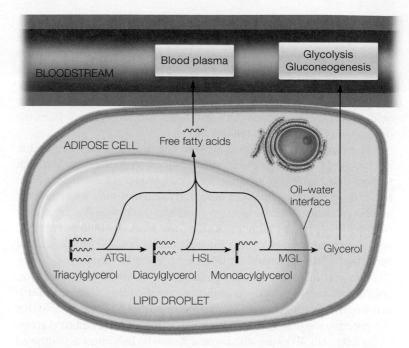

▲ **FIGURE 16.10 Mobilization of adipose cell triacylglycerols by lipolysis.** Three lipases act sequentially to hydrolyze triacylglycerol (TG) to glycerol and free fatty acids (FFA). These enzymes act at the oil–water interface of the lipid droplet. FFA are exported to the blood plasma, where they are bound to albumin for transport to liver and other tissues for subsequent oxidation. Glycerol is released to the blood to be taken up by liver cells, where it serves as a gluconeogenic substrate. ATGL, adipose triglyceride lipase; HSL, hormone-sensitive lipase; MGL, monoacylglycerol lipase.

activation of HSL and ATGL. Depending on the physiological state, glucagon, epinephrine, parathyroid hormone, thyrotropin, or adrenocorticotropin can all stimulate lipolysis. In adipose tissue, the primary hormonal effects are mediated by epinephrine in stress situations and by glucagon during fasting.

The free fatty acid hydrolysis products exit the adipocyte by passive diffusion and find their way to the blood plasma, where the fatty acids become bound to **albumin.** Each molecule of albumin can bind up to 10 molecules of free fatty acid, although the actual amount bound is usually far lower. Fatty acids are released from albumin and taken up by tissues largely by passive diffusion, so that fatty acid uptake into cells is driven primarily by concentration. Most of the glycerol released to the bloodstream is taken up by liver cells, where it serves as a gluconeogenic substrate, leading to the production of glucose (see Figure 16.6).

16.2 Fatty Acid Oxidation

In the preceding sections, we have discussed how fat is obtained from the diet and distributed to the tissues. We are now ready to learn how these fatty acids are oxidized to yield energy.

Early Experiments

The nature of the pathway by which fatty acids are oxidized was revealed beginning in 1904, in a brilliant series of experiments by the German chemist Franz Knoop. The experiments involved the first known use of metabolic tracers, more than 40 years before radioactive tracers became available. Knoop fed dogs a series of fatty acids in which the terminal methyl group was derivatized with a phenyl group. The expectation was that these analogs would follow metabolic pathways similar to those used for oxidizing normal fatty acids. Knoop found that when the derivatized fatty acid had an even-numbered carbon chain, the final breakdown product, recovered from urine, was phenylacetic acid. When the derivatized fatty acid had an odd-numbered chain, the product was benzoic acid (**FIGURE 16.11**).

These results led Knoop to propose that fatty acids are oxidized in a stepwise fashion, with initial attack on carbon 3 (the β-carbon with respect to the carboxyl group). This attack would release the terminal two carbons, and the remainder of the fatty acid molecule could undergo another oxidation. Release of a two-carbon fragment would occur at each step in the oxidation. With the analogs, the process would be repeated until the remaining acid, either phenylacetic or benzoic acid, could not be further metabolized and would be excreted in the urine.

The next major development came in the 1940s, when Luis Leloir and Albert Lehninger independently demonstrated fatty acid oxidation in cell-free liver homogenates. Lehninger showed that ATP was essential for this process, suggesting that ATP somehow activates the carboxyl group of the fatty acid. Working with Eugene Kennedy, Lehninger also showed that the process occurs in mitochondria and that it releases two-carbon fragments that are oxidized in the citric acid cycle. In Munich, Feodor Lynen demonstrated that the ATP-dependent activation esterifies the fatty acid carboxyl group with the thiol group of coenzyme A, and it was later

shown that all of the intermediates in the subsequent oxidative reactions are fatty acyl-CoA thioesters. Thus, by the mid-1950s, the basic outlines of the fatty acid oxidation pathway were clear. As we will discuss in the next sections, the pathway consists of activation of the carboxyl group, transport into the mitochondrial matrix, and stepwise oxidation of the carbon chain, two carbons at a time, from the end containing the carboxyl group. This process of fatty acid oxidation is summarized in **FIGURE 16.12**.

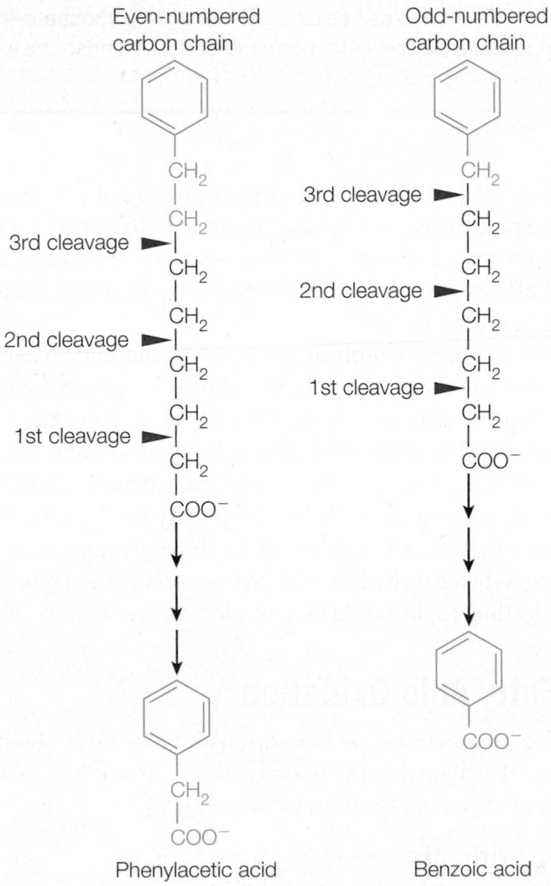

▲ **FIGURE 16.11 Oxidation of phenyl derivatives of fatty acids in Knoop's experiment.** Red triangles represent presumed sites of cleavage of these model fatty acids.

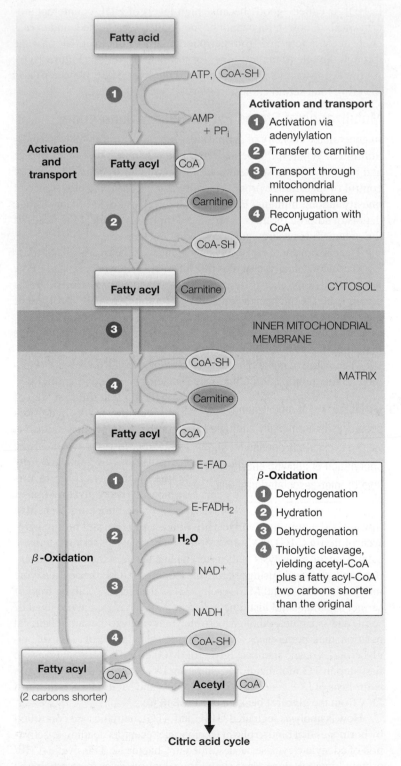

▲ **FIGURE 16.12** Overview of the fatty acid oxidation pathway.

Fatty Acid Activation and Transport into Mitochondria

Fatty acids arise in the cytosol from three processes: de novo biosynthesis (as discussed later in this chapter); hydrolysis of triacylglycerols stored in the cell; and transport of fatty acids from fat depots from elsewhere in the body. These fatty acids must be transported into the mitochondrial matrix for oxidation. Because the inner membrane is impermeable to free long-chain fatty acids and acyl-CoAs, a specific transport system comes into play. That transport system operates hand in hand with the metabolic activation needed to initiate the β-oxidation pathway. Fatty acids with 10 or fewer carbons can simply diffuse through the inner mitochondrial membrane, bypassing the need for a specific transport system.

A series of **acyl-CoA synthetases,** specific for short-chain, medium-chain, or long-chain fatty acids, catalyze formation of the fatty acyl thioester conjugate with coenzyme A (step ❶ in the upper part of Figure 16.12):

$$\text{R—COO}^- + \text{ATP} + \text{CoA—SH} \underset{\text{synthetase}}{\overset{\text{acyl-CoA}}{\rightleftharpoons}}$$

Fatty acid   Coenzyme A

$$\text{R—C(=O)—S—CoA} + \text{AMP} + \text{PP}_i \qquad \Delta G°' \approx -15 \text{ kJ/mol}$$

Fatty acyl thioester

The long-chain enzyme, which plays the predominant role in initiating fatty acid oxidation, acts on fatty acids with chain lengths of 10 to 20 carbons; the medium-chain enzyme acts on 4- to 12-carbon chains; and the short-chain enzyme prefers acetate and propionate. The synthetase specific for long-chain acids is a membrane-bound enzyme, found in both the endoplasmic reticulum and the outer mitochondrial membrane; the short-chain and medium-chain enzymes are found primarily in the mitochondrial matrix.

Chemically, the energy-rich thioester link in long-chain fatty acyl-CoAs is identical to that of acetyl-CoA (see Chapter 13). Recall that pyruvate oxidation provides the energy to drive acetyl-CoA formation in the pyruvate dehydrogenase reaction. The acyl-CoA synthetases, on the other hand, use a two-step mechanism involving cleavage of ATP to drive the endergonic thioester formation (**FIGURE 16.13**). First comes activation of the carboxyl group by ATP to

● **CONCEPT** Fatty acids are activated for oxidation by ATP-dependent acylation of coenzyme A.

give a **fatty acyl adenylate,** with concomitant release of pyrophosphate. Next, the activated carboxyl group is attacked by the nucleophilic thiol group of CoA, thereby displacing AMP and forming the fatty acyl-CoA derivative. Carboxyl groups of amino acids are activated for protein synthesis in similar fashion.

Although each fatty acyl-CoA, like ATP itself, is an energy-rich compound ($\Delta G°'$ of hydrolysis ~ -30 kJ/mol), cleavage of ATP to AMP ($\Delta G°' = -45.6$ kJ/mol; Figure 3.7) provides the driving force for formation of the fatty acyl-CoA. The reaction is made essentially irreversible because of the active pyrophosphatase present in most cells:

$$\text{PP}_i + \text{H}_2\text{O} \rightleftharpoons 2\text{P}_i \qquad \Delta G°' = -19.2 \text{ kJ/mol}$$

Thus, the overall reaction (the sum of the two previous reactions) proceeds far in the direction of completion, with a net $\Delta G°'$ of about -35 kJ/mol. Because ATP is hydrolyzed to AMP rather than ADP, activation of each fatty acid requires the equivalent of two ATPs.

Long-chain fatty acyl-CoAs are formed on the outer mitochondrial membrane (**FIGURE 16.14**). Hence, they must move through the inner

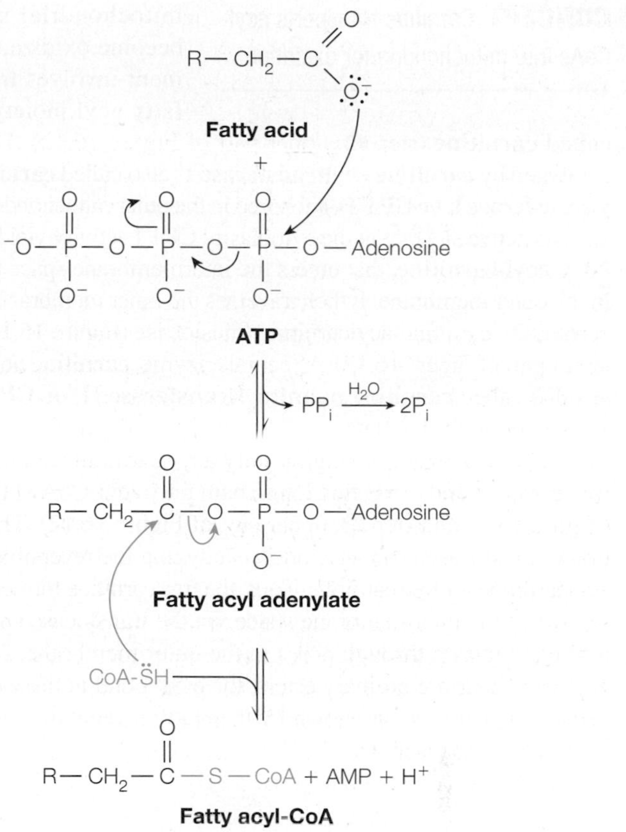

▲ **FIGURE 16.13 Mechanism of acyl-CoA synthetase reactions.** The figure shows reversible formation of the activated fatty acyl adenylate, nucleophilic attack by the thiol sulfur of CoA-SH on the activated carboxyl group, and the quasi-irreversible pyrophosphatase reaction, which draws the overall reaction toward fatty acyl-CoA.

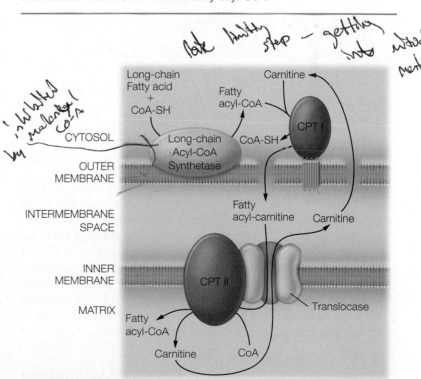

▲ **FIGURE 16.14 The carnitine acyltransferase system, for transport of long-chain fatty acyl-CoAs into mitochondria.**

CONCEPT Carnitine transports acyl-CoAs into mitochondria for oxidation.

mitochondrial membrane to become oxidized. This movement involves transfer of the fatty acyl moiety to a carrier called **carnitine** (step ❷, upper part of Figure 16.12). The reaction is catalyzed by **carnitine acyltransferase I** (also called **carnitine palmitoyltransferase I, or CPT I**), anchored in the outer mitochondrial membrane with its active site facing the cytoplasm. CPT I activity yields a derivative, **fatty acyl-carnitine,** that enters the intermembrane space through pores in the outer membrane. It then traverses the inner membrane via a specific carrier, the carnitine-acylcarnitine translocase (Figure 16.14 and step ❸, upper part of Figure 16.12). A second enzyme, **carnitine acyltransferase II** (also called **carnitine palmitoyltransferase II, or CPT II**), loosely associated with the matrix side of the inner membrane, completes the transfer process by exchanging fatty acyl-carnitine and CoA, releasing free carnitine and producing long-chain fatty acyl-CoA within the matrix (Figure 16.14 and step ❹, upper part of Figure 16.12). The inner membrane translocase is an antiporter, catalyzing the reversible exchange of acylcarnitine for free carnitine. Thus, the free carnitine formed in the matrix returns to the intermembrane space via the translocase and then moves to the cytoplasm through pores in the outer membrane. Although fatty acyl-carnitines are ordinary esters, the ester bond in these compounds is somewhat activated, as shown by the ready reversibility of the carnitine acyltransferase reactions.

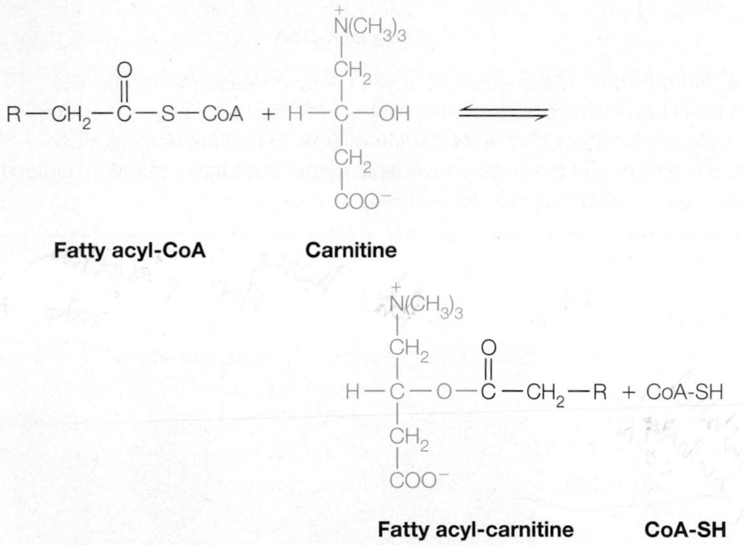

What is the point of this rather complex shuttling process? It exists to regulate fatty acid oxidation, preventing the futile cycle that would occur if oxidation and resynthesis were taking place at the same time. Carnitine acyltransferase I is strongly inhibited by **malonyl-CoA,** the first committed intermediate in fatty acid synthesis (page 533). CPT II is insensitive to malonyl-CoA. Indeed, entry of fatty acids into mitochondria is rate-limiting for β-oxidation and is the primary point of regulation. Thus, conditions in the cell that favor fatty acid synthesis prevent the transfer of fatty acyl moieties to their intracellular sites of oxidation and, hence, prevent that oxidation.

The β-Oxidation Pathway

Once inside the mitochondrial matrix, fatty acyl-CoAs are oxidized as predicted by Knoop, with oxidation of the β-carbon in a series of steps in which the fatty acyl chain is shortened by two carbons at a time. Release of each two-carbon fragment, as acetyl-CoA, involves four reactions (**FIGURE 16.15** and steps ❶–❹ in the lower part of Figure 16.12).

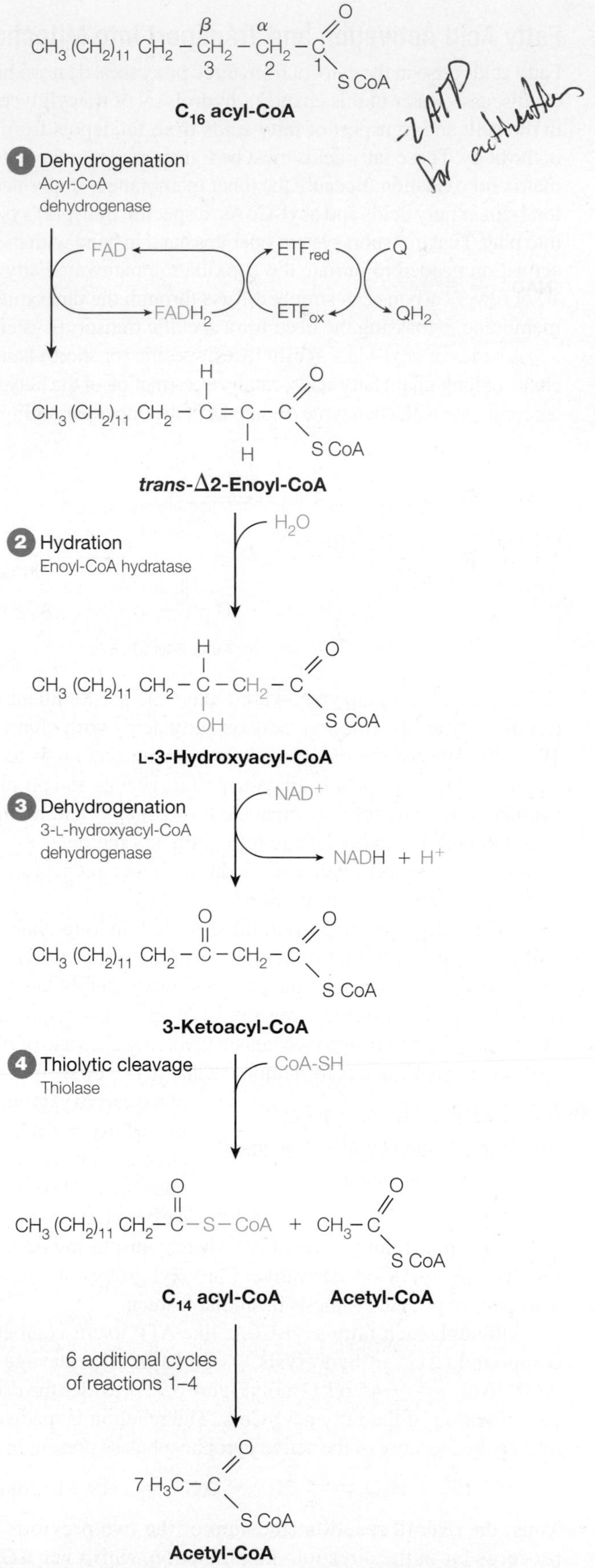

▲ **FIGURE 16.15 Outline of the β-oxidation of fatty acids.** In the diagram, a 16-carbon saturated fatty acyl-CoA (palmitoyl-CoA) undergoes seven cycles of oxidation to yield eight molecules of acetyl-CoA. These reactions correspond to steps ❶–❹ in the lower part of Figure 16.12.

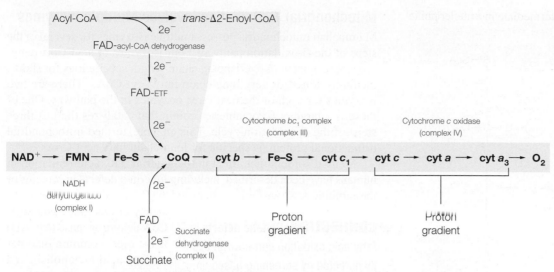

▲ **FIGURE 16.16 Fate of reducing equivalents derived from fatty acyl-CoA dehydrogenation.** Enzyme-bound FAD becomes reduced and then transfers its electrons to ETF, which in turn passes them to coenzyme Q, which is also a collection point for electrons from NADH dehydrogenase and succinate dehydrogenase.

bond into a carbonyl compound, and like most enzymes that catalyze this type of oxidation, acyl-CoA dehydrogenase uses a tightly bound FAD prosthetic group.

As shown in **FIGURE 16.16**, the enzyme-bound $FADH_2$ that is formed contributes a pair of electrons to a shuttle protein, the **electron-transferring flavoprotein (ETF)**. These electrons are passed in turn to coenzyme Q via ETF-Q oxidoreductase, an integral membrane protein, and are then shuttled along the respiratory chain, yielding ATP via oxidative phosphorylation. In this respect, ETF-Q oxidoreductase is comparable to NADH dehydrogenase and succinate dehydrogenase. All three are flavoproteins that transfer electrons to the mobile electron carrier coenzyme Q (see Figure 14.12).

The pathway is cyclic in that each 4-reaction step ends with formation of an acyl-CoA, shortened by two carbons, which undergoes the same process in the next step, or cycle. For example, 1 mole of palmitoyl-CoA, derived from a 16-carbon fatty acid, undergoes seven cycles of oxidation to give 8 moles of acetyl-CoA. Each cycle releases 1 two-carbon unit, concomitant with 2 two-electron oxidation–reduction reactions. Because each cycle results in oxidation of the β-carbon (the α-carbon remains at the methylene oxidation state), the pathway is called β**-oxidation.**

Mechanistically, this pathway is remarkably similar to that used to oxidize succinate in the citric acid cycle (compare with Figure 13.4). As shown in Figure 16.15, each cycle in the oxidation of a saturated fatty acyl-CoA involves the following reactions: ❶ dehydrogenation to give an enoyl derivative; ❷ hydration of the resultant double bond, with the β-carbon undergoing hydroxylation; ❸ dehydrogenation of the hydroxyl group (oxidation to a ketone); and ❹ cleavage by attack of a second molecule of coenzyme A on the β-carbon, to release acetyl-CoA and a fatty acyl-CoA two carbons shorter than the original substrate. Oxidation of unsaturated fatty acyl-CoAs is slightly different, as will be discussed shortly.

Acetyl-CoA from β-oxidation enters the citric acid cycle, where it is oxidized to CO_2 in the same fashion as the acetyl-CoA derived from the oxidation of pyruvate. Like the citric acid cycle, β-oxidation generates reduced electron carriers, whose reoxidation in the mitochondria generates ATP via oxidative phosphorylation from ADP. Now let us describe the individual reactions in detail (refer to Figure 16.15).

Reaction ❶: The Initial Dehydrogenation

The first reaction is catalyzed by an **acyl-CoA dehydrogenase,** which catalyzes the removal of two hydrogen atoms from the α- and β-carbons to give a *trans* α, β-unsaturated acyl-CoA (*trans*-2-enoyl-CoA) as the product. This oxidation introduces a conjugated double

Reactions ❷ and ❸: Hydration and Dehydrogenation

Like succinate oxidation, an initial FAD-dependent fatty acyl-CoA oxidation is followed by hydration and an NAD^+-dependent dehydrogenation. In β-oxidation, the latter two reactions are catalyzed by **enoyl-CoA hydratase** and **L-3-hydroxyacyl-CoA dehydrogenase,** respectively (Figure 16.15). Both reactions are stereospecific. Because carbon 3 is β with respect to the carboxyl carbon, the products of these two reactions are sometimes called L-β-hydroxyacyl-CoA and β-ketoacyl-CoA, respectively. Hence the term β-*oxidation*.

Reaction ❹: Thiolytic Cleavage

The fourth and last reaction in each cycle of the β-oxidation pathway involves attack of the nucleophilic thiol sulfur of coenzyme A on the electron-poor keto (β) carbon of 3-ketoacyl-CoA, with cleavage of the α—β bond and release of acetyl-CoA. The other product is a shortened fatty acyl-CoA, ready to begin a new cycle of oxidation:

3-Ketoacyl-CoA

Acyl-CoA   **Acetyl-CoA**

Because this reaction involves cleavage by a thiol, it is referred to as a **thiolytic cleavage,** by analogy with hydrolysis, which involves cleavage by water. The enzyme is commonly called β**-ketothiolase** or simply **thiolase.** An essential nucleophilic cysteine thiol group on the enzyme (E-SH) attacks the substrate, with formation of an acyl-enzyme intermediate and acetyl-CoA in a retro-Claisen reaction (see Chapter 11, page 350). Free CoA-SH then attacks the intermediate in a nucleophilic acyl substitution reaction.

Mechanism of thiolytic cleavage

As noted earlier, the overall oxidation pathway as just described is applicable to the most abundant fatty acids—those that contain even numbers of carbon atoms and are fully saturated. Shortly, we shall

● **CONCEPT** Fatty acids are oxidized by repeated cycles of dehydrogenation, hydration, dehydrogenation, and thiolytic cleavage, with each cycle yielding acetyl-CoA and a fatty acyl-CoA shorter by two carbons than the input acyl-CoA.

describe the variations in this pathway that permit oxidation of other fatty acids. For the saturated, even-chain fatty acyl-CoAs, oxidation simply proceeds stepwise, with two carbons lost as acetyl-CoA after each cycle. For the C_{16} palmitoyl-CoA, the example shown in Figure 16.15, the first cycle yields acetyl-CoA plus the C_{14} myristoyl-CoA. A second cycle, acting on the latter substrate, yields acetyl-CoA plus the C_{12} lauroyl-CoA. In the seventh and last cycle, the 3-hydroxyacyl-CoA dehydrogenase reaction yields acetoacetyl-CoA. Thiolytic cleavage of this substrate yields 2 moles of acetyl-CoA (see below). Thus, the oxidation of 1 mole of palmitic acid involves six successive cycles, each of which yields 1 mole of acetyl-CoA, and a seventh cycle, which yields 2 moles. Other saturated even-chain fatty acids are degraded identically. For example, stearic acid oxidation involves eight cycles, with two acetyl-CoAs resulting from the last cycle.

The final thiolytic cleavage

Mitochondrial β-Oxidation Involves Multiple Isozymes

Mammalian mitochondria possess multiple isozymes for several of the steps of the β-oxidation pathway. Four isozymes of acyl-CoA dehydrogenase exist with overlapping chain-length specificities for short-, medium-, long-, or very long-chain fatty acyl-CoAs. There are two isozymes for each of the next three enzymes in the pathway. One of these is a large $\alpha_4\beta_4$ octameric protein that catalyzes the last three steps of the β-oxidation cycle. This enzyme, termed mitochondrial trifunctional protein, is specific for long-chain fatty acyl-CoAs.

Many disorders that affect mitochondrial fatty acid oxidation in humans have been described, including inherited defects in proteins of the carnitine acyltransferase system and the β-oxidation enzymes. Defi-

● **CONNECTION** Genetic defects in fatty acid oxidation can now be easily detected by screening based on mass spectrometry.

ciency of medium-chain acyl-CoA dehydrogenase (MCAD) is the most common disorder of fatty acid metabolism and has been associated with some cases of sudden infant death syndrome (SIDS). Children with fatty acid oxidation disorders usually present within the first year of life with recurring episodes of fatty liver (steatosis), high blood levels of fatty acid intermediates, and hypoglycemia, and may also exhibit skeletal and cardiac myopathies. Improvements in mass spectrometry (see Tools of Biochemistry 11A) have facilitated the screening of newborn infants using blood spots, allowing affected infants to be detected before the onset of symptoms. This is important because many of these disorders can be managed by relatively simple dietary means.

Energy Yield from Fatty Acid Oxidation

We can now write a balanced equation for the overall degradation of palmitoyl-CoA to 8 moles of acetyl-CoA:

$$\text{Palmitoyl-CoA} + 7\,\text{CoA-SH} + 7\,\text{FAD} + 7\,\text{NAD}^+ + 7\,\text{H}_2\text{O} \longrightarrow$$
$$8\,\text{acetyl-CoA} + 7\,\text{FADH}_2 + 7\,\text{NADH} + 7\,\text{H}^+$$

Each of the products is metabolized exactly as described earlier for oxidation of carbohydrates. Acetyl-CoA is catabolized via the citric acid cycle, and $FADH_2$ and NADH transfer electrons to the respiratory chain through ETF (page 527) and complex I, respectively (Figure 16.16). Thus, we can easily compute the metabolic energy yield from fatty acid oxidation in terms of moles of ATP synthesized from ADP. Recall from Chapter 14 that the P/O ratios for oxidation of flavoproteins and NADH are ~1.5 and ~2.5, respectively, and oxidation of acetyl-CoA in one turn of the citric acid cycle yields 10 ATPs. The following summation, using palmitate as an example, gives the total energy yield:

Reaction	ATP Yield
Activation of palmitate to palmitoyl-CoA	−2
Oxidation of 8 acetyl-CoA	$8 \times 10 = 80$
Oxidation of 7 $FADH_2$	$7 \times 1.5 = 10.5$
Oxidation of 7 NADH	$7 \times 2.5 = 17.5$
Net: Palmitate $\longrightarrow$ CO_2 + H_2O	106

From this, you can calculate the ATP yield per carbon oxidized to CO_2 as 106/16, or about 6.6. The corresponding value for glucose is 5–5.3 (30–32 ATPs formed per 6 carbons oxidized). Thus, the energy yield from fat oxidation is higher than that from oxidation of the less highly reduced carbohydrate, whether measured on a weight basis (see Chapter 11, page 351) or a molar basis.

Oxidation of Unsaturated Fatty Acids

Recall from Chapter 10 that many fatty acids in natural lipids are unsaturated; that is, they contain one or more double bonds (see Table

● **CONCEPT** Two enzymes, enoyl-CoA isomerase and 2,4-dienoyl-CoA reductase, play essential roles in the oxidation of unsaturated fatty acids.

10.1). Because these bonds are usually in the *cis* configuration in natural lipids, they cannot be hydrated by enoyl-CoA hydratase, which acts only on *trans* compounds. Two additional enzymes, **enoyl-CoA isomerase** and **2,4-dienoyl-CoA reductase,** must come into play for unsaturated fatty acids to be oxidized. The isomerase acts upon monounsaturated fatty acids, such as the 18-carbon Δ9 compound, oleic acid, which contains a *cis* double bond between carbons 9 and 10. Oleic acid is activated, transported into mitochondria, and carried through three cycles of β-oxidation, just as are the saturated fatty acids. The product of the third cycle is the CoA ester of a 12-carbon fatty acid with a *cis* double bond between carbons 3 and 4. Not only is the double bond in the wrong configuration to be hydrated, but it is also in the wrong position. The enoyl-CoA isomerase enzyme converts this *cis*-3-enoyl-CoA to the more stable, conjugated *trans*-2-enoyl-CoA, which can then be acted on by enoyl-CoA hydratase. This hydration and all subsequent reactions are identical to those already described for saturated fatty acids.

The other auxiliary enzyme, 2,4-dienoyl-CoA reductase, comes into play during the oxidation of polyunsaturated fatty acids, such as linoleic acid (18:2c Δ9,12). This 18-carbon fatty acid contains *cis* double bonds between carbons 9 and 10 and between carbons 12 and 13. As shown in **FIGURE 16.17**, linoleoyl-CoA undergoes three cycles of β-oxidation, just as does oleoyl-CoA, to give a C_{12} acyl-CoA with *cis* double bonds between carbons 3 and 4 and between 6 and 7. Enoyl-CoA isomerase converts the Δ3 *cis* double bond to a Δ2 *trans* double bond. Next follows hydration, dehydrogenation, and thiolytic cleavage to give acetyl-CoA plus a 10-carbon enoyl-CoA, unsaturated between carbons 4 and 5. Action of acyl-CoA dehydrogenase yields a dienoyl-CoA, unsaturated at C4—C5 and at C2—C3. The NADPH-dependent 2,4-dienoyl-CoA reductase converts this to a C_{10} *cis*-Δ3-enoyl-CoA. The enoyl-CoA isomerase comes into play once more, generating a *trans*-Δ2-enoyl-CoA, which undergoes the remaining cycles of β-oxidation normally.

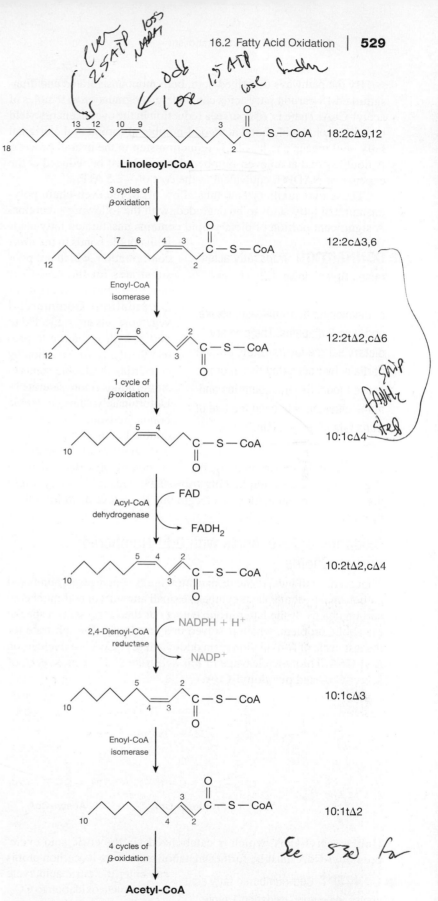

▲ **FIGURE 16.17** *β-oxidation pathway for polyunsaturated fatty acids.* This example, using linoleoyl-CoA, shows sites of action of enoyl-CoA isomerase and 2,4-dienoyl-CoA reductase, enzymes specific to unsaturated fatty acid oxidation.

By the pathways described here, both monounsaturated and diunsaturated 18-carbon fatty acids can undergo degradation to 9 moles of acetyl-CoA. There is, of course, a reduction in the overall energy yield because each double bond at an <u>odd-numbered</u> carbon in the original fatty acid means one <u>less FAD</u> reduction step in the overall process. A double bond at an <u>even-numbered</u> carbon must be reduced at the expense of <u>NADPH</u>, equivalent to the cost of ~2.5 ATP.

These two auxiliary enzymes allow all of the even-chain polyunsaturated fatty acids to be degraded, with the following exception. A significant portion of dietary lipid contains unsaturated fatty acids with double bonds in the *trans* configuration, which are poor substrates for the acyl-CoA dehydrogenase reaction of β-oxidation. Commercial vegetable oils are subjected to partial hydrogenation to protect them against oxidation by converting the double bonds to single bonds (more saturated). Unfortunately, some *cis* double bonds are isomerized to *trans* double bonds during this process, and *trans* fatty acids can thus be quite abundant in margarine and cooking oils. Because there is growing evidence implicating *trans* fatty acids in coronary artery disease, food manufacturers are beginning to remove them from their products.

● **CONNECTION** *Trans* fatty acids raise LDL and lower HDL cholesterol levels and increase your risk of developing heart disease, stroke, and type 2 diabetes. The primary dietary source for *trans* fats are the partially hydrogenated oils in processed food. Several countries and many cities have banned the use of *trans* fats in restaurants.

Oxidation of Fatty Acids with Odd-Numbered Carbon Chains

Though most of the fatty acids in natural lipids contain even-numbered carbon chains, some plants synthesize small amounts of odd-numbered carbon chains. If the latter group enters our diet, it presents a special metabolic problem, which is solved in a novel way. The substrate for the last cycle of β-oxidation of an odd-chain acyl-CoA is a five-carbon acyl-CoA. Thiolytic cleavage of this substrate yields 1 mole each of acetyl-CoA and **propionyl-CoA.**

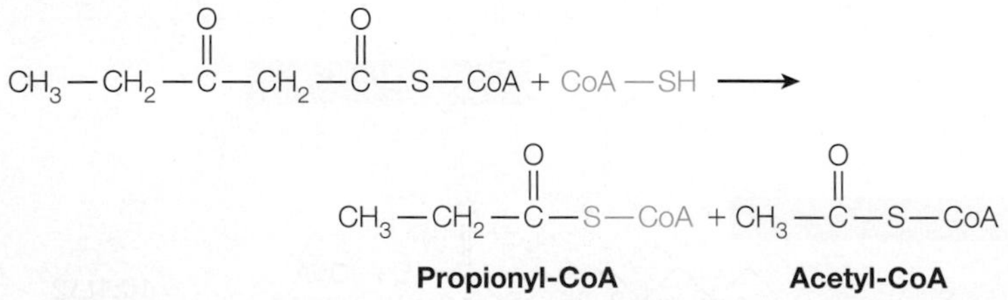

Unlike acetyl-CoA, which is catabolized via the citric acid cycle, propionyl-CoA must be further metabolized before its carbon atoms

● **CONCEPT** Odd-numbered fatty acid chains yield upon oxidation 1 mole of propionyl-CoA, whose conversion to succinyl-CoA involves a biotin-dependent carboxylation and a coenzyme B₁₂-dependent rearrangement.

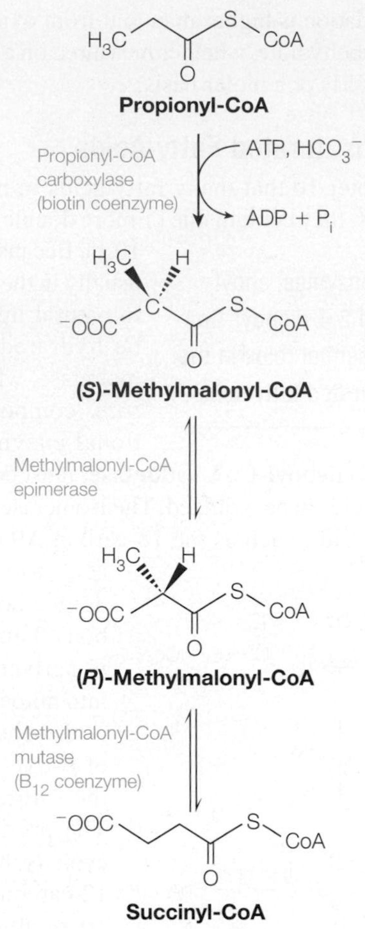

can enter the citric acid cycle for complete oxidation to CO_2. This involves first the ATP-dependent carboxylation of propionyl-CoA, catalyzed by the biotin-containing enzyme **propionyl-CoA carboxylase** (**FIGURE 16.18**). The product,

▲ **FIGURE 16.18** Pathway for catabolism of propionyl-CoA.

(S)-methylmalonyl-CoA (D-methylmalonyl-CoA), then undergoes epimerization to its (R) stereoisomer by action of methylmalonyl-CoA epimerase. Next, this branched-chain acyl-CoA derivative is converted to the corresponding straight-chain compound, which happens to be succinyl-CoA, by an unusual reaction. The enzyme, methylmalonyl-CoA mutase, requires a cofactor derived from vitamin B_{12}. However, because B_{12} coenzymes are also involved in amino acid metabolism, we reserve study of this coenzyme for Chapter 18.

Propionyl-CoA is also produced during catabolism of isoleucine and valine (see Chapter 18). Inability to catabolize propionyl-CoA properly has severe consequences in humans. If there is defective activity of methylmalonyl-CoA mutase or of the synthesis of the B_{12} coenzyme, (R)-methylmalonyl-CoA accumulates and exits from cells as methylmalonic acid. This process causes a severe acidosis (lowering of blood pH) and also damages the central nervous system. This rare condition, called **methylmalonic acidemia,** is usually fatal in early life. The disease can sometimes be treated by administering large doses of vitamin B_{12} and a diet restricted in protein. In these cases, the mutation decreases the affinity of the mutase for its B_{12} coenzyme, and the enzyme can be induced to function if the coenzyme concentration can be increased substantially.

Control of Fatty Acid Oxidation

In most cells, fatty acid oxidation is controlled by the availability of substrates for oxidation, the fatty acids themselves. In animals, this availability is controlled in turn by the hormonal control of fat mobilization in adipocytes. Because the function of adipose tissue is to store fat for use in other cells, it makes good metabolic sense for breakdown and release of this stored fat to be regulated by hormones, which are extracellular messengers. Recall from Section 16.1 that triacylglycerol lipase (hormone-sensitive lipase) activity is regulated by hormonally initiated regulatory cascades involving cyclic AMP. The action of glucagon or epinephrine causes fat breakdown and release, which leads ultimately to fatty acid accumulation in other cells. Also, as noted on page 526, malonyl-CoA provides another important regulatory mechanism, by inhibiting fatty acyl-CoA movement into mitochondria by the acyl-carnitine shuttle. As we shall soon see, malonyl-CoA levels are also hormonally controlled, providing tight coordination between fatty acid oxidation and fatty acid synthesis.

Ketogenesis

Thus far we have focused on only one metabolic fate of acetyl-CoA: its oxidation to CO_2 in the citric acid cycle. In the next section, we will learn about the role of acetyl-CoA in the biosynthesis of fatty acids. But

● **CONCEPT** When carbohydrate catabolism is limited, acetyl-CoA is converted to ketone bodies, mainly acetoacetate and β-hydroxybutyrate— important metabolic fuels in some circumstances.

another major pathway comes into play in mitochondria (primarily in liver) when acetyl-CoA accumulates beyond its capacity to be oxidized or used for fatty acid synthesis. That pathway is called **ketogenesis,** and it leads to a class of compounds called **ketone bodies.**

During fasting or starvation, when carbohydrate intake is too low, oxaloacetate levels fall so that flux through citrate synthase is impaired, causing acetyl-CoA levels to rise. Under these conditions, 2 moles of acetyl-CoA undergo a reversal of the thiolase reaction

● **CONNECTION** 3-Hydroxy-3-methylglutaryl-CoA (HMG-CoA), an intermediate in ketogenesis, is also an intermediate in cholesterol biosynthesis.

to give acetoacetyl-CoA (**FIGURE 16.19**). Acetoacetyl-CoA can react in turn with a third mole of acetyl-CoA to give **3-hydroxy-3-methylglutaryl-CoA (HMG-CoA),** catalyzed by **HMG-CoA synthase.** In mitochondria, HMG-CoA is

acted on by **HMG-CoA lyase** to yield **acetoacetate** plus acetyl-CoA. Acetoacetate undergoes either NADH-dependent reduction to give D-β-hydroxybutyrate or, in very small amounts, spontaneous decarboxylation to acetone.

Collectively, acetoacetate, acetone, and β-hydroxybutyrate are called ketone bodies, even though the last compound does not contain a keto carbonyl group. Liver also produces free acetate, by direct hydrolysis of acetyl-CoA. Acetoacetate and β-hydroxybutyrate can be utilized by peripheral tissues as an alternative fuel under ketogenic conditions. Extrahepatic tissues take up acetate from the blood and transport it into mitochondria, where it is converted back to acetyl-CoA in an ATP-dependent reaction catalyzed by **acetyl-CoA synthetase:**

$$\text{acetate} + \text{ATP} + \text{CoA-SH} \rightleftharpoons \text{acetyl-CoA} + \text{AMP} + \text{PP}_i$$

Acetyl-CoA is then oxidized through the citric acid cycle for ATP production.

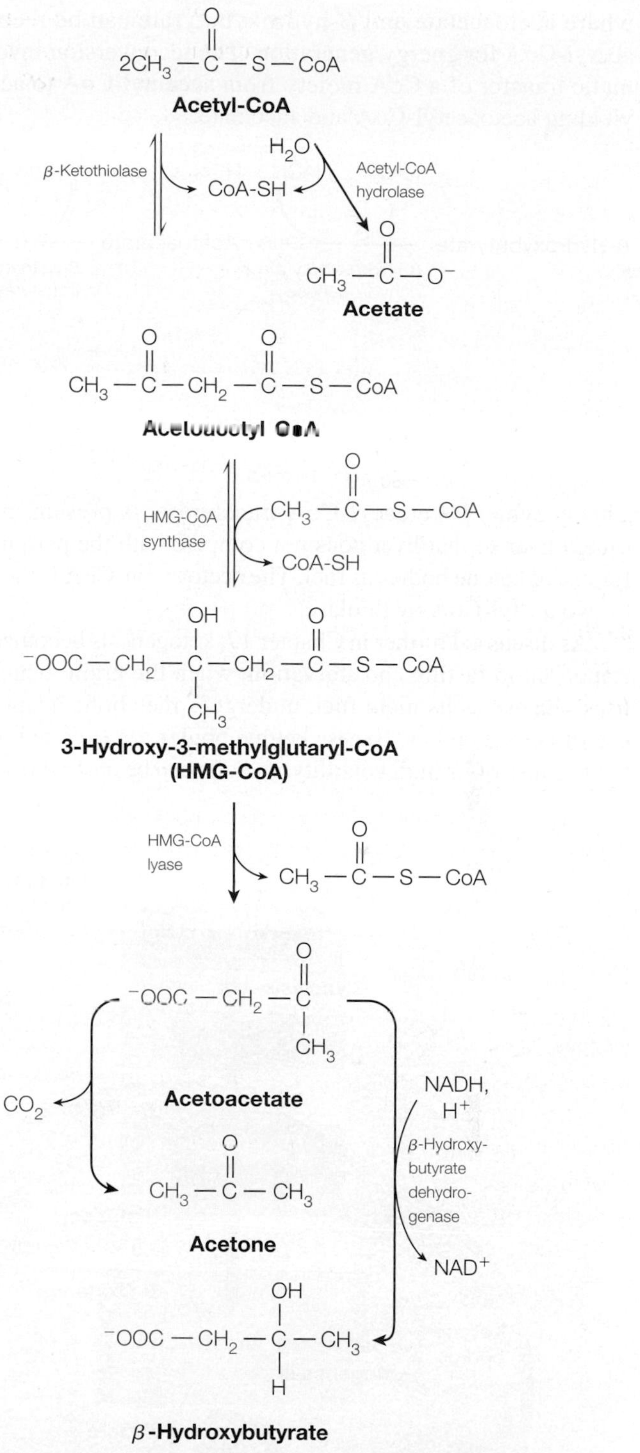

▲ **FIGURE 16.19 Biosynthesis of ketone bodies in the liver.** The three water-soluble compounds commonly called ketone bodies are boxed. Acetone is formed in very small quantities by nonenzymatic decarboxylation of acetoacetate. Acetate is also produced and released by liver for utilization by peripheral tissues.

In some circumstances, ketogenesis can be considered an "overflow pathway." As already noted, it is stimulated when acetyl-CoA accumulates because of deficient carbohydrate intake. Ketogenesis occurs primarily in liver because of the high levels of HMG-CoA synthase in that tissue. Ketone bodies are transported from liver to other tissues, where acetoacetate and β-hydroxybutyrate can be reconverted to acetyl-CoA for energy generation. The reconversion involves enzymatic transfer of a CoA moiety from succinyl-CoA to acetoacetate, yielding acetoacetyl-CoA and succinate.

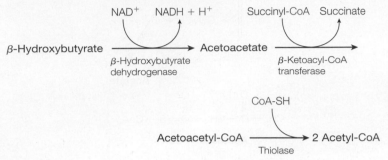

This enzyme, β-ketoacyl-CoA transferase, is present in all tissues *except* liver so that liver does not compete with the peripheral tissues for use of ketone bodies as fuel. The acetoacetyl-CoA is then converted to two acetyl-CoA by thiolase.

As discussed further in Chapter 17, ketogenesis becomes extremely important in fasting and starvation, when the brain, which normally uses glucose as its main fuel, undergoes metabolic adaptation to the use of ketone bodies. Excess ketone bodies are excreted in the urine, but because of its high volatility, acetone can be detected on the breath.

● **CONNECTION** A healthy person can go into mild ketosis from fasting (e.g., skipping breakfast) or from maintaining a low-carbohydrate diet. A sweet and fruity odor can often be smelled on the breath of a person in ketosis due to the volatile acetone.

● **CONNECTION** Excess production of ketone bodies contributes to acidosis and other complications of diabetes.

Under normal conditions, certain other tissues, particularly heart, derive much of their energy by metabolizing ketone bodies produced in the liver. In untreated diabetes, where tissues are unable to efficiently use glucose, ketone bodies are produced in excess of the capacity of peripheral tissues to use them, a condition referred to as **ketosis.** Blood levels of ketone bodies in these patients can exceed 100 mg/dL and high levels of the weak acids acetoacetate and β-hydroxybutyrate lower blood pH (**acidosis**). This **ketoacidosis** is a diagnostic feature of diabetes.

16.3 Fatty Acid Biosynthesis

Thus far, we have focused on those pathways of lipid metabolism that generate energy—that is, the catabolic pathways. We are now ready to learn something about the anabolic processes of lipid metabolism, the biosynthesis of fatty acids and triacylglycerols. Along the way, we will discuss the control of fatty acid synthesis and how the catabolic and anabolic processes are coordinately regulated.

Relationship of Fatty Acid Synthesis to Carbohydrate Metabolism

We have noted that the vast majority of the stored fuel in most animal cells is in the form of fat. However, a large proportion of the caloric intake of many animal diets—certainly most human diets—is carbohydrate. Because carbohydrate storage reserves are strictly limited, there must be efficient mechanisms for conversion of carbohydrate to fat. In this section, our primary focus is on fatty acid synthesis.

As schematized in **FIGURE 16.20**, a central metabolite is acetyl-CoA, which comes both from pyruvate in the pyruvate dehydrogenase reaction and from fatty acid β-oxidation. Acetyl-CoA is, in turn, converted in the cytosol to fatty acids. Thus, acetyl-CoA is derived from both fat breakdown and carbohydrate breakdown, and is also the major fat precursor. However, in animals *acetyl-CoA cannot undergo net conversion to carbohydrate*. This is because of the virtual irreversibility of the pyruvate dehydrogenase reaction. As noted in Chapter 13, the glyoxylate cycle in plants and some microorganisms permits a bypass of this step, with net conversion of acetyl-CoA

▲ **FIGURE 16.20 Acetyl-CoA is a key intermediate between fat and carbohydrate metabolism.** Arrows identify major routes of formation or utilization of acetyl-CoA. Citrate serves as a carrier to transport acetyl units from the mitochondrion to the cytosol for fatty acid synthesis. Note that acetyl-CoA is readily converted into fatty acids, but acetyl-CoA cannot undergo net conversion to carbohydrate.

● **CONCEPT** Animals readily convert carbohydrate to fat but cannot carry out net conversion of fat to carbohydrate.

to gluconeogenic precursors. However, *in animals the conversion of carbohydrate to fat is unidirectional.* Moreover, although fatty acid synthesis is regulated, the total capacity for fat storage is not.

Early Studies of Fatty Acid Synthesis

Early in the twentieth century, when it became evident that most fatty acids in lipids contain even-numbered chains, it was reasonable to expect that the biosynthetic process would involve some stepwise addition of activated two-carbon fragments, in the same sense that oxidation proceeds two carbons at a time. Indeed, this process was demonstrated experimentally in the 1940s, in one of the first metabolic experiments using isotopic tracers. David Rittenberg and Konrad Bloch fed mice acetate labeled with the stable isotopes carbon-13, ^{13}C, and deuterium, ^{2}H ($C^{2}H_{3}^{13}COO^{-}$) and found both isotopes incorporated into fatty acids. This experiment showed that both carbons of acetate were used in the synthesis of fatty acids.

Once the β-oxidation pathway had been discovered, it was generally thought that fatty acid synthesis would proceed simply by a reversal of its degradation pathway. However, when biochemists began to fractionate enzyme systems capable of synthesizing fatty acids, they found that the activities of β-oxidation were lacking from their purified fractions. Today we know that, although the chemistries of fatty acid synthesis and degradation are similar, the pathways differ in the enzymes involved, acyl group carriers, stereochemistry of the intermediates, electron carriers, intracellular location, and regulation. Most importantly, there is just enough difference in the chemistry of fatty acid synthesis and degradation to ensure that both pathways are exergonic, and thus unidirectional, in their respective directions. Indeed, fatty acid metabolism is one of the best examples of the statement that anabolic pathways are never the simple reversal of catabolic pathways.

● **CONCEPT** Fatty acid synthesis occurs through intermediates similar to those of fatty acid oxidation, but with differences in electron carriers, carboxyl group activation, stereochemistry, and cellular location.

The overall process of fatty acid synthesis is similar in all prokaryotic and eukaryotic systems analyzed to date. Three separate enzyme systems catalyze, respectively, (1) biosynthesis of the 16-carbon fatty acid palmitate from acetyl-CoA, (2) chain elongation starting from palmitate, and (3) desaturation. In eukaryotic cells the first pathway occurs in the cytosol, chain elongation occurs both in mitochondria and in the endoplasmic reticulum, and desaturation occurs in the endoplasmic reticulum.

Biosynthesis of Palmitate from Acetyl-CoA

As outlined in **FIGURE 16.21**, the chemistry of fatty acid synthesis is remarkably similar to that of fatty acid oxidation run in reverse. The synthetic process comprises stepwise additions of two-carbon units, with each step proceeding via condensation, reduction, dehydration, and another reduction. The major distinctions are (1) the need for an activated intermediate, *malonyl-CoA,* at each two-carbon addition step, (2) the nature of the acyl group carrier, and (3) the use of

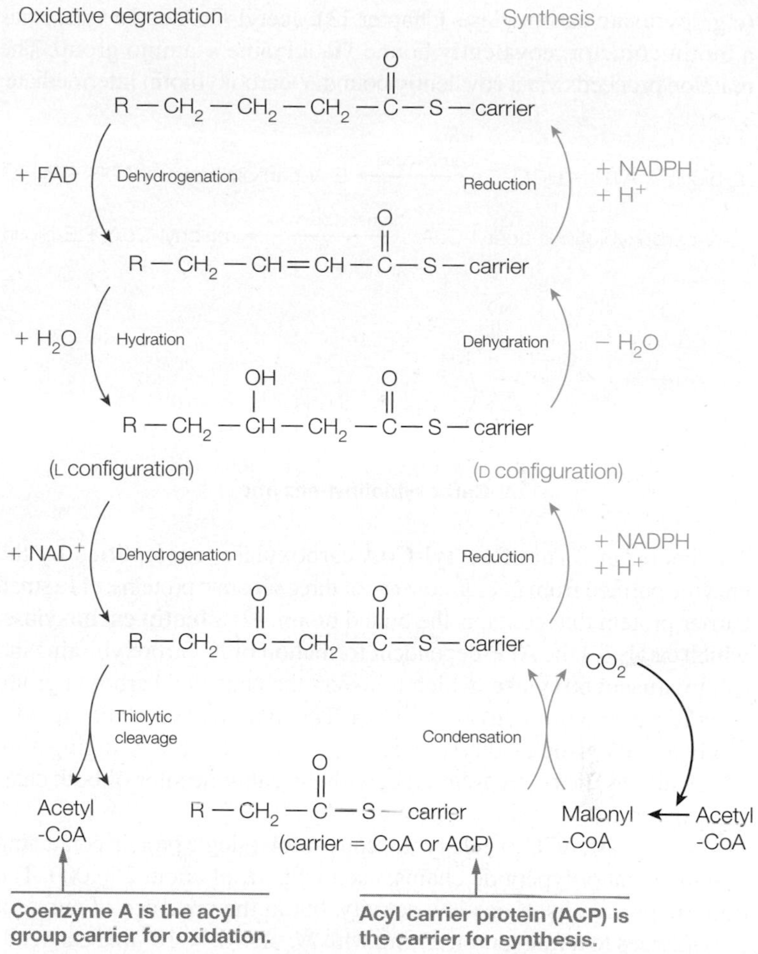

▲ **FIGURE 16.21 Chemical similarities between oxidation and synthesis of a fatty acid.** The figure shows a single cycle of oxidation (down) or addition (up) of a two-carbon fragment.

NADPH-requiring enzymes in the reductive reactions. Details of these and the other reactions follow.

Synthesis of Malonyl-CoA

The first committed step in fatty acid biosynthesis is the formation of malonyl-CoA from acetyl-CoA and bicarbonate, catalyzed by **acetyl-CoA carboxylase (ACC).**

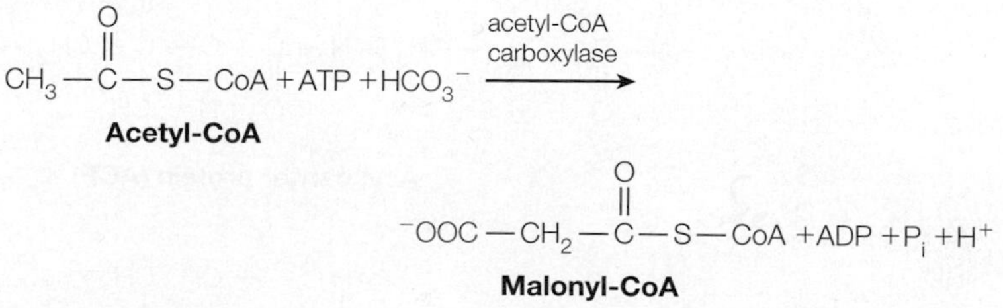

Like other committed steps in biosynthetic pathways, this reaction, driven by ATP hydrolysis, is so exergonic as to be virtually irreversible. This chemistry makes the entire pathway thermodynamically favorable and thus subject to regulation (recall from Chapter 11 that regulation can be imposed only on pathways that are displaced from equilibrium). Similar to other enzymes catalyzing carboxylation reactions (e.g., pyruvate carboxylase, Chapter 13), acetyl-CoA carboxylase has a biotin cofactor, covalently bound via a lysine ε-amino group. The reaction proceeds via a covalently bound N-carboxybiotin intermediate.

$$\text{E-biotin} + \text{ATP} + \text{HCO}_3^- \xrightarrow[\text{carboxylase}]{\text{biotin}} \text{E-}N\text{-carboxybiotin} + \text{ADP} + \text{P}_i$$

$$\text{E-}N\text{-carboxybiotin} + \text{acetyl-CoA} \xrightarrow{\text{transcarboxylase}} \text{malonyl-CoA} + \text{E-biotin}$$

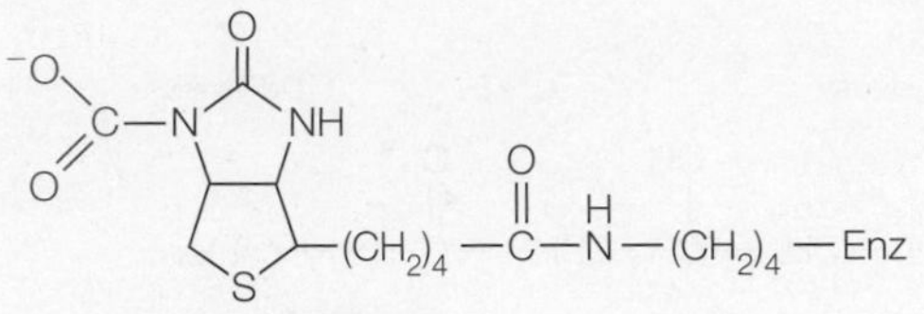

N-Carboxybiotinyl-enzyme

The bacterial form of acetyl-CoA carboxylase, exemplified by the enzyme purified from *E. coli*, consists of three separate proteins: (1) a small carrier protein that contains the bound biotin, (2) a **biotin carboxylase,** which catalyzes the ATP-dependent formation of N-carboxybiotin, and (3) a **transcarboxylase,** which transfers the activated carboxyl group from N-carboxybiotin to acetyl-CoA. The hydrocarbon chains in both biotin and its associated lysine residue act as a flexible swinging arm, which allows the biotin to interact with the catalytic sites of both catalytic subunits.

By contrast, ACC in eukaryotes consists of a single protein containing two identical polypeptide chains, each with M_r of about 250,000. The dimeric protein itself has low activity, but in the presence of citrate it polymerizes to a filamentous form, with M_r of 4-8 $\times$ 10^6 that can readily be visualized in the electron microscope. The equilibrium between inactive protein dimers and the active filamentous form, and its control by metabolic intermediates, represents an important mechanism for regulating fatty acid biosynthesis. (The regulation of ACC and fatty acid synthesis will be discussed after we describe the rest of the pathway.)

Malonyl-CoA to Palmitate

Recall that all of the intermediates in fatty acid oxidation are activated via their linkage to a carrier molecule, coenzyme A. A similar activation is involved in fatty acid synthesis, but the carrier is different. It is a small protein (77 residues in *E. coli*) called **acyl carrier protein (ACP).** The chemistry of activation is identical to that in acyl-CoAs. Indeed, ACP uses an identical phosphopantetheine moiety with its reactive sulfhydryl group as that found in CoA. In ACP, the phosphopantetheine moiety is covalently linked to a serine group in the polypeptide (**FIGURE 16.22**). All of the chemistry catalyzed by fatty acid synthase occurs on substrates attached to ACP via a thioester linkage. As we shall see, the phosphopantetheine moiety acts as a swinging arm to transfer the acyl group between the active sites of the complex.

To begin the synthesis of a new palmitate molecule, the fatty acid synthase must first be charged with starter substrates. The acetyl moiety of acetyl-CoA is loaded onto ACP in a reaction catalyzed by **malonyl/acetyl-CoA-ACP transacylase (MAT)** (**FIGURE 16.23** reaction ❶a). The acetyl group is then transferred to a Cys-SH in the active site of the **β-ketoacyl-ACP synthase (KS),** giving acetyl-KS, reaction ❶b. The phosphopantetheine moiety of ACP is now available to be charged with the second substrate, malonyl-CoA, also catalyzed by MAT, reaction ❷. Because the energy-rich thioester bonds in acyl-CoAs and acyl-ACPs are identical, these transacylase reactions are readily reversible. The fatty acid synthase is now activated for the first cycle of chain elongation. In each cycle, a primer substrate (an acyl group on KS) is condensed with an extender molecule (a malonyl group on ACP). The synthetic cycle proceeds via *condensation, reduction, dehydration,* and *reduction.*

● **CONCEPT** Malonyl-CoA represents an activated source of two-carbon fragments for fatty acid biosynthesis, with the loss of CO_2 driving C—C bond formation.

For the first cycle of synthesis (**FIGURE 16.24**, reactions ❶–❹), we start with 1 mole each of malonyl-ACP and acetyl-KS, and in four reactions we generate 1 mole of butyryl-ACP. These are the reactions that resemble the reactions (in reverse) of fatty acid oxidation (see Figure 16.21). The key carbon–carbon bond forming reaction is a Claisen-type condensation (see Figure 11.8, page 349) between acetyl-KS and malonyl-ACP, reaction ❶. This reaction, catalyzed by KS, involves decarboxylation of the malonyl moiety to give a nucleophilic enolate ion, which attacks the electrophilic acetyl thioester on KS. Breakdown of the tetrahedral intermediate involves elimination of the KS cysteine thiol, freeing it

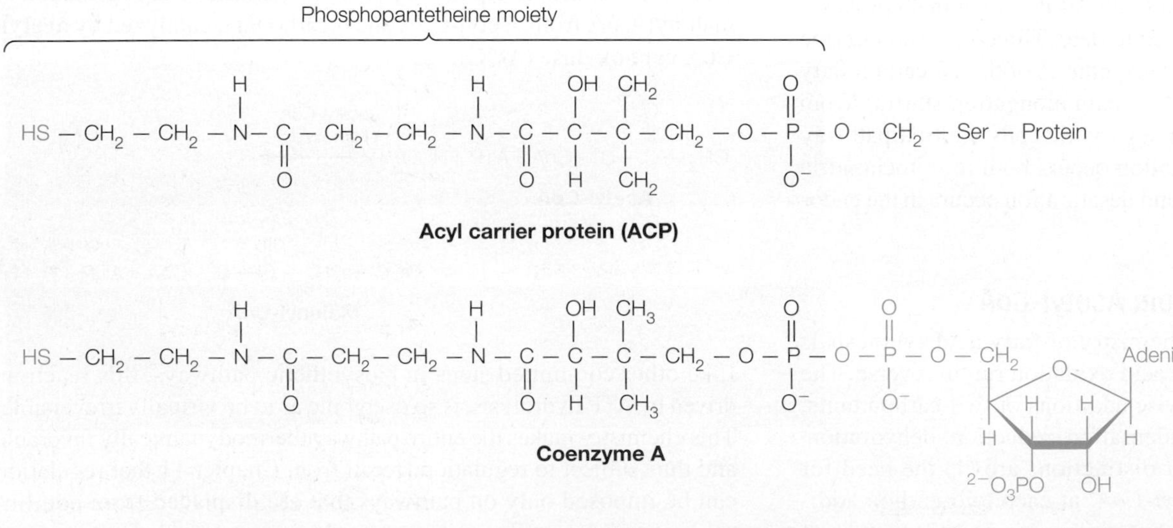

Phosphopantetheine moiety

Acyl carrier protein (ACP)

Coenzyme A

◀ **FIGURE 16.22** Phosphopantetheine as the reactive unit in ACP and CoA.

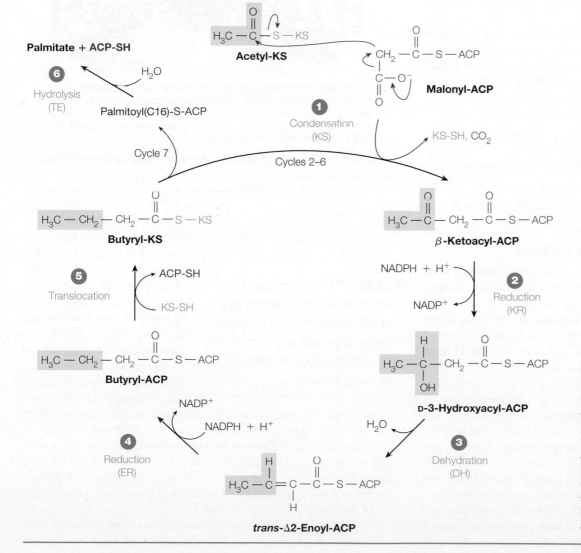

FIGURE 16.23 Malonyl/acetyl-CoA-ACP transacylase (MAT) loads fatty acid synthase with substrates. The acyl groups (acetyl or malonyl) are transferred from the SH group of CoA to the SH group of the phosphopantetheine moiety of acyl carrier protein (ACP). The reactions shown here produce acetyl-KS and malonyl-ACP, which are used in the remaining reactions of the cycle.

for attachment of the elongated acyl group at the end of the cycle, reaction ❺. The condensation product, a β-ketoacyl-ACP thioester, is next reduced to a D-β-hydroxyacyl-ACP in an NADPH-dependent reaction catalyzed by **β-ketoacyl-ACP reductase (KR),** reaction ❷. Dehydration of the D-β-hydroxyacyl-ACP, catalyzed by **β-hydroxyacyl-ACP dehydrase (DH),** reaction ❸, yields a *trans*-2-enoyl-ACP, which

undergoes a second NADPH-dependent reduction, catalyzed by **enoyl-ACP reductase (ER),** reaction ❹, to yield butyryl-ACP in the first cycle of synthesis. To end the first cycle, the butyryl group is translocated from ACP to the Cys-SH of KS, reaction ❺, and ACP is charged with a second malonyl-CoA. To start the second cycle, butyryl-KS reacts with another molecule of malonyl-ACP, and the product of the

FIGURE 16.24 Synthesis of palmitate, starting with malonyl-ACP and acetyl-KS. The first cycle of four reactions generates butyryl-ACP. Following translocation from ACP, butyryl-KS reacts with a second molecule of malonyl-ACP, leading to a second cycle of two-carbon addition. A total of seven such cycles generates palmitoyl-ACP. Hydrolysis of this product releases palmitate. KS, β-ketoacyl-ACP synthase; KR, β-ketoacyl-ACP reductase; DH, β-hydroxyacyl-ACP dehydrase; ER, enoyl-ACP reductase; TE, thioesterase.

second cycle is hexanoyl-ACP. The same pattern continues until the product of the seventh cycle, palmitoyl-ACP, undergoes hydrolysis to yield palmitate and free ACP. This final step is catalyzed by **thioesterase (TE)**, reaction ⑥.

What is the molecular logic of using malonyl-ACP as a donor of an acetyl unit? The condensation of two activated acetyl units is normally quite endergonic. A comparable reaction in reverse—namely, the thio-lytic cleavage of acetoacetyl-CoA—is strongly exergonic. However, the carboxyl group of malonyl-ACP is a good leaving group because the β-carbonyl group can act as an electron acceptor during the decar-boxylation. This decarboxylation makes the condensation reaction exergonic. Ultimately, it is ATP hydrolysis that drives this otherwise endergonic condensation reaction because ATP participated in the original synthesis of malonyl-CoA from acetyl-CoA (page 533). This condensation process explains the early observation that bicarbonate is not incorporated into the final product. Rather, all of the carbons in fatty acids come from acetate.

Like most biosynthetic pathways, this one requires both *energy* (as ATP) and *reducing equivalents* (as NADPH). The quantitative require-ments can be seen from the stoichiometry of the complete seven-cycle process:

$$Acetyl\text{-}CoA + 7\ malonyl\text{-}CoA + 14\ NADPH + 14\ H^+ \longrightarrow$$
$$palmitate + 7\ CO_2 + 14\ NADP^+ + 8\ CoA\text{-}SH + 6\ H_2O$$

Although one H_2O is released in each cycle, a net of only six H_2O is produced in the complete process because one H_2O is used to hydro-lyze the thioester linkage to release free palmitate (reaction 6 in Figure 16.24). To see the ATP requirement, we must consider the synthesis of the 7 moles of malonyl-CoA:

$$7\ Acetyl\text{-}CoA + 7\ CO_2 + 7\ ATP \longrightarrow$$
$$7\ malonyl\text{-}CoA + 7\ ADP + 7\ P_i + 7\ H^+$$

Hence, the following equation describes the overall process.

$$8\ Acetyl\text{-}CoA + 7\ ATP + 14\ NADPH + 7\ H^+ \longrightarrow$$
$$palmitate + 14\ NADP^+ + 8\ CoA\text{-}SH + 7\ ADP + 7\ P_i + 6\ H_2O$$

Multifunctional Proteins in Fatty Acid Synthesis

The fatty acid synthesis reaction pathway is essentially identical in all known organisms, but the enzymology involved is startlingly variable. Fatty acid synthesis was first worked out in *E. coli,* and it was discov-ered that the reactions are catalyzed by seven distinct monofunctional enzymes, which can be separately purified. This same organization is found in all plants and most bacteria, and is called **type II fatty acid synthesis.** In animals and lower eukaryotes, as well as a few bacteria, all of the fatty acid synthesis activities are associated in a single multifunctional enzyme referred to as a **megasynthase,** or **type I fatty acid synthase (FAS).** The animal enzyme is a homodi-mer of 273,000 Da subunits (0.54 MDa total). Each subunit of the dimeric mammalian FAS is folded into seven distinct domains, and each domain carries out a specific function in the reaction sequence. Six of these domains carry active sites for the chemical steps, and the seventh is the ACP domain with its phosphopantetheine moiety— truly, a multifunctional protein. The order of these domains along the polypeptide chain is illustrated in **FIGURE 16.25(a)**. Note the locations of the KS and ACP domains—at opposite ends of the polypeptide chain.

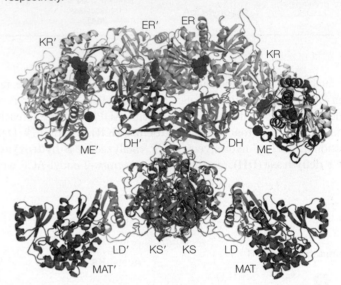

(a) Arrangement of domains on the 273,000-dalton polypeptide chain. KS ketoacyl-ACP synthase; MAT malonyl/acetyl transacylase; DH hydroxyacyl-ACP dehydrase; ER enoyl-ACP reductase; KR ketoacyl-ACP reductase; ACP acyl carrier protein; TE thioesterase. The KR domain is actually composed of two discontinuous regions split by the ER domain. LD linker domain; ME = noncatalytic structural domain. The cysteine thiol and phosphopantetheine thiol groups are shown attached to KS and ACP, respectively.

(b) A cartoon representation of the X-ray structure of the porcine FAS homodimer (PDB ID: 2vz9). Each domain is abbreviated and colored as in panel a; domains of the second subunit are indicated by a prime ('). The ACP anchor site is indicated by a red sphere, and NADP bound at the KR and ER active sites is shown in blue.

▲ **FIGURE 16.25** Structure of the mammalian "megasynthase" fatty acid synthase complex.

The sulfhydryl groups of these two domains must come into close proximity for the condensation step (reaction 1 in Figure 16.24). The X-ray structure of the porcine FAS (Figure 16.25(b)) reveals how this might occur. The two subunits form an intertwined X-structure on which two fatty acids can be synthesized simultaneously. The domains that catalyze the condensing reaction are found in the lower arms of the structure, whereas the domains catalyzing the next three reactions are found in the upper arms. The ACP and thioesterase (TE) domains are not visible in this structure, but their attachment site (C-terminus of the KR domain) is known. During the reaction cycle, the phosphopantetheine moiety attached to the ACP domain serves as a swinging arm for bringing acyl groups into contact with all of the active sites of the complex.

It has long been known that intermediates between acetyl-CoA and palmitate do not accumulate in cells that are synthesizing fatty

● **CONCEPT** In eukaryotes, fatty acid synthesis is carried out by a multifunctional enzyme (megasyn-thase), organized so that its various catalytic functions are executed by separate domains.

acids—the process is extremely fast. The basis for this is clear because all the intermediates are covalently bound to the ACP domain of a multifunctional protein. This arrangement ensures that substrates need not seek catalytic sites by simple diffusion. Indeed, starting from acetyl-CoA and malonyl-CoA, eukaryotic megasynthases can synthesize palmitate in less than a second. Multifunctional proteins have now been described in most major metabolic processes, although few other enzymes carry as many as four activities on one polypeptide chain.

Although all of these fatty acid synthase systems use essentially the same chemistry to build fatty acids, differences in their enzymology provide the possibility that specific inhibitors might be developed as antimicrobial drugs. For example, triclosan is widely used as an antibacterial agent. Triclosan is a potent inhibitor of bacterial enoyl-ACP reductase (ER) (reaction 4 in Figure 16.24). Unfortunately, the widespread use of triclosan in toothpastes, underarm deodorants, antiseptic soaps, plastic kitchenware, and many other household products has led to

⚫ **CONNECTION** Triclosan, a potent inhibitor of bacterial fatty acid synthase, is added as an antibacterial agent in many household products. However, its overuse has generated bacterial species resistant to its effects, causing the U.S. Food and Drug Administration in 2016 to ban the use of triclosan in antibacterial soaps.

the development of triclosan resistance in several bacterial species, including *E. coli, Staphylococcus aureus, Pseudomonas aeruginosa,* and *Salmonella enterica.* Drug resistance is an increasingly common problem associated with the overuse of antibiotics.

Triclosan

Transport of Acetyl Units and Reducing Equivalents into the Cytosol

Because acetyl-CoA is generated in the mitochondrial matrix, it must be transported to the cytosol for use in fatty acid synthesis. Like longer-chain acyl-CoAs, acetyl-CoA cannot penetrate the inner membrane. A shuttle system is used, which is interesting both because it provides a control mechanism for fatty acid synthesis and because it generates much of the NADPH needed for the process. The shuttle involves citrate, which is formed in mitochondria from acetyl-CoA and oxaloacetate, in the first step of the citric acid cycle (step ❶ in **FIGURE 16.26**). When citrate is being generated in excess of the amount needed for oxidation in the citric acid cycle, it is transported through

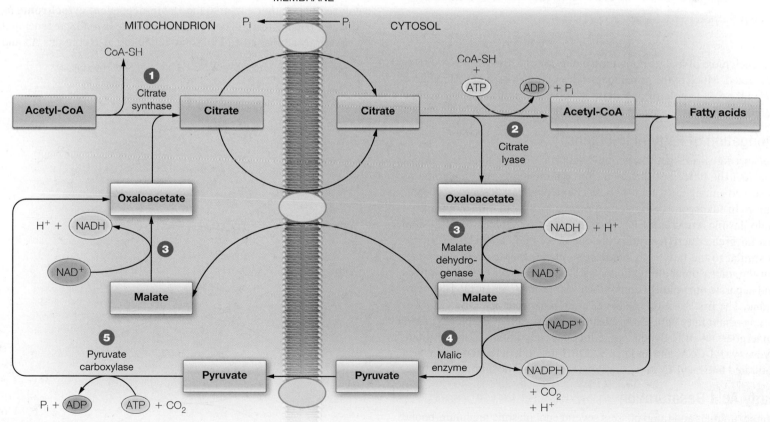

▲ **FIGURE 16.26 Transport of acetyl units and reducing equivalents used in fatty acid synthesis.** This diagram shows the shuttle mechanism for transferring acetyl units and reducing equivalents from mitochondria to the cytosol, for use in fatty acid synthesis. The transporter that carries citrate through the mitochondrial inner membrane is an antiporter, which, for every molecule of citrate exported to the cytosol, one molecule of another negatively charged molecule (either P_i or malate) must be carried into the mitochondrion. Malate that is not exchanged generates some of the NADPH for fatty acid synthesis, through action of the malic enzyme. Purple ovals represent transport systems located in the mitochondrial membrane.

● **CONCEPT** Citrate serves as a carrier of two-carbon fragments from mitochondria to cytosol for fatty acid biosynthesis.

the mitochondrial membrane to the cytosol. There it is acted on by **citrate lyase** to regenerate acetyl-CoA and oxaloacetate at the expense of one ATP, step ❷:

$$\text{Citrate} + \text{ATP} + \text{CoA-SH} \longrightarrow$$
$$\text{acetyl-CoA} + \text{ADP} + P_i + \text{oxaloacetate}$$

Oxaloacetate cannot directly return to the mitochondrial matrix because the inner membrane lacks a transporter for this compound. First, it is reduced by a cytosolic malate dehydrogenase to malate in step ❸, and some malate is oxidatively decarboxylated by the malic enzyme to give pyruvate in step ❹. (Note that the malic enzyme is working here opposite to the direction shown in Chapter 13, page 442). However, some of the malate formed returns to the mitochondrion in exchange for citrate.

$$\text{Oxaloacetate} + \text{NADH} + \text{H}^+ \longrightarrow \text{malate} + \text{NAD}^+$$
$$\text{Malate} + \text{NADP}^+ + \text{H}_2\text{O} \longrightarrow \text{pyruvate} + \text{HCO}_3^- + \text{NADPH} + \text{H}^+$$

The resultant pyruvate is transported back into mitochondria, where it is reconverted to oxaloacetate by pyruvate carboxylase in step ❺ (Chapter 13, page 442).

$$\text{Pyruvate} + \text{HCO}_3^- + \text{ATP} \longrightarrow \text{oxaloacetate} + \text{ADP} + P_i + 2\,\text{H}^+$$

The net reaction catalyzed by these three enzymes is as follows:

$$\text{NADP}^+ + \text{NADH} + \text{ATP} + \text{H}_2\text{O} \longrightarrow$$
$$\text{NADPH} + \text{NAD}^+ + \text{ADP} + P_i + 2\,\text{H}^+$$

For each mole of malate remaining in the cytosol, 1 mole of NADPH is generated. Much of the remainder of the 14 moles of NADPH required to synthesize 1 mole of palmitate is generated in the cytosol via the pentose phosphate pathway (Chapter 12).

Elongation of Fatty Acid Chains

Because fatty acid synthase action leads primarily to palmitate, we must consider the processes that lead from palmitate to give the variations observed among fatty acids in both chain length and degree of unsaturation. In eukaryotic cells, elongation occurs in both mitochondria and endoplasmic reticulum (ER). The latter, so-called microsomal, system has far greater activity and is the one described here. The chemistry is similar to the fatty acid synthase sequence that leads to palmitate, but the microsomal elongation system involves acyl-CoA derivatives and separate enzymes bound to the cytoplasmic face of the ER membrane. The first reaction is a condensation between malonyl-CoA and a long-chain fatty acyl-CoA substrate. The resultant β-ketoacyl-CoA undergoes NADPH-dependent reduction, dehydration of the resultant hydroxyacyl-CoA, and another NADPH-dependent reduction to give a saturated fatty acyl-CoA two carbons longer than the original substrate.

Fatty Acid Desaturation

Higher animals and fungi possess several endoplasmic reticulum-bound acyl-CoA desaturases that catalyze the production of mono- and polyunsaturated fatty acids. The first *cis*-double bond is always introduced between carbons 9 and 10, counting from the carboxyl end of the fatty acid. Additional double bonds are introduced toward the carboxyl end at three-carbon intervals so that the double bonds are separated by a methylene group and therefore are not conjugated. The most common

monounsaturated fatty acids in animal lipids are oleic acid, an 18:1c∆9 acid, and palmitoleic acid, a 16:1c∆9 compound (see Table 10.1). These compounds are synthesized from stearate and palmitate, respectively, by a microsomal ∆9 desaturating system called **stearoyl-CoA desaturase.** The overall reaction for stearoyl-CoA desaturation is as follows:

$$\text{Stearoyl-CoA (18:0)} + \text{NADH} + \text{H}^+ + \text{O}_2 \longrightarrow$$
$$\text{oleoyl-CoA (18:1c∆9)} + \text{NAD}^+ + 2\,\text{H}_2\text{O}$$

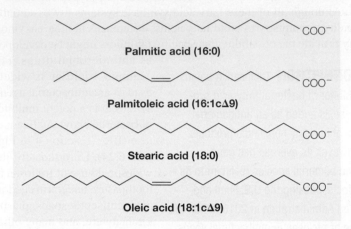

Palmitic acid (16:0)

Palmitoleic acid (16:1c∆9)

Stearic acid (18:0)

Oleic acid (18:1c∆9)

The desaturase uses molecular oxygen as the oxidant in this dehydrogenation in which an active site Fe²⁺ is oxidized to Fe³⁺. The oxidized desaturase is regenerated by reduced cytochrome b_5 in a reaction catalyzed by another enzyme, the flavin-dependent **cytochrome b_5 reductase** (**FIGURE 16.27**). The oxidized reductase is finally regenerated by NADH. Mammalian endoplasmic reticulum also contains ∆5 and

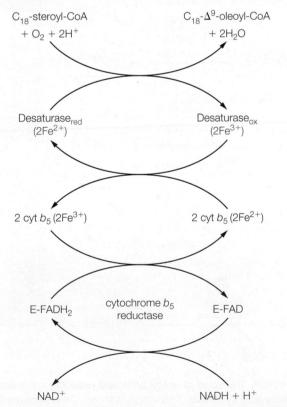

▲ **FIGURE 16.27 Fatty acid ∆9 desaturation system.** The path of electron flow is from bottom to top as the fatty acid and NADH are oxidized, and O_2 is reduced to H_2O. ∆5 and ∆6 desaturases use the same mechanism.

Δ6 desaturases, which function similarly to the Δ9 desaturating system.

Animals are unable to introduce double bonds beyond Δ9 (i.e., between C10 and the methyl carbon) in the fatty acid chain. Hence, they cannot synthesize either linoleic acid (18:2cΔ9,12) or linolenic acid (18:3cΔ9,12,15). These are called **essential fatty acids** because they are required lipid components that must be provided in the diet from plants, which do possess Δ12 and Δ15 desaturases.* After ingestion by mammals, linoleic acid and linolenic acid become substrates for further desaturation and elongation reactions. Linolenic acid is the precursor for omega-3 PUFAs EPA (20:5cΔ5,8,11,14,17) and DHA (22:6cΔ4,7,10,13,16,19) (see page 522). Dietary linoleic acid is the precursor for **arachidonic acid** (20:4cΔ5,8,11,14), which in turn serves as precursor to a class of compounds called the *eicosanoids*. As discussed later in this chapter, eicosanoids include two important classes of metabolic regulators, the prostaglandins and the thromboxanes.

Control of Fatty Acid Synthesis

To a large extent, fatty acid biosynthesis is controlled by hormonal mechanisms. Much of the fatty acid synthesis in animals takes place in adipose tissue, where fat is being stored for release and transport to other tissues on demand, to help meet their energy needs. As extracellular messengers, hormones are well suited to these interorgan regulatory roles.

FIGURE 16.28 summarizes the major regulatory processes of fatty acid synthesis in animal cells. Insulin acts in several ways to stimulate synthesis and storage of fatty acids in adipose tissue and liver. One of its effects is to increase glucose entry into cells by stimulating translocation of the glucose transporter to the plasma membrane. This increases flux through glycolysis and the pyruvate dehydrogenase reaction, which provides acetyl-CoA for fatty acid synthesis. Insulin also activates the pyruvate dehydrogenase complex by stimulating its dephosphorylation to the active form (see Section 13.5).

Another site for regulation is the transfer of acetyl units from the mitochondrial matrix to the cytosol, where fatty acid synthesis occurs. Citrate lyase, a key enzyme in this process (page 538), is activated by phosphorylation. Insulin and other growth factors stimulate this activation through the PI3K pathway (more about this in Chapter 17).

The first enzyme whose action is committed to fatty acid synthesis is acetyl-CoA carboxylase (ACC; page 533). Activities of this enzyme are quite low in starved animals, reflecting its regulation by allosteric and covalent modification

● **CONCEPT** Acetyl-CoA carboxylase is the committed enzyme and major control point for fatty acid synthesis.

mechanisms. Phosphorylation of acetyl-CoA carboxylase causes its inactivation. Two protein kinases are known to phosphorylate ACC:

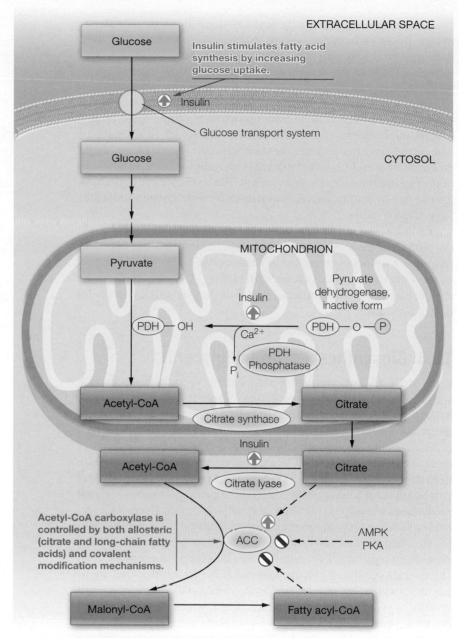

▲ **FIGURE 16.28 Regulation of fatty acid synthesis in animal cells.** The rate-limiting enzyme, acetyl-CoA carboxylase (ACC), is controlled by both allosteric (citrate and long-chain fatty acids) and covalent modification mechanisms. Phosphorylation by AMP-activated protein kinase (AMPK) or cyclic AMP-dependent protein kinase (PKA) inactivates ACC. Insulin stimulates fatty acid synthesis by increasing glucose uptake and increasing flux through pyruvate dehydrogenase to produce acetyl-CoA. The dephosphorylated form of PDH is the enzymatically active form.

AMP-activated protein kinase (AMPK) and cyclic AMP-dependent protein kinase (PKA). The AMPK system acts as a sensor of cellular energy status—it is activated by increases in the cellular AMP:ATP ratio (more about this in Chapter 17). Thus, under conditions of low energy charge, such as would occur during fasting or starvation, activated AMPK switches off fatty acid synthesis by inhibiting ACC (Figure 16.28). ACC activity is also under hormonal control. Glucagon and epinephrine activate PKA, which phosphorylates and inhibits the enzyme.

---

*Fish are also good sources of linoleic acid and linolenic acid, but they also cannot synthesize these essential fatty acids—like all animals, fish must obtain them from their diet.

Citrate and long-chain fatty acyl-CoAs are allosteric modulators of acetyl-CoA carboxylase (Figure 16.28). As noted earlier, ACC must undergo a reversible polymerization in order to be active. Long-chain fatty acyl-CoAs at low levels prevent polymerization, thereby inactivating the enzyme and providing feedback inhibition of the pathway. Citrate is an allosteric activator of ACC, stimulating its polymerization. As the carrier of acetyl units from mitochondria (see Figure 16.20), cytoplasmic citrate levels rise when mitochondrial acetyl-CoA and ATP concentrations increase. Thus, the same molecule (citrate) serves as the precursor of acetyl-CoA for fatty acid biosynthesis and as an activator of the rate-limiting step for this process. The levels of fatty acyl-CoAs are lowered by insulin, another mechanism by which insulin stimulates fatty acid synthesis.

Fatty acid degradation is also regulated. Recall that malonyl-CoA, the product of the acetyl-CoA carboxylase reaction, inhibits fatty acid oxidation by blocking carnitine acyltransferase I, the first committed step in fatty acid synthesis (page 526). Malonyl-CoA thus provides an elegant mechanism to coordinately regulate the synthesis and degradation of fatty acids.

16.4 Biosynthesis of Triacylglycerols

Fatty acyl-CoAs and glycerol-3-phosphate serve as the major precursors to triacylglycerols and, as we shall see in Part II of this chapter, to the glycerophospholipids used to build membranes. In many cases, the glycerol backbone is the limiting substrate. Glycerol-3-phosphate is derived either from the reduction of the glycolytic intermediate dihydroxyacetone phosphate (DHAP), catalyzed by **glycerol-3-phosphate dehydrogenase** or from the ATP-dependent phosphorylation of glycerol by **glycerol kinase** (**FIGURE 16.29**). The pathway involving DHAP predominates in adipose tissue because adipocytes lack glycerol kinase. During fasting or starvation conditions when glycolysis is reduced, adipocytes, hepatocytes, and cancer cells can synthesize glycerol-3-phosphate from pyruvate. This pathway, termed **glyceroneogenesis**, is an abbreviated version of gluconeogenesis (Chapter 13) that uses pyruvate carboxylase, phosphoenolpyruvate carboxykinase (PEPCK), and reversal of glycolytic steps to produce DHAP, which is then reduced to glycerol-3-phosphate by glycerol-3-phosphate dehydrogenase.

Whatever the source, glycerol-3-phosphate undergoes two successive enzymatic esterifications with fatty acyl-CoAs to yield **diacylglycerol-3-phosphate,** also called **phosphatidic acid.** The first esterification is catalyzed by **glycerophosphate acyltransferase** (GPAT) and, as

● **CONNECTION** In addition to its role in triacylglycerol synthesis, phosphatidic acid is also a key precursor to glycerophospholipids, found primarily as components of membranes.

the first committed step of the cycle, is an important site of regulation. The pathway to triacylglycerols is completed by hydrolytic removal of the phosphate, followed by transfer of another fatty acyl moiety from an acyl-CoA.

As we have noted, triacylglycerols represent the major form in which energy can be stored. Normally in an adult animal, synthesis and degradation are balanced so that there is no net change in the total body amount of triacylglycerols. If dietary intake exceeds caloric needs, then proteins, carbohydrate, or fat can each readily provide acetyl-CoA to drive the synthesis of fatty acids and triacylglycerols. On the other hand, fat reserves allow animals to go for rather long

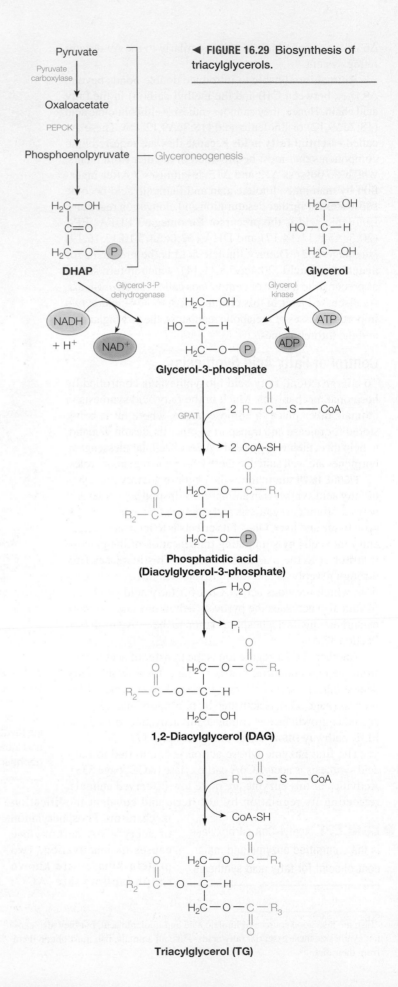

◀ **FIGURE 16.29** Biosynthesis of triacylglycerols.

times without eating and still maintain adequate energy levels. Such fasting does generate some metabolic stresses, though, as we describe further in Chapter 17.

Hibernating animals have adapted remarkably well to cope with such stresses. For example, bears store huge amounts of fat just before beginning a hibernation that may last as long as seven months. During this period, all of the bear's energy comes from breakdown of the stored fat. Moreover, the bear excretes so little water that the water released from fat oxidation meets the animal's needs. Similarly, the glycerol released from triacylglycerols provides a source of gluconeogenic precursors.

PART II: METABOLISM OF MEMBRANE LIPIDS, STEROIDS, AND OTHER COMPLEX LIPIDS

So far, we have been concerned primarily with the energetic aspects of lipid metabolism—synthesis and oxidation of fatty acids, interorgan transport of lipoproteins, and metabolism of triacylglycerols. In addition to their roles in energy storage, lipids function as membrane components and as biological regulators. Our attention now shifts to the roles played by these more complex lipids, as well as pathways for their synthesis and degradation.

Part II of this chapter focuses on the following major classes of lipids: *glycerophospholipids* (also called *phosphoglycerides*), which are primarily membrane components but which also play some specialized regulatory roles; *sphingolipids,* which in animals are found abundantly in nervous tissue; *steroids* and other *isoprenoid* compounds, which function as hormones, vitamins, and membrane constituents; and *eicosanoids,* a class of biological regulators synthesized from arachidonic acid. Two major topics involving lipids and regulation are discussed later in this book: the actions of steroid hormones (in Chapter 20) and the second-messenger regulatory role of inositol phospholipids (in Chapter 20).

16.5 Glycerophospholipids

The most abundant phospholipids are those derived from glycerol. These **glycerophospholipids** are found primarily as components of membranes. Membrane phospholipids are also metabolic precursors to various regulatory elements of signal transduction pathways. In animals, phospholipids also participate in the transport of triacylglycerols and cholesterol, as discussed earlier, by forming the surface of lipoproteins. In addition, phospholipids play specific roles in processes as diverse as blood clotting and lung function. Pathways of glycerophospholipid synthesis are outlined in **FIGURE 16.30**. In the following

section, we will limit our discussion to the metabolic strategies used to synthesize the glycerophospholipids.

Most eukaryotic cells contain six classes of glycerophospholipids—phosphatidylethanolamine (PE), phosphatidylglycerol (PG), cardiolipin (CL), phosphatidylserine (PS), phosphatidylcholine (PC), and phosphatidylinositol (PI), with PC and PE being the most abundant. Most bacterial membranes contain only three phospholipids in significant amounts, PE, PG, and CL. It is clear from Figure 16.30 that the major precursor to all glycerophospholipids is phosphatidic acid, which is synthesized from glycerol-3-phosphate, as we saw in

● **CONCEPT** Phosphatidic acid is a key intermediate in glycerophospholipid biosynthesis.

Figure 16.32. From phosphatidic acid, the pathways to PE, PS, PG, and CL are virtually identical in eukaryotes and bacteria.

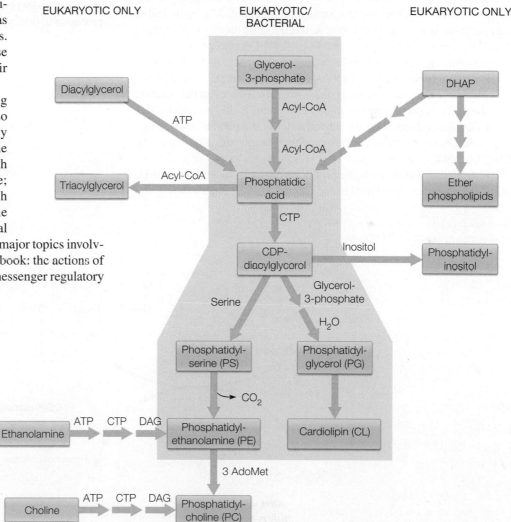

▲ **FIGURE 16.30 Pathways in glycerophospholipid biosynthesis.** The major phospholipids found in membranes are shown in green. Pathways found in both bacterial and eukaryotic cells are highlighted in light purple. Other reactions are confined to eukaryotic cells. DHAP = dihydroxyacetone phosphate; DAG = diacylglycerol; AdoMet = *S*-adenosylmethionine.

As we learned in Chapter 10, glycerophospholipids are composed of a diacylglycerol moiety linked to a hydrophilic head group via a phosphodiester linkage.

A glycerophospholipid

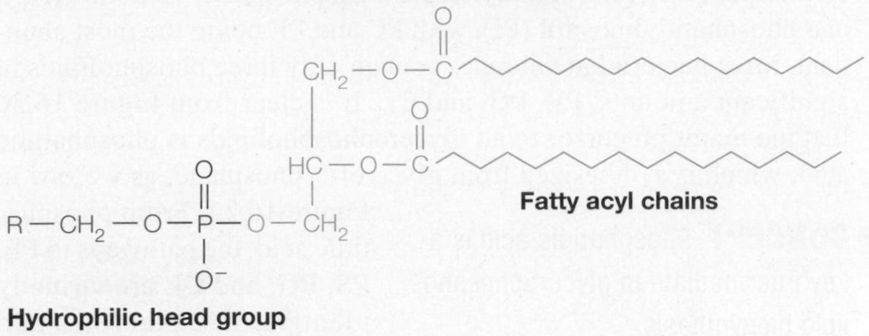

Fatty acyl chains

Hydrophilic head group

Diacylglycerol and the head groups are both alcohols, and one or the other must be metabolically activated to achieve this coupling. The energy cofactor for phospholipid biosynthesis is cytidine triphosphate (CTP), whose role is similar to that of UTP in polysaccharide synthesis (Chapter 9). The predominant pathway involves activation of the diacylglycerol moiety, starting from phosphatidic acid. Mechanistically, this reaction is reminiscent of the activation of glucose-1-phosphate by UTP to yield UDP-glucose (Section 12.8). The phosphoryl oxygen of phosphatidic acid attacks the α phosphorus atom of CTP to form **CDP-diacylglycerol** and pyrophosphate (**FIGURE 16.31**). The $\Delta G^{\circ\prime}$ of this phosphoanhydride exchange is nearly zero. However, the reaction is pulled to completion by the highly exergonic hydrolysis of the PP$_i$.

CDP-diacylglycerol is now activated (CMP is a good leaving group) for nucleophilic attack by the various polar head groups, leading to PS, PE, PG, and CL (Figure 16.30).

Phosphatidylcholine can be produced by three successive methylations of phosphatidylethanolamine (Figure 16.30). The methyl group

● **CONCEPT** Metabolic activation of phospholipid precursors is carried out by reaction with CTP.

donor in this pathway is an activated derivative of methionine, **S-adenosyl-L-methionine** (AdoMet). The chemistry of this "universal" methyl donor will be discussed in Chapter 18.

Finally, eukaryotic cells possess alternative pathways that start with the free bases choline and ethanolamine, which arise through the turnover of preexisting phospholipids. These "salvage pathways" lead to PC and PE, respectively (see end-of-chapter Problem 32).

16.6 Sphingolipids

Recall from Chapter 10 that sphingolipids are derivatives of the base *sphingosine*. Plant sphingolipids contain a slightly different form of this compound, called **phytosphingosine.** The sphingolipids include *ceramide* (*N*-acylsphingosine), *sphingomyelin* (*N*-acylsphingosine phosphorylcholine), and a family of carbohydrate-containing sphingolipids called neutral and acidic **glycosphingolipids;** the latter substances include *cerebrosides* and *gangliosides* (which also contain sialic acid). Ceramide serves as the precursor to both sphingomyelin and the glycosphingolipids.

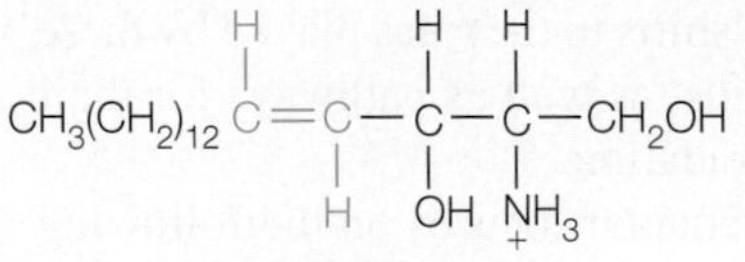

Sphingosine

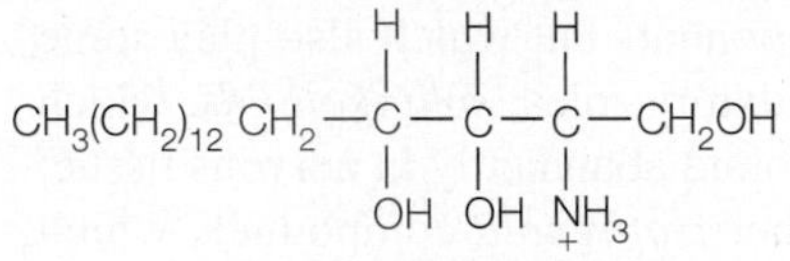

Phytosphingosine

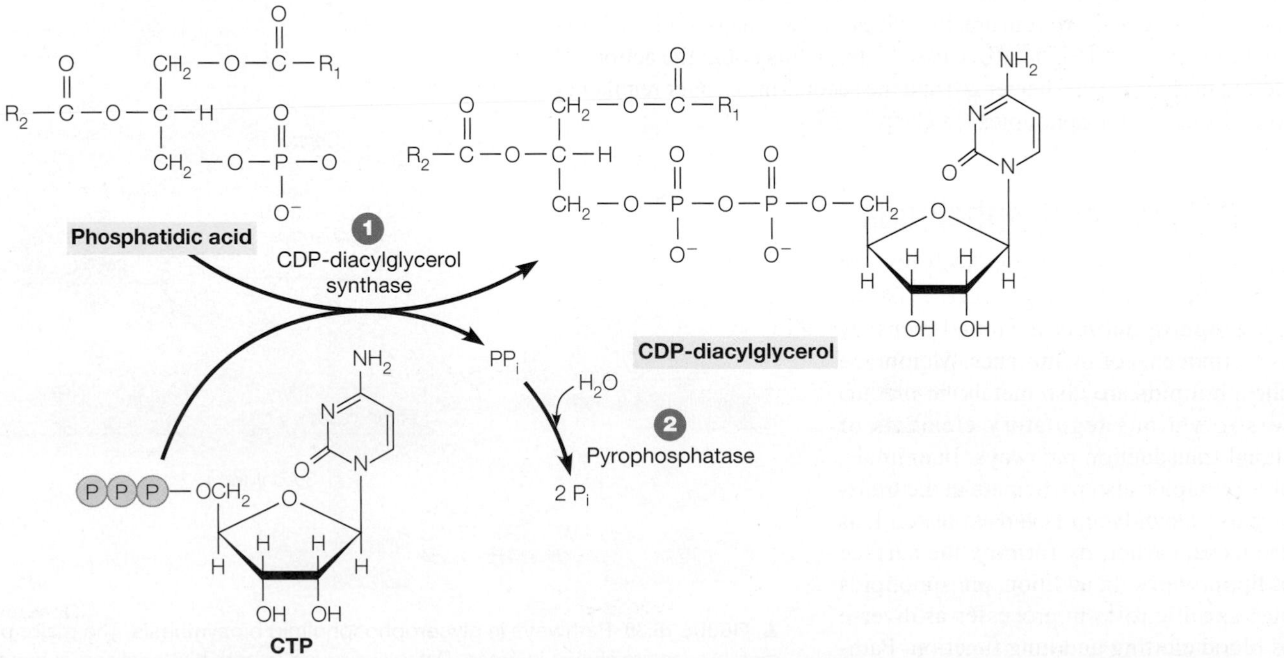

▲ **FIGURE 16.31 Activation of phosphatidic acid.** The reaction of CTP with phosphatidic acid is catalyzed by CDP-diacylglycerol synthase (reaction 1). This reaction is drawn to the right by the enzymatic hydrolysis of pyrophosphate, catalyzed by the ubiquitous pyrophosphatase (reaction 2).

CONNECTION Genetic defects in glycosphingolipid catabolism cause breakdown intermediates to accumulate in nervous tissue, with severe consequences. In Tay-Sachs disease, the ganglioside GM2 accumulates, causing mental retardation, blindness, and early death.

CONCEPT Nucleotide-linked sugars and glycosyltransferases are involved in glycosphingolipid biosynthesis.

16.7 Steroid Metabolism

We turn now to an extraordinarily large and diverse group of lipids, the **isoprenoids,** or **terpenes.** These compounds are built up from one or more five-carbon activated derivatives of **isoprene.** Steroids are the most well-known members of the diverse isoprenoid family, but also included in this family are bile acids; the lipid-soluble vitamins; the dolichol and undecaprenol phosphates we encountered in glycoprotein synthesis; phytol, the long-chain alcohol in chlorophyll; **gibberellins,** a family of plant growth hormones; insect juvenile hormones; the major components of rubber; coenzyme Q; and many more compounds.

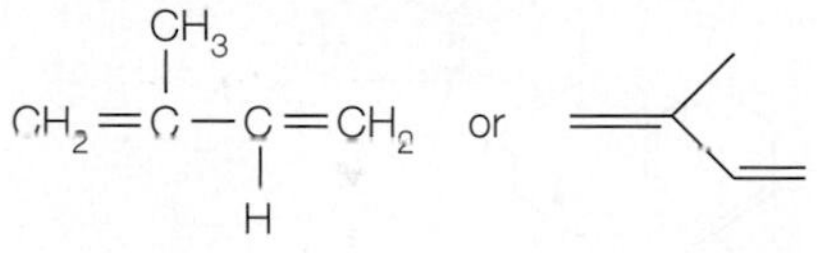

Isoprene

Much of our discussion of isoprenoids focuses on a single steroid compound: cholesterol. As discussed in Chapter 10, this lipid is a major component of animal cell membranes, where it participates in modulation of membrane fluidity. In animals it also serves as precursor to all of the steroids, to vitamin D, and to the bile acids, which aid in fat digestion. And as we discussed at the start of this chapter, there is intense medical interest in cholesterol because of the relationships among diet, blood cholesterol levels, atherosclerosis, and heart disease. These biological relationships, coupled with the complex stereochemistry of its structure and the elegance of its biosynthetic pathway from a single low-molecular-weight precursor, have focused attention on this compound ever since its first isolation from gallstones in 1784.

In animals the pathway to ceramide starts from palmitoyl-CoA and serine (**FIGURE 16.32**). After reduction of the palmitoyl keto group, the amino group of the sphingosine base unit is acylated and desaturated to give a ceramide. Transfer of a phosphocholine unit from phosphatidylcholine yields sphingomyelin plus diacylglycerol.

The pathways leading to glycosphingolipids are more numerous, but the metabolic strategies are comparable to those we have encountered before in synthesis of the oligosaccharide chains of glycoproteins (see Chapter 9). The pathways involve the stepwise addition of monosaccharide units, using nucleotide-linked sugars as the activated biosynthetic substrates (e.g., UDP-glucose) and with ceramide as the initial monosaccharide acceptor (Figure 16.32).

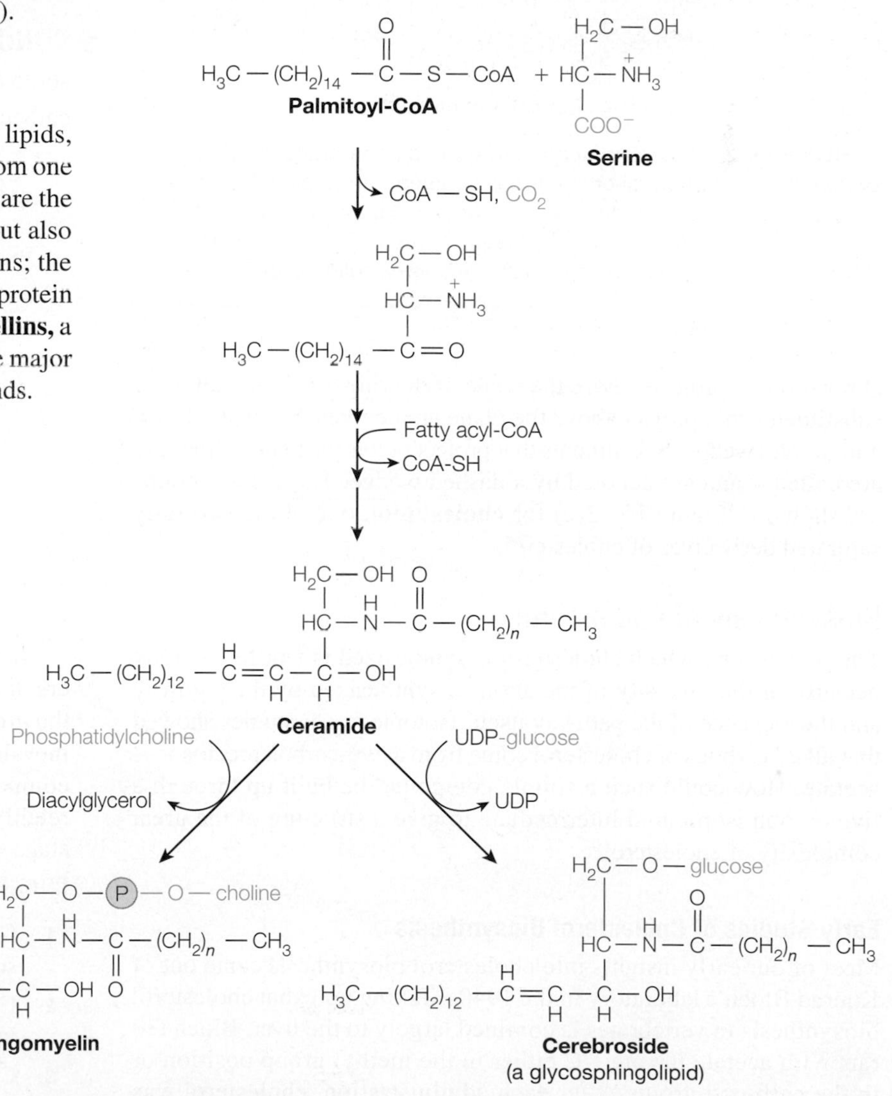

Steroids: Some Structural Considerations

Steroids constitute a class of lipids that are derivatives based on a saturated tetracyclic hydrocarbon structure (**FIGURE 16.33**). Note the letters used to denote the four rings—A, B, C, and D, with D being the five-membered ring—and the carbon numbering system. Cholesterol differs from the basic ring system in having an aliphatic chain at C-17, axial methyl groups at C-10 and C-13, a double bond in ring B, and a hydroxyl group in ring A. The alcoholic functional group and the carbon chain at C-17 make cholesterol a **sterol,** which is the generic term used to identify steroid alcohols.

Much of the cholesterol in lipoproteins and intracellular storage droplets is esterified at this position with a long-chain fatty acid, which makes the resultant cholesterol ester much more hydrophobic than cholesterol itself.

During most of our treatment of steroid metabolism, we shall use structural representations, as in Figure 16.33(c), instead of three-dimensional configurational models. By convention, the methyl group

▲ **FIGURE 16.32 Biosynthesis of sphingolipids in animal cells.** The enzymes catalyzing these reactions are localized on the cytoplasmic face of the endoplasmic reticulum.

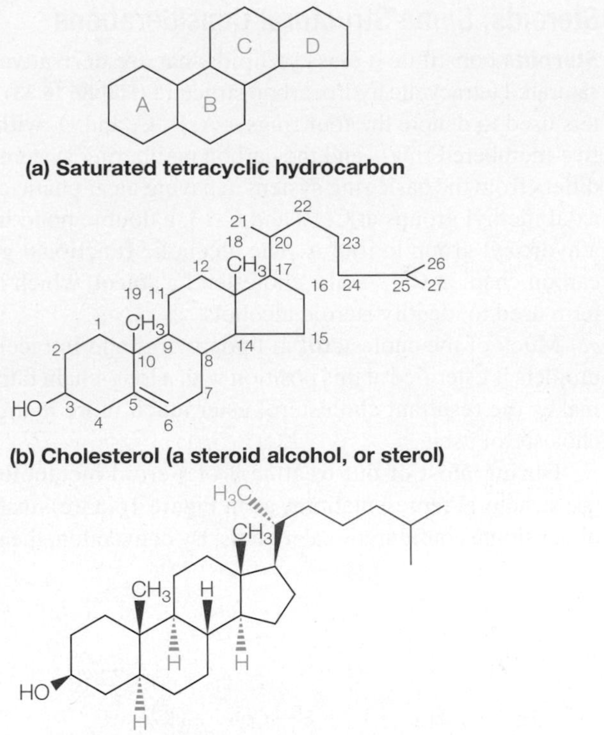

(a) Saturated tetracyclic hydrocarbon

(b) Cholesterol (a steroid alcohol, or sterol)

(c) Stereochemistry of cholestanol

▲ **FIGURE 16.33** Ring identification system (a), carbon numbering system (b) and structural conventions (c) used for steroids. Structural conventions, with cholestanol as the example. α substituents project below the plane of the steroid ring system (blue dashed wedge), and β substituents project above that plane (magenta solid wedge).

at position 10 projects *above* the plane of the rings. This and all other substituents that project above the plane are denoted β and are drawn with a solid wedge. Substituents that project *below* the plane of the ring are called α and are denoted by a dashed wedge. These conventions are shown in Figure 16.33(c) for **cholestanol,** one of the two fully saturated derivatives of cholesterol.

Biosynthesis of Cholesterol

The pathway by which cholesterol is synthesized is worthy of study because of the diversity of metabolites synthesized by the pathway and the elegance of the pathway itself. Isotopic tracer studies showed that all 27 carbons of cholesterol come from a two-carbon precursor—acetate. How could such a simple compound be built up through a five-carbon isoprenoid intermediate to give a structure of the great complexity of cholesterol?

Early Studies of Cholesterol Biosynthesis

Most of our early insights into cholesterol biosynthesis came out of Konrad Bloch's laboratory in the 1940s. Taking note that cholesterol biosynthesis in vertebrates is confined largely to the liver, Bloch fed rats with acetate having ^{14}C either in the methyl group position or in the carboxyl group. After each administration, cholesterol was isolated from the liver and subjected to chemical degradation, with radioactive counting of the fragments. This procedure established the

pattern shown below, in which each carbon of cholesterol was found to originate from either the methyl carbon (blue) or the carboxyl carbon (magenta) of acetate (actually, acetyl-CoA).

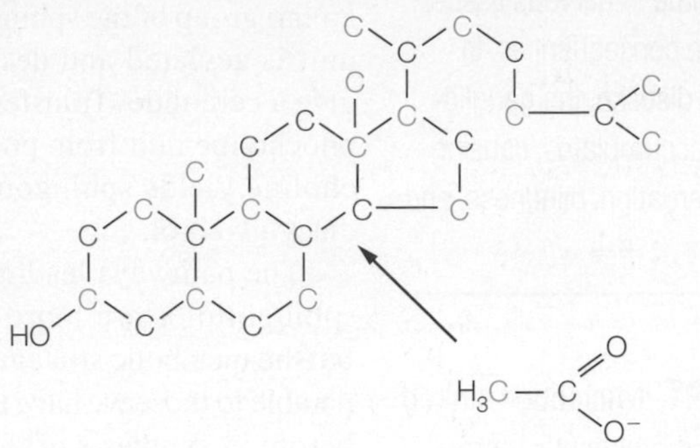

Other early insights came from the realization that the five carbons of isoprene could be derived metabolically from three molecules of acetate, and the prediction that cholesterol was a product of the cyclization of the linear C_{30} hydrocarbon **squalene.** Squalene contains six isoprene units (delineated by magenta marks on the structures below), and its configuration makes it a plausible steroid precursor.

● **CONCEPT** Cholesterol, the precursor to all steroids, derives all of its carbon atoms from acetate.

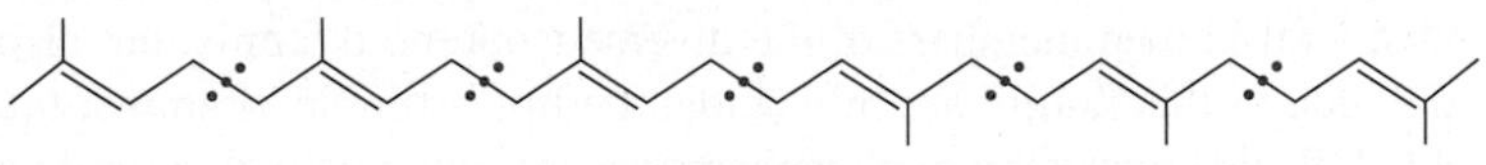

Squalene

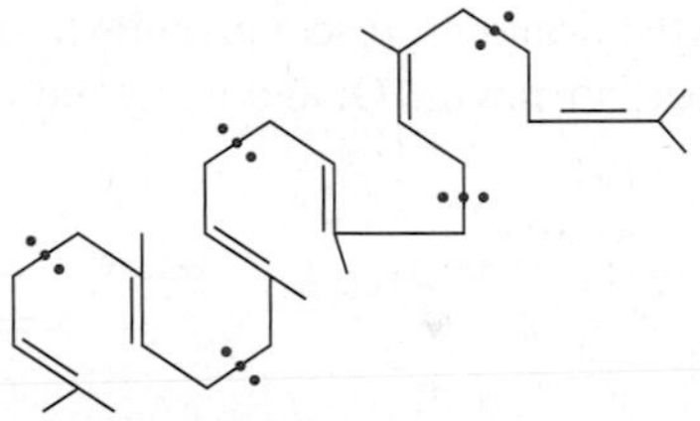

Postulated precyclization configuration of squalene

In 1956, another important development occurred, when Karl Folkers discovered that a C_6 organic acid, **mevalonic acid,** could permit the growth of certain acetate-requiring bacteria. Folkers showed that mevalonic acid was readily converted to an activated C_5 isoprenoid compound, **isopentenyl pyrophosphate.** In animals, mevalonate is readily converted to squalene. Once this had been established, the stage was set for considering cholesterol biosynthesis as three distinct processes.

1. Conversion of C_2 fragments (acetate) to a C_6 isoprenoid precursor (mevalonate)

2. Conversion of six C_6 mevalonates, via activated C_5 isoprenoid intermediates, to the C_{30} squalene

3. Cyclization of squalene and its transformation to the C_{27} cholesterol

Now let us consider some of the details of these three processes.

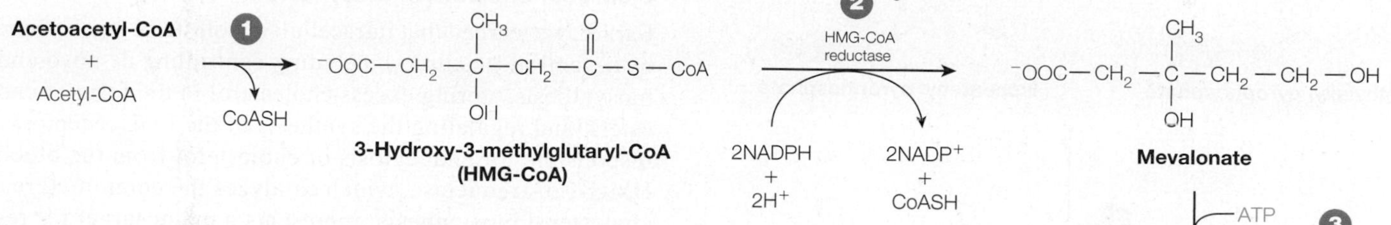

Lipitor inhibits ✎

Stage 1: Formation of Mevalonate

The first part of the pathway is identical to reactions used in ketogenesis (Figure 16.19), although it occurs in a different cell compartment. Ketogenesis occurs in mitochondria, whereas cholesterol biosynthesis occurs in the cytosol and the endoplasmic reticulum (ER).

Stage 1 begins with condensation of two molecules of acetyl-CoA to give acetoacetyl-CoA. **FIGURE 16.34** shows the rest of this stage. In step ❶, acetoacetyl-CoA reacts with a third molecule of acetyl-CoA to give 3-hydroxy-3-methylglutaryl-CoA (HMG-CoA). Recall from Figure 16.19 that during ketogenesis, HMG-CoA cleaves to give acetoacetate plus acetyl-CoA in the mitochondrial matrix. However, the HMG-CoA lyase that accomplishes this cleavage is missing from the ER, where cholesterol biosynthesis begins. Instead, **HMG-CoA reductase,** an integral membrane protein in the ER, catalyzes the reduction of HMG-CoA to mevalonate (step ❷). This multistep reaction requires two equivalents of NADPH (four electrons) to reduce the thioester to an alcohol. This is the major step that regulates the overall pathway of cholesterol biosynthesis.

● **CONCEPT** Hydroxymethylglutaryl-CoA reductase, which catalyzes an early reaction in cholesterol biosynthesis, is the major control point for the overall process.

Stage 2: Synthesis of Squalene from Mevalonate

The next several reactions occur in the cytosol. First, mevalonate is activated by three successive phosphorylations (steps ❸, ❹, and ❺). The first two are simple nucleophilic substitutions on the γ-phosphate of ATP. The third phosphorylation, at position 3, sets the stage for a spontaneous decarboxylation to give the five-carbon isopentenyl pyrophosphate (IPP) in step ❻.

IPP isomerase catalyzes the isomerization of one molecule of the resulting isopentenyl pyrophosphate to the C_5 **dimethylallyl pyrophosphate** (step ❼). The latter compound, as shown in **FIGURE 16.35**, reacts with a second molecule of isopentenyl pyrophosphate to give the C_{10} **geranyl pyrophosphate** (step ❶), and still another molecule of isopentenyl pyrophosphate reacts with this product to give the C_{15} **farnesyl pyrophosphate** (step ❷).

The final reactions of squalene synthesis are catalyzed by **squalene synthase,** which is bound to membranes of the endoplasmic reticulum. This enzyme uses carbocation intermediates to join two molecules of farnesyl pyrophosphate in head-to-head fashion. Following pyrophosphate elimination and rearrangement, the enzyme-bound intermediate is reduced by NADPH to yield the C_{30} squalene (step ❸).

Stage 3: Cyclization of Squalene to Lanosterol and Its Conversion to Cholesterol

All subsequent reactions occur in the endoplasmic reticulum. The cyclization of squalene to lanosterol and the conversion of lanosterol to cholesterol are summarized in **FIGURE 16.36**. The formation of lanosterol, which has the four-ring sterol nucleus, occurs in three steps. A series of about 19 reactions follows, involving double-bond reductions and three demethylations to yield cholesterol.

● **CONCEPT** Cyclization of squalene, a C_{30} hydrocarbon, creates the four-ring sterol nucleus.

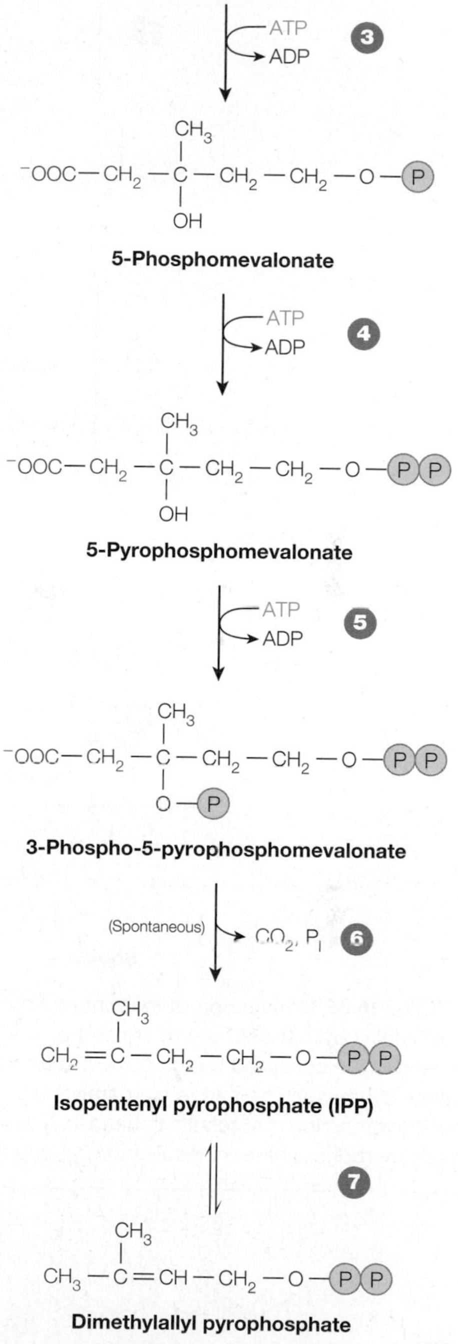

▲ **FIGURE 16.34** Biosynthesis of mevalonate and conversion to isopentenyl pyrophosphate and dimethylallyl pyrophosphate. The two carbons of the third acetyl group are shown in magenta.

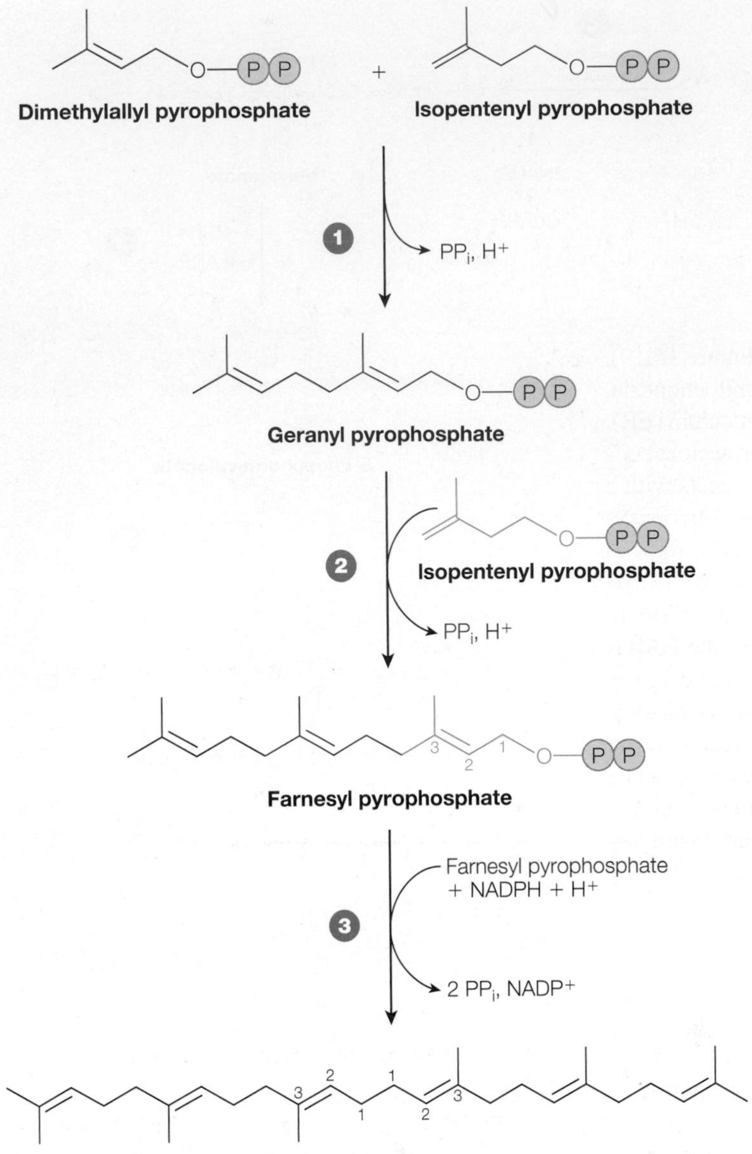

▲ **FIGURE 16.35 Conversion of isopentenyl pyrophosphate and dimethylallyl pyrophosphate to squalene.** The first two head-to-tail condensations are catalyzed by the same prenyltransferase (farnesyl pyrophosphate synthase). The third step, catalyzed by squalene synthase, is a complex reaction that results in head-to-head condensation of two farnesyl pyrophosphate molecules to give squalene.

Control of Cholesterol Biosynthesis

Earlier, we learned that intracellular cholesterol levels are regulated by several mechanisms, including controlling de novo cholesterol biosynthesis, storing excess cholesterol in the form of cholesterol esters, and regulating the synthesis of the LDL receptor, which is responsible for endocytosis of cholesterol from the bloodstream. HMG-CoA reductase, which catalyzes the committed reaction in cholesterol biosynthesis, represents a major target for regulation of the overall pathway (see Figure 16.9). It has long been known from feeding studies that dietary cholesterol efficiently reduces the endogenous synthesis of cholesterol. But this regulation is not a simple case of feedback inhibition of the rate-limiting enzyme. The cholesterol pathway presents two complications. First, the end product—cholesterol—is located entirely within the membrane. How does the cell monitor the levels of a membrane component? Second, the pathway from mevalonate produces several other important products besides cholesterol, including geranyl and farnesyl pyrophosphate for prenylation of proteins and synthesis of ubiquinone and dolichol. How does the cell coordinate the synthesis of all of these products?

Control of HMG-CoA reductase occurs at both transcriptional and posttranscriptional levels, in a process that employs a set of protein–protein interactions that take place within the ER membrane. Mammalian HMG-CoA reductase is anchored in the ER membrane, by a hydrophobic N-terminal domain composed of eight membrane-spanning segments. The central players in this elegant regulatory mechanism are **Insigs** (Insulin-induced growth response genes), **SREBPs** (sterol regulatory element binding proteins), and **Scap** (SREBP cleavage-activating protein) (**FIGURE 16.37**). Like HMG-CoA reductase, all of these regulatory proteins are anchored in the ER membrane by multiple transmembrane segments. We now understand that Insigs control cholesterol synthesis through sterol-induced protein–protein interactions within the ER membrane.

As implied by their name (sterol regulatory element binding proteins), SREBPs are transcription factors that bind to the promoters of genes required to produce cholesterol, including the HMG-CoA reductase gene. Conversion of acetyl-CoA to cholesterol requires more than 20 enzymes, all of whose genes are activated by SREBP binding. SREBP also activates transcription of the gene for the LDL receptor, which mediates uptake of dietary cholesterol. However, as shown in Figure 16.37, SREBPs are synthesized as integral membrane proteins in the ER. When cells are depleted of cholesterol, Scap then

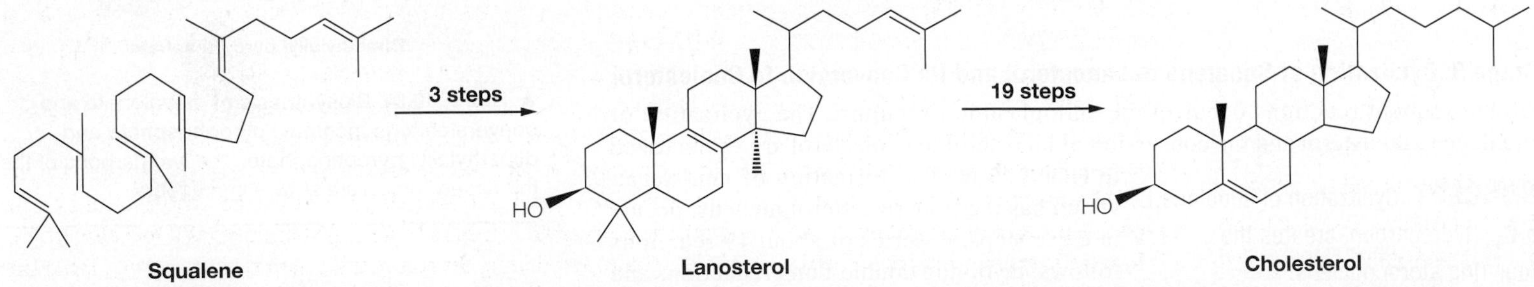

▲ **FIGURE 16.36 Conversion of squalene to cholesterol.**

escorts SREBPs to the Golgi, where they are proteolytically processed to yield active transcription factors that enter the nucleus. When sterols accumulate in ER membranes (either from cellular uptake or from de novo biosynthesis), Scap binds cholesterol, causing a conformational change in Scap that promotes Insig binding. This inhibits translocation of Scap–SREBP complexes to the Golgi, blocking the proteolytic activation of SREBPs. This causes a decrease in the transcription of SREBP target genes, leading to lower rates of cholesterol synthesis and uptake.

Mammalian HMG-CoA reductase is also regulated posttranscriptionally by controlled proteolysis. When sterols accumulate in the ER membrane, HMG-CoA reductase is rapidly degraded ($t_{1/2} < 1$ hr) by the ubiquitin-proteasome pathway (more about this in Chapter 18). Membrane sterols bind to HMG-CoA reductase, causing it to bind to a population of Insigs that are associated with a ubiquitination complex. This protein-protein interaction, stimulated by sterol binding, leads to proteolysis of HMG-CoA reductase.

Insig proteins thus integrate both the transcriptional and posttranscriptional regulatory mechanisms. In both processes, sterol binding to a sterol-sensing domain (in HMG-CoA reductase or Scap) stimulates Insig binding. For HMG-CoA reductase, Insig binding means ubiquitination and destruction. For Scap, Insig binding turns off a transcriptional activation pathway. Together, these mechanisms ensure fine control over cellular cholesterol metabolism.

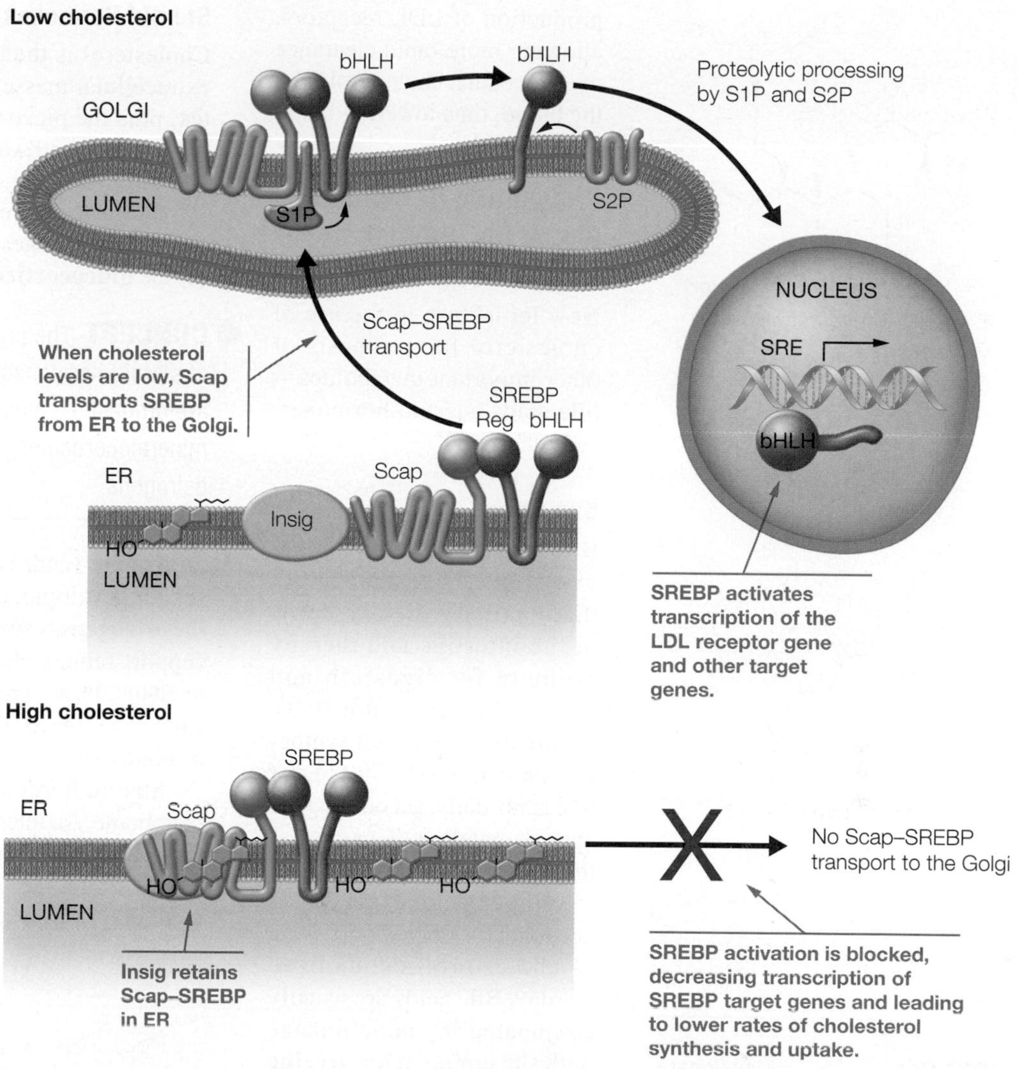

Finally, in some tissues, HMG-CoA reductase is subject to short-term regulation by reversible phosphorylation/dephosphorylation. Phosphorylation is catalyzed by AMP-activated protein kinase (AMPK). As we shall see in Chapter 17, the AMPK system acts as a sensor of cellular energy status—it is activated by increases in the cellular AMP:ATP ratio. Cholesterol biosynthesis is a particularly expensive pathway, requiring 36 moles of ATP and 16 moles of NADPH per mole of cholesterol. Thus, under conditions of low energy charge, activated AMPK switches off cholesterol synthesis by inhibiting HMG-CoA reductase. The enzyme is reactivated by dephosphorylation, catalyzed by a type 2A protein phosphatase.

In vertebrates, cholesterol homeostasis is maintained by a mechanism that coordinates dietary intake of cholesterol, rate of endogenous cholesterol synthesis in the liver (and to a lesser extent in the intestine), and rate of cholesterol use by cells. However, when these regulatory mechanisms break down in disease or are overwhelmed by excessive dietary cholesterol, hypercholesterolemia, and eventually atherosclerosis, can result. Once the rate-limiting role of HMG-CoA reductase in cholesterol biosynthesis was understood, specific inhibitors were sought as a therapeutic approach to lowering blood cholesterol levels. Several compounds were discovered, collectively called **statins,** that act by competitively inhibiting HMG-CoA reductase. Shown on the next page are the structures of several widely used statins, including the fungal polyketides lovastatin and simvastatin and the synthetic atorvastatin (Lipitor™). Each statin carries a mevalonate-like moiety (blue), explaining the competitive nature of its activity. Inhibition of HMG-CoA reductase depresses de novo cholesterol biosynthesis and, hence, intracellular cholesterol levels. This in turn leads to increased

● **CONNECTION** Statins have proven to be spectacularly effective in the treatment of hypercholesterolemia and are among the most widely prescribed drugs in the United States, Canada, and other developed nations.

Low cholesterol

GOLGI

LUMEN

bHLH

bHLH

Proteolytic processing by S1P and S2P

S1P

S2P

When cholesterol levels are low, Scap transports SREBP from ER to the Golgi.

Scap–SREBP transport

SREBP Reg bHLH

ER

Insig

Scap

HO

LUMEN

NUCLEUS

SRE

bHLH

SREBP activates transcription of the LDL receptor gene and other target genes.

High cholesterol

SREBP

ER

Scap

HO

HO

HO

LUMEN

Insig retains Scap–SREBP in ER

No Scap–SREBP transport to the Golgi

SREBP activation is blocked, decreasing transcription of SREBP target genes and leading to lower rates of cholesterol synthesis and uptake.

▲ **FIGURE 16.37 Insig-mediated regulation of SREBP activation.** When cholesterol levels in the cell are low, Scap transports SREBP to the Golgi. Proteolytic processing by membrane-bound proteases S1P and S2P releases SREBP's transcription factor domain (bHLH), which enters the nucleus and binds to sterol regulatory elements (SRE) in the promoters of target genes, stimulating their transcription. High levels of cholesterol block this process by Insig-mediated retention of Scap-SREBP in the ER.

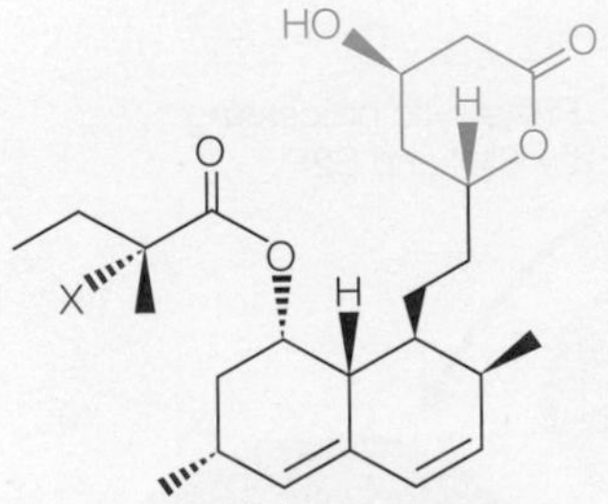

X = H   **Lovastatin (Mevacor)**
X = CH₃ **Simvastatin (Zocor)**

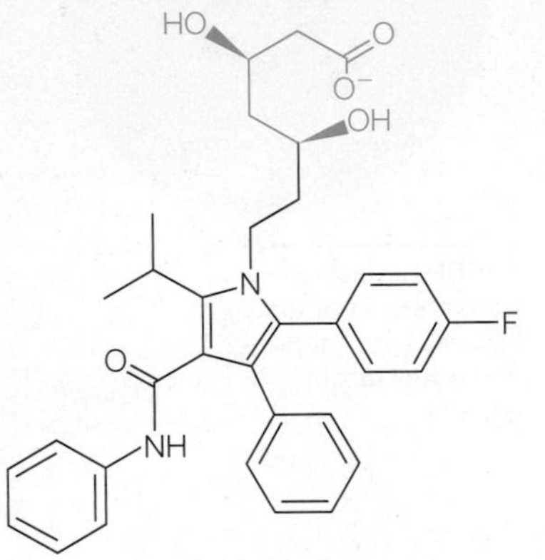

Atorvastatin (Lipitor)

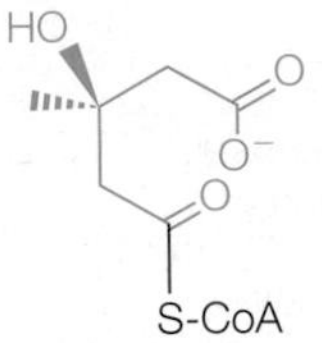

HMG-CoA 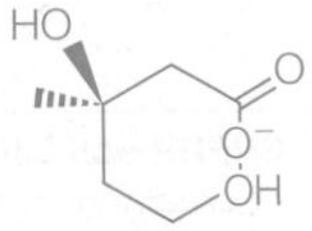   **Mevalonate**

Glycine

Glycocholate

Taurine

Taurocholate

production of LDL receptors, allowing more rapid clearance of extracellular cholesterol from the blood, thus lowering blood cholesterol levels.

Cholesterol Derivatives: Bile Acids, Steroid Hormones, and Vitamin D

Now let us turn to the use of cholesterol for synthesis of other important metabolites—bile acids, steroid hormones, and vitamin D.

Bile Acids

Bile acids are steroid derivatives with detergent properties that emulsify dietary lipids in the intestine and thereby promote fat digestion and absorption (see Figure 16.3). A normal human adult synthesizes about 400 to 500 mg of bile acids daily, accounting for approximately 90% of cholesterol catabolism. By contrast, steroid hormone synthesis accounts for only about 50 mg of cholesterol metabolized per day. Bile acids are usually conjugated in amide linkage with the amino acids **glycine** or **taurine**, giving compounds called **bile salts.** The cholic acid conjugates **glycocholate** and **taurocholate** are major bile salts in mammals.

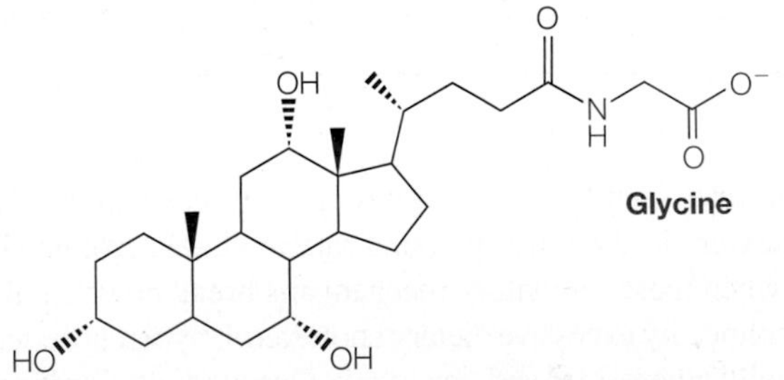

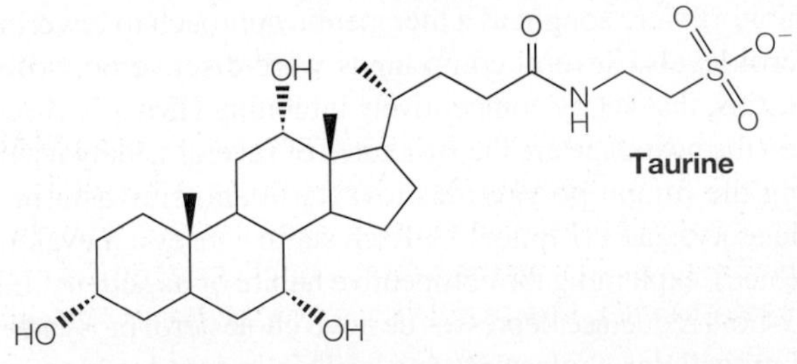

Steroid Hormones

Cholesterol is the biosynthetic source of all steroid hormones, the extracellular messengers secreted by the gonads and the adrenal cortex, plus the placenta in pregnant females. Here we summarize the biosynthetic pathways to steroid hormones; their actions are discussed in Chapter 20.

There are five major classes of steroid hormones: (1) the **progestins** (progesterone), which regulate events during pregnancy; (2) the **glucocorticoids** (cortisol and corticosterone), which promote gluconeogenesis and, in pharmacological doses, suppress inflammation reactions; (3) the **mineralocorticoids** (aldosterone), which regulate ion balance by promoting reabsorption of K^+, Na^+, Cl^-, and HCO_3^- in the kidney; (4) the **androgens** (androstenedione and testosterone), which promote male sexual development and maintain male sex characteristics; and (5) the **estrogens** (estrone and estradiol), or female sex hormones, which support female characteristics. Most of these hormones are shown in **FIGURE 16.38**, which also summarizes their routes of synthesis. In each case, the side chain in cholesterol is either greatly shortened or nonexistent.

● **CONCEPT** The principal categories of steroid hormones in vertebrates are progestins, glucocorticoids, mineralocorticoids, androgens, and estrogens.

Steroid hormone biosynthesis begins in the inner mitochondrial membrane. An integral membrane cytochrome P450 called **cholesterol side-chain cleavage enzyme** hydroxylates the side chain of cholesterol at C-20 and C-22 and cleaves it, to yield **pregnenolone,** the precursor to all other steroid hormones.

Cholesterol

Pregnenolone

Progesterone

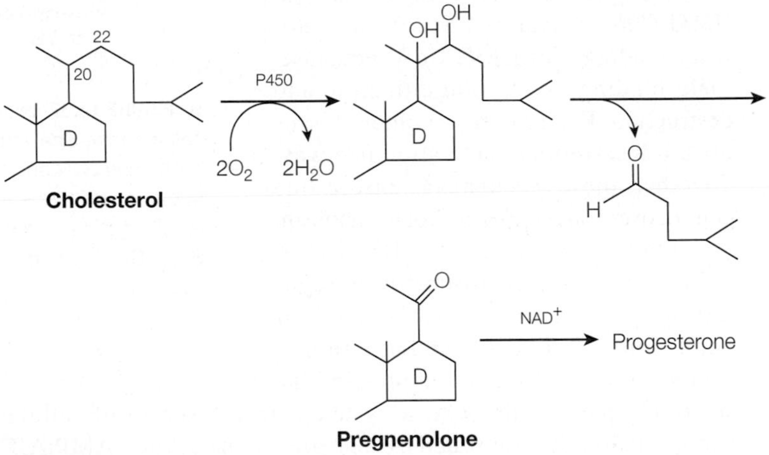

Pregnenolone then moves to the endoplasmic reticulum, where a series of dehydrogenations and hydroxylations lead to all of the other steroid hormones, as shown in Figure 16.38. All of these hydroxylations are catalyzed by cytochrome P450 enzymes, typically obtaining the necessary reducing equivalents from NADPH (see Section 14.9).

Vitamin D

Vitamin D acts analogously to a steroid hormone and is involved in the regulation of calcium and phosphorus metabolism. The most abundant

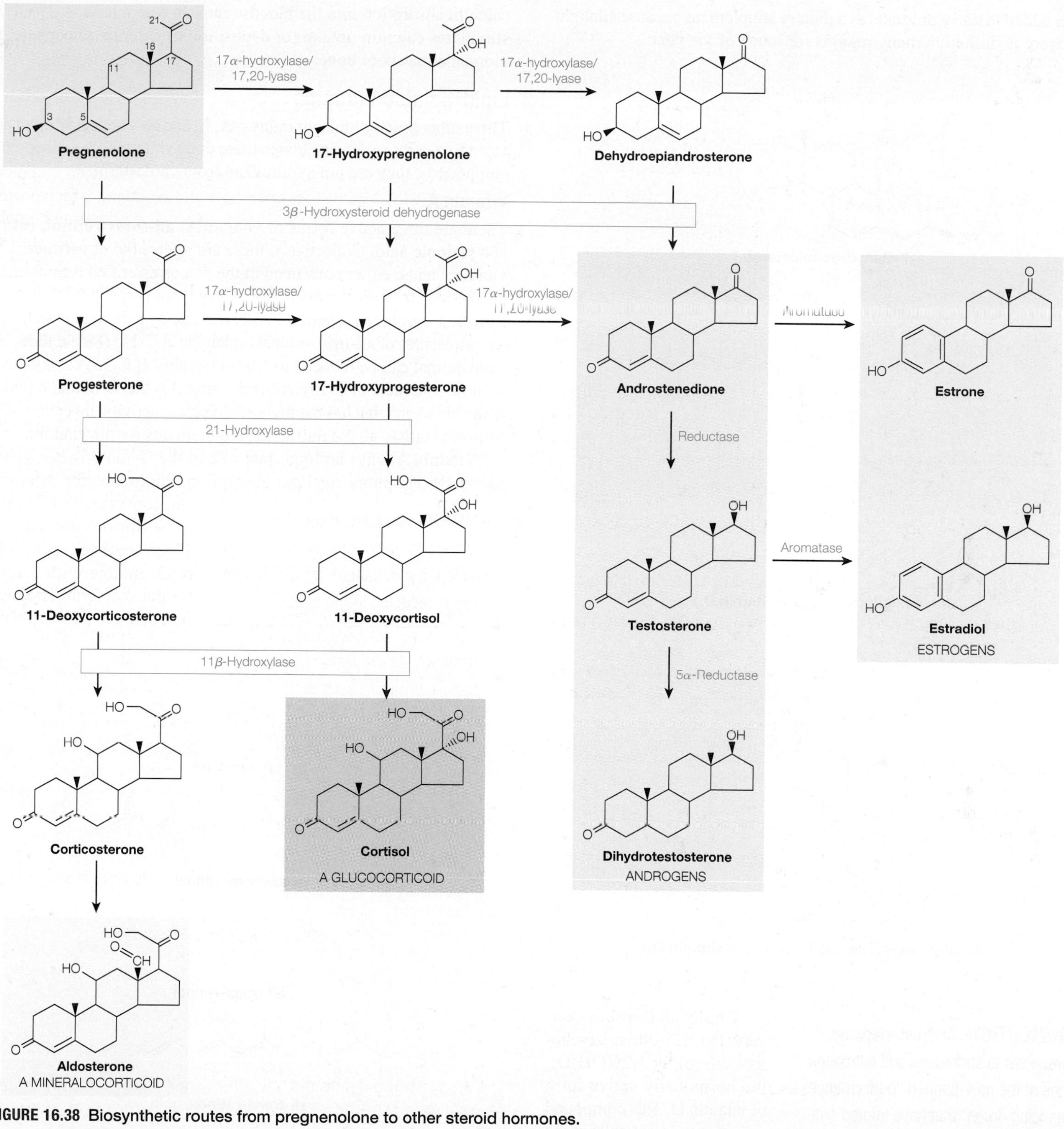

▲ **FIGURE 16.38** Biosynthetic routes from pregnenolone to other steroid hormones.

form of this lipid-soluble vitamin is vitamin D_3, or **cholecalciferol.** This is not truly a vitamin because it is not required in the diet. Instead,

● **CONCEPT** Pregnenolone is an intermediate en route from cholesterol to all other known steroid compounds.

it arises by synthesis from 7-dehydrocholesterol, an intermediate in cholesterol biosynthesis (see next page).

In skin cells, 7-dehydrocholesterol undergoes ultraviolet photolysis to give cholecalciferol. Because the UV rays come from sunlight, insufficient sunlight exposure can cause a deficiency of vitamin D_3 and result in the bone malformation known as **rickets.** Mild deficiency can lead to **osteoporosis,** loss of calcium from the bones. Vitamin D_3 is often added to dairy products as a dietary supplement because sunlight exposure is limited in many regions for much of the year.

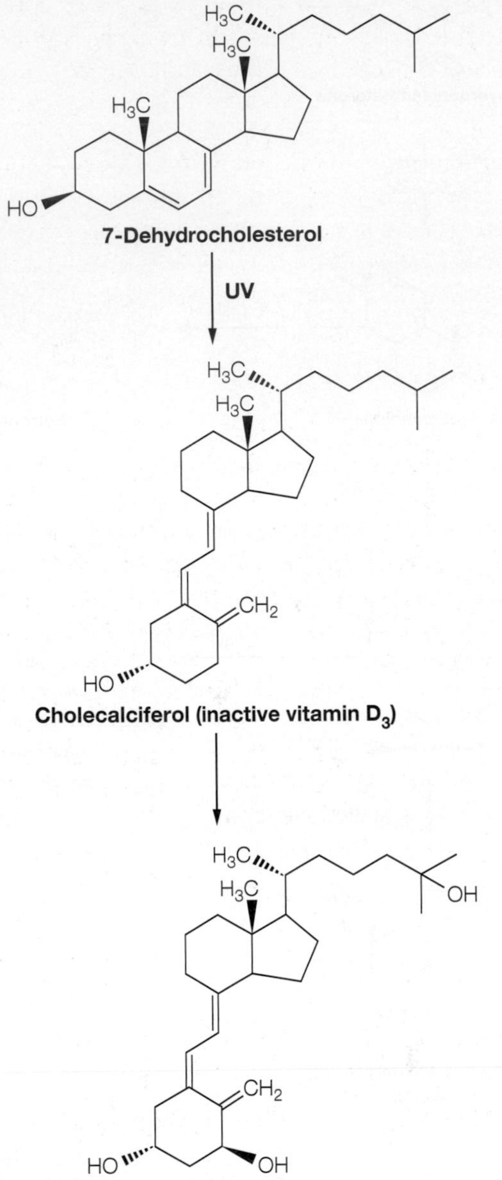

7-Dehydrocholesterol

UV

Cholecalciferol (inactive vitamin D₃)

1,25-Dihydroxycholecalciferol (active vitamin D₃)

● **CONNECTION** Androstenedione, a precursor to androgens and estrogens, is one of the now-banned "performance-enhancing drugs" that have tainted professional sports.

● **CONCEPT** 1,25-Dihydroxychole-calciferol controls bone metabolism by regulating intestinal absorption of calcium.

Cholecalciferol is converted to 1,25-dihydroxycho-lecalciferol, or 1,25(OH)D₃, the hormonally active form of vitamin D. This compound migrates to target cells in the intestine and in osteoblasts (bone cells), where it binds to protein receptors that migrate to the cell nucleus. In intestine, the hormone–receptor complex stimulates transcription, resulting in synthesis of a protein that stimulates

calcium absorption into the bloodstream. In osteoblasts, 1,25(OH)D₃ stimulates calcium uptake for deposition as calcium phosphate, the inorganic matrix of bone.

Lipid-Soluble Vitamins

Three other lipid-soluble vitamins—A, E, and K—are made up of activated five-carbon units. Although these three vitamins are all isoprenoid compounds, they are not synthesized from cholesterol.

Vitamin A

There are three active forms of vitamin A: **all-*trans*-retinol, retinal,** and **retinoic acid.** Collectively, these are referred to as **retinoids.** The vitamin can be either consumed in the diet as esterified retinol or biosynthesized from **β-carotene,** a plant isoprenoid especially abundant in carrots. β-carotene is cleaved in the intestine by a dioxygenase to form two molecules of all-*trans*-retinal (aldehyde at C-15) (**FIGURE 16.39**). All-*trans*-retinal can be reduced to retinol (alcohol at C-15) or oxidized to retinoic acid (acid at C-15). All-*trans*-retinol is the form that circulates in the blood and that has the highest biological activity. Ingestion of this form will satisfy all the nutritional requirements for the vitamin.

Vitamin A plays an important role in the visual process, serving as the chromophore for light absorption. Photoreceptor cells (rods and cones) are rich in **opsin,** a membrane protein with **11-*cis*-retinal** in a Schiff base linkage with a lysine residue that forms **rhodopsin** in rods or **photopsin** in

● **CONCEPT** Isomerization of a protein-bound form of vitamin A in the retina is the mechanism by which light energy is received in the eye.

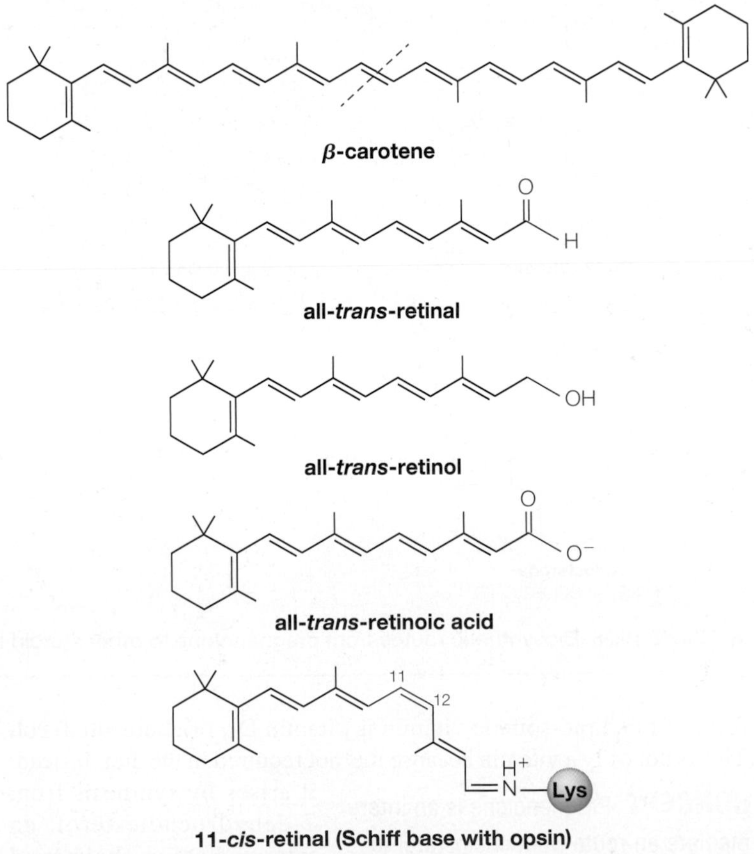

β-carotene

all-*trans*-retinal

all-*trans*-retinol

all-*trans*-retinoic acid

11-*cis*-retinal (Schiff base with opsin)

▲ **FIGURE 16.39** Retinoids.

cones (Figure 16.39). Absorption of a photon of light by 11-*cis*-retinal causes photoisomerization, which triggers a chain of events leading to neural excitation. This process involves cyclic GMP and a G protein called **transducin.** Further details of this process are presented in Chapter 20.

Retinoids (mainly all-*trans*- and 9-*cis*-retinoic acid) also play important roles as developmental regulators, acting much like steroid hormones (see Chapter 20). They interact with specific receptor proteins in the cell nucleus. The ligand–receptor complexes bind to specific DNA sequences, where they control the transcription of genes involved in embryonic development, reproduction, postnatal growth, immune responses, and differentiation of epithelia. Indeed, the earliest effect of vitamin A deficiency is a keratinization of epithelial tissues of the respiratory and urogenital tracts, in which the columnar epithelia become replaced by squamous epithelium. Lesions in the eyes (e.g., night blindness) occur much later in vitamin A deficiency.

Vitamin E

Vitamin E, also called **α-tocopherol,** was originally recognized in nutritional studies as an agent that prevented sterility in rats. The vitamin appears to play an antioxidant role, particularly in preventing the attack of peroxides on unsaturated fatty acids in membrane lipids. Additional biological roles seem likely, since vitamin E deficiency results in symptoms such as erythrocyte hemolysis and neuromuscular dysfunction that are not relieved by other antioxidants.

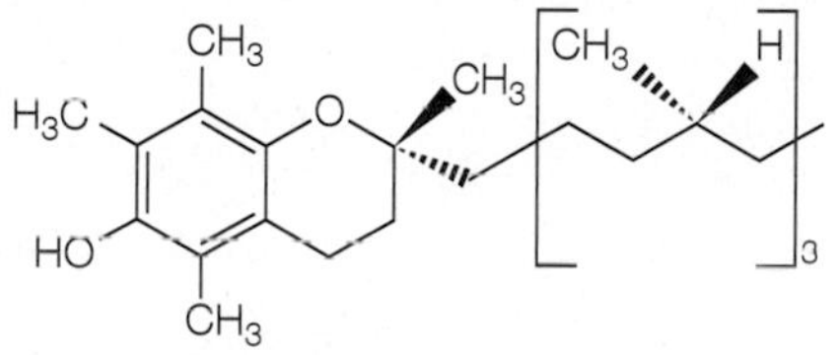

α-Tocopherol (vitamin E)

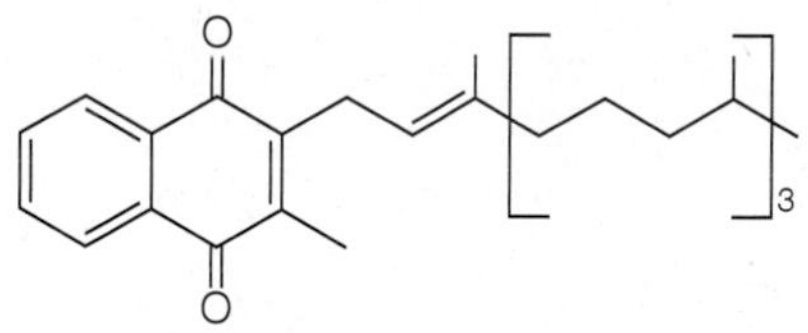

Phylloquinone (vitamin K₁)

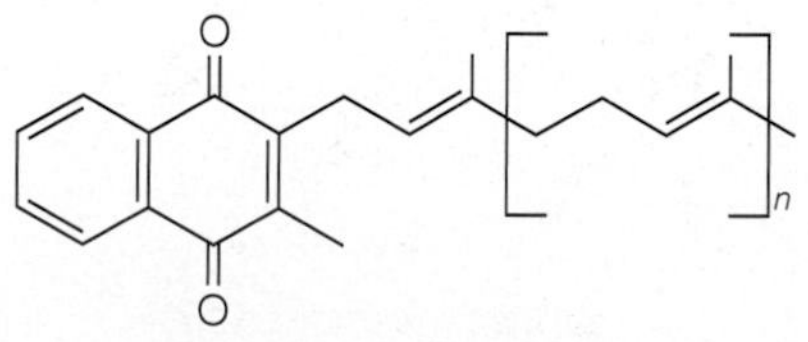

Menaquinone (vitamin K₂)

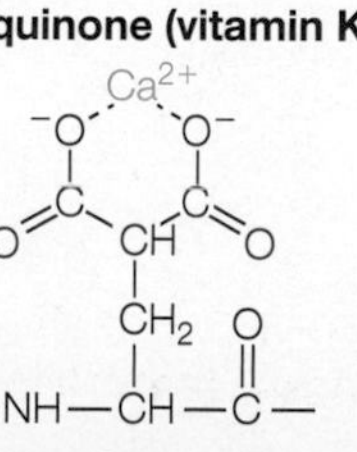

A γ-carboxyglutamate residue complexed with calcium

Vitamin K

Vitamin K was originally discovered as a lipid-soluble substance involved in blood coagulation. Vitamin K_1, or **phylloquinone,** is found in plants; the quinone portion of this molecule has a largely saturated side chain. Another form of the vitamin, vitamin K_2, or **menaquinone,** is found largely in animals and bacteria. Menaquinone has a partly unsaturated side chain. In animals, vitamin K_2 is essential for the carboxylation of glutamate residues in certain proteins, to give **γ-carboxyglutamate.** This modification allows the protein to bind calcium, an essential event in the blood-clotting cascade (discussed in Chapter 8). Newborn children routinely receive vitamin K injections because most of our vitamin K comes from intestinal bacteria, which have not yet colonized the guts of newborns.

16.8 Eicosanoids: Prostaglandins, Thromboxanes, and Leukotrienes

We turn finally to the **eicosanoids,** a family of locally acting signaling molecules. These lipids are distinguished by their potent physiological properties, low levels in tissues, rapid metabolic turnover, and common metabolic origin. The most important of these compounds are the **prostaglandins;** also included are the **thromboxanes** and **leukotrienes.** Collectively, these compounds are called eicosanoids because of their common origin from C_{20} polyunsaturated fatty acids, the eicosaenoic acids, particularly arachidonic acid, which is all-*cis*-5,8,11,14-eicosatetraenoic acid. Recall that arachidonic acid is synthesized from linolenic acid (Section 16.3). **FIGURE 16.40** summarizes the biosynthesis of these compounds from arachidonic acid.

● **CONCEPT** The biologically active eicosanoids, derived from arachidonic acid, include prostaglandins, thromboxanes, and leukotrienes. They are short-lived, locally acting signaling molecules.

LTA₄ and other leukotrienes

Arachidonic acid

PGH synthase (COX)

PGH₂ and other prostaglandins

TxA₂ and other thromboxanes

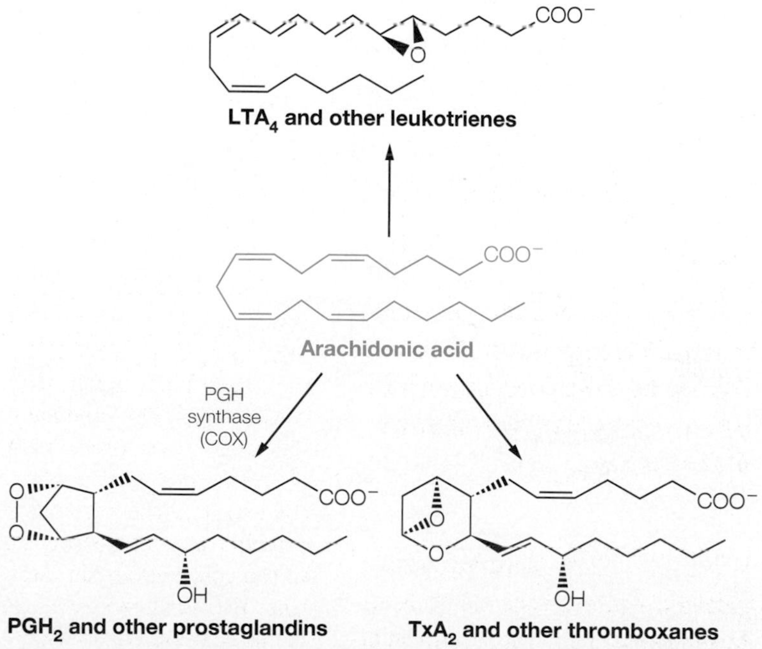

▲ **FIGURE 16.40** Eicosanoid biosynthesis.

The key step in the synthesis of prostaglandins is catalyzed by **PGH synthase,** a bifunctional heme-containing endoplasmic reticulum membrane enzyme. The two enzyme activities, **cyclooxygenase** and **peroxidase,** act on free arachidonate to generate PGH_2 (Figure 16.40). Mammalian cells contain two distinct forms of PGH synthase, called PGHS-1 and PGHS-2 (or COX-1 and COX-2, where COX stands for cyclooxygenase). COX-1 is constitutively expressed in most tissues and is responsible for the physiological production of prostaglandins. COX-2 is induced by cytokines, mitogens, and endotoxins in inflammatory cells and is responsible for the elevated production of prostaglandins during inflammation.

● **CONNECTION** The cyclooxygenase reaction, one of the first steps in eicosanoid synthesis, is the target site for aspirin action.

The anti-inflammatory activity of aspirin has been known ever since it was isolated by the German chemical company Bayer in 1897. But its mechanism of action remained a mystery until 1971, when it was discovered that aspirin inhibits one of the enzymes in prostaglandin biosynthesis, PGH synthase. This inhibition is now known as the major

site of action of aspirin and other nonsteroidal anti-inflammatory drugs (NSAIDs). Both COX isoforms are covalently modified, and hence inactivated, by reaction with aspirin (acetylsalicylic acid). As shown, aspirin acetylates a specific serine residue, which in turn blocks access of the fatty acid substrate to the cyclooxygenase active site.

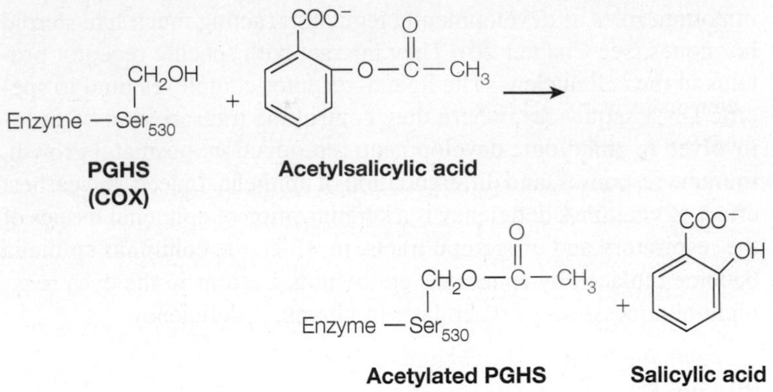

Summary

- **Triacylglycerols** are the main form for storage of biological energy. Low-density lipoproteins represent the major vehicle for transport of cholesterol to peripheral tissues. Cholesterol levels in blood are regulated through control of synthesis of LDL receptors, involved in cellular uptake of LDL by endocytosis. Faulty control of LDL levels contributes to the development of **atherosclerotic plaque.** Fat depots are mobilized by enzymatic hydrolysis of triacylglycerols to fatty acids plus glycerol in a process hormonally controlled via cyclic AMP. (Section 16.1)

- Most **fatty acid degradation** occurs through **β-oxidation.** β-oxidation is a mitochondrial process that involves stepwise oxidation and removal of two-carbon fragments as acetyl-CoA. Under conditions in which further oxidation of acetyl-CoA through the citric acid cycle is limited, acetyl-CoA is used to synthesize **ketone bodies,** which are excellent energy substrates for brain and heart. (Section 16.2)

- **Fatty acid biosynthesis** occurs via the stepwise addition of two-carbon fragments, in a process that superficially resembles a reversal of β-oxidation. Metabolic activation involves acyl carrier protein, and the reductive power comes from NADPH. In eukaryotic cells, the seven enzyme activities are linked covalently on multifunctional enzymes or multienzyme complexes called **megasynthases.** Elegant mechanisms have evolved to coordinately regulate the synthesis and degradation of fatty acids. Triacylglycerols are synthesized by straightforward pathways in which acyl groups in fatty acyl-CoAs are transferred to the hydroxyl groups of glycerol-3-phosphate and diacylglycerol. (Sections 16.3-16.4)

- **Glycerophospholipids,** the predominant membrane lipids, are synthesized by routes that start from phosphatidic acid and intermediates activated by reaction with cytidine triphosphate. (Section 16.5)

- **Sphingolipids** are assembled from ceramide and successive sugar additions involving glycosyltransferases and nucleotide-linked sugars. Enzymatic defects in the turnover of these compounds causes accumulation of intermediates in nervous tissue, with severe consequences. (Section 16.6)

- All **steroid** compounds—and indeed all isoprenoid compounds—are synthesized from acetate. The pathway proceeds through the six-carbon mevalonic acid and involves C_5, C_{10}, C_{15}, and C_{30} (squalene) intermediates. Cyclization of squalene leads to **cholesterol,** the precursor for all bile acids and steroid hormones. All steroid hormone synthesis proceeds from cholesterol through pregnenolone. (Section 16.7)

- Arachidonic acid is the precursor to physiologically potent, locally acting hormones that include **prostaglandins, thromboxanes,** and **leukotrienes.** Although actions of these biological regulators are not yet understood at the molecular level, metabolism of these **eicosanoids** presents therapeutic targets for drugs used to control inflammation, blood clotting, and gastric secretion and to manipulate reproductive processes in various ways. (Section 16.8)

Problems

Enhanced by Mastering Chemistry for Biochemistry | Mastering Chemistry for Biochemistry provides select end-of-chapter problems and feedback enriched tutorial problems, animations, and interactive figures to deepen your understanding of complex topics while practicing problem solving.

Answers to red problems are available in the Answer Appendix.

1. Calculate the ATP yield from oxidation of palmitic acid, taking into account the energy needed to activate the fatty acid and transport it into mitochondria. Do the same for stearic acid, linoleic acid, and oleic acid.

2. If palmitic acid is subjected to complete combustion in a bomb calorimeter, one can calculate a standard free energy of combustion of 9788 kJ/mol. From the ATP yield of palmitate oxidation, what is the metabolic efficiency of the biological oxidation, in terms of kilojoules saved as ATP per kilojoule released? (Ignore the cost of fatty acid activation.)

3. Calculate the number of ATPs generated by the complete metabolic oxidation of tripalmitin (tripalmitoylglycerol). Hydrolysis of the triacylglycerol occurs at the cell surface. Consider the energy yield from catabolism of glycerol, as well as from the fatty acids. Calculate the ATP yield per carbon atom oxidized, and compare it with the energy yield from glucose.

4. Write a balanced equation for the *complete* metabolic oxidation of each of the following. Include O_2, ADP, and P_i as reactants and ATP, CO_2, and H_2O as products.
   (a) Stearic acid
   (b) Oleic acid
   (c) Palmitic acid
   (d) Linoleic acid

5. Calculate the number of ATPs generated from the metabolic oxidation of the four carbons of acetoacetyl-CoA to CO_2. Now consider the homolog derived from oxidation of an odd-numbered carbon chain, namely, propionoacetyl-CoA. Calculate the net ATP yield from oxidation of the five carbons of this compound to CO_2.

6. Under conditions where ketone bodies are being produced in the liver, how many ATPs can be produced from a molecule of palmitic acid if all resulting molecules of acetyl-CoA are converted into β-hydroxybutyrate?

7. β-hydroxybutyrate dehydrogenase catalyzes the last step in ketogenesis, the reduction of acetoacetate to β-hydroxybutyrate.
   (a) Write a balanced equation for this reaction.
   (b) Calculate the standard free energy ($\Delta G°'$) associated with this reaction, using the data in Table 14.1. Is this reaction favorable in the direction written under standard conditions? If not, what cellular conditions would be required for this reaction to continuously produce β-hydroxybutyrate at a high rate?

8. 2-Bromopalmitoyl-CoA inhibits the oxidation of palmitoyl-CoA by isolated mitochondria but has no effect on the oxidation of palmitoylcarnitine. What is the most likely site of inhibition by 2-bromopalmitoyl-CoA?

9. When the identical subunits of chicken liver fatty acid synthase are dissociated in vitro, all of the activities can be detected in the separated subunits except for the β-ketoacyl synthase reaction and the overall synthesis of palmitate. Explain these observations.

10. Mammals cannot undergo *net* synthesis of carbohydrate from acetyl-CoA, but the carbons of acetyl-CoA can be incorporated into glucose and amino acids. Present pathways by which this could come about.

11. Describe a pathway whereby some of the carbon from a fatty acid with an odd-numbered carbon chain could undergo a net conversion to carbohydrate.

12. As low-carbohydrate diets have experienced a dramatic increase in popularity, arguments have been made that glucose can be made from odd-chain fatty acids. Based on the metabolism of such molecules, what quantity of a C-19 fatty acid would be required to produce 1 g of glucose? Considering that odd-chain fatty acids make up approximately 1% of the fat in our diet, what quantity of fatty acids would be needed to produce 1 g of glucose?

13. How many tritium atoms (^{3}H) are incorporated into palmitate when fatty acid synthesis is carried out in vitro with the following labeled substrate?

$$^-OOC - C^3H_2 - \overset{\overset{\displaystyle O}{\|}}{C} - S - CoA$$

14. A dialyzed pigeon liver extract will catalyze the conversion of acetyl-CoA to palmitate and CoASH if supplied with Mg^{2+}, NADPH, ATP, HCO_3^-, and citrate.
    (a) If $H^{14}CO_3^-$ is supplied, what compounds will become labeled (permanently or transiently) during the course of the reaction? In what compounds will ^{14}C accumulate?
    (b) Explain the role of citrate in this reaction.

15. The following link carbohydrate metabolism with lipid biosynthesis:
    (a) How many molecules of glucose are required to provide the carbon for synthesis of one molecule of palmitate?
    (b) How many molecules of glucose are required if all of the glucose first proceeds through the pentose phosphate pathway before proceeding through the rest of glycolysis on its way to pyruvate?

16. What would be the effect on fatty acid synthesis of an increase in intramitochondrial oxaloacetate level? Briefly explain your answer.

17. Glucagon secretion causes inhibition of intracellular acetyl-CoA carboxylase activity by several mechanisms. Name all you can think of.

18. Identify and briefly discuss each mechanism ensuring against simultaneous fatty acid synthesis and oxidation in the same cell.

19. Discuss the metabolic rationale for phosphorylation of acetyl-CoA carboxylase by AMP-activated protein kinase (AMPK) and cyclic AMP–dependent protein kinase (PKA).

20. Describe the probable effect in adipocytes of insulin-stimulated uptake of glucose into these cells.

21. Phosphatidylserine (PS) is considered to be an intermediate in the biosynthesis of phosphatidylethanolamine (PE) in *E. coli*, yet PS is not found in appreciable amounts among *E. coli* membrane phospholipids. Because PS must be present in the membrane to serve as an intermediate, how might you explain its failure to accumulate to a significant extent? What kinds of experiments could test your proposed explanation?

22. What would you expect to happen to levels of mevalonate in human plasma if an individual were to go from a meat-containing diet to a vegetarian diet?

23. Write a balanced equation for the synthesis of *sn*-1-stearoyl-2-oleoylglycerophosphorylserine, starting with glycerol, the fatty acids involved, and serine.

24. If mevalonate labeled with ^{14}C in the carboxyl carbon were administered to rats, which carbons of cholesterol would become labeled?

25. Which step in lipid metabolism would you expect to be affected by 3,4-dihydroxybutyl-1-phosphonic acid (shown here)? Explain your answer.

$$\begin{array}{c} CH_2OH \\ | \\ HO-C-H \qquad\qquad O \\ | \qquad\qquad\qquad \| \\ CH_2-CH_2-P-O^- \\ | \\ O^- \end{array}$$

26. Identify a pathway for utilization of the four carbons of acetoacetate in cholesterol biosynthesis. Carry your pathway as far as the rate-determining reaction in cholesterol biosynthesis.

27. Explain why a deficiency of steroid 21-hydroxylase leads to excessive production of sex steroids (androgens and estrogens).

28. *cis*-Vaccenate is an 18-carbon unsaturated fatty acid abundant in *E. coli* membrane lipids. Propose a metabolic route for synthesis of this fatty acid, in light of the fact that stearic acid, the C_{18} saturated analogous fatty acid, is virtually absent from *E. coli* lipids.

29. Briefly describe how cyclic AMP controls phospholipid synthesis.

30. In the experiment shown below, fibroblasts obtained from a normal subject (closed symbols) or from a patient homozygous for familial hypercholesterolemia (FH Homozygote) (open symbols) were grown in monolayer cultures. At time zero, the medium was replaced with fresh medium depleted of lipoproteins, and HMG-CoA reductase activity was measured in extracts prepared at the indicated times (panel a). Twenty-four hours after addition of the lipoprotein-deficient medium, human LDL was added to the cells at the indicated levels, and HMG-CoA reductase activity was measured at the indicated time.

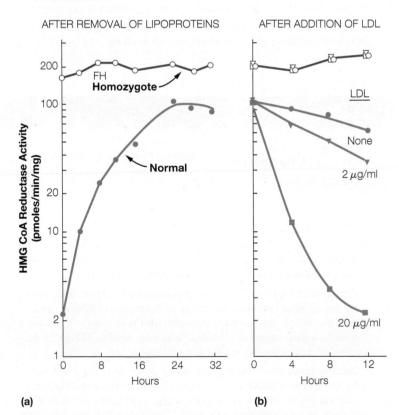

(a)

(b)

Based on your understanding of HMG-CoA reductase regulation, explain the following results:
(a) When cultured in the presence of LDL, normal cells showed low activity of HMG-CoA reductase. After removal of lipoproteins, including LDL, HMG-CoA reductase activities increased some 50- to 100-fold in

normal cells (panel a). This high level of enzyme activity was rapidly suppressed upon addition of LDL back to normal cells (panel b).

(b) Cells from FH individuals showed high levels of reductase activity, whether cultured in the presence or absence of LDL.

31. Thyroid-stimulating hormone (TSH) has been shown to stimulate cholesterol synthesis in liver. Zhang et al, in *J. Lipid Res.* (2105) 56, 963–971 investigated the mechanism of this effect by treating mouse hepatocytes (liver cells) with TSH. The data below show the ratio of phosphorylated HMG-CoA reductase to total HMG-CoA reductase in these cells following treatment with 0, 1, or 4 μM bovine TSH (bTSH). Based on these data and your knowledge of the regulation of HMG-CoA reductase, propose a mechanism that explains the stimulatory effect of TSH on cholesterol synthesis.

32. In addition to the pathway described in Figure 16.31, eukaryotes can also synthesize phosphatidylcholine and phosphatidylethanolamine, starting with free choline or ethanolamine, respectively. In this salvage pathway, the phosphodiester linkage in the glycerophospholipid is formed by activating the –OH of the head group instead of the diacylglycerol moiety. Propose a pathway for the synthesis of phosphatidylcholine from free choline using this salvage route.

33. During constant flight, hummingbirds expend about 2.9 kJ/hr, relying on fat oxidation as an energy source.

(a) Calculate the grams of fat necessary to sustain a 20-hr nonstop flight.

(b) If the birds relied on carbohydrate for this energy, how much additional weight would they need to have to start the flight?

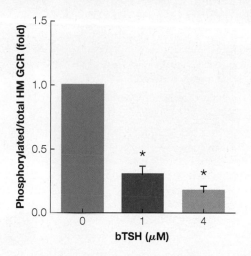

 References

For a list of references related to this chapter, see Appendix II.

A complex regulatory system has evolved to coordinate and integrate the control of energy metabolism. Malfunction of a single component can disrupt this delicate balance. The mouse on the left is unable to produce the hormone leptin and so grows to three times the weight of its normal littermate.

Interorgan and Intracellular Coordination of Energy Metabolism in Vertebrates

OUR PRESENTATION OF intermediary metabolism in earlier chapters has placed primary emphasis on the cell—its reactions, its individual enzymes, and its particular metabolites. In this chapter we integrate these individual pathways in two ways. First, we review the metabolic profiles of the major organs in vertebrates: the fuels they use, the fuels they generate, and the way the organs interact under stress to maintain appropriate energy balance. Second, we describe how these interactions are controlled in large part by hormonal signals, some of which have already been introduced. The molecular mechanisms of hormonal signal transduction will be discussed in detail in Chapter 20.

As we consider metabolic integration, keep in mind the major metabolic control mechanisms first introduced in Chapter 11 (Section 11.5). We have seen in the previous five chapters that metabolism

is controlled to a great extent by the availability of substrates for specific metabolic pathways. In general, substrate concentrations within cells fall below saturating levels for the enzymes that metabolize them. Therefore, fluxes through particular enzymes vary as the concentrations of their substrates vary. A good example is the metabolic adaptation that occurs during a marathon run. Once the glycogen stores in liver and muscle are exhausted, flux through glycolysis decreases in muscle, not for any

● **CONCEPT** Metabolite concentrations represent a significant intracellular control mechanism.

hormonal reason but simply because glucose phosphates are less available. Hormonal

adjustments do occur to allow increased use of fatty acids in muscle, but the primary factor determining which metabolic pathway becomes functional in the cell—and thus which substrates are catabolized for energy—is the concentration of each of the usable substrates.

Overlaid on top of these substrate-level regulatory mechanisms operating within a eukaryotic cell are

● **CONCEPT** Hormonal signals are an extracellular control mechanism that coordinates metabolism throughout the entire animal.

hormonal signals that coordinate metabolism throughout the entire animal. These extracellular messengers interact with specific receptors on target organs, resulting in specific metabolic changes in the target cell. Hormonal regulation may involve control of enzyme synthesis or enzyme activity.

17.1 Interdependence of the Major Organs in Vertebrate Fuel Metabolism

In this section, we look at metabolism not as the activities of a selected group of enzymes in just one cell, but as a set of interconnected chemical reactions in a complex multicellular animal. We emphasize the specialized roles that each of the major organs—brain, muscle, liver, adipose tissue, and heart—plays in fuel metabolism, and we describe the varying relationships among these organs as the animal encounters different physiological conditions.

Fuel Inputs and Outputs

In a differentiated organism, each tissue must be provided with fuels that it can use, in amounts sufficient to meet its own energy needs and to perform its specialized roles. The kidney, for example, must generate ATP for the osmotic work of transporting solutes against a concentration gradient for excretion. Muscle must generate ATP for the mechanical work of contraction, and particularly in heart muscle, this energy supply must be continuous. The liver generates ATP for biosynthetic purposes, whether for plasma protein synthesis, cholesterol generation, fatty acid synthesis, gluconeogenesis, or the production of urea for nitrogen excretion. Energy production must meet widely varying needs, depending on level of exertion, composition of fuel molecules in the diet, time since last feeding, and so forth. For example, in humans the daily caloric needs may vary by four-fold, depending in part on the level of exertion—from 1500 to 6000 kcal/day (6000 to 25,000 kJ/day) in an average-sized human.

Metabolic Division of Labor Among the Major Organs

The major organs involved in fuel metabolism vary in their levels of specific enzymes, so that each organ is specialized for the storage,

● **CONCEPT** The major fuel reserves are triacylglycerols (adipose tissue and liver) and glycogen (muscle and liver).

use, and generation of different fuels. The major fuel reserves are *triacylglycerols*, stored primarily in adipose tissue; *protein*, most of which exists in skeletal muscle; and *glycogen*, which is stored in both liver and muscle. In general, an organ specialized to *produce* a particular fuel lacks the enzymes to *use* that fuel. For example, liver is a major producer of ketone bodies, but little catabolism of ketone bodies occurs in the liver. Now let us review how the mobilization of each reserve is controlled and how the organs involved communicate with each other to meet the energy needs of the animal. This information is summarized in **FIGURE 17.1** and **TABLE 17.1**.

Brain

The brain is the most fastidious, and one of the most voracious, of the organs. It must generate ATP in large quantities to drive the ion pumps that maintain the membrane potentials essential for transmission of nerve impulses (see Chapter 10). Under normal conditions, the brain meets its prodigious energy requirement strictly by using glucose, which amounts to about 60% of the glucose utilization of a human at rest. The brain's need for about 120 grams of glucose per day is equivalent to 1760 kJ—about 15% of the total energy consumed by one person. The brain's quantitative requirement for glucose remains quite constant, even when an animal is at rest or asleep. Furthermore, the brain is a highly aerobic organ, and its metabolism demands some 20% of the total oxygen consumed by a human. Because the brain has no significant glycogen or other fuel reserves, the supply of both oxygen and glucose cannot be interrupted, even for a short time. Otherwise, irreversible brain damage results. However, the brain can adapt during fasting or starvation to use ketone bodies (see Section 16.2) instead of glucose as a major fuel.

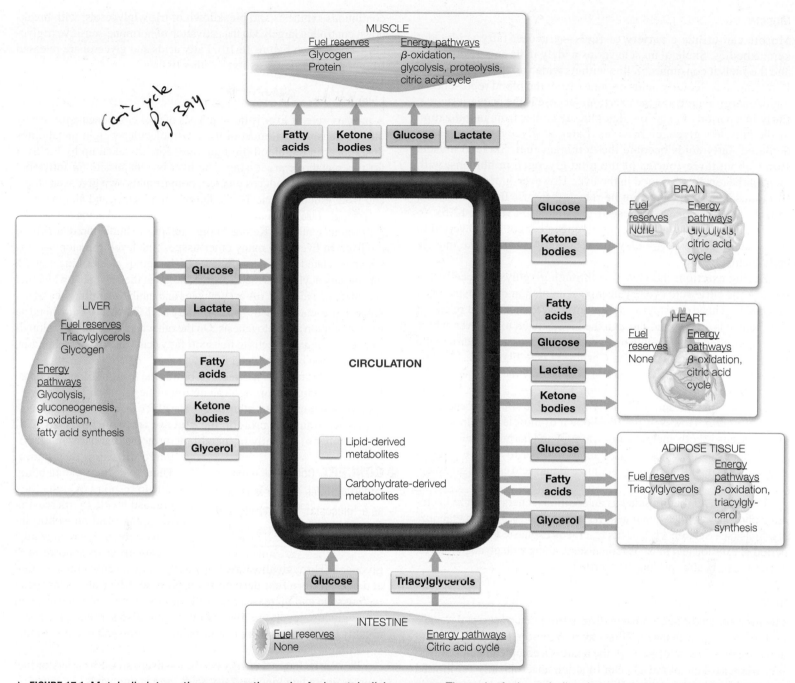

Cori cycle
Pg 394

▲ FIGURE 17.1 Metabolic interactions among the major fuel-metabolizing organs. The major fuel metabolites imported and exported by each organ are shown, along with the fuel stores and major energy pathways that take place in each organ. Lipid-derived metabolites are highlighted in yellow; carbohydrate-derived metabolites are highlighted in blue.

TABLE 17.1 Profiles of the major vertebrate organs in fuel metabolism

Tissue	Fuel Reserves	Preferred Fuel	Fuel Sources Exported
Brain	None	Glucose (ketone bodies during starvation)	None
Skeletal muscle (resting)	Glycogen, protein	Fatty acids	None
Skeletal muscle (during exertion)	None	Glucose	Lactate
Heart muscle	None	Fatty acids	None
Adipose tissue	Triacylglycerols	Fatty acids	Fatty acids, glycerol
Liver	Glycogen, triacylglycerols	Glucose, fatty acids, amino acids	Fatty acids, glucose, ketone bodies

Muscle

Muscle can utilize a variety of fuels—glucose, fatty acids, and ketone bodies. Skeletal muscle varies widely in its energy demands and the fuels it consumes, in line with its wide variations in activity. In resting muscle, fatty acids obtained from the blood represent the major energy source; during exertion, glucose is the primary source. Early in a period of exertion, that glucose comes from mobilization of the muscle's glycogen reserves. Later, as glycogen reserves are depleted, fatty acids become the dominant fuel. Skeletal muscle stores about three-fourths of the total glycogen in humans, with most of the remainder stored in the liver. However, glucose produced from muscle glycogen cannot be released for use by other tissues. Muscle has little or no glucose-6-phosphatase (Section 12.5), so glucose phosphates derived from glycogen cannot be converted to free glucose for export—they are retained for use by the muscle cells themselves.

During exertion, the flux rate through glycolysis exceeds that through the citric acid cycle, causing accumulation of pyruvate from glycolysis. This excess pyruvate is converted to lactate and released. This lactate is transported through the bloodstream to the liver, where it is reconverted through gluconeogenesis to glucose, for return to the muscle and other tissues by the Cori cycle (see Figure 12.12). The major fate of lactate during exertion is uptake by the heart for use as fuel, via oxidation to CO_2 (see Section 12.3).

Muscle contains another readily mobilizable source of energy—its own protein. However, the breakdown of muscle protein to meet energy needs is both energetically wasteful and harmful to an animal, which needs its muscles to move about in order to survive. Protein breakdown is regulated so as to minimize amino acid catabolism except in starvation.

Finally, recall that muscle has an additional energy reserve in creatine phosphate, which generates ATP without the need for metabolizing fuels (see Section 11.4). This reserve is exhausted very early in a period of exertion and must be replenished, along with glycogen stores, as muscle rests after prolonged exertion.

Heart

The metabolism of heart muscle differs from that of skeletal muscle in three important respects. First, work output is far more constant than in skeletal muscle. Second, the heart is a completely aerobic tissue, whereas skeletal muscle can function anaerobically for limited periods. Mitochondria are much more densely packed in heart than in other cells, making up nearly half the volume of a heart cell. Third, the heart contains negligible energy reserves as glycogen or lipid, although there is a small amount of creatine phosphate. Therefore, the supply of both oxygen and fuels from the blood must be continuous to meet the unending energy demands of the heart. The heart uses a variety of fuels—mainly fatty acids but also glucose, lactate, and ketone bodies.

Adipose Tissue

Adipose tissue represents the major fuel reserve for an animal. The total stored triacylglycerols amount to some 555,000 kJ (133,000 Cal) in an average-sized human (see Section 16.1). This is enough fuel, metabolic complications aside, to sustain life for a couple of months in the absence of further caloric intake. The adipocyte is designed for continuous synthesis and breakdown of triacylglycerols, with breakdown controlled largely via the activation of hormone-sensitive triglyceride lipases (see Figure 16.10). Fatty acids and glycerol are released from the adipocyte for export to other tissues.

Liver

A primary role of liver is the synthesis of fuel components for use by other organs. In fact, most of the low-molecular-weight metabolites that appear in the blood through digestion are taken up by the liver for this metabolic processing. The liver is a major site for fatty acid synthesis. It also produces glucose, both from its own glycogen stores and from gluconeogenesis, the latter using lactate and alanine from muscle, glycerol from adipose tissue, and the amino acids not needed for protein synthesis. Ketone bodies are also manufactured largely in the liver. In liver (and many other tissues) the level of malonyl-CoA, which is related to the energy status of the cell, is a determinant of the fate of fatty acyl-CoAs (see Section 16.2). On the one hand, when fuel is abundant, malonyl-CoA accumulates and inhibits carnitine acyltransferase I, preventing the transport of fatty acyl-CoAs into mitochondria for β-oxidation and ketogenesis. On the other hand, shrinking malonyl-CoA pools signal the cells to transport fatty acids into the mitochondria, for generation of energy and fuels.

An important role of liver is to buffer the level of blood glucose. It does this largely through the action of hexokinase IV, the liver-specific isozyme, with a high $K_{0.5}$ for glucose, and partly through a high-K_M transport protein, the **glucose transporter (GLUT2),** one member of a family of membrane proteins that carry out facilitated diffusion of glucose. Thus, liver is unique in being able to respond to high blood glucose levels by increasing the uptake and phosphorylation of glucose, which results eventually in its deposition as glycogen.

● **CONCEPT** One of the most important roles of liver is to serve as a "glucostat," monitoring and stabilizing blood glucose levels.

Through allosteric mechanisms, liver senses the fed state and acts to store fuel derived from glucose. Liver also senses the fasting state and increases the synthesis and export of glucose when blood glucose levels are low. (Other organs also sense the fed state, notably the pancreas, which adjusts its glucagon and insulin outputs accordingly.)

To meet its internal energy needs, the liver can use a variety of fuel sources, including glucose, fatty acids, and amino acids.

Blood

All of the organs we have discussed are connected by the bloodstream, which transports what may be one organ's waste product but another organ's fuel (for example, transport of ketone bodies from the liver to the brain). Blood also transports oxygen from lungs to tissues, enabling exergonic oxidative pathways to occur, followed by transport of the resultant CO_2 back to the lungs for exhalation, as described in Chapter 7. And, as described in Chapter 16, the lipoprotein components of blood plasma (e.g., chylomicrons and very low-density lipoproteins) play indispensable roles in transporting lipids. Of course, blood is also the medium of transport of hormonal signals from one tissue to another and for excreted metabolic end products, such as urea, via the kidneys.

In terms of the blood's own energy metabolism, the most prominent pathway is glycolysis in the erythrocyte. Blood cells constitute nearly half the volume of blood, and erythrocytes constitute more than 99% of blood cells. Mammalian erythrocytes contain no mitochondria and depend exclusively on anaerobic glycolysis to meet their limited energy needs.

17.2 Hormonal Regulation of Fuel Metabolism

In animals, it is supremely important to maintain blood glucose levels within rather narrow limits, particularly for proper functioning of the nervous system. Of course, blood glucose levels vary, depending on nutritional status. Between meals, the normal level in humans is about 80 mg per 100 mL of blood, or 4.4 mM. Shortly after a meal, that level might rise to 120 mg per 100 mL (6.6 mM). In response, homeostatic mechanisms promote uptake of glucose into cells and its use by tissues. Later, when glucose levels begin to fall, other mechanisms promote glucose release from liver glycogen stores as well as gluconeogenesis, so that the normal level is maintained. The liver plays a central role in these homeostatic mechanisms. However, hormonal regulation also plays a critical role. Although we discuss molecular mechanisms of hormone action in detail in Chapter 20, it is appropriate here to discuss, at the physiological level, some of the hormones involved in fuel metabolism.

Actions of the Major Hormones

The most important hormone promoting glucose uptake and use is insulin. The hormones glucagon and epinephrine have the opposite effect, increasing blood glucose levels. The major effects of these three hormones are summarized in **TABLE 17.2**. **FIGURE 17.2** illustrates the interplay between insulin and glucagon, which are both produced in the pancreas. Epinephrine, released from the adrenal gland

● **CONCEPT** Maintenance of blood glucose within narrow limits is critical to brain function because under normal conditions the brain meets its prodigious energy requirement strictly by using glucose.

● **CONCEPT** The key hormones regulating glucose metabolism are insulin, which promotes glucose uptake and utilization, and glucagon and epinephrine, which increase the release of glucose into circulation.

TABLE 17.2 Major hormones controlling fuel metabolism in mammals

Hormone	Biochemical Actions	Enzyme Target	Physiological Actions
Insulin	↑ Glucose uptake (muscle, adipose tissue)	GLUT4	Signals fed state:
	↑ Glycolysis (liver, muscle)	PFK-1 (via PFK-2/FBPase-2)	↓ Blood glucose level
	↑ Acetyl-CoA production (liver, muscle)	Pyruvate dehydrogenase complex	↑ Fuel storage
	↑ Glycogen synthesis (liver, muscle)	Glycogen synthase	↑ Cell growth and differentiation
	↑ Triacylglycerol synthesis (liver)	Acetyl-CoA carboxylase	
	↓ Gluconeogenesis (liver)	FBPase-1 (via PFK-2/FBPase-2)	
	↓ Lipolysis		
	↓ Protein degradation		
	↑ Protein, DNA, RNA synthesis		
Glucagon	↑ cAMP level (liver, adipose tissue)		Signals fasting state:
	↑ Glycogenolysis (liver)	Glycogen phosphorylase	↑ Glucose release from liver
	↓ Glycogen synthesis (liver)	Glycogen synthase	↑ Blood glucose level
	↑ Triacylglycerol hydrolysis and mobilization (adipose tissue)	Hormone-sensitive lipase, adipose triglyceride lipase	↑ Ketone bodies as alternative fuel for brain
	↑ Gluconeogenesis (liver)	FBPase-1 (via PFK-2/FBPase-2), pyruvate kinase, PEPCK	
	↓ Glycolysis (liver)	PFK-1 (via PFK-2/FBPase-2)	
	↑ Ketogenesis (liver)	Acetyl-CoA carboxylase	
Epinephrine	↑ cAMP level (muscle)		Signals stress:
	↑ Triacylglycerol mobilization (adipose tissue)	Hormone-sensitive lipase, adipose triglyceride lipase	
	↑ Glycogenolysis (liver, muscle)	Glycogen phosphorylase	↑ Glucose release from liver
	↓ Glycogen synthesis (liver, muscle)	Glycogen synthase	↑ Blood glucose level
	↑ Glycolysis (muscle)	Glycogen phosphorylase, providing increased glucose	

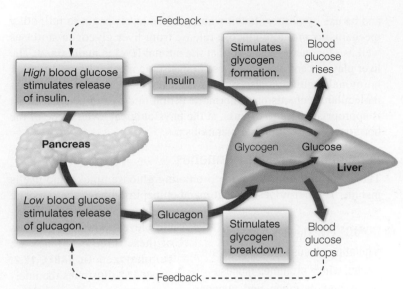

▲ **FIGURE 17.2 Aspects of the control of blood glucose levels by pancreatic secretion of insulin and glucagon.** Conditions resulting from high glucose levels are shown in blue, and those from low glucose levels in red. Dashed lines indicate feedback of blood glucose levels to the pancreas.

during the "fight-or-flight" response, has shorter-lived effects on fuel metabolism than insulin or glucagon.

Insulin

Insulin is a 5.8-kilodalton protein (see Figures 5.16 and 5.20) that is synthesized in the pancreas. The pancreas has both **endocrine cells,** which secrete hormones directly into the bloodstream, and **exocrine cells,** which secrete zymogen precursors of digestive enzymes into the upper small intestine. The endocrine tissue, which takes the form of cell clusters known as *islets of Langerhans,* contains at least five different cell types, each specialized for synthesis of one hormone. Insulin is synthesized in the β (or B) cells, which sense glucose levels and secrete insulin in response to increased levels of blood glucose. The β cells sense and respond to increased blood glucose by taking up and catabolizing glucose, which results in increased intracellular ATP levels. Increased intracellular ATP causes closure of ATP-gated K^+ channels and depolarization of the plasma membrane (see Section 10.8). Voltage-gated Ca^{2+} channels open in response to this membrane depolarization, leading to increased cytosolic $[Ca^{2+}]$, which finally triggers exocytosis of insulin granules.

The simplest way to describe the several actions of insulin is to say that *insulin is a signal that indicates the fed state* and thereby promotes (1) uptake of fuel substrates into some cells, (2) storage of fuels (lipids and glycogen), and (3) biosynthesis of macromolecules (nucleic acids and protein). As summarized in Table 17.2, specific effects include increased uptake of glucose in muscle and adipose tissue; activation of glycolysis in liver; increased synthesis of fatty acids and triacylglycerols in liver and adipose tissue; inhibition of gluconeogenesis in liver; increased glycogen synthesis in liver and muscle; increased uptake of amino acids into muscle with consequent activation of muscle protein synthesis; and inhibition of protein degradation. Because of its promotion of biosynthesis, it is appropriate to consider insulin a growth hormone.

The mechanism by which insulin stimulates glucose uptake into muscle and adipose cells is an area of intense investigation. One important action involves the glucose transporter, GLUT4, expressed by muscle and adipose tissue. An important consequence of glucose uptake in adipocytes is its conversion to glycerol-3-phosphate, which is required, along with fatty acids, for triacylglycerol synthesis (see Figure 16.29).

Glucagon

Glucagon is a 3.5-kilodalton polypeptide hormone synthesized by α cells of the islets of Langerhans in the pancreas. These endocrine cells sense the blood glucose concentration and release glucagon in response to low levels (Figure 17.2).

The primary target of glucagon is the liver, and its principal effect is to increase cyclic AMP levels in liver cells, as schematized in **FIGURE 17.3.** The resultant metabolic cascades, discussed in Chapter 12, promote glycogenolysis and inhibit glycogen synthesis. Additionally, cAMP inhibits glycolysis and activates gluconeogenesis by activating the hydrolysis of fructose-2,6-bisphosphate (see Figure 12.17). Glucagon also brings about inhibition of pyruvate kinase (PK) in the liver, causing phosphoenolpyruvate (PEP) to accumulate. The level of pyruvate decreases, both because its synthesis from PEP is blocked and because it continues to be converted to PEP (via the pyruvate carboxylase and phosphoenolpyruvate carboxykinase reactions; see Figure 12.14). Although accumulation of PEP is slight, it suffices to promote gluconeogenesis, while inhibition of pyruvate kinase diminishes the

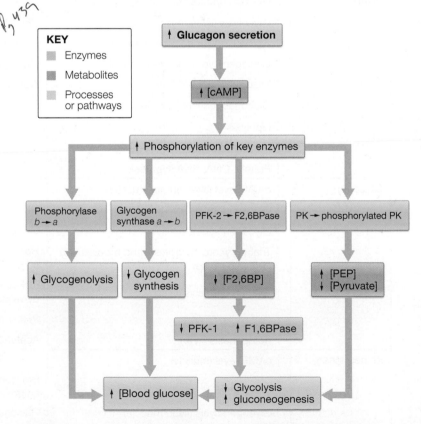

▲ **FIGURE 17.3 Actions of glucagon in liver that lead to a rise in blood glucose.** Brackets indicate concentration; ↑ and ↓ indicate increase or decrease in metabolite level.

glycolytic flux rate. The result of all these changes in the liver mediated by glucagon and increased cyclic AMP—increased glycogenolysis and gluconeogenesis, and decreased glycolysis and glycogen synthesis—is to increase blood glucose levels.

Glucagon also raises cAMP levels in adipose tissue. There, the chief effect of cAMP is to promote triacylglycerol mobilization via phosphorylation of hormone-sensitive lipase, yielding glycerol and fatty acids (see Figure 16.10).

Epinephrine

When released from adrenal medulla in response to low blood glucose levels, epinephrine (a catecholamine) interacts with second-messenger systems in many tissues, with varied effects. In muscle, epinephrine activates adenylate cyclase, with concomitant activation of glycogenolysis and inhibition of glycogen synthesis (see Chapter 12). Triacylglycerol breakdown in adipose tissue is also stimulated, providing fuel for the muscle tissue. Epinephrine also exerts effects on the pancreas, inhibiting insulin secretion and stimulating glucagon secretion. These effects tend to increase glucose production and release by the liver. The net result is to increase blood glucose levels. Unlike glucagon, catecholamines like epinephrine have short-lived metabolic effects. As discussed in Chapter 12, epinephrine action on skeletal and heart muscle cells is a crucial part of the "fight-or-flight" response. Note that epinephrine also functions as a neurotransmitter (as we will see in Chapter 20).

Epinephrine

Coordination of Energy Homeostasis

All organisms, indeed all cells, must balance the ingestion and absorption of fuel molecules with the metabolism and storage of these nutrients to meet immediate, as well as long-term, energy needs. Maintenance of this balance is termed **energy homeostasis,** and a complex regulatory system has evolved to coordinate and integrate these processes. This regulatory system comprises a large number of components, but two key protein kinases, **AMPK** and **mTOR,** play central roles in orchestrating the metabolic activity of mammalian cells. Although these two protein kinases evolved to allow the cell to sense and respond to changes in its energy status, they also play a role in insulin signaling. A third highly conserved family of enzymes, the **sirtuins,** are also emerging as important regulators of energy metabolism.

● **CONCEPT** AMPK and mTOR protein kinases play central roles in orchestrating the metabolic activity of mammalian cells.

AMP-Activated Protein Kinase (AMPK)

We have seen AMPK (AMP-activated protein kinase) several times in the preceding chapters. This serine/threonine protein kinase, found in all eukaryotes, is activated when the energy charge of the cell is low (i.e., high AMP/ATP ratio), such as occurs during nutrient starvation or hypoxia. Once activated, AMPK initiates a signaling process that conserves cellular energy by stimulating pathways that lead to ATP production while inhibiting pathways that utilize ATP. AMPK is

● **CONCEPT** AMPK is activated when the energy charge of the cell is low.

a heterotrimer, composed of a catalytic subunit (α) and two regulatory subunits (β and γ). Activation of the catalytic α subunit involves the binding of AMP to four nucleotide-binding sites in the γ subunit (**FIGURE 17.4**). Activation of AMPK also requires phosphorylation of a specific threonine residue in the α subunit, catalyzed by several upstream protein kinases.

AMPK phosphorylates multiple substrates that enhance energy-producing pathways, including targets that stimulate glycolysis in heart and mitochondrial biogenesis. Activation of AMPK also stimulates glucose uptake. Before stimulation, the GLUT4 glucose transporter is not present on the cell surface; rather, it is localized in vesicles in the cytosol. The protein is translocated to the cell surface in response to signaling by AMPK, where it facilitates glucose uptake. Insulin also stimulates translocation of GLUT4 to the plasma membrane. At the

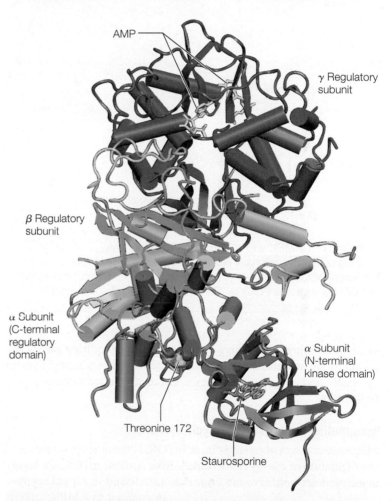

▲ **FIGURE 17.4 Mammalian AMP-activated protein kinase (AMPK).** This figure shows the X-ray crystal structure of the heterotrimeric enzyme (PDB ID: 2y94). The catalytic α subunit is composed of two domains, an N-terminal kinase domain (dark blue) and a C-terminal regulatory domain (cyan). The kinase active site is marked by a bound inhibitor, staurosporine (orange). Phosphorylated threonine 172 is shown in space-filling representation. The regulatory β subunit (green) is normally 272 amino acids long, but only its C-terminal 85 residues are included in this structure. The regulatory γ subunit (red) is composed of four nucleotide-binding sites. Two of these sites are filled with AMP molecules (yellow).

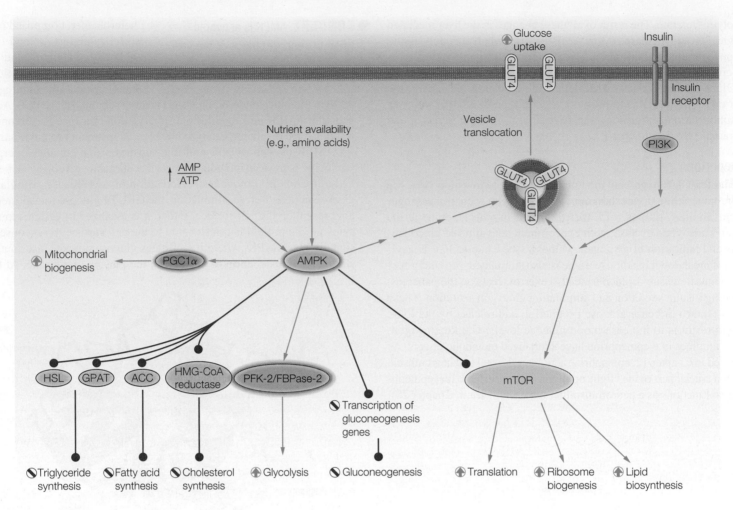

▲ **FIGURE 17.5 AMPK and mTOR signaling pathways.** Green arrows indicate activation; red balls indicate inhibition. For example, phosphorylation of ACC by AMPK inhibits fatty acid synthesis; phosphorylation of PGC-1α activates PGC-1α. Some of the metabolic responses mediated by AMPK and mTOR are tissue-specific. Refer to text for details. ACC, acetyl-CoA carboxylase; GLUT4, glucose transporter; GPAT, glycerophosphate acyltransferase; HSL, hormone-sensitive lipase; PGC-1α, PPARγ coactivator 1α; PI3K, phosphoinositide 3-kinase.

same time, other AMPK targets inhibit energy-requiring pathways, including hepatic gluconeogenesis, fatty acid synthesis, triacylglycerol synthesis, and cholesterol synthesis. These effects are summarized in **FIGURE 17.5.**

Mammalian Target of Rapamycin (mTOR)

Mammalian target of rapamycin, or mTOR, is the other main player in the regulation of energy homeostasis. Like AMPK, mTOR is a highly conserved serine/threonine protein kinase found in all eukaryotes. In contrast to AMPK, mTOR is active under nutrient-rich conditions and inactive under nutrient-poor conditions. Activated mTOR promotes anabolic processes, including cell proliferation, protein synthesis, and lipid biosynthesis. Rapamycin, produced by a *Streptomyces* bacterium, allosterically inhibits mTOR.

● **CONNECTION** mTOR was discovered during biochemical studies with the bacterial natural product, rapamycin, a potent immunosuppressant. mTOR inhibitors such as rapamycin can suppress the immune system by disrupting the cell cycle of lymphocytes and other proliferating immune cells.

Rapamycin (Sirolimus)

mTOR activity is regulated by a number of upstream inputs from a wide variety of environmental signals, including energy status, nutrient availability (e.g., amino acids), and growth factors (e.g., insulin and epidermal growth factor). When insulin binds to its plasma membrane receptor, the tyrosine kinase activity of the receptor is

● **CONCEPT** mTOR, in contrast to AMPK, is active under nutrient-rich conditions and inactive under nutrient-poor conditions.

activated, initiating a phosphorylation cascade (Figure 17.5). Through the action of **phosphoinositide 3-kinase (PI3K)** and several other protein kinases not shown here, this cascade activates mTOR. Insulin binding also stimulates GLUT4 translocation to the plasma membrane, accounting for insulin-stimulated glucose transport in muscle cells. We will discuss the insulin signal transduction pathway in more detail in Chapter 20.

The mechanism by which mTOR senses energy status involves AMPK. Recall that AMPK is activated during nutrient starvation, directly sensing the adenylate energy charge (AMP/ATP ratio). Activated AMPK then phosphorylates mTOR and other targets, which leads to inhibition of mTOR function.

In summary, AMPK and mTOR play opposing roles in controlling the metabolic activity of cells in response to intracellular and extracellular signals that report on the energy status of the individual cell and the organism as a whole. These two protein kinases, and the nutrient signaling pathways they control, are evolutionarily conserved from yeast to humans.

Sirtuins

Sirtuins are a highly conserved family of protein deacetylases. These enzymes catalyze the deacetylation of acetylated lysine residues in target proteins. Recall from Section 8.9 that acetylation of lysine residues is a

● **CONCEPT** Sirtuins are a highly conserved family of NAD$^+$-dependent protein deacetylases.

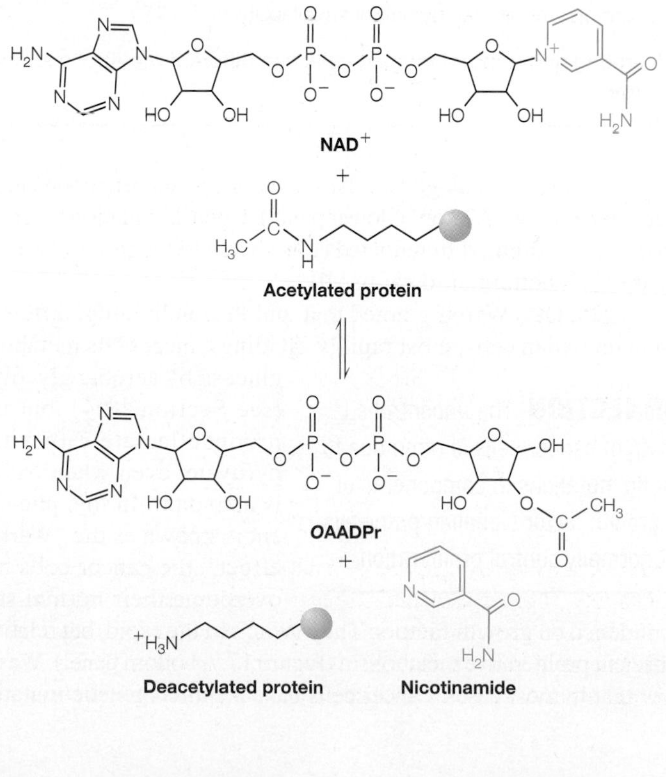

common covalent modification of proteins, catalyzed by a family of protein acetyltransferases. Reversal of this modification requires an enzyme-catalyzed deacetylation. Of the three types of protein deacetylases found in nature, sirtuins are unique in that they require NAD$^+$ for their deacetylation activity. NAD$^+$ functions not as a redox cofactor in this complex reaction, but rather as a substrate that is cleaved to nicotinamide and 2'-O-acetyl-ADP-ribose (OAADPr).

NAD$^+$

+

Acetylated protein

OAADPr

+

Deacetylated protein   Nicotinamide

● **CONNECTION** Recent studies have revealed that precise control of cellular NAD$^+$ levels is important for prolonging both health and life span. Sirtuins are central players in this regulation of the aging process.

● **CONCEPT** Sirtuins act as metabolic sensors of the cellular redox state and are activated at high NAD$^+$/NADH ratio.

teins have been found to be acetylated in mammalian cells, and the list of pathways known to be regulated by SIRT-mediated deacetylation is expanding rapidly. For most protein targets, deacetylation increases the activity of the target protein. The deacetylase activity of sirtuins is sensitive to changes in the cellular NAD$^+$ levels, being enhanced at high NAD$^+$/NADH ratios. Thus, sirtuins act as metabolic sensors of the cellular redox state.

Sirtuins are part of an intricate regulatory system that controls flux through fuel utilization pathways in response to the dietary availability of alternative fuels. For example, in the next section, we will see that under fasting conditions, mammals initiate a reprogramming of their metabolic systems across several tissue types. This response includes increasing the rate of glucose synthesis in the liver and kidney and increasing the utilization of fatty acids as a fuel source in peripheral tissues. One of the most important participants in this metabolic reprogramming is the **peroxisome proliferator-activated receptor-γ coactivator 1α (PGC-1α)**. PGC-1α was initially discovered as a coactivator of the transcription factor peroxisome proliferator-activated receptor-γ (PPARγ), but PGC-1α binds to and stimulates the transcriptional activity of several transcription factors.

The transcriptional coactivator function of PGC-1α is sensitive to its acetylation status, and PGC-1α can be deacetylated by SIRT1. Fasting (low nutrients) results in a higher NAD$^+$/NADH ratio, activating the sirtuins. Deacetylation of PGC-1α by SIRT1 causes upregulation of its coactivator function (**FIGURE 17.6**). In liver, upregulation of PGC-1α

● **CONCEPT** Deacetylation of PGC-1α by SIRT1 initiates transcriptional responses that lead to increased oxidation of fatty acids and decreased utilization of glucose.

function stimulates gluconeogenesis by activating the transcription of several key genes. In skeletal muscle and heart, the transcriptional responses mediated by PGC-1α lead to increased oxidation of fatty acids and decreased utilization

of glucose. Deacetylated PGC-1α also coactivates the transcription of nuclear genes that encode subunits of the mitochondrial respiratory chain (Chapter 14) as well as genes that encode components of the mitochondrial gene expression machinery. This enhanced mitochondrial biogenesis increases the capacity of the cell for fatty acid oxidation and is a critical part of the PGC-1α-dependent metabolic reprogramming that occurs in heart and skeletal muscle when carbohydrate fuel is scarce.

PGC-1α is capable of integrating multiple signals that monitor the cellular energy state. Recall that AMPK also stimulates mitochondrial

Sirtuins are named after the founding member of the family, yeast Sir2 (silent information regulator 2), which deacetylates histones to silence (turn off) transcription of genes involved in sexual reproduction in yeast. Mammals possess seven sirtuins (SIRT1-7), which differ in their cellular localization (nucleus, mitochondria, cytoplasm) and in their protein targets. Sirtuins act on many proteins besides histones—more than 2000 pro-

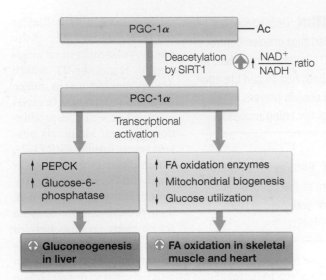

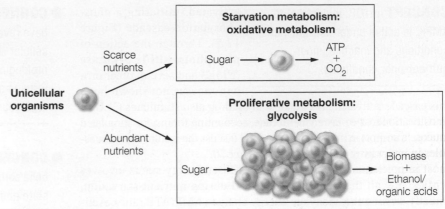

Proliferative metabolism, in both unicellular and multicellular organisms, relies on glycolysis, a rapid but relatively inefficient process for generating ATP. This "fermentative" metabolism requires abundant nutrients. When nutrients are scarce, unicellular organisms adapt to a starvation metabolism, characterized by the slower, but more efficient, oxidative metabolism.

▲ **FIGURE 17.6 PGC-1α and SIRT1 control the reprogramming of fuel utilization pathways in response to fasting.** A high $NAD^+/NADH$ ratio, in response to low nutrients (fasting), activates SIRT1 to deacetylate PGC-1α, upregulating its transcriptional coactivator function. Tissue-specific transcriptional activation programs result in increased gluconeogenesis (liver) and increased fatty acid oxidation (skeletal and heart muscle).

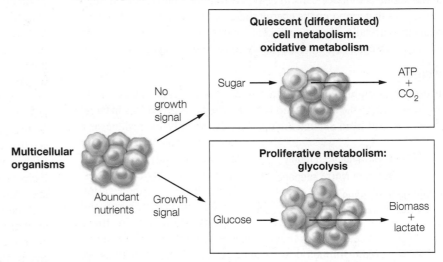

This same efficient oxidative metabolism is used by nondividing, differentiated mammalian cells. Nutrient abundance is rarely an issue in multicellular organisms so that the switch between proliferative and quiescent metabolism is determined by the presence or absence of appropriate growth factors, rather than by nutrient availability.

▲ **FIGURE 17.7 Proliferating and nonproliferating cells use different metabolic strategies to generate energy.**

biogenesis. AMPK phosphorylates PGC-1α, which also causes its activation (see Figure 17.5). It appears that phosphorylation primes PGC-1α for subsequent deacetylation by SIRT1. Thus, PGC-1α senses both the AMP/ATP ratio (via AMPK) and the $NAD^+/NADH$ ratio (via SIRT1).

Endocrine Regulation of Energy Homeostasis

AMPK, mTOR, and sirtuins all evolved in unicellular organisms as part of a control system that sensed the nutrient supply and initiated appropriate metabolic responses. If nutrients are abundant, the cells take up the fuels and metabolize them via glycolysis, a rapid, but relatively inefficient process (**FIGURE 17.7**, bottom of upper panel). This *proliferative metabolism* provides the building blocks and free energy needed to produce biomass (new cells) during exponential growth. When nutrients are scarce, the cells adapt to a *starvation metabolism* (Figure 17.7, top of upper panel). Biomass production ceases, and the cells switch to a slower, but more efficient, oxidative metabolism in order to extract maximum energy from the limiting nutrients. During the evolution of metazoans (multicellular organisms), this control system grew in complexity, responding to new inputs and developing new outputs. In contrast to unicellular organisms, most of the cells in multicellular organisms are bathed in a relatively constant supply of nutrients via the circulatory system, and this supply often exceeds the levels needed to support cell growth and replication. Growth control in metazoans thus evolved to occur at the levels of nutrient intake, transport, and utilization (metabolism). Most mammalian cells therefore exhibit a strict dependence on growth factors (hormones) to switch from a quiescent, differentiated state to a proliferative state. Nondividing, differentiated mammalian cells typically use oxidative metabolism (aerobic glycolysis + citric acid cycle) to

metabolize glucose, analogous to the starvation metabolism of unicellular organisms (Figure 17.7, top of lower panel). Upon stimulation by growth factors (e.g., insulin), differentiated cells switch to the faster glycolysis (Figure 17.7, bottom of lower panel).

In 1925, Otto Warburg noted that, unlike nondividing, differentiated mammalian cells, most rapidly dividing cancer cells metabolize glucose by aerobic glycolysis (see Section 12.4), but they produce lactate rather than pyruvate, even when oxygen is abundant. In this phenomenon, known as the "Warburg effect," the cancer cells have overcome their normal strict dependence on growth factors. They switch to the rapid, but relatively inefficient proliferative metabolism (Figure 17.7, bottom panel). We now know that in most cases, cancer cells have acquired genetic mutations

● **CONNECTION** The uncontrolled growth of cancer cells is often due to genetic mutations in components of the growth factor signaling pathways that normally control proliferation.

in components of the growth factor signaling pathways that normally control proliferation. We will see several examples in Chapter 20.

In mammals, the brain coordinates whole-body energy homeostasis. The brain receives information about the quality and quantity of nutrients being consumed, the levels of fuels already present in the blood, and the amounts of energy present in various storage reserves in the body. These fuel and hormonal signals converge in neurons in the **arcuate nucleus** of the hypothalamus, where appetite, and thus food intake, is controlled. Not surprisingly, AMPK and mTOR play central roles in integrating these signals in the hypothalamus (**FIGURE 17.8**). The most important endocrine regulators of food intake are **insulin** and **leptin,** both of which inhibit food intake, and **ghrelin** and **adiponectin,** both of which promote food intake. Each of these acts via specific receptors on particular cells in the arcuate nucleus to initiate signaling pathways that converge on AMPK and mTOR. Hypothalamic AMPK is activated by ghrelin and adiponectin in response to low levels of nutrients such as glucose, branched-chain amino acids, and free fatty

acids. Activated AMPK then promotes food intake. Insulin and leptin activate mTOR in response to ample nutrient levels, resulting in the inhibition of food intake.

Leptin is a peptide hormone released by adipocytes (an **adipokine**) when fat stores are adequate. Leptin binding to a specific receptor in neurons in the arcuate nucleus of the hypothalamus activates mTOR and inhibits AMPK function, suppressing food intake (Figure 17.8).

● **CONCEPT** Leptin functions as a "lipostat," sensing the amount of fat stored in the adipocytes—when fat stores are adequate, leptin controls feeding behavior to limit fat intake.

Dramatic evidence that leptin is a negative regulator of food intake comes from genetic studies with mice. Leptin is encoded by the *OB* gene in mice (*OB,* for *obese*). Mice bearing two defective alleles of the *OB* gene (*ob/ob*) grow to body weights as much as three times normal (see chapter opener, p. 556). The leptin receptor is encoded by the *DB* gene in mice; mice with two defective alleles of the *DB* gene (*db/db*) are obese and develop diabetes. Leptin evidently functions as a "lipostat," sensing the amount of fat stored in the adipocytes. When fat stores are adequate, leptin levels are high, and the signaling system controls feeding behavior to limit fat deposition. During starvation, leptin levels decline, which promotes feeding and fat storage within the adipocyte. *ob/ob* mice, lacking functional leptin, act as if perpetually starved, and their overeating makes them obese; injections of leptin lower their feeding rates and cause them to lose weight dramatically. Obese humans are different from obese mice, however, in that they contain high levels of leptin. Current research is aimed at the premise that these individuals are somehow unresponsive to normal leptin signaling.

Ghrelin, a small peptide hormone (28 amino acids) produced in cells lining the stomach, is a hunger-stimulating signal. Its levels increase before meals and decrease after meals. Ghrelin binds to specific receptors in the arcuate nucleus of the hypothalamus and promotes food intake by activating AMPK.

Adiponectin, another adipokine produced by adipocytes, circulates in higher concentrations than most hormones, but its levels are lower in obese individuals than in lean individuals. Adiponectin levels are also lower in people with type 2 diabetes. Adiponectin binds to specific receptors and stimulates food intake by activating AMPK in hypothalamus (Figure 17.8).

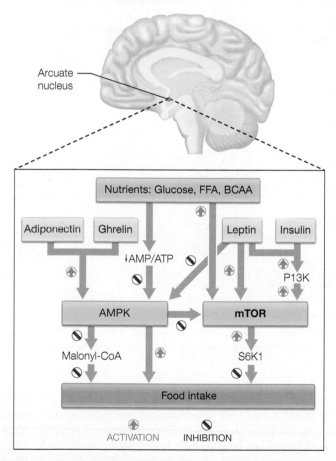

▲ **FIGURE 17.8 Fuel and hormonal control of appetite in the arcuate nucleus of the hypothalamus.** Activation of AMPK promotes food intake; inhibition of AMPK ultimately suppresses food intake. Green arrows indicate stimulatory effects; red bars indicate inhibitory effects. Insulin acts through the phosphoinositide 3-kinase (PI3K) cascade to inhibit food intake. Leptin has both direct and indirect effects (through PI3K and AMPK) on mTOR. Adiponectin and ghrelin stimulate food intake by activating AMPK. AMPK inhibits production of malonyl-CoA by ACC (see Figure 17.5). Malonyl-CoA inhibits food intake, so decreased malonyl-CoA stimulates food intake. S6K, ribosomal protein S6 kinase; BCAA, branched-chain amino acids; FFA, free fatty acids.

17.3 Responses to Metabolic Stress: Starvation, Diabetes

An excellent way to understand how the interorgan and hormonal relationships we have discussed actually integrate fuel metabolism is to examine the effects of metabolic stress. In this section, we consider two examples—prolonged fasting, in which the intake of fuel substrates is inadequate; and **diabetes mellitus,** in which a functional insufficiency of insulin impairs the ability of the body to use glucose, even when the sugar is present in abundance.

First, let us review how glucose levels are maintained during normal feeding cycles (**FIGURE 17.9**). The blood glucose elevation occurring shortly after a carbohydrate-containing meal stimulates the secretion of insulin from the pancreas and suppresses the secretion of glucagon. Together these effects promote uptake of glucose into the liver, stimulate glycogen synthesis, and suppress glycogen breakdown. Flux

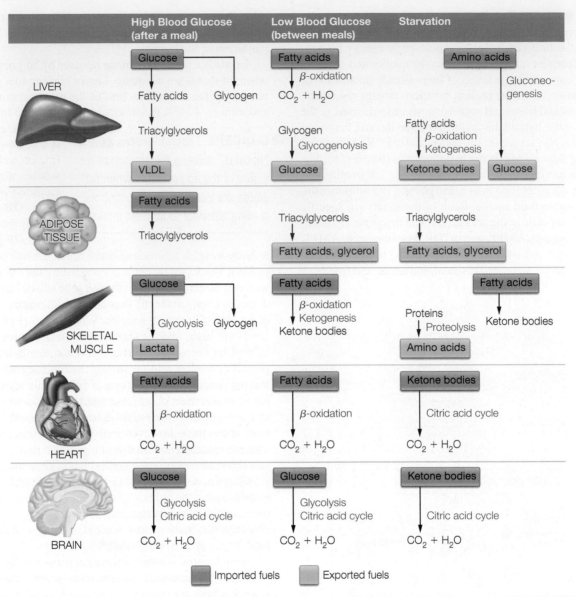

▲ **FIGURE 17.9 Major events in the storage, retrieval, and use of fuels in the fed and unfed states and in early starvation.** Purple indicates fuels imported into the tissue; green indicates fuels exported from tissue.

through hexokinase increases in response to elevated glucose levels, providing substrates for glycogen synthesis. In addition, activation of acetyl-CoA carboxylase in the liver stimulates fatty acid synthesis, with subsequent transport to adipose tissue as triacylglycerols in very low-density lipoproteins (VLDL). There, the increased levels of glycolytic intermediates and fatty acids stimulate triacylglycerol synthesis. The liver-derived VLDL also delivers fatty acids to the heart for β-oxidation. Increased glucose uptake into skeletal muscle not only provides fuel, but also increases levels of substrates for glycogen synthesis in that tissue. The brain, which has no significant glycogen or other fuel reserves, relies exclusively on blood glucose during normal feeding cycles.

Several hours later, when blood glucose levels begin to fall, the above events are reversed. Insulin secretion from the pancreas slows

and glucagon secretion increases. This promotes glycogen mobilization in liver via the cAMP-dependent cascade mechanisms that activate glycogen phosphorylase and inactivate glycogen synthase. Triacylglycerol breakdown in adipocytes is activated as well, via the action of hormone-sensitive lipase, generating fatty acids for use as fuel by liver and muscle. At the same time, the decrease in insulin levels reduces glucose use by muscle, liver, and adipose tissue. Consequently, nearly all the glucose produced in the liver is exported to the blood and is available for use by the brain.

Starvation

Suppose that food intake is denied not just for a few hours, as just described, but for many days. Given that a 70-kg human can store at most the equivalent of 6800 kJ of energy as glycogen (see Section

16.1), this source of blood glucose will be exhausted in just a few hours. Because it is critical for brain function that blood glucose levels be maintained near 4.4 mM, the organism adapts metabolically to increase the use of fuels other than carbohydrate, primarily fat.

Before we discuss the metabolic adjustments involved, let us recall the other major energy stores: about 555,000 kJ as triacylglycerol, largely in adipose tissue, and 100,000 kJ as mobilizable proteins, largely in muscle. These stores provide sufficient energy to permit survival for up to several months. However, use of these stores presents problems. Triacylglycerol mobilization generates metabolic fuel largely in the form of acetyl-CoA, whose further oxidation in the citric acid cycle requires oxaloacetate. Recall from Section 13.8 that oxaloacetate and other citric acid cycle intermediates are used in other metabolic reactions and must be replenished via anaplerotic pathways. The most important of these processes is the pyruvate carboxylase reaction, with most of the pyruvate coming from carbohydrate catabolism. However, when carbohydrate availability is limited, the resupply of citric acid cycle intermediates is limited, and flux through the cycle may be reduced.

During carbohydrate limitation, citric acid cycle intermediates can be provided from other sources. For example, the glycerol released from lipolysis can be used, but it is not produced in amounts adequate to maintain levels of citric acid cycle intermediates. Alternatively, these intermediates can be produced from protein catabolism and transamination. However, this process is energetically wasteful and has the undesirable effect of wasting the muscle and weakening the fasting subject. Nevertheless, proteolysis is accelerated in muscle during the first few days of starvation (Figure 17.9) because amino acids for protein synthesis are not present in sufficient amounts to counterbalance protein breakdown, which continues at normal rates. A major fate of the released amino acids is transport to the liver for gluconeogenesis, as the body attempts to cope with the absence of glycogen stores by synthesizing its own glucose. During this time, the liver and muscle are shifting to fatty acids as the dominant fuels for their own use (Figure 17.9).

Meanwhile, the increased use of carbon for gluconeogenesis diminishes the amount of oxaloacetate available to combine with acetyl-CoA in the citric acid cycle. Because fat breakdown has been

● **CONCEPT** Metabolic adaptations promote alternative fuel use during starvation so that glucose homeostasis is maintained for several weeks.

activated, both acetyl-CoA and reduced electron carriers accumulate in the liver to the point that the acetyl-CoA cannot all be oxidized, and ketone bodies begin to accumulate (Figure 17.9). Accumulation of acetoacetate and β-hydroxybutyrate increases flux through the reactions that catabolize these ketone bodies. Thus, the brain adapts to reduced glucose levels by increasing the use of ketone bodies as alternative energy substrates. This trend continues for the duration of starvation. On the third day, the brain derives about one-third of its energy needs from ketone bodies; by day 40, that usage has increased to two-thirds. This adaptation reduces the need for gluconeogenesis and spares the mobilization of muscle protein. In fact, the loss of muscle protein *decreases* by about four-fold late in starvation—from about 75 grams consumed per day on day 3 to about 20 grams per day on day 40. The metabolic

changes accompanying starvation compromise the organism's abilities to respond to further stresses, such as extreme cold or infection. However, the adaptations do allow life to continue for many weeks without food intake, the total period being determined largely by the size of the fat deposits.

Diabetes

In starvation, glucose utilization is abnormally low because of inadequate glucose supplies. In **diabetes mellitus,** glucose utilization is also abnormally low, but the reason in this case is that the hormonal stimulus to glucose utilization—namely, insulin—is defective. As a result, glucose is actually present in excessive amounts. The consequences of insulin deficiency are comparable to those of starvation in revealing important aspects of interorgan metabolic relationships.

Diabetes is a major public health issue, having reached epidemic proportions in the United States and around the world. It is esti-

● **CONCEPT** Diabetes results either from insulin deficiency or from defects in the insulin response mechanism.

mated that more than 12% of the adult population in the United States is afflicted with this disease. Diabetes is not a single disease, but rather a family of diseases. **Type 1 diabetes,** formerly called insulin-dependent diabetes, or juvenile diabetes because of its typical early onset, often involves autoimmune destruction of the β cells of the pancreas, which can be caused by various factors, including viral infection. Some forms of type 1 diabetes have a genetic origin. Mutations in insulin structure can render the hormone inactive, and other mutations cause defects in the conversion of preproinsulin or proinsulin to the active hormone (see Figure 5.20). Either way, type 1 diabetes is characterized by an actual deficiency in insulin and can be treated by administration of insulin.

Type 2 diabetes, formerly called adult-onset diabetes, obesity-related diabetes, or non-insulin-dependent diabetes mellitus, is characterized by *insulin resistance*—patients cannot respond to therapeutic doses of insulin. Type 2 diabetes accounts for more than 95% of people with diabetes.

Unfortunately, despite an intense research effort extending back nearly 100 years, the specific defects that lead to insulin resistance in type 2 diabetes remain undefined. However, several clues are begin-

● **CONNECTION** Metabolic syndrome—defined by abdominal obesity, hypertension, high blood sugar, and insulin resistance—often precedes both cardiovascular disease and type 2 diabetes.

ning to shed some light on the disease. The first is that most people with type 2 diabetes are also obese. In fact, obesity is so closely associated with insulin resistance that there must be a mechanistic link. Like diabetes, the prevalence of obesity has increased dramatically in the United States

since the 1970s, with 36% of adults classified as obese in 2014. The second clue comes from the close relationship between type 2 diabetes and **metabolic syndrome.** Metabolic syndrome, which afflicts some 50 million Americans, is defined by abdominal obesity, hypertension, high blood sugar, and, most importantly, insulin resistance. These metabolic abnormalities often precede both cardiovascular disease and diabetes. Common to both obesity and metabolic syndrome is excess fuel intake and abnormal accumulation of lipid in "ectopic sites," primarily liver and skeletal muscle. Excess lipids initially accumulate in adipose cells, increasing their size. Eventually, fuel intake exceeds the storage capacity of adipose tissue, and excess lipids are shunted to ectopic sites.

There is growing evidence that this abnormal lipid accumulation causes insulin resistance by affecting downstream signaling pathways. Two related mechanisms have gained wide support as potential causes of the disease. The **lipid overload** hypothesis states that when fat accumulates in muscle cells, it blocks the insulin signaling pathway that normally stimulates translocation of GLUT4 (the major glucose transporter in muscle) to the plasma membrane (see Figure 17.5). Thus, insulin no longer efficiently stimulates glucose transport; that is, the cell is insulin resistant. The **inflammation** hypothesis states that as adipose cells increase in size with excess lipids, they secrete inflammatory adipokines and cytokines, including TNF-α, interleukins, and resistin. These cytokines bind their receptors in peripheral tissues such as muscle and interfere with insulin signaling, causing insulin resistance. Thus, although type 2 diabetes has long been characterized by defects in carbohydrate metabolism, abnormal lipid metabolism may be at the root of the disease.

● **CONCEPT** Two related mechanisms have been proposed as causes of type 2 diabetes: the lipid overload hypothesis and the inflammation hypothesis.

Whatever the cause of the functional insulin deficiency, diabetes can truly be called "starvation in the midst of plenty." The insufficient production of insulin or the failure of insulin to act normally in promoting glucose utilization, with resultant glucose accumulation in the blood, starves the cells of nutrients and promotes metabolic responses similar to those of fasting (**FIGURE 17.10**).

● **CONCEPT** Diabetes can be thought of as "starvation in the midst of plenty" because cells are unable to utilize the glucose that accumulates in the blood.

Liver cells attempt to generate more glucose by stimulating gluconeogenesis. Most of the substrates come from amino acids, which in turn come largely from degradation of muscle proteins. Glucose cannot be reused for resynthesis of amino acids or of fatty acids, so a person with diabetes may lose weight even while consuming what would normally be adequate calories in the diet.

● **CONNECTION** A special danger with diabetic persons experiencing ketoacidosis is that they may lose consciousness. This, coupled with a sweet organic odor on the breath, may give the impression that they are intoxicated, when in fact their lives are in jeopardy.

As cells attempt to generate usable energy sources, triacylglycerol reserves are mobilized in response to the abnormally low insulin-to-glucagon ratio. Fatty acid oxidation is elevated, with concomitant generation of acetyl-CoA. Flux through the citric acid cycle may decrease because of the accumulation

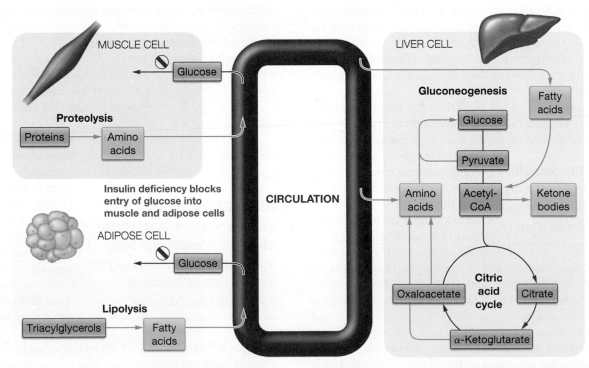

▲ **FIGURE 17.10 The metabolic abnormalities in diabetes.** The insulin deficiency blocks the uptake of glucose into muscle and adipose tissue and reduces glucose catabolism in all tissues. Proteolysis in muscle and lipolysis in adipose tissue are enhanced. In the liver, gluconeogenesis from amino acids and citric acid cycle intermediates is stimulated as the cells attempt to remedy the perceived lack of usable glucose, and fatty acid oxidation and ketogenesis are also increased. Green indicates pathways activated; pink indicates pathways diminished.

of reduced electron carriers and/or oxaloacetate limitation. In liver, both effects accelerate ketone body formation, generating increased levels of organic acids in the blood (ketosis). These acids can lower the blood pH from the normal value of 7.4 to 6.8 or lower (ketoacidosis). Decarboxylation of acetoacetate, which is stimulated at low pH, generates acetone, which can be smelled on the breath of patients with severe ketoacidosis.

The excessive concentrations of glucose in body fluids generate other metabolic problems, quite different from anything seen in starvation. At blood glucose levels above 10 mM, the kidney can no longer reabsorb all of the glucose out of the blood filtrate, and glucose is spilled into the urine, sometimes in amounts approaching 100 grams per day. In fact, the Latin name *diabetes mellitus* literally means "honey-sweet urine." Glucose excretion creates an osmotic load, which causes large amounts of water to be excreted as well, and under these conditions the kidney cannot reabsorb most of this water. Indeed, the earliest indications of diabetes are often frequent and excessive urination, coupled with excessive thirst. Long before biochemistry was a science, the loss of nutrients, excessive urination, and breakdown of fat and protein were recognized as hallmarks of diabetes.

● **CONNECTION** To understand the causes and treatment of diabetes is to understand the metabolism of carbohydrates, lipids, and proteins, as well as mechanisms of hormonal control.

As early as the first century A.D., diabetes was described as "the flesh and bones running together into urine." It took another 1800 years or so before Israel Kleiner in New York and Frederick Banting, Charles Best, James Collip, and John Macleod in Toronto discovered that extracts of dog pancreas possessed the ability to lower glucose levels and restore health to children and young adults suffering from diabetes. These studies culminated in the identification of insulin as the active component in 1922.

In type 1 diabetes, the metabolic imbalance is usually more severe and difficult to control than in the milder and more common type 2 diabetes. The latter can often be controlled by exercise and dietary restriction of carbohydrate, whereas treatment for type 1 diabetes involves daily self-injection of insulin. For many years this insulin was purified from bovine pancreas, and its high cost, coupled with occasional problems resulting from the minor structural differences between human and bovine insulin, led the fledgling biotechnology industry to attempt to produce human insulin through recombinant DNA techniques. In the late 1970s, the gene for human insulin was cloned into *E. coli* in a form that allowed it to be expressed, and in 1982 cloned human insulin became the first recombinant DNA product to be approved for human use.

● **CONNECTION** Human insulin was the first recombinant DNA product to be approved for human use.

Summary

- Each organ or tissue of a multicellular organism has a distinctive profile of metabolic activities that allows it to serve its specialized functions. These tissues must remain in constant communication to maintain homeostasis. In vertebrates, the most essential element of this homeostasis is maintenance of constant blood glucose levels, primarily for proper brain function. (Section 17.1)

- The actions of three hormones—**insulin, glucagon,** and **epinephrine**—play the dominant roles in glucose homeostasis. Insulin signals the fed state and promotes glucose utilization and synthesis of energy storage compounds. Glucagon acts primarily upon liver cells, increasing blood glucose by several mechanisms involving cyclic AMP. Epinephrine has similar effects on muscle cells. (Section 17.2)

- Energy homeostasis, maintaining the balance of fuel intake with the metabolism and storage of nutrients to meet energy needs, is coordinated by a complex intracellular regulatory system. Two protein kinases, **AMPK** and **mTOR,** play central roles in orchestrating the metabolic activity of mammalian cells. The **sirtuins,** a highly conserved family of NAD^+-dependent protein deacetylases, act as metabolic sensors of the cellular redox state. (Section 17.2)

- The response to metabolic stresses such as starvation and diabetes reveals the interorgan and hormonal relationships that integrate fuel metabolism in mammals. (Section 17.3)

Problems

Enhanced by
Mastering Chemistry
for Biochemistry

Mastering Chemistry for Biochemistry provides select end-of-chapter problems and feedback-enriched tutorial problems, animations, and interactive figures to deepen your understanding of complex topics while practicing problem solving.

Answers to red problems are available in the Answer Appendix.

1. On your way to class this morning, you stop at your favorite coffeehouse and grab a caffé mocha and a blueberry scone. The caffé mocha contains 260 Calories and the scone contains 460 Calories.
   (a) What fraction of your recommended daily caloric intake did you just consume? Use the estimated calorie requirements for your age and gender at www.nhlbi.nih.gov/health/public/heart/obesity/wecan/healthy-weight-basics/balance.htm.
   (b) Convert the calories you consumed for breakfast into kilojoules.

2. Marathon runners preparing for a race engage in "carb loading" to maximize their carbohydrate reserves. This involves eating large quantities of starchy foods. Why is starch preferable to candy or sugar-rich foods?

3. Supposing that an average human consumes energy at the rate of 1500 kcal/day at rest and that long-distance running consumes energy at 10 times that rate, how long would the glycogen reserves last during a marathon run?

4. What proportion of the total energy consumption supports brain function in an average resting human? What proportion in a human running in a marathon?

5. Proteolysis increases during the early phases of fasting, but later it decreases as the body adapts to using alternative energy sources. Given that feedback control mechanisms have not been described for intracellular proteases, how might you explain these apparent changes in protease activity?

6. Shortly after a typical meal, your blood glucose will rise from its fasting level of 4.4 mM to 6.6 mM.
   (a) How many grams of glucose does this increase represent? (An average adult male weighing 70 kg has a blood volume of about 5 L.)
   (b) Estimate the change in velocity for the liver hexokinase isozyme (IV) and the muscle hexokinase isozyme (I) that the fasting versus fed glucose concentrations would produce. Refer to the substrate–velocity data on p. 379 of Chapter 12.

7. Glucose has been found to react nonenzymatically with hemoglobin, through Schiff base formation between C-1 of glucose and the amino termini of the β chains. How might this finding be applied in monitoring diabetic patients?

8. Ketone bodies are exported from liver for use by other tissues. Because many tissues can synthesize ketone bodies, what enzymatic property of liver might contribute to its special ability to export these compounds?

9. Adipose tissue cannot resynthesize triacylglycerols from glycerol released during lipolysis (fat breakdown). Why not? Describe the metabolic route that is used to generate a glycerol compound for triacylglycerol synthesis.

10. (a) Briefly describe the relationship between intracellular malonyl-CoA levels in the liver and the control of ketogenesis.
    (b) Describe how the action of hexokinase IV helps the liver to buffer the level of blood glucose.

11. The action of glucagon on liver cells leads to inhibition of pyruvate kinase. What is the most probable mechanism for this effect?

12. AMPK and mTOR can both be considered intracellular signal integrators. Explain this definition.

13. Pancreatic β cells secrete insulin in response to increased blood glucose. This process requires the catabolism of glucose to pyruvate via glycolysis, producing ATP. which initiates the exocytosis of insulin as described in Section 17.2. However, it is known that mitochondrial pyruvate metabolism is also critical for glucose-stimulated insulin secretion from pancreatic β cells. Ferdaoussi et al. in *J. Clin. Invest.* **125**, 3847–3860 (2015) investigated a pathway that amplifies glucose-stimulated insulin secretion. In these studies, they infused β cells with various metabolites (see panel a on the next page) and measured the exocytosis response. You will need to recall the various routes by which pyruvate can be metabolized in mitochondria (see Chapter 13) to address this problem.
    (a) Panels b, c, and d show the effects of treatment with the citric acid cycle intermediates isocitrate and α-ketoglutarate (α-KG). Describe the effect (amplify, inhibit, no effect) of each of these metabolites on glucose-stimulated exocytosis.
    (b) Panels e and f show the effects of treatment with either NADPH (in either a 1:10 or a 10:1 molar ratio with $NADP^+$), or NADH (in a 10:1 molar ratio with NAD^+). Describe the effect (amplify, inhibit, no effect) of each of these treatments on glucose-stimulated exocytosis.
    (c) In other experiments not shown here, the authors discovered that cytosolic isocitrate dehydrogenase (see Chapter 13, p. 435) is required for amplification of the glucose-stimulated exocytosis. Based on your analysis of the data, explain the role of cytosolic isocitrate dehydrogenase in this pathway.
    (d) The authors also observed that glucose stimulation causes an increase in reduced glutathione (GSH) and that this increase was dependent on the activity of cytosolic isocitrate dehydrogenase. They then showed that GSH amplifies the glucose-stimulated exocytosis from β cells (panel g). Propose a metabolic pathway that connects all of these results, starting with pyruvate and ending with GSH.

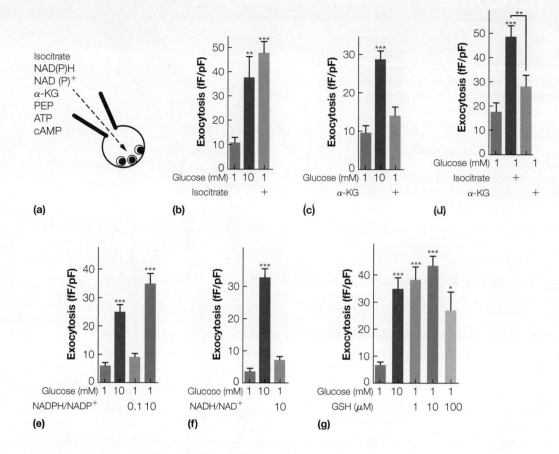

References

For a list of references related to this chapter, see Appendix II.

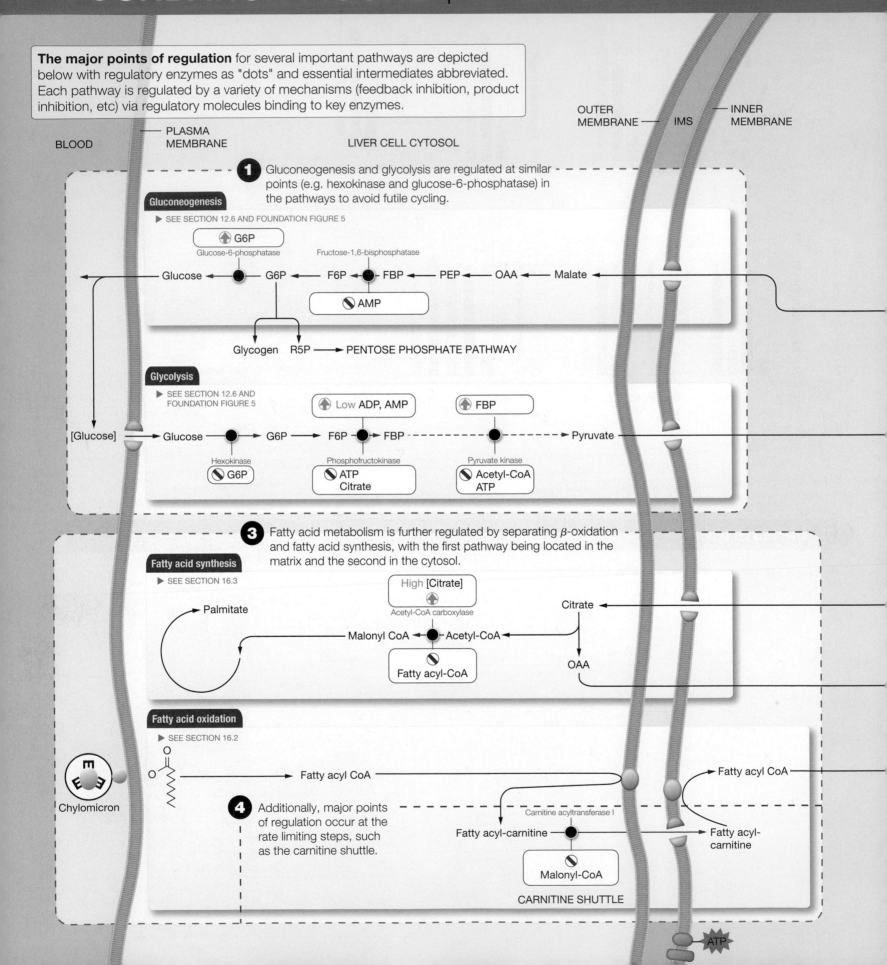

The major points of regulation for several important pathways are depicted below with regulatory enzymes as "dots" and essential intermediates abbreviated. Each pathway is regulated by a variety of mechanisms (feedback inhibition, product inhibition, etc) via regulatory molecules binding to key enzymes.

BLOOD

PLASMA MEMBRANE

LIVER CELL CYTOSOL

OUTER MEMBRANE — IMS — INNER MEMBRANE

1 Gluconeogenesis and glycolysis are regulated at similar points (e.g. hexokinase and glucose-6-phosphatase) in the pathways to avoid futile cycling.

Gluconeogenesis

▶ SEE SECTION 12.6 AND FOUNDATION FIGURE 5

⬆ G6P
Glucose-6-phosphatase

Fructose-1,6-bisphosphatase

Glucose ← G6P ← F6P ← FBP ← PEP ← OAA ← Malate

🚫 AMP

Glycogen ← R5P → PENTOSE PHOSPHATE PATHWAY

Glycolysis

▶ SEE SECTION 12.6 AND FOUNDATION FIGURE 5

⬆ Low ADP, AMP   ⬆ FBP

[Glucose] → Glucose → G6P → F6P → FBP ----- → Pyruvate

Hexokinase   Phosphofructokinase   Pyruvate kinase

🚫 G6P   🚫 ATP Citrate   🚫 Acetyl-CoA ATP

3 Fatty acid metabolism is further regulated by separating β-oxidation and fatty acid synthesis, with the first pathway being located in the matrix and the second in the cytosol.

Fatty acid synthesis

▶ SEE SECTION 16.3

High [Citrate]
⬆
Acetyl-CoA carboxylase

Palmitate

Malonyl CoA ← Acetyl-CoA ← Citrate

🚫 Fatty acyl-CoA

OAA

Fatty acid oxidation

▶ SEE SECTION 16.2

Chylomicron

O → Fatty acyl CoA ----- Fatty acyl CoA

Fatty acyl CoA

4 Additionally, major points of regulation occur at the rate limiting steps, such as the carnitine shuttle.

Carnitine acyltransferase I

Fatty acyl-carnitine ← ● → Fatty acyl-carnitine

🚫 Malonyl-CoA

CARNITINE SHUTTLE

ATP

Mastering **Chemistry** for Biochemistry

Mastering Chemistry for Biochemistry provides select end-of-chapter problems and feedback-enriched tutorial problems, animations, and interactive figures to depen your understanding of complex topics while practicing problem solving.

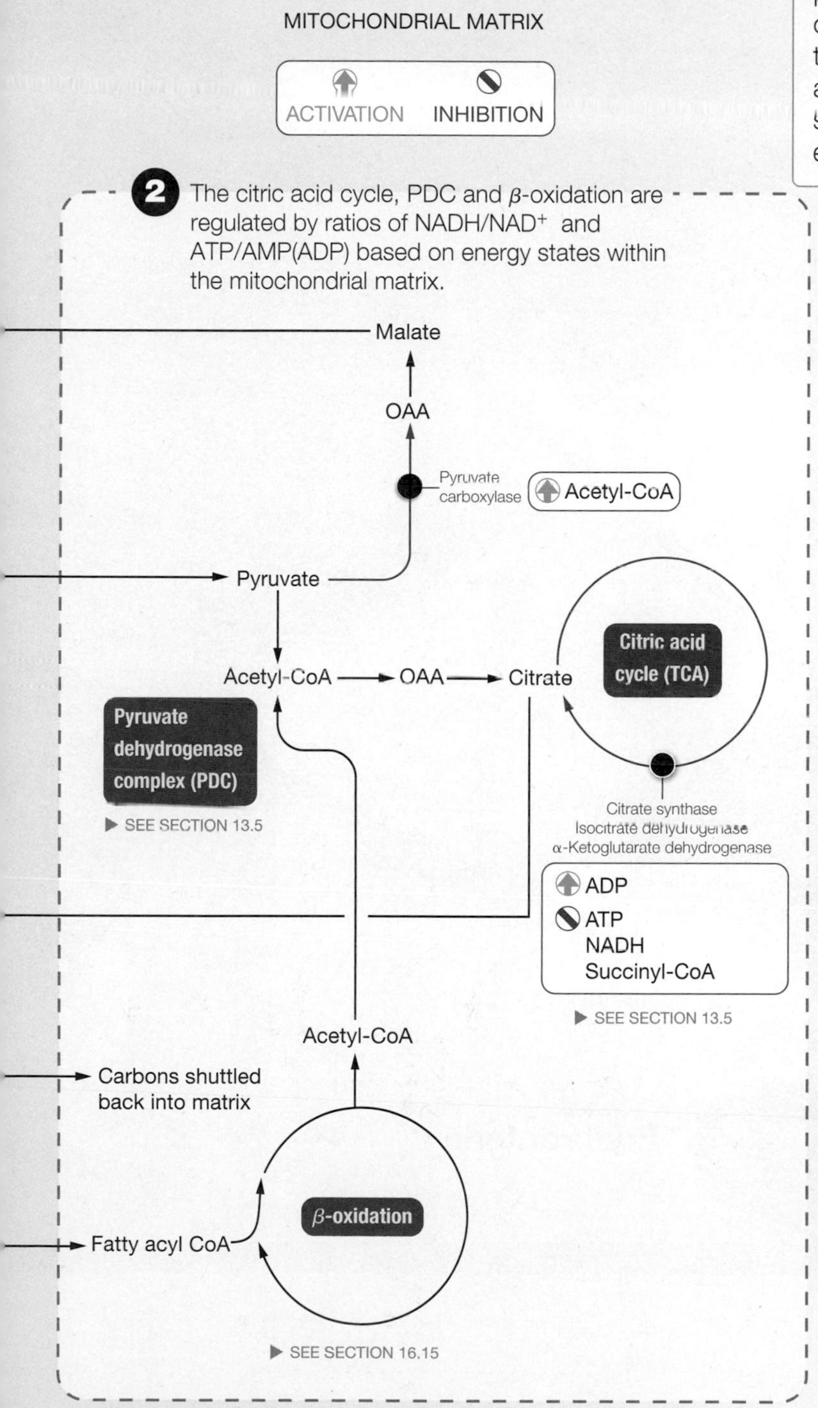

The table below summarizes the coordinated regulation of many pathways under "high energy conditions." Here this indicates a situation where ATP, NADH and other metabolic intermediates accumulate as products from glycolysis, the TCA cycle and β-oxidation. When the concentration of these molecules increases beyond the needs of the cell these pathways will be down-regulated, while energy storing pathways are upregulated at key enzymes. The opposite, "low energy conditions" shown at bottom results in coordinated regulation that reverses the effects seen under "high energy conditions".

MITOCHONDRIAL MATRIX

⬆ ACTIVATION   ⊘ INHIBITION

2 The citric acid cycle, PDC and β-oxidation are regulated by ratios of NADH/NAD$^+$ and ATP/AMP(ADP) based on energy states within the mitochondrial matrix.

Malate

↑

OAA

↑

● Pyruvate carboxylase   ⬆ Acetyl-CoA

Pyruvate

↓

Acetyl-CoA → OAA → Citrate

Citric acid cycle (TCA)

Pyruvate dehydrogenase complex (PDC)

▶ SEE SECTION 13.5

Citrate synthase
Isocitrate dehydrogenase
α-Ketoglutarate dehydrogenase

⬆ ADP
⊘ ATP
NADH
Succinyl-CoA

▶ SEE SECTION 13.5

Acetyl-CoA

↑

Carbons shuttled back into matrix

↑

Fatty acyl CoA →

β-oxidation

▶ SEE SECTION 16.15

High Energy Conditions	Pathways or enzymes upregulated	Pathways or enzymes downregulated
High [ATP]/([ADP] + [AMP])	Gluconeogenesis	Glycolysis PDC
High [NADH]/[NAD$^+$]	Minor effects	PDC TCA cycle β-oxidation
High [G6P]	Gluconeogenesis	Glycolysis
High [ACETYL CoA]	Gluconeogenesis (via PEPCK and PC)	PDC β-oxidation Glycolysis (via PK)
High [MALONYL CoA]	Fatty acid synthesis	Fatty acid oxidation (via carnitine shuttle)
High CITRATE	Fatty acid synthesis (via acetyl-CoA carboxylase)	Glycolysis (via PFK)

Low Energy Conditions	Pathways or enzymes upregulated	Pathways or enzymes downregulated
Low [ATP]/([ADP] + [AMP])	Glycolysis PDC TCA cycle AMPK	Gluconeogenesis
Low [NADH]/[NAD$^+$]	PDC TCA cycle β-oxidation Sirtuins	Minor effects
Low [G6P]	Glycolysis	Gluconeogenesis
Low [ACETYL CoA]	PDC β-oxidation	Minor effects
Low [MALONYL CoA]	Fatty acid oxidation (via acyltransferase I)	Fatty acid synthesis
Low CITRATE	Minor effects	Fatty acid synthesis (via acetyl-CoA carboxylase)

Sepiapterin

Pteridine

Leukopterin

Biopterin

Isoxanthopterin

Erythropterin

The pigments in butterfly wings are based on a class of nitrogen-rich heterocylic compounds called pteridines. In fact, pteridines are named after the Greek *pteron* (wing). Pteridine is also a component of folic acid, a central coenzyme in amino acid metabolism.

Amino Acid and Nitrogen Metabolism

18

THUS FAR, OUR STUDY of metabolism has concerned itself primarily with compounds that can be degraded completely to carbon dioxide and water—in other words, compounds containing only carbon, hydrogen, and oxygen. In this chapter and the next, we turn to the metabolism of nitrogen-containing compounds—amino acids and their derivatives, nucleotides, and the polymeric nucleic acids and proteins (**FIGURE 18.1**). Unifying principles of amino acid and nitrogen metabolism are presented in this chapter, and nucleotide metabolism is covered in Chapter 19. This chapter describes how cells assimilate nitrogen, common routes for utilizing and excreting ammonia, and coenzymes used in nitrogen metabolism. We will outline the metabolism of the 20 standard amino acids, focusing on the fates and sources of their carbon skeletons. Our approach is to organize these amino acids into families that are metabolically related. Finally, we will mention some of the major roles of amino acids as precursors to hormones, vitamins, coenzymes, porphyrins, pigments, and neurotransmitters.

Chapters 18 and 19 also reveal how much we have learned from naturally occurring human mutations, as well as from mutations

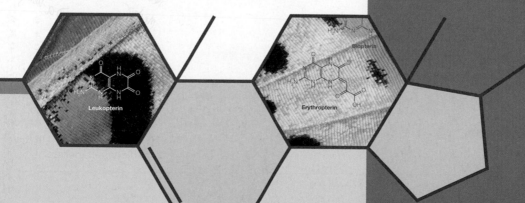

Leukopterin

Biopterin

Erythropterin

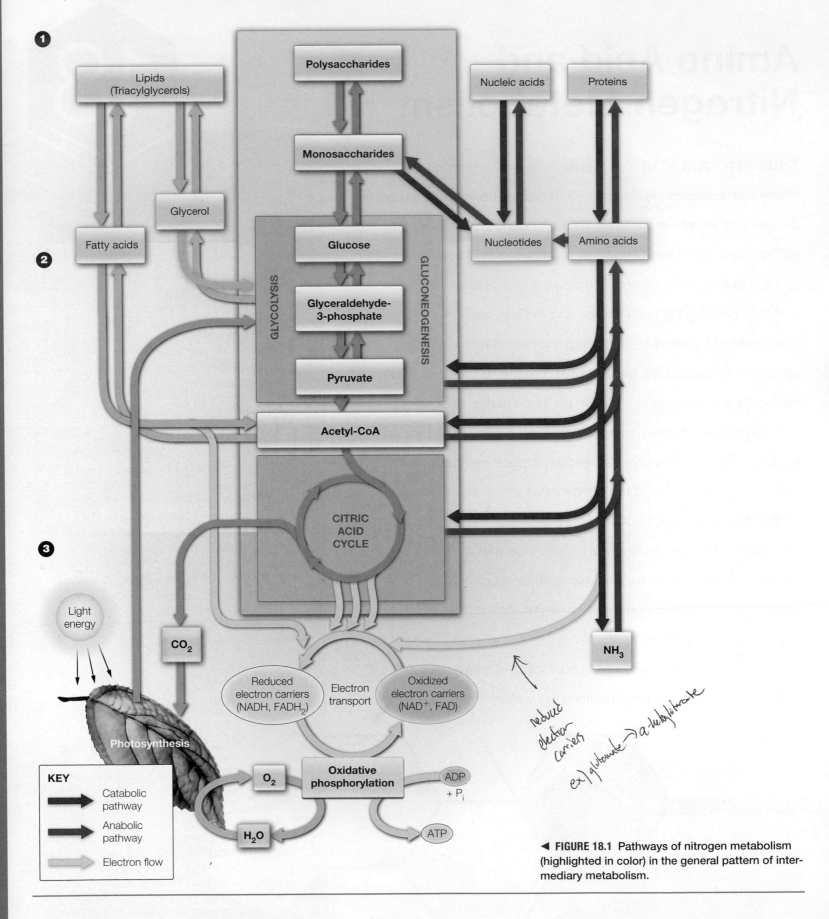

◀ **FIGURE 18.1** Pathways of nitrogen metabolism (highlighted in color) in the general pattern of intermediary metabolism.

generated in the laboratory, in cultured cells, or in bacteria. Whereas a mutation that inactivates an enzyme

● **CONNECTION** The clinical consequences of mutations that affect amino acid or nucleotide metabolism have greatly enhanced our understanding of human biochemistry.

in one of the central energy-generating or energy-storing pathways is likely to be lethal and, hence, not

observed in living individuals, mutations that affect amino acid or nucleotide metabolism are often not lethal and *are* found in living humans. The clinical consequences of these mutations are often quite severe and tragic, but these inherited metabolic diseases have greatly enhanced our understanding of human biochemistry.

18.1 Utilization of Inorganic Nitrogen: The Nitrogen Cycle

For many organisms, growth and reproduction are limited by the availability of utilizable nitrogen, which in turn is limited by the abilities of organisms to utilize different inorganic forms of nitrogen. All organisms

● **CONCEPT** Few organisms can use the N_2 in air, and many soils are poor in nitrate. Thus, nitrogen bioavailability limits growth for most organisms.

can convert ammonia (NH_3) to organic nitrogen compounds—that is, substances containing $C—N$ bonds. However, not all organisms can synthesize ammonia from the far more abundant forms of inorganic

nitrogen—dinitrogen gas (N_2), the most abundant component of the Earth's atmosphere, and nitrate ion (NO_3^-), a soil constituent essential for the growth of most plants. The reduction of N_2 to NH_3, termed **biological nitrogen fixation,** is carried out only by certain prokaryotes called **diazotrophs.** The reduction of NO_3^- to NH_3, by contrast, is widespread among both plants and microorganisms.

As in the consideration of any limited resource, it is useful to think about nitrogen metabolism in terms of an economy—a **nitrogen economy**—that focuses on questions of supply, demand, turnover, reuse, growth, and maintenance of a steady state. Within the biosphere, a balance is maintained between total inorganic and total organic forms

of nitrogen. The conversion of inorganic to organic nitrogen, which starts with nitrogen fixation or nitrate reduction, is counterbalanced by processes that return inorganic nitrogen to the biosphere (**FIGURE 18.2**). Catabolism of proteins and nucleic acids and other macromolecules yields ammonia and various organic nitrogenous end products, which can in turn be oxidized by various bacteria. These oxidations generate biological energy, just as other organisms derive energy from oxidation of carbohydrate or fat to CO_2. In the process, inorganic nitrate and nitrite are produced. **Denitrification,** the catabolism of ammonia to N_2, is a process that is carried out by other bacteria called **denitrifying bacteria.**

Biological Nitrogen Fixation

Although nitrogen gas makes up about 80% of the Earth's atmosphere, its reduction to ammonia occurs in relatively few living systems—some free-living soil bacteria, such as *Klebsiella* and *Azotobacter;* photosynthetic cyanobacteria; some archaea; and symbiotic nodules on the roots of **leguminous plants,** such as beans or alfalfa, that have been infected with certain diazotrophs, notably of the genus *Rhizobium* (**FIGURE 18.3**). The infecting bacterium assumes a modified form, called a **bacteroid,** inside the cells of infected plants. The nodules contain an abundant protein called **leghemoglobin,** which maintains an anaerobic environment by binding any O_2 that finds its way into the nodule—this is critical because the nitrogen-fixing enzymes are extremely sensitive to oxygen. This symbiosis allows the host plant to grow without an exogenous nitrogen source.

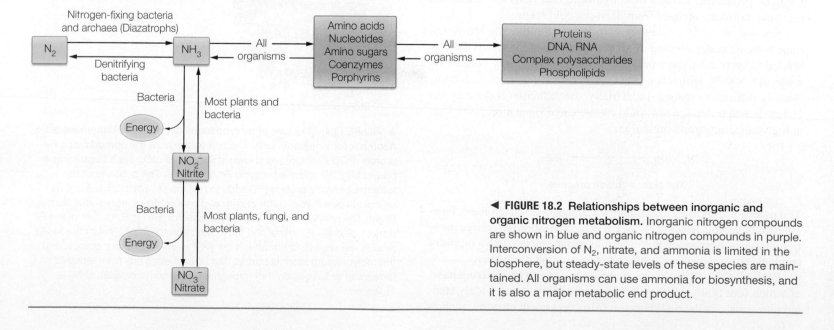

◄ **FIGURE 18.2 Relationships between inorganic and organic nitrogen metabolism.** Inorganic nitrogen compounds are shown in blue and organic nitrogen compounds in purple. Interconversion of N_2, nitrate, and ammonia is limited in the biosphere, but steady-state levels of these species are maintained. All organisms can use ammonia for biosynthesis, and it is also a major metabolic end product.

▲ **FIGURE 18.3 The site of nitrogen fixation in symbiotic root nodules.** This root of a soybean plant is infected by nitrogen-fixing bacteria of the genus *Rhizobium*.

● **CONNECTION** Crops with nitrogen-fixing ability, such as clovers, soybeans, or alfalfa, can be rotated with non-nitrogen-fixing crops, to replenish the soil with nitrogen.

Some trees, such as alder, also form nitrogen-fixing nodules and thus have the capacity to fix nitrogen. Recent discoveries have revealed an impressive diversity of diazotrophs, including hyperthermophilic methane-producing archaea from hydrothermal vents and anaerobic methane-oxidizing archaea from deep-sea cold seeps.

Because nitrogen availability is the factor limiting the fertility of most soils, an understanding of biological nitrogen fixation is directly related to increasing the world's food supply. The triply bonded N_2 molecule, $N \equiv N$, with a bond energy of about 940 kJ/mol, is extraordinarily difficult to reduce. Industrially, the reduction is done by the Haber–Bosch process, a low-yield catalytic hydrogenation carried out at high temperature and pressure.

$$N_2 + 3H_2 \xrightarrow[\text{450°C, 270 atm}]{\text{catalyst}} 2NH_3$$

The Haber–Bosch process

This process is used in the manufacture of ammonia-based fertilizers. Interest in the molecular details of biological nitrogen fixation has derived partly from hopes of supplanting this energy-intensive process with a means of ammonia production that can take place under milder conditions.

Biological N_2 reduction is catalyzed by the enzyme **nitrogenase,** of which four types are known. The most abundant and widely studied nitrogenase is the molybdenum (Mo)-dependent enzyme, such as

that found in *Azotobacter vinelandii*. The stoichiometry of the overall reaction is as follows:

$$N_2 + 8\,H^+ + 16\,MgATP + 8\,e^- \longrightarrow$$
$$2\,NH_3 + H_2 + 16\,MgADP + 16\,P_i$$

Nitrogen fixation is a very expensive process requiring hydrolysis of two ATPs per electron transferred. The ATP is generated through energy-yielding pathways of the organism, primarily carbohydrate catabolism. Although a total of eight electrons are required, reduction of N_2 to $2NH_3$ is a six-electron process. The other two electrons are "wasted" in the formation of H_2, a by-product of nitrogen reduction. Electrons for N_2 reduction are derived from low-potential carriers, either reduced ferredoxin or flavodoxin, a low-potential flavoprotein.

The Mo-dependent nitrogenase consists of two separate metalloproteins (**FIGURE 18.4**). One protein—called **molybdenum–iron (MoFe) protein, dinitrogenase,** or **component I**—catalyzes the reduction of

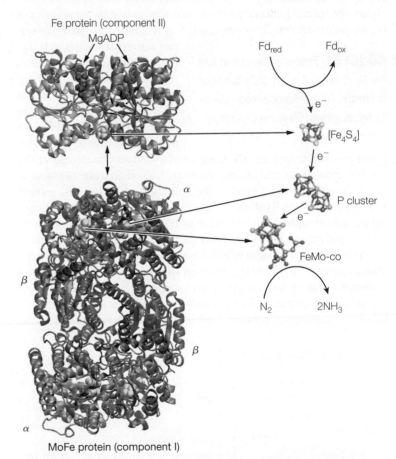

▲ **FIGURE 18.4 Structure of molybdenum-dependent nitrogenase from** *Azotobacter vinelandii*. **Left:** The two subunits of the homodimeric Fe protein (PDB ID: 1fp6) are shown in shades of pink, each containing a bound MgADP and the bridging Fe_4S_4 cluster. The subunits of the $\alpha_2\beta_2$ tetramer of MoFe protein (PDB ID: 1m1n) are shown in blue. Each $\alpha\beta$ unit binds one P iron–sulfur cluster and one FeMo-co iron–sulfur cluster. **Right:** The relative positions and structures of the Fe_4S_4 cluster of the Fe protein, and the P cluster and the FeMo cofactor (FeMo-co) of the MoFe protein are shown. Sulfur atoms are yellow, iron atoms are orange, and the molybdenum atom is purple. The flow of electrons from reduced ferredoxin or flavodoxin (Fd) through the iron–sulfur clusters to N_2 is indicated.

N_2. The other—called **iron (Fe) protein, dinitrogenase reductase,** or **component II**—transfers electrons and protons, one at a time, to the MoFe protein, in a process coupled to the hydrolysis of two MgATPs. Both proteins contain iron–sulfur clusters, and MoFe protein also contains molybdenum, in the form of a tightly bound **iron–molybdenum cofactor** (FeMo-co). FeMo-co, with its nine sulfurs, seven irons, and one molybdenum, is one of the largest and most complex metal centers in biological systems. N_2 binds to this cofactor during its reduction, although the precise mode of binding is not yet known. As shown in Figure 18.4, electrons flow from reduced ferredoxin or flavodoxin to the Fe_4S_4 complex in the Fe protein, and the hydrolysis of bound ATP somehow drives the electrons to the P cluster in the MoFe protein and then to FeMo-co. These three clusters are sufficiently close together in the complex to allow facile electron transfer.

Nitrate Utilization

The ability to reduce nitrate to ammonia is common to virtually all plants, fungi, and bacteria (Figure 18.2). The first step, reduction of nitrate (+5 oxidation state) to nitrite (+3 oxidation state) is catalyzed by **nitrate reductase.** The eukaryotic enzyme contains bound FAD, molybdenum, and a cytochrome b_5. The enzyme carries out the overall reaction:

$$NO_3^- + NAD(P)H + H^+ \longrightarrow NO_2^- + NAD(P)^+ + H_2O$$

The electrons are transferred from NADH or NADPH to enzyme-bound FAD, then to cytochrome b_5, then to molybdenum, and finally to the substrate.

Reduction of nitrite to ammonia is carried out in three steps

$$NO_2^- \longrightarrow NO^- \longrightarrow NH_2OH \longrightarrow NH_3$$

by one enzyme, **nitrite reductase.** Higher plants, algae, and cyanobacteria use ferredoxin as the electron donor in this six-electron reaction.

18.2 Utilization of Ammonia: Biogenesis of Organic Nitrogen

Although plants, animals, and bacteria derive their nitrogen from different sources, virtually all organisms share a few common routes for utilization of inorganic nitrogen in the form of ammonia. Ammonia in high concentrations is quite toxic, but at lower levels it is a central metabolite, serving as substrate for four enzymes that convert it to various organic nitrogen compounds (**FIGURE 18.5**). At physiological pH the dominant ionic species is ammonium ion, NH_4^+ ($pK_a = 9.2$). However, the four reactions involve the unshared electron pair of NH_3, which is therefore the reactive species.

● **CONCEPT** Several ubiquitous enzymes use ammonia as a substrate for synthesis of glutamate, glutamine, asparagine, or carbamoyl phosphate.

All organisms assimilate ammonia via reactions leading to glutamate, glutamine, asparagine, and **carbamoyl phosphate.**

$$H_2N-\overset{\overset{\displaystyle O}{\|}}{C}-O-\overset{\overset{\displaystyle O}{\|}}{\underset{\underset{\displaystyle O^-}{|}}{P}}-O^-$$

Carbamoyl phosphate

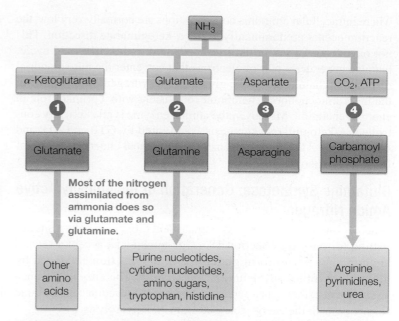

▲ **FIGURE 18.5 Reactions in assimilation of ammonia and major fates of the fixed nitrogen.** ❶ Glutamate dehydrogenase; ❷ Glutamine synthetase; ❸ Asparagine synthetase; ❹ Carbamoyl phosphate synthetase. The carbon skeletons that accept the ammonia in these four reactions are in green.

Carbamoyl phosphate is used only in the biosynthesis of arginine, urea, and the pyrimidine nucleotides. Thus, most of the nitrogen that finds its way from ammonia to amino acids and other nitrogenous compounds does so via the three amino acids glutamate, glutamine, and asparagine. The α-amino nitrogen of glutamate and the side-chain *amide* nitrogen of glutamine are both primary sources of N in biosynthetic pathways.

Glutamate Dehydrogenase: Reductive Amination of α-Ketoglutarate

Glutamate dehydrogenase (Figure 18.5, reaction ❶) catalyzes the reductive amination of α-ketoglutarate, forming glutamate:

$$
\begin{array}{c}
COO^- \\
| \\
CH_2 \\
| \\
CH_2 \\
| \\
C{=}O \\
| \\
COO^-
\end{array}
\;+\; NH_3 \;+\; NAD(P)H \;+\; 2H^+ \;\rightleftharpoons
$$

α-Ketoglutarate

$$
\begin{array}{c}
COO^- \\
| \\
CH_2 \\
| \\
CH_2 \\
| \\
H-C-\overset{+}{N}H_3 \\
| \\
COO^-
\end{array}
\;+\; H_2O \;+\; NAD(P)^+
$$

Glutamate

The reaction is reversible, but in most bacteria and many plants it acts primarily in the direction of glutamate formation. In animal cells, where intracellular ammonia concentrations are normally very low, the reaction occurs predominantly in the α-ketoglutarate direction. This not only feeds an important intermediate into the citric acid cycle, but also generates a pair of electrons that can enter the mitochondrial respiratory chain. In animals, glutamate dehydrogenase is located in the inner mitochondrial membrane, consistent with a primary role in energy generation. Moreover, the animal enzyme is allosterically controlled; α-ketoglutarate synthesis is inhibited by GTP and ATP and stimulated by ADP. Thus, the enzyme is activated under conditions of low energy charge.

Glutamine Synthetase: Generation of Biologically Active Amide Nitrogen

Whether formed by action of glutamate dehydrogenase or by transamination (discussed in Section 18.4), glutamate can accept a second ammonia moiety to form glutamine in the reaction catalyzed by **glutamine synthetase** (Figure 18.5, reaction ❷). This enzyme is named a *synthetase*, rather than a *synthase*, because the reaction couples bond formation with the energy released from ATP hydrolysis.

The glutamine synthetase reaction occurs via an acyl phosphate intermediate. ATP phosphorylates the δ-carboxylate of glutamate to give a carboxylic-phosphoric acid anhydride (γ-glutamyl phosphate) which undergoes nucleophilic attack by the nitrogen of ammonia to give the amide product, glutamine.

Glutamate   **γ-Glutamyl phosphate**

Glutamine

Asparagine synthetase (Figure 18.5, reaction ❸) catalyzes a reaction comparable to that of glutamine synthetase, forming asparagine from aspartate and either NH_3 or the amide of glutamine as nitrogen donor. Although asparagine synthetase is found in most organisms, it accounts for much less ammonia assimilation than glutamine synthetase.

Carbamoyl Phosphate Synthetase: Generation of an Intermediate for Arginine and Pyrimidine Synthesis

The final route for assimilating ammonia goes through carbamoyl phosphate (Figure 18.5, reaction ❹). The enzyme responsible is **carbamoyl phosphate synthetase (CPS).** Either ammonia or glutamine can serve as the nitrogen donor:

$$NH_3 + HCO_3^- + 2\ ATP \longrightarrow \text{carbamoyl phosphate} + 2\ ADP + P_i$$

$$Glutamine + H_2O + HCO_3^- + 2\ ATP \longrightarrow$$
$$\text{carbamoyl phosphate} + 2\ ADP + P_i + glutamate$$

The bacterial enzyme can catalyze both reactions, although glutamine is the preferred substrate. Eukaryotic cells contain two forms of the enzyme. CPS I, localized in mitochondria, has a preference for ammonia as substrate and is used in the arginine biosynthetic pathway and the urea cycle (see Section 18.5). CPS II, present in the cytosol, has a strong preference for glutamine. CPS II is inhibited by uridine triphosphate (UTP), consistent with its involvement in pyrimidine nucleotide biosynthesis (as we will see in Chapter 19).

18.3 The Nitrogen Economy and Protein Turnover

Thus far, we have discussed the supply side of the nitrogen economy—the routes by which inorganic nitrogen is converted to ammonia and how this ammonia is utilized in the biosynthesis of organic nitrogen compounds such as amino acids and nucleotides. We are now ready to discuss the demand side of the nitrogen economy—the turnover and reuse of this limited resource.

Metabolic Consequences of the Absence of Nitrogen Storage Compounds

Early biochemists and physiologists believed that the proteins of an adult animal were quite stable, while proteins in the diet were immediately metabolized to provide energy and the end products excreted. This dogma was challenged in the 1930s by Rudolf Schoenheimer. Schoenheimer had escaped Hitler's Germany and joined the Department of Biochemistry at Columbia University, where Harold Urey had discovered deuterium a few years earlier. This environment inspired Schoenheimer to explore the use of isotopic tracers to study metabolism in whole animals (see Tools of Biochemistry 11B). Schoenheimer's group synthesized ^{15}N-labeled tyrosine and discovered that following administration of this tracer to a rat, only about 50% of the ^{15}N was recovered in the urine. Most of the remainder was incorporated into tissue proteins. Importantly, only a small amount of the ^{15}N found in the tissue proteins was attached to the original tyrosine carbon skeleton—*most had been incorporated into other amino acids.* Finally, they observed that an equivalent amount of unlabeled protein nitrogen was excreted, keeping the rat in **nitrogen equilibrium.**

Whereas carbohydrates and lipids can be stored for mobilization as needed by an organism for energy generation or for biosynthesis, most organisms past the embryonic state have no polymeric nitrogen compounds whose function is to be stored and released on demand. The lack of such nitrogen depots imposes special requirements on organisms, particularly because of the limited availability of utilizable nitrogen. Animals must continually replenish nitrogen supplies through the diet to replace nitrogen lost through catabolism. In much of the world, protein-rich foods cannot be produced in sufficient quantity to meet the nutritional needs of humans and domestic animals. When dietary protein is insufficient, proteins synthesized for other purposes, mostly muscle proteins, are broken down and not replaced.

● **CONCEPT** Most organisms lack nitrogen storage depots.

● **CONNECTION** A healthy 70-kg adult requires about 8 g of nitrogen per day in his or her diet to remain in nitrogen balance. This corresponds to a daily intake of about 52 g of protein.

Just as we can think of a nitrogen economy for the biosphere, we can see it also in relation to individual organisms. Under optimal conditions, animals maintain nitrogen intake and excretion at equivalent rates. A well-nourished adult is said to be in nitrogen equilibrium or **normal nitrogen balance** if the daily intake of nitrogen through the diet is equal to that lost through urinary excretion and other processes, such as ammonia loss in sweat.

Protein Turnover

Proteins are subject to continuous biosynthesis and degradation, a process called **protein turnover.** As revealed in Schoenheimer's ^{15}N experiments, many of the amino acids released during protein turnover are reutilized in the synthesis of new proteins. Individual proteins exhibit tremendous variability in their metabolic lifetimes, from a few minutes to many months. In the rat, the average protein has a half-life of 1 or 2 days. Proteins that are secreted into an extracellular environment, such as digestive enzymes or polypeptide hormones, turn over quite rapidly, whereas proteins that play a predominantly structural role, such as collagen of connective tissue, are much more stable metabolically. Enzymes catalyzing rate-determining steps in metabolic pathways are typically short-lived. Indeed, for many enzymes the rate of breakdown is an important regulatory factor in controlling intracellular enzyme levels.

Like all other intracellular constituents, proteins are subjected to a barrage of environmental influences, primarily reactive oxygen species (see Chapter 14), which can affect their structure, conformation, and biological activity. The capacity of proteins to repair the resulting damage is limited. Thus, protein turnover serves as a quality control system in which modified proteins are degraded and replaced. We now know that proteolysis requires an input of energy; this is a surprising finding given that amide bond hydrolysis is exergonic. This energy requirement reflects the fact that the process is nonrandom and highly regulated. Indeed, protein molecules that have become chemically altered are preferentially degraded. A certain chemical change may mark a protein molecule, targeting it for degradation by a proteolytic enzyme that specifically recognizes the marker.

● **CONCEPT** All proteins are in a constant state of turnover, for replacement of damaged proteins and for biological regulation

Over the past three decades, it has become clear that the process of limited proteolysis—regulated cleavage of a few specific peptide bonds in a protein—has a host of functions, including regulation of gene expression, response to environmental stress, and participation in cell-signaling pathways. Of great current interest is the involvement of selective proteolytic reactions in signaling pathways leading to *apoptosis,* a process in normal development in which certain cells, having fulfilled their function in differentiation, undergo a programmed death. Here we concern ourselves with those aspects of protein turnover that relate specifically to amino acid metabolism: to identifying major classes of intracellular proteases, and to describing some of the structural features that mark certain proteins for degradation.

Intracellular Proteases and Sites of Turnover

Eukaryotes possess several types of intracellular proteases. **Cathepsins** are contained in lysosomes, which form by budding from the Golgi complex. Lysosomes are essentially bags of digestive enzymes, containing nucleases, lipases, and glycosidases as well as the cathepsin proteases. The lysosomal system functions primarily in the proteolysis of extracytoplasmic proteins, which enter the cell via endocytosis and are degraded within the vacuolar lumen. Lysosomes play various cellular roles: secretion of digestive enzymes, digestion of organelles destined for destruction, digestion of food particles or bacteria engulfed by phagocytosis, and the nonselective engulfment and degradation of bulk cellular constituents (**autophagy**).

In addition to cathepsins, eukaryotic cells also possess many nonlysosomal proteases, including a large (2.5 megadalton) multisubunit ATP-dependent protease called the **proteasome** (FIGURE 18.6). The

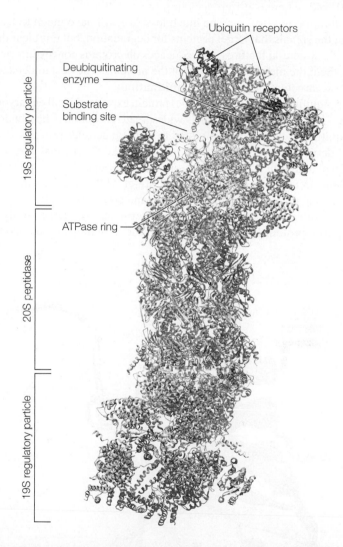

▲ **FIGURE 18.6 Structure of the human proteasome.** The proteasome consists of a 28-subunit core particle (also known as the 20S particle) capped on one or both ends by a 19-subunit regulatory particle (also known as the 19S regulatory particle). The proteolytic active sites are located within the large internal space (~100 × 60 Å) of the 20S core particle. The 19S regulatory particle controls substrate entry into the 20S core particle. Proteins tagged with the protein ubiquitin pass through the tube in an ATP-dependent fashion. This structure was generated by averaging multiple cryo-electron micrographs of the human 26S proteasome. Image produced from PDB ID: 5t0c by David Taylor, Univ. of Texas, Austin.

proteasome, as we shall see, degrades proteins that have been modified by the attachment of the small protein **ubiquitin.** In contrast to lysosomal enzymes, which are usually safely sequestered in their vesicles, any protease activity free in normal cytosol must be under strict control, so as to attack only those proteins whose destruction is needed—damaged, mutant, or otherwise dispensable proteins. The identification, or marking, of those proteins whose degradation suits the interests of the cell involves various tagging schemes described in the next section.

Chemical Signals for Turnover—Ubiquitination

The in vivo turnover rates for different proteins vary by as much as 1000-fold, whereas differences in protein stability, as measured by denaturation in vitro, may be much less. We still have much to learn about the signals that target proteins for degradation, but it is clear that specific structural features and sequences on proteins convey information about the metabolic stability of the proteins. The best understood of these structural features is **ubiquitination.**

Ubiquitin is a small (76-residue) protein expressed in all eukaryotic cells—it derives its name from its widespread (ubiquitous) distribution. Ubiquitin is covalently conjugated to specific cellular proteins in an ATP-dependent reaction, which condenses the C-terminal carboxyl group of ubiquitin with specific lysine amino groups on target proteins, forming an *isopeptide* bond. Expenditure of ATP energy ensures that the tagging reaction is both irreversible and specific.

The selection of target proteins is determined by a large family of ubiquitin–protein ligases (E3 ligases). The human genome encodes

~600 of these ligases. In addition to the initial conjugation of ubiquitin to a target protein, E3 ligases can also catalyze the successive addition of a ubiquitin moiety to a previously conjugated ubiquitin, forming a polyubiquitin chain in which a Lys of ubiquitin forms an isopeptide bond with the C-terminal Gly carboxyl group of the succeeding ubiquitin. Because of the central role of ubiquitin E3 ligases in determining the specificity and selectivity of the proteasome system, these proteins are implicated in a number of disease states, including neurodegenerative disorders, inflammatory diseases, muscle wasting disorders, and cancer.

Polyubiquitin chains serve as recognition markers for the proteasome, and such tagged proteins dock at the proteasome via specific ubiquitin receptors. The proteasome contains a molecular motor, a hexameric ring of ATPase subunits (Figure 18.6) that uses the energy of ATP hydrolysis to unfold and translocate the target proteins into the peptidase. These substrates are subsequently degraded inside the proteasome to short peptides. The polyubiquitin chains are not degraded, but instead are released and disassembled by deubiquitinating enzymes in the lid of the proteasome. The free ubiquitin monomers can then be reused.

Polyubiquitin chains also have nonproteasomal functions involved in the regulation of many critical processes, including cell-cycle progression, cholesterol synthesis, inflammation, response to hypoxia, and apoptosis. As we have seen for other regulatory systems, ubiquitination is reversible. Ubiquitin removal is catalyzed by a family of **deubiquitinating enzymes** (Dubs); the human genome encodes ~95 distinct Dubs.

Although bacteria lack the 26S proteasome, they do possess functionally similar ATP-dependent proteolytic assemblies known as the **Lon** and **Clp proteases.** Both of these proteins are barrel-shaped, with the protease active sites located in a central cavity, like the 26S proteasome of eukaryotes. Homologs of Lon are found in archaea and in mitochondria of eukaryotes.

18.4 Coenzymes Involved in Nitrogen Metabolism

Before presenting in detail the metabolism of amino acids and nucleotides, as we do in this chapter and the next, we should consider three families of coenzymes that play major roles in amino acid and/or nucleotide metabolism. Although all have been mentioned previously in other chapters, we shall consider their actions in detail here. These cofactors include (1) pyridoxal phosphate, the cofactor for transamination and many other reactions of amino acid metabolism; (2) the folic acid coenzymes, which transfer single-carbon functional groups in synthesizing nucleotides and certain amino acids; and (3) the B_{12}, or cobalamin, coenzymes, which participate in the synthesis of methionine and, as noted in Chapter 16, the catabolism of methylmalonyl-CoA.

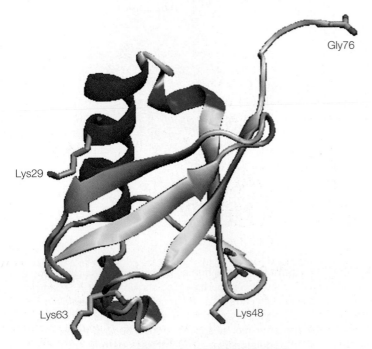

Human ubiquitin (PDB ID: 1ubi)

● **CONNECTION** Marginal vitamin B_6 deficiency occurs frequently in humans and is associated with coronary artery disease, stroke, and an elevated risk of Alzheimer's disease.

Pyridoxal Phosphate

Vitamin B_6 was discovered in the 1930s as the result of nutritional studies with rats fed vitamin-free diets. The vitamin as originally isolated is **pyridoxine,** named from its structural similarity to pyridine. Pyridoxine contains a hydroxymethyl group at position 4 of the pyridine ring. However, in the active coenzyme, this group has been oxidized to an aldehyde, and the hydroxymethyl group at position 5 is phosphorylated. Pyridoxal phosphate (which we shall abbreviate PLP) is the predominant coenzyme form, with pyridoxamine phosphate (PMP) being an intermediate in transamination reactions.

Pyridoxine

Pyridoxal phosphate (PLP)

Pyridoxamine phosphate (PMP)

Pyridoxal phosphate (PLP) is a remarkably versatile coenzyme. In addition to its involvement in transamination reactions, PLP serves as a coenzyme for the majority of enzymes that catalyze some chemical change at the α-, β-, or γ-carbons of the common amino acids, including decarboxylations, eliminations, racemizations, and retro-aldol reactions. All of these pyridoxal phosphate–requiring enzymes act via the formation of a Schiff base between the amino acid and coenzyme. This is the same chemical strategy used in the fructose-1,6-bisphosphate aldolase reaction (see Figure 12.5), in which a cationic imine (the Schiff base) lowers the energy barrier to the reaction. The ability of PLP to form a stable Schiff base is the key to its versatility in enzyme-catalyzed reactions.

Although pyridoxal phosphate is the coenzyme for all of these reactions, the reactive species is not the aldehyde group but rather an aldimine, formed between the coenzyme and an ε-amino group of a lysine residue in the active site of the enzyme (below). A hydrogen bond between the phenolic proton and the imino nitrogen of the lysine residue favors a planar structure between the aldimine and the aromatic pyridine ring.

Enzyme-bound pyridoxal phosphate

We now know that all pyridoxal phosphate–requiring enzymes act via the formation of a Schiff base between the amino acid substrate and enzyme-bound coenzyme, displacing the lysine amino group. Recall from Section 12.2 that a Schiff base is a nucleophilic addition product between an amino group and a carbonyl group. The planarity of the structure results in a large conjugated π molecular orbital system, which is essential for catalysis.

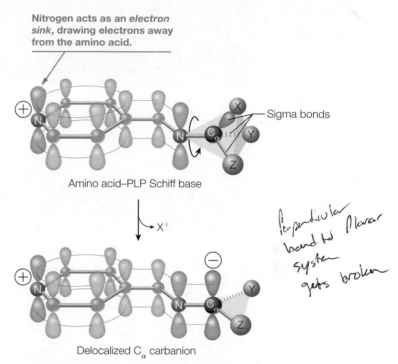

π **Molecular orbital system of PLP-amino acid Schiff base**

The most important catalytic feature of the coenzyme is the electrophilic nitrogen of the pyridine ring, which acts as an *electron sink,* drawing electrons away from the amino acid and labilizing one of the three σ bonds on the α-carbon. When an amino acid forms an imine with PLP, and the pyridinium N is protonated, all three σ bonds on the α-carbon become electron-deficient and are susceptible to heterolytic cleavage. The σ bond that is aligned perpendicular to the plane of the π molecular orbital system is the one cleaved, and this is determined by the angle of rotation of the C_α—N bond, specified by interactions with the enzyme active site (see above). *All of the known reactions of PLP enzymes can be described mechanistically in the same way:* formation of a planar Schiff base or aldimine intermediate, followed by bond cleavage and formation of a resonance-stabilized carbanion with a quinonoid structure, as shown in **FIGURE 18.7**. Stabilizing the carbanion intermediate that results from bond cleavage is the other important function of the electrophilic nitrogen of the pyridine ring. Depending on the bond labilized, formation of the aldimine can lead to transamination (as detailed in Figure 18.7), to decarboxylation, to racemization, or to retroaldol cleavage.

● **CONCEPT** All pyridoxal phosphate reactions involve initial Schiff base formation, followed by bond labilization caused by electron withdrawal to the coenzyme's pyridine ring.

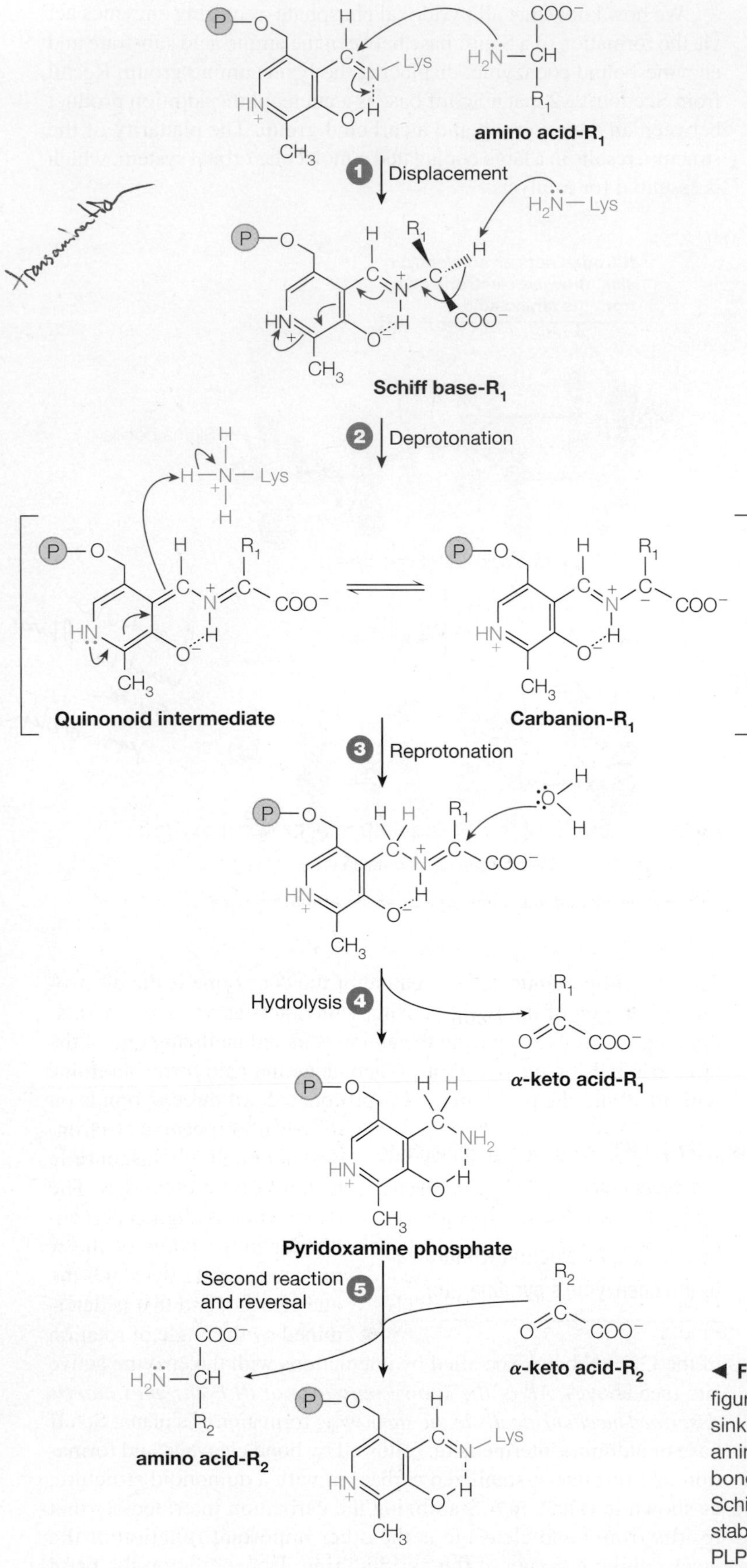

Folic Acid Coenzymes and One-Carbon Metabolism

Discovery and Chemistry of Folic Acid

Coenzymes derived from the vitamin **folic acid** participate in the generation and utilization of single-carbon functional groups—methyl, methylene, and formyl. The vitamin was discovered by the British physician Lucy Wills in the 1930s, when she found that people with a certain type of **megaloblastic anemia** could be cured by treatment with yeast or liver extracts. The condition is characterized, like all anemias, by reduced levels of erythrocytes. The cells that remain are characteristically large and immature, suggesting a role for the vitamin in cell proliferation and/or maturation. Isolation and structural identification of the vitamin required several more years due to its natural low abundance. Esmond Snell and Herschel Mitchell, working with Roger Williams at the University of Texas at Austin, had to process *four tons* of spinach leaves to obtain a few hundred micrograms of the active component. They named it folic acid, from the Latin *folium* for leaf. Folic acid and its many derivatives are referred to as **folates.**

Naturally occurring folates are formed from three distinct moieties: (1) a bicyclic, heterocyclic **pteridine** ring; (2) **p-aminobenzoic acid (PABA),** which is itself required for the growth of many bacteria; and (3) a "tail" of glutamate residues, ranging from three to eight or more residues. These residues are linked to one another, not by the familiar peptide bond but rather by an amide bond between the γ-carboxyl group of the first glutamate and the α-amino group of the next. These three moieties are shown in the overall structure of the active form of the coenzyme, **tetrahydrofolate (THF):**

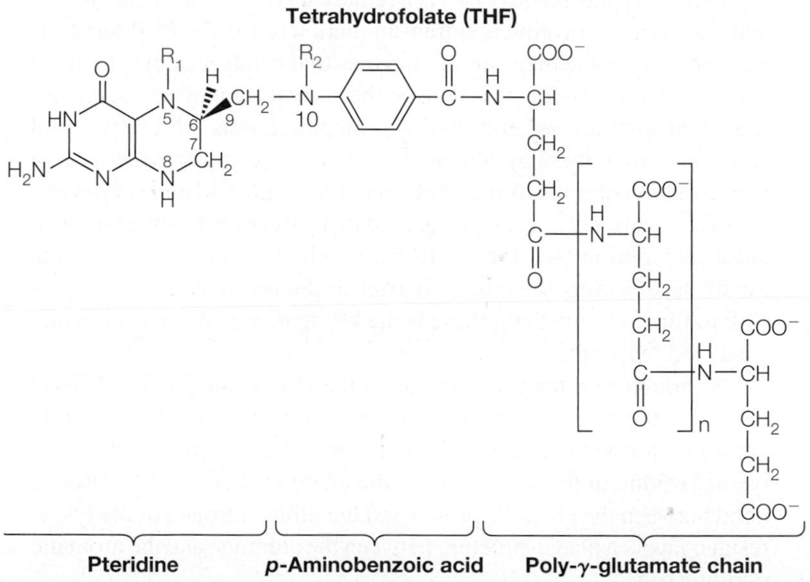

◀ FIGURE 18.7 Involvement of pyridoxal phosphate in transamination. The figure shows the action of the positively charged pyridinium ion as an electron sink. ❶ Amino acid R_1 reacts with the enzyme-bound PLP, displacing the lysine amino group. ❷ Base-catalyzed deprotonation (cleavage of the labile C—H σ bond that is perpendicular to the plane of the π molecular orbital system of the Schiff base intermediate) leads to formation of a carbanion, which is resonance stabilized by interconversion with a quinonoid intermediate. ❸ Reprotonation on the PLP carbon results in tautomerization of the imine C—N bond. ❹ Hydrolysis via a carbinolamine intermediate yields an α-keto acid R_1 and pyridoxamine phosphate. ❺ The transamination is completed by reaction with a second α-keto acid (R_2) and conversion, by reversal of steps 1–4, to enzyme-bound PLP and amino acid R_2.

● **CONNECTION** The pteridine ring is also found in a large class of biological pigments. Insect wings and eyes contain pteridine pigments, as does the skin of amphibians and fish. Butterfly wings are particularly abundant in pteridines and were the first source from which any such compounds were identified structurally. These compounds are named after the Greek *pteron* ("wing").

The N-5 and N-10 positions can carry single carbon units, designated here as R_1 and R_2, for donation in biosynthetic pathways, which we will discuss in more detail later.

Conversion of Folic Acid to Tetrahydrofolate

The folic acid we obtain from our diet has an oxidized pteridine ring. Once inside a cell, it is converted to active forms by two successive reductions. Both reactions are catalyzed by the NADPH-specific enzyme **dihydrofolate reductase (DHFR).** The first reduction yields **7,8-dihydrofolate,** and the second reduction yields **5,6,7,8-tetrahydrofolate (THF).**

Reduction of the 5–6 double bond generates a new chiral center at C-6; the 6*S*-isomer of tetrahydrofolate is the naturally occurring form used by enzymes.

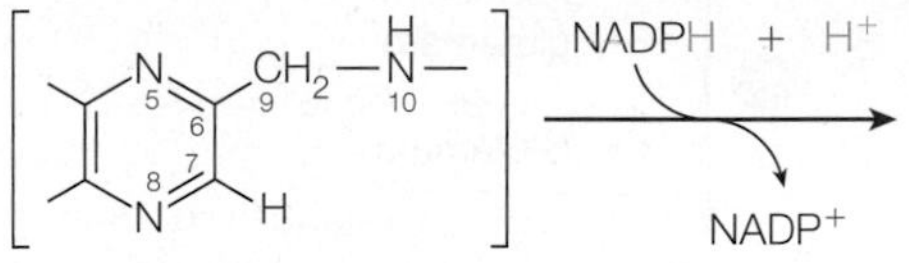

Folate (partial structure)

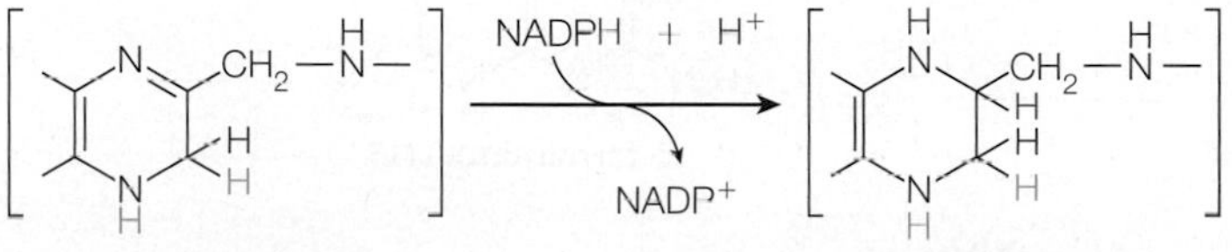

7,8-Dihydrofolate (DHF)   **5,6,7,8-Tetrahydrofolate (THF)**

Dihydrofolate reductase is the target for action of a number of clinically useful **antimetabolites.** An antimetabolite is a synthetic compound, usually a structural analog of a normal metabolite, that interferes with the utilization of the metabolite to which it is related structurally. As early as 1948, two analogs of folate—**aminopterin** and **methotrexate**—had been synthesized and found to induce remissions in acute leukemias.

● **CONNECTION** Dihydrofolate reductase is the target for a number of useful anticancer, antibacterial, and antiparasitic drugs.

These compounds inhibit dihydrofolate reductase, binding to the enzyme at least 1000-fold more tightly than the normal substrates do. Thus, these analogs block the utilization of folate and dihydrofolate. We now know that their effectiveness derives from the involvement of dihydrofolate reductase in the biosynthesis of thymine nucleotides and, hence, of DNA. Inhibiting DNA synthesis blocks the proliferation of cancer cells, as discussed further in Chapter 19.

Folate analogs such as methotrexate have been used in treating many different cancers in addition to leukemia. Other clinically useful dihydrofolate reductase inhibitors show selectivity among species-specific forms of the enzyme. **Trimethoprim** specifically inhibits bacterial dihydrofolate reductases and is widely used to treat bacterial infections, and **pyrimethamine** shows similar specificity against the enzyme of protozoal origin.

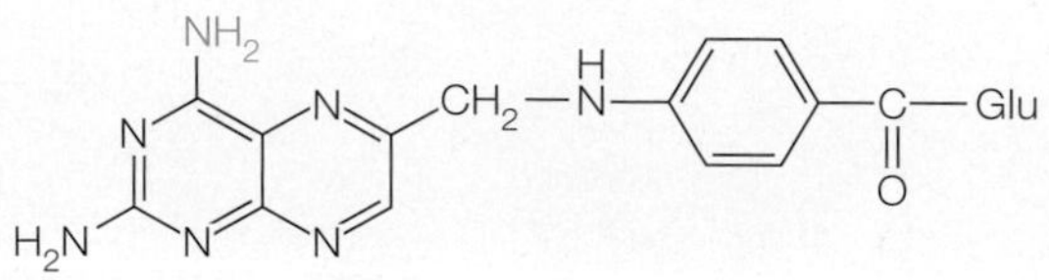

Aminopterin (4-aminofolate)

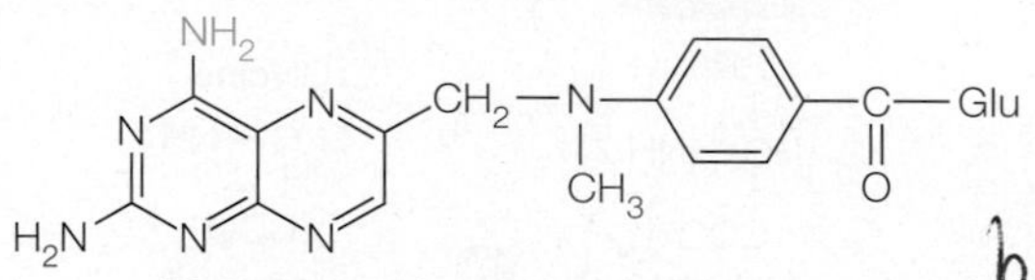

Methotrexate (4-amino-10-methylfolate)

binds tighter to DHFR 1000x more effectively to folate (competitive inhibitor) anti-metabolite

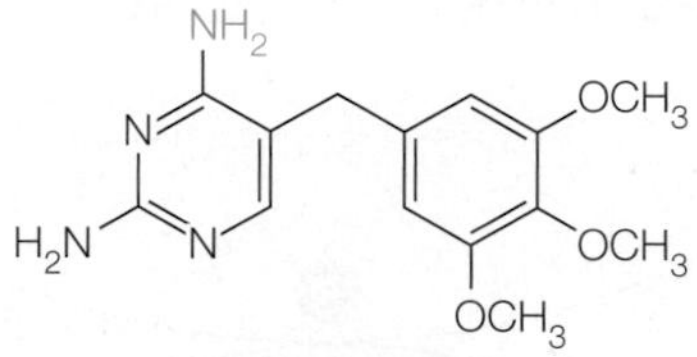

Trimethoprim

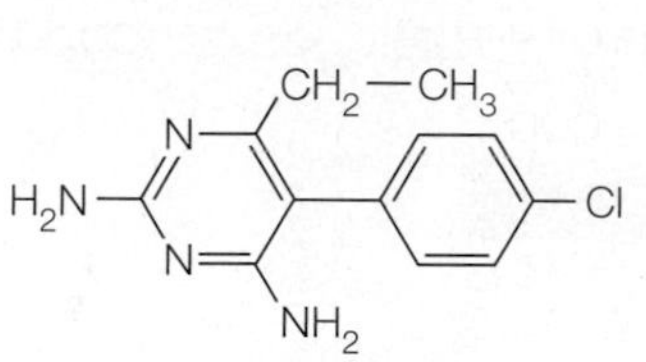

Pyrimethamine

Tetrahydrofolate in the Metabolism of One-Carbon Units

The coenzymatic function of tetrahydrofolate (THF) is the mobilization and utilization of single-carbon functional groups (one-carbon units).

● **CONCEPT** Tetrahydrofolate coenzymes transfer and interconvert one-carbon units at the methyl, methylene, and formyl oxidation levels.

These reactions are involved in the metabolism of serine, glycine, methionine, and histidine, among the amino acids, and in the biosynthesis of purine nucleotides and the methyl group of thymine.

Tetrahydrofolate carries one-carbon units at the methyl, methylene, and formyl oxidation levels, equivalent in oxidation level to methanol, formaldehyde, and formic acid, respectively. One-carbon groups on THF can be carried on N-5 or N-10, or bridged between N-5 and N-10. The THF derivatives are named according to the oxidation state of the one-carbon unit and the nitrogen positions to which it is attached. Thus, **5,10-methylenetetrahydrofolate** (5,10-methylene-THF) carries a methylene group ($-CH_2-$) attached to N–5 and N–10:

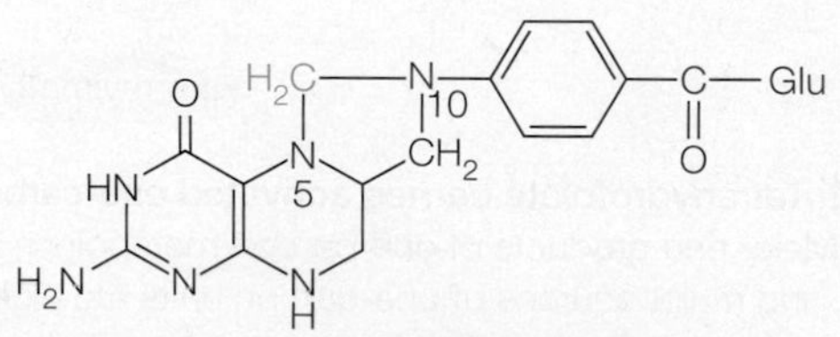

5,10-methylenetetrahydrofolate

One-carbon units attached to tetrahydrofolate are activated for the formation of new carbon bonds in various biosynthetic reactions (**FIGURE 18.8**). The most reduced form, **5-methyltetrahydrofolate**

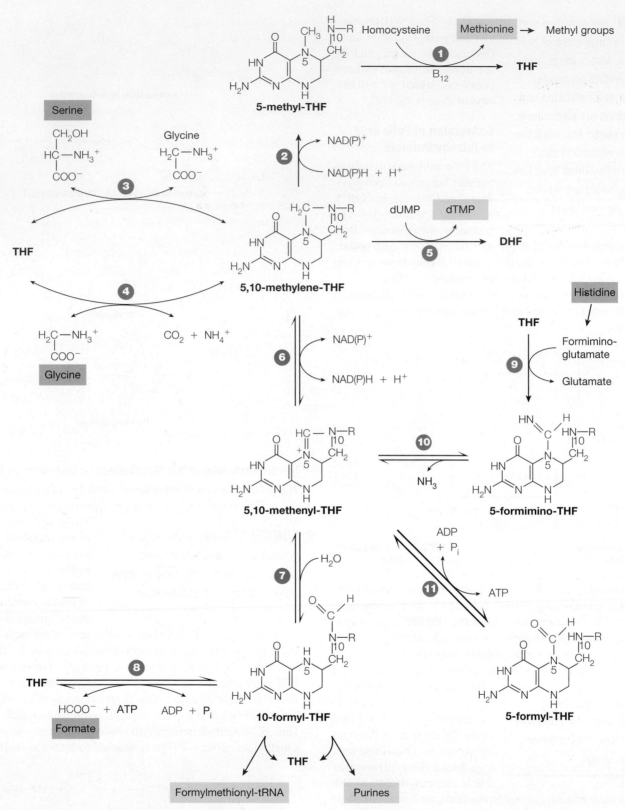

▲ **FIGURE 18.8 Tetrahydrofolate carries activated one-carbon units for biosynthesis.** Major end products of one-carbon metabolism are highlighted in pink, and major sources of one-carbon units are highlighted in orange. The enzymes involved are ❶ homocysteine methyltransferase (also called methionine synthase; uses methyl-B$_{12}$ as cofactor), ❷ methylenetetrahydrofolate reductase, ❸ serine hydroxymethyltransferase, ❹ glycine cleavage system, ❺ thymidylate synthase, ❻ methylenetetrahydrofolate dehydrogenase, ❼ methenyltetrahydrofolate cyclohydrolase, ❽ 10-formyltetrahydrofolate synthetase, ❾ glutamate formiminotransferase, ❿ 5-formiminotetrahydrofolate cyclodeaminase, and ⑪ 5-formyltetrahydrofolate cycloligase (also called methenyltetrahydrofolate synthetase). THF = tetrahydrofolate, DHF = dihydrofolate, R = PABA-glutamate.

(5-methyl-THF), donates its one-carbon unit to just one acceptor—homocysteine, forming the terminal C—S bond of methionine (reaction ❶ in Figure 18.8). The one-carbon unit carried by 5,10-methylenetetrahydrofolate is used to form new C—C bonds, as in reactions ❸, ❹, and ❺. The most oxidized form, **10-formyltetrahydrofolate** (10-formyl-THF), forms new C—N bonds.

As shown in Figure 18.8, one-carbon units derived from THF coenzymes are used in many biosynthetic processes. In the synthesis of thymine nucleotides, catalyzed by **thymidylate synthase** (reaction ❺), the THF coenzyme serves both as a one-carbon donor and as a source of reducing power. Because this enzyme generates the methyl group of thymine from 5,10-methylene-THF, it catalyzes both a one-carbon transfer and a reduction (see Figure 19.18). The electrons come from the reduced pteridine ring, to give dihydrofolate as a product. Although dihydrofolate reductase can act on either folate or dihydrofolate, the reduction of dihydrofolate is more significant in vivo than that of folate because of the need for constant regeneration of tetrahydrofolate from dihydrofolate produced in the thymidylate synthase reaction. This is the basis for the use of dihydrofolate reductase inhibitors in the treatment of cancer and infectious diseases (more about this in Section 19.5).

Folic Acid in the Prevention of Heart Disease and Birth Defects

In the mid-1990s, a series of clinical reports described correlations between folate deficiencies and increased risk of myocardial infarction. The same studies revealed that individuals at risk for heart attack also showed abnormally high levels of serum homocysteine. The simplest interpretation is that in folate-deficient individuals, decreased levels of tetrahydrofolate cofactors limit metabolic flux through the methionine synthase reaction (reaction ❶, Figure 18.8), with consequent accumulation of homocysteine, the substrate for this enzyme. Elevated plasma homocysteine (**homocysteinemia**) is now considered to be a major independent risk factor for many types of cardiovascular disease, including coronary artery disease, stroke, and peripheral vascular occlusive disease. Homocysteine is presumed to be the toxic metabolite responsible for damage to the heart, although the mechanisms are not known. Indeed, some studies have failed to show a correlation between folate status and heart disease. However, folate deficiencies have other recently recognized biological consequences, including abnormally high levels of uracil in DNA. As we discuss in Chapter 19, this phenomenon, which can lead to chromosome breakage, is a consequence of limitation of the biosynthesis of thymine nucleotides.

● **CONNECTION** Folic acid deficiency increases the risk of cardiovascular disease and birth defects in humans.

Folate deficiency during embryogenesis causes a number of birth defects, including those affecting the craniofacies (e.g., cleft palate), the neural tube (e.g., anencephaly and spina bifida), and the heart. Consequently, women are urged to take folic acid supplements throughout their pregnancies, but especially in the early stages, when the fetal nervous system develops most rapidly. Maternal supplementation with folic acid is known to reduce the incidence of neural tube defects (NTDs) by as much as 70%. Because the window during which folic acid supplementation is effective generally occurs before a woman discovers she is pregnant, the United States began fortifying enriched flour and other cereal grain products with folic acid to ensure that women of childbearing age have adequate folate levels. The prevalence of NTDs has indeed declined in the United States since 1998, when fortification became mandatory.

B₁₂ Coenzymes

Vitamin B_{12} was discovered through studies of a formerly incurable disease, pernicious anemia. This condition begins with a megaloblastic anemia, which is virtually identical to that seen in folate deficiency but which leads to an irreversible degeneration of the nervous system if untreated. In 1926 two Harvard physicians, George Minot and William Murphy, found that symptoms of the disease could be alleviated by feeding patients large amounts of raw liver. The active material in the liver, which was named vitamin B_{12}, was present in exceedingly small amounts, so many years passed until sufficient material had been isolated for characterization. In England in 1956, Dorothy Hodgkin and her colleagues used X-ray crystallography to complete the structure determination for this active substance. Hodgkin was awarded the Nobel Prize for this work.

The structure of vitamin B_{12} is shown in **FIGURE 18.9**. The metal cobalt is coordinated with a tetrapyrrole ring system, called a **corrin** ring, which is similar to the porphyrin ring of heme compounds. The cobalt is also linked to a heterocyclic base, 5,6-dimethylbenzimidazole (DMB). Because of the presence of cobalt and many amide nitrogens, B_{12} compounds are called **cobamides** or, more commonly but less accurately, **cobalamins.** In the vitamin as isolated, the sixth coordination position of the cobalt is occupied by cyanide ion, but this ion is introduced during isolation. In the two known coenzymatically active forms of B_{12}, the

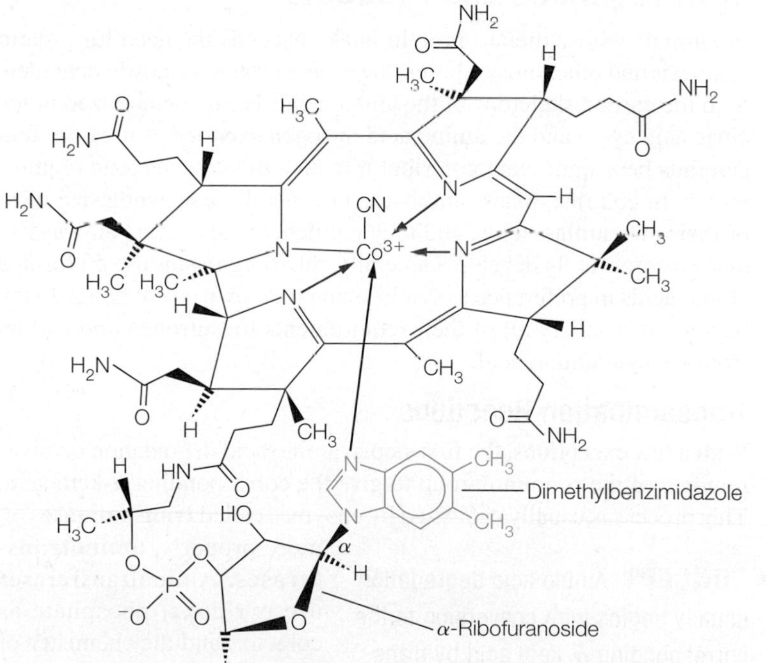

▲ **FIGURE 18.9 Structure of vitamin B₁₂.** The molecule shown here is the cyanide-containing form originally isolated (cyanocobalamin). In cells, a water molecule or hydroxyl group takes the place of CN, forming the precursor to the coenzyme forms of B₁₂. The corrin ring is shown in magenta. 5,6-Dimethylbenzimidazole (DMB), which is linked to the cobalt, is shown in blue.

cyanide ion is replaced by either a methyl or a 5′-deoxyadenosyl group. **Methylcobalamin,** or methyl-B$_{12}$, is used in the methionine synthase reaction (reaction ❶ Figure 18.8). **5′-Deoxyadenosylcobalamin** is used by a number of enzymes, including methylmalonyl-CoA mutase (Figure 16.18, p. 530).

B$_{12}$ Coenzymes and Pernicious Anemia

Recall that vitamin B$_{12}$ was isolated as a factor that could cure pernicious anemia, which suggested that the disease is caused by B$_{12}$ deficiency. In fact, pernicious anemia is a disease of the stomach. Gastric

● **CONCEPT** Pernicious anemia is caused by deficiency of a glycoprotein needed for intestinal absorption of vitamin B$_{12}$, leading to intracellular deficiencies of B$_{12}$ coenzymes.

tissue secretes a glycoprotein called **intrinsic factor,** which complexes with ingested B$_{12}$ in the digestive tract and promotes its efficient absorption through the terminal portion of the small intestine into the bloodstream. Pernicious anemia results from insufficient secretion of intrinsic factor. This is usually caused by an autoimmune process in which the body destroys the gastric lining cells that produce intrinsic factor. Indeed, patients who undergo surgical removal of the stomach for cancer or other problems can also develop the symptoms of pernicious anemia. The uncomplexed vitamin can be absorbed, but so poorly that massive doses must be administered to cure or prevent the disease.

18.5 Amino Acid Degradation and Metabolism of Nitrogenous End Products

In animals whose dietary protein intake exceeds the need for protein synthesis and other biosyntheses, the excess protein is mostly degraded, with the carbon skeletons of the amino acids being metabolized in the citric acid cycle and the amino acid nitrogen excreted as urea. Protein can thus be a significant contributor to an animal's energetic requirements. In contrast, plants and bacteria generally can synthesize most of their own amino acids, and they regulate the anabolic pathways so that excesses rarely develop. Generally, microorganisms use preformed amino acids in preference to synthesizing their own, even though many bacteria can satisfy all of their requirements for nitrogen *and* carbon from a single amino acid.

Transamination Reactions

With a few exceptions, the first step in amino acid degradation involves removal of the α-amino group to give the corresponding α-keto acid. This process is usually catalyzed by enzymes called **transaminases** or, more properly, **aminotrans-**

● **CONCEPT** Amino acid degradation usually begins with conversion to the corresponding α-keto acid by transamination or oxidative deamination.

ferases. Aminotransferases use pyridoxal phosphate as cofactor, and the chemistry of transamination was shown in Figure 18.7. Transamination plays a central role in amino acid metabolism in that it provides a route for redistributing amino acid nitrogen. Because of the key role of glutamate in ammonia assimilation, it is a star player in transamination. In other words, glutamate is an abundant product of ammonia assimilation, and transamination uses glutamate nitrogen to synthesize other amino acids.

As shown here, transamination involves transfer of the α-amino group, usually of glutamate, to an α-keto acid, with formation of the corresponding amino acid plus the α-keto derivative of glutamate, which is α-ketoglutarate.

Glutamate   α-Keto acid   α-Ketoglutarate   α-Amino acid

Transamination reactions have equilibrium constants close to unity. Therefore, the direction in which a particular transamination proceeds is controlled in large part by the intracellular concentrations of substrates and products. This means that transamination can be used not only for amino acid synthesis but also for degradation of amino acids that accumulate in excess of need. In degradation, the aminotransferase works in concert with glutamate dehydrogenase, as exemplified by the degradation of alanine:

Alanine + α-ketoglutarate $\xrightarrow{\text{Aminotransferase}}$ pyruvate + glutamate

Glutamate + NAD$^+$ + H$_2$O $\xrightarrow{\text{Glutamate dehydrogenase}}$

$$\alpha\text{-ketoglutarate} + \text{NADH} + \overset{+}{\text{N}}\text{H}_4$$

Net: Alanine + NAD$^+$ + H$_2$O $\longrightarrow$ pyruvate + NADH + $\overset{+}{\text{N}}$H$_4$

● **CONCEPT** Transamination is the reversible transfer of an amino group from an α-amino acid to an α-keto acid, with pyridoxal phosphate as a coenzyme.

● **CONNECTION** Serum glutamate-oxaloacetate transaminase (SGOT) and serum glutamate-pyruvate transaminase (SGPT) are important in the clinical diagnosis of human disease. Abundant in heart and in liver, these enzymes are released as part of the cell injury that occurs in myocardial infarction, infectious hepatitis, or other damage to either organ. Assays of these enzyme activities in blood serum can be used both in diagnosis and in monitoring the progress of a patient during treatment.

The net process is the deamination of the α-amino acid (here, alanine) to the corresponding α-keto acid (here, pyruvate) plus ammonia. Thus, we see transamination as a mechanism for amino acid synthesis *or* degradation. Because the amino acids within a cell are rarely present in the proportions needed to synthesize the specific proteins of that cell, transamination plays an important role in bringing the amino acid composition into line with the organism's needs. It also participates in funneling excess amino acids toward catabolism and energy generation.

Most aminotransferases use glutamate/α-ketoglutarate as one of the two α-amino/α-keto acid pairs involved. For example, SGOT catalyzes the interconversion of oxaloacetate and aspartate, and SGPT catalyzes the interconversion of pyruvate and alanine. But both are coupled to the glutamate/α-ketoglutarate pair:

$$\text{Glutamate} + \text{oxaloacetate} \underset{}{\overset{\text{SGOT}}{\rightleftharpoons}} \text{α-ketoglutarate} + \text{aspartate}$$

$$\text{Glutamate} + \text{pyruvate} \underset{}{\overset{\text{SGPT}}{\rightleftharpoons}} \text{α-ketoglutarate} + \text{alanine}$$

Once the nitrogen has been removed from an amino acid, the carbon skeleton can, depending on the physiological state of the organism, either proceed toward oxidation in the citric acid cycle or be used for biosynthesis of carbohydrate. The individual pathways are presented in Section 18.6.

Detoxification and Excretion of Ammonia

Although ammonia is a universal participant in amino acid synthesis and degradation, its accumulation in abnormal concentrations has toxic consequences. Therefore, cells undergoing active amino acid catabolism must be able to detoxify and/or excrete ammonia as fast as it is generated. For most aquatic animals, which can take in and pass out unlimited quantities of water, ammonia simply dissolves in the water and diffuses away. Because terrestrial animals must conserve water, they convert ammonia to a form that can be excreted without large water losses. Birds, terrestrial reptiles, and insects convert most of their excess ammonia to **uric acid,** an oxidized purine. Because uric acid is quite insoluble, it precipitates and can be excreted without a large water loss and without building up osmotic pressure. Most mammals excrete the bulk of their nitrogen in the form of **urea.** Urea is highly soluble and, lacking ionizable groups, does not affect the pH when it accumulates, as does ammonia.

● **CONCEPT** Animals have evolved pathways, adapted to their lifestyles, for excretion of ammonia, uric acid, or urea as the major nitrogenous end product.

● **CONNECTION** Humans have used seabird excrement (guano) as a soil amendment for at least 1500 years. The high content of uric acid, ammonium oxalate, and phosphates makes for a rich fertilizer.

Uric acid   **Urea**

Transport of Ammonia to the Liver

All animal organs degrade amino acids and produce ammonia. Two mechanisms are involved in transporting this ammonia from other tissues to liver for its eventual conversion to urea. Most tissues use glutamine synthetase to convert ammonia to the nontoxic, and electrically neutral, glutamine (**FIGURE 18.10**; see also Figure 18.5). The glutamine is then transported in the blood to the liver, where it is cleaved hydrolytically by glutaminase, yielding glutamate and ammonia.

Muscle, which derives most of its energy from glycolysis, uses a different route, the **glucose–alanine cycle** (Figure 18.10). Glycolysis generates pyruvate, which undergoes transamination with glutamate to give alanine and α-ketoglutarate. The glutamate in turn has acquired its nitrogen from ammonia, via glutamate dehydrogenase. The resultant alanine is transported to the liver, where it loses its nitrogen

● **CONCEPT** The glucose–alanine cycle removes toxic ammonia from muscle. Glutamine synthetase and glutaminase do the same for most other tissues.

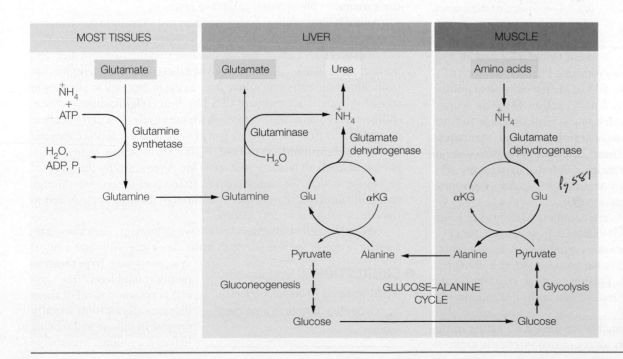

◀ **FIGURE 18.10 Transport of ammonia to the liver for urea synthesis.** The carrier is glutamine in most tissues, but it is alanine in muscle. αKG, α-ketoglutarate.

by a reversal of the previous processes. This reversal yields ammonia for urea synthesis, plus pyruvate. The pyruvate undergoes gluconeogenesis to give glucose, which is released to the blood for transport back to the muscle or for nourishment of the brain. This cyclic process helps muscle get rid of ammonia, with the carbon from the pyruvate being returned to the liver for gluconeogenesis.

The Krebs–Henseleit Urea Cycle

Urea is synthesized almost exclusively in the liver and then transported to the kidneys for excretion. The pathway, which is cyclic, was discovered by Hans Krebs and Kurt Henseleit in 1932, five years before the other cycle for which Krebs is famous. Krebs and Henseleit were investigating the pathway by adding possible precursors to liver slices and then measuring the amount of urea produced. When arginine was added, urea was produced in 30-fold molar excess over the amount of arginine administered. Similar results were seen if either of two structurally related amino acids, **ornithine** or **citrulline,** was substituted for arginine. Because these three amino acids seemed to function catalytically to promote urea synthesis, Krebs and Henseleit proposed the existence of a cyclic pathway.

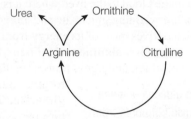

That proposal was correct, as was subsequently confirmed by isolation of the enzymes involved and identification of the biosynthetic route to ornithine, which begins the pathway. These details are shown in **FIGURE 18.11**. Ornithine serves as a "carrier," upon which are assembled the carbon and nitrogen atoms that will eventually constitute urea. The source of the carbon and one nitrogen atom in urea is carbamoyl phosphate, synthesized from NH_4^+ and HCO_3^- (reaction ❶) by carbamoyl phosphate synthetase I (CPS I; Section 18.2). Carbamoyl phosphate reacts with ornithine, via the enzyme **ornithine transcarbamoylase,** to give citrulline (reaction ❷). The second nitrogen comes from aspartate, which reacts with citrulline to form **argininosuccinate,** through the action of **argininosuccinate synthetase** (reaction ❸). Next, **argininosuccinase** cleaves argininosuccinate in a β-elimination reaction to give arginine and fumarate (reaction ❹). Arginine is cleaved hydrolytically by **arginase,** to regenerate ornithine and yield one molecule of urea (reaction ❺).

> **CONCEPT** Urea is synthesized by an energy-requiring cyclic pathway that begins and ends with ornithine.

The reactions of the urea cycle are compartmentalized in mitochondria and cytosol of liver cells. Glutamate dehydrogenase, the citric acid cycle enzymes, carbamoyl phosphate synthetase I, and ornithine carbamoyltransferase are localized in the mitochondrion, and the rest of the cycle occurs in the cytosol. This means that, in order for the cycle to proceed, ornithine must be transported into mitochondria, and citrulline exported to the cytosol.

The enzyme arginase is responsible for the cyclic nature of the urea biosynthetic pathway. Virtually all organisms synthesize arginine

> **CONNECTION** The capacity to synthesize arginase develops in frogs at the same time that they undergo metamorphosis from the tadpole stage to the adult animal. Because the tadpole lives in water, it can excrete ammonia. The adult frog, being adapted to a terrestrial lifestyle, develops the ability to synthesize urea.

from ornithine by the reactions shown in Figure 18.11. However, only **ureotelic** organisms (those excreting most of their nitrogen as urea) contain arginase, and, hence, only those organisms carry out the cyclic pathway.

As already noted, one nitrogen atom in urea comes from aspartate. This atom is derived from ammonia, which is transferred to glutamate via the glutamate dehydrogenase reaction, then to aspartate by transamination. Note from the bottom of Figure 18.11 that a second cycle is used to maintain carbon balance by conversion of the fumarate produced from argininosuccinate cleavage to oxaloacetate in the citric acid cycle and then back to aspartate by transamination.

The net reaction for one turn of the urea cycle is as follows:

$$HCO_3^- + NH_4^+ + 3ATP + \text{aspartate} + 2H_2O \longrightarrow$$
$$\text{urea} + 2ADP + 2P_i + AMP + PP_i + \text{fumarate}$$

Two molecules of ATP are required to reconvert AMP to ATP, so really four (not three) high-energy phosphates are consumed in each turn of the cycle. Thus, the synthesis of this excretion product is energetically expensive. Excess ammonia is also produced when the animal is forced to catabolize amino acids in muscle as energy sources, for example, during fasting. The urea cycle is also important for maintaining pH balance in mammals. HCO_3^-, a major product of protein catabolism, is titrated by protons from NH_4^+ in the carbamoyl phosphate synthetase reaction.

> **CONNECTION** Adult humans produce about 30 g of urea each day. This expensive pathway consumes up to 50% of the ATP produced in liver.

Animals have evolved both long-term and short-term mechanisms to regulate flux through the urea cycle. The levels of the four urea cycle enzymes and CPS I are increased in animals on high-protein diets and are decreased in animals fed protein-free diets. This long-term mechanism allows the animal to adjust its urea cycle capacity to changes in its diet. Allosteric activation of CPS I by **N-acetylglutamate** provides short-term regulation of flux through the urea cycle. N-acetylglutamate is synthesized from acetyl-CoA and glutamate in a reaction catalyzed by **N-acetylglutamate synthase** (Figure 18.11, upper right). Mitochondrial levels of N-acetylglutamate are determined by the levels of glutamate, which rise with increased amino acid breakdown through transamination reactions. Thus, flux through the urea cycle is tied to the rate of amino acid degradation.

Several inherited disorders are known in humans, each caused by a defect in one of the five enzymes in the urea cycle. Although clinical symptoms vary, **hyperammonemia** (high blood NH_4^+ levels) is common to all of these diseases. Symptoms usually present in infants and include lethargy, vomiting, irreversible

> **CONNECTION** At least five different diseases in humans result from inherited defects in urea cycle enzymes.

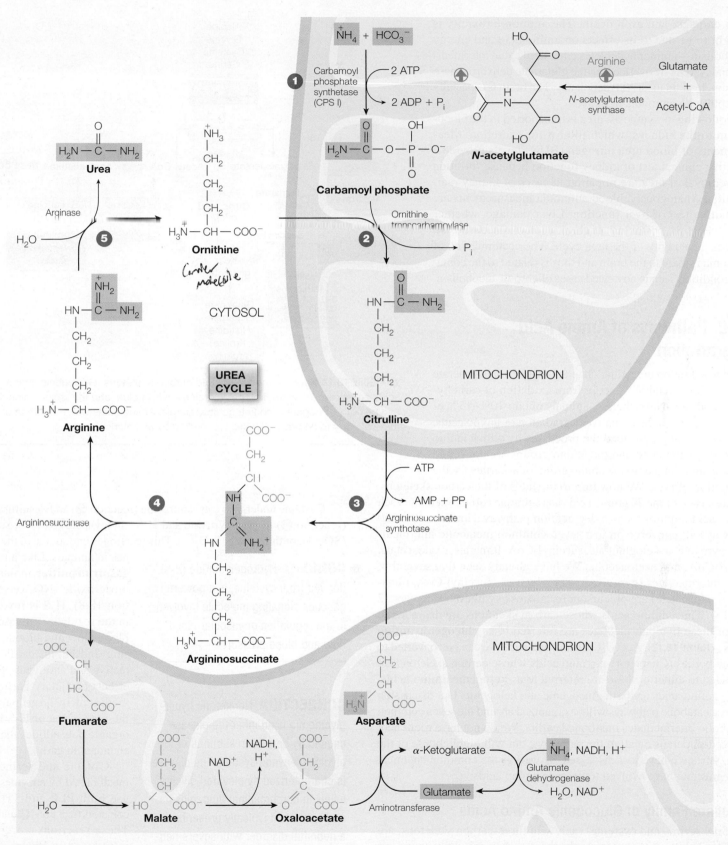

▲ **FIGURE 18.11 The Krebs–Henseleit urea cycle.** Urea (upper left) contains a carbon and a nitrogen (orange) derived from carbamoyl phosphate and a nitrogen (purple) derived from aspartate. NH_4^+ and HCO_3^-, the ultimate sources of these atoms, were incorporated in turn through the actions of carbamoyl phosphate synthetase (upper right) and glutamate dehydrogenase (lower right). Glutamate can serve directly as the source of some urea nitrogen. *N*-acetylglutamate, synthesized from glutamate and acetyl-CoA, allosterically activates carbamoyl phosphate synthetase.

brain damage, and even death. The ammonia toxicity is thought to be due to its effects on amino acid and energy metabolism—ammonia depletes citric acid cycle intermediates and NADH by overloading the glutamate dehydrogenase reaction, leading to an ATP deficiency. The brain is especially susceptible to ammonia toxicity.

Following its synthesis, urea is transported in the bloodstream to the kidneys, which filter it for excretion. Measurements of blood urea nitrogen (BUN) levels provide a sensitive clinical test of kidney function because filtration and removal of urea are impaired in cases of kidney malfunction. Analogously, blood ammonia measurements are a sensitive test of liver function. Liver damage, whether acute (hepatitis, poisoning) or chronic (alcoholic cirrhosis), reduces the activity of the urea cycle. The accumulation of ammonia is toxic to the brain and thus is related to the comatose condition seen in advanced cases of chronic alcoholism.

18.6 Pathways of Amino Acid Degradation

We learned in the preceding chapters that animals generate most of their metabolic energy from oxidation of carbohydrates and fats. Nevertheless, animals obtain 10%–15% of their energy from the oxidative degradation of amino acids. We have already described the processes by which amino acids are deaminated and the amino groups are converted either to ammonia or to the amino group of aspartate for the production of urea. We now turn to the fates of the carbon skeletons. Because each of the 20 amino acids has a unique carbon skeleton, each amino acid requires its own degradation pathway. However, these 20 pathways all converge on just seven common metabolic intermediates: pyruvate, α-ketoglutarate, succinyl-CoA, fumarate, oxaloacetate, acetyl-CoA, and acetoacetate. We have already seen that several of these intermediates (pyruvate, α-ketoglutarate, succinyl-CoA, fumarate, oxaloacetate) are precursors for glucose synthesis (Chapters 12 and 13). Thus, amino acids whose carbon skeletons are degraded to one of these five intermediates are referred to as **glucogenic amino acids** (FIGURE 18.12). Acetyl-CoA and acetoacetate can be converted to ketone bodies (Chapter 16); amino acids whose carbon skeletons are degraded to either of these are referred to as **ketogenic amino acids.** Some amino acids are both glucogenic and ketogenic. Our discussion of these catabolic pathways will be organized around the seven common metabolic intermediates mentioned earlier. We do not have room here to give full details on all of the pathways, but will highlight use of the coenzymes we have just discussed, as well as the common chemical strategies that have evolved to degrade amino acids.

Pyruvate Family of Glucogenic Amino Acids

Alanine, serine, and cysteine, each with three-carbon skeletons, are converted in one or two steps to the three-carbon metabolite pyruvate (FIGURE 18.13). Five of these reactions use pyridoxal phosphate, illustrating the versatility of this coenzyme. For example, alanine is transaminated to pyruvate (reaction ❻) using the PLP-dependent chemistry we described in Section 18.4. Serine is dehydrated and deaminated to pyruvate by another PLP-dependent enzyme (reaction ❺).

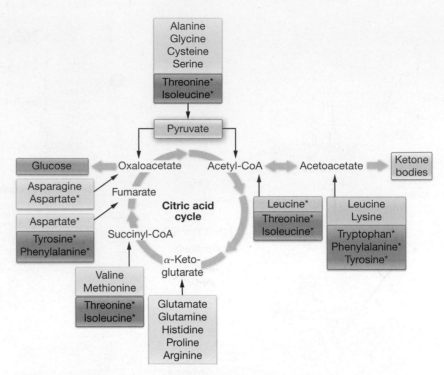

▲ **FIGURE 18.12 Fates of the amino acid carbon skeletons.** Glucogenic amino acids are shown in yellow, ketogenic amino acids in blue, and the amino acids that can be both glucogenic and ketogenic in purple. Amino acids with more than one route of entry to central pathways are marked by an asterisk.

Cysteine undergoes transamination (reaction ❼) and desulfuration (reaction ❽) to give pyruvate and H_2S, thiocyanate (SCN^-), sulfite (SO_3^{2-}), or thiosulfate ($S_2O_3^{2-}$). This reaction is one source of the H_2S gas in animals. Like another **gasotransmitter** molecule, nitric oxide (NO; see Section 18.8), H_2S is involved in the regulation of vascular blood flow and blood pressure. Indeed, the cardioprotective and antihypertensive effects of dietary garlic are mediated in large part by the production of H_2S from organic polysulfides that are abundant in garlic.

Glycine and serine are metabolized by enzymes that use both PLP and THF as cofactors (reactions ❸ and ❹). These reactions were first introduced in Figure 18.8. The hydroxymethyl group of serine (C3) comes from 5,10-methylene-THF, which might derive from glycine cleavage (reaction ❸) or some other one-carbon donor (see Figure 18.8).

● **CONCEPT** Hydrogen sulfide (H_2S), derived from cysteine, is a powerful gaseous signaling molecule involved in the regulation of vascular blood flow and blood pressure.

● **CONNECTION** Nonketotic hyperglycinemia is an inherited disease caused by defects in subunits of the glycine cleavage system. The disease is characterized by elevated glycine levels in cerebrospinal fluid, plasma, and urine, and typically presents as a neonatal disorder with severe neurological symptoms, including mental retardation.

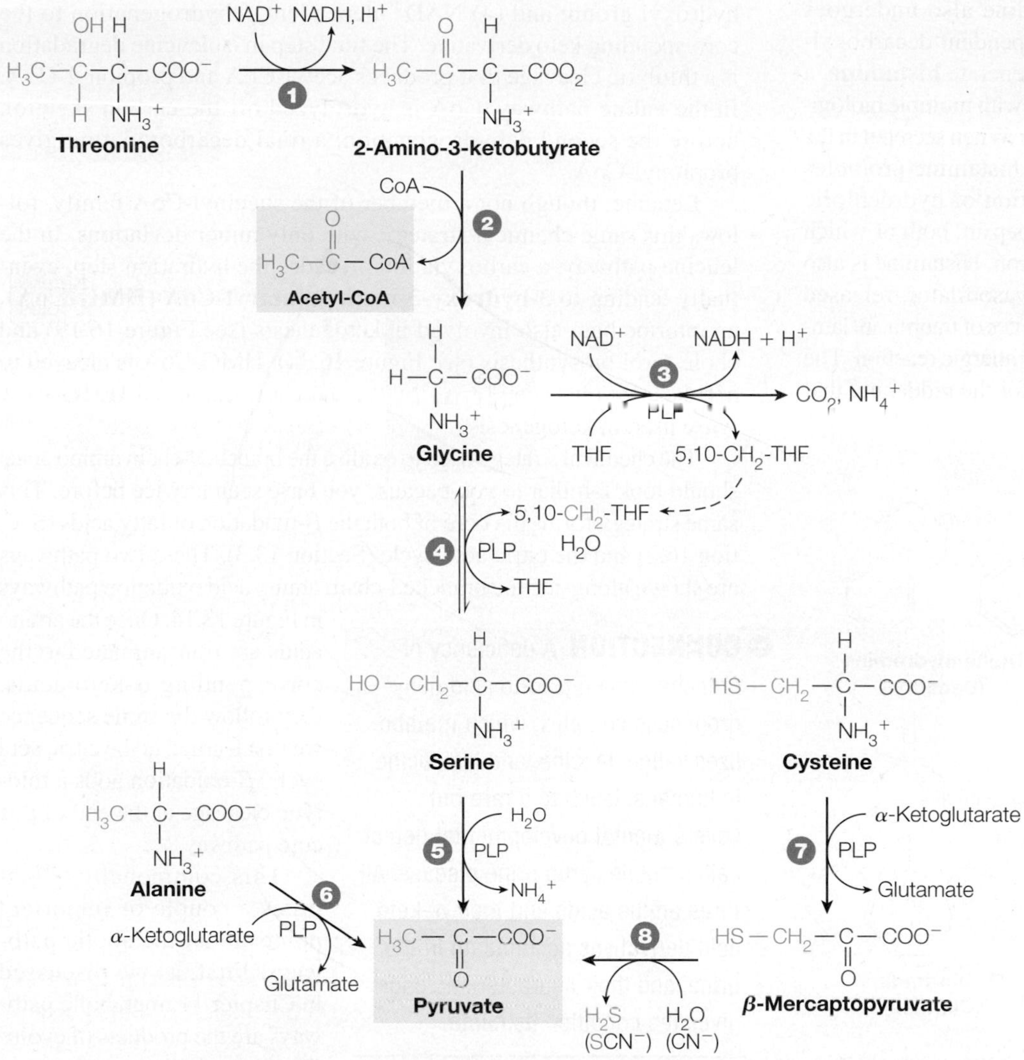

▲ **FIGURE 18.13 Alanine, cysteine, glycine, serine, and threonine are degraded to pyruvate in mammals.** In humans, threonine is degraded by a different pathway, to succinyl-CoA. All of these enzymes are located in mitochondria, except for the cytoplasmic serine dehydratase (reaction 5). See text for descriptions of the other reactions.

● **CONNECTION** Asparaginase is an interesting example of enzyme-based chemotherapy. The *E. coli* enzyme is widely used in the treatment of childhood acute lymphoblastic leukemia. Normal and malignant lymphocytes depend on the uptake of asparagine from the blood for growth—asparaginase depletes circulating asparagine, and this leads to complete remission in many cases.

Oxaloacetate Family of Glucogenic Amino Acids

The four-carbon skeletons of asparagine and aspartate are converted to oxaloacetate in a simple pathway (see below). **Asparaginase** catalyzes hydrolytic cleavage of the asparagine amide to aspartate and ammonium. Aspartate is then transaminated directly to oxaloacetate. Recall that aspartate can alternatively be converted to fumarate as part of the urea cycle (Figure 18.11).

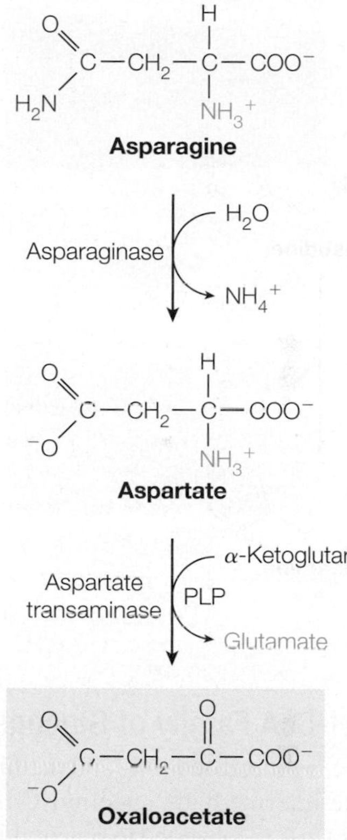

α-Ketoglutarate Family of Glucogenic Amino Acids

The carbon skeletons of glutamine, proline, arginine, and histidine are all degraded to glutamate (see below). Glutamine is hydrolyzed to glutamate by *glutaminase*, which in animals participates in transporting ammonia to the liver (see Figure 18.10). Finally, glutamate is oxidatively deaminated to α-ketoglutarate by glutamate dehydrogenase, which was discussed in depth in Section 18.2.

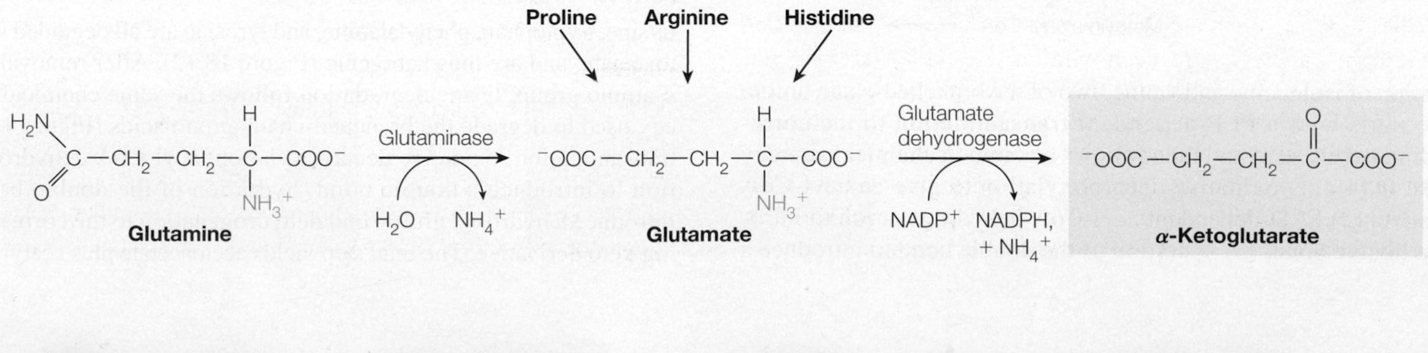

● **CONNECTION** Histamine is produced from decarboxylation of histidine. A large number of **antihistamines,** such as diphenylhydramine and desloratadine, are used to treat allergies and other inflammations. Typically, these drugs prevent the binding of histamine to its receptors.

Histidine also undergoes a PLP-dependent decarboxylation to generate **histamine,** a substance with multiple biological actions. When secreted in the stomach, histamine promotes the secretion of hydrochloric acid and pepsin, both of which aid digestion. Histamine is also a potent vasodilator, released locally in sites of trauma, inflammation, or allergic reaction. The local enlargement of blood capillaries is the basis for the reddening that occurs in inflamed tissues.

hydroxyl group; and (4) NAD^+-dependent dehydrogenation to the corresponding keto derivative. The final step in isoleucine degradation is a thiolytic cleavage that produces acetyl-CoA and propionyl-CoA. In the valine pathway, CoA is hydrolyzed off the carbon skeleton before the second dehydrogenation; a final decarboxylation gives propionyl-CoA.

Leucine, though not a member of the succinyl-CoA family, follows this same chemical strategy, with only minor deviations. In the leucine pathway, a carboxylation precedes the hydration step, eventually leading to **3-hydroxy-3-methylglutaryl-CoA (HMG-CoA),** an intermediate also involved in ketogenesis (see Figure 16.19) and cholesterol biosynthesis (see Figure 16.34). HMG-CoA is cleaved to acetoacetate plus acetyl-CoA by the same mitochondrial **HMG-CoA lyase** used in ketogenesis.

The chemical strategy used to oxidize the branched-chain amino acids should look familiar to you because you have seen it twice before. This same strategy forms the core of both the β-oxidation of fatty acids (Section 16.2) and the citric acid cycle (Section 13.3). These two pathways are shown alongside the branched-chain amino acid oxidation pathways in Figure 18.14. Once the amino acids are transaminated to the corresponding α-keto acids, they follow the same sequence we first learned in the citric acid cycle. β-oxidation adds a thiolytic cleavage at the end of this core pathway.

● **CONNECTION** A deficiency of branched-chain α-keto acid dehydrogenase complex, which metabolizes valine, leucine, and isoleucine in humans, leads to a rare but severe mental developmental defect called maple syrup urine disease. All three amino acids and their α-keto acid derivatives accumulate in the urine, and their characteristic odor gives the condition its name.

This commonality illustrates a couple of important points about metabolic pathways. First, as we discussed in Chapter 11, metabolic pathways are the products of evolution, being built from enzymes and pathways that may have had other functions initially. It is likely that the branched-chain amino acid oxidation pathways arose from the evolutionarily ancient citric acid cycle pathway. Second, a cell can "economize" by using the same enzyme to metabolize several related substrates. For example, the oxidative decarboxylation of all three α-keto acid derivatives of the branched-chain amino acids (Figure 18.14) is carried out by the mitochondrial **branched-chain α-keto acid dehydrogenase complex.** This enzyme has the same E_1–E_2–E_3 multienzyme structure and mechanism of the pyruvate dehydrogenase and α-ketoglutarate dehydrogenase complexes (Sections 13.2–13.3). Indeed, the E_3 subunits are identical in all three complexes from a given species.

Succinyl-CoA Family of Glucogenic Amino Acids

Isoleucine, valine, threonine, and methionine are degraded to the citric acid cycle intermediate, succinyl-CoA, by way of propionyl-CoA. Propionyl-CoA is converted to succinyl-CoA by the same B_{12}-dependent pathway we introduced in Section 16.2 as a key step in the oxidation of odd-chain fatty acids.

Oxidation of isoleucine and valine, two of the branched-chain amino acids, starts with a PLP-dependent transamination to the corresponding α-keto acid and then follows a common chemical strategy (**FIGURE 18.14**): (1) oxidative decarboxylation to give an acyl-CoA derivative; (2) FAD-dependent acyl-CoA dehydrogenation to introduce a double bond; (3) hydration of the double bond to introduce a

Acetoacetate/Acetyl-CoA Family of Ketogenic Amino Acids

Lysine, tryptophan, phenylalanine, and tyrosine are all degraded to acetoacetate, and are thus ketogenic (Figure 18.12). After removal of its ε-amino group, lysine degradation follows the same chemical strategy used to degrade the branched-chain amino acids (Figure 18.14): transamination; oxidative decarboxylation; acyl-CoA dehydrogenation to introduce a double bond; hydration of the double bond to introduce a hydroxyl group; and dehydrogenation to the corresponding keto derivative. The final step yields acetoacetate plus acetyl-CoA.

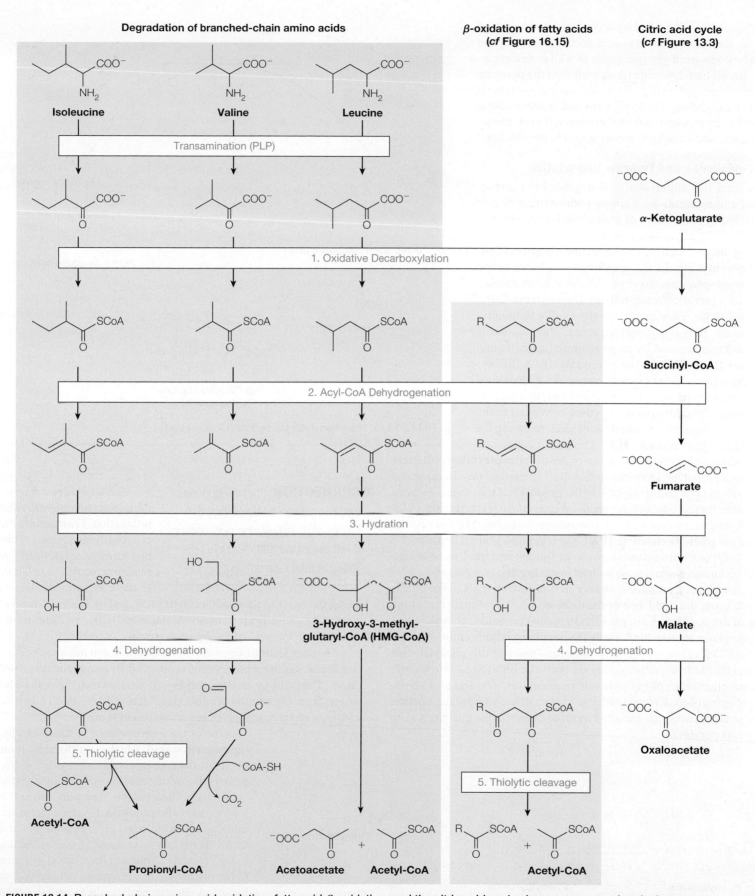

▲ **FIGURE 18.14** Branched-chain amino acid oxidation, fatty acid β-oxidation, and the citric acid cycle share a common chemical strategy.

Tryptophan is degraded by a complex pathway that yields alanine in addition to acetoacetate, so tryptophan can be considered gluconeogenic as well as ketogenic.

Recall that threonine (discussed with the pyruvate family) and isoleucine (discussed with the succinyl-CoA family) also yield acetyl-CoA and/or acetoacetate, and like tryptophan, are thus considered both gluconeogenic and ketogenic amino acids (Figure 18.12).

Phenylalanine and Tyrosine Degradation

The final two amino acids are degraded to acetoacetate and fumarate by a single pathway that begins with the hydroxylation of phenylalanine to tyrosine, catalyzed by **phenylalanine hydroxylase.** This interesting enzyme belongs to the **aromatic amino acid hydroxylase** family, along with tyrosine hydroxylase and tryptophan hydroxylase. All three hydroxylases require a pterin cofactor, **tetrahydrobiopterin** (BH_4) (**FIGURE 18.15**). Though structurally similar to tetrahydrofolate, BH_4 is not derived from folic acid but is instead synthesized by most mammalian cells and tissues from GTP. In the phenylalanine hydroxylase reaction, one O atom becomes the hydroxyl of tyrosine and the other hydroxylates BH_4 to a carbinolamine. The carbinolamine is then converted to the quinonoid form of 7,8-dihydrobiopterin, releasing the second oxygen atom as H_2O. The pterin coenzyme is regenerated through the action of the **dihydropteridine reductase** (analogous, but not identical, to dihydrofolate reductase). Because one atom from O_2 is incorporated into the product and the other is reduced to water, these hydroxylases are *monooxygenases* (see Section 14.9).

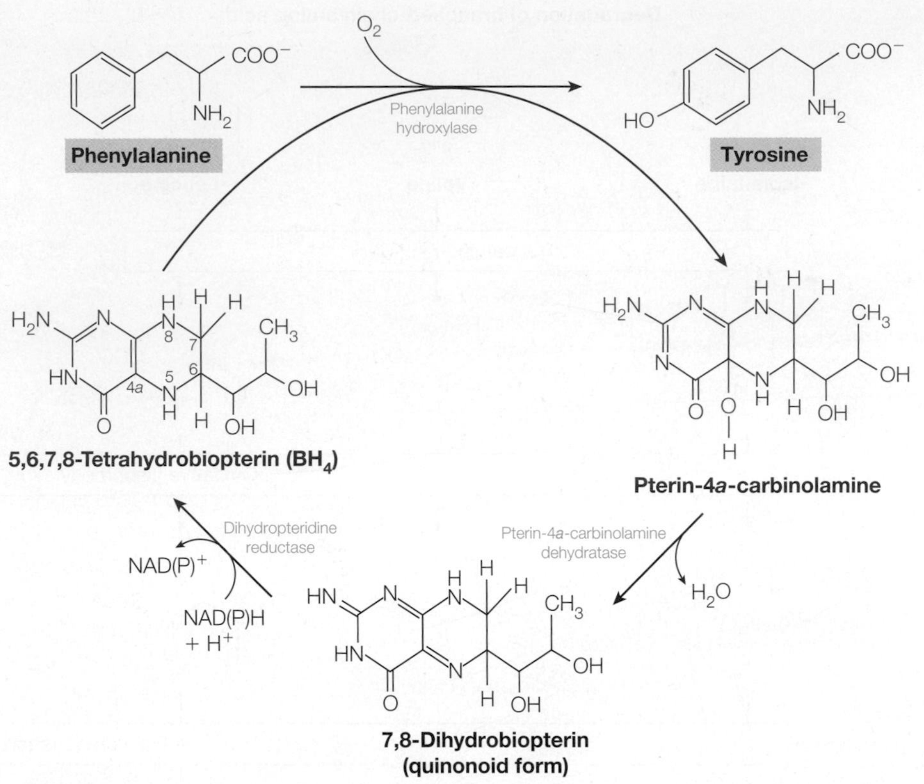

A hereditary deficiency of phenylalanine hydroxylase is responsible for **phenylketonuria (PKU),** a condition that afflicts about 1 in 10,000 newborn infants in western Europe and the United States. PKU is an autosomal recessive trait, meaning that two parents heterozygous for the trait have 1 chance in 4 of having a phenylketonuric child. From the incidence of the disease, we can estimate that about 2% of the population are carriers. In phenylketonuria, phenylalanine accumulates to very high levels (**hyperphenylalaninemia**) because of the block in conversion to tyrosine, and much of this phenylalanine is metabolized via pathways that are normally little used—particularly transamination to phenylpyruvate (a phenyl*ketone*), and also subsequent conversion of phenylpyruvate to phenyllactate and phenylacetate. These compounds are excreted in urine in enormous quantities (1 to 2 grams per day).

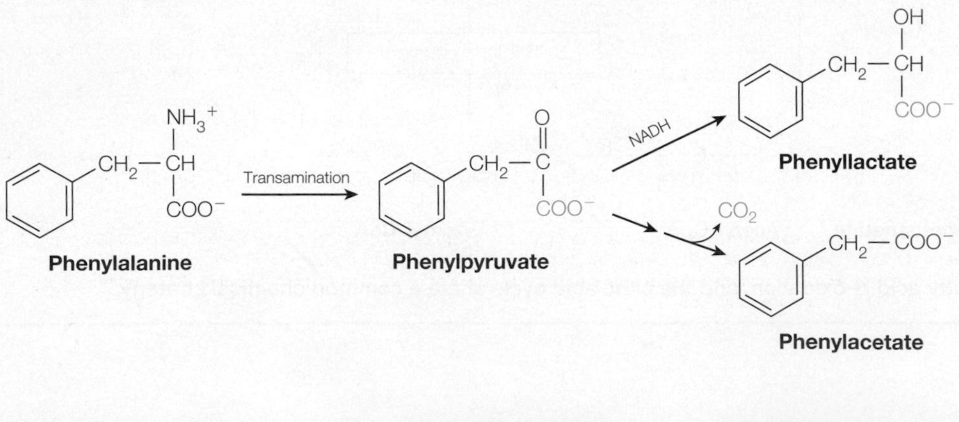

▲ **FIGURE 18.15** The phenylalanine hydroxylation system.

● **CONNECTION** Phenylketonuria (PKU) is caused by hereditary deficiency of phenylalanine hydroxylase. If left untreated, PKU leads to profound mental retardation.

If undetected and untreated, PKU leads to profound mental retardation. Fortunately, PKU can readily be detected at birth, and most hospitals carry out routine screening of newborns. If the condition is detected early, the onset of retardation can be prevented by feeding for several years a synthetic diet low in phenylalanine and rich in tyrosine, to allow normal development of the nervous system.

The degradation of tyrosine to acetoacetate and fumarate occurs in a complex multistep process that begins with its transamination (**FIGURE 18.16**). The pathway involves two *dioxygenases,* which incorporate both atoms from O_2 into the product (see Section 14.9), and ends with a cleavage reaction that gives acetoacetate and fumarate.

One of the intermediates in this pathway is **homogentisic acid,** which is oxidized by **homogentisate dioxygenase.** A hereditary deficiency of this enzyme in humans causes a condition that was known for centuries as the "dark urine disease" but is now called **alkaptonuria.** Homogentisate accumulates and is excreted in large amounts in the urine; its oxidation on standing causes the urine to become dark. Although the clinical symptoms of the disease are not severe, it is of considerable historical interest. Early in the twentieth century, Sir Archibald Garrod examined pedigrees of the families of afflicted individuals, and in 1909 he

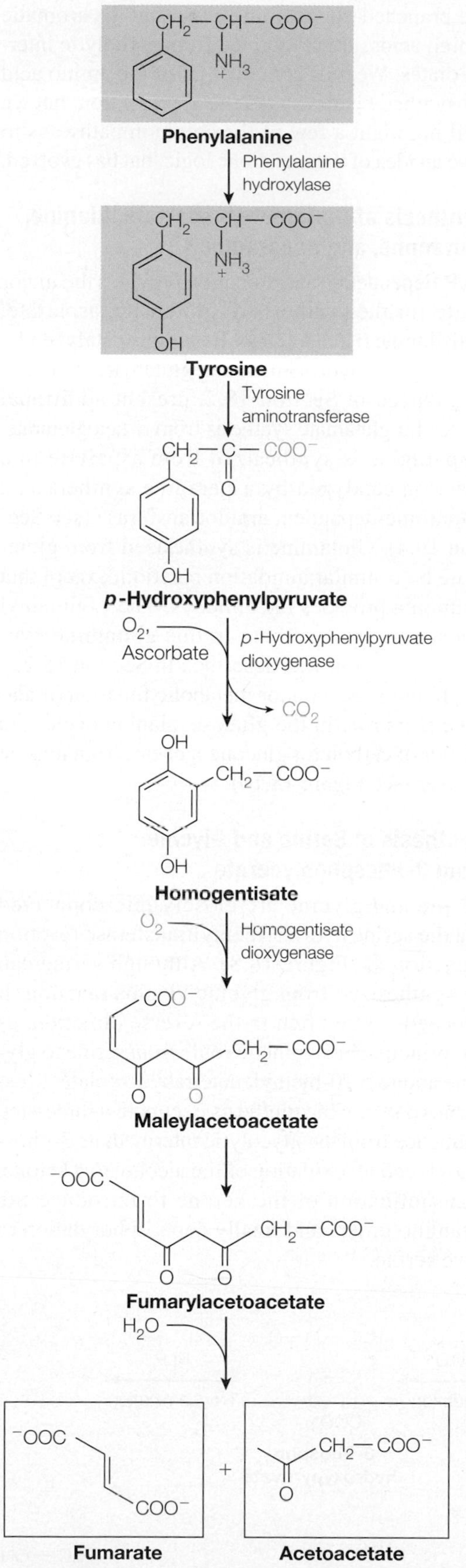

Phenylalanine

Phenylalanine hydroxylase

Tyrosine

Tyrosine aminotransferase

p-Hydroxyphenylpyruvate

O_2, Ascorbate

p-Hydroxyphenylpyruvate dioxygenase

CO_2

Homogentisate

O_2

Homogentisate dioxygenase

Maleylacetoacetate

Fumarylacetoacetate

H_2O

Fumarate   **Acetoacetate**

▲ **FIGURE 18.16** Catabolism of phenylalanine and tyrosine to fumarate and acetoacetate.

● **CONNECTION** Alkaptonuria, caused by hereditary deficiency of homogentisate dioxygenase, was described by Garrod in 1909. His study of this disease in humans led to the concept of inheritable metabolic diseases.

wrote, "We may further conceive that the splitting of the benzene ring in normal metabolism is the work of a special enzyme, that in congenital alcaptonuria this enzyme is wanting, whilst in disease its working may be partially or even completely inhibited." In other words, Garrod proposed that one gene encodes one enzyme, long before the chemical nature of either genes or enzymes was known, and developed the concept of inheritable metabolic diseases.

18.7 Amino Acid Biosynthesis

Biosynthetic Capacities of Organisms

Organisms vary widely in their ability to synthesize amino acids. Many bacteria and most plants can synthesize all of their nitrogenous metabolites starting from a single nitrogen source, such as ammonia or nitrate. However, many microorganisms will use a preformed amino acid, when available, in preference to synthesizing that amino acid. Sometimes preformed amino acids are required. For example, over the course of evolution, the genus *Lactobacillus* has lost many biosynthetic capacities because they grow in milk, a very nutrient-rich environment. Some *Lactobacillus* species must therefore be provided with most of the 20 amino acids to be grown in the laboratory. Mammals are intermediate, being able to biosynthesize about half of the amino acids in quantities needed for growth and for maintenance of normal nitrogen balance.

● **CONCEPT** Essential amino acids cannot be biosynthesized in adequate amounts and must be provided in the diet.

The amino acids that must be provided in the diet to meet an animal's metabolic needs are called **essential amino acids** (**TABLE 18.1**). Those that need not be provided because they can be biosynthesized in adequate amounts are called **nonessential amino acids.** In general, the essential amino acids include those with complex structures, including aromatic rings and hydrocarbon side chains. The nonessential amino acids include those that are readily synthesized from abundant metabolites, such as intermediates in glycolysis or the citric acid cycle.

Although dieticians recommend a protein intake of 50 to 100 grams per day or more, a human can do quite well on a diet containing as little as 20 grams per day, if that protein is of high nutritional quality—that is, if it contains adequate proportions of essential amino acids. In general, the more closely the amino acid composition of ingested protein resembles the amino acid composition of the animal eating the protein, the higher the nutritional quality of that protein. For humans, mammalian protein is of the highest nutritional quality, followed by fish and poultry, and then by fruits and vegetables. (In this context, nutritional quality refers only to the single criterion of essential amino acid content.) Plant proteins in particular are often deficient in lysine, methionine, or tryptophan. However, a vegetarian diet provides adequate protein if it contains a variety of protein sources, with a deficiency in one source being compensated for by excess in another source.

TABLE 18.1 Nutritional requirements for amino acids in mammals
Essential
Arginine,* histidine, isoleucine, leucine, lysine, methionine,* phenylalanine, threonine, tryptophan, valine
Nonessential
Alanine, asparagine, aspartate, cysteine, glutamate, glutamine, glycine, proline, serine, tyrosine**

* Although mammals can synthesize arginine and methionine, the use of these amino acids for the production of urea and methyl groups, respectively, is greater than the capacity of their biosynthetic pathways.
** Tyrosine is considered nonessential because mammals can produce it during phenylalanine degradation via the phenylalanine hydroxylase reaction.

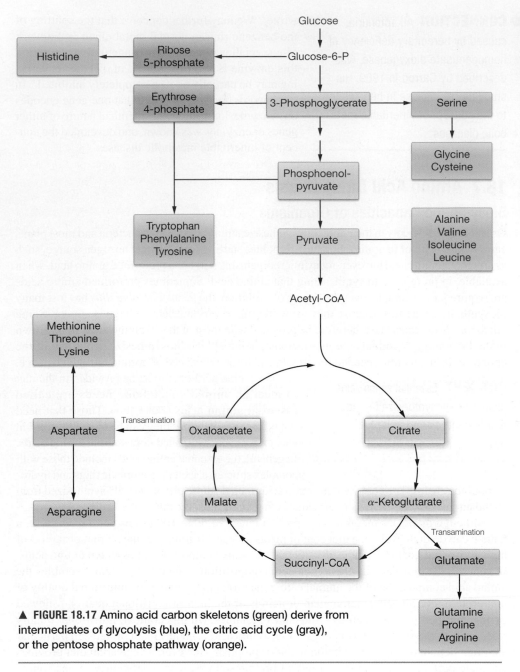

▲ **FIGURE 18.17 Amino acid carbon skeletons (green) derive from intermediates of glycolysis (blue), the citric acid cycle (gray), or the pentose phosphate pathway (orange).**

the branched-chain amino acids and the aromatic amino acids, are all derived from glycolytic intermediates. We will not cover all of the amino acid biosynthetic pathways in this final section, but we will highlight a few of the common pathways to give an idea of the metabolic logic that has evolved.

Synthesis of Glutamate, Aspartate, Alanine, Glutamine, and Asparagine

PLP-dependent transamination provides the major route for the synthesis of glutamate, aspartate, and alanine (**FIGURE 18.18**). Reactions catalyzed by glutamate dehydrogenase and glutamate synthase, introduced in Section 18.2, present additional routes for glutamate synthesis from α-ketoglutarate. Asparagine is synthesized from aspartate in a reaction catalyzed by asparagine synthetase, a glutamine-dependent amidotransferase (see Section 19.4). Glutamine is synthesized from glutamate by a similar amidation reaction, except that ammonia provides the amide N. The γ-glutamyl phosphate intermediate in this glutamine synthetase reaction was described in Section 18.2.

In animals, a major metabolic function of alanine is its role in the glucose–alanine cycle as a carrier of carbon for gluconeogenesis from muscle to liver (see Figure 18.10).

Synthesis of Serine and Glycine from 3-Phosphoglycerate

Serine and glycine are closely interconnected via the serine hydroxymethyltransferase reaction (reaction ❸, Figure 18.8). Although serine can be synthesized from glycine via this reaction, it proceeds more often in the reverse direction, as the principal biosynthetic route *from* serine to glycine and to 5,10-methylenetetrahydrofolate. Most de novo serine biosynthesis occurs in a three-step sequence from the glycolytic intermediate 3-phosphoglycerate: oxidation of the alcohol to a ketone, transamination of the ketone to introduce the α-amino group, and finally dephosphorylation to give serine.

Amino Acid Biosynthetic Pathways

All amino acids can be synthesized from intermediates in glycolysis, the pentose phosphate pathway, or the citric acid cycle (**FIGURE 18.17**). About half are biosynthesized more or less directly from intermediates in the citric acid cycle or from pyruvate. We include in this family glutamate, aspartate, and alanine, which can be formed directly by transamination from α-ketoglutarate, oxaloacetate, and pyruvate, respectively. The family also includes glutamine and asparagine, which are formed directly from glutamate and aspartate, respectively; and proline and arginine, which are formed in short pathways from glutamate. Recall that arginine is produced from ornithine in the urea cycle (Figure 18.11); that ornithine comes from glutamate. Aspartate is also the starting point for threonine, methionine, and lysine. The carbon skeletons of serine, glycine, cysteine, and histidine, as well as

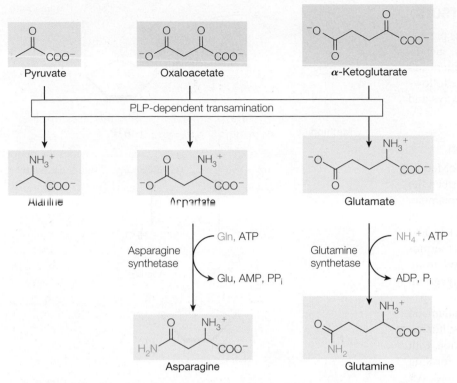

▲ **FIGURE 18.18 Synthesis of alanine, aspartate, glutamate, asparagine, and glutamine.**

● **CONCEPT** Serine is involved in glycine, phospholipid, and cysteine synthesis. Glycine is active in biosynthesis of purine nucleotides and porphyrins.

● **CONCEPT** Serine and glycine are both major contributors to the pool of activated one-carbon groups, in the form of 5,10-methylenetetrahydrofolate.

In bacteria and plants, the first committed step to serine synthesis, the NAD-dependent oxidation of 3-phosphoglycerate, is feedback inhibited by L-serine.

Serine is quite active metabolically; we have already considered its roles in the biosynthesis of phospholipids (Chapter 16) and its contribution of activated one-carbon units to the pool of tetrahydrofolate coenzymes (Section 18.4). In addition, serine provides the carbon skeleton for cysteine (Figure 18.17). Glycine also plays multiple roles, including

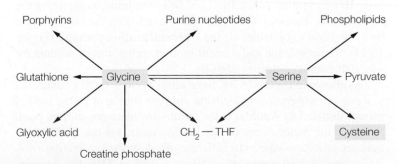

▲ **FIGURE 18.19 Metabolic interconversions and fates of serine and glycine.**

contributions to the one-carbon pool and as a precursor to glutathione, to purine nucleotides (see Chapter 19), and to porphyrins. **FIGURE 18.19** summarizes the metabolic fates of glycine and serine.

Synthesis of Valine, Leucine, and Isoleucine from Pyruvate

The branched-chain amino acids valine, leucine, and isoleucine are essential for mammals, and they are synthesized primarily in plant and bacterial cells. Furthermore, none of these amino acids is known to play significant metabolic roles other than as protein constituents and as substrates for their own degradation. The pathways involved are complex, and they are shown here only in outline in order to emphasize the way these pathways are regulated.

Valine, leucine, and isoleucine are structurally related, and they share certain reactions and enzymes in their biosynthetic pathways (**FIGURE 18.20**). The last four reactions in valine and isoleucine biosynthesis are catalyzed by the same four enzymes. Valine biosynthesis begins with transfer of a two-carbon fragment from hydroxyethyl thiamine pyrophosphate (TPP) to pyruvate. The two-carbon fragment derives from a second molecule of pyruvate in a TPP-dependent reaction similar to that catalyzed by pyruvate decarboxylase (Section 12.3). A similar TPP-dependent transfer of a two-carbon unit to α-ketobutyrate begins the pathway to isoleucine. The keto acid analog of valine is the input for a four-step pathway to leucine. In bacteria, each of these three amino acids controls its own synthesis by feedback inhibition of a different enzyme. In fact, the concept of allosteric control was developed largely in studies on the inhibition of threonine dehydratase by isoleucine.

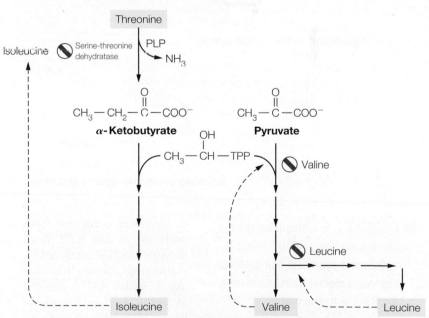

▲ **FIGURE 18.20 Biosynthesis of isoleucine, leucine, and valine.** After the serine–threonine dehydratase reaction, one set of enzymes catalyzes the comparable reactions in valine and isoleucine synthesis. In bacteria, each end product regulates its own synthesis by inhibiting a specific enzyme.

18.8 Amino Acids as Biosynthetic Precursors

Beyond their incorporation into proteins, amino acids serve as precursors for a tremendous variety of other important metabolites, such as polyamines, glutathione, methyl groups, heme, neurotransmitters and other signaling molecules, and nucleotides (Chapter 19). In the following section, we highlight some of the most important pathways and products of amino acid metabolism.

S-Adenosylmethionine and Biological Methylation

In Section 16.5 we introduced S-adenosyl-L-methionine (AdoMet) as a metabolically activated form of methionine, when we described the biosynthesis of phosphatidylcholine from phosphatidylethanolamine. AdoMet is a "universal" methyl donor, involved in numerous methyl group transfer reactions in lipid, protein, amino acid, and nucleic acid metabolism. It has an unstable **sulfonium ion,** which has a high thermodynamic tendency to transfer its strongly electrophilic methyl group to nucleophiles and lose its charge. The other product of methyl group transfer is **S-adenosyl-L-homocysteine** (AdoHcy).

● **CONCEPT** S-Adenosyl-L-methionine is a "universal" methyl donor, involved in numerous methyl group transfer reactions in lipid, protein, amino acid, and nucleic acid metabolism.

Note that the sulfonium ion is a chiral center—the sulfur has a pyramidal configuration, with the lone electron pair forming the absent fourth ligand. All known AdoMet-dependent methyltransferases are specific for the S isomer (shown below).

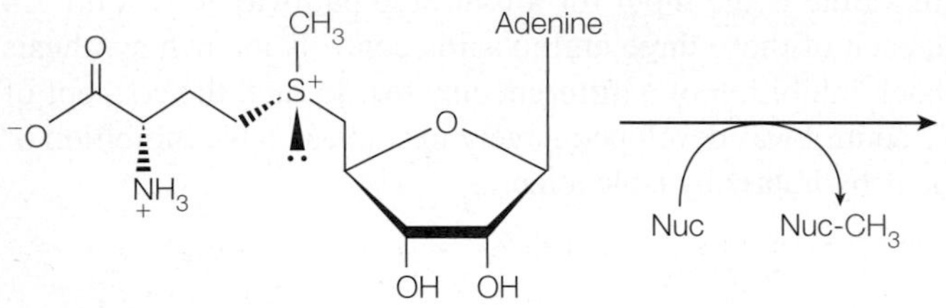

S-Adenosyl-L-methionine (AdoMet)

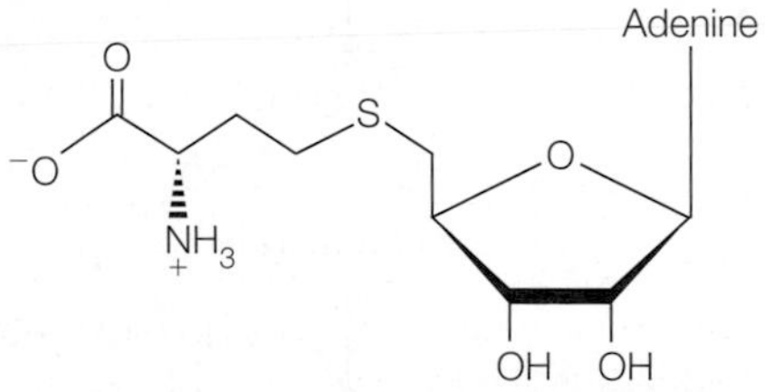

S-Adenosyl-L-homocysteine (AdoHcy)

● **CONCEPT** 5-Methyltetrahydrofolate transfers a methyl group in methionine synthesis, but all other biological methyl transfers involve S-adenosylmethionine.

AdoMet is formed from methionine and ATP in an unusual reaction catalyzed by methionine adenosyltransferase, in which ATP is cleaved to yield inorganic triphosphate (PPP_i) plus an adenosyl moiety linked directly via the ribose C5 to the methionine sulfur. The triphosphate is hydrolyzed by the enzyme to pyrophosphate (PP_i) and orthophosphate (P_i), drawing the reaction to completion.

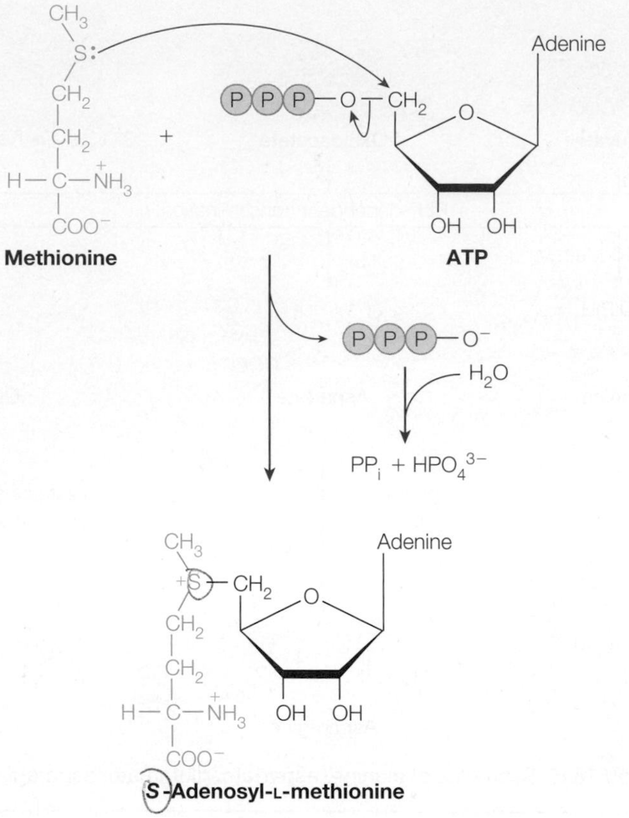

Methionine   **ATP**

$PP_i + HPO_4^{3-}$

S-Adenosyl-L-methionine

The central metabolic role of AdoMet can be appreciated if you keep in mind that, except for a few reactions in bacterial metabolism, *the only known methyl group transfer that does not involve AdoMet is the synthesis of methionine itself.* As noted in Section 18.4, a methyl group is generated de novo through the reduction of 5,10-methylenetetrahydrofolate to 5-methyltetrahydrofolate (5-methyl-THF), catalyzed by a flavin-dependent methylenetetrahydrofolate reductase (MTHFR). The MTHFR reaction in mammalian liver uses NADPH as electron donor and consequently is physiologically irreversible due to the high cytosolic NADPH/NADP$^+$ ratio and the large standard free energy change for the reduction of 5,10-methylene-THF. The methyl group is then transferred to homocysteine to yield methionine via methyl-B$_{12}$ and the action of methionine synthase (reaction ❸, **FIGURE 18.21**). This methyl group is then activated for methyl transfers by the ATP-dependent conversion of methionine to AdoMet. S-adenosylhomocysteine, formed from AdoMet-dependent transmethylations, is hydrolyzed to yield adenosine and homocysteine by S-adenosylhomocysteine hydrolase. Homocysteine is remethylated to methionine, completing the *methyl cycle.* Homocysteine can be removed from the methyl cycle by conversion to cysteine, via the **transsulfuration** pathway (Figure 18.21). Remethylation and transsulfuration each normally account for ~50% of homocysteine metabolism.

As shown in Figure 18.21, the methyl cycle is tightly connected to a *one-carbon cycle*—indeed, the ultimate source of all the methyl groups donated by AdoMet is the tetrahydrofolate one-carbon pool. Serine is the major donor of one-carbon units, via the PLP-dependent serine hydroxymethyltransferase, ❶, described in Section 18.4. Reduction of 5,10-methylene-THF to 5-methyl-THF, ❷, commits this one-carbon unit to methyl group biogenesis because methionine synthase is the only enzyme known to utilize 5-methyl-THF in

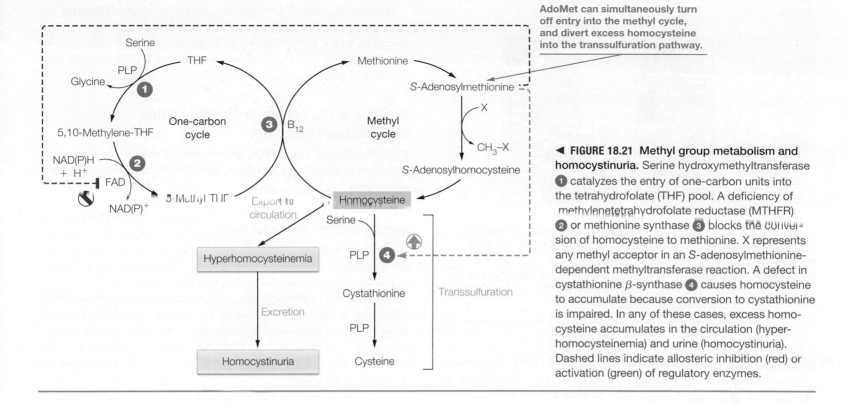

◄ FIGURE 18.21 Methyl group metabolism and homocystinuria. Serine hydroxymethyltransferase ❶ catalyzes the entry of one-carbon units into the tetrahydrofolate (THF) pool. A deficiency of methylenetetrahydrofolate reductase (MTHFR) ❷ or methionine synthase ❸ blocks the conversion of homocysteine to methionine. X represents any methyl acceptor in an *S*-adenosylmethionine-dependent methyltransferase reaction. A defect in cystathionine β-synthase ❹ causes homocysteine to accumulate because conversion to cystathionine is impaired. In any of these cases, excess homocysteine accumulates in the circulation (hyperhomocysteinemia) and urine (homocystinuria). Dashed lines indicate allosteric inhibition (red) or activation (green) of regulatory enzymes.

eukaryotes. Thus, if the methionine synthase reaction is blocked for some reason (e.g., a vitamin B_{12} deficiency), 5-methyl-THF accumulates, at the expense of depleted pools of the other THF coenzymes. This is referred to as a "methyl trap."

As you might expect, regulatory mechanisms have evolved to maintain balance between the supply of one-carbon units and the production of methyl groups in the form of AdoMet. Because the MTHFR reaction catalyzes the committed step in methyl group biogenesis, the regulation of MTHFR is crucial for one-carbon metabolism in all organisms. Eukaryotic MTHFRs are feedback inhibited by AdoMet, a key regulatory feature that prevents 5,10-methylene-THF depletion when AdoMet levels are adequate (reaction ❷, Figure 18.21). The other important regulatory site is the first step of the transsulfuration pathway, where AdoMet is a positive allosteric effector of cystathionine β-synthase (reaction ❹, Figure 18.21). Thus, under conditions of adequate methionine and methyl groups, AdoMet simultaneously turns off entry of one-carbon units into the methyl cycle and diverts excess homocysteine into the transsulfuration pathway.

In humans, a genetic deficiency of cystathionine β-synthase, (reaction ❹, Figure 18.21) leads to a condition called **homocystinuria,** in which homocysteine accumulates, as evidenced by excessive urinary excretion of homocystine (the oxidized disulfide derivative of homocysteine). Homocystinuria results in severe mental retardation, damage to blood vessels, and dislocation of the lens of the eye. In some patients, homocystinuria can be treated with vitamin B_6 (pyridoxine). Cystathionine β-synthase is a PLP-dependent enzyme (Figure 18.21). B_6-responsive patients typically express mutant cystathionine β-synthase enzymes with two- to five-fold lower affinity for PLP. These "K_M mutants" can often be rescued in vivo by administration of high doses of vitamin B_6 (see end-of-chapter Problem 2).

After cystathionine β-synthase deficiency was described, it was found that similar symptoms resulted from deficiencies of either of two related enzymes— methionine synthase, ❸, or 5,10-methylenetetrahydrofolate reductase, ❷ (Figure 18.21). Deficiency of methionine synthase causes megaloblastic anemia and homocysteinemia but is very rare in humans. MTHFR deficiency, however, is the most common inborn error of folate metabolism. Although a few cases of severe MTHFR deficiency are known, causing homocystinuria, much more common are patients who exhibit only moderate deficiency of MTHFR enzyme activity. In 1995, Rima Rozen and colleagues at McGill University in Montreal identified a $C \rightarrow T$ substitution in exon 4 of the human MTHFR gene, producing an Ala $\rightarrow$ Val substitution at position 222 in the catalytic domain of the enzyme (**FIGURE 18.22**). This missense, or nonsynonymous, codon change is quite common in North America, with up to 45% of the Caucasian population being heterozygous (carrying one T allele and one C allele), and 10%–15% being homozygous for the T allele. When a particular mutation occurs relatively frequently ($>$1% of the alleles in a population), it is called a *polymorphism*. Because this genetic change involves only a single nucleotide, it is called a single-nucleotide polymorphism, or SNP (see Section 21.4). Patients who are homozygous for this $C \rightarrow T$ polymorphism in MTHFR exhibit mild homocysteinemia, but only if they also have low folate and riboflavin nutritional status. Heterozygotes generally have normal plasma homocysteine levels.

Biochemical analysis of the valine variant of the enzyme (product of the T allele) has provided a molecular explanation for the clinical findings. Compared to the normal alanine-containing enzyme (product of the C allele), the valine variant is *thermolabile*—it is completely inactivated at 46 °C and is less stable than the Ala enzyme even at 37 °C. Lymphocytes taken from TT homozygotes have only ~50%

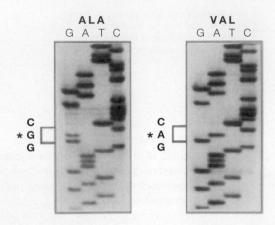

▲ **FIGURE 18.22 DNA sequencing reveals a common polymorphism in human methylenetetrahydrofolate reductase (MTHFR).** The DNA sequence was obtained from the template strands of the MTHFR gene from two individuals, one carrying the C allele (left) and the other the T allele (right). The corresponding coding strands would encode a GCC codon (Ala) or GTC codon (Val), respectively.

● **CONNECTION** Homocysteinemia can result from genetic defects in cystathionine β-synthase, methionine synthase, or 5,10-methylenetetrahydrofolate reductase, as well as from dietary deficiency of folic acid, vitamin B₆, or vitamin B₁₂.

of the MTHFR enzymatic activity of CC homozygotes in vitro assays. This decrease in enzyme activity leads to lower 5-methyl-THF, restricting the remethylation of homocysteine (Figure 18.21). Thus, homocysteine accumulates in patients homozygous for the T allele of MTHFR.

Precursor Functions of Glutamate

In addition to providing a source of α-ketoglutarate for the citric acid cycle (Section 18.2) and a source of nitrogen (via transamination) for other amino acids, glutamate serves as a precursor for the biosynthesis of numerous important metabolites, including the **polyamines**, glutathione, and the neurotransmitter **α-aminobutyric acid (GABA).**

Glutamate

γ-Aminobutyric acid

Putrescine (a polyamine)

Glutathione (γ-glutamylcysteinylglycine)

Finally, glutamate is involved in the synthesis, via an ATP-dependent conjugating system, of the polyglutamate tails of folic acid and its coenzymes (see Section 18.4). As in glutathione synthesis, the glutamate residues in the polyglutamate tails are linked through their γ-carboxyl groups.

Arginine Is the Precursor for Nitric Oxide and Creatine Phosphate

Beginning in the late 1980s, an unexpected role for arginine was described as a precursor to a novel second messenger and neurotransmitter. This novel regulator has been identified as a gas, the free radical **nitric oxide** (NO·), which is produced from arginine in an unusual reaction that also yields citrulline (**FIGURE 18.23**).

● **CONCEPT** Nitric oxide (NO·), derived from arginine, is a powerful gasotransmitter involved in the regulation of vascular blood flow and blood pressure.

Nitric oxide synthase (NOS), which catalyzes a two-step NADPH-dependent oxidation of arginine by O₂, contains bound FMN, FAD, heme iron, and tetrahydrobiopterin, the same cofactor involved in the synthesis of tyrosine from phenylalanine (see Figure 18.15). The physiological role and mechanism of action of this gasotransmitter are described in Chapter 20.

Arginine is also the precursor to the energy storage compound creatine phosphate (see Figure 3.7). The guanidino group is transferred to glycine, releasing the rest of the arginine molecule as ornithine (Figure 18.23). Arginine can be regenerated from ornithine via the urea cycle (see Figure 18.11). The N-methyl group of creatine derives from S-adenosylmethionine. Creatine kinase catalyzes the phosphorylation of creatine to creatine phosphate. Because creatine kinase is very low in liver, creatine is released into the bloodstream. Tissues with high energy demand, such as muscle and brain, take up creatine from blood and phosphorylate it to creatine phosphate.

Both creatine and creatine phosphate spontaneously cyclize to creatinine (Figure 18.23), which is excreted in the urine (1–2 g per day in a normal adult). A typical Western diet can supply about half of this lost creatine; the other half is from endogenous biosynthesis, and it has been estimated that adult humans synthesize ~4–8 mmol creatine per day. Indeed, the methylation of guanidinoacetic acid to creatine consumes more AdoMet than all other methylation reactions combined.

Tryptophan and Tyrosine Are Precursors of Neurotransmitters and Biological Regulators

Many amino acids and their metabolites participate in signal transduction processes—in hormonal control and in synaptic transmission of nervous impulses. As introduced in Chapter 10 and discussed further in Chapter 20, these two roles are comparable in that a low-molecular-weight substance released from one cell migrates to a target cell, where it interacts with specific receptors in the target cell membrane. The difference is that neurotransmission involves movement across a synapse, between two adjacent cells, whereas hormonal transmission occurs over a distance, with the hormonal messenger being transported through the bloodstream to the effector cell.

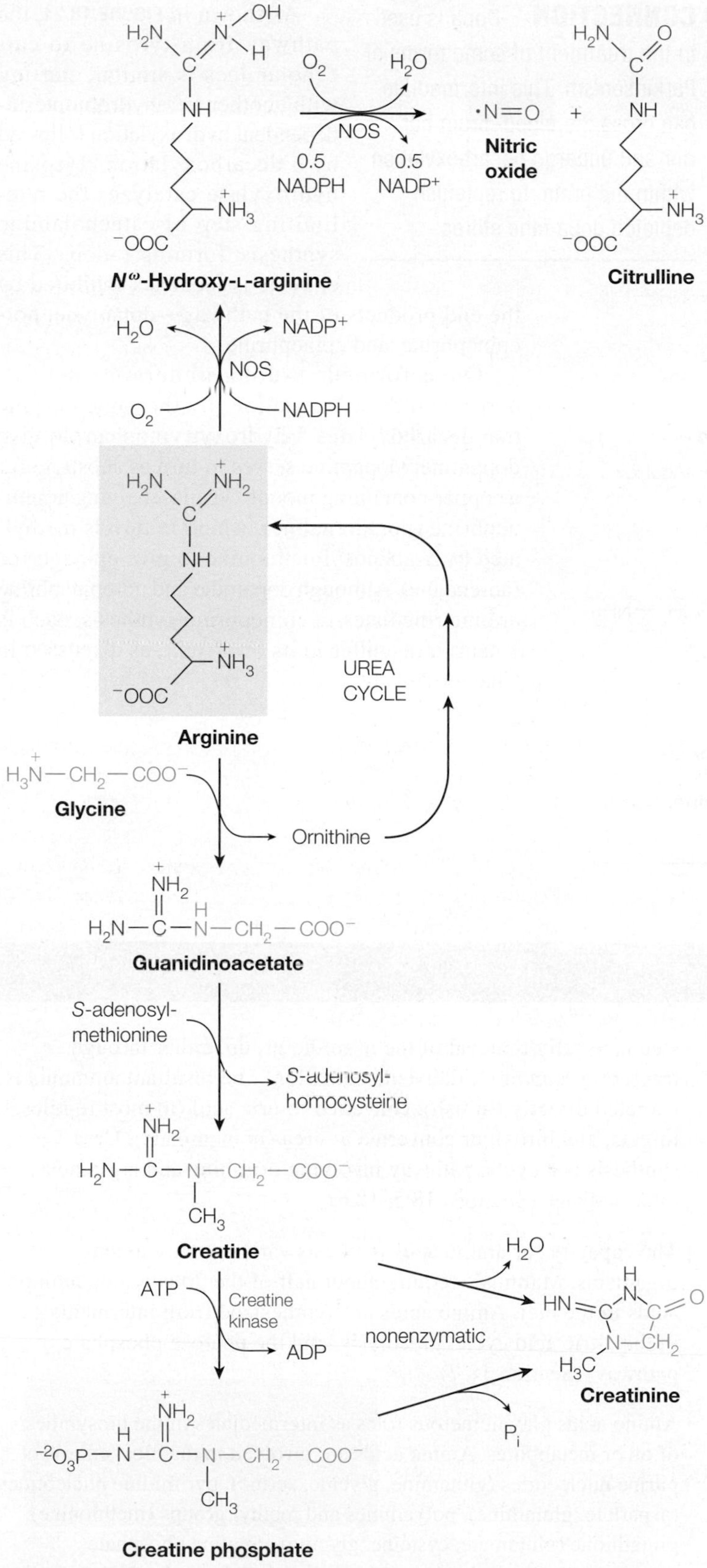

▲ **FIGURE 18.23 Biosynthesis of nitric oxide and creatine phosphate from arginine.** NOS, nitric oxide synthase.

● **CONCEPT** Glutamate, tyrosine, glycine, and tryptophan serve as neurotransmitters or precursors to neurotransmitters.

Among canonical amino acids that serve directly as neurotransmitters are glycine and glutamate. As noted previously, GABA, the decarboxylation product of glutamate, is also a neurotransmitter. Several aromatic amino acid metabolites also function in neurotransmission. They include histamine, derived from histidine (Section 18.6); **serotonin** (5-hydroxytryptamine), derived from tryptophan; and the **catecholamines**—epinephrine, **dopamine,** and **norepinephrine**—derived from tyrosine.

The pathway to serotonin begins with hydroxylation of tryptophan by a tetrahydrobiopterin-dependent aromatic amino acid hydroxylase, similar to phenylalanine hydroxylase (see Figure 18.15). This reaction is followed by a PLP-dependent decarboxylation to yield serotonin.

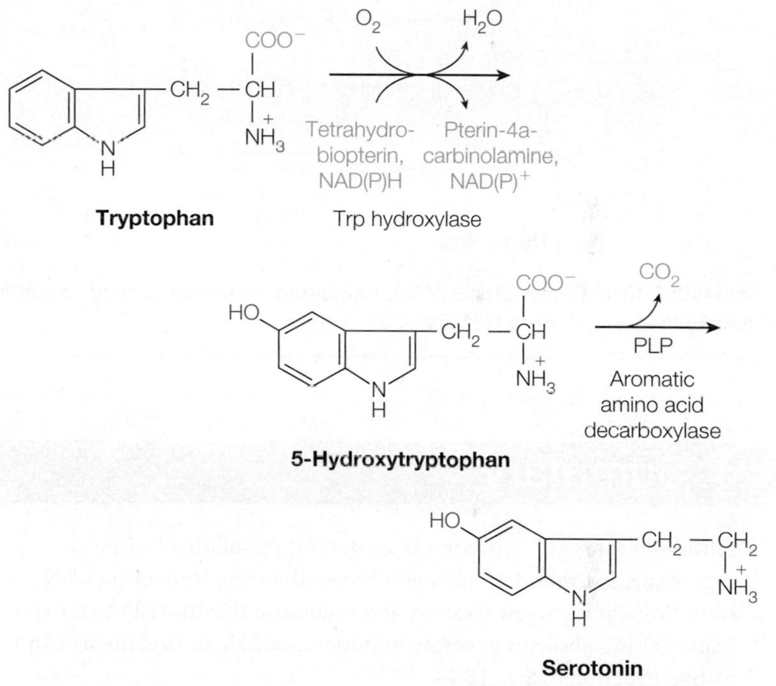

● **CONCEPT** Tryptophan hydroxylase and tyrosine hydroxylase are both tetrahydrobiopterin-dependent monooxygenases and are mechanistically and structurally related to phenylalanine hydroxylase.

Serotonin plays multiple regulatory roles in the nervous system, including neurotransmission. It is produced in the pineal gland, where it serves as precursor to **melatonin** (*O*-methyl-*N*-acetylserotonin). The pineal gland is known to regulate light–dark cycles in animals, and the levels of serotonin and melatonin undergo cyclic variations in phase with these cycles. Many long-distance airline passengers take melatonin pills to escape jet lag by resetting their biological clocks. Serotonin is also a potent vasoconstrictor, which helps regulate blood pressure.

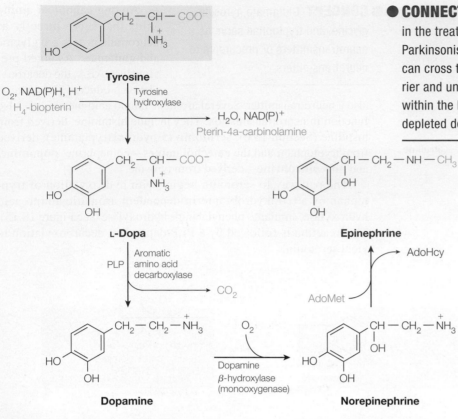

● CONNECTION L-Dopa is used in the treatment of some forms of Parkinsonism. This intermediate can cross the blood–brain barrier and undergo decarboxylation within the brain, to replenish depleted dopamine stores.

As shown in **FIGURE 18.24**, the pathway from tyrosine to catecholamines is similar, starting with another tetrahydrobiopterin-dependent hydroxylation followed by a decarboxylation. Tyrosine hydroxylase catalyzes the rate-limiting step of catecholamine synthesis, forming L-dopa. This enzyme is feedback inhibited by the end products of the pathway—dopamine, norepinephrine, and epinephrine.

Once formed, L-dopa undergoes a PLP-dependent decarboxylation (by the same enzyme that decarboxylates 5-hydroxytryptophan) to give dopamine. Dopamine serves in turn as substrate for a copper-containing monooxygenase, giving norepinephrine (noradrenaline), which in turn is methylated by *S*-adenosylmethionine to give epinephrine (adrenaline). Although dopamine and norepinephrine are intermediates in epinephrine synthesis, each is a neurotransmitter in its own right, as discussed in Chapter 20.

▲ **FIGURE 18.24** Biosynthesis of the catecholamines—dopamine, norepinephrine, and epinephrine—from tyrosine.

 Summary

- Although inorganic nitrogen is abundant, metabolism of most organisms is limited by nitrogen bioavailability. Reduction of N_2 in biological nitrogen fixation and reduction of nitrate in plant and bacterial metabolism generate ammonia, which all organisms can utilize (Sections 18.1–18.2).

- Proteins are in a continual state of turnover and replacement. The turnover is partly for replacement of damaged proteins and partly as the result of normal cellular regulatory mechanisms. Most amino acids released by protein turnover are reutilized for protein synthesis. (Section 18.3)

- Transamination and numerous additional reactions undergone by amino acids use pyridoxal phosphate as a coenzyme. Tetrahydrofolate binds one-carbon units at three different oxidation states, interconverts them, and transfers them in the synthesis of purine nucleotides, thymidine nucleotides, and several amino acids. B_{12} coenzymes include methylcobalamin, which participates in methionine biosynthesis, and 5′-deoxyadenosylcobalamin, the coenzyme for methylmalonyl-CoA mutase. Folate metabolism presents various chemotherapeutic targets, and folate and B_{12} deficiencies both have important clinical consequences. (Section 18.4)

- When amino acids are degraded, either for catabolism of an oversupply or when needed for energy generation, the first step is usually removal of the α-amino group, either through transamination or oxidative deamination. The resultant ammonia is excreted directly (in fish), converted to uric acid (in most reptiles, insects, and birds), or converted to urea (in mammals). Urea synthesis is a cyclic pathway involving ornithine and arginine as intermediates (Sections 18.5–18.6).

- The capacity for amino acid synthesis varies greatly among organisms. Mammals require about half of the 20 common amino acids in the diet. Amino acids are synthesized from intermediates in the citric acid cycle, glycolysis, and the pentose phosphate pathway. (Section 18.7)

- Amino acids play numerous roles as intermediates in the biosynthesis of other metabolites. Amino acids are involved in the biosynthesis of purine nucleotides (glutamine, glycine, serine), pyrimidine nucleotides (aspartate, glutamine), polyamines and methyl groups (methionine), glutathione (glutamate, cysteine, glycine), creatine phosphate (arginine), neurotransmitters (tyrosine, tryptophan, glutamate, arginine), lignin, aromatic compounds and pigments (phenylalanine), hormones (tyrosine, histidine), porphyrins (glycine and glutamate in plants), and other amino acids. The roles of amino acids as neurotransmitters and neurotransmitter precursors are particularly important, as are their roles in porphyrin synthesis. (Section 18.8)

Problems

Enhanced by
Mastering Chemistry
for Biochemistry

Mastering Chemistry for Biochemistry provides select end-of-chapter problems and feedback-enriched tutorial problems, animations, and interactive figures to deepen your understanding of complex topics while practicing problem solving.

Answers to red problems are available in the Answer Appendix.

1. Identify the most likely additional substrates, products, and coenzymes for each reaction in the following imaginary pathway.

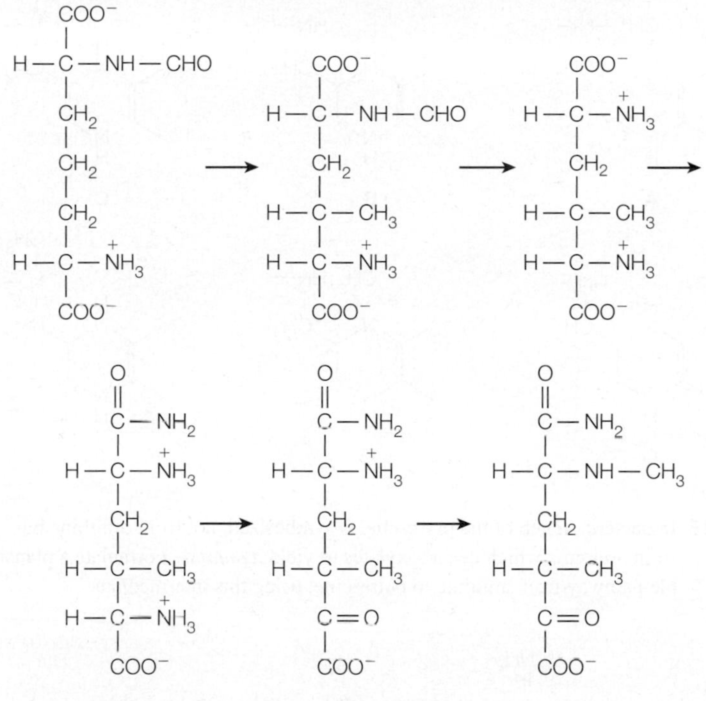

2. The following diagram shows the biosynthesis of B_{12} coenzymes, starting with the vitamin. DMB is dimethylbenzimidazole.

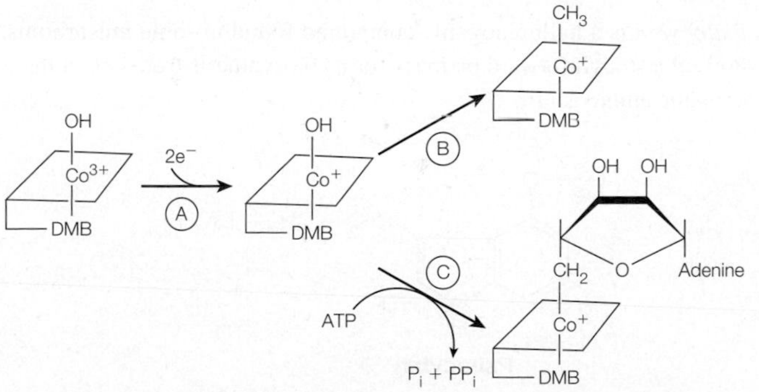

(a) What one additional substrate or cofactor is required by enzyme B?
(b) Genetic deficiency in animals of enzyme C would result in excessive urinary excretion of what compound?
(c) Some forms of the condition described in (b) can be successfully treated by injection of rather massive doses of vitamin B_{12}. What kind of genetic alteration in the enzyme would be consistent with this result?
(d) Genetic deficiency in animals of enzyme B will result in excessive urinary excretion of what amino acid?

3. Using the principles described in the text regarding pyridoxal phosphate mechanisms, propose a mechanism for the reaction catalyzed by serine hydroxymethyltransferase.

4. Use numbers 1 to 5 to identify each carbon atom in the product of this reaction. What coenzyme is involved?

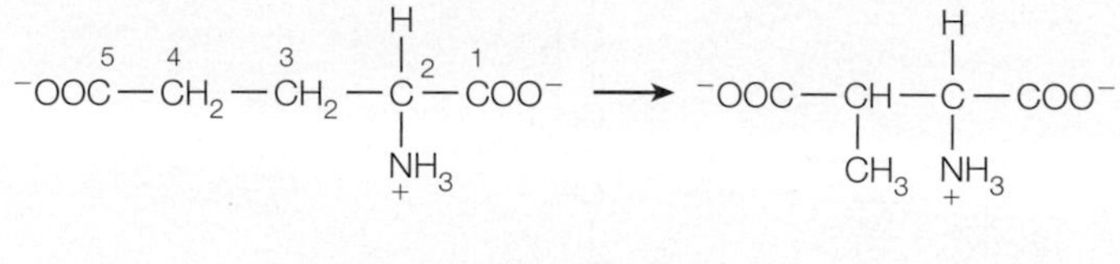

5. The precise mechanism of ammonia toxicity to the brain is not known. Speculate on a possible mechanism, based on possible effects of ammonia on levels of key intermediates in energy generation.

6. Mutants of *Neurospora crassa* that lack carbamoyl phosphate synthetase I (CPS I) require arginine in the medium in order to grow, whereas mutants that lack carbamoyl-phosphate synthetase II (CPS II) require a pyrimidine, such as uracil. A priori, one would expect the active CPS II in the arginine mutants to provide sufficient carbamoyl phosphate for arginine synthesis, and the active CPS I in the pyrimidine mutants to "feed" the pyrimidine pathway. Explain these observations.

7. Indicate whether each of the following statements is true or false, and briefly explain your answer.
   (a) In general, the metabolic oxidation of protein in mammals is less efficient, in terms of energy conserved, than the metabolic oxidation of carbohydrate or fat.
   (b) Given that the nitrogen of glutamate can be redistributed by transamination, glutamate should be a good supplement for nutritionally poor proteins.
   (c) Arginine is a nonessential amino acid for mammals because the enzymes of arginine synthesis are abundant in liver.
   (d) Alanine is an essential amino acid because it is a constituent of every protein.

8. Although proteins differ in nitrogen content because of differences in their amino acid compositions, the N content of protein in a typical Western diet averages 16% (w/w). Using this average value, calculate your daily dietary protein requirement to maintain nitrogen balance assuming you excrete 120 mg of N/kg body weight each day.

9. Write a series of balanced equations and a summary equation for the reactions of the glucose–alanine cycle.

10. Consider the following questions about glutamate dehydrogenase.
   (a) The reaction as shown on page 581 has NH_3 as a reactant, instead of NH_4^+, which is far more abundant at physiological pH. Why is NH_3 preferred?
   (b) Glutamate dehydrogenase has a K_M for ammonia (NH_3) of ~1 mM. However, at physiological pH, the dominant ionic species is ammonium ion, NH_4^+ ($pK_a = 9.2$). Calculate the velocity (as a fraction of V_{max}) that would be achieved by glutamate dehydrogenase if the total intracellular ammonia concentration ($NH_3 + NH_4^+$) is 100 μM (approximate physiological concentration). Assume a mitochondrial matrix pH of 8.0.
   (c) The thermodynamic equilibrium for the reaction greatly favors α-ketoglutarate reduction; yet in mitochondria the enzyme acts primarily to oxidize glutamate to α-ketoglutarate. Explain.
   (d) Propose a reasonable mechanism for this reaction.

11. Explain the basis for the following statement: As a coenzyme, pyridoxal phosphate is covalently bound to enzymes with which it functions; yet during catalysis the coenzyme is not covalently bound.

12. Lysine degradation requires removal of two amino groups. Removal of its ε-amino group gives α-aminoadipic semialdehyde. This product is then degraded to acetoacetate by the same chemical strategy used to degrade the branched-chain amino acids. Draw the proposed intermediates in this pathway (*Hint*: see Figure 18.14).

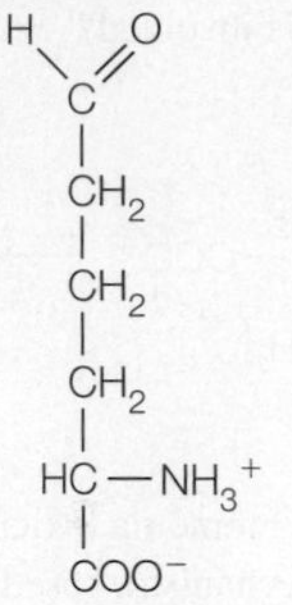

α-Aminoadipic semialdehyde

13. Suppose that you wanted to determine the metabolic half-life of glutamine synthetase in HeLa cells (a line of human tumor cells) growing in tissue culture. Describe how this could be done experimentally.

14. Folic acid is synthesized in bacteria as dihydrofolate, in a pathway starting from guanosine triphosphate. In this pathway, C-8 is lost as formate. From the structural similarities between guanine and pterin, predict which carbon and nitrogen atoms of GTP are the precursors to N-1, C-2, C-4, N-5, C-7, N-8, and C-9 of dihydrofolate.

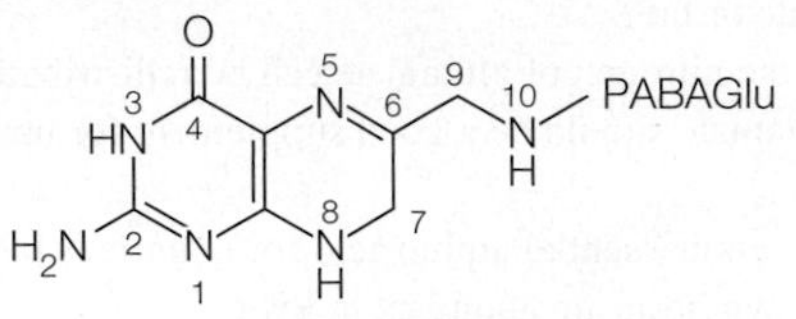

Dihydrofolate

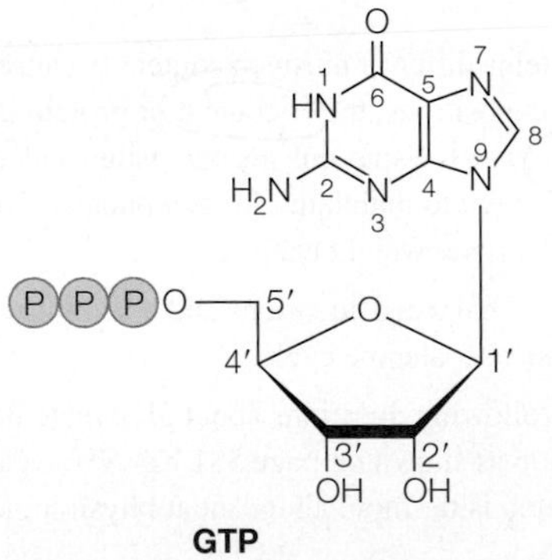

GTP

15. A clinical test sometimes used to diagnose folate deficiency or B_{12} deficiency is a histidine tolerance test, where one injects a large dose of histidine into the bloodstream and then carries out a series of biochemical determinations. What histidine metabolite would you expect to accumulate in a folate- or B_{12}-deficient patient, and why?

16. The mitochondrial form of carbamoyl phosphate synthetase is allosterically activated by *N*-acetylglutamate. Briefly describe a rationale for this effect.

17. Which folate structure (from the list below)
   (a) is the substrate for the enzyme that is inhibited by methotrexate and trimethoprim?
   (b) has the most highly oxidized one-carbon substituent?
   (c) is used in the conversion of serine to glycine?
   (d) transfers its one-carbon substituent to a B_{12} coenzyme? What amino acid is synthesized as the end result of this reaction?
   (e) is the coenzyme for the thymidylate synthase reaction?
   (f) is not known to exist in nature?
   (g) is used in purine nucleotide synthesis?

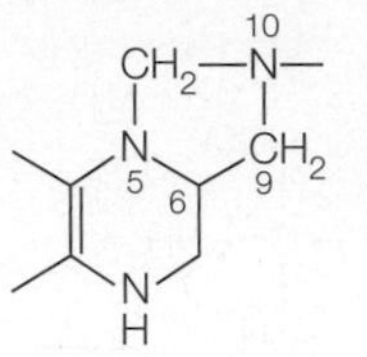

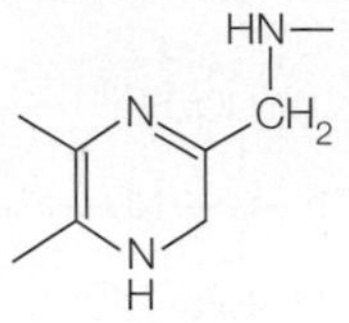

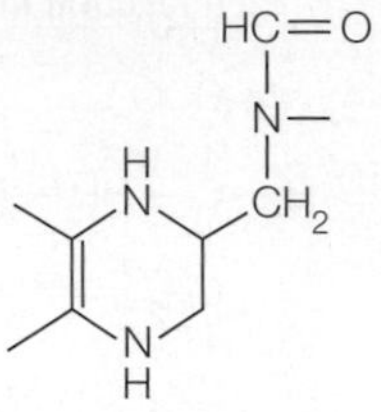

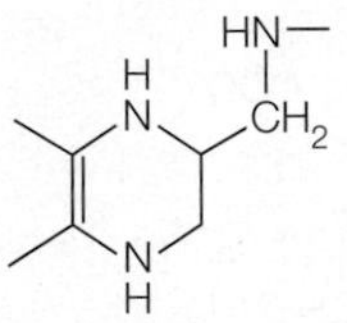

18. In bacteria, much of the putrescine is synthesized, not from ornithine but from arginine, which decarboxylates to yield *agmatine*. Formulate a plausible pathway from arginine to putrescine, using this intermediate.

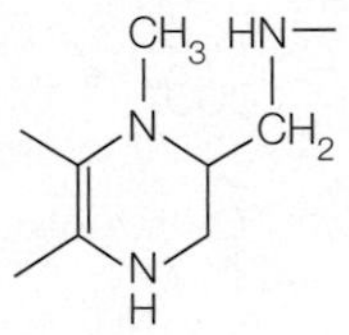

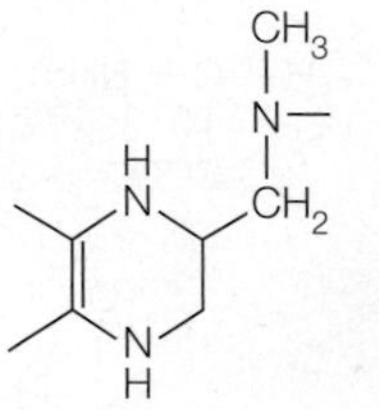

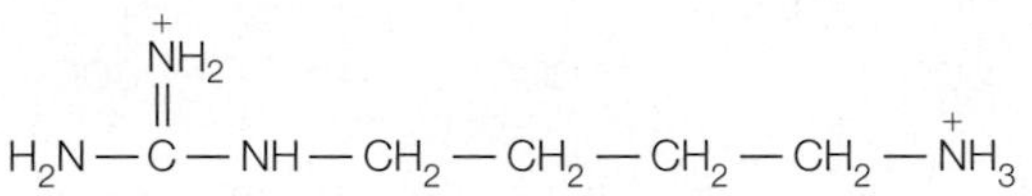

Agmatine

19. *Psilocybin* is a hallucinogenic compound found in some mushrooms. Present a straightforward pathway for its biosynthesis from one of the aromatic amino acids.

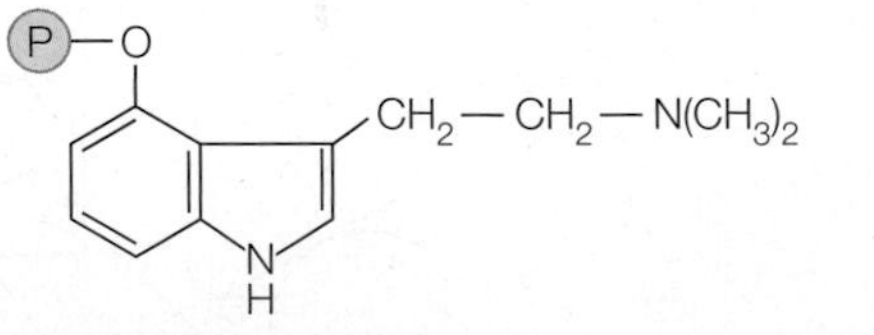

Psilocybin

20. One can identify phenylketonurics and PKU carriers (heterozygotes) by means of a phenylalanine tolerance test. One injects a large dose of phenylalanine into the bloodstream and measures its clearance from the blood by measuring serum phenylalanine levels at regular intervals. Sketch curves showing relative blood phenylalanine concentration versus time that you would expect to be displayed by (a) a PKU patient, (b) a heterozygote, and (c) a normal individual. What kind of tolerance test could you devise to distinguish between PKU resulting from either phenylalanine hydroxylase deficiency or dihydropteridine reductase deficiency?

21. (a) Formaldehyde reacts nonenzymatically with tetrahydrofolate to generate 5,10-methylenetetrahydrofolate. [^{14}C]Formaldehyde can be used to prepare serine labeled in the β-carbon. What else would be needed?

(b) [^{14}C]Serine, prepared as described above, is useful for many things, but you would probably not want to use it for studies on protein synthesis because it would label nucleic acids, carbohydrates, and lipids, as well as proteins. Indicate how each of these classes of compounds could become labeled by this precursor.

22. If oxidation of acetyl-CoA yields 10 ATPs per mole through the citric acid cycle, how many ATPs will be derived from the complete metabolic oxidation of 1 mole of alanine in a mammal? Would the corresponding energy yield in a fish be higher or lower? Why? How much energy would be derived from the metabolic oxidation of 1 mole of isoleucine to CO_2, H_2O, and NH_3? Of tyrosine?

23. Proline betaine is a putative osmoprotectant in plants and bacteria, helping to prevent dehydration of cells.

Propose a plausible pathway for biosynthesis of this compound.

24. Most bacterial mutants that require isoleucine for growth also require valine. Why? Which enzyme or reaction would be defective in a mutant requiring only isoleucine (not valine) for growth?

25. The structure shown below is an intermediate in the synthesis of which biogenic amine? Use arrows to show how the next intermediate in this reaction is formed, and draw the structure of that intermediate.

26. Why is phenylketonuria resulting from dihydropteridine reductase deficiency a more serious disorder than PKU resulting from phenylalanine hydroxylase deficiency?

27. Methyl-labeled [^{14}C]methionine at a specific activity of 2.0 millicuries per millimole was injected into rats. Six hours later the rats were killed. Phosphatidylcholine was isolated from the liver and found to have a specific activity of 1.5 millicuries per millimole. Calculate the proportions of phosphatidylcholine synthesized by the phosphatidylserine pathway and by the pathway starting from free choline. What further information would you need for your calculated values to reflect the true rates of these processes?

28. Propose a mechanism for the TPP-dependent step in isoleucine biosynthesis (see Figure 18.20). Draw the predicted product of this reaction.

References

For a list of references related to this chapter, see Appendix II.

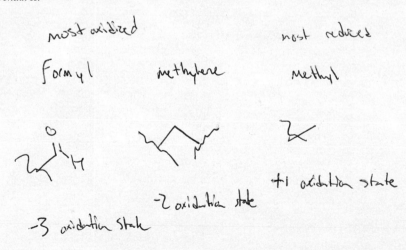

Gout, a disease of purine nucleotide metabolism. Uric acid, an oxidized purine, is rather insoluble, and its accumulation can cause it to crystallize (crystals shown here), often in joints of the large toe. This causes a painful arthritis.

Nucleotide Metabolism

19

WE HAVE ENCOUNTERED nucleotides repeatedly during our exploration of biochemistry. They serve as precursors to nucleic acids, as critical players in energy metabolism, as carriers of activated metabolites for biosynthesis (such as nucleoside diphosphate sugars), as structural moieties of coenzymes, and, finally, as metabolic regulators and signaling molecules (notably, cyclic AMP). In this chapter we discuss metabolism of purine and pyrimidine nucleotides, and we explore the regulation of these processes—particularly critical in pathways leading to DNA replication. We shall see that nucleotide metabolic enzymes are important targets for treating infectious diseases, cancer, and other ailments, and we shall explore the metabolic consequences of certain heritable alterations of nucleotide metabolism.

Before starting this chapter, you may find it useful to review the information on nucleotide structure in Chapter 4. You should also be aware of the distinction between nucleosides and nucleotides. On complete hydrolysis, one mole of *nucleoside* yields one mole each of a sugar and a heterocyclic base (or *nucleobase*), whereas one mole of *nucleotide* yields one mole each of a sugar and a base, plus one or more moles of inorganic phosphate. A mole of *mononucleotide* contains only one mole each of base and sugar, but it may contain more than one phosphate. If it contains, for example, three phosphates, it is called a *nucleoside triphosphate*. The *deoxyribonucleotides,*

which are used in DNA synthesis, are formed from *ribonucleotides* (RNA constituents) by pathways discussed later in this chapter.

19.1 Outlines of Pathways in Nucleotide Metabolism

Here we introduce the major pathways involved in nucleotide biosynthesis and degradation, showing the pathways in brief outline form. Details will be filled in as we proceed.

Biosynthetic Routes: De Novo and Salvage Pathways

We concern ourselves first with metabolism of the *purine* nucleotides, containing adenine or guanine, and next with the *pyrimidine* nucleotides, containing cytosine, uracil, or thymine. Most organisms can synthesize both purine and pyrimidine nucleotides from low-molecular-weight precursors—amino acids, sugar phosphates, CO_2, and so forth. These

are called **de novo** pathways, meaning they are newly synthesized. These pathways are largely conserved throughout the biological world. **FIGURE 19.1** summarizes the metabolic network of de novo pathways and interconversions that lead to the major ribonucleoside and deoxyribonucleoside triphosphates.

Most organisms can also synthesize nucleotides from preformed nucleosides or nucleobases that become available either in the diet or from intracellular nucleic acid turnover. These are called **salvage** pathways because they involve the use of preformed components. A few organisms, including

● **CONCEPT** Nucleotides arise through de novo synthesis from low-molecular-weight precursors or through salvage of nucleosides or nucleobases.

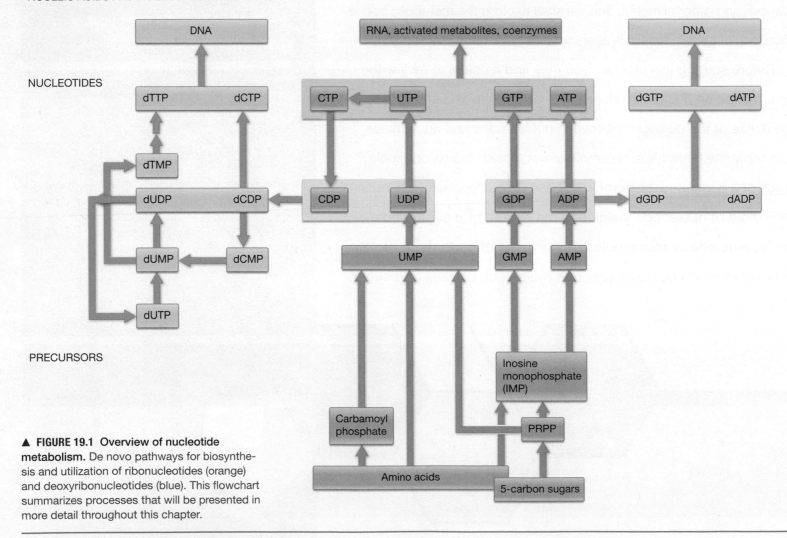

▲ **FIGURE 19.1 Overview of nucleotide metabolism.** De novo pathways for biosynthesis and utilization of ribonucleotides (orange) and deoxyribonucleotides (blue). This flowchart summarizes processes that will be presented in more detail throughout this chapter.

protozoans like the malaria parasite *Plasmodium falciparum,* lack the de novo biosynthetic pathways and are absolutely dependent on salvage pathways for nucleotide synthesis. This biochemical distinction may yield strategies for treating parasitic diseases.

Nucleic Acid Degradation and the Importance of Nucleotide Salvage

Because salvage, or reuse, of purine and pyrimidine bases involves molecules released by nucleic acid degradation, we briefly consider these degradative processes here (**FIGURE 19.2**). Degradation can occur intracellularly (through the turnover of unstable mRNA species or as a result of DNA damage), as a result of cell death or, in animals, through digestion of nucleic acids in the diet.

In animals, the extracellular hydrolysis of ingested nucleic acids represents the major route by which bases and nucleosides become available. The breakdown processes are comparable to those involved in protein digestion. Cleavage processes begin at internal linkages—in this case, phosphodiester bonds. Catalysis occurs via **endonucleases,** such as pancreatic ribonuclease or deoxyribonuclease, which function to digest nucleic acids in the small intestine. Endonucleolytic cleavages yield oligonucleotides, which are then cleaved by **exonucleases** (at linkages near the ends of molecules), in particular by nonspecific enzymes called **phosphodiesterases.** The products of phosphodiesterase action are mononucleotides—nucleoside 5′– or 3′–monophosphates, depending on the specificities of the enzymes involved. Nucleotides can then be cleaved hydrolytically, by a group of phosphomonoesterases called **nucleotidases,** to yield orthophosphate plus the corresponding nucleoside. The most common route for cleavage of a nucleoside to the free base involves the action of a **nucleoside phosphorylase.** Like glycogen phosphorylase,

nucleoside phosphorylases cleave a glycosidic bond by phosphorolysis—addition across the bond of the elements of phosphoric acid, yielding the corresponding base plus ribose-1-phosphate (or deoxyribose-1-phosphate if the substrate is a deoxyribonucleoside).

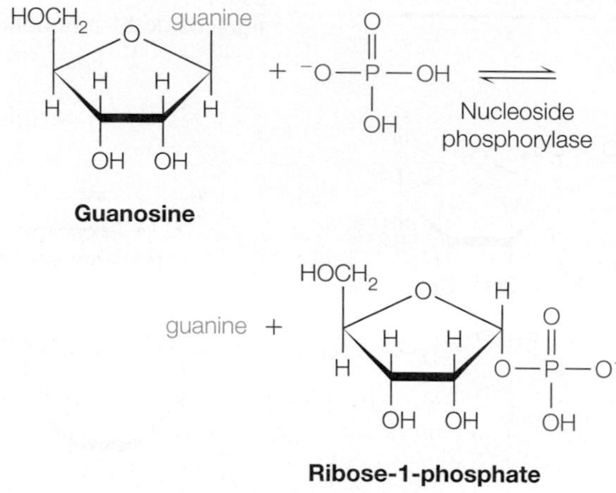

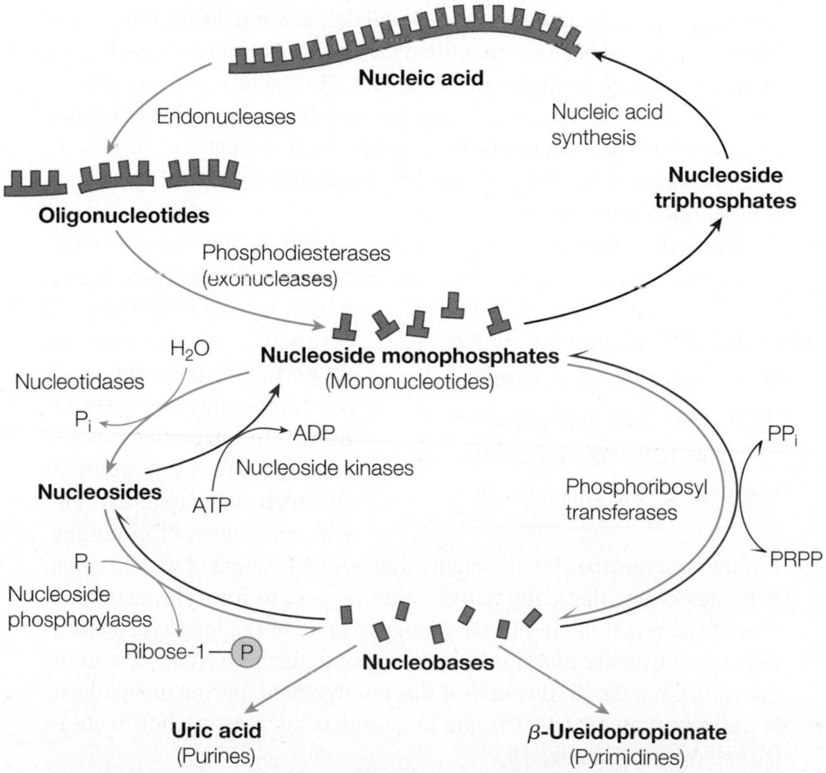

Because nucleoside phosphorylase reactions are readily reversible, these enzymes can also catalyze the first reaction in a salvage biosynthetic pathway. When that occurs, the product nucleoside can be phosphorylated by ATP, through the action of a **nucleoside kinase** (see Figure 19.2). The nucleoside kinases and phosphorylases are widely, but not universally, distributed. For example, animal cells contain neither a guanosine kinase nor a uridine phosphorylase, although these enzymes are found in other organisms. Finally, when the purine or pyrimidine bases accumulate beyond the point where they can be used biosynthetically, they are subject to further degradation, as shown in Figure 19.2.

PRPP, a Central Metabolite in De Novo and Salvage Pathways

As shown in Figure 19.2, salvage pathways to nucleotides can begin with either nucleosides or nucleobases. Utilization of nucleobases requires an activated sugar-phosphate donor, **5-phosphoribosyl-α-D-1-pyrophosphate (PRPP).** Also involved in amino acid biosynthesis, PRPP is a key intermediate in both salvage and de novo pathways. It is formed by the action of **PRPP synthetase,** which activates C1 of ribose-5-phosphate by transferring to it the pyrophosphate moiety of ATP:

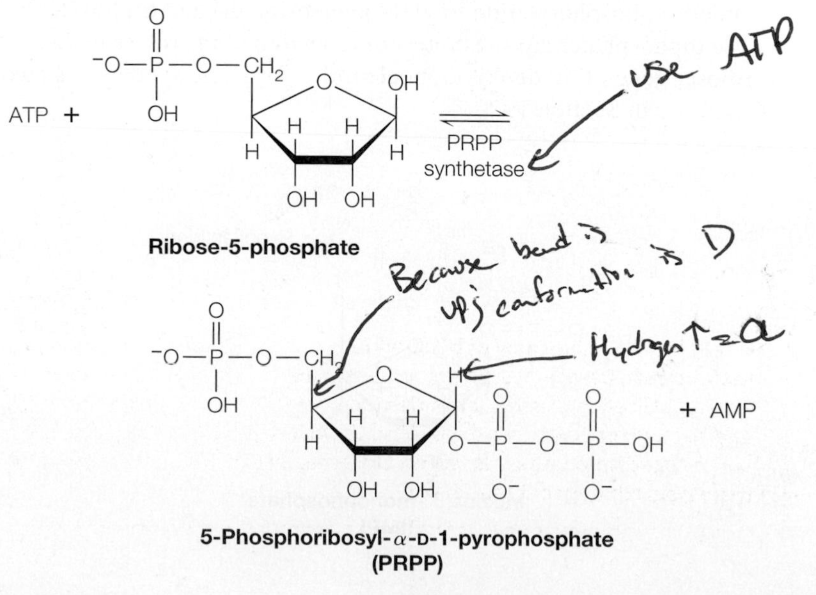

FIGURE 19.2 Reutilization of purine and pyrimidine bases and nucleosides. The figure shows relationships between nucleic acid catabolism (blue) and resynthesis of nucleotides by salvage pathways (magenta).

● **CONCEPT** PRPP is an activated ribose-5-phosphate derivative used in both salvage and de novo nucleotide biosynthetic pathways.

A phosphoribosyltransferase reaction catalyzes the reversible transfer of a free base to the ribose of PRPP, displacing pyrophosphate and producing a nucleoside 5′-monophosphate.

PRPP

GMP

Because phosphoribosyltransferase reactions are reversible, they could be used in nucleotide degradation. However, pyrophosphate is readily hydrolyzed to orthophosphate in most tissues, which drives these reactions to the right and ensures that they are used primarily for nucleotide biosynthesis.

19.2 De Novo Biosynthesis of Purine Ribonucleotides

As we mentioned previously, organisms vary in their capacity for nucleotide synthesis by salvage pathways. However, for the great majority of organisms that possess de novo pathways for nucleotide biosynthesis, those pathways are almost identical. Purine nucleotides are formed by a 10-step pathway leading from PRPP to a branch point, **inosine 5′-monophosphate,** or IMP, a nucleotide for which the purine base is **hypoxanthine.** Separate short pathways lead from IMP to the major purine nucleotides, adenosine 5′-monophosphate (AMP) and guanosine 5′-monophosphate (GMP). These nucleotides undergo phosphorylation to yield nucleoside di- and triphosphates. The diphosphates are substrates for reduction of the ribose to deoxyribose, generating deoxyribonucleotides for DNA synthesis, as we shall see in Section 19.5.

Inosine 5′-monophosphate (IMP)

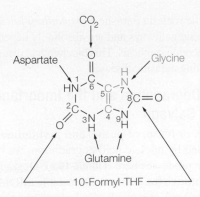

▲ **FIGURE 19.3 Low-molecular-weight precursors to the purine ring.** The source of each atom in the ring, as established with isotopic tracer analysis of uric acid synthesis.

By contrast, pyrimidine nucleotides are formed by an unbranched pathway—six reactions leading to uridine 5′-monophosphate (UMP), followed by two phosphorylations yielding uridine triphosphate (UTP) and then a one-step conversion to cytidine triphosphate (CTP). Just as seen with purines, the substrates for reduction to deoxyribonucleotides are the ribonucleoside diphosphates, UDP and CDP. We shall return to the pyrimidine nucleotides in Section 19.4.

Synthesis of the Purine Ring

An early clue to the nature of the purine synthesis pathway came from the realization that birds excrete most of their excess nitrogen compounds in the form of uric acid, an oxidized purine. Various isotopically labeled metabolites were fed to pigeons, and uric acid was isolated from the droppings and subjected to chemical degradation. Determination of label in the C and N atoms identified the metabolic source of each atom in the purine ring, as shown in **FIGURE 19.3**. This pattern plus isolation of the individual enzymes eventually yielded the entire pathway, shown in **FIGURE 19.4** as it occurs in vertebrates. Minor changes are seen in other organisms, but the series of intermediates shown in Figure 19.4 is essentially universal.

More important than recalling every detail of this complex pathway is understanding the involvement of specific reaction types, which we see both here and elsewhere in metabolism. Of particular importance are reactions catalyzed by **glutamine amidotransferases** and tetrahydrofolate-dependent **formyltransferases.** The amide nitrogen of glutamine is rather nonreactive, but in conjunction with ATP breakdown, this atom becomes biosynthetically active. With respect to formyltransferases, two of the reactions in purine synthesis involve the tetrahydrofolate-dependent transfer of a single-carbon group, reactions that were introduced in Chapter 18. Because of this involvement, purine biosynthesis is an important target of folate antagonists such as methotrexate or trimethoprim (Chapter 18).

● **CONCEPT** Purines are synthesized at the nucleotide level, starting with PRPP conversion to phosphoribosylamine, followed by stepwise ring assembly on the amino group.

As noted earlier, the overall pathway (Figure 19.4) is a 10-step reaction sequence leading from PRPP to inosine monophosphate (IMP). Reaction ❶ is catalyzed by an amidotransferase, with the amide

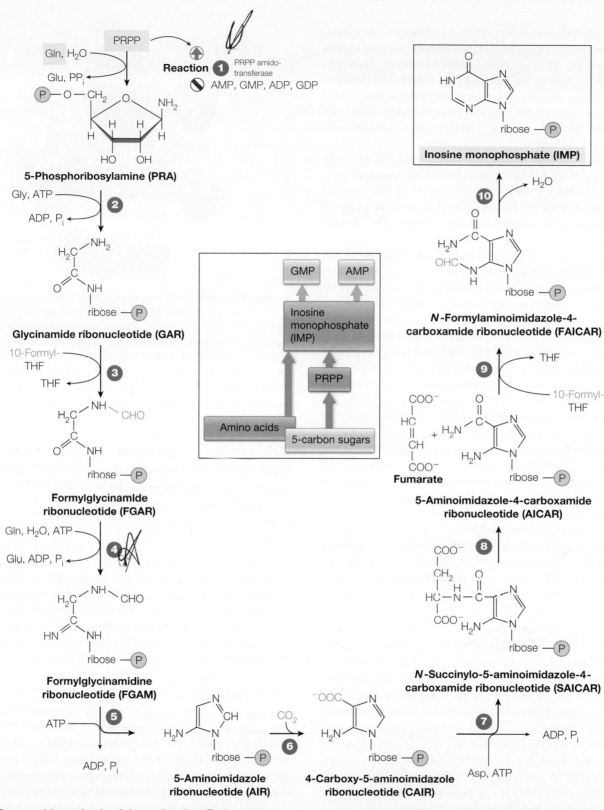

▲ **FIGURE 19.4 De novo biosynthesis of the purine ring.** Each enzyme is numbered in relation to the overall sequence. Hence, E1 catalyzes the first reaction, PRPP amidotransferase, which converts PRPP to phospho-ribosylamine. Subsequent enzymes are not named, for simplicity. In this and in subsequent figures, ribose—P refers to the ribose-5-phosphate moiety in a nucleotide. The encircled P represents a phosphoryl group. The inset shows that part of the Figure 19.1 flowchart that is represented by this pathway.

nitrogen of glutamine being transferred to C1 of ribose-5-phosphate. Unlike other glutamine amidotransferases, this enzyme does not require ATP. That is because the other substrate, PRPP, has already been activated in the PRPP synthetase reaction, and the weakly nucleophilic amide nitrogen of glutamine readily displaces the pyrophosphate at C1 of PRPP. This reaction yields the simplest possible nucleotide, with the ribose-phosphate linked to —NH$_2$. The incorporated N atom will become N9 of the purine ring.

Reaction ❷ incorporates two carbon atoms and one nitrogen atom of the growing purine ring through the ATP-dependent condensation of the carboxyl group of glycine with the N atom of phosphoribosylamine (PRA). ATP activates the carboxyl group of glycine by formation of an acyl-phosphate intermediate, preparing it for attack by the weakly nucleophilic amino group of PRA. The incorporated glycine molecule will form C4, C5, and N7 of the purine ring. Next, in reaction ❸, the first of two tetrahydrofolate-dependent formyltransferases transfers a single-carbon group from 10-formyltetrahydrofolate. Now the five atoms that will form the five-membered ring of the purine moiety are in place. In reaction ❹, an ATP-dependent glutamine amidotransferase begins the assembly of the six-membered ring. The product of reaction 4 is formylglycin-amidine ribonucleotide, or FGAM. Reaction ❺ is an ATP-dependent closure of the five-membered ring in which the formyl oxygen is activated by the γ-phosphate of ATP for nucleophilic attack by N1 of FGAM. This is followed by reaction ❻, a CO$_2$-dependent carboxylation reaction, distinctive in that it does not require biotin. In reaction ❼, ATP drives the formation of an amide link between the newly formed carboxyl group and the amino group of aspartate. In reaction ❽ an α,β-elimination releases the four carbons of aspartate as fumarate, with the amino nitrogen of Asp now part of the six-membered ring. Where have you seen this kind of reaction before? (See Problem 13 at the end of this chapter.)

Another formyltransferase follows in reaction ❾, transferring a one-carbon group from 10-formyl-THF to the nearly complete six-membered ring. This is followed by reaction ❿ in which an intramolecular condensation completes the synthesis of the purine ring in the form of IMP.

Enzyme Organization in the Purine Biosynthetic Pathway

In earlier chapters, we have discussed the metabolic advantages of enzyme organization, specifically, juxtaposing enzymes that act in the same process, to facilitate the flow of intermediates through metabolic pathways. Examples include the pyruvate dehydrogenase complex, the fatty acid synthase multienzyme protein, and the mitochondrial respiratory complexes. Such complexes facilitate multistep reaction pathways either by effecting direct transfer of intermediates from one active site to the next or by maintaining high local concentrations of intermediates so that enzymes can operate at or near V_{max}.

In vertebrates, the purine synthetic enzymes display two organizational features: (1) multifunctional enzymes and (2) reversible enzyme complex formation. The active sites for the second, third, and fifth enzymes in the pathway (E2/E3/E5) are located on the same protein, and E6/E7 and E9/E10 are both bifunctional proteins. Hence, the 10 reactions leading from PRPP to IMP are carried out at catalytic sites on just six proteins. Those six proteins associate noncovalently and reversibly to form a multienzyme complex, called the **purinosome.**

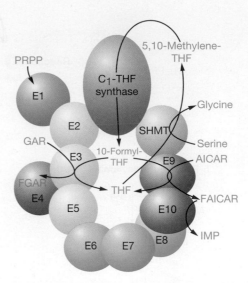

▲ **FIGURE 19.5 Model of the purinosome.** This schematic diagram shows the association of the six proteins involved in converting PRPP to IMP, plus the two enzymes that provide 10-formyl-THF, the one-carbon donor for E3 and E9. Together these two enzymes channel one-carbon units from serine to 10-formyl-THF for the formyltransferase reactions. Each *protein* has a distinctive color; for example, the yellow color identifies the single protein that contains three different enzymatic activities.

● **CONCEPT** Organizing enzymes to place active sites for sequential reactions into close contact facilitates the flow of material through multistep reaction pathways.

Associated with this complex, as shown in **FIGURE 19.5**, are two enzymes of single-carbon metabolism, serine hydroxymethyltransferase (SHMT) and a multifunctional enzyme of tetrahydrofolate metabolism (C-1 THF synthase). These two enzymes combine to produce 10-formyl-THF at its major sites of utilization, the catalytic sites of E3 and E9. Association of the eight proteins is reversible, and signaling processes that promote association and dissociation in living cells are under active investigation. Recent work shows that assembly of the purinosome and flux through the purine synthesis pathway are controlled by mTOR, the central growth regulator that we introduced in Chapter 17.

Synthesis of ATP and GTP from Inosine Monophosphate

IMP represents a branch point in purine nucleotide biosynthesis, being converted in separate short pathways to adenosine 5′-monophosphate (AMP) and guanosine 5′-monophosphate (GMP), as shown in **FIGURE 19.6**.

En route to GMP, IMP undergoes an NAD$^+$-linked reaction in ❶A in which the hypoxanthine ring undergoes hydration followed by dehydrogenation. This yields **xanthosine monophosphate** (XMP), a nucleotide for which the purine base is **xanthine.** In reaction ❷A, XMP undergoes a glutamine amidotransferase reaction, yielding GMP.

● **CONCEPT** IMP, the first fully formed purine nucleotide, lies at a branch point between adenine and guanine nucleotide biosynthesis.

On the other side, IMP reacts with aspartate in reaction ❶B to yield **adenylosuccinate,** followed by an α,β-elimination reaction in ❷B to give AMP plus fumarate, almost identical to the eighth

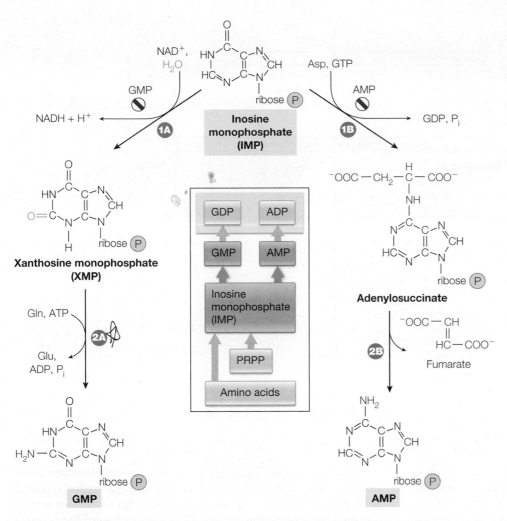

▲ **FIGURE 19.6** Pathways from inosine monophosphate (IMP) to GMP and AMP. The inset shows that part of the Figure 19.1 flowchart represented by this pathway.

and ninth reactions in the synthesis of IMP. Indeed, the same enzyme (adenylosuccinate lyase) catalyzes both elimination reactions.

Xanthine

Activities at the branch point are coordinated in at least two ways to balance the cell's needs for adenine and guanine nucleotides. First, ATP is required in the pathway leading to GMP, while GTP is the energy cofactor in the first reaction from IMP to AMP. Second, GMP allosterically inhibits the first reaction from IMP to GMP, while AMP inhibits the first reaction from IMP to AMP. These effects all tend to promote AMP synthesis when GMP is abundant, and vice versa.

AMP and GMP, along with ADP and GDP, also regulate the pathway to IMP, through allosteric inhibition of both PRPP synthetase and PRPP amidotransferase (the enzyme catalyzing the first reaction in Figure 19.4). In addition, PRPP allosterically activates PRPP

amidotransferase. This latter regulatory effect has medical significance, described in Section 19.3.

Nucleotides are active in metabolism primarily as their respective nucleoside triphosphates (NTPs). AMP and GMP are converted first to their respective diphosphates by specific ATP-dependent kinases. Adenylate kinase, which catalyzes the reversible phosphorylation of AMP, plays an important role in energy metabolism. When ATP levels drop dangerously low, the reaction can run backward from the direction shown, yielding one ATP and one AMP.

$$GMP + ATP \xrightleftharpoons{\text{Guanylate kinase}} GDP + ADP$$

$$AMP + ATP \xrightleftharpoons{\text{Adenylate kinase}} 2ADP$$

Converting ADP to ATP occurs primarily through energy metabolism—substrate-level phosphorylation in glycolysis and the citric acid cycle, oxidative phosphorylation in aerobic organisms, and photophosphorylation in plants. ATP is the most abundant of the four ribonucleoside triphosphates, or rNTPs—ATP, CTP, GTP, and UTP—and it serves as the principal phosphate donor for converting GDP, CDP, and UDP to their respective triphosphates, through the action of a widely distributed enzyme, **nucleoside diphosphate kinase.** This enzyme catalyzes the reversible transfer of phosphate from any common nucleoside triphosphate to any common diphosphate, as shown for GDP.

$$GDP + ATP \xrightleftharpoons{\text{NDP kinase}} GTP + ADP \quad \Delta G°' = 0$$

The enzyme acts similarly on deoxyribonucleoside diphosphates. As noted earlier, the enzyme is readily reversible, and because ATP is by far the most abundant nucleoside triphosphate, it tends to drive, by mass action, the conversion of the other common ribo- and deoxyribonucleoside diphosphates to their respective triphosphate forms at the expense of ATP.

19.3 Purine Catabolism and Its Medical Significance

Having seen how purine nucleotides are synthesized, we turn now to their breakdown, an area of metabolism that has considerable medical significance.

Uric Acid, a Primary End Product

Degradation of nucleic acids, either by digestion or as a result of cell damage or death, yields nucleoside monophosphates. Generally, these NMPs are acted upon by nucleotidases to yield nucleosides. In purine metabolism, the substrates for catabolism include not only AMP and GMP, but also the biosynthetic intermediates IMP and XMP. Purine nucleosides are then acted upon by **purine nucleoside phosphorylase** (PNP), as shown in **FIGURE 19.7**, to yield nucleobases and ribose-1-phosphate. Adenosine is not a substrate for this enzyme, so it undergoes deamination to give inosine, prior to PNP action.

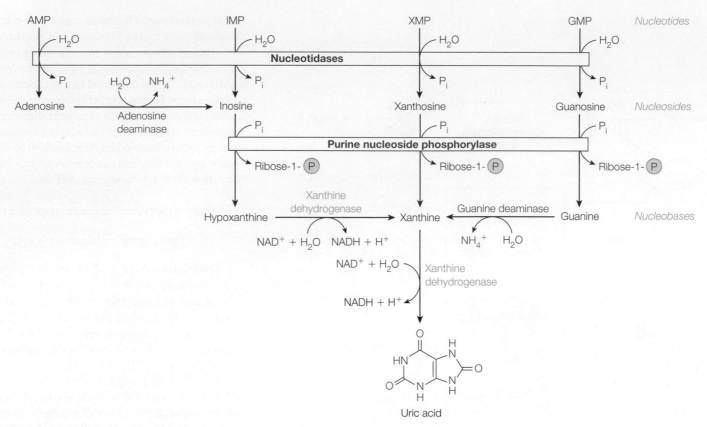

▲ **FIGURE 19.7** Catabolism of purine nucleosides to uric acid.

Further degradation of guanine involves hydrolytic deamination, yielding xanthine. Xanthine is converted to uric acid by **xanthine dehydrogenase,** an enzyme of broad specificity for oxidation of heterocyclic rings, sometimes called xanthine oxidase. Oxidation of the substrate occurs via a **molybdopterin** center, somewhat akin to the molybdenum cofactor in photosynthesis. Electrons are transferred via iron-sulfur clusters and FAD, ultimately to NAD^+. Xanthine dehydrogenase also carries out the oxidation of hypoxanthine to xanthine, as shown in Figure 19.7.

Purine catabolism in humans and the large primates, birds, reptiles, and insects ends with uric acid, which is excreted. However, most animals possess enzymes to further oxidize the purine ring, ultimately to CO_2. The insolubility of uric acid has important human health implications, as we discuss next.

Medical Abnormalities of Purine Catabolism
Gout

Uric acid and its salts are quite insoluble. This property is advantageous in birds, reptiles, and insects because it allows disposition of excess nitrogen with virtually no loss of water; the waste material is excreted essentially as uric acid crystals. This is important during embryonic development in the egg, which is a closed system, where osmotic pressure cannot be allowed to build up. Because the insolubility of uric acid can present difficulties in mammalian metabolism, most mammals have an active enzyme, urate oxidase, which converts uric acid to allantoin, a soluble derivative. However, some 8 to 24 million years ago, during hominid evolution, humans and great apes lost the gene for this enzyme, leading to approximately tenfold elevations of blood uric acid levels relative to those in other mammals.

What advantage might this evolutionary adaptation serve? For one thing, uric acid is a powerful scavenger of free radicals, providing protection against oxidative damage. Uric acid readily undergoes a one-electron reduction, meaning that it can convert, for example, hydroxyl radical to peroxide, for disposition via catalase. However, an unfortunate consequence of the evolutionary pressure to elevate uric acid levels is a predisposition in humans for crystalline uric acid deposition. Indeed, in North America and Europe, about 3 in 1000 individuals suffer from **hyperuricemia,** chronic elevation of blood uric acid to the extent that it crystallizes as sodium urate in the synovial fluid of joints. This condition, known clinically as **gout,** causes inflammation, resulting in a painful arthritis, which if untreated leads to severe degeneration of the joints.

Gout is actually a family of diseases. Some forms result from a kidney disorder that interferes with filtration of urate from the blood. Various biochemical abnormalities are known, including a deficiency of the purine salvage enzyme, **hypoxanthine-guanine phosphoribosyltransferase** (HGPRT). This enzyme salvages both hypoxanthine and guanine (adenine is salvaged by a separate phosphoribosyltransferase).

$$\text{Hypoxanthine} + \text{PRPP} \rightleftharpoons \text{IMP} + \text{PP}_i$$

$$\text{Guanine} + \text{PRPP} \xrightleftharpoons{\text{HGPRT}} \text{GMP} + \text{PP}_i$$

How might an HGPRT deficiency result in chronic uric acid elevation? Two factors contribute, as shown in **FIGURE 19.8**. Decreased flux through the HGPRT reaction causes PRPP to accumulate. PRPP is an allosteric activator of PRPP amidotransferase, the enzyme catalyzing the first committed step in de novo purine synthesis. Hence, PRPP accumulation increases flux through the IMP synthesis pathway, resulting

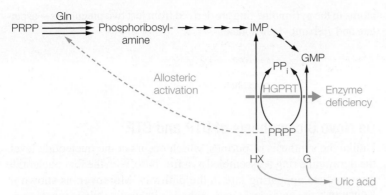

▲ **FIGURE 19.8 Uric acid accumulation as a consequence of HGPRT deficiency.** Excess HX and G (red) are catabolized to uric acid, and excess PRPP (red) allosterically activates purine de novo synthesis, generating purine nucleotides to accumulate in excess of their metabolic needs.

in more purine nucleotides than the cell can absorb, with the excess undergoing degradation to uric acid. In addition, the HGPRT deficiency causes the substrates, hypoxanthine and guanine, to accumulate, and this stimulates their catabolism to uric acid.

Many cases of gout are successfully treated with **allopurinol,** a hypoxanthine analog in which the N7 and C8 positions are interchanged. Xanthine dehydrogenase oxidizes allopurinol as shown, yielding **alloxanthine,** which remains tightly bound to the enzyme. This blocks conversion of hypoxanthine and xanthine to uric acid, and these soluble purine bases are more readily excreted than is uric acid. Allopurinol represents an example of *suicide inhibition*—conversion of an analog into an active inhibitor by the target enzyme itself.

● **CONNECTION** Allopurinol, a hypoxanthine analog, is one of the most successful drugs used to treat gout.

OH

Allopurinol
(enol form)

OH

Hypoxanthine
(enol form)

Xanthine dehydrogenase

OH

HO

Alloxanthine
(enol form)

OH

HO

Xanthine
(enol form)

Lesch–Nyhan Syndrome

Analyses of patients with gout resulting from HGPRT deficiency reveal that in most cases the *HGPRT* gene has undergone a missense mutation,

yielding a mutant form of the enzyme with low residual catalytic activity. Far more serious is a condition called **Lesch–Nyhan syndrome,** first described in 1964 by a medical student (Lesch) and his faculty mentor (Nyhan). This condition results from null mutations, such as chain terminators or frameshifts (Chapter 7), which result in a complete absence of HGPRT activity. Because the gene for HGPRT lies on the X chromosome, this condition affects males almost exclusively. Patients have gout so severe that they cannot be treated with allopurinol. They also display a range of behavioral symptoms, including learning disability, motor disability, and hostile or aggressive behavior, often self-directed. In the most extreme cases, patients nibble off the ends of their fingertips or, if restrained, their lips, causing severe self-mutilation. Nyhan has described this behavior as "nail biting with the volume turned up." The biochemical basis for the behavioral pattern is unknown, but it is clear that all of the behavioral manifestations, as well as the gout, derive from the single well-characterized enzyme deficiency affecting HGPRT.

● **CONNECTION** Null mutations in the gene for HGPRT have far more serious clinical consequences than do mutations that leave some residual enzyme activity.

Severe Combined Immunodeficiency Disease

Severe combined immunodeficiency disease (SCID) is a rare hereditary condition marked by an inability to react to an immune challenge. Neither B cells nor T cells in the immune system can proliferate in response to a challenge (Chapter 7), and as a result, individuals with this condition are susceptible, often fatally, to infectious diseases. This disease was made famous by the case of "bubble boy" David Vetter, who lived all his life in a sterile environment. In 1972, it was found that some cases of this condition resulted from a hereditary deficiency of adenosine deaminase (ADA; see Figure 19.7).

How might ADA deficiency affect immune cell proliferation? The most likely explanation (there are others) stems from the fact that ADA acts upon deoxyadenosine, as well as adenosine. When ADA is absent, deoxyadenosine accumulates and is converted to dATP by salvage pathways. As we discuss in Section 19.5, dATP is an allosteric inhibitor of **ribonucleotide reductase,** the enzyme that begins the flow of nucleotides into DNA, by reducing the four ribonucleoside diphosphates (rNDPs) to the corresponding deoxyribonucleoside diphosphates (dNDPs). Hence, although dATP accumulates to high levels, the other dNTPs disappear, and the DNA replication needed to support cell proliferation is blocked. Another heritable immunodeficiency is caused by a lack of purine nucleoside phosphorylase (page 617). In this condition, in which only T-cell proliferation is affected, dGTP is the principal nucleotide to accumulate.

ADA deficiency was one of the first genetic diseases to be treated by gene therapy, in 1990. The gene for ADA was spliced into the genome of a retrovirus, which had been engineered to be nonpathogenic. ADA was expressed from the retroviral genome. As of 2016, a total of 18 patients had been treated successfully. A more conventional way to treat the disease involves simply administering periodic injections of the purified enzyme linked to an inert support. Although

● **CONNECTION** Severe combined immunodeficiency disease was one of the earliest genetic diseases to be successfully treated by enzyme replacement therapy.

enzyme replacement therapies for most enzyme deficiencies require that the missing enzyme become established within cells, ADA can act effectively within the extracellular environment.

19.4 Pyrimidine Ribonucleotide Metabolism

● **CONCEPT** Pyrimidine biosynthesis occurs by an unbranched pathway, at the nucleobase level, with UTP serving both as an end product and a precursor to CTP.

Because of the simpler structure of the pyrimidine ring relative to the purine ring, the pyrimidine synthetic pathway is simpler than the purine pathway. In fact, all six

atoms in the pyrimidine ring are derived from just two precursors—aspartate and carbamoyl phosphate.

De Novo Biosynthesis of UTP and CTP

Unlike the synthesis of purines, which occurs at the nucleotide level, the pyrimidine ring is assembled as a free base, with the first nucleotide intermediate appearing late in the pathway. Moreover, as shown in **FIGURE 19.9**, the pathway is unbranched, with uridine triphosphate serving both as an end product and as a precursor to cytidine triphosphate.

Pyrimidine synthesis begins with the synthesis of carbamoyl phosphate, catalyzed by **carbamoyl phosphate synthetase** (reaction ❶ in Figure 19.9; see also Chapter 18). The details of this four-step process are shown in the figure at the top of the next page.

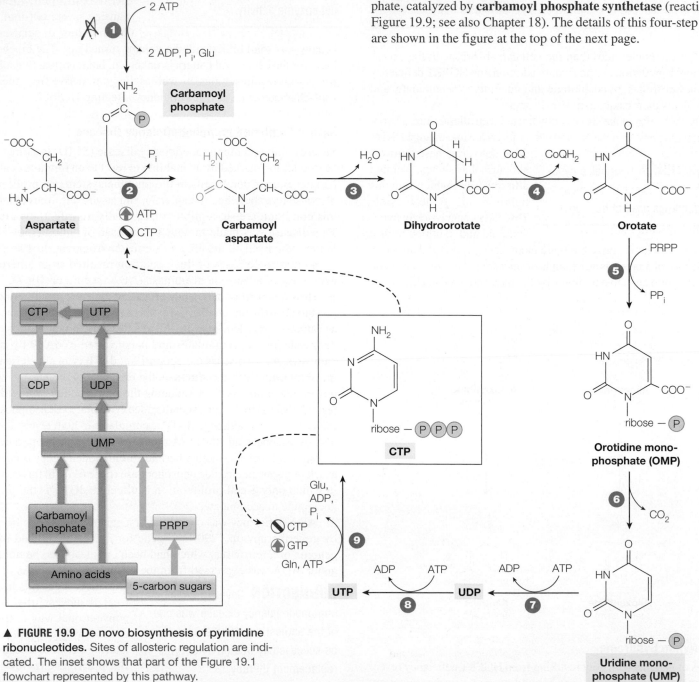

▲ **FIGURE 19.9 De novo biosynthesis of pyrimidine ribonucleotides.** Sites of allosteric regulation are indicated. The inset shows that part of the Figure 19.1 flowchart represented by this pathway.

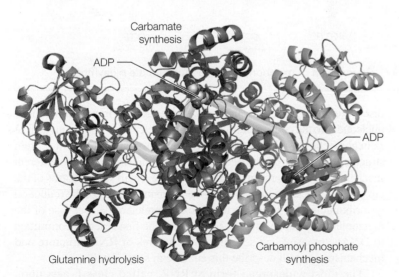

Bicarbonate   **Carboxyphosphate**   **Carbamate**   **Carbamoyl phosphate**

This amidotransferase reaction joins ammonia (released from glutamine) with bicarbonate and phosphate from ATP. Step ❶ activates a carboxyl group by linking it with phosphate derived from ATP. Step ❷ hydrolyzes glutamine, with the resulting ammonia attacking the activated carboxyl group and forming carbamate (step ❸). In step ❹ a second ATP activates the carboxyl group of carbamate, yielding carbamoyl phosphate.

In bacteria, carbamoyl phosphate synthetase is a heterodimeric protein (**FIGURE 19.10**). The small subunit (orange in the figure) hydrolyzes glutamine and transfers ammonia to a catalytic site in the large subunit. The structure of the protein reveals a tunnel that connects three active sites in the enzyme, allowing for channeling of unstable intermediates from site to site, as we have discussed for other multifunctional enzymes, such as tryptophan synthetase (Chapter 18) and the purinosome, discussed in Section 19.2.

Let's return now to the overall pyrimidine synthesis pathway shown in Figure 19.9. In reaction ❷, catalyzed by **aspartate transcarbamoylase** (ATCase), the α-amino group of aspartate nucleophilically attacks the activated carboxyl of carbamoyl phosphate, yielding carbamoyl aspartate. This undergoes intramolecular ring closure in reaction ❸, catalyzed by **dihydroorotase.** These first three reactions occur in the cytosol. Next follows reaction ❹, a mitochondrial NAD^+-linked dehydrogenation, to give

orotate, which is actually a pyrimidine base. The enzyme, **dihydroorotate dehydrogenase,** is a flavoprotein, which transfers electrons from NADH to flavin nucleotides to coenzyme Q within the mitochondrion. The rest of the pathway occurs in the cytosol of eukaryotic cells.

Orotate is the substrate for a phosphoribosyltransferase in reaction ❺, yielding **orotidine 5′-monophosphate** (OMP). Decarboxylation of this intermediate in reaction ❻ yields uridine monophosphate. Subsequent phosphorylations catalyzed by **UMP kinase** in reaction ❼ and nucleoside diphosphate kinase in reaction ❽ yield UTP. As noted above, UTP is an end product, but it is also a substrate in reaction ❾ for a glutamine-dependent amidotranferase, **CTP synthetase,** to yield CTP. CTP synthetase is allosterically inhibited by CTP itself and activated by GTP, two effects that help to maintain favorable intracellular levels of the four ribonucleoside triphosphates.

Glutamine-Dependent Amidotransferases

Examine the CTP synthetase reaction in Figure 19.9. This is one of five glutamine-dependent amidotransferase reactions we have seen in purine and pyrimidine biosynthesis (what are the other four?). Each of these reactions is thought to occur by the same mechanism, shown below, in which the glutamine amide group (unreactive, nonnucleophilic) is activated by hydrolysis at the active site to deliver nucleophilic ammonia in high local concentrations sufficient to amidate the specific substrate, in this case the carbonyl carbon of an amide. ATP drives the reaction by phosphorylating the substrate, which generates a good leaving group upon attack by the nucleophilic NH_3. These enzymes are typically heterodimers, with one subunit containing the glutaminase activity and the other catalyzing the specific amidation reaction. Carbamoyl phosphate synthetase, which we saw in Figure 19.10, illustrates this enzyme architecture. The glutaminase domain is evolutionarily conserved among this family of enzymes.

▲ **FIGURE 19.10 Channeling in carbamoyl phosphate synthetase.** The figure shows the crystal structure of the *E. coli* heterodimeric enzyme (PDB ID: 1c30). An interior tunnel (yellow) connects the active site of the small-subunit glutaminase (orange) with two active sites in the large subunit, each of which contains a bound ADP in space-filling configuration. The N-terminal domain of the large subunit (blue) contains the active site for subreactions 1 and 3, while the final synthesis of carbamoyl phosphate occurs at a site in the C-terminus of the large subunit (cyan).

Multifunctional Enzymes in Eukaryotic Pyrimidine Metabolism

Although the eight reactions leading to UTP and CTP are identical throughout the biological world, there are some major differences in the enzymes involved. ATCase of bacteria was one of the earliest known allosteric enzymes, being activated by ATP and inhibited by CTP (Chapter 8). The *E. coli* enzyme contains six each of two subunits, arranged as two catalytic trimers and three regulatory dimers. By contrast, the eukaryotic ATCase is a multifunctional protein that catalyzes the first three reactions of the pathway (reactions ❶, ❷, and ❸ in Figure 19.9). This multifunctional enzyme is called the CAD protein—an acronym combining the first letters of the names of each of the three enzymes—carbamoyl phosphate synthetase, aspartate transcarbamoylase, and dihydroorotase.

Eukaryotic cells also combine catalytic sites for reactions ❺ and ❻ (Figure 19.9) into a bifunctional enzyme, called UMP synthase. In humans a deficiency of this enzyme causes a condition called **hereditary orotic aciduria.** The enzyme deficiency causes orotic acid to accumulate, resulting in high urinary excretion of this intermediate. The resulting pyrimidine nucleotide deficiency causes a megaloblastic anemia that cannot be cured by administration of folic acid or vitamin B_{12}.

19.5 Deoxyribonucleotide Metabolism

Most cells contain 5 to 10 times as much RNA as DNA. Moreover, as we have seen, ribonucleotides have multiple metabolic roles, while the sole function of deoxyribonucleotides is to serve as constituents of DNA. Accordingly, most of the carbon that flows through nucleotide biosynthetic pathways goes into ribonucleoside triphosphate (rNTP) pools, and intracellular rNTP pools are often orders of magnitude larger than dNTP pools. Typically, intracellular ATP concentration may be 2 to 5 mM and the other rNTPs somewhat lower, while levels of the dNTPs may be 10 to 100 μM, and much lower in non-proliferating cells. However, those nucleotides that do fill the dNTP pools are of paramount importance to the life of the cell. **FIGURE 19.11** summarizes pathways of dNTP synthesis.

● **CONCEPT** Partly because of the multiple metabolic functions played by ribonucleotides and the limited function of deoxyribonucleotides as DNA precursors, intracellular rNTP pools are far larger than those of dNTPs.

Recalling that DNA differs chemically from RNA in the nature of the sugar and in the identity of one of the pyrimidine bases, we can focus our discussion of dNTP biosynthesis on two specific processes—the origin of deoxyribose and the origin of the thymine methyl group. Both of these processes occur at the nucleotide level, and both are of interest from the standpoint of regulation and as targets for drugs used to treat cancer and infectious diseases.

Reduction of Ribonucleotides to Deoxyribonucleotides

Peter Reichard showed that the immediate precursor to a deoxyribonucleotide is a ribonucleotide, meaning that the hydroxyl at C-2′ of the ribonucleotide ribose becomes replaced by a hydrogen atom. This reaction is catalyzed by **ribonucleotide reductase** (RNR), which was

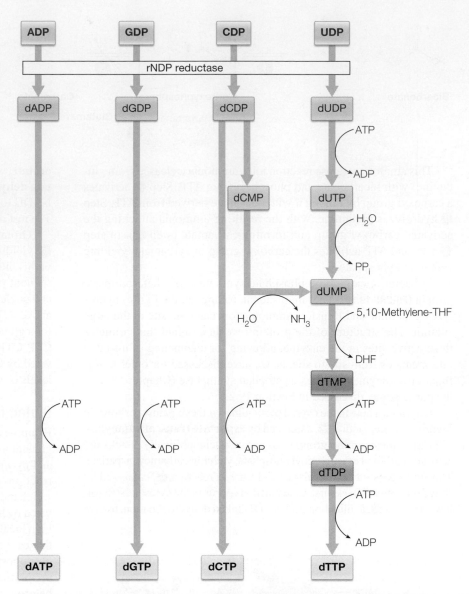

▲ **FIGURE 19.11 Overview of deoxyribonucleoside triphosphate biosynthesis.**

discovered by Reichard. A single enzyme reduces all four common ribonucleotides to the corresponding deoxyribonucleotides, using a novel free radical mechanism. Evolution has created three quite different protein structures for this enzyme, used in different organisms, and three different mechanisms for generating the radical. However, all known forms of the enzyme use the same fundamental mechanism to catalyze the reduction of a ribonucleotide substrate to a deoxyribonucleotide. Because of the metabolic importance of RNR (it catalyzes the first reaction committed to DNA synthesis) and several novel features of RNR structure and mechanism, we will describe this enzyme in some detail.

The most widespread form of RNR, called class I, acts upon ribonucleoside diphosphate substrates (rNDPs) and reduces them to the corresponding deoxyribonucleoside diphosphates (dNDPs). The catalytically essential free radical is on a specific tyrosine residue, where it is stabilized by two ferric ions linked to an oxygen

● **CONCEPT** Deoxyribonucleotides arise via reduction of ribonucleotides in which a free radical process replaces –OH at C-2′ of a ribonucleotide with –H, with retention of configuration.

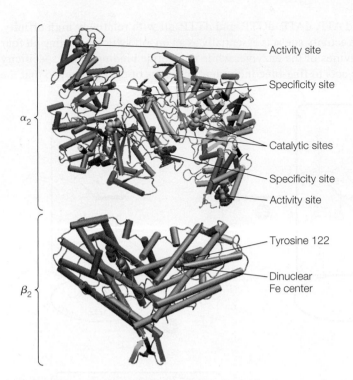

▲ FIGURE 19.12 Structure of *E. coli* ribonucleotide reductase. Crystal structures of the two homodimeric proteins. The two α subunits of the large (R1) protein are in blue and green, and the two β subunits of the small (R2) protein are in lavender and pink. Each of the two β subunits contains a tyrosine radical (Tyr–O·) stabilized by an iron-bridged oxygen atom $(Fe^{3+}—O—Fe^{3+})$. PDB ID: 1pfr. Each of the two α subunits contains a catalytic site and two distinct allosteric control sites, the *activity site* and the *specificity site*. Each catalytic site contains three essential thiol groups (not shown). PDB ID: 3r1r and 4r1r.

atom; some class I RNRs use manganese instead of iron. Class II RNRs use a B_{12} coenzyme to generate a radical, while class III enzymes use *S*-adenosylmethionine to create a specific glycine radical on the protein. Class III enzymes, found only in anaerobic organisms, are distinctive also in acting upon ribonucleoside triphosphate substrates. Here we consider only class I enzymes.

RNR Structure and Mechanism

Class I RNR, as studied in *E. coli*, yeast, and human cells, is an $α_2β_2$ heterodimer, as schematized in **FIGURE 19.12**. In *E. coli* the large protein, called R1, consists of two 87-kDa α polypeptide chains, and the small protein, R2, consists of two 43-kDa β chains. The tyrosine radical (Tyr-122 in *E. coli*) is located in the small (R2) protein, some 35 Å from the catalytic site in the large (E1) protein. Also in the R1 protein are two different allosteric control sites and a site for interaction with an external reductant.

Our understanding of the RNR mechanism is based on the following observations. (1) Radiolabeling studies show that the bond linking the ribose C-3′ to H is cleaved during the reaction. (2) The reaction proceeds with retention of configuration at C-2′, which rules out simple displacement of the 2′ hydroxyl by a hydride ion. (3) The catalytic site thiol groups undergo oxidation during the reaction. (4) The radical at Tyr-122 (*E. coli* RNR) is essential for catalysis. This was shown originally by the fact that hydroxyurea, an RNR inhibitor,

acts by destroying the radical. Tyr-122 lies some 35 Å from the catalytic site, meaning that a long-range electron transport system must link the radical to the active site.

$$HO—NH—\overset{\overset{\textstyle O}{\|}}{C}—NH_2 \qquad \text{Hydroxyurea}$$

All these observations suggest a mechanism for *E. coli* RNR outlined in **FIGURE 19.13**. In step ❶ Cys-439 in the catalytic site is converted to a **thiyl radical** by loss of an electron in a *long-range proton-coupled electron transport* process that results in reduction of the Tyr-122 radical on the β subunit. In step ❷ the thiyl radical abstracts a hydrogen atom from C-3′ of the substrate, creating a substrate radical. Glu-448 in the active site abstracts a proton from the substrate radical. This is followed in step ❸ by loss of a hydroxide ion (water) from C-2′ and migration of the radical to C-2′. Reduction at C-2′ by the redox-active cysteines (225 and 462) generates a disulfide radical anion in step ❹, which then reduces C-3′ in step ❺. The proton-coupled electron transport process of step 1 is then reversed in steps ❻ and ❼, regenerating the radical at Tyr-122 of the small subunit and completing the reduction of the substrate. At this point, the redox-active thiols in the catalytic site (Cys-225 and Cys-462) are in the disulfide form and next undergo reduction by disulfide exchange with another pair of redox-active thiols (step 8). These in turn are reduced by an external electron carrier (shown as Grx; see the next section). Finally, the product dNDP dissociates to regenerate enzyme for another cycle (not shown).

The amazing long-range proton-coupled electron transport process that results in radical propagation from Tyr-122 in $β_2$ to Cys-439 in $α_2$ is but the first known example of what turns out to be a widespread process in enzyme chemistry. Other enzymes that use this process include the cytochrome P450 monooxygenases (Chapter 14) and the oxygen-evolving complex of photosystem II in photosynthesis (Chapter 15).

Source of Electrons for Ribonucleotide Reduction

The electrons for ribonucleotide reduction come ultimately from NADPH, but they are shuttled through either one of two low-molecular-weight proteins that contain redox-active thiols. The first of these protein cofactors, thioredoxin (Trx), was introduced in Chapter 15 in connection with its function in photosynthesis; this small protein plays multiple functions in redox metabolism. **FIGURE 19.14** (red arrows) shows the pathway by which thioredoxin participates in ribonucleotide reduction. The disulfide form of the protein is reduced by the flavoprotein enzyme **thioredoxin reductase.** Reduced thioredoxin then transfers its electrons to the redox site on the enzyme.

An alternative electron carrier, **glutaredoxin (Grx),** was discovered when *E. coli* mutants lacking thioredoxin were found to be viable. Grx can be reduced nonenzymatically by glutathione. The electron transport process linking Grx to RNR is shown by blue arrows in Figure 19.14.

Regulation of Ribonucleotide Reductase Activity

Because deoxyribonucleotides are used only for DNA synthesis and because a single enzyme system is used to reduce all four ribonucleotide substrates, regulation of both the *activity* and the *specificity* of ribonucleotide reductase is essential to ensure that optimally balanced pools of DNA precursors are maintained. To some extent, the intracellular concentrations of the rNDP substrates participate in this control because all four compete for binding to the same active site. A finer level of regulation is exercised through binding of dNTPs to two classes of allosteric control sites—the *activity sites,* which bind ATP and dATP with relatively low affinity, and the *specificity sites,* which bind ATP, dATP, dGTP, and dTTP, all with relatively high affinity. The activity site acts essentially as an on–off switch, affecting all four activities of the enzyme, while nucleotide binding in the specificity site acts to fine-tune individual activities of the enzyme, so that the

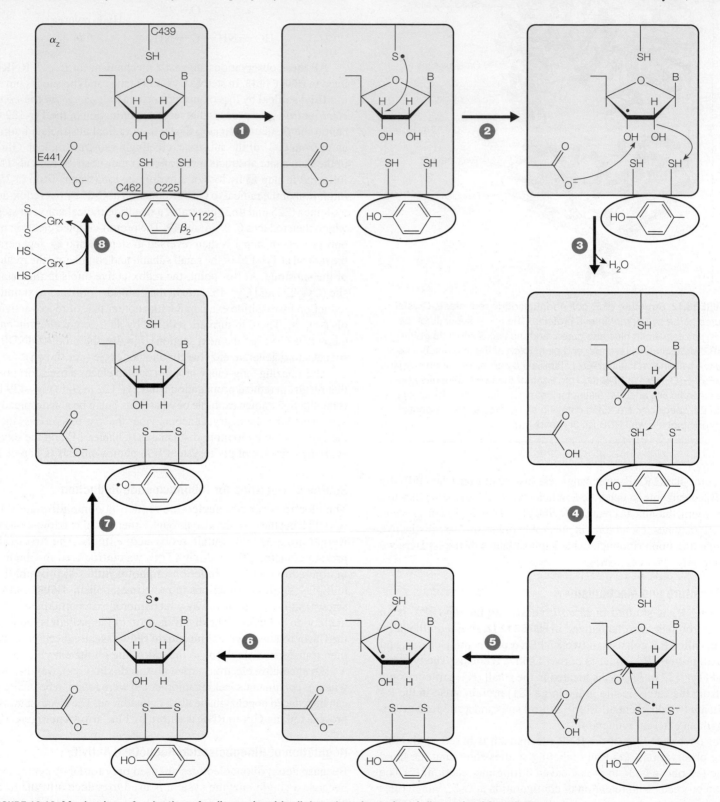

▲ **FIGURE 19.13 Mechanism of reduction of a ribonucleoside diphosphate by a class I ribonucleotide reductase.**
Blue represents the catalytic site in R1, and light yellow represents the tyrosyl radical in R2. Residue numbering is for the *E. coli* RNR. B is any nucleobase.

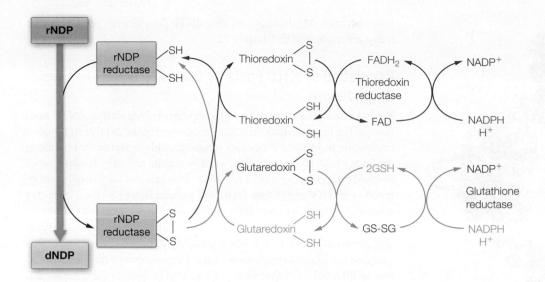

◄ **FIGURE 19.14 Reductive electron transport sequences in the action of ribonucleotide reductase.** The enzyme glutathione reductase was introduced in Chapter 12. Either thioredoxin or glutaredoxin in the dithiol form can reduce the oxidized form of ribonucleotide reductase.

four deoxyribonucleotides are produced at rates corresponding to the base composition of the DNA of the organism. As an example of the fine tuning, dTTP bound in the specificity site stimulates reduction of GDP and inhibits reduction of CDP and UDP. In this case, a pyrimidine is helping to maintain a balance by increasing purine production, while blocking the synthesis of more pyrimidine dNTPs. **TABLE 19.1** summarizes the regulatory effects.

Structural relationships among the allosteric and catalytic sites are shown in **FIGURE 19.15**, which is a schematic model of α_2, the large subunit. The specificity sites are located relatively close to the catalytic sites, and they transmit information to the catalytic site through a flexible domain called loop 2. The activity sites are located in an N-terminal domain called an ATP cone.

As noted earlier, the active form of RNR is a dimer of α_2 and β_2 subunits. Regulation at the activity site involves a novel allosteric process, as shown in **FIGURE 19.16** for the *E. coli* enzyme. In the presence of dATP bound at the activity site, the enzyme forms an $\alpha_4\beta_4$ tetramer. This can be visualized in the electron microscope, and several biophysical investigations establish that the tetramer has an $\alpha_2\beta_2\alpha_2\beta_2$ ring structure. In that structure, the ATP-cone domain of α_2 buries a portion of the β_2 subunit, which effectively blocks the long-range proton-coupled electron transport chain essential to initiate catalysis. ATP, which does not promote formation of the tetramer, competes with dATP for binding at the activity site and, hence, activates the enzyme

when it is present at sufficient concentrations. Human RNR behaves differently; addition of dATP to active enzyme creates an inactive α_6 circular hexamer.

Inhibition of rNDP reductase by dATP helps to explain the immune deficiency syndrome that arises from a mutation affecting adenosine deaminase (page 619). Proliferation of antibody-forming white blood

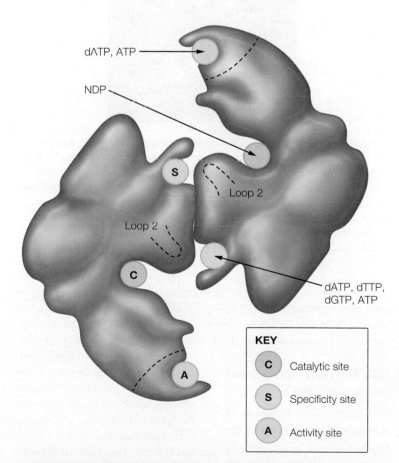

▲ **FIGURE 19.15 Structural relationships among catalytic and allosteric regulatory sites in the RNR large subunit.**

TABLE 19.1 Regulation of the activities of ribonucleotide reductase

Nucleotide Bound in			
Activity Site	Specificity Site	Activates Reduction of	Inhibits Reduction of
ATP	ATP or dATP	CDP, UDP	
ATP	dTTP	GDP	CDP, UDP
ATP	dGTP	ADP	CDP, UDP[a]
dATP	Any effector		ADP, GDP, CDP, UDP

[a]dGTP binding inhibits the reduction of pyrimidine nucleotides by the mammalian enzyme but not by the *E. coli* enzyme.

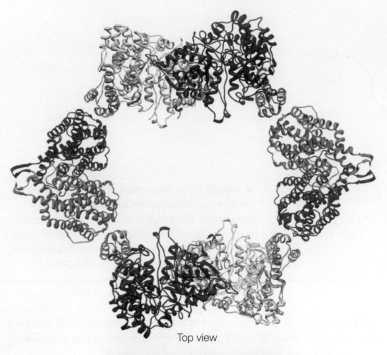

Top view

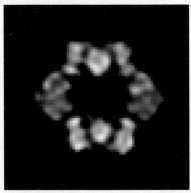

▲ **FIGURE 19.16 Structural basis for RNR inhibition by dATP.** Above, a top view of an $\alpha_4\beta_4$ tetramer formed in the presence of 50 μM dATP. Each of the two dimeric R1 proteins shows two α subunits, one in light blue and one in dark blue, while each of the two R2 dimers shows one red and one orange β_2 subunit. Regions that bind dATP and interact with β are green. Below is an image-processed representation of the tetramer as viewed in the electron microscope.

cells is essential in mounting an immune response. Accumulation of dATP when the conversion of deoxyadenosine to deoxyinosine is blocked inhibits DNA replication by depriving DNA polymerase of three of its essential precursors, so that cells of the immune system cannot proliferate.

● **CONNECTION** Ribonucleotide reductase inhibition as a result of dATP accumulation caused by adenosine deaminase deficiency can explain the block to immune cell proliferation in severe combined immunodeficiency disease.

The importance of RNR regulation can be seen in the effects of mutations affecting the allosteric sites. Both in bacteria and in mammalian cells, mutations affecting either activity-site or specificity-site control lead to abnormal dNTP pool sizes in the mutant cells, and these conditions often increase spontaneous

mutation rates. Mechanisms relating dNTP pool sizes to genomic instability are discussed in Chapter 22.

Regulation of dNTP Pools by Selective dNTP Degradation

We have seen the complexity of mechanisms regulating dNTP pool sizes at the level of deoxyribonucleotide synthesis, catalyzed by ribonucleotide reductase. Processes of comparable complexity operate at the level of dNTP *degradation,* as shown quite recently. Initial studies focused on a protein called SAMDH1. This protein was initially recognized as an HIV restriction factor—a protein in white blood cells that allows them to resist virus infection. In 2011, SAMDH1 was shown to be a *deoxyribonucleoside triphosphatase,* an enzyme that hydrolyzes any common dNTP to the corresponding deoxyribonucleoside plus inorganic triphosphate. Action of the enzyme drives dNTP levels so low in infected cells that viral reverse transcription (see Chapter 22) is inhibited. More recently, the enzyme has been shown to be widely distributed in mammalian tissues and to play a role in regulating relative dNTP pool sizes. Comparable to RNR, the enzyme acts on all four common dNTPs and is controlled by allosteric mechanisms currently being investigated.

Biosynthesis of Thymine Deoxyribonucleotides

Once the four deoxyribonucleoside diphosphates (dNDPs) have been created through RNR action, three of them—dADP, dCDP, and dGDP—are converted directly to the respective dNTPs by nucleoside diphosphate kinase. dUDP can be a precursor to deoxythymidine triphosphate, as indicated in **FIGURE 19.17**. However, in most cells, most of the thymidine triphosphate that is formed comes from a pathway involving deoxycytidine nucleotides. Note that we use the terms *thymidine* and *deoxythymidine* interchangeably. That is because thymine *ribonucleotides* are not normal metabolites, so the nucleoside containing thymine and deoxyribose need not be specifically identified as a deoxyribonucleoside.

Like dADP, dCDP, and dGDP, dUDP is also a substrate for nucleoside diphosphate kinase. The resultant dUTP is rapidly cleaved to the monophosphate, dUMP, by **dUTPase,** an enzyme that helps to minimize the accumulation of uracil in DNA (see Chapter 23). dUMP, the immediate precursor to thymidine monophosphate, dTMP, is also derived from dCDP. A reaction similar to that of adenylate kinase converts dCDP to dCMP and dCTP.

$$2\ \text{dCDP} \rightleftharpoons \text{dCMP} + \text{dCTP}$$

dCTP is used directly as a DNA precursor, while dCMP undergoes hydrolytic deamination by **dCMP deaminase** to form dUMP. (In *E. coli* and some other bacteria, the deamination occurs at the triphosphate level.)

However it is formed, dUMP serves as the immediate precursor to thymidine monophosphate (dTMP) through the action of **thymidylate synthase** (TS). This enzyme transfers a one-carbon unit at the methylene oxidation level and reduces it to a methyl group as part of the overall reaction (see Chapter 18). The one-carbon donor is 5,10-methylenetetrahydrofolate, which in this unusual reaction also serves as an electron donor, reducing the methylene group to methyl and oxidizing the folate cofactor to dihydrofolate, DHF. As shown in

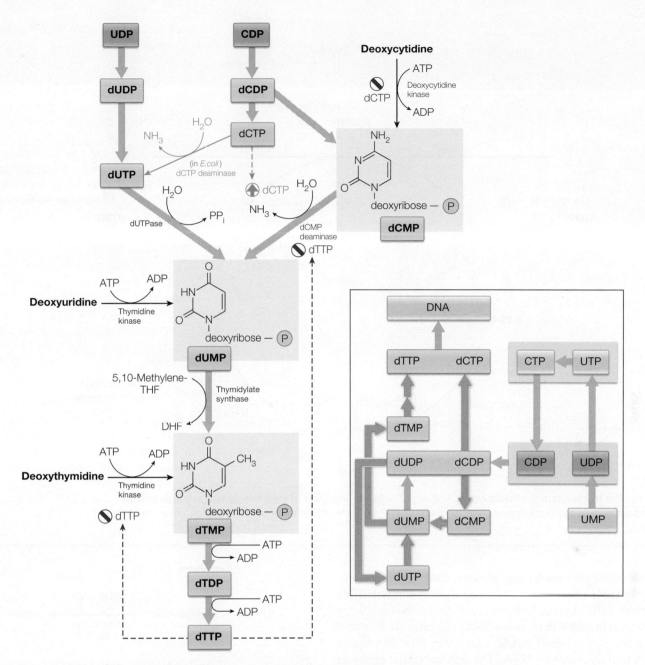

▲ **FIGURE 19.17 Salvage and de novo synthetic pathways to thymidine triphosphate.** The de novo pathways begin with UDP or CDP, shown at the top. Dashed arrows identify sites of allosteric control. The inset shows that part of the Figure 19.1 flow chart represented by this pathway.

FIGURE 19.18, the cofactor is then reduced by dihydrofolate reductase (DHFR). The resultant THF must acquire another methylene group, to begin the process anew. Most commonly, the one-carbon group arises via the serine hydroxymethyltransferase reaction. Inhibition of any one of the enzymes in this three-reaction cycle blocks the whole cycle. This is the basis for the antiproliferative effects of dihydrofolate reductase inhibitors, such as methotrexate or trimethoprim, which we mentioned in Chapter 18. In proliferating cells, inhibiting DHFR causes all reduced folate cofactors to accumulate as dihydrofolate, through this cycle, and THF needed for other reactions is not available. When TS is not active, reduced folates remain reduced. dTMP, once formed by TS, undergoes two successive phosphorylations, yielding dTTP.

Salvage Routes to Deoxyribonucleotides

● **CONNECTION** Severe genetic diseases, involving either the nervous system or muscle function, arise in humans when mitochondrial deoxyribonucleotide salvage enzymes are deficient and mitochondria in affected tissues cannot maintain adequate amounts of mitochondrial DNA.

As mentioned previously in this chapter, purine salvage pathways often involve phosphoribosyltransferase reactions and PRPP. After phosphorylation to ribonucleoside diphosphates, the purine rNDPs can enter deoxyribonucleotide metabolism via ribonucleotide reductase. Pyrimidine salvage to deoxyribonucleotides generally involves

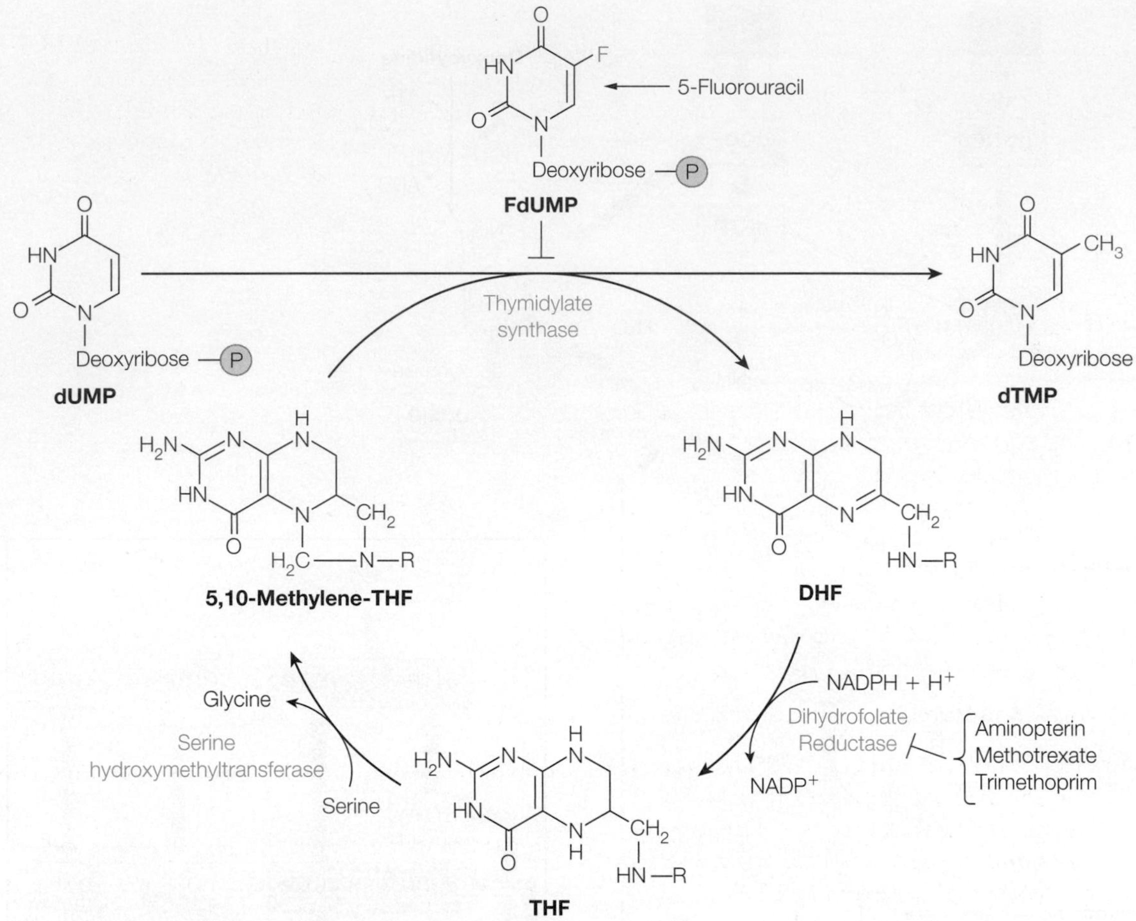

▲ **FIGURE 19.18 The thymidylate synthesis cycle.** The cyclic pathway, involving thymidylate synthase, dihydrofolate reductase, and serine hydroxymethyltransferase, converts dUMP to dTMP. Inhibition of any of the three enzymes blocks dTMP synthesis.

ATP-dependent deoxyribonucleoside kinases. Cells and organisms vary widely in their contents of ribo- and deoxyribonucleoside kinases. Human cells contain four different deoxyribonucleoside kinases. As shown in **FIGURE 19.19**, two of these enzymes are localized in the cytosol and two in mitochondria, where they provide a source of precursors to mitochondrial DNA. The mitochondrial enzymes have broad specificity, so that the two enzymes can provide sufficient pools of all four dNTPs. The importance of the mitochondrial deoxyribonucleoside kinases is underscored by a family of diseases called **mitochondrial DNA depletion syndromes.** Genetic deficiency in humans of either thymidine kinase 2 or deoxyguanosine kinase causes severe reduction in mtDNA copy number in affected tissues, leading to defective mitochondrial function and causing progressive muscle weakness or nervous system dysfunction.

Throughout the rest of this chapter, we will be mentioning nucleoside analogs that are used in treating HIV-AIDS or cancer. Invariably, these **pro-drugs** must be converted to deoxyribonucleotides to be effective, and this has focused attention on the deoxyribonucleoside kinases. While effective, sometimes pro-drugs have serious side effects. For example, a side effect of **3-azido-2′,3′-dideoxythymidine** (AZT or zidovudine), the first drug to receive approval for treating human immunodeficiency virus (HIV) infections, is cardiotoxicity—damage to the heart muscle.

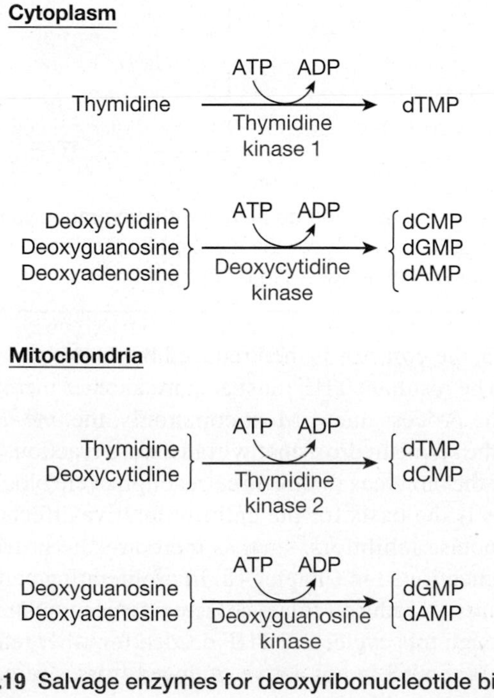

▲ **FIGURE 19.19 Salvage enzymes for deoxyribonucleotide biosynthesis in human cells.**

**3′-Azido-2′,3′-dideoxythymidine
(AZT or Zidovudine)**

AZT is converted to the triphosphate, which is an inhibitor of HIV reverse transcriptase (Chapter 22). This conversion occurs primarily in mitochondria, through the action of thymidine kinase 2. The triphosphate of AZT is also an alternative substrate for mitochondrial DNA polymerase, and this may explain its cardiotoxicity. However, AZT is also an inhibitor of thymidine kinase 2, and its effect on mitochondrial function may be related to depletion of mitochondrial dTTP pools. Current research is aimed at developing effective analogs whose metabolic activation does not occur in mitochondria.

Thymidylate Synthase: A Target Enzyme for Chemotherapy

A goal of **chemotherapy**—the treatment of diseases with chemical agents—is to exploit a biochemical difference between the pathological process and surrounding normal tissue, so as to interfere selectively with the disease process while normal tissue remains unaffected. Many therapeutically effective antimetabolites were developed by chance, through testing analogs of normal metabolites. The effectiveness of these agents is often compromised by unanticipated side effects, incomplete selectivity, and the development of resistance to the agent. An active area of current biochemical pharmacology is rational drug design—the design of specific inhibitors based on knowledge of the molecular structure of the site to which the inhibitor binds and the mechanism of action of the target molecule. For drugs whose target is an enzyme, the researcher must know the three-dimensional structure of the enzyme and its mechanism of action. Obtaining this information involves a fusion of bioorganic chemistry, structural biology, and site-directed mutagenesis. Thymidylate synthase (TS) provides an excellent example of the utility of these approaches.

● **CONCEPT** Inhibition of thymidylate synthase is an approach to cancer chemotherapy, by causing specific inhibition of DNA synthesis.

Because TS participates only in the synthesis of a deoxyribonucleotide, any disease that involves uncontrolled cell proliferation can in principle be treated with a TS inhibitor. Blocking the production of a DNA precursor should inhibit DNA replication with minimal effects on cells that are not dividing. Thus, cancer and a range of infectious diseases should be amenable to treatment by TS inhibitors.

None of this was recognized in the mid-1950s. In fact, TS had not yet been discovered. It was known, however, that certain tumor cells took up and metabolized uracil much more rapidly than normal cells. Without knowing the metabolic fates of uracil in detail, Charles Heidelberger hoped to kill tumor cells selectively by treatment with analogs that would block uracil metabolism in tumor cells. To that end, he synthesized **5-fluorouracil** (FUra) and its deoxyribonucleoside,

5-fluorodeoxyuridine (FdUrd). Both compounds were found to be potent inhibitors of DNA synthesis. FUra and FdUrd are pro-drugs—their action as inhibitors requires their intracellular conversion to **5-fluorodeoxyuridine monophosphate** (FdUMP), a dUMP analog that exerts biological activity, as an irreversible inhibitor of TS.

**5-Fluorouracil
(FUra)**

Salvage pathways

**5-Fluorodeoxyuridine
monophosphate
(FdUMP)**

**5-Fluorodeoxyuridine
(FdUrd)**

Both fluorouracil and fluorodeoxyuridine are used in cancer treatment. However, these antimetabolites are not completely selective in their effects. For example, fluorouracil can be incorporated into RNA by salvage routes normally used for uracil, thereby interfering with transcription in both cancer and normal cells. A detailed understanding of the active site of TS could lead to the design of completely specific enzyme inhibitors.

Analysis of the binding of FdUMP to TS has opened the door to understanding the enzyme's reaction mechanism and the structure of the active site. FdUMP is a true **mechanism-based inhibitor** in that irreversible binding occurs only in the presence of the other substrate, 5,10-methylenetetrahydrofolate. Binding of the coenzyme induces a conformational change in the active site that duplicates early steps in the catalytic reaction and leads to irreversible FdUMP binding. Proteolytic digestion of the ternary complex containing FdUMP, methylene-THF, and enzyme led to isolation of a peptide fragment of the enzyme containing both the inhibitor and the coenzyme.

Methylene-THF
(partial structure)

**Ternary complex between FdUMP,
5,10-methylene-THF, and
thymidylate synthase**

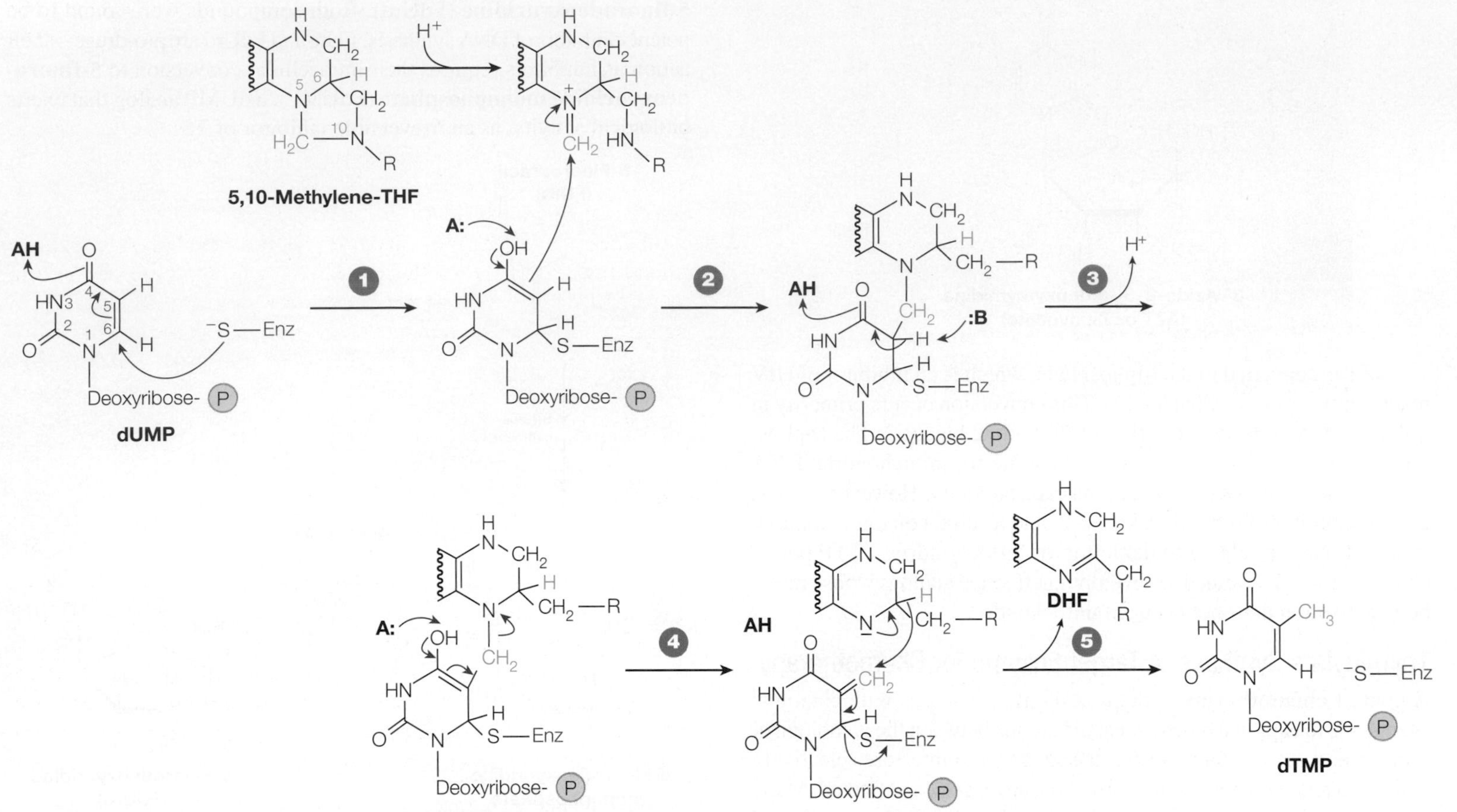

▲ **FIGURE 19.20 Mechanism of the reaction catalyzed by thymidylate synthase.** A and B represent an active site general acid and base, respectively. DHF, dihydrofolate; THF, tetrahydrofolate.

Eventually it was shown that FdUMP was linked to the methylene carbon of the THF coenzyme through C-5 of the pyrimidine ring and to the enzyme through a cysteine sulfur covalently bonded to C-6 of the pyrimidine. The structure of this ternary complex suggested that the enzymatic reaction begins with nucleophilic attack by the cysteine thiol on C-6 of the dUMP substrate, leading to the proposed mechanism shown in **FIGURE 19.20**. We don't describe the mechanism in detail here, but one point is of crucial importance. In step ❹ a hydrogen atom is transferred as a hydride ion from the reduced ring of tetrahydrofolate to the pyrimidine ring of dUMP. When tetrahydrofolate was prepared with tritium at C-6, the label was transferred quantitatively to dUMP, showing how methylene-THF serves as both a one-carbon donor and a reductive cofactor. This mechanism helps to explain the selective antiproliferative properties of dihydrofolate reductase inhibitors, such as methotrexate.

Crystallization of *E. coli* TS as a complex with FdUMP and 5,10-methylene-THF provided strong support for the proposed TS mechanism (**FIGURE 19.21**). Analysis of these complexes confirmed the conformations of substrate and cofactor bound in the active site and the identification of active-site residues in contact with these ligands. In these ternary complex structures, FdUMP is covalently bound to cysteine 146 (*E. coli* numbering) through C-6 and to the folate cofactor through C-5. Tyrosine 94 is thought to be the general base that abstracts the proton from C-5 in step ❸ (Figure 19.20), and glutamate 58 is the general acid that transfers a proton to and from the pyrimidine ring in steps ❶ through ❹.

The crystal structure of the human enzyme was reported in 1991, and this enzyme has been the focus of further drug development. Analysis of substrate-binding interactions has led to design and synthesis of folate cofactor analogs (**antifolates**) that compete effectively with 5,10-methylene-THF for binding to TS (K_i values as low as 0.4 nM). Three such inhibitors are shown here.

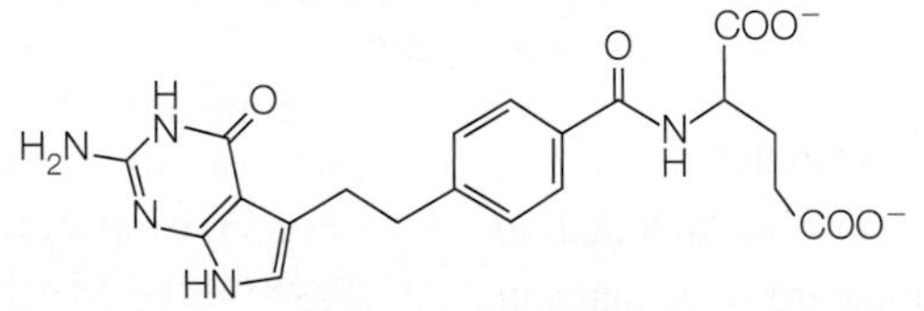

Pemetrexed (Alimta)

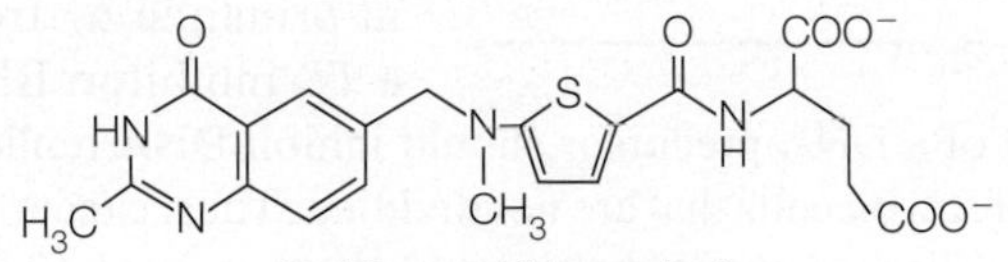

Raltitrexed (Tomudex)

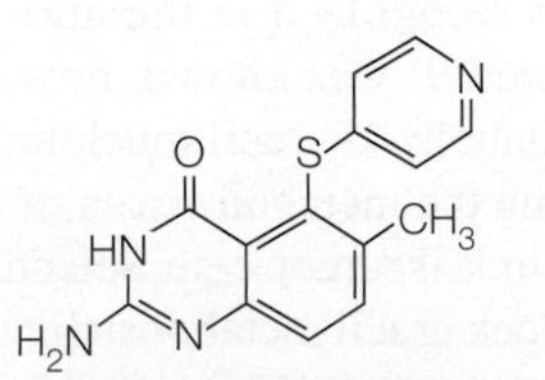

Nolatrexed (Thymitaq)

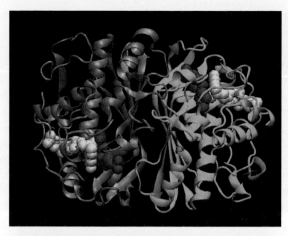

(a) Structure of the homodimer with bound folate cofactor (yellow) and FdUMP (red).

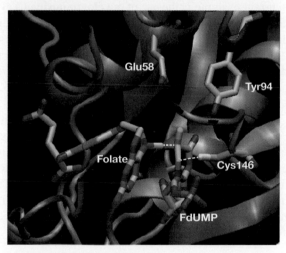

(b) Active-site region of the orange subunit in (a) showing the ternary complex formed between 5,10-methylene-THF, F-dUMP, and Cys-146. The dashed lines indicate covalent bonds of the ternary complex. The close proximity of Glu 58 (general acid) and Tyr-94 (general base) is also evident.

▲ **FIGURE 19.21 Structure of the homodimeric thymidylate synthase from *E. coli.*** (PDB ID: 1tls and 1tsn)

● **CONNECTION** Useful drugs are being developed as a result of detailed models of the active sites and mechanisms of target enzymes.

Pemetrexed and raltitrexed are both used clinically to treat certain types of cancer. The same type of approach is underway in dozens of laboratories and companies, focused on drug targets that include membrane-bound or intracellular receptor proteins and nucleic acids, as well as enzymes. The development of specific HIV protease inhibitors as anti-AIDS drugs is an example of the utility of this approach.

Diseases other than cancer are also susceptible to attack by inhibition of thymidylate synthase. For example, parasitic protozoans, such as those that cause malaria, synthesize an unusual form of TS—a bifunctional enzyme, with both thymidylate synthase and dihydrofolate reductase activities. Knowing the structure of this

enzyme's active site should allow development of inhibitors that would block this novel enzyme but not thymidylate synthase of the animal or human host.

An additional opportunity to exploit TS as a chemotherapeutic target came to light in 2002, when several microorganisms were found to contain a completely different form of thymidylate synthase, which shows no sequence homology with "classical" TS. The novel enzyme uses 5,10-methylene-THF as the source of the dTMP methyl group, but it does not oxidize the cofactor to DHF. Instead, the electrons to reduce the methylene to methyl come from NADPH, which reduces an enzyme-bound flavin cofactor, FAD. About 30% of known microorganisms, including several human pathogens, contain this novel flavin-dependent TS. Investigators are trying to develop specific inhibitors of the novel enzyme, which would not affect human TS and hence, could serve as effective antibiotics.

19.6 Virus-Directed Alterations of Nucleotide Metabolism

That viruses can redirect the metabolism of their host cells first came to light in 1957, through studies of nucleotide biosynthesis in *E. coli* bacteria infected with bacteriophage T4 and its close relatives, T2 and T6. Gerry Wyatt and Seymour Cohen had shown in 1952 that the DNA of these bacterial viruses contains no cytosine but instead contains **5-hydroxymethylcytosine,** with most of the hydroxymethyl groups further modified by glycosylation with glucose.

Cytosine   **5-Hydroxy-methylcytosine**   **α-Glucosyl-5-hydroxy-methylcytosine**

The cytosine modifications were shown to occur at the nucleotide level, through the action of virus-coded enzymes (**FIGURE 19.22**). The principal enzymes in bacteria infected by phages T2, T4, and T6 include a **dCTPase,** which cleaves dCTP to dCMP; a **dCMP hydroxymethyltransferase,** which transfers a one-carbon group from 5,10-methylene-THF to dCMP without reducing that single-carbon group; and a **deoxyribonucleoside monophosphate kinase,** which can phosphorylate 5-hydroxymethyl-dCMP and also phosphorylates dTMP and dGMP. dCMP hydroxymethyltransferase uses the same mechanism as thymidylate synthase, but it does not reduce the methylene group. Phosphorylation of 5-hydroxymethyl-dCDP to the triphosphate is catalyzed by nucleoside diphosphate kinase of the host cell. The glucosylation reactions occur after the modified nucleotide has been incorporated into DNA. The T4 phage genome encodes two UDP-glucose-dependent **glucosyltransferase** enzymes, one of which transfers glucose in the α configuration and one in the β. In addition, the viral genome encodes nucleases that specifically attack unmodified, cytosine-containing DNA. This process helps to abolish expression of host-cell genes, and it also provides a source of precursors for viral DNA synthesis. In 2009, 5-hydroxymethylcytosine

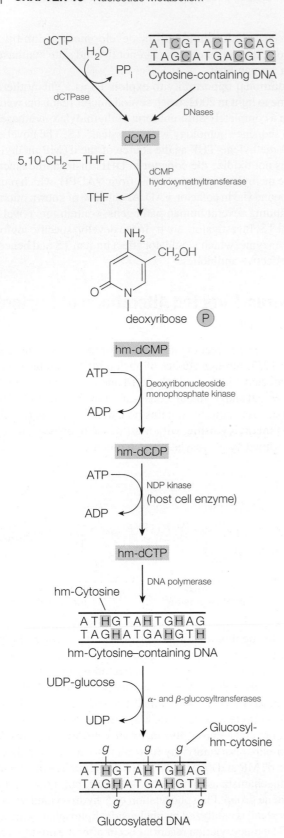

▲ **FIGURE 19.22 Metabolic pathways leading to nucleotide modifications in *E. coli* infected by phages T2, T4, or T6.** Virus-coded enzymes are shown in red. hm, hydroxymethyl.

was discovered as a minor component of eukaryotic DNAs, where it is synthesized by a completely different route and plays a different function (Chapter 26).

Although base modifications are unusual, several other bacteriophages have similar modifications. One group of phages infecting *Bacillus subtilis* contains uracil substituted for thymine; another contains 5-hydroxymethyluracil. A phage infecting *Xanthomonas oryzae* contains 5-methylcytosine completely substituted for cytosine. In each of these cases, the virus creates the DNA modifications through synthesis of virus-coded enzymes that create novel metabolic pathways in the cells that they infect, just as is shown here for T4 and its relatives.

Although plant and animal viruses do not contain extensive base modifications, virus-coded enzymes are often produced to help the infected cell augment its synthesis of nucleic acid precursors. If the virus-coded enzyme differs sufficiently from its host-cell counterpart, it can be used as a chemotherapeutic target to treat viral infections. The best current example is the use of **acycloguanosine (Acyclovir)** and its relative **Ganciclovir** in treating herpes virus infections. The herpes viruses, such as herpes simplex, are large DNA viruses whose

● **CONNECTION** Virus-coded enzymes with expanded substrate specificity can serve as targets for chemotherapy of viral diseases.

genomes encode several novel enzymes, including thymidine kinase (HSV-TK). This virus-coded enzyme differs from mammalian cell TK in having extremely broad substrate specificity, extending to purine nucleosides as well as thymidine. Nucleoside analogs like Acyclovir and Ganciclovir are readily phosphorylated by HSV-TK, and the resultant triphosphates interfere with viral DNA replication. Uninfected cells, being unable to phosphorylate the analog, are unaffected.

Acycloguanosine (Acyclovir)   **Ganciclovir**

19.7 Other Medically Useful Analogs

As we are seeing, nucleotide metabolism provides a variety of biochemical targets for disease treatment and prevention. Some, like 5-fluorouracil, arose through serendipity. Others, like Pemetrexed, were developed in rational fashion. An excellent example of the former is **6-mercaptopurine** (6-MP), synthesized in the late 1940s by Gertrude Elion and George Hitchings in the hope that it might interfere with nucleic acid metabolism in a useful way. Indeed it does, and 6-MP is still one of the most useful drugs in treating some forms of leukemia. The analog is an excellent substrate for HGPRT, and the resultant thiolated purine nucleotides interfere with nucleic acid metabolism in several ways. The Elion–Hitchings partnership was extremely fruitful. They and their colleagues at the pharmaceutical company Burroughs-Wellcome developed several other drugs we have mentioned, including allopurinol, azidothymidine, and acyclovir. Elion and Hitchings were awarded a share of the 1988 Nobel Prize in Physiology or Medicine.

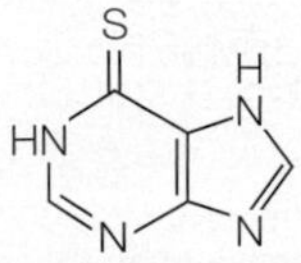

S
6-Mercaptopurine

 CONNECTION HIV infections can be treated with pyrimidine nucleoside analogs that can be converted to dNTPs, but that prevent replication after incorporation into DNA because of the absence of a 3′ hydroxyl terminus for chain extension.

Another family of nucleoside analogs, following the success of AZT in treating HIV-AIDS, was also synthesized as pro-drugs—analogs that could be metabolized by normal kinases to the triphosphate level and that would then interfere with viral DNA replication. These analogs, including **2′,3′-dideoxycytidine** (ddC), **2′,3′-dideoxyinosine** (ddI), **3′-thiacytidine** (3TC), and **2′,3′-didehydro-3′-deoxythymidine** (d4T), can all be incorporated, as the corresponding nucleotide, into DNA, but after incorporation will block chain extension because of the absence of a 3′ hydroxyl group to which the next nucleotide

can be attached. All four analogs have been approved for treatment of HIV-AIDS, and both AZT and 3TC are components, along with HIV protease inhibitors, of the three-drug "cocktails" responsible for long-term remissions of HIV infections.

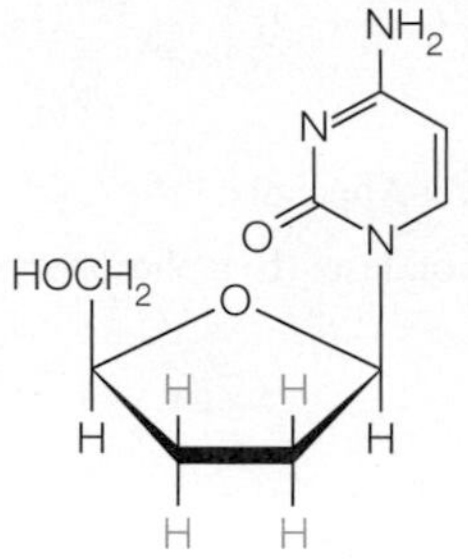

2′,3′-Dideoxycytidine (ddC)

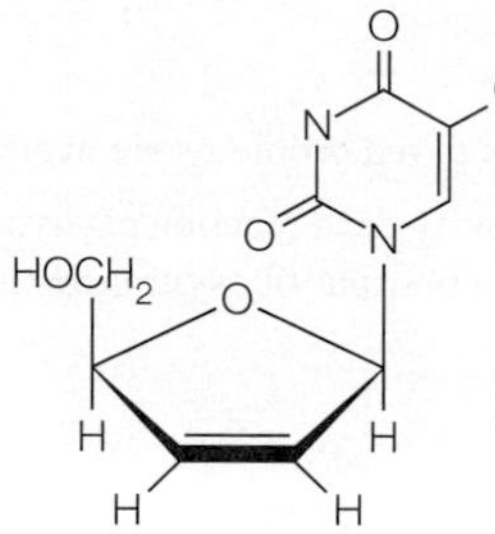

2′,3′-Didehydro-3′-deoxythymidine (d4T)

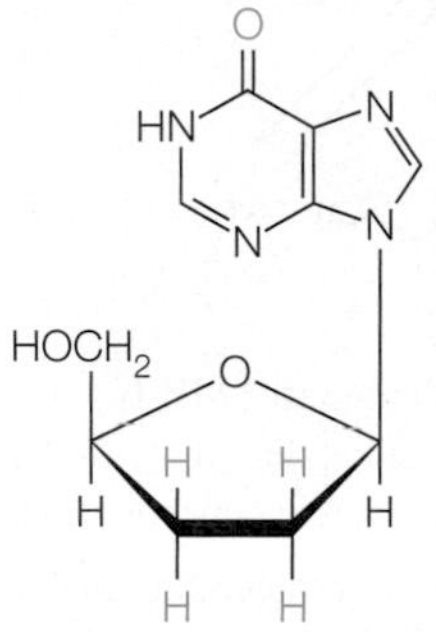

2′,3′-Dideoxyinosine (ddI)

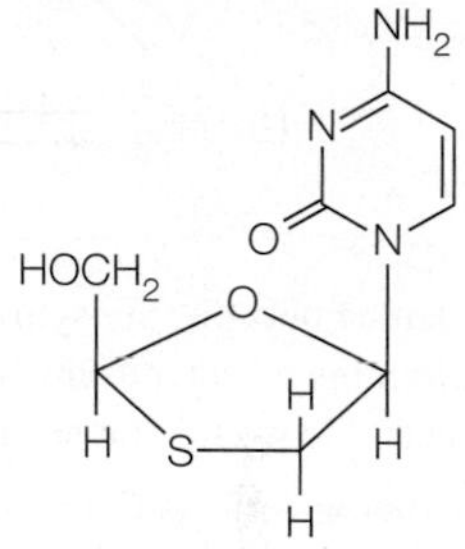

3′-Thiacytidine (3TC)

Summary

- Purine and pyrimidine nucleotides arise within cells from nucleic acid breakdown, from reuse (or **salvage**) of preformed nucleosides or nucleobases, or from **de novo** biosynthesis. (Sections 19.1 and 19.2)

- Purine nucleotides are formed at the nucleotide level, in a 10-step pathway that leads from **PRPP** to **inosine monophosphate (IMP).** Beyond the branch point (formation of IMP), separate pathways lead to adenine and guanine nucleotides. Purine catabolism yields uric acid, an insoluble compound that is formed in excess in a variety of disease states. (Section 19.2)

- Pyrimidines are synthesized at the base level, with conversion to a nucleotide occurring late in the pathway. An unbranched pathway leads to both UTP and CTP. (Section 19.4)

- In most organisms, the ribonucleoside diphosphates are substrates for reduction of the ribose sugar in situ, yielding deoxyribonucleoside diphosphates, which in turn lead to the four dNTP DNA precursors. **Ribonucleotide reductase** is an important control site, inasmuch as it represents the first metabolic reaction committed to DNA synthesis. (Section 19.5)

- Biosynthesis of thymine nucleotides involves transfer of the methylene group of 5,10-methylenetetrahydrofolate to a deoxyuridine nucleotide, followed by reduction of the methylene group. Reactions of deoxyribonucleotide biosynthesis are target sites for enzyme inhibitors that have found use as anticancer, antimicrobial, antiviral, and antiparasitic drugs. (Sections 19.6, 19.7)

Problems

Enhanced by
Mastering Chemistry
for Biochemistry

Mastering Chemistry for Biochemistry provides select end-of-chapter problems and feedback-enriched tutorial problems, animations, and interactive figures to deepen your understanding of complex topics while practicing problem solving.

Answers to red problems are available in the Answer Appendix.

1. Identify each reaction catalyzed by (a) a nucleotidase; (b) a phosphorylase; (c) a phosphoribosyltransferase.

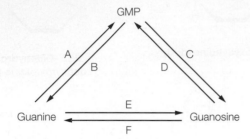

2. Adenine phosphoribosyltransferase converts adenine to AMP. If you were to determine whether a similar reaction converts adenine directly to dAMP, what metabolite would you need to find in cells at appreciable concentrations?

3. Predict the effects of the following compounds on intracellular nucleoside triphosphate levels. For each answer, plot percent initial nucleotide level as a function of time after administration of the agent, for each of the four nucleotides (semiquantitatively). Consider not only the primary effect of the compound but also any indirect effects on allosteric enzymes caused by nucleotide accumulation or depletion.
   (a) Effect of thymidine on dNTP levels
   (b) Effect of trimethoprim on bacterial rNTP levels
   (c) Effect of fluorodeoxyuridine on dNTP levels
   (d) Effect of hydroxyurea on dNTP levels

4. Radioactive uracil can be used to label all of the pyrimidine residues in DNA. Using either names or structures, present pathways for the conversion of uracil to dTTP and to dCTP. For each reaction, show the involvement of cofactors, and identify sites of allosteric regulation.

5. Similarly, hypoxanthine (HX) can be used to label purine residues. As in Problem 4, write reactions showing the conversion of hypoxanthine to dATP and dGTP.

6. Leukemia is a neoplastic (cancerous) proliferation of white blood cells. Clinicians are currently testing deoxycoformycin, an adenosine deaminase inhibitor, as a possible antileukemic agent. Why might one expect this therapy to be effective?

7. The text states that a side effect of 5-fluorouracil therapy is its incorporation into RNA. Show pathways by which 5-fluorouracil could be converted both to FdUMP and to an immediate RNA precursor.

8. A classic way to isolate thymidylate synthase–negative mutants of bacteria is to treat a growing culture with thymidine and trimethoprim. Most of the cells are killed, and the survivors are greatly enriched in thymidylate synthase–negative mutants.
   (a) What phenotype would allow you to identify these mutants?
   (b) What is the biochemical rationale for the selection? (That is, why are the mutants not killed under these conditions?)
   (c) How would the procedure need to be modified to select mammalian cell mutants defective in thymidylate synthase?

9. As stated in Chapter 23, mammalian cells can become resistant to the lethal action of methotrexate by the selective survival of cells containing increases in dihydrofolate reductase gene copy number, so that intracellular levels of the enzyme become very high. What other biochemical or genetic changes in cells could cause them to become resistant to methotrexate?

10. As stated in the text, bacteriophages have been discovered with the following base substitutions in their DNA:
   (a) dUMP completely substituting for dTMP
   (b) 5-hydroxymethyl-dUMP completely substituting for dTMP
   (c) 5-methyl-dCMP completely substituting for dCMP

   For any one of these cases, formulate a set of virus-coded enzyme activities that could lead to the observed substitution. Write a balanced equation for each reaction you propose.

11. 6-Diazo-5-oxonorleucine (DON) irreversibly inhibits glutamine-dependent amidotransferases.

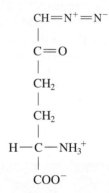

   (a) Which intermediate or intermediates in purine nucleotide biosynthesis would you expect to accumulate in DON-treated cells?
   (b) Which participants in pyrimidine synthesis would you expect to accumulate?
   (c) Speculate on the mechanism by which DON inhibits these enzymes.

12. Consider either the purinosome or a multifunctional protein such as the CAD protein, and propose one or two experiments to determine whether the enzyme "channels" substrates through a multistep reaction sequence. Define the term *metabolite channeling*.

13. Write balanced equations for the three known reactions that transfer an amino group to a substrate by condensation with aspartate to give an intermediate that then undergoes an α,β-elimination to give the product plus fumarate.

14. CTP synthetase catalyzes the glutamine-dependent conversion of UTP to CTP. The enzyme is allosterically inhibited by the product, CTP. Mammalian cells defective in this allosteric inhibition are found to have a complex phenotype: They require thymidine in the growth medium, they have unbalanced nucleotide pools, and they have an elevated spontaneous mutation rate. Explain the likely basis for these observations.

15. The purinosome contains enzymes that convert the serine hydroxymethyl group to the formyl group of 10-formyltetrahydrofolate. Write a balanced equation for each reaction in this conversion.

16. Sulfonamide drugs like sulfanilamide inhibit tetrahydrofolate biosynthesis. What intermediate or intermediates in purine synthesis would you expect to accumulate in sulfanilamide-treated bacteria? Suppose that a cultured mammalian cell line was treated with sulfanilamide. What intermediates might accumulate in these cells?

17. The text states that ATP is synthesized primarily by energy metabolism, whereas other nucleoside triphosphates are formed from the action of nucleoside diphosphate kinase. What additional pathway exists for GTP synthesis?

18. The text describes a form of gout that results from HGPRT deficiency. Propose one or two additional enzyme abnormalities that might similarly lead to hyperuricemia.

19. CTP synthetase is allosterically activated by GTP. What function might this play in the cell?

20. As stated in the text, adenosine deaminase deficiency can be treated by injection of the stabilized enzyme. Why might this treatment be effective, while injection of the missing enzyme is ineffective in other conditions, such as HGPRT deficiency or a deficiency of one of the mitochondrial deoxyribonucleoside kinases?

21. Consider the large number of therapeutically effective nucleoside analogs. Why do you suppose that these drugs are administered as nucleosides, rather than as nucleotides, which would be more readily converted to the active form of the drug?

22. In the pyrimidine degradative pathway, all pyrimidines undergo conversion to uracil, which undergoes an NADPH-dependent reduction. Show plausible reactions leading from cytidine, cytosine, and uridine to uracil, and write a plausible structure for the product of uracil reduction (dihydrouracil).

23. The text states that in *E. coli* ribonucleotide reductase Tyrosine-122 was identified as the source of the catalytically essential free radical. Describe experimental evidence that would support this conclusion.

24. Cell cultures can be *synchronized*, or brought into the same phase of the cell cycle, by various means. For example, adding thymidine to a cell culture causes all cells to become arrested early in S phase. What is the mechanism by which thymidine treatment blocks DNA replication?

25. Oxidation of DNA bases is mutagenic. An oxidized form of guanine, 8-oxoguanine, can base-pair with adenine, leading to errors in DNA replication. A gene called *mutT* acts to minimize the tendency of A-oxoG base pairs to form. The *mutT* gene product was purified and shown to be an enzyme. Speculate on the reaction catalyzed by this enzyme.

26. The paper cited in References by Franzolin et al. (2013) presents evidence that dNTP pool sizes are controlled not only at the level of dNTP *synthesis* (RNR), but also at the level of dNTP *breakdown*, the latter catalyzed by an enzyme called SAMHD1. Speculate on the reaction catalyzed by SAMHD1. Speculate on allosteric control mechanisms that might regulate SAMHD1 activity.

27. Gemcitabine is a deoxycytidine analog used in treating pancreatic and ovarian cancer. The drug requires metabolic activation after uptake into cells. Speculate on the pathway for activation and the identity of the target enzyme.

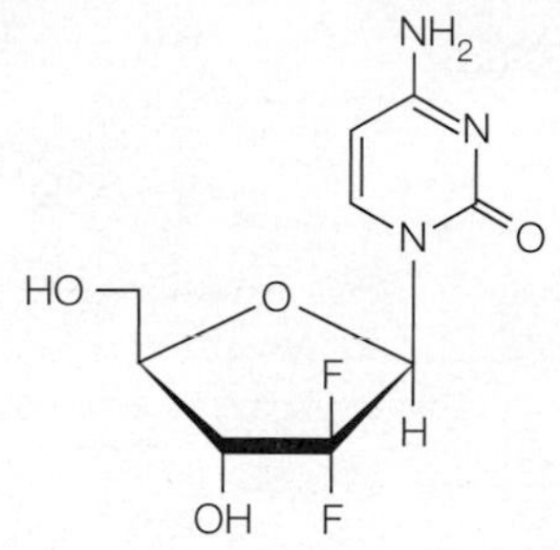

References

For a list of references related to this chapter, see Appendix II.

Human prostate cancer cells, visualized by scanning electron micros-copy. Cancer often results from mutations that change proteins involved in signaling. About half of all human tumors contain mutations affecting p53, a protein that has been called a "guardian of the genome." Why is this an appropriate description?

Mechanisms of Signal Transduction

20

IN PREVIOUS CHAPTERS—particularly Chapters 11, 12, and 17—we have discussed the actions of hormones in regulating metabolism. In this chapter, we focus on *mechanisms* of hormone action, and we will see that hormones represent but one class of extracellular chemical messengers. Hormones are molecules that are released from controller cells in the endocrine system and travel to target cells, usually some distance away, where they interact with specific receptors in the target cell. This process is known as *signal transduction.* In a broader sense, signal transduction refers to reception of an environmental stimulus by a cell, leading to metabolic change that adapts the cell to that stimulus. The interaction of a hormone with its receptor stimulates or inhibits specific metabolic events within the target cell. The examples described in earlier chapters involve the effects of epinephrine and glucagon on metabolic cascades controlling the breakdown or synthesis of glycogen in animal tissues. In fact, these processes yielded our earliest understanding of molecular mechanisms in signal transduction, and these hormones, together with insulin, will receive the most attention here.

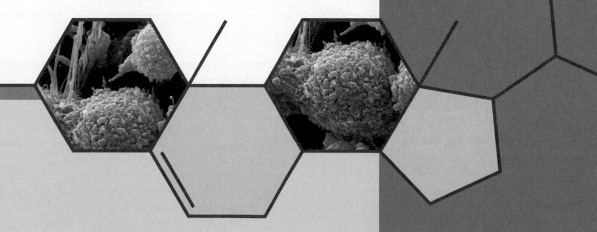

Early research on hormones in animals revealed little about how they act, but it did show fundamental similarities among different hormones. First, they are secreted by specific tissues, the **endocrine glands.** Second, they are secreted directly into the bloodstream rather than being excreted through ducts or stored in bladders. Thus, the response to a hormonal signal comes as a direct and rapid result of its secretion. **FIGURE 20.1** shows the locations of the major endocrine organs in the human body.

Hormones usually stimulate metabolic activities in tissues remote from the secretory organ. They are active at exceedingly low concentrations, in the micromolar to picomolar ranges. Furthermore, many hormones are metabolized rapidly, so their effects are often short-lived, allowing rapid adaptations to metabolic changes. The prostaglandins, which we introduced in Chapter 16, are also involved in signal transduction. These mediators act like hormones but are distinctive in their extreme metabolic lability, their synthesis in many cell types instead of just one endocrine gland, their lower target organ specificity, and their actions primarily on cells near those from which they were secreted.

● **CONCEPT** Signal transduction involves cell-to-cell communication, via neurotransmitters, hormones, growth factors, and pheromones.

Hormones comprise but one class of extra-cellular messengers. Other such messengers include **pheromones,** which are transmitted between cells of different individuals; **neurotransmitters,** which act immediately across a synaptic junction from their sites of release; and **growth factors,** which are continuously growth-stimulating

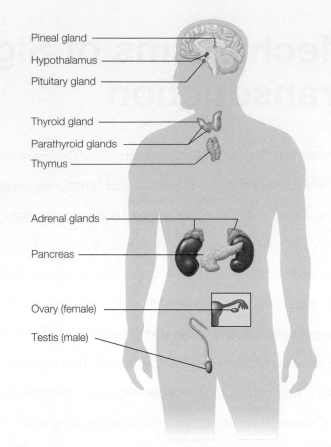

▲ **FIGURE 20.1 The major human endocrine glands and their central nervous system control centers.** Some other tissues also produce hormones, such as the lining of parts of the gastrointestinal tract.

rather than being short-lived in response to a burst of secretion. Another class of signaling agents, the **cytokines,** bind to specific receptors and stimulate cell growth and differentiation in the immune response. These agents are similar in that they are released from a controller cell and exert their effects on a target cell through interaction with specific receptors in the target cell.

20.1 An Overview of Hormone Action

Given the potent activities of hormones when secreted at minute levels, how can we understand their rapid and immediate actions? We have already mentioned in Chapter 12 the "fight-or-flight" response, in which epinephrine secretion leads to massive breakdown of muscle glycogen to glucose 1-phosphate, beginning within seconds. Until the 1950s, it was thought that hormones enter cells and act directly on target enzymes. However, investigations in the laboratories of Earl Sutherland and Edwin Krebs showed that epinephrine does *not* enter cells, as it must if a rate-limiting enzyme is to be activated directly.

Instead, as shown in Figure 12.26, epinephrine binds to a macromolecular receptor at the cell surface and stimulates the formation of cyclic AMP, which acts as a *second messenger* within the cell, influencing the phosphorylation of target enzymes. The hormone itself is the first messenger. Today we know that hormones act by binding to specific receptors, whether those receptors are located inside the target cell or on the cell surface. The distribution of specific receptors on specific cell types determines how hormones, secreted into the bloodstream, affect only specific target tissues. For example, the preferential action of glucagon in stimulating glycogenolysis in liver derives from the density of glucagon receptors on the surface of liver cells. Second messengers are often used to transmit the message to the target metabolic pathway, though not all hormone actions involve a second messenger.

Chemical Nature of Hormones and Other Signaling Agents

Chemically, the hormones in vertebrate metabolism include (1) *peptides* or polypeptides, such as insulin or glucagon; (2) *steroids,* including glucocorticoids and the sex hormones; and (3) *amino acid derivatives,* including the catecholamines, such as epinephrine, and thyroxine, derived from tyrosine. Hormonal mechanisms include (1) enzyme activation or inhibition via second messengers, as noted for epinephrine and glucagon; (2) stimulation of the synthesis of target proteins, through activation of specific genes; and (3) selective increases in the cellular uptake of certain metabolites. Among this last category are some receptors that serve directly as ion channels, with hormone binding causing a conformational change that opens the channel.

● **CONCEPT** Hormone action can influence (1) enzyme activity (via second messengers), (2) the synthesis of specific proteins, or (3) membrane permeability to ions or small metabolites.

The receptor to which a hormone binds may be located either in the plasma membrane or inside the cell. Most hormones interacting with *intracellular receptors* (also called nuclear receptors) exert their effects at the gene level. The hormone–receptor complex migrates to the nucleus, where it interacts with specific DNA sites and affects rates of transcription of neighboring genes. These hormones include steroids, thyroid hormones, and the hormonal forms of vitamin D. In addition, **retinoids,** derived from retinoic acid (related to vitamin A), exert regulatory effects in embryonic development, through interactions with intracellular receptors (see Chapter 16).

We recognize three major classes of *membrane-bound receptors.* The first class interacts with G proteins and influences the synthesis of second messengers, like those introduced in Chapter 12. The second class includes ion channels—comparable to the nicotinic acetylcholine receptor (page 657). Peptide hormones and epinephrine act primarily through these two classes of receptors. The third, exemplified by the insulin receptor, is a transmembrane protein with a ligand-binding site on the extracellular side and a catalytic domain on the cytosolic side. In receptors of this class, that catalyst is a protein kinase, which is stimulated by hormone binding to the extracellular domain to phosphorylate tyrosine residues, or occasionally serine or threonine, on target proteins.

● **CONCEPT** Membrane receptors include (1) proteins that influence second-messenger synthesis, (2) ion channels, and (3) proteins with intrinsic enzyme activity.

A major biochemical reaction in signal transduction is protein phosphorylation. This reaction was discovered in the late 1950s by Edwin Krebs and Edmond Fischer, in the sequence of reversible protein phosphorylations during the epinephrine-induced glycogenolytic cascade, as introduced in Chapter 12. At that time, no one had predicted the extent to which protein phosphorylation would turn out to dominate cell-signaling mechanisms. More than 500 different protein kinases have now been shown to exist in human cells, all of them

● **CONCEPT** The end result of many signal transduction events is the phosphorylation or dephosphorylation of target proteins.

related in terms of amino acid sequence. The importance of protein phosphorylation was formally acknowledged with award of the 1992 Nobel Prize for Medicine or Physiology to Krebs and Fischer. More recent work has uncovered a host of specific protein *phosphatases* that are also controlled by cell-signaling mechanisms. Also, as noted in Chapter 17, protein acetylation is becoming recognized as a reversible protein modification, possibly comparable to protein phosphorylation in how widely it is used.

Hierarchical Nature of Hormonal Control

Hormonal regulation involves a hierarchy of cell types acting on each other either to stimulate or to modulate the release and action of a hormone. In vertebrates the secretion of hormones from endocrine cells is stimulated by chemical signals from regulatory cells that occupy a higher position in this hierarchy. **FIGURE 20.2** summarizes the hierarchy that controls the secretion and action of glucagon, epinephrine, and insulin. Hormonal action is ultimately controlled by the central nervous system. The master coordinator in mammals is the **hypothalamus,** a specialized center of the brain. The hypothalamus receives and processes sensory inputs from the environment via the central nervous system. In response, it produces a number of hypothalamic hormones,

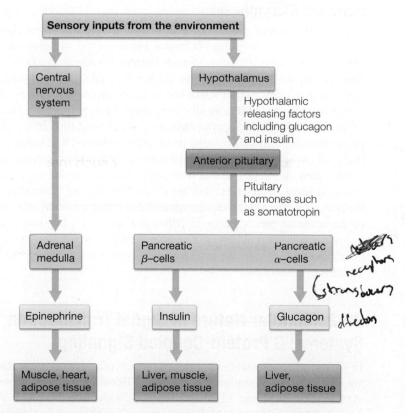

▲ **FIGURE 20.2 Hierarchical nature of hormone action in vertebrates.** The pituitary, which is under hypothalamic control, is the first target for most hormones, including glucagon and insulin. Pituitary hormones then act on secondary targets, the hormone products of which collectively influence ultimate organs and tissues. Neural stimulation of the adrenal medulla controls the release of epinephrine.

● **CONCEPT** Specific hormone-releasing factors from the hypothalamus control the release—and thus the action—of other hormones.

● **CONCEPT** The central nervous system transmits signals to the hypothalamus, which produces releasing factors that act upon endocrine glands to control the secretion of hormones with specific metabolic effects on target tissues.

some of them called **releasing factors.** These factors act on the pituitary, which is located just beneath the hypothalamus. Releasing factors stimulate the anterior portion of the pituitary to release specific hormones. Other hypothalamic hormones inhibit the secretion of particular pituitary hormones. Some pituitary hormones stimulate target tissue directly. For example, **prolactin** stimulates mammary glands to produce milk. However, most pituitary hormones act on endocrine glands that occupy an intermediate, or secondary, position in the hierarchy, stimulating them to produce hormones that exert the ultimate actions on target tissues. Pituitary hormones that act on other endocrine glands are called **tropic hormones** or **tropins.** An example is **somatotropin,** a peptide secreted from the anterior pituitary, which stimulates islet cells in the pancreas to release insulin (β cells) or glucagon (α cells), which then act on a number of tissues, including liver and muscle.

Hormone Biosynthesis

The biosyntheses of steroid hormones (Chapter 16) and of catecholamines and thyroid hormones (Chapter 18) occur via straightforward metabolic pathways. However, peptide hormone synthesis is more complex. Nearly all peptide hormones are synthesized as inactive precursors and then converted to active hormones by proteolytic processing. Recall from Chapter 5 that insulin, consisting of two polypeptide chains of 21 and 30 residues, is synthesized by processing the 105-residue preproinsulin to the 80-residue proinsulin, followed by cleavage to the final two-chain hormone. All known polypeptide hormones are synthesized in "prepro" form, with an N-terminal "signal sequence" and additional sequence(s) that are cleaved out during maturation of the hormone. The signal sequence facilitates transmembrane transport of the hormone precursor, as described in Chapter 25. In some cases, a single precursor polypeptide sequence contains two or more distinct hormones, which may be produced in different tissues by different patterns of proteolysis.

20.2 Modular Nature of Signal Transduction Systems: G Protein-Coupled Signaling

In Chapter 12 we described how glycogen synthesis and breakdown are regulated by epinephrine or glucagon. Recall that a three-protein system is used (Figure 12.26)—a *receptor,* which interacts with a *G protein* (a *transducer*), which then interacts with adenylate cyclase (an *effector*). Activation of adenylate cyclase stimulates the synthesis of cyclic AMP (the second messenger), which promotes numerous metabolic responses. In this section, we see that this three-component system is but one example of a great variety of similar modular systems; this variety gives great flexibility among cells and tissues in response to hormonal stimulation.

Receptors
Receptors as Defined by Interactions with Drugs

Before we knew anything about the molecular nature of hormone receptors, those that respond to epinephrine—**adrenergic receptors**—had been categorized pharmacologically in terms of the response of cells or tissues to epinephrine analogs. The term *adrenergic* is based on **adrenaline,** an older term for epinephrine. Epinephrine analogs, such as **isoproterenol** or **propranolol,** could be categorized as epinephrine **agonists** (agents that act similarly to epinephrine), or **antagonists,** those that block the action of epinephrine. On one hand, an agonist is an analog that binds to a receptor and stimulates a response similar to that of the natural hormone. An antagonist, on the other hand, binds the receptor but has no effect other than to block productive binding by the hormone itself.

Studies of many catecholamine agonists and antagonists originally revealed in vertebrates four types of receptors, each of which has a distinctive pattern of response to these analogs. Although several more are now recognized, the basic four are called α_1-, α_2-, β_1-, and β_2-adrenergic receptors. These receptors, when stimulated, have specific and diverse physiological effects in different tissues, some of which are summarized in **TABLE 20.1**. Propranolol is one of the class of drugs called beta-blockers. When it was introduced in 1962, long before receptors had been isolated or described biochemically, this epinephrine antagonist revolutionized the treatment of heart conditions, particularly angina pectoris. Note from Table 20.1 that stimulation of β_1 or β_2 receptors controls heartbeat rate, contractility, and expelled blood volume.

● **CONNECTION** Propranolol, the first epinephrine analog in clinical use, is one of a class of beta-blockers, which antagonize β receptors and are used to treat a variety of heart disorders.

Isoproterenol

Epinephrine

Propranolol

Receptors and Adenylate Cyclase as Distinct Components of Signal Transduction Systems

Because adenylate cyclase, which synthesizes cAMP, is located in the plasma membrane, early workers thought that the adrenergic receptor *is* adenylate cyclase. Two observations suggested otherwise. First, hormones other than epinephrine were found to activate adenylate cyclase; more than a dozen are now known, including glucagon. Adenylate cyclase seemed unlikely to have that many hormone-binding sites. Second, the binding of catecholamines to the α_2 class of receptors was found to *inhibit* adenylate cyclase, suggesting that different kinds of proteins interact with adenylate cyclase to produce distinct metabolic effects.

TABLE 20.1 Some biological actions associated with adrenergic receptors

Receptor Class	Action
α_1	Smooth muscle contraction in blood vessels and skin, gastrointestinal system, kidney, and urethral sphincter
	Increased sweat gland secretion
α_2	Decreased glucagon and insulin release from pancreas
	Contraction of sphincters in the gastrointestinal tract
β_1	Increased heart rate, contraction, and expelled fluid volume
	Increased renin secretion from kidney
β_2	Smooth muscle relaxation in gastrointestinal tract, lung bronchia
	Increased lipolysis
	Increased glycogenolysis and gluconeogenesis
	Decreased histamine release from mast cells
	Increased anabolism in skeletal muscle

Both of these observations indicated that the receptor and adenylate cyclase are distinct proteins. In fact, the resolution of β-adrenergic receptors from adenylate cyclase, observed experimentally in 1977, was important because it showed this hormonal response system to be far more flexible and versatile than previously thought. A variety of hormones could exert a multitude of biological effects through a common mechanism, namely, activation or inhibition of cAMP synthesis. The diversity of signals and responses was built into both the diversity of receptors and the diversity of enzymes in target cells whose activities could be increased or inhibited by cAMP-stimulated phosphorylation. It was soon learned that transduction of the hormonal signal to adenylate

● **CONCEPT** Hormones that act through second messengers involve a three-protein module—receptor, transducer (G protein), and effector (adenylate cyclase or related enzyme).

cyclase involved a third class of proteins—the G proteins, which we also introduced in Chapter 12. Martin Rodbell, who showed that receptors are distinct from adenylate cyclase, and Alfred Gilman, who discovered G proteins, were recognized for these discoveries with the 1994 Nobel Prize for Medicine or Physiology.

Structural Analysis of G Protein-Coupled Receptors

Because receptors are embedded in the membrane and are present in minute amounts, isolating them in quantities sufficient for biochemical analysis was difficult. Cloning the receptor genes was crucial for elucidating the complete amino acid sequence. Among the many receptor proteins whose amino acid sequences have now been determined—including both α_2- and β_2-adrenergic receptors—there are some remarkable structural similarities. The proteins are of comparable size, with 415 to 480 residues, including seven conserved regions that are rich in hydrophobic amino acids. These represent regions of α helix that are embedded in the membrane and linked by hydrophilic loops, projecting into both the extracellular environment and the cytosol, with the recognition site for the signaling agent on the extracellular side. The amino acid sequence of the β_2-adrenergic receptor β_2AR is depicted in **FIGURE 20.3**, along with the membrane-spanning domains. Since these receptor molecules snake back and

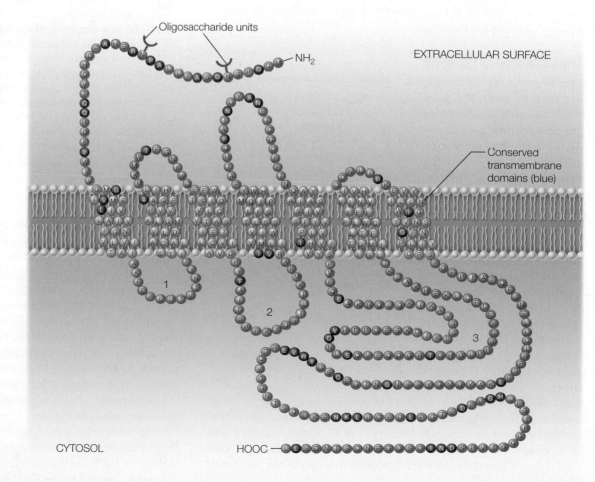

▶ **FIGURE 20.3** Amino acid sequence of the human β_2-adrenergic receptor. The seven conserved transmembrane domains are shown in blue. Note also the three extracellular and three cytoplasmic loops and the two N-linked oligosaccharide units on the extracellular side (bound to asparagine residues). Interaction of the receptor with G proteins is controlled in part by reversible phosphorylation of serine and threonine residues near the C-terminus. The amino acids colored black are different in the hamster β_2-adrenergic receptor sequence.

● **CONNECTION** G protein-coupled receptors are the targets for nearly half of all non-antibiotic prescription drugs.

forth through the membrane, they sometimes are called *serpentine* receptors. Because nearly all of these receptors function in concert with G proteins, they are also called **G protein-coupled receptors** (GPCRs). It is estimated that half of all prescription, non-antibiotic drugs target members of this receptor class. The potential for future drug discovery is enormous because analysis of the human genome suggests that at least 800 different GPCRs exist, many of which are potential or actual drug targets.

Isolation of GPCRs for structural analysis was challenging because of their association with membranes. The first GPCR to be structurally characterized was rhodopsin in 2000. However, rhodopsin participates in light sensing in vision rather than in hormone action (see Chapter 17). The structure of the β_2-adrenergic receptor (β_2AR) was reported in 2007, and several have been described since then. **FIGURE 20.4** shows β_2AR in complex with carazolol, a β-adrenergic antagonist. The figure shows a striking similarity between the structures of rhodopsin and β_2AR, in the placement of the seven transmembrane helices and in the binding site for ligand (carazolol for β_2AR, 11-*cis*-retinal for rhodopsin). A significant difference is the involvement of a helical region of one of the extracellular domains of β_2AR in binding the ligand.

Carazolol

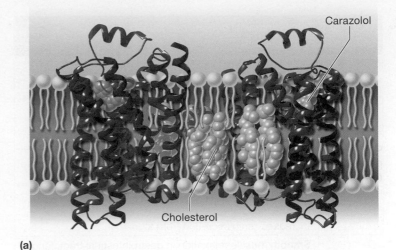

(a)

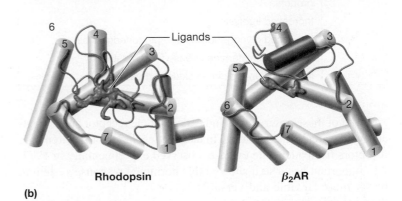

(b)

▲ **FIGURE 20.4 Structure of the human β_2 adrenergic receptor.** (a) A model of two receptor molecules embedded in the membrane and joined by cholesterol molecules (in yellow). The ligand carazolol is shown in green. (b) Comparison of the top (extracellular) views of rhodopsin (PDB ID: 1f88) and the β_2 adrenergic receptor, showing similarities in arrangement of the transmembrane helices. Each ligand (11-*cis*-retinal in rhodopsin, carazolol in β_2AR) is shown in green. A helical region of an extracellular domain, which forms part of the binding pocket for epinephrine, is also shown in red.

Transducers: G Proteins

The G proteins, so named because of their ability to bind guanine nucleotides, are the second component of the receptor–transducer–effector signaling system first described for the β-adrenergic response. In 1971, guanosine triphosphate (GTP) was found to be required for the activation of adenylate cyclase by β-adrenergic agonists, and late in the decade the basis for this requirement emerged: GTP-binding membrane proteins interact with receptor systems that activate or inhibit adenylate cyclase. Of the several known G proteins, the two best characterized are G_s, a family of G proteins involved in *stimulating* adenylate cyclase, and G_i, a closely related family involved in responses that *inhibit* adenylate cyclase. Although both types of G proteins interact with other receptors as well (and with target proteins other than adenylate cyclase), we first describe their functions in terms of the adrenergic receptors.

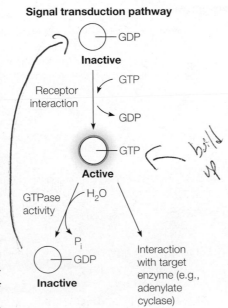

Signal transduction pathway

Actions of G Proteins

The G proteins are membrane proteins that in the *inactive* state bind guanosine diphosphate (GDP). Recall from Chapter 12 that a hormone response leading to stimulation of adenylate cyclase—the binding of extracellular hormone or agonist to a receptor, typically a β-adrenergic receptor—causes a conformational change that stimulates the receptor to interact with a nearby molecule of G_s. This in turn stimulates an exchange of bound GDP for GTP—that is, the dissociation of GDP from G_s, to be replaced by GTP. G_s is thereby converted to a protein that activates adenylate cyclase, producing cAMP from ATP. Cyclic AMP synthesis results in the activation of cAMP-dependent protein kinase (protein kinase A), with consequent phosphorylation of target proteins, such as phosphorylase b kinase in cells that activate glycogen phosphorolysis.

To summarize, this signal transduction pathway involves (1) hormone binding to receptor; (2) receptor interaction with G_s, stimulating release of GDP and the association of GTP with G_s; (3) stimulation of adenylate cyclase by the GTP-bound G_s; (4) stimulation by cAMP of protein phosphorylation; and (5) stimulation or inhibition of metabolic reactions.

The initial exchange reaction, stimulated by hormone binding to the receptor (**FIGURE 20.5**), is usually assisted by one of a class of proteins called **guanine nucleotide exchange factors** (GEFs). Continued activation of G_s depends on the presence of bound GTP. The hormonal response is limited by the presence of a slow GTPase activity inherent to the G protein. Thus, bound GTP is slowly hydrolyzed to GDP, with concomitant loss of the ability to stimulate adenylate cyclase. This process, like the initial activation, is protein-assisted, being helped by a **GTPase-activating protein** (GAP). The G_i protein functions similarly, but in response to extracellular signals whose response is the *inhibition* of adenylate cyclase, typically α_2 agonists. Here the binding of GTP provokes an inhibitory interaction of G_i with adenylate cyclase, which decreases the synthesis of cAMP.

Structure of G Proteins

G_s, G_i, and other G proteins have an $\alpha\beta\gamma$ heterotrimeric structure (Figure 20.5) consisting of a 39- to 46-kilodalton α subunit, a 37-kilodalton β subunit, and an 8-kilodalton γ subunit. The human genome encodes at least 24 different α proteins, 5 β, and 6 γ, allowing for a great variety of different heterotrimeric G proteins. In most of these, the γ subunit is **prenylated;** that is, it contains a covalently bound C_{20} isoprenoid moiety at the C-terminal cysteine, which helps anchor the protein in the membrane and may facilitate protein–protein interactions. The α subunit is **myristylated** in two other G proteins, G_i and G_o, and palmitylated in G_s. The myristyl or palmityl moiety is bound via amide linkage with an N-terminal cysteine. The guanine nucleotide-binding site and its associated GTPase activity are both located on the α subunit. A hormonal stimulus causes the exchange of GDP for GTP and the dissociation of the G protein, with the α–GTP complex moving along the membrane until it encounters a molecule of adenylate cyclase or other effector molecule. The slow GTPase activity mentioned earlier eventually reconverts α–GTP to α–GDP, and the α–GDP complex dissociates from adenylate cyclase and rejoins the $\beta\gamma$ complex.

● **CONCEPT** G proteins are activated when GTP displaces GDP bound to the *A* subunit and the complex dissociates from $\beta\gamma$. Hormone action is limited by slow hydrolysis of the bound GTP.

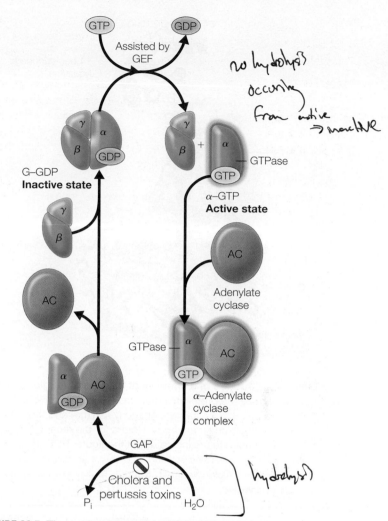

▲ **FIGURE 20.5 The cycle of G protein dissociation and reassociation.** α, β, and γ are the three subunits of the G protein. The active form is the α–GTP complex (orange highlight), while the inactive GDP complexes are shown without highlights. The sites where the pertussis and cholera toxins act are also shown. GEF, guanine nucleotide exchange factor; GAP, GTPase-activating protein.

Consequences of Blocking GTPase

Blocking the GTPase activity shows just how important it is in controlling the hormone response. Several bacterial toxins have G proteins as their biological targets. The toxin of *Vibrio cholerae* is an enzyme with the ability to cleave NAD^+ and transfer its ADP-ribose moiety to a specific site in the α subunit of G_s. This modification of G_s inhibits its GTPase activity and converts the α subunit to an irreversible activator of adenylate cyclase.

$$NAD^+ + \alpha_s \longrightarrow \text{nicotinamide} + \text{ADP-ribosyl-}\alpha_s$$

In the intestine, the cAMP that accumulates promotes the uncontrollable secretion of water and Na^+ and is responsible for the severe diarrhea and resulting dehydration and loss of salt that accompany cholera.

● **CONNECTION** Bacterial toxins responsible for serious infectious diseases target G proteins.

A component of the toxin of *Bordetella pertussis,* which causes whooping cough, has a similar effect on the α subunit of the G_i protein, with different physiological effects—namely, lowered blood glucose and hypersensitivity to histamine.

The Versatility of G Proteins

The G protein mechanism is used in many signal transduction pathways. Reassortment of the several α, β, and γ proteins means that a large number of different G proteins exist, giving great flexibility in response to this signal transduction element. Interaction with target enzymes is a function of the α subunits. Some interact with adenylate cyclase, some with ion channels, and some with phospholipases. One large subfamily of G proteins, called G_{olf}, is present in olfactory cells

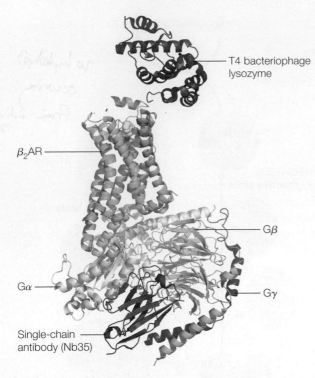

▲ **FIGURE 20.6 Structure of a complex of the β_2 adrenergic receptor with G_s.** To stabilize the proteins and aid in their crystallization, the receptor protein was fused to T4 bacteriophage lysozyme, and the G protein was fused to a single-chain antibody. Colors for the receptor and each of the G protein subunits are indicated on the figure. PDB ID: 3sn6. Data from Rasmussen, S. G. F., and 19 coauthors (2011) Crystal structure of the β_2 adrenergic receptor G_s protein complex. Nature 477:549–557. The first structural analysis of a receptor-G protein complex.

in the nose and functions with a large number of receptors involved in the sensory reception of odors.

Interaction of GPCRs with G Proteins

How does a GPCR interact with an associated G protein? The crystal structure of the β_2-adrenergic receptor complexed with G_s, determined in 2011 and presented in **FIGURE 20.6**, shows that the α subunit of G_s is in direct contact with the receptor. Binding significantly altered the positions of two of the transmembrane helices of the receptor and also displaced the α-helical domain of the α subunit of G_s. Because analyses of this type will reveal molecular mechanisms of signal transduction, this contribution was recognized with the award of the 2012 Nobel Prize in Physiology or Medicine to Robert Lefkowitz and Brian Kobilka.

G Proteins in the Visual Process

There are remarkable similarities between the actions of G proteins in transmitting hormonal signals and their actions in the transmission of signals from light. Much of our understanding of G proteins in hormonal signal transduction came from studies of a G protein called **transducin** in the visual process. The extracellular stimulus and the biochemical end point are quite different in vision from those in hormone action, but the transmembrane signaling processes are almost identical.

As mentioned in Chapter 16, the extracellular signal in vision is a photon of light, and the membrane receptor is **rhodopsin,** an abundant membrane protein in the outer segment of rod cells in the retina. A photochemical change in the structure of rhodopsin causes it to activate transducin so that it binds GTP. Green arrows denote activation.

The transducin–GTP complex activates a specific **phosphodiesterase,** which cleaves a cyclic nucleotide, **guanosine 3,5-monophosphate** (cyclic GMP, or cGMP). Cleavage of cGMP, in turn, stimulates intracellular reactions that generate a visual signal to the brain. Thus, the stimulated *hydrolysis* of cGMP is the visual analog of the stimulated *synthesis* of cAMP in β-adrenergic responses.

Effectors

Although G proteins have several targets involved in signal transduction, we focus here on adenylate cyclase (AC) because it is involved in adrenergic signaling. As noted earlier, AC catalyzes the conversion of ATP to cAMP plus pyrophosphate. Mammalian cells contain 10 AC isoforms that are regulated by G proteins, each consisting of two transmembrane domains (M1 and M2) and two homologous cytoplasmic domains (C1α and C2α). **FIGURE 20.7(a)** shows the structure of the cytoplasmic domains crystallized in the presence of the α subunit of G_s and **forskolin,** a plant diterpene that activates all but one of these adenylate cyclases. The action of forskolin, as shown partly by this structure, is to draw the two cytoplasmic domains together in a catalytically active conformation. Other studies indicate that the binding site for α_s is on the opposite side of the pseudosymmetric catalytic domain, where it appears to prevent association of the catalytic domains (not shown). Both regulatory subunits bind to sites remote from the active site of the enzyme, indicating that complex allosteric processes must be involved in regulating AC activity.

The adenylate cyclase reaction activates the 3′ hydroxyl of ATP for nucleophilic attack on the α (inner) phosphorus to create a phosphodiester bond, with pyrophosphate as a leaving group. The reaction is similar to that catalyzed by DNA polymerases,

Cyclic GMP

5′ GMP

Forskolin

α-Adenylate cyclase complex

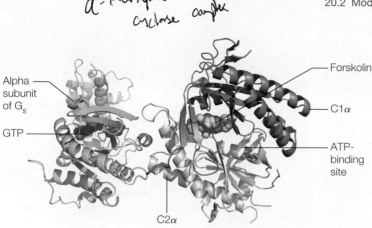

Alpha
subunit
of G_s

GTP

Forskolin

C1α

ATP-
binding
site

C2α

(a) The C1α and C2α catalytic domains (tan and green) were crystallized as a complex with forskolin (yellow) and α_s, the α subunit of G_s (turquoise). The catalytic site where ATP is bound consists of residues from both domains. GTP bound to α_s is shown as well (blue and red colored atoms).

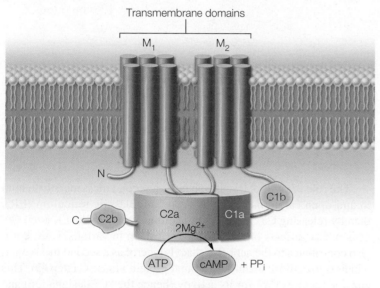

Transmembrane domains

M_1   M_2

N

C1b

C   C2b   C2a   C1a

2Mg²⁺

ATP   cAMP + PP_i

(b) Schematic diagram showing relationships of the catalytic domains to the transmembrane helical regions.

◀ **FIGURE 20.7 Crystal structure of an adenylate cyclase catalytic domain. (a)** Structure of the protein. **(b)** Schematic diagram showing relationships of the catalytic domains to the transmembrane helical regions.

→ see pg 406

Second Messengers

Cyclic AMP

We described in Chapter 12 the second messenger function of cAMP, particularly its effects on protein kinase A, or cAMP-dependent protein kinase. Cyclic AMP also has regulatory effects at the gene level. Protein kinase A, after cAMP activation, phosphorylates a protein called CREB (**cAMP response element binding protein**), and the resulting phosphorylated protein controls the transcription of genes, including those encoding particular receptors. Some of these actions represent the adaptation of a cell to action of a hormone. In addition, cAMP functions as a second messenger in the actions of signaling agents other than epinephrine and glucagon, including dopamine, β-corticotropin (ACTH), histamine, serotonin, and prostaglandins.

With this diversity of signaling agents acting through cAMP, how do we explain the specificity of hormone action? Part of the answer lies in the distribution of hormone receptors among tissues. Glucagon receptors, for example, are located in liver and adipose tissue, which explains the preferential effects of glucagon on these tissues. Also, the distribution of stimulatory and inhibitory G proteins within different cells determines whether binding of a hormone to that cell will increase or decrease the intracellular cAMP concentration. More recently, a class of proteins, called AKAPs (**A kinase anchoring proteins**) was discovered; these are bound to specific sites within a cell and are controlled by localized pools of cAMP, thereby accounting for differential effects of cAMP within the same cell. Among the 30 or so human AKAPs are forms associated with microtubules, ion channels, or mitochondria; these bind protein kinase A at those specific sites, thus localizing the effects of cAMP within a single cell.

Cyclic GMP and Nitric Oxide

Cyclic AMP was the earliest second messenger known but is far from the only one. Here we mention cyclic GMP (cGMP) and the phosphoinositide system (Chapter 16). Much interest has been focused on cGMP, particularly with respect to its role in nitric oxide (NO·) metabolism. NO· is a gaseous signaling molecule, synthesized from arginine (Chapter 18), which plays important regulatory roles. NO· was originally identified as an agent in vasodilation of endothelial vascular cells and the underlying smooth muscle. Signals that decrease blood pressure and inhibit platelet aggregation use NO· as an intermediary. In inflammatory and immune responses, an inducible form of nitric oxide synthase produces NO· at levels sufficient to be toxic to pathogenic organisms. NO· also regulates neurotransmission in the central nervous system.

The nitric oxide synthase in endothelial vascular cells is acutely sensitive to calcium ion concentration; activation of the enzyme by Ca²⁺

● **CONNECTION** Nitric oxide regulates blood pressure, platelet aggregation, neurotransmission, and aspects of the immune response.

except that the AC reaction is intramolecular. Strategically placed Asp residues in the active site (not shown), plus the requirement for two metal ions, support a mechanism for adenylate cyclase quite similar to the two-metal mechanism now widely accepted for DNA polymerases (Chapter 22).

causes NO· to accumulate. Because NO· is a very small molecule, it diffuses rapidly into neighboring cells, where it exerts its control by binding to ferrous ion in a soluble form of guanylate cyclase and stimulating cGMP formation. In other words, guanylate cyclase is an intracellular receptor for NO·. Guanylate cyclase may not be the only regulatory target for NO·, but cGMP elevation is probably the major cellular effect of NO· release.

NO· is unstable, so its effects are short-lived. Because of its function in stimulating vasodilation, NO· plays a role in stimulating the erection of the penis. The drug sildenafil (^{TM}Viagra) counteracts erectile dysfunction by inhibiting cGMP phosphodiesterase, thus increasing the metabolic half-life of cGMP.

Sildenafil (Viagra)

Many cells contain a cGMP-stimulated protein kinase that, unlike the cAMP-activated enzyme, contains both catalytic and regulatory domains on one polypeptide chain of a homodimeric protein. Our

● **CONNECTION** Sildenafil acts by inhibiting cyclic GMP phosphodiesterase, thereby prolonging the vasodilatory action of cGMP.

understanding of the roles of cGMP in signal transduction has been achieved only more recently because its intracellular concentrations are 10- to 100-fold lower than those of cAMP.

Although the discovery of NO· as a signaling molecule was unexpected, it is not the only gaseous signaling agent. Both hydrogen sulfide (H_2S) and carbon monoxide (CO), released in small, subtoxic doses, also have anti-inflammatory and vasodilatory actions similar to those of NO·, although different mechanisms are involved. H_2S, produced by desulfhydration of cysteine (Chapter 18), activates an ATP-sensitive potassium channel in smooth muscle cells. Less is known about actions of CO.

Phosphoinositides

Although similar in many respects to the adenylate cyclase system, the **phosphoinositide system** is distinctive in that the hormonal stimulus activates a reaction that generates *two* second messengers, both derived from a specific lipid in the phosphoinositide family. **Phosphatidylinositol 4,5-bisphosphate (PIP$_2$)** is a membrane-associated storage form for two second messengers. As shown in **FIGURE 20.8**, the binding of an agonist to a receptor (step ❶) stimulates a G protein to bind GTP (step ❷), just as occurs during the adrenergic response. However, this G protein activates not adenylate cyclase but a different membrane-bound enzyme, **phospholipase C,** which, as shown, cleaves PIP$_2$ to yield two products (step ❸), **sn-1,2-diacylglycerol (DAG)** and **inositol 1,4,5-trisphosphate (InsP$_3$).** Both DAG and InsP$_3$ act as second messengers. Therefore, the cleavage of PIP$_2$ by phospholipase C is the functional equivalent of the synthesis of cAMP by adenylate cyclase.

Phosphatidylinositol 4,5-bisphosphate (PIP$_2$)

Phospholipase C

sn-1,2-Diacylglycerol (DAG)

Inositol 1,4,5-trisphosphate (InsP$_3$)

The role of inositol trisphosphate as a second messenger is to bind to and open calcium channels in the endoplasmic reticulum (ER), thereby releasing Ca^{2+} from its intracellular stores in the ER (step ❹). This release affects intracellular metabolism in various ways, but it also contributes to the action of diacylglycerol as a second messenger, which is to activate membrane-bound protein kinase C (step ❺). This enzyme requires Ca^{2+} for its activity (hence the "C" designation) and a *phospholipid* (specifically, phosphatidylserine). Diacylglycerol, the other second messenger, stimulates protein kinase C activity by greatly increasing the affinity of the enzyme for Ca^{2+}. This requirement is specific for the *sn*-1,2-DAG; neither the 1,3- nor the 2,3-isomer is active. The enzyme phosphorylates specific serine and threonine residues in target proteins (step ❻). As with cAMP-stimulated protein kinase, the specific cellular responses to protein kinase C activation, such as the phosphorylation of calmodulin, (steps ❼ and ❽), depend on the ensemble of target proteins that become phosphorylated in a given cell. Other known target proteins include the insulin receptor, β-adrenergic receptor, glucose transporter, HMG-CoA reductase, cytochrome P450, and tyrosine hydroxylase.

We now briefly consider the metabolism of inositol trisphosphate (InsP$_3$) after it is released from PIP$_2$. Three sequential hydrolytic steps yield inositol, which is then reincorporated into phosphatidylinositol, as discussed in Chapter 16, to regenerate PIP and PIP$_2$. The last hydrolytic step, in which inositol monophosphate is hydrolyzed to inositol, is specifically inhibited by **lithium ion** (Li^+). Blocking this step inhibits the resynthesis of InsP$_3$ by depleting the cell of inositol. This may be related to the action of Li^+ in the treatment of bipolar disorder.

$$\text{Inositol monophosphate} + H_2O \longrightarrow \text{inositol} + P_i$$

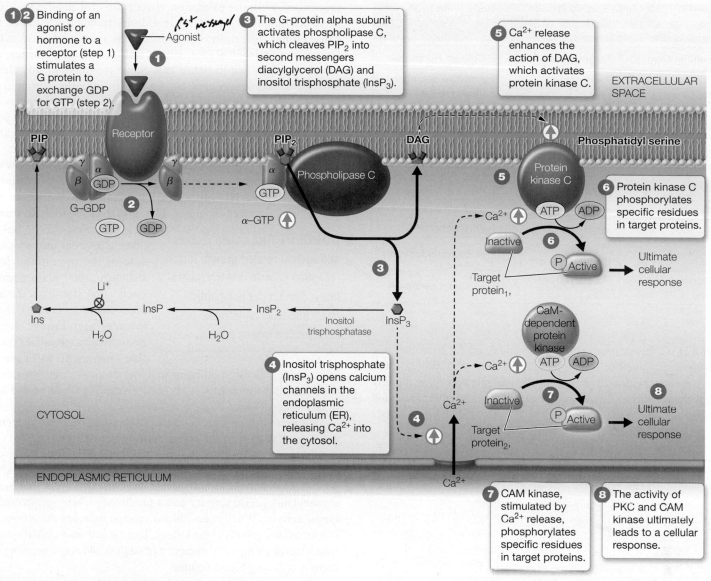

1 2 Binding of an agonist or hormone to a receptor (step 1) stimulates a G protein to exchange GDP for GTP (step 2).

3 The G-protein alpha subunit activates phospholipase C, which cleaves PIP₂ into second messengers diacylglycerol (DAG) and inositol trisphosphate (InsP₃).

5 Ca²⁺ release enhances the action of DAG, which activates protein kinase C.

6 Protein kinase C phosphorylates specific residues in target proteins.

4 Inositol trisphosphate (InsP₃) opens calcium channels in the endoplasmic reticulum (ER), releasing Ca²⁺ into the cytosol.

7 CAM kinase, stimulated by Ca²⁺ release, phosphorylates specific residues in target proteins.

8 The activity of PKC and CAM kinase ultimately leads to a cellular response.

▲ **FIGURE 20.8 Signal transduction pathways involving phospho-inositide turnover.** DAG, *sn*-1,2-diacylglycerol; Ins, inositol; InsP, inositol monophosphate; PIP, phosphatidylinositol-4-phosphate; PIP₂, phosphatidylinositol 4,5-bisphosphate; InsP₃, inositol 1,4,5-trisphosphate; InsP₂, inositol 1,4-bisphosphate. Most of the effects of calcium ion (Ca²⁺) result from its binding to calmodulin (CaM). A23187 is a calcium ionophore, which can be used experimentally to release calcium from intracellular stores. The release of Ca²⁺ stimulates both protein kinase C and calmodulin kinase. Green arrows denote activation.

Because many metabolic processes are controlled by calcium fluxes and by the phosphorylation of specific proteins, the phosphoinositide system has great versatility as a control mechanism. The fact that a cell can use either DAG or InsP₃, or both mechanisms, from a single extracellular stimulus further increases this versatility. Some of the processes controlled by the phosphoinositide system are listed in **TABLE 20.2**.

Several observations imply a role for the phosphoinositide system not only in metabolic regulation but also in the control of cellular growth. First is the activity of a group of natural products called **phorbol esters,** part of whose structure resembles that of DAG (shown in red). These compounds are called **tumor promoters.** Not carcinogenic by themselves, they stimulate the formation of tumors when

● **CONCEPT** Second messengers include cAMP, cGMP, InsP₃, and DAG.

● **CONNECTION** Wound healing is controlled in part by lipid signaling.

applied along with a carcinogen to experimental animals. Some phorbol esters have been found to activate protein kinase C independently of DAG. This finding is consistent with the hypothesis that protein kinase C activation is part of the normal growth control process that becomes perturbed in tumorigenesis.

To recapitulate: Cyclic AMP was the earliest known second messenger. However, several comparable second messengers are now known, including cGMP,

TABLE 20.2 Some cellular processes controlled by the phospho-inositide second-messenger system

Extracellular Signal	Target Tissue	Cellular Response
Acetylcholine	Pancreas Pancreas (islet cells) Smooth muscle	Amylase secretion Insulin release Contraction
Vasopressin	Liver	Glycogenolysis
Thrombin	Blood platelets	Platelet aggregation
Antigens	Lymphoblasts Mast cells	DNA synthesis Histamine secretion
Growth factors	Fibroblasts	DNA synthesis
Spermatozoa	Eggs (sea urchin)	Fertilization
Light	Photoreceptors (*Limulus*)	Phototransduction
Thyrotropin-releasing hormone	Pituitary anterior lobe	Prolactin secretion

InsP$_3$, and DAG. In addition, phosphoinositides are far from the only phospholipids involved in cell signaling. For example, lysophosphatidic acid interacts with a different G protein-coupled receptor, thereby regulating the growth and development of cell types involved in wound healing.

A phorbol ester,
1-*O*-tetradecanoylphorbol-13-acetate

sn-1,2-Diacylglycerol
(DAG)

20.3 Receptor Tyrosine Kinases and Insulin Signaling

We turn now to a family of related membrane receptors that are distinctive because they have a single membrane-spanning domain, a ligand-binding site on the extracellular part of the molecule, and an intrinsic protein tyrosine kinase activity in the cytoplasmic domain. These include the **insulin receptor** (IR); receptors for related peptide growth factors, including those for **epidermal growth factor** (EGF), **platelet-derived growth factor** (PDGF), **colony-stimulating factor 1** (CSF-1), **nerve growth factor** (NGF), **fibroblast growth factor** (FGF), and a peptide **insulin-like growth factor 1** (IGF-1). These receptor tyrosine kinases (RTKs) represent a family of related proteins because the tyrosine kinase domains share homologous amino acid

sequences. With the exception of receptors for insulin and IGF-1, these are monomeric proteins that dimerize when stimulated by an extracellular ligand, thus stimulating the tyrosine kinase activity. Humans contain 58 RTKs, which fall into 20 subfamilies, some of which are shown in **FIGURE 20.9**. Some members of this class act through second messengers, but most act by initiating a protein phosphorylation cascade, similar to what we have seen in glycogenolysis. Receptors in this class include those for many peptides that control cell growth and differentiation, as well as insulin (which can also be considered a growth factor).

Structural studies on the EGF receptor have suggested how extracellular ligand binding stimulates protein tyrosine kinase activity. As schematized in **FIGURE 20.10**, ligand binding stimulates dimerization of the monomeric receptor in step ❶ to give a surprisingly asymmetric structure. Parts of the polypeptide chains near the cytoplasmic side of the membrane form a structure, the "juxtamembrane latch," that stabilizes the asymmetric dimer. One of the two intracellular kinase domains allosterically stimulates the tyrosine kinase activity of the other domain, step ❷. Both domains undergo extensive autophosphorylation, step ❸. Similar processes occur in the dimerization and activation of other RTKs.

● **CONCEPT** The insulin receptor and several related growth factor receptors contain one transmembrane domain per polypeptide chain and have an intrinsic protein tyrosine kinase activity.

Some more distantly related membrane receptors have other enzyme activities. Proteins of the **transforming growth factor β** (TGF-β) family bind to a receptor that has a protein serine/threonine kinase activity (like protein kinase A). **Atrial natriuretic factor (ANF),** a peptide that controls blood volume, binds to a receptor that has both a guanylate cyclase activity and a predicted protein serine/threonine kinase activity. When excess blood volume stretches the atrium, ANF is released and travels to the kidney. There it activates cGMP synthesis, which increases the renal excretion of sodium and accompanying water, thereby reducing blood volume.

Once an agonist has bound and the protein tyrosine kinase has become activated, proteins are recruited to the intracellular domains of the RTK by virtue of containing either **SH2** (S̲rc h̲omology) domains or **PTB** (p̲hosphotyrosine b̲inding) domains. Proteins associating with the phosphotyrosine domain may be either signaling agents or adaptor proteins, which create binding sites for downstream signaling proteins. **FIGURE 20.11** depicts schematically three pathways following activation of the insulin receptor. A prominent substrate for the insulin receptor kinase is **insulin receptor substrate-1** (IRS-1), a protein that binds to the kinase domain of the insulin receptor via its SH2 domains. First, the phosphorylation of IRS-1 leads to activation of **phosphoinositide 3-kinase,** (PI3K) which converts phosphatidylinositol 4,5-bisphosphate (PIP$_2$) to phosphatidylinositol 3,4,5-trisphosphate (PIP$_3$), a second messenger with a variety of intracellular targets. In the response to insulin, one effect of PIP$_3$ is activation of **protein kinase B** (PKB), by way of its attachment to the membrane. PKB then phosphorylates another protein kinase, glycogen synthase kinase (GSK3), which inactivates it, thereby blocking phosphorylation and inactivation of glycogen synthase. Therefore, as a result of insulin binding to its receptor, glycogen synthase cannot be inactivated, ensuring that glycogen synthesis continues.

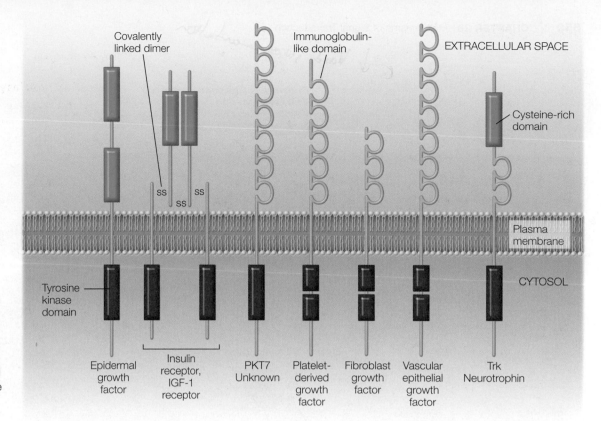

Covalently linked dimer

Immunoglobulin-like domain

EXTRACELLULAR SPACE

Cysteine-rich domain

ss   ss
    ss

Plasma membrane

CYTOSOL

Tyrosine kinase domain

Epidermal growth factor

Insulin receptor, IGF-1 receptor

PKT7 Unknown

Platelet-derived growth factor

Fibroblast growth factor

Vascular epithelial growth factor

Trk Neurotrophin

► **FIGURE 20.9** Representative receptor tyrosine kinase families. Each protein has one or two intracellular kinase domains (in red) and various conserved extracellular domains, as indicated. The family containing the insulin receptor is distinctive because it is a covalently linked dimer.

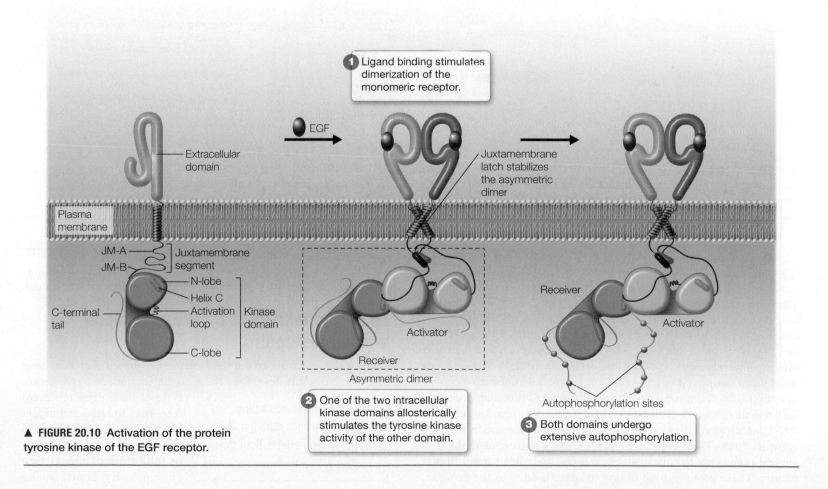

1 Ligand binding stimulates dimerization of the monomeric receptor.

EGF

Juxtamembrane latch stabilizes the asymmetric dimer

Extracellular domain

Plasma membrane

JM-A
JM-B   Juxtamembrane segment

N-lobe
Helix C
C-terminal tail   Activation loop   Kinase domain
C-lobe

Receiver

Activator

Receiver

Activator

Asymmetric dimer

Autophosphorylation sites

2 One of the two intracellular kinase domains allosterically stimulates the tyrosine kinase activity of the other domain.

3 Both domains undergo extensive autophosphorylation.

▲ **FIGURE 20.10** Activation of the protein tyrosine kinase of the EGF receptor.

↓ blood glucose concentration

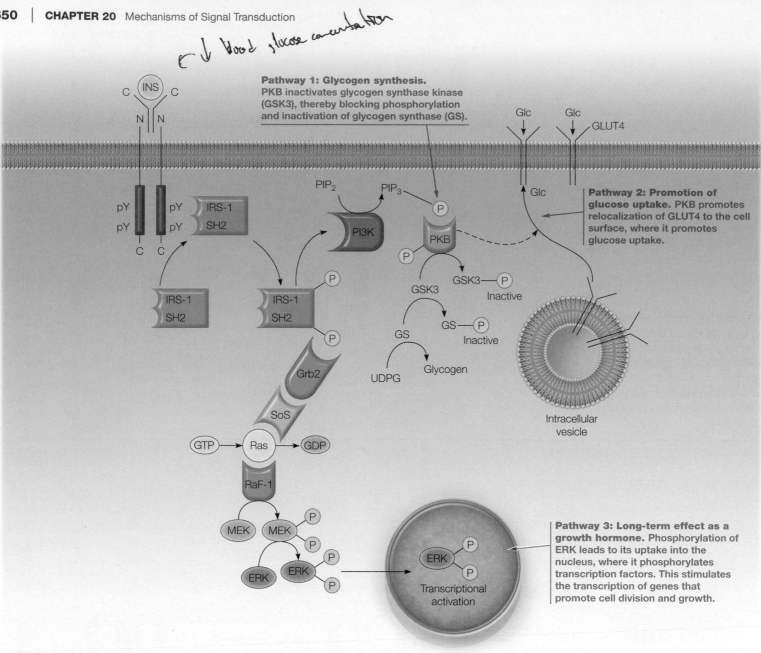

Pathway 1: Glycogen synthesis.
PKB inactivates glycogen synthase kinase (GSK3), thereby blocking phosphorylation and inactivation of glycogen synthase (GS).

Pathway 2: Promotion of glucose uptake. PKB promotes relocalization of GLUT4 to the cell surface, where it promotes glucose uptake.

Pathway 3: Long-term effect as a growth hormone. Phosphorylation of ERK leads to its uptake into the nucleus, where it phosphorylates transcription factors. This stimulates the transcription of genes that promote cell division and growth.

▲ **FIGURE 20.11** Signaling pathways involving the insulin receptor, illustrative of RTKs. INS, insulin. GLUT4 is one member of a family of glucose transporters. IRS, insulin receptor substrate; PKB, protein kinase B; GSK, glycogen synthase kinase; GLUT4, glucose transporter 4.

A second result of insulin binding is the action of PKB in promoting relocalization of the glucose transporter GLUT4 from internal membrane vesicles to the cell surface, where it promotes glucose uptake. The third effect of insulin is its long-term effect as a growth hormone, and this results from establishment of another phosphorylation cascade. Figure 20.11 shows that phosphorylated IRS-1 interacts with an adaptor protein called Grb2, which in turn activates Sos, which next binds to Ras (see page 654 for more details about Ras). Sos activates Ras by promoting the binding of GTP, coupled with the dissociation of bound GDP (comparable to the activation of G proteins, discussed earlier in this chapter). GTP-bound Ras activates a protein kinase called Raf-1, which then activates MEK, another protein kinase. MEK in turn activates ERK. Phosphorylation of ERK leads to its uptake, through nuclear pores, into the nucleus, where it phosphorylates transcription factors. These proteins bind in turn to specific sites on the genome,

promoting the transcription of genes whose action promotes cell division and growth. The phosphorylation cascade shown for insulin is similar to that involved in the actions of other peptide growth factors.

20.4 Hormones and Gene Expression: Nuclear Receptors

Hormonal effects occurring via G protein-coupled receptors tend to last only a short time. Like the epinephrine-induced glycogenolytic cascade, they represent responses to rapid and urgent physiological demands, and they involve the activation or inhibition of preexisting enzymes. By contrast, the

● **CONCEPT** Hormones acting through nuclear receptors generally have longer-lived effects than those interacting with membrane receptors.

TABLE 20.3 Target organs for steroid and thyroid hormones and major proteins whose synthesis is affected

Hormone Class	Target Organ	Protein[a]
Glucocorticoids	Liver	Tyrosine aminotransferase
		Tryptophan oxygenase
		α Fetoprotein ($\downarrow$)
		Metallothionein
	Retina	Glutamine synthetase
	Kidney	Phosphoenolpyruvate carboxykinase
	Oviduct	Ovalbumin
	Pituitary	Pro-opiomelanocortin (peptide hormone precursor)
Estrogens	Oviduct	Ovalbumin
		Lysozyme
	Liver	Vitellogenin
		Apo-VLDL
Progesterone	Oviduct	Ovalbumin
		Avidin
	Uterus	Uteroglobin
Androgens	Prostate	Aldolase
	Kidney	β-Glucuronidase
	Oviduct	Albumin
1,25-Dihydroxyvitamin D$_3$ Thyroxine	Intestine	Calcium-binding protein
	Liver	Carbamoyl phosphate synthetase
	Pituitary	Malic enzyme
		Growth hormone
		Prolactin ($\downarrow$)

[a]Synthesis of each indicated protein is increased by the hormone, except for the two identified by ($\downarrow$)

Nuclear receptors exist at levels of only about 10^4 molecules per cell, which makes their purification difficult. However, because they bind to hormones quite tightly, these proteins can be purified by affinity chromatography. Sequence analysis has revealed structural similarities among this class of receptors, and the construction of hybrid receptors by recombinant DNA technology has led to identification of domains of function within the receptor molecule. Each receptor protein within this family contains a central conserved domain of about 80 residues, which is involved in DNA binding (**FIGURE 20.12**). On the N-terminal side of this domain is a region essential to transcriptional activation. Toward the C-terminus are domains involved in hormone binding, protein dimerization, and transcriptional activation.

● **CONCEPT** The family of steroid receptors contains a conserved, zinc-containing DNA-binding sequence and a C-terminal hormone-binding domain.

All of the known receptors in this family contain bound zinc ion (Zn^{2+}), which is essential for DNA binding, and the DNA-binding sequences show a completely conserved distribution of cysteine residues. These observations suggest that the zinc ions are complexed by the cysteine sulfurs in a pattern akin to the "zinc finger" structural motif associated with a number of other eukaryotic transcriptional regulatory proteins (see Chapter 26). **FIGURE 20.13** illustrates this binding schematically.

The utility of a set of long-term-acting regulators is evident from a couple of examples. Estrogens and progesterone regulate the female reproductive cycle. In humans, these hormones interact over a four-week cycle to prepare the uterus for implantation of a fertilized ovum. Proliferation of the endometrium, the epithelial lining of the uterus, is the major event, and it requires new protein synthesis and increased blood flow to the uterus. These processes stop when a pituitary signal triggers decreased release of the hormones, causing cells in the uterine lining to slough off and the beginning of menstrual bleeding.

The actions of glucocorticoids are comparable in that control of the synthesis of particular proteins allows for long-term metabolic adaptation. Whereas estrogens exercise control of reproductive metabolism over a several-week period, the secretion of glucocorticoids provides a way to adapt to longer-term stress. This adaptation involves the stimulation of gluconeogenesis and the synthesis of a variety of proteins, including some that counteract the effects of inflammation. Unlike estrogens, which act chiefly in reproductive tissues, the glucocorticoids influence cells in a wide variety of target tissues.

Steroid hormone receptors are target sites for several drugs. **Tamoxifen** binds to estrogen receptors but does not activate estrogen-responsive genes. The growth of some breast tumor cells is activated by estrogen. Treating these patients with tamoxifen after surgery or chemotherapy often antagonizes estrogen binding in residual tumor cells and retards their growth. However, patients taking tamoxifen after breast cancer surgery must be monitored carefully because there is also an increased risk of uterine cancer. **RU486,** which was developed in France, binds to progesterone receptors and blocks the events essential

● **CONNECTION** Tamoxifen effectively treats breast cancers characterized by high densities of estrogen receptors.

effects of steroid hormones involve longer-term changes related to growth and differentiation of cells. Steroids and related hormones (thyroid, vitamin D, and retinoic acid hormones) act intracellularly. By virtue of their hydrophobic nature, they traverse the plasma membrane and exert their effects within the cell—actually, within the nucleus, where they control the activities of specific genes. In most cases, target genes are activated. **TABLE 20.3** lists several proteins whose synthesis is regulated by these hormones.

These regulatory effects occur at the level of transcription of steroid-responsive genes. Steroid and related hormones act by binding in the cytosol to specific receptor proteins, which undergo subunit structure change under the influence of the hormone. Binding in the cytosol is followed by movement of the hormone–receptor complex into the nucleus, where the complex interacts with specific DNA sites called **hormone-responsive elements** (HREs). Binding of the complex to DNA affects the transcription rates of nearby genes. Because of their site of action, members of this protein family are also called nuclear receptors.

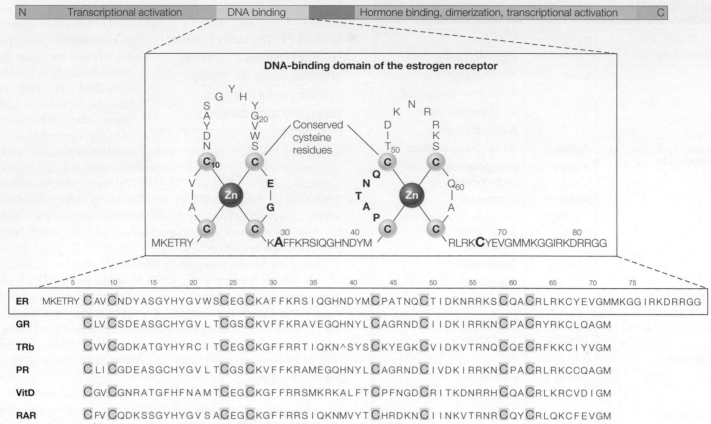

Structural domains within estrogen receptor

| N | Transcriptional activation | DNA binding | Hormone binding, dimerization, transcriptional activation | C |

DNA-binding domain of the estrogen receptor

Conserved cysteine residues

DNA-binding domain sequences of related human receptors, with the conserved cysteine residues highlighted.

ER	MKETRY C AV C NDYASGYHYGVWS C EG C KAFFKRS I QGHNDYM C PATNQ C T I DKNRRKS C QA C RLRKCYEVGMMKGG I RKDRRGG
GR	C LV C SDEASGCHYGV L T C GS C KVFFKRAVEGQHNYL C AGRND C I I DK I RRKN C PA C RYRKCLQAGM
TRb	C VV C GDKATGYHYRC I T C EG C KGFFRRT I QKN^SYS C KYEGK C V I DKVTRNQ C QE C RFKKC I YVGM
PR	C L I C GDEASGCHYGV L T C GS C KVFFKRAMEGQHNYL C AGRND C I VDK I RRKN C PA C RLRKCCQAGM
VitD	C GV C GNRATGFHFNAM T C EG C KGFFRRSMKRKALFT C PFNGD C R I TKDNRRH C QA C RLKRCVD I GM
RAR	C FV C QDKSSGYHYGV S A C EG C KGFFRRS I QKNMVYT C HRDKN C I I NKVTRNR C QY C RLQKCFEVGM

▲ **FIGURE 20.12 The conserved DNA-binding domain in steroid receptors.** At the top are structural domains within steroid receptors, illustrated for the estrogen receptor. In the center is the DNA-binding domain of the estrogen receptor, showing conserved cysteine residues that contact the bound zinc ions. At the bottom are the DNA-binding domain sequences of related human receptors, with the conserved cysteine residues highlighted. ER, estrogen receptor; GR, glucocorticoid receptor; Trb, thyroid hormone receptor; PR, progesterone receptor; VitD, vitamin D receptor; RAR, retinoic acid receptor.

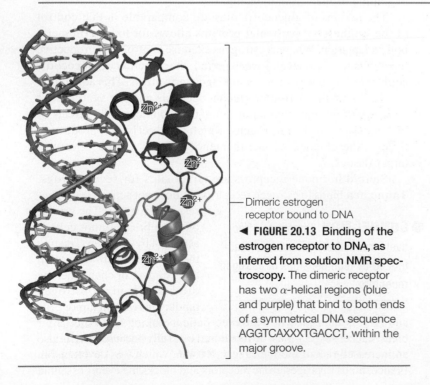

— Dimeric estrogen receptor bound to DNA

◄ **FIGURE 20.13 Binding of the estrogen receptor to DNA, as inferred from solution NMR spectroscopy.** The dimeric receptor has two α-helical regions (blue and purple) that bind to both ends of a symmetrical DNA sequence AGGTCAXXXTGACCT, within the major groove.

to implant a fertilized ovum in the uterus. Hence, RU486 is an effective contraceptive agent, even when taken after intercourse.

Tamoxifen

RU486

20.5 Signal Transduction, Growth Control, and Cancer

Intense current research focuses on genetic differences between cancer cells and the normal cells from which they are derived. These investigations have revealed, in a wide variety of tumor cells, mutationally altered forms or levels of proteins involved in signal transduction—including altered protein kinases, G proteins, nuclear receptors, growth factors, and growth factor receptors. Some tumor cells contain a normal signal transduction protein but in excessive amounts. Genes responsible for such alterations are called **oncogenes.** Investigations of protein products of oncogenes, termed **oncoproteins,** have illuminated the roles of the normal forms of these proteins in regulating cell metabolism and growth and have spotlighted how normal control mechanisms go awry in a cancer cell.

Viral and Cellular Oncogenes

Two developments are noteworthy in the history of the study of cancer, one involving tumor viruses and the other involving genetic analysis of human tumors. Regarding the first development, it has long been known that certain viruses cause cancer in infected animals. The first known tumor virus was **Rous sarcoma virus,** discovered in 1911 by Peyton Rous, who showed that it caused tumors in chickens.

Whether a virus contains RNA (like Rous sarcoma virus) or DNA, certain features are common in viral infections leading to cancer. First, cells become **transformed;** they lose normal growth control mechanisms,

and in cell culture they continue to proliferate under conditions that arrest the growth of normal cells. Second, the transformed cells are themselves tumorigenic; that is, injecting them into animals causes tumors. Third, part or all of the viral genome becomes linearly inserted into chromosomes of the transformed cells. For RNA viruses like Rous sarcoma virus, the viral genome must be converted to double-stranded DNA before this insertion can occur. The viral enzyme that synthesizes DNA from a single-stranded RNA template is called **reverse transcriptase,** and viruses containing this enzyme are called **retroviruses** (see Chapter 22). The product of a reverse transcriptase reaction is called **complementary DNA,** or **cDNA,** because it is complementary in sequence to its RNA template.

A number of nontumorigenic mutants of Rous sarcoma virus exist. Mapping mutations in these strains identified *src,* the viral oncogene responsible for transforming infected cells. Some of these mutants contain extensive deletions, which permitted Raymond Erikson to use nucleic acid hybridization techniques and, in 1978, to clone a cDNA corresponding to the viral *src* gene. Two surprising findings emerged. First, expression of the cloned gene yielded an enzyme with a protein tyrosine kinase activity. Thus, a specific enzyme activity, which might be associated with signal transduction, was also associated with the oncogene product. Second, further analysis showed that sequences corresponding to the viral *src* gene were present in normal cells. This finding suggested that viral oncogenes had their origins in normal cellular genes, or vice versa. One way to explain the transfer of an oncogene, or oncogene precursor, from cells to viruses is to postulate a rare genome excision event, as depicted in **FIGURE 20.14**.

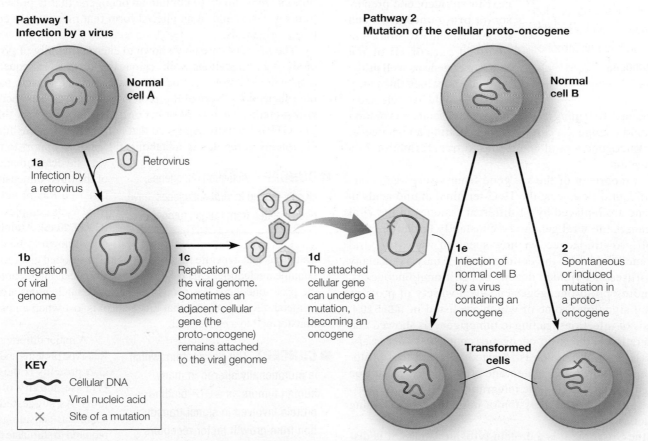

Pathway 1
Infection by a virus

Normal cell A

1a Infection by a retrovirus

Retrovirus

1b Integration of viral genome

1c Replication of the viral genome. Sometimes an adjacent cellular gene (the proto-oncogene) remains attached to the viral genome

1d The attached cellular gene can undergo a mutation, becoming an oncogene

Pathway 2
Mutation of the cellular proto-oncogene

Normal cell B

1e Infection of normal cell B by a virus containing an oncogene

2 Spontaneous or induced mutation in a proto-oncogene

Transformed cells

KEY
〜 Cellular DNA
— Viral nucleic acid
✕ Site of a mutation

▲ **FIGURE 20.14 Pathways by which proto-oncogenes can become oncogenes.** A proto-oncogene is a normal cellular gene that can be converted to an oncogene and cause transformation to a cancer cell. This process can occur in two ways: (1) infection by a virus, which integrates into a chromosomal site next to a proto-oncogene and carries that gene along in its own genome when the virus replicates, or (2) mutation of the cellular proto-oncogene. In the first case, once cellular DNA becomes part of a viral genome, it can undergo mutation that converts the proto-oncogene to an oncogene. The oncogene can then cause transformation when this virus infects another cell.

TABLE 20.4 Oncogene products as elements of signal transduction pathways

Signal Transduction Element	Oncogene	Isolated from	Gene Product
Growth factors	sis	Retrovirus	Platelet-derived growth factor
Growth factor receptors	erbB, neu	Retrovirus	Epidermal growth factor receptor
	fms	Retrovirus	Colony-stimulating factor 1 receptor
	trk	Tumor	Nerve growth factor receptor
	ros	Retrovirus	Insulin receptor
	kit	Retrovirus	PDGF receptor
	flg	Retrovirus	Fibroblast growth factor receptor
Intracellular transducers	src	Retrovirus	Protein tyrosine kinase
	abl	Retrovirus	Protein tyrosine kinase
	raf	Retrovirus	Protein serine kinase
	gsp	Tumor	G protein α subunit
	ras	Tumor, retrovirus	GTP/GDP-binding protein
Nuclear transcription factors	jun	Retrovirus	Transcription factor (AP-1)
	fos	Retrovirus	Transcription factor (AP-1)
	myc	Tumor, retrovirus	Transcription factor
	erbA	Retrovirus	Thyroid receptor

Source: Data from J. D. Watson, M. Gilman, J. Witkowski, and M. Zoller, *Recombinant DNA*, 2nd ed. (New York: Scientific American Books, 1992), p. 339.

● **CONCEPT** Viral oncogenes are errant cellular proto-oncogenes, mostly encoding signal transduction elements that have been taken into viral genomes and have undergone subsequent mutations.

If an infection had caused the insertion of the viral genome next to an oncogene precursor (or **proto-oncogene**), and if a subsequent excision event removed part or all of the proto-oncogene as well as the viral genome, then this faulty excision would have created a novel viral genome, containing a cellular gene. Subsequent evolution of the virus could change the cellular gene, creating an oncogene. Action of the oncogene would contribute to transformation in a subsequent infection.

Sequence comparison of the *src* gene from viruses and cells revealed significant differences; the 19 C-terminal amino acids in the cellular gene are replaced by 11 different residues in the viral gene, which causes the viral gene to be constantly turned on, thus promoting cell growth. Hence, we now speak of *v-src*, the viral form of the gene, and *c-src*, the cellular form. Analysis of many other tumor viruses has yielded dozens of additional oncogenes. The corresponding proto-oncogenes encode a variety of proteins involved in cell signaling, some of which are listed in **TABLE 20.4**. Further analysis of infections leading to tumorigenesis showed that mutational alteration of the proto-oncogene is not always necessary. In some cases, the viral genome is inserted adjacent to a proto-oncogene. Elements of the viral genome stimulate transcription of the DNA sequences flanking the integration site. Thus, tumorigenesis can result from overexpression of normal genes encoding signal transduction machinery.

Although the Src protein is a protein tyrosine kinase, it is distinct from the receptor tyrosine kinases in that it is located in the cytoplasm rather than at the membrane. By convention, the name of the gene (*src*) is italicized, whereas the name of the corresponding protein (Src) is not.

Oncogenes in Human Tumors

How else, other than from tumor viruses, might oncogenes come into existence in human cells? In the late 1980s, Robert Weinberg and others pursued the isolation and analysis of transforming genes from human tumors. Weinberg isolated DNA from bladder cancer tissue and used it to **transfect** normal mouse fibroblasts (connective tissue precursor cells). That is, DNA was introduced into these cells, and transformed cells were isolated after outgrowth of the cells. DNA recovered from transformed cells contained human sequences. After additional rounds of transfection, the human DNA associated with the transformed mouse fibroblasts was sequenced. The transforming gene was nearly identical with a previously described oncogene from *H*arvey *rat* *s*arcoma virus, called the **H-ras** gene. Sequence analysis showed the *H-ras* gene sequence to be identical to *c-ras,* its counterpart in untransformed cells, with but a single difference—a mutation in the twelfth codon that changed a glycine codon in *c-ras* to a valine codon in the oncogene isolated from tumor tissue. Thus, human tumors were shown to contain an oncogene that is present in some tumor viruses, and in an altered form that presumably contributed to tumorigenesis.

The *ras* genes are now known to encode a family of proteins—all of about 21 kilodaltons, with regions homologous to sequences in the α subunit of G proteins. Like the α subunit, the Ras proteins bind guanine nucleotides. Normal Ras proteins possess a GTPase activity, as do G-α proteins, whereas most *ras* oncogene proteins lack this activity. The GTPase activity suggested that normal Ras proteins function like G proteins in regulating metabolism. Lending support to this model was the determination in 1988 of the three-dimensional structure of a Ras protein, crystallized as its complex with GDP (**FIGURE 20.15**). Amino acid residues known to be changed in mutations that generate *ras* oncogenes are positioned close to the bound guanine nucleotide. This positioning suggests that interactions between the proto-oncogene Ras protein and guanine nucleotides are important to metabolic control and that this control is lost when a normal cell is transformed to a cancer cell.

● **CONCEPT** Activated oncogenes, closely related to viral oncogenes, have been isolated from human tumors.

● **CONCEPT** The Ras protein, which is mutationally altered in many human tumors, is a GTP-binding protein involved in signal transduction from growth factor receptors on the plasma membrane to specific gene activation events in the nucleus.

A major difference between Ras-type proteins and the related G-α proteins is the far higher GTPase activity of G-α proteins. As suggested earlier, a set of Ras-activating proteins is required to stimulate the GTPase activity of Ras. The basis for this difference was seen in 1994, with the first structural determination of a G-α protein. G-α,

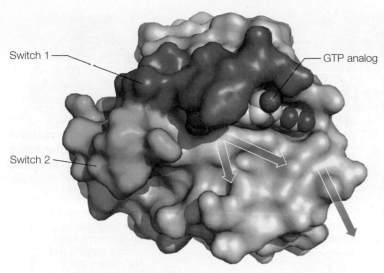

Switch 1

GTP analog

Switch 2

(a) The complex of human *c-H-ras* protein with a nonhydrolyzable GTP analog (red, gray, and blue).

Most prominent changes are in switch 1 and switch 2, both of which are in close contact with bound guanine nucleotide.

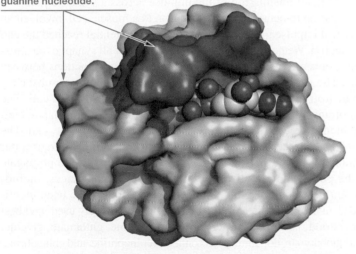

(b) The Ras-GDP complex. Large conformational changes in the protein result from the displacement of GDP by GTP.

▲ **FIGURE 20.15** **Structural basis for Ras activation by GTP.** Most prominent changes are in the regions called switch 1 (blue) and switch 2 (green). PDB ID: 5p21 and 4q21.

but not Ras, proteins contain a conserved arginine residue (R178), which stabilizes the transition state for GTP hydrolysis by interacting with the phosphates of bound GTP.

Many other genetic alterations have been detected in tumor tissue. Some mutations lie in **tumor suppressor genes.** Unlike proto-oncogenes, these are genes that in the normal form suppress tumorigenesis. Loss of normal gene function, however, as in a deletion, leads to tumor formation because tumor suppression is now deficient. The most prominent tumor suppressor gene encodes a protein called **p53** (a protein of 53 kilodaltons). Loss of p53 function leads to tumorigenesis, and at least half of all human tumors examined display p53 gene mutations. p53

is a DNA-binding protein that regulates metabolic processes following DNA-damaging events. If the damage is moderate, then the effect is to retard progression through the cell cycle from G1 into S phase until the damage has been repaired. If the damage is more extensive, then p53 signals that the cell shall undergo **apoptosis,** or programmed cell death. The loss of such a checkpoint would involve inappropriate cell growth, contributing to carcinogenesis. Binding to specific DNA sequences is critical for p53 to function properly, as shown by the structure of a p53–DNA complex, and in particular by the fact that those amino acids in closest contact with DNA are the ones most often mutated in the *p53* gene analyzed in human tumors (**FIGURE 20.16**).

● **CONCEPT** Human tumors contain a series of mutations, affecting both signal transduction components and tumor suppressor genes and gene products.

The Cancer Genome Mutational Landscape

Given what we now know about cell growth control and the biochemical functions of oncoproteins and tumor suppressor proteins in cell signaling, what are we learning about the biology of cancer? Large-scale sequence analysis of DNA from human tumors yields what is called the "mutational landscape of the cancer genome." What has been found is that some tumors have as few as 10 genes that have undergone mutations, while others have as many as 1000. The average colorectal cancer cell has about 80 mutations in protein-coding genes. Other tumors, such as breast, show a completely different mutational landscape, with a different spectrum of mutations. What is found is that the "driver" mutations, those occurring most frequently in different tumors and thought to be related to carcinogenesis, are mutations that affect components of cell-signaling pathways. For example, genes most

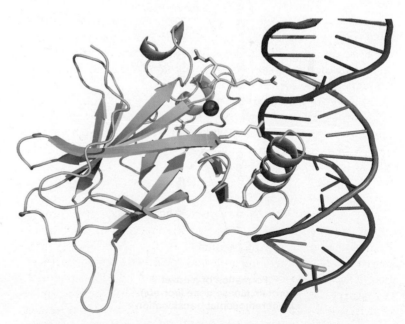

▲ **FIGURE 20.16** **Structure of the p53–DNA complex.** This ribbon drawing shows the DNA-binding domain of one subunit of the homotetrameric p53 (cyan) complexed with an oligonucleotide pair containing the p53 binding site (dark blue and orange). A bound zinc ion is shown in red. Shown in yellow are the six amino acid residues most often changed in mutant p53 proteins. PDB ID: 1tup.

frequently mutated in colorectal cancer are those encoding Ras, p53, and phosphoinositide 3-kinase.

What is the source of the several dozen mutations in the genome of a cancer cell? Much of the genetic abnormality results from unrepaired DNA damage, a topic to which we return in Chapter 23. However, statistical analysis of the genome sequences of tumor cells suggests that many of the mutations in a cancer genome arise in random fashion—a concept which, if correct, makes more distant the time when we will really understand cancer.

At the same time, some notable successes have been achieved, based on what has been learned. In 90% of chronic myelogenous leukemia cases, a chromosomal translocation has occurred, linking portions of chromosomes 9 and 22. This translocation fuses parts of two oncogenes, *c-abl* and

● **CONNECTION** A typical cancer cell has about 80 mutations in protein-coding genes but may have as few as 10 and as many as 1000.

bcr. The *c-abl* gene encodes a protein tyrosine kinase related to *src*. The gene fusion creates a novel protein kinase, which is over-expressed and stimulates growth-promoting pathways. The drug Gleevec™ is a specific inhibitor of this abnormal protein kinase and is a highly specific and effective treatment for this form of leukemia. Another drug, Herceptin, is similarly effective in treating specific forms of breast cancer. These examples suggest that the more we learn about genetic and biochemical changes in individual cancers, the better position science will be in to devise specific therapies.

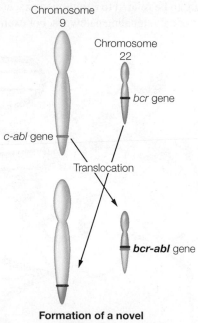

Chromosome 9

Chromosome 22

bcr gene

c-abl gene

Translocation

bcr-abl gene

Formation of a novel protein kinase gene (*bcr-abl*) by chromosomal translocation

20.6 Neurotransmission

In Chapter 10, we described the propagation of an action potential along a neuron as a nervous impulse is transmitted. Recall that this involves a wave of depolarization as ion channels open to admit extracellular sodium and to allow the efflux of intracellular potassium. It was long

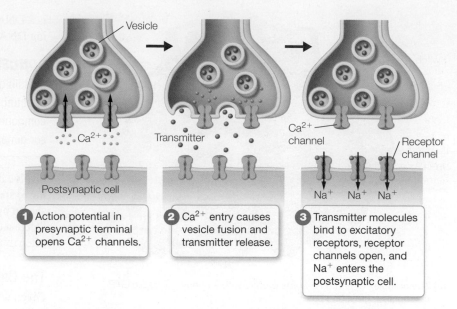

Vesicle

Ca²⁺

Postsynaptic cell

1 Action potential in presynaptic terminal opens Ca²⁺ channels.

Transmitter

2 Ca²⁺ entry causes vesicle fusion and transmitter release.

Ca²⁺ channel

Receptor channel

Na⁺ Na⁺ Na⁺

3 Transmitter molecules bind to excitatory receptors, receptor channels open, and Na⁺ enters the postsynaptic cell.

▲ **FIGURE 20.17** Transmission of a neural impulse across a synapse, such as a cholinergic synapse.

thought that transmission of the impulse across a synaptic junction, from neuron to neuron or from a neuron to a muscle cell, involved an electrical impulse triggered when the action potential reached the end of a neuron. We now know that fewer than 1% of all synaptic transmission events are mediated electrically, through gap junctions between cells. The vast majority involve the release of a chemical substance, the neurotransmitter, from the upstream, or **presynaptic,** cell. The neurotransmitter then diffuses through an intercellular synaptic cleft and binds to receptors in the downstream, or **postsynaptic,** cell. The postsynaptic receptor is a ligand-gated ion channel; neurotransmitter binding triggers events that allow the nervous impulse to propagate in the postsynaptic cell. About 100 different substances, including amino acids, biogenic amines, and peptides, have been identified as neurotransmitters in the brain. The most widely used, and best understood, neurotransmitters are acetylcholine, glutamate, glycine, γ-aminobutyrate, dopamine, serotonin, norepinephrine, and epinephrine.

The Cholinergic Synapse

Our earliest biochemical understanding of neurotransmission involved the action of acetylcholine as the transmitter at the neuromuscular junction, where nervous stimulation of a muscle cell triggers its contraction. These studies generated the scheme for neurotransmission shown in **FIGURE 20.17**. When acetylcholine is the neurotransmitter, this event is called a **cholinergic synapse;** when dopamine is the neurotransmitter, it is known as a **dopaminergic synapse.** Acetylcholine is synthesized by the coenzyme A-dependent acetylation of choline, which in turn was formed by the breakdown of phosphatidylcholine.

The synaptic cleft is about 20 nm wide. The neurotransmitter is stored in vesicles within a bulb at the end of the presynaptic cell, with about 5000 molecules per vesicle. Arrival of the action potential opens voltage-gated calcium channels, triggering an influx of Ca²⁺ to the bulb, step **1**. This in turn causes the vesicles to fuse with the plasma membrane at the end of the bulb, spilling acetylcholine into the extracellular space of the synaptic cleft, step **2**. The neurotransmitter diffuses through the cleft and binds to receptors in the postsynaptic cell membrane, step **3**. This in turn triggers the opening of channels, allowing an influx of Na^+ and generating an action potential. If the postsynaptic cell is a muscle cell, then the action potential opens Ca^{2+} channels, and the resulting rise in intracellular Ca^{2+} concentration triggers contraction. In the meantime, on the presynaptic side, some of the released acetylcholine is taken back into the cell, and an acetylcholine transporter protein refills vesicles with the neurotransmitter, exchanging protons for acetylcholine with the help of a **vacuolar ATPase.** The remainder of the acetylcholine in the cleft is inactivated by hydrolysis, terminating the signaling event. The entire process occurs about 1000 times per second. Similar events occur during neuron-to-neuron transmission, except that here the increase in Na^+ in the postsynaptic cell triggers the establishment of an action potential and its transmission down that cell.

Structural studies of a receptor called the **nicotinic acetylcholine receptor,** because it binds the alkaloid nicotine, show it to be composed of five subunits, two of which are identical and each of which has five α-helical regions that span the membrane. As shown in **FIGURE 20.18,** the subunits interact to form a channel through the membrane. Reconstitution of this multisubunit protein into membrane vesicles shows that the addition of acetylcholine causes ions to flow through it. Thus, this receptor is a gated channel that undergoes a conformational change and opens a pore in response to binding of the neurotransmitter. Our understanding of the structure and function of this receptor originally involved the isolation of the receptor from organs of the electric eel (*Electrophorus*) or electric ray (*Torpedo*). Both of these animals have the receptor densely packed in an organ, called the electroplaque, which allows each animal to shock its prey by generating potentials of several hundred volts.

Within the intersynaptic cleft, acetylcholine is rapidly hydrolyzed by acetylcholinesterase, thereby destroying excess neurotransmitter and restoring the resting potential in the postsynaptic cell membrane. Acetylcholinesterase is the target for several organophosphate compounds such as the nerve gas **sarin,** which was developed as a chemical warfare agent (and used in 2013 during the civil war in Syria), which causes paralysis by reacting with an active site serine and irreversibly inactivating acetylcholinesterase (Figure 8.32). Other agents act by binding to the receptor itself. **Tubocurarine,** one such agent, blocks the channel in the closed position and is considered a receptor antagonist; this is in contrast to **nicotine,** which activates the receptor and is an agonist. Tubocurarine is derived from curare, originally used by aboriginal people to poison arrow tips and paralyze human or animal prey. In the hands of qualified medical personnel, tubocurarine is a useful muscle relaxant.

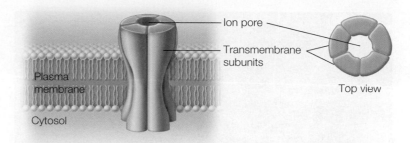

(a) Schematic model of the nicotinic acetylcholine receptor. Five subunits combine to form a transmembrane structure with an ion pore in the center.

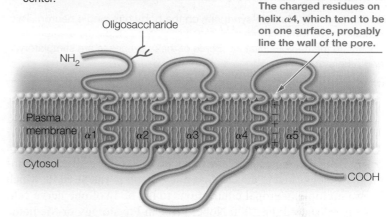

(b) Structure of an individual subunit. There are four different kinds of subunits, but their sequences are all similar, and each individual subunit has the kind of structure depicted here. Five α helices (α1 to α5) in each subunit traverse the membrane.

▲ **FIGURE 20.18** The nicotinic acetylcholine receptor.

transmission and transmission at the neuromuscular junction. Some interneuronal transmission events, however, are excitatory, and others are inhibitory. In excitatory transmission, binding the transmitter at the postsynaptic receptor causes an influx of Na^+, which depolarizes the membrane and stimulates transmission of the action potential. γ-Aminobutyric acid (GABA) is the principal inhibitory neurotransmitter. Binding GABA triggers an influx of Cl^-, which causes hyperpolarization and inhibits the propagation of the action potential.

$$H_3C-\overset{\overset{O}{\|}}{C}-O-CH_2-CH_2-\overset{+}{N}(CH_3)_3$$

Acetylcholine

$$\Big\downarrow\quad H_2O$$

Acetylcholinesterase

$$\searrow CH_3COO^-$$

$$HO-CH_2-CH_2-\overset{+}{N}(CH_3)_3$$

Choline

Tubocurarine, an antagonist

Nicotine, an agonist

Fast and Slow Synaptic Transmission

For some time, it was thought that the model of synaptic transmission obtained with acetylcholine receptors described all neuron-to-neuron

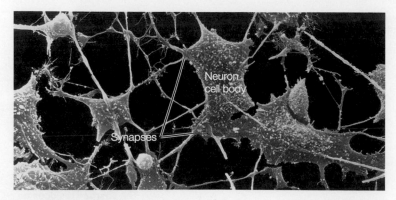

▲ **FIGURE 20.19 Multiple synapses on the body of a single neuron.** This scanning electron micrograph demonstrates the complexity of interconnection in the nervous system. Some of these synapses are stimulatory, and others are inhibitory.

The situation is considerably more complex, however. Paul Greengard has pointed out that each of the 100 billion cells in the human brain communicates directly with about 1000 other cells. The complexity is suggested in the electron micrograph in **FIGURE 20.19**, which shows multiple neuronal connections to the body of one nerve cell. Greengard shared the 2000 Nobel Prize in Physiology or Medicine with Arvid Carlsson and Eric Kandel for their discoveries regarding slow synaptic transmission events, which occur with a time scale of hundreds of milliseconds to several minutes. In these cases, neurotransmitter binding stimulates intracellular metabolic events comparable to those resulting from the actions of hormones. Some of these events are mediated through second messengers, such as cAMP, cGMP, and phosphoinositides.

Actions of Specific Neurotransmitters

Most of the fast excitatory synapses in the brain use glutamate as the neurotransmitter, and most of the fast inhibitory synapses use GABA. Excessive firing of glutamatergic synapses, with consequent damage to the central nervous system, can result from ingesting too much glutamate. For this reason monosodium glutamate, found in soy sauce, was removed from baby food formulas some years ago, where it was being used as a flavor enhancer. It was removed because the developing nervous system is particularly susceptible to this kind of damage.

Both glutamate and GABA participate in slow synapses as well, and it now seems clear that biogenic amines and peptide neurotransmitters participate only in slow synaptic transmission. Of particular interest are synapses for which dopamine is the neurotransmitter. Arvid Carlsson and others showed that several important psychiatric disorders involve abnormalities in dopamine signaling, including Parkinsonism, schizophrenia, drug addiction, and attention-deficit/hyperactivity disorder (ADHD). Parkinsonism involves the death of dopamine-producing nerve cells in a portion of the brain called the **substantia nigra.** Success in treating some forms of Parkinsonism has resulted from treatment with massive doses of L-dopa, which can traverse a permeability blockade called the **blood–brain barrier** and then undergo decarboxylation within the brain to replenish depleted dopamine stores.

The relationship with schizophrenia is seen because some effective antischizophrenic drugs antagonize the effect of dopamine at one or more classes of receptors by binding to the receptor and blocking the binding of dopamine. The drug **chlorpromazine** is not closely related to dopamine structurally, but it effectively antagonizes dopamine receptor binding. However, several drugs of abuse, including **mescaline** and **amphetamine,** are closely related to dopamine, and they act as dopamine agonists, binding to dopamine and mimicking its effects. Dopamine is considered a "pleasure agent," and many abused drugs stimulate the firing of dopaminergic synapses, either by raising dopamine levels or by mimicking the natural neurotransmitter. In a recent study, a targeted deletion in the mouse of D4, one of the four classes of dopamine receptors, caused the animals to become hypersensitive to ethanol, cocaine, and methamphetamine.

Synapses using serotonin as the neurotransmitter are also involved in the pathophysiology related to schizophrenia. **Lysergic acid diethylamide** (LSD) is an indole derivative, like serotonin. LSD acts as a serotonin agonist, binding to a class of serotonin receptors and mimicking its effect.

● **CONNECTION** Many drugs of abuse act through interactions with dopamine or serotonin receptors.

Finally, **Ritalin,** used to treat ADHD, promotes the release of dopamine. This stimulant effect is paradoxical because the drug is used to calm hyperactive children. Recent research suggests, however, that its calming effect may result more from elevating levels of serotonin.

Not all drugs of abuse involve slow synaptic transmission, however, and understanding the actions of dopamine and serotonin is only a small part of grasping the complexities of schizophrenia. Recent work has implicated glutamate receptors as well in the control of addictive behavior. Inhibiting glutamate neurotransmission in rats modulates the compulsive drug-seeking behavior after an initial drug experience; neuroscientists are excited at the prospect of developing therapies involving glutamate antagonists, which might increase the likelihood that an addict will remain "clean" after treatment. Glutamate receptors are also strongly linked to the action of phencyclidine (PCP, or "angel dust"). PCP blocks glutamate binding to the N-methyl-D-aspartate (NMDA) class of glutamate receptors, thereby inducing a schizophrenia-like state thought until recently to result from decreased glutamatergic neurotransmission. However, neuroscientists have found that lowering brain glutamate levels with another drug greatly diminishes the effectiveness of PCP; such an approach might therefore aid in treating schizophrenia.

● **CONNECTION** Many drugs of abuse act as agonists or antagonists of central nervous system neurotransmitters.

Ritalin
(methyl phenidylacetate)

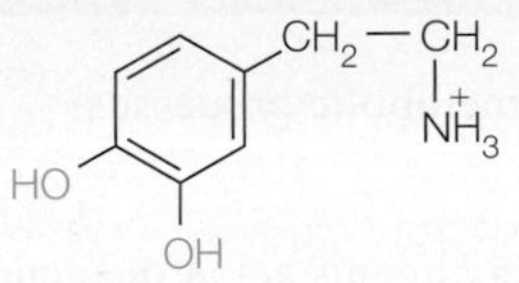

Dopamine

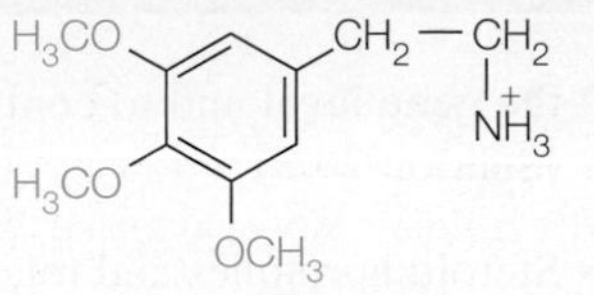

Mescaline

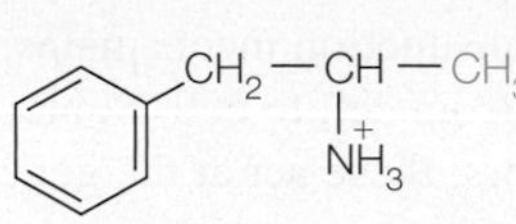

Amphetamine

Dopamine agonists

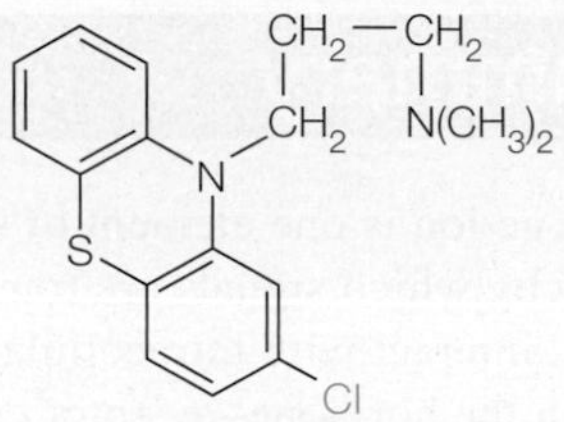

Chlorpromazine,
a dopamine antagonist

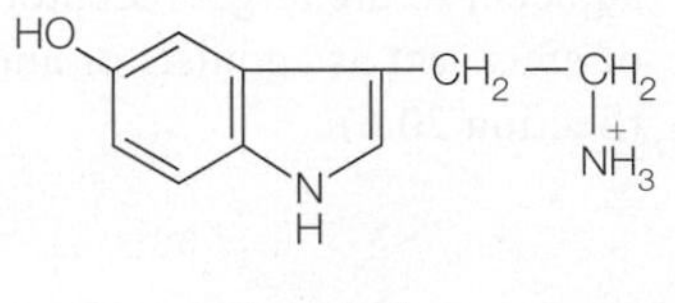

Serotonin

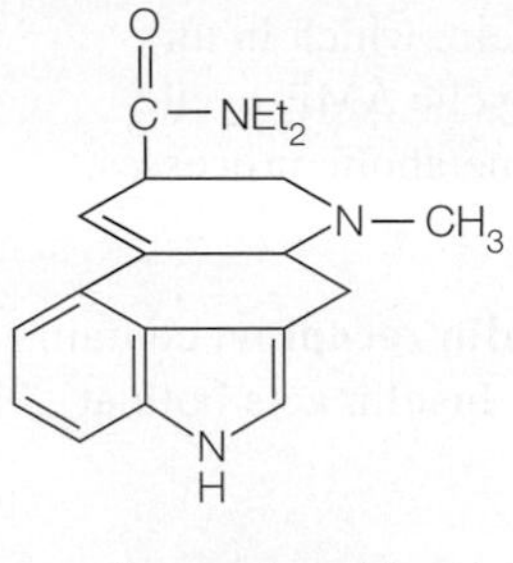

Lysergic acid
diethylamide,
a serotonin agonist

Drugs That Act in the Synaptic Cleft

So far, the psychopharmacological agents we have discussed act primarily as receptor agonists and antagonists. Other important classes of drugs affect neurotransmitter metabolism in the synaptic cleft. Catecholamines are catabolized in the cleft either by methylation (**catecholamine O-methyltransferase,** COMT) or by oxidation (**monoamine oxidase,** MAO). These enzymes limit the biological effects of catecholamine neurotransmitters, just as acetylcholinesterase limits the firing of cholinergic neurons. A number of drugs used to treat depression are inhibitors of either COMT or MAO, and their action increases the effective amounts of neurotransmitter by limiting its breakdown. A more recently developed drug, **fluoxetine,** marketed as Prozac™, acts as a selective reuptake inhibitor. Secreted neurotransmitter has three possible fates—binding to postsynaptic receptors, catabolism in the cleft, or reuptake into the presynaptic cell for re-packaging into storage vesicles. Prozac selectively blocks the reuptake of serotonin, thereby increasing the amount that reaches the postsynaptic side and potentiating serotonergic synapses. Originally marketed as an antidepressant drug, Prozac has found use against a range of psychiatric disorders.

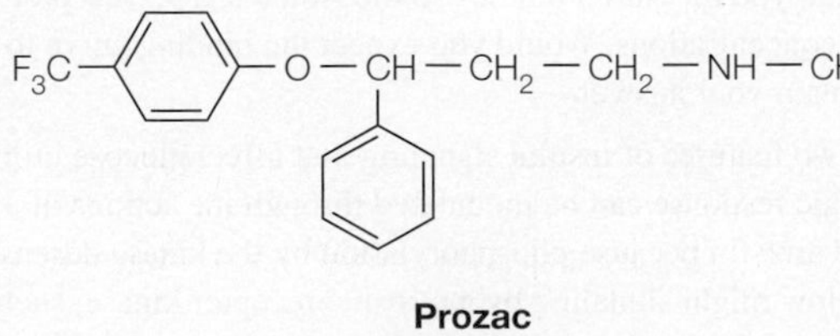

Prozac
(fluoxetine)

Peptide Neurotransmitters and Neurohormones

Finally, a number of peptides, including somatostatin, neurotensin, and the **enkephalins,** act as neurotransmitters. The enkephalins, along with **β-endorphin,** also function as **neurohormones,** acting to modify the ways in which nerve cells respond to transmitters. The endorphins were discovered in the 1970s, as the result of Solomon Snyder's efforts to understand the effects of opiate drugs, such as **morphine.** After detecting morphine receptors in the human brain, Snyder realized that the receptors must have a natural endogenous ligand, since morphine is derived from a nonhuman source, the poppy. This work led to the isolation of the endorphins, small peptides that function as natural analgesics. The modification of neural signals by these substances appears to be responsible for the insensitivity to pain that is experienced under conditions of great stress or shock. The effectiveness of opiate analgesics such as morphine is a consequence, perhaps accidental, of the neurohormones recognizing these opiates despite being structurally different from neurohormones.

Summary

- Hormone action is one element of signal transduction mechanisms, processes by which signals are transmitted from cell to cell. Some hormones interact with intracellular receptors; these act at the gene level, with the hormone–receptor complex affecting transcription of specific genes in the target tissue. Other hormones interact with plasma membrane receptors. (Section 20.1)

- Plasma membrane receptors comprise part of a modular system, in which the receptor interacts with a G protein, which in turn interacts with a second messenger, such as cyclic AMP, cyclic GMP, or phosphoinositides, to affect target metabolic processes. (Section 20.2)

- Another class of receptors, notably the **insulin receptor,** contains an intrinsic protein tyrosine kinase activity. Insulin acts both at

the gene level and to control target metabolic processes. (Section 20.3)

- Steroid hormones and related signaling agents act at the gene level, through binding to receptors that controls transcription of target genes (Section 20.4).

- Aberrant signal transduction processes are associated with cancer. (Section 20.5)

- Neurotransmission is a form of signal transduction in which the signaling agent, a neurotransmitter, is released immediately adjacent to the target receptor. Many psychiatric drugs and drugs of abuse act as agonists or antagonists of neurotransmitters. (Section 20.6)

Problems

Enhanced by
Mastering Chemistry
for Biochemistry

Mastering Chemistry for Biochemistry provides select end-of-chapter problems and feedback-enriched tutorial problems, animations, and interactive figures to deepen your understanding of complex topics while practicing problem solving.

Answers to red problems are available in the Answer Appendix.

1. Caffeine is an inhibitor of cyclic AMP phosphodiesterase. How would drinking several cups of coffee affect muscle function? How might it affect lipid metabolism?

2. List two or three factors that make it advantageous for peptide hormones to be synthesized as inactive prohormones that are activated by proteolytic cleavage.

3. Describe a mechanism by which a steroid hormone might act to increase intracellular levels of cyclic AMP.

4. Signaling molecules interact with cells through specific macromolecular receptors. For each of the four receptors identified below, list all characteristics, by number, which accurately describe that receptor.
   (a) An adrenergic receptor
   (b) A steroid receptor
   (c) The LDL receptor
   (d) The insulin receptor
      (1) Located at the cell surface
      (2) Associated with the protein clathrin
      (3) Ligand binding stimulates the activity of phospholipase C
      (4) A transmembrane protein
      (5) A DNA-binding protein
      (6) Located in the cell interior
      (7) Receptor–ligand complex moves to the lysosome
      (8) Receptor–ligand complex becomes concentrated in the nucleus
      (9) Receptor activation can inhibit the synthesis of glycogen
      (10) The hormone–receptor complex activates specific gene transcription
      (11) Internalization decreases the synthesis of cholesterol esters
      (12) Action of this receptor diminishes the synthesis and activity of β-hydroxy-β-methylglutaryl-CoA reductase (HMG-CoA reductase)
      (13) This receptor activates its own synthesis
      (14) Biological activity of this receptor involves interaction with guanine nucleotide–binding proteins

   (15) This receptor has a protein kinase activity
   (16) Not known to act through a second messenger

5. Upon activation by a receptor, a G protein exchanges bound GDP for GTP, rather than phosphorylating GDP that is already bound. Similarly, the α subunit–GTP complex has a slow GTPase activity that hydrolyzes bound GTP, rather than exchanging it for GDP. Describe experimental evidence that would be consistent with these conclusions.

6. Answer each of the following questions in about one sentence:
   (a) What is the biochemical basis for the action of Viagra?
   (b) What is the biochemical basis for the action of Prozac?
   (c) Oral administration of S-adenosylmethionine has been reported to be effective in treating depression. Suggest a possible explanation.

7. Write a balanced equation for the hydrolysis of cGMP, catalyzed by cGMP phosphodiesterase. Would you expect an inhibitor of this enzyme to potentiate or antagonize the action of Viagra? Explain.

8. Lithium ion inhibits the synthesis of inositol trisphosphate by inhibiting a reaction in the breakdown of inositol trisphosphate (see page 646). Explain this apparent paradox.

9. Suppose that you measured binding to the isolated EGF receptor of EGF at various concentrations. Would you expect the binding curve to be hyperbolic? Explain your answer.

10. Describe two features of insulin signaling that affect glucose utilization. A β-adrenergic response can be modulated through the actions of a receptor kinase and arrestin because phosphorylation by the kinase desensitizes the receptor. How might signaling by a tyrosine receptor kinase, such as the insulin receptor, be modulated?

11. Suppose that a G protein undergoes a mutation that allows the exchange of bound GTP for GDP to occur in the absence of G protein binding to a receptor. How might this mutation affect signaling involving a GPCR? Which subunit of the G protein is most likely affected by the mutation?

12. Studies on cAMP actions in cultured cells usually involve adding to the cell culture not cAMP, but dibutyryl cAMP (see structure). Why is this structural modification necessary? How could you test the premise that di-Bu-cAMP has the same biochemical effects as cAMP?

Dibutyryl cAMP

13. Bisphenol A is widely used as a building block in polymer synthesis and is found in the polycarbonate hard plastics of reusable drink containers, DVDs, cell phones, and other consumer goods. Bisphenol A is reported to have estrogenic activity, and its widespread occurrence in our environment is a potential concern. Describe one or two biochemical experiments that could be done to compare the activity of bisphenol A with that of its estra-diol, its structural relative.

Bisphenol A

14. In early studies of adrenergic signaling, it was thought that the epinephrine receptor and adenylate cyclase were one and the same protein. What kind of evidence would prove otherwise?

15. Mutations that inactivate p53 have a recessive phenotype, whereas mutations affecting Ras are dominant. Explain the difference.

16. GTPγS is a nonhydrolyzable analog of GTP. Suppose this compound were added to a cell-free system containing active components of an adrenergic

signaling system. What consequences would you expect? What would be the effects on cAMP levels?

GTPγS

17. Suppose that you had a monoclonal antibody that recognized phosphotyrosine. How would you expect that antibody to affect insulin signaling?

18. Pertussis toxin ADP ribosylates the α subunit of G_i. How would you expect this toxin to affect blood pressure?

19. β-adrenergic receptors are subject to phosphorylation at several serine residues by **β-adrenergic receptor kinase,** and this is followed by binding to the phosphorylated sites by a protein called **arrestin.** What do you think might be the purpose of these modifications?

20. Explain the effectiveness of Gleevec in treating chronic myelogenous leukemia.

21. Herceptin is an antibody that is used to treat certain forms of breast cancer by binding to a class of estrogen receptors. What is the basis for its effectiveness in treating certain forms of breast cancer?

22. How did analysis of the mechanism of chymotrypsin action help lead to understanding of the action of the paralytic action of the nerve poison Sarin?

Pg 640
last paragraph

Sarin

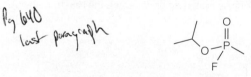

References

For a list of references related to this chapter, see Appendix II.

INTEGRATING SIGNALING AND METABOLISM

This figure shows how different inputs, each acting via a different signaling mechanism, can regulate metabolism in a liver cell. This is one example of how metabolism can be adjusted in response to different conditions by different signaling mechanisms. Similar kinds of signal integration lead to the appropriate regulation of all the activities in a cell, and coordination of functions across the entire organism.

- The dual function enzyme 6-phosphofructo-2-kinase/fructose-2,6-bisphosphatase (PF-K2/FBPase-2) is a key regulator of glycolysis and gluconeogenesis (see Section 12.6)
- Blood glucose levels are maintained by signaling cascades initiated by the hormones insulin and glucagon (see Chapters 17 and 20), which regulate PF-K2/FBPase-2 activity by dephosphorylation and phosphorylation, respectively (Figure 12.17).
- In addition, transcriptional regulation of PF-K2/FBPase-2 (and other glycolytic enzymes) by the HIF-1α transcription factor allows for increased glycolysis during hypoxia.

1 In hypoxic conditions, signaling through a receptor tyrosine kinase (RTK) pathway leads to inactivation of the machinery that degrades HIF-1α.

HIF-1α PATHWAY Increases glucose utilization in hypoxic conditions

Hypoxia-inducible factor 1 (HIF-1α) is a transcription factor that is normally broken down by ubiquitination and proteosomal degradation. However, in hypoxic conditions, signaling through a receptor tyrosine kinase (RTK) pathway leads to inactivation of the machinery that degrades HIF-1α. This pathway is particularly relevant in the formation of tumors, when cancer cells do not have sufficient blood supply, and overcome these hypoxic conditions by increasing glycolysis via the HIF-1α pathway.

▶ SECTION 13.7

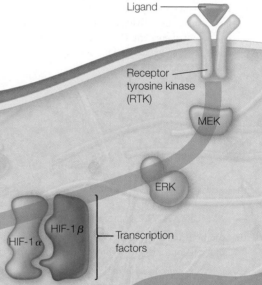

Ligand

Receptor tyrosine kinase (RTK)

MEK

ERK

HIF-1β — HIF-1α — Transcription factors

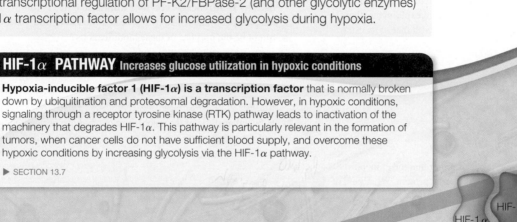

2 HIF-1α is transported into the nucleus.

3 HIF-1α increases transcription of its target genes, some of which are enzymes of glycolysis.

DNA (gene)

HIF-1 transcription factors

Transcription

mRNA

Nucleus

Nuclear export

mRNA processing

Folding

Translation

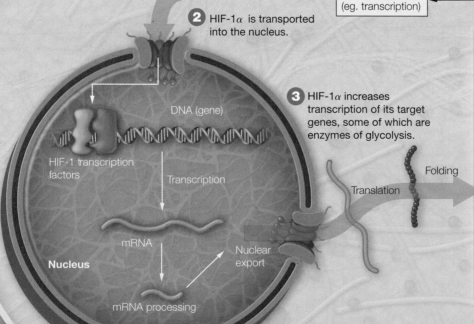

Other effects (eg. transcription)

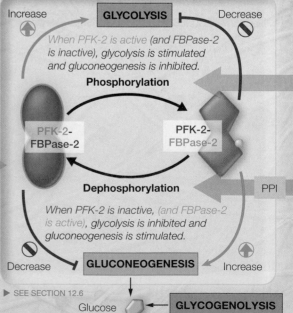

Increase — **GLYCOLYSIS** — Decrease

When PFK-2 is active (and FBPase-2 is inactive), glycolysis is stimulated and gluconeogenesis is inhibited.

Phosphorylation

PFK-2-FBPase-2

PFK-2-FBPase-2

Dephosphorylation

PPI

When PFK-2 is inactive, (and FBPase-2 is active), glycolysis is inhibited and gluconeogenesis is stimulated.

Decrease — **GLUCONEOGENESIS** — Increase

▶ SEE SECTION 12.6

Glucose — **GLYCOGENOLYSIS**

Mastering Chemistry for Biochemistry

Mastering Chemistry for Biochemistry provides select end-of-chapter problems and feedback-enriched tutorial problems, animations, and interactive figures to deepen your understanding of complex topics while practicing problem solving.

INSULIN SIGNALING PATHWAY Increases glucose transport into cells and decreases blood glucose levels

When blood glucose levels are high, the pancreatic β-cells secrete the hormone insulin. Insulin binds to its receptor on the liver cell, causing autophosphorylation and activation of the receptor. Activation of the insulin signaling pathway leads to activation of two main proteins– pAKT, and Ras. Dephosphorylated PFK-2-FBPase-2 increases glycolysis while decreasing gluconeogenesis. These two activities of insulin (increased glucose transport into cells, and increased glucose utilization) result in an overall decrease in the levels of blood glucose.

▶ SEE CHAPTER 12, CHAPTER 17, AND SECTION 20.3

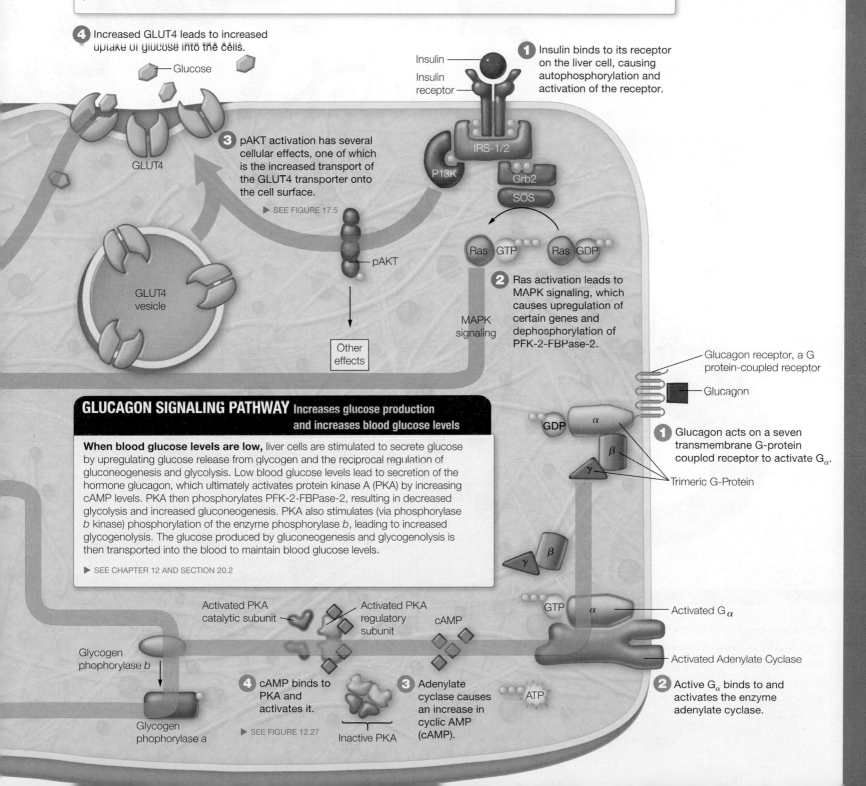

4 Increased GLUT4 leads to increased uptake of glucose into the cells.

— Glucose

GLUT4

3 pAKT activation has several cellular effects, one of which is the increased transport of the GLUT4 transporter onto the cell surface.

▶ SEE FIGURE 17.5

GLUT4 vesicle

— pAKT

Other effects

Insulin —

Insulin receptor

1 Insulin binds to its receptor on the liver cell, causing autophosphorylation and activation of the receptor.

IRS-1/2

P13K

Grb2

SOS

Ras GTP   Ras GDP

2 Ras activation leads to MAPK signaling, which causes upregulation of certain genes and dephosphorylation of PFK-2-FBPase-2.

MAPK signaling

Glucagon receptor, a G protein-coupled receptor

— Glucagon

GLUCAGON SIGNALING PATHWAY Increases glucose production and increases blood glucose levels

When blood glucose levels are low, liver cells are stimulated to secrete glucose by upregulating glucose release from glycogen and the reciprocal regulation of gluconeogenesis and glycolysis. Low blood glucose levels lead to secretion of the hormone glucagon, which ultimately activates protein kinase A (PKA) by increasing cAMP levels. PKA then phosphorylates PFK-2-FBPase-2, resulting in decreased glycolysis and increased gluconeogenesis. PKA also stimulates (via phosphorylase b kinase) phosphorylation of the enzyme phosphorylase b, leading to increased glycogenolysis. The glucose produced by gluconeogenesis and glycogenolysis is then transported into the blood to maintain blood glucose levels.

▶ SEE CHAPTER 12 AND SECTION 20.2

GDP α

β

γ

1 Glucagon acts on a seven transmembrane G-protein coupled receptor to activate G_α.

Trimeric G-Protein

γ β

GTP α — Activated G_α

Activated PKA catalytic subunit

Activated PKA regulatory subunit

cAMP

— Activated Adenylate Cyclase

Glycogen phophorylase b

4 cAMP binds to PKA and activates it.

▶ SEE FIGURE 12.27

Inactive PKA

3 Adenylate cyclase causes an increase in cyclic AMP (cAMP).

ATP

2 Active G_α binds to and activates the enzyme adenylate cyclase.

Glycogen phophorylase a

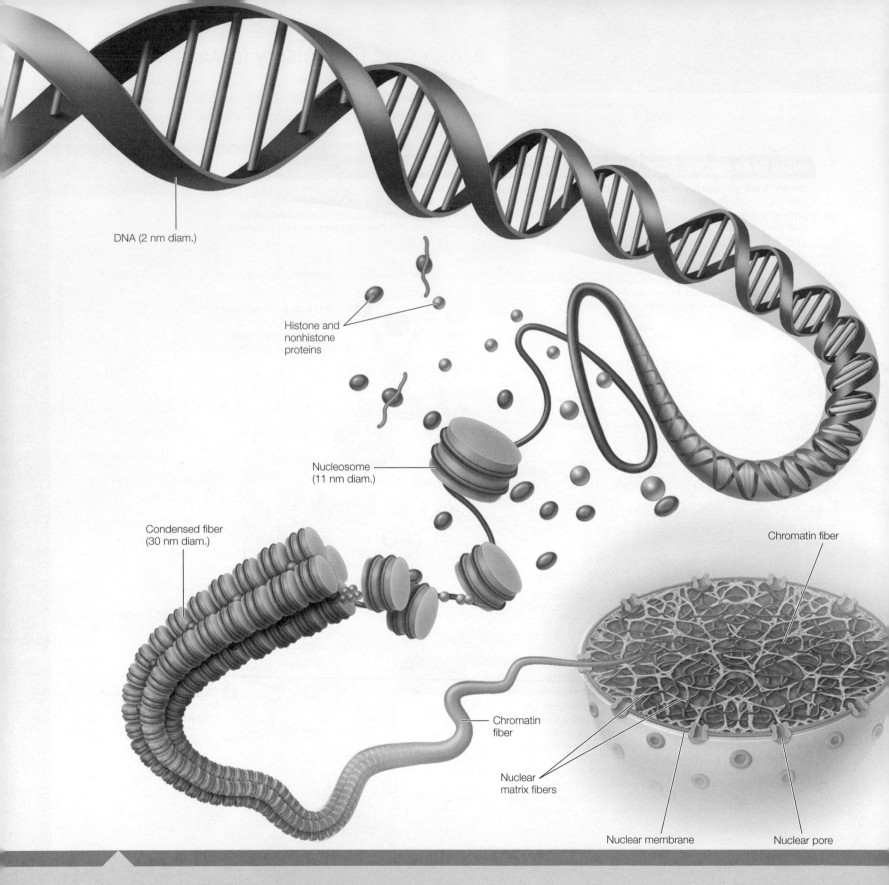

DNA (2 nm diam.)

Histone and
nonhistone
proteins

Nucleosome
(11 nm diam.)

Condensed fiber
(30 nm diam.)

Chromatin fiber

Chromatin
fiber

Nuclear
matrix fibers

Nuclear membrane

Nuclear pore

Packaging eukaryotic DNA within the nucleus. In order to fit within the
nucleus, DNA is wrapped about histone core particles, as described in
this chapter, forming a fiber some 11 nm in diameter. This fiber is further
condensed by winding into a greatly thickened and shortened structure,
which undergoes additional winding and compaction until it can fit within
the nucleus.

Genes, Genomes, and Chromosomes

IN THIS LAST major section of the book, we are concerned with the storage, retrieval, processing, and transmission of biological information—processes we might call information metabolism (to distinguish it from intermediary metabolism). In intermediary metabolism, all the information specifying the nature of a chemical reaction lies within the three-dimensional structure of the enzyme involved. That structure determines which substrates are bound and which reactions are catalyzed. Of course, all metabolic reactions are controlled ultimately by genetic information, which specifies the structures and properties of enzymes. However, the reactions we encounter from here onward are distinguished by the direct involvement of genetic information—specifically, the requirement for a *template*, which functions along with enzymes to specify the reaction catalyzed. The biological templates, nucleic acids, generally play a passive role, determining which substrates are bound and leaving catalysis to enzymes, although ribozymes—RNA enzymes—present an important exception to this generalization (Chapter 8).

The basic processes—DNA replication, transcription, and translation—are described in detail in subsequent chapters. Here, however, we focus on the organization and nature of the genes that carry and transmit that information, as well as the genomes and chromosomes within which individual genes reside.

As defined in Chapter 4, a gene is a chromosomal segment that either encodes a single polypeptide chain or RNA molecule or plays a regulatory function. Although we might call this section of our discipline genetic biochemistry, it may now be more appropriate to call it *genomic* biochemistry. That is because the powerful methods used to determine the complete nucleotide sequences of hundreds of organisms allow us to think of biological processes in a more global sense. Whereas the sequencing and cloning of individual genes, and the analysis of cloned genes and recombinant enzymes, allow insight into metabolism at the level of individual or grouped reactions, we can now think of cellular and organismal function in terms of the coordinated expression of large blocks of genes—an integrated system.

21.1 Bacterial and Viral Genomes

With the exception of RNA viruses, the genomes of all organisms are made up of DNA sequences containing just four different nucleotides (plus small amounts of a few modified nucleotides, as in transfer RNA or 5-methylcytosine in DNA). However, there is enormous variability in genome size and in the physical state of the genome within its intracellular milieu. Single-celled organisms have relatively simple metabolic needs and so can thrive with as few as several hundred genes.

Viral Genomes

Viruses, which coopt the metabolic machinery of the cells they infect, are even simpler. On the one hand, the smallest DNA viruses have a dozen genes or fewer, whereas some RNA viruses have only four. On the other hand, *Mimivirus*, which was discovered in 1992, has a larger genome than those of many bacteria. The more recently discovered *pandoraviruses* are even larger. **TABLE 21.1** lists features of the genomes of several viruses that are either useful in research or related to important diseases, or both. The *E. coli* genome is listed for comparison.

Bacterial Genomes—The Nucleoid

All cells, whether bacterial, archaeal, or eukaryotic—as well as viruses—must be able to pack gigantic nucleic acid molecules into severely limited space. For example, the *E. coli* genome, a DNA molecule more than 1 mm in length, is packed into a cell no longer than 5 μm. Moreover, the packing must be carried out in an organized way, so that the molecule can undergo rapid replication on demand, and individual genes can be accessible to RNA polymerase and gene-regulatory factors. Bacterial and eukaryotic cells solve the problem in quite different ways, representing one of the fundamental distinctions between prokaryotes and eukaryotes (see the opening illustration and **FIGURE 21.1**).

In bacteria, chromosomal compaction occurs largely by negative supercoiling and by organizing the genome into separate loops of supercoiled DNA, with each loop bound to protein. Typically, each loop is some 50 kbp in length, as schematized in Figure 21.1. This compacted

TABLE 21.1 Properties of some bacterial and viral genomes

Organism or Virus	Genome Size (bp or bases)	Number of Genes	Physical Nature of the Genome
Escherichia coli	4,639,221	~4400	Circular duplex
Bacteriophage T4	168,889	~175	Linear duplex, circularly permuted*
Bacteriophage T7	39,936	~35	Linear duplex, small repetition each end
Bacteriophage λ	48,502	~50	Linear duplex, single-stranded ends
Mimivirus	1,181,404	979	Linear duplex
Influenza virus	~13,500	12	Single-stranded RNA
Human immunodeficiency virus (HIV)	9,749	23	Single-stranded RNA
Variola (smallpox)	~186,000	~195	Linear duplex
Bacteriophage φX174	5,387	11	Circular single-stranded DNA
Bacteriophage M13	6,407	11	Circular single-stranded DNA
Simian virus 40	5,226	6	Circular duplex DNA
Tobacco mosaic virus	~6,400	4	Single-stranded RNA
Zika virus	10,794	13	Single-stranded RNA

*Circularly permuted means that all genomes have the same linear sequence of genes, but the beginning and ending points vary among different genomes.

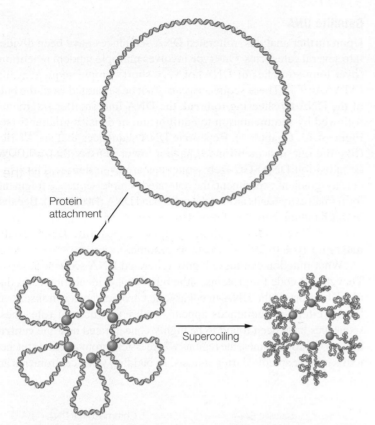

Protein
attachment

Supercoiling

(a) Independent domains of supercoiling, each stabilized by binding to protein.

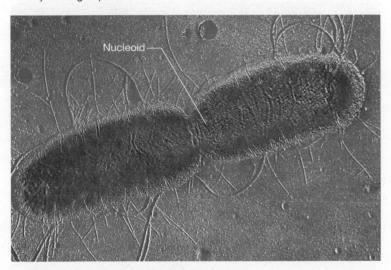

Nucleoid

(b) Electron micrograph of a dividing E. coli cell.

▲ **FIGURE 21.1 Structure of a bacterial nucleoid.** The nucleoid is the green and orange structure in the cell interior.

structure, called a **nucleoid,** exists within the cytosol of a bacterial cell, with a small number of attachment points to the membrane. See also Figure 1.9. As schematized in the chapter-opening illustration, eukaryotic genomes are organized quite differently, with DNA being confined to the nucleus and wrapped around the outside of protein assemblies. We return to this topic on page 671.

21.2 Eukaryotic Genomes

Genome Sizes

The genomes of the more complex multicellular organisms are far larger than those of bacteria or viruses. As **FIGURE 21.2** shows, however, there is no simple relationship between genome size and organismal complexity. The human genome, consisting of about three billion base

● **CONCEPT** Most eukaryotes require—and have—much larger genomes than eubacteria or archaea.

pairs, is dwarfed by the genomes of some amphibians and some plants, which may be as much as 50-fold larger than those of humans. Because there is no basis for concluding that a bean plant is 50 times more complex than a human being, a logical conclusion is that eukaryotic genomes must contain considerable DNA that does not code for proteins or for the RNA machinery of protein synthesis. In the following sections, we describe several kinds of noncoding DNA sequences in eukaryotic genomes.

An obvious difference between bacterial and eukaryotic genomes is the diploid nature of eukaryotic genomes; bacteria and archaea have but one chromosome per cell, whereas eukaryotic cells usually have two copies of each chromosome (with the exception of sex chromosomes—for example, the X and Y chromosomes in humans).

Once we became aware of the great sizes of many eukaryotic genomes, additional complexity came into focus. The preliminary base sequence of the human genome, reported in 2001, underscored this complexity because the number of sequences that could represent protein-coding genes was far lower than the total potential coding capacity (page 681). Indeed, the complexity of eukaryotic genomes had already been recognized for some years, beginning with the finding

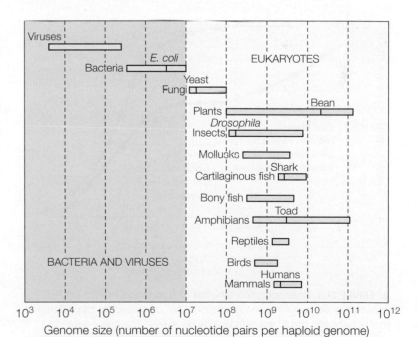

▲ **FIGURE 21.2 Genome size.** The bars show the range of haploid genome sizes for different groups of organisms. A few specific organisms are marked with vertical lines—for example, 3×10^9 for humans and 4.6×10^6 for E. coli. Note that the genome size scale is logarithmic and that many organisms have larger genomes than humans.

that many sequences are repeated multiple times in the genome (see below), even though most protein-coding genes are represented but once. Also, the discovery that most eukaryotic protein-coding genes are interrupted by noncoding segments (*introns*; see Chapter 7) further underscored the complexity of these genomes. Here we discuss some of the noncoding sequences in eukaryotic genomes.

Repetitive Sequences

The earliest indication that eukaryotic chromosomes might contain noncoding DNA came in 1968, when Roy Britten and David Kohne developed a technique for analyzing DNA reassociation kinetics. In this method, the total DNA from an organism is cut by shear forces into pieces about 300 base pairs long. The fragments are heated to cause strand separation, then slowly cooled, to allow strands of complementary sequence to reassociate. Sequences that are present in multiple copies reassociate relatively quickly, whereas single sequences reassociate slowly because their abundance is low. When Britten and Kohne analyzed bovine DNA using this method, they were surprised to find that almost half the DNA reassociated much more rapidly than expected for single-copy sequences in the large genome (**FIGURE 21.3**). To account for this rapid reassociation, they concluded that some DNA sequences are reiterated 10^5 to 10^6 times in one cell. Practically all of the DNA of *E. coli* is single copy, whereas only about half of most mammalian DNAs and one-third of plant DNAs fall into this category.

● **CONCEPT** Repetitive DNA sequences in eukaryotic genomes include satellite DNAs and scattered duplicate sequences.

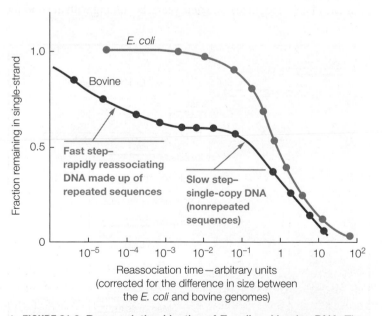

▲ **FIGURE 21.3 Reassociation kinetics of *E. coli* and bovine DNA.** The abscissa corresponds to reassociation time, corrected for the difference in size between the *E. coli* and bovine genomes. The curve for *E. coli* corresponds to that expected for a collection of single-copy genes in a genome of the *E. coli* size—4.67×10^6 bp. The curve for bovine DNA exhibits two steps in reassociation. A slow step corresponds to single-copy DNA (nonrepeated sequences). A faster step corresponds to rapidly reassociating DNA made up of repeated sequences. Many classes of repeated DNA are represented in this phase of the reassociation.

Satellite DNA

Upon further analysis, reiterated DNA sequences have been divided into several categories. One type involves multiple tandem repetitions (over long stretches of DNA) of very short, simple sequences like (ATAAACT)*n*. These sequences can often be separated from the bulk of the DNA by shearing to break the DNA into smaller fragments, followed by sedimentation to equilibrium in density gradients (see Figure 4.15, Chapter 4). Repetitive DNA sequences that are AT-rich (like the one just mentioned) have a lower density than average-composition DNA; GC-rich sequences are even denser. Thus, in a density-gradient experiment, the reiterated, simple-sequence fragments form small extra bands around the main-band DNA (**FIGURE 21.4**). Because of this banding, clusters of repeated sequences are sometimes referred to as **satellite DNAs.** In higher eukaryotes, satellite DNA usually makes up 10% to 20% of the total genome.

What function can these highly reiterated DNA sequences serve? They do not code for proteins, although recent evidence indicates that much or most of the DNA in eukaryotic chromosomes is transcribed. Some reiterated sequences appear to play a structural role; these sequences have been found to be highly concentrated near the **centromeres** of chromosomes, the regions where sister chromatids are attached to the mitotic spindle during mitosis. In budding yeast centromeres are

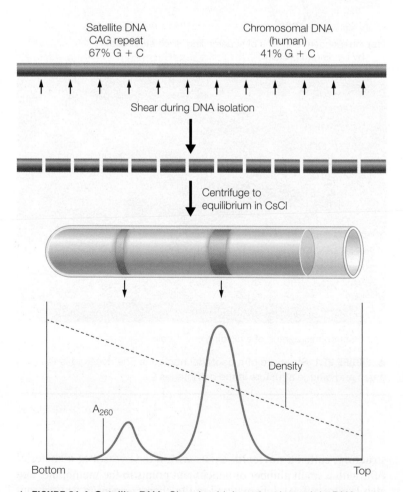

▲ **FIGURE 21.4 Satellite DNA.** Shearing high-molecular-weight DNA yields short fragments that have distinctive densities and can be separated in a CsCl gradient.

AT-rich segments about 125 base pairs long, although most other organisms have shorter repeating sequences in centromeric DNA. Centromeres serve as binding sites for proteins that attach the spindle fibers when paired chromosomes are drawn to opposite poles of a cell during mitosis.

Duplications of Functional Genes

There are other classes of DNA sequences with varying degrees of repetition. Some represent duplications of functional genes, and in some cases the repetitiveness plays a useful role by allowing high levels of much-needed transcripts to be produced. Examples include the genes for ribosomal RNAs (up to several thousand copies may be present) and tRNA genes (hundreds of copies of each type are often found). The cell's continual need for large quantities of ribosomes and tRNAs for translation is met by having multiple copies of these genes. The same is true for the genes for some much-used proteins, such as the histones that bind to eukaryotic DNA to form chromatin (see page 672). As pointed out in Chapter 23, even genes that are normally single copy are sometimes amplified, either in response to environmental stress or in special tissues during embryonic development.

Alu Elements

Other kinds of repeated DNA sequences exist that do not code for proteins but whose true function remains mysterious. These sequences are often scattered throughout the genome rather than being clustered like the satellite DNAs. One of the most common such families in mammals consists of the so-called *Alu sequences.* There are more than one million copies of these sequences (each about 300 bp long) in the human genome. Their name reflects the existence of a single site for the restriction endonuclease *Alu*I in most members of this class. The *Alu* sequences can be transcribed into RNA, although they are not known to be translated. *Alu* sequences are examples of **SINES** (short interspersed elements). The human genome also contains up to one million **LINES** (long interspersed elements) sequences that can be as long as 10 kbp. SINES and LINES are both thought to be transposable elements, capable of movement from place to place within the genome (see Chapter 23).

● **CONCEPT** Repeated sequences make up a significant fraction of the human genome, yet the functions of many of these sequences are as yet unknown.

Introns

Another reason for the large size of eukaryotic genomes is that most eukaryotic genes are interrupted by introns. In Chapter 7 we pointed out that the coding regions of most eukaryotic genes—exons—are interrupted by noncoding regions—introns. Recall that the β globin gene consists of three exons interrupted by two noncoding intron regions. This kind of structure is common in eukaryotes and is often more extreme than in this β globin example. **FIGURE 21.5** depicts analysis of the gene for *hexon*, the major structural protein of adenovirus, a small DNA virus. In this experiment, an *Eco*RI restriction fragment containing the cloned hexon gene was denatured and subjected to annealing conditions in the presence of purified hexon mRNA. The genomic DNA pairs to the mRNA along the exons, but the introns, which contain no counterpart in the mRNA molecule,

● **CONCEPT** Most eukaryotic genes contain introns.

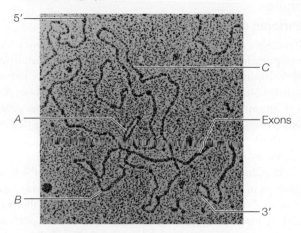

(a) Map of the hexon gene, showing exons 1–4 (blue) and introns A, B, and C (gray).

(b) Electron micrograph of a hybrid molecule formed by renaturing an *Eco*RI fragment containing the hexon gene with purified adenovirus mRNA.

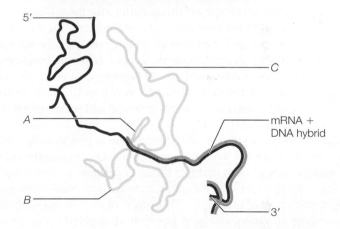

(c) Diagram showing how the intron regions loop out in R loops in the DNA–mRNA hybrid. The RNA is shown in red, the DNA exons in blue, and the single-strand DNA introns in gray.

▲ **FIGURE 21.5 Exon–intron structure of the adenovirus gene for hexon.** Data are from S. M. Berget, SC. Moore, and P. A. Sharp (1977) *Proc. Natl. Acad. Sci. USA* 74:3171–3175.

are looped out as single-stranded DNA, to form so-called **R loops**—DNA that has no RNA counterpart. Figure 21.5(b) shows an electron micrograph of one such molecule, and the tracings in Figure 21.5(c) identify the single-stranded regions (introns A, B, and C), as well as the thicker DNA–RNA hybrid region, which shows locations of the exons.

Sequence analysis shows that introns are present in most eukaryotic structural genes and frequently exceed exons in total length. Lower eukaryotes like yeast usually have many fewer introns, and their genome size is correspondingly smaller.

As mentioned in Chapter 7, the function of introns is not yet wholly understood. They do serve as loci for genetic recombination, allowing functional parts of proteins to be interchanged in evolution, a process called exon shuffling. Such loci also allow eukaryotes to make variants of a protein from a single gene, by splicing different exons together. This **alternative splicing** is discussed in Chapter 24.

Gene Families

Multiple Variants of a Gene

Despite alternative splicing, in many cases variants of genes for the same type of protein are expressed in different tissues or at different stages in development. We encountered an example in Chapter 7, where the embryonic (ζ and ε), fetal (α and γ), and adult (α, β, and δ) globins of mammals were described. For each of these proteins there exists a complete gene in every cell of the animal. As discussed in Chapter 7, these related DNA sequences arose through gene duplication followed by separate lines of evolution.

Figure 7.35 (Chapter 7) depicts the clusters of genes for the α and β classes of globins in humans. Each of these genes has the kind of exon–intron structure shown in Figure 21.5. In addition, the genes themselves are separated by long stretches of noncoding DNA. Some portions of these intervening regions must contain control signals because the expression of the globin genes is under complex and subtle regulation. In the first place, the globin gene clusters are found in all human cells, but they are expressed only in **erythropoietic** cells, those that give rise to red blood cells. Furthermore, as we saw in Chapter 7, the expression of each variant is strictly constrained to certain developmental stages. In the early embryo, for example, only the ζ and ε genes are transcribed; all other globin genes are turned off. As development proceeds, transcription switches first to the fetal α and γ genes, and at about the time of birth the adult β variant begins to dominate and transcription of γ ceases (see Figure 7.35).

Many other gene families exist. Some seem to play developmental roles rather like those of the globin gene family. Others, like the immunoglobulin genes, exist in multiple forms to generate immense amino acid sequence diversity within a common protein structural framework, as discussed in detail in Chapter 23. In each case, it seems likely that the members of a particular gene family have evolved by successive duplications of an original, ancestral gene.

Pseudogenes

Gene families often include one or more **pseudogenes,** which until recently were thought to be nonexpressed genes. Pseudogenes can bear strong sequence similarity to expressed genes, from which they undoubtedly evolved. Because their gene products are nonfunctional, pseudogenes are no longer under strong selective control in evolution. In a sense, it does not matter what happens to them, so they can accumulate mutations that would be selected against in functional genes. Until recently it was thought that pseudogenes play no significant biological function—that they are evolutionary remnants. However, recent research indicates that this view may require modification. Examples of pseudogenes can be seen in Figure 7.35, which shows the arrangement of α and β globin gene variants in human DNA.

In summary, a number of quite different explanations combine to account for the very large size of the genomes of eukaryotic organisms. At the same time, we still find it hard to rationalize the extreme variations in amount of DNA that are sometimes observed even between closely related organisms. For example, among the amphibians alone there is more than a 100-fold range of genome sizes. The function, if any, of such enormous variation is still obscure, which suggests that we still do not understand some fundamental features of the eukaryotic genome.

The ENCODE Project and the Concept of "Junk DNA"

The finding that the human genome contains only about 21,000 protein-coding genes, as well as the difficulty in assigning functions to highly repetitive noncoding DNA sequences, led to the idea that a large fraction of eukaryotic genomes is "junk DNA," with no functional significance. To test this idea, a consortium of laboratories analyzed the human genome in the "ENCODE Project" (<u>E</u>ncyclopedia <u>o</u>f <u>D</u>NA <u>E</u>lements), using a range of techniques to assess biological function. In 2012, the consortium simultaneously published 30 papers. The essence of these studies was that, in contrast to the junk DNA concept, at least 80% of the human genome plays some biological function. However, the nature of many of those functions remains to be established.

21.3 Physical Organization of Eukaryotic Genes: Chromosomes and Chromatin

The Nucleus

Unlike the bacterial chromosome (nucleoid), the chromosomes of eukaryotes are not normally found free in the cytoplasm. In nondividing cells, the chromosomes are segregated within the nucleus (see the opening illustration) as an entangled mass of fibers of **chromatin**—the DNA–protein complex that comprises the chromosome. The chromosomes are held in place by the **nuclear envelope,** a membrane bilayer studded with **nuclear pores**—openings, about 9 μm in diameter, that permit free diffusion of small molecules between nucleus and cytoplasm, but are selective in allowing transport of RNA and protein molecules. The pores are lined with a huge multiprotein assembly called the **nuclear pore complex** (**FIGURE 21.6**), an assembly of 500 to 1000 protein molecules (**nucleoporins**) that allows selective transport of RNA and protein molecules into and out of the nucleus. Two classes of helper proteins, called **exportins** and **importins,** are also involved in these processes.

Because protein synthesis in eukaryotes occurs outside the nucleus, transcription and translation cannot be directly coupled, as they are in bacteria (Chapter 25). Therefore, messenger RNA must be exported from the nucleus for translation. As described in Chapter 24, the primary transcript undergoes posttranscriptional processing before it is exported. During mitosis, the nuclear envelope breaks down, and the diploid chromosomes condense into compact structures like the one shown in **FIGURE 21.7**. In this electron micrograph, two newly replicated chromosomes are joined at the *centromere,* which becomes linked to the spindle fibers as chromosome pairs are drawn apart in the latter stage of mitosis.

Not seen in the electron micrograph of Figure 21.7 are the *telomeres*—structures at each end of eukaryotic chromosomes that protect the DNA from degradation and ensure that each chromosome is completely copied during replication. A telomere is a simple tandemly repeated sequence that may be several kbp in length.

● **CONCEPT** The multiple eukaryotic chromosomes are contained within the nucleus, except during mitosis when the nuclear envelope disintegrates.

Side view   **Cytoplasmic view**

25°   65°

Cargo in transit

(a) Image based upon cryoelectron microscopy.

Cytoplasmic filaments

Nuclear envelope

Nuclear basket

POMs
Coat nucleoporins
Adaptor nucleoporins
Channel nucleoporins
FG repeats

(b) Model of the complex. The four concentric circles that line the pore are composed of integral pore membrane proteins (POMs), coat nucleoporins, adaptor nucleoporins, and channel nucleoporins. Natively unfolded phenylalanine-glycine (FG) repeats of several nucleoporins comprise a transport barrier in the central channel and are depicted as a transparent plug.

▲ **FIGURE 21.6** A schematic model of the nuclear pore complex. From A. Hoelz, E. W. Debler, and G. Blobel (2011) *Ann. Rev. Biochem.* 80:613–643. Reprinted with permission from *Annual Reviews*.

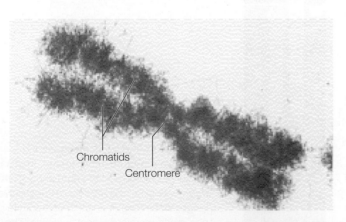

Chromatids
Centromere

▲ **FIGURE 21.7 A mitotic chromosome.** An electron microscope image of a human chromosome during the metaphase stage of mitosis. The constriction at the centromere and the lengthwise division into sister chromatids are clearly visible. The hairy-looking surface consists of loops of highly coiled chromatin.

Usually one strand is G-rich, and, as pointed out in Chapter 4, the dGMP residues often form a G-quadruplex structure. In humans, the telomeric repeat sequence is 5′-TTAGGG. **FIGURE 21.8** depicts a spread of human metaphase chromosomes showing centromeres (pink) and telomeres (green). The color marking was carried out with **fluorescent in situ hybridization (FISH;** see page 678). We discuss telomeres and their function in stabilizing chromosome ends in Chapter 22.

Chromatin

As noted earlier, DNA is tremendously compacted when folded into a cell. The diploid DNA content of a human cell is more than 6×10^9 bp, corresponding to a total length of about 2 m. Somehow, all of this DNA must be packed into a nucleus about $10\,\mu m$ (10^{-5} m) in diameter. As with bacterial and archaeal cells, eukaryotes also use extensive negative supercoiling as a primary means of achieving compaction. Unlike the bacterial nucleoid, however, eukaryotic DNA is packed in a different way, along with a set of special proteins, to form the protein–DNA complex called chromatin. The most abundant proteins in chromatin are a set of five highly conserved, low-molecular-weight, strongly basic proteins, the **histones.** Chromatin also includes a much larger number of nonhistone proteins, which are present in smaller amounts and some of which participate in gene expression and its control.

● **CONCEPT** The chromatin of eukaryotes consists of DNA complexed with histones and nonhistone proteins.

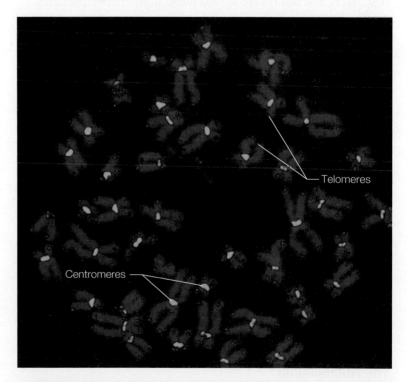

Telomeres

Centromeres

▲ **FIGURE 21.8** Mitotic human chromosomes in metaphase, stained separately by fluorescent in situ hybridization (page 678) for telomeres (red) and centromeres (green).

TABLE 21.2 Properties of the major histone types

Histone Type	Molecular Weight	Number of Amino Acid Residues	mol % Lys	mol % Arg	Role
H1	22,550	244	29.5	1.3	Associated with linker DNA; helps form higher-order structure
H2A	13,960	129	10.9	9.3	Two of each form the histone octamer core of the nucleosome
H2B	13,774	125	16.0	6.4	
H3	15,273	135	9.6	13.3	
H4	11,236	102	10.8	13.7	

Note: All data are for calf thymus histones, except for H1, which is from rabbit.

Histones and Nonhistone Chromosomal Proteins

Properties of the five major types of histones are outlined in **TABLE 21.2**. All histones are small, highly basic proteins rich in lysine and arginine. Histones are the fundamental building blocks of chromatin structure. The amino acid sequences of some have been strikingly well conserved throughout evolution. All known histone H4 molecules, for example, contain exactly 102 amino acid residues. Histone H4 shows only two substitutions between humans and peas and only eight substitutions between humans and yeast. The nucleoids of bacterial cells also have proteins associated with DNA, but these proteins are quite different from the histones and do not form a comparable chromatin structure. Some archaea have been shown to contain a chromatin-like structure. In all kinds of eukaryotic nuclei, from yeast to human, the histones are present in an amount of about 1 gram per gram of DNA, and histones H2A, H2B, H3, and H4 are always found in equimolar quantities.

The histones are accompanied by a much more diverse group of nonhistone chromosomal proteins. Total amounts of these proteins vary greatly from one cell type to another, ranging from about 0.05 to 1 gram per gram of DNA. They include a variety of proteins, such as polymerases and other nuclear enzymes, nuclear receptor proteins, and transcription factors. It is possible to count, on two-dimensional electrophoretic gels, approximately 1000 different nonhistone chromosomal proteins in a typical eukaryotic nucleus. Among the most abundant are topoisomerases and a class called **SMC proteins** (for <u>s</u>tructural <u>m</u>aintenance of <u>c</u>hromosomes). The major proteins in this class are **cohesins,** which help to hold sister chromatids together immediately after replication and continue to hold them together until chromosomes condense at metaphase, and **condensins,** which are essential to chromosome condensation as cells enter mitosis.

As early as 1888, the German chemist Albrecht Kossel isolated histones from nuclei and recognized them as basic substances that would bind to the nucleic acid. Histones were, in fact, the first class of proteins to be recognized. However, their precise role was not understood until about 1974. Then, research in several laboratories showed that these proteins combine in a specific way to form the **nucleosome,** a repeating element of chromatin structure.

The Nucleosome

If naked DNA (i.e., DNA containing no bound proteins) is partially digested with a nonspecific endonuclease such as micrococcal nuclease, which cuts double strands randomly, a broad smear of polynucleotide fragments is produced, as shown by gel electrophoresis. When chromatin is treated the same way, however, the DNA is cleaved in a specific, nonrandom way. On a polyacrylamide gel, the DNA from nuclease-digested chromatin shows a series of bands that are multiples of approximately 200 base pairs (**FIGURE 21.9**). This indicates that the nuclease can find easy access to the DNA only at regularly spaced points.

Electron micrographs of extended chromatin fibers reveal a regular "beaded" pattern in the chromatin structure, with one bead about every 200 bp. Other experiments revealed that if nuclease digestion of chromatin is continued, it slows down and nearly stops when about 30% of the DNA has been consumed. The remaining protected DNA is present in particles corresponding to the beads seen in the electron micrographs. These particles, called nucleosomes (or more precisely, *nucleosomal core particles*), have a simple, definite composition that is practically invariant over the entire eukaryotic realm. They always contain 146 bp

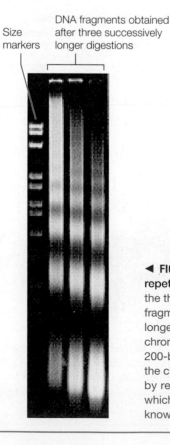

Size markers

DNA fragments obtained after three successively longer digestions

◄ **FIGURE 21.9 Evidence suggesting a repetitive structure in chromatin.** In this gel, the three columns on the right show DNA fragments obtained after three successively longer digestions of chicken erythrocyte chromatin by micrococcal nuclease. The 200-base-pair length difference between the chromatin fragments was estimated by reference to the left-hand column, which displays DNA restriction fragments of known molecular weights.

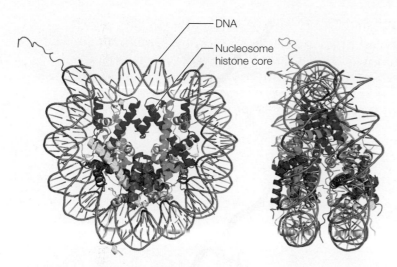

▲ **FIGURE 21.10 Structure of the nucleosome core particle as revealed by X-ray diffraction.** A high-resolution (2.8 Å) model of the nucleosome core particle with DNA wrapped around it. Two views perpendicular to the twofold axis are shown. Histone molecules are colored as follows: H2A, yellow; H2B, red; H3, blue; H4, green. The N-terminal tails of the histones are not completely resolved. PDB ID: 1ao1.

of DNA, wrapped about an octamer of histone molecules—two each of H2A, H2B, H3, and H4. This composition explains why these four histones are found in equal amounts in chromatin. Both nucleosomes and nucleosome histone cores have been crystallized, and X-ray diffraction studies have revealed the structure shown in **FIGURE 21.10**. The DNA lies on the surface of the octamer and makes about 1.7 left-hand solenoidal superhelical turns about it. The structure of the octamer provides a left-hand helical "ramp" on which the DNA is bound. The high-resolution data now available on the histone octamer reveal a commonality in histone structure—the *histone fold*—that was not evident from the examination of sequences alone. It suggests an early common ancestor for these proteins.

● **CONCEPT** The basic repeating structure in chromatin is the nucleosome, in which nearly two superhelical turns of DNA are wrapped about an octamer of histones.

Although the nucleosome itself is a nearly invariant structure in eukaryotes, the way in which nucleosomes are spaced along the DNA varies considerably among organisms and even among tissues in the same organism. The length of DNA between nucleosomes may vary from about 20 bp to over 100 bp. Exactly what determines the arrangement of nucleosomes along the DNA is still not completely understood. However, it is now clear that at least some nucleosomes occupy defined positions along the DNA. The implications of this finding are discussed later. The internucleosomal, or *linker,* DNA is occupied by the H1-type (very lysine-rich) histones and nonhistone proteins. **FIGURE 21.11** provides an overall schematic view of the fundamental elements of chromatin structure.

Histone molecules in chromatin are subject to numerous post-translational modifications, including the acetylation and methylation of lysine ε-amino groups, as well as phosphorylation and ubiquitylation of other amino acids. These modifications help to regulate gene

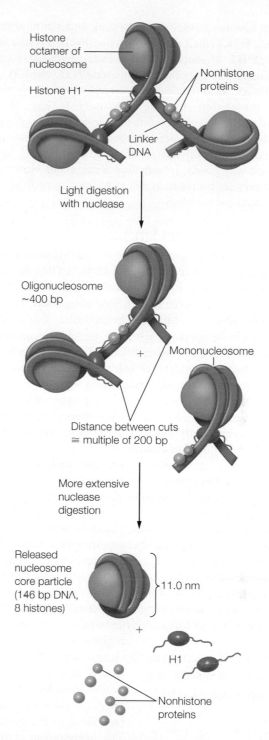

▲ **FIGURE 21.11 The elements of chromatin structure.** At the top is our current understanding of the extended structure of a chromatin fiber. Light digestion with nuclease releases first mononucleosomes and oligonucleosomes. Then, as linker DNA is further digested, nonhistone proteins and H1 are released, to yield the core particle, whose structure is shown in Figure 21.10.

expression. In order to be transcribed, DNA must at least transiently undergo dissociation from the histone core, and histone amino acid modifications affect the strength of the interactions between DNA and the histone core. Most of the modifications occur at specific amino acid residues on H3 or H4. Modifications are identified in terms of the histone, modified amino acid, and nature of the modification. Thus, H3K27me3 refers to a trimethylated lysine at position 27 in histone H3. Modifications associated with *heterochromatin,* which is transcriptionally inactive, include H3K9me3, H3K27me3, and H4K20me3. Modifications associated with actively transcribed regions of the genome include H3K4me3, H3Kac, H3K36me3, H4Kac, and uH2B, where ac refers to acetylation; ubiquitylation and acetylation occur at many sites, whereas most methylations occur at well-defined sites. Modifications of this kind can bring about stable changes in the way genetic information is expressed. How these modification patterns are transmitted when chromosomes divide at mitosis is a major unsolved problem.

Higher-Order Chromatin Structure in the Nucleus

Wrapping DNA about histone cores to form nucleosomes accomplishes part of the compaction necessary to fit eukaryotic DNA into the nucleus because it shortens the strand severalfold. Much of the chromatin in the nucleus, however, is even more highly compacted. The next stage in compaction involves folding the beaded fiber into a thicker fiber like that shown in the opening illustration. These fibers are about 30 nm in diameter and must be further folded on themselves to make the even thicker chromatin fibers visible in metaphase chromosomes (see Figure 21.7). Additional modifications occur because chromatin exists in interphase nuclei as two major forms—**euchromatin,** which is transcriptionally active, and the much thicker fibers of **heterochromatin,** which is transcriptionally inactive. Evidently, heterochromatin has undergone further rounds of compaction.

During mitosis, chromatin of both types undergoes further processing and thickening, to yield the compact and organized structures such as the metaphase chromosome shown in Figure 21.7. However, when viewing a nucleus at interphase, such as the one schematized in the opening illustration, one is struck by the evident lack of structural organization, with chromatin fibers seemingly being distributed at random. On the other hand, a recently developed technique called **chromosome conformation capture** has established a much higher degree of order than expected from microscopic observation of interphase chromosomes. In this technique (**FIGURE 21.12**), chromatin fragments are treated with formaldehyde, which reversibly cross-links regions of chromatin that are in close contact. Next, the preparation is digested with a restriction endonuclease that generates DNA cohesive or "sticky" ends (see Tools of Biochemistry 4B). The preparation is diluted and subjected to DNA-renaturing conditions and DNA ligase treatment, so that crosslinked fragments undergo ligation. At that point, the cross-linking is reversed and the ligated DNA molecules are subjected to sequence analysis. Applications of this technology demonstrate "chromosome neighborhoods"—regions of DNA that may be far apart in terms of the linear DNA sequence of a chromosome are in fact close together in the three-dimensional space within the nucleus. Often, the chromosome folding demonstrated by this technology brings into close association a particular gene and a sequence responsible for regulating expression of that gene. More recent analysis demonstrates two distinct chromosome folding patterns that occur as intermediates in chromosome condensation as cells prepare for mitosis.

● **CONCEPT** The fiber formed by the nucleosomes is folded in vivo to form higher-order chromatin structure.

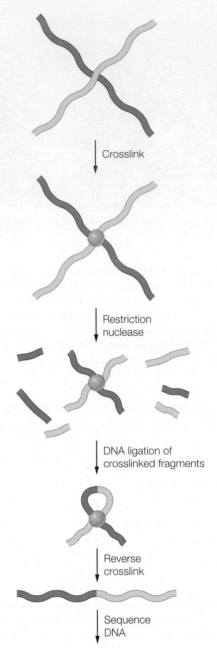

▲ **FIGURE 21.12 Chromosome conformation capture,** a technique to identify chromosome "neighborhoods," regions that are in close contact in three-dimensional space. The violet and green denote regions of chromatin from the same chromosome.

21.4 Nucleotide Sequence Analysis of Genomes

How is the complete nucleotide sequence of a genome determined? Our ability to map and determine the nucleotide sequence of genomes depended crucially on the existence of **restriction endonucleases,** which we identified in Chapter 4 as enzymes that catalyze double-strand DNA cleavage in a sequence-specific fashion. Before discussing the sequence determination of large genomes, we digress here to describe these remarkable enzymes and the biological processes in which they are involved—**host-induced restriction and modification.** These processes, which we describe in the following section, help to protect bacteria against foreign invaders, such as viruses.

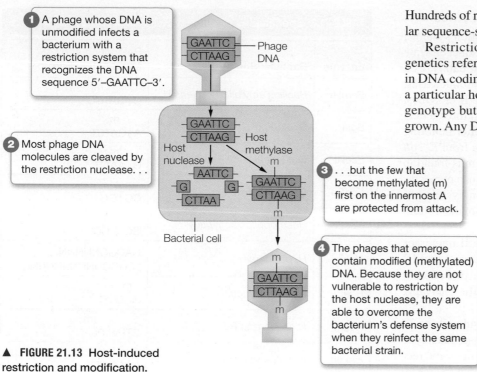

1 A phage whose DNA is unmodified infects a bacterium with a restriction system that recognizes the DNA sequence 5'–GAATTC–3'.

2 Most phage DNA molecules are cleaved by the restriction nuclease. . .

3 . . .but the few that become methylated (m) first on the innermost A are protected from attack.

4 The phages that emerge contain modified (methylated) DNA. Because they are not vulnerable to restriction by the host nuclease, they are able to overcome the bacterium's defense system when they reinfect the same bacterial strain.

▲ **FIGURE 21.13** Host-induced restriction and modification.

Hundreds of restriction endonucleases are now known to catalyze similar sequence-specific cleavages, some with cytosine as the target base.

Restriction and modification are **epigenetic** phenomena; epigenetics refers to heritable characteristics that do not involve changes in DNA coding capacity. The ability of a phage to escape restriction in a particular host bacterium depends not on any changes in the phage's genotype but on the host strain in which that phage was previously grown. Any DNA within the cell is subject to restriction and modification, including transforming and plasmid DNAs, as well as chromosomal or phage DNAs.

The importance of these developments was that, for the first time, scientists could isolate homogeneous DNA fragments of precisely defined length by treating DNA with a restriction nuclease in vitro and then resolving the fragments in the digest on an electrophoretic gel. These developments gave rise to gene cloning, as discussed in Tools of Biochemistry 4B. With respect to genome analysis, restriction fragments resolved by electrophoresis (**FIGURE 21.14**) can be arranged in order, so as to give physical maps of DNA molecules. The maps are called **restriction maps** because they show the physical locations of restriction sites.

Restriction and Modification

Our ability to clone, map, and sequence genes, as noted earlier, depended on enzymes that cleave DNA at specific sites. These capabilities came from our understanding of *host-induced restriction and modification.* Investigation of these processes, begun in Switzerland in the 1960s, shows dramatically how discoveries of great practical utility can arise from untargeted basic research. The earliest work focused on bacteriophage λ, a DNA virus that infects *E. coli*. In many infections, the DNA of the invading virus was seen to quickly break down, with a productive infection not being established; in other words, the viral genome was *restricted*. In some infections, however, the viral genome was protected from restriction; it became *modified* and maintained its capability to direct a round of viral growth.

Restriction–modification systems are widespread among bacteria. Some are encoded by chromosomal genes and some by plasmids. Each system consists of two enzymes—a DNA endonuclease and a DNA methylase. In 1970, Hamilton Smith observed that a particular restriction nuclease catalyzed double-strand cleavage of DNA within a specific short nucleotide sequence. Shortly thereafter, it was found that modification has the same sequence specificity, as schematized in **FIGURE 21.13**. In this example, one nucleotide within a six-nucleotide sequence is the substrate for a specific DNA methylase. When that site becomes methylated, by transfer of a methyl group from *S*-adenosylmethionine to an adenine within the site, the DNA becomes resistant to cleavage by a nuclease that recognizes the same hexanucleotide sequence; when that site is unmethylated, the DNA is susceptible to attack at that site.

▶ **FIGURE 21.14** **Fragmentation of bacteriophage λ DNA with restriction endonucleases** *Eco*RI **or** *Bam*HI. Mapping the restriction sites also requires data from partial digestion or with digestion of the DNA with *Eco*RI and *Bam*HI together (not shown).

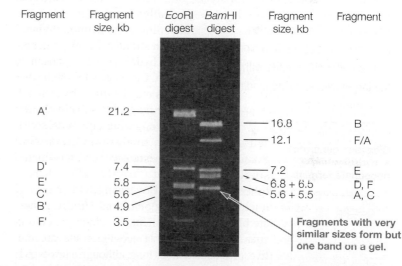

Fragment	Fragment size, kb	*Eco*RI digest	*Bam*HI digest	Fragment size, kb	Fragment
A'	21.2				
				16.8	B
				12.1	F/A
D'	7.4			7.2	E
E'	5.8			6.8 + 6.5	D, F
C'	5.6			5.6 + 5.5	A, C
B'	4.9				
F'	3.5				

Fragments with very similar sizes form but one band on a gel.

(a) Experimental determination of fragmentation patterns, resulting from enzymatic digestion of the 48.5-kb linear DNA molecule from phage λ. Restriction digests are subjected to agarose gel electrophoresis, and the fragments are visualized under ultraviolet light after staining the gel with ethidium bromide, a fluorescent dye.

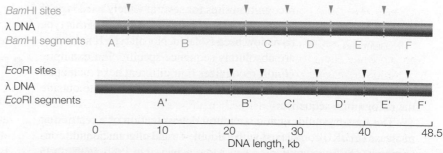

(b) Restriction maps of cleavage sites for each enzyme on the DNA molecule. By convention, the fragments are assigned letters as shown. In this experiment the 12.1-kb fragment in the *Bam*HI digest resulted from linking the A and F (terminal) fragments by base pairing at their cohesive ends (Chapter 23).

Properties of Restriction and Modification Enzymes

We recognize three major types of restriction–modification systems, called types I, II, and III. Each system consists of two distinct enzyme activities: a DNA methylase and an endonuclease that catalyzes a double-stranded DNA break. The sequence-specific endonucleases, those most widely used in molecular biology, are type II enzymes. Regardless of type, the enzymes are named with the first three letters denoting the bacterial species of origin and a fourth letter denoting an individual strain. For example, the restriction system from *E. coli* strain K is called *Eco*K. If more than one enzyme system is found in a given strain, the different enzymes are designated by Roman numerals. For instance, *Eco*RI is one of two known restriction systems in *E. coli* strain R, and *Hind*III is one of three enzymes from the d strain of *Haemophilus influenzae.*

Among the three types of restriction systems, type II nucleases are the only ones that cleave *within* the target site. Types I and III enzymes recognize specific sequences, but they cleave outside of those sequences, so they cannot be used for sequence-specific cleavage. Therefore, we focus our discussion on type II systems.

Most of the type II enzymes are homodimers, with subunits of 30 to 40 kilodaltons. A divalent cation is required for cleavage. Each type II nuclease has a counterpart methylase, which binds to the same recognition sequence and methylates one nucleotide within that sequence. A **hemimethylated** DNA (with methyl group on one strand only) is a preferred substrate for the methylase but not for the nuclease, which generally cleaves only when the recognition site is unmethylated on both strands. Cleavage generates 3′ hydroxyl and 5′ phosphate termini. Cleavage sites on the two strands may be offset by as much as four nucleotides (as in *Eco*RI) or more, giving cuts with short, self-complementary, single-stranded termini. Some enzymes cleave to give a 5′-terminated single-stranded end ("overhang"), whereas others generate a 3′ overhang. Other type II nucleases, including *Sma*I and *Hind*II, generate blunt-ended fragments in which the cutting sites are not offset. Most recognition sites are four, five, or six nucleotides long, although a few type II enzymes recognize an eight-nucleotide sequence. Most show twofold rotational sequence symmetry, suggesting that the two enzyme subunits are also arranged symmetrically. **TABLE 21.3** lists the recognition sites for several widely used type II nucleases. Several hundred enzymes of this type have now been isolated. Not all type II nucleases are absolutely sequence specific. For example, *Hind*II recognizes four different hexanucleotide sequences, and some enzymes (such as *Hga*I) cleave at a site outside the recognition sequence.

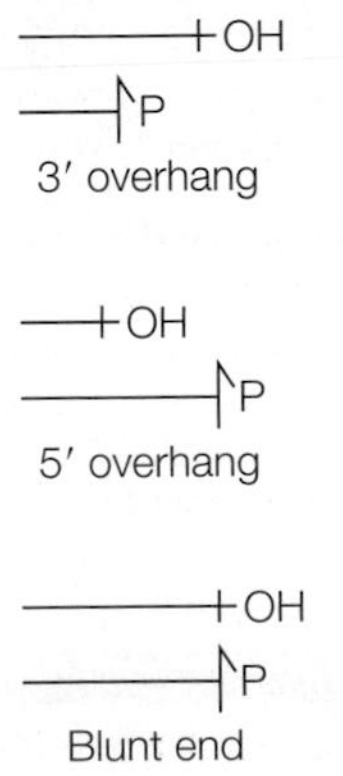

The first crystallographic structural determination of a restriction nuclease (*Eco*RI) complexed with a double-strand oligonucleotide containing its DNA recognition sequence was achieved in 1986. **FIGURE 21.15** shows one polypeptide subunit of the dimeric enzyme in contact with its DNA recognition sequence. The DNA is bound in a cleft, and the

● **CONCEPT** The restriction enzymes most useful to biologists cleave both DNA strands within the target sequence, depending on base methylation.

Cleavage generates 3′ hydroxyl and 5′ phosphate termini:

——————+OH
——→P
3′ overhang

——————+OH
——————→P
5′ overhang

——————+OH
——————→P
Blunt end

TABLE 21.3 Specificities of some type II restriction systems

Enzyme	Bacterial Source	Restriction and Modification Site[a]
*Bam*HI	*Bacillus amyloliquefaciens* H	m G↓GATCC
*Bgl*II	*B. globiggi*	m A↓GATCT
*Eco*RI	*Escherichia coli* RY13	m G↓AATTC
*Eco*RII	*E. coli* R245	m CC↓GG
*Hae*III	*Haemophilus aegyptius*	m GG ↓ CC
*Hga*I	*H. gallinarum*	GACGCNNNNN ↓ CTGCGNNNNNNNNNN ↓
*Hha*I	*H. haemolyticus*	m GCG↓C
*Hind*II	*H. influenzae* Rd	m GTPy↓PuAC
*Hind*III	*H. influenzae* Rd	m A↓AGCTT
*Hinf*I	*H. influenzae* Rf	G↓ANTC
*Hpa*I	*H. parainfluenzae*	GTT↓AAC
*Hpa*II	*H. parainfluenzae*	m C↓CGG
*Msp*I	*Moraxella* sp.	C↓CGG
*Not*I	*Nocardia rubra*	GC↓GGCCGC
*Ple*I	*Pseudomonas lemoignei*	GAGTCNNNN ↓ CTCAGNNNN ↓
*Pst*I	*Providencia stuartii*	CTGCA↓G
*Sal*I	*Streptomyces albus* G	G↓TCGAC
*Sma*I	*Serratia marcescens* Sb	m CCC↓GGG
*Xba*I	*Xanthomonas badrii*	T↓CTAGA

[a]The methylated base in each site, where known, is identified with the letter "m." All sequences read 5′ to 3′, left to right. The cleavage on the opposite strand in each case can be inferred from the symmetry of the site (except for *Hga*I and *Ple*I, each of which has an asymmetric site). Pu = purine, Py = pyrimidine, N = any base.

protein has an N-terminal "arm" that wraps around the DNA. Sequence specificity is maintained by 12 hydrogen bonds, which link the purine residues in the site to a glutamate and two arginine residues (not shown in the figure). Binding of the DNA alters its structure to generate "kinks"; the sequences immediately flanking the six-nucleotide cutting site (GAATTC) adopt the A duplex conformation, and the B structure is retained within the cutting site. The other subunit, not shown in the figure, contacts the substrate identically, accounting for the ability of the enzyme to catalyze symmetrical cleavages within the

● **CONCEPT** Modification methylases swing the target DNA base totally out of the helix in order to act upon it.

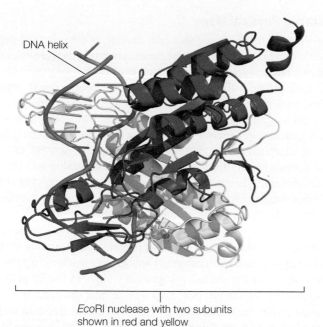

FIGURE 21.15 Structure of the *Eco*RI nuclease complexed with its DNA substrate.

▲ **FIGURE 21.15 Structure of the *Eco*RI nuclease complexed with its DNA substrate.** The DNA helix is shown in blue, while the two subunits of the protein are shown in red and yellow, respectively. Note the "kink" in the DNA structure, resulting from the fact that the enzyme binds the central six-base-pair cutting site in the B conformation, while the flanking sequences are bound as A-form DNA. Note also the N-terminal "arm" on each protein subunit, which wraps around the DNA. PDB ID: 1eri. Based on J. A. McClarin et al., *Science* (1986) 234:1526–1541.

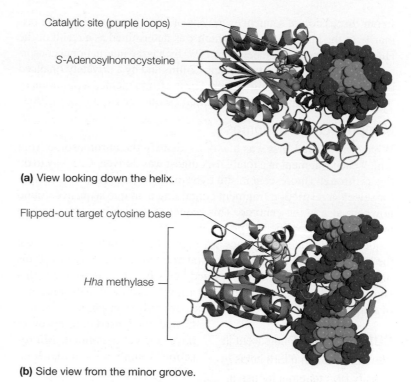

(a) View looking down the helix.

(b) Side view from the minor groove.

▲ **FIGURE 21.16 Structure of a complex of a type II DNA methylase with DNA.** The structure is a ternary complex containing *Hha* methylase from *Haemophilus haemolyticus,* DNA, and *S*-adenosylhomocysteine. The loops containing the catalytic site are in purple, and the rest of the enzyme is in orange. *S*-Adenosylhomocysteine is in yellow, the DNA backbone is blue, and the bases are green. In both views, the flipped-out target cytosine base is clearly visible. PDB ID: 1mht. Based on S. Klimasauskas, S. Kumar, R. J. Roberts, and X. Cheng, *Cell* (1994) 76:357–369.

cutting site. Subsequent crystallographic analyses indicate that type II restriction nucleases use varying strategies to recognize specific DNA sequences.

Structural studies on type II DNA methylases have been informative as well. For example, structural studies on the *Hha* DNA methylase showed that the bases undergoing methylation rotate completely out of the DNA duplex and into a catalytic pocket within the enzyme, where methylation occurs (**FIGURE 21.16**). Other enzymes that act upon specific DNA bases have since been shown to similarly flip the target base, including both DNA methylases, as shown here, and glycosylases, enzymes involved in DNA repair (Chapter 23).

Methylation is not the only form of modification. A widely distributed restriction-modification system uses **phosphorothioation** as the basis for modification. In this case, a sulfur atom replaces one of the oxygen atoms in the modified phosphate group.

Determining Genome Nucleotide Sequences

The ability to map restriction cleavage sites within a genome made it possible to determine the complete nucleotide sequence of small genomes. Once a restriction map of the genome is available, the genome is simply fragmented by restriction cleavage into relatively small pieces, each of which can be cloned into a suitable vector for subsequent sequence analysis (Tools of Biochemistry 4B). The collection of cloned genomic fragments is called a **library** because it is a collection of separate information-containing units. Each "book" in the library is sequenced, with the final sequence being reassembled on the basis of the restriction map. Generally, digestion must be carried out with two or more restriction nucleases, so that the identification

of overlap regions will make it possible to align adjacent fragments. Even simpler is "shotgun sequencing," a process that yields a random set of fragments, arising either from partial digestion of the genome by a restriction nuclease or by mechanical shear. As a result, different clones in the library contain some identical sequences, representing regions of overlap between contiguous segments. Overlapping regions can be identified on the basis of restriction fragment length patterns, which allow all of the fragments to be aligned properly. Alternatively, the fragments can be sequenced individually and then aligned by use of computer programs that look for regions of sequence identity. These means were used to sequence the 5386 bases in the φX174 bacteriophage genome as early as 1977, with the 16,569 base pairs in human mitochondrial DNA being identified a few years later. A shotgun approach yielded the first sequence of a free-living organism, the bacterium *Haemophilus influenzae,* in 1995. This genome, consisting of 1,830,137 base pairs, encodes about 1740 proteins. Many other bacterial genomes have since been sequenced by comparable approaches.

These simple approaches were impractical, however, when applied to the much larger eukaryotic genomes. Even the unicellular yeast, *Saccharomyces cerevisiae,* contains a genome of 12 million base pairs, distributed among 16 chromosomes. Much more daunting is the human genome, with more than 3 billion base pairs in 24 chromosomes (22 autosomes plus the X and Y sex chromosomes). Different approaches were needed, both for aligning genome fragments of contiguous sequence ("contigs") and for assigning them to specific chromosomes.

In our discussion of sequence analysis of large genomes, we refer primarily to the human genome, which was determined as a result of the Human Genome Project, begun in 1990 by a large international consortium eventually directed by Francis Collins and by a privately financed program directed by Craig Venter. These efforts yielded a preliminary "draft sequence" in 2000, with a "final" sequence following in 2003.

Mapping Large Genomes

What was needed first was a way to identify the chromosome from which each fragment in a total DNA digest was derived. One way to do this is through fluorescent in situ hybridization (FISH). The investigator selects a restriction fragment containing a unique sequence, found nowhere else in the genome. This sequence, typically a few hundred nucleotides, is amplified by **polymerase chain reaction** (PCR), a technique that selectively amplifies any region of a genome, provided that the base sequences flanking the region of interest are known (see Tools of Biochemistry 21A). The amplified DNA fragment is tagged with a fluorescent dye, then denatured and allowed to anneal in the presence of a preparation of human chromosomes arrested in metaphase, where replicated condensed chromosomes have not yet separated. Microscopic visualization under light of the appropriate wavelength for the fluorescent dye identifies the chromosome to which the probe has become annealed (**FIGURE 21.17**).

● **CONNECTION** Fluorescent in *situ* hybridization (FISH) helps to identify DNA features for use in genetic counseling.

FISH was used to label the telomeres and centromeres shown in Figure 21.8. FISH has additional uses, including finding specific features in DNA for use in genetic counseling and medicine.

Sequence analysis of the human genome thus becomes 24 separate projects, each focused on one chromosome. Each chromosome is still enormous, though, and additional landmarks are needed that make it possible to properly place fragments of sequence information as they become available. What might those sites be? Classical genetics used visually distinguishable markers, such as eye color in *Drosophila,* and mapping was done by analysis of the results of genetic crosses.

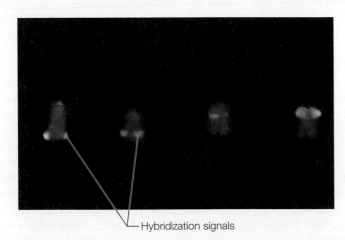

└─ Hybridization signals

▲ **FIGURE 21.17 Mapping genes by fluorescent in situ hybridization.** Four different DNA probes representing different genes were each tagged with a fluorescent dye and hybridized to human chromosome 21 in metaphase, showing hybridization signals (in yellow) at four distinct locations along the chromosome. Because metaphase chromosomes are made up of two identical sister chromatids, each probe produced a pair of signals.

Generating Physical Maps

What we need for sequence analysis, though, is not a map of genetic distances determined by biological measurements, which are essentially statistical, but a physical map, in which markers can be placed according to their physical distance (i.e., number of kilobase pairs) from one another. Such markers are also indispensable for genetic analysis. Well before investigators had proposed sequencing the 3-billion-base-pair human genome, geneticists were trying to locate within the genome genes in which mutations give rise to inherited diseases. For example, cystic fibrosis results from a deficiency or abnormality in a chloride channel protein. When the gene for this protein was mapped in 1989, it was hoped that the information might soon lead to a cure for this disease. Unfortunately, that has not yet happened. However, identification of the gene and gene product related to a disease does help immeasurably in understanding the disease, seeking prevention, and guiding therapy.

The key to locating disease genes was to have markers on a physical genome map and then to use association studies in family pedigrees to find markers on the physical map that lie near the gene of interest. But what might those markers be? In 1980, David Botstein realized that random variations among human genome sequences could lead to the gain or loss of restriction cleavage sites, and that the resulting **polymorphisms**—variable sequences within the same region—could provide the markers needed to construct physical maps.

Consider the four-base sequence 5′-GATC, the recognition site for the restriction endonuclease *Mbo*I. On average this site will occur at intervals of 256 base pairs (1/4 × 1/4 × 1/4 × 1/4). Suppose that individual A has two homologous chromosomes identical within this region, containing *Mbo*I cleavage sites 650 base pairs apart. Both chromosomes contain an additional *Mbo*I site, spaced 250 nucleotides from the 5′ end of this region, as drawn in **FIGURE 21.18**. Now suppose that a second individual, B,

● **CONCEPT** Sequencing large genomes requires prior construction of a physical map of the genome.

● **CONNECTION** Restriction length polymorphisms can be used to locate genes on the human genome that, when mutated, contribute to specific disease states

has a **restriction fragment length polymorphism** (RFLP) within this region. One chromosome is identical to that just described, while the other has 250 base pairs from the 5′ end, not GATC, but GTTC. *Mbo*I will no longer cleave at this site. Now, if the genome of A is digested with *Mbo*I, this region will yield a 250-base-pair fragment and a 400-bp fragment, while the genome of B will yield the same 250- and 400-bp fragments from one chromosome and a 650-bp fragment from the chromosome not containing the internal *Mbo*I site. How can these fragments be detected against the enormous background resulting from *Mbo*I cleavage throughout the entire genome? A powerful technique called **Southern blotting,** or **Southern transfer,** after its inventor, Edwin Southern, makes this analysis possible.

The Principle of Southern Analysis

Suppose we want to examine the *Mbo*I fragments in the region shown in Figure 21.18. If we simply digest total human DNA with *Mbo*I and resolve the fragments by agarose gel electrophoresis, and then examine the pattern of fragments on the gel by the standard method—ethidium bromide staining followed by examining its fluorescence under ultraviolet light—then the pattern will be a smear. There are far too many fragments, with each one present in minute amounts,

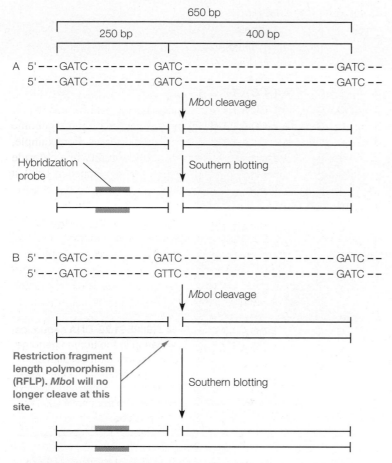

▲ **FIGURE 21.18** Analysis of a restriction fragment length polymorphism. See the text for additional details. Each dashed line represents one DNA strand from one chromosome. Each solid line represents a DNA strand in a restriction fragment. The thick yellow line represents a hybridization probe—a labeled oligonucleotide complementary to part of the sequence toward the left end of each molecule as drawn.

for any one fragment to be visualized. In the Southern technique (**FIGURE 21.19**) the contents of the gel are transferred, or "blotted," to a sheet of nitrocellulose, under denaturing conditions.

The initial Southern technique literally involved blotting; absorbent paper was placed between the electrophoretic gel and a sheet of nitrocellulose. Nitrocellulose binds tightly to single-stranded DNA, so what results is a replica of the agarose gel, with all DNA fragments tightly bound. Next you prepare a single-stranded DNA fragment identical to part of the region of interest, made highly radioactive by the incorporation of ^{32}P. This is incubated with the nitrocellulose sheet under annealing conditions, allowing the radiolabeled DNA "probe" to find complementary sequences and yield renatured radioactive double-strand DNA molecules. After this, the nitrocellulose is washed to remove unbound probe and is subjected to autoradiography. Only those fragments in the DNA digest containing sequences complementary to those of the probe will be detected because of the bound ^{32}P. More recently, use of radiolabeled hybridization probes has been supplanted by use of fluorescent nucleotide analogs. This allows the gels to be examined after transfer under UV light, without the need for hazardous radioisotopes. Also, the transfer from agarose gel to nitrocellulose is now done through application of an electric current (sometimes called *electroblotting*).

In the example shown in Figure 21.18, suppose that the labeled probe is complementary to a unique sequence in the 250-bp fragment.

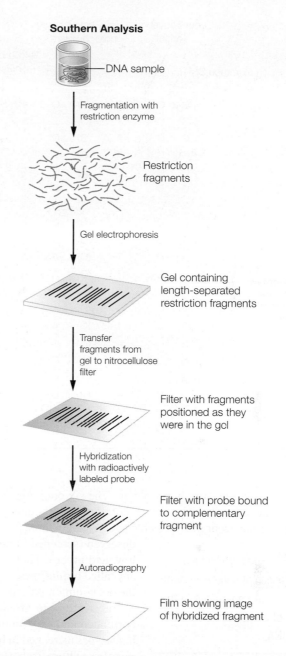

▲ **FIGURE 21.19** The principle of Southern transfer and hybridization. Reproduced from N. G. Cooper, 1994 (ed.), *The Human Genome Project. Deciphering the Blueprint of Heredity*, University Science Books, Mill Valley, CA, p. 63.

Then Southern analysis of an *Mbo*I digest will show a 250-bp fragment from the DNA of individual A and fragments of 250 and 650 bp in the digest from individual B.

How are the radiolabeled probes produced? One method involves the automated chemical synthesis of oligonucleotides in the presence of ^{32}P, or a fluorescent-tagged dNTP. More often, probes are prepared by polymerase chain reaction (Tools of Biochemistry 21A).

Polymorphisms within genomes include not only single-base changes, as shown in Figure 21.18, but also insertions, deletions, and repetitions of short sequences. **FIGURE 21.20** shows typical variation in a 5-kbp region of the genome for 10 individuals, carrying 20 distinct copies of the human genome. Shown are 10 single-nucleotide polymorphisms (SNPs), an insertion–deletion polymorphism (indel), and

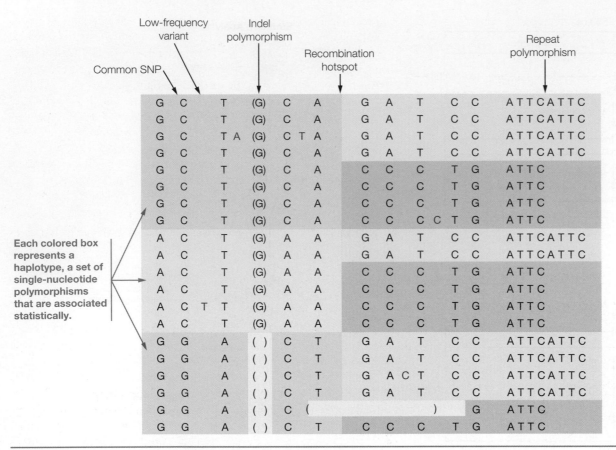

◀ **FIGURE 21.20 DNA sequence variation in the human genome.** See the text for a description. In the column labeled "Indel polymorphism" ( ) represents deletion of a G more frequently seen at that position.

a tetranucleotide repeat polymorphism. The six common polymorphisms on the left side are strongly correlated. Although these six could in principle occur in 2^6 possible patterns, only three patterns are observed (screened in pink, yellow, and green). These patterns are called **haplotypes.** Similarly, the six common polymorphisms on the right side are strongly correlated and exist in only two haplotypes (screened in blue and purple). There is little or no correlation between the two groups of polymorphisms because there is a "hotspot," or region of high genetic recombination frequency between them.

● **CONCEPT** Southern blotting allows detection of minute amounts of specific DNA sequences in the presence of a vast excess of non-specific DNA.

Southern Transfer and DNA Fingerprinting

Southern transfer and analysis has many applications, some of which are discussed elsewhere in this book. Best known to the general public is DNA fingerprinting in forensic analysis. Restriction fragment length polymorphisms (RFLPs, or "ruflups"), such as the variant sequences shown in Figures 21.18 and 21.20, exist throughout the genome. No two individuals, with the exception of identical twins, have the same pattern of RFLPs. Thus, a crime investigator who carries out Southern analysis using probes covering sufficient regions of the genome can generate a "DNA fingerprint" that establishes the identity of an individual with far more confidence than can be done from the prints on our fingertips. Moreover, the power of PCR to amplify minute amounts of DNA means that DNA fingerprinting can be carried out with miniscule amounts of material left at a crime scene, such as the DNA in a single human hair.

● **CONNECTION** DNA fingerprinting has many uses other than crime detection, including determining paternity, checking for food adulteration, and analyzing ancient DNA.

Most of the RFLPs used in forensic analysis are not single-base changes, but short-tandem repeats (*repeat polymorphisms*), such as that shown in the blue portion of Figure 21.20. **FIGURE 21.21** shows the results of a typical analysis.

Since DNA fingerprinting became routine in forensics laboratories in the late 1990s, many innocent criminal defendants have been acquitted and hundreds of previously convicted individuals have been exonerated. The large number of those freed, often after spending years in prison or having been sentenced to death, may yield satisfaction in the fact that molecular biology has contributed so successfully to the criminal justice system; it also evokes a good deal of sadness for the years and lives unjustly sacrificed owing to deficiencies in that same system.

Locating Genes on the Human Genome

The large number of known polymorphic regions in the human genome—those characterized by the presence of one or more RFLPs—provides many markers comparable to those of classical genetics. Just as alleles of a gene can be recognized by observable phenotypes—eye color, for example—so can alleles at polymorphic regions be recognized by the presence or absence of specific restriction sites. This allows disease genes to be mapped in terms of their proximity to a known polymorphic region. Earlier we mentioned mapping the mutations responsible for cystic fibrosis. Another example is Huntington's disease, an inherited neurological disease that took the life of folk singer Woody Guthrie. By carrying out RFLP analysis on the DNA of members of a family afflicted with

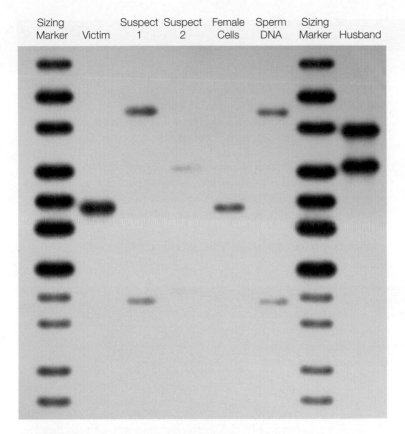

Sizing Marker | Victim | Suspect 1 | Suspect 2 | Female Cells | Sperm DNA | Sizing Marker | Husband

▲ **FIGURE 21.21 Results of a DNA forensic analysis in a rape investigation, involving repeat polymorphisms.** Each suspect contains two variable-length segments of this particular repeat, as does the victim's husband, so the fingerprint for each displays two bands when probed by Southern analysis with a radiolabeled oligonucleotide containing the repeated sequence. The fingerprint of suspect 1 corresponds to the pattern obtained from sperm in the victim's body.

this disease, those polymorphic regions, and specific sequence variants, that remain associated with susceptibility to the disease can be identified in several individuals. By these means, the physical location of the gene responsible for Huntington's disease was located, so that it could be sequenced, cloned, and investigated, with an eye toward identifying the mutant protein responsible for the disease and ways that this information could be used to seek treatment options. Thus, RFLPs have shown great value both in mapping the human genome and in seeking cures for genetic diseases based on an analysis of the responsible genes. Unfortunately, until a cure is found for Huntington's disease, the detection of a Huntington's marker in an individual's genome is devastating news because it means that the person is certain to be affected later in life.

Sequence Analysis Using Artificial Chromosomes

Although recent advances in DNA sequencing technology now permit many thousands of nucleotides to be sequenced in one run, the original Sanger method using fluorescent-tagged dideoxynucleotides was limited to several hundred nucleotides per sequencing run (see Tools of Biochemistry 4B). Because a human chromosome might contain 100 million base pairs or more, there was a need for sequence analysis in DNA segments much smaller than individual chromosomes, but also much larger than typical restriction fragments. Yeast artificial chromosomes (YACs) were useful for this purpose. A YAC is a DNA construct that contains a centromere, a DNA replication origin, telomeres at each end of the linear structure, a

selectable marker such as a drug resistance gene, and a cloning site. Inserts as large as 1000 kbp can be cloned into these constructs, which, after reintroduction into yeast cells, are maintained as the cell divides, just like natural chromosomes. A technique called **pulsed field gel electrophoresis,** which involves alternating voltage gradients during electrophoresis, allows resolution of the much larger DNA fragments resulting from analysis of these long DNA inserts. This approach subdivided each large sequencing project into sequence analysis of individual YACs, followed by assembly of sequences for a given chromosome. This effective, but laborious, procedure has now been supplanted by "second-generation" instruments that allow much more rapid DNA sequence analysis.

Size of the Human Genome

With all of these technologies in place, and with the aid of hundreds of workers, primarily from 20 institutions designated by the NIH as genome centers, and with Craig Venter's privately financed effort, the entire human genome sequence, comprising some three billion base pairs, was determined. A surprise, given the large amount of DNA in the genome, was the relatively small number of genes that are expressed as proteins. From the size of the genome and the amount represented by noncoding structures such as satellite sequences, it was expected that the genome would contain about 100,000 genes. However, analysis of the draft sequence suggested that the true number of genes, identified as open reading frames with appropriate punctuation for initiation and termination of translation (Chapter 25), was closer to 30,000. Subsequent refinement of the sequence led the number of putative genes further downward, to about 21,000. However, phenomena such as alternative splicing and posttranslational modification allow more than one protein product to be encoded by one gene. Nevertheless, the fact is that humans maintain complexity and diversity using far fewer protein-coding genes than had originally been expected. Note, however, that results previously mentioned from the ENCODE Project suggest that most of the noncoding region of the human genome may play some functions. Most of these functions are yet to be identified (page 670).

● **CONCEPT** The human genome contains between 20,000 and 25,000 genes, far fewer than predicted from the amount of DNA in a human cell.

From the first human sequence that was determined and from the analysis of RFLPs, it turns out that any two humans show differences in DNA sequence at about one base in 1000. In other words, any two humans, regardless of ethnic origin, are about 99.9% identical, genetically speaking.

The cost of large-scale genome sequence analysis continues to drop as technological innovations in sequencing methods keep making the process faster and more efficient. The government-sponsored Human Genome Project cost about $400 million. In 1999, sequence analysis of one million base pairs of DNA cost about U.S. $20,000; by 2010 that cost was 20 cents and dropping. By mid-2010, Illumina, the designer of one of the "second-generation" sequence technologies (Tools of Biochemistry 4A), was offering complete sequence determination of one individual for $9500, and by 2013 that cost had dropped by an additional threefold. By the time you read this book, the cost will probably be below $1000. Although whole-genome sequencing has not yet become routine in clinical medicine, human genome analysis has become a cottage industry. Companies such as 23andMe will analyze DNA in a cheek swab, looking for disease-related genetic markers such as those for cystic fibrosis. Analysis of this type is not without its problems, as discussed in the references at the end of this chapter.

Summary

- Bacterial, viral, and eukaryotic genomes vary in unexpected ways in size and complexity. Viral genomes, either RNA or DNA, encode as few as three proteins or as many as 175, or even more. Bacterial genomes usually consist of one circular double-strand DNA molecule encoding about 1500 to 2000 proteins, while the human genome encodes about 21,000 proteins, considerably fewer than predicted from the amount of DNA in the genome. Most of the DNA in a bacterial genome encodes proteins or RNA or serves a regulatory function. Most eukaryotic cells are diploid, with two of each chromosome. An exception is the sex chromosomes, where a male human typically has one X and one Y chromosome. (Sections 21.1 and 21.2)

- Chromosomal DNA is compacted by folding about **histone** octamers, to form **nucleosome** core particles. These particles plus linker DNA make up chromatin, which refolds to form 30-nm fibers. Further compaction processes fold these fibers into the highly condensed form seen in **chromatin.** (Section 21.3)

- Sequence determination of complex genomes, notably those of humans, depended crucially on the development of new technologies, including use of type II restriction endonucleases, fluorescent in situ hybridization (FISH), Southern transfer, **polymerase chain reaction** (PCR), and the use of yeast artificial chromosomes (YACs) as vectors for cloning huge amounts of DNA. (Section 21.4)

Problems

Enhanced by
Mastering Chemistry
for Biochemistry

Mastering Chemistry for Biochemistry provides select end-of-chapter problems and feedback-enriched tutorial problems, animations, and interactive figures to deepen your understanding of complex topics while practicing problem solving.

Answers to red problems are available in the Answer Appendix.

1. The exponential nature of PCR allows spectacular increases in the abundance of a DNA sequence being amplified. Consider a 10-kbp DNA sequence in a genome of 10^{10} base pairs. What fraction of the genome does this sequence represent? That is, what is the fractional abundance of this sequence in this genome? Calculate the fractional abundance of this target sequence after 10, 15, and 20 cycles of PCR, starting with DNA representing the whole genome and assuming that no other sequences in the genome undergo amplification in the process.

2. Of the restriction enzymes listed in Table 21.3, which enzymes generate flush, or blunt-ended, fragments? Of those that recognize offset sites and generate staggered cuts, which of these cuts cannot be converted to flush ends by the action of DNA polymerase? Why?

3. The following diagram shows one-half of a restriction site.
   (a) Draw the other half.

$$\text{G A C G T C}$$

   (b) Use heavy arrows (↑↓) to identify type II cleavage sites that would yield blunt-ended duplex DNA products.
   (c) Use light arrows (↑↓) to identify type II cleavage sites yielding staggered cuts that could be converted directly to recombinant DNA molecules by DNA ligase, with no other enzymes involved.
   (d) If this were the recognition site for a type I restriction endonuclease, where would the duplex be cut?
   (e) If DNA sequences were completely random, how large an interval (in kilobase pairs) would you expect on average between identical copies of this sequence in DNA?

4. pBR322 DNA (4.36 kb; see Figure 4A.3) was cleaved with *Hin*dIII nuclease and ligated to a *Hin*dIII digest of human mitochondrial DNA. One recombinant plasmid DNA was analyzed by gel electrophoresis of restriction cleavage fragments, with the following results: lane A = *Eco*RI-treated recombinant, lane B = *Hin*dIII-treated vector, and lane C = *Hin*dIII-treated recombinant.

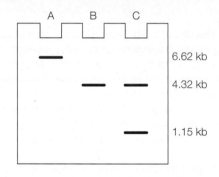

(a) Why was the recombinant plasmid treated with *Eco*RI to determine its size?
(b) How might you explain the discrepancy between the size of the recombinant molecule and the sum of the sizes of the *Hin*dIII cleavage fragments?
(c) Draw a diagram of the recombinant showing the locations of the *Hin*dIII cleavage sites.

5. A small DNA molecule was cleaved with several different restriction nucleases, and the size of each fragment was determined by gel electrophoresis. The following data were obtained.

Enzyme	Fragment Size (kb)
*Eco*RI	1.3, 1.3
*Hpa*II	2.6
*Hin*dIII	2.6
*Eco*RI + *Hpa*II	1.3, 0.8, 0.5
*Eco*RI + *Hin*dIII	0.6, 0.7, 1.3

(a) Is the original molecule linear or circular?
(b) Draw a map of restriction sites (showing distances between sites) that is consistent with the data given.
(c) How many additional maps are compatible with the data?
(d) What would have to be done to locate the cleavage sites unambiguously with respect to each other?

6. The average human chromosome contains about 1×10^8 bp of DNA.
   (a) If each base pair has a mass of about 660 daltons and there are about 2 g of protein (histones plus nonhistones) per gram of DNA, how much does such a chromosome weigh (in grams)?
   (b) If the DNA were extended, how long would it be?
   (c) An actual chromosome is about 5 mm long. What is the approximate compaction ratio?
   (d) You have about 4×10^{12} cells in your body. If you have 46 chromosomes in each cell, what is the approximate extended length of *all* of your DNA? For comparison, the distance from the earth to the sun is about 1.5×10^8 km.

7. Forming nucleosomes and wrapping them into a 30-nm fiber provide part of the compaction of DNA in chromatin. If the fiber contains about six nucleosomes per 10 nm of length, what is the approximate compaction ratio achieved? Compare this answer with the one from Problem 6(c).

8. Huntington's disease is an inherited neurological ailment with a variable age of onset (see Chapter 22). A protein called huntingtin has a sequence of repeated glutamine residues, all encoded by CAG. The number of repeated CAG triplets is expanded in Huntington's disease, apparently as the result of replication errors. The age of onset is related to the number of CAG triplets in the repeat region; the more glutamine codons, the earlier the onset. Describe experimental evidence, using Southern blotting, that is consistent with this finding.

9. A sample of chromatin was partially digested by the enzyme Staphylococcal nuclease. The DNA fragments from this digestion were purified and resolved on a polyacrylamide gel. A set of DNA restriction fragments was used as markers. Distances of migration (*d*) are given below in cm. From these data, estimate the nucleosome repeat distance in the chromatin.

Marker DNA Fragment		Chromatin DNA Fragment
Size (bp)	*d* (cm)	*d* (cm)
94	40	30.
145	34.2	19.2
263	25.2	14.4
498	16.7	11.5
794	11.5	—

10. Suppose that you wished to determine the number of pseudogenes related to a particular gene in an organism whose complete genome had not yet been sequenced. How might you do this experimentally?

11. Some viruses, such as SV40, contain closed circular DNAs carrying nucleosomes. If the SV40 viral chromosome is treated with topoisomerase, and the histone is then removed, it is still supercoiled. If histones are removed *before* topoisomerase treatment, however, the DNA is relaxed. Explain.

12. Histone genes are unusual among eukaryotic genes because they do not have introns, and histone mRNAs do not have poly(A) tails (see Chapter 24). Moreover, in almost all eukaryotes, histone genes are arranged in multiple tandem domains, each domain carrying one copy of each of the five histone genes. Explain these features in terms of the special requirements for histone synthesis.

13. Consider the λ bacteriophage DNA molecule as shown in Figure 21.14. Total digestion with the two restriction endonucleases used does not allow unambiguous placement of restriction sites, as shown in the map. Describe restriction patterns—either from partial digestion or from digestion with both *Bam*HI and *Eco*RI—that would help to establish the specific map of restriction sites shown in Figure 21.14(b).

14. RFLPs can include sequence changes created by single-base change, single-base insertion, single-base deletion, or tetranucleotide repeat, as shown in Figure 21.20. For each polymorphism, show a specific sequence change that would alter the restriction fragmentation pattern at that site and describe how the pattern would change—that is, addition or loss of a site for a specific restriction endonuclease. Refer to Table 21.3 for the cleavage sites of several common restriction nucleases.

15. Why do you suppose that forensic DNA analysis relies principally on short tandem repeats (repeat polymorphisms) rather than single-nucleotide polymorphisms, such as that described on page 678 and in Figure 21.20?

16. DNA methylation is considered an epigenetic phenomenon because the pattern of methylation can be inherited. Briefly discuss the question of whether histone modifications in chromatin represent an epigenetic phenomenon.

17.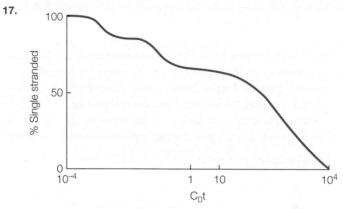

DNA renaturation curves occasionally show three distinct phases of renaturation. In this graph, DNA renaturation is plotted against C_0t (initial concentration times time of renaturation—essentially a measure of relative renaturation time). See Figure 21.3.
   (a) Identify each part of this plot that corresponds to reannealing of (1) unique sequences, (2) moderately repetitive sequences, and (3) highly repetitive sequences.
   (b) Suppose that you cloned a single-copy gene, such as the gene for dihydrofolate reductase (DHFR), into a plasmid vector and subjected it to renaturation analysis. Sketch the curve you might expect.
   (c) Suppose that you used reverse transcriptase (Chapter 24) to copy the ovalbumin mRNA and cloned this *complementary DNA* (cDNA) into a plasmid vector. Would you expect this cDNA to reanneal (1) more slowly, (2) more rapidly, or (3) at the same rate as genomic DNA? Briefly explain your answer.

18. Refer to Figure 19.22 (page 632), which describes the base modifications of bacteriophage T4 DNA, and briefly describe some issues that must be dealt with in preparing a restriction map of T4 DNA.

19. The restriction endonuclease *Not*I recognizes the octanucleotide sequence GCGGCCGC. Calculate the expected number of *Not*I cleavage sites in the bacteriophage λ genome, a linear DNA duplex 48.5 kbp in length with a (G + C) content of 50%.

20. Huntington's disease is a hereditary central nervous system disorder characterized by tandem repeats of the sequence 5'-CAG-3' in the gene that encodes a protein called huntingtin. The disease is progressive from generation to generation, meaning that in later generations the number of CAG repeats increases and the age of onset of symptoms decreases. Refer to Figure 21.4 and describe the sort of evidence supporting the generational increase in the number of CAG repeats. Note: This will be different from the kind of evidence asked for in problem 8.

 References

For a list of references related to this chapter, see Appendix II.

As we discuss in Chapter 4, gene cloning by recombinant DNA techniques revolutionized biology in the mid-1970s because it allows an investigator to isolate and amplify individual genes for the analysis of their sequence, expression, and regulation. Cloning requires living cells, into which DNA molecules must be introduced for amplification. An equally revolutionary technique, **polymerase chain reaction (PCR),** was invented by Kary Mullis in 1983. PCR amplifies exceedingly small amounts of DNA in vitro, without prior transfer into living cells. This technique has facilitated the analysis of eukaryotic genes because it avoids some of the tedium involved in cloning DNA from very large genomes. In addition, the technique has dozens of practical applications.

Recall from Chapter 4 that DNA polymerase catalyzes the addition of a deoxyribonucleoside triphosphate to a preexisting 3′-hydroxyl terminus of a growing daughter DNA strand (the primer). A template DNA strand is required, which instructs the polymerase regarding the correct nucleotide to insert at each step (see Figure 4.32). The reaction is presented in detail in Chapter 22, but is shown in outline form here.

3′-pApTpTpCpApApGpApGpG.... + dTTP ⟶
5′-pTpApApG-OH

3′-pApTpTpCpApApGpApGpG....
5′-pTpApApGpT-OH

Most DNA polymerases also contain an associated 3′-exonuclease. This activity helps the enzyme to proofread errors by removing nucleotides that occasionally are incorporated in non-Watson/Crick fashion.

PCR requires knowledge of the sequences that flank the region to be amplified. Oligonucleotides complementary to these sequences are produced by automated chemical synthesis and are used as primers in a special series of DNA polymerase–catalyzed reactions (**FIGURE 21A.1**). First, DNA containing the sequences to be amplified is heat-denatured at 94–96 °C and then annealed to the primers, which are present in excess (steps ❶ and ❷). Annealing is carried out at 50–65°, depending on the base composition of the DNAs involved. Next, polymerase chain extension is carried out from the 3′ termini of the primers (step ❸). Then a second cycle of heat denaturation, annealing, and primer extension is carried out. Using a thermostable form of DNA polymerase, such as **Taq polymerase,** from an organism that lives at high temperatures (in hot springs) avoids the need to add more polymerase at each cycle because the enzyme is not inactivated at DNA-denaturing temperature. Step ❸ is typically carried out at 72°, the temperature optimum for *Taq* polymerase. This cycle is repeated 30 or more times

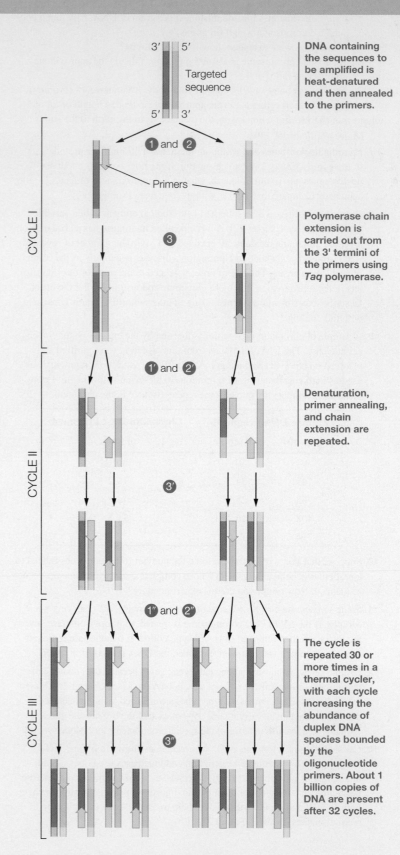

▶ FIGURE 21A.1 **Three cycles of the polymerase chain reaction.** A segment within the region shown in blue is amplified by use of primers (green) that are complementary to the ends of the blue segment. The amplification process is exponential.

DNA containing the sequences to be amplified is heat-denatured and then annealed to the primers.

Polymerase chain extension is carried out from the 3' termini of the primers using *Taq* polymerase.

Denaturation, primer annealing, and chain extension are repeated.

The cycle is repeated 30 or more times in a thermal cycler, with each cycle increasing the abundance of duplex DNA species bounded by the oligonucleotide primers. About 1 billion copies of DNA are present after 32 cycles.

in an automated temperature-regulating device (a *thermal cycler*), with each cycle increasing the abundance of duplex DNA species bounded by the oligonucleotide primers. Two such molecules have been formed by the end of cycle III, and the number doubles with each successive cycle. About 1 billion copies are present after 32 cycles (2^n to be exact, where n is the number of cycles).

Many applications of this technique have been developed, including forensic analysis, in which DNA can be amplified from exceedingly small samples of biological material (e.g., blood, semen, or hair) for the identification of suspects in criminal cases or of fathers in paternity cases. PCR-based forensic analysis focuses on parts of the genome with variable numbers of tetranucleotide repeats (see Figure 21.20). Some parts of the human genome have a particular tetranucleotide sequence repeated variable numbers of times up to 30. Analyses of these regions by PCR can determine the number of repeats in one individual's genome. By analyzing several of these regions, an individual can be identified with great confidence, beginning with a DNA sample as small as one nanogram (i.e., 10^{-9} g).

The field of "molecular anthropology" has developed from PCR and sequence analysis of human mitochondrial DNA, with the results used to formulate models of human evolution. Mitochondrial DNA is useful, both because it is small (rapidly sequenced) and because it undergoes mutation more rapidly than nuclear DNA. This means that the amount of sequence variation is a more sensitive indicator of evolutionary time than is seen in nuclear DNA. A comparable field of "molecular paleontology" exists, with minute amounts of DNA being extracted from long-preserved biological samples, such as organisms frozen in ice or insects trapped in amber. The recent sequence analysis of DNA from minute bone fragments of Neanderthals and other archaic humans has generated new insights into human evolution. (This

application required modification of the original method so that only one strand is amplified; (can you tell why?) In the same way, PCR is also used to diagnose microbial or viral infections and in prenatal diagnosis of genetic diseases.

PCR does suffer some limitations. For example, because some thermostable DNA polymerases, such as the commonly used *Taq* polymerase, lack a proofreading $3'$ exonuclease, the DNA synthesis in PCR can be relatively inaccurate. This is not usually a problem if one wants to sequence the PCR product because errors are uniformly distributed over the length of DNA being amplified, with the error frequency at each site being too low to affect sequencing operations. Also, proofreading polymerases from thermophilic organisms are now available. However, precautions must be taken if you want to clone PCR products with the assurance that a natural sequence is being cloned. Another problem is the great sensitivity of the technique, which can cause amplification of minute amounts of DNA contaminants in the sample.

A number of variants on the original PCR technique have been devised. In quantitative PCR or real-time PCR, one of the nucleotides has a fluorescent label on it, which makes it possible to follow the reaction in real time and relate the data to the abundance of the target DNA in the original sample. Another variation of the technique, called RT-PCR, can be used to monitor gene expression by analyzing the levels of a particular mRNA species. The sample is treated first with reverse transcriptase, an RNA-dependent DNA polymerase (Chapter 22), which converts an mRNA molecule to a DNA of complementary sequence, which can then be amplified by PCR.

References

For a list of references related to this chapter, see Appendix II.

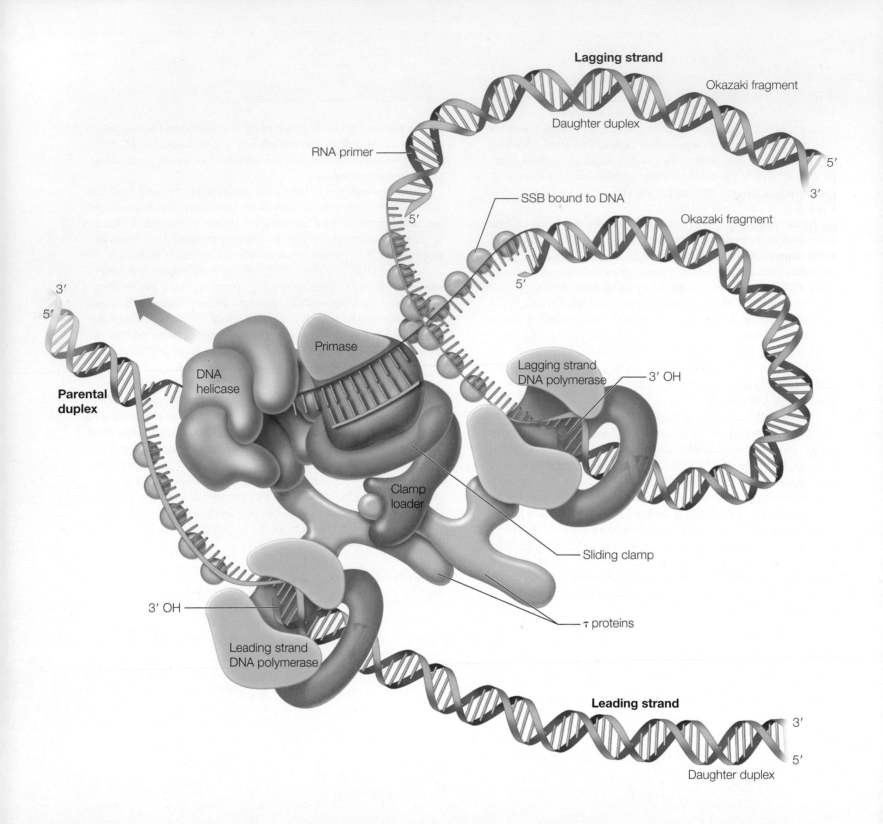

Lagging strand

Okazaki fragment

Daughter duplex

RNA primer

SSB bound to DNA

Okazaki fragment

5′

3′

5′

Primase

Lagging strand
DNA polymerase

3′ OH

3′

5′

DNA
helicase

**Parental
duplex**

Clamp
loader

Sliding clamp

τ proteins

3′ OH

Leading strand
DNA polymerase

Leading strand

3′

5′

Daughter duplex

DNA replication is catalyzed by a dynamic multiprotein machine called
the replisome. DNA polymerase, which catalyzes phosphodiester bond
formation, is but one of the many proteins that must act in concert to
faithfully copy each of the two DNA strands in a single replication fork.

DNA Replication

OUR FOCUS FOR the remainder of this book is on the biochemistry of informational macromolecules—processes in genome replication, genome maintenance, and gene expression. We discuss DNA replication in this chapter. Next, Chapter 23 focuses on DNA repair, recombination, and genome rearrangement. In Chapter 24 we turn to the readout of information encoded in DNA—transcription and RNA processing. Chapter 25 deals with protein synthesis—how information encoded in the four-letter nucleic acid language is translated into the 20-letter amino acid language, and how proteins undergo processing and trafficking to their destinations as mature proteins, in the correct intracellular or extracellular location. Finally, Chapter 26 revisits these process-oriented chapters to consider how gene expression is regulated.

● CONCEPT The biosynthesis of nucleic acids and proteins is carried out through the processes of replication, transcription, and translation. DNA repair, recombination, and rearrangement play essential functions in stabilizing and diversifying the genome.

DNA replication is an amazing process. From Chapter 4, you know that DNA strands must unwind, after which incoming nucleotides must be fitted to template strands and connected covalently to growing strands, followed by rewinding of the new duplex

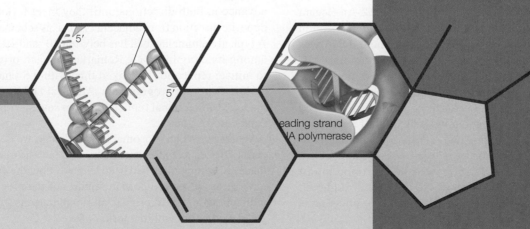

5′

5′

eading strand
A polymerase

molecules. The whole process occurs with hundreds of nucleotides incorporated per second, and with the replication machinery making errors only once in about one billion incorporation events.

22.1 Early Insights into DNA Replication

In 1953, Watson and Crick predicted three central features of DNA replication. First was their explicit prediction that DNA replication

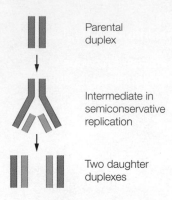

Parental duplex

Intermediate in semiconservative replication

Two daughter duplexes

is semiconservative—that each of the two identical daughter DNA molecules contains one parental strand and one newly synthesized strand. This prediction was confirmed in 1958 by Meselson and Stahl (Figure 4.14). Not explicitly required by the model, but implied, was the idea that unwinding of the parental strand and synthesis of new DNA occur simultaneously, in the same microenvironment. In other words, replication occurs at a **fork** in which parental strands are unwinding and daughter strands are undergoing elongation, as suggested by the diagram. Also suggested by the Watson–Crick model was the premise that replication begins at one or more fixed sites— **replication origins**—on a chromosome.

The first images of replicating DNA were consistent with the idea that replication occurs within a fork. In *E. coli* the chromosome is one circular DNA molecule about 1.3 mm long. In 1963, John Cairns was able to visualize *E. coli* chromosomes in the act of replication. What he did was to label replicating DNA by growing the bacteria in medium containing [³H]thymidine. With great care he teased out a few of the immensely long molecules without breaking them and deposited them on a membrane. Exposure of the membrane to X-ray film showed structures like that presented in **FIGURE 22.1**. The circular molecule shown contains two Y-shaped structures, either or both of which could be a replication fork. This *theta structure* (from the Greek letter θ) does not permit one to conclude whether replication is unidirectional, with one Y junction representing the site of replication initiation, or whether it is bidirectional, with two forks that are moving from a fixed origin halfway between the two Y's.

The two alternatives are shown at the top of **FIGURE 22.2**. An elegant variation of Cairns's experiment established that for *E. coli*, replication is bidirectional from one fixed origin. In this experiment, DNA was radiolabeled by growing cells in the presence of [³H]thymidine. Just as cells were terminating one round of replication and initiating the next, the specific radioactivity of the thymidine was increased severalfold, so that areas of active DNA synthesis could be distinguished in an autoradiogram by increased darkening of the film. If replication were unidirectional, then termination and reinitiation would have to occur at the same site. Examination of individual DNA molecules, such as that depicted in Figure 22.2, showed that termination of one round of replication and initiation of the next occurred about 180° apart from each other, which could occur only if replication is bidirectional.

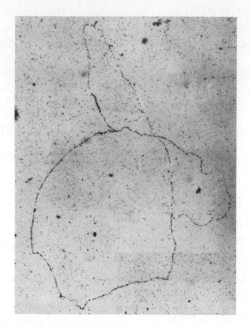

▲ **FIGURE 22.1** Autoradiogram of a replicating *E. coli* chromosome after two generations of growth in [³H]thymidine.

What about eukaryotic cells, which have huge linear molecules in their chromosomes? Typically, a mammalian cell in culture replicates all of its DNA in an eight-hour S phase (as compared to 40 minutes for doubling of the *E. coli* chromosome). Autoradiographic and other evidence shows that individual DNA chains in eukaryotic cells grow an order of magnitude more slowly than bacterial chromosomes— about 100 nucleotides per second.

● **CONCEPT** In bacteria, DNA replication initiates from a fixed origin and proceeds in both directions from that origin.

Autoradiographic studies reveal that replication initiates bidirectionally on linear eukaryotic chromosomes, but from numerous fixed origins on each chromosome. Forks advance in both directions until they meet forks proceeding in the opposite direction from adjacent origins, as schematized in **FIGURE 22.3**. A typical mammalian cell has between 10^3 and 10^4 origins distributed among its chromosomes. Remarkably, each origin is programmed to initiate replication at a fixed time within S phase; in other words, "origin firing" is programmed. Archaeal DNA replication evidently involves bidirectional replication from several origins on a single circular chromosome.

Eukaryotic chromosomal DNA is linear; we deal with events at the ends of these molecules later in this chapter. The main point, however, is that even before much had been learned about the enzymology of DNA replication, several important features of the process were evident— bidirectionality, fixed origins, and simultaneous strand unwinding and chain elongation within a fork.

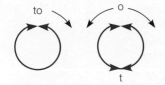

Unidirectional   Bidirectional

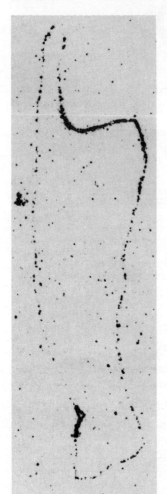

◀ **FIGURE 22.2 Demonstration of bidirectional replication by autoradiography.** This *E. coli* chromosome was labeled by long-term growth in [³H]thymidine, and the specific activity was increased fourfold just as a culture completed one round of replication and commenced another. The diagrams show the labeling patterns that would be expected for unidirectional replication, where the terminus (t) and origin (o) are adjacent, and for bidirectional replication, where they are on opposite sides of the chromosome. The data support the bidirectional replication model.

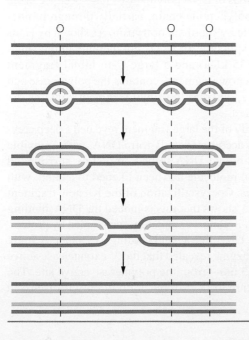

◀ **FIGURE 22.3 Bidirectional replication from several fixed origins (o) on a linear eukaryotic chromosome.**

22.2 DNA Polymerases: Enzymes Catalyzing Polynucleotide Chain Elongation

The biochemical elucidation of DNA replication began in the mid-1950s, with Arthur Kornberg's discovery of DNA polymerase. Kornberg carried out the chemical synthesis of deoxyribonucleoside 5′-triphosphates (dNTPs; unknown at that time) and isolated an enzyme from *E. coli* that incorporated these nucleotides into DNA. The enzyme required added DNA, plus Mg^{2+}. As mentioned in Chapter 4, two DNA molecules are required for the reaction—the *template* and the *primer*—with the primer becoming covalently extended from its 3′ hydroxyl group and the template specifying the nucleotide sequence of the product DNA strand. Also, the polarity of the product DNA strand is opposite from that of the template strand, as expected from the antiparallel orientation of DNA strands in the Watson–Crick model.

● **CONCEPT** DNA polymerase catalyzes nucleophilic attack by the 3′ hydroxyl at the primer terminus upon the α-phosphate of an incoming dNTP, base-paired with its template.

As shown in **FIGURE 22.4**, the DNA polymerase reaction involves nucleophilic attack by the 3′ hydroxyl group of the primer terminus on the α-phosphate of the deoxyribonucleotide substrate (step ②), which leads to covalent bond formation. The 3′ hydroxyl

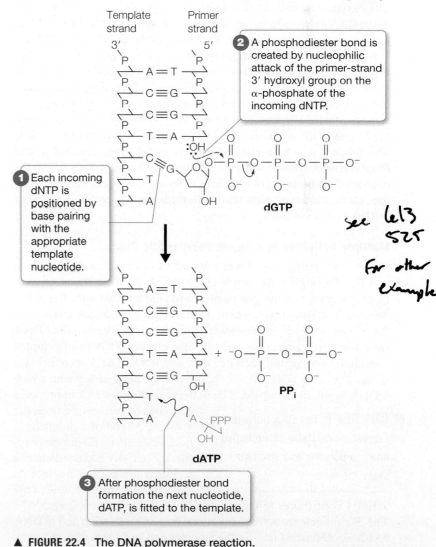

see 613 525 for other examples

▲ **FIGURE 22.4 The DNA polymerase reaction.**

is a weak nucleophile, but the pyrophosphate is a good leaving group, thus facilitating this nucleophilic substitution. The reaction is readily reversible. In cells and in crude preparations, the reaction is drawn to the right by pyrophosphatase, which hydrolyzes pyrophosphate to two orthophosphates ($PP_i + H_2O \rightarrow 2 P_i$). Hence, two energy-rich phosphates are expended per nucleotide incorporated.

Structure and Activities of DNA Polymerase I

The DNA polymerase discovered by Kornberg in *E. coli* was later shown to be one of five different DNA polymerases expressed in most bacterial cells. The Kornberg enzyme is now called DNA polymerase I. We shall use this term henceforth and introduce the other polymerases later in this chapter.

DNA Substrates for the Polymerase Reaction

DNA polymerase I requires both template DNA and a primer, either DNA or RNA. In vitro, these roles can be played either by two distinct nucleic acids or by one molecule. As shown in **FIGURE 22.5(c)**, a single-stranded DNA with self-complementary sequences can fold itself into a **hairpin** or **stem-loop** structure whose 3′ end can be extended by polymerase, using the 5′ end as the template. The figure shows other primer–template substrates that DNA polymerase can act upon in vitro. The enzyme can copy around a circular single-stranded template, such as the DNA extracted from small bacteriophages, like ϕX174 or M13, as long as a primer is present, but it cannot join the ends [Figure 22.5(a)]. When the template is linear, polymerase copies only to the 5′ end of the template, and then it dissociates [Figure 22.5(b) and (c)]. In similar fashion, the enzyme can fill in a gap [Figure 22.5(d)], dissociating when the gap is reduced to a nick. Under some conditions, the enzyme can also extend from a 3′ hydroxyl group at a nick. Typically, when this occurs, the 5′ end of the preexisting DNA is displaced in advance of the nick. This process, which generates a single-stranded "flap," is called **strand displacement** synthesis [Figure 22.5(e)]. Under conditions where the displaced strand is degraded, there is no net DNA accumulation, and the reaction is called **nick translation** (distinct from *translation,* the mRNA-coded synthesis of proteins).

Multiple Activities in a Single Polypeptide Chain

When DNA polymerase I was purified from *E. coli,* it was found to consist of a single polypeptide chain ($M_r = 103,000$). In addition to its polymerase activity, the purified enzyme has two nuclease activities—a **3′ exonuclease,** which degrades single-stranded DNA from the 3′ end, and a **5′ exonuclease,** which degrades base-paired DNA from the 5′ terminus. The enzyme also cleaves RNA from a duplex containing one strand each of DNA and RNA. The 3′ exonuclease serves a "proofreading" function, improving the accuracy with which a DNA template is copied. The activity will remove an improperly base-paired nucleotide from the growing 3′ end of a polydeoxynucleotide chain, giving the polymerase activity a second chance to insert the correct nucleotide specified by the template. This activity contributes to the fidelity of DNA replication (page 707). The 5′ exonuclease activity functions both in replication and in DNA repair, as described later in this chapter and in Chapter 23.

● **CONCEPT** The DNA polymerase I molecule contains three active sites: a polymerase and two exonucleases.

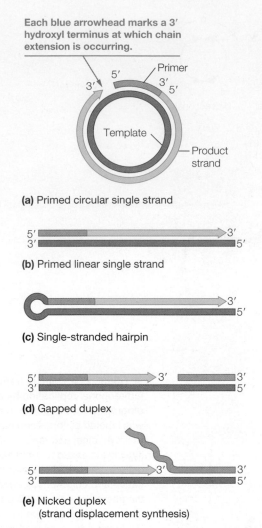

Each blue arrowhead marks a 3′ hydroxyl terminus at which chain extension is occurring.

(a) Primed circular single strand

(b) Primed linear single strand

(c) Single-stranded hairpin

(d) Gapped duplex

(e) Nicked duplex (strand displacement synthesis)

▲ **FIGURE 22.5** DNA substrates that can be acted upon by purified DNA polymerase.

Structure of DNA Polymerase I

The three catalytic activities of DNA polymerase I have been localized to regions of the long polypeptide chain, partially through limited proteolysis of the enzyme by subtilisin or trypsin. As shown by Hans Klenow, this treatment splits the 103-kDa polypeptide into a small N-terminal fragment (35 kDa) and a large C-terminal fragment (68 kDa). The large "Klenow fragment" contains the polymerase and 3′ exonuclease domains, whereas the small fragment contains the 5′ exonuclease domain.

Crystallographic study of the large fragment revealed a deep crevice, just large enough to accommodate B-form DNA, with a flexible subdomain that allows bound DNA to be completely surrounded (**FIGURE 22.6**). The protein molecule has been likened to a hand, with palm, thumb, and fingers. Cocrystallization of the Klenow fragment with a short duplex DNA shows that this is indeed the DNA-binding site and that bound DNA is almost completely surrounded by protein, which wraps around the DNA like a hand holding a cylinder. A study of mutant forms of the enzyme revealed that the 3′ exonuclease active site is quite far—about 3 nm—from the polymerase active site. The significance of this feature to DNA replication fidelity is discussed later in this chapter.

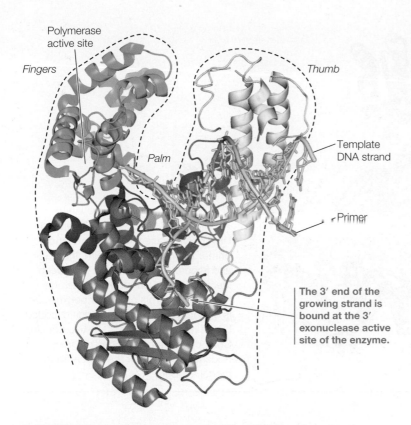

▲ **FIGURE 22.6** **The Klenow fragment of *E. coli* DNA polymerase I.** Ribbon backbone representation of the Klenow fragment complexed with DNA, in an editing configuration. That is, the 3′ end of the growing strand (light green) is bound at the 3′ exonuclease active site of the enzyme (orange). The locations of the polymerase and 3′ exonuclease active sites have been identified by site-directed mutagenesis. The template strand is shown in turquoise. PDB ID: 1kln.

Discovery of Additional DNA Polymerases

For his discovery of DNA polymerase, Arthur Kornberg was awarded the 1959 Nobel Prize in Physiology or Medicine. Much, however, remained to be learned about DNA replication. In particular, if polymerase adds nucleotides only in a 5′ → 3′ direction, how is the complementary chain extended, if both DNA chains are extended in the same replication fork? Moreover, V_{max} for DNA polymerase I in vitro is only about 20 nucleotides per second, while DNA strands during replication in living bacteria must grow at several hundred nucleotides per second to replicate the entire *E. coli* genome in 40 minutes. How is one to account for these discrepancies?

In 1969, John Cairns isolated an *E. coli* mutant that was deficient in DNA polymerase I activity. This made it possible to detect two additional DNA polymerase activities in *E. coli* cells. These were named DNA polymerases II and III. (Recall that the original Kornberg polymerase is DNA polymerase I.) Three decades later, two additional polymerases were described in *E. coli,* named IV and V (see Chapter 23).

DNA polymerase III was assigned the major role in nucleotide incorporation during replication. What evidence supports this conclusion? First, polymerase III has a high V_{max} (250 to 1000 nucleotides per second), consistent with the high replicative DNA chain growth

rate. Second, there are 10 or fewer molecules of the enzyme per cell, as expected if it functions only at replication forks. Third, and most definitive, is the existence of temperature-sensitive (*ts*) mutant cells that contain a thermolabile form of DNA polymerase III and in which DNA replication in vivo is blocked at high temperature. Because a single mutation both inactivates polymerase III *and* causes inhibition of DNA replication, the same mutation must be responsible for both effects, confirming that polymerase III plays an essential role in DNA replication. The gene encoding the catalytic subunit of polymerase III was named *polC*, with the genes for polymerases I and II being named *polA* and *polB*, respectively.

What do these developments tell us about the physiological roles of the other DNA polymerases? In the original *polA* mutant described by Cairns, DNA replication occurred normally. However, the bacteria were abnormally sensitive to DNA-damaging agents such as ultraviolet light, suggesting a role for polymerase I in DNA repair. Further study of the Cairns mutant did reveal an additional role in replication. Although lacking the *polymerase* activity of DNA polymerase I, the mutant bacteria did synthesize the small N-terminal fragment of the protein, which contains the 5′ exonuclease activity. Still later it was found that mutants lacking this 5′ exonuclease activity are also defective in DNA replication. Thus, it was established that two polymerases, I and III, play essential roles in DNA replication. Polymerase II is inducible by DNA-damaging events, as are IV and V, so it is thought to function in DNA repair.

Structure and Mechanism of DNA Polymerases

Although there is considerable diversity in the primary structure of DNA polymerases, all enzymes in this family show features in common. Like polymerase I, all of the polymerases described so far have a structure described for the Klenow fragment, with domains identified as "palm," "thumb," and "fingers." **FIGURE 22.7** shows these features for a polymerase holoenzyme—the DNA polymerase encoded by bacteriophage T7.

Analyses of DNA–enzyme complexes, conserved amino acid residues, and targeted mutations indicate that the polymerase active site lies in the palm domain and that the 3′ exonuclease site lies at the base of the palm. High-resolution polymerase structures show two magnesium ions, bound to nucleotide phosphates and to conserved aspartate residues already known to be essential for catalysis. These features support a general polymerase mechanism proposed by Thomas Steitz (**FIGURE 22.8**). In that mechanism, one metal ion polarizes the hydroxyl group at the 3′ primer terminus, facilitating nucleophilic attack of that moiety upon the α-phosphate of the dNTP substrate. Both metals stabilize a trigonal bipyramidal transition state, in which the α-phosphorus atom is linked to five oxygens, and the second metal facilitates leaving by the pyrophosphate. Extensive contacts occur between the enzyme and the DNA minor groove— contacts that could occur only with a properly base-paired duplex (not shown). Also, structures of polymerase–DNA complexes show that DNA near the primer terminus adopts a conformation more like that of A form than B form and that this conformational change facilitates the minor groove interactions. Moreover, the incoming dNTP was shown to fit snugly into a pocket that favors correct base pairing with the template. Thus, these structures reveal both how the reaction is catalyzed and how the enzyme copies its template DNA with high accuracy.

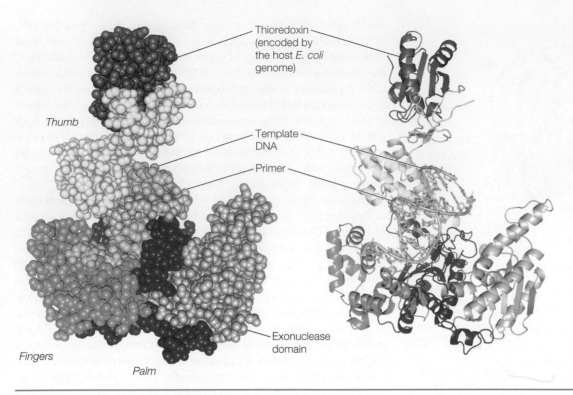

◀ **FIGURE 22.7 Structure of T7 phage DNA polymerase with template and primer bound.** Note the palm, thumb, and fingers domains, which have been identified in all DNA polymerases analyzed to date. This particular polymerase is distinctive in having a subunit encoded by the host *E. coli* genome—thioredoxin. Color coding for the ribbon diagram is identical to that for the space-filling model. PDB ID: 1t7p.

22.3 Other Proteins at the Replication Fork

As noted earlier, DNA polymerase III was identified as the major replicative polymerase from the analysis of temperature-sensitive mutants in the structural gene for polymerase III. Genetic analysis of a great many *ts* mutations in *E. coli* revealed the existence of additional proteins playing essential roles in DNA replication.

Genetic Maps of *E. coli* and Bacteriophage T4

FIGURE 22.9 shows part of the *E. coli* genetic map, identifying genes known to participate, either directly or indirectly, in DNA replication. The large number of these genes underscores the complexities of DNA replication, and we shall refer frequently to genes shown on this map. The genes were mapped by analysis of bacterial conjugation. As shown by Joshua Lederberg, bacteria can carry out sexual reproduction under certain conditions, and the sexual partners can undergo genetic recombination. Analysis of recombination frequencies between genetic markers allows placement of a gene on a map such as that shown in Figure 22.9.

By convention, each gene has an italicized lower-case three-letter designation that refers either to the gene product or to the phenotype

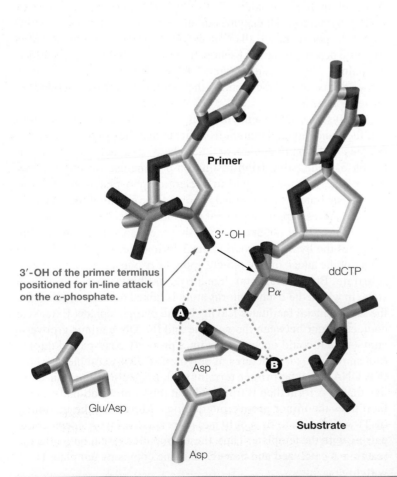

◀ **FIGURE 22.8 The two-metal mechanism of the DNA polymerase reaction.** This mechanism was suggested originally from the structure of the T7 phage DNA polymerase–substrate complex. Two Mg^{2+} ions in the polymerase active site are contacted by two conserved aspartate residues. Both metal ions ligate the α-phosphate of the incoming dNTP (ddCTP, an analog of dCTP), while metal ion B also ligates the β- and γ-phosphates. The 3'-OH of the primer terminus is positioned for an in-line attack on the α-phosphate, followed by the release of pyrophosphate. Atoms are shown in standard colors; the metal ion coordination is represented as dotted lines.

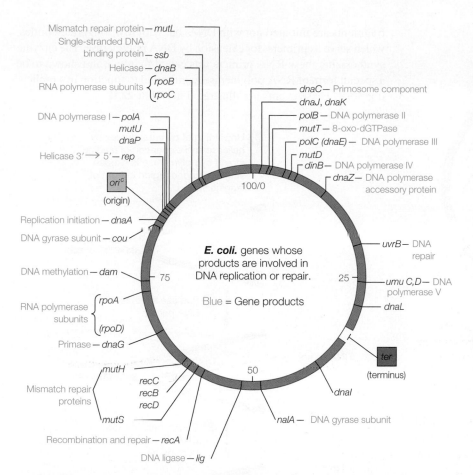

▲ **FIGURE 22.9 Partial genetic map of *E. coli*.** Genes whose products are involved in DNA replication or repair are shown. The names of gene products are given in blue. The *dna* genes play essential roles in DNA replication. The *mut* genes specify proteins that, when mutated, can cause elevated rates of spontaneous mutation. Initiation and termination of genome replication occur at the *ori* and *ter* sites, respectively. The numbers refer to times, in minutes, of chromosome transfer from one cell to another during conjugation between mating bacteria.

T4 map (**FIGURE 22.11**) was identified as the structural gene for a virus-coded DNA polymerase. Figure 22.11 shows locations on the T4 genome of 15 genes whose products play essential roles in DNA replication. Three of these gene products are enzymes that synthesize the modified DNA base 5-hydroxymethylcytosine (Chapter 19). However, the remaining proteins replicate the T4 genome, through mechanisms similar to those in cellular DNA replication. Analysis of these proteins and their interactions continues to enlighten us about the protein chemistry of DNA replication, well into the twenty-first century.

Replication Proteins in Addition to DNA Polymerase

Although the discovery of DNA polymerase revealed the biochemical process by which DNA chains are elongated in replication, important questions remained. How are two antiparallel DNA chains extended in the same fork if polymerase can extend in only one direction? How is the synthesis of new chains initiated if polymerase is capable only of adding nucleotides to preexisting 3′ termini? How are parental DNA chains unwound?

Discontinuous DNA Synthesis

The first two questions were answered primarily through the work of Reiji Okazaki, who proposed that DNA replication is discontinuous. One parental strand (the **leading strand**) could be extended continuously, with polymerase moving 5′ to 3′ (from the 5′ terminus to the 3′ terminus) in the direction of fork movement. Synthesis on the **lagging strand** would be discontinuous; chain extension along the leading strand would expose the single-stranded template on the lagging strand. This template could be copied in short fragments, with polymerase moving *opposite* to the

of mutants defective in that gene, followed by a capital letter denoting the order of discovery of the gene or gene product. For example, the structural genes encoding DNA polymerases I, II, and III are *polA*, *polB*, and *polC*, respectively. Other genes were named from the defective phenotype. For example, all genes designated *dna* were originally identified from temperature-sensitive (*ts*) mutations affecting DNA replication. For instance, the *dnaG* gene product was found to be **primase.** Gene *products*, which are usually proteins, are conventionally given the corresponding nonitalicized and capitalized designations. Thus, DnaG is another designation for primase, the product of the *dnaG* gene. Similarly, PolC is identical to DNA polymerase III. Different conventions are used for naming eukaryotic genes and gene products.

Bacteriophage T4, which infects *E. coli*, was an additional source of early information about proteins essential to replication (**FIGURE 22.10**). This virus contains a large duplex DNA genome, some 169 kbp in length, about 5% of the size of the *E. coli* genome. Phage T4 encodes nearly all of its own replicative proteins and enzymes. T4 was the first biological system in which replication-defective mutants were described and in which the function of DNA polymerase was confirmed. The latter event occurred in 1965, when gene 43 on the

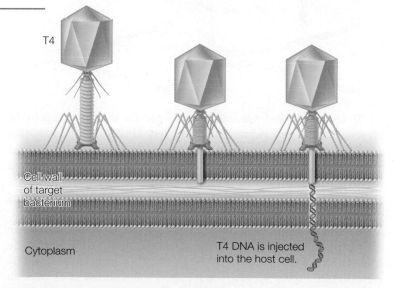

▲ **FIGURE 22.10 Bacteriophage T4.** DNA is packed into the head of this virus particle. Infection occurs when the tail fibers attach to an *E. coli* cell; then the tail contracts, and the DNA is injected through the tail and baseplate into the interior of the infected cell.

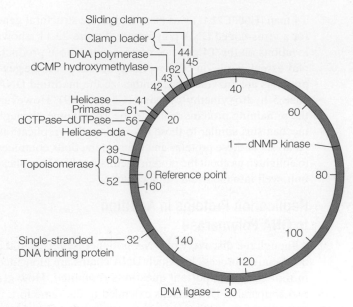

▲ **FIGURE 22.11 Partial map of the T4 genome.** Genes whose products participate in DNA metabolism are shown on the outside of the circular map. The inner numbers represent distances in kbp from the reference point (0), which represents the divide between two genes *rIIA* and *rIIB* (not shown).

direction of fork movement. Thus, lagging-strand synthesis would occur in short pieces. These short pieces could then be attached to high-molecular-weight DNA by the enzyme **DNA ligase.** See below and also Figure 22.13.

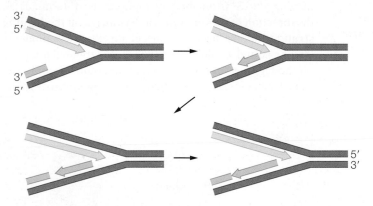

To test this model, Okazaki grew *E. coli* for short intervals in the presence of [³H]thymidine and found that ³H was incorporated into small DNA fragments (later called Okazaki fragments). Next, excess nonradioactive thymidine was added. After further growth, Okazaki found that the radiolabel, initially found in low-molecular-weight DNA, had moved into high-molecular-weight DNA. This suggested that nucleotides were initially incorporated into small pieces of DNA, which later became high-molecular-weight DNA. DNA ligase, an enzyme that closes nicks in double-stranded DNA, was thought to be responsible for this process. In phage T4, DNA ligase is encoded by gene 30 (see Figure 22.11). When *E. coli* was infected at 42 °C with a *ts* gene 30 T4 mutant, Okazaki fragments accumulated and were not incorporated into high-molecular-weight DNA until the temperature of the culture was reduced. This result supported the postulated function of DNA ligase in processing Okazaki fragments.

If DNA polymerase can only extend preexisting chains, then how can these short fragments be initiated? It turns out that Okazaki

fragments are initiated not with DNA, but with *RNA* oligonucleotides, which serve as primers, for extension by DNA polymerase. The enzyme synthesizing these RNA primers was called **primase** and shown to be a special form of RNA polymerase, configured to function in a replication fork. These events are illustrated in **FIGURE 22.12**.

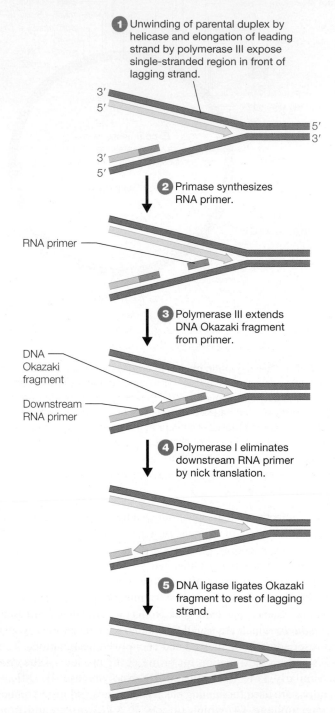

▲ **FIGURE 22.12 A model for DNA replication involving discontinuous synthesis on the lagging strand.** A short RNA primer (orange) initiates the synthesis of low-molecular-weight DNA fragments on the lagging strand (blue). RNA primers are replaced by deoxyribonucleotides (see text), and the fragments are sealed to high-molecular-weight DNA by DNA ligase. Each arrowhead on a nucleic acid strand identifies a 3′ hydroxyl group that can undergo polymerase-catalyzed chain extension. The replication fork is moving to the right.

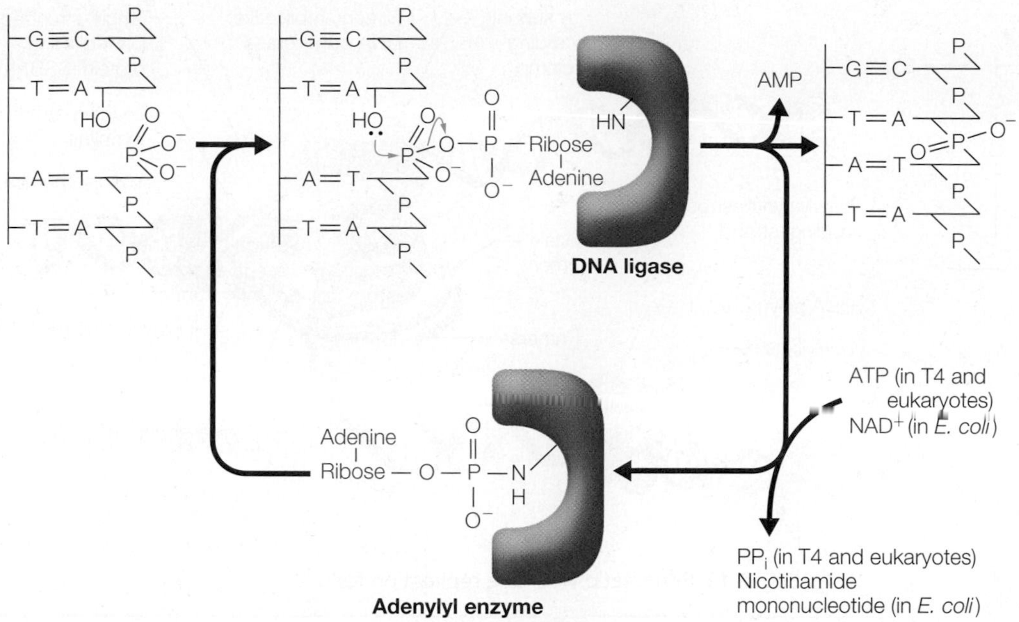

◀ **FIGURE 22.13 The reaction catalyzed by DNA ligase.** The adenylyl enzyme shows a covalent bond between a lysine amino group and the AMP phosphate.

In order for DNA ligase to seal a nick, the nick must contain $3'$ hydroxyl and $5'$ phosphoryl termini, and the nucleotides being linked must be adjacent in a duplex structure and properly base-paired. DNA ligase is activated by adenylylation of a lysine residue in the active site (**FIGURE 22.13**). The enzyme then transfers the adenylyl (AMP) moiety to the $5'$ terminal phosphate of the DNA substrate, thereby activating it for nucleophilic attack by the $3'$ hydroxyl group, with formation of a phosphodiester bond and displacement of AMP. The T4 phage enzyme uses ATP to adenylylate the enzyme, as do eukaryotic DNA ligases, but the enzyme from *E. coli* and other bacteria uses NAD^+. Instead of being a redox cofactor, the dinucleotide in this enzyme is cleaved, yielding adenylylated enzyme plus nicotinamide mononucleotide (NMN).

Although the leading strand can be elongated continuously, as shown in Figure 22.12, recent data indicate that in vivo both strands are elongated discontinuously. This observation may result from a DNA repair process, but the mechanism is not known.

RNA Primers

The enzyme that synthesizes RNA primers is a special RNA polymerase called primase. Unlike DNA polymerases, primase does not require a primer. In *E. coli,* primase is the product of the *dnaG* gene. Priming involves pairing a ribonucleoside $5'$-triphosphate opposite a deoxyribonucleotide residue in template DNA, followed by sequential ribonucleotide additions to the $3'$ hydroxyl termini, just as occurs with DNA polymerases. At some point, RNA synthesis stops, and DNA polymerase continues to extend from the $3'$ hydroxyl terminus of the RNA primer, but now deoxyribonucleotides are incorporated. In T4 and *E. coli* DNA replication, the primer lengths are 5 and 11 nucleotides, respectively.

For Okazaki fragments to be ligated to high-molecular-weight DNA, the RNA primers must be excised and replaced with corresponding deoxyribonucleotides. In *E. coli* DNA polymerase I is involved, through its nick translation activity. The removal of ribonucleotides

from the $5'$ end of the primer, by the $5'$ exonuclease activity of polymerase I, is coordinated with their replacement by deoxyribonucleotides (see **FIGURE 22.14**). An alternative means for digesting RNA primers is **ribonuclease H,** an enzyme that specifically hydrolyzes RNA base-paired with DNA.

● **CONCEPT** Primase, a special class of RNA polymerase, synthesizes short RNA molecules as primers for lagging-strand DNA replication.

Proteins at the Replication Fork

Now we can consider all of the proteins in a replication fork, as well as the functions of each. **FIGURE 22.15**, an idealized fork, includes proteins we have discussed—DNA polymerases, primase, and ligase. Note that the replicative polymerase is dimeric, with one polymerase molecule each assigned to the leading and lagging strands. The primase is complexed with a **helicase,** one of several enzymes that can use the energy of ATP hydrolysis to drive parental strand unwinding. The helicase–primase complex is called a **primosome.** Attached to each polymerase molecule is a circular **sliding clamp,** which keeps polymerase bound to DNA over thousands of catalytic cycles. As DNA unwinding exposes single-stranded template DNA, that DNA is coated with **single-stranded DNA-binding protein** (SSB), which holds DNA in an extended conformation so that it can base-pair efficiently with incoming nucleotides. Finally, because unwinding the parental duplex creates torsional stress, a **topoisomerase** moves ahead of the fork, acting to relieve that stress. Ultimately, a topoisomerase introduces negative supercoiling into replicated DNA. Not shown in Figure 22.15 is the **clamp-loading complex,** which both fastens and unfastens the sliding clamp to DNA. The complex of proteins supporting replication is called the **replisome.** Unlike a static complex, however, such as pyruvate dehydrogenase, the individual proteins within the replisome are in constant motion, dissociating from the complex and reassociating as parts of a dynamic machine.

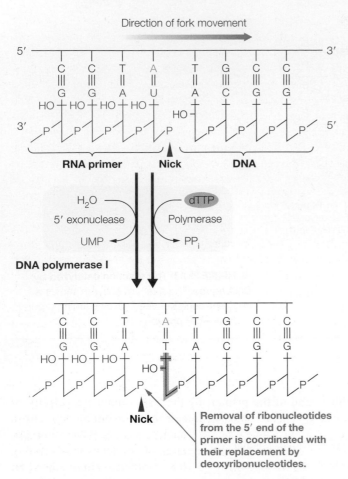

Direction of fork movement

RNA primer   **Nick**   **DNA**

H_2O

5′ exonuclease   dTTP   Polymerase

UMP PP_i

DNA polymerase I

Nick

Removal of ribonucleotides from the 5′ end of the primer is coordinated with their replacement by deoxyribonucleotides.

(a) Base-paired UMP in the RNA primer is replaced by dTMP in the growing DNA chain. The template DNA is the lagging strand.

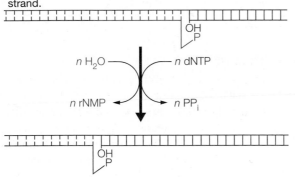

$n\,H_2O$ n dNTP

n rNMP $n\,PP_i$

(b) Nick translation proceeding in cycles of ribonucleotide excision and deoxyribonucleotide replacement.

▲ **FIGURE 22.14** Nick translation in the removal of RNA primers by coordinated action of the 5′ exonuclease and polymerase activities of DNA polymerase I.

The DNA Polymerase III Holoenzyme

The *E. coli polC* gene encodes a single polypeptide chain ($M_r = 130{,}000$). This protein has an intrinsic polymerase activity, but it is quite low. Within cells, the PolC protein achieves a high V_{max} as part of a multiprotein aggregate called the **DNA polymerase III holoenzyme.** As shown in **FIGURE 22.16**, the holoenzyme contains 10 different polypeptide chains, each identified with a Greek letter.

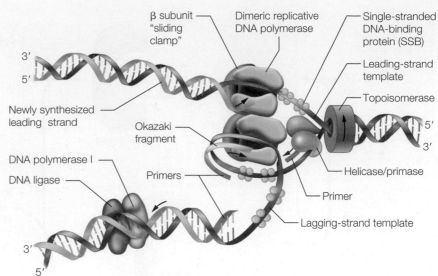

▲ **FIGURE 22.15** Schematic view of a replication fork.

The α, ε, and θ subunits make up the "core polymerase," where α is the *polC* gene product (i.e., the protein with polymerase activity) and ε has an editing 3′ exonuclease activity. The function of θ is unknown, but it may act to improve the catalytic efficiency of α. The τ protein dimerizes the holoenzyme, holding leading- and lagging-strand polymerases together, thereby ensuring that both DNA strands are elongated at the replication fork, even though the lagging-strand polymerase moves in the direction opposite that of fork movement. χ mediates the switch from RNA primers to DNA.

● **CONCEPT** DNA polymerase III holoenzyme, a complex bacterial enzyme containing at least 10 subunits, plays the predominant role in chain elongation during replication.

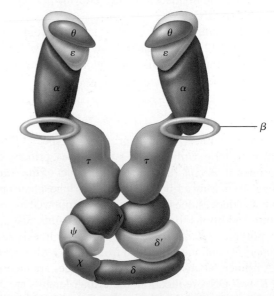

▲ **FIGURE 22.16** Subunit structure of the *E. coli* DNA polymerase III holoenzyme. Greek letter designations for the subunits are identified in the text.

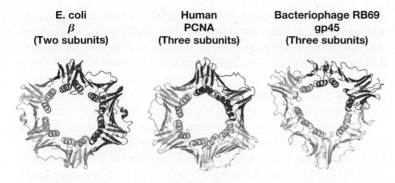

E. coli
β
(Two subunits)

Human
PCNA
(Three subunits)

Bacteriophage RB69
gp45
(Three subunits)

▲ **FIGURE 22.17 Structure of the sliding clamp.** Left, the *E coli* β protein (PDB ID: 2pol); center, human PCNA (proliferating cell nuclear antigen; PDB ID: 1axc); right, gp45, the sliding clamp of the T4-related phage, RB69 (PDB ID: 1b77. Each protein forms a "doughnut" that can completely surround double-stranded DNA and thus keep polymerase associated with its DNA templates. The α helices on the inner surface of the subunit contact DNA but do not bind tightly enough to retard the movement of the protein. The *E. coli* protein has two identical subunits with three DNA-associating domains, while the human and RB69 proteins have three subunits, each with two DNA-associating domains. Data from DNA Repair 8:570–578 (2009), L. B. Bloom, Loading clamps for DNA replication and repair.

Sliding Clamp

The β subunit is essential for the *processivity* of DNA polymerase. That is, β is needed to keep DNA polymerase bound to its template through many cycles of nucleotide addition. Without it, the core polymerase would remain bound only long enough to extend a primer strand by 10–20 nucleotides. However, β tethers the enzyme to DNA, allowing it to incorporate several thousand nucleotides per binding event. As a result, β converts DNA polymerase III from a highly *distributive* enzyme, which incorporates just a few nucleotides per binding event, to a highly *processive* enzyme, which remains bound through many incorporation cycles. β is a circular molecule with a 3.5-nm opening, capable of completely surrounding double-stranded DNA. As shown in **FIGURE 22.17**, six α-helical domains face the interior of the circle, with the hydrophobic residues in these helices having little attraction for DNA. Thus, the molecule acts as a *sliding clamp,* permitting polymerase to slide readily along DNA, but not to dissociate. The structure is remarkably well conserved evolutionarily, even though most other forms of this protein are trimers of dimers, not, as in *E. coli,* dimers of trimers.

● **CONCEPT** Single-stranded DNA-binding proteins are essential in DNA replication, repair, and recombination because they facilitate both DNA denaturation and renaturation.

Clamp Loading Complex

How does a circular molecule wrap itself around DNA to begin processive synthesis? That is the function of the remaining five proteins (γ, δ, δ', χ, and ψ in Figure 22.16), which form the γ complex. Also called the **clamp loader,** it contains three copies of the γ protein and one copy each of the related δ and δ'

subunits. The χ and ψ proteins are also considered part of the clamp loader complex, but they do not participate directly.

FIGURE 22.18(a) shows schematically how the clamp loader operates. Exchange of bound ADP for ATP increases the affinity of the complex for a closed clamp. Upon binding, the clamp loader opens the clamp. Binding of the template-primer DNA leads to ATP hydrolysis, which helps eject the clamp loader and close the clamp. Figure 22.18(b) shows the structure of the γ complex from three different organisms. The figure does not show DNA binding, either in (a) or (b). Structural and functional domains are conserved from one organism to another, even though the bacterial complex contains three proteins (3γ, δ, and δ'), the eukaryotic complex contains five different proteins, and the T4 phage complex contains just two proteins.

The event shown in Figure 22.18(a) needs to occur only once per round of replication on the leading strand. On the lagging strand, however, polymerase must rebind at initiation of the synthesis of each Okazaki fragment and must dissociate when the 5' end of the preexisting daughter DNA strand is reached. In *E. coli* Okazaki fragments are 1 to 2 kb, and DNA chains are extended at about 800 nucleotides per second. Consequently, the clamp loading and unloading cycle must occur almost once every second, and it must do so with the lagging-strand core polymerase unit remaining bound to its leading-strand partner, at the fork.

Single-Stranded DNA-Binding Proteins: Maintaining Optimal Template Conformation

One of the earliest replication proteins to be identified, other than DNA polymerase itself, was a T4 phage protein, gp32, the product of gene 32, called *single-stranded DNA-binding protein* (SSB). Analysis of the purified gp32 showed that it binds specifically to

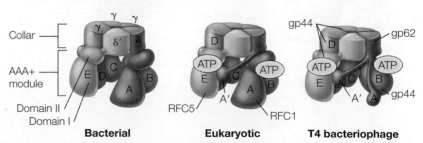

Clamp loader

(a) Scheme for action of the *E. coli* clamp loader

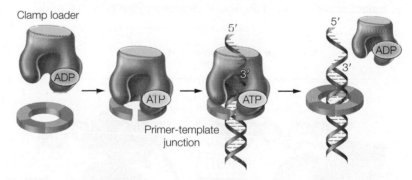

Collar

AAA+ module

Domain II
Domain I

Bacterial

Eukaryotic

T4 bacteriophage

(b) Structures of three clamp-loading complexes

▲ **FIGURE 22.18 Scheme for action of the *E. coli* clamp loader (a) and structures of three clamp-loading complexes (b).** Modified from *Science* 334:1675–1680 (2011), B. A. Kelch, D. L. Makino, M. O'Donnell, and J. Kuriyan. How a DNA polymerase clamp loader opens a sliding clamp.

single-stranded DNA. Moreover, binding is strongly *cooperative,* meaning that the protein is far more likely to bind to DNA adjacent to a site already occupied than to an isolated site. In other words, the binding of one gp32 molecule facilitates the binding of others, and the protein tends to bind in clusters. Thus, gp32 promotes the denaturation of DNA. Although it does not initiate denaturation, its presence lowers the melting temperature of DNA by as much as 40 °C.

The role of gp32 is to keep the template in an extended, single-stranded conformation, with the purine and pyrimidine bases exposed so that they can base-pair readily with incoming nucleotides. This function is essential for DNA repair and genetic recombination, as well as for replication. Given that all three processes also involve the re-formation of duplex structures, with one strand each of parental and daughter DNA, it is significant that gp32 facilitates the renaturation of single-stranded DNA, as well as strand separation.

SSB proteins are widely distributed. The *E. coli* SBB protein (encoded by the *ssb* gene) also binds cooperatively to single-stranded DNA. However, the mechanism of binding is quite different from that in T4. In *E. coli,* DNA is wrapped about the outer surface of the tetrameric SSB protein. Moreover, under some conditions, the SSB protein binding shows *negative* cooperativity. It may be that the protein binds in different modes, depending on whether it is participating in replication, DNA repair, or recombination.

In eukaryotic cells, a heterotrimeric protein called replication factor A (RFA) serves the role of SSB in DNA replication. This protein undergoes phosphorylation during S phase or after DNA damage, suggesting a role in the cellular coordination of DNA metabolism.

Helicases: Unwinding DNA Ahead of the Fork

Single-stranded DNA-binding proteins stabilize single-stranded DNA but cannot actively unwind duplex DNA strands. That is the job of the helicase proteins. They catalyze the ATP-dependent unwinding of double-stranded DNA. *E. coli* cells contain at least a half dozen different helicases, some of which participate in DNA repair and some in bacterial conjugation. The principal helicase in DNA replication is DnaB (the protein product of the *dnaB* gene), which interacts with DnaG and other proteins to form the primosome (see Figure 22.15). The comparable roles in T4 DNA replication are played by gp41 (helicase) and gp61 (primase).

All known helicases are multimeric proteins. Most are homodimers, but a few, like DnaB, are homohexamers. In vitro, each helicase binds initially to single-stranded DNA, adjacent to a duplex region, and proceeds in a fixed direction ($5' \rightarrow 3'$ or $3' \rightarrow 5'$), displacing the unbound DNA strand as it moves and with ATP hydrolysis coupled to movement. Although the helicases are homo-oligomers, they are structurally asymmetric when bound to DNA. In the dimeric Rep helicase of *E. coli,* for example, the ATP-binding and DNA-binding properties of the two subunits are quite different. This suggests a rolling or "hand-over-hand" mechanism, in which each subunit alternates between tight-binding and loose-binding conformations, depending on whether ATP or ADP is bound at a given instant (**FIGURE 22.19**). The primosome interacts with the τ and χ subunits of the γ complex to coordinate the switch from primase to polymerase on the lagging strand.

Two inherited human diseases, Werner's syndrome and Bloom's syndrome, result from helicase defects. Both conditions increase a person's susceptibility to cancer, and Werner's syndrome patients also undergo premature aging, usually going gray in their twenties, developing cataracts, and dying of natural causes before age 50. Mapping

● **CONCEPT** Helicases are multimeric proteins that bind preferentially to one strand of a DNA duplex and use the energy of ATP hydrolysis to actively unwind the duplex.

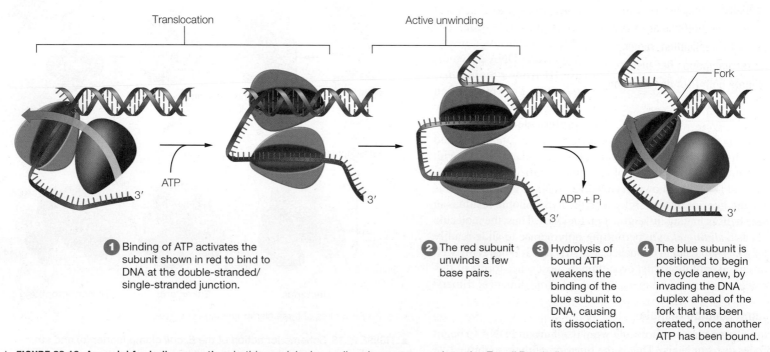

① Binding of ATP activates the subunit shown in red to bind to DNA at the double-stranded/single-stranded junction.

② The red subunit unwinds a few base pairs.

③ Hydrolysis of bound ATP weakens the binding of the blue subunit to DNA, causing its dissociation.

④ The blue subunit is positioned to begin the cycle anew, by invading the DNA duplex ahead of the fork that has been created, once another ATP has been bound.

▲ **FIGURE 22.19 A model for helicase action.** In this model a homodimeric enzyme, such as the *E. coli* Rep helicase, shows $3' \rightarrow 5'$ polarity. By alternating in their binding to DNA, the two subunits cause the enzyme to "roll" counterclockwise as shown in this model, unwinding the duplex as it moves.

● **CONNECTION** Studies on helicase mechanisms have illuminated the clinical problems in Bloom's syndrome and Werner's syndrome, both of which involve defective helicases.

the defective genes responsible for these conditions revealed in both cases that the genes encode proteins related to the *E. coli recQ* gene product. In *E. coli* this helicase participates in a homologous recombination pathway and may be involved in the resumption of DNA replication after repair of radiation-induced DNA damage. Like the RecQ protein, the Werner's syndrome protein has a $3' \rightarrow 5'$ helicase activity. These findings provide intriguing clues to understanding relationships between genomic instability and both cancer and aging.

Topoisomerases: Relieving Torsional Stress

Bidirectional replication of the circular *E. coli* chromosome unwinds about 100,000 base pairs per minute. Were there not some mechanism for relieving the resulting torsional stress, then DNA ahead of the fork would become overwound as DNA at the fork became unwound, and replication could not be sustained. Topoisomerases, a group of enzymes that can interconvert different topological isomers of DNA (see Chapter 4), provide a "swivel" mechanism for relieving this stress. Topoisomerase action is demonstrated most simply in vitro by the relaxation of supercoiled DNA. If supercoiled DNA is incubated with a purified topoisomerase, then the intermediate stages in conversion of the supercoiled substrate to relaxed circular DNA can be observed by gel electrophoresis (**FIGURE 22.20**). This analysis reveals the existence of two general classes of topoisomerases—type I enzymes, which change the linking number in units of 1, and type II enzymes, which change the linking number in units of 2. See Chapter 4 for a definition of linking number.

Actions of Type I and Type II Topoisomerases

A type I topoisomerase breaks just one strand of the duplex (**FIGURE 22.21**). The enzyme remains covalently attached to the 5' end of the broken strand by forming a phosphodiester bond between the 5' phosphate and a tyrosine hydroxyl. The 3' end is then free to rotate (by one turn in the example shown). The hydroxyl group on the 3' end then attacks the activated, covalently bound 5' phosphate, closing the nick—in fact, *E. coli* type I topoisomerase was originally called nicking–closing enzyme. As a result, the linking number has been changed by 1. Eukaryotic topoisomerase I acts similarly, but the 3' end, not the 5' end, is immobilized during the reaction.

By contrast, a type II topoisomerase catalyzes a double-strand break, and the unbroken part of the duplex passes through the gap that is created (**FIGURE 22.22**). The most thoroughly studied type II topoisomerase is an *E. coli* enzyme also called **DNA gyrase,** because it can not only relax a supercoiled molecule but also introduce negative superhelical turns into DNA. ATP hydrolysis is required for both activities of most type II enzymes. DNA gyrase is a tetramer, with two A and two B subunits. The A subunits bind and cleave DNA, while the B subunits carry out the energy transduction resulting from ATP hydrolysis.

As shown in Figure 22.22, gyrase action begins with DNA wrapping about the enzyme. The A subunit cleaves both DNA strands and immobilizes them, and both strands of the duplex pass through the opening. Next, both strands of the duplex

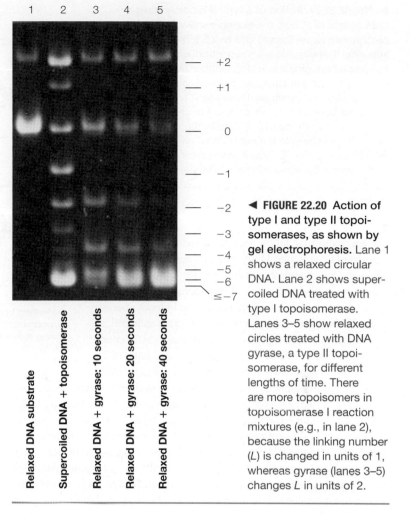

◄ **FIGURE 22.20 Action of type I and type II topoisomerases, as shown by gel electrophoresis.** Lane 1 shows a relaxed circular DNA. Lane 2 shows supercoiled DNA treated with type I topoisomerase. Lanes 3–5 show relaxed circles treated with DNA gyrase, a type II topoisomerase, for different lengths of time. There are more topoisomers in topoisomerase I reaction mixtures (e.g., in lane 2), because the linking number (L) is changed in units of 1, whereas gyrase (lanes 3–5) changes L in units of 2.

Lane labels (left to right):
- Relaxed DNA substrate
- Supercoiled DNA + topoisomerase
- Relaxed DNA + gyrase: 10 seconds
- Relaxed DNA + gyrase: 20 seconds
- Relaxed DNA + gyrase: 40 seconds

Band position markers: +2, +1, 0, −1, −2, −3, −4, −5, −6, ≤−7

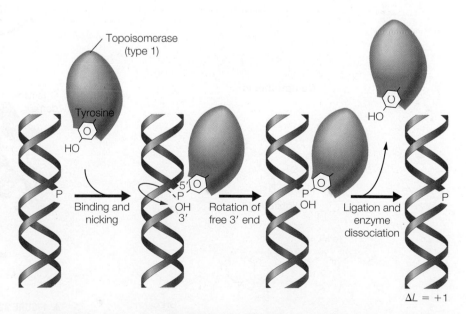

▲ **FIGURE 22.21 Action of a type I topoisomerase.** See text for details. The linking number is increased by 1 in the example shown (an underwound DNA). It would decrease by one, by essentially the same mechanism, in an overwound DNA.

Figure 22.21 labels: Topoisomerase (type 1), Tyrosine, HO, P; Binding and nicking; Rotation of free 3' end; Ligation and enzyme dissociation; $\Delta L = +1$

▶ **FIGURE 22.22 Action of a type II topoisomerase.**
DNA gyrase of *E. coli;* the example shown is a tetrameric protein with two A (green) and two B (orange and blue) subunits. The enzyme is shown introducing two negative turns and changing the linking number from +1 to −1. The enzyme catalyzes a double-stranded break, and the two DNA ends are bound by A subunits, which move the DNA ends apart, so that the unbroken duplex can pass through the gap. Resealing converts the positive supertwist to a negative one, giving the overall molecule a ΔL of −2. Type II topoisomerases can relax underwound duplexes by reversing the same pathway.

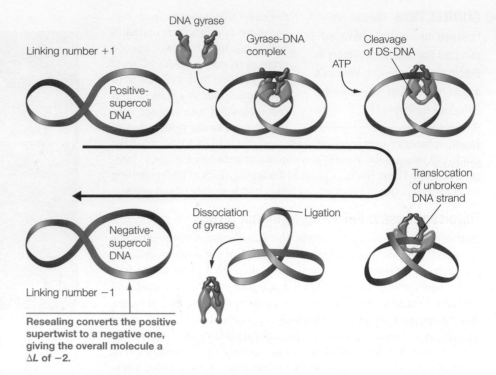

Resealing converts the positive supertwist to a negative one, giving the overall molecule a ΔL of −2.

are resealed and the enzyme dissociates. In the example shown, a circular DNA with one positive supercoil is converted to a circular DNA with one negative supercoil. Thus, the linking number is changed by 2, the distinction between a type I and a type II topoisomerase. In Figure 22.20, note that DNA treated with a type I topoisomerase (lane 2) shows twice as many intermediates as identical DNA treated with a type II enzyme (lanes 3–5), because the type I enzyme changes linking number in units of 1.

In the crystal structure of human topoisomerase I, as shown in **FIGURE 22.23**, the enzyme completely wraps around its DNA substrate. Most of the DNA–protein contacts involve the DNA sugar–phosphate backbone rather than the bases, so DNA is bound as an undistorted B-form helix. Because of the involvement of topoisomerases in DNA replication, a variety of topoisomerase inhibitors have been developed as anticancer drugs and antibacterial agents. Analysis of mutants resistant to the anticancer drug **camptothecin** (CPT) reveals where the inhibitor is bound; this information may lead to the design of more effective inhibitors.

Camptothecin

Nalidixic Acid

Novobiocin

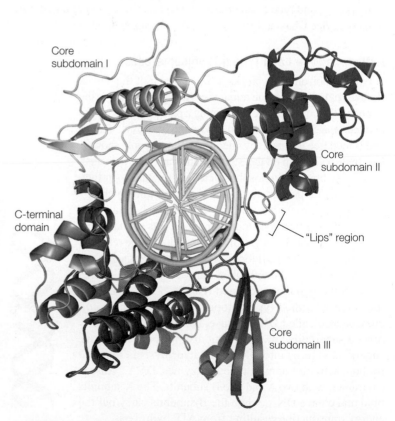

▲ **FIGURE 22.23 Crystal structure of human topoisomerase I in complex with a 22-bp DNA duplex.** End-on view of the topoisomerase-DNA complex. Most of the sites of camptothecin (CPT)-resistant mutations lie near the DNA-binding site. PDB ID: 1a35.

The Four Topoisomerases of *E. coli*

E. coli contains four different topoisomerases. The terminology is a bit confusing because the enzymes named topoisomerase I and topoisomerase III are both type I topoisomerases, whereas topoisomerase II (also called DNA gyrase) and topoisomerase IV are both type II topoisomer-

● **CONCEPT** Type I topoisomerases break and reseal one DNA strand, whereas type II topoisomerases break and reseal both strands; hence, type I and type II enzymes change the DNA linking number in units of 1 and 2, respectively.

● **CONNECTION** Topoisomerases are targets for several antibiotics and anticancer agents in current clinical use.

ases. Of these four enzymes, DNA gyrase plays the dominant role during replicative chain elongation, both in relieving stress ahead of the fork and in introducing negative supercoils into newly synthesized DNA. We know of this role primarily through studies of the properties of gyrase inhibitors. The gyrase A subunit is the target for the binding of **nalidixic acid,** a compound long known to inhibit DNA replication. Another replication inhibitor, **novobiocin,** binds to the B subunit and inhibits ATP cleavage. Inhibitors such as these are useful as antibacterial drugs.

Topoisomerase IV plays a critical role in the completion of a round of replication. Type II topoisomerases catalyze a variety of topological interconversions, including knotting, unknotting, **catenation** (linking), and **decatenation** (unlinking) of circular DNAs, as shown in **FIGURE 22.24**.

A circular DNA nearing the end of a round of replication will generate two interlinked circles, so the action of a type II topoisomerase is necessary for separating the newly replicated molecules. As two forks approach each other at the replication terminus, steric barriers ultimately interfere with the unwinding activities of topoisomerases

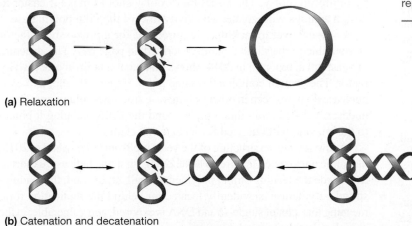

(a) Relaxation

(b) Catenation and decatenation

(c) Knotting and unknotting

▲ **FIGURE 22.24** The types of topological interconversions catalyzed by type II topoisomerases.

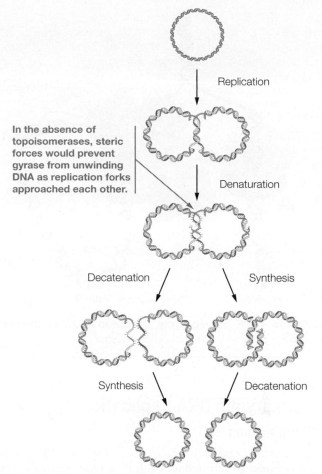

In the absence of topoisomerases, steric forces would prevent gyrase from unwinding DNA as replication forks approached each other.

Replication

Denaturation

Decatenation

Synthesis

Synthesis

Decatenation

▲ **FIGURE 22.25 Topoisomerase action in termination of replication.** Topoisomerase IV allows the circles to decatenate. It is not known whether decatenation occurs before or after the completion of replication. Both possibilities are shown here.

ahead of the two forks (**FIGURE 22.25**). At this stage, the two incompletely replicated chromosomes are still interlinked. Topoisomerase IV is then necessary for the decatenation process, as shown in Figure 22.25. Recent evidence suggests a role for topoisomerase III in this process as well.

A Model of the Replisome

Strictly speaking, topoisomerases are not components of the replisome because they act at a distance from the fork. However, now that we have discussed the major proteins involved in bacterial fork propagation, and as a prelude to discussing comparable eukaryotic proteins, we present in **FIGURE 22.26** a model of the *E. coli* replisome, highlighting the relationships among the constituent proteins.

Recently, it has become possible to image the *E. coli* replisome in vivo, using fluorescent-tagged proteins and single-molecule techniques. These experiments indicate that the replisome contains three molecules of DNA polymerase, not just the two shown in Figure 22.26. Only two of the three molecules are linked to sliding clamps, but three copies of τ are present, suggesting that the unlinked polymerase is "waiting its turn" to have a clamp loaded, so that it can catalyze synthesis of the next Okazaki fragment.

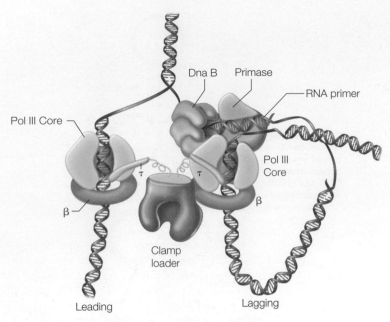

▲ **FIGURE 22.26 The *E. coli* replisome.** The clamp loader, via the τ protein, serves as a link to dimerize the pol III holoenzyme.

22.4 Eukaryotic DNA Replication

DNA Polymerases

The mechanism of replicative DNA chain elongation has remained remarkably constant throughout evolution. One striking difference between prokaryotic and eukaryotic replication is the requirement for three different DNA polymerases to propagate a eukaryotic replication fork. Early fractionation of DNA polymerases from yeast or mammalian cells yielded five different enzymes ($\alpha - \varepsilon$), three of which are involved in nuclear DNA replication (α, δ, and ε), one in mitochondrial DNA replication (γ) and one in repair (β). Later work revealed the existence of at least nine additional DNA polymerases in human cells, most of which participate in specialized repair processes. Although there is some disagreement in the literature, most evidence assigns Pol ε the major role in replicative leading-strand synthesis and Pol δ the major role in lagging-strand chain elongation. Pol α has both polymerase and primase activities, and it participates in the synthesis and processing of RNA primers.

The first eukaryotic DNA polymerase to be described structurally was the human mitochondrial pol γ, reported in 2009. The structure shows the sites of several mutations known to be responsible for human genetic diseases. For example, the W748S mutation is associated with a condition called Alpers syndrome, which causes both cerebral cortical atrophy and liver failure in children. In addition, pol γ is responsible

● **CONNECTION** The structure of human DNA polymerase γ leads to greater understanding of hereditary disorders of mitochondrial function and the toxicity of nucleoside antiviral drugs.

for the toxicity of several drugs used in HIV therapy. As pointed out in Chapter 19, nucleoside analogs such as dideoxycytidine act by conversion to their 5′ triphosphates, which interfere with HIV reverse transcriptase. Several of these analogs, which undergo conversion to triphosphates within

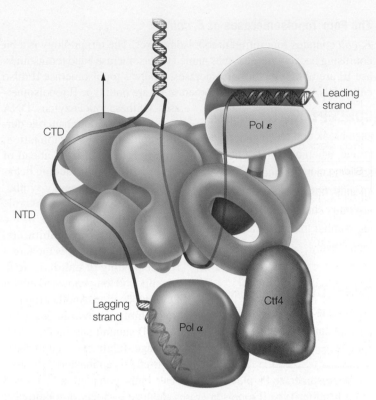

▲ **FIGURE 22.27 Model of the yeast replisome, based on high-resolution electron microscopy.** CTD and NTD, C- and N-terminal domains, respectively, of the multisubunit helicase; Ctf4, a protein that binds together several subunits of the replisome. Source: J. Sun et al. (2015) *Nature Str, Mol. Biol.* 22:976–982.

the mitochondrion, also interfere with mitochondrial DNA replication, by inhibiting pol γ. The structure of pol γ should make it easier to design analogs with greater selectivity toward the viral polymerase.

Pol ε is novel in lacking a requirement for a processivity factor comparable to the sliding clamp described on page 697. The structure of yeast pol ε, reported in 2014, shows that it has a built-in processivity factor. The protein contains the same palm, fingers, thumb, and exonuclease domains seen in other polymerase structures, plus a previously undescribed *P domain* that wraps around the DNA, keeping it bound to the enzyme without need for an external clamp.

A low-resolution structure of the yeast replisome, described in 2015, has some unexpected features (**FIGURE 22.27**). First, the helicase, expected to precede the leading-strand polymerase, trails it. Second, the leading-strand polymerase is evidently located far behind the replication fork, meaning that a lot of single-strand DNA is exposed, and vulnerable, in the complex. In this image the lagging-strand polymerase is not identified.

Other Eukaryotic Replication Proteins

Most of what we know about the functions of individual proteins in eukaryotic DNA replication has come from in vitro systems that replicate an easily handled small replication substrate—the circular duplex DNA from the tumor virus SV40. These studies, plus earlier work on the phage T4 system and later work with yeast, reveal striking uniformity in the kinds of proteins at replication forks and the biochemical functions of each, as shown in **TABLE 22.1**. One noteworthy difference is that eukaryotes require two enzymes to remove RNA primers—the

TABLE 22.1 Proteins That Carry Out Analogous Functions in DNA Replication

Function	*E. coli*	Phage T4	SV40/Human	Yeast
DNA polymerase	Pol III core enzyme	gp43	Pol δ, Pol ε	Pol δ, Pol ε
Primase	DnaG	gp61	Pol α	Pol α
Helicase	DnaB	gp41	SV40 T antigen	MCM proteins
Proofreading	ε subunit of Pol III holoenzyme	gp43	Pol δ	Pol δ, Pol ε 3′ exonuclease
Sliding clamp	β subunit of Pol III holoenzyme	gp45	PCNA	PCNA
Clamp loader	γ complex of Pol III holoenzyme	gp44/62	Replication factor C	Replication factor C
Single-stranded DNA-binding protein	SSB	gp32	Replication protein A	Replication protein A
RNA primer removal	Pol I, RNase H	*E. coli* Pol I T4 RNase H	Pol δ, FEN 1	Pol δ, FEN 1

5′ exonuclease activity of Pol δ and the 5′ "flap" endonuclease activity of FEN 1 (flap endonuclease), which removes a single-stranded end like that shown in Figure 22.5.

Replication of Chromatin

Unlike bacterial and bacteriophage systems, the eukaryotic replisome has to deal with chromatin. Chromatin must be dismantled in advance of replication forks and reassembled on daughter DNA strands after a fork has passed through. **FIGURE 22.28** summarizes our current understanding.

Nucleosomes are disassembled ahead of the replication fork and then reassembled on both daughter strands. Both preexisting and newly synthesized histones are used in the new nucleosomes, evidently with random deposition on the daughter strands. Old and new histone molecules mix on the daughter strands, but the process is not entirely random. Tetramers containing two H3 and two H4 molecules each tend to remain intact, as do dimers containing one each of H2A and H2B. This result conforms with results of in vitro studies, where (H3/H4)$_2$ tetramers and H2A/H2B dimers are stable when released from nucleosomes, but the octamer is not.

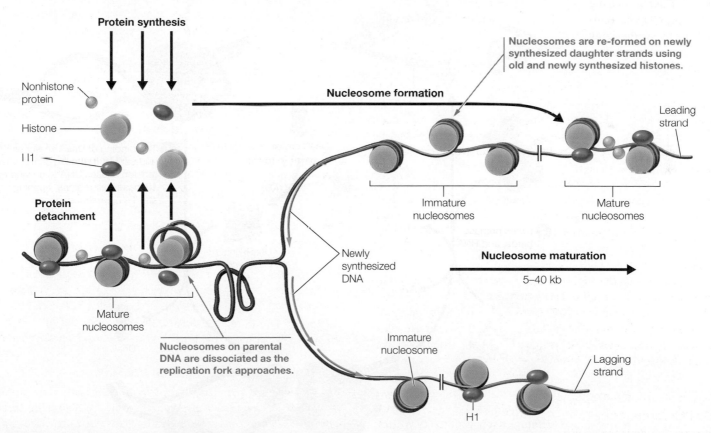

▲ **FIGURE 22.28 Model for chromatin replication.** Nucleosomes on parental DNA are dissociated as the replication fork approaches and are re-formed on newly synthesized daughter strands, with both old and newly synthesized histones being used. Maturation occurs slowly, with full organization not being reestablished until many kb behind the moving fork. For simplicity, this figure does not show proteins of the replisome.

How is the precise arrangement of nucleosomes and nonhistone proteins reestablished after replication? In particular, how are histone modifications reestablished? The exact processes are still not understood, but a class of proteins called *histone chaperones* facilitates the association of histones with nascent DNA.

22.5 Initiation of DNA Replication

Replication proceeds from fixed origins, so there are two requirements for initiation: a nucleotide sequence that specifically binds initiation proteins and a mechanism that generates a primer terminus to which nucleotides can be added by DNA polymerase. A number of phage, bacterial, plasmid, and organelle replication origins have been isolated by gene cloning, and their nucleotide sequences have been determined. In general, these origins include repeated sequences of either identical polarity (**direct repeats**) or opposite polarity (**inverted repeats**). This finding suggests that initiation proteins bind in multiple copies.

The two most straightforward ways to generate a primer terminus at the origin are to nick a strand of the parental duplex, exposing a 3′ hydroxyl terminus, and to unwind the parental duplex and synthesize an RNA primer to expose a 3′ hydroxyl *ribonucleotide* terminus. Whereas some small phages use a nicking process, duplex DNA replication occurs by synthesizing RNA primers, not by nicking the parental duplex.

Initiation of *E. coli* DNA Replication at *ori*^{c}

In *E. coli* initiation of replication is reasonably well understood because the origin sequence has been cloned into plasmids whose replication from the origin can then be studied in vitro. This origin sequence, called *ori*^{c}, is 245 base pairs long. It contains four repeats of a nine-base-pair sequence that binds an initiation protein, the *dnaA* gene product. To the left of these sites, as shown in **FIGURE 22.29**, are three direct repeats of a 13-base-pair sequence that is rich in A and T, which facilitates

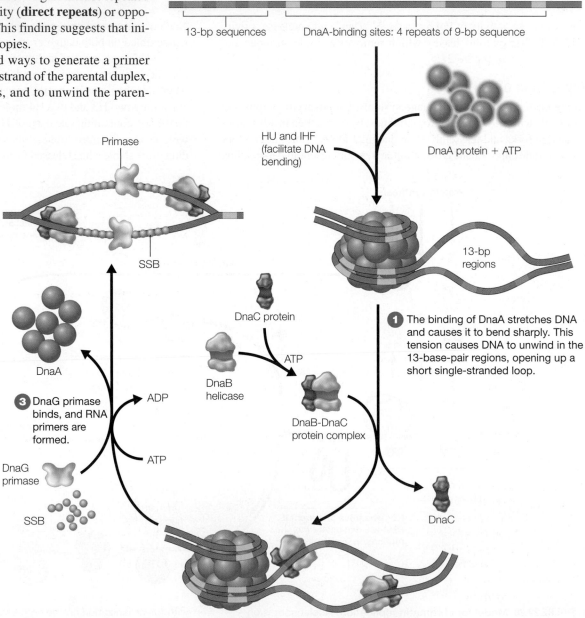

▶ **FIGURE 22.29 A model for the initiation of *E. coli* DNA replication at *ori*^{c}.** HU and IHF are double-stranded DNA-binding proteins that help DNA bend at the origin.

1 The binding of DnaA stretches DNA and causes it to bend sharply. This tension causes DNA to unwind in the 13-base-pair regions, opening up a short single-stranded loop.

2 DnaB helicase binds in both forks of the loop, and the helicase activity further unwinds this structure.

3 DnaG primase binds, and RNA primers are formed.

strand separation. The sequence also contains binding sites for several basic proteins (HU and IHF) that facilitate DNA bending, an important step in the sequence leading to initiation.

Step ❶ is the binding of 10 to 20 molecules of a complex of DnaA protein and ATP. The protein is activated for this step by reacting with the phospholipid cardiolipin, a process that may represent coordination between DNA replication and membrane growth. The binding of DnaA, as well as some basic proteins, stretches DNA and causes it to bend sharply, creating negative superhelical tension. In turn, this tension causes DNA to unwind in the 13-base-pair regions, opening up a short single-stranded loop. Aided by DnaC, another initiation protein, the DnaB helicase, in step ❷, binds in both forks of this loop, and the helicase activity further unwinds this structure. In step ❸, and subsequent steps (not shown), DnaG primase binds, and RNA primers are formed (RNA polymerase may also be involved). The RNA primers are extended by DNA polymerase III on both the leading and lagging strands, and the two forks in the initiation complex mature, with a leading and a lagging strand in each fork. This may represent a general mechanism for the initiation of chromosomal DNA replication.

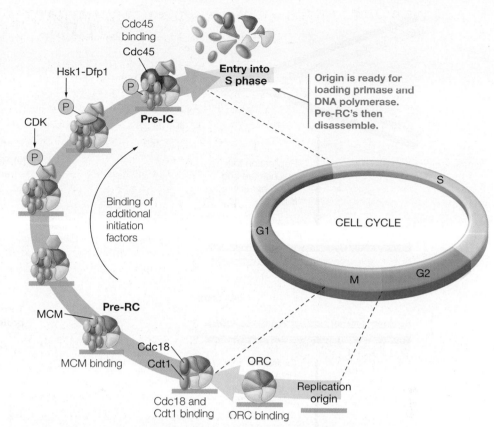

▲ **FIGURE 22.30** Preparing a fission yeast replication origin for initiation.

Initiation of Eukaryotic Replication

While most prokaryotic organisms have a single origin of replication on a single chromosome, a typical eukaryotic cell has thousands of replication origins distributed over multiple chromosomes. The replication origins become active, or "fire," at different times within S phase, and not all origins have to fire within a single cell cycle. Moreover, there is the problem of "licensing" each origin—namely, the process by which initiation at that origin is controlled so that it fires not more than once per cell cycle, to ensure that all of the genome is replicated once and only once.

Early insight into yeast replication origins came from the identification of *autonomously replicating sequences* (ARSs), which are sequences essential for the replication of plasmids after being introduced into yeast cells. An ARS is typically several hundred base pairs long, with subsequences carrying copies of the 11-base consensus sequence 5′ TTTTATATTTT 3′. The AT-rich composition suggests that, as with *E. coli ori^c*, initiation involves strand unwinding at the origin.

The complexity of eukaryotic initiation can be appreciated by considering the number of proteins involved in preparing origins to fire in the fission yeast, *Schizosaccharomyces pombe,* and the time needed for this process. **FIGURE 22.30** shows that this process begins in mitosis, hours before the onset of S phase, with the binding of a six-protein origin replication complex (ORC). This prepares the site to bind two more proteins, Cdc18 and Cdt1. These proteins then recruit the six Mcm proteins, which constitute a putative helicase for unwinding DNA strands at the origin, to form a prereplicative complex (pre-RC). Licensing occurs during assembly of the pre-RC, by an undefined mechanism. Several more initiation factors bind, some of which undergo phosphorylation by a *cyclin-dependent kinase* (CDK) or Hsk1-Dfp1—thought to be required for loading Cdc45 onto the origin. This process completes

● **CONCEPT** Protein binding, causing DNA to bend at the origin, initiates duplex DNA replication. Stress from bending causes nearby DNA to unwind, and primosomes assemble in the forks, forming RNA primers that are extended by DNA polymerases.

formation of the preinitiation complex (Pre-IC). This prepares the origin for loading primase and DNA polymerase, which occurs in S phase, when replication actually begins. Once primase and polymerase have been loaded, all pre-RCs must be disassembled to ensure that each licensed origin fires only once.

22.6 Replication of Linear Genomes

Thus far, we have discussed in detail only the replication of circular DNA genomes. Linear genomes, including those of several viruses, as well as the chromosomes of eukaryotic cells, face the additional problem of how to complete the replication of the lagging strand (**FIGURE 22.31**). Excision of an RNA primer from the 5′ end of a linear molecule would leave a gap that cannot be filled by DNA polymerase action, because there is no 3′ primer terminus to extend. If this DNA could not be replicated, then the chromosome would shorten a bit with each round of replication.

Linear Virus Genome Replication

Viruses reveal at least three strategies for dealing with this problem. Phages T4 and T7 exhibit **terminal redundancy,** the duplication of a small part of the genome at each end of the chromosome. Thus, recombination can occur at the incompletely replicated ends of two nascent

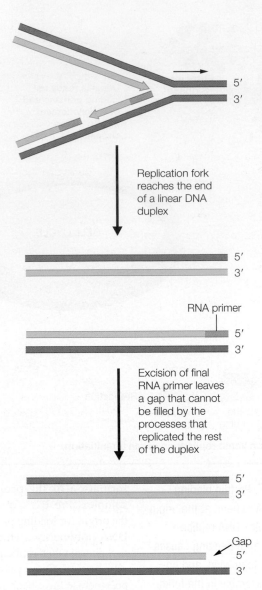

▲ **FIGURE 22.31** The problem of completing the 5′ end in copying a linear DNA molecule.

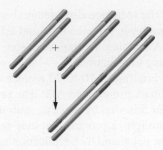

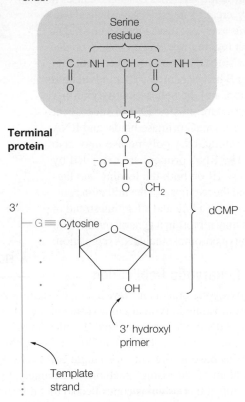

(a) Recombination between redundant chromosome ends.

(b) Using a protein serine hydroxyl as a primer.

▲ **FIGURE 22.32** Strategies for replicating linear genomes.

DNA molecules without the loss of genetic information [**FIGURE 22.32(a)**]. This process is repeated in subsequent rounds of replication until the end-to-end linear aggregate (called a **concatemer**) is more than 20 times the length of a single phage chromosome. A virally specified nuclease then cuts this giant DNA into genome-length pieces where it is packaged into phage heads.

Bacteriophage ϕ 29 and the adenoviruses have evolved a different strategy. The genomes of these viruses contain inverted repeat sequences at the ends. Replication begins at one end of the linear duplex, with a protein called **terminal protein** serving as the primer. This protein in adenovirus reacts with dCTP to form a dCMP residue covalently linked through its phosphate to a serine residue [Figure 22.32(b)]. The dCMP serves as primer for the replication of the 3′-terminated strand, with the 5′ end being displaced as a single strand.

Poxviruses, such as smallpox virus or the closely related vaccinia, use a third mechanism. The two strands at each end of this linear

genome are covalently linked together. When approached by a replication fork, this structure helps move the leading strand *around* the link between the strands, so that the final lagging-strand primer can be replaced by DNA as in a circular duplex (not shown).

Telomerase

For the linear DNA molecules in eukaryotes, the end replication problem has been solved by the addition of **telomeres** at the ends of each chromosome. As mentioned in Chapter 21, telomeric DNA consists of simple tandemly repeated sequences. Typically, one strand is G-rich, and the other is C-rich. The G-rich strand forms a 3′-terminal overhang, up to 15 residues in length. These sequences are repeatedly added to the 3′ termini of chromosomal DNAs by the enzyme **telomerase**

● **CONCEPT** Telomerase adds repeated short DNA segments to the ends of chromosomes.

(FIGURE 22.33) discovered by Carol Greider and Elizabeth Blackburn. These telomeres provide room for a primer to bind and initiate lagging-strand synthesis on the other strand, maintaining the approximate length of the chromosome and preventing the loss of coding sequences.

Telomerase can add nucleotides without the use of a DNA primer because each telomerase molecule has an essential RNA oligonucleotide that is complementary to the telomeric sequence being synthesized and thus acts as a template. Telomerase may be an evolutionary relic of a ribozyme that once catalyzed DNA synthesis, a process long ago taken over by wholly protein polymerases. Because of its RNA-templated DNA synthesis, the protein portion of the enzyme, without its associated RNA, is called more descriptively TERT (<u>t</u>elomerase <u>r</u>everse <u>t</u>ranscriptase).

Telomeres and telomerase have wide-ranging significance in addition to their practical function in preventing the shortening of chromosomes. A strong correlation exists between aging and cell senescence, on the one hand, and low levels of telomerase, on the

● **CONNECTION** The development of telomerase inhibitors is being sought as a means of limiting the life span of cancer cells, which have high telomerase activities.

● **CONNECTION** Life adversity in childhood predicts telomere shortening in later life, with possible effects on life span.

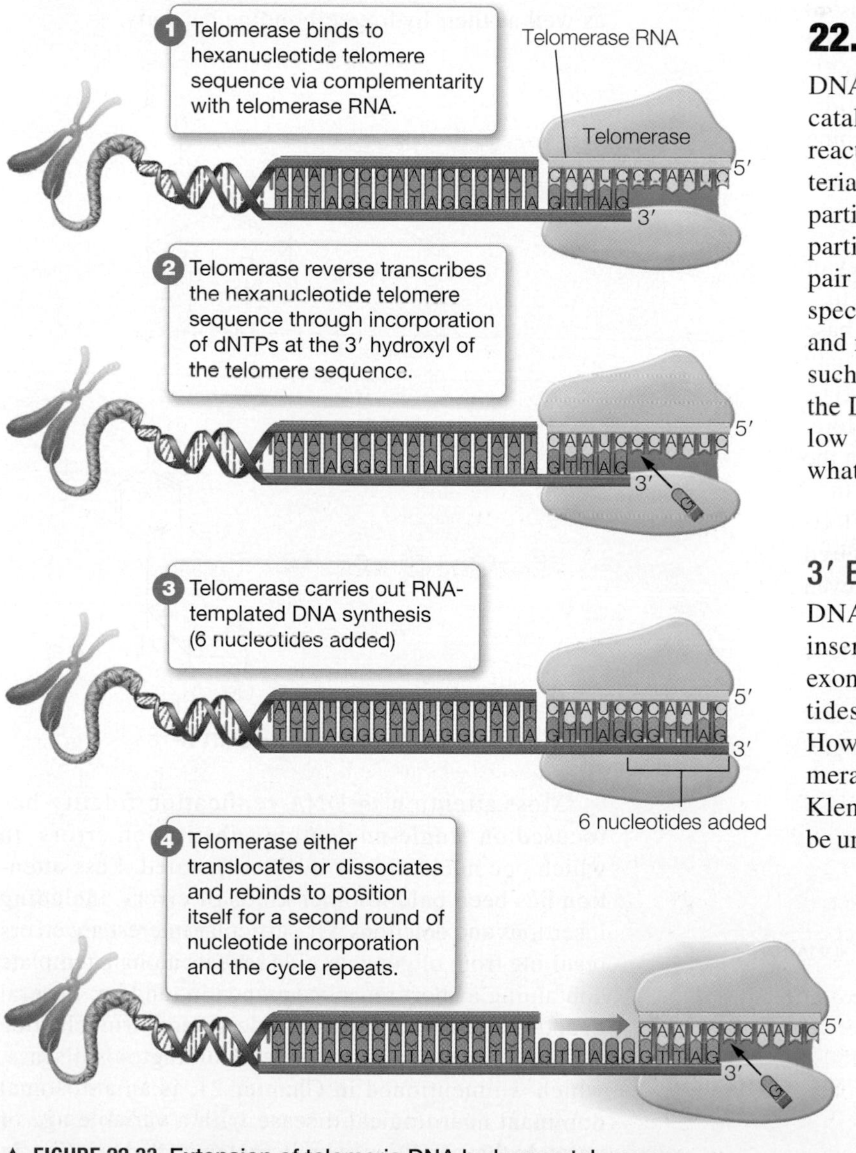

other. Telomeres shorten with age, both in organisms and in cultured cells. Chronic stress and other aging-related conditions are also known to shorten telomeres. Life adversity in childhood (e.g., fetal alcohol syndrome or malnutrition) has a particularly strong effect on telomere length in adulthood. Conversely, cells in culture can be "immortalized" by introducing active telomerase genes. Under these conditions, telomeres do not undergo shortening over extended periods of growth and division. These observations, together with the discovery that malignant tumor cells invariably have high levels of telomerase, have spurred intense interest in telomerase inhibition as a possible cancer therapy.

22.7 Fidelity of DNA Replication

DNA replication is among the most accurate of all known enzyme-catalyzed reactions, all the more remarkable given the speed of the reaction—almost 1000 nucleotides incorporated per second in bacteria. From spontaneous mutation rates, which for all positions in a particular gene amount to about 10^{-6} per generation, the chance that a particular nucleotide will be copied incorrectly is about 10^{-9} per base pair per round of replication. About two orders of magnitude of this specificity comes from **mismatch repair,** a process that recognizes and removes mismatched nucleotides and other aberrant structures, such as looped-out nucleotides (see Chapter 23). Still, this means that the DNA polymerase reaction per se can have an error frequency as low as 10^{-7} errors per nucleotide incorporated, and in fact, that is what is observed.

3′ Exonucleolytic Proofreading

DNA polymerase accuracy is determined at both the nucleotide insertion step and the 3′ exonucleolytic proofreading step. The 3′-exonuclease is positioned to recognize and remove mispaired nucleotides before the polymerase activity can add the next nucleotide. However, crystallographic analysis shows that for most DNA polymerases the exonuclease site is far from the polymerase site. In the Klenow fragment, for example (Figure 22.6), eight base pairs must be unwound to move the 3′-terminal nucleotide in the daughter strand from the polymerase site to the exonuclease site. How, then, is a mismatched nucleotide recognized? How is a high polymerization rate maintained if each newly incorporated nucleotide must be unwound from its template in order to be inspected?

Experiments using template–primer oligonucleotide pairs with deliberate mismatches show that the K_M for chain extension from a mismatch is about one thousandfold higher than extension from a correctly matched terminus; misalignment makes a mismatched 3′-terminal nucleotide a poor substrate. At physiological dNTP concentrations, this means

1. Telomerase binds to hexanucleotide telomere sequence via complementarity with telomerase RNA.

Telomerase RNA

Telomerase

2. Telomerase reverse transcribes the hexanucleotide telomere sequence through incorporation of dNTPs at the 3′ hydroxyl of the telomere sequence.

3. Telomerase carries out RNA-templated DNA synthesis (6 nucleotides added)

6 nucleotides added

4. Telomerase either translocates or dissociates and rebinds to position itself for a second round of nucleotide incorporation and the cycle repeats.

▲ **FIGURE 22.33 Extension of telomeric DNA by human telomerase.**

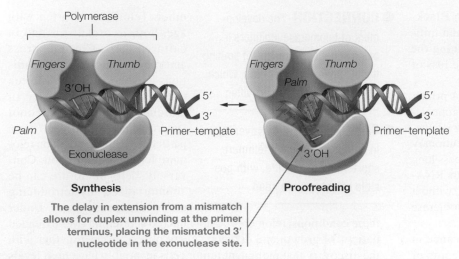

Synthesis

Proofreading

The delay in extension from a mismatch allows for duplex unwinding at the primer terminus, placing the mismatched 3′ nucleotide in the exonuclease site.

▲ **FIGURE 22.34** Kinetic basis for preferential excision of mismatched nucleotides by a 3′-exonuclease site distant from the polymerase site.

that extension from a mismatch is very slow, and the delay in extension gives time for partial unwinding of the template-primer, a process that places the primer in the exonuclease site (**FIGURE 22.34**). Hence, the 3′-exonuclease activity has no specificity for a mismatched nucleotide; rather, a mismatched nucleotide has a far higher probability of reaching that site than does a matched nucleotide.

Polymerase Insertion Specificity

Implicit in the Watson–Crick model was the idea that the accuracy of DNA replication is governed by the hydrogen bonds that link A to T and G to C. However, the free energy of forming a Watson–Crick base pair differs from that for forming a mismatch (i.e., A-C or G-T) by only 4–13 kJ/mol. This corresponds to only a 100- to 1000-fold difference in binding affinity between correct and incorrect base pairs. Therefore, if DNA polymerase were completely passive, only incorporating the nucleotide that spontaneously associates with a template base, then errors would be made approximately 0.1%–1.0% of the time, reflecting the relative abundance of correctly and incorrectly base-paired structures. These error rates are far higher than what is observed, even when proofreading is not occurring.

Pre-steady-state kinetic analysis of DNA polymerase reactions shows that dNTP binding to a polymerase-DNA complex precedes a major conformational change that must occur before the formation and breakage of chemical bonds. A complete kinetic scheme for a single-nucleotide addition is as follows.

1   **2**   **3**

$$E \leqq DNA \underset{k_{-1}}{\overset{k_1}{\rightleftharpoons}} E{\cdot}DNA \underset{k_{-2}}{\overset{\overset{dNTP}{k_2}}{\rightleftharpoons}} E{\cdot}DNA{\cdot}dNTP \underset{k_{-3}}{\overset{k_3}{\rightleftharpoons}} E^*{\cdot}DNA{\cdot}dNTP$$

Catalysis Step

$$\underset{k_{-4}}{\overset{k_4}{\rightleftharpoons}} \;\; \textbf{4}$$

5

$$E{\cdot}DNA_{+1} \underset{k_{-6}}{\overset{k_6}{\rightleftharpoons}} E{\cdot}DNA_{+1}{\cdot}PP_i \underset{k_{-5}}{\overset{k_5}{\rightleftharpoons}} E^*{\cdot}DNA_{+1}{\cdot}PP_i$$

Data from Joyce, C. M., and S. J. Benkovic (2004) DNA polymerase fidelity: Kinetics, structure, and checkpoints. *Biochemistry* 43:14317–14324.

For most polymerases, discrimination occurs at both steps **2** (dNTP binding) and **3** (the conformational change that occurs prior to phosphodiester bond formation in step **4**). The bond formation step is not rate-limiting and does not contribute to discrimination. Evidence suggests that the conformational change occurring with a misaligned dNTP distorts catalytic residues on the enzyme, so that dissociation of the misaligned dNTP becomes more likely.

Is hydrogen-bonding specificity sufficient to account for insertion specificity? Eric Kool tested this idea by synthesizing dNTP analogs that are geometrically equivalent to natural dNTPs, but that lack hydrogen-bonding atoms (see the dTTP analog). Although these analogs are incorporated at slower rates, they show discrimination comparable to natural dNTPs; for example, the analog shown can compete with dTTP for incorporation opposite a dAMP residue in template DNA. Thus, specificity in DNA polymerase insertion reactions results from the shapes of the substrate molecules as well as their hydrogen-bonding capacity.

dTTP

Geometric analog of dTTP

Most attention in DNA replication fidelity has focused on single-nucleotide substitution errors, in which one nucleotide is misincorporated. Less attention has been paid to other kinds of errors, including insertions and deletions. Of particular interest are errors resulting from oligonucleotide repeats along a template containing a short repeat sequence in tandem. Several genetic diseases involve multiple tandem trinucleotide repeats within an affected gene. Huntington's disease, which we mentioned in Chapter 21, is an autosomal dominant neurological disease with a variable age of onset. In this condition, a brain protein called **huntingtin** contains a stretch of consecutive glutamine residues,

3′....G—T—C—G—T—C—G—T—C—G—T—C—G—T—C—G—T—C—G—T—C—G—T—C....

5′....C—A—G—C—A—G—C—A—G—C—A—G C—A—G—C—A—G—C—A—G—OH

(looped-out single-stranded region)
C   G
A   A
G   C
C   G
A

each encoded by 5′ CAG 3′. Normal individuals contain between 6 and 31 glutamines at this site, whereas affected individuals contain as many as 80 glutamines or more. The number of repeats is thought to increase if a product DNA strand slips during an interval between successive nucleotide additions, forming a looped-out structure replication intermediate. In the example shown above, the product DNA strand (red), after completion of replication, would have three more glutamine codons than the template (blue). In Huntington's disease, the array of glutamine codons tends to become longer from one generation to the next, with a correspondingly earlier age of onset. The longer the array, the earlier the age of onset of this invariably fatal disease.

● **CONNECTION** Huntington's disease is an example of genetic disorders resulting from DNA polymerase slippage errors that yield the tandem repetition of short oligonucleotide sequences in DNA.

DNA Precursor Metabolism and Genomic Stability

Accurate DNA replication depends on maintaining balanced pools of the four dNTPs. The importance of pool balance was stressed in Chapter 19, where we described the complex allosteric control of ribonucleotide reductase. If one dNTP is present in excess at a replication site, then it can force errors by either of the two mechanisms shown in **FIGURE 22.35**: (1) It can compete with a less abundant nucleotide for incorrect insertion to create a mismatch, or (2) By a "next nucleotide effect" it can drive correct insertion opposite a template base just past a site where a mismatch has occurred and where exonucleolytic proofreading hasn't taken place.

Are dNTP imbalances responsible for genomic instability in living cells? It has long been known, for example, that the frequency of spontaneous mutations is much higher for the mitochondrial genome than it is for the nuclear genome. A large number of genetic diseases result from mutations either in the mitochondrial genome (see Chapter 14) or in nuclear genes that affect mitochondrial metabolism. One such disease is *mitochondrial neurogastrointestinal encephalomyopathy* (MNGIE), a condition that results in age-related accumulation of mitochondrial mutations, with impairment of several normal functions. The disease results from a mutation in the nuclear gene for thymidine phosphorylase, an enzyme that catalyzes thymidine catabolism (see Chapter 19):

$$\text{Thymidine} + P_i \rightarrow \text{Thymine} + \text{deoxyribose-1-phosphate}$$

A deficiency of this enzyme causes thymidine to accumulate, and its uptake into mitochondria leads to an excess of dTTP, which causes GC-to-AT transitions by forcing the formation of G-T base pairs during replication of mtDNA. The dTTP excess also leads to a dCTP deficiency, which contributes to mutagenesis. What might be a mechanism for the dTTP-induced dCTP depletion?

Ribonucleotide Incorporation and Genomic Stability

Until recently, all investigations of DNA polymerase fidelity have focused on the incorporation of incorrect deoxyribonucleotides. However, yeast replicative DNA polymerases have been shown to incorporate *ribo*nucleotides at significant rates. Given the discrimination ratios seen in vitro, as well as the fact that intracellular ribonucleoside triphosphate concentrations are much higher than those of dNTPs, it is estimated that 10,000 ribonucleotide molecules might be incorporated into yeast DNA in each round of replication. If unrepaired, ribonucleotide incorporation at this level is mutagenic. Repair involves the action of RNase H2, a ribonuclease specific for RNA in DNA–RNA hybrid molecules. RNase H2 acts preferentially at sites where a single ribonucleotide has been incorporated. If

● **CONNECTION** Either spontaneous or induced mutations in mitochondrial DNA can result from mutagenic accumulation of dNTPs.

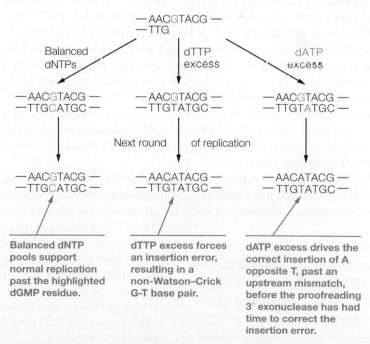

▲ **FIGURE 22.35 How dNTP pool imbalance can lead to replication errors.** In both examples shown here, the dNTP imbalance causes a GC-to-AT transition mutation.

● **CONNECTION** The importance of repairing misincorporation of ribonucleotides into DNA is underscored by the severe clinical consequences of RNase H2 deficiency.

these ribonucleotides are not excised, then the cleavage of DNA during topoisomerase I action in regions of short repeated sequences leads to short deletions. A rare neuroinflammatory disorder in humans called Aicardi-Goutières syndrome is associated with a deficiency of RNase H2.

22.8 RNA Viruses: The Replication of RNA Genomes

We conclude Chapter 22 with a brief discussion of the replication of RNA viral genomes. Almost all known plant viruses contain RNA instead of DNA, as do several bacteriophages and many important animal viruses, including poliovirus, influenza, Ebola, and Zika viruses. The **retroviruses,** which are responsible for many tumors and for HIV-AIDS, also contain RNA genomes.

RNA-Dependent RNA Replicases

Most RNA viruses contain a genome consisting of a single molecule of single-stranded RNA. That RNA is usually the "sense" or "plus" strand for expression of genetic information. In other words, the RNA molecule that passes from the viral particle into the infected cell can serve directly as a messenger RNA, without first requiring the synthesis of complementary-strand RNA. One of the early products of translation of this input genome is the enzyme **replicase,** or RNA-dependent RNA polymerase. Replicase copies the input RNA (the plus strand), starting from the 3' end. Thus, the new strand (minus strand) is laid down from its 5' end to its 3' end, the same direction in which DNA polymerases function (see the accompanying figure). The newly replicated minus strands then serve as templates for the synthesis of plus strands, which are packaged into progeny **virions,** or virus particles. More complex mechanisms come into play when the RNA genome is double-stranded or segmented (three or four separate RNA molecules) or when the virion RNA is itself the minus strand (these latter viruses are called **negative-strand viruses**).

The known RNA replicases lack proofreading activity, so viral RNA replication is more error prone than DNA replication, and RNA viruses mutate and evolve far more rapidly than the organisms they infect. These characteristics are related to viral pathogenesis, in part because a virus population infecting a plant or an animal can undergo change so rapidly that it can evade or counteract the host's defense mechanisms.

● **CONCEPT** The rapid mutation rates of RNA viruses are largely due to RNA replication mechanisms that do not involve proofreading for mismatched nucleotides.

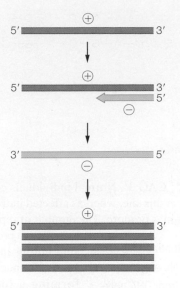

Replication of Retroviral Genomes

Different strategies for genome replication are involved in the action of retroviruses, so named because of the presence of RNA-dependent DNA polymerase, or **reverse transcriptase.** In this class of viruses, the single-stranded RNA genome achieves latency—the ability to persist in a host cell for a long period without pathological effects—by making a DNA copy of itself and inserting that copy into the host-cell genome. The DNA copy is made by reverse transcriptase, a multifunctional enzyme that is packaged in virions and enters the infected cell along with the viral genome. As shown in **FIGURE 22.36**, reverse transcriptase uses viral RNA as a template for synthesizing a complementary DNA strand, with a specific transfer RNA molecule serving as the primer (step ❶). An RNase H activity of the enzyme then partially digests the RNA (step ❷), and the structure circularizes by DNA–RNA base pairing (step ❸). The nascent DNA chain is extended around the circle, and the tRNA primer is removed by RNase H activity (step ❹). Strand displacement synthesis then occurs, with the RNA strand being displaced (step ❺). The resultant double-stranded DNA circle then recombines with a site on a chromosomal DNA and is linearly inserted into that DNA in the process (step ❻). Under these conditions, the integrated proviral genome can persist in a noninfectious state for many years, with most of its own genes turned off. Environmental stresses, still undetermined, can trigger excision of the integrated viral genome and the return of the virus to an infectious state.

As discussed when we introduced azidothymidine (AZT) in Chapter 19, the reverse transcriptase of human immunodeficiency virus (HIV), the virus that causes AIDS, is an obvious target for antiviral therapy. Like RNA-dependent RNA replicases, reverse transcriptases lack proofreading activity. Hence, the fidelity of reverse transcriptase is much lower than that of other DNA polymerases, causing HIV to have a high mutation rate. This presents a challenge, both in designing reverse transcriptase inhibitors as antiviral drugs and in designing effective antibodies against the virus.

● **CONNECTION** A major problem in treating diseases caused by RNA viruses such as HIV is the high frequency with which the virus mutates, thus leading to drug resistance.

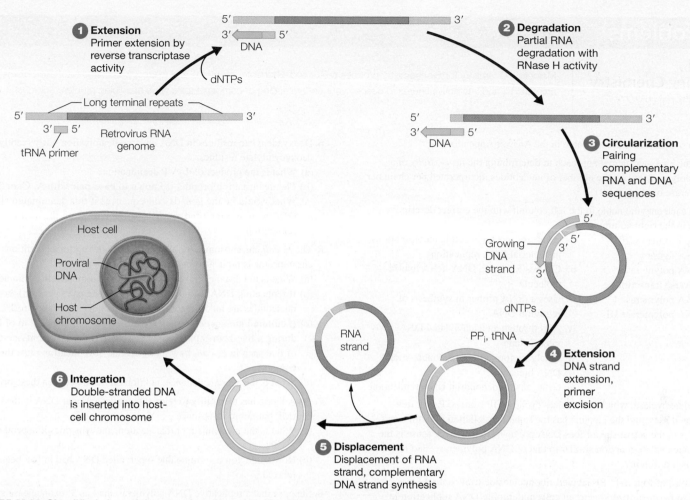

① Extension
Primer extension by reverse transcriptase activity

5′ ▭▭▭▭▭▭▭▭▭▭▭ 3′
3′ ◀▭▭ 5′
DNA

dNTPs

Long terminal repeats

5′ ▭▭▭▭▭▭▭▭▭▭▭ 3′
3′ ▭ 5′
tRNA primer   Retrovirus RNA genome

② Degradation
Partial RNA degradation with RNase H activity

5′ ▭▭▭▭▭▭▭▭▭▭▭ 3′
3′ ◀▭▭ 5′
DNA

③ Circularization
Pairing complementary RNA and DNA sequences

Growing DNA strand

dNTPs

PPᵢ, tRNA

④ Extension
DNA strand extension, primer excision

Host cell

Proviral DNA

Host chromosome

RNA strand

⑥ Integration
Double-stranded DNA is inserted into host-cell chromosome

⑤ Displacement
Displacement of RNA strand, complementary DNA strand synthesis

▲ **FIGURE 22.36 Simplified view of the retrovirus life cycle.** The RNA genome contains long terminal repeats, to one of which a transfer RNA molecule binds. Primer extension, partial RNA digestion, and circularization generate a substrate for extensive DNA synthesis. Ultimately, a circular duplex DNA molecule is formed, and it is the likely substrate for integration into a host chromosome.

Summary

- DNA replication initiates from fixed **replication origins** and proceeds semiconservatively and bidirectionally. (Section 22.1)

- DNA replication is catalyzed by DNA polymerases. DNA polymerases are enzymes that catalyze the template-directed incorporation of deoxyribonucleotides, moving in a 5′-to-3′ direction. Bacterial cells contain 5 different DNA polymerases, and eukaryotic cells at least 15. Not all DNA polymerases are involved in DNA replication. (Section 22.2)

- Proteins in addition to DNA polymerase are needed to replicate entire chromosomes. These additional proteins include **DNA ligase**, **single-strand DNA-binding protein**, **primase**, **sliding clamp**, **clamp loaders**, **helicases**, and **topoisomerases**. Topoisomerases also participate in termination of replication. (Sections 22.3 and 22.4)

- Initiation of DNA replication begins with duplex DNA unwinding to expose single-stranded templates. Initiation involves the synthesis of RNA primers that are extended by DNA polymerases and then removed. Special mechanisms, notably **telomerase** action, are required to complete replication of both DNA strands. (Sections 22.5 and 22.6)

- Several processes contribute to the high accuracy of DNA replication, including 3′ exonucleolytic proofreading, polymerase insertion specificity, and maintenance of balanced DNA precursor pools. (Section 22.7)

- RNA virus genomes are replicated by RNA-templated RNA polymerases or **reverse transcriptase**, depending on whether the viral genome is RNA or DNA. Viral genome replication is highly inaccurate because it is not proofread. (Section 22.8)

Problems

Enhanced by
Mastering Chemistry
for Biochemistry

Mastering Chemistry for Biochemistry provides select end-of-chapter problems and feedback-enriched tutorial problems, animations, and interactive figures to deepen your understanding of complex topics while practicing problem solvings.

Answers to red problems are available in the Answer Appendix.

1. Describe an experimental approach to determining the *processivity* of a DNA polymerase (i.e., the number of nucleotides incorporated per chain per polymerase binding event).

2. Match each enzyme name in the left column with the correct descriptive phrase in the right column.

   (a) Topoisomerase II
   (b) DNA ligase
   (c) DNA polymerase γ
   (d) Reverse transcriptase
   (e) DNA polymerase I
   (f) DNA polymerase III

   i. Catalyzes most nucleotide incorporations in bacterial DNA replication
   ii. Cleaves RNA in a DNA–RNA hybrid molecule
   iii. Uses a tRNA primer in synthesis of retroviral DNA
   iv. Acts through an adenylylated DNA intermediate
   v. Catalyzes formation of a double-strand DNA break
   vi. Catalyzes mitochondrial DNA replication

3. Adenylate cyclase, which synthesizes cyclic AMP from ATP, requires two metal ions, and the enzyme has the same constellation of amino acid residues in the active site as does DNA polymerase I. In what sense is the adenylate cyclase reaction similar to that of DNA polymerase, and in what sense is it different?

4. A mixture of four α-[^{32}P]–labeled ribonucleoside triphosphates was added to permeabilized bacterial cells undergoing DNA replication in the presence of an RNA polymerase inhibitor, and incorporation into high-molecular-weight material was followed over time, as shown in the accompanying graph. After 10 minutes of incubation, a 1000-fold excess of unlabeled ribonucleoside triphosphates was added, with the results shown in the graph.

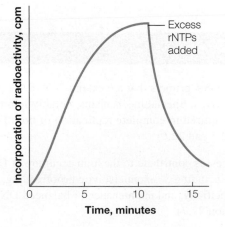

(a) Why was the excess of unlabeled rNTPs added?
(b) How could you tell that radioactivity is being incorporated as ribonucleotides rather than as an alternative such as reduction to deoxyribonucleotides, followed by incorporation?
(c) What does this experiment tell you about the process of DNA replication?

5. Deoxyadenylate residues in DNA undergo deamination fairly readily, as do deoxycytidylate residues.
   (a) What is the product of dAMP deamination?
   (b) The deamination product is known to base-pair with A, C, or T. What would be the genetic consequences if this deaminated site in DNA were not repaired and if it paired with C on the next round of replication?

6. The *E. coli* chromosome is 1.28 mm long. Under optimal conditions, the chromosome is replicated in 40 minutes.
   (a) What is the distance traversed by one replication fork in 1 minute?
   (b) If replicating DNA is in the B form (10.4 base pairs per turn), how many nucleotides are incorporated in 1 minute in one replication fork?
   (c) If cultured human cells (such as HeLa cells) replicate 1.2 m of DNA during a five-hour S phase and at a rate of fork movement one-tenth of that seen in *E. coli,* how many origins of replication must the cells contain?
   (d) What is the average distance, in kilobase pairs, between these origins?

7. DNA ligase has the ability to relax supercoiled circular DNA in the presence of AMP but not in its absence.
   (a) What is the mechanism of this reaction, and why does it depend on AMP?
   (b) How could you determine that supercoiled DNA had in fact been relaxed?

8. Suppose that a replicative DNA polymerase had its 3′ exonuclease site 1.5 nm from the polymerase site, rather than the 3.0 nm seen in Klenow fragment. How would this change affect the fidelity of the enzyme? Why? Describe an additional change that could give this enzyme the same fidelity as Klenow fragment while retaining the 1.5-nm inter-site distance.

9. Although DNA polymerases require both a template and a primer, the following single-stranded polynucleotide was found to serve as a substrate for DNA polymerase in the absence of any additional DNA.

   3′ HO-ATGGGCTCATAGCCGGAGCCCTAACC-
   GTAGACCACGAATAGCATTAGG-p 5′

   What is the structure of the product of this reaction?

10. The 3′-exonuclease activity of *E. coli* DNA polymerase I was found to show no discrimination between correctly and incorrectly base-paired nucleotides at the 3′-terminus; properly and improperly base-paired nucleotides are cleaved at equal rates there. How can this observation be reconciled with the fact that the 3′-exonuclease activity increases the accuracy with which template DNA is copied?

11. 2′,3′-Dideoxyinosine has been approved as an anti-HIV drug (Chapter 19). Propose a mechanism by which it might block the growth of the HIV virus.

12. What information would you need to tell whether the leading and lagging strands in yeast are replicated with equal fidelity?

13. It takes 40 minutes to completely replicate the *E. coli* chromosome, even in an optimally nourished cell. However, bacterial cells can divide as frequently as every 20 minutes. How can cells divide more rapidly, apparently, than their DNA can be copied?

14. 5-Bromouracil (BU) resembles thymine sufficiently that BU base-pairs readily with adenine in a DNA helix, and it can readily substitute for thymine in DNA replication. However, its electron distribution allows it to resemble cytosine sufficiently that BU can pair with guanine. Show how a few rounds of replication in the presence of BU could convert an A-T base pair to G-C, and identify the products of each round of replication.

15. The oxidation of a DNA-guanine residue to 8-oxoguanine (8-oxoG) is mutagenic because 8-oxoG pairs readily with adenine in the next round of replication. Show a pathway beginning with the oxidation of G in a G-C base pair and leading to a mutant base pair, and identify the products of each round of replication.

16. In mammalian cells, genes that are expressed in a particular cell are reported to undergo replication during the first half of S phase, and genes not expressed in that cell are replicated in the latter half of S phase. Briefly describe an experiment that could lead to this conclusion. You might consider approaches that involve 5-bromouracil incorporation.

17. DNA precursor imbalances are mutagenic. For example, if dGTP accumulates, it can compete with dATP for incorporation opposite dTMP in the template, leading to a transition mutation. Investigators have shown that a modest *balanced* increase in all four dNTPs, three- or fourfold, stimulated

mutagenesis out of proportion to the dNTP pool change. Describe a mechanism by which this could occur.

18. Bacteriophage T4 has a linear double-stranded DNA genome, yet mapping many mutations, as shown in Figure 22.11, generates a circular linkage map. How might you explain this discrepancy?

19. In the mitochondrial disease MNGIE (see page 709) a deficiency of thymidine phosphorylase causes dTTP to accumulate in mitochondria. Describe the mechanism by which this occurs. A secondary effect is depletion of mitochondrial dCTP pools. Describe a plausible mechanism for this effect.

20. A major difficulty in preparing vaccines against viral diseases is the rapidity with which the virus mutates to vaccine resistance. What is the likely mechanism for this hypermutability? Would you expect this problem to affect DNA and RNA viruses equally? Explain.

21. Bacteriophage T4 mutants defective in genetic recombination usually show defective DNA replication as well. What specific defect might you expect to see? Describe a plausible explanation for this effect.

22. Before about 1960 it wasn't clear whether both replicating DNA strands were extended in the same fork. Did the Cairns experiment (Figure 22.1) shed light on this question? Why or why not?

References

For a list of references related to this chapter, see Appendix II.

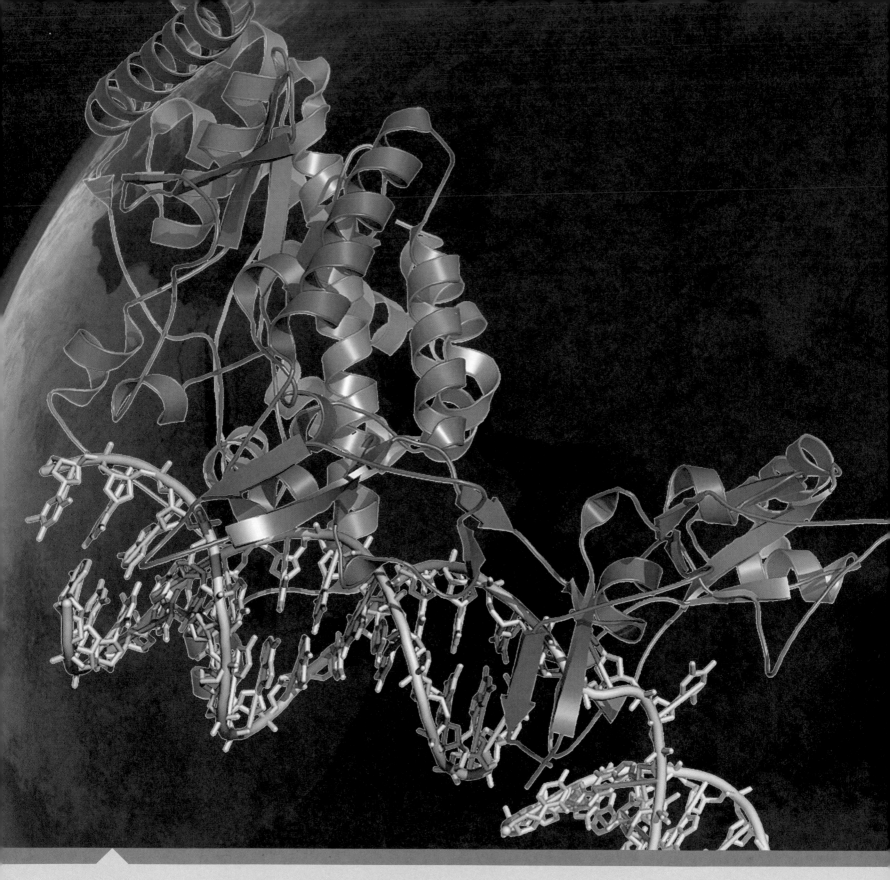

Ultraviolet rays in sunlight are among the most potent environmental DNA-damaging agents. The protein XPC, shown here in green, recognizes damaged DNA and begins a process that removes damaged nucleotides. Humans with a mutation lacking this protein are extraordinarily sensitive to UV light and develop skin cancers.

DNA Repair, Recombination, and Rearrangement

23

OUR FOCUS NOW shifts from DNA as a *template* for its own replication to DNA as a *substrate,* in processes that we can call *information remodeling.* DNA is well suited for its role in storing genetic information because of its chemical stability. The fact that DNA from prehistoric human tissues can be isolated and sequenced attests to this stability. However, like all biomolecules, DNA is continuously exposed to damaging agents, including radiation, environmental chemicals, and endogenous agents such as reactive oxygen species (ROS). In addition, chemical change is introduced as errors during normal DNA replication. For DNA to function as a storehouse of genetic information, the cell must efficiently repair damage suffered by DNA. We will see that genetic recombination—the breakage and rejoining of DNA molecules, originally studied as a process ensuring biological diversity—is also an integral part of some repair mechanisms. Finally, we briefly consider the plasticity of the genome in terms of the ability of DNA segments to move from point to point within a genome or from genome to genome, or to undergo selective amplification.

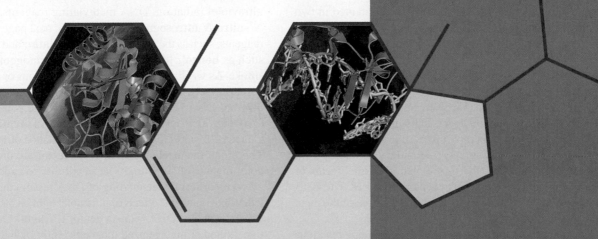

The processes we consider in this chapter include (1) metabolic responses to DNA structural damage, principally mutagenesis and repair; (2) recombination, whereby the contents of a genome are redistributed (e.g., during meiosis); (3) gene rearrangements, including transpositions of DNA segments from one chromosomal integration site into another and the joining of DNA segments from distant parts of a genome; (4) diversification of the immune response by recombination; and (5) gene amplification, an increase in the copy number of individual segments of DNA, which occurs both as a normal developmental process and as a response to environmental stresses (**FIGURE 23.1**). Collectively, these processes are essential to the survival of cells. In broader terms, recombination and gene rearrangements (including gene duplication) are the source of most of the genetic variability in a population of cells or organisms and, along with mutation, form the basis for evolutionary change.

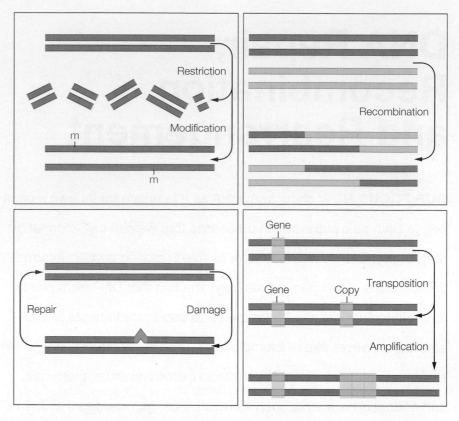

▲ **FIGURE 23.1 A summary of the major processes in information remodeling.** Restriction and modification were introduced in Chapter 21, and the other processes are presented in this chapter.

23.1 DNA Repair

Types and Consequences of DNA Damage

Unprogrammed chemical changes occur in all biological macromolecules because of either environmental damage or errors in synthesis. For most biomolecules, including RNA, protein, and membrane lipids, the effects of these changes are minimized by turnover and replacement of the altered molecules. DNA is distinctive, however, in that its information content must be transmitted virtually intact from one cell to another during cell division or the reproduction of an organism. Thus, DNA requires metabolic stability. This stability is maintained in two ways—by a highly accurate replication process and by mechanisms for correcting genetic mistakes when DNA suffers damage. In Chapter 22 we described mechanisms used to ensure high replication fidelity. Here we discuss the kinds of environmental and endogenous damage that occur and the processes that repair the damage. **FIGURE 23.2** identifies the major kinds of DNA damage that arise through either replication errors or the actions of environmental agents. The figure also identifies each of the major processes that can repair particular types of damage. And because unrepaired damage is mutagenic and cancer arises from genomic instability (Chapter 20), the figure also identifies forms of cancer that result from defects in particular repair processes. The 2015 Nobel Prize in Chemistry was awarded to pioneers in three of the most important DNA repair processes.

Some of the most prominent endogenous DNA-damaging reactions include (1) depurination, resulting from cleavage of the glycosidic

● **CONNECTION** One of the principal causes of cancer is genomic instability resulting from the defective repair of DNA damage.

bond between deoxyribose and a purine base; (2) deamination, usually the hydrolytic conversion of a DNA–cytosine residue to uracil; (3) oxidation, notably the oxidation of guanine to 8-oxoguanine or of thymine to thymine glycol and (4) nonenzymatic methylation by *S*-adenosylmethionine. **FIGURE 23.3** shows the frequency with which these reactions occur in mammalian cells.

Environmental DNA-damaging agents include ionizing radiation; ultraviolet radiation; DNA methylating reagents, such as ***N*-methyl-*N*'-nitro-*N*-nitrosoguanidine** (MNNG; next page); DNA cross-linking reagents, such as the anticancer drug **cisplatin;** and bulky hydrocarbons, such as **benzo[*a*]pyrene,** one of the carcinogenic agents in tobacco smoke. As we discuss the repair of spontaneous or environmental DNA damage, we will see that some repair processes are accurate (i.e., the original DNA sequence is restored), while some lead to mutations (i.e., the repair process is inaccurate). As mentioned earlier and in Chapter 20, cancer results from an accumulation of somatic cell mutations. Therefore, the mechanisms of DNA repair are under intense study as determinants of an animal's susceptibility to cancer.

Our earliest understanding of DNA repair came from study of the lethal and mutagenic effects of ultraviolet light on bacteria or viruses. How do we know that DNA is the target for UV light? Organisms were found to be killed or mutagenized most efficiently by light of wavelength 260 nm, the absorption maximum for DNA. Analysis of the **photoproducts,**

Type of damage

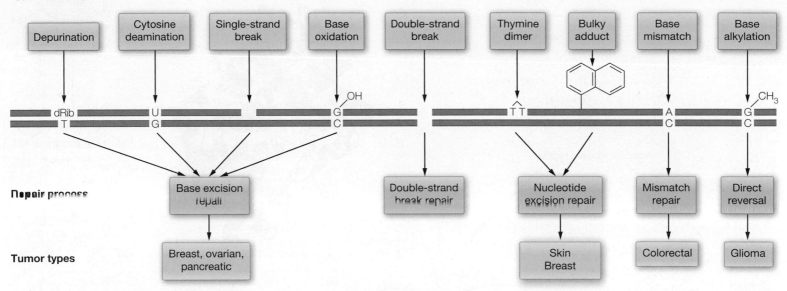

▲ **FIGURE 23.2 Types of DNA damage and repair processes.** Tumor types resulting from deficiencies in specific repair processes are identified at the bottom of the figure.

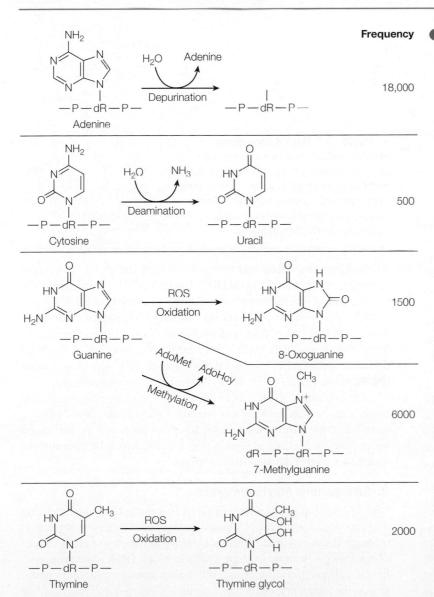

Frequency

Depurination (Adenine → Adenine)	18,000
Deamination (Cytosine → Uracil)	500
Oxidation (Guanine → 8-Oxoguanine)	1500
Methylation (Guanine → 7-Methylguanine)	6000
Oxidation (Thymine → Thymine glycol)	2000

● **CONCEPT** Cyclobutane thymine dimer is the most lethal photoproduct in UV light–irradiated DNA, and 6–4 dimer is the most mutagenic.

or altered DNA constituents after irradiation, showed the most prominent to be intra-strand dimers consisting of two bases (usually thymine) joined either by a cyclobutane ring involving carbons 5 and 6 or by a bond from C4 of one pyrimidine to C6 of the next (**FIGURE 23.4**). These **thymine dimers** were identified early as biologically significant photoprod-ucts because the relative abundance of thymine dimers in irradiated DNA correlated most closely with death or mutagenesis of irradi-ated phages or bacte-ria. Thus, the ability of an organism to survive ultraviolet irradiation was partially due to its ability to remove thy-mine dimers from its DNA. The cyclobutane structure draws the adjacent thymine resi-dues together, distort-ing the helix in a way that blocks replication past this site, while the 6–4 dimer is strongly mutagenic because it induces inaccurate replication.

N-Methyl-*N*'-nitro-*N*-nitrosoguanidine **(MNNG)**

Cisplatin

Benzo(*a*)pyrene

◀ **FIGURE 23.3 Endogenous DNA-damaging reactions.** The approx-imate frequency of each reaction, in number of lesions per mam-malian cell per day, is indicated. Data from Friedberg, E. C., G. C. Walker, W. Siede, R. D. Wood, R. A. Schultz, and T. Ellenberger (2006) *DNA Repair and Mutagenesis*, Second Edition, ASM Press, Washington, DC.

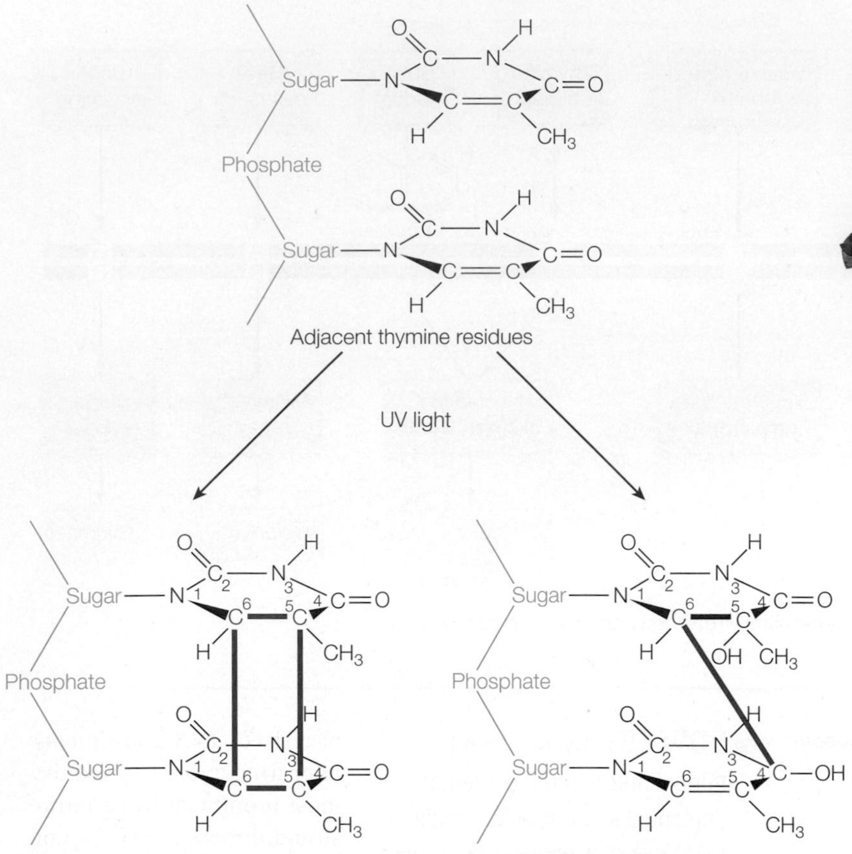

(a) Cyclobutane thymine dimer

(b) 6–4 photoproduct

▲ **FIGURE 23.4** Structures of cyclobutane and pyrimidine dimer photoproducts.

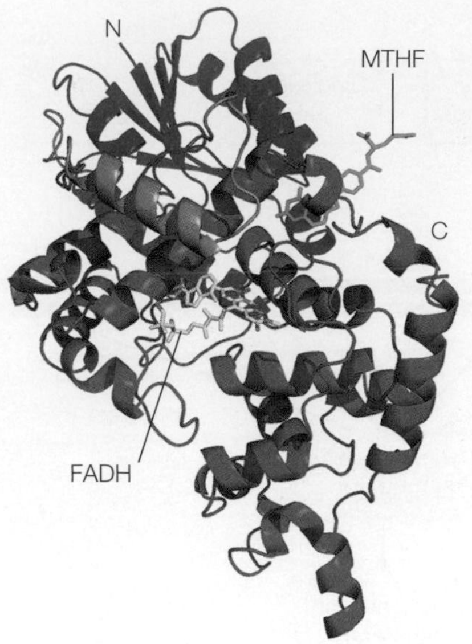

(a)

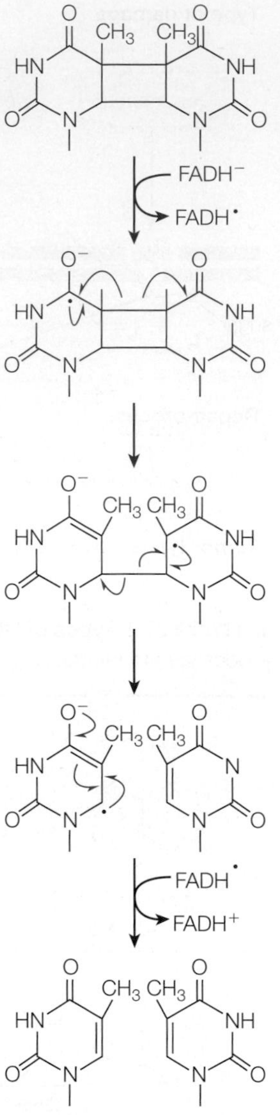

(b)

▲ **FIGURE 23.5** The thymine dimer photolyase. (a) Structure of the *E. coli* enzyme, showing distinct N-terminal (blue) and C-terminal (red) domains, with a linker in orange and showing both bound folate and flavin cofactors. PDB ID: 1dnp. (b) The likely reaction pathway.

Direct Repair of Damaged DNA Bases: Photoreactivation and Alkyltransferases

Of the half dozen well-understood DNA repair processes, most involve removal of the damaged nucleotides, along with several adjacent residues, followed by replacement of the excised region using information encoded in the complementary (undamaged) strand. However, at least two processes involve reactions that *directly change* the damaged bases, rather than removing them.

● **CONCEPT** DNA can be repaired directly, by changing a damaged base to a normal one, or indirectly, by replacing a DNA segment containing the damaged nucleotide.

Photoreactivation

In some organisms, not including mammals, DNA damage can be repaired directly, by visible light irradiation at about 370 nm, in a process called **photoreactivation.** This involves the action of the enzyme **DNA photolyase.** The enzyme binds to DNA in a light-independent process, specifically at the site of pyrimidine dimers. In the presence of visible-wavelength light, the bonds linking the pyrimidine rings are broken, after which the enzyme can dissociate in the dark.

Clues to the photolyase mechanism have come with the finding that the enzyme contains two chromophores. One chromophore is bound flavin adenine dinucleotide, deprotonated and in the reduced state (FADH$^-$); the second in many photolyases is 5,10-methenyltetrahydrofolate (introduced in Chapter 18). Mechanistic studies suggest a process akin to photosynthesis, with the second chromophore functioning as a light-harvesting factor and transmitting light energy by fluorescent resonance energy transfer to FADH$^-$. Hence, FADH$^-$ functions like the photochemical reaction center, transferring an electron to the dimer and breaking the pyrimidine–pyrimidine bonds by a free radical mechanism. The crystal structure of the *E. coli* photolyase shows 5,10-methenyltetrahydrofolate bound at the surface, between the N-terminal and C-terminal domains, with FADH$^-$ bound deeply within the C-terminal domain (**FIGURE 23.5a**). Figure 23.5b shows the probable reaction pathway.

Although photolyase has been detected in numerous eukaryotic organisms, human cells do not contain a photolyase. Nor do frogs: thinning of the earth's ozone layer has been proposed as one reason for population declines in certain frog species, which lack photolyase and, hence, suffer damage from sunlight.

O^6-Alkylguanine Alkyltransferase

Like UV irradiation, exposure of DNA to a methylating or ethylating reagent yields modified DNA bases, some of which cause mutagenesis or death if not repaired. Some alkylating agents are used in cancer chemotherapy because of their ability to block DNA replication and, hence, cell proliferation. Some alkylating agents are used as mutagens in the laboratory, including *N*-methyl-*N'*-nitro-*N*-nitrosoguanidine

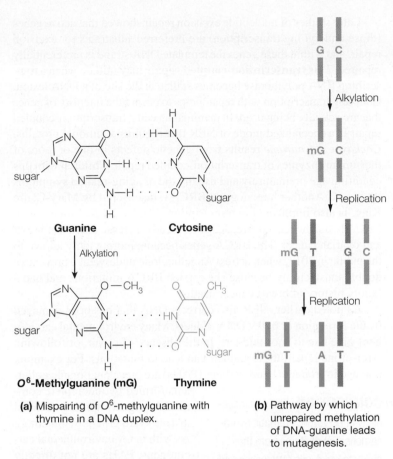

(a) Mispairing of O^6-methylguanine with thymine in a DNA duplex.

(b) Pathway by which unrepaired methylation of DNA-guanine leads to mutagenesis.

▲ **FIGURE 23.6** Genetic consequences of DNA–guanine methylation.

(MNNG, page 717) and others shown here. MNNG and other alkylating agents react primarily with purines. The most highly mutagenic of these products, O^6-**methylguanine,** has a high probability of pairing with thymine when the modified strand replicates (**FIGURE 23.6**). Thus, alkylation of a DNA-guanine stimulates a GC → AT transition mutation (see Figure 23.6b) where mG is a methylguanine residue.

Methylnitrosourea (MNU)

Ethylmethanesulfonate (EMS)

Repair of this type of damage involves action of O^6-**alkylguanine alkyltransferase,** which transfers a methyl or ethyl group from an O^6-methylguanine or O^6-ethylguanine residue to a cysteine residue in the active site of the protein. Remarkably, this "enzyme," which is widely distributed in bacteria and eukaryotes, can function only once. Having become alkylated, it cannot remove the alkyl group, and the protein molecule cannot repeat the process. However, the alkylated form of the protein

● **CONCEPT** Direct repair enzymes include photolyase, which uses light energy to pyrimidine dimers, and alkyltransferases, "enzymes" that are inactivated after just one catalytic cycle.

is a transcriptional activator, which stimulates transcription of the gene encoding the alkyltransferase. It thereby allows the cell to adapt to alkylation damage by using the alkylated protein to signal the cell to produce more of the protein needed to repair the damage.

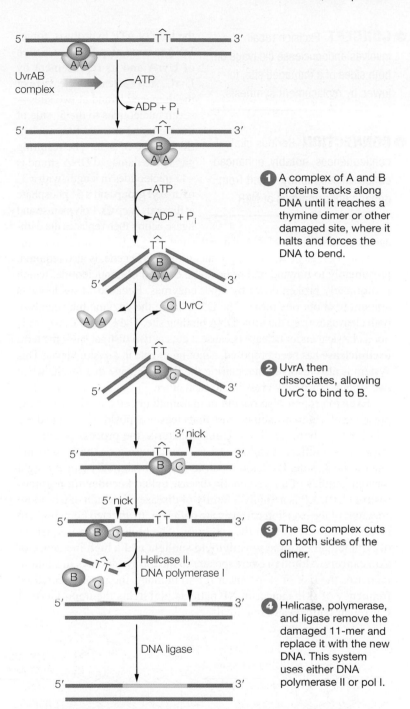

1. A complex of A and B proteins tracks along DNA until it reaches a thymine dimer or other damaged site, where it halts and forces the DNA to bend.

2. UvrA then dissociates, allowing UvrC to bind to B.

3. The BC complex cuts on both sides of the dimer.

4. Helicase, polymerase, and ligase remove the damaged 11-mer and replace it with the new DNA. This system uses either DNA polymerase II or pol I.

▲ **FIGURE 23.7** Excision repair of thymine dimers by the UvrABC excinuclease of *E. coli.* Data from A. Sancar and J. E. Hearst, *Science* (1993) 259:1415–1420. AAAS.

Nucleotide Excision Repair: Excinucleases

Nucleotide excision repair (NER) is an enzyme system capable of repairing thymine dimers created in DNA by UV irradiation. Unlike photoreactivation, this process can take place in the dark. The enzyme system involved, which in *E. coli* includes the products of genes *uvrA, uvrB,* and *uvrC,* also acts upon a wide range of other DNA lesions that may be quite bulky, such as those created by large alkyl or aryl groups that distort the DNA double helix. Similar systems exist in mammalian cells and in yeast, so the process is probably universal.

As shown in **FIGURE 23.7** for *E. coli,* the three-subunit UvrABC enzyme recognizes a lesion (a thymine dimer in the example shown) and, with

● **CONCEPT** Excision repair involves endonuclease cleavage on both sides of a damaged site, followed by replacement synthesis.

● **CONNECTION** Serious clinical consequences, notably, enhanced cancer susceptibility, result from genetic abnormalities of NER.

the help of ATP hydrolysis, forces DNA to bend, step ❶. Dissociation of UvrA and its replacement by UvrC, step ❷, leads to cleavage of the damaged strand at two sites—seven nucleotides to the 5′ side of the damaged site and four nucleotides to the 3′ side. The resulting gap in the damaged DNA strand is ~11 nucleotides in length, with a 3′ hydroxyl group and a 5′ phosphate at the ends, step ❸. Polymerase and ligase action then replaces the damaged 11-mer with new DNA, using the undamaged DNA strand as a template, step ❹. Helicase II, the product of the *uvrD* gene, is also required, presumably to unwind and remove the excised oligonucleotide, which is ultimately broken down by other enzymes. Because 11 nucleotides amount to about one turn of the DNA helix, the enzyme may catalyze both cleavages from the same DNA binding site. The UvrABC enzyme is not a classical endonuclease because it cuts at two distinct sites; the term **excinuclease** has been proposed, denoting its role in *exci*sion repair. This system is also involved in repairing the DNA damage that results when two strands covalently crosslink to each other.

Excision repair also occurs in mammalian cells, as shown by the presence of a human excinuclease that cleaves at positions −22 and +6 relative to a thymine dimer. Unlike bacteria, the process in humans requires two different endonucleases—one cutting on the 5′ side and one on the 3′ side. Excision repair in humans originally came to light through studies of a rare genetic disease called **xeroderma pigmentosum (XP)**. XP is actually a family of diseases in which one or more enzymes of the excision pathway are deficient. In affected humans there is no known way to treat the condition. The biological consequences of XP include extreme sensitivity to sunlight and a high incidence of skin cancers. Although overexposure to the ultraviolet rays in sunlight increases the risk of skin cancer for all humans, the greatly increased frequency of skin cancer in XP patients highlights the importance of UV repair pathways to mammals.

Later studies of nucleotide excision repair showed that active genes (those undergoing transcription) are preferred substrates for excision repair, and within these genes the template DNA strand is preferentially repaired. This **transcription-coupled repair** may initiate when a transcribing RNA polymerase becomes stalled at the site of a DNA lesion. Coupling transcription with repair helps to ensure the integrity of genes that are actually being used. In mammalian cells, transcription-coupled repair is a specialized mode of NER that requires additional proteins. *Cockayne's syndrome* results from genetic defects in one or more of the human enzymes of transcription-coupled repair. Children with this condition age prematurely and usually die of aging-related symptoms by age 12. Another human gene, BRCA1, discovered by Mary-Claire King, is also implicated in transcription-coupled NER. Mutations in BRCA1, and the related BRCA2 gene, greatly increase the risk of breast and ovarian cancer. The BRCA genes became more widely known to the public in 2013, when actress Angelina Jolie underwent a preventive double mastectomy because she carried BRCA mutations and had a family history of breast cancer deaths.

As noted earlier, NER also corrects DNA that has been damaged by the formation of bulky DNA adducts. Many environmental carcinogens give rise to such adducts. In the absence of repair, or following error-prone repair, these adducts can lead to mutations. For example, polycyclic aromatic hydrocarbons (PAHs) are prevalent organic pollutants formed by incomplete combustion (e.g., by burning charcoal in a backyard grill or by smoking). As with many environmental carcinogens, PAHs are not directly reactive with DNA. Indeed, their lack of chemical reactivity accounts for their persistence in the environment. However, PAHs can undergo *metabolic activation* (i.e., *biotransformation*), thus leading to reactive electrophilic intermediates that can bind to DNA and other macromolecules. These reactions can occur in the liver and in other organs. The best-characterized route for the metabolic activation of PAHs is the pathway for benzo[*a*]pyrene metabolism (**FIGURE 23.8**). First, oxygenation catalyzed by a cytochrome P450 generates an epoxide intermediate. This epoxide is hydrolyzed, in a reaction catalyzed by *epoxide hydrolase*, to give a dihydrodiol product. Finally, a second cytochrome P450-catalyzed oxygenation produces the highly reactive benzo[*a*]pyrene 7,8-dihydrodiol-9,10-epoxide (BPDE). BPDE is highly reactive with DNA, forming bulky covalent adducts primarily, but not exclusively, with guanine.

● **CONNECTION** The carcinogenicity of polycyclic aromatic hydrocarbons (PAHs) stems from their ability to react with DNA bases and form mutagenic intermediates.

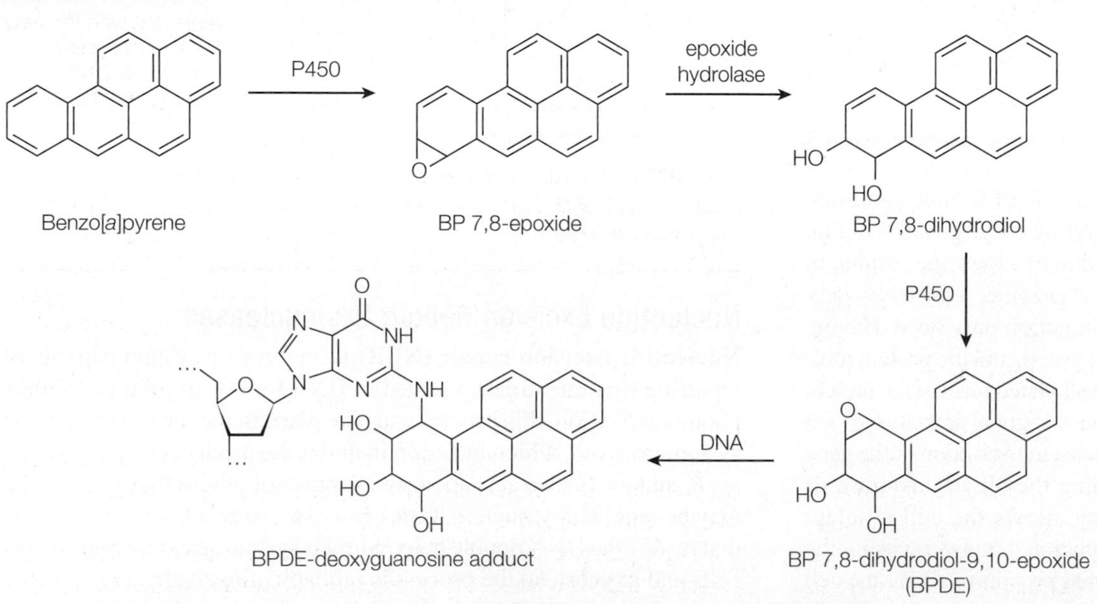

Benzo[a]pyrene → (P450) → BP 7,8-epoxide → (epoxide hydrolase) → BP 7,8-dihydrodiol → (P450) → BP 7,8-dihydrodiol-9,10-epoxide (BPDE) → (DNA) → BPDE-deoxyguanosine adduct

◄ **FIGURE 23.8** Metabolic activation of a carcinogenic polycyclic aromatic hydrocarbon, followed by reaction of the activated dihydrodiol epoxide with a DNA dGMP residue. Some H atoms in the structures are omitted for simplicity.

Base Excision Repair: DNA *N*-Glycosylases

In addition to NER, there is another form of excision repair, **base excision repair (BER).** Like NER, BER also removes one or more nucleotides from a site of base damage. However, this process begins with enzymatic cleavage of the glycosidic bond between the damaged base and deoxyribose.

Replacement of Uracil in DNA by BER

One of the best-understood BER systems scans DNA to remove uracil. Uracil can base-pair with adenine in a DNA duplex, and DNA polymer-

● **CONCEPT** A base excision repair process removes uracil residues in DNA, whether they arose through the deamination of cytosine residues or the incorporation of deoxyuridine nucleotides instead of thymidine nucleotides.

ases can readily accept deoxyuridine triphosphate as a substrate in place of thymidine triphosphate. Yet cells possess an elaborate two-stage process that prevents dUMP residues from accumulating in DNA. The first stage, described in Chapter 19, involves cleavage by dUTPase of dUTP to dUMP, thereby minimizing the dUTP pool and its use as a replication substrate. The second stage involves **uracil-DNA *N*-glycosylase (Ung),** an enzyme that removes any dUMP residues in DNA that might have arisen either through deamination of a dCMP residue or through incorporation of a dUTP that escaped the action of dUTPase.

As shown in **FIGURE 23.9**, Ung hydrolytically cleaves the glycosidic bond between N-1 of uracil and C-1′ of deoxyribose. This yields free uracil and DNA with an **apyrimidinic site** (i.e., a sugar residue lacking an attached pyrimidine). Another enzyme, **apyrimidinic endonuclease** (AP endonuclease), recognizes this site and cleaves the phosphodiester bond on the 5′ side of the deoxyribose moiety. This is followed in bacteria by the nick translation activity of DNA polymerase I (Figure 22.14), which inserts dTTP as a replacement for the dUMP that was removed, displacing the deoxyribose phosphate residue in the AP site. This is removed by a **deoxyribose-5′-phosphatase.** The resulting nick is sealed by DNA ligase.

Ung, like all DNA glycosylases examined to date, acts by flipping the target base out of the DNA duplex and binding it in a pocket where cleavage occurs. Structural analysis of human UNG shows that the pocket is small enough to exclude purine bases. More important, as shown in **FIGURE 23.10**, the pocket excludes thymine because of negative steric interaction between the thymine methyl group at C5 and Tyr-147. (By convention, eukaryotic genes and gene products are denoted with capital letters; thus, *UNG* is the human gene that encodes the protein UNG. In bacteria, gene *ung* encodes protein Ung.)

Why do cells go to all this trouble just to replace a nucleotide that does not affect the information encoded in DNA? Almost certainly, U-A base pairs are not the true target of this DNA repair system. Uracil residues in DNA also arise through spontaneous deamination of cytosine residues (Figure 23.3). This alteration *does* change the genetic sense because it converts a G-C base pair to a G-U pair, and in a subsequent round of replication the U-containing strand would give rise to an A-T base pair—a GC-to-AT **transition** (change of a purine-pyrimidine pair to another purine-pyrimidine). The uracil repair system prevents this mutation but does not discriminate between uracils paired with adenines or with guanines. Consistent with this model is the **hypermutable** phenotype displayed by mutant

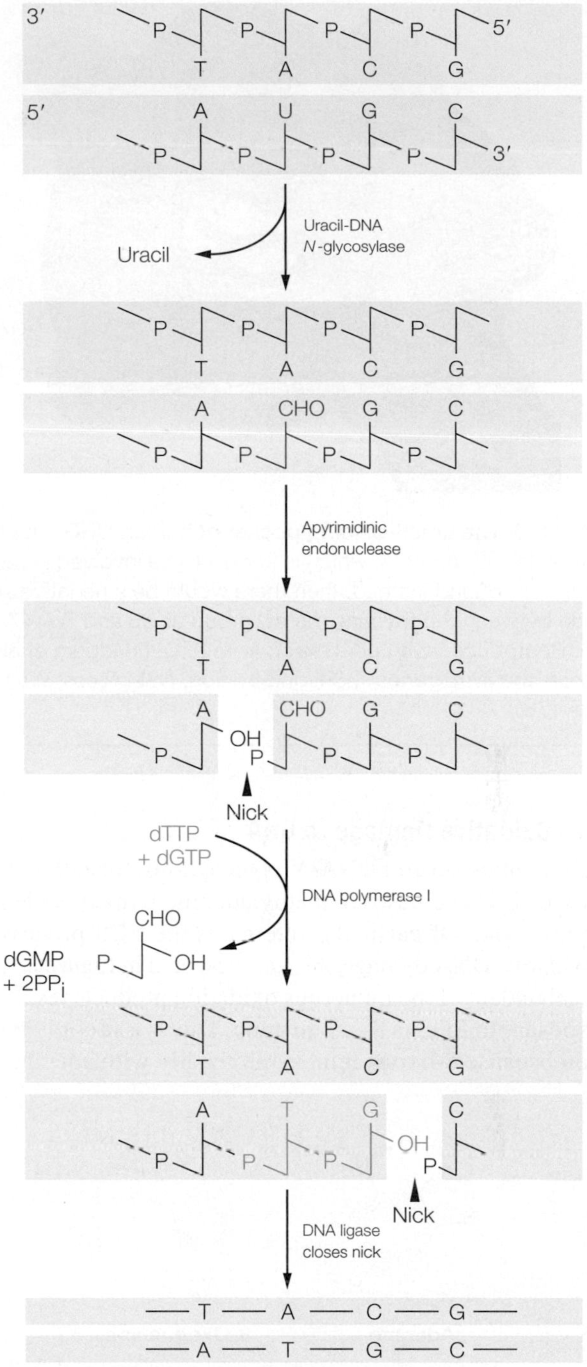

▲ **FIGURE 23.9 Action of the DNA uracil repair system.** Uracil-DNA *N*-glycosylase (Ung) removes uracil, leaving an apyrimidinic site. A specific endonuclease recognizes this site and cleaves on the 5′ side. DNA polymerase I replaces the missing nucleotide, leaving deoxyribose-5-phosphate on the 5′ side of the nick. This is removed hydrolytically, and DNA ligase seals the nick.

bacteria lacking an active Ung. Such strains exhibit elevated rates of spontaneous mutagenesis, resulting from the accumulation of DNA dUMP residues paired with dGMP.

The replacement of U by T in DNA is one example of BER. Another example, described next, involves repair of damage caused by reactive oxygen species.

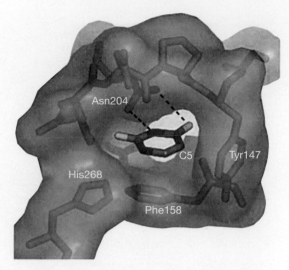

▲ **FIGURE 23.10 The uracil-binding pocket of human UNG.** The bound uracil interacts with Asn204, which is known to be involved in catalysis. If thymine were bound instead, then there would be a negative steric interaction between the thymine methyl group at C5 and Tyr147. PDB ID: 1ugh. Reproduced with permission from E. C. Friedberg et al. (2006) *DNA Repair and Mutagenesis*, Second Edition, ASM Press, Washington, DC, page 176.

Repair of Oxidative Damage to DNA

Most cells contain several DNA-*N*-glycosylases, including those specific for the alkylated bases *N*-methyladenine, 3-methyladenine, and 7-methylguanine. Of particular interest is the BER process used to repair oxidative DNA damage. As was indicated in Figure 23.2, one of the most abundant of the numerous oxidation products resulting from DNA exposure to ROS is 8-oxoguanine. This is a strongly mutagenic alteration because 8-oxoguanine pairs readily with adenine.

Adenine   8-Oxoguanine

Oxidation of a G paired with C gives rise to a C-oxoG base pair. A subsequent round of replication readily gives rise to an A-oxoG pair, which then becomes A-T in the next round. Hence, the C-oxoG mispair can be an intermediate in a **transversion** mutation C-G → C-oxoG → A-oxoG → A-T. (A *transversion* changes a pyrimidine-purine base pair to a purine-pyrimidine pair.) In *E. coli*, three genes encode proteins that minimize oxidative mutagenesis: *mutM, mutY,* and *mutT*. (The "mut" designation indicates that mutations in any one of these genes have the effect of increasing the spontaneous <u>mutation</u> rate.) All three gene products have mammalian homologs. MutM and MutY are both DNA-base glycosylases. MutM (or its mammalian homolog, OGG1), which we describe here, cleaves 8-oxoguanine from DNA to initiate BER. The repair process, similar to DNA-uracil repair, results in replacement of oxoG by G.

Like UNG, described earlier, human OGG1 DNA glycosylase distinguishes target oxoG bases from nontarget bases such as guanine by excluding nontarget bases from the active site where cleavage occurs. Structural analysis of human OGG1 revealed that 8-oxoguanine is able to enter the binding pocket, while guanine is completely excluded due to both attractive and repulsive forces (**FIGURE 23.11**).

Mismatch Repair

Mismatches, or non-Watson–Crick base pairs in a DNA duplex, can arise through replication errors, through deamination of 5-methylcytosine in

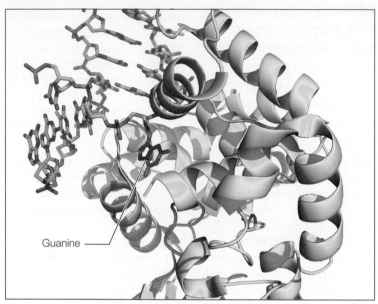

G complex

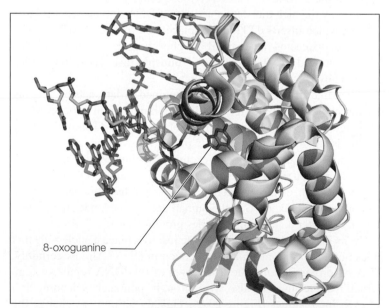

oxoG complex

▲ **FIGURE 23.11 Structure of human OGG1 with guanine (top) and 8-oxoguanine (bottom) bound near the catalytic pocket.** Guanine (in magenta) is flipped out of the DNA helix but excluded from the active site, while 8-oxoguanine is bound in the active site. PDB ID: 1yqk, 1yqr. Based on A. Banerjee, W. Yang, M. Karplus, and G. L. Verdine (2005) *Nature* 434, 612–618.

DNA to yield thymine, or through recombination between DNA segments that are not completely homologous. In addition, mismatches result when DNA polymerase encounters a short repeated sequence and slides along its template, creating short loops or bulges in duplex DNA (see Figure 22.7). We best understand the correction of replication errors, so that is what we describe here.

If DNA polymerase introduces an incorrect nucleotide, creating a non-Watson–Crick base pair, the error is usually corrected by the polymerase-associated 3′ exonuclease activity (Chapter 22). If the error is not corrected immediately, the fully replicated DNA will contain a mismatch at that site. This error can be corrected by **mismatch repair.** In *E. coli*, the proteins that participate include the products of genes *mutH, mutL,* and *mutS,* plus an enzyme called helicase II.

The mismatch correction system scans newly replicated DNA, looking for both mismatched bases and single-base insertions or deletions. MutS binds to DNA at the site of the mismatch, followed by the binding of MutL and then MutH. MutS is a "motor protein," which uses the energy of ATP hydrolysis to pull DNA from both directions until it reaches the site at which repair is to begin. When it finds an appropriate signal, part of one strand containing the mismatched region is cut out and replaced (**FIGURE 23.12**).

● **CONCEPT** The mismatch repair system in bacteria uses DNA methylation to identify the strand that has a mispaired nucleotide.

How does the mismatch repair system recognize the correct strand to repair? If it chose either strand randomly, it would choose incorrectly half the time and there would be negligible gain in replication accuracy. The mismatch repair enzymes identify the newly replicated strand because for a short period that DNA is unmethylated. In *E. coli,* the sequence –GATC– is crucial because that is the site methylated soon after replication, by the product of the *dam* gene (DNA adenine methylase). The mismatch repair enzymes look for –GATC– sequences that are not methylated. Recognition of an unmethylated GATC can target that strand for mismatch correction at a site as far as 1 kbp between the mismatch and the GATC site, in either direction. Once the methylation system has acted on all GATC sites in the daughter strand, it is too late for the mismatch repair system to recognize the more recently synthesized DNA strand, and any advantage in total DNA replication fidelity is lost. When the system functions properly, it has the effect of increasing overall replication fidelity by about 100-fold—from about 1 error in 10^7 base pairs replicated to about 1 in 10^9 or better.

As shown in Figure 23.12, the MutHLS complex moves along DNA in both directions until it encounters the nearest 5′-GATC sequence. An endonuclease activity of MutH then cleaves on the 5′ side of the G in the unmethylated strand. At that point helicase II unwinds the DNA, moving back past the mismatch, followed by an exonuclease that digests the displaced single strand. The resultant gap is filled by DNA polymerase III holoenzyme and DNA ligase, working in concert with SSB.

A similar mismatch repair system exists in eukaryotic cells. The mismatch recognition step is somewhat more complex, however, because three different MutS homologs (MSH proteins) are involved—MSH2, MSH3, and MSH6. These three proteins form heterodimers, with specificities for different types of mismatches. Still unknown is the mechanism by which eukaryotic systems recognize and initiate repair on the newly replicated DNA strand; it is clear that DNA methylation is not involved. Recently, it has been proposed that ribonucleotides incorporated by imperfect specificity of DNA polymerase (Chapter 22)

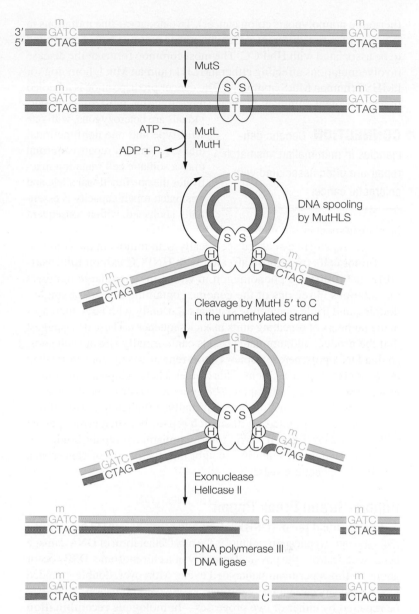

▲ **FIGURE 23.12 Methyl-directed mismatch repair in *E. coli*.** The newly replicated daughter strand (lavender) contains a T mismatched to G in the template strand (blue). The mismatch repair system identifies the daughter strand because it is not yet methylated. Thus, this system must function before the newly replicated daughter strand becomes methylated, through action of the Dam methylase on the A residue in the GATC sequence.

could provide such a signal. In the parental strands, any ribonucleotide incorporation would have been repaired.

Just as is seen with bacteria, mutations in eukaryotic genes that control mismatch repair confer a mutator phenotype, raising spontaneous mutation rates at all loci. How do such mutations affect the biology of human cells? As mentioned in Chapter 20, the progression of a normal cell to a cancer cell involves the accumulation of multiple mutations in tumor cell precursors. Cells lacking mismatch repair activity do have elevated mutation rates, so it was a logical development when mutations in mismatch repair genes were found in tumor cells from individuals with a heritable cancer predisposition called HNPCC (heritable nonpolyposis colon cancer). To date, germ-line mutations in the genes for five different mismatch repair proteins have been found to be associated with HNPCC. The most common forms of the disease involve mutations affecting either hMLH1 (human Mut L homolog) or hMHS2 (human MutS homolog). The cancer predisposition is inherited in an autosomal dominant fashion, suggesting that most affected individuals are heterozygous, with one

● **CONNECTION** Genetic deficiencies in mammalian mismatch repair are often associated with colorectal cancer.

wild-type and one nonfunctional allele. Mismatch repair is normal until a somatic cell mutation inactivates the one functional allele and mismatch repair capacity is essentially abolished, with a consequent increase in spontaneous mutagenesis. It is unclear why mutations affecting mismatch repair are associated specifically with tumors in the colon.

Tumor cells from those affected with HNPCC exhibit **microsatellite instability**—a phenomenon in which there are large numbers of mutations in regions of the genome containing repeats of single-, double-, and triple-nucleotide sequences, usually with large increases in the numbers of repeating units in such sequences. These data suggest that the product and template strands can normally slip at such sites, so that DNA polymerase copies a short repeating sequence more than once, or else skips a segment. This creates a heteroduplex with a short loop, as we showed in Chapter 22 (Figure 22.7) for replication errors leading to Huntington's disease. Normally, a replication error of this type would be corrected by mismatch repair, but such errors persist and accumulate in a cell lacking normal mismatch repair. Studies of this type have given scientists insight into the nature of cancer as a progressive genetic disease.

Double-Strand Break Repair

A double-strand break (DSB), a lesion that can be caused by ionizing radiation or by replication stalling, is the most lethal form of DNA damage because it destroys the physical integrity of a chromosome. DSBs occur about 50 times per mammalian cell cycle. Moreover, double-stranded DNA breaks occur naturally during meiotic recombination. DSBs can be repaired by either of two processes—**homologous recombination** (HR), using sequence information from an undamaged sister chromatid, or **nonhomologous end joining** (NHEJ). (From the standpoint of evolution, *homologous* means "derived from a common ancestor," but we use it in a broader sense, meaning "having similar sequences.")

NHEJ is more efficient, but if DNA ends are not rejoined at the precise sites where breakage occurred, genetic information becomes lost or scrambled. By contrast, HR normally repairs the broken site precisely, but it can occur only during the S or G_2 cell-cycle phases, when a homologous chromosome is available. Both processes begin with several signaling proteins associating at the severed ends. A principal player in vertebrates is a protein kinase called ATM (ataxia telangiectasia mutated).

An early event in both HR and NHEJ is phosphorylation of a variant form of H2 histone, H2AX, in nucleosomes near the break. This may be part of a chromatin remodeling process that helps to move core particles out of the way, exposing DNA ends for processing.

Double-strand break repair is better understood for homologous recombination, and that is what we outline here. Although it is simple to visualize conceptually, as shown in **FIGURE 23.13**, HR is a complex process, involving extensive signaling reactions as part of the **DNA**

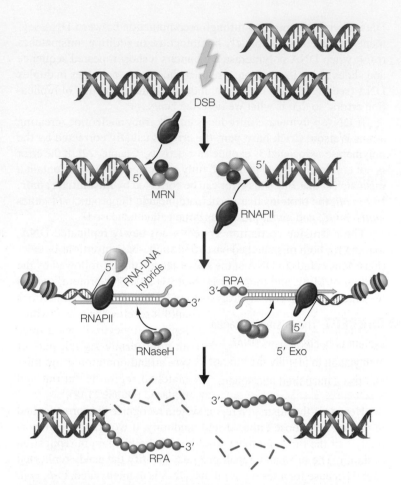

▲ **FIGURE 23.13** A pathway for homologous recombination (HR) to repair DNA double-strand breaks. RNAPII is RNA polymerase II. **For description see text.** Based on C. Ohle et al (2016) *Cell* 167, 1001–1013.

damage response (see page 725). As studied in yeast, a signaling complex called ATM activates a nuclease in another complex called MRN, as well as signaling downstream effectors via an associated protein kinase activity. The MRN-associated nuclease trims away 5′ DNA ends, leaving 3′-terminated single-stranded ends, which become coated with RPA (replication protein A; single-strand binding protein). The BRCA2 protein (page 720) is involved in this process.

● **CONNECTION** Genetic defects involving double-stranded break repair, particularly in genes BRCA1 and BRCA2, are risk factors for breast or ovarian cancer.

RNA polymerase II transcribes the single-stranded DNA ends, and ribonuclease H degrades the RNA in the hybrid. RPA then coats the 3′ DNA ends. Not shown is the binding of Rad51, a protein that scans the single-stranded DNA ends and the homologous undamaged chromosome, seeking regions of sequence homology, thereby facilitating base pairing and subsequent ligation. Rad51 is the eukaryotic counterpart of bacterial RecA, whose action is described on page 728.

Human mutations are known that affect several of the proteins involved in DSB repair. As mentioned previously, mutations in either BRCA1 or BRCA2 are risk factors for developing breast or ovarian

● **CONCEPT** Double-strand DNA breaks can be repaired either by homologous recombination (HR) or by nonhomologous end joining (NHEJ), a process that does not require DNA sequence homology at the ends being joined.

cancer. BRCA2 mutations are also associated with Fanconi's anemia, a rare condition characterized by bone marrow failure. ATM received its name from the deficiency syndrome ataxia telangiectasia, which involves premature aging, cerebellar degeneration, and enhanced cancer susceptibility.

Daughter-Strand Gap Repair

Bacteria also use HR as a way to repair DNA damage. However, because bacteria have but a single chromosome, the undamaged duplex needed as a repair template is the replicated portion of an incompletely replicated chromosome. This process, called **daughter-strand gap repair,** comes into play if DNA damage is sufficient to exceed the cell's capacity for photoreactivation or excision repair. The process is schematized in **FIGURE 23.14**.

When a replisome encounters a thymine dimer or a bulky lesion, it cannot replicate past this site. A recombination-related process can take place, not to repair the damage, but to allow replication to continue, saving the damaged site for subsequent repair by another process. Daughter-strand gap repair depends on the RecA protein mentioned earlier. RecA, like its eukaryotic counterpart, Rad51, catalyzes **strand pairing,** or **strand assimilation**—the joining of two different DNAs by homologous base pairing.

Translesion Synthesis and the DNA Damage Response

With the exception of direct-reversal repair processes like photoreactivation, DNA repair requires the action of a polymerase to fill in a single-strand gap in the DNA duplex. If the gap contains a damaged structure, like a thymine dimer or a bulky lesion, a replicative polymerase is unable to copy past the damaged site, and the repair process aborts. A new set of damage-tolerant DNA polymerases, inducible by DNA damage, comes into play. Recall from Chapter 22 that bacteria contain five DNA polymerases, only two of which are involved in replication, while eukaryotic cells have as many as nine damage-inducible polymerases. Most of these enzymes have low fidelity, partly because they lack 3′ exonucleolytic proofreading and partly because their catalytic sites are larger and more open than those of high-fidelity replicative polymerases. The enzymes are inducible as a result of DNA damage, but the control mechanisms differ markedly between bacterial and eukaryotic organisms.

The **SOS response** in bacteria is a metabolic alarm system that helps the cell to save itself in the presence of potentially lethal stresses. Inducers of the SOS response include DNA damaging agents such as ultraviolet irradiation, thymine starvation, and inactivation of genes essential to DNA replication. Responses include mutagenesis, filamentation (in which cells elongate by growth but don't divide), and activated excision repair. Mutagenesis occurs because, under SOS conditions, the gaps that are formed are filled by inaccurate DNA polymerases. In fact, this process is the principal pathway by which ultraviolet light stimulates mutagenesis in bacteria.

Induction of the SOS response by UV light or other DNA damage activates transcription of about 40 genes in *E. coli.* We discuss the mechanism of transcriptional activation in Chapter 26. Among these SOS-inducible genes are those encoding DNA polymerases

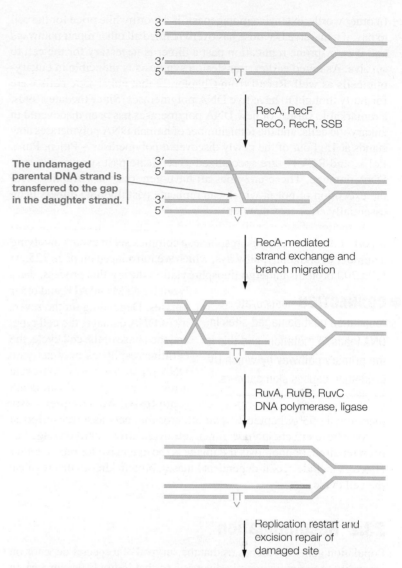

▲ **FIGURE 23.14 Daughter-strand gap repair.** The undamaged parental DNA strand is transferred to the gap in the daughter strand, formed by the inability of DNA polymerase to replicate past it. The remaining steps occur by mechanisms similar or identical to those in homologous recombination. The functions of RuvA, B, and C are discussed in Section 23.3.

● **CONCEPT** DNA can be repaired after replication, either by recombination or by inducible error-prone repair. Both processes require RecA.

IV and V. Both of these enzymes are highly error-prone. After SOS activation, the processing of two gene products (UmuC and UmuD) gives rise to a pol V molecule. This associates with RecA bound to ATP, giving the active form of the enzyme. When the replisome containing pol III holoenzyme stalls at a damage site, pol V replaces it and binds to the sliding clamp. Once the damaged site has been copied (inaccurately), pol V somehow steps aside, and pol III holoenzyme completes the replication process. Pol IV functions similarly, but we know less about it.

Considering that most mutations are probably deleterious, what is the advantage to the cell of accumulating mutations during DNA repair? Probably none, except that the alternative would be for the cell to die.

In other words, extensive mutagenesis is a worthwhile price for the cell to pay, if a massive UV dose has overwhelmed all other repair pathways and if error-prone replication past a dimer is necessary for the cell to survive. As noted earlier, translesion synthesis is inducible in eukaryotic cells as well. Recall from Chapter 22 that eukaryotic cells were formerly thought to have five DNA polymerases. Since the late 1990s, a remarkable variety of new DNA polymerases has been discovered in eukaryotic cells, and the total number of human DNA polymerases now stands at 15. Four of the newly discovered polymerases—Pol η, Pol ι, Pol κ, and REV1—are specialized to replicate past specific types of DNA damage. These enzymes all have generally low fidelity, with the exception of pol η, which can replicate past thymine dimers in an essentially error-free manner.

Induction of some or all of the repair polymerases in eukaryotic cells is part of the DNA damage response, a complex set of events involving a number of signaling pathways, which we introduced on page 725. At least 700 proteins undergo phosphorylation during this process, catalyzed by ATM and ATR and other kinases. Depending on the severity of DNA damage, the cell type, and the phase of the cell cycle, the cellular responses may activate DNA repair pathways, cell-cycle arrest, or programmed cell death (**apoptosis**). We have previously mentioned p53 (Chapter 20), an effector that acts as a transcription factor whose effects include either cell-cycle arrest, until damage has been repaired, or apoptosis if damage is too extensive for repair. Other effectors include cyclin-dependent kinases, protein kinases that regulate the cell cycle.

● **CONNECTION** Inaccurate replication past damaged sites in DNA leads to mutation, and this is the primary pathway by which UV irradiation causes skin cancers.

23.2 Recombination

Population genetics teaches us that the survival of a species depends on its ability to maintain genetic diversity, so that individuals can vary in their ability to respond to unforeseen environmental pressures. Diversity is maintained through mutation, which alters single genes or small groups of genes in an individual, and recombination, which redistributes the contents of a genome among various progeny during reproduction.

As initially described by cytogeneticists, recombination results from crossing over between paired homologous chromosomes during meiosis in eukaryotes. Recombination, however, encompasses more processes and biological functions than those involved in sexual reproduction. Strictly speaking, **recombination** is any process that involves the formation of new DNA from distinct DNA molecules, such that genetic information from each parental DNA molecule is present in the new molecules. The double-strand break repair and daughter-strand gap repair processes described earlier are forms of recombination. So, too, is the formation of recombinant genotypes in bacterial conjugation—and, of course, recombinant DNA molecules constructed in vitro.

● **CONCEPT** Recombination is any process that creates end-to-end joining from two different DNA molecules.

We begin our discussion with **site-specific recombination,** which was initially studied as a model for homologous recombination; it is important for the insight it provides into infection of cells by some viruses, including retroviruses.

Site-Specific Recombination

Many DNA viruses maintain a long-term relationship with the cells that they infect by inserting their genome into one or more chromosomes of the infected cell. Retroviruses carry out a similar process, which was outlined for HIV in Chapters 20 and 22; in this process, the RNA genome makes a DNA copy, which can then become inserted into host-cell chromosomes. The mechanism of viral genome integration is best understood for bacteriophage λ, and that is what we describe here.

Phage T4 is a *lytic* phage. After infecting a bacterial cell, a cycle of virus multiplication always occurs, followed by lysis of the cell, which releases hundreds of new viral particles. Phage λ, by contrast, is a *temperate* phage, which does not always kill its host. Infection can lead either to a lytic growth cycle, like T4, or a *lysogenic* response, in which the viral chromosome becomes linearly inserted into the bacterial chromosome; once inserted, it can remain in a quiescent state, called a *prophage,* for many generations. A change in environmental conditions can trigger the expression of genes that direct excision of the viral chromosome from that of the host and turn on genes leading to virus multiplication and eventual lysis of the cell. Understanding these processes contributed importantly to our understanding of transcriptional regulation, which we will discuss in Chapter 26.

The phage λ genome, as it exists in phage particles, is a linear duplex DNA molecule with short single-stranded tails at each 5' end. These are called *cohesive ends,* or sticky ends, because they are complementary in sequence. Hence, the molecule can circularize by base pairing between the ends. The resultant nicks are closed by DNA ligase. In an infection leading to lysogeny, the covalently closed λ circle becomes linearly inserted into the *E. coli* host-cell chromosome, as schematized in **FIGURE 23.15**. A virally encoded protein called *integrase* recognizes

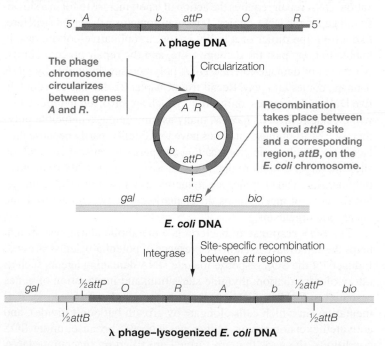

▲ **FIGURE 23.15 Site-specific recombination establishes lysogeny in bacteriophage λ.** The phage chromosome circularizes between genes A and R, and recombination takes place between the viral *attP* site and a corresponding region, *attB,* on the *E. coli* chromosome between two genes called *gal* and *bio*. Integrase carries out the site-specific recombination event. A bacterial protein (not shown) also participates. *O* and *b* are additional markers on the phage chromosome.

specific sequences in the viral and bacterial chromosomes, which are brought together by their joint attraction for integrase.

Unlike homologous recombination, which requires several hundred base pairs of sequence homology in order to occur, site-specific recombination requires fewer than two dozen homologous base pairs. In the case of phage λ (see Figure 23.15), sequences in the phage and bacterial genomes, called *attP* and *attB,* respectively, bind to specific sites on the integrase molecule, which catalyzes strand breaking and resealing reactions that linearly insert the viral chromosome into the chromosome of the infected *E. coli* cell. In this process, the sites of DNA recognition, breakage, and joining are brought about by specific DNA–protein interactions rather than by the DNA–DNA interactions via extensive sequence homology that categorize homologous recombination.

Homologous Recombination

Breaking and Joining of Chromosomes

The most straightforward way to accomplish recombination is to break and rejoin DNA molecules. However, if recombination occurs this way, the sites of breakage must be precisely the same on both recombining chromosomes for intact genes to be regenerated. Some researchers favored alternative mechanisms, but in 1961 Matthew Meselson and Jean Weigle showed that recombination in fact does occur via breakage and rejoining of chromosomes. The demonstration involved a Meselson–Stahl type of experiment (see Figure 4.17). *E. coli* was infected with two genetically distinct λ phage populations, one of which had been density labeled by growth in ^{13}C–^{15}N medium. The phage particles resulting from this mixed infection were centrifuged to equilibrium in a cesium chloride gradient. Phages with recombinant genotypes were recovered from all parts of the gradient, whereas nonrecombinant phages were uniformly light or heavy. This result means that recombinant phages contained DNA derived from both parents, which could have occurred only by breaking and rejoining DNA strands.

Recombination is stimulated by processes that nick or break DNA strands, such as thymidine starvation or UV irradiation. This suggested that single-stranded DNA, or free DNA ends, play a role in initiating recombination.

Models for Recombination

In 1964, Robin Holliday proposed a model for homologous recombination between duplex DNA molecules. Detailed in **FIGURE 23.16**, Holliday's model continues to dominate thinking and experiments about this process.

Holliday proposed that recombination begins with nicking at the same site on two paired chromosomes (step **1**). Partial unwinding of the duplexes is followed by **strand invasion,** in which a free single-strand 3′ end from one duplex pairs with its unbroken complementary strand in the other duplex, and vice versa (step **2**). Enzymatic ligation generates a crossed-strand intermediate, called a **Holliday junction** (step **3**). The crossed-strand structure can move in either direction by duplex unwinding and rewinding (branch migration, step **4**). The Holliday junction "resolves" itself into two unbroken duplexes by a process of strand breaking and rejoining. The process leading to recombination begins with isomerization of the Holliday structure (step **5**), followed by strand breakage, so that the strands that break (in step **9**) are those that were *not* broken in step 1. Resolution of the resulting structure (steps **10** and **11**) generates two chromosomes recombinant for DNA flanking the region and each containing a heteroduplex region (each strand from a different parental molecule). However, if the original

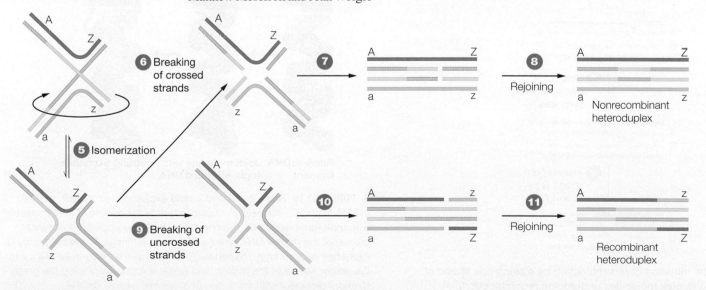

▲ **FIGURE 23.16 The Holliday model for homologous recombination.** A, a, Z, and z are genetic markers.

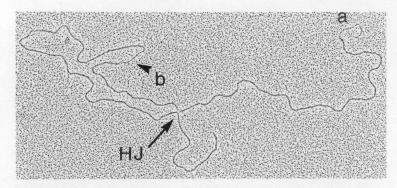

▲ **FIGURE 23.17** Electron microscopic visualization of a Holliday junction. This junction was created by recombination between two plasmid DNA molecules, a and b. Science VU/H. Potter-D. Dressler/Visuals Unlimited, Inc.

crossed strands (those that *were* broken in step 1) break and rejoin (steps ⑥, ⑦ and ⑧), the products are nonrecombinant duplexes, each containing a heteroduplex region (i.e., nonrecombinant with respect to the outside markers A and Z).

Considerable evidence now supports the central tenets of the Holliday model, including electron microscopic visualization of Holliday junctions (**FIGURE 23.17**). However, a major problem with the Holliday model was the requirement that two duplex DNAs undergo single-strand nicking at precisely the same points. How might that problem be resolved? Matthew Meselson and Charles Radding proposed that recombination could start with a single nick. The 3′ end of the nick invades the unbroken duplex and initiates strand displacement synthesis, with the displaced single strand then pairing with its partner in the unbroken strand, as indicated in **FIGURE 23.18**.

● **CONCEPT** Homologous recombination breaks and rejoins chromosomes.

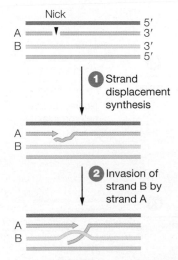

▲ **FIGURE 23.18** Initiation of recombination by nicking one strand of one of the two duplex molecules undergoing recombination.

Proteins Involved in Homologous Recombination

The Holliday and Meselson–Radding models explained most of the existing data on homologous recombination between paired chromosomes, particularly as studied in lower eukaryotes such as yeast. Moreover, the models could easily be adapted to explain daughter-strand gap repair or the recombination that occurs in bacteria after transformation or conjugation. Some of the proteins thought to participate, notably DNA polymerase, DNA ligase, and single-strand DNA-binding protein, had been characterized and shown to participate in recombination.

What other proteins might participate in recombination if the models are largely correct? To answer this question, we turn back to *E. coli* and its phages and the characteristics of bacterial mutants defective in recombination. Mutations conferring a recombination-defective (rec^-) phenotype map in several loci, and two important proteins are responsible for most bacterial recombination events. One of these gene products, the RecA protein, was introduced earlier. The other protein is called exonuclease V, or the RecBCD nuclease.

RecA is a multifunctional protein with molecular weight of about 38,000. As we will see in Chapter 26, it participates in regulating the SOS response. In recombination, it promotes the ATP-dependent pairing of homologous strands, as mentioned earlier in connection with daughter-strand gap repair.

As schematized in **FIGURE 23.19** (left panel), the process begins with a reaction between RecA and single-stranded (ss) DNA, to give a RecA-ssDNA filament. RecA wraps about ssDNA as a multisubunit right-handed helix, with six RecA monomers per turn. Once ssDNA is bound, the filament searches double-stranded (ds) DNA, seeking sequences

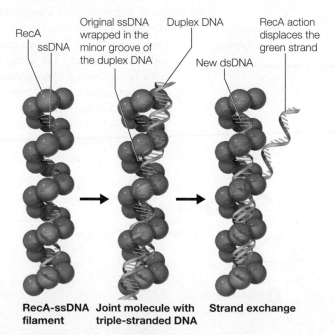

▲ **FIGURE 23.19** RecA-mediated strand exchange. At the left, a RecA-ssDNA filament, with ssDNA shown in red. In the middle, a joint molecule, with triple-stranded DNA; the original ssDNA is wrapped in the minor groove of the duplex DNA (yellow and green strands). At the right, strand exchange is occurring. The red ssDNA is complementary in sequence to the yellow strand of the duplex, and RecA action is displacing the green strand, coincident with formation of a new red-yellow dsDNA.

complementary to those in the single strand already bound. In this process, dsDNA is also taken up within the filament, giving a "joint molecule" (Figure 23.19, center panel). Binding to dsDNA requires ATP, but that ATP need not undergo hydrolysis. By contrast, movement of the ssDNA-protein complex with respect to the dsDNA *does* require ATP hydrolysis. During this process, dsDNA becomes underwound and stretched to about 1.5 times its normal length. The complex moves in a 5′-to-3′ direction along the initially bound ssDNA. During the movement, a triple-stranded structure transiently forms, with ssDNA (the red strand in Figure 23.19) wound in the minor groove of the dsDNA. The ssDNA continually tests the antiparallel strand in dsDNA (yellow) for sequence complementarity. It is not clear how this occurs, but evidently short oligonucleotide sequences swing out from the duplex structure and can pair with ssDNA if sequence complementarity is found.

Once complementarity is established, branch migration occurs, with simultaneous strand exchange. A duplex is formed between the red strand (ssDNA originally) and the yellow strand (complementary to the red strand), while the displaced (green) strand is spooled out, away from the complex. ATP hydrolysis during this period may promote rotation of the DNA within the filament, facilitating the release of the displaced strand.

> ● **CONCEPT** RecA, a multifunctional bacterial enzyme, uses ATP to promote the pairing of homologous DNA sequences.

Recombination occurs preferentially at or near particular DNA sequences. In *E. coli*, recombination is favored near a particular octanucleotide sequence, 5′-GCTGGTCC, called Chi (for <u>c</u>rossover <u>h</u>otspot <u>i</u>nstigator). How does this site act to stimulate recombination? The RecBCD protein, a multifunctional heterotrimeric enzyme encoded by the *recB, recC,* and *recD* genes, displays sequence specificity for Chi. This enzyme binds at a double-strand break on duplex DNA and uses two helicase activities—RecB and RecD—to unwind and partially degrade the DNA.

As shown from the crystal structure of a RecBCD-DNA complex, both helicases are in contact with DNA (**FIGURE 23.20**). The RecD helicase activity is higher than that of RecB, so as the protein moves, the 3′ end is displaced as a single-stranded loop ahead of RecB and becomes coated with SSB protein. Both strands are degraded by associated nucleases, but because of the differential speeds of the protein motors, more of the 3′ end is saved as the loop. When the enzyme reaches Chi, the protein pauses briefly, and a sequence-specific interaction causes RecBCD to switch speeds and change its preferred polarity of DNA degradation. As RecB moves faster, the 3′-terminated loop is reeled in by RecB and coated with RecA. An associated nuclease releases the bound 3′ end (not shown), freeing it for strand invasion of a neighboring duplex.

Once a Holliday junction is formed, branch migration is essential for eventual formation of recombinant structures, as was shown in Figure 23.16. This is largely the responsibility of three other proteins. In *E. coli*, these three proteins are products of the *ruvA, ruvB,* and *ruvC* genes. RuvA is a DNA-binding protein, whose specificity directs it toward the four-stranded Holliday structure. RuvB protein is an ATP-requiring motor protein, which binds to two opposed arms of the junction. In the model shown in **FIGURE 23.21**, which is based on crystal structures of the isolated

> ● **CONCEPT** RecBCD, a multifunctional enzyme, unwinds and rewinds DNA, with one strand being unwound more rapidly and converted to a single-stranded 3′ end.

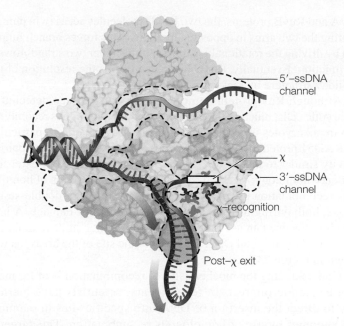

▲ **FIGURE 23.20** A model of the *E. coli* RecBCD protein in complex with DNA, based on the crystal structure. PDB ID: 1w36. Each DNA strand is in a separate channel after rewinding, with the 3′ end coming in contact with a Chi recognition domain. DNA after Chi recognition exits through a separate channel (purple shading). RecB, pink; RecC, blue; RecD, green. Amino acid side chains involved in Chi recognition are in red and green. Based on L. Yang et al (2012) *Proc. Natl. Acad. Sci.* USA 109: 8907–8912.

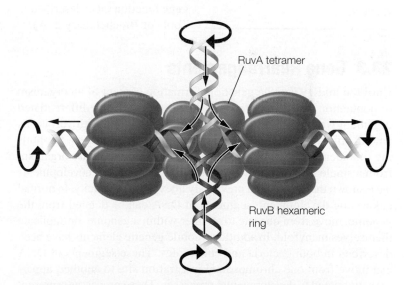

▲ **FIGURE 23.21** A model for the RuvA-RuvB-Holliday junction structure. This is based on crystal structures of RuvA and RuvB. (PDB ID: 3uwx). Branch migration is believed to involve spooling of DNA to the left and right through the RuvB twin pumps, with the upper and lower arms being drawn into the center and eventually out through the pumps. *Science* 18 October 1996: Vol. 274 no. 5286 pp. 415-421 DOI: 10.1126/science .274.5286.415. John B. Rafferty et al. Courtesy of Peter Artymiuk, Krebs Institute, Sheffield.

RuvA and RuvB proteins, the two RuvB molecules act as twin pumps, rotating the two arms in opposite directions. This forces branch migration by driving the rotational movement of the other two strands toward the junction. Eventually, RuvC binds and begins the resolution of the Holliday structure by nicking two strands.

Although Ruv protein homologs have not yet been detected in eukaryotic cells, much of the biochemistry of homologous recombination in eukaryotes is similar to what is described here. In particular, the RAD51 protein of both human cells and yeast has a strand-pairing activity similar to that of RecA, and the two proteins show extensive sequence homology. As noted earlier, an essential function of homologous recombination in eukaryotic cells is the repair of double-strand breaks, both during meiosis and as a result of DNA damage. A broken chromosome can use the sequence information in its homolog to reconstruct the original DNA sequence at the site of the break, as was shown in Figure 23.14.

Understanding the biochemistry of recombination is of far more than academic interest. In recent years, scientists have learned how to direct the insertion of DNA into specific sites in mammalian genomes by use of homologous recombination. This targeted insertion allows the creation of a "knockout mouse," in which any desired gene can be inactivated to allow investigation of the biological function of the deleted gene (see Tools of Biochemistry 23A). The recent development of CRISPR-Cas9 technology has made it even simpler to obliterate any desired gene function (also described in Tools of Biochemistry 23A).

● **CONNECTION** The ability to knock out the expression of any gene in a mouse by targeted homologous recombination has led to important insights into human diseases including cancer, obesity, heart disease, diabetes, arthritis, aging, and Parkinson's disease.

23.3 Gene Rearrangements

Until the mid-1970s, the genetic information content of an organism or population was considered to be static. All cells of a differentiated organism were thought to have identical genomes, with variations among different cells arising at the level of gene expression. Supporting this idea was the possibility of cloning a multicellular organism from a single differentiated cell. However, more recent developments have shown that DNA has a previously unexpected plasticity. In normal eukaryotic development, segments of DNA can be deleted from the genome, move from one site to another within a genome, or duplicate themselves manyfold. In addition, mobile genetic elements have been described in both bacteria and eukaryotes. These segments of DNA can move from one chromosomal integration site to another, apparently unrelated to developmental processes. These processes represent a specialized form of recombination.

Actually, the plasticity of DNA was predicted, but by few scientists. Barbara McClintock's work on maize genetics, starting in the 1940s, led her to postulate genetic regulatory mechanisms effected through the action of mobile genetic elements. However, three decades passed before the physical demonstration of such elements in bacteria focused attention on McClintock's pioneering work. In the remainder of this chapter, we discuss three important aspects of genome plasticity: the

genetic basis for antibody variability in vertebrates, gene transposition, and gene amplification.

Immunoglobulin Synthesis: Generating Antibody Diversity

Recall from Chapter 7 that antibodies are proteins manufactured by vertebrate immune systems that aid in defense against infectious agents and other substances foreign to the animal. The immune response, resulting from the introduction of an antigen, elicits the production of several highly specific antibodies. It is estimated that a human is capable of synthesizing more than 10 million distinct antibodies. Most of this great diversity is generated through the action of precisely controlled gene rearrangements, involving but a small fraction of the coding capacity of the genome. These rearrangements occur during differentiation of many individual clones of cells, each clone specialized for the synthesis of one and only one antibody. Other large protein families, notably, T-cell receptors (also involved in immune responses), are diversified by similar mechanisms. The immune response involves the proliferation of clones of cells that produce antibodies specialized to bind the specific antigen, or immunogen, provoking that response. This clonal expansion allows large-scale production of the specific antibodies needed to combat infection or another challenge.

To see how immunological diversity is generated, let us consider one type of antibody, the immunoglobulin G (IgG) class. Recall from Chapter 7 that these proteins consist of two heavy chains and two light chains. Each chain comprises two distinct segments—a domain of variable polypeptide sequences and a constant domain, which is invariant among different IgG light or heavy chains. We focus on the light chains and in particular the κ class of light chains. (Another class, λ, has somewhat different sequences in its constant region, but its development involves similar mechanisms.) Much of what we know about this process comes from the laboratories of Susumu Tonegawa, Lee Hood, and Philip Leder.

FIGURE 23.22 shows the organization of the precursor genes to κ chains in germ-line cells, undifferentiated for antibody formation, and the rearrangements leading to one such gene in a differentiated antibody-producing cell. The light chains are encoded by DNA sequences that are noncontiguous in the genome of undifferentiated cells but are all in the same chromosome. These sequences are called V (variable), C (constant), and J (joining). The human genome contains about 300 different V sequences, each of which encodes the first 95 amino acids of the variable region; four different J sequences, each of which encodes the last 12 residues of the variable region and which joins it to the constant region; and one C sequence, which encodes the constant region. The V sequences in an embryonic cell, each preceded by a leader sequence containing a transcriptional activator that is not expressed, form a tight cluster; the J sequences form another cluster some distance away; and the C sequence follows shortly after the J cluster. Each J sequence is flanked by nonexpressed spacer sequences.

In the differentiation of one antibody-forming clone of cells, a recombinational event links one of the approximately 300 V sequences with one of the four J sequences. All of the DNA lying between these two chosen sequences is spliced out and deleted in this rearrangement, disappearing from all progeny of this cell line. Any upstream V sequences (on the 5' side, to the left in Figure 23.22) and downstream

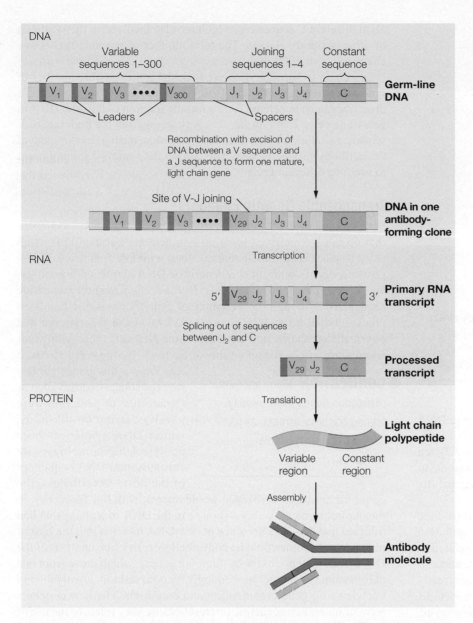

▲ **FIGURE 23.22** Gene rearrangements in antibody gene maturation. The rearrangements of C, V, and J sequences produce one mature κ light chain gene, and the transcription, processing, and translation of this gene produce an antibody κ light chain protein.

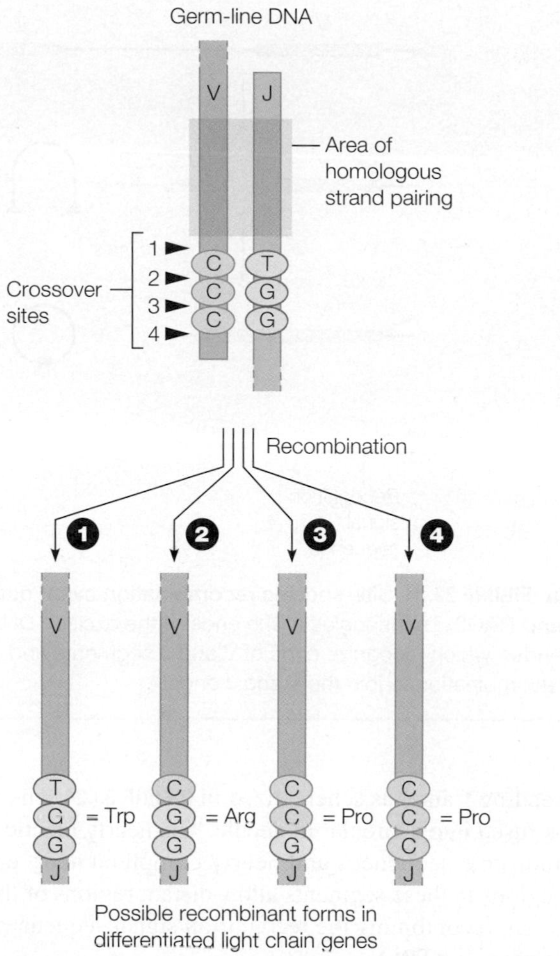

▲ **FIGURE 23.23 Generation of diversity by variability in V-J joining.** Four crossover events are possible at a V-J junction, giving rise to codons for any three possible amino acids in this example. Only one DNA strand is shown.

J sequences (on the 3′ side, to the right) remain in these cells but are not used in antibody synthesis.

Additional diversity is provided by the way in which the V and J sequences recombine. The cutting and splicing can occur within the terminal trinucleotide sequences of V and J in any way that yields one trinucleotide sequence in the spliced product (**FIGURE 23.23**). This increases the total number of different light chain sequences by about 2.5 (the average number of different amino acids encoded by four random triplets). Thus, the total number of possible light chain sequences that can be formed from 300 V sequences and four J sequences is about 3000 (300 × 4 × 2.5).

Related DNA sequences are found to the 3′ side of each V sequence and to the 5′ side of each J sequence, and they represent recognition

sites for the enzymes involved in the joining reaction. Those sequences, which are called recognition signal sequences, are as follows:

$$5′......V....CACAGTG....12\ bases....ACAAAAAC.....3′$$
$$3′......J....GTGTCAC.....23\ bases....TGTTTTTG......5′$$

Note that the complementary regions of these sequences are inverted repeats.

Recombination begins with two proteins called RAG1 and RAG2. These proteins begin the process by catalyzing double-strand breaks between two recognition signal sequences and the respective V and J coding sequences. DNA repair proteins process the double-strand DNA breaks, with the V and J sequences fused in the appropriate

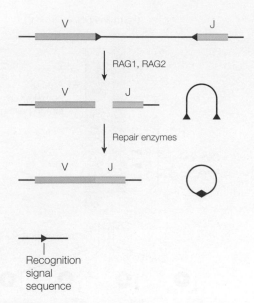

▲ **FIGURE 23.24 Site-specific recombination event catalyzed by RAG1 and RAG2.** The triangles at the ends of the excised DNA are "signal ends," which recognize ends of V and J segments and participate in recombination to join the V and J ends.

reading frame, as schematized in **FIGURE 23.24**. The intervening DNA is fused into a circular molecule. The nearly identical seven-base palindromic sequences and nearly complementary eight-base AT-rich regions in these segments allow distant regions of the chromosome to align, recombining the recognition signal sequences and excising the intervening DNA.

Somatic hypermutation is another mechanism to introduce sequence diversity into antibody chains. The spontaneous mutation rate becomes very high, but specifically within sequences related to antibody formation. The cytosine residues in these sequences undergo a high rate of deamination, leaving uracil residues paired incorrectly with guanine. Upon replication, these mutant strands pair with adenine, completing the transition from G-C base pairs to A-T. The enzyme involved, **activation-induced deoxycytidine deaminase** (AID), evidently acts at paused transcription complexes, with the nontemplate DNA strand serving as the deamination substrate. Why this process occurs only within antibody-forming genes is unclear.

The final step in producing a light chain polypeptide is the joining of the C and J segments (see Figure 23.22). This occurs not at the DNA level but at the level of mRNA. As discussed in Chapters 7 and 24, eukaryotic gene expression usually involves cutting and splicing of an mRNA precursor, thus excising sequences that are not represented in the final gene product. In this case, transcription yields an RNA molecule extending from the 5′ side of the V gene that is spliced to J to the 3′ side of C. Depending on which J region has been spliced to V in this cell, the RNA excised during splicing may contain sequences corresponding to other J regions.

● **CONCEPT** The diversity of the immune response involves recombination among thousands of different DNA sequences, thus yielding a vast array of antibodies.

Heavy chains are formed similarly—from V sequences, J sequences, and a class of sequences called D. In addition, there are eight different C sequences, which are also involved in the synthesis of other antibody classes. The total number of possible IgG heavy chains is about 5000. Because any light chain can combine with any heavy chain to form a complete IgG, the total possible number of IgG molecules is 3000 × 5000, or 1.5×10^7. In this way, enormous diversity can be generated from a minute fraction of the total DNA in germ-line cells. Even further diversity arises from the high rate of V sequence alteration by somatic hypermutation, during development of the antibody-producing cell. This allows the same V-J joining event to produce different IgGs.

Transposable Genetic Elements

In this section, we discuss transposable genetic elements—sequences that do not have a fixed location in a genome but can move from place to place within the genome, albeit with low frequency. Transposition occurs without the benefit of DNA sequence homology, but the enzymes catalyzing transposition recognize short nucleotide sequences. Although the existence of gene transposition had been predicted by Barbara McClintock's work on maize genetics, the first physical characterization of transposable elements arose from studies of antibiotic-resistant strains of bacteria. By the early 1970s, it

● **CONNECTION** Many antibiotic resistance genes move readily among bacterial strains, as passengers on transposable genetic elements.

was known that genes conferring resistance to drugs such as tetracycline or penicillin were usually carried on plasmids, whose DNA sequences bore no detectable homology with chromosomal DNA sequences of the host. Nevertheless, the

genes for antibiotic resistance would appear, with low frequency, in the chromosome of the bacterium or in the DNA of a phage that had infected that cell. The presence of new DNA inserted into the host or phage chromosome could be confirmed by restriction analysis of the DNA. The existence of these "jumping genes," which move from one chromosome to another in seemingly random fashion, greatly altered our views on gene organization and evolution. The new concepts were of more than academic interest because they relate to the use of antibiotics to treat bacterial infections—specifically, to the ease with which populations of antibiotic-resistant bacteria can arise.

Transposable elements have been demonstrated in many eukaryotes, including maize, *Drosophila,* and yeast. Here we concentrate on bacteria, whose physical structures and transposition mechanisms are best understood. We first point out several distinctions between bacterial transposition and other recombinational processes. First, transposition does not require extensive DNA sequence homology. Second, transposition occurs normally in a *recA*⁻ host, confirming that extensive homologous DNA strand pairing is not involved. Third, DNA synthesis is involved in bacterial transposition. Transposition always involves duplication of the target site, the short sequence (3–12 base pairs) at which the transposable element is inserted. Sometimes the transposable element is itself replicated, with one copy being deposited in the new sequence and one remaining in the donor sequence. Finally, transposable elements can restructure a host chromosome.

A bacterial transposable element, or **transposon,** is a length of DNA flanked by short repeated sequences. Depending on whether these repeats are oriented in the same direction (direct repeats) or in opposite directions (inverted repeats), recombination between them

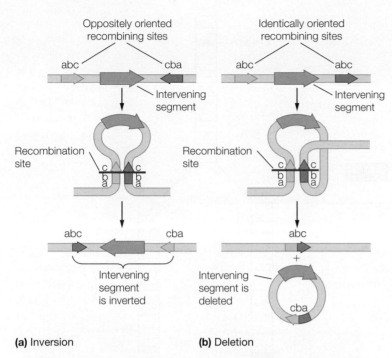

(a) Inversion

(b) Deletion

▲ **FIGURE 23.25 Genome rearrangements that can be promoted by homologous recombination between two copies of the same transposable element.** Depending on the orientation of the two copies, either (a) inversion or (b) deletion can result.

can yield either a deletion or an inversion, as shown in **FIGURE 23.25**. In the laboratory, insertional inactivation of genes is useful for isolating mutants defective in specific functions.

We recognize two types of transposon—simple and complex. In a simple transposon, or **insertion sequence,** the DNA between the two flanking repeats encodes a single protein, called **transposase,** the enzyme catalyzing the reactions of transposition. A complex transposon encodes a transposase plus additional proteins, which may include drug resistance elements. Transposase initiates its action by recognizing a short sequence in target DNA and making a staggered cut at each end of the sequence (**FIGURE 23.26**, step ❶). The transposon is then inserted, and the resultant gaps are filled by DNA polymerase and DNA ligase in order to generate the flanking repeat sequences, step ❷.

In some cases, both the transposon and the target sequences are replicated. A transient intermediate in this *replicative* transposition contains a tandem repeat of the entire transposon. Another enzyme, **resolvase,** catalyzes a site-specific recombinational event, which results in single copies of the transposon, one in its original site and the other in the site to which the sequence was transposed.

Retroviruses

Gene transposition in eukaryotic systems presents some similarities to and some differences from transposition in bacteria. The first major distinction is that integration and excision are distinct processes in eukaryotes. Thus, the

transposable element can be isolated in free form, often as a double-stranded circular DNA. Second, replication of that DNA often involves the synthesis of an RNA intermediate. Both of these properties are seen in the retroviruses of vertebrates, perhaps the most widely studied class of eukaryotic transposable elements. As we noted in Chapter 22, these RNA viruses use reverse transcriptase to synthesize a circular duplex DNA, which can integrate into many sites of the host-cell chromosome. The integrated retroviral genome bears remarkable resemblance to a bacterial complex transposon. The prototypical retroviral genome has three structural genes—*gag,* which encodes a polyprotein that undergoes cleavage to give virion core proteins; *pol,* which encodes the viral polymerase, or reverse transcriptase; and *env,* the major glycoprotein of the viral envelope. Flanking these structural genes are two direct repeats, the **long terminal repeats (LTRs)** of about 250 to 1400 base pairs each. Each LTR is flanked in turn by short inverted repeat sequences, 5 to 13 base pairs in length. Integration occurs by a mechanism that duplicates the target site, so that the integrated viral gene, called a **provirus,** is flanked by direct repeats of 5 to 13 base pairs each of host-cell DNA.

● **CONCEPT** Retroviral genomes and eukaryotic transposable elements have sequence similarities to each other and to bacterial transposons.

Just as bacterial transposons can carry passenger genes, so also can retroviruses. We discussed in Chapter 20 the fact that tumor viruses carry a gene, in altered form, that in its original form in a host cell was involved in signal transduction; the protein tyrosine kinase of Rous sarcoma virus was the earliest recognized oncogene product. If the genetic rearrangements that created the virus substituted the proto-oncogene for an essential gene of the virus, as schematized in **FIGURE 23.27c**, then the virus is defective and can multiply only if the missing function is provided, for example, by coinfection with a *helper virus,* a related retrovirus that provides the missing function(s).

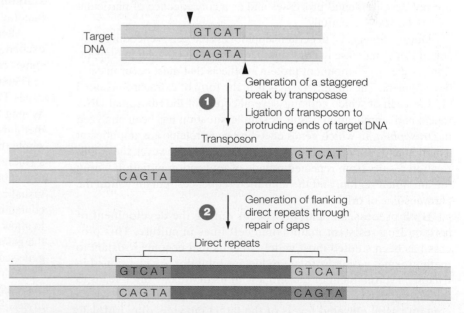

▲ **FIGURE 23.26 Action of transposase to generate direct repeats during the insertion of a transposon or an insertion sequence.**

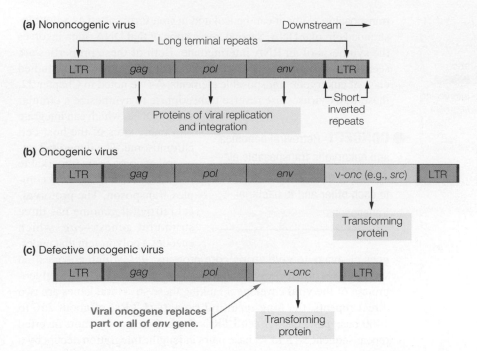

(a) Nononcogenic virus

Downstream →

Long terminal repeats

Proteins of viral replication and integration

Short inverted repeats

(b) Oncogenic virus

Transforming protein

(c) Defective oncogenic virus

Viral oncogene replaces part or all of *env* gene.

Transforming protein

▲ **FIGURE 23.27 Structure of retroviral genomes in the integrated state. (a)** A nononcogenic virus. **(b)** An oncogenic virus such as Rous sarcoma virus, showing the viral oncogene downstream (rightward) from the viral replication genes. **(c)** A defective oncogenic virus with the viral oncogene replacing part or all of a gene (*env*) essential to viral replication. In each case the LTRs are direct repeats, flanked by short inverted repeats.

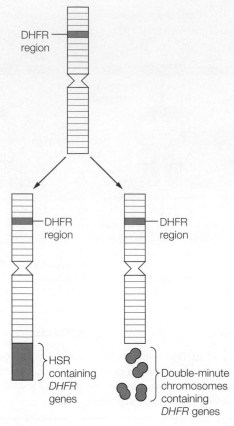

▲ **FIGURE 23.28** Two modes of gene amplification leading to drug resistance.

Gene Amplification

The final process we discuss is the selective amplification of specific regions of the genome, principally in eukaryotic cells. This occurs in normal developmental processes and as a consequence of particular metabolic stress situations.

During oögenesis, in certain amphibians the genes encoding ribosomal RNAs increase in copy number by some 2000-fold, in preparation for the large amount of protein synthesis that must occur in early development. The amplified DNA is in the form of extrachromosomal circles, each of which contains several copies of the ribosomal DNA repeat and a replication origin. A similar situation has been analyzed in *Drosophila,* in which genes encoding egg proteins are amplified at a particular developmental stage. In the latter case, however, the amplification results from repeated rounds of replication initiation within the amplified region, and the amplified sequences remain within the chromosome of origin.

Both processes apparently occur during the development of certain drug-resistant mammalian cell lines in culture. This process has been studied most widely in cells that become resistant to methotrexate, a dihydrofolate reductase inhibitor. As mentioned in Chapter 19, treatment of leukemia with methotrexate often leads to the emergence of drug-resistant leukemic cell populations, which contain vastly elevated levels of the target enzyme, dihydrofolate reductase (DHFR). As shown originally by Robert Schimke, overproduction of the enzyme usually results from specific amplification of a large DNA segment that includes the *DHFR* gene. In one process, tandem duplication of the DNA segment generates a giant chromosome with multiple gene copies, in what is called a **homogeneously staining region** (HSR), because it lacks the typical chromosome banding pattern.

Alternatively, a DNA segment containing the *DHFR* gene can be excised, apparently by a recombinational process, to form minichromosomes called **double-minute chromosomes,** as shown schematically in **FIGURE 23.28**. Some resistant cells contain both types of amplified genes. Double-minute chromosomes are maintained within a cell only as long as selective pressure is maintained by growth of the cell in methotrexate. However, the chromosomally amplified phenotype is stable through many generations of cell growth. **FIGURE 23.29** shows a fluorescence micrograph of metaphase chromosomes from a stably amplified Chinese hamster ovary cell line. DHFR sequences were visualized by in situ hybridization with a fluorescent-tagged DNA containing DHFR sequences. This technique is sufficiently sensitive to allow detection of single-copy sequences (white arrows). Note also the giant chromosomes containing many gene-equivalents of *DHFR* gene sequences.

Amplification of genes under selective conditions has been widely observed—for example, in development of pesticide-resistant forms of insects. The mechanism of amplification is not yet clear. However, evidence such as that of Figure 23.29 shows that the amplified sequences are on the same chromosome as the original single-copy gene site, but some

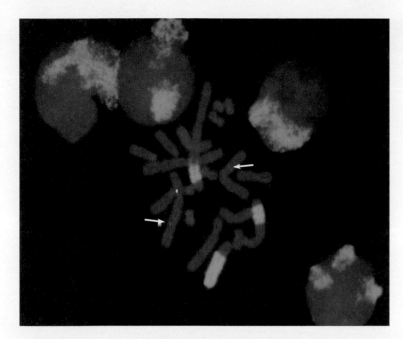

▲ **FIGURE 23.29 Structural changes to the chromosome that accompany dihydrofolate reductase gene amplification.** The micrograph shows metaphase chromosomes from Chinese hamster ovary cells that are highly resistant to methotrexate. Chromosomal DNA was subjected to hybridization in situ with a fluorescence-labeled *DHFR* gene probe. White arrows point to single-copy genes. The amplified chromosomal sequences are on a giant form of the chromosome that also contains one of the original single-copy sequences.

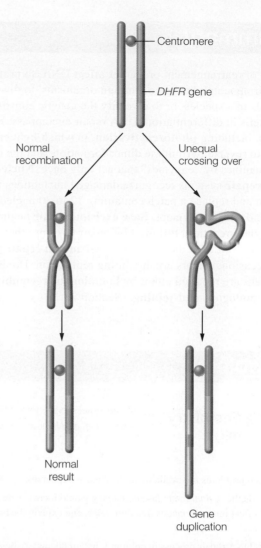

▲ **FIGURE 23.30 Unequal crossing over, as a mechanism to explain early steps in gene amplification.**

● **CONCEPT** Gene amplification generates multiple copies of DNA sequences at a separate site on the same chromosome. Recombination among homologous segments yields extrachromosomal amplified sequences.

distance away from that site. Such structures could arise either through recombination with unequal sister-chromatid exchange, schematized in **FIGURE 23.30**, or through a conservative transposition. Later, homologous recombination within an amplified region can lead to excision of sequences containing one or more amplified sequences. Such elements probably represent the double-minute chromosomes.

The presence of selective pressure, such as the continuous presence of methotrexate, promotes specifically the survival of cells that can respond to that pressure, for instance, by overproducing DHFR. Once two or more copies of the gene are present on a chromosome, additional copies can be generated by further recombinational events or by abnormalities of replication. Resistance is thus developed in stepwise fashion and occurs over many generations of growth. These findings have practical significance because cancer chemotherapy often involves long-term treatment with low doses of an antimetabolite—precisely the conditions most likely to nullify the effect of treatment

by generating drug-resistant cells. Findings on gene amplification not only have changed the way in which anticancer drugs are administered

● **CONNECTION** Understanding mechanisms of gene amplification has led to significant changes in the ways anticancer drugs are administered.

but also have been extended to the tumorigenic process itself. Investigators have found that specific oncogenes become amplified during the clinical progression of certain human tumors. Thus, gene amplification is seen as a mechanism in normal development, in cellular adaptation to stress, and in abnormal developmental processes. Also, the gene duplications that have occurred during evolution (see Chapter 7) probably have taken place by similar pathways.

Summary

• A variety of rearrangement processes affect DNA, to protect it from environmental damage or foreign organisms, to diversify the individuals in a species, or to diversify the genetic constitution of somatic cells in differentiation. DNA repair encompasses several processes, including **photoreactivation**, in which light energy is captured to reverse pyrimidine dimer formation, and the removal of alkylguanines by "enzymes" that act only once. **Nucleotide excision repair** systems recognize damaged nucleotides and cleave out and replace a patch containing 12 to 30 nucleotides that flanks the damaged segment. **Base excision repair** begins with an N-glycosylase reaction, followed by repair enzymes that excise one or more nucleotides at the abasic site. **Mismatch repair** systems correct occasional errors arising during replication. Double-strand DNA breaks are repaired either by **homologous recombination** or by **nonhomologous end-joining**. (Section 23.1)

• Genetic recombination between homologous DNA segments involves helicase-catalyzed duplex unwinding, site-specific DNA cutting, duplex-strand invasion by single-stranded 3′ hydroxyl termini, chain extension, branch migration, and resolution of Holliday structures. **Strand invasion** catalyzed by RecA in bacteria and RAD51 in eukaryotes is integral to this process. Some recombinational events, such as integration of bacteriophage λ, are site-specific and governed by DNA–protein interactions. (Section 23.2)

• A range of gene rearrangements includes the joining reactions in differentiation of the immune response, gene transposition, retroviral genome integration, and gene amplification. Gene amplification probably occurs by a recombinational mechanism and can either be a normal developmental process or occur in response to a specific environmental stress. (Section 23.3)

Problems

Enhanced by
Mastering Chemistry
for Biochemistry

Mastering chemistry for Biochemistry provides select end-of-chapter problems and feedback-enriched tutorial problems, animations, and interactive figures to deepen your understanding of complex topics while practicing problem solving.

Answers to red problems are available in the Answer Appendix.

1. Predict whether a *dam* methylase deficiency would increase, decrease, or have no effect on spontaneous mutation rates, and explain the basis for your prediction.

2. For each DNA repair process in column I, list *all* characteristics from column II that correctly describe that process.

I
(a) Nucleotide excision repair
(b) Photoreactivation
(c) Base excision repair
(d) Recombinational repair
(e) SOS-driven error-prone repair
(f) Alkyltransferase repair
(g) Mismatch repair
(h) Double-strand break repair

II
1. RecA protein participates.
2. Damaged nucleotides are removed by nick translation.
3. A free radical mechanism is involved.
4. The repair enzyme functions only once.
5. The key enzyme contains a bound folate cofactor.
6. No bases or nucleotides are removed from the DNA.
7. Deficiency of this enzyme in humans greatly increases the risk of skin cancer.
8. This system is chiefly responsible for the mutagenic effect of ultraviolet light.
9. This process begins with cleavage of two phosphodiester bonds.
10. This process begins up to 1 kbp away from the site to be repaired.
11. DNA ligase catalyzes the final reaction.
12. This process also occurs in meiotic recombination.
13. Replication fork regression might occur during this process.

3. For each of the following characteristics, list all of the bases to which they apply.

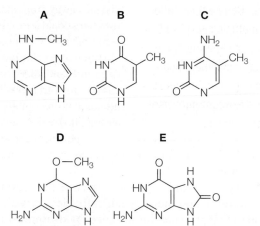

(a) A signal that identifies a parental DNA strand in the MutH,L,S mismatch correction system
(b) Most likely to be involved in cyclobutane dimer formation after ultraviolet irradiation of DNA
(c) A methylated base found immediately to the 5′ side of dGMP residues in eukaryotic DNA
(d) Created by treating DNA with alkylating agents that transfer methyl groups and repaired by an "enzyme" that functions only once in its lifetime
(e) Created by AdoMet-dependent methylation of a nucleotide residue in DNA

(f) A substrate for deamination at the DNA level, which would lead to a GC → AT transition

(g) A mutagenic base that can arise in DNA through ROS action

4. Homologous recombination in *E. coli* forms heteroduplex regions of DNA containing mismatched bases. Why are these mismatches not eliminated by the mismatch repair system?

5. Deficiencies in the activity of either dUTPase or DNA ligase stimulate recombination. Why?

6. Suppose that you want to study retroviral integration mechanisms by determining the nucleotide sequence at the integration site—several dozen nucleotides on each side of the viral–cellular DNA junction. Describe how to isolate DNA containing a junction site in amounts sufficient for sequence analysis.

7. Analysis of p53 gene mutations in human tumors shows that a large propor tion of these mutations involve GC → AT transitions originating at sites of DNA methylation. Propose a model to explain preferential mutagenesis of this type at these sites.

8. Recent data indicate, as mentioned in the text, that the signal for eukaryotic mismatch repair, for identifying the strand to be repaired, is incorporated ribonucleotides that have not yet been replaced by deoxyribonucleotides. Previous studies implicated nicks in the most recently synthesized DNA as the signal identifying the strand to be repaired. If DNA nicks comprise the signal, would you expect the nicked strand or the intact strand to be the strand to be repaired? Briefly explain your answer.

9. Identify and briefly describe three of the processes by which deamination of DNA cytosine residues by AID could lead to mutagenesis.

10. In what ways can insertion of a transposon affect the expression of genes in the neighborhood of the insertion site?

11. Briefly explain how integration of a retroviral genome could activate transcription of genes adjacent to the integration site.

12. Outline a pathway for mutagenesis, comparable to that shown on page 722, resulting from a *mutM* gene mutation.

13. Write a balanced equation for the hydrolytic deamination of a DNA-5-methylcytosine residue.

14. There is evidence that some oxidative damage to DNA occurs at the nucleotide level, with oxidation of a nucleotide, followed by incorporation of the damaged nucleotide into DNA.

(a) Describe a pathway by which this could occur.

(b) Propose one or more experiments to test whether your proposed pathway does occur.

15. Consider the Meselson–Weigle experiment (page 727), which established that recombination occurs via the breaking and joining of DNA strands. For the experiment outlined in the text, construct a graph showing the distribution through a CsCl gradient of parental genotypes and recombinant genotypes. Your graph should show density on the horizontal axis and phage titer (infectious particles/ml) on the vertical axis.

16. A mammalian cell line was cultured for many generations in the presence of methotrexate, whose concentration in the culture medium was steadily increased. After maximum resistance to methotrexate had been achieved, the cells were transferred to drug-free medium for many more generations. At the beginning of the experiment, dihydrofolate reductase represented 0.1% of the soluble protein in the cell; at maximum resistance it was 10.0%; and at the conclusion of the experiment, 1.5%. Describe in qualitative terms the processes involved in generating maximum resistance.

17. Consider the two processes involved in repairing double-strand DNA breaks. Which process—homologous recombination or NHEJ—is more likely to restore the original gene sequences and why?

18. Of the four DNA bases, why is guanine most likely to be modified after exposure to polycyclic hydrocarbons such as benzo(*a*)pyrene, as shown in Figure 23.8? (*Hint:* Consider the pK_a values of the four DNA bases.)

References

For a list of references related to this chapter, see Appendix II.

The value of mutant organisms as probes for metabolic processes has been evident since the early 1940s, when Beadle and Tatum used X-ray-induced mutants of *Neurospora crassa* to define pathways of amino acid biosynthesis (see Chapter 11). The power of selection techniques (drug resistance, auxotrophy), coupled with the ability to work with huge numbers of microbial cells, allowed investigators to isolate mutants affecting the expression of almost any gene. An example is the analysis of *E. coli* mutations affecting lactose metabolism, which led to the operon model of gene regulation (Chapter 26).

The introduction of recombinant DNA technology, coupled with increased understanding of homologous recombination processes, allowed investigators to target mutations to defined regions of the genome. A vector could be designed with sequences homologous to the chromosomal site in which exogenous DNA was to be inserted. Depending on the design of the vector, recombination between vector sequences and homologous chromosomal sequences could lead either to insertion of new sequences into the genome or excision of chromosomal sequences adjacent to the targeting sequence.

Extension of these approaches to eukaryotes was challenging because of the much longer generation times, the large numbers of organisms needed, and the diploid nature of the organism, which required the same event to occur in two homologous chromosomes. Nevertheless, Mario Capecchi and Oliver Smithies achieved success using mouse embryonic stem cells and powerful selection techniques. One approach, illustrated in **FIGURE 23A.1**, uses a neomycin resistance gene (*neo*^{r}) and the gene for hypoxanthine guanine phosphoribosyl-transferase (*hprt*, see Chapter 19) as selectable markers for the targeted deletion of gene 8. The targeting vector contains *neo*^{R} substituted for gene 8, which encodes HPRT. The targeting vector is microinjected into embryonic stem cells. Recombinant cells are selected by their resistance to G418, a relative of neomycin, and by the absence of HPRT

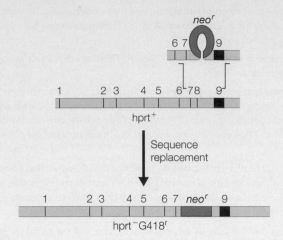

▲ **FIGURE 23A.1 Targeted insertion of the *neo*^{R} gene into the chromosomal *hprt* gene.** Courtesy of M. R. Capecchi (2007) Nobel Lecture, © The Nobel Foundation.

activity, which is monitored by resistance to 6-thioguanine. Those rare cells that are resistant to both G418 and 6-thioguanine are injected into an early mouse embryo, which is surgically implanted into a foster mother. The offspring are mosaics because they contain cells derived both from the manipulated stem cells and from the recipient embryo. Breeding these chimeras with wild-type mice leads to homozygous transgenic animals, showing that the transgene was transmitted through the germ line.

Although the procedure sounds complicated, approaches such as that described here have been used to "knock out" several thousand individual genes of the mouse, with subsequent analysis of the functions of the deleted genes. However, since 2013 a far simpler approach has been introduced, and that approach, called CRISPR-Cas9, is creating a revolution in the biological sciences. The success of this approach provides dramatic evidence of the rewards of serendipity in scientific research. Only a brief description of the phenomenon is possible here.

Many eubacteria and most archaea were found to contain within the genome a region containing clustered repetitive interspersed repeat sequences—a short sequence repeated several times, with each pair of repeats separated by a stretch of nonhomologous DNA (hence, the acronym CRISPR). This represents a bacterial defense mechanism, comparable to restriction-modification systems. When a bacterial culture is attacked, for example by a bacteriophage, the surviving cells place a distinctive part of the attacker's genome between two of the clustered repeat sequences. If the surviving bacteria or their descendants are reinfected by the same phage or other agent, the stored sequence in the CRISPR array serves as a reminder of the previous attack, triggering a chain of events that inactivates the attacker by introducing a double-strand break in its genome. FIGURE 23A.2 represents the process as it occurs in *Streptococcus thermophilus*. Although the process appears complex, the result is straightforward; the activated Cas9 protein catalyzes a site-specific double-strand DNA cleavage. By designing an appropriate crRNA sequence, the investigator can engineer Cas to catalyze a DNA double-strand break at any desired site in the genome of a target organism. The severed ends can be rejoined to the original chromosome or to any DNA molecule introduced into the cell by transformation or microinjection. The attachment can occur by either homologous recombination or nonhomologous end joining. So CRISPR technology can be used for gene knockout, as described, or for site-directed mutagenesis, or for insertion of new DNA into a chromosome.

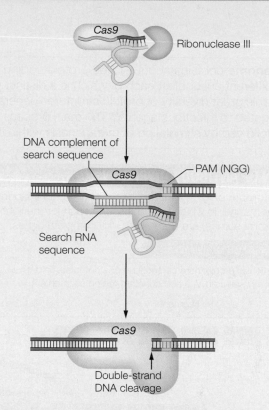

▲ **FIGURE 23A.2 CRISPR-Cas9 system from Streptococcus thermophilus.** Cas9 and associated proteins transcribe a genomic sequence in the CRISPR array that was deposited in a previous infection (not shown). This RNA pairs with a homologous DNA strand near PAM site (Promoter adjacent motif). This triggers Cas9 to catalyze a double-strand DNA cleavage, thereby inactivating the invader. Based on E. S. Lander (2016) *Cell* 164:18–28.

The human genome contains about 21,000 protein-coding genes. Yet the human immune system is capable of specifying the structures of about 10^{11} different antibodies, each of which is a distinct protein molecule composed of two light polypeptide chains and two heavy chains. How is this great diversity of protein structures generated from a limited genetic resource, so that highly specific antibodies can be formed and targeted to infectious agents? The main process involves random reassortment of diverse chromosomal DNA segments as shown below, followed by stimulation of mutagenesis within the spliced antibody genes.

1 GENERATION OF ANTIBODY DIVERSITY Gene segments are assembled into complete genes for each light and heavy chain.

The variable domain of each light and heavy chain of an antibody molecule contributes to antigen binding. The variability of these protein domains originates in the manner in which the genes, one each for the light and heavy chains, are assembled. Gene segments, from a very large selection available on chromosomes, are assembled into complete genes.

LIGHT-CHAIN LOCUS: V–J recombination

Within any particular β cell, there can only be one complete light chain gene, formed by splicing together randomly selected V, J and constant region segments.

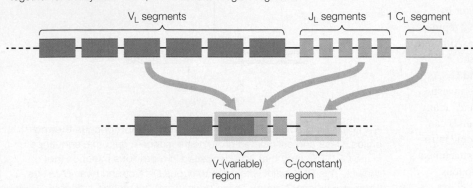

Somatic hypermutation is another mechanism to introduce sequence diversity into antibody chains. The spontaneous mutation rate becomes very high, but particularly within the variable sequences that make up the antigen binding sites.

▶ SEE SECTION 23.3 FOR REFS

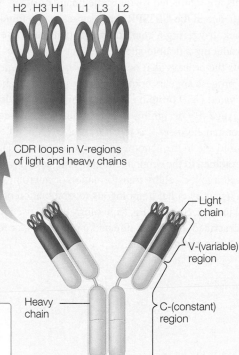

CDR loops in V-regions of light and heavy chains

Light chain

V-(variable) region

Heavy chain

C-(constant) region

Antibody schematic structure

▶ THE STRUCTURE OF ANTIBODIES IS DESCRIBED IN DETAIL IN SECTION 7.3. ANTIBODY:ANTIGEN INTERACTIONS ARE DESCRIBED IN SECTION 7.4.

HEAVY-CHAIN LOCUS: V-D-J recombination

Similarly, only one complete heavy chain gene is assembled using V, D, J and constant region segments for heavy chains, which lie on another chromosome.

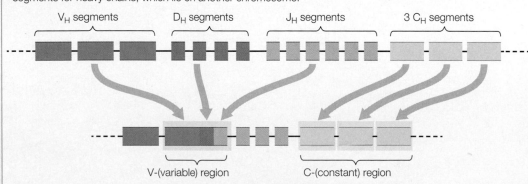

THE CLONAL SELECTION THEORY OF THE ADAPTIVE IMMUNE RESPONSE

Every B cell produces an antibody with unique antigen binding specificity when the assembled light and heavy chain genes are expressed. The combinatorial effect of bringing together the V_L and V_H domains (variable regions on the light and heavy chains, respectively) to form the antigen binding sites contributes to further diversity.

Mastering Chemistry for Biochemistry

Mastering Chemistry for Biochemistry provides select end-of-chapter problems and feedback-enriched tutorial problems, animations, and interactive figures to deepen your understanding of complex topics while practicing problem solving.

THE CLONAL SELECTION THEORY (CONT'D)

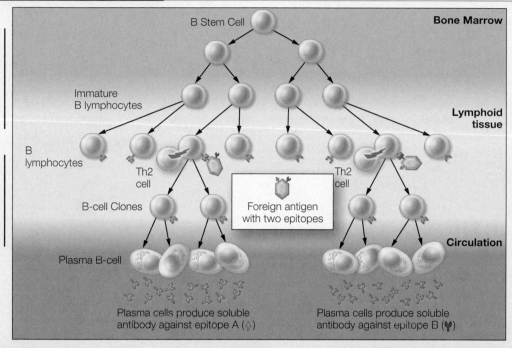

Stem cells in the bone marrow (B stem cells) differentiate and migrate to the lymphoid tissue. Each of the differentiated cells (B lymphocytes) synthesizes a unique kind of antibody, which it carries on its surface.

When the B lymphocytes encounter antigens (shown as an orange hexagon; middle), the B cells that carry antibodies to the antigenic determinants are stimulated by helper T cells ("Th2"cell) to multiply, forming B-cell clones. In this case, either the "red" or "green" epitope on the foreign antigen. Th2 cells stimulate the B-lymphocytes by secretion of interleukin-2 (lightning bolt symbol).

Some of the cloned B cells, called plasma cells, produce soluble antibodies, with each clone producing antibody against a single antigenic determinant (epitope).

▶ SEE FIGURE 7.2 FOR DETAILS

② ANTIBODIES IN DISEASE TREATMENT Specific cellular targets and antibody-drug conjugates

Some antibodies act as therapeutics by binding to specific cellular targets, which are associated with disease, and interrupting their function. For example, Herceptin® binds to HER2 receptors in (some) breast cancers and disrupts HER2 signaling.

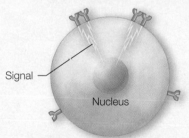

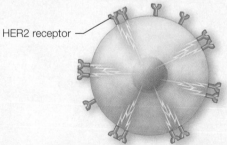

HER2-normal breast/stomach cancer cell

HER2 receptor dimerizes, then sends signals telling cells to grow and divide

HER2+ breast/stomach cancer cell

Overexpression of HER2 receptors leads to increased signaling and unregulated cell growth (cancer).

Herceptin attaches to HER2 receptors and disrupts dimerization, which disrupts signaling

A different strategy for the design of antibody-based drugs is that of antibody-drug conjugates. An ADC links the targeting function of an antibody to the cell-killing function of a potent toxin that is activated upon uptake by the target cell. For example, Adcetris® comprises an anti-CD30 monoclonal antibody attached by a protease-cleavable linker to a microtubule disrupting agent, monomethyl auristatin E (MMAE). The ADC is designed to be stable in the bloodstream but to release MMAE upon internalization into CD30-expressing tumor cells (e.g., Hodgkin lymphoma). ▶ SEE FIGURE 7.11 FOR DETAILS

① Adcetris® binds to CD30 and the ADC-CD30 complex enters the cell.

② MMAE is released and binds to tubulin, disrupting the microtubule network.

③ The cell cycle is arrested and apoptosis occurs.

The deadly nightshade (*Amanita phalloides*), one of the most poison-ous mushrooms known, contains α-amanitin, which blocks eukaryotic messenger RNA synthesis because of its potent inhibition of RNA polymerase II.

Transcription and Posttranscriptional Processing

WE TURN NOW to *transcription,* the process by which information stored in the nucleotide sequence of DNA is read out through the template-dependent synthesis of polyribonucleotides. Mechanistically, transcription is similar to DNA replication, particularly in the use of nucleoside triphosphate substrates and the template-directed growth of nucleic acid chains in a $5' \rightarrow 3'$ direction. There are, however, two major differences. First, with a few exceptions, only one strand of the DNA duplex is used as a template and transcribed for any particular gene, and second, only part of the entire genome is transcribed in any one cell. The importance of understanding transcription lies in regulation—the processes used to select particular genes and template strands for transcription. This selection, in large part, governs the metabolic capabilities of a cell. These regulatory mechanisms chiefly operate at the levels of initiation and termination of transcription, through the actions of proteins that contact DNA in a highly site-specific manner. **FIGURE 24.1** gives a preview of what we have learned about transcriptional regulation through the analysis

of DNA–protein interactions. We discuss this topic both here and in Chapter 26.

The products of transcriptive RNA synthesis are rarely used directly. In eukaryotic cells transcription products are subject to further processing, including trimming, cutting, splicing, and modifications at the 3′ and 5′ ends. Comparable processing occurs for prokaryotic ribosomal and transfer RNAs. Therefore, we also discuss posttranscriptional processing in this chapter. We are concerned primarily with the synthesis and processing of messenger RNA, ribosomal RNA, and transfer RNA, the species most directly involved in gene expression. However, some RNA transcripts serve additional functions, including catalysis (Chapters 8 and 25) and the telomerase reaction (Chapter 22). In Chapter 26, we will discuss other RNA species and functions—RNA editing, riboswitches, small interfering RNAs, microRNAs, and noncoding RNAs.

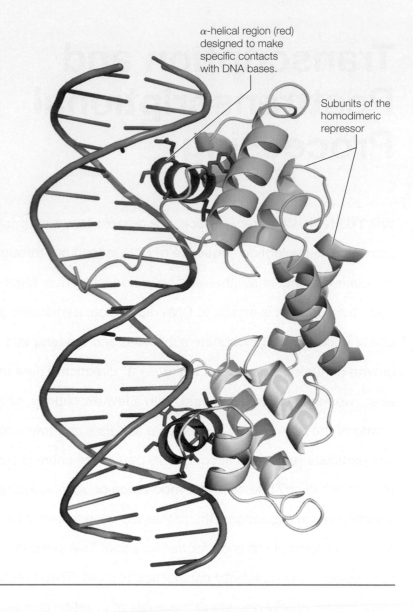

α-helical region (red) designed to make specific contacts with DNA bases.

Subunits of the homodimeric repressor

▶ **FIGURE 24.1 Regulation of transcription by DNA–protein interactions.** Gene expression is controlled by proteins that recognize particular DNA sequences and bind at those sites. Shown here is a regulatory protein (a *repressor*) of bacteriophage λ, which binds to a specific *operator* site on λ phage DNA and regulates the viral reproductive cycle by controlling the transcription of genes adjacent to this site. Yellow and blue depict subunits of the homodimeric repressor, and red depicts α-helical regions of the repressor, designed to make specific contacts with DNA bases. PBD ID: 1lmb

24.1 DNA as the Template for RNA Synthesis

The idea that RNA molecules are complementary copies of DNA templates originated from genetic analysis. In France's Pasteur Institute, François Jacob, Jacques Monod, and Andre Lwoff were studying genes controlling the utilization of lactose in *E. coli* and the control of lysogeny in bacteriophage λ. Recall from Chapter 23 that lysogeny is the state of gene repression after insertion of a circular bacteriophage genome into the chromosome of a host bacterium. Both the lactose and phage λ systems are described more fully in Chapter 26. Here we focus upon lactose utilization.

● **CONCEPT** Our original concepts of RNA synthesis came from genetic studies that predicted the existence of messenger RNA.

The Predicted Existence of Messenger RNA

Lactose utilization in *E. coli* was known to be controlled by three enzymes, whose genes are adjacent on the chromosome. One of these enzymes is **β-galactosidase,** which hydrolyzes lactose and other β-galactosides. When bacteria are grown in a medium containing both glucose and lactose, the genes controlling lactose utilization are silent, and glucose is preferentially metabolized. Once the glucose supply is exhausted, the three enzymes of lactose utilization are rapidly synthesized, allowing the bacteria to take up and metabolize lactose. This enzyme **induction** is rapidly reversed upon adding glucose back to the medium, whereupon the β-galactosidase activity reverts to a low level. The rapid changes in β-galactosidase-forming capacity suggested that the template for synthesizing this enzyme is metabolically unstable—synthesized rapidly on demand and degraded when the stimulus to induction is removed. Ribosomal and transfer RNAs are metabolically stable. Hence, these species were unlikely to be intermediates in information transfer. What might these observations tell us about RNA synthesis?

Jacob and Monod analyzed *E. coli* mutants that displayed faulty control over induction of the lactose-utilizing enzymes. Some expressed all three genes at high levels even when lactose or a similar inducer was absent, and others could not produce any of the enzymes, even after the addition of lactose to a bacterial culture. Based on these studies, as well as parallel work by Lwoff with phage λ, in 1961 Jacob and Monod proposed a unifying hypothesis of gene regulation in which

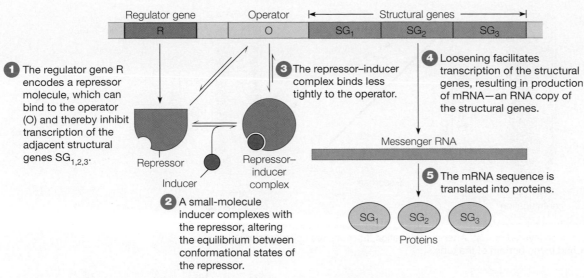

▲ FIGURE 24.2 The operon model, as proposed in 1961 by Jacob and Monod.

1. The regulator gene R encodes a repressor molecule, which can bind to the operator (O) and thereby inhibit transcription of the adjacent structural genes SG$_{1,2,3}$.

Repressor

Inducer

2. A small-molecule inducer complexes with the repressor, altering the equilibrium between conformational states of the repressor.

Repressor–inducer complex

3. The repressor–inducer complex binds less tightly to the operator.

4. Loosening facilitates transcription of the structural genes, resulting in production of mRNA—an RNA copy of the structural genes.

Messenger RNA

5. The mRNA sequence is translated into proteins.

SG$_1$ SG$_2$ SG$_3$

Proteins

transcription, or copying information encoded in DNA, was regulated specifically at the level of initiation. Hypothetical regulatory elements called *repressors* and *operators* controlled the synthesis of other hypothetical entities called *messenger RNAs* (mRNAs). mRNA was postulated to be a complementary copy of the DNA that encompassed a set of structural genes, which encode proteins, as schematized in **FIGURE 24.2**. A set of contiguous genes plus adjacent regulatory elements that control their expression was termed an **operon,** and the Jacob–Monod hypothesis thus became known as the **operon model.**

Jacob and Monod predicted several characteristics of the hypothetical mRNA. First, they predicted a high rate of mRNA synthesis followed by rapid degradation, which would explain the fast turn-on of the genes after induction by lactose and turn-off in the presence of another preferentially used sugar. Second, because of rapid synthesis and degradation, they expected mRNA to accumulate rapidly but not to high steady-state levels. Third, because they thought that the messenger was a copy of two or more contiguous genes, they expected it to be fairly large and part of a heterogeneous size class of RNA molecules. Finally, if the mRNA was a complementary copy of DNA, its nucleotide sequence should be identical to a portion of one of the DNA strands.

● **CONCEPT** Bacterial genetics predicted messenger RNA to be a collection of metabolically active RNAs that are present in low abundance, heterogeneous in size, and complementary in sequence to portions of the DNA genome.

T2 Bacteriophage and the Demonstration of Messenger RNA

The first physical demonstration of mRNA came from work with the closely related T2 and T4 bacteriophages. Infection by these large DNA viruses arrests all expression of host-cell genes, and no significant accumulation of RNA can be detected after infection. However, in 1956 the use of radioisotopes led to detection of a distinctive RNA in T2-infected *E. coli*. When infected cultures were labeled for short periods (a 3- or 4-minute "pulse") with [^{32}P]orthophosphate, about 2% of the total RNA became radioactive. This radiolabeled RNA had

two properties that led to its eventual identification as viral mRNA. First, it was metabolically unstable; after termination of the labeling interval, this RNA rapidly lost radioactivity, as if the labeled species were being degraded. Second, the RNA seemed to be a product of viral DNA metabolism because its nucleotide composition resembled that of T2 DNA; the radiolabeled RNA was rich in adenine and uracil and low in guanine and cytosine; T2 DNA has a high A + T content.

Additional evidence came from sucrose gradient centrifugation of the rapidly labeled phage RNA, which showed that the labeled material sediments heterogeneously and distinctly from any of the known rRNA or tRNA species. In the experiment of **FIGURE 24.3(a)**, an ultracentrifuge tube contains buffer with a linear gradient of sucrose concentration. The RNA solution being analyzed is carefully layered onto the top of the gradient. The sucrose minimizes convection, so that in a centrifugal field molecules sediment as discrete bands, at rates related to their size and shape. This differs from equilibrium gradient centrifugation, as in the Meselson–Stahl experiment (Chapter 4), where molecules are separated on the basis of density differences (mass per unit volume).

In the experiment of Figure 24.3(a), fractions obtained after centrifugation were analyzed for their ultraviolet light absorption, which measures RNA concentration, and radioactivity, which measures the relative amount of RNA synthesized during the pulse. The UV absorbance profile shows two large peaks at 16S and 23S, and a smaller peak at 4S. (S stands for "Svedberg units," a measure of sedimentation rate; it is related to molecular size and shape.) The 16S and 23S peaks represent rRNA, and the 4S peak is mostly tRNA, which is a much smaller molecule. The radioactivity was distributed heterogeneously throughout the gradient, suggesting that the labeled RNA is a collection of molecules of varying sizes, none of which was present in great abundance.

Benjamin Hall and Sol Spiegelman established that this phage RNA is a viral gene product when they carried out the first DNA–RNA hybridization experiment, showing that the labeled RNA was complementary in sequence to phage DNA. DNA and RNA were heated and slow-cooled together, so that RNA could base-pair with DNA of complementary nucleotide sequence. The earliest experiments were based on the fact that RNA is denser than DNA. (Why is this so?) Thus, a DNA–RNA hybrid could be detected in equilibrium gradient centrifugation as a species of intermediate density containing label derived from both DNA and RNA. Such a hybrid was formed when T2 RNA was heated and slowly cooled along with T2 DNA, but not when the DNA came from *E. coli*. Additional density-labeling experiments established that

● **CONCEPT** The ability of T2 phage RNA to hybridize with T2 DNA and to associate with ribosomes made before infection was evidence for the existence of mRNA.

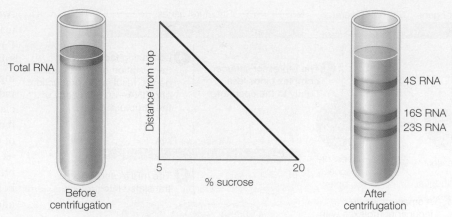

(a) A centrifuge tube containing a linear 5% to 20% gradient of sucrose with an RNA preparation layered on top. After centrifugation, fractions are collected through a hole punched in the bottom of the tube.

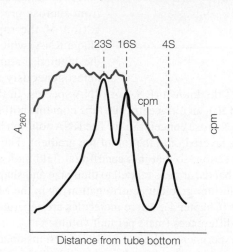

(b) Sedimentation profile of total and pulse-labeled RNAs in T2 phage-infected *E. coli*. Total concentration of RNA in each fraction (black) was determined by ultraviolet absorbance (A_{260}). The radioactivity profile (red) shows the heterogeneous size distribution of RNA molecules synthesized during the pulse.

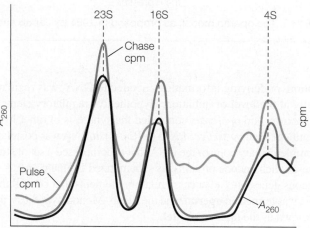

(c) Pulse-labeled RNA species in uninfected bacteria and their fate in a chase. The orange line represents RNA molecules labeled in a 3-minute pulse. The blue line depicts an identical culture with the label chased by 0.7 generation of growth in nonradioactive medium, and the black line is the A_{260} profile, representing ribosomal and transfer RNA species as in panel (b).

▲ **FIGURE 24.3 Demonstration of mRNA by pulse labeling and sedimentation.** Panel (a) demonstrates the technique of gradient centrifugation, while panels (b) and (c) depict the results of analysis of pulse-labeled RNA preparations.

---

phage proteins were synthesized on ribosomes that had been formed before infection. These experiments showed the ribosome to be a non-specific workbench on which any protein could be assembled, depending on which template became associated with that workbench.

RNA Dynamics in Uninfected Cells

The experiments described in the previous section supported the existence of mRNA in phage-infected *E. coli*. What about uninfected bacteria? Spiegelman and his colleagues showed that pulse-labeled RNA from uninfected *E. coli* hybridized to *E. coli* DNA. At very short labeling intervals, the sedimentation pattern showed incorporation into both rRNA and tRNA species *and* a heterogeneously sedimenting species (Figure 24.3(c)). After a "chase," when the radiolabeled phosphate was removed and incubation was continued, the radioactivity profile followed the absorbance profile, showing that all RNA species had become labeled to equivalent specific radioactivities. This finding is consistent

with the postulated short lifetime of mRNA ($t_{1/2} = 2$–3 minutes). mRNA would reach its maximal radioactivity within just a few minutes, but during the chase, mRNA turnover would release nucleotides that could flow into stable RNA molecules. Because those stable RNA molecules do not turn over, label accumulates, and the fraction of total label in the stable RNA species continues to increase. Consistent with this idea, Spiegelman also showed that highly labeled ribosomal and transfer RNAs hybridize to *E. coli* DNA, demonstrating that all three major classes of RNA are synthesized from template DNA strands.

● **CONCEPT** The demonstration of mRNA in *E. coli* involved a classic "pulse-chase" experiment, in which incorporation of a radioisotope into a metabolic species takes place during a short time interval; then the fate of that species is followed after removal of the radioisotope.

As noted previously, the earliest DNA–RNA hybridization experiments involved gradient centrifugation, a laborious technique. Spiegelman and his colleagues made the important discovery that single-stranded DNA binds irreversibly to membrane filters made of material such as nitrocellulose. This technique allowed rapid hybridization analysis of a large number of samples because a radiolabeled RNA

● **CONNECTION** The discovery that single-stranded DNA, immobilized on a membrane filter, can hybridize with complementary-sequence RNA or DNA underlies forensic DNA analysis and gene-expression analysis by microarray technology.

could hybridize to denatured DNA immobilized on a nitrocellulose filter. After washing unbound radioactivity off the filter, the extent of hybridization could be determined simply by counting the radioactivity of the filter. The same principle—immobilization of DNA on nitrocellulose

followed by the analysis of bound radioactivity—underlies Southern blotting (described in Chapter 21), which is now more widely used to analyze gene organization and expression. These developments also led to microarray technology, in which thousands of DNA–RNA hybridization reactions are analyzed on a single DNA chip (see Tools of Biochemistry 24A; see also Figure 1.11).

24.2 Enzymology of RNA Synthesis: RNA Polymerase

As discussed earlier, RNA synthesis involves copying a template DNA strand. However, the earliest known enzyme capable of RNA synthesis in vitro did *not* require a template. This RNA-synthesizing enzyme, called **polynucleotide phosphorylase,** was discovered in the 1950s. The enzyme was quite different from DNA polymerase. The enzyme required no template, and it used ribonucleoside *diphosphates* (rNDPs) as substrates to produce a random-sequence polynucleotide whose base composition matched the nucleotide composition of the reaction medium.

$$n \text{ rNDP} \rightleftharpoons (\text{rNMP})_n + n \text{ P}_i$$

Initially, it was thought that polynucleotide phosphorylase might be the major RNA-synthesizing enzyme, but the lack of a template requirement was puzzling, as was the apparent absence of the

● **CONCEPT** Polynucleotide phosphorylase catalyzes the reversible, template-independent synthesis of random-sequence polyribonucleotides.

enzyme in eukaryotic cells. Ultimately, polynucleotide phosphorylase turned out to play no role in RNA synthesis in vivo but instead was found to participate in the

degradation of bacterial mRNAs. The enzyme was of great value, however, in the synthesis of polynucleotides used as templates for in vitro protein synthesis, when the genetic code was being elucidated (see Chapter 25).

Investigators continued to search for an enzyme that would copy a DNA template in vitro. In 1960, such an enzyme was discovered almost simultaneously in four different laboratories. The enzyme, **DNA-directed RNA polymerase,** resembled DNA polymerases in the nature of the reaction catalyzed.

$$n \text{ (ATP + GTP + CTP + UTP)} \xrightarrow{\text{Mg}^{2+}, \text{DNA}}$$
$$(\text{AMP} - \text{GMP} - \text{CMP} - \text{UMP})_n + n \text{ PP}_i$$

The reaction product is a complementary RNA copy of the DNA template.

Biological Role of RNA Polymerase

In bacteria, a single RNA polymerase catalyzes the synthesis of mRNA, rRNA, and tRNA. This was shown in experiments with **rifampicin,** an antibiotic that inhibits RNA polymerase in vitro and blocks the synthesis of mRNA, rRNA, and tRNA in vivo. Rifampicin-resistant mutants of *E. coli* were found both to contain a rifampicin-resistant form

● **CONNECTION** Rifampicin is a widely used antibiotic that specifically targets bacterial RNA polymerase.

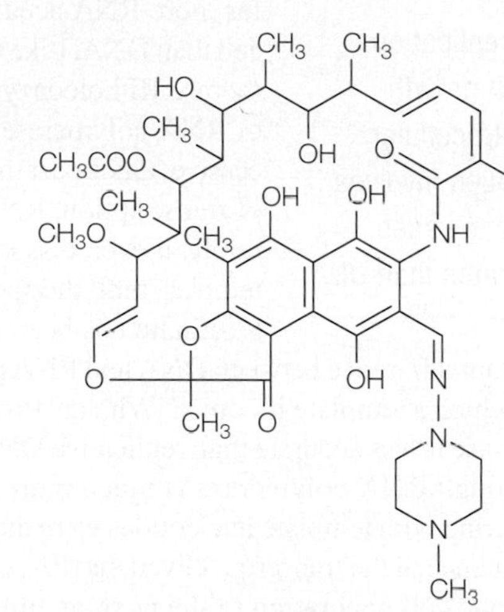

of RNA polymerase and to be capable of synthesizing all three RNA classes in vivo in the presence of rifampicin. Because a single mutation affects both the RNA polymerase and the synthesis of all RNA types in vivo, RNA polymerase must be the one enzyme catalyzing all forms of transcription in bacteria.

Rifampicin

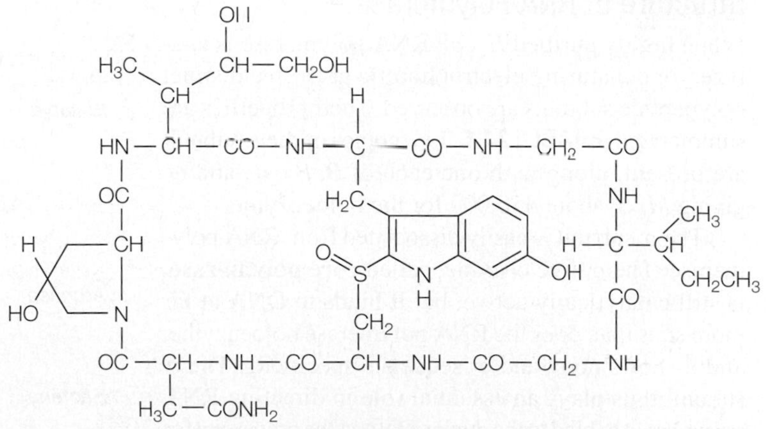

α-Amanitin

In contrast, eukaryotes contain three distinct RNA polymerases, one each for synthesis of rRNA, mRNA, and small RNAs (tRNA plus the 5S species of rRNA)—RNA polymerases I, II, and III, respectively. The existence of separate enzymes was revealed partly because they differ in their sensitivity to inhibition by ***α*-amanitin,** a toxin from the poisonous *Amanita* mushroom. RNA polymerase II is inhibited at low concentrations of

● **CONNECTION** α-Amanitin, the toxic agent in one of the most deadly poisonous mushrooms, acts by inhibiting RNA synthesis and, hence, by blocking all protein synthesis.

α-amanitin, RNA polymerase III is inhibited only at high concentrations, and RNA polymerase I is quite resistant.

Because DNA polymerases and RNA polymerases catalyze similar reactions, it is worthwhile to compare some of their kinetic features. The k_{cat} for *E. coli* DNA polymerase III holoenzyme, at about 500 to 1000 nucleotides per second, is much higher than k_{cat} for purified RNA polymerase—50 nucleotides per second, which is about the same as the rate of transcription in vivo. Although there are only about 10 molecules of DNA polymerase III per *E. coli* cell, there are some 2000 molecules of RNA polymerase, of which half might be involved in transcription at any instant. Why is this significant? As seen in Chapter 22, replicative DNA chain growth is rapid but takes place at few sites, whereas transcription is much slower but occurs at many sites. The result is that

● **CONCEPT** DNA replication involves rapid chain growth occurring at few intracellular sites, and transcription involves slower growth at many sites. More RNA accumulates than DNA.

far more RNA accumulates in the cell than DNA. Like the DNA polymerase III holoenzyme, the action of RNA polymerase is highly processive. Once past the initial stages of transcription, RNA polymerase rarely, if ever, dissociates from the template until the specific signal to terminate has been reached.

Another important difference between DNA and RNA polymerases is the accuracy with which a template is copied. With an error rate of about 10^{-5}, RNA polymerase is less accurate than replicative DNA polymerase holoenzymes, although RNA polymerase is much more accurate than expected if base-pairing with template nucleotides were the sole determinant of the base sequence of the transcript. Given that RNA does not carry information from one cell generation to the next, an ultrahigh-fidelity template-copying mechanism is evidently not needed.

Structure of RNA Polymerase

When highly purified *E. coli* RNA polymerase is analyzed in denaturing electrophoretic gels, five distinct polypeptide subunits are observed. Their properties are summarized in **TABLE 24.1**. Two copies of the α subunit are present, along with one each of β, β', σ, and ω, giving M_r of about 450,000 for the holoenzyme.

The σ subunit is easily dissociated from RNA polymerase. The σ-free enzyme, called **core polymerase,** is still catalytically active, but it binds to DNA at far more sites than does the RNA polymerase holoenzyme, and it shows no strand or sequence specificity. The σ subunit thus plays an essential role in directing RNA polymerase to bind to the duplex DNA at the proper site for initiation—the **promoter** site—and to select the correct strand as the template for transcription. The addition of σ to core polymerase reduces the affinity of the enzyme for *nonpromoter* sites by about 10^4, thereby increasing the enzyme's specificity for binding to promoters.

These discoveries about σ suggested that gene expression could be regulated by having core polymerase interact with different forms of σ, which would in turn direct the

TABLE 24.1 Subunit composition of *E. coli* RNA polymerase

Subunit	M_r	Number per Enzyme Molecule	Function
α	36.5 kDa	2	Chain initiation, interaction with regulatory proteins and upstream promoter elements
β	151.0 kDa	1	Chain initiation and elongation
β'	155.0 kDa	1	DNA binding
σ	70.0 kDa[a]	1	Promoter recognition
ω	11.0 kDa	1	Promotion of enzyme assembly

[a]The 70-kDa σ subunit is one of several alternative σ subunits.

holoenzyme to different promoters. In fact, this does occur. In one example, an *E. coli* culture is stressed by a sudden temperature increase. In these *heat-shocked* cells, a new form of σ appears and directs the modified RNA polymerase to a different set of promoters, thereby activating transcription of a block of genes called heat-shock genes. The most abundant σ in *E. coli,* and the one that will frame our discussions, is called σ^{70} because of its 70-kDa molecular weight. Subunit σ^{70} is one of seven different σ factors known in *E. coli,* each designed to direct RNA polymerase to a functionally related set of genes.

Eukaryotic RNA polymerases show similarity with the bacterial enzymes, but they have a more complex subunit structure. RNA polymerase II (pol II) from yeast has 12 subunits. The common ancestry among RNA polymerases is evident, however, as shown in **FIGURE 24.4**. For example, the two largest subunits in the three eukaryotic multisubunit

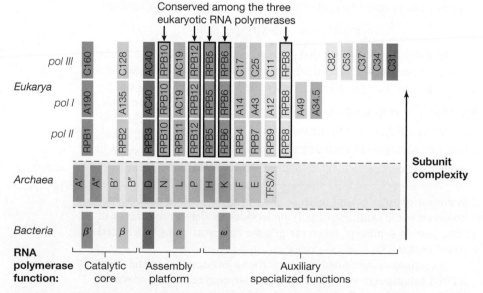

▲ **FIGURE 24.4 RNA polymerase subunit structures in the three domains of life.** The subunits are arranged by function rather than by size. Homologous subunits are color-coded. Subunits enclosed in a box are conserved among the three eukaryotic RNA polymerases. Only core enzyme subunits are shown. Data from F. Werner (2007) Structure and function of archaeal RNA polymerases, *Molecular Microbiology* 65:1395–1404.

polymerases are related to β and β' of the bacterial enzyme. The figure also shows the subunit composition of archaeal RNA polymerases, which reveal them to be much more closely related to eukaryotic than to bacterial polymerases. Eukaryotic RNA polymerases have no direct counterpart to bacterial σ. Instead, a series of proteins called **transcription factors** functions in comparable fashion, helping to direct RNA polymerase to promoter sites, forming an initiation complex (page 753).

Although the multisubunit motif for RNA polymerases is the dominant structural theme, it is not universal. Exceptions include single-subunit polymerases encoded by some bacteriophage genomes, including T7 and SP6, and a mitochondrial RNA polymerase found in vertebrates. Plant chloroplasts also contain an organelle-specific RNA polymerase as well as two additional RNA polymerases, IV and V. These enzymes are involved in RNA-directed DNA methylation, a process in gene silencing (Chapter 26).

The crystal structures of eukaryotic and bacterial RNA polymerases were first determined in 2001 and 2002 in the laboratories of Roger Kornberg and Seth Darst, respectively. **FIGURE 24.5(a)** shows the structure of *E. coli* RNA polymerase holoenzyme (containing sigma), reported in 2013, and Figure 24.5(b) shows a cutaway view of yeast RNA polymerase II, illustrating structural features common to all of the multisubunit RNA polymerases. We shall describe the relationship between these structures and the RNA polymerase reaction presently.

24.3 Mechanism of Transcription in Bacteria

Like DNA replication and protein synthesis, transcription occurs in three distinct phases—initiation, elongation, and termination. Initiation and termination signals in the DNA sequence punctuate the genetic message by directing RNA polymerase to particular genes and by specifying where transcription will start, where it will stop, and which DNA strand will be transcribed. The signals involve both instructions encoded in DNA base sequences and interactions between DNA and proteins other than RNA polymerase. We focus first on bacterial RNA polymerases, exemplified by the widely studied *E. coli* enzyme, but the basic mechanics of transcription are similar in all organisms. Actions of the eukaryotic enzymes are discussed later.

Initiation of Transcription: Interactions with Promoters

Promoters, the DNA sites at which RNA polymerase binds and initiates transcription, were originally identified from DNA base-sequence analysis upstream (on the 5′ side) from sites at which transcription begins. This analysis for several *E. coli* genes identified two conserved adenine- and thymine-rich sequences centered at about 10 and 35 nucleotides to the 5′ side of the transcriptional start site, or, at positions −10 and −35, where +1 represents the first DNA nucleotide to be transcribed (**FIGURE 24.6**). There was some sequence variation among the promoters analyzed, but a *consensus sequence* emerged—meaning those nucleotides found most often at each position in each site—within this conserved region.

Biochemical analysis by a technique called **footprinting** confirmed that RNA polymerase binds tightly to DNA sequences immediately upstream from the transcription start site and that the −10 and −35 "boxes" are in close contact with the bound enzyme. In footprinting, a cloned segment of DNA is made radioactive by use of a kinase enzyme

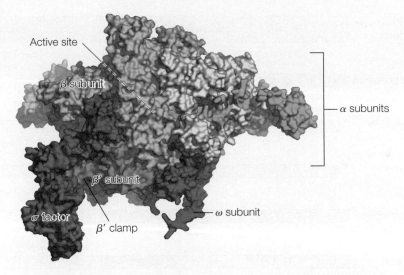

(a) *E. coli* holoenzyme. Some functional domains are indicated. For other views, see http://www.jbc.org/content/suppl/2013/02/06/M112.430900.DC1/jbc.M112.430900-1.mov. PDB ID: 4igc.

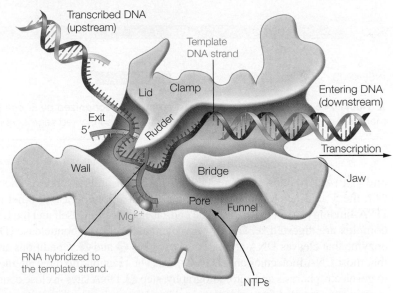

(b) Cutaway view of the yeast RNA polymerase II elongation complex. Cut surfaces are in light gray. Nontemplate strand (blue) is not shown where it is disordered. The 3′ end of the growing RNA chain is adjacent to one of two catalytically essential Mg^{2+} ions.

▲ **FIGURE 24.5 Crystal structure of RNA polymerase.** (a) *E. coli* holoenzyme. α subunits in yellow and green; β, cyan; β_1, pink; ω, gray; σ^{70}, orange. Some functional domains are indicated. PDB ID: 4igc. Data from K. S. Murakami (2013), X-ray Crystal Structure of *Escherichia coli* RNA Polymerase σ70 Holoenzyme, *J. Biol. Chem.* 288:9126–9134; (b) Cutaway view of the yeast RNA polymerase II elongation complex. Labeled parts of the structure are described in the text. Cut surfaces are in light gray. Template DNA strand is in purple. Nontemplate strand (blue) is not shown where it is disordered. RNA that is hybridized to the template strand is in orange. The 3′ end of the growing RNA chain is adjacent to one of two catalytically essential Mg^{2+} ions. PDB ID: 1l6h.

trigonal bipyramidal

two metal catalysis

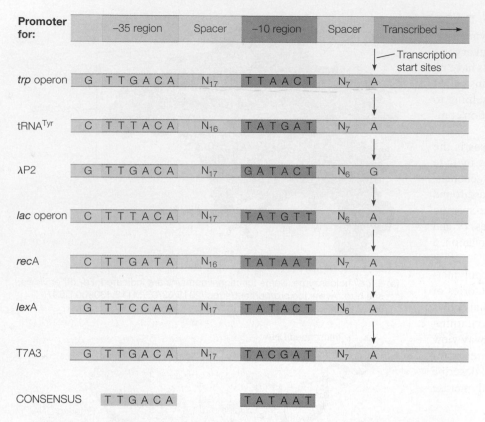

▲ **FIGURE 24.6 Conserved sequences in promoters recognized by *E. coli* RNA polymerase.** Lengths of spacer regions between the two conserved segments are shown.

and [^{32}P]ATP to phosphorylate the DNA molecule at its 5′ end. As shown in **FIGURE 24.7**, the 5′ end-labeled DNA is mixed with RNA polymerase or other site-specific DNA-binding protein of interest. The end-labeled DNA by itself and the DNA–protein complex are digested separately with pancreatic deoxyribonuclease (DNase I), an enzyme that cleaves DNA at random sites, steps ❷ and ❷′. Conditions are chosen so that most DNA molecules are cleaved once. The reaction mixtures are then subjected to gel electrophoresis and autoradiography, step ❸. DNA sites in close contact with the protein are protected from cleavage, and bands representing those sites are not seen.

The search process by which RNA polymerase finds a promoter site embedded in many kilobases of DNA is not well understood. σ is required, for core polymerase binds weakly to DNA with no specificity whatsoever. The initial encounter between RNA polymerase holoenzyme and a promoter generates a **closed-promoter complex.** Whereas DNA strands unwind later in transcription, no unwinding is detectable in a closed-promoter complex. Binding is primarily electrostatic and of low affinity. Footprinting studies show that polymerase is in contact with DNA from about nucleotide −55 to −5.

● **CONCEPT** Transcription begins with sequence-specific interaction between RNA polymerase and a promoter site, where duplex unwinding and template strand selection occur.

Next, RNA polymerase unwinds several base pairs of DNA, from about −12 to −1, giving an **open-promoter complex,** so-called because it binds DNA whose strands are open, or unwound. The open-promoter complex is extremely stable; it forms with a K_a as high as 10^{12} M^{-1}. Structural analysis of DNA complexed with σ shows specific interactions between the protein and DNA bases in the −10 box. DNA melting is an essential feature of promoter recognition.

Next, a Mg^{2+}-dependent isomerization occurs, giving a modified form of the open-promoter complex with the unwound DNA region now extending from −12 to +2. DNA

is bent in this structure, as shown in **FIGURE 24.8**. Note also the positions of the major promoter elements (−10 and −35) and the bound Mg^{2+} at the catalytic site.

Initiation and Elongation: Incorporation of Ribonucleotides

RNA chain growth begins with binding of the template-specified rNTP, followed by binding of the second nucleotide and its fitting to the template. Nucleophilic attack by the 3′ hydroxyl of the first nucleotide on the α (innermost) phosphorus of the second nucleotide generates the first

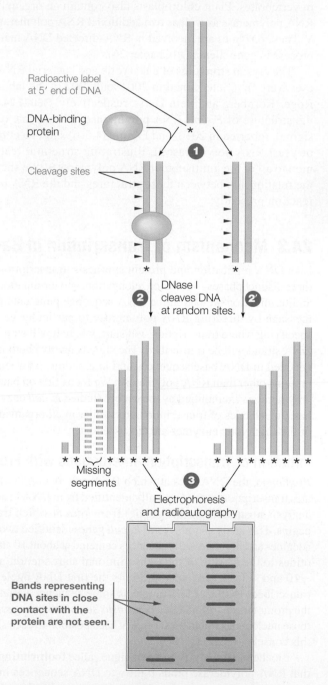

▲ **FIGURE 24.7 Footprinting to identify binding sites on DNA for site-specific DNA-binding proteins.**

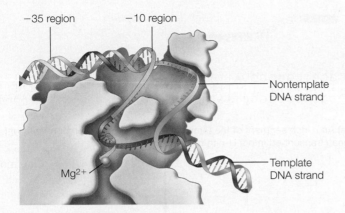

▲ **FIGURE 24.8 DNA bending in an open-promoter complex.** This image is taken from the crystal structure of *Thermus aquaticus* RNA polymerase. Parts of β have been removed to show interior structure. Note the sharp bend in the template strand. The first two ribonucleotides to be incorporated (i and i + 1) are shown close to a catalytically essential Mg^{2+}. Much of the α and β′ subunits are rendered transparent. PDB ID: 1I9u. Data from *Science* 296:1285–1290, K. S. Murakami, S. et al. (2002) Taq RNA polymerase.

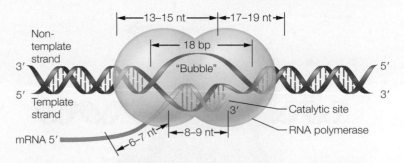

▲ **FIGURE 24.9 The transcription bubble.** This idealized view was originally determined by indirect means, such as the reactivity of transcription complexes with reagents specific for nucleotides in single-stranded conformation. Nt = nucleotide.

phosphodiester bond and leaves an intact triphosphate moiety at the 5′ position of the first nucleotide. Nucleotide incorporation occurs via the two-metal mechanism described for DNA polymerases in Chapter 22. The transcript is bound unstably during the first several phosphodiester bond-forming reactions, as shown by the fact that most initiations are abortive, with release of oligonucleotides two to nine residues long. The basis for this low efficiency of initiation is not fully understood.

During incorporation of the first 10 nucleotides, the σ subunit dissociates from the transcription complex, and the remainder of the transcription process is catalyzed by the core polymerase. Once σ has dissociated, the elongation complex becomes quite stable. Transcription, as studied in vitro, can no longer be inhibited by adding rifampicin, and virtually all transcription events proceed to completion.

During *elongation*, the core enzyme moves along the duplex DNA template. As it moves, it simultaneously unwinds the DNA, exposing a single-stranded template for base pairing with incoming nucleotides and with the nascent transcript (the most recently synthesized RNA); it rewinds the template behind the 3′ end of the growing RNA chain, as suggested in **FIGURE 24.9**. In this model, about 18 base pairs of DNA are unwound at any given time, forming a moving "transcription bubble." As one base pair becomes unwound in advance of the 3′ end of the nascent RNA strand, one base pair becomes rewound near the trailing end of the RNA polymerase molecule. About nine base pairs of the 3′ end of the nascent transcript are hybridized to the template DNA strand. During elongation, RNA polymerase functions as a true molecular motor. Techniques for analyzing single complexes show that RNA polymerase generates forces exceeding those of well-studied cytoskeletal motor proteins, such as myosin and kinesin.

Refer back to Figure 24.5(b) for a stylized picture of an elongation complex. Although this model was derived from the structure of yeast RNA polymerase II, it shows generally applicable features of transcriptional elongation, including the large "wall," which forces DNA to bend, almost at right angles. In this schematic illustration, the direction of polymerase motion is left to right as shown. DNA entering

the enzyme is gripped by protein "jaws" (upper jaw not shown in this cutaway model). The 3′ end of the growing RNA is adjacent to one of the catalytically essential Mg^{2+} ions. The wall forces the DNA to turn. rNTPs probably enter the active site, as shown, through a funnel structure and pore. The 5′ end of the growing RNA chain is diverted from the DNA template by a protein loop called the rudder, which limits the length of RNA hybridized to template DNA. The rudder and lid, which guide the exit of RNA, emanate from a large clamp that swings from back to front, as shown, over the catalytic site and contributes to binding nucleic acids, and hence, to the high processivity of transcription.

The concept of a transcription bubble as a central intermediate in transcription, as depicted in Figure 24.9, suggests that the enzyme moves along the DNA template in register with the growing RNA transcript, with the footprint advancing by one base pair for each ribonucleotide incorporated into the transcript. However, footprinting of numerous initiation and elongation complexes has shown that the enzyme often advances discontinuously, holding its position for several cycles of nucleotide addition and then jumping forward by several base pairs along the template. These and other observations suggest that the means of RNA polymerase translocation is fundamentally different from the continuous movement implied by a picture of the transcription bubble. It has long been known that some DNA sequences are difficult to transcribe, and RNA polymerase "pauses" when it reaches such a site in vitro, often sitting at the same site for several seconds before transcription is resumed. At such sites, RNA polymerase often translocates backward, and in the process the 3′ end of the nascent transcript is displaced from the catalytic site of the enzyme. This gives a 3′ "tail," which may be several nucleotides long and is not base-paired to the template, protruding downstream of the enzyme (**FIGURE 24.10**). For transcription to resume, an RNA 3′ end must be positioned in the active site. At this point, the enzyme may either cleave the single-stranded RNA end or move forward without new nucleotide incorporation, thereby repositioning the RNA 3′ nucleotide in duplex conformation. These pausing, backtracking, and cleavage reactions are thought to help control the transcription rate and to enhance transcriptional fidelity by removing mismatched or damaged ribonucleotides.

Punctuation of Transcription: Termination

Because of the great stability of transcription complexes, termination of transcription, with release of the nascent transcript, is an involved

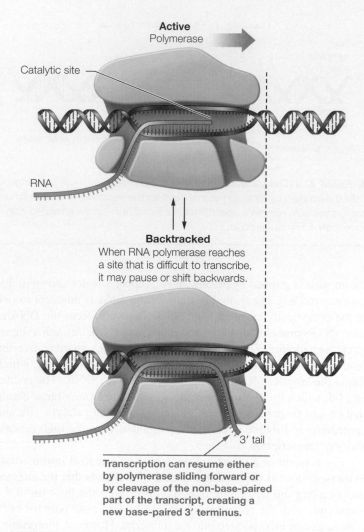

FIGURE 24.10 Backtracking in an elongation complex. Above, the 3′ terminus of the transcript is in the catalytic site. Below, the enzyme has slipped backward, leaving the 3′ transcript terminus at the end of a non–base-paired RNA tail, several nucleotides long. Data from E. Nudler (2012) RNA polymerase backtracking in gene regulation and genome instability. *Cell* 149;1438–1445.

process. In bacteria, we recognize two distinct types of termination events—those that depend on the action of a protein **termination factor** called ρ (rho), and those that are ρ factor–independent.

Factor-Independent Termination

Sequencing the 3′ ends of genes that terminate in a factor-independent manner reveals two structural features shared by many such genes and is illustrated in **FIGURE 24.11**: (1) two symmetrical GC-rich segments that in the transcript have the potential to form a stem–loop structure, and (2) a downstream run of four to eight A residues.

● **CONCEPT** DNA sequences that promote factor-independent termination include a GC-rich region that forms a stem–loop followed by a run of 4 to 8 A residues.

These features suggest the following as elements of the termination mechanism. First, RNA polymerase slows down, or pauses, when it reaches the first GC-rich segment, because the stability of G-C base pairs

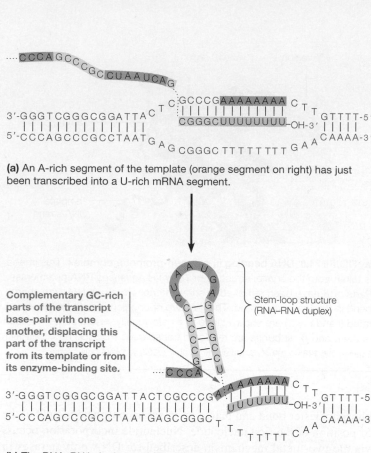

(a) An A-rich segment of the template (orange segment on right) has just been transcribed into a U-rich mRNA segment.

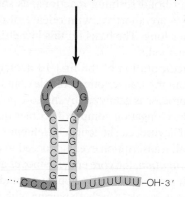

Complementary GC-rich parts of the transcript base-pair with one another, displacing this part of the transcript from its template or from its enzyme-binding site.

Stem-loop structure (RNA–RNA duplex)

(b) The RNA–RNA duplex, stabilized by G–C base pairs (yellow), eliminates some of the base pairing between template and transcript.

(c) The unstable A–U bonds linking transcript to template hybrid dissociate, releasing the transcript.

FIGURE 24.11 A model for factor-independent termination of transcription.

makes the template hard to unwind. In vitro, RNA polymerase does pause, sometimes for as long as several minutes, at a GC-rich segment. Second, pausing gives time for the complementary GC-rich parts of the nascent transcript to base-pair with one another, thereby displacing this part of the transcript from its template or from its enzyme binding site. Hence, the ternary complex of RNA polymerase, DNA template, and RNA is weakened. Further weakening, leading to dissociation, occurs when the A-rich segment is transcribed to give a series of A–U bonds (very weak) linking transcript to template.

The actual mechanism of termination is more complex than just described, in part because DNA sequences both upstream and downstream from the regions shown in Figure 24.11 also influence termination efficiency.

Factor-Dependent Termination

Factor-dependent termination sites are less frequent, and the process requires an additional protein factor, called ρ. The ρ protein, a

● **CONCEPT** In factor-dependent termination, ρ protein acts as an RNA–DNA helicase, unwinding the template–transcript duplex and facilitating release of the transcript.

hexamer composed of identical subunits, was originally discovered in studies on termination of λ phage DNA transcription in vitro. This protein, which has been characterized as an RNA–DNA helicase, contains a nucleoside triphosphatase activity that is activated by binding to polynucle-

otides. Apparently, ρ acts by binding to the nascent transcript at a C-rich site near the 3′ end, when RNA polymerase has paused (**FIGURE 24.12**). Then ρ moves along the transcript toward the 3′ end, with the helicase activity unwinding the 3′ end of the transcript from the template (and/or the RNA polymerase molecule) and causing it to be released. Release involves an additional protein, NusA (not shown).

24.4 Transcription in Eukaryotic Cells

Transcription in eukaryotes is more complex than in bacteria and their phages. Not only is there much more discrimination in what is to be transcribed and what is not, but this transcription must be precisely programmed during development and tissue differentiation. Furthermore, the transcription machinery must deal with the complex levels of structure in eukaryotic chromatin. Reflecting this complexity is the fact that eukaryotic cells have three different RNA polymerases, each with a specialized function (we leave aside the two additional polymerases

● **CONCEPT** Eukaryotic cells have three RNA polymerases, each requiring additional protein factors to initiate transcription.

in plants). For each polymerase, several proteins must assemble at promoters and other upstream sites on the template DNA, along with RNA polymerase, in order to form a functional transcrip-

tion complex. None of the three polymerases has a direct counterpart to the σ factor of bacterial complexes. However, all three require a set of transcription factors that play roles comparable to that of σ, in addition to proteins that might be specialized for the transcription of a particular gene. By convention, transcription factors are named TFI, TFII, or TFIII, depending on whether they function with RNA polymerase I, II, or III, respectively. Within one class of transcription factors, each individual factor is identified with a letter; thus, TFIIA is one of several transcription factors functioning with RNA polymerase II. The ensemble of proteins required to form an initiation complex is conserved among polymerases I, II, and III.

Other differences from bacteria pertain to the fact that bacterial genomes are organized into blocks of functionally related genes—operons, such as the lactose operon mentioned earlier in this chapter—which are co-transcribed to give multicistronic mRNAs, while eukaryotic genes

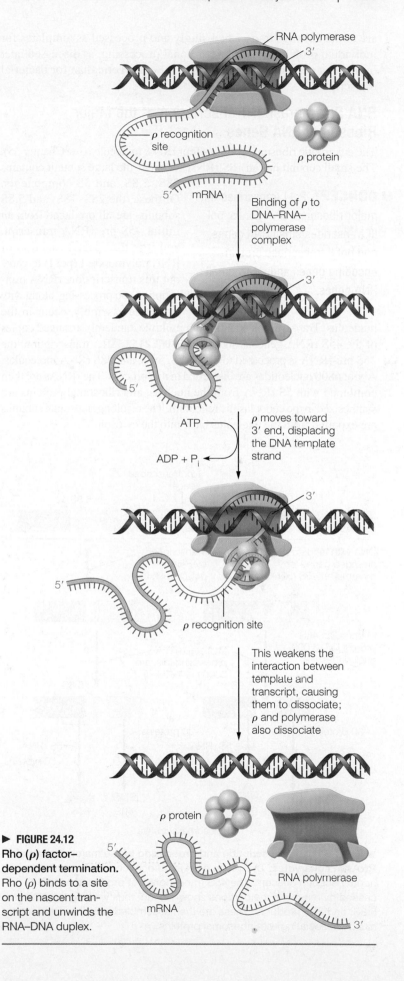

▶ **FIGURE 24.12**
Rho (ρ) factor-dependent termination. Rho (ρ) binds to a site on the nascent transcript and unwinds the RNA–DNA duplex.

are almost always transcribed singly and processed as templates for individual proteins. Posttranscriptional processing, as discussed later in this chapter, is far more complex for eukaryotic than for bacterial transcripts.

RNA Polymerase I: Transcription of the Major Ribosomal RNA Genes

The eukaryotic ribosome contains four rRNA molecules (see Chapter 25). The small subunit has an 18S rRNA, whereas the large subunit contains 28S, 5.8S, and 5S components. Of these, the 28S, 18S, and 5.8S subunits are all produced from an initial 45S pre-rRNA transcript, and it is the special function of RNA polymerase I (pol I) to carry out this transcription. rRNA transcription and processing, along with ribosome assembly, occur in the nucleolus. Transcription occurs from multiple, tandemly arranged copies of the 45S rRNA gene, as shown in **FIGURE 24.13**. After transcription, the 45S pre-rRNA is processed to yield 18S, 5.8S, and 28S rRNA molecules. About 6800 nucleotides are discarded in this process. The rRNAs are then combined with 5S rRNA from the nucleus, and ribosomal proteins are synthesized in the endoplasmic reticulum. The resulting ribosomal subunits are exported from the nucleolus back into the cytosol.

● **CONCEPT** Pol I transcribes the major ribosomal RNA genes; pol III transcribes small RNA genes; and pol II transcribes protein-encoding genes and a few small RNA genes.

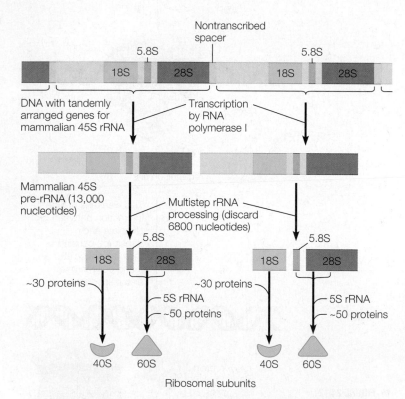

DNA with tandemly arranged genes for mammalian 45S rRNA

Transcription by RNA polymerase I

Mammalian 45S pre-rRNA (13,000 nucleotides)

Multistep rRNA processing (discard 6800 nucleotides)

~30 proteins
5S rRNA
~50 proteins

~30 proteins
5S rRNA
~50 proteins

40S 60S   40S 60S

Ribosomal subunits

▲ **FIGURE 24.13 Transcription and processing of the major ribosomal RNAs in eukaryotes.** The genes exist in tandem copies, separated by nontranscribed spacers. The 45S transcripts first produced are processed by removing the portions shown in tan, thus yielding the 18S, 5.8S, and 28S products. These are then assembled into ribosomal subunits by associating with ribosomal proteins.

RNA Polymerase III: Transcription of Small RNA Genes

RNA polymerase III (pol III) is the largest and most complex of the eukaryotic RNA polymerases (see Figure 24.4). The major targets for pol III are the genes for all the tRNAs and for the ribosomal 5S rRNA. Like the major ribosomal genes described in the previous section, these small genes are present in multiple copies, but they are usually not grouped together in tandem arrays, nor are they localized in one region of the nucleus. Rather, they are scattered over the genome and throughout the nucleus.

Of all the genes transcribed by pol III, the most thoroughly studied are those for 5S rRNA. At least three transcription factors, TFIIIA, –B, and –C, are needed for the expression of these genes. TFIIIB and TFIIIC participate in transcribing tRNA genes as well, but TFIIIA is specific only for the 5S genes. TFIIIA is an example of an abundant class of sequence-specific DNA-binding proteins, in which metal-binding *zinc fingers* make contact with and identify DNA sequences (**FIGURE 24.14**). This class of proteins contains conserved histidine and cysteine residues, which complex with Zn^{2+}. This DNA-binding protein motif was mentioned in Chapter 20 as a structural element in steroid hormone receptors.

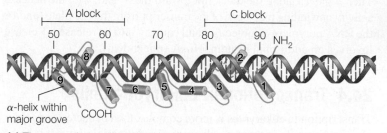

(a) The transcription factor TFIIIA binds to the 5S RNA gene via zinc fingers inserted into the major groove. The two major recognition regions, A block and C block, are contacted by fingers 7–9 and 1–3, respectively.

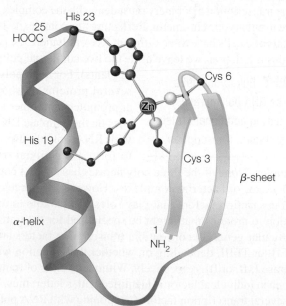

(b) Structure of a synthetic polypeptide that contains the zinc finger motif. The α-helix lies within the DNA major groove, as shown in panel (a). The two histidine and two cysteine residues that coordinate the zinc are shown in detail.

▲ **FIGURE 24.14 Zinc fingers.** Data from M. S. Lee et al. (1989) *Science* 245:635–637.

RNA Polymerase II: Transcription of Structural Genes

All of the structural genes (those encoding proteins) in the eukaryotic cell are transcribed by RNA polymerase II (pol II). This enzyme also transcribes some of the small nuclear RNAs involved in splicing (discussed in Section 24.5). Like other RNA polymerases, pol II is a complex multisubunit enzyme. However, not even its 12 subunits are sufficient to allow pol II to initiate transcription on a eukaryotic promoter. Because the expression of many eukaryotic genes is either tissue-specific or developmental stage-specific or both, eukaryotic promoter structure is far more complex than that of bacteria. Protein factors in addition to RNA polymerase are required for promoter recognition, recruitment of RNA polymerase to a promoter, and generation of an active elongation complex. A typical initiation complex contains 60 proteins in addition to RNA polymerase. Of these, about half form a preinitiation complex that assists the enzyme in recognizing and binding to a promoter. The remainder are involved in regulation.

A typical eukaryotic promoter contains an initiator region (Inr) with the sequence YYANWYY, where N is any nucleotide, Y is a pyrimidine (C or T), W is either A or T, and N represents the +1 initiation site. A counterpart to the bacterial −10 region, called the *TATA box,* positioned between −20 and −30, has the sequence TATA-AAA. Upstream from that box are arrayed additional control elements, including the *CAAT box* (GGCCAATCT), the *GC box* (GGGCGG), and *Octamer* (ATTTGCAT). Additional regulatory sites may exist several kbp upstream from the initiation site; these are called **enhancer** regions. Although these far upstream activator sites are involved in transcriptional regulation, they are not considered part of the promoter itself. **FIGURE 24.15** shows the locations of these elements in several well-studied eukaryotic promoters.

RNA polymerase II interacts with several general transcription factors, including TBP (TATA box-binding protein) and TFIIA, -B, -E, -F, and H. The formation of initiation and elongation complexes has been studied in detail with yeast RNA polymerase II by use of crystallography and cryo-electron microscopy. A remarkable feature of the preinitiation complex, as shown by Roger Kornberg's laboratory, is the fact that DNA is bound in this complex but is not in direct contact with RNA polymerase (**FIGURE 24.16**).

TFIIB plays an important role in converting the initial closed-promoter complex to an open-promoter complex. The process, summarized below, was described and illustrated by A. Cheung and P. Kramer (2012) Cell:149,1431–1437.

In forming the closed-promoter complex, TBP first binds to DNA and bends it by 90 degrees. The C-terminal domain of TFIIB binds to TBP and flanking DNA regions. The N-terminal domain recruits RNA polymerase to promoter DNA near the transcription start site, forming the closed-promoter complex. Next, a TFIIB element called the B-linker opens DNA before the transcription start site, leading to an open-promoter complex. ATP is required for this process. TFIIB threads the template DNA strand into the active center. For this TFIIB uses another structural element, the B-reader, which consists of a helix followed by a mobile loop. Next, DNA is scanned for an initiator (Inr) motif near the start site; DNA "scrunching," movements of the nontemplate strand, facilitate this process. Following this, the first two ribonucleotide substrates are positioned opposite Inr, and the first

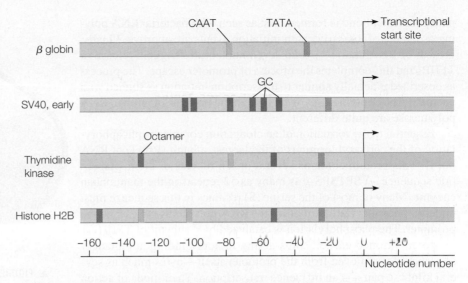

▲ **FIGURE 24.15 Structures of some eukaryotic promoters.** The colored boxes represent different regulatory elements: orange, TATA box; blue, GC box; yellow, CAAT box; purple, Octamer. Based in part on Genes IV, B. Lewin, Oxford: Oxford University Press, 1990.

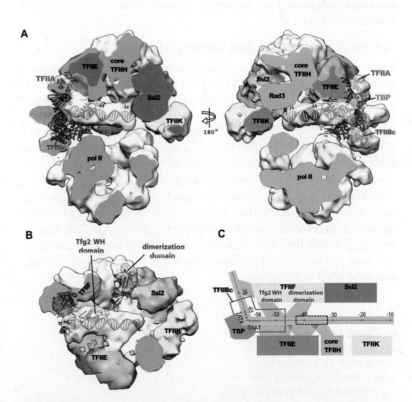

▲ **FIGURE 24.16 Structure of the yeast RNA polymerase II transcription preinitiation complex.** Cutaway section through the complex as visualized by cryoelectron microscopy. Cut surfaces are shown in gray. The complex is divided into two parts—pol II below and the transcription factors (IIA, IIE, etc.) above. Ss12 is a helicase subunit of TFIIH, and TBP is the TATA box binding protein subunit of the TFIIB C-terminal domain. From K. Murakami et al. (2013) *Science* 349, 1238724. DOI:10.1126/science.1238724. Courtesy of Roger D. Kornberg.

phosphodiester bond is formed. Just as seen with bacterial RNA polymerase, most of the early chain initiation events are abortive. Finally, growth of RNA chains beyond seven nucleotides triggers release of TFIIB, and this completes the process of promoter escape. The process as described is actually similar to transcription initiation as studied with bacterial RNA polymerases, even though the proteins other than RNA polymerase are quite different.

Essential to the formation of an elongation complex is phosphorylation of the carboxyl terminus of the largest subunit (Rpb I) of RNA pol II. This protein contains several dozen repeats of the heptapeptide sequence –YSPTSPS–, as many as 52 repeats in the mammalian enzyme. Many or most of the serine (S) residues in this sequence must undergo phosphorylation in order for RNA polymerase to escape the promoter. The phosphorylation is catalyzed by a subunit of TFIIH.

As mentioned earlier, trans-acting factors binding at enhancer sequences far removed from the promoter itself—by as much as several kilobase pairs—can influence transcription. Their mode of action appears to involve DNA looping, perhaps mediated by nucleosomes, which can bring enhancer-bound proteins into close physical contact with proteins bound to the promoter. Some transcription factors can bind in either promoter or enhancer regions. These can act as intermediates between activators or repressors bound to enhancer regions and the core transcription complex as schematized in **FIGURE 24.17**. Also involved in communication between upstream control elements and proteins bound at the promoter is a multiprotein complex called **mediator** (Chapter 26).

Chromatin Structure and Transcription

The complex interplay of transcription factors and polymerases we have described occurs not on naked DNA but on chromatin. The chromatin structure presents two major problems: First, how can the transcription factors and initiation complex bind to DNA in the presence of nucleosomes? Second, how can the actively transcribing polymerase pass through arrays of nucleosomes? This is an area of intense research interest, which we treat in more detail in Chapter 26. Here we can point out that transcription often initiates in chromatin regions containing nuclease-accessible sites—regions in which the DNA in isolated chromatin is readily cleaved, as if it were not complexed with histones at those sites, granting ready access to RNA polymerase.

● **CONCEPT** Nuclease-accessible (open) sites disrupt chromatin to allow initiation.

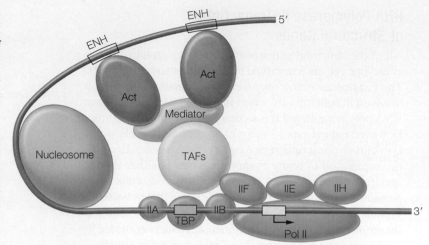

▲ **FIGURE 24.17** A schematic representation of DNA looping as a process to bring enhancer-bound (ENH) activator (Act) proteins into contact with trans-acting factors (TAFs) associated with the core transcription complex.

How are accessible sites established in previously unresponsive genes? In some cases, such as the globin genes, it seems that the chromatin structure is rearranged at the time of replication. In other instances, protein factors seem able to interfere with chromatin structure at specific loci, opening hypersensitive sites. In either case, the clearance of histones from nuclease-susceptible sites involves the action of **chromatin remodeling factors.** These are proteins that enable promoter regions to be able to accept the complex and bulky machinery depicted in Figure 24.16. These complexes, which we discuss in Chapter 26, require ATP hydrolysis to somehow "open" nucleosomes transiently, to allow transcription complexes to form.

Another and perhaps equally important role is played by histone acetyltransferases and deacetylases. Histones of the nucleosome core are subject to acetylation at specific lysine residues in the N-terminal tails (see **FIGURE 24.18**); chromatin is subject to other

▶ **FIGURE 24.18** Acetylation of core histones. The general structure of each of the four core histones involves a helical "histone-fold" domain plus an unstructured, highly basic N-terminal domain. Acetylation in nuclei occurs exclusively in the N-terminal domains, at the highly conserved sites indicated in red. Data from J. C. Hansen, C. Tse, and A. P. Wolffe, *Biochemistry* (1997) 37:17637–17641.

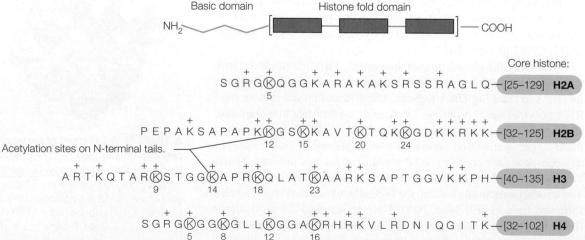

modifications as well, as mentioned in Chapter 21. High levels of acetylation are correlated with high transcriptional activity, and low acetylation with low activity. Chemically, this makes sense: neutralization of histone basic residues by acetylation would loosen ionic interactions between histones and DNA in chromatin. A number of proteins recruited to the initiation complex by activators and trans-acting factors have histone acetylase activity. The fact that specific transcription factors are involved in this process may provide the long-sought explanation for how the chromatin of *specific* genes can be targeted for disruption.

Transcriptional Elongation

Formation of the open-promoter complex (see Figure 24.16) is followed, in the presence of rNTPs and ATP, by melting of a short region of DNA and initiation of transcription. As noted previously, the C-terminal tail of the Rpb1 subunit of pol II becomes highly phosphorylated, leading to promoter release, and elongation begins, with a helicase activity clearing the way. A number of the core transcription factors are released, and pol II, together with TFIIF, moves along the DNA. A residual complex, containing TBP, TFIIA, TAFs, and probably activator proteins, remains at the start site, ready to initiate another round.

At this point, the polymerase also acquires several *elongation factors.* Some of these factors assist the enzyme in traversing pause sites in the DNA. As previously noted for bacterial RNA polymerase,

● **CONCEPT** Pol II, with the aid of other proteins, can transcribe through nucleosome arrays.

transcription is relatively slow and interrupted by frequent pauses, especially in T-rich regions. Elongation factors assist the enzyme in passing such sites. Nucleosomes form even larger obstacles to the progress of an RNA polymerase II along the DNA, as mentioned previously.

Just how pol II transcribes through nucleosomes is still something of a mystery. Do the nucleosomes unfold and re-form as the polymerase passes? Are they temporarily displaced? Current evidence favors temporary displacement, but the issue is far from settled.

Termination of Transcription

The termination of mRNA transcription is also different in eukaryotes. Whereas the bacterial RNA polymerase recognizes terminator signals, which sometimes function with the aid of the ρ protein, the eukaryotic polymerase II usually continues to transcribe well past the end of the gene. In doing so, it passes through one or more TTATTT signals, which lie beyond the 3′ end of the coding region (**FIGURE 24.19**). The pre-mRNA, carrying this signal as AAUAAA, is then cleaved by a special endonuclease that recognizes the signal and cuts at a site 11 to 30 residues downstream of it. At this point, a tail of polyriboadenylic acid, poly(A), as many as 300 bases long, is added by sequential ATP incorporation, catalyzed by a special nontemplate-directed enzyme, **poly(A) polymerase.** The functions of the poly(A) tails of eukaryotic mRNAs include mRNA stabilization and facilitation of transport from nucleus to cytoplasm. We know that they cannot be essential for all messages because some mRNAs (e.g., most histone mRNAs in higher eukaryotes) do not have them. However, poly(A) tails relate to message stability, for tail-less messages typically have much shorter lifetimes in the nucleus. Recent evidence indicates that the specific TTATTT

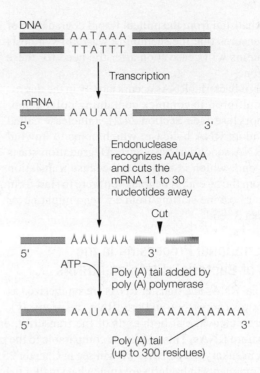

▲ **FIGURE 24.19 Termination of transcription in eukaryotes: Addition of poly(A) tails.** There is a TTATTT sequence near the 3′ end of most eukaryotic genes. When the complementary strand is transcribed to AAUAAA, it provides a signal for endonuclease cleavage and poly(A) tail addition.

signals used for termination vary in different tissues, so that the 3′ end of an mRNA is partly tissue-specific. The functional significance of this variation is not yet known.

24.5 Posttranscriptional Processing

Bacterial mRNA Turnover

Critical to mRNA metabolism in eukaryotes are events occurring *after* transcription—events that are necessary for messages to move from the nucleus to their sites of utilization in the cytosol. We discuss these events later in this chapter. In bacteria, by contrast, mRNAs are available for use in protein synthesis immediately. In fact, a nascent mRNA can serve as a template for translation at its 5′ end while still in the process of being synthesized toward the 3′ end. That is, transcription is coupled directly to translation.

The major posttranscriptional event in metabolism of bacterial mRNA is its own degradation, which in most cases is rapid. A few bacterial mRNAs, notably those encoding outer membrane proteins, are long-lived; however, many bacterial messages have half-lives of only 2 to 3 minutes. This short life span means that genes being expressed must be transcribed continuously and that many mRNA molecules are translated only a few times. Although this might seem wasteful, it is consistent with bacterial lifestyles, which necessitate rapid adaptation to environmental changes. Earlier we noted the selective advantage to bacteria of expressing the genes for lactose utilization only when an inducer is present. By the same token, it would be wasteful for the cell to continue producing these proteins after lactose or a related sugar was exhausted from the milieu. Rapid degradation of *lac* mRNA ensures that the seemingly wasteful synthesis of these proteins will cease soon after the need for these proteins is gone.

Turnover of bacterial RNAs occurs largely in the **degradosome,** a multiprotein complex containing polynucleotide phosphorylase (see Section 24.2), a nuclease called RNase E, and an RNA helicase, which helps to unwind regions of RNA secondary structure. Degradation starts from the 5′ end, which is important because translation also starts from the 5′ end. If degradation were to start from the 3′ end, a ribosome starting from a 5′ end might never reach an intact 3′ end.

Posttranscriptional Processing in the Synthesis of Bacterial rRNAs and tRNAs

Both ribosomal RNAs and transfer RNAs are synthesized as larger transcripts (pre-rRNA and pre-tRNA, respectively), which undergo cleavage at both ends of the transcript, en route to becoming mature RNAs. This process is comparable to the processing of pre-rRNA in eukaryotic cells. As we will see in Chapter 25, however, the rRNA components in bacteria are somewhat smaller than in eukaryotes—23S, 16S, and 5S. The total amount of DNA encoding these rRNAs and tRNAs accounts for less than 1% of the *E. coli* genome, but because of the instability of mRNA, rRNA and tRNA constitute about 98% of the total RNA in a bacterial cell. Transcription of rRNA genes is extremely efficient when cells are growing rapidly. The intracellular concentrations of ribonucleoside triphosphates are important control elements here; ATP, whose level is high in rapidly growing cells, activates rRNA gene transcription by stabilizing the relevant open-promoter complexes.

rRNA Processing

The *E. coli* genome contains seven different operons for rRNA species. Each one encodes, in a single 30S transcript, sequences for one copy each of 16S, 23S, and 5S rRNAs (**FIGURE 24.20**). Because the three species are used in equal amounts, the logic of this organization is apparent. Less obvious is that each transcript also includes sequences for one to four tRNA molecules, which vary among the seven different operons. This interspersion of rRNA and tRNA sequences may represent a means of coordinating the rates of synthesis of these RNAs.

Processing of the 30S pre-rRNA probably begins with cleavage within two stem–loop structures by RNase III, a nuclease specific for double-stranded RNA. One double-strand cut in each of two giant stem–loop regions releases precursors to 16S and 23S rRNAs. RNase E is similarly involved in the processing of 5S rRNA. Further maturation steps require the presence of particular ribosomal proteins, which begin to assemble on the precursor RNAs while transcription is still in progress. The embedded tRNA sequences are processed to give mature tRNAs, along the same routes used for other tRNA species, as discussed next.

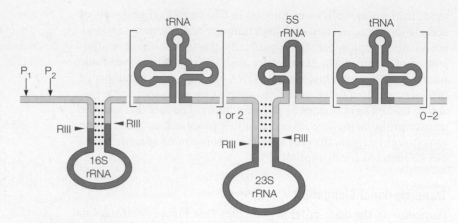

▲ **FIGURE 24.20** Structure of *E. coli* 30S pre-rRNA. Sequences complementary to two promoter sites (P₁ and P₂), RNase III cleavage sites (RIII) that release 16S and 23S species, and the locations of tRNA sequences embedded within the transcript are shown.

tRNA Processing

Aside from the tRNAs embedded in pre-rRNA transcripts, the other tRNAs are synthesized in transcripts that contain one to seven tRNAs each, all surrounded by lengthy flanking sequences. The maturation steps are summarized in **FIGURE 24.21**, using as an example the well-studied case of the *E. coli* tyrosine tRNA species (tRNA^Tyr).

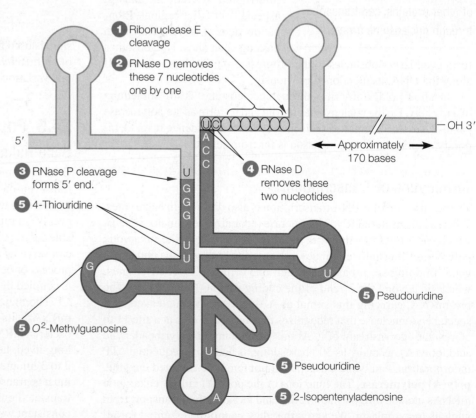

▲ **FIGURE 24.21** Modification steps (①–④) that occur in the maturation of *E. coli* tRNA^Tyr from its transcript and modified bases (step ⑤) seen in the mature tRNA. The tRNA sequence is shown in blue.

Maturation (step ❶) starts with RNase E, which cleaves next to a stem–loop structure on the 3' side of the tRNA sequence. This is followed by the action of **ribonuclease D** (step ❷), which carries out exonucleolytic cleavage to a point two nucleotides removed from the CCA sequence at the 3' end. Next, the 5' end is formed by the action of **ribonuclease P** (step ❸), which cleaves to leave a phosphate on the 5' terminal G. This enzyme creates the 5' terminus of all tRNA molecules. It is unclear what structural features are recognized by RNase P, because different sequences are contained in the cleavage sites. As pointed out in Chapter 8, ribonuclease P was one of the first identified ribozymes. The enzyme consists of one RNA molecule of 377 nucleotides and one protein molecule with M_r of about 20,000. Both components are necessary for full catalytic activity, but the RNA molecule alone can catalyze accurate cleavage.

> ● **CONCEPT** Bacterial transcripts undergo posttranscriptional processing, involving both endonucleolytic and exonucleolytic cleavage.

Once the proper 5' terminus has been formed, ribonuclease D removes the remaining two nucleotides from the 3' end (step ❹). Should excessive "nibbling" occur through faulty control of RNase D activity, there is an enzyme (CCA nucleotidyltransferase) that will restore the CCA end to any tRNA in a nontemplated fashion. This enzyme specifically recognizes the 3' terminus of tRNAs that lack the CCA end and catalyzes sequential reactions with a CTP, another CTP, and an ATP. Note that the CCA end is encoded by every tRNA gene, so the nucleotidyltransferase is essentially a repair enzyme.

> ● **CONCEPT** Posttranscriptional processing in bacteria involves cleavage of the primary transcript, modification of bases (in tRNA synthesis), and nontranscriptive nucleotide addition.

Creation of the modified bases (see Chapters 4 and 25) occurs at the final stage, including methylations, thiolations, and reduction of uracil to dihydrouracil. In the specific example shown, the modifications include formation of two pseudouridines, one 2-isopentenyladenosine, one O^2-methylguanosine, and one 4-thiouridine (step ❺). These modifications serve to stabilize the tRNA molecules against intracellular degradation, and in some cases they promote translational fidelity. The modifications are not essential for tRNA function, however, because many tRNAs lacking the modifications are fully active in vitro. Pathways for eukaryotic tRNA synthesis are similar, including the involvement of ribonuclease P. In the yeast *Saccharomyces cerevisiae* the average tRNA molecule has more than 12 of its 75–80 bases modified.

Processing of Eukaryotic mRNA

Bacterial and eukaryotic cells differ significantly in the ways that mRNAs for protein-coding genes are produced and processed. Recall that bacterial mRNAs are synthesized at the nucleoid in direct contact with the cytosol and are *immediately* available for translation. A specific nucleotide sequence at the 5' end recognizes a site on the bacterial rRNA, allowing attachment of the ribosome and initiation of translation, often even before transcription of the message is completed. Hence, there is little or no posttranscriptional processing of bacterial mRNAs.

In eukaryotes, mRNA is produced in the nucleus and must be exported to the cytosol for translation. Furthermore, the initial product of transcription (*pre-mRNA*) includes all of the introns and substantial flanking regions; the introns must be removed before correct translation can occur. For these reasons, eukaryotic mRNA requires extensive processing before it can be used as a template. This processing, including polyadenylylation at the 3' end (Section 24.4), takes place while pre-mRNA is still in the nucleus.

Capping

The first modification occurs at the 5' end of the pre-mRNA. First, one phosphate is removed hydrolytically from the triphosphate moiety at the 5' terminal nucleotide. Next, the resulting 5' diphosphate end attacks the α (inner) phosphate of a GTP molecule; in essence, the guanine nucleotide is added in *reverse* orientation ($5' \rightarrow 5'$). Together with the first two nucleotides of the chain, it forms a *cap* (**FIGURE 24.22**). The cap is further modified by the addition of methyl groups to the N-7 position of the guanine and to one or two sugar hydroxyl groups of the cap nucleotides. This cap structure positions the mRNA on the ribosome for translation, and also it probably contributes to stabilization of the message.

Splicing

After being capped, the pre-mRNA becomes complexed with a number of *small nuclear ribonucleoprotein particles* (*snRNPs*, often called "snurps"), which are themselves complexes of *small nuclear RNAs* (*snRNAs*) and special splicing proteins. The snRNAs are all less than 300 nucleotides long. The snRNP–pre-mRNA complex is called a **spliceosome,** and it is here that the most elegant part of the processing takes place—the cutting and splicing that is necessary to excise introns from the pre-mRNA and join the ends of the two exons. In forming a spliceosome, snRNAs recognize and bind intron–exon splice sites by means of complementary sequences

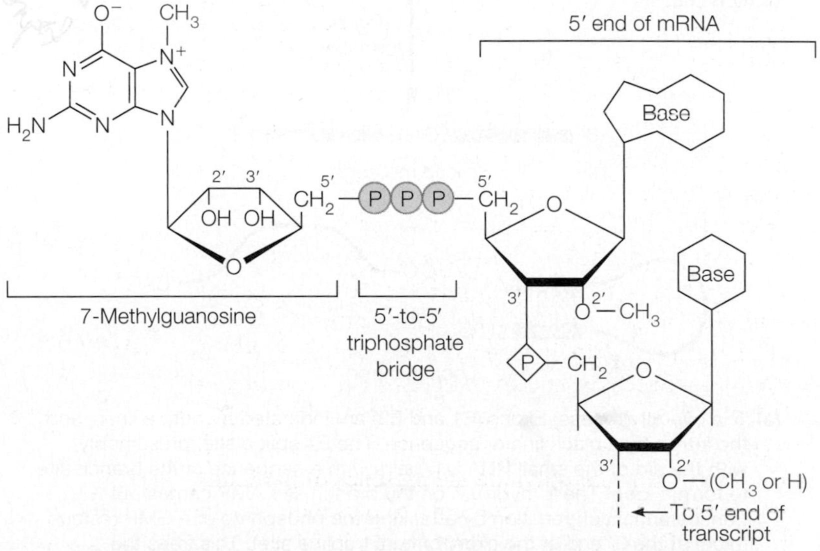

▲ **FIGURE 24.22 Structure of a processed mRNA 5' end.** Details of the 5' cap region are shown. Methyl groups that are added are in red.

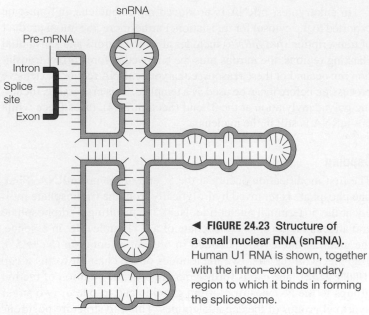

▲ **FIGURE 24.23 Structure of a small nuclear RNA (snRNA).** Human U1 RNA is shown, together with the intron–exon boundary region to which it binds in forming the spliceosome.

(**FIGURE 24.23**). Precise recognition of splice sequences is essential because even a single-base error would disrupt the sense of the genetic message. A schematic view of the chemistry of splicing is shown in **FIGURE 24.24**.

Excision of a single intron involves assembling and disassembling a spliceosome. **FIGURE 24.25** depicts the overall process. The sequence begins with binding of the U1 snRNP to the G site at the 5′ end of the intron (step ❶). The U2 snRNP then binds at the branch site (step ❷). With continued assembly of the spliceosome, including the addition of several more snRNPs, the lariat loop in the intron is formed and the two exons are joined (steps ❸–❺). Catalytically essential Mg^{2+} ions are located in the RNA of U6 snRNP, implying that RNA plays a catalytic role. Splicing has now been accomplished, and the products—a ligated mRNA and a looped intron—are released (step ❻). As the spliceosome disintegrates, the looped intron is released (step ❼) and degraded (step ❽), and the mRNA is exported from the nucleus.

Early steps in spliceosome assembly require more proteins and sites than are indicated in Figure 24.25. Kinetic experiments indicate that all of the early reactions in spliceosome assembly are reversible,

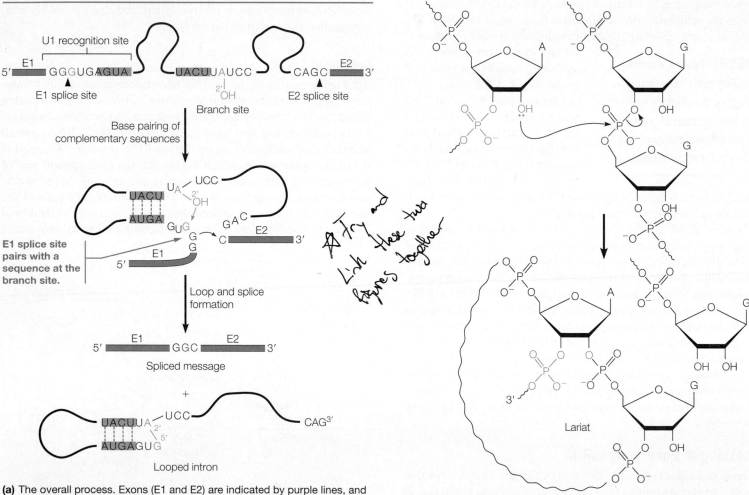

(a) The overall process. Exons (E1 and E2) are indicated by purple lines, and the intron by a black line or sequence. The E1 splice site, presumably with the aid of the small RNA U1, pairs with a sequence at the branch site to form a loop. The 2′ hydroxyl on the branch site AMP carries out a transesterification reaction by attacking the phosphate of a GMP residue (blue) at the 3′ end of the exon 1 (the E1 splice site). This frees the adjacent G (red) to attack with its 3′ hydroxyl the phosphate 5′ to the C at the 5′ end of exon 2. The products are a spliced message and a looped intron "lariat" structure, which is then degraded.

(b) The first transesterification reaction. The second reaction (not shown) involves nucleophilic attack of the GMP 3′-hydroxyl group (in red) upon the phosphate 5′ to a CMP residue, as seen schematically in part a.

▲ **FIGURE 24.24 A schematic view of the mechanism of mRNA splicing.**

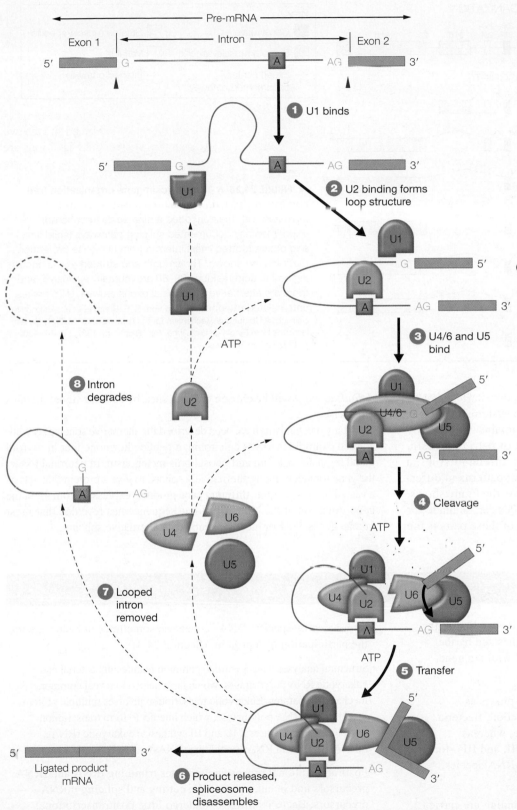

▲ **FIGURE 24.25 The overall process of splicing.** The pre-mRNA plus assorted snRNPs assemble and disassemble a spliceosome, which carries out the splicing reaction. The snRNPs are designated U1, U2, and so on. U1 is bound in step ❶, which together with U2 binding (step ❷) leads to a looped structure. Factors U4/6 and U5 then bind (step ❸), and cleavage and transfer then occur (steps ❹ and ❺). The spliceosome disassembles, releasing the ligated product (step ❻) and the looped intron (step ❼). This is degraded into small oligonucleotides (step ❽).

with the process being driven by the irreversibility of the late steps. In 2017, the structure of the catalytically activated spliceosomes of yeast and human were solved by cryo-electron microscopy, yielding important mechanistic understanding of this complex process. The structure of the human spliceosome is depicted on the cover of this book.

Because of the importance of accurate splicing for correct expression of genetic information, splicing errors are responsible for many genetic diseases. As many as 15% of all known genetic diseases arise from splicing errors. In some forms of thalassemia, a family of diseases arising from defective synthesis of hemoglobin chains (Chapter 7), mutations have been found

● **CONNECTION** Defects in pre-mRNA splicing are responsible for about 15% of known genetic diseases.

in both the 5′ and 3′ splice sites of both genes for the β chain of human hemoglobin. Usually, an incorrect mRNA chain is formed, which leads to premature termination of translation of the message. Another example of defective splicing is Hutchinson–Gilford progeria syndrome, a disease of accelerated aging in which afflicted individuals die in their teens from aging-related symptoms such as cardiovascular disease. The disease affects a nuclear envelope protein called Lamin A. The mutation activates a cryptic donor splicing site in the gene for a lamin A precursor, leading to accumulation of a truncated form of the protein.

Alternative Splicing

Once investigators discovered and described mRNA splicing, they were surprised to learn that the same pre-mRNA can undergo splicing in several different ways. The existence of **alternative splicing** means that different combinations of exons from the same gene can be processed into different mature mRNAs and then undergo translation into quite different proteins in different tissues or at different developmental stages of the same organism. Alternative splicing made it less surprising when

● **CONCEPT** Alternative splicing allows one gene to specify more than one protein.

the Human Genome Project revealed the existence of far fewer genes than had been expected given the size of the genome and the complexity of *Homo sapiens*. Alternative splicing greatly enlarges the repertoire of proteins that can be encoded by a genome.

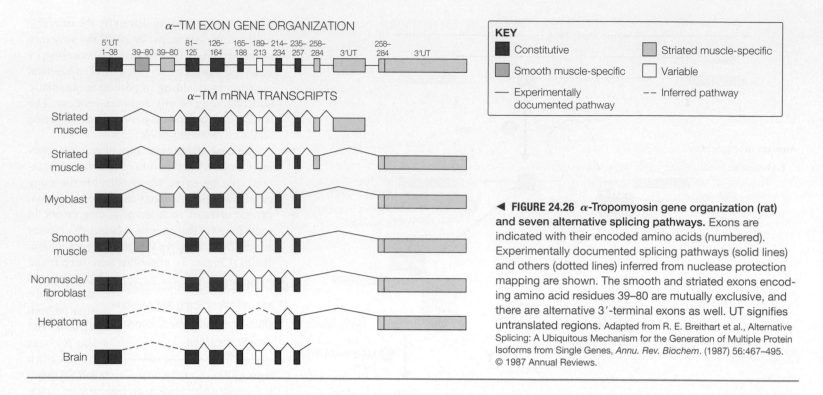

α–TM EXON GENE ORGANIZATION

α–TM mRNA TRANSCRIPTS

Striated muscle

Striated muscle

Myoblast

Smooth muscle

Nonmuscle/ fibroblast

Hepatoma

Brain

KEY

- ■ Constitutive
- ▨ Smooth muscle-specific
- ▨ Striated muscle-specific
- □ Variable
- — Experimentally documented pathway
- – – Inferred pathway

◀ **FIGURE 24.26** α-Tropomyosin gene organization (rat) and seven alternative splicing pathways. Exons are indicated with their encoded amino acids (numbered). Experimentally documented splicing pathways (solid lines) and others (dotted lines) inferred from nuclease protection mapping are shown. The smooth and striated exons encoding amino acid residues 39–80 are mutually exclusive, and there are alternative 3′-terminal exons as well. UT signifies untranslated regions. Adapted from R. E. Breithart et al., Alternative Splicing: A Ubiquitous Mechanism for the Generation of Multiple Protein Isoforms from Single Genes, *Annu. Rev. Biochem.* (1987) 56:467–495. © 1987 Annual Reviews.

A dramatic example of alternative splicing is shown in **FIGURE 24.26**. The protein α-*tropomyosin* is used in contractile systems in various cell types. Apparently, the need for functional domains coded for by different exons differs among the various uses of α-tropomyosin. Rather than having different genes expressed in different tissues, a single gene is employed, but the specific splicing patterns in different tissues provide a variety of α-tropomyosins. As the figure shows, there are two positions at which alternative choices can be made for which exon to splice in. The 3′ member of each of these pairs is the

default exon; it will be chosen unless a specific cellular signal dictates otherwise.

Many mechanisms have been described for alternative splicing. A well-studied example involves calcitonin, a peptide hormone that in thyroid gland regulates calcium and phosphorus metabolism. In neuronal tissue, the gene for calcitonin is alternatively spliced to give a protein that acts as a vasodilator—two quite different gene products expressed from the same gene. Analysis of the human genome sequence has revealed that most human genes, 90% or more, are subject to alternative splicing.

Summary

- The discovery of messenger RNA led to our understanding of the enzymology of RNA synthesis. All RNA is synthesized by the template-dependent copying of one DNA strand within a gene, catalyzed by RNA polymerase. (Section 24.1)

- RNA polymerases use 5′-ribonucleoside triphosphates as substrates, and they transcribe in a 5′ → 3′ direction. Bacteria synthesize all RNA classes with one polymerase, whereas eukaryotic cells have different polymerases—I, II, and III—for synthesis of ribosomal, messenger, and transfer RNA species, respectively. (Section 24.2)

- Strand selection and duplex unwinding and rewinding are carried out by RNA polymerase in conjunction with other proteins. The enzyme binds at a promoter site, by formation of specific DNA–protein contacts, largely involving the enzyme's σ subunit in bacteria. Most initiations are abortive, but after a productive initiation involving different factors in bacteria and eukaryotes, elongation continues. In bacteria this is carried out by the core polymerase, $\alpha_2\beta\beta'\omega$. Transcription is highly processive and is

terminated by specific DNA sequences, sometimes in bacteria with the participation of ρ protein. (Section 24.3)

- Structural analyses have identified common features in bacterial and eukaryotic RNA polymerases, which have helped to reveal common mechanistic features. Eukaryotic transcription involves multiple proteins in addition to RNA polymerase, which interact to form transcription complexes. Polymerases I, II, and III synthesize eukaryotic rRNAs, mRNAs, and small RNAs (including tRNAs), respectively. (Section 24.4)

- Posttranscriptional processing includes trimming tRNA and rRNA precursors and in eukaryotic cells cutting and splicing mRNA precursors. Bacterial mRNAs undergo little posttranscriptional processing, whereas eukaryotic messages are extensively processed, with polyadenylylation at the 3′ end, capping at the 5′ end, and splicing throughout the gene. Splicing is carried out by small nuclear ribonucleoprotein particles guided by base sequence interactions. Alternative splicing expands the information content of a genome by directing different mRNA splicing patterns in different tissues and developmental stages. (Section 24.5)

Problems

Enhanced by
Mastering Chemistry
for Biochemistry

Mastering Chemistry for Biochemistry provides select end-of-chapter problems and feedback-enriched tutorial problems, animations, and interactive figures to deepen your understanding of complex topics while practicing problem solving.

Answers to red problems are available in the Answer Appendix.

1. Outline an experimental approach to determining the average chain growth rate for transcription in vivo. Chain growth rate is the number of nucleotides polymerized per minute per RNA chain.

2. Outline an experimental approach to determining the average RNA chain growth rate during transcription of a cloned gene in vitro.

3. Measurements of RNA chain growth rates are often led astray by the phenomenon of *pausing,* in which an RNA polymerase molecule stops transcription when it reaches certain sites, for intervals that may be as long as several seconds. How might pausing be detected?

4. Suppose you want to study the transcription in vitro of one particular gene in a DNA molecule that contains several genes and promoters. Without adding specific regulatory proteins, how might you stimulate transcription from the gene of interest relative to the transcription of the other genes on your DNA template? To make all of the complexes identical, you would like to arrest all transcriptional events at the same position on the DNA template before isolating the complex. How might you do this?

5. The *tac* promoter, an artificial promoter made from portions of the *trp* and *lacUV5* promoters, has been introduced into a plasmid. It is a hybrid of the *lac* and *trp* (tryptophan) promoters, containing the −35 region of one and the −10 region of the other. This promoter directs transcription initiation more efficiently than either the *trp* or *lac* promoters. Why?

6. Explain the basis for the following statement: Transcription of two genes on a plasmid can occur without the concomitant action of a topoisomerase, but only if those two genes are oriented in opposite directions.

7. Some years ago, it was suggested that the function of the poly(A) tail on a eukaryotic message may be to "ticket" the message. That is, each time the message is used, one or more residues is removed, and the message is degraded after the tail is shortened below a critical length. Suggest an experiment to test this hypothesis.

8. For the original detection of DNA–RNA hybrid molecules, as described on page 745, the DNA–RNA hybrid was detected in a CsCl equilibrium gradient. Why are RNA and DNA–RNA hybrids denser than double-stranded DNA?

9. Shown below is an R loop prepared for electron microscopy by annealing a purified eukaryotic messenger RNA with DNA from a genomic clone containing the full-length gene corresponding to the mRNA.

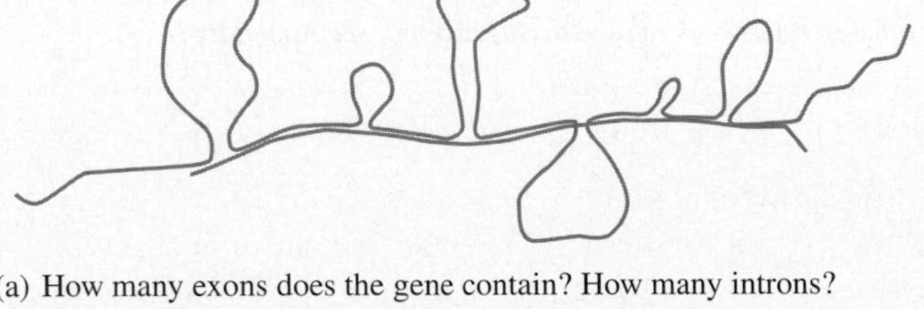

(a) How many exons does the gene contain? How many introns?
(b) Where in this structure would you expect to find a 5′,5′-internucleotide bond? Where would you expect to find a polyadenylic acid sequence?

10. Introns in protein-coding genes of some eukaryotes are rarely shorter than 65 nucleotides long. What might be a rationale for this limitation?

11. Heparin is a polyanionic polysaccharide that blocks initiation by RNA polymerase by virtue of its binding to double-stranded DNA. But heparin inhibits only when added before the onset of transcription, and not if added after transcription begins. Explain this difference.

12. Estimate the time needed for *E. coli* RNA polymerase at 37 °C to transcribe the entire gene for a 50-kilodalton protein. What assumption or assumptions must be made for this estimate to be accurate?

13. Describe how RNA polymerase backtracking could function to increase the fidelity of transcription.

14. Is RNA polymerase saturated with substrates in vivo? Describe experiments that might indicate whether RNA polymerase is operating at V_{max} with respect to its nucleotide substrates.

15. As discussed in the text, promoters were originally identified as consensus sequences upstream from transcriptional start sites. What additional evidence might support the assignment of these sequences as parts of promoters?

16. In this chapter and elsewhere, we have described two types of ultracentrifugation experiments—sucrose-gradient centrifugation and equilibrium density-gradient centrifugation. Briefly discuss these procedures with respect to the physical bases on which molecular species are separated in each method, the kinds of isotopic compounds used in the analysis, and the biological processes that have been or can be analyzed with each.

17. About 98% of the *E. coli* genome codes for proteins, yet mRNA, the template for protein synthesis, comprises only about 2% of the total RNA in the cell. Explain this apparent discrepancy.

18. Briefly explain why RNA-seq gives more information about the transcriptome than does microarray analysis. Read Tools of Biochemistry 24A before answering this question.

19. Why does it make biological sense for RNA synthesis *in vivo* to be less accurate than DNA synthesis?

20. Consider the −10 region sequences for the bacterial promoters in Figure 24.6. Why is the consensus sequence more active in promoting transcription than any of the sequences associated with actual promoters?

References

For a list of references related to this chapter, see Appendix II.

The finding in the 1960s that single-stranded DNA binds irreversibly to membrane filters, and the development of recombinant DNA technology in the 1970s, led to a number of techniques for analyzing gene expression (i.e., measuring the levels of transcripts of particular genes in living cells). RNA could be radiolabeled in vivo and hybridized to gene-specific DNA—a cloned gene or a restriction fragment—and the bound radioactivity analyzed by autoradiography or in a liquid scintillation counter. Several techniques, such as Northern analysis, were based on these developments. Northern analysis is comparable to Southern analysis, except that in Northern analysis RNA is hybridized to DNA restriction fragments separated by gel electrophoresis and then immobilized on a filter. However, such approaches allow the analysis of just one or a few genes in each experiment. With the availability of complete genome sequences, it became desirable to analyze levels of transcripts from many genes in a single experiment, that is, patterns of gene expression, which could be compared under different physiological conditions. Microarray technology makes this kind of analysis possible.

In a microarray experiment, minute amounts of gene-specific DNAs—usually several thousand—are immobilized on a substrate, such as glass or a membrane filter. The gene-specific DNAs are either cloned cDNAs or synthetic oligonucleotides. Using robotic technology, the investigator "prints" single-stranded DNAs onto the substrate, which may be a microscope slide, suitably coated to bind the applied DNAs. The DNAs are printed as a large array, which allows the investigator to identify each gene from its position on the array. The DNAs are fixed irreversibly on the substrate, so that the "DNA chip" can be used repeatedly by stripping the annealed RNA targets off the chip after each experiment.

Typically, a microarray experiment involves the comparison of gene-expression profiles under different conditions—comparing a tumor with the tissue of origin, for example, or comparing a hormone-stimulated tissue with unstimulated tissue. The investigator wishes to learn which genes are activated under the conditions being analyzed and which are repressed. Total mRNA is isolated from each tissue or cell culture and converted to a population of cDNAs using reverse transcriptase. During the enzymatic synthesis of the cDNAs, one of the deoxyribonucleoside triphosphates is tagged with a fluorescent dye. Typically, the reference sample is labeled with a red fluorophore, and the test sample is labeled with a green fluorophore. After cDNA synthesis is complete, the two samples are mixed and subjected to annealing conditions in the presence of the microarray. Unhybridized cDNAs are washed off, and the array is then scanned. Scanning at wavelengths corresponding to emission maxima of the fluorophores reveals which transcripts are more abundant in the test than in the reference (more green fluorescence) and which are less abundant (more red fluorescence). Analysis of the image reveals which genes were stimulated and which repressed under the conditions being tested. An example of a microarray analysis was shown in Figure 1.11 (page 14).

Microarray technology has numerous applications in addition to measuring patterns of gene expression. For example, by using arrayed oligonucleotides representing different mutant forms of a gene of interest, we can carry out DNA–DNA hybridization on the gene chip and identify mutations or single-nucleotide polymorphisms in biological samples.

The advent of "second-generation" DNA sequencing technology, which we mentioned in Chapter 4, has led to improved methods for transcriptome analysis, of which the most important is RNA-seq. In this method, RNA populations are reverse-transcribed by reverse transcriptase, and the resultant cDNA molecules are subjected to DNA sequence analysis in parallel. Hence, every RNA molecule in a population is sequenced, not only those molecules that can hybridize with the immobilized cDNAs or chemically synthesized oligonucleotides.

In a typical RNA-seq protocol the RNA population being analyzed is treated with poly(A) polymerase (page 757), which gives each RNA molecule a 3′ poly(A) tail. These molecules are then captured by hybridization to poly(dT)-coated magnetic beads and converted to double-stranded DNA by reverse transcriptase. These molecules are then fitted with adaptor oligonucleotides at each end and amplified by PCR. The amplified DNA population is then subjected to second-generation sequence analysis in parallel. Typically one microgram of total RNA suffices for the total analysis.

As mentioned earlier, microarray technology is used primarily for analysis of gene expression. RNA-seq can also be used to analyze differential gene-expression profiles, but it has additional capabilities. For example, because every expressed sequence is determined, RNA-seq can identify intron-exon junctions. RNA-seq can also be used to analyze patterns of noncoding RNA synthesis. Finally, RNA-seq is not limited, as is microarray analysis, by the population of cDNAs or oligonucleotides selected for immobilization. All expressed RNAs are captured.

References

For a list of references related to this chapter, see Appendix II.

So far, the techniques we have discussed for identifying and characterizing binding sites for proteins such as RNA polymerase and repressors have involved single binding sites, whereas regulatory proteins such as nuclear hormone receptors act at multiple binding sites. Moreover, techniques such as footprinting are carried out in vitro, although our main interest is characterizing attachment sites for DNA-binding proteins in intact cells. Chromatin immunoprecipitation (ChIP) makes it possible to identify in vivo binding sites on a genome-wide basis.

The principle of ChIP is that any DNA-binding protein can be covalently attached to its DNA binding site(s) in vivo by use of a cross-linking reagent that can penetrate cell membranes and react covalently with both protein and DNA in a reversible manner. Formaldehyde is most commonly used, as shown in **FIGURE 24B.1**. After whole cells are treated with formaldehyde, chromatin is isolated and subjected to sonic oscillation under conditions that reduce the length of each DNA molecule to fragments several hundred base pairs long. The mixture is treated with antibody to the protein of interest, and the immunoprecipitated DNA–protein complexes are collected. At this point, the cross-links are broken, and the DNA that was precipitated along with the protein is subjected to sequence analysis. Originally, this was done most often by PCR amplification of the DNA followed by conventional sequence analysis. This approach, however, makes it possible only to analyze known and suspected DNA sequences. An alternate approach involves cloning all DNA fragments in the mixture and then using PCR primers corresponding to flanking sequences on the vector, followed by sequence analysis of each

clone. With the advent of microarray technology (Tools of Biochemistry 24A), it became possible to screen the DNA fragments against a DNA microarray containing hundreds or thousands of DNA sequences. This technique is called ChIP-chip, because the immunoprecipitated DNA fragments are identified on a gene chip. A still more recent innovation, called ChIP-seq, sequences all of the immunoprecipitated DNA in parallel using a next-generation sequencing technology that permits simultaneous sequence analysis of hundreds or thousands of DNA molecules.

References

For a list of references related to this chapter, see Appendix II.

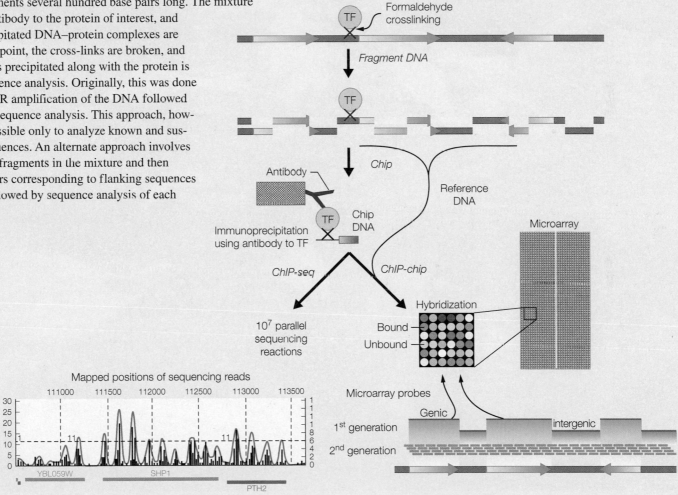

▲ **FIGURE 24B.1 Chromatin immunoprecipitation.** A transcription factor (TF) is crosslinked to the DNA sites to which it binds in chromatin. After fragmentation and immunoprecipitation using antibody to TF, the TF-bound DNA fragments are either identified by microarray analysis, in which the DNA is bound to a red fluorophore and subjected to hybridization analysis with an array of genomic fragments (ChIP-chip), or subjected to massive parallel sequence analysis (ChIP-seq). Data from B. J. Venters and B. F. Pugh (2009) How eukaryotic genes are transcribed, *Critical Reviews in Biochemistry and Molecular Biology* 44:117–141.

Many antibiotics act by inhibiting steps in protein synthesis. There is much criticism of the large-scale administration of antibiotics to livestock, particularly in feedlots. What is the rationale for this use of antibiotics, and what is the basis for criticism of the practice?

Information Decoding: Translation and Posttranslational Protein Processing

WE NOW TURN to what may be the most complex process in biological information transfer—the decoding of genetic messages in the 4-letter language of nucleic acids into polypeptides expressed in the 20-letter language of proteins. In DNA replication, transcription, and reverse transcription, information transfer is guided strictly by Watson–Crick base pairing between the template nucleic acid and the product, whether the template is DNA or RNA. By contrast, when a messenger RNA sequence directs the synthesis of a specific protein, base-sequence complementarity is still crucially involved, but a more complex overall process converts information encoded in a nucleotide sequence to information expressed as a specific sequence of the 20 common amino acids found in proteins.

In terms of the number of components involved—ribosomal RNAs (rRNAs) and proteins, transfer RNAs (tRNAs), amino acid activating enzymes, and soluble protein factors—and the variety of different proteins in each cell, protein synthesis is among the most complex of all metabolic processes, and it certainly involves the dominant portion of a cell's metabolic effort. In a logarithmically growing bacterial cell, as much as 90% of the total metabolic effort may be devoted to protein biosynthesis, with the metabolic machinery for translation accounting for 35% of the cell's dry weight.

Protein synthesis includes not only translation, which yields a specific amino acid sequence, but also posttranslational processing and trafficking because each protein must be properly modified and transported to its ultimate intracellular or extracellular destination. We have already seen, for example, that protein processing often involves cleavage, as in the conversion of preproinsulin to insulin, and modification of individual amino acids, as in the hydroxylation of proline residues in collagen synthesis or the phosphorylation or acetylation of specific amino acid residues. Here we briefly consider protein trafficking, the process by which mature (or maturing) proteins are moved to their ultimate destinations, whether inside or outside the cell.

25.1 An Overview of Translation

We introduced translation in Chapters 4 and 5. Figure 4.34 shows, for example, how translation involves the movement of a ribosome with respect to an mRNA molecule, three nucleotides at a time, with each trinucleotide mRNA sequence pairing with a tRNA charged with a specific amino acid, and with the polypeptide chain growing stepwise, one amino acid per step, from the N-terminus to the C-terminus. **FIGURE 25.1** shows a somewhat more detailed picture of this process and serves as an overview of translation. Figure 5.17 presents the genetic code, which shows how each of the 64 possible trinucleotide sequences corresponds to one of the 20 possible amino acids. In this chapter we expand on both the process of protein synthesis and the elucidation and nature of the genetic code.

In 1958, several years before elucidation of the genetic code or the demonstration of messenger RNA, Francis Crick predicted the existence of *adaptor molecules,* each of which would help translate the genetic message by binding to a specific amino acid

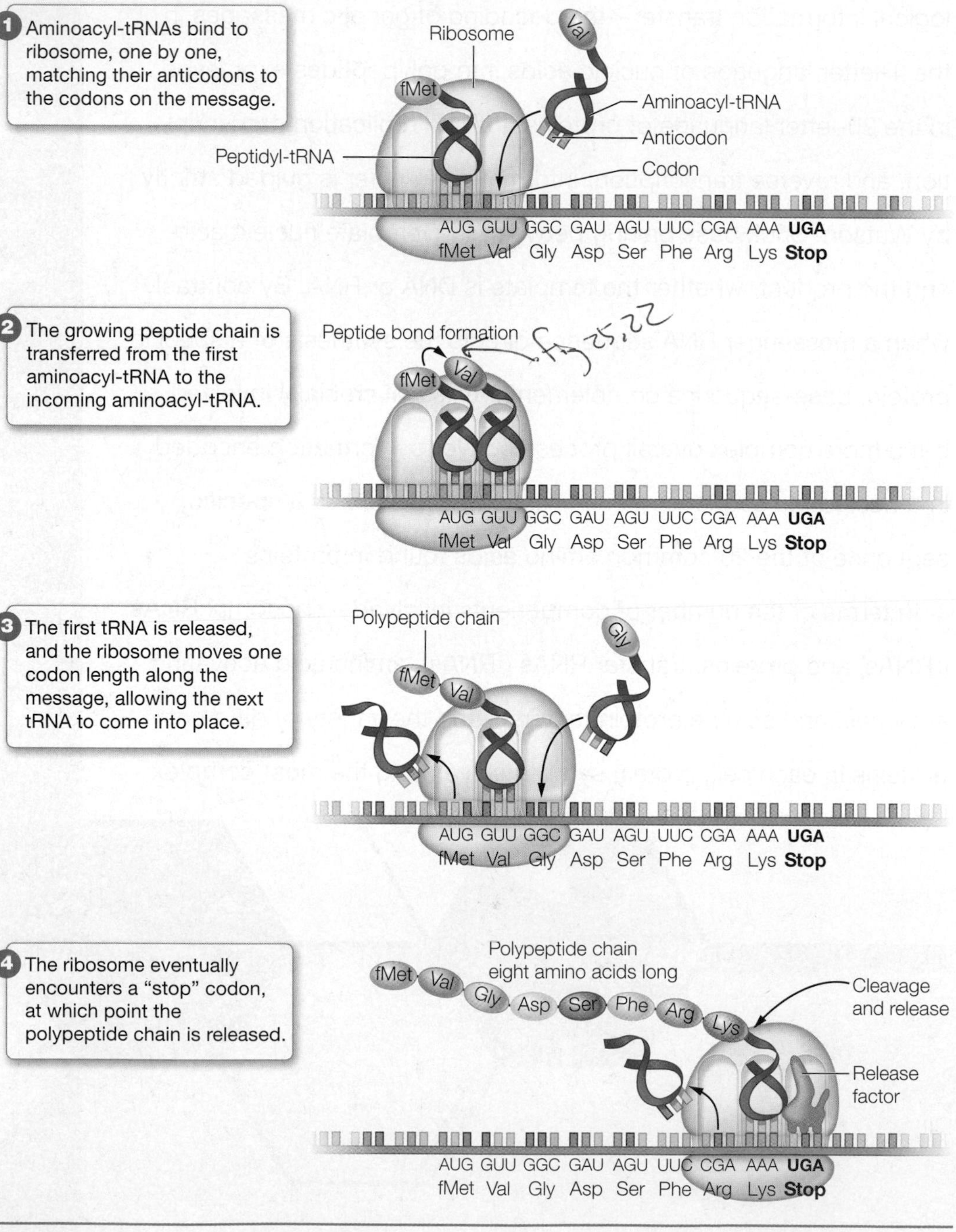

① Aminoacyl-tRNAs bind to ribosome, one by one, matching their anticodons to the codons on the message.

② The growing peptide chain is transferred from the first aminoacyl-tRNA to the incoming aminoacyl-tRNA.

③ The first tRNA is released, and the ribosome moves one codon length along the message, allowing the next tRNA to come into place.

④ The ribosome eventually encounters a "stop" codon, at which point the polypeptide chain is released.

▶ **FIGURE 25.1 Translation of an RNA message into a protein.** As the ribosome moves along the mRNA, it accepts specific aminoacyl tRNAs in succession, selecting them by matching the trinucleotide anticodon on the tRNA to the trinucleotide codon on the RNA message. See text for details.

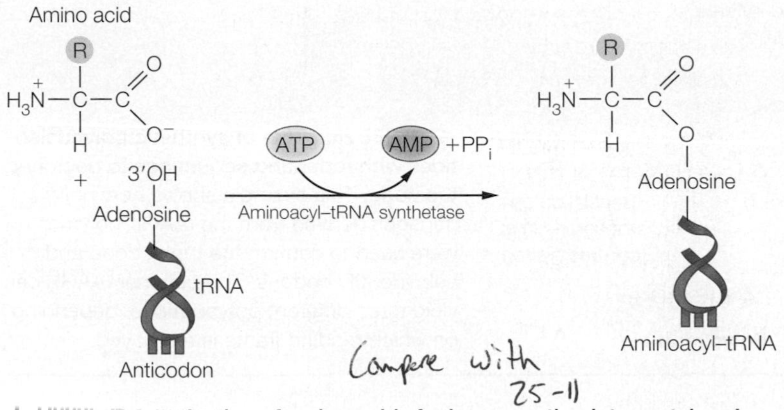

▲ FIGURE 25.2 Activation of amino acids for incorporation into proteins. A specific enzyme, aminoacyl-tRNA synthetase, recognizes both a particular amino acid and a tRNA carrying the corresponding anticodon. This synthetase catalyzes the formation of an aminoacyl tRNA, with accompanying hydrolysis of one ATP to AMP.

and linking it to a molecular code word in the translation machinery. These adaptors turned out to be transfer RNAs. As discussed in Chapter 4, each tRNA molecule is 75–80 nucleotides long (some as large as 93 nucleotides), folded by intramolecular hydrogen bonding into a three-loop structure. Each tRNA molecule is designed to bind one of the 20 amino acids through the specificity of an amino acid-activating

● **CONCEPT** tRNAs are the adaptor molecules that match amino acids to codons.

enzyme, more properly called an *aminoacyl-tRNA synthetase.* The overall reaction is shown in **FIGURE 25.2** and described in detail on page 775.

Each tRNA contains, in a region known as the **anticodon loop,** a trinucleotide sequence called the **anticodon** that is complementary to the appropriate trinucleotide codon in the message. Thus, the whole set of tRNAs contained in a cell comprises a molecular dictionary for the translation—it defines the correspondences between words in the 4-letter nucleic acid language (gene nucleotide sequence) and words in the 20-letter amino acid language (protein amino acid sequence).

The mRNA is bound to a ribosome, as shown in Figure 25.1. The aminoacyl tRNAs also bind here, one by one, matching their anticodons to the codons on the message, as shown in Figure 25.1 (step ❶). The growing peptide chain is transferred from the tRNA to which it is bound to the incoming aminoacyl-tRNA (step ❷). The first tRNA is then released, and the ribosome moves one codon length along the message, allowing the next tRNA to come into place, carrying *its* amino acid (step ❸). Again, energy from high-energy phosphate hydrolysis is expended at each step in the movement. As the ribosome moves along the mRNA, it eventually encounters a "stop" codon, at which point the polypeptide chain is released. Step ❹ shows a completed, albeit short, protein. In every cell, of every kind of organism, this remarkable machinery translates the information coded in thousands of different genes into thousands of different proteins. The mRNA message is always read in the $5' \rightarrow 3'$ direction, and the polypeptide chain is synthesized starting with its N-terminal residue.

The simple picture of translation presented so far leaves a host of questions unanswered. How are specific tRNAs matched to specific

● **CONCEPT** Messenger RNA is read $5' \rightarrow 3'$. Polypeptide synthesis begins at the N-terminus.

amino acids? How does the ribosome attach to the mRNA and move along it? How does it catalyze peptide bond formation? How does it start and stop translation correctly? How does it avoid making mistakes? Where does the energy for all of this activity come from? To answer such questions, we must dissect the whole process of translation, with careful examination of each of its parts. First, let us consider the genetic code in more detail.

25.2 The Genetic Code

We introduced the genetic code in Chapters 4 and 5. Here, we describe how the code was deciphered. By the late 1950s, it was generally accepted that a protein's amino acid sequence was encoded by the sequence of bases in a nucleic acid template. A triplet code seemed most likely, with three nucleotides specifying one amino acid. A doublet code wouldn't work because there are only 16 possible dinucleotide sequences (4×4), and we need at least 20 code words if each amino acid is to have its own. So, a triplet code seemed the simplest. With 64 possible trinucleotides ($4 \times 4 \times 4$), it was likely that some amino acids had more than one code word. Genetic experiments supported the idea of a triplet code, as well as a code that is *nonoverlapping* and *unpunctuated*:

$$-UCG-GGA-AAC-UCC-UCA-$$
$$-aa1 \quad -aa2 \quad -aa3 \quad -aa4 \quad -aa5$$

How the Code Was Deciphered

Biochemical elucidation of the code began in 1961 when Marshall Nirenberg and Heinrich Matthaei used artificial RNA templates for in vitro protein synthesis. Recall from Chapter 24 that the enzyme polynucleotide phosphorylase catalyzes the nontemplate-dependent synthesis, from a mixture of ribonucleoside diphosphates, of a random-sequence RNA, whose nucleotide composition matches that of the reaction mixture. Nirenberg and Matthaei polymerized UDP with the enzyme to synthesize polyU, a polyribonucleotide containing only UMP residues. When this artificial RNA was placed in a cell-free system containing a bacterial extract, ATP, GTP, and the 20 canonical amino acids (i.e., those commonly found in proteins), the product was a polypeptide containing only phenylalanine. Thus, the genetic code word for phenylalanine was a specific sequence of UMP residues—three if we are really dealing with a triplet code. In short order, polyC was shown to encode only proline and polyA only lysine. Other experiments with random-sequence polymers containing two or more different nucleotides established the nucleotide *composition* of most codons, but not their *sequences* (see Problem 1 at the end of this chapter).

Two approaches led to the identification of codon sequences. First, H. Gobind Khorana synthesized polyribonucleotides of regular repeating sequence. For example, the polymer UCUCUCUC . . . was shown to direct the synthesis of an alternating copolymer, Ser-Leu-Ser-Leu-Ser-Leu . . . If the code is triplet and nonoverlapping, then UCU encodes either serine or leucine and CUC encodes the other amino acid. Because serine was previously determined to have a codon with 2 Us and 1 C, and leucine a codon with 2 Cs and 1 U, this established UCU

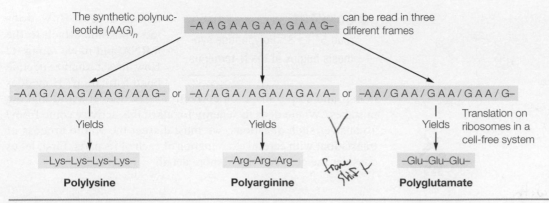

The synthetic polynucleotide (AAG)$_n$

–AAGAAGAAGAAG–

can be read in three different frames

–AAG/AAG/AAG/AAG– or –A/AGA/AGA/AGA/A– or –AA/GAA/GAA/GAA/G–

Yields | Yields | Yields

–Lys–Lys–Lys–Lys– | –Arg–Arg–Arg– | –Glu–Glu–Glu–

Polylysine | **Polyarginine** | **Polyglutamate**

Translation on ribosomes in a cell-free system

frame shift

◀ **FIGURE 25.3 Use of synthetic polynucleotides with repeating sequences to decipher the code.** This example shows how polypeptides derived from the (AAG)$_n$ polymer were used to confirm the triplet code and help identify codons. The polymer (AAG)$_n$ can yield three different polypeptides, depending on which reading frame is employed.

as a serine codon and CUC as a leucine codon. When a trinucleotide was used as the repeating unit, a different result was seen, as shown in **FIGURE 25.3**. The polymer AAGAAGAAG… directed synthesis of three homopolypeptides—polyLys, polyArg, and polyGlu. This experiment didn't give codon sequences, but it did establish the triplet and non-overlapping nature of the code. Here the nature of the product was set by the initial **reading frame**—the trinucleotide sequence chosen for the first amino acid incorporation event. If GAA was selected, for example, then every subsequent codon would also be GAA, making all amino acids in the product identical. The experiment did establish GAA, AGA, and AAG as codons for the three amino acids but could not directly assign each amino acid to one particular codon.

Experiments of this kind identified many code words, but in 1964 Philip Leder and Marshall Nirenberg discovered that synthetic trinucleotides would bind to ribosomes and direct the binding of specific tRNAs. For example, UUU and UUC stimulated the binding of phenylalanine tRNAs to ribosomes, and CCC and CCU stimulated the binding of proline tRNA. Such experiments not only yielded codon assignments, but they provided unequivocal evidence for the *redundancy* of the code, because several different codons were found to correspond to a single amino acid. By the combined use of these techniques, the entire genetic code was established within a few years after the demonstration of polyU-directed incorporation of Phe.

Features of the Code

In the genetic code (**FIGURE 25.4**), 61 of the 64 trinucleotides are called "sense" codons because they code for one amino acid. The remaining three are called "nonsense" codons because normally they do *not* code for an amino acid (with some exceptions; see **TABLE 25.1** and the accompanying discussion). When a ribosome encounters a nonsense codon (UAG, UAA, or UGA) in the correct reading frame, there is no aminoacyl-tRNA in the cell containing a matching anticodon, so translation ceases. As we will see later, these codons are used as part of the normal machinery for terminating translation of a message. The code is *degenerate* (or redundant) in the sense that most amino acids have more than one codon, but *unambiguous,* in the sense that each particular trinucleotide encodes one and only one amino acid. There are some exceptions to this generalization, as summarized in Table 25.1, so the genetic code is almost, but not quite, universal.

The genetic code was assigned using in vitro systems and synthetic mRNA templates, so how can we be sure that these codon assignments

are valid for the translation of messages in living cells? Some of the validation came from amino acid sequence analysis of mutant human hemoglobins (Chapter 7). Most of the amino acid sequence changes

			Second position			
		U	C	A	G	
First position (5' end)	U	UUU ⎫ Phe UUC ⎭ UUA ⎫ Leu UUG ⎭	UCU ⎫ UCC ⎪ Ser UCA ⎪ UCG ⎭	UAU ⎫ Tyr UAC ⎭ UAA Stop UAG Stop	UGU ⎫ Cys UGC ⎭ UGA Stop UGG Trp	U C A G
	C	CUU ⎫ CUC ⎪ Leu CUA ⎪ CUG ⎭	CCU ⎫ CCC ⎪ Pro CCA ⎪ CCG ⎭	CAU ⎫ His CAC ⎭ CAA ⎫ Gln CAG ⎭	CGU ⎫ CGC ⎪ Arg CGA ⎪ CGG ⎭	U C A G
	A	AUU ⎫ AUC ⎪ Ile AUA ⎭ AUG Met	ACU ⎫ ACC ⎪ Thr ACA ⎪ ACG ⎭	AAU ⎫ Asn AAC ⎭ AAA ⎫ Lys AAG ⎭	AGU ⎫ Ser AGC ⎭ AGA ⎫ Arg AGG ⎭	U C A G
	G	GUU ⎫ GUC ⎪ Val GUA ⎪ GUG ⎭	GCU ⎫ GCC ⎪ Ala GCA ⎪ GCG ⎭	GAU ⎫ Asp GAC ⎭ GAA ⎫ Glu GAG ⎭	GGU ⎫ GGC ⎪ Gly GGA ⎪ GGG ⎭	U C A G

Third position (3' end)

▲ **FIGURE 25.4 The genetic code, as used in most organisms.** Chain termination (stop) codons are shown in red, and the usual start codon AUG is dark green. Rarely, bacteria use other start codons besides AUG; these are shown in light green. When AUG is used as a start codon in bacteria, it codes for *N*-formylmethionine (fMet); when not used as a start codon, it codes for methionine (Met). AUG always codes for methionine in eukaryotes. Exceptions to these codon assignments are listed in Table 25.1.

TABLE 25.1 Modifications of the genetic code

Codon	Usual Use	Alternate Use	Where Alternate Use Occurs
AGA AGG	Arg	Stop, Ser	Some animal mitochondria, some protozoans
AUA	Ile	Met	Mitochondria
CGG	Arg	Trp	Plant mitochondria
CUU CUC CUA CUG	Leu	Thr	Yeast mitochondria
AUU	Ile	Start (N-fMet)	Some bacteria
GUG	Val	Start	
UUG	Leu	Start	
UAA	Stop	Glu	Some protozoans
UAG	Stop	Pyrrolysine	Various archaea
		Glu	Some protozoans
UGA	Stop	Trp	Mitochondria, mycoplasmas
		Selenocysteine	Widespread[a]
		Selenocysteine and Cys	*Euplotes*

[a]Depends on context of message and other factors

● **CONCEPT** The genetic code is almost, but not quite, universal.

could be accounted for by substitution mutations involving a single base, the most frequent spontaneous mutation. For example, the Glu → Val substitution seen in sickle-cell hemoglobin could be accounted for by changing a GAA Glu codon to a GUA Val codon, or GAG to GUG. In subsequent years, alignment of amino acid sequences for purified proteins with nucleotide sequences in the corresponding genes has provided conclusive evidence for the validity of the code.

Deviations from the Genetic Code

Why has the genetic code remained almost unchanged over so vast an evolutionary span? Perhaps it is simply because even small codon changes could be devastating. A single codon change could alter the sequence of nearly every protein made by an organism. Some of these changes would almost certainly have lethal effects. Therefore, codon changes have been opposed by intense selective pressure during evolution. They represent changes in the most basic rules of the game. Yet significant deviations do occur, most notably the differences in the mitochondrial code and coding for the "21st and 22nd" amino acids"—namely, selenocysteine and pyrrolysine (see Figure 5.8). A significant change in the mitochondrial code, as shown in Table 25.1, is the change in AUA from an isoleucine to a methionine codon. Mitochondrial proteins contain methionine in higher abundance than do proteins in other cell compartments, but the evolutionary significance of this codon change is not clear.

Selenocysteine (Sec, 21st amino acid) and pyrrolysine (Pyl, 22nd amino acid) are translated differently. Both use codons that are otherwise used in translation termination—UGA for Sec and UAG for Pyl. A special tRNA, tRNA^{Sec}, is a substrate for a seryl-tRNA^{Sec} synthetase, which charges serine directly, to give Ser-tRNA^{Sec}. (Note the convention; Ser refers to the amino acid bound, and superscript Sec denotes the amino acid corresponding to the anticodon on that tRNA molecule.) The tRNA-linked Ser is then converted to Sec by a two-step process beginning with phosphorylation of the serine hydroxyl group. The resultant Sec-tRNA^{Sec} responds to a UGA codon. For a particular UGA to be translated as Sec rather than read as a stop codon, that UGA must have a special Sec Insertion Sequence (SECIS) nearby, usually in the 3'-untranslated region of the mRNA transcript (3'UTR). Although insertion of selenocysteine is rather rare among proteins, the human proteome does include 25 selenoproteins. As mentioned in Chapter 15, some of these proteins participate in oxidant protection.

Pyl, by contrast, has a much narrower distribution, having been found so far in only about 1% of all examined organisms, mostly methanogenic archaea. Pyl is converted directly to a pyrrolysyl-tRNA by its own amino acyl-tRNA synthetase. The resultant Pyl-tRNA^{Pyl} has an anticodon that pairs with UAG, normally used in chain termination. So far, it is not clear whether any of the UAGs in these genomes are read as stop codons or whether all encode Pyl.

Finally, although we say that the code is unambiguous, at least one gene in *E. crassus* (a ciliated protozoan of the genus *Euplotes*) has UGA triplets that encode both cysteine and selenocysteine. Sequence context is key to ensuring correct insertional specificity. Since this organism also uses UGA as a tryptophan codon in its mitochondria, UGA plays multiple roles.

The Wobble Hypothesis

Most amino acids in the codon table (Figure 25.4) are characterized by the first two codon letters. For example, all four Pro codons start with CC, and all four Val codons start with GU. Thus, redundancy is usually expressed in the third letter—ACU, ACC, ACA, and ACG all code for Thr. Soon after the code was deciphered, it was observed that a single tRNA may recognize several different codons. The multiple recognition always involves the 3' residue of the codon, which corresponds to the 5' residue of the anticodon.

● **CONCEPT** The code is redundant. Several codons may correspond to a single amino acid, sometimes via wobble in the 5' anticodon position.

In 1966, Francis Crick proposed that the 5' base of the anticodon was capable of "wobble" in its position during translation, allowing it to make alternative (non-Watson–Crick) hydrogen-bonding arrangements with several different codon bases. An example is shown in **FIGURE 25.5**. G in the 5' anticodon position can pair with either C or U in the codon, depending on the orientation of the pair. Considering both base-pairing possibilities and the observed selectivity of tRNAs, Crick proposed the set of "wobble rules" listed in **TABLE 25.2**. This hypothesis nicely explains the frequently observed degeneracy in the 3' site of the codon. The rather uncommon nucleoside, *inosine* (I; Chapter 19), is found in a number of anticodons in the 5' position, where it can pair with A, U, or C.

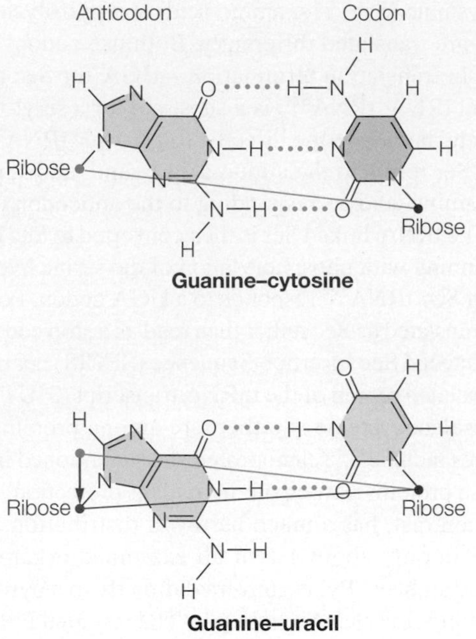

Guanine–cytosine

Guanine–uracil

▲ **FIGURE 25.5 The wobble hypothesis.** As an example, we show how the anticodon base G can pair with either C or U in a codon. Movement ("wobble") of the base in the 5′ anticodon position is necessary for this to occur (see arrow).

Not all cases of multiple codon use involve the translation of a single tRNA using wobble. For example, four of the six Leu codons begin with CU and in principle could be translated by two different tRNAs, using wobble. However, the remaining two codons, UUA and UUG, will require a different anticodon, such as 3′-AAU-5′, which could translate both codons. In fact, *E. coli* contains five different leucine tRNAs, and multiple **isoaccepting tRNAs**—tRNAs accepting and translating the same amino acid—are common.

tRNA Abundance and Codon Bias

What we have learned about tRNA structure and the basis for degeneracy of the code has considerable practical significance, particularly for those wishing to express recombinant eukaryotic proteins in bacteria. Consider *E. coli,* a frequently used host for recombinant

gene expression. Of the six arginine codons, two (AGA and AGG) are rarely used in *E.coli* mRNAs, meaning that each of these triplets represents fewer than 1% of the arginine codons in the entire genome. The intracellular concentration of the tRNA with the anticodon, 3′-TCI-5′, which can translate these two rare codons, is also quite low. This *codon bias* means that a recombinant gene with more abundant representation of these codons may be poorly expressed after transfer into *E. coli* because a tRNA to translate that codon is present in low abundance. This situation can be remedied by site-directed mutagenesis of the recombinant genes to change these rare codons to arginine codons that are more abundant, and more efficiently translated, in the *E. coli* environment. An alternative approach is to engineer the *E. coli* host to overexpress the rare tRNA, so that the codons AGA and AGG can be efficiently translated.

● **CONNECTION** Overexpression of recombinant proteins often requires adjustment of the gene to be expressed, resulting in the use of codons for which corresponding tRNAs are abundant in the host cell.

A code that has synonymous codons with similar structures means that many mutations involving single-base changes are silent because a codon change (e.g., UUA → UUG) doesn't change the sense of the genetic message (i.e., it still codes for Leu). Not only are many single-base changes silent, but many more are conservative, in the sense that a mutation may substitute a structurally similar amino acid that can be tolerated by the protein with no loss of function. For example, each of the six leucine codons can be converted to a codon for the closely related valine by a single-base change. This suggests that the code has evolved to maximize genetic stability. You might ask yourself: how many of the possible single-base changes involving the six leucine codons yield another leucine codon? How many yield a conservative Leu → Ile substitution? How many Ile → Val? The genetic code is buffered against change.

Punctuation: Stopping and Starting

Because the mRNA is invariably longer than the sequence that is to be translated, specific start and stop signals are required to begin and end translation. In almost all organisms, UAA, UAG, and UGA are used for stop signals and do not code for any amino acid (with the exceptions discussed above). A stop signal indicates that translation is to terminate and that the polypeptide product is to be released by the ribosome. Three stop signals are more than is absolutely necessary, so these codons are also used for designating amino acids in mitochondria and in other special cases (see Table 25.1).

Although nature has been generous in designating stop signals, it has been less generous in apportioning starts. The start signal commonly used in translation is AUG, which also serves as the single methionine codon. On page 775 we discuss how the translation machinery distinguishes between an AUG that is intended as an initiating codon and an AUG that specifies an internal methionine residue. As shown in Figure 25.4, GUG (Val), UUG (Leu), and AUU (Ile) are occasionally used as initiator codons in bacteria.

● **CONCEPT** Messenger RNAs contain translational start and stop signals.

TABLE 25.2 Base-pairing capabilities in wobble pairs		
Base at 5′ Position in Anticodon		**Base at 3′ Position in Codon**
G	pairs with	C or U
C	pairs with	G
A	pairs with	U
U	pairs with	A or G
I	pairs with	A, U, or C

25.3 The Major Participants in Translation: mRNA, tRNA, and Ribosomes

Messenger RNA

As we indicated in Chapter 24, eukaryotic mRNAs are quite different from prokaryotic mRNAs. Many or most bacterial mRNAs are *polycistronic*—that is, they encode two or more polypeptide chains. This means that the mRNA sequence must be punctuated, so that translation of the RNA corresponding to each gene is controlled by its own initiation and termination signals. Eukaryotic messages, by contrast, almost always encode just one protein, and the mRNA structure is the result of posttranscriptional processing that is far more extensive than that seen in bacterial systems.

As a good example of a bacterial mRNA, consider that produced by transcription of the *E. coli lac* operon, which was introduced in Chapter 24 and receives further attention in Chapter 26. This group of three linked genes—*lacZ, lacY,* and *lacA*—controls the utilization of lactose and related sugars by bacteria. As shown in **FIGURE 25.6**, these three genes are expressed as a single mRNA molecule some 5300 nucleotides in length. Within this mRNA are three **open reading frames,** corresponding to the *lacZ, Y,* and *A* genes. An open reading frame is a sequence within an mRNA, bounded by start and stop codons, that can be continuously translated. Each open reading frame has its own start and stop signals, and you can see that these signals vary somewhat. There is extra, untranslated RNA between the reading frames and at the ends. The regions 5′ to each start signal contain sequences rich in A and G, which help to align the mRNA on the ribosome so that translation can begin at the proper points and in the correct reading frame. Such attachment sequences, found on all bacterial mRNAs, are called *Shine–Dalgarno sequences,* after John Shine and Lynn Dalgarno, who first described them. A Shine–Dalgarno sequence can base-pair with a sequence contained in 16S rRNA, to properly align the mRNA on the ribosome for starting translation. The different attachment sequences appear to have different affinities for ribosomes. In the *lac* operon (Figure 25.6), for example, *lacZ* is translated much more frequently than *lacY* or *lacA*. The mRNA produced from the *lac* operon has all the basic elements necessary for its function: sequences to align it properly on the ribosome and sequences that start and stop translation at the proper points.

● **CONCEPT** Shine–Dalgarno sequences help align mRNAs on ribosomes to properly start translation in bacteria.

Transfer RNA

Any cell, bacterial, archaeal, or eukaryotic, contains a battery of different tRNA molecules sufficient to incorporate all 20 amino acids into protein. This does not mean that there need be as many tRNA types as there are codons, because some tRNAs can recognize more than one codon, when the difference is in the third, or wobble, position. However, *E. coli* has 88 different tRNA genes—more than enough to translate all 61 sense codons, with some amino acids having multiple tRNAs.

tRNA was the first natural polynucleotide sequence to be determined, in a pioneering study of yeast tRNA^{Ala} by Robert Holley in 1965. Since then, thousands of tRNAs have been sequenced.

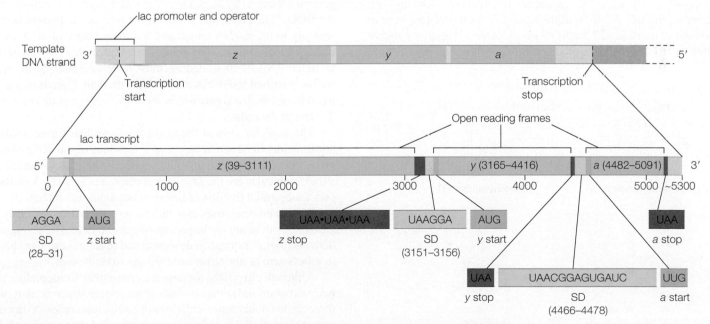

▲ **FIGURE 25.6 The *lac* operon mRNA.** The mRNA for the *E. coli lac* operon is about 5300 nucleotides long and contains open reading frames for the *lacZ, lacY,* and *lacA* genes, each flanked appropriately by start, stop, and Shine–Dalgarno (SD) sequences.

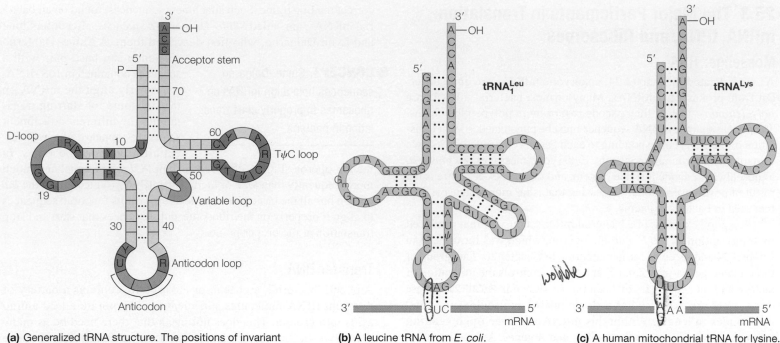

(a) Generalized tRNA structure. The positions of invariant and rarely varied bases are shown in purple. Regions in the D loop and the variable loop that can contain different numbers of nucleotides are shown in blue. The anticodon is shown in orange.

(b) A leucine tRNA from *E. coli*.

(c) A human mitochondrial tRNA for lysine. Code for bases: Y = pyrimidine, R = purine, ψ = pseudouridine, T = ribothymidine, and D = dihydrouridine (see Figure 25.8).

▲ **FIGURE 25.7** Structure of tRNAs.

All have the general base-pairing structure shown schematically in **FIGURE 25.7(a)** and have similar sequences of 70 to 80 nucleotides or more (see also Chapter 4). There is, however, considerable variation in detail, as shown in the examples in Figure 25.7(b) and (c). Furthermore, the tRNAs are unique among RNA molecules in their high content of unusual and modified bases, three of which are shown in **FIGURE 25.8**. Biosynthesis of the modified bases always occurs

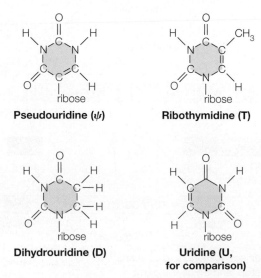

Pseudouridine (ψ)

Ribothymidine (T)

Dihydrouridine (D)

Uridine (U, for comparison)

▲ **FIGURE 25.8** A sampling of the modified and unusual bases found in tRNAs.

posttranscriptionally. Consider the types of reactions that could bring about each of the modifications shown in Figure 25.8.

Cloverleaf models of the kind shown in Figure 25.7 highlight the general pattern of hydrogen bonding and denote the functional parts of the tRNA. The *anticodon triplet* in the loop at the bottom is complementary to the mRNA codon and will base-pair with it. Because the codon and anticodon, when paired, constitute a short stretch of double-stranded RNA, their directions must be antiparallel. In Figure 25.7, we have written the tRNA molecules with their 5′ ends to the left, so the mRNAs in this figure are in the unconventional orientation, with 3′ ends to the right.

The *acceptor stem* at the top of the cloverleaf figure is where the amino acid will be attached, at the 3′ terminus of the tRNA. This stem always has the sequence 5′...CCA–OH 3′. Other common features of tRNA molecules are the *D loop* and the *TΨC* loop, regions that contain a substantial fraction of invariant positions and frequently contain modified or unusual bases as well. The so-called *variable loop* is indeed variable, both in nucleotide composition and in length, as Figure 25.7 demonstrates. Presumably, the conserved structural elements play roles in attachment of aminoacylated tRNAs to ribosomes.

Although cloverleaf models are convenient for depicting the primary structure and some elements of secondary structure, they are poor three-dimensional representations of tRNA molecules. X-ray diffraction studies of tRNA molecules have revealed the complex molecular shape shown in **FIGURE 25.9** and Figure 4.25. Based on figures like these, a tRNA molecule looks rather like a hand-held drill or soldering gun. The anticodon loop is at the bottom of the grip, and the acceptor stem is at the working tip. The *D* loop and the *T Ψ C* loop are folded

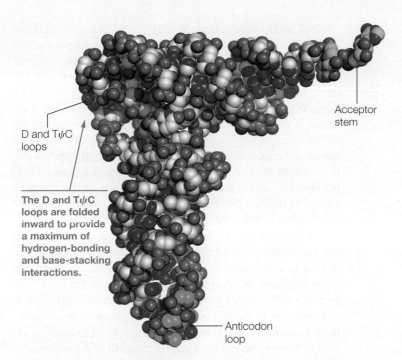

▲ **FIGURE 25.9** Space-filling model of yeast phenylalanine tRNA derived from X-ray diffraction studies. Carbon atoms are colored gray, except in the anticodon loop, where they are green, and in the 3′ acceptor stem, where they are cyan. PDB ID: 4tna

● **CONCEPT** All tRNAs share a general common structure that includes an anticodon loop, which pairs with codons, and an acceptor stem, to which the amino acid is attached.

inward in a complex fashion near the top of the grip, to provide a maximum of hydrogen-bonding and base-stacking interactions. Some of the hydrogen-bonding patterns required to produce this folding are rather unusual (**FIGURE 25.10**), including some where a single base is paired with two other bases. The three-dimensional shapes of the tRNAs are highly conserved, even though the primary structures vary, probably so that each tRNA can fit equally well onto the ribosome and carry out its function.

Aminoacyl-tRNA Synthetases: The First Step in Protein Synthesis

Amino acids are attached to tRNAs by a covalent bond between the carboxylate of the amino acid and the ribose 3′ hydroxyl group of the invariant 3′ terminal adenosine residue on the tRNA. Pairing of the correct amino acid residues and the tRNAs is accomplished by the *aminoacyl-tRNA synthetases* (abbreviated aaRS; see Figure 25.2). There are 21 aaRSs in *E. coli,* each of which recognizes one amino acid and one or more tRNAs. Lysine is unique in having two aaRSs. The reaction linking the two molecules, shown in **FIGURE 25.11**, consists of two steps. In step ❶, the amino acid, which is bound to the aaRS, is activated by ATP to form an **aminoacyl adenylate.** While still bound to the enzyme, this intermediate reacts with one of the correct tRNAs to form the covalent bond and release AMP (step ❷).

▲ **FIGURE 25.10** Unusual base pairings in tRNA. All are from the yeast tRNA^Phe shown in Figure 25.9. The bases prefixed by m are methylated at the carbon atom corresponding to the superscript. Numbers following the letters designating bases show the position in the sequence.

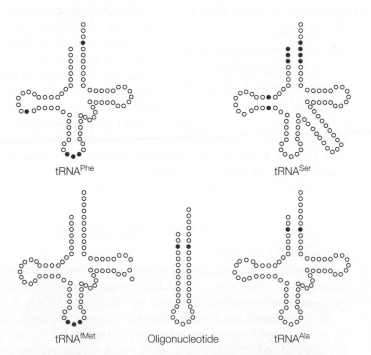

▲ **FIGURE 25.11 Formation of aminoacyl-tRNAs by aminoacyl-tRNA synthetase.** In step **1** the amino acid is accepted by the synthetase and is adenylylated, with the aminoacyl adenylate remaining bound to the enzyme. In step **2** the proper tRNA is accepted by the synthetase, and the amino acid residue is transferred to the 3′ OH of the 3′-terminal residue of the tRNA (class II enzymes) or to the 2′ OH, followed by isomerization to the 3′ aminoacyl-tRNA (class I enzymes). For class I enzymes, the 2′ OH of the 3′-terminal AMP residue is the nucleophile for reaction 2.

● **CONCEPT** Amino acids are coupled to their appropriate tRNAs by aminoacyl-tRNA synthetases (aaRSs).

There are two general classes of aminoacyl-tRNA synthetases (I and II). Their active sites are completely different, and the two classes bind their cognate tRNAs from opposite sides. Furthermore, the class I enzymes tend to function as monomers, whereas the class II enzymes function as dimers or tetramers. Moreover, the enzymes differ mechanistically. Class II enzymes link the aminoacyl moiety in the aminoacyl adenylate intermediate directly to the 3′ hydroxyl in the tRNA acceptor, while class I enzymes synthesize first a 2′-aminoacyl-tRNA intermediate, which then undergoes intramolecular transesterification, giving the 3′-aminoacyl-tRNA product.

The reasons for these mechanistic differences are unknown, but they may reflect the use of some amino acids in proteins before others in the very early evolution of protein synthesis.

You might expect that the aaRS would identify the correct tRNA on the basis of its anticodon, but recent studies indicate that the identification process is more complex, and various tRNA nucleotides act as *identity elements*. Indeed, changing a single base pair (between residues 3 and 70 in the acceptor stem) of tRNA^{Cys} or tRNA^{Phe} to the G-U pair found in tRNA^{Ala} causes the alanine synthetase to accept the tRNA^{Cys} or tRNA^{Phe} and couple it to alanine. Other tRNAs appear to be recognized by their aaRSs at many different locations (**FIGURE 25.12**). No simple rule has emerged, although identity elements are clustered in the anticodon loop and the acceptor stem. The importance of identity elements is dramatically illustrated by the fact that the yeast tRNA^{Ala} depicted in Figure 25.12 can be trimmed to just a single-hairpin molecule, as shown, and the molecule can be efficiently and accurately aminoacylated as long as a critical G-U base pair (shown in red) is present.

Aminoacyl-tRNA synthetases contribute to the fidelity of translation through a process akin to proofreading by DNA polymerases. In the instant between formation of an enzyme-bound aminoacyl adenylate and its conversion to aminoacyl-tRNA, the enzyme can sense the improper fit of the amino acid side chain and hydrolyze the intermediate before the condensation reaction that links the amino acid carboxyl group to the tRNA 3′ hydroxyl. Moreover, even if the wrong aminoacyl-tRNA is synthesized, the enzyme has a short time in which it can identify the mischarged amino acid as incorrect and

tRNA^{Phe} tRNA^{Ser}

tRNA^{fMet} Oligonucleotide   tRNA^{Ala}

▲ **FIGURE 25.12 Major "identity elements" in some tRNAs.** Red circles represent the positions that have been shown to identify the tRNA to its cognate synthetase. Shown also is a synthetic polynucleotide containing the G-U alanine identity element (in red), which is a good substrate for alanyl-tRNA synthetase.

Data from L. Schulman and J. Abelson, (1988) Recent excitement in understanding transfer RNA identity. *Science* 240:1591–1592.

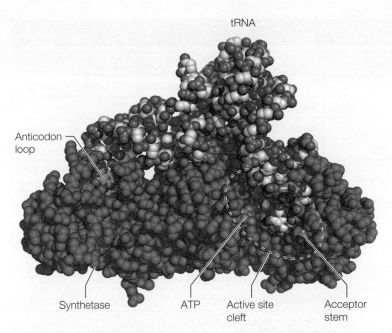

▲ **FIGURE 25.13** Crystal structure of the *E. coli* glutaminyl-tRNA synthetase coupled with its tRNA and ATP. The tRNA is represented by a model colored as in Figure 25.9, with the protein in magenta. The ATP (dark green) and the 3′ acceptor stem of the tRNA fit into a deep cleft in the synthetase. This cleft will also accommodate the amino acid. This is a monomeric class I synthetase. PDB ID: 1gtr.

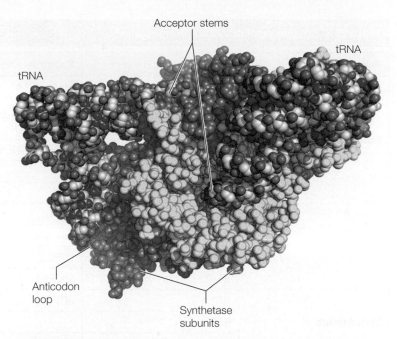

▲ **FIGURE 25.14** Yeast aspartyl-tRNA synthetase complexed with two molecules of tRNA^{Asp}. The two subunits of the enzyme homodimer are in blue and yellow. PDB ID: 1asy.

hydrolyze it before it can be released to participate in translation. In these ways, aminoacyl-tRNAs contribute to an overall error frequency for protein synthesis of about 10^{-4}. This is less accurate than DNA replication ($\sim 10^{-7}$), but the consequences of error are much lower, too, because the error is not propagated to the next generations.

Insight into the recognition of tRNAs by their synthetases has been provided by crystallographic analysis of the complexes formed. **FIGURE 25.13** shows the structure of a class I synthetase-tRNA complex involving *E. coli* glutaminyl-tRNA and its cognate synthetase. As shown in the figure, the tRNA lies across the protein, making a number of specific contacts, including crucial ones in the anticodon region and in the acceptor stem. Both of these regions are distorted in the complex, with the acceptor stem being elongated and inserted into the active site pocket. This pocket is formed by a common protein structural motif called the *dinucleotide fold,* which frequently acts as a nucleotide-binding region. In this case it also binds the ATP required for acylation. It provides a binding site for glutamine as well. Thus, all three participants in the reactions are grouped close together.

Similar interactions are seen with a class II synthetase. **FIGURE 25.14** shows the dimeric yeast aspartyl-tRNA synthetase complexed with two molecules of tRNA^{Asp}. Note that the opposite side of the tRNA is bound to the enzyme from that seen with class I synthetases. That is, the acceptor stem points to the right in Figure 25.14 (class II), whereas it points to the left in Figure 25.13 (class I). Only one of the two tRNA molecules is bound in a catalytically productive conformation.

The polypeptide chains of aaRS molecules range from about 500 to several thousand amino acid residues, with corresponding structural diversity. Within each structure is a conserved catalytic domain

120 to 130 residues in length. Charles Carter and colleagues designed and expressed genes for these domains, yielding polypeptides called urzymes (meaning original or primordial enzymes). Remarkably, these incomplete aaRS molecules catalyze amino acid acylation with rate enhancements some 10^6-fold over the uncatalyzed reaction (still at least one thousandfold lower than those for fully evolved aaRSs). Carter argues that urzymes represent intermediate stages in the evolution of aaRS molecules and, hence, that primordial evolution involved RNA and peptide interactions rather than a purely RNA-based process, as postulated by the "RNA world" model (Chapter 1).

In higher eukaryotes, most aaRSs are "moonlighting proteins"— proteins that evolved initially to function in protein synthesis, but that in further evolution acquired additional purposes. In humans, aaRS molecules are involved in functions as diverse as autoimmunity, control of apoptosis, regulation of rRNA synthesis, vascular development, and coordination of the DNA damage response. In all cases studied, the catalytic machinery for aminoacyl-tRNA synthesis has remained undisturbed, and evolutionary modifications to convey additional functions occur elsewhere on the protein molecule.

The Ribosome and Its Associated Factors

We have now described two of the participants that must be brought together to carry out translation—the mRNA and the set of tRNAs charged with the appropriate amino acids. The actors are in the wings, and all that is needed is a proper director and a stage on which the events can unfold. Both are provided by the ribosome, and the typical cell requires many. An *E. coli* cell, for example, contains as many as 20,000 ribosomes, accounting for about 25% of the dried cell mass. Thus, a cell devotes a large part of its energy to producing ribosomes and to using them in protein synthesis.

TABLE 25.3 Soluble protein factors in translation

Function	Factor (bacteria)	Factor (eukaryotes)	Role in Translation
Initiation	IF1	eIF1, eIF1A	Promotes dissociation of preexisting 70S or 80S ribosome
	IF2	eIF2, eIF2B	Helps attach initiator tRNA
	IF3	eIF3, eIF4C	Similar to IF1; prepares mRNA for ribosome binding
		eIF4A, eIF4B, eIF4F	Same as eIF1, eIF1A
		eIF5	Helps dissociate eIF2, eIF3, eIF4C
		eIF6	Helps dissociate 60S subunit from inactive ribosomes
Elongation	EF-Tu	eEF1α	Helps deliver aminoacyl-tRNA to ribosomes
	EF-Ts	eEF1$\beta\gamma$	Helps recharge EF-Tu with GTP
	EF-G	eEF2	Facilitates translocation
	EF-P	a/eIF5A	Helps translate consecutive proline codons
Termination	RF1	eRF	Release factor (UAA,UAG)
	RF2		Release factor (UAA, UGA)
	RF3		A GTPase that promotes release

Soluble Protein Factors in Translation

Before describing ribosomes in detail, however, we mention one more set of participants, whose functions will be described later. These are the soluble proteins that participate in the three stages of translation—initiation factors, elongation factors, and release factors. **TABLE 25.3** introduces these factors as initially studied in bacteria, as well as their eukaryotic counterparts. We shall refer back to the information in this table as we discuss mechanisms in translation.

Components of Ribosomes

The ribosome is a large ribonucleoprotein particle containing 60–70% RNA and 30–40% protein. Ribosomes and their subunits are characterized according to their sedimentation rates (S; see Chapter 24, Section 24.1) during ultracentrifugation. The individual bacterial ribosome has a sedimentation coefficient of 70S, consistent with a molecular mass of about 2.5×10^6 Da. Eukaryotic ribosomes are somewhat larger, with a sedimentation coefficient of 80S and a molecular mass of 4.2×10^6 Da. When isolated ribosomes are placed in a buffer containing Mg^{2+} at low concentration, they dissociate into two smaller subunits. As shown in **FIGURE 25.15**, bacterial 70S ribosomes dissociate into 30S and 50S subunits. We shall see later that dissociation and reassociation of these subunits occur during translation. Figure 25.15 also shows the number of RNA and protein components in each subunit. Note that the 50S bacterial subunit contains two rRNA molecules (5S and 23S) and 34 different proteins, while the 30S subunit contains just one rRNA (16S) and 21 proteins—all different from those in the 50S subunit. Proteins from the small subunit are called S1, S2, S3, …, and S21, while those from the large subunit are called L1, L2, L3, …, and L34. All proteins are present in one copy per ribosome, except for L12, which is present in four copies. Eukaryotic ribosomes are significantly larger, with larger rRNAs and more proteins. Here we shall be discussing primarily bacterial ribosomes, whose structures and functions are known in greater detail.

Once the complexity of the ribosome was revealed, particularly the large numbers of proteins in each subunit, it seemed a daunting task to determine the structure of the particle and to understand the function of each protein. However, as early as 1968, Peter Traub and Masayasu Nomura learned that they could reassemble 30S ribosomal subunits from the separated RNA and protein components. The product, when combined with 50S subunits, was active for in vitro protein synthesis. An obligatory order of assembly was seen, with some proteins being incorporated only after the binding of certain other proteins. The ability to assemble ribosomes in vitro made it possible to analyze the function of individual ribosomal proteins because ribosomal subunits could be assembled with specific proteins missing, followed by functional analysis of these deliberately altered particles. Later work established that the ribosomal assembly pathway as it occurs in vivo is similar but with some signficant differences.

● **CONCEPT** Despite their complexity, ribosomal subunits can be assembled in vitro.

Sequence analyses of the corresponding proteins in the ribosomes of different organisms reveal considerable evolutionary conservation. Thus, the ribosome is a complex object that evolved early in the history of life and has remained relatively unchanged. Although the ribosomes of eukaryotes differ significantly from those of bacteria, the evolutionary continuity is clear. The sequences of many rRNAs tell the same story. Indeed, because of their relatively slow evolutionary rates of change, rRNAs are useful as evolutionary yardsticks over vast phylogenetic distances. In fact, it was sequence analysis of 16S rRNAs that led Carl Woese to propose the existence of a third domain of life, the archaea.

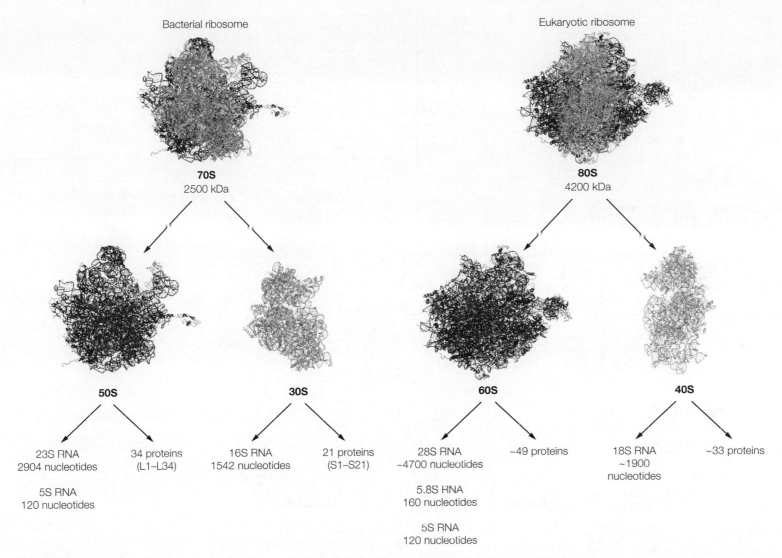

▲ **FIGURE 25.15 Components of bacterial and eukaryotic ribosomes.** Bacterial and eukaryotic ribosomes are assembled along the same structural plan, with eukaryotic ribosomes being somewhat larger and more complex. The molecular size and shape of each subunit were determined by X-ray crystallography.

Ribosomal RNA Structure

When the sequences of 16S rRNAs were originally determined, they were found to contain many regions of self-complementarity, which are capable of forming double-helical segments. A pattern like that shown in **FIGURE 25.16** may seem so complex as to appear almost arbitrary, but comparison with other, even distantly related, 16S RNA sequences shows that the potentially double-stranded regions are highly conserved. Indeed, the secondary structure seems more highly conserved than the primary structure because compensatory mutations are often found in double-helical regions that function to maintain base pairing. A schematic illustration like that in Figure 25.16 is analogous to the cloverleaf visualization of a tRNA (Figure 25.7). The actual rRNA is folded into a three-dimensional structure, just as is the tRNA. In the case of the ribosomal subunit, however, the structure is further complicated by the presence of ribosomal proteins bound to the RNA. However, the pattern shown in Figure 25.16 faithfully describes the secondary structure of 16S rRNA. 23S rRNA

has a comparable secondary structure but is actually more complex, reflecting its larger size.

Internal Structure of the Ribosome

Although electron microscopic images of intact ribosomes and their subunits were obtained some time ago, high resolution was difficult to achieve because of the necessity of staining or shadowing the particles. Nor could such techniques hope to tell us how the proteins and RNA were arranged inside the ribosome.

Attempts to crystallize ribosomes began in the 1970s, but the size and complexity of these particles frustrated early attempts. A key to the success of crystallization efforts, particularly in the laboratory of Ada Yonath, was the use of extremophilic bacteria (archaea) as the source material. Although the *E. coli* ribosome was the most thoroughly studied to that point, the archaeal thermophile *Thermus thermophilus* and the halophile *Haloarcula marismortui* yielded the best crystals.

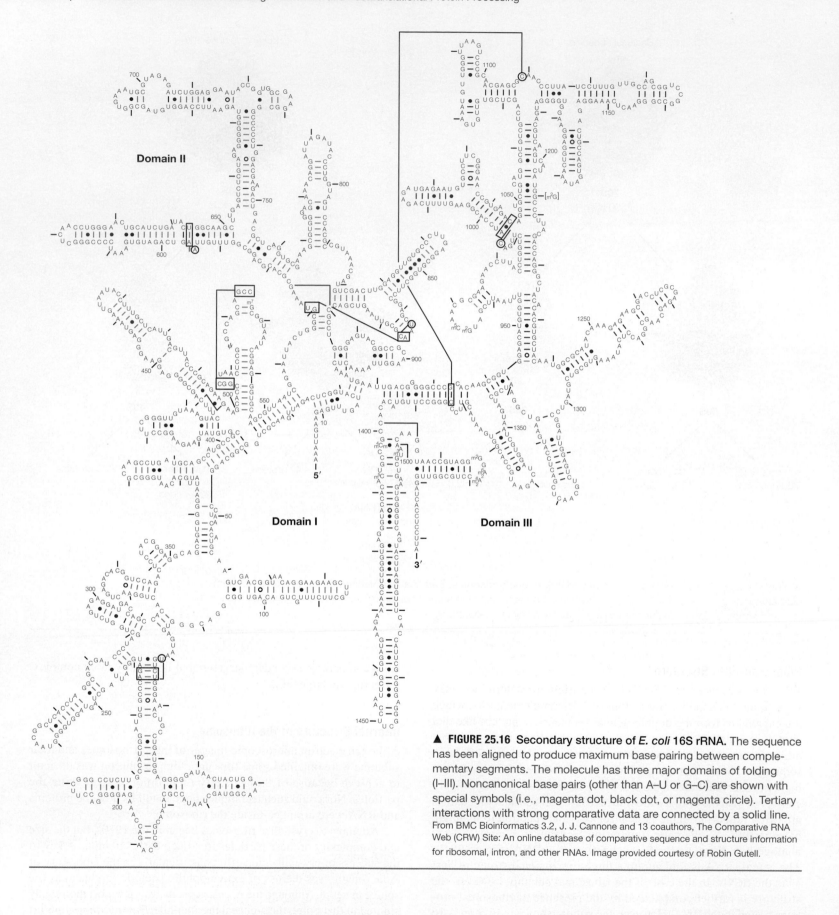

▲ **FIGURE 25.16 Secondary structure of *E. coli* 16S rRNA.** The sequence has been aligned to produce maximum base pairing between complementary segments. The molecule has three major domains of folding (I–III). Noncanonical base pairs (other than A–U or G–C) are shown with special symbols (i.e., magenta dot, black dot, or magenta circle). Tertiary interactions with strong comparative data are connected by a solid line. From BMC Bioinformatics 3.2, J. J. Cannone and 13 coauthors, The Comparative RNA Web (CRW) Site: An online database of comparative sequence and structure information for ribosomal, intron, and other RNAs. Image provided courtesy of Robin Gutell.

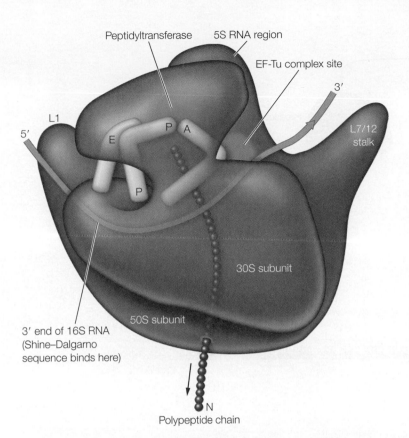

▲ **FIGURE 25.17 A model of the 70S ribosome based on early structural data.** This model shows all three tRNA binding sites (E, P, and A) occupied simultaneously, which does not normally occur. In this view, the 30S subunit is in front of the 50S subunit. The peptidyltransferase site is behind the site indicated, on the 50S subunit. Locations of some large-subunit proteins are shown (L1 and L7/12), as are locations of other components in translation (FF-Tu, Shine–Dalgarno sequence, peptidyltransferase, and 5S rRNA).

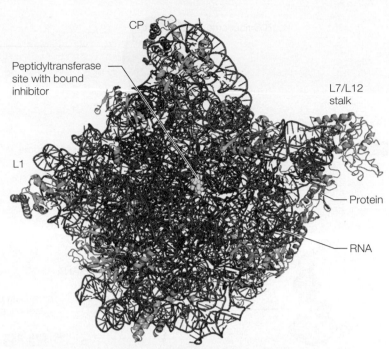

▲ **FIGURE 25.18 A high-resolution model of the 50S ribosomal subunit with RNA in raspberry and protein in green.** This view shows the two stalks (L1 and L7/L12) and a central protuberance (CP), seen also in early electron micrographs. The placement of these structural features confirmed the correctness of the models derived from electron microscopy. The peptidyltransferase site, in yellow, is identified from the binding of an inhibitor. PDB ID: 2qa4.

Because the structure of the ribosome is well conserved evolutionarily, these organisms made satisfactory models.

FIGURE 25.17 shows a model of the 70S ribosome based on the first medium-resolution structures, in the late 1990s. Critical features known already or shown by this model are that the ribosome has three tRNA binding sites (E, P, and A), that mRNA binding and decoding occur on the 30S subunit, that aminoacyl-tRNAs fill the gap between the 30S and 50S subunits, that the newly synthesized polypeptide chain exits the ribosome through a tunnel in the 50S subunit, and that the peptidyltransferase reaction, which creates peptide bonds, occurs at a site on the 50S subunit.

Within a year after the medium-resolution structure was published, Thomas Steitz's laboratory published a high-resolution structure of the 50S subunit from *H. marismortui,* and shortly afterward Venki Ramakrishnan and colleagues described the 30S subunit from *T. thermophilus,* followed by its complete 70S ribosome. **FIGURE 25.18** shows the Steitz 50S structure, and **FIGURE 25.19** shows the Ramakrishnan 70S structure. A striking feature of both structures is that the peptidyltransferase site (colored yellow in Figures 25.18 and 25.19) lies far from any protein (green in both figures). This structural work established conclusively that the ribosome is a ribozyme. We shall return to this reaction and to the question of how this structural work illuminated ribosome function. Yonath, Steitz, and Ramakrishnan shared the 2009 Nobel Prize in Chemistry for their contributions to ribosome structure and function.

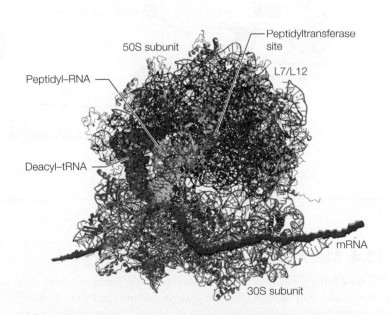

▲ **FIGURE 25.19 Crystal structure of the 70S ribosome, with mRNA and tRNA bound.** The 30S subunit is orange (RNA) and blue (protein), and the 50S subunit is raspberry (RNA) and green (protein). Two bound tRNAs can be seen—namely, peptidyl-tRNA in turquoise and deacylated tRNA in purple. mRNA (a bound oligonucleotide) is shown in red. The peptidyl-transferase site is yellow. PDB ID: 2j00(30S-1), 2j01 (50S-1), 2j02 (30S-2), and 2j03 (50S-2).

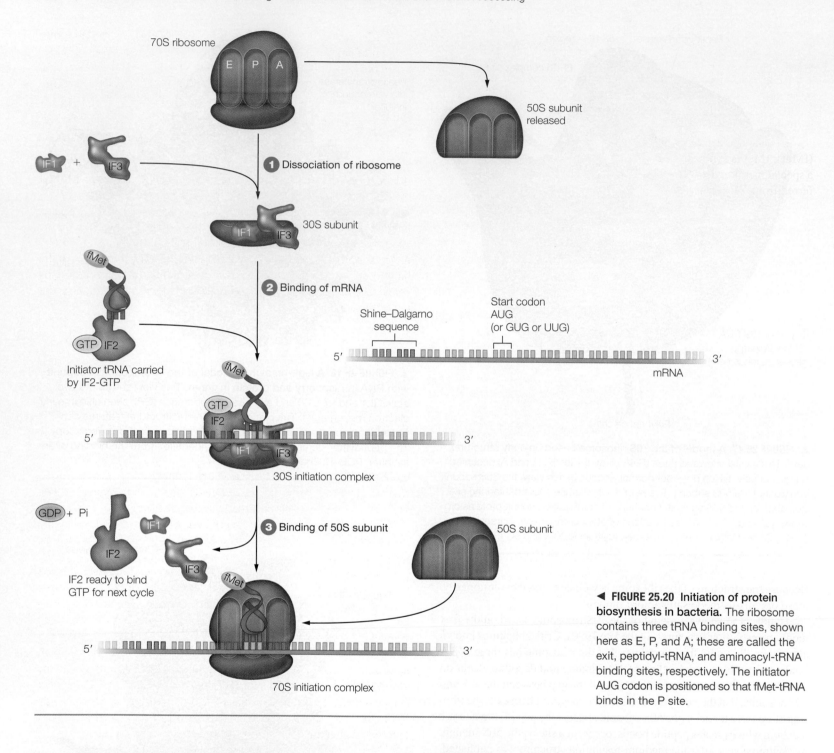

◄ **FIGURE 25.20 Initiation of protein biosynthesis in bacteria.** The ribosome contains three tRNA binding sites, shown here as E, P, and A; these are called the exit, peptidyl-tRNA, and aminoacyl-tRNA binding sites, respectively. The initiator AUG codon is positioned so that fMet-tRNA binds in the P site.

25.4 Mechanism of Translation

We now have identified all of the major participants in translation: mRNA, charged tRNAs, soluble protein factors, and the ribosome, where the actual translation events occur. Just as in transcription, we can divide translation into three stages: *initiation, elongation,* and *termination.* When we describe these steps here, we focus mainly on bacteria and archaea, where our understanding is most detailed. Significant, though not fundamental, differences in eukaryotic protein synthesis are also discussed.

Each step in translation requires a number of specific postteins that interact with the major participants listed above. These proteins

● **CONCEPT** Translation involves three steps—initiation, elongation, and termination—each aided by soluble protein factors.

are referred to as *initiation factors* (IFs), *elongation factors* (EFs), and *release factors* (RFs). These factors, together with some of their properties and functions, are listed in Table 25.3.

Initiation

The initiation of translation is schematized in **FIGURE 25.20**. During initiation, the mRNA and *initiator tRNA* first bind to a free 30S subunit, and then the 50S subunit is added to form the entire complex. In most cases, the initiation codon to which the initiator tRNA binds is AUG, which is also used as an internal methionine codon. The initiator AUG is distinguished from internal AUG methionine codons by the presence of an upstream Shine–Dalgarno sequence, which binds to a complementary sequence in 16S rRNA, thereby positioning the 30S subunit at the initiator AUG.

In bacteria, the initiator tRNA is charged with **N-formylmethionine (fMet)**. fMet is synthesized at the acyl-tRNA level by formylation of a special methionyl-tRNA (Met-tRNA^{fMet}). The formyl group is transferred from N^{10}-formyltetrahydrofolate to a charged tRNA.

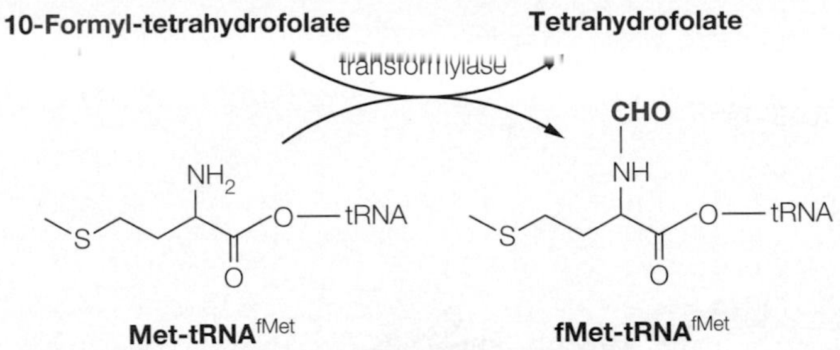

10-Formyl-tetrahydrofolate **Tetrahydrofolate**

transformylase

CHO

NH$_2$ NH

O—tRNA   O—tRNA

O   O

Met-tRNA^{fMet}   **fMet-tRNA^{fMet}**

fMet-tRNA^{fMet} is the only charged tRNA that can bind the 30S subunit on its own; all subsequent charged tRNAs require the fully assembled 70S ribosome. Therefore, most (not all) bacterial proteins are synthesized with the same N-terminal residue, N-formylmethionine. In most cases, the formyl group is removed during chain elongation. For many proteins the methionine itself is also cleaved off later.

The binding of mRNA and initiator tRNA to a free 30S subunit also requires binding of the three **initiation factors** (IF1, IF2, and IF3). IF1 and IF3 promote the dissociation of preexisting 70S ribosomes, thereby producing the free 30S subunits needed for initiation (Figure 25.20, step ❶). The third factor, IF2, binds a molecule of GTP; it delivers the charged initiator tRNA to the 30S subunit. IF2 is a G protein, similar to those involved in signal transduction. At about the same time that the IF2–fMet-tRNA^{fMet} complex binds the 30S subunit, the mRNA also is bound (step ❷). Although the order of these additions is still uncertain, IF2–GTP is absolutely required for the binding of the first (initiator) tRNA. With the binding of the initiator tRNA and the mRNA, formation of the *30S initiation complex* is complete. The initiation complex has high affinity for a 50S subunit and binds one from the available pool (step ❸), with simultaneous release of IF3. Initiation ultimately results in formation of a *70S initiation complex,* which consists of a complete ribosome bound to mRNA and to the charged initiator tRNA.

The mRNA attaches to the 30S subunit near the 5′ end of the message, which is appropriate because all messages are translated in the 5′ → 3′ direction. As mentioned above, an AUG initiation codon is identified by the presence of an upstream Shine–Dalgarno sequence, which is complementary to the sequence, 3′…UCCUCC…5′ in 16S rRNA. This rRNA sequence will pair with any Shine–Dalgarno sequence on mRNA. The pairing aligns the message correctly for the start of translation. In particular, it places the initiator codon next to the P site, one of three tRNA binding sites in the ribosome (see Figure 25.20 and below).

● **CONCEPT** In initiation, the correct attachment of mRNA to the ribosome is determined by binding of the Shine–Dalgarno sequence to a sequence on the 16S rRNA.

Translation cannot proceed until the 50S subunit has bound to the 30S initiation complex. The ribosome has three sites for tRNA binding, called the P (peptidyl) site, the A (aminoacyl) site, and the E (exit) site. The AUG initiator codon with its bound fMet-tRNA^{fMet} aligns with the P site. At this point, the GTP molecule carried by IF2 is hydrolyzed, and IF2–GDP, P$_i$, and IF1 are all released. The complete 70S initiation complex is now ready to accept a second charged tRNA and begin elongation of the protein chain.

The locations of the A, P, and E binding sites for tRNAs were originally established by chemical cross-linking but are now confirmed by X-ray crystallography (Figure 25.19). The anticodon ends of the tRNA molecules contact the 30S subunit, whereas the acceptor ends interact specifically with the 50S subunit. All of the ribosomal proteins contacted lie in the cavity between the 30S and 50S subunits. The tRNA molecules are oriented with their anticodons reaching the mRNA at the bottom of the cavity close to the 30S subunit and their acceptor ends contacting the peptidyltransferase region on the 50S subunit, near the top of the cavity.

Elongation

Growth of the polypeptide chain on the ribosome occurs in a cyclic process; a single round is illustrated in **FIGURE 25.21**. In this particular example, the fifth amino acid from the N-terminus is being linked to the sixth. However, all cycles are the same until a termination signal is reached.

At the beginning of each cycle, the nascent polypeptide chain is attached to a tRNA in the P (peptidyl) site, and the A (aminoacyl) and E (exit) sites are empty. Aligned with the A site is the mRNA codon corresponding to the *next* amino acid to be incorporated. In step ❶ the charged (aminoacylated) tRNA is escorted to the A site in a complex with the elongation factor EF-Tu (a protein), which also carries a molecule of GTP. (Note the parallel to IF2–GTP here.) EF-Tu plays an active role in ensuring that the correct aminoacyl-tRNA is fitted to its codon. The aminoacyl-tRNA becomes distorted in its complex with EF-Tu. Initial binding puts the tRNA anticodon loop into the *decoding center,* a site on the 30S subunit, with the acceptor stem near the EF-Tu site. Nucleotides in the decoding center probe the major groove of the anticodon loop, specifically in positions 1 and 2. This structural work (not shown) confirmed the wobble hypothesis by showing that codon–anticodon fitting is more stringent in the first two positions. GTP hydrolysis by EF-Tu results in conformational changes that move the aminoacyl-tRNA entirely into the A site and cause dissociation of EF-Tu itself. The EF-Tu–GTP complex is then regenerated by the subsidiary cycle shown in the upper right area in Figure 25.21. After the charged tRNA is in place, it is checked both before and after the GTP hydrolysis and rejected if incorrect.

The next, and crucial, step is peptide bond formation (Figure 25.21, step ❷). The polypeptide chain that was attached to the tRNA in the P site is now transferred to the amino group of the amino acid carried by the tRNA in the A site. This step is catalyzed by *peptidyltransferase,* an integral part of the 50S subunit. As mentioned previously, the structure determination of the 50S subunit established conclusively that catalysis is carried out by the RNA portion of the subunit; the ribosome is indeed a ribozyme.

Analysis of the 50S subunit in the Steitz laboratory showed that a conserved AMP residue (number 2451 in *E. coli*) exists in an environment that makes the purine ring unusually basic, probably resulting from hydrogen bonding to a nearby GMP. This suggests a process in which N3 abstracts a proton from the amino group of the aminoacyl-tRNA,

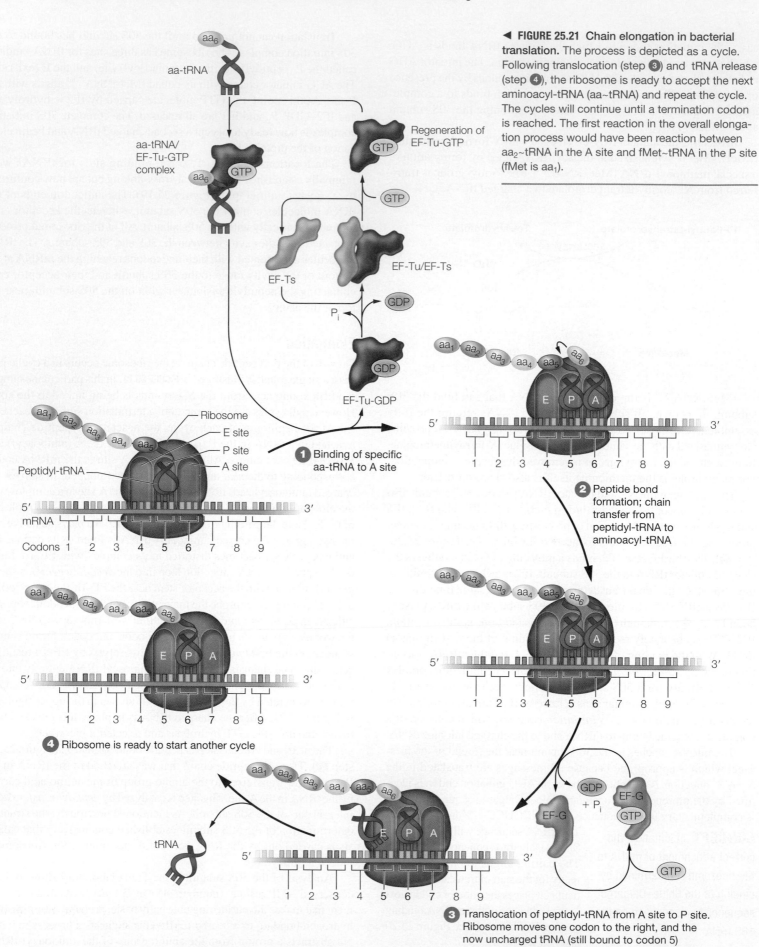

◀ **FIGURE 25.21 Chain elongation in bacterial translation.** The process is depicted as a cycle. Following translocation (step ❸) and tRNA release (step ❹), the ribosome is ready to accept the next aminoacyl-tRNA (aa~tRNA) and repeat the cycle. The cycles will continue until a termination codon is reached. The first reaction in the overall elongation process would have been reaction between aa_2~tRNA in the A site and fMet~tRNA in the P site (fMet is aa_1).

aa-tRNA

aa-tRNA/
EF-Tu-GTP
complex

Regeneration of
EF-Tu-GTP

EF-Ts

EF-Tu/EF-Ts

GTP

GTP

GDP

P_i

EF-Tu·GDP

Ribosome
E site
P site
A site

Peptidyl-tRNA

mRNA

Codons

❶ Binding of specific
aa-tRNA to A site

❷ Peptide bond
formation; chain
transfer from
peptidyl-tRNA to
aminoacyl-tRNA

GDP
+ P_i

EF-G

EF-G
GTP

GTP

❹ Ribosome is ready to start another cycle

tRNA

❸ Translocation of peptidyl-tRNA from A site to P site.
Ribosome moves one codon to the right, and the
now uncharged tRNA (still bound to codon 5)
moves from P site to E site

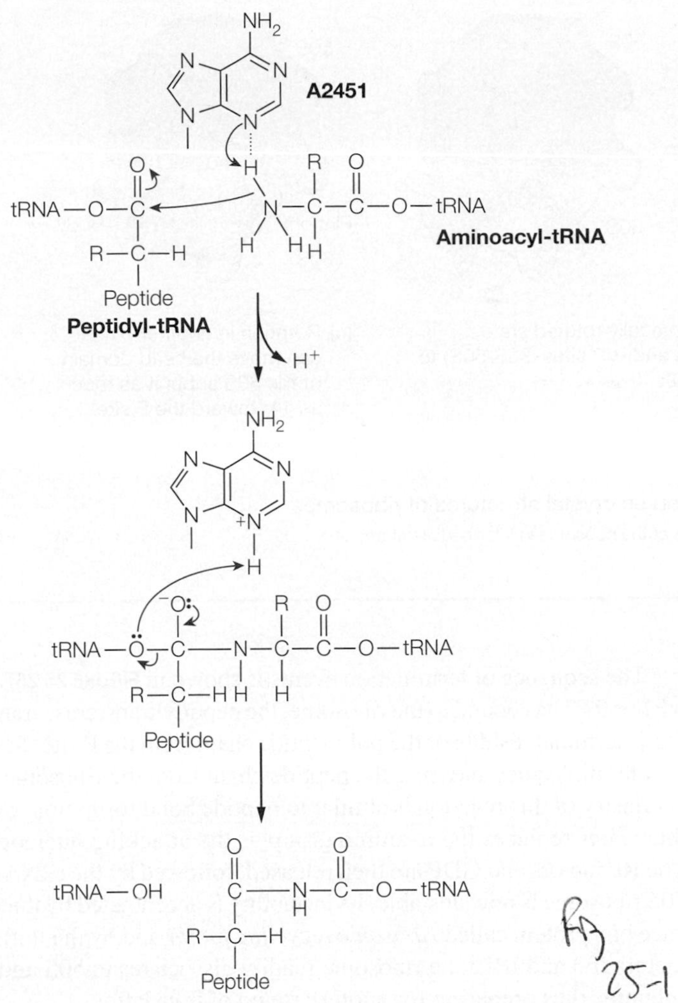

▲ **FIGURE 25.22 A mechanism for peptidyltransferase involving A2451 (*E. coli*) as a general base.** This mechanism is based on the structure of the 50S subunit. Data from P. Nissen et al., (2000) The Structural Basis of Ribosome Activity in Peptide Bond Synthesis. *Science* 289:920–930.

converting the amino group to a better nucleophile that attacks the carboxyl carbon of the C-terminal amino acid, linked to peptidyl-tRNA, as shown in **FIGURE 25.22**. The protonated N3 then stabilizes a tetrahedral carbon intermediate by binding to the oxyanion. Next the proton is transferred to the peptidyl-tRNA 3′ hydroxyl as the newly formed

● **CONCEPT** In elongation, the growing peptide chain at the P site is transferred to the newly arrived aminoacyl-tRNA in the A site. Translocation then moves this tRNA to the P site and the previous tRNA to the E site.

peptide deacylates. Concomitant with this transfer is a switch from the simple P and A states to hybrid states, in which the acceptor ends of the two tRNA molecules move into the leftward positions while the codon ends remain fixed as before. These hybrid sites are indicated as E/P and P/A in **FIGURE 25.23**. Although other mechanisms have been proposed, this can be considered the first half of the *translocation* step (step ❸ in Figure 25.21).

During translation, the ribosome is "ratcheted" along the mRNA molecule by a process that rotates the two ribosomal subunits with respect to each other. Based on structural analysis of the ribosome in intermediate states of rotation, the ratcheting process can be shown schematically as in **FIGURE 25.24**.

At this point, the E and P sites are occupied, but A is empty. As the deacylated tRNA is released from E (step ❹, Figure 25.21), the affinity of A increases greatly and it accepts the aminoacyl-tRNA dictated by the next codon. A cycle of elongation is now complete. All is as it was at the start, except that now:

1. The polypeptide chain has grown by one residue.

2. The ribosome has moved along the mRNA by three nucleotide residues—one codon.

3. At least two molecules of GTP have been hydrolyzed.

The whole process is repeated again and again until a termination signal is reached, with the newly synthesized polypeptide chain being forced to exit the ribosome through the tunnel mentioned previously.

Termination

The completion of polypeptide synthesis is signaled by the translocation of one of the *stop codons* (UAA, UAG, or UGA) into the A site. Because there are no tRNAs that recognize these codons under normal circumstances, termination of the chain does not involve the binding of a tRNA. Instead, protein *release factors* participate in the termination process. The three release factors found in bacteria are listed in Table 25.3. Two of these factors can bind to the ribosome when a stop codon occupies the A site—namely,

● **CONCEPT** Termination requires protein release factors that recognize stop codons.

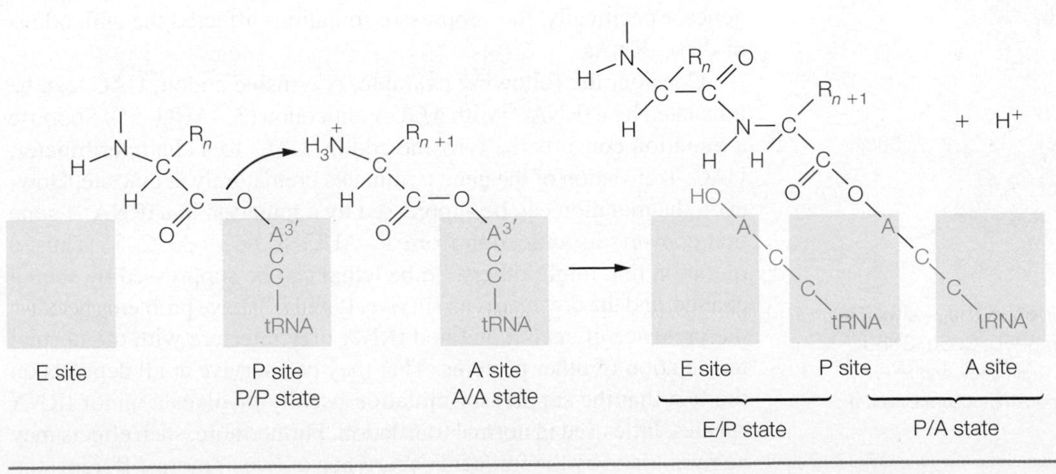

◄ **FIGURE 25.23 Hybrid tRNA-binding sites during translocation.**

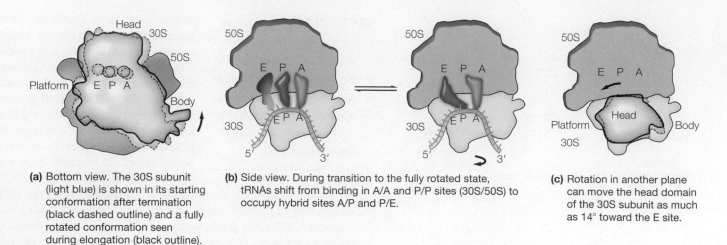

(a) Bottom view. The 30S subunit (light blue) is shown in its starting conformation after termination (black dashed outline) and a fully rotated conformation seen during elongation (black outline).

(b) Side view. During transition to the fully rotated state, tRNAs shift from binding in A/A and P/P sites (30S/50S) to occupy hybrid sites A/P and P/E.

(c) Rotation in another plane can move the head domain of the 30S subunit as much as 14° toward the E site.

▲ **FIGURE 25.24** A schematic view of ribosome subunit rotational motions, based on crystal structures of ribosomes in intermediate states. Data from W. Zhang, J. A. Dunkle, and J. H. D. Cate, (2009) Structures of the ribosome in intermediate states of ratcheting. *Science* 325:1014–1017.

RF1 recognizes UAA and UAG, and RF2 recognizes UAA and UGA. The third factor, RF3, is a GTPase that appears to stimulate the release process, via GTP binding and hydrolysis. Structural analysis of the ribosome complexed with RF1 shows that the release factor interacts directly with a termination codon and the decoding center (**FIGURE 25.25**).

Large subunit

tRNA in the E site

tRNA in the P site

50S

RF1 in the decoding site

mRNA

30S

Small subunit

▲ **FIGURE 25.25** Interaction of RF1 wth a stop codon in the A site. The 70S ribosome in complex with RF1. PDB ID: 3O5a, 3O5b, 3O5c, 3O5d. Data from Laurberg, M., H. Asahara, A. Korostelev, J. Zhu, S. Trakhanov, and H. F. Noller, (2008) Structural basis for translation termination on the 70S ribosome. *Nature* 454:852–857.

The sequence of termination events is shown in **FIGURE 25.26**. After RF1 or RF2 has bound to the ribosome, the peptidyltransferase transfers the C-terminal residue of the polypeptide chain from the P site tRNA to a water molecule, releasing the peptide chain from the ribosome. The chemistry of this reaction is similar to peptide bond formation, except that water replaces the α-amino group as the attacking nucleophile. The RF factors and GDP are then released, followed by the tRNA. The 70S ribosome is now unstable. Its instability is accentuated by the presence of a protein called *ribosome recycling factor* and by the initiation factors IF3 and IF1. The ribosome readily dissociates to 50S and 30S subunits, thus preparing for another round of translation.

Suppression of Nonsense Mutations

Understanding the process of termination helped explain some puzzling observations concerning nonsense mutations. Recall from Chapter 5 that a *nonsense mutation* is one in which a codon for some amino acid has been mutated into a stop codon, so that translation of the polypeptide chain terminates prematurely. These mutations were originally discovered because their phenotypic expression could be *suppressed* by a class of mutations located in other genes; that is, wild-type function was restored by a second mutation at a different site. Upon examination, the suppressors of nonsense mutations were found to lie in tRNA genes. Specifically, the suppressor mutations affected the anticodons of these tRNAs.

Consider the following example. A tyrosine codon, UAC, can be translated by a tRNA^Tyr with a GUA anticodon (3'–AUG–5'). Suppose a mutation converts the tyrosine codon, UAC, to a chain terminator, UAG. Translation of the gene terminates prematurely at that site. However, the mutation can be suppressed by a mutation in a tRNA^Tyr gene that converts its anticodon from 3'–AUG–5' to 3'–AUC–5'. Thus, a mutation that might otherwise be lethal can be suppressed by such a change, and the organism can survive. It will still have problems because the presence of such a mutated tRNA may interfere with the normal termination of other proteins. That they can survive at all depends on the fact that the suppressor mutation usually involves a minor tRNA species, little used in normal translation. Furthermore, such effects may be minimized by the frequent occurrence of two or more different stop

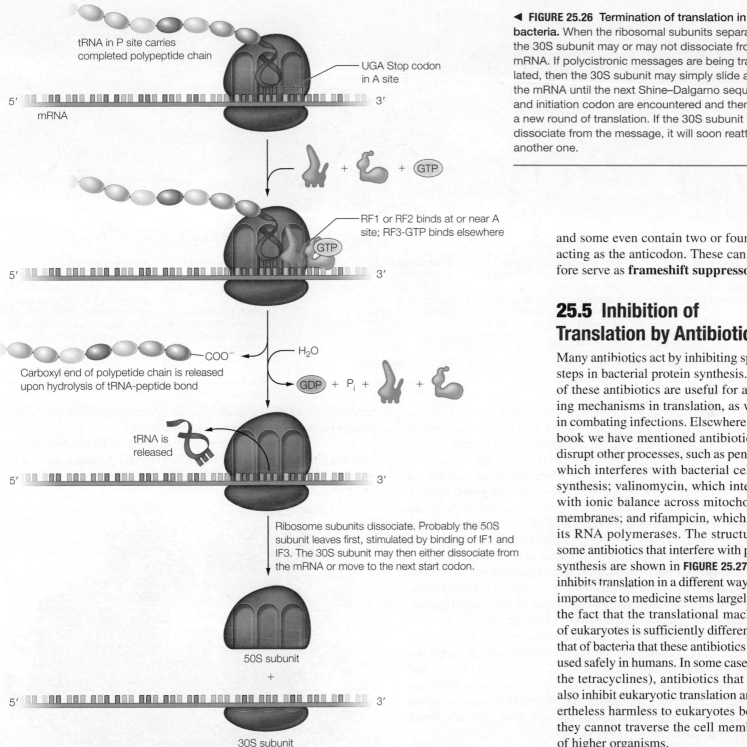

tRNA in P site carries completed polypeptide chain

UGA Stop codon in A site

5′ ——— 3′

mRNA

+ + GTP

RF1 or RF2 binds at or near A site; RF3-GTP binds elsewhere

GTP

5′ ——— 3′

COO⁻ H₂O

Carboxyl end of polypetide chain is released upon hydrolysis of tRNA-peptide bond

GDP + Pᵢ + +

tRNA is released

5′ ——— 3′

Ribosome subunits dissociate. Probably the 50S subunit leaves first, stimulated by binding of IF1 and IF3. The 30S subunit may then either dissociate from the mRNA or move to the next start codon.

50S subunit

+

5′ ——— 3′

30S subunit

▲ **FIGURE 25.26 Termination of translation in bacteria.** When the ribosomal subunits separate, the 30S subunit may or may not dissociate from its mRNA. If polycistronic messages are being translated, then the 30S subunit may simply slide along the mRNA until the next Shine–Dalgarno sequence and initiation codon are encountered and then begin a new round of translation. If the 30S subunit does dissociate from the message, it will soon reattach to another one.

and some even contain two or four bases acting as the anticodon. These can therefore serve as **frameshift suppressors.**

25.5 Inhibition of Translation by Antibiotics

Many antibiotics act by inhibiting specific steps in bacterial protein synthesis. Some of these antibiotics are useful for analyzing mechanisms in translation, as well as in combating infections. Elsewhere in this book we have mentioned antibiotics that disrupt other processes, such as penicillin, which interferes with bacterial cell wall synthesis; valinomycin, which interferes with ionic balance across mitochondrial membranes; and rifampicin, which inhibits RNA polymerases. The structures of some antibiotics that interfere with protein synthesis are shown in **FIGURE 25.27**. Each inhibits translation in a different way. Their importance to medicine stems largely from the fact that the translational machinery of eukaryotes is sufficiently different from that of bacteria that these antibiotics can be used safely in humans. In some cases (e.g., the tetracyclines), antibiotics that would also inhibit eukaryotic translation are nevertheless harmless to eukaryotes because they cannot traverse the cell membranes of higher organisms.

A major problem with the therapeutic use of antibiotics is that microorganisms can develop resistance to many of them. An important example is erythromycin resistance. The erythromycin-binding site on the bacterial ribosome includes a specific region of the 23S rRNA, and binding of the antibiotic can be inhibited by an enzyme that methylates a specific adenine residue in this region. Because many such resistance genes are carried on plasmids, which are easily transferred from

● **CONCEPT** The effects of nonsense mutations can be suppressed by suppressor mutations, in which a tRNA mutates to recognize a stop codon and inserts an amino acid instead.

signals in tandem in mRNAs. Even if the first stop codon is suppressed, the "emergency brake" still holds. As shown in Figure 25.6, the termination signal for *lacZ* is three stop codons in a row.

● **CONNECTION** Several antibiotics in clinical use act by inhibiting translation in bacterial cells.

Suppressor mutations are by no means confined to the correction of nonsense mutations. Some mutated tRNAs correct missense mutations,

Tetracycline: Inhibits the binding of aminoacyl–tRNAs to the ribosome, blocking continued translation

Streptomycin: Interferes with normal pairing between aminoacyl–tRNAs and message codons, causing misreading and thereby producing aberrant proteins

Erythromycin: Binds to a specific site on the 23S RNA and blocks elongation by interfering with the translocation step

Chloramphenicol: Blocks elongation by acting as a competitive inhibitor for the peptidyltransferase complex. The amide link (in blue) resembles a peptide bond.

Puromycin: Causes premature chain termination. The red portion of the molecule resembles the 3′–end of the aminoacylated tRNA. It will enter the A site and transfer to the growing chain, causing premature chain release.

▲ **FIGURE 25.27** Some antibiotics that act by interfering with protein biosynthesis.

● **CONNECTION** The indiscriminate use of antibiotics, both to treat infectious diseases and in the livestock industry, promotes the emergence of resistant bacterial strains and compromises the effectiveness of these drugs.

bacterium to bacterium, the frequency with which antibiotic-resistant strains arise is far higher than if the elements were carried on chromosomal genes. Another problem is the widespread use of antibiotics in animal husbandry, not to cure an infection, but to suppress any possible infections for improving animal weight maintenance and preventing the spread of infection from animal to animal, under the crowded conditions in feedlots. That may be fine in the short run, but the increased emergence of antibiotic-resistant strains makes many question the wisdom of this practice.

Structural studies on ribosomes have stimulated the development of new classes of antimicrobial agents to which resistance might not so readily develop. In the same sense that knowledge of enzyme and receptor structure makes it possible to design entirely new inhibitors with therapeutic properties, the ribosome, because of its many activities and its structural conservation among bacteria, is an attractive target for drug development.

25.6 Translation in Eukaryotes

The mechanism for translating mRNA into protein in eukaryotic cells is basically the same as in bacteria. In eukaryotes, virtually all mRNAs

● **CONCEPT** In eukaryotes, translational initiation is more complex and requires more protein factors than in bacteria.

are monocistronic. Moreover, the ribosome, while of the same overall shape, is larger and more complex (**FIGURE 25.28**). There are many more soluble protein

factors, as we saw in Table 25.3, but the functions performed are comparable to those we have discussed for bacteria.

The most significant differences are in initiation mechanisms. Aside from the greater complexity of the ribosome and the larger number of soluble protein factors, the major differences are (1) that the 5′ end

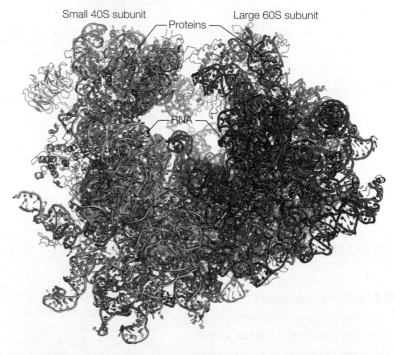

▲ **FIGURE 25.28** The yeast 80S ribosome. Data from coordinates published by A. Ben-Shem, L. Jenner, G. Yusupova, and M. Yusupov, (2010) Crystal Structure of the Eukaryotic Ribosome. *Science* 330:1203–1209. PDB ID: 3O2z, 3O30, 3O5, and 3O5H to PDB ID: 4v7r.

▲ **FIGURE 25.29 ADP-ribosylated diphthamide derivative of histidine in eEF2.** Synthesis of this derivative of a modified histidine in eEF2 using NAD^+ is catalyzed by diphtheria toxin. eEF2 is inactivated, and protein synthesis is therefore blocked. ADP ribose from NAD^+ is shown in blue. Diphthamide is in black.

of a message is sensed not by a Shine–Dalgarno sequence, but by the 7-methylguanine cap, and (2) the N-terminal amino acid, inserted at the initiator AUG, is Met, not *N*-fMet. After detecting the 5′ cap, the ribosomal 40S subunit then scans along the mRNA (an ATP-dependent process) until the first AUG is found. At this point, the initiation factors are released, and the 60S subunit is attached to begin translation.

● **CONNECTION** Diphtheria toxin is toxic because it inhibits protein synthesis via its ADP-ribosylation and inactivation of eEF2.

Cycloheximide

A number of the common inhibitors of bacterial translation are also effective in eukaryotic cells. They include pactamycin, tetracycline, and puromycin. There are also inhibitors that are effective *only* in eukaryotes, such as *cycloheximide* and *diphtheria toxin*. Cycloheximide inhibits the translocation activity in the eukaryotic ribosome and is often used in biochemical studies when processes must be studied in the absence of protein synthesis. Diphtheria toxin is an enzyme, coded for by a temperate bacteriophage gene in the bacterium *Corynebacterium diphtheriae*. It catalyzes a reaction in which NAD^+ adds an *ADP-ribose* group to a specially modified histidine in the translocation factor eEF2, the eukaryotic equivalent of EF-G (**FIGURE 25.29**). Because the toxin is a catalyst, minute amounts can irreversibly block a cell's protein synthetic machinery; pure diphtheria toxin is one of the deadliest substances known.

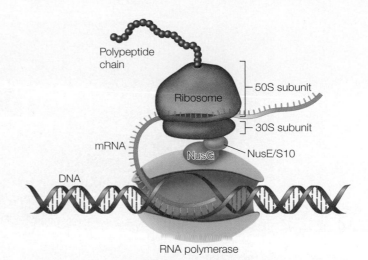

▲ **FIGURE 25.30 Coupling of transcription to translation in *E. coli* via the interaction between NusE and NusG.** Data from Roberts, J. W., (2010) Syntheses that stay together. *Science* 328:436–437.

25.7 Rate of Translation; Polyribosomes

Translation is a rapid process in bacteria. At 37 °C an *E. coli* ribosome can synthesize a 300-residue polypeptide chain in about 20 seconds. This means that a single ribosome passes through about 15 codons (i.e., 45 nucleotides) in each second. This rate is almost exactly the same as our best estimates of the rate of bacterial *transcription,* which means that mRNA can be translated as fast as it is transcribed. This is not a coincidence. Recent studies with *E. coli* show that a ribosomal protein, NusE, interacts in the cell with an RNA polymerase component, NusG, and that through this interaction transcription and translation are physically coupled, as shown in **FIGURE 25.30**, with the rate of transcription being controlled by the rate of translation. Direct coupling of this type cannot occur in eukaryotic cells because the two processes take place in separate cellular compartments.

The rate of 15 codons translated per second represents the growth of individual polypeptide chains but does not account for the total rate of protein synthesis in the cell because many ribosomes may be simultaneously translating a given message. In fact, if we were to carefully lyse *E. coli* cells, we would observe **polyribosomes** (also called polysomes) like those shown in **FIGURE 25.31**. As soon as one ribosome has moved clear of the 5′ region of the mRNA, another attaches. Under some conditions, as many as 50 ribosomes may be packed onto an mRNA, with one finishing translation every few seconds. Because each *E. coli* cell contains 15,000 ribosomes or more, all of them operating at full capacity can synthesize about 750 polypeptide chains of 300 residues each second.

25.8 The Final Stages in Protein Synthesis: Folding and Covalent Modification

The polypeptide chain that emerges from the ribosome is not a completed, functional protein. As described in Chapter 6, it must fold into its tertiary structure, and it may have to associate with other subunits. In some cases, disulfide bonds must be formed and other covalent

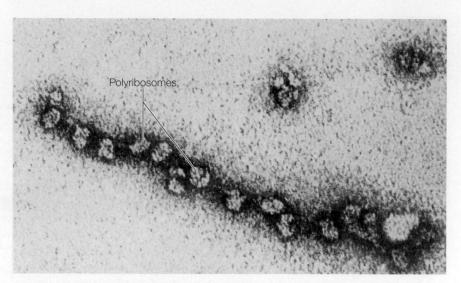

◀ FIGURE 25.31 Polyribosomes.

(a) Electron micrograph showing *E. coli* polyribosomes. The ribosomes are closely clustered on an mRNA molecule.

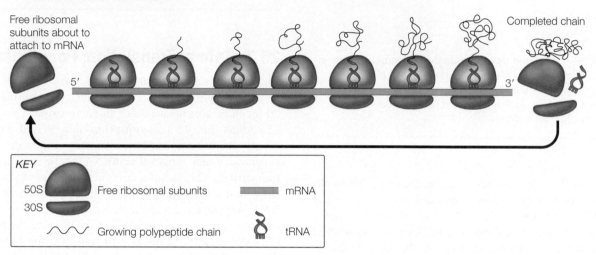

(b) Schematic picture of a polyribosome like that shown in **(a)**. Imagine that each ribosome moves from left to right.

modifications, such as the hydroxylation of specific prolines and lysines, must take place. Complexing with carbohydrate or lipid occurs after translation. In addition, many proteins are subjected to specific proteolytic cleavage to remove portions of the nascent chain.

Chain Folding

The cell need not wait until the entire chain is released from the ribosome to commence its finishing touches. The first portion of the nascent chain (about 30 residues) is constrained from folding as it passes through the tunnel in the ribosome. However, changes begin almost as soon as the N-terminal end emerges from the ribosome. There is good evidence that folding into the tertiary structure starts during translation and is nearly complete by the time the chain is released. For example, antibodies to *E. coli* β-galactosidase that recognize the tertiary folding of the molecule will attach to polyribosomes synthesizing this protein. This enzyme displays catalytic activity only as a tetramer. It has been demonstrated that nascent β-galactosidase chains, still attached to ribosomes, can associate with free subunits to form a functional

tetramer. Thus, even quaternary structure can be partially established before synthesis is complete.

This behavior should not be surprising, if we recall (from Chapters 5 and 6) that formation of the secondary, tertiary, and quaternary levels of protein structure is thermodynamically favored. However, as we have seen in Chapter 6, in some cases this spontaneous folding must be aided by chaperone proteins.

Covalent Modification

Some of the covalent modifications of polypeptide chains also occur during translation. We mentioned earlier that the *N*-formyl group is removed from the initial *N*-fMet of most bacterial proteins. A specific *deformylase* catalyzes this reaction. Deformylation happens almost as soon as the N-terminus emerges from the ribosome. Removal of the N-terminal methionine itself can also be an early event, but whether or not it happens apparently depends on the cotranslational folding of the chain. Presumably, in some cases this residue is "tucked away" and protected from proteolysis in the folded structure.

● **CONCEPT** Translation is immediately followed by various kinds of protein processing, including chain folding, covalent modification, and directed transport.

Some bacterial (and many eukaryotic) proteins experience much more significant proteolytic modifications. These proteins are usually the ones that are going to be exported from the cell or are destined for membrane or organelle locations. We discuss the more complicated eukaryotic protein processing in the next section and concentrate here on what happens in bacteria.

Bacterial proteins that are destined for membrane insertion or for secretion (**translocation** across the cell membrane) are characterized by hydrophobic **signal sequences** or **leader sequences** in the N-terminal regions. After the protein has passed through the membrane, the leader sequence is cleaved off at a predetermined point.

A current model for translocation in bacteria is shown in **FIGURE 25.32**. In most cases, the protein to be translocated (the pro-protein) is first complexed in the cytoplasm with a chaperone—the SecB protein in the example shown (step ❶). This complexing keeps the protein from folding prematurely, which would prevent it from passing through a secretory pore in the membrane. This pore is composed of a heterotrimeric protein made up of SecE, SecY, and SecG—the "SecYEG translocon." The secretory pore is also a target for a fourth protein component, SecA (step ❷). SecA is an ATPase, and both ATP hydrolysis and the electrochemical potential gradient across the membrane help drive translocation. Structural analysis of the SecA protein with and without bound adenine nucleotide suggests a mechanism comparable to that of DNA-dependent helicases in driving the protein through the membrane. After the pro-protein has

been translocated, the leader peptide is cleaved off by a membrane-associated peptidase, and the protein can fold (step ❸). The cleavage site usually lies between a small amino acid (often G or A) and an acidic or a basic one.

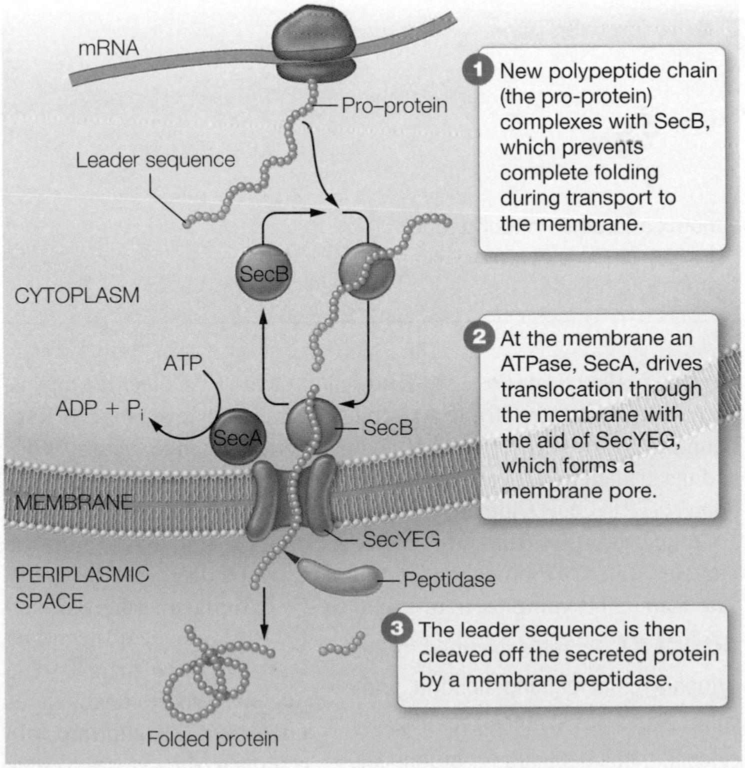

25.9 Protein Targeting in Eukaryotes

The eukaryotic cell is a multicompartmental structure. Each of its several organelles requires different proteins, only a few of which are synthesized within the organelles themselves. Most mitochondrial and chloroplast proteins, for example, are encoded by the nuclear genome and synthesized in the cytoplasm. They must be distinguished from other newly synthesized proteins and selectively transported to their appropriate addresses. Other new proteins are destined for export out of the cell or into vesicles like lysosomes. The diversity of destinations for different proteins implies the existence of a complex system for labeling and sorting newly synthesized proteins and ensuring that they end up in their proper places. And, as seen with bacteria, there must be a process by which protein molecules, which may be hydrophilic, engage the hydrophobic membrane and find a way either to pass through or, as in the case of integral membrane proteins, to become embedded within the membrane.

Proteins Synthesized in the Cytoplasm

Proteins destined for the cytoplasm and those to be incorporated into mitochondria, chloroplasts, or nuclei are synthesized on polyribosomes free in the cytoplasm. The proteins targeted to organelles, as initially synthesized, contain specific signal sequences. These sequences probably aid in membrane insertion, but they also signal that these polypeptides will interact with a particular class of chaperones. These chaperones are members of the "heat-shock" Hsp70 family, and they act to ensure that the newly synthesized protein remains unfolded and is delivered to a receptor site on the organelle membrane. The unfolded protein then passes through membranes, through gates containing transport proteins that discriminate among proteins destined for the lumen, the membranes, or an organelle matrix. If it passes into an organelle matrix, the protein may be taken up by intraorganelle chaperones for final folding. The N-terminal targeting sequence is also cleaved off during this transport.

Several pathways for the transport of proteins into mitochondria are shown schematically in **FIGURE 25.33**. In the best-known "presequence" pathway (in blue), the Hsp70-bound protein attaches via a basic N-terminal signal sequence to a receptor protein as part of a structure called the TOM complex (*t*ranslocation of *o*uter *m*embrane). An ATP-dependent reaction releases the protein from the receptor and inserts it into a pore (another part of the TOM complex; not shown). The signal sequence then interacts with the TIM complex (*t*ranslocation of *i*nner *m*embrane), in the inner membrane. The electrochemical gradient across the inner membrane pulls the signal sequence through. A mitochondrial Hsp70 binds to the protein as it becomes exposed within the mitochondrial matrix and, in another energy-dependent reaction, pulls the rest of the protein through. The signal sequence is removed by a specific protease (MPP, *m*atrix *p*rocessing *p*eptidase) within the matrix. This process pulls the protein through both outer and inner mitochondrial membranes. The other pathways shown are for proteins destined for the outer membrane, the inner membrane, or the intermembrane space.

mRNA

Pro–protein

Leader sequence

❶ New polypeptide chain (the pro-protein) complexes with SecB, which prevents complete folding during transport to the membrane.

CYTOPLASM

SecB

ATP

ADP + P_i

SecA

SecB

❷ At the membrane an ATPase, SecA, drives translocation through the membrane with the aid of SecYEG, which forms a membrane pore.

MEMBRANE

SecYEG

PERIPLASMIC SPACE

Peptidase

❸ The leader sequence is then cleaved off the secreted protein by a membrane peptidase.

Folded protein

▲ **FIGURE 25.32 A model for protein secretion by bacteria.**

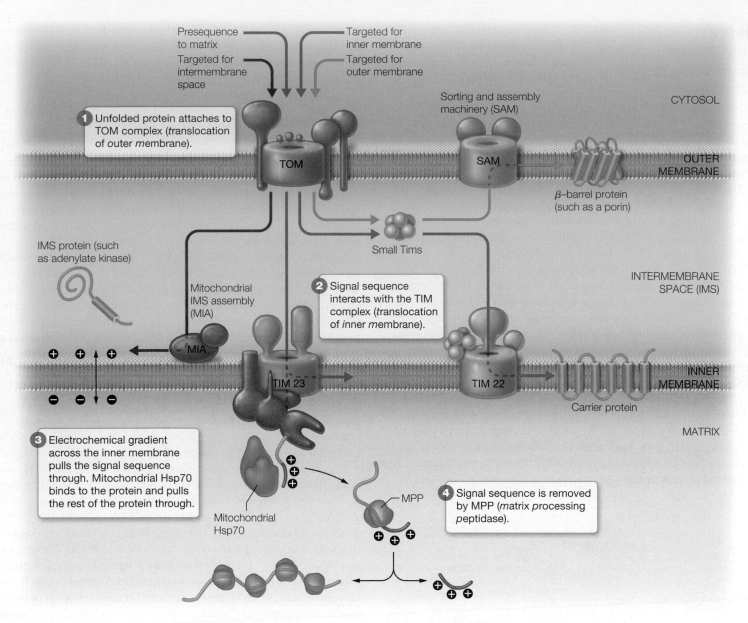

▲ **FIGURE 25.33** Transport of newly synthesized mitochondrial proteins into the matrix and other mitochondrial sites. Based on J. Dudek, P. Rehling, and M. van der Laan, (2012) Mitochondrial protein import: Common principles and physiological networks. *Biochim. Biophys. Acta* 1833:274–285.

A quite different process occurs in nuclear transport. Originally it was thought that nuclear proteins simply diffused into the nucleus through the nuclear pores and were then bound to chromatin. It turns out, however, that the nuclear pores are complex gates rather than open channels (see Figure 21.7). Proteins destined for the nucleus contain *nuclear localization sequences* (NLS) that help these proteins select the nucleus as their destination. Nuclear localization sequences can be found anywhere within the polypeptide sequence, not only the N-terminus. Moreover, the NLS is not removed as a consequence of transport. This is important because the nuclear membrane breaks down in each cell division cycle, and each nuclear protein must be re-transported into the nucleus after the nuclear envelope is reestablished.

The nuclear localization signal on a prospective cargo protein interacts with a protein called **importin,** which carries the protein through the nuclear pore complex. Energy for transport is provided by a monomeric

G protein called **Ran** (*Ra*s-related *n*uclear protein). Like other G proteins, such as Ras (Chapter 20), Ran becomes activated by exchange of Ran-bound GDP for GTP, by a guanine nucleotide exchange factor (GEF), and inactivated by GTPase-activating protein (GAP), which hydrolyzes bound GTP to GDP. Ultimately, GTP cleavage drives the transport process.

Once the importin–cargo protein complex has passed into the nucleus, Ran–GTP binds to that complex, which displaces the cargo. The Ran–GTP complex is returned to the cytoplasm, where bound GTP is converted to GDP. Ran–GDP returns to the nucleus, where it exchanges GDP for GTP, and importin returns to the cytosol to seek a new NLS-containing cargo protein.

● **CONCEPT** Proteins destined for the cytoplasm, nuclei, mitochondria, and chloroplasts are synthesized in the cytosol; those destined for organelles have specific targeting sequences.

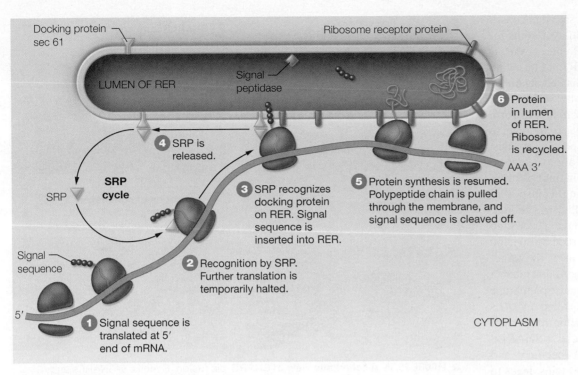

▲ **FIGURE 25.34** The sequence of events in the synthesis of proteins on the rough endoplasmic reticulum (**RER**). The time sequence of events is from left to right.

Proteins Synthesized on the Rough Endoplasmic Reticulum

Proteins destined for cellular membranes, lysosomes, or extracellular transport use a quite different distribution system. The key structures in this system are the **rough endoplasmic reticulum** (RER) and the Golgi complex. The RER is a network of membrane-enclosed spaces within the cytoplasm, continuous with the nuclear envelope. The RER membrane is heavily coated on the outer, cytosolic surface with polyribosomes; this coating is what gives the membrane its rough appearance. The Golgi complex resembles the RER in that it is a stack of thin, membrane-bound sacs. However, the Golgi sacs are not interconnected, nor do they carry polyribosomes on their surfaces. The role of the Golgi complex is to act as a "switching center" for proteins with various destinations.

Proteins that are to be directed to their destinations via the Golgi complex are synthesized by polyribosomes associated with the RER. Synthesis actually begins in the cytoplasm (**FIGURE 25.34**, step ❶).

● **CONCEPT** Proteins destined for cell membranes, lysosomes, or export are synthesized on the rough endoplasmic reticulum, then modified and transported via the Golgi apparatus.

The first sequence to be synthesized is an N-terminal *signal sequence,* part of a mechanism for attaching the ribosome and nascent protein to the RER. *Signal recognition particles (SRPs),* containing several proteins and a small (7S) RNA, recognize the signal sequences of the appropriate nascent proteins and bind to them as they are being extruded from the ribosomes (step ❷).

The SRP has two functions. First, its binding temporarily halts translation, so that no more than the N-terminal signal sequence extends from the ribosome. This pause prevents completion of the protein in the wrong place—that is, in the cytosol—and also inhibits premature folding of the polypeptide chain. Thus, the SRP is acting as a kind of chaperone. The second function of the SRP is to recognize a docking protein in the RER membrane. This is the trimeric Sec61 complex, homologous to the bacterial SecYEG. The docking protein binds the ribosome to the RER, and the signal sequence is inserted into the RER membrane (step ❸). The SRP is then released (step ❹), allowing translation to resume (step ❺). The protein being synthesized is actually *pulled* through the membrane by an ATP-dependent process. Before translation is complete, signal sequences are cleaved from some proteins by an RER-associated protease. These proteins are released into the lumen of the RER and further transported (step ❻). Proteins that will remain in the endoplasmic reticulum have resistant signal peptides and thereby remain anchored to the RER membrane.

Role of the Golgi Complex

The proteins that enter the lumen of the RER undergo the first stages of glycosylation at this point. Vesicles carrying these proteins then bud off the RER and move to the Golgi complex (**FIGURE 25.35**). Here the carbohydrate moieties of glycoproteins are completed, and a final

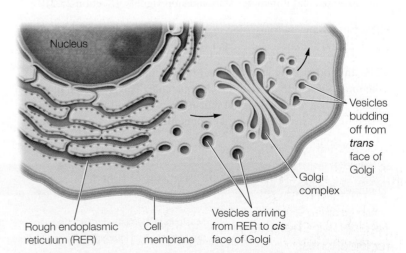

▲ **FIGURE 25.35** Transfer from the rough endoplasmic reticulum (**RER**) to the Golgi complex. Note that vesicles bud off the RER and move to the *cis* face of the Golgi. Primary lysosomal vesicles bud from the *trans* portion of the Golgi.

sorting occurs. The multiple membrane sacs that constitute the Golgi complex represent a multilayer arena for these processes. Vesicles from the RER enter at the *cis* face of the Golgi (that closest to the RER) and fuse with the Golgi membrane. Proteins are then passed, again via vesicles, to the intermediate layers. Finally, vesicles bud off from the *trans* face of the Golgi complex to form lysosomes, peroxisomes, or glyoxysomes or to travel to the plasma membrane. All of this transport of vesicles, from the RER to the *cis* face of the Golgi, to successive levels of the Golgi, and on to their final destinations, requires high specificity in targeting. Transport of vesicles to the wrong destinations would cause cellular chaos. This sorting is accomplished by having each kind of protein cargo packed in a vesicle marked by specific vesicle membrane proteins. In some cases, the target membranes contain complementary proteins that interact with the vesicle membrane proteins and cause membrane fusion. These complementary pairs are called *SNARE*s (*soluble N*-ethylmaleimide-sensitive factor *a*ttachment protein *r*eceptors)—v-SNARES on vesicles, t-SNARES on target membranes. The interaction of specific v- and t-SNARES, aided by cytosolic fusion proteins, leads to fusion of the vesicle and target membranes and delivery of the cargo (see **FIGURE 25.36**).

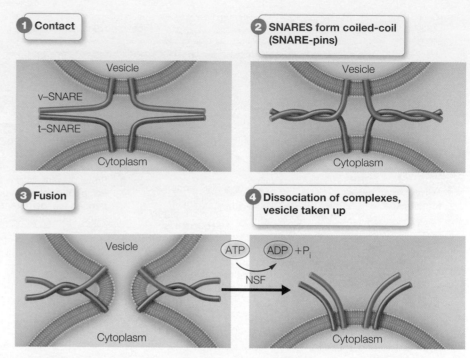

▲ **FIGURE 25.36 A schematic view of SNARE–pin fusion.** Specific v-SNAREs and t-SNAREs dictate interaction and form coiled-coil structures. After fusion, these are broken up by the factor NSF, using the energy of ATP hydrolysis.

Summary

- Translation occurs via the movement of a ribosome in a 5′-to-3′ direction along messenger RNA, with amino acids selected by pairing of tRNA **anticodons** with mRNA codons. (Section 25.1)

- The connection between mRNA sequence and protein sequence is dictated by the genetic code. The code is almost, but not quite, uniform in all organisms. The code is redundant, with multiple codons for most amino acids, and it includes start and stop signals. (Section 25.2)

- The participants in translation include mRNAs, at least one tRNA for each amino acid, ribosomes, and several soluble factors. (Section 25.3)

- The anticodon for each amino acid is a trinucleotide sequence, the anticodon, one on each tRNA molecule. The ribosome moves along mRNA with sequential pairing between codon and anticodon. Specific trinucleotide sequences punctuate the genetic message.

- An RNA component of the large ribosomal subunit provides the catalytic site for peptide bond formation. (Section 25.4)

- Much has been learned about the mechanism of translation from the study of certain antibiotics that inhibit specific steps in the process. (Section 25.5)

- Translation in eukaryotes differs from that in bacteria in small but significant ways, such as the absence of polycistronic mRNAs in eukaryotes. (Section 25.6)

- The rate of translation is maximized by use of **polyribosomes**—mRNA molecules simultaneously binding multiple translating ribosomes. (Section 25.7)

- Simultaneously with translation, nascent polypeptide chains are undergoing folding, posttranslational modification, and transport through membranes en route to their final intracellular or extracellular destinations. (Sections 25.8 and 25.9)

Problems

Enhanced by
Mastering Chemistry
for Biochemistry

Mastering Chemistry for Biochemistry provides select end-of-chapter problems and feedback-enriched tutorial problems, animations, and interactive figures to deepen your understanding of complex topics while practicing problem solving.

Answers to red problems are available in the Answer Appendix.

1. A random-sequence polyribonucleotide produced by polynucleotide phosphorylase, with CDP and ADP in a 5:1 molar ratio stimulated the incorporation of proline, histidine, threonine, glutamine, asparagine, and lysine in a cell-free translation system in the following proportions: 100, 23.4, 20, 3.3, 3.3, and 1.0, respectively. What does this experiment reveal about the nucleotide composition of coding triplets for these six amino acids?

2. The following polynucleotide was synthesized and used as a template for peptide synthesis in a cell-free system from *E. coli*.

…AUAUAUAUAUAUAU…

What polypeptide would you expect to be produced? What information would this give you about the code?

3. If the same polynucleotide described in Problem 2 is used with a *mitochondria-derived* cell-free protein-synthesizing system, the product is

…Met–Tyr–Met–Tyr–Met–Tyr…

What does this say about differences between the mitochondrial and bacterial codes?

4. When polynucleotides are synthesized with repeating triplets of nucleotide residues, one, two, or three different polypeptide chains will be produced in cell-free synthesis.
(a) Explain why these different results are possible.
(b) Predict polypeptides produced when the following are used with an *E. coli* system: $(GUA)_n$, $(UUA)_n$.

5. Although the Shine–Dalgarno sequences vary considerably in different genes, they include examples like GAGGGG that could serve as code—in this case, for Glu–Gly. Does this imply that the sequence Glu–Gly cannot ever occur in a protein, lest it be read as a Shine–Dalgarno sequence? Speculate.

6. According to wobble rules, what codons should be recognized by the following anticodons? What amino acid residues do these correspond to?
(a) $5'$ — ICC — $3'$ (b) $5'$ — GCU — $3'$

7. In the early days of ribosome research, before the exact role of ribosomes was clear, a researcher made the following observation. She could find, in sedimentation experiments on bacterial lysates, not only 30S, 50S, and 70S particles, but also some particles that sedimented at about 100S and 130S. When she treated such a mixture with EDTA, everything dissociated to 30S and 50S particles. Upon adding divalent ions, she could regain 70S particles but never 100S or 130S particles.
(a) Suggest what the 100S and 130S particles might represent, in light of current knowledge of protein synthesis. What important discovery did the researcher miss?
(b) Why do you think reassociation to 100S and 130S particles did not work?

8. The E site may not require codon recognition. Why?

9. What is the minimum number of tRNA molecules that a cell must contain in order to translate all 61 sense codons?

10. Suppose that the probability of making a mistake in translation at each translational step is a small number, δ. Show that the probability, p, that a given protein molecule, containing n residues, will be completely error-free is $(1 - \delta)^n$.

11. Assume that the translational error frequency, δ, is 1×10^{-4}.
(a) Calculate the probability of making a perfect protein of 100 residues.
(b) Repeat for a 1000-residue protein.

12. An important validation of the genetic code occurred when George Streisinger determined the amino acid sequence of bacteriophage T4 lysozyme and of mutants induced by **proflavin**, a dye with a planar structure that can intercalate (fit) between successive base pairs in DNA and induce *frameshift* mutations—that is, mutations involving additions or deletions of a single base. Streisinger and colleagues found that a particular single-base insertion mutation could be suppressed, with wild-type function restored, by a mutation that evidently involved a single-base deletion at a nearby site. Shown below are portions of the amino acid sequence of wild-type T4 lysozyme and the putative double mutant. The remaining parts of the sequence were unchanged. Identify mRNA sequences that could encode each of these

amino acid sequences and determine whether your codon assignments are consistent with the genetic code.

Wild-type …Lys–Ser–Pro–Ser–Leu–Asn–Ala…
Double mutant …Lys–Val–His–His–Leu–Met–Ala…

13. In another such analysis, the Streisinger group found that the altered amino acid sequence in the affected part of the double mutation involved addition of one amino acid, as shown below. Explain how this result is compatible with the codon assignments determined in vitro.

Wild-type …Lys–Ser–Pro–Ser–Leu–Asn–Ala…
Double mutant …Lys–Ser–Val–His–His–Leu–Met–Ala…

14. Assuming that glucose is metabolized to CO_2 as an energy source, how many amino acid residues can be incorporated into a protein molecule for each glucose consumed by a cell? Is this a maximum or minimum?

15. The antibacterial protein colicin E3 is an effective inhibitor of protein synthesis in bacteria. This protein is a nuclease, specifically attacking a phosphodiester bond near the 3′ end of the 16S RNA. Suggest a mechanism for the effect of colicin E3 on translation.

16. Compare and contrast the metabolic pathways leading to thymine in DNA and thymine as a modified base in tRNA.

17. The earliest work on the genetic code established UUU, CCC, and AAA as the codons for Phe, Pro, and Lys, respectively. Can you think of a reason why polyG was not used as a translation template in these experiments?

18. The genetic code is thought to have evolved to maximize genetic stability by minimizing the effect on protein function of most substitution mutations (single-base changes). We will use the six arginine codons to test this idea. Consider all of the substitutions that could affect all of the six arginine codons.
(a) How many total mutations are possible?
(b) How many of these mutations are "silent," in the sense that the mutant codon is changed to another Arg codon?
(c) How many of these mutations are conservative, in the sense that an Arg codon is changed to a functionally similar Lys codon?

19. Ribosomal proteins have high pI values. Why is this advantageous for ribosome stability?

20. The 5′ sequence for the mRNA for *E. coli* ribosomal L10 protein is shown below. Identify the Shine–Dalgarno sequence and the initiator codon.

5′ — CUACCAGGAGCAAAGCUAAUGGCUUUA — 3′

21. Chaperones are generally thought to facilitate protein folding. What additional functions do mitochondrial chaperones perform?

22. Suppression of a nonsense mutation involves a change in nucleotide sequence of a tRNA molecule.
(a) What part of the tRNA molecule is changed?
(b) How might this affect translation globally within a cell?
(c) Why does such a mutation usually not have deleterious effects?

23. Suppose that a gene underwent a mutation that changed a GAA codon to UAA.
(a) Name the amino acid encoded by the original triplet.
(b) Identify a tRNA anticodon that could translate the nonsense UAA triplet.
(c) What other amino acid could be encoded by the mutant tRNA?

24. A nonsense mutation is a substitution mutation that creates a chain-terminating codon in the mRNA corresponding to the mutant gene. Identify three substitution mutations that could change a tryptophan codon to a nonsense triplet.

25. A nonsense codon can be suppressed. Suppression is a mutation at an unlinked site that restores wild-type function at the mutant site without changing the original mutation. Suppression of nonsense mutations was found to alter the structures of particular tRNAs. Consider a nonsense mutation resulting from alteration of a trp codon and identify a specific mutation in a tRNA-encoding gene that could result in translation of the nonsense codon.

References

For a list of references related to this chapter, see Appendix II.

In 1990, attempts to use genetic engineering to increase pigmentation in petunias failed spectacularly, but the results opened the door to discovery of RNA interference, a major process in gene regulation (see page 816).

Regulation of Gene Expression

26

A KEY TO understanding life processes is learning how the expression of genes is controlled. Why is hemoglobin present only in red blood cells, and why are digestive enzymes like trypsin synthesized only in the pancreas? How does the addition of glucose to a bacterial culture immediately shut off β-galactosidase synthesis? What factors enable stem cells to express genes that are permanently repressed in most cells? These are a few questions whose answers demand that we understand gene regulation.

As we mentioned in Chapter 24, seminal insights into genetic control were gained in 1960, when François Jacob and Jacques Monod proposed the operon model of gene regulation, based on their genetic analysis of lactose utilization in *E. coli*. Contributing equally to Jacob and Monod's success were Andre Lwoff's parallel investigations of the genetic regulation of reproduction of the temperate bacteriophage λ in *E. coli*. Based on the similarities in regulatory phenotypes between these two quite different systems, Jacob and Monod proposed that regulation of gene expression occurs primarily at the level of transcription and specifically at the level of

transcriptional initiation. Their model was largely correct, but further research in subsequent years has shown that regulation can occur at any stage in the expression of a gene. For example, we have seen in Chapter 23 that regulation can occur at the level of gene copy number, when environmental stresses cause amplification of genes whose products deal with that stress. Chapter 18 mentioned controlled protein degradation in regulating gene expression during the cell division cycle. Later in this chapter we discuss regulatory RNA molecules, including short interfering RNA.

Jacob and Monod were correct, however; most regulation does occur at the transcriptional level, although important translational control processes are now known. We will begin with bacterial systems, which provide historical context; then we will move to the more complex eukaryotic regulatory processes. Along the way, we will present some examples of regulatory processes that occur at other levels of gene expression, particularly translation. We will also discuss recent discoveries that have identified the functions of small RNA molecules in gene regulation.

26.1 Regulation of Transcription in Bacteria

Our understanding of gene regulation in bacteria and their phages has developed largely from analysis of a few "systems." (By system we mean a set of functionally related genes under some coordinated control process.) The systems to be described include (1) the lactose operon, which was introduced in Chapter 24 during our discussion of transcription; (2) bacteriophage λ, which was introduced in Chapter 23 in connection with recombination; (3) the tryptophan operon, which includes structural genes for the five enzymes in tryptophan synthesis; and (4) the SOS "regulon," also mentioned in Chapter 23—a group of unlinked bacterial genes that help a bacterial cell recover from DNA damage. As you read these pages, keep in mind that the systems selected are *examples* and that additional mechanisms are involved in regulating other systems.

The Lactose Operon— Earliest Insights into Transcriptional Regulation

Recall from Chapter 24 that lactose utilization in *E. coli* is controlled by three contiguous genes—*lacZ* (β-galactosidase, a hydrolytic enzyme), *lacY* (β-galactoside permease, a transport protein), and *lacA* (thiogalactoside transacetylase, an enzyme whose function is still unknown). In the presence of an inducer, all three proteins begin to accumulate simultaneously but at different rates. Lactose itself leads to induction of the lactose operon, but the true intracellular inducer is **allolactose,** Galβ (1 → 6) Glc, a minor product of β-galactosidase action. To study the *lac* operon

Lactose

Allolactose

Isopropyl β-thiogalactoside

in the laboratory, one often uses a synthetic inducer such as **isopropyl β-thiogalactoside (IPTG),** which induces the lactose operon but is not cleaved by β-galactosidase. Hence, its concentration does not change during an experiment.

A mutation in a structural gene—*lacZ*, for example—can inactivate its product (β-galactosidase) without affecting the expression of the other two genes. However, mutations in the regulatory regions mapping *outside* genes *lacZ*, *lacY*, and *lacA* can affect the expression of all three structural genes. In their early work, Jacob and Monod recognized two distinct mutant phenotypes—**constitutive,** in which all three gene products are synthesized at high levels even when an inducer is absent, and **noninducible,** in which all three enzyme activities remain low even after the addition of an inducer. These mutations mapped in two sites, termed *lacO* and *lacI*. Importantly, Jacob and Monod were able to establish dominance relationships involving regulatory mutations by using interrupted bacterial mating to create partial diploids, as we will discuss shortly.

The original Jacob–Monod model for gene regulation, based on the lactose system, was presented in Figure 24.2. A more contemporary map of the *lac* operon is shown in **FIGURE 26.1**. As Jacob and Monod correctly proposed, transcription of the three structural genes is initiated near an adjacent site, called the **operator** (the *lacO* region). Transcription yields a single **polycistronic messenger RNA,** that is, an RNA molecule that includes the coding sequences for all three proteins. (The term **cistron** has genetic significance. For our purposes, it is a region of a genome that encodes one polypeptide chain.)

The *lacI* gene product is a protein **repressor,** which in the active form binds to the operator, blocking transcription (**FIGURE 26.2(a)**). The repressor protein also has a binding site for an inducer. Binding of IPTG or allolactose at this site inactivates the repressor by vastly decreasing its affinity for DNA (Figure 26.2(b)). This repressor inactivation stimulates transcription of *lacZ, Y,* and *A* because dissociation of the repressor–inducer complex from the operator removes a steric block to binding of RNA polymerase at the initiation site. Thus, the introduction of lactose to the cell activates synthesis of the gene products involved in its catabolism by removing a barrier to their transcription. This mode of regulation is essentially negative because the active regulatory element (the repressor) is an *inhibitor* of transcription. Positive control, in which the active element *stimulates* transcription, was discovered

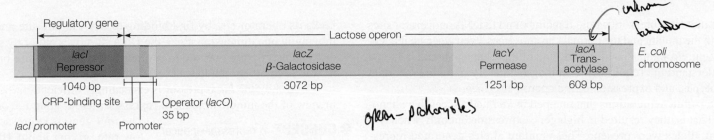

unknown function

open-prokaryotes

▲ **FIGURE 26.1** **A map of the lactose operon.** The cAMP receptor protein (CRP) site is the binding site for cAMP receptor protein, a regulatory factor (page 802). The promoter is a binding site for RNA polymerase. Adjacent to the promoter is the operator, a binding site for the *lac* repressor. Synthesis of the repressor initiates at its own promoter (*lacI* promoter). Additional repressor-binding sites exist—82 nucleotides upstream and 432 nucleotides downstream, respectively, from the transcriptional start point (not shown).

later. This involves the CRP site shown in Figure 26.1, and it will be discussed shortly.

Essential for developing the operon model was the ability to analyze regulation in partial diploids—bacteria containing one complete chromosome plus part of another, transferred by conjugation. One copy of the operon resides in the chromosome, while another lies in an incomplete chromosome, introduced into a cell as part of the bacterial mating process. In this technique, bacterial pairs undergoing conjugation are separated by shear forces while the process is under way, disrupting chromosome transfer. Jacob and Monod systematically created and analyzed partial diploids for different pairs of mutations, as well as for each mutation paired with the wild-type allele.

Noninducible mutations that mapped in *lacI* had a dominant phenotype; that is, expression of the structural genes was low when both wild-type and mutant alleles of *lacI* were present. Jacob and Monod proposed that the mutant alleles give rise to mutant repressors that are unable to bind an inducer.

● **CONCEPT** The phenotypes of partial diploids involving *lac* regulatory genes gave indispensable clues to mechanisms of transcriptional regulation.

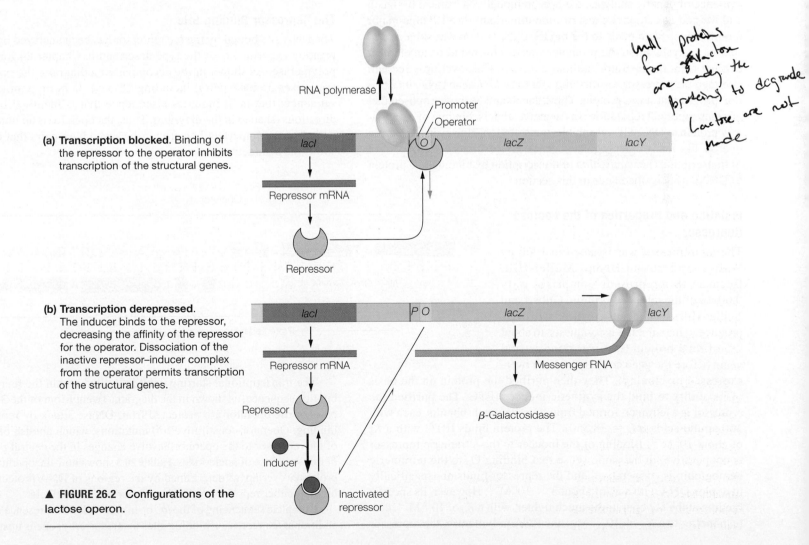

RNA polymerase

Promoter
Operator

(a) Transcription blocked. Binding of the repressor to the operator inhibits transcription of the structural genes.

lacI O *lacZ* *lacY*

Repressor mRNA

Repressor

(b) Transcription derepressed. The inducer binds to the repressor, decreasing the affinity of the repressor for the operator. Dissociation of the inactive repressor–inducer complex from the operator permits transcription of the structural genes.

lacI P O *lacZ* *lacY*

Repressor mRNA

Messenger RNA

Repressor

β-Galactosidase

Inducer

Inactivated repressor

▲ **FIGURE 26.2** **Configurations of the lactose operon.**

until proteins for galactose are made; the proteins to degrade lactose are not made

These mutant repressors would remain bound to DNA at operator sites on both the mutant and the normal chromosomes, even when an inducer is present. Problems 6 and 9 at the end of this chapter should help you to understand the relationships among regulatory mutations, mutant phenotypes, and expression of *lac* operon enzymes.

Constitutive mutations that mapped in *lacI* had a recessive phenotype. That is, they resulted in high gene expression but only when two mutant alleles were present. These mutant alleles generated repressors that were defective in operator binding and thus could not turn off gene expression. Such mutations are recessive because a normal repressor in the same cytosol can bind to all operators and inhibit transcription. These observations showed that repressor mutations are **trans-dominant,** meaning that the *lacI* gene product encoded by one genome can affect gene expression from other genomes in the same cell. This finding led to the conclusion that the repressor is a diffusible product, capable of acting on any DNA site in the cell to which it could bind. Conversely, the constitutive mutations mapping in *lacO* had a **cis-dominant** effect. That is, in a cell with one wild-type operator and one mutant operator, only the genes on the same chromosome as the mutant operator were expressed constitutively. Given that a protein would be capable of diffusing through the cytosol and acting on other chromosomes, this finding suggested that the operator does not encode a gene product.

For a molecular mechanism based almost entirely on the indirect evidence of genetic analysis, the operon model as advanced by Jacob and Monod has stood the test of time remarkably well. Three major modifications were made to the model as the system was subjected to further analysis. First, the promoter was discovered to be an element distinct from the operator (although the two sites overlap). Second, although the repressor was first thought to be *lacI*-gene RNA, its isolation proved that it is a protein. Third, Jacob and Monod proposed that all transcriptional regulation was negative; that is, binding of a regulatory protein to DNA always inhibits transcription. However, the lactose operon, like many other regulated genes, also exhibits *positive* control of transcription (i.e., *activation* of transcription by binding of a protein to DNA), as described later in this section.

Isolation and Properties of the Lactose Repressor

The *lac* repressor was isolated in 1966 by Walter Gilbert and Benno Müller-Hill. Because this repressor comprises only 0.001% of the total cell protein, Gilbert and Müller-Hill used mutants engineered to overproduce it, maximizing its synthesis, to about 2% of total protein (this was done several years before the gene could have been overexpressed by cloning). They then purified the protein on the basis of its ability to bind the synthetic inducer IPTG. The purified *lac* repressor is a tetramer, formed from four identical subunits, each with 360 amino acids ($M_r = 38,350$). The protein binds IPTG with a K_a of about 10^6M^{-1}. Binding of the inducer to the tetrameric repressor is cooperative, in the same sense that binding O_2 to the tetrameric hemoglobin is cooperative, and the repressor binds nonspecifically to duplex DNA with a K_a of about $3 \times 10^6 \text{M}^{-1}$. However, its specific binding at the *lac* operator is much tighter, with a K_a of 10^{13}M^{-1}, corresponding to a K_d of 10^{-13}M. Like RNA polymerase, the repressor

seeks its operator site by first binding to DNA at any site and then moving in one dimension along the DNA. It moves either by sliding or by transfer from one site to another, if the two sites are brought next to each other on adjacent loops of DNA.

Control by the *lac* repressor is exceedingly efficient, particularly in view of the minute amount of repressor present in an *E. coli* cell.

● **CONCEPT** A repressor–inducer system provides negative control of the *lac* operon. The repressor binds to the operator, interfering with transcription initiation. An inducer binds to the repressor, reducing its affinity for the operator.

The *lacI* gene is expressed at a low rate, giving about 10 molecules of repressor tetramer per cell. Although this value corresponds to a concentration of only about 10^{-8} M, it is several orders of magnitude higher than the *dissociation* constant. That is, in a *noninduced* cell the operator is bound by the repressor more than 99.9% of the time—hence, the very low levels of *lac* operon proteins in uninduced cells (less than one molecule per cell). However, binding of the inducer decreases the affinity of the repressor-inducer complex for the operator by many orders of magnitude. Under these conditions, nonspecific binding of the repressor–inducer complex at other DNA sites becomes significant, so that in *induced* cells the operator is occupied by the repressor less than 5% of the time.

The Repressor Binding Site

The DNA site bound by the *lac* repressor has been analyzed by footprinting experiments of the type described in Chapter 24 for RNA polymerase. As shown in the accompanying diagram, the operator comprises 35 base pairs, including 28 base pairs of symmetrical sequence; that is, it includes a sequence that is identical in both directions (shaded in the diagram). Thus, the operator is an imperfect palindrome—imperfect because of the seven base pairs that do not show this symmetry.

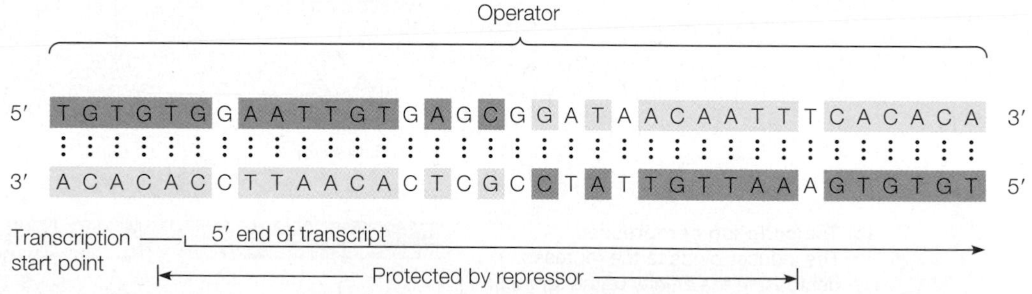

Operator

5′ TGTGTGGAATTGTGAGCGGATAACAATTTCACACA 3′

3′ ACACACCTTAACACTCGCCTATTGTTAAAGTGTGT 5′

Transcription start point ⌐ 5′ end of transcript

|← Protected by repressor →|

The transcriptional starting point is located within the repressor-binding sequence, as shown in the diagram. Twenty-four of the 35 base pairs of the operator are protected from DNase attack by repressor binding. Operator-constitutive (o^c) mutations, which abolish binding of the repressor to the operator, involve changes in the central portion of this sequence of nucleotides. **FIGURE 26.3** shows how the operator and promoter overlap, as determined by the regions of DNA protected by binding either repressor or RNA polymerase, respectively.

Complete sequencing of the *lac* operon revealed the presence of two additional *lac* repressor-binding sites located nearby; one is upstream,

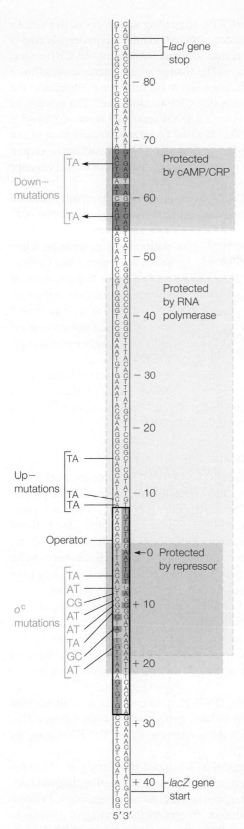

◀ **FIGURE 26.3** **The 122-base *lac* regulatory region.** As determined by DNase I footprinting, the binding sites for repressor and RNA polymerase, as well as for CRP (discussed shortly), are boxed. The palindromic areas of the operator are shaded in green. The CRP binding site also contains an inverted repeat sequence, and this is shaded in orange. The sequence changes encountered in promoter and operator mutations are shown (e.g., GC→TA at position 66, which decreases *lac* expression), as well as the start codon for *lacZ* gene translation and the stop codon for the *lacI* gene. Mutations labeled o^c are constitutive because repressor binding by inducer is decreased. Nucleotides are numbered from 0 at the *lac* transcription start point (the first mRNA nucleotide is +1).

Confirmation came with the crystal structure determination of the *lac* repressor. The tetrameric protein consists of two dimeric units, joined by a hinge region. Each dimer binds DNA separately, suggesting that the tetrameric protein binds to both the +11 and −82 sites, thereby creating a DNA loop of 93 base pairs. **FIGURE 26.4** shows a dimeric form of the repressor protein in complex with operator DNA. The DNA-binding domain of the protein is an α-helical region that contacts bases within the major groove of the operator DNA. This helical binding motif has been seen in other sequence-specific DNA-binding proteins, as we shall discuss shortly.

The structure of the tetrameric repressor protein suggests the mechanism of induction. The repressor is an allosteric protein which, upon binding inducer, significantly increases the angle at which two monomeric units in a dimer are oriented with respect to one another, as shown by the transparent parts of the structure in Figure 26.4. The conformational change in DNA binding drives the DNA-binding helices apart by 3.5 Å, so that they can no longer contact DNA-binding sites, as they must in order to bind tightly and maintain repression.

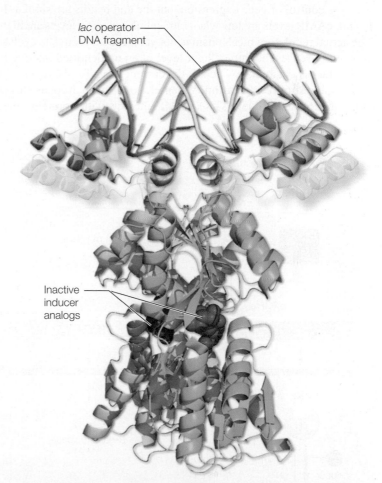

▲ **FIGURE 26.4** **Structure of the *lac* repressor.** This image shows a dimeric form of the repressor complexed with a *lac* operator DNA fragment. The two polypeptide chains are in green and aqua. At the junction between the C- and N-terminal domains of the two polypeptide chains is bound one molecule per chain of an inactive inducer analog, ONPF (o-nitrophenylfucoside), in red and lavender. The transparent portion indicates how the DNA-binding helices move apart from one another as a result of inducer binding. PDB ID: 1efa and 1tif.

centered at position −82, and one is downstream, within the *lacZ* gene itself, centered at position +432 (the original operator is centered at +11). Genetic analysis indicated that the upstream site participates in *lac* operon regulation; mutations affecting the upstream site led to partial repression of the operon.

Regulation of the *lac* Operon by Glucose: A Positive Control System

The *lac* repressor–operator system keeps the operon turned *off* in the absence of utilizable β-galactosides. An overlapping regulatory system, summarized in **FIGURE 26.5**, turns the operon *on* only when alternative energy sources are unavailable. *E. coli* has long been known to use glucose in preference to most other energy substrates. When grown in a medium containing both glucose and lactose, the cells metabolize glucose exclusively until the supply is exhausted. Then growth slows, and the lactose operon becomes activated in preparation for continued growth using lactose. This phenomenon, now known to involve a transcriptional *activation* mechanism, was originally called glucose repression or catabolite repression. Transcriptional activation occurs when glucose levels are low and control is exerted through intracellular levels of cyclic AMP.

Recall from Chapter 12 that in animal cells a rise in cAMP levels stimulates catabolic enzymes, which increase the levels of energy substrates. Those effects are mediated metabolically through hormonal signals and triggering of metabolic cascades. In bacteria, the activation involves control of gene expression, but the end results are similar. In *E. coli,* cAMP levels are low when intracellular glucose levels are high. The actual regulatory mechanism is not yet known. Adenylate cyclase apparently senses the intracellular level of an unidentified intermediate in glucose catabolism—hence, the current name for the regulatory process, **catabolite activation.** When glucose levels drop, as shown in Figure 26.5, cAMP levels rise. cAMP interacts with a protein called

cAMP receptor protein (CRP), triggering activation of the lactose operon. CRP protein is a dimer, with two identical polypeptide chains each with 210 amino acid residues. When it binds cAMP, CRP undergoes a conformational change. The change greatly increases its affinity for certain DNA sites, including a site in the *lac* operon adjacent to the RNA polymerase binding site (the CRP-binding site). The binding of cAMP–CRP at this site protects a DNA sequence from −68 to −55, as shown in Figure 26.3. This binding facilitates transcription of the *lac* operon by stimulating the binding of RNA polymerase to form a closed-promoter complex or by increasing the rate of open-promoter complex formation.

Our understanding of CRP action is still incomplete, partly because the cAMP–CRP complex activates several different gene systems in *E. coli,* all of which are involved with energy generation. They include operons for utilization of other sugars, such as galactose, maltose, arabinose, and sorbitol, as well as several amino acids. Among the operons that have been analyzed, the DNA-binding site of the cAMP-activated dimer varies considerably with respect to the transcriptional start point; this suggests that regulatory mechanisms involving this protein are complex.

> ● **CONCEPT** Cyclic AMP receptor protein (CRP) provides positive control of *lac* and several other catabolite-repressible operons. The cAMP–CRP complex binds at the *lac* promoter when glucose levels are low, and it facilitates initiation of transcription.

The CRP–DNA Complex

The structure of the CRP–cAMP–DNA complex, as revealed by X-ray crystallography (**FIGURE 26.6**), shows how the protein binds to DNA. Each CRP subunit contains a characteristic pair of α-helices, which are joined by a turn. One helix of each pair, shown as perpendicular to the plane of the figure, lies within the major groove of DNA. This **helix–turn–helix** structural motif, which was observed at about the same time as the structures of the λ phage Cro and cI repressors (see Figure 26.8), is found in several DNA-binding regulatory proteins, suggesting common evolutionary origins for this family of proteins. Helix–turn–helix was the first known sequence-specific DNA-binding structural motif, recognized earlier than the zinc finger motif mentioned in Chapters 20 and 24.

Analysis of the DNA–protein complex also shows that CRP induces DNA to bend quite sharply when it binds. This bending can be seen in the structure of the CRP–DNA complex shown in Figure 26.6.

The operon model as advanced from analysis of the *lac* system offered a paradigm of gene regulation, both in bacteria and in more complex multicellular organisms. Positive and negative regulation effected by the binding of proteins to specific regulatory DNA sites, and modulation of DNA binding by small molecules, were powerful concepts that found validation in numerous analyses of other genetic systems. However, in formulating these concepts, early investigators were aided by parallel investigations in other biological systems, in particular, as noted earlier, parallel investigations with bacteriophage λ.

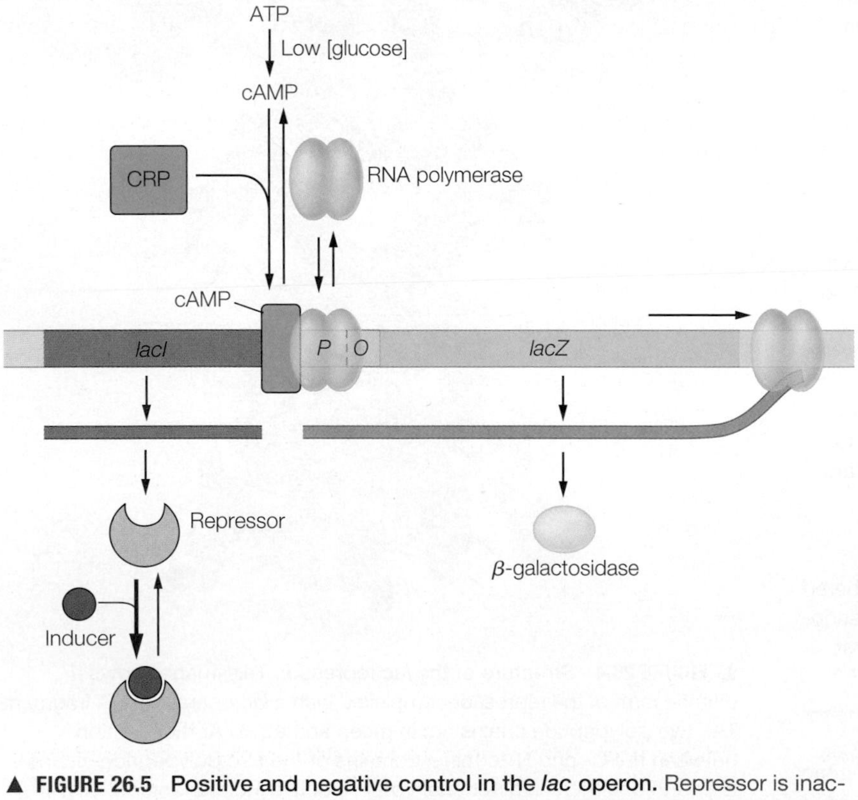

▲ **FIGURE 26.5 Positive and negative control in the *lac* operon.** Repressor is inactivated by binding inducer, and cAMP receptor protein (CRP) is activated by binding cyclic AMP. Binding of the CRP–cAMP complex to DNA facilitates initiation of transcription by RNA polymerase.

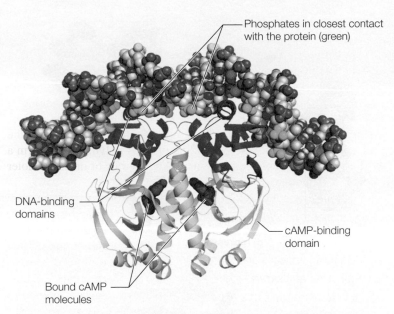

Phosphates in closest contact
with the protein (green)

DNA-binding
domains

cAMP-binding
domain

Bound cAMP
molecules

▲ **FIGURE 26.6 A transcriptional regulatory protein.** CRP, the cAMP receptor protein, binds to a site on the promoter for the *E. coli lac* operon in the presence of cyclic AMP and activates transcription of the operon. DNA is bent as a consequence of its binding to the CRP–cAMP complex. The DNA-binding domains of the protein—two α-helices that contact DNA bases in the major groove (see page 804)—are in purple. The cAMP-binding domain is in aqua, and two bound cAMP molecules are in red. On the DNA molecules, those phosphates that are in closest contact with the protein are in green. Those phosphates not in direct contact with DNA, on the outer edge of the bend, are in yellow. PDB ID: 1cgp.

Some Other Bacterial Transcriptional Regulatory Systems: Variations on a Theme

In this section, we describe how studies on phage λ reinforced the operon model developed with the *lac* system, and we also discuss

significant ways in which studies on λ and other systems confirmed and extended the operon concept.

Bacteriophage λ: Multiple Operators, Dual Repressors, Interspersed Promoters and Operators

The phage λ system is more complex than the *lac* operon. However, λ displayed mutants with regulatory phenotypes comparable to those in the *lac* system, and these similarities, along with Lwoff's investigations, helped guide Jacob and Monod toward their model. The complexity arises because the phage genome is much larger than the 6-kbp *lac* operon and because regulation involves far more than a simple on–off switch. As discussed in Chapter 23 (see Figure 23.16), infection of a bacterial cell by this virus can lead either to lytic growth, with the viral reproduction program activated, or to lysogeny, where most viral genes are turned off and the circularized chromosome becomes integrated into the genome of the host. Once integrated, the lysogenic state must be maintained until an environmental stimulus triggers viral chromosome excision and a cycle of lytic growth. In addition, lytic growth and viral reproduction involve a precisely timed developmental program in which, for example, the viral genome replicates early in the cycle, followed by synthesis of viral structural proteins, followed in turn by assembly of intact viral particles and lysis of the host cell.

The essence of the decision between lytic and lysogenic infection is controlled by protein–DNA interactions within a 75-base-pair region of the phage genome called $O_R P_R$ (**FIGURE 26.7**). Within this region multiple operators and promoters are interspersed, meaning that repressors and RNA polymerase bind the phage genome at overlapping sites. Three 17-base-pair regions, called $O_R 1$, $O_R 2$, and $O_R 3$ are binding sites for *two* different repressors, called cI and Cro. The binding sites are not identical, so they vary in their affinity for each repressor. The −35 and −10 boxes for two promoters lie within this region, and from their locations it is evident that binding sites for RNA polymerase and the two repressors are interspersed and overlapping. Depending on the occupancy of the three operators by either cI or Cro protein, transcription

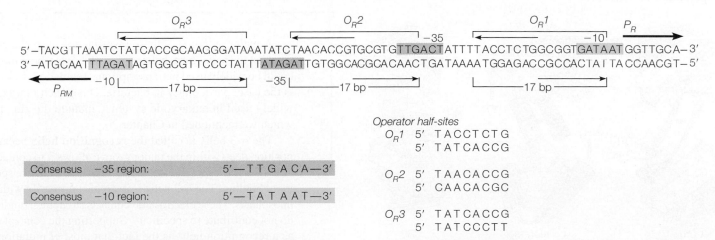

▲ **FIGURE 26.7 The λ phage $O_R P_R$ region.** The upper diagram shows the nucleotide sequence of the $O_R P_R$ region, including the three repressor-binding sites ($O_R 1$, $O_R 2$, *and* $O_R 3$), the rightward promoter P_R and the leftward promoter P_{RM}, and the −35 and −10 regions for the two promoters (−35 regions are shaded purple and −10 regions are

shaded blue). The thin horizontal arrows mark the operator half-sites. The thick arrows represent transcription initiation sites. Below are the consensus sequences for the −35 and −10 regions and a chart that shows the partial homology among the operator half-sites (that is, the repressor-binding sites).

Consensus −35 region: 5′—T T G A C A—3′

Consensus −10 region: 5′—T A T A A T—3′

Operator half-sites

$O_R 1$ 5′ T A C C T C T G
        5′ T A T C A C C G

$O_R 2$ 5′ T A A C A C C G
        5′ C A A C A C G C

$O_R 3$ 5′ T A T C A C C G
        5′ T A T C C C T T

can occur either rightward from the promoter P_R or leftward from a promoter called P_{RM}. Among other things, this system acts to ensure that *cI,* the gene for the major repressor, is transcribed at a high rate when lysogeny (repression of the viral genome) is being established—the lysogenic response is competing with a lytic response. However, once lysogeny is established and the lytic genes have been turned off, much less cI repressor is needed to maintain the lysogenic state. An additional pattern of regulated gene expression comes into play when environmental changes within a lysogenized bacterial cell cause the dormant prophage to excise itself from the bacterial chromosome and begin a cycle of lytic growth.

It was structural analysis of the λ Cro repressor that first suggested the helix–turn–helix motif as the mechanism by which gene-regulatory proteins bind with specificity and high affinity to their target sites. Structures of the λ cI repressor and the *E. coli* CRP, which we mentioned earlier, supported this model. As seen in **FIGURE 26.8**, each protein has a pair of similar α-helices in each of two identical subunits—the helix–turn–helix motif that we have mentioned. Model building with DNA suggested that one of the two helices lies across the opening of the major groove in DNA, providing mostly electrostatic interactions to promote DNA binding. The other helix lies deep within the major groove and forms specific interactions between amino acid residues on the protein and DNA bases within the major groove sites with varying affinities, leading to varying occupancy of each binding site by each repressor under different physiological conditions. The cI repressor also serves, under certain conditions, as a transcriptional activator, promoting the expression of some genes while repressing transcription of other genes.

Amino acid sequence homologies among these regions are shown for Cro and cI in **FIGURE 26.9**. Note that the sequences, though similar, are not identical. If they were identical, we would not be able to explain how Cro and cI repressors differ in their relative affinities for different operators. The cI repressor contains an additional binding determinant—a pair of "arms," or short polypeptide segments that extend from helix 1 and can be seen in Figure 24.1 (page 744) extending around the helix and establishing contacts on the other side of the DNA duplex. These arms probably explain why cI binds more tightly to its operators than does Cro.

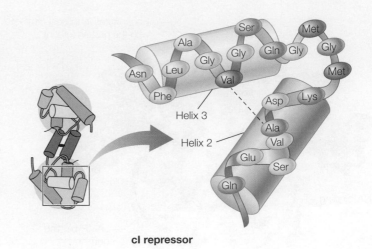

cI repressor

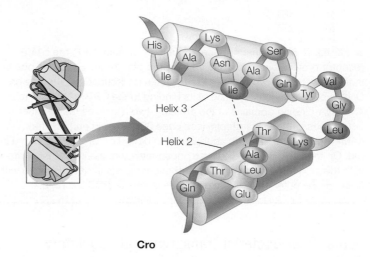

Cro

▲ **FIGURE 26.9 Conserved residues in the DNA-binding helices of** λ **cI repressor and Cro.** Conservative substitutions are shown in blue and identities in green. In both proteins, the alanine in helix 2 contacts a residue in helix 3, which helps to position the helices with respect to each other.

A major outcome of these investigations was the conclusion that the helix–turn–helix is an evolved structural motif that could be widely used in transcriptional regulation. Indeed that is the case, particularly in bacteria. Other such motifs, more widely used in eukaryotic systems, include the zinc finger, which we mentioned in Chapter 20.

The α-3 helix is called the **recognition helix** because its position deep within the major groove allows it to contact specific DNA bases and hence to determine the sequence specificity of binding. The α-2 helix is in contact primarily with DNA phosphates. These electrostatic contacts strengthen binding but do not contribute to specificity. Supporting the concept of α-3 as a recognition helix is the fact that most *cI* mutations that reduce the specific binding of the repressor to operator DNA alter the amino acid sequence in this region of the protein.

Crystallographic analysis of the respective DNA–protein complexes explains how Cro and cI can bind to the same operator sites with different binding affinities. As shown in

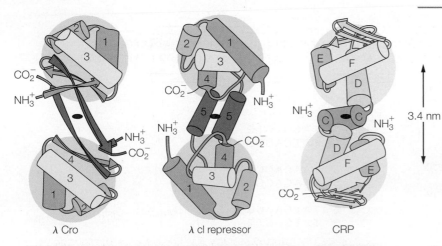

▲ **FIGURE 26.8 DNA-binding faces of** λ **Cro,** λ **cI repressor, and CRP, showing the helix–turn–helix motif.** The motif involves helices 2 and 3 in Cro and cI and helices E and F in CRP. The black ellipses mark centers of symmetry.

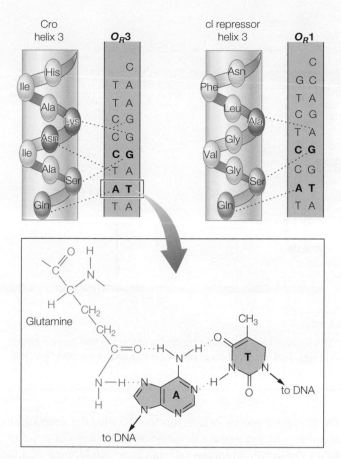

▲ **FIGURE 26.10** **Specific amino acid–nucleotide contacts for cI and Cro repressors.** The conserved residues (green) bind to nucleotides common to all of the operators, and unique residues (blue) bind to non-conserved nucleotides in the operators. Also shown is the structure of a glutamine residue in contact with an A-T base pair. Note that the two repressors are shown binding to different operators.

● **CONCEPT** The helix–turn–helix motif is widely used in bacterial transcriptional regulatory proteins. Specific contacts are made between DNA major groove bases and amino acids in a recognition helix.

FIGURE 26.10, the residues common to both proteins are in contact with DNA sequence elements common to all of the operators. In both proteins, a glutamine residue interacts with one A-T base pair. The cI repressor establishes specificity through a contact in O_R1 with a unique alanine residue, whereas Cro can be in contact with three specific base pairs in O_R3 with unique asparagine and lysine residues. Also, because the two α-3 helices lie closer together in Cro (2.9 nm) than in cI (3.4 nm), the orientations of these helices with respect to the major grooves of operator DNA are quite different.

The SOS Regulon: Activation of Multiple Operons by a Common Set of Environmental Signals

In order for the lysogenic state of λ phage to be broken, initiating prophage excision and a cycle of lytic growth, the λ cI repressor must be inactivated. How does this happen? Various DNA-damaging treatments are known to induce λ prophages, including ultraviolet irradiation,

inhibition of DNA replication, and chemical damage to DNA. Evidently, the virus finds it advantageous to leave a damaged cell, like "rats leaving a sinking ship." Because of the similarity in genetic control between the λ and *lac* systems, investigators sought a small molecule, perhaps a nucleotide, that would accumulate after these treatments and might be the ligand that binds to cI and inactivates it. Surprisingly, the λ repressor was found to be inactivated by a quite different mechanism—proteolytic cleavage. Analysis of this cleavage reaction revealed the SOS system described in Chapter 23 as one of the elements in error-prone DNA repair, in which the genes are controlled by a single repressor–operator system. Such a set of unlinked genes, regulated by a common mechanism, is called a **regulon**. The heat-shock genes, all activated by a transient temperature rise, comprise another regulon.

The control elements in the *E. coli* SOS regulon are the products of the *lexA* and *recA* genes. We have encountered RecA protein before, in its role of stimulating DNA strand pairing during recombination. Remarkably, this small protein has an enzymatic activity in addition to the activities involved in recombination. When bound to single-stranded DNA, it can stimulate proteolytic cleavage of the proteins encoded

● **CONCEPT** The SOS regulon is activated by DNA damage, which stimulates RecA to cause proteolytic cleavage of the LexA and λ cI repressors.

by *cI, lexA,* and *umuD.* LexA is a repressor that binds to at least two dozen operators scattered about the *E. coli* genome (**FIGURE 26.11**). Each operator controls the transcription of one or more proteins that help the cell respond after

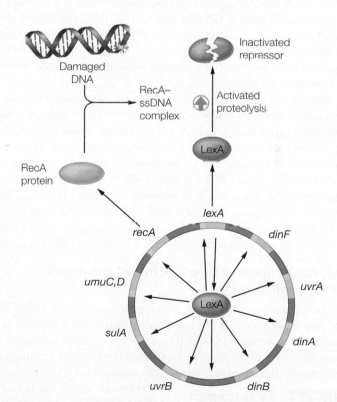

▲ **FIGURE 26.11** **The SOS regulon.** The figure shows locations on the *E. coli* chromosome of some of the genes controlled by the LexA repressor (in blue) and their LexA-sensitive operators (in yellow). LexA repressor (purple) is inactivated by proteolysis, which is enhanced by a complex of RecA protein (blue) and single-stranded DNA.

environmental damage that might harm the genetic apparatus. About 40 known proteins are induced as a result of LexA inactivation. These proteins include the gene products of *uvrA* and *uvrB*, involved in excision repair; *umuC,D*, the genes for the error-prone DNA polymerase V; *sulA*, involved in cell division control; *dinA*, the structural gene for DNA polymerase II; *recA* itself; *lexA* itself; *dinB*, the structural gene for error-prone DNA polymerase IV; and *dinF* (function unknown).

In a healthy cell, *lexA* and *recA* are expressed at low levels, with sufficient LexA protein to turn off the expression of the other SOS genes completely. LexA protein does not completely abolish transcription of either *lexA* or *recA*. The trigger that activates the SOS system after damage is single-stranded DNA. As we have seen, UV irradiation generates gapped DNA structures, and so do other conditions that induce the SOS system. RecA binding within a gap activates LexA proteolysis by a mechanism that is not yet clear. Intracellular levels of LexA decrease, removing the LexA barrier to *recA* transcription. RecA protein accumulates in large amounts. Simultaneously, cleavage of the LexA protein activates transcription of all genes under *lexA* control. Once damaged DNA has been repaired, single-strand gaps in DNA have been filled, RecA is no longer activated, and expression of the SOS genes is limited. However, in a λ lysogen, cleavage of the λ cI repressor leads to irreversible change because that activates prophage excision and phage reproduction, as discussed earlier.

Sequencing of LexA-sensitive operators has yielded a consensus sequence, with seven highly conserved bases in a 20-base-pair region. However, in some LexA-sensitive genes this sequence is located quite differently with respect to the transcriptional start site. The location of each LexA operator, and its similarity to the consensus sequence, control the time after DNA damage at which a particular gene begins to be expressed. Thus, the SOS response appears to be a coordinated series of events, controlled in part by the nature and severity of the initial DNA-damaging event. Other stress situations, such as abrupt changes in growth temperature, involve activation of different regulons.

Biosynthetic Operons: Ligand-Activated Repressors and Attenuation

The lactose operon is involved with catabolism of a substrate. Therefore, the gene products are not needed unless the substrate is also present to be consumed. A different situation is encountered with genes whose products catalyze biosynthesis—those of an amino acid, for example. Because biosynthesis consumes energy, it is to the cell's advantage to use the preformed amino acid, if it is available. Therefore, the regulatory goal is to repress gene activity, by turning *off* the synthesis of enzymes in the pathway when the end product is available. Regulation

● **CONCEPT** The *trp* repressor inhibits tryptophan synthesis by binding as the repressor–ligand complex to the *trp* operator, blocking transcription.

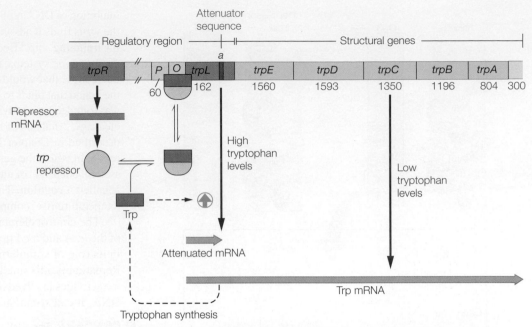

▲ **FIGURE 26.12** The *trp* operon. The figure shows regulation by the *trp* repressor and by attenuation. The attenuator site *trpa* is shown in red. The length of each gene and regulatory region is given in base pairs.

of the *E. coli trp* operon, which controls the final five reactions in the biosynthesis of tryptophan, demonstrates two ways of accomplishing this shutdown: a repressor design in which binding of a small-molecule ligand *activates* the repressor, rather than inactivating it, and premature termination of transcription.

The *trp* operon consists of five adjacent structural genes whose transcription is controlled from a common promoter–operator regulatory region (**FIGURE 26.12**). The *trp* repressor, a 58-kilodalton protein encoded by the nonadjacent *trpR* gene, binds a low-molecular-weight ligand, the amino acid tryptophan. However, in this case the protein–ligand complex is the *active* form of the protein, which binds to the operator and blocks transcription. When intracellular tryptophan levels decrease, the ligand–protein complex dissociates and the free protein ("aporepressor") leaves the operator, so that transcription is activated. If we call lactose an inducer in a catabolic system, it is appropriate to call tryptophan a **corepressor** in this anabolic system. The crystal structure of the *trp* repressor–DNA complex shows a helix–turn–helix motif, comparable to that seen with the λ cI, λ Cro, and *lac* repressors; binding tryptophan to this protein reorients the helices to activate binding to DNA.

The *trp* operon has an additional regulatory feature, now known to be involved in controlling numerous biosynthetic operons. Charles Yanofsky found that activities of the *trp* enzymes varied over a 600-fold range under different physiological conditions, more than could be accounted for by a repressor–operator mechanism alone. Analysis revealed a second mechanism, called *attenuation,* that involves early termination of *trp* operon transcription under conditions of tryptophan abundance. Note from Figure 26.12 a 162-nucleotide sequence called *trpL*, the *trp* leader region. A site called *a,* the **attenuator,** is 133 nucleotides from the 5′ end of the *trpL* sequence. When tryptophan levels are high, transcription terminates at *a,* to give a truncated 133-nucleotide transcript rather than the complete 7000-nucleotide *trp* mRNA. The structural genes are not transcribed, so tryptophan is not synthesized.

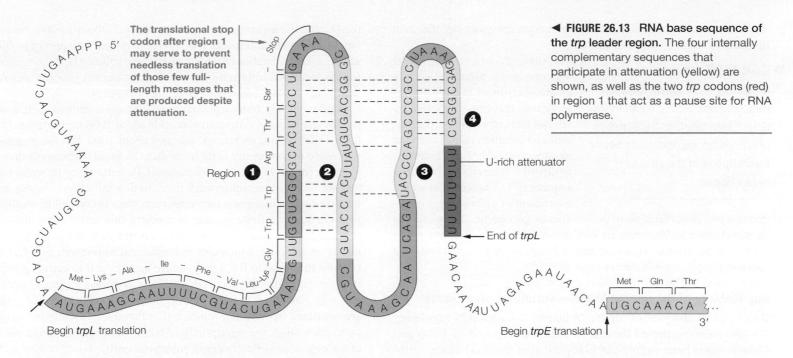

The translational stop codon after region 1 may serve to prevent needless translation of those few full-length messages that are produced despite attenuation.

◀ **FIGURE 26.13** RNA base sequence of the *trp* leader region. The four internally complementary sequences that participate in attenuation (yellow) are shown, as well as the two *trp* codons (red) in region 1 that act as a pause site for RNA polymerase.

Critical to understanding the mechanism of attenuation is the presence of four oligonucleotide sequences in the *trp* leader region that are capable of base pairing to form stem-loop structures in the RNA transcript (**FIGURE 26.13**). In the most stable conformation (**FIGURE 26.14(a)**), region 1 pairs with 2, and region 3 pairs with 4, to give two stem loops. The 3–4 structure, being followed by eight U's, is an efficient transcription terminator because it resembles the factor-independent terminator structure shown in Figure 24.11.

When tryptophan levels are low (Figure 26.14(b)), formation of the 3–4 stem-loop is inhibited, and termination does not occur at the attenuation site. Note that region 1 contains two tryptophan codons (see Figure 26.13). In bacteria, translation is coupled to transcription, so a ribosome can begin translating a message from its 5' end while the message is still being synthesized at its 3' end. In this case, the ribosome stalls when it reaches the two tryptophan codons, because there is insufficient tryptophanyl-tRNA to translate them. The presence of the bulky ribosome prevents region 1 from base pairing with 2, leaving region 2 free to base-pair with 3. Once region 3 is unavailable to base-pair with 4, the 3–4 stem-loop transcriptional terminator cannot form, and the entire message is synthesized. Conversely, when tryptophan is abundant (Figure 26.14(c)), the ribosome does not stall, thereby occluding region 2 and allowing the 3–4 stem-loop structure to form, which leads to transcription termination on the 3 side of 3–4.

Neither the *trpR* system nor the attenuator is simply an on–off system. Both respond in graded fashion to the intracellular tryptophan level. Even though both systems are controlled by the same signal, the action of two distinct control systems greatly extends the possible range of transcription rates of the *trp* operon, giving maximum efficiency to regulation of these genes. At low tryptophan concentration, the repressor–operator interaction is the principal regulatory

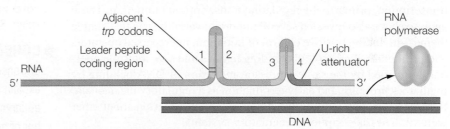

(a) Most stable conformation for leader mRNA.

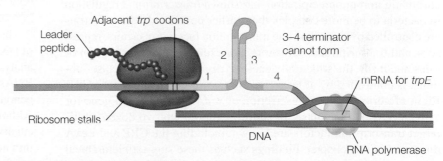

(b) Conformation for leader mRNA at low tryptophan levels.

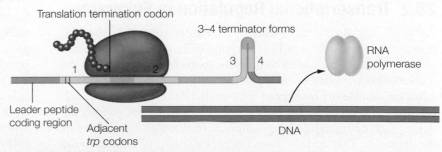

(c) Conformation for leader mRNA at high tryptophan levels.

▲ **FIGURE 26.14** Mechanism of attenuation in the *trp* operon.

mechanism, whereas the effects of attenuation are more significant at moderate to high tryptophan levels.

● **CONCEPT** Attenuation is a regulatory mechanism in which ribosome positioning on an mRNA determines whether transcription of an operon will terminate before transcription of the structural genes begins.

Although the foregoing model was originally proposed simply by inspection of the *trp* leader sequence, it is now supported by several lines of evidence. Significant confirmation comes from the existence in other attenuation-controlled operons of "stalling sequences"—sequences at which movement of a ribosome is inhibited at a low concentration of the product of the operon. These include bacterial operons for synthesis of leucine, with four adjacent leucine codons in the leader sequence, and of histidine, with seven histidine codons.

Applicability of the Operon Model—Variations on a Theme

Biochemical analyses of the *lac,* λ phage, *trp,* and SOS regulatory systems have confirmed the central features proposed by Jacob and Monod—that gene expression is regulated at the level of transcription and that specific protein–DNA interactions control the rate of transcription, primarily by regulating transcriptional initiation. These analyses have also revealed several important variations on that simple theme—including positive control of initiation, interspersed operators and promoters, dual proteins binding to the same site, multiple operons controlled by the same repressor, induction of DNA bending by regulatory proteins, and early termination as a regulatory mechanism. References in Appendix II will direct you to information about other well-studied transcriptional regulatory systems.

Although the operon concept is well established, more recent analysis by genomic techniques, including microarray analysis and chromatin immunoprecipitation, has shown transcriptional regulation in bacteria to be more complex than earlier postulated. Many operons are controlled by more than one transcription factor—σ factors, repressors, and transcriptional activators. For example, the transcription of *nrdA* and *nrdB,* the structural genes for ribonucleotide reductase subunits, is controlled by at least five different proteins, including CRP. Eight proteins control transcription of *sodA,* the structural gene for Mn-superoxide dismutase. In addition, more than two dozen multi-target transcription factors are known, including the CRP and LexA discussed in this chapter. Findings such as these suggest hierarchical networks of transcriptional regulation.

26.2 Transcriptional Regulation in Eukaryotes

As we already discovered in Chapter 24, transcription and its regulation are far more complex in eukaryotic cells than in bacteria and viruses. Much of the added complexity stems from the fact that the transcription template in eukaryotes is chromatin, not naked DNA. Also, the complexity associated with multicellularity and highly differentiated states demands higher orders of regulation. Among the eukaryotes, complexity of regulation increases with complexity of the organism. That complexity is not necessarily proportional to the number of protein-coding genes in a genome. Consider the nematode worm *Caenorhabditis elegans,* whose genome contains about 20,000 genes, close to the number of protein-coding genes in humans, while

the *Drosophila melanogaster* genome contains about 14,000. No one would argue that humans are comparable in genetic complexity to flies and worms. As mentioned earlier, alternative splicing helps to account for the increased information content of the human genome, despite surprisingly small differences in numbers of genes.

Differences in transcriptional regulation contribute as well. Both worms and flies are estimated to encode about 1000 transcription factors, while the human genome encodes about 3000. But the resultant difference in complexity is far more than threefold because regulation by transcription factors is *combinatorial.* Because many different factors are used in assembling each transcription initiation complex and because each transcription factor can participate in controlling multiple genes, a nearly infinite number of transcription factor combinations is possible. By contrast, bacteria have about a half dozen RNA polymerase σ subunits, which can be considered as transcription factors because they bind to RNA polymerase and direct it to certain groups of promoters. Any single bacterial RNA polymerase molecule associates with just one σ factor at a time. So, even though bacterial genes are regulated by multiple transcription factors, as mentioned in the previous section, the multiplicity of factors controlling transcription of a single gene is far greater in eukaryotic cells.

Additional complexity is brought about by the more recently discovered small regulatory RNA molecules, called microRNA, or miRNA (page 816). As we see later in this chapter, the number of microRNA

● **CONCEPT** A large number of transcription factors helps to explain the complexity of higher eukaryotes, even though the number of genes in a genome may be smaller than originally expected from the size of the genome.

molecules encoded by a genome is closely related to the complexity of the organism. Despite the complexity of eukaryotic transcriptional regulation, the model put forth a half century ago for bacterial transcriptional regulation applies also to eukaryotic cells; gene regulation is expressed largely at the level of transcriptional initiation.

In Chapter 21 we mentioned the ENCODE Project (Encyclopedia of DNA Elements). Recently developed techniques such as microarray analysis, chromatin immunoprecipitation, rapid DNA sequencing, and RNAseq have led to the finding that as much as 80% of the human genome is transcribed, even though only about 2% encodes proteins. Although transcriptional initiation is the primary known process in eukaryotic gene regulation, a complete understanding will need to take into account the functions of DNA sequences recently thought to be nonfunctional.

Chromatin and Transcription

Recall from Chapter 21 that chromatin consists of individual units—nucleosomes—with each nucleosome containing 200 or more base pairs of DNA, 147 of which are wrapped about a core of eight histone proteins. Within these 147 base pairs, the central 80 or so are organized by a heterotetramer of histones H3 and H4. About 40 base pairs on each side are more loosely associated with two H2A/H2B dimers. Spacer DNA, which is not associated with the core histones, associates with histone H1 and other proteins. The core histones (H2A, H2B, H3, and H4) each have a common helical core structure, called the *histone fold,* joined to a relatively unstructured N-terminal tail. Specific amino acid residues within the tails are subject to modification—acetylation, methylation (mono-, di-, and tri-) ubiquitylation, sumoylation,

ADP-ribosylation, and phosphorylation. Acetylation usually occurs on lysine amino groups. By neutralizing charge, acetylation loosens the ionic bonds that link histones to DNA, and generally activates the target nucleosome as a transcription template, whereas deacetylation, conversely, tends to inhibit transcription. The action of *histone deacetylase* (HDAC), which hydrolytically removes acetyl groups, is associated with transcriptional inhibition. A protein called polycomb repressive complex (PRC) recruits HDAC to nucleosomes.

Chromatin, as noted earlier, exists within cells either in a highly compacted form, *heterochromatin,* or a more loosely structured form, *euchromatin.* Heterochromatin is transcriptionally inactive, and genes in particular tissues, which are permanently silenced in those tissues as a result of differentiation, are usually found in heterochromatin. The repressed chromatin structure is often associated with methylation of histone H3 at lysine 9 (H3K9). Methylation at this site leads to binding of HP1 (Heterochromatin Protein 1), which further contributes to compaction. Repression is also induced by other histone modifications, such as methylation of H3K27 or ubiquitylation of H3K119. Whether these modifications are causes or effects of the repressive structure is not yet clear. By contrast, the H3K4 methylation mark is a transcriptional activating signal. Many biochemists believe in a "histone code," which would relate specific chromatin modifications to specific regulatory effects in a predictable and rational way, as seen with the genetic code. So far, however, the histone code has proved to be elusive.

In order for a gene to be transcribed, the promoter must be relatively clear of nucleosomes. In other words, it should lie in a nucleosome-free region (NFR). To some extent, clearing a promoter of histones is a function of chromatin remodeling complexes (page 810). **FIGURE 26.15** shows schematic representations of chromatin in repressed and transcriptionally active configurations. Not all histone modifications are shown.

● **CONCEPT** Several factors, including chromatin remodeling and histone modifications, combine to render eukaryotic transcription start sites relatively free of nucleosomes.

An additional event in preparing chromatin for transcription initiation is replacement of histones H3 and H2A with modified forms of these histones, called H3.3 and H2A.Z. ChIP-seq analysis (described in Tools of Biochemistry 24.B) shows that H2A.Z is specifically associated with nucleosomes near the transcription start site, evidently being replaced once transcription is under way (**FIGURE 26.16**). Note that the transcription start site (TSS) is in a nucleosome-free region.

Transcriptional Control Sites and Genes

As in bacterial gene expression, transcription in eukaryotes is regulated by the interplay of *trans*-acting proteins that interact with *cis*-acting sites on the DNA template. *Cis*-acting sites within the promoter include the NFR (nucleosome-free region), TSS (transcription start site), Inr (initiation region), TATA box, and BRE (TFIIB recognition element). This list is not exhaustive. Sites farther upstream are generally called *enhancers* in metazoans, and *UAS* (upstream activator sites) or *URS* (upstream repressor sites) in yeast. Because of the large number of proteins bound in each initiation complex, there is tremendous variability among different genes in promoter and enhancer sequences.

Trans-acting factors include not only general transcription factors (Chapter 24), but chromatin remodeling factors, *histone chaperones* (which facilitate chromatin reformation once a transcription complex

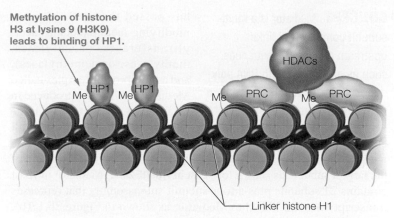

Methylation of histone H3 at lysine 9 (H3K9) leads to binding of HP1.

Linker histone H1

Repressed. Repression is maintained partly by histone modifications, partly by linker histones H1 (red ovals), partly by HP1 (see text), and partly by binding of polycomb repressive complex (PRC), which recruits histone deacetylases (HDAC).

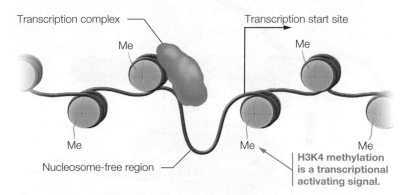

Transcription complex

Transcription start site

Me

Me

Me

Me

Me

Nucleosome-free region

H3K4 methylation is a transcriptional activating signal.

Active. Transcriptionally active chromatin displays a nucleosome-free region downstream of the transcription complex (blue oval). Nucleosomes in green contain modified histones associated with transcription initiation.

▲ **FIGURE 26.15** Schematic representations of transcriptionally repressed (top) and active (bottom) chromatin.

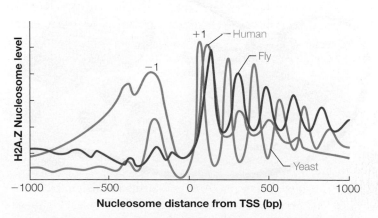

▲ **FIGURE 26.16** Distribution of H2A.Z-containing nucleosomes near a transcription start site (TSS). In human, *Drosophila,* and yeast cells, H2A.Z is most abundant in nucleosomes at positions +1 and −1 relative to the TSS. These data were obtained by chromatin immunoprecipitation (see Tools of Biochemistry 24B).

● **CONCEPT** Mediator is a multi-subunit complex that links upstream regulatory sequences, such as enhancers, with RNA polymerase II and general transcription factors at the promoter site.

has passed through), histone-modifying factors (such as acetyltransferases, deacetylases, methylases, or demethylases), and a multiprotein complex called Mediator, which we introduced in Chapter 24. Mediator is a large multisubunit complex found in all eukaryotes, which links upstream activating (or repressing) elements with RNA polymerase II—specifically, with the C-terminal domain of the largest subunit, as shown for yeast in **FIGURE 26.17**. Mediator in yeast contains 21 subunits plus a four-subunit sub-assembly that represses transcription when linked to Mediator, as shown in Figure 26.17(b).

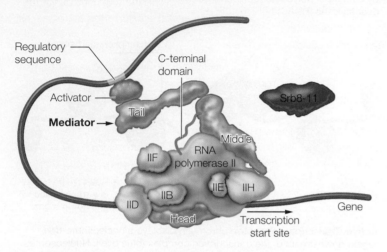

(a) Activation. Mediator has three distinct structural domains – head, tail, and middle. The tail domain interacts with transcriptional activators (in brown) and links Mediator (blue) with RNA polymerase (green) and general transcription factors (yellow).

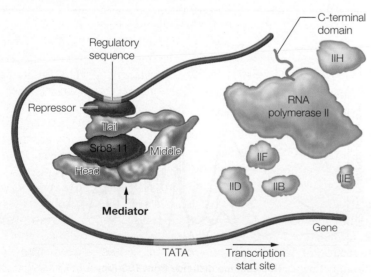

(b) Repression. The four-subunit complex containing Srb8, Srb9, Srb10, and Srb11, when bound to Mediator, prevents its interaction with RNA polymerase II and the basal transcription machinery.

▲ **FIGURE 26.17** Mediator as a bridge between gene-specific regulatory factors and the general transcription machinery at the pol II promoter.

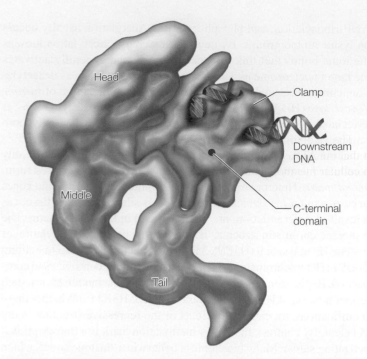

▲ **FIGURE 26.18** The structure of yeast Mediator in complex with RNA polymerase II. The structure of Mediator (dark blue) was determined by image processing of cryoelectron micrographs, and that of RNA polymerase II (green) from its crystal structure. The red dot marks the location where the C-terminal domain (CTD) of the largest pol II subunit (the one that undergoes phosphorylation) emerges from the surface of the enzyme.

Mediator's large size can be seen in **FIGURE 26.18**, which shows image-processed electron micrographs of the complex in association with RNA polymerase II.

Chromatin Remodeling Complexes

Chromatin remodeling complexes couple the energy of ATP hydrolysis to changes in chromatin structure, usually resulting in transcriptional activation. In most cases, the energy of ATP hydrolysis is coupled to translocation of the histone core particle along DNA, with the core histones either sliding along DNA or being transiently displaced. Four families of chromatin remodelers are particularly well studied, classified in terms of the structure of the ATPase-containing subunit. **FIGURE 26.19** shows the domain organization of the ATPase-containing subunit in each of the four described families.

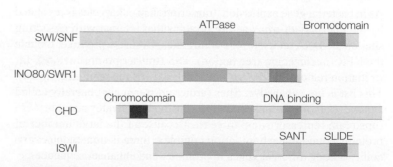

▲ **FIGURE 26.19** Four chromatin remodeler families and conserved domains of the ATPase-containing subunit.

The ATPase subunit of the SWI/SNF family contains a **bromodomain** (named after the *Drosophila* BRM gene), a domain that interacts specifically with acetylated lysines. This draws the complex to chromatin that has already become partially activated by histone acetylation. Recent evidence indicates that RSC, a member of this family, completely dissociates histones from DNA, beginning with disruption of DNA–histone bonds by the remodeler, followed by ATP-dependent translocation.

The catalytic ATPase subunit in the INO80 family is distinctive in that the ATPase domain is split. This complex plays a broader role in cellular metabolism than the other remodelers in that it is found in DNA repair complexes and at sites of resolution of stalled replication forks. A principal action of this remodeler is replacement in chromatin of canonical histones with histone variants, specifically, exchange of H2A for H2A.Z, which occurs near TSS (see Figure 26.15(b)).

The ATPase subunit of the CHD remodeler family contains a **chromodomain,** which interacts specifically with methylated histones, particularly the transcriptionally activating H3K4 that we mentioned earlier. Although CHD has been found to be associated with transcriptionally active chromatin, its specific function has not been identified.

● **CONCEPT** Nucleosome remodeling complexes use the energy of ATP hydrolysis to move nucleosomes out of the way for transcription initiation, but they have other roles as well.

However, interest has focused on the relationship between one member of this family and the ability of embryonic stem cells to maintain the pluripotent state, that is, the ability to develop into any kind of differentiated cell. So the current picture of this family of remodelers is that, by helping to maintain all chromatin in an open conformation, CHD helps the cell to retain the ability to express any combination of genes needed for a specific developmental pathway.

Unlike the three families of histone remodeling complexes just described, the ISWI family of remodelers is associated with transcriptional repression. Although little is known about the function of this family, recent evidence in *Drosophila* identifies a role in maintaining the higher-order structure of the male X chromosome.

● **CONNECTION** Learning the function of the CHD family of nucleosome remodelers is relevant to understanding the pluripotent state of embryonic stem cells.

Transcription Initiation

As noted earlier, most sequence-specific regulatory sequences, such as nuclear receptor binding sites, are found some distance upstream from transcription start sites, and Mediator plays a key role in connecting these sites to the downstream promoters. It is estimated that each eukaryotic gene contains about five specific regulatory sites. Because of the large number of sequence-specific regulatory proteins and their uses in controlling multiple genes, this complexity is probably necessary to prevent the accidental expression of a gene that might result if only one or two specific regulatory events needed to occur.

The action of chromatin remodeling complexes, the modifications of histones near promoter sites, the binding of upstream sequence-specific transcriptional activators, and their relationship to Mediator set the stage for final assembly of a transcription complex: binding of RNA polymerase II and the several general transcription factors that we identified in Chapter 24, including TATA box-binding protein and TFII proteins A,

B, D, E, F, and H. TFIIH contains at least two enzymatic activities—an ATP-dependent helicase that unwinds template DNA strands to expose the transcription template, and a protein kinase that converts the initiation complex to an elongation complex by phosphorylating serine residues in the C-terminal domain of the largest pol II subunit.

Recent studies with ChIP techniques indicate that RNA polymerase II is bound at most genes, regardless of whether the genes are expressed. This finding suggests that binding of general transcription factors, not binding of pol II itself, is rate-limiting for initiation. Although these events are largely independent of the gene being transcribed, there is some gene specificity. For example, Rap1 is a protein that binds directly to TFIID, and this interaction helps to drive the transcription of the highly expressed genes for ribosomal protein synthesis. In any event, the sequence of binding general transcription factors is, as we indicated in Chapter 24, leading to a preinitiation complex.

● **CONCEPT** Most, but not all, events involving general transcription factor binding at promoters are nonspecific with regard to the gene being activated.

As is also seen with bacterial transcription, many eukaryotic initiation complexes pause soon after initiation. This may involve a site for specific regulation because several protein factors are known both to promote and to overcome pausing. One such positive regulator is called P-TEFb (*positive transcriptional elongation factor* b). This protein phosphorylates the second serine in each of the heptapeptide carboxy terminal repeats in RNA polymerase II—perhaps augmenting the activity of the TFIIH protein kinase, which acts at the fifth serine residue in each repeat.

Regulation of the Elongation Cycle by RNA Polymerase Phosphorylation

The heptad repeats in CTD, the C-terminal domain of the large RNA polymerase subunit, are unphosphorylated when the enzyme interacts with Mediator for placement at the promoter. The phosphorylation status of these repeats changes with time after initiation, in a functionally significant way that suggests a "CTD phosphorylation code." Proteins that interact with the CTD can be classified as code "writers" (serine kinases), "readers" (proteins that interact with specific phosphorylation patterns), or "erasers" (serine phosphatases). Among the proteins binding early are those involved in synthesizing the RNA 5' methyl-G cap (Chapter 24). Other CTD readers are enzymes that change the histone modification status, and these patterns change as the predominant CTD phosphorylation changes from Ser-5 to Ser-2 as termination approaches. **FIGURE 26.20** shows the predominant histone and CTD modification patterns as transcription proceeds. The late modifications help to recruit protein factors involved in transcription termination.

● **CONCEPT** The shifting pattern of CTD phosphorylation as RNA polymerase moves through a gene controls events such as mRNA capping, histone modifications, and the recruitment of transcription termination factors.

Time-dependent changes in CTD phosphorylation are of intense interest to investigators trying to understand the "histone code," mentioned earlier as a presumed relationship between specific histone modification patterns and functional consequences, such as activation or repression of specific genes, or *epigenetic* phenomena.

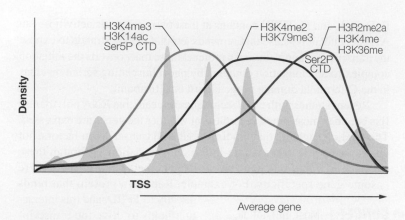

▲ **FIGURE 26.20** Histone and pol II CTD modifications as a function of gene position. The nucleosome distribution relative to the TSS is shown in gray. The blue, black, and red traces represent genome-wide histone and CTD modification patterns for the respective modifications shown.

26.3 DNA Methylation, Gene Silencing, and Epigenetics

So far, our discussion of gene regulation has focused on reversible processes in which transcription can be either turned on or off. In eukaryotes there is a need for differentiated cells to permanently turn off or silence most genes. The process best understood for gene silencing is DNA methylation.

DNA Methylation in Eukaryotes

As we discussed in Chapter 21, DNA methylation in bacteria is well understood. The target base for methylation is usually adenine, and the processes involving methylation include restriction/modification and methyl-directed mismatch repair. The situation in eukaryotes is quite different. The only eukaryotic DNA base to undergo methylation is cytosine (although N-methyladenine has been recently detected in some eukaryotic DNAs). In eukaryotes, DNA mismatch repair does not involve methylation, and there is no known eukaryotic counterpart to restriction and modification. Instead, DNA methylation participates in processes involving gene silencing.

Those cytosine residues undergoing methylation in eukaryotes are usually C's that are immediately 5′ to G's, that is, C's in a CpG dinucleotide. CpG is relatively underrepresented in eukaryotic genomes. However, there are regions of eukaryotic genomes in which CpG dinucleotides are found at the statistically expected frequency. These regions are called **CpG islands.** CpG islands are generally longer than about 500 base pairs, and they have a total GC content of 55% or greater. Generally, C's in CpG islands are undermethylated, while most CpG methylation occurs in CpG-poor regions of the genome. For unknown reasons, this distribution of methylated sites is often reversed in cancer cells. Interest is focused on whether this is related to altered patterns of gene expression seen in cancer, such as decreased expression of tumor suppressor genes. 5-MethylC residues in DNA undergo deamination more readily than C. Because deamination converts a G-C or G-mC base pair to A–T, interest focuses on whether deamination-induced mutagenesis is partially responsible for the finding that cancer cells have elevated mutation rates relative to normal cells.

● **CONNECTION** DNA methylation patterns are altered in cancer, and analyses of the alterations should enhance our understanding of carcinogenesis.

DNA methylation patterns are heritable, making DNA methylation the best understood example of an **epigenetic** process. Epigenetics refers to the heritable transmission of a gene-expression pattern that does not involve a change in the DNA coding capacity. The pattern of methylation in a particular genome is established early in embryonic development. Mammalian cells contain three different DNA-cytosine methyltransferase enzymes—Dnmt1, Dnmt3a, and Dnmt3b. De novo methylation, early in development, is carried out by Dnmt3a and 3b. How these enzymes generate specific methylation patterns is not yet clear. However, once an embryonic methylation pattern is established, that pattern can be faithfully reproduced by the maintenance methylase, Dnmt1, which tracks with the replication apparatus and methylates every C in the daughter strand that was methylated in the parental strand. This process is illustrated in **FIGURE 26.21**.

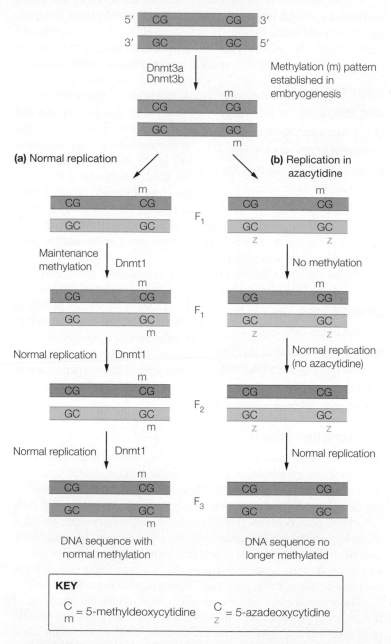

▲ **FIGURE 26.21** De novo and maintenance methylation of DNA and the effect of 5-azacytidine on DNA methylation. F1, F2, and F3 refer to generations of growth.

A key finding leading to understanding this phenomenon is the effect of **5-azacytidine** on maintenance methylation. Azacytidine is a cytidine analog that can be metabolized to the 5-aza analog of dCTP and become incorporated into DNA. However, because the six-membered ring of 5-azacytidine contains N instead of C at position 5, it cannot undergo stable methylation. Hence, in the presence of 5-azacytidine, a 5-mCpG dinucleotide is converted to CpG in three rounds of replication. This treatment was found to activate previously repressed genes, a finding that constituted important early evidence that correlated DNA methylation with gene silencing. For example, adult bone marrow cells were found to reactivate the synthesis of fetal hemoglobin, which is normally repressed after birth.

● **CONCEPT** DNA methylation can lead to permanent gene inactivation by processes involving chromatin structure and histone modification.

5-Azacytidine

5-Azadeoxycytidine triphosphate

More recent evidence points to a natural role of DNA demethylation in activation of repressed genes. In 2010, two laboratories reported that deamination of DNA-methylcytosine bases by activation-induced deoxycytidine deaminase (AID), involved in antibody maturation (Chapter 23), also plays a role in the induced dedifferentiation of somatic cells to induced pluripotent cells, capable of differentiation in culture to any state. This is an important development in any attempt to use stem cell biology for therapeutic purposes.

There does not seem to be a single mechanism accounting for gene repression as a consequence of methylation. In some cases, binding of transcription factors to methylated DNA is inhibited. In other cases, the effects on histone modifications promote transcriptional inactivation. For example, some H3K4 *protein* methyltransferases, expected to cause transcriptional activation, are unable to modify histones that are associated in chromatin with DNA containing CpG islands.

DNA Methylation and Gene Silencing

Whatever the mechanism in terms of effects on chromatin structure, it is clear that DNA-cytosine methylation is responsible for permanent gene silencing. Two such phenomena are well established—*X chromosome inactivation* and establishment and maintenance of

gene imprinting. During embryonic development in mammals, one of the two X chromosomes in female cells is permanently inactivated by DNA-cytosine methylation. As described above, this involves de novo methylation, followed by maintenance methylation throughout life. The significance of this modification is that the level of expression of genes carried on the X chromosome is then approximately equal for both male and female cells. In most mammals, the choice of a chromosome to inactivate is random. Gene imprinting is similar in that it involves permanent gene inactivation during embryonic development. For some genes only a single parental allele is expressed—some from the father, some from the mother. Again, DNA methylation is responsible for shutting off expression of the imprinted gene.

Imprinting is affected in a rare condition called *Prader–Willi syndrome.* In this condition, a region of paternal chromosome 15 is deleted or unexpressed. This region includes a gene that controls imprinting; normally, paternal genes from this region are expressed, while maternal genes are silenced. The defect in imprinting means that, while most people have one working copy of each gene in this region, affected individuals have none. The condition is characterized by short stature, obesity, and delayed puberty. A different condition, *Angelman syndrome,* arises when the same region of chromosome 15 is affected in the maternally derived genetic material. Individuals with this condition exhibit a range of symptoms, including intellectual and developmental disability and sleep disturbance.

● **CONNECTION** Genetic abnormalities in DNA methylation have developmental consequences in humans.

● **CONCEPT** DNA CpG methylation is responsible for at least two developmental gene inactivation processes—X chromosome inactivation and gene imprinting.

Genomic Distribution of Methylated Cytosines

Recently, the availability of second- and third generation DNA sequencing procedures, which allow rapid generation of vast amounts of sequence data, has made it possible to analyze the entire human "DNA methylome," that is, the distribution of 5-methylcytosine bases throughout the genome. The approach is to treat DNA with sodium bisulfite, which quantitatively deaminates cytosine to uracil, but has little effect on methylcytosine. This is followed by high-throughput sequence analysis. An intriguing recent finding is that embryonic stem cells have about one quarter of their methylcytosines in a non-CpG context. When embryonic stem cells underwent differentiation, the non-CpG methylation disappeared. Moreover, it is known that certain cell lines can be induced to return to a pluripotent state (able to differentiate into any state); these induced pluripotent cell lines were found once again to contain a large proportion of non-CpG methylated sites. These findings provide important clues to the potential use of embryonic stem cells in treating diseases such as diabetes or Parkinsonism. Recent technical improvements allow determination of DNA base sequence and cytosine methylation status in a single operation, without a need for bisulfite treatment.

Other Proposed Epigenetic Phenomena

5-Hydroxymethylcytosine

In 2009, it was reported that DNAs from some cells, mostly in nervous tissue, contain significant proportions not only of 5-methylcytosine, but also of *5-hydroxymethylcytosine*. As we noted in Chapter 19, bacteriophage T4 and its relatives have 5-hydroxymethylcytosine completely substituted for cytosine in their DNA. In phages, the modification occurs at the nucleotide level. The situation in mammalian cells is quite different; hydroxymethylcytosine arises through oxidation of methyl groups in 5-methylcytosine. More recently, smaller amounts of 5-formylcytosine and 5-carboxycytosine have been detected, suggesting that 5-hydroxymethylcytosine is an intermediate in an oxidative pathway for DNA demethylation.

Oxidative pathway for demethylation of 5-methylcytosine

The finding that hydroxymethylcytosine is particularly abundant in DNA from nervous tissue suggests a possible epigenetic role for this modification independent of that for 5-methylcytosine, but so far, evidence for this idea is scant.

Chromatin Histone Modifications

Intense interest is focusing on other possible mechanisms of epigenetic inheritance, including processes involving chromatin histone side-chain modification and noncoding RNA molecules. In particular, the premise that modification of histone amino acid residues is an epigenetic phenomenon is widely accepted. There is no doubt that histone side-chain modifications affect gene expression and that no changes in DNA base sequence are involved. The issue is whether patterns of histone modification are inherited from one cell generation to the next. As stated in a recent article, "The problem with this characterization is that overwhelmingly, experiments have shown it to be false: histone modifications are *not* maintained as cells divide" (Ptashne, 2013, p. 1); however, see the recent article by De and Kassis, cited at the end of this book. Understanding processes by which histone modification patterns are sustained through time is one of the most important research frontiers in biology.

The other issue regarding epigenetics and DNA methylation arises from the fact that several organisms, including the widely studied *Drosophila melanogaster* and *Caenorhabditis elegans,* contain no detectable methylated DNA bases. Do such organisms have undiscovered epigenetic processes?

26.4 Regulation of Translation

As we have mentioned, the Jacob–Monod paradigm dominated our thinking for many years about gene expression and its control. We have seen how much of the operon model applies also to the more complex processes of eukaryotic gene expression. At the same time, we have encountered many striking differences—mRNA splicing, mRNA capping and tailing, multiple transcription factors in initiation complexes, chromatin as the transcription template, chromatin remodeling, the involvement of Mediator, control by phosphorylation of RNA polymerase, and the absence of multicistronic mRNAs, to cite a few examples. Equally striking are recent discoveries of important regulatory processes acting at the translational level. We present examples of these regulatory processes first for bacteria, then for eukaryotes. Some of these RNA regulatory phenomena, such as RNA interference, operate at the level of mRNA degradation and are treated in a separate section (26.5). Other processes, including riboswitches (Section 26.6) and RNA editing (Section 26.7), are also treated separately.

Regulation of Bacterial Translation

Bacterial translation is regulated by at least three mechanisms—ribosome occlusion due to mRNA tertiary structure, translational repression caused by protein binding to mRNA, and actions of regulatory RNA molecules, which base-pair with mRNA molecules. The example we describe here falls in this last category.

The idea that RNA molecules themselves could serve as genetic regulators has been known since the early 1980s, but the broad significance of this mechanism has become evident only recently. An early finding was the discovery of antisense regulation, in which mRNA translation is blocked by a complementary RNA. An example involves *ompC* and *ompF,* two genes that encode outer membrane proteins in *E. coli.* These genes are osmoregulated: Cells respond to growth in a medium of high osmolarity by shutting down the synthesis of OmpF protein and activating the synthesis of OmpC, so that the total amount of protein is constant and the cell's internal environment is maintained. The postulated mechanism of the *ompF* shutoff is shown in **FIGURE 26.22**. High osmolarity in some way triggers synthesis of an antisense RNA, the product of the *micF* gene. This RNA is partly complementary to sequences in the 5′ end of *ompF* mRNA. The *micF* RNA inactivates the *ompF* message by annealing to it and forming a duplex RNA in vivo. The translational initiation sequences of OmpF mRNA, which must be single-stranded to direct translation, are included in this duplex. Incorporation of the translation initiation sequences into a duplex is responsible for blocking translation of the message. Another gene regulated by antisense RNA is *crp,* the structural gene for cAMP receptor protein (not shown).

More recently, analysis of whole-genome sequences followed by confirming experimental evidence has revealed that regulation by small RNAs (now called sRNAs) is much more widespread than originally realized. In *E. coli* alone, some 80 sRNAs have been discovered. Some, like the MicF RNA mentioned above, bind to just one target sequence. Others bind several different sequences. Some, like the antisense examples mentioned here, act by sequence-specific binding to RNA. Others act by binding specific target proteins. Most of the regulatory effects mentioned here are exerted at the level of translational initiation. Again, this is most economical. Regulation by RNAs instead of proteins makes

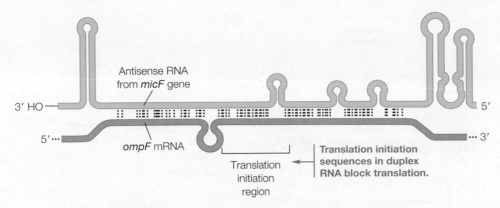

▲ **FIGURE 26.22** Inactivation of *ompF* mRNA by pairing with antisense RNA from the *micF* gene. A change in osmolarity stimulates transcription of the *micF* gene. The transcript is largely complementary to a region in *ompF* RNA that includes the translational start site. Hairpin loops within the sequences allow base pairing between complementary regions on the two mRNAs; in this way, both transcripts are prevented from serving as templates for protein synthesis.

good metabolic sense because the energy-using reactions of translation can be avoided.

Synthesis of designed antisense RNAs is being widely used as an approach to **gene knockdown,** when an investigator wishes to block expression of a specific gene without the extensive efforts involved in creating knockout organisms by specific gene interruption. A synthetic oligonucleotide can be introduced into cells, usually following treatment to transiently increase membrane permeability, with a sequence targeted to the gene whose expression is to be inhibited. Often the knockdown reagent is not an RNA molecule but an RNA analog, one that has been modified so as to avoid enzymatic degradation within the cell. A popular family of antisense reagents is the *morpholinos,* oligomers or polymers that use the same bases as found in natural RNA, but that have morpholine instead of ribose and a phosphorodiamidate bond to link adjacent "nucleotide" units; see the illustration.

Antisense analogs such as morpholinos are being developed as therapeutic reagents, where a target nucleic acid can be identified, such as a viral genome, to become bound and thereby inactivated. The stumbling block to developing this approach is the difficulty of producing reagents in a form that is permeable to cell membranes—something that is far easier to do with cells in culture than with cells in living organisms.

Regulation of Eukaryotic Translation

Comparable translational regulatory processes occur in bacterial and eukaryotic cells, although there are some significant differences, such as control by phosphorylation of initiation factors (see below). Most striking, however, are parallel developments in bacterial and eukaryotic cells regarding the regulatory roles of noncoding RNAs. In bacteria, these RNA molecules are generally short, as implied by the name sRNA. By contrast, this family of RNA molecules in eukaryotes is larger, being normally over 200 nucleotide residues in length; these are usually called long noncoding RNAs, or lncRNAs. We discuss these molecules after presenting a better-understood example of translational regulation.

Phosphorylation of Eukaryotic Initiation Factors

A number of the soluble protein translation factors in eukaryotes are subject to control by phosphorylation. For example, eIF2 is a G protein involved in binding Met-tRNA to the P site as an essential step in initiation (Table 25.3). Four different protein kinases are known to phosphorylate the α subunit of eIF2 at Ser-51. Phosphorylation of eIF2 increases its affinity for its guanine nucleotide exchange factor, eIF2B, leading to formation of an inactive eIF2B-eIF2-GDP complex, thereby shutting down translation initiation. One of the four relevant protein kinases is a heme-regulated inhibitor kinase (HRI), which is activated under conditions of heme deprivation, thereby shutting off synthesis of hemoglobin when the heme cofactor is scarce. This effect is important in reticulocytes (immature erythrocytes), which are enucleated but which have ample mRNA

● **CONCEPT** Phosphorylation of initiation factors is commonly used for translation-level control in eukaryotes.

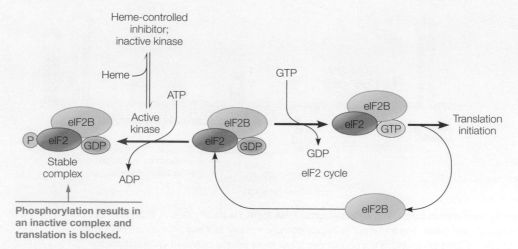

▲ **FIGURE 26.23** **Regulation of translation in erythropoietic cells by heme levels.** If heme levels fall, the heme-controlled kinase becomes active and phosphorylates eIF2 (red arrow). This blocks further translation by tying this factor into a stable complex with eIF2B. When heme levels are adequate, the kinase is inhibited, and eIF2 is available for translation initiation.

for globin synthesis. As shown in **FIGURE 26.23**, the biosynthesis of hemoglobin is regulated by this process, and, hence, the translation of globin mRNA is shut off unless adequate heme is available to complex with the protein.

Long Noncoding RNAs

Because a large proportion of a mammalian genome is transcribed and yet only a small portion codes for proteins, mammalian transcriptomes must contain large numbers of noncoding RNA molecules. Although some of these molecules may represent transcriptional "noise," results of the ENCODE Project suggest that many of the newly discovered transcripts are, in fact, functional. Indeed, for some of these ncRNAs that have been characterized, there is evidence that they may be used in an antisense regulatory mechanism, as we have seen for sRNAs in bacteria. Unlike bacteria, however, the eukaryotic ncRNAs analyzed to date appear to act at the transcriptional level. Transcription of a noncoding region in the vicinity of a protein-coding gene can inhibit transcription of that gene by interfering with the binding of transcription factors. An interesting example is transcription of human *DHFR*, the structural gene for dihydrofolate reductase. In this case, a noncoding RNA forms a triplex structure with the *DHFR* promoter, which results in disruption of the preinitiation complex. Of course, this pushes the regulatory issue back one step; what controls expression of this regulatory ncRNA?

So far there are few indications that ncRNAs act at the translational level, for example, by occluding ribosome binding sites as has been seen for bacterial sRNAs. However, the phenomenon of RNA interference, which we consider next, establishes that noncoding RNAs in eukaryotes regulate translation by controlling *degradation* of mRNA molecules.

26.5 RNA Interference

A series of striking discoveries, beginning in the late 1990s, revealed unexpected processes involving RNA-based genetic regulatory and cellular defense mechanisms. The original observations stemmed from attempts to use genetic engineering to increase purple pigmentation in petunias by introducing into purple flowers additional genes encoding enzymes of the pigment synthesis pathway. Surprisingly, the resultant transgenic plants were not purple; flowers either had highly variegated pigmentation or were white. It appeared that the pigment-forming genes had somehow switched each other off. At first it was thought that this could be resulting from formation of antisense RNA, which would pair with normal-sense RNA, creating nontranslatable double-stranded RNAs. Further experiments, primarily by Craig Mello and Andrew Fire, led to a different conclusion and to the discovery of **RNA interference**, or RNAi. The term covers two distinct processes, both of which involve small RNA molecules, some 21 to 24 nucleotides in length. One class of these molecules, called **microRNA**, or miRNA, is involved in gene regulation, while **small interfering RNAs**, or siRNAs, are formed primarily as a cellular defense mechanism.

MicroRNAs

MicroRNAs are specific gene-regulatory products; it is estimated that nearly one-third of all human genes are regulated by miRNA molecules. The number of specific miRNA molecules in cells of a particular organism is related to the evolutionary complexity of the organism. In a recent study, the total numbers of known miRNAs were reported as 677 for humans and 491 for the mouse. By contrast, *Drosophila* had 147, while the sponge had only eight. miRNAs are derived from partially palindromic RNA molecules. As shown in **FIGURE 26.24**, these precursor molecules result from transcription by RNA polymerase II, followed by 5′ capping and 3′ polyadenylylation, just as in mRNA synthesis, yielding a primary transcript (pri-miRNA). Ends of the molecule meet to give a stem-loop hairpin, and complementary sequences pair up. Processing begins in the nucleus, with cleavage by an enzyme complex called Drosha, some 22 nucleotides from the stem-loop junction, to give a partial hairpin with a short 3′ overhang (pre-miRNA). This is recognized by an exportin complex (XPO5 in the figure), which carries the miRNA precursor to the cytoplasm, where it is bound by another complex called Dicer. In conjunction with another protein called Argonaute, Dicer degrades one strand of the partial duplex miRNA precursor and transfers the remaining single strand, which is now a

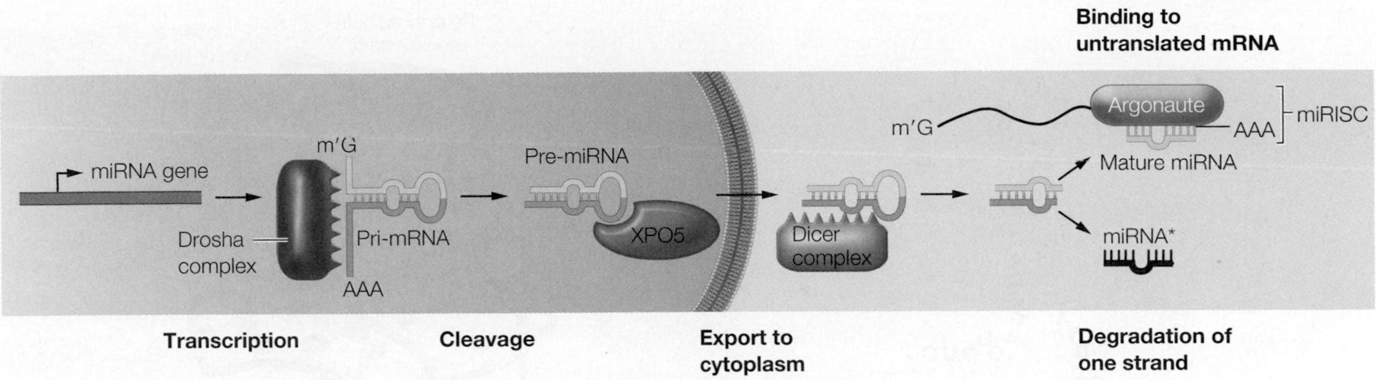

▲ FIGURE 26.24 Biogenesis of miRNA. The mRNA sequence selected for processing can be in either an exon or an intron. For further details, see text.

completed miRNA molecule, along with the Argonaute protein, to still another complex, called RISC (RNA-induced silencing complex). RISC then binds to the 3′ untranslated sequences of target mRNA molecules, using sequence complementarity with the miRNA as a guide. The region along which sequence complementarity is sought is about seven nucleotides. If the miRNA sequence is completely complementary to that of the target mRNA within those seven nucleotides, that mRNA is completely degraded (not shown in the figure). This process is catalytic in that once an mRNA has been degraded, RISC can seek further targets.

If the miRNA–mRNA sequences match only partially, then activity of the mRNA is slowed by various mechanisms. There may be inhibition of binding of translational initiation factors, or ribosome stalling, or activation of enzymes that remove the 3′ polyA tail. There may not always be immediate mRNA degradation, but its translational activity is slowed.

Eventually, the inhibited mRNA is transferred to a cytosolic site called the **P-body** (P for processing), where it is sequestered from ribosomes, and hence, is translationally inactive. The P-body is the site for ultimate degradation of all mRNA molecules, whether or not they arrived there accompanied by RISC. Because RISC can be reused, a single miRNA molecule can participate in regulating as many as several hundred different mRNAs. Also, regulation by miRNAs can be combinatorial: The 3′ UTR of an mRNA can be bound by two or more miRNAs, with each binding of a second miRNA yielding further repression of the translational activity of that mRNA.

● **CONCEPT** Small RNAs produced from processing of double-stranded RNAs are widely used in gene regulation (miRNAs) and as a defense mechanism (siRNAs).

Small Interfering RNAs

RNAi as a defense mechanism normally arises as a result of virus infection of a cell. If the virus has an RNA genome but is not a retrovirus, perfectly matched double-stranded RNA molecules arise in the infected cell as intermediates in viral genome replication. These dsRNAs are cleaved in the cytosol by Dicer, giving a series of perfectly matched double-stranded RNA molecules, each about 23 base pairs long, called siRNAs (small interfering RNAs). These are targeted to viral RNA molecules, and because the nucleotide match

is perfect, the RNA molecules are cleaved as indicated in our discussion of miRNA synthesis. This process is particularly effective in plants, whose cells are connected by fine channels. Hence, the RNA interference activity can be spread from cell to cell, leading an entire plant to become virus-resistant, even though only a few cells might have been infected initially.

Although we have much to learn about particular aspects of miRNA regulation, we do know that the synthesis of specific miRNAs is tissue- and developmental-stage-specific. Initial transcription of the pri-miRNA molecule is subject to the same regulatory processes as those we have discussed for mRNA biosynthesis; each following step is controlled, although many details remain to be elucidated.

With respect to miRNA function, we know that these molecules are involved in regulating processes as diverse as cellular proliferation, control of development, apoptosis, homeostasis, and tumorigenesis. There is tremendous interest in harnessing miRNA biology to treat or prevent diseases such as cancer. Although this field is in its infancy, RNAi has found widespread use in the laboratory for specific gene knockdown. The most complete and specific and permanent way to study gene function is by ablation of that gene. However, use of RNAi, like that of morpholinos, is a much faster, albeit less complete and less specific, way to achieve the same goal. A short hairpin RNA (shRNA) can be prepared by chemical synthesis and introduced into target cells. These hairpin molecules are metabolized identically to siRNAs and lead to degradation of the target mRNA.

26.6 Riboswitches

The versatility of RNA molecules was highlighted by the discovery of RNA interference, coming on top of the discovery of catalytic RNAs and regulation by noncoding RNAs. An additional regulatory function came with the fairly recent discovery of **riboswitches**—mRNA molecules whose translation is controlled by specific binding of a metabolic end product of the pathway in which that mRNA is involved. These molecules were

● **CONCEPT** A riboswitch is an mRNA molecule that has a specific binding site for a metabolic end product, which, when bound, blocks either transcription or translation of downstream genes.

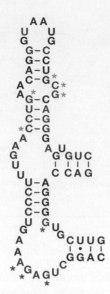

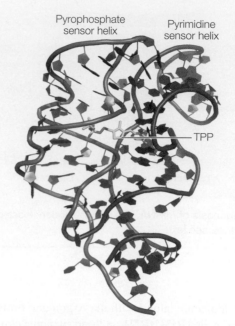

Pyrophosphate sensor helix

Pyrimidine sensor helix

TPP

(a) Secondary structure diagram for the TPP-binding domain of the riboswitch. Residues involved in pyrophosphate binding are shown with green asterisks and those in thiamine binding with red asterisks. Conserved nucleotides are shown in red.

(b) Crystal structure of the TPP riboswitch, with bound TPP shown in yellow and the "sensor helices," which bind portions of the TPP molecule as shown.

▲ **FIGURE 26.25** Structure of the thiamine pyrophosphate riboswitch from *Arabidopsis thaliana*. PDB ID: 2cky.

originally discovered in bacteria and have recently been found in some plants and fungi as well.

A riboswitch usually has an **aptamer** at or near its 5′ end. In the laboratory an aptamer is a specific binding site created from an oligonucleotide or polynucleotide sequence. With the discovery of riboswitches, in about 2002, it was found that aptamers of great binding specificity had also been created by evolution. As noted above, a riboswitch RNA has a binding site for a particular metabolite at the 5′ end of an mRNA that encodes an enzyme in the metabolic pathway leading to that metabolite. Many riboswitches contain sites for binding nucleotides or coenzymes—thiamine pyrophosphate, flavin adenine dinucleotide, *S*-adenosylmethionine, and so on. The first identified riboswitch has a binding site for adenosylcobalamin (B_{12} cofactor). Specificity of the binding sites is high, as shown by the finding that a riboswitch for *S*-adenosylmethionine was shown to bind that molecule at least 100-fold more tightly than the closely related *S*-adenosylhomocysteine. **FIGURE 26.25** shows the crystal structure of a riboswitch binding site for thiamine pyrophosphate. Although this image does not show the specific binding interactions, it is evident that a complex folding pattern has evolved to completely enfold the target molecule. The dissociation constant for the complex shown is about 50 nM. It is essential that riboswitches bind their target molecules tightly, because they are controlling the biosynthesis of nucleotides and coenzymes that exist within cells at very low concentrations. In each case, binding of the target molecule has the effect of shutting off the expression of genes involved in synthesis of the target molecule. This can occur at the level of either transcription or translation.

26.7 RNA Editing

RNA editing was another unanticipated form of gene regulation when it was first described in 1986. In this process, the nucleotide sequence of an RNA molecule is actually changed posttranscriptionally. It was found that in some protozoans the mRNAs for some mitochondrial proteins were modified by insertion and occasional removal of UMP residues. In some cases, as many as half of the nucleotides in the mature mRNA molecule were U's that had been inserted. Obviously, the complete sense of the genetic message is changed. However, the process is not random. It is directed by *guide RNAs*, which have 5′ ends complementary to one end of the molecule to be edited, followed by a sequence of nucleotides identical to that which is to be inserted. Each insertion step of a single nucleotide involves cleavage, insertion, and re-ligation of the RNA chain. The biological significance of this seemingly wasteful process is still not known.

RNA editing in mammalian cells is much different, simply involving enzymatic deamination of selected AMP or CMP RNA nucleotides. As many as 1000 mammalian genes may be subject to RNA editing. The deaminase enzymes involved recognize a double-stranded RNA that is formed between the site that is to be edited and a complementary sequence elsewhere in the molecule, typically in a downstream intron. A good example is editing of the mRNA for apolipoprotein B in synthesis of human lipoproteins (Chapter 16). As shown in **FIGURE 26.26**, apoproteins B-48 and B-100 are expressed from the same gene. In liver, translation of the mRNA yields a 100-kDa protein, Apo B-100, which is a VLDL component. In intestine, however, a specific CAA

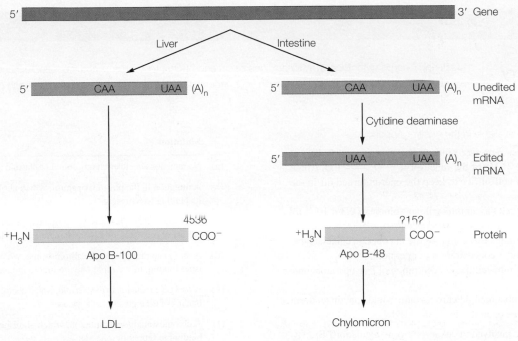

▲ **FIGURE 26.26** RNA editing of the lipoprotein B gene transcript.

in the mRNA is deaminated to give UAA, a translational stop codon. This yields a 48-kDa protein, Apo B-48, with quite different properties, which becomes a component of chylomicrons.

Editing is used by lymphocytes as a defense against HIV infection. These cells carry out large-scale deamination of CMP residues to UMP, using an enzyme similar to AID, the activation-induced DNA-deoxycytidine deaminase mentioned previously. Because the RNA in this case is the genome of the virus, the extensive deamination creates

● **CONNECTION** RNA editing represents a battleground in the relationship between HIV and its host cell.

multiple mutations, which cripple or kill the virus. However, the virus has evolved a defense mechanism of its own—a protein inhibitor of the deaminase, which is carried into the infected cell along with the viral genome.

Summary

- In bacteria, gene expression is regulated primarily at the level of transcription. Control is exercised by the binding of repressors and transcriptional activators at promoter sites, thereby influencing the binding of RNA polymerase. Many or most genes are expressed as operons, in which as many as a dozen proteins or more can be translated from the same multicistronic mRNA. (Section 26.1)

- Gene regulation in eukaryotes also occurs at the level of transcription initiation, but regulation is much more complex. As many as 50 proteins assemble at a promoter and upstream enhancer in order for RNA polymerase II to bind and transcribe a gene. These proteins include sequence-specific activators, bound at enhancers; Mediator, which connects enhancers to the transcription machinery; chromatin remodeling complexes, which clear the transcription start region for binding RNA polymerase; and general transcription factors. Time-dependent changes in the phosphorylation pattern of the RNA polymerase II C-terminal domain control the timing of events in transcription. (Section 26.2)

- Methylation of DNA cytosine residues leads to gene silencing. DNA methylation patterns are heritable, thanks to DNA cytosine methylases that track with the replisome. (Section 26.3)

- Although most gene regulation occurs at the level of transcription, important translational regulatory processes are known as well. (Section 26.4)

- Important RNA-based regulatory processes have recently been discovered, notably, long noncoding RNAs and **RNA interference** (RNAi). In RNAi processing of small RNAs to short duplex RNAs and then to single-stranded **microRNAs** is now recognized as an important translational regulatory mechanism. (Section 26.5)

- **Riboswitches** are RNAs that control biosynthetic pathways by specific and tight binding of metabolic end products of those pathways. (Section 26.6)

- RNA editing causes changes in RNA base sequence, which affect the way an mRNA is translated. (Section 26.7)

26. 1, 2, 15, 19, 24

Problems

Enhanced by
Mastering Chemistry
for Biochemistry

Mastering Chemistry for Biochemistry provides select end-of-chapter problems and feedback-enriched tutorial problems, animations, and interactive figures to deepen your understanding of complex topics while practicing problem solving.

Answers to red problems are available in the Answer Appendix.

1. The active form of lactose repressor binds to the operator with a dissociation constant of 10^{-13} M for the reaction $R + O \rightleftharpoons RO$. About 10 molecules per *E. coli* cell suffice to keep the operon turned off in the absence of inducer.
   (a) If the average *E. coli* cell has an intracellular volume of 0.3×10^{-12} mL, calculate the approximate intracellular concentration of repressor.
   (b) If the average cell contains two copies of the *lac* operon, calculate the approximate intracellular concentration of operators.
   (c) Calculate the average intracellular concentration of *free* operators under these conditions.
   (d) Explain how a cell with a haploid chromosome could contain an average of two copies of the *lac* operon.

2. Is attenuation likely to be involved in eukaryotic gene regulation? Briefly explain your answer.

3. What are the major differences between an operon and a regulon?

4. Suppose you want to study the transcription in vitro of one particular gene in a DNA molecule that contains several genes and promoters. Without adding specific regulatory proteins, how might you stimulate transcription from the gene of interest relative to the transcription of the other genes on your DNA template? To make all of the complexes identical, you would like to arrest all transcriptional events at the same position on the DNA template before isolating the complex. How might you do this?

5. For some time, it was not clear whether *lac* repressor inhibits *lac* operon transcription by inhibiting the binding of RNA polymerase to its promoter or by allowing transcription initiation but blocking elongation past the site of bound repressor. How might you distinguish between these possibilities?

6. A *lac* operon containing one mutation was cloned into a plasmid, which was introduced by transformation into a bacterium containing a wild-type *lac* operon. The three genes of the chromosomal operon were rendered non-inducible in the presence of the plasmid.
   (a) What kind of mutation in the plasmid operon could have this effect?
   (b) Suppose the result of transformation was to cause the three plasmid *lac* genes to be expressed constitutively, at a high level. What type of plasmid gene mutation could have this result?

7. Why would phage λ need to synthesize more *cI* repressor during establishment of lysogeny, early in infection, rather than in maintenance of the lysogenic state?

8. Repressors are inactivated either by interaction with a small-molecule inducer or by proteolytic cleavage. Why is it advantageous for a repressor like the *lac* repressor to be inactivated by binding to allolactose rather than by proteolytic cleavage?

9. Partial diploid forms of *E. coli* were created, each of which contained a complete lactose operon at its normal chromosomal site and the regulatory sequences only *(lacI, P, lacO)* on a plasmid. Predict the effect of each mutation on the activity of β-galactosidase before and after the addition of inducer. Use −, +, or + + to indicate approximate activity levels. Briefly explain the basis for each of your predictions.

Mutation	Before	After
(a) No mutations in either chromosomal or plasmid genes		
(b) A mutation in the plasmid operator, which abolishes its binding to repressor		
(c) A mutation in the chromosomal promoter, which reduces affinity of promoter for RNA polymerase by 10-fold		
(d) A *lacI* gene mutation in the chromosome, which abolishes binding of the *lacI* gene product to inducer		
(e) A *lacI* gene mutation in the plasmid, which abolishes binding of the repressor to the inducer		
(f) A chromosomal *lacO* mutation, which abolishes its binding to repressor		
(g) A mutation in the gene for CRP, which abolishes its binding to cyclic AMP		

10. What type of mutation of the *lac* repressor might be both constitutive and *trans*-dominant?

11. It has been proposed that thiogalactoside transacetylase (LacA in the lactose operon) plays a role in detoxification—ridding the cell of potentially toxic β-galactosides by acetylating them to inhibit their reuptake after their diffusion out of the cell. How might you test this proposal?

12. Riboswitches are generally considered to have been discovered in about 2002. But a comparable regulatory process was described much earlier, when Nomura et al. (*Proc. Natl. Acad. Sci. USA* 77:7084 (1980)) described the regulation of ribosomal protein synthesis carried out by binding of ribosomal proteins to their own mRNAs (Figure 26.25). Was the riboswitch actually discovered years earlier? Discuss similarities and differences in control of ribosomal protein synthesis and riboswitch regulation as discussed in this chapter text.

13. Not long ago investigators were surprised to learn that more than 95% of a mammalian genome is transcribed, even though less than 2% encodes proteins. What kind of evidence could be used to determine the percentage of the genome that is transcribed?

14. What do studies on attenuation tell us about mechanisms of transcription termination in bacteria?

15. In eukaryotic transcription, what is the function of a histone chaperone? Of a chromatin remodeling complex? Of Mediator?

16. Why does histone deacetylase action tend to repress transcription?

17. Refer to Figure 26.16, which shows the distribution of histone H2A.Z on nucleosomes near a transcription start site. What experimental technique would have been used to generate the data for this figure? Briefly describe the operation of this technique.

18. 5-Azacytidine is a reagent that suppresses DNA methylation through its conversion to 5-azadeoxycytidine triphosphate. Outline a metabolic pathway leading from Aza-C to Aza-dCTP.

19. Briefly explain why CpG islands might have come to be underrepresented in eukaryotic genomes.

20. Explain how a gene knockdown reagent, such as RNAi or a morpholino, could interfere with expression of a gene without affecting the rate of transcription of the target gene.

21. Briefly explain how heme regulates the expression of globin genes in cells that synthesize hemoglobin.

22. What condition must be fulfilled in order for histone modification to be considered an epigenetic phenomenon? Does DNA methylation meet that criterion? Explain.

23. When binding of sequence-specific regulatory proteins such as *lac* repressor or CRP to its respective DNA binding site was studied in vitro, the rates of association were much higher than predicted by collision theory. Propose an explanation for this observation and a simple experimental test of your hypothesis.

24. Cancer is thought to result in part from an increase in the spontaneous mutation rate in precancerous cells. At the same time, the DNA methylation pattern changes during oncogenic transformation. Identify a process by which DNA methylation might lead to increased mutagenesis.

References

For a list of references related to this chapter, see Appendix II.

comp DNA
& RNA polymerase

FOUNDATION FIGURE | Information Flow in Biological Systems

In the Central Dogma of molecular biology, genetic information stored in DNA is passed through an RNA intermediate before being expressed as protein. In replication, DNA is used as a template for new DNA synthesis prior to cell division.

1 Replication

2 Transcription

3 Translation

DNA

RNA

Protein

In transcription, DNA is used as a template for the synthesis of an RNA copy of the gene. In translation, the genetic code found in the sequence of bases along the mRNA is interpreted and used to drive the synthesis of protein.

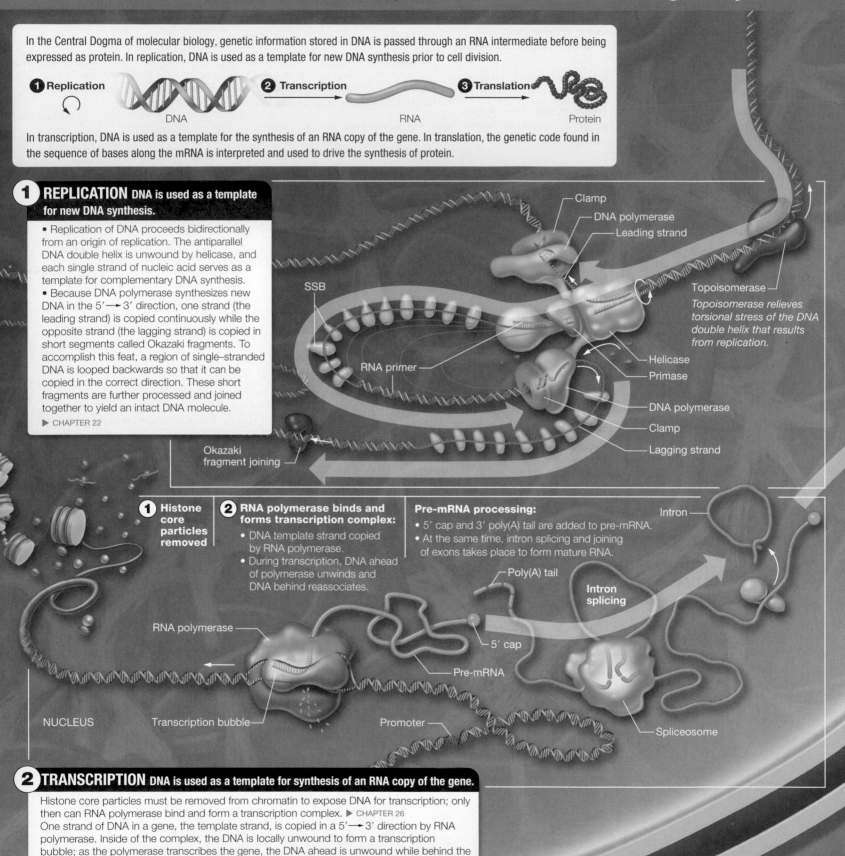

1 REPLICATION DNA is used as a template for new DNA synthesis.

- Replication of DNA proceeds bidirectionally from an origin of replication. The antiparallel DNA double helix is unwound by helicase, and each single strand of nucleic acid serves as a template for complementary DNA synthesis.
- Because DNA polymerase synthesizes new DNA in the 5′⟶3′ direction, one strand (the leading strand) is copied continuously while the opposite strand (the lagging strand) is copied in short segments called Okazaki fragments. To accomplish this feat, a region of single–stranded DNA is looped backwards so that it can be copied in the correct direction. These short fragments are further processed and joined together to yield an intact DNA molecule.

▶ CHAPTER 22

Clamp
DNA polymerase
Leading strand
Topoisomerase

Topoisomerase relieves torsional stress of the DNA double helix that results from replication.

SSB

RNA primer

Helicase
Primase
DNA polymerase
Clamp
Lagging strand

Okazaki fragment joining

1 Histone core particles removed

2 RNA polymerase binds and forms transcription complex:
- DNA template strand copied by RNA polymerase.
- During transcription, DNA ahead of polymerase unwinds and DNA behind reassociates.

Pre-mRNA processing:
- 5′ cap and 3′ poly(A) tail are added to pre-mRNA.
- At the same time, intron splicing and joining of exons takes place to form mature RNA.

Intron

Poly(A) tail

Intron splicing

RNA polymerase

5′ cap

Pre-mRNA

NUCLEUS

Transcription bubble

Promoter

Spliceosome

2 TRANSCRIPTION DNA is used as a template for synthesis of an RNA copy of the gene.

Histone core particles must be removed from chromatin to expose DNA for transcription; only then can RNA polymerase bind and form a transcription complex. ▶ CHAPTER 26

One strand of DNA in a gene, the template strand, is copied in a 5′⟶3′ direction by RNA polymerase. Inside of the complex, the DNA is locally unwound to form a transcription bubble; as the polymerase transcribes the gene, the DNA ahead is unwound while behind the polymerase the DNA strands reassociate. Regulation of gene expression occurs at the level of every process shown here. Perhaps most important is the involvement of proteins, including repressors and activators, that bind near transcription start sites (promoters) and either stimulate or inhibit transcriptional initiation.

▶ CHAPTER 24

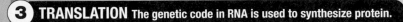

❸ TRANSLATION The genetic code in RNA is used to synthesize protein.

Beginning at the 5' end of the mRNA, ribosomes start translating the genetic message at the initiator codon. Charged transfer RNA (tRNA) molecules bind to the ribosome, and protein synthesis continues in an N⟶C direction. The newly synthesized protein emerges from the ribosome and begins the process of folding into its native structure.

▶ CHAPTER 25

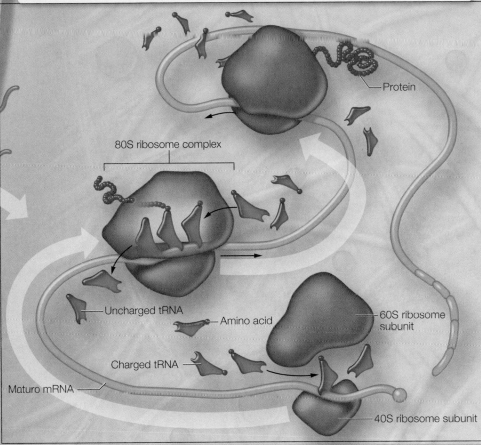

Nuclear pore

Protein

80S ribosome complex

Uncharged tRNA

Amino acid

60S ribosome subunit

Charged tRNA

Maturo mRNA

40S ribosome subunit

CYTOPLASM

EUKARYOTES VS. PROKARYOTES

• Replication, transcription, and translation **all occur in the same cell compartment in prokaryotes** as there is no nuclear membrane present.

• Antibacterial agents specifically target bacteria by taking advantage of the differences between prokaryotes and eukaryotes.

▶ CHAPTER 25

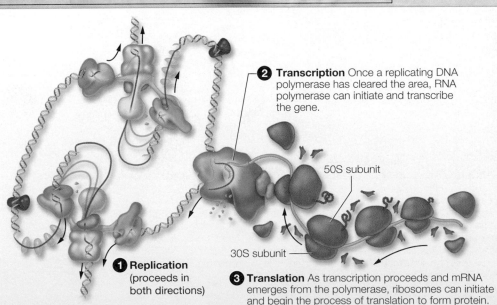

❷ Transcription Once a replicating DNA polymerase has cleared the area, RNA polymerase can initiate and transcribe the gene.

50S subunit

30S subunit

❶ Replication (proceeds in both directions)

❸ Translation As transcription proceeds and mRNA emerges from the polymerase, ribosomes can initiate and begin the process of translation to form protein.

CHAPTER 2

1. **(a)** Equation 2.3 predicts that the interaction energy between the ions will be greater in the lower dielectric medium; thus, the attraction between the Na^+ and Cl^- will be greater in pentane.
   (b) Since the length of the ionic bonds is 2×1.16 Å in each case (i.e., the value of r is the same for Ca-F and Na-F), Equation 2.3 predicts that the interaction energy between the ion will be greater as the values of q increase; thus, the attraction between the Ca^{2+} ($q = 2$) and F^- ($q = -1$) will be greater than the attraction between the Na^+ ($q = 1$) and F^- ($q = -1$).
   (c) Ca^{2+} will be bound more tightly by a $-COOH$ group that is fully deprotonated. At pH $= 3$ the $-COOH$ form will predominate. At pH $= 4.2$ the $-COOH$ and $-COO^-$ forms will be in equal concentration. At pH 8 the $-COO^-$ will predominate; thus, expect greatest Ca^{2+} binding at pH $= 8$.

4. **(a)** HCl is a strong acid; thus, $[HCl] \approx [H^+]$ and pH $= -\log[H^+] = -\log[0.35] = 0.456$
   (b) Acetic acid is a weak acid with $K_a = 1.74 \times 10^{-5}$; thus, use ICE table to solve this problem:

$$H_2O + CH_3COOH \rightleftharpoons H_3O^+ + CH_3COO^-$$

	CH_3COOH	H^+	CH_3COO^-
Initial	0.35 M	0 M	0 M
Change	$-x$	$+x$	$+x$
Equilibrium	$0.35 - x$ M	x M	x M

$$K_a = \frac{[CH_3COO^-][H_3O^+]}{[CH_3COOH]} = \frac{x^2}{[CH_3COOH] - x}$$

Assume $[CH_3COO^-] = [H_3O^+]$ and $[CH_3COOH] \gg x$; therefore:

$$1.74 \times 10^{-5} = \frac{x^2}{0.35 - x}$$

$x = 2.47 \times 10^{-3} = [H_3O^+]$

pH $= -\log(2.47 \times 10^{-3}) = 2.61$

This answer verifies the initial assumption that $[CH_3COOH] \gg x$.
   (c) Here $[CH_3COOH] \gg x$ cannot be assumed, so use ICE table approach with the quadratic equation to solve this problem:

$$K_a = \frac{x^2}{0.035 - x}$$

Rearrange to
$$0 = x^2 + K_a x - 0.035 K_a = x^2 + (1.74 \times 10^{-5})x - (6.09 \times 10^{-7})$$

Solve using the quadratic equation:

$$x = \frac{-(1.74 \times 10^{-5}) \pm \sqrt{(1.74 \times 10^{-5})^2 - (4 \times -6.09 \times 10^{-7})}}{2}$$

$x = 7.717 \times 10^{-4} = [H^+]$; thus, pH $= -\log[7.717 \times 10^{-4}] = 3.11$

5. **(a)** See Table 2.6, which indicates that NH_4^+ is a weak acid with $K_a = 5.62 \times 10^{-10}$; thus, use ICE table to solve this problem (note that the "initial" conditions here are hypothetical; we imagine a starting concentration of NH_4^+ of 1 M and assume that the final concentration of H^+ from dissociation of NH_4^+ will be significantly greater than 10^{-7} M, which results from the autolysis of water):

$$HA \rightleftharpoons H^+ + A^-$$

	NH_4^+	H^+	NH_3
Initial	1 M	~0 M	0 M
Change	$-x$	$+x$	$+x$
Equilibrium	$1 - x$ M	x M	x M

$$K_a = \frac{x^2}{1 - x} = 5.62 \times 10^{-10}$$

Solve for $x = 2.371 \times 10^{-5} = [H^+]$; thus pH $= -\log(2.371 \times 10^{-5}) = 4.63$
   (b) NH_4^+ is a weak acid with $pK_a = 9.25$. Here, NaOH is consuming H^+ from the NH_4^+ thus, use this alternate version of the ICE table to solve for $[HA]$ and $[A^-]$ after addition of NaOH (a source of ^-OH). Note: the activity of H_2O is assumed to be unity (see Equation 2.7), so it does not appear in these calculations.

$$HA + {^-}OH \rightleftharpoons A^- + H_2O$$

	NH_4^+	^-OH	NH_3
Initial	0.040 mol	0.010 mol	~0
Change	-0.010 mol	-0.010 mol	$+0.010$ mol
Equilibrium	0.030 mol	0	0.010 mol

Solve for $[H^+]$ using the Henderson-Hasselbalch equation:

$$pH = pK_a + \log\frac{[A^-]}{[HA]} = 9.25 + \log\left(\frac{\left(\frac{0.010 \text{ mol}}{0.050 \text{ L}}\right)}{\left(\frac{0.030 \text{ mol}}{0.050 \text{ L}}\right)}\right) = 9.25 + -0.477 = 8.77$$

   (c) Solve as in part (b)

$$HA + {^-}OH \rightleftharpoons A^- + H_2O$$

	NH_4^+	^-OH	NH_3
Initial	0.040 mol	0.030 mol	0
Change	-0.030 mol	-0.030 mol	$+0.030$ mol
Equilibrium	0.010 mol	0	0.030 mol

Solve for $[H^+]$ using the Henderson-Hasselbalch equation:

$$pH = pK_a + \log\frac{[A^-]}{[HA]} = 9.25 + \log\left(\frac{\left(\frac{0.030 \text{ mol}}{0.070 \text{ L}}\right)}{\left(\frac{0.010 \text{ mol}}{0.070 \text{ L}}\right)}\right)$$

$$= 9.25 + 0.477 = 9.73$$

9. **(a)** $H_2PO_4^- + OH^- \rightleftharpoons HPO_4^{-2} + H_2O$
   $pK_a = 6.86$
   pH $= 7.00$
   $[H_2PO_4^-] = 0.1$ M

$$pH = pK_a + \log\left[\frac{A^-}{HA}\right] = 7.0$$

$$7.0 = 6.86 + \log\left[\frac{A^-}{0.1}\right]$$

$$0.14 = \log\left[\frac{A^-}{0.1}\right]$$

$[A^-] = 0.138$ M

A-1

9 (b) $[H_2PO_4^-] + [HPO_4^{-2}] = 0.3\ M$

$[H_2PO_4^-] = 0.3\ M - [HPO_4^{-2}]$

$$7.00 = pK_a + \log\left[\frac{HPO_4^{-2}}{H_2PO_4^-}\right] = 6.86 + \log\left[\frac{0.3 - H_2PO_4^-}{H_2PO_4^-}\right]$$

$$\log\left[\frac{0.3 - H_2PO_4^-}{H_2PO_4^-}\right] = 0.14$$

$$10^{0.14} = \left[\frac{0.3 - H_2PO_4^-}{H_2PO_4^-}\right] = 1.38$$

$$1.38 = \frac{0.3 - x}{x}$$

solve : x

$x = 0.126\ M = [KH_2PO_4]$

$[Na_2HPO_4] = 0.174\ M$

10. Formic acid is a weak acid with $pK_a = 3.75$. Since the formate buffer is at pH = pK_a, [HA] = [A$^-$] initially. Here, KOH is consuming H$^+$ from the HCOOH thus, use this alternate version of the ICE table to solve for [HA] and [A$^-$] after addition of KOH (a source of $^-$OH). Note: the activity of H_2O is assumed to be unity (see Equation 2.7), so it does not appear in these calculations.

$$HA + {}^-OH \rightleftharpoons A^- + H_2O$$

	HCOOH	$^-$OH	HCOO$^-$
Initial	0.025 mol	0.005 mol	0.025 mol
Change	−0.005 mol	−0.005 mol	+0.005 mol
Equilibrium	0.020 mol	0	0.030 mol

Solve for [H$^+$] using the Henderson-Hasselbalch equation:

$$pH = pK_a + \log\frac{[A^-]}{[HA]} = 3.75 + \log\left(\frac{\left(\frac{0.030\ mol}{0.505\ L}\right)}{\left(\frac{0.020\ mol}{0.505\ L}\right)}\right)$$

$$= 3.75 + 0.176 = 3.93$$

14. Protein molecules in aqueous solution become increasingly protonated as the pH decreases. Thus, proteins become more positively charged because carboxylic acids become *less negatively charged* as pH drops, whereas amines become *more positively charged*. Proteins become increasingly deprotonated as pH increases. Thus, proteins will become more negatively charged as pH increases, because acidic groups become more negatively charged while the basic groups become less positively charged.

15. (a) Species III is the isoelectric species (not net charge); thus, the pK_as that describe ionization equilibria that include this species will be used to calculate the pI. Since Species I is not one of these (and Species I is present at an insignificant concentration), we can ignore it for this simple case (i.e., a molecule with only three ionizable groups).

$$pI = \left(\frac{pK_{a1} + pK_{a2}}{2}\right) = \left(\frac{8.99 + 12.5}{2}\right) = 10.75$$

(b) At pH = 9.20 the deprotonation of the α–carboxylic acid will be essentially 100% ($pK_a = 1.82$ is >7 pH units below the pH 9.20; thus, the deprotonated form will predominate by 7 orders of magnitude). Thus, it is safe to assume a charge of −1 on the α–carboxylate at pH 9.20.

At pH = 9.20 the deprotonation of the α–amino group will be closer to 50% ($pK_a = 8.99$ is close to the pH 9.20; thus, the [HA] and [A$^-$] will be within a factor of ten). The fractional charge on the

α–amino group can be calculated using the Henderson-Hasselbalch equation:

$$9.20 = 8.99 + \log\left(\frac{[A^-]}{[HA]}\right) \longrightarrow \frac{[A^-]}{[HA]} = 1.62$$

Thus, at pH = 9.2, for every 1 mol of HA, there are 1.62 mol of A$^-$. With this information the mole fractions of A$^-$ (i.e., $-NH_2$) and HA (i.e., $-NH_3^+$) can be calculated:

$$\text{mole fraction A}^- = \frac{\text{mol A}^-}{\text{mol A}^- + \text{mol HA}} = \frac{1.62}{1.62 + 1} = 0.618$$

$$\text{mole fraction HA} = \frac{\text{mol HA}}{\text{mol A}^- + \text{mol HA}} = \frac{1}{1.62 + 1} = 0.382$$

Since the A$^-$ form is uncharged (i.e., $-NH_2$), it will not contribute to the overall molecular charge. However, at pH 9.20, 38.2% of the α–amino groups are in the positively charged ($-NH_3^+$) form. Thus, the average charge on this group is +0.382 at pH 9.20.

At pH = 9.20 the deprotonation of the side chain guanidinium group will be negligible ($pK_a = 12.5$ is more than 3 pH units above the pH 9.20; thus, the [HA] form will predominate by three orders of magnitude). Thus, it is safe to assume a charge of +1 on the side chain guanidinium group at pH 9.20.

The average molecular charge at pH 9.20 is the sum of the average charges on each ionizable group: $(-1) + (+0.382) + (+1) = +0.382$.

(c) Yes. When pH < pI the molecule is predicted to carry a positive charge.

19. (a) The enzyme will be more active at pH = 3.5 because the −COOH group will be more protonated at that pH (protonation is favored when pH < pK_a).

(b) When pH = pK_a the ionizable group is 50% protonated; thus, 50% of the enzymes will be in an active state.

(c) The enzyme will be 78% active when 78% of the −COOH groups are protonated, or when the mole fraction of −COOH vs, −COO$^-$ is 0.78. The ratio of COOH: COO$^-$ can be calculated by setting mol COOH to an arbitrary value and solving for mol COO$^-$. For this purpose setting mol COOH = 1 is convenient:

$$\text{mole fraction COOH} = \frac{\text{mol COOH}}{\text{mol COO}^- + \text{mol COOH}}$$

$$= \frac{1}{x + 1} = 0.78 \longrightarrow x = 0.282$$

Now use the Henderson-Hasselbalch equation to solve for pH:

$$pH = 4.07 + \log\left(\frac{[0.282]}{[1]}\right) = 4.07 + -0.55 = 3.52$$

20. (a) At pH = 7.20 [H_3PO_4] and [PO_4^{3-}] will be negligible. The pK_a of H_3PO_4 = 2.14, which is ~ five pH units below 7.20. Thus, the Henderson-Hasselbalch equation predicts that [$H_2PO_4^-$] > [H_3PO_4] by five orders of magnitude (~100,000:1) at pH 7.20. Likewise, the pK_a of HPO_4^{2-} = 12.4, which is ~ five pH units above 7.20. Thus, the Henderson-Hasselbalch equation predicts that [HPO_4^{2-}] > [PO_4^{3-}] by five orders of magnitude (~100,000:1) at pH 7.20.

(b) At pH = 7.20 the predominant phosphate species are $H_2PO_4^-$ (the conjugate acid in this equilibrium) and HPO_4^{2-} (the conjugate base). Thus, as HCl is generated it will be neutralized by the conjugate base HPO_4^{2-} to produce more conjugate acid $H_2PO_4^-$:

$$A^- + HCl \rightleftharpoons HA + Cl^-$$

	HPO_4^{2-}	H^+	$H_2PO_4^-$
Initial	$17.2\ 10^{-6}$ mol	$3.80\ 10^{-6}$ mol	$7.85\ 10^{-6}$ mol
Change	$-3.80\ 10^{-6}$ mol	$-3.80\ 10^{-6}$ mol	$+3.80\ 10^{-6}$ mol
Equilibrium	$13.4\ 10^{-6}$ mol	0	$11.65\ 10^{-6}$ mol

To complete the ICE table, you need to solve for initial concentrations of phosphate species using the Henderson-Hasselbalch equation:

$$pH = pK_a + \log\frac{[A^-]}{[HA]} \longrightarrow 7.20 = 6.86 + \log\left(\frac{[HPO_4^{2-}]}{[H_2PO_4^-]}\right) \longrightarrow$$

$$10^{0.34} = 2.188 = \frac{[HPO_4^{2-}]}{[H_2PO_4^-]}$$

$$\text{mole fraction } HPO_4^{2-} = \frac{\text{mol } HPO_4^{2-}}{\text{mol } HPO_4^{2-} + \text{mol } H_2PO_4^-}$$

$$= \frac{2.188}{2.188 + 1} = 0.686$$

$$\text{mole fraction } H_2PO_4^- = 0.314$$

Thus, the initial moles of HPO_4^{2-} =
$(0.00100 \text{ L}) \times (0.0250 \text{ M}) \times (0.686) = 17.2 \; 10^{-6} \text{ mol}$
Likewise, the initial moles of $H_2PO_4^-$ =
$(0.00100 \text{ L}) \times (0.0250 \text{ M}) \times (0.314) = 7.85 \; 10^{-6} \text{ mol}$
Once the ICE table is completed, you can plug the final values ("equilibrium" line of the ICE table) for the phosphate species back into the Henderson-Hasselbalch equation to solve for pH:

$$\text{final pH} = 6.86 + \log\left(\frac{\dfrac{13.4 \times 10^{-6} \text{ mol}}{0.00100 \text{ L}}}{\dfrac{11.65 \times 10^{-6} \text{ mol}}{0.00100 \text{ L}}}\right) = 6.92$$

21. Assuming RNaseA functions in cells with a pH ~ 7.4, and it must bind to a highly negatively-charged RNA molecule to perform its function, it is more likely to have a pI of 9.2 than 5.0. At pH = 7.4 a protein with a pI = 5.0 will carry excess negative charge (because pH > pI); thus, we would expect this protein to repel an RNA molecule. At pH = 7.4 a protein with a pI = 9.2 will carry excess positive charge (because pH < pI); thus, we would expect this protein to bind favorably an RNA molecule.

CHAPTER 3

1. (a) $\Delta G° = \Delta H° - T\Delta S°$

$$= \left(-280\frac{kJ}{mol}\right) - (298 \; K)\left(-0.790\frac{kJ}{mol \times K}\right) = -44.6\frac{kJ}{mol}$$

(b) For *unfolding* to be favorable, *folding* must be unfavorable. Thus, unfolding will occur when $\Delta G°$ for folding is > 0. The values of $\Delta H°$ and $\Delta S°$ for the folding process are both negative. Thus, *folding* is predicted to become unfavorable at T increases (see Table 3.3), or, as T increases *unfolding* (the reverse of the process as written) will become favorable. The temperature at which $\Delta G° = 0$ is called the "melting temperature", or T_m. At T above the T_m unfolding will be favorable; thus, finding the value of T_m will provide the temperature above which $\Delta G° > 0$ and unfolding will be favorable:

$$\Delta G° = \Delta H° - T\Delta S°$$

$$0 = \left(-280\frac{kJ}{mol}\right) - (T)\left(-0.790\frac{kJ}{mol \times K}\right)$$

$$-354.4 \; K = -T$$

$T = 354.4 \; K$ or $81.3 \; °C$.

(c) The ratio of unfolded to folded protein is given in a mass action expression, which is equivalent to an equilibrium constant (K) at some temperature T. Thus, we can use a second expression for $\Delta G°$ (Equation 3.21) to solve for the T at which the ratio unfolded:folded is 1:5 (note: in the first equation K refers to the equilibrium constant; later, K appears as the temperature unit "Kelvin").

$$\Delta G° = -RT \ln K = -RT \ln\left(\frac{[\text{folded}]}{[\text{unfolded}]}\right) = \Delta H° - T\Delta S°$$

Dividing by $(-RT)$ gives

$$\ln\left(\frac{[\text{folded}]}{[\text{unfolded}]}\right) = \left(-\frac{\Delta H°}{RT}\right) + \left(\frac{\Delta S°}{R}\right)$$

$$\ln\left(\frac{5}{1}\right) = \left(-\frac{-280\dfrac{kJ}{mol}}{\left(0.008314 \; J\dfrac{kJ}{mol \times K}\right)T}\right) + \left(\frac{-0.790\dfrac{kJ}{mol \times K}}{\left(0.008314\dfrac{kJ}{mol \times K}\right)}\right)$$

$$1.609 = \left(\frac{3.368 \times 10^4}{\dfrac{T}{(K)}}\right) + (-95.02)$$

$$T = \left(\frac{3.368 \times 10^4}{(1.609) + (95.02)}\right)K = 348.5K = 75.4 \; °C$$

4. (a) $\Delta G° = \Delta H° - T\Delta S°$
$\quad = (-2816 \text{ kJ mol}^{-1}) - (310 \text{ K})(0.181 \text{ kJ K}^{-1} \text{ mol}^{-1})$
$\quad = -2872.1 \text{ kJ mol}^{-1}$

(b) From Table 3.5:
$ADP + P_i + H^+ \longrightarrow ATP + H_2O \qquad \Delta G = 32.2 \text{ kJ mol}^{-1}$
$(32.2 \text{ kJ mol}^{-1}) \times (32 \text{ mol ATP}) = 1030.4 \text{ kJ}$
$C_6H_{12}O_6 + 6O_2 + 32ADP + 32P_i + 32H^+ \longrightarrow$
$\qquad\qquad\qquad\qquad\qquad\qquad\qquad 6CO_2 + 38H_2O + 32ATP$

$\Delta G = (-2872.1 \text{ kJ}) + (1030.4 \text{ kJ})$
$\quad = -1841.7 \text{ kJ}$ for the stoichiometry as written (per one mol glucose or per 32 mol ATP)

(c) % efficiency $= |\Delta G°_{\text{ATP synthesis}}/\Delta G°_{\text{total available}}| \times 100\%$
$\quad = (1030.4/2872.1) \times 100$
$\quad = 35.9\%$

5. (a) glucose $+ P_i \longrightarrow$ G6P $+ H_2O \qquad \Delta G°' = 13.8 \text{ kJ mol}^{-1}$
$K_{eq} = ([G6P] \times [H_2O]/[\text{glucose}] \times [P_i]) = e^{-\Delta G°/RT}$
Note: in the biochemical standard state, the activity of H_2O is assigned a value of 1.
$([G6P] \times (1))/((0.005) \times (0.005)) = e^{(-13.8)/(0.008314) \times (310)}$
$[G6P] = 0.000025 \times e^{-5.36}$
$[G6P] = 1.2 \times 10^{-7} \text{ M}$

(b) $ATP + H_2O \longrightarrow ADP + P_i + H^+ \qquad \Delta G° = -32.2 \text{ kJ mol}^{-1}$
$\underline{\text{Glucose} + P_i \longrightarrow \text{G6P} + H_2O \qquad \Delta G° = +13.8 \text{ kJ mol}^{-1}}$
$ATP + \text{glucose} \longrightarrow ADP + G6P + H^+ \qquad \Delta G° = -18.4 \text{ kJ mol}^{-1}$

(c) $K_{eq} = ([G6P] \times [ADP] \times [H^+])/([ATP] \times [\text{glucose}]) = e^{-\Delta G°/RT}$
Note: the activity of H^+ is referenced to a biochemical standard state concentration of 1×10^{-7} M.

$K_{eq} = ([G6P] \times 0.001 \times 10^{-0.4})/(0.003 \times 0.005) =$
$e^{-(-18.4)/(0.008314) \times (310)}$
$[G6P] = (0.000015 \times e^{7.14})/(0.001 \times 10^{-0.4}) = (0.0168)/(0.000398)$
$\quad = 42.2 \text{ M}$

This G6P concentration is never reached because G6P is continuously consumed by other reactions, and so the reaction never reaches true thermodynamic equilibrium.

7. (a) ΔS must be positive because the increase in available states corresponds to an increase in entropy.

(b) Since $\Delta G = \Delta H - T\Delta S$, a positive ΔS yields a negative contribution to ΔG (T is always a positive number). Thus, for proteins to be stable, which requires ΔG for the above to be positive, denaturation must involve a large positive ΔH and/or an additional negative contribution to ΔS. As we shall see in Chapter 6, both occur.

11. **(a)** See problem 10. Plot $\ln K_w$ vs. $1/T$ and fit a line to the points. The slope will correspond to $-\Delta H°/R$.

$1/T$	$\ln K_w$
0.00366	−34.4
0.00336	−32.2
0.00330	−31.9
0.00323	−31.3

$\ln K_w = (-7093.9 \text{ K})(1/\text{T}) - 8.426$

$\text{slope} = -7093.9 \text{ K} = -\Delta H°/R$

$-7093.9 \text{ K} = -\Delta H°/(0.008314 \text{ kJ mol}^{-1} \text{ K}^{-1})$

$\Delta H° = 59.0 \text{ kJ mol}^{-1}$

(b) $\ln K = -\Delta H°/RT + \Delta S°/R$

$K = K_w = 10^{-14}$

$\Delta S° = [\ln K + (\Delta H°/RT)] \times R$

$\Delta S° = [-32.2 + (59.0 \text{ kJ mol}^{-1}/(0.008314 \text{ kJ mol}^{-1} \text{ K}^{-1} \times 298 \text{ K}))] \times 0.008314 \text{ kJ mol}^{-1} \text{ K}^{-1}$

$\Delta S° = -0.0700 \text{ kJ mol}^{-1} \text{ K}^{-1}$ or $-70.0 \text{ J mol}^{-1} \text{ K}^{-1}$

13. To be favorable the reaction must have $\Delta G < 0$.

$0 > \Delta G = \Delta G°' + RT \ln\left(\dfrac{[\text{oxaloacetate}][\text{NADH}][\text{H}^+]}{[\text{malate}][\text{NAD}^+]}\right)$

$0 > +29.7\dfrac{\text{kJ}}{\text{mol}} + \left(0.008314\dfrac{\text{kJ}}{\text{mol} \times \text{K}}\right)(310 \text{ K})$

$\ln\left(\dfrac{[\text{oxaloacetate}][0.0003][1]}{[0.0004][0.020]}\right)$

$-11.5 > \ln\left(\dfrac{[\text{oxaloacetate}][0.0003][1]}{[0.0004][0.020]}\right) = \ln(37.5[\text{oxaloacetate}])$

$e^{-11.5} = 9.89 \times 10^{-6} > 37.5[\text{oxaloacetate}]$

$2.63 \times 10^{-7} > [\text{oxaloacetate}]$

Thus, the reaction is unfavorable under these conditions when [oxaloacetate} exceeds 2.63×10^{-7} M.

17. **(a)** As written 4 reactants combine to produce 3 products (there are 2 mol H^+ as written, thus there are 4 reactants). Typically, producing fewer products than reactants would require a loss of entropy; however, in this case on of the products is a gas (and none of the reactants are gases). Gas formation is typically associated with a significant increase in entropy. Thus, the prediction that the entropy change is > 0 is reasonable.

(b)

$\Delta G = \Delta G°' + RT \ln Q = \Delta G°' + RT\ln\left(\dfrac{[\text{ethanol}][\text{NAD}^+][\text{CO}_2]}{[\text{pyruvate}][\text{NADH}][\text{H}^+]^2}\right)$

$-38.3\dfrac{\text{kJ}}{\text{mol}} = -64.4\dfrac{\text{kJ}}{\text{mol}} +$

$RT\ln\left(\dfrac{(\text{ethanol})\left(\dfrac{0.000350 \text{ M}}{1 \text{ M}}\right)\left(\dfrac{15 \text{ torr}}{750 \text{ torr}}\right)}{\left(\dfrac{62 \times 10^{-6} \text{ M}}{1 \text{ M}}\right)\left(\dfrac{15 \times 10^{-6} \text{ M}}{1 \text{ M}}\right)\left(\dfrac{10^{-7.4} \text{ M}}{10^{-7.0} \text{ M}}\right)^2}\right)$

$-38.3\dfrac{\text{kJ}}{\text{mol}} = -64.4\dfrac{\text{kJ}}{\text{mol}} + \left(0.008314\dfrac{\text{kJ}}{\text{mol}}\right)(310\text{K})$

$\ln\left(\dfrac{(\text{ethanol}) \times (7.0 \times 10^{-6})}{1.47 \times 10^{-10}}\right)$

$+26.1\dfrac{\text{kJ}}{\text{mol}} = 2.577\dfrac{\text{kJ}}{\text{mol}}\ln[(\text{ethanol}) \times (4.75 \times 10^4)]$

$e^{\left(\dfrac{26.1\frac{\text{kJ}}{\text{mol}}}{2.577\frac{\text{kJ}}{\text{mol}}}\right)}$

$= [(\text{ethanol}) \times (4.75 \times 10^4)] = 2.50 \times 10^4$

$(\text{ethanol}) = \dfrac{2.50 \times 10^4}{4.75 \times 10^4} = 0.527$

Thus, [ethanol] = 0.527 M.

(c) The activity of H^+ appears in the denominator of the Q term. As pH decreases, $[\text{H}^+]$ increases, and the value for the activity of H^+ would increase. This would make the value of $RT \ln Q$ less positive, which in turn makes the value of ΔG less positive (i.e., more negative or more thermodynamically favorable). Thus, decreasing pH is expected to make the reaction as written more favorable. This is also predicted by applying Le Chatelier's Principle to the chemical equation as written: increasing $[\text{H}^+]$ would drive the reaction to the right (more favorable).

(d) The activity of CO_2 appears in the numerator of the Q term. As CO_2 concentration increases, the value of $RT\ln Q$ becomes more positive, which in turn makes the value of ΔG more positive (i.e., less thermodynamically favorable). Thus, increasing CO_2 concentration is expected to make the reaction as written less favorable. This is also predicted by applying Le Chatelier's Principle to the chemical equation as written: increasing CO_2 concentration would drive the reaction to the left (less favorable).

(e) $\Delta G°'$ is a constant that refers only to the free energy of the equilibrium state compared to the standard state.

---

18. $\Delta G = \Delta G°' + RT\ln Q = \Delta G°' + RT\ln\left(\dfrac{[\text{pyruvate}]^2[\text{ATP}]^2[\text{H}_2\text{O}]^2[\text{NADH}]^2[\text{H}^+]^2}{[\text{glucose}][\text{ADP}]^2[\text{P}_i]^2[\text{NAD}^+]^2}\right)$

Evaluation of $\Delta G°'$ requires combination of $\Delta G°'$ values for ATP hydrolysis (given) and glucose oxidation (which can be calculated from the $E°'$ info provided):

glucose	$\longrightarrow$	$2\text{pyruvate} + 6\text{H}^+ + 4e^-$	this is the e^- donor with $E°' = -0.590$ V
$2\text{NAD}^+ + 2\text{H}^+ + 4e^-$	$\longrightarrow$	2NADH	this is the e^- acceptor with $E°' = -0.315$ V
NET: $\text{glucose} + 2\text{NAD}^+$	$\longrightarrow$	$2\text{pyruvate} + 2\text{NADH} + 4\text{H}^+$	$\Delta E°' = (-0.315 \text{ V}) - (-0.590 \text{ V}) = +0.275$ V

Thus, $\Delta G°' = -n\text{F}\Delta E°' = -(4)(96.5 \text{ kJ/mol} \times \text{V})(+0.275\text{V}) = -106.2 \text{ kJ/mol}$

glucose $+ 2\text{NAD}^+$	$\longrightarrow$	$2\text{pyruvate} + 2\text{NADH} + 4\text{H}^+$		$\Delta G°' = -106.2 \text{ kJ/mol}$
$2\text{ADP} + 2\text{P}_i + 2\text{H}^+$	$\longrightarrow$	$2\text{ATP} + 2\text{H}_2\text{O}$		$\Delta G°' = +64.4 \text{ kJ/mol}$
NET: $\text{glucose} + 2\text{NAD}^+ + 2\text{ADP} + 2\text{P}_i$	$\longrightarrow$	$2\text{pyruvate} + 2\text{ATP} + 2\text{H}_2\text{O} + 2\text{NADH} + 2\text{H}^+$	$\Delta G°' = -41.8 \text{ kJ/mol}$

$\Delta G = -41.8\dfrac{\text{kJ}}{\text{mol}} + RT\ln\left(\dfrac{(62 \times 10^{-6})^2(0.0031)^2(1)^2(15 \times 10^{-6})^2(10^{-0.4})^2}{(0.0051)(2.2 \times 10^{-4})^2(0.0059)^2(3.5 \times 10^{-4})^2}\right)$

$\Delta G = -41.8\dfrac{\text{kJ}}{\text{mol}} + \left(0.008314\dfrac{\text{kJ}}{\text{mol}}\right)(310 \text{ K})\ln(1.25 \times 10^{-3}) = -59.0\dfrac{\text{kJ}}{\text{mol}}$

20. This uses a logic similar to that in Problem 1c. In this case, since the problem is asking for fraction unfolded, we can use the reverse process: *folded protein* ⇌ *unfolded protein*. Thus, we will be reversing the signs of the given thermodynamic parameters as they apply to the process of folding: *unfolded protein* ⇌ *folded protein*.

$$\Delta G° = -RT \ln K = -RT \ln\left(\frac{[\text{unfolded}]}{[\text{folded}]}\right) = \Delta H° - T\Delta S°$$

As in Problem 1, dividing by ($-RT$) gives

$$\ln\left(\frac{[\text{unfolded}]}{[\text{folded}]}\right) = \left(-\frac{\Delta H°}{RT}\right) + \left(\frac{\Delta S°}{R}\right)$$

$$\ln\left(\frac{[\text{unfolded}]}{[\text{folded}]}\right) = \left(-\frac{+280\frac{\text{kJ}}{\text{mol}}}{\left(0.008314\frac{\text{kJ}}{\text{mol} \times \text{K}}\right) \times (315\text{K})}\right)$$
$$+ \left(\frac{+0.790\frac{\text{kJ}}{\text{mol} \times \text{K}}}{\left(0.008314\frac{\text{kJ}}{\text{mol} \times \text{K}}\right)}\right)$$

$$\ln\left(\frac{[\text{unfolded}]}{[\text{folded}]}\right) = (-106.9) + (95.02) = -11.88$$

$$\frac{[\text{unfolded}]}{[\text{folded}]} = e^{-11.88} = 6.928 \times 10^{-6}$$

This shows that for every mol of folded protein there is 6.9×10^{-6} mol of unfolded protein. In summary, the fraction of unfolded ribonuclease at 42 °C is insignificant.

CHAPTER 4

1.

b Adenosine	b Uridine
c Cytidine 5′-monophosphate	e Deoxycytidine 5′-triphosphate
a Guanine	d Deoxyuridine
a Thymine	c Guanosine 5′-diphosphate
d Deoxyguanosine	e Thymidine triphosphate

2. A nucleoside triphosphate yields upon complete hydrolysis one nucleobase, one sugar, and three phosphates. A trinucleotide yields three bases, three sugars, and at least two phosphates. Other answers are possible.

3. (a)
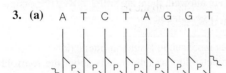
A T C T A G G T

(b) ACCTAGAT or dACCTAGAT

4. (a) and **(b)**

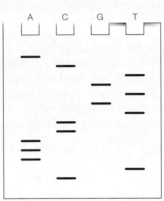

(c) The non-template DNA strand has the same base sequence as the RNA transcribed from that DNA molecule (thymine instead of uracil).

5. If cytosine is at 23%, then so is guanine. The total, 46%, is subtracted to give A/T content of 54%. If A = T and G = C, then A is 27% and so is T.

12. *Neurospora crassa* DNA has a higher GC content than human DNA, so its T_m is greater.

15. The GGG at the 3′ terminus can pair with the internal CCC, thereby providing a duplex segment that can bind DNA polymerase, a 3′ hydroxyl terminus (GGG-OH) and a template for nucleotide addition (GATTTGGACACAGT-5′).

16. The 3′ GGG-OH sequence could serve as a primer, if properly aligned with the CCC internal sequence, yielding a gel pattern that would yield bands corresponding to the 14 nucleotides at the 5′ end of the oligonucleotide

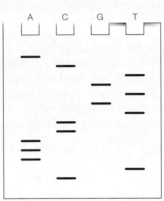

20. 5′ CCAGATAAA**GGG**AGATTCCAG 3′

CHAPTER 5

3. (a) D-Valine   **(b)** D-Threonine   **(c)** D-Glutamine

5. (a) The curve will exhibit inflections corresponding to two groups titrating near 4 (carboxyl terminus and glutamic side chain), one group near 7 (histidine), two near 9 (N-terminus and lysine), one near 10 (tyrosine), and one near 12 (arginine).

(b) At pH = 1 the overall charge will be +3.998; at pH = 5 the overall charge will be +2.104; at pH = 11 the overall charge will be −1.827.

(c) pI = 8.813

6. (a) SYSMEHFR, WGKPV.

(b) SYSM*, EHFRWGKPV; M* = homoserine lactone.

(c) SYS, MEHFRWGKP, V.

8. (a) ~−1 @ pH = 7, and ~−4 @ pH = 12.

(b) Cyanogen bromide: 2 peptides (SEPIM* + APVEYPK); trypsin: no cleavage; chymotrypsin: 2 peptides (SEPI + MAPVEYPK; no cleavage after the Y because there is a C-terminal P residue).

(c) At pH = 7 SEPI will carry a (−) charge and MAPVEYPK will carry a charge ~0. Thus, these two peptides can be separated by electrophoresis, or by chromatography on an anion exchange column (column carries a (+) charge) at pH 7 (SEPI will stick to the column while MAPVEYPK flows through).

11. (a) Two disulfides.

(b) One disulfide between Cys1-Cys3 and the second between Cys11-Cys15

15. (a) The primary amine of lysine, and the N-terminal amine.

(b) The thiolate of cysteine.

(c) The hydroxyl groups of serine, threonine, and tyrosine.

16. (a) The (+) charge on amines is neutralized by acetylation; thus, there is a reduction in (+) charge when a protein is treated with acetic anhydride. At pH = 7 the thiol of cysteine is mostly protonated (uncharged); thus, addition of negatively-charged carboxymethyl groups will result in increased ($-$) charge density. Phosphorylation at pH $= 7$ will increase ($-$) charge density since hydroxyl groups are uncharged at physiological pH.

(b) pI will decrease after acetylation of amines (Lys), or carboxymethylation of thiols (Cys), or phosphorylation of Ser, Thr, Tyr because the modified protein will have greater ($-$) charge density relative to the unmodified protein. Thus, it is necessary to decrease pH to achieve the ($+$) charge density that will counterbalance this increased ($-$) charge density.

18. Since the mutations replace nonionizable side chains with ionizable side chains, we do not need to know anything about the charge state of the Gln and Phe that are replaced (since they carry no charge in the wt protein). We can then assume that any difference in charge between the wt and the mutant at pH $= 5.5$ is due to the presence of partial charges on the mutant side chains. Using the Henderson–Hasselbalch equation with the pK_as given in the problem, the charges on the His and Glu side chains can be calculated at pH $= 5.5$ (the pI of the wild-type protein; see Problem 15b at the end of Chapter 2). This is done because the charge on the wt protein is zero when pH $= 5.5$. This analysis yields an overall charge on the mutant of $+0.559$ at pH $=$ pI (see below). If a protein is positively charged the pH must be BELOW the pI; thus, the pI of the mutant is greater than 5.5.

At pH $= 5.5$ the deprotonation of the carboxylic acid on the Glu side chain will be between 0 and 50% ($pK_a = 6.0$ is above the pH by less than one pH unit). This can be confirmed as follows:

$$5.5 = 6.0 + \log\left(\frac{[A^-]}{[HA]}\right) = 6.0 + \log\left(\frac{[-COO^-]}{[-COOH]}\right)$$

$$\longrightarrow \frac{[-COO^-]}{[-COOH]} = 0.316$$

Thus, at pH $= 5.5$, for every 1 mol of $-COOH$, there is 0.316 mol of $-COO^-$. With this information the mole fractions of conjugate base "A^-" (i.e., $-COO^-$) and conjugate acid "HA" (i.e., $-COOH$) can be calculated:

$$\bar{x}_{A^-} = \text{mole fraction } A^- = \frac{\text{mol } A^-}{\text{mol } A^- + \text{mol } HA} = \frac{0.316}{0.316 + 1} = 0.240$$

$$\bar{x}_{HA} = \text{mole fraction } HA = \frac{\text{mol } HA}{\text{mol } A^- + \text{mol } HA} = \frac{1}{0.316 + 1} = 0.760$$

For a carboxylic acid it is the $-COO^-$ form that carries charge; thus, the (mole fraction of the charged species) multiplied by (the charge on that species) gives the overall charge (in this case "HA" has a charge of zero, so whatever the value of mole fraction for HA, it contributes zero to the overall charge on the side chain):

$$\text{total charge} = (\bar{x}_{A^-} \times (-1)) + (\bar{x}_{HA} \times (0))$$
$$= (0.240 \times (-1)) + (0.760 \times (0)) = -0.240$$

At pH $= 5.5$ the deprotonation of the imidazole on the His side chain will be between 0 and 50% ($pK_a = 6.1$ is above pH 5.5 by less than 1 pH unit). This can be confirmed as follows:

$$5.5 = 6.1 + \log\left(\frac{[A^-]}{[HA]}\right) = 6.1 + \log\left(\frac{[-N:]}{[-NH^+]}\right)$$

$$\longrightarrow \frac{[-N:]}{[-NH^+]} = 0.251$$

Thus, at pH $= 5.5$, for every 1 mol of $-NH^+$, there is 0.251 mol of $-N:$. As was done above, the mole fractions of conjugate base "A^-" (i.e., $-N:$) and conjugate acid "HA" (i.e., $-NH^+$) can be calculated:

$$\bar{x}_{A^-} = \frac{\text{mol } A^-}{\text{mol } A^- + \text{mol } HA} = \frac{0.251}{0.251 + 1} = 0.201$$

$$\bar{x}_{HA} = \frac{\text{mol } HA}{\text{mol } A^- + \text{mol } HA} = \frac{1}{0.251 + 1} = 0.799$$

For the imidazole side chain of His, it is the conjugate acid (i.e., the protonated $-NH^+$ form) that carries charge; thus:

$$\text{total charge} = (\bar{x}_{A^-} \times (0)) + (\bar{x}_{HA} \times (+1))$$
$$= (0.201 \times (0)) + (0.799 \times (+1)) = +0.799$$

The average molecular charge on the mutant at pH 5.5 is the sum of the average charges on each mutant side chain:

$$(-0.240) + (+0.799) = +0.559.$$

20. The expect score for the shorter sequence is more likely to be closer to one (i.e., the shorter sequence is more likely to match to many entries in the database, making these matches less statistically significant).

21. **(a)** Insulin binds receptors that are accessed via the circulatory system; thus, direct injection introduces insulin to its sites of action. More importantly, insulin (and other proteins) would not survive transit through the stomach and gut- environments designed to break down protein "nutrients."

(b) Peptides containing D-amino acids are less likely to be recognized as substrates by proteolytic enzymes in the gut.

(c) Peptides typically carry charged groups (i.e., N- or C-termini and some side chains) that reduce transport across the hydrophobic core of the membrane bilayer.

(d) Modification of the N- and C-termini (by, respectively, acetylation and amidation) will reduce the charge density on peptides. This would make them more likely to cross membranes.

23. At pH $= 7$ a DNA-binding protein is expected to carry excess ($+$) charge that will lead to favorable binding to the ($-$) charge on the DNA molecule (e.g., histones are highly positively charged). A positively-charged protein will bind to a column carrying excess ($-$) charge. Thus, at pH $= 7$, CM-cellulose is expected to bind to a DNA-binding protein.

CHAPTER 6

2. (5 possible random disulfides) $\times$ (3 possible random disulfides) $\times$ (1 possible disulfide) $= 15$ possible random combinations of disulfides $\rightarrow$ 1 native combination/15 possible $= 6.67\%$ (1 in 15) native disulfides would form randomly.

3. **(a)** A hydrophobic side chain at every 3rd or 4th residue would give an extended hydrophobic face along the helix. Four such helices could be arranged so that the hydrophobic side chains would all point toward the center of the bundle and would pack together there. This would give a stabilizing hydrophobic core.

(b) A proline at this point would break the helix near the Fe_2 binding sites. This would probably mean that Fe_2 could not be bound, and the mutant protein would be nonfunctional.

6. **(a)** $3^{200} = 2.7 \times 10^{95}$

(b) Not all of these conformations will be sterically possible. But even if only 0.1% of these are allowed, there are still 2.7×10^{92} possible conformations-a very large number.

10. **(a)** C_8 or D_4.

(b) D_4, because it involves more subunit–subunit interactions.

(c) Both. There must be heterologous interactions about the fourfold axis and isologous interactions about the two-fold axes.

11. The favorable intramolecular ionic or H-bonding interactions in a folded protein replace interactions between solvent (water) and the ionic species (or H-bond donors and acceptors) in the unfolded state. The favorable ΔH obtained by formation of intramolecular bonds in the folded protein is offset by the energy required to break many interactions with solvent going from the unfolded to the folded state.

13. In the absence of BME, a single band of $M_W \approx 70,000$ is obtained. This suggests, but does not prove, that there are two identical, noncovalently linked subunits. However, the addition of BME removes this band and gives two bands of $\approx 30,000$ and $\approx 40,000$, respectively. The sum of these is 70,000, strongly suggesting that the native molecule contains four subunits, two of $\approx 30,000$, two of $\approx 40,000$. The 30,000 and 40,000 units are paired by at least one disulfide bond.

16. **(a)** No. DNA is charged and therefore polar, so most of the DNA-binding helix is likely to be composed of polar residues that interact with

either the DNA or the solvent (this is supported by sequence analysis of the bHLH family of DNA binding proteins).

(b) The N-terminus is interacting with the DNA. Two reasons might be given for this: (1) the α-amino group of the N-terminus is positively charged and will interact favorably with the negative charge on the phosphodiester backbone of the DNA; (2) this orientation also situates the "partial positive" end of the helical macrodipole for favorable electrostatic interactions with the negatively charged phosphodiester backbone of the DNA.

17. Consider how the mutation would affect the relative free energies in the folded and unfolded states of the protein. Because the mutation is on the surface, loss of hydrophobic contacts in the protein core is not an issue and the amino acid is likely to be interacting with solvent to similar extents in both folded and unfolded states; thus, effects on ΔH are predicted to be minimal. For the same reason the $\Delta S_{solvent}$ is likely to be small because side chain solvation is predicted to be similar in both the folded and unfolded states (see Equation 6.3). $\Delta S_{protein}$ changes the most due to the conformational flexibility of Gly compared to Pro. Gly will stabilize both the folded and the unfolded states; however, it stabilizes the unfolded state more due to the dramatic increase in conformational entropy of the unfolded state as a result of this mutation. The stabilization of the unfolded state for the mutant means that $\Delta G_{unfolding\,(wt)} > \Delta G_{unfolding\,(mutant)}$; thus, the mutation is destabilizing.

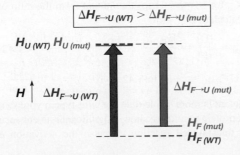

18. The orientation of four contiguous residues is required to initiate one turn of α-helical structure (Figure 6.4a). Likewise a minimum of three contiguous residues must be ordered to initiate a turn (Figure 6.20), which could then initiate H-bonding between antiparallel strands. Nucleation of an antiparallel sheet can therefore be faster than nucleation of a helix. The initiation of a parallel sheet requires a noncontiguous sequence to form H-bonding interactions. Because the effective concentration of contiguous residues is higher than that of noncontiguous residues, the nucleation of a parallel sheet will be significantly slower.

20. Chaperones are meant to prevent irreversible protein aggregation of unfolded proteins, which is mediated by the intermolecular association of hydrophobic surfaces. A folded protein will minimize solvent-exposed hydrophobic surface area. Thus, as proteins unfold more hydrophobic surface area will be exposed, triggering recognition by chaperones.

22. Due to loss of van der Waals interactions in the folded state, $\Delta H_{F \rightarrow U\,(WT)} > \Delta H_{F \rightarrow U\,(mut)}$ (i.e., $\Delta H_{F \rightarrow U\,(WT)}$ will be more (+) and $\Delta H_{F \rightarrow U\,(mut)}$ will be less (+) for the process of unfolding):

This enthalpic effect makes $\Delta G_{F \rightarrow U}$ more unfavorable for WT vs. the mutant (i.e., the WT is more stable to unfolding than the mutant).

$\Delta S_{solvent}$ is likely to be less unfavorable for the mutant because the smaller Ala side chain is predicted to have less clathrate structure in the unfolded state than the WT Leu side chain. All the clathrate structure will be lost to bulk solvent in the folded state. Thus, $\Delta S_{solvent\,F \rightarrow U\,(WT)} < \Delta S_{solvent\,F \rightarrow U\,(mut)}$ (i.e., $\Delta S_{solvent\,F \rightarrow U\,(WT)}$ will be more ($-$) and $\Delta S_{solvent\,F \rightarrow U\,(mut)}$ will be less ($-$) for the process of unfolding):

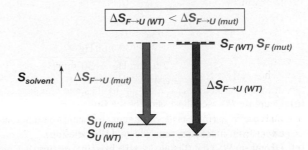

This entropic effect makes $\Delta G_{F \rightarrow U}$ more unfavorable for WT vs. the mutant (i.e., the WT is more stable to unfolding than the mutant).

$\Delta S_{protein\,F \rightarrow U\,(WT)} \approx \Delta S_{protein\,F \rightarrow U\,(mut)}$ because differences in conformational entropy around ϕ (Phi) and ψ (Psi) angles are small for Leu vs. Ala (see Figures 6.9 and 6.10). Thus, this entropic effect is not predicted to be significantly different for WT vs. mutant.

The differences in ΔH and $\Delta S_{solvent}$ rationalized above both predict that the mutant will be less stable to denaturation than the WT protein.

24. $\Delta S° = \Delta S°_{peptide} + \Delta S°_{solvent}$. $\Delta S°_{peptide}$ will be affected by the Gly content and will therefore be greatest for adaH2 over Top7. Given the similarities in size and structure, it is reasonable to assume that $S°_{peptide}$ in the folded state will be similar for adaH2 and Top7. In the unfolded state $S°_{peptide}$ will be greater for adaH2 vs. Top7 due to the greater Gly content in adaH2. Thus, $\Delta S°_{peptide}$ for folding is predicted to be more negative for adaH2 than for Top7:

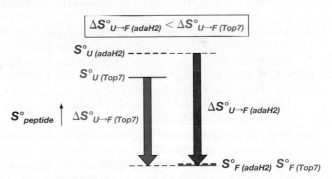

Given this prediction that the conformational entropy change $\Delta S°_{peptide\,(adaH2)} < \Delta S°_{peptide\,(Top7)}$, and the data in the table that show the overall entropy change $\Delta S°_{adaH2} > \Delta S°_{Top7}$ (i.e., $-53\ \text{J}/(\text{mol} \times \text{K}) > -58\ \text{J}/(\text{mol*K})$), it follows that $\Delta S°_{solvent\,(adaH2)}$ must be $> \Delta S°_{solvent\,(Top7)}$. A greater value for $\Delta S°_{solvent}$ correlates with greater release of clathrate structure to the bulk solvent, which in turn correlates with the amount of hydrophobic surface area that is buried. Thus, adaH2 is predicted to bury more hydrophobic surface area upon folding that Top7.

CHAPTER 7

2. In the former case, the binding epitope is composed of regions of the protein that are distant in primary sequence, but close in the folded structure (a "discontinuous epitope"). In the latter case, the binding epitope is composed of contiguous amino acid sequence (a "continuous epitope").

4. (a) ~2.6 mm Hg. This can be calculated from any line of data in the table. For example:

$$Y_{O_2} = \frac{P_{O_2}}{P_{50} + P_{O_2}}$$

$$0.161 = \frac{0.5}{x + 0.5}$$

$$x = 2.6 \text{ mm Hg}$$

(b) ~92 %

$$Y_{O_2} = \frac{P_{O_2}}{P_{50} + P_{O_2}}$$

$$x = \frac{30 \text{ mm Hg}}{2.6 \text{ mm Hg} + 30 \text{ mm Hg}}$$

$$x = 0.92 = 92\%$$

5. (a) Decrease P_{50} (see discussion of the Bohr effect).

(b) Increase P_{50} (increased carbamate formation and increased Bohr effect from increased carbonic acid dissociation).

(c) Decrease P_{50} (the T state is stabilized by quaternary interactions, which are disrupted when the subunits dissociate; thus, the R state is favored when subunits dissociate).

(d) Decrease P_{50} (2,3-BPG binding stabilizes the T state; thus, a reduction in 2,3-BPG binding will promote the R state and reduce P_{50}).

7. In the absence of O_2 from the lungs, the residual O_2 bound to hemoglobin is used for continued energy production in respiring cells (such as cardiac and skeletal muscle). Underwater, the crocodile continues to generate CO_2 from metabolic activity. The CO_2 is rapidly converted to HCO_3^- by carbonic anhydrase. As $[HCO_3^-]$ increases, more will bind to hemoglobin, shifting it to greater T-state conformation. This favors O_2 release; thus, as the crocodile stays underwater its hemoglobin delivers most of the bound O_2 as a result of increased binding of HCO_3^- to the T conformation.

8. (a) Because H146 lies in the α/β interface (see Figure 7.23), mutation should be expected to interfere with the T $\rightleftharpoons$ R transition by favoring the R sate. The effect is to increase affinity. This mutation is known; it is hemoglobin Hiroshima.

(b) Because F8 is involved in heme binding (see Figure 7.25), the heme should be unstable. Leucine will not ligate to the heme iron.

In each case a single base change could give rise to the mutation.

12. (a) Yes, because the linkage between subunits is such that forcing one pair of helices apart favors moving the other pair apart (to maintain the favorable electrostatic interactions shown in the figure), making O_2 binding easier in the second pair.

(b) Probably. Deprotonation of His 13 destroys the salt bridge, allowing easier opening of the O_2 binding site.

(c) The molecule would exhibit higher O_2 affinity, and probably display reduced cooperativity because the O_2 binding sites would be opened further. It is also possible that the whole structure would become unstable.

13. The negative charge on the Asp side chain can form salt bridges with the other (+)-charged side chains in the BPG–binding pocket and stabilize the T-state. In essence, the Asp side chain is mimicking the (−) charge on BPG.

15. (a) The disulfide can only form in the oxy state (i.e., the R state); thus, disulfide-bonded HbR is predicted to favor the R state and exhibit increased O_2 binding affinity.

(b) The Bohr effect is reduced in HbR compared to HbA because a covalent disulfide (which stabilizes the R state) is stronger than the Asp-His salt bridge that stabilizes the T state.

(c) Adaption to altitude will occur more slowly for the HbR individual because a higher concentration of 2,3-BPG will be required to drive disulfide-bonded HbR into the T state. More time is require to produce higher [2,3-BPG]. This would not be beneficial to a mountaineer.

(d) For the HbR individual:

$$\Delta Y_{O_2} = Y_{O_2}(\text{lungs}) - Y_{O_2}(\text{capillaries})$$
$$= \left(\frac{(48)^{1.2}}{(14)^{1.2} + (48)^{1.2}} \right) - \left(\frac{(15)^{1.2}}{(14)^{1.2} + (15)^{1.2}} \right) = 0.814 - 0.521 = 0.293$$

For the HbA individual:

$$\Delta Y_{O_2} = Y_{O_2}(\text{lungs}) - Y_{O_2}(\text{capillaries})$$
$$= \left(\frac{(48)^{3.2}}{(28)^{3.2} + (48)^{3.2}} \right) - \left(\frac{(15)^{3.2}}{(28)^{3.2} + (15)^{3.2}} \right) = 0.849 - 0.119 = 0.730$$

Much less O_2 is released from the HbR compared to the HbA.

19. (a) Release of the myosin headpiece from the thin filament requires ATP binding. Until ATP binds, the myosin-actin cross-bridge will remain intact, thereby preventing extension of the sarcomeres.

(b) Without the action of the Ca^{2+} transporter, Ca^{2+} will leak across the membrane from the side of high Ca^{2+} concentration (in the transverse tubule) to the side of lower Ca^{2+} concentration (inside the sarcomere). As $[Ca^{2+}]$ increases, it will bind TnC, thereby stimulating myosin binding to actin. The lack of ATP will result in a persistent cross-bridge (see part [a]), characteristic of the *rigor* state.

(c) Decomposition includes cleavage of actin and myosin by intracellular proteases (e.g., enzymes such as trypsin and chymotrypsin described in Chapter 5).

CHAPTER 8

2. The rate enhancement can be evaluated using Equation 8.16:

$$\text{rate enhancement} = \left(\frac{A_{cat}}{A_{non}} \right) e^{\left(\frac{\Delta\Delta G^{\circ\ddagger}}{RT} \right)}$$

$$= \left(\frac{A_{cat}}{A_{non}} \right) e^{\left(\frac{(125-46)\frac{kJ}{mol}}{\left(0.008314 \frac{kJ}{mol * K}\right)(294K)} \right)} = \left(\frac{A_{cat}}{A_{non}} \right) e^{(32.32)}$$

$$\text{rate enhancement} = \left(\frac{A_{cat}}{A_{non}} \right) (1.09 \times 10^{14})$$

Assuming the ratio A_{cat}: A_{non} is 1, the expected rate enhancement is ~1.1×10^{14}.

3. The Asp must be protonated to act as a general acid catalyst; thus, activity will be higher when pH < pK_a and lower when pH > pK_a. At pH = pK_a expect 50% of maximal activity because the Asp will be 50% protonated.

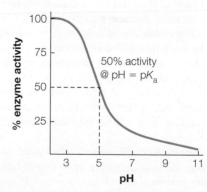

4. Find K_{eq} for F $\rightleftharpoons$ U for the mutant from the ratio of rate constants; then, calculate $\Delta G^{\circ\prime}_{F \to U}$:

$$K_{eq} = \frac{[U]}{[F]} = \frac{[k_{F \to U}]}{[k_{U \to F}]} = \frac{3.62 \times 10^{-5} \text{ s}^{-1}}{255 \text{ s}^{-1}} = 1.42 \times 10^{-7}$$

$$\Delta G^{\circ\prime}_{F \to U} = -RT\ln(1.42 \times 10^{-7}) = +39.1 \frac{kJ}{mol}$$

The mutant is more stable than the wild-type myoglobin.

6. Such a protein would not show a significant rate enhancement compared to the uncatalyzed reaction, because the activation energies ($\Delta G^{\circ\ddagger}$)

would be identical for both the catalyzed and the uncatalyzed reactions. Equation 8.16 predicts that for a value of $\Delta\Delta G^{\circ\ddagger} = 0$, the rate enhancement is A_{cat}/A_{non}.

9. (a) E270 acts as a GBC in step one and as a GAC in step two.

   (b) R145 provides specific ion–ion interactions with the C-terminal carboxylate of the substrate. This confers specificity for cleavage of the C-terminal residue from the peptide substrate.

11. (a) If we take k_{cat}/K_M ratios, we find that PAPAF would be digested most rapidly and PAPAG most slowly.

   (b) A hydrophobic residue C-terminal to the bond cleaved seems to be favored. Elastase always requires a small residue (like Ala) to the N-terminal side.

   (c) Serine and histidine (i.e., key components of the serine protease catalytic triad).

12. (a) The enzyme must be stable both to the presence of detergents and to moderately high temperatures.

   (b) Replace the methionine, by site-directed mutagenesis, with another residue. Because methionine is quite hydrophobic, a hydrophobic replacement would seem appropriate. A single base change in the Met codon could yield Phe, Leu, Ile, or Val.

   (c) The oxyanion is formed after S binds; thus, for a mutation of a residue that only interacts with the oxyanion intermediate one would not expect K_M to change significantly (it does not, according to Wells et al.); however, k_{cat} should be reduced due to the loss of enthalpic stabilization of the transition state (it is—by factor of 2500).

   (d) Using Equation 8.16, and assuming that the value of A is the same for both mutants:

   $$\text{rate enhancement} = \frac{k_{cat\,N155}}{k_{cat\,T155}} = \frac{(A_{N155})e^{\left(\frac{-\Delta G^{\circ\ddagger}_{N155}}{RT}\right)}}{(A_{T155})e^{\left(\frac{-\Delta G^{\circ\ddagger}_{T155}}{RT}\right)}} = \left(\frac{A_{N155}}{A_{T155}}\right)e^{\left(\frac{\Delta\Delta G^{\circ\ddagger}}{RT}\right)}$$
   $$= 2500$$

   Solving for $\Delta\Delta G^{\circ\ddagger}$ gives $\Delta\Delta G^{\circ\ddagger} = 20$ kJ/mol, which corresponds to the stabilization provided by the N155 H-bond.

   (e) The dielectric constant, ε, is lower in the enzyme active site than it is in water; thus, Coulomb's law (Equation 2.2) predicts a stronger interaction between the H-bond donor and acceptor.

   $$F = k\left(\frac{q_1 q_2}{\varepsilon r^2}\right)$$

16. (a) Enalaprilat is too polar to cross membranes; whereas enalapril is less polar and can cross membranes to get from the gut to circulation.

   (b) The inhibited reaction will have an apparent $K_M = \alpha K_M$, and the observed velocity will be 10% of the uninhibited reaction under the conditions given in the problem. These two ideas can be expressed mathematically in this way:

   $$v_{inhibited} = \frac{V_{max}[S]}{\alpha K_M + [S]} = \frac{0.1 V_{max}[S]}{K_M + [S]} \text{ this can be rearranged to give:}$$

   $$10 = \frac{\alpha K_M + [S]}{K_M + [S]}$$

   solve for α using the given values of [S] and K_M:

   $$\alpha = 66.25 = 1 + \frac{[I]}{K_I} \text{ solve for } K_I$$

   $$K_I = \frac{2.4 \times 10^{-9}\,\text{M}}{66.25} = 3.7 \times 10^{-11}\,\text{M}$$

17. (a) $k_{cat} = V_{max}/[E_{total}] = (234.7\ \mu\text{mol ml}^{-1}\,\text{s}^{-1})/(0.003\ \mu\text{mol ml}^{-1})$
       $= 78200\ \text{s}^{-1}$.

   (b) Use V_{max} above and any line of data in the table to solve Michaelis-Menten equation for K_M:

   $$v_0 = \frac{V_{max}[S]}{K_M + [S]}$$

$$92.0\ \mu\text{mol mL}^{-1}\text{s}^{-1} = \frac{(234.7\ \mu\text{mol mL}^{-1}\text{s}^{-1})[80\ \mu\text{M}]}{K_M + [80\ \mu\text{M}]}$$

$K_M = 124\ \mu\text{M}$.

   (c) For uncompetitive inhibition, the y-intercept of a lineweaver-Burk plot is α'/V_{max} solve for α'

   $$0.352\ \mu\text{mol}^{-1}\text{mLs} = \frac{\alpha'}{(234.7\ \mu\text{mol mL}^{-1}\text{s}^{-1})} \rightarrow \alpha' = 82.6$$

   $$\alpha' = 82.6 = 1 + \frac{[I]}{K'_I}$$

   solve for K'_I

   $$K'_I = \frac{12 \times 10^{-5}\,\text{M}}{81.6} = 1.45 \times 10^{-7}\,\text{M}$$

19. (a) Activation of the molecule by trypsin will be greatly reduced due to removal of the R that trypsin binds in its active site.

   (b) If activation occurs, the N-terminal peptide will no longer be constrained by an S—S bond and may be released.

   (c) Automodification of π-chymotrypsin to α-chymotrypsin could be reduced, although this mutation is relatively conservative (loss of a methyl group); thus, the effect on activation might be slight.

22. The T state of ATCase is favored at low [S] and the R state is favored at high [S]. CTP stabilizes the T state, while ATP stabilizes the R state. Thus, CTP is predicted to bind preferentially to T-state ATCase, which will predominate when [S] is lower, and ATP is predicted to bind preferentially to R-state ATCase, which will predominate when [S] is higher.

23. (a) E117 acts as a GAC in step one and as a GBC in step three; Y409 acts as a covalent catatlyst in step 2; D149 acts as GBC in step 2 and a GAC in step 3.

   (b) E117 appears to play a key role in the catalytic mechanism as a GABC. Thus, the E117A mutation is expected to result in a significant reduction of k_{cat} (since the Ala side chain cannot function as a GABC). Effects on K_M are more difficult to predict given the information in the scheme. Equation 8.24 predicts that a reduction in k_{cat} might also reduce the value of K_M when k_{-1} is not $\gg k_{cat}$.

   (c) During the catalytic cycle C1 of the saccharide transitions from an sp^3 center to an sp^2 center and back. The differences in bonding geometry during catalysis could result in the carboxylate group on C1 moving closer to the R374 side chain, thereby increasing the enthalpic stabilization of the transition state. This is reminiscent of the stabilization provided by the oxyanion hole in the serine protease mechanism as the amide oxygen changes from sp^2 to sp^3 (see Figure 8.17).

24. (a) The slope of a Lineweaver-Burk plot is K_M/V_{max}. $V_{max} = k_{cat}[E]_{total} = 26.8\ \text{s}^{-1} \times 0.021\ \text{mM} = 0.563\ \text{mM s}^{-1}$.

   $$51.2\ s = \frac{K_M}{V_{max}} = \frac{K_M}{0.563\ \text{mMs}^{-1}}$$

   $K_M = 28.8\ \text{mM}$

   (b) The slope of the Lineweaver-Burk plot is $(\alpha K_M)/V_{max}$. Thus,

   $$198.8s = \frac{\alpha K_M}{V_{max}} = \frac{\alpha(28.8\ \text{mM})}{0.563\ \text{mMs}^{-1}}$$

   $$\alpha = 3.88 = 1 + \frac{[I]}{K_I} = 1 + \frac{0.040\ \text{mM}}{K_I}$$

   $K_I = 0.0139\ \text{mM}$ or $13.9\ \mu\text{M}$

CHAPTER 9

1. (a)

   (b)

   α-D-Xylofuranose

(c)

(d)

4. A type O individual does not produce the antigenic A or B oligosaccharides and, hence, can donate blood to any recipient without stimulating an immune reaction. A type AB individual carries both A and B oligosaccharides, and therefore will not recognize either A or B as foreign in donated blood, so no immune reaction will be stimulated as the result of a blood transfusion from any donor.

6. Reducing: maltose, cellobiose, lactose, gentiobiose. Nonreducing: trehalose, sucrose

12. **(a)** 1 **(b)** 6 **(c)** 1 **(d)** 2,3 **(e)** 4,6

13. The influenza hemagglutinin on the virus particle binds to the surface of any cell that contains sialic acid. When it binds to the surface of erythrocytes, this causes the cells to agglutinate, or to aggregate or clump.

15. Mannose and galactose, no; they differ in configuration at both C2 and C4. Allose and altrose, yes; they differ only at C2. Gulose and talose, no; they differ at C2 and C3. Ribose and arabinose, yes; they differ only at C2.

17. A branched polymer has far more terminal glucose residues than an unbranched polymer of equivalent molecular weight, thereby allowing a larger number of terminal glucose residues to be mobilized in the face of an energy demand. It seems likely that the enzyme and the polymer coevolved to meet the potential need for rapid mobilization.

19. Dextran sucrase uses the bond energy stored in the glycosidic link between glucose and fructose to drive formation of the glucose-glucose bond; therefore, ATP is not needed to make the reaction exergonic.

23. Transfusion of type AB blood would introduce both A and B tetrasaccharides, which would be recognized as foreign and trigger an immune reaction in recipients lacking either the A saccharide (type B) or B (type A) or both (type O). By contrast the type AB blood would contain both A and B and would not recognize these tetrasaccharides as foreign.

CHAPTER 10

2. **(a** Fatty acid, long-chain alcohol.
   (b) Glycerol, fatty acid.
   (c) Carbohydrate, long-chain alcohol.

4. **(a)** (30 Å to span bilayer)/(1.5 Å rise per residue in α helix) = 20 residues.
   (b) (30 Å to span bilayer)/(3.3 Å rise per residue in β strand) = 9 residues.
   (c) The helical conformation spanning the bilayer satisfies all the main-chain H-bond donors and acceptors within the helix, whereas the strand would be unable to form intramolecular H-bonds to satisfy main-chain H-bond donors and acceptors.
   (d) The sequence MVGALLLLVVALGIGILFM is 19 residues long, is very hydrophobic, and has a number of helix-forming residues. Thus, it is a likely sequence to span the bilayer.

5. Use Equation 10.3 to solve parts (a) and (b):
   (a) The concentration ratio is 10^5.

$$\Delta G = RT \ln\left(\frac{C_{in}}{C_{out}}\right) + zF\Delta\psi$$

$$= \left(0.008314\frac{kJ}{mol \cdot K}\right)(310\ K)\ln\left(\frac{0.100}{0.000001}\right) + 0 = +29.7\frac{kJ}{mol}$$

   (b) $\Delta G = RT\ln\left(\frac{C_{in}}{C_{out}}\right) + zF\Delta\psi = \left(0.008314\frac{kJ}{mol \cdot K}\right)$

$$(310\ K)\ln\left(\frac{0.100}{0.000001}\right) + (+1)\left(96.5\frac{kJ}{mol \cdot V}\right)(+0.070\ V)$$

$$\Delta G = +36.4\frac{kJ}{mol}$$

(c) Because ΔG for ATP hydrolysis is -55.1 kJ/mol under intracellular conditions (see below), in either case, hydrolysis of 1 mole of ATP would suffice to transport 1 mole of Na^+.

$$\Delta G = \Delta G^{\circ\prime} + RT\ln\left(\frac{\frac{[ADP]}{1M}\frac{[P_i]}{1M}\frac{[H^+]}{10^{-7}M}}{\frac{[ATP]}{1M}[H_2O]}\right)$$

$$\Delta G = -32.2\frac{kJ}{mol} + \left(0.008314\frac{kJ}{mol*K}\right)$$

$$(310K)\ln\left(\frac{[0.000310][0.00510][10^{-0.4}]}{[0.00460][1]}\right)$$

$$\Delta G = -32.2\frac{kJ}{mol} + -22.9\frac{kJ}{mol} = -55.1\frac{kJ}{mol}$$

8. Peptide hormones are generally water-soluble proteins/oligopeptides that would not easily cross the outer membrane of a cell. Peptides vary in polarity, but water-soluble peptides are typically polar and composed of residues with polar and/or charged side chains. Steroid hormones are generally hydrophobic, carry no charged groups, and can cross the outer membrane bilayer by non-mediated diffusion.

9. The calcium ion is *maintained* at this concentration difference when ΔG for transport = 0. From Equation (10.3):

$$\Delta G = 0 = RT\ln\left(\frac{C_{in}}{C_{out}}\right) + zF\Delta\Psi = \left(0.008314\frac{kJ}{mol \cdot K}\right)$$

$$(310\ K)\ln\left(\frac{1000}{1}\right) + (+2)\left(96.5\frac{kJ}{mol \cdot V}\right)(\Psi_{in} - \Psi_{out})$$

$$0 = \left(17.8\frac{kJ}{mol}\right) + \left(193\frac{kJ}{mol \cdot V}\right)(\Psi_{in} - \Psi_{out})$$

$$(\Psi_{in} - \Psi_{out}) = -0.0922\ V = -92.2\ mV$$

Since $(\Psi_{in} - \Psi_{out})$ is < 0, the inside of the organelle must be negative (by 92.2 mV) with respect to the outside.

12. To maintain fluidity at lower temperature, the membranes of cells grown at 25 °C would have
    (i) shorter fatty acids with
    (ii) a greater number of *cis* double bonds compared to the membranes of cells grown at 37 °C.

15. **(a)** This process comprises two parts, (1) the transport "reaction" and (2) the phosphorylation reaction:

glucose$_{out}$ → glucose$_{in}$
glucose$_{in}$ + ATP → G6P + ADP

The phosphorylation reaction effectively reduces the [glucose$_{in}$] concentration by converting it to G6P. In terms of Le Chatelier's principle, glucose$_{in}$ is being removed from the product side of the transport "reaction" and that drives the transport process to the left. Note that G6P is not transported across the plasma membrane; thus, once glucose is converted to G6P it is effectively sequestered inside the cell (until the phosphoryl group is hydrolyzed, at which point the glucose could be exported).

(b) The maximum value of [G6P]$_{in}$ will be achieved when ΔG for the glucose phosphorylation reaction inside the cell is zero (i.e., when the thermodynamic driving force goes from favorable (<0) to zero). If we assume the concentrations of free glucose outside and inside the cell are equal (i.e., 5 mM), the theoretical maximum value of [G6P]$_{in}$ = 313 M. This absurdly high value is never achieved, as G6P is rapidly consumed in other metabolic reactions (see Chapter 12).

Initially, we need to calculate a value for $\Delta G^{\circ\prime}$ for the coupled processes:

ATP + H_2O	→	ADP + P_i + H^+	$\Delta G^{\circ\prime} = -32.2$ kJ/mol
glucose$_{in}$ + P_i	→	G6P$_{in}$ + H_2O	$\Delta G^{\circ\prime} = +13.8$ kJ/mol
NET: glucose$_{in}$ + ATP	→	G6P$_{in}$ + ADP + H^+	$\Delta G^{\circ\prime} = -18.4$ kJ/mol

$$\Delta G = 0 = \Delta G^{\circ\prime} + RT\ln Q = -18.4\frac{kJ}{mol} + RT\ln\left(\frac{(ADP)(G6P)(H^+)}{(ATP)(glucose)}\right)$$

$$e^{\left(\frac{18.4\frac{kJ}{mol}}{0.008314\frac{kJ}{mol\cdot K}(310K)}\right)} = e^{\ln\left(\frac{(0.00015)(G6P)(0.631)}{(0.0047)(0.005)}\right)}$$

$e^{7.139} = 1260 = 4.03\,(G6P)$ Solving for [G6P] gives: [G6P] = 313 M

16. (a) The pH inside = 8.2 and outside pH = 7.4. Equation 10.3 applies:

$$\Delta G = RT\ln\left(\frac{[H^+]_{in}}{[H^+]_{out}}\right) + zF\Delta\Psi = \left(0.008314\frac{kJ}{mol\cdot K}\right)$$

$$(310\,K)\ln\left(\frac{10^{-1.2}}{10^{-0.4}}\right) + (+1)\left(96.5\frac{kJ}{mol\cdot V}\right)(-0.168\,V)$$

$$\Delta G = \left(-4.75\frac{kJ}{mol}\right) + \left(-16.2\frac{kJ}{mol}\right) = -21.0\frac{kJ}{mol}$$

(b) $ADP + P_i + 3H^+_{out} \longrightarrow ATP + H_2O + 2H^+_{in}$

(c) Equation 10.6 applies because ATP synthesis is coupled to H^+ transport:

$$\Delta G = \Delta G^{\circ\prime} + RT\ln\left(\frac{[ATP][H_2O][H^+]^2_{in}}{[ADP][P_i][H^+]^3_{out}}\right) + nzF\Delta\Psi$$

$$-11.7\frac{kJ}{mol} = +32.2\frac{kJ}{mol} + RT\ln\left(\frac{[0.0027][1][0.0631]^2}{[ADP][0.0052][0.398]^3}\right)$$

$$+ (3)(+1)\left(96.5\frac{kJ}{mol\cdot V}\right)(-0.168\,V)$$

$$\left(-11.7\frac{kJ}{mol}\right) = +\left(-32.2\frac{kJ}{mol}\right) + \left(+48.6\frac{kj}{mol}\right)$$

$$= +4.736\frac{kJ}{mol} = RT\ln\left(\frac{0.0328}{(ADP)}\right)$$

$$e^{\left(\frac{4.736\frac{kJ}{mol}}{0.008314\frac{kJ}{mol\cdot K}(310K)}\right)} = e^{1.838} = e^{\ln\left(\frac{0.0328}{(ADP)}\right)}$$

$$6.28 = \left(\frac{0.0328}{(ADP)}\right)$$

[ADP] = 0.0052 M or 5.2 mM.

18. Make a mutant Lep which substitutes non-charged residues for the (+) charged side chains in the loop, and put (+) charged side chains in terminal positions. If the inside-positive rule applies, the mutant ought to have the reversed orientation in the membrane.

CHAPTER 11

1. (a) 0.67 (b) 1.0 (c) 0.69 (d) 0.71 (e) 0.72
2. (a) >1 (b) <1
3. (a) 0 (b) 1 (c) 4
6. $\Delta G = \Delta G^{\circ\prime} + RT\ln([ADP][P_i]/[ATP])$

$$= -32200\,J/mol + (8.315\,J/mol\cdot K)(310\,K)\ln(1\times10^{-3}\,M)$$
$$(1\times10^{-3}\,M)/(3\times10^{-3}\,M)$$
$$= -32200\,J/mol -20638\,J/mol$$
$$= -52838\,J/mol = -52.84\,kJ/mol$$

10. (a) $K = e^{\left(\frac{-\Delta G^{\circ\prime}}{RT}\right)} = e^{\left(\frac{16700\,J/mol}{(8.315\,J/mol\cdot K)(298\,K)}\right)}$

$$= 845 = \frac{[ADP][glucose\text{-}6\text{-}phosphate]}{[ATP][glucose]}$$

Since [ATP] = [ADP]: [glucose-6-P]/[glucose] = 845 under physiological conditions.

(b) K = [glucose][P_i]/[glucose-6-P] = 262
[glucose]/[glucose-6-P] = 262/(1 × 10⁻³ M) = 262,000

(c) Any condition where the [glucose]/[glucose-6-P] ratio is between 262,000 and 0.0012 (1/845)

(d) The relative concentrations of allosteric regulators that exert kinetic control over the enzymes in the opposing pathways.

12. (a) First, calculate the volume of the mitochondrion in L:
Vol of mitochondrion (1st calculate in cm³ = ml): vol of cylinder
$= V = \pi r^2 h = \pi(0.3\times10^{-4}\,cm)^2(1.5\times10^{-4}\,cm) =$
$4.25\times10^{-13}\,cm^3 = 4.25\times10^{-13}\,ml = 4.25\times10^{-16}\,L$
Citrate synthase is present at 1 μM = 1 × 10⁻⁶ M:
thus 1 × 10⁻⁶ mol/L × 4.2 × 10⁻¹⁶ L = 4.2 × 10⁻²² mol
(4.2 × 10⁻²² mol) (6.02 × 10²³ molecules/mol) = 253 molecules citrate synthase in a mitochondrion

(b) Diameter of single spherical citrate synthase molecule = 10 nm = 1 × 10⁻⁶ cm
Volume of single spherical citrate synthase molecule:
$$V = \frac{4}{3}\pi r^3 = \frac{4}{3}\pi(0.5\times10^{-6}\,cm)^3 = \frac{4}{3}\pi(1.25\times10^{-19}\,cm^3) =$$
$5.24\times10^{-19}\,cm^3 = 5.24\times10^{-19}\,ml = 5.24\times10^{-22}\,L$

Total volume occupied by 253 molecules = 5.24 × 10⁻²² L × 253 = 1.32 × 10⁻¹⁹ L; which is 3 × 10³ less than the volume of the mitochondrion, so this is a reasonable answer.

15. (a) 20%
(b) 1.5 × 10⁶ molecules/min/cell
(c) Prepare an acid extract of the cells (e.g., 5% trichloroacetic acid), separate the nucleotides by ion-exchange HPLC, and determine in the dTTP fraction its radioactivity and its mass, the latter from UV absorbance.

CHAPTER 12

1. $\Delta G = \Delta G^{\circ\prime} + RT\ln([FBP][ADP]/[F6P][ATP])$. With the concentrations given, $\Delta G = \Delta G^{\circ\prime}$, since the natural log term is zero. Therefore, the reaction is neither more nor less exergonic in muscle than under standard conditions.
4. (a) 4 (2 each from glucose and fructose).
(b) 5 (3 from G6P + 2 from fructose).
5. C-3 or C-4. Both become C-1 of pyruvate, which is lost as CO_2 in the pyruvate decarboxylase reaction.
7. Glyceraldehyde-3-phosphate dehydrogenase. The acyl arsenate analog of 1,3-bisphosphoglycerate spontaneously hydrolyzes.
8. From the creatine phosphate stores in the muscle:
creatine phosphate + ADP ⇌ ATP + creatine (catalyzed by muscle creatine kinase) (see Section 11.4)

13.
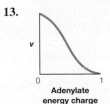

14. The two Mg^{2+} ions neutralize the negative charges on the substrate and lower the pK_a of the C-2 proton, making it easier to abstract.
16. (a) 2DHAP ⟶ F1,6BP ⟶ F6P ⟶ G6P ⟶ G1P →1ATP
UDPG ⟶ glucosyl residue
(b) 1.
(c) 7 (see Figure 12.10).
(d) 1.
17. (a) 2 generated (glycolysis).
(b) 6 consumed (gluconeogenesis requires ATP).
18. (a) because lactate is the only substrate that must go through the biotin-requiring pyruvate carboxylase reaction.
19. None. The CO_2 that is fixed comes off in the PEP carboxykinase reaction.
20. (a) Fructose-6-phosphate + ATP ⟶ fructose-2,6-bisphosphate + ADP
(b) 2 Oxaloacetate + 2ATP + 2GTP + 2NADH + 2H⁺ + 4H₂O ⟶
glucose + 2CO₂ + 2NAD⁺ + 2ADP + 2GDP + 4P_i

(c) Glucose + ATP + UTP $\longrightarrow$ UDP-Glc + PP$_i$ + ADP

(d) 2Glycerol + 2ATP + 2NAD$^+$ + 2H$_2$O $\longrightarrow$
        glucose + 2ADP + 2NADH + 2H$^+$ + 2P$_i$

(e) 2Malate + 2ATP + 2GTP + 3H$_2$O $\longrightarrow$
     glucose-6-phosphate + 2CO$_2$ + 2ADP + 2GDP + 3P$_i$

24. AMP activation of glycogen phosphorylase *b*; glucose-6-P activation of glycogen synthase *b*.

28. (a) G6P + 2NADP$^+$ + H$_2$O $\longrightarrow$ R5P + CO$_2$ + 2NADPH + 2H$^+$

 (b) G6P + P$_i$ + 12NADP$^+$ + 7H$_2$O $\longrightarrow$ 6CO$_2$ + 12NADPH + 12H$^+$

29. C-1 and C-3 of fructose-6-phosphate should be labeled. Erythrose-4-phosphate should be unlabeled.

33. (a) You will not be able to find a value for the $\Delta G^{\circ\prime}$ for the dephosphorylation of phosphorylated Ser-14 listed anywhere. However, phosphorylated serine is a simple phosphate ester, similar to that in glycerol-3-phosphate, for which a $\Delta G^{\circ\prime}$ value for hydrolysis does exist (see Figure 3.7). This value, -9.2 kJ/mol, would be a pretty good estimate for the $\Delta G^{\circ\prime}$ for the dephosphorylation of phosphorylated Ser-14.

 (b) In order to estimate $\Delta G^{\circ\prime}$ for this reaction, you need to calculate the difference in the free energies of ATP versus the phosphorylated serine residue. You can look up the $\Delta G^{\circ\prime}$ for ATP hydrolysis (-32.2 kJ/mol in Table 3.5), and use the estimate for dephosphorylation of phosphorylated serine from part a (-9 kJ/mol). Thus, a good approximation of the $\Delta G^{\circ\prime}$ for the phosphorylation of Ser-14 would be: -32.2 kJ/mol $- (-9$ kJ/mol) $= -23.2$ kJ/mol

 (c) Yes. Certainly both reactions occur with large negative free energy changes under standard conditions, and probably even more so under cellular conditions. And for a regulatory process to be physiologically useful, both directions must be thermodynamically favorable under cellular conditions.

CHAPTER 13

2. C-1: all released as CO$_2$. C-2 and C-3: all retained in oxaloacetate.

3. First turn: one-quarter. Second turn: three-eighths.

4. The action of pyruvate carboxylase on the labeled pyruvate would yield oxaloacetate labeled such that, when these carbons proceed through the citric acid cycle, C-5 of isocitrate would be labeled.

5. Addition to isolated glyoxysomes of citrate, isocitrate, glyoxylate, malate, or oxaloacetate would stimulate succinate formation out of proportion to the amount added.

6. C-3 and C-4, since these become the carboxyl group of pyruvate, which is lost in the pyruvate dehydrogenase reaction.

12. Step 1: The decarboxylation step is metabolically irreversible since the CO$_2$ product diffuses away from the enzyme, and does not rebind to any significant extent. The other 4 reactions are transfer reactions or redox reactions that are easily reversed.

14. (a) 0.08   (b) No, because it catalyzes a freely reversible reaction.

15.

17. 2 acetyl-CoA + 2NAD$^+$ + E-FAD + 3H$_2$O $\longrightarrow$ oxaloacetate + 2NADH + E-FADH$_2$ + 2CoA-SH + 4H$^+$

19. Metabolic reactions in Chapters 12 and 13 that form or break C—C bonds:

Reaction	Is the bond α or β to carbonyl?	Use TPP?
Aldolase	β	No
Pyruvate decarboxylase	α	TPP
Pyruvate carboxylase	β	No
PEPCK	β	No
6-Phosphogluconate dehydrogenase	β	No
Transketolase	α	TPP
Transaldolase	β	No
Pyruvate dehydrogenase	α	TPP
Citrate synthase	β	No
Isocitrate dehydrogenase	β	No
α-Ketoglutarate dehydrogenase	α	TPP
PEP carboxylase	β	No
Malic enzyme	β	No
Isocitrate lyase	β	No
Malate synthase	β	No

When the C—C bond formation or cleavage is between the carbonyl and the α carbon, TPP is used. When the bond formation or cleavage is between α and β carbons, TPP is never used. C—C bond formation or cleavage involves carbanion intermediates that must be stabilized by resonance. For $\alpha - \beta$ reactions, the necessary resonance stabilization occurs via formation of a Schiff base. When the C—C bond is between the carbonyl and the α carbon, the carbanion cannot be stabilized by resonance. These reactions require TPP in which the thiazole ring forms the resonance-stabilized carbanion by acting as an electron sink (see Figure 12.35). We see the same situation in decarboxylations. β-keto acids can stabilize the carbanion transition state that develops during decarboxylation. α-Keto acids cannot; they require TPP — the thiazole ring stabilizes the carbanion intermediate through resonance (see Fig. 13.11).

20. [NAD$^+$]/[NADH] should be high, so that it can promote the oxidation of substrates, e.g., malate + NAD$^+$ $\rightleftharpoons$ oxaloacetate + NADH + H$^+$. Conversely, since NADPH and NADP$^+$ usually promote reduction of substrates, we expect [NADP$^+$]/[NADPH] to be low.

CHAPTER 14

1. $+28.95$ kJ/mol—from the equation $\Delta G^{\circ\prime} = -nF\Delta E^{\circ\prime}$.
$\Delta E^{\circ\prime} = -0.17 - (-0.32)$.

3. (a) β-hydroxybutyrate $\longrightarrow$ NADH $\longrightarrow$ complex I $\longrightarrow$ CoQ $\longrightarrow$ complex III $\longrightarrow$ cytochrome *c*

 (b) 2, because cytochrome *c* oxidase is bypassed.

 (c) Because NADH cannot freely enter the mitochondrion.

 (d) To block cytochrome oxidase, so that electrons exit the chain at cytochrome *c*.

 (e) β-hydroxybutyrate $+2$ cyt c_{ox} + 2 ADP + 2P$_i$ + 4H$^+$
     $\longrightarrow$ acetoacetate + 2 cyt c_{red} + 2ATP + 2H$_2$O

 (f) -53.8 kJ/mol β-hydroxybutyrate (calculated as in Problem 2).

6. $+67.6$ kJ/mol (solved as in Problem 1). Minimum [NAD$^+$]/[NADH] ratio $= 2.5 \times 10^{10}$, calculated from equation 3.23 (Chapter 3):

$$\Delta G = \Delta G^{\circ\prime} + RT \ln\left(\frac{[\text{fumarate}][\text{NADH}]}{[\text{succinate}][\text{NAD}^+]}\right)$$

where $\Delta G < 0$.

8. 1.8×10^{17}. Calculate $\Delta G^{\circ\prime}$ as in Problem 1, and then apply $\Delta G^{\circ\prime} = -RT \ln K_{eq}$. Use $E^{\circ\prime} = +0.30$ V for H_2O_2 (see Table 14.1).

10. Because the energy not used for ATP was dissipated as heat, and the subjects developed uncontrollable fevers.

11. 1.

16. (a) $\Delta G^{\circ\prime} = -nF\Delta E^{\circ\prime} = -2(96485)(0.82 - 0.03) = -151$ kJ/mol
   (b) For protons (equation 14.5, p. 466):
   $\Delta G^{\circ\prime} = 2.3RT\Delta pH + F\Delta\Psi = 24.9$ kJ/mol protons.
   $151/24.9 = 6$ protons maximum
   (c) Complexes III and IV

18. (a) A, 1; B, 2; C, 3; D, 4; E, 5.
   (b) 3 divided by 4 (ratio of uptakes in presence and absence of ADP, with substrate present for both measurements).
   (c) 5; O_2 depleted.
   (d) 4, ADP level low because ATP level is high
   (e) 3; rapid ATP production and turnover demand rapid and continuous O_2 uptake.

20. Because the electrical component ($\Delta\Psi$) makes a greater contribution to the protonmotive force than the pH gradient (ΔpH) (see eqn. 14.5), valinomycin would be expected to be a more effective uncoupler than nigericin.

21. (a)
$$\Delta G = RT \ln\left(\frac{[Ca^{2+}]_m}{[Ca^{2+}]_c}\right) + zF\Delta\psi$$

$$0 = RT \ln\left(\frac{[Ca^{2+}]_m}{[Ca^{2+}]_c}\right) + (2)(96485 \text{ J/V}\cdot\text{mol})(-0.18 \text{ V})$$

$$(2578 \text{ J/mol}) \ln\left(\frac{[Ca^{2+}]_m}{[Ca^{2+}]_c}\right) = 34735 \text{ J/mol}$$

$$\ln\left(\frac{[Ca^{2+}]_m}{[Ca^{2+}]_c}\right) = 13.47$$

$$\left(\frac{[Ca^{2+}]_m}{[Ca^{2+}]_c}\right) = e^{13.47} = 7.1 \times 10^5$$

   (b) If $[Ca^{2+}]_c = 10^{-7}$ M, and $\left(\frac{[Ca^{2+}]_m}{[Ca^{2+}]_c}\right)$
   $= 7.1 \times 10^5$, then at equilibrium
   $[Ca^{2+}]_m = (10^{-7}\text{M})(7.1 \times 10^5) = 7.1 \times 10^{-2}$ M

This value for matrix calcium concentration (71 mM) is incompatible with mitochondrial integrity, and much higher than experimentally determined values. This discrepancy is explained by the existence of Ca^{2+} efflux pathways (Na^+/Ca^{2+} exchanger and H^+/Ca^{2+} antiporter). These oppose the tendency of Ca^{2+} to reach electrochemical equilibrium, resulting in a $\left(\frac{[Ca^{2+}]_m}{[Ca^{2+}]_c}\right)$ ratio closer to 100.

CHAPTER 15

1. $\Delta G^{\circ} = 164$ kJ/mol. An einstein of 700 nm light would yield 171 kJ.

2. 14.7%.

5. Assuming 48 photons per mole of hexose, the leaf could theoretically produce 0.0263 mol, or 4.73 g, of hexose in 1 hour. It will, in fact, produce only a small fraction of this, for not all photons are absorbed, nor do all absorbed photons serve to pass electrons through the photosynthetic pathway.

6. (a) No, because plastoquinones are not involved in this process.
   (b) Addition of ferricyanide as an electron donor allows a Hill reaction.

8. Carbon 3.

9. Competitive.

12. (a) Because both O_2 and CO_2 are being consumed and produced by the opposing processes of photorespiration and photosynthesis, a steady-state ratio will be attained.
   (b) The relative affinity of rubisco for CO_2 and O_2.

15. Compound 7 acts at photosystem II (PSII), most likely at the oxygen-evolving complex (OEC). PSI is not a target, since electron transport from DCPIP to MV was not significantly affected. The most pronounced inhibitory effect was on electron transport from water to DCBQ and water to SiMo, both PSII partial reactions involving OEC. The DPC to DCPIP reaction was partially inhibited, suggesting that compound 7 also inhibits PSII between P_{680} and Q_B.

CHAPTER 16

1. Palmitic acid, 106 ATP; stearic acid, 120 ATP; linoleic acid, 116 ATP; oleic acid, 118.5 ATP.

2. 36%.

3. 336.5 (106 from each palmitate, 18.5 from glycerol). 336.5 ATP/51 carbons $= 6.6$ ATPs per carbon atom (5.3 ATPs per carbon atom for glucose).

7. (a) Acetoacetate $+$ NADH $+$ H$^+$ $\rightleftharpoons$ β-hydroxybutyrate $+$ NAD$^+$
   (b) $E^{\circ\prime}$
   Acetoacetate $+$ 2H$^+$ $+$ 2e$^-$ $\rightleftharpoons$ β-hydroxybutyrate -0.35 (Acceptor)
   NAD$^+$ $+$ 2H$^+$ $+$ 2e$^-$ $\rightleftharpoons$ NADH $+$ H$^+$ -0.32 (Donor)
   $\Delta G^{\circ\prime} = -nF\Delta E^{\circ\prime}$ $\Delta E^{\circ\prime} = E^{\circ\prime}_{Acc} - E^{\circ\prime}_{Don} = -0.35 - (-0.32) = -0.03\text{V}$
   $\Delta G^{\circ\prime} = (-2)(96480 \text{ J/V}\cdot\text{mol})(-0.03\text{V}) = 5789 \text{ J/mol} = 5.8 \text{ kJ/mol}$
   Positive $\Delta G^{\circ\prime}$ indicates that reaction is not favorable under standard conditions. Removal of product (β-hydroxybutyrate) by subsequent reactions would allow the reaction to continuously produce β-hydroxybutyrate at a high rate.

8. Carnitine acyltransferase I. If inhibition occurred at a later step, then palmitoylcarnitine oxidation, as well as that of palmitoyl-CoA, would be inhibited.

9. The acyl-ACP produced by one subunit undergoes the next round of reductive two-carbon addition on the other subunit.

11. Propionyl-CoA $\rightarrow$ methylmalonyl-CoA $\rightarrow$ succinyl-CoA $\rightarrow$ oxaloacetate $\rightarrow$ PEP $\rightarrow$ glucose

13. 7 (during each cycle, one tritium atom is lost at the 3-hydroxyacyl-ACP dehydratase step, and the resulting double bond is reduced by unlabeled NADPH).

16. Increase in citrate levels would increase generation of acetyl-CoA in cytosol, hence stimulating fatty acid synthesis.

19. This could be a way for a cell to inhibit fatty acid synthesis under conditions where substrates are needed for oxidation, to provide ATP.

22. Substitution of vegetable fats for animal fats could decrease cholesterol levels. This would ultimately decrease inhibition of HMG-CoA reductase levels by cholesterol, which could result in increased mevalonate levels.

24. None, since the carboxyl C is lost in conversion to the isopentenyl pyrophosphate intermediate.

26. Acetoacetate $+$ succinyl-CoA $\rightarrow$ acetoacetyl-CoA $+$ succinate
   acetoacetyl-CoA $+$ acetyl-CoA $\rightarrow$ HMGCoA $+$ CoASH
   HMGCoA $+$ 2NADPH $+$ 2H$^+$ $\rightarrow$ mevalonate $+$ 2NADP$^+$ $+$ CoASH

27. By shutting down the pathway leading to aldosterone, this deficiency increases the supply of progesterone available for conversion to sex steroids.

29. Cyclic AMP promotes triacylglycerol breakdown, through activation of hormone-sensitive lipase. This probably increases intracellular levels of diacylglycerol, which could in turn increase flux through the last reactions in the salvage pathways to PE and PC (see Figure 16.30).

31. bTSH treatment leads to dephosphorylation of HMG-CoA reductase (decreases the ratio of phosphorylated to total HMG-CoA reductase). HMG-CoA reductase is inhibited in some tissues (including liver) by AMPK-dependent phosphorylation. bTSH could either inhibit AMPK, or stimulate a type 2A protein phosphatase, leading to dephosphorylation, and thus activation, of HMG-CoA reductase, resulting in stimulation of cholesterol synthesis.

33. **(a)** For a 20-hr flight, 58 kJ are required. The complete oxidation of fat yields 37 kJ/gm, so 1.5 gm of fat is required.

(b) Carbohydrate yields 17 kJ/gm, so 3.4 gm of glycogen is required. However, the hydrophilic nature of glycogen requires that about 2 gm of water is bound per gm of glycogen stored, so the total weight increase is $3.4 + 6.8 = 10.2$ gm.

CHAPTER 17

1. **(a)** www.nhlbi.nih.gov/health/public/heart/obesity/wecan/healthy-weight-basics/balance.htm lists the average calorie requirement for a moderately active 20-year old female at 2000–2200 calories per day. Using 2100 Cal, the caffé mocha (260 Cal) and scone (460 Cal) represent 720 /2100 × 100% = 34% of your recommended daily caloric intake.

(b) $720 \text{ Cal} \times \dfrac{4.184 \text{ kJ}}{1 \text{ Cal}} = 3012 \text{ kJ}$

4. About 28% at rest, about one-tenth that value during a marathon.

6. **(a)** $\dfrac{2.2 \text{ mmol}}{\text{L}} \times 5 \text{ L} \times \dfrac{1 \text{ mol}}{10^3 \text{ mmol}} \times \dfrac{180 \text{ gm}}{\text{mol}} = 1.98 \text{ gm}$

(b) In liver, hexokinase IV velocity would increase from ~50% V_{max} to ~60% V_{max}. In muscle, hexokinase I is already operating at V_{max} even at very low glucose concentrations, so the velocity would not increase in this tissue.

9. Adipose tissue lacks glycerol kinase. Glycolysis generates dihydroxyacetone phosphate, which is reduced to glycerol 3-phosphate (see Chapter 16).

10. **(a)** Malonyl-CoA at high levels inhibits carnitine acyltransferase I, and this inhibits ketogenesis by blocking the transport of fatty acids into mitochondria, both for β-oxidation and for ketogenesis.

(b) The high $K_{0.5}$ (half saturation concentration) of hexokinase IV, a liver-specific enzyme, allows the liver to control the rate of glucose phosphorylation over a wide range of glucose concentrations. Accumulation of glucose-6-phosphate activates the *b* form of glycogen synthase and promotes glycogen deposition. By several mechanisms, the liver also senses when blood glucose levels are low and mobilizes its glycogen reserves accordingly (see Chapter 12).

11. Phosphorylation of pyruvate kinase by cyclic AMP–dependent protein kinase. The phosphorylated form of the enzyme is far less active than the dephosphorylated form (see Chapter 13).

13. **(a)** Isocitrate amplifies glucose-stimulated exocytosis (panel b: 1 mM glucose + isocitrate gives higher exocytosis than even 10 mM glucose alone). α-KG has no effect on glucose-stimulated exocytosis (panels c,d: compare 1 mM glucose $\pm$ α-KG).

(b) NADPH amplifies glucose-stimulated exocytosis, but only when NADPH is in excess over NADP$^+$ (panel e). NADH has no effect (panel f).

(c) The isocitrate effect (panels b,d) is mimicked by NADPH (panel e). Unlike the mitochondrial isocitrate dehydrogenase, the cytoplasmic isocitrate dehydrogenase is NADP-specific. The data suggest that the cytoplasmic isocitrate dehydrogenase oxidizes isocitrate, producing NADPH that then stimulates glucose-stimulated exocytosis.

(d) As described in Chapter 12, glutathione is a major protective mechanism against oxidative stress. In this process, GSH is oxidized to GSSG. The protective GSH is regenerated by glutathione reductase, which requires NADPH as a reductant. These data suggest the following metabolic pathway: pyruvate is converted in mitochondria to oxaloacetate (OAA) via pyruvate carboxylase (Chapter 13). OAA is then converted to isocitrate via the 1st two reactions of the citric acid cycle. Isocitrate is transported into the cytoplasm where it is oxidized to α-KG via the NADP-specific cytoplasmic isocitrate dehydrogenase, producing NADPH. The high NADPH/NADP$^+$ ratio maintains glutathione in its reduced form, GSH.

CHAPTER 18

1. 5'-Deoxyadenosyl-B$_{12}$;tetrahydrofolate;ATP + glutamine;α-ketoglutarate + pyridoxal phosphate; S-adenosylmethionine.

2. **(a)** 5-Methyltetrahydrofolate.

(b) Methylmalonate.

(c) Decreased affinity of enzyme C for B$_{12}$ coenzyme (K$_M$ mutant).

(d) Homocysteine (or homocystine).

6. CPS I is in mitochondria, and CPS II is in cytosol. Apparently, carbamoyl phosphate cannot cross the mitochondrial membrane, so the carbamoyl phosphate that is formed in mitochondria can be used only for arginine synthesis, and that formed in cytosol is used only for pyrimidine synthesis.

7. **(a)** True.

(b) False.

(c) False.

(d) False.

8. A N content of 16% (w/w) equates to:

$$\dfrac{1 \text{ g protein}}{0.16 \text{ g N}} = \dfrac{6.25 \text{ g protein}}{\text{g N}}$$

To replace 120 mg N/kg body weight:

$$\dfrac{0.12 \text{ g protein}}{\text{kg per day}} \times \dfrac{6.25 \text{ g protein}}{\text{g N}} \times 70 \text{ kg} = 5.25 \text{ g protein per day}$$

11. PLP forms a covalent Schiff base between the aldehyde carbon of the coenzyme and the ε-amino group of a lysine residue in the active site of the enzyme. Obviously, this bond must be broken for the coenzyme to form a Schiff base with an amino acid substrate.

12. α-Aminoadipic semialdedhyde is first oxidized to the acid, and then transaminated to remove the α-amino group. The remainder of the pathway follows the degradation of leucine in Fig. 18.14, except that an extra decarboxylation is necessary to remove the terminal carboxyl group: oxidative decarboxylation, acyl-CoA dehydrogenation, decarboxylation, dehydration, dehydrogenation, acetylation, and finally cleavage to give acetoacetate + acetyl- CoA.

14.

Dihydrofolate	GTP
N-1	N-3
C-2	C-2
C-4	C-6
N-5	N-7
C-7	C-1'
N-8	N-9
C-9	C-3'

15. Formiminoglutamate, because the next reaction in its catabolism requires tetrahydrofolate.

17. **(a)** B. **(b)** C. **(c)** D. **(d)** E, methionine. **(e)** A. **(f)** F. **(g)** C.

24. Because the same enzymes are involved in comparable steps of both isoleucine and valine biosynthesis. Threonine dehydratase.

26. Because a pteridine reductase deficiency would impair all tetrahydrobiopterin-dependent reactions, which include the synthesis of catecholamines, serotonin, and nitric oxide, as well as tyrosine.

CHAPTER 19

3. **(a)** 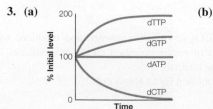 **(b)**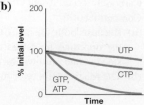

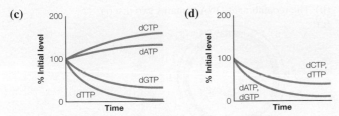

(a) Thymidine causes dTTP to accumulate via salvage synthesis. dTTP activates GDP reduction and inhibits CDP reduction by ribonucleotide reductase, causing secondary effects on the dGTP and dCTP pools.

(b) Trimethoprim inhibits dihydrofolate reductase, which depletes the cell of tetrahydrofolate cofactors needed for purine nucleotide biosynthesis. An immediate effect upon pyrimidine ribonucleotide pools is not expected, so minimal changes are seen in UTP and CTP pools.

(c) Fluorodeoxyuridine forms FdUMP, which inhibits thymidylate synthase and causes dTTP to become depleted. Because dTTP is the prime allosteric activator for GDP reduction by ribonucleotide reductase, dGTP pools decline secondarily. dCTP and dATP might accumulate because allosteric inhibitors of CDP and ADP reduction are depleted.

(d) By inhibiting all four reactions catalyzed by ribonucleotide reductase, hydroxyurea causes all four dNTP pools to decline. In some cells the dGTP and dATP pools are small, so they decline rapidly as the result of continuing DNA synthesis.

6. Deoxycoformycin might lead to adenosine and deoxyadenosine accumulation, since their deamination would be inhibited, as in adenosine deaminase deficiency. This could cause dATP to accumulate, thereby inhibiting all four activities of ribonucleotide reductase.

8. (a) Either a deficiency of thymidylate synthase activity, measured in a cell-free extract, or a growth requirement of cells for thymidine would confirm that cells are deficient in thymidylate synthase.

(b) Because the mutants have negligible flux through the thymidylate synthase reaction because of the enzyme deficiency, they do not deplete their intracellular tetrahydrofolate coenzyme pools, which remain available for functions other than dTMP synthesis. The added thymidine satisfies the growth requirement for the mutants.

(c) Use methotrexate instead of trimethoprim for the selection, because trimethoprim is an ineffective inhibitor of mammalian dihydrofolate reductases (see Chapter 18).

12. Channeling refers to a metabolic pathway in which each intermediate is preferentially bound to the next enzyme in the pathway, as opposed to free diffusion into the surrounding medium. A classic way to test the possibility of channeling is to look for evidence of a facilitated multistep pathway. What is seen is that intermediates in a pathway accumulate only to a low level, because they are rapidly transferred to the next enzyme in a sequence. Also the product of a reaction sequence reaches its maximal rate of synthesis faster than in a nonchanneled pathway, because each intermediate need not diffuse to find the next enzyme and becomes rapidly bound to that enzyme. In other words, the "transient time," needed for the product to reach its maximal rate of formation, is shortened in a channeled system. Another test is to add one of the intermediates to a system carrying out the pathway. If the pathway is indeed channeled, then the added intermediate will not be incorporated into the product; it will not mix with the pool of the intermediate in the channeled pathway.

14. Uncontrolled conversion of UTP to CTP elevates pools of cytidine and deoxycytidine nucleotides, while pools of uridine and thymidine nucleotide pools are diminished. The depletion of endogenous thymidine nucleotides explains the growth requirement for exogenous thymidine,

and the perturbed dNTP pool imbalance (dCTP/dTTP pool ratio is elevated) causes DNA replication errors, principally C incorporated opposite A, that lead to mutations.

16. Intermediates that react with 10-formyl-THF would accumulate because the synthesis of folate coenzymes is inhibited. In purine nucleotide synthesis this would be glycinamide ribonucleotide (substrate for reaction 3) and 5-aminoimidzole-4-carboxamide ribonucleotide (substrate for reaction 9). In addition to these intermediates, glycine and deoxyuridine monophosphate might be expected to accumulate in sulfonamide–treated cells. However, mammalian cells do not synthesize tetrahydrofolate, so they may not be affected so long as required substances, particularly folate, are present in the medium.

20. Injected adenosine deaminase need not be taken up into cells in order to be effective because the substrates, adenosine and deoxyadenosine, are uncharged and, when they accumulate, can efflux from cells and undergo deamination in the extracellular space and then be taken up by cells. The other enzymes mentioned can only function within the cell or organelle.

23. What was done was to mutate the gene for the *E. coli* R2 protein, so that the codon for Tyr-122, the suspected critical Tyr, was changed to a phenylalanine codon. The mutant enzyme was purified as a recombinant protein and found to be completely inactive. In addition it did not contain a free radical when analyzed by electron paramagnetic spectroscopy.

24. Thymidine added to cell culture medium is taken up into cells and converted by salvage pathways to thymidine triphosphate. dTTP accumulation blocks DNA synthesis reversibly by allosteric inhibition of CDP reduction by ribonucleotide reductase. So thymidine-treated cells progress through the cell cycle until they reach S phase, where they are blocked because a deficiency of dCTP makes DNA synthesis impossible. Eventually, when all cells have become blocked in early S phase, the cells are transferred to thymidine-free medium, whereupon all of the cells begin DNA synthesis in synchrony and remain synchronized for at least one complete cell cycle.

25. The *mutT* gene product was shown to be a nucleotidase. Guanine can undergo oxidation either as a DNA base, 8-oxoguanine, or as a DNA precursor, 8-oxo-dGTP. The *mutT* enzyme was shown to hydrolyze 8-oxo-dGTP to 8-oxo-dGMP plus pyrophosphate, thereby preventing the incorporation of the oxidized guanine into DNA.

26. One can readily predict that SAMHD1 catalyzes dNTP breakdown, and in fact the enzyme was shown to hydrolyse each of the four dNTPs to the respective deoxyribonucleoside plus tripolyphosphate, PPP$_i$. Allosteric control could be predicted to involve stimulation by dNTPs, and in fact the enzyme was shown to be activated by dATP or dTTP in physiological range.

CHAPTER 20

1. Inhibiting cAMP breakdown would increase cAMP levels, which in turn would promote both glycogenolysis and lipolysis. Muscle function should be enhanced as long as blood glucose elevation persists. (Not covered in the chapter, but relevant: caffeine acts through adenosine receptors at low levels as well as through inhibition of cAMP phosphodiesterase at higher levels. Both effects tend to increase intracellular [cAMP].)

2. (a) Suppression of the hormonal signal until it is needed.

(b) Storage of the hormone to ensure availability when it is needed.

(c) Signal sequence on prohormone molecule could direct the protein to the correct subcellular site, then be cleaved off when no longer needed.

(d) Additional polypeptide sequence may promote correct folding of the polypeptide chain.

4. (a) 1, 4, 9, 14
   (b) 5, 6, 8, 10, 16
   (c) 1, 2, 7, 12, 16
   (d) 1, 4, 10, 15
6. (a) Viagra inhibits cyclic GMP phosphodiesterase, so cGMP levels remain high, thereby prolonging the effects of nitric oxide on regional blood flow.
   (b) Prozac selectively inhibits reuptake of serotonin by presynaptic neurons, thereby increasing the amount of the neurotransmitter that reaches the postsynaptic cell, stimulating synaptic transmission in areas related to regulation of mood.
   (c) If AdoMet could cross the blood-brain barrier, it might conceivably increase methylation of norepinephrine to give epinephrine. Alternatively, AdoMet could act through folate metabolism, increasing the pool of labile methyl groups (Chapter 17) or it could affect membrane fluidity through effects upon phospholipid metabolism (Chapter 16). Because all possible mechanisms are highly speculative, perhaps one should view the initial report with skepticism.
7. cGMP + H_2O $\longrightarrow$ 5'-GMP. Because Viagra acts by inhibiting this reaction, any other compound acting similarly should potentiate the effect of Viagra.
14. Protein fractionation, with a demonstration that epinephrine-binding activity and adenylate cyclase activity could be physically separated from one another was, in fact, what was done.
16. So long as GTPγS could substitute for GTP in the G protein-catalyzed exchange reaction, its effects should be similar or identical to those of GTP, and cAMP levels should increase. The difference is that the analog is not hydrolyzed, so stimulation of adenylate cyclase is lost much more slowly.
18. This would inactivate G_i, blocking the inhibition of adenylate cyclase and allowing cAMP to accumulate, with a corresponding increase in blood pressure.
20. As described on page 656, chronic myelogenous leukemia is associated with a chromosomal translocation, which generates a hybrid gene that encodes a novel protein kinase, which is inhibited by Gleevec. Presumably the novel protein kinase is involved in abnormal signaling pathways, which in turn are associated with abnormal growth control associated with the disease. Treatment with Gleevec inhibits these abnormal signaling pathways.

CHAPTER 21

1. Fractional abundance F = 10 kbp × 10^3 bp/kpb/10^{10} bp = 10^{-6}
   For PCR each cycle doubles the target sequence. Amplification = 2^n where n is the number of cycles.
   After 10 cycles $F = 2^{10} \times 1/(2^{10} + 10^{-6}) = 0.00102$
   After 15 cycles $F = 2^{15} \times 1/(2^{15} + 10^{-6}) = 0.0317$
   After 20 cycles $F = 2^{20} \times 1/(2^{20} + 10^{-6}) = 0.511$
3. (a), (b), (c)

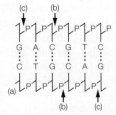

   (d) Cleavage would occur outside this site (up to 1 kb away).
   (e) A particular hexanucleotide sequence would be present on average at intervals of $1/4^6$, or once in 4.096 kbp.
4. (a) Relationships between DNA fragment size and electrophoretic mobility are based upon migration rate of linear molecules. Circularization and supercoiling make it impossible to compare migration rates with those of linear standards.

(b) The recombinant DNA contains two inserts.
(c)

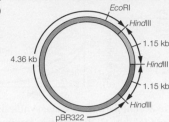

5. (a) Circular, because if linear, HpaII or HindIII would each have yielded two fragments.
   (b)

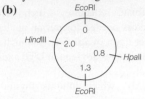

   (c) Fifteen if four possible locations for HpaII and four for HindIII and the sites are independent
   (d) Cleave with HpaII plus HindIII; possibly cleavage with additional restriction enzymes would be necessary.
6. (a) 6.6×10^2 g/mol × 10^8 bp ÷ 6.02×10^{23} molecules/mol = 1.1×10^{-13} g DNA/chromosome. Therefore, if the chromosome contains 2 g protein per gram DNA, then 1.1×10^{-13} g DNA/chromosome + 2.2×10^{-13} g protein/chromosome = 3.3×10^{-13} or 0.33 picogram total weight of one chromosome.
   (b) 10^8 bp × 3.4 Å/ bp × 10^{-8} cm/Å = 3.4 cm
   (c) 3.4×10^{-2} m/ 5×10^{-6} m = 6800
   (d) 46 chrom./cell × 4×10^{12} cells × 3.4×10^{-2} m/chromosome = 6.24×10^{12} m
7. If we take the average nucleosome (including linker) to contain about 200 bp, the 30-nm fiber has 6 × 200 bp/10 nm = 120 bp/nm
   DNA has 10 bp/3.4 nm = 2.94 bp/nm
   Compaction ratio is 120/2.94 = 40.8
   Since the metaphase chromosome has an overall compaction ratio of about 7000, the 30-nm fiber must undergo further rounds of folding upon itself.
17. (a)

   (b) The line labeled (b)
   (c) The genomic DNA would renature more slowly than cDNA because it has greater complexity among the 300-bp fragments used for the analysis.
20. Analysis of DNA satellite bands by DNA shearing followed by equilibrium gradient centrifugation would reveal the size of the satellite band to increase with the number of generations in which the family had been afflicted.

CHAPTER 22

2. (a) v (b) iv (c) vi (d) iii (e) ii (f) i
3. Similar in that it uses a nucleoside triphosphate as a substrate, with cleavage between the α and β phosphates providing the energy to drive the reaction. Different in that the adenylate cyclase catalyzes an

intramolecular esterification reaction, while DNA polymerase catalyzes a bimolecular reaction.

5. (a) Deoxyinosine monophosphate (dIMP)

   (b) dIMP would most likely be misread as dGMP, so that dCMP would be incorporated opposite dIMP, and the ultimate result would be conversion of an A-T base pair to a G-C pair.

6. (a) Replication rate = 1.28 mm/40 min. = 0.032 mm/min. = 32 μm/min.
   Replication involves 2 forks, so replication rate/fork = 32/2 = 16 μm/min.

   (b) bases/min. = 10.4 bases/turn $\times$ 1 turn/3.4 nm $\times$ 16 μm/min/fork $\times$ 2 strands/fork $\times$ 10^3 nm/μm = 97,882 bases/minute

   (c) 1.2 m/5 hr $\times$ 1 hr/60 min $\times$ 10^3 nm/mm = 4 mm/min.
   #forks = (4 mm/min/1.6 μm/min) $\times$ 103 μm/mm = 2500 forks
   2500 forks/2 forks/origin = 1250 replication origins

   (d) 10.4 base pairs/turn $\times$ 1 turn/3.4 nm $\times$ 1.2 $\times$ 10^9 nm = 3.67 $\times$ 10^9 base pairs. 3.67 $\times$ 10^9 base pairs/1250 origins = 2936 kilobase pairs (given that this number is derived from estimates, a more realistic answer would be 2700–3300).

9.
```
  /A—T—A\
 G          CTCGGGATTGGCATCTGGTGCTTATCGTAATCC-OH
 |          ||||||||||||||||||||||||||||||||||
 C          GAGCCCTAACCGTAGACCACGAATAGCATTAGG-p 5′
  \C—G/
```

Note. The 3′-exonuclease removes the mismatched A and T before polymerase action begins. Incorporated nucleotides are shown in italics.

11. Dideoxyinosine (ddI), as an uncharged nucleoside, is taken up into the cell and converted to the corresponding dideoxynucleoside 5′ triphosphate, ddITP. Incorporation of this nucleotide in place of dGTP would block further chain elongation because of the absence of a 3′ hydroxyl terminus. The HIV reverse transcriptase lacks a 3′-exonuclease, so it incorporates ddITP more readily than do the cellular replicative polymerases, causing replication of the viral genome to be selectively inhibited.

14. BU substitutes for T, opposite template A, in the first round of replication. BU is occasionally misread as C in the second round, inserting G. G templates C in the third round, converting an A-T pair to G-C.

A-T $\longrightarrow$ A-BU $\longrightarrow$ G-BU $\longrightarrow$ G-C
$\quad$ + $\quad\quad$ + $\quad\quad$ +
$\quad$ A-T $\quad\quad$ A-T $\quad\quad$ A-BU

15. G-C $\xrightarrow{\text{ROS}}$ oxoG-C $\longrightarrow$ oxoG-A $\longrightarrow$ T-A
$\quad\quad\quad\quad\quad$ + $\quad\quad\quad$ +
$\quad\quad\quad\quad$ G-C $\quad\quad$ oxoG-C

19. In the mitochondrial disease MNGIE (see page 709) a deficiency of thymidine phosphorylase causes dTTP to accumulate in mitochondria. Describe the mechanism by which this occurs. A secondary effect is depletion of mitochondrial dCTP pools. Describe a plausible mechanism for this effect.

Thymidine phosphorylase (TP) breaks down thymidine:

Thymidine + P_i $\rightarrow$ thymine + deoxyribose-1-phosphate

A TP deficiency would cause thymidine to accumulate, leading in turn to dTTP accumulation through salvage pathways from thymidine. dTTP is an allosteric inhibitor of CDP reduction by ribonucleotide reductase, so dTTP accumulation would inhibit CDP reduction, leading to dCTP pool depletion.

CHAPTER 23

2. (a) 7, 9, 11 (b) 3, 5, 6 (c) 2, 11 (d) 1, 11, 13 (e) 1, 8, 11
   (f) 4, 6 (g) 10, 11 (h) 11, 12

3. (a) A (b) B (c) C (d) D (e) A, C (f) C (g) E

4. Because both DNA strands are methylated and thereby identified as "correct"

5. dUTPase deficiency increases dUTP incorporation into DNA and increases subsequent excision repair. Ligase deficiency increases the mean lifetime of Okazaki fragments. Both conditions increase the number of single-strand interruptions, the structural features that stimulate recombination.

10. By inserting within a gene, the transposon would interrupt expression of the gene. By inserting into a regulatory region, the result could be either enhanced or inhibited expression of the gene, depending on the nature of the regulatory region. Depending on whether there are promoters in the transposon that read outward and cause transcription in the neighborhood, there could be effects on the transcription of neighboring genes.

12. G-C $\xrightarrow{\text{ROS}}$ 8-oxoG-C $\longrightarrow$ 8-oxoG-A $\longrightarrow$ T-A
$\quad\quad\quad\quad\quad\quad\quad\quad$ + G-C $\quad\quad$ + 8-oxoG-C

13. DNA-5mC + H_2O $\rightleftharpoons$ DNA-thymine + NH_3

18. The scheme in Figure 23.8 shows a guanine amino nitrogen initiating a nucleophilic attack on an epoxide carbon. The attacking nitrogen must be unprotonated to serve as a nucleophile. Among the nucleic acid bases guanine has the lowest pK_a on its amino-N, and, hence, is most likely to be unprotonated at physiological pH.

CHAPTER 24

5. Because it uses the -35 region of the *trp* promoter and the -10 region of the *lacUV5* promoter, the *tac* promoter more closely resembles the consensus -35 and -10 sequences than does either *trp* or *lac*. *tac* has 12 out of 12 identities with the consensus sequences, whereas *lac* has 9 and *trp* has 8.

8. The oxygen atom at the 2′ position of ribose makes a ribonucleotide molecule slightly heavier than a deoxyribonucleotide molecule, while the volumes occupied are essentially the same. CsCl gradients separate on the basis of density (mass/volume, see Chapter 4), so a molecule made up of ribonucleotides has a slightly, but significantly, higher density than a molecule made up of deoxyribonucleotides.

9. (a) seven exons, six introns

   (b) The 5′, 5′-internucleotide bond would be at the 5′ end, where capping occurs. The poly(A) sequence would be at the 3′ end—at the right end of the figure, where the RNA product does not anneal to the DNA template because of the absence of a complementary sequence.

10. The ribonucleoprotein machinery needed to recognize the two splice junctions and the lariat site probably cannot be assembled if the junctions are too close together.

11. Once the transcription complex is formed, it is extremely stable, and inaccessible to heparin.

12. A 50-kDa protein would contain about 450 amino acid residues, and therefore would be encoded by a gene containing 450 $\times$ 3 or 1350 nucleotides. If V_{max} for RNA polymerase is 50 nucleotides per second, this gene could be transcribed in 27 seconds, assuming that the gene contains no introns (highly unlikely in a bacterial gene). Also the time needed for initiation and formation of a stable transcription complex would lengthen the time needed. Pausing within the gene could lengthen the time needed for transcription of the gene by an unknown amount.

13. The choice to backtrack must be more likely if a mistake has been made, probably because of kinetic partitioning: if *forward* translocation slows down, the *reverse* process will occur more often. Just as we have seen with the 3′ to 5′ exonuclease activity of DNA polymerases, the main fidelity gain probably comes from slow extension of a mismatch

(increased K_M for extension from a mismatch). The backtracking activity allows the polymerase to make a second attempt instead of just falling off. Because the probability of repeating an error is low, the second attempt to transcribe that same stretch of DNA has a high probability of transcribing it accurately.

15. Important evidence involved footprinting analysis, to show that RNA polymerase binds tightly to DNA in the vicinity of these consensus sequences. Also, mutations engineered in the site under study should alter (usually decrease) the rate of transcription of adjacent genes. Finally, there are methods (S1 mapping, not described in this book), which allow the investigator to identify transcription start sites. These should be close to the consensus sequences.

17. mRNA is unstable so does not accumulate in bacteria. Ribosomal and transfer RNAs are metabolically stable, so they accumulate to high steady-state levels. Also the amount of mRNA used in one translation complex is small in comparison to the amounts of tRNAs and rRNAs, particularly if several ribosomes are involved in simultaneously translating from a single mRNA molecule.

19. The information encoded in DNA is transmitted from generation to generation, whereas the information encoded in RNA is used only in the generation in which it is synthesized. There is an energy cost in maintaining ultrahigh-fidelity nucleic acid biosynthesis.

CHAPTER 25

5. The Shine-Dalgarno sequence, like the AUG that is read uniquely as fMet, occurs near the 5′ end of a message. The implication is that, in forming an initiation complex, only a region near the 5′ end of an mRNA molecule can bind a 30S ribosomal subunit. However, because Shine-Dalgarno sequences and AUG codons do appear in the interior of multicistronic messages, there may be additional signals to identify the 5′ end of an mRNA sequence. The Shine-Dalgarno sequence at a translational initiation site is recognized by a 30S ribosomal subunit, while an internal site is recognized by a 70S ribosome, which may be incapable of base-pairing with that sequence.

Also, because GAG and GGG represent the least frequently used codons for Glu and Gly in *E. coli*, the premise of the question may be correct; the sequence GAGGGG may occur infrequently at interior Glu-Gly coding sequences.

7. (a) The simplest explanation for particles of 100S and 130S is that they represent dimers and trimers of 70S particles (S increases with molecular mass, but in a nonlinear fashion). From what we know now, these represented small oligoribosomes bound to mRNA (*i.e.*, two and three 70S ribosomes bound to one mRNA molecule). The experimenter missed discovering polyribosomes as the active particulate structure in protein synthesis.

(b) Dissociation would have produced 30S and 50S ribosomal subunits plus mRNA. The mRNA would have already been fragmented if only 100S and 130S particles had been seen, representing oligoribosome dimers and trimers. mRNA is unstable and probably would have further degraded, particularly if traces of ribonuclease were present, so that, while 70S ribosomes could have reformed, there would be no mRNA fragments long enough to bind two or more 70 ribosomes. Perhaps this result means that 70S ribosomes can form in the absence of mRNA.

10. If Å is the probability that an error occurs at each step, then $1 - Å$ is the probability that an error has not been made in any step. The probability that no error has been made in n steps, then, is $(1 - Å)^n$. This is the probability that a polypeptide chain containing n amino acid residues has been synthesized with no errors because the probability of a series of independent events is the product of the individual probabilities.

11. (a) $(1 - 0.0001)^{100} = 0.990$

(b) $(1 - 0.0001)^{1000} = 0.904$; In other words, nearly ten percent of the completed polypeptide chains would contain one or more errors.

12. Data are from E. Terzaghi, Y. Okada, G. Streisinger, J. Emrich, M. Inouye, and A. Tsugita (1966) *Proc. Natl. Acad. Sci. USA* 56: 500–507. The assumption is that the first mutation involved a one-base insertion or deletion that disrupted the reading frame and that the second mutation restored the reading frame by a one-base deletion or insertion, respectively, with the intervening amino acid sequence representing acceptable missense.

13. Data are from Y. Okada, E. Terzaghi, G. Streisinger, J. Emrich, M. Inouye, and A. Tsugita (1966) *Proc. Natl. Acad. Sci. USA* 56:1692–1698. The assumption is that the first mutation involved a two-base insertion and that the second mutation restored the reading frame with a one-base insertion, with the one-amino acid lengthening of the polypeptide chain representing acceptable missense.

Wild-type sequence

| ... | Lys | Ser | Pro | Ser | Leu | Asn | Ala | ... |

... AAᴳ/ᴬ AGU CCA UCA CUU AAU GC . | ...

+GU or UG   +G or A

... AAᴳ/ᴬ AGU GUC CAU CAC UUA AUᴳ/ᴬ GC .

| ... | Lys | Ser | Val | His | His | Leu | Met | Ala |

Double mutant sequence

15. Cutting a phosphodiester bond near the 3′ end of 16S rRNA might (and does) remove the sequence that binds the Shine-Dalgarno sequence on the messenger RNA.

18. (a) Each nucleotide in each codon can undergo three substitutions; therefore 6 Arg codons × 3 nucleotides/codon × 3 substitutions/nucleotide = 54 possible substitution mutations

(b) (c)

Codon	Number substitutions to Arg codon	Number substitutions to Lys codon
CGU	3	0
CGC	3	0
CGA	4	0
CGG	4	0
AGA	2	1
AGG	2	1

21. Mitochondrial chaperones, such as Hsp70, act to ensure that proteins destined for mitochondria remain unfolded so that they can pass through membranes en route to their destination within the mitochondrion.

25. Example: The tRNA anticodon sequence for 5′-UGG-3′ is 3′-ACC-5′. Consider a nonsense mutation resulting from a GC → AT transition that changes an mRNA trp codon from 5′-UGG-3′ to 5′-UAG-3′. This nonsense mutation could be suppressed by a GC → AT transition in the gene for a glutamate tRNA that changes the anticodon from 5′-CUC-3′ to 5′-CUA-3′. This corresponds to a 5′-GAG-3′ codon, which would result in the original tryptophan codon being translated as glutamate. A 5′-CUA-3′ anticodon will bind to a 5′-UAG-3′ codon and will result in the nonsense codon being translated as glutamate.

CHAPTER 26

1. **(a)** $K_d = 10^{-13}$

   Repressor [R] = 10 molecules per *E. coli* cell keep operon turned off in the absence of inducer

   Volume of one *E. coli* cell = 0.3×10^{-12} ml

   $$\frac{(10 \text{ molecules/cell})/6.02 \times 10^{23} \text{ molecules/mol}}{0.3 \times 10^{-15} \text{L/cell}} = 5.5 \times 10^{-8} \text{M}$$

   (b) $\dfrac{(2 \text{ molecules/cell})/6.02 \times 10^{23} \text{ molecules/mol}}{0.3 \times 10^{-15} \text{ L/cell}} = 1.1 \times 10^{-8} \text{M}$

   (c) $K_d = \dfrac{[R][O]}{[RO]} = 10^{-13} \text{M} [R] = 5.5 \times 10^{-8}$

   $M[O]_{total} = 1.1 \times 10^{-8} M$

   $[RO]_{bound} + [O]_{free} = [O]_{total} = 1.1 \times 10^{-8} M$

   $[RO]_{bound} = -[O]_{free} + 1.1 \times 10^{-8} M$

   $[RO]_{bound} + [R]_{free} = 5.5 \times 10^{-8} M$

   $[R]_{free} = -[RO]_{bound} + 5.5 \times 10^{-8} M$

   $[R]_{free} \times [O]_{free} = (-[RO]_{bound} + 5.5 \times 10^{-8} M) \times [O]_{free}$

   $\qquad = ([O]_{free} - 1.1 \times 10^{-8} M + 5.5 \times 10^{-8} M)[O_{free}]$

   $K_d = \dfrac{[R_{free}][O_{free}]}{-[O_{free}] + 1.1 \times 10^{-8} M} = 10^{-13} M$

   $([O]_{free} - 1.1 \times 10^{-8} M + 5.5 \times 10^{-8} M) \times [O]_{free}$

   $\qquad = (-[O]_{free} + 1.1 \times 10^{-8} M) \times 10^{-13}$

   $[O]_{free}^2 + 4.4 \times 10^{-8} M \times [O]_{free} - 1.1 \times 10^{-21} M^2$

   $[O]_{free} = 2.5 \times 10^{-14} M$

   (d) At any given time an average cell in a rapidly growing culture is midway through duplicating its chromosome, so there is more than one copy of the operon.

2. It's unlikely because in eukaryotes transcription occurs in the nucleus and translation in the cytosol. Hence, direct coupling between transcription and translation, essential for attenuation, is absent.

3. An operon is a set of physically linked genes under common regulatory processes, while a regulon is a set of unlinked genes, dispersed about the genome, that is controlled by common regulatory processes.

4. A dinucleotide, complementary to the first two template nucleotides, can bypass the first nucleotide incorporation step and permit efficient transcription at low nucleotide concentrations. Then one can arrest all transcription events at one nucleotide by knowing the nucleotide sequence of the gene being transcribed and adding only two or three rNTPs, instead of all four, with the missing nucleotide being the one

that would have been inserted at the site at which you desire to arrest transcription.

6. **(a)** The plasmid *laci* gene encodes a repressor that cannot bind inducer.
   (b) The plasmid operator cannot bind repressor.

9.

	Before	After	Comment
(a)	−	+ +	Wild-type regulation
(b)	−	+	Excess repressor may bind chromosomal operator and prevent full expression upon induction
(c)	−	+	Reduction in RNA polymerase binding to promoter decreases *lac* operon transcription
(d)	−	−	Chromosome-coded *lacI* repressor binds irreversibly to both chromosomal and plasmid operators
(e)	−	−	Plasmid-coded *lacI* repressor binds irreversibly to both chromosomal and plasmid operators
(f)	+ +	+ +	Transcription is constitutive from the chromosomal *lacZ* gene
(g)	−	+	The stimulation of *lacZ* transcription by CRP-cAMP complex is abolished

10. A mutation that abolishes binding to operator. Subunits of the mutant repressor could interact with those of the wild-type repressor synthesized in the same cell, to form mixed dimers with reduced affinity for operator DNA.

14. Analysis of nucleotide sequences in genes subject to attenuation control confirmed the hypothesis that formation of a stem-loop structure followed by a uracil-rich sequence is an essential structural feature of a transcriptional termination site—at least, for those genes subject to attenuation control.

16. Removal of acetyl groups from lysine residues in histones yields positively charged free amino groups, which form ionic interactions with negatively charged DNA phosphates. A more significant effect, not discussed in this chapter, is that deacetylases may allow for more stable supramolecular structures via inter-nucleosome interaction. Moreover, deacetylated histones may recruit gene-silencing proteins.

18. 5-Azacytidine → 5-Azacytidine 5′-monophosphate → 5-Azacytidine diphosphate → 5-Azadeoxycytidine diphosphate → 5-Azadeoxycytidine triphosphate

    The first two reactions are catalyzed by ATP-dependent kinases. The third reaction is catalyzed by ribonucleotide reductase, and the fourth reaction is catalyzed by nucleoside diphosphate kinase.

22. Epigenetics refers to heritable changes in gene expression that do not involve a change in coding capacity of the gene whose expression is changed. Until recently (2017) it had not been established that histone modifications are transmitted from one generation to the next. Because histone modifications do affect gene expression, and with recent evidence that histone modification patterns are heritable, it is appropriate to consider histone modifications as epigenetic. With respect to DNA methylation, the coding properties of cytosine and 5-methylcytosine are identical, and the mechanism of intergenerational transmission of the methyl mark is known, making cytosine methylation the classic example of an epigenetic phenomenon.

CHAPTER 1

Costanza, M., *et al.* (2016) A global genetic interaction network maps a wiring diagram of cellular function. *Science* 353:1381. A one-page summary of a much longer article available on line.

Jasry, B. R., and D. Kennedy, eds. (2001) The Human Genome. *Science* 291:1148–1432. A special issue of *Science*, reporting and analyzing the near completion of the sequence determination of human DNA.

Jasry, B. R., and L. Roberts, eds. (2003) Building on the DNA revolution. *Science* 300:277–296. A series of articles in a special issue of *Science* commemorating the fiftieth anniversary of the Watson–Crick discovery.

Kornberg, A. (1987) The two cultures: Chemistry and biology. *Biochemistry* 26:6888–6891. The author argues eloquently that biologists should learn more chemistry.

Koshland, D. E. (2002) The seven pillars of life. *Science* 295:2215–2216. A two-page essay outlining seven distinctive attributes of living matter.

Lander, E. S. (2011) Initial impact of the sequencing of the human genome. *Nature* 470:187–197. The impact of the human genome sequence upon human biomedicine and the prospective impact of genomics upon medicine.

CHAPTER 2

Noncovalent Interactions

Creighton, T. E. (2010) *The Physical and Chemical Basis of Molecular Biology.* UK: Helvetian Press. See Chapters 2 and 3 for more detailed discussion of noncovalent interactions and acid–base chemistry.

Leckband, D., and J. Israelachvili (2001) Intermolecular forces in biology. *Quart. Rev. Biophys.* 34:105–267. Extensive review of the theory for predicting forces and the practice of measuring forces in biological systems.

van Holde, K. E., W. C. Johnson, and P. S. Ho (2006) *Principles of Physical Biochemistry* (2nd ed.). Upper Saddle River, NJ: Prentice-Hall. Covers most of the topics in this chapter in considerably more depth.

Water

Moore, F. G., and G. L. Richmond (2008) Integration or Segregation: How do Molecules Behave at Oil/Water Interfaces? *Accts. of Chem. Res.* 41:739–748. A detailed study of interfacial regions between water and nonpolar fluids.

Tanford, C. (1980) *The Hydrophobic Effect. Formation of Micelles and Biological Membranes.* New York: Wiley. A classic study of hydrophobicity.

Ionic Equilibria

Phillips, R., J. Kondev, and J. Theriot (2009) *Physical Biology of the Cell.* New York: Garland Science. Chapter 9 has a good discussion of water and electrostatics in ionic solutions.

Tossell, J. A. (2006) H_2CO_3 and its oligomers: Structures, stabilities, vibrational and NMR spectra, and acidities. *Inorganic Chemistry* 45:5061–5970. Computational analysis of stability and pK_a of H_2CO_3.

Electrophoresis

Hames, B. D., and D. Rickwood, eds. (1981) *Gel Electrophoresis of Proteins.* IRL Press, Oxford, Washington, DC, and Rickwood, D., and B. D. Hames, eds.

(1982) *Gel Electrophoresis of Nucleic Acids.* Oxford: IRL Press. These two volumes are useful laboratory manuals for gel electrophoresis techniques.

Osterman, L. A. (1984) *Methods of Protein and Nucleic Acids Research,* Vol. 1, Parts 1 and 2. New York: Springer-Verlag. A comprehensive summary of electrophoresis and isoelectric focusing.

CHAPTER 3

Chapter 3 presents an abbreviated treatment of thermodynamics. For the student who wishes a more rigorous background in this field and more information about its applications to biochemistry, we recommend the following books:

Dill, K. A., and S. Bromberg (2010) *Molecular Driving Forces* (2nd ed.). Garland Science, New York. An excellent resource for those desiring a clear and comprehensive presentation of thermodynamic principles.

Phillips, R., J. Kondev, and J. Theriot (2009) *Physical Biology of the Cell.* Garland Science, New York Chapters 5 and 6 provide greater detail on many of the topics covered here.

van Holde, K. E., W. C. Johnson, and P. S. Ho (2006) *Principles of Physical Biochemistry* (2nd ed.). Prentice Hall, Upper Saddle River, N.J. Chapters 2–4 extend the applications of thermodynamics to biochemistry.

For a sophisticated discussion of the effect of ionic conditions on the free energy changes in phosphate ester hydrolysis, see the following article:

Alberty, R. A. (1992) Equilibrium calculations on systems of biochemical reactions at specified pH and pMg. *Biophys. Chem.* 42:117–131.

For a more rigorous definition of the biochemical standard state and standard free energies, see these sources:

Alberty, R. A., A. Cornish-Bowden, R. N. Goldberg, G. G. Hammes, K. Tipton, and H. V. Westerhoff (2011) Recommendations for terminology and databases for biochemical thermodynamics. *Biophys. Chem.* 155:89–103.

Frey, P., and A. Arabshahi (1995) Standard free energy change for the hydrolysis of the α, β-phosphoanhydride bridge in ATP. *Biochemistry* 34:11307–11310.

Lundblad, R. L., and F. M. MacDonald (eds.) (2010) *Handbook of Biochemistry and Molecular Biology* (4th ed.). CRC Press, Boca Raton, FL.

Méndez, E. (2008) Biochemical thermodynamics under near physiological conditions. *Biochem. Mol. Biol. Educ.* 36:116–119.

CHAPTER 4

Bacolla, A., and R. D. Wells (2004) Non-B DNA conformations, genomic rearrangements, and human diseases. *J. Biol. Chem.* 279:47411–47414. A mini-review dealing with unconventional DNA structures.

Bochman, M. L., K. Paeschke, and V. A. Zakian (2012) DNA secondary structures: Stability and function of G-quadruplex structures. *Nature Rev. Gen.* 13:770–780. Chemistry and biology of this unusual DNA structure.

Deweese, J. E., M. A. Osheroff, and N. Osheroff (2009) DNA topology and topoisomerases. Teaching a "knotty" subject. *Biochem. Mol. Biol. Education* 37:2–10. A clearly written short review, with discussion of topoisomerases as drug targets.

Joyce, G. F. (2002) The antiquity of RNA-based evolution. *Nature* 418:214–221. Thoughts about a primordial RNA world.

Judson, H. F. (1979) *The Eighth Day of Creation.* New York: Simon & Schuster. A fascinating and dramatic account of the development of modern ideas about molecular genetics, often through the eyes and in the words of the major participants.

Meselson, M., and F. Stahl (1958) The replication of DNA in *Escherichia coli. Proc. Natl. Acad. Sci. USA* 44:671–682. An example of a beautifully designed and executed experiment.

Mukherjee, S. (2016) *The Gene: An Intimate History*. Simon & Schuster. A book-length history of genetics written by a cancer researcher.

Olby, R. (2009) *Francis Crick, Hunter of Life's Secrets*. Cold Spring Harbor Laboratory Press. A distinguished science historian records the discovery of the structure of DNA and other aspects of the life of this brilliant scientist.

Pääbo, S. (2015) The diverse origins of the human gene pool. *Nature Rev. Genet.* 16:313–314. A summary of what has been learned about human evolution from sequence analysis of ancient DNA.

Watson, J. D., and F. H. C. Crick (1953) Molecular structure of nucleic acids. A structure for deoxyribose nucleic acid. *Nature* 171:737–738. Two pages that shook the world of biology.

Tools of Biochemistry 4A Manipulating DNA

Goodwin, S., J. D. McPherson, and W. R. McCombie (2016) Coming of age: Ten years of next-generation sequencing technologies. *Nature Rev. Genet.* 13:233–251. Technical advances allow complete genome sequences from single cells.

Green, M. R., and P. J. Sambrook (2012) *Molecular Cloning, A Laboratory Manual, Volumes 1–3,* Fourth Edition. Cold Spring Harbor, NY: Cold Spring Harbor Laboratory. The definitive laboratory handbook of molecular biological methods.

Strano, M. S. (2012) Functional DNA origami devices *Science* 338:890–891. Synthetic DNA devices may have practical applications.

Tools of Biochemistry 4B An Introduction to X-Ray Diffraction

Callaway, E. (2015) The Revolution will not be crystallized. *Nature* 525, 172–174. A brief introduction to cryo-EM.

van Holde, K. E., W. C. Johnson, and P. S. Ho (2006) *Principles of Physical Biochemistry* (2nd ed., Chapter 6). Upper Saddle River, NJ: Pearson/Prentice Hall. A more detailed treatment of X-ray diffraction of biopolymers.

CHAPTER 5

Brändén, C., and J. Tooze (1999) *Introduction to Protein Structure* (2nd ed.). Garland, New York. Contains much information on all levels of structure. Excellent illustrations.

Crooks, D. E., G. Hon, J. M. Chandonia, and S. E. Brenner (2004) Weblogo: A sequence logo generator. *Genome Res.* 14:1188–1190. (See: **weblogo.berkeley.edu**)

Lander, E. S., et al. (2001) Initial sequencing and analysis of the human genome. *Nature* 409:860–921.

Liljas, A., L. Liljas, J. Piskur, G. Lindblom, P. Nissen, and M. Kjeldgaard (2009) *Textbook of Structural Biology*. World Scientific Publishing, Singapore. Brief treatment on basics of protein structure; but gives a broad and reasonably detailed overview of protein structures and functions.

Petsko, G. A., and D. Ringe (2004) *Protein Structure and Function*. New Science Press, London. Concise and clearly written. Excellent illustrations complete with Protein Data Bank ID codes.

Roberts, L. (2001) Controversial from the start. *Science* 291:1182–1188. A review of some controversies surrounding the human genome sequencing project.

Rose, G. D., A. R. Geselowitz, G. J. Lesser, R. H. Lee, and M. H. Zehfus (1985) Hydrophobicity of amino acid residues in globular proteins. *Science* 229:834–838.

Turanov, A. A., A. V. Lobanov, D. E. Fomenko, H. G. Morrison, M. L. Sogin, L. A. Klobutcher, D. L. Hatfiled, and V. M. Gladyshev (2009) Genetic code supports targeted insertion of two amino acids. *Science* 323:259–261.

Venter, J. C., et al. (2001) The sequence of the human genome. *Science* 291:1304–1351.

Ye, J., S. McGinnis, and T. L. Madden (2006) BLAST: Improvements for better sequence analysis. *Nucleic Acids Res.* 34:6–9.

URLs for access to public sequence databases, sequence alignment tools, and other protein analysis tools (e.g., mass and/or pI calculations, DNA sequence translation, etc.):

GOLD: www.genomesonline.org

GenBank:

www.ncbi.nlm.nih.gov/Genbank/index.html

www.ncbi.nlm.nih.gov/Genomes/index.html

KEGG:

http://www.kegg.jp/kegg/

BLAST:

blast.ncbi.nlm.nih.gov/Blast.cgi

Proteomics tools: www.expasy.ch

Tools of Biochemistry 5A Protein Expression and Purification

Roe, S. (ed.) (2001) *Protein Purification Techniques: A Practical Approach* (2nd ed.). Oxford University Press, Oxford.

Rosenberg, I. M. (2005) *Protein Analysis and Purification: Benchtop Techniques* (2nd ed.). Birkhauser, Boston.

Tools of Biochemistry 5B Mass, Sequence, and Amino Acid Analyses of Purified Proteins

Cañas, B., D. López-Ferrer, A. Ramos-Fernández, E. Camafeita, and E. Calvo (2006) Mass spectrometry technologies for proteomics. *Brief. Funct. Genom. Proteom.* 4:295–320.

Dunn, M. J. (2000) Studying heart disease using the proteomic approach. *Drug Discov. Today* 5:76–84.

Goh, W. W. B., Y. H. Lee, R. M. Zubaidah, J. Jin, D. Dong, Q. Lin, M. C. M. Chung, and L. Wong (2011) Network-based pipeline for analyzing MS data: An application toward liver cancer. *J. Proteome Res.* 10:2261–2272.

Graves, P. R., and T. A. Haystead (2002) Molecular biologist's guide to proteomics. *Microbiol. Mol. Biol. Rev.* 66: 39–63.

Spacil, Z., S. Elliott, L. Reeber, M. H. Gelb, C. R. Scott, and F. Turecek (2011) Comparative triplex tandem mass spectrometry assays of lysosomal enzyme activities in dried blood spots using fast liquid chromatography: Application to newborn screening of Pompe, Fabry, and Hurler diseases. *Anal. Chem.* 83:4822–4828.

Sutton, C. W., N. Rustogi, C. Gurkan, A. Scally, M. A. Loizidou, A. Hadjisavvas, and K. Kyriacou (2010) Quantitative proteomic profiling of matched normal and tumor breast tissues. *J. Proteome Res.* 9:3891–3902.

Thomas, J. J., R. Bakhtiar, and G. Suizdak (2000) Mass spectrometry in viral proteomics. *Acc. Chem. Res.* 33:179–187.

CHAPTER 6

General

Creighton, T. E. (2010) *The Biophysical Chemistry of Nucleic Acids and Proteins*. Helvetian Press, UK.

Liljas, A., L. Liljas, J. Piskur, G. Lindblom, P. Nissen, and M. Kjeldgaard (2009) *Textbook of Structural Biology*. World Scientific Publishing, Singapore.

Links to Protein Rendering and Viewing Software

http://pymol.org/educational/

http://spdbv.vital-it.ch/

http://www.rcsb.org/pdb/static.do?p=software/software_links/molecular_graphics.html

Databases of Domain Structure and Classification

Class Architecture Topology Homologous Superfamily (CATH): http://www .cathdb.info/

Structural Classification of Proteins (SCOP): http://scop2.mrc-lmb.cam.ac.uk/

Protein Folding and Stability

Baase, W. A., L. Liu, D. E. Tronrud, and B. W. Matthews (2010) Lessons from the lysozyme of phage t4. *Prot. Sci.* 19:631–641.

Dill, K. A., and H. S. Chan (1997) From Levinthal to pathways to funnels. *Nature Struct. Biol.* 4:10–19.

Onuchic, J. N., and P. G. Wolynes (2004) Theory of protein folding. *Curr. Opin. Struct. Biol.* 14:70–75.

Chaperones

Kim, Y. E., M. S. Hipp, A. Bracher, M. Hayer-Hartl, and F. U. Hartl (2013) Molecular chaperone functions in protein folding and proteostasis. *Annu. Rev. Biochem.* 82:323–355. doi: 10.1146/annurev-biochem-060208-092442.

Lindberg, I., J. Shorter, R. L. Wiseman, F. Chiti, C. A. Dickey, and P. J. McLean (2015) Chaperones in neurodegeneration. *J. Neurosci.*, 35:13853–13859; DOI: https://doi.org/10.1523/JNEUROSCI.2600-15.2015

Balchin, D., M. Hayer-Hartl, and F. U. Hartl (2016) In vivo aspects of protein folding and quality control. *Science* 353:42. doi: 10.1126/science.aac4354.

Prediction of Protein Structure

Rost, B. (2009) Prediction of protein structure in 1D-Secondary structure, membrane regions and solvent accessibility. In J. Gu and P. E. Bourne, eds., *Structural Bioinfomatics* (2nd ed.). Wiley-Blackwell, Hoboken, NJ.

Access to several secondary structure prediction programs is available from: www.expasy.ch

Tertiary Structure Prediction:

Adams P. D., D. Baker, A. T. Brunger, R. Das, F. DiMaio, R. J. Read, D. C. Richardson, J. S. Richardson, and T. C. Terwilliger (2013) Advances, interactions, and future developments in the CNS, Phenix, and Rosetta structural biology software systems. *Annu. Rev. Biophys.* 42:265–287. doi: 10.1146/annurev-biophys-083012-130253.

Misfolding Disease

Knowles, T. P. J., M. Vendruscolo, and C. M. Dobson (2014) The amyloid state and its association with protein misfolding diseases. *Nature Rev. Mol. Cell Biol.* 15:384–396. DOI: 10.1038/nrm3810.

Valastyan, J. S., and S. Lindquist (2014) Mechanisms of protein-folding diseases at a glance. *Dis. Mod. & Mech.*, 7:9–14. http://doi.org/10.1242/dmm.013474

Tools of Biochemistry 6A Spectroscopic Methods for Studying Macromolecular Conformation in Solution

Campbell, I. D., and R. A. Dwek (1984) *Biological Spectroscopy.* Benjamin/ Cummings, Menlo Park, CA.

Cavanagh, J., W. J. Fairbrother, A. G. Palmer, M. Rance, and N. J. Skelton (2007) *Protein NMR Spectroscopy: Principles and Practice.* Academic Press, San Diego, CA. A comprehensive and detailed treatise.

Johnson, W. C., Jr. (1990) Protein secondary structure and circular dichroism: A practical guide. *Proteins Struct. Funct. Genet.* 7:205–214.

Tsien, R. Y. (2009) Constructing and exploiting the fluorescent protein paintbox (Nobel Lecture). *Angew. Chem. Int. Ed.* 48:5612–5626.

Tools of Biochemistry 6B Mass, Sequence, and Amino Acid Analyses of Purified Proteins

Hames, B. D., and D. Rickwood, eds. (1990) *Gel Electrophoresis of Proteins,* 2nd ed. IRL Press, Oxford, Washington, DC.

van Holde, K. E., W. C. Johnson, and P. S. Ho (2006) *Principles of Physical Biochemistry,* 2nd ed. Prentice Hall, Upper Saddle River, NJ.

ImageJ software (public domain): https://imagej.nih.gov/ij/

CHAPTER 7

The Immune Response

Excler, J. L., M. L. Robb, and J. H. Kim (2015) Prospects for a globally effective HIV-1 vaccine. *Vaccine* 33:D4–12 (doi: 10.1016/j.vaccine.2105.03.059).

Chaplin, D. D. (2010) Overview of the immune response. *J. Allergy Clin. Immunol.* 125:S3–23.

Antibody-Based Therapeutics

Harris, T. J., and C. G. Drake (2013) Primer on tumor immunology and cancer immunotherapy. *J. Immunother. Cancer* 1:12–20 (doi: 10.1186/2051-1426-1-12)

van Gils, M. J., and R. W. Sanders (2013) Broadly neutralizing antibodies against HIV-1: Templates for a vaccine. *Virology* 435:46–56 (doi:10.1016/j.virol.2012.10.004).

Allosteric Models

Barrick, D., N. T. Ho, V. Simplaceanu, F. Dahlquist, and C. Ho (1997) A test of the role of the proximal histidines in the Perutz model for cooperativity in haemoglobin. *Nature Struct. Biol.* 4:78–83.

Eaton, W. A., E. R. Henry, J. Hofrichter, S. Bettati, C. Viappiani, and A. Mozzarelli (2007) Evolution of allosteric models for haemoglobin. *IUBMB Life* 59:586–599.

Koshland, D. E., G. Nemethy, and D. Filmer (1966) Comparison of experimental binding data and theoretical models in proteins containing subunits. *Biochemistry* 5:365–385.

Monod, J., J. Wyman, and J. P. Changeux (1965) On the nature of allosteric transitions: A plausible model. *J. Mol. Biol.* 12:88–118.

Perutz, M. F., A. J. Wilkinson, M. Paoli, and G. G. Dodson (1998) The stereochemical mechanism of cooperative effects in hemoglobin revisited. *Annu. Rev. Biophys. Biomol. Struct.* 27:1–34.

Evolution of Globin Proteins and Theories of Protein Evolution

Hardison, R. C. (2012) Evolution of hemoglobin and its genes. *Cold Spring Harb. Perspect. Med.* 2:a011627 (doi: 10.1101cshperspect.a011627).

Harms, M. J., and J. W. Thornton (2013) Evolutionary biochemistry: Revealing the historical and physical causes of protein properties. *Nature Rev. Genet.* 14:559–571 (doi: 10.1038/nrg3540).

Storz, J. F. (2016) Gene duplication and evolutionary innovations in hemoglobin-oxygen transport. Physiology 31:223–232 (doi:10.1152/physiol.00060.2015).

Variant Hemoglobins and Hemoglobin Pathologies

Hosseini, P., S. Z. Abidi, E. Du, D. P. Papageorgiou, Y. Choi, Y. Park, J. M. Higgins, G. J. Kato, S. Suresh, M. Dao, Z. Yaqoob, and P. T. C. So (2016) Cellular normoxic biophysical markers of hydroxyurea treatment in sickle cell disease. *Proc. Natl Acad. Sci. (USA)* 113:9527–9532 (doi.org/10.1073/pnas.1610435113).

Schechter, A. N. (2008) Hemoglobin research and the origins of molecular medicine. *Blood* 112:3927–3938.

See also the HbVar database for human hemoglobin variants and thalassemias: http://globin.bx.psu.edu/hbvar/menu.html

Motor Proteins

Cooke, R. (2004) The sliding filament model: 1972–2004. *J. Gen. Physiol.* 123:643–656.

Bhabha, G., G. T. Johnson, C. M. Schroeder, and R. D. Vale (2016) How dynein moves along microtubules. *Trends Biochem. Sci.* 41:94–105 (doi: 10.1016/j.tibs.2015.11.004).

See also the following animations created by G. Johnson:
http://www.scripps.edu/cb/milligan/research/movies/myosin_text.html
http:// http://www.scripps.edu/milligan/projects.html
https://valelab.ucsf.edu/molecular-animations/

CHAPTER 8

General

Copeland, R. (2005) *Evaluation of Enzyme Inhibitors in Drug Discovery.* John Wiley and Sons, Inc., Hoboken, NJ. Gives many examples of inhibitory compounds that are used as drugs.

Cuesta, S. M., N. Furnham, S. A. Rahman, I. Sillitoe, and J. M. Thornton (2014) The evolution of enzyme finction in the isomerases. *Curr. Op. Struct. Biol.*, 26:121–130.

Fersht, A. (1999) *Structure and Mechanism in Protein Science.* W. H. Freeman and Co., New York. A fine treatise on almost all aspects of enzymology.

Park, J. O., S. A. Rubin, Y.-F. Xu, D. Amador-Noguez, J. Fan, T. Shlomi, and J. D. Rabinowitz (2016) Metabolite concentrations, fluxes and free energies imply efficient enzyme usage. *Nature Chem. Biol.*, 12:482–489.

Enzyme Databases

BRENDA http://www.brenda-enzymes.info
ExPASy http://expasy.org/enzyme

Enzyme Mechanisms and Kinetics

Amyes, T. L., and J. P. Richard (2013) Specificity in transition state binding: The Pauling model revisited. *Biochemistry* 52:2021–2035.

Herschlag, D., and A. Natarajan (2013) Fundamental challenges in mechanistic enzymology: Progress toward understanding the rate enhancements of enzymes. *Biochemistry* 52:2050–2067.

Hanoian, P., C. T. Liu, S. Hammes-Schiffer, and S. Benkovic (2015) Perspectives on electrostatics and conformational motions in enzyme catalysis. *Acc. Chem. Res.*, 48:482–489.

Johnson, K. A. (2013) A century of enzyme kinetic analysis, 1913 to 2013. *FEBS Letters* 587:2753–2766.

Lysozyme and Serine Proteases

Vocadlo, D. J., G. J. Davies, R. Laine, and S. G. Withers (2001) Catalysis by hen egg-white lysozyme proceeds via a covalent intermediate. *Nature* 412:835–838.

Wilmouth, R. C., K. Edman, R. Neutze, P. A. Wright, I. J. Clifton, T. R. Schneider, C. J. Schofield, and J. Hajdu (2001) X-ray snapshots of serine protease catalysis reveal a tetrahedral intermediate. *Nature Struct. Biol.* 8:689–694.

Ribozymes and DNAzymes

Zhou, C., J. L. Avins, P. C. Klauser, B. M. Brandsen, Y. Lee, and S. K. Silverman (2016) DNA-catalyzed amide hydrolysis. *J. Am. Chem. Soc.* 138:2106–2109.

Ponce-Salvatierra, A., K. Wawrzyniak-Turek, U. Steuerwald, C. Höbartner, and V. Pena (2016) Crystal structure of a DNA catalyst *Nature* 529:231–234.

Silverman, S. K. (2015) Pursuing DNA catalysts for protein modification. *Acc. Chem. Res.*, 48:1369–1379.

Regulation of Enzyme Activity

Cárdenas, M. L. (2013) Michaelis and Menten and the long road to the discovery of cooperativity. *FEBS Letters* 587:2767–2771.

Cockrell, G. M., Y. Zheng, W. Guo, A. W. Peterson, and E. R. Kantrowitz, (2013) New paradigm for allosteric regulation of *Escherichia coli* aspartate transcarbamoylase. *Biochemistry* 52:8036–8047.

Goodey, N. M., and S. J. Benkovic (2008) Allosteric regulation and catalysis emerge via a common route. *Nature Chem. Biol.* 8:474–482.

Swain, J. F., and L. M. Gierasch (2006) The changing landscape of protein allostery. *Curr. Op. Struct. Biol.* 16:102–108.

Tools of Biochemistry 8A How to Measure the Rates of Enzyme-Catalyzed Reactions

Cleland, W. W. (2003) The use of isotope effects to determine enzyme mechanism. *J. Biol. Chem.* 278:51975–51984.

Fersht, A. (1999) *Structure and Mechanism in Protein Science.* W. H. Freeman and Co., New York.

Johnson, K. A. (2003) Introduction to kinetic analysis of enzyme systems. In *Kinetic Analysis of Macromolecules: A Practical Approach,* K. A. Johnson (ed.). Oxford University Press, New York.

Reyes, A. C., T. L. Amyes, and J. P. Richard (2016) Structure–reactivity effects on intrinsic primary kinetic isotope effects for hydride transfer catalyzed by glycerol-3-phosphate dehydrogenase. *J. Am. Chem. Soc.* 138:14526–14529.

CHAPTER 9

Dalziel, M., M. Crispin, C. N. Scanlan, N. Zitzmann, and R. A. Dwek (2014) Emerging principles for the therapeutic exploitation of glycosylation. *Science* 343:37. Summary of a review available at http://.dx.doi.org/10:1126/science.1235681.

Finklestein, J. (2007) Glycochemistry and glycobiology. *Nature* 446:999. Introduction to a special series of review articles on contemporary carbohydrate biochemistry.

Kim, J. H., and fourteen coauthors (2013) Mechanism-based covalent neuraminidase inhibitors with broad-spectrum influenza antiviral activity. *Science* 340:71–75. Improving on Tamiflu.

Laughlin, S. T., and C. R. Bertozzi (2009) Imaging the glycome. *Proc. Natl. Acad. Sci. U.S.A.* 106:12–17. Emerging technology for visualizing specific glycans in living cells.

Shieh, P., M. S. Siegrist, A. J. Cullen, and C. R. Bertozzi (2014) Imaging bacterial peptidoglycan with near-infrared fluorogenic azide probes. *Proc. Natl. Acad. Sci. USA* 111:5456–5461. Visualization of peptidoglycan in living cells.

Seeberger, P. H., and D. B. Werz (2007) Synthesis and medical applications of oligosaccharides. *Nature* 446:1046–1051. Sequence analysis and synthesis of complex carbohydrates both involve special problems.

Taylor, M. E., and K. Drickamer (2011) *Introduction to Glycobiology,* Third Edition. Oxford, UK: Oxford University Press. A book-length review.

Tynas, A., M. Banzhaf, C. A. Gross, and W. Vollmer (2011) From the regulation of peptidoglycan synthesis to bacterial growth and morphology. *Nat. Rev. Microbiol.* 10:123–136.

Varki, A. (2007) Glycan-based interactions involving vertebrate sialic-acid-recognizing proteins. *Nature* 446:1023–1029. Roles of sialic acid in cell recognition.

CHAPTER 10

General

Engleman, D. M. (2005) Membranes are more mosaic than fluid. *Nature* 438:578–580. This issue of *Nature* includes several review articles on membranes.

Membrane Asymmetry and Structure

Daleke, D. L. (2007) Phospholipid flippases. *J. Biol. Chem.* 282:821–825.

Hartlova, A., L. Cerveny, M. Hubalek, Z. Krocova, and J. Stulik (2010) Membrane rafts: A potential gateway for bacterial entry into host cells. *Microbiol. Immunol.* 54:237–245.

Lingwood, D., and K. Simons (2010) Lipid rafts as a membrane-organizing principle. *Science* 327:46–50.

Sham, L.-T., E. K. Butler, M. D. Lebar, D. Kahne, T. G. Bernhardt, and N. Ruiz (2014) MurJ is the flippase of lipid-linked precursors for peptidoglycan biogenesis. *Science* 345:220–222.

Membrane Proteins

McMorran, L. M., D. J. Brockwell, and S. E. Radford (2014) Mechanistic studies of the biogenesis and folding of outer membrane proteins *in vitro* and *in vivo*: What have we learned to date? *Arch. Biochem. Biophys.* 546:265–280.

Popot, J. L., and D. M. Engleman (2016) Membranes do not tell proteins how to fold. *Biochemistry* 55:5–18.

Translocon Structure and Function

Becker, T., S. Bhushan, A. Jarasch, J.-P. Armache, S. Funes, F. Jossinet, J. Gumbart, T. Mielke, O. Berninghausen, K. Schulten, E. Westhof, R. Gilmore, E. C. Mandon, and R. Beckmann (2009) Structure of monomeric yeast and mammalian Sec61 complexes interacting with the translating ribosome. *Science* 326:1369–1373.

Egea, P. F., and R. M. Stroud (2010) Lateral opening of a translocon upon entry of protein suggests the mechanism of insertion into membranes. *Proc. Natl. Acad. Sci. USA* 107:17182–17187.

Reithlinger, J. H., C. Yim, S. Kim, H. Lee, and H. Kim (2014) Structural and functional profiling of the lateral gate of the Sec61 translocon. *J. Biol. Chem.* 289:15845–15855.

Voohees, R. M., and R. S. Hegde (2016) Structure of the Sec61 channel opened by a signal sequence. *Science* 351:88–91.

Transport across Membranes

Catterall, W. A. (2010) Ion channel voltage sensors: Structure, function and pathophysiology. *Neuron* 67:915–928.

Cole, S. P. C. (2014) Multidrug Resistance Protein 1 (MRP1, ABCC1), a "Multitasking" ATP-binding Cassette (ABC) Transporter *J. Biol. Chem.* 289:30880–30888.

Fletcher, J. I., M. Haber, M. J. Henderson, and M. D. Norris (2010) ABC transporters in cancer: More than just drug efflux pumps. *Nat. Rev. Cancer* 10:147–156.

Gouaux, E., and R. MacKinnon (2005) Principles of selective ion transport in channels and pumps. *Science* 310:1461–1465.

Jensen, M. Ø., V. Jogini, D. W. Borhani, A. E. Leffler, R. O. Dror, and D. E. Shaw (2012) Mechanism of voltage gating in potassium channels. *Science* 336:229–233.

S. W. Lockless (2015) Determinants of cation transport selectivity: Equilibrium binding and transport kinetics. *J. Gen. Physiol.* 146:3–13.

CHAPTER 11

Metabolic Design Principles

Atkinson, D. E. (1977) *Cellular Energy Metabolism and Its Regulation.* New York: Academic Press. This excellent book lays out the bioenergetic foundations of metabolic processes.

Brosnan, J. T. (2005) Metabolic design principles: Chemical and physical determinants of cell chemistry. *Adv. Enzyme Regul.* 45:27–36. This essay, based largely on concepts in Atkinson's book, discusses how the design of metabolic systems follows logically from chemical and physical constraints.

Experimental Techniques in the Study of Metabolism

Cunningham, R. E. (2010) Overview of flow cytometry and fluorescent probes for flow cytometry. *Methods Mol. Biol.* 588:319–326. The power of fluorescence-activated cell sorting.

Freifelder, D. (1982) *Physical Biochemistry,* 2nd ed. San Francisco: W. H. Freeman. Chapter 5 of this book presents a clear description of techniques in radioactive labeling and counting.

Liu, X., and J. W. Locasale (2017) Metabolomics: A Primer. *Trends Biochem. Sci.* 42, 274–284.

Shulman, R. G., and D. L. Rothman (2001) ^{13}C NMR of intermediary metabolism: Implications for systemic physiology. *Annu. Rev. Physiol.* 63:15–48. The power of NMR in the study of metabolism.

Tsien, R. Y. (2009) Constructing and exploiting the fluorescent protein paintbox (Nobel Lecture). *Angew. Chem. Int. Ed. Engl.* 48:5612–5626. This review by the 2008 Chemistry Nobel Laureate describes another powerful technique for noninvasive metabolic monitoring of individual cells.

Compartmentation and Intracellular Enzyme Organization

Dzeja, P. P., and A. Terzic (2003) Phosphotransfer networks and cellular energetics. *J. Exp. Biol.* 206:2039–2047. A contemporary account of the bioenergetic role of creatine phosphate, emphasizing the importance of compartmentation as a metabolic control phenomenon.

Goodsell, D. S. (1991) Inside a living cell. *Trends Biochem. Sci.* 16:203–206. Classic drawings of the interior of a bacterial cell, based on physical information about the sizes, shapes, and distribution of cellular constituents.

Ovádi, J., and V. Saks (2004) On the origin of intracellular compartmentation and organized metabolic systems. *Mol. Cell. Biochem.* 256–257:5–12. A concise review of evidence for the organization of sequential metabolic pathways, including both membranous complexes and complexes involving soluble enzymes.

Enzyme Control and Metabolic Regulation

Fell, D. (1997) *Understanding the Control of Metabolism.* London: Portland Press Ltd. This book goes into depth on most of the topics of this chapter, including metabolic control analysis.

Newsholme, E. A., R. A. J. Challiss, and B. Crabtree (1984) Substrate cycles: Their role in improving sensitivity in metabolic control. *Trends Biochem. Sci.* 9:277–280. A brief but lucid discussion of substrate cycle control, with several examples.

CHAPTER 12

Glycolysis

Kresge, N., R. D. Simoni, and R. L. Hill (2005) Otto Fritz Meyerhof and the elucidation of the glycolytic pathway. *J. Biol. Chem.* 280:e3. A review that summarizes the classic papers from Meyerhof and coworkers.

Ovadi, J., and P. A. Srere (2000) Macromolecular compartmentation and channeling. *Int. Rev. Cytol.* 192:255–280. A review written by two of the pioneers of metabolic compartmentation.

Glycogen Metabolism

Chen, Y.-T. (2001) Glycogen storage diseases. In: *The Metabolic and Molecular Bases of Inherited Disease,* edited by C. R. Scriver, A. L. Beaudet, W. S. Sly, D. Valle, B. Childs, K. W. Kinzler, and B. Vogelstein, Vol. I, Ch. 71, pp. 1521–1551. New York: McGraw-Hill. A chapter in the four-volume treatise considered the most authoritative reference on heritable metabolic human diseases.

Johnson, L. N. (2009) The regulation of protein phosphorylation. *Biochem. Soc. Trans.* 37:627–641. An excellent review that describes recent as well as historical work on protein phosphorylation, including the classical studies on glycogen phosphorylase.

Leloir, L. F. (1983) Long ago and far away. *Annu. Rev. Biochem.* 52:1–16. A personal reminiscence, describing the author's Nobel Prize–winning role in the discovery of nucleotide-linked sugars and the mechanism of glycogen synthesis.

Regulation of Carbohydrate Metabolism

Agius, L. (2008) Glucokinase and molecular aspects of liver glycogen metabolism. *Biochem. J.* 414:1–18.

Bocarsly, M. E., E. S. Powell, N. M. Avena, and B. G. Hoebel (2010) High-fructose corn syrup causes characteristics of obesity in rats: Increased body weight, body fat and triglyceride levels. *Pharmacol. Biochem. Behav.* 97:101–106. Experimental evidence that supports a link between high-fructose corn syrup and obesity and type 2 diabetes.

Brosnan, J. T. (1999) Comments on metabolic needs for glucose and the role of gluconeogenesis. *Eur. J. Clin. Nutr.* 53 Suppl 1;S107–S111.

Sprang, S. R., S. G. Withers, E. J. Goldsmith, R. J. Fletterick, and N. B. Madsen (1991) Structural basis for the activation of glycogen phosphorylase b by adenosine monophosphate. *Science* 254:1367–1371. One of a series of reports describing the crystal structure of glycogen phosphorylase in activated and inactivated states.

Vander Heiden, M. G., L. C. Cantley, and C. B. Thompson (2009) Understanding the Warburg effect: The metabolic requirements of cell proliferation. *Science* 324:1029–1033. A short review that describes the role of aerobic glycolysis in cancer.

Evolution of Carbohydrate Metabolic Pathways

Martin, W., J. Baross, D. Kelley, and M. J. Russell (2008) Hydrothermal vents and the origin of life. *Nat. Rev. Microbiol.* 6:805–814. The chemistry found at hydrothermal vents provides clues about the kinds of reactions that might have initiated the chemistry of life.

Pentose Phosphate Pathway and Oxidative Stress

Sies, H. (1999) Glutathione and its role in cellular functions. *Free Radic. Biol. Med.* 27:916–921. A review of the chemistry and biochemistry of an important biological reductant.

Stincone, A., A. Prigione, T. Cramer, M. M. C. Wamelink, K. Campbell, E. Cheung, V. Olin-Sandoval, N.-M. Grüning, A. Krüger, M. Tauqeer Alam, M. A. Keller, M. Breitenbach, K. M. Brindle, J. D. Rabinowitz, and M. Ralser (2015) The return of metabolism: biochemistry and physiology of the pentose phosphate pathway. *Biological Reviews* 90, 927–963

CHAPTER 13

Regulation of the Citric Acid Cycle

Gray, L. R., S. C. Tompkins, and E. B. Taylor (2014) Regulation of pyruvate metabolism and human disease. *Cellular and Molecular Life Sciences* 71, 2577–2604.

Maj, M. C., J. M. Cameron, and B. H. Robinson (2006) Pyruvate dehydrogenase phosphatase deficiency: Orphan disease or an under-diagnosed condition? *Mol. Cell. Endocrinol.* 249:1–9.

Enzymes of the Citric Acid Cycle and Related Pathways

Kaelin, W. G., Jr., and C. B. Thompson (2010) Q&A: Cancer: Clues from cell metabolism. *Nature* 465:562–564. A short article that describes how mutations in metabolic enzymes, including citric acid cycle enzymes, can lead to cancer.

Mesecar, A. D., and D. E. Koshland Jr. (2000) A new model for protein stereospecificity. *Nature* 403:614–615. A concise illustration of the four-point location model to explain stereospecific binding to enzymes.

Patel, M. S., N. S. Nemeria, W. Furey, and F. Jordan (2014) The Pyruvate Dehydrogenase Complexes: Structure-based Function and Regulation. *J. Biol. Chem.* 289, 16615–16623

Perham, R. N. (2000) Swinging arms and swinging domains in multifunctional enzymes: Catalytic machines for multistep reactions. *Annu. Rev. Biochem.* 69:961–1004. An excellent review of multienzyme complexes, including those that use lipoic acid and biotin.

Srere, P. A., A. D. Sherry, C. R. Malloy, and B. Sumegi (1997) Channelling in the Krebs tricarboxylic acid cycle. In *Channelling in Intermediary Metabolism*, L. Agius, and H. S. A. Sherratt, eds., Vol. IX, pp. 201–217. Portland Press Ltd., London. A general review of the concept of metabolons and channeling in metabolic pathways.

Van Vranken, J. G., U. Na, D. R. Winge, and J. Rutter, (2015) Protein-mediated assembly of succinate dehydrogenase and its cofactors. *Crit. Rev. Biochem. Mol. Biol.* 50:168–180. A review on the biochemistry and medical connections of this key respiratory enzyme.

Zhou, Z. H., D. B. McCarthy, C. M. O'Connor, L. J. Reed, and J. K. Stoops (2001) The remarkable structural and functional organization of the eukaryotic pyruvate dehydrogenase complexes. *Proc. Natl. Acad. Sci.* USA 98:14802–14807.

Experimental Background of the Citric Acid Cycle

Krebs, H. A. (1970) The history of the tricarboxylic acid cycle. *Perspect. Biol. Med.* 14:154–170. A historical account by the man responsible for most of the history.

Snell, E. E. (1993) From bacterial nutrition to enzyme structure: A personal odyssey. *Annu Rev Biochem* 62:1–27. A memoir by one of the scientists most intimately involved in discoveries of vitamins and coenzymes.

Sumegi, B., A. D. Sherry, and C. R. Malloy (1990) Channeling of TCA cycle intermediates in cultured *Saccharomyces cerevisiae*. *Biochemistry* 29:9106–9110. A paper that describes ^{13}C-NMR studies that revealed nonrandom labeling of the symmetrical succinate and fumarate intermediates, suggesting substrate channeling in the cycle.

Wu, F., and S. Minteer (2015) Krebs Cycle Metabolon: Structural Evidence of Substrate Channeling Revealed by Cross-Linking and Mass Spectrometry. *Angewandte Chemie International Edition* 54, 1851–1854. This paper confirms the existence of a multienzyme complex of citric acid cycle enzymes that could support substrate channeling.

The Glyoxylate Cycle

Eastmond, P. J., and I. A. Graham (2001) Re-examining the role of the glyoxylate cycle in oilseeds. *Trends Plant Sci.* 6:72–78. An *Arabidopsis* mutant reveals an anaeroplotic function of the glyoxylate cycle.

CHAPTER 14

Historical Background

Saier, M. H., Jr. (1997) Peter Mitchell and his chemiosmotic theories. *ASM News* 63:13–21. A short scientific biography of the biochemist who proposed the proton gradient as the driving force for ATP synthesis.

Mitochondrial Structure and Function

Scheffler, I. E. (2008) *Mitochondria.* Wiley-Liss, Hoboken, NJ. An up-to-date compendium of mitochondrial structure, function, genetics, and evolution.

Mechanisms in Electron Transport

Letts, J. A., and L. A. Sazanov (2015) Gaining mass: the structure of respiratory complex I — from bacterial towards mitochondrial versions. *Curr. Opin. Struc. Biol.* 33, 135–145. A review of the structure and function of complex I.

Sun, F., Q. Zhou, X. Pang, Y. Xu, and Z. Rao (2013) Revealing various coupling of electron transfer and proton pumping in mitochondrial respiratory chain. *Curr. Opin. Struc. Biol.* 23, 526–538. A review of the structure and proton pumping mechanisms of the mitochondrial electron transfer complexes.

Mechanisms in Oxidative Phosphorylation

Boyer, P. D. (1997) The ATP synthase—A splendid molecular machine. *Annu. Rev. Biochem.* 66:717–750. A mechanistic analysis of the function of F_1F_0 ATP synthase by the person who predicted the correct mechanism of ATP synthesis and did the crucial early experiments.

Maldonado, E. N., and J. J. Lemasters (2014) ATP/ADP ratio, the missed connection between mitochondria and the Warburg effect. *Mitochondrion* 19, 78–84. A minireview on how the adenine nucleotide transporter (ANT) and the mitochondrial membrane potential maintain a high cytoplasmic ATP/ADP ratio and how the Warburg effect controls this ratio in cancer cells.

Nicholls, D. G., and S. J. Ferguson (2002) *Bioenergetics 3,* Academic Press, London. An excellent source on the thermodynamics and mechanisms of chemiosmosis and redox chemistry.

Noji, H., and M. Yoshida (2001) The rotary machine in the cell, ATP synthase. *J. Biol. Chem.* 276:1665–1668. A minireview on the experimental evidence for rotation of the complex in the membrane.

Walker, J. E. (2013) The ATP synthase: the understood, the uncertain and the unknown. *Biochem Soc Trans* 41, 1–16. A review of the structure and function of the F_1F_0 ATP synthase.

Watt, I. N., M. G. Montgomery, M. J. Runswick, A. G. Leslie, and J. E. Walker (2010) Bioenergetic cost of making an adenosine triphosphate molecule in animal mitochondria. *Proc. Natl. Acad. Sci. USA* 107:16823–16827. The structure of the bovine mitochondrial c-ring reveals the stoichiometry of ATP synthesis in higher eukaryotes.

Mitochondrial Genetics, Diseases, and Evolution

Lane, N. (2005) *Power, Sex, Suicide. Mitochondria and the Meaning of Life.* Oxford University Press, Oxford. This excellent book presents the case for the central role played by mitochondria in the evolution of eukaryotic cells, and the consequences for human disease and aging.

Nunnari, J., and A. Suomalainen (2012) Mitochondria: In sickness and in health. *Cell* 148:1145–1159.

Vafai, S. B., and V. K. Mootha (2012) Mitochondrial disorders as windows into an ancient organelle. *Nature* 491:374–383. This review of human mitochondrial disorders connects the biochemistry, genetics, and evolution of mitochondria.

Oxygen Metabolism

Dickinson, B. C., and C. J. Chang (2011) Chemistry and biology of reactive oxygen species in signaling or stress responses. *Nat. Chem. Biol.* 7:504–511.

CHAPTER 15

General

Blankenship, R. E. (2007) *Molecular Mechanisms of Photosynthesis.* John Wiley & Sons, Ltd., Chi-chester. A concise, but complete, introduction to the history, chemistry, mechanisms, physiology, and evolution of photosynthetic systems.

Bowsher, C., M. W. Steer, and A. K. Tobin (2008) *Plant Biochemistry.* Garland Science, New York. An introduction to all aspects of the biochemistry of plants.

Croce, R., and H. van Amerongen (2014) Natural strategies for photosynthetic light harvesting. *Nat. Chem. Biol.* 10, 492–501

Evolution of Photosynthesis

Allen, J. F., and W. Martin (2007) Evolutionary biology: Out of thin air. *Nature* 445:610–612. A concise speculation about when and how cells first learned to split water to make oxygen.

Leslie, M. (2009) On the origin of photosynthesis. *Science* 323:1286–1287.

Light Reactions

Barber, J. (2008) Photosynthetic generation of oxygen. *Philos. Trans. R. Soc. Lond. B. Biol. Sci.* 363:2665–2674. A review of the structure and function of photosystem II.

McEvoy, J. P., and G. W. Brudvig (2006) Water-splitting chemistry of photosystem II. *Chem. Rev.* 106:4455–4483. A detailed review of the chemistry and structure of the manganese cluster of the oxygen-evolving complex.

Rochaix, J.-D. (2011) Regulation of photosynthetic electron transport. *Biochim. Biophys. Acta-Bioenergetics* 1807:375–383.

Structures

Amunts, A., and N. Nelson (2009) Plant photosystem I design in the light of evolution. *Structure* 17:637–650.

Guskov, A., J. Kern, A. Gabdulkhakov, M. Broser, A. Zouni, and W. Saenger (2009) Cyanobacterial photosystem II at 2.9-A resolution and the role of quinones, lipids, channels and chloride. *Nat. Struct. Mol. Biol.* 16:334–342. Describes the structure of the *Thermosynechococcus elongatus* PSII complex illustrated in Figure 15.13.

Carbon Reactions and Photorespiration

Benson, A. A. (2002) Following the path of carbon in photosynthesis: A personal story. *Photosyn. Res.* 73:29–49. A history of the elucidation of the dark reactions by one of the participants.

Whitney, S. M., R. L. Houtz, and H. Alonso (2011) Advancing our understanding and capacity to engineer nature's CO_2-sequestering enzyme, Rubisco. *Plant Physiol* 155:27–35.

CHAPTER 16

Lipid and Lipoprotein Metabolism in Animals

Goldstein, J. L., and M. S. Brown (2015) A Century of Cholesterol and Coronaries: From Plaques to Genes to Statins. *Cell* 161, 161–172. A short account of the discovery and actions of the LDL receptor, written by its discoverers.

Havel, R. J., and J. P. Kane (2001) Introduction: Structure and metabolism of plasma lipoproteins. In *The Metabolic and Molecular Bases of Inherited Disease,* C. R. Scriver, A. L. Beaudet, W. S. Sly, D. Valle, B. Childs, K. W. Kinzler, and B. Vogelstein, eds., Vol. II, Ch. 114, pp. 2705–2716, McGraw-Hill, New York. The first in a series of 10 chapters dealing with clinical disorders of lipid and lipoprotein metabolism.

Walther, T. C., and R. V. Farese (2012) Lipid droplets and cellular lipid metabolism. *Annual Review of Biochemistry* 81:687–714.

Fatty Acid Metabolism

Grininger, M. (2014) Perspectives on the evolution, assembly and conformational dynamics of fatty acid synthase type I (FAS I) systems. *Curr. Opin. Struct. Biol.* 25, 49–56. This review is based on recent structures of mammalian, fungal and bacterial megasynthases.

Saggerson, D. (2008) Malonyl-CoA, a key signaling molecule in mammalian cells. *Ann. Rev. Nutr.* 28:253–272. Reviews the regulation of fatty acid metabolism.

Biochemical Insights into Obesity

Samuel, V. T., and G. I. Shulman (2012) Mechanisms for Insulin Resistance: Common Threads and Missing Links. *Cell* 148, 852-871. This review discusses current thinking on the role of disordered lipid metabolism in diabetes and obesity.

Phospholipid Metabolism

Vance, J. E. (2015) Phospholipid Synthesis and Transport in Mammalian Cells. *Traffic* 16, 1–18.

Sphingolipids

Olson, D. K., F. Fröhlich, R. V. Farese Jr, and T. C. Walther (2016) Taming the sphinx: Mechanisms of cellular sphingolipid homeostasis. *Biochimica et Biophysica Acta (BBA)—Molecular and Cell Biology of Lipids* 1861, 784–792. This review discusses the biosynthetic pathways and regulation of sphingolipids.

Gravel, R. A., M. M. Kaback, R. L. Proia, K. Sandhoff, K. Suzuki, and K. Suzuki (2001) The G_{M2} gangliosidoses. In *The Metabolic and Molecular Bases of Inherited Disease*, C. R. Scriver, A. L. Beaudet, W. S. Sly, D. Valle, B. Childs, K. W. Kinzler, and B. Vogelstein, eds., Vol. III, Ch. 153, pp. 3827–3876, McGraw-Hill, New York. This chapter from a four-volume series on inherited metabolic disorders covers many of the lipid storage diseases.

Steroids and Isoprenoids

Sharpe, L. J., E. C. L. Cook, N. Zelcer, and A. J. Brown (2014) The UPS and downs of cholesterol homeostasis. *Trends Biochem. Sci.* 39, 527–535. This minreview discusses the role of the ubiquitin proteasome system in regulation of cholesterol metabolism.

Sharpe, L. J., and A. J. Brown (2013) Controlling cholesterol synthesis beyond 3-hydroxy-3-methylglutaryl-CoA reductase (HMGCR). *J. Biol. Chem. 288*, 18707–18715. Highlights some of the regulatory mechanisms affecting other enzymes in the cholesterol synthesis pathway.

Lipid-Soluble Vitamins

Card, D. J., R. Gorska, J. Cutler, and D. J. Harrington (2014) Vitamin K metabolism: Current knowledge and future research. *Mol. Nutr. Food Res.* 58, 1590–1600.

Christakos, S., P. Dhawan, A. Verstuyf, L. Verlinden, and G. Carmeliet (2016) Vitamin D: Metabolism, Molecular Mechanism of Action, and Pleiotropic Effects. *Physiol Rev* 96, 365–408. This comprehensive review highlights the roles of vitamin D in bone, skin, the immune system, cardiovascular health, and cancer.

von Lintig, J. (2010) Colors with functions: Elucidating the biochemical and molecular basis of carotenoid metabolism. *Annu. Rev. Nutr.* 30:35–56. Summarizes the pathways of vitamin A synthesis from carotenoids in animals and some of the physiological functions of the vitamin.

Zingg, J.-M. (2015) Vitamin E: A Role in Signal Transduction. *Annual Review of Nutrition* 35, 135–173.

Eicosanoids

Shimizu, T. (2009) Lipid mediators in health and disease: Enzymes and receptors as therapeutic targets for the regulation of immunity and inflammation. *Annu. Rev. Pharmacol. Toxicol.* 49:123–150. Covers prostaglandins, leukotrienes, platelet-activating factor, lysophosphatidic acid, sphingosine 1-phosphate, and other lipid mediators efficiently and completely.

CHAPTER 17

Hormonal Regulation of Fuel Metabolism

Roh, E., D. K. Song, and M.-S. Kim (2016) Emerging role of the brain in the homeostatic regulation of energy and glucose metabolism. *Exp Mol Med* 48, e216.

AMPK, mTOR, and Sirtuins

Albert, V., and M. N. Hall (2015) mTOR signaling in cellular and organismal energetics. *Current Opinion in Cell Biology* 33, 55–66.

Baeza, J., M. J. Smallegan, and J. M. Denu (2016) Mechanisms and Dynamics of Protein Acetylation in Mitochondria. *Trends Biochem. Sci.* 41, 231–244. The review highlights the role of reversible acetylation in the regulation of mitochondrial energy metabolism.

Bonkowski, M. S., and D. A. Sinclair (2016) Slowing ageing by design: the rise of NAD^+ and sirtuin-activating compounds. *Nat Rev Mol Cell Biol* 17, 679–690. This article reviews the connections between calorie restriction, sirtuins, and aging in mammals.

Weikel, K. A., N. B. Ruderman, and J. M. Cacicedo (2016) Unraveling the actions of AMP-activated protein kinase in metabolic diseases: Systemic to molecular insights. *Metabolism* 65, 634–645. A comprehensive review of AMPK function.

Diabetes and Obesity

Friedman, J. M. (2010) A tale of two hormones. *Nat. Med.* 16:1100–1106. A brief history of the discoveries of insulin and leptin, written by the discoverer of leptin.

Samuel, V. T., and G. I. Shulman (2016) The pathogenesis of insulin resistance: integrating signaling pathways and substrate flux. *The J. Clin. Invest.* 126, 12–22. A detailed review of the lipid overload hypothesis.

CHAPTER 18

Inorganic Nitrogen Fixation

Hu, Y., and M. W. Ribbe (2015) Nitrogenase and homologs. *J Biol Inorg Chem* 20, 435–445. An overview of the three classes of nitrogenase known to date.

Udvardi, M., and P. S. Poole (2013) Transport and Metabolism in Legume-Rhizobia Symbioses. *Annu. Rev. Plant Biol.* 64, 781–805. A contemporary view of the complexities of symbiotic nitrogen fixation.

Urea Cycle

Erez, A. (2013) Argininosuccinic aciduria: from a monogenic to a complex disorder. *Genet Med* 15, 251–257. This minireview describes how single-gene defects often lead to complex phenotypes, using a urea cycle disorder as an example.

Diez-Fernandez, C., J. Gallego, J. Haberle, J. Cervera, and V. Rubio (2015) The Study of Carbamoyl Phosphate Synthetase 1 Deficiency Sheds Light on the Mechanism for Switching On/Off the Urea Cycle. *J Genet Genomics* 42, 249–260. An excellent example of investigating the biochemistry of naturally occurring human mutations to advance our understanding of normal regulatory processes.

Protein Turnover

Kleiger, G., and T. Mayor, Perilous journey: a tour of the ubiquitin-proteasome system. *Trends Cell Biol.* 24, 352–359.

Schoenheimer, R., S. Ratner, and D. Rittenberg (1939) Studies in protein metabolism. VII. The metabolism of tyrosine. *J. Biol. Chem.* 127:333–344. This classic paper describes one of the earliest uses of an isotopic tracer to study metabolism in whole animals (indeed, this experiment was performed on a single rat!).

Folate and B_{12} Coenzymes

Bridwell-Rabb, J., and C. L. Drennan (2017) Vitamin B_{12} in the spotlight again. *Curr. Opin. Chem. Biol.* 37, 63–70. An excellent recent review of B_{12} mechanisms, focused on the chemistry and structures of enzymes involved.

Ducker, G. S., and J. D. Rabinowitz (2017) One-Carbon Metabolism in Health and Disease. *Cell Metab* 25, 27–42.

Tibbetts, A. S., and D. R. Appling (2010) Compartmentation of mammalian folate-mediated one-carbon metabolism. *Ann. Rev. Nutr.* 30:57–81. Several of these reactions are catalyzed by multifunctional proteins or multienzyme complexes, and this has implications for optimal therapeutic use of folate antimetabolites.

Amino Acid Metabolism

Brosnan, M. E., L. MacMillan, J. R. Stevens, and J. T. Brosnan (2015) Division of labour: how does folate metabolism partition between one-carbon metabolism and amino acid oxidation? *Biochem J* 472, 135–146.

Burrage, L. C., S. C. S. Nagamani, P. M. Campeau, and B. H. Lee (2014) Branched-chain amino acid metabolism: from rare Mendelian diseases to more common disorders. *Hum. Mol. Genet.* 23, R1–R8. This review describes the biochemistry and clinical consequences of deficiency of the branched-chain α-keto acid dehydrogenase.

Garrod, A. E. (1909) *Inborn Errors of Metabolism. The Croonian lectures delivered before the Royal College of Physicians of London, in June, 1908.* London: Frowde, Hodder & Stoughton. In this classic text, Garrod develops his concept of inheritable metabolic diseases, based on his study of alkaptonuria.

Paul, B. D., and S. H. Snyder (2015) H₂S: A Novel Gasotransmitter that Signals by Sulfhydration. *Trends Biochem. Sci.* 40, 687–700.

Roberts, K. M., and P. F. Fitzpatrick (2013) Mechanisms of Tryptophan and Tyrosine Hydroxylase. *IUBMB Life* 65, 350–357.

Rozen, R. (2001) Polymorphisms of folate and cobalamin metabolism. In *Homocysteine in Health and Disease,* R. Carmel and D. W. Jacobsen, eds., Ch. 22, pp. 259–269. Cambridge, UK: Cambridge University Press. This chapter from a compendium on homocysteine metabolism describes biochemical and genetic aspects of methylenetetrahydrofolate reductase.

Scriver, C. R., and S. Kaufman (2001) Hyperphenylalaninemia: Phenylalanine hydroxylase deficiency. In *The Metabolic and Molecular Bases of Inherited Disease,* C. R. Scriver, A. L. Beaudet, W. S. Sly, D. Valle, B. Childs, K. W. Kinzler, and B. Vogelstein, eds., Vol. II, Ch. 77, pp. 1667–1734. New York: McGraw-Hill. This is the first of 14 chapters in this compendium that describe heritable metabolic disorders of amino acid metabolism.

Wang, R. (2014) Gasotransmitters: growing pains and joys. *Trends Biochem. Sci.* 39, 227–232. This review covers the function of nitric oxide and hydrogen sulfide, and discusses other potential members of this growing family.

CHAPTER 19

Ando, N., and eight coauthors (2015) Allosteric inhibition of human ribonucleotide reductase by dATP entails the stabilization of a hexamer. *Biochemistry* 53:373–381. Structural mechanism of allosteric inhibition.

Blakley, R. L., and S. J. Benkovic (1984) *Folates and Pterins,* Vol. 1. Academic Press, New York. An older but still useful book on folate cofactors and enzymes.

Elion, G. B. (1989) The purine path to chemotherapy. *Science* 244:41–47. Elion's Nobel Prize address, which described the development of allopurinol, acyclovir, 6-thioguanine, and other therapeutically valuable purine analogs.

Finer-Moore, J. S., D. V. Santi, and R. M. Stroud (2003) Lessons and conclusions from dissecting the mechanism of a bisubstrate enzyme: Thymidylate synthase mutagenesis, function, and structure. *Biochemistry* 42:248–256. Protein chemistry of an important drug target.

Franzolin, E., G. Pontarin, C. Rampazzo, C. Miazzi, P. Ferraro, P. Reichard, and V. Bianchi (2013) The deoxynucleotide triphosphohydrolase SAMDH1 is a major regulator of DNA precursor pools in mammalian cells. *Proc. Natl. Acad. Sci. USA* 110:14272–14277. New evidence for dNTP breakdown as a site for controlling intracellular dNTP pools.

Hershfield, M. S., and B. S. Mitchell (2001) Immunodeficiency diseases caused by adenosine deaminase deficiency and purine nucleoside phosphorylase deficiency. In *The Metabolic and Molecular Bases of Inherited Disease*, C. R. Scriver, A. L. Beaudet, W. S. Sly, D. Valle, B. Childs, K. W. Kinzler, and B. Vogelstein, eds., Vol. II, Ch. 109, pp. 2585–2625. New York: McGraw-Hill.

Lee, H., J. Hanes, and K. A. Johnson (2003) Toxicity of nucleoside analogues used to treat AIDS and the selectivity of the mitochondrial DNA polymerase. *Biochemistry* 42:14711–14719. How to minimize toxicity of HIV-AIDS drugs.

Mathews, C. K. (2015) Deoxyribonucleotide metabolism, mutagenesis, and cancer. *Nature Reviews Cancer* 15:528–539. How nucleotide metabolic defects affect mutation rates.

Nordlund, P., and P. Reichard (2006) Ribonucleotide reductases. *Annu. Rev. Biochem.* 75:681–706. A comprehensive review, focusing on the evolutionary significance of the existence of widely divergent classes of this important enzyme.

Pedley, A. M., and S. J. Benkovic (2017) A new view into the regulation of purine metabolism: The purinosome. *Trends in Biochem. Sci.* 42:141–154. A recent review about an important metabolon.

Oda, M., Y. Satta, O. Takenaka, and N. Takahata (2002) Loss of urate oxidase activity in hominoids and its evolutionary implications. *Mol. Biol. Evol.* 19:640–653. Molecular genetic analysis of the urate oxidase gene during hominid evolution.

CHAPTER 20

Black, J. (1989) Drugs from emasculated hormones: The principle of syntopic antagonism. *Science* 245:486–493. Black's Nobel Prize address, which describes the development of drugs that are adrenergic receptor antagonists.

Hill, B. G., B. P. Dranka, S. M. Bailey, J. R. Lancaster, Jr., and V. M. Darley-Usmar (2010) What part of NO don't you understand? Some answers to the cardinal questions in nitric oxide biology. *J. Biol. Chem.* 285:19699–19704.

Kandel, E., and L. Squire (2000) Neuroscience: Breaking down scientific barriers to the study of brain and mind. *Science* 290:1113–1120. Kandel's Nobel Prize lecture published alongside lectures by fellow Nobelists Arvid Carlsson and Paul Greengard.

Lemmon, M. A., and J. Schlessinger (2010) Cell signaling by receptor tyrosine kinases. *Cell* 141:1117–1134. A comprehensive recent review.

Pert, C. B., and S. H. Snyder (1973) Opiate receptor: Demonstration in nervous tissue. *Science* 179:1011–1014. The discovery of opiate receptors.

Rasmussen, S. G. F., and 19 coauthors (2011) Crystal structure of the β₂ adrenergic receptor G₅ protein complex. *Nature* 477:549–557. The first structural analysis of a receptor-G protein complex.

Reinhardt, H. C., and B. Schumacher (2012) The p53 network: Cellular and systemic DNA damage responses in aging and cancer. *Trends in Genetics* 28:128–136. Complexities in the actions of the best-known tumor suppressor.

Scott, J. D., and T. Pawson (2009) Cell signaling in space and time: Where proteins come together and when they're apart. *Science* 326:1220–1224. A recent review by leaders in understanding A kinase anchoring proteins.

Sunahara, R. K., and P. A. Insel (2016) The molecular pharmacology of G protein signaling then and now: A tribute to Alfred G. Glman. *Mol. Pharmacol.* 89:585–592. Gilman was centrally involved in many important advances in cell signaling.

Vogelstein, B., N. Papadopoulos, V. E. Velculesco, S. Zhou, L. A. Diaz, Jr., and K. W. Kinzler (2013) Cancer genome landscapes. *Science* 339:1546–1558. The complexity of cancer as revealed from DNA sequencing in individual tumors. Part of a special issue of *Science* on cancer genomics.

CHAPTER 21

Altshuler, D., M. J. Daly, and E. S. Lander (2008) Genetic mapping in human disease. *Science* 322:881–888. A review of the methods used to map human disease-related genes.

Bonev, B., and G. Cavalli (2016) Organization and function of the 3D genome. *Nature Rev. Genetics* 17:661–678. How genome architecture affects gene expression and evolution.

Ecker, J. R. (2012) Serving up a genome feast. *Nature* 489:52–53. Introduction to a series of short articles that present the essence of the report of the ENCODE Project.

Gibson, D. G., and 23 coauthors (2010) Creation of a bacterial cell controlled by a chemically synthesized genome. *Science* 329:52–56. A big step in the creation of "synthetic life," from the J. Craig Venter Institute.

Koboldt, D. C., K. M. Steinberg, D. E. Larson, R. K. Wilson, and E. R. Mardis (2013) The next-generation sequencing revolution and its impact on genomics. *Cell* 155:27–38. More about genomics than about the actual details of new sequencing techniques.

Kosinski, J., and 11 coauthors (2016) Molecular architecture of the inner ring scaffold of the human nuclear pore complex. *Science* 352, 363–365. One of two articles in this issue of *Science* to describe high-resolution structural analysis of the nuclear pore complex.

Krietenstein, N., M. Wal, S. Watanabe, B. Park, A. L. Peterson, B. F. Pugh, and P. Korber (2016) Genomic nucleosome organization reconstituted with pure proteins. *Cell* 167:700–721. The ultimate test of a structure determination is reassembly of the structure from its constituent parts.

Lander, E. S. (2011) Initial impact of sequencing of the human genome. *Nature* 470:187–197. A look back after 10 years.

Pombo, A., and N. Dillon (2015) Three-dimensional genome architecture: players and mechanisms. *Nature Rev. Mol. Biol.* 16:245–257. New approaches to chromosome organization.

Roach, J. C., and 14 coauthors (2010) Analysis of genetic inheritance in a family quartet by whole-genome sequencing. *Science* 328:636–639. Sequence analysis of closely related family members is an effective way to identify sequencing errors and identify disease genes.

Servick, K. (2015) Can 23andMe have it all? *Science* 349:1472–1477. One of several articles in a special section on corporate approaches to human molecular genetics.

Starr, D. (2016) When DNA is lying. *Science* 351:1133–1136. Part of a special section of *Science* on Forensics.

Tools of Biochemistry 21A Polymerase Chain Reaction

Green, M. R., and J. Sambrook (2012) In vitro amplification of DNA by the polymerase chain reaction. Chapter in *Molecular Cloning: A Laboratory Manual,* Fourth Edition, Cold Spring Harbor, NY: Cold Spring Harbor Press. Specific instructions for using PCR.

Meyer, M., and 33 coauthors (2012) A high-coverage genome sequence from an archaic Denisovan individual *Science* 338:222–226. A spectacular demonstration of the amount of DNA sequence information that can be obtained from minute amounts of ancient tissue.

Mullis, K. B. (1997) *Nobel Lectures, Chemistry, 1991–1995,* B. G. Malmström, ed., Singapore: World Scientific Publishing Co. Available online at:

http://nobelprize.org/nobel_prizes/chemistry/laureates/1993/mullis-lecture.html

CHAPTER 22

Blackburn, E. H. (2009) http://www.nobelprize.org/nobel_prizes/medicine/laureates/2009/blackburn-lecture.html. Telomeres and telomerase: The means to the end. Elizabeth Blackburn's Nobel lecture.

Bleichert, F., M. R. Botchan, and J. M. Berger (2017) Mechanisms for initiating cellular DNA replication. *Science* 355:811. Distinct initiation mechanisms between bacteria and archaea/eukaryotes.

Boye, E., and B. Grallert (2009) In DNA replication, the early bird INSERTes the worm. *Cell* 136:812–814. A nice two-page summary of the complexity of eukaryotic replication initiation.

Burgers, P. M. J., D. Dordenin, and T. A. Kunkel (2016) Who is leading the replication fork, Pol ε or Pol δ? *Mol. Cell* 61:16–19. A brief discussion of a controversy in polymerase biology.

Ganal, R. A., G. O. Bylund, and E. Johansson (2015) Switching between polymerase and exonuclease sites in DNA polymerase ε. *Nucl. Ac. Res.* 43:932–942. Structure-function analysis in a eukaryotic DNA polymerase.

Kelch, B. A., D. L. Makino, M. O'Donnell, and J. Kuriyan (2011) How a DNA polymerase clamp loader opens a sliding clamp. *Science* 334:1675–1680. The mechanism is based on crystallographic analysis of clamp loaders..

López, V., M. L. Martinez-Robles, P. Hernández, D. B. Krimer, and J. B. Schvartzman (2012) Topo IV is the topoisomerase that knots and unknots sister duplexes during DNA replication. *Nucl. Ac. Res.* 40:3563–3573. Nice treatment of the topological complexities of DNA replication.

Mathews, C. K. (2014) Deoxyribonucleotides as genetic and metabolic regulators. *FASEB J* 28:3832–3840. A review of relationships between nucleotide metabolism and DNA replication fidelity.

Pterman, E., A. Gemmill, D. Karasek, D. Weir, N. E. Adler, A. A. Prather, and E. Epel (2016) Lifespan adversity and later adulthood telomere length in the nationally representative U.S. health and retirement survey. *Proc Natl, Acad. Sci USA* E6335–E6342. A large controlled study reveals how lifespan adversity in early life affects the telomeres.

Sun, J., Y. Shi, R. E. Georgescu, Z. Yuan, B. T. Chait, H. Li, and M. E. O'Donnell (2015) The architecture of a eukaryotic replisome. *Nature Str, Mol, Biol.* 22:976–982. A single-particle electron microscopic analysis of the yeast replication complex.

Symanski, M. R., V. B. Kuznetsov, C. Shumate, Q. Meng, Y.-S. Lee, G. Patel, S. Patel, an Y. W. Yin (2015) Structural basis for processivity and antiviral drug toxicity in human mitochondrial DNA replicase. *EMBO J* 34:1845–1985. Structural analysis of human mitochondrial DNA polymerase has a large potential biomedical payoff.

Wang, J. C. (2009) *Untangling the Double Helix: DNA Entanglement and the Action of the DNA Topoisomerases.* 233 pages. Cold Spring Harbor, NY: Cold Spring Harbor Press. Up-to-date presentation of topoisomerases and DNA tertiary structure by the discoverer of topoisomerases.

Yao, N. Y., and M. E. O'Donnell (2017) DNA replication: How does a sliding clamp slide? *Current Bol.* 27:R172–R190. Insights revealed from use of biophysical techniques.

Yeeles, J. T. P., A. Janska, A. Early, and J. F. X. Diffley (2017) How the eukaryotic replisome achieves rapid and efficient DNA replication. *Molecular Cell* 65:106–116. One of two companion papers describing reconstitution of a eukaryotic repisome from purified components.

CHAPTER 23

Bell, J. C., and S. C. Kowalczykowski (2016) RecA: Regulation and mechanism of a molecular search engine. *Trends in Biochem. Res.* 41:491–507. How a protein scans DNA searching for sequence homology.

Capecchi, M. R. (2007) Gene targeting, 1977–present. http://nobelprize.org/nobel_prizes/medicine/laureates/2007capecchi lecture.html. A description of gene targeting for specific gene inactivation.

Chapman, J. R., M. R. G. Taylor, and S. J. Boulton (2012) Playing the end game: DNA double-strand break repair pathway choice. *Mol. Cell* 47:497–510. Recent review of both pathways of double-strand break repair.

Cleaver, J. E., E. T. Lam, and I. Revet (2009) Disorders of nucleotide excision repair: The genetic and molecular basis of heterogeneity. *Nat. Rev. Genet.* 10:756–768. Xeroderma pigmentosum is not the only disease resulting from faulty NER.

Doudna, J. A., and E. Charpentier (2014) The new frontier of genome editing with CRISPR-Cas9. *Science* 346:1258096(2014).DOI:10.1126/science.1258096. One of several recent reviews by two of the central figures. A one-page summary of the on-line version is at *Science* 346, 1077 (2014).

Friedberg, E. C., G. C. Walker, W. Siede, R. D. Wood, R. A. Schultz, and T. Ellenberger (2006) *DNA Repair and Mutagenesis,* 2nd ed. Washington, DC: ASM Press. The most authoritative contemporary book-length review on DNA repair.

Haber, J. E. (2015) Deciphering the DNA damage response. *Cell* 160, 1193–1195. An appreciation of the 2015 Lasker Awards, given in recognition of the elucidation of the DNA damage response.

Holloman, M. K. (2011) Unraveling the mechanism of BRCA2 in homologous recombination. *Nature Str. and Mol. Biol.* 18:748–754. Because of the status of BRCA2 mutations as breast cancer risk factors, action of the gene product is of special interest.

Kunkel, T. A. (2015) Celebrating DNA's repair crew. *Cell* 162:1201–1203. A brief appreciation of the three DNA repair processes that were recognized with the 2015 Nobel Prize in Chemistry.

Lander, E. S. (2016) The heroes of CRISPR. *Cell* 164:18–27. A readable account of a series of serendipitous findings in several laboratories that triggered a revolution in biology.

McClintock, B. (1984) The significance of responses of the genome to challenge. *Science* 226:792–801. McClintock's Nobel Prize address, giving the history of the first description of mobile genetic elements.

Montano, S. P., and P. A. Rice (2011) Moving DNA around: DNA transposition and retroviral integration. *Curr. Opin. Str. Biol.* 21:370–378. Recent review of mobile genetic elements.

Schlissel, M. S., D. Schulz, and C. Vetterman (2009) A histone code for regulating V(D)J recombination. *Mol. Cell* 34:639–640. A recent minireview of antibody gene rearrangements.

CHAPTER 24

Cheung, A. C. M., and P. Cramer (2012) A movie of RNA polymerase II transcription. *Cell* 149:1431–1437. A description of the process with reference to a spectacular movie. http://dx/doi.org/10.1016/j.cell.2012.06.006

Egloff, S., M. Dienstbier, and S. Murphy (2012) Updating the RNA polymerase CTD code: Adding gene-specific layers. *Trends Genet.* 28:333–341. The phosphorylation state of RNA polymerase II C-terminal domain controls numerous functions.

Haag, J. R., and C. S. Pikaard (2011) Multisubunit RNA polymerases IV and V: Purveyors of non-coding RNA for plant gene silencing. *Nat. Rev. Mol. Cell Biol.* 12:483–492. Review of the plant RNA polymerases not described in this chapter.

Kornblitt, A. R., I. E. Schor, M. Alló, G. Dujardin, E. Petrillo, and M. J. Muñoz (2013) Alternative splicing: A pivotal step between eukaryotic transcription and translation. *Nat. Rev. Mol. Cell Biol.* 13:153–165. Timely review of alternative splicing.

Ma, J., and M. D. Wang (2014) RNA polymerase is a powerful torsional motor. *Cell Cycle* 13:337–338. A new experimental approach to understanding mechanical properties of RNA polymerase.

Matera, A. G., and Z. Wang (2014) A day in the life of the spliceosome. *Nature Rev. Mol. Cell Biol.* 15:108–121. A recent review of spliceosome structure and function.

Murakami, K., (2013) X-ray crystal structure of *Escherichia coli* RNA polymerase σ^{70} holoenzyme. *J. Biol. Chem.* 288:9126–9134. The first structure of a bacterial RNA polymerase holoenzyme.

Murakami, K., and 11 coauthors (2013) Architecture of an RNA polymerase II transcription pre-initiation complex. *Science* 342:709. Summary of a multipronged structural analysis, with reference to the full paper.

Nudler, E. (2012) RNA polymerase backtracking in gene regulation and genome instability. *Cell* 149:1438–1445. Backtracking serves multiple functions.

Padgett, R. A. (2012) New connections between splicing and human disease. *Trends Genet.* 28:147–154. Many genetic diseases in humans are caused by defective splicing.

Petesch, S. J., and J. T. Lis (2012) Overcoming the nucleosome barrier during transcription elongation. *Trends Genet.* 28:285–294. A review of one of the most difficult problems in understanding eukaryotic transcription.

Vannini, A., and P. Cramer (2012) Conservation between the RNA polymerase I, II, and III transcription initiation machines. *Mol. Cell* 45:439–446. Fundamental similarities in the actions of the three eukaryotic RNA polymerases.

Werner, F., and D. Grohmann (2011) Evolution of multisubunit RNA polymerases in the three domains of life. *Nat. Rev. Microbiol.* 9:85–98. Comparison of bacterial, archaeal, and eukaryotic RNA polymerases.

Yang, C., R. Wan, R. Bai, G. Huang, and Y. Shi (2017) Structure of a yeast step II catalytically activated spliceosome. *Science* 355, 149–155. A Chinese laboratory shocked the world with a complete spliceosome structure.

Tools of Biochemistry 24A Analyzing the Transcriptome

Ioannidis, J. P. A., and 15 coauthors (2009) Repeatability of published microarray gene expression analyses. *Nat. Genet.* 41:149–155. Analyses of data in 18 articles emphasize the importance of adequate controls, statistical analysis, and experimental details in generating and publishing meaningful microarray data.

Mittal, V. (2001) DNA array technology. In Sambrook, J., and D. W. Russell, eds., *Molecular Cloning: A Laboratory Manual,* 3rd ed. Vol. 3, pp. A.10.1–A.10.19. Cold Spring Harbor, NY: Cold Spring Harbor Laboratory. Straightforward description of the technology.

Wang, Z., M. Gerstein, and M. Snyder (2009) RNA-Seq: A revolutionary tool for transcriptomics. *Nature Rev. Genet.* 10:57–63. An informative review.

Tools of Biochemistry 24B Chromatin Immunoprecipitation

Furey, T. S. (2012) ChIP-seq and beyond: New and improved methodologies to detect and characterize protein-DNA interactions. *Nat. Rev Genet.*13:840–847. A recent review with references to the earlier literature.

Rhee, H. S., and B. F. Pugh (2012) Comprehensive genome-wide protein-DNA interactions detected at single-nucleotide resolution. *Cell* 147:1408–1419. Recent improvements in ChIP techniques.

CHAPTER 25

Becker, T., L. Böttinger, and N. Pfanner (2011) Mitochondrial protein import: From transport pathways to an integrated network. *Trends Biochem. Sci.* 37:85–91. A review of five pathways of protein transport through mitochondrial membranes.

Lajoie, M. J., D. Söll, and G. M. Church (2016) Overcoming challenges in engineering the genetic code. *J. Mol. Biol.* 428:1004–1021. Recoding—the repurposing of genetic codons—is a powerful strategy for enhancing genomes with functions not commonly found in nature.

Li, L., C. Francklyn, and C. W. Carter, Jr. (2013) Aminoacylating urzymes challenge the RNA world hypothesis. *J. Biol. Chem.* 288:26856–25863. Surprising catalytic efficiency of aminoacyl-tRNA synthetase fragments.

Mulder, A. M., C. Yoshioka, A. H. Beck, A. E. Bunner, R. A. Milligan, C. S. Potter, B. Carragher, and J. R. Williamson (2010) Visualizing ribosome biogenesis: Parallel assembly pathways for the 30S subunit. *Science* 330:673–677. Evidence for major distinctions between ribosome assembly pathways in vitro and in vivo.

Pulk, A., and J. H. D. Cate (2013) Control of ribosomal subunit rotation by elongation factor G. *Science* 340:1235790–1–7. Structure of G and its mechanistic significance.

Ramakrishnan, V. (2009) Decoding the genetic message: The 3D version. 2009 Nobel Lecture. http://nobelprize.org/nobel_prizes/chemistry/laureates/2009/ramakrishnan-lecture.html. This wide-ranging lecture includes a movie showing ribosomal movements during translation.

Roberts, J. W. (2010) Syntheses that stay together. *Science* 328:436–437. A summary of two papers that uncovered the transcription–translation coupling mechanism.

Schmeing, T. M., R. M. Voorhees, A. C. Kelley, Y-G. Gao, F. V. Murphy IV, J. R. Weir, and V. Ramakrishnan (2009) The crystal structure of the ribosome bound to EF-Tu and aminoacyl-tRNA. *Science* 326:688–694. This and a companion paper from the same laboratory present a structural analysis of events in translation.

Steitz, T. A. (2009) From understanding ribosome structure and function to new antibiotics. 2009 Nobel Lecture in Chemistry. http://nobelprize.org/nobel_prizes/chemistry/laureates/2009/steitz-lecture.html. Steitz's Nobel lecture covers known antibiotic action from a structural standpoint and describes new antibiotics emerging from structural work.

Yuan, J., P. O'Donoghue, A. Ambrogelly, S. Gundllapalli, R. L. Sherrer, S. Paliora, M. Siminovic, and D. Söll (2010) Distinct genetic code expansion strategies for selenocysteine and pyrrolysine are reflected in different aminoacyl-tRNA formation systems. *FEBS Letters* 584:342–349. Coding for the 21st and 22nd amino acids.

Zhu, J., A. Korostelev, L. Lancaster, and H. F. Noller (2012) Crystal structures of 70S ribosomes bound to release factors RF1, RF2, and RF3. *Curr. Opin. Str. Biol* 22:733–742. Structural insights into polypeptide chain termination.

Zimmerman, E., and A. Yonath (2009) Biological implications of the ribosome's stunning stereochemistry. *ChemBioChem* 10:63–72. A minireview coauthored by one of the 2009 Nobel laureates honored for ribosome structure determination.

CHAPTER 26

Baylin, S. B. (2016) Jacob, Monod, the Lac operon, and the PaJaMa experiment—Gene expression circuitry changing the face of cancer research. *Cancer Res.* 76:2060–2062. How half-century-old science is impacting current cancer research.

Gurtan, A. M, and P. A. Sharp (2013) The role of miRNAs in regulating gene expression networks. J. Mol. Biol. 425:3582–3600. A comprehensive, fairly recent review.

Harmston, N., and B. Lenhard (2013) Chromatin and epigenetic features of long-range gene regulation. *Nucl. Ac. Res.* 41:7185–7199. Histone modifications and enhancers.

Ishihama, A. (2010) Prokaryotic gene regulation: Multifactor promoters, multitarget regulators and hierarchic networks. *FEMS Microbiol. Rev.* 34:628–645. Complexities of prokaryotic transcription revealed by genomic analysis.

Jones, P. A. (2012) Functions of DNA methylation: Islands, start sites, gene bodies, and beyond. *Nature Rev. Genet.* 13:484–492. Nice review of DNA methylation.

Malik, S., and R. G. Roeder (2016) Mediator: A drawbridge across the enhancer-promoter divide. *Molecular Cell* 64:433–434. A recent minireview that summaries two longer research papers in the same issue of the journal.

Nishikura, K. (2016) A-to-I editing of coding and noncoding RNAs by ADARs. *Nature Revs. Mol. Cell Biol.* 17:83–86.

Ptashne, M. (2014) The chemistry of regulation of genes and other things. *J. Biol. Chem.* 289:5417–5435. A semi-autobiographical article, in which Ptashne outlines his contributions and those of others to gene regulation in phage λ and in yeast.

Rasmussen, K. D., and K. Helin (2016) Role of TET enzymes in DNA methylation, development, and cancer. *Genes and Development* 30:733–750. Important aspects of DNA methylation and hydroxymethylation.

Serganov, A., and E. Nudler (2013) A decade of riboswitches. *Cell* 152:17–24.

Venters, B. J., and B. F. Pugh (2009) How eukaryotic genes are transcribed. *Crit. Rev. Biochem. Mol. Biol.* 44:117–141. A comprehensive and readable review.

Workman, J. L. (2016) It takes teamwork to modify chromatin. *Science* 351:667. A timely one-page microreview.

Zentner, G. E., and S. Henikoff (2013) Regulation of nucleosome dynamics by histone modifications. *Nature Str. and Mol. Biol.* 20:259–266.

Credits

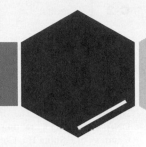

PHOTO CREDITS

CHAPTER 1   **page 2:** Comaniciu Dan/123RF. **page 5:** DEA/G. DAGLI ORTI/De Agostini/Getty Images. **page 6:** A. Barrington Brown/Science Source. **page 9:** Don W. Fawcett/Science Source. **page 14:** Juergen Berger/Science Source. **page 15 top right:** Don W. Fawcett/Science Source. **page 15 bottom right:** Biophoto Associates/Science Source. **page 16 left:** Alfred Pasieka/Science Source. **page 16 right:** SPL/Science Source. **page 17:** From: A global genetic interaction network maps a wiring diagram of cellular function. M. Constanzo et al. *Science.* 2016 Sep 23;353(6306). pii: aaf1420, Fig. 1.

CHAPTER 2   **page 18:** Niyazz/Shutterstock. **page 22:** MarcelClemens/Shutterstock. **page 30:** D. W. Deamer/P. B. Armstrong/University of California, Davis. **page 45 top:** Jarrod Erbe/Shutterstock. **page 45 bottom left:** U.S. Food and Drug Administration.

CHAPTER 3   **page 48:** Elodie Giuge/Moment open/Getty Images. **page 48 inset:** Anita P Peppers/Fotolia.

CHAPTER 4   **page 72:** Permission Used by Nadrian C. Seeman. **page 74:** Biology Pics/Science Source. **page 81:** Science Source. **page 87 left:** Dr. Gopal Murti/Science Source. **page 87 right:** Howard Hughes Medical Institute. **page 89:** James C. Wang MCB Emeritus Faculty. **page 101 left:** © 2004 Chengde Mao. **page 101 right:** Prof. Nadrian C. Seeman, Department of Chemistry, New York University. **page 106:** P. Shing Ho Dept of Biochemistry and Molecular Biology College of Natural Sciences Colorado State University. **page 107:** Wah Chiu/Baylor College of Medicine.

CHAPTER 5   **page 108:** Michael J. Klein, M.D./Cultura RM Exclusive/Getty Images. **page 137:** Cory Hamada.

CHAPTER 6   **page 144:** Copyright © 2017 Copyright Clearance Center, Inc. All Rights Reserved. **page 155:** J. Gross/Biozentrum, University of Basel/Science Source. page 171: Adapted from Proceedings of the National Academy of Science of the United States of America 99: 9196–9201, J.L. Jimenez, E.J. Nettleton, M. Bouchard, C.V. Robinson, C.M. Dobson, and H.R. Saibil, *The protofilament structure of insulin amyloid fibrils.* © 2002 National Academy of Sciences, USA. **page 173:** Science Source. **page 180:** William Horton/Portland Shriners Research Center/Oregon Health & Science University. **page 181:** Nathan Shaner/Roger Tsien Lab/Composite by Paul Steinbach.

CHAPTER 7   **page 190:** Dlumen/istock/Getty Images. **page 199 top left:** Courtesy of Hans Gelderblom, Robert Koch Institute, Berlin. **page 199 top center:** C. Goldsmith, P. Feorino, E. L. Palmer, and W. R. McManus/Centers for Disease Control. **page 200:** Eye of Science/Science Source. **page 219 top right:** Science Source. **page 219 bottom right:** Wellems TE and Josephs R (1979). Crystallization of deoxyhemoglobin S by fiber alignment and fusion. *J Mol Biol* 135:651–674. **page 220 top left:** Bridget Carragher. **page 220 top center:** Bridget Carragher. **page 221:** Courtesy of Roger Craig. **page 222:**

Courtesy of Thomas D. Pollard, Yale University. **page 224 top left:** Mediscan/Alamy Stock Photo. **page 224 top right:** Mary C. Reedy. **page 231 left:** Department of Autoimmunology, Statens Serum Institut, Copenhagen, Denmark. nhe@ssi.dk. **page 231 right:** Dr. Alexey Khodjakov/Science Source.

CHAPTER 9   **page 293 left:** Biophoto Associates/Science Source. **page 293 center:** Dr. Lloyd M. Beidler/Science Source. **page 293 right:** MedImage/Science Source. **page 295:** Biophoto Associates/Science Source. **page 296:** Republished with permission of Elsevier Science and Technology Journals, from *Electron Microscopy Reviews*, Joseph A. Buckwalter, Lawrence C. Rosenberg, 1(1):87–112, 1988 permission conveyed through Copyright Clearance Center, Inc.

CHAPTER 10   **page 309:** Steve Gschmeissner/Science Source. **page 320 bottom left:** Courtesy of Marjorie Longo.

CHAPTER 11   **page 340:** MedImage/Science Source. **page 361:** The Rockefeller University Press. *The Journal of Cell Biology*, 1979, 82:114–139, J. J. Wolosewick and K. R. Porter, Microtrabecular lattice of the cytoplasmic ground substance: Artifact or reality.

CHAPTER 12   **page 374:** Michael Interisano/Design Pics Inc/Alamy Stock Photo. **page 388:** © 1999 National Academy of Sciences, U.S.A.

CHAPTER 13   **page 420:** National Institutes of Health Office of History. **page 424:** Professors Pietro M. Motta/Tomonori Naguro/Science Source. **page 427:** *The remarkable structural and functional organization of the eukaryotic pyruvate dehydrogenase complexes* Z. Hong Zhou, Diane B. McCarthy, Catherine M. O'Connor, Lester J. Reed, and James K. Stoops. Proc Natl Acad Sci U S A. Dec 18, 2001; 98(26);14802–14807. doi:10.1073/pnas.011597698 © 2001 National Academy of Sciences, U.S.A.

CHAPTER 14   **page 450:** Boris Jordan Photography/Moment Open/Getty Images. **page 454:** Keith R. Porter/Science Source. **page 468:** Alex Tzagoloff. **page 471:** *Journal of Biological Chemistry* 276:1665–1668, H. Noji and M. Yoshida, The rotary machine in the cell, ATP synthase. © 2001 The American Society for Biochemistry and Molecular Biology. All rights reserved. **page 474:** Dean Appling.

CHAPTER 15   **page 486:** Mythja/Shutterstock. **page 491:** Biophoto Associates/Science Source. **page 491:** Biophoto Associates/Science Source.

CHAPTER 16   **page 512:** Steve Byland/Fotolia. **page 520:** Courtesy of R. G. W. Anderson, M. S. Brown, and J. L. Goldstein. **page 521 left, center:** RCSB Protein Data Bank, http://www.rcsb.org/pdb/home/home.do. Molecule of the Month, illustrations by David Goodsell. **page 521 right:** Courtesy of John Heuser.

CHAPTER 17   **page 556:** ORNL/Science Source.

CHAPTER 18   **page 576:** Sukpaiboonwat/Shutterstock. **page 580:** Dan Guravich/Science Source. **page 604:** Rowena Matthews.

CHAPTER 19   **page 610:** Scenics & Science/Alamy Stock Photo. **page 610 inset:** Joloei/Shutterstock. **page 626:** Figure provided courtesy of Edward Brignole and Catherine L. Drennan.

CHAPTER 20   **page 636:** Dr. Gopal Murti/Science Source. **page 658:** SPL/Science Source.

CHAPTER 21   **page 667:** CNRI/Science Source. **page 669:** Phillip A. Sharp and The National Academy of Sciences. **page 671 bottom left:** Courtesy of G. F. Bahr, Armed Forces Institute of Pathology. **page 671 bottom right:** Courtesy of Peter Lansdorp, University of British Columbia, Canada and University Medical Center, Groningen, The Netherlands. **page 672:** Ken van Holde. **page 675:** Catherine Z. Mathew. **page 677:** Reproduced from *The Human Genome Project: Deciphering the Blueprint of Heredity*, N.G. Cooper, ed., p. 112. Copyright 1994 University Science Books, Mill Valley, CA. **page 681:** Martin Shields/Alamy Stock Photo.

CHAPTER 22   **page 688:** *Cold Spring Harbor Symposia on Quantitative Biology* 28:44, J. Cairns. 1963 Cold Spring Harbor Laboratory Press. **page 689:** Reprinted from *Journal of Molecular Biology* 74, 599–604, R. L. Rodriguez, M. S. Dalbey, and C. I. Davern © 1973, with permission from Elsevier. **page 699:** From *Science* 206:1081–1083, P. O. Brown and N. R. Cozzarelli, A sign inversion mechanism for enzymatic supercoiling of DNA. 1979. Reprinted with permission from AAAS and Pat Brown.

CHAPTER 23   **page 714:** NASA Earth Observatory. **page 728:** Courtesy of R. B. Inman. **page 735:** Courtesy of Professor Joyce L. Hamlin.

CHAPTER 24   **page 742:** Damnederangel/iStock/Getty images.

CHAPTER 25   **page 766:** Monty Rakusen/Cultura/Getty Images. **page 790:** Dr. Barbara Hamkalo.

CHAPTER 26   **page 796:** Juliet photography/Shutterstock. **page 796 inset:** From Napoli, C., Lemieux, C., and Jorgensen, R. (1990) The Plant Cell 2:279–289.

TEXT CREDITS

CHAPTER 1   **page 3:** From The two cultures: Chemistry and biology. *Biochemistry* 26:6888–6891. by Arthur Kornberg. Published by ACS publications, © 1987. **page 4** From Wöhler to Berzelius (1828), Published by J. M. McBride, 2003. **page 13 Figure 1.10:** Adapted from *Microbe Magazine*, January 2008, p. 17, © 2008 American Society for Microbiology. Used with permission.

CHAPTER 2   **page 22 Table 2.1:** Data from Advances in Protein Chemistry 39:125–189, S. K. Burley and G. A. Petsko, Weakly polar interactions in proteins.

CHAPTER 4   **page 82 Figure 4.11a, c:** Illustration, Irving Geis. Image from the Irving Geis Collection/Howard Hughes Medical Institute. Rights owned by HHMI. **page 86 Figure 4.17:** Data from R.E. Dickerson, *Sci. Am.* December 1983, pp. 100–104. **page 101 Figure 4.A5:** Nadrian C. Seeman, Nanomaterials Based on DNA, *Annu. Rev. Biochem.* 2010. 79:65–87. **page 103 Figure 4.A7:** Data from Dr. Robert H. Lyons, The University of Michigan DNA Sequencing Core. **page 105 Figure 4.B6:** Courtesy of P.Shing Ho, Colorado State University.

CHAPTER 5 **page 115 Figure 5.6:** Data from *Advances in Protein Chemistry* 17:303–390, D. B. Wetlaufer, Ultraviolet spectra of proteins and amino acids. © 1962. **page 143 Figure 5.B6a.c:** Courtesy of Jack Benner. **page 143 Figure 5.B6d:** Based on data from UniProt database, http://www.uniprot.org/uniprot/?query=UniProt&sort=score, retrieved from http://www.expasy.org, courtesy of Swiss Institute of Bioinformatics, covered under http://creativecommons.org/licenses/by-nd/2.5.

CHAPTER 6 **page 151 Figure 6.9:** Plots courtesy S. A. Hollingsworth and P. A. Karplus, Oregon State University. **page 151 Table 6.2:** Data from *Protein Science* 18:1321–1325 (2009), S. A. Hollingsworth, D. S. Berkholz, and P. A. Karplus, On the occurrence of linear groups in proteins. **page 152 Figure 6.10:** Plots courtesy S. A. Hollingsworth and P. A. Karplus, Oregon State University. **page 152 Figure 6.11:** Plots courtesy S.A. Hollingsworth and P. A. Karplus, Oregon State University. **page 153 Table 6.3:** Data from *Journal of Chemical Information and Modeling* 50:690–700, J. M. Otaki, M. Tsutsumi, T. Gotoh, and H. Yamamoto, Secondary structure characterization based on amino acid composition and availability in proteins. **page 162 Figure 6.22b:** Data from *Biochemistry* 4:2159–2174 (1965), A. Ginsburg and W. R. Carroll, Some specific ion effects on the conformation and thermal stability of ribonuclease. **page 165 Figure 6.26:** Data from *European Journal of Biochemistry* (1971) 23:401–411 J. P. Vincent, R. Chicheportiche, and M. Lazdunski, The conformational properties of the basic pancreatic trypsin-inhibitor. **page 165 Figure 6.27:** Images in panel (b) courtesy of J. S. Olson (Rice University). **page 166 Figure 6.28:** Images courtesy of J. S. Olson (Rice University). **page 169 Figure 6.30a:** *Nature Reviews Drug Discovery* by Nature Publishing Group. Reproduced with permission of Nature Publishing Group in the format Republish in a book via Copyright Clearance Center. **page 169 Figure 6.30b:** *Nature structural & molecular biology* by Nature Publishing Group. Reproduced with permission of Nature Publishing Group in the format Republish in a book via Copyright Clearance Center. **page 169 Figure 6.31d:** *Nature* by Nature Publishing Group. Reproduced with permission of Nature Publishing Group in the format Republish in a book via Copyright Clearance Center. **page 169 Figure 6.31e:** Source: Adapted from *Trends in Biochemical Science* 23:68–73, W. S. Netzer and F. U. Hartl, Protein folding in the cytosol: Chaperonin-dependent and -independent mechanisms. © 1998. **page 171 Figure 6.32c:** *Nature* by Nature Publishing Group. Reproduced with permission of Nature Publishing Group in the format Republish in a book via Copyright Clearance Center. **page 171 Figure 6.32d:** Source: Robert Tycko, National Institutes of Health, Bethesda, MD. **page 172 Figure 6.33:** Republished with permission of Annual Reviews, from *Annual Review of Biochemistry* 77:363–382, R. Das and D. Baker, Macromolecular modeling with Rosetta. © 2008 Annual Reviews.; permission conveyed through Copyright Clearance Center, Inc. **page 173 Figure 6.35b:** Appling, D. R., Mathews, C. K., & Anthony-Cahill, S. J. (2016). *Biochemistry: Concepts and Connections.* New Jersey: Pearson Education. **page 183 Figure 6A.9:**

Source: Courtesy of S. Delbecq and R. Klevit, University of Washington. **page 188 Figure FF.6.1:** Plots courtesy S.A. Hollingsworth and P. A. Karplus, Oregon State University.

CHAPTER 7 **page 199 Figure 7.11:** Source: Reproduced with permission from *Accounts of Chemical Research* 41:98–107, R. V. Chari, *Targeted cancer therapy: Conferring specificity to cytotoxic drugs.* © 2008 American Chemical Society. **page 203 Figure 7.17:** Source: Courtesy of John S. Olson, Rice University. **page 214 Figure 7.30:** Source: Data from British Medical Bulletin 32:209–212, J. V. Kilmartin, Interaction of haemoglobin with protons, CO2 and 2, 3-disphosphoglycerate. © 1976. **page 217 Figure 7.33:** Source: Data from British Medical Bulletin 32:282–287, W. G. Wood, *Haemoglobin synthesis during human fetal development.* © 1976 Oxford University Press.

CHAPTER 8 **page 261 Figure 8.33:** From: Figure 1 in S.R. Stone and J.F. Morrison, "Kinetic Mechanism of the Reaction Catalyzed by Dihydrofolate Reductase from Escherichia coli" *Biochemistry*, 1982, vol 21: pages 3757–3765.

CHAPTER 9 **page 300 Figure 9.25:** Images of the H1N1 Influenza Virus, Centers for Disease Control and Prevention. http://www.cdc.gov/h1n1flu/images.htm.

CHAPTER 10 **page 310 Figure 10.5c:** Source: Generated from data published in Heller, H., Schaefer, M., and K. Schulten (1993) "Molecular dynamics simulation of a bilayer of 200 lipids in the gel and the liquid-crystal phases." *J. Phys. Chem.* 97:8343–8360. **page 311 Table 10.4:** Data from C. Tanford (1973) The Hydrophobic Effect. Wiley, New York. **page 313 Figure 10.T5:** Adapted from Annual Review of Biochemistry 41:731, G. Guidotti, Membrane proteins. © 1972 Annual Reviews. **page 314 Figure 10.11a:** Generated from computational modeling data published in Heller, H., Schaefer, M., and K. Schulten (1993) "Molecular dynamics simulation of a bilayer of 200 lipids in the gel and the liquid-crystal phases." *J. Phys. Chem.* 97:8343–8360. **page 323 Figure 10.T6:** Data from M. K. Jain and R. C. Wagner (1980) *Introduction to Biological Membranes.* Wiley, New York. **page 326 Figure 10.28c:** Based on Federation of the European Biochemical Societies, FEBS Letters 555:72–78, P. Agre and D. Kozono, *Water channels: Molecular mechanisms for human disease.*

CHAPTER 11 **page 369 Figure 11.A2:** Courtesy of Stefano Tiziani, University of Texas at Austin. **page 371 Figure 11.B1a–d:** Courtesy of Dean Sherry, Craig Malloy and Jimin Ren of University of Texas-Southwestern Medical Center.

CHAPTER 13 **page 427 Figure 13.5b–e:** Source: *The Journal of Biological Chemistry* 276:38329–38336, L. Reed, A trail of research from lipoic acid to ?-keto acid dehydrogenase complexes. Reprinted with permission. © 2001. The American Society for Biochemistry and Molecular Biology. **page 437 bottom left:** Source: Data from Roberts and Company Publishers as seen in *The Organic Chemistry of Biological Pathways* by John McMurry and Tadhg Begley.

CHAPTER 14 **page 471 Figure 14.22a:** Source: Data from *Cell* 93:1117–1124, R. Yasuda, H. Noji, K. Kinosita Jr., and Y. Masasuke, F1-ATPase is a highly efficient molecular motor that rotates with discrete 120° steps. **page 478 Figure 14.28b:** Data for (b) from *Biochimica et Biophysica Acta* 1658:80–88, S. DiMauro, Mitochondrial

diseases. **page 482 bottom left:** Cordes et al (*J. Biol. Chem.* 291:14274-14284, 2016).

CHAPTER 15 **page 503:** John F. Allen.

CHAPTER 16 **page 521 Figure 16.8a–b:** Adapted from RCSB Protein Data Bank, http://www.rcsb.org/pdb/home/home.do. Molecule of the Month, illustrations by David Goodsell. http://creativecommons.org/licenses/by/4.0/.

CHAPTER 17 **page 573:** The journal of clinical investigation by American Society for Clinical Investigation Reproduced with permission of American Society for Clinical Investigation the format Republish in a book via Copyright Clearance Center.

CHAPTER 18 **page 583 Figure 18.6:** Source: Courtesy of M. E. Matyskiela, G. C. Lander, and A. Martin, Univ. of California, Berkeley.

CHAPTER 19 **page 626 Figure 19.16:** Source: From N. Ando et al (2011) Proc. Natl. Acad. Sci. USA 108:21046–21051.

CHAPTER 20 **page 644 Figure 20.6:** Source: Data from Rasmussen, S. G. F., and 19 coauthors (2011) Crystal structure of the b2 adrenergic receptor GS protein complex. *Nature* 477:549–557. The first structural analysis of a receptor-G protein complex. **page 654 Figure 20.T4:** Source: Data from J. D. Watson, M. Gilman, J. Witkowski, and M. Zoller, *Recombinant DNA*, 2nd ed. (New York: Scientific American Books, 1992), p. 339.

CHAPTER 21 **page 669 Figure 21.5a, c:** Data are from Berget, S. M., C. Moore, and P. A. Sharp (1977) Proc. Natl. Acad. Sci. USA 74:3171–3175. **page 671 Figure 21.6:** From Hoelz, A., E. W. Debler, and G. Blobel (2011) Ann. Rev. Biochem. 80:613–643, Reprinted with permission from Annual Reviews. **page 677 Figure 21.15:** Based on J. A. McClarin et al., *Science* (1986) 234:1526–1541. **page 677 Figure 21.16:** Based on S. Klimasauskas, S. Kumar, R. J. Roberts, and X. Cheng, *Cell* (1994) 76:357–369. **page 679 Figure 21.19:** Reproduced from N. G. Cooper, 1994 (ed.) *The Human Genome Project. Deciphering the Blueprint of Heredity*, University Science Books, Mill Valley, CA, p. 63.

CHAPTER 22 **page 697 Figure 22.18b:** Modified from *Science* 334:1675–1680 (2011), B. A. Kelch, D. L. Makino, M. O'Donnell, and J. Kuriyan. How a DNA polymerase clamp loader opens a sliding clamp.

CHAPTER 23 **page 717 Figure 23.3:** Data from Friedberg, E. C., G. C. Walker, W. Siede, R. D. Wood, R. A. Schultz, and T. Ellenberger (2006) DNA Repair and Mutagenesis, 2nd. ed. ASM Press, Washington, D.C. **page 719 Figure 23.7:** Data from A. Sancar and J. E. Hearst, *Science* (1993) 259:1415–1420. AAAS. **page 722 Figure 23.11:** Based on A. Banerjee, W. Yang, M. Karplus, and G.L. Verdine (2005) *Nature* 434, 612–618. **page 729 Figure 23.20:** Based on L. Yang et al (2012) Proc. Natl. Acad. Sci. USA 109:8907–8912. **page 729 Figure 23.21:** Courtesy of BioMed Central. **page 738 Figure 23A.1:** Courtesy of Mario R. Capecchi.

CHAPTER 24 **page 754 Figure 24.14:** Data from M. S. Lee et al (1989) *Science* 245:635–637. **page 755 Figure 24.15:** Based in part upon Genes IV, B. Lewin, Oxford: Oxford University Press, 1990. **page 755 Figure 24.16:** From K. Murakami et al (2013) *Science* 349, 1238724. DOI:10.1126/science.1238724. Courtesy of Roger D. Kornberg. **page 756 Figure 24.18:** Data from J. C. Hansen, C. Tse, and A. P. Wolffe, Biochemistry (1997)

37:17637–17641. **page 765 Figure 24B.1:** Data from B. J. Venters and B. F. Pugh (2009) How eukaryotic genes are transcribed, *Critical Reviews in Biochemistry and Molecular Biology* 44:117–141.

CHAPTER 25 **page 776 Figure 25.12:** Data from L. Schulman and J. Abelson, (1988) Recent excitement in understanding transfer RNA identity. *Science* 240:1591–1592. **page 780 Figure 25.16:** From *BMC Bioinformatics* 3.2, J. J. Cannone and 13 coauthors, The Comparative RNA Web (CRW) Site: An online database of comparative sequence and structure information for ribos omal, intron, and other RNAs. Image provided courtesy of Robin Gutell. **page 785 Figure 25.22:** Data from P. Nissen et al, (2000) The Structural Basis of Ribosome Activity in Peptide Bond Synthesis. *Science* 289:920–930. **page 786 Figure 25.24:** Data from W. Zhang, J. A. Dunkle, and J. H. D. Cate, (2009) Structures of the ribosome in intermediate states of ratcheting. *Science* 325:1014–1017. **page 786 Figure 25.25:** Data from Laurberg, M., H. Asahara, A. Korostelev, J. Zhu, S. Trakhanov, and H. F. Noller, (2008) Structural basis for translation termination on the 70S ribosome. *Nature* 454:852–857. **page 788 Figure 25.28:** Data from coordinates published by A. Ben-Shem, L. Jenner, G. Yusupova, and M. Yusupov, (2010) Crystal Structure of the Eukaryotic Ribosome. *Science* 330:1203–1209. PDB ID 302Z, 3030, 305, and 305H. **page 789 Figure 25.30:** Data from Roberts, J. W. (2010) Syntheses that stay together. *Science* 328:436–437. **page 792 Figure 25.33:** Based on J. Dudek, P. Rehling, and M. van der Laan, (2012) Mitochondrial protein import: Common principles and physiological networks. *Biochim. Biophys. Acta* 1833:274–285.

CHAPTER 26 **page 814:** *Ptashne*, 2013, p. 1.

Index

I-27

Useful Equations

Henderson–Hasselbalch equation	$pH = pK_a + \log([A^-]/[HA])$
Michaelis–Menten equation	$\nu = V_{max}[S]/(K_M + [S])$
Free energy change under non-standard-state conditions	$\Delta G = \Delta G^\circ + RT \ln([C][D]/[A][B])$
Free energy change and standard reduction potential	$\Delta G^{\circ\prime} = -nF\Delta E^{\circ\prime}$
Reduction potentials in a redox reaction	$\Delta E^{\circ\prime} = E^{\circ\prime}(\text{acceptor}) - E^{\circ\prime}(\text{donor})$
Proton motive force	$\Delta p = \Delta\Psi - 2.3RT\,\Delta pH/F$
Passive diffusion of a charged species	$\Delta G = G_2 - G_1 = RT \ln(C_2/C_1) + ZF\Delta\Psi$

Common Abbreviations Used by Biochemists

Ab	antibody		dTTP	thymidine triphosphate
Ac-CoA	acetyl-coenzyme A		E	reduction potential
ACP	acyl carrier protein		EF	elongation factor
ADH	alcohol dehydrogenase		EGF	epidermal growth factor
AdoMet	*S*-adenosylmethionine		EPR	electron paramagnetic resonance
ADP	adenosine diphosphate		ER	endoplasmic reticulum
Ag	antigen		F	phenylalanine
AIDS	acquired immune deficiency syndrome		F	Faraday constant
Ala	alanine		F_{ab}	antibody molecule fragment that binds antigen
AMP	adenosine monophosphate		FAD	flavin adenine dinucleotide
Arg	arginine		$FADH_2$	reduced flavin adenine dinucleotide
ARS	autonomously replicating sequence		FBP	fructose-1,6-bisphosphate
Asn	asparagine		FBPase	fructose bisphosphatase
Asp	aspartic acid		Fd	ferredoxin
atm	atmosphere		fMet	*N*-formylmethionine
ATP	adenosine triphosphate		FMN	flavin mononucleotide
bp	base pair		F1P	fructose-1-phosphate
BPG	bisphosphoglycerate		F6P	fructose-6-phosphate
cal	calorie		G	Gibbs free energy
cAMP	cyclic 3′,5′-adenosine monophosphate		GABA	γ-aminobutyric acid
CD	circular dichroism		Gal	galactose
cDNA	complementary DNA		GAP	glyceraldehyde-3-phosphate
CDP	cytidine diphosphate		GC-MS	gas chromatography-mass spectrometry
Chl	chlorophyll		GDP	guanosine diphosphate
CMP	cytidine monophosphate		Glc	glucose
CoA or CoA-SH	coenzyme A		Gln	glutamine
CoQ	coenzyme Q		Glu	glutamic acid
cpm	counts per minute		Gly	glycine
CRP	cAMP receptor protein (catabolite activator protein)		GMP	guanosine monophosphate
CTP	cytidine triphosphate		G1P	glucose-1-phosphate
Cys	cysteine		GS	glutamine synthetase
d	deoxy		GSH	glutathione (reduced glutathione)
Da	dalton		G6P	glucose-6-phosphate
dd	dideoxy		GSSG	glutathione disulfide (oxidized glutathione)
DEAE	diethylaminoethyl		GTP	guanosine triphosphate
DHAP	dihydroxyacetone phosphate		h	hour
DHF	dihydrofolate		h	Planck's constant
DHFR	dihydrofolate reductase		Hb	hemoglobin
DNA	deoxyribonucleic acid		HDL	high-density lipoprotein
DNP	dinitrophenol		HIV	human immunodeficiency virus
dopa	dihydroxyphenylalanine		hnRNA	heterogeneous nuclear RNA
dTDP	thymidine diphosphate		HPLC	high-pressure (or high-performance) liquid chromatography
dTMP	thymidine monophosphate or thymidylate			